MATHEMATICAL METHODS FOR PHYSICS AND ENGINEERING

The third edition of this highly acclaimed undergraduate textbook is suitable for teaching all the mathematics ever likely to be needed for an undergraduate course in any of the physical sciences. As well as lucid descriptions of all the topics covered and many worked examples, it contains more than 800 exercises. A number of additional topics have been included and the text has undergone significant reorganisation in some areas. New stand-alone chapters:

- give a systematic account of the 'special functions' of physical science
- cover an extended range of practical applications of complex variables including WKB methods and saddle-point integration techniques
- provide an introduction to quantum operators.

Further tabulations, of relevance in statistics and numerical integration, have been added. In this edition, all 400 odd-numbered exercises are provided with complete worked solutions in a separate manual, available to both students and their teachers; these are in addition to the hints and outline answers given in the main text. The even-numbered exercises have no hints, answers or worked solutions and can be used for unaided homework; full solutions to them are available to instructors on a password-protected website.

KEN RILEY read mathematics at the University of Cambridge and proceeded to a Ph.D. there in theoretical and experimental nuclear physics. He became a research associate in elementary particle physics at Brookhaven, and then, having taken up a lectureship at the Cavendish Laboratory, Cambridge, continued this research at the Rutherford Laboratory and Stanford; in particular he was involved in the experimental discovery of a number of the early baryonic resonances. As well as having been Senior Tutor at Clare College, where he has taught physics and mathematics for over 40 years, he has served on many committees concerned with the teaching and examining of these subjects at all levels of tertiary and undergraduate education. He is also one of the authors of *200 Puzzling Physics Problems*.

MICHAEL HOBSON read natural sciences at the University of Cambridge, specialising in theoretical physics, and remained at the Cavendish Laboratory to complete a Ph.D. in the physics of star-formation. As a research fellow at Trinity

Hall, Cambridge and subsequently an advanced fellow of the Particle Physics and Astronomy Research Council, he developed an interest in cosmology, and in particular in the study of fluctuations in the cosmic microwave background. He was involved in the first detection of these fluctuations using a ground-based interferometer. He is currently a University Reader at the Cavendish Laboratory, his research interests include both theoretical and observational aspects of cosmology, and he is the principal author of *General Relativity: An Introduction for Physicists*. He is also a Director of Studies in Natural Sciences at Trinity Hall and enjoys an active role in the teaching of undergraduate physics and mathematics.

STEPHEN BENCE obtained both his undergraduate degree in Natural Sciences and his Ph.D. in Astrophysics from the University of Cambridge. He then became a Research Associate with a special interest in star-formation processes and the structure of star-forming regions. In particular, his research concentrated on the physics of jets and outflows from young stars. He has had considerable experience of teaching mathematics and physics to undergraduate and pre-universtiy students.

MATHEMATICAL METHODS FOR PHYSICS AND ENGINEERING

THIRD EDITION

K. F. RILEY, M. P. HOBSON and S. J. BENCE

CAMBRIDGE
UNIVERSITY PRESS

CAMBRIDGE
UNIVERSITY PRESS

University Printing House, Cambridge CB2 8BS, United Kingdom

Cambridge University Press is a part of the University of Cambridge.

It furthers the University's mission by disseminating knowledge in the pursuit of education, learning and research at the highest international levels of excellence.

www.cambridge.org
Information on this title: www.cambridge.org/9780521861533

© K. F. Riley, M. P. Hobson and S. J. Bence 2006

First edition © Cambridge University Press 1998
Reprinted 1998 (with minor corrections), 2000 (twice), 2001
Second edition © Ken Riley, Mike Hobson, Stephen Bence 2002
Reprinted (with corrections) 2003, 2004
Reprinted 2005
Third edition 2006
14th printing 2016

Printed in the United Kingdom by TJ International Ltd. Padstow Cornwall

A catalogue record for this publication is available from the British Library

ISBN 978-0-521-86153-3 Hardback
ISBN 978-0-521-67971-8 Paperback

Contents

Areas and volumes; masses, centres of mass and centroids; Pappus' theorems; moments of inertia; mean values of functions

Change of variables in double integrals; evaluation of the integral $I = \int_{-\infty}^{\infty} e^{-x^2}\, dx$; change of variables in triple integrals; general properties of Jacobians

Scalar product; vector product; scalar triple product; vector triple product

I am the very Model for a Student Mathematical

I am the very model for a student mathematical;
I've information rational, and logical and practical.
I know the laws of algebra, and find them quite symmetrical,
And even know the meaning of 'a variate antithetical'.

I'm extremely well acquainted, with all things mathematical.
I understand equations, both the simple and quadratical.
About binomial theorems I'm teeming with a lot o'news,
With many cheerful facts about the square of the hypotenuse.

I'm very good at integral and differential calculus,
And solving paradoxes that so often seem to rankle us.
In short in matters rational, and logical and practical,
I am the very model for a student mathematical.

I know the singularities of equations differential,
And some of these are regular, but the rest are quite essential.
I quote the results of giants; with Euler, Newton, Gauss, Laplace,
And can calculate an orbit, given a centre, force and mass.

I can reconstruct equations, both canonical and formal,
And write all kinds of matrices, orthogonal, real and normal.
I show how to tackle problems that one has never met before,
By analogy or example, or with some clever metaphor.

I seldom use equivalence to help decide upon a class,
But often find an integral, using a contour o'er a pass.
In short in matters rational, and logical and practical,
I am the very model for a student mathematical.

When you have learnt just what is meant by 'Jacobian' and 'Abelian';
When you at sight can estimate, for the modal, mean and median;
When describing normal subgroups is much more than recitation;
When you understand precisely what is 'quantum excitation';

When you know enough statistics that you can recognise RV;
When you have learnt all advances that have been made in SVD;
And when you can spot the transform that solves some tricky PDE,
You will feel no better student has ever sat for a degree.

Your accumulated knowledge, whilst extensive and exemplary,
Will have only been brought down to the beginning of last century,
But still in matters rational, and logical and practical,
You'll be the very model of a student mathematical.

KFR, with apologies to W.S. Gilbert

Preface to the third edition

As is natural, in the four years since the publication of the second edition of this book we have somewhat modified our views on what should be included and how it should be presented. In this new edition, although the range of topics covered has been extended, there has been no significant shift in the general level of difficulty or in the degree of mathematical sophistication required. Further, we have aimed to preserve the same style of presentation as seems to have been well received in the first two editions. However, a significant change has been made to the format of the chapters, specifically to the way that the exercises, together with their hints and answers, have been treated; the details of the change are explained below.

The two major chapters that are new in this third edition are those dealing with 'special functions' and the applications of complex variables. The former presents a systematic account of those functions that appear to have arisen in a more or less haphazard way as a result of studying particular physical situations, and are deemed 'special' for that reason. The treatment presented here shows that, in fact, they are nearly all particular cases of the hypergeometric or confluent hypergeometric functions, and are special only in the sense that the parameters of the relevant function take simple or related values.

The second new chapter describes how the properties of complex variables can be used to tackle problems arising from the description of physical situations or from other seemingly unrelated areas of mathematics. To topics treated in earlier editions, such as the solution of Laplace's equation in two dimensions, the summation of series, the location of zeros of polynomials and the calculation of inverse Laplace transforms, has been added new material covering Airy integrals, saddle-point methods for contour integral evaluation, and the WKB approach to asymptotic forms.

Other new material includes a stand-alone chapter on the use of coordinate-free operators to establish valuable results in the field of quantum mechanics; amongst the physical topics covered are angular momentum and uncertainty principles. There are also significant additions to the treatment of numerical integration. In particular, Gaussian quadrature based on Legendre, Laguerre, Hermite and Chebyshev polynomials is discussed, and appropriate tables of points and weights are provided.

We now turn to the most obvious change to the format of the book, namely the way that the exercises, hints and answers are treated. The second edition of *Mathematical Methods for Physics and Engineering* carried more than twice as many exercises, based on its various chapters, as did the first. In its preface we discussed the general question of how such exercises should be treated but, in the end, decided to provide hints and outline answers to all problems, as in the first edition. This decision was an uneasy one as, on the one hand, it did not allow the exercises to be set as totally unaided homework that could be used for assessment purposes but, on the other, it did not give a full explanation of how to tackle a problem when a student needed explicit guidance or a model answer.

In order to allow both of these educationally desirable goals to be achieved, we have, in this third edition, completely changed the way in which this matter is handled. A large number of exercises have been included in the penultimate subsections of the appropriate, sometimes reorganised, chapters. Hints and outline answers are given, as previously, in the final subsections, *but only for the odd-numbered exercises*. This leaves all even-numbered exercises free to be set as unaided homework, as described below.

For the four hundred plus **odd-numbered** exercises, *complete* solutions are available, to both students and their teachers, in the form of a separate manual, *Student Solutions Manual for Mathematical Methods for Physics and Engineering* (Cambridge: Cambridge University Press, 2006); the hints and outline answers given in this main text are brief summaries of the model answers given in the manual. There, each original exercise is reproduced and followed by a fully worked solution. For those original exercises that make internal reference to this text or to other (even-numbered) exercises not included in the solutions manual, the questions have been reworded, usually by including additional information, so that the questions can stand alone.

In many cases, the solution given in the manual is even fuller than one that might be expected of a good student that has understood the material. This is because we have aimed to make the solutions instructional as well as utilitarian. To this end, we have included comments that are intended to show how the plan for the solution is formulated and have given the justifications for particular intermediate steps (something not always done, even by the best of students). We have also tried to write each individual substituted formula in the form that best indicates how it was obtained, before simplifying it at the next or a subsequent stage. Where several lines of algebraic manipulation or calculus are needed to obtain a final result, they are normally included in full; this should enable the student to determine whether an incorrect answer is due to a misunderstanding of principles or to a technical error.

The remaining four hundred or so **even-numbered** exercises have no hints or answers, outlined or detailed, available for general access. They can therefore be used by instructors as a basis for setting unaided homework. Full solutions to these exercises, in the same general format as those appearing in the manual (though they may contain references to the main text or to other exercises), are available without charge to accredited teachers as downloadable pdf files on the password-protected website http://www.cambridge.org/9780521679718. Teachers wishing to have access to the website should contact solutions@cambridge.org for registration details.

In all new publications, errors and typographical mistakes are virtually unavoidable, and we would be grateful to any reader who brings instances to our attention. Retrospectively, we would like to record our thanks to Reinhard Gerndt, Paul Renteln and Joe Tenn for making us aware of some errors in the second edition. Finally, we are extremely grateful to Dave Green for his considerable and continuing advice concerning LATEX.

Ken Riley, Michael Hobson,
Cambridge, 2006

Preface to the second edition

Since the publication of the first edition of this book, both through teaching the material it covers and as a result of receiving helpful comments from colleagues, we have become aware of the desirability of changes in a number of areas. The most important of these is that the mathematical preparation of current senior college and university entrants is now less thorough than it used to be. To match this, we decided to include a preliminary chapter covering areas such as polynomial equations, trigonometric identities, coordinate geometry, partial fractions, binomial expansions, necessary and sufficient condition and proof by induction and contradiction.

Whilst the general level of what is included in this second edition has not been raised, some areas have been expanded to take in topics we now feel were not adequately covered in the first. In particular, increased attention has been given to non-square sets of simultaneous linear equations and their associated matrices. We hope that this more extended treatment, together with the inclusion of singular value matrix decomposition, will make the material of more practical use to engineering students. In the same spirit, an elementary treatment of linear recurrence relations has been included. The topic of normal modes has been given a small chapter of its own, though the links to matrices on the one hand, and to representation theory on the other, have not been lost.

Elsewhere, the presentation of probability and statistics has been reorganised to give the two aspects more nearly equal weights. The early part of the probability chapter has been rewritten in order to present a more coherent development based on Boolean algebra, the fundamental axioms of probability theory and the properties of intersections and unions. Whilst this is somewhat more formal than previously, we think that it has not reduced the accessibility of these topics and hope that it has increased it. The scope of the chapter has been somewhat extended to include all physically important distributions and an introduction to cumulants.

Statistics now occupies a substantial chapter of its own, one that includes systematic discussions of estimators and their efficiency, sample distributions and t- and F-tests for comparing means and variances. Other new topics are applications of the chi-squared distribution, maximum-likelihood parameter estimation and least-squares fitting. In other chapters we have added material on the following topics: curvature, envelopes, curve-sketching, more refined numerical methods for differential equations and the elements of integration using Monte Carlo techniques.

Over the last four years we have received somewhat mixed feedback about the number of exercises at the ends of the various chapters. After consideration, we decided to increase the number substantially, partly to correspond to the additional topics covered in the text but mainly to give both students and their teachers a wider choice. There are now nearly 800 such exercises, many with several parts. An even more vexed question has been whether to provide hints and answers to all the exercises or just to 'the odd-numbered' ones, as is the normal practice for textbooks in the United States, thus making the remainder more suitable for setting as homework. In the end, we decided that hints and outline solutions should be provided for all the exercises, in order to facilitate independent study while leaving the details of the calculation as a task for the student.

In conclusion, we hope that this edition will be thought by its users to be 'heading in the right direction' and would like to place on record our thanks to all who have helped to bring about the changes and adjustments. Naturally, those colleagues who have noted errors or ambiguities in the first edition and brought them to our attention figure high on the list, as do the staff at The Cambridge University Press. In particular, we are grateful to Dave Green for continued LaTeX advice, Susan Parkinson for copy-editing the second edition with her usual keen eye for detail and flair for crafting coherent prose and Alison Woollatt for once again turning our basic LaTeX into a beautifully typeset book. Our thanks go to all of them, though of course we accept full responsibility for any remaining errors or ambiguities, of which, as with any new publication, there are bound to be some.

On a more personal note, KFR again wishes to thank his wife Penny for her unwavering support, not only in his academic and tutorial work, but also in their joint efforts to convert time at the bridge table into 'green points' on their record. MPH is once more indebted to his wife, Becky, and his mother, Pat, for their tireless support and encouragement above and beyond the call of duty. MPH dedicates his contribution to this book to the memory of his father, Ronald Leonard Hobson, whose gentle kindness, patient understanding and unbreakable spirit made all things seem possible.

Ken Riley, Michael Hobson
Cambridge, 2002

Preface to the first edition

A knowledge of mathematical methods is important for an increasing number of university and college courses, particularly in physics, engineering and chemistry, but also in more general science. Students embarking on such courses come from diverse mathematical backgrounds, and their core knowledge varies considerably. We have therefore decided to write a textbook that assumes knowledge only of material that can be expected to be familiar to all the current generation of students starting physical science courses at university. In the United Kingdom this corresponds to the standard of Mathematics A-level, whereas in the United States the material assumed is that which would normally be covered at junior college.

Starting from this level, the first six chapters cover a collection of topics with which the reader may already be familiar, but which are here extended and applied to typical problems encountered by first-year university students. They are aimed at providing a common base of general techniques used in the development of the remaining chapters. Students who have had additional preparation, such as Further Mathematics at A-level, will find much of this material straightforward.

Following these opening chapters, the remainder of the book is intended to cover at least that mathematical material which an undergraduate in the physical sciences might encounter up to the end of his or her course. The book is also appropriate for those beginning graduate study with a mathematical content, and naturally much of the material forms parts of courses for mathematics students. Furthermore, the text should provide a useful reference for research workers.

The general aim of the book is to present a topic in three stages. The first stage is a qualitative introduction, wherever possible from a physical point of view. The second is a more formal presentation, although we have deliberately avoided strictly mathematical questions such as the existence of limits, uniform convergence, the interchanging of integration and summation orders, etc. on the grounds that 'this is the real world; it must behave reasonably'. Finally a worked example is presented, often drawn from familiar situations in physical science and engineering. These examples have generally been fully worked, since, in the authors' experience, partially worked examples are unpopular with students. Only in a few cases, where trivial algebraic manipulation is involved, or where repetition of the main text would result, has an example been left as an exercise for the reader. Nevertheless, a number of exercises also appear at the end of each chapter, and these should give the reader ample opportunity to test his or her understanding. Hints and answers to these exercises are also provided.

With regard to the presentation of the mathematics, it has to be accepted that many equations (especially partial differential equations) can be written more compactly by using subscripts, e.g. u_{xy} for a second partial derivative, instead of the more familiar $\partial^2 u/\partial x \partial y$, and that this certainly saves typographical space. However, for many students, the labour of mentally unpacking such equations is sufficiently great that it is not possible to think of an equation's physical interpretation at the same time. Consequently, wherever possible we have decided to write out such expressions in their more obvious but longer form.

During the writing of this book we have received much help and encouragement from various colleagues at the Cavendish Laboratory, Clare College, Trinity Hall and Peterhouse. In particular, we would like to thank Peter Scheuer, whose comments and general enthusiasm proved invaluable in the early stages. For reading sections of the manuscript, for pointing out misprints and for numerous useful comments, we thank many of our students and colleagues at the University of Cambridge. We are especially grateful to Chris Doran, John Huber, Garth Leder, Tom Körner and, not least, Mike Stobbs, who, sadly, died before the book was completed. We also extend our thanks to the University of Cambridge and the Cavendish teaching staff, whose examination questions and lecture hand-outs have collectively provided the basis for some of the examples included. Of course, any errors and ambiguities remaining are entirely the responsibility of the authors, and we would be most grateful to have them brought to our attention.

We are indebted to Dave Green for a great deal of advice concerning typesetting in LaTeX and to Andrew Lovatt for various other computing tips. Our thanks also go to Anja Visser and Graça Rocha for enduring many hours of (sometimes heated) debate. At Cambridge University Press, we are very grateful to our editor Adam Black for his help and patience and to Alison Woollatt for her expert typesetting of such a complicated text. We also thank our copy-editor Susan Parkinson for many useful suggestions that have undoubtedly improved the style of the book.

Finally, on a personal note, KFR wishes to thank his wife Penny, not only for a long and happy marriage, but also for her support and understanding during his recent illness – and when things have not gone too well at the bridge table! MPH is indebted both to Rebecca Morris and to his parents for their tireless support and patience, and for their unending supplies of tea. SJB is grateful to Anthony Gritten for numerous relaxing discussions about J. S. Bach, to Susannah Ticciati for her patience and understanding, and to Kate Isaak for her calming late-night e-mails from the USA.

Ken Riley, Michael Hobson and Stephen Bence
Cambridge, 1997

1

Preliminary algebra

This opening chapter reviews the basic algebra of which a working knowledge is presumed in the rest of the book. Many students will be familiar with much, if not all, of it, but recent changes in what is studied during secondary education mean that it cannot be taken for granted that they will already have a mastery of all the topics presented here. The reader may assess which areas need further study or revision by attempting the exercises at the end of the chapter. The main areas covered are polynomial equations and the related topic of partial fractions, curve sketching, coordinate geometry, trigonometric identities and the notions of proof by induction or contradiction.

1.1 Simple functions and equations

It is normal practice when starting the mathematical investigation of a physical problem to assign an algebraic symbol to the quantity whose value is sought, either numerically or as an explicit algebraic expression. For the sake of definiteness, in this chapter we will use x to denote this quantity most of the time. Subsequent steps in the analysis involve applying a combination of known laws, consistency conditions and (possibly) given constraints to derive one or more equations satisfied by x. These equations may take many forms, ranging from a simple polynomial equation to, say, a partial differential equation with several boundary conditions. Some of the more complicated possibilities are treated in the later chapters of this book, but for the present we will be concerned with techniques for the solution of relatively straightforward algebraic equations.

1.1.1 Polynomials and polynomial equations

Firstly we consider the simplest type of equation, a *polynomial equation*, in which a *polynomial* expression in x, denoted by $f(x)$, is set equal to zero and thereby forms an equation which is satisfied by particular values of x, called the *roots* of the equation:

$$f(x) = a_n x^n + a_{n-1} x^{n-1} + \cdots + a_1 x + a_0 = 0. \tag{1.1}$$

Here n is an integer > 0, called the *degree* of both the polynomial and the equation, and the known coefficients $a_0, a_1, \ldots, a_n$ are real quantities with $a_n \neq 0$.

Equations such as (1.1) arise frequently in physical problems, the coefficients a_i being determined by the physical properties of the system under study. What is needed is to find some or all of the roots of (1.1), i.e. the x-values, α_k, that satisfy $f(\alpha_k) = 0$; here k is an index that, as we shall see later, can take up to n different values, i.e. $k = 1, 2, \ldots, n$. The roots of the polynomial equation can equally well be described as the zeros of the polynomial. When they are *real*, they correspond to the points at which a graph of $f(x)$ crosses the x-axis. Roots that are complex (see chapter 3) do not have such a graphical interpretation.

For polynomial equations containing powers of x greater than x^4 general methods do not exist for obtaining explicit expressions for the roots α_k. Even for $n = 3$ and $n = 4$ the prescriptions for obtaining the roots are sufficiently complicated that it is usually preferable to obtain exact or approximate values by other methods. Only for $n = 1$ and $n = 2$ can closed-form solutions be given. These results will be well known to the reader, but they are given here for the sake of completeness. For $n = 1$, (1.1) reduces to the *linear* equation

$$a_1 x + a_0 = 0; \tag{1.2}$$

the solution (root) is $\alpha_1 = -a_0/a_1$. For $n = 2$, (1.1) reduces to the *quadratic* equation

$$a_2 x^2 + a_1 x + a_0 = 0; \tag{1.3}$$

the two roots α_1 and α_2 are given by

$$\alpha_{1,2} = \frac{-a_1 \pm \sqrt{a_1^2 - 4a_2 a_0}}{2a_2}. \tag{1.4}$$

When discussing specifically quadratic equations, as opposed to more general polynomial equations, it is usual to write the equation in one of the two notations

$$ax^2 + bx + c = 0, \qquad ax^2 + 2bx + c = 0, \tag{1.5}$$

with respective explicit pairs of solutions

$$\alpha_{1,2} = \frac{-b \pm \sqrt{b^2 - 4ac}}{2a}, \qquad \alpha_{1,2} = \frac{-b \pm \sqrt{b^2 - ac}}{a}. \tag{1.6}$$

Of course, these two notations are entirely equivalent and the only important point is to associate each form of answer with the corresponding form of equation; most people keep to one form, to avoid any possible confusion.

If the value of the quantity appearing under the square root sign is positive then both roots are real; if it is negative then the roots form a complex conjugate pair, i.e. they are of the form $p \pm iq$ with p and q real (see chapter 3); if it has zero value then the two roots are equal and special considerations usually arise.

Thus linear and quadratic equations can be dealt with in a cut-and-dried way. We now turn to methods for obtaining partial information about the roots of higher-degree polynomial equations. In some circumstances the knowledge that an equation has a root lying in a certain range, or that it has no real roots at all, is all that is actually required. For example, in the design of electronic circuits it is necessary to know whether the current in a proposed circuit will break into spontaneous oscillation. To test this, it is sufficient to establish whether a certain polynomial equation, whose coefficients are determined by the physical parameters of the circuit, has a root with a positive real part (see chapter 3); complete determination of all the roots is not needed for this purpose. If the complete set of roots of a polynomial equation is required, it can usually be obtained to any desired accuracy by numerical methods such as those described in chapter 27.

There is no explicit step-by-step approach to finding the roots of a general polynomial equation such as (1.1). In most cases analytic methods yield only information *about* the roots, rather than their exact values. To explain the relevant techniques we will consider a particular example, 'thinking aloud' on paper and expanding on special points about methods and lines of reasoning. In more routine situations such comment would be absent and the whole process briefer and more tightly focussed.

Example: the cubic case

Let us investigate the roots of the equation

$$g(x) = 4x^3 + 3x^2 - 6x - 1 = 0 \tag{1.7}$$

or, in an alternative phrasing, investigate the zeros of $g(x)$. We note first of all that this is a *cubic* equation. It can be seen that for x large and positive $g(x)$ will be large and positive and, equally, that for x large and negative $g(x)$ will be large and negative. Therefore, intuitively (or, more formally, by continuity) $g(x)$ must cross the x-axis at least once and so $g(x) = 0$ must have at least one real root. Furthermore, it can be shown that if $f(x)$ is an nth-degree polynomial then the graph of $f(x)$ must cross the x-axis an even or odd number of times as x varies between $-\infty$ and $+\infty$, according to whether n itself is even or odd. Thus a polynomial of odd degree always has at least one real root, but one of even degree may have no real root. A small complication, discussed later in this section, occurs when repeated roots arise.

Having established that $g(x) = 0$ has at least one real root, we may ask how many real roots it *could* have. To answer this we need one of the fundamental theorems of algebra, mentioned above:

An *n*th-degree polynomial equation has exactly *n* roots.

It should be noted that this does not imply that there are *n real* roots (only that there are not more than *n*); some of the roots may be of the form $p + iq$.

To make the above theorem plausible and to see what is meant by repeated roots, let us suppose that the *n*th-degree polynomial equation $f(x) = 0$, (1.1), has *r* roots $\alpha_1, \alpha_2, \ldots, \alpha_r$, considered distinct for the moment. That is, we suppose that $f(\alpha_k) = 0$ for $k = 1, 2, \ldots, r$, so that $f(x)$ vanishes only when x is equal to one of the *r* values α_k. But the same can be said for the function

$$F(x) = A(x - \alpha_1)(x - \alpha_2) \cdots (x - \alpha_r), \tag{1.8}$$

in which A is a non-zero constant; $F(x)$ can clearly be multiplied out to form a polynomial expression.

We now call upon a second fundamental result in algebra: that if two polynomial functions $f(x)$ and $F(x)$ have equal values for *all* values of x, then their coefficients are equal on a term-by-term basis. In other words, we can equate the coefficients of each and every power of x in the two expressions (1.8) and (1.1); in particular we can equate the coefficients of the highest power of x. From this we have $Ax^r \equiv a_n x^n$ and thus that $r = n$ and $A = a_n$. As r is both equal to n and to the number of roots of $f(x) = 0$, we conclude that the *n*th-degree polynomial $f(x) = 0$ has n roots. (Although this line of reasoning may make the theorem plausible, it does not constitute a proof since we have not shown that it is permissible to write $f(x)$ in the form of equation (1.8).)

We next note that the condition $f(\alpha_k) = 0$ for $k = 1, 2, \ldots, r$, could also be met if (1.8) were replaced by

$$F(x) = A(x - \alpha_1)^{m_1}(x - \alpha_2)^{m_2} \cdots (x - \alpha_r)^{m_r}, \tag{1.9}$$

with $A = a_n$. In (1.9) the m_k are integers ≥ 1 and are known as the multiplicities of the roots, m_k being the multiplicity of α_k. Expanding the right-hand side (RHS) leads to a polynomial of degree $m_1 + m_2 + \cdots + m_r$. This sum must be equal to n. Thus, if any of the m_k is greater than unity then the number of *distinct* roots, r, is less than n; the total number of roots remains at n, but one or more of the α_k counts more than once. For example, the equation

$$F(x) = A(x - \alpha_1)^2(x - \alpha_2)^3(x - \alpha_3)(x - \alpha_4) = 0$$

has exactly seven roots, α_1 being a double root and α_2 a triple root, whilst α_3 and α_4 are unrepeated (*simple*) roots.

We can now say that our particular equation (1.7) has either one or three real roots but in the latter case it may be that not all the roots are distinct. To decide how many real roots the equation has, we need to anticipate two ideas from the

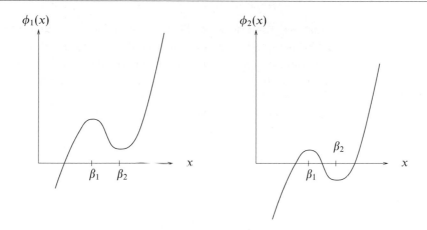

Figure 1.1   Two curves $\phi_1(x)$ and $\phi_2(x)$, both with zero derivatives at the same values of x, but with different numbers of real solutions to $\phi_i(x) = 0$.

next chapter. The first of these is the notion of the derivative of a function, and the second is a result known as Rolle's theorem.

The *derivative* $f'(x)$ of a function $f(x)$ measures the slope of the tangent to the graph of $f(x)$ at that value of x (see figure 2.1 in the next chapter). For the moment, the reader with no prior knowledge of calculus is asked to accept that the derivative of ax^n is nax^{n-1}, so that the derivative $g'(x)$ of the curve $g(x) = 4x^3 + 3x^2 - 6x - 1$ is given by $g'(x) = 12x^2 + 6x - 6$. Similar expressions for the derivatives of other polynomials are used later in this chapter.

Rolle's theorem states that if $f(x)$ has equal values at two different values of x then at some point between these two x-values its derivative is equal to zero; i.e. the tangent to its graph is parallel to the x-axis at that point (see figure 2.2).

Having briefly mentioned the derivative of a function and Rolle's theorem, we now use them to establish whether $g(x)$ has one or three real zeros. If $g(x) = 0$ does have three real roots α_k, i.e. $g(\alpha_k) = 0$ for $k = 1, 2, 3$, then it follows from Rolle's theorem that between any consecutive pair of them (say α_1 and α_2) there must be some real value of x at which $g'(x) = 0$. Similarly, there must be a further zero of $g'(x)$ lying between α_2 and α_3. Thus a *necessary* condition for three real roots of $g(x) = 0$ is that $g'(x) = 0$ itself has two real roots.

However, this condition on the number of roots of $g'(x) = 0$, whilst necessary, is not *sufficient* to guarantee three real roots of $g(x) = 0$. This can be seen by inspecting the cubic curves in figure 1.1. For each of the two functions $\phi_1(x)$ and $\phi_2(x)$, the derivative is equal to zero at both $x = \beta_1$ and $x = \beta_2$. Clearly, though, $\phi_2(x) = 0$ has three real roots whilst $\phi_1(x) = 0$ has only one. It is easy to see that the crucial difference is that $\phi_1(\beta_1)$ and $\phi_1(\beta_2)$ have the same sign, whilst $\phi_2(\beta_1)$ and $\phi_2(\beta_2)$ have opposite signs.

It will be apparent that for some equations, $\phi(x) = 0$ say, $\phi'(x)$ equals zero at a value of x for which $\phi(x)$ is also zero. Then the graph of $\phi(x)$ just touches the x-axis. When this happens the value of x so found is, in fact, a double real root of the polynomial equation (corresponding to one of the m_k in (1.9) having the value 2) and must be counted twice when determining the number of real roots.

Finally, then, we are in a position to decide the number of real roots of the equation

$$g(x) = 4x^3 + 3x^2 - 6x - 1 = 0.$$

The equation $g'(x) = 0$, with $g'(x) = 12x^2 + 6x - 6$, is a quadratic equation with explicit solutions[§]

$$\beta_{1,2} = \frac{-3 \pm \sqrt{9 + 72}}{12},$$

so that $\beta_2 = -1$ and $\beta_1 = \frac{1}{2}$. The corresponding values of $g(x)$ are $g(\beta_2) = 4$ and $g(\beta_1) = -\frac{11}{4}$, which are of opposite sign. This indicates that $4x^3 + 3x^2 - 6x - 1 = 0$ has three real roots, one lying in the range $-1 < x < \frac{1}{2}$ and the others one on each side of that range.

The techniques we have developed above have been used to tackle a cubic equation, but they can be applied to polynomial equations $f(x) = 0$ of degree greater than 3. However, much of the analysis centres around the equation $f'(x) = 0$ and this itself, being then a polynomial equation of degree 3 or more, either has no closed-form general solution or one that is complicated to evaluate. Thus the amount of information that can be obtained about the roots of $f(x) = 0$ is correspondingly reduced.

A more general case

To illustrate what can (and cannot) be done in the more general case we now investigate as far as possible the real roots of

$$f(x) = x^7 + 5x^6 + x^4 - x^3 + x^2 - 2 = 0.$$

The following points can be made.

(i) This is a seventh-degree polynomial equation; therefore the number of real roots is 1, 3, 5 or 7.
(ii) $f(0)$ is negative whilst $f(\infty) = +\infty$, so there must be at least one positive root. It also follows that the total number of real roots to the right of $x = 0$ must be odd. Further, since the overall total number of roots must be odd, the number to the left must be even (0, 2, 4 or 6).

[§] The two roots β_1, β_2 are written as $\beta_{1,2}$. By convention β_1 refers to the upper symbol in $\pm$, β_2 to the lower symbol.

(iii) The equation $f'(x) = 0$ can be written as $x(7x^5 + 30x^4 + 4x^2 - 3x + 2) = 0$ and thus $x = 0$ is a root. The derivative of $f'(x)$, denoted by $f''(x)$, equals $42x^5 + 150x^4 + 12x^2 - 6x + 2$. That $f'(x)$ is zero whilst $f''(x)$ is positive at $x = 0$ indicates (subsection 2.1.8) that $f(x)$ has a minimum there.

This is about all that can be deduced by *simple* analytic methods in this case, although some further progress can be made in the ways indicated in exercise 1.3.

There are, in fact, more sophisticated tests that examine the relative signs of successive terms in an equation such as (1.1), and in quantities derived from them, to place limits on the numbers and positions of roots. But they are not prerequisites for the remainder of this book and will not be pursued further here.

We conclude this section with a worked example which demonstrates that the practical application of the ideas developed so far can be both short and decisive.

▶*For what values of k, if any, does*

$$f(x) = x^3 - 3x^2 + 6x + k = 0$$

have three real roots?

Firstly we study the equation $f'(x) = 0$, i.e. $3x^2 - 6x + 6 = 0$. This is a quadratic equation but, using (1.6), because $6^2 < 4 \times 3 \times 6$, it can have no real roots. Therefore, it follows immediately that $f(x)$ has no maximum or minimum; consequently $f(x) = 0$ cannot have more than one real root, whatever the value of k. ◀

1.1.2 Factorising polynomials

In the previous subsection we saw how a polynomial with r given distinct zeros α_k could be constructed as the product of factors containing those zeros:

$$\begin{aligned} f(x) &= a_n(x - \alpha_1)^{m_1}(x - \alpha_2)^{m_2} \cdots (x - \alpha_r)^{m_r} \\ &= a_n x^n + a_{n-1}x^{n-1} + \cdots + a_1 x + a_0, \end{aligned} \tag{1.10}$$

with $m_1 + m_2 + \cdots + m_r = n$, the degree of the polynomial. It will cause no loss of generality in what follows to suppose that all the zeros are simple, i.e. all $m_k = 1$ and $r = n$, and this we will do.

Sometimes it is desirable to be able to reverse this process, in particular when one exact zero has been found by some method and the remaining zeros are to be investigated. Suppose that we have located one zero, α; it is then possible to write (1.10) as

$$f(x) = (x - \alpha)f_1(x), \tag{1.11}$$

where $f_1(x)$ is a polynomial of degree $n-1$. How can we find $f_1(x)$? The procedure is much more complicated to describe in a general form than to carry out for an equation with given numerical coefficients a_i. If such manipulations are too complicated to be carried out mentally, they could be laid out along the lines of an algebraic 'long division' sum. However, a more compact form of calculation is as follows. Write $f_1(x)$ as

$$f_1(x) = b_{n-1}x^{n-1} + b_{n-2}x^{n-2} + b_{n-3}x^{n-3} + \cdots + b_1 x + b_0.$$

Substitution of this form into (1.11) and subsequent comparison of the coefficients of x^p for $p = n, n-1, \ldots, 1, 0$ with those in the second line of (1.10) generates the series of equations

$$b_{n-1} = a_n,$$
$$b_{n-2} - \alpha b_{n-1} = a_{n-1},$$
$$b_{n-3} - \alpha b_{n-2} = a_{n-2},$$
$$\vdots$$
$$b_0 - \alpha b_1 = a_1,$$
$$-\alpha b_0 = a_0.$$

These can be solved successively for the b_j, starting either from the top or from the bottom of the series. In either case the final equation used serves as a check; if it is not satisfied, at least one mistake has been made in the computation – or α is not a zero of $f(x) = 0$. We now illustrate this procedure with a worked example.

▶Determine by inspection the simple roots of the equation
$$f(x) = 3x^4 - x^3 - 10x^2 - 2x + 4 = 0$$
and hence, by factorisation, find the rest of its roots.

From the pattern of coefficients it can be seen that $x = -1$ is a solution to the equation. We therefore write

$$f(x) = (x+1)(b_3 x^3 + b_2 x^2 + b_1 x + b_0),$$

where

$$b_3 = 3,$$
$$b_2 + b_3 = -1,$$
$$b_1 + b_2 = -10,$$
$$b_0 + b_1 = -2,$$
$$b_0 = 4.$$

These equations give $b_3 = 3, b_2 = -4, b_1 = -6, b_0 = 4$ (check) and so

$$f(x) = (x+1)f_1(x) = (x+1)(3x^3 - 4x^2 - 6x + 4).$$

We now note that $f_1(x) = 0$ if x is set equal to 2. Thus $x - 2$ is a factor of $f_1(x)$, which therefore can be written as

$$f_1(x) = (x - 2)f_2(x) = (x - 2)(c_2 x^2 + c_1 x + c_0)$$

with

$$c_2 = 3,$$
$$c_1 - 2c_2 = -4,$$
$$c_0 - 2c_1 = -6,$$
$$-2c_0 = 4.$$

These equations determine $f_2(x)$ as $3x^2 + 2x - 2$. Since $f_2(x) = 0$ is a quadratic equation, its solutions can be written explicitly as

$$x = \frac{-1 \pm \sqrt{1 + 6}}{3}.$$

Thus the four roots of $f(x) = 0$ are $-1, 2, \frac{1}{3}(-1 + \sqrt{7})$ and $\frac{1}{3}(-1 - \sqrt{7})$. ◄

1.1.3 Properties of roots

From the fact that a polynomial equation can be written in any of the alternative forms

$$f(x) = a_n x^n + a_{n-1} x^{n-1} + \cdots + a_1 x + a_0 = 0,$$
$$f(x) = a_n (x - \alpha_1)^{m_1} (x - \alpha_2)^{m_2} \cdots (x - \alpha_r)^{m_r} = 0,$$
$$f(x) = a_n (x - \alpha_1)(x - \alpha_2) \cdots (x - \alpha_n) = 0,$$

it follows that it must be possible to express the coefficients a_i in terms of the roots α_k. To take the most obvious example, comparison of the constant terms (formally the coefficient of x^0) in the first and third expressions shows that

$$a_n(-\alpha_1)(-\alpha_2) \cdots (-\alpha_n) = a_0,$$

or, using the product notation,

$$\prod_{k=1}^{n} \alpha_k = (-1)^n \frac{a_0}{a_n}. \tag{1.12}$$

Only slightly less obvious is a result obtained by comparing the coefficients of x^{n-1} in the same two expressions of the polynomial:

$$\sum_{k=1}^{n} \alpha_k = -\frac{a_{n-1}}{a_n}. \tag{1.13}$$

Comparing the coefficients of other powers of x yields further results, though they are of less general use than the two just given. One such, which the reader may wish to derive, is

$$\sum_{j=1}^{n} \sum_{k>j}^{n} \alpha_j \alpha_k = \frac{a_{n-2}}{a_n}. \tag{1.14}$$

9

In the case of a quadratic equation these root properties are used sufficiently often that they are worth stating explicitly, as follows. If the roots of the quadratic equation $ax^2 + bx + c = 0$ are α_1 and α_2 then

$$\alpha_1 + \alpha_2 = -\frac{b}{a},$$

$$\alpha_1 \alpha_2 = \frac{c}{a}.$$

If the alternative standard form for the quadratic is used, b is replaced by $2b$ in both the equation and the first of these results.

▶*Find a cubic equation whose roots are* $-4, 3$ *and* 5.

From results (1.12) – (1.14) we can compute that, arbitrarily setting $a_3 = 1$,

$$-a_2 = \sum_{k=1}^{3} \alpha_k = 4, \qquad a_1 = \sum_{j=1}^{3} \sum_{k>j}^{3} \alpha_j \alpha_k = -17, \qquad a_0 = (-1)^3 \prod_{k=1}^{3} \alpha_k = 60.$$

Thus a possible cubic equation is $x^3 + (-4)x^2 + (-17)x + (60) = 0$. Of course, any multiple of $x^3 - 4x^2 - 17x + 60 = 0$ will do just as well. ◀

1.2 Trigonometric identities

So many of the applications of mathematics to physics and engineering are concerned with periodic, and in particular sinusoidal, behaviour that a sure and ready handling of the corresponding mathematical functions is an essential skill. Even situations with no obvious periodicity are often expressed in terms of periodic functions for the purposes of analysis. Later in this book whole chapters are devoted to developing the techniques involved, but, as a necessary prerequisite, we here establish (or remind the reader of) some standard identities with which he or she should be fully familiar, so that the manipulation of expressions containing sinusoids becomes automatic and reliable. So as to emphasise the angular nature of the argument of a sinusoid we will denote it in this section by θ rather than x.

1.2.1 Single-angle identities

We give without proof the basic identity satisfied by the sinusoidal functions $\sin \theta$ and $\cos \theta$, namely

$$\cos^2 \theta + \sin^2 \theta = 1. \tag{1.15}$$

If $\sin \theta$ and $\cos \theta$ have been defined geometrically in terms of the coordinates of a point on a circle, a reference to the name of Pythagoras will suffice to establish this result. If they have been defined by means of series (with θ expressed in radians) then the reader should refer to Euler's equation (3.23) on page 93, and note that $e^{i\theta}$ has unit modulus if θ is real.

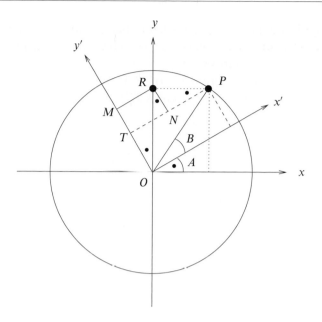

Figure 1.2 Illustration of the compound-angle identities. Refer to the main text for details.

Other standard single-angle formulae derived from (1.15) by dividing through by various powers of $\sin\theta$ and $\cos\theta$ are

$$1 + \tan^2\theta = \sec^2\theta, \tag{1.16}$$

$$\cot^2\theta + 1 = \mathrm{cosec}^2\theta. \tag{1.17}$$

1.2.2 Compound-angle identities

The basis for building expressions for the sinusoidal functions of compound angles are those for the sum and difference of just two angles, since all other cases can be built up from these, in principle. Later we will see that a study of complex numbers can provide a more efficient approach in some cases.

To prove the basic formulae for the sine and cosine of a compound angle $A + B$ in terms of the sines and cosines of A and B, we consider the construction shown in figure 1.2. It shows two sets of axes, Oxy and $Ox'y'$, with a common origin but rotated with respect to each other through an angle A. The point P lies on the unit circle centred on the common origin O and has coordinates $\cos(A + B), \sin(A + B)$ with respect to the axes Oxy and coordinates $\cos B, \sin B$ with respect to the axes $Ox'y'$.

Parallels to the axes Oxy (dotted lines) and $Ox'y'$ (broken lines) have been drawn through P. Further parallels (MR and RN) to the $Ox'y'$ axes have been

11

drawn through R, the point $(0, \sin(A+B))$ in the Oxy system. That all the angles marked with the symbol $\bullet$ are equal to A follows from the simple geometry of right-angled triangles and crossing lines.

We now determine the coordinates of P in terms of lengths in the figure, expressing those lengths in terms of both sets of coordinates:

(i) $\cos B = x' = TN + NP = MR + NP$

$$= OR \sin A + RP \cos A = \sin(A+B) \sin A + \cos(A+B) \cos A;$$

(ii) $\sin B = y' = OM - TM = OM - NR$

$$= OR \cos A - RP \sin A = \sin(A+B) \cos A - \cos(A+B) \sin A.$$

Now, if equation (i) is multiplied by $\sin A$ and added to equation (ii) multiplied by $\cos A$, the result is

$$\sin A \cos B + \cos A \sin B = \sin(A+B)(\sin^2 A + \cos^2 A) = \sin(A+B).$$

Similarly, if equation (ii) is multiplied by $\sin A$ and subtracted from equation (i) multiplied by $\cos A$, the result is

$$\cos A \cos B - \sin A \sin B = \cos(A+B)(\cos^2 A + \sin^2 A) = \cos(A+B).$$

Corresponding graphically based results can be derived for the sines and cosines of the difference of two angles; however, they are more easily obtained by setting B to $-B$ in the previous results and remembering that $\sin B$ becomes $-\sin B$ whilst $\cos B$ is unchanged. The four results may be summarised by

$$\sin(A \pm B) = \sin A \cos B \pm \cos A \sin B \tag{1.18}$$

$$\cos(A \pm B) = \cos A \cos B \mp \sin A \sin B. \tag{1.19}$$

Standard results can be deduced from these by setting one of the two angles equal to π or to $\pi/2$:

$$\sin(\pi - \theta) = \sin\theta, \quad \cos(\pi - \theta) = -\cos\theta, \tag{1.20}$$

$$\sin\left(\tfrac{1}{2}\pi - \theta\right) = \cos\theta, \quad \cos\left(\tfrac{1}{2}\pi - \theta\right) = \sin\theta. \tag{1.21}$$

From these basic results many more can be derived. An immediate deduction, obtained by taking the ratio of the two equations (1.18) and (1.19) and then dividing both the numerator and denominator of this ratio by $\cos A \cos B$, is

$$\tan(A \pm B) = \frac{\tan A \pm \tan B}{1 \mp \tan A \tan B}. \tag{1.22}$$

One application of this result is a test for whether two lines on a graph are orthogonal (perpendicular); more generally, it determines the angle between them. The standard notation for a straight-line graph is $y = mx + c$, in which m is the slope of the graph and c is its intercept on the y-axis. It should be noted that the slope m is also the tangent of the angle the line makes with the x-axis.

Consequently the angle θ_{12} between two such straight-line graphs is equal to the difference in the angles they individually make with the x-axis, and the tangent of that angle is given by (1.22):

$$\tan \theta_{12} = \frac{\tan \theta_1 - \tan \theta_2}{1 + \tan \theta_1 \tan \theta_2} = \frac{m_1 - m_2}{1 + m_1 m_2}. \tag{1.23}$$

For the lines to be orthogonal we must have $\theta_{12} = \pi/2$, i.e. the final fraction on the RHS of the above equation must equal ∞, and so

$$m_1 m_2 = -1. \tag{1.24}$$

A kind of inversion of equations (1.18) and (1.19) enables the sum or difference of two sines or cosines to be expressed as the product of two sinusoids; the procedure is typified by the following. Adding together the expressions given by (1.18) for $\sin(A + B)$ and $\sin(A - B)$ yields

$$\sin(A + B) + \sin(A - B) = 2 \sin A \cos B.$$

If we now write $A + B = C$ and $A - B = D$, this becomes

$$\sin C + \sin D = 2 \sin \left(\frac{C + D}{2} \right) \cos \left(\frac{C - D}{2} \right). \tag{1.25}$$

In a similar way each of the following equations can be derived:

$$\sin C - \sin D = 2 \cos \left(\frac{C + D}{2} \right) \sin \left(\frac{C - D}{2} \right), \tag{1.26}$$

$$\cos C + \cos D = 2 \cos \left(\frac{C + D}{2} \right) \cos \left(\frac{C - D}{2} \right), \tag{1.27}$$

$$\cos C - \cos D = -2 \sin \left(\frac{C + D}{2} \right) \sin \left(\frac{C - D}{2} \right). \tag{1.28}$$

The minus sign on the right of the last of these equations should be noted; it may help to avoid overlooking this 'oddity' to recall that if $C > D$ then $\cos C < \cos D$.

1.2.3 Double- and half-angle identities

Double-angle and half-angle identities are needed so often in practical calculations that they should be committed to memory by any physical scientist. They can be obtained by setting B equal to A in results (1.18) and (1.19). When this is done, and use made of equation (1.15), the following results are obtained:

$$\sin 2\theta = 2 \sin \theta \cos \theta, \tag{1.29}$$

$$\cos 2\theta = \cos^2 \theta - \sin^2 \theta$$

$$= 2 \cos^2 \theta - 1$$

$$= 1 - 2 \sin^2 \theta, \tag{1.30}$$

$$\tan 2\theta = \frac{2 \tan \theta}{1 - \tan^2 \theta}. \tag{1.31}$$

A further set of identities enables sinusoidal functions of θ to be expressed in terms of polynomial functions of a variable $t = \tan(\theta/2)$. They are not used in their primary role until the next chapter, but we give a derivation of them here for reference.

If $t = \tan(\theta/2)$, then it follows from (1.16) that $1+t^2 = \sec^2(\theta/2)$ and $\cos(\theta/2) = (1 + t^2)^{-1/2}$, whilst $\sin(\theta/2) = t(1 + t^2)^{-1/2}$. Now, using (1.29) and (1.30), we may write:

$$\sin \theta = 2 \sin \frac{\theta}{2} \cos \frac{\theta}{2} = \frac{2t}{1 + t^2}, \tag{1.32}$$

$$\cos \theta = \cos^2 \frac{\theta}{2} - \sin^2 \frac{\theta}{2} = \frac{1 - t^2}{1 + t^2}, \tag{1.33}$$

$$\tan \theta = \frac{2t}{1 - t^2}. \tag{1.34}$$

It can be further shown that the derivative of θ with respect to t takes the algebraic form $2/(1 + t^2)$. This completes a package of results that enables expressions involving sinusoids, particularly when they appear as integrands, to be cast in more convenient algebraic forms. The proof of the derivative property and examples of use of the above results are given in subsection 2.2.7.

We conclude this section with a worked example which is of such a commonly occurring form that it might be considered a standard procedure.

▶ *Solve for θ the equation*

$$a \sin \theta + b \cos \theta = k,$$

where a, b and k are given real quantities.

To solve this equation we make use of result (1.18) by setting $a = K \cos \phi$ and $b = K \sin \phi$ for suitable values of K and ϕ. We then have

$$k = K \cos \phi \sin \theta + K \sin \phi \cos \theta = K \sin(\theta + \phi),$$

with

$$K^2 = a^2 + b^2 \quad \text{and} \quad \phi = \tan^{-1} \frac{b}{a}.$$

Whether ϕ lies in $0 \leq \phi \leq \pi$ or in $-\pi < \phi < 0$ has to be determined by the individual signs of a and b. The solution is thus

$$\theta = \sin^{-1} \left(\frac{k}{K} \right) - \phi,$$

with K and ϕ as given above. Notice that the inverse sine yields two values in the range 0 to 2π and that there is no real solution to the original equation if $|k| > |K| = (a^2 + b^2)^{1/2}$. ◀

1.3 Coordinate geometry

We have already mentioned the standard form for a straight-line graph, namely

$$y = mx + c, \tag{1.35}$$

representing a linear relationship between the independent variable x and the dependent variable y. The slope m is equal to the tangent of the angle the line makes with the x-axis whilst c is the intercept on the y-axis.

An alternative form for the equation of a straight line is

$$ax + by + k = 0, \tag{1.36}$$

to which (1.35) is clearly connected by

$$m = -\frac{a}{b} \quad \text{and} \quad c = -\frac{k}{b}.$$

This form treats x and y on a more symmetrical basis, the intercepts on the two axes being $-k/a$ and $-k/b$ respectively.

A power relationship between two variables, i.e. one of the form $y = Ax^n$, can also be cast into straight-line form by taking the logarithms of both sides. Whilst it is normal in mathematical work to use natural logarithms (to base e, written $\ln x$), for practical investigations logarithms to base 10 are often employed. In either case the form is the same, but it needs to be remembered which has been used when recovering the value of A from fitted data. In the mathematical (base e) form, the power relationship becomes

$$\ln y = n \ln x + \ln A. \tag{1.37}$$

Now the slope gives the power n, whilst the intercept on the $\ln y$ axis is $\ln A$, which yields A, either by exponentiation or by taking antilogarithms.

The other standard coordinate forms of two-dimensional curves that students should know and recognise are those concerned with the *conic sections* – so called because they can all be obtained by taking suitable sections across a (double) cone. Because the conic sections can take many different orientations and scalings their general form is complex,

$$Ax^2 + By^2 + Cxy + Dx + Ey + F = 0, \tag{1.38}$$

but each can be represented by one of four generic forms, an ellipse, a parabola, a hyperbola or, the degenerate form, a pair of straight lines. If they are reduced to their standard representations, in which their axes of symmetry are made to coincide with the coordinate axes, the first three take the forms

$$\frac{(x-\alpha)^2}{a^2} + \frac{(y-\beta)^2}{b^2} = 1 \qquad \text{(ellipse)}, \tag{1.39}$$

$$(y-\beta)^2 = 4a(x-\alpha) \qquad \text{(parabola)}, \tag{1.40}$$

$$\frac{(x-\alpha)^2}{a^2} - \frac{(y-\beta)^2}{b^2} = 1 \qquad \text{(hyperbola)}. \tag{1.41}$$

Here, (α, β) gives the position of the 'centre' of the curve, usually taken as the origin $(0,0)$ when this does not conflict with any imposed conditions. The parabola equation given is that for a curve symmetric about a line parallel to the x-axis. For one symmetrical about a parallel to the y-axis the equation would read $(x-\alpha)^2 = 4a(y-\beta)$.

Of course, the circle is the special case of an ellipse in which $b = a$ and the equation takes the form

$$(x-\alpha)^2 + (y-\beta)^2 = a^2. \tag{1.42}$$

The distinguishing characteristic of this equation is that when it is expressed in the form (1.38) the coefficients of x^2 and y^2 are equal and that of xy is zero; this property is not changed by any reorientation or scaling and so acts to identify a general conic as a circle.

Definitions of the conic sections in terms of geometrical properties are also available; for example, a parabola can be defined as the locus of a point that is always at the same distance from a given straight line (the *directrix*) as it is from a given point (the *focus*). When these properties are expressed in Cartesian coordinates the above equations are obtained. For a circle, the defining property is that all points on the curve are a distance a from (α, β); (1.42) expresses this requirement very directly. In the following worked example we derive the equation for a parabola.

▶*Find the equation of a parabola that has the line $x = -a$ as its directrix and the point $(a, 0)$ as its focus.*

Figure 1.3 shows the situation in Cartesian coordinates. Expressing the defining requirement that PN and PF are equal in length gives

$$(x+a) = [(x-a)^2 + y^2]^{1/2} \quad \Rightarrow \quad (x+a)^2 = (x-a)^2 + y^2$$

which, on expansion of the squared terms, immediately gives $y^2 = 4ax$. This is (1.40) with α and β both set equal to zero. ◀

Although the algebra is more complicated, the same method can be used to derive the equations for the ellipse and the hyperbola. In these cases the distance from the fixed point is a definite fraction, e, known as the *eccentricity*, of the distance from the fixed line. For an ellipse $0 < e < 1$, for a circle $e = 0$, and for a hyperbola $e > 1$. The parabola corresponds to the case $e = 1$.

16

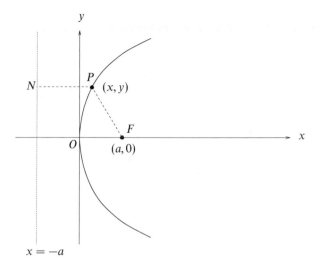

Figure 1.3   Construction of a parabola using the point $(a,0)$ as the focus and the line $x = -a$ as the directrix.

The values of a and b (with $a \geq b$) in equation (1.39) for an ellipse are related to e through

$$e^2 = \frac{a^2 - b^2}{a^2}$$

and give the lengths of the semi-axes of the ellipse. If the ellipse is centred on the origin, i.e. $\alpha = \beta = 0$, then the focus is $(-ae, 0)$ and the directrix is the line $x = -a/e$.

For each conic section curve, although we have two variables, x and y, they are not independent, since if one is given then the other can be determined. However, determining y when x is given, say, involves solving a quadratic equation on each occasion, and so it is convenient to have *parametric* representations of the curves. A parametric representation allows each point on a curve to be associated with a unique value of a *single* parameter t. The simplest parametric representations for the conic sections are as given below, though that for the hyperbola uses hyperbolic functions, not formally introduced until chapter 3. That they do give valid parameterizations can be verified by substituting them into the standard forms (1.39)–(1.41); in each case the standard form is reduced to an algebraic or trigonometric identity.

$$
\begin{aligned}
x &= \alpha + a \cos \phi, & y &= \beta + b \sin \phi & \text{(ellipse)}, \\
x &= \alpha + at^2, & y &= \beta + 2at & \text{(parabola)}, \\
x &= \alpha + a \cosh \phi, & y &= \beta + b \sinh \phi & \text{(hyperbola)}.
\end{aligned}
$$

As a final example illustrating several topics from this section we now prove the well-known result that the angle subtended by a diameter at any point on a circle is a right angle.

▶ *Taking the diameter to be the line joining $Q = (-a, 0)$ and $R = (a, 0)$ and the point P to be any point on the circle $x^2 + y^2 = a^2$, prove that angle QPR is a right angle.*

If P is the point (x, y), the slope of the line QP is

$$m_1 = \frac{y - 0}{x - (-a)} = \frac{y}{x + a}.$$

That of RP is

$$m_2 = \frac{y - 0}{x - (a)} = \frac{y}{x - a}.$$

Thus

$$m_1 m_2 = \frac{y^2}{x^2 - a^2}.$$

But, since P is on the circle, $y^2 = a^2 - x^2$ and consequently $m_1 m_2 = -1$. From result (1.24) this implies that QP and RP are orthogonal and that QPR is therefore a right angle. Note that this is true for *any* point P on the circle. ◀

1.4 Partial fractions

In subsequent chapters, and in particular when we come to study integration in chapter 2, we will need to express a function $f(x)$ that is the ratio of two polynomials in a more manageable form. To remove some potential complexity from our discussion we will assume that all the coefficients in the polynomials are real, although this is not an essential simplification.

The behaviour of $f(x)$ is crucially determined by the location of the zeros of its denominator, i.e. if $f(x)$ is written as $f(x) = g(x)/h(x)$ where both $g(x)$ and $h(x)$ are polynomials,[§] then $f(x)$ changes extremely rapidly when x is close to those values α_i that are the roots of $h(x) = 0$. To make such behaviour explicit, we write $f(x)$ as a sum of terms such as $A/(x - \alpha)^n$, in which A is a constant, α is one of the α_i that satisfy $h(\alpha_i) = 0$ and n is a positive integer. Writing a function in this way is known as expressing it in *partial fractions*.

Suppose, for the sake of definiteness, that we wish to express the function

$$f(x) = \frac{4x + 2}{x^2 + 3x + 2}$$

[§] It is assumed that the ratio has been reduced so that $g(x)$ and $h(x)$ do not contain any common factors, i.e. there is no value of x that makes both vanish at the same time. We may also assume without any loss of generality that the coefficient of the highest power of x in $h(x)$ has been made equal to unity by, if necessary, dividing both numerator and denominator by the coefficient of this highest power.

in partial fractions, i.e. to write it as

$$f(x) = \frac{g(x)}{h(x)} = \frac{4x+2}{x^2+3x+2} = \frac{A_1}{(x-\alpha_1)^{n_1}} + \frac{A_2}{(x-\alpha_2)^{n_2}} + \cdots.$$
(1.43)

The first question that arises is that of how many terms there should be on the right-hand side (RHS). Although some complications occur when $h(x)$ has repeated roots (these are considered below) it is clear that $f(x)$ only becomes infinite at the *two* values of x, α_1 and α_2, that make $h(x) = 0$. Consequently the RHS can only become infinite at the same two values of x and therefore contains only two partial fractions – these are the ones shown explicitly. This argument can be trivially extended (again temporarily ignoring the possibility of repeated roots of $h(x)$) to show that if $h(x)$ is a polynomial of degree n then there should be n terms on the RHS, each containing a different root α_i of the equation $h(\alpha_i) = 0$.

A second general question concerns the appropriate values of the n_i. This is answered by putting the RHS over a common denominator, which will clearly have to be the product $(x-\alpha_1)^{n_1}(x-\alpha_2)^{n_2}\cdots$. Comparison of the highest power of x in this new RHS with the same power in $h(x)$ shows that $n_1 + n_2 + \cdots = n$. This result holds whether or not $h(x) = 0$ has repeated roots and, although we do not give a rigorous proof, strongly suggests the following correct conclusions.

- The number of terms on the RHS is equal to the number of *distinct* roots of $h(x) = 0$, each term having a different root α_i in its denominator $(x-\alpha_i)^{n_i}$.
- If α_i is a multiple root of $h(x) = 0$ then the value to be assigned to n_i in (1.43) is that of m_i when $h(x)$ is written in the product form (1.9). Further, as discussed on p. 23, A_i has to be replaced by a polynomial of degree $m_i - 1$. This is also formally true for non-repeated roots, since then both m_i and n_i are equal to unity.

Returning to our specific example we note that the denominator $h(x)$ has zeros at $x = \alpha_1 = -1$ and $x = \alpha_2 = -2$; these x-values are the simple (non-repeated) roots of $h(x) = 0$. Thus the partial fraction expansion will be of the form

$$\frac{4x+2}{x^2+3x+2} = \frac{A_1}{x+1} + \frac{A_2}{x+2}.$$
(1.44)

We now list several methods available for determining the coefficients A_1 and A_2. We also remind the reader that, as with all the explicit examples and techniques described, these methods are to be considered as models for the handling of any ratio of polynomials, with or without characteristics that make it a special case.

(i) The RHS can be put over a common denominator, in this case $(x+1)(x+2)$, and then the coefficients of the various powers of x can be equated in the numerators on both sides of the equation. This leads to

$$4x + 2 = A_1(x + 2) + A_2(x + 1),$$
$$4 = A_1 + A_2 \qquad 2 = 2A_1 + A_2.$$

Solving the simultaneous equations for A_1 and A_2 gives $A_1 = -2$ and $A_2 = 6$.

(ii) A second method is to substitute two (or more generally n) different values of x into each side of (1.44) and so obtain two (or n) simultaneous equations for the two (or n) constants A_i. To justify this practical way of proceeding it is necessary, strictly speaking, to appeal to method (i) above, which establishes that there are unique values for A_1 and A_2 valid for all values of x. It is normally very convenient to take zero as one of the values of x, but of course any set will do. Suppose in the present case that we use the values $x = 0$ and $x = 1$ and substitute in (1.44). The resulting equations are

$$\frac{2}{2} = \frac{A_1}{1} + \frac{A_2}{2},$$
$$\frac{6}{6} = \frac{A_1}{2} + \frac{A_2}{3},$$

which on solution give $A_1 = -2$ and $A_2 = 6$, as before. The reader can easily verify that any other pair of values for x (except for a pair that includes α_1 or α_2) gives the same values for A_1 and A_2.

(iii) The very reason why method (ii) fails if x is chosen as one of the roots α_i of $h(x) = 0$ can be made a basis for determining the values of the A_i corresponding to non-multiple roots, without having to solve simultaneous equations. The method is conceptually more difficult than the other methods presented here, and needs results from the theory of complex variables (chapter 24) to justify it. However, we give a practical 'cookbook' recipe for determining the coefficients.

   (a) To determine the coefficient A_k, imagine the denominator $h(x)$ written as the product $(x - \alpha_1)(x - \alpha_2) \cdots (x - \alpha_n)$, with any m-fold repeated root giving rise to m factors in parentheses.
   (b) Now set x equal to α_k and evaluate the expression obtained after omitting the factor that reads $\alpha_k - \alpha_k$.
   (c) Divide the value so obtained into $g(\alpha_k)$; the result is the required coefficient A_k.

For our specific example we find that in step (a) that $h(x) = (x + 1)(x + 2)$ and that, when evaluating A_1, step (b) yields $-1 + 2$, i.e. 1. Since $g(-1) = 4(-1) + 2 = -2$, step (c) gives A_1 as $(-2)/(1)$, i.e in agreement with our other evaluations. In a similar way A_2 is evaluated as $(-6)/(-1) = 6$.

Thus any one of the methods listed above shows that

$$\frac{4x+2}{x^2+3x+2} = \frac{-2}{x+1} + \frac{6}{x+2}.$$

The best method to use in any particular circumstance will depend on the complexity, in terms of the degrees of the polynomials and the multiplicities of the roots of the denominator, of the function being considered and, to some extent, on the individual inclinations of the student; some prefer lengthy but straightforward solution of simultaneous equations, whilst others feel more at home carrying through shorter but more abstract calculations in their heads.

1.4.1 Complications and special cases

Having established the basic method for partial fractions, we now show, through further worked examples, how some complications are dealt with by extensions to the procedure. These extensions are introduced one at a time, but of course in any practical application more than one may be involved.

The degree of the numerator is greater than or equal to that of the denominator

Although we have not specifically mentioned the fact, it will be apparent from trying to apply method (i) of the previous subsection to such a case, that if the degree of the numerator (m) is not less than that of the denominator (n) then the ratio of two polynomials cannot be expressed in partial fractions.

To get round this difficulty it is necessary to start by dividing the denominator $h(x)$ into the numerator $g(x)$ to obtain a further polynomial, which we will denote by $s(x)$, together with a function $t(x)$ that *is* a ratio of two polynomials for which the degree of the numerator is less than that of the denominator. The function $t(x)$ *can* therefore be expanded in partial fractions. As a formula,

$$f(x) = \frac{g(x)}{h(x)} = s(x) + t(x) \equiv s(x) + \frac{r(x)}{h(x)}. \tag{1.45}$$

It is apparent that the polynomial $r(x)$ is the *remainder* obtained when $g(x)$ is divided by $h(x)$, and, in general, will be a polynomial of degree $n-1$. It is also clear that the polynomial $s(x)$ will be of degree $m-n$. The actual division process can be set out as an algebraic long division sum but is probably more easily handled by writing (1.45) in the form

$$g(x) = s(x)h(x) + r(x) \tag{1.46}$$

or, more explicitly, as

$$g(x) = (s_{m-n}x^{m-n} + s_{m-n-1}x^{m-n-1} + \cdots + s_0)h(x) + (r_{n-1}x^{n-1} + r_{n-2}x^{n-2} + \cdots + r_0) \tag{1.47}$$

and then equating coefficients.

We illustrate this procedure with the following worked example.

▶*Find the partial fraction decomposition of the function*
$$f(x) = \frac{x^3 + 3x^2 + 2x + 1}{x^2 - x - 6}.$$

Since the degree of the numerator is 3 and that of the denominator is 2, a preliminary long division is necessary. The polynomial $s(x)$ resulting from the division will have degree $3 - 2 = 1$ and the remainder $r(x)$ will be of degree $2 - 1 = 1$ (or less). Thus we write

$$x^3 + 3x^2 + 2x + 1 = (s_1 x + s_0)(x^2 - x - 6) + (r_1 x + r_0).$$

From equating the coefficients of the various powers of x on the two sides of the equation, starting with the highest, we now obtain the simultaneous equations

$$1 = s_1,$$
$$3 = s_0 - s_1,$$
$$2 = -s_0 - 6s_1 + r_1,$$
$$1 = -6s_0 + r_0.$$

These are readily solved, in the given order, to yield $s_1 = 1$, $s_0 = 4$, $r_1 = 12$ and $r_0 = 25$. Thus $f(x)$ can be written as

$$f(x) = x + 4 + \frac{12x + 25}{x^2 - x - 6}.$$

The last term can now be decomposed into partial fractions as previously. The zeros of the denominator are at $x = 3$ and $x = -2$ and the application of any method from the previous subsection yields the respective constants as $A_1 = 12\frac{1}{5}$ and $A_2 = -\frac{1}{5}$. Thus the final partial fraction decomposition of $f(x)$ is

$$x + 4 + \frac{61}{5(x - 3)} - \frac{1}{5(x + 2)}. ◀$$

Factors of the form $a^2 + x^2$ in the denominator

We have so far assumed that the roots of $h(x) = 0$, needed for the factorisation of the denominator of $f(x)$, can always be found. In principle they always can but in some cases they are not real. Consider, for example, attempting to express in partial fractions a polynomial ratio whose denominator is $h(x) = x^3 - x^2 + 2x - 2$. Clearly $x = 1$ is a zero of $h(x)$, and so a first factorisation is $(x - 1)(x^2 + 2)$. However we cannot make any further progress because the factor $x^2 + 2$ cannot be expressed as $(x - \alpha)(x - \beta)$ for any real α and β.

Complex numbers are introduced later in this book (chapter 3) and, when the reader has studied them, he or she may wish to justify the procedure set out below. It can be shown to be equivalent to that already given, but the zeros of $h(x)$ are now allowed to be complex and terms that are complex conjugates of each other are combined to leave only real terms.

Since quadratic factors of the form $a^2 + x^2$ that appear in $h(x)$ cannot be reduced to the product of two linear factors, partial fraction expansions including them need to have numerators in the corresponding terms that are not simply constants

A_i but linear functions of x, i.e. of the form $B_i x + C_i$. Thus, in the expansion, linear terms (first-degree polynomials) in the denominator have constants (zero-degree polynomials) in their numerators, whilst quadratic terms (second-degree polynomials) in the denominator have linear terms (first-degree polynomials) in their numerators. As a symbolic formula, the partial fraction expansion of

$$\frac{g(x)}{(x - \alpha_1)(x - \alpha_2) \cdots (x - \alpha_p)(x^2 + a_1^2)(x^2 + a_2^2) \cdots (x^2 + a_q^2)}$$

should take the form

$$\frac{A_1}{x - \alpha_1} + \frac{A_2}{x - \alpha_2} + \cdots + \frac{A_p}{x - \alpha_p} + \frac{B_1 x + C_1}{x^2 + a_1^2} + \frac{B_2 x + C_2}{x^2 + a_2^2} + \cdots + \frac{B_q x + C_q}{x^2 + a_q^2}.$$

Of course, the degree of $g(x)$ must be less than $p + 2q$; if it is not, an initial division must be carried out as demonstrated earlier.

Repeated factors in the denominator

Consider trying (incorrectly) to expand

$$f(x) = \frac{x - 4}{(x + 1)(x - 2)^2}$$

in partial fraction form as follows:

$$\frac{x - 4}{(x + 1)(x - 2)^2} = \frac{A_1}{x + 1} + \frac{A_2}{(x - 2)^2}.$$

Multiplying both sides of this supposed equality by $(x + 1)(x - 2)^2$ produces an equation whose LHS is linear in x, whilst its RHS is quadratic. This is clearly wrong and so an expansion in the above form cannot be valid. The correction we must make is very similar to that needed in the previous subsection, namely that since $(x - 2)^2$ is a quadratic polynomial the numerator of the term containing it must be a first-degree polynomial, and not simply a constant.

The correct form for the part of the expansion containing the repeated root is therefore $(Bx + C)/(x - 2)^2$. Using this form and either of methods (i) and (ii) for determining the constants gives the full partial fraction expansion as

$$\frac{x - 4}{(x + 1)(x - 2)^2} = -\frac{5}{9(x + 1)} + \frac{5x - 16}{9(x - 2)^2},$$

as the reader may verify.

Since any term of the form $(Bx + C)/(x - \alpha)^2$ can be written as

$$\frac{B(x - \alpha) + C + B\alpha}{(x - \alpha)^2} = \frac{B}{x - \alpha} + \frac{C + B\alpha}{(x - \alpha)^2},$$

and similarly for multiply repeated roots, an alternative form for the part of the partial fraction expansion containing a repeated root α is

$$\frac{D_1}{x - \alpha} + \frac{D_2}{(x - \alpha)^2} + \cdots + \frac{D_p}{(x - \alpha)^p}. \tag{1.48}$$

In this form, all x-dependence has disappeared from the numerators but at the expense of $p-1$ additional terms; the total number of constants to be determined remains unchanged, as it must.

When describing possible methods of determining the constants in a partial fraction expansion, we noted that method (iii), p. 20, which avoids the need to solve simultaneous equations, is restricted to terms involving non-repeated roots. In fact, it can be applied in repeated-root situations, when the expansion is put in the form (1.48), but only to find the constant in the term involving the largest inverse power of $x - \alpha$, i.e. D_p in (1.48).

We conclude this section with a more protracted worked example that contains all three of the complications discussed.

▶ *Resolve the following expression $F(x)$ into partial fractions:*

$$F(x) = \frac{x^5 - 2x^4 - x^3 + 5x^2 - 46x + 100}{(x^2 + 6)(x - 2)^2}.$$

We note that the degree of the denominator (4) is not greater than that of the numerator (5), and so we must start by dividing the latter by the former. It follows, from the difference in degrees and the coefficients of the highest powers in each, that the result will be a linear expression $s_1 x + s_0$ with the coefficient s_1 equal to 1. Thus the numerator of $F(x)$ must be expressible as

$$(x + s_0)(x^4 - 4x^3 + 10x^2 - 24x + 24) + (r_3 x^3 + r_2 x^2 + r_1 x + r_0),$$

where the second factor in parentheses is the denominator of $F(x)$ written as a polynomial. Equating the coefficients of x^4 gives $-2 = -4 + s_0$ and fixes s_0 as 2. Equating the coefficients of powers less than 4 gives equations involving the coefficients r_i as follows:

$$-1 = -8 + 10 + r_3,$$
$$5 = -24 + 20 + r_2,$$
$$-46 = 24 - 48 + r_1,$$
$$100 = 48 + r_0.$$

Thus the remainder polynomial $r(x)$ can be constructed and $F(x)$ written as

$$F(x) = x + 2 + \frac{-3x^3 + 9x^2 - 22x + 52}{(x^2 + 6)(x - 2)^2} \equiv x + 2 + f(x).$$

The polynomial ratio $f(x)$ can now be expressed in partial fraction form, noting that its denominator contains both a term of the form $x^2 + a^2$ and a repeated root. Thus

$$f(x) = \frac{Bx + C}{x^2 + 6} + \frac{D_1}{x - 2} + \frac{D_2}{(x - 2)^2}.$$

We could now put the RHS of this equation over the common denominator $(x^2 + 6)(x - 2)^2$ and find B, C, D_1 and D_2 by equating coefficients of powers of x. It is quicker, however, to use methods (iii) and (ii). Method (iii) gives D_2 as $(-24 + 36 - 44 + 52)/(4 + 6) = 2$. We choose to evaluate the other coefficients by method (ii), and setting $x = 0$, $x = 1$ and

$x = -1$ gives respectively

$$\frac{52}{24} = \frac{C}{6} - \frac{D_1}{2} + \frac{2}{4},$$

$$\frac{36}{7} = \frac{B+C}{7} - D_1 + 2,$$

$$\frac{86}{63} = \frac{C-B}{7} - \frac{D_1}{3} + \frac{2}{9}.$$

These equations reduce to

$$4C - 12D_1 = 40,$$
$$B + C - 7D_1 = 22,$$
$$-9B + 9C - 21D_1 = 72,$$

with solution $B = 0$, $C = 1$, $D_1 = -3$.

Thus, finally, we may rewrite the original expression $F(x)$ in partial fractions as

$$F(x) = x + 2 + \frac{1}{x^2 + 6} - \frac{3}{x-2} + \frac{2}{(x-2)^2}. \quad \blacktriangleleft$$

1.5 Binomial expansion

Earlier in this chapter we were led to consider functions containing powers of the sum or difference of two terms, e.g. $(x - \alpha)^m$. Later in this book we will find numerous occasions on which we wish to write such a product of repeated factors as a polynomial in x or, more generally, as a sum of terms each of which contains powers of x and α separately, as opposed to a power of their sum or difference.

To make the discussion general and the result applicable to a wide variety of situations, we will consider the general expansion of $f(x) = (x+y)^n$, where x and y may stand for constants, variables or functions and, for the time being, n is a positive integer. It may not be obvious what form the general expansion takes but some idea can be obtained by carrying out the multiplication explicitly for small values of n. Thus we obtain successively

$$(x + y)^1 = x + y,$$
$$(x + y)^2 = (x + y)(x + y) = x^2 + 2xy + y^2,$$
$$(x + y)^3 = (x + y)(x^2 + 2xy + y^2) = x^3 + 3x^2y + 3xy^2 + y^3,$$
$$(x + y)^4 = (x + y)(x^3 + 3x^2y + 3xy^2 + y^3) = x^4 + 4x^3y + 6x^2y^2 + 4xy^3 + y^4.$$

This does not *establish* a general formula, but the regularity of the terms in the expansions and the suggestion of a pattern in the coefficients indicate that a general formula for power n will have $n + 1$ terms, that the powers of x and y in every term will add up to n and that the coefficients of the first and last terms will be unity whilst those of the second and penultimate terms will be n.

In fact, the general expression, the *binomial expansion* for power n, is given by

$$(x+y)^n = \sum_{k=0}^{k=n} {}^nC_k x^{n-k} y^k, \tag{1.49}$$

where ${}^nC_k$ is called the *binomial coefficient* and is expressed in terms of factorial functions by $n!/[k!(n-k)!]$. Clearly, simply to make such a statement does not constitute proof of its validity, but, as we will see in subsection 1.5.2, (1.49) can be *proved* using a method called induction. Before turning to that proof, we investigate some of the elementary properties of the binomial coefficients.

1.5.1 Binomial coefficients

As stated above, the binomial coefficients are defined by

$$ {}^nC_k \equiv \frac{n!}{k!(n-k)!} \equiv \binom{n}{k} \qquad \text{for } 0 \le k \le n, \tag{1.50}$$

where in the second identity we give a common alternative notation for ${}^nC_k$. Obvious properties include

(i) ${}^nC_0 = {}^nC_n = 1$,
(ii) ${}^nC_1 = {}^nC_{n-1} = n$,
(iii) ${}^nC_k = {}^nC_{n-k}$.

We note that, for any given n, the largest coefficient in the binomial expansion is the middle one ($k = n/2$) if n is even; the middle two coefficients ($k = \frac{1}{2}(n \pm 1)$) are equal largest if n is odd. Somewhat less obvious is the result

$$\begin{aligned}
{}^nC_k + {}^nC_{k-1} &= \frac{n!}{k!(n-k)!} + \frac{n!}{(k-1)!(n-k+1)!} \\
&= \frac{n![(n+1-k)+k]}{k!(n+1-k)!} \\
&= \frac{(n+1)!}{k!(n+1-k)!} = {}^{n+1}C_k.
\end{aligned} \tag{1.51}$$

An equivalent statement, in which k has been redefined as $k+1$, is

$$ {}^nC_k + {}^nC_{k+1} = {}^{n+1}C_{k+1}. \tag{1.52}$$

1.5.2 Proof of the binomial expansion

We are now in a position to *prove* the binomial expansion (1.49). In doing so, we introduce the reader to a procedure applicable to certain types of problems and known as the *method of induction*. The method is discussed much more fully in subsection 1.7.1.

We start by *assuming* that (1.49) is true for some positive integer $n = N$. We now proceed to show that this implies that it must also be true for $n = N+1$, as follows:

$$(x + y)^{N+1} = (x + y) \sum_{k=0}^{N} {}^N C_k x^{N-k} y^k$$

$$= \sum_{k=0}^{N} {}^N C_k x^{N+1-k} y^k + \sum_{k=0}^{N} {}^N C_k x^{N-k} y^{k+1}$$

$$= \sum_{k=0}^{N} {}^N C_k x^{N+1-k} y^k + \sum_{j=1}^{N+1} {}^N C_{j-1} x^{(N+1)-j} y^j,$$

where in the first line we have used the assumption and in the third line have moved the second summation index by unity, by writing $k + 1 = j$. We now separate off the first term of the first sum, ${}^N C_0 x^{N+1}$, and write it as ${}^{N+1} C_0 x^{N+1}$; we can do this since, as noted in (i) following (1.50), ${}^n C_0 = 1$ for every n. Similarly, the last term of the second summation can be replaced by ${}^{N+1} C_{N+1} y^{N+1}$.

The remaining terms of each of the two summations are now written together, with the summation index denoted by k in both terms. Thus

$$(x + y)^{N+1} = {}^{N+1} C_0 x^{N+1} + \sum_{k=1}^{N} \left({}^N C_k + {}^N C_{k-1} \right) x^{(N+1)-k} y^k + {}^{N+1} C_{N+1} y^{N+1}$$

$$= {}^{N+1} C_0 x^{N+1} + \sum_{k-1}^{N} {}^{N+1} C_k x^{(N+1)-k} y^k + {}^{N+1} C_{N+1} y^{N+1}$$

$$= \sum_{k=0}^{N+1} {}^{N+1} C_k x^{(N+1)-k} y^k.$$

In going from the first to the second line we have used result (1.51). Now we observe that the final overall equation is just the original assumed result (1.49) but with $n = N + 1$. Thus it has been shown that if the binomial expansion is *assumed* to be true for $n = N$, then it can be *proved* to be true for $n = N + 1$. But it holds trivially for $n = 1$, and therefore for $n = 2$ also. By the same token it is valid for $n = 3, 4, \ldots$, and hence is established for all positive integers n.

1.6 Properties of binomial coefficients

1.6.1 Identities involving binomial coefficients

There are many identities involving the binomial coefficients that can be derived directly from their definition, and yet more that follow from their appearance in the binomial expansion. Only the most elementary ones, given earlier, are worth committing to memory but, as illustrations, we now derive two results involving sums of binomial coefficients.

The first is a further application of the method of induction. Consider the proposal that, for any $n \geq 1$ and $k \geq 0$,

$$\sum_{s=0}^{n-1} {}^{k+s}C_k = {}^{n+k}C_{k+1}.$$
(1.53)

Notice that here n, the number of terms in the sum, is the parameter that varies, k is a fixed parameter, whilst s is a summation index and does not appear on the RHS of the equation.

Now we suppose that the statement (1.53) about the value of the sum of the binomial coefficients ${}^kC_k, {}^{k+1}C_k, \ldots, {}^{k+n-1}C_k$ is true for $n = N$. We next write down a series with an extra term and determine the implications of the supposition for the new series:

$$\sum_{s=0}^{N+1-1} {}^{k+s}C_k = \sum_{s=0}^{N-1} {}^{k+s}C_k + {}^{k+N}C_k$$
$$= {}^{N+k}C_{k+1} + {}^{N+k}C_k$$
$$= {}^{N+k+1}C_{k+1}.$$

But this is just proposal (1.53) with n now set equal to $N + 1$. To obtain the last line, we have used (1.52), with n set equal to $N + k$.

It only remains to consider the case $n = 1$, when the summation only contains one term and (1.53) reduces to

$${}^kC_k = {}^{1+k}C_{k+1}.$$

This is trivially valid for any k since both sides are equal to unity, thus completing the proof of (1.53) for all positive integers n.

The second result, which gives a formula for combining terms from two sets of binomial coefficients in a particular way (a kind of 'convolution', for readers who are already familiar with this term), is derived by applying the binomial expansion directly to the identity

$$(x + y)^p (x + y)^q \equiv (x + y)^{p+q}.$$

Written in terms of binomial expansions, this reads

$$\sum_{s=0}^{p} {}^pC_s x^{p-s} y^s \sum_{t=0}^{q} {}^qC_t x^{q-t} y^t = \sum_{r=0}^{p+q} {}^{p+q}C_r x^{p+q-r} y^r.$$

We now equate coefficients of $x^{p+q-r} y^r$ on the two sides of the equation, noting that on the LHS all combinations of s and t such that $s + t = r$ contribute. This gives as an identity that

$$\sum_{t=0}^{r} {}^pC_{r-t} {}^qC_t = {}^{p+q}C_r = \sum_{t=0}^{r} {}^pC_t {}^qC_{r-t}.$$
(1.54)

28

We have specifically included the second equality to emphasise the symmetrical nature of the relationship with respect to p and q.

Further identities involving the coefficients can be obtained by giving x and y special values in the defining equation (1.49) for the expansion. If both are set equal to unity then we obtain (using the alternative notation so as to produce familiarity with it)

$$\binom{n}{0} + \binom{n}{1} + \binom{n}{2} + \cdots + \binom{n}{n} = 2^n, \tag{1.55}$$

whilst setting $x = 1$ and $y = -1$ yields

$$\binom{n}{0} - \binom{n}{1} + \binom{n}{2} - \cdots + (-1)^n \binom{n}{n} = 0. \tag{1.56}$$

1.6.2 Negative and non-integral values of n

Up till now we have restricted n in the binomial expansion to be a positive integer. Negative values can be accommodated, but only at the cost of an infinite series of terms rather than the finite one represented by (1.49). For reasons that are intuitively sensible and will be discussed in more detail in chapter 4, very often we require an expansion in which, at least ultimately, successive terms in the infinite series decrease in magnitude. For this reason, if $|x| > |y|$ we consider $(x + y)^{-m}$, where m itself is a positive integer, in the form

$$(x + y)^n = (x + y)^{-m} = x^{-m} \left(1 + \frac{y}{x} \right)^{-m}.$$

Since the ratio y/x is less than unity, terms containing higher powers of it will be small in magnitude, whilst raising the unit term to any power will not affect its magnitude. If $|y| > |x|$ the roles of the two must be interchanged.

We can now state, but will not explicitly prove, the form of the binomial expansion appropriate to negative values of n (n equal to $-m$):

$$(x + y)^n = (x + y)^{-m} = x^{-m} \sum_{k=0}^{\infty} {}^{-m}C_k \left(\frac{y}{x} \right)^k, \tag{1.57}$$

where the hitherto undefined quantity ${}^{-m}C_k$, which appears to involve factorials of negative numbers, is given by

$$ {}^{-m}C_k = (-1)^k \frac{m(m+1)\cdots(m+k-1)}{k!} = (-1)^k \frac{(m+k-1)!}{(m-1)!k!} = (-1)^k \, {}^{m+k-1}C_k. \tag{1.58}$$

The binomial coefficient on the extreme right of this equation has its normal meaning and is well defined since $m + k - 1 \geq k$.

Thus we have a definition of binomial coefficients for negative integer values of n in terms of those for positive n. The connection between the two may not be obvious, but they are both formed in the same way in terms of recurrence relations. Whatever the sign of n, the series of coefficients ^nC_k can be generated by starting with $^nC_0 = 1$ and using the recurrence relation

$$^nC_{k+1} = \frac{n-k}{k+1}\,^nC_k. \tag{1.59}$$

The difference is that for positive integer n the series terminates when $k = n$, whereas for negative n there is no such termination – in line with the infinite series of terms in the corresponding expansion.

Finally we note that, in fact, equation (1.59) generates the appropriate coefficients for all values of n, positive or negative, integer or non-integer, with the obvious exception of the case in which $x = -y$ and n is negative. For non-integer n the expansion does not terminate, even if n is positive.

1.7 Some particular methods of proof

Much of the mathematics used by physicists and engineers is concerned with obtaining a particular value, formula or function from a given set of data and stated conditions. However, just as it is essential in physics to formulate the basic laws and so be able to set boundaries on what can or cannot happen, so it is important in mathematics to be able to state general propositions about the outcomes that are or are not possible. To this end one attempts to establish theorems that state in as general a way as possible mathematical results that apply to particular types of situation. We conclude this introductory chapter by describing two methods that can sometimes be used to prove particular classes of theorems.

The two general methods of proof are known as proof by induction (which has already been met in this chapter) and proof by contradiction. They share the common characteristic that at an early stage in the proof an assumption is made that a particular (unproven) statement is true; the consequences of that assumption are then explored. In an inductive proof the conclusion is reached that the assumption is self-consistent and has other equally consistent but broader implications, which are then applied to establish the general validity of the assumption. A proof by contradiction, however, establishes an internal inconsistency and thus shows that the assumption is unsustainable; the natural consequence of this is that the negative of the assumption is established as true.

Later in this book use will be made of these methods of proof to explore new territory, e.g. to examine the properties of vector spaces, matrices and groups. However, at this stage we will draw our illustrative and test examples from earlier sections of this chapter and other topics in elementary algebra and number theory.

1.7.1 Proof by induction

The proof of the binomial expansion given in subsection 1.5.2 and the identity established in subsection 1.6.1 have already shown the way in which an inductive proof is carried through. They also indicated the main limitation of the method, namely that only an initially supposed result can be proved. Thus the method of induction is of no use for *deducing* a previously unknown result; a putative equation or result has to be arrived at by some other means, usually by noticing patterns or by trial and error using simple values of the variables involved. It will also be clear that propositions that can be proved by induction are limited to those containing a parameter that takes a range of integer values (usually infinite).

For a proposition involving a parameter n, the five steps in a proof using induction are as follows.

(i) Formulate the supposed result for general n.

(ii) Suppose (i) to be true for $n = N$ (or more generally for all values of $n \leq N$; see below), where N is restricted to lie in the stated range.

(iii) Show, using only proven results and supposition (ii), that proposition (i) is true for $n = N + 1$.

(iv) Demonstrate directly, and without any assumptions, that proposition (i) is true when n takes the lowest value in its range.

(v) It then follows from (iii) and (iv) that the proposition is valid for all values of n in the stated range.

It should be noted that, although many proofs at stage (iii) require the validity of the proposition only for $n = N$, some require it for all n less than or equal to N – hence the form of inequality given in parentheses in the stage (ii) assumption.

To illustrate further the method of induction, we now apply it to two worked examples; the first concerns the sum of the squares of the first n natural numbers.

> ▶ *Prove that the sum of the squares of the first n natural numbers is given by*
>
> $$\sum_{r=1}^{n} r^2 = \tfrac{1}{6}n(n+1)(2n+1). \qquad (1.60)$$

As previously we start by assuming the result is true for $n = N$. Then it follows that

$$\begin{aligned}
\sum_{r=1}^{N+1} r^2 &= \sum_{r=1}^{N} r^2 + (N+1)^2 \\
&= \tfrac{1}{6}N(N+1)(2N+1) + (N+1)^2 \\
&= \tfrac{1}{6}(N+1)[N(2N+1) + 6N + 6] \\
&= \tfrac{1}{6}(N+1)[(2N+3)(N+2)] \\
&= \tfrac{1}{6}(N+1)[(N+1)+1][2(N+1)+1].
\end{aligned}$$

This is precisely the original assumption, but with N replaced by $N + 1$. To complete the proof we only have to verify (1.60) for $n = 1$. This is trivially done and establishes the result for all positive n. The same and related results are obtained by a different method in subsection 4.2.5. ◀

Our second example is somewhat more complex and involves two nested proofs by induction: whilst trying to establish the main result by induction, we find that we are faced with a second proposition which itself requires an inductive proof.

> ▶Show that $Q(n) = n^4 + 2n^3 + 2n^2 + n$ is divisible by 6 (without remainder) for all positive integer values of n.

Again we start by assuming the result is true for some particular value N of n, whilst noting that it is trivially true for $n = 0$. We next examine $Q(N + 1)$, writing each of its terms as a binomial expansion:

$$Q(N + 1) = (N + 1)^4 + 2(N + 1)^3 + 2(N + 1)^2 + (N + 1)$$
$$= (N^4 + 4N^3 + 6N^2 + 4N + 1) + 2(N^3 + 3N^2 + 3N + 1)$$
$$+ 2(N^2 + 2N + 1) + (N + 1)$$
$$= (N^4 + 2N^3 + 2N^2 + N) + (4N^3 + 12N^2 + 14N + 6).$$

Now, by our assumption, the group of terms within the first parentheses in the last line is divisible by 6 and clearly so are the terms $12N^2$ and 6 within the second parentheses. Thus it comes down to deciding whether $4N^3 + 14N$ is divisible by 6 – or equivalently, whether $R(N) = 2N^3 + 7N$ is divisible by 3.

To settle this latter question we try using a second inductive proof and assume that $R(N)$ is divisible by 3 for $N = M$, whilst again noting that the proposition is trivially true for $N = M = 0$. This time we examine $R(M + 1)$:

$$R(M + 1) = 2(M + 1)^3 + 7(M + 1)$$
$$= 2(M^3 + 3M^2 + 3M + 1) + 7(M + 1)$$
$$= (2M^3 + 7M) + 3(2M^2 + 2M + 3)$$

By assumption, the first group of terms in the last line is divisible by 3 and the second group is patently so. We thus conclude that $R(N)$ is divisible by 3 for all $N \geq M$, and taking $M = 0$ shows that it is divisible by 3 for all N.

We can now return to the main proposition and conclude that since $R(N) = 2N^3 + 7N$ is divisible by 3, $4N^3 + 12N^2 + 14N + 6$ is divisible by 6. This in turn establishes that the divisibility of $Q(N + 1)$ by 6 follows from the assumption that $Q(N)$ divides by 6. Since $Q(0)$ clearly divides by 6, the proposition in the question is established for all values of n. ◀

1.7.2 Proof by contradiction

The second general line of proof, but again one that is normally only useful when the result is already suspected, is proof by contradiction. The questions it can attempt to answer are only those that can be expressed in a proposition that is either true or false. Clearly, it could be argued that any mathematical result can be so expressed but, if the proposition is no more than a guess, the chances of success are negligible. Valid propositions containing even modest formulae are either the result of true inspiration or, much more normally, yet another reworking of an old chestnut!

The essence of the method is to exploit the fact that mathematics is required to be self-consistent, so that, for example, two calculations of the same quantity, starting from the same given data but proceeding by different methods, must give the same answer. Equally, it must not be possible to follow a line of reasoning and draw a conclusion that contradicts either the input data or any other conclusion based upon the same data.

It is this requirement on which the method of proof by contradiction is based. The crux of the method is to assume that the proposition to be proved is *not* true, and then use this incorrect assumption and 'watertight' reasoning to draw a conclusion that contradicts the assumption. The only way out of the self-contradiction is then to conclude that the assumption was indeed false and therefore that the proposition is true.

It must be emphasised that once a (false) contrary assumption has been made, every subsequent conclusion in the argument *must* follow of necessity. Proof by contradiction fails if at any stage we have to admit 'this may or may not be the case'. That is, each step in the argument must be a *necessary* consequence of results that precede it (taken together with the assumption), rather than simply a *possible* consequence.

It should also be added that if no contradiction can be found using sound reasoning based on the assumption then no conclusion can be drawn about either the proposition or its negative and some other approach must be tried.

We illustrate the general method with an example in which the mathematical reasoning is straightforward, so that attention can be focussed on the structure of the proof.

> ►*A rational number r is a fraction $r = p/q$ in which p and q are integers with q positive. Further, r is expressed in its lowest terms, any integer common factor of p and q having been divided out.*
> *Prove that the square root of an integer m cannot be a rational number, unless the square root itself is an integer.*

We begin by supposing that the stated result is *not* true and that we *can* write an equation

$$\sqrt{m} = r = \frac{p}{q} \quad \text{for integers } m, p, q \text{ with } q \neq 1.$$

It then follows that $p^2 = mq^2$. But, since r is expressed in its lowest terms, p and q, and hence p^2 and q^2, have no factors in common. However, m is an integer; this is only possible if $q = 1$ and $p^2 = m$. This conclusion contradicts the requirement that $q \neq 1$ and so leads to the conclusion that it was wrong to suppose that $\sqrt{m}$ can be expressed as a non-integer rational number. This completes the proof of the statement in the question. ◄

Our second worked example, also taken from elementary number theory, involves slightly more complicated mathematical reasoning but again exhibits the structure associated with this type of proof.

> ►*The prime integers p_i are labelled in ascending order, thus $p_1 = 1$, $p_2 = 2$, $p_5 = 7$, etc. Show that there is no largest prime number.*

Assume, on the contrary, that there is a largest prime and let it be p_N. Consider now the number q formed by multiplying together all the primes from p_1 to p_N and then adding one to the product, i.e.

$$q = p_1 p_2 \cdots p_N + 1.$$

By our assumption p_N is the largest prime, and so no number can have a prime factor greater than this. However, for every prime p_i, $i = 1, 2, \ldots, N$, the quotient q/p_i has the form $M_i + (1/p_i)$ with M_i an integer and $1/p_i$ non-integer. This means that q/p_i cannot be an integer and so p_i cannot be a divisor of q.

Since q is not divisible by any of the (assumed) finite set of primes, it must be itself a prime. As q is also clearly greater than p_N, we have a contradiction. This shows that our assumption that there is a largest prime integer must be false, and so it follows that there is no largest prime integer.

It should be noted that the given construction for q does not generate all the primes that actually exist (e.g. for $N = 3, q = 7$ rather than the next actual prime value of 5, is found), but this does not matter for the purposes of our proof by contradiction. ◄

1.7.3 Necessary and sufficient conditions

As the final topic in this introductory chapter, we consider briefly the notion of, and distinction between, necessary and sufficient conditions in the context of proving a mathematical proposition. In ordinary English the distinction is well defined, and that distinction is maintained in mathematics. However, in the authors' experience students tend to overlook it and assume (wrongly) that, having proved that the validity of proposition A implies the truth of proposition B, it follows by 'reversing the argument' that the validity of B automatically implies that of A.

As an example, let proposition A be that an integer N is divisible without remainder by 6, and proposition B be that N is divisible without remainder by 2. Clearly, if A is true then it follows that B is true, i.e. A is a sufficient condition for B; it is not however a necessary condition, as is trivially shown by taking N as 8. Conversely, the same value of N shows that whilst the validity of B is a necessary condition for A to hold, it is not sufficient.

An alternative terminology to 'necessary' and 'sufficient' often employed by mathematicians is that of 'if' and 'only if', particularly in the combination 'if and only if' which is usually written as IFF or denoted by a double-headed arrow $\Longleftrightarrow$. The equivalent statements can be summarised by

A if B	A is true if B is true *or*	$B \Longrightarrow A$,
	B is a sufficient condition for A	$B \Longrightarrow A$,
A only if B	A is true only if B is true *or*	$A \Longrightarrow B$,
	B is a necessary consequence of A	$A \Longrightarrow B$,

A IFF B A is true if and only if B is true *or* $B \iff A$,
        A and B necessarily imply each other $B \iff A$.

Although at this stage in the book we are able to employ for illustrative purposes only simple and fairly obvious results, the following example is given as a model of how necessary and sufficient conditions should be proved. The essential point is that for the second part of the proof (whether it be the 'necessary' part or the 'sufficient' part) one needs to start again from scratch; more often than not, the lines of the second part of the proof will *not* be simply those of the first written in reverse order.

> ►*Prove that (A) a function $f(x)$ is a quadratic polynomial with zeros at $x = 2$ and $x = 3$ if and only if (B) the function $f(x)$ has the form $\lambda(x^2 - 5x + 6)$ with λ a non-zero constant.*

(1) Assume A, i.e. that $f(x)$ *is* a quadratic polynomial with zeros at $x = 2$ and $x = 3$. Let its form be $ax^2 + bx + c$ with $a \neq 0$. Then we have

$$4a + 2b + c = 0,$$
$$9a + 3b + c = 0,$$

and subtraction shows that $5a + b = 0$ and $b = -5a$. Substitution of this into the first of the above equations gives $c = -4a - 2b = -4a + 10a = 6a$. Thus, it follows that

$$f(x) = a(x^2 - 5x + 6) \quad \text{with} \quad a \neq 0,$$

and establishes the 'A only if B' part of the stated result.

(2) Now assume that $f(x)$ *has* the form $\lambda(x^2 - 5x + 6)$ with λ a non-zero constant. Firstly we note that $f(x)$ is a quadratic polynomial, and so it only remains to prove that its zeros occur at $x = 2$ and $x = 3$. This could be done by straightforward evaluation of $f(2)$ and $f(3)$, but, to demonstrate the method, we proceed using a technique known as *completing the square*. Omitting the non-zero overall multiplier λ, we write $f(x) = 0$ as

$$x^2 - 5x + (\tfrac{5}{2})^2 - (\tfrac{5}{2})^2 + 6 = 0,$$
$$(x - \tfrac{5}{2})^2 = \tfrac{1}{4},$$
$$x - \tfrac{5}{2} = \pm\tfrac{1}{2}.$$

The two roots of $f(x) = 0$ are therefore $x = 2$ and $x = 3$; these x-values give the zeros of $f(x)$. This establishes the second ('A if B') part of the result. Thus we have shown that the assumption of either condition implies the validity of the other and the proof is complete. ◄

It should be noted that the propositions have to be carefully and precisely formulated. If, for example, the word 'quadratic' were omitted from A, statement B would still be a sufficient condition for A but not a necessary one, since $f(x)$ could then be $x^3 - 4x^2 + x + 6$ and A would not require B. Omitting the constant λ from the stated form of $f(x)$ in B has the same effect. Conversely, if A were to state that $f(x) = 3(x - 2)(x - 3)$ then B would be a necessary condition for A but not a sufficient one.

1.8 Exercises

Polynomial equations

1.1   Continue the investigation of equation (1.7), namely

$$g(x) = 4x^3 + 3x^2 - 6x - 1,$$

as follows.

(a) Make a table of values of $g(x)$ for integer values of x between -2 and 2. Use it and the information derived in the text to draw a graph and so determine the roots of $g(x) = 0$ as accurately as possible.

(b) Find one accurate root of $g(x) = 0$ by inspection and hence determine precise values for the other two roots.

(c) Show that $f(x) = 4x^3 + 3x^2 - 6x + k = 0$ has only one real root unless $-5 \le k \le \frac{7}{4}$.

1.2   Determine how the number of real roots of the equation

$$g(x) = 4x^3 - 17x^2 + 10x + k = 0$$

depends upon k. Are there any cases for which the equation has exactly two distinct real roots?

1.3   Continue the analysis of the polynomial equation

$$f(x) = x^7 + 5x^6 + x^4 - x^3 + x^2 - 2 = 0,$$

investigated in subsection 1.1.1, as follows.

(a) By writing the fifth-degree polynomial appearing in the expression for $f'(x)$ in the form $7x^5 + 30x^4 + a(x - b)^2 + c$, show that there is in fact only one positive root of $f(x) = 0$.

(b) By evaluating $f(1)$, $f(0)$ and $f(-1)$, and by inspecting the form of $f(x)$ for negative values of x, determine what you can about the positions of the real roots of $f(x) = 0$.

1.4   Given that $x = 2$ is one root of

$$g(x) = 2x^4 + 4x^3 - 9x^2 - 11x - 6 = 0,$$

use factorisation to determine how many real roots it has.

1.5   Construct the quadratic equations that have the following pairs of roots:
(a) $-6, -3$; (b) $0, 4$; (c) $2, 2$; (d) $3 + 2i, 3 - 2i$, where $i^2 = -1$.

1.6   Use the results of (i) equation (1.13), (ii) equation (1.12) and (iii) equation (1.14) to prove that if the roots of $3x^3 - x^2 - 10x + 8 = 0$ are α_1, α_2 and α_3 then

(a) $\alpha_1^{-1} + \alpha_2^{-1} + \alpha_3^{-1} = 5/4$,

(b) $\alpha_1^2 + \alpha_2^2 + \alpha_3^2 = 61/9$,

(c) $\alpha_1^3 + \alpha_2^3 + \alpha_3^3 = -125/27$.

(d) Convince yourself that eliminating (say) α_2 and α_3 from (i), (ii) and (iii) does *not* give a simple explicit way of finding α_1.

Trigonometric identities

1.7   Prove that

$$\cos \frac{\pi}{12} = \frac{\sqrt{3} + 1}{2\sqrt{2}}$$

by considering

(a) the sum of the sines of $\pi/3$ and $\pi/6$,
(b) the sine of the sum of $\pi/3$ and $\pi/4$.

1.8 The following exercises are based on the half-angle formulae.

(a) Use the fact that $\sin(\pi/6) = 1/2$ to prove that $\tan(\pi/12) = 2 - \sqrt{3}$.
(b) Use the result of (a) to show further that $\tan(\pi/24) = q(2 - q)$ where $q^2 = 2 + \sqrt{3}$.

1.9 Find the real solutions of

(a) $3\sin\theta - 4\cos\theta = 2$,
(b) $4\sin\theta + 3\cos\theta = 6$,
(c) $12\sin\theta - 5\cos\theta = -6$.

1.10 If $s = \sin(\pi/8)$, prove that

$$8s^4 - 8s^2 + 1 = 0,$$

and hence show that $s = [(2 - \sqrt{2})/4]^{1/2}$.

1.11 Find all the solutions of

$$\sin\theta + \sin 4\theta = \sin 2\theta + \sin 3\theta$$

that lie in the range $-\pi < \theta \leq \pi$. What is the multiplicity of the solution $\theta = 0$?

Coordinate geometry

1.12 Obtain in the form (1.38) the equations that describe the following:

(a) a circle of radius 5 with its centre at $(1, -1)$;
(b) the line $2x + 3y + 4 = 0$ and the line orthogonal to it which passes through $(1, 1)$;
(c) an ellipse of eccentricity 0.6 with centre $(1, 1)$ and its major axis of length 10 parallel to the y-axis.

1.13 Determine the forms of the conic sections described by the following equations:

(a) $x^2 + y^2 + 6x + 8y = 0$;
(b) $9x^2 - 4y^2 - 54x - 16y + 29 = 0$;
(c) $2x^2 + 2y^2 + 5xy - 4x + y - 6 = 0$;
(d) $x^2 + y^2 + 2xy - 8x + 8y = 0$.

1.14 For the ellipse

$$\frac{x^2}{a^2} + \frac{y^2}{b^2} = 1$$

with eccentricity e, the two points $(-ae, 0)$ and $(ae, 0)$ are known as its foci. Show that the sum of the distances from *any* point on the ellipse to the foci is $2a$. (The constancy of the sum of the distances from two fixed points can be used as an alternative defining property of an ellipse.)

Partial fractions

1.15 Resolve the following into partial fractions using the three methods given in section 1.4, verifying that the same decomposition is obtained by each method:

(a) $\dfrac{2x + 1}{x^2 + 3x - 10}$, (b) $\dfrac{4}{x^2 - 3x}$.

1.16   Express the following in partial fraction form:

$$\text{(a) } \frac{2x^3 - 5x + 1}{x^2 - 2x - 8}, \qquad \text{(b) } \frac{x^2 + x - 1}{x^2 + x - 2}.$$

1.17   Rearrange the following functions in partial fraction form:

$$\text{(a) } \frac{x - 6}{x^3 - x^2 + 4x - 4}, \qquad \text{(b) } \frac{x^3 + 3x^2 + x + 19}{x^4 + 10x^2 + 9}.$$

1.18   Resolve the following into partial fractions in such a way that x does not appear in any numerator:

$$\text{(a) } \frac{2x^2 + x + 1}{(x - 1)^2(x + 3)}, \qquad \text{(b) } \frac{x^2 - 2}{x^3 + 8x^2 + 16x}, \qquad \text{(c) } \frac{x^3 - x - 1}{(x + 3)^3(x + 1)}.$$

Binomial expansion

1.19   Evaluate those of the following that are defined: (a) $^5C_3$, (b) $^3C_5$, (c) $^{-5}C_3$, (d) $^{-3}C_5$.

1.20   Use a binomial expansion to evaluate $1/\sqrt{4.2}$ to five places of decimals, and compare it with the accurate answer obtained using a calculator.

Proof by induction and contradiction

1.21   Prove by induction that

$$\sum_{r=1}^{n} r = \tfrac{1}{2}n(n + 1) \qquad \text{and} \qquad \sum_{r=1}^{n} r^3 = \tfrac{1}{4}n^2(n + 1)^2.$$

1.22   Prove by induction that

$$1 + r + r^2 + \cdots + r^k + \cdots + r^n = \frac{1 - r^{n+1}}{1 - r}.$$

1.23   Prove that $3^{2n} + 7$, where n is a non-negative integer, is divisible by 8.

1.24   If a sequence of terms, u_n, satisfies the recurrence relation $u_{n+1} = (1 - x)u_n + nx$, with $u_1 = 0$, show, by induction, that, for $n \geq 1$,

$$u_n = \frac{1}{x}[nx - 1 + (1 - x)^n].$$

1.25   Prove by induction that

$$\sum_{r=1}^{n} \frac{1}{2^r} \tan\left(\frac{\theta}{2^r}\right) = \frac{1}{2^n} \cot\left(\frac{\theta}{2^n}\right) - \cot\theta.$$

1.26   The quantities a_i in this exercise are all positive real numbers.

(a) Show that

$$a_1 a_2 \leq \left(\frac{a_1 + a_2}{2}\right)^2.$$

(b) Hence prove, by induction on m, that

$$a_1 a_2 \cdots a_p \leq \left(\frac{a_1 + a_2 + \cdots + a_p}{p}\right)^p,$$

where $p = 2^m$ with m a positive integer. Note that each increase of m by unity doubles the number of factors in the product.

1.27   Establish the values of k for which the binomial coefficient ${}^pC_k$ is divisible by p when p is a prime number. Use your result and the method of induction to prove that $n^p - n$ is divisible by p for all integers n and all prime numbers p. Deduce that $n^5 - n$ is divisible by 30 for any integer n.

1.28   An arithmetic progression of integers a_n is one in which $a_n = a_0 + nd$, where a_0 and d are integers and n takes successive values $0, 1, 2, \ldots$.

   (a) Show that if any one term of the progression is the cube of an integer then so are infinitely many others.
   (b) Show that no cube of an integer can be expressed as $7n + 5$ for some positive integer n.

1.29   Prove, by the method of contradiction, that the equation

$$x^n + a_{n-1}x^{n-1} + \cdots + a_1 x + a_0 = 0,$$

in which all the coefficients a_i are integers, cannot have a rational root, unless that root is an integer. Deduce that any integral root must be a divisor of a_0 and hence find all rational roots of

   (a) $x^4 + 6x^3 + 4x^2 + 5x + 4 = 0,$
   (b) $x^4 + 5x^3 + 2x^2 - 10x + 6 = 0.$

Necessary and sufficient conditions

1.30   Prove that the equation $ax^2 + bx + c = 0$, in which a, b and c are real and $a > 0$, has two real distinct solutions IFF $b^2 > 4ac$.

1.31   For the real variable x, show that a sufficient, but not necessary, condition for $f(x) = x(x + 1)(2x + 1)$ to be divisible by 6 is that x is an integer.

1.32   Given that at least one of a and b, and at least one of c and d, are non-zero, show that $ad = bc$ is both a necessary and sufficient condition for the equations

$$ax + by = 0,$$
$$cx + dy = 0,$$

to have a solution in which at least one of x and y is non-zero.

1.33   The coefficients a_i in the polynomial $Q(x) = a_4 x^4 + a_3 x^3 + a_2 x^2 + a_1 x$ are all integers. Show that $Q(n)$ is divisible by 24 for all integers $n \geq 0$ if and only if all of the following conditions are satisfied:
   (i) $2a_1 + a_3$ is divisible by 4;
   (ii) $a_4 + a_2$ is divisible by 12;
   (iii) $a_4 + a_3 + a_2 + a_1$ is divisible by 24.

1.9 Hints and answers

1.1   (b) The roots are $1, \frac{1}{8}(-7 + \sqrt{33}) = -0.1569, \frac{1}{8}(-7 - \sqrt{33}) = -1.593$. (c) -5 and $\frac{7}{4}$ are the values of k that make $f(-1)$ and $f(\frac{1}{2})$ equal to zero.

1.3   (a) $a = 4$, $b = \frac{3}{8}$ and $c = \frac{23}{16}$ are all positive. Therefore $f'(x) > 0$ for all $x > 0$.
   (b) $f(1) = 5$, $f(0) = -2$ and $f(-1) = 5$, and so there is at least one root in each of the ranges $0 < x < 1$ and $-1 < x < 0$. $(x^7 + 5x^6) + (x^4 - x^3) + (x^2 - 2)$ is positive definite for $-5 < x < -\sqrt{2}$. There are therefore no roots in this range, but there must be one to the left of $x = -5$.

1.5   (a) $x^2 + 9x + 18 = 0$; (b) $x^2 - 4x = 0$; (c) $x^2 - 4x + 4 = 0$; (d) $x^2 - 6x + 13 = 0$.

1.7   (a) Use $\sin(\pi/4) = 1/\sqrt{2}$. (b) Use results (1.20) and (1.21).

1.9   (a) $1.339, -2.626$. (b) No solution because $6^2 > 4^2 + 3^2$. (c) $-0.0849, -2.276$.

1.11   Show that the equation is equivalent to $\sin(5\theta/2)\sin(\theta)\sin(\theta/2) = 0$.
Solutions are $-4\pi/5, -2\pi/5, 0, 2\pi/5, 4\pi/5, \pi$. The solution $\theta = 0$ has multiplicity 3.

1.13   (a) A circle of radius 5 centred on $(-3, -4)$.
(b) A hyperbola with 'centre' $(3, -2)$ and 'semi-axes' 2 and 3.
(c) The expression factorises into two lines, $x + 2y - 3 = 0$ and $2x + y + 2 = 0$.
(d) Write the expression as $(x + y)^2 = 8(x - y)$ to see that it represents a parabola passing through the origin, with the line $x + y = 0$ as its axis of symmetry.

1.15   (a) $\dfrac{5}{7(x-2)} + \dfrac{9}{7(x+5)}$,   (b) $-\dfrac{4}{3x} + \dfrac{4}{3(x-3)}$.

1.17   (a) $\dfrac{x+2}{x^2+4} - \dfrac{1}{x-1}$,   (b) $\dfrac{x+1}{x^2+9} + \dfrac{2}{x^2+1}$.

1.19   (a) 10, (b) not defined, (c) -35, (d) -21.

1.21   Look for factors common to the $n = N$ sum and the additional $n = N + 1$ term, so as to reduce the sum for $n = N + 1$ to a single term.

1.23   Write 3^{2n} as $8m - 7$.

1.25   Use the half-angle formulae of equations (1.32) to (1.34) to relate functions of $\theta/2^k$ to those of $\theta/2^{k+1}$.

1.27   Divisible for $k = 1, 2, \ldots, p - 1$. Expand $(n + 1)^p$ as $n^p + \sum_1^{p-1} {}^pC_k n^k + 1$. Apply the stated result for $p = 5$. Note that $n^5 - n = n(n-1)(n+1)(n^2+1)$; the product of any three consecutive integers must divide by both 2 and 3.

1.29   By assuming $x = p/q$ with $q \neq 1$, show that a fraction $-p^n/q$ is equal to an integer $a_{n-1}p^{n-1} + \cdots + a_1 pq^{n-2} + a_0 q^{n-1}$. This is a contradiction, and is only resolved if $q = 1$ and the root is an integer.
(a) The only possible candidates are $\pm 1, \pm 2, \pm 4$. None is a root.
(b) The only possible candidates are $\pm 1, \pm 2, \pm 3, \pm 6$. Only -3 is a root.

1.31 $f(x)$ can be written as $x(x + 1)(x + 2) + x(x + 1)(x - 1)$. Each term consists of the product of three consecutive integers, of which one must therefore divide by 2 and (a different) one by 3. Thus each term separately divides by 6, and so therefore does $f(x)$. Note that if x is the root of $2x^3 + 3x^2 + x - 24 = 0$ that lies near the non-integer value $x = 1.826$, then $x(x + 1)(2x + 1) = 24$ and therefore divides by 6.

1.33   Note that, e.g., the condition for $6a_4 + a_3$ to be divisible by 4 is the same as the condition for $2a_4 + a_3$ to be divisible by 4.
For the necessary (only if) part of the proof set $n = 1, 2, 3$ and take integer combinations of the resulting equations.
For the sufficient (if) part of the proof use the stated conditions to prove the proposition by induction. Note that $n^3 - n$ is divisible by 6 and that $n^2 + n$ is even.

2

Preliminary calculus

This chapter is concerned with the formalism of probably the most widely used mathematical technique in the physical sciences, namely the calculus. The chapter divides into two sections. The first deals with the process of differentiation and the second with its inverse process, integration. The material covered is essential for the remainder of the book and serves as a reference. Readers who have previously studied these topics should ensure familiarity by looking at the worked examples in the main text and by attempting the exercises at the end of the chapter.

2.1 Differentiation

Differentiation is the process of determining how quickly or slowly a function varies, as the quantity on which it depends, its *argument*, is changed. More specifically it is the procedure for obtaining an expression (numerical or algebraic) for the rate of change of the function with respect to its argument. Familiar examples of rates of change include acceleration (the rate of change of velocity) and chemical reaction rate (the rate of change of chemical composition). Both acceleration and reaction rate give a measure of the change of a quantity with respect to time. However, differentiation may also be applied to changes with respect to other quantities, for example the change in pressure with respect to a change in temperature.

Although it will not be apparent from what we have said so far, differentiation is in fact a limiting process, that is, it deals only with the infinitesimal change in one quantity resulting from an infinitesimal change in another.

2.1.1 Differentiation from first principles

Let us consider a function $f(x)$ that depends on only one variable x, together with numerical constants, for example, $f(x) = 3x^2$ or $f(x) = \sin x$ or $f(x) = 2 + 3/x$.

41

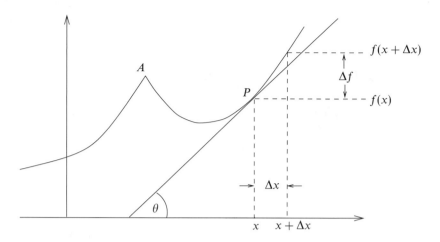

Figure 2.1   The graph of a function $f(x)$ showing that the gradient or slope of the function at P, given by $\tan\theta$, is approximately equal to $\Delta f/\Delta x$.

Figure 2.1 shows an example of such a function. Near any particular point, P, the value of the function changes by an amount Δf, say, as x changes by a small amount Δx. The slope of the tangent to the graph of $f(x)$ at P is then approximately $\Delta f/\Delta x$, and the change in the value of the function is $\Delta f = f(x + \Delta x) - f(x)$. In order to calculate the true value of the gradient, or *first derivative*, of the function at P, we must let Δx become infinitesimally small. We therefore define the first derivative of $f(x)$ as

$$f'(x) \equiv \frac{df(x)}{dx} \equiv \lim_{\Delta x \to 0} \frac{f(x + \Delta x) - f(x)}{\Delta x}, \tag{2.1}$$

provided that the limit exists. The limit will depend in almost all cases on the value of x. If the limit does exist at a point $x = a$ then the function is said to be differentiable at a; otherwise it is said to be non-differentiable at a. The formal concept of a limit and its existence or non-existence is discussed in chapter 4; for present purposes we will adopt an intuitive approach.

In the definition (2.1), we allow Δx to tend to zero from either positive or negative values and require the same limit to be obtained in both cases. A function that is differentiable at a is necessarily continuous at a (there must be no jump in the value of the function at a), though the converse is not necessarily true. This latter assertion is illustrated in figure 2.1: the function is continuous at the 'kink' A but the two limits of the gradient as Δx tends to zero from positive or negative values are different and so the function is not differentiable at A.

It should be clear from the above discussion that near the point P we may approximate the change in the value of the function, Δf, that results from a small change Δx in x by

$$\Delta f \approx \frac{df(x)}{dx}\Delta x. \tag{2.2}$$

As one would expect, the approximation improves as the value of Δx is reduced. In the limit in which the change Δx becomes infinitesimally small, we denote it by the *differential dx*, and (2.2) reads

$$df = \frac{df(x)}{dx}dx. \tag{2.3}$$

This *equality* relates the infinitesimal change in the function, df, to the infinitesimal change dx that causes it.

So far we have discussed only the first derivative of a function. However, we can also define the *second derivative* as the gradient of the gradient of a function. Again we use the definition (2.1) but now with $f(x)$ replaced by $f'(x)$. Hence the second derivative is defined by

$$f''(x) \equiv \lim_{\Delta x \to 0} \frac{f'(x + \Delta x) - f'(x)}{\Delta x}, \tag{2.4}$$

provided that the limit exists. A physical example of a second derivative is the second derivative of the distance travelled by a particle with respect to time. Since the first derivative of distance travelled gives the particle's velocity, the second derivative gives its acceleration.

We can continue in this manner, the nth derivative of the function $f(x)$ being defined by

$$f^{(n)}(x) \equiv \lim_{\Delta x \to 0} \frac{f^{(n-1)}(x + \Delta x) - f^{(n-1)}(x)}{\Delta x}. \tag{2.5}$$

It should be noted that with this notation $f'(x) \equiv f^{(1)}(x)$, $f''(x) \equiv f^{(2)}(x)$, etc., and that formally $f^{(0)}(x) \equiv f(x)$.

All this should be familiar to the reader, though perhaps not with such formal definitions. The following example shows the differentiation of $f(x) = x^2$ from first principles. In practice, however, it is desirable simply to remember the derivatives of standard functions; the techniques given in the remainder of this section can be applied to find more complicated derivatives.

▶ *Find from first principles the derivative with respect to x of $f(x) = x^2$.*

Using the definition (2.1),

$$f'(x) = \lim_{\Delta x \to 0} \frac{f(x + \Delta x) - f(x)}{\Delta x}$$

$$= \lim_{\Delta x \to 0} \frac{(x + \Delta x)^2 - x^2}{\Delta x}$$

$$= \lim_{\Delta x \to 0} \frac{2x\Delta x + (\Delta x)^2}{\Delta x}$$

$$= \lim_{\Delta x \to 0} (2x + \Delta x).$$

As Δx tends to zero, $2x + \Delta x$ tends towards $2x$, hence

$$f'(x) = 2x. \blacktriangleleft$$

Derivatives of other functions can be obtained in the same way. The derivatives of some simple functions are listed below (note that a is a constant):

$$\frac{d}{dx}(x^n) = nx^{n-1}, \qquad \frac{d}{dx}(e^{ax}) = ae^{ax}, \qquad \frac{d}{dx}(\ln ax) = \frac{1}{x},$$

$$\frac{d}{dx}(\sin ax) = a\cos ax, \qquad \frac{d}{dx}(\cos ax) = -a\sin ax, \qquad \frac{d}{dx}(\sec ax) = a\sec ax\tan ax,$$

$$\frac{d}{dx}(\tan ax) = a\sec^2 ax, \qquad \frac{d}{dx}(\csc ax) = -a\,\csc ax\cot ax,$$

$$\frac{d}{dx}(\cot ax) = -a\,\csc^2 ax, \qquad \frac{d}{dx}\left(\sin^{-1}\frac{x}{a}\right) = \frac{1}{\sqrt{a^2 - x^2}},$$

$$\frac{d}{dx}\left(\cos^{-1}\frac{x}{a}\right) = \frac{-1}{\sqrt{a^2 - x^2}}, \qquad \frac{d}{dx}\left(\tan^{-1}\frac{x}{a}\right) = \frac{a}{a^2 + x^2}.$$

Differentiation from first principles emphasises the definition of a derivative as the gradient of a function. However, for most practical purposes, returning to the definition (2.1) is time consuming and does not aid our understanding. Instead, as mentioned above, we employ a number of techniques, which use the derivatives listed above as 'building blocks', to evaluate the derivatives of more complicated functions than hitherto encountered. Subsections 2.1.2–2.1.7 develop the methods required.

2.1.2 Differentiation of products

As a first example of the differentiation of a more complicated function, we consider finding the derivative of a function $f(x)$ that can be written as the product of two other functions of x, namely $f(x) = u(x)v(x)$. For example, if $f(x) = x^3 \sin x$ then we might take $u(x) = x^3$ and $v(x) = \sin x$. Clearly the separation is not unique. (In the given example, possible alternative break-ups would be $u(x) = x^2$, $v(x) = x \sin x$, or even $u(x) = x^4 \tan x$, $v(x) = x^{-1} \cos x$.)

The purpose of the separation is to split the function into two (or more) parts, of which we know the derivatives (or at least we can evaluate these derivatives more easily than that of the whole). We would gain little, however, if we did not know the relationship between the derivative of f and those of u and v. Fortunately, they are very simply related, as we shall now show.

Since $f(x)$ is written as the product $u(x)v(x)$, it follows that

$$f(x + \Delta x) - f(x) = u(x + \Delta x)v(x + \Delta x) - u(x)v(x)$$
$$= u(x + \Delta x)[v(x + \Delta x) - v(x)] + [u(x + \Delta x) - u(x)]v(x).$$

From the definition of a derivative (2.1),

$$\frac{df}{dx} = \lim_{\Delta x \to 0} \frac{f(x + \Delta x) - f(x)}{\Delta x}$$
$$= \lim_{\Delta x \to 0} \left\{ u(x + \Delta x) \left[\frac{v(x + \Delta x) - v(x)}{\Delta x} \right] + \left[\frac{u(x + \Delta x) - u(x)}{\Delta x} \right] v(x) \right\}.$$

In the limit $\Delta x \to 0$, the factors in square brackets become dv/dx and du/dx (by the definitions of these quantities) and $u(x + \Delta x)$ simply becomes $u(x)$. Consequently we obtain

$$\frac{df}{dx} = \frac{d}{dx}[u(x)v(x)] = u(x)\frac{dv(x)}{dx} + \frac{du(x)}{dx}v(x). \tag{2.6}$$

In primed notation and without writing the argument x explicitly, (2.6) is stated concisely as

$$f' = (uv)' = uv' + u'v. \tag{2.7}$$

This is a general result obtained without making any assumptions about the specific forms f, u and v, other than that $f(x) = u(x)v(x)$. In words, the result reads as follows. *The derivative of the product of two functions is equal to the first function times the derivative of the second plus the second function times the derivative of the first.*

▶ *Find the derivative with respect to x of $f(x) = x^3 \sin x$.*

Using the product rule, (2.6),

$$\frac{d}{dx}(x^3 \sin x) = x^3 \frac{d}{dx}(\sin x) + \frac{d}{dx}(x^3) \sin x$$
$$= x^3 \cos x + 3x^2 \sin x. \blacktriangleleft$$

The product rule may readily be extended to the product of three or more functions. Considering the function

$$f(x) = u(x)v(x)w(x) \tag{2.8}$$

and using (2.6), we obtain, as before omitting the argument,

$$\frac{df}{dx} = u\frac{d}{dx}(vw) + \frac{du}{dx}vw.$$

Using (2.6) again to expand the first term on the RHS gives the complete result

$$\frac{d}{dx}(uvw) = uv\frac{dw}{dx} + u\frac{dv}{dx}w + \frac{du}{dx}vw \tag{2.9}$$

or

$$(uvw)' = uvw' + uv'w + u'vw. \tag{2.10}$$

It is readily apparent that this can be extended to products containing any number n of factors; the expression for the derivative will then consist of n terms with the prime appearing in successive terms on each of the n factors in turn. This is probably the easiest way to recall the product rule.

2.1.3 The chain rule

Products are just one type of complicated function that we may encounter in differentiation. Another is the function of a function, e.g. $f(x) = (3+x^2)^3 = u(x)^3$, where $u(x) = 3 + x^2$. If Δf, Δu and Δx are small finite quantities, it follows that

$$\frac{\Delta f}{\Delta x} = \frac{\Delta f}{\Delta u}\frac{\Delta u}{\Delta x};$$

As the quantities become infinitesimally small we obtain

$$\frac{df}{dx} = \frac{df}{du}\frac{du}{dx}. \tag{2.11}$$

This is the *chain rule*, which we must apply when differentiating a function of a function.

▶*Find the derivative with respect to x of $f(x) = (3+x^2)^3$.*

Rewriting the function as $f(x) = u^3$, where $u(x) = 3 + x^2$, and applying (2.11) we find

$$\frac{df}{dx} = 3u^2\frac{du}{dx} = 3u^2\frac{d}{dx}(3+x^2) = 3u^2 \times 2x = 6x(3+x^2)^2. \blacktriangleleft$$

Similarly, the derivative with respect to x of $f(x) = 1/v(x)$ may be obtained by rewriting the function as $f(x) = v^{-1}$ and applying (2.11):

$$\frac{df}{dx} = -v^{-2}\frac{dv}{dx} = -\frac{1}{v^2}\frac{dv}{dx}. \tag{2.12}$$

The chain rule is also useful for calculating the derivative of a function f with respect to x when both x and f are written in terms of a variable (or parameter), say t.

46

▶*Find the derivative with respect to x of $f(t) = 2at$, where $x = at^2$.*

We could of course substitute for t and then differentiate f as a function of x, but in this case it is quicker to use

$$\frac{df}{dx} = \frac{df}{dt}\frac{dt}{dx} = 2a\frac{1}{2at} = \frac{1}{t},$$

where we have used the fact that

$$\frac{dt}{dx} = \left(\frac{dx}{dt}\right)^{-1}. \ \blacktriangleleft$$

2.1.4 Differentiation of quotients

Applying (2.6) for the derivative of a product to a function $f(x) = u(x)[1/v(x)]$, we may obtain the derivative of the quotient of two factors. Thus

$$f' = \left(\frac{u}{v}\right)' = u\left(\frac{1}{v}\right)' + u'\left(\frac{1}{v}\right) = u\left(-\frac{v'}{v^2}\right) + \frac{u'}{v},$$

where (2.12) has been used to evaluate $(1/v)'$ This can now be rearranged into the more convenient and memorisable form

$$f' = \left(\frac{u}{v}\right)' = \frac{vu' - uv'}{v^2}. \tag{2.13}$$

This can be expressed in words as *the derivative of a quotient is equal to the bottom times the derivative of the top minus the top times the derivative of the bottom, all over the bottom squared.*

▶*Find the derivative with respect to x of $f(x) = \sin x/x$.*

Using (2.13) with $u(x) = \sin x$, $v(x) = x$ and hence $u'(x) = \cos x$, $v'(x) = 1$, we find

$$f'(x) = \frac{x\cos x - \sin x}{x^2} = \frac{\cos x}{x} - \frac{\sin x}{x^2}. \ \blacktriangleleft$$

2.1.5 Implicit differentiation

So far we have only differentiated functions written in the form $y = f(x)$. However, we may not always be presented with a relationship in this simple form. As an example consider the relation $x^3 - 3xy + y^3 = 2$. In this case it is not possible to rearrange the equation to give y as a function of x. Nevertheless, by differentiating term by term with respect to x (*implicit differentiation*), we can find the derivative of y.

▶*Find dy/dx if $x^3 - 3xy + y^3 = 2$.*

Differentiating each term in the equation with respect to x we obtain

$$\frac{d}{dx}(x^3) - \frac{d}{dx}(3xy) + \frac{d}{dx}(y^3) = \frac{d}{dx}(2),$$

$$\Rightarrow \quad 3x^2 - \left(3x\frac{dy}{dx} + 3y\right) + 3y^2\frac{dy}{dx} = 0,$$

where the derivative of $3xy$ has been found using the product rule. Hence, rearranging for dy/dx,

$$\frac{dy}{dx} = \frac{y - x^2}{y^2 - x}.$$

Note that dy/dx is a function of both x and y and cannot be expressed as a function of x only. ◀

2.1.6 Logarithmic differentiation

In circumstances in which the variable with respect to which we are differentiating is an exponent, taking logarithms and then differentiating implicitly is the simplest way to find the derivative.

▶*Find the derivative with respect to x of $y = a^x$.*

To find the required derivative we first take logarithms and then differentiate implicitly:

$$\ln y = \ln a^x = x \ln a \quad \Rightarrow \quad \frac{1}{y}\frac{dy}{dx} = \ln a.$$

Now, rearranging and substituting for y, we find

$$\frac{dy}{dx} = y \ln a = a^x \ln a. \blacktriangleleft$$

2.1.7 Leibnitz' theorem

We have discussed already how to find the derivative of a product of two or more functions. We now consider *Leibnitz' theorem*, which gives the corresponding results for the higher derivatives of products.

Consider again the function $f(x) = u(x)v(x)$. We know from the product rule that $f' = uv' + u'v$. Using the rule once more for each of the products, we obtain

$$f'' = (uv'' + u'v') + (u'v' + u''v)$$
$$= uv'' + 2u'v' + u''v.$$

Similarly, differentiating twice more gives

$$f''' = uv''' + 3u'v'' + 3u''v' + u'''v,$$
$$f^{(4)} = uv^{(4)} + 4u'v''' + 6u''v'' + 4u'''v' + u^{(4)}v.$$

The pattern emerging is clear and strongly suggests that the results generalise to

$$f^{(n)} = \sum_{r=0}^{n} \frac{n!}{r!(n-r)!} u^{(r)} v^{(n-r)} = \sum_{r=0}^{n} {}^{n}C_r u^{(r)} v^{(n-r)}, \qquad (2.14)$$

where the fraction $n!/[r!(n-r)!]$ is identified with the binomial coefficient ${}^{n}C_r$ (see chapter 1). To *prove* that this is so, we use the method of induction as follows. Assume that (2.14) is valid for n equal to some integer N. Then

$$f^{(N+1)} = \sum_{r=0}^{N} {}^{N}C_r \frac{d}{dx} \left(u^{(r)} v^{(N-r)} \right)$$

$$= \sum_{r=0}^{N} {}^{N}C_r [u^{(r)} v^{(N-r+1)} + u^{(r+1)} v^{(N-r)}]$$

$$= \sum_{s=0}^{N} {}^{N}C_s u^{(s)} v^{(N+1-s)} + \sum_{s=1}^{N+1} {}^{N}C_{s-1} u^{(s)} v^{(N+1-s)},$$

where we have substituted summation index s for r in the first summation, and for $r+1$ in the second. Now, from our earlier discussion of binomial coefficients, equation (1.51), we have

$$^{N}C_s + {}^{N}C_{s-1} = {}^{N+1}C_s$$

and so, after separating out the first term of the first summation and the last term of the second, obtain

$$f^{(N+1)} = {}^{N}C_0 u^{(0)} v^{(N+1)} + \sum_{s=1}^{N} {}^{N+1}C_s u^{(s)} v^{(N+1-s)} + {}^{N}C_N u^{(N+1)} v^{(0)}.$$

But ${}^{N}C_0 = 1 = {}^{N+1}C_0$ and ${}^{N}C_N = 1 = {}^{N+1}C_{N+1}$, and so we may write

$$f^{(N+1)} = {}^{N+1}C_0 u^{(0)} v^{(N+1)} + \sum_{s=1}^{N} {}^{N+1}C_s u^{(s)} v^{(N+1-s)} + {}^{N+1}C_{N+1} u^{(N+1)} v^{(0)}$$

$$= \sum_{s=0}^{N+1} {}^{N+1}C_s u^{(s)} v^{(N+1-s)}.$$

This is just (2.14) with n set equal to $N+1$. Thus, assuming the validity of (2.14) for $n = N$ implies its validity for $n = N + 1$. However, when $n = 1$ equation (2.14) is simply the product rule, and this we have already proved directly. These results taken together establish the validity of (2.14) for all n and prove Leibnitz' theorem.

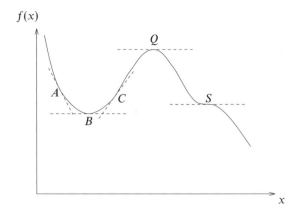

$f(x)$

Figure 2.2 A graph of a function, $f(x)$, showing how differentiation corresponds to finding the gradient of the function at a particular point. Points B, Q and S are stationary points (see text).

> ▶ *Find the third derivative of the function $f(x) = x^3 \sin x$.*

Using (2.14) we immediately find

$$f'''(x) = 6 \sin x + 3(6x) \cos x + 3(3x^2)(-\sin x) + x^3(-\cos x)$$
$$= 3(2 - 3x^2) \sin x + x(18 - x^2) \cos x. \ ◄$$

2.1.8 Special points of a function

We have interpreted the derivative of a function as the gradient of the function at the relevant point (figure 2.1). If the gradient is zero for some particular value of x then the function is said to have a *stationary point* there. Clearly, in graphical terms, this corresponds to a horizontal tangent to the graph.

Stationary points may be divided into three categories and an example of each is shown in figure 2.2. Point B is said to be a *minimum* since the function *increases* in value in both directions away from it. Point Q is said to be a *maximum* since the function *decreases* in both directions away from it. Note that B is not the overall minimum value of the function and Q is not the overall maximum; rather, they are a local minimum and a local maximum. Maxima and minima are known collectively as *turning points*.

The third type of stationary point is the *stationary point of inflection*, S. In this case the function falls in the positive x-direction and rises in the negative x-direction so that S is neither a maximum nor a minimum. Nevertheless, the gradient of the function is zero at S, i.e. the graph of the function is flat there, and this justifies our calling it a stationary point. Of course, a point at which the gradient of the function is zero but the function rises in the positive x-direction and falls in the negative x-direction is also a stationary point of inflection.

The above distinction between the three types of stationary point has been made rather descriptively. However, it is possible to define and distinguish stationary points mathematically. From their definition as points of zero gradient, all stationary points must be characterised by $df/dx = 0$. In the case of the minimum, B, the slope, i.e. df/dx, changes from negative at A to positive at C through zero at B. Thus df/dx is increasing and so the second derivative d^2f/dx^2 must be positive. Conversely, at the maximum, Q, we must have that d^2f/dx^2 is negative.

It is less obvious, but intuitively reasonable, that at S, d^2f/dx^2 is zero. This may be inferred from the following observations. To the left of S the curve is concave upwards so that df/dx is increasing with x and hence $d^2f/dx^2 > 0$. To the right of S, however, the curve is concave downwards so that df/dx is decreasing with x and hence $d^2f/dx^2 < 0$.

In summary, at a stationary point $df/dx = 0$ and

(i) for a minimum, $d^2f/dx^2 > 0$,

(ii) for a maximum, $d^2f/dx^2 < 0$,

(iii) for a stationary point of inflection, $d^2f/dx^2 = 0$ and d^2f/dx^2 changes sign through the point.

In case (iii), a stationary point of inflection, in order that d^2f/dx^2 changes sign through the point we normally require $d^3f/dx^3 \neq 0$ at that point. This simple rule can fail for some functions, however, and in general if the first non-vanishing derivative of $f(x)$ at the stationary point is $f^{(n)}$ then if n is even the point is a maximum or minimum and if n is odd the point is a stationary point of inflection. This may be seen from the Taylor expansion (see equation (4.17)) of the function about the stationary point, but it is not proved here.

> ▶ *Find the positions and natures of the stationary points of the function*
> $$f(x) = 2x^3 - 3x^2 - 36x + 2.$$

The first criterion for a stationary point is that $df/dx = 0$, and hence we set

$$\frac{df}{dx} = 6x^2 - 6x - 36 = 0,$$

from which we obtain

$$(x - 3)(x + 2) = 0.$$

Hence the stationary points are at $x = 3$ and $x = -2$. To determine the nature of the stationary point we must evaluate d^2f/dx^2:

$$\frac{d^2f}{dx^2} = 12x - 6.$$

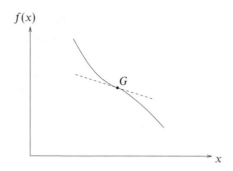

Figure 2.3   The graph of a function $f(x)$ that has a general point of inflection at the point G.

Now, we examine each stationary point in turn. For $x = 3$, $d^2f/dx^2 = 30$. Since this is positive, we conclude that $x = 3$ is a minimum. Similarly, for $x = -2$, $d^2f/dx^2 = -30$ and so $x = -2$ is a maximum. ◄

So far we have concentrated on stationary points, which are defined to have $df/dx = 0$. We have found that at a stationary point of inflection d^2f/dx^2 is also zero and changes sign. This naturally leads us to consider points at which d^2f/dx^2 is zero and changes sign but at which df/dx is *not*, in general, zero. Such points are called *general points of inflection* or simply *points of inflection*. Clearly, a stationary point of inflection is a special case for which df/dx is also zero. At a general point of inflection the graph of the function changes from being concave upwards to concave downwards (or vice versa), but the tangent to the curve at this point need not be horizontal. A typical example of a general point of inflection is shown in figure 2.3.

The determination of the stationary points of a function, together with the identification of its zeros, infinities and possible asymptotes, is usually sufficient to enable a graph of the function showing most of its significant features to be sketched. Some examples for the reader to try are included in the exercises at the end of this chapter.

2.1.9 Curvature of a function

In the previous section we saw that at a point of inflection of the function $f(x)$, the second derivative d^2f/dx^2 changes sign and passes through zero. The corresponding graph of f shows an inversion of its curvature at the point of inflection. We now develop a more quantitative measure of the curvature of a function (or its graph), which is applicable at general points and not just in the neighbourhood of a point of inflection.

As in figure 2.1, let θ be the angle made with the x-axis by the tangent at a

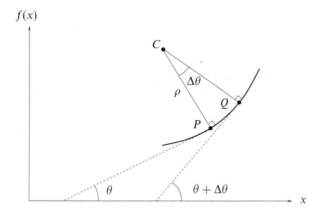

Figure 2.4   Two neighbouring tangents to the curve $f(x)$ whose slopes differ by $\Delta\theta$. The angular separation of the corresponding radii of the circle of curvature is also $\Delta\theta$.

point P on the curve $f = f(x)$, with $\tan\theta = df/dx$ evaluated at P. Now consider also the tangent at a neighbouring point Q on the curve, and suppose that it makes an angle $\theta + \Delta\theta$ with the x-axis, as illustrated in figure 2.4.

It follows that the corresponding normals at P and Q, which are perpendicular to the respective tangents, also intersect at an angle $\Delta\theta$. Furthermore, their point of intersection, C in the figure, will be the position of the centre of a circle that approximates the arc PQ, at least to the extent of having the same tangents at the extremities of the arc. This circle is called the *circle of curvature*.

For a finite arc PQ, the lengths of CP and CQ will not, in general, be equal, as they would be if $f = f(x)$ *were* in fact the equation of a circle. But, as Q is allowed to tend to P, i.e. as $\Delta\theta \to 0$, they do become equal, their common value being ρ, the radius of the circle, known as the *radius of curvature*. It follows immediately that the curve and the circle of curvature have a common tangent at P and lie on the same side of it. The reciprocal of the radius of curvature, ρ^{-1}, defines the *curvature* of the function $f(x)$ at the point P.

The radius of curvature can be defined more mathematically as follows. The length Δs of arc PQ is approximately equal to $\rho\Delta\theta$ and, in the limit $\Delta\theta \to 0$, this relationship defines ρ as

$$\rho = \lim_{\Delta\theta\to 0} \frac{\Delta s}{\Delta\theta} = \frac{ds}{d\theta}. \tag{2.15}$$

It should be noted that, as s increases, θ may increase or decrease according to whether the curve is locally concave upwards (i.e. shaped as if it were near a minimum in $f(x)$) or concave downwards. This is reflected in the sign of ρ, which therefore also indicates the position of the curve (and of the circle of curvature) relative to the common tangent, above or below. Thus a negative value of ρ indicates that the curve is locally concave downwards and that the tangent lies above the curve.

We next obtain an expression for ρ, not in terms of s and θ but in terms of x and $f(x)$. The expression, though somewhat cumbersome, follows from the defining equation (2.15), the defining property of θ that $\tan\theta = df/dx \equiv f'$ and the fact that the rate of change of arc length with x is given by

$$\frac{ds}{dx} = \left[1 + \left(\frac{df}{dx}\right)^2\right]^{1/2}. \tag{2.16}$$

This last result, simply quoted here, is proved more formally in subsection 2.2.13.

From the chain rule (2.11) it follows that

$$\rho = \frac{ds}{d\theta} = \frac{ds}{dx}\frac{dx}{d\theta}. \tag{2.17}$$

Differentiating both sides of $\tan\theta = df/dx$ with respect to x gives

$$\sec^2\theta\,\frac{d\theta}{dx} = \frac{d^2f}{dx^2} \equiv f'',$$

from which, using $\sec^2\theta = 1 + \tan^2\theta = 1 + (f')^2$, we can obtain $dx/d\theta$ as

$$\frac{dx}{d\theta} = \frac{1 + \tan^2\theta}{f''} = \frac{1 + (f')^2}{f''}. \tag{2.18}$$

Substituting (2.16) and (2.18) into (2.17) then yields the final expression for ρ,

$$\rho = \frac{\left[1 + (f')^2\right]^{3/2}}{f''}. \tag{2.19}$$

It should be noted that the quantity in brackets is always positive and that its three-halves root is also taken as positive. The sign of ρ is thus solely determined by that of d^2f/dx^2, in line with our previous discussion relating the sign to whether the curve is concave or convex upwards. If, as happens at a point of inflection, d^2f/dx^2 is zero then ρ is formally infinite and the curvature of $f(x)$ is zero. As d^2f/dx^2 changes sign on passing through zero, both the local tangent and the circle of curvature change from their initial positions to the opposite side of the curve.

> ▶*Show that the radius of curvature at the point (x, y) on the ellipse*
>
> $$\frac{x^2}{a^2} + \frac{y^2}{b^2} = 1$$
>
> *has magnitude $(a^4 y^2 + b^4 x^2)^{3/2}/(a^4 b^4)$ and the opposite sign to y. Check the special case $b = a$, for which the ellipse becomes a circle.*

Differentiating the equation of the ellipse with respect to x gives

$$\frac{2x}{a^2} + \frac{2y}{b^2}\frac{dy}{dx} = 0$$

and so

$$\frac{dy}{dx} = -\frac{b^2 x}{a^2 y}.$$

A second differentiation, using (2.13), then yields

$$\frac{d^2 y}{dx^2} = -\frac{b^2}{a^2}\left(\frac{y - xy'}{y^2}\right) = -\frac{b^4}{a^2 y^3}\left(\frac{y^2}{b^2} + \frac{x^2}{a^2}\right) = -\frac{b^4}{a^2 y^3},$$

where we have used the fact that (x, y) lies on the ellipse. We note that d^2y/dx^2, and hence ρ, has the opposite sign to y^3 and hence to y. Substituting in (2.19) gives for the magnitude of the radius of curvature

$$|\rho| = \left|\frac{\left[1 + b^4 x^2/(a^4 y^2)\right]^{3/2}}{-b^4/(a^2 y^3)}\right| = \frac{(a^4 y^2 + b^4 x^2)^{3/2}}{a^4 b^4}.$$

For the special case $b = a$, $|\rho|$ reduces to $a^{-2}(y^2 + x^2)^{3/2}$ and, since $x^2 + y^2 = a^2$, this in turn gives $|\rho| = a$, as expected. ◀

The discussion in this section has been confined to the behaviour of curves that lie in one plane; examples of the application of curvature to the bending of loaded beams and to particle orbits under the influence of a central forces can be found in the exercises at the ends of later chapters. A more general treatment of curvature in three dimensions is given in section 10.3, where a vector approach is adopted.

2.1.10 Theorems of differentiation

Rolle's theorem

Rolle's theorem (figure 2.5) states that if a function $f(x)$ is continuous in the range $a \le x \le c$, is differentiable in the range $a < x < c$ and satisfies $f(a) = f(c)$ then for at least one point $x = b$, where $a < b < c$, $f'(b) = 0$. Thus Rolle's theorem states that for a well-behaved (continuous and differentiable) function that has the same value at two points either there is at least one stationary point between those points or the function is a constant between them. The validity of the theorem is immediately apparent from figure 2.5 and a full analytic proof will not be given. The theorem is used in deriving the mean value theorem, which we now discuss.

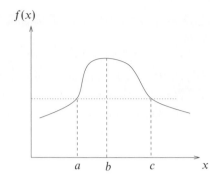

Figure 2.5   The graph of a function $f(x)$, showing that if $f(a) = f(c)$ then at one point at least between $x = a$ and $x = c$ the graph has zero gradient.

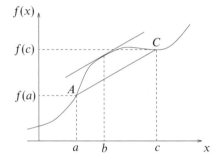

Figure 2.6   The graph of a function $f(x)$; at some point $x = b$ it has the same gradient as the line AC.

Mean value theorem

The mean value theorem (figure 2.6) states that if a function $f(x)$ is continuous in the range $a \leq x \leq c$ and differentiable in the range $a < x < c$ then

$$f'(b) = \frac{f(c) - f(a)}{c - a},\tag{2.20}$$

for at least one value b where $a < b < c$. Thus the mean value theorem states that for a well-behaved function the gradient of the line joining two points on the curve is equal to the slope of the tangent to the curve for at least one intervening point.

The proof of the mean value theorem is found by examination of figure 2.6, as follows. The equation of the line AC is

$$g(x) = f(a) + (x - a)\frac{f(c) - f(a)}{c - a},$$

and hence the difference between the curve and the line is

$$h(x) = f(x) - g(x) = f(x) - f(a) - (x - a)\frac{f(c) - f(a)}{c - a}.$$

Since the curve and the line intersect at A and C, $h(x) = 0$ at both of these points. Hence, by an application of Rolle's theorem, $h'(x) = 0$ for at least one point b between A and C. Differentiating our expression for $h(x)$, we find

$$h'(x) = f'(x) - \frac{f(c) - f(a)}{c - a},$$

and hence at b, where $h'(x) = 0$,

$$f'(b) = \frac{f(c) - f(a)}{c - a}.$$

Applications of Rolle's theorem and the mean value theorem

Since the validity of Rolle's theorem is intuitively obvious, given the conditions imposed on $f(x)$, it will not be surprising that the problems that can be solved by applications of the theorem alone are relatively simple ones. Nevertheless we will illustrate it with the following example.

> ▶ *What semi-quantitative results can be deduced by applying Rolle's theorem to the following functions $f(x)$, with a and c chosen so that $f(a) = f(c) = 0$? (i) $\sin x$, (ii) $\cos x$, (iii)$x^2 - 3x + 2$, (iv) $x^2 + 7x + 3$, (v) $2x^3 - 9x^2 - 24x + k$.*

(i) If the consecutive values of x that make $\sin x = 0$ are $\alpha_1, \alpha_2, \ldots$ (actually $x = n\pi$, for any integer n) then Rolle's theorem implies that the derivative of $\sin x$, namely $\cos x$, has at least one zero lying between each pair of values α_i and α_{i+1}.

(ii) In an exactly similar way, we conclude that the derivative of $\cos x$, namely $- \sin x$, has at least one zero lying between consecutive pairs of zeros of $\cos x$. These two results taken together (but neither separately) imply that $\sin x$ and $\cos x$ have interleaving zeros.

(iii) For $f(x) = x^2 - 3x + 2$, $f(a) = f(c) = 0$ if a and c are taken as 1 and 2 respectively. Rolle's theorem then implies that $f'(x) = 2x - 3 = 0$ has a solution $x = b$ with b in the range $1 < b < 2$. This is obviously so, since $b = 3/2$.

(iv) With $f(x) = x^2 + 7x + 3$, the theorem tells us that if there are two roots of $x^2 + 7x + 3 = 0$ then they have the root of $f'(x) = 2x + 7 = 0$ lying between them. Thus if there are any (real) roots of $x^2 + 7x + 3 = 0$ then they lie one on either side of $x = -7/2$. The actual roots are $(-7 \pm \sqrt{37})/2$.

(v) If $f(x) = 2x^3 - 9x^2 - 24x + k$ then $f'(x) = 0$ is the equation $6x^2 - 18x - 24 = 0$, which has solutions $x = -1$ and $x = 4$. Consequently, if α_1 and α_2 are two different roots of $f(x) = 0$ then at least one of -1 and 4 must lie in the open interval α_1 to α_2. If, as is the case for a certain range of values of k, $f(x) = 0$ has three roots, α_1, α_2 and α_3, then $\alpha_1 < -1 < \alpha_2 < 4 < \alpha_3$.

In each case, as might be expected, the application of Rolle's theorem does no more than focus attention on particular ranges of values; it does not yield precise answers. ◄

Direct verification of the mean value theorem is straightforward when it is applied to simple functions. For example, if $f(x) = x^2$, it states that there is a value b in the interval $a < b < c$ such that

$$c^2 - a^2 = f(c) - f(a) = (c - a)f'(b) = (c - a)2b.$$

This is clearly so, since $b = (a + c)/2$ satisfies the relevant criteria.

As a slightly more complicated example we may consider a cubic equation, say $f(x) = x^3 + 2x^2 + 4x - 6 = 0$, between two specified values of x, say 1 and 2. In this case we need to verify that there is a value of x lying in the range $1 < x < 2$ that satisfies

$$18 - 1 = f(2) - f(1) = (2 - 1)f'(x) = 1(3x^2 + 4x + 4).$$

This is easily done, either by evaluating $3x^2 + 4x + 4 - 17$ at $x = 1$ and at $x = 2$ and checking that the values have opposite signs or by solving $3x^2 + 4x + 4 - 17 = 0$ and showing that one of the roots lies in the stated interval.

The following applications of the mean value theorem establish some general inequalities for two common functions.

► *Determine inequalities satisfied by $\ln x$ and $\sin x$ for suitable ranges of the real variable x.*

Since for positive values of its argument the derivative of $\ln x$ is x^{-1}, the mean value theorem gives us

$$\frac{\ln c - \ln a}{c - a} = \frac{1}{b}$$

for some b in $0 < a < b < c$. Further, since $a < b < c$ implies that $c^{-1} < b^{-1} < a^{-1}$, we have

$$\frac{1}{c} < \frac{\ln c - \ln a}{c - a} < \frac{1}{a},$$

or, multiplying through by $c - a$ and writing $c/a = x$ where $x > 1$,

$$1 - \frac{1}{x} < \ln x < x - 1.$$

Applying the mean value theorem to $\sin x$ shows that

$$\frac{\sin c - \sin a}{c - a} = \cos b$$

for some b lying between a and c. If a and c are restricted to lie in the range $0 \leq a < c \leq \pi$, in which the cosine function is monotonically decreasing (i.e. there are no turning points), we can deduce that

$$\cos c < \frac{\sin c - \sin a}{c - a} < \cos a. ◄$$

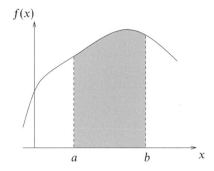

Figure 2.7   An integral as the area under a curve.

2.2 Integration

The notion of an integral as the area under a curve will be familiar to the reader. In figure 2.7, in which the solid line is a plot of a function $f(x)$, the shaded area represents the quantity denoted by

$$I = \int_a^b f(x)\,dx. \tag{2.21}$$

This expression is known as the *definite integral* of $f(x)$ between the *lower limit* $x = a$ and the *upper limit* $x = b$, and $f(x)$ is called the *integrand*.

2.2.1 Integration from first principles

The definition of an integral as the area under a curve is not a formal definition, but one that can be readily visualised. The formal definition of I involves subdividing the finite interval $a \le x \le b$ into a large number of subintervals, by defining intermediate points ξ_i such that $a = \xi_0 < \xi_1 < \xi_2 < \cdots < \xi_n = b$, and then forming the sum

$$S = \sum_{i=1}^n f(x_i)(\xi_i - \xi_{i-1}), \tag{2.22}$$

where x_i is an arbitrary point that lies in the range $\xi_{i-1} \le x_i \le \xi_i$ (see figure 2.8). If now n is allowed to tend to infinity in any way whatsoever, subject only to the restriction that the length of every subinterval ξ_{i-1} to ξ_i tends to zero, then S might, or might not, tend to a unique limit, I. If it does then the definite integral of $f(x)$ between a and b is defined as having the value I. If no unique limit exists the integral is undefined. For continuous functions and a finite interval $a \le x \le b$ the existence of a unique limit is assured and the integral is guaranteed to exist.

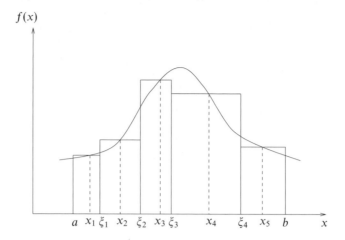

Figure 2.8   The evaluation of a definite integral by subdividing the interval $a \leq x \leq b$ into subintervals.

▶Evaluate from first principles the integral $I = \int_0^b x^2 \, dx$.

We first approximate the area under the curve $y = x^2$ between 0 and b by n rectangles of equal width h. If we take the value at the lower end of each subinterval (in the limit of an infinite number of subintervals we could equally well have chosen the value at the upper end) to give the height of the corresponding rectangle, then the area of the kth rectangle will be $(kh)^2 h = k^2 h^3$. The total area is thus

$$A = \sum_{k=0}^{n-1} k^2 h^3 = (h^3)\tfrac{1}{6}n(n-1)(2n-1),$$

where we have used the expression for the sum of the squares of the natural numbers derived in subsection 1.7.1. Now $h = b/n$ and so

$$A = \left(\frac{b^3}{n^3}\right)\frac{n}{6}(n-1)(2n-1) = \frac{b^3}{6}\left(1 - \frac{1}{n}\right)\left(2 - \frac{1}{n}\right).$$

As $n \to \infty$, $A \to b^3/3$, which is thus the value I of the integral. ◀

Some straightforward properties of definite integrals that are almost self-evident are as follows:

$$\int_a^b 0 \, dx = 0, \qquad \int_a^a f(x) \, dx = 0, \tag{2.23}$$

$$\int_a^c f(x) \, dx = \int_a^b f(x) \, dx + \int_b^c f(x) \, dx, \tag{2.24}$$

$$\int_a^b [f(x) + g(x)] \, dx = \int_a^b f(x) \, dx + \int_a^b g(x) \, dx. \tag{2.25}$$

60

Combining (2.23) and (2.24) with c set equal to a shows that

$$\int_a^b f(x)\,dx = -\int_b^a f(x)\,dx. \tag{2.26}$$

2.2.2 Integration as the inverse of differentiation

The definite integral has been defined as the area under a curve between two fixed limits. Let us now consider the integral

$$F(x) = \int_a^x f(u)\,du \tag{2.27}$$

in which the lower limit a remains fixed but the upper limit x is now variable. It will be noticed that this is essentially a restatement of (2.21), but that the variable x in the integrand has been replaced by a new variable u. It is conventional to rename the *dummy variable* in the integrand in this way in order that the same variable does not appear in both the integrand and the integration limits.

It is apparent from (2.27) that $F(x)$ is a continuous function of x, but at first glance the definition of an integral as the area under a curve does not connect with our assertion that integration is the inverse process to differentiation. However, by considering the integral (2.27) and using the elementary property (2.24), we obtain

$$F(x + \Delta x) = \int_a^{x+\Delta x} f(u)\,du$$

$$= \int_a^x f(u)\,du + \int_x^{x+\Delta x} f(u)\,du$$

$$= F(x) + \int_x^{x+\Delta x} f(u)\,du.$$

Rearranging and dividing through by Δx yields

$$\frac{F(x + \Delta x) - F(x)}{\Delta x} = \frac{1}{\Delta x}\int_x^{x+\Delta x} f(u)\,du.$$

Letting $\Delta x \to 0$ and using (2.1) we find that the LHS becomes dF/dx, whereas the RHS becomes $f(x)$. The latter conclusion follows because when Δx is small the value of the integral on the RHS is approximately $f(x)\Delta x$, and in the limit $\Delta x \to 0$ no approximation is involved. Thus

$$\frac{dF(x)}{dx} = f(x), \tag{2.28}$$

or, substituting for $F(x)$ from (2.27),

$$\frac{d}{dx}\left[\int_a^x f(u)\,du\right] = f(x).$$

From the last two equations it is clear that integration can be considered as the inverse of differentiation. However, we see from the above analysis that the lower limit a is arbitrary and so differentiation does not have a *unique* inverse. Any function $F(x)$ obeying (2.28) is called an *indefinite integral* of $f(x)$, though any two such functions can differ by at most an arbitrary additive constant. Since the lower limit is arbitrary, it is usual to write

$$F(x) = \int^x f(u)\,du \tag{2.29}$$

and explicitly include the arbitrary constant only when evaluating $F(x)$. The evaluation is conventionally written in the form

$$\int f(x)\,dx = F(x) + c \tag{2.30}$$

where c is called the *constant of integration*. It will be noticed that, in the absence of any integration limits, we use the same symbol for the arguments of both f and F. This can be confusing, but is sufficiently common practice that the reader needs to become familiar with it.

We also note that the definite integral of $f(x)$ between the fixed limits $x = a$ and $x = b$ can be written in terms of $F(x)$. From (2.27) we have

$$\int_a^b f(x)\,dx = \int_{x_0}^b f(x)\,dx - \int_{x_0}^a f(x)\,dx$$
$$= F(b) - F(a), \tag{2.31}$$

where x_0 is *any* third fixed point. Using the notation $F'(x) = dF/dx$, we may rewrite (2.28) as $F'(x) = f(x)$, and so express (2.31) as

$$\int_a^b F'(x)\,dx = F(b) - F(a) \equiv [F]_a^b.$$

In contrast to differentiation, where repeated applications of the product rule and/or the chain rule will always give the required derivative, it is not always possible to find the integral of an arbitrary function. Indeed, in most real physical problems exact integration cannot be performed and we have to revert to numerical approximations. Despite this cautionary note, it is in fact possible to integrate many simple functions and the following subsections introduce the most common types. Many of the techniques will be familiar to the reader and so are summarised by example.

2.2.3 Integration by inspection

The simplest method of integrating a function is by inspection. Some of the more elementary functions have well-known integrals that should be remembered. The reader will notice that these integrals are precisely the inverses of the derivatives found near the end of subsection 2.1.1. A few are presented below, using the form given in (2.30):

$$\int a\, dx = ax + c, \qquad \int ax^n\, dx = \frac{ax^{n+1}}{n+1} + c,$$

$$\int e^{ax}\, dx = \frac{e^{ax}}{a} + c, \qquad \int \frac{a}{x}\, dx = a\ln x + c,$$

$$\int a\cos bx\, dx = \frac{a\sin bx}{b} + c, \qquad \int a\sin bx\, dx = \frac{-a\cos bx}{b} + c,$$

$$\int a\tan bx\, dx = \frac{-a\ln(\cos bx)}{b} + c, \qquad \int a\cos bx \sin^n bx\, dx = \frac{a\sin^{n+1} bx}{b(n+1)} + c,$$

$$\int \frac{a}{a^2+x^2}\, dx = \tan^{-1}\left(\frac{x}{a}\right) + c, \qquad \int a\sin bx \cos^n bx\, dx = \frac{-a\cos^{n+1} bx}{b(n+1)} + c,$$

$$\int \frac{-1}{\sqrt{a^2-x^2}}\, dx = \cos^{-1}\left(\frac{x}{a}\right) + c, \qquad \int \frac{1}{\sqrt{a^2-x^2}}\, dx = \sin^{-1}\left(\frac{x}{a}\right) + c,$$

where the integrals that depend on n are valid for all $n \neq -1$ and where a and b are constants. In the two final results $|x| \leq a$.

2.2.4 Integration of sinusoidal functions

Integrals of the type $\int \sin^n x\, dx$ and $\int \cos^n x\, dx$ may be found by using trigonometric expansions. Two methods are applicable, one for odd n and the other for even n. They are best illustrated by example.

▶ Evaluate the integral $I = \int \sin^5 x\, dx$.

Rewriting the integral as a product of $\sin x$ and an even power of $\sin x$, and then using the relation $\sin^2 x = 1 - \cos^2 x$ yields

$$I = \int \sin^4 x \sin x\, dx$$

$$= \int (1 - \cos^2 x)^2 \sin x\, dx$$

$$= \int (1 - 2\cos^2 x + \cos^4 x) \sin x\, dx$$

$$= \int (\sin x - 2\sin x \cos^2 x + \sin x \cos^4 x)\, dx$$

$$= -\cos x + \tfrac{2}{3}\cos^3 x - \tfrac{1}{5}\cos^5 x + c,$$

where the integration has been carried out using the results of subsection 2.2.3. ◀

▶ *Evaluate the integral* $I = \int \cos^4 x \, dx$.

Rewriting the integral as a power of $\cos^2 x$ and then using the double-angle formula $\cos^2 x = \frac{1}{2}(1 + \cos 2x)$ yields

$$I = \int (\cos^2 x)^2 \, dx = \int \left(\frac{1 + \cos 2x}{2} \right)^2 dx$$

$$= \int \frac{1}{4}(1 + 2\cos 2x + \cos^2 2x) \, dx.$$

Using the double-angle formula again we may write $\cos^2 2x = \frac{1}{2}(1 + \cos 4x)$, and hence

$$I = \int \left[\frac{1}{4} + \frac{1}{2}\cos 2x + \frac{1}{8}(1 + \cos 4x) \right] dx$$

$$= \frac{1}{4}x + \frac{1}{4}\sin 2x + \frac{1}{8}x + \frac{1}{32}\sin 4x + c$$

$$= \frac{3}{8}x + \frac{1}{4}\sin 2x + \frac{1}{32}\sin 4x + c. \blacktriangleleft$$

2.2.5 Logarithmic integration

Integrals for which the integrand may be written as a fraction in which the numerator is the derivative of the denominator may be evaluated using

$$\int \frac{f'(x)}{f(x)} \, dx = \ln f(x) + c. \tag{2.32}$$

This follows directly from the differentiation of a logarithm as a function of a function (see subsection 2.1.3).

▶ *Evaluate the integral*

$$I = \int \frac{6x^2 + 2\cos x}{x^3 + \sin x} \, dx.$$

We note first that the numerator can be factorised to give $2(3x^2 + \cos x)$, and then that the quantity in brackets is the derivative of the denominator. Hence

$$I = 2 \int \frac{3x^2 + \cos x}{x^3 + \sin x} \, dx = 2\ln(x^3 + \sin x) + c. \blacktriangleleft$$

2.2.6 Integration using partial fractions

The method of partial fractions was discussed at some length in section 1.4, but in essence consists of the manipulation of a fraction (here the integrand) in such a way that it can be written as the sum of two or more simpler fractions. Again we illustrate the method by an example.

▶*Evaluate the integral*

$$I = \int \frac{1}{x^2 + x}\, dx.$$

We note that the denominator factorises to give $x(x + 1)$. Hence

$$I = \int \frac{1}{x(x + 1)}\, dx.$$

We now separate the fraction into two partial fractions and integrate directly:

$$I = \int \left(\frac{1}{x} - \frac{1}{x + 1} \right) dx = \ln x - \ln(x + 1) + c = \ln \left(\frac{x}{x + 1} \right) + c. \blacktriangleleft$$

2.2.7 Integration by substitution

Sometimes it is possible to make a substitution of variables that turns a complicated integral into a simpler one, which can then be integrated by a standard method. There are many useful substitutions and knowing which to use is a matter of experience. We now present a few examples of particularly useful substitutions.

▶*Evaluate the integral*

$$I = \int \frac{1}{\sqrt{1 - x^2}}\, dx.$$

Making the substitution $x = \sin u$, we note that $dx = \cos u\, du$, and hence

$$I = \int \frac{1}{\sqrt{1 - \sin^2 u}} \cos u\, du = \int \frac{1}{\sqrt{\cos^2 u}} \cos u\, du = \int du = u + c.$$

Now substituting back for u,

$$I = \sin^{-1} x + c.$$

This corresponds to one of the results given in subsection 2.2.3. ◀

Another particular example of integration by substitution is afforded by integrals of the form

$$I = \int \frac{1}{a + b \cos x}\, dx \quad \text{or} \quad I = \int \frac{1}{a + b \sin x}\, dx. \tag{2.33}$$

In these cases, making the substitution $t = \tan(x/2)$ yields integrals that can be solved more easily than the originals. Formulae expressing $\sin x$ and $\cos x$ in terms of t were derived in equations (1.32) and (1.33) (see p. 14), but before we can use them we must relate dx to dt as follows.

Since

$$\frac{dt}{dx} = \frac{1}{2}\sec^2\frac{x}{2} = \frac{1}{2}\left(1 + \tan^2\frac{x}{2}\right) = \frac{1 + t^2}{2},$$

the required relationship is

$$dx = \frac{2}{1 + t^2}\,dt. \tag{2.34}$$

> ▶Evaluate the integral
>
> $$I = \int \frac{2}{1 + 3\cos x}\,dx.$$

Rewriting $\cos x$ in terms of t and using (2.34) yields

$$I = \int \frac{2}{1 + 3\left[(1 - t^2)(1 + t^2)^{-1}\right]}\left(\frac{2}{1 + t^2}\right)dt$$

$$= \int \frac{2(1 + t^2)}{1 + t^2 + 3(1 - t^2)}\left(\frac{2}{1 + t^2}\right)dt$$

$$= \int \frac{2}{2 - t^2}\,dt = \int \frac{2}{(\sqrt{2} - t)(\sqrt{2} + t)}\,dt$$

$$= \int \frac{1}{\sqrt{2}}\left(\frac{1}{\sqrt{2} - t} + \frac{1}{\sqrt{2} + t}\right)dt$$

$$= -\frac{1}{\sqrt{2}}\ln(\sqrt{2} - t) + \frac{1}{\sqrt{2}}\ln(\sqrt{2} + t) + c$$

$$= \frac{1}{\sqrt{2}}\ln\left[\frac{\sqrt{2} + \tan(x/2)}{\sqrt{2} - \tan(x/2)}\right] + c. \blacktriangleleft$$

Integrals of a similar form to (2.33), but involving $\sin 2x$, $\cos 2x$, $\tan 2x$, $\sin^2 x$, $\cos^2 x$ or $\tan^2 x$ instead of $\cos x$ and $\sin x$, should be evaluated by using the substitution $t = \tan x$. In this case

$$\sin x = \frac{t}{\sqrt{1 + t^2}}, \quad \cos x = \frac{1}{\sqrt{1 + t^2}} \quad \text{and} \quad dx = \frac{dt}{1 + t^2}. \tag{2.35}$$

A final example of the evaluation of integrals using substitution is the method of completing the square (cf. subsection 1.7.3).

►*Evaluate the integral*

$$I = \int \frac{1}{x^2 + 4x + 7}\, dx.$$

We can write the integral in the form

$$I = \int \frac{1}{(x+2)^2 + 3}\, dx.$$

Substituting $y = x + 2$, we find $dy = dx$ and hence

$$I = \int \frac{1}{y^2 + 3}\, dy.$$

Hence, by comparison with the table of standard integrals (see subsection 2.2.3)

$$I = \frac{\sqrt{3}}{3} \tan^{-1}\left(\frac{y}{\sqrt{3}}\right) + c = \frac{\sqrt{3}}{3} \tan^{-1}\left(\frac{x+2}{\sqrt{3}}\right) + c. \;\blacktriangleleft$$

2.2.8 Integration by parts

Integration by parts is the integration analogy of product differentiation. The principle is to break down a complicated function into two functions, at least one of which can be integrated by inspection. The method in fact relies on the result for the differentiation of a product. Recalling from (2.6) that

$$\frac{d}{dx}(uv) = u\frac{dv}{dx} + \frac{du}{dx}v,$$

where u and v are functions of x, we now integrate to find

$$uv = \int u\frac{dv}{dx}\, dx + \int \frac{du}{dx} v\, dx.$$

Rearranging into the standard form for integration by parts gives

$$\int u\frac{dv}{dx}\, dx = uv - \int \frac{du}{dx} v\, dx. \tag{2.36}$$

Integration by parts is often remembered for practical purposes in the form *the integral of a product of two functions is equal to {the first times the integral of the second} minus the integral of {the derivative of the first times the integral of the second}*. Here, u is 'the first' and dv/dx is 'the second'; clearly the integral v of 'the second' must be determinable by inspection.

►*Evaluate the integral* $I = \int x \sin x\, dx.$

In the notation given above, we identify x with u and $\sin x$ with dv/dx. Hence $v = -\cos x$ and $du/dx = 1$ and so using (2.36)

$$I = x(-\cos x) - \int (1)(-\cos x)\, dx = -x\cos x + \sin x + c. \;\blacktriangleleft$$

The separation of the functions is not always so apparent, as is illustrated by the following example.

▶Evaluate the integral $I = \int x^3 e^{-x^2}\, dx$.

Firstly we rewrite the integral as

$$I = \int x^2 \left(x e^{-x^2} \right) dx.$$

Now, using the notation given above, we identify x^2 with u and xe^{-x^2} with dv/dx. Hence $v = -\frac{1}{2}e^{-x^2}$ and $du/dx = 2x$, so that

$$I = -\tfrac{1}{2}x^2 e^{-x^2} - \int (-x)e^{-x^2}\, dx = -\tfrac{1}{2}x^2 e^{-x^2} - \tfrac{1}{2}e^{-x^2} + c. \blacktriangleleft$$

A trick that is sometimes useful is to take '1' as one factor of the product, as is illustrated by the following example.

▶Evaluate the integral $I = \int \ln x\, dx$.

Firstly we rewrite the integral as

$$I = \int (\ln x)\, 1\, dx.$$

Now, using the notation above, we identify $\ln x$ with u and 1 with dv/dx. Hence we have $v = x$ and $du/dx = 1/x$, and so

$$I = (\ln x)(x) - \int \left(\frac{1}{x} \right) x\, dx = x \ln x - x + c. \blacktriangleleft$$

It is sometimes necessary to integrate by parts more than once. In doing so, we may occasionally re-encounter the original integral I. In such cases we can obtain a linear algebraic equation for I that can be solved to obtain its value.

▶Evaluate the integral $I = \int e^{ax} \cos bx\, dx$.

Integrating by parts, taking e^{ax} as the first function, we find

$$I = e^{ax} \left(\frac{\sin bx}{b} \right) - \int a e^{ax} \left(\frac{\sin bx}{b} \right) dx,$$

where, for convenience, we have omitted the constant of integration. Integrating by parts a second time,

$$I = e^{ax} \left(\frac{\sin bx}{b} \right) - a e^{ax} \left(\frac{-\cos bx}{b^2} \right) + \int a^2 e^{ax} \left(\frac{-\cos bx}{b^2} \right) dx.$$

Notice that the integral on the RHS is just $-a^2/b^2$ times the original integral I. Thus

$$I = e^{ax} \left(\frac{1}{b} \sin bx + \frac{a}{b^2} \cos bx \right) - \frac{a^2}{b^2} I.$$

Rearranging this expression to obtain I explicitly and including the constant of integration we find

$$I = \frac{e^{ax}}{a^2 + b^2}(b \sin bx + a \cos bx) + c. \tag{2.37}$$

Another method of evaluating this integral, using the exponential of a complex number, is given in section 3.6. ◀

2.2.9 Reduction formulae

Integration using reduction formulae is a process that involves first evaluating a simple integral and then, in stages, using it to find a more complicated integral.

> ▶ Using integration by parts, find a relationship between I_n and I_{n-1} where
>
> $$I_n = \int_0^1 (1 - x^3)^n \, dx$$
>
> and n is any positive integer. Hence evaluate $I_2 = \int_0^1 (1 - x^3)^2 \, dx$.

Writing the integrand as a product and separating the integral into two we find

$$I_n = \int_0^1 (1 - x^3)(1 - x^3)^{n-1} \, dx$$

$$= \int_0^1 (1 - x^3)^{n-1} \, dx - \int_0^1 x^3(1 - x^3)^{n-1} \, dx.$$

The first term on the RHS is clearly I_{n-1} and so, writing the integrand in the second term on the RHS as a product,

$$I_n = I_{n-1} - \int_0^1 (x)x^2(1 - x^3)^{n-1} \, dx.$$

Integrating by parts we find

$$I_n = I_{n-1} + \left[\frac{x}{3n}(1 - x^3)^n \right]_0^1 - \int_0^1 \frac{1}{3n}(1 - x^3)^n \, dx$$

$$= I_{n-1} + 0 - \frac{1}{3n}I_n,$$

which on rearranging gives

$$I_n = \frac{3n}{3n + 1}I_{n-1}.$$

We now have a relation connecting successive integrals. Hence, if we can evaluate I_0, we can find I_1, I_2 etc. Evaluating I_0 is trivial:

$$I_0 = \int_0^1 (1 - x^3)^0 \, dx = \int_0^1 dx = [x]_0^1 = 1.$$

Hence

$$I_1 = \frac{(3 \times 1)}{(3 \times 1) + 1} \times 1 = \frac{3}{4}, \qquad I_2 = \frac{(3 \times 2)}{(3 \times 2) + 1} \times \frac{3}{4} = \frac{9}{14}.$$

Although the first few I_n could be evaluated by direct multiplication, this becomes tedious for integrals containing higher values of n; these are therefore best evaluated using the reduction formula. ◀

2.2.10 Infinite and improper integrals

The definition of an integral given previously does not allow for cases in which either of the limits of integration is infinite (an *infinite integral*) or for cases in which $f(x)$ is infinite in some part of the range (an *improper integral*), e.g. $f(x) = (2 - x)^{-1/4}$ near the point $x = 2$. Nevertheless, modification of the definition of an integral gives infinite and improper integrals each a meaning.

In the case of an integral $I = \int_a^b f(x)\,dx$, the infinite integral, in which b tends to ∞, is defined by

$$I = \int_a^\infty f(x)\,dx = \lim_{b \to \infty} \int_a^b f(x)\,dx = \lim_{b \to \infty} F(b) - F(a).$$

As previously, $F(x)$ is the indefinite integral of $f(x)$ and $\lim_{b \to \infty} F(b)$ means the limit (or value) that $F(b)$ approaches as $b \to \infty$; it is evaluated *after* calculating the integral. The formal concept of a limit will be introduced in chapter 4.

▶ *Evaluate the integral*

$$I = \int_0^\infty \frac{x}{(x^2 + a^2)^2}\,dx.$$

Integrating, we find $F(x) = -\frac{1}{2}(x^2 + a^2)^{-1} + c$ and so

$$I = \lim_{b \to \infty} \left[\frac{-1}{2(b^2 + a^2)} \right] - \left(\frac{-1}{2a^2} \right) = \frac{1}{2a^2}. \quad \blacktriangleleft$$

For the case of improper integrals, we adopt the approach of excluding the unbounded range from the integral. For example, if the integrand $f(x)$ is infinite at $x = c$ (say), $a \le c \le b$ then

$$\int_a^b f(x)\,dx = \lim_{\delta \to 0} \int_a^{c-\delta} f(x)\,dx + \lim_{\epsilon \to 0} \int_{c+\epsilon}^b f(x)\,dx.$$

▶ *Evaluate the integral $I = \int_0^2 (2 - x)^{-1/4}\,dx$.*

Integrating directly,

$$I = \lim_{\epsilon \to 0} \left[-\tfrac{4}{3}(2 - x)^{3/4} \right]_0^{2-\epsilon} = \lim_{\epsilon \to 0} \left[-\tfrac{4}{3}\epsilon^{3/4} \right] + \tfrac{4}{3}2^{3/4} = \left(\tfrac{4}{3} \right) 2^{3/4}. \quad \blacktriangleleft$$

2.2.11 Integration in plane polar coordinates

In plane polar coordinates ρ, ϕ, a curve is defined by its distance ρ from the origin as a function of the angle ϕ between the line joining a point on the curve to the origin and the x-axis, i.e. $\rho = \rho(\phi)$. The area of an element is given by

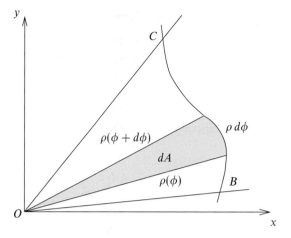

Figure 2.9 Finding the area of a sector OBC defined by the curve $\rho(\phi)$ and the radii OB, OC, at angles to the x-axis ϕ_1, ϕ_2 respectively.

$dA = \frac{1}{2}\rho^2 \, d\phi$, as illustrated in figure 2.9, and hence the total area between two angles ϕ_1 and ϕ_2 is given by

$$A = \int_{\phi_1}^{\phi_2} \frac{1}{2}\rho^2 \, d\phi. \tag{2.38}$$

An immediate observation is that the area of a circle of radius a is given by

$$A = \int_0^{2\pi} \frac{1}{2}a^2 \, d\phi = \left[\frac{1}{2}a^2\phi\right]_0^{2\pi} = \pi a^2.$$

▶ *The equation in polar coordinates of an ellipse with semi-axes a and b is*

$$\frac{1}{\rho^2} = \frac{\cos^2 \phi}{a^2} + \frac{\sin^2 \phi}{b^2}.$$

Find the area A of the ellipse.

Using (2.38) and symmetry, we have

$$A = \frac{1}{2}\int_0^{2\pi} \frac{a^2 b^2}{b^2 \cos^2 \phi + a^2 \sin^2 \phi} \, d\phi = 2a^2 b^2 \int_0^{\pi/2} \frac{1}{b^2 \cos^2 \phi + a^2 \sin^2 \phi} \, d\phi.$$

To evaluate this integral we write $t = \tan \phi$ and use (2.35):

$$A = 2a^2 b^2 \int_0^\infty \frac{1}{b^2 + a^2 t^2} \, dt = 2b^2 \int_0^\infty \frac{1}{(b/a)^2 + t^2} \, dt.$$

Finally, from the list of standard integrals (see subsection 2.2.3),

$$A = 2b^2 \left[\frac{1}{(b/a)} \tan^{-1} \frac{t}{(b/a)}\right]_0^\infty = 2ab\left(\frac{\pi}{2} - 0\right) = \pi ab. \blacktriangleleft$$

2.2.12 Integral inequalities

Consider the functions $f(x)$, $\phi_1(x)$ and $\phi_2(x)$ such that $\phi_1(x) \le f(x) \le \phi_2(x)$ for all x in the range $a \le x \le b$. It immediately follows that

$$\int_a^b \phi_1(x)\,dx \le \int_a^b f(x)\,dx \le \int_a^b \phi_2(x)\,dx, \tag{2.39}$$

which gives us a way of estimating an integral that is difficult to evaluate explicitly.

▶*Show that the value of the integral*

$$I = \int_0^1 \frac{1}{(1+x^2+x^3)^{1/2}}\,dx$$

lies between 0.810 and 0.882.

We note that for x in the range $0 \le x \le 1$, $0 \le x^3 \le x^2$. Hence

$$(1+x^2)^{1/2} \le (1+x^2+x^3)^{1/2} \le (1+2x^2)^{1/2},$$

and so

$$\frac{1}{(1+x^2)^{1/2}} \ge \frac{1}{(1+x^2+x^3)^{1/2}} \ge \frac{1}{(1+2x^2)^{1/2}}.$$

Consequently,

$$\int_0^1 \frac{1}{(1+x^2)^{1/2}}\,dx \ge \int_0^1 \frac{1}{(1+x^2+x^3)^{1/2}}\,dx \ge \int_0^1 \frac{1}{(1+2x^2)^{1/2}}\,dx,$$

from which we obtain

$$\left[\ln(x+\sqrt{1+x^2})\right]_0^1 \ge I \ge \left[\tfrac{1}{\sqrt{2}}\ln\left(x+\sqrt{\tfrac{1}{2}+x^2}\right)\right]_0^1$$

$$0.8814 \ge I \ge 0.8105$$

$$0.882 \ge I \ge 0.810.$$

In the last line the calculated values have been rounded to three significant figures, one rounded up and the other rounded down so that the proved inequality cannot be unknowingly made invalid. ◀

2.2.13 Applications of integration

Mean value of a function

The mean value m of a function between two limits a and b is defined by

$$m = \frac{1}{b-a}\int_a^b f(x)\,dx. \tag{2.40}$$

The mean value may be thought of as the height of the rectangle that has the same area (over the same interval) as the area under the curve $f(x)$. This is illustrated in figure 2.10.

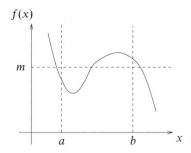

Figure 2.10   The mean value m of a function.

▶ *Find the mean value m of the function $f(x) = x^2$ between the limits $x = 2$ and $x = 4$.*

Using (2.40),

$$m - \frac{1}{4-2} \int_2^4 x^2 \, dx = \frac{1}{2} \left[\frac{x^3}{3} \right]_2^4 = \frac{1}{2} \left(\frac{4^3}{3} - \frac{2^3}{3} \right) = \frac{28}{3}. \quad ◀$$

Finding the length of a curve

Finding the area between a curve and certain straight lines provides one example of the use of integration. Another is in finding the length of a curve. If a curve is defined by $y = f(x)$ then the distance along the curve, Δs, that corresponds to small changes Δx and Δy in x and y is given by

$$\Delta s \approx \sqrt{(\Delta x)^2 + (\Delta y)^2}; \tag{2.41}$$

this follows directly from Pythagoras' theorem (see figure 2.11). Dividing (2.41) through by Δx and letting $\Delta x \to 0$ we obtain[§]

$$\frac{ds}{dx} = \sqrt{1 + \left(\frac{dy}{dx} \right)^2}. \tag{2.42}$$

Clearly the total length s of the curve between the points $x = a$ and $x = b$ is then given by integrating both sides of the equation:

$$s = \int_a^b \sqrt{1 + \left(\frac{dy}{dx} \right)^2} \, dx. \tag{2.43}$$

[§] Instead of considering small changes Δx and Δy and letting these tend to zero, we could have derived (2.42) by considering infinitesimal changes dx and dy from the start. After writing $(ds)^2 = (dx)^2 + (dy)^2$, (2.42) may be deduced by using the formal device of dividing through by dx. Although not mathematically rigorous, this method is often used and generally leads to the correct result.

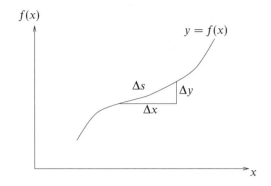

Figure 2.11 The distance moved along a curve, Δs, corresponding to the small changes Δx and Δy.

In plane polar coordinates,

$$ds = \sqrt{(dr)^2 + (r\, d\phi)^2} \quad \Rightarrow \quad s = \int_{r_1}^{r_2} \sqrt{1 + r^2 \left(\frac{d\phi}{dr}\right)^2}\, dr.$$

(2.44)

▶ *Find the length of the curve $y = x^{3/2}$ from $x = 0$ to $x = 2$.*

Using (2.43) and noting that $dy/dx = \frac{3}{2}\sqrt{x}$, the length s of the curve is given by

$$s = \int_0^2 \sqrt{1 + \tfrac{9}{4}x}\, dx$$

$$= \left[\tfrac{2}{3} \left(\tfrac{4}{9}\right) \left(1 + \tfrac{9}{4}x\right)^{3/2} \right]_0^2 = \tfrac{8}{27} \left[\left(1 + \tfrac{9}{4}x\right)^{3/2} \right]_0^2$$

$$= \tfrac{8}{27} \left[\left(\tfrac{11}{2}\right)^{3/2} - 1 \right]. \quad ◀$$

Surfaces of revolution

Consider the surface S formed by rotating the curve $y = f(x)$ about the x-axis (see figure 2.12). The surface area of the 'collar' formed by rotating an element of the curve, ds, about the x-axis is $2\pi y\, ds$, and hence the total surface area is

$$S = \int_a^b 2\pi y\, ds.$$

Since $(ds)^2 = (dx)^2 + (dy)^2$ from (2.41), the total surface area between the planes $x = a$ and $x = b$ is

$$S = \int_a^b 2\pi y \sqrt{1 + \left(\frac{dy}{dx}\right)^2}\, dx.$$

(2.45)

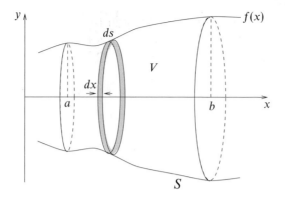

Figure 2.12   The surface and volume of revolution for the curve $y = f(x)$.

▶ *Find the surface area of a cone formed by rotating about the x-axis the line $y = 2x$ between $x = 0$ and $x = h$.*

Using (2.45), the surface area is given by

$$S = \int_0^h (2\pi)2x \sqrt{1 + \left[\frac{d}{dx}(2x)\right]^2} \, dx$$

$$= \int_0^h 4\pi x \left(1 + 2^2\right)^{1/2} dx = \int_0^h 4\sqrt{5}\pi x \, dx$$

$$= \left[2\sqrt{5}\pi x^2\right]_0^h = 2\sqrt{5}\pi(h^2 - 0) = 2\sqrt{5}\pi h^2. \; ◀$$

We note that a surface of revolution may also be formed by rotating a line about the y-axis. In this case the surface area between $y = a$ and $y = b$ is

$$S = \int_a^b 2\pi x \sqrt{1 + \left(\frac{dx}{dy}\right)^2} \, dy. \tag{2.46}$$

Volumes of revolution

The volume V enclosed by rotating the curve $y = f(x)$ about the x-axis can also be found (see figure 2.12). The volume of the disc between x and $x + dx$ is given by $dV = \pi y^2 \, dx$. Hence the total volume between $x = a$ and $x = b$ is

$$V = \int_a^b \pi y^2 \, dx. \tag{2.47}$$

▶*Find the volume of a cone enclosed by the surface formed by rotating about the x-axis the line $y = 2x$ between $x = 0$ and $x = h$.*

Using (2.47), the volume is given by

$$V = \int_0^h \pi(2x)^2 \, dx = \int_0^h 4\pi x^2 \, dx$$

$$= \left[\tfrac{4}{3}\pi x^3\right]_0^h = \tfrac{4}{3}\pi(h^3 - 0) = \tfrac{4}{3}\pi h^3. \blacktriangleleft$$

As before, it is also possible to form a volume of revolution by rotating a curve about the y-axis. In this case the volume enclosed between $y = a$ and $y = b$ is

$$V = \int_a^b \pi x^2 \, dy. \tag{2.48}$$

2.3 Exercises

2.1   Obtain the following derivatives from first principles:

(a) the first derivative of $3x + 4$;
(b) the first, second and third derivatives of $x^2 + x$;
(c) the first derivative of $\sin x$.

2.2   Find from first principles the first derivative of $(x+3)^2$ and compare your answer with that obtained using the chain rule.

2.3   Find the first derivatives of

(a) $x^2 \exp x$, (b) $2\sin x \cos x$, (c) $\sin 2x$, (d) $x \sin ax$,
(e) $(\exp ax)(\sin ax)\tan^{-1} ax$, (f) $\ln(x^a + x^{-a})$,
(g) $\ln(a^x + a^{-x})$, (h) x^x.

2.4   Find the first derivatives of

(a) $x/(a + x)^2$, (b) $x/(1 - x)^{1/2}$, (c) $\tan x$, as $\sin x / \cos x$,
(d) $(3x^2 + 2x + 1)/(8x^2 - 4x + 2)$.

2.5   Use result (2.12) to find the first derivatives of

(a) $(2x + 3)^{-3}$, (b) $\sec^2 x$, (c) $\operatorname{cosech}^3 3x$, (d) $1/\ln x$, (e) $1/[\sin^{-1}(x/a)]$.

2.6   Show that the function $y(x) = \exp(-|x|)$ defined by

$$y(x) = \begin{cases} \exp x & \text{for } x < 0, \\ 1 & \text{for } x = 0, \\ \exp(-x) & \text{for } x > 0, \end{cases}$$

is *not* differentiable at $x = 0$. Consider the limiting process for both $\Delta x > 0$ and $\Delta x < 0$.

2.7   Find dy/dx if $x = (t - 2)/(t + 2)$ and $y = 2t/(t + 1)$ for $-\infty < t < \infty$. Show that it is always non-negative, and make use of this result in sketching the curve of y as a function of x.

2.8   If $2y + \sin y + 5 = x^4 + 4x^3 + 2\pi$, show that $dy/dx = 16$ when $x = 1$.

2.9   Find the second derivative of $y(x) = \cos[(\pi/2) - ax]$. Now set $a = 1$ and verify that the result is the same as that obtained by first setting $a = 1$ and simplifying $y(x)$ before differentiating.

2.10   The function $y(x)$ is defined by $y(x) = (1 + x^m)^n$.

(a) Use the chain rule to show that the first derivative of y is $nmx^{m-1}(1 + x^m)^{n-1}$.

(b) The binomial expansion (see section 1.5) of $(1 + z)^n$ is

$$(1 + z)^n = 1 + nz + \frac{n(n-1)}{2!}z^2 + \cdots + \frac{n(n-1)\cdots(n-r+1)}{r!}z^r + \cdots .$$

Keeping only the terms of zeroth and first order in dx, apply this result twice to derive result (a) from first principles.

(c) Expand y in a series of powers of x before differentiating term by term. Show that the result is the series obtained by expanding the answer given for dy/dx in (a).

2.11   Show by differentiation and substitution that the differential equation

$$4x^2 \frac{d^2 y}{dx^2} - 4x \frac{dy}{dx} + (4x^2 + 3)y = 0$$

has a solution of the form $y(x) = x^n \sin x$, and find the value of n.

2.12   Find the positions and natures of the stationary points of the following functions:

(a) $x^3 - 3x + 3$; (b) $x^3 - 3x^2 + 3x$; (c) $x^3 + 3x + 3$;
(d) $\sin ax$ with $a \neq 0$; (e) $x^5 + x^3$; (f) $x^5 - x^3$.

2.13   Show that the lowest value taken by the function $3x^4 + 4x^3 - 12x^2 + 6$ is -26.

2.14   By finding their stationary points and examining their general forms, determine the range of values that each of the following functions $y(x)$ can take. In each case make a sketch-graph incorporating the features you have identified.

(a) $y(x) = (x - 1)/(x^2 + 2x + 6)$.
(b) $y(x) = 1/(4 + 3x - x^2)$.
(c) $y(x) = (8 \sin x)/(15 + 8 \tan^2 x)$.

2.15   Show that $y(x) = xa^{2x} \exp x^2$ has no stationary points other than $x = 0$, if $\exp(-\sqrt{2}) < a < \exp(\sqrt{2})$.

2.16   The curve $4y^3 = a^2(x + 3y)$ can be parameterised as $x = a \cos 3\theta$, $y = a \cos \theta$.

(a) Obtain expressions for dy/dx (i) by implicit differentiation and (ii) in parameterised form. Verify that they are equivalent.

(b) Show that the only point of inflection occurs at the origin. Is it a stationary point of inflection?

(c) Use the information gained in (a) and (b) to sketch the curve, paying particular attention to its shape near the points $(-a, a/2)$ and $(a, -a/2)$ and to its slope at the 'end points' (a, a) and $(-a, -a)$.

2.17   The parametric equations for the motion of a charged particle released from rest in electric and magnetic fields at right angles to each other take the forms

$$x = a(\theta - \sin \theta), \qquad y = a(1 - \cos \theta).$$

Show that the tangent to the curve has slope $\cot(\theta/2)$. Use this result at a few calculated values of x and y to sketch the form of the particle's trajectory.

2.18   Show that the maximum curvature on the catenary $y(x) = a \cosh(x/a)$ is $1/a$. You will need some of the results about hyperbolic functions stated in subsection 3.7.6.

2.19   The curve whose equation is $x^{2/3} + y^{2/3} = a^{2/3}$ for positive x and y and which is completed by its symmetric reflections in both axes is known as an astroid. Sketch it and show that its radius of curvature in the first quadrant is $3(axy)^{1/3}$.

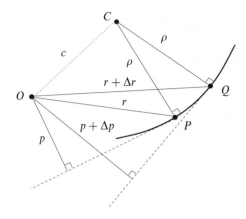

Figure 2.13   The coordinate system described in exercise 2.20.

2.20   A two-dimensional coordinate system useful for orbit problems is the tangential-polar coordinate system (figure 2.13). In this system a curve is defined by r, the distance from a fixed point O to a general point P of the curve, and p, the perpendicular distance from O to the tangent to the curve at P. By proceeding as indicated below, show that the radius of curvature, ρ, at P can be written in the form $\rho = r\, dr/dp$.

Consider two neighbouring points, P and Q, on the curve. The normals to the curve through those points meet at C, with (in the limit $Q \to P$) $CP = CQ = \rho$. Apply the cosine rule to triangles OPC and OQC to obtain two expressions for c^2, one in terms of r and p and the other in terms of $r + \Delta r$ and $p + \Delta p$. By equating them and letting $Q \to P$ deduce the stated result.

2.21   Use Leibnitz' theorem to find

(a) the second derivative of $\cos x \sin 2x$,
(b) the third derivative of $\sin x \ln x$,
(c) the fourth derivative of $(2x^3 + 3x^2 + x + 2) \exp 2x$.

2.22   If $y = \exp(-x^2)$, show that $dy/dx = -2xy$ and hence, by applying Leibnitz' theorem, prove that for $n \geq 1$

$$y^{(n+1)} + 2xy^{(n)} + 2ny^{(n-1)} = 0.$$

2.23   Use the properties of functions at their turning points to do the following:

(a) By considering its properties near $x = 1$, show that $f(x) = 5x^4 - 11x^3 + 26x^2 - 44x + 24$ takes negative values for some range of x.
(b) Show that $f(x) = \tan x - x$ cannot be negative for $0 \leq x < \pi/2$, and deduce that $g(x) = x^{-1} \sin x$ decreases monotonically in the same range.

2.24   Determine what can be learned from applying Rolle's theorem to the following functions $f(x)$: (a) e^x; (b) $x^2 + 6x$; (c) $2x^2 + 3x + 1$; (d) $2x^2 + 3x + 2$; (e) $2x^3 - 21x^2 + 60x + k$. (f) If $k = -45$ in (e), show that $x = 3$ is one root of $f(x) = 0$, find the other roots, and verify that the conclusions from (e) are satisfied.

2.25   By applying Rolle's theorem to $x^n \sin nx$, where n is an arbitrary positive integer, show that $\tan nx + x = 0$ has a solution α_1 with $0 < \alpha_1 < \pi/n$. Apply the theorem a second time to obtain the nonsensical result that there is a real α_2 in $0 < \alpha_2 < \pi/n$, such that $\cos^2(n\alpha_2) = -n$. Explain why this incorrect result arises.

2.26   Use the mean value theorem to establish bounds in the following cases.

(a) For $-\ln(1-y)$, by considering $\ln x$ in the range $0 < 1 - y < x < 1$.
(b) For $e^y - 1$, by considering $e^x - 1$ in the range $0 < x < y$.

2.27   For the function $y(x) = x^2 \exp(-x)$ obtain a simple relationship between y and dy/dx and then, by applying Leibnitz' theorem, prove that

$$xy^{(n+1)} + (n + x - 2)y^{(n)} + ny^{(n-1)} = 0.$$

2.28   Use Rolle's theorem to deduce that, if the equation $f(x) = 0$ has a repeated root x_1, then x_1 is also a root of the equation $f'(x) = 0$.

(a) Apply this result to the 'standard' quadratic equation $ax^2 + bx + c = 0$, to show that a necessary condition for equal roots is $b^2 = 4ac$.
(b) Find all the roots of $f(x) = x^3 + 4x^2 - 3x - 18 = 0$, given that one of them is a repeated root.
(c) The equation $f(x) = x^4 + 4x^3 + 7x^2 + 6x + 2 = 0$ has a repeated integer root. How many real roots does it have altogether?

2.29   Show that the curve $x^3 + y^3 - 12x - 8y - 16 = 0$ touches the x-axis.
2.30   Find the following indefinite integrals:

(a) $\int (4 + x^2)^{-1}\, dx$; (b) $\int (8 + 2x - x^2)^{-1/2}\, dx$ for $2 \le x \le 4$;
(c) $\int (1 + \sin\theta)^{-1}\, d\theta$; (d) $\int (x\sqrt{1-x})^{-1}\, dx$ for $0 < x \le 1$.

2.31   Find the indefinite integrals J of the following ratios of polynomials:

(a) $(x + 3)/(x^2 + x - 2)$;
(b) $(x^3 + 5x^2 + 8x + 12)/(2x^2 + 10x + 12)$;
(c) $(3x^2 + 20x + 28)/(x^2 + 6x + 9)$;
(d) $x^3/(a^8 + x^8)$.

2.32   Express $x^2(ax + b)^{-1}$ as the sum of powers of x and another integrable term, and hence evaluate

$$\int_0^{b/a} \frac{x^2}{ax + b}\, dx.$$

2.33   Find the integral J of $(ax^2 + bx + c)^{-1}$, with $a \ne 0$, distinguishing between the cases (i) $b^2 > 4ac$, (ii) $b^2 < 4ac$ and (iii) $b^2 = 4ac$.
2.34   Use logarithmic integration to find the indefinite integrals J of the following:

(a) $\sin 2x/(1 + 4\sin^2 x)$;
(b) $e^x/(e^x - e^{-x})$;
(c) $(1 + x\ln x)/(x\ln x)$;
(d) $[x(x^n + a^n)]^{-1}$.

2.35   Find the derivative of $f(x) = (1 + \sin x)/\cos x$ and hence determine the indefinite integral J of $\sec x$.
2.36   Find the indefinite integrals, J, of the following functions involving sinusoids:

(a) $\cos^5 x - \cos^3 x$;
(b) $(1 - \cos x)/(1 + \cos x)$;
(c) $\cos x \sin x/(1 + \cos x)$;
(d) $\sec^2 x/(1 - \tan^2 x)$.

2.37   By making the substitution $x = a\cos^2\theta + b\sin^2\theta$, evaluate the definite integrals J between limits a and b $(> a)$ of the following functions:

(a) $[(x - a)(b - x)]^{-1/2}$;
(b) $[(x - a)(b - x)]^{1/2}$;

(c) $[(x-a)/(b-x)]^{1/2}$.

2.38 Determine whether the following integrals exist and, where they do, evaluate them:

(a) $\int_0^\infty \exp(-\lambda x)\,dx$; (b) $\int_{-\infty}^\infty \dfrac{x}{(x^2+a^2)^2}\,dx$;

(c) $\int_1^\infty \dfrac{1}{x+1}\,dx$; (d) $\int_0^1 \dfrac{1}{x^2}\,dx$;

(e) $\int_0^{\pi/2} \cot\theta\,d\theta$; (f) $\int_0^1 \dfrac{x}{(1-x^2)^{1/2}}\,dx$.

2.39 Use integration by parts to evaluate the following:

(a) $\int_0^y x^2 \sin x\,dx$; (b) $\int_1^y x\ln x\,dx$;

(c) $\int_0^y \sin^{-1} x\,dx$; (d) $\int_1^y \ln(a^2+x^2)/x^2\,dx$.

2.40 Show, using the following methods, that the indefinite integral of $x^3/(x+1)^{1/2}$ is

$$J = \tfrac{2}{35}(5x^3 - 6x^2 + 8x - 16)(x+1)^{1/2} + c.$$

(a) Repeated integration by parts.
(b) Setting $x+1 = u^2$ and determining dJ/du as $(dJ/dx)(dx/du)$.

2.41 The gamma function $\Gamma(n)$ is defined for all $n > -1$ by

$$\Gamma(n+1) = \int_0^\infty x^n e^{-x}\,dx.$$

Find a recurrence relation connecting $\Gamma(n+1)$ and $\Gamma(n)$.

(a) Deduce (i) the value of $\Gamma(n+1)$ when n is a non-negative integer, and (ii) the value of $\Gamma\left(\tfrac{7}{2}\right)$, given that $\Gamma\left(\tfrac{1}{2}\right) = \sqrt{\pi}$.
(b) Now, taking factorial m for *any* m to be defined by $m! = \Gamma(m+1)$, evaluate $\left(-\tfrac{3}{2}\right)!$.

2.42 Define $J(m,n)$, for non-negative integers m and n, by the integral

$$J(m,n) = \int_0^{\pi/2} \cos^m\theta \sin^n\theta\,d\theta.$$

(a) Evaluate $J(0,0)$, $J(0,1)$, $J(1,0)$, $J(1,1)$, $J(m,1)$, $J(1,n)$.
(b) Using integration by parts, prove that, for m and n both > 1,

$$J(m,n) = \frac{m-1}{m+n}J(m-2,n) \quad\text{and}\quad J(m,n) = \frac{n-1}{m+n}J(m,n-2).$$

(c) Evaluate (i) $J(5,3)$, (ii) $J(6,5)$ and (iii) $J(4,8)$.

2.43 By integrating by parts twice, prove that I_n as defined in the first equality below for positive integers n has the value given in the second equality:

$$I_n = \int_0^{\pi/2} \sin n\theta \cos\theta\,d\theta = \frac{n - \sin(n\pi/2)}{n^2 - 1}.$$

2.44 Evaluate the following definite integrals:

(a) $\int_0^\infty xe^{-x}\,dx$; (b) $\int_0^1 [(x^3+1)/(x^4+4x+1)]\,dx$;

(c) $\int_0^{\pi/2}[a + (a-1)\cos\theta]^{-1}\,d\theta$ with $a > \tfrac{1}{2}$; (d) $\int_{-\infty}^\infty (x^2+6x+18)^{-1}\,dx$.

2.45   If J_r is the integral

$$\int_0^\infty x^r \exp(-x^2)\,dx$$

show that

(a) $J_{2r+1} = (r!)/2$,
(b) $J_{2r} = 2^{-r}(2r-1)(2r-3)\cdots(5)(3)(1)\,J_0$.

2.46   Find positive constants a, b such that $ax \le \sin x \le bx$ for $0 \le x \le \pi/2$. Use this inequality to find (to two significant figures) upper and lower bounds for the integral

$$I = \int_0^{\pi/2} (1 + \sin x)^{1/2}\,dx.$$

Use the substitution $t = \tan(x/2)$ to evaluate I exactly.

2.47   By noting that for $0 \le \eta \le 1$, $\eta^{1/2} \ge \eta^{3/4} \ge \eta$, prove that

$$\frac{2}{3} \le \frac{1}{a^{5/2}} \int_0^a (a^2 - x^2)^{3/4}\,dx \le \frac{\pi}{4}.$$

2.48   Show that the total length of the astroid $x^{2/3} + y^{2/3} = a^{2/3}$, which can be parameterised as $x = a\cos^3\theta$, $y = a\sin^3\theta$, is $6a$.

2.49   By noting that $\sinh x < \frac{1}{2}e^x < \cosh x$, and that $1 + z^2 < (1+z)^2$ for $z > 0$, show that, for $x > 0$, the length L of the curve $y = \frac{1}{2}e^x$ measured from the origin satisfies the inequalities $\sinh x < L < x + \sinh x$.

2.50   The equation of a cardioid in plane polar coordinates is

$$\rho = a(1 - \sin\phi).$$

Sketch the curve and find (i) its area, (ii) its total length, (iii) the surface area of the solid formed by rotating the cardioid about its axis of symmetry and (iv) the volume of the same solid.

2.4 Hints and answers

2.1   (a) 3; (b) $2x + 1$, 2, 0; (c) $\cos x$.

2.3   Use: the product rule in (a), (b), (d) and (e)[3 factors]; the chain rule in (c), (f) and (g); logarithmic differentiation in (g) and (h).
(a) $(x^2 + 2x)\exp x$; (b) $2(\cos^2 x - \sin^2 x) = 2\cos 2x$;
(c) $2\cos 2x$; (d) $\sin ax + ax\cos ax$;
(e) $(a\exp ax)[(\sin ax + \cos ax)\tan^{-1} ax + (\sin ax)(1 + a^2x^2)^{-1}]$;
(f) $[a(x^a - x^{-a})]/[x(x^a + x^{-a})]$; (g) $[(a^x - a^{-x})\ln a]/(a^x + a^{-x})$; (h) $(1 + \ln x)x^x$.

2.5   (a) $-6(2x + 3)^{-4}$; (b) $2\sec^2 x\tan x$; (c) $-9\operatorname{cosech}^3 3x\coth 3x$;
(d) $-x^{-1}(\ln x)^{-2}$; (e) $-(a^2 - x^2)^{-1/2}[\sin^{-1}(x/a)]^{-2}$.

2.7   Calculate dy/dt and dx/dt and divide one by the other. $(t + 2)^2/[2(t + 1)^2]$. Alternatively, eliminate t and find dy/dx by implicit differentiation.

2.9 $-\sin x$ in both cases.

2.11   The required conditions are $8n - 4 = 0$ and $4n^2 - 8n + 3 = 0$; both are satisfied by $n = \frac{1}{2}$.

2.13   The stationary points are the zeros of $12x^3 + 12x^2 - 24x$. The lowest stationary value is -26 at $x = -2$; other stationary values are 6 at $x = 0$ and 1 at $x = 1$.

2.15   Use logarithmic differentiation. Set $dy/dx = 0$, obtaining $2x^2 + 2x\ln a + 1 = 0$.

2.17   See figure 2.14.

2.19 $\dfrac{dy}{dx} = -\left(\dfrac{y}{x}\right)^{1/3}$; $\dfrac{d^2y}{dx^2} = \dfrac{a^{2/3}}{3x^{4/3}y^{1/3}}.$

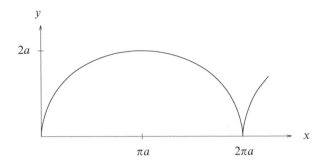

Figure 2.14   The solution to exercise 2.17.

2.21   (a) $2(2 - 9\cos^2 x)\sin x$; (b) $(2x^{-3} - 3x^{-1})\sin x - (3x^{-2} + \ln x)\cos x$; (c) $8(4x^3 + 30x^2 + 62x + 38)\exp 2x$.

2.23   (a) $f(1) = 0$ whilst $f'(1) \neq 0$, and so $f(x)$ must be negative in some region with $x = 1$ as an endpoint.
        (b) $f'(x) = \tan^2 x > 0$ and $f(0) = 0$; $g'(x) = (-\cos x)(\tan x - x)/x^2$, which is never positive in the range.

2.25   The false result arises because $\tan nx$ is not differentiable at $x = \pi/(2n)$, which lies in the range $0 < x < \pi/n$, and so the conditions for applying Rolle's theorem are not satisfied.

2.27   The relationship is $x\, dy/dx = (2 - x)y$.

2.29   By implicit differentiation, $y'(x) = (3x^2 - 12)/(8 - 3y^2)$, giving $y'(\pm 2) = 0$. Since $y(2) = 4$ and $y(-2) = 0$, the curve touches the x-axis at the point $(-2, 0)$.

2.31   (a) Express in partial fractions; $J = \frac{1}{3}\ln[(x - 1)^4/(x + 2)] + c$.
        (b) Divide the numerator by the denominator and express the remainder in partial fractions; $J = x^2/4 + 4\ln(x + 2) - 3\ln(x + 3) + c$.
        (c) After division of the numerator by the denominator, the remainder can be expressed as $2(x + 3)^{-1} - 5(x + 3)^{-2}$; $J = 3x + 2\ln(x + 3) + 5(x + 3)^{-1} + c$.
        (d) Set $x^4 = u$; $J = (4a^4)^{-1}\tan^{-1}(x^4/a^4) + c$.

2.33   Writing $b^2 - 4ac$ as $\Delta^2 > 0$, or $4ac - b^2$ as $\Delta'^2 > 0$:
        (i) $\Delta^{-1}\ln[(2ax + b - \Delta)/(2ax + b + \Delta)] + k$;
        (ii) $2\Delta'^{-1}\tan^{-1}[(2ax + b)/\Delta'] + k$;
        (iii) $-2(2ax + b)^{-1} + k$.

2.35 $f'(x) = (1 + \sin x)/\cos^2 x = f(x)\sec x$; $J = \ln(f(x)) + c = \ln(\sec x + \tan x) + c$.

2.37   Note that $dx = 2(b - a)\cos\theta\sin\theta\, d\theta$.
        (a) π; (b) $\pi(b - a)^2/8$; (c) $\pi(b - a)/2$.

2.39   (a) $(2 - y^2)\cos y + 2y\sin y - 2$; (b) $[(y^2\ln y)/2] + [(1 - y^2)/4]$;
        (c) $y\sin^{-1} y + (1 - y^2)^{1/2} - 1$;
        (d) $\ln(a^2 + 1) - (1/y)\ln(a^2 + y^2) + (2/a)[\tan^{-1}(y/a) - \tan^{-1}(1/a)]$.

2.41 $\Gamma(n + 1) = n\Gamma(n)$; (a) (i) $n!$, (ii) $15\sqrt{\pi}/8$; (b) $-2\sqrt{\pi}$.

2.43   By integrating twice, recover a multiple of I_n.

2.45 $J_{2r+1} = rJ_{2r-1}$ and $2J_{2r} = (2r - 1)J_{2r-2}$.

2.47   Set $\eta = 1 - (x/a)^2$ throughout, and $x = a\sin\theta$ in one of the bounds.

2.49 $L = \int_0^x \left(1 + \frac{1}{4}\exp 2x\right)^{1/2} dx$.

Complex numbers and hyperbolic functions

This chapter is concerned with the representation and manipulation of complex numbers. Complex numbers pervade this book, underscoring their wide application in the mathematics of the physical sciences. The application of complex numbers to the description of physical systems is left until later chapters and only the basic tools are presented here.

3.1 The need for complex numbers

Although complex numbers occur in many branches of mathematics, they arise most directly out of solving polynomial equations. We examine a specific quadratic equation as an example.

Consider the quadratic equation

$$z^2 - 4z + 5 = 0. \tag{3.1}$$

Equation (3.1) has two solutions, z_1 and z_2, such that

$$(z - z_1)(z - z_2) = 0. \tag{3.2}$$

Using the familiar formula for the roots of a quadratic equation, (1.4), the solutions z_1 and z_2, written in brief as $z_{1,2}$, are

$$z_{1,2} = \frac{4 \pm \sqrt{(-4)^2 - 4(1 \times 5)}}{2}$$

$$= 2 \pm \frac{\sqrt{-4}}{2}. \tag{3.3}$$

Both solutions contain the square root of a negative number. However, it is not true to say that there are no solutions to the quadratic equation. The *fundamental theorem of algebra* states that a quadratic equation will always have two solutions and these are in fact given by (3.3). The second term on the RHS of (3.3) is called an *imaginary* term since it contains the square root of a negative number;

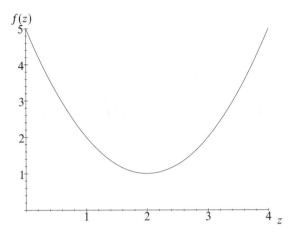

Figure 3.1   The function $f(z) = z^2 - 4z + 5$.

the first term is called a *real* term. The full solution is the sum of a real term and an imaginary term and is called a *complex number*. A plot of the function $f(z) = z^2 - 4z + 5$ is shown in figure 3.1. It will be seen that the plot does not intersect the z-axis, corresponding to the fact that the equation $f(z) = 0$ has no purely real solutions.

The choice of the symbol z for the quadratic variable was not arbitrary; the conventional representation of a complex number is z, where z is the sum of a real part x and i times an imaginary part y, i.e.

$$z = x + iy,$$

where i is used to denote the square root of -1. The real part x and the imaginary part y are usually denoted by $\mathrm{Re}\, z$ and $\mathrm{Im}\, z$ respectively. We note at this point that some physical scientists, engineers in particular, use j instead of i. However, for consistency, we will use i throughout this book.

In our particular example, $\sqrt{-4} = 2\sqrt{-1} = 2i$, and hence the two solutions of (3.1) are

$$z_{1,2} = 2 \pm \frac{2i}{2} = 2 \pm i.$$

Thus, here $x = 2$ and $y = \pm 1$.

For compactness a complex number is sometimes written in the form

$$z = (x, y),$$

where the components of z may be thought of as coordinates in an xy-plot. Such a plot is called an *Argand diagram* and is a common representation of complex numbers; an example is shown in figure 3.2.

84

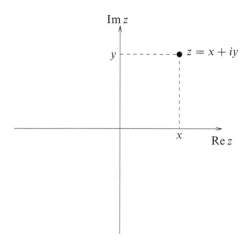

Figure 3.2   The Argand diagram.

Our particular example of a quadratic equation may be generalised readily to polynomials whose highest power (degree) is greater than 2, e.g. cubic equations (degree 3), quartic equations (degree 4) and so on. For a general polynomial $f(z)$, of degree n, the fundamental theorem of algebra states that the equation $f(z) = 0$ will have exactly n solutions. We will examine cases of higher-degree equations in subsection 3.4.3.

The remainder of this chapter deals with: the algebra and manipulation of complex numbers; their polar representation, which has advantages in many circumstances; complex exponentials and logarithms; the use of complex numbers in finding the roots of polynomial equations; and hyperbolic functions.

3.2 Manipulation of complex numbers

This section considers basic complex number manipulation. Some analogy may be drawn with vector manipulation (see chapter 7) but this section stands alone as an introduction.

3.2.1 Addition and subtraction

The addition of two complex numbers, z_1 and z_2, in general gives another complex number. The real components and the imaginary components are added separately and in a like manner to the familiar addition of real numbers:

$$z_1 + z_2 = (x_1 + iy_1) + (x_2 + iy_2) = (x_1 + x_2) + i(y_1 + y_2),$$

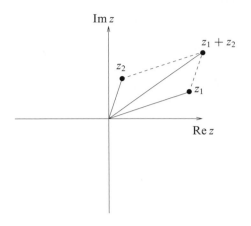

Figure 3.3   The addition of two complex numbers.

or in component notation

$$z_1 + z_2 = (x_1, y_1) + (x_2, y_2) = (x_1 + x_2, y_1 + y_2).$$

The Argand representation of the addition of two complex numbers is shown in figure 3.3.

By straightforward application of the commutativity and associativity of the real and imaginary parts separately, we can show that the addition of complex numbers is itself commutative and associative, i.e.

$$z_1 + z_2 = z_2 + z_1,$$

$$z_1 + (z_2 + z_3) = (z_1 + z_2) + z_3.$$

Thus it is immaterial in what order complex numbers are added.

▶*Sum the complex numbers* $1 + 2i$, $3 - 4i$, $-2 + i$.

Summing the real terms we obtain

$$1 + 3 - 2 = 2,$$

and summing the imaginary terms we obtain

$$2i - 4i + i = -i.$$

Hence

$$(1 + 2i) + (3 - 4i) + (-2 + i) = 2 - i. \blacktriangleleft$$

The subtraction of complex numbers is very similar to their addition. As in the case of real numbers, if two identical complex numbers are subtracted then the result is zero.

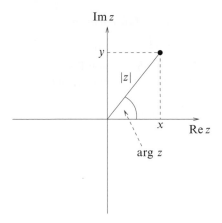

Figure 3.4   The modulus and argument of a complex number.

3.2.2 Modulus and argument

The modulus of the complex number z is denoted by $|z|$ and is defined as

$$|z| = \sqrt{x^2 + y^2}. \tag{3.4}$$

Hence the modulus of the complex number is the distance of the corresponding point from the origin in the Argand diagram, as may be seen in figure 3.4.

The argument of the complex number z is denoted by arg z and is defined as

$$\arg z = \tan^{-1}\left(\frac{y}{x}\right). \tag{3.5}$$

It can be seen that arg z is the angle that the line joining the origin to z on the Argand diagram makes with the positive x-axis. The anticlockwise direction is taken to be positive by convention. The angle arg z is shown in figure 3.4. Account must be taken of the signs of x and y individually in determining in which quadrant arg z lies. Thus, for example, if x and y are both negative then arg z lies in the range $-\pi < \arg z < -\pi/2$ rather than in the first quadrant $(0 < \arg z < \pi/2)$, though both cases give the same value for the ratio of y to x.

▶*Find the modulus and the argument of the complex number $z = 2 - 3i$.*

Using (3.4), the modulus is given by

$$|z| = \sqrt{2^2 + (-3)^2} = \sqrt{13}.$$

Using (3.5), the argument is given by

$$\arg z = \tan^{-1}\left(-\tfrac{3}{2}\right).$$

The two angles whose tangents equal -1.5 are -0.9828 rad and 2.1588 rad. Since $x = 2$ and $y = -3$, z clearly lies in the fourth quadrant; therefore arg $z = -0.9828$ is the appropriate answer. ◀

3.2.3 Multiplication

Complex numbers may be multiplied together and in general give a complex number as the result. The product of two complex numbers z_1 and z_2 is found by multiplying them out in full and remembering that $i^2 = -1$, i.e.

$$
\begin{aligned}
z_1 z_2 &= (x_1 + iy_1)(x_2 + iy_2) \\
&= x_1 x_2 + ix_1 y_2 + iy_1 x_2 + i^2 y_1 y_2 \\
&= (x_1 x_2 - y_1 y_2) + i(x_1 y_2 + y_1 x_2).
\end{aligned}
\tag{3.6}
$$

▶*Multiply the complex numbers* $z_1 = 3 + 2i$ *and* $z_2 = -1 - 4i$.

By direct multiplication we find

$$
\begin{aligned}
z_1 z_2 &= (3 + 2i)(-1 - 4i) \\
&= -3 - 2i - 12i - 8i^2 \\
&= 5 - 14i. \ \blacktriangleleft
\end{aligned}
\tag{3.7}
$$

The multiplication of complex numbers is both commutative and associative, i.e.

$$
z_1 z_2 = z_2 z_1,
\tag{3.8}
$$

$$
(z_1 z_2) z_3 = z_1 (z_2 z_3).
\tag{3.9}
$$

The product of two complex numbers also has the simple properties

$$
|z_1 z_2| = |z_1||z_2|,
\tag{3.10}
$$

$$
\arg(z_1 z_2) = \arg z_1 + \arg z_2.
\tag{3.11}
$$

These relations are derived in subsection 3.3.1.

▶*Verify that (3.10) holds for the product of* $z_1 = 3 + 2i$ *and* $z_2 = -1 - 4i$.

From (3.7)

$$
|z_1 z_2| = |5 - 14i| = \sqrt{5^2 + (-14)^2} = \sqrt{221}.
$$

We also find

$$
|z_1| = \sqrt{3^2 + 2^2} = \sqrt{13},
$$
$$
|z_2| = \sqrt{(-1)^2 + (-4)^2} = \sqrt{17},
$$

and hence

$$
|z_1||z_2| = \sqrt{13}\sqrt{17} = \sqrt{221} = |z_1 z_2|. \ \blacktriangleleft
$$

We now examine the effect on a complex number z of multiplying it by ± 1 and $\pm i$. These four multipliers have modulus unity and we can see immediately from (3.10) that multiplying z by another complex number of unit modulus gives a product with the same modulus as z. We can also see from (3.11) that if we

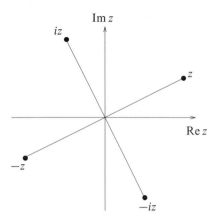

Figure 3.5   Multiplication of a complex number by ± 1 and $\pm i$.

multiply z by a complex number then the argument of the product is the sum of the argument of z and the argument of the multiplier. Hence multiplying z by unity (which has argument zero) leaves z unchanged in both modulus and argument, i.e. z is completely unaltered by the operation. Multiplying by -1 (which has argument π) leads to rotation, through an angle π, of the line joining the origin to z in the Argand diagram. Similarly, multiplication by i or $-i$ leads to corresponding rotations of $\pi/2$ or $-\pi/2$ respectively. This geometrical interpretation of multiplication is shown in figure 3.5.

▶ *Using the geometrical interpretation of multiplication by i, find the product $i(1 - i)$.*

The complex number $1 - i$ has argument $-\pi/4$ and modulus $\sqrt{2}$. Thus, using (3.10) and (3.11), its product with i has argument $+\pi/4$ and unchanged modulus $\sqrt{2}$. The complex number with modulus $\sqrt{2}$ and argument $+\pi/4$ is $1 + i$ and so

$$i(1 - i) = 1 + i,$$

as is easily verified by direct multiplication. ◀

The division of two complex numbers is similar to their multiplication but requires the notion of the complex conjugate (see the following subsection) and so discussion is postponed until subsection 3.2.5.

3.2.4 Complex conjugate

If z has the convenient form $x + iy$ then the complex conjugate, denoted by z^{*}, may be found simply by changing the sign of the imaginary part, i.e. if $z = x + iy$ then $z^{*} = x - iy$. More generally, we may define the complex conjugate of z as the (complex) number having the same magnitude as z that when multiplied by

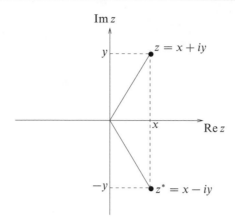

Figure 3.6   The complex conjugate as a mirror image in the real axis.

z leaves a real positive result, i.e. the product has a real positive part and no imaginary one.

In the case where z can be written in the form $x + iy$ it is easily verified, by direct multiplication of the components, that the product zz^* gives a real result:

$$zz^* = (x + iy)(x - iy) = x^2 - ixy + ixy - i^2 y^2 = x^2 + y^2 = |z|^2.$$

Complex conjugation corresponds to a reflection of z in the real axis of the Argand diagram, as may be seen in figure 3.6.

▶*Find the complex conjugate of $z = a + 2i + 3ib$.*

The complex number is written in the standard form

$$z = a + i(2 + 3b);$$

then, replacing i by $-i$, we obtain

$$z^* = a - i(2 + 3b). ◀$$

In some cases, however, it may not be simple to rearrange the expression for z into the standard form $x + iy$. Nevertheless, given two complex numbers, z_1 and z_2, it is straightforward to show that the complex conjugate of their sum (or difference) is equal to the sum (or difference) of their complex conjugates, i.e. $(z_1 \pm z_2)^* = z_1^* \pm z_2^*$. Similarly, it may be shown that the complex conjugate of the product (or quotient) of z_1 and z_2 is equal to the product (or quotient) of their complex conjugates, i.e. $(z_1 z_2)^* = z_1^* z_2^*$ and $(z_1/z_2)^* = z_1^*/z_2^*$.

Using these results, it can be deduced that, no matter how complicated the expression, its complex conjugate may *always* be found by replacing every i by $-i$. To apply this rule, however, we must always ensure that all complex parts are first written out in full, so that no i's are hidden.

90

▶ *Find the complex conjugate of the complex number* $z = w^{(3y+2ix)}$, *where* $w = x + 5i$.

Although we do not discuss complex powers until section 3.5, the simple rule given above still enables us to find the complex conjugate of z.

In this case w itself contains real and imaginary components and so must be written out in full, i.e.

$$z = w^{3y+2ix} = (x + 5i)^{3y+2ix}.$$

Now we can replace each i by $-i$ to obtain

$$z^* = (x - 5i)^{(3y-2ix)}.$$

It can be shown that the product zz^* is real, as required. ◀

The following properties of the complex conjugate are easily proved and others may be derived from them. If $z = x + iy$ then

$$(z^*)^* = z, \tag{3.12}$$

$$z + z^* = 2\,\mathrm{Re}\,z = 2x, \tag{3.13}$$

$$z - z^* = 2i\,\mathrm{Im}\,z = 2iy, \tag{3.14}$$

$$\frac{z}{z^*} = \left(\frac{x^2 - y^2}{x^2 + y^2}\right) + i\left(\frac{2xy}{x^2 + y^2}\right). \tag{3.15}$$

The derivation of this last relation relies on the results of the following subsection.

3.2.5 Division

The division of two complex numbers z_1 and z_2 bears some similarity to their multiplication. Writing the quotient in component form we obtain

$$\frac{z_1}{z_2} = \frac{x_1 + iy_1}{x_2 + iy_2}. \tag{3.16}$$

In order to separate the real and imaginary components of the quotient, we multiply both numerator and denominator by the complex conjugate of the denominator. By definition, this process will leave the denominator as a real quantity. Equation (3.16) gives

$$\frac{z_1}{z_2} = \frac{(x_1 + iy_1)(x_2 - iy_2)}{(x_2 + iy_2)(x_2 - iy_2)} = \frac{(x_1x_2 + y_1y_2) + i(x_2y_1 - x_1y_2)}{x_2^2 + y_2^2}$$

$$= \frac{x_1x_2 + y_1y_2}{x_2^2 + y_2^2} + i\frac{x_2y_1 - x_1y_2}{x_2^2 + y_2^2}.$$

Hence we have separated the quotient into real and imaginary components, as required.

In the special case where $z_2 = z_1^*$, so that $x_2 = x_1$ and $y_2 = -y_1$, the general result reduces to (3.15).

> ►*Express z in the form $x + iy$, when*
> $$z = \frac{3 - 2i}{-1 + 4i}.$$

Multiplying numerator and denominator by the complex conjugate of the denominator we obtain

$$z = \frac{(3 - 2i)(-1 - 4i)}{(-1 + 4i)(-1 - 4i)} = \frac{-11 - 10i}{17}$$
$$= -\frac{11}{17} - \frac{10}{17}i. \blacktriangleleft$$

In analogy to (3.10) and (3.11), which describe the multiplication of two complex numbers, the following relations apply to their division:

$$\left| \frac{z_1}{z_2} \right| = \frac{|z_1|}{|z_2|}, \tag{3.17}$$

$$\arg\left(\frac{z_1}{z_2} \right) = \arg z_1 - \arg z_2. \tag{3.18}$$

The proof of these relations is left until subsection 3.3.1.

3.3 Polar representation of complex numbers

Although considering a complex number as the sum of a real and an imaginary part is often useful, sometimes the *polar representation* proves easier to manipulate. This makes use of the complex exponential function, which is defined by

$$e^z = \exp z \equiv 1 + z + \frac{z^2}{2!} + \frac{z^3}{3!} + \cdots. \tag{3.19}$$

Strictly speaking it is the function $\exp z$ that is defined by (3.19). The number e is the value of $\exp(1)$, i.e. it is just a number. However, it may be shown that e^z and $\exp z$ are equivalent when z is real and rational and mathematicians then *define* their equivalence for irrational and complex z. For the purposes of this book we will not concern ourselves further with this mathematical nicety but, rather, assume that (3.19) is valid for all z. We also note that, using (3.19), by multiplying together the appropriate series we may show that (see chapter 24)

$$e^{z_1}e^{z_2} = e^{z_1 + z_2}, \tag{3.20}$$

which is analogous to the familiar result for exponentials of real numbers.

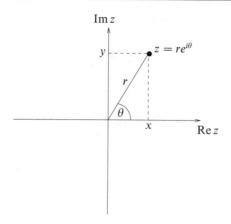

Figure 3.7   The polar representation of a complex number.

From (3.19), it immediately follows that for $z = i\theta$, θ real,

$$e^{i\theta} = 1 + i\theta - \frac{\theta^2}{2!} - \frac{i\theta^3}{3!} + \cdots \tag{3.21}$$

$$= 1 - \frac{\theta^2}{2!} + \frac{\theta^4}{4!} - \cdots + i\left(\theta - \frac{\theta^3}{3!} + \frac{\theta^5}{5!} - \cdots\right) \tag{3.22}$$

and hence that

$$e^{i\theta} = \cos\theta + i\sin\theta, \tag{3.23}$$

where the last equality follows from the series expansions of the sine and cosine functions (see subsection 4.6.3). This last relationship is called *Euler's equation*. It also follows from (3.23) that

$$e^{in\theta} = \cos n\theta + i\sin n\theta$$

for all n. From Euler's equation (3.23) and figure 3.7 we deduce that

$$re^{i\theta} = r(\cos\theta + i\sin\theta)$$
$$= x + iy.$$

Thus a complex number may be represented in the polar form

$$z = re^{i\theta}. \tag{3.24}$$

Referring again to figure 3.7, we can identify r with $|z|$ and θ with arg z. The simplicity of the representation of the modulus and argument is one of the main reasons for using the polar representation. The angle θ lies conventionally in the range $-\pi < \theta \leq \pi$, but, since rotation by θ is the same as rotation by $2n\pi + \theta$, where n is any integer,

$$re^{i\theta} \equiv re^{i(\theta + 2n\pi)}.$$

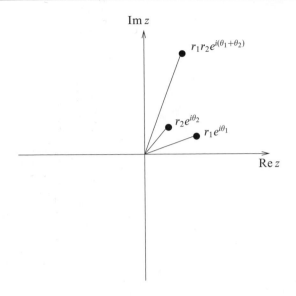

Figure 3.8 The multiplication of two complex numbers. In this case r_1 and r_2 are both greater than unity.

The algebra of the polar representation is different from that of the real and imaginary component representation, though, of course, the results are identical. Some operations prove much easier in the polar representation, others much more complicated. The best representation for a particular problem must be determined by the manipulation required.

3.3.1 Multiplication and division in polar form

Multiplication and division in polar form are particularly simple. The product of $z_1 = r_1 e^{i\theta_1}$ and $z_2 = r_2 e^{i\theta_2}$ is given by

$$z_1 z_2 = r_1 e^{i\theta_1} r_2 e^{i\theta_2}$$
$$= r_1 r_2 e^{i(\theta_1 + \theta_2)}. \tag{3.25}$$

The relations $|z_1 z_2| = |z_1||z_2|$ and $\arg(z_1 z_2) = \arg z_1 + \arg z_2$ follow immediately. An example of the multiplication of two complex numbers is shown in figure 3.8.

Division is equally simple in polar form; the quotient of z_1 and z_2 is given by

$$\frac{z_1}{z_2} = \frac{r_1 e^{i\theta_1}}{r_2 e^{i\theta_2}} = \frac{r_1}{r_2} e^{i(\theta_1 - \theta_2)}. \tag{3.26}$$

The relations $|z_1/z_2| = |z_1|/|z_2|$ and $\arg(z_1/z_2) = \arg z_1 - \arg z_2$ are again

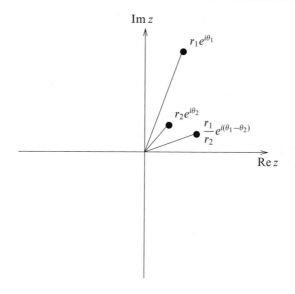

Im z

$r_1 e^{i\theta_1}$

$r_2 e^{i\theta_2}$

$\dfrac{r_1}{r_2} e^{i(\theta_1 - \theta_2)}$

Re z

Figure 3.9   The division of two complex numbers As in the previous figure, r_1 and r_2 are both greater than unity.

immediately apparent. The division of two complex numbers in polar form is shown in figure 3.9.

3.4 de Moivre's theorem

We now derive an extremely important theorem. Since $\left(e^{i\theta}\right)^n = e^{in\theta}$, we have

$$(\cos\theta + i\sin\theta)^n = \cos n\theta + i\sin n\theta, \tag{3.27}$$

where the identity $e^{in\theta} = \cos n\theta + i\sin n\theta$ follows from the series definition of $e^{in\theta}$ (see (3.21)). This result is called *de Moivre's theorem* and is often used in the manipulation of complex numbers. The theorem is valid for all n whether real, imaginary or complex.

There are numerous applications of de Moivre's theorem but this section examines just three: proofs of trigonometric identities; finding the nth roots of unity; and solving polynomial equations with complex roots.

3.4.1 *Trigonometric identities*

The use of de Moivre's theorem in finding trigonometric identities is best illustrated by example. We consider the expression of a multiple-angle function in terms of a polynomial in the single-angle function, and its converse.

95

▶*Express* $\sin 3\theta$ *and* $\cos 3\theta$ *in terms of powers of* $\cos \theta$ *and* $\sin \theta$.

Using de Moivre's theorem,

$$\cos 3\theta + i \sin 3\theta = (\cos \theta + i \sin \theta)^3$$
$$= (\cos^3 \theta - 3 \cos \theta \sin^2 \theta) + i(3 \sin \theta \cos^2 \theta - \sin^3 \theta). \tag{3.28}$$

We can equate the real and imaginary coefficients separately, i.e.

$$\cos 3\theta = \cos^3 \theta - 3 \cos \theta \sin^2 \theta$$
$$= 4 \cos^3 \theta - 3 \cos \theta \tag{3.29}$$

and

$$\sin 3\theta = 3 \sin \theta \cos^2 \theta - \sin^3 \theta$$
$$= 3 \sin \theta - 4 \sin^3 \theta. \blacktriangleleft$$

This method can clearly be applied to finding power expansions of $\cos n\theta$ and $\sin n\theta$ for any positive integer n.

The converse process uses the following properties of $z = e^{i\theta}$,

$$z^n + \frac{1}{z^n} = 2 \cos n\theta, \tag{3.30}$$

$$z^n - \frac{1}{z^n} = 2i \sin n\theta. \tag{3.31}$$

These equalities follow from simple applications of de Moivre's theorem, i.e.

$$z^n + \frac{1}{z^n} = (\cos \theta + i \sin \theta)^n + (\cos \theta + i \sin \theta)^{-n}$$
$$= \cos n\theta + i \sin n\theta + \cos(-n\theta) + i \sin(-n\theta)$$
$$= \cos n\theta + i \sin n\theta + \cos n\theta - i \sin n\theta$$
$$= 2 \cos n\theta$$

and

$$z^n - \frac{1}{z^n} = (\cos \theta + i \sin \theta)^n - (\cos \theta + i \sin \theta)^{-n}$$
$$= \cos n\theta + i \sin n\theta - \cos n\theta + i \sin n\theta$$
$$= 2i \sin n\theta.$$

In the particular case where $n = 1$,

$$z + \frac{1}{z} = e^{i\theta} + e^{-i\theta} = 2 \cos \theta, \tag{3.32}$$

$$z - \frac{1}{z} = e^{i\theta} - e^{-i\theta} = 2i \sin \theta. \tag{3.33}$$

> ►*Find an expression for* $\cos^3 \theta$ *in terms of* $\cos 3\theta$ *and* $\cos \theta$.

Using (3.32),

$$\cos^3 \theta = \frac{1}{2^3} \left(z + \frac{1}{z} \right)^3$$

$$= \frac{1}{8} \left(z^3 + 3z + \frac{3}{z} + \frac{1}{z^3} \right)$$

$$= \frac{1}{8} \left(z^3 + \frac{1}{z^3} \right) + \frac{3}{8} \left(z + \frac{1}{z} \right).$$

Now using (3.30) and (3.32), we find

$$\cos^3 \theta = \tfrac{1}{4} \cos 3\theta + \tfrac{3}{4} \cos \theta. \; ◄$$

This result happens to be a simple rearrangement of (3.29), but cases involving larger values of n are better handled using this direct method than by rearranging polynomial expansions of multiple-angle functions.

3.4.2 Finding the nth roots of unity

The equation $z^2 = 1$ has the familiar solutions $z = \pm 1$. However, now that we have introduced the concept of complex numbers we can solve the general equation $z^n = 1$. Recalling the fundamental theorem of algebra, we know that the equation has n solutions. In order to proceed we rewrite the equation as

$$z^n = e^{2ik\pi},$$

where k is any integer. Now taking the nth root of each side of the equation we find

$$z = e^{2ik\pi/n}.$$

Hence, the solutions of $z^n = 1$ are

$$z_{1,2,...,n} = 1, \;\; e^{2i\pi/n}, \;\; ..., \;\; e^{2i(n-1)\pi/n},$$

corresponding to the values $0, 1, 2, ..., n-1$ for k. Larger integer values of k do not give new solutions, since the roots already listed are simply cyclically repeated for $k = n, n+1, n+2$, etc.

> ►*Find the solutions to the equation* $z^3 = 1$.

By applying the above method we find

$$z = e^{2ik\pi/3}.$$

Hence the three solutions are $z_1 = e^{0i} = 1$, $z_2 = e^{2i\pi/3}$, $z_3 = e^{4i\pi/3}$. We note that, as expected, the next solution, for which $k = 3$, gives $z_4 = e^{6i\pi/3} = 1 = z_1$, so that there are only three separate solutions. ◄

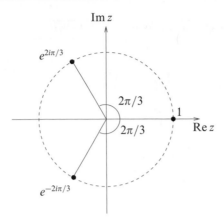

Figure 3.10   The solutions of $z^3 = 1$.

Not surprisingly, given that $|z^3| = |z|^3$ from (3.10), all the roots of unity have unit modulus, i.e. they all lie on a circle in the Argand diagram of unit radius. The three roots are shown in figure 3.10.

The cube roots of unity are often written 1, ω and ω^2. The properties $\omega^3 = 1$ and $1 + \omega + \omega^2 = 0$ are easily proved.

3.4.3 Solving polynomial equations

A third application of de Moivre's theorem is to the solution of polynomial equations. Complex equations in the form of polynomial relationships must first be solved for z, or for powers of z, using the same methods as those employed to find the roots of real polynomial equations. The complex roots may then be deduced.

▶*Solve the equation $z^6 - z^5 + 4z^4 - 6z^3 + 2z^2 - 8z + 8 = 0$.*

We first factorise to give

$$(z^3 - 2)(z^2 + 4)(z - 1) = 0.$$

Hence $z^3 = 2$ or $z^2 = -4$ or $z = 1$. The solutions to the quadratic equation are $z = \pm 2i$; to find the complex cube roots, we first write the equation in the form

$$z^3 = 2 = 2e^{2ik\pi},$$

where k is any integer. If we now take the cube root, we get

$$z = 2^{1/3}e^{2ik\pi/3}.$$

To avoid the duplication of solutions, we use the fact that $-\pi < \arg z \le \pi$ and find

$$z_1 = 2^{1/3},$$

$$z_2 = 2^{1/3}e^{2\pi i/3} = 2^{1/3}\left(-\frac{1}{2} + \frac{\sqrt{3}}{2}i\right),$$

$$z_3 = 2^{1/3}e^{-2\pi i/3} = 2^{1/3}\left(-\frac{1}{2} - \frac{\sqrt{3}}{2}i\right).$$

The complex numbers z_1, z_2 and z_3, together with $z_4 = 2i$, $z_5 = -2i$ and $z_6 = 1$ are the solutions to the original polynomial equation.

As expected from the fundamental theorem of algebra, we find that the total number of complex roots (six, in this case) is equal to the largest power of z in the polynomial. ◀

A useful result is that the roots of a polynomial with real coefficients occur in conjugate pairs (i.e. if z_1 is a root, then z_1^* is a second distinct root, unless z_1 is real). This may be proved as follows. Let the polynomial equation of which z is a root be

$$a_n z^n + a_{n-1}z^{n-1} + \cdots + a_1 z + a_0 = 0.$$

Taking the complex conjugate of this equation,

$$a_n^*(z^*)^n + a_{n-1}^*(z^*)^{n-1} + \cdots + a_1^* z^* + a_0^* = 0.$$

But the a_n are real, and so z^* satisfies

$$a_n(z^*)^n + a_{n-1}(z^*)^{n-1} + \cdots + a_1 z^* + a_0 = 0,$$

and is also a root of the original equation.

3.5 Complex logarithms and complex powers

The concept of a complex exponential has already been introduced in section 3.3, where it was assumed that the definition of an exponential as a series was valid for complex numbers as well as for real numbers. Similarly we can define the logarithm of a complex number and we can use complex numbers as exponents.

Let us denote the natural logarithm of a complex number z by $w = \mathrm{Ln}\, z$, where the notation Ln will be explained shortly. Thus, w must satisfy

$$z = e^w.$$

Using (3.20), we see that

$$z_1 z_2 = e^{w_1} e^{w_2} = e^{w_1 + w_2},$$

and taking logarithms of both sides we find

$$\mathrm{Ln}\,(z_1 z_2) = w_1 + w_2 = \mathrm{Ln}\, z_1 + \mathrm{Ln}\, z_2, \tag{3.34}$$

which shows that the familiar rule for the logarithm of the product of two real numbers also holds for complex numbers.

We may use (3.34) to investigate further the properties of $\operatorname{Ln} z$. We have already noted that the argument of a complex number is multivalued, i.e. $\arg z = \theta + 2n\pi$, where n is any integer. Thus, in polar form, the complex number z should strictly be written as

$$z = re^{i(\theta + 2n\pi)}.$$

Taking the logarithm of both sides, and using (3.34), we find

$$\operatorname{Ln} z = \ln r + i(\theta + 2n\pi), \tag{3.35}$$

where $\ln r$ is the natural logarithm of the real positive quantity r and so is written normally. Thus from (3.35) we see that $\operatorname{Ln} z$ is itself multivalued. To avoid this multivalued behaviour it is conventional to define another function $\ln z$, the *principal value* of $\operatorname{Ln} z$, which is obtained from $\operatorname{Ln} z$ by restricting the argument of z to lie in the range $-\pi < \theta \leq \pi$.

▶*Evaluate* $\operatorname{Ln}(-i)$.

By rewriting $-i$ as a complex exponential, we find

$$\operatorname{Ln}(-i) = \operatorname{Ln}\left[e^{i(-\pi/2 + 2n\pi)}\right] = i(-\pi/2 + 2n\pi),$$

where n is any integer. Hence $\operatorname{Ln}(-i) = -i\pi/2,\ 3i\pi/2,\ \ldots$ We note that $\ln(-i)$, the principal value of $\operatorname{Ln}(-i)$, is given by $\ln(-i) = -i\pi/2$. ◀

If z and t are both complex numbers then the zth power of t is defined by

$$t^z = e^{z\operatorname{Ln} t}.$$

Since $\operatorname{Ln} t$ is multivalued, so too is this definition.

▶*Simplify the expression* $z = i^{-2i}$.

Firstly we take the logarithm of both sides of the equation to give

$$\operatorname{Ln} z = -2i\operatorname{Ln} i.$$

Now inverting the process we find

$$e^{\operatorname{Ln} z} = z = e^{-2i\operatorname{Ln} i}.$$

We can write $i = e^{i(\pi/2 + 2n\pi)}$, where n is any integer, and hence

$$\operatorname{Ln} i = \operatorname{Ln}\left[e^{i(\pi/2 + 2n\pi)}\right]$$
$$= i\left(\pi/2 + 2n\pi\right).$$

We can now simplify z to give

$$i^{-2i} = e^{-2i \times i(\pi/2 + 2n\pi)}$$
$$= e^{(\pi + 4n\pi)},$$

which, perhaps surprisingly, is a real quantity rather than a complex one. ◀

Complex powers and the logarithms of complex numbers are discussed further in chapter 24.

3.6 Applications to differentiation and integration

We can use the exponential form of a complex number together with de Moivre's theorem (see section 3.4) to simplify the differentiation of trigonometric functions.

▶Find the derivative with respect to x of $e^{3x} \cos 4x$.

We could differentiate this function straightforwardly using the product rule (see subsection 2.1.2). However, an alternative method in this case is to use a complex exponential. Let us consider the complex number

$$z = e^{3x}(\cos 4x + i \sin 4x) = e^{3x}e^{4ix} = e^{(3+4i)x},$$

where we have used de Moivre's theorem to rewrite the trigonometric functions as a complex exponential. This complex number has $e^{3x} \cos 4x$ as its real part. Now, differentiating z with respect to x we obtain

$$\frac{dz}{dx} = (3 + 4i)e^{(3+4i)x} = (3 + 4i)e^{3x}(\cos 4x + i \sin 4x), \tag{3.36}$$

where we have again used de Moivre's theorem. Equating real parts we then find

$$\frac{d}{dx}\left(e^{3x} \cos 4x\right) = e^{3x}(3 \cos 4x - 4 \sin 4x).$$

By equating the imaginary parts of (3.36), we also obtain, as a bonus,

$$\frac{d}{dx}\left(e^{3x} \sin 4x\right) = e^{3x}(4 \cos 4x + 3 \sin 4x). \blacktriangleleft$$

In a similar way the complex exponential can be used to evaluate integrals containing trigonometric and exponential functions.

▶Evaluate the integral $I = \int e^{ax} \cos bx \, dx$.

Let us consider the integrand as the real part of the complex number

$$e^{ax}(\cos bx + i \sin bx) = e^{ax}e^{ibx} = e^{(a+ib)x},$$

where we use de Moivre's theorem to rewrite the trigonometric functions as a complex exponential. Integrating we find

$$\int e^{(a+ib)x} \, dx = \frac{e^{(a+ib)x}}{a + ib} + c$$

$$= \frac{(a - ib)e^{(a+ib)x}}{(a - ib)(a + ib)} + c$$

$$= \frac{e^{ax}}{a^2 + b^2}\left(ae^{ibx} - ibe^{ibx}\right) + c, \tag{3.37}$$

where the constant of integration c is in general complex. Denoting this constant by $c = c_1 + ic_2$ and equating real parts in (3.37) we obtain

$$I = \int e^{ax} \cos bx \, dx = \frac{e^{ax}}{a^2 + b^2}(a \cos bx + b \sin bx) + c_1,$$

which agrees with result (2.37) found using integration by parts. Equating imaginary parts in (3.37) we obtain, as a bonus,

$$J = \int e^{ax} \sin bx \, dx = \frac{e^{ax}}{a^2 + b^2}(a \sin bx - b \cos bx) + c_2. \blacktriangleleft$$

3.7 Hyperbolic functions

The *hyperbolic functions* are the complex analogues of the trigonometric functions. The analogy may not be immediately apparent and their definitions may appear at first to be somewhat arbitrary. However, careful examination of their properties reveals the purpose of the definitions. For instance, their close relationship with the trigonometric functions, both in their identities and in their calculus, means that many of the familiar properties of trigonometric functions can also be applied to the hyperbolic functions. Further, hyperbolic functions occur regularly, and so giving them special names is a notational convenience.

3.7.1 Definitions

The two fundamental hyperbolic functions are $\cosh x$ and $\sinh x$, which, as their names suggest, are the hyperbolic equivalents of $\cos x$ and $\sin x$. They are defined by the following relations:

$$\cosh x = \tfrac{1}{2}(e^x + e^{-x}), \tag{3.38}$$

$$\sinh x = \tfrac{1}{2}(e^x - e^{-x}). \tag{3.39}$$

Note that $\cosh x$ is an even function and $\sinh x$ is an odd function. By analogy with the trigonometric functions, the remaining hyperbolic functions are

$$\tanh x = \frac{\sinh x}{\cosh x} = \frac{e^x - e^{-x}}{e^x + e^{-x}}, \tag{3.40}$$

$$\operatorname{sech} x = \frac{1}{\cosh x} = \frac{2}{e^x + e^{-x}}, \tag{3.41}$$

$$\operatorname{cosech} x = \frac{1}{\sinh x} = \frac{2}{e^x - e^{-x}}, \tag{3.42}$$

$$\coth x = \frac{1}{\tanh x} = \frac{e^x + e^{-x}}{e^x - e^{-x}}. \tag{3.43}$$

All the hyperbolic functions above have been defined in terms of the real variable x. However, this was simply so that they may be plotted (see figures 3.11–3.13); the definitions are equally valid for any complex number z.

3.7.2 Hyperbolic–trigonometric analogies

In the previous subsections we have alluded to the analogy between trigonometric and hyperbolic functions. Here, we discuss the close relationship between the two groups of functions.

Recalling (3.32) and (3.33) we find

$$\cos ix = \tfrac{1}{2}(e^x + e^{-x}),$$

$$\sin ix = \tfrac{1}{2}i(e^x - e^{-x}).$$

102

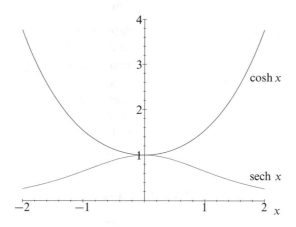

Figure 3.11   Graphs of cosh x and sechx.

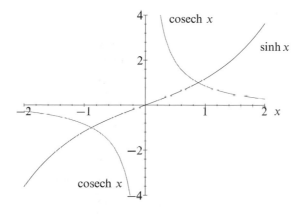

Figure 3.12   Graphs of sinh x and cosechx.

Hence, by the definitions given in the previous subsection,

$$\cosh x = \cos ix,$$ (3.44)

$$i \sinh x = \sin ix,$$ (3.45)

$$\cos x = \cosh ix,$$ (3.46)

$$i \sin x = \sinh ix.$$ (3.47)

These useful equations make the relationship between hyperbolic and trigono-

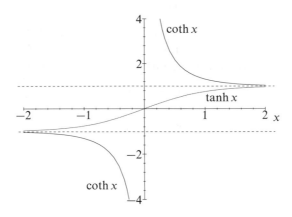

Figure 3.13   Graphs of $\tanh x$ and $\coth x$.

metric functions transparent. The similarity in their calculus is discussed further in subsection 3.7.6.

3.7.3 Identities of hyperbolic functions

The analogies between trigonometric functions and hyperbolic functions having been established, we should not be surprised that all the trigonometric identities also hold for hyperbolic functions, with the following modification. Wherever $\sin^2 x$ occurs it must be replaced by $-\sinh^2 x$, and vice versa. Note that this replacement is necessary even if the $\sin^2 x$ is hidden, e.g. $\tan^2 x = \sin^2 x / \cos^2 x$ and so must be replaced by $(-\sinh^2 x / \cosh^2 x) = -\tanh^2 x$.

▶ *Find the hyperbolic identity analogous to* $\cos^2 x + \sin^2 x = 1$.

Using the rules stated above $\cos^2 x$ is replaced by $\cosh^2 x$, and $\sin^2 x$ by $-\sinh^2 x$, and so the identity becomes

$$\cosh^2 x - \sinh^2 x = 1.$$

This can be verified by direct substitution, using the definitions of $\cosh x$ and $\sinh x$; see (3.38) and (3.39). ◀

Some other identities that can be proved in a similar way are

$$\text{sech}^2 x = 1 - \tanh^2 x, \tag{3.48}$$
$$\text{cosech}^2 x = \coth^2 x - 1, \tag{3.49}$$
$$\sinh 2x = 2 \sinh x \cosh x, \tag{3.50}$$
$$\cosh 2x = \cosh^2 x + \sinh^2 x. \tag{3.51}$$

3.7.4 Solving hyperbolic equations

When we are presented with a hyperbolic equation to solve, we may proceed by analogy with the solution of trigonometric equations. However, it is almost always easier to express the equation directly in terms of exponentials.

▶ *Solve the hyperbolic equation* $\cosh x - 5\sinh x - 5 = 0$.

Substituting the definitions of the hyperbolic functions we obtain
$$\tfrac{1}{2}(e^x + e^{-x}) - \tfrac{5}{2}(e^x - e^{-x}) - 5 = 0.$$
Rearranging, and then multiplying through by $-e^x$, gives in turn
$$-2e^x + 3e^{-x} - 5 = 0$$
and
$$2e^{2x} + 5e^x - 3 = 0.$$
Now we can factorise and solve:
$$(2e^x - 1)(e^x + 3) = 0.$$
Thus $e^x = 1/2$ or $e^x = -3$. Hence $x = -\ln 2$ or $x = \ln(-3)$. The interpretation of the logarithm of a negative number has been discussed in section 3.5. ◀

3.7.5 Inverses of hyperbolic functions

Just like trigonometric functions, hyperbolic functions have inverses. If $y = \cosh x$ then $x = \cosh^{-1} y$, which serves as a definition of the inverse. By using the fundamental definitions of hyperbolic functions, we can find closed-form expressions for their inverses. This is best illustrated by example.

▶ *Find a closed-form expression for the inverse hyperbolic function* $y = \sinh^{-1} x$.

First we write x as a function of y, i.e.
$$y = \sinh^{-1} x \quad \Rightarrow \quad x = \sinh y.$$
Now, since $\cosh y = \tfrac{1}{2}(e^y + e^{-y})$ and $\sinh y = \tfrac{1}{2}(e^y - e^{-y})$,
$$e^y = \cosh y + \sinh y$$
$$= \sqrt{1 + \sinh^2 y} + \sinh y$$
$$= \sqrt{1 + x^2} + x,$$
and hence
$$y = \ln(\sqrt{1 + x^2} + x). \quad ◀$$

In a similar fashion it can be shown that
$$\cosh^{-1} x = \ln(x \pm \sqrt{x^2 - 1}).$$

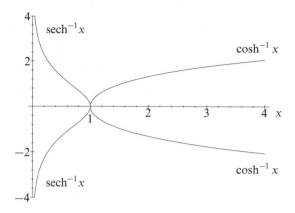

Figure 3.14   Graphs of $\cosh^{-1} x$ and $\operatorname{sech}^{-1}x$.

▶*Find a closed-form expression for the inverse hyperbolic function $y = \tanh^{-1} x$.*

First we write x as a function of y, i.e.

$$y = \tanh^{-1} x \quad \Rightarrow \quad x = \tanh y.$$

Now, using the definition of $\tanh y$ and rearranging, we find

$$x = \frac{e^y - e^{-y}}{e^y + e^{-y}} \quad \Rightarrow \quad (x+1)e^{-y} = (1-x)e^y.$$

Thus, it follows that

$$e^{2y} = \frac{1+x}{1-x} \quad \Rightarrow \quad e^y = \sqrt{\frac{1+x}{1-x}},$$

$$y = \ln \sqrt{\frac{1+x}{1-x}},$$

$$\tanh^{-1} x = \frac{1}{2} \ln \left(\frac{1+x}{1-x} \right). \blacktriangleleft$$

Graphs of the inverse hyperbolic functions are given in figures 3.14–3.16.

3.7.6 Calculus of hyperbolic functions

Just as the identities of hyperbolic functions closely follow those of their trigono-metric counterparts, so their calculus is similar. The derivatives of the two basic

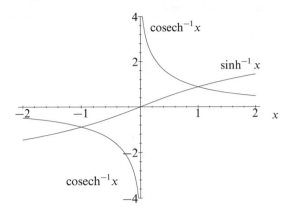

Figure 3.15   Graphs of $\sinh^{-1} x$ and $\operatorname{cosech}^{-1}x$.

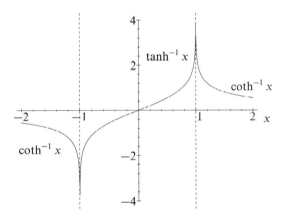

Figure 3.16   Graphs of $\tanh^{-1} x$ and $\coth^{-1} x$.

hyperbolic functions are given by

$$\frac{d}{dx}(\cosh x) = \sinh x, \tag{3.52}$$

$$\frac{d}{dx}(\sinh x) = \cosh x. \tag{3.53}$$

They may be deduced by considering the definitions (3.38), (3.39) as follows.

▶*Verify the relation $(d/dx)\cosh x = \sinh x$.*

Using the definition of $\cosh x$,

$$\cosh x = \tfrac{1}{2}(e^x + e^{-x}),$$

and differentiating directly, we find

$$\frac{d}{dx}(\cosh x) = \tfrac{1}{2}(e^x - e^{-x})$$
$$= \sinh x. \; ◀$$

Clearly the integrals of the fundamental hyperbolic functions are also defined by these relations. The derivatives of the remaining hyperbolic functions can be derived by product differentiation and are presented below only for completeness.

$$\frac{d}{dx}(\tanh x) = \operatorname{sech}^2 x, \tag{3.54}$$

$$\frac{d}{dx}(\operatorname{sech} x) = -\operatorname{sech} x \tanh x, \tag{3.55}$$

$$\frac{d}{dx}(\operatorname{cosech} x) = -\operatorname{cosech} x \coth x, \tag{3.56}$$

$$\frac{d}{dx}(\coth x) = -\operatorname{cosech}^2 x. \tag{3.57}$$

The inverse hyperbolic functions also have derivatives, which are given by the following:

$$\frac{d}{dx}\left(\cosh^{-1}\frac{x}{a}\right) = \frac{\pm 1}{\sqrt{x^2 - a^2}}, \tag{3.58}$$

$$\frac{d}{dx}\left(\sinh^{-1}\frac{x}{a}\right) = \frac{1}{\sqrt{x^2 + a^2}}, \tag{3.59}$$

$$\frac{d}{dx}\left(\tanh^{-1}\frac{x}{a}\right) = \frac{a}{a^2 - x^2}, \quad \text{for } x^2 < a^2, \tag{3.60}$$

$$\frac{d}{dx}\left(\coth^{-1}\frac{x}{a}\right) = \frac{-a}{x^2 - a^2}, \quad \text{for } x^2 > a^2. \tag{3.61}$$

These may be derived from the logarithmic form of the inverse (see subsection 3.7.5).

▶ *Evaluate* $(d/dx)\sinh^{-1} x$ *using the logarithmic form of the inverse.*

From the results of section 3.7.5,

$$\frac{d}{dx}\left(\sinh^{-1} x\right) = \frac{d}{dx}\left[\ln\left(x + \sqrt{x^2 + 1}\right)\right]$$

$$= \frac{1}{x + \sqrt{x^2 + 1}}\left(1 + \frac{x}{\sqrt{x^2 + 1}}\right)$$

$$= \frac{1}{x + \sqrt{x^2 + 1}}\left(\frac{\sqrt{x^2 + 1} + x}{\sqrt{x^2 + 1}}\right)$$

$$= \frac{1}{\sqrt{x^2 + 1}}. \blacktriangleleft$$

3.8 Exercises

3.1   Two complex numbers z and w are given by $z = 3 + 4i$ and $w = 2 - i$. On an Argand diagram, plot

(a) $z + w$, (b) $w - z$, (c) wz, (d) z/w,
(e) $z^*w + w^*z$, (f) w^2, (g) $\ln z$, (h) $(1 + z + w)^{1/2}$.

3.2   By considering the real and imaginary parts of the product $e^{i\theta}e^{i\phi}$ prove the standard formulae for $\cos(\theta + \phi)$ and $\sin(\theta + \phi)$.

3.3   By writing $\pi/12 = (\pi/3) - (\pi/4)$ and considering $e^{i\pi/12}$, evaluate $\cot(\pi/12)$.

3.4   Find the locus in the complex z-plane of points that satisfy the following equations.

(a) $z - c = \rho\left(\dfrac{1 + it}{1 - it}\right)$, where c is complex, ρ is real and t is a real parameter that varies in the range $-\infty < t < \infty$.

(b) $z = a + bt + ct^2$, in which t is a real parameter and a, b, and c are complex numbers with b/c real.

3.5   Evaluate

(a) $\operatorname{Re}(\exp 2iz)$, (b) $\operatorname{Im}(\cosh^2 z)$, (c) $(-1 + \sqrt{3}i)^{1/2}$,
(d) $|\exp(i^{1/2})|$, (e) $\exp(i^3)$, (f) $\operatorname{Im}(2^{i+3})$, (g) i^i, (h) $\ln[(\sqrt{3} + i)^3]$.

3.6   Find the equations in terms of x and y of the sets of points in the Argand diagram that satisfy the following:

(a) $\operatorname{Re} z^2 = \operatorname{Im} z^2$;
(b) $(\operatorname{Im} z^2)/z^2 = -i$;
(c) $\arg[z/(z - 1)] = \pi/2$.

3.7   Show that the locus of all points $z = x + iy$ in the complex plane that satisfy

$$|z - ia| = \lambda|z + ia|, \quad \lambda > 0,$$

is a circle of radius $|2\lambda a/(1 - \lambda^2)|$ centred on the point $z = ia[(1 + \lambda^2)/(1 - \lambda^2)]$. Sketch the circles for a few typical values of λ, including $\lambda < 1$, $\lambda > 1$ and $\lambda = 1$.

3.8   The two sets of points $z = a$, $z = b$, $z = c$, and $z = A$, $z = B$, $z = C$ are the corners of two similar triangles in the Argand diagram. Express in terms of $a, b, \ldots, C$

(a) the equalities of corresponding angles, and
(b) the constant ratio of corresponding sides,

in the two triangles.

By noting that any complex quantity can be expressed as

$$z = |z| \exp(i \arg z),$$

deduce that

$$a(B - C) + b(C - A) + c(A - B) = 0.$$

3.9   For the real constant a find the loci of all points $z = x + iy$ in the complex plane that satisfy

(a) $\operatorname{Re}\left\{ \ln\left(\dfrac{z - ia}{z + ia} \right) \right\} = c,$ $c > 0,$

(b) $\operatorname{Im}\left\{ \ln\left(\dfrac{z - ia}{z + ia} \right) \right\} = k,$ $0 \le k \le \pi/2.$

Identify the two families of curves and verify that in case (b) all curves pass through the two points $\pm ia$.

3.10   The most general type of transformation between one Argand diagram, in the z-plane, and another, in the Z-plane, that gives one and only one value of Z for each value of z (and conversely) is known as the *general bilinear transformation* and takes the form

$$z = \frac{aZ + b}{cZ + d}.$$

(a) Confirm that the transformation from the Z-plane to the z-plane is also a general bilinear transformation.
(b) Recalling that the equation of a circle can be written in the form

$$\left| \frac{z - z_1}{z - z_2} \right| = \lambda, \qquad \lambda \ne 1,$$

show that the general bilinear transformation transforms circles into circles (or straight lines). What is the condition that z_1, z_2 and λ must satisfy if the transformed circle is to be a straight line?

3.11   Sketch the parts of the Argand diagram in which

(a) $\operatorname{Re} z^2 < 0, \; |z^{1/2}| \le 2;$
(b) $0 \le \arg z^* \le \pi/2;$
(c) $|\exp z^3| \to 0$ as $|z| \to \infty.$

What is the area of the region in which all three sets of conditions are satisfied?

3.12   Denote the nth roots of unity by $1, \omega_n, \omega_n^2, \ldots, \omega_n^{n-1}$.

(a) Prove that

(i) $\displaystyle\sum_{r=0}^{n-1} \omega_n^r = 0,$ (ii) $\displaystyle\prod_{r=0}^{n-1} \omega_n^r = (-1)^{n+1}.$

(b) Express $x^2 + y^2 + z^2 - yz - zx - xy$ as the product of two factors, each linear in x, y and z, with coefficients dependent on the third roots of unity (and those of the x terms arbitrarily taken as real).

110

3.13 Prove that $x^{2m+1} - a^{2m+1}$, where m is an integer ≥ 1, can be written as

$$x^{2m+1} - a^{2m+1} = (x - a) \prod_{r=1}^{m} \left[x^2 - 2ax \cos\left(\frac{2\pi r}{2m+1}\right) + a^2 \right].$$

3.14 The complex position vectors of two parallel interacting equal fluid vortices moving with their axes of rotation always perpendicular to the z-plane are z_1 and z_2. The equations governing their motions are

$$\frac{dz_1^*}{dt} = -\frac{i}{z_1 - z_2}, \qquad \frac{dz_2^*}{dt} = -\frac{i}{z_2 - z_1}.$$

Deduce that (a) $z_1 + z_2$, (b) $|z_1 - z_2|$ and (c) $|z_1|^2 + |z_2|^2$ are all constant in time, and hence describe the motion geometrically.

3.15 Solve the equation

$$z^7 - 4z^6 + 6z^5 - 6z^4 + 6z^3 - 12z^2 + 8z + 4 = 0,$$

(a) by examining the effect of setting z^3 equal to 2, and then
(b) by factorising and using the binomial expansion of $(z + a)^4$.

Plot the seven roots of the equation on an Argand plot, exemplifying that complex roots of a polynomial equation always occur in conjugate pairs if the polynomial has real coefficients.

3.16 The polynomial $f(z)$ is defined by

$$f(z) = z^5 - 6z^4 + 15z^3 - 34z^2 + 36z - 48.$$

(a) Show that the equation $f(z) - 0$ has roots of the form $z = \lambda i$, where λ is real, and hence factorize $f(z)$.
(b) Show further that the cubic factor of $f(z)$ can be written in the form $(z + a)^3 + b$, where a and b are real, and hence solve the equation $f(z) = 0$ completely.

3.17 The binomial expansion of $(1 + x)^n$, discussed in chapter 1, can be written for a positive integer n as

$$(1 + x)^n = \sum_{r=0}^{n} {}^nC_r x^r,$$

where ${}^nC_r = n!/[r!(n-r)!]$.

(a) Use de Moivre's theorem to show that the sum

$$S_1(n) = {}^nC_0 - {}^nC_2 + {}^nC_4 - \cdots + (-1)^m \, {}^nC_{2m}, \qquad n - 1 \leq 2m \leq n,$$

has the value $2^{n/2} \cos(n\pi/4)$.
(b) Derive a similar result for the sum

$$S_2(n) = {}^nC_1 - {}^nC_3 + {}^nC_5 - \cdots + (-1)^m \, {}^nC_{2m+1}, \qquad n - 1 \leq 2m + 1 \leq n,$$

and verify it for the cases $n = 6$, 7 and 8.

3.18 By considering $(1 + \exp i\theta)^n$, prove that

$$\sum_{r=0}^{n} {}^nC_r \cos r\theta = 2^n \cos^n(\theta/2) \cos(n\theta/2),$$

$$\sum_{r=0}^{n} {}^nC_r \sin r\theta = 2^n \cos^n(\theta/2) \sin(n\theta/2),$$

where ${}^nC_r = n!/[r!(n-r)!]$.

3.19   Use de Moivre's theorem with $n = 4$ to prove that

$$\cos 4\theta = 8\cos^4\theta - 8\cos^2\theta + 1,$$

and deduce that

$$\cos\frac{\pi}{8} = \left(\frac{2 + \sqrt{2}}{4}\right)^{1/2}.$$

3.20   Express $\sin^4\theta$ entirely in terms of the trigonometric functions of multiple angles and deduce that its average value over a complete cycle is $\frac{3}{8}$.

3.21   Use de Moivre's theorem to prove that

$$\tan 5\theta = \frac{t^5 - 10t^3 + 5t}{5t^4 - 10t^2 + 1},$$

where $t = \tan\theta$. Deduce the values of $\tan(n\pi/10)$ for $n = 1, 2, 3, 4$.

3.22   Prove the following results involving hyperbolic functions.

(a) That

$$\cosh x - \cosh y = 2\sinh\left(\frac{x+y}{2}\right)\sinh\left(\frac{x-y}{2}\right).$$

(b) That, if $y = \sinh^{-1} x$,

$$(x^2 + 1)\frac{d^2 y}{dx^2} + x\frac{dy}{dx} = 0.$$

3.23   Determine the conditions under which the equation

$$a\cosh x + b\sinh x = c, \qquad c > 0,$$

has zero, one, or two real solutions for x. What is the solution if $a^2 = c^2 + b^2$?

3.24   Use the definitions and properties of hyperbolic functions to do the following:

(a) Solve $\cosh x = \sinh x + 2\,\text{sech}\,x$.
(b) Show that the real solution x of $\tanh x = \text{cosech}\,x$ can be written in the form $x = \ln(u + \sqrt{u})$. Find an explicit value for u.
(c) Evaluate $\tanh x$ when x is the real solution of $\cosh 2x = 2\cosh x$.

3.25   Express $\sinh^4 x$ in terms of hyperbolic cosines of multiples of x, and hence find the real solutions of

$$2\cosh 4x - 8\cosh 2x + 5 = 0.$$

3.26   In the theory of special relativity, the relationship between the position and time coordinates of an event, as measured in two frames of reference that have parallel x-axes, can be expressed in terms of hyperbolic functions. If the coordinates are x and t in one frame and x' and t' in the other, then the relationship take the form

$$x' = x\cosh\phi - ct\sinh\phi,$$
$$ct' = -x\sinh\phi + ct\cosh\phi.$$

Express x and ct in terms of x', ct' and ϕ and show that

$$x^2 - (ct)^2 = (x')^2 - (ct')^2.$$

3.27   A closed barrel has as its curved surface the surface obtained by rotating about the x-axis the part of the curve

$$y = a[2 - \cosh(x/a)]$$

lying in the range $-b \le x \le b$, where $b < a \cosh^{-1} 2$. Show that the total surface area, A, of the barrel is given by

$$A = \pi a[9a - 8a \exp(-b/a) + a \exp(-2b/a) - 2b].$$

3.28   The principal value of the logarithmic function of a complex variable is defined to have its argument in the range $-\pi < \arg z \le \pi$. By writing $z = \tan w$ in terms of exponentials show that

$$\tan^{-1} z = \frac{1}{2i} \ln \left(\frac{1 + iz}{1 - iz} \right).$$

Use this result to evaluate

$$\tan^{-1} \left(\frac{2\sqrt{3} - 3i}{7} \right).$$

3.9 Hints and answers

3.1   (a) $5 + 3i$; (b) $-1 - 5i$; (c) $10 + 5i$; (d) $2/5 + 11i/5$; (e) 4; (f) $3 - 4i$;
        (g) $\ln 5 + i[\tan^{-1}(4/3) + 2n\pi]$; (h) $\pm(2.521 + 0.595i)$.
3.3   Use $\sin \pi/4 = \cos \pi/4 = 1/\sqrt{2}$, $\sin \pi/3 = \sqrt{3}/2$ and $\cos \pi/3 = 1/2$.
        $\cot \pi/12 = 2 + \sqrt{3}$.
3.5   (a) $\exp(-2y) \cos 2x$; (b) $(\sin 2y \sinh 2x)/2$; (c) $\sqrt{2} \exp(\pi i/3)$ or $\sqrt{2} \exp(4\pi i/3)$;
        (d) $\exp(1/\sqrt{2})$ or $\exp(-1/\sqrt{2})$; (e) $0.540 - 0.841i$; (f) $8 \sin(\ln 2) = 5.11$;
        (g) $\exp(-\pi/2 - 2\pi n)$; (h) $\ln 8 + i(6n + 1/2)\pi$.
3.7   Starting from $|x + iy - ia| = \lambda|x + iy + ia|$, show that the coefficients of x and y are equal, and write the equation in the form $x^2 + (y - \alpha)^2 = r^2$.
3.9   (a) Circles enclosing $z = -ia$, with $\lambda = \exp c > 1$.
        (b) The condition is that $\arg[(z - ia)/(z + ia)] = k$. This can be rearranged to give $a(z + z^*) = (a^2 - |z|^2) \tan k$, which becomes in x, y coordinates the equation of a circle with centre $(-a \cot k, 0)$ and radius $a \operatorname{cosec} k$.
3.11   All three conditions are satisfied in $3\pi/2 \le \theta \le 7\pi/4$, $|z| \le 4$; area $= 2\pi$.
3.13   Denoting $\exp[2\pi i/(2m + 1)]$ by Ω, express $x^{2m+1} - a^{2m+1}$ as a product of factors like $(x - a\Omega^r)$ and then combine those containing Ω^r and Ω^{2m+1-r}. Use the fact that $\Omega^{2m+1} = 1$.
3.15   The roots are $2^{1/3} \exp(2\pi ni/3)$ for $n = 0, 1, 2$; $1 \pm 3^{1/4}$; $1 \pm 3^{1/4}i$.
3.17   Consider $(1 + i)^n$. (b) $S_2(n) = 2^{n/2} \sin(n\pi/4)$. $S_2(6) = -8$, $S_2(7) = -8$, $S_2(8) = 0$.
3.19   Use the binomial expansion of $(\cos \theta + i \sin \theta)^4$.
3.21   Show that $\cos 5\theta = 16c^5 - 20c^3 + 5c$, where $c = \cos \theta$, and correspondingly for $\sin 5\theta$. Use $\cos^{-2} \theta = 1 + \tan^2 \theta$. The four required values are
        $[(5 - \sqrt{20})/5]^{1/2}$, $(5 - \sqrt{20})^{1/2}$, $[(5 + \sqrt{20})/5]^{1/2}$, $(5 + \sqrt{20})^{1/2}$.
3.23   Reality of the root(s) requires $c^2 + b^2 \ge a^2$ and $a + b > 0$. With these conditions, there are two roots if $a^2 > b^2$, but only one if $b^2 > a^2$.
        For $a^2 = c^2 + b^2$, $x = \frac{1}{2} \ln[(a - b)/(a + b)]$.
3.25   Reduce the equation to $16 \sinh^4 x = 1$, yielding $x = \pm 0.481$.

113

3.27   Show that $ds = (\cosh x/a)\, dx$;
        curved surface area $= \pi a^2 [8 \sinh(b/a) - \sinh(2b/a)] - 2\pi ab$.
        flat ends area $= 2\pi a^2 [4 - 4\cosh(b/a) + \cosh^2(b/a)]$.

4

Series and limits

4.1 Series

Many examples exist in the physical sciences of situations where we are presented with a *sum of terms* to evaluate. For example, we may wish to add the contributions from successive slits in a diffraction grating to find the total light intensity at a particular point behind the grating.

A series may have either a finite or infinite number of terms. In either case, the sum of the first N terms of a series (often called a partial sum) is written

$$S_N = u_1 + u_2 + u_3 + \cdots + u_N,$$

where the terms of the series u_n, $n = 1, 2, 3, \ldots, N$ are numbers, that may in general be complex. If the terms are complex then S_N will in general be complex also, and we can write $S_N = X_N + iY_N$, where X_N and Y_N are the partial sums of the real and imaginary parts of each term separately and are therefore real. If a series has only N terms then the partial sum S_N is of course the sum of the series. Sometimes we may encounter series where each term depends on some variable, x, say. In this case the partial sum of the series will depend on the value assumed by x. For example, consider the infinite series

$$S(x) = 1 + x + \frac{x^2}{2!} + \frac{x^3}{3!} + \cdots .$$

This is an example of a power series; these are discussed in more detail in section 4.5. It is in fact the Maclaurin expansion of $\exp x$ (see subsection 4.6.3). Therefore $S(x) = \exp x$ and, of course, varies according to the value of the variable x. A series might just as easily depend on a complex variable z.

A general, random sequence of numbers can be described as a series and a sum of the terms found. However, for cases of practical interest, there will usually be

115

some sort of relationship between successive terms. For example, if the nth term of a series is given by

$$u_n = \frac{1}{2^n},$$

for $n = 1, 2, 3, \ldots, N$ then the sum of the first N terms will be

$$S_N = \sum_{n=1}^{N} u_n = \frac{1}{2} + \frac{1}{4} + \frac{1}{8} + \cdots + \frac{1}{2^N}. \tag{4.1}$$

It is clear that the sum of a finite number of terms is always finite, provided that each term is itself finite. It is often of practical interest, however, to consider the sum of a series with an infinite number of finite terms. The sum of an infinite number of terms is best defined by first considering the partial sum of the first N terms, S_N. If the value of the partial sum S_N tends to a finite limit, S, as N tends to infinity, then the series is said to converge and its sum is given by the limit S. In other words, the sum of an infinite series is given by

$$S = \lim_{N \to \infty} S_N,$$

provided the limit exists. For complex infinite series, if S_N approaches a limit $S = X + iY$ as $N \to \infty$, this means that $X_N \to X$ and $Y_N \to Y$ separately, i.e. the real and imaginary parts of the series are each convergent series with sums X and Y respectively.

However, not all infinite series have finite sums. As $N \to \infty$, the value of the partial sum S_N may diverge: it may approach $+\infty$ or $-\infty$, or oscillate finitely or infinitely. Moreover, for a series where each term depends on some variable, its convergence can depend on the value assumed by the variable. Whether an infinite series converges, diverges or oscillates has important implications when describing physical systems. Methods for determining whether a series converges are discussed in section 4.3.

4.2 Summation of series

It is often necessary to find the sum of a finite series or a convergent infinite series. We now describe arithmetic, geometric and arithmetico-geometric series, which are particularly common and for which the sums are easily found. Other methods that can sometimes be used to sum more complicated series are discussed below.

4.2.1 Arithmetic series

An *arithmetic series* has the characteristic that the difference between successive terms is constant. The sum of a general arithmetic series is written

$$S_N = a + (a + d) + (a + 2d) + \cdots + [a + (N-1)d] = \sum_{n=0}^{N-1}(a + nd).$$

Rewriting the series in the opposite order and adding this term by term to the original expression for S_N, we find

$$S_N = \frac{N}{2}[a + a + (N-1)d] = \frac{N}{2}(\text{first term} + \text{last term}). \qquad (4.2)$$

If an infinite number of such terms are added the series will increase (or decrease) indefinitely; that is to say, it diverges.

▶Sum the integers between 1 and 1000 inclusive.

This is an arithmetic series with $a = 1$, $d = 1$ and $N = 1000$. Therefore, using (4.2) we find

$$S_N = \frac{1000}{2}(1 + 1000) = 500500,$$

which can be checked directly only with considerable effort. ◀

4.2.2 Geometric series

Equation (4.1) is a particular example of a *geometric series*, which has the characteristic that the ratio of successive terms is a constant (one-half in this case). The sum of a geometric series is in general written

$$S_N = a + ar + ar^2 + \cdots + ar^{N-1} = \sum_{n=0}^{N-1} ar^n,$$

where a is a constant and r is the ratio of successive terms, the *common ratio*. The sum may be evaluated by considering S_N and rS_N:

$$S_N = a + ar + ar^2 + ar^3 + \cdots + ar^{N-1},$$
$$rS_N = ar + ar^2 + ar^3 + ar^4 + \cdots + ar^{N}.$$

If we now subtract the second equation from the first we obtain

$$(1 - r)S_N = a - ar^N,$$

and hence

$$S_N = \frac{a(1 - r^N)}{1 - r}. \qquad (4.3)$$

For a series with an infinite number of terms and $|r| < 1$, we have $\lim_{N \to \infty} r^N = 0$, and the sum tends to the limit

$$S = \frac{a}{1 - r}. \tag{4.4}$$

In (4.1), $r = \frac{1}{2}$, $a = \frac{1}{2}$, and so $S = 1$. For $|r| \geq 1$, however, the series either diverges or oscillates.

> ▶*Consider a ball that drops from a height of 27 m and on each bounce retains only a third of its kinetic energy; thus after one bounce it will return to a height of 9 m, after two bounces to 3 m, and so on. Find the total distance travelled between the first bounce and the Mth bounce.*

The total distance travelled between the first bounce and the Mth bounce is given by the sum of $M - 1$ terms:

$$S_{M-1} = 2(9 + 3 + 1 + \cdots) = 2 \sum_{m=0}^{M-2} \frac{9}{3^m}$$

for $M > 1$, where the factor 2 is included to allow for both the upward and the downward journey. Inside the parentheses we clearly have a geometric series with first term 9 and common ratio $1/3$ and hence the distance is given by (4.3), i.e.

$$S_{M-1} = 2 \times \frac{9 \left[1 - \left(\frac{1}{3} \right)^{M-1} \right]}{1 - \frac{1}{3}} = 27 \left[1 - \left(\frac{1}{3} \right)^{M-1} \right],$$

where the number of terms N in (4.3) has been replaced by $M - 1$. ◀

4.2.3 Arithmetico-geometric series

An arithmetico-geometric series, as its name suggests, is a combined arithmetic and geometric series. It has the general form

$$S_N = a + (a + d)r + (a + 2d)r^2 + \cdots + [a + (N - 1)d] r^{N-1} = \sum_{n=0}^{N-1} (a + nd)r^n,$$

and can be summed, in a similar way to a pure geometric series, by multiplying by r and subtracting the result from the original series to obtain

$$(1 - r)S_N = a + rd + r^2 d + \cdots + r^{N-1} d - [a + (N - 1)d] r^N.$$

Using the expression for the sum of a geometric series (4.3) and rearranging, we find

$$S_N = \frac{a - [a + (N - 1)d] r^N}{1 - r} + \frac{rd(1 - r^{N-1})}{(1 - r)^2}.$$

For an infinite series with $|r| < 1$, $\lim_{N \to \infty} r^N = 0$ as in the previous subsection, and the sum tends to the limit

$$S = \frac{a}{1 - r} + \frac{rd}{(1 - r)^2}. \tag{4.5}$$

As for a geometric series, if $|r| \geq 1$ then the series either diverges or oscillates.

▶*Sum the series*

$$S = 2 + \frac{5}{2} + \frac{8}{2^2} + \frac{11}{2^3} + \cdots .$$

This is an infinite arithmetico-geometric series with $a = 2$, $d = 3$ and $r = 1/2$. Therefore, from (4.5), we obtain $S = 10$. ◀

4.2.4 The difference method

The difference method is sometimes useful in summing series that are more complicated than the examples discussed above. Let us consider the general series

$$\sum_{n=1}^{N} u_n = u_1 + u_2 + \cdots + u_N.$$

If the terms of the series, u_n, can be expressed in the form

$$u_n = f(n) - f(n-1)$$

for some function $f(n)$ then its (partial) sum is given by

$$S_N = \sum_{n=1}^{N} u_n = f(N) - f(0).$$

This can be shown as follows. The sum is given by

$$S_N = u_1 + u_2 + \cdots + u_N$$

and since $u_n = f(n) - f(n-1)$, it may be rewritten

$$S_N = [f(1) - f(0)] + [f(2) - f(1)] + \cdots + [f(N) - f(N-1)].$$

By cancelling terms we see that

$$S_N = f(N) - f(0).$$

▶*Evaluate the sum*

$$\sum_{n=1}^{N} \frac{1}{n(n+1)}.$$

Using partial fractions we find

$$u_n = - \left(\frac{1}{n+1} - \frac{1}{n} \right).$$

Hence $u_n = f(n) - f(n-1)$ with $f(n) = -1/(n+1)$, and so the sum is given by

$$S_N = f(N) - f(0) = -\frac{1}{N+1} + 1 = \frac{N}{N+1}. \quad ◀$$

The difference method may be easily extended to evaluate sums in which each term can be expressed in the form

$$u_n = f(n) - f(n - m),\tag{4.6}$$

where m is an integer. By writing out the sum to N terms with each term expressed in this form, and cancelling terms in pairs as before, we find

$$S_N = \sum_{k=1}^{m} f(N - k + 1) - \sum_{k=1}^{m} f(1 - k).$$

▶ *Evaluate the sum*

$$\sum_{n=1}^{N} \frac{1}{n(n + 2)}.$$

Using partial fractions we find

$$u_n = -\left[\frac{1}{2(n + 2)} - \frac{1}{2n}\right].$$

Hence $u_n = f(n) - f(n - 2)$ with $f(n) = -1/[2(n + 2)]$, and so the sum is given by

$$S_N = f(N) + f(N - 1) - f(0) - f(-1) = \frac{3}{4} - \frac{1}{2}\left(\frac{1}{N + 2} + \frac{1}{N + 1}\right). \blacktriangleleft$$

In fact the difference method is quite flexible and may be used to evaluate sums even when each term cannot be expressed as in (4.6). The method still relies, however, on being able to write u_n in terms of a single function such that most terms in the sum cancel, leaving only a few terms at the beginning and the end. This is best illustrated by an example.

▶ *Evaluate the sum*

$$\sum_{n=1}^{N} \frac{1}{n(n + 1)(n + 2)}.$$

Using partial fractions we find

$$u_n = \frac{1}{2(n + 2)} - \frac{1}{n + 1} + \frac{1}{2n}.$$

Hence $u_n = f(n) - 2f(n - 1) + f(n - 2)$ with $f(n) = 1/[2(n + 2)]$. If we write out the sum, expressing each term u_n in this form, we find that most terms cancel and the sum is given by

$$S_N = f(N) - f(N - 1) - f(0) + f(-1) = \frac{1}{4} + \frac{1}{2}\left(\frac{1}{N + 2} - \frac{1}{N + 1}\right). \blacktriangleleft$$

120

4.2.5 Series involving natural numbers

Series consisting of the natural numbers 1, 2, 3, ..., or the square or cube of these numbers, occur frequently and deserve a special mention. Let us first consider the sum of the first N natural numbers,

$$S_N = 1 + 2 + 3 + \cdots + N = \sum_{n=1}^{N} n.$$

This is clearly an arithmetic series with first term $a = 1$ and common difference $d = 1$. Therefore, from (4.2), $S_N = \frac{1}{2}N(N+1)$.

Next, we consider the sum of the squares of the first N natural numbers:

$$S_N = 1^2 + 2^2 + 3^2 + \ldots + N^2 = \sum_{n=1}^{N} n^2,$$

which may be evaluated using the difference method. The nth term in the series is $u_n = n^2$, which we need to express in the form $f(n) - f(n-1)$ for some function $f(n)$. Consider the function

$$f(n) = n(n+1)(2n+1) \quad \Rightarrow \quad f(n-1) = (n-1)n(2n-1).$$

For this function $f(n) - f(n-1) = 6n^2$, and so we can write

$$u_n = \frac{1}{6}[f(n) - f(n-1)].$$

Therefore, by the difference method,

$$S_N = \frac{1}{6}[f(N) - f(0)] = \frac{1}{6}N(N+1)(2N+1).$$

Finally, we calculate the sum of the cubes of the first N natural numbers,

$$S_N = 1^3 + 2^3 + 3^3 + \cdots + N^3 = \sum_{n=1}^{N} n^3,$$

again using the difference method. Consider the function

$$f(n) = [n(n+1)]^2 \quad \Rightarrow \quad f(n-1) = [(n-1)n]^2,$$

for which $f(n) - f(n-1) = 4n^3$. Therefore we can write the general nth term of the series as

$$u_n = \frac{1}{4}[f(n) - f(n-1)],$$

and using the difference method we find

$$S_N = \frac{1}{4}[f(N) - f(0)] = \frac{1}{4}N^2(N+1)^2.$$

Note that this is the square of the sum of the natural numbers, i.e.

$$\sum_{n=1}^{N} n^3 = \left(\sum_{n=1}^{N} n\right)^2.$$

►*Sum the series*

$$\sum_{n=1}^{N}(n+1)(n+3).$$

The nth term in this series is

$$u_n = (n+1)(n+3) = n^2 + 4n + 3,$$

and therefore we can write

$$\sum_{n=1}^{N}(n+1)(n+3) = \sum_{n=1}^{N}(n^2 + 4n + 3)$$

$$= \sum_{n=1}^{N}n^2 + 4\sum_{n=1}^{N}n + \sum_{n=1}^{N}3$$

$$= \tfrac{1}{6}N(N+1)(2N+1) + 4 \times \tfrac{1}{2}N(N+1) + 3N$$

$$= \tfrac{1}{6}N(2N^2 + 15N + 31). ◄$$

4.2.6 Transformation of series

A complicated series may sometimes be summed by transforming it into a familiar series for which we already know the sum, perhaps a geometric series or the Maclaurin expansion of a simple function (see subsection 4.6.3). Various techniques are useful, and deciding which one to use in any given case is a matter of experience. We now discuss a few of the more common methods.

The differentiation or integration of a series is often useful in transforming an apparently intractable series into a more familiar one. If we wish to differentiate or integrate a series that already depends on some variable then we may do so in a straightforward manner.

►*Sum the series*

$$S(x) = \frac{x^4}{3(0!)} + \frac{x^5}{4(1!)} + \frac{x^6}{5(2!)} + \cdots.$$

Dividing both sides by x we obtain

$$\frac{S(x)}{x} = \frac{x^3}{3(0!)} + \frac{x^4}{4(1!)} + \frac{x^5}{5(2!)} + \cdots,$$

which is easily differentiated to give

$$\frac{d}{dx}\left[\frac{S(x)}{x}\right] = \frac{x^2}{0!} + \frac{x^3}{1!} + \frac{x^4}{2!} + \frac{x^5}{3!} + \cdots.$$

Recalling the Maclaurin expansion of $\exp x$ given in subsection 4.6.3, we recognise that the RHS is equal to $x^2 \exp x$. Having done so, we can now integrate both sides to obtain

$$S(x)/x = \int x^2 \exp x \, dx.$$

Integrating the RHS by parts we find

$$S(x)/x = x^2 \exp x - 2x \exp x + 2 \exp x + c,$$

where the value of the constant of integration c can be fixed by the requirement that $S(x)/x = 0$ at $x = 0$. Thus we find that $c = -2$ and that the sum is given by

$$S(x) = x^3 \exp x - 2x^2 \exp x + 2x \exp x - 2x. \ \blacktriangleleft$$

Often, however, we require the sum of a series that does not depend on a variable. In this case, in order that we may differentiate or integrate the series, we define a function of some variable x such that the value of this function is equal to the sum of the series for some particular value of x (usually at $x = 1$).

▶Sum the series

$$S = 1 + \frac{2}{2} + \frac{3}{2^2} + \frac{4}{2^3} + \cdots.$$

Let us begin by defining the function

$$f(x) = 1 + 2x + 3x^2 + 4x^3 + \cdots,$$

so that the sum $S = f(1/2)$. Integrating this function we obtain

$$\int f(x) \, dx = x + x^2 + x^3 + \cdots,$$

which we recognise as an infinite geometric series with first term $a = x$ and common ratio $r = x$. Therefore, from (4.4), we find that the sum of this series is $x/(1 - x)$. In other words

$$\int f(x) \, dx = \frac{x}{1 - x},$$

so that $f(x)$ is given by

$$f(x) = \frac{d}{dx} \left(\frac{x}{1 - x} \right) = \frac{1}{(1 - x)^2}.$$

The sum of the original series is therefore $S = f(1/2) = 4$. ◀

Aside from differentiation and integration, an appropriate substitution can sometimes transform a series into a more familiar form. In particular, series with terms that contain trigonometric functions can often be summed by the use of complex exponentials.

▶Sum the series

$$S(\theta) = 1 + \cos \theta + \frac{\cos 2\theta}{2!} + \frac{\cos 3\theta}{3!} + \cdots.$$

Replacing the cosine terms with a complex exponential, we obtain

$$S(\theta) = \operatorname{Re} \left\{ 1 + \exp i\theta + \frac{\exp 2i\theta}{2!} + \frac{\exp 3i\theta}{3!} + \cdots \right\}$$

$$= \operatorname{Re} \left\{ 1 + \exp i\theta + \frac{(\exp i\theta)^2}{2!} + \frac{(\exp i\theta)^3}{3!} + \cdots \right\}.$$

Again using the Maclaurin expansion of $\exp x$ given in subsection 4.6.3, we notice that

$$S(\theta) = \text{Re}\,[\exp(\exp i\theta)] = \text{Re}\,[\exp(\cos\theta + i\sin\theta)]$$
$$= \text{Re}\,\{[\exp(\cos\theta)][\exp(i\sin\theta)]\} = [\exp(\cos\theta)]\text{Re}\,[\exp(i\sin\theta)]$$
$$= [\exp(\cos\theta)][\cos(\sin\theta)].\ \blacktriangleleft$$

4.3 Convergence of infinite series

Although the sums of some commonly occurring infinite series may be found, the sum of a general infinite series is usually difficult to calculate. Nevertheless, it is often useful to know whether the partial sum of such a series converges to a limit, even if the limit cannot be found explicitly. As mentioned at the end of section 4.1, if we allow N to tend to infinity, the partial sum

$$S_N = \sum_{n=1}^{N} u_n$$

of a series may tend to a definite limit (i.e. the sum S of the series), or increase or decrease without limit, or oscillate finitely or infinitely.

To investigate the convergence of any given series, it is useful to have available a number of tests and theorems of general applicability. We discuss them below; some we will merely state, since once they have been stated they become almost self-evident, but are no less useful for that.

4.3.1 Absolute and conditional convergence

Let us first consider some general points concerning the convergence, or otherwise, of an infinite series. In general an infinite series $\sum u_n$ can have complex terms, and in the special case of a real series the terms can be positive or negative. From any such series, however, we can always construct another series $\sum |u_n|$ in which each term is simply the modulus of the corresponding term in the original series. Then each term in the new series will be a positive real number.

If the series $\sum |u_n|$ converges then $\sum u_n$ also converges, and $\sum u_n$ is said to be *absolutely convergent*, i.e. the series formed by the absolute values is convergent. For an absolutely convergent series, the terms may be reordered without affecting the convergence of the series. However, if $\sum |u_n|$ diverges whilst $\sum u_n$ converges then $\sum u_n$ is said to be *conditionally convergent*. For a conditionally convergent series, rearranging the order of the terms can affect the behaviour of the sum and, hence, whether the series converges or diverges. In fact, a theorem due to Riemann shows that, by a suitable rearrangement, a conditionally convergent series may be made to converge to any arbitrary limit, or to diverge, or to oscillate finitely or infinitely! Of course, if the original series $\sum u_n$ consists only of positive real terms and converges then automatically it is absolutely convergent.

4.3.2 Convergence of a series containing only real positive terms

As discussed above, in order to test for the absolute convergence of a series $\sum u_n$, we first construct the corresponding series $\sum |u_n|$ that consists only of real positive terms. Therefore in this subsection we will restrict our attention to series of this type.

We discuss below some tests that may be used to investigate the convergence of such a series. Before doing so, however, we note the following *crucial consideration*. In all the tests for, or discussions of, the convergence of a series, it is not what happens in the first ten, or the first thousand, or the first million terms (or any other finite number of terms) that matters, but what happens *ultimately*.

Preliminary test

A necessary *but not sufficient* condition for a series of real positive terms $\sum u_n$ to be convergent is that the term u_n tends to zero as n tends to infinity, i.e. we require

$$\lim_{n \to \infty} u_n = 0.$$

If this condition is not satisfied then the series must diverge. Even if it is satisfied, however, the series may still diverge, and further testing is required.

Comparison test

The comparison test is the most basic test for convergence. Let us consider two series $\sum u_n$ and $\sum v_n$ and suppose that we *know* the latter to be convergent (by some earlier analysis, for example). Then, if each term u_n in the first series is less than or equal to the corresponding term v_n in the second series, for all n greater than some fixed number N that will vary from series to series, then the original series $\sum u_n$ is also convergent. In other words, if $\sum v_n$ is convergent and

$$u_n \leq v_n \qquad \text{for } n > N,$$

then $\sum u_n$ converges.

However, if $\sum v_n$ diverges and $u_n \geq v_n$ for all n greater than some fixed number then $\sum u_n$ diverges.

► *Determine whether the following series converges:*

$$\sum_{n=1}^{\infty} \frac{1}{n! + 1} = \frac{1}{2} + \frac{1}{3} + \frac{1}{7} + \frac{1}{25} + \cdots. \tag{4.7}$$

Let us compare this series with the series

$$\sum_{n=0}^{\infty} \frac{1}{n!} = \frac{1}{0!} + \frac{1}{1!} + \frac{1}{2!} + \frac{1}{3!} + \cdots = 2 + \frac{1}{2!} + \frac{1}{3!} + \cdots, \tag{4.8}$$

which is merely the series obtained by setting $x = 1$ in the Maclaurin expansion of $\exp x$ (see subsection 4.6.3), i.e.

$$\exp(1) = e = 1 + \frac{1}{1!} + \frac{1}{2!} + \frac{1}{3!} + \cdots .$$

Clearly this second series is convergent, since it consists of only positive terms and has a finite sum. Thus, since each term u_n in the series (4.7) is less than the corresponding term $1/n!$ in (4.8), we conclude from the comparison test that (4.7) is also convergent. ◄

D'Alembert's ratio test

The ratio test determines whether a series converges by comparing the relative magnitudes of successive terms. If we consider a series $\sum u_n$ and set

$$\rho = \lim_{n \to \infty} \left(\frac{u_{n+1}}{u_n} \right), \qquad (4.9)$$

then if $\rho < 1$ the series is convergent; if $\rho > 1$ the series is divergent; if $\rho = 1$ then the behaviour of the series is undetermined by this test.

To prove this we observe that if the limit (4.9) is less than unity, i.e. $\rho < 1$ then we can find a value r in the range $\rho < r < 1$ and a value N such that

$$\frac{u_{n+1}}{u_n} < r,$$

for all $n > N$. Now the terms u_n of the series that follow u_N are

$$u_{N+1}, \qquad u_{N+2}, \qquad u_{N+3}, \qquad \ldots,$$

and each of these is less than the corresponding term of

$$r u_N, \qquad r^2 u_N, \qquad r^3 u_N, \qquad \ldots . \qquad (4.10)$$

However, the terms of (4.10) are those of a geometric series with a common ratio r that is less than unity. This geometric series consequently converges and therefore, by the comparison test discussed above, so must the original series $\sum u_n$. An analogous argument may be used to prove the divergent case when $\rho > 1$.

►*Determine whether the following series converges:*

$$\sum_{n=0}^{\infty} \frac{1}{n!} = \frac{1}{0!} + \frac{1}{1!} + \frac{1}{2!} + \frac{1}{3!} + \cdots = 2 + \frac{1}{2!} + \frac{1}{3!} + \cdots .$$

As mentioned in the previous example, this series may be obtained by setting $x = 1$ in the Maclaurin expansion of $\exp x$, and hence we know already that it converges and has the sum $\exp(1) = e$. Nevertheless, we may use the ratio test to confirm that it converges. Using (4.9), we have

$$\rho = \lim_{n \to \infty} \left[\frac{n!}{(n+1)!} \right] = \lim_{n \to \infty} \left(\frac{1}{n+1} \right) = 0 \qquad (4.11)$$

and since $\rho < 1$, the series converges, as expected. ◄

Ratio comparison test

As its name suggests, the ratio comparison test is a combination of the ratio and comparison tests. Let us consider the two series $\sum u_n$ and $\sum v_n$ and assume that we know the latter to be convergent. It may be shown that if

$$\frac{u_{n+1}}{u_n} \leq \frac{v_{n+1}}{v_n}$$

for all n greater than some fixed value N then $\sum u_n$ is also convergent.
Similarly, if

$$\frac{u_{n+1}}{u_n} \geq \frac{v_{n+1}}{v_n}$$

for all sufficiently large n, and $\sum v_n$ diverges then $\sum u_n$ also diverges.

▶Determine whether the following series converges:

$$\sum_{n=1}^{\infty} \frac{1}{(n!)^2} = 1 + \frac{1}{2^2} + \frac{1}{6^2} + \cdots .$$

In this case the ratio of successive terms, as n tends to infinity, is given by

$$R = \lim_{n \to \infty} \left[\frac{n!}{(n+1)!} \right]^2 = \lim_{n \to \infty} \left(\frac{1}{n+1} \right)^2 ,$$

which is less than the ratio seen in (4.11). Hence, by the ratio comparison test, the series converges. (It is clear that this series could also be found to be convergent using the ratio test.) ◀

Quotient test

The quotient test may also be considered as a combination of the ratio and comparison tests. Let us again consider the two series $\sum u_n$ and $\sum v_n$, and define ρ as the limit

$$\rho = \lim_{n \to \infty} \left(\frac{u_n}{v_n} \right) . \tag{4.12}$$

Then, it can be shown that:

   (i) if $\rho \neq 0$ but is finite then $\sum u_n$ and $\sum v_n$ either both converge or both diverge;

  (ii) if $\rho = 0$ and $\sum v_n$ converges then $\sum u_n$ converges;

 (iii) if $\rho = \infty$ and $\sum v_n$ diverges then $\sum u_n$ diverges.

> ►*Given that the series $\sum_{n=1}^{\infty} 1/n$ diverges, determine whether the following series converges:*
>
> $$\sum_{n=1}^{\infty} \frac{4n^2 - n - 3}{n^3 + 2n}. \tag{4.13}$$

If we set $u_n = (4n^2 - n - 3)/(n^3 + 2n)$ and $v_n = 1/n$ then the limit (4.12) becomes

$$\rho = \lim_{n \to \infty} \left[\frac{(4n^2 - n - 3)/(n^3 + 2n)}{1/n} \right] = \lim_{n \to \infty} \left[\frac{4n^3 - n^2 - 3n}{n^3 + 2n} \right] = 4.$$

Since ρ is finite but non-zero and $\sum v_n$ diverges, from (i) above $\sum u_n$ must also diverge. ◄

Integral test

The integral test is an extremely powerful means of investigating the convergence of a series $\sum u_n$. Suppose that there exists a function $f(x)$ which monotonically decreases for x greater than some fixed value x_0 and for which $f(n) = u_n$, i.e. the value of the function at integer values of x is equal to the corresponding term in the series under investigation. Then it can be shown that, if the limit of the integral

$$\lim_{N \to \infty} \int^N f(x)\, dx$$

exists, the series $\sum u_n$ is convergent. Otherwise the series diverges. Note that the integral defined here has no lower limit; the test is sometimes stated with a lower limit, equal to unity, for the integral, but this can lead to unnecessary difficulties.

> ►*Determine whether the following series converges:*
>
> $$\sum_{n=1}^{\infty} \frac{1}{(n - 3/2)^2} = 4 + 4 + \frac{4}{9} + \frac{4}{25} + \cdots.$$

Let us consider the function $f(x) = (x - 3/2)^{-2}$. Clearly $f(n) = u_n$ and $f(x)$ monotonically decreases for $x > 3/2$. Applying the integral test, we consider

$$\lim_{N \to \infty} \int^N \frac{1}{(x - 3/2)^2}\, dx = \lim_{N \to \infty} \left(\frac{-1}{N - 3/2} \right) = 0.$$

Since the limit exists the series converges. Note, however, that if we had included a lower limit, equal to unity, in the integral then we would have run into problems, since the integrand diverges at $x = 3/2$. ◄

The integral test is also useful for examining the convergence of the Riemann zeta series. This is a special series that occurs regularly and is of the form

$$\sum_{n=1}^{\infty} \frac{1}{n^p}.$$

It converges for $p > 1$ and diverges if $p \le 1$. These convergence criteria may be derived as follows.

Using the integral test, we consider

$$\lim_{N\to\infty}\int^N \frac{1}{x^p}dx = \lim_{N\to\infty}\left(\frac{N^{1-p}}{1-p}\right),$$

and it is obvious that the limit tends to zero for $p > 1$ and to ∞ for $p \le 1$.

Cauchy's root test

Cauchy's root test may be useful in testing for convergence, especially if the nth terms of the series contains an nth power. If we define the limit

$$\rho = \lim_{n\to\infty}(u_n)^{1/n},$$

then it may be proved that the series $\sum u_n$ converges if $\rho < 1$. If $\rho > 1$ then the series diverges. Its behaviour is undetermined if $\rho = 1$.

▶Determine whether the following series converges:

$$\sum_{n=1}^{\infty}\left(\frac{1}{n}\right)^n = 1 + \frac{1}{4} + \frac{1}{27} + \cdots.$$

Using Cauchy's root test, we find

$$\rho = \lim_{n\to\infty}\left(\frac{1}{n}\right) = 0,$$

and hence the series converges. ◀

Grouping terms

We now consider the Riemann zeta series, mentioned above, with an alternative proof of its convergence that uses the method of grouping terms. In general there are better ways of determining convergence, but the grouping method may be used if it is not immediately obvious how to approach a problem by a better method.

First consider the case where $p > 1$, and group the terms in the series as follows:

$$S_N = \frac{1}{1^p} + \left(\frac{1}{2^p} + \frac{1}{3^p}\right) + \left(\frac{1}{4^p} + \cdots + \frac{1}{7^p}\right) + \cdots.$$

Now we can see that each bracket of this series is less than each term of the geometric series

$$S_N = \frac{1}{1^p} + \frac{2}{2^p} + \frac{4}{4^p} + \cdots.$$

This geometric series has common ratio $r = \left(\frac{1}{2}\right)^{p-1}$; since $p > 1$, it follows that $r < 1$ and that the geometric series converges. Then the comparison test shows that the Riemann zeta series also converges for $p > 1$.

The divergence of the Riemann zeta series for $p \leq 1$ can be seen by first considering the case $p = 1$. The series is

$$S_N = 1 + \frac{1}{2} + \frac{1}{3} + \frac{1}{4} + \cdots,$$

which does *not* converge, as may be seen by bracketing the terms of the series in groups in the following way:

$$S_N = \sum_{n=1}^{N} u_n = 1 + \left(\frac{1}{2}\right) + \left(\frac{1}{3} + \frac{1}{4}\right) + \left(\frac{1}{5} + \frac{1}{6} + \frac{1}{7} + \frac{1}{8}\right) + \cdots.$$

The sum of the terms in each bracket is $\geq \frac{1}{2}$ and, since as many such groupings can be made as we wish, it is clear that S_N increases indefinitely as N is increased.

Now returning to the case of the Riemann zeta series for $p < 1$, we note that each term in the series is greater than the corresponding one in the series for which $p = 1$. In other words $1/n^p > 1/n$ for $n > 1$, $p < 1$. The comparison test then shows us that the Riemann zeta series will diverge for all $p \leq 1$.

4.3.3 Alternating series test

The tests discussed in the last subsection have been concerned with determining whether the series of real positive terms $\sum |u_n|$ converges, and so whether $\sum u_n$ is absolutely convergent. Nevertheless, it is sometimes useful to consider whether a series is merely convergent rather than absolutely convergent. This is especially true for series containing an infinite number of both positive and negative terms. In particular, we will consider the convergence of series in which the positive and negative terms alternate, i.e. an *alternating series*.

An alternating series can be written as

$$\sum_{n=1}^{\infty} (-1)^{n+1} u_n = u_1 - u_2 + u_3 - u_4 + u_5 - \cdots,$$

with all $u_n \geq 0$. Such a series can be shown to converge provided (i) $u_n \to 0$ as $n \to \infty$ and (ii) $u_n < u_{n-1}$ for all $n > N$ for some finite N. If these conditions are not met then the series oscillates.

To prove this, suppose for definiteness that N is odd and consider the series starting at u_N. The sum of its first $2m$ terms is

$$S_{2m} = (u_N - u_{N+1}) + (u_{N+2} - u_{N+3}) + \cdots + (u_{N+2m-2} - u_{N+2m-1}).$$

By condition (ii) above, all the parentheses are positive, and so S_{2m} increases as m increases. We can also write, however,

$$S_{2m} = u_N - (u_{N+1} - u_{N+2}) - \cdots - (u_{N+2m-3} - u_{N+2m-2}) - u_{N+2m-1},$$

and since each parenthesis is positive, we must have $S_{2m} < u_N$. Thus, since S_{2m}

is always less than u_N for all m and $u_n \to 0$ as $n \to \infty$, the alternating series converges. It is clear that an analogous proof can be constructed in the case where N is even.

►*Determine whether the following series converges:*

$$\sum_{n=1}^{\infty}(-1)^{n+1}\frac{1}{n} = 1 - \frac{1}{2} + \frac{1}{3} - \cdots .$$

This alternating series clearly satisfies conditions (i) and (ii) above and hence converges. However, as shown above by the method of grouping terms, the corresponding series with all positive terms is divergent. ◄

4.4 Operations with series

Simple operations with series are fairly intuitive, and we discuss them here only for completeness. The following points apply to both finite and infinite series unless otherwise stated.

(i) If $\sum u_n = S$ then $\sum k u_n = kS$ where k is any constant.

(ii) If $\sum u_n = S$ and $\sum v_n = T$ then $\sum (u_n + v_n) = S + T$.

(iii) If $\sum u_n = S$ then $a + \sum u_n = a + S$. A simple extension of this trivial result shows that the removal or insertion of a finite number of terms anywhere in a series does not affect its convergence.

(iv) If the infinite series $\sum u_n$ and $\sum v_n$ are both absolutely convergent then the series $\sum w_n$, where

$$w_n = u_1 v_n + u_2 v_{n-1} + \cdots + u_n v_1,$$

is also absolutely convergent. The series $\sum w_n$ is called the *Cauchy product* of the two original series. Furthermore, if $\sum u_n$ converges to the sum S and $\sum v_n$ converges to the sum T then $\sum w_n$ converges to the sum ST.

(v) It is not true in general that term-by-term differentiation or integration of a series will result in a new series with the same convergence properties.

4.5 Power series

A power series has the form

$$P(x) = a_0 + a_1 x + a_2 x^2 + a_3 x^3 + \cdots ,$$

where a_0, a_1, a_2, a_3 etc. are constants. Such series regularly occur in physics and engineering and are useful because, for $|x| < 1$, the later terms in the series may become very small and be discarded. For example the series

$$P(x) = 1 + x + x^2 + x^3 + \cdots ,$$

although in principle infinitely long, in practice may be simplified if x happens to have a value small compared with unity. To see this note that $P(x)$ for $x = 0.1$ has the following values: 1, if just one term is taken into account; 1.1, for two terms; 1.11, for three terms; 1.111, for four terms, etc. If the quantity that it represents can only be measured with an accuracy of two decimal places, then all but the first three terms may be ignored, i.e. when $x = 0.1$ or less

$$P(x) = 1 + x + x^2 + O(x^3) \approx 1 + x + x^2.$$

This sort of approximation is often used to simplify equations into manageable forms. It may seem imprecise at first but is perfectly acceptable insofar as it matches the experimental accuracy that can be achieved.

The symbols O and $\approx$ used above need some further explanation. They are used to compare the behaviour of two functions when a variable upon which both functions depend tends to a particular limit, usually zero or infinity (and obvious from the context). For two functions $f(x)$ and $g(x)$, with g positive, the formal *definitions* of the above symbols are as follows:

(i) If there exists a constant k such that $|f| \le kg$ as the limit is approached then $f = O(g)$.
(ii) If as the limit of x is approached f/g tends to a limit l, where $l \neq 0$, then $f \approx lg$. The statement $f \approx g$ means that the ratio of the two sides tends to unity.

4.5.1 Convergence of power series

The convergence or otherwise of power series is a crucial consideration in practical terms. For example, if we are to use a power series as an approximation, it is clearly important that it tends to the precise answer as more and more terms of the approximation are taken. Consider the general power series

$$P(x) = a_0 + a_1 x + a_2 x^2 + \cdots.$$

Using d'Alembert's ratio test (see subsection 4.3.2), we see that $P(x)$ converges absolutely if

$$\rho = \lim_{n \to \infty} \left| \frac{a_{n+1}}{a_n} x \right| = |x| \lim_{n \to \infty} \left| \frac{a_{n+1}}{a_n} \right| < 1.$$

Thus the convergence of $P(x)$ depends upon the value of x, i.e. there is, in general, a range of values of x for which $P(x)$ converges, an *interval of convergence*. Note that at the limits of this range $\rho = 1$, and so the series may converge or diverge. The convergence of the series at the end-points may be determined by substituting these values of x into the power series $P(x)$ and testing the resulting series using any applicable method (discussed in section 4.3).

> ▶*Determine the range of values of x for which the following power series converges:*
> $$P(x) = 1 + 2x + 4x^2 + 8x^3 + \cdots.$$

By using the interval-of-convergence method discussed above,

$$\rho = \lim_{n\to\infty} \left| \frac{2^{n+1}}{2^n} x \right| = |2x|,$$

and hence the power series will converge for $|x| < 1/2$. Examining the end-points of the interval separately, we find

$$P(1/2) = 1 + 1 + 1 + \cdots,$$
$$P(-1/2) = 1 - 1 + 1 - \cdots.$$

Obviously $P(1/2)$ diverges, while $P(-1/2)$ oscillates. Therefore $P(x)$ is not convergent at either end-point of the region but is convergent for $-1 < x < 1$. ◀

The convergence of power series may be extended to the case where the parameter z is complex. For the power series

$$P(z) = a_0 + a_1 z + a_2 z^2 + \cdots,$$

we find that $P(z)$ converges if

$$\rho = \lim_{n\to\infty} \left| \frac{a_{n+1}}{a_n} z \right| = |z| \lim_{n\to\infty} \left| \frac{a_{n+1}}{a_n} \right| < 1.$$

We therefore have a range in $|z|$ for which $P(z)$ converges, i.e. $P(z)$ converges for values of z lying within a circle in the Argand diagram (in this case centred on the origin of the Argand diagram). The radius of the circle is called the *radius of convergence*: if z lies inside the circle, the series will converge whereas if z lies outside the circle, the series will diverge; if, though, z lies on the circle then the convergence must be tested using another method. Clearly the radius of convergence R is given by $1/R = \lim_{n\to\infty} |a_{n+1}/a_n|$.

> ▶*Determine the range of values of z for which the following complex power series converges:*
> $$P(z) = 1 - \frac{z}{2} + \frac{z^2}{4} - \frac{z^3}{8} + \cdots.$$

We find that $\rho = |z/2|$, which shows that $P(z)$ converges for $|z| < 2$. Therefore the circle of convergence in the Argand diagram is centred on the origin and has a radius $R = 2$. On this circle we must test the convergence by substituting the value of z into $P(z)$ and considering the resulting series. On the circle of convergence we can write $z = 2 \exp i\theta$. Substituting this into $P(z)$, we obtain

$$P(z) = 1 - \frac{2 \exp i\theta}{2} + \frac{4 \exp 2i\theta}{4} - \cdots$$
$$= 1 - \exp i\theta + [\exp i\theta]^2 - \cdots,$$

which is a complex infinite geometric series with first term $a = 1$ and common ratio

$r = -\exp i\theta$. Therefore, on the the circle of convergence we have

$$P(z) = \frac{1}{1 + \exp i\theta}.$$

Unless $\theta = \pi$ this is a finite complex number, and so $P(z)$ converges at all points on the circle $|z| = 2$ except at $\theta = \pi$ (i.e. $z = -2$), where it diverges. Note that $P(z)$ is just the binomial expansion of $(1 + z/2)^{-1}$, for which it is obvious that $z = -2$ is a singular point. In general, for power series expansions of complex functions about a given point in the complex plane, the circle of convergence extends as far as the nearest singular point. This is discussed further in chapter 24. ◀

Note that the centre of the circle of convergence does not necessarily lie at the origin. For example, applying the ratio test to the complex power series

$$P(z) = 1 + \frac{z-1}{2} + \frac{(z-1)^2}{4} + \frac{(z-1)^3}{8} + \cdots,$$

we find that for it to converge we require $|(z-1)/2| < 1$. Thus the series converges for z lying within a circle of radius 2 centred on the point $(1,0)$ in the Argand diagram.

4.5.2 Operations with power series

The following rules are useful when manipulating power series; they apply to power series in a real or complex variable.

(i) If two power series $P(x)$ and $Q(x)$ have regions of convergence that overlap to some extent then the series produced by taking the sum, the difference or the product of $P(x)$ and $Q(x)$ converges in the common region.

(ii) If two power series $P(x)$ and $Q(x)$ converge for all values of x then one series may be substituted into the other to give a third series, which also converges for all values of x. For example, consider the power series expansions of $\sin x$ and e^x given below in subsection 4.6.3,

$$\sin x = x - \frac{x^3}{3!} + \frac{x^5}{5!} - \frac{x^7}{7!} + \cdots$$

$$e^x = 1 + x + \frac{x^2}{2!} + \frac{x^3}{3!} + \frac{x^4}{4!} + \cdots,$$

both of which converge for all values of x. Substituting the series for $\sin x$ into that for e^x we obtain

$$e^{\sin x} = 1 + x + \frac{x^2}{2!} - \frac{3x^4}{4!} - \frac{8x^5}{5!} + \cdots,$$

which also converges for all values of x.

If, however, either of the power series $P(x)$ and $Q(x)$ has only a limited region of convergence, or if they both do so, then further care must be taken when substituting one series into the other. For example, suppose $Q(x)$ converges for all x, but $P(x)$ only converges for x within a finite range. We may substitute

We may follow a similar procedure to obtain a Taylor series about an arbitrary point $x = a$.

▶ *Expand $f(x) = \cos x$ as a Taylor series about $x = \pi/3$.*

As in the above example, it is easily shown that the nth derivative of $f(x)$ is given by

$$f^{(n)}(x) = \cos\left(x + \frac{n\pi}{2}\right).$$

Therefore the remainder after expanding $f(x)$ as an $(n-1)$th-order polynomial about $x = \pi/3$ is given by

$$R_n(x) = \frac{(x - \pi/3)^n}{n!} \cos\left(\xi + \frac{n\pi}{2}\right),$$

where ξ lies in the range $[\pi/3, x]$. The modulus of the cosine term is always less than or equal to unity, and so $|R_n(x)| < |(x - \pi/3)^n|/n!$. As in the previous example, $\lim_{n \to \infty} R_n(x) = 0$ for any particular value of x, and so $\cos x$ can be represented by an infinite Taylor series about $x = \pi/3$.

Evaluating the function and its derivatives at $x = \pi/3$ we obtain

$$f(\pi/3) = \cos(\pi/3) = 1/2,$$
$$f'(\pi/3) = \cos(5\pi/6) = -\sqrt{3}/2,$$
$$f''(\pi/3) = \cos(4\pi/3) = -1/2,$$

and so on. Thus the Taylor series expansion of $\cos x$ about $x = \pi/3$ is given by

$$\cos x = \frac{1}{2} - \frac{\sqrt{3}}{2}(x - \pi/3) - \frac{1}{2}\frac{(x - \pi/3)^2}{2!} + \cdots . \blacktriangleleft$$

4.6.2 Approximation errors in Taylor series

In the previous subsection we saw how to represent a function $f(x)$ by an infinite power series, which is exactly equal to $f(x)$ for all x within the interval of convergence of the series. However, in physical problems we usually do not want to have to sum an infinite number of terms, but prefer to use only a finite number of terms in the Taylor series to *approximate* the function in some given range of x. In this case it is desirable to know what is the maximum possible error associated with the approximation.

As given in (4.18), a function $f(x)$ can be represented by a finite $(n-1)$th-order power series together with a remainder term such that

$$f(x) = f(a) + (x - a)f'(a) + \frac{(x - a)^2}{2!}f''(a) + \cdots + \frac{(x - a)^{n-1}}{(n-1)!}f^{(n-1)}(a) + R_n(x),$$

where

$$R_n(x) = \frac{(x - a)^n}{n!}f^{(n)}(\xi)$$

and ξ lies in the range $[a, x]$. $R_n(x)$ is the remainder term, and represents the error in approximating $f(x)$ by the above $(n-1)$th-order power series. Since the exact value of ξ that satisfies the expression for $R_n(x)$ is not known, an upper limit on the error may be found by differentiating $R_n(x)$ with respect to ξ and equating the derivative to zero in the usual way for finding maxima.

▶*Expand $f(x) = \cos x$ as a Taylor series about $x = 0$ and find the error associated with using the approximation to evaluate $\cos(0.5)$ if only the first two non-vanishing terms are taken. (Note that the Taylor expansions of trigonometric functions are only valid for angles measured in radians.)*

Evaluating the function and its derivatives at $x = 0$, we find

$$f(0) = \cos 0 = 1,$$
$$f'(0) = -\sin 0 = 0,$$
$$f''(0) = -\cos 0 = -1,$$
$$f'''(0) = \sin 0 = 0.$$

So, for small $|x|$, we find from (4.18)

$$\cos x \approx 1 - \frac{x^2}{2}.$$

Note that since $\cos x$ is an even function, its power series expansion contains only even powers of x. Therefore, in order to estimate the error in this approximation, we must consider the term in x^4, which is the next in the series. The required derivative is $f^{(4)}(x)$ and this is (by chance) equal to $\cos x$. Thus, adding in the remainder term $R_4(x)$, we find

$$\cos x = 1 - \frac{x^2}{2} + \frac{x^4}{4!} \cos \xi,$$

where ξ lies in the range $[0, x]$. Thus, the maximum possible error is $x^4/4!$, since $\cos \xi$ cannot exceed unity. If $x = 0.5$, taking just the first two terms yields $\cos(0.5) \approx 0.875$ with a predicted error of less than 0.002 60. In fact $\cos(0.5) = 0.877\,58$ to 5 decimal places. Thus, to this accuracy, the true error is 0.002 58, an error of about 0.3%. ◀

4.6.3 Standard Maclaurin series

It is often useful to have a readily available table of Maclaurin series for standard elementary functions, and therefore these are listed below.

$$\sin x = x - \frac{x^3}{3!} + \frac{x^5}{5!} - \frac{x^7}{7!} + \cdots \quad \text{for } -\infty < x < \infty,$$

$$\cos x = 1 - \frac{x^2}{2!} + \frac{x^4}{4!} - \frac{x^6}{6!} + \cdots \quad \text{for } -\infty < x < \infty,$$

$$\tan^{-1} x = x - \frac{x^3}{3} + \frac{x^5}{5} - \frac{x^7}{7} + \cdots \quad \text{for } -1 < x < 1,$$

$$e^x = 1 + x + \frac{x^2}{2!} + \frac{x^3}{3!} + \frac{x^4}{4!} + \cdots \quad \text{for } -\infty < x < \infty,$$

$$\ln(1 + x) = x - \frac{x^2}{2} + \frac{x^3}{3} - \frac{x^4}{4} + \cdots \quad \text{for } -1 < x \le 1,$$

$$(1 + x)^n = 1 + nx + n(n - 1)\frac{x^2}{2!} + n(n - 1)(n - 2)\frac{x^3}{3!} + \cdots \quad \text{for } -\infty < x < \infty.$$

These can all be derived by straightforward application of Taylor's theorem to the expansion of a function about $x = 0$.

4.7 Evaluation of limits

The idea of the limit of a function $f(x)$ as x approaches a value a is fairly intuitive, though a strict definition exists and is stated below. In many cases the limit of the function as x approaches a will be simply the value $f(a)$, but sometimes this is not so. Firstly, the function may be undefined at $x = a$, as, for example, when

$$f(x) = \frac{\sin x}{x},$$

which takes the value $0/0$ at $x = 0$. However, the limit as x approaches zero does exist and can be evaluated as unity using l'Hôpital's rule below. Another possibility is that even if $f(x)$ is defined at $x = a$ its value may not be equal to the limiting value $\lim_{x \to a} f(x)$. This can occur for a discontinuous function at a point of discontinuity. The strict definition of a limit is that *if* $\lim_{x \to a} f(x) = l$ *then for any number* ϵ *however small, it must be possible to find a number* η *such that* $|f(x) - l| < \epsilon$ *whenever* $|x - a| < \eta$. In other words, as x becomes arbitrarily close to a, $f(x)$ becomes arbitrarily close to its limit, l. To remove any ambiguity, it should be stated that, in general, the number η will depend on both ϵ and the form of $f(x)$.

The following observations are often useful in finding the limit of a function.

(i) A limit may be $\pm \infty$. For example as $x \to 0$, $1/x^2 \to \infty$.

(ii) A limit may be approached from below or above and the value may be different in each case. For example consider the function $f(x) = \tan x$. As x tends to $\pi/2$ from below $f(x) \to \infty$, but if the limit is approached from above then $f(x) \to -\infty$. Another way of writing this is

$$\lim_{x \to \frac{\pi}{2}^-} \tan x = \infty, \qquad \lim_{x \to \frac{\pi}{2}^+} \tan x = -\infty.$$

(iii) It may ease the evaluation of a limit if the function under consideration is split into a sum, product or quotient. Provided that in each case a limit exists, the rules for evaluating the original limit are as follows.

(a) $\lim_{x \to a} \{f(x) + g(x)\} = \lim_{x \to a} f(x) + \lim_{x \to a} g(x)$.

(b) $\lim_{x \to a} \{f(x)g(x)\} = \lim_{x \to a} f(x) \lim_{x \to a} g(x)$.

(c) $\lim_{x \to a} \dfrac{f(x)}{g(x)} = \dfrac{\lim_{x \to a} f(x)}{\lim_{x \to a} g(x)}$, provided that

  the numerator and denominator are
  not both equal to zero or infinity.

Examples of cases (a)–(c) are discussed below.

141

▶*Evaluate the limits*

$$\lim_{x\to 1}(x^2 + 2x^3), \qquad \lim_{x\to 0}(x\cos x), \qquad \lim_{x\to \pi/2}\frac{\sin x}{x}.$$

Using (a) above,

$$\lim_{x\to 1}(x^2 + 2x^3) = \lim_{x\to 1} x^2 + \lim_{x\to 1} 2x^3 = 3.$$

Using (b),

$$\lim_{x\to 0}(x\cos x) = \lim_{x\to 0} x \lim_{x\to 0} \cos x = 0 \times 1 = 0.$$

Using (c),

$$\lim_{x\to \pi/2}\frac{\sin x}{x} = \frac{\lim_{x\to \pi/2} \sin x}{\lim_{x\to \pi/2} x} = \frac{1}{\pi/2} = \frac{2}{\pi}. \ \blacktriangleleft$$

(iv) Limits of functions of x that contain exponents that themselves depend on x can often be found by taking logarithms.

▶*Evaluate the limit*

$$\lim_{x\to\infty}\left(1 - \frac{a^2}{x^2}\right)^{x^2}.$$

Let us define

$$y = \left(1 - \frac{a^2}{x^2}\right)^{x^2}$$

and consider the logarithm of the required limit, i.e.

$$\lim_{x\to\infty}\ln y = \lim_{x\to\infty}\left[x^2 \ln\left(1 - \frac{a^2}{x^2}\right)\right].$$

Using the Maclaurin series for $\ln(1 + x)$ given in subsection 4.6.3, we can expand the logarithm as a series and obtain

$$\lim_{x\to\infty}\ln y = \lim_{x\to\infty}\left[x^2\left(-\frac{a^2}{x^2} - \frac{a^4}{2x^4} + \cdots\right)\right] = -a^2.$$

Therefore, since $\lim_{x\to\infty}\ln y = -a^2$ it follows that $\lim_{x\to\infty} y = \exp(-a^2)$. ◀

(v) L'Hôpital's rule may be used; it is an extension of (iii)(c) above. In cases where both numerator and denominator are zero or both are infinite, further consideration of the limit must follow. Let us first consider $\lim_{x\to a} f(x)/g(x)$, where $f(a) = g(a) = 0$. Expanding the numerator and denominator as Taylor series we obtain

$$\frac{f(x)}{g(x)} = \frac{f(a) + (x - a)f'(a) + [(x - a)^2/2!]f''(a) + \cdots}{g(a) + (x - a)g'(a) + [(x - a)^2/2!]g''(a) + \cdots}.$$

However, $f(a) = g(a) = 0$ so

$$\frac{f(x)}{g(x)} = \frac{f'(a) + [(x - a)/2!]f''(a) + \cdots}{g'(a) + [(x - a)/2!]g''(a) + \cdots}.$$

Therefore we find

$$\lim_{x \to a} \frac{f(x)}{g(x)} = \frac{f'(a)}{g'(a)},$$

provided $f'(a)$ and $g'(a)$ are not themselves both equal to zero. If, however, $f'(a)$ and $g'(a)$ *are* both zero then the same process can be applied to the ratio $f'(x)/g'(x)$ to yield

$$\lim_{x \to a} \frac{f(x)}{g(x)} = \frac{f''(a)}{g''(a)},$$

provided that at least one of $f''(a)$ and $g''(a)$ is non-zero. If the original limit does exist then it can be found by repeating the process as many times as is necessary for the ratio of corresponding nth derivatives not to be of the indeterminate form $0/0$, i.e.

$$\lim_{x \to a} \frac{f(x)}{g(x)} = \frac{f^{(n)}(a)}{g^{(n)}(a)}.$$

▶*Evaluate the limit*

$$\lim_{x \to 0} \frac{\sin x}{x}.$$

We first note that if $x = 0$, both numerator and denominator are zero. Thus we apply l'Hôpital's rule: differentiating, we obtain

$$\lim_{x \to 0}(\sin x / x) = \lim_{x \to 0}(\cos x / 1) = 1. \ \blacktriangleleft$$

So far we have only considered the case where $f(a) = g(a) = 0$. For the case where $f(a) = g(a) = \infty$ we may still apply l'Hôpital's rule by writing

$$\lim_{x \to a} \frac{f(x)}{g(x)} = \lim_{x \to a} \frac{1/g(x)}{1/f(x)},$$

which is now of the form $0/0$ at $x = a$. Note also that l'Hôpital's rule is still valid for finding limits as $x \to \infty$, i.e. when $a = \infty$. This is easily shown by letting $y = 1/x$ as follows:

$$\lim_{x \to \infty} \frac{f(x)}{g(x)} = \lim_{y \to 0} \frac{f(1/y)}{g(1/y)}$$

$$= \lim_{y \to 0} \frac{-f'(1/y)/y^2}{-g'(1/y)/y^2}$$

$$= \lim_{y \to 0} \frac{f'(1/y)}{g'(1/y)}$$

$$= \lim_{x \to \infty} \frac{f'(x)}{g'(x)}.$$

143

Summary of methods for evaluating limits

To find the limit of a continuous function $f(x)$ at a point $x = a$, simply substitute the value a into the function noting that $\frac{0}{\infty} = 0$ and that $\frac{\infty}{0} = \infty$. The only difficulty occurs when either of the expressions $\frac{0}{0}$ or $\frac{\infty}{\infty}$ results. In this case differentiate top and bottom and try again. Continue differentiating until the top and bottom limits are no longer both zero or both infinity. If the undetermined form $0 \times \infty$ occurs then it can always be rewritten as $\frac{0}{0}$ or $\frac{\infty}{\infty}$.

4.8 Exercises

4.1   Sum the even numbers between 1000 and 2000 inclusive.

4.2   If you invest £1000 on the first day of each year, and interest is paid at 5% on your balance at the end of each year, how much money do you have after 25 years?

4.3   How does the convergence of the series

$$\sum_{n=r}^{\infty} \frac{(n-r)!}{n!}$$

depend on the integer r?

4.4   Show that for testing the convergence of the series

$$x + y + x^2 + y^2 + x^3 + y^3 + \cdots,$$

where $0 < x < y < 1$, the D'Alembert ratio test fails but the Cauchy root test is successful.

4.5   Find the sum S_N of the first N terms of the following series, and hence determine whether the series are convergent, divergent or oscillatory:

$$\text{(a) } \sum_{n=1}^{\infty} \ln\left(\frac{n+1}{n}\right), \qquad \text{(b) } \sum_{n=0}^{\infty}(-2)^n, \qquad \text{(c) } \sum_{n=1}^{\infty} \frac{(-1)^{n+1}n}{3^n}.$$

4.6   By grouping and rearranging terms of the absolutely convergent series

$$S = \sum_{n=1}^{\infty} \frac{1}{n^2},$$

show that

$$S_{\mathrm{o}} = \sum_{\substack{n=1 \\ n \text{ odd}}}^{\infty} \frac{1}{n^2} = \frac{3S}{4}.$$

4.7   Use the difference method to sum the series

$$\sum_{n=2}^{N} \frac{2n-1}{2n^2(n-1)^2}.$$

144

4.8   The $N + 1$ complex numbers ω_m are given by $\omega_m = \exp(2\pi i m / N)$, for $m = 0, 1, 2, \ldots, N$.

(a) Evaluate the following:

$$\text{(i) } \sum_{m=0}^{N} \omega_m, \quad \text{(ii) } \sum_{m=0}^{N} \omega_m^2, \quad \text{(iii) } \sum_{m=0}^{N} \omega_m x^m.$$

(b) Use these results to evaluate:

$$\text{(i) } \sum_{m=0}^{N} \left[\cos \left(\frac{2\pi m}{N} \right) - \cos \left(\frac{4\pi m}{N} \right) \right], \quad \text{(ii) } \sum_{m=0}^{3} 2^m \sin \left(\frac{2\pi m}{3} \right).$$

4.9   Prove that

$$\cos \theta + \cos(\theta + \alpha) + \cdots + \cos(\theta + n\alpha) = \frac{\sin \frac{1}{2}(n+1)\alpha}{\sin \frac{1}{2}\alpha} \cos(\theta + \tfrac{1}{2} n\alpha).$$

4.10   Determine whether the following series converge (θ and p are positive real numbers):

$$\text{(a) } \sum_{n=1}^{\infty} \frac{2 \sin n\theta}{n(n+1)}, \quad \text{(b) } \sum_{n=1}^{\infty} \frac{2}{n^2}, \quad \text{(c) } \sum_{n=1}^{\infty} \frac{1}{2n^{1/2}},$$

$$\text{(d) } \sum_{n=2}^{\infty} \frac{(-1)^n (n^2 + 1)^{1/2}}{n \ln n}, \quad \text{(e) } \sum_{n=1}^{\infty} \frac{n^p}{n!}.$$

4.11   Find the real values of x for which the following series are convergent:

$$\text{(a) } \sum_{n=1}^{\infty} \frac{x^n}{n+1}, \quad \text{(b) } \sum_{n-1}^{\infty} (\sin x)^n, \quad \text{(c) } \sum_{n=1}^{\infty} n^x,$$

$$\text{(d) } \sum_{n-1}^{\infty} e^{nx}, \quad \text{(e) } \sum_{n=2}^{\infty} (\ln n)^x.$$

4.12   Determine whether the following series are convergent:

$$\text{(a) } \sum_{n=1}^{\infty} \frac{n^{1/2}}{(n+1)^{1/2}}, \quad \text{(b) } \sum_{n=1}^{\infty} \frac{n^2}{n!}, \quad \text{(c) } \sum_{n=1}^{\infty} \frac{(\ln n)^n}{n^{n/2}}, \quad \text{(d) } \sum_{n=1}^{\infty} \frac{n^n}{n!}.$$

4.13   Determine whether the following series are absolutely convergent, convergent or oscillatory:

$$\text{(a) } \sum_{n=1}^{\infty} \frac{(-1)^n}{n^{5/2}}, \quad \text{(b) } \sum_{n=1}^{\infty} \frac{(-1)^n (2n+1)}{n}, \quad \text{(c) } \sum_{n=0}^{\infty} \frac{(-1)^n |x|^n}{n!},$$

$$\text{(d) } \sum_{n=0}^{\infty} \frac{(-1)^n}{n^2 + 3n + 2}, \quad \text{(e) } \sum_{n=1}^{\infty} \frac{(-1)^n 2^n}{n^{1/2}}.$$

4.14   Obtain the positive values of x for which the following series converges:

$$\sum_{n=1}^{\infty} \frac{x^{n/2} e^{-n}}{n}.$$

4.15   Prove that

$$\sum_{n=2}^{\infty} \ln\left[\frac{n^r + (-1)^n}{n^r}\right]$$

is absolutely convergent for $r = 2$, but only conditionally convergent for $r = 1$.

4.16   An extension to the proof of the integral test (subsection 4.3.2) shows that, if $f(x)$ is positive, continuous and monotonically decreasing, for $x \geq 1$, and the series $f(1) + f(2) + \cdots$ is convergent, then its sum does not exceed $f(1) + L$, where L is the integral

$$\int_1^{\infty} f(x)\, dx.$$

Use this result to show that the sum $\zeta(p)$ of the Riemann zeta series $\sum n^{-p}$, with $p > 1$, is not greater than $p/(p-1)$.

4.17   Demonstrate that rearranging the order of its terms can make a conditionally convergent series converge to a different limit by considering the series $\sum(-1)^{n+1}n^{-1} = \ln 2 = 0.693$. Rearrange the series as

$$S = \tfrac{1}{1} + \tfrac{1}{3} - \tfrac{1}{2} + \tfrac{1}{5} + \tfrac{1}{7} - \tfrac{1}{4} + \tfrac{1}{9} + \tfrac{1}{11} - \tfrac{1}{6} + \tfrac{1}{13} + \cdots$$

and group each set of three successive terms. Show that the series can then be written

$$\sum_{m=1}^{\infty} \frac{8m-3}{2m(4m-3)(4m-1)},$$

which is convergent (by comparison with $\sum n^{-2}$) and contains only positive terms. Evaluate the first of these and hence deduce that S is not equal to $\ln 2$.

4.18   Illustrate result (iv) of section 4.4, concerning Cauchy products, by considering the double summation

$$S = \sum_{n=1}^{\infty} \sum_{r=1}^{n} \frac{1}{r^2(n+1-r)^3}.$$

By examining the points in the nr-plane over which the double summation is to be carried out, show that S can be written as

$$S = \sum_{n=r}^{\infty} \sum_{r=1}^{\infty} \frac{1}{r^2(n+1-r)^3}.$$

Deduce that $S \leq 3$.

4.19   A Fabry–Pérot interferometer consists of two parallel heavily silvered glass plates; light enters normally to the plates, and undergoes repeated reflections between them, with a small transmitted fraction emerging at each reflection. Find the intensity of the emerging wave, $|B|^2$, where

$$B = A(1-r)\sum_{n=0}^{\infty} r^n e^{in\phi},$$

with r and ϕ real.

146

4.20 Identify the series

$$\sum_{n=1}^{\infty} \frac{(-1)^{n+1} x^{2n}}{(2n-1)!},$$

and then, by integration and differentiation, deduce the values S of the following series:

(a) $\sum_{n=1}^{\infty} \frac{(-1)^{n+1} n^2}{(2n)!},$ (b) $\sum_{n=1}^{\infty} \frac{(-1)^{n+1} n}{(2n+1)!},$

(c) $\sum_{n=1}^{\infty} \frac{(-1)^{n+1} n \pi^{2n}}{4^n (2n-1)!},$ (d) $\sum_{n=0}^{\infty} \frac{(-1)^n (n+1)}{(2n)!}.$

4.21 Starting from the Maclaurin series for $\cos x$, show that

$$(\cos x)^{-2} = 1 + x^2 + \frac{2x^4}{3} + \cdots.$$

Deduce the first three terms in the Maclaurin series for $\tan x$.

4.22 Find the Maclaurin series for:

(a) $\ln\left(\frac{1+x}{1-x}\right),$ (b) $(x^2 + 4)^{-1},$ (c) $\sin^2 x.$

4.23 Writing the nth derivative of $f(x) = \sinh^{-1} x$ as

$$f^{(n)}(x) = \frac{P_n(x)}{(1+x^2)^{n-1/2}},$$

where $P_n(x)$ is a polynomial (of degree $n-1$), show that the $P_n(x)$ satisfy the recurrence relation

$$P_{n+1}(x) - (1+x^2) P_n'(x) - (2n-1) x P_n(x).$$

Hence generate the coefficients necessary to express $\sinh^{-1} x$ as a Maclaurin series up to terms in x^5.

4.24 Find the first three non-zero terms in the Maclaurin series for the following functions:

(a) $(x^2 + 9)^{-1/2},$ (b) $\ln[(2 + x)^3],$ (c) $\exp(\sin x),$
(d) $\ln(\cos x),$ (e) $\exp[-(x-a)^{-2}],$ (f) $\tan^{-1} x.$

4.25 By using the logarithmic series, prove that if a and b are positive and nearly equal then

$$\ln \frac{a}{b} \simeq \frac{2(a-b)}{a+b}.$$

Show that the error in this approximation is about $2(a-b)^3/[3(a+b)^3]$.

4.26 Determine whether the following functions $f(x)$ are (i) continuous, and (ii) differentiable at $x = 0$:

(a) $f(x) = \exp(-|x|)$;
(b) $f(x) = (1 - \cos x)/x^2$ for $x \neq 0$, $f(0) = \frac{1}{2}$;
(c) $f(x) = x \sin(1/x)$ for $x \neq 0$, $f(0) = 0$;
(d) $f(x) = [4 - x^2]$, where $[y]$ denotes the integer part of y.

4.27 Find the limit as $x \to 0$ of $[\sqrt{1+x^m} - \sqrt{1-x^m}]/x^n$, in which m and n are positive integers.

4.28 Evaluate the following limits:

$$\text{(a) } \lim_{x \to 0} \frac{\sin 3x}{\sinh x}, \qquad \text{(b) } \lim_{x \to 0} \frac{\tan x - \tanh x}{\sinh x - x},$$

$$\text{(c) } \lim_{x \to 0} \frac{\tan x - x}{\cos x - 1}, \qquad \text{(d) } \lim_{x \to 0} \left(\frac{\cosec x}{x^3} - \frac{\sinh x}{x^5} \right).$$

4.29   Find the limits of the following functions:

(a) $\dfrac{x^3 + x^2 - 5x - 2}{2x^3 - 7x^2 + 4x + 4}$,   as $x \to 0$, $x \to \infty$ and $x \to 2$;

(b) $\dfrac{\sin x - x \cosh x}{\sinh x - x}$,   as $x \to 0$;

(c) $\displaystyle\int_x^{\pi/2} \left(\frac{y \cos y - \sin y}{y^2} \right) dy$,   as $x \to 0$.

4.30   Use Taylor expansions to three terms to find approximations to (a) $\sqrt[4]{17}$, and (b) $\sqrt[3]{26}$.

4.31   Using a first-order Taylor expansion about $x = x_0$, show that a better approximation than x_0 to the solution of the equation

$$f(x) = \sin x + \tan x = 2$$

is given by $x = x_0 + \delta$, where

$$\delta = \frac{2 - f(x_0)}{\cos x_0 + \sec^2 x_0}.$$

(a) Use this procedure twice to find the solution of $f(x) = 2$ to six significant figures, given that it is close to $x = 0.9$.

(b) Use the result in (a) to deduce, to the same degree of accuracy, one solution of the quartic equation

$$y^4 - 4y^3 + 4y^2 + 4y - 4 = 0.$$

4.32   Evaluate

$$\lim_{x \to 0} \left[\frac{1}{x^3} \left(\cosec x - \frac{1}{x} - \frac{x}{6} \right) \right].$$

4.33   In quantum theory, a system of oscillators, each of fundamental frequency v and interacting at temperature T, has an average energy $\bar{E}$ given by

$$\bar{E} = \frac{\sum_{n=0}^{\infty} nhv e^{-nx}}{\sum_{n=0}^{\infty} e^{-nx}},$$

where $x = hv/kT$, h and k being the Planck and Boltzmann constants, respectively. Prove that both series converge, evaluate their sums, and show that at high temperatures $\bar{E} \approx kT$, whilst at low temperatures $\bar{E} \approx hv \exp(-hv/kT)$.

4.34   In a very simple model of a crystal, point-like atomic ions are regularly spaced along an infinite one-dimensional row with spacing R. Alternate ions carry equal and opposite charges $\pm e$. The potential energy of the ith ion in the electric field due to another ion, the jth, is

$$\frac{q_i q_j}{4\pi \epsilon_0 r_{ij}},$$

where q_i, q_j are the charges on the ions and r_{ij} is the distance between them.

Write down a series giving the total contribution V_i of the ith ion to the overall potential energy. Show that the series converges, and, if V_i is written as

$$V_i = \frac{\alpha e^2}{4\pi \epsilon_0 R},$$

find a closed-form expression for α, the Madelung constant for this (unrealistic) lattice.

4.35 One of the factors contributing to the high relative permittivity of water to static electric fields is the permanent electric dipole moment, p, of the water molecule. In an external field E the dipoles tend to line up with the field, but they do not do so completely because of thermal agitation corresponding to the temperature, T, of the water. A classical (non-quantum) calculation using the Boltzmann distribution shows that the average polarisability per molecule, α, is given by

$$\alpha = \frac{p}{E}(\coth x - x^{-1}),$$

where $x = pE/(kT)$ and k is the Boltzmann constant.

At ordinary temperatures, even with high field strengths (10^4 V m^{-1} or more), $x \ll 1$. By making suitable series expansions of the hyperbolic functions involved, show that $\alpha = p^2/(3kT)$ to an accuracy of about one part in $15x^{-2}$.

4.36 In quantum theory, a certain method (the Born approximation) gives the (so-called) amplitude $f(\theta)$ for the scattering of a particle of mass m through an angle θ by a uniform potential well of depth V_0 and radius b (i.e. the potential energy of the particle is $-V_0$ within a sphere of radius b and zero elsewhere) as

$$f(0) = \frac{2mV_0}{\hbar^2 K^3}(\sin Kb - Kb\cos Kb).$$

Here $\hbar$ is the Planck constant divided by 2π, the energy of the particle is $\hbar^2 k^2/(2m)$ and K is $2k\sin(\theta/2)$.

Use l'Hôpital's rule to evaluate the amplitude at low energies, i.e. when k and hence K tend to zero, and so determine the low-energy total cross-section.

[Note: the differential cross-section is given by $|f(\theta)|^2$ and the total cross section by the integral of this over all solid angles, i.e. $2\pi \int_0^\pi |f(\theta)|^2 \sin\theta \, d\theta$.]

4.9 Hints and answers

4.1 Write as $2(\sum_{n=1}^{1000} n - \sum_{n=1}^{499} n) = 751\,500$.

4.3 Divergent for $r < 1$; convergent for $r \geq 2$.

4.5 (a) $\ln(N+1)$, divergent; (b) $\frac{1}{3}[1-(-2)^n]$, oscillates infinitely; (c) Add $\frac{1}{3}S_N$ to the S_N series; $\frac{3}{16}[1-(-3)^{-N}] + \frac{3}{4}N(-3)^{-N-1}$, convergent to $\frac{3}{16}$.

4.7 Write the nth term as the difference between two consecutive values of a partial-fraction function of n. The sum equals $\frac{1}{2}(1 - N^{-2})$.

4.9 Sum the geometric series with rth term $\exp[i(\theta + r\alpha)]$. Its real part is

$$\{\cos\theta - \cos[(n+1)\alpha + \theta] - \cos(\theta - \alpha) + \cos(\theta + n\alpha)\}/4\sin^2(\alpha/2),$$

which can be reduced to the given answer.

4.11 (a) $-1 \leq x < 1$; (b) all x except $x = (2n \pm 1)\pi/2$; (c) $x < -1$; (d) $x < 0$; (e) always divergent. Clearly divergent for $x > -1$. For $-X = x < -1$, consider

$$\sum_{k=1}^{\infty} \sum_{n=M_{k-1}+1}^{M_k} \frac{1}{(\ln M_k)^X},$$

where $\ln M_k = k$ and note that $M_k - M_{k-1} = e^{-1}(e-1)M_k$; hence show that the series diverges.

4.13 (a) Absolutely convergent, compare with exercise 4.10(b). (b) Oscillates finitely. (c) Absolutely convergent for all x. (d) Absolutely convergent; use partial fractions. (e) Oscillates infinitely.

4.15   Divide the series into two series, n odd and n even. For $r = 2$ both are absolutely convergent, by comparison with $\sum n^{-2}$. For $r = 1$ neither series is convergent, by comparison with $\sum n^{-1}$. However, the sum of the two is convergent, by the alternating sign test or by showing that the terms cancel in pairs.

4.17   The first term has value 0.833 and all other terms are positive.

4.19 $|A|^2(1-r)^2/(1+r^2-2r\cos\phi)$.

4.21   Use the binomial expansion and collect terms up to x^4. Integrate both sides of the displayed equation. $\tan x = x + x^3/3 + 2x^5/15 + \cdots$.

4.23   For example, $P_5(x) = 24x^4 - 72x^2 + 9$. $\sinh^{-1} x = x - x^3/6 + 3x^5/40 - \cdots$.

4.25   Set $a = D + \delta$ and $b = D - \delta$ and use the expansion for $\ln(1 \pm \delta/D)$.

4.27   The limit is 0 for $m > n$, 1 for $m = n$, and ∞ for $m < n$.

4.29   (a) $-\frac{1}{2}, \frac{1}{2}, \infty$; (b) -4; (c) $-1 + 2/\pi$.

4.31   (a) First approximation 0.886452; second approximation 0.886287. (b) Set $y = \sin x$ and re-express $f(x) = 2$ as a polynomial equation. $y = \sin(0.886287) = 0.774730$.

4.33   If $S(x) = \sum_{n=0}^{\infty} e^{-nx}$ evaluate $S(x)$ and consider $dS(x)/dx$. $E = h\nu[\exp(h\nu/kT) - 1]^{-1}$.

4.35   The series expansion is $\dfrac{px}{E}\left(\dfrac{1}{3} - \dfrac{x^2}{45} + \cdots\right)$.

5

Partial differentiation

In chapter 2, we discussed functions f of only one variable x, which were usually written $f(x)$. Certain constants and parameters may also have appeared in the definition of f, e.g. $f(x) = ax + 2$ contains the constant 2 and the parameter a, but only x was considered as a variable and only the derivatives $f^{(n)}(x) = d^n f/dx^n$ were defined.

However, we may equally well consider functions that depend on more than one variable, e.g. the function $f(x, y) = x^2 + 3xy$, which depends on the two variables x and y. For any pair of values x, y, the function $f(x, y)$ has a well-defined value, e.g. $f(2, 3) = 22$. This notion can clearly be extended to functions dependent on more than two variables. For the n-variable case, we write $f(x_1, x_2, \ldots, x_n)$ for a function that depends on the variables $x_1, x_2, \ldots, x_n$. When $n = 2$, x_1 and x_2 correspond to the variables x and y used above.

Functions of one variable, like $f(x)$, can be represented by a graph on a plane sheet of paper, and it is apparent that functions of two variables can, with little effort, be represented by a surface in three-dimensional space. Thus, we may also picture $f(x, y)$ as describing the variation of height with position in a mountainous landscape. Functions of many variables, however, are usually very difficult to visualise and so the preliminary discussion in this chapter will concentrate on functions of just two variables.

5.1 Definition of the partial derivative

It is clear that a function $f(x, y)$ of two variables will have a gradient in all directions in the xy-plane. A general expression for this rate of change can be found and will be discussed in the next section. However, we first consider the simpler case of finding the rate of change of $f(x, y)$ in the positive x- and y-directions. These rates of change are called the *partial derivatives* with respect to x and y respectively, and they are extremely important in a wide range of physical applications.

For a function of two variables $f(x, y)$ we may define the derivative with respect to x, for example, by saying that it is that for a one-variable function when y is held fixed and treated as a constant. To signify that a derivative is with respect to x, but at the same time to recognize that a derivative with respect to y also exists, the former is denoted by $\partial f / \partial x$ and is the *partial derivative of $f(x, y)$ with respect to x*. Similarly, the partial derivative of f with respect to y is denoted by $\partial f / \partial y$.

To define formally the partial derivative of $f(x, y)$ with respect to x, we have

$$\frac{\partial f}{\partial x} = \lim_{\Delta x \to 0} \frac{f(x + \Delta x, y) - f(x, y)}{\Delta x}, \tag{5.1}$$

provided that the limit exists. This is much the same as for the derivative of a one-variable function. The other partial derivative of $f(x, y)$ is similarly defined as a limit (provided it exists):

$$\frac{\partial f}{\partial y} = \lim_{\Delta y \to 0} \frac{f(x, y + \Delta y) - f(x, y)}{\Delta y}. \tag{5.2}$$

It is common practice in connection with partial derivatives of functions involving more than one variable to indicate those variables that are held constant by writing them as subscripts to the derivative symbol. Thus, the partial derivatives defined in (5.1) and (5.2) would be written respectively as

$$\left(\frac{\partial f}{\partial x} \right)_y \quad \text{and} \quad \left(\frac{\partial f}{\partial y} \right)_x.$$

In this form, the subscript shows explicitly which variable is to be kept constant. A more compact notation for these partial derivatives is f_x and f_y. However, it is extremely important when using partial derivatives to remember which variables are being held constant and it is wise to write out the partial derivative in explicit form if there is any possibility of confusion.

The extension of the definitions (5.1), (5.2) to the general n-variable case is straightforward and can be written formally as

$$\frac{\partial f(x_1, x_2, \ldots, x_n)}{\partial x_i} = \lim_{\Delta x_i \to 0} \frac{[f(x_1, x_2, \ldots, x_i + \Delta x_i, \ldots, x_n) - f(x_1, x_2, \ldots, x_i, \ldots, x_n)]}{\Delta x_i},$$

provided that the limit exists.

Just as for one-variable functions, second (and higher) partial derivatives may be defined in a similar way. For a two-variable function $f(x, y)$ they are

$$\frac{\partial}{\partial x} \left(\frac{\partial f}{\partial x} \right) = \frac{\partial^2 f}{\partial x^2} = f_{xx}, \qquad \frac{\partial}{\partial y} \left(\frac{\partial f}{\partial y} \right) = \frac{\partial^2 f}{\partial y^2} = f_{yy},$$

$$\frac{\partial}{\partial x} \left(\frac{\partial f}{\partial y} \right) = \frac{\partial^2 f}{\partial x \partial y} = f_{xy}, \qquad \frac{\partial}{\partial y} \left(\frac{\partial f}{\partial x} \right) = \frac{\partial^2 f}{\partial y \partial x} = f_{yx}.$$

Only three of the second derivatives are independent since the relation

$$\frac{\partial^2 f}{\partial x \partial y} = \frac{\partial^2 f}{\partial y \partial x},$$

is always obeyed, provided that the second partial derivatives are continuous at the point in question. This relation often proves useful as a labour-saving device when evaluating second partial derivatives. It can also be shown that for a function of n variables, $f(x_1, x_2, \ldots, x_n)$, under the same conditions,

$$\frac{\partial^2 f}{\partial x_i \partial x_j} = \frac{\partial^2 f}{\partial x_j \partial x_i}.$$

▶ *Find the first and second partial derivatives of the function*

$$f(x, y) = 2x^3 y^2 + y^3.$$

The first partial derivatives are

$$\frac{\partial f}{\partial x} = 6x^2 y^2, \qquad \frac{\partial f}{\partial y} = 4x^3 y + 3y^2,$$

and the second partial derivatives are

$$\frac{\partial^2 f}{\partial x^2} = 12xy^2, \qquad \frac{\partial^2 f}{\partial y^2} = 4x^3 + 6y, \qquad \frac{\partial^2 f}{\partial x \partial y} = 12x^2 y, \qquad \frac{\partial^2 f}{\partial y \partial x} = 12x^2 y,$$

the last two being equal, as expected. ◀

5.2 The total differential and total derivative

Having defined the (first) partial derivatives of a function $f(x, y)$, which give the rate of change of f along the positive x- and y-axes, we consider next the rate of change of $f(x, y)$ in an arbitrary direction. Suppose that we make simultaneous small changes Δx in x and Δy in y and that, as a result, f changes to $f + \Delta f$. Then we must have

$$\begin{aligned}
\Delta f &= f(x + \Delta x, y + \Delta y) - f(x, y) \\
&= f(x + \Delta x, y + \Delta y) - f(x, y + \Delta y) + f(x, y + \Delta y) - f(x, y) \\
&= \left[\frac{f(x + \Delta x, y + \Delta y) - f(x, y + \Delta y)}{\Delta x} \right] \Delta x + \left[\frac{f(x, y + \Delta y) - f(x, y)}{\Delta y} \right] \Delta y.
\end{aligned}$$

$$(5.3)$$

In the last line we note that the quantities in brackets are very similar to those involved in the definitions of partial derivatives (5.1), (5.2). For them to be strictly equal to the partial derivatives, Δx and Δy would need to be infinitesimally small. But even for finite (but not too large) Δx and Δy the approximate formula

$$\Delta f \approx \frac{\partial f(x, y)}{\partial x} \Delta x + \frac{\partial f(x, y)}{\partial y} \Delta y \tag{5.4}$$

can be obtained. It will be noticed that the first bracket in (5.3) actually approximates to $\partial f(x, y + \Delta y)/\partial x$ but that this has been replaced by $\partial f(x, y)/\partial x$ in (5.4). This approximation clearly has the same degree of validity as that which replaces the bracket by the partial derivative.

How valid an approximation (5.4) is to (5.3) depends not only on how small Δx and Δy are but also on the magnitudes of higher partial derivatives; this is discussed further in section 5.7 in the context of Taylor series for functions of more than one variable. Nevertheless, letting the small changes Δx and Δy in (5.4) become infinitesimal, we can define the *total differential df* of the function $f(x, y)$, without any approximation, as

$$df = \frac{\partial f}{\partial x}dx + \frac{\partial f}{\partial y}dy. \tag{5.5}$$

Equation (5.5) can be extended to the case of a function of n variables, $f(x_1, x_2, \ldots, x_n)$;

$$df = \frac{\partial f}{\partial x_1}dx_1 + \frac{\partial f}{\partial x_2}dx_2 + \cdots + \frac{\partial f}{\partial x_n}dx_n. \tag{5.6}$$

▶ *Find the total differential of the function* $f(x, y) = y \exp(x + y)$.

Evaluating the first partial derivatives, we find

$$\frac{\partial f}{\partial x} = y \exp(x + y), \quad \frac{\partial f}{\partial y} = \exp(x + y) + y \exp(x + y).$$

Applying (5.5), we then find that the total differential is given by

$$df = [y \exp(x + y)]dx + [(1 + y) \exp(x + y)]dy. \ ◀$$

In some situations, despite the fact that several variables x_i, $i = 1, 2, \ldots, n$, appear to be involved, effectively only one of them is. This occurs if there are subsidiary relationships constraining all the x_i to have values dependent on the value of one of them, say x_1. These relationships may be represented by equations that are typically of the form

$$x_i = x_i(x_1), \qquad i = 2, 3, \ldots, n. \tag{5.7}$$

In principle f can then be expressed as a function of x_1 alone by substituting from (5.7) for $x_2, x_3, \ldots, x_n$, and then the *total derivative* (or simply the derivative) of f with respect to x_1 is obtained by ordinary differentiation.

Alternatively, (5.6) can be used to give

$$\frac{df}{dx_1} = \frac{\partial f}{\partial x_1} + \left(\frac{\partial f}{\partial x_2}\right)\frac{dx_2}{dx_1} + \cdots + \left(\frac{\partial f}{\partial x_n}\right)\frac{dx_n}{dx_1}. \tag{5.8}$$

It should be noted that the LHS of this equation is the total derivative df/dx_1, whilst the partial derivative $\partial f/\partial x_1$ forms only a part of the RHS. In evaluating this partial derivative account must be taken only of *explicit* appearances of x_1 in the function f, and *no* allowance must be made for the knowledge that changing x_1 necessarily changes $x_2, x_3, \ldots, x_n$. The contribution from these latter changes is precisely that of the remaining terms on the RHS of (5.8). Naturally, what has been shown using x_1 in the above argument applies equally well to any other of the x_i, with the appropriate consequent changes.

▶*Find the total derivative of $f(x, y) = x^2 + 3xy$ with respect to x, given that $y = \sin^{-1} x$.*

We can see immediately that

$$\frac{\partial f}{\partial x} = 2x + 3y, \qquad \frac{\partial f}{\partial y} = 3x, \qquad \frac{dy}{dx} = \frac{1}{(1 - x^2)^{1/2}}$$

and so, using (5.8) with $x_1 = x$ and $x_2 = y$,

$$\frac{df}{dx} = 2x + 3y + 3x \frac{1}{(1 - x^2)^{1/2}}$$

$$= 2x + 3 \sin^{-1} x + \frac{3x}{(1 - x^2)^{1/2}}.$$

Obviously the same expression would have resulted if we had substituted for y from the start, but the above method often produces results with reduced calculation, particularly in more complicated examples. ◀

5.3 Exact and inexact differentials

In the last section we discussed how to find the total differential of a function, i.e. its infinitesimal change in an arbitrary direction, in terms of its gradients $\partial f / \partial x$ and $\partial f / \partial y$ in the x- and y- directions (see (5.5)). Sometimes, however, we wish to reverse the process and find the function f that differentiates to give a known differential. Usually, finding such functions relies on inspection and experience.

As an example, it is easy to see that the function whose differential is $df = x \, dy + y \, dx$ is simply $f(x, y) = xy + c$, where c is a constant. Differentials such as this, which integrate directly, are called *exact differentials*, whereas those that do not are *inexact differentials*. For example, $x \, dy + 3y \, dx$ is not the straightforward differential of any function (see below). Inexact differentials can be made exact, however, by multiplying through by a suitable function called an integrating factor. This is discussed further in subsection 14.2.3.

▶*Show that the differential $x \, dy + 3y \, dx$ is inexact.*

On the one hand, if we integrate with respect to x we conclude that $f(x, y) = 3xy + g(y)$, where $g(y)$ is any function of y. On the other hand, if we integrate with respect to y we conclude that $f(x, y) = xy + h(x)$ where $h(x)$ is any function of x. These conclusions are inconsistent for any and every choice of $g(y)$ and $h(x)$, and therefore the differential is inexact. ◀

It is naturally of interest to investigate which properties of a differential make it exact. Consider the general differential containing two variables,

$$df = A(x, y)\, dx + B(x, y)\, dy.$$

We see that

$$\frac{\partial f}{\partial x} = A(x, y), \qquad \frac{\partial f}{\partial y} = B(x, y)$$

and, using the property $f_{xy} = f_{yx}$, we therefore require

$$\frac{\partial A}{\partial y} = \frac{\partial B}{\partial x}. \tag{5.9}$$

This is in fact both a necessary and a sufficient condition for the differential to be exact.

▶ *Using (5.9) show that $x\, dy + 3y\, dx$ is inexact.*

In the above notation, $A(x, y) = 3y$ and $B(x, y) = x$ and so

$$\frac{\partial A}{\partial y} = 3, \qquad \frac{\partial B}{\partial x} = 1.$$

As these are not equal it follows that the differential is inexact. ◀

Determining whether a differential containing many variable $x_1, x_2, \ldots, x_n$ is exact is a simple extension of the above. A differential containing many variables can be written in general as

$$df = \sum_{i=1}^{n} g_i(x_1, x_2, \ldots, x_n)\, dx_i$$

and will be exact if

$$\frac{\partial g_i}{\partial x_j} = \frac{\partial g_j}{\partial x_i} \qquad \text{for all pairs } i, j. \tag{5.10}$$

There will be $\frac{1}{2}n(n-1)$ such relationships to be satisfied.

▶ *Show that*
$$(y + z)\, dx + x\, dy + x\, dz$$
is an exact differential.

In this case, $g_1(x, y, z) = y + z$, $g_2(x, y, z) = x$, $g_3(x, y, z) = x$ and hence $\partial g_1 / \partial y = 1 = \partial g_2 / \partial x$, $\partial g_3 / \partial x = 1 = \partial g_1 / \partial z$, $\partial g_2 / \partial z = 0 = \partial g_3 / \partial y$; therefore, from (5.10), the differential is exact. As mentioned above, it is sometimes possible to show that a differential is exact simply by finding by inspection the function from which it originates. In this example, it can be seen easily that $f(x, y, z) = x(y + z) + c$. ◀

5.4 Useful theorems of partial differentiation

So far our discussion has centred on a function $f(x, y)$ dependent on two variables, x and y. Equally, however, we could have expressed x as a function of f and y, or y as a function of f and x. To emphasise the point that all the variables are of equal standing, we now replace f by z. This does not imply that x, y and z are coordinate positions (though they might be). Since x is a function of y and z, it follows that

$$dx = \left(\frac{\partial x}{\partial y}\right)_z dy + \left(\frac{\partial x}{\partial z}\right)_y dz \tag{5.11}$$

and similarly, since $y = y(x, z)$,

$$dy = \left(\frac{\partial y}{\partial x}\right)_z dx + \left(\frac{\partial y}{\partial z}\right)_x dz. \tag{5.12}$$

We may now substitute (5.12) into (5.11) to obtain

$$dx = \left(\frac{\partial x}{\partial y}\right)_z \left(\frac{\partial y}{\partial x}\right)_z dx + \left[\left(\frac{\partial x}{\partial y}\right)_z \left(\frac{\partial y}{\partial z}\right)_x + \left(\frac{\partial x}{\partial z}\right)_y\right] dz. \tag{5.13}$$

Now if we hold z constant, so that $dz = 0$, we obtain the *reciprocity relation*

$$\left(\frac{\partial x}{\partial y}\right)_z = \left(\frac{\partial y}{\partial x}\right)_z^{-1},$$

which holds provided both partial derivatives exist and neither is equal to zero. Note, further, that this relationship only holds when the variable being kept constant, in this case z, is the same on both sides of the equation.

Alternatively we can put $dx = 0$ in (5.13). Then the contents of the square brackets also equal zero, and we obtain the *cyclic relation*

$$\left(\frac{\partial y}{\partial z}\right)_x \left(\frac{\partial z}{\partial x}\right)_y \left(\frac{\partial x}{\partial y}\right)_z = -1,$$

which holds unless any of the derivatives vanish. In deriving this result we have used the reciprocity relation to replace $(\partial x/\partial z)_y^{-1}$ by $(\partial z/\partial x)_y$.

5.5 The chain rule

So far we have discussed the differentiation of a function $f(x, y)$ with respect to its variables x and y. We now consider the case where x and y are themselves functions of another variable, say u. If we wish to find the derivative df/du, we could simply substitute in $f(x, y)$ the expressions for $x(u)$ and $y(u)$ and then differentiate the resulting function of u. Such substitution will quickly give the desired answer in simple cases, but in more complicated examples it is easier to make use of the total differentials described in the previous section.

From equation (5.5) the total differential of $f(x, y)$ is given by

$$df = \frac{\partial f}{\partial x} dx + \frac{\partial f}{\partial y} dy,$$

but we now note that by using the formal device of dividing through by du this immediately implies

$$\frac{df}{du} = \frac{\partial f}{\partial x}\frac{dx}{du} + \frac{\partial f}{\partial y}\frac{dy}{du}, \tag{5.14}$$

which is called the *chain rule* for partial differentiation. This expression provides a direct method for calculating the total derivative of f with respect to u and is particularly useful when an equation is expressed in a parametric form.

> ► *Given that $x(u) = 1 + au$ and $y(u) = bu^3$, find the rate of change of $f(x, y) = xe^{-y}$ with respect to u.*

As discussed above, this problem could be addressed by substituting for x and y to obtain f as a function only of u and then differentiating with respect to u. However, using (5.14) directly we obtain

$$\frac{df}{du} = (e^{-y})a + (-xe^{-y})3bu^2,$$

which on substituting for x and y gives

$$\frac{df}{du} = e^{-bu^3}(a - 3bu^2 - 3bau^3). \quad ◄$$

Equation (5.14) is an example of the chain rule for a function of two variables each of which depends on a single variable. The chain rule may be extended to functions of many variables, each of which is itself a function of a variable u, i.e. $f(x_1, x_2, x_3, \ldots, x_n)$, with $x_i = x_i(u)$. In this case the chain rule gives

$$\frac{df}{du} = \sum_{i=1}^{n} \frac{\partial f}{\partial x_i}\frac{dx_i}{du} = \frac{\partial f}{\partial x_1}\frac{dx_1}{du} + \frac{\partial f}{\partial x_2}\frac{dx_2}{du} + \cdots + \frac{\partial f}{\partial x_n}\frac{dx_n}{du}. \tag{5.15}$$

5.6 Change of variables

It is sometimes necessary or desirable to make a change of variables during the course of an analysis, and consequently to have to change an equation expressed in one set of variables into an equation using another set. The same situation arises if a function f depends on one set of variables x_i, so that $f = f(x_1, x_2, \ldots, x_n)$ but the x_i are themselves functions of a further set of variables u_j and given by the equations

$$x_i = x_i(u_1, u_2, \ldots, u_m). \tag{5.16}$$

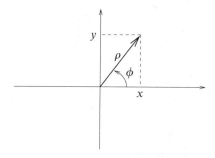

Figure 5.1   The relationship between Cartesian and plane polar coordinates.

For each different value of i, x_i will be a different function of the u_j. In this case the chain rule (5.15) becomes

$$\frac{\partial f}{\partial u_j} = \sum_{i=1}^{n} \frac{\partial f}{\partial x_i} \frac{\partial x_i}{\partial u_j}, \qquad j = 1, 2, \ldots, m, \tag{5.17}$$

and is said to express a *change of variables*. In general the number of variables in each set need not be equal, i.e. m need not equal n, but if both the x_i and the u_i are sets of independent variables then $m = n$.

▶ *Plane polar coordinates, ρ and ϕ, and Cartesian coordinates, x and y, are related by the expressions*

$$x = \rho \cos \phi, \qquad y = \rho \sin \phi,$$

as can be seen from figure 5.1. An arbitrary function $f(x, y)$ can be re-expressed as a function $g(\rho, \phi)$. Transform the expression

$$\frac{\partial^2 f}{\partial x^2} + \frac{\partial^2 f}{\partial y^2}$$

into one in ρ and ϕ.

We first note that $\rho^2 = x^2 + y^2$, $\phi = \tan^{-1}(y/x)$. We can now write down the four partial derivatives

$$\frac{\partial \rho}{\partial x} = \frac{x}{(x^2 + y^2)^{1/2}} = \cos \phi, \qquad \frac{\partial \phi}{\partial x} = \frac{-(y/x^2)}{1 + (y/x)^2} = -\frac{\sin \phi}{\rho},$$

$$\frac{\partial \rho}{\partial y} = \frac{y}{(x^2 + y^2)^{1/2}} = \sin \phi, \qquad \frac{\partial \phi}{\partial y} = \frac{1/x}{1 + (y/x)^2} = \frac{\cos \phi}{\rho}.$$

Thus, from (5.17), we may write

$$\frac{\partial}{\partial x} = \cos\phi\frac{\partial}{\partial\rho} - \frac{\sin\phi}{\rho}\frac{\partial}{\partial\phi}, \qquad \frac{\partial}{\partial y} = \sin\phi\frac{\partial}{\partial\rho} + \frac{\cos\phi}{\rho}\frac{\partial}{\partial\phi}.$$

Now it is only a matter of writing

$$\begin{aligned}
\frac{\partial^2 f}{\partial x^2} &= \frac{\partial}{\partial x}\left(\frac{\partial f}{\partial x}\right) = \frac{\partial}{\partial x}\left(\frac{\partial}{\partial x}\right)f \\
&= \left(\cos\phi\frac{\partial}{\partial\rho} - \frac{\sin\phi}{\rho}\frac{\partial}{\partial\phi}\right)\left(\cos\phi\frac{\partial}{\partial\rho} - \frac{\sin\phi}{\rho}\frac{\partial}{\partial\phi}\right)g \\
&= \left(\cos\phi\frac{\partial}{\partial\rho} - \frac{\sin\phi}{\rho}\frac{\partial}{\partial\phi}\right)\left(\cos\phi\frac{\partial g}{\partial\rho} - \frac{\sin\phi}{\rho}\frac{\partial g}{\partial\phi}\right) \\
&= \cos^2\phi\frac{\partial^2 g}{\partial\rho^2} + \frac{2\cos\phi\sin\phi}{\rho^2}\frac{\partial g}{\partial\phi} - \frac{2\cos\phi\sin\phi}{\rho}\frac{\partial^2 g}{\partial\phi\partial\rho} \\
&\quad + \frac{\sin^2\phi}{\rho}\frac{\partial g}{\partial\rho} + \frac{\sin^2\phi}{\rho^2}\frac{\partial^2 g}{\partial\phi^2}
\end{aligned}$$

and a similar expression for $\partial^2 f/\partial y^2$,

$$\begin{aligned}
\frac{\partial^2 f}{\partial y^2} &= \left(\sin\phi\frac{\partial}{\partial\rho} + \frac{\cos\phi}{\rho}\frac{\partial}{\partial\phi}\right)\left(\sin\phi\frac{\partial}{\partial\rho} + \frac{\cos\phi}{\rho}\frac{\partial}{\partial\phi}\right)g \\
&= \sin^2\phi\frac{\partial^2 g}{\partial\rho^2} - \frac{2\cos\phi\sin\phi}{\rho^2}\frac{\partial g}{\partial\phi} + \frac{2\cos\phi\sin\phi}{\rho}\frac{\partial^2 g}{\partial\phi\partial\rho} \\
&\quad + \frac{\cos^2\phi}{\rho}\frac{\partial g}{\partial\rho} + \frac{\cos^2\phi}{\rho^2}\frac{\partial^2 g}{\partial\phi^2}.
\end{aligned}$$

When these two expressions are added together the change of variables is complete and we obtain

$$\frac{\partial^2 f}{\partial x^2} + \frac{\partial^2 f}{\partial y^2} = \frac{\partial^2 g}{\partial\rho^2} + \frac{1}{\rho}\frac{\partial g}{\partial\rho} + \frac{1}{\rho^2}\frac{\partial^2 g}{\partial\phi^2}. \quad \blacktriangleleft$$

5.7 Taylor's theorem for many-variable functions

We have already introduced Taylor's theorem for a function $f(x)$ of one variable, in section 4.6. In an analogous way, the Taylor expansion of a function $f(x, y)$ of two variables is given by

$$\begin{aligned}
f(x, y) = f(x_0, y_0) &+ \frac{\partial f}{\partial x}\Delta x + \frac{\partial f}{\partial y}\Delta y \\
&+ \frac{1}{2!}\left[\frac{\partial^2 f}{\partial x^2}(\Delta x)^2 + 2\frac{\partial^2 f}{\partial x\partial y}\Delta x\Delta y + \frac{\partial^2 f}{\partial y^2}(\Delta y)^2\right] + \cdots, \quad (5.18)
\end{aligned}$$

where $\Delta x = x - x_0$ and $\Delta y = y - y_0$, and all the derivatives are to be evaluated at (x_0, y_0).

▶*Find the Taylor expansion, up to quadratic terms in $x-2$ and $y-3$, of $f(x,y) = y\exp xy$ about the point $x = 2$, $y = 3$.*

We first evaluate the required partial derivatives of the function, i.e.

$$\frac{\partial f}{\partial x} = y^2 \exp xy, \qquad \frac{\partial f}{\partial y} = \exp xy + xy \exp xy,$$

$$\frac{\partial^2 f}{\partial x^2} = y^3 \exp xy, \qquad \frac{\partial^2 f}{\partial y^2} = 2x \exp xy + x^2 y \exp xy,$$

$$\frac{\partial^2 f}{\partial x \partial y} = 2y \exp xy + xy^2 \exp xy.$$

Using (5.18), the Taylor expansion of a two-variable function, we find

$$f(x,y) \approx e^6 \Big\{ 3 + 9(x-2) + 7(y-3)$$

$$+ (2!)^{-1} \left[27(x-2)^2 + 48(x-2)(y-3) + 16(y-3)^2 \right] \Big\}. \blacktriangleleft$$

It will be noticed that the terms in (5.18) containing first derivatives can be written as

$$\frac{\partial f}{\partial x}\Delta x + \frac{\partial f}{\partial y}\Delta y = \left(\Delta x \frac{\partial}{\partial x} + \Delta y \frac{\partial}{\partial y} \right) f(x,y),$$

where both sides of this relation should be evaluated at the point (x_0, y_0). Similarly the terms in (5.18) containing second derivatives can be written as

$$\frac{1}{2!} \left[\frac{\partial^2 f}{\partial x^2}(\Delta x)^2 + 2\frac{\partial^2 f}{\partial x \partial y}\Delta x \Delta y + \frac{\partial^2 f}{\partial y^2}(\Delta y)^2 \right] = \frac{1}{2!} \left(\Delta x \frac{\partial}{\partial x} + \Delta y \frac{\partial}{\partial y} \right)^2 f(x,y), \tag{5.19}$$

where it is understood that the partial derivatives resulting from squaring the expression in parentheses act only on $f(x, y)$ and its derivatives, and not on Δx or Δy; again both sides of (5.19) should be evaluated at (x_0, y_0). It can be shown that the higher-order terms of the Taylor expansion of $f(x, y)$ can be written in an analogous way, and that we may write the full Taylor series as

$$f(x,y) = \sum_{n=0}^{\infty} \frac{1}{n!} \left[\left(\Delta x \frac{\partial}{\partial x} + \Delta y \frac{\partial}{\partial y} \right)^n f(x,y) \right]_{x_0,y_0}$$

where, as indicated, all the terms on the RHS are to be evaluated at (x_0, y_0).

The most general form of Taylor's theorem, for a function $f(x_1, x_2, \ldots, x_n)$ of n variables, is a simple extension of the above. Although it is not necessary to do so, we may think of the x_i as coordinates in n-dimensional space and write the function as $f(\mathbf{x})$, where $\mathbf{x}$ is a vector from the origin to $(x_1, x_2, \ldots, x_n)$. Taylor's theorem then becomes

$$f(\mathbf{x}) = f(\mathbf{x}_0) + \sum_i \frac{\partial f}{\partial x_i} \Delta x_i + \frac{1}{2!} \sum_i \sum_j \frac{\partial^2 f}{\partial x_i \partial x_j} \Delta x_i \Delta x_j + \cdots,$$

(5.20)

where $\Delta x_i = x_i - x_{i_0}$ and the partial derivatives are evaluated at $(x_{1_0}, x_{2_0}, \ldots, x_{n_0})$. For completeness, we note that in this case the full Taylor series can be written in the form

$$f(\mathbf{x}) = \sum_{n=0}^{\infty} \frac{1}{n!} \left[(\Delta \mathbf{x} \cdot \nabla)^n f(\mathbf{x}) \right]_{\mathbf{x}=\mathbf{x}_0},$$

where ∇ is the vector differential operator del, to be discussed in chapter 10.

5.8 Stationary values of many-variable functions

The idea of the *stationary points* of a function of just one variable has already been discussed in subsection 2.1.8. We recall that the function $f(x)$ has a stationary point at $x = x_0$ if its gradient df/dx is zero at that point. A function may have any number of stationary points, and their nature, i.e. whether they are maxima, minima or stationary points of inflection, is determined by the value of the second derivative at the point. A stationary point is

  (i) a minimum if $d^2 f/dx^2 > 0$;
 (ii) a maximum if $d^2 f/dx^2 < 0$;
(iii) a stationary point of inflection if $d^2 f/dx^2 = 0$ and changes sign through the point.

We now consider the stationary points of functions of more than one variable; we will see that partial differential analysis is ideally suited to the determination of the position and nature of such points. It is helpful to consider first the case of a function of just two variables but, even in this case, the general situation is more complex than that for a function of one variable, as can be seen from figure 5.2.

This figure shows part of a three-dimensional model of a function $f(x, y)$. At positions P and B there are a peak and a bowl respectively or, more mathematically, a local maximum and a local minimum. At position S the gradient in any direction is zero but the situation is complicated, since a section parallel to the plane $x = 0$ would show a maximum, but one parallel to the plane $y = 0$ would show a minimum. A point such as S is known as a *saddle point*. The orientation of the 'saddle' in the xy-plane is irrelevant; it is as shown in the figure solely for ease of discussion. For any saddle point the function increases in some directions away from the point but decreases in other directions.

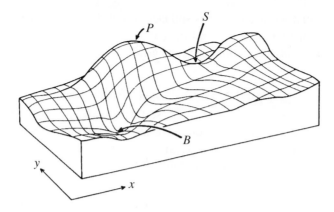

Figure 5.2 Stationary points of a function of two variables. A minimum occurs at B, a maximum at P and a saddle point at S.

For functions of two variables, such as the one shown, it should be clear that a necessary condition for a stationary point (maximum, minimum or saddle point) to occur is that

$$\frac{\partial f}{\partial x} = 0 \quad \text{and} \quad \frac{\partial f}{\partial y} = 0. \tag{5.21}$$

The vanishing of the partial derivatives in directions parallel to the axes is enough to ensure that the partial derivative in any arbitrary direction is also zero. The latter can be considered as the superposition of two contributions, one along each axis; since both contributions are zero, so is the partial derivative in the arbitrary direction. This may be made more precise by considering the total differential

$$df = \frac{\partial f}{\partial x}dx + \frac{\partial f}{\partial y}dy.$$

Using (5.21) we see that although the infinitesimal changes dx and dy can be chosen independently the change in the value of the infinitesimal function df is always zero at a stationary point.

We now turn our attention to determining the nature of a stationary point of a function of two variables, i.e. whether it is a maximum, a minimum or a saddle point. By analogy with the one-variable case we see that $\partial^2 f/\partial x^2$ and $\partial^2 f/\partial y^2$ must both be positive for a minimum and both be negative for a maximum. However these are not sufficient conditions since they could also be obeyed at complicated saddle points. What is important for a minimum (or maximum) is that the second partial derivative must be positive (or negative) in all directions, not just in the x- and y- directions.

To establish just what constitutes sufficient conditions we first note that, since f is a function of two variables and $\partial f/\partial x = \partial f/\partial y = 0$, a Taylor expansion of the type (5.18) about the stationary point yields

$$f(x, y) - f(x_0, y_0) \approx \frac{1}{2!} \left[(\Delta x)^2 f_{xx} + 2\Delta x \Delta y f_{xy} + (\Delta y)^2 f_{yy} \right],$$

where $\Delta x = x - x_0$ and $\Delta y = y - y_0$ and where the partial derivatives have been written in more compact notation. Rearranging the contents of the bracket as the weighted sum of two squares, we find

$$f(x, y) - f(x_0, y_0) \approx \frac{1}{2} \left[f_{xx} \left(\Delta x + \frac{f_{xy}\Delta y}{f_{xx}} \right)^2 + (\Delta y)^2 \left(f_{yy} - \frac{f_{xy}^2}{f_{xx}} \right) \right].$$
(5.22)

For a minimum, we require (5.22) to be positive for all Δx and Δy, and hence $f_{xx} > 0$ and $f_{yy} - (f_{xy}^2/f_{xx}) > 0$. Given the first constraint, the second can be written $f_{xx}f_{yy} > f_{xy}^2$. Similarly for a maximum we require (5.22) to be negative, and hence $f_{xx} < 0$ and $f_{xx}f_{yy} > f_{xy}^2$. For minima and maxima, symmetry requires that f_{yy} obeys the same criteria as f_{xx}. When (5.22) is negative (or zero) for some values of Δx and Δy but positive (or zero) for others, we have a saddle point. In this case $f_{xx}f_{yy} < f_{xy}^2$. In summary, all stationary points have $f_x = f_y = 0$ and they may be classified further as

(i) minima if both f_{xx} and f_{yy} are positive *and* $f_{xy}^2 < f_{xx}f_{yy}$,

(ii) maxima if both f_{xx} and f_{yy} are negative *and* $f_{xy}^2 < f_{xx}f_{yy}$,

(iii) saddle points if f_{xx} and f_{yy} have opposite signs *or* $f_{xy}^2 > f_{xx}f_{yy}$.

Note, however, that if $f_{xy}^2 = f_{xx}f_{yy}$ then $f(x, y) - f(x_0, y_0)$ can be written in one of the four forms

$$\pm \frac{1}{2} \left(\Delta x |f_{xx}|^{1/2} \pm \Delta y |f_{yy}|^{1/2} \right)^2.$$

For some choice of the ratio $\Delta y/\Delta x$ this expression has zero value, showing that, for a displacement from the stationary point in this particular direction, $f(x_0 + \Delta x, y_0 + \Delta y)$ does not differ from $f(x_0, y_0)$ to second order in Δx and Δy; in such situations further investigation is required. In particular, if f_{xx}, f_{yy} and f_{xy} are all zero then the Taylor expansion has to be taken to a higher order. As examples, such extended investigations would show that the function $f(x, y) = x^4 + y^4$ has a minimum at the origin but that $g(x, y) = x^4 + y^3$ has a saddle point there.

> ▶Show that the function $f(x,y) = x^3 \exp(-x^2 - y^2)$ has a maximum at the point $(\sqrt{3/2}, 0)$, a minimum at $(-\sqrt{3/2}, 0)$ and a stationary point at the origin whose nature cannot be determined by the above procedures.

Setting the first two partial derivatives to zero to locate the stationary points, we find

$$\frac{\partial f}{\partial x} = (3x^2 - 2x^4) \exp(-x^2 - y^2) = 0, \tag{5.23}$$

$$\frac{\partial f}{\partial y} = -2yx^3 \exp(-x^2 - y^2) = 0. \tag{5.24}$$

For (5.24) to be satisfied we require $x = 0$ or $y = 0$ and for (5.23) to be satisfied we require $x = 0$ or $x = \pm\sqrt{3/2}$. Hence the stationary points are at $(0,0)$, $(\sqrt{3/2}, 0)$ and $(-\sqrt{3/2}, 0)$. We now find the second partial derivatives:

$$f_{xx} = (4x^5 - 14x^3 + 6x) \exp(-x^2 - y^2),$$
$$f_{yy} = x^3(4y^2 - 2) \exp(-x^2 - y^2),$$
$$f_{xy} = 2x^2 y(2x^2 - 3) \exp(-x^2 - y^2).$$

We then substitute the pairs of values of x and y for each stationary point and find that at $(0,0)$

$$f_{xx} = 0, \quad f_{yy} = 0, \quad f_{xy} = 0$$

and at $(\pm\sqrt{3/2}, 0)$

$$f_{xx} = \mp 6\sqrt{3/2} \exp(-3/2), \quad f_{yy} = \mp 3\sqrt{3/2} \exp(-3/2), \quad f_{xy} = 0.$$

Hence, applying criteria (i)–(iii) above, we find that $(0,0)$ is an undetermined stationary point, $(\sqrt{3/2}, 0)$ is a maximum and $(-\sqrt{3/2}, 0)$ is a minimum. The function is shown in figure 5.3. ◀

Determining the nature of stationary points for functions of a general number of variables is considerably more difficult and requires a knowledge of the eigenvectors and eigenvalues of matrices. Although these are not discussed until chapter 8, we present the analysis here for completeness. The remainder of this section can therefore be omitted on a first reading.

For a function of n real variables, $f(x_1, x_2, \ldots, x_n)$, we require that, at all stationary points,

$$\frac{\partial f}{\partial x_i} = 0 \quad \text{for all } x_i.$$

In order to determine the nature of a stationary point, we must expand the function as a Taylor series about the point. Recalling the Taylor expansion (5.20) for a function of n variables, we see that

$$\Delta f = f(\mathbf{x}) - f(\mathbf{x}_0) \approx \frac{1}{2} \sum_i \sum_j \frac{\partial^2 f}{\partial x_i \partial x_j} \Delta x_i \Delta x_j. \tag{5.25}$$

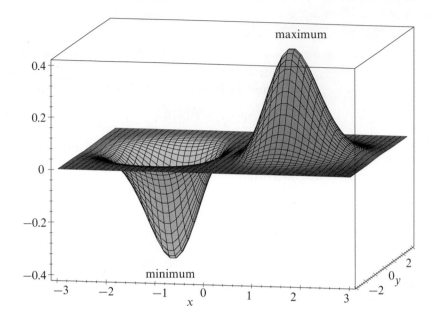

Figure 5.3   The function $f(x, y) = x^3 \exp(-x^2 - y^2)$.

If we define the matrix M to have elements given by

$$M_{ij} = \frac{\partial^2 f}{\partial x_i \partial x_j},$$

then we can rewrite (5.25) as

$$\Delta f = \tfrac{1}{2} \Delta \mathbf{x}^{\mathrm{T}} \mathsf{M} \Delta \mathbf{x}, \tag{5.26}$$

where $\Delta \mathbf{x}$ is the column vector with the Δx_i as its components and $\Delta \mathbf{x}^{\mathrm{T}}$ is its transpose. Since M is real and symmetric it has n real eigenvalues λ_r and n orthogonal eigenvectors $\mathbf{e}_r$, which after suitable normalisation satisfy

$$\mathsf{M}\mathbf{e}_r = \lambda_r \mathbf{e}_r, \qquad \mathbf{e}_r^{\mathrm{T}} \mathbf{e}_s = \delta_{rs},$$

where the *Kronecker delta*, written δ_{rs}, equals unity for $r = s$ and equals zero otherwise. These eigenvectors form a basis set for the n-dimensional space and we can therefore expand $\Delta \mathbf{x}$ in terms of them, obtaining

$$\Delta \mathbf{x} = \sum_r a_r \mathbf{e}_r,$$

166

where the a_r are coefficients dependent upon Δx. Substituting this into (5.26), we find

$$\Delta f = \tfrac{1}{2}\Delta x^T M \Delta x = \tfrac{1}{2}\sum_r \lambda_r a_r^2.$$

Now, for the stationary point to be a minimum, we require $\Delta f = \tfrac{1}{2}\sum_r \lambda_r a_r^2 > 0$ for all sets of values of the a_r, and therefore all the eigenvalues of M to be greater than zero. Conversely, for a maximum we require $\Delta f = \tfrac{1}{2}\sum_r \lambda_r a_r^2 < 0$, and therefore all the eigenvalues of M to be less than zero. If the eigenvalues have mixed signs, then we have a saddle point. Note that the test may fail if some or all of the eigenvalues are equal to zero and all the non-zero ones have the same sign.

▶*Derive the conditions for maxima, minima and saddle points for a function of two real variables, using the above analysis.*

For a two-variable function the matrix M is given by

$$M = \begin{pmatrix} f_{xx} & f_{xy} \\ f_{yx} & f_{yy} \end{pmatrix}.$$

Therefore its eigenvalues satisfy the equation

$$\begin{vmatrix} f_{xx} - \lambda & f_{xy} \\ f_{xy} & f_{yy} - \lambda \end{vmatrix} = 0.$$

Hence

$$(f_{xx} - \lambda)(f_{yy} - \lambda) - f_{xy}^2 = 0$$

$$\Rightarrow \quad f_{xx}f_{yy} - (f_{xx} + f_{yy})\lambda + \lambda^2 - f_{xy}^2 = 0$$

$$\Rightarrow \quad 2\lambda = (f_{xx} + f_{yy}) \pm \sqrt{(f_{xx} + f_{yy})^2 - 4(f_{xx}f_{yy} - f_{xy}^2)},$$

which by rearrangement of the terms under the square root gives

$$2\lambda = (f_{xx} + f_{yy}) \pm \sqrt{(f_{xx} - f_{yy})^2 + 4f_{xy}^2}.$$

Now, that M is real and symmetric implies that its eigenvalues are real, and so for both eigenvalues to be positive (corresponding to a minimum), we require f_{xx} and f_{yy} positive and also

$$f_{xx} + f_{yy} > \sqrt{(f_{xx} + f_{yy})^2 - 4(f_{xx}f_{yy} - f_{xy}^2)},$$

$$\Rightarrow \quad f_{xx}f_{yy} - f_{xy}^2 > 0.$$

A similar procedure will find the criteria for maxima and saddle points. ◀

5.9 Stationary values under constraints

In the previous section we looked at the problem of finding stationary values of a function of two or more variables when all the variables may be independently varied. However, it is often the case in physical problems that not all the variables used to describe a situation are in fact independent, i.e. some relationship between the variables must be satisfied. For example, if we walk through a hilly landscape and we are constrained to walk along a path, we will never reach the highest peak on the landscape unless the path happens to take us to it. Nevertheless, we can still find the highest point that we have reached during our journey.

We first discuss the case of a function of just two variables. Let us consider finding the maximum value of the differentiable function $f(x, y)$ subject to the constraint $g(x, y) = c$, where c is a constant. In the above analogy, $f(x, y)$ might represent the height of the land above sea-level in some hilly region, whilst $g(x, y) = c$ is the equation of the path along which we walk.

We could, of course, use the constraint $g(x, y) = c$ to substitute for x or y in $f(x, y)$, thereby obtaining a new function of only one variable whose stationary points could be found using the methods discussed in subsection 2.1.8. However, such a procedure can involve a lot of algebra and becomes very tedious for functions of more than two variables. A more direct method for solving such problems is the *method of Lagrange undetermined multipliers*, which we now discuss.

To maximise f we require

$$df = \frac{\partial f}{\partial x} dx + \frac{\partial f}{\partial y} dy = 0.$$

If dx and dy were independent, we could conclude $f_x = 0 = f_y$. However, here they are not independent, but constrained because g is constant:

$$dg = \frac{\partial g}{\partial x} dx + \frac{\partial g}{\partial y} dy = 0.$$

Multiplying dg by an as yet unknown number λ and adding it to df we obtain

$$d(f + \lambda g) = \left(\frac{\partial f}{\partial x} + \lambda \frac{\partial g}{\partial x} \right) dx + \left(\frac{\partial f}{\partial y} + \lambda \frac{\partial g}{\partial y} \right) dy = 0,$$

where λ is called a *Lagrange undetermined multiplier*. In this equation dx and dy are to be independent and arbitrary; we must therefore choose λ such that

$$\frac{\partial f}{\partial x} + \lambda \frac{\partial g}{\partial x} = 0, \tag{5.27}$$

$$\frac{\partial f}{\partial y} + \lambda \frac{\partial g}{\partial y} = 0. \tag{5.28}$$

These equations, together with the constraint $g(x, y) = c$, are sufficient to find the three unknowns, i.e. λ and the values of x and y at the stationary point.

> ▶ *The temperature of a point (x, y) on a unit circle is given by $T(x, y) = 1 + xy$. Find the*
> *temperature of the two hottest points on the circle.*

We need to maximise $T(x, y)$ subject to the constraint $x^2 + y^2 = 1$. Applying (5.27) and (5.28), we obtain

$$y + 2\lambda x = 0, \qquad (5.29)$$

$$x + 2\lambda y = 0. \qquad (5.30)$$

These results, together with the original constraint $x^2 + y^2 = 1$, provide three simultaneous equations that may be solved for λ, x and y.

From (5.29) and (5.30) we find $\lambda = \pm 1/2$, which in turn implies that $y = \mp x$. Remembering that $x^2 + y^2 = 1$, we find that

$$y = x \;\Rightarrow\; x = \pm\frac{1}{\sqrt{2}}, \quad y = \pm\frac{1}{\sqrt{2}}$$

$$y = -x \;\Rightarrow\; x = \mp\frac{1}{\sqrt{2}}, \quad y = \pm\frac{1}{\sqrt{2}}.$$

We have not yet determined which of these stationary points are maxima and which are minima. In this simple case, we need only substitute the four pairs of x- and y- values into $T(x, y) = 1 + xy$ to find that the maximum temperature on the unit circle is $T_{\text{max}} = 3/2$ at the points $y = x = \pm 1/\sqrt{2}$. ◀

The method of Lagrange multipliers can be used to find the stationary points of functions of more than two variables, subject to several constraints, provided that the number of constraints is smaller than the number of variables. For example, if we wish to find the stationary points of $f(x, y, z)$ subject to the constraints $g(x, y, z) = c_1$ and $h(x, y, z) = c_2$, where c_1 and c_2 are constants, then we proceed as above, obtaining

$$\frac{\partial}{\partial x}(f + \lambda g + \mu h) = \frac{\partial f}{\partial x} + \lambda\frac{\partial g}{\partial x} + \mu\frac{\partial h}{\partial x} = 0,$$

$$\frac{\partial}{\partial y}(f + \lambda g + \mu h) = \frac{\partial f}{\partial y} + \lambda\frac{\partial g}{\partial y} + \mu\frac{\partial h}{\partial y} = 0, \qquad (5.31)$$

$$\frac{\partial}{\partial z}(f + \lambda g + \mu h) = \frac{\partial f}{\partial z} + \lambda\frac{\partial g}{\partial z} + \mu\frac{\partial h}{\partial z} = 0.$$

We may now solve these three equations, together with the two constraints, to give λ, μ, x, y and z.

> ▶ *Find the stationary points of $f(x,y,z) = x^3 + y^3 + z^3$ subject to the following constraints:*
>
> (i) $g(x,y,z) = x^2 + y^2 + z^2 = 1$;
> (ii) $g(x,y,z) = x^2 + y^2 + z^2 = 1$ *and* $h(x,y,z) = x + y + z = 0$.

Case (i). Since there is only one constraint in this case, we need only introduce a single Lagrange multiplier to obtain

$$\frac{\partial}{\partial x}(f + \lambda g) = 3x^2 + 2\lambda x = 0,$$

$$\frac{\partial}{\partial y}(f + \lambda g) = 3y^2 + 2\lambda y = 0, \qquad (5.32)$$

$$\frac{\partial}{\partial z}(f + \lambda g) = 3z^2 + 2\lambda z = 0.$$

These equations are highly symmetrical and clearly have the solution $x = y = z = -2\lambda/3$. Using the constraint $x^2 + y^2 + z^2 = 1$ we find $\lambda = \pm\sqrt{3}/2$ and so stationary points occur at

$$x = y = z = \pm\frac{1}{\sqrt{3}}. \qquad (5.33)$$

In solving the three equations (5.32) in this way, however, we have implicitly assumed that x, y and z are non-zero. However, it is clear from (5.32) that any of these values can equal zero, with the exception of the case $x = y = z = 0$ since this is prohibited by the constraint $x^2 + y^2 + z^2 = 1$. We must consider the other cases separately.

If $x = 0$, for example, we require

$$3y^2 + 2\lambda y = 0,$$
$$3z^2 + 2\lambda z = 0,$$
$$y^2 + z^2 = 1.$$

Clearly, we require $\lambda \neq 0$, otherwise these equations are inconsistent. If neither y nor z is zero we find $y = -2\lambda/3 = z$ and from the third equation we require $y = z = \pm 1/\sqrt{2}$. If $y = 0$, however, then $z = \pm 1$ and, similarly, if $z = 0$ then $y = \pm 1$. Thus the stationary points having $x = 0$ are $(0, 0, \pm 1)$, $(0, \pm 1, 0)$ and $(0, \pm 1/\sqrt{2}, \pm 1/\sqrt{2})$. A similar procedure can be followed for the cases $y = 0$ and $z = 0$ respectively and, in addition to those already obtained, we find the stationary points $(\pm 1, 0, 0)$, $(\pm 1/\sqrt{2}, 0, \pm 1/\sqrt{2})$ and $(\pm 1/\sqrt{2}, \pm 1/\sqrt{2}, 0)$.

Case (ii). We now have two constraints and must therefore introduce two Lagrange multipliers to obtain (cf. (5.31))

$$\frac{\partial}{\partial x}(f + \lambda g + \mu h) = 3x^2 + 2\lambda x + \mu = 0, \qquad (5.34)$$

$$\frac{\partial}{\partial y}(f + \lambda g + \mu h) = 3y^2 + 2\lambda y + \mu = 0, \qquad (5.35)$$

$$\frac{\partial}{\partial z}(f + \lambda g + \mu h) = 3z^2 + 2\lambda z + \mu = 0. \qquad (5.36)$$

These equations are again highly symmetrical and the simplest way to proceed is to subtract (5.35) from (5.34) to obtain

$$3(x^2 - y^2) + 2\lambda(x - y) = 0$$
$$\Rightarrow \quad 3(x + y)(x - y) + 2\lambda(x - y) = 0. \qquad (5.37)$$

This equation is clearly satisfied if $x = y$; then, from the second constraint, $x + y + z = 0$, we find $z = -2x$. Substituting these values into the first constraint, $x^2 + y^2 + z^2 = 1$, we obtain

$$x = \pm\frac{1}{\sqrt{6}}, \qquad y = \pm\frac{1}{\sqrt{6}}, \qquad z = \mp\frac{2}{\sqrt{6}}. \qquad (5.38)$$

Because of the high degree of symmetry amongst the equations (5.34)–(5.36), we may obtain by inspection two further relations analogous to (5.37), one containing the variables y, z and the other the variables x, z. Assuming $y = z$ in the first relation and $x = z$ in the second, we find the stationary points

$$x = \pm\frac{1}{\sqrt{6}}, \qquad y = \mp\frac{2}{\sqrt{6}}, \qquad z = \pm\frac{1}{\sqrt{6}} \qquad (5.39)$$

and

$$x = \mp\frac{2}{\sqrt{6}}, \qquad y = \pm\frac{1}{\sqrt{6}}, \qquad z = \pm\frac{1}{\sqrt{6}}. \qquad (5.40)$$

We note that in finding the stationary points (5.38)–(5.40) we did not need to evaluate the Lagrange multipliers λ and μ explicitly. This is not always the case, however, and in some problems it may be simpler to begin by finding the values of these multipliers.

Returning to (5.37) we must now consider the case where $x \neq y$; then we find

$$3(x + y) + 2\lambda = 0. \qquad (5.41)$$

However, in obtaining the stationary points (5.39), (5.40), we did *not* assume $x = y$ but only required $x = z$ and $y = z$ respectively. It is clear that $x \neq y$ at these stationary points, and it can be shown that they do indeed satisfy (5.41). Similarly, several stationary points for which $x \neq z$ or $y \neq z$ have already been found.

Thus we need to consider further only two cases, $x = y = z$, and x, y and z are all different. The first is clearly prohibited by the constraint $x + y + z = 0$. For the second case, (5.41) must be satisfied, together with the analogous equations containing y, z and x, z respectively, i.e.

$$3(x + y) + 2\lambda = 0,$$
$$3(y + z) + 2\lambda = 0,$$
$$3(x + z) + 2\lambda = 0.$$

Adding these three equations together and using the constraint $x + y + z = 0$ we find $\lambda = 0$. However, for $\lambda = 0$ the equations are inconsistent for non-zero x, y and z. Therefore all the stationary points have already been found and are given by (5.38)–(5.40). ◄

The method may be extended to functions of any number n of variables subject to any smaller number m of constraints. This means that effectively there are $n - m$ independent variables and, as mentioned above, we could solve by substitution and then by the methods of the previous section. However, for large n this becomes cumbersome and the use of Lagrange undetermined multipliers is a useful simplification.

►*A system contains a very large number N of particles, each of which can be in any of R energy levels with a corresponding energy E_i, $i = 1, 2, \ldots, R$. The number of particles in the ith level is n_i and the total energy of the system is a constant, E. Find the distribution of particles amongst the energy levels that maximises the expression*

$$P = \frac{N!}{n_1! n_2! \cdots n_R!},$$

subject to the constraints that both the number of particles and the total energy remain constant, i.e.

$$g = N - \sum_{i=1}^{R} n_i = 0 \quad \text{and} \quad h = E - \sum_{i=1}^{R} n_i E_i = 0.$$

The way in which we proceed is as follows. In order to maximise P, we must minimise its denominator (since the numerator is fixed). Minimising the denominator is the same as minimising the logarithm of the denominator, i.e.

$$f = \ln(n_1! n_2! \cdots n_R!) = \ln(n_1!) + \ln(n_2!) + \cdots + \ln(n_R!).$$

Using Stirling's approximation, $\ln(n!) \approx n \ln n - n$, we find that

$$f = n_1 \ln n_1 + n_2 \ln n_2 + \cdots + n_R \ln n_R - (n_1 + n_2 + \cdots + n_R)$$

$$= \left(\sum_{i=1}^{R} n_i \ln n_i \right) - N.$$

It has been assumed here that, for the desired distribution, all the n_i are large. Thus, we now have a function f subject to two constraints, $g = 0$ and $h = 0$, and we can apply the Lagrange method, obtaining (cf. (5.31))

$$\frac{\partial f}{\partial n_1} + \lambda \frac{\partial g}{\partial n_1} + \mu \frac{\partial h}{\partial n_1} = 0,$$

$$\frac{\partial f}{\partial n_2} + \lambda \frac{\partial g}{\partial n_2} + \mu \frac{\partial h}{\partial n_2} = 0,$$

$$\vdots$$

$$\frac{\partial f}{\partial n_R} + \lambda \frac{\partial g}{\partial n_R} + \mu \frac{\partial h}{\partial n_R} = 0.$$

Since all these equations are alike, we consider the general case

$$\frac{\partial f}{\partial n_k} + \lambda \frac{\partial g}{\partial n_k} + \mu \frac{\partial h}{\partial n_k} = 0,$$

for $k = 1, 2, \ldots, R$. Substituting the functions f, g and h into this relation we find

$$\frac{n_k}{n_k} + \ln n_k + \lambda(-1) + \mu(-E_k) = 0,$$

which can be rearranged to give

$$\ln n_k = \mu E_k + \lambda - 1,$$

and hence

$$n_k = C \exp \mu E_k.$$

We now have the general form for the distribution of particles amongst energy levels, but in order to determine the two constants μ, C we recall that

$$\sum_{k=1}^{R} C \exp \mu E_k = N$$

and

$$\sum_{k=1}^{R} C E_k \exp \mu E_k = E.$$

This is known as the Boltzmann distribution and is a well-known result from statistical mechanics. ◄

5.10 Envelopes

As noted at the start of this chapter, many of the functions with which physicists, chemists and engineers have to deal contain, in addition to constants and one or more variables, quantities that are normally considered as parameters of the system under study. Such parameters may, for example, represent the capacitance of a capacitor, the length of a rod, or the mass of a particle – quantities that are normally taken as fixed for any particular physical set-up. The corresponding variables may well be time, currents, charges, positions and velocities. However, the parameters *could* be varied and in this section we study the effects of doing so; in particular we study how the form of dependence of one variable on another, typically $y = y(x)$, is affected when the value of a parameter is changed in a smooth and continuous way. In effect, we are making the parameter into an additional variable.

As a particular parameter, which we denote by α, is varied over its permitted range, the shape of the plot of y against x will change, usually, but not always, in a smooth and continuous way. For example, if the muzzle speed v of a shell fired from a gun is increased through a range of values then its height–distance trajectories will be a series of curves with a common starting point that are essentially just magnified copies of the original; furthermore the curves do not cross each other. However, if the muzzle speed is kept constant but θ, the angle of elevation of the gun, is increased through a series of values, the corresponding trajectories do not vary in a monotonic way. When θ has been increased beyond $45°$ the trajectories then do cross some of the trajectories corresponding to $\theta < 45°$. The trajectories for $\theta > 45°$ all lie within a curve that touches each individual trajectory at one point. Such a curve is called the *envelope* to the set of trajectory solutions; it is to the study of such envelopes that this section is devoted.

For our general discussion of envelopes we will consider an equation of the form $f = f(x, y, \alpha) = 0$. A function of three Cartesian variables, $f = f(x, y, \alpha)$, is defined at all points in $xy\alpha$-space, whereas $f = f(x, y, \alpha) = 0$ is a *surface* in this space. A plane of constant α, which is parallel to the xy-plane, cuts such

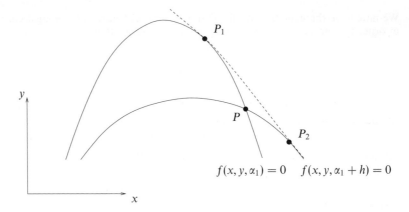

$$f(x, y, \alpha_1) = 0 \qquad f(x, y, \alpha_1 + h) = 0$$

Figure 5.4 Two neighbouring curves in the xy-plane of the family $f(x, y, \alpha) = 0$ intersecting at P. For fixed α_1, the point P_1 is the limiting position of P as $h \to 0$. As α_1 is varied, P_1 delineates the envelope of the family (broken line).

a surface in a curve. Thus different values of the parameter α correspond to different curves, which can be plotted in the xy-plane. We now investigate how the *envelope equation* for such a family of curves is obtained.

5.10.1 Envelope equations

Suppose $f(x, y, \alpha_1) = 0$ and $f(x, y, \alpha_1 + h) = 0$ are two neighbouring curves of a family for which the parameter α differs by a small amount h. Let them intersect at the point P with coordinates x, y, as shown in figure 5.4. Then the envelope, indicated by the broken line in the figure, touches $f(x, y, \alpha_1) = 0$ at the point P_1, which is defined as the limiting position of P when α_1 is fixed but $h \to 0$. The full envelope is the curve traced out by P_1 as α_1 changes to generate successive members of the family of curves. Of course, for any finite h, $f(x, y, \alpha_1 + h) = 0$ is one of these curves and the envelope touches it at the point P_2.

We are now going to apply Rolle's theorem, see subsection 2.1.10, with the parameter α as the independent variable and x and y fixed as constants. In this context, the two curves in figure 5.4 can be thought of as the projections onto the xy-plane of the planar curves in which the *surface* $f = f(x, y, \alpha) = 0$ meets the planes $\alpha = \alpha_1$ and $\alpha = \alpha_1 + h$.

Along the normal to the page that passes through P, as α changes from α_1 to $\alpha_1 + h$ the value of $f = f(x, y, \alpha)$ will depart from zero, because the normal meets the surface $f = f(x, y, \alpha) = 0$ only at $\alpha = \alpha_1$ and at $\alpha = \alpha_1 + h$. However, at these end points the values of $f = f(x, y, \alpha)$ will both be zero, and therefore equal. This allows us to apply Rolle's theorem and so to conclude that for some θ in the range $0 \le \theta \le 1$ the partial derivative $\partial f(x, y, \alpha_1 + \theta h)/\partial \alpha$ is zero. When

h is made arbitrarily small, so that $P \rightarrow P_1$, the three defining equations reduce to two, which define the envelope point P_1:

$$f(x, y, \alpha_1) = 0 \quad \text{and} \quad \frac{\partial f(x, y, \alpha_1)}{\partial \alpha} = 0. \tag{5.42}$$

In (5.42) both the function and the gradient are evaluated at $\alpha = \alpha_1$. The equation of the envelope $g(x, y) = 0$ is found by eliminating α_1 between the two equations.

As a simple example we will now solve the problem which when posed mathematically reads 'calculate the envelope appropriate to the family of straight lines in the xy-plane whose points of intersection with the coordinate axes are a fixed distance apart'. In more ordinary language, the problem is about a ladder leaning against a wall.

> ►*A ladder of length L stands on level ground and can be leaned at any angle against a vertical wall. Find the equation of the curve bounding the vertical area below the ladder.*

We take the ground and the wall as the x- and y-axes respectively. If the foot of the ladder is a from the foot of the wall and the top is b above the ground then the straight-line equation of the ladder is

$$\frac{x}{a} + \frac{y}{b} = 1,$$

where a and b are connected by $a^2 + b^2 = L^2$. Expressed in standard form with only one independent parameter, a, the equation becomes

$$f(x, y, a) = \frac{x}{a} + \frac{y}{(L^2 - a^2)^{1/2}} - 1 = 0. \tag{5.43}$$

Now, differentiating (5.43) with respect to a and setting the derivative $\partial f / \partial a$ equal to zero gives

$$-\frac{x}{a^2} + \frac{ay}{(L^2 - a^2)^{3/2}} = 0;$$

from which it follows that

$$a = \frac{L x^{1/3}}{(x^{2/3} + y^{2/3})^{1/2}} \quad \text{and} \quad (L^2 - a^2)^{1/2} = \frac{L y^{1/3}}{(x^{2/3} + y^{2/3})^{1/2}}.$$

Eliminating a by substituting these values into (5.43) gives, for the equation of the envelope of all possible positions on the ladder,

$$x^{2/3} + y^{2/3} = L^{2/3}.$$

This is the equation of an astroid (mentioned in exercise 2.19), and, together with the wall and the ground, marks the boundary of the vertical area below the ladder. ◄

Other examples, drawn from both geometry and and the physical sciences, are considered in the exercises at the end of this chapter. The shell trajectory problem discussed earlier in this section is solved there, but in the guise of a question about the water bell of an ornamental fountain.

5.11 Thermodynamic relations

Thermodynamic relations provide a useful set of physical examples of partial differentiation. The relations we will derive are called *Maxwell's thermodynamic relations*. They express relationships between four thermodynamic quantities describing a unit mass of a substance. The quantities are the pressure P, the volume V, the thermodynamic temperature T and the entropy S of the substance. These four quantities are not independent; any two of them can be varied independently, but the other two are then determined.

The first law of thermodynamics may be expressed as

$$dU = T\,dS - P\,dV, \tag{5.44}$$

where U is the internal energy of the substance. Essentially this is a conservation of energy equation, but we shall concern ourselves, not with the physics, but rather with the use of partial differentials to relate the four basic quantities discussed above. The method involves writing a total differential, dU say, in terms of the differentials of two variables, say X and Y, thus

$$dU = \left(\frac{\partial U}{\partial X}\right)_Y dX + \left(\frac{\partial U}{\partial Y}\right)_X dY, \tag{5.45}$$

and then using the relationship

$$\frac{\partial^2 U}{\partial X \partial Y} = \frac{\partial^2 U}{\partial Y \partial X}$$

to obtain the required Maxwell relation. The variables X and Y are to be chosen from P, V, T and S.

▶*Show that $(\partial T/\partial V)_S = -(\partial P/\partial S)_V$.*

Here the two variables that have to be held constant, in turn, happen to be those whose differentials appear on the RHS of (5.44). And so, taking X as S and Y as V in (5.45), we have

$$T\,dS - P\,dV = dU = \left(\frac{\partial U}{\partial S}\right)_V dS + \left(\frac{\partial U}{\partial V}\right)_S dV,$$

and find directly that

$$\left(\frac{\partial U}{\partial S}\right)_V = T \quad \text{and} \quad \left(\frac{\partial U}{\partial V}\right)_S = -P.$$

Differentiating the first expression with respect to V and the second with respect to S, and using

$$\frac{\partial^2 U}{\partial V \partial S} = \frac{\partial^2 U}{\partial S \partial V},$$

we find the Maxwell relation

$$\left(\frac{\partial T}{\partial V}\right)_S = -\left(\frac{\partial P}{\partial S}\right)_V. \quad ◀$$

▶*Show that $(\partial S / \partial V)_T = (\partial P / \partial T)_V$.*

Applying (5.45) to dS, with independent variables V and T, we find

$$dU = T\, dS - P\, dV = T \left[\left(\frac{\partial S}{\partial V} \right)_T dV + \left(\frac{\partial S}{\partial T} \right)_V dT \right] - P\, dV.$$

Similarly applying (5.45) to dU, we find

$$dU = \left(\frac{\partial U}{\partial V} \right)_T dV + \left(\frac{\partial U}{\partial T} \right)_V dT.$$

Thus, equating partial derivatives,

$$\left(\frac{\partial U}{\partial V} \right)_T = T \left(\frac{\partial S}{\partial V} \right)_T - P \quad \text{and} \quad \left(\frac{\partial U}{\partial T} \right)_V = T \left(\frac{\partial S}{\partial T} \right)_V.$$

But, since

$$\frac{\partial^2 U}{\partial T \partial V} = \frac{\partial^2 U}{\partial V \partial T}, \quad \text{i.e.} \quad \frac{\partial}{\partial T} \left(\frac{\partial U}{\partial V} \right)_T = \frac{\partial}{\partial V} \left(\frac{\partial U}{\partial T} \right)_V,$$

it follows that

$$\left(\frac{\partial S}{\partial V} \right)_T + T \frac{\partial^2 S}{\partial T \partial V} - \left(\frac{\partial P}{\partial T} \right)_V = \frac{\partial}{\partial V} \left[T \left(\frac{\partial S}{\partial T} \right)_V \right]_T = T \frac{\partial^2 S}{\partial V \partial T}.$$

Thus finally we get the Maxwell relation

$$\left(\frac{\partial S}{\partial V} \right)_T = \left(\frac{\partial P}{\partial T} \right)_V. \blacktriangleleft$$

The above derivation is rather cumbersome, however, and a useful trick that can simplify the working is to define a new function, called a *potential*. The internal energy U discussed above is one example of a potential but three others are commonly defined and they are described below.

▶*Show that $(\partial S / \partial V)_T = (\partial P / \partial T)_V$ by considering the potential $U - ST$.*

We first consider the differential $d(U - ST)$. From (5.5), we obtain

$$d(U - ST) = dU - S\, dT - T\, dS = -S\, dT - P\, dV$$

when use is made of (5.44). We rewrite $U - ST$ as F for convenience of notation; F is called the *Helmholtz potential*. Thus

$$dF = -S\, dT - P\, dV,$$

and it follows that

$$\left(\frac{\partial F}{\partial T} \right)_V = -S \quad \text{and} \quad \left(\frac{\partial F}{\partial V} \right)_T = -P.$$

Using these results together with

$$\frac{\partial^2 F}{\partial T \partial V} = \frac{\partial^2 F}{\partial V \partial T},$$

we can see immediately that

$$\left(\frac{\partial S}{\partial V} \right)_T = \left(\frac{\partial P}{\partial T} \right)_V,$$

which is the same Maxwell relation as before. ◀

Although the Helmholtz potential has other uses, in this context it has simply provided a means for a quick derivation of the Maxwell relation. The other Maxwell relations can be derived similarly by using two other potentials, the *enthalpy*, $H = U + PV$, and the *Gibbs free energy*, $G = U + PV - ST$ (see exercise 5.25).

5.12 Differentiation of integrals

We conclude this chapter with a discussion of the differentiation of integrals. Let us consider the indefinite integral (cf. equation (2.30))

$$F(x,t) = \int f(x,t)\, dt,$$

from which it follows immediately that

$$\frac{\partial F(x,t)}{\partial t} = f(x,t).$$

Assuming that the second partial derivatives of $F(x,t)$ are continuous, we have

$$\frac{\partial^2 F(x,t)}{\partial t \partial x} = \frac{\partial^2 F(x,t)}{\partial x \partial t},$$

and so we can write

$$\frac{\partial}{\partial t}\left[\frac{\partial F(x,t)}{\partial x}\right] = \frac{\partial}{\partial x}\left[\frac{\partial F(x,t)}{\partial t}\right] = \frac{\partial f(x,t)}{\partial x}.$$

Integrating this equation with respect to t then gives

$$\frac{\partial F(x,t)}{\partial x} = \int \frac{\partial f(x,t)}{\partial x}\, dt. \tag{5.46}$$

Now consider the definite integral

$$I(x) = \int_{t=u}^{t=v} f(x,t)\, dt$$
$$= F(x,v) - F(x,u),$$

where u and v are constants. Differentiating this integral with respect to x, and using (5.46), we see that

$$\frac{dI(x)}{dx} = \frac{\partial F(x,v)}{\partial x} - \frac{\partial F(x,u)}{\partial x}$$
$$= \int^v \frac{\partial f(x,t)}{\partial x}\, dt - \int^u \frac{\partial f(x,t)}{\partial x}\, dt$$
$$= \int_u^v \frac{\partial f(x,t)}{\partial x}\, dt.$$

This is *Leibnitz' rule* for differentiating integrals, and basically it states that for constant limits of integration the order of integration and differentiation can be reversed.

In the more general case where the limits of the integral are themselves functions of x, it follows immediately that

$$I(x) = \int_{t=u(x)}^{t=v(x)} f(x,t)\, dt$$
$$= F(x, v(x)) - F(x, u(x)),$$

which yields the partial derivatives

$$\frac{\partial I}{\partial v} = f(x, v(x)), \qquad \frac{\partial I}{\partial u} = -f(x, u(x)).$$

Consequently

$$\frac{dI}{dx} = \left(\frac{\partial I}{\partial v}\right)\frac{dv}{dx} + \left(\frac{\partial I}{\partial u}\right)\frac{du}{dx} + \frac{\partial I}{\partial x}$$

$$= f(x, v(x))\frac{dv}{dx} - f(x, u(x))\frac{du}{dx} + \frac{\partial}{\partial x}\int_{u(x)}^{v(x)} f(x,t)dt$$

$$= f(x, v(x))\frac{dv}{dx} - f(x, u(x))\frac{du}{dx} + \int_{u(x)}^{v(x)} \frac{\partial f(x,t)}{\partial x}dt, \qquad (5.47)$$

where the partial derivative with respect to x in the last term has been taken inside the integral sign using (5.46). This procedure is valid because $u(x)$ and $v(x)$ are being held constant in this term.

►Find the derivative with respect to x of the integral

$$I(x) = \int_{x}^{x^2} \frac{\sin xt}{t}\, dt.$$

Applying (5.47), we see that

$$\frac{dI}{dx} = \frac{\sin x^3}{x^2}(2x) - \frac{\sin x^2}{x}(1) + \int_{x}^{x^2} \frac{t \cos xt}{t} dt$$

$$= \frac{2\sin x^3}{x} - \frac{\sin x^2}{x} + \left[\frac{\sin xt}{x}\right]_{x}^{x^2}$$

$$= 3\frac{\sin x^3}{x} - 2\frac{\sin x^2}{x}$$

$$= \frac{1}{x}(3\sin x^3 - 2\sin x^2). ◄$$

5.13 Exercises

5.1   Using the appropriate properties of ordinary derivatives, perform the following.

    (a) Find all the first partial derivatives of the following functions $f(x, y)$:
        (i) x^2y, (ii) $x^2 + y^2 + 4$, (iii) $\sin(x/y)$, (iv) $\tan^{-1}(y/x)$,
        (v) $r(x, y, z) = (x^2 + y^2 + z^2)^{1/2}$.
    (b) For (i), (ii) and (v), find $\partial^2 f/\partial x^2$, $\partial^2 f/\partial y^2$ and $\partial^2 f/\partial x\partial y$.
    (c) For (iv) verify that $\partial^2 f/\partial x\partial y = \partial^2 f/\partial y\partial x$.

5.2   Determine which of the following are exact differentials:

    (a) $(3x + 2)y\, dx + x(x + 1)\, dy$;
    (b) $y\tan x\, dx + x\tan y\, dy$;
    (c) $y^2(\ln x + 1)\, dx + 2xy\ln x\, dy$;
    (d) $y^2(\ln x + 1)\, dy + 2xy\ln x\, dx$;
    (e) $[x/(x^2 + y^2)]\, dy - [y/(x^2 + y^2)]\, dx$.

5.3   Show that the differential

$$df = x^2\, dy - (y^2 + xy)\, dx$$

    is not exact, but that $dg = (xy^2)^{-1}df$ is exact.

5.4   Show that

$$df = y(1 + x - x^2)\, dx + x(x + 1)\, dy$$

    is not an exact differential.
        Find the differential equation that a function $g(x)$ must satisfy if $d\phi = g(x)df$ is to be an exact differential. Verify that $g(x) = e^{-x}$ is a solution of this equation and deduce the form of $\phi(x, y)$.

5.5   The equation $3y = z^3 + 3xz$ defines z implicitly as a function of x and y. Evaluate all three second partial derivatives of z with respect to x and/or y. Verify that z is a solution of

$$x\frac{\partial^2 z}{\partial y^2} + \frac{\partial^2 z}{\partial x^2} = 0.$$

5.6   A possible equation of state for a gas takes the form

$$PV = RT\exp\left(-\frac{\alpha}{VRT}\right),$$

    in which α and R are constants. Calculate expressions for

$$\left(\frac{\partial P}{\partial V}\right)_T, \qquad \left(\frac{\partial V}{\partial T}\right)_P, \qquad \left(\frac{\partial T}{\partial P}\right)_V,$$

    and show that their product is -1, as stated in section 5.4.

5.7   The function $G(t)$ is defined by

$$G(t) = F(x, y) = x^2 + y^2 + 3xy,$$

    where $x(t) = at^2$ and $y(t) = 2at$. Use the chain rule to find the values of (x, y) at which $G(t)$ has stationary values as a function of t. Do any of them correspond to the stationary points of $F(x, y)$ as a function of x and y?

5.8   In the xy-plane, new coordinates s and t are defined by

$$s = \tfrac{1}{2}(x + y), \qquad t = \tfrac{1}{2}(x - y).$$

    Transform the equation

$$\frac{\partial^2 \phi}{\partial x^2} - \frac{\partial^2 \phi}{\partial y^2} = 0$$

    into the new coordinates and deduce that its general solution can be written

$$\phi(x, y) = f(x + y) + g(x - y),$$

    where $f(u)$ and $g(v)$ are arbitrary functions of u and v, respectively.

5.9   The function $f(x, y)$ satisfies the differential equation

$$y\frac{\partial f}{\partial x} + x\frac{\partial f}{\partial y} = 0.$$

By changing to new variables $u = x^2 - y^2$ and $v = 2xy$, show that f is, in fact, a function of $x^2 - y^2$ only.

5.10   If $x = e^u \cos\theta$ and $y = e^u \sin\theta$, show that

$$\frac{\partial^2\phi}{\partial u^2} + \frac{\partial^2\phi}{\partial\theta^2} = (x^2 + y^2)\left(\frac{\partial^2 f}{\partial x^2} + \frac{\partial^2 f}{\partial y^2}\right),$$

where $f(x, y) = \phi(u, \theta)$.

5.11   Find and evaluate the maxima, minima and saddle points of the function

$$f(x, y) = xy(x^2 + y^2 - 1).$$

5.12   Show that

$$f(x, y) = x^3 - 12xy + 48x + by^2, \qquad b \neq 0,$$

has two, one, or zero stationary points, according to whether $|b|$ is less than, equal to, or greater than 3.

5.13   Locate the stationary points of the function

$$f(x, y) = (x^2 - 2y^2)\exp[-(x^2 + y^2)/a^2],$$

where a is a non-zero constant.
   Sketch the function along the x- and y-axes and hence identify the nature and values of the stationary points.

5.14   Find the stationary points of the function

$$f(x, y) = x^3 + xy^2 - 12x - y^2$$

and identify their natures.

5.15   Find the stationary values of

$$f(x, y) = 4x^2 + 4y^2 + x^4 - 6x^2y^2 + y^4$$

and classify them as maxima, minima or saddle points. Make a rough sketch of the contours of f in the quarter plane $x, y \geq 0$.

5.16   The temperature of a point (x, y, z) on the unit sphere is given by

$$T(x, y, z) = 1 + xy + yz.$$

By using the method of Lagrange multipliers, find the temperature of the hottest point on the sphere.

5.17   A rectangular parallelepiped has all eight vertices on the ellipsoid

$$x^2 + 3y^2 + 3z^2 = 1.$$

Using the symmetry of the parallelepiped about each of the planes $x = 0$, $y = 0$, $z = 0$, write down the surface area of the parallelepiped in terms of the coordinates of the vertex that lies in the octant $x, y, z \geq 0$. Hence find the maximum value of the surface area of such a parallelepiped.

5.18   Two horizontal corridors, $0 \leq x \leq a$ with $y \geq 0$, and $0 \leq y \leq b$ with $x \geq 0$, meet at right angles. Find the length L of the longest ladder (considered as a stick) that may be carried horizontally around the corner.

5.19   A barn is to be constructed with a uniform cross-sectional area A throughout its length. The cross-section is to be a rectangle of wall height h (fixed) and width w, surmounted by an isosceles triangular roof that makes an angle θ with

181

the horizontal. The cost of construction is α per unit height of wall and β per unit (slope) length of roof. Show that, irrespective of the values of α and β, to minimise costs w should be chosen to satisfy the equation

$$w^4 = 16A(A - wh),$$

and θ made such that $2 \tan 2\theta = w/h$.

5.20 Show that the envelope of all concentric ellipses that have their axes along the x- and y-coordinate axes, and that have the sum of their semi-axes equal to a constant L, is the same curve (an astroid) as that found in the worked example in section 5.10.

5.21 Find the area of the region covered by points on the lines

$$\frac{x}{a} + \frac{y}{b} = 1,$$

where the sum of any line's intercepts on the coordinate axes is fixed and equal to c.

5.22 Prove that the envelope of the circles whose diameters are those chords of a given circle that pass through a fixed point on its circumference, is the cardioid

$$r = a(1 + \cos \theta).$$

Here a is the radius of the given circle and (r, θ) are the polar coordinates of the envelope. Take as the system parameter the angle ϕ between a chord and the polar axis from which θ is measured.

5.23 A water feature contains a spray head at water level at the centre of a round basin. The head is in the form of a small hemisphere perforated by many evenly distributed small holes, through which water spurts out at the same speed, v_0, in all directions.

(a) What is the shape of the 'water bell' so formed?

(b) What must be the minimum diameter of the bowl if no water is to be lost?

5.24 In order to make a focussing mirror that concentrates parallel axial rays to one spot (or conversely forms a parallel beam from a point source), a parabolic shape should be adopted. If a mirror that is part of a circular cylinder or sphere were used, the light would be spread out along a curve. This curve is known as a *caustic* and is the envelope of the rays reflected from the mirror. Denoting by θ the angle which a typical incident axial ray makes with the normal to the mirror at the place where it is reflected, the geometry of reflection (the angle of incidence equals the angle of reflection) is shown in figure 5.5.

Show that a parametric specification of the caustic is

$$x = R \cos \theta \left(\tfrac{1}{2} + \sin^2 \theta \right), \qquad y = R \sin^3 \theta,$$

where R is the radius of curvature of the mirror. The curve is, in fact, part of an epicycloid.

5.25 By considering the differential

$$dG = d(U + PV - ST),$$

where G is the Gibbs free energy, P the pressure, V the volume, S the entropy and T the temperature of a system, and given further that the internal energy U satisfies

$$dU = T \, dS - P \, dV,$$

derive a Maxwell relation connecting $(\partial V / \partial T)_P$ and $(\partial S / \partial P)_T$.

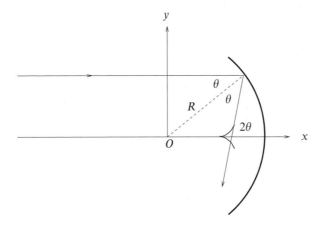

Figure 5.5   The reflecting mirror discussed in exercise 5.24.

5.26   Functions $P(V, T)$, $U(V, T)$ and $S(V, T)$ are related by

$$T \, dS = dU + P \, dV,$$

where the symbols have the same meaning as in the previous question. The pressure P is known from experiment to have the form

$$P = \frac{T^4}{3} + \frac{T}{V},$$

in appropriate units. If

$$U = \alpha V T^4 + \beta T,$$

where α, β, are constants (or, at least, do not depend on T or V), deduce that α must have a specific value, but that β may have any value. Find the corresponding form of S.

5.27   As in the previous two exercises on the thermodynamics of a simple gas, the quantity $dS = T^{-1}(dU + P \, dV)$ is an exact differential. Use this to prove that

$$\left(\frac{\partial U}{\partial V} \right)_T = T \left(\frac{\partial P}{\partial T} \right)_V - P.$$

In the van der Waals model of a gas, P obeys the equation

$$P = \frac{RT}{V - b} - \frac{a}{V^2},$$

where R, a and b are constants. Further, in the limit $V \to \infty$, the form of U becomes $U = cT$, where c is another constant. Find the complete expression for $U(V, T)$.

5.28   The entropy $S(H, T)$, the magnetisation $M(H, T)$ and the internal energy $U(H, T)$ of a magnetic salt placed in a magnetic field of strength H, at temperature T, are connected by the equation

$$T \, dS = dU - H \, dM.$$

183

By considering $d(U - TS - HM)$ prove that

$$\left(\frac{\partial M}{\partial T}\right)_H = \left(\frac{\partial S}{\partial H}\right)_T.$$

For a particular salt,

$$M(H, T) = M_0[1 - \exp(-\alpha H/T)].$$

Show that if, at a fixed temperature, the applied field is increased from zero to a strength such that the magnetization of the salt is $\frac{3}{4}M_0$, then the salt's entropy *decreases* by an amount

$$\frac{M_0}{4\alpha}(3 - \ln 4).$$

5.29   Using the results of section 5.12, evaluate the integral

$$I(y) = \int_0^\infty \frac{e^{-xy}\sin x}{x}\, dx.$$

Hence show that

$$J = \int_0^\infty \frac{\sin x}{x}\, dx = \frac{\pi}{2}.$$

5.30   The integral

$$\int_{-\infty}^\infty e^{-\alpha x^2}\, dx$$

has the value $(\pi/\alpha)^{1/2}$. Use this result to evaluate

$$J(n) = \int_{-\infty}^\infty x^{2n} e^{-x^2}\, dx,$$

where n is a positive integer. Express your answer in terms of factorials.

5.31   The function $f(x)$ is differentiable and $f(0) = 0$. A second function $g(y)$ is defined by

$$g(y) = \int_0^y \frac{f(x)\, dx}{\sqrt{y - x}}.$$

Prove that

$$\frac{dg}{dy} = \int_0^y \frac{df}{dx} \frac{dx}{\sqrt{y - x}}.$$

For the case $f(x) = x^n$, prove that

$$\frac{d^n g}{dy^n} = 2(n!)\sqrt{y}.$$

5.32   The functions $f(x, t)$ and $F(x)$ are defined by

$$f(x, t) = e^{-xt},$$
$$F(x) = \int_0^x f(x, t)\, dt.$$

Verify, by explicit calculation, that

$$\frac{dF}{dx} = f(x, x) + \int_0^x \frac{\partial f(x, t)}{\partial x}\, dt.$$

184

5.33   If

$$I(\alpha) = \int_0^1 \frac{x^\alpha - 1}{\ln x} dx, \qquad \alpha > -1,$$

what is the value of $I(0)$? Show that

$$\frac{d}{d\alpha} x^\alpha = x^\alpha \ln x,$$

and deduce that

$$\frac{d}{d\alpha} I(\alpha) = \frac{1}{\alpha + 1}.$$

Hence prove that $I(\alpha) = \ln(1 + \alpha)$.

5.34   Find the derivative, with respect to x, of the integral

$$I(x) = \int_x^{3x} \exp xt \, dt.$$

5.35   The function $G(t, \xi)$ is defined for $0 \le t \le \pi$ by

$$G(t, \xi) = \begin{cases} -\cos t \sin \xi & \text{for } \xi \le t, \\ \sin t \cos \xi & \text{for } \xi > t. \end{cases}$$

Show that the function $x(t)$ defined by

$$x(t) = \int_0^\pi G(t, \xi) f(\xi) \, d\xi$$

satisfies the equation

$$\frac{d^2 x}{dt^2} + x = f(t),$$

where $f(t)$ can be *any* arbitrary (continuous) function. Show further that $x(0) = [dx/dt]_{t=\pi} = 0$, again for any $f(t)$, but that the *value* of $x(\pi)$ does depend upon the form of $f(t)$.

[The function $G(t, \xi)$ is an example of a Green's function, an important concept in the solution of differential equations and one studied extensively in later chapters.]

5.14 Hints and answers

5.1   (a) (i) $2xy, x^2$; (ii) $2x, 2y$; (iii) $y^{-1} \cos(x/y), (-x/y^2) \cos(x/y)$;
          (iv) $-y/(x^2 + y^2), x/(x^2 + y^2)$; (v) $x/r, y/r, z/r$.
       (b) (i) $2y, 0, 2x$; (ii) $2, 2, 0$; (v) $(y^2 + z^2)r^{-3}, (x^2 + z^2)r^{-3}, -xyr^{-3}$.
       (c) Both second derivatives are equal to $(y^2 - x^2)(x^2 + y^2)^{-2}$.

5.3 $2x \ne -2y - x$. For g, both sides of equation (5.9) equal y^{-2}.

5.5 $\partial^2 z/\partial x^2 = 2xz(z^2 + x)^{-3}, \partial^2 z/\partial x \partial y = (z^2 - x)(z^2 + x)^{-3}, \partial^2 z/\partial y^2 = -2z(z^2 + x)^{-3}$.

5.7 $(0, 0)$, $(a/4, -a)$ and $(16a, -8a)$. Only the saddle point at $(0, 0)$.

5.9   The transformed equation is $2(x^2 + y^2)\partial f/\partial v = 0$; hence f does not depend on v.

5.11   Maxima, equal to $1/8$, at $\pm(1/2, -1/2)$, minima, equal to $-1/8$, at $\pm(1/2, 1/2)$, saddle points, equalling 0, at $(0, 0), (0, \pm 1), (\pm 1, 0)$.

5.13   Maxima equal to $a^2 e^{-1}$ at $(\pm a, 0)$, minima equal to $-2a^2 e^{-1}$ at $(0, \pm a)$, saddle point equalling 0 at $(0, 0)$.

5.15   Minimum at $(0, 0)$; saddle points at $(\pm 1, \pm 1)$. To help with sketching the contours, determine the behaviour of $g(x) = f(x, x)$.

5.17   The Lagrange multiplier method gives $z = y = x/2$, for a maximal area of 4.

5.19   The cost always includes $2\alpha h$, which can therefore be ignored in the optimisation. With Lagrange multiplier λ, $\sin\theta = \lambda w/(4\beta)$ and $\beta\sec\theta - \frac{1}{2}\lambda w\tan\theta = \lambda h$, leading to the stated results.

5.21   The envelope of the lines $x/a + y/(c-a) - 1 = 0$, as a is varied, is $\sqrt{x} + \sqrt{y} = \sqrt{c}$. Area $= c^2/6$.

5.23   (a) Using $\alpha = \cot\theta$, where θ is the initial angle a jet makes with the vertical, the equation is $f(z, \rho, \alpha) = z - \rho\alpha + [g\rho^2(1+\alpha^2)/(2v_0^2)]$, and setting $\partial f/\partial\alpha = 0$ gives $\alpha = v_0^2/(g\rho)$. The water bell has a parabolic profile $z = v_0^2/(2g) - g\rho^2/(2v_0^2)$.
        (b) Setting $z = 0$ gives the minimum diameter as $2v_0^2/g$.

5.25   Show that $(\partial G/\partial P)_T = V$ and $(\partial G/\partial T)_P = -S$. From each result, obtain an expression for $\partial^2 G/\partial T\partial P$ and equate these, giving $(\partial V/\partial T)_P = -(\partial S/\partial P)_T$.

5.27   Find expressions for $(\partial S/\partial V)_T$ and $(\partial S/\partial T)_V$, and equate $\partial^2 S/\partial V\partial T$ with $\partial^2 S/\partial T\partial V$. $U(V, T) = cT - aV^{-1}$.

5.29 $dI/dy = -\mathrm{Im}[\int_0^\infty \exp(-xy + ix)\,dx] = -1/(1+y^2)$. Integrate dI/dy from 0 to ∞. $I(\infty) = 0$ and $I(0) = J$.

5.31   Integrate the RHS of the equation by parts, before differentiating with respect to y. Repeated application of the method establishes the result for all orders of derivative.

5.33 $I(0) = 0$; use Leibnitz' rule.

5.35   Write $x(t) = -\cos t \int_0^t \sin\xi\, f(\xi)\,d\xi - \sin t \int_t^\pi \cos\xi\, f(\xi)\,d\xi$ and differentiate each term as a product to obtain dx/dt. Obtain d^2x/dt^2 in a similar way. Note that integrals that have equal lower and upper limits have value zero. The value of $x(\pi)$ is $\int_0^\pi \sin\xi\, f(\xi)\,d\xi$.

186

6

Multiple integrals

For functions of several variables, just as we may consider derivatives with respect to two or more of them, so may the integral of the function with respect to more than one variable be formed. The formal definitions of such multiple integrals are extensions of that for a single variable, discussed in chapter 2. We first discuss double and triple integrals and illustrate some of their applications. We then consider changing the variables in multiple integrals and discuss some general properties of Jacobians.

6.1 Double integrals

For an integral involving two variables – a double integral – we have a function, $f(x, y)$ say, to be integrated with respect to x and y between certain limits. These limits can usually be represented by a closed curve C bounding a region R in the xy-plane. Following the discussion of single integrals given in chapter 2, let us divide the region R into N subregions ΔR_p of area ΔA_p, $p = 1, 2, \ldots, N$, and let (x_p, y_p) be any point in subregion ΔR_p. Now consider the sum

$$S = \sum_{p-1}^{N} f(x_p, y_p) \Delta A_p,$$

and let $N \to \infty$ as each of the areas $\Delta A_p \to 0$. If the sum S tends to a unique limit, I, then this is called the *double integral of $f(x, y)$ over the region R* and is written

$$I = \int_R f(x, y) \, dA, \tag{6.1}$$

where dA stands for the element of area in the xy-plane. By choosing the subregions to be small rectangles each of area $\Delta A = \Delta x \Delta y$, and letting both Δx

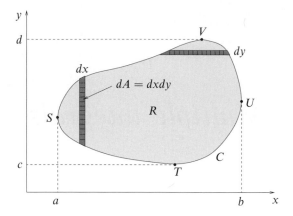

Figure 6.1   A simple curve C in the xy-plane, enclosing a region R.

and $\Delta y \to 0$, we can also write the integral as

$$I = \iint_R f(x, y)\, dx\, dy, \tag{6.2}$$

where we have written out the element of area explicitly as the product of the two coordinate differentials (see figure 6.1).

Some authors use a single integration symbol whatever the dimension of the integral; others use as many symbols as the dimension. In different circumstances both have their advantages. We will adopt the convention used in (6.1) and (6.2), that as many integration symbols will be used as differentials *explicitly* written.

The form (6.2) gives us a clue as to how we may proceed in the evaluation of a double integral. Referring to figure 6.1, the limits on the integration may be written as an equation $c(x, y) = 0$ giving the boundary curve C. However, an explicit statement of the limits can be written in two distinct ways.

One way of evaluating the integral is first to sum up the contributions from the small rectangular elemental areas in a horizontal strip of width dy (as shown in the figure) and then to combine the contributions of these horizontal strips to cover the region R. In this case, we write

$$I = \int_{y=c}^{y=d} \left\{ \int_{x=x_1(y)}^{x=x_2(y)} f(x, y)\, dx \right\} dy, \tag{6.3}$$

where $x = x_1(y)$ and $x = x_2(y)$ are the equations of the curves TSV and TUV respectively. This expression indicates that first $f(x, y)$ is to be integrated with respect to x (treating y as a constant) between the values $x = x_1(y)$ and $x = x_2(y)$ and then the result, considered as a function of y, is to be integrated between the limits $y = c$ and $y = d$. Thus the double integral is evaluated by expressing it in terms of two single integrals called *iterated* (or *repeated*) integrals.

An alternative way of evaluating the integral, however, is first to sum up the contributions from the elemental rectangles arranged into *vertical* strips and then to combine these vertical strips to cover the region R. We then write

$$I = \int_{x=a}^{x=b} \left\{ \int_{y=y_1(x)}^{y=y_2(x)} f(x,y)\,dy \right\} dx, \tag{6.4}$$

where $y = y_1(x)$ and $y = y_2(x)$ are the equations of the curves STU and SVU respectively. In going to (6.4) from (6.3), we have essentially interchanged the order of integration.

In the discussion above we assumed that the curve C was such that any line parallel to either the x- or y-axis intersected C at most twice. In general, provided $f(x,y)$ is continuous everywhere in R and the boundary curve C has this simple shape, the same result is obtained irrespective of the order of integration. In cases where the region R has a more complicated shape, it can usually be subdivided into smaller simpler regions R_1, R_2 etc. that satisfy this criterion. The double integral over R is then merely the sum of the double integrals over the subregions.

▶*Evaluate the double integral*

$$I = \iint_R x^2 y\,dx\,dy,$$

where R is the triangular area bounded by the lines $x = 0$, $y = 0$ and $x + y = 1$. Reverse the order of integration and demonstrate that the same result is obtained.

The area of integration is shown in figure 6.2. Suppose we choose to carry out the integration with respect to y first. With x fixed, the range of y is 0 to $1 - x$. We can therefore write

$$I = \int_{x=0}^{x=1} \left\{ \int_{y=0}^{y=1-x} x^2 y\,dy \right\} dx$$

$$= \int_{x=0}^{x=1} \left[\frac{x^2 y^2}{2} \right]_{y=0}^{y=1-x} dx = \int_0^1 \frac{x^2(1-x)^2}{2}\,dx = \frac{1}{60}.$$

Alternatively, we may choose to perform the integration with respect to x first. With y fixed, the range of x is 0 to $1 - y$, so we have

$$I = \int_{y=0}^{y=1} \left\{ \int_{x=0}^{x=1-y} x^2 y\,dx \right\} dy$$

$$= \int_{y=0}^{y=1} \left[\frac{x^3 y}{3} \right]_{x=0}^{x=1-y} dx = \int_0^1 \frac{(1-y)^3 y}{3}\,dy = \frac{1}{60}.$$

As expected, we obtain the same result irrespective of the order of integration. ◀

We may avoid the use of braces in expressions such as (6.3) and (6.4) by writing (6.4), for example, as

$$I = \int_a^b dx \int_{y_1(x)}^{y_2(x)} dy\, f(x,y),$$

where it is understood that each integral symbol acts on everything to its right,

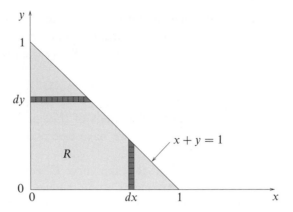

Figure 6.2   The triangular region whose sides are the axes $x = 0$, $y = 0$ and the line $x + y = 1$.

and that the order of integration is from right to left. So, in this example, the integrand $f(x, y)$ is first to be integrated with respect to y and then with respect to x. With the double integral expressed in this way, we will no longer write the independent variables explicitly in the limits of integration, since the differential of the variable with respect to which we are integrating is always adjacent to the relevant integral sign.

Using the order of integration in (6.3), we could also write the double integral as

$$I = \int_c^d dy \int_{x_1(y)}^{x_2(y)} dx\, f(x, y).$$

Occasionally, however, interchange of the order of integration in a double integral is not permissible, as it yields a different result. For example, difficulties might arise if the region R were unbounded with some of the limits infinite, though in many cases involving infinite limits the same result is obtained whichever order of integration is used. Difficulties can also occur if the integrand $f(x, y)$ has any discontinuities in the region R or on its boundary C.

6.2 Triple integrals

The above discussion for double integrals can easily be extended to triple integrals. Consider the function $f(x, y, z)$ defined in a closed three-dimensional region R. Proceeding as we did for double integrals, let us divide the region R into N subregions ΔR_p of volume ΔV_p, $p = 1, 2, \ldots, N$, and let (x_p, y_p, z_p) be any point in the subregion ΔR_p. Now we form the sum

$$S = \sum_{p=1}^N f(x_p, y_p, z_p)\Delta V_p,$$

and let $N \to \infty$ as each of the volumes $\Delta V_p \to 0$. If the sum S tends to a unique limit, I, then this is called the *triple integral of $f(x, y, z)$ over the region R* and is written

$$I = \int_R f(x, y, z) \, dV, \tag{6.5}$$

where dV stands for the element of volume. By choosing the subregions to be small cuboids, each of volume $\Delta V = \Delta x \Delta y \Delta z$, and proceeding to the limit, we can also write the integral as

$$I = \iiint_R f(x, y, z) \, dx \, dy \, dz, \tag{6.6}$$

where we have written out the element of volume explicitly as the product of the three coordinate differentials. Extending the notation used for double integrals, we may write triple integrals as three iterated integrals, for example,

$$I = \int_{x_1}^{x_2} dx \int_{y_1(x)}^{y_2(x)} dy \int_{z_1(x,y)}^{z_2(x,y)} dz \, f(x, y, z),$$

where the limits on each of the integrals describe the values that x, y and z take on the boundary of the region R. As for double integrals, in most cases the order of integration does not affect the value of the integral.

We can extend these ideas to define multiple integrals of higher dimensionality in a similar way.

6.3 Applications of multiple integrals

Multiple integrals have many uses in the physical sciences, since there are numerous physical quantities which can be written in terms of them. We now discuss a few of the more common examples.

6.3.1 Areas and volumes

Multiple integrals are often used in finding areas and volumes. For example, the integral

$$A = \int_R dA = \iint_R dx \, dy$$

is simply equal to the area of the region R. Similarly, if we consider the surface $z = f(x, y)$ in three-dimensional Cartesian coordinates then the volume under this surface that stands vertically above the region R is given by the integral

$$V = \int_R z \, dA = \iint_R f(x, y) \, dx \, dy,$$

where volumes above the xy-plane are counted as positive, and those below as negative.

191

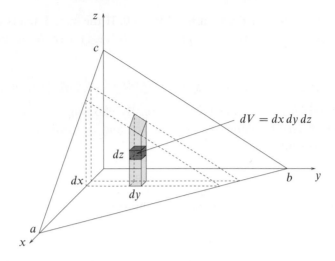

Figure 6.3   The tetrahedron bounded by the coordinate surfaces and the plane $x/a + y/b + z/c = 1$ is divided up into vertical slabs, the slabs into columns and the columns into small boxes.

►*Find the volume of the tetrahedron bounded by the three coordinate surfaces $x = 0$, $y = 0$ and $z = 0$ and the plane $x/a + y/b + z/c = 1$.*

Referring to figure 6.3, the elemental volume of the shaded region is given by $dV = z\,dx\,dy$, and we must integrate over the triangular region R in the xy-plane whose sides are $x = 0$, $y = 0$ and $y = b - bx/a$. The total volume of the tetrahedron is therefore given by

$$V = \iint_R z\,dx\,dy = \int_0^a dx \int_0^{b-bx/a} dy\, c\left(1 - \frac{y}{b} - \frac{x}{a}\right)$$

$$= c \int_0^a dx \left[y - \frac{y^2}{2b} - \frac{xy}{a} \right]_{y=0}^{y=b-bx/a}$$

$$= c \int_0^a dx \left(\frac{bx^2}{2a^2} - \frac{bx}{a} + \frac{b}{2} \right) = \frac{abc}{6}. \;\blacktriangleleft$$

Alternatively, we can write the volume of a three-dimensional region R as

$$V = \int_R dV = \iiint_R dx\,dy\,dz, \tag{6.7}$$

where the only difficulty occurs in setting the correct limits on each of the integrals. For the above example, writing the volume in this way corresponds to dividing the tetrahedron into elemental boxes of volume $dx\,dy\,dz$ (as shown in figure 6.3); integration over z then adds up the boxes to form the shaded column in the figure. The limits of integration are $z = 0$ to $z = c\left(1 - y/b - x/a\right)$, and

192

the total volume of the tetrahedron is given by

$$V = \int_0^a dx \int_0^{b-bx/a} dy \int_0^{c\left(1-y/b-x/a\right)} dz, \qquad (6.8)$$

which clearly gives the same result as above. This method is illustrated further in the following example.

> ▶Find the volume of the region bounded by the paraboloid $z = x^2 + y^2$ and the plane $z = 2y$.

The required region is shown in figure 6.4. In order to write the volume of the region in the form (6.7), we must deduce the limits on each of the integrals. Since the integrations can be performed in any order, let us first divide the region into vertical slabs of thickness dy perpendicular to the y-axis, and then as shown in the figure we cut each slab into horizontal strips of height dz, and each strip into elemental boxes of volume $dV = dx\,dy\,dz$. Integrating first with respect to x (adding up the elemental boxes to get a horizontal strip), the limits on x are $x = -\sqrt{z-y^2}$ to $x = \sqrt{z-y^2}$. Now integrating with respect to z (adding up the strips to form a vertical slab) the limits on z are $z = y^2$ to $z = 2y$. Finally, integrating with respect to y (adding up the slabs to obtain the required region), the limits on y are $y = 0$ and $y = 2$, the solutions of the simultaneous equations $z = 0^2 + y^2$ and $z = 2y$. So the volume of the region is

$$V = \int_0^2 dy \int_{y^2}^{2y} dz \int_{-\sqrt{z-y^2}}^{\sqrt{z-y^2}} dx = \int_0^2 dy \int_{y^2}^{2y} dz\, 2\sqrt{z-y^2}$$

$$= \int_0^2 dy\, \left[\tfrac{4}{3}(z-y^2)^{3/2}\right]_{z=y^2}^{z=2y} = \int_0^2 dy\, \tfrac{4}{3}(2y-y^2)^{3/2}.$$

The integral over y may be evaluated straightforwardly by making the substitution $y = 1 + \sin u$, and gives $V = \pi/2$. ◀

In general, when calculating the volume (area) of a region, the volume (area) elements need not be small boxes as in the previous example, but may be of any convenient shape. The latter is usually chosen to make evaluation of the integral as simple as possible.

6.3.2 Masses, centres of mass and centroids

It is sometimes necessary to calculate the mass of a given object having a non-uniform density. Symbolically, this mass is given simply by

$$M = \int dM,$$

where dM is the element of mass and the integral is taken over the extent of the object. For a solid three-dimensional body the element of mass is just $dM = \rho\,dV$, where dV is an element of volume and ρ is the variable density. For a laminar body (i.e. a uniform sheet of material) the element of mass is $dM = \sigma\,dA$, where σ is the mass per unit area of the body and dA is an area element. Finally, for a body in the form of a thin wire we have $dM = \lambda\,ds$, where λ is the mass per

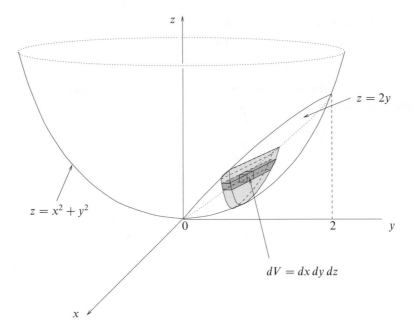

Figure 6.4 The region bounded by the paraboloid $z = x^2 + y^2$ and the plane $z = 2y$ is divided into vertical slabs, the slabs into horizontal strips and the strips into boxes.

unit length and ds is an element of arc length along the wire. When evaluating the required integral, we are free to divide up the body into mass elements in the most convenient way, provided that over each mass element the density is approximately constant.

> ▶ *Find the mass of the tetrahedron bounded by the three coordinate surfaces and the plane* $x/a + y/b + z/c = 1$, *if its density is given by* $\rho(x, y, z) = \rho_0(1 + x/a)$.

From (6.8), we can immediately write down the mass of the tetrahedron as

$$M = \int_R \rho_0 \left(1 + \frac{x}{a}\right) dV = \int_0^a dx\, \rho_0 \left(1 + \frac{x}{a}\right) \int_0^{b-bx/a} dy \int_0^{c(1-y/b-x/a)} dz,$$

where we have taken the density outside the integrations with respect to z and y since it depends only on x. Therefore the integrations with respect to z and y proceed exactly as they did when finding the volume of the tetrahedron, and we have

$$M = c\rho_0 \int_0^a dx \left(1 + \frac{x}{a}\right) \left(\frac{bx^2}{2a^2} - \frac{bx}{a} + \frac{b}{2}\right). \tag{6.9}$$

We could have arrived at (6.9) more directly by dividing the tetrahedron into triangular slabs of thickness dx perpendicular to the x-axis (see figure 6.3), each of which is of constant density, since ρ depends on x alone. A slab at a position x has volume $dV = \frac{1}{2}c(1 - x/a)(b - bx/a)\,dx$ and mass $dM = \rho\,dV = \rho_0(1 + x/a)\,dV$. Integrating over x we again obtain (6.9). This integral is easily evaluated and gives $M = \frac{5}{24}abc\rho_0$. ◀

194

The coordinates of the centre of mass of a solid or laminar body may also be written as multiple integrals. The centre of mass of a body has coordinates $\bar{x}$, $\bar{y}$, $\bar{z}$ given by the three equations

$$\bar{x}\int dM = \int x\,dM$$

$$\bar{y}\int dM = \int y\,dM$$

$$\bar{z}\int dM = \int z\,dM,$$

where again dM is an element of mass as described above, x, y, z are the coordinates of the centre of mass of the element dM and the integrals are taken over the entire body. Obviously, for any body that lies entirely in, or is symmetrical about, the xy-plane (say), we immediately have $\bar{z} = 0$. For completeness, we note that the three equations above can be written as the single vector equation (see chapter 7)

$$\bar{\mathbf{r}} = \frac{1}{M}\int \mathbf{r}\,dM,$$

where $\bar{\mathbf{r}}$ is the position vector of the body's centre of mass with respect to the origin, $\mathbf{r}$ is the position vector of the centre of mass of the element dM and $M = \int dM$ is the total mass of the body. As previously, we may divide the body into the most convenient mass elements for evaluating the necessary integrals, provided each mass element is of constant density.

We further note that the coordinates of the *centroid* of a body are defined as those of its centre of mass if the body had uniform density.

> ▶*Find the centre of mass of the solid hemisphere bounded by the surfaces $x^2 + y^2 + z^2 = a^2$ and the xy-plane, assuming that it has a uniform density ρ.*

Referring to figure 6.5, we know from symmetry that the centre of mass must lie on the z-axis. Let us divide the hemisphere into volume elements that are circular slabs of thickness dz parallel to the xy-plane. For a slab at a height z, the mass of the element is $dM = \rho\,dV = \rho\pi(a^2 - z^2)\,dz$. Integrating over z, we find that the z-coordinate of the centre of mass of the hemisphere is given by

$$\bar{z}\int_0^a \rho\pi(a^2 - z^2)\,dz = \int_0^a z\rho\pi(a^2 - z^2)\,dz.$$

The integrals are easily evaluated and give $\bar{z} = 3a/8$. Since the hemisphere is of uniform density, this is also the position of its centroid. ◀

6.3.3 Pappus' theorems

The theorems of Pappus (which are about seventeen centuries old) relate centroids to volumes of revolution and areas of surfaces, discussed in chapter 2, and may be useful for finding one quantity given another that can be calculated more easily.

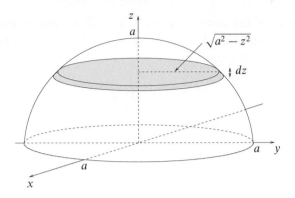

Figure 6.5   The solid hemisphere bounded by the surfaces $x^2 + y^2 + z^2 = a^2$ and the xy-plane.

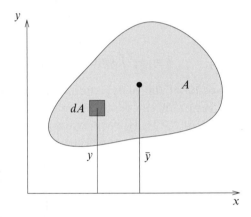

Figure 6.6   An area A in the xy-plane, which may be rotated about the x-axis to form a volume of revolution.

If a plane area is rotated about an axis that does not intersect it then the solid so generated is called a *volume of revolution*. *Pappus' first theorem* states that the volume of such a solid is given by the plane area A multiplied by the distance moved by its centroid (see figure 6.6). This may be proved by considering the definition of the centroid of the plane area as the position of the centre of mass if the density is uniform, so that

$$\bar{y} = \frac{1}{A} \int y \, dA.$$

Now the volume generated by rotating the plane area about the x-axis is given by

$$V = \int 2\pi y \, dA = 2\pi \bar{y} A,$$

which is the area multiplied by the distance moved by the centroid.

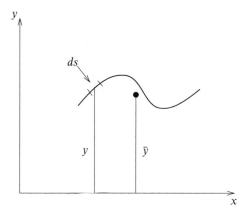

Figure 6.7 A curve in the xy-plane, which may be rotated about the x-axis to form a surface of revolution.

Pappus' second theorem states that if a plane curve is rotated about a coplanar axis that does not intersect it then the area of the *surface of revolution* so generated is given by the length of the curve L multiplied by the distance moved by its centroid (see figure 6.7). This may be proved in a similar manner to the first theorem by considering the definition of the centroid of a plane curve,

$$\bar{y} = \frac{1}{L} \int y \, ds,$$

and noting that the surface area generated is given by

$$S = \int 2\pi y \, ds = 2\pi \bar{y} L,$$

which is equal to the length of the curve multiplied by the distance moved by its centroid.

▶ *A semicircular uniform lamina is freely suspended from one of its corners. Show that its straight edge makes an angle of 23.0° with the vertical.*

Referring to figure 6.8, the suspended lamina will have its centre of gravity C vertically below the suspension point and its straight edge will make an angle $\theta = \tan^{-1}(d/a)$ with the vertical, where $2a$ is the diameter of the semicircle and d is the distance of its centre of mass from the diameter.

Since rotating the lamina about the diameter generates a sphere of volume $\frac{4}{3}\pi a^3$, Pappus' first theorem requires that

$$\tfrac{4}{3}\pi a^3 = 2\pi d \times \tfrac{1}{2}\pi a^2.$$

Hence $d = \frac{4a}{3\pi}$ and $\theta = \tan^{-1}\frac{4}{3\pi} = 23.0°$. ◀

197

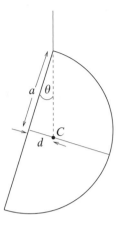

Figure 6.8   Suspending a semicircular lamina from one of its corners.

6.3.4 Moments of inertia

For problems in rotational mechanics it is often necessary to calculate the moment of inertia of a body about a given axis. This is defined by the multiple integral

$$I = \int l^2 \, dM,$$

where l is the distance of a mass element dM from the axis. We may again choose mass elements convenient for evaluating the integral. In this case, however, in addition to elements of constant density we require all parts of each element to be at approximately the same distance from the axis about which the moment of inertia is required.

> ▶ *Find the moment of inertia of a uniform rectangular lamina of mass M with sides a and b about one of the sides of length b.*

Referring to figure 6.9, we wish to calculate the moment of inertia about the y-axis. We therefore divide the rectangular lamina into elemental strips parallel to the y-axis of width dx. The mass of such a strip is $dM = \sigma b \, dx$, where σ is the mass per unit area of the lamina. The moment of inertia of a strip at a distance x from the y-axis is simply $dI = x^2 \, dM = \sigma b x^2 \, dx$. The total moment of inertia of the lamina about the y-axis is therefore

$$I = \int_0^a \sigma b x^2 \, dx = \frac{\sigma b a^3}{3}.$$

Since the total mass of the lamina is $M = \sigma a b$, we can write $I = \frac{1}{3} M a^2$. ◀

198

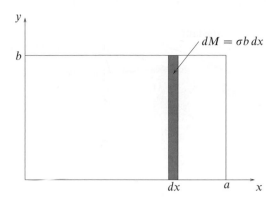

Figure 6.9   A uniform rectangular lamina of mass M with sides a and b can be divided into vertical strips.

6.3.5 Mean values of functions

In chapter 2 we discussed average values for functions of a single variable. This is easily extended to functions of several variables. Let us consider, for example, a function $f(x, y)$ defined in some region R of the xy-plane. Then the average value $\bar{f}$ of the function is given by

$$\bar{f} \int_R dA = \int_R f(x, y) \, dA. \tag{6.10}$$

This definition is easily extended to three (and higher) dimensions; if a function $f(x, y, z)$ is defined in some three-dimensional region of space R then the average value $\bar{f}$ of the function is given by

$$\bar{f} \int_R dV = \int_R f(x, y, z) \, dV. \tag{6.11}$$

▶ *A tetrahedron is bounded by the three coordinate surfaces and the plane $x/a + y/b + z/c = 1$ and has density $\rho(x, y, z) = \rho_0(1 + x/a)$. Find the average value of the density.*

From (6.11), the average value of the density is given by

$$\bar{\rho} \int_R dV = \int_R \rho(x, y, z) \, dV.$$

Now the integral on the LHS is just the volume of the tetrahedron, which we found in subsection 6.3.1 to be $V = \frac{1}{6}abc$, and the integral on the RHS is its mass $M = \frac{5}{24}abc\rho_0$, calculated in subsection 6.3.2. Therefore $\bar{\rho} = M/V = \frac{5}{4}\rho_0$. ◀

6.4 Change of variables in multiple integrals

It often happens that, either because of the form of the integrand involved or because of the boundary shape of the region of integration, it is desirable to

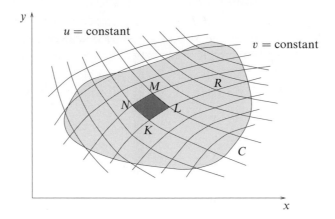

Figure 6.10   A region of integration R overlaid with a grid formed by the family of curves $u = $ constant and $v = $ constant. The parallelogram $KLMN$ defines the area element dA_{uv}.

express a multiple integral in terms of a new set of variables. We now consider how to do this.

6.4.1 Change of variables in double integrals

Let us begin by examining the change of variables in a double integral. Suppose that we require to change an integral

$$I = \iint_R f(x, y)\, dx\, dy,$$

in terms of coordinates x and y, into one expressed in new coordinates u and v, given in terms of x and y by differentiable equations $u = u(x, y)$ and $v = v(x, y)$ with inverses $x = x(u, v)$ and $y = y(u, v)$. The region R in the xy-plane and the curve C that bounds it will become a new region R' and a new boundary C' in the uv-plane, and so we must change the limits of integration accordingly. Also, the function $f(x, y)$ becomes a new function $g(u, v)$ of the new coordinates.

Now the part of the integral that requires most consideration is the area element. In the xy-plane the element is the rectangular area $dA_{xy} = dx\, dy$ generated by constructing a grid of straight lines parallel to the x- and y- axes respectively. Our task is to determine the corresponding area element in the uv-coordinates. In general the corresponding element dA_{uv} will not be the same shape as dA_{xy}, but this does not matter since all elements are infinitesimally small and the value of the integrand is considered constant over them. Since the sides of the area element are infinitesimal, dA_{uv} will in general have the shape of a parallelogram. We can find the connection between dA_{xy} and dA_{uv} by considering the grid formed by the family of curves $u = $ constant and $v = $ constant, as shown in figure 6.10. Since v

is constant along the line element KL, the latter has components $(\partial x/\partial u)\, du$ and $(\partial y/\partial u)\, du$ in the directions of the x- and y-axes respectively. Similarly, since u is constant along the line element KN, the latter has corresponding components $(\partial x/\partial v)\, dv$ and $(\partial y/\partial v)\, dv$. Using the result for the area of a parallelogram given in chapter 7, we find that the area of the parallelogram $KLMN$ is given by

$$dA_{uv} = \left| \frac{\partial x}{\partial u}\, du \frac{\partial y}{\partial v}\, dv - \frac{\partial x}{\partial v}\, dv \frac{\partial y}{\partial u}\, du \right|$$

$$= \left| \frac{\partial x}{\partial u} \frac{\partial y}{\partial v} - \frac{\partial x}{\partial v} \frac{\partial y}{\partial u} \right|\, du\, dv.$$

Defining the *Jacobian* of x, y with respect to u, v as

$$J = \frac{\partial(x, y)}{\partial(u, v)} \equiv \frac{\partial x}{\partial u} \frac{\partial y}{\partial v} - \frac{\partial x}{\partial v} \frac{\partial y}{\partial u},$$

we have

$$dA_{uv} = \left| \frac{\partial(x, y)}{\partial(u, v)} \right|\, du\, dv.$$

The reader acquainted with determinants will notice that the Jacobian can also be written as the 2×2 determinant

$$J = \frac{\partial(x, y)}{\partial(u, v)} = \begin{vmatrix} \dfrac{\partial x}{\partial u} & \dfrac{\partial y}{\partial u} \\[2mm] \dfrac{\partial x}{\partial v} & \dfrac{\partial y}{\partial v} \end{vmatrix}.$$

Such determinants can be evaluated using the methods of chapter 8.

So, in summary, the relationship between the size of the area element generated by dx, dy and the size of the corresponding area element generated by du, dv is

$$dx\, dy = \left| \frac{\partial(x, y)}{\partial(u, v)} \right|\, du\, dv.$$

This equality should be taken as meaning that when transforming from coordinates x, y to coordinates u, v, the area element $dx\, dy$ should be replaced by the expression on the RHS of the above equality. Of course, the Jacobian can, and in general will, vary over the region of integration. We may express the double integral in either coordinate system as

$$I = \iint_R f(x, y)\, dx\, dy = \iint_{R'} g(u, v) \left| \frac{\partial(x, y)}{\partial(u, v)} \right|\, du\, dv. \tag{6.12}$$

When evaluating the integral in the new coordinate system, it is usually advisable to sketch the region of integration R' in the uv-plane.

> ▶*Evaluate the double integral*
> $$I = \iint_R \left(a + \sqrt{x^2 + y^2}\right) \, dx \, dy,$$
> *where R is the region bounded by the circle* $x^2 + y^2 = a^2$.

In Cartesian coordinates, the integral may be written

$$I = \int_{-a}^{a} dx \int_{-\sqrt{a^2-x^2}}^{\sqrt{a^2-x^2}} dy \left(a + \sqrt{x^2 + y^2}\right),$$

and can be calculated directly. However, because of the circular boundary of the integration region, a change of variables to plane polar coordinates ρ, ϕ is indicated. The relationship between Cartesian and plane polar coordinates is given by $x = \rho \cos \phi$ and $y = \rho \sin \phi$. Using (6.12) we can therefore write

$$I = \iint_{R'} (a + \rho) \left| \frac{\partial(x, y)}{\partial(\rho, \phi)} \right| d\rho \, d\phi,$$

where R' is the rectangular region in the $\rho\phi$-plane whose sides are $\rho = 0$, $\rho = a$, $\phi = 0$ and $\phi = 2\pi$. The Jacobian is easily calculated, and we obtain

$$J = \frac{\partial(x, y)}{\partial(\rho, \phi)} = \begin{vmatrix} \cos \phi & \sin \phi \\ -\rho \sin \phi & \rho \cos \phi \end{vmatrix} = \rho(\cos^2 \phi + \sin^2 \phi) = \rho.$$

So the relationship between the area elements in Cartesian and in plane polar coordinates is

$$dx \, dy = \rho \, d\rho \, d\phi.$$

Therefore, when expressed in plane polar coordinates, the integral is given by

$$I = \iint_{R'} (a + \rho)\rho \, d\rho \, d\phi$$
$$= \int_0^{2\pi} d\phi \int_0^a d\rho \, (a + \rho)\rho = 2\pi \left[\frac{a\rho^2}{2} + \frac{\rho^3}{3} \right]_0^a = \frac{5\pi a^3}{3}. \quad ◀$$

6.4.2 Evaluation of the integral $I = \int_{-\infty}^{\infty} e^{-x^2} \, dx$

By making a judicious change of variables, it is sometimes possible to evaluate an integral that would be intractable otherwise. An important example of this method is provided by the evaluation of the integral

$$I = \int_{-\infty}^{\infty} e^{-x^2} \, dx.$$

Its value may be found by first constructing I^2, as follows:

$$I^2 = \int_{-\infty}^{\infty} e^{-x^2} \, dx \int_{-\infty}^{\infty} e^{-y^2} \, dy = \int_{-\infty}^{\infty} dx \int_{-\infty}^{\infty} dy \, e^{-(x^2+y^2)}$$
$$= \iint_R e^{-(x^2+y^2)} \, dx \, dy,$$

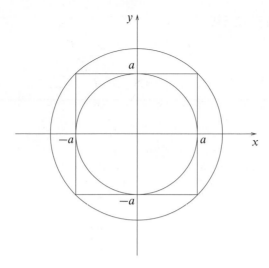

Figure 6.11   The regions used to illustrate the convergence properties of the integral $I(a) = \int_{-a}^{a} e^{-x^2} \, dx$ as $a \to \infty$.

where the region R is the whole xy-plane. Then, transforming to plane polar coordinates, we find

$$I^2 = \iint_{R'} e^{-\rho^2} \rho \, d\rho \, d\phi = \int_0^{2\pi} d\phi \int_0^{\infty} d\rho \, \rho e^{-\rho^2} = 2\pi \left[-\tfrac{1}{2} e^{-\rho^2} \right]_0^{\infty} = \pi.$$

Therefore the original integral is given by $I = \sqrt{\pi}$. Because the integrand is an even function of x, it follows that the value of the integral from 0 to ∞ is simply $\sqrt{\pi}/2$.

We note, however, that, unlike in all the previous examples, the regions of integration R and R' are both infinite in extent (i.e. unbounded). It is therefore prudent to derive this result more rigorously; this we do by considering the integral

$$I(a) = \int_{-a}^{a} e^{-x^2} \, dx.$$

We then have

$$I^2(a) = \iint_R e^{-(x^2+y^2)} \, dx \, dy,$$

where R is the square of side $2a$ centred on the origin. Referring to figure 6.11, since the integrand is always positive the value of the integral taken over the square lies between the value of the integral taken over the region bounded by the inner circle of radius a and the value of the integral taken over the outer circle of radius $\sqrt{2}a$. Transforming to plane polar coordinates as above, we may

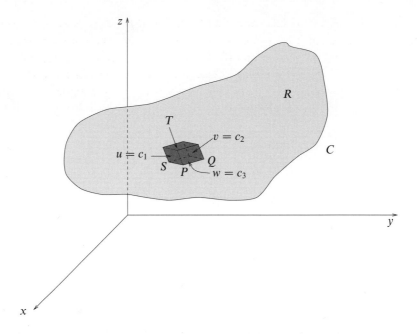

Figure 6.12   A three-dimensional region of integration R, showing an element of volume in u, v, w coordinates formed by the coordinate surfaces $u = $ constant, $v = $ constant, $w = $ constant.

evaluate the integrals over the inner and outer circles respectively, and we find

$$\pi \left(1 - e^{-a^2}\right) < I^2(a) < \pi \left(1 - e^{-2a^2}\right).$$

Taking the limit $a \to \infty$, we find $I^2(a) \to \pi$. Therefore $I = \sqrt{\pi}$, as we found previously. Substituting $x = \sqrt{\alpha} y$ shows that the corresponding integral of $\exp(-\alpha x^2)$ has the value $\sqrt{\pi/\alpha}$. We use this result in the discussion of the normal distribution in chapter 30.

6.4.3 Change of variables in triple integrals

A change of variable in a triple integral follows the same general lines as that for a double integral. Suppose we wish to change variables from x, y, z to u, v, w. In the x, y, z coordinates the element of volume is a cuboid of sides dx, dy, dz and volume $dV_{xyz} = dx\, dy\, dz$. If, however, we divide up the total volume into infinitesimal elements by constructing a grid formed from the coordinate surfaces $u = $ constant, $v = $ constant and $w = $ constant, then the element of volume dV_{uvw} in the new coordinates will have the shape of a parallelepiped whose faces are the coordinate surfaces and whose edges are the curves formed by the intersections of these surfaces (see figure 6.12). Along the line element PQ the coordinates v and

w are constant, and so PQ has components $(\partial x/\partial u)\, du$, $(\partial y/\partial u)\, du$ and $(\partial z/\partial u)\, du$ in the directions of the x-, y- and z- axes respectively. The components of the line elements PS and ST are found by replacing u by v and w respectively.

The expression for the volume of a parallelepiped in terms of the components of its edges with respect to the x-, y- and z-axes is given in chapter 7. Using this, we find that the element of volume in u, v, w coordinates is given by

$$dV_{uvw} = \left| \frac{\partial(x, y, z)}{\partial(u, v, w)} \right| du\, dv\, dw,$$

where the Jacobian of x, y, z with respect to u, v, w is a short-hand for a 3×3 determinant:

$$\frac{\partial(x, y, z)}{\partial(u, v, w)} \equiv \begin{vmatrix} \dfrac{\partial x}{\partial u} & \dfrac{\partial y}{\partial u} & \dfrac{\partial z}{\partial u} \\[2mm] \dfrac{\partial x}{\partial v} & \dfrac{\partial y}{\partial v} & \dfrac{\partial z}{\partial v} \\[2mm] \dfrac{\partial x}{\partial w} & \dfrac{\partial y}{\partial w} & \dfrac{\partial z}{\partial w} \end{vmatrix}.$$

So, in summary, the relationship between the elemental volumes in multiple integrals formulated in the two coordinate systems is given in Jacobian form by

$$dx\, dy\, dz = \left| \frac{\partial(x, y, z)}{\partial(u, v, w)} \right| du\, dv\, dw,$$

and we can write a triple integral in either set of coordinates as

$$I = \iiint_R f(x, y, z)\, dx\, dy\, dz = \iiint_{R'} g(u, v, w) \left| \frac{\partial(x, y, z)}{\partial(u, v, w)} \right| du\, dv\, dw.$$

▶ *Find an expression for a volume element in spherical polar coordinates, and hence calculate the moment of inertia about a diameter of a uniform sphere of radius a and mass M.*

Spherical polar coordinates r, θ, ϕ are defined by

$$x = r \sin \theta \cos \phi, \quad y = r \sin \theta \sin \phi, \quad z = r \cos \theta$$

(and are discussed fully in chapter 10). The required Jacobian is therefore

$$J = \frac{\partial(x, y, z)}{\partial(r, \theta, \phi)} = \begin{vmatrix} \sin \theta \cos \phi & \sin \theta \sin \phi & \cos \theta \\ r \cos \theta \cos \phi & r \cos \theta \sin \phi & -r \sin \theta \\ -r \sin \theta \sin \phi & r \sin \theta \cos \phi & 0 \end{vmatrix}.$$

The determinant is most easily evaluated by expanding it with respect to the last column (see chapter 8), which gives

$$J = \cos \theta (r^2 \sin \theta \cos \theta) + r \sin \theta (r \sin^2 \theta)$$
$$= r^2 \sin \theta (\cos^2 \theta + \sin^2 \theta) = r^2 \sin \theta.$$

Therefore the volume element in spherical polar coordinates is given by

$$dV = \frac{\partial(x, y, z)}{\partial(r, \theta, \phi)} dr\, d\theta\, d\phi = r^2 \sin \theta\, dr\, d\theta\, d\phi,$$

which agrees with the result given in chapter 10.

If we place the sphere with its centre at the origin of an x, y, z coordinate system then its moment of inertia about the z-axis (which is, of course, a diameter of the sphere) is

$$I = \int \left(x^2 + y^2\right) dM = \rho \int \left(x^2 + y^2\right) dV,$$

where the integral is taken over the sphere, and ρ is the density. Using spherical polar coordinates, we can write this as

$$I = \rho \iiint_V \left(r^2 \sin^2 \theta\right) r^2 \sin \theta \, dr \, d\theta \, d\phi$$

$$= \rho \int_0^{2\pi} d\phi \int_0^{\pi} d\theta \, \sin^3 \theta \int_0^a dr \, r^4$$

$$= \rho \times 2\pi \times \tfrac{4}{3} \times \tfrac{1}{5} a^5 = \tfrac{8}{15} \pi a^5 \rho.$$

Since the mass of the sphere is $M = \tfrac{4}{3} \pi a^3 \rho$, the moment of inertia can also be written as $I = \tfrac{2}{5} Ma^2$. ◄

6.4.4 General properties of Jacobians

Although we will not prove it, the general result for a change of coordinates in an n-dimensional integral from a set x_i to a set y_j (where i and j both run from 1 to n) is

$$dx_1 \, dx_2 \cdots dx_n = \left| \frac{\partial(x_1, x_2, \ldots, x_n)}{\partial(y_1, y_2, \ldots, y_n)} \right| dy_1 \, dy_2 \cdots dy_n,$$

where the n-dimensional Jacobian can be written as an $n \times n$ determinant (see chapter 8) in an analogous way to the two- and three-dimensional cases.

For readers who already have sufficient familiarity with matrices (see chapter 8) and their properties, a fairly compact proof of some useful general properties of Jacobians can be given as follows. Other readers should turn straight to the results (6.16) and (6.17) and return to the proof at some later time.

Consider three sets of variables x_i, y_i and z_i, with i running from 1 to n for each set. From the chain rule in partial differentiation (see (5.17)), we know that

$$\frac{\partial x_i}{\partial z_j} = \sum_{k=1}^{n} \frac{\partial x_i}{\partial y_k} \frac{\partial y_k}{\partial z_j}. \tag{6.13}$$

Now let A, B and C be the matrices whose ijth elements are $\partial x_i / \partial y_j$, $\partial y_i / \partial z_j$ and $\partial x_i / \partial z_j$ respectively. We can then write (6.13) as the matrix product

$$c_{ij} = \sum_{k=1}^{n} a_{ik} b_{kj} \qquad \text{or} \qquad \mathsf{C} = \mathsf{AB}. \tag{6.14}$$

We may now use the general result for the determinant of the product of two matrices, namely $|\mathsf{AB}| = |\mathsf{A}||\mathsf{B}|$, and recall that the Jacobian

$$J_{xy} = \frac{\partial(x_1, \ldots, x_n)}{\partial(y_1, \ldots, y_n)} = |\mathsf{A}|, \tag{6.15}$$

and similarly for J_{yz} and J_{xz}. On taking the determinant of (6.14), we therefore obtain

$$J_{xz} = J_{xy} J_{yz}$$

or, in the usual notation,

$$\frac{\partial(x_1,\ldots,x_n)}{\partial(z_1,\ldots,z_n)} = \frac{\partial(x_1,\ldots,x_n)}{\partial(y_1,\ldots,y_n)} \frac{\partial(y_1,\ldots,y_n)}{\partial(z_1,\ldots,z_n)}. \tag{6.16}$$

As a special case, if the set z_i is taken to be identical to the set x_i, and the obvious result $J_{xx} = 1$ is used, we obtain

$$J_{xy} J_{yx} = 1$$

or, in the usual notation,

$$\frac{\partial(x_1,\ldots,x_n)}{\partial(y_1,\ldots,y_n)} = \left[\frac{\partial(y_1,\ldots,y_n)}{\partial(x_1,\ldots,x_n)}\right]^{-1}. \tag{6.17}$$

The similarity between the properties of Jacobians and those of derivatives is apparent, and to some extent is suggested by the notation. We further note from (6.15) that since $|\mathsf{A}| = |\mathsf{A}^\mathsf{T}|$, where A^T is the transpose of A, we can interchange the rows and columns in the determinantal form of the Jacobian without changing its value.

6.5 Exercises

6.1 Identify the curved wedge bounded by the surfaces $y^2 = 4ax$, $x + z = a$ and $z = 0$, and hence calculate its volume V.

6.2 Evaluate the volume integral of $x^2 + y^2 + z^2$ over the rectangular parallelepiped bounded by the six surfaces $x = \pm a$, $y = \pm b$ and $z = \pm c$.

6.3 Find the volume integral of $x^2 y$ over the tetrahedral volume bounded by the planes $x = 0$, $y = 0$, $z = 0$, and $x + y + z = 1$.

6.4 Evaluate the surface integral of $f(x, y)$ over the rectangle $0 \le x \le a$, $0 \le y \le b$ for the functions

$$\text{(a) } f(x, y) = \frac{x}{x^2 + y^2}, \qquad \text{(b) } f(x, y) = (b - y + x)^{-3/2}.$$

6.5 Calculate the volume of an ellipsoid as follows:

(a) Prove that the area of the ellipse

$$\frac{x^2}{a^2} + \frac{y^2}{b^2} = 1$$

is πab.

(b) Use this result to obtain an expression for the volume of a slice of thickness dz of the ellipsoid

$$\frac{x^2}{a^2} + \frac{y^2}{b^2} + \frac{z^2}{c^2} = 1.$$

Hence show that the volume of the ellipsoid is $4\pi abc/3$.

6.6   The function

$$\Psi(r) = A\left(2 - \frac{Zr}{a}\right)e^{-Zr/2a}$$

gives the form of the quantum-mechanical wavefunction representing the electron in a hydrogen-like atom of atomic number Z, when the electron is in its first allowed spherically symmetric excited state. Here r is the usual spherical polar coordinate, but, because of the spherical symmetry, the coordinates θ and ϕ do not appear explicitly in Ψ. Determine the value that A (assumed real) must have if the wavefunction is to be correctly normalised, i.e. if the volume integral of $|\Psi|^2$ over all space is to be equal to unity.

6.7   In quantum mechanics the electron in a hydrogen atom in some particular state is described by a wavefunction Ψ, which is such that $|\Psi|^2\,dV$ is the probability of finding the electron in the infinitesimal volume dV. In spherical polar coordinates $\Psi = \Psi(r, \theta, \phi)$ and $dV = r^2 \sin\theta\,dr\,d\theta\,d\phi$. Two such states are described by

$$\Psi_1 = \left(\frac{1}{4\pi}\right)^{1/2}\left(\frac{1}{a_0}\right)^{3/2} 2e^{-r/a_0},$$

$$\Psi_2 = -\left(\frac{3}{8\pi}\right)^{1/2}\sin\theta\,e^{i\phi}\left(\frac{1}{2a_0}\right)^{3/2}\frac{re^{-r/2a_0}}{a_0\sqrt{3}}.$$

(a) Show that each Ψ_i is normalised, i.e. the integral over all space $\int |\Psi|^2\,dV$ is equal to unity – physically, this means that the electron must be somewhere.
(b) The (so-called) dipole matrix element between the states 1 and 2 is given by the integral

$$p_x = \int \Psi_1^* qr \sin\theta \cos\phi\,\Psi_2\,dV,$$

where q is the charge on the electron. Prove that p_x has the value $-2^7 qa_0/3^5$.

6.8   A planar figure is formed from uniform wire and consists of two equal semicircular arcs, each with its own closing diameter, joined so as to form a letter 'B'. The figure is freely suspended from its top left-hand corner. Show that the straight edge of the figure makes an angle θ with the vertical given by $\tan\theta = (2 + \pi)^{-1}$.

6.9   A certain torus has a circular vertical cross-section of radius a centred on a horizontal circle of radius $c\ (> a)$.

(a) Find the volume V and surface area A of the torus, and show that they can be written as

$$V = \frac{\pi^2}{4}(r_o^2 - r_i^2)(r_o - r_i), \qquad A = \pi^2(r_o^2 - r_i^2),$$

where r_o and r_i are, respectively, the outer and inner radii of the torus.
(b) Show that a vertical circular cylinder of radius c, coaxial with the torus, divides A in the ratio

$$\pi c + 2a \ : \ \pi c - 2a.$$

6.10   A thin uniform circular disc has mass M and radius a.

(a) Prove that its moment of inertia about an axis perpendicular to its plane and passing through its centre is $\frac{1}{2}Ma^2$.
(b) Prove that the moment of inertia of the same disc about a diameter is $\frac{1}{4}Ma^2$.

This is an example of the general result for planar bodies that the moment of inertia of the body about an axis perpendicular to the plane is equal to the sum of the moments of inertia about two perpendicular axes lying in the plane; in an obvious notation

$$I_z = \int r^2 \, dm = \int (x^2 + y^2) \, dm = \int x^2 \, dm + \int y^2 \, dm = I_y + I_x.$$

6.11 In some applications in mechanics the moment of inertia of a body about a single point (as opposed to about an axis) is needed. The moment of inertia, I, about the origin of a uniform solid body of density ρ is given by the volume integral

$$I = \int_V (x^2 + y^2 + z^2) \rho \, dV.$$

Show that the moment of inertia of a right circular cylinder of radius a, length $2b$ and mass M about its centre is

$$M \left(\frac{a^2}{2} + \frac{b^2}{3} \right).$$

6.12 The shape of an axially symmetric hard-boiled egg, of uniform density ρ_0, is given in spherical polar coordinates by $r = a(2 - \cos\theta)$, where θ is measured from the axis of symmetry.

(a) Prove that the mass M of the egg is $M = \frac{40}{3}\pi\rho_0 a^3$.

(b) Prove that the egg's moment of inertia about its axis of symmetry is $\frac{342}{175} M a^2$.

6.13 In spherical polar coordinates r, θ, ϕ the element of volume for a body that is symmetrical about the polar axis is $dV = 2\pi r^2 \sin\theta \, dr \, d\theta$, whilst its element of surface area is $2\pi r \sin\theta [(dr)^2 + r^2(d\theta)^2]^{1/2}$. A particular surface is defined by $r = 2a \cos\theta$, where a is a constant and $0 \le \theta \le \pi/2$. Find its total surface area and the volume it encloses, and hence identify the surface.

6.14 By expressing both the integrand and the surface element in spherical polar coordinates, show that the surface integral

$$\int \frac{x^2}{x^2 + y^2} \, dS$$

over the surface $x^2 + y^2 = z^2$, $0 \le z \le 1$, has the value $\pi/\sqrt{2}$.

6.15 By transforming to cylindrical polar coordinates, evaluate the integral

$$I = \int \int \int \ln(x^2 + y^2) \, dx \, dy \, dz$$

over the interior of the conical region $x^2 + y^2 \le z^2$, $0 \le z \le 1$.

6.16 Sketch the two families of curves

$$y^2 = 4u(u - x), \qquad y^2 = 4v(v + x),$$

where u and v are parameters.

By transforming to the uv-plane, evaluate the integral of $y/(x^2 + y^2)^{1/2}$ over the part of the quadrant $x > 0$, $y > 0$ that is bounded by the lines $x = 0$, $y = 0$ and the curve $y^2 = 4a(a - x)$.

6.17 By making two successive simple changes of variables, evaluate

$$I = \int \int \int x^2 \, dx \, dy \, dz$$

209

over the ellipsoidal region

$$\frac{x^2}{a^2} + \frac{y^2}{b^2} + \frac{z^2}{c^2} \le 1.$$

6.18 Sketch the domain of integration for the integral

$$I = \int_0^1 \int_{x=y}^{1/y} \frac{y^3}{x} \exp[y^2(x^2 + x^{-2})]\, dx\, dy$$

and characterise its boundaries in terms of new variables $u = xy$ and $v = y/x$. Show that the Jacobian for the change from (x, y) to (u, v) is equal to $(2v)^{-1}$, and hence evaluate I.

6.19 Sketch the part of the region $0 \le x$, $0 \le y \le \pi/2$ that is bounded by the curves $x = 0$, $y = 0$, $\sinh x \cos y = 1$ and $\cosh x \sin y = 1$. By making a suitable change of variables, evaluate the integral

$$I = \int \int (\sinh^2 x + \cos^2 y) \sinh 2x \sin 2y \, dx\, dy$$

over the bounded subregion.

6.20 Define a coordinate system u, v whose origin coincides with that of the usual x, y system and whose u-axis coincides with the x-axis, whilst the v-axis makes an angle α with it. By considering the integral $I = \int \exp(-r^2)\, dA$, where r is the radial distance from the origin, over the area defined by $0 \le u < \infty$, $0 \le v < \infty$, prove that

$$\int_0^\infty \int_0^\infty \exp(-u^2 - v^2 - 2uv \cos \alpha)\, du\, dv = \frac{\alpha}{2 \sin \alpha}.$$

6.21 As stated in section 5.11, the first law of thermodynamics can be expressed as

$$dU = T\, dS - P\, dV.$$

By calculating and equating $\partial^2 U/\partial Y\, \partial X$ and $\partial^2 U/\partial X \partial Y$, where X and Y are an unspecified pair of variables (drawn from P, V, T and S), prove that

$$\frac{\partial(S, T)}{\partial(X, Y)} = \frac{\partial(V, P)}{\partial(X, Y)}.$$

Using the properties of Jacobians, deduce that

$$\frac{\partial(S, T)}{\partial(V, P)} = 1.$$

6.22 The distances of the variable point P, which has coordinates x, y, z, from the fixed points $(0, 0, 1)$ and $(0, 0, -1)$ are denoted by u and v respectively. New variables ξ, η, ϕ are defined by

$$\xi = \tfrac{1}{2}(u + v), \qquad \eta = \tfrac{1}{2}(u - v),$$

and ϕ is the angle between the plane $y = 0$ and the plane containing the three points. Prove that the Jacobian $\partial(\xi, \eta, \phi)/\partial(x, y, z)$ has the value $(\xi^2 - \eta^2)^{-1}$ and that

$$\int \int \int_{\text{all space}} \frac{(u - v)^2}{uv} \exp\left(-\frac{u + v}{2}\right) dx\, dy\, dz = \frac{16\pi}{3e}.$$

6.23 This is a more difficult question about 'volumes' in an increasing number of dimensions.

(a) Let R be a real positive number and define K_m by

$$K_m = \int_{-R}^{R} \left(R^2 - x^2\right)^m dx.$$

Show, using integration by parts, that K_m satisfies the recurrence relation

$$(2m + 1)K_m = 2mR^2 K_{m-1}.$$

(b) For integer n, define $I_n = K_n$ and $J_n = K_{n+1/2}$. Evaluate I_0 and J_0 directly and hence prove that

$$I_n = \frac{2^{2n+1}(n!)^2 R^{2n+1}}{(2n+1)!} \quad \text{and} \quad J_n = \frac{\pi(2n+1)! R^{2n+2}}{2^{2n+1} n!(n+1)!}.$$

(c) A sequence of functions $V_n(R)$ is defined by

$$V_0(R) = 1,$$

$$V_n(R) = \int_{-R}^{R} V_{n-1}\left(\sqrt{R^2 - x^2}\right) dx, \qquad n \geq 1.$$

Prove by induction that

$$V_{2n}(R) = \frac{\pi^n R^{2n}}{n!}, \qquad V_{2n+1}(R) = \frac{\pi^n 2^{2n+1} n! R^{2n+1}}{(2n+1)!}.$$

(d) For interest,

  (i) show that $V_{2n+2}(1) < V_{2n}(1)$ and $V_{2n+1}(1) < V_{2n-1}(1)$ for all $n \geq 3$;
  (ii) hence, by explicitly writing out $V_k(R)$ for $1 \leq k \leq 8$ (say), show that the 'volume' of the totally symmetric solid of unit radius is a maximum in five dimensions.

6.6 Hints and answers

6.1   For integration order z, y, x, the limits are $(0, a - x)$, $(-\sqrt{4ax}, \sqrt{4ax})$ and $(0, a)$. For integration order y, x, z, the limits are $(-\sqrt{4ax}, \sqrt{4ax})$, $(0, a - z)$ and $(0, a)$. $V = 16a^3/15$.

6.3 $1/360$.

6.5   (a) Evaluate $\int 2b[1 - (x/a)^2]^{1/2} dx$ by setting $x = a\cos\phi$;
      (b) $dV = \pi \times a[1 - (z/c)^2]^{1/2} \times b[1 - (z/c)^2]^{1/2} dz$.

6.7   Write $\sin^3 \theta$ as $(1 - \cos^2 \theta)\sin\theta$ when integrating $|\Psi_2|^2$.

6.9   (a) $V = 2\pi c \times \pi a^2$ and $A = 2\pi a \times 2\pi c$. Setting $r_o = c + a$ and $r_i = c - a$ gives the stated results. (b) Show that the centre of gravity of either half is $2a/\pi$ from the cylinder.

6.11 Transform to cylindrical polar coordinates.

6.13 $4\pi a^2$; $4\pi a^3/3$; a sphere.

6.15 The volume element is $\rho\, d\phi\, d\rho\, dz$. The integrand for the final z-integration is given by $2\pi[(z^2 \ln z) - (z^2/2)]$; $I = -5\pi/9$.

6.17 Set $\xi = x/a, \eta = y/b, \zeta = z/c$ to map the ellipsoid onto the unit sphere, and then change from (ξ, η, ζ) coordinates to spherical polar coordinates; $I = 4\pi a^3 bc/15$.

6.19 Set $u = \sinh x \cos y$ and $v = \cosh x \sin y$; $J_{xy,uv} = (\sinh^2 x + \cos^2 y)^{-1}$ and the integrand reduces to $4uv$ over the region $0 \leq u \leq 1, 0 \leq v \leq 1$; $I = 1$.

6.21 Terms such as $T\partial^2 S/\partial Y \partial X$ cancel in pairs. Use equations (6.17) and (6.16).

6.23 (c) Show that the two expressions mutually support the integration formula given for computing a volume in the next higher dimension.
      (d)(ii) $2, \pi, 4\pi/3, \pi^2/2, 8\pi^2/15, \pi^3/6, 16\pi^3/105, \pi^4/24$.

7

Vector algebra

This chapter introduces space vectors and their manipulation. Firstly we deal with the description and algebra of vectors, then we consider how vectors may be used to describe lines and planes and finally we look at the practical use of vectors in finding distances. Much use of vectors will be made in subsequent chapters; this chapter gives only some basic rules.

7.1 Scalars and vectors

The simplest kind of physical quantity is one that can be completely specified by its magnitude, a single number, together with the units in which it is measured. Such a quantity is called a *scalar* and examples include temperature, time and density.

A *vector* is a quantity that requires both a magnitude (≥ 0) and a direction in space to specify it completely; we may think of it as an arrow in space. A familiar example is force, which has a magnitude (strength) measured in newtons and a direction of application. The large number of vectors that are used to describe the physical world include velocity, displacement, momentum and electric field. Vectors are also used to describe quantities such as angular momentum and surface elements (a surface element has an area and a direction defined by the normal to its tangent plane); in such cases their definitions may seem somewhat arbitrary (though in fact they are standard) and not as physically intuitive as for vectors such as force. A vector is denoted by bold type, the convention of this book, or by underlining, the latter being much used in handwritten work.

This chapter considers basic vector algebra and illustrates just how powerful vector analysis can be. All the techniques are presented for three-dimensional space but most can be readily extended to more dimensions.

Throughout the book we will represent a vector in diagrams as a line together with an arrowhead. We will make no distinction between an arrowhead at the

212

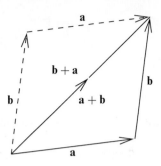

Figure 7.1   Addition of two vectors showing the commutation relation. We make no distinction between an arrowhead at the end of the line and one along the line's length, but rather use that which gives the clearer diagram.

end of the line and one along the line's length but, rather, use that which gives the clearer diagram. Furthermore, even though we are considering three-dimensional vectors, we have to draw them in the plane of the paper. It should not be assumed that vectors drawn thus are coplanar, unless this is explicitly stated.

7.2 Addition and subtraction of vectors

The *resultant* or *vector sum* of two displacement vectors is the displacement vector that results from performing first one and then the other displacement, as shown in figure 7.1; this process is known as vector addition. However, the principle of addition has physical meaning for vector quantities other than displacements; for example, if two forces act on the same body then the resultant force acting on the body is the vector sum of the two. The addition of vectors only makes physical sense if they are of a like kind, for example if they are both forces acting in three dimensions. It may be seen from figure 7.1 that vector addition is commutative, i.e.

$$\mathbf{a} + \mathbf{b} = \mathbf{b} + \mathbf{a}. \tag{7.1}$$

The generalisation of this procedure to the addition of three (or more) vectors is clear and leads to the associativity property of addition (see figure 7.2), e.g.

$$\mathbf{a} + (\mathbf{b} + \mathbf{c}) = (\mathbf{a} + \mathbf{b}) + \mathbf{c}. \tag{7.2}$$

Thus, it is immaterial in what order any number of vectors are added.

The subtraction of two vectors is very similar to their addition (see figure 7.3), that is,

$$\mathbf{a} - \mathbf{b} = \mathbf{a} + (-\mathbf{b})$$

where $-\mathbf{b}$ is a vector of equal magnitude but exactly opposite direction to vector $\mathbf{b}$.

213

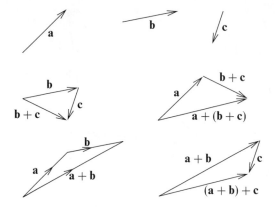

Figure 7.2 Addition of three vectors showing the associativity relation.

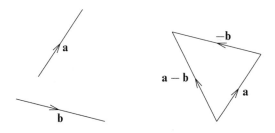

Figure 7.3 Subtraction of two vectors.

The subtraction of two equal vectors yields the zero vector, **0**, which has zero magnitude and no associated direction.

7.3 Multiplication by a scalar

Multiplication of a vector by a scalar (not to be confused with the 'scalar product', to be discussed in subsection 7.6.1) gives a vector in the same direction as the original but of a proportional magnitude. This can be seen in figure 7.4. The scalar may be positive, negative or zero. It can also be complex in some applications. Clearly, when the scalar is negative we obtain a vector pointing in the opposite direction to the original vector. Multiplication by a scalar is associative, commutative and distributive over addition. These properties may be summarised for arbitrary vectors **a** and **b** and arbitrary scalars λ and μ by

$$(\lambda\mu)\mathbf{a} = \lambda(\mu\mathbf{a}) = \mu(\lambda\mathbf{a}), \tag{7.3}$$

$$\lambda(\mathbf{a} + \mathbf{b}) = \lambda\mathbf{a} + \lambda\mathbf{b}, \tag{7.4}$$

$$(\lambda + \mu)\mathbf{a} = \lambda\mathbf{a} + \mu\mathbf{a}. \tag{7.5}$$

214

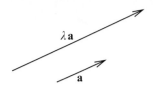

Figure 7.4   Scalar multiplication of a vector (for $\lambda > 1$).

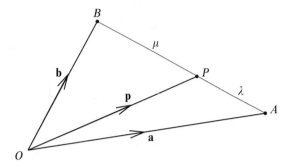

Figure 7.5   An illustration of the ratio theorem. The point P divides the line segment AB in the ratio $\lambda : \mu$.

Having defined the operations of addition, subtraction and multiplication by a scalar, we can now use vectors to solve simple problems in geometry.

►*A point P divides a line segment AB in the ratio $\lambda : \mu$ (see figure 7.5). If the position vectors of the points A and B are $\mathbf{a}$ and $\mathbf{b}$, respectively, find the position vector of the point P.*

As is conventional for vector geometry problems, we denote the vector from the point A to the point B by $\mathbf{AB}$. If the position vectors of the points A and B, relative to some origin O, are $\mathbf{a}$ and $\mathbf{b}$, it should be clear that $\mathbf{AB} = \mathbf{b} - \mathbf{a}$.

Now, from figure 7.5 we see that one possible way of reaching the point P from O is first to go from O to A and to go along the line AB for a distance equal to the the fraction $\lambda/(\lambda + \mu)$ of its total length. We may express this in terms of vectors as

$$\mathbf{OP} = \mathbf{p} = \mathbf{a} + \frac{\lambda}{\lambda + \mu}\mathbf{AB}$$

$$= \mathbf{a} + \frac{\lambda}{\lambda + \mu}(\mathbf{b} - \mathbf{a})$$

$$= \left(1 - \frac{\lambda}{\lambda + \mu}\right)\mathbf{a} + \frac{\lambda}{\lambda + \mu}\mathbf{b}$$

$$= \frac{\mu}{\lambda + \mu}\mathbf{a} + \frac{\lambda}{\lambda + \mu}\mathbf{b}, \tag{7.6}$$

which expresses the position vector of the point P in terms of those of A and B. We would, of course, obtain the same result by considering the path from O to B and then to P. ◄

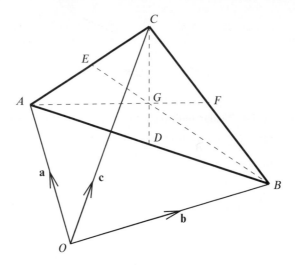

Figure 7.6   The centroid of a triangle. The triangle is defined by the points A, B and C that have position vectors $\mathbf{a}$, $\mathbf{b}$ and $\mathbf{c}$. The broken lines CD, BE, AF connect the vertices of the triangle to the mid-points of the opposite sides; these lines intersect at the centroid G of the triangle.

Result (7.6) is a version of the *ratio theorem* and we may use it in solving more complicated problems.

▶ *The vertices of triangle ABC have position vectors $\mathbf{a}$, $\mathbf{b}$ and $\mathbf{c}$ relative to some origin O (see figure 7.6). Find the position vector of the centroid G of the triangle.*

From figure 7.6, the points D and E bisect the lines AB and AC respectively. Thus from the ratio theorem (7.6), with $\lambda = \mu = 1/2$, the position vectors of D and E relative to the origin are

$$\mathbf{d} = \tfrac{1}{2}\mathbf{a} + \tfrac{1}{2}\mathbf{b},$$

$$\mathbf{e} = \tfrac{1}{2}\mathbf{a} + \tfrac{1}{2}\mathbf{c}.$$

Using the ratio theorem again, we may write the position vector of a general point on the line CD that divides the line in the ratio $\lambda : (1 - \lambda)$ as

$$\mathbf{r} = (1 - \lambda)\mathbf{c} + \lambda\mathbf{d},$$
$$= (1 - \lambda)\mathbf{c} + \tfrac{1}{2}\lambda(\mathbf{a} + \mathbf{b}), \tag{7.7}$$

where we have expressed $\mathbf{d}$ in terms of $\mathbf{a}$ and $\mathbf{b}$. Similarly, the position vector of a general point on the line BE can be expressed as

$$\mathbf{r} = (1 - \mu)\mathbf{b} + \mu\mathbf{e},$$
$$= (1 - \mu)\mathbf{b} + \tfrac{1}{2}\mu(\mathbf{a} + \mathbf{c}). \tag{7.8}$$

Thus, at the intersection of the lines CD and BE we require, from (7.7), (7.8),

$$(1 - \lambda)\mathbf{c} + \tfrac{1}{2}\lambda(\mathbf{a} + \mathbf{b}) = (1 - \mu)\mathbf{b} + \tfrac{1}{2}\mu(\mathbf{a} + \mathbf{c}).$$

By equating the coefficients of the vectors $\mathbf{a}$, $\mathbf{b}$, $\mathbf{c}$ we find

$$\lambda = \mu, \qquad \tfrac{1}{2}\lambda = 1 - \mu, \qquad 1 - \lambda = \tfrac{1}{2}\mu.$$

These equations are consistent and have the solution $\lambda = \mu = 2/3$. Substituting these values into either (7.7) or (7.8) we find that the position vector of the centroid G is given by

$$\mathbf{g} = \tfrac{1}{3}(\mathbf{a} + \mathbf{b} + \mathbf{c}). \blacktriangleleft$$

7.4 Basis vectors and components

Given any three different vectors $\mathbf{e}_1$, $\mathbf{e}_2$ and $\mathbf{e}_3$, which do not all lie in a plane, it is possible, in a three-dimensional space, to write any other vector in terms of scalar multiples of them:

$$\mathbf{a} = a_1\mathbf{e}_1 + a_2\mathbf{e}_2 + a_3\mathbf{e}_3. \tag{7.9}$$

The three vectors $\mathbf{e}_1$, $\mathbf{e}_2$ and $\mathbf{e}_3$ are said to form a *basis* (for the three-dimensional space); the scalars a_1, a_2 and a_3, which may be positive, negative or zero, are called the *components* of the vector $\mathbf{a}$ with respect to this basis. We say that the vector has been *resolved* into components.

Most often we shall use basis vectors that are mutually perpendicular, for ease of manipulation, though this is not necessary. In general, a basis set must

(i) have as many basis vectors as the number of dimensions (in more formal language, the basis vectors must span the space) and

(ii) be such that no basis vector may be described as a sum of the others, or, more formally, the basis vectors must be *linearly independent*. Putting this mathematically, in N dimensions, we require

$$c_1\mathbf{e}_1 + c_2\mathbf{e}_2 + \cdots + c_N\mathbf{e}_N \neq \mathbf{0},$$

for any set of coefficients $c_1, c_2, \ldots, c_N$ except $c_1 = c_2 = \cdots = c_N = 0$.

In this chapter we will only consider vectors in three dimensions; higher dimensionality can be achieved by simple extension.

If we wish to label points in space using a Cartesian coordinate system (x, y, z), we may introduce the unit vectors $\mathbf{i}$, $\mathbf{j}$ and $\mathbf{k}$, which point along the positive x-, y- and z- axes respectively. A vector $\mathbf{a}$ may then be written as a sum of three vectors, each parallel to a different coordinate axis:

$$\mathbf{a} = a_x\mathbf{i} + a_y\mathbf{j} + a_z\mathbf{k}. \tag{7.10}$$

A vector in three-dimensional space thus requires three components to describe fully both its direction and its magnitude. A displacement in space may be thought of as the sum of displacements along the x-, y- and z- directions (see figure 7.7). For brevity, the components of a vector $\mathbf{a}$ with respect to a particular coordinate system are sometimes written in the form (a_x, a_y, a_z). Note that the

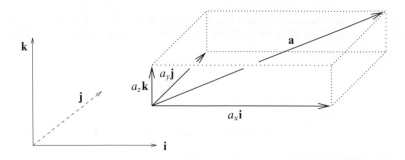

Figure 7.7   A Cartesian basis set. The vector **a** is the sum of $a_x\mathbf{i}$, $a_y\mathbf{j}$ and $a_z\mathbf{k}$.

basis vectors **i**, **j** and **k** may themselves be represented by $(1,0,0)$, $(0,1,0)$ and $(0,0,1)$ respectively.

We can consider the addition and subtraction of vectors in terms of their components. The sum of two vectors **a** and **b** is found by simply adding their components, i.e.

$$\mathbf{a} + \mathbf{b} = a_x\mathbf{i} + a_y\mathbf{j} + a_z\mathbf{k} + b_x\mathbf{i} + b_y\mathbf{j} + b_z\mathbf{k}$$
$$= (a_x + b_x)\mathbf{i} + (a_y + b_y)\mathbf{j} + (a_z + b_z)\mathbf{k}, \tag{7.11}$$

and their difference by subtracting them,

$$\mathbf{a} - \mathbf{b} = a_x\mathbf{i} + a_y\mathbf{j} + a_z\mathbf{k} - (b_x\mathbf{i} + b_y\mathbf{j} + b_z\mathbf{k})$$
$$= (a_x - b_x)\mathbf{i} + (a_y - b_y)\mathbf{j} + (a_z - b_z)\mathbf{k}. \tag{7.12}$$

▶*Two particles have velocities* $\mathbf{v}_1 = \mathbf{i} + 3\mathbf{j} + 6\mathbf{k}$ *and* $\mathbf{v}_2 = \mathbf{i} - 2\mathbf{k}$, *respectively. Find the velocity* **u** *of the second particle relative to the first.*

The required relative velocity is given by

$$\mathbf{u} = \mathbf{v}_2 - \mathbf{v}_1 = (1 - 1)\mathbf{i} + (0 - 3)\mathbf{j} + (-2 - 6)\mathbf{k}$$
$$= -3\mathbf{j} - 8\mathbf{k}. ◀$$

7.5 Magnitude of a vector

The magnitude of the vector **a** is denoted by $|\mathbf{a}|$ or a. In terms of its components in three-dimensional Cartesian coordinates, the magnitude of **a** is given by

$$a \equiv |\mathbf{a}| = \sqrt{a_x^2 + a_y^2 + a_z^2}. \tag{7.13}$$

Hence, the magnitude of a vector is a measure of its length. Such an analogy is useful for displacement vectors but magnitude is better described, for example, by 'strength' for vectors such as force or by 'speed' for velocity vectors. For instance,

218

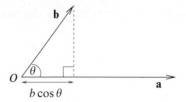

Figure 7.8   The projection of **b** onto the direction of **a** is $b\cos\theta$. The scalar product of **a** and **b** is $ab\cos\theta$.

in the previous example, the speed of the second particle relative to the first is given by

$$u = |\mathbf{u}| = \sqrt{(-3)^2 + (-8)^2} = \sqrt{73}.$$

A vector whose magnitude equals unity is called a *unit vector*. The unit vector in the direction **a** is usually notated **â** and may be evaluated as

$$\hat{\mathbf{a}} = \frac{\mathbf{a}}{|\mathbf{a}|}. \tag{7.14}$$

The unit vector is a useful concept because a vector written as $\lambda\hat{\mathbf{a}}$ then has magnitude λ and direction **â**. Thus magnitude and direction are explicitly separated.

7.6 Multiplication of vectors

We have already considered multiplying a vector by a scalar. Now we consider the concept of multiplying one vector by another vector. It is not immediately obvious what the product of two vectors represents and in fact two products are commonly defined, the *scalar product* and the *vector product* As their names imply, the scalar product of two vectors is just a number, whereas the vector product is itself a vector. Although neither the scalar nor the vector product is what we might normally think of as a product, their use is widespread and numerous examples will be described elsewhere in this book.

7.6.1 Scalar product

The scalar product (or dot product) of two vectors **a** and **b** is denoted by $\mathbf{a} \cdot \mathbf{b}$ and is given by

$$\mathbf{a} \cdot \mathbf{b} \equiv |\mathbf{a}||\mathbf{b}| \cos\theta, \quad 0 \le \theta \le \pi, \tag{7.15}$$

where θ is the angle between the two vectors, placed 'tail to tail' or 'head to head'. Thus, the value of the scalar product $\mathbf{a} \cdot \mathbf{b}$ equals the magnitude of **a** multiplied by the projection of **b** onto **a** (see figure 7.8).

From (7.15) we see that the scalar product has the particularly useful property that

$$\mathbf{a} \cdot \mathbf{b} = 0 \qquad (7.16)$$

is a necessary and sufficient condition for $\mathbf{a}$ to be perpendicular to $\mathbf{b}$ (unless either of them is zero). It should be noted in particular that the Cartesian basis vectors $\mathbf{i}$, $\mathbf{j}$ and $\mathbf{k}$, being mutually orthogonal unit vectors, satisfy the equations

$$\mathbf{i} \cdot \mathbf{i} = \mathbf{j} \cdot \mathbf{j} = \mathbf{k} \cdot \mathbf{k} = 1, \qquad (7.17)$$

$$\mathbf{i} \cdot \mathbf{j} = \mathbf{j} \cdot \mathbf{k} = \mathbf{k} \cdot \mathbf{i} = 0. \qquad (7.18)$$

Examples of scalar products arise naturally throughout physics and in particular in connection with energy. Perhaps the simplest is the work done $\mathbf{F} \cdot \mathbf{r}$ in moving the point of application of a constant force $\mathbf{F}$ through a displacement $\mathbf{r}$; notice that, as expected, if the displacement is perpendicular to the direction of the force then $\mathbf{F} \cdot \mathbf{r} = 0$ and no work is done. A second simple example is afforded by the potential energy $-\mathbf{m} \cdot \mathbf{B}$ of a magnetic dipole, represented in strength and orientation by a vector $\mathbf{m}$, placed in an external magnetic field $\mathbf{B}$.

As the name implies, the scalar product has a magnitude but no direction. The scalar product is commutative and distributive over addition:

$$\mathbf{a} \cdot \mathbf{b} = \mathbf{b} \cdot \mathbf{a} \qquad (7.19)$$

$$\mathbf{a} \cdot (\mathbf{b} + \mathbf{c}) = \mathbf{a} \cdot \mathbf{b} + \mathbf{a} \cdot \mathbf{c}. \qquad (7.20)$$

▶*Four non-coplanar points A, B, C, D are positioned such that the line AD is perpendicular to BC and BD is perpendicular to AC. Show that CD is perpendicular to AB.*

Denote the four position vectors by $\mathbf{a}$, $\mathbf{b}$, $\mathbf{c}$, $\mathbf{d}$. As none of the three pairs of lines actually intersect, it is difficult to indicate their orthogonality in the diagram we would normally draw. However, the orthogonality can be expressed in vector form and we start by noting that, since $AD \perp BC$, it follows from (7.16) that

$$(\mathbf{d} - \mathbf{a}) \cdot (\mathbf{c} - \mathbf{b}) = 0.$$

Similarly, since $BD \perp AC$,

$$(\mathbf{d} - \mathbf{b}) \cdot (\mathbf{c} - \mathbf{a}) = 0.$$

Combining these two equations we find

$$(\mathbf{d} - \mathbf{a}) \cdot (\mathbf{c} - \mathbf{b}) = (\mathbf{d} - \mathbf{b}) \cdot (\mathbf{c} - \mathbf{a}),$$

which, on multiplying out the parentheses, gives

$$\mathbf{d} \cdot \mathbf{c} - \mathbf{a} \cdot \mathbf{c} - \mathbf{d} \cdot \mathbf{b} + \mathbf{a} \cdot \mathbf{b} = \mathbf{d} \cdot \mathbf{c} - \mathbf{b} \cdot \mathbf{c} - \mathbf{d} \cdot \mathbf{a} + \mathbf{b} \cdot \mathbf{a}.$$

Cancelling terms that appear on both sides and rearranging yields

$$\mathbf{d} \cdot \mathbf{b} - \mathbf{d} \cdot \mathbf{a} - \mathbf{c} \cdot \mathbf{b} + \mathbf{c} \cdot \mathbf{a} = 0,$$

which simplifies to give

$$(\mathbf{d} - \mathbf{c}) \cdot (\mathbf{b} - \mathbf{a}) = 0.$$

From (7.16), we see that this implies that CD is perpendicular to AB. ◀

If we introduce a set of basis vectors that are mutually orthogonal, such as $\mathbf{i}$, $\mathbf{j}$, $\mathbf{k}$, we can write the components of a vector $\mathbf{a}$, with respect to that basis, in terms of the scalar product of $\mathbf{a}$ with each of the basis vectors, i.e. $a_x = \mathbf{a} \cdot \mathbf{i}$, $a_y = \mathbf{a} \cdot \mathbf{j}$ and $a_z = \mathbf{a} \cdot \mathbf{k}$. In terms of the components a_x, a_y and a_z the scalar product is given by

$$\mathbf{a} \cdot \mathbf{b} = (a_x\mathbf{i} + a_y\mathbf{j} + a_z\mathbf{k}) \cdot (b_x\mathbf{i} + b_y\mathbf{j} + b_z\mathbf{k}) = a_xb_x + a_yb_y + a_zb_z, \qquad (7.21)$$

where the cross terms such as $a_x\mathbf{i} \cdot b_y\mathbf{j}$ are zero because the basis vectors are mutually perpendicular; see equation (7.18). It should be clear from (7.15) that the value of $\mathbf{a} \cdot \mathbf{b}$ has a geometrical definition and that this value is independent of the actual basis vectors used.

▶ *Find the angle between the vectors* $\mathbf{a} = \mathbf{i} + 2\mathbf{j} + 3\mathbf{k}$ *and* $\mathbf{b} = 2\mathbf{i} + 3\mathbf{j} + 4\mathbf{k}$.

From (7.15) the cosine of the angle θ between $\mathbf{a}$ and $\mathbf{b}$ is given by

$$\cos\theta = \frac{\mathbf{a} \cdot \mathbf{b}}{|\mathbf{a}||\mathbf{b}|}.$$

From (7.21) the scalar product $\mathbf{a} \cdot \mathbf{b}$ has the value

$$\mathbf{a} \cdot \mathbf{b} = 1 \times 2 + 2 \times 3 + 3 \times 4 = 20,$$

and from (7.13) the lengths of the vectors are

$$|\mathbf{a}| = \sqrt{1^2 + 2^2 + 3^2} = \sqrt{14} \qquad \text{and} \qquad |\mathbf{b}| = \sqrt{2^2 + 3^2 + 4^2} = \sqrt{29}.$$

Thus,

$$\cos\theta = \frac{20}{\sqrt{14}\sqrt{29}} \approx 0.9926 \qquad \Rightarrow \qquad \theta = 0.12 \text{ rad.} \blacktriangleleft$$

We can see from the expressions (7.15) and (7.21) for the scalar product that if θ is the angle between $\mathbf{a}$ and $\mathbf{b}$ then

$$\cos\theta = \frac{a_x}{a}\frac{b_x}{b} + \frac{a_y}{a}\frac{b_y}{b} + \frac{a_z}{a}\frac{b_z}{b}$$

where a_x/a, a_y/a and a_z/a are called the *direction cosines* of $\mathbf{a}$, since they give the cosine of the angle made by $\mathbf{a}$ with each of the basis vectors. Similarly b_x/b, b_y/b and b_z/b are the direction cosines of $\mathbf{b}$.

If we take the scalar product of any vector $\mathbf{a}$ with itself then clearly $\theta = 0$ and from (7.15) we have

$$\mathbf{a} \cdot \mathbf{a} = |\mathbf{a}|^2.$$

Thus the magnitude of $\mathbf{a}$ can be written in a coordinate-independent form as $|\mathbf{a}| = \sqrt{\mathbf{a} \cdot \mathbf{a}}$.

Finally, we note that the scalar product may be extended to vectors with complex components if it is redefined as

$$\mathbf{a} \cdot \mathbf{b} = a_x^* b_x + a_y^* b_y + a_z^* b_z,$$

where the asterisk represents the operation of complex conjugation. To accom-

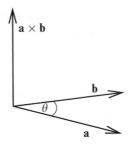

Figure 7.9   The vector product. The vectors **a**, **b** and **a** × **b** form a right-handed set.

modate this extension the commutation property (7.19) must be modified to read

$$\mathbf{a} \cdot \mathbf{b} = (\mathbf{b} \cdot \mathbf{a})^{*}. \tag{7.22}$$

In particular it should be noted that $(\lambda\mathbf{a}) \cdot \mathbf{b} = \lambda^{*}\mathbf{a} \cdot \mathbf{b}$, whereas $\mathbf{a} \cdot (\lambda\mathbf{b}) = \lambda\mathbf{a} \cdot \mathbf{b}$. However, the magnitude of a complex vector is still given by $|\mathbf{a}| = \sqrt{\mathbf{a} \cdot \mathbf{a}}$, since $\mathbf{a} \cdot \mathbf{a}$ is always real.

7.6.2 Vector product

The vector product (or cross product) of two vectors **a** and **b** is denoted by $\mathbf{a} \times \mathbf{b}$ and is defined to be a vector of magnitude $|\mathbf{a}||\mathbf{b}| \sin\theta$ in a direction perpendicular to both **a** and **b**;

$$|\mathbf{a} \times \mathbf{b}| = |\mathbf{a}||\mathbf{b}| \sin\theta.$$

The direction is found by 'rotating' **a** into **b** through the smallest possible angle. The sense of rotation is that of a right-handed screw that moves forward in the direction $\mathbf{a} \times \mathbf{b}$ (see figure 7.9). Again, θ is the angle between the two vectors placed 'tail to tail' or 'head to head'. With this definition **a**, **b** and $\mathbf{a} \times \mathbf{b}$ form a right-handed set. A more directly usable description of the relative directions in a vector product is provided by a right hand whose first two fingers and thumb are held to be as nearly mutually perpendicular as possible. If the first finger is pointed in the direction of the first vector and the second finger in the direction of the second vector, then the thumb gives the direction of the vector product.

The vector product may be shown (with a little work) to be distributive over addition, but *anticommutative* and *non-associative*:

$$(\mathbf{a} + \mathbf{b}) \times \mathbf{c} = (\mathbf{a} \times \mathbf{c}) + (\mathbf{b} \times \mathbf{c}), \tag{7.23}$$

$$\mathbf{b} \times \mathbf{a} = -(\mathbf{a} \times \mathbf{b}), \tag{7.24}$$

$$(\mathbf{a} \times \mathbf{b}) \times \mathbf{c} \neq \mathbf{a} \times (\mathbf{b} \times \mathbf{c}). \tag{7.25}$$

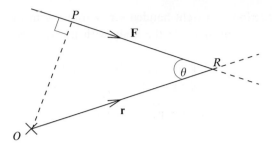

Figure 7.10 The moment of the force **F** about O is $\mathbf{r} \times \mathbf{F}$. The cross represents the direction of $\mathbf{r} \times \mathbf{F}$, which is perpendicularly into the plane of the paper.

From its definition, we see that the vector product has the very useful property that if $\mathbf{a} \times \mathbf{b} = \mathbf{0}$ then **a** is parallel or antiparallel to **b** (unless either of them is zero). We also note that

$$\mathbf{a} \times \mathbf{a} = \mathbf{0}. \tag{7.26}$$

▶ *Show that if* $\mathbf{a} = \mathbf{b} + \lambda \mathbf{c}$, *for some scalar* λ, *then* $\mathbf{a} \times \mathbf{c} = \mathbf{b} \times \mathbf{c}$.

From (7.23) we have

$$\mathbf{a} \times \mathbf{c} = (\mathbf{b} + \lambda \mathbf{c}) \times \mathbf{c} = \mathbf{b} \times \mathbf{c} + \lambda \mathbf{c} \times \mathbf{c}.$$

However, from (7.26), $\mathbf{c} \times \mathbf{c} = \mathbf{0}$ and so

$$\mathbf{a} \times \mathbf{c} = \mathbf{b} \times \mathbf{c}. \tag{7.27}$$

We note in passing that the fact that (7.27) is satisfied does *not* imply that $\mathbf{a} = \mathbf{b}$. ◀

An example of the use of the vector product is that of finding the area, A, of a parallelogram with sides **a** and **b**, using the formula

$$A = |\mathbf{a} \times \mathbf{b}|. \tag{7.28}$$

Another example is afforded by considering a force **F** acting through a point R, whose vector position relative to the origin O is **r** (see figure 7.10). Its *moment* or *torque* about O is the strength of the force times the perpendicular distance OP, which numerically is just $Fr \sin \theta$, i.e. the magnitude of $\mathbf{r} \times \mathbf{F}$. Furthermore, the sense of the moment is clockwise about an axis through O that points perpendicularly into the plane of the paper (the axis is represented by a cross in the figure). Thus the moment is completely represented by the vector $\mathbf{r} \times \mathbf{F}$, in both magnitude and spatial sense. It should be noted that the same vector product is obtained wherever the point R is chosen, so long as it lies on the line of action of **F**.

Similarly, if a solid body is rotating about some axis that passes through the origin, with an angular velocity ω then we can describe this rotation by a vector $\boldsymbol{\omega}$ that has magnitude ω and points along the axis of rotation. The direction of $\boldsymbol{\omega}$

is the forward direction of a right-handed screw rotating in the same sense as the body. The velocity of any point in the body with position vector $\mathbf{r}$ is then given by $\mathbf{v} = \boldsymbol{\omega} \times \mathbf{r}$.

Since the basis vectors $\mathbf{i}$, $\mathbf{j}$, $\mathbf{k}$ are mutually perpendicular unit vectors, forming a right-handed set, their vector products are easily seen to be

$$\mathbf{i} \times \mathbf{i} = \mathbf{j} \times \mathbf{j} = \mathbf{k} \times \mathbf{k} = \mathbf{0}, \tag{7.29}$$

$$\mathbf{i} \times \mathbf{j} = -\mathbf{j} \times \mathbf{i} = \mathbf{k}, \tag{7.30}$$

$$\mathbf{j} \times \mathbf{k} = -\mathbf{k} \times \mathbf{j} = \mathbf{i}, \tag{7.31}$$

$$\mathbf{k} \times \mathbf{i} = -\mathbf{i} \times \mathbf{k} = \mathbf{j}. \tag{7.32}$$

Using these relations, it is straightforward to show that the vector product of two general vectors $\mathbf{a}$ and $\mathbf{b}$ is given in terms of their components with respect to the basis set $\mathbf{i}$, $\mathbf{j}$, $\mathbf{k}$, by

$$\mathbf{a} \times \mathbf{b} = (a_y b_z - a_z b_y)\mathbf{i} + (a_z b_x - a_x b_z)\mathbf{j} + (a_x b_y - a_y b_x)\mathbf{k}. \tag{7.33}$$

For the reader who is familiar with determinants (see chapter 8), we record that this can also be written as

$$\mathbf{a} \times \mathbf{b} = \begin{vmatrix} \mathbf{i} & \mathbf{j} & \mathbf{k} \\ a_x & a_y & a_z \\ b_x & b_y & b_z \end{vmatrix}.$$

That the cross product $\mathbf{a} \times \mathbf{b}$ is perpendicular to both $\mathbf{a}$ and $\mathbf{b}$ can be verified in component form by forming its dot products with each of the two vectors and showing that it is zero in both cases.

▶*Find the area A of the parallelogram with sides $\mathbf{a} = \mathbf{i} + 2\mathbf{j} + 3\mathbf{k}$ and $\mathbf{b} = 4\mathbf{i} + 5\mathbf{j} + 6\mathbf{k}$.*

The vector product $\mathbf{a} \times \mathbf{b}$ is given in component form by

$$\mathbf{a} \times \mathbf{b} = (2 \times 6 - 3 \times 5)\mathbf{i} + (3 \times 4 - 1 \times 6)\mathbf{j} + (1 \times 5 - 2 \times 4)\mathbf{k}$$
$$= -3\mathbf{i} + 6\mathbf{j} - 3\mathbf{k}.$$

Thus the area of the parallelogram is

$$A = |\mathbf{a} \times \mathbf{b}| = \sqrt{(-3)^2 + 6^2 + (-3)^2} = \sqrt{54}. \blacktriangleleft$$

7.6.3 Scalar triple product

Now that we have defined the scalar and vector products, we can extend our discussion to define products of three vectors. Again, there are two possibilities, the *scalar triple product* and the *vector triple product*.

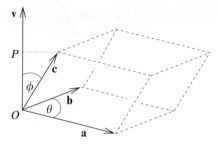

Figure 7.11   The scalar triple product gives the volume of a parallelepiped.

The scalar triple product is denoted by

$$[\mathbf{a}, \mathbf{b}, \mathbf{c}] \equiv \mathbf{a} \cdot (\mathbf{b} \times \mathbf{c})$$

and, as its name suggests, it is just a number. It is most simply interpreted as the volume of a parallelepiped whose edges are given by $\mathbf{a}$, $\mathbf{b}$ and $\mathbf{c}$ (see figure 7.11). The vector $\mathbf{v} = \mathbf{a} \times \mathbf{b}$ is perpendicular to the base of the solid and has magnitude $v = ab \sin \theta$, i.e. the area of the base. Further, $\mathbf{v} \cdot \mathbf{c} = vc \cos \phi$. Thus, since $c \cos \phi = OP$ is the vertical height of the parallelepiped, it is clear that $(\mathbf{a} \times \mathbf{b}) \cdot \mathbf{c} =$ area of the base $\times$ perpendicular height $=$ volume. It follows that, if the vectors $\mathbf{a}$, $\mathbf{b}$ and $\mathbf{c}$ are coplanar, $\mathbf{a} \cdot (\mathbf{b} \times \mathbf{c}) = 0$.

Expressed in terms of the components of each vector with respect to the Cartesian basis set $\mathbf{i}$, $\mathbf{j}$, $\mathbf{k}$ the scalar triple product is

$$\mathbf{a} \cdot (\mathbf{b} \times \mathbf{c}) = a_x(b_y c_z - b_z c_y) + a_y(b_z c_x - b_x c_z) + a_z(b_x c_y - b_y c_x),$$
(7.34)

which can also be written as a determinant:

$$\mathbf{a} \cdot (\mathbf{b} \times \mathbf{c}) = \begin{vmatrix} a_x & a_y & a_z \\ b_x & b_y & b_z \\ c_x & c_y & c_z \end{vmatrix}.$$

By writing the vectors in component form, it can be shown that

$$\mathbf{a} \cdot (\mathbf{b} \times \mathbf{c}) = (\mathbf{a} \times \mathbf{b}) \cdot \mathbf{c},$$

so that the dot and cross symbols can be interchanged without changing the result. More generally, the scalar triple product is unchanged under cyclic permutation of the vectors $\mathbf{a}, \mathbf{b}, \mathbf{c}$. Other permutations simply give the negative of the original scalar triple product. These results can be summarised by

$$[\mathbf{a}, \mathbf{b}, \mathbf{c}] = [\mathbf{b}, \mathbf{c}, \mathbf{a}] = [\mathbf{c}, \mathbf{a}, \mathbf{b}] = -[\mathbf{a}, \mathbf{c}, \mathbf{b}] = -[\mathbf{b}, \mathbf{a}, \mathbf{c}] = -[\mathbf{c}, \mathbf{b}, \mathbf{a}].$$
(7.35)

▶*Find the volume V of the parallelepiped with sides* $\mathbf{a} = \mathbf{i} + 2\mathbf{j} + 3\mathbf{k}$, $\mathbf{b} = 4\mathbf{i} + 5\mathbf{j} + 6\mathbf{k}$ *and* $\mathbf{c} = 7\mathbf{i} + 8\mathbf{j} + 10\mathbf{k}$.

We have already found that $\mathbf{a} \times \mathbf{b} = -3\mathbf{i} + 6\mathbf{j} - 3\mathbf{k}$, in subsection 7.6.2. Hence the volume of the parallelepiped is given by

$$V = |\mathbf{a} \cdot (\mathbf{b} \times \mathbf{c})| = |(\mathbf{a} \times \mathbf{b}) \cdot \mathbf{c}|$$
$$= |(-3\mathbf{i} + 6\mathbf{j} - 3\mathbf{k}) \cdot (7\mathbf{i} + 8\mathbf{j} + 10\mathbf{k})|$$
$$= |(-3)(7) + (6)(8) + (-3)(10)| = 3. \;\blacktriangleleft$$

A useful formula that can be proved using the properties of the scalar triple product is Lagrange's identity (see exercise 7.9), which reads

$$(\mathbf{a} \times \mathbf{b}) \cdot (\mathbf{c} \times \mathbf{d}) \equiv (\mathbf{a} \cdot \mathbf{c})(\mathbf{b} \cdot \mathbf{d}) - (\mathbf{a} \cdot \mathbf{d})(\mathbf{b} \cdot \mathbf{c}). \tag{7.36}$$

7.6.4 Vector triple product

By the vector triple product of three vectors $\mathbf{a}$, $\mathbf{b}$, $\mathbf{c}$ we mean the vector $\mathbf{a} \times (\mathbf{b} \times \mathbf{c})$. Clearly, $\mathbf{a} \times (\mathbf{b} \times \mathbf{c})$ is perpendicular to $\mathbf{a}$ and lies in the plane of $\mathbf{b}$ and $\mathbf{c}$ and so can be expressed in terms of them (see (7.37) below). We note, from (7.25), that the vector triple product is not associative, i.e. $\mathbf{a} \times (\mathbf{b} \times \mathbf{c}) \neq (\mathbf{a} \times \mathbf{b}) \times \mathbf{c}$.

Two useful formulae involving the vector triple product are

$$\mathbf{a} \times (\mathbf{b} \times \mathbf{c}) = (\mathbf{a} \cdot \mathbf{c})\mathbf{b} - (\mathbf{a} \cdot \mathbf{b})\mathbf{c}, \tag{7.37}$$
$$(\mathbf{a} \times \mathbf{b}) \times \mathbf{c} = (\mathbf{a} \cdot \mathbf{c})\mathbf{b} - (\mathbf{b} \cdot \mathbf{c})\mathbf{a}, \tag{7.38}$$

which may be derived by writing each vector in component form (see exercise 7.8). It can also be shown that for any three vectors $\mathbf{a}$, $\mathbf{b}$, $\mathbf{c}$,

$$\mathbf{a} \times (\mathbf{b} \times \mathbf{c}) + \mathbf{b} \times (\mathbf{c} \times \mathbf{a}) + \mathbf{c} \times (\mathbf{a} \times \mathbf{b}) = \mathbf{0}.$$

7.7 Equations of lines, planes and spheres

Now that we have described the basic algebra of vectors, we can apply the results to a variety of problems, the first of which is to find the equation of a line in vector form.

7.7.1 Equation of a line

Consider the line passing through the fixed point A with position vector $\mathbf{a}$ and having a direction $\mathbf{b}$ (see figure 7.12). It is clear that the position vector $\mathbf{r}$ of a general point R on the line can be written as

$$\mathbf{r} = \mathbf{a} + \lambda\mathbf{b}, \tag{7.39}$$

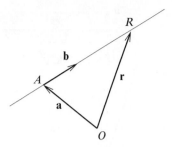

Figure 7.12   The equation of a line. The vector **b** is in the direction AR and λ**b** is the vector from A to R.

since R can be reached by starting from O, going along the translation vector **a** to the point A on the line and then adding some multiple λ**b** of the vector **b**. Different values of λ give different points R on the line.

Taking the components of (7.39), we see that the equation of the line can also be written in the form

$$\frac{x - a_x}{b_x} = \frac{y - a_y}{b_y} = \frac{z - a_z}{b_z} = \text{constant}. \tag{7.40}$$

Taking the vector product of (7.39) with **b** and remembering that $\mathbf{b} \times \mathbf{b} = \mathbf{0}$ gives an alternative equation for the line

$$(\mathbf{r} - \mathbf{a}) \times \mathbf{b} = \mathbf{0}.$$

We may also find the equation of the line that passes through two fixed points A and C with position vectors **a** and **c**. Since AC is given by $\mathbf{c} - \mathbf{a}$, the position vector of a general point on the line is

$$\mathbf{r} = \mathbf{a} + \lambda(\mathbf{c} - \mathbf{a}).$$

7.7.2 Equation of a plane

The equation of a plane through a point A with position vector **a** and perpendicular to a unit position vector $\hat{\mathbf{n}}$ (see figure 7.13) is

$$(\mathbf{r} - \mathbf{a}) \cdot \hat{\mathbf{n}} = 0. \tag{7.41}$$

This follows since the vector joining A to a general point R with position vector **r** is $\mathbf{r} - \mathbf{a}$; **r** will lie in the plane if this vector is perpendicular to the normal to the plane. Rewriting (7.41) as $\mathbf{r} \cdot \hat{\mathbf{n}} = \mathbf{a} \cdot \hat{\mathbf{n}}$, we see that the equation of the plane may also be expressed in the form $\mathbf{r} \cdot \hat{\mathbf{n}} = d$, or in component form as

$$lx + my + nz = d, \tag{7.42}$$

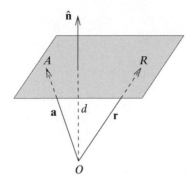

Figure 7.13   The equation of the plane is $(\mathbf{r} - \mathbf{a}) \cdot \hat{\mathbf{n}} = 0$.

where the unit normal to the plane is $\hat{\mathbf{n}} = l\mathbf{i} + m\mathbf{j} + n\mathbf{k}$ and $d = \mathbf{a} \cdot \hat{\mathbf{n}}$ is the perpendicular distance of the plane from the origin.

The equation of a plane containing points $\mathbf{a}$, $\mathbf{b}$ and $\mathbf{c}$ is

$$\mathbf{r} = \mathbf{a} + \lambda(\mathbf{b} - \mathbf{a}) + \mu(\mathbf{c} - \mathbf{a}).$$

This is apparent because starting from the point $\mathbf{a}$ in the plane, all other points may be reached by moving a distance along each of two (non-parallel) directions in the plane. Two such directions are given by $\mathbf{b} - \mathbf{a}$ and $\mathbf{c} - \mathbf{a}$. It can be shown that the equation of this plane may also be written in the more symmetrical form

$$\mathbf{r} = \alpha\mathbf{a} + \beta\mathbf{b} + \gamma\mathbf{c},$$

where $\alpha + \beta + \gamma = 1$.

▶Find the direction of the line of intersection of the two planes $x + 3y - z = 5$ and $2x - 2y + 4z = 3$.

The two planes have normal vectors $\mathbf{n}_1 = \mathbf{i} + 3\mathbf{j} - \mathbf{k}$ and $\mathbf{n}_2 = 2\mathbf{i} - 2\mathbf{j} + 4\mathbf{k}$. It is clear that these are not parallel vectors and so the planes must intersect along some line. The direction $\mathbf{p}$ of this line must be parallel to both planes and hence perpendicular to both normals. Therefore

$$\begin{aligned}
\mathbf{p} &= \mathbf{n}_1 \times \mathbf{n}_2 \\
&= [(3)(4) - (-2)(-1)]\,\mathbf{i} + [(-1)(2) - (1)(4)]\,\mathbf{j} + [(1)(-2) - (3)(2)]\,\mathbf{k} \\
&= 10\mathbf{i} - 6\mathbf{j} - 8\mathbf{k}. \ ◀
\end{aligned}$$

7.7.3   Equation of a sphere

Clearly, the defining property of a sphere is that all points on it are equidistant from a fixed point in space and that the common distance is equal to the radius of the sphere. This is easily expressed in vector notation as

$$|\mathbf{r} - \mathbf{c}|^2 = (\mathbf{r} - \mathbf{c}) \cdot (\mathbf{r} - \mathbf{c}) = a^2, \tag{7.43}$$

where $\mathbf{c}$ is the position vector of the centre of the sphere and a is its radius.

> ►*Find the radius ρ of the circle that is the intersection of the plane $\hat{\mathbf{n}} \cdot \mathbf{r} = p$ and the sphere of radius a centred on the point with position vector $\mathbf{c}$.*

The equation of the sphere is

$$|\mathbf{r} - \mathbf{c}|^2 = a^2, \tag{7.44}$$

and that of the circle of intersection is

$$|\mathbf{r} - \mathbf{b}|^2 = \rho^2, \tag{7.45}$$

where $\mathbf{r}$ is restricted to lie in the plane and $\mathbf{b}$ is the position of the circle's centre.

As $\mathbf{b}$ lies on the plane whose normal is $\hat{\mathbf{n}}$, the vector $\mathbf{b} - \mathbf{c}$ must be parallel to $\hat{\mathbf{n}}$, i.e. $\mathbf{b} - \mathbf{c} = \lambda\hat{\mathbf{n}}$ for some λ. Further, by Pythagoras, we must have $\rho^2 + |\mathbf{b} - \mathbf{c}|^2 = a^2$. Thus $\lambda^2 = a^2 - \rho^2$.

Writing $\mathbf{b} = \mathbf{c} + \sqrt{a^2 - \rho^2}\,\hat{\mathbf{n}}$ and substituting in (7.45) gives

$$r^2 - 2\mathbf{r} \cdot \left(\mathbf{c} + \sqrt{a^2 - \rho^2}\,\hat{\mathbf{n}}\right) + c^2 + 2(\mathbf{c} \cdot \hat{\mathbf{n}})\sqrt{a^2 - \rho^2} + a^2 - \rho^2 = \rho^2,$$

whilst, on expansion, (7.44) becomes

$$r^2 - 2\mathbf{r} \cdot \mathbf{c} + c^2 = a^2.$$

Subtracting these last two equations, using $\hat{\mathbf{n}} \cdot \mathbf{r} = p$ and simplifying yields

$$p - \mathbf{c} \cdot \hat{\mathbf{n}} = \sqrt{a^2 - \rho^2}.$$

On rearrangement, this gives ρ as $\sqrt{a^2 - (p - \mathbf{c} \cdot \hat{\mathbf{n}})^2}$, which places obvious geometrical constraints on the values $a, \mathbf{c}, \hat{\mathbf{n}}$ and p can take if a real intersection between the sphere and the plane is to occur. ◄

7.8 Using vectors to find distances

This section deals with the practical application of vectors to finding distances. Some of these problems are extremely cumbersome in component form, but they all reduce to neat solutions when general vectors, with no explicit basis set, are used. These examples show the power of vectors in simplifying geometrical problems.

7.8.1 Distance from a point to a line

Figure 7.14 shows a line having direction $\mathbf{b}$ that passes through a point A whose position vector is $\mathbf{a}$. To find the *minimum distance* d of the line from a point P whose position vector is $\mathbf{p}$, we must solve the right-angled triangle shown. We see that $d = |\mathbf{p} - \mathbf{a}|\sin\theta$; so, from the definition of the vector product, it follows that

$$d = |(\mathbf{p} - \mathbf{a}) \times \hat{\mathbf{b}}|.$$

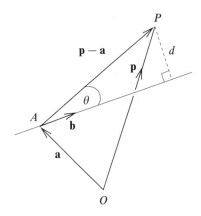

Figure 7.14   The minimum distance from a point to a line.

▶*Find the minimum distance from the point P with coordinates* $(1, 2, 1)$ *to the line* $\mathbf{r} = \mathbf{a} + \lambda\mathbf{b}$, *where* $\mathbf{a} = \mathbf{i} + \mathbf{j} + \mathbf{k}$ *and* $\mathbf{b} = 2\mathbf{i} - \mathbf{j} + 3\mathbf{k}$.

Comparison with (7.39) shows that the line passes through the point $(1, 1, 1)$ and has direction $2\mathbf{i} - \mathbf{j} + 3\mathbf{k}$. The unit vector in this direction is

$$\hat{\mathbf{b}} = \frac{1}{\sqrt{14}}(2\mathbf{i} - \mathbf{j} + 3\mathbf{k}).$$

The position vector of P is $\mathbf{p} = \mathbf{i} + 2\mathbf{j} + \mathbf{k}$ and we find

$$(\mathbf{p} - \mathbf{a}) \times \hat{\mathbf{b}} = \frac{1}{\sqrt{14}} [\mathbf{j} \times (2\mathbf{i} - \mathbf{j} + 3\mathbf{k})]$$

$$= \frac{1}{\sqrt{14}}(3\mathbf{i} - 2\mathbf{k}).$$

Thus the minimum distance from the line to the point P is $d = \sqrt{13/14}$. ◀

7.8.2 Distance from a point to a plane

The minimum distance d from a point P whose position vector is $\mathbf{p}$ to the plane defined by $(\mathbf{r} - \mathbf{a}) \cdot \hat{\mathbf{n}} = 0$ may be deduced by finding any vector from P to the plane and then determining its component in the normal direction. This is shown in figure 7.15. Consider the vector $\mathbf{a} - \mathbf{p}$, which is a particular vector from P to the plane. Its component normal to the plane, and hence its distance from the plane, is given by

$$d = (\mathbf{a} - \mathbf{p}) \cdot \hat{\mathbf{n}}, \tag{7.46}$$

where the sign of d depends on which side of the plane P is situated.

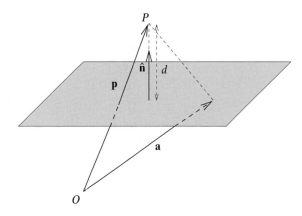

Figure 7.15   The minimum distance d from a point to a plane.

▶*Find the distance from the point P with coordinates $(1, 2, 3)$ to the plane that contains the points A, B and C having coordinates $(0, 1, 0)$, $(2, 3, 1)$ and $(5, 7, 2)$.*

Let us denote the position vectors of the points A, B, C by $\mathbf{a}$, $\mathbf{b}$, $\mathbf{c}$. Two vectors in the plane are

$$\mathbf{b} - \mathbf{a} = 2\mathbf{i} + 2\mathbf{j} + \mathbf{k} \qquad \text{and} \qquad \mathbf{c} - \mathbf{a} = 5\mathbf{i} + 6\mathbf{j} + 2\mathbf{k},$$

and hence a vector normal to the plane is

$$\mathbf{n} = (2\mathbf{i} + 2\mathbf{j} + \mathbf{k}) \times (5\mathbf{i} + 6\mathbf{j} + 2\mathbf{k}) = -2\mathbf{i} + \mathbf{j} + 2\mathbf{k},$$

and its unit normal is

$$\hat{\mathbf{n}} = \frac{\mathbf{n}}{|\mathbf{n}|} = \tfrac{1}{3}(-2\mathbf{i} + \mathbf{j} + 2\mathbf{k}).$$

Denoting the position vector of P by $\mathbf{p}$, the minimum distance from the plane to P is given by

$$
\begin{aligned}
d &= (\mathbf{a} - \mathbf{p}) \cdot \hat{\mathbf{n}} \\
&= (-\mathbf{i} - \mathbf{j} - 3\mathbf{k}) \cdot \tfrac{1}{3}(-2\mathbf{i} + \mathbf{j} + 2\mathbf{k}) \\
&= \tfrac{2}{3} - \tfrac{1}{3} - 2 = -\tfrac{5}{3}.
\end{aligned}
$$

If we take P to be the origin O, then we find $d = \tfrac{1}{3}$, i.e. a positive quantity. It follows from this that the original point P with coordinates $(1, 2, 3)$, for which d was negative, is on the opposite side of the plane from the origin. ◀

7.8.3 Distance from a line to a line

Consider two lines in the directions $\mathbf{a}$ and $\mathbf{b}$, as shown in figure 7.16. Since $\mathbf{a} \times \mathbf{b}$ is by definition perpendicular to both $\mathbf{a}$ and $\mathbf{b}$, the unit vector normal to both these lines is

$$\hat{\mathbf{n}} = \frac{\mathbf{a} \times \mathbf{b}}{|\mathbf{a} \times \mathbf{b}|}.$$

231

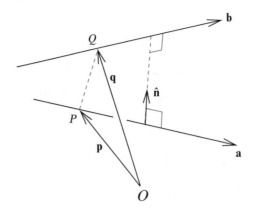

Figure 7.16   The minimum distance from one line to another.

If $\mathbf{p}$ and $\mathbf{q}$ are the position vectors of any two points P and Q on different lines then the vector connecting them is $\mathbf{p} - \mathbf{q}$. Thus, the minimum distance d between the lines is this vector's component along the unit normal, i.e.

$$d = |(\mathbf{p} - \mathbf{q}) \cdot \hat{\mathbf{n}}|.$$

▶*A line is inclined at equal angles to the x-, y- and z-axes and passes through the origin. Another line passes through the points $(1, 2, 4)$ and $(0, 0, 1)$. Find the minimum distance between the two lines.*

The first line is given by

$$\mathbf{r}_1 = \lambda(\mathbf{i} + \mathbf{j} + \mathbf{k}),$$

and the second by

$$\mathbf{r}_2 = \mathbf{k} + \mu(\mathbf{i} + 2\mathbf{j} + 3\mathbf{k}).$$

Hence a vector normal to both lines is

$$\mathbf{n} = (\mathbf{i} + \mathbf{j} + \mathbf{k}) \times (\mathbf{i} + 2\mathbf{j} + 3\mathbf{k}) = \mathbf{i} - 2\mathbf{j} + \mathbf{k},$$

and the unit normal is

$$\hat{\mathbf{n}} = \frac{1}{\sqrt{6}}(\mathbf{i} - 2\mathbf{j} + \mathbf{k}).$$

A vector between the two lines is, for example, the one connecting the points $(0, 0, 0)$ and $(0, 0, 1)$, which is simply $\mathbf{k}$. Thus it follows that the minimum distance between the two lines is

$$d = \frac{1}{\sqrt{6}}|\mathbf{k} \cdot (\mathbf{i} - 2\mathbf{j} + \mathbf{k})| = \frac{1}{\sqrt{6}}. \blacktriangleleft$$

7.8.4 *Distance from a line to a plane*

Let us consider the line $\mathbf{r} = \mathbf{a} + \lambda\mathbf{b}$. This line will intersect any plane to which it is not parallel. Thus, if a plane has a normal $\hat{\mathbf{n}}$ then the minimum distance from the line to the plane is zero unless

$$\mathbf{b} \cdot \hat{\mathbf{n}} = 0,$$

in which case the distance, d, will be

$$d = |(\mathbf{a} - \mathbf{r}) \cdot \hat{\mathbf{n}}|,$$

where $\mathbf{r}$ is any point in the plane.

> ▶ *A line is given by* $\mathbf{r} = \mathbf{a} + \lambda \mathbf{b}$, *where* $\mathbf{a} = 5\mathbf{i} + 7\mathbf{j} + 9\mathbf{k}$ *and* $\mathbf{b} = 4\mathbf{i} + 5\mathbf{j} + 6\mathbf{k}$. *Find the coordinates of the point* P *at which the line intersects the plane*
> $$x + 2y + 3z = 6.$$

A vector normal to the plane is

$$\mathbf{n} = \mathbf{i} + 2\mathbf{j} + 3\mathbf{k},$$

from which we find that $\mathbf{b} \cdot \mathbf{n} \neq 0$. Thus the line does indeed intersect the plane. To find the point of intersection we merely substitute the x-, y- and z- values of a general point on the line into the equation of the plane, obtaining

$$5 + 4\lambda + 2(7 + 5\lambda) + 3(9 + 6\lambda) = 6 \quad \Rightarrow \quad 46 + 32\lambda = 6.$$

This gives $\lambda = -\frac{5}{4}$, which we may substitute into the equation for the line to obtain $x = 5 - \frac{5}{4}(4) = 0$, $y = 7 - \frac{5}{4}(5) = \frac{3}{4}$ and $z = 9 - \frac{5}{4}(6) = \frac{3}{2}$. Thus the point of intersection is $(0, \frac{3}{4}, \frac{3}{2})$. ◀

7.9 Reciprocal vectors

The final section of this chapter introduces the concept of reciprocal vectors, which have particular uses in crystallography.

The two sets of vectors $\mathbf{a}$, $\mathbf{b}$, $\mathbf{c}$ and $\mathbf{a}'$, $\mathbf{b}'$, $\mathbf{c}'$ are called *reciprocal sets* if

$$\mathbf{a} \cdot \mathbf{a}' = \mathbf{b} \cdot \mathbf{b}' = \mathbf{c} \cdot \mathbf{c}' = 1 \tag{7.47}$$

and

$$\mathbf{a}' \cdot \mathbf{b} = \mathbf{a}' \cdot \mathbf{c} = \mathbf{b}' \cdot \mathbf{a} = \mathbf{b}' \cdot \mathbf{c} = \mathbf{c}' \cdot \mathbf{a} = \mathbf{c}' \cdot \mathbf{b} = 0. \tag{7.48}$$

It can be verified (see exercise 7.19) that the reciprocal vectors of $\mathbf{a}$, $\mathbf{b}$ and $\mathbf{c}$ are given by

$$\mathbf{a}' = \frac{\mathbf{b} \times \mathbf{c}}{\mathbf{a} \cdot (\mathbf{b} \times \mathbf{c})}, \tag{7.49}$$

$$\mathbf{b}' = \frac{\mathbf{c} \times \mathbf{a}}{\mathbf{a} \cdot (\mathbf{b} \times \mathbf{c})}, \tag{7.50}$$

$$\mathbf{c}' = \frac{\mathbf{a} \times \mathbf{b}}{\mathbf{a} \cdot (\mathbf{b} \times \mathbf{c})}, \tag{7.51}$$

where $\mathbf{a} \cdot (\mathbf{b} \times \mathbf{c}) \neq 0$. In other words, reciprocal vectors only exist if $\mathbf{a}$, $\mathbf{b}$ and $\mathbf{c}$ are not coplanar. Moreover, if $\mathbf{a}$, $\mathbf{b}$ and $\mathbf{c}$ are mutually orthogonal unit vectors then $\mathbf{a}' = \mathbf{a}$, $\mathbf{b}' = \mathbf{b}$ and $\mathbf{c}' = \mathbf{c}$, so that the two systems of vectors are identical.

▶*Construct the reciprocal vectors of* $\mathbf{a} = 2\mathbf{i}$, $\mathbf{b} = \mathbf{j} + \mathbf{k}$, $\mathbf{c} = \mathbf{i} + \mathbf{k}$.

First we evaluate the triple scalar product:

$$\mathbf{a} \cdot (\mathbf{b} \times \mathbf{c}) = 2\mathbf{i} \cdot [(\mathbf{j} + \mathbf{k}) \times (\mathbf{i} + \mathbf{k})]$$
$$= 2\mathbf{i} \cdot (\mathbf{i} + \mathbf{j} - \mathbf{k}) = 2.$$

Now we find the reciprocal vectors:

$$\mathbf{a}' = \tfrac{1}{2}(\mathbf{j} + \mathbf{k}) \times (\mathbf{i} + \mathbf{k}) = \tfrac{1}{2}(\mathbf{i} + \mathbf{j} - \mathbf{k}),$$
$$\mathbf{b}' = \tfrac{1}{2}(\mathbf{i} + \mathbf{k}) \times 2\mathbf{i} = \mathbf{j},$$
$$\mathbf{c}' = \tfrac{1}{2}(2\mathbf{i}) \times (\mathbf{j} + \mathbf{k}) = -\mathbf{j} + \mathbf{k}.$$

It is easily verified that these reciprocal vectors satisfy their defining properties (7.47), (7.48). ◀

We may also use the concept of reciprocal vectors to define the components of a vector $\mathbf{a}$ with respect to basis vectors $\mathbf{e}_1$, $\mathbf{e}_2$, $\mathbf{e}_3$ that are not mutually orthogonal. If the basis vectors are of unit length and mutually orthogonal, such as the Cartesian basis vectors $\mathbf{i}$, $\mathbf{j}$, $\mathbf{k}$, then (see the text preceeding (7.21)) the vector $\mathbf{a}$ can be written in the form

$$\mathbf{a} = (\mathbf{a} \cdot \mathbf{i})\mathbf{i} + (\mathbf{a} \cdot \mathbf{j})\mathbf{j} + (\mathbf{a} \cdot \mathbf{k})\mathbf{k}.$$

If the basis is not orthonormal, however, then this is no longer true. Nevertheless, we may write the components of $\mathbf{a}$ with respect to a non-orthonormal basis $\mathbf{e}_1$, $\mathbf{e}_2$, $\mathbf{e}_3$ in terms of its reciprocal basis vectors $\mathbf{e}_1'$, $\mathbf{e}_2'$, $\mathbf{e}_3'$, which are defined as in (7.49)–(7.51). If we let

$$\mathbf{a} = a_1\mathbf{e}_1 + a_2\mathbf{e}_2 + a_3\mathbf{e}_3,$$

then the scalar product $\mathbf{a} \cdot \mathbf{e}_1'$ is given by

$$\mathbf{a} \cdot \mathbf{e}_1' = a_1\mathbf{e}_1 \cdot \mathbf{e}_1' + a_2\mathbf{e}_2 \cdot \mathbf{e}_1' + a_3\mathbf{e}_3 \cdot \mathbf{e}_1' = a_1,$$

where we have used the relations (7.48). Similarly, $a_2 = \mathbf{a} \cdot \mathbf{e}_2'$ and $a_3 = \mathbf{a} \cdot \mathbf{e}_3'$; so now

$$\mathbf{a} = (\mathbf{a} \cdot \mathbf{e}_1')\mathbf{e}_1 + (\mathbf{a} \cdot \mathbf{e}_2')\mathbf{e}_2 + (\mathbf{a} \cdot \mathbf{e}_3')\mathbf{e}_3. \tag{7.52}$$

7.10 Exercises

7.1 Which of the following statements about general vectors $\mathbf{a}$, $\mathbf{b}$ and $\mathbf{c}$ are true?

(a) $\mathbf{c} \cdot (\mathbf{a} \times \mathbf{b}) = (\mathbf{b} \times \mathbf{a}) \cdot \mathbf{c}$.
(b) $\mathbf{a} \times (\mathbf{b} \times \mathbf{c}) = (\mathbf{a} \times \mathbf{b}) \times \mathbf{c}$.
(c) $\mathbf{a} \times (\mathbf{b} \times \mathbf{c}) = (\mathbf{a} \cdot \mathbf{c})\mathbf{b} - (\mathbf{a} \cdot \mathbf{b})\mathbf{c}$.
(d) $\mathbf{d} = \lambda\mathbf{a} + \mu\mathbf{b}$ implies $(\mathbf{a} \times \mathbf{b}) \cdot \mathbf{d} = 0$.
(e) $\mathbf{a} \times \mathbf{c} = \mathbf{b} \times \mathbf{c}$ implies $\mathbf{c} \cdot \mathbf{a} - \mathbf{c} \cdot \mathbf{b} = c|\mathbf{a} - \mathbf{b}|$.
(f) $(\mathbf{a} \times \mathbf{b}) \times (\mathbf{c} \times \mathbf{b}) = \mathbf{b}[\mathbf{b} \cdot (\mathbf{c} \times \mathbf{a})]$.

7.2 A unit cell of diamond is a cube of side A, with carbon atoms at each corner, at the centre of each face and, in addition, at positions displaced by $\frac{1}{4}A(\mathbf{i}+\mathbf{j}+\mathbf{k})$ from each of those already mentioned; $\mathbf{i}$, $\mathbf{j}$, $\mathbf{k}$ are unit vectors along the cube axes. One corner of the cube is taken as the origin of coordinates. What are the vectors joining the atom at $\frac{1}{4}A(\mathbf{i}+\mathbf{j}+\mathbf{k})$ to its four nearest neighbours? Determine the angle between the carbon bonds in diamond.

7.3 Identify the following surfaces:

(a) $|\mathbf{r}| = k$; (b) $\mathbf{r} \cdot \mathbf{u} = l$; (c) $\mathbf{r} \cdot \mathbf{u} = m|\mathbf{r}|$ for $-1 \le m \le +1$;
(d) $|\mathbf{r} - (\mathbf{r} \cdot \mathbf{u})\mathbf{u}| = n$.

Here k, l, m and n are fixed scalars and $\mathbf{u}$ is a fixed unit vector.

7.4 Find the angle between the position vectors to the points $(3, -4, 0)$ and $(-2, 1, 0)$ and find the direction cosines of a vector perpendicular to both.

7.5 A, B, C and D are the four corners, in order, of one face of a cube of side 2 units. The opposite face has corners E, F, G and H, with AE, BF, CG and DH as parallel edges of the cube. The centre O of the cube is taken as the origin and the x-, y- and z-axes are parallel to AD, AE and AB, respectively. Find the following:

(a) the angle between the face diagonal AF and the body diagonal AG;
(b) the equation of the plane through B that is parallel to the plane CGE;
(c) the perpendicular distance from the centre J of the face $BCGF$ to the plane OCG;
(d) the volume of the tetrahedron $JOCG$.

7.6 Use vector methods to prove that the lines joining the mid-points of the opposite edges of a tetrahedron $OABC$ meet at a point and that this point bisects each of the lines.

7.7 The edges OP, OQ and OR of a tetrahedron $OPQR$ are vectors $\mathbf{p}$, $\mathbf{q}$ and $\mathbf{r}$, respectively, where $\mathbf{p} = 2\mathbf{i} + 4\mathbf{j}$, $\mathbf{q} = 2\mathbf{i} - \mathbf{j} + 3\mathbf{k}$ and $\mathbf{r} = 4\mathbf{i} - 2\mathbf{j} + 5\mathbf{k}$. Show that OP is perpendicular to the plane containing OQR. Express the volume of the tetrahedron in terms of $\mathbf{p}$, $\mathbf{q}$ and $\mathbf{r}$ and hence calculate the volume.

7.8 Prove, by writing it out in component form, that

$$(\mathbf{a} \times \mathbf{b}) \times \mathbf{c} = (\mathbf{a} \cdot \mathbf{c})\mathbf{b} - (\mathbf{b} \cdot \mathbf{c})\mathbf{a},$$

and deduce the result, stated in equation (7.25), that the operation of forming the vector product is non-associative.

7.9 Prove Lagrange's identity, i.e.

$$(\mathbf{a} \times \mathbf{b}) \cdot (\mathbf{c} \times \mathbf{d}) = (\mathbf{a} \cdot \mathbf{c})(\mathbf{b} \cdot \mathbf{d}) - (\mathbf{a} \cdot \mathbf{d})(\mathbf{b} \cdot \mathbf{c}).$$

7.10 For four arbitrary vectors $\mathbf{a}$, $\mathbf{b}$, $\mathbf{c}$ and $\mathbf{d}$, evaluate

$$(\mathbf{a} \times \mathbf{b}) \times (\mathbf{c} \times \mathbf{d})$$

in two different ways and so prove that

$$\mathbf{a}[\mathbf{b}, \mathbf{c}, \mathbf{d}] - \mathbf{b}[\mathbf{c}, \mathbf{d}, \mathbf{a}] + \mathbf{c}[\mathbf{d}, \mathbf{a}, \mathbf{b}] - \mathbf{d}[\mathbf{a}, \mathbf{b}, \mathbf{c}] = 0.$$

Show that this reduces to the normal Cartesian representation of the vector $\mathbf{d}$, i.e. $d_x\mathbf{i} + d_y\mathbf{j} + d_z\mathbf{k}$, if $\mathbf{a}, \mathbf{b}$ and $\mathbf{c}$ are taken as $\mathbf{i}$, $\mathbf{j}$ and $\mathbf{k}$, the Cartesian base vectors.

7.11 Show that the points $(1, 0, 1)$, $(1, 1, 0)$ and $(1, -3, 4)$ lie on a straight line. Give the equation of the line in the form

$$\mathbf{r} = \mathbf{a} + \lambda\mathbf{b}.$$

7.12 The plane P_1 contains the points A, B and C, which have position vectors $\mathbf{a} = -3\mathbf{i} + 2\mathbf{j}$, $\mathbf{b} = 7\mathbf{i} + 2\mathbf{j}$ and $\mathbf{c} = 2\mathbf{i} + 3\mathbf{j} + 2\mathbf{k}$, respectively. Plane P_2 passes through A and is orthogonal to the line BC, whilst plane P_3 passes through B and is orthogonal to the line AC. Find the coordinates of $\mathbf{r}$, the point of intersection of the three planes.

7.13 Two planes have non-parallel unit normals $\hat{\mathbf{n}}$ and $\hat{\mathbf{m}}$ and their closest distances from the origin are λ and μ, respectively. Find the vector equation of their line of intersection in the form $\mathbf{r} = v\mathbf{p} + \mathbf{a}$.

7.14 Two fixed points, A and B, in three-dimensional space have position vectors $\mathbf{a}$ and $\mathbf{b}$. Identify the plane P given by

$$(\mathbf{a} - \mathbf{b}) \cdot \mathbf{r} = \tfrac{1}{2}(a^2 - b^2),$$

where a and b are the magnitudes of $\mathbf{a}$ and $\mathbf{b}$.

Show also that the equation

$$(\mathbf{a} - \mathbf{r}) \cdot (\mathbf{b} - \mathbf{r}) = 0$$

describes a sphere S of radius $|\mathbf{a} - \mathbf{b}|/2$. Deduce that the intersection of P and S is also the intersection of two spheres, centred on A and B, and each of radius $|\mathbf{a} - \mathbf{b}|/\sqrt{2}$.

7.15 Let O, A, B and C be four points with position vectors $\mathbf{0}$, $\mathbf{a}$, $\mathbf{b}$ and $\mathbf{c}$, and denote by $\mathbf{g} = \lambda\mathbf{a} + \mu\mathbf{b} + v\mathbf{c}$ the position of the centre of the sphere on which they all lie.

(a) Prove that λ, μ and v simultaneously satisfy

$$(\mathbf{a} \cdot \mathbf{a})\lambda + (\mathbf{a} \cdot \mathbf{b})\mu + (\mathbf{a} \cdot \mathbf{c})v = \tfrac{1}{2}a^2$$

and two other similar equations.

(b) By making a change of origin, find the centre and radius of the sphere on which the points $\mathbf{p} = 3\mathbf{i} + \mathbf{j} - 2\mathbf{k}$, $\mathbf{q} = 4\mathbf{i} + 3\mathbf{j} - 3\mathbf{k}$, $\mathbf{r} = 7\mathbf{i} - 3\mathbf{k}$ and $\mathbf{s} = 6\mathbf{i} + \mathbf{j} - \mathbf{k}$ all lie.

7.16 The vectors $\mathbf{a}$, $\mathbf{b}$ and $\mathbf{c}$ are coplanar and related by

$$\lambda\mathbf{a} + \mu\mathbf{b} + v\mathbf{c} = 0,$$

where λ, μ, v are not all zero. Show that the condition for the points with position vectors $\alpha\mathbf{a}$, $\beta\mathbf{b}$ and $\gamma\mathbf{c}$ to be collinear is

$$\frac{\lambda}{\alpha} + \frac{\mu}{\beta} + \frac{v}{\gamma} = 0.$$

7.17 Using vector methods:

(a) Show that the line of intersection of the planes $x + 2y + 3z = 0$ and $3x + 2y + z = 0$ is equally inclined to the x- and z-axes and makes an angle $\cos^{-1}(-2/\sqrt{6})$ with the y-axis.

(b) Find the perpendicular distance between one corner of a unit cube and the major diagonal not passing through it.

7.18 Four points X_i, $i = 1, 2, 3, 4$, taken for simplicity as all lying within the octant $x, y, z \geq 0$, have position vectors $\mathbf{x}_i$. Convince yourself that the direction of vector $\mathbf{x}_n$ lies within the sector of space defined by the directions of the other three vectors if

$$\min_{\text{over } j} \left[\frac{\mathbf{x}_i \cdot \mathbf{x}_j}{|\mathbf{x}_i||\mathbf{x}_j|} \right],$$

considered for $i = 1, 2, 3, 4$ in turn, takes its maximum value for $i = n$, i.e. n equals that value of i for which the largest of the set of angles which $\mathbf{x}_i$ makes with the other vectors, is found to be the lowest. Determine whether any of the four

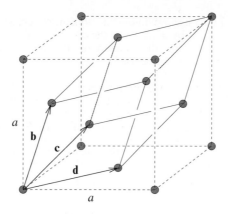

Figure 7.17   A face-centred cubic crystal.

points with coordinates

$$X_1 = (3, 2, 2), \quad X_2 = (2, 3, 1), \quad X_3 = (2, 1, 3), \quad X_4 = (3, 0, 3)$$

lies within the tetrahedron defined by the origin and the other three points.

7.19   The vectors **a**, **b** and **c** are not coplanar. The vectors **a**′, **b**′ and **c**′ are the associated reciprocal vectors. Verify that the expressions (7.49)–(7.51) define a set of reciprocal vectors **a**′, **b**′ and **c**′ with the following properties:

(a) $\mathbf{a}' \cdot \mathbf{a} = \mathbf{b}' \cdot \mathbf{b} = \mathbf{c}' \cdot \mathbf{c} = 1$;
(b) $\mathbf{a}' \cdot \mathbf{b} = \mathbf{a}' \cdot \mathbf{c} = \mathbf{b}' \cdot \mathbf{a}$ etc $= 0$;
(c) $[\mathbf{a}', \mathbf{b}', \mathbf{c}'] = 1/[\mathbf{a}, \mathbf{b}, \mathbf{c}]$;
(d) $\mathbf{a} = (\mathbf{b}' \times \mathbf{c}')/[\mathbf{a}', \mathbf{b}', \mathbf{c}']$.

7.20   Three non-coplanar vectors **a**, **b** and **c**, have as their respective reciprocal vectors the set **a**′, **b**′ and **c**′. Show that the normal to the plane containing the points $k^{-1}\mathbf{a}$, $l^{-1}\mathbf{b}$ and $m^{-1}\mathbf{c}$ is in the direction of the vector $k\mathbf{a}' + l\mathbf{b}' + m\mathbf{c}'$.

7.21   In a crystal with a face-centred cubic structure, the basic cell can be taken as a cube of edge a with its centre at the origin of coordinates and its edges parallel to the Cartesian coordinate axes; atoms are sited at the eight corners and at the centre of each face. However, other basic cells are possible. One is the rhomboid shown in figure 7.17, which has the three vectors **b**, **c** and **d** as edges.

(a)   Show that the volume of the rhomboid is one-quarter that of the cube.

(b)   Show that the angles between pairs of edges of the rhomboid are 60° and that the corresponding angles between pairs of edges of the rhomboid defined by the reciprocal vectors to **b**, **c**, **d** are each 109.5°. (This rhomboid can be used as the basic cell of a body-centred cubic structure, more easily visualised as a cube with an atom at each corner and one at its centre.)

(c)   In order to use the Bragg formula, $2d \sin \theta = n\lambda$, for the scattering of X-rays by a crystal, it is necessary to know the perpendicular distance d between successive planes of atoms; for a given crystal structure, d has a particular value for each set of planes considered. For the face-centred cubic structure find the distance between successive planes with normals in the **k**, **i** + **j** and **i** + **j** + **k** directions.

237

7.22   In subsection 7.6.2 we showed how the moment or torque of a force about an axis could be represented by a vector in the direction of the axis. The magnitude of the vector gives the size of the moment and the sign of the vector gives the sense. Similar representations can be used for angular velocities and angular momenta.

(a) The magnitude of the angular momentum about the origin of a particle of mass m moving with velocity $\mathbf{v}$ on a path that is a perpendicular distance d from the origin is given by $m|\mathbf{v}|d$. Show that if $\mathbf{r}$ is the position of the particle then the vector $\mathbf{J} = \mathbf{r} \times m\mathbf{v}$ represents the angular momentum.

(b) Now consider a rigid collection of particles (or a solid body) rotating about an axis through the origin, the angular velocity of the collection being represented by $\boldsymbol{\omega}$.

(i) Show that the velocity of the ith particle is

$$\mathbf{v}_i = \boldsymbol{\omega} \times \mathbf{r}_i$$

and that the total angular momentum $\mathbf{J}$ is

$$\mathbf{J} = \sum_i m_i [r_i^2 \boldsymbol{\omega} - (\mathbf{r}_i \cdot \boldsymbol{\omega})\mathbf{r}_i].$$

(ii) Show further that the component of $\mathbf{J}$ along the axis of rotation can be written as $I\omega$, where I, the moment of inertia of the collection about the axis or rotation, is given by

$$I = \sum_i m_i \rho_i^2.$$

Interpret ρ_i geometrically.

(iii) Prove that the total kinetic energy of the particles is $\frac{1}{2}I\omega^2$.

7.23   By proceeding as indicated below, prove the *parallel axis theorem*, which states that, for a body of mass M, the moment of inertia I about any axis is related to the corresponding moment of inertia I_0 about a parallel axis that passes through the centre of mass of the body by

$$I = I_0 + Ma_\perp^2,$$

where $a_\perp$ is the perpendicular distance between the two axes. Note that I_0 can be written as

$$\int (\hat{\mathbf{n}} \times \mathbf{r}) \cdot (\hat{\mathbf{n}} \times \mathbf{r}) \, dm,$$

where $\mathbf{r}$ is the vector position, relative to the centre of mass, of the infinitesimal mass dm and $\hat{\mathbf{n}}$ is a unit vector in the direction of the axis of rotation. Write a similar expression for I in which $\mathbf{r}$ is replaced by $\mathbf{r}' = \mathbf{r} - \mathbf{a}$, where $\mathbf{a}$ is the vector position of any point on the axis to which I refers. Use Lagrange's identity and the fact that $\int \mathbf{r} \, dm = \mathbf{0}$ (by the definition of the centre of mass) to establish the result.

7.24   Without carrying out any further integration, use the results of the previous exercise, the worked example in subsection 6.3.4 and exercise 6.10 to prove that the moment of inertia of a uniform rectangular lamina, of mass M and sides a and b, about an axis perpendicular to its plane and passing through the point $(\alpha a/2, \beta b/2)$, with $-1 \le \alpha, \beta \le 1$, is

$$\frac{M}{12}[a^2(1 + 3\alpha^2) + b^2(1 + 3\beta^2)].$$

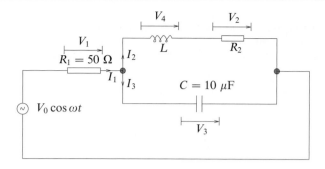

Figure 7.18   An oscillatory electric circuit. The power supply has angular frequency $\omega = 2\pi f = 400\pi \text{ s}^{-1}$.

7.25   Define a set of (non-orthogonal) base vectors $\mathbf{a} = \mathbf{j} + \mathbf{k}$, $\mathbf{b} = \mathbf{i} + \mathbf{k}$ and $\mathbf{c} = \mathbf{i} + \mathbf{j}$.

(a) Establish their reciprocal vectors and hence express the vectors $\mathbf{p} = 3\mathbf{i} - 2\mathbf{j} + \mathbf{k}$, $\mathbf{q} = \mathbf{i} + 4\mathbf{j}$ and $\mathbf{r} = -2\mathbf{i} + \mathbf{j} + \mathbf{k}$ in terms of the base vectors $\mathbf{a}$, $\mathbf{b}$ and $\mathbf{c}$.
(b) Verify that the scalar product $\mathbf{p} \cdot \mathbf{q}$ has the same value, -5, when evaluated using either set of components.

7.26   Systems that can be modelled as damped harmonic oscillators are widespread; pendulum clocks, car shock absorbers, tuning circuits in television sets and radios, and collective electron motions in plasmas and metals are just a few examples.

In all these cases, one or more variables describing the system obey(s) an equation of the form

$$\ddot{x} + 2\gamma \dot{x} + \omega_0^2 x = P \cos \omega t,$$

where $\dot{x} = dx/dt$, etc. and the inclusion of the factor 2 is conventional. In the steady state (i.e. after the effects of any initial displacement or velocity have been damped out) the solution of the equation takes the form

$$x(t) = A \cos(\omega t + \phi).$$

By expressing each term in the form $B \cos(\omega t + \epsilon)$, and representing it by a vector of magnitude B making an angle ϵ with the x-axis, draw a closed vector diagram, at $t = 0$, say, that is equivalent to the equation.

(a) Convince yourself that whatever the value of ω (> 0) ϕ must be negative ($-\pi < \phi \leq 0$) and that

$$\phi = \tan^{-1}\left(\frac{-2\gamma\omega}{\omega_0^2 - \omega^2}\right).$$

(b) Obtain an expression for A in terms of P, ω_0 and ω.

7.27   According to alternating current theory, the currents and potential differences in the components of the circuit shown in figure 7.18 are determined by Kirchhoff's laws and the relationships

$$I_1 = \frac{V_1}{R_1}, \quad I_2 = \frac{V_2}{R_2}, \quad I_3 = i\omega C V_3, \quad V_4 = i\omega L I_2.$$

The factor $i = \sqrt{-1}$ in the expression for I_3 indicates that the phase of I_3 is 90° ahead of V_3. Similarly the phase of V_4 is 90° ahead of I_2.

Measurement shows that V_3 has an amplitude of $0.661V_0$ and a phase of $+13.4°$ relative to that of the power supply. Taking $V_0 = 1 \text{ V}$, and using a series of vector plots for potential differences and currents (they could all be on the same plot if suitable scales were chosen), determine all unknown currents and potential differences and find values for the inductance of L and the resistance of R_2.

[Scales of $1\,\text{cm} = 0.1\,\text{V}$ for potential differences and $1\,\text{cm} = 1\,\text{mA}$ for currents are convenient.]

7.11 Hints and answers

7.1   (c), (d) and (e).

7.3   (a) A sphere of radius k centred on the origin; (b) a plane with its normal in the direction of $\mathbf{u}$ and at a distance l from the origin; (c) a cone with its axis parallel to $\mathbf{u}$ and of semiangle $\cos^{-1} m$; (d) a circular cylinder of radius n with its axis parallel to $\mathbf{u}$.

7.5   (a) $\cos^{-1} \sqrt{2/3}$; (b) $z - x = 2$; (c) $1/\sqrt{2}$; (d) $\frac{1}{3}\frac{1}{2}(\mathbf{c} \times \mathbf{g}) \cdot \mathbf{j} = \frac{1}{3}$.

7.7   Show that $\mathbf{q} \times \mathbf{r}$ is parallel to $\mathbf{p}$; volume $= \frac{1}{3} \left[\frac{1}{2}(\mathbf{q} \times \mathbf{r}) \cdot \mathbf{p} \right] = \frac{5}{3}$.

7.9   Note that $(\mathbf{a} \times \mathbf{b}) \cdot (\mathbf{c} \times \mathbf{d}) = \mathbf{d} \cdot [(\mathbf{a} \times \mathbf{b}) \times \mathbf{c}]$ and use the result for a triple vector product to expand the expression in square brackets.

7.11   Show that the position vectors of the points are linearly dependent; $\mathbf{r} = \mathbf{a} + \lambda \mathbf{b}$ where $\mathbf{a} = \mathbf{i} + \mathbf{k}$ and $\mathbf{b} = -\mathbf{j} + \mathbf{k}$.

7.13   Show that $\mathbf{p}$ must have the direction $\hat{\mathbf{n}} \times \hat{\mathbf{m}}$ and write $\mathbf{a}$ as $x\hat{\mathbf{n}} + y\hat{\mathbf{m}}$. By obtaining a pair of simultaneous equations for x and y, prove that $x = (\lambda - \mu\hat{\mathbf{n}} \cdot \hat{\mathbf{m}})/[1 - (\hat{\mathbf{n}} \cdot \hat{\mathbf{m}})^2]$ and that $y = (\mu - \lambda\hat{\mathbf{n}} \cdot \hat{\mathbf{m}})/[1 - (\hat{\mathbf{n}} \cdot \hat{\mathbf{m}})^2]$.

7.15   (a) Note that $|\mathbf{a} - \mathbf{g}|^2 = R^2 = |\mathbf{0} - \mathbf{g}|^2$, leading to $\mathbf{a} \cdot \mathbf{a} = 2\mathbf{a} \cdot \mathbf{g}$.
        (b) Make $\mathbf{p}$ the new origin and solve the three simultaneous linear equations to obtain $\lambda = 5/18$, $\mu = 10/18$, $\nu = -3/18$, giving $\mathbf{g} = 2\mathbf{i} - \mathbf{k}$ and a sphere of radius $\sqrt{5}$ centred on $(5, 1, -3)$.

7.17   (a) Find two points on both planes, say $(0, 0, 0)$ and $(1, -2, 1)$, and hence determine the direction cosines of the line of intersection; (b) $\left(\frac{2}{3}\right)^{1/2}$.

7.19   For (c) and (d), treat $(\mathbf{c} \times \mathbf{a}) \times (\mathbf{a} \times \mathbf{b})$ as a triple vector product with $\mathbf{c} \times \mathbf{a}$ as one of the three vectors.

7.21   (b) $\mathbf{b}' = a^{-1}(-\mathbf{i} + \mathbf{j} + \mathbf{k})$, $\mathbf{c}' = a^{-1}(\mathbf{i} - \mathbf{j} + \mathbf{k})$, $\mathbf{d}' = a^{-1}(\mathbf{i} + \mathbf{j} - \mathbf{k})$; (c) $a/2$ for direction $\mathbf{k}$; successive planes through $(0, 0, 0)$ and $(a/2, 0, a/2)$ give a spacing of $a/\sqrt{8}$ for direction $\mathbf{i} + \mathbf{j}$; successive planes through $(-a/2, 0, 0)$ and $(a/2, 0, 0)$ give a spacing of $a/\sqrt{3}$ for direction $\mathbf{i} + \mathbf{j} + \mathbf{k}$.

7.23   Note that $a^2 - (\hat{\mathbf{n}} \cdot \mathbf{a})^2 = a_\perp^2$.

7.25 $\mathbf{p} = -2\mathbf{a} + 3\mathbf{b}$, $\mathbf{q} = \frac{3}{2}\mathbf{a} - \frac{3}{2}\mathbf{b} + \frac{5}{2}\mathbf{c}$ and $\mathbf{r} = 2\mathbf{a} - \mathbf{b} - \mathbf{c}$. Remember that $\mathbf{a} \cdot \mathbf{a} = \mathbf{b} \cdot \mathbf{b} = \mathbf{c} \cdot \mathbf{c} = 2$ and $\mathbf{a} \cdot \mathbf{b} = \mathbf{a} \cdot \mathbf{c} = \mathbf{b} \cdot \mathbf{c} = 1$.

7.27   With currents in mA and potential differences in volts:
$I_1 = (7.76, -23.2°)$, $I_2 = (14.36, -50.8°)$, $I_3 = (8.30, 103.4°)$;
$V_1 = (0.388, -23.2°)$, $V_2 = (0.287, -50.8°)$, $V_4 = (0.596, 39.2°)$;
$L = 33$ mH, $R_2 = 20\ \Omega$.

8

Matrices and vector spaces

In the previous chapter we defined a *vector* as a geometrical object which has both a magnitude and a direction and which may be thought of as an arrow fixed in our familiar three-dimensional space, a space which, if we need to, we define by reference to, say, the fixed stars. This geometrical definition of a vector is both useful and important since it is *independent* of any coordinate system with which we choose to label points in space.

In most specific applications, however, it is necessary at some stage to choose a coordinate system and to break down a vector into its *component vectors* in the directions of increasing coordinate values. Thus for a particular Cartesian coordinate system (for example) the component vectors of a vector $\mathbf{a}$ will be $a_x\mathbf{i}$, $a_y\mathbf{j}$ and $a_z\mathbf{k}$ and the complete vector will be

$$\mathbf{a} = a_x\mathbf{i} + a_y\mathbf{j} + a_z\mathbf{k}. \tag{8.1}$$

Although we have so far considered only real three-dimensional space, we may extend our notion of a vector to more abstract spaces, which in general can have an arbitrary number of dimensions N. We may still think of such a vector as an 'arrow' in this abstract space, so that it is again *independent* of any (N-dimensional) coordinate system with which we choose to label the space. As an example of such a space, which, though abstract, has very practical applications, we may consider the description of a mechanical or electrical system. If the state of a system is uniquely specified by assigning values to a set of N variables, which could be angles or currents, for example, then that state can be represented by a vector in an N-dimensional space, the vector having those values as its components.

In this chapter we first discuss general *vector spaces* and their properties. We then go on to discuss the transformation of one vector into another by a linear operator. This leads naturally to the concept of a *matrix*, a two-dimensional array of numbers. The properties of matrices are then discussed and we conclude with a discussion of how to use these properties to solve systems of linear equations. The application of matrices to the study of oscillations in physical systems is taken up in chapter 9.

8.1 Vector spaces

A set of objects (vectors) $\mathbf{a}$, $\mathbf{b}$, $\mathbf{c}$, ... is said to form a *linear vector space V* if:

(i) the set is closed under commutative and associative addition, so that

$$\mathbf{a} + \mathbf{b} = \mathbf{b} + \mathbf{a}, \tag{8.2}$$

$$(\mathbf{a} + \mathbf{b}) + \mathbf{c} = \mathbf{a} + (\mathbf{b} + \mathbf{c}); \tag{8.3}$$

(ii) the set is closed under multiplication by a scalar (any complex number) to form a new vector $\lambda\mathbf{a}$, the operation being both distributive and associative so that

$$\lambda(\mathbf{a} + \mathbf{b}) = \lambda\mathbf{a} + \lambda\mathbf{b}, \tag{8.4}$$

$$(\lambda + \mu)\mathbf{a} = \lambda\mathbf{a} + \mu\mathbf{a}, \tag{8.5}$$

$$\lambda(\mu\mathbf{a}) = (\lambda\mu)\mathbf{a}, \tag{8.6}$$

where λ and μ are arbitrary scalars;

(iii) there exists a *null vector* $\mathbf{0}$ such that $\mathbf{a} + \mathbf{0} = \mathbf{a}$ for all $\mathbf{a}$;

(iv) multiplication by unity leaves any vector unchanged, i.e. $1 \times \mathbf{a} = \mathbf{a}$;

(v) all vectors have a corresponding *negative vector* $-\mathbf{a}$ such that $\mathbf{a} + (-\mathbf{a}) = \mathbf{0}$. It follows from (8.5) with $\lambda = 1$ and $\mu = -1$ that $-\mathbf{a}$ is the same vector as $(-1) \times \mathbf{a}$.

We note that if we restrict all scalars to be real then we obtain a *real vector space* (an example of which is our familiar three-dimensional space); otherwise, in general, we obtain a *complex vector space*. We note that it is common to use the terms 'vector space' and 'space', instead of the more formal 'linear vector space'.

The *span* of a set of vectors $\mathbf{a}, \mathbf{b}, \dots, \mathbf{s}$ is defined as the set of all vectors that may be written as a linear sum of the original set, i.e. all vectors

$$\mathbf{x} = \alpha\mathbf{a} + \beta\mathbf{b} + \cdots + \sigma\mathbf{s} \tag{8.7}$$

that result from the infinite number of possible values of the (in general complex) scalars $\alpha, \beta, \dots, \sigma$. If $\mathbf{x}$ in (8.7) is equal to $\mathbf{0}$ for some choice of $\alpha, \beta, \dots, \sigma$ (not *all* zero), i.e. if

$$\alpha\mathbf{a} + \beta\mathbf{b} + \cdots + \sigma\mathbf{s} = \mathbf{0}, \tag{8.8}$$

then the set of vectors $\mathbf{a}, \mathbf{b}, \dots, \mathbf{s}$, is said to be *linearly dependent*. In such a set at least one vector is redundant, since it can be expressed as a linear sum of the others. If, however, (8.8) is not satisfied by *any* set of coefficients (other than the trivial case in which all the coefficients are zero) then the vectors are *linearly independent*, and no vector in the set can be expressed as a linear sum of the others.

If, in a given vector space, there exist sets of N linearly independent vectors, but no set of $N + 1$ linearly independent vectors, then the vector space is said to be N-dimensional. (In this chapter we will limit our discussion to vector spaces of finite dimensionality; spaces of infinite dimensionality are discussed in chapter 17.)

8.1.1 Basis vectors

If V is an N-dimensional vector space then *any* set of N linearly independent vectors $e_1, e_2, \ldots, e_N$ forms a *basis* for V. If x is an arbitrary vector lying in V then the set of $N + 1$ vectors $x, e_1, e_2, \ldots, e_N$, must be *linearly dependent* and therefore such that

$$\alpha e_1 + \beta e_2 + \cdots + \sigma e_N + \chi x = 0, \tag{8.9}$$

where the coefficients $\alpha, \beta, \ldots, \chi$ are not all equal to 0, and in particular $\chi \neq 0$. Rearranging (8.9) we may write x as a linear sum of the vectors e_i as follows:

$$x = x_1 e_1 + x_2 e_2 + \cdots + x_N e_N = \sum_{i=1}^{N} x_i e_i, \tag{8.10}$$

for some set of coefficients x_i that are simply related to the original coefficients, e.g. $x_1 = -\alpha/\chi$, $x_2 = -\beta/\chi$, etc. Since any x lying in the span of V can be expressed in terms of the *basis* or *base vectors* e_i, the latter are said to form a *complete* set. The coefficients x_i are the *components* of x with respect to the e_i-basis. These components are *unique*, since if both

$$x = \sum_{i=1}^{N} x_i e_i \quad \text{and} \quad x = \sum_{i=1}^{N} y_i e_i,$$

then

$$\sum_{i=1}^{N} (x_i - y_i) e_i = 0, \tag{8.11}$$

which, since the e_i are linearly independent, has only the solution $x_i = y_i$ for all $i = 1, 2, \ldots, N$.

From the above discussion we see that *any* set of N linearly independent vectors can form a basis for an N-dimensional space. If we choose a different set e_i', $i = 1, \ldots, N$ then we can write x as

$$x = x_1' e_1' + x_2' e_2' + \cdots + x_N' e_N' = \sum_{i=1}^{N} x_i' e_i'. \tag{8.12}$$

We reiterate that the vector $\mathbf{x}$ (a geometrical entity) is independent of the basis – it is only the components of $\mathbf{x}$ that depend on the basis. We note, however, that given a set of vectors $\mathbf{u}_1, \mathbf{u}_2, \ldots, \mathbf{u}_M$, where $M \neq N$, in an N-dimensional vector space, then *either* there exists a vector that cannot be expressed as a linear combination of the $\mathbf{u}_i$ *or*, for some vector that can be so expressed, the components are not unique.

8.1.2 The inner product

We may usefully add to the description of vectors in a vector space by defining the *inner product* of two vectors, denoted in general by $\langle \mathbf{a} | \mathbf{b} \rangle$, which is a scalar function of $\mathbf{a}$ and $\mathbf{b}$. The scalar or dot product, $\mathbf{a} \cdot \mathbf{b} \equiv |\mathbf{a}||\mathbf{b}| \cos \theta$, of vectors in real three-dimensional space (where θ is the angle between the vectors), was introduced in the last chapter and is an example of an inner product. In effect the notion of an inner product $\langle \mathbf{a} | \mathbf{b} \rangle$ is a generalisation of the dot product to more abstract vector spaces. Alternative notations for $\langle \mathbf{a} | \mathbf{b} \rangle$ are $(\mathbf{a}, \mathbf{b})$, or simply $\mathbf{a} \cdot \mathbf{b}$.

The inner product has the following properties:

(i) $\langle \mathbf{a} | \mathbf{b} \rangle = \langle \mathbf{b} | \mathbf{a} \rangle^*$,
(ii) $\langle \mathbf{a} | \lambda \mathbf{b} + \mu \mathbf{c} \rangle = \lambda \langle \mathbf{a} | \mathbf{b} \rangle + \mu \langle \mathbf{a} | \mathbf{c} \rangle$.

We note that in general, for a complex vector space, (i) and (ii) imply that

$$\langle \lambda \mathbf{a} + \mu \mathbf{b} | \mathbf{c} \rangle = \lambda^* \langle \mathbf{a} | \mathbf{c} \rangle + \mu^* \langle \mathbf{b} | \mathbf{c} \rangle, \tag{8.13}$$

$$\langle \lambda \mathbf{a} | \mu \mathbf{b} \rangle = \lambda^* \mu \langle \mathbf{a} | \mathbf{b} \rangle. \tag{8.14}$$

Following the analogy with the dot product in three-dimensional real space, two vectors in a general vector space are defined to be *orthogonal* if $\langle \mathbf{a} | \mathbf{b} \rangle = 0$. Similarly, the *norm* of a vector $\mathbf{a}$ is given by $\|\mathbf{a}\| = \langle \mathbf{a} | \mathbf{a} \rangle^{1/2}$ and is clearly a generalisation of the length or modulus $|\mathbf{a}|$ of a vector $\mathbf{a}$ in three-dimensional space. In a general vector space $\langle \mathbf{a} | \mathbf{a} \rangle$ can be positive or negative; however, we shall be primarily concerned with spaces in which $\langle \mathbf{a} | \mathbf{a} \rangle \geq 0$ and which are thus said to have a *positive semi-definite norm*. In such a space $\langle \mathbf{a} | \mathbf{a} \rangle = 0$ implies $\mathbf{a} = \mathbf{0}$.

Let us now introduce into our N-dimensional vector space a basis $\hat{\mathbf{e}}_1, \hat{\mathbf{e}}_2, \ldots, \hat{\mathbf{e}}_N$ that has the desirable property of being *orthonormal* (the basis vectors are mutually orthogonal and each has unit norm), i.e. a basis that has the property

$$\langle \hat{\mathbf{e}}_i | \hat{\mathbf{e}}_j \rangle = \delta_{ij}. \tag{8.15}$$

Here δ_{ij} is the *Kronecker delta* symbol (of which we say more in chapter 26) and has the properties

$$\delta_{ij} = \begin{cases} 1 & \text{for } i = j, \\ 0 & \text{for } i \neq j. \end{cases}$$

In the above basis we may express any two vectors **a** and **b** as

$$\mathbf{a} = \sum_{i=1}^{N} a_i \hat{\mathbf{e}}_i \quad \text{and} \quad \mathbf{b} = \sum_{i=1}^{N} b_i \hat{\mathbf{e}}_i.$$

Furthermore, *in such an orthonormal basis* we have, for any **a**,

$$\langle \hat{\mathbf{e}}_j | \mathbf{a} \rangle = \sum_{i=1}^{N} \langle \hat{\mathbf{e}}_j | a_i \hat{\mathbf{e}}_i \rangle = \sum_{i=1}^{N} a_i \langle \hat{\mathbf{e}}_j | \hat{\mathbf{e}}_i \rangle = a_j. \tag{8.16}$$

Thus the components of **a** are given by $a_i = \langle \hat{\mathbf{e}}_i | \mathbf{a} \rangle$. Note that this is *not* true unless the basis is orthonormal. We can write the inner product of **a** and **b** in terms of their components in an orthonormal basis as

$$\langle \mathbf{a} | \mathbf{b} \rangle = \langle a_1 \hat{\mathbf{e}}_1 + a_2 \hat{\mathbf{e}}_2 + \cdots + a_N \hat{\mathbf{e}}_N | b_1 \hat{\mathbf{e}}_1 + b_2 \hat{\mathbf{e}}_2 + \cdots + b_N \hat{\mathbf{e}}_N \rangle$$

$$= \sum_{i=1}^{N} a_i^* b_i \langle \hat{\mathbf{e}}_i | \hat{\mathbf{e}}_i \rangle + \sum_{i=1}^{N} \sum_{j \neq i}^{N} a_i^* b_j \langle \hat{\mathbf{e}}_i | \hat{\mathbf{e}}_j \rangle$$

$$= \sum_{i=1}^{N} a_i^* b_i,$$

where the second equality follows from (8.14) and the third from (8.15). This is clearly a generalisation of the expression (7.21) for the dot product of vectors in three-dimensional space.

We may generalise the above to the case where the base vectors $\mathbf{e}_1, \mathbf{e}_2, \ldots, \mathbf{e}_N$ are *not* orthonormal (or orthogonal). In general we can define the N^2 numbers

$$G_{ij} = \langle \mathbf{e}_i | \mathbf{e}_j \rangle. \tag{8.17}$$

Then, if $\mathbf{a} = \sum_{i=1}^{N} a_i \mathbf{e}_i$ and $\mathbf{b} = \sum_{i=1}^{N} b_i \mathbf{e}_i$, the inner product of **a** and **b** is given by

$$\langle \mathbf{a} | \mathbf{b} \rangle = \left\langle \sum_{i=1}^{N} a_i \mathbf{e}_i \,\middle|\, \sum_{j=1}^{N} b_j \mathbf{e}_j \right\rangle$$

$$= \sum_{i=1}^{N} \sum_{j=1}^{N} a_i^* b_j \langle \mathbf{e}_i | \mathbf{e}_j \rangle$$

$$= \sum_{i=1}^{N} \sum_{j=1}^{N} a_i^* G_{ij} b_j. \tag{8.18}$$

We further note that from (8.17) and the properties of the inner product we require $G_{ij} = G_{ji}^*$. This in turn ensures that $\|\mathbf{a}\| = \langle \mathbf{a} | \mathbf{a} \rangle$ is real, since then

$$\langle \mathbf{a} | \mathbf{a} \rangle^* = \sum_{i=1}^{N} \sum_{j=1}^{N} a_i G_{ij}^* a_j^* = \sum_{j=1}^{N} \sum_{i=1}^{N} a_j^* G_{ji} a_i = \langle \mathbf{a} | \mathbf{a} \rangle.$$

8.1.3 Some useful inequalities

For a set of objects (vectors) forming a linear vector space in which $\langle \mathbf{a}|\mathbf{a}\rangle \geq 0$ for all $\mathbf{a}$, the following inequalities are often useful.

(i) *Schwarz's inequality* is the most basic result and states that

$$|\langle \mathbf{a}|\mathbf{b}\rangle| \leq \|\mathbf{a}\|\,\|\mathbf{b}\|, \tag{8.19}$$

where the equality holds when $\mathbf{a}$ is a scalar multiple of $\mathbf{b}$, i.e. when $\mathbf{a} = \lambda\mathbf{b}$. It is important here to distinguish between the *absolute value* of a scalar, $|\lambda|$, and the *norm* of a vector, $\|\mathbf{a}\|$. Schwarz's inequality may be proved by considering

$$
\begin{aligned}
\|\mathbf{a} + \lambda\mathbf{b}\|^2 &= \langle \mathbf{a} + \lambda\mathbf{b}|\mathbf{a} + \lambda\mathbf{b}\rangle \\
&= \langle \mathbf{a}|\mathbf{a}\rangle + \lambda\langle \mathbf{a}|\mathbf{b}\rangle + \lambda^*\langle \mathbf{b}|\mathbf{a}\rangle + \lambda\lambda^*\langle \mathbf{b}|\mathbf{b}\rangle.
\end{aligned}
$$

If we write $\langle \mathbf{a}|\mathbf{b}\rangle$ as $|\langle \mathbf{a}|\mathbf{b}\rangle|e^{i\alpha}$ then

$$\|\mathbf{a} + \lambda\mathbf{b}\|^2 = \|\mathbf{a}\|^2 + |\lambda|^2\|\mathbf{b}\|^2 + \lambda|\langle \mathbf{a}|\mathbf{b}\rangle|e^{i\alpha} + \lambda^*|\langle \mathbf{a}|\mathbf{b}\rangle|e^{-i\alpha}.$$

However, $\|\mathbf{a} + \lambda\mathbf{b}\|^2 \geq 0$ for all λ, so we may choose $\lambda = re^{-i\alpha}$ and require that, for all r,

$$0 \leq \|\mathbf{a} + \lambda\mathbf{b}\|^2 = \|\mathbf{a}\|^2 + r^2\|\mathbf{b}\|^2 + 2r|\langle \mathbf{a}|\mathbf{b}\rangle|.$$

This means that the quadratic equation in r formed by setting the RHS equal to zero must have no real roots. This, in turn, implies that

$$4|\langle \mathbf{a}|\mathbf{b}\rangle|^2 \leq 4\|\mathbf{a}\|^2\|\mathbf{b}\|^2,$$

which, on taking the square root (all factors are necessarily positive) of both sides, gives Schwarz's inequality.

(ii) The *triangle inequality* states that

$$\|\mathbf{a} + \mathbf{b}\| \leq \|\mathbf{a}\| + \|\mathbf{b}\| \tag{8.20}$$

and may be derived from the properties of the inner product and Schwarz's inequality as follows. Let us first consider

$$\|\mathbf{a} + \mathbf{b}\|^2 = \|\mathbf{a}\|^2 + \|\mathbf{b}\|^2 + 2\,\mathrm{Re}\,\langle \mathbf{a}|\mathbf{b}\rangle \leq \|\mathbf{a}\|^2 + \|\mathbf{b}\|^2 + 2|\langle \mathbf{a}|\mathbf{b}\rangle|.$$

Using Schwarz's inequality we then have

$$\|\mathbf{a} + \mathbf{b}\|^2 \leq \|\mathbf{a}\|^2 + \|\mathbf{b}\|^2 + 2\|\mathbf{a}\|\,\|\mathbf{b}\| = (\|\mathbf{a}\| + \|\mathbf{b}\|)^2,$$

which, on taking the square root, gives the triangle inequality (8.20).

(iii) *Bessel's inequality* requires the introduction of an orthonormal basis $\hat{\mathbf{e}}_i$, $i = 1, 2, \ldots, N$ into the N-dimensional vector space; it states that

$$\|\mathbf{a}\|^2 \geq \sum_i |\langle \hat{\mathbf{e}}_i|\mathbf{a}\rangle|^2, \tag{8.21}$$

where the equality holds if the sum includes all N basis vectors. If not all the basis vectors are included in the sum then the inequality results (though of course the equality remains if those basis vectors omitted all have $a_i = 0$). Bessel's inequality can also be written

$$\langle \mathbf{a}|\mathbf{a} \rangle \geq \sum_i |a_i|^2,$$

where the a_i are the components of $\mathbf{a}$ in the orthonormal basis. From (8.16) these are given by $a_i = \langle \hat{\mathbf{e}}_i|\mathbf{a} \rangle$. The above may be proved by considering

$$\left\| \mathbf{a} - \sum_i \langle \hat{\mathbf{e}}_i|\mathbf{a} \rangle \hat{\mathbf{e}}_i \right\|^2 = \left\langle \mathbf{a} - \sum_i \langle \hat{\mathbf{e}}_i|\mathbf{a} \rangle \hat{\mathbf{e}}_i \middle| \mathbf{a} - \sum_j \langle \hat{\mathbf{e}}_j|\mathbf{a} \rangle \hat{\mathbf{e}}_j \right\rangle.$$

Expanding out the inner product and using $\langle \hat{\mathbf{e}}_i|\mathbf{a} \rangle^* = \langle \mathbf{a}|\hat{\mathbf{e}}_i \rangle$, we obtain

$$\left\| \mathbf{a} - \sum_i \langle \hat{\mathbf{e}}_i|\mathbf{a} \rangle \hat{\mathbf{e}}_i \right\|^2 = \langle \mathbf{a}|\mathbf{a} \rangle - 2 \sum_i \langle \mathbf{a}|\hat{\mathbf{e}}_i \rangle \langle \hat{\mathbf{e}}_i|\mathbf{a} \rangle + \sum_i \sum_j \langle \mathbf{a}|\hat{\mathbf{e}}_i \rangle \langle \hat{\mathbf{e}}_j|\mathbf{a} \rangle \langle \hat{\mathbf{e}}_i|\hat{\mathbf{e}}_j \rangle.$$

Now $\langle \hat{\mathbf{e}}_i|\hat{\mathbf{e}}_j \rangle = \delta_{ij}$, since the basis is orthonormal, and so we find

$$0 \leq \left\| \mathbf{a} - \sum_i \langle \hat{\mathbf{e}}_i|\mathbf{a} \rangle \hat{\mathbf{e}}_i \right\|^2 = \|\mathbf{a}\|^2 - \sum_i |\langle \hat{\mathbf{e}}_i|\mathbf{a} \rangle|^2,$$

which is Bessel's inequality.

We take this opportunity to mention also

(iv) the *parallelogram equality*

$$\|\mathbf{a} + \mathbf{b}\|^2 + \|\mathbf{a} - \mathbf{b}\|^2 = 2 \left(\|\mathbf{a}\|^2 + \|\mathbf{b}\|^2 \right), \tag{8.22}$$

which may be proved straightforwardly from the properties of the inner product.

8.2 Linear operators

We now discuss the action of *linear operators* on vectors in a vector space. A linear operator $\mathcal{A}$ associates with every vector $\mathbf{x}$ another vector

$$\mathbf{y} = \mathcal{A}\mathbf{x},$$

in such a way that, for two vectors $\mathbf{a}$ and $\mathbf{b}$,

$$\mathcal{A}(\lambda\mathbf{a} + \mu\mathbf{b}) = \lambda\mathcal{A}\mathbf{a} + \mu\mathcal{A}\mathbf{b},$$

where λ, μ are scalars. We say that $\mathcal{A}$ 'operates' on $\mathbf{x}$ to give the vector $\mathbf{y}$. We note that the action of $\mathcal{A}$ is *independent* of any basis or coordinate system and may be thought of as 'transforming' one geometrical entity (i.e. a vector) into another.

If we now introduce a basis $\mathbf{e}_i$, $i = 1, 2, \ldots, N$, into our vector space then the action of $\mathcal{A}$ on each of the basis vectors is to produce a linear combination of the latter; this may be written as

$$\mathcal{A}\mathbf{e}_j = \sum_{i=1}^{N} A_{ij}\mathbf{e}_i, \tag{8.23}$$

where A_{ij} is the ith component of the vector $\mathcal{A}\mathbf{e}_j$ in this basis; collectively the numbers A_{ij} are called the components of the linear operator in the $\mathbf{e}_i$-basis. *In this basis we can express the relation* $\mathbf{y} = \mathcal{A}\mathbf{x}$ in component form as

$$\mathbf{y} = \sum_{i=1}^{N} y_i\mathbf{e}_i = \mathcal{A}\left(\sum_{j=1}^{N} x_j\mathbf{e}_j\right) = \sum_{j=1}^{N} x_j \sum_{i=1}^{N} A_{ij}\mathbf{e}_i,$$

and hence, in purely component form, in this basis we have

$$y_i = \sum_{j=1}^{N} A_{ij}x_j. \tag{8.24}$$

If we had chosen a different basis $\mathbf{e}'_i$, in which the components of $\mathbf{x}$, $\mathbf{y}$ and $\mathcal{A}$ are x'_i, y'_i and A'_{ij} respectively then the geometrical relationship $\mathbf{y} = \mathcal{A}\mathbf{x}$ would be represented in this new basis by

$$y'_i = \sum_{j=1}^{N} A'_{ij}x'_j.$$

We have so far assumed that the vector $\mathbf{y}$ is in the same vector space as $\mathbf{x}$. If, however, $\mathbf{y}$ belongs to a different vector space, which may in general be M-dimensional ($M \neq N$) then the above analysis needs a slight modification. By introducing a basis set $\mathbf{f}_i$, $i = 1, 2, \ldots, M$, into the vector space to which $\mathbf{y}$ belongs we may generalise (8.23) as

$$\mathcal{A}\mathbf{e}_j = \sum_{i=1}^{M} A_{ij}\mathbf{f}_i,$$

where the components A_{ij} of the linear operator $\mathcal{A}$ relate to both of the bases $\mathbf{e}_j$ and $\mathbf{f}_i$.

8.2.1 Properties of linear operators

If $\mathbf{x}$ is a vector and $\mathcal{A}$ and $\mathcal{B}$ are two linear operators then it follows that

$$(\mathcal{A} + \mathcal{B})\mathbf{x} = \mathcal{A}\mathbf{x} + \mathcal{B}\mathbf{x},$$
$$(\lambda\mathcal{A})\mathbf{x} = \lambda(\mathcal{A}\mathbf{x}),$$
$$(\mathcal{A}\mathcal{B})\mathbf{x} = \mathcal{A}(\mathcal{B}\mathbf{x}),$$

where in the last equality we see that the action of two linear operators in succession is associative. The product of two linear operators is not in general commutative, however, so that in general $\mathcal{A}\mathcal{B}\mathbf{x} \neq \mathcal{B}\mathcal{A}\mathbf{x}$. In an obvious way we define the null (or zero) and identity operators by

$$\mathcal{O}\mathbf{x} = \mathbf{0} \quad \text{and} \quad \mathcal{I}\mathbf{x} = \mathbf{x},$$

for any vector $\mathbf{x}$ in our vector space. Two operators $\mathcal{A}$ and $\mathcal{B}$ are equal if $\mathcal{A}\mathbf{x} = \mathcal{B}\mathbf{x}$ for all vectors $\mathbf{x}$. Finally, if there exists an operator $\mathcal{A}^{-1}$ such that

$$\mathcal{A}\mathcal{A}^{-1} = \mathcal{A}^{-1}\mathcal{A} = \mathcal{I}$$

then $\mathcal{A}^{-1}$ is the *inverse* of $\mathcal{A}$. Some linear operators do not possess an inverse and are called *singular*, whilst those operators that do have an inverse are termed *non-singular*.

8.3 Matrices

We have seen that in a particular basis $\mathbf{e}_i$ both vectors and linear operators can be described in terms of their components with respect to the basis. These components may be displayed as an array of numbers called a *matrix*. In general, if a linear operator $\mathcal{A}$ transforms vectors from an N-dimensional vector space, for which we choose a basis $\mathbf{e}_j$, $j = 1, 2, \ldots, N$, into vectors belonging to an M-dimensional vector space, with basis $\mathbf{f}_i$, $i = 1, 2, \ldots, M$, then we may represent the operator $\mathcal{A}$ by the matrix

$$\mathsf{A} = \begin{pmatrix} A_{11} & A_{12} & \ldots & A_{1N} \\ A_{21} & A_{22} & \ldots & A_{2N} \\ \vdots & \vdots & \ddots & \vdots \\ A_{M1} & A_{M2} & \ldots & A_{MN} \end{pmatrix}. \tag{8.25}$$

The *matrix elements* A_{ij} are the components of the linear operator with respect to the bases $\mathbf{e}_j$ and $\mathbf{f}_i$; the component A_{ij} of the linear operator appears in the ith row and jth column of the matrix. The array has M rows and N columns and is thus called an $M \times N$ matrix. If the dimensions of the two vector spaces are the same, i.e. $M = N$ (for example, if they are the same vector space) then we may represent $\mathcal{A}$ by an $N \times N$ or *square* matrix of *order* N. The component A_{ij}, which in general may be complex, is also denoted by $(\mathsf{A})_{ij}$.

In a similar way we may denote a vector $\mathbf{x}$ in terms of its components x_i in a basis $\mathbf{e}_i$, $i = 1, 2, \ldots, N$, by the array

$$\mathsf{x} = \begin{pmatrix} x_1 \\ x_2 \\ \vdots \\ x_N \end{pmatrix},$$

which is a special case of (8.25) and is called a *column matrix* (or conventionally, and slightly confusingly, a *column vector* or even just a *vector* – strictly speaking the term 'vector' refers to the geometrical entity $\mathbf{x}$). The column matrix x can also be written as

$$\mathsf{x} = (x_1 \quad x_2 \quad \cdots \quad x_N)^{\mathrm{T}},$$

which is the *transpose* of a *row matrix* (see section 8.6).

We note that in a different basis $\mathbf{e}'_i$ the vector $\mathbf{x}$ would be represented by a *different* column matrix containing the components x'_i in the new basis, i.e.

$$\mathsf{x}' = \begin{pmatrix} x'_1 \\ x'_2 \\ \vdots \\ x'_N \end{pmatrix}.$$

Thus, we use x and x' to denote different column matrices which, in different bases $\mathbf{e}_i$ and $\mathbf{e}'_i$, represent the *same* vector $\mathbf{x}$. In many texts, however, this distinction is not made and $\mathbf{x}$ (rather than x) is equated to the corresponding column matrix; if we regard $\mathbf{x}$ as the geometrical entity, however, this can be misleading and so we explicitly make the distinction. A similar argument follows for linear operators; the same linear operator $\mathcal{A}$ is described in different bases by different matrices A and A', containing different matrix elements.

8.4 Basic matrix algebra

The basic algebra of matrices may be deduced from the properties of the linear operators that they represent. In a given basis the action of two linear operators $\mathcal{A}$ and $\mathcal{B}$ on an arbitrary vector $\mathbf{x}$ (see the beginning of subsection 8.2.1), when written in terms of components using (8.24), is given by

$$\sum_j (\mathsf{A} + \mathsf{B})_{ij} x_j = \sum_j A_{ij} x_j + \sum_j B_{ij} x_j,$$

$$\sum_j (\lambda \mathsf{A})_{ij} x_j = \lambda \sum_j A_{ij} x_j,$$

$$\sum_j (\mathsf{AB})_{ij} x_j = \sum_k A_{ik} (\mathsf{Bx})_k = \sum_j \sum_k A_{ik} B_{kj} x_j.$$

Now, since $\mathbf{x}$ is arbitrary, we can immediately deduce the way in which matrices are added or multiplied, i.e.

$$(\mathsf{A} + \mathsf{B})_{ij} = A_{ij} + B_{ij}, \tag{8.26}$$

$$(\lambda \mathsf{A})_{ij} = \lambda A_{ij}, \tag{8.27}$$

$$(\mathsf{AB})_{ij} = \sum_k A_{ik} B_{kj}. \tag{8.28}$$

We note that a matrix element may, in general, be complex. We now discuss matrix addition and multiplication in more detail.

8.4.1 Matrix addition and multiplication by a scalar

From (8.26) we see that the sum of two matrices, $\mathsf{S} = \mathsf{A} + \mathsf{B}$, is the matrix whose elements are given by

$$S_{ij} = A_{ij} + B_{ij}$$

for every pair of subscripts i, j, with $i = 1, 2, \ldots, M$ and $j = 1, 2, \ldots, N$. For example, if A and B are 2×3 matrices then $\mathsf{S} = \mathsf{A} + \mathsf{B}$ is given by

$$
\begin{aligned}
\begin{pmatrix} S_{11} & S_{12} & S_{13} \\ S_{21} & S_{22} & S_{23} \end{pmatrix} &= \begin{pmatrix} A_{11} & A_{12} & A_{13} \\ A_{21} & A_{22} & A_{23} \end{pmatrix} + \begin{pmatrix} B_{11} & B_{12} & B_{13} \\ B_{21} & B_{22} & B_{23} \end{pmatrix} \\
&= \begin{pmatrix} A_{11} + B_{11} & A_{12} + B_{12} & A_{13} + B_{13} \\ A_{21} + B_{21} & A_{22} + B_{22} & A_{23} + B_{23} \end{pmatrix}.
\end{aligned} \tag{8.29}
$$

Clearly, for the sum of two matrices to have any meaning, the matrices must have the same dimensions, i.e. both be $M \times N$ matrices.

From definition (8.29) it follows that $\mathsf{A} + \mathsf{B} = \mathsf{B} + \mathsf{A}$ and that the sum of a number of matrices can be written unambiguously without bracketing, i.e. matrix addition is *commutative* and *associative*.

The difference of two matrices is defined by direct analogy with addition. The matrix $\mathsf{D} = \mathsf{A} - \mathsf{B}$ has elements

$$D_{ij} = A_{ij} - B_{ij}, \quad \text{for } i = 1, 2, \ldots, M, \ j = 1, 2, \ldots, N. \tag{8.30}$$

From (8.27) the product of a matrix A with a scalar λ is the matrix with elements λA_{ij}, for example

$$
\lambda \begin{pmatrix} A_{11} & A_{12} & A_{13} \\ A_{21} & A_{22} & A_{23} \end{pmatrix} = \begin{pmatrix} \lambda A_{11} & \lambda A_{12} & \lambda A_{13} \\ \lambda A_{21} & \lambda A_{22} & \lambda A_{23} \end{pmatrix}. \tag{8.31}
$$

Multiplication by a scalar is distributive and associative.

▶ *The matrices* A, B *and* C *are given by*

$$A = \begin{pmatrix} 2 & -1 \\ 3 & 1 \end{pmatrix}, \qquad B = \begin{pmatrix} 1 & 0 \\ 0 & -2 \end{pmatrix}, \qquad C = \begin{pmatrix} -2 & 1 \\ -1 & 1 \end{pmatrix}.$$

Find the matrix D = A + 2B − C.

$$
\begin{aligned}
D &= \begin{pmatrix} 2 & -1 \\ 3 & 1 \end{pmatrix} + 2\begin{pmatrix} 1 & 0 \\ 0 & -2 \end{pmatrix} - \begin{pmatrix} -2 & 1 \\ -1 & 1 \end{pmatrix} \\
&= \begin{pmatrix} 2 + 2 \times 1 - (-2) & -1 + 2 \times 0 - 1 \\ 3 + 2 \times 0 - (-1) & 1 + 2 \times (-2) - 1 \end{pmatrix} = \begin{pmatrix} 6 & -2 \\ 4 & -4 \end{pmatrix}. \quad ◀
\end{aligned}
$$

From the above considerations we see that the set of all, in general complex, $M \times N$ matrices (with fixed M and N) forms a linear vector space of dimension MN. One basis for the space is the set of $M \times N$ matrices $\mathsf{E}^{(p,q)}$ with the property that $E_{ij}^{(p,q)} = 1$ if $i = p$ and $j = q$ whilst $E_{ij}^{(p,q)} = 0$ for all other values of i and j, i.e. each matrix has only one non-zero entry, which equals unity. Here the pair (p, q) is simply a label that picks out a particular one of the matrices $\mathsf{E}^{(p,q)}$, the total number of which is MN.

8.4.2 Multiplication of matrices

Let us consider again the 'transformation' of one vector into another, $\mathbf{y} = \mathcal{A}\mathbf{x}$, which, from (8.24), may be described in terms of components with respect to a particular basis as

$$y_i = \sum_{j=1}^{N} A_{ij} x_j \qquad \text{for } i = 1, 2, \dots, M. \tag{8.32}$$

Writing this in matrix form as $\mathbf{y} = \mathsf{A}\mathbf{x}$ we have

$$
\begin{pmatrix} y_1 \\ \boxed{y_2} \\ \vdots \\ y_M \end{pmatrix} = \begin{pmatrix} A_{11} & A_{12} & \dots & A_{1N} \\ \boxed{A_{21}} & \boxed{A_{22}} & \dots & \boxed{A_{2N}} \\ \vdots & \vdots & \ddots & \vdots \\ A_{M1} & A_{M2} & \dots & A_{MN} \end{pmatrix} \begin{pmatrix} \boxed{x_1} \\ \boxed{x_2} \\ \vdots \\ \boxed{x_N} \end{pmatrix} \tag{8.33}
$$

where we have highlighted with boxes the components used to calculate the element y_2: using (8.32) for $i = 2$,

$$y_2 = A_{21}x_1 + A_{22}x_2 + \cdots + A_{2N}x_N.$$

All the other components y_i are calculated similarly.

If instead we operate with $\mathcal{A}$ on a basis vector $\mathbf{e}_j$ having all components zero except for the jth, which equals unity, then we find

$$\mathsf{A}e_j = \begin{pmatrix} A_{11} & A_{12} & \cdots & A_{1N} \\ A_{21} & A_{22} & \cdots & A_{2N} \\ \vdots & \vdots & \ddots & \vdots \\ A_{M1} & A_{M2} & \cdots & A_{MN} \end{pmatrix} \begin{pmatrix} 0 \\ 0 \\ \vdots \\ 1 \\ \vdots \\ 0 \end{pmatrix} = \begin{pmatrix} A_{1j} \\ A_{2j} \\ \vdots \\ A_{Mj} \end{pmatrix},$$

and so confirm our identification of the matrix element A_{ij} as the ith component of $\mathsf{A}e_j$ in this basis.

From (8.28) we can extend our discussion to the product of two matrices $\mathsf{P} = \mathsf{AB}$, where P is the matrix of the quantities formed by the operation of the rows of A on the columns of B, treating each column of B in turn as the vector $\mathbf{x}$ represented in component form in (8.32). It is clear that, for this to be a meaningful definition, the number of columns in A must equal the number of rows in B. Thus the product AB of an $M \times N$ matrix A with an $N \times R$ matrix B is itself an $M \times R$ matrix P, where

$$P_{ij} = \sum_{k=1}^{N} A_{ik} B_{kj} \quad \text{for } i = 1, 2, \dots, M, \quad j = 1, 2, \dots, R.$$

For example, $\mathsf{P} = \mathsf{AB}$ may be written in matrix form

$$\begin{pmatrix} P_{11} & P_{12} \\ P_{21} & P_{22} \end{pmatrix} = \begin{pmatrix} A_{11} & A_{12} & A_{13} \\ A_{21} & A_{22} & A_{23} \end{pmatrix} \begin{pmatrix} B_{11} & B_{12} \\ B_{21} & B_{22} \\ B_{31} & B_{32} \end{pmatrix}$$

where

$$P_{11} = A_{11} B_{11} + A_{12} B_{21} + A_{13} B_{31},$$
$$P_{21} = A_{21} B_{11} + A_{22} B_{21} + A_{23} B_{31},$$
$$P_{12} = A_{11} B_{12} + A_{12} B_{22} + A_{13} B_{32},$$
$$P_{22} = A_{21} B_{12} + A_{22} B_{22} + A_{23} B_{32}.$$

Multiplication of more than two matrices follows naturally and is associative. So, for example,

$$\mathsf{A}(\mathsf{BC}) \equiv (\mathsf{AB})\mathsf{C}, \tag{8.34}$$

provided, of course, that all the products are defined.

As mentioned above, if A is an $M \times N$ matrix and B is an $N \times M$ matrix then two product matrices are possible, i.e.

$$\mathsf{P} = \mathsf{AB} \quad \text{and} \quad \mathsf{Q} = \mathsf{BA}.$$

These are clearly not the same, since P is an $M \times M$ matrix whilst Q is an $N \times N$ matrix. Thus, particular care must be taken to write matrix products in the intended order; P = AB but Q = BA. We note in passing that A^2 means AA, A^3 means A(AA) = (AA)A etc. Even if both A and B are square, in general

$$AB \neq BA, \tag{8.35}$$

i.e. the multiplication of matrices is not, in general, commutative.

▶ *Evaluate* P = AB *and* Q = BA *where*

$$A = \begin{pmatrix} 3 & 2 & -1 \\ 0 & 3 & 2 \\ 1 & -3 & 4 \end{pmatrix}, \quad B = \begin{pmatrix} 2 & -2 & 3 \\ 1 & 1 & 0 \\ 3 & 2 & 1 \end{pmatrix}.$$

As we saw for the 2×2 case above, the element P_{ij} of the matrix P = AB is found by mentally taking the 'scalar product' of the ith row of A with the jth column of B. For example, $P_{11} = 3 \times 2 + 2 \times 1 + (-1) \times 3 = 5$, $P_{12} = 3 \times (-2) + 2 \times 1 + (-1) \times 2 = -6$, etc. Thus

$$P = AB = \begin{pmatrix} 3 & 2 & -1 \\ 0 & 3 & 2 \\ 1 & -3 & 4 \end{pmatrix} \begin{pmatrix} 2 & -2 & 3 \\ 1 & 1 & 0 \\ 3 & 2 & 1 \end{pmatrix} = \begin{pmatrix} 5 & -6 & 8 \\ 9 & 7 & 2 \\ 11 & 3 & 7 \end{pmatrix},$$

and, similarly,

$$Q = BA = \begin{pmatrix} 2 & -2 & 3 \\ 1 & 1 & 0 \\ 3 & 2 & 1 \end{pmatrix} \begin{pmatrix} 3 & 2 & -1 \\ 0 & 3 & 2 \\ 1 & -3 & 4 \end{pmatrix} = \begin{pmatrix} 9 & -11 & 6 \\ 3 & 5 & 1 \\ 10 & 9 & 5 \end{pmatrix}.$$

These results illustrate that, in general, two matrices do not commute. ◀

The property that matrix multiplication is distributive over addition, i.e. that

$$(A + B)C = AC + BC \tag{8.36}$$

and

$$C(A + B) = CA + CB, \tag{8.37}$$

follows directly from its definition.

8.4.3 The null and identity matrices

Both the null matrix and the identity matrix are frequently encountered, and we take this opportunity to introduce them briefly, leaving their uses until later. The *null* or *zero* matrix 0 has all elements equal to zero, and so its properties are

$$A0 = 0 = 0A,$$

$$A + 0 = 0 + A = A.$$

The *identity* matrix I has the property

$$\mathsf{AI} = \mathsf{IA} = \mathsf{A}.$$

It is clear that, in order for the above products to be defined, the identity matrix must be square. The $N \times N$ identity matrix (often denoted by I_N) has the form

$$\mathsf{I}_N = \begin{pmatrix} 1 & 0 & \cdots & 0 \\ 0 & 1 & & \vdots \\ \vdots & & \ddots & 0 \\ 0 & \cdots & 0 & 1 \end{pmatrix}.$$

8.5 Functions of matrices

If a matrix A is *square* then, as mentioned above, one can define *powers* of A in a straightforward way. For example $\mathsf{A}^2 = \mathsf{AA}$, $\mathsf{A}^3 = \mathsf{AAA}$, or in the general case

$$\mathsf{A}^n = \mathsf{AA} \cdots \mathsf{A} \qquad (n \text{ times}),$$

where n is a positive integer. Having defined powers of a square matrix A, we may construct *functions* of A of the form

$$\mathsf{S} = \sum_n a_n \mathsf{A}^n,$$

where the a_k are simple scalars and the number of terms in the summation may be finite or infinite. In the case where the sum has an infinite number of terms, the sum has meaning only if it converges. A common example of such a function is the *exponential* of a matrix, which is defined by

$$\exp \mathsf{A} = \sum_{n=0}^{\infty} \frac{\mathsf{A}^n}{n!}. \tag{8.38}$$

This definition can, in turn, be used to define other functions such as $\sin \mathsf{A}$ and $\cos \mathsf{A}$.

8.6 The transpose of a matrix

We have seen that the components of a linear operator in a given coordinate system can be written in the form of a matrix A. We will also find it useful, however, to consider the different (but clearly related) matrix formed by interchanging the rows and columns of A. The matrix is called the *transpose* of A and is denoted by A^T.

> ►*Find the transpose of the matrix*
> $$A = \begin{pmatrix} 3 & 1 & 2 \\ 0 & 4 & 1 \end{pmatrix}.$$

By interchanging the rows and columns of A we immediately obtain

$$A^T = \begin{pmatrix} 3 & 0 \\ 1 & 4 \\ 2 & 1 \end{pmatrix}. \blacktriangleleft$$

It is obvious that if A is an $M \times N$ matrix then its transpose A^T is a $N \times M$ matrix. As mentioned in section 8.3, the transpose of a column matrix is a row matrix and vice versa. An important use of column and row matrices is in the representation of the inner product of two real vectors in terms of their components in a given basis. This notion is discussed fully in the next section, where it is extended to complex vectors.

The transpose of the product of two matrices, $(AB)^T$, is given by the product of their transposes taken in the reverse order, i.e.

$$(AB)^T = B^T A^T. \tag{8.39}$$

This is proved as follows:

$$(AB)^T_{ij} = (AB)_{ji} = \sum_k A_{jk} B_{ki}$$
$$= \sum_k (A^T)_{kj} (B^T)_{ik} = \sum_k (B^T)_{ik} (A^T)_{kj} = (B^T A^T)_{ij},$$

and the proof can be extended to the product of several matrices to give

$$(ABC \cdots G)^T = G^T \cdots C^T B^T A^T.$$

8.7 The complex and Hermitian conjugates of a matrix

Two further matrices that can be derived from a given general $M \times N$ matrix are the *complex conjugate*, denoted by A^*, and the *Hermitian conjugate*, denoted by $A^\dagger$.

The complex conjugate of a matrix A is the matrix obtained by taking the complex conjugate of each of the elements of A, i.e.

$$(A^*)_{ij} = (A_{ij})^*.$$

Obviously if a matrix is *real* (i.e. it contains only real elements) then $A^* = A$.

▶ *Find the complex conjugate of the matrix*
$$A = \begin{pmatrix} 1 & 2 & 3i \\ 1+i & 1 & 0 \end{pmatrix}.$$

By taking the complex conjugate of each element we obtain immediately

$$A^* = \begin{pmatrix} 1 & 2 & -3i \\ 1-i & 1 & 0 \end{pmatrix}. \blacktriangleleft$$

The Hermitian conjugate, or *adjoint*, of a matrix A is the transpose of its complex conjugate, or equivalently, the complex conjugate of its transpose, i.e.

$$A^\dagger = (A^*)^T = (A^T)^*.$$

We note that if A is real (and so $A^* = A$) then $A^\dagger = A^T$, and taking the Hermitian conjugate is equivalent to taking the transpose. Following the previous line of argument for the transpose of the product of several matrices, the Hermitian conjugate of such a product can be shown to be given by

$$(AB \cdots G)^\dagger = G^\dagger \cdots B^\dagger A^\dagger. \tag{8.40}$$

▶ *Find the Hermitian conjugate of the matrix*
$$A - \begin{pmatrix} 1 & 2 & 3i \\ 1+i & 1 & 0 \end{pmatrix}.$$

Taking the complex conjugate of A and then forming the transpose we find

$$A^\dagger = \begin{pmatrix} 1 & 1-i \\ 2 & 1 \\ -3i & 0 \end{pmatrix}.$$

We obtain the same result, of course, if we first take the transpose of A and then take the complex conjugate. ◀

An important use of the Hermitian conjugate (or transpose in the real case) is in connection with the inner product of two vectors. Suppose that in a given orthonormal basis the vectors **a** and **b** may be represented by the column matrices

$$a = \begin{pmatrix} a_1 \\ a_2 \\ \vdots \\ a_N \end{pmatrix} \quad \text{and} \quad b = \begin{pmatrix} b_1 \\ b_2 \\ \vdots \\ b_N \end{pmatrix}. \tag{8.41}$$

Taking the Hermitian conjugate of **a**, to give a row matrix, and multiplying (on the right) by b we obtain

$$\mathbf{a}^\dagger \mathbf{b} = (a_1^* \; a_2^* \; \cdots \; a_N^*) \begin{pmatrix} b_1 \\ b_2 \\ \vdots \\ b_N \end{pmatrix} = \sum_{i=1}^N a_i^* b_i, \tag{8.42}$$

which is the expression for the inner product $\langle \mathbf{a} | \mathbf{b} \rangle$ in that basis. We note that for real vectors (8.42) reduces to $\mathbf{a}^{\mathrm{T}} \mathbf{b} = \sum_{i=1}^N a_i b_i$.

If the basis $\mathbf{e}_i$ is *not* orthonormal, so that, in general,

$$\langle \mathbf{e}_i | \mathbf{e}_j \rangle = G_{ij} \neq \delta_{ij},$$

then, from (8.18), the scalar product of $\mathbf{a}$ and $\mathbf{b}$ in terms of their components with respect to this basis is given by

$$\langle \mathbf{a} | \mathbf{b} \rangle = \sum_{i=1}^N \sum_{j=1}^N a_i^* G_{ij} b_j = \mathbf{a}^\dagger \mathsf{G} \mathbf{b},$$

where G is the $N \times N$ matrix with elements G_{ij}.

8.8 The trace of a matrix

For a given matrix A, in the previous two sections we have considered various other matrices that can be derived from it. However, sometimes one wishes to derive a single number from a matrix. The simplest example is the *trace* (or *spur*) of a square matrix, which is denoted by $\mathrm{Tr}\,\mathsf{A}$. This quantity is defined as the sum of the diagonal elements of the matrix,

$$\mathrm{Tr}\,\mathsf{A} = A_{11} + A_{22} + \cdots + A_{NN} = \sum_{i=1}^N A_{ii}. \tag{8.43}$$

It is clear that taking the trace is a linear operation so that, for example,

$$\mathrm{Tr}(\mathsf{A} \pm \mathsf{B}) = \mathrm{Tr}\,\mathsf{A} \pm \mathrm{Tr}\,\mathsf{B}.$$

A very useful property of traces is that the trace of the product of two matrices is independent of the order of their multiplication; this results holds whether or not the matrices commute and is proved as follows:

$$\mathrm{Tr}\,\mathsf{AB} = \sum_{i=1}^N (\mathsf{AB})_{ii} = \sum_{i=1}^N \sum_{j=1}^N A_{ij} B_{ji} = \sum_{i=1}^N \sum_{j=1}^N B_{ji} A_{ij} = \sum_{j=1}^N (\mathsf{BA})_{jj} = \mathrm{Tr}\,\mathsf{BA}. \tag{8.44}$$

The result can be extended to the product of several matrices. For example, from (8.44), we immediately find

$$\mathrm{Tr}\,\mathsf{ABC} = \mathrm{Tr}\,\mathsf{BCA} = \mathrm{Tr}\,\mathsf{CAB},$$

which shows that the trace of a multiple product is invariant under cyclic permutations of the matrices in the product. Other easily derived properties of the trace are, for example, $\mathrm{Tr}\,A^T = \mathrm{Tr}\,A$ and $\mathrm{Tr}\,A^\dagger = (\mathrm{Tr}\,A)^*$.

8.9 The determinant of a matrix

For a given matrix A, the determinant det A (like the trace) is a single number (or algebraic expression) that depends upon the elements of A. Also like the trace, the determinant is defined only for *square* matrices. If, for example, A is a 3×3 matrix then its determinant, of *order* 3, is denoted by

$$
\det A = |A| = \begin{vmatrix} A_{11} & A_{12} & A_{13} \\ A_{21} & A_{22} & A_{23} \\ A_{31} & A_{32} & A_{33} \end{vmatrix}. \tag{8.45}
$$

In order to calculate the value of a determinant, we first need to introduce the notions of the *minor* and the *cofactor* of an element of a matrix. (We shall see that we can use the cofactors to write an order-3 determinant as the weighted sum of three order-2 determinants, thereby simplifying its evaluation.) The minor M_{ij} of the element A_{ij} of an $N \times N$ matrix A is the determinant of the $(N-1) \times (N-1)$ matrix obtained by removing all the elements of the ith row and jth column of A; the associated cofactor, C_{ij}, is found by multiplying the minor by $(-1)^{i+j}$.

▶*Find the cofactor of the element A_{23} of the matrix*

$$
A = \begin{pmatrix} A_{11} & A_{12} & A_{13} \\ A_{21} & A_{22} & A_{23} \\ A_{31} & A_{32} & A_{33} \end{pmatrix}.
$$

Removing all the elements of the second row and third column of A and forming the determinant of the remaining terms gives the minor

$$
M_{23} = \begin{vmatrix} A_{11} & A_{12} \\ A_{31} & A_{32} \end{vmatrix}.
$$

Multiplying the minor by $(-1)^{2+3} = (-1)^5 = -1$ gives

$$
C_{23} = - \begin{vmatrix} A_{11} & A_{12} \\ A_{31} & A_{32} \end{vmatrix}. ◀
$$

We now define a determinant as *the sum of the products of the elements of any row or column and their corresponding cofactors*, e.g. $A_{21}C_{21} + A_{22}C_{22} + A_{23}C_{23}$ or $A_{13}C_{13} + A_{23}C_{23} + A_{33}C_{33}$. Such a sum is called a *Laplace expansion*. For example, in the first of these expansions, using the elements of the second row of the determinant defined by (8.45) and their corresponding cofactors, we write $|A|$ as the Laplace expansion

$$|A| = A_{21}(-1)^{(2+1)}M_{21} + A_{22}(-1)^{(2+2)}M_{22} + A_{23}(-1)^{(2+3)}M_{23}$$

$$= -A_{21}\begin{vmatrix} A_{12} & A_{13} \\ A_{32} & A_{33} \end{vmatrix} + A_{22}\begin{vmatrix} A_{11} & A_{13} \\ A_{31} & A_{33} \end{vmatrix} - A_{23}\begin{vmatrix} A_{11} & A_{12} \\ A_{31} & A_{32} \end{vmatrix}.$$

We will see later that the value of the determinant is independent of the row or column chosen. Of course, we have not yet determined the value of $|A|$ but, rather, written it as the weighted sum of three determinants of order 2. However, applying again the definition of a determinant, we can evaluate each of the order-2 determinants.

▶ *Evaluate the determinant*

$$\begin{vmatrix} A_{12} & A_{13} \\ A_{32} & A_{33} \end{vmatrix}.$$

By considering the products of the elements of the first row in the determinant, and their corresponding cofactors, we find

$$\begin{vmatrix} A_{12} & A_{13} \\ A_{32} & A_{33} \end{vmatrix} = A_{12}(-1)^{(1+1)}|A_{33}| + A_{13}(-1)^{(1+2)}|A_{32}|$$

$$= A_{12}A_{33} - A_{13}A_{32},$$

where the values of the order-1 determinants $|A_{33}|$ and $|A_{32}|$ are defined to be A_{33} and A_{32} respectively. It must be remembered that the determinant is *not* the same as the modulus, e.g. det $(-2) = |-2| = -2$, not 2. ◀

We can now combine all the above results to show that the value of the determinant (8.45) is given by

$$|A| = -A_{21}(A_{12}A_{33} - A_{13}A_{32}) + A_{22}(A_{11}A_{33} - A_{13}A_{31})$$

$$- A_{23}(A_{11}A_{32} - A_{12}A_{31}) \tag{8.46}$$

$$= A_{11}(A_{22}A_{33} - A_{23}A_{32}) + A_{12}(A_{23}A_{31} - A_{21}A_{33})$$

$$+ A_{13}(A_{21}A_{32} - A_{22}A_{31}), \tag{8.47}$$

where the final expression gives the form in which the determinant is usually remembered and is the form that is obtained immediately by considering the Laplace expansion using the first row of the determinant. The last equality, which essentially rearranges a Laplace expansion using the second row into one using the first row, supports our assertion that the value of the determinant is unaffected by which row or column is chosen for the expansion.

> ►*Suppose the rows of a real 3×3 matrix A are interpreted as the components in a given basis of three (three-component) vectors* **a**, **b** *and* **c**. *Show that one can write the determinant of A as*
> $$|A| = \mathbf{a} \cdot (\mathbf{b} \times \mathbf{c}).$$

If one writes the rows of A as the components in a given basis of three vectors **a**, **b** and **c**, we have from (8.47) that

$$|A| = \begin{vmatrix} a_1 & a_2 & a_3 \\ b_1 & b_2 & b_3 \\ c_1 & c_2 & c_3 \end{vmatrix} = a_1(b_2 c_3 - b_3 c_2) + a_2(b_3 c_1 - b_1 c_3) + a_3(b_1 c_2 - b_2 c_1).$$

From expression (7.34) for the scalar triple product given in subsection 7.6.3, it follows that we may write the determinant as

$$|A| = \mathbf{a} \cdot (\mathbf{b} \times \mathbf{c}). \tag{8.48}$$

In other words, $|A|$ is the volume of the parallelepiped defined by the vectors **a**, **b** and **c**. (One could equally well interpret the *columns* of the matrix A as the components of three vectors, and result (8.48) would still hold.) This result provides a more memorable (and more meaningful) expression than (8.47) for the value of a 3×3 determinant. Indeed, using this geometrical interpretation, we see immediately that, if the vectors $\mathbf{a}_1, \mathbf{a}_2, \mathbf{a}_3$ are not linearly independent then the value of the determinant vanishes: $|A| = 0$. ◄

The evaluation of determinants of order greater than 3 follows the same general method as that presented above, in that it relies on successively reducing the order of the determinant by writing it as a Laplace expansion. Thus, a determinant of order 4 is first written as a sum of four determinants of order 3, which are then evaluated using the above method. For higher-order determinants, one cannot write down directly a simple geometrical expression for $|A|$ analogous to that given in (8.48). Nevertheless, it is still true that if the rows or columns of the $N \times N$ matrix A are interpreted as the components in a given basis of N (N-component) vectors $\mathbf{a}_1, \mathbf{a}_2, \ldots, \mathbf{a}_N$, then the determinant $|A|$ vanishes if these vectors are not all linearly independent.

8.9.1 Properties of determinants

A number of properties of determinants follow straightforwardly from the definition of det A; their use will often reduce the labour of evaluating a determinant. We present them here without specific proofs, though they all follow readily from the alternative form for a determinant, given in equation (26.29) on page 942, and expressed in terms of the Levi–Civita symbol ϵ_{ijk} (see exercise 26.9).

(i) *Determinant of the transpose.* The transpose matrix A^T (which, we recall, is obtained by interchanging the rows and columns of A) has the same determinant as A itself, i.e.

$$|A^T| = |A|. \tag{8.49}$$

It follows that *any* theorem established for the rows of A will apply to the columns as well, and vice versa.

(ii) *Determinant of the complex and Hermitian conjugate.* It is clear that the matrix A^* obtained by taking the complex conjugate of each element of A has the determinant $|A^*| = |A|^*$. Combining this result with (8.49), we find that

$$|A^\dagger| = |(A^*)^T| = |A^*| = |A|^*. \tag{8.50}$$

(iii) *Interchanging two rows or two columns.* If two rows (columns) of A are interchanged, its determinant changes sign but is unaltered in magnitude.

(iv) *Removing factors.* If all the elements of a single row (column) of A have a common factor, λ, then this factor may be removed; the value of the determinant is given by the product of the remaining determinant and λ. Clearly this implies that if all the elements of any row (column) are zero then $|A| = 0$. It also follows that if every element of the $N \times N$ matrix A is multiplied by a constant factor λ then

$$|\lambda A| = \lambda^N |A|. \tag{8.51}$$

(v) *Identical rows or columns.* If any two rows (columns) of A are identical or are multiples of one another, then it can be shown that $|A| = 0$.

(vi) *Adding a constant multiple of one row (column) to another.* The determinant of a matrix is unchanged in value by adding to the elements of one row (column) any fixed multiple of the elements of another row (column).

(vii) *Determinant of a product.* If A and B are square matrices of the same order then

$$|AB| = |A||B| = |BA|. \tag{8.52}$$

A simple extension of this property gives, for example,

$$|AB \cdots G| = |A||B| \cdots |G| = |A||G| \cdots |B| = |A \cdots GB|,$$

which shows that the determinant is invariant under permutation of the matrices in a multiple product.

There is no explicit procedure for using the above results in the evaluation of any given determinant, and judging the quickest route to an answer is a matter of experience. A general guide is to try to reduce all terms but one in a row or column to zero and hence in effect to obtain a determinant of smaller size. The steps taken in evaluating the determinant in the example below are certainly not the fastest, but they have been chosen in order to illustrate the use of most of the properties listed above.

▶️*Evaluate the determinant*

$$|A| = \begin{vmatrix} 1 & 0 & 2 & 3 \\ 0 & 1 & -2 & 1 \\ 3 & -3 & 4 & -2 \\ -2 & 1 & -2 & -1 \end{vmatrix}.$$

Taking a factor 2 out of the third column and then adding the second column to the third gives

$$|A| = 2 \begin{vmatrix} 1 & 0 & 1 & 3 \\ 0 & 1 & -1 & 1 \\ 3 & -3 & 2 & -2 \\ -2 & 1 & -1 & -1 \end{vmatrix} = 2 \begin{vmatrix} 1 & 0 & 1 & 3 \\ 0 & 1 & 0 & 1 \\ 3 & -3 & -1 & -2 \\ -2 & 1 & 0 & -1 \end{vmatrix}.$$

Subtracting the second column from the fourth gives

$$|A| = 2 \begin{vmatrix} 1 & 0 & 1 & 3 \\ 0 & 1 & 0 & 0 \\ 3 & -3 & -1 & 1 \\ -2 & 1 & 0 & -2 \end{vmatrix}.$$

We now note that the second row has only one non-zero element and so the determinant may conveniently be written as a Laplace expansion, i.e.

$$|A| = 2 \times 1 \times (-1)^{2+2} \begin{vmatrix} 1 & 1 & 3 \\ 3 & -1 & 1 \\ -2 & 0 & -2 \end{vmatrix} = 2 \begin{vmatrix} 4 & 0 & 4 \\ 3 & -1 & 1 \\ -2 & 0 & -2 \end{vmatrix},$$

where the last equality follows by adding the second row to the first. It can now be seen that the first row is minus twice the third, and so the value of the determinant is zero, by property (v) above. ◀

8.10 The inverse of a matrix

Our first use of determinants will be in defining the *inverse* of a matrix. If we were dealing with ordinary numbers we would consider the relation P = AB as equivalent to B = P/A, provided that A ≠ 0. However, if A, B and P are matrices then this notation does not have an obvious meaning. What we really want to know is whether an explicit formula for B can be obtained in terms of A and P. It will be shown that this is possible for those cases in which $|A| \neq 0$. A square matrix whose determinant is zero is called a *singular* matrix; otherwise it is *non-singular*. We will show that if A is non-singular we can define a matrix, denoted by A^{-1} and called the *inverse* of A, which has the property that if AB = P then $B = A^{-1}P$. In words, B can be obtained by multiplying P from the left by A^{-1}. Analogously, if B is non-singular then, by multiplication from the right, $A = PB^{-1}$.

It is clear that

$$AI = A \quad \Rightarrow \quad I = A^{-1}A, \tag{8.53}$$

where I is the unit matrix, and so $A^{-1}A = I = AA^{-1}$. These statements are equivalent to saying that if we first multiply a matrix, B say, by A and then multiply by the inverse A^{-1}, we end up with the matrix we started with, i.e.

$$A^{-1}AB = B. \tag{8.54}$$

This justifies our use of the term inverse. It is also clear that the inverse is only defined for square matrices.

So far we have only defined what we mean by the inverse of a matrix. Actually finding the inverse of a matrix A may be carried out in a number of ways. We will show that one method is to construct first the matrix C containing the cofactors of the elements of A, as discussed in section 8.9. Then the required inverse A^{-1} can be found by forming the transpose of C and dividing by the determinant of A. Thus the elements of the inverse A^{-1} are given by

$$(A^{-1})_{ik} = \frac{(C)_{ik}^{T}}{|A|} = \frac{C_{ki}}{|A|}. \tag{8.55}$$

That this procedure does indeed result in the inverse may be seen by considering the components of $A^{-1}A$, i.e.

$$(A^{-1}A)_{ij} = \sum_k (A^{-1})_{ik}(A)_{kj} = \sum_k \frac{C_{ki}}{|A|} A_{kj} = \frac{|A|}{|A|} \delta_{ij}. \tag{8.56}$$

The last equality in (8.56) relies on the property

$$\sum_k C_{ki}A_{kj} = |A|\delta_{ij}; \tag{8.57}$$

this can be proved by considering the matrix A' obtained from the original matrix A when the ith column of A is replaced by one of the other columns, say the jth. Thus A' is a matrix with two identical columns and so has zero determinant. However, replacing the ith column by another does not change the cofactors C_{ki} of the elements in the ith column, which are therefore the same in A and A'. Recalling the Laplace expansion of a determinant, i.e.

$$|A| = \sum_k A_{ki}C_{ki},$$

we obtain

$$0 = |A'| = \sum_k A'_{ki}C'_{ki} = \sum_k A_{kj}C_{ki}, \quad i \neq j,$$

which together with the Laplace expansion itself may be summarised by (8.57).

It is immediately obvious from (8.55) that the inverse of a matrix is not defined if the matrix is singular (i.e. if $|A| = 0$).

▶ *Find the inverse of the matrix*

$$A = \begin{pmatrix} 2 & 4 & 3 \\ 1 & -2 & -2 \\ -3 & 3 & 2 \end{pmatrix}.$$

We first determine $|A|$:

$$|A| = 2[-2(2) - (-2)3] + 4[(-2)(-3) - (1)(2)] + 3[(1)(3) - (-2)(-3)]$$
$$= 11. \tag{8.58}$$

This is non-zero and so an inverse matrix can be constructed. To do this we need the matrix of the cofactors, C, and hence C^T. We find

$$C = \begin{pmatrix} 2 & 4 & -3 \\ 1 & 13 & -18 \\ -2 & 7 & -8 \end{pmatrix} \quad \text{and} \quad C^T = \begin{pmatrix} 2 & 1 & -2 \\ 4 & 13 & 7 \\ -3 & -18 & -8 \end{pmatrix},$$

and hence

$$A^{-1} = \frac{C^T}{|A|} = \frac{1}{11} \begin{pmatrix} 2 & 1 & -2 \\ 4 & 13 & 7 \\ -3 & -18 & -8 \end{pmatrix}. \blacktriangleleft \tag{8.59}$$

For a 2×2 matrix, the inverse has a particularly simple form. If the matrix is

$$A = \begin{pmatrix} A_{11} & A_{12} \\ A_{21} & A_{22} \end{pmatrix}$$

then its determinant $|A|$ is given by $|A| = A_{11}A_{22} - A_{12}A_{21}$, and the matrix of cofactors is

$$C = \begin{pmatrix} A_{22} & -A_{21} \\ -A_{12} & A_{11} \end{pmatrix}.$$

Thus the inverse of A is given by

$$A^{-1} = \frac{C^T}{|A|} = \frac{1}{A_{11}A_{22} - A_{12}A_{21}} \begin{pmatrix} A_{22} & -A_{12} \\ -A_{21} & A_{11} \end{pmatrix}. \tag{8.60}$$

It can be seen that the transposed matrix of cofactors for a 2×2 matrix is the same as the matrix formed by swapping the elements on the leading diagonal (A_{11} and A_{22}) and changing the signs of the other two elements (A_{12} and A_{21}). This is completely general for a 2×2 matrix and is easy to remember.

The following are some further useful properties related to the inverse matrix and may be straightforwardly derived.

(i) $(A^{-1})^{-1} = A$.

(ii) $(A^T)^{-1} = (A^{-1})^T$.

(iii) $(A^\dagger)^{-1} = (A^{-1})^\dagger$.

(iv) $(AB)^{-1} = B^{-1}A^{-1}$.

(v) $(AB\cdots G)^{-1} = G^{-1}\cdots B^{-1}A^{-1}$.

▶*Prove the properties* (i)–(v) *stated above.*

We begin by writing down the fundamental expression defining the inverse of a non-singular square matrix A:

$$AA^{-1} = I = A^{-1}A. \tag{8.61}$$

Property (i). This follows immediately from the expression (8.61).

Property (ii). Taking the transpose of each expression in (8.61) gives

$$(AA^{-1})^T = I^T = (A^{-1}A)^T.$$

Using the result (8.39) for the transpose of a product of matrices and noting that $I^T = I$, we find

$$(A^{-1})^T A^T = I = A^T (A^{-1})^T.$$

However, from (8.61), this implies $(A^{-1})^T = (A^T)^{-1}$ and hence proves result (ii) above.

Property (iii). This may be proved in an analogous way to property (ii), by replacing the transposes in (ii) by Hermitian conjugates and using the result (8.40) for the Hermitian conjugate of a product of matrices.

Property (iv). Using (8.61), we may write

$$(AB)(AB)^{-1} = I = (AB)^{-1}(AB),$$

From the left-hand equality it follows, by multiplying on the left by A^{-1}, that

$$A^{-1}AB(AB)^{-1} = A^{-1}I \quad \text{and hence} \quad B(AB)^{-1} = A^{-1}.$$

Now multiplying on the left by B^{-1} gives

$$B^{-1}B(AB)^{-1} = B^{-1}A^{-1},$$

and hence the stated result.

Property (v). Finally, result (iv) may extended to case (v) in a straightforward manner. For example, using result (iv) twice we find

$$(ABC)^{-1} = (BC)^{-1}A^{-1} = C^{-1}B^{-1}A^{-1}. \blacktriangleleft$$

We conclude this section by noting that the determinant $|A^{-1}|$ of the inverse matrix can be expressed very simply in terms of the determinant $|A|$ of the matrix itself. Again we start with the fundamental expression (8.61). Then, using the property (8.52) for the determinant of a product, we find

$$|AA^{-1}| = |A||A^{-1}| = |I|.$$

It is straightforward to show by Laplace expansion that $|I| = 1$, and so we arrive at the useful result

$$|A^{-1}| = \frac{1}{|A|}. \tag{8.62}$$

8.11 The rank of a matrix

The *rank* of a general $M \times N$ matrix is an important concept, particularly in the solution of sets of simultaneous linear equations, to be discussed in the next section, and we now discuss it in some detail. Like the trace and determinant, the rank of matrix A is a single number (or algebraic expression) that depends on the elements of A. Unlike the trace and determinant, however, the rank of a matrix can be defined even when A is not square. As we shall see, there are two *equivalent* definitions of the rank of a general matrix.

Firstly, the rank of a matrix may be defined in terms of the *linear independence* of vectors. Suppose that the columns of an $M \times N$ matrix are interpreted as the components in a given basis of N (M-component) vectors $\mathbf{v}_1, \mathbf{v}_2, \ldots, \mathbf{v}_N$, as follows:

$$
A = \begin{pmatrix} \uparrow & \uparrow & & \uparrow \\ \mathbf{v}_1 & \mathbf{v}_2 & \cdots & \mathbf{v}_N \\ \downarrow & \downarrow & & \downarrow \end{pmatrix}.
$$

Then the *rank* of A, denoted by rank A or by $R(A)$, is defined as the number of *linearly independent* vectors in the set $\mathbf{v}_1, \mathbf{v}_2, \ldots, \mathbf{v}_N$, and equals the dimension of the vector space spanned by those vectors. Alternatively, we may consider the rows of A to contain the components in a given basis of the M (N-component) vectors $\mathbf{w}_1, \mathbf{w}_2, \ldots, \mathbf{w}_M$ as follows:

$$
A = \begin{pmatrix} \leftarrow & \mathbf{w}_1 & \rightarrow \\ \leftarrow & \mathbf{w}_2 & \rightarrow \\ & \vdots & \\ \leftarrow & \mathbf{w}_M & \rightarrow \end{pmatrix}.
$$

It may then be shown[§] that the rank of A is also equal to the number of linearly independent vectors in the set $\mathbf{w}_1, \mathbf{w}_2, \ldots, \mathbf{w}_M$. From this definition it is should be clear that the rank of A is unaffected by the exchange of two rows (or two columns) or by the multiplication of a row (or column) by a constant. Furthermore, suppose that a constant multiple of one row (column) is added to another row (column): for example, we might replace the row $\mathbf{w}_i$ by $\mathbf{w}_i + c\mathbf{w}_j$. This also has no effect on the number of linearly independent rows and so leaves the rank of A unchanged. We may use these properties to evaluate the rank of a given matrix.

A second (equivalent) definition of the rank of a matrix may be given and uses the concept of *submatrices*. A submatrix of A is any matrix that can be formed from the elements of A by ignoring one, or more than one, row or column. It

---

[§] For a fuller discussion, see, for example, C. D. Cantrell, *Modern Mathematical Methods for Physicists and Engineers* (Cambridge: Cambridge University Press, 2000), chapter 6.

may be shown that the rank of a general $M \times N$ matrix is equal to the size of the largest square submatrix of A whose determinant is non-zero. Therefore, if a matrix A has an $r \times r$ submatrix S with $|S| \neq 0$, but no $(r+1) \times (r+1)$ submatrix with non-zero determinant then the rank of the matrix is r. From either definition it is clear that the rank of A is less than or equal to the smaller of M and N.

▶Determine the rank of the matrix

$$A = \begin{pmatrix} 1 & 1 & 0 & -2 \\ 2 & 0 & 2 & 2 \\ 4 & 1 & 3 & 1 \end{pmatrix}.$$

The largest possible square submatrices of A must be of dimension 3×3. Clearly, A possesses four such submatrices, the determinants of which are given by

$$\begin{vmatrix} 1 & 1 & 0 \\ 2 & 0 & 2 \\ 4 & 1 & 3 \end{vmatrix} = 0, \qquad \begin{vmatrix} 1 & 1 & -2 \\ 2 & 0 & 2 \\ 4 & 1 & 1 \end{vmatrix} = 0,$$

$$\begin{vmatrix} 1 & 0 & -2 \\ 2 & 2 & 2 \\ 4 & 3 & 1 \end{vmatrix} = 0, \qquad \begin{vmatrix} 1 & 0 & -2 \\ 0 & 2 & 2 \\ 1 & 3 & 1 \end{vmatrix} = 0.$$

(In each case the determinant may be evaluated as described in subsection 8.9.1.)

The next largest square submatrices of A are of dimension 2×2. Consider, for example, the 2×2 submatrix formed by ignoring the third row and the third and fourth columns of A; this has determinant

$$\begin{vmatrix} 1 & 1 \\ 2 & 0 \end{vmatrix} = 1 \times 0 - 2 \times 1 = -2.$$

Since its determinant is non-zero, A is of rank 2 and we need not consider any other 2×2 submatrix. ◀

In the special case in which the matrix A is a *square $N \times N$* matrix, by comparing either of the above definitions of rank with our discussion of determinants in section 8.9, we see that $|A| = 0$ unless the rank of A is N. In other words, A is *singular* unless $R(A) = N$.

8.12 Special types of square matrix

Matrices that are square, i.e. $N \times N$, are very common in physical applications. We now consider some special forms of square matrix that are of particular importance.

8.12.1 Diagonal matrices

The unit matrix, which we have already encountered, is an example of a *diagonal* matrix. Such matrices are characterised by having non-zero elements only on the *leading diagonal*, i.e. only elements A_{ij} with $i = j$ may be non-zero. For example,

$$A = \begin{pmatrix} 1 & 0 & 0 \\ 0 & 2 & 0 \\ 0 & 0 & -3 \end{pmatrix},$$

is a 3×3 diagonal matrix. Such a matrix is often denoted by $A = \text{diag}\,(1, 2, -3)$. By performing a Laplace expansion, it is easily shown that the determinant of an $N \times N$ diagonal matrix is equal to the product of the diagonal elements. Thus, if the matrix has the form $A = \text{diag}(A_{11}, A_{22}, \ldots, A_{NN})$ then

$$|A| = A_{11}A_{22}\cdots A_{NN}. \tag{8.63}$$

Moreover, it is also straightforward to show that the inverse of A is also a diagonal matrix given by

$$A^{-1} = \text{diag}\left(\frac{1}{A_{11}}, \frac{1}{A_{22}}, \ldots, \frac{1}{A_{NN}}\right).$$

Finally, we note that, if two matrices A and B are *both* diagonal then they have the useful property that their product is commutative:

$$AB = BA.$$

This is *not* true for matrices in general.

8.12.2 Lower and upper triangular matrices

A square matrix A is called *lower triangular* if all the elements *above* the principal diagonal are zero. For example, the general form for a 3×3 lower triangular matrix is

$$A = \begin{pmatrix} A_{11} & 0 & 0 \\ A_{21} & A_{22} & 0 \\ A_{31} & A_{32} & A_{33} \end{pmatrix},$$

where the elements A_{ij} may be zero or non-zero. Similarly an *upper triangular* square matrix is one for which all the elements *below* the principal diagonal are zero. The general 3×3 form is thus

$$A = \begin{pmatrix} A_{11} & A_{12} & A_{13} \\ 0 & A_{22} & A_{23} \\ 0 & 0 & A_{33} \end{pmatrix}.$$

By performing a Laplace expansion, it is straightforward to show that, in the general $N \times N$ case, the determinant of an upper or lower triangular matrix is equal to the product of its diagonal elements,

$$|A| = A_{11}A_{22}\cdots A_{NN}. \tag{8.64}$$

Clearly result (8.63) for diagonal matrices is a special case of this result. Moreover, it may be shown that the inverse of a non-singular lower (upper) triangular matrix is also lower (upper) triangular.

8.12.3 Symmetric and antisymmetric matrices

A square matrix A of order N with the property $A = A^T$ is said to be *symmetric*. Similarly a matrix for which $A = -A^T$ is said to be *anti-* or *skew*-symmetric and its diagonal elements $a_{11}, a_{22}, \ldots, a_{NN}$ are necessarily zero. Moreover, if A is (anti-)symmetric then so too is its inverse A^{-1}. This is easily proved by noting that if $A = \pm A^T$ then

$$(A^{-1})^T = (A^T)^{-1} = \pm A^{-1}.$$

Any $N \times N$ matrix A can be written as the sum of a symmetric and an antisymmetric matrix, since we may write

$$A = \tfrac{1}{2}(A + A^T) + \tfrac{1}{2}(A - A^T) = B + C,$$

where clearly $B = B^T$ and $C = -C^T$. The matrix B is therefore called the symmetric part of A, and C is the antisymmetric part.

▶*If A is an $N \times N$ antisymmetric matrix, show that $|A| = 0$ if N is odd.*

If A is antisymmetric then $A^T = -A$. Using the properties of determinants (8.49) and (8.51), we have

$$|A| = |A^T| = |-A| = (-1)^N |A|.$$

Thus, if N is odd then $|A| = -|A|$, which implies that $|A| = 0$. ◀

8.12.4 Orthogonal matrices

A non-singular matrix with the property that its transpose is also its inverse,

$$A^T = A^{-1}, \tag{8.65}$$

is called an *orthogonal matrix*. It follows immediately that the inverse of an orthogonal matrix is also orthogonal, since

$$(A^{-1})^T = (A^T)^{-1} = (A^{-1})^{-1}.$$

Moreover, since for an orthogonal matrix $A^T A = I$, we have

$$|A^T A| = |A^T||A| = |A|^2 = |I| = 1.$$

Thus the determinant of an orthogonal matrix must be $|A| = \pm 1$.

An orthogonal matrix represents, in a particular basis, a linear operator that leaves the norms (lengths) of real vectors unchanged, as we will now show.

Suppose that $\mathbf{y} = \mathcal{A}\mathbf{x}$ is represented in some coordinate system by the matrix equation $y = Ax$; then $\langle \mathbf{y}|\mathbf{y}\rangle$ is given in this coordinate system by

$$y^T y = x^T A^T A x = x^T x.$$

Hence $\langle \mathbf{y}|\mathbf{y}\rangle = \langle \mathbf{x}|\mathbf{x}\rangle$, showing that the action of a linear operator represented by an orthogonal matrix does not change the norm of a real vector.

8.12.5 Hermitian and anti-Hermitian matrices

An *Hermitian* matrix is one that satisfies $A = A^\dagger$, where $A^\dagger$ is the Hermitian conjugate discussed in section 8.7. Similarly if $A^\dagger = -A$, then A is called *anti-Hermitian*. A real (anti-)symmetric matrix is a special case of an (anti-)Hermitian matrix, in which all the elements of the matrix are real. Also, if A is an (anti-)Hermitian matrix then so too is its inverse A^{-1}, since

$$(A^{-1})^\dagger = (A^\dagger)^{-1} = \pm A^{-1}.$$

Any $N \times N$ matrix A can be written as the sum of an Hermitian matrix and an anti-Hermitian matrix, since

$$A = \tfrac{1}{2}(A + A^\dagger) + \tfrac{1}{2}(A - A^\dagger) = B + C,$$

where clearly $B = B^\dagger$ and $C = -C^\dagger$. The matrix B is called the Hermitian part of A, and C is called the anti-Hermitian part.

8.12.6 Unitary matrices

A *unitary matrix* A is defined as one for which

$$A^\dagger = A^{-1}. \tag{8.66}$$

Clearly, if A is real then $A^\dagger = A^T$, showing that a real orthogonal matrix is a special case of a unitary matrix, one in which all the elements are real. We note that the inverse A^{-1} of a unitary matrix is also unitary, since

$$(A^{-1})^\dagger = (A^\dagger)^{-1} = (A^{-1})^{-1}.$$

Moreover, since for a unitary matrix $A^\dagger A = I$, we have

$$|A^\dagger A| = |A^\dagger||A| = |A|^*|A| = |I| = 1.$$

Thus the determinant of a unitary matrix has unit modulus.

A unitary matrix represents, in a particular basis, a linear operator that leaves the norms (lengths) of complex vectors unchanged. If $\mathbf{y} = \mathcal{A}\mathbf{x}$ is represented in some coordinate system by the matrix equation $y = Ax$ then $\langle \mathbf{y}|\mathbf{y}\rangle$ is given in this coordinate system by

$$y^\dagger y = x^\dagger A^\dagger A x = x^\dagger x.$$

Hence $\langle \mathbf{y}|\mathbf{y}\rangle = \langle \mathbf{x}|\mathbf{x}\rangle$, showing that the action of the linear operator represented by a unitary matrix does not change the norm of a complex vector. The action of a unitary matrix on a complex column matrix thus parallels that of an orthogonal matrix acting on a real column matrix.

8.12.7 Normal matrices

A final important set of special matrices consists of the *normal* matrices, for which

$$AA^\dagger = A^\dagger A,$$

i.e. a normal matrix is one that commutes with its Hermitian conjugate.

We can easily show that Hermitian matrices and unitary matrices (or symmetric matrices and orthogonal matrices in the real case) are examples of normal matrices. For an Hermitian matrix, $A = A^\dagger$ and so

$$AA^\dagger = AA = A^\dagger A.$$

Similarly, for a unitary matrix, $A^{-1} = A^\dagger$ and so

$$AA^\dagger = AA^{-1} = A^{-1}A = A^\dagger A.$$

Finally, we note that, if A is normal then so too is its inverse A^{-1}, since

$$A^{-1}(A^{-1})^\dagger = A^{-1}(A^\dagger)^{-1} = (A^\dagger A)^{-1} = (AA^\dagger)^{-1} = (A^\dagger)^{-1}A^{-1} = (A^{-1})^\dagger A^{-1}.$$

This broad class of matrices is important in the discussion of eigenvectors and eigenvalues in the next section.

8.13 Eigenvectors and eigenvalues

Suppose that a linear operator $\mathcal{A}$ transforms vectors $\mathbf{x}$ in an N-dimensional vector space into other vectors $\mathcal{A}\mathbf{x}$ in the same space. The possibility then arises that there exist vectors $\mathbf{x}$ each of which is transformed by $\mathcal{A}$ into a multiple of itself. Such vectors would have to satisfy

$$\mathcal{A}\mathbf{x} = \lambda\mathbf{x}. \tag{8.67}$$

Any non-zero vector $\mathbf{x}$ that satisfies (8.67) for some value of λ is called an *eigenvector* of the linear operator $\mathcal{A}$, and λ is called the corresponding *eigenvalue*. As will be discussed below, in general the operator $\mathcal{A}$ has N independent eigenvectors $\mathbf{x}^i$, with eigenvalues λ_i. The λ_i are not necessarily all distinct.

If we choose a particular basis in the vector space, we can write (8.67) in terms of the components of $\mathcal{A}$ and $\mathbf{x}$ with respect to this basis as the matrix equation

$$A\mathbf{x} = \lambda\mathbf{x}, \tag{8.68}$$

where A is an $N \times N$ matrix. The column matrices x that satisfy (8.68) obviously represent the eigenvectors **x** of $\mathcal{A}$ in our chosen coordinate system. Conventionally, these column matrices are also referred to as the *eigenvectors of the matrix* A.[§] Clearly, if x is an eigenvector of A (with some eigenvalue λ) then any scalar multiple μx is also an eigenvector with the same eigenvalue. We therefore often use *normalised* eigenvectors, for which

$$x^\dagger x = 1$$

(note that $x^\dagger x$ corresponds to the inner product $\langle x|x \rangle$ in our basis). Any eigenvector x can be normalised by dividing all its components by the scalar $(x^\dagger x)^{1/2}$.

As will be seen, the problem of finding the eigenvalues and corresponding eigenvectors of a square matrix A plays an important role in many physical investigations. Throughout this chapter we denote the *i*th eigenvector of a square matrix A by x^i and the corresponding eigenvalue by λ_i. This superscript notation for eigenvectors is used to avoid any confusion with components.

▶*A non-singular matrix* A *has eigenvalues* λ_i *and eigenvectors* x^i. *Find the eigenvalues and eigenvectors of the inverse matrix* A^{-1}.

The eigenvalues and eigenvectors of A satisfy

$$Ax^i = \lambda_i x^i.$$

Left-multiplying both sides of this equation by A^{-1}, we find

$$A^{-1}Ax^i = \lambda_i A^{-1}x^i.$$

Since $A^{-1}A = I$, on rearranging we obtain

$$A^{-1}x^i = \frac{1}{\lambda_i}x^i.$$

Thus, we see that A^{-1} has the *same* eigenvectors x^i as does A, but the corresponding eigenvalues are $1/\lambda_i$. ◀

In the remainder of this section we will discuss some useful results concerning the eigenvectors and eigenvalues of certain special (though commonly occurring) square matrices. The results will be established for matrices whose elements may be complex; the corresponding properties for real matrices may be obtained as special cases.

8.13.1 Eigenvectors and eigenvalues of a normal matrix

In subsection 8.12.7 we defined a normal matrix A as one that commutes with its Hermitian conjugate, so that

$$A^\dagger A = AA^\dagger.$$

[§] In this context, when referring to linear combinations of eigenvectors x we will normally use the term 'vector'.

We also showed that both Hermitian and unitary matrices (or symmetric and orthogonal matrices in the real case) are examples of normal matrices. We now discuss the properties of the eigenvectors and eigenvalues of a normal matrix.

If x is an eigenvector of a normal matrix A with corresponding eigenvalue λ then $Ax = \lambda x$, or equivalently,

$$(A - \lambda I)x = 0. \tag{8.69}$$

Denoting $B = A - \lambda I$, (8.69) becomes $Bx = 0$ and, taking the Hermitian conjugate, we also have

$$(Bx)^\dagger = x^\dagger B^\dagger = 0. \tag{8.70}$$

From (8.69) and (8.70) we then have

$$x^\dagger B^\dagger Bx = 0. \tag{8.71}$$

However, the product $B^\dagger B$ is given by

$$B^\dagger B = (A - \lambda I)^\dagger (A - \lambda I) = (A^\dagger - \lambda^* I)(A - \lambda I) = A^\dagger A - \lambda^* A - \lambda A^\dagger + \lambda\lambda^*.$$

Now since A is normal, $AA^\dagger = A^\dagger A$ and so

$$B^\dagger B = AA^\dagger - \lambda^* A - \lambda A^\dagger + \lambda\lambda^* = (A - \lambda I)(A - \lambda I)^\dagger = BB^\dagger,$$

and hence B is also normal. From (8.71) we then find

$$x^\dagger B^\dagger Bx = x^\dagger BB^\dagger x = (B^\dagger x)^\dagger B^\dagger x = 0,$$

from which we obtain

$$B^\dagger x = (A^\dagger - \lambda^* I)x = 0.$$

Therefore, for a normal matrix A, *the eigenvalues of $A^\dagger$ are the complex conjugates of the eigenvalues of A.*

Let us now consider two eigenvectors x^i and x^j of a normal matrix A corresponding to two *different* eigenvalues λ_i and λ_j. We then have

$$Ax^i = \lambda_i x^i, \tag{8.72}$$
$$Ax^j = \lambda_j x^j. \tag{8.73}$$

Multiplying (8.73) on the left by $(x^i)^\dagger$ we obtain

$$(x^i)^\dagger Ax^j = \lambda_j (x^i)^\dagger x^j. \tag{8.74}$$

However, on the LHS of (8.74) we have

$$(x^i)^\dagger A = (A^\dagger x^i)^\dagger = (\lambda_i^* x^i)^\dagger = \lambda_i (x^i)^\dagger, \tag{8.75}$$

where we have used (8.40) and the property just proved for a normal matrix to write $A^\dagger x^i = \lambda_i^* x^i$. From (8.74) and (8.75) we have

$$(\lambda_i - \lambda_j)(x^i)^\dagger x^j = 0. \tag{8.76}$$

Thus, *if $\lambda_i \neq \lambda_j$ the eigenvectors x^i and x^j must be orthogonal*, i.e. $(x^i)^\dagger x^j = 0$.

It follows immediately from (8.76) that if all N eigenvalues of a normal matrix A are distinct then all N eigenvectors of A are mutually orthogonal. If, however, two or more eigenvalues are the same then further consideration is required. An eigenvalue corresponding to two or more different eigenvectors (i.e. they are not simply multiples of one another) is said to be *degenerate*. Suppose that λ_1 is k-fold degenerate, i.e.

$$Ax^i = \lambda_1 x^i \quad \text{for } i = 1, 2, \ldots, k, \tag{8.77}$$

but that it is different from any of λ_{k+1}, λ_{k+2}, etc. Then any linear combination of these x^i is also an eigenvector with eigenvalue λ_1, since, for $z = \sum_{i=1}^{k} c_i x^i$,

$$Az \equiv A \sum_{i=1}^{k} c_i x^i = \sum_{i=1}^{k} c_i A x^i = \sum_{i=1}^{k} c_i \lambda_1 x^i = \lambda_1 z. \tag{8.78}$$

If the x^i defined in (8.77) are not already mutually orthogonal then we can construct new eigenvectors z^i that are orthogonal by the following procedure:

$$z^1 = x^1,$$
$$z^2 = x^2 - \left[(\hat{z}^1)^\dagger x^2\right] \hat{z}^1,$$
$$z^3 = x^3 - \left[(\hat{z}^2)^\dagger x^3\right] \hat{z}^2 - \left[(\hat{z}^1)^\dagger x^3\right] \hat{z}^1,$$
$$\vdots$$
$$z^k = x^k - \left[(\hat{z}^{k-1})^\dagger x^k\right] \hat{z}^{k-1} - \cdots - \left[(\hat{z}^1)^\dagger x^k\right] \hat{z}^1.$$

In this procedure, known as *Gram–Schmidt orthogonalisation*, each new eigenvector z^i is normalised to give the unit vector $\hat{z}^i$ before proceeding to the construction of the next one (the normalisation is carried out by dividing each element of the vector z^i by $[(z^i)^\dagger z^i]^{1/2}$). Note that each factor in brackets $(\hat{z}^m)^\dagger x^n$ is a scalar product and thus only a number. It follows that, as shown in (8.78), each vector z^i so constructed is an eigenvector of A with eigenvalue λ_1 and will remain so on normalisation. It is straightforward to check that, provided the previous new eigenvectors have been normalised as prescribed, each z^i is orthogonal to all its predecessors. (In practice, however, the method is laborious and the example in subsection 8.14.1 gives a less rigorous but considerably quicker way.)

Therefore, even if A has some degenerate eigenvalues we can *by construction* obtain a set of N mutually orthogonal eigenvectors. Moreover, it may be shown (although the proof is beyond the scope of this book) that these eigenvectors are *complete* in that they form a basis for the N-dimensional vector space. As a result any arbitrary vector y can be expressed as a linear combination of the eigenvectors x^i:

$$y = \sum_{i=1}^{N} a_i x^i, \qquad (8.79)$$

where $a_i = (x^i)^\dagger y$. Thus, the eigenvectors form an orthogonal basis for the vector space. By normalising the eigenvectors so that $(x^i)^\dagger x^i = 1$ this basis is made orthonormal.

▶Show that a normal matrix A can be written in terms of its eigenvalues λ_i and orthonormal eigenvectors x^i as

$$A = \sum_{i=1}^{N} \lambda_i x^i (x^i)^\dagger. \qquad (8.80)$$

The key to proving the validity of (8.80) is to show that both sides of the expression give the same result when acting on an arbitary vector y. Since A is normal, we may expand y in terms of the eigenvectors x^i, as shown in (8.79). Thus, we have

$$Ay = A \sum_{i=1}^{N} a_i x^i = \sum_{i=1}^{N} a_i \lambda_i x^i.$$

Alternatively, the action of the RHS of (8.80) on y is given by

$$\sum_{i=1}^{N} \lambda_i x^i (x^i)^\dagger y = \sum_{i=1}^{N} a_i \lambda_i x^i,$$

since $a_i = (x^i)^\dagger y$. We see that the two expressions for the action of each side of (8.80) on y are identical, which implies that this relationship is indeed correct. ◀

8.13.2 Eigenvectors and eigenvalues of Hermitian and anti-Hermitian matrices

For a normal matrix we showed that if $Ax = \lambda x$ then $A^\dagger x = \lambda^* x$. However, if A is also Hermitian, $A = A^\dagger$, it follows necessarily that $\lambda = \lambda^*$. Thus, the eigenvalues of an Hermitian matrix are real, a result which may be proved directly.

▶Prove that the eigenvalues of an Hermitian matrix are real.

For any particular eigenvector x^i, we take the Hermitian conjugate of $Ax^i = \lambda_i x^i$ to give

$$(x^i)^\dagger A^\dagger = \lambda_i^* (x^i)^\dagger. \qquad (8.81)$$

Using $A^\dagger = A$, since A is Hermitian, and multiplying on the right by x^i, we obtain

$$(x^i)^\dagger A x^i = \lambda_i^* (x^i)^\dagger x^i. \qquad (8.82)$$

But multiplying $Ax^i = \lambda_i x^i$ through on the left by $(x^i)^\dagger$ gives

$$(x^i)^\dagger A x^i = \lambda_i (x^i)^\dagger x^i.$$

Subtracting this from (8.82) yields

$$0 = (\lambda_i^* - \lambda_i)(x^i)^\dagger x^i.$$

But $(x^i)^\dagger x^i$ is the modulus squared of the non-zero vector x^i and is thus non-zero. Hence λ_i^* must equal λ_i and thus be real. The same argument can be used to show that the eigenvalues of a real symmetric matrix are themselves real. ◄

The importance of the above result will be apparent to any student of quantum mechanics. In quantum mechanics the eigenvalues of operators correspond to measured values of observable quantities, e.g. energy, angular momentum, parity and so on, and these clearly must be real. If we use Hermitian operators to formulate the theories of quantum mechanics, the above property guarantees physically meaningful results.

Since an Hermitian matrix is also a normal matrix, its eigenvectors are orthogonal (or can be made so using the Gram–Schmidt orthogonalisation procedure). Alternatively we can prove the orthogonality of the eigenvectors directly.

► *Prove that the eigenvectors corresponding to different eigenvalues of an Hermitian matrix are orthogonal.*

Consider two unequal eigenvalues λ_i and λ_j and their corresponding eigenvectors satisfying

$$A x^i = \lambda_i x^i, \tag{8.83}$$
$$A x^j = \lambda_j x^j. \tag{8.84}$$

Taking the Hermitian conjugate of (8.83) we find $(x^i)^\dagger A^\dagger = \lambda_i^*(x^i)^\dagger$. Multiplying this on the right by x^j we obtain

$$(x^i)^\dagger A^\dagger x^j = \lambda_i^*(x^i)^\dagger x^j,$$

and similarly multiplying (8.84) through on the left by $(x^i)^\dagger$ we find

$$(x^i)^\dagger A x^j = \lambda_j(x^i)^\dagger x^j.$$

Then, since $A^\dagger = A$, the two left-hand sides are equal and, because the λ_i are real, on subtraction we obtain

$$0 = (\lambda_i - \lambda_j)(x^i)^\dagger x^j.$$

Finally we note that $\lambda_i \neq \lambda_j$ and so $(x^i)^\dagger x^j = 0$, i.e. the eigenvectors x^i and x^j are orthogonal. ◄

In the case where some of the eigenvalues are equal, further justification of the orthogonality of the eigenvectors is needed. The Gram–Schmidt orthogonalisation procedure discussed above provides a proof of, and a means of achieving, orthogonality. The general method has already been described and we will not repeat it here.

We may also consider the properties of the eigenvalues and eigenvectors of an anti-Hermitian matrix, for which $A^\dagger = -A$ and thus

$$A A^\dagger = A(-A) = (-A)A = A^\dagger A.$$

Therefore matrices that are anti-Hermitian are also normal and so have mutually orthogonal eigenvectors. The properties of the eigenvalues are also simply deduced, since if $Ax = \lambda x$ then

$$\lambda^* x = A^\dagger x = -Ax = -\lambda x.$$

Hence $\lambda^* = -\lambda$ and so λ must be *pure imaginary* (or *zero*). In a similar manner to that used for Hermitian matrices, these properties may be proved directly.

8.13.3 Eigenvectors and eigenvalues of a unitary matrix

A unitary matrix satisfies $A^\dagger = A^{-1}$ and is also a normal matrix, with mutually orthogonal eigenvectors. To investigate the eigenvalues of a unitary matrix, we note that if $Ax = \lambda x$ then

$$x^\dagger x = x^\dagger A^\dagger A x = \lambda^* \lambda x^\dagger x,$$

and we deduce that $\lambda\lambda^* = |\lambda|^2 = 1$. Thus, the eigenvalues of a unitary matrix have unit modulus.

8.13.4 Eigenvectors and eigenvalues of a general square matrix

When an $N \times N$ matrix is not normal there are no general properties of its eigenvalues and eigenvectors; in general it is not possible to find any orthogonal set of N eigenvectors or even to find *pairs* of orthogonal eigenvectors (except by chance in some cases). While the N non-orthogonal eigenvectors are usually linearly independent and hence form a basis for the N-dimensional vector space, this is not necessarily so. It may be shown (although we will not prove it) that any $N \times N$ matrix with *distinct* eigenvalues has N linearly independent eigenvectors, which therefore form a basis for the N-dimensional vector space. If a general square matrix has degenerate eigenvalues, however, then it may or may not have N linearly independent eigenvectors. A matrix whose eigenvectors are not linearly independent is said to be *defective*.

8.13.5 Simultaneous eigenvectors

We may now ask under what conditions two different normal matrices can have a common set of eigenvectors. The result – that they do so if, and only if, they commute – has profound significance for the foundations of quantum mechanics.

To prove this important result let A and B be two $N \times N$ normal matrices and x^i be the ith eigenvector of A corresponding to eigenvalue λ_i, i.e.

$$Ax^i = \lambda_i x^i \quad \text{for} \quad i = 1, 2, \dots, N.$$

For the present we assume that the eigenvalues are all different.

(i) First suppose that A and B commute. Now consider

$$ABx^i = BAx^i = B\lambda_i x^i = \lambda_i Bx^i,$$

where we have used the commutativity for the first equality and the eigenvector property for the second. It follows that $A(Bx^i) = \lambda_i(Bx^i)$ and thus that Bx^i is an eigenvector of A corresponding to eigenvalue λ_i. But the eigenvector solutions of $(A - \lambda_i I)x^i = 0$ are unique to within a scale factor, and we therefore conclude that

$$Bx^i = \mu_i x^i$$

for some scale factor μ_i. However, this is just an eigenvector equation for B and shows that x^i is an eigenvector of B, in addition to being an eigenvector of A. By reversing the roles of A and B, it also follows that every eigenvector of B is an eigenvector of A. Thus the two sets of eigenvectors are identical.

(ii) Now suppose that A and B have all their eigenvectors in common, a typical one x^i satisfying both

$$Ax^i = \lambda_i x^i \quad \text{and} \quad Bx^i = \mu_i x^i.$$

As the eigenvectors span the N-dimensional vector space, any arbitrary vector x in the space can be written as a linear combination of the eigenvectors,

$$x = \sum_{i=1}^{N} c_i x^i.$$

Now consider both

$$ABx = AB \sum_{i=1}^{N} c_i x^i - A \sum_{i=1}^{N} c_i \mu_i x^i = \sum_{i=1}^{N} c_i \lambda_i \mu_i x^i,$$

and

$$BAx = BA \sum_{i=1}^{N} c_i x^i = B \sum_{i=1}^{N} c_i \lambda_i x^i = \sum_{i=1}^{N} c_i \mu_i \lambda_i x^i.$$

It follows that ABx and BAx are the same for any arbitrary x and hence that

$$(AB - BA)x = 0$$

for all x. That is, A and B *commute*.

This completes the proof that a necessary and sufficient condition for two normal matrices to have a set of eigenvectors in common is that they commute. It should be noted that if an eigenvalue of A, say, is degenerate then not all of its possible sets of eigenvectors will also constitute a set of eigenvectors of B. However, provided that by taking linear combinations one set of joint eigenvectors can be found, the proof is still valid and the result still holds.

When extended to the case of Hermitian operators and continuous eigenfunctions (sections 17.2 and 17.3) the connection between commuting matrices and a set of common eigenvectors plays a fundamental role in the postulatory basis of quantum mechanics. It draws the distinction between commuting and non-commuting observables and sets limits on how much information about a system can be known, even in principle, at any one time.

8.14 Determination of eigenvalues and eigenvectors

The next step is to show how the eigenvalues and eigenvectors of a given $N \times N$ matrix A are found. To do this we refer to (8.68) and as in (8.69) rewrite it as

$$\mathsf{A}\mathsf{x} - \lambda \mathsf{I}\mathsf{x} = (\mathsf{A} - \lambda \mathsf{I})\mathsf{x} = 0. \tag{8.85}$$

The slight rearrangement used here is to write x as $\mathsf{I}\mathsf{x}$, where I is the unit matrix of order N. The point of doing this is immediate since (8.85) now has the form of a homogeneous set of simultaneous equations, the theory of which will be developed in section 8.18. What will be proved there is that the equation $\mathsf{B}\mathsf{x} = 0$ only has a non-trivial solution x if $|\mathsf{B}| = 0$. Correspondingly, therefore, we must have in the present case that

$$|\mathsf{A} - \lambda \mathsf{I}| = 0, \tag{8.86}$$

if there are to be non-zero solutions x to (8.85).

Equation (8.86) is known as the *characteristic equation* for A and its LHS as the *characteristic* or *secular determinant* of A. The equation is a polynomial of degree N in the quantity λ. The N roots of this equation λ_i, $i = 1, 2, \ldots, N$, give the eigenvalues of A. Corresponding to each λ_i there will be a column vector x^i, which is the ith eigenvector of A and can be found by using (8.68).

It will be observed that when (8.86) is written out as a polynomial equation in λ, the coefficient of $-\lambda^{N-1}$ in the equation will be simply $A_{11} + A_{22} + \cdots + A_{NN}$ relative to the coefficient of λ^N. As discussed in section 8.8, the quantity $\sum_{i=1}^{N} A_{ii}$ is the *trace* of A and, from the ordinary theory of polynomial equations, will be equal to the sum of the roots of (8.86):

$$\sum_{i=1}^{N} \lambda_i = \mathrm{Tr}\,\mathsf{A}. \tag{8.87}$$

This can be used as one check that a computation of the eigenvalues λ_i has been done correctly. Unless equation (8.87) is satisfied by a computed set of eigenvalues, they have not been calculated correctly. However, that equation (8.87) is satisfied is a necessary, but not sufficient, condition for a correct computation. An alternative proof of (8.87) is given in section 8.16.

▶*Find the eigenvalues and normalised eigenvectors of the real symmetric matrix*

$$\mathsf{A} = \begin{pmatrix} 1 & 1 & 3 \\ 1 & 1 & -3 \\ 3 & -3 & -3 \end{pmatrix}.$$

Using (8.86),

$$\begin{vmatrix} 1 - \lambda & 1 & 3 \\ 1 & 1 - \lambda & -3 \\ 3 & -3 & -3 - \lambda \end{vmatrix} = 0.$$

Expanding out this determinant gives

$$(1 - \lambda)\,[(1 - \lambda)(-3 - \lambda) - (-3)(-3)] + 1\,[(-3)(3) - 1(-3 - \lambda)]$$
$$+ 3\,[1(-3) - (1 - \lambda)(3)] = 0,$$

which simplifies to give

$$(1 - \lambda)(\lambda^2 + 2\lambda - 12) + (\lambda - 6) + 3(3\lambda - 6) = 0,$$
$$\Rightarrow \quad (\lambda - 2)(\lambda - 3)(\lambda + 6) = 0.$$

Hence the roots of the characteristic equation, which are the eigenvalues of A, are $\lambda_1 = 2$, $\lambda_2 = 3$, $\lambda_3 = -6$. We note that, as expected,

$$\lambda_1 + \lambda_2 + \lambda_3 = -1 = 1 + 1 - 3 = A_{11} + A_{22} + A_{33} = \text{Tr A}.$$

For the first root, $\lambda_1 = 2$, a suitable eigenvector $\mathbf{x}^1$, with elements x_1, x_2, x_3, must satisfy $A\mathbf{x}^1 = 2\mathbf{x}^1$ or, equivalently,

$$x_1 + x_2 + 3x_3 = 2x_1,$$
$$x_1 + x_2 - 3x_3 = 2x_2, \qquad (8.88)$$
$$3x_1 - 3x_2 - 3x_3 = 2x_3.$$

These three equations are consistent (to ensure this was the purpose in finding the particular values of λ) and yield $x_3 = 0$, $x_1 = x_2 = k$, where k is any non-zero number. A suitable eigenvector would thus be

$$\mathbf{x}^1 = (k \quad k \quad 0)^{\text{T}}.$$

If we apply the normalisation condition, we require $k^2 + k^2 + 0^2 = 1$ or $k = 1/\sqrt{2}$. Hence

$$\mathbf{x}^1 = \left(\frac{1}{\sqrt{2}} \quad \frac{1}{\sqrt{2}} \quad 0 \right)^{\text{T}} = \frac{1}{\sqrt{2}}(1 \quad 1 \quad 0)^{\text{T}}.$$

Repeating the last paragraph, but with the factor 2 on the RHS of (8.88) replaced successively by $\lambda_2 = 3$ and $\lambda_3 = -6$, gives two further normalised eigenvectors

$$\mathbf{x}^2 = \frac{1}{\sqrt{3}}(1 \quad -1 \quad 1)^{\text{T}}, \qquad \mathbf{x}^3 = \frac{1}{\sqrt{6}}(1 \quad -1 \quad -2)^{\text{T}}. \blacktriangleleft$$

In the above example, the three values of λ are all different and A is a real symmetric matrix. Thus we expect, and it is easily checked, that the three eigenvectors are mutually orthogonal, i.e.

$$\left(\mathbf{x}^1\right)^{\text{T}}\mathbf{x}^2 = \left(\mathbf{x}^1\right)^{\text{T}}\mathbf{x}^3 = \left(\mathbf{x}^2\right)^{\text{T}}\mathbf{x}^3 = 0.$$

It will be apparent also that, as expected, the normalisation of the eigenvectors has no effect on their orthogonality.

8.14.1 Degenerate eigenvalues

We return now to the case of degenerate eigenvalues, i.e. those that have two or more associated eigenvectors. We have shown already that it is always possible to construct an orthogonal set of eigenvectors for a normal matrix, see subsection 8.13.1, and the following example illustrates one method for constructing such a set.

►*Construct an orthonormal set of eigenvectors for the matrix*
$$A = \begin{pmatrix} 1 & 0 & 3 \\ 0 & -2 & 0 \\ 3 & 0 & 1 \end{pmatrix}.$$

We first determine the eigenvalues using $|A - \lambda I| = 0$:

$$0 = \begin{vmatrix} 1-\lambda & 0 & 3 \\ 0 & -2-\lambda & 0 \\ 3 & 0 & 1-\lambda \end{vmatrix} = -(1-\lambda)^2(2+\lambda) + 3(3)(2+\lambda)$$

$$= (4-\lambda)(\lambda+2)^2.$$

Thus $\lambda_1 = 4$, $\lambda_2 = -2 = \lambda_3$. The eigenvector $\mathbf{x}^1 = (x_1 \quad x_2 \quad x_3)^{\mathrm{T}}$ is found from

$$\begin{pmatrix} 1 & 0 & 3 \\ 0 & -2 & 0 \\ 3 & 0 & 1 \end{pmatrix} \begin{pmatrix} x_1 \\ x_2 \\ x_3 \end{pmatrix} = 4 \begin{pmatrix} x_1 \\ x_2 \\ x_3 \end{pmatrix} \quad \Rightarrow \quad \mathbf{x}^1 = \frac{1}{\sqrt{2}} \begin{pmatrix} 1 \\ 0 \\ 1 \end{pmatrix}.$$

A general column vector that is orthogonal to $\mathbf{x}^1$ is

$$\mathbf{x} = (a \quad b \quad -a)^{\mathrm{T}}, \tag{8.89}$$

and it is easily shown that

$$A\mathbf{x} = \begin{pmatrix} 1 & 0 & 3 \\ 0 & -2 & 0 \\ 3 & 0 & 1 \end{pmatrix} \begin{pmatrix} a \\ b \\ -a \end{pmatrix} = -2 \begin{pmatrix} a \\ b \\ -a \end{pmatrix} = -2\mathbf{x}.$$

Thus $\mathbf{x}$ is a eigenvector of A with associated eigenvalue -2. It is clear, however, that there is an infinite set of eigenvectors $\mathbf{x}$ all possessing the required property; the geometrical analogue is that there are an infinite number of corresponding vectors $\mathbf{x}$ lying in the plane that has $\mathbf{x}^1$ as its normal. We do require that the two remaining eigenvectors are orthogonal to one another, but this still leaves an infinite number of possibilities. For $\mathbf{x}^2$, therefore, let us choose a simple form of (8.89), suitably normalised, say,

$$\mathbf{x}^2 = (0 \quad 1 \quad 0)^{\mathrm{T}}.$$

The third eigenvector is then specified (to within an arbitrary multiplicative constant) by the requirement that it must be orthogonal to $\mathbf{x}^1$ and $\mathbf{x}^2$; thus $\mathbf{x}^3$ may be found by evaluating the vector product of $\mathbf{x}^1$ and $\mathbf{x}^2$ and normalising the result. This gives

$$\mathbf{x}^3 = \frac{1}{\sqrt{2}}(-1 \quad 0 \quad 1)^{\mathrm{T}},$$

to complete the construction of an orthonormal set of eigenvectors. ◄

8.15 Change of basis and similarity transformations

Throughout this chapter we have considered the vector $\mathbf{x}$ as a geometrical quantity that is independent of any basis (or coordinate system). If we introduce a basis $\mathbf{e}_i$, $i = 1, 2, \ldots, N$, into our N-dimensional vector space then we may write

$$\mathbf{x} = x_1\mathbf{e}_1 + x_2\mathbf{e}_2 + \cdots + x_N\mathbf{e}_N,$$

and represent $\mathbf{x}$ in this basis by the column matrix

$$\mathbf{x} = (x_1 \quad x_2 \quad \cdots \quad x_n)^{\mathrm{T}},$$

having components x_i. We now consider how these components change as a result of a prescribed change of basis. Let us introduce a new basis $\mathbf{e}'_i$, $i = 1, 2, \ldots, N$, which is related to the old basis by

$$\mathbf{e}'_j = \sum_{i=1}^{N} S_{ij}\mathbf{e}_i, \tag{8.90}$$

the coefficient S_{ij} being the ith component of $\mathbf{e}'_j$ with respect to the old (unprimed) basis. For an arbitrary vector $\mathbf{x}$ it follows that

$$\mathbf{x} = \sum_{i=1}^{N} x_i\mathbf{e}_i = \sum_{j=1}^{N} x'_j\mathbf{e}'_j = \sum_{j=1}^{N} x'_j \sum_{i=1}^{N} S_{ij}\mathbf{e}_i.$$

From this we derive the relationship between the components of $\mathbf{x}$ in the two coordinate systems as

$$x_i = \sum_{j=1}^{N} S_{ij}x'_j,$$

which we can write in matrix form as

$$\mathbf{x} = \mathsf{S}\mathbf{x}' \tag{8.91}$$

where S is the *transformation matrix* associated with the change of basis.

Furthermore, since the vectors $\mathbf{e}'_j$ are linearly independent, the matrix S is non-singular and so possesses an inverse S^{-1}. Multiplying (8.91) on the left by S^{-1} we find

$$\mathbf{x}' = \mathsf{S}^{-1}\mathbf{x}, \tag{8.92}$$

which relates the components of $\mathbf{x}$ in the new basis to those in the old basis. Comparing (8.92) and (8.90) we note that the components of $\mathbf{x}$ transform inversely to the way in which the basis vectors $\mathbf{e}_i$ themselves transform. This has to be so, as the vector $\mathbf{x}$ itself must remain unchanged.

We may also find the transformation law for the components of a linear operator under the same change of basis. Now, the operator equation $\mathbf{y} = \mathcal{A}\mathbf{x}$ (which is basis independent) can be written as a matrix equation in each of the two bases as

$$\mathbf{y} = \mathsf{A}\mathbf{x}, \qquad \mathbf{y}' = \mathsf{A}'\mathbf{x}'. \tag{8.93}$$

But, using (8.91), we may rewrite the first equation as

$$\mathsf{S}\mathbf{y}' = \mathsf{A}\mathsf{S}\mathbf{x}' \quad \Rightarrow \quad \mathbf{y}' = \mathsf{S}^{-1}\mathsf{A}\mathsf{S}\mathbf{x}'.$$

Comparing this with the second equation in (8.93) we find that the components of the linear operator $\mathcal{A}$ transform as

$$A' = S^{-1}AS. \qquad (8.94)$$

Equation (8.94) is an example of a *similarity transformation* – a transformation that can be particularly useful in converting matrices into convenient forms for computation.

Given a square matrix A, we may interpret it as representing a linear operator $\mathcal{A}$ in a given basis $\mathbf{e}_i$. From (8.94), however, we may also consider the matrix $A' = S^{-1}AS$, for any non-singular matrix S, as representing the same linear operator $\mathcal{A}$ but in a new basis $\mathbf{e}'_j$, related to the old basis by

$$\mathbf{e}'_j = \sum_i S_{ij}\mathbf{e}_i.$$

Therefore we would expect that any property of the matrix A that represents some (basis-independent) property of the linear operator $\mathcal{A}$ will also be shared by the matrix A'. We list these properties below.

(i) If $A = I$ then $A' = I$, since, from (8.94),

$$A' = S^{-1}IS = S^{-1}S = I. \qquad (8.95)$$

(ii) The value of the determinant is unchanged:

$$|A'| = |S^{-1}AS| = |S^{-1}||A||S| = |A||S^{-1}||S| = |A||S^{-1}S| = |A|. \qquad (8.96)$$

(iii) The characteristic determinant and hence the eigenvalues of A' are the same as those of A: from (8.86),

$$\begin{aligned} |A' - \lambda I| &= |S^{-1}AS - \lambda I| = |S^{-1}(A - \lambda I)S| \\ &= |S^{-1}||S||A - \lambda I| = |A - \lambda I|. \end{aligned} \qquad (8.97)$$

(iv) The value of the trace is unchanged: from (8.87),

$$\begin{aligned} \operatorname{Tr} A' &= \sum_i A'_{ii} = \sum_i \sum_j \sum_k (S^{-1})_{ij} A_{jk} S_{ki} \\ &= \sum_i \sum_j \sum_k S_{ki}(S^{-1})_{ij} A_{jk} = \sum_j \sum_k \delta_{kj} A_{jk} = \sum_j A_{jj} \\ &= \operatorname{Tr} A. \end{aligned} \qquad (8.98)$$

An important class of similarity transformations is that for which S is a unitary matrix; in this case $A' = S^{-1}AS = S^{\dagger}AS$. Unitary transformation matrices are particularly important, for the following reason. If the original basis $\mathbf{e}_i$ is orthonormal and the transformation matrix S is unitary then

$$\langle e_i'|e_j'\rangle = \left\langle \sum_k S_{ki} e_k \middle| \sum_r S_{rj} e_r \right\rangle$$

$$= \sum_k S_{ki}^* \sum_r S_{rj} \langle e_k|e_r\rangle$$

$$= \sum_k S_{ki}^* \sum_r S_{rj} \delta_{kr} = \sum_k S_{ki}^* S_{kj} = (S^\dagger S)_{ij} = \delta_{ij},$$

showing that the new basis is also orthonormal.

Furthermore, in addition to the properties of general similarity transformations, for unitary transformations the following hold.

(i) If A is Hermitian (anti-Hermitian) then A′ is Hermitian (anti-Hermitian), i.e. if $A^\dagger = \pm A$ then

$$(A')^\dagger = (S^\dagger A S)^\dagger = S^\dagger A^\dagger S = \pm S^\dagger A S = \pm A'. \tag{8.99}$$

(ii) If A is unitary (so that $A^\dagger - A^{-1}$) then A′ is unitary, since

$$(A')^\dagger A' = (S^\dagger A S)^\dagger (S^\dagger A S) = S^\dagger A^\dagger S S^\dagger A S = S^\dagger A^\dagger A S$$

$$= S^\dagger I S = I. \tag{8.100}$$

8.16 Diagonalisation of matrices

Suppose that a linear operator $\mathcal{A}$ is represented in some basis e_i, $i = 1, 2, \ldots, N$, by the matrix A. Consider a new basis x^j given by

$$x^j = \sum_{i=1}^N S_{ij} e_i,$$

where the x^j are chosen to be the eigenvectors of the linear operator $\mathcal{A}$, i.e.

$$\mathcal{A} x^j = \lambda_j x^j. \tag{8.101}$$

In the new basis, $\mathcal{A}$ is represented by the matrix $A' = S^{-1}AS$, which has a particularly simple form, as we shall see shortly. The element S_{ij} of S is the ith component, in the old (unprimed) basis, of the jth eigenvector x^j of A, i.e. the columns of S are the eigenvectors of the matrix A:

$$S = \begin{pmatrix} \uparrow & \uparrow & & \uparrow \\ x^1 & x^2 & \cdots & x^N \\ \downarrow & \downarrow & & \downarrow \end{pmatrix},$$

that is, $S_{ij} = (\mathbf{x}^j)_i$. Therefore A$'$ is given by

$$(\mathsf{S}^{-1}A\mathsf{S})_{ij} = \sum_k \sum_l (\mathsf{S}^{-1})_{ik} A_{kl} S_{lj}$$

$$= \sum_k \sum_l (\mathsf{S}^{-1})_{ik} A_{kl} (\mathbf{x}^j)_l$$

$$= \sum_k (\mathsf{S}^{-1})_{ik} \lambda_j (\mathbf{x}^j)_k$$

$$= \sum_k \lambda_j (\mathsf{S}^{-1})_{ik} S_{kj} = \lambda_j \delta_{ij}.$$

So the matrix A$'$ is diagonal with the eigenvalues of $\mathcal{A}$ as the diagonal elements, i.e.

$$\mathsf{A}' = \begin{pmatrix} \lambda_1 & 0 & \cdots & 0 \\ 0 & \lambda_2 & & \vdots \\ \vdots & & \ddots & 0 \\ 0 & \cdots & 0 & \lambda_N \end{pmatrix}.$$

Therefore, given a matrix A, if we construct the matrix S that has the eigenvectors of A as its columns then the matrix A$' = \mathsf{S}^{-1}A\mathsf{S}$ is diagonal and has the eigenvalues of A as its diagonal elements. Since we require S to be non-singular ($|\mathsf{S}| \neq 0$), the N eigenvectors of A must be linearly independent and form a basis for the N-dimensional vector space. It may be shown that *any matrix with distinct eigenvalues* can be diagonalised by this procedure. If, however, a general square matrix has degenerate eigenvalues then it may, or may not, have N linearly independent eigenvectors. If it does not then it *cannot* be diagonalised.

For normal matrices (which include Hermitian, anti-Hermitian and unitary matrices) the N eigenvectors are indeed linearly independent. Moreover, when normalised, these eigenvectors form an *orthonormal* set (or can be made to do so). Therefore the matrix S with these normalised eigenvectors as columns, i.e. whose elements are $S_{ij} = (\mathbf{x}^j)_i$, has the property

$$(\mathsf{S}^\dagger\mathsf{S})_{ij} = \sum_k (\mathsf{S}^\dagger)_{ik}(\mathsf{S})_{kj} = \sum_k S^*_{ki} S_{kj} = \sum_k (\mathbf{x}^i)^*_k (\mathbf{x}^j)_k = (\mathbf{x}^i)^\dagger \mathbf{x}^j = \delta_{ij}.$$

Hence S is unitary ($\mathsf{S}^{-1} = \mathsf{S}^\dagger$) and the original matrix A can be diagonalised by

$$\mathsf{A}' = \mathsf{S}^{-1}\mathsf{A}\mathsf{S} = \mathsf{S}^\dagger\mathsf{A}\mathsf{S}.$$

Therefore, any normal matrix A can be diagonalised by a similarity transformation using a *unitary* transformation matrix S.

▶Diagonalise the matrix

$$A = \begin{pmatrix} 1 & 0 & 3 \\ 0 & -2 & 0 \\ 3 & 0 & 1 \end{pmatrix}.$$

The matrix A is symmetric and so may be diagonalised by a transformation of the form $A' = S^\dagger AS$, where S has the normalised eigenvectors of A as its columns. We have already found these eigenvectors in subsection 8.14.1, and so we can write straightaway

$$S = \frac{1}{\sqrt{2}} \begin{pmatrix} 1 & 0 & -1 \\ 0 & \sqrt{2} & 0 \\ 1 & 0 & 1 \end{pmatrix}.$$

We note that although the eigenvalues of A are degenerate, its three eigenvectors are linearly independent and so A can still be diagonalised. Thus, calculating $S^\dagger AS$ we obtain

$$S^\dagger AS = \frac{1}{2} \begin{pmatrix} 1 & 0 & 1 \\ 0 & \sqrt{2} & 0 \\ -1 & 0 & 1 \end{pmatrix} \begin{pmatrix} 1 & 0 & 3 \\ 0 & -2 & 0 \\ 3 & 0 & 1 \end{pmatrix} \begin{pmatrix} 1 & 0 & -1 \\ 0 & \sqrt{2} & 0 \\ 1 & 0 & 1 \end{pmatrix}$$

$$= \begin{pmatrix} 4 & 0 & 0 \\ 0 & -2 & 0 \\ 0 & 0 & -2 \end{pmatrix},$$

which is diagonal, as required, and has as its diagonal elements the eigenvalues of A. ◀

If a matrix A is diagonalised by the similarity transformation $A' = S^{-1}AS$, so that $A' = \mathrm{diag}(\lambda_1, \lambda_2, \ldots, \lambda_N)$, then we have immediately

$$\mathrm{Tr}\, A' = \mathrm{Tr}\, A = \sum_{i=1}^{N} \lambda_i, \tag{8.102}$$

$$|A'| = |A| = \prod_{i=1}^{N} \lambda_i, \tag{8.103}$$

since the eigenvalues of the matrix are unchanged by the transformation. Moreover, these results may be used to prove the rather useful *trace formula*

$$|\exp A| = \exp(\mathrm{Tr}\, A), \tag{8.104}$$

where the exponential of a matrix is as defined in (8.38).

▶Prove the trace formula (8.104).

At the outset, we note that for the similarity transformation $A' = S^{-1}AS$, we have

$$(A')^n = (S^{-1}AS)(S^{-1}AS) \cdots (S^{-1}AS) = S^{-1}A^nS.$$

Thus, from (8.38), we obtain $\exp A' = S^{-1}(\exp A)S$, from which it follows that $|\exp A'| =$

$|\exp \mathsf{A}|$. Moreover, by choosing the similarity transformation so that it diagonalises A, we have $\mathsf{A}' = \mathrm{diag}(\lambda_1, \lambda_2, \ldots, \lambda_N)$, and so

$$|\exp \mathsf{A}| = |\exp \mathsf{A}'| = |\exp[\mathrm{diag}(\lambda_1, \lambda_2, \ldots, \lambda_N)]| = |\mathrm{diag}(\exp \lambda_1, \exp \lambda_2, \ldots, \exp \lambda_N)| = \prod_{i=1}^{N} \exp \lambda_i.$$

Rewriting the final product of exponentials of the eigenvalues as the exponential of the sum of the eigenvalues, we find

$$|\exp \mathsf{A}| = \prod_{i=1}^{N} \exp \lambda_i = \exp\left(\sum_{i=1}^{N} \lambda_i\right) = \exp(\mathrm{Tr}\,\mathsf{A}),$$

which gives the trace formula (8.104). ◄

8.17 Quadratic and Hermitian forms

Let us now introduce the concept of quadratic forms (and their complex analogues, Hermitian forms). A quadratic form Q is a scalar function of a real vector $\mathbf{x}$ given by

$$Q(\mathbf{x}) = \langle \mathbf{x} | \mathcal{A}\,\mathbf{x} \rangle, \tag{8.105}$$

for some real linear operator $\mathcal{A}$. In any given basis (coordinate system) we can write (8.105) in matrix form as

$$Q(\mathbf{x}) = \mathbf{x}^{\mathrm{T}} \mathsf{A} \mathbf{x}, \tag{8.106}$$

where A is a real matrix. In fact, as will be explained below, we need only consider the case where A is symmetric, i.e. $\mathsf{A} = \mathsf{A}^{\mathrm{T}}$. As an example in a three-dimensional space,

$$Q = \mathbf{x}^{\mathrm{T}} \mathsf{A} \mathbf{x} = \begin{pmatrix} x_1 & x_2 & x_3 \end{pmatrix} \begin{pmatrix} 1 & 1 & 3 \\ 1 & 1 & -3 \\ 3 & -3 & -3 \end{pmatrix} \begin{pmatrix} x_1 \\ x_2 \\ x_3 \end{pmatrix}$$

$$= x_1^2 + x_2^2 - 3x_3^2 + 2x_1 x_2 + 6x_1 x_3 - 6x_2 x_3. \tag{8.107}$$

It is reasonable to ask whether a quadratic form $Q = \mathbf{x}^{\mathrm{T}} \mathsf{M} \mathbf{x}$, where M is any (possibly non-symmetric) real square matrix, is a more general definition. That this is not the case may be seen by expressing M in terms of a symmetric matrix $\mathsf{A} = \frac{1}{2}(\mathsf{M} + \mathsf{M}^{\mathrm{T}})$ and an antisymmetric matrix $\mathsf{B} = \frac{1}{2}(\mathsf{M} - \mathsf{M}^{\mathrm{T}})$ such that $\mathsf{M} = \mathsf{A} + \mathsf{B}$. We then have

$$Q = \mathbf{x}^{\mathrm{T}} \mathsf{M} \mathbf{x} = \mathbf{x}^{\mathrm{T}} \mathsf{A} \mathbf{x} + \mathbf{x}^{\mathrm{T}} \mathsf{B} \mathbf{x}. \tag{8.108}$$

However, Q is a scalar quantity and so

$$Q = Q^{\mathrm{T}} = (\mathbf{x}^{\mathrm{T}} \mathsf{A} \mathbf{x})^{\mathrm{T}} + (\mathbf{x}^{\mathrm{T}} \mathsf{B} \mathbf{x})^{\mathrm{T}} = \mathbf{x}^{\mathrm{T}} \mathsf{A}^{\mathrm{T}} \mathbf{x} + \mathbf{x}^{\mathrm{T}} \mathsf{B}^{\mathrm{T}} \mathbf{x} = \mathbf{x}^{\mathrm{T}} \mathsf{A} \mathbf{x} - \mathbf{x}^{\mathrm{T}} \mathsf{B} \mathbf{x}. \tag{8.109}$$

Comparing (8.108) and (8.109) shows that $\mathbf{x}^{\mathrm{T}} \mathsf{B} \mathbf{x} = 0$, and hence $\mathbf{x}^{\mathrm{T}} \mathsf{M} \mathbf{x} = \mathbf{x}^{\mathrm{T}} \mathsf{A} \mathbf{x}$, i.e. Q is unchanged by considering only the symmetric part of M. Hence, with no loss of generality, we may assume $A = A^T$ in (8.106).

From its definition (8.105), Q is clearly a basis- (i.e. coordinate-) independent quantity. Let us therefore consider a new basis related to the old one by an orthogonal transformation matrix S, the components in the two bases of any vector $\mathbf{x}$ being related (as in (8.91)) by $\mathbf{x} = S\mathbf{x}'$ or, equivalently, by $\mathbf{x}' = S^{-1}\mathbf{x} = S^T\mathbf{x}$. We then have

$$Q = \mathbf{x}^T A\mathbf{x} = (\mathbf{x}')^T S^T A S\mathbf{x}' = (\mathbf{x}')^T A'\mathbf{x}',$$

where (as expected) the matrix describing the linear operator $\mathcal{A}$ in the new basis is given by $A' = S^T A S$ (since $S^T = S^{-1}$). But, from the last section, if we choose as S the matrix whose columns are the *normalised* eigenvectors of A then $A' = S^T A S$ is diagonal with the eigenvalues of A as the diagonal elements. (Since A is symmetric, its normalised eigenvectors are orthogonal, or can be made so, and hence S is orthogonal with $S^{-1} = S^T$.)

In the new basis

$$Q = \mathbf{x}^T A\mathbf{x} = (\mathbf{x}')^T \Lambda\mathbf{x}' = \lambda_1 x_1'^2 + \lambda_2 x_2'^2 + \cdots + \lambda_N x_N'^2, \tag{8.110}$$

where $\Lambda = \operatorname{diag}(\lambda_1, \lambda_2, \ldots, \lambda_N)$ and the λ_i are the eigenvalues of A. It should be noted that Q contains no cross-terms of the form $x_1' x_2'$.

▶*Find an orthogonal transformation that takes the quadratic form (8.107) into the form*

$$\lambda_1 x_1'^2 + \lambda_2 x_2'^2 + \lambda_3 x_3'^2.$$

The required transformation matrix S has the *normalised* eigenvectors of A as its columns. We have already found these in section 8.14, and so we can write immediately

$$S = \frac{1}{\sqrt{6}} \begin{pmatrix} \sqrt{3} & \sqrt{2} & 1 \\ \sqrt{3} & -\sqrt{2} & -1 \\ 0 & \sqrt{2} & -2 \end{pmatrix},$$

which is easily verified as being orthogonal. Since the eigenvalues of A are $\lambda = 2$, 3, and -6, the general result already proved shows that the transformation $\mathbf{x} = S\mathbf{x}'$ will carry (8.107) into the form $2x_1'^2 + 3x_2'^2 - 6x_3'^2$. This may be verified most easily by writing out the inverse transformation $\mathbf{x}' = S^{-1}\mathbf{x} = S^T\mathbf{x}$ and substituting. The inverse equations are

$$\begin{aligned} x_1' &= (x_1 + x_2)/\sqrt{2}, \\ x_2' &= (x_1 - x_2 + x_3)/\sqrt{3}, \\ x_3' &= (x_1 - x_2 - 2x_3)/\sqrt{6}. \end{aligned} \tag{8.111}$$

If these are substituted into the form $Q = 2x_1'^2 + 3x_2'^2 - 6x_3'^2$ then the original expression (8.107) is recovered. ◀

In the definition of Q it was assumed that the components x_1, x_2, x_3 and the matrix A were real. It is clear that in this case the quadratic form $Q \equiv \mathbf{x}^T A\mathbf{x}$ is real also. Another, rather more general, expression that is also real is the *Hermitian form*

$$H(\mathbf{x}) \equiv \mathbf{x}^\dagger A\mathbf{x}, \tag{8.112}$$

where A is Hermitian (i.e. $A^\dagger = A$) and the components of $\mathbf{x}$ may now be complex. It is straightforward to show that H is real, since

$$H^* = (H^T)^* = \mathbf{x}^\dagger A^\dagger \mathbf{x} = \mathbf{x}^\dagger A\mathbf{x} = H.$$

With suitable generalisation, the properties of quadratic forms apply also to Hermitian forms, but to keep the presentation simple we will restrict our discussion to quadratic forms.

A special case of a quadratic (Hermitian) form is one for which $Q = \mathbf{x}^T A\mathbf{x}$ is greater than zero for all column matrices $\mathbf{x}$. By choosing as the basis the eigenvectors of A we have Q in the form

$$Q = \lambda_1 x_1^2 + \lambda_2 x_2^2 + \lambda_3 x_3^2.$$

The requirement that $Q > 0$ for all $\mathbf{x}$ means that all the eigenvalues λ_i of A must be positive. A symmetric (Hermitian) matrix A with this property is called *positive definite*. If, instead, $Q \geq 0$ for all $\mathbf{x}$ then it is possible that some of the eigenvalues are zero, and A is called *positive semi-definite*.

8.17.1 The stationary properties of the eigenvectors

Consider a quadratic form, such as $Q(\mathbf{x}) = \langle \mathbf{x} | \mathcal{A} \mathbf{x} \rangle$, equation (8.105), in a fixed basis. As the vector $\mathbf{x}$ is varied, through changes in its three components x_1, x_2 and x_3, the value of the quantity Q also varies. Because of the homogeneous form of Q we may restrict any investigation of these variations to vectors of unit length (since multiplying any vector $\mathbf{x}$ by any scalar k simply multiplies the value of Q by a factor k^2).

Of particular interest are any vectors $\mathbf{x}$ that make the value of the quadratic form a maximum or minimum. A necessary, but not sufficient, condition for this is that Q is stationary with respect to small variations $\Delta \mathbf{x}$ in $\mathbf{x}$, whilst $\langle \mathbf{x} | \mathbf{x} \rangle$ is maintained at a constant value (unity).

In the chosen basis the quadratic form is given by $Q = \mathbf{x}^T A\mathbf{x}$ and, using Lagrange undetermined multipliers to incorporate the variational constraints, we are led to seek solutions of

$$\Delta[\mathbf{x}^T A\mathbf{x} - \lambda(\mathbf{x}^T\mathbf{x} - 1)] = 0. \tag{8.113}$$

This may be used directly, together with the fact that $(\Delta\mathbf{x}^T)A\mathbf{x} = \mathbf{x}^T A \Delta\mathbf{x}$, since A is symmetric, to obtain

$$A\mathbf{x} = \lambda\mathbf{x} \tag{8.114}$$

as the necessary condition that x must satisfy. If (8.114) is satisfied for some eigenvector x then the value of $Q(x)$ is given by

$$Q = x^T A x = x^T \lambda x = \lambda. \tag{8.115}$$

However, if x and y are eigenvectors corresponding to different eigenvalues then they are (or can be chosen to be) orthogonal. Consequently the expression $y^T A x$ is necessarily zero, since

$$y^T A x = y^T \lambda x = \lambda y^T x = 0. \tag{8.116}$$

Summarising, those column matrices x of unit magnitude that make the quadratic form Q stationary are eigenvectors of the matrix A, and the stationary value of Q is then equal to the corresponding eigenvalue. It is straightforward to see from the proof of (8.114) that, conversely, any eigenvector of A makes Q stationary.

Instead of maximising or minimising $Q = x^T A x$ subject to the constraint $x^T x = 1$, an equivalent procedure is to extremise the function

$$\lambda(x) = \frac{x^T A x}{x^T x}.$$

▶*Show that if $\lambda(x)$ is stationary then x is an eigenvector of A and $\lambda(x)$ is equal to the corresponding eigenvalue.*

We require $\Delta\lambda(x) = 0$ with respect to small variations in x. Now

$$\Delta\lambda = \frac{1}{(x^T x)^2} \left[(x^T x) \left(\Delta x^T A x + x^T A \Delta x \right) - x^T A x \left(\Delta x^T x + x^T \Delta x \right) \right]$$

$$= \frac{2\Delta x^T A x}{x^T x} - 2 \left(\frac{x^T A x}{x^T x} \right) \frac{\Delta x^T x}{x^T x},$$

since $x^T A \Delta x = (\Delta x^T) A x$ and $x^T \Delta x = (\Delta x^T) x$. Thus

$$\Delta\lambda = \frac{2}{x^T x} \Delta x^T [A x - \lambda(x) x].$$

Hence, if $\Delta\lambda = 0$ then $A x = \lambda(x) x$, i.e. x is an eigenvector of A with eigenvalue $\lambda(x)$. ◀

Thus the eigenvalues of a symmetric matrix A are the values of the function

$$\lambda(x) = \frac{x^T A x}{x^T x}$$

at its stationary points. The eigenvectors of A lie along those directions in space for which the quadratic form $Q = x^T A x$ has stationary values, given a fixed magnitude for the vector x. Similar results hold for Hermitian matrices.

8.17.2 Quadratic surfaces

The results of the previous subsection may be turned round to state that the surface given by

$$x^T Ax = \text{constant} = 1 \text{ (say)} \tag{8.117}$$

and called a *quadratic surface*, has stationary values of its radius (i.e. origin–surface distance) in those directions that are along the eigenvectors of A. More specifically, in three dimensions the quadratic surface $x^T Ax = 1$ has its principal axes along the three mutually perpendicular eigenvectors of A, and the squares of the corresponding principal radii are given by λ_i^{-1}, $i = 1, 2, 3$. As well as having this stationary property of the radius, a *principal axis* is characterised by the fact that any section of the surface perpendicular to it has some degree of symmetry about it. If the eigenvalues corresponding to any two principal axes are degenerate then the quadratic surface has rotational symmetry about the third principal axis and the choice of a pair of axes perpendicular to that axis is not uniquely defined.

> ▶*Find the shape of the quadratic surface*
> $$x_1^2 + x_2^2 - 3x_3^2 + 2x_1 x_2 + 6x_1 x_3 - 6x_2 x_3 = 1.$$

If, instead of expressing the quadratic surface in terms of x_1, x_2, x_3, as in (8.107), we were to use the new variables x_1', x_2', x_3' defined in (8.111), for which the coordinate axes are along the three mutually perpendicular eigenvector directions $(1, 1, 0)$, $(1, -1, 1)$ and $(1, -1, -2)$, then the equation of the surface would take the form (see (8.110))

$$\frac{x_1'^2}{(1/\sqrt{2})^2} + \frac{x_2'^2}{(1/\sqrt{3})^2} - \frac{x_3'^2}{(1/\sqrt{6})^2} = 1.$$

Thus, for example, a section of the quadratic surface in the plane $x_3' = 0$, i.e. $x_1 - x_2 - 2x_3 = 0$, is an ellipse, with semi-axes $1/\sqrt{2}$ and $1/\sqrt{3}$. Similarly a section in the plane $x_1' = x_1 + x_2 = 0$ is a hyperbola. ◀

Clearly the simplest three-dimensional situation to visualise is that in which all the eigenvalues are positive, since then the quadratic surface is an ellipsoid.

8.18 Simultaneous linear equations

In physical applications we often encounter sets of simultaneous linear equations. In general we may have M equations in N unknowns $x_1, x_2, \ldots, x_N$ of the form

$$
\begin{aligned}
A_{11}x_1 + A_{12}x_2 + \cdots + A_{1N}x_N &= b_1, \\
A_{21}x_1 + A_{22}x_2 + \cdots + A_{2N}x_N &= b_2, \\
&\;\;\vdots \\
A_{M1}x_1 + A_{M2}x_2 + \cdots + A_{MN}x_N &= b_M,
\end{aligned}
\tag{8.118}
$$

where the A_{ij} and b_i have known values. If all the b_i are zero then the system of equations is called *homogeneous*, otherwise it is *inhomogeneous*. Depending on the given values, this set of equations for the N unknowns $x_1, x_2, \ldots, x_N$ may have either a unique solution, no solution or infinitely many solutions. Matrix analysis may be used to distinguish between the possibilities. The set of equations may be expressed as a single matrix equation $\mathsf{A}\mathbf{x} = \mathbf{b}$, or, written out in full, as

$$
\begin{pmatrix}
A_{11} & A_{12} & \cdots & A_{1N} \\
A_{21} & A_{22} & \cdots & A_{2N} \\
\vdots & \vdots & \ddots & \vdots \\
A_{M1} & A_{M2} & \cdots & A_{MN}
\end{pmatrix}
\begin{pmatrix}
x_1 \\
x_2 \\
\vdots \\
x_N
\end{pmatrix}
=
\begin{pmatrix}
b_1 \\
b_2 \\
\vdots \\
b_M
\end{pmatrix}.
$$

8.18.1 The range and null space of a matrix

As we discussed in section 8.2, we may interpret the matrix equation $\mathsf{A}\mathbf{x} = \mathbf{b}$ as representing, in some basis, the linear transformation $\mathcal{A}\mathbf{x} = \mathbf{b}$ of a vector $\mathbf{x}$ in an N-dimensional vector space V into a vector $\mathbf{b}$ in some other (in general different) M-dimensional vector space W.

In general the operator $\mathcal{A}$ will map *any* vector in V into some particular *subspace* of W, which may be the entire space. This subspace is called the *range* of $\mathcal{A}$ (or A) and its dimension is equal to the *rank* of A. Moreover, if $\mathcal{A}$ (and hence A) is *singular* then there exists some subspace of V that is mapped onto the zero vector $\mathbf{0}$ in W; that is, any vector $\mathbf{y}$ that lies in the subspace satisfies $\mathcal{A}\mathbf{y} = \mathbf{0}$. This subspace is called the *null space* of A and the dimension of this null space is called the *nullity* of A. We note that the matrix A *must* be singular if $M \neq N$ and *may* be singular even if $M = N$.

The dimensions of the range and the null space of a matrix are related through the fundamental relationship

$$\text{rank } \mathsf{A} + \text{nullity } \mathsf{A} = N, \tag{8.119}$$

where N is the number of original unknowns $x_1, x_2, \ldots, x_N$.

▶*Prove the relationship (8.119).*

As discussed in section 8.11, if the columns of an $M \times N$ matrix A are interpreted as the components, in a given basis, of N (M-component) vectors $\mathbf{v}_1, \mathbf{v}_2, \ldots, \mathbf{v}_N$ then rank A is equal to the number of linearly independent vectors in this set (this number is also equal to the dimension of the vector space spanned by these vectors). Writing (8.118) in terms of the vectors $\mathbf{v}_1, \mathbf{v}_2, \ldots, \mathbf{v}_N$, we have

$$x_1\mathbf{v}_1 + x_2\mathbf{v}_2 + \cdots + x_N\mathbf{v}_N = \mathbf{b}. \tag{8.120}$$

From this expression, we immediately deduce that the range of A is merely the span of the vectors $\mathbf{v}_1, \mathbf{v}_2, \ldots, \mathbf{v}_N$ and hence has dimension $r = \text{rank A}$.

If a vector $\mathbf{y}$ lies in the null space of A then $\mathcal{A}\,\mathbf{y} = \mathbf{0}$, which we may write as

$$y_1\mathbf{v}_1 + y_2\mathbf{v}_2 + \cdots + y_N\mathbf{v}_N = \mathbf{0}. \tag{8.121}$$

As just shown above, however, only r $(\leq N)$ of these vectors are linearly independent. By renumbering, if necessary, we may assume that $\mathbf{v}_1, \mathbf{v}_2, \ldots, \mathbf{v}_r$ form a linearly independent set; the remaining vectors, $\mathbf{v}_{r+1}, \mathbf{v}_{r+2}, \ldots, \mathbf{v}_N$, can then be written as a linear superposition of $\mathbf{v}_1, \mathbf{v}_2, \ldots, \mathbf{v}_r$. We are therefore free to choose the $N - r$ coefficients $y_{r+1}, y_{r+2}, \ldots, y_N$ arbitrarily and (8.121) will still be satisfied for some set of r coefficients $y_1, y_2, \ldots, y_r$ (which are not all zero). The dimension of the null space is therefore $N - r$, and this completes the proof of (8.119). ◄

Equation (8.119) has far-reaching consequences for the existence of solutions to sets of simultaneous linear equations such as (8.118). As mentioned previously, these equations may have *no solution*, a *unique solution* or *infinitely many solutions*. We now discuss these three cases in turn.

No solution

The system of equations possesses no solution unless $\mathbf{b}$ lies in the range of $\mathcal{A}$; in this case (8.120) will be satisfied for some $x_1, x_2, \ldots, x_N$. This in turn requires the set of vectors $\mathbf{b}, \mathbf{v}_1, \mathbf{v}_2, \ldots, \mathbf{v}_N$ to have the same span (see (8.8)) as $\mathbf{v}_1, \mathbf{v}_2, \ldots, \mathbf{v}_N$. In terms of matrices, this is equivalent to the requirement that the matrix A and the *augmented matrix*

$$\mathsf{M} = \begin{pmatrix} A_{11} & A_{12} & \ldots & A_{1N} & b_1 \\ A_{21} & A_{22} & \ldots & A_{2N} & b_1 \\ \vdots & & \ddots & & \vdots \\ A_{M1} & A_{M2} & \ldots & A_{MN} & b_M \end{pmatrix}$$

have the *same* rank r. If this condition is satisfied then $\mathbf{b}$ does lie in the range of $\mathcal{A}$, and the set of equations (8.118) will have either a unique solution or infinitely many solutions. If, however, A and M have different ranks then there will be no solution.

A unique solution

If $\mathbf{b}$ lies in the range of $\mathcal{A}$ and if $r = N$ then all the vectors $\mathbf{v}_1, \mathbf{v}_2, \ldots, \mathbf{v}_N$ in (8.120) are linearly independent and the equation has a *unique solution* $x_1, x_2, \ldots, x_N$.

Infinitely many solutions

If $\mathbf{b}$ lies in the range of $\mathcal{A}$ and if $r < N$ then only r of the vectors $\mathbf{v}_1, \mathbf{v}_2, \ldots, \mathbf{v}_N$ in (8.120) are linearly independent. We may therefore choose the coefficients of $n - r$ vectors in an arbitrary way, while still satisfying (8.120) for some set of coefficients $x_1, x_2, \ldots, x_N$. There are therefore *infinitely many solutions*, which span an $(n-r)$-dimensional vector space. We may also consider this space of solutions in terms of the null space of A: if $\mathbf{x}$ is some vector satisfying $\mathcal{A}\mathbf{x} = \mathbf{b}$ and $\mathbf{y}$ is *any* vector in the null space of $\mathcal{A}$ (i.e. $\mathcal{A}\,\mathbf{y} = \mathbf{0}$) then

$$\mathcal{A}\,(\mathbf{x} + \mathbf{y}) = \mathcal{A}\,\mathbf{x} + \mathcal{A}\,\mathbf{y} = \mathcal{A}\,\mathbf{x} + \mathbf{0} = \mathbf{b},$$

and so $\mathbf{x} + \mathbf{y}$ is also a solution. Since the null space is $(n - r)$-dimensional, so too is the space of solutions.

We may use the above results to investigate the special case of the solution of a *homogeneous* set of linear equations, for which $\mathbf{b} = \mathbf{0}$. Clearly the set *always* has the trivial solution $x_1 = x_2 = \cdots = x_n = 0$, and if $r = N$ this will be the only solution. If $r < N$, however, there are infinitely many solutions; they form the null space of A, which has dimension $n - r$. In particular, we note that if $M < N$ (i.e. there are fewer equations than unknowns) then $r < N$ automatically. Hence a set of *homogeneous* linear equations with fewer equations than unknowns *always* has infinitely many solutions.

8.18.2 N simultaneous linear equations in N unknowns

A special case of (8.118) occurs when $M = N$. In this case the matrix A is *square* and we have the same number of equations as unknowns. Since A is square, the condition $r = N$ corresponds to $|A| \neq 0$ and the matrix A is *non-singular*. The case $r < N$ corresponds to $|A| = 0$, in which case A is *singular*.

As mentioned above, the equations will have a solution provided b lies in the range of A. If this is true then the equations will possess a unique solution when $|A| \neq 0$ or infinitely many solutions when $|A| = 0$. There exist several methods for obtaining the solution(s). Perhaps the most elementary method is *Gaussian elimination*; this method is discussed in subsection 27.3.1, where we also address numerical subtleties such as equation interchange (pivoting). In this subsection, we will outline three further methods for solving a square set of simultaneous linear equations.

Direct inversion

Since A is square it will possess an inverse, provided $|A| \neq 0$. Thus, if A is non-singular, we immediately obtain

$$\mathbf{x} = \mathsf{A}^{-1}\mathbf{b} \tag{8.122}$$

as the unique solution to the set of equations. However, if $\mathbf{b} = 0$ then we see immediately that the set of equations possesses only the trivial solution $\mathbf{x} = 0$. The direct inversion method has the advantage that, once A^{-1} has been calculated, one may obtain the solutions $\mathbf{x}$ corresponding to different vectors $\mathbf{b}_1$, $\mathbf{b}_2$, ... on the RHS, with little further work.

▶*Show that the set of simultaneous equations*

$$2x_1 + 4x_2 + 3x_3 = 4,$$
$$x_1 - 2x_2 - 2x_3 = 0, \qquad (8.123)$$
$$-3x_1 + 3x_2 + 2x_3 = -7,$$

has a unique solution, and find that solution.

The simultaneous equations can be represented by the matrix equation $Ax = b$, i.e.

$$\begin{pmatrix} 2 & 4 & 3 \\ 1 & -2 & -2 \\ -3 & 3 & 2 \end{pmatrix} \begin{pmatrix} x_1 \\ x_2 \\ x_3 \end{pmatrix} = \begin{pmatrix} 4 \\ 0 \\ -7 \end{pmatrix}.$$

As we have already shown that A^{-1} exists and have calculated it, see (8.59), it follows that $x = A^{-1}b$ or, more explicitly, that

$$\begin{pmatrix} x_1 \\ x_2 \\ x_3 \end{pmatrix} = \frac{1}{11} \begin{pmatrix} 2 & 1 & -2 \\ 4 & 13 & 7 \\ -3 & -18 & -8 \end{pmatrix} \begin{pmatrix} 4 \\ 0 \\ -7 \end{pmatrix} = \begin{pmatrix} 2 \\ -3 \\ 4 \end{pmatrix}. \qquad (8.124)$$

Thus the unique solution is $x_1 = 2$, $x_2 = -3$, $x_3 = 4$. ◀

LU decomposition

Although conceptually simple, finding the solution by calculating A^{-1} can be computationally demanding, especially when N is large. In fact, as we shall now show, it is not necessary to perform the full inversion of A in order to solve the simultaneous equations $Ax = b$. Rather, we can perform a *decomposition* of the matrix into the product of a square *lower triangular* matrix L and a square *upper triangular* matrix U, which are such that

$$A = LU, \qquad (8.125)$$

and then use the fact that triangular systems of equations can be solved very simply.

We must begin, therefore, by finding the matrices L and U such that (8.125) is satisfied. This may be achieved straightforwardly by writing out (8.125) in component form. For illustration, let us consider the 3×3 case. It is, in fact, always possible, and convenient, to take the diagonal elements of L as unity, so we have

$$
A = \begin{pmatrix} 1 & 0 & 0 \\ L_{21} & 1 & 0 \\ L_{31} & L_{32} & 1 \end{pmatrix} \begin{pmatrix} U_{11} & U_{12} & U_{13} \\ 0 & U_{22} & U_{23} \\ 0 & 0 & U_{33} \end{pmatrix}
$$
$$
= \begin{pmatrix} U_{11} & U_{12} & U_{13} \\ L_{21}U_{11} & L_{21}U_{12} + U_{22} & L_{21}U_{13} + U_{23} \\ L_{31}U_{11} & L_{31}U_{12} + L_{32}U_{22} & L_{31}U_{13} + L_{32}U_{23} + U_{33} \end{pmatrix} \qquad (8.126)
$$

The nine unknown elements of L and U can now be determined by equating the nine elements of (8.126) to those of the 3×3 matrix A. This is done in the particular order illustrated in the example below.

Once the matrices L and U have been determined, one can use the decomposition to solve the set of equations $Ax = b$ in the following way. From (8.125), we have $LUx = b$, but this can be written as *two* triangular sets of equations

$$Ly = b \quad \text{and} \quad Ux = y,$$

where y is another column matrix to be determined. One may easily solve the first triangular set of equations for y, which is then substituted into the second set. The required solution x is then obtained readily from the second triangular set of equations. We note that, as with direct inversion, once the LU decomposition has been determined, one can solve for various RHS column matrices b_1, b_2, ... , with little extra work.

▶*Use LU decomposition to solve the set of simultaneous equations (8.123).*

We begin the determination of the matrices L and U by equating the elements of the matrix in (8.126) with those of the matrix

$$A = \begin{pmatrix} 2 & 4 & 3 \\ 1 & -2 & -2 \\ -3 & 3 & 2 \end{pmatrix}.$$

This is performed in the following order:

1st row:	$U_{11} = 2$,	$U_{12} = 4$,	$U_{13} = 3$
1st column:	$L_{21}U_{11} = 1$,	$L_{31}U_{11} = -3$	$\Rightarrow L_{21} = \frac{1}{2}, L_{31} = -\frac{3}{2}$
2nd row:	$L_{21}U_{12} + U_{22} = -2$	$L_{21}U_{13} + U_{23} = -2$	$\Rightarrow U_{22} = -4, U_{23} = -\frac{7}{2}$
2nd column:	$L_{31}U_{12} + L_{32}U_{22} = 3$		$\Rightarrow L_{32} = -\frac{9}{4}$
3rd row:	$L_{31}U_{13} + L_{32}U_{23} + U_{33} = 2$		$\Rightarrow U_{33} = -\frac{11}{8}$

Thus we may write the matrix A as

$$A = LU = \begin{pmatrix} 1 & 0 & 0 \\ \frac{1}{2} & 1 & 0 \\ -\frac{3}{2} & -\frac{9}{4} & 1 \end{pmatrix} \begin{pmatrix} 2 & 4 & 3 \\ 0 & -4 & -\frac{7}{2} \\ 0 & 0 & -\frac{11}{8} \end{pmatrix}.$$

We must now solve the set of equations $Ly = b$, which read

$$\begin{pmatrix} 1 & 0 & 0 \\ \frac{1}{2} & 1 & 0 \\ -\frac{3}{2} & -\frac{9}{4} & 1 \end{pmatrix} \begin{pmatrix} y_1 \\ y_2 \\ y_3 \end{pmatrix} = \begin{pmatrix} 4 \\ 0 \\ -7 \end{pmatrix}.$$

Since this set of equations is triangular, we quickly find

$$y_1 = 4, \quad y_2 = 0 - (\tfrac{1}{2})(4) = -2, \quad y_3 = -7 - (-\tfrac{3}{2})(4) - (-\tfrac{9}{4})(-2) = -\tfrac{11}{2}.$$

These values must then be substituted into the equations $Ux = y$, which read

$$\begin{pmatrix} 2 & 4 & 3 \\ 0 & -4 & -\frac{7}{2} \\ 0 & 0 & -\frac{11}{8} \end{pmatrix} \begin{pmatrix} x_1 \\ x_2 \\ x_3 \end{pmatrix} = \begin{pmatrix} 4 \\ -2 \\ -\frac{11}{2} \end{pmatrix}.$$

This set of equations is also triangular, and we easily find the solution

$$x_1 = 2, \quad x_2 = -3, \quad x_3 = 4,$$

which agrees with the result found above by direct inversion. ◀

We note, in passing, that one can calculate both the inverse and the determinant of A from its LU decomposition. To find the inverse A^{-1}, one solves the system of equations $Ax = b$ repeatedly for the N different RHS column matrices $b = e_i$, $i = 1, 2, \ldots, N$, where e_i is the column matrix with its ith element equal to unity and the others equal to zero. The solution x in each case gives the corresponding column of A^{-1}. Evaluation of the determinant $|A|$ is much simpler. From (8.125), we have

$$|A| = |LU| = |L||U|. \tag{8.127}$$

Since L and U are triangular, however, we see from (8.64) that their determinants are equal to the products of their diagonal elements. Since $L_{ii} = 1$ for all i, we thus find

$$|A| = U_{11}U_{22} \cdots U_{NN} = \prod_{i=1}^{N} U_{ii}.$$

As an illustration, in the above example we find $|A| = (2)(-4)(-11/8) = 11$, which, as it must, agrees with our earlier calculation (8.58).

Finally, we note that if the matrix A is symmetric and positive semi-definite then we can decompose it as

$$A = LL^{\dagger}, \tag{8.128}$$

where L is a lower triangular matrix whose diagonal elements are *not*, in general, equal to unity. This is known as a *Cholesky decomposition* (in the special case where A is real, the decomposition becomes $A = LL^T$). The reason that we cannot set the diagonal elements of L equal to unity in this case is that we require the same number of independent elements in L as in A. The requirement that the matrix be positive semi-definite is easily derived by considering the Hermitian form (or quadratic form in the real case)

$$x^{\dagger}Ax = x^{\dagger}LL^{\dagger}x = (L^{\dagger}x)^{\dagger}(L^{\dagger}x).$$

Denoting the column matrix $L^{\dagger}x$ by y, we see that the last term on the RHS is $y^{\dagger}y$, which must be greater than or equal to zero. Thus, we require $x^{\dagger}Ax \geq 0$ for any arbitrary column matrix x, and so A must be positive semi-definite (see section 8.17).

We recall that the requirement that a matrix be positive semi-definite is equivalent to demanding that all the eigenvalues of A are positive or zero. If one of the eigenvalues of A is zero, however, then from (8.103) we have $|A| = 0$ and so A is *singular*. Thus, if A is a non-singular matrix, it must be *positive definite* (rather than just positive semi-definite) in order to perform the Cholesky decomposition (8.128). In fact, in this case, the inability to find a matrix L that satisfies (8.128) implies that A cannot be positive definite.

The Cholesky decomposition can be applied in an analogous way to the *LU* decomposition discussed above, but we shall not explore it further.

Cramer's rule

An alternative method of solution is to use *Cramer's rule*, which also provides some insight into the nature of the solutions in the various cases. To illustrate this method let us consider a set of three equations in three unknowns,

$$
\begin{aligned}
A_{11}x_1 + A_{12}x_2 + A_{13}x_3 &= b_1, \\
A_{21}x_1 + A_{22}x_2 + A_{23}x_3 &= b_2, \\
A_{31}x_1 + A_{32}x_2 + A_{33}x_3 &= b_3,
\end{aligned}
\tag{8.129}
$$

which may be represented by the matrix equation $Ax = b$. We wish either to find the solution(s) x to these equations or to establish that there are no solutions. From result (vi) of subsection 8.9.1, the determinant $|A|$ is unchanged by adding to its first column the combination

$$
\frac{x_2}{x_1} \times \text{(second column of } |A|) + \frac{x_3}{x_1} \times \text{(third column of } |A|).
$$

We thus obtain

$$
|A| = \begin{vmatrix} A_{11} & A_{12} & A_{13} \\ A_{21} & A_{22} & A_{23} \\ A_{31} & A_{32} & A_{33} \end{vmatrix} = \begin{vmatrix} A_{11} + (x_2/x_1)A_{12} + (x_3/x_1)A_{13} & A_{12} & A_{13} \\ A_{21} + (x_2/x_1)A_{22} + (x_3/x_1)A_{23} & A_{22} & A_{23} \\ A_{31} + (x_2/x_1)A_{32} + (x_3/x_1)A_{33} & A_{32} & A_{33} \end{vmatrix},
$$

which, on substituting b_i/x_1 for the ith entry in the first column, yields

$$
|A| = \frac{1}{x_1} \begin{vmatrix} b_1 & A_{12} & A_{13} \\ b_2 & A_{22} & A_{23} \\ b_3 & A_{32} & A_{33} \end{vmatrix} = \frac{1}{x_1}\Delta_1.
$$

The determinant Δ_1 is known as a *Cramer determinant*. Similar manipulations of the second and third columns of $|A|$ yield x_2 and x_3, and so the full set of results reads

$$
x_1 = \frac{\Delta_1}{|A|}, \qquad x_2 = \frac{\Delta_2}{|A|}, \qquad x_3 = \frac{\Delta_3}{|A|},
\tag{8.130}
$$

where

$$
\Delta_1 = \begin{vmatrix} b_1 & A_{12} & A_{13} \\ b_2 & A_{22} & A_{23} \\ b_3 & A_{32} & A_{33} \end{vmatrix}, \qquad \Delta_2 = \begin{vmatrix} A_{11} & b_1 & A_{13} \\ A_{21} & b_2 & A_{23} \\ A_{31} & b_3 & A_{33} \end{vmatrix}, \qquad \Delta_3 = \begin{vmatrix} A_{11} & A_{12} & b_1 \\ A_{21} & A_{22} & b_2 \\ A_{31} & A_{32} & b_3 \end{vmatrix}.
$$

It can be seen that each Cramer determinant Δ_i is simply $|A|$ but with column i replaced by the RHS of the original set of equations. If $|A| \neq 0$ then (8.130) gives the unique solution. The proof given here appears to fail if any of the solutions x_i is zero, but it can be shown that result (8.130) is valid even in such a case.

▶ *Use Cramer's rule to solve the set of simultaneous equations (8.123).*

Let us again represent these simultaneous equations by the matrix equation $\mathsf{Ax} = \mathsf{b}$, i.e.

$$
\begin{pmatrix} 2 & 4 & 3 \\ 1 & -2 & -2 \\ -3 & 3 & 2 \end{pmatrix} \begin{pmatrix} x_1 \\ x_2 \\ x_3 \end{pmatrix} = \begin{pmatrix} 4 \\ 0 \\ -7 \end{pmatrix}.
$$

From (8.58), the determinant of A is given by $|\mathsf{A}| = 11$. Following the discussion given above, the three Cramer determinants are

$$
\Delta_1 = \begin{vmatrix} 4 & 4 & 3 \\ 0 & -2 & -2 \\ -7 & 3 & 2 \end{vmatrix}, \quad \Delta_2 = \begin{vmatrix} 2 & 4 & 3 \\ 1 & 0 & -2 \\ -3 & -7 & 2 \end{vmatrix}, \quad \Delta_3 = \begin{vmatrix} 2 & 4 & 4 \\ 1 & -2 & 0 \\ -3 & 3 & -7 \end{vmatrix}.
$$

These may be evaluated using the properties of determinants listed in subsection 8.9.1 and we find $\Delta_1 = 22$, $\Delta_2 = -33$ and $\Delta_3 = 44$. From (8.130) the solution to the equations (8.123) is given by

$$
x_1 = \frac{22}{11} = 2, \quad x_2 = \frac{-33}{11} = -3, \quad x_3 = \frac{44}{11} = 4,
$$

which agrees with the solution found in the previous example. ◀

At this point it is useful to consider each of the three equations (8.129) as representing a plane in three-dimensional Cartesian coordinates. Using result (7.42) of chapter 7, the sets of components of the vectors normal to the planes are (A_{11}, A_{12}, A_{13}), (A_{21}, A_{22}, A_{23}) and (A_{31}, A_{32}, A_{33}), and using (7.46) the perpendicular distances of the planes from the origin are given by

$$
d_i = \frac{b_i}{\left(A_{i1}^2 + A_{i2}^2 + A_{i3}^2\right)^{1/2}} \quad \text{for } i = 1, 2, 3.
$$

Finding the solution(s) to the simultaneous equations above corresponds to finding the point(s) of intersection of the planes.

If there is a unique solution the planes intersect at only a single point. This happens if their normals are linearly independent vectors. Since the rows of A represent the directions of these normals, this requirement is equivalent to $|\mathsf{A}| \neq 0$. If $\mathsf{b} = (0 \ \ 0 \ \ 0)^{\mathrm{T}} = 0$ then all the planes pass through the origin and, since there is only a single solution to the equations, the origin is that solution.

Let us now turn to the cases where $|\mathsf{A}| = 0$. The simplest such case is that in which all three planes are parallel; this implies that the normals are all parallel and so A is of rank 1. Two possibilities exist:

(i) the planes are coincident, i.e. $d_1 = d_2 = d_3$, in which case there is an infinity of solutions;
(ii) the planes are not all coincident, i.e. $d_1 \neq d_2$ and/or $d_1 \neq d_3$ and/or $d_2 \neq d_3$, in which case there are no solutions.

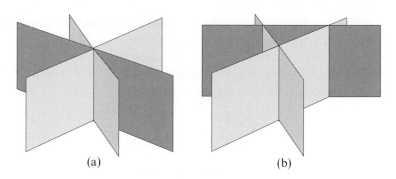

(a)   (b)

Figure 8.1   The two possible cases when A is of rank 2. In both cases all the normals lie in a horizontal plane but in (a) the planes all intersect on a single line (corresponding to an infinite number of solutions) whilst in (b) there are no common intersection points (no solutions).

It is apparent from (8.130) that case (i) occurs when all the Cramer determinants are zero and case (ii) occurs when at least one Cramer determinant is non-zero.

The most complicated cases with $|A| = 0$ are those in which the normals to the planes themselves lie in a plane but are not parallel. In this case A has rank 2. Again two possibilities exist and these are shown in figure 8.1. Just as in the rank-1 case, if all the Cramer determinants are zero then we get an infinity of solutions (this time on a line). Of course, in the special case in which $b = 0$ (and the system of equations is homogeneous), the planes all pass through the origin and so they must intersect on a line through it. If at least one of the Cramer determinants is non-zero, we get no solution.

These rules may be summarised as follows.

(i) $|A| \neq 0$, $b \neq 0$: The three planes intersect at a single point that is not the origin, and so there is only one solution, given by both (8.122) and (8.130).

(ii) $|A| \neq 0$, $b = 0$: The three planes intersect at the origin only and there is only the trivial solution, $x = 0$.

(iii) $|A| = 0$, $b \neq 0$, Cramer determinants all zero: There is an infinity of solutions either on a line if A is rank 2, i.e. the cofactors are not all zero, or on a plane if A is rank 1, i.e. the cofactors are all zero.

(iv) $|A| = 0$, $b \neq 0$, Cramer determinants not all zero: No solutions.

(v) $|A| = 0$, $b = 0$: The three planes intersect on a line through the origin giving an infinity of solutions.

8.18.3 Singular value decomposition

There exists a very powerful technique for dealing with a simultaneous set of linear equations $Ax = b$, such as (8.118), which may be applied *whether or not* the number of simultaneous equations M is equal to the number of unknowns N. This technique is known as *singular value decomposition* (SVD) and is the method of choice in analysing *any* set of simultaneous linear equations.

We will consider the general case, in which A is an $M \times N$ (complex) matrix. Let us suppose we can write A as the product[‡]

$$A = USV^\dagger, \tag{8.131}$$

where the matrices U, S and V have the following properties.

(i) The square matrix U has dimensions $M \times M$ and is *unitary*.
(ii) The matrix S has dimensions $M \times N$ (the same dimensions as those of A) and is *diagonal* in the sense that $S_{ij} = 0$ if $i \neq j$. We denote its diagonal elements by s_i for $i = 1, 2, \ldots, p$, where $p = \min(M, N)$; these elements are termed the *singular values* of A.
(iii) The square matrix V has dimensions $N \times N$ and is *unitary*.

We must now determine the elements of these matrices in terms of the elements of A. From the matrix A, we can construct two square matrices: $A^\dagger A$ with dimensions $N \times N$ and $AA^\dagger$ with dimensions $M \times M$. Both are clearly *Hermitian*. From (8.131), and using the fact that U and V are unitary, we find

$$A^\dagger A = VS^\dagger U^\dagger USV^\dagger = VS^\dagger SV^\dagger \tag{8.132}$$

$$AA^\dagger = USV^\dagger VS^\dagger U^\dagger = USS^\dagger U^\dagger, \tag{8.133}$$

where $S^\dagger S$ and $SS^\dagger$ are diagonal matrices with dimensions $N \times N$ and $M \times M$ respectively. The first p elements of each diagonal matrix are s_i^2, $i = 1, 2, \ldots, p$, where $p = \min(M, N)$, and the rest (where they exist) are zero.

These two equations imply that both $V^{-1}A^\dagger AV \left(= V^{-1}A^\dagger A(V^\dagger)^{-1}\right)$ and, by a similar argument, $U^{-1}AA^\dagger U$, must be diagonal. From our discussion of the diagonalisation of Hermitian matrices in section 8.16, we see that the columns of V must therefore be the normalised eigenvectors v^i, $i = 1, 2, \ldots, N$, of the matrix $A^\dagger A$ and the columns of U must be the normalised eigenvectors u^j, $j = 1, 2, \ldots, M$, of the matrix $AA^\dagger$. Moreover, the singular values s_i must satisfy $s_i^2 = \lambda_i$, where the λ_i are the eigenvalues of the smaller of $A^\dagger A$ and $AA^\dagger$. Clearly, the λ_i are also some of the eigenvalues of the larger of these two matrices, the remaining ones being equal to zero. Since each matrix is Hermitian, the λ_i are real and the singular values s_i may be taken as real and non-negative. Finally, to make the decomposition (8.131) unique, it is customary to arrange the singular values in decreasing order of their values, so that $s_1 \geq s_2 \geq \cdots \geq s_p$.

[‡] The proof that such a decomposition always exists is beyond the scope of this book. For a full account of SVD one might consult, for example, G. H. Golub and C. F. Van Loan, *Matrix Computations*, 3rd edn (Baltimore MD: Johns Hopkins University Press, 1996).

> ►*Show that, for $i = 1, 2, \ldots, p$, $\mathsf{A}v^i = s_i u^i$ and $\mathsf{A}^\dagger u^i = s_i v^i$, where $p = \min(M, N)$.*

Post-multiplying both sides of (8.131) by V, and using the fact that V is unitary, we obtain

$$\mathsf{AV} = \mathsf{US}.$$

Since the columns of V and U consist of the vectors v^i and u^j respectively and S has only diagonal non-zero elements, we find immediately that, for $i = 1, 2, \ldots, p$,

$$\mathsf{A}v^i = s_i u^i. \tag{8.134}$$

Moreover, we note that $\mathsf{A}v^i = 0$ for $i = p+1, p+2, \ldots, N$.

Taking the Hermitian conjugate of both sides of (8.131) and post-multiplying by U, we obtain

$$\mathsf{A}^\dagger \mathsf{U} = \mathsf{VS}^\dagger = \mathsf{VS}^\mathsf{T},$$

where we have used the fact that U is unitary and S is real. We then see immediately that, for $i = 1, 2, \ldots, p$,

$$\mathsf{A}^\dagger u^i = s_i v^i. \tag{8.135}$$

We also note that $\mathsf{A}^\dagger u^i = 0$ for $i = p+1, p+2, \ldots, M$. Results (8.134) and (8.135) are useful for investigating the properties of the SVD. ◄

The decomposition (8.131) has some advantageous features for the analysis of sets of simultaneous linear equations. These are best illustrated by writing the decomposition (8.131) in terms of the vectors u^i and v^i as

$$\mathsf{A} = \sum_{i=1}^{p} s_i u^i (v^i)^\dagger,$$

where $p = \min(M, N)$. It may be, however, that some of the singular values s_i are *zero*, as a result of degeneracies in the set of M linear equations $\mathsf{A}x = b$. Let us suppose that there are r non-zero singular values. Since our convention is to arrange the singular values in order of decreasing size, the non-zero singular values are s_i, $i = 1, 2, \ldots, r$, and the zero singular values are $s_{r+1}, s_{r+2}, \ldots, s_p$. Therefore we can write A as

$$\mathsf{A} = \sum_{i=1}^{r} s_i u^i (v^i)^\dagger. \tag{8.136}$$

Let us consider the action of (8.136) on an arbitrary vector x. This is given by

$$\mathsf{A}x = \sum_{i=1}^{r} s_i u^i (v^i)^\dagger x.$$

Since $(v^i)^\dagger x$ is just a number, we see immediately that the vectors u^i, $i = 1, 2, \ldots, r$, must span the *range* of the matrix A; moreover, these vectors form an orthonormal basis for the range. Further, since this subspace is r-dimensional, we have rank $\mathsf{A} = r$, i.e. the rank of A is equal to the number of non-zero singular values.

The SVD is also useful in characterising the null space of A. From (8.119), we already know that the null space must have dimension $N - r$; so, if A has r

non-zero singular values s_i, $i = 1, 2, \ldots, r$, then from the worked example above we have

$$A v^i = 0 \qquad \text{for } i = r+1, r+2, \ldots, N.$$

Thus, the $N - r$ vectors v^i, $i = r+1, r+2, \ldots, N$, form an orthonormal basis for the null space of A.

▶ *Find the singular value decompostion of the matrix*

$$A = \begin{pmatrix} 2 & 2 & 2 & 2 \\ \frac{17}{10} & \frac{1}{10} & -\frac{17}{10} & -\frac{1}{10} \\ \frac{3}{5} & \frac{9}{5} & -\frac{3}{5} & -\frac{9}{5} \end{pmatrix}. \tag{8.137}$$

The matrix A has dimension 3×4 (i.e. $M = 3$, $N = 4$), and so we may construct from it the 3×3 matrix $AA^\dagger$ and the 4×4 matrix $A^\dagger A$ (in fact, since A is real, the Hermitian conjugates are just transposes). We begin by finding the eigenvalues λ_i and eigenvectors u^i of the smaller matrix $AA^\dagger$. This matrix is easily found to be given by

$$AA^\dagger = \begin{pmatrix} 16 & 0 & 0 \\ 0 & \frac{29}{5} & \frac{12}{5} \\ 0 & \frac{12}{5} & \frac{36}{5} \end{pmatrix},$$

and its characteristic equation reads

$$\begin{vmatrix} 16 - \lambda & 0 & 0 \\ 0 & \frac{29}{5} - \lambda & \frac{12}{5} \\ 0 & \frac{12}{5} & \frac{36}{5} - \lambda \end{vmatrix} = (16 - \lambda)(36 - 13\lambda + \lambda^2) = 0.$$

Thus, the eigenvalues are $\lambda_1 = 16$, $\lambda_2 = 9$, $\lambda_3 = 4$. Since the singular values of A are given by $s_i = \sqrt{\lambda_i}$ and the matrix S in (8.131) has the same dimensions as A, we have

$$S = \begin{pmatrix} 4 & 0 & 0 & 0 \\ 0 & 3 & 0 & 0 \\ 0 & 0 & 2 & 0 \end{pmatrix}, \tag{8.138}$$

where we have arranged the singular values in order of decreasing size. Now the matrix U has as its columns the normalised eigenvectors u^i of the 3×3 matrix $AA^\dagger$. These normalised eigenvectors correspond to the eigenvalues of $AA^\dagger$ as follows:

$$\lambda_1 = 16 \quad \Rightarrow \quad u^1 = (1 \quad 0 \quad 0)^T$$
$$\lambda_2 = 9 \quad \Rightarrow \quad u^2 = (0 \quad \tfrac{3}{5} \quad \tfrac{4}{5})^T$$
$$\lambda_3 = 4 \quad \Rightarrow \quad u^3 = (0 \quad -\tfrac{4}{5} \quad \tfrac{3}{5})^T,$$

and so we obtain the matrix

$$U = \begin{pmatrix} 1 & 0 & 0 \\ 0 & \frac{3}{5} & -\frac{4}{5} \\ 0 & \frac{4}{5} & \frac{3}{5} \end{pmatrix}. \tag{8.139}$$

The columns of the matrix V in (8.131) are the normalised eigenvectors of the 4×4 matrix $A^\dagger A$, which is given by

$$A^\dagger A = \frac{1}{4} \begin{pmatrix} 29 & 21 & 3 & 11 \\ 21 & 29 & 11 & 3 \\ 3 & 11 & 29 & 21 \\ 11 & 3 & 21 & 29 \end{pmatrix}.$$

We already know from the above discussion, however, that the non-zero eigenvalues of this matrix are *equal* to those of $AA^\dagger$ found above, and that the remaining eigenvalue is *zero*. The corresponding normalised eigenvectors are easily found:

$$\lambda_1 = 16 \quad \Rightarrow \quad v^1 = \tfrac{1}{2}(1 \quad 1 \quad 1 \quad 1)^T$$

$$\lambda_2 = 9 \quad \Rightarrow \quad v^2 = \tfrac{1}{2}(1 \quad 1 \quad -1 \quad -1)^T$$

$$\lambda_3 = 4 \quad \Rightarrow \quad v^3 = \tfrac{1}{2}(-1 \quad 1 \quad 1 \quad -1)^T$$

$$\lambda_4 = 0 \quad \Rightarrow \quad v^4 = \tfrac{1}{2}(1 \quad -1 \quad 1 \quad -1)^T$$

and so the matrix V is given by

$$V = \frac{1}{2} \begin{pmatrix} 1 & 1 & -1 & 1 \\ 1 & 1 & 1 & -1 \\ 1 & -1 & 1 & 1 \\ 1 & -1 & -1 & -1 \end{pmatrix}. \tag{8.140}$$

Alternatively, we could have found the first three columns of V by using the relation (8.135) to obtain

$$v^i = \frac{1}{s_i} A^\dagger u^i \qquad \text{for } i = 1, 2, 3.$$

The fourth eigenvector could then be found using the Gram–Schmidt orthogonalisation procedure. We note that if there were more than one eigenvector corresponding to a zero eigenvalue then we would need to use this procedure to orthogonalise these eigenvectors before constructing the matrix V.

Collecting our results together, we find the SVD of the matrix A:

$$A = USV^\dagger = \begin{pmatrix} 1 & 0 & 0 \\ 0 & \tfrac{3}{5} & -\tfrac{4}{5} \\ 0 & \tfrac{4}{5} & \tfrac{3}{5} \end{pmatrix} \begin{pmatrix} 4 & 0 & 0 & 0 \\ 0 & 3 & 0 & 0 \\ 0 & 0 & 2 & 0 \end{pmatrix} \begin{pmatrix} \tfrac{1}{2} & \tfrac{1}{2} & \tfrac{1}{2} & \tfrac{1}{2} \\ \tfrac{1}{2} & \tfrac{1}{2} & -\tfrac{1}{2} & -\tfrac{1}{2} \\ -\tfrac{1}{2} & \tfrac{1}{2} & \tfrac{1}{2} & -\tfrac{1}{2} \\ \tfrac{1}{2} & -\tfrac{1}{2} & \tfrac{1}{2} & \tfrac{1}{2} \end{pmatrix};$$

this can be verified by direct multiplication. ◄

Let us now consider the use of SVD in solving a set of M simultaneous linear equations in N unknowns, which we write again as $Ax = b$. Firstly, consider the solution of a homogeneous set of equations, for which $b = 0$. As mentioned previously, if A is square and non-singular (and so possesses no zero singular values) then the equations have the unique trivial solution $x = 0$. Otherwise, *any* of the vectors v^i, $i = r + 1, r + 2, \ldots, N$, or any linear combination of them, will be a solution.

In the inhomogeneous case, where b is not a zero vector, the set of equations will possess solutions if b lies in the range of A. To investigate these solutions, it is convenient to introduce the $N \times M$ matrix $\overline{S}$, which is constructed by taking the transpose of S in (8.131) and replacing each non-zero singular value s_i on the diagonal by $1/s_i$. It is clear that, with this construction, $S\overline{S}$ is an $M \times M$ diagonal matrix with diagonal entries that equal unity for those values of j for which $s_j \neq 0$, and zero otherwise.

Now consider the vector

$$\hat{x} = V\overline{S}U^\dagger b. \tag{8.141}$$

305

Using the unitarity of the matrices U and V, we find that

$$A\hat{x} - b = US\overline{S}U^\dagger b - b = U(S\overline{S} - I)U^\dagger b. \tag{8.142}$$

The matrix $(S\overline{S} - I)$ is diagonal and the jth element on its leading diagonal is non-zero (and equal to -1) only when $s_j = 0$. However, the jth element of the vector $U^\dagger b$ is given by the scalar product $(u^j)^\dagger b$; if b lies in the range of A, this scalar product can be non-zero only if $s_j \neq 0$. Thus the RHS of (8.142) must equal zero, and so $\hat{x}$ given by (8.141) is a solution to the equations $Ax = b$. We may, however, add to this solution *any* linear combination of the $N - r$ vectors v^i, $i = r+1, r+2, \ldots, N$, that form an orthonormal basis for the null space of A; thus, in general, there exists an infinity of solutions (although it is straightforward to show that (8.141) is the solution vector of shortest length). The only way in which the solution (8.141) can be *unique* is if the rank r equals N, so that the matrix A does not possess a null space; this only occurs if A is square and non-singular.

If b does not lie in the range of A then the set of equations $Ax = b$ does not have a solution. Nevertheless, the vector (8.141) provides the closest possible 'solution' in a least-squares sense. In other words, although the vector (8.141) does not exactly solve $Ax = b$, it is the vector that minimises the *residual*

$$\epsilon = |Ax - b|,$$

where here the vertical lines denote the absolute value of the quantity they contain, not the determinant. This is proved as follows.

Suppose we were to add some arbitrary vector x' to the vector $\hat{x}$ in (8.141). This would result in the addition of the vector $b' = Ax'$ to $A\hat{x} - b$; b' is clearly in the range of A since any part of x' belonging to the null space of A contributes nothing to Ax'. We would then have

$$|A\hat{x} - b + b'| = |(US\overline{S}U^\dagger - I)b + b'|$$
$$= |U[(S\overline{S} - I)U^\dagger b + U^\dagger b']|$$
$$= |(S\overline{S} - I)U^\dagger b + U^\dagger b'|; \tag{8.143}$$

in the last line we have made use of the fact that the length of a vector is left unchanged by the action of the unitary matrix U. Now, the jth component of the vector $(S\overline{S} - I)U^\dagger b$ will only be non-zero when $s_j = 0$. However, the jth element of the vector $U^\dagger b'$ is given by the scalar product $(u^j)^\dagger b'$, which is non-zero only if $s_j \neq 0$, since b' lies in the range of A. Thus, as these two terms only contribute to (8.143) for two disjoint sets of j-values, its minimum value, as x' is varied, occurs when $b' = 0$; this requires $x' = 0$.

▶*Find the solution(s) to the set of simultaneous linear equations $Ax = b$, where A is given by (8.137) and $b = (1 \quad 0 \quad 0)^T$.*

To solve the set of equations, we begin by calculating the vector given in (8.141),

$$x = V\overline{S}U^\dagger b,$$

where U and V are given by (8.139) and (8.140) respectively and $\overline{S}$ is obtained by taking the transpose of S in (8.138) and replacing all the non-zero singular values s_i by $1/s_i$. Thus, $\overline{S}$ reads

$$\overline{S} = \begin{pmatrix} \frac{1}{4} & 0 & 0 \\ 0 & \frac{1}{3} & 0 \\ 0 & 0 & \frac{1}{2} \\ 0 & 0 & 0 \end{pmatrix}.$$

Substituting the appropriate matrices into the expression for x we find

$$x = \tfrac{1}{8}(1 \quad 1 \quad 1 \quad 1)^T. \tag{8.144}$$

It is straightforward to show that this solves the set of equations $Ax = b$ exactly, and so the vector $b = (1 \quad 0 \quad 0)^T$ must lie in the range of A. This is, in fact, immediately clear, since $b = u^1$. The solution (8.144) is *not*, however, unique. There are three non-zero singular values, but $N = 4$. Thus, the matrix A has a one-dimensional null space, which is 'spanned' by v^4, the fourth column of V, given in (8.140). The solutions to our set of equations, consisting of the sum of the exact solution and *any* vector in the null space of A, therefore lie along the line

$$x = \tfrac{1}{8}(1 \quad 1 \quad 1 \quad 1)^T + \alpha(1 \quad -1 \quad 1 \quad -1)^T,$$

where the parameter α can take any real value. We note that (8.144) is the point on this line that is closest to the origin. ◄

8.19 Exercises

8.1   Which of the following statements about linear vector spaces are true? Where a statement is false, give a counter-example to demonstrate this.

(a) Non-singular $N \times N$ matrices form a vector space of dimension N^2.
(b) Singular $N \times N$ matrices form a vector space of dimension N^2.
(c) Complex numbers form a vector space of dimension 2.
(d) Polynomial functions of x form an infinite-dimensional vector space.
(e) Series $\{a_0, a_1, a_2, \ldots, a_N\}$ for which $\sum_{n=0}^{N} |a_n|^2 = 1$ form an N-dimensional vector space.
(f) Absolutely convergent series form an infinite-dimensional vector space.
(g) Convergent series with terms of alternating sign form an infinite-dimensional vector space.

8.2   Evaluate the determinants

$$\text{(a)} \quad \begin{vmatrix} a & h & g \\ h & b & f \\ g & f & c \end{vmatrix}, \qquad \text{(b)} \quad \begin{vmatrix} 1 & 0 & 2 & 3 \\ 0 & 1 & -2 & 1 \\ 3 & -3 & 4 & -2 \\ -2 & 1 & -2 & 1 \end{vmatrix}$$

and

$$\text{(c)} \quad \begin{vmatrix} gc & ge & a+ge & gb+ge \\ 0 & b & b & b \\ c & e & e & b+e \\ a & b & b+f & b+d \end{vmatrix}.$$

8.3 Using the properties of determinants, solve with a minimum of calculation the following equations for x:

$$
\text{(a)} \quad \begin{vmatrix} x & a & a & 1 \\ a & x & b & 1 \\ a & b & x & 1 \\ a & b & c & 1 \end{vmatrix} = 0, \qquad \text{(b)} \quad \begin{vmatrix} x+2 & x+4 & x-3 \\ x+3 & x & x+5 \\ x-2 & x-1 & x+1 \end{vmatrix} = 0.
$$

8.4 Consider the matrices

$$
\text{(a)} \quad \mathsf{B} = \begin{pmatrix} 0 & -i & i \\ i & 0 & -i \\ -i & i & 0 \end{pmatrix}, \qquad \text{(b)} \quad \mathsf{C} = \frac{1}{\sqrt{8}} \begin{pmatrix} \sqrt{3} & -\sqrt{2} & -\sqrt{3} \\ 1 & \sqrt{6} & -1 \\ 2 & 0 & 2 \end{pmatrix}.
$$

Are they (i) real, (ii) diagonal, (iii) symmetric, (iv) antisymmetric, (v) singular, (vi) orthogonal, (vii) Hermitian, (viii) anti-Hermitian, (ix) unitary, (x) normal?

8.5 By considering the matrices

$$
\mathsf{A} = \begin{pmatrix} 1 & 0 \\ 0 & 0 \end{pmatrix}, \qquad \mathsf{B} = \begin{pmatrix} 0 & 0 \\ 3 & 4 \end{pmatrix},
$$

show that $\mathsf{AB} = 0$ does *not* imply that either A or B is the zero matrix, but that it does imply that at least one of them is singular.

8.6 This exercise considers a crystal whose unit cell has base vectors that are not necessarily mutually orthogonal.

(a) The basis vectors of the unit cell of a crystal, with the origin O at one corner, are denoted by $\mathbf{e}_1$, $\mathbf{e}_2$, $\mathbf{e}_3$. The matrix G has elements G_{ij}, where $G_{ij} = \mathbf{e}_i \cdot \mathbf{e}_j$ and H_{ij} are the elements of the matrix $\mathsf{H} \equiv \mathsf{G}^{-1}$. Show that the vectors $\mathbf{f}_i = \sum_j H_{ij} \mathbf{e}_j$ are the reciprocal vectors and that $H_{ij} = \mathbf{f}_i \cdot \mathbf{f}_j$.

(b) If the vectors $\mathbf{u}$ and $\mathbf{v}$ are given by

$$
\mathbf{u} = \sum_i u_i \mathbf{e}_i, \qquad \mathbf{v} = \sum_i v_i \mathbf{f}_i,
$$

obtain expressions for $|\mathbf{u}|$, $|\mathbf{v}|$, and $\mathbf{u} \cdot \mathbf{v}$.

(c) If the basis vectors are each of length a and the angle between each pair is $\pi/3$, write down G and hence obtain H.

(d) Calculate (i) the length of the normal from O onto the plane containing the points $p^{-1}\mathbf{e}_1$, $q^{-1}\mathbf{e}_2$, $r^{-1}\mathbf{e}_3$, and (ii) the angle between this normal and $\mathbf{e}_1$.

8.7 Prove the following results involving Hermitian matrices:

(a) If A is Hermitian and U is unitary then $\mathsf{U}^{-1}\mathsf{A}\mathsf{U}$ is Hermitian.

(b) If A is anti-Hermitian then $i\mathsf{A}$ is Hermitian.

(c) The product of two Hermitian matrices A and B is Hermitian if and only if A and B commute.

(d) If S is a real antisymmetric matrix then $\mathsf{A} = (\mathsf{I} - \mathsf{S})(\mathsf{I} + \mathsf{S})^{-1}$ is orthogonal. If A is given by

$$
\mathsf{A} = \begin{pmatrix} \cos\theta & \sin\theta \\ -\sin\theta & \cos\theta \end{pmatrix}
$$

then find the matrix S that is needed to express A in the above form.

(e) If K is skew-hermitian, i.e. $\mathsf{K}^\dagger = -\mathsf{K}$, then $\mathsf{V} = (\mathsf{I} + \mathsf{K})(\mathsf{I} - \mathsf{K})^{-1}$ is unitary.

8.8 A and B are real non-zero 3×3 matrices and satisfy the equation

$$
(\mathsf{AB})^\mathsf{T} + \mathsf{B}^{-1}\mathsf{A} = 0.
$$

(a) Prove that if B is orthogonal then A is antisymmetric.

308

(b) Without assuming that B is orthogonal, prove that A is singular.

8.9   The *commutator* $[X, Y]$ of two matrices is defined by the equation

$$[X, Y] = XY - YX.$$

Two anticommuting matrices A and B satisfy

$$A^2 = I, \qquad B^2 = I, \qquad [A, B] = 2iC.$$

(a) Prove that $C^2 = I$ and that $[B, C] = 2iA$.
(b) Evaluate $[[[A, B], [B, C]], [A, B]]$.

8.10   The four matrices S_x, S_y, S_z and I are defined by

$$S_x = \begin{pmatrix} 0 & 1 \\ 1 & 0 \end{pmatrix}, \qquad S_y = \begin{pmatrix} 0 & -i \\ i & 0 \end{pmatrix},$$

$$S_z = \begin{pmatrix} 1 & 0 \\ 0 & -1 \end{pmatrix}, \qquad I = \begin{pmatrix} 1 & 0 \\ 0 & 1 \end{pmatrix},$$

where $i^2 = -1$. Show that $S_x^2 = I$ and $S_x S_y = iS_z$, and obtain similar results by permutting x, y and z. Given that $\mathbf{v}$ is a vector with Cartesian components (v_x, v_y, v_z), the matrix $S(\mathbf{v})$ is defined as

$$S(\mathbf{v}) = v_x S_x + v_y S_y + v_z S_z.$$

Prove that, for general non-zero vectors $\mathbf{a}$ and $\mathbf{b}$,

$$S(\mathbf{a})S(\mathbf{b}) = \mathbf{a} \cdot \mathbf{b}\, I + i\, S(\mathbf{a} \times \mathbf{b}).$$

Without further calculation, deduce that $S(\mathbf{a})$ and $S(\mathbf{b})$ commute if and only if $\mathbf{a}$ and $\mathbf{b}$ are parallel vectors.

8.11   A general triangle has angles α, β and γ and corresponding opposite sides a, b and c. Express the length of each side in terms of the lengths of the other two sides and the relevant cosines, writing the relationships in matrix and vector form, using the vectors having components a, b, c and $\cos\alpha, \cos\beta, \cos\gamma$. Invert the matrix and hence deduce the cosine-law expressions involving α, β and γ.

8.12   Given a matrix

$$A = \begin{pmatrix} 1 & \alpha & 0 \\ \beta & 1 & 0 \\ 0 & 0 & 1 \end{pmatrix},$$

where α and β are non-zero complex numbers, find its eigenvalues and eigenvectors. Find the respective conditions for (a) the eigenvalues to be real and (b) the eigenvectors to be orthogonal. Show that the conditions are jointly satisfied if and only if A is Hermitian.

8.13   Using the Gram–Schmidt procedure:

(a) construct an orthonormal set of vectors from the following:

$$x_1 = (0 \quad 0 \quad 1 \quad 1)^T, \qquad x_2 = (1 \quad 0 \quad -1 \quad 0)^T,$$
$$x_3 = (1 \quad 2 \quad 0 \quad 2)^T, \qquad x_4 = (2 \quad 1 \quad 1 \quad 1)^T;$$

(b) find an orthonormal basis, within a four-dimensional Euclidean space, for the subspace spanned by the three vectors $(1 \quad 2 \quad 0 \quad 0)^{\mathrm{T}}$, $(3 \quad -1 \quad 2 \quad 0)^{\mathrm{T}}$ and $(0 \quad 0 \quad 2 \quad 1)^{\mathrm{T}}$.

8.14   If a unitary matrix U is written as $A + iB$, where A and B are Hermitian with non-degenerate eigenvalues, show the following:

(a) A and B commute;
(b) $A^2 + B^2 = I$;
(c) The eigenvectors of A are also eigenvectors of B;
(d) The eigenvalues of U have unit modulus (as is necessary for any unitary matrix).

8.15   Determine which of the matrices below are mutually commuting, and, for those that are, demonstrate that they have a complete set of eigenvectors in common:

$$A = \begin{pmatrix} 6 & -2 \\ -2 & 9 \end{pmatrix}, \qquad B = \begin{pmatrix} 1 & 8 \\ 8 & -11 \end{pmatrix},$$

$$C = \begin{pmatrix} -9 & -10 \\ -10 & 5 \end{pmatrix}, \qquad D = \begin{pmatrix} 14 & 2 \\ 2 & 11 \end{pmatrix}.$$

8.16   Find the eigenvalues and a set of eigenvectors of the matrix

$$\begin{pmatrix} 1 & 3 & -1 \\ 3 & 4 & -2 \\ -1 & -2 & 2 \end{pmatrix}.$$

Verify that its eigenvectors are mutually orthogonal.

8.17   Find three real orthogonal column matrices, each of which is a simultaneous eigenvector of

$$A = \begin{pmatrix} 0 & 0 & 1 \\ 0 & 1 & 0 \\ 1 & 0 & 0 \end{pmatrix} \qquad \text{and} \qquad B = \begin{pmatrix} 0 & 1 & 1 \\ 1 & 0 & 1 \\ 1 & 1 & 0 \end{pmatrix}.$$

8.18   Use the results of the first worked example in section 8.14 to evaluate, without repeated matrix multiplication, the expression $A^6 x$, where $x = (2 \quad 4 \quad -1)^{\mathrm{T}}$ and A is the matrix given in the example.

8.19   Given that A is a real symmetric matrix with normalised eigenvectors e^i, obtain the coefficients α_i involved when column matrix x, which is the solution of

$$Ax - \mu x = v, \qquad\qquad (*)$$

is expanded as $x = \sum_i \alpha_i e^i$. Here μ is a given constant and v is a given column matrix.

(a) Solve (*) when

$$A = \begin{pmatrix} 2 & 1 & 0 \\ 1 & 2 & 0 \\ 0 & 0 & 3 \end{pmatrix},$$

$\mu = 2$ and $v = (1 \quad 2 \quad 3)^{\mathrm{T}}$.

(b) Would (*) have a solution if $\mu = 1$ and (i) $v = (1 \quad 2 \quad 3)^{\mathrm{T}}$, (ii) $v = (2 \quad 2 \quad 3)^{\mathrm{T}}$?

8.20   Demonstrate that the matrix

$$A = \begin{pmatrix} 2 & 0 & 0 \\ -6 & 4 & 4 \\ 3 & -1 & 0 \end{pmatrix}$$

is defective, i.e. does not have three linearly independent eigenvectors, by showing the following:

(a) its eigenvalues are degenerate and, in fact, all equal;
(b) any eigenvector has the form $(\mu \quad (3\mu - 2v) \quad v)^{\mathsf{T}}$.
(c) if two pairs of values, μ_1, v_1 and μ_2, v_2, define two independent eigenvectors v_1 and v_2, then *any* third similarly defined eigenvector v_3 can be written as a linear combination of v_1 and v_2, i.e.

$$v_3 = av_1 + bv_2,$$

where

$$a = \frac{\mu_3 v_2 - \mu_2 v_3}{\mu_1 v_2 - \mu_2 v_1} \quad \text{and} \quad b = \frac{\mu_1 v_3 - \mu_3 v_1}{\mu_1 v_2 - \mu_2 v_1}.$$

Illustrate (c) using the example $(\mu_1, v_1) = (1, 1)$, $(\mu_2, v_2) = (1, 2)$ and $(\mu_3, v_3) = (0, 1)$.
   Show further that any matrix of the form

$$\begin{pmatrix} 2 & 0 & 0 \\ 6n - 6 & 4 - 2n & 4 - 4n \\ 3 - 3n & n - 1 & 2n \end{pmatrix}$$

is defective, with the same eigenvalues and eigenvectors as A.

8.21   By finding the eigenvectors of the Hermitian matrix

$$H = \begin{pmatrix} 10 & 3i \\ -3i & 2 \end{pmatrix},$$

construct a unitary matrix U such that $U^{\dagger}HU = \Lambda$, where Λ is a real diagonal matrix.

8.22   Use the stationary properties of quadratic forms to determine the maximum and minimum values taken by the expression

$$Q = 5x^2 + 4y^2 + 4z^2 + 2xz + 2xy$$

on the unit sphere, $x^2 + y^2 + z^2 = 1$. For what values of x, y and z do they occur?

8.23   Given that the matrix

$$A = \begin{pmatrix} 2 & -1 & 0 \\ -1 & 2 & -1 \\ 0 & -1 & 2 \end{pmatrix}$$

has two eigenvectors of the form $(1 \quad y \quad 1)^{\mathsf{T}}$, use the stationary property of the expression $J(x) = x^{\mathsf{T}}Ax/(x^{\mathsf{T}}x)$ to obtain the corresponding eigenvalues. Deduce the third eigenvalue.

8.24   Find the lengths of the semi-axes of the ellipse

$$73x^2 + 72xy + 52y^2 = 100,$$

and determine its orientation.

8.25   The equation of a particular conic section is

$$Q \equiv 8x_1^2 + 8x_2^2 - 6x_1x_2 = 110.$$

Determine the type of conic section this represents, the orientation of its principal axes, and relevant lengths in the directions of these axes.

8.26   Show that the quadratic surface

$$5x^2 + 11y^2 + 5z^2 - 10yz + 2xz - 10xy = 4$$

is an ellipsoid with semi-axes of lengths 2, 1 and 0.5. Find the direction of its longest axis.

8.27   Find the direction of the axis of symmetry of the quadratic surface

$$7x^2 + 7y^2 + 7z^2 - 20yz - 20xz + 20xy = 3.$$

8.28   For the following matrices, find the eigenvalues and sufficient of the eigenvectors to be able to describe the quadratic surfaces associated with them:

$$\text{(a)} \begin{pmatrix} 5 & 1 & -1 \\ 1 & 5 & 1 \\ -1 & 1 & 5 \end{pmatrix}, \quad \text{(b)} \begin{pmatrix} 1 & 2 & 2 \\ 2 & 1 & 2 \\ 2 & 2 & 1 \end{pmatrix}, \quad \text{(c)} \begin{pmatrix} 1 & 2 & 1 \\ 2 & 4 & 2 \\ 1 & 2 & 1 \end{pmatrix}.$$

8.29   This exercise demonstrates the reverse of the usual procedure of diagonalising a matrix.

(a) Rearrange the result $A' = S^{-1}AS$ of section 8.16 to express the original matrix A in terms of the unitary matrix S and the diagonal matrix A'. Hence show how to construct a matrix A that has given eigenvalues and given (orthogonal) column matrices as its eigenvectors.

(b) Find the matrix that has as eigenvectors $(1 \quad 2 \quad 1)^T$, $(1 \quad -1 \quad 1)^T$ and $(1 \quad 0 \quad -1)^T$, with corresponding eigenvalues λ, μ and ν.

(c) Try a particular case, say $\lambda = 3$, $\mu = -2$ and $\nu = 1$, and verify by explicit solution that the matrix so found does have these eigenvalues.

8.30   Find an orthogonal transformation that takes the quadratic form

$$Q \equiv -x_1^2 - 2x_2^2 - x_3^2 + 8x_2x_3 + 6x_1x_3 + 8x_1x_2$$

into the form

$$\mu_1 y_1^2 + \mu_2 y_2^2 - 4y_3^2,$$

and determine μ_1 and μ_2 (see section 8.17).

8.31   One method of determining the nullity (and hence the rank) of an $M \times N$ matrix A is as follows.

• Write down an augmented transpose of A, by adding on the right an $N \times N$ unit matrix and thus producing an $N \times (M + N)$ array B.
• Subtract a suitable multiple of the first row of B from each of the other lower rows so as to make $B_{i1} = 0$ for $i > 1$.
• Subtract a suitable multiple of the second row (or the uppermost row that does not start with M zero values) from each of the other lower rows so as to make $B_{i2} = 0$ for $i > 2$.
• Continue in this way until all remaining rows have zeros in the first M places. The number of such rows is equal to the nullity of A, and the N rightmost entries of these rows are the components of vectors that span the null space. They can be made orthogonal if they are not so already.

Use this method to show that the nullity of

$$A = \begin{pmatrix} -1 & 3 & 2 & 7 \\ 3 & 10 & -6 & 17 \\ -1 & -2 & 2 & -3 \\ 2 & 3 & -4 & 4 \\ 4 & 0 & -8 & -4 \end{pmatrix}$$

is 2 and that an orthogonal base for the null space of A is provided by any two column matrices of the form $(2 + \alpha_i \quad -2\alpha_i \quad 1 \quad \alpha_i)^T$, for which the α_i $(i = 1, 2)$ are real and satisfy $6\alpha_1\alpha_2 + 2(\alpha_1 + \alpha_2) + 5 = 0$.

8.32   Do the following sets of equations have non-zero solutions? If so, find them.

(a) $3x + 2y + z = 0,$ $\qquad x - 3y + 2z = 0,$ $\qquad 2x + y + 3z = 0.$
(b) $2x = b(y + z),$ $\qquad x = 2a(y - z),$ $\qquad x = (6a - b)y - (6a + b)z.$

8.33   Solve the simultaneous equations

$$2x + 3y + z = 11,$$
$$x + y + z = 6,$$
$$5x - y + 10z = 34.$$

8.34   Solve the following simultaneous equations for x_1, x_2 and x_3, using matrix methods:

$$x_1 + 2x_2 + 3x_3 = 1,$$
$$3x_1 + 4x_2 + 5x_3 = 2,$$
$$x_1 + 3x_2 + 4x_3 = 3.$$

8.35   Show that the following equations have solutions only if $\eta = 1$ or 2, and find them in these cases:

$$x + y + z = 1,$$
$$x + 2y + 4z = \eta,$$
$$x + 4y + 10z = \eta^2.$$

8.36   Find the condition(s) on α such that the simultaneous equations

$$x_1 + \alpha x_2 = 1,$$
$$x_1 - x_2 + 3x_3 = -1,$$
$$2x_1 - 2x_2 + \alpha x_3 = -2$$

have (a) exactly one solution, (b) no solutions, or (c) an infinite number of solutions; give all solutions where they exist.

8.37   Make an LU decomposition of the matrix

$$A = \begin{pmatrix} 3 & 6 & 9 \\ 1 & 0 & 5 \\ 2 & -2 & 16 \end{pmatrix}$$

and hence solve $Ax = b$, where (i) $b = (21 \quad 9 \quad 28)^T$, (ii) $b = (21 \quad 7 \quad 22)^T$.

8.38   Make an LU decomposition of the matrix

$$A = \begin{pmatrix} 2 & -3 & 1 & 3 \\ 1 & 4 & -3 & -3 \\ 5 & 3 & -1 & -1 \\ 3 & -6 & -3 & 1 \end{pmatrix}.$$

Hence solve $Ax = b$ for (i) $b = (-4 \quad 1 \quad 8 \quad -5)^T$, (ii) $b = (-10 \quad 0 \quad -3 \quad -24)^T$. Deduce that $\det A = -160$ and confirm this by direct calculation.

8.39   Use the Cholesky separation method to determine whether the following matrices are positive definite. For each that is, determine the corresponding lower diagonal matrix L:

$$A = \begin{pmatrix} 2 & 1 & 3 \\ 1 & 3 & -1 \\ 3 & -1 & 1 \end{pmatrix}, \qquad B = \begin{pmatrix} 5 & 0 & \sqrt{3} \\ 0 & 3 & 0 \\ \sqrt{3} & 0 & 3 \end{pmatrix}.$$

8.40   Find the equation satisfied by the squares of the singular values of the matrix associated with the following over-determined set of equations:

$$2x + 3y + z = 0$$
$$x - y - z = 1$$
$$2x + y = 0$$
$$2y + z = -2.$$

Show that one of the singular values is close to zero. Determine the two larger singular values by an appropriate iteration process and the smallest one by indirect calculation.

8.41   Find the SVD of

$$A = \begin{pmatrix} 0 & -1 \\ 1 & 1 \\ -1 & 0 \end{pmatrix},$$

showing that the singular values are $\sqrt{3}$ and 1.

8.42   Find the SVD form of the matrix

$$A = \begin{pmatrix} 22 & 28 & -22 \\ 1 & -2 & -19 \\ 19 & -2 & -1 \\ -6 & 12 & 6 \end{pmatrix}.$$

Use it to determine the best solution x of the equation $Ax = b$ when (i) $b = (6 \quad -39 \quad 15 \quad 18)^T$, (ii) $b = (9 \quad -42 \quad 15 \quad 15)^T$, showing that (i) has an exact solution, but that the best solution to (ii) has a residual of $\sqrt{18}$.

8.43   Four experimental measurements of particular combinations of three physical variables, x, y and z, gave the following inconsistent results:

$$13x + 22y - 13z = 4,$$
$$10x - 8y - 10z = 44,$$
$$10x - 8y - 10z = 47,$$
$$9x - 18y - 9z = 72.$$

Find the SVD best values for x, y and z. Identify the null space of A and hence obtain the general SVD solution.

8.20 Hints and answers

8.1   (a) False. O_N, the $N \times N$ null matrix, is *not* non-singular.

(b) False. Consider the sum of $\begin{pmatrix} 1 & 0 \\ 0 & 0 \end{pmatrix}$ and $\begin{pmatrix} 0 & 0 \\ 0 & 1 \end{pmatrix}$.

(c) True.

(d) True.

(e) False. Consider $b_n = a_n + a_n$ for which $\sum_{n=0}^{N} |b_n|^2 = 4 \neq 1$, or note that there is no zero vector with unit norm.

(f) True.

(g) False. Consider the two series defined by

$$a_0 = \tfrac{1}{2}, \qquad a_n = 2(-\tfrac{1}{2})^n \quad \text{for} \quad n \geq 1; \qquad b_n = -(-\tfrac{1}{2})^n \quad \text{for} \quad n \geq 0.$$

The series that is the sum of $\{a_n\}$ and $\{b_n\}$ does not have alternating signs and so closure does not hold.

8.3   (a) $x = a$, b or c; (b) $x = -1$; the equation is linear in x.

8.5   Use the property of the determinant of a matrix product.

8.7   (d) $S = \begin{pmatrix} 0 & -\tan(\theta/2) \\ \tan(\theta/2) & 0 \end{pmatrix}$.

        (e) Note that $(I + K)(I - K) = I - K^2 = (I - K)(I + K)$.

8.9   (b) $32iA$.

8.11 $a = b\cos\gamma + c\cos\beta$, and cyclic permutations; $a^2 = b^2 + c^2 - 2bc\cos\alpha$, and cyclic permutations.

8.13   (a) $2^{-1/2}(0 \quad 0 \quad 1 \quad 1)^T$, $6^{-1/2}(2 \quad 0 \quad -1 \quad 1)^T$,
           $39^{-1/2}(-1 \quad 6 \quad -1 \quad 1)^T$, $13^{-1/2}(2 \quad 1 \quad 2 \quad -2)^T$.
        (b) $5^{-1/2}(1 \quad 2 \quad 0 \quad 0)^T$, $(345)^{-1/2}(14 \quad -7 \quad 10 \quad 0)^T$,
           $(18\,285)^{-1/2}(-56 \quad 28 \quad 98 \quad 69)^T$.

8.15   C does not commute with the others; A, B and D have $(1 \quad -2)^T$ and $(2 \quad 1)^T$ as common eigenvectors.

8.17   For A : $(1 \quad 0 \quad -1)^T$, $(1 \quad \alpha_1 \quad 1)^T$, $(1 \quad \alpha_2 \quad 1)^T$.
        For B : $(1 \quad 1 \quad 1)^T$, $(\beta_1 \quad \gamma_1 \quad -\beta_1 - \gamma_1)^T$, $(\beta_2 \quad \gamma_2 \quad -\beta_2 - \gamma_2)^T$.
        The α_i, β_i and γ_i are arbitrary.
        Simultaneous and orthogonal: $(1 \quad 0 \quad -1)^T$, $(1 \quad 1 \quad 1)^T$, $(1 \quad -2 \quad 1)^T$.

8.19 $\alpha_j = (v \cdot e^{j*})/(\lambda_j - \mu)$, where λ_j is the eigenvalue corresponding to e^j.

        (a) $x = (2 \quad 1 \quad 3)^T$.
        (b) Since μ is equal to one of A's eigenvalues λ_j, the equation only has a solution if $v \cdot e^{j*} = 0$; (i) no solution; (ii) $x = (1 \quad 1 \quad 3/2)^T$.

8.21 $U = (10)^{-1/2}(1, 3i; 3i, 1)$, $\Lambda = (1, 0; 0, 11)$.

8.23 $J = (2y^2 - 4y + 4)/(y^2 + 2)$, with stationary values at $y = \pm\sqrt{2}$ and corresponding eigenvalues $2 \mp \sqrt{2}$. From the trace property of A, the third eigenvalue equals 2.

8.25   Ellipse; $\theta - \pi/4$, $a = \sqrt{22}$; $\theta = 3\pi/4$, $b = \sqrt{10}$.

8.27   The direction of the eigenvector having the unrepeated eigenvalue is $(1, 1, -1)/\sqrt{3}$.

8.29   (a) $A = SA'S^\dagger$, where S is the matrix whose columns are the eigenvectors of the matrix A to be constructed, and $A' = \text{diag}(\lambda, \mu, \nu)$.
        (b) $A = (\lambda + 2\mu + 3\nu, 2\lambda - 2\mu, \lambda + 2\mu - 3\nu; 2\lambda - 2\mu, 4\lambda + 2\mu, 2\lambda - 2\mu;$
               $\lambda + 2\mu - 3\nu, 2\lambda - 2\mu, \lambda + 2\mu + 3\nu)$.
        (c) $\frac{1}{3}(1, 5, -2; 5, 4, 5; -2, 5, 1)$.

8.31   The null space is spanned by $(2 \quad 0 \quad 1 \quad 0)^T$ and $(1 \quad -2 \quad 0 \quad 1)^T$.

8.33 $x = 3$, $y = 1$, $z = 2$.

8.35   First show that A is singular. $\eta = 1$, $x = 1 + 2z$, $y = -3z$; $\eta = 2$, $x = 2z$, $y = 1 - 3z$.

8.37 $L = (1, 0, 0; \frac{1}{3}, 1, 0; \frac{2}{3}, 3, 1)$, $U = (3, 6, 9; 0, -2, 2; 0, 0, 4)$.

        (i) $x = (-1 \quad 1 \quad 2)^T$. (ii) $x = (-3 \quad 2 \quad 2)^T$.

8.39   A is not positive definite, as L_{33} is calculated to be $\sqrt{-6}$.
        $B = LL^T$, where the non-zero elements of L are
        $L_{11} = \sqrt{5}$, $L_{31} = \sqrt{3/5}$, $L_{22} = \sqrt{3}$, $L_{33} = \sqrt{12/5}$.

8.41

$$A^\dagger A = \begin{pmatrix} 2 & 1 \\ 1 & 2 \end{pmatrix}, \quad U = \frac{1}{\sqrt{6}}\begin{pmatrix} -1 & \sqrt{3} & \sqrt{2} \\ 2 & 0 & \sqrt{2} \\ -1 & -\sqrt{3} & \sqrt{2} \end{pmatrix}, \quad V = \frac{1}{\sqrt{2}}\begin{pmatrix} 1 & 1 \\ 1 & -1 \end{pmatrix}.$$

8.43   The singular values are $12\sqrt{6}, 0, 18\sqrt{3}$ and the calculated best solution is $x = 1.71$, $y = -1.94$, $z = -1.71$. The null space is the line $x = z$, $y = 0$ and the general SVD solution is $x = 1.71 + \lambda$, $y = -1.94$, $z = -1.71 + \lambda$.

9

Normal modes

Any student of the physical sciences will encounter the subject of oscillations on many occasions and in a wide variety of circumstances, for example the voltage and current oscillations in an electric circuit, the vibrations of a mechanical structure and the internal motions of molecules. The matrices studied in the previous chapter provide a particularly simple way to approach what may appear, at first glance, to be difficult physical problems.

We will consider only systems for which a position-dependent potential exists, i.e., the potential energy of the system in any particular configuration depends upon the coordinates of the configuration, which need not be be lengths, however; the potential must *not* depend upon the time derivatives (generalised velocities) of these coordinates. So, for example, the potential $-q\mathbf{v} \cdot \mathbf{A}$ used in the Lagrangian description of a charged particle in an electromagnetic field is excluded. A further restriction that we place is that the potential has a local minimum at the equilibrium point; physically, this is a necessary and sufficient condition for stable equilibrium. By suitably defining the origin of the potential, we may take its value at the equilibrium point as zero.

We denote the coordinates chosen to describe a configuration of the system by q_i, $i = 1, 2, \ldots, N$. The q_i need not be distances; some could be angles, for example. For convenience we can define the q_i so that they are all zero at the equilibrium point. The instantaneous velocities of various parts of the system will depend upon the time derivatives of the q_i, denoted by $\dot{q}_i$. For small oscillations the velocities will be linear in the $\dot{q}_i$ and consequently the total kinetic energy T will be quadratic in them – and will include cross terms of the form $\dot{q}_i \dot{q}_j$ with $i \neq j$. The general expression for T can be written as the quadratic form

$$T = \sum_i \sum_j a_{ij} \dot{q}_i \dot{q}_j = \dot{\mathbf{q}}^{\mathrm{T}} \mathsf{A} \dot{\mathbf{q}}, \tag{9.1}$$

where $\dot{\mathbf{q}}$ is the column vector $(\dot{q}_1 \quad \dot{q}_2 \quad \cdots \quad \dot{q}_N)^{\mathrm{T}}$ and the $N \times N$ matrix A is real and may be chosen to be symmetric. Furthermore, A, like any matrix corresponding to a kinetic energy, is positive definite; that is, whatever non-zero real values the $\dot{q}_i$ take, the quadratic form (9.1) has a value > 0.

Turning now to the potential energy, we may write its value for a configuration q by means of a Taylor expansion about the origin $q = 0$,

$$V(\mathsf{q}) = V(0) + \sum_i \frac{\partial V(0)}{\partial q_i} q_i + \frac{1}{2}\sum_i \sum_j \frac{\partial^2 V(0)}{\partial q_i \partial q_j} q_i q_j + \cdots.$$

However, we have chosen $V(0) = 0$ and, since the origin is an equilibrium point, there is no force there and $\partial V(0)/\partial q_i = 0$. Consequently, to second order in the q_i we also have a quadratic form, but in the coordinates rather than in their time derivatives:

$$V = \sum_i \sum_j b_{ij} q_i q_j = \mathsf{q}^\mathsf{T} \mathsf{B} \mathsf{q}, \tag{9.2}$$

where B is, or can be made, symmetric. In this case, and in general, the requirement that the potential is a minimum means that the potential matrix B, like the kinetic energy matrix A, is real and positive definite.

9.1 Typical oscillatory systems

We now introduce particular examples, although the results of this section are general, given the above restrictions, and the reader will find it easy to apply the results to many other instances.

Consider first a uniform rod of mass M and length l, attached by a light string also of length l to a fixed point P and executing small oscillations in a vertical plane. We choose as coordinates the angles θ_1 and θ_2 shown, with exaggerated magnitude, in figure 9.1. In terms of these coordinates the centre of gravity of the rod has, to *first order* in the θ_i, a velocity component in the x-direction equal to $l\dot{\theta}_1 + \frac{1}{2}l\dot{\theta}_2$ and in the y-direction equal to zero. Adding in the rotational kinetic energy of the rod about its centre of gravity we obtain, to second order in the $\dot{\theta}_i$,

$$T \approx \frac{1}{2}Ml^2(\dot{\theta}_1^2 + \frac{1}{4}\dot{\theta}_2^2 + \dot{\theta}_1\dot{\theta}_2) + \frac{1}{24}Ml^2\dot{\theta}_2^2$$

$$= \frac{1}{6}Ml^2\left(3\dot{\theta}_1^2 + 3\dot{\theta}_1\dot{\theta}_2 + \dot{\theta}_2^2\right) = \frac{1}{12}Ml^2\dot{\mathsf{q}}^\mathsf{T}\begin{pmatrix} 6 & 3 \\ 3 & 2 \end{pmatrix}\dot{\mathsf{q}}, \tag{9.3}$$

where $\dot{\mathsf{q}}^\mathsf{T} = (\dot{\theta}_1 \quad \dot{\theta}_2)$. The potential energy is given by

$$V = Mlg\left[(1 - \cos\theta_1) + \frac{1}{2}(1 - \cos\theta_2)\right] \tag{9.4}$$

so that

$$V \approx \frac{1}{4}Mlg(2\theta_1^2 + \theta_2^2) = \frac{1}{12}Mlg\mathsf{q}^\mathsf{T}\begin{pmatrix} 6 & 0 \\ 0 & 3 \end{pmatrix}\mathsf{q}, \tag{9.5}$$

where g is the acceleration due to gravity and $\mathsf{q} = (\theta_1 \quad \theta_2)^\mathsf{T}$; (9.5) is valid to second order in the θ_i.

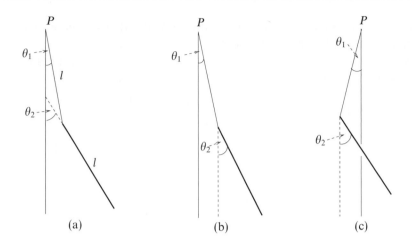

Figure 9.1   A uniform rod of length l attached to the fixed point P by a light string of the same length: (a) the general coordinate system; (b) approximation to the normal mode with lower frequency; (c) approximation to the mode with higher frequency.

With these expressions for T and V we now apply the conservation of energy,

$$\frac{d}{dt}(T + V) = 0, \tag{9.6}$$

assuming that there are no external forces other than gravity. In matrix form (9.6) becomes

$$\frac{d}{dt}(\dot{q}^T A\dot{q} + q^T Bq) = \ddot{q}^T A\dot{q} + \dot{q}^T A\ddot{q} + \dot{q}^T Bq + q^T B\dot{q} = 0,$$

which, using $A = A^T$ and $B = B^T$, gives

$$2\dot{q}^T(A\ddot{q} + Bq) = 0.$$

We will assume, although it is not clear that this gives the only possible solution, that the above equation implies that the coefficient of each $\dot{q}_i$ is separately zero. Hence

$$A\ddot{q} + Bq = 0. \tag{9.7}$$

For a rigorous derivation Lagrange's equations should be used, as in chapter 22.

Now we search for sets of coordinates q that *all* oscillate with the same period, i.e. the total motion repeats itself *exactly* after a *finite* interval. Solutions of this form will satisfy

$$q = x \cos \omega t; \tag{9.8}$$

the relative values of the elements of x in such a solution will indicate how each

318

coordinate is involved in this special motion. In general there will be N values of ω if the matrices A and B are $N \times N$ and these values are known as *normal frequencies* or *eigenfrequencies*.

Putting (9.8) into (9.7) yields

$$-\omega^2 Ax + Bx = (B - \omega^2 A)x = 0. \tag{9.9}$$

Our work in section 8.18 showed that this can have non-trivial solutions only if

$$|B - \omega^2 A| = 0. \tag{9.10}$$

This is a form of characteristic equation for B, except that the unit matrix I has been replaced by A. It has the more familiar form if a choice of coordinates is made in which the kinetic energy T is a simple sum of squared terms, i.e. it has been diagonalised, and the scale of the new coordinates is then chosen to make each diagonal element unity.

However, even in the present case, (9.10) can be solved to yield ω_k^2 for $k = 1, 2, \ldots, N$, where N is the order of A and B. The values of ω_k can be used with (9.9) to find the corresponding column vector x^k and the initial (stationary) physical configuration that, on release, will execute motion with period $2\pi/\omega_k$.

In equation (8.76) we showed that the eigenvectors of a real symmetric matrix were, except in the case of degeneracy of the eigenvalues, mutually orthogonal. In the present situation an analogous, but not identical, result holds. It is shown in section 9.3 that if x^1 and x^2 are two eigenvectors satisfying (9.9) for different values of ω^2 then they are orthogonal in the sense that

$$(x^2)^T A x^1 = 0 \qquad \text{and} \qquad (x^2)^T B x^1 = 0.$$

The direct 'scalar product' $(x^2)^T x^1$, formally equal to $(x^2)^T I x^1$, is not, in general, equal to zero.

Returning to the suspended rod, we find from (9.10)

$$\left| \frac{Mlg}{12} \begin{pmatrix} 6 & 0 \\ 0 & 3 \end{pmatrix} - \frac{\omega^2 M l^2}{12} \begin{pmatrix} 6 & 3 \\ 3 & 2 \end{pmatrix} \right| = 0.$$

Writing $\omega^2 l/g = \lambda$, this becomes

$$\begin{vmatrix} 6 - 6\lambda & -3\lambda \\ -3\lambda & 3 - 2\lambda \end{vmatrix} = 0 \quad \Rightarrow \quad \lambda^2 - 10\lambda + 6 = 0,$$

which has roots $\lambda = 5 \pm \sqrt{19}$. Thus we find that the two normal frequencies are given by $\omega_1 = (0.641g/l)^{1/2}$ and $\omega_2 = (9.359g/l)^{1/2}$. Putting the lower of the two values for ω^2, namely $(5 - \sqrt{19})g/l$, into (9.9) shows that for this mode

$$x_1 : x_2 = 3(5 - \sqrt{19}) : 6(\sqrt{19} - 4) = 1.923 : 2.153.$$

This corresponds to the case where the rod and string are almost straight out, i.e. they almost form a simple pendulum. Similarly it may be shown that the higher frequency corresponds to a solution where the string and rod are moving with opposite phase and $x_1 : x_2 = 9.359 : -16.718$. The two situations are shown in figure 9.1.

In connection with quadratic forms it was shown in section 8.17 how to make a change of coordinates such that the matrix for a particular form becomes diagonal. In exercise 9.6 a method is developed for diagonalising simultaneously two quadratic forms (though the transformation matrix may not be orthogonal). If this process is carried out for A and B in a general system undergoing stable oscillations, the kinetic and potential energies in the new variables η_i take the forms

$$T = \sum_i \mu_i \dot{\eta}_i^2 = \dot{\eta}^{\mathsf{T}} \mathsf{M} \dot{\eta}, \quad \mathsf{M} = \mathrm{diag}\,(\mu_1, \mu_2, \ldots, \mu_N), \tag{9.11}$$

$$V = \sum_i v_i \eta_i^2 = \eta^{\mathsf{T}} \mathsf{N} \eta, \quad \mathsf{N} = \mathrm{diag}\,(v_1, v_2 \ldots, v_N), \tag{9.12}$$

and the equations of motion are the *uncoupled* equations

$$\mu_i \ddot{\eta}_i + v_i \eta_i = 0, \quad i = 1, 2, \ldots, N. \tag{9.13}$$

Clearly a simple renormalisation of the η_i can be made that reduces all the μ_i in (9.11) to unity. When this is done the variables so formed are called *normal coordinates* and equations (9.13) the *normal equations*.

When a system is executing one of these simple harmonic motions it is said to be in a *normal mode*, and once started in such a mode it will repeat its motion exactly after each interval of $2\pi/\omega_i$. Any arbitrary motion of the system may be written as a superposition of the normal modes, and each component mode will execute harmonic motion with the corresponding eigenfrequency; however, unless by chance the eigenfrequencies are in integer relationship, the system will never return to its initial configuration after any finite time interval.

As a second example we will consider a number of masses coupled together by springs. For this type of situation the potential and kinetic energies are automatically quadratic functions of the coordinates and their derivatives, provided the elastic limits of the springs are not exceeded, and the oscillations do not have to be vanishingly small for the analysis to be valid.

> ▶*Find the normal frequencies and modes of oscillation of three particles of masses m, μm, m connected in that order in a straight line by two equal light springs of force constant k. This arrangement could serve as a model for some linear molecules, e.g. CO_2.*

The situation is shown in figure 9.2; the coordinates of the particles, x_1, x_2, x_3, are measured from their equilibrium positions, at which the springs are neither extended nor compressed.

The kinetic energy of the system is simply

$$T = \tfrac{1}{2} m \left(\dot{x}_1^2 + \mu \dot{x}_2^2 + \dot{x}_3^2 \right),$$

320

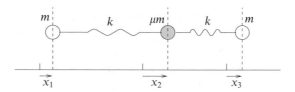

Figure 9.2 Three masses m, μm and m connected by two equal light springs of force constant k.

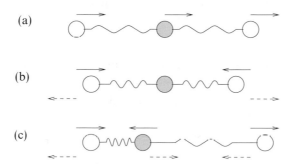

Figure 9.3 The normal modes of the masses and springs of a linear molecule such as CO_2. (a) $\omega^2 = 0$; (b) $\omega^2 = k/m$; (c) $\omega^2 = [(\mu + 2)/\mu](k/m)$.

whilst the potential energy stored in the springs is

$$V = \tfrac{1}{2}k\left[(x_2 - x_1)^2 + (x_3 - x_2)^2\right].$$

The kinetic- and potential-energy symmetric matrices are thus

$$A = \frac{m}{2}\begin{pmatrix} 1 & 0 & 0 \\ 0 & \mu & 0 \\ 0 & 0 & 1 \end{pmatrix}, \qquad B = \frac{k}{2}\begin{pmatrix} 1 & -1 & 0 \\ -1 & 2 & -1 \\ 0 & -1 & 1 \end{pmatrix}.$$

From (9.10), to find the normal frequencies we have to solve $|B - \omega^2 A| = 0$. Thus, writing $m\omega^2/k = \lambda$, we have

$$\begin{vmatrix} 1-\lambda & -1 & 0 \\ -1 & 2-\mu\lambda & -1 \\ 0 & -1 & 1-\lambda \end{vmatrix} = 0,$$

which leads to $\lambda = 0$, 1 or $1 + 2/\mu$. The corresponding eigenvectors are respectively

$$\mathsf{x}^1 = \frac{1}{\sqrt{3}}\begin{pmatrix} 1 \\ 1 \\ 1 \end{pmatrix}, \qquad \mathsf{x}^2 = \frac{1}{\sqrt{2}}\begin{pmatrix} 1 \\ 0 \\ -1 \end{pmatrix}, \qquad \mathsf{x}^3 = \frac{1}{\sqrt{2 + (4/\mu^2)}}\begin{pmatrix} 1 \\ -2/\mu \\ 1 \end{pmatrix}.$$

The physical motions associated with these normal modes are illustrated in figure 9.3. The first, with $\lambda = \omega = 0$ and all the x_i equal, merely describes bodily translation of the whole system, with no (i.e. zero-frequency) internal oscillations.

In the second solution the central particle remains stationary, $x_2 = 0$, whilst the other two oscillate with equal amplitudes in antiphase with each other. This motion, which has frequency $\omega = (k/m)^{1/2}$, is illustrated in figure 9.3(b).

The final and most complicated of the three normal modes has angular frequency $\omega = \{[(\mu + 2)/\mu](k/m)\}^{1/2}$, and involves a motion of the central particle which is in antiphase with that of the two outer ones and which has an amplitude $2/\mu$ times as great. In this motion (see figure 9.3(c)) the two springs are compressed and extended in turn. We also note that in the second and third normal modes the centre of mass of the molecule remains stationary. ◄

9.2 Symmetry and normal modes

It will have been noticed that the system in the above example has an obvious symmetry under the interchange of coordinates 1 and 3: the matrices A and B, the equations of motion and the normal modes illustrated in figure 9.3 are all unaltered by the interchange of x_1 and $-x_3$. This reflects the more general result that for each physical symmetry possessed by a system, there is at least one normal mode with the same symmetry.

The general question of the relationship between the symmetries possessed by a physical system and those of its normal modes will be taken up more formally in chapter 29 where the representation theory of groups is considered. However, we can show here how an appreciation of a system's symmetry properties will sometimes allow its normal modes to be guessed (and then verified), something that is particularly helpful if the number of coordinates involved is greater than two and the corresponding eigenvalue equation (9.10) is a cubic or higher-degree polynomial equation.

Consider the problem of determining the normal modes of a system consisting of four equal masses M at the corners of a square of side $2L$, each pair of masses being connected by a light spring of modulus k that is unstretched in the equilibrium situation. As shown in figure 9.4, we introduce Cartesian coordinates x_n, y_n, with $n = 1, 2, 3, 4$, for the positions of the masses and denote their displacements from their equilibrium positions $\mathbf{R}_n$ by $\mathbf{q}_n = x_n\mathbf{i} + y_n\mathbf{j}$. Thus

$$\mathbf{r}_n = \mathbf{R}_n + \mathbf{q}_n \quad \text{with} \quad \mathbf{R}_n = \pm L\mathbf{i} \pm L\mathbf{j}.$$

The coordinates for the system are thus $x_1, y_1, x_2, \ldots, y_4$ and the kinetic energy matrix A is given trivially by $M\mathsf{I}_8$, where I_8 is the 8×8 identity matrix.

The potential energy matrix B is much more difficult to calculate and involves, for each pair of values m, n, evaluating the quadratic approximation to the expression

$$b_{mn} = \tfrac{1}{2}k \left(|\mathbf{r}_m - \mathbf{r}_n| - |\mathbf{R}_m - \mathbf{R}_n|\right)^2.$$

Expressing each $\mathbf{r}_i$ in terms of $\mathbf{q}_i$ and $\mathbf{R}_i$ and making the normal assumption that

322

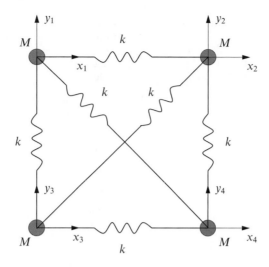

Figure 9.4 The arrangement of four equal masses and six equal springs discussed in the text. The coordinate systems x_n, y_n for $n = 1, 2, 3, 4$ measure the displacements of the masses from their equilibrium positions.

$|\mathbf{R}_m - \mathbf{R}_n| \gg |\mathbf{q}_m - \mathbf{q}_n|$, we obtain b_{mn} ($= b_{nm}$):

$$b_{mn} = \tfrac{1}{2}k \left[|(\mathbf{R}_m - \mathbf{R}_n) + (\mathbf{q}_m - \mathbf{q}_n)| - |\mathbf{R}_m - \mathbf{R}_n| \right]^2$$

$$= \tfrac{1}{2}k \left\{ \left[|\mathbf{R}_m - \mathbf{R}_n|^2 + 2(\mathbf{q}_m - \mathbf{q}_n) \cdot (\mathbf{R}_M - \mathbf{R}_n) + |\mathbf{q}_m - \mathbf{q}_n)|^2 \right]^{1/2} - |\mathbf{R}_m - \mathbf{R}_n| \right\}^2$$

$$= \tfrac{1}{2}k |\mathbf{R}_m - \mathbf{R}_n|^2 \left\{ \left[1 + \frac{2(\mathbf{q}_m - \mathbf{q}_n) \cdot (\mathbf{R}_M - \mathbf{R}_n)}{|\mathbf{R}_m - \mathbf{R}_n|^2} + \cdots \right]^{1/2} - 1 \right\}^2$$

$$\approx \tfrac{1}{2}k \left\{ \frac{(\mathbf{q}_m - \mathbf{q}_n) \cdot (\mathbf{R}_M - \mathbf{R}_n)}{|\mathbf{R}_m - \mathbf{R}_n|} \right\}^2.$$

This final expression is readily interpretable as the potential energy stored in the spring when it is extended by an amount equal to the component, along the equilibrium direction of the spring, of the relative displacement of its two ends.

Applying this result to each spring in turn gives the following expressions for the elements of the potential matrix.

m	n	$2b_{mn}/k$
1	2	$(x_1 - x_2)^2$
1	3	$(y_1 - y_3)^2$
1	4	$\tfrac{1}{2}(-x_1 + x_4 + y_1 - y_4)^2$
2	3	$\tfrac{1}{2}(x_2 - x_3 + y_2 - y_3)^2$
2	4	$(y_2 - y_4)^2$
3	4	$(x_3 - x_4)^2.$

The potential matrix is thus constructed as

$$B = \frac{k}{4} \begin{pmatrix} 3 & -1 & -2 & 0 & 0 & 0 & -1 & 1 \\ -1 & 3 & 0 & 0 & 0 & -2 & 1 & -1 \\ -2 & 0 & 3 & 1 & -1 & -1 & 0 & 0 \\ 0 & 0 & 1 & 3 & -1 & -1 & 0 & -2 \\ 0 & 0 & -1 & -1 & 3 & 1 & -2 & 0 \\ 0 & -2 & -1 & -1 & 1 & 3 & 0 & 0 \\ -1 & 1 & 0 & 0 & -2 & 0 & 3 & -1 \\ 1 & -1 & 0 & -2 & 0 & 0 & -1 & 3 \end{pmatrix}.$$

To solve the eigenvalue equation $|B - \lambda A| = 0$ directly would mean solving an eighth-degree polynomial equation. Fortunately, we can exploit intuition and the symmetries of the system to obtain the eigenvectors and corresponding eigenvalues without such labour.

Firstly, we know that bodily translation of the whole system, without any internal vibration, must be possible and that there will be two independent solutions of this form, corresponding to translations in the x- and y- directions. The eigenvector for the first of these (written in row form to save space) is

$$x^{(1)} = (1 \quad 0 \quad 1 \quad 0 \quad 1 \quad 0 \quad 1 \quad 0)^T.$$

Evaluation of $Bx^{(1)}$ gives

$$Bx^{(1)} = (0 \quad 0 \quad 0 \quad 0 \quad 0 \quad 0 \quad 0 \quad 0)^T,$$

showing that $x^{(1)}$ is a solution of $(B - \omega^2 A)x = 0$ corresponding to the eigenvalue $\omega^2 = 0$, whatever form Ax may take. Similarly,

$$x^{(2)} = (0 \quad 1 \quad 0 \quad 1 \quad 0 \quad 1 \quad 0 \quad 1)^T$$

is a second eigenvector corresponding to the eigenvalue $\omega^2 = 0$.

The next intuitive solution, again involving no internal vibrations, and, therefore, expected to correspond to $\omega^2 = 0$, is pure rotation of the whole system about its centre. In this mode each mass moves perpendicularly to the line joining its position to the centre, and so the relevant eigenvector is

$$x^{(3)} = \frac{1}{\sqrt{2}}(1 \quad 1 \quad 1 \quad -1 \quad -1 \quad 1 \quad -1 \quad -1)^T.$$

It is easily verified that $Bx^{(3)} = 0$ thus confirming both the eigenvector and the corresponding eigenvalue. The three non-oscillatory normal modes are illustrated in diagrams (a)–(c) of figure 9.5.

We now come to solutions that do involve real internal oscillations, and, because of the four-fold symmetry of the system, we expect one of them to be a mode in which all the masses move along radial lines – the so-called 'breathing

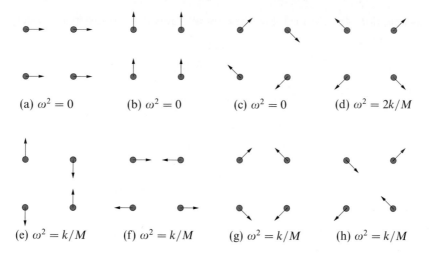

(a) $\omega^2 = 0$ (b) $\omega^2 = 0$ (c) $\omega^2 = 0$ (d) $\omega^2 = 2k/M$

(e) $\omega^2 = k/M$ (f) $\omega^2 = k/M$ (g) $\omega^2 = k/M$ (h) $\omega^2 = k/M$

Figure 9.5   The displacements and frequencies of the eight normal modes of the system shown in figure 9.4. Modes (a), (b) and (c) are not true oscillations: (a) and (b) are purely translational whilst (c) is a mode of bodily rotation. Mode (d), the 'breathing mode', has the highest frequency and the remaining four, (e)–(h), of lower frequency, are degenerate.

mode'. Expressing this motion in coordinate form gives as the fourth eigenvector

$$x^{(4)} = \frac{1}{\sqrt{2}}(-1 \quad 1 \quad 1 \quad 1 \quad -1 \quad -1 \quad 1 \quad 1)^{\mathrm{T}}.$$

Evaluation of $Bx^{(4)}$ yields

$$Bx^{(4)} = \frac{k}{4\sqrt{2}}(-8 \quad 8 \quad 8 \quad 8 \quad -8 \quad -8 \quad 8 \quad -8)^{\mathrm{T}} = 2kx^{(4)},$$

i.e. a multiple of $x^{(4)}$, confirming that it is indeed an eigenvector. Further, since $Ax^{(4)} = Mx^{(4)}$, it follows from $(B - \omega^2 A)x = 0$ that $\omega^2 = 2k/M$ for this normal mode. Diagram (d) of the figure illustrates the corresponding motions of the four masses.

As the next step in exploiting the symmetry properties of the system we note that, because of its reflection symmetry in the x-axis, the system is invariant under the double interchange of y_1 with $-y_3$ and y_2 with $-y_4$. This leads us to try an eigenvector of the form

$$x^{(5)} = (0 \quad \alpha \quad 0 \quad \beta \quad 0 \quad -\alpha \quad 0 \quad -\beta)^{\mathrm{T}}.$$

Substituting this trial vector into $(B - \omega^2 A)x = 0$ gives, of course, eight simultaneous equations for α and β, but they are all equivalent to just two, namely

$$\alpha + \beta = 0,$$

$$5\alpha + \beta = \frac{4M\omega^2}{k}\alpha;$$

these have the solution $\alpha = -\beta$ and $\omega^2 = k/M$. The latter thus gives the frequency of the mode with eigenvector

$$\mathsf{x}^{(5)} = (0 \quad 1 \quad 0 \quad -1 \quad 0 \quad -1 \quad 0 \quad 1)^{\mathrm{T}}.$$

Note that, in this mode, when the spring joining masses 1 and 3 is most stretched, the one joining masses 2 and 4 is at its most compressed. Similarly, based on reflection symmetry in the y-axis,

$$\mathsf{x}^{(6)} = (1 \quad 0 \quad -1 \quad 0 \quad -1 \quad 0 \quad 1 \quad 0)^{\mathrm{T}}$$

can be shown to be an eigenvector corresponding to the same frequency. These two modes are shown in diagrams (e) and (f) of figure 9.5.

This accounts for six of the expected eight modes, and the other two could be found by considering motions that are symmetric about both diagonals of the square or are invariant under successive reflections in the x- and y- axes. However, since A is a multiple of the unit matrix, and since we know that $(\mathsf{x}^{(j)})^{\mathrm{T}}\mathsf{A}\mathsf{x}^{(i)} = 0$ if $i \neq j$, we can find the two remaining eigenvectors more easily by requiring them to be orthogonal to each of those found so far.

Let us take the next (seventh) eigenvector, $\mathsf{x}^{(7)}$, to be given by

$$\mathsf{x}^{(7)} = (a \quad b \quad c \quad d \quad e \quad f \quad g \quad h)^{\mathrm{T}}.$$

Then orthogonality with each of the $\mathsf{x}^{(n)}$ for $n = 1, 2, \ldots, 6$ yields six equations satisfied by the unknowns $a, b, \ldots, h$. As the reader may verify, they can be reduced to the six simple equations

$$a + g = 0, \quad d + f = 0, \quad a + f = d + g,$$
$$b + h = 0, \quad c + e = 0, \quad b + c = e + h.$$

With six homogeneous equations for eight unknowns, effectively separated into two groups of four, we may pick one in each group arbitrarily. Taking $a = b = 1$ gives $d = e = 1$ and $c = f = g = h = -1$ as a solution. Substitution of

$$\mathsf{x}^{(7)} = (1 \quad 1 \quad -1 \quad 1 \quad 1 \quad -1 \quad -1 \quad -1)^{\mathrm{T}}.$$

into the eigenvalue equation checks that it is an eigenvector and shows that the corresponding eigenfrequency is given by $\omega^2 = k/M$.

We now have the eigenvectors for seven of the eight normal modes and the eighth can be found by making it simultaneously orthogonal to each of the other seven. It is left to the reader to show (or verify) that the final solution is

$$\mathsf{x}^{(8)} = (1 \quad -1 \quad 1 \quad 1 \quad -1 \quad -1 \quad -1 \quad 1)^{\mathrm{T}}$$

and that this mode has the same frequency as three of the other modes. The general topic of the degeneracy of normal modes is discussed in chapter 29. The movements associated with the final two modes are shown in diagrams (g) and (h) of figure 9.5; this figure summarises all eight normal modes and frequencies.

Although this example has been lengthy to write out, we have seen that the actual calculations are quite simple and provide the full solution to what is formally a matrix eigenvalue equation involving 8×8 matrices. It should be noted that our exploitation of the intrinsic symmetries of the system played a crucial part in finding the correct eigenvectors for the various normal modes.

9.3 Rayleigh–Ritz method

We conclude this chapter with a discussion of the Rayleigh–Ritz method for estimating the eigenfrequencies of an oscillating system. We recall from the introduction to the chapter that for a system undergoing small oscillations the potential and kinetic energy are given by

$$V = \mathsf{q}^{\mathsf{T}}\mathsf{B}\mathsf{q} \qquad \text{and} \qquad T - \dot{\mathsf{q}}^{\mathsf{T}}\mathsf{A}\dot{\mathsf{q}},$$

where the components of q are the coordinates chosen to represent the configuration of the system and A and B are symmetric matrices (or may be chosen to be such). We also recall from (9.9) that the normal modes x^i and the eigenfrequencies ω_i are given by

$$(\mathsf{B} - \omega_i^2 \mathsf{A})\mathsf{x}^i = 0. \tag{9.14}$$

It may be shown that the eigenvectors x^i corresponding to different normal modes are linearly independent and so form a complete set. Thus, any coordinate vector q can be written $\mathsf{q} = \sum_j c_j \mathsf{x}^j$. We now consider the value of the generalised quadratic form

$$\lambda(\mathsf{x}) = \frac{\mathsf{x}^{\mathsf{T}}\mathsf{B}\mathsf{x}}{\mathsf{x}^{\mathsf{T}}\mathsf{A}\mathsf{x}} = \frac{\sum_m (\mathsf{x}^m)^{\mathsf{T}} c_m^* \mathsf{B} \sum_i c_i \mathsf{x}^i}{\sum_j (\mathsf{x}^j)^{\mathsf{T}} c_j^* \mathsf{A} \sum_k c_k \mathsf{x}^k},$$

which, since both numerator and denominator are positive definite, is itself non-negative. Equation (9.14) can be used to replace $\mathsf{B}\mathsf{x}^i$, with the result that

$$\lambda(\mathsf{x}) = \frac{\sum_m (\mathsf{x}^m)^{\mathsf{T}} c_m^* \mathsf{A} \sum_i \omega_i^2 c_i \mathsf{x}^i}{\sum_j (\mathsf{x}^j)^{\mathsf{T}} c_j^* \mathsf{A} \sum_k c_k \mathsf{x}^k}$$

$$= \frac{\sum_m (\mathsf{x}^m)^{\mathsf{T}} c_m^* \sum_i \omega_i^2 c_i \mathsf{A}\mathsf{x}^i}{\sum_j (\mathsf{x}^j)^{\mathsf{T}} c_j^* \mathsf{A} \sum_k c_k \mathsf{x}^k}. \tag{9.15}$$

Now the eigenvectors x^i obtained by solving $(\mathsf{B} - \omega^2 \mathsf{A})\mathsf{x} = 0$ are not mutually orthogonal unless either A or B is a multiple of the unit matrix. However, it may be shown that they do possess the desirable properties

$$(\mathbf{x}^j)^{\mathrm{T}}\mathsf{A}\mathbf{x}^i = 0 \quad \text{and} \quad (\mathbf{x}^j)^{\mathrm{T}}\mathsf{B}\mathbf{x}^i = 0 \quad \text{if } i \neq j. \tag{9.16}$$

This result is proved as follows. From (9.14) it is clear that, for general i and j,

$$(\mathbf{x}^j)^{\mathrm{T}}(\mathsf{B} - \omega_i^2\mathsf{A})\mathbf{x}^i = 0. \tag{9.17}$$

But, by taking the transpose of (9.14) with i replaced by j and recalling that A and B are real and symmetric, we obtain

$$(\mathbf{x}^j)^{\mathrm{T}}(\mathsf{B} - \omega_j^2\mathsf{A}) = 0.$$

Forming the scalar product of this with $\mathbf{x}^i$ and subtracting the result from (9.17) gives

$$(\omega_j^2 - \omega_i^2)(\mathbf{x}^j)^{\mathrm{T}}\mathsf{A}\mathbf{x}^i = 0.$$

Thus, for $i \neq j$ and non-degenerate eigenvalues ω_i^2 and ω_j^2, we have that $(\mathbf{x}^j)^{\mathrm{T}}\mathsf{A}\mathbf{x}^i = 0$, and substituting this into (9.17) immediately establishes the corresponding result for $(\mathbf{x}^j)^{\mathrm{T}}\mathsf{B}\mathbf{x}^i$. Clearly, if either A or B is a multiple of the unit matrix then the eigenvectors are mutually orthogonal in the normal sense. The orthogonality relations (9.16) are derived again, and extended, in exercise 9.6.

Using the first of the relationships (9.16) to simplify (9.15), we find that

$$\lambda(\mathbf{x}) = \frac{\sum_i |c_i|^2\omega_i^2(\mathbf{x}^i)^{\mathrm{T}}\mathsf{A}\mathbf{x}^i}{\sum_k |c_k|^2(\mathbf{x}^k)^{\mathrm{T}}\mathsf{A}\mathbf{x}^k}. \tag{9.18}$$

Now, if ω_0^2 is the lowest eigenfrequency then $\omega_i^2 \geq \omega_0^2$ for all i and, further, since $(\mathbf{x}^i)^{\mathrm{T}}\mathsf{A}\mathbf{x}^i \geq 0$ for all i the numerator of (9.18) is $\geq \omega_0^2 \sum_i |c_i|^2(\mathbf{x}^i)^{\mathrm{T}}\mathsf{A}\mathbf{x}^i$. Hence

$$\lambda(\mathbf{x}) \equiv \frac{\mathbf{x}^{\mathrm{T}}\mathsf{B}\mathbf{x}}{\mathbf{x}^{\mathrm{T}}\mathsf{A}\mathbf{x}} \geq \omega_0^2, \tag{9.19}$$

for any $\mathbf{x}$ whatsoever (whether $\mathbf{x}$ is an eigenvector or not). Thus we are able to estimate the lowest eigenfrequency of the system by evaluating λ for a variety of vectors $\mathbf{x}$, the components of which, it will be recalled, give the ratios of the coordinate amplitudes. This is sometimes a useful approach if many coordinates are involved and direct solution for the eigenvalues is not possible.

An additional result is that the maximum eigenfrequency ω_m^2 may also be estimated. It is obvious that if we replace the statement '$\omega_i^2 \geq \omega_0^2$ for all i' by '$\omega_i^2 \leq \omega_m^2$ for all i', then $\lambda(\mathbf{x}) \leq \omega_m^2$ for any $\mathbf{x}$. Thus $\lambda(\mathbf{x})$ always lies between the lowest and highest eigenfrequencies of the system. Furthermore, $\lambda(\mathbf{x})$ has a *stationary* value, equal to ω_k^2, when $\mathbf{x}$ is the kth eigenvector (see subsection 8.17.1).

▶ *Estimate the eigenfrequencies of the oscillating rod of section 9.1.*

Firstly we recall that

$$A = \frac{Ml^2}{12} \begin{pmatrix} 6 & 3 \\ 3 & 2 \end{pmatrix} \quad \text{and} \quad B = \frac{Mlg}{12} \begin{pmatrix} 6 & 0 \\ 0 & 3 \end{pmatrix}.$$

Physical intuition suggests that the slower mode will have a configuration approximating that of a simple pendulum (figure 9.1), in which $\theta_1 = \theta_2$, and so we use this as a *trial vector*. Taking $x = (\theta \;\; \theta)^T$,

$$\lambda(x) = \frac{x^T B x}{x^T A x} = \frac{3Mlg\theta^2/4}{7Ml^2\theta^2/6} = \frac{9g}{14l} = 0.643 \frac{g}{l},$$

and we conclude from (9.19) that the lower (angular) frequency is $\leq (0.643g/l)^{1/2}$. We have already seen on p. 319 that the true answer is $(0.641g/l)^{1/2}$ and so we have come very close to it.

Next we turn to the higher frequency. Here, a typical pattern of oscillation is not so obvious but, rather preempting the answer, we try $\theta_2 = -2\theta_1$; we then obtain $\lambda = 9g/l$ and so conclude that the higher eigenfrequency $\geq (9g/l)^{1/2}$. We have already seen that the exact answer is $(9.359g/l)^{1/2}$ and so again we have come close to it. ◀

A simplified version of the Rayleigh–Ritz method may be used to estimate the eigenvalues of a symmetric (or in general Hermitian) matrix B, the eigenvectors of which will be mutually orthogonal. By repeating the calculations leading to (9.18), A being replaced by the unit matrix I, it is easily verified that if

$$\lambda(x) = \frac{x^T B x}{x^T x}$$

is evaluated for *any* vector x then

$$\lambda_1 \leq \lambda(x) \leq \lambda_m,$$

where $\lambda_1, \lambda_2 \ldots, \lambda_m$ are the eigenvalues of B in order of increasing size. A similar result holds for Hermitian matrices.

9.4 Exercises

9.1 Three coupled pendulums swing perpendicularly to the horizontal line containing their points of suspension, and the following equations of motion are satisfied:

$$-m\ddot{x}_1 = cmx_1 + d(x_1 - x_2),$$
$$-M\ddot{x}_2 = cMx_2 + d(x_2 - x_1) + d(x_2 - x_3),$$
$$-m\ddot{x}_3 = cmx_3 + d(x_3 - x_2),$$

where x_1, x_2 and x_3 are measured from the equilibrium points; m, M and m are the masses of the pendulum bobs; and c and d are positive constants. Find the normal frequencies of the system and sketch the corresponding patterns of oscillation. What happens as $d \to 0$ or $d \to \infty$?

9.2 A double pendulum, smoothly pivoted at A, consists of two light rigid rods, AB and BC, each of length l, which are smoothly jointed at B and carry masses m and αm at B and C respectively. The pendulum makes small oscillations in one plane under gravity. At time t, AB and BC make angles $\theta(t)$ and $\phi(t)$, respectively, with the downward vertical. Find quadratic expressions for the kinetic and potential energies of the system and hence show that the normal modes have angular frequencies given by

$$\omega^2 = \frac{g}{l}\left[1 + \alpha \pm \sqrt{\alpha(1+\alpha)}\right].$$

For $\alpha = 1/3$, show that in one of the normal modes the mid-point of BC does not move during the motion.

9.3   Continue the worked example, modelling a linear molecule, discussed at the end of section 9.1, for the case in which $\mu = 2$.

    (a) Show that the eigenvectors derived there have the expected orthogonality properties with respect to both A and B.

    (b) For the situation in which the atoms are released from rest with initial displacements $x_1 = 2\epsilon$, $x_2 = -\epsilon$ and $x_3 = 0$, determine their subsequent motions and maximum displacements.

9.4   Consider the circuit consisting of three equal capacitors and two different inductors shown in the figure. For charges Q_i on the capacitors and currents I_i

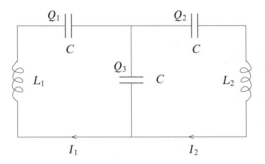

through the components, write down Kirchhoff's law for the total voltage change around each of two complete circuit loops. Note that, to within an unimportant constant, the conservation of current implies that $Q_3 = Q_1 - Q_2$. Express the loop equations in the form given in (9.7), namely

$$A\ddot{Q} + BQ = 0.$$

Use this to show that the normal frequencies of the circuit are given by

$$\omega^2 = \frac{1}{CL_1L_2}\left[L_1 + L_2 \pm (L_1^2 + L_2^2 - L_1L_2)^{1/2}\right].$$

Obtain the same matrices and result by finding the total energy stored in the various capacitors (typically $Q^2/(2C)$) and in the inductors (typically $LI^2/2$).

    For the special case $L_1 = L_2 = L$ determine the relevant eigenvectors and so describe the patterns of current flow in the circuit.

9.5   It is shown in physics and engineering textbooks that circuits containing capacitors and inductors can be analysed by replacing a capacitor of capacitance C by a 'complex impedance' $1/(i\omega C)$ and an inductor of inductance L by an impedance $i\omega L$, where ω is the angular frequency of the currents flowing and $i^2 = -1$.

    Use this approach and Kirchhoff's circuit laws to analyse the circuit shown in the figure and obtain three linear equations governing the currents I_1, I_2 and I_3. Show that the only possible frequencies of self-sustaining currents satisfy either

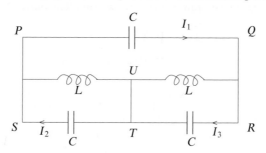

(a) $\omega^2 LC = 1$ or (b) $3\omega^2 LC = 1$. Find the corresponding current patterns and, in each case, by identifying parts of the circuit in which no current flows, draw an equivalent circuit that contains only one capacitor and one inductor.

9.6   *The simultaneous reduction to diagonal form of two real symmetric quadratic forms.*
Consider the two real symmetric quadratic forms $u^T A u$ and $u^T B u$, where u^T stands for the row matrix $(x \quad y \quad z)$, and denote by u^n those column matrices that satisfy

$$B u^n = \lambda_n A u^n, \tag{E9.1}$$

in which n is a label and the λ_n are real, non-zero and all different.

(a) By multiplying (E9.1) on the left by $(u^m)^T$, and the transpose of the corresponding equation for u^m on the right by u^n, show that $(u^m)^T A u^n = 0$ for $n \neq m$.
(b) By noting that $A u^n = (\lambda_n)^{-1} B u^n$, deduce that $(u^m)^T B u^n = 0$ for $m \neq n$.
(c) It can be shown that the u^n are linearly independent; the next step is to construct a matrix P whose columns are the vectors u^n.
(d) Make a change of variables $u = Pv$ such that $u^T A u$ becomes $v^T C v$, and $u^T B u$ becomes $v^T D v$. Show that C and D are diagonal by showing that $c_{ij} = 0$ if $i \neq j$, and similarly for d_{ij}.

Thus $u = Pv$ or $v = P^{-1} u$ reduces both quadratics to diagonal form.
    To summarise, the method is as follows:

(a) find the λ_n that allow (E9.1) a non-zero solution, by solving $|B - \lambda A| = 0$;
(b) for each λ_n construct u^n;
(c) construct the non-singular matrix P whose columns are the vectors u^n;
(d) make the change of variable $u = Pv$.

9.7   (*It is recommended that the reader does not attempt this question until exercise 9.6 has been studied.*)
    If, in the pendulum system studied in section 9.1, the string is replaced by a second rod identical to the first then the expressions for the kinetic energy T and the potential energy V become (to second order in the θ_i)

$$T \approx Ml^2 \left(\tfrac{8}{3}\dot\theta_1^2 + 2\dot\theta_1\dot\theta_2 + \tfrac{2}{3}\dot\theta_2^2 \right),$$
$$V \approx Mgl \left(\tfrac{3}{2}\theta_1^2 + \tfrac{1}{2}\theta_2^2 \right).$$

Determine the normal frequencies of the system and find new variables ξ and η that will reduce these two expressions to diagonal form, i.e. to

$$a_1\dot\xi^2 + a_2\dot\eta^2 \qquad \text{and} \qquad b_1\xi^2 + b_2\eta^2.$$

9.8   (*It is recommended that the reader does not attempt this question until exercise 9.6 has been studied.*)
Find a real linear transformation that simultaneously reduces the quadratic forms

$$3x^2 + 5y^2 + 5z^2 + 2yz + 6zx - 2xy,$$

$$5x^2 + 12y^2 + 8yz + 4zx$$

to diagonal form.

9.9   Three particles of mass m are attached to a light horizontal string having fixed ends, the string being thus divided into four equal portions each of length a and under a tension T. Show that for small transverse vibrations the amplitudes x^i of the normal modes satisfy $\mathsf{B}x = (ma\omega^2/T)x$, where B is the matrix

$$\begin{pmatrix} 2 & -1 & 0 \\ -1 & 2 & -1 \\ 0 & -1 & 2 \end{pmatrix}.$$

Estimate the lowest and highest eigenfrequencies using trial vectors $(3 \quad 4 \quad 3)^{\mathsf{T}}$ and $(3 \quad -4 \quad 3)^{\mathsf{T}}$. Use also the exact vectors $\left(1 \quad \sqrt{2} \quad 1\right)^{\mathsf{T}}$ and $\left(1 \quad -\sqrt{2} \quad 1\right)^{\mathsf{T}}$ and compare the results.

9.10   Use the Rayleigh–Ritz method to estimate the lowest oscillation frequency of a heavy chain of N links, each of length a ($= L/N$), which hangs freely from one end. Consider simple calculable configurations such as all links but one vertical, or all links collinear, etc.

9.5 Hints and answers

9.1   See figure 9.6.
9.3   (b) $x_1 = \epsilon(\cos \omega t + \cos \sqrt{2}\omega t)$, $x_2 = -\epsilon \cos \sqrt{2}\omega t$, $x_3 = \epsilon(-\cos \omega t + \cos \sqrt{2}\omega t)$. At various times the three displacements will reach $2\epsilon, \epsilon, 2\epsilon$ respectively. For example, x_1 can be written as $2\epsilon \cos[(\sqrt{2}-1)\omega t/2] \cos[(\sqrt{2}+1)\omega t/2]$, i.e. an oscillation of angular frequency $(\sqrt{2}+1)\omega/2$ and modulated amplitude $2\epsilon \cos[(\sqrt{2}-1)\omega t/2]$; the amplitude will reach 2ϵ after a time $\approx 4\pi/[\omega(\sqrt{2}-1)]$.

9.5   As the circuit loops contain no voltage sources, the equations are homogeneous, and so for a non-trivial solution the determinant of coefficients must vanish.
(a) $I_1 = 0$, $I_2 = -I_3$; no current in PQ; equivalent to two separate circuits of capacitance C and inductance L.
(b) $I_1 = -2I_2 = -2I_3$; no current in TU; capacitance $3C/2$ and inductance $2L$.

9.7 $\omega = (2.634g/l)^{1/2}$ or $(0.3661g/l)^{1/2}$; $\theta_1 = \xi + \eta$, $\theta_2 = 1.431\xi - 2.097\eta$.
9.9   Estimated, $10/17 < Ma\omega^2/T < 58/17$; exact, $2 - \sqrt{2} \le Ma\omega^2/T \le 2 + \sqrt{2}$.

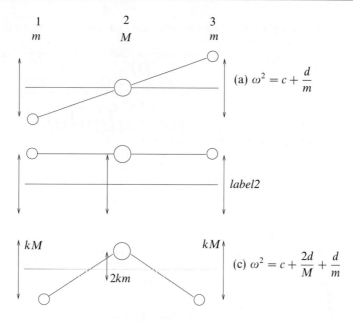

Figure 9.6   The normal modes, as viewed from above, of the coupled pendulums in example 9.1.

333

10

Vector calculus

In chapter 7 we discussed the algebra of vectors, and in chapter 8 we considered how to transform one vector into another using a linear operator. In this chapter and the next we discuss the calculus of vectors, i.e. the differentiation and integration both of vectors describing particular bodies, such as the velocity of a particle, and of vector fields, in which a vector is defined as a function of the coordinates throughout some volume (one-, two- or three-dimensional). Since the aim of this chapter is to develop methods for handling multi-dimensional physical situations, we will assume throughout that the functions with which we have to deal have sufficiently amenable mathematical properties, in particular that they are continuous and differentiable.

10.1 Differentiation of vectors

Let us consider a vector $\mathbf{a}$ that is a function of a scalar variable u. By this we mean that with each value of u we associate a vector $\mathbf{a}(u)$. For example, in Cartesian coordinates $\mathbf{a}(u) = a_x(u)\mathbf{i} + a_y(u)\mathbf{j} + a_z(u)\mathbf{k}$, where $a_x(u)$, $a_y(u)$ and $a_z(u)$ are scalar functions of u and are the components of the vector $\mathbf{a}(u)$ in the x-, y- and z- directions respectively. We note that if $\mathbf{a}(u)$ is continuous at some point $u = u_0$ then this implies that each of the Cartesian components $a_x(u)$, $a_y(u)$ and $a_z(u)$ is also continuous there.

Let us consider the derivative of the vector function $\mathbf{a}(u)$ with respect to u. The derivative of a vector function is defined in a similar manner to the ordinary derivative of a scalar function $f(x)$ given in chapter 2. The small change in the vector $\mathbf{a}(u)$ resulting from a small change Δu in the value of u is given by $\Delta \mathbf{a} = \mathbf{a}(u + \Delta u) - \mathbf{a}(u)$ (see figure 10.1). The derivative of $\mathbf{a}(u)$ with respect to u is defined to be

$$\frac{d\mathbf{a}}{du} = \lim_{\Delta u \to 0} \frac{\mathbf{a}(u + \Delta u) - \mathbf{a}(u)}{\Delta u}, \tag{10.1}$$

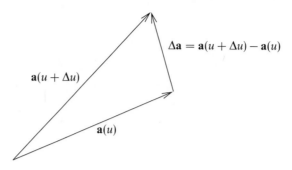

Figure 10.1 A small change in a vector $\mathbf{a}(u)$ resulting from a small change in u.

assuming that the limit exists, in which case $\mathbf{a}(u)$ is said to be differentiable at that point. Note that $d\mathbf{a}/du$ is also a vector, which is not, in general, parallel to $\mathbf{a}(u)$. In Cartesian coordinates, the derivative of the vector $\mathbf{a}(u) = a_x\mathbf{i} + a_y\mathbf{j} + a_z\mathbf{k}$ is given by

$$\frac{d\mathbf{a}}{du} = \frac{da_x}{du}\mathbf{i} + \frac{da_y}{du}\mathbf{j} + \frac{da_z}{du}\mathbf{k}.$$

Perhaps the simplest application of the above is to finding the velocity and acceleration of a particle in classical mechanics. If the time-dependent position vector of the particle with respect to the origin in Cartesian coordinates is given by $\mathbf{r}(t) = x(t)\mathbf{i} + y(t)\mathbf{j} + z(t)\mathbf{k}$ then the velocity of the particle is given by the vector

$$\mathbf{v}(t) = \frac{d\mathbf{r}}{dt} = \frac{dx}{dt}\mathbf{i} + \frac{dy}{dt}\mathbf{j} + \frac{dz}{dt}\mathbf{k}.$$

The direction of the velocity vector is along the tangent to the path $\mathbf{r}(t)$ at the instantaneous position of the particle, and its magnitude $|\mathbf{v}(t)|$ is equal to the speed of the particle. The acceleration of the particle is given in a similar manner by

$$\mathbf{a}(t) = \frac{d\mathbf{v}}{dt} = \frac{d^2x}{dt^2}\mathbf{i} + \frac{d^2y}{dt^2}\mathbf{j} + \frac{d^2z}{dt^2}\mathbf{k}.$$

▶*The position vector of a particle at time t in Cartesian coordinates is given by $\mathbf{r}(t) = 2t^2\mathbf{i} + (3t-2)\mathbf{j} + (3t^2-1)\mathbf{k}$. Find the speed of the particle at $t = 1$ and the component of its acceleration in the direction $\mathbf{s} = \mathbf{i} + 2\mathbf{j} + \mathbf{k}$.*

The velocity and acceleration of the particle are given by

$$\mathbf{v}(t) = \frac{d\mathbf{r}}{dt} = 4t\mathbf{i} + 3\mathbf{j} + 6t\mathbf{k},$$

$$\mathbf{a}(t) = \frac{d\mathbf{v}}{dt} = 4\mathbf{i} + 6\mathbf{k}.$$

335

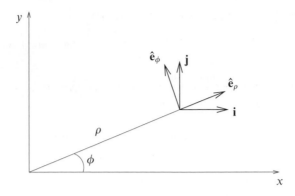

Figure 10.2   Unit basis vectors for two-dimensional Cartesian and plane polar coordinates.

The speed of the particle at $t = 1$ is simply

$$|\mathbf{v}(1)| = \sqrt{4^2 + 3^2 + 6^2} = \sqrt{61}.$$

The acceleration of the particle is constant (i.e. independent of t), and its component in the direction $\mathbf{s}$ is given by

$$\mathbf{a} \cdot \hat{\mathbf{s}} = \frac{(4\mathbf{i} + 6\mathbf{k}) \cdot (\mathbf{i} + 2\mathbf{j} + \mathbf{k})}{\sqrt{1^2 + 2^2 + 1^2}} = \frac{5\sqrt{6}}{3}. \; \blacktriangleleft$$

Note that in the case discussed above $\mathbf{i}$, $\mathbf{j}$ and $\mathbf{k}$ are fixed, time-independent basis vectors. This may not be true of basis vectors in general; when we are not using Cartesian coordinates the basis vectors themselves must also be differentiated. We discuss basis vectors for non-Cartesian coordinate systems in detail in section 10.10. Nevertheless, as a simple example, let us now consider two-dimensional plane polar coordinates ρ, ϕ.

Referring to figure 10.2, imagine holding ϕ fixed and moving radially outwards, i.e. in the direction of increasing ρ. Let us denote the unit vector in this direction by $\hat{\mathbf{e}}_\rho$. Similarly, imagine keeping ρ fixed and moving around a circle of fixed radius in the direction of increasing ϕ. Let us denote the unit vector tangent to the circle by $\hat{\mathbf{e}}_\phi$. The two vectors $\hat{\mathbf{e}}_\rho$ and $\hat{\mathbf{e}}_\phi$ are the basis vectors for this two-dimensional coordinate system, just as $\mathbf{i}$ and $\mathbf{j}$ are basis vectors for two-dimensional Cartesian coordinates. All these basis vectors are shown in figure 10.2.

An important difference between the two sets of basis vectors is that, while $\mathbf{i}$ and $\mathbf{j}$ are constant in magnitude *and direction*, the vectors $\hat{\mathbf{e}}_\rho$ and $\hat{\mathbf{e}}_\phi$ have constant magnitudes but their directions change as ρ and ϕ vary. Therefore, when calculating the derivative of a vector written in polar coordinates we must also differentiate the basis vectors. One way of doing this is to express $\hat{\mathbf{e}}_\rho$ and $\hat{\mathbf{e}}_\phi$

in terms of **i** and **j**. From figure 10.2, we see that

$$\hat{\mathbf{e}}_\rho = \cos\phi\,\mathbf{i} + \sin\phi\,\mathbf{j},$$
$$\hat{\mathbf{e}}_\phi = -\sin\phi\,\mathbf{i} + \cos\phi\,\mathbf{j}.$$

Since **i** and **j** are constant vectors, we find that the derivatives of the basis vectors $\hat{\mathbf{e}}_\rho$ and $\hat{\mathbf{e}}_\phi$ with respect to t are given by

$$\frac{d\hat{\mathbf{e}}_\rho}{dt} = -\sin\phi\frac{d\phi}{dt}\,\mathbf{i} + \cos\phi\frac{d\phi}{dt}\,\mathbf{j} = \dot{\phi}\,\hat{\mathbf{e}}_\phi, \qquad (10.2)$$

$$\frac{d\hat{\mathbf{e}}_\phi}{dt} = -\cos\phi\frac{d\phi}{dt}\,\mathbf{i} - \sin\phi\frac{d\phi}{dt}\,\mathbf{j} = -\dot{\phi}\,\hat{\mathbf{e}}_\rho, \qquad (10.3)$$

where the overdot is the conventional notation for differentiation with respect to time.

> ► *The position vector of a particle in plane polar coordinates is* $\mathbf{r}(t) = \rho(t)\hat{\mathbf{e}}_\rho$. *Find expressions for the velocity and acceleration of the particle in these coordinates.*

Using result (10.4) below, the velocity of the particle is given by

$$\mathbf{v}(t) = \dot{\mathbf{r}}(t) = \dot{\rho}\,\hat{\mathbf{e}}_\rho + \rho\,\dot{\hat{\mathbf{e}}}_\rho = \dot{\rho}\,\hat{\mathbf{e}}_\rho + \rho\dot{\phi}\,\hat{\mathbf{e}}_\phi,$$

where we have used (10.2). In a similar way its acceleration is given by

$$\mathbf{a}(t) = \frac{d}{dt}(\dot{\rho}\,\hat{\mathbf{e}}_\rho + \rho\dot{\phi}\,\hat{\mathbf{e}}_\phi)$$
$$= \ddot{\rho}\,\hat{\mathbf{e}}_\rho + \dot{\rho}\,\dot{\hat{\mathbf{e}}}_\rho + \rho\dot{\phi}\,\dot{\hat{\mathbf{e}}}_\phi + \rho\ddot{\phi}\,\hat{\mathbf{e}}_\phi + \dot{\rho}\dot{\phi}\,\hat{\mathbf{e}}_\phi$$
$$= \ddot{\rho}\,\hat{\mathbf{e}}_\rho + \dot{\rho}(\dot{\phi}\,\hat{\mathbf{e}}_\phi) + \rho\dot{\phi}(-\dot{\phi}\,\hat{\mathbf{e}}_\rho) + \rho\ddot{\phi}\,\hat{\mathbf{e}}_\phi + \dot{\rho}\dot{\phi}\,\hat{\mathbf{e}}_\phi$$
$$= (\ddot{\rho} - \rho\dot{\phi}^2)\,\hat{\mathbf{e}}_\rho + (\rho\ddot{\phi} + 2\dot{\rho}\dot{\phi})\,\hat{\mathbf{e}}_\phi.$$

Here we have used (10.2) and (10.3). ◄

10.1.1 Differentiation of composite vector expressions

In composite vector expressions each of the vectors or scalars involved may be a function of some scalar variable u, as we have seen. The derivatives of such expressions are easily found using the definition (10.1) and the rules of ordinary differential calculus. They may be summarised by the following, in which we assume that **a** and **b** are differentiable vector functions of a scalar u and that ϕ is a differentiable scalar function of u:

$$\frac{d}{du}(\phi\mathbf{a}) = \phi\frac{d\mathbf{a}}{du} + \frac{d\phi}{du}\mathbf{a}, \qquad (10.4)$$

$$\frac{d}{du}(\mathbf{a}\cdot\mathbf{b}) = \mathbf{a}\cdot\frac{d\mathbf{b}}{du} + \frac{d\mathbf{a}}{du}\cdot\mathbf{b}, \qquad (10.5)$$

$$\frac{d}{du}(\mathbf{a}\times\mathbf{b}) = \mathbf{a}\times\frac{d\mathbf{b}}{du} + \frac{d\mathbf{a}}{du}\times\mathbf{b}. \qquad (10.6)$$

The order of the factors in the terms on the RHS of (10.6) is, of course, just as important as it is in the original vector product.

> ►*A particle of mass m with position vector* **r** *relative to some origin O experiences a force* **F**, *which produces a torque (moment)* **T** = **r** × **F** *about O. The angular momentum of the particle about O is given by* **L** = **r** × m**v**, *where* **v** *is the particle's velocity. Show that the rate of change of angular momentum is equal to the applied torque.*

The rate of change of angular momentum is given by

$$\frac{d\mathbf{L}}{dt} = \frac{d}{dt}(\mathbf{r} \times m\mathbf{v}).$$

Using (10.6) we obtain

$$\frac{d\mathbf{L}}{dt} = \frac{d\mathbf{r}}{dt} \times m\mathbf{v} + \mathbf{r} \times \frac{d}{dt}(m\mathbf{v})$$

$$= \mathbf{v} \times m\mathbf{v} + \mathbf{r} \times \frac{d}{dt}(m\mathbf{v})$$

$$= \mathbf{0} + \mathbf{r} \times \mathbf{F} = \mathbf{T},$$

where in the last line we use Newton's second law, namely $\mathbf{F} = d(m\mathbf{v})/dt$. ◄

If a vector **a** is a function of a scalar variable s that is itself a function of u, so that $s = s(u)$, then the chain rule (see subsection 2.1.3) gives

$$\frac{d\mathbf{a}(s)}{du} = \frac{ds}{du}\frac{d\mathbf{a}}{ds}. \tag{10.7}$$

The derivatives of more complicated vector expressions may be found by repeated application of the above equations.

One further useful result can be derived by considering the derivative

$$\frac{d}{du}(\mathbf{a} \cdot \mathbf{a}) = 2\mathbf{a} \cdot \frac{d\mathbf{a}}{du};$$

since $\mathbf{a} \cdot \mathbf{a} = a^2$, where $a = |\mathbf{a}|$, we see that

$$\mathbf{a} \cdot \frac{d\mathbf{a}}{du} = 0 \quad \text{if } a \text{ is constant.} \tag{10.8}$$

In other words, if a vector $\mathbf{a}(u)$ has a constant magnitude as u varies then it is perpendicular to the vector $d\mathbf{a}/du$.

10.1.2 Differential of a vector

As a final note on the differentiation of vectors, we can also define the *differential* of a vector, in a similar way to that of a scalar in ordinary differential calculus. In the definition of the vector derivative (10.1), we used the notion of a small change $\Delta\mathbf{a}$ in a vector $\mathbf{a}(u)$ resulting from a small change Δu in its argument. In the limit $\Delta u \to 0$, the change in **a** becomes infinitesimally small, and we denote it by the differential $d\mathbf{a}$. From (10.1) we see that the differential is given by

$$d\mathbf{a} = \frac{d\mathbf{a}}{du} du. \tag{10.9}$$

Note that the differential of a vector is also a vector. As an example, the infinitesimal change in the position vector of a particle in an infinitesimal time dt is

$$d\mathbf{r} = \frac{d\mathbf{r}}{dt}\,dt = \mathbf{v}\,dt,$$

where $\mathbf{v}$ is the particle's velocity.

10.2 Integration of vectors

The integration of a vector (or of an expression involving vectors that may itself be either a vector or scalar) with respect to a scalar u can be regarded as the inverse of differentiation. We must remember, however, that

(i) the integral has the same nature (vector or scalar) as the integrand,
(ii) the constant of integration for indefinite integrals must be of the same nature as the integral.

For example, if $\mathbf{a}(u) = d[\mathbf{A}(u)]/du$ then the indefinite integral of $\mathbf{a}(u)$ is given by

$$\int \mathbf{a}(u)\,du = \mathbf{A}(u) + \mathbf{b},$$

where $\mathbf{b}$ is a constant vector. The definite integral of $\mathbf{a}(u)$ from $u = u_1$ to $u = u_2$ is given by

$$\int_{u_1}^{u_2} \mathbf{a}(u)\,du = \mathbf{A}(u_2) - \mathbf{A}(u_1).$$

▶ *A small particle of mass m orbits a much larger mass M centred at the origin O. According to Newton's law of gravitation, the position vector* $\mathbf{r}$ *of the small mass obeys the differential equation*

$$m\frac{d^2\mathbf{r}}{dt^2} = -\frac{GMm}{r^2}\hat{\mathbf{r}}.$$

Show that the vector $\mathbf{r} \times d\mathbf{r}/dt$ *is a constant of the motion.*

Forming the vector product of the differential equation with $\mathbf{r}$, we obtain

$$\mathbf{r} \times \frac{d^2\mathbf{r}}{dt^2} = -\frac{GM}{r^2}\mathbf{r} \times \hat{\mathbf{r}}.$$

Since $\mathbf{r}$ and $\hat{\mathbf{r}}$ are collinear, $\mathbf{r} \times \hat{\mathbf{r}} = \mathbf{0}$ and therefore we have

$$\mathbf{r} \times \frac{d^2\mathbf{r}}{dt^2} = \mathbf{0}. \tag{10.10}$$

However,

$$\frac{d}{dt}\left(\mathbf{r} \times \frac{d\mathbf{r}}{dt}\right) = \mathbf{r} \times \frac{d^2\mathbf{r}}{dt^2} + \frac{d\mathbf{r}}{dt} \times \frac{d\mathbf{r}}{dt} = \mathbf{0},$$

339

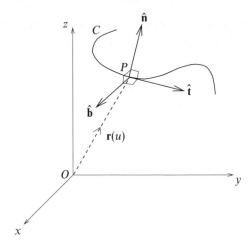

Figure 10.3   The unit tangent $\hat{\mathbf{t}}$, normal $\hat{\mathbf{n}}$ and binormal $\hat{\mathbf{b}}$ to the space curve C at a particular point P.

since the first term is zero by (10.10), and the second is zero because it is the vector product of two parallel (in this case identical) vectors. Integrating, we obtain the required result

$$\mathbf{r} \times \frac{d\mathbf{r}}{dt} = \mathbf{c}, \tag{10.11}$$

where $\mathbf{c}$ is a constant vector.

As a further point of interest we may note that in an infinitesimal time dt the change in the position vector of the small mass is $d\mathbf{r}$ and the element of area swept out by the position vector of the particle is simply $dA = \frac{1}{2}|\mathbf{r} \times d\mathbf{r}|$. Dividing both sides of this equation by dt, we conclude that

$$\frac{dA}{dt} = \frac{1}{2}\left|\mathbf{r} \times \frac{d\mathbf{r}}{dt}\right| = \frac{|\mathbf{c}|}{2},$$

and that the physical interpretation of the above result (10.11) is that the position vector $\mathbf{r}$ of the small mass sweeps out equal areas in equal times. This result is in fact valid for motion under any force that acts along the line joining the two particles. ◄

10.3 Space curves

In the previous section we mentioned that the velocity vector of a particle is a tangent to the curve in space along which the particle moves. We now give a more complete discussion of curves in space and also a discussion of the geometrical interpretation of the vector derivative.

A curve C in space can be described by the vector $\mathbf{r}(u)$ joining the origin O of a coordinate system to a point on the curve (see figure 10.3). As the parameter u varies, the end-point of the vector moves along the curve. In Cartesian coordinates,

$$\mathbf{r}(u) = x(u)\mathbf{i} + y(u)\mathbf{j} + z(u)\mathbf{k},$$

where $x = x(u)$, $y = y(u)$ and $z = z(u)$ are the *parametric* equations of the curve.

340

This parametric representation can be very useful, particularly in mechanics when the parameter may be the time t. We can, however, also represent a space curve by $y = f(x)$, $z = g(x)$, which can be easily converted into the above parametric form by setting $u = x$, so that

$$\mathbf{r}(u) = u\mathbf{i} + f(u)\mathbf{j} + g(u)\mathbf{k}.$$

Alternatively, a space curve can be represented in the form $F(x, y, z) = 0$, $G(x, y, z) = 0$, where each equation represents a surface and the curve is the intersection of the two surfaces.

A curve may sometimes be described in parametric form by the vector $\mathbf{r}(s)$, where the parameter s is the *arc length* along the curve measured from a fixed point. Even when the curve is expressed in terms of some other parameter, it is straightforward to find the arc length between any two points on the curve. For the curve described by $\mathbf{r}(u)$, let us consider an infinitesimal vector displacement

$$d\mathbf{r} = dx\,\mathbf{i} + dy\,\mathbf{j} + dz\,\mathbf{k}$$

along the curve. The square of the infinitesimal distance moved is then given by

$$(ds)^2 = d\mathbf{r} \cdot d\mathbf{r} - (dx)^2 + (dy)^2 \mid (dz)^2,$$

from which it can be shown that

$$\left(\frac{ds}{du}\right)^2 = \frac{d\mathbf{r}}{du} \cdot \frac{d\mathbf{r}}{du}.$$

Therefore, the arc length between two points on the curve $\mathbf{r}(u)$, given by $u = u_1$ and $u = u_2$, is

$$s = \int_{u_1}^{u_2} \sqrt{\frac{d\mathbf{r}}{du} \cdot \frac{d\mathbf{r}}{du}} \, du. \tag{10.12}$$

▶ *A curve lying in the xy-plane is given by $y = y(x)$, $z = 0$. Using (10.12), show that the arc length along the curve between $x = a$ and $x = b$ is given by $s = \int_a^b \sqrt{1 + y'^2}\,dx$, where $y' = dy/dx$.*

Let us first represent the curve in parametric form by setting $u = x$, so that

$$\mathbf{r}(u) = u\mathbf{i} + y(u)\mathbf{j}.$$

Differentiating with respect to u, we find

$$\frac{d\mathbf{r}}{du} = \mathbf{i} + \frac{dy}{du}\mathbf{j},$$

from which we obtain

$$\frac{d\mathbf{r}}{du} \cdot \frac{d\mathbf{r}}{du} = 1 + \left(\frac{dy}{du}\right)^2.$$

341

Therefore, remembering that $u = x$, from (10.12) the arc length between $x = a$ and $x = b$ is given by

$$s = \int_a^b \sqrt{\frac{d\mathbf{r}}{du} \cdot \frac{d\mathbf{r}}{du}}\, du = \int_a^b \sqrt{1 + \left(\frac{dy}{dx}\right)^2}\, dx.$$

This result was derived using more elementary methods in chapter 2. ◄

If a curve C is described by $\mathbf{r}(u)$ then, by considering figures 10.1 and 10.3, we see that, at any given point on the curve, $d\mathbf{r}/du$ is a vector tangent to C at that point, in the direction of increasing u. In the special case where the parameter u is the arc length s along the curve then $d\mathbf{r}/ds$ is a *unit* tangent vector to C and is denoted by $\hat{\mathbf{t}}$.

The rate at which the unit tangent $\hat{\mathbf{t}}$ changes with respect to s is given by $d\hat{\mathbf{t}}/ds$, and its magnitude is defined as the *curvature* κ of the curve C at a given point,

$$\kappa = \left|\frac{d\hat{\mathbf{t}}}{ds}\right| = \left|\frac{d^2\mathbf{r}}{ds^2}\right|.$$

We can also define the quantity $\rho = 1/\kappa$, which is called the *radius of curvature*.

Since $\hat{\mathbf{t}}$ is of constant (unit) magnitude, it follows from (10.8) that it is perpendicular to $d\hat{\mathbf{t}}/ds$. The unit vector in the direction perpendicular to $\hat{\mathbf{t}}$ is denoted by $\hat{\mathbf{n}}$ and is called the *principal normal* at the point. We therefore have

$$\frac{d\hat{\mathbf{t}}}{ds} = \kappa\,\hat{\mathbf{n}}. \tag{10.13}$$

The unit vector $\hat{\mathbf{b}} = \hat{\mathbf{t}} \times \hat{\mathbf{n}}$, which is perpendicular to the plane containing $\hat{\mathbf{t}}$ and $\hat{\mathbf{n}}$, is called the *binormal* to C. The vectors $\hat{\mathbf{t}}$, $\hat{\mathbf{n}}$ and $\hat{\mathbf{b}}$ form a right-handed rectangular cooordinate system (or *triad*) at any given point on C (see figure 10.3). As s changes so that the point of interest moves along C, the triad of vectors also changes.

The rate at which $\hat{\mathbf{b}}$ changes with respect to s is given by $d\hat{\mathbf{b}}/ds$ and is a measure of the *torsion* τ of the curve at any given point. Since $\hat{\mathbf{b}}$ is of constant magnitude, from (10.8) it is perpendicular to $d\hat{\mathbf{b}}/ds$. We may further show that $d\hat{\mathbf{b}}/ds$ is also perpendicular to $\hat{\mathbf{t}}$, as follows. By definition $\hat{\mathbf{b}} \cdot \hat{\mathbf{t}} = 0$, which on differentiating yields

$$\begin{aligned}
0 &= \frac{d}{ds}\left(\hat{\mathbf{b}} \cdot \hat{\mathbf{t}}\right) = \frac{d\hat{\mathbf{b}}}{ds} \cdot \hat{\mathbf{t}} + \hat{\mathbf{b}} \cdot \frac{d\hat{\mathbf{t}}}{ds} \\
&= \frac{d\hat{\mathbf{b}}}{ds} \cdot \hat{\mathbf{t}} + \hat{\mathbf{b}} \cdot \kappa\,\hat{\mathbf{n}} \\
&= \frac{d\hat{\mathbf{b}}}{ds} \cdot \hat{\mathbf{t}},
\end{aligned}$$

where we have used the fact that $\hat{\mathbf{b}} \cdot \hat{\mathbf{n}} = 0$. Hence, since $d\hat{\mathbf{b}}/ds$ is perpendicular to both $\hat{\mathbf{b}}$ and $\hat{\mathbf{t}}$, we must have $d\hat{\mathbf{b}}/ds \propto \hat{\mathbf{n}}$. The constant of proportionality is $-\tau$, so we finally obtain

$$\frac{d\hat{\mathbf{b}}}{ds} = -\tau\,\hat{\mathbf{n}}. \tag{10.14}$$

Taking the dot product of each side with $\hat{\mathbf{n}}$, we see that the torsion of a curve is given by

$$\tau = -\hat{\mathbf{n}} \cdot \frac{d\hat{\mathbf{b}}}{ds}.$$

We may also define the quantity $\sigma = 1/\tau$, which is called the *radius of torsion*.

Finally, we consider the derivative $d\hat{\mathbf{n}}/ds$. Since $\hat{\mathbf{n}} = \hat{\mathbf{b}} \times \hat{\mathbf{t}}$ we have

$$\begin{aligned}
\frac{d\hat{\mathbf{n}}}{ds} &= \frac{d\hat{\mathbf{b}}}{ds} \times \hat{\mathbf{t}} + \hat{\mathbf{b}} \times \frac{d\hat{\mathbf{t}}}{ds} \\
&= -\tau\,\hat{\mathbf{n}} \times \hat{\mathbf{t}} + \hat{\mathbf{b}} \times \kappa\,\hat{\mathbf{n}} \\
&= \tau\,\hat{\mathbf{b}} - \kappa\,\hat{\mathbf{t}}. \tag{10.15}
\end{aligned}$$

In summary, $\hat{\mathbf{t}}$, $\hat{\mathbf{n}}$ and $\hat{\mathbf{b}}$ and their derivatives with respect to s are related to one another by the relations (10.13), (10.14) and (10.15), the *Frenet–Serret formulae*,

$$\frac{d\hat{\mathbf{t}}}{ds} = \kappa\,\hat{\mathbf{n}}, \qquad \frac{d\hat{\mathbf{n}}}{ds} = \tau\,\hat{\mathbf{b}} - \kappa\,\hat{\mathbf{t}}, \qquad \frac{d\hat{\mathbf{b}}}{ds} = -\tau\,\hat{\mathbf{n}}. \tag{10.16}$$

▶*Show that the acceleration of a particle travelling along a trajectory $\mathbf{r}(t)$ is given by*

$$\mathbf{a}(t) = \frac{dv}{dt}\,\hat{\mathbf{t}} + \frac{v^2}{\rho}\,\hat{\mathbf{n}},$$

where v is the speed of the particle, $\hat{\mathbf{t}}$ is the unit tangent to the trajectory, $\hat{\mathbf{n}}$ is its principal normal and ρ is its radius of curvature.

The velocity of the particle is given by

$$\mathbf{v}(t) = \frac{d\mathbf{r}}{dt} = \frac{d\mathbf{r}}{ds}\frac{ds}{dt} = \frac{ds}{dt}\,\hat{\mathbf{t}},$$

where ds/dt is the speed of the particle, which we denote by v, and $\hat{\mathbf{t}}$ is the unit vector tangent to the trajectory. Writing the velocity as $\mathbf{v} = v\,\hat{\mathbf{t}}$, and differentiating once more with respect to time t, we obtain

$$\mathbf{a}(t) = \frac{d\mathbf{v}}{dt} = \frac{dv}{dt}\,\hat{\mathbf{t}} + v\frac{d\hat{\mathbf{t}}}{dt};$$

but we note that

$$\frac{d\hat{\mathbf{t}}}{dt} = \frac{ds}{dt}\frac{d\hat{\mathbf{t}}}{ds} = v\kappa\,\hat{\mathbf{n}} = \frac{v}{\rho}\,\hat{\mathbf{n}}.$$

Therefore, we have

$$\mathbf{a}(t) = \frac{dv}{dt}\,\hat{\mathbf{t}} + \frac{v^2}{\rho}\,\hat{\mathbf{n}}.$$

This shows that in addition to an acceleration dv/dt along the tangent to the particle's trajectory, there is also an acceleration v^2/ρ in the direction of the principal normal. The latter is often called the *centripetal* acceleration. ◀

Finally, we note that a curve $\mathbf{r}(u)$ representing the trajectory of a particle may sometimes be given in terms of some parameter u that is not necessarily equal to the time t but is functionally related to it in some way. In this case the velocity of the particle is given by

$$\mathbf{v} = \frac{d\mathbf{r}}{dt} = \frac{d\mathbf{r}}{du}\frac{du}{dt}.$$

Differentiating again with respect to time gives the acceleration as

$$\mathbf{a} = \frac{d\mathbf{v}}{dt} = \frac{d}{dt}\left(\frac{d\mathbf{r}}{du}\frac{du}{dt}\right) = \frac{d^2\mathbf{r}}{du^2}\left(\frac{du}{dt}\right)^2 + \frac{d\mathbf{r}}{du}\frac{d^2u}{dt^2}.$$

10.4 Vector functions of several arguments

The concept of the derivative of a vector is easily extended to cases where the vectors (or scalars) are functions of more than one independent scalar variable, $u_1, u_2, \ldots, u_n$. In this case, the results of subsection 10.1.1 are still valid, except that the derivatives become partial derivatives $\partial\mathbf{a}/\partial u_i$ defined as in ordinary differential calculus. For example, in Cartesian coordinates,

$$\frac{\partial\mathbf{a}}{\partial u_r} = \frac{\partial a_x}{\partial u_r}\mathbf{i} + \frac{\partial a_y}{\partial u_r}\mathbf{j} + \frac{\partial a_z}{\partial u_r}\mathbf{k}.$$

In particular, (10.7) generalises to the chain rule of partial differentiation discussed in section 5.5. If $\mathbf{a} = \mathbf{a}(u_1, u_2, \ldots, u_n)$ and each of the u_i is also a function $u_i(v_1, v_2, \ldots, v_n)$ of the variables v_i then, generalising (5.17),

$$\frac{\partial\mathbf{a}}{\partial v_i} = \frac{\partial\mathbf{a}}{\partial u_1}\frac{\partial u_1}{\partial v_i} + \frac{\partial\mathbf{a}}{\partial u_2}\frac{\partial u_2}{\partial v_i} + \cdots + \frac{\partial\mathbf{a}}{\partial u_n}\frac{\partial u_n}{\partial v_i} = \sum_{j=1}^{n}\frac{\partial\mathbf{a}}{\partial u_j}\frac{\partial u_j}{\partial v_i}. \tag{10.17}$$

A special case of this rule arises when $\mathbf{a}$ is an explicit function of some variable v, as well as of scalars $u_1, u_2, \ldots, u_n$ that are themselves functions of v; then we have

$$\frac{d\mathbf{a}}{dv} = \frac{\partial\mathbf{a}}{\partial v} + \sum_{j=1}^{n}\frac{\partial\mathbf{a}}{\partial u_j}\frac{\partial u_j}{\partial v}. \tag{10.18}$$

We may also extend the concept of the differential of a vector given in (10.9) to vectors dependent on several variables $u_1, u_2, \ldots, u_n$:

$$d\mathbf{a} = \frac{\partial\mathbf{a}}{\partial u_1}du_1 + \frac{\partial\mathbf{a}}{\partial u_2}du_2 + \cdots + \frac{\partial\mathbf{a}}{\partial u_n}du_n = \sum_{j=1}^{n}\frac{\partial\mathbf{a}}{\partial u_j}du_j. \tag{10.19}$$

As an example, the infinitesimal change in an electric field $\mathbf{E}$ in moving from a position $\mathbf{r}$ to a neighbouring one $\mathbf{r} + d\mathbf{r}$ is given by

$$d\mathbf{E} = \frac{\partial\mathbf{E}}{\partial x}dx + \frac{\partial\mathbf{E}}{\partial y}dy + \frac{\partial\mathbf{E}}{\partial z}dz. \tag{10.20}$$

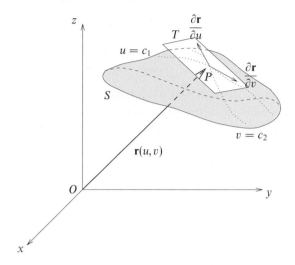

Figure 10.4 The tangent plane T to a surface S at a particular point P; $u = c_1$ and $v = c_2$ are the coordinate curves, shown by dotted lines, that pass through P. The broken line shows some particular parametric curve $\mathbf{r} = \mathbf{r}(\lambda)$ lying in the surface.

10.5 Surfaces

A surface S in space can be described by the vector $\mathbf{r}(u,v)$ joining the origin O of a coordinate system to a point on the surface (see figure 10.4). As the parameters u and v vary, the end-point of the vector moves over the surface. This is very similar to the parametric representation $\mathbf{r}(u)$ of a curve, discussed in section 10.3, but with the important difference that we require *two* parameters to describe a surface, whereas we need only one to describe a curve.

In Cartesian coordinates the surface is given by

$$\mathbf{r}(u,v) = x(u,v)\mathbf{i} + y(u,v)\mathbf{j} + z(u,v)\mathbf{k},$$

where $x = x(u,v)$, $y = y(u,v)$ and $z = z(u,v)$ are the parametric equations of the surface. We can also represent a surface by $z = f(x,y)$ or $g(x,y,z) = 0$. Either of these representations can be converted into the parametric form in a similar manner to that used for equations of curves. For example, if $z = f(x,y)$ then by setting $u = x$ and $v = y$ the surface can be represented in parametric form by

$$\mathbf{r}(u,v) = u\mathbf{i} + v\mathbf{j} + f(u,v)\mathbf{k}.$$

Any curve $\mathbf{r}(\lambda)$, where λ is a parameter, on the surface S can be represented by a pair of equations relating the parameters u and v, for example $u = f(\lambda)$ and $v = g(\lambda)$. A parametric representation of the curve can easily be found by straightforward substitution, i.e. $\mathbf{r}(\lambda) = \mathbf{r}(u(\lambda), v(\lambda))$. Using (10.17) for the case where the vector is a function of a single variable λ so that the LHS becomes a total derivative, the tangent to the curve $\mathbf{r}(\lambda)$ at any point is given by

$$\frac{d\mathbf{r}}{d\lambda} = \frac{\partial \mathbf{r}}{\partial u}\frac{du}{d\lambda} + \frac{\partial \mathbf{r}}{\partial v}\frac{dv}{d\lambda}. \tag{10.21}$$

The two curves $u = $ constant and $v = $ constant passing through any point P on S are called *coordinate curves*. For the curve $u = $ constant, for example, we have $du/d\lambda = 0$, and so from (10.21) its tangent vector is in the direction $\partial \mathbf{r}/\partial v$. Similarly, the tangent vector to the curve $v = $ constant is in the direction $\partial \mathbf{r}/\partial u$.

If the surface is smooth then at any point P on S the vectors $\partial \mathbf{r}/\partial u$ and $\partial \mathbf{r}/\partial v$ are linearly independent and define the *tangent plane* T at the point P (see figure 10.4). A vector normal to the surface at P is given by

$$\mathbf{n} = \frac{\partial \mathbf{r}}{\partial u} \times \frac{\partial \mathbf{r}}{\partial v}. \tag{10.22}$$

In the neighbourhood of P, an infinitesimal vector displacement $d\mathbf{r}$ is written

$$d\mathbf{r} = \frac{\partial \mathbf{r}}{\partial u} du + \frac{\partial \mathbf{r}}{\partial v} dv.$$

The *element of area* at P, an infinitesimal parallelogram whose sides are the coordinate curves, has magnitude

$$dS = \left| \frac{\partial \mathbf{r}}{\partial u} du \times \frac{\partial \mathbf{r}}{\partial v} dv \right| = \left| \frac{\partial \mathbf{r}}{\partial u} \times \frac{\partial \mathbf{r}}{\partial v} \right| du\,dv = |\mathbf{n}|\,du\,dv. \tag{10.23}$$

Thus the total area of the surface is

$$A = \iint_R \left| \frac{\partial \mathbf{r}}{\partial u} \times \frac{\partial \mathbf{r}}{\partial v} \right| du\,dv = \iint_R |\mathbf{n}|\,du\,dv, \tag{10.24}$$

where R is the region in the uv-plane corresponding to the range of parameter values that define the surface.

> ▶ *Find the element of area on the surface of a sphere of radius a, and hence calculate the total surface area of the sphere.*

We can represent a point $\mathbf{r}$ on the surface of the sphere in terms of the two parameters θ and ϕ:

$$\mathbf{r}(\theta, \phi) = a\sin\theta\cos\phi\,\mathbf{i} + a\sin\theta\sin\phi\,\mathbf{j} + a\cos\theta\,\mathbf{k},$$

where θ and ϕ are the polar and azimuthal angles respectively. At any point P, vectors tangent to the coordinate curves $\theta = $ constant and $\phi = $ constant are

$$\frac{\partial \mathbf{r}}{\partial \theta} = a\cos\theta\cos\phi\,\mathbf{i} + a\cos\theta\sin\phi\,\mathbf{j} - a\sin\theta\,\mathbf{k},$$

$$\frac{\partial \mathbf{r}}{\partial \phi} = -a\sin\theta\sin\phi\,\mathbf{i} + a\sin\theta\cos\phi\,\mathbf{j}.$$

A normal $\mathbf{n}$ to the surface at this point is then given by

$$\mathbf{n} = \frac{\partial \mathbf{r}}{\partial \theta} \times \frac{\partial \mathbf{r}}{\partial \phi} = \begin{vmatrix} \mathbf{i} & \mathbf{j} & \mathbf{k} \\ a\cos\theta\cos\phi & a\cos\theta\sin\phi & -a\sin\theta \\ -a\sin\theta\sin\phi & a\sin\theta\cos\phi & 0 \end{vmatrix}$$

$$= a^2 \sin\theta(\sin\theta\cos\phi\,\mathbf{i} + \sin\theta\sin\phi\,\mathbf{j} + \cos\theta\,\mathbf{k}),$$

which has a magnitude of $a^2 \sin\theta$. Therefore, the element of area at P is, from (10.23),

$$dS = a^2 \sin\theta\, d\theta\, d\phi,$$

and the total surface area of the sphere is given by

$$A = \int_0^\pi d\theta \int_0^{2\pi} d\phi\, a^2 \sin\theta = 4\pi a^2.$$

This familiar result can, of course, be proved by much simpler methods! ◀

10.6 Scalar and vector fields

We now turn to the case where a particular scalar or vector quantity is defined not just at a point in space but continuously as a *field* throughout some region of space R (which is often the whole space). Although the concept of a field is valid for spaces with an arbitrary number of dimensions, in the remainder of this chapter we will restrict our attention to the familiar three-dimensional case. A *scalar field* $\phi(x, y, z)$ associates a scalar with each point in R, while a *vector field* $\mathbf{a}(x, y, z)$ associates a vector with each point. In what follows, we will assume that the variation in the scalar or vector field from point to point is both continuous and differentiable in R.

Simple examples of scalar fields include the pressure at each point in a fluid and the electrostatic potential at each point in space in the presence of an electric charge. Vector fields relating to the same physical systems are the velocity vector in a fluid (giving the local speed and direction of the flow) and the electric field.

With the study of continuously varying scalar and vector fields there arises the need to consider their derivatives and also the integration of field quantities along lines, over surfaces and throughout volumes in the field. We defer the discussion of line, surface and volume integrals until the next chapter, and in the remainder of this chapter we concentrate on the definitions of vector differential operators and their properties.

10.7 Vector operators

Certain differential operations may be performed on scalar and vector fields and have wide-ranging applications in the physical sciences. The most important operations are those of finding the *gradient* of a scalar field and the *divergence* and *curl* of a vector field. It is usual to define these operators from a strictly mathematical point of view, as we do below. In the following chapter, however, we will discuss their geometrical definitions, which rely on the concept of integrating vector quantities along lines and over surfaces.

Central to all these differential operations is the vector operator ∇, which is called *del* (or sometimes *nabla*) and in Cartesian coordinates is defined by

$$\nabla \equiv \mathbf{i}\frac{\partial}{\partial x} + \mathbf{j}\frac{\partial}{\partial y} + \mathbf{k}\frac{\partial}{\partial z}. \tag{10.25}$$

The form of this operator in non-Cartesian coordinate systems is discussed in sections 10.9 and 10.10.

10.7.1 Gradient of a scalar field

The *gradient* of a scalar field $\phi(x, y, z)$ is defined by

$$\text{grad } \phi = \nabla \phi = \mathbf{i}\frac{\partial \phi}{\partial x} + \mathbf{j}\frac{\partial \phi}{\partial y} + \mathbf{k}\frac{\partial \phi}{\partial z}. \tag{10.26}$$

Clearly, $\nabla \phi$ is a vector field whose x-, y- and z- components are the first partial derivatives of $\phi(x, y, z)$ with respect to x, y and z respectively. Also note that the vector field $\nabla \phi$ should not be confused with the vector operator $\phi \nabla$, which has components $(\phi \, \partial/\partial x, \phi \, \partial/\partial y, \phi \, \partial/\partial z)$.

▶*Find the gradient of the scalar field $\phi = xy^2 z^3$.*

From (10.26) the gradient of ϕ is given by

$$\nabla \phi = y^2 z^3 \mathbf{i} + 2xyz^3 \mathbf{j} + 3xy^2 z^2 \mathbf{k}. \ ◀$$

The gradient of a scalar field ϕ has some interesting geometrical properties. Let us first consider the problem of *calculating the rate of change of ϕ in some particular direction*. For an infinitesimal vector displacement $d\mathbf{r}$, forming its scalar product with $\nabla \phi$ we obtain

$$\nabla \phi \cdot d\mathbf{r} = \left(\mathbf{i}\frac{\partial \phi}{\partial x} + \mathbf{j}\frac{\partial \phi}{\partial y} + \mathbf{k}\frac{\partial \phi}{\partial z} \right) \cdot (\mathbf{i}\, dx + \mathbf{j}\, dy + \mathbf{k}\, dx),$$

$$= \frac{\partial \phi}{\partial x}\, dx + \frac{\partial \phi}{\partial y}\, dy + \frac{\partial \phi}{\partial z}\, dz,$$

$$= d\phi, \tag{10.27}$$

which is the infinitesimal change in ϕ in going from position $\mathbf{r}$ to $\mathbf{r} + d\mathbf{r}$. In particular, if $\mathbf{r}$ depends on some parameter u such that $\mathbf{r}(u)$ defines a space curve

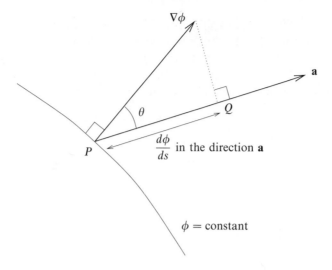

Figure 10.5   Geometrical properties of $\nabla\phi$. PQ gives the value of $d\phi/ds$ in the direction **a**.

then the total derivative of ϕ with respect to u along the curve is simply

$$\frac{d\phi}{du} = \nabla\phi \cdot \frac{d\mathbf{r}}{du}. \tag{10.28}$$

In the particular case where the parameter u is the arc length s along the curve, the total derivative of ϕ with respect to s along the curve is given by

$$\frac{d\phi}{ds} = \nabla\phi \cdot \hat{\mathbf{t}}, \tag{10.29}$$

where $\hat{\mathbf{t}}$ is the unit tangent to the curve at the given point, as discussed in section 10.3.

In general, the rate of change of ϕ with respect to the distance s in a particular direction **a** is given by

$$\frac{d\phi}{ds} = \nabla\phi \cdot \hat{\mathbf{a}} \tag{10.30}$$

and is called the directional derivative. Since $\hat{\mathbf{a}}$ is a unit vector we have

$$\frac{d\phi}{ds} = |\nabla\phi| \cos\theta$$

where θ is the angle between $\hat{\mathbf{a}}$ and $\nabla\phi$ as shown in figure 10.5. Clearly $\nabla\phi$ lies in the direction of the fastest increase in ϕ, and $|\nabla\phi|$ is the largest possible value of $d\phi/ds$. Similarly, the largest rate of decrease of ϕ is $d\phi/ds = -|\nabla\phi|$ in the direction of $-\nabla\phi$.

> ►*For the function $\phi = x^2 y + yz$ at the point $(1, 2, -1)$, find its rate of change with distance in the direction $\mathbf{a} = \mathbf{i} + 2\mathbf{j} + 3\mathbf{k}$. At this same point, what is the greatest possible rate of change with distance and in which direction does it occur?*

The gradient of ϕ is given by (10.26):

$$\nabla\phi = 2xy\mathbf{i} + (x^2 + z)\mathbf{j} + y\mathbf{k},$$
$$= 4\mathbf{i} + 2\mathbf{k} \quad \text{at the point } (1, 2, -1).$$

The unit vector in the direction of $\mathbf{a}$ is $\hat{\mathbf{a}} = \frac{1}{\sqrt{14}}(\mathbf{i} + 2\mathbf{j} + 3\mathbf{k})$, so the rate of change of ϕ with distance s in this direction is, using (10.30),

$$\frac{d\phi}{ds} = \nabla\phi \cdot \hat{\mathbf{a}} = \frac{1}{\sqrt{14}}(4 + 6) = \frac{10}{\sqrt{14}}.$$

From the above discussion, at the point $(1, 2, -1)$ $d\phi/ds$ will be greatest in the direction of $\nabla\phi = 4\mathbf{i} + 2\mathbf{k}$ and has the value $|\nabla\phi| = \sqrt{20}$ in this direction. ◄

We can extend the above analysis to find the rate of change of a vector field (rather than a scalar field as above) in a particular direction. The scalar differential operator $\hat{\mathbf{a}} \cdot \nabla$ can be shown to give the rate of change with distance in the direction $\hat{\mathbf{a}}$ of the quantity (vector or scalar) on which it acts. In Cartesian coordinates it may be written as

$$\hat{\mathbf{a}} \cdot \nabla = a_x \frac{\partial}{\partial x} + a_y \frac{\partial}{\partial y} + a_z \frac{\partial}{\partial z}. \tag{10.31}$$

Thus we can write the infinitesimal change in an electric field in moving from $\mathbf{r}$ to $\mathbf{r} + d\mathbf{r}$ given in (10.20) as $d\mathbf{E} = (d\mathbf{r} \cdot \nabla)\mathbf{E}$.

A second interesting geometrical property of $\nabla\phi$ may be found by considering the surface defined by $\phi(x, y, z) = c$, where c is some constant. If $\hat{\mathbf{t}}$ is a unit tangent to this surface at some point then clearly $d\phi/ds = 0$ in this direction and from (10.29) we have $\nabla\phi \cdot \hat{\mathbf{t}} = 0$. In other words, $\nabla\phi$ *is a vector normal to the surface* $\phi(x, y, z) = c$ *at every point*, as shown in figure 10.5. If $\hat{\mathbf{n}}$ is a unit normal to the surface in the direction of increasing $\phi(x, y, z)$, then the gradient is sometimes written

$$\nabla\phi \equiv \frac{\partial\phi}{\partial n}\hat{\mathbf{n}}, \tag{10.32}$$

where $\partial\phi/\partial n \equiv |\nabla\phi|$ is the rate of change of ϕ in the direction $\hat{\mathbf{n}}$ and is called the *normal derivative*.

> ►*Find expressions for the equations of the tangent plane and the line normal to the surface $\phi(x, y, z) = c$ at the point P with coordinates x_0, y_0, z_0. Use the results to find the equations of the tangent plane and the line normal to the surface of the sphere $\phi = x^2 + y^2 + z^2 = a^2$ at the point $(0, 0, a)$.*

A vector normal to the surface $\phi(x, y, z) = c$ at the point P is simply $\nabla\phi$ evaluated at that point; we denote it by $\mathbf{n}_0$. If $\mathbf{r}_0$ is the position vector of the point P relative to the origin,

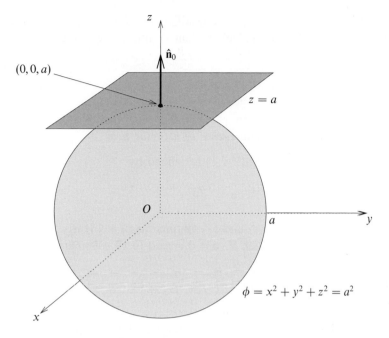

Figure 10.6   The tangent plane and the normal to the surface of the sphere $\phi = x^2 + y^2 + z^2 = a^2$ at the point $\mathbf{r}_0$ with coordinates $(0, 0, a)$.

and $\mathbf{r}$ is the position vector of any point on the tangent plane, then the vector equation of the tangent plane is, from (7.41),

$$(\mathbf{r} - \mathbf{r}_0) \cdot \mathbf{n}_0 = 0.$$

Similarly, if $\mathbf{r}$ is the position vector of any point on the straight line passing through P (with position vector $\mathbf{r}_0$) in the direction of the normal $\mathbf{n}_0$ then the vector equation of this line is, from subsection 7.7.1,

$$(\mathbf{r} - \mathbf{r}_0) \times \mathbf{n}_0 = \mathbf{0}.$$

For the surface of the sphere $\phi = x^2 + y^2 + z^2 = a^2$,

$$\nabla \phi = 2x\mathbf{i} + 2y\mathbf{j} + 2z\mathbf{k}$$
$$= 2a\mathbf{k} \quad \text{at the point } (0, 0, a).$$

Therefore the equation of the tangent plane to the sphere at this point is

$$(\mathbf{r} - \mathbf{r}_0) \cdot 2a\mathbf{k} = 0.$$

This gives $2a(z - a) = 0$ or $z = a$, as expected. The equation of the line normal to the sphere at the point $(0, 0, a)$ is

$$(\mathbf{r} - \mathbf{r}_0) \times 2a\mathbf{k} = \mathbf{0},$$

which gives $2ay\mathbf{i} - 2ax\mathbf{j} = \mathbf{0}$ or $x = y = 0$, i.e. the z-axis, as expected. The tangent plane and normal to the surface of the sphere at this point are shown in figure 10.6. ◀

Further properties of the gradient operation, which are analogous to those of the ordinary derivative, are listed in subsection 10.8.1 and may be easily proved.

In addition to these, we note that the gradient operation also obeys the chain rule as in ordinary differential calculus, i.e. if ϕ and ψ are scalar fields in some region R then

$$\nabla \left[\phi(\psi)\right] = \frac{\partial \phi}{\partial \psi} \nabla \psi.$$

10.7.2 Divergence of a vector field

The *divergence* of a vector field $\mathbf{a}(x, y, z)$ is defined by

$$\operatorname{div} \mathbf{a} = \nabla \cdot \mathbf{a} = \frac{\partial a_x}{\partial x} + \frac{\partial a_y}{\partial y} + \frac{\partial a_z}{\partial z}, \tag{10.33}$$

where a_x, a_y and a_z are the x-, y- and z- components of $\mathbf{a}$. Clearly, $\nabla \cdot \mathbf{a}$ is a scalar field. Any vector field $\mathbf{a}$ for which $\nabla \cdot \mathbf{a} = 0$ is said to be *solenoidal*.

▶ *Find the divergence of the vector field* $\mathbf{a} = x^2 y^2 \mathbf{i} + y^2 z^2 \mathbf{j} + x^2 z^2 \mathbf{k}$.

From (10.33) the divergence of $\mathbf{a}$ is given by

$$\nabla \cdot \mathbf{a} = 2xy^2 + 2yz^2 + 2x^2 z = 2(xy^2 + yz^2 + x^2 z). \blacktriangleleft$$

We will discuss fully the geometric definition of divergence and its physical meaning in the next chapter. For the moment, we merely note that the divergence can be considered as a quantitative measure of how much a vector field diverges (spreads out) or converges at any given point. For example, if we consider the vector field $\mathbf{v}(x, y, z)$ describing the local velocity at any point in a fluid then $\nabla \cdot \mathbf{v}$ is equal to the net rate of outflow of fluid per unit volume, evaluated at a point (by letting a small volume at that point tend to zero).

Now if some vector field $\mathbf{a}$ is itself derived from a scalar field via $\mathbf{a} = \nabla \phi$ then $\nabla \cdot \mathbf{a}$ has the form $\nabla \cdot \nabla \phi$ or, as it is usually written, $\nabla^2 \phi$, where ∇^2 (del squared) is the scalar differential operator

$$\nabla^2 \equiv \frac{\partial^2}{\partial x^2} + \frac{\partial^2}{\partial y^2} + \frac{\partial^2}{\partial z^2}. \tag{10.34}$$

$\nabla^2 \phi$ is called the *Laplacian* of ϕ and appears in several important partial differential equations of mathematical physics, discussed in chapters 20 and 21.

▶ *Find the Laplacian of the scalar field* $\phi = xy^2 z^3$.

From (10.34) the Laplacian of ϕ is given by

$$\nabla^2 \phi = \frac{\partial^2 \phi}{\partial x^2} + \frac{\partial^2 \phi}{\partial y^2} + \frac{\partial^2 \phi}{\partial z^2} = 2xz^3 + 6xy^2 z. \blacktriangleleft$$

10.7.3 Curl of a vector field

The *curl* of a vector field $\mathbf{a}(x, y, z)$ is defined by

$$\operatorname{curl} \mathbf{a} = \nabla \times \mathbf{a} = \left(\frac{\partial a_z}{\partial y} - \frac{\partial a_y}{\partial z} \right) \mathbf{i} + \left(\frac{\partial a_x}{\partial z} - \frac{\partial a_z}{\partial x} \right) \mathbf{j} + \left(\frac{\partial a_y}{\partial x} - \frac{\partial a_x}{\partial y} \right) \mathbf{k},$$

where a_x, a_y and a_z are the x-, y- and z- components of $\mathbf{a}$. The RHS can be written in a more memorable form as a determinant:

$$\nabla \times \mathbf{a} = \begin{vmatrix} \mathbf{i} & \mathbf{j} & \mathbf{k} \\ \dfrac{\partial}{\partial x} & \dfrac{\partial}{\partial y} & \dfrac{\partial}{\partial z} \\ a_x & a_y & a_z \end{vmatrix}, \tag{10.35}$$

where it is understood that, on expanding the determinant, the partial derivatives in the second row act on the components of $\mathbf{a}$ in the third row. Clearly, $\nabla \times \mathbf{a}$ is itself a vector field. Any vector field $\mathbf{a}$ for which $\nabla \times \mathbf{a} = \mathbf{0}$ is said to be *irrotational*.

▶ *Find the curl of the vector field* $\mathbf{a} = x^2 y^2 z^2 \mathbf{i} + y^2 z^2 \mathbf{j} + x^2 z^2 \mathbf{k}$.

The curl of $\mathbf{a}$ is given by

$$\nabla \times \mathbf{a} = \begin{vmatrix} \mathbf{i} & \mathbf{j} & \mathbf{k} \\ \dfrac{\partial}{\partial x} & \dfrac{\partial}{\partial y} & \dfrac{\partial}{\partial z} \\ x^2 y^2 z^2 & y^2 z^2 & x^2 z^2 \end{vmatrix} = -2 \left[y^2 z \mathbf{i} + (xz^2 - x^2 y^2 z) \mathbf{j} + x^2 yz^2 \mathbf{k} \right]. \; ◀$$

For a vector field $\mathbf{v}(x, y, z)$ describing the local velocity at any point in a fluid, $\nabla \times \mathbf{v}$ is a measure of the angular velocity of the fluid in the neighbourhood of that point. If a small paddle wheel were placed at various points in the fluid then it would tend to rotate in regions where $\nabla \times \mathbf{v} \neq \mathbf{0}$, while it would not rotate in regions where $\nabla \times \mathbf{v} = \mathbf{0}$.

Another insight into the physical interpretation of the curl operator is gained by considering the vector field $\mathbf{v}$ describing the velocity at any point in a rigid body rotating about some axis with angular velocity ω. If $\mathbf{r}$ is the position vector of the point with respect to some origin on the axis of rotation then the velocity of the point is given by $\mathbf{v} = \boldsymbol{\omega} \times \mathbf{r}$. Without any loss of generality, we may take $\boldsymbol{\omega}$ to lie along the z-axis of our coordinate system, so that $\boldsymbol{\omega} = \omega \mathbf{k}$. The velocity field is then $\mathbf{v} = -\omega y \mathbf{i} + \omega x \mathbf{j}$. The curl of this vector field is easily found to be

$$\nabla \times \mathbf{v} = \begin{vmatrix} \mathbf{i} & \mathbf{j} & \mathbf{k} \\ \dfrac{\partial}{\partial x} & \dfrac{\partial}{\partial y} & \dfrac{\partial}{\partial z} \\ -\omega y & \omega x & 0 \end{vmatrix} = 2\omega \mathbf{k} = 2\boldsymbol{\omega}. \tag{10.36}$$

$$\nabla(\phi + \psi) = \nabla\phi + \nabla\psi$$
$$\nabla \cdot (\mathbf{a} + \mathbf{b}) = \nabla \cdot \mathbf{a} + \nabla \cdot \mathbf{b}$$
$$\nabla \times (\mathbf{a} + \mathbf{b}) = \nabla \times \mathbf{a} + \nabla \times \mathbf{b}$$
$$\nabla(\phi\psi) = \phi\nabla\psi + \psi\nabla\phi$$
$$\nabla(\mathbf{a} \cdot \mathbf{b}) = \mathbf{a} \times (\nabla \times \mathbf{b}) + \mathbf{b} \times (\nabla \times \mathbf{a}) + (\mathbf{a} \cdot \nabla)\mathbf{b} + (\mathbf{b} \cdot \nabla)\mathbf{a}$$
$$\nabla \cdot (\phi\mathbf{a}) = \phi\nabla \cdot \mathbf{a} + \mathbf{a} \cdot \nabla\phi$$
$$\nabla \cdot (\mathbf{a} \times \mathbf{b}) = \mathbf{b} \cdot (\nabla \times \mathbf{a}) - \mathbf{a} \cdot (\nabla \times \mathbf{b})$$
$$\nabla \times (\phi\mathbf{a}) = \nabla\phi \times \mathbf{a} + \phi\nabla \times \mathbf{a}$$
$$\nabla \times (\mathbf{a} \times \mathbf{b}) = \mathbf{a}(\nabla \cdot \mathbf{b}) - \mathbf{b}(\nabla \cdot \mathbf{a}) + (\mathbf{b} \cdot \nabla)\mathbf{a} - (\mathbf{a} \cdot \nabla)\mathbf{b}$$

Table 10.1 Vector operators acting on sums and products. The operator ∇ is defined in (10.25); ϕ and ψ are scalar fields, $\mathbf{a}$ and $\mathbf{b}$ are vector fields.

Therefore the curl of the velocity field is a vector equal to twice the angular velocity vector of the rigid body about its axis of rotation. We give a full geometrical discussion of the curl of a vector in the next chapter.

10.8 Vector operator formulae

In the same way as for ordinary vectors (chapter 7), for vector operators certain identities exist. In addition, we must consider various relations involving the action of vector operators on sums and products of scalar and vector fields. Some of these relations have been mentioned earlier, but we list all the most important ones here for convenience. The validity of these relations may be easily verified by direct calculation (a quick method of deriving them using tensor notation is given in chapter 26).

Although some of the following vector relations are expressed in Cartesian coordinates, it may be proved that they are all independent of the choice of coordinate system. This is to be expected since grad, div and curl all have clear geometrical definitions, which are discussed more fully in the next chapter and which do not rely on any particular choice of coordinate system.

10.8.1 Vector operators acting on sums and products

Let ϕ and ψ be scalar fields and $\mathbf{a}$ and $\mathbf{b}$ be vector fields. Assuming these fields are differentiable, the action of grad, div and curl on various sums and products of them is presented in table 10.1.

These relations can be proved by direct calculation.

▶*Show that*

$$\nabla \times (\phi\mathbf{a}) = \nabla\phi \times \mathbf{a} + \phi\nabla \times \mathbf{a}.$$

The x-component of the LHS is

$$\frac{\partial}{\partial y}(\phi a_z) - \frac{\partial}{\partial z}(\phi a_y) = \phi\frac{\partial a_z}{\partial y} + \frac{\partial \phi}{\partial y}a_z - \phi\frac{\partial a_y}{\partial z} - \frac{\partial \phi}{\partial z}a_y,$$

$$= \phi\left(\frac{\partial a_z}{\partial y} - \frac{\partial a_y}{\partial z}\right) + \left(\frac{\partial \phi}{\partial y}a_z - \frac{\partial \phi}{\partial z}a_y\right),$$

$$= \phi(\nabla \times \mathbf{a})_x + (\nabla\phi \times \mathbf{a})_x,$$

where, for example, $(\nabla\phi \times \mathbf{a})_x$ denotes the x-component of the vector $\nabla\phi \times \mathbf{a}$. Incorporating the y- and z- components, which can be similarly found, we obtain the stated result. ◀

Some useful special cases of the relations in table 10.1 are worth noting. If $\mathbf{r}$ is the position vector relative to some origin and $r = |\mathbf{r}|$, then

$$\nabla\phi(r) = \frac{d\phi}{dr}\,\hat{\mathbf{r}},$$

$$\nabla \cdot [\phi(r)\mathbf{r}] = 3\phi(r) + r\frac{d\phi(r)}{dr},$$

$$\nabla^2\phi(r) = \frac{d^2\psi(r)}{dr^2} + \frac{2}{r}\frac{d\phi(r)}{dr},$$

$$\nabla \times [\phi(r)\mathbf{r}] = \mathbf{0}.$$

These results may be proved straightforwardly using Cartesian coordinates but far more simply using spherical polar coordinates, which are discussed in subsection 10.9.2. Particular cases of these results are

$$\nabla r = \hat{\mathbf{r}}, \qquad \nabla \cdot \mathbf{r} = 3, \qquad \nabla \times \mathbf{r} = \mathbf{0},$$

together with

$$\nabla\left(\frac{1}{r}\right) = -\frac{\hat{\mathbf{r}}}{r^2},$$

$$\nabla \cdot \left(\frac{\hat{\mathbf{r}}}{r^2}\right) = -\nabla^2\left(\frac{1}{r}\right) = 4\pi\delta(r),$$

where $\delta(r)$ is the Dirac delta function, discussed in chapter 13. The last equation is important in the solution of certain partial differential equations and is discussed further in chapter 20.

10.8.2 Combinations of grad, div and curl

We now consider the action of two vector operators in succession on a scalar or vector field. We can immediately discard four of the nine obvious combinations of grad, div and curl, since they clearly do not make sense. If ϕ is a scalar field and

a is a vector field, these four combinations are grad(grad ϕ), div(div **a**), curl(div **a**) and grad(curl **a**). In each case the second (outer) vector operator is acting on the wrong type of field, i.e. scalar instead of vector or vice versa. In grad(grad ϕ), for example, grad acts on grad ϕ, which is a vector field, but we know that grad only acts on scalar fields (although in fact we will see in chapter 26 that we can form the *outer product* of the del operator with a vector to give a tensor, but that need not concern us here).

Of the five valid combinations of grad, div and curl, two are identically zero, namely

$$\text{curl grad } \phi = \nabla \times \nabla \phi = \mathbf{0}, \tag{10.37}$$

$$\text{div curl } \mathbf{a} = \nabla \cdot (\nabla \times \mathbf{a}) = 0. \tag{10.38}$$

From (10.37), we see that if **a** is derived from the gradient of some scalar function such that $\mathbf{a} = \nabla \phi$ then it is necessarily irrotational ($\nabla \times \mathbf{a} = 0$). We also note that if **a** is an irrotational vector field then another irrotational vector field is $\mathbf{a} + \nabla \phi + \mathbf{c}$, where ϕ is any scalar field and **c** is a constant vector. This follows since

$$\nabla \times (\mathbf{a} + \nabla \phi + \mathbf{c}) = \nabla \times \mathbf{a} + \nabla \times \nabla \phi = \mathbf{0}.$$

Similarly, from (10.38) we may infer that if **b** is the curl of some vector field **a** such that $\mathbf{b} = \nabla \times \mathbf{a}$ then **b** is solenoidal ($\nabla \cdot \mathbf{b} = 0$). Obviously, if **b** is solenoidal and **c** is any constant vector then $\mathbf{b} + \mathbf{c}$ is also solenoidal.

The three remaining combinations of grad, div and curl are

$$\text{div grad } \phi = \nabla \cdot \nabla \phi = \nabla^2 \phi = \frac{\partial^2 \phi}{\partial x^2} + \frac{\partial^2 \phi}{\partial y^2} + \frac{\partial^2 \phi}{\partial z^2}, \tag{10.39}$$

$$\text{grad div } \mathbf{a} = \nabla(\nabla \cdot \mathbf{a}),$$

$$= \left(\frac{\partial^2 a_x}{\partial x^2} + \frac{\partial^2 a_y}{\partial x \partial y} + \frac{\partial^2 a_z}{\partial x \partial z} \right) \mathbf{i} + \left(\frac{\partial^2 a_x}{\partial y \partial x} + \frac{\partial^2 a_y}{\partial y^2} + \frac{\partial^2 a_z}{\partial y \partial z} \right) \mathbf{j}$$

$$+ \left(\frac{\partial^2 a_x}{\partial z \partial x} + \frac{\partial^2 a_y}{\partial z \partial y} + \frac{\partial^2 a_z}{\partial z^2} \right) \mathbf{k}, \tag{10.40}$$

$$\text{curl curl } \mathbf{a} = \nabla \times (\nabla \times \mathbf{a}) = \nabla(\nabla \cdot \mathbf{a}) - \nabla^2 \mathbf{a}, \tag{10.41}$$

where (10.39) and (10.40) are expressed in Cartesian coordinates. In (10.41), the term $\nabla^2 \mathbf{a}$ has the linear differential operator ∇^2 acting on a vector (as opposed to a scalar as in (10.39)), which of course consists of a sum of unit vectors multiplied by components. Two cases arise.

(i) If the unit vectors are constants (i.e. they are independent of the values of the coordinates) then the differential operator gives a non-zero contribution only when acting upon the components, the unit vectors being merely multipliers.

(ii) If the unit vectors vary as the values of the coordinates change (i.e. are not constant in direction throughout the whole space) then the derivatives of these vectors appear as contributions to $\nabla^2 \mathbf{a}$.

Cartesian coordinates are an example of the first case in which each component satisfies $(\nabla^2 \mathbf{a})_i = \nabla^2 a_i$. In this case (10.41) can be applied to each component separately:

$$[\nabla \times (\nabla \times \mathbf{a})]_i = [\nabla(\nabla \cdot \mathbf{a})]_i - \nabla^2 a_i. \tag{10.42}$$

However, cylindrical and spherical polar coordinates come in the second class. For them (10.41) is still true, but the further step to (10.42) cannot be made.

More complicated vector operator relations may be proved using the relations given above.

▶*Show that*
$$\nabla \cdot (\nabla \phi \times \nabla \psi) = 0,$$
where ϕ and ψ are scalar fields.

From the previous section we have
$$\nabla \cdot (\mathbf{a} \times \mathbf{b}) = \mathbf{b} \cdot (\nabla \times \mathbf{a}) - \mathbf{a} \cdot (\nabla \times \mathbf{b}).$$
If we let $\mathbf{a} = \nabla \phi$ and $\mathbf{b} = \nabla \psi$ then we obtain
$$\nabla \cdot (\nabla \phi \times \nabla \psi) = \nabla \psi \cdot (\nabla \times \nabla \phi) - \nabla \phi \cdot (\nabla \times \nabla \psi) = 0, \tag{10.43}$$
since $\nabla \times \nabla \phi = 0 = \nabla \times \nabla \psi$, from (10.37). ◀

10.9 Cylindrical and spherical polar coordinates

The operators we have discussed in this chapter, i.e. grad, div, curl and ∇^2, have all been defined in terms of Cartesian coordinates, but for many physical situations other coordinate systems are more natural. For example, many systems, such as an isolated charge in space, have spherical symmetry and spherical polar coordinates would be the obvious choice. For axisymmetric systems, such as fluid flow in a pipe, cylindrical polar coordinates are the natural choice. The physical laws governing the behaviour of the systems are often expressed in terms of the vector operators we have been discussing, and so it is necessary to be able to express these operators in these other, non-Cartesian, coordinates. We first consider the two most common non-Cartesian coordinate systems, i.e. cylindrical and spherical polars, and go on to discuss general curvilinear coordinates in the next section.

10.9.1 Cylindrical polar coordinates

As shown in figure 10.7, the position of a point in space P having Cartesian coordinates x, y, z may be expressed in terms of cylindrical polar coordinates

ρ, ϕ, z, where

$$x = \rho \cos \phi, \qquad y = \rho \sin \phi, \qquad z = z, \tag{10.44}$$

and $\rho \geq 0, 0 \leq \phi < 2\pi$ and $-\infty < z < \infty$. The position vector of P may therefore be written

$$\mathbf{r} = \rho \cos \phi \, \mathbf{i} + \rho \sin \phi \, \mathbf{j} + z \, \mathbf{k}. \tag{10.45}$$

If we take the partial derivatives of $\mathbf{r}$ with respect to ρ, ϕ and z respectively then we obtain the three vectors

$$\mathbf{e}_\rho = \frac{\partial \mathbf{r}}{\partial \rho} = \cos \phi \, \mathbf{i} + \sin \phi \, \mathbf{j}, \tag{10.46}$$

$$\mathbf{e}_\phi = \frac{\partial \mathbf{r}}{\partial \phi} = -\rho \sin \phi \, \mathbf{i} + \rho \cos \phi \, \mathbf{j}, \tag{10.47}$$

$$\mathbf{e}_z = \frac{\partial \mathbf{r}}{\partial z} = \mathbf{k}. \tag{10.48}$$

These vectors lie in the directions of increasing ρ, ϕ and z respectively but are not all of unit length. Although $\mathbf{e}_\rho$, $\mathbf{e}_\phi$ and $\mathbf{e}_z$ form a useful set of basis vectors in their own right (we will see in section 10.10 that such a basis is sometimes the *most* useful), it is usual to work with the corresponding *unit* vectors, which are obtained by dividing each vector by its modulus to give

$$\hat{\mathbf{e}}_\rho = \mathbf{e}_\rho = \cos \phi \, \mathbf{i} + \sin \phi \, \mathbf{j}, \tag{10.49}$$

$$\hat{\mathbf{e}}_\phi = \frac{1}{\rho} \mathbf{e}_\phi = -\sin \phi \, \mathbf{i} + \cos \phi \, \mathbf{j}, \tag{10.50}$$

$$\hat{\mathbf{e}}_z = \mathbf{e}_z = \mathbf{k}. \tag{10.51}$$

These three unit vectors, like the Cartesian unit vectors $\mathbf{i}$, $\mathbf{j}$ and $\mathbf{k}$, form an orthonormal triad at each point in space, i.e. the basis vectors are mutually orthogonal and of unit length (see figure 10.7). Unlike the fixed vectors $\mathbf{i}$, $\mathbf{j}$ and $\mathbf{k}$, however, $\hat{\mathbf{e}}_\rho$ and $\hat{\mathbf{e}}_\phi$ change direction as P moves.

The expression for a general infinitesimal vector displacement $d\mathbf{r}$ in the position of P is given, from (10.19), by

$$\begin{aligned} d\mathbf{r} &= \frac{\partial \mathbf{r}}{\partial \rho} d\rho + \frac{\partial \mathbf{r}}{\partial \phi} d\phi + \frac{\partial \mathbf{r}}{\partial z} dz \\ &= d\rho \, \mathbf{e}_\rho + d\phi \, \mathbf{e}_\phi + dz \, \mathbf{e}_z \\ &= d\rho \, \hat{\mathbf{e}}_\rho + \rho \, d\phi \, \hat{\mathbf{e}}_\phi + dz \, \hat{\mathbf{e}}_z. \end{aligned} \tag{10.52}$$

This expression illustrates an important difference between Cartesian and cylindrical polar coordinates (or non-Cartesian coordinates in general). In Cartesian coordinates, the distance moved in going from x to $x + dx$, with y and z held constant, is simply $ds = dx$. However, in cylindrical polars, if ϕ changes by $d\phi$, with ρ and z held constant, then the distance moved is *not* $d\phi$, but $ds = \rho \, d\phi$.

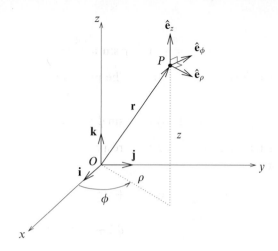

Figure 10.7   Cylindrical polar coordinates ρ, ϕ, z.

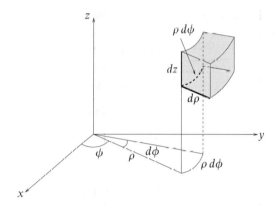

Figure 10.8   The element of volume in cylindrical polar coordinates is given by $\rho\, d\rho\, d\phi\, dz$.

Factors, such as the ρ in $\rho\, d\phi$, that multiply the coordinate differentials to give distances are known as *scale factors*. From (10.52), the scale factors for the ρ-, ϕ- and z- coordinates are therefore 1, ρ and 1 respectively.

The magnitude ds of the displacement $d\mathbf{r}$ is given in cylindrical polar coordinates by

$$(ds)^2 = d\mathbf{r} \cdot d\mathbf{r} = (d\rho)^2 + \rho^2 (d\phi)^2 + (dz)^2,$$

where in the second equality we have used the fact that the basis vectors are orthonormal. We can also find the volume element in a cylindrical polar system (see figure 10.8) by calculating the volume of the infinitesimal parallelepiped

$$
\nabla\Phi = \frac{\partial\Phi}{\partial\rho}\hat{\mathbf{e}}_\rho + \frac{1}{\rho}\frac{\partial\Phi}{\partial\phi}\hat{\mathbf{e}}_\phi + \frac{\partial\Phi}{\partial z}\hat{\mathbf{e}}_z
$$

$$
\nabla\cdot\mathbf{a} = \frac{1}{\rho}\frac{\partial}{\partial\rho}(\rho a_\rho) + \frac{1}{\rho}\frac{\partial a_\phi}{\partial\phi} + \frac{\partial a_z}{\partial z}
$$

$$
\nabla\times\mathbf{a} = \frac{1}{\rho}
\begin{vmatrix}
\hat{\mathbf{e}}_\rho & \rho\hat{\mathbf{e}}_\phi & \hat{\mathbf{e}}_z \\
\dfrac{\partial}{\partial\rho} & \dfrac{\partial}{\partial\phi} & \dfrac{\partial}{\partial z} \\
a_\rho & \rho a_\phi & a_z
\end{vmatrix}
$$

$$
\nabla^2\Phi = \frac{1}{\rho}\frac{\partial}{\partial\rho}\left(\rho\frac{\partial\Phi}{\partial\rho}\right) + \frac{1}{\rho^2}\frac{\partial^2\Phi}{\partial\phi^2} + \frac{\partial^2\Phi}{\partial z^2}
$$

Table 10.2   Vector operators in cylindrical polar coordinates; Φ is a scalar field and $\mathbf{a}$ is a vector field.

defined by the vectors $d\rho\,\hat{\mathbf{e}}_\rho$, $\rho\,d\phi\,\hat{\mathbf{e}}_\phi$ and $dz\,\hat{\mathbf{e}}_z$:

$$
dV = |d\rho\,\hat{\mathbf{e}}_\rho\cdot(\rho\,d\phi\,\hat{\mathbf{e}}_\phi\times dz\,\hat{\mathbf{e}}_z)| = \rho\,d\rho\,d\phi\,dz,
$$

which again uses the fact that the basis vectors are orthonormal. For a simple coordinate system such as cylindrical polars the expressions for $(ds)^2$ and dV are obvious from the geometry.

We will now express the vector operators discussed in this chapter in terms of cylindrical polar coordinates. Let us consider a vector field $\mathbf{a}(\rho,\phi,z)$ and a scalar field $\Phi(\rho,\phi,z)$, where we use Φ for the scalar field to avoid confusion with the azimuthal angle ϕ. We must first write the vector field in terms of the basis vectors of the cylindrical polar coordinate system, i.e.

$$
\mathbf{a} = a_\rho\,\hat{\mathbf{e}}_\rho + a_\phi\,\hat{\mathbf{e}}_\phi + a_z\,\hat{\mathbf{e}}_z,
$$

where a_ρ, a_ϕ and a_z are the components of $\mathbf{a}$ in the ρ-, ϕ- and z- directions respectively. The expressions for grad, div, curl and ∇^2 can then be calculated and are given in table 10.2. Since the derivations of these expressions are rather complicated we leave them until our discussion of general curvilinear coordinates in the next section; the reader could well postpone examination of these formal proofs until some experience of using the expressions has been gained.

▶Express the vector field $\mathbf{a} = yz\,\mathbf{i} - y\,\mathbf{j} + xz^2\,\mathbf{k}$ in cylindrical polar coordinates, and hence calculate its divergence. Show that the same result is obtained by evaluating the divergence in Cartesian coordinates.

The basis vectors of the cylindrical polar coordinate system are given in (10.49)–(10.51). Solving these equations simultaneously for $\mathbf{i}$, $\mathbf{j}$ and $\mathbf{k}$ we obtain

$$
\mathbf{i} = \cos\phi\,\hat{\mathbf{e}}_\rho - \sin\phi\,\hat{\mathbf{e}}_\phi
$$

$$
\mathbf{j} = \sin\phi\,\hat{\mathbf{e}}_\rho + \cos\phi\,\hat{\mathbf{e}}_\phi
$$

$$
\mathbf{k} = \hat{\mathbf{e}}_z.
$$

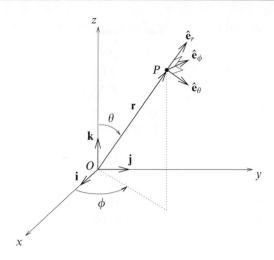

Figure 10.9   Spherical polar coordinates r, θ, ϕ.

Substituting these relations and (10.44) into the expression for $\mathbf{a}$ we find

$$\mathbf{a} = z\rho \sin \phi \, (\cos \phi \, \hat{\mathbf{e}}_\rho - \sin \phi \, \hat{\mathbf{e}}_\phi) - \rho \sin \phi \, (\sin \phi \, \hat{\mathbf{e}}_\rho + \cos \phi \, \hat{\mathbf{e}}_\phi) + z^2 \rho \cos \phi \, \hat{\mathbf{e}}_z$$
$$= (z\rho \sin \phi \cos \phi - \rho \sin^2 \phi) \, \hat{\mathbf{e}}_\rho - (z\rho \sin^2 \phi + \rho \sin \phi \cos \phi) \, \hat{\mathbf{e}}_\phi + z^2 \rho \cos \phi \, \hat{\mathbf{e}}_z.$$

Substituting into the expression for $\nabla \cdot \mathbf{a}$ given in table 10.2,

$$\nabla \cdot \mathbf{a} = 2z \sin \phi \cos \phi - 2 \sin^2 \phi - 2z \sin \phi \cos \phi - \cos^2 \phi + \sin^2 \phi + 2z\rho \cos \phi$$
$$= 2z\rho \cos \phi - 1.$$

Alternatively, and much more quickly in this case, we can calculate the divergence directly in Cartesian coordinates. We obtain

$$\nabla \cdot \mathbf{a} = \frac{\partial a_x}{\partial x} + \frac{\partial a_y}{\partial y} + \frac{\partial a_z}{\partial z} = 2zx - 1,$$

which on substituting $x = \rho \cos \phi$ yields the same result as the calculation in cylindrical polars. ◄

Finally, we note that similar results can be obtained for (two-dimensional) polar coordinates in a plane by omitting the z-dependence. For example, $(ds)^2 = (d\rho)^2 + \rho^2 (d\phi)^2$, while the element of volume is replaced by the element of area $dA = \rho \, d\rho \, d\phi$.

10.9.2 Spherical polar coordinates

As shown in figure 10.9, the position of a point in space P, with Cartesian coordinates x, y, z, may be expressed in terms of spherical polar coordinates r, θ, ϕ, where

$$x = r \sin \theta \cos \phi, \qquad y = r \sin \theta \sin \phi, \qquad z = r \cos \theta, \tag{10.53}$$

and $r \geq 0$, $0 \leq \theta \leq \pi$ and $0 \leq \phi < 2\pi$. The position vector of P may therefore be written as

$$\mathbf{r} = r \sin \theta \cos \phi \, \mathbf{i} + r \sin \theta \sin \phi \, \mathbf{j} + r \cos \theta \, \mathbf{k}.$$

If, in a similar manner to that used in the previous section for cylindrical polars, we find the partial derivatives of $\mathbf{r}$ with respect to r, θ and ϕ respectively and divide each of the resulting vectors by its modulus then we obtain the unit basis vectors

$$\hat{\mathbf{e}}_r = \sin \theta \cos \phi \, \mathbf{i} + \sin \theta \sin \phi \, \mathbf{j} + \cos \theta \, \mathbf{k},$$
$$\hat{\mathbf{e}}_\theta = \cos \theta \cos \phi \, \mathbf{i} + \cos \theta \sin \phi \, \mathbf{j} - \sin \theta \, \mathbf{k},$$
$$\hat{\mathbf{e}}_\phi = - \sin \phi \, \mathbf{i} + \cos \phi \, \mathbf{j}.$$

These unit vectors are in the directions of increasing r, θ and ϕ respectively and are the orthonormal basis set for spherical polar coordinates, as shown in figure 10.9.

A general infinitesimal vector displacement in spherical polars is, from (10.19),

$$d\mathbf{r} = dr \, \hat{\mathbf{e}}_r + r \, d\theta \, \hat{\mathbf{e}}_\theta + r \sin \theta \, d\phi \, \hat{\mathbf{e}}_\phi; \tag{10.54}$$

thus the scale factors for the r-, θ- and ϕ- coordinates are 1, r and $r \sin \theta$ respectively. The magnitude ds of the displacement $d\mathbf{r}$ is given by

$$(ds)^2 = d\mathbf{r} \cdot d\mathbf{r} = (dr)^2 + r^2 (d\theta)^2 + r^2 \sin^2 \theta (d\phi)^2,$$

since the basis vectors form an orthonormal set. The element of volume in spherical polar coordinates (see figure 10.10) is the volume of the infinitesimal parallelepiped defined by the vectors $dr \, \hat{\mathbf{e}}_r$, $r \, d\theta \, \hat{\mathbf{e}}_\theta$ and $r \sin \theta \, d\phi \, \hat{\mathbf{e}}_\phi$ and is given by

$$dV = |dr \, \hat{\mathbf{e}}_r \cdot (r \, d\theta \, \hat{\mathbf{e}}_\theta \times r \sin \theta \, d\phi \, \hat{\mathbf{e}}_\phi)| = r^2 \sin \theta \, dr \, d\theta \, d\phi,$$

where again we use the fact that the basis vectors are orthonormal. The expressions for $(ds)^2$ and dV in spherical polars can be obtained from the geometry of this coordinate system.

We will now express the standard vector operators in spherical polar coordinates, using the same techniques as for cylindrical polar coordinates. We consider a scalar field $\Phi(r, \theta, \phi)$ and a vector field $\mathbf{a}(r, \theta, \phi)$. The latter may be written in terms of the basis vectors of the spherical polar coordinate system as

$$\mathbf{a} = a_r \, \hat{\mathbf{e}}_r + a_\theta \, \hat{\mathbf{e}}_\theta + a_\phi \, \hat{\mathbf{e}}_\phi,$$

where a_r, a_θ and a_ϕ are the components of $\mathbf{a}$ in the r-, θ- and ϕ- directions respectively. The expressions for grad, div, curl and ∇^2 are given in table 10.3. The derivations of these results are given in the next section.

As a final note, we mention that, in the expression for $\nabla^2 \Phi$ given in table 10.3,

$$\nabla\Phi = \frac{\partial\Phi}{\partial r}\hat{\mathbf{e}}_r + \frac{1}{r}\frac{\partial\Phi}{\partial\theta}\hat{\mathbf{e}}_\theta + \frac{1}{r\sin\theta}\frac{\partial\Phi}{\partial\phi}\hat{\mathbf{e}}_\phi$$

$$\nabla\cdot\mathbf{a} = \frac{1}{r^2}\frac{\partial}{\partial r}(r^2 a_r) + \frac{1}{r\sin\theta}\frac{\partial}{\partial\theta}(\sin\theta\, a_\theta) + \frac{1}{r\sin\theta}\frac{\partial a_\phi}{\partial\phi}$$

$$\nabla\times\mathbf{a} = \frac{1}{r^2\sin\theta}\begin{vmatrix} \hat{\mathbf{e}}_r & r\hat{\mathbf{e}}_\theta & r\sin\theta\,\hat{\mathbf{e}}_\phi \\ \dfrac{\partial}{\partial r} & \dfrac{\partial}{\partial\theta} & \dfrac{\partial}{\partial\phi} \\ a_r & ra_\theta & r\sin\theta\, a_\phi \end{vmatrix}$$

$$\nabla^2\Phi = \frac{1}{r^2}\frac{\partial}{\partial r}\left(r^2\frac{\partial\Phi}{\partial r}\right) + \frac{1}{r^2\sin\theta}\frac{\partial}{\partial\theta}\left(\sin\theta\frac{\partial\Phi}{\partial\theta}\right) + \frac{1}{r^2\sin^2\theta}\frac{\partial^2\Phi}{\partial\phi^2}$$

Table 10.3   Vector operators in spherical polar coordinates; Φ is a scalar field and $\mathbf{a}$ is a vector field.

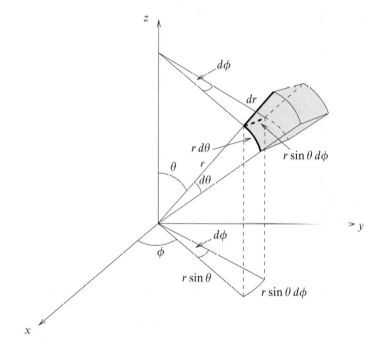

Figure 10.10   The element of volume in spherical polar coordinates is given by $r^2\sin\theta\, dr\, d\theta\, d\phi$.

we can rewrite the first term on the RHS as follows:

$$\frac{1}{r^2}\frac{\partial}{\partial r}\left(r^2\frac{\partial\Phi}{\partial r}\right) = \frac{1}{r}\frac{\partial^2}{\partial r^2}(r\Phi),$$

which can often be useful in shortening calculations.

10.10 General curvilinear coordinates

As indicated earlier, the contents of this section are more formal and technically complicated than hitherto. The section could be omitted until the reader has had some experience of using its results.

Cylindrical and spherical polars are just two examples of what are called *general curvilinear coordinates*. In the general case, the position of a point P having Cartesian coordinates x, y, z may be expressed in terms of three curvilinear coordinates u_1, u_2, u_3, where

$$x = x(u_1, u_2, u_3), \qquad y = y(u_1, u_2, u_3), \qquad z = z(u_1, u_2, u_3),$$

and similarly

$$u_1 = u_1(x, y, z), \qquad u_2 = u_2(x, y, z), \qquad u_3 = u_3(x, y, z).$$

We assume that all these functions are continuous, differentiable and have a single-valued inverse, except perhaps at or on certain isolated points or lines, so that there is a one-to-one correspondence between the x, y, z and u_1, u_2, u_3 systems. The u_1-, u_2- and u_3- coordinate curves of a general curvilinear system are analogous to the x-, y- and z- axes of Cartesian coordinates. The surfaces $u_1 = c_1$, $u_2 = c_2$ and $u_3 = c_3$, where c_1, c_2, c_3 are constants, are called the *coordinate surfaces* and each pair of these surfaces has its intersection in a curve called a *coordinate curve* or *line* (see figure 10.11).

If at each point in space the three coordinate surfaces passing through the point meet at right angles then the curvilinear coordinate system is called *orthogonal*. For example, in spherical polars $u_1 = r$, $u_2 = \theta$, $u_3 = \phi$ and the three coordinate surfaces passing through the point (R, Θ, Φ) are the sphere $r = R$, the circular cone $\theta = \Theta$ and the plane $\phi = \Phi$, which intersect at right angles at that point. Therefore spherical polars form an orthogonal coordinate system (as do cylindrical polars) .

If $\mathbf{r}(u_1, u_2, u_3)$ is the position vector of the point P then $\mathbf{e}_1 = \partial \mathbf{r}/\partial u_1$ is a vector tangent to the u_1-curve at P (for which u_2 and u_3 are constants) in the direction of increasing u_1. Similarly, $\mathbf{e}_2 = \partial \mathbf{r}/\partial u_2$ and $\mathbf{e}_3 = \partial \mathbf{r}/\partial u_3$ are vectors tangent to the u_2- and u_3- curves at P in the direction of increasing u_2 and u_3 respectively. Denoting the lengths of these vectors by h_1, h_2 and h_3, the *unit* vectors in each of these directions are given by

$$\hat{\mathbf{e}}_1 = \frac{1}{h_1} \frac{\partial \mathbf{r}}{\partial u_1}, \qquad \hat{\mathbf{e}}_2 = \frac{1}{h_2} \frac{\partial \mathbf{r}}{\partial u_2}, \qquad \hat{\mathbf{e}}_3 = \frac{1}{h_3} \frac{\partial \mathbf{r}}{\partial u_3},$$

where $h_1 = |\partial \mathbf{r}/\partial u_1|$, $h_2 = |\partial \mathbf{r}/\partial u_2|$ and $h_3 = |\partial \mathbf{r}/\partial u_3|$.

The quantities h_1, h_2, h_3 are the scale factors of the curvilinear coordinate system. The element of distance associated with an infinitesimal change du_i in one of the coordinates is $h_i \, du_i$. In the previous section we found that the scale

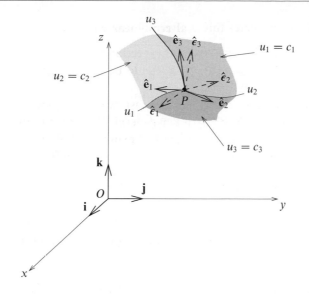

Figure 10.11   General curvilinear coordinates.

factors for cylindrical and spherical polar coordinates were

$$\text{for cylindrical polars} \quad h_\rho = 1, \quad h_\phi = \rho, \quad h_z = 1,$$
$$\text{for spherical polars} \quad h_r = 1, \quad h_\theta = r, \quad h_\phi = r \sin \theta.$$

Although the vectors $\mathbf{e}_1$, $\mathbf{e}_2$, $\mathbf{e}_3$ form a perfectly good basis for the curvilinear coordinate system, it is usual to work with the corresponding unit vectors $\hat{\mathbf{e}}_1$, $\hat{\mathbf{e}}_2$, $\hat{\mathbf{e}}_3$. For an orthogonal curvilinear coordinate system these unit vectors form an orthonormal basis.

An infinitesimal vector displacement in general curvilinear coordinates is given by, from (10.19),

$$d\mathbf{r} = \frac{\partial \mathbf{r}}{\partial u_1} \, du_1 + \frac{\partial \mathbf{r}}{\partial u_2} \, du_2 + \frac{\partial \mathbf{r}}{\partial u_3} \, du_3 \tag{10.55}$$

$$= du_1 \, \mathbf{e}_1 + du_2 \, \mathbf{e}_2 + du_3 \, \mathbf{e}_3 \tag{10.56}$$

$$= h_1 \, du_1 \, \hat{\mathbf{e}}_1 + h_2 \, du_2 \, \hat{\mathbf{e}}_2 + h_3 \, du_3 \, \hat{\mathbf{e}}_3. \tag{10.57}$$

In the case of *orthogonal* curvilinear coordinates, where the $\hat{\mathbf{e}}_i$ are mutually perpendicular, the element of arc length is given by

$$(ds)^2 = d\mathbf{r} \cdot d\mathbf{r} = h_1^2 (du_1)^2 + h_2^2 (du_2)^2 + h_3^2 (du_3)^2. \tag{10.58}$$

The volume element for the coordinate system is the volume of the infinitesimal parallelepiped defined by the vectors $(\partial \mathbf{r} / \partial u_i) \, du_i = du_i \, \mathbf{e}_i = h_i \, du_i \, \hat{\mathbf{e}}_i$, for $i = 1, 2, 3$.

For orthogonal coordinates this is given by

$$dV = |du_1 \, \mathbf{e}_1 \cdot (du_2 \, \mathbf{e}_2 \times du_3 \, \mathbf{e}_3)|$$
$$= |h_1 \, \hat{\mathbf{e}}_1 \cdot (h_2 \, \hat{\mathbf{e}}_2 \times h_3 \, \hat{\mathbf{e}}_3)| \, du_1 \, du_2 \, du_3$$
$$= h_1 h_2 h_3 \, du_1 \, du_2 \, du_3.$$

Now, in addition to the set $\{\hat{\mathbf{e}}_i\}$, $i = 1, 2, 3$, there exists another useful set of three unit basis vectors at P. Since ∇u_1 is a vector normal to the surface $u_1 = c_1$, a unit vector in this direction is $\hat{\boldsymbol{\epsilon}}_1 = \nabla u_1/|\nabla u_1|$. Similarly, $\hat{\boldsymbol{\epsilon}}_2 = \nabla u_2/|\nabla u_2|$ and $\hat{\boldsymbol{\epsilon}}_3 = \nabla u_3/|\nabla u_3|$ are unit vectors normal to the surfaces $u_2 = c_2$ and $u_3 = c_3$ respectively.

Therefore at each point P in a curvilinear coordinate system, there exist, in general, two sets of unit vectors: $\{\hat{\mathbf{e}}_i\}$, tangent to the coordinate curves, and $\{\hat{\boldsymbol{\epsilon}}_i\}$, normal to the coordinate surfaces. A vector $\mathbf{a}$ can be written in terms of either set of unit vectors:

$$\mathbf{a} = a_1 \hat{\mathbf{e}}_1 + a_2 \hat{\mathbf{e}}_2 + a_3 \hat{\mathbf{e}}_3 = A_1 \hat{\boldsymbol{\epsilon}}_1 + A_2 \hat{\boldsymbol{\epsilon}}_2 + A_3 \hat{\boldsymbol{\epsilon}}_3,$$

where a_1, a_2, a_3 and A_1, A_2, A_3 are the components of $\mathbf{a}$ in the two systems. It may be shown that the two bases become identical if the coordinate system is orthogonal.

Instead of the *unit* vectors discussed above, we could instead work directly with the two sets of vectors $\{\mathbf{e}_i = \partial \mathbf{r}/\partial u_i\}$ and $\{\boldsymbol{\epsilon}_i = \nabla u_i\}$, which are not, in general, of unit length. We can then write a vector $\mathbf{a}$ as

$$\mathbf{a} = \alpha_1 \mathbf{e}_1 + \alpha_2 \mathbf{e}_2 + \alpha_3 \mathbf{e}_3 = \beta_1 \boldsymbol{\epsilon}_1 + \beta_2 \boldsymbol{\epsilon}_2 + \beta_3 \boldsymbol{\epsilon}_3,$$

or more explicitly as

$$\mathbf{a} = \alpha_1 \frac{\partial \mathbf{r}}{\partial u_1} + \alpha_2 \frac{\partial \mathbf{r}}{\partial u_2} + \alpha_3 \frac{\partial \mathbf{r}}{\partial u_3} = \beta_1 \nabla u_1 + \beta_2 \nabla u_2 + \beta_3 \nabla u_3,$$

where α_1, α_2, α_3 and β_1, β_2, β_3 are called the *contravariant* and *covariant* components of $\mathbf{a}$ respectively. A more detailed discussion of these components, in the context of tensor analysis, is given in chapter 26. The (in general) non-unit bases $\{\mathbf{e}_i\}$ and $\{\boldsymbol{\epsilon}_i\}$ are often the most natural bases in which to express vector quantities.

▶ *Show that $\{\mathbf{e}_i\}$ and $\{\boldsymbol{\epsilon}_i\}$ are reciprocal systems of vectors.*

Let us consider the scalar product $\mathbf{e}_i \cdot \boldsymbol{\epsilon}_j$; using the Cartesian expressions for $\mathbf{r}$ and ∇, we obtain

$$\mathbf{e}_i \cdot \boldsymbol{\epsilon}_j = \frac{\partial \mathbf{r}}{\partial u_i} \cdot \nabla u_j$$

$$= \left(\frac{\partial x}{\partial u_i} \mathbf{i} + \frac{\partial y}{\partial u_i} \mathbf{j} + \frac{\partial z}{\partial u_i} \mathbf{k} \right) \cdot \left(\frac{\partial u_j}{\partial x} \mathbf{i} + \frac{\partial u_j}{\partial y} \mathbf{j} + \frac{\partial u_j}{\partial z} \mathbf{k} \right)$$

$$= \frac{\partial x}{\partial u_i} \frac{\partial u_j}{\partial x} + \frac{\partial y}{\partial u_i} \frac{\partial u_j}{\partial y} + \frac{\partial z}{\partial u_i} \frac{\partial u_j}{\partial z} = \frac{\partial u_j}{\partial u_i}.$$

In the last step we have used the chain rule for partial differentiation. Therefore $\mathbf{e}_i \cdot \boldsymbol{\epsilon}_j = 1$ if $i = j$, and $\mathbf{e}_i \cdot \boldsymbol{\epsilon}_j = 0$ otherwise. Hence $\{\mathbf{e}_i\}$ and $\{\boldsymbol{\epsilon}_j\}$ are reciprocal systems of vectors. ◀

We now derive expressions for the standard vector operators in *orthogonal* curvilinear coordinates. Despite the useful properties of the non-unit bases discussed above, the remainder of our discussion in this section will be in terms of the unit basis vectors $\{\hat{\mathbf{e}}_i\}$. The expressions for the vector operators in cylindrical and spherical polar coordinates given in tables 10.2 and 10.3 respectively can be found from those derived below by inserting the appropriate scale factors.

Gradient

The change $d\Phi$ in a scalar field Φ resulting from changes du_1, du_2, du_3 in the coordinates u_1, u_2, u_3 is given by, from (5.5),

$$d\Phi = \frac{\partial \Phi}{\partial u_1} du_1 + \frac{\partial \Phi}{\partial u_2} du_2 + \frac{\partial \Phi}{\partial u_3} du_3.$$

For orthogonal curvilinear coordinates u_1, u_2, u_3 we find from (10.57), and comparison with (10.27), that we can write this as

$$d\Phi = \nabla\Phi \cdot d\mathbf{r}, \tag{10.59}$$

where $\nabla\Phi$ is given by

$$\nabla\Phi = \frac{1}{h_1} \frac{\partial \Phi}{\partial u_1} \hat{\mathbf{e}}_1 + \frac{1}{h_2} \frac{\partial \Phi}{\partial u_2} \hat{\mathbf{e}}_2 + \frac{1}{h_3} \frac{\partial \Phi}{\partial u_3} \hat{\mathbf{e}}_3. \tag{10.60}$$

This implies that the del operator can be written as

$$\nabla = \frac{\hat{\mathbf{e}}_1}{h_1} \frac{\partial}{\partial u_1} + \frac{\hat{\mathbf{e}}_2}{h_2} \frac{\partial}{\partial u_2} + \frac{\hat{\mathbf{e}}_3}{h_3} \frac{\partial}{\partial u_3}$$

when acting on scalar (but not vector) quantities. See also, Section 26.20

▶*Show that for orthogonal curvilinear coordinates* $\nabla u_i = \hat{\mathbf{e}}_i/h_i$. *Hence show that the two sets of vectors* $\{\hat{\mathbf{e}}_i\}$ *and* $\{\hat{\boldsymbol{\epsilon}}_i\}$ *are identical in this case.*

Letting $\Phi = u_i$ in (10.60) we find immediately that $\nabla u_i = \hat{\mathbf{e}}_i/h_i$. Therefore $|\nabla u_i| = 1/h_i$, and so $\hat{\boldsymbol{\epsilon}}_i = \nabla u_i/|\nabla u_i| = h_i \nabla u_i = \hat{\mathbf{e}}_i$. ◀

Divergence

In order to derive the expression for the divergence of a vector field in orthogonal curvilinear coordinates, we must first write the vector field in terms of the basis vectors of the coordinate system:

$$\mathbf{a} = a_1 \hat{\mathbf{e}}_1 + a_2 \hat{\mathbf{e}}_2 + a_3 \hat{\mathbf{e}}_3.$$

The divergence is then given by

$$\nabla \cdot \mathbf{a} = \frac{1}{h_1 h_2 h_3} \left[\frac{\partial}{\partial u_1}(h_2 h_3 a_1) + \frac{\partial}{\partial u_2}(h_3 h_1 a_2) + \frac{\partial}{\partial u_3}(h_1 h_2 a_3) \right]. \tag{10.61}$$

▶*Prove the expression for $\nabla \cdot \mathbf{a}$ in orthogonal curvilinear coordinates.*

Let us consider the sub-expression $\nabla \cdot (a_1 \hat{\mathbf{e}}_1)$. Now $\hat{\mathbf{e}}_1 = \hat{\mathbf{e}}_2 \times \hat{\mathbf{e}}_3 = h_2 \nabla u_2 \times h_3 \nabla u_3$. Therefore

$$\nabla \cdot (a_1 \hat{\mathbf{e}}_1) = \nabla \cdot (a_1 h_2 h_3 \nabla u_2 \times \nabla u_3),$$
$$= \nabla(a_1 h_2 h_3) \cdot (\nabla u_2 \times \nabla u_3) + a_1 h_2 h_3 \nabla \cdot (\nabla u_2 \times \nabla u_3).$$

However, $\nabla \cdot (\nabla u_2 \times \nabla u_3) = 0$, from (10.43), so we obtain

$$\nabla \cdot (a_1 \hat{\mathbf{e}}_1) = \nabla(a_1 h_2 h_3) \cdot \left(\frac{\hat{\mathbf{e}}_2}{h_2} \times \frac{\hat{\mathbf{e}}_3}{h_3} \right) = \nabla(a_1 h_2 h_3) \cdot \frac{\hat{\mathbf{e}}_1}{h_2 h_3};$$

letting $\Phi = a_1 h_2 h_3$ in (10.60) and substituting into the above equation, we find

$$\nabla \cdot (a_1 \hat{\mathbf{e}}_1) = \frac{1}{h_1 h_2 h_3} \frac{\partial}{\partial u_1} (a_1 h_2 h_3).$$

Repeating the analysis for $\nabla \cdot (a_2 \hat{\mathbf{e}}_2)$ and $\nabla \cdot (a_3 \hat{\mathbf{e}}_3)$, and adding the results we obtain (10.61), as required. ◀

Laplacian

In the expression for the divergence (10.61), let

$$\mathbf{a} = \nabla \Phi = \frac{1}{h_1} \frac{\partial \Phi}{\partial u_1} \hat{\mathbf{e}}_1 + \frac{1}{h_2} \frac{\partial \Phi}{\partial u_2} \hat{\mathbf{e}}_2 + \frac{1}{h_3} \frac{\partial \Phi}{\partial u_3} \hat{\mathbf{e}}_3,$$

where we have used (10.60). We then obtain

$$\nabla^2 \Phi = \frac{1}{h_1 h_2 h_3} \left[\frac{\partial}{\partial u_1} \left(\frac{h_2 h_3}{h_1} \frac{\partial \Phi}{\partial u_1} \right) + \frac{\partial}{\partial u_2} \left(\frac{h_3 h_1}{h_2} \frac{\partial \Phi}{\partial u_2} \right) + \frac{\partial}{\partial u_3} \left(\frac{h_1 h_2}{h_3} \frac{\partial \Phi}{\partial u_3} \right) \right],$$

which is the expression for the Laplacian in orthogonal curvilinear coordinates.

Curl

The curl of a vector field $\mathbf{a} = a_1 \hat{\mathbf{e}}_1 + a_2 \hat{\mathbf{e}}_2 + a_3 \hat{\mathbf{e}}_3$ in orthogonal curvilinear coordinates is given by

$$\nabla \times \mathbf{a} = \frac{1}{h_1 h_2 h_3} \begin{vmatrix} h_1 \hat{\mathbf{e}}_1 & h_2 \hat{\mathbf{e}}_2 & h_3 \hat{\mathbf{e}}_3 \\ \dfrac{\partial}{\partial u_1} & \dfrac{\partial}{\partial u_2} & \dfrac{\partial}{\partial u_3} \\ h_1 a_1 & h_2 a_2 & h_3 a_3 \end{vmatrix}. \tag{10.62}$$

▶*Prove the expression for $\nabla \times \mathbf{a}$ in orthogonal curvilinear coordinates.*

Let us consider the sub-expression $\nabla \times (a_1 \hat{\mathbf{e}}_1)$. Since $\hat{\mathbf{e}}_1 = h_1 \nabla u_1$ we have

$$\nabla \times (a_1 \hat{\mathbf{e}}_1) = \nabla \times (a_1 h_1 \nabla u_1),$$
$$= \nabla(a_1 h_1) \times \nabla u_1 + a_1 h_1 \nabla \times \nabla u_1.$$

But $\nabla \times \nabla u_1 = 0$, so we obtain

$$\nabla \times (a_1 \hat{\mathbf{e}}_1) = \nabla(a_1 h_1) \times \frac{\hat{\mathbf{e}}_1}{h_1}.$$

$$
\nabla\Phi \;=\; \frac{1}{h_1}\frac{\partial\Phi}{\partial u_1}\hat{\mathbf{e}}_1 + \frac{1}{h_2}\frac{\partial\Phi}{\partial u_2}\hat{\mathbf{e}}_2 + \frac{1}{h_3}\frac{\partial\Phi}{\partial u_3}\hat{\mathbf{e}}_3
$$

$$
\nabla\cdot\mathbf{a} \;=\; \frac{1}{h_1 h_2 h_3}\left[\frac{\partial}{\partial u_1}(h_2 h_3 a_1) + \frac{\partial}{\partial u_2}(h_3 h_1 a_2) + \frac{\partial}{\partial u_3}(h_1 h_2 a_3)\right]
$$

$$
\nabla\times\mathbf{a} \;=\; \frac{1}{h_1 h_2 h_3}\begin{vmatrix} h_1\hat{\mathbf{e}}_1 & h_2\hat{\mathbf{e}}_2 & h_3\hat{\mathbf{e}}_3 \\[2pt] \dfrac{\partial}{\partial u_1} & \dfrac{\partial}{\partial u_2} & \dfrac{\partial}{\partial u_3} \\[2pt] h_1 a_1 & h_2 a_2 & h_3 a_3 \end{vmatrix}
$$

$$
\nabla^2\Phi \;=\; \frac{1}{h_1 h_2 h_3}\left[\frac{\partial}{\partial u_1}\left(\frac{h_2 h_3}{h_1}\frac{\partial\Phi}{\partial u_1}\right) + \frac{\partial}{\partial u_2}\left(\frac{h_3 h_1}{h_2}\frac{\partial\Phi}{\partial u_2}\right) + \frac{\partial}{\partial u_3}\left(\frac{h_1 h_2}{h_3}\frac{\partial\Phi}{\partial u_3}\right)\right]
$$

Table 10.4   Vector operators in orthogonal curvilinear coordinates u_1, u_2, u_3. Φ is a scalar field and $\mathbf{a}$ is a vector field.

Letting $\Phi = a_1 h_1$ in (10.60) and substituting into the above equation, we find

$$
\nabla\times(a_1\hat{\mathbf{e}}_1) = \frac{\hat{\mathbf{e}}_2}{h_3 h_1}\frac{\partial}{\partial u_3}(a_1 h_1) - \frac{\hat{\mathbf{e}}_3}{h_1 h_2}\frac{\partial}{\partial u_2}(a_1 h_1).
$$

The corresponding analysis of $\nabla\times(a_2\hat{\mathbf{e}}_2)$ produces terms in $\hat{\mathbf{e}}_3$ and $\hat{\mathbf{e}}_1$, whilst that of $\nabla\times(a_3\hat{\mathbf{e}}_3)$ produces terms in $\hat{\mathbf{e}}_1$ and $\hat{\mathbf{e}}_2$. When the three results are added together, the coefficients multiplying $\hat{\mathbf{e}}_1$, $\hat{\mathbf{e}}_2$ and $\hat{\mathbf{e}}_3$ are the same as those obtained by writing out (10.62) explicitly, thus proving the stated result. ◄

The general expressions for the vector operators in orthogonal curvilinear coordinates are shown for reference in table 10.4. The explicit results for cylindrical and spherical polar coordinates, given in tables 10.2 and 10.3 respectively, are obtained by substituting the appropriate set of scale factors in each case.

A discussion of the expressions for vector operators in tensor form, which are valid even for non-orthogonal curvilinear coordinate systems, is given in chapter 26.

10.11 Exercises

10.1   Evaluate the integral

$$
\int \left[\mathbf{a}(\dot{\mathbf{b}}\cdot\mathbf{a} + \mathbf{b}\cdot\dot{\mathbf{a}}) + \dot{\mathbf{a}}(\mathbf{b}\cdot\mathbf{a}) - 2(\dot{\mathbf{a}}\cdot\mathbf{a})\mathbf{b} - \dot{\mathbf{b}}|\mathbf{a}|^2\right]\,dt
$$

in which $\dot{\mathbf{a}}$, $\dot{\mathbf{b}}$ are the derivatives of $\mathbf{a}$, $\mathbf{b}$ with respect to t.

10.2   At time $t = 0$, the vectors $\mathbf{E}$ and $\mathbf{B}$ are given by $\mathbf{E} = \mathbf{E}_0$ and $\mathbf{B} = \mathbf{B}_0$, where the unit vectors, $\mathbf{E}_0$ and $\mathbf{B}_0$ are fixed and orthogonal. The equations of motion are

$$
\frac{d\mathbf{E}}{dt} = \mathbf{E}_0 + \mathbf{B}\times\mathbf{E}_0,
$$

$$
\frac{d\mathbf{B}}{dt} = \mathbf{B}_0 + \mathbf{E}\times\mathbf{B}_0.
$$

Find $\mathbf{E}$ and $\mathbf{B}$ at a general time t, showing that after a long time the directions of $\mathbf{E}$ and $\mathbf{B}$ have almost interchanged.

10.3   The general equation of motion of a (non-relativistic) particle of mass m and charge q when it is placed in a region where there is a magnetic field $\mathbf{B}$ and an electric field $\mathbf{E}$ is

$$m\ddot{\mathbf{r}} = q(\mathbf{E} + \dot{\mathbf{r}} \times \mathbf{B});$$

here $\mathbf{r}$ is the position of the particle at time t and $\dot{\mathbf{r}} = d\mathbf{r}/dt$, etc. Write this as three separate equations in terms of the Cartesian components of the vectors involved.

For the simple case of crossed uniform fields $\mathbf{E} = E\mathbf{i}$, $\mathbf{B} = B\mathbf{j}$, in which the particle starts from the origin at $t = 0$ with $\dot{\mathbf{r}} = v_0\mathbf{k}$, find the equations of motion and show the following:

(a) if $v_0 = E/B$ then the particle continues its initial motion;
(b) if $v_0 = 0$ then the particle follows the space curve given in terms of the parameter ξ by

$$x = \frac{mE}{B^2q}(1 - \cos\xi), \qquad y = 0, \qquad z = \frac{mE}{B^2q}(\xi - \sin\xi).$$

Interpret this curve geometrically and relate ξ to t. Show that the total distance travelled by the particle after time t is given by

$$\frac{2E}{B}\int_0^t \left|\sin\frac{Bqt'}{2m}\right| dt'.$$

10.4   Use vector methods to find the maximum angle to the horizontal at which a stone may be thrown so as to ensure that it is always moving away from the thrower.

10.5   If two systems of coordinates with a common origin O are rotating with respect to each other, the measured accelerations differ in the two systems. Denoting by $\mathbf{r}$ and $\mathbf{r}'$ position vectors in frames $OXYZ$ and $OX'Y'Z'$, respectively, the connection between the two is

$$\ddot{\mathbf{r}}' = \ddot{\mathbf{r}} + \dot{\boldsymbol{\omega}} \times \mathbf{r} + 2\boldsymbol{\omega} \times \dot{\mathbf{r}} + \boldsymbol{\omega} \times (\boldsymbol{\omega} \times \mathbf{r}),$$

where $\boldsymbol{\omega}$ is the angular velocity vector of the rotation of $OXYZ$ with respect to $OX'Y'Z'$ (taken as fixed). The third term on the RHS is known as the Coriolis acceleration, whilst the final term gives rise to a centrifugal force.

Consider the application of this result to the firing of a shell of mass m from a stationary ship on the steadily rotating earth, working to the first order in ω ($= 7.3 \times 10^{-5}\,\mathrm{rad\,s^{-1}}$). If the shell is fired with velocity $\mathbf{v}$ at time $t = 0$ and only reaches a height that is small compared with the radius of the earth, show that its acceleration, as recorded on the ship, is given approximately by

$$\ddot{\mathbf{r}} = \mathbf{g} - 2\boldsymbol{\omega} \times (\mathbf{v} + \mathbf{g}t),$$

where $m\mathbf{g}$ is the weight of the shell measured on the ship's deck.

The shell is fired at another stationary ship (a distance $\mathbf{s}$ away) and $\mathbf{v}$ is such that the shell would have hit its target had there been no Coriolis effect.

(a) Show that without the Coriolis effect the time of flight of the shell would have been $\tau = -2\mathbf{g} \cdot \mathbf{v}/g^2$.
(b) Show further that when the shell actually hits the sea it is off-target by approximately

$$\frac{2\tau}{g^2}[(\mathbf{g} \times \boldsymbol{\omega}) \cdot \mathbf{v}](\mathbf{g}\tau + \mathbf{v}) - (\boldsymbol{\omega} \times \mathbf{v})\tau^2 - \frac{1}{3}(\boldsymbol{\omega} \times \mathbf{g})\tau^3.$$

(c) Estimate the order of magnitude Δ of this miss for a shell for which the initial speed v is $300\ \mathrm{m\,s^{-1}}$, firing close to its maximum range ($\mathbf{v}$ makes an angle of $\pi/4$ with the vertical) in a northerly direction, whilst the ship is stationed at latitude $45°$ North.

10.6　Prove that for a space curve $\mathbf{r} = \mathbf{r}(s)$, where s is the arc length measured along the curve from a fixed point, the triple scalar product

$$\left(\frac{d\mathbf{r}}{ds} \times \frac{d^2\mathbf{r}}{ds^2}\right) \cdot \frac{d^3\mathbf{r}}{ds^3}$$

at any point on the curve has the value $\kappa^2\tau$, where κ is the curvature and τ the torsion at that point.

10.7　For the twisted space curve $y^3 + 27axz - 81a^2y = 0$, given parametrically by

$$x = au(3 - u^2), \qquad y = 3au^2, \qquad z = au(3 + u^2),$$

show that the following hold:

(a) $ds/du = 3\sqrt{2}a(1+u^2)$, where s is the distance along the curve measured from the origin;
(b) the length of the curve from the origin to the Cartesian point $(2a, 3a, 4a)$ is $4\sqrt{2}a$;
(c) the radius of curvature at the point with parameter u is $3a(1 + u^2)^2$;
(d) the torsion τ and curvature κ at a general point are equal;
(e) any of the Frenet–Serret formulae that you have not already used directly are satisfied.

10.8　The shape of the curving slip road joining two motorways, that cross at right angles and are at vertical heights $z = 0$ and $z = h$, can be approximated by the space curve

$$\mathbf{r} = \frac{\sqrt{2h}}{\pi} \ln\cos\left(\frac{z\pi}{2h}\right)\mathbf{i} + \frac{\sqrt{2h}}{\pi} \ln\sin\left(\frac{z\pi}{2h}\right)\mathbf{j} + z\mathbf{k}.$$

Show that the radius of curvature ρ of the slip road is $(2h/\pi)\operatorname{cosec}(z\pi/h)$ at height z and that the torsion $\tau = -1/\rho$. To shorten the algebra, set $z = 2h\theta/\pi$ and use θ as the parameter.

10.9　In a magnetic field, field lines are curves to which the magnetic induction $\mathbf{B}$ is everywhere tangential. By evaluating $d\mathbf{B}/ds$, where s is the distance measured along a field line, prove that the radius of curvature at any point on a line is given by

$$\rho = \frac{B^3}{|\mathbf{B} \times (\mathbf{B} \cdot \nabla)\mathbf{B}|}.$$

10.10　Find the areas of the given surfaces using parametric coordinates.

(a) Using the parameterisation $x = u\cos\phi$, $y = u\sin\phi$, $z = u\cot\Omega$, find the sloping surface area of a right circular cone of semi-angle Ω whose base has radius a. Verify that it is equal to $\frac{1}{2} \times$ perimeter of the base $\times$ slope height.
(b) Using the same parameterization as in (a) for x and y, and an appropriate choice for z, find the surface area between the planes $z = 0$ and $z = Z$ of the paraboloid of revolution $z = \alpha(x^2 + y^2)$.

10.11　Parameterising the hyperboloid

$$\frac{x^2}{a^2} + \frac{y^2}{b^2} - \frac{z^2}{c^2} = 1$$

by $x = a\cos\theta\sec\phi$, $y = b\sin\theta\sec\phi$, $z = c\tan\phi$, show that an area element on its surface is

$$dS = \sec^2\phi \left[c^2\sec^2\phi \left(b^2\cos^2\theta + a^2\sin^2\theta\right) + a^2b^2\tan^2\phi\right]^{1/2} d\theta\,d\phi.$$

371

Use this formula to show that the area of the curved surface $x^2 + y^2 - z^2 = a^2$ between the planes $z = 0$ and $z = 2a$ is

$$\pi a^2 \left(6 + \frac{1}{\sqrt{2}} \sinh^{-1} 2\sqrt{2} \right).$$

10.12   For the function

$$z(x, y) = (x^2 - y^2)e^{-x^2 - y^2},$$

find the location(s) at which the steepest gradient occurs. What are the magnitude and direction of that gradient? The algebra involved is easier if plane polar coordinates are used.

10.13   Verify by direct calculation that

$$\nabla \cdot (\mathbf{a} \times \mathbf{b}) = \mathbf{b} \cdot (\nabla \times \mathbf{a}) - \mathbf{a} \cdot (\nabla \times \mathbf{b}).$$

10.14   In the following exercises, $\mathbf{a}$, $\mathbf{b}$ and $\mathbf{c}$ are vector fields.

(a) Simplify

$$\nabla \times \mathbf{a}(\nabla \cdot \mathbf{a}) \;+\; \mathbf{a} \times [\nabla \times (\nabla \times \mathbf{a})] \;+\; \mathbf{a} \times \nabla^2 \mathbf{a}.$$

(b) By explicitly writing out the terms in Cartesian coordinates, prove that

$$[\mathbf{c} \cdot (\mathbf{b} \cdot \nabla) - \mathbf{b} \cdot (\mathbf{c} \cdot \nabla)] \, \mathbf{a} = (\nabla \times \mathbf{a}) \cdot (\mathbf{b} \times \mathbf{c}).$$

(c) Prove that $\mathbf{a} \times (\nabla \times \mathbf{a}) = \nabla(\tfrac{1}{2}a^2) - (\mathbf{a} \cdot \nabla)\mathbf{a}$.

10.15   Evaluate the Laplacian of the function

$$\psi(x, y, z) = \frac{zx^2}{x^2 + y^2 + z^2}$$

(a) directly in Cartesian coordinates, and (b) after changing to a spherical polar coordinate system. Verify that, as they must, the two methods give the same result.

10.16   Verify that (10.42) is valid for each component separately when $\mathbf{a}$ is the Cartesian vector $x^2 y\,\mathbf{i} + xyz\,\mathbf{j} + z^2 y\,\mathbf{k}$, by showing that each side of the equation is equal to $z\,\mathbf{i} + (2x + 2z)\,\mathbf{j} + x\,\mathbf{k}$.

10.17   The (Maxwell) relationship between a time-independent magnetic field $\mathbf{B}$ and the current density $\mathbf{J}$ (measured in SI units in $\mathrm{A\,m^{-2}}$) producing it,

$$\nabla \times \mathbf{B} = \mu_0 \mathbf{J},$$

can be applied to a long cylinder of conducting ionised gas which, in cylindrical polar coordinates, occupies the region $\rho < a$.

(a) Show that a uniform current density $(0, C, 0)$ and a magnetic field $(0, 0, B)$, with B constant $(= B_0)$ for $\rho > a$ and $B = B(\rho)$ for $\rho < a$, are consistent with this equation. Given that $B(0) = 0$ and that $\mathbf{B}$ is continuous at $\rho = a$, obtain expressions for C and $B(\rho)$ in terms of B_0 and a.

(b) The magnetic field can be expressed as $\mathbf{B} = \nabla \times \mathbf{A}$, where $\mathbf{A}$ is known as the vector potential. Show that a suitable $\mathbf{A}$ that has only one non-vanishing component, $A_\phi(\rho)$, can be found, and obtain explicit expressions for $A_\phi(\rho)$ for both $\rho < a$ and $\rho > a$. Like $\mathbf{B}$, the vector potential is continuous at $\rho = a$.

(c) The gas pressure $p(\rho)$ satisfies the hydrostatic equation $\nabla p = \mathbf{J} \times \mathbf{B}$ and vanishes at the outer wall of the cylinder. Find a general expression for p.

10.18   Evaluate the Laplacian of a vector field using two different coordinate systems as follows.

(a) For cylindrical polar coordinates ρ, ϕ, z, evaluate the derivatives of the three unit vectors with respect to each of the coordinates, showing that only $\partial\hat{\mathbf{e}}_\rho/\partial\phi$ and $\partial\hat{\mathbf{e}}_\phi/\partial\phi$ are non-zero.

  (i) Hence evaluate $\nabla^2\mathbf{a}$ when $\mathbf{a}$ is the vector $\hat{\mathbf{e}}_\rho$, i.e. a vector of unit magnitude everywhere directed radially outwards and expressed by $a_\rho = 1$, $a_\phi = a_z = 0$.

  (ii) Note that it is trivially obvious that $\nabla \times \mathbf{a} = \mathbf{0}$ and hence that equation (10.41) requires that $\nabla(\nabla \cdot \mathbf{a}) = \nabla^2\mathbf{a}$.

  (iii) Evaluate $\nabla(\nabla \cdot \mathbf{a})$ and show that the latter equation holds, but that

$$[\nabla(\nabla \cdot \mathbf{a})]_\rho \neq \nabla^2 a_\rho.$$

(b) Rework the same problem in Cartesian coordinates (where, as it happens, the algebra is more complicated).

10.19 Maxwell's equations for electromagnetism in free space (i.e. in the absence of charges, currents and dielectric or magnetic media) can be written

(i) $\nabla \cdot \mathbf{B} = 0$,   (ii) $\nabla \cdot \mathbf{E} = 0$,

(iii) $\nabla \times \mathbf{E} + \dfrac{\partial\mathbf{B}}{\partial t} = \mathbf{0}$,   (iv) $\nabla \times \mathbf{B} - \dfrac{1}{c^2}\dfrac{\partial\mathbf{E}}{\partial t} = \mathbf{0}$.

A vector $\mathbf{A}$ is defined by $\mathbf{B} = \nabla \times \mathbf{A}$, and a scalar ϕ by $\mathbf{E} = -\nabla\phi - \partial\mathbf{A}/\partial t$. Show that if the condition

(v) $\nabla \cdot \mathbf{A} + \dfrac{1}{c^2}\dfrac{\partial\phi}{\partial t} = 0$

is imposed (this is known as choosing the Lorentz gauge), then $\mathbf{A}$ and ϕ satisfy wave equations as follows:

(vi) $\nabla^2\phi - \dfrac{1}{c^2}\dfrac{\partial^2\phi}{\partial t^2} = 0$,

(vii) $\nabla^2\mathbf{A} - \dfrac{1}{c^2}\dfrac{\partial^2\mathbf{A}}{\partial t^2} = \mathbf{0}$.

The reader is invited to proceed as follows.

(a) Verify that the expressions for $\mathbf{B}$ and $\mathbf{E}$ in terms of $\mathbf{A}$ and ϕ are consistent with (i) and (iii).

(b) Substitute for $\mathbf{E}$ in (ii) and use the derivative with respect to time of (v) to eliminate $\mathbf{A}$ from the resulting expression. Hence obtain (vi).

(c) Substitute for $\mathbf{B}$ and $\mathbf{E}$ in (iv) in terms of $\mathbf{A}$ and ϕ. Then use the gradient of (v) to simplify the resulting equation and so obtain (vii).

10.20 In a description of the flow of a very viscous fluid that uses spherical polar coordinates with axial symmetry, the components of the velocity field $\mathbf{u}$ are given in terms of the *stream function* ψ by

$$u_r = \frac{1}{r^2 \sin\theta}\frac{\partial\psi}{\partial\theta}, \qquad u_\theta = \frac{-1}{r\sin\theta}\frac{\partial\psi}{\partial r}.$$

Find an explicit expression for the differential operator E defined by

$$E\psi = -(r\sin\theta)(\nabla \times \mathbf{u})_\phi.$$

The stream function satisfies the equation of motion $E^2\psi = 0$ and, for the flow of a fluid past a sphere, takes the form $\psi(r, \theta) = f(r)\sin^2\theta$. Show that $f(r)$ satisfies the (ordinary) differential equation

$$r^4 f^{(4)} - 4r^2 f'' + 8rf' - 8f = 0.$$

10.21 Paraboloidal coordinates u, v, ϕ are defined in terms of Cartesian coordinates by

$$x = uv \cos \phi, \qquad y = uv \sin \phi, \qquad z = \tfrac{1}{2}(u^2 - v^2).$$

Identify the coordinate surfaces in the u, v, ϕ system. Verify that each coordinate surface (u = constant, say) intersects every coordinate surface on which one of the other two coordinates (v, say) is constant. Show further that the system of coordinates is an orthogonal one and determine its scale factors. Prove that the u-component of $\nabla \times \mathbf{a}$ is given by

$$\frac{1}{(u^2 + v^2)^{1/2}} \left(\frac{a_\phi}{v} + \frac{\partial a_\phi}{\partial v} \right) - \frac{1}{uv} \frac{\partial a_v}{\partial \phi}.$$

10.22 Non-orthogonal curvilinear coordinates are difficult to work with and should be avoided if at all possible, but the following example is provided to illustrate the content of section 10.10.

In a new coordinate system for the region of space in which the Cartesian coordinate z satisfies $z \geq 0$, the position of a point $\mathbf{r}$ is given by (α_1, α_2, R), where α_1 and α_2 are respectively the cosines of the angles made by $\mathbf{r}$ with the x- and y-coordinate axes of a Cartesian system and $R = |\mathbf{r}|$. The ranges are $-1 \leq \alpha_i \leq 1$, $0 \leq R < \infty$.

(a) Express $\mathbf{r}$ in terms of α_1, α_2, R and the unit Cartesian vectors $\mathbf{i}, \mathbf{j}, \mathbf{k}$.
(b) Obtain expressions for the vectors $\mathbf{e}_i \, (= \partial \mathbf{r}/\partial \alpha_1, \ldots)$ and hence show that the scale factors h_i are given by

$$h_1 = \frac{R(1 - \alpha_2^2)^{1/2}}{(1 - \alpha_1^2 - \alpha_2^2)^{1/2}}, \qquad h_2 = \frac{R(1 - \alpha_1^2)^{1/2}}{(1 - \alpha_1^2 - \alpha_2^2)^{1/2}}, \qquad h_3 = 1.$$

(c) Verify formally that the system is not an orthogonal one.
(d) Show that the volume element of the coordinate system is

$$dV = \frac{R^2 \, d\alpha_1 \, d\alpha_2 \, dR}{(1 - \alpha_1^2 - \alpha_2^2)^{1/2}},$$

and demonstrate that this is always less than or equal to the corresponding expression for an orthogonal curvilinear system.
(e) Calculate the expression for $(ds)^2$ for the system, and show that it differs from that for the corresponding orthogonal system by

$$\frac{2\alpha_1 \alpha_2 R^2}{1 - \alpha_1^2 - \alpha_2^2} \, d\alpha_1 d\alpha_2.$$

10.23 Hyperbolic coordinates u, v, ϕ are defined in terms of Cartesian coordinates by

$$x = \cosh u \cos v \cos \phi, \qquad y = \cosh u \cos v \sin \phi, \qquad z = \sinh u \sin v.$$

Sketch the coordinate curves in the $\phi = 0$ plane, showing that far from the origin they become concentric circles and radial lines. In particular, identify the curves $u = 0$, $v = 0$, $v = \pi/2$ and $v = \pi$. Calculate the tangent vectors at a general point, show that they are mutually orthogonal and deduce that the appropriate scale factors are

$$h_u = h_v = (\cosh^2 u - \cos^2 v)^{1/2}, \qquad h_\phi = \cosh u \cos v.$$

Find the most general function $\psi(u)$ of u only that satisfies Laplace's equation $\nabla^2 \psi = 0$.

10.24 In a Cartesian system, A and B are the points $(0, 0, -1)$ and $(0, 0, 1)$ respectively. In a new coordinate system a general point P is given by (u_1, u_2, u_3) with $u_1 = \tfrac{1}{2}(r_1 + r_2)$, $u_2 = \tfrac{1}{2}(r_1 - r_2)$, $u_3 = \phi$; here r_1 and r_2 are the distances AP and BP and ϕ is the angle between the plane ABP and $y = 0$.

(a) Express z and the perpendicular distance ρ from P to the z-axis in terms of u_1, u_2, u_3.

(b) Evaluate $\partial x/\partial u_i$, $\partial y/\partial u_i$, $\partial z/\partial u_i$, for $i = 1, 2, 3$.

(c) Find the Cartesian components of $\hat{\mathbf{u}}_j$ and hence show that the new coordinates are mutually orthogonal. Evaluate the scale factors and the infinitesimal volume element in the new coordinate system.

(d) Determine and sketch the forms of the surfaces $u_i = $ constant.

(e) Find the most general function f of u_1 only that satisfies $\nabla^2 f = 0$.

10.12 Hints and answers

10.1   Group the term so that they form the total derivatives of compound vector expressions. The integral has the value $\mathbf{a} \times (\mathbf{a} \times \mathbf{b}) + \mathbf{h}$.

10.3   For crossed uniform fields, $\ddot{x} + (Bq/m)^2 x = q(E - Bv_0)/m$, $\ddot{y} = 0$, $m\dot{z} = qBx + mv_0$;
(b) $\xi = Bqt/m$; the path is a cycloid in the plane $y = 0$; $ds = [(dx/dt)^2 + (dz/dt)^2]^{1/2}\, dt$.

10.5 $\mathbf{g} = \ddot{\mathbf{r}}' - \boldsymbol{\omega} \times (\boldsymbol{\omega} \times \mathbf{r})$, where $\ddot{\mathbf{r}}'$ is the shell's acceleration measured by an observer fixed in space. To first order in ω, the direction of $\mathbf{g}$ is radial, i.e. parallel to $\ddot{\mathbf{r}}'$.

(a) Note that $\mathbf{s}$ is orthogonal to $\mathbf{g}$.

(b) If the actual time of flight is T, use $(\mathbf{s} + \Delta) \cdot \mathbf{g} = 0$ to show that

$$T \approx \tau(1 + 2g^{-2}(\mathbf{g} \times \boldsymbol{\omega}) \cdot \mathbf{v} + \cdots).$$

In the Coriolis terms, it is sufficient to put $T \approx \tau$.

(c) For this situation $(\mathbf{g} \times \boldsymbol{\omega}) \cdot \mathbf{v} = 0$ and $\boldsymbol{\omega} \times \mathbf{v} = 0$; $\tau \approx 43$ s and $\Delta = 10$–15 m to the East.

10.7   (a) Evaluate $(d\mathbf{r}/du) \cdot (d\mathbf{r}/du)$.

(b) Integrate the previous result between $u = 0$ and $u = 1$.

(c) $\hat{\mathbf{t}} = [\sqrt{2}(1 + u^2)]^{-1}[(1 - u^2)\mathbf{i} + 2u\mathbf{j} + (1 + u^2)\mathbf{k}]$. Use $d\hat{\mathbf{t}}/ds = (d\hat{\mathbf{t}}/du)/(ds/du)$; $\rho^{-1} = |d\hat{\mathbf{t}}/ds|$.

(d) $\hat{\mathbf{n}} = (1 + u^2)^{-1}[-2u\mathbf{i} + (1 - u^2)\mathbf{j}]$. $\hat{\mathbf{b}} = [\sqrt{2}(1 + u^2)]^{-1}[(u^2 - 1)\mathbf{i} - 2u\mathbf{j} + (1 + u^2)\mathbf{k}]$. Use $d\hat{\mathbf{b}}/ds = (d\hat{\mathbf{b}}/du)/(ds/du)$ and show that this equals $-[3a(1 + u^2)^2]^{-1}\hat{\mathbf{n}}$.

(e) Show that $d\hat{\mathbf{n}}/ds = \tau(\hat{\mathbf{b}} - \hat{\mathbf{t}}) = -2[3\sqrt{2}a(1 + u^2)^3]^{-1}[(1 - u^2)\mathbf{i} + 2u\mathbf{j}]$.

10.9   Note that $d\mathbf{B} = (d\mathbf{r} \cdot \nabla)\mathbf{B}$ and that $\mathbf{B} = B\hat{\mathbf{t}}$, with $\hat{\mathbf{t}} = d\mathbf{r}/ds$. Obtain $(\mathbf{B} \cdot \nabla)\mathbf{B}/B = \hat{\mathbf{t}}(dB/ds) + \hat{\mathbf{n}}(B/\rho)$ and then take the vector product of $\hat{\mathbf{t}}$ with this equation.

10.11   To integrate $\sec^2 \phi(\sec^2 \phi + \tan^2 \phi)^{1/2}\, d\phi$ put $\tan \phi = 2^{-1/2} \sinh \psi$.

10.13   Work in Cartesian coordinates, regrouping the terms obtained by evaluating the divergence on the LHS.

10.15   (a) $2z(x^2 + y^2 + z^2)^{-3}[(y^2 + z^2)(y^2 + z^2 - 3x^2) - 4x^4]$; (b) $2r^{-1} \cos \theta (1 - 5 \sin^2 \theta \cos^2 \phi)$; both are equal to $2zr^{-4}(r^2 - 5x^2)$.

10.17   Use the formulae given in table 10.2.

(a) $C = -B_0/(\mu_0 a)$; $B(\rho) = B_0 \rho/a$.

(b) $B_0 \rho^2/(3a)$ for $\rho < a$, and $B_0[\rho/2 - a^2/(6\rho)]$ for $\rho > a$.

(c) $[B_0^2/(2\mu_0)][1 - (\rho/a)^2]$.

10.19   Recall that $\nabla \times \nabla \phi = \mathbf{0}$ for any scalar ϕ and that $\partial/\partial t$ and ∇ act on different variables.

10.21   Two sets of paraboloids of revolution about the z-axis and the sheaf of planes containing the z-axis. For constant u, $-\infty < z < u^2/2$; for constant v, $-v^2/2 < z < \infty$. The scale factors are $h_u = h_v = (u^2 + v^2)^{1/2}$, $h_\phi = uv$.

10.23   The tangent vectors are as follows: for $u = 0$, the line joining $(1, 0, 0)$ and $(-1, 0, 0)$; for $v = 0$, the line joining $(1, 0, 0)$ and $(\infty, 0, 0)$; for $v = \pi/2$, the line $(0, 0, z)$; for $v = \pi$, the line joining $(-1, 0, 0)$ and $(-\infty, 0, 0)$.
$\psi(u) = 2 \tan^{-1} e^u + c$, derived from $\partial[\cosh u (\partial \psi / \partial u)] / \partial u = 0$.

11

Line, surface and volume integrals

In the previous chapter we encountered continuously varying scalar and vector fields and discussed the action of various differential operators on them. In addition to these differential operations, the need often arises to consider the integration of field quantities along lines, over surfaces and throughout volumes. In general the integrand may be scalar or vector in nature, but the evaluation of such integrals involves their reduction to one or more scalar integrals, which are then evaluated. In the case of surface and volume integrals this requires the evaluation of double and triple integrals (see chapter 6).

11.1 Line integrals

In this section we discuss *line* or *path integrals*, in which some quantity related to the field is integrated between two given points in space, A and B, along a prescribed curve C that joins them. In general, we may encounter line integrals of the forms

$$\int_C \phi \, d\mathbf{r}, \qquad \int_C \mathbf{a} \cdot d\mathbf{r}, \qquad \int_C \mathbf{a} \times d\mathbf{r}, \qquad (11.1)$$

where ϕ is a scalar field and $\mathbf{a}$ is a vector field. The three integrals themselves are respectively vector, scalar and vector in nature. As we will see below, in physical applications line integrals of the second type are by far the most common.

The formal definition of a line integral closely follows that of ordinary integrals and can be considered as the limit of a sum. We may divide the path C joining the points A and B into N small line elements $\Delta\mathbf{r}_p$, $p = 1, \ldots, N$. If (x_p, y_p, z_p) is any point on the line element $\Delta\mathbf{r}_p$ then the second type of line integral in (11.1), for example, is defined as

$$\int_C \mathbf{a} \cdot d\mathbf{r} = \lim_{N \to \infty} \sum_{p=1}^{N} \mathbf{a}(x_p, y_p, z_p) \cdot \Delta\mathbf{r}_p,$$

where it is assumed that all $|\Delta\mathbf{r}_p| \to 0$ as $N \to \infty$.

Each of the line integrals in (11.1) is evaluated over some curve C that may be either open (A and B being distinct points) or closed (the curve C forms a loop, so that A and B are coincident). In the case where C is closed, the line integral is written $\oint_C$ to indicate this. The curve may be given either parametrically by $\mathbf{r}(u) = x(u)\mathbf{i} + y(u)\mathbf{j} + z(u)\mathbf{k}$ or by means of simultaneous equations relating x, y, z for the given path (in Cartesian coordinates). A full discussion of the different representations of space curves was given in section 10.3.

In general, the value of the line integral depends not only on the end-points A and B but also on the path C joining them. For a closed curve we must also specify the direction around the loop in which the integral is taken. It is usually taken to be such that a person walking around the loop C in this direction always has the region R on his/her left; this is equivalent to traversing C in the anticlockwise direction (as viewed from above).

11.1.1 Evaluating line integrals

The method of evaluating a line integral is to reduce it to a set of scalar integrals. It is usual to work in Cartesian coordinates, in which case $d\mathbf{r} = dx\,\mathbf{i} + dy\,\mathbf{j} + dz\,\mathbf{k}$. The first type of line integral in (11.1) then becomes simply

$$\int_C \phi\,d\mathbf{r} = \mathbf{i} \int_C \phi(x, y, z)\,dx + \mathbf{j} \int_C \phi(x, y, z)\,dy + \mathbf{k} \int_C \phi(x, y, z)\,dz.$$

The three integrals on the RHS are ordinary scalar integrals that can be evaluated in the usual way once the path of integration C has been specified. Note that in the above we have used relations of the form

$$\int \phi\,\mathbf{i}\,dx = \mathbf{i} \int \phi\,dx,$$

which is allowable since the Cartesian unit vectors are of constant magnitude and direction and hence may be taken out of the integral. If we had been using a different coordinate system, such as spherical polars, then, as we saw in the previous chapter, the unit basis vectors would not be constant. In that case the basis vectors could not be factorised out of the integral.

The second and third line integrals in (11.1) can also be reduced to a set of scalar integrals by writing the vector field $\mathbf{a}$ in terms of its Cartesian components as $\mathbf{a} = a_x\mathbf{i} + a_y\mathbf{j} + a_z\mathbf{k}$, where a_x, a_y, a_z are each (in general) functions of x, y, z. The second line integral in (11.1), for example, can then be written as

$$\int_C \mathbf{a} \cdot d\mathbf{r} = \int_C (a_x\mathbf{i} + a_y\mathbf{j} + a_z\mathbf{k}) \cdot (dx\,\mathbf{i} + dy\,\mathbf{j} + dz\,\mathbf{k})$$
$$= \int_C (a_x\,dx + a_y\,dy + a_z\,dz)$$
$$= \int_C a_x\,dx + \int_C a_y\,dy + \int_C a_z\,dz. \tag{11.2}$$

A similar procedure may be followed for the third type of line integral in (11.1), which involves a cross product.

Line integrals have properties that are analogous to those of ordinary integrals. In particular, the following are useful properties (which we illustrate using the second form of line integral in (11.1) but which are valid for all three types).

(i) Reversing the path of integration changes the sign of the integral. If the path C along which the line integrals are evaluated has A and B as its end-points then

$$\int_A^B \mathbf{a} \cdot d\mathbf{r} = - \int_B^A \mathbf{a} \cdot d\mathbf{r}.$$

This implies that if the path C is a loop then integrating around the loop in the opposite direction changes the sign of the integral.

(ii) If the path of integration is subdivided into smaller segments then the sum of the separate line integrals along each segment is equal to the line integral along the whole path. So, if P is any point on the path of integration that lies between the path's end-points A and B then

$$\int_A^B \mathbf{a} \cdot d\mathbf{r} = \int_A^P \mathbf{a} \cdot d\mathbf{r} + \int_P^B \mathbf{a} \cdot d\mathbf{r}.$$

▶ *Evaluate the line integral $I = \int_C \mathbf{a} \cdot d\mathbf{r}$, where $\mathbf{a} = (x+y)\mathbf{i} + (y-x)\mathbf{j}$, along each of the paths in the xy-plane shown in figure 11.1, namely*

(i) *the parabola $y^2 = x$ from $(1,1)$ to $(4,2)$,*
(ii) *the curve $x = 2u^2 + u + 1$, $y = 1 + u^2$ from $(1,1)$ to $(4,2)$,*
(iii) *the line $y = 1$ from $(1,1)$ to $(4,1)$, followed by the line $x = 4$ from $(4,1)$ to $(4,2)$.*

Since each of the paths lies entirely in the xy-plane, we have $d\mathbf{r} = dx\,\mathbf{i} + dy\,\mathbf{j}$. We can therefore write the line integral as

$$I = \int_C \mathbf{a} \cdot d\mathbf{r} = \int_C [(x+y)\,dx + (y-x)\,dy]. \tag{11.3}$$

We must now evaluate this line integral along each of the prescribed paths.

Case (i). Along the parabola $y^2 = x$ we have $2y\,dy = dx$. Substituting for x in (11.3) and using just the limits on y, we obtain

$$I = \int_{(1,1)}^{(4,2)} [(x+y)\,dx + (y-x)\,dy] = \int_1^2 [(y^2+y)2y + (y-y^2)]\,dy = 11\tfrac{1}{3}.$$

Note that we could just as easily have substituted for y and obtained an integral in x, which would have given the same result.

Case (ii). The second path is given in terms of a parameter u. We could eliminate u between the two equations to obtain a relationship between x and y directly and proceed as above, but it is usually quicker to write the line integral in terms of the parameter u. Along the curve $x = 2u^2 + u + 1$, $y = 1 + u^2$ we have $dx = (4u+1)\,du$ and $dy = 2u\,du$.

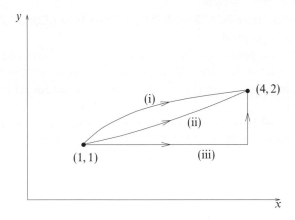

Figure 11.1   Different possible paths between the points $(1, 1)$ and $(4, 2)$.

Substituting for x and y in (11.3) and writing the correct limits on u, we obtain

$$I = \int_{(1,1)}^{(4,2)} [(x + y) \, dx + (y - x) \, dy]$$

$$= \int_0^1 [(3u^2 + u + 2)(4u + 1) - (u^2 + u)2u] \, du = 10\tfrac{2}{3}.$$

Case (iii). For the third path the line integral must be evaluated along the two line segments separately and the results added together. First, along the line $y = 1$ we have $dy = 0$. Substituting this into (11.3) and using just the limits on x for this segment, we obtain

$$\int_{(1,1)}^{(4,1)} [(x + y) \, dx + (y - x) \, dy] = \int_1^4 (x + 1) \, dx = 10\tfrac{1}{2}.$$

Next, along the line $x = 4$ we have $dx = 0$. Substituting this into (11.3) and using just the limits on y for this segment, we obtain

$$\int_{(4,1)}^{(4,2)} [(x + y) \, dx + (y - x) \, dy] = \int_1^2 (y - 4) \, dy = -2\tfrac{1}{2}.$$

The value of the line integral along the whole path is just the sum of the values of the line integrals along each segment, and is given by $I = 10\tfrac{1}{2} - 2\tfrac{1}{2} = 8$. ◄

When calculating a line integral along some curve C, which is given in terms of x, y and z, we are sometimes faced with the problem that the curve C is such that x, y and z are not single-valued functions of one another over the entire length of the curve. This is a particular problem for closed loops in the xy-plane (and also for some open curves). In such cases the path may be subdivided into shorter line segments along which one coordinate is a single-valued function of the other two. The sum of the line integrals along these segments is then equal to the line integral along the entire curve C. A better solution, however, is to represent the curve in a parametric form $\mathbf{r}(u)$ that is valid for its entire length.

> ▶Evaluate the line integral $I = \oint_C x\,dy$, where C is the circle in the xy-plane defined by
> $x^2 + y^2 = a^2$, $z = 0$.

Adopting the usual convention mentioned above, the circle C is to be traversed in the anticlockwise direction. Taking the circle as a whole means x is not a single-valued function of y. We must therefore divide the path into two parts with $x = +\sqrt{a^2 - y^2}$ for the semicircle lying to the right of $x = 0$, and $x = -\sqrt{a^2 - y^2}$ for the semicircle lying to the left of $x = 0$. The required line integral is then the sum of the integrals along the two semicircles. Substituting for x, it is given by

$$I = \oint_C x\,dy = \int_{-a}^{a} \sqrt{a^2 - y^2}\,dy + \int_{a}^{-a} \left(-\sqrt{a^2 - y^2}\right)\,dy$$
$$= 4\int_0^a \sqrt{a^2 - y^2}\,dy = \pi a^2.$$

Alternatively, we can represent the entire circle parametrically, in terms of the azimuthal angle ϕ, so that $x = a\cos\phi$ and $y = a\sin\phi$ with ϕ running from 0 to 2π. The integral can therefore be evaluated over the whole circle at once. Noting that $dy = a\cos\phi\,d\phi$, we can rewrite the line integral completely in terms of the parameter ϕ and obtain

$$I = \oint_C x\,dy = a^2 \int_0^{2\pi} \cos^2\phi\,d\phi = \pi a^2. \blacktriangleleft$$

11.1.2 Physical examples of line integrals

There are many physical examples of line integrals, but perhaps the most common is the expression for the total work done by a force $\mathbf{F}$ when it moves its point of application from a point A to a point B along a given curve C. We allow the magnitude and direction of $\mathbf{F}$ to vary along the curve. Let the force act at a point $\mathbf{r}$ and consider a small displacement $d\mathbf{r}$ along the curve; then the small amount of work done is $dW = \mathbf{F} \cdot d\mathbf{r}$, as discussed in subsection 7.6.1 (note that dW can be either positive or negative). Therefore, the total work done in traversing the path C is

$$W_C = \int_C \mathbf{F} \cdot d\mathbf{r}.$$

Naturally, other physical quantities can be expressed in such a way. For example, the electrostatic potential energy gained by moving a charge q along a path C in an electric field $\mathbf{E}$ is $-q \int_C \mathbf{E} \cdot d\mathbf{r}$. We may also note that Ampère's law concerning the magnetic field $\mathbf{B}$ associated with a current-carrying wire can be written as

$$\oint_C \mathbf{B} \cdot d\mathbf{r} = \mu_0 I,$$

where I is the current enclosed by a closed path C traversed in a right-handed sense with respect to the current direction.

Magnetostatics also provides a physical example of the third type of line integral in (11.1). If a loop of wire C carrying a current I is placed in a magnetic field $\mathbf{B}$ then the force $d\mathbf{F}$ on a small length $d\mathbf{r}$ of the wire is given by $d\mathbf{F} = I \, d\mathbf{r} \times \mathbf{B}$, and so the total (vector) force on the loop is

$$\mathbf{F} = I \oint_C d\mathbf{r} \times \mathbf{B}.$$

11.1.3 Line integrals with respect to a scalar

In addition to those listed in (11.1), we can form other types of line integral, which depend on a particular curve C but for which we integrate with respect to a scalar du, rather than the vector differential $d\mathbf{r}$. This distinction is somewhat arbitrary, however, since we can always rewrite line integrals containing the vector differential $d\mathbf{r}$ as a line integral with respect to some scalar parameter. If the path C along which the integral is taken is described parametrically by $\mathbf{r}(u)$ then

$$d\mathbf{r} = \frac{d\mathbf{r}}{du} \, du,$$

and the second type of line integral in (11.1), for example, can be written as

$$\int_C \mathbf{a} \cdot d\mathbf{r} = \int_C \mathbf{a} \cdot \frac{d\mathbf{r}}{du} \, du.$$

A similar procedure can be followed for the other types of line integral in (11.1).

Commonly occurring special cases of line integrals with respect to a scalar are

$$\int_C \phi \, ds, \qquad \int_C \mathbf{a} \, ds,$$

where s is the arc length along the curve C. We can always represent C parametrically by $\mathbf{r}(u)$, and from section 10.3 we have

$$ds = \sqrt{\frac{d\mathbf{r}}{du} \cdot \frac{d\mathbf{r}}{du}} \, du.$$

The line integrals can therefore be expressed entirely in terms of the parameter u and thence evaluated.

▶ *Evaluate the line integral* $I = \int_C (x-y)^2 \, ds$, *where* C *is the semicircle of radius a running from* $A = (a, 0)$ *to* $B = (-a, 0)$ *and for which* $y \geq 0$.

The semicircular path from A to B can be described in terms of the azimuthal angle ϕ (measured from the x-axis) by

$$\mathbf{r}(\phi) = a \cos \phi \, \mathbf{i} + a \sin \phi \, \mathbf{j},$$

where ϕ runs from 0 to π. Therefore the element of arc length is given, from section 10.3, by

$$ds = \sqrt{\frac{d\mathbf{r}}{d\phi} \cdot \frac{d\mathbf{r}}{d\phi}} \, d\phi = a(\cos^2 \phi + \sin^2 \phi) \, d\phi = a \, d\phi.$$

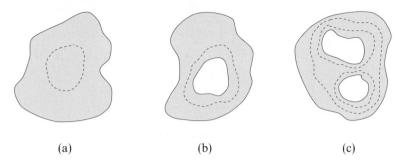

(a)   (b)   (c)

Figure 11.2   (a) A simply connected region; (b) a doubly connected region; (c) a triply connected region.

Since $(x - y)^2 = a^2(1 - \sin 2\phi)$, the line integral becomes

$$I = \int_C (x - y)^2 \, ds = \int_0^\pi a^3(1 - \sin 2\phi) \, d\phi = \pi a^3. \ \blacktriangleleft$$

As discussed in the previous chapter, the expression (10.58) for the square of the element of arc length in three-dimensional orthogonal curvilinear coordinates u_1, u_2, u_3 is

$$(ds)^2 = h_1^2 \, (du_1)^2 + h_2^2 \, (du_2)^2 + h_3^2 \, (du_3)^2,$$

where h_1, h_2, h_3 are the scale factors of the coordinate system. If a curve C in three dimensions is given parametrically by the equations $u_i = u_i(\lambda)$ for $i = 1, 2, 3$ then the element of arc length along the curve is

$$ds = \sqrt{h_1^2 \left(\frac{du_1}{d\lambda} \right)^2 + h_2^2 \left(\frac{du_2}{d\lambda} \right)^2 + h_3^2 \left(\frac{du_3}{d\lambda} \right)^2} \, d\lambda.$$

11.2 Connectivity of regions

In physical systems it is usual to define a scalar or vector field in some region R. In the next and some later sections we will need the concept of the *connectivity* of such a region in both two and three dimensions.

We begin by discussing planar regions. A plane region R is said to be *simply connected* if every simple closed curve within R can be continuously shrunk to a point without leaving the region (see figure 11.2(a)). If, however, the region R contains a hole then there exist simple closed curves that cannot be shrunk to a point without leaving R (see figure 11.2(b)). Such a region is said to be doubly connected, since its boundary has two distinct parts. Similarly, a region with $n - 1$ holes is said to be *n-fold connected*, or *multiply connected* (the region in figure 11.2(c) is triply connected).

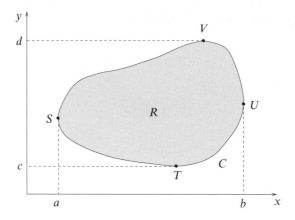

Figure 11.3   A simply connected region R bounded by the curve C.

These ideas can be extended to regions that are not planar, such as general three-dimensional surfaces and volumes. The same criteria concerning the shrinking of closed curves to a point also apply when deciding the connectivity of such regions. In these cases, however, the curves must lie in the surface or volume in question. For example, the interior of a torus is not simply connected, since there exist closed curves in the interior that cannot be shrunk to a point without leaving the torus. The region between two concentric spheres of different radii is simply connected.

11.3 Green's theorem in a plane

In subsection 11.1.1 we considered (amongst other things) the evaluation of line integrals for which the path C is closed and lies entirely in the xy-plane. Since the path is closed it will enclose a region R of the plane. We now discuss how to express the line integral around the loop as a double integral over the enclosed region R.

Suppose the functions $P(x, y)$, $Q(x, y)$ and their partial derivatives are single-valued, finite and continuous inside and on the boundary C of some simply connected region R in the xy-plane. *Green's theorem in a plane* (sometimes called the divergence theorem in two dimensions) then states

$$\oint_C (P\,dx + Q\,dy) = \iint_R \left(\frac{\partial Q}{\partial x} - \frac{\partial P}{\partial y} \right) dx\,dy, \qquad (11.4)$$

and so relates the line integral around C to a double integral over the enclosed region R. This theorem may be proved straightforwardly in the following way. Consider the simply connected region R in figure 11.3, and let $y = y_1(x)$ and

$y = y_2(x)$ be the equations of the curves STU and SVU respectively. We then write

$$\iint_R \frac{\partial P}{\partial y}\, dx\, dy = \int_a^b dx \int_{y_1(x)}^{y_2(x)} dy\, \frac{\partial P}{\partial y} = \int_a^b dx \Big[P(x, y) \Big]_{y=y_1(x)}^{y=y_2(x)}$$

$$= \int_a^b \Big[P(x, y_2(x)) - P(x, y_1(x)) \Big]\, dx$$

$$= -\int_a^b P(x, y_1(x))\, dx - \int_b^a P(x, y_2(x))\, dx = -\oint_C P\, dx.$$

If we now let $x = x_1(y)$ and $x = x_2(y)$ be the equations of the curves TSV and TUV respectively, we can similarly show that

$$\iint_R \frac{\partial Q}{\partial x}\, dx\, dy = \int_c^d dy \int_{x_1(y)}^{x_2(y)} dx\, \frac{\partial Q}{\partial x} = \int_c^d dy \Big[Q(x, y) \Big]_{x=x_1(y)}^{x=x_2(y)}$$

$$= \int_c^d \Big[Q(x_2(y), y) - Q(x_1(y), y) \Big]\, dy$$

$$= \int_d^c Q(x_1, y)\, dy + \int_c^d Q(x_2, y)\, dy - \oint_C Q\, dy.$$

Subtracting these two results gives Green's theorem in a plane.

▶ *Show that the area of a region R enclosed by a simple closed curve C is given by $A = \frac{1}{2}\oint_C (x\, dy - y\, dx) = \oint_C x\, dy = -\oint_C y\, dx$. Hence calculate the area of the ellipse $x = a\cos\phi$, $y = b\sin\phi$.*

In Green's theorem (11.4) put $P = -y$ and $Q = x$; then

$$\oint_C (x\, dy - y\, dx) = \iint_R (1 + 1)\, dx\, dy = 2 \iint_R dx\, dy = 2A.$$

Therefore the area of the region is $A = \frac{1}{2}\oint_C (x\, dy - y\, dx)$. Alternatively, we could put $P = 0$ and $Q = x$ and obtain $A = \oint_C x\, dy$, or put $P = -y$ and $Q = 0$, which gives $A = -\oint_C y\, dx$.
The area of the ellipse $x = a\cos\phi$, $y = b\sin\phi$ is given by

$$A = \frac{1}{2}\oint_C (x\, dy - y\, dx) = \frac{1}{2}\int_0^{2\pi} ab(\cos^2\phi + \sin^2\phi)\, d\phi$$

$$= \frac{ab}{2}\int_0^{2\pi} d\phi = \pi ab. \blacktriangleleft$$

It may further be shown that Green's theorem in a plane is also valid for multiply connected regions. In this case, the line integral must be taken over all the distinct boundaries of the region. Furthermore, each boundary must be traversed in the positive direction, so that a person travelling along it in this direction always has the region R on their left. In order to apply Green's theorem

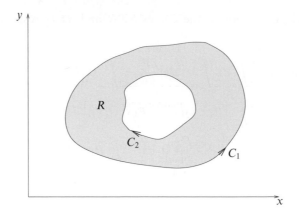

Figure 11.4 A doubly connected region R bounded by the curves C_1 and C_2.

to the region R shown in figure 11.4, the line integrals must be taken over both boundaries, C_1 and C_2, in the directions indicated, and the results added together.

We may also use Green's theorem in a plane to investigate the path independence (or not) of line integrals when the paths lie in the xy-plane. Let us consider the line integral

$$I = \int_A^B (P\,dx + Q\,dy).$$

For the line integral from A to B to be independent of the path taken, it must have the same value along any two arbitrary paths C_1 and C_2 joining the points. Moreover, if we consider as the path the closed loop C formed by $C_1 - C_2$ then the line integral around this loop must be zero. From Green's theorem in a plane, (11.4), we see that a *sufficient* condition for $I = 0$ is that

$$\frac{\partial P}{\partial y} = \frac{\partial Q}{\partial x}, \tag{11.5}$$

throughout some simply connected region R containing the loop, where we assume that these partial derivatives are continuous in R.

It may be shown that (11.5) is also a *necessary* condition for $I = 0$ and is equivalent to requiring $P\,dx + Q\,dy$ to be an exact differential of some function $\phi(x, y)$ such that $P\,dx + Q\,dy = d\phi$. It follows that $\int_A^B (P\,dx + Q\,dy) = \phi(B) - \phi(A)$ and that $\oint_C (P\,dx + Q\,dy)$ around any closed loop C in the region R is identically zero. These results are special cases of the general results for paths in three dimensions, which are discussed in the next section.

> ►*Evaluate the line integral*
>
> $$I = \oint_C [(e^x y + \cos x \sin y)\, dx + (e^x + \sin x \cos y)\, dy],$$
>
> *around the ellipse* $x^2/a^2 + y^2/b^2 = 1$.

Clearly, it is not straightforward to calculate this line integral directly. However, if we let

$$P = e^x y + \cos x \sin y \qquad \text{and} \qquad Q = e^x + \sin x \cos y,$$

then $\partial P/\partial y = e^x + \cos x \cos y = \partial Q/\partial x$, and so $P\, dx + Q\, dy$ is an exact differential (it is actually the differential of the function $f(x,y) = e^x y + \sin x \sin y$). From the above discussion, we can conclude immediately that $I = 0$. ◄

11.4 Conservative fields and potentials

So far we have made the point that, in general, the value of a line integral between two points A and B depends on the path C taken from A to B. In the previous section, however, we saw that, for paths in the xy-plane, line integrals whose integrands have certain properties are independent of the path taken. We now extend that discussion to the full three-dimensional case.

For line integrals of the form $\int_C \mathbf{a} \cdot d\mathbf{r}$, there exists a class of vector fields for which the line integral between two points is *independent* of the path taken. Such vector fields are called *conservative*. A vector field $\mathbf{a}$ that has continuous partial derivatives in a simply connected region R is conservative if, and only if, any of the following is true.

(i) The integral $\int_A^B \mathbf{a} \cdot d\mathbf{r}$, where A and B lie in the region R, is independent of the path from A to B. Hence the integral $\oint_C \mathbf{a} \cdot d\mathbf{r}$ around any closed loop in R is zero.

(ii) There exists a single-valued function ϕ of position such that $\mathbf{a} = \nabla\phi$.

(iii) $\nabla \times \mathbf{a} = \mathbf{0}$.

(iv) $\mathbf{a} \cdot d\mathbf{r}$ is an exact differential.

The validity or otherwise of any of these statements implies the same for the other three, as we will now show.

First, let us assume that (i) above is true. If the line integral from A to B is independent of the path taken between the points then its value must be a function only of the positions of A and B. We may therefore write

$$\int_A^B \mathbf{a} \cdot d\mathbf{r} = \phi(B) - \phi(A), \tag{11.6}$$

which defines a single-valued scalar function of position ϕ. If the points A and B are separated by an infinitesimal displacement $d\mathbf{r}$ then (11.6) becomes

$$\mathbf{a} \cdot d\mathbf{r} = d\phi,$$

which shows that we require $\mathbf{a} \cdot d\mathbf{r}$ to be an exact differential: condition (iv). From (10.27) we can write $d\phi = \nabla\phi \cdot d\mathbf{r}$, and so we have

$$(\mathbf{a} - \nabla\phi) \cdot d\mathbf{r} = 0.$$

Since $d\mathbf{r}$ is arbitrary, we find that $\mathbf{a} = \nabla\phi$; this immediately implies $\nabla \times \mathbf{a} = \mathbf{0}$, condition (iii) (see (10.37)).

Alternatively, if we suppose that there exists a single-valued function of position ϕ such that $\mathbf{a} = \nabla\phi$ then $\nabla \times \mathbf{a} = \mathbf{0}$ follows as before. The line integral around a closed loop then becomes

$$\oint_C \mathbf{a} \cdot d\mathbf{r} = \oint_C \nabla\phi \cdot d\mathbf{r} = \oint d\phi.$$

Since we defined ϕ to be single-valued, this integral is zero as required.

Now suppose $\nabla \times \mathbf{a} = \mathbf{0}$. From Stoke's theorem, which is discussed in section 11.9, we immediately obtain $\oint_C \mathbf{a} \cdot d\mathbf{r} = 0$; then $\mathbf{a} = \nabla\phi$ and $\mathbf{a} \cdot d\mathbf{r} = d\phi$ follow as above.

Finally, let us suppose $\mathbf{a} \cdot d\mathbf{r} = d\phi$. Then immediately we have $\mathbf{a} = \nabla\phi$, and the other results follow as above.

▶*Evaluate the line integral* $I = \int_A^B \mathbf{a} \cdot d\mathbf{r}$, *where* $\mathbf{a} = (xy^2 + z)\mathbf{i} + (x^2y + 2)\mathbf{j} + x\mathbf{k}$, A *is the point* (c, c, h) *and* B *is the point* $(2c, c/2, h)$, *along the different paths*

(i) C_1, *given by* $x = cu$, $y = c/u$, $z = h$,
(ii) C_2, *given by* $2y = 3c - x$, $z = h$.

Show that the vector field $\mathbf{a}$ *is in fact conservative, and find* ϕ *such that* $\mathbf{a} = \nabla\phi$.

Expanding out the integrand, we have

$$I = \int_{(c,c,h)}^{(2c,c/2,h)} \left[(xy^2 + z)\,dx + (x^2y + 2)\,dy + x\,dz \right], \tag{11.7}$$

which we must evaluate along each of the paths C_1 and C_2.

(i) Along C_1 we have $dx = c\,du$, $dy = -(c/u^2)\,du$, $dz = 0$, and on substituting in (11.7) and finding the limits on u, we obtain

$$I = \int_1^2 c\left(h - \frac{2}{u^2}\right)\,du = c(h - 1).$$

(ii) Along C_2 we have $2\,dy = -dx$, $dz = 0$ and, on substituting in (11.7) and using the limits on x, we obtain

$$I = \int_c^{2c} \left(\tfrac{1}{2}x^3 - \tfrac{9}{4}cx^2 + \tfrac{9}{4}c^2x + h - 1 \right)\,dx = c(h - 1).$$

Hence the line integral has the same value along paths C_1 and C_2. Taking the curl of $\mathbf{a}$, we have

$$\nabla \times \mathbf{a} = (0 - 0)\mathbf{i} + (1 - 1)\mathbf{j} + (2xy - 2xy)\mathbf{k} = \mathbf{0},$$

so $\mathbf{a}$ is a conservative vector field, and the line integral between two points must be

388

independent of the path taken. Since $\mathbf{a}$ is conservative, we can write $\mathbf{a} = \nabla \phi$. Therefore, ϕ must satisfy

$$\frac{\partial \phi}{\partial x} = xy^2 + z,$$

which implies that $\phi = \frac{1}{2}x^2y^2 + zx + f(y, z)$ for some function f. Secondly, we require

$$\frac{\partial \phi}{\partial y} = x^2y + \frac{\partial f}{\partial y} = x^2y + 2,$$

which implies $f = 2y + g(z)$. Finally, since

$$\frac{\partial \phi}{\partial z} = x + \frac{\partial g}{\partial z} = x,$$

we have $g = \text{constant} = k$. It can be seen that we have explicitly constructed the function $\phi = \frac{1}{2}x^2y^2 + zx + 2y + k$. ◀

The quantity ϕ that figures so prominently in this section is called the *scalar potential function* of the conservative vector field $\mathbf{a}$ (which satisfies $\nabla \times \mathbf{a} = \mathbf{0}$), and is unique up to an arbitrary additive constant. Scalar potentials that are multi-valued functions of position (but in simple ways) are also of value in describing some physical situations, the most obvious example being the scalar magnetic potential associated with a current-carrying wire. When the integral of a field quantity around a closed loop is considered, provided the loop does not enclose a net current, the potential is single-valued and all the above results still hold. If the loop does enclose a net current, however, our analysis is no longer valid and extra care must be taken.

If, instead of being conservative, a vector field $\mathbf{b}$ satisfies $\nabla \cdot \mathbf{b} = 0$ (i.e. $\mathbf{b}$ is solenoidal) then it is both possible and useful, for example in the theory of electromagnetism, to define a *vector field* $\mathbf{a}$ such that $\mathbf{b} = \nabla \times \mathbf{a}$. It may be shown that such a vector field $\mathbf{a}$ always exists. Further, if $\mathbf{a}$ is one such vector field then $\mathbf{a}' = \mathbf{a} + \nabla \psi + \mathbf{c}$, where ψ is any scalar function and $\mathbf{c}$ is any constant vector, also satisfies the above relationship, i.e. $\mathbf{b} = \nabla \times \mathbf{a}'$. This was discussed more fully in subsection 10.8.2.

11.5 Surface integrals

As with line integrals, integrals over surfaces can involve vector and scalar fields and, equally, can result in either a vector or a scalar. The simplest case involves entirely scalars and is of the form

$$\int_S \phi \, dS. \tag{11.8}$$

As analogues of the line integrals listed in (11.1), we may also encounter surface integrals involving vectors, namely

$$\int_S \phi \, d\mathbf{S}, \qquad \int_S \mathbf{a} \cdot d\mathbf{S}, \qquad \int_S \mathbf{a} \times d\mathbf{S}. \tag{11.9}$$

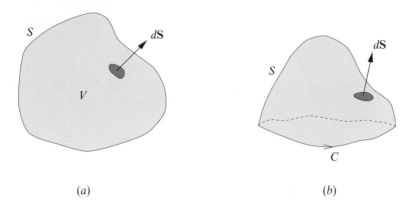

Figure 11.5 (a) A closed surface and (b) an open surface. In each case a normal to the surface is shown: $d\mathbf{S} = \hat{\mathbf{n}}\,dS$.

All the above integrals are taken over some surface S, which may be either open or closed, and are therefore, in general, double integrals. Following the notation for line integrals, for surface integrals over a closed surface $\int_S$ is replaced by $\oint_S$.

The vector differential $d\mathbf{S}$ in (11.9) represents a vector area element of the surface S. It may also be written $d\mathbf{S} = \hat{\mathbf{n}}\,dS$, where $\hat{\mathbf{n}}$ is a unit normal to the surface at the position of the element and dS is the scalar area of the element used in (11.8). The convention for the direction of the normal $\hat{\mathbf{n}}$ to a surface depends on whether the surface is open or closed. A closed surface, see figure 11.5(a), does not have to be simply connected (for example, the surface of a torus is not), but it does have to enclose a volume V, which may be of infinite extent. The direction of $\hat{\mathbf{n}}$ is taken to point outwards from the enclosed volume as shown. An open surface, see figure 11.5(b), spans some perimeter curve C. The direction of $\hat{\mathbf{n}}$ is then given by the right-hand sense with respect to the direction in which the perimeter is traversed, i.e. follows the right-hand screw rule discussed in subsection 7.6.2. An open surface does not have to be simply connected but for our purposes it must be two-sided (a Möbius strip is an example of a one-sided surface).

The formal definition of a surface integral is very similar to that of a line integral. We divide the surface S into N elements of area ΔS_p, $p = 1, 2, \ldots, N$, each with a unit normal $\hat{\mathbf{n}}_p$. If (x_p, y_p, z_p) is any point in ΔS_p then the second type of surface integral in (11.9), for example, is defined as

$$\int_S \mathbf{a} \cdot d\mathbf{S} = \lim_{N \to \infty} \sum_{p=1}^{N} \mathbf{a}(x_p, y_p, z_p) \cdot \hat{\mathbf{n}}_p \Delta S_p,$$

where it is required that all $\Delta S_p \to 0$ as $N \to \infty$.

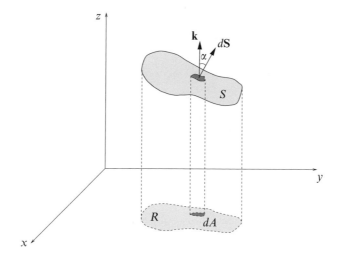

Figure 11.6 A surface S (or part thereof) projected onto a region R in the xy-plane; dS is a surface element.

11.5.1 Evaluating surface integrals

We now consider how to evaluate surface integrals over some general surface. This involves writing the scalar area element dS in terms of the coordinate differentials of our chosen coordinate system. In some particularly simple cases this is very straightforward. For example, if S is the surface of a sphere of radius a (or some part thereof) then using spherical polar coordinates θ, ϕ on the sphere we have $dS = a^2 \sin\theta \, d\theta \, d\phi$. For a general surface, however, it is not usually possible to represent the surface in a simple way in any particular coordinate system. In such cases, it is usual to work in Cartesian coordinates and consider the projections of the surface onto the coordinate planes.

Consider a surface (or part of a surface) S as in figure 11.6. The surface S is projected onto a region R of the xy-plane, so that an element of surface area dS projects onto the area element dA. From the figure, we see that $dA = |\cos\alpha| \, dS$, where α is the angle between the unit vector $\mathbf{k}$ in the z-direction and the unit normal $\hat{\mathbf{n}}$ to the surface at P. So, at any given point of S, we have simply

$$dS = \frac{dA}{|\cos\alpha|} = \frac{dA}{|\hat{\mathbf{n}} \cdot \mathbf{k}|}.$$

Now, if the surface S is given by the equation $f(x, y, z) = 0$ then, as shown in subsection 10.7.1, the unit normal at any point of the surface is given by $\hat{\mathbf{n}} = \nabla f / |\nabla f|$ evaluated at that point, cf. (10.32). The scalar element of surface area then becomes

$$dS = \frac{dA}{|\hat{\mathbf{n}} \cdot \mathbf{k}|} = \frac{|\nabla f| \, dA}{\nabla f \cdot \mathbf{k}} = \frac{|\nabla f| \, dA}{\partial f / \partial z}, \qquad (11.10)$$

where $|\nabla f|$ and $\partial f/\partial z$ are evaluated on the surface S. We can therefore express any surface integral over S as a double integral over the region R in the xy-plane.

▶Evaluate the surface integral $I = \int_S \mathbf{a} \cdot d\mathbf{S}$, where $\mathbf{a} = x\mathbf{i}$ and S is the surface of the hemisphere $x^2 + y^2 + z^2 = a^2$ with $z \geq 0$.

The surface of the hemisphere is shown in figure 11.7. In this case dS may be easily expressed in spherical polar coordinates as $dS = a^2 \sin\theta\, d\theta\, d\phi$, and the unit normal to the surface at any point is simply $\hat{\mathbf{r}}$. On the surface of the hemisphere we have $x = a\sin\theta\cos\phi$ and so

$$\mathbf{a} \cdot d\mathbf{S} = x\,(\mathbf{i} \cdot \hat{\mathbf{r}})\, dS = (a\sin\theta\cos\phi)(\sin\theta\cos\phi)(a^2\sin\theta\, d\theta\, d\phi).$$

Therefore, inserting the correct limits on θ and ϕ, we have

$$I = \int_S \mathbf{a} \cdot d\mathbf{S} = a^3 \int_0^{\pi/2} d\theta\, \sin^3\theta \int_0^{2\pi} d\phi\, \cos^2\phi = \frac{2\pi a^3}{3}.$$

We could, however, follow the general prescription above and project the hemisphere S onto the region R in the xy-plane that is a circle of radius a centred at the origin. Writing the equation of the surface of the hemisphere as $f(x, y) = x^2 + y^2 + z^2 - a^2 = 0$ and using (11.10), we have

$$I = \int_S \mathbf{a} \cdot d\mathbf{S} = \int_S x\,(\mathbf{i} \cdot \hat{\mathbf{r}})\, dS = \int_R x\,(\mathbf{i} \cdot \hat{\mathbf{r}}) \frac{|\nabla f|\, dA}{\partial f/\partial z}.$$

Now $\nabla f = 2x\mathbf{i} + 2y\mathbf{j} + 2z\mathbf{k} = 2\mathbf{r}$, so on the surface S we have $|\nabla f| = 2|\mathbf{r}| = 2a$. On S we also have $\partial f/\partial z = 2z = 2\sqrt{a^2 - x^2 - y^2}$ and $\mathbf{i} \cdot \hat{\mathbf{r}} = x/a$. Therefore, the integral becomes

$$I = \iint_R \frac{x^2}{\sqrt{a^2 - x^2 - y^2}}\, dx\, dy.$$

Although this integral may be evaluated directly, it is quicker to transform to plane polar coordinates:

$$I = \iint_{R'} \frac{\rho^2 \cos^2\phi}{\sqrt{a^2 - \rho^2}}\, \rho\, d\rho\, d\phi$$

$$= \int_0^{2\pi} \cos^2\phi\, d\phi \int_0^a \frac{\rho^3\, d\rho}{\sqrt{a^2 - \rho^2}}.$$

Making the substitution $\rho = a\sin u$, we finally obtain

$$I = \int_0^{2\pi} \cos^2\phi\, d\phi \int_0^{\pi/2} a^3 \sin^3 u\, du = \frac{2\pi a^3}{3}. \quad ◀$$

In the above discussion we assumed that any line parallel to the z-axis intersects S only once. If this is not the case, we must split up the surface into smaller surfaces S_1, S_2 etc. that are of this type. The surface integral over S is then the sum of the surface integrals over S_1, S_2 and so on. This is always necessary for closed surfaces.

Sometimes we may need to project a surface S (or some part of it) onto the zx- or yz-plane, rather than the xy-plane; for such cases, the above analysis is easily modified.

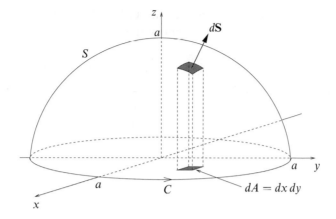

Figure 11.7   The surface of the hemisphere $x^2 + y^2 + z^2 = a^2$, $z \geq 0$.

11.5.2 Vector areas of surfaces

The vector area of a surface S is defined as

$$\mathbf{S} = \int_S d\mathbf{S},$$

where the surface integral may be evaluated as above.

▶ *Find the vector area of the surface of the hemisphere $x^2 + y^2 + z^2 = a^2$ with $z \geq 0$.*

As in the previous example, $d\mathbf{S} = a^2 \sin\theta \, d\theta \, d\phi \, \hat{\mathbf{r}}$ in spherical polar coordinates. Therefore the vector area is given by

$$\mathbf{S} = \iint_S a^2 \sin\theta \, \hat{\mathbf{r}} \, d\theta \, d\phi.$$

Now, since $\hat{\mathbf{r}}$ varies over the surface S, it also must be integrated. This is most easily achieved by writing $\hat{\mathbf{r}}$ in terms of the constant Cartesian basis vectors. On S we have

$$\hat{\mathbf{r}} = \sin\theta \cos\phi \, \mathbf{i} + \sin\theta \sin\phi \, \mathbf{j} + \cos\theta \, \mathbf{k},$$

so the expression for the vector area becomes

$$\mathbf{S} = \mathbf{i} \left(a^2 \int_0^{2\pi} \cos\phi \, d\phi \int_0^{\pi/2} \sin^2\theta \, d\theta \right) + \mathbf{j} \left(a^2 \int_0^{2\pi} \sin\phi \, d\phi \int_0^{\pi/2} \sin^2\theta \, d\theta \right)$$

$$+ \mathbf{k} \left(a^2 \int_0^{2\pi} d\phi \int_0^{\pi/2} \sin\theta \cos\theta \, d\theta \right)$$

$$= \mathbf{0} + \mathbf{0} + \pi a^2 \mathbf{k} = \pi a^2 \mathbf{k}.$$

Note that the magnitude of $\mathbf{S}$ is the projected area of the hemisphere onto the xy-plane, and not the surface area of the hemisphere. ◀

393

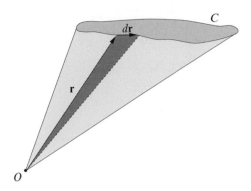

Figure 11.8 The conical surface spanning the perimeter C and having its vertex at the origin.

The hemispherical shell discussed above is an example of an open surface. For a closed surface, however, the vector area is always zero. This may be seen by projecting the surface down onto each Cartesian coordinate plane in turn. For each projection, every positive element of area on the upper surface is cancelled by the corresponding negative element on the lower surface. Therefore, each component of $\mathbf{S} = \oint_S d\mathbf{S}$ vanishes.

An important corollary of this result is that the vector area of an open surface depends only on its perimeter, or boundary curve, C. This may be proved as follows. If surfaces S_1 and S_2 have the same perimeter then $S_1 - S_2$ is a closed surface, for which

$$\oint d\mathbf{S} = \int_{S_1} d\mathbf{S} - \int_{S_2} d\mathbf{S} = \mathbf{0}.$$

Hence $\mathbf{S}_1 = \mathbf{S}_2$. Moreover, we may derive an expression for the vector area of an open surface S solely in terms of a line integral around its perimeter C. Since we may choose any surface with perimeter C, we will consider a cone with its vertex at the origin (see figure 11.8). The vector area of the elementary triangular region shown in the figure is $d\mathbf{S} = \frac{1}{2}\mathbf{r} \times d\mathbf{r}$. Therefore, the vector area of the cone, and hence of *any* open surface with perimeter C, is given by the line integral

$$\mathbf{S} = \frac{1}{2}\oint_C \mathbf{r} \times d\mathbf{r}.$$

For a surface confined to the xy-plane, $\mathbf{r} = x\mathbf{i} + y\mathbf{j}$ and $d\mathbf{r} = dx\,\mathbf{i} + dy\,\mathbf{j}$, and we obtain for this special case that the area of the surface is given by $A = \frac{1}{2}\oint_C(x\,dy - y\,dx)$, as we found in section 11.3.

> ►*Find the vector area of the surface of the hemisphere $x^2 + y^2 + z^2 = a^2$, $z \geq 0$, by evaluating the line integral $\mathbf{S} = \frac{1}{2} \oint_C \mathbf{r} \times d\mathbf{r}$ around its perimeter.*

The perimeter C of the hemisphere is the circle $x^2 + y^2 = a^2$, on which we have

$$\mathbf{r} = a\cos\phi\,\mathbf{i} + a\sin\phi\,\mathbf{j}, \qquad d\mathbf{r} = -a\sin\phi\,d\phi\,\mathbf{i} + a\cos\phi\,d\phi\,\mathbf{j}.$$

Therefore the cross product $\mathbf{r} \times d\mathbf{r}$ is given by

$$\mathbf{r} \times d\mathbf{r} = \begin{vmatrix} \mathbf{i} & \mathbf{j} & \mathbf{k} \\ a\cos\phi & a\sin\phi & 0 \\ -a\sin\phi\,d\phi & a\cos\phi\,d\phi & 0 \end{vmatrix} = a^2(\cos^2\phi + \sin^2\phi)\,d\phi\,\mathbf{k} = a^2\,d\phi\,\mathbf{k},$$

and the vector area becomes

$$\mathbf{S} = \tfrac{1}{2}a^2\mathbf{k} \int_0^{2\pi} d\phi = \pi a^2\,\mathbf{k}. \ \blacktriangleleft$$

11.5.3 Physical examples of surface integrals

There are many examples of surface integrals in the physical sciences. Surface integrals of the form (11.8) occur in computing the total electric charge on a surface or the mass of a shell, $\int_S \rho(\mathbf{r})\,dS$, given the charge or mass density $\rho(\mathbf{r})$. For surface integrals involving vectors, the second form in (11.9) is the most common. For a vector field $\mathbf{a}$, the surface integral $\int_S \mathbf{a} \cdot d\mathbf{S}$ is called the *flux* of $\mathbf{a}$ through S. Examples of physically important flux integrals are numerous. For example, let us consider a surface S in a fluid with density $\rho(\mathbf{r})$ that has a velocity field $\mathbf{v}(\mathbf{r})$. The mass of fluid crossing an element of surface area $d\mathbf{S}$ in time dt is $dM = \rho\mathbf{v} \cdot d\mathbf{S}\,dt$. Therefore the *net* total mass flux of fluid crossing S is $M = \int_S \rho(\mathbf{r})\mathbf{v}(\mathbf{r}) \cdot d\mathbf{S}$. As another example, the electromagnetic flux of energy out of a given volume V bounded by a surface S is $\oint_S (\mathbf{E} \times \mathbf{H}) \cdot d\mathbf{S}$.

The solid angle, to be defined below, subtended at a point O by a surface (closed or otherwise) can also be represented by an integral of this form, although it is not strictly a flux integral (unless we imagine isotropic rays radiating from O). The integral

$$\Omega = \int_S \frac{\mathbf{r} \cdot d\mathbf{S}}{r^3} = \int_S \frac{\hat{\mathbf{r}} \cdot d\mathbf{S}}{r^2}, \tag{11.11}$$

gives the *solid angle Ω subtended at O by a surface S* if $\mathbf{r}$ is the position vector measured from O of an element of the surface. A little thought will show that (11.11) takes account of all three relevant factors: the size of the element of surface, its inclination to the line joining the element to O and the distance from O. Such a general expression is often useful for computing solid angles when the three-dimensional geometry is complicated. Note that (11.11) remains valid when the surface S is not convex and when a single ray from O in certain directions would cut S in more than one place (but we exclude multiply connected regions).

In particular, when the surface is closed $\Omega = 0$ if O is outside S and $\Omega = 4\pi$ if O is an interior point.

Surface integrals resulting in vectors occur less frequently. An example is afforded, however, by the total resultant force experienced by a body immersed in a stationary fluid in which the hydrostatic pressure is given by $p(\mathbf{r})$. The pressure is everywhere inwardly directed and the resultant force is $\mathbf{F} = - \oint_S p \, d\mathbf{S}$, taken over the whole surface.

11.6 Volume integrals

Volume integrals are defined in an obvious way and are generally simpler than line or surface integrals since the element of volume dV is a scalar quantity. We may encounter volume integrals of the forms

$$\int_V \phi \, dV, \qquad \int_V \mathbf{a} \, dV. \tag{11.12}$$

Clearly, the first form results in a scalar, whereas the second form yields a vector. Two closely related physical examples, one of each kind, are provided by the total mass of a fluid contained in a volume V, given by $\int_V \rho(\mathbf{r}) \, dV$, and the total linear momentum of that same fluid, given by $\int_V \rho(\mathbf{r})\mathbf{v}(\mathbf{r}) \, dV$ where $\mathbf{v}(\mathbf{r})$ is the velocity field in the fluid. As a slightly more complicated example of a volume integral we may consider the following.

▶*Find an expression for the angular momentum of a solid body rotating with angular velocity* $\boldsymbol{\omega}$ *about an axis through the origin.*

Consider a small volume element dV situated at position $\mathbf{r}$; its linear momentum is $\rho \, dV \dot{\mathbf{r}}$, where $\rho = \rho(\mathbf{r})$ is the density distribution, and its angular momentum about O is $\mathbf{r} \times \rho \dot{\mathbf{r}} \, dV$. Thus for the whole body the angular momentum $\mathbf{L}$ is

$$\mathbf{L} = \int_V (\mathbf{r} \times \dot{\mathbf{r}})\rho \, dV.$$

Putting $\dot{\mathbf{r}} = \boldsymbol{\omega} \times \mathbf{r}$ yields

$$\mathbf{L} = \int_V [\mathbf{r} \times (\boldsymbol{\omega} \times \mathbf{r})] \, \rho \, dV = \int_V \boldsymbol{\omega} r^2 \rho \, dV - \int_V (\mathbf{r} \cdot \boldsymbol{\omega})\mathbf{r}\rho \, dV. \blacktriangleleft$$

The evaluation of the first type of volume integral in (11.12) has already been considered in our discussion of multiple integrals in chapter 6. The evaluation of the second type of volume integral follows directly since we can write

$$\int_V \mathbf{a} \, dV = \mathbf{i} \int_V a_x \, dV + \mathbf{j} \int_V a_y \, dV + \mathbf{k} \int_V a_z \, dV, \tag{11.13}$$

where a_x, a_y, a_z are the Cartesian components of $\mathbf{a}$. Of course, we could have written $\mathbf{a}$ in terms of the basis vectors of some other coordinate system (e.g. spherical polars) but, since such basis vectors are not, in general, constant, they

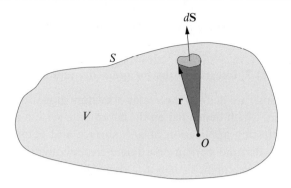

Figure 11.9   A general volume V containing the origin and bounded by the closed surface S.

cannot be taken out of the integral sign as in (11.13) and must be included as part of the integrand.

11.6.1 Volumes of three-dimensional regions

As discussed in chapter 6, the volume of a three-dimensional region V is simply $V = \int_V dV$, which may be evaluated directly once the limits of integration have been found. However, the volume of the region obviously depends only on the surface S that bounds it. We should therefore be able to express the volume V in terms of a surface integral over S. This is indeed possible, and the appropriate expression may derived as follows. Referring to figure 11.9, let us suppose that the origin O is contained within V. The volume of the small shaded cone is $dV = \frac{1}{3} \mathbf{r} \cdot d\mathbf{S}$; the total volume of the region is thus given by

$$ V = \frac{1}{3} \oint_S \mathbf{r} \cdot d\mathbf{S}. $$

It may be shown that this expression is valid even when O is not contained in V. Although this surface integral form is available, in practice it is usually simpler to evaluate the volume integral directly.

▶ Find the volume enclosed between a sphere of radius a centred on the origin and a circular cone of half-angle α with its vertex at the origin.

The element of vector area $d\mathbf{S}$ on the surface of the sphere is given in spherical polar coordinates by $a^2 \sin\theta \, d\theta \, d\phi \, \hat{\mathbf{r}}$. Now taking the axis of the cone to lie along the z-axis (from which θ is measured) the required volume is given by

$$ V = \frac{1}{3} \oint_S \mathbf{r} \cdot d\mathbf{S} = \frac{1}{3} \int_0^{2\pi} d\phi \int_0^{\alpha} a^2 \sin\theta \, \mathbf{r} \cdot \hat{\mathbf{r}} \, d\theta $$

$$ = \frac{1}{3} \int_0^{2\pi} d\phi \int_0^{\alpha} a^3 \sin\theta \, d\theta = \frac{2\pi a^3}{3}(1 - \cos\alpha). \blacktriangleleft $$

397

11.7 Integral forms for grad, div and curl

In the previous chapter we defined the vector operators grad, div and curl in purely mathematical terms, which depended on the coordinate system in which they were expressed. An interesting application of line, surface and volume integrals is the expression of grad, div and curl in coordinate-free, geometrical terms. If ϕ is a scalar field and $\mathbf{a}$ is a vector field then it may be shown that at any point P

$$\nabla\phi = \lim_{V \to 0} \left(\frac{1}{V} \oint_S \phi \, d\mathbf{S} \right) \qquad (11.14)$$

$$\nabla \cdot \mathbf{a} = \lim_{V \to 0} \left(\frac{1}{V} \oint_S \mathbf{a} \cdot d\mathbf{S} \right) \qquad (11.15)$$

$$\nabla \times \mathbf{a} = \lim_{V \to 0} \left(\frac{1}{V} \oint_S d\mathbf{S} \times \mathbf{a} \right) \qquad (11.16)$$

where V is a small volume enclosing P and S is its bounding surface. Indeed, we may consider these equations as the (geometrical) *definitions* of grad, div and curl. An alternative, but equivalent, geometrical definition of $\nabla \times \mathbf{a}$ at a point P, which is often easier to use than (11.16), is given by

$$(\nabla \times \mathbf{a}) \cdot \hat{\mathbf{n}} = \lim_{A \to 0} \left(\frac{1}{A} \oint_C \mathbf{a} \cdot d\mathbf{r} \right), \qquad (11.17)$$

where C is a plane contour of area A enclosing the point P and $\hat{\mathbf{n}}$ is the unit normal to the enclosed planar area.

It may be shown, *in any coordinate system*, that all the above equations are consistent with our definitions in the previous chapter, although the difficulty of proof depends on the chosen coordinate system. The most general coordinate system encountered in that chapter was one with orthogonal curvilinear coordinates u_1, u_2, u_3, of which Cartesians, cylindrical polars and spherical polars are all special cases. Although it may be shown that (11.14) leads to the usual expression for grad in curvilinear coordinates, the proof requires complicated manipulations of the derivatives of the basis vectors with respect to the coordinates and is not presented here. In Cartesian coordinates, however, the proof is quite simple.

▶*Show that the geometrical definition of* grad *leads to the usual expression for* $\nabla\phi$ *in Cartesian coordinates.*

Consider the surface S of a small rectangular volume element $\Delta V = \Delta x \, \Delta y \, \Delta z$ that has its faces parallel to the x, y, and z coordinate surfaces; the point P (see above) is at one corner. We must calculate the surface integral (11.14) over each of its six faces. Remembering that the normal to the surface points outwards from the volume on each face, the two faces with $x = $ constant have areas $\Delta\mathbf{S} = -\mathbf{i} \, \Delta y \, \Delta z$ and $\Delta\mathbf{S} = \mathbf{i} \, \Delta y \, \Delta z$ respectively. Furthermore, over each small surface element, we may take ϕ to be constant, so that the net contribution to the surface integral from these two faces is, to first order in Δx,

$$[(\phi + \Delta\phi) - \phi]\,\Delta y\,\Delta z\,\mathbf{i} = \left(\phi + \frac{\partial\phi}{\partial x}\Delta x - \phi\right)\Delta y\,\Delta z\,\mathbf{i}$$

$$= \frac{\partial\phi}{\partial x}\Delta x\,\Delta y\,\Delta z\,\mathbf{i}.$$

The surface integral over the pairs of faces with $y = $ constant and $z = $ constant respectively may be found in a similar way, and we obtain

$$\oint_S \phi\,d\mathbf{S} = \left(\frac{\partial\phi}{\partial x}\mathbf{i} + \frac{\partial\phi}{\partial y}\mathbf{j} + \frac{\partial\phi}{\partial z}\mathbf{k}\right)\Delta x\,\Delta y\,\Delta z.$$

Therefore $\nabla\phi$ at the point P is given by

$$\nabla\phi = \lim_{\Delta x,\Delta y,\Delta z \to 0}\left[\frac{1}{\Delta x\,\Delta y\,\Delta z}\left(\frac{\partial\phi}{\partial x}\mathbf{i} + \frac{\partial\phi}{\partial y}\mathbf{j} + \frac{\partial\phi}{\partial z}\mathbf{k}\right)\Delta x\,\Delta y\,\Delta z\right]$$

$$= \frac{\partial\phi}{\partial x}\mathbf{i} + \frac{\partial\phi}{\partial y}\mathbf{j} + \frac{\partial\phi}{\partial z}\mathbf{k}.\; \blacktriangleleft$$

We now turn to (11.15) and (11.17). These geometrical definitions may be shown straightforwardly to lead to the usual expressions for div and curl in orthogonal curvilinear coordinates.

▶ *By considering the infinitesimal volume element $dV = h_1 h_2 h_3\,\Delta u_1\,\Delta u_2\,\Delta u_3$ shown in figure 11.10, show that (11.15) leads to the usual expression for $\nabla\cdot\mathbf{a}$ in orthogonal curvilinear coordinates.*

Let us write the vector field in terms of its components with respect to the basis vectors of the curvilinear coordinate system as $\mathbf{a} = a_1\hat{\mathbf{e}}_1 + a_2\hat{\mathbf{e}}_2 + a_3\hat{\mathbf{e}}_3$. We consider first the contribution to the RHS of (11.15) from the two faces with $u_1 = $ constant, i.e. $PQRS$ and the face opposite it (see figure 11.10). Now, the volume element is formed from the orthogonal vectors $h_1\,\Delta u_1\,\hat{\mathbf{e}}_1$, $h_2\,\Delta u_2\,\hat{\mathbf{e}}_2$ and $h_3\,\Delta u_3\,\hat{\mathbf{e}}_3$ at the point P and so for $PQRS$ we have

$$\Delta\mathbf{S} = h_2 h_3\,\Delta u_2\,\Delta u_3\,\hat{\mathbf{e}}_3 \times \hat{\mathbf{e}}_2 = -h_2 h_3\,\Delta u_2\,\Delta u_3\,\hat{\mathbf{e}}_1.$$

Reasoning along the same lines as in the previous example, we conclude that the contribution to the surface integral of $\mathbf{a}\cdot d\mathbf{S}$ over $PQRS$ and its opposite face taken together is given by

$$\frac{\partial}{\partial u_1}(\mathbf{a}\cdot\Delta\mathbf{S})\,\Delta u_1 = \frac{\partial}{\partial u_1}(a_1 h_2 h_3)\,\Delta u_1\,\Delta u_2\,\Delta u_3.$$

The surface integrals over the pairs of faces with $u_2 = $ constant and $u_3 = $ constant respectively may be found in a similar way, and we obtain

$$\oint_S \mathbf{a}\cdot d\mathbf{S} = \left[\frac{\partial}{\partial u_1}(a_1 h_2 h_3) + \frac{\partial}{\partial u_2}(a_2 h_3 h_1) + \frac{\partial}{\partial u_3}(a_3 h_1 h_2)\right]\Delta u_1\,\Delta u_2\,\Delta u_3.$$

Therefore $\nabla\cdot\mathbf{a}$ at the point P is given by

$$\nabla\cdot\mathbf{a} = \lim_{\Delta u_1,\Delta u_2,\Delta u_3 \to 0}\left[\frac{1}{h_1 h_2 h_3\,\Delta u_1\,\Delta u_2\,\Delta u_3}\oint_S \mathbf{a}\cdot d\mathbf{S}\right]$$

$$= \frac{1}{h_1 h_2 h_3}\left[\frac{\partial}{\partial u_1}(a_1 h_2 h_3) + \frac{\partial}{\partial u_2}(a_2 h_3 h_1) + \frac{\partial}{\partial u_3}(a_3 h_1 h_2)\right].\; \blacktriangleleft$$

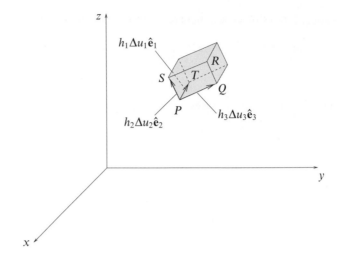

Figure 11.10 A general volume ΔV in orthogonal curvilinear coordinates u_1, u_2, u_3. PT gives the vector $h_1 \Delta u_1 \hat{\mathbf{e}}_1$, PS gives $h_2 \Delta u_2 \hat{\mathbf{e}}_2$ and PQ gives $h_3 \Delta u_3 \hat{\mathbf{e}}_3$.

▶By considering the infinitesimal planar surface element $PQRS$ in figure 11.10, show that (11.17) leads to the usual expression for $\nabla \times \mathbf{a}$ in orthogonal curvilinear coordinates.

The planar surface $PQRS$ is defined by the orthogonal vectors $h_2 \Delta u_2 \hat{\mathbf{e}}_2$ and $h_3 \Delta u_3 \hat{\mathbf{e}}_3$ at the point P. If we traverse the loop in the direction $PSRQ$ then, by the right-hand convention, the unit normal to the plane is $\hat{\mathbf{e}}_1$. Writing $\mathbf{a} = a_1 \hat{\mathbf{e}}_1 + a_2 \hat{\mathbf{e}}_2 + a_3 \hat{\mathbf{e}}_3$, the line integral around the loop in this direction is given by

$$\oint_{PSRQ} \mathbf{a} \cdot d\mathbf{r} = a_2 h_2 \, \Delta u_2 + \left[a_3 h_3 + \frac{\partial}{\partial u_2}(a_3 h_3) \, \Delta u_2 \right] \Delta u_3$$

$$- \left[a_2 h_2 + \frac{\partial}{\partial u_3}(a_2 h_2) \, \Delta u_3 \right] \Delta u_2 - a_3 h_3 \, \Delta u_3$$

$$= \left[\frac{\partial}{\partial u_2}(a_3 h_3) - \frac{\partial}{\partial u_3}(a_2 h_2) \right] \Delta u_2 \, \Delta u_3.$$

Therefore from (11.17) the component of $\nabla \times \mathbf{a}$ in the direction $\hat{\mathbf{e}}_1$ at P is given by

$$(\nabla \times \mathbf{a})_1 = \lim_{\Delta u_2, \Delta u_3 \to 0} \left[\frac{1}{h_2 h_3 \, \Delta u_2 \, \Delta u_3} \oint_{PSRQ} \mathbf{a} \cdot d\mathbf{r} \right]$$

$$= \frac{1}{h_2 h_3} \left[\frac{\partial}{\partial u_2}(h_3 a_3) - \frac{\partial}{\partial u_3}(h_2 a_2) \right].$$

The other two components are found by cyclically permuting the subscripts 1, 2, 3. ◀

Finally, we note that we can also write the ∇^2 operator as a surface integral by setting $\mathbf{a} = \nabla \phi$ in (11.15), to obtain

$$\nabla^2 \phi = \nabla \cdot \nabla \phi = \lim_{V \to 0} \left(\frac{1}{V} \oint_S \nabla \phi \cdot d\mathbf{S} \right).$$

11.8 Divergence theorem and related theorems

The divergence theorem relates the total flux of a vector field out of a closed surface S to the integral of the divergence of the vector field over the enclosed volume V; it follows almost immediately from our geometrical definition of divergence (11.15).

Imagine a volume V, in which a vector field $\mathbf{a}$ is continuous and differentiable, to be divided up into a large number of small volumes V_i. Using (11.15), we have for each small volume

$$(\nabla \cdot \mathbf{a})V_i \approx \oint_{S_i} \mathbf{a} \cdot d\mathbf{S},$$

where S_i is the surface of the small volume V_i. Summing over i we find that contributions from surface elements interior to S cancel since each surface element appears in two terms with opposite signs, the outward normals in the two terms being equal and opposite. Only contributions from surface elements that are also parts of S survive. If each V_i is allowed to tend to zero then we obtain the *divergence theorem*,

$$\int_V \nabla \cdot \mathbf{a}\, dV = \oint_S \mathbf{a} \cdot d\mathbf{S}. \tag{11.18}$$

We note that the divergence theorem holds for both simply and multiply connected surfaces, provided that they are closed and enclose some non-zero volume V. The divergence theorem may also be extended to tensor fields (see chapter 26).

The theorem finds most use as a tool in formal manipulations, but sometimes it is of value in transforming surface integrals of the form $\int_S \mathbf{a} \cdot d\mathbf{S}$ into volume integrals or vice versa. For example, setting $\mathbf{a} = \mathbf{r}$ we immediately obtain

$$\int_V \nabla \cdot \mathbf{r}\, dV = \int_V 3\, dV = 3V = \oint_S \mathbf{r} \cdot d\mathbf{S},$$

which gives the expression for the volume of a region found in subsection 11.6.1. The use of the divergence theorem is further illustrated in the following example.

> ► *Evaluate the surface integral* $I = \int_S \mathbf{a} \cdot d\mathbf{S}$, *where* $\mathbf{a} = (y - x)\mathbf{i} + x^2 z\, \mathbf{j} + (z + x^2)\mathbf{k}$ *and* S *is the open surface of the hemisphere* $x^2 + y^2 + z^2 = a^2$, $z \geq 0$.

We could evaluate this surface integral directly, but the algebra is somewhat lengthy. We will therefore evaluate it by use of the divergence theorem. Since the latter only holds for closed surfaces enclosing a non-zero volume V, let us first consider the closed surface $S' = S + S_1$, where S_1 is the circular area in the xy-plane given by $x^2 + y^2 \leq a^2$, $z = 0$; S' then encloses a hemispherical volume V. By the divergence theorem we have

$$\int_V \nabla \cdot \mathbf{a}\, dV = \oint_{S'} \mathbf{a} \cdot d\mathbf{S} = \int_S \mathbf{a} \cdot d\mathbf{S} + \int_{S_1} \mathbf{a} \cdot d\mathbf{S}.$$

Now $\nabla \cdot \mathbf{a} = -1 + 0 + 1 = 0$, so we can write

$$\int_S \mathbf{a} \cdot d\mathbf{S} = -\int_{S_1} \mathbf{a} \cdot d\mathbf{S}.$$

401

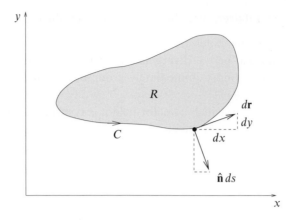

Figure 11.11 A closed curve C in the xy-plane bounding a region R. Vectors tangent and normal to the curve at a given point are also shown.

The surface integral over S_1 is easily evaluated. Remembering that the normal to the surface points outward from the volume, a surface element on S_1 is simply $d\mathbf{S} = -\mathbf{k}\,dx\,dy$. On S_1 we also have $\mathbf{a} = (y-x)\mathbf{i} + x^2\,\mathbf{k}$, so that

$$I = -\int_{S_1} \mathbf{a} \cdot d\mathbf{S} = \iint_R x^2\,dx\,dy,$$

where R is the circular region in the xy-plane given by $x^2 + y^2 \le a^2$. Transforming to plane polar coordinates we have

$$I = \iint_{R'} \rho^2 \cos^2\phi \; \rho\,d\rho\,d\phi = \int_0^{2\pi} \cos^2\phi\,d\phi \int_0^a \rho^3\,d\rho = \frac{\pi a^4}{4}. \; \blacktriangleleft$$

It is also interesting to consider the two-dimensional version of the divergence theorem. As an example, let us consider a two-dimensional planar region R in the xy-plane bounded by some closed curve C (see figure 11.11). At any point on the curve the vector $d\mathbf{r} = dx\,\mathbf{i} + dy\,\mathbf{j}$ is a tangent to the curve and the vector $\hat{\mathbf{n}}\,ds = dy\,\mathbf{i} - dx\,\mathbf{j}$ is a normal pointing out of the region R. If the vector field $\mathbf{a}$ is continuous and differentiable in R then the two-dimensional divergence theorem in Cartesian coordinates gives

$$\iint_R \left(\frac{\partial a_x}{\partial x} + \frac{\partial a_y}{\partial y} \right) dx\,dy = \oint \mathbf{a} \cdot \hat{\mathbf{n}}\,ds = \oint_C (a_x\,dy - a_y\,dx).$$

Letting $P = -a_y$ and $Q = a_x$, we recover Green's theorem in a plane, which was discussed in section 11.3.

11.8.1 Green's theorems

Consider two scalar functions ϕ and ψ that are continuous and differentiable in some volume V bounded by a surface S. Applying the divergence theorem to the vector field $\phi\nabla\psi$ we obtain

$$\oint_S \phi\nabla\psi \cdot d\mathbf{S} = \int_V \nabla \cdot (\phi\nabla\psi)\, dV$$

$$= \int_V \left[\phi\nabla^2\psi + (\nabla\phi) \cdot (\nabla\psi)\right]\, dV. \qquad (11.19)$$

Reversing the roles of ϕ and ψ in (11.19) and subtracting the two equations gives

$$\oint_S (\phi\nabla\psi - \psi\nabla\phi) \cdot d\mathbf{S} = \int_V (\phi\nabla^2\psi - \psi\nabla^2\phi)\, dV. \qquad (11.20)$$

Equation (11.19) is usually known as Green's first theorem and (11.20) as his second. Green's second theorem is useful in the development of the Green's functions used in the solution of partial differential equations (see chapter 21).

11.8.2 Other related integral theorems

There exist two other integral theorems which are closely related to the divergence theorem and which are of some use in physical applications. If ϕ is a scalar field and $\mathbf{b}$ is a vector field and both ϕ and $\mathbf{b}$ satisfy our usual differentiability conditions in some volume V bounded by a closed surface S then

$$\int_V \nabla\phi\, dV = \oint_S \phi\, d\mathbf{S}, \qquad (11.21)$$

$$\int_V \nabla \times \mathbf{b}\, dV = \oint_S d\mathbf{S} \times \mathbf{b}. \qquad (11.22)$$

▶ *Use the divergence theorem to prove (11.21).*

In the divergence theorem (11.18) let $\mathbf{a} = \phi\mathbf{c}$, where $\mathbf{c}$ is a constant vector. We then have

$$\int_V \nabla \cdot (\phi\mathbf{c})\, dV = \oint_S \phi\mathbf{c} \cdot d\mathbf{S}.$$

Expanding out the integrand on the LHS we have

$$\nabla \cdot (\phi\mathbf{c}) = \phi\nabla \cdot \mathbf{c} + \mathbf{c} \cdot \nabla\phi = \mathbf{c} \cdot \nabla\phi,$$

since $\mathbf{c}$ is constant. Also, $\phi\mathbf{c} \cdot d\mathbf{S} = \mathbf{c} \cdot \phi d\mathbf{S}$, so we obtain

$$\int_V \mathbf{c} \cdot (\nabla\phi)\, dV = \oint_S \mathbf{c} \cdot \phi\, d\mathbf{S}.$$

Since $\mathbf{c}$ is constant we may take it out of both integrals to give

$$\mathbf{c} \cdot \int_V \nabla\phi\, dV = \mathbf{c} \cdot \oint_S \phi\, d\mathbf{S},$$

and since $\mathbf{c}$ is arbitrary we obtain the stated result (11.21). ◀

Equation (11.22) may be proved in a similar way by letting $\mathbf{a} = \mathbf{b} \times \mathbf{c}$ in the divergence theorem, where $\mathbf{c}$ is again a constant vector.

11.8.3 Physical applications of the divergence theorem

The divergence theorem is useful in deriving many of the most important partial differential equations in physics (see chapter 20). The basic idea is to use the divergence theorem to convert an integral form, often derived from observation, into an equivalent differential form (used in theoretical statements).

> ▶ *For a compressible fluid with time-varying position-dependent density $\rho(\mathbf{r}, t)$ and velocity field $v(\mathbf{r}, t)$, in which fluid is neither being created nor destroyed, show that*
>
> $$\frac{\partial \rho}{\partial t} + \nabla \cdot (\rho \mathbf{v}) = 0.$$

For an arbitrary volume V in the fluid, the conservation of mass tells us that the rate of increase or decrease of the mass M of fluid in the volume must equal the net rate at which fluid is entering or leaving the volume, i.e.

$$\frac{dM}{dt} = -\oint_S \rho \mathbf{v} \cdot d\mathbf{S},$$

where S is the surface bounding V. But the mass of fluid in V is simply $M = \int_V \rho \, dV$, so we have

$$\frac{d}{dt} \int_V \rho \, dV + \oint_S \rho \mathbf{v} \cdot d\mathbf{S} = 0.$$

Taking the derivative inside the first integral on the RHS and using the divergence theorem to rewrite the second integral, we obtain

$$\int_V \frac{\partial \rho}{\partial t} \, dV + \int_V \nabla \cdot (\rho \mathbf{v}) \, dV = \int_V \left[\frac{\partial \rho}{\partial t} + \nabla \cdot (\rho \mathbf{v}) \right] dV = 0.$$

Since the volume V is arbitrary, the integrand (which is assumed continuous) must be identically zero, so we obtain

$$\frac{\partial \rho}{\partial t} + \nabla \cdot (\rho \mathbf{v}) = 0.$$

This is known as the *continuity equation*. It can also be applied to other systems, for example those in which ρ is the density of electric charge or the heat content, etc. For the flow of an incompressible fluid, $\rho = $ constant and the continuity equation becomes simply $\nabla \cdot \mathbf{v} = 0$. ◀

In the previous example, we assumed that there were no sources or sinks in the volume V, i.e. that there was no part of V in which fluid was being created or destroyed. We now consider the case where a finite number of *point* sources and/or sinks are present in an incompressible fluid. Let us first consider the simple case where a single source is located at the origin, out of which a quantity of fluid flows radially at a rate Q (m^3 s^{-1}). The velocity field is given by

$$\mathbf{v} = \frac{Q\mathbf{r}}{4\pi r^3} = \frac{Q\hat{\mathbf{r}}}{4\pi r^2}.$$

Now, for a sphere S_1 of radius r centred on the source, the flux across S_1 is

$$\oint_{S_1} \mathbf{v} \cdot d\mathbf{S} = |\mathbf{v}| 4\pi r^2 = Q.$$

Since $\mathbf{v}$ has a singularity at the origin it is not differentiable there, i.e. $\nabla \cdot \mathbf{v}$ is not defined there, but at all other points $\nabla \cdot \mathbf{v} = 0$, as required for an incompressible fluid. Therefore, from the divergence theorem, for any closed surface S_2 that does not enclose the origin we have

$$\oint_{S_2} \mathbf{v} \cdot d\mathbf{S} = \int_V \nabla \cdot \mathbf{v}\, dV = 0.$$

Thus we see that the surface integral $\oint_S \mathbf{v} \cdot d\mathbf{S}$ has value Q or zero depending on whether or not S encloses the source. In order that the divergence theorem is valid for *all* surfaces S, irrespective of whether they enclose the source, we write

$$\nabla \cdot \mathbf{v} = Q\delta(\mathbf{r}),$$

where $\delta(\mathbf{r})$ is the three-dimensional Dirac delta function. The properties of this function are discussed fully in chapter 13, but for the moment we note that it is defined in such a way that

$$\delta(\mathbf{r} - \mathbf{a}) = 0 \qquad \text{for } \mathbf{r} \neq \mathbf{a},$$

$$\int_V f(\mathbf{r})\delta(\mathbf{r} - \mathbf{a})\, dV = \begin{cases} f(\mathbf{a}) & \text{if } \mathbf{a} \text{ lies in } V \\ 0 & \text{otherwise} \end{cases}$$

for any well-behaved function $f(\mathbf{r})$. Therefore, for any volume V containing the source at the origin, we have

$$\int_V \nabla \cdot \mathbf{v}\, dV = Q \int_V \delta(\mathbf{r})\, dV = Q,$$

which is consistent with $\oint_S \mathbf{v} \cdot d\mathbf{S} = Q$ for a closed surface enclosing the source. Hence, by introducing the Dirac delta function the divergence theorem can be made valid even for non-differentiable point sources.

The generalisation to several sources and sinks is straightforward. For example, if a source is located at $\mathbf{r} = \mathbf{a}$ and a sink at $\mathbf{r} = \mathbf{b}$ then the velocity field is

$$\mathbf{v} = \frac{(\mathbf{r} - \mathbf{a})Q}{4\pi|\mathbf{r} - \mathbf{a}|^3} - \frac{(\mathbf{r} - \mathbf{b})Q}{4\pi|\mathbf{r} - \mathbf{b}|^3}$$

and its divergence is given by

$$\nabla \cdot \mathbf{v} = Q\delta(\mathbf{r} - \mathbf{a}) - Q\delta(\mathbf{r} - \mathbf{b}).$$

Therefore, the integral $\oint_S \mathbf{v} \cdot d\mathbf{S}$ has the value Q if S encloses the source, $-Q$ if S encloses the sink and 0 if S encloses neither the source nor sink or encloses them both. This analysis also applies to other physical systems – for example, in electrostatics we can regard the sources and sinks as positive and negative point charges respectively and replace $\mathbf{v}$ by the electric field $\mathbf{E}$.

11.9 Stokes' theorem and related theorems

Stokes' theorem is the 'curl analogue' of the divergence theorem and relates the integral of the curl of a vector field over an open surface S to the line integral of the vector field around the perimeter C bounding the surface.

Following the same lines as for the derivation of the divergence theorem, we can divide the surface S into many small areas S_i with boundaries C_i and unit normals $\hat{\mathbf{n}}_i$. Using (11.17), we have for each small area

$$(\nabla \times \mathbf{a}) \cdot \hat{\mathbf{n}}_i \, S_i \approx \oint_{C_i} \mathbf{a} \cdot d\mathbf{r}.$$

Summing over i we find that on the RHS all parts of all interior boundaries that are not part of C are included twice, being traversed in opposite directions on each occasion and thus contributing nothing. Only contributions from line elements that are also parts of C survive. If each S_i is allowed to tend to zero then we obtain Stokes' theorem,

$$\int_S (\nabla \times \mathbf{a}) \cdot d\mathbf{S} = \oint_C \mathbf{a} \cdot d\mathbf{r}. \tag{11.23}$$

We note that Stokes' theorem holds for both simply and multiply connected open surfaces, provided that they are two-sided. Stokes' theorem may also be extended to tensor fields (see chapter 26).

Just as the divergence theorem (11.18) can be used to relate volume and surface integrals for certain types of integrand, Stokes' theorem can be used in evaluating surface integrals of the form $\oint_S (\nabla \times \mathbf{a}) \cdot d\mathbf{S}$ as line integrals or vice versa.

▶ *Given the vector field* $\mathbf{a} = y\,\mathbf{i} - x\,\mathbf{j} + z\,\mathbf{k}$, *verify Stokes' theorem for the hemispherical surface* $x^2 + y^2 + z^2 = a^2$, $z \geq 0$.

Let us first evaluate the surface integral

$$\int_S (\nabla \times \mathbf{a}) \cdot d\mathbf{S}$$

over the hemisphere. It is easily shown that $\nabla \times \mathbf{a} = -2\,\mathbf{k}$, and the surface element is $d\mathbf{S} = a^2 \sin\theta \, d\theta \, d\phi \, \hat{\mathbf{r}}$ in spherical polar coordinates. Therefore

$$\int_S (\nabla \times \mathbf{a}) \cdot d\mathbf{S} = \int_0^{2\pi} d\phi \int_0^{\pi/2} d\theta \, \left(-2a^2 \sin\theta\right) \hat{\mathbf{r}} \cdot \mathbf{k}$$

$$= -2a^2 \int_0^{2\pi} d\phi \int_0^{\pi/2} \sin\theta \left(\frac{z}{a}\right) d\theta$$

$$= -2a^2 \int_0^{2\pi} d\phi \int_0^{\pi/2} \sin\theta \cos\theta \, d\theta = -2\pi a^2.$$

We now evaluate the line integral around the perimeter curve C of the surface, which is the circle $x^2 + y^2 = a^2$ in the xy-plane. This is given by

$$\oint_C \mathbf{a} \cdot d\mathbf{r} = \oint_C (y\,\mathbf{i} - x\,\mathbf{j} + z\,\mathbf{k}) \cdot (dx\,\mathbf{i} + dy\,\mathbf{j} + dz\,\mathbf{k})$$

$$= \oint_C (y\,dx - x\,dy).$$

Using plane polar coordinates, on C we have $x = a\cos\phi$, $y = a\sin\phi$ so that $dx = -a\sin\phi\,d\phi$, $dy = a\cos\phi\,d\phi$, and the line integral becomes

$$\oint_C (y\,dx - x\,dy) = -a^2 \int_0^{2\pi} (\sin^2\phi + \cos^2\phi)\,d\phi = -a^2 \int_0^{2\pi} d\phi = -2\pi a^2.$$

Since the surface and line integrals have the same value, we have verified Stokes' theorem in this case. ◀

The two-dimensional version of Stokes' theorem also yields Green's theorem in a plane. Consider the region R in the xy-plane shown in figure 11.11, in which a vector field $\mathbf{a}$ is defined. Since $\mathbf{a} = a_x\,\mathbf{i} + a_y\,\mathbf{j}$, we have $\nabla \times \mathbf{a} = (\partial a_y/\partial x - \partial a_x/\partial y)\,\mathbf{k}$, and Stokes' theorem becomes

$$\iint_R \left(\frac{\partial a_y}{\partial x} - \frac{\partial a_x}{\partial y} \right) dx\,dy = \oint_C (a_x\,dx + a_y\,dy).$$

Letting $P = a_x$ and $Q = a_y$ we recover Green's theorem in a plane, (11.4).

11.9.1 Related integral theorems

As for the divergence theorem, there exist two other integral theorems that are closely related to Stokes' theorem. If ϕ is a scalar field and $\mathbf{b}$ is a vector field, and both ϕ and $\mathbf{b}$ satisfy our usual differentiability conditions on some two-sided open surface S bounded by a closed perimeter curve C, then

$$\int_S d\mathbf{S} \times \nabla\phi = \oint_C \phi\,d\mathbf{r}, \tag{11.24}$$

$$\int_S (d\mathbf{S} \times \nabla) \times \mathbf{b} = \oint_C d\mathbf{r} \times \mathbf{b}. \tag{11.25}$$

▶ *Use Stokes' theorem to prove (11.24).*

In Stokes' theorem, (11.23), let $\mathbf{a} = \phi\mathbf{c}$, where $\mathbf{c}$ is a constant vector. We then have

$$\int_S [\nabla \times (\phi\mathbf{c})] \cdot d\mathbf{S} = \oint_C \phi\mathbf{c} \cdot d\mathbf{r}. \tag{11.26}$$

Expanding out the integrand on the LHS we have

$$\nabla \times (\phi\mathbf{c}) = \nabla\phi \times \mathbf{c} + \phi\nabla \times \mathbf{c} = \nabla\phi \times \mathbf{c},$$

since $\mathbf{c}$ is constant, and the scalar triple product on the LHS of (11.26) can therefore be written

$$[\nabla \times (\phi\mathbf{c})] \cdot d\mathbf{S} = (\nabla\phi \times \mathbf{c}) \cdot d\mathbf{S} = \mathbf{c} \cdot (d\mathbf{S} \times \nabla\phi).$$

Substituting this into (11.26) and taking $\mathbf{c}$ out of both integrals because it is constant, we find

$$\mathbf{c} \cdot \int_S d\mathbf{S} \times \nabla\phi = \mathbf{c} \cdot \oint_C \phi\, d\mathbf{r}.$$

Since $\mathbf{c}$ is an arbitrary constant vector we therefore obtain the stated result (11.24). ◀

Equation (11.25) may be proved in a similar way, by letting $\mathbf{a} = \mathbf{b} \times \mathbf{c}$ in Stokes' theorem, where $\mathbf{c}$ is again a constant vector. We also note that by setting $\mathbf{b} = \mathbf{r}$ in (11.25) we find

$$\int_S (d\mathbf{S} \times \nabla) \times \mathbf{r} = \oint_C d\mathbf{r} \times \mathbf{r}.$$

Expanding out the integrand on the LHS gives

$$(d\mathbf{S} \times \nabla) \times \mathbf{r} = d\mathbf{S} - d\mathbf{S}(\nabla \cdot \mathbf{r}) = d\mathbf{S} - 3\, d\mathbf{S} = -2\, d\mathbf{S}.$$

Therefore, as we found in subsection 11.5.2, the vector area of an open surface S is given by

$$\mathbf{S} = \int_S d\mathbf{S} = \frac{1}{2} \oint_C \mathbf{r} \times d\mathbf{r}.$$

11.9.2 Physical applications of Stokes' theorem

Like the divergence theorem, Stokes' theorem is useful in converting integral equations into differential equations.

▶*From Ampère's law, derive Maxwell's equation in the case where the currents are steady, i.e.* $\nabla \times \mathbf{B} - \mu_0 \mathbf{J} = \mathbf{0}$.

Ampère's rule for a distributed current with current density $\mathbf{J}$ is

$$\oint_C \mathbf{B} \cdot d\mathbf{r} = \mu_0 \int_S \mathbf{J} \cdot d\mathbf{S},$$

for any circuit C bounding a surface S. Using Stokes' theorem, the LHS can be transformed into $\int_S (\nabla \times \mathbf{B}) \cdot d\mathbf{S}$; hence

$$\int_S (\nabla \times \mathbf{B} - \mu_0 \mathbf{J}) \cdot d\mathbf{S} = 0$$

for *any* surface S. This can only be so if $\nabla \times \mathbf{B} - \mu_0 \mathbf{J} = \mathbf{0}$, which is the required relation. Similarly, from Faraday's law of electromagnetic induction we can derive Maxwell's equation $\nabla \times \mathbf{E} = -\partial \mathbf{B}/\partial t$. ◀

In subsection 11.8.3 we discussed the flow of an incompressible fluid in the presence of several sources and sinks. Let us now consider *vortex* flow in an incompressible fluid with a velocity field

$$\mathbf{v} = \frac{1}{\rho}\hat{\mathbf{e}}_\phi,$$

in cylindrical polar coordinates ρ, ϕ, z. For this velocity field $\nabla \times \mathbf{v}$ equals zero everywhere except on the axis $\rho = 0$, where $\mathbf{v}$ has a singularity. Therefore $\oint_C \mathbf{v} \cdot d\mathbf{r}$ equals zero for any path C that does not enclose the vortex line on the axis and 2π if C does enclose the axis. In order for Stokes' theorem to be valid for all paths C, we therefore set

$$\nabla \times \mathbf{v} = 2\pi \delta(\rho),$$

where $\delta(\rho)$ is the Dirac delta function, to be discussed in subsection 13.1.3. Now, since $\nabla \times \mathbf{v} = \mathbf{0}$, except on the axis $\rho = 0$, there exists a scalar potential ψ such that $\mathbf{v} = \nabla \psi$. It may easily be shown that $\psi = \phi$, the azimuthal angle. Therefore, if C does not enclose the axis then

$$\oint_C \mathbf{v} \cdot d\mathbf{r} = \oint d\phi = 0,$$

and if C does enclose the axis,

$$\oint_C \mathbf{v} \cdot d\mathbf{r} = \Delta\phi = 2\pi n,$$

where n is the number of times we traverse C. Thus ϕ is a multivalued potential.

Similar analyses are valid for other physical systems – for example, in magnetostatics we may replace the vortex lines by current-carrying wires and the velocity field $\mathbf{v}$ by the magnetic field $\mathbf{B}$.

11.10 Exercises

11.1   The vector field $\mathbf{F}$ is defined by

$$\mathbf{F} = 2xz\mathbf{i} + 2yz^2\mathbf{j} + (x^2 + 2y^2z - 1)\mathbf{k}.$$

Calculate $\nabla \times \mathbf{F}$ and deduce that $\mathbf{F}$ can be written $F = \nabla\phi$. Determine the form of ϕ.

11.2   The vector field $\mathbf{Q}$ is defined by

$$\mathbf{Q} = \left[3x^2(y+z) + y^3 + z^3\right]\mathbf{i} + \left[3y^2(z+x) + z^3 + x^3\right]\mathbf{j} + \left[3z^2(x+y) + x^3 + y^3\right]\mathbf{k}.$$

Show that $\mathbf{Q}$ is a conservative field, construct its potential function and hence evaluate the integral $J = \int \mathbf{Q} \cdot d\mathbf{r}$ along any line connecting the point A at $(1, -1, 1)$ to B at $(2, 1, 2)$.

11.3 $\mathbf{F}$ is a vector field $xy^2\mathbf{i} + 2\mathbf{j} + x\mathbf{k}$, and L is a path parameterised by $x = ct$, $y = c/t$, $z = d$ for the range $1 \leq t \leq 2$. Evaluate (a) $\int_L \mathbf{F}\,dt$, (b) $\int_L \mathbf{F}\,dy$ and (c) $\int_L \mathbf{F} \cdot d\mathbf{r}$.

11.4   By making an appropriate choice for the functions $P(x, y)$ and $Q(x, y)$ that appear in Green's theorem in a plane, show that the integral of $x - y$ over the upper half of the unit circle centred on the origin has the value $-\frac{2}{3}$. Show the same result by direct integration in Cartesian coordinates.

11.5   Determine the point of intersection P, in the first quadrant, of the two ellipses

$$\frac{x^2}{a^2} + \frac{y^2}{b^2} = 1 \quad \text{and} \quad \frac{x^2}{b^2} + \frac{y^2}{a^2} = 1.$$

Taking $b < a$, consider the contour L that bounds the area in the first quadrant that is common to the two ellipses. Show that the parts of L that lie along the coordinate axes contribute nothing to the line integral around L of $x\,dy - y\,dx$. Using a parameterisation of each ellipse similar to that employed in the example in section 11.3, evaluate the two remaining line integrals and hence find the total area common to the two ellipses.

11.6 By using parameterisations of the form $x = a\cos^n\theta$ and $y = a\sin^n\theta$ for suitable values of n, find the area bounded by the curves

$$x^{2/5} + y^{2/5} = a^{2/5} \quad \text{and} \quad x^{2/3} + y^{2/3} = a^{2/3}.$$

11.7 Evaluate the line integral

$$I = \oint_C \left[y(4x^2 + y^2)\,dx + x(2x^2 + 3y^2)\,dy \right]$$

around the ellipse $x^2/a^2 + y^2/b^2 = 1$.

11.8 Criticise the following 'proof' that $\pi = 0$.

(a) Apply Green's theorem in a plane to the functions $P(x, y) = \tan^{-1}(y/x)$ and $Q(x, y) = \tan^{-1}(x/y)$, taking the region R to be the unit circle centred on the origin.

(b) The RHS of the equality so produced is

$$\int\int_R \frac{y - x}{x^2 + y^2}\,dx\,dy,$$

which, either from symmetry considerations or by changing to plane polar coordinates, can be shown to have zero value.

(c) In the LHS of the equality, set $x = \cos\theta$ and $y = \sin\theta$, yielding $P(\theta) = \theta$ and $Q(\theta) = \pi/2 - \theta$. The line integral becomes

$$\int_0^{2\pi} \left[\left(\frac{\pi}{2} - \theta\right)\cos\theta - \theta\sin\theta \right]\,d\theta,$$

which has the value 2π.

(d) Thus $2\pi = 0$ and the stated result follows.

11.9 A single-turn coil C of arbitrary shape is placed in a magnetic field $\mathbf{B}$ and carries a current I. Show that the couple acting upon the coil can be written as

$$\mathbf{M} = I \int_C (\mathbf{B} \cdot \mathbf{r})\,d\mathbf{r} - I \int_C \mathbf{B}(\mathbf{r} \cdot d\mathbf{r}).$$

For a planar rectangular coil of sides $2a$ and $2b$ placed with its plane vertical and at an angle ϕ to a uniform horizontal field $\mathbf{B}$, show that $\mathbf{M}$ is, as expected, $4abBI\cos\phi\,\mathbf{k}$.

11.10 Find the vector area $\mathbf{S}$ of the part of the curved surface of the hyperboloid of revolution

$$\frac{x^2}{a^2} - \frac{y^2 + z^2}{b^2} = 1$$

that lies in the region $z \geq 0$ and $a \leq x \leq \lambda a$.

11.11 An axially symmetric solid body with its axis AB vertical is immersed in an incompressible fluid of density ρ_0. Use the following method to show that, whatever the shape of the body, for $\rho = \rho(z)$ in cylindrical polars the Archimedean upthrust is, as expected, $\rho_0 g V$, where V is the volume of the body.

Express the vertical component of the resultant force on the body, $-\int p\,d\mathbf{S}$, where p is the pressure, in terms of an integral; note that $p = -\rho_0 g z$ and that for an annular surface element of width dl, $\mathbf{n} \cdot \mathbf{n}_z\,dl = -d\rho$. Integrate by parts and use the fact that $\rho(z_A) = \rho(z_B) = 0$.

11.12   Show that the expression below is equal to the solid angle subtended by a rectangular aperture, of sides $2a$ and $2b$, at a point on the normal through its centre, and at a distance c from the aperture:

$$\Omega = 4 \int_0^b \frac{ac}{(y^2 + c^2)(y^2 + c^2 + a^2)^{1/2}} \, dy.$$

By setting $y = (a^2 + c^2)^{1/2} \tan \phi$, change this integral into the form

$$\int_0^{\phi_1} \frac{4ac \cos \phi}{c^2 + a^2 \sin^2 \phi} \, d\phi,$$

where $\tan \phi_1 = b/(a^2 + c^2)^{1/2}$, and hence show that

$$\Omega = 4 \tan^{-1} \left[\frac{ab}{c(a^2 + b^2 + c^2)^{1/2}} \right].$$

11.13   A vector field $\mathbf{a}$ is given by $-zxr^{-3}\mathbf{i} - zyr^{-3}\mathbf{j} + (x^2+y^2)r^{-3}\mathbf{k}$, where $r^2 = x^2+y^2+z^2$. Establish that the field is conservative (a) by showing that $\nabla \times \mathbf{a} = \mathbf{0}$, and (b) by constructing its potential function ϕ.

11.14   A vector field $\mathbf{a}$ is given by $(z^2 + 2xy)\mathbf{i} + (x^2 + 2yz)\mathbf{j} + (y^2 + 2zx)\mathbf{k}$. Show that $\mathbf{a}$ is conservative and that the line integral $\int \mathbf{a} \cdot d\mathbf{r}$ along any line joining $(1,1,1)$ and $(1,2,2)$ has the value 11.

11.15   A force $\mathbf{F}(\mathbf{r})$ acts on a particle at $\mathbf{r}$. In which of the following cases can $\mathbf{F}$ be represented in terms of a potential? Where it can, find the potential.

(a) $\mathbf{F} = F_0 \left[\mathbf{i} - \mathbf{j} - \dfrac{2(x-y)}{a^2} \mathbf{r} \right] \exp\left(-\dfrac{r^2}{a^2} \right)$;

(b) $\mathbf{F} = \dfrac{F_0}{a} \left[z\mathbf{k} + \dfrac{(x^2 + y^2 - a^2)}{a^2} \mathbf{r} \right] \exp\left(-\dfrac{r^2}{a^2} \right)$;

(c) $\mathbf{F} = F_0 \left[\mathbf{k} + \dfrac{a(\mathbf{r} \times \mathbf{k})}{r^2} \right]$.

11.16   One of Maxwell's electromagnetic equations states that all magnetic fields $\mathbf{B}$ are solenoidal (i.e. $\nabla \cdot \mathbf{B} = 0$). Determine whether each of the following vectors could represent a real magnetic field; where it could, try to find a suitable vector potential $\mathbf{A}$, i.e. such that $\mathbf{B} = \nabla \times \mathbf{A}$. (Hint: seek a vector potential that is parallel to $\nabla \times \mathbf{B}$.):

(a) $\dfrac{B_0 b}{r^3} [(x - y)z\,\mathbf{i} + (x - y)z\,\mathbf{j} + (x^2 - y^2)\,\mathbf{k}]$ in Cartesians with $r^2 = x^2 + y^2 + z^2$;

(b) $\dfrac{B_0 b^3}{r^3} [\cos\theta\ \cos\phi\,\hat{\mathbf{e}}_r - \sin\theta\ \cos\phi\,\hat{\mathbf{e}}_\theta + \sin 2\theta\ \sin\phi\,\hat{\mathbf{e}}_\phi]$ in spherical polars;

(c) $B_0 b^2 \left[\dfrac{z\rho}{(b^2 + z^2)^2}\,\hat{\mathbf{e}}_\rho + \dfrac{1}{b^2 + z^2}\,\hat{\mathbf{e}}_z \right]$ in cylindrical polars.

11.17   The vector field $\mathbf{f}$ has components $y\mathbf{i} - x\mathbf{j} + \mathbf{k}$ and γ is a curve given parametrically by

$$\mathbf{r} = (a - c + c\cos\theta)\mathbf{i} + (b + c\sin\theta)\mathbf{j} + c^2\theta\mathbf{k}, \quad 0 \le \theta \le 2\pi.$$

Describe the shape of the path γ and show that the line integral $\int_\gamma \mathbf{f} \cdot d\mathbf{r}$ vanishes. Does this result imply that $\mathbf{f}$ is a conservative field?

11.18   A vector field $\mathbf{a} = f(r)\mathbf{r}$ is spherically symmetric and everywhere directed away from the origin. Show that $\mathbf{a}$ is irrotational, but that it is also solenoidal only if $f(r)$ is of the form Ar^{-3}.

11.19   Evaluate the surface integral $\int \mathbf{r} \cdot d\mathbf{S}$, where $\mathbf{r}$ is the position vector, over that part of the surface $z = a^2 - x^2 - y^2$ for which $z \geq 0$, by each of the following methods.

(a) Parameterise the surface as $x = a \sin\theta \cos\phi$, $y = a \sin\theta \sin\phi$, $z = a^2 \cos^2\theta$, and show that

$$\mathbf{r} \cdot d\mathbf{S} = a^4(2 \sin^3\theta \cos\theta + \cos^3\theta \sin\theta) \, d\theta \, d\phi.$$

(b) Apply the divergence theorem to the volume bounded by the surface and the plane $z = 0$.

11.20   Obtain an expression for the value ϕ_P at a point P of a scalar function ϕ that satisfies $\nabla^2\phi = 0$, in terms of its value and normal derivative on a surface S that encloses it, by proceeding as follows.

(a) In Green's second theorem, take ψ at any particular point Q as $1/r$, where r is the distance of Q from P. Show that $\nabla^2\psi = 0$, except at $r = 0$.

(b) Apply the result to the doubly connected region bounded by S and a small sphere Σ of radius δ centred on P.

(c) Apply the divergence theorem to show that the surface integral over Σ involving $1/\delta$ vanishes, and prove that the term involving $1/\delta^2$ has the value $4\pi\phi_P$.

(d) Conclude that

$$\phi_P = -\frac{1}{4\pi} \int_S \phi \frac{\partial}{\partial n}\left(\frac{1}{r}\right) dS + \frac{1}{4\pi} \int_S \frac{1}{r} \frac{\partial\phi}{\partial n} \, dS.$$

This important result shows that the value at a point P of a function ϕ that satisfies $\nabla^2\phi = 0$ everywhere within a closed surface S that encloses P may be expressed *entirely* in terms of its value and normal derivative on S. This matter is taken up more generally in connection with Green's functions in chapter 21 and in connection with functions of a complex variable in section 24.10.

11.21   Use result (11.21), together with an appropriately chosen scalar function ϕ, to prove that the position vector $\bar{\mathbf{r}}$ of the centre of mass of an arbitrarily shaped body of volume V and uniform density can be written

$$\bar{\mathbf{r}} = \frac{1}{V} \oint_S \tfrac{1}{2} r^2 \, d\mathbf{S}.$$

11.22   A rigid body of volume V and surface S rotates with angular velocity $\boldsymbol{\omega}$. Show that

$$\boldsymbol{\omega} = -\frac{1}{2V} \oint_S \mathbf{u} \times d\mathbf{S},$$

where $\mathbf{u}(\mathbf{x})$ is the velocity of the point $\mathbf{x}$ on the surface S.

11.23   Demonstrate the validity of the divergence theorem:

(a) by calculating the flux of the vector

$$\mathbf{F} = \frac{\alpha\mathbf{r}}{(r^2 + a^2)^{3/2}}$$

through the spherical surface $|\mathbf{r}| = \sqrt{3}a$;

(b) by showing that

$$\nabla \cdot \mathbf{F} = \frac{3\alpha a^2}{(r^2 + a^2)^{5/2}}$$

and evaluating the volume integral of $\nabla \cdot \mathbf{F}$ over the interior of the sphere $|\mathbf{r}| = \sqrt{3}a$. The substitution $r = a \tan\theta$ will prove useful in carrying out the integration.

11.24   Prove equation (11.22) and, by taking $\mathbf{b} = zx^2\mathbf{i} + zy^2\mathbf{j} + (x^2 - y^2)\mathbf{k}$, show that the two integrals

$$I = \int x^2 \, dV \quad \text{and} \quad J = \int \cos^2 \theta \sin^3 \theta \cos^2 \phi \, d\theta \, d\phi,$$

both taken over the unit sphere, must have the same value. Evaluate both directly to show that the common value is $4\pi/15$.

11.25   In a uniform conducting medium with unit relative permittivity, charge density ρ, current density $\mathbf{J}$, electric field $\mathbf{E}$ and magnetic field $\mathbf{B}$, Maxwell's electromagnetic equations take the form (with $\mu_0 \epsilon_0 = c^{-2}$)

(i) $\nabla \cdot \mathbf{B} = 0$,   (ii) $\nabla \cdot \mathbf{E} = \rho/\epsilon_0$,
(iii) $\nabla \times \mathbf{E} + \dot{\mathbf{B}} = \mathbf{0}$,   (iv) $\nabla \times \mathbf{B} - (\dot{\mathbf{E}}/c^2) = \mu_0 \mathbf{J}$.

The density of stored energy in the medium is given by $\frac{1}{2}(\epsilon_0 E^2 + \mu_0^{-1} B^2)$. Show that the rate of change of the total stored energy in a volume V is equal to

$$-\int_V \mathbf{J} \cdot \mathbf{E} \, dV - \frac{1}{\mu_0} \oint_S (\mathbf{E} \times \mathbf{B}) \cdot d\mathbf{S},$$

where S is the surface bounding V.

[The first integral gives the ohmic heating loss, whilst the second gives the electromagnetic energy flux out of the bounding surface. The vector $\mu_0^{-1}(\mathbf{E} \times \mathbf{B})$ is known as the Poynting vector.]

11.26   A vector field $\mathbf{F}$ is defined in cylindrical polar coordinates ρ, θ, z by

$$\mathbf{F} = F_0 \left(\frac{x \cos \lambda z}{a} \mathbf{i} + \frac{y \cos \lambda z}{a} \mathbf{j} + (\sin \lambda z)\mathbf{k} \right) \equiv \frac{F_0 \rho}{a}(\cos \lambda z)\mathbf{e}_\rho + F_0(\sin \lambda z)\mathbf{k},$$

where $\mathbf{i}, \mathbf{j}$ and $\mathbf{k}$ are the unit vectors along the Cartesian axes and $\mathbf{e}_\rho$ is the unit vector $(x/\rho)\mathbf{i} + (y/\rho)\mathbf{j}$.

(a) Calculate, as a surface integral, the flux of $\mathbf{F}$ through the closed surface bounded by the cylinders $\rho = a$ and $\rho = 2a$ and the planes $z = \pm a\pi/2$.
(b) Evaluate the same integral using the divergence theorem.

11.27   The vector field $\mathbf{F}$ is given by

$$\mathbf{F} = (3x^2 yz + y^3 z + xe^{-x})\mathbf{i} + (3xy^2 z + x^3 z + ye^x)\mathbf{j} + (x^3 y + y^3 x + xy^2 z^2)\mathbf{k}.$$

Calculate (a) directly, and (b) by using Stokes' theorem the value of the line integral $\int_L \mathbf{F} \cdot d\mathbf{r}$, where L is the (three-dimensional) closed contour $OABCDEO$ defined by the successive vertices $(0,0,0)$, $(1,0,0)$, $(1,0,1)$, $(1,1,1)$, $(1,1,0)$, $(0,1,0)$, $(0,0,0)$.

11.28   A vector force field $\mathbf{F}$ is defined in Cartesian coordinates by

$$\mathbf{F} = F_0 \left[\left(\frac{y^3}{3a^3} + \frac{y}{a}e^{xy/a^2} + 1 \right)\mathbf{i} + \left(\frac{xy^2}{a^3} + \frac{x+y}{a}e^{xy/a^2} \right)\mathbf{j} + \frac{z}{a}e^{xy/a^2}\mathbf{k} \right].$$

Use Stokes' theorem to calculate

$$\oint_L \mathbf{F} \cdot d\mathbf{r},$$

where L is the perimeter of the rectangle $ABCD$ given by $A = (0,1,0)$, $B = (1,1,0)$, $C = (1,3,0)$ and $D = (0,3,0)$.

11.11 Hints and answers

11.1   Show that $\nabla \times \mathbf{F} = \mathbf{0}$. The potential $\phi_F(\mathbf{r}) = x^2 z + y^2 z^2 - z$.

11.3   (a) $c^3 \ln 2 \, \mathbf{i} + 2 \mathbf{j} + (3c/2)\mathbf{k}$; (b) $(-3c^4/8)\mathbf{i} - c\,\mathbf{j} - (c^2 \ln 2)\mathbf{k}$; (c) $c^4 \ln 2 - c$.

11.5   For P, $x = y = ab/(a^2 + b^2)^{1/2}$. The relevant limits are $0 \leq \theta_1 \leq \tan^{-1}(b/a)$ and $\tan^{-1}(a/b) \leq \theta_2 \leq \pi/2$. The total common area is $4ab \tan^{-1}(b/a)$.

11.7   Show that, in the notation of section 11.3, $\partial Q/\partial x - \partial P/\partial y = 2x^2$; $I = \pi a^3 b/2$.

11.9 $\mathbf{M} = I \int_C \mathbf{r} \times (d\mathbf{r} \times \mathbf{B})$. Show that the horizontal sides in the first term and the whole of the second term contribute nothing to the couple.

11.11   Note that, if $\hat{\mathbf{n}}$ is the outward normal to the surface, $\hat{\mathbf{n}}_z \cdot \hat{\mathbf{n}} \, dl$ is equal to $-d\rho$.

11.13   (b) $\phi = c + z/r$.

11.15   (a) Yes, $F_0(x - y)\exp(-r^2/a^2)$; (b) yes, $-F_0[(x^2 + y^2)/(2a)]\exp(-r^2/a^2)$; (c) no, $\nabla \times \mathbf{F} \neq \mathbf{0}$.

11.17   A spiral of radius c with its axis parallel to the z-direction and passing through (a, b). The pitch of the spiral is $2\pi c^2$. No, because (i) γ is not a closed loop and (ii) the line integral must be zero for *every* closed loop, not just for a particular one. In fact $\nabla \times \mathbf{f} = -2\mathbf{k} \neq \mathbf{0}$ shows that $\mathbf{f}$ is not conservative.

11.19   (a) $d\mathbf{S} = (2a^3 \cos\theta \sin^2\theta \cos\phi \, \mathbf{i} + 2a^3 \cos\theta \sin^2\theta \sin\phi \, \mathbf{j} + a^2 \cos\theta \sin\theta \, \mathbf{k}) \, d\theta \, d\phi$.
        (b) $\nabla \cdot \mathbf{r} = 3$; over the plane $z = 0$, $\mathbf{r} \cdot d\mathbf{S} = 0$.
        The necessarily common value is $3\pi a^4/2$.

11.21   Write $\mathbf{r}$ as $\nabla(\frac{1}{2}r^2)$.

11.23   The answer is $3\sqrt{3}\pi a/2$ in each case.

11.25   Identify the expression for $\nabla \cdot (\mathbf{E} \times \mathbf{B})$ and use the divergence theorem.

11.27   (a) The successive contributions to the integral are:
        $1 - 2e^{-1}$, 0, $2 + \frac{1}{2}e$, $-\frac{7}{3}$, $-1 + 2e^{-1}$, $-\frac{1}{2}$.
        (b) $\nabla \times \mathbf{F} = 2xyz^2\mathbf{i} - y^2z^2\mathbf{j} + ye^x\mathbf{k}$. Show that the contour is equivalent to the sum of two plane square contours in the planes $z = 0$ and $x = 1$, the latter being traversed in the negative sense. Integral $= \frac{1}{6}(3e - 5)$.

12

Fourier series

We have already discussed, in chapter 4, how complicated functions may be expressed as power series. However, this is not the only way in which a function may be represented as a series, and the subject of this chapter is the expression of functions as a sum of sine and cosine terms. Such a representation is called a *Fourier series*. Unlike Taylor series, a Fourier series can describe functions that are not everywhere continuous and/or differentiable. There are also other advantages in using trigonometric terms. They are easy to differentiate and integrate, their moduli are easily taken and each term contains only one characteristic frequency. This last point is important because, as we shall see later, Fourier series are often used to represent the response of a system to a periodic input, and this response often depends directly on the frequency content of the input. Fourier series are used in a wide variety of such physical situations, including the vibrations of a finite string, the scattering of light by a diffraction grating and the transmission of an input signal by an electronic circuit.

12.1 The Dirichlet conditions

We have already mentioned that Fourier series may be used to represent some functions for which a Taylor series expansion is not possible. The particular conditions that a function $f(x)$ must fulfil in order that it may be expanded as a Fourier series are known as the *Dirichlet conditions*, and may be summarised by the following four points:

(i) the function must be periodic;
(ii) it must be single-valued and continuous, except possibly at a finite number of finite discontinuities;
(iii) it must have only a finite number of maxima and minima within one period;
(iv) the integral over one period of $|f(x)|$ must converge.

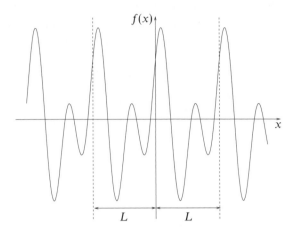

Figure 12.1   An example of a function that may be represented as a Fourier series without modification.

If the above conditions are satisfied then the Fourier series converges to $f(x)$ at all points where $f(x)$ is continuous. The convergence of the Fourier series at points of discontinuity is discussed in section 12.4. The last three Dirichlet conditions are almost always met in real applications, but not all functions are periodic and hence do not fulfil the first condition. It may be possible, however, to represent a non-periodic function as a Fourier series by manipulation of the function into a periodic form. This is discussed in section 12.5. An example of a function that may, without modification, be represented as a Fourier series is shown in figure 12.1.

We have stated without proof that any function that satisfies the Dirichlet conditions may be represented as a Fourier series. Let us now show why this is a plausible statement. We require that any reasonable function (one that satisfies the Dirichlet conditions) can be expressed as a linear sum of sine and cosine terms. We first note that we cannot use just a sum of sine terms since sine, being an odd function (i.e. a function for which $f(-x) = -f(x)$), cannot represent even functions (i.e. functions for which $f(-x) = f(x)$). This is obvious when we try to express a function $f(x)$ that takes a non-zero value at $x = 0$. Clearly, since $\sin nx = 0$ for all values of n, we cannot represent $f(x)$ at $x = 0$ by a sine series. Similarly odd functions cannot be represented by a cosine series since cosine is an even function. Nevertheless, it is possible to represent *all* odd functions by a sine series and *all* even functions by a cosine series. Now, since all functions may be written as the sum of an odd and an even part,

$$f(x) = \tfrac{1}{2}[f(x) + f(-x)] + \tfrac{1}{2}[f(x) - f(-x)]$$
$$= f_{\text{even}}(x) + f_{\text{odd}}(x),$$

416

we can write any function as the sum of a sine series and a cosine series.

All the terms of a Fourier series are mutually orthogonal, i.e. the integrals, over one period, of the product of any two terms have the following properties:

$$\int_{x_0}^{x_0+L} \sin\left(\frac{2\pi rx}{L}\right) \cos\left(\frac{2\pi px}{L}\right) dx = 0 \quad \text{for all } r \text{ and } p, \tag{12.1}$$

$$\int_{x_0}^{x_0+L} \cos\left(\frac{2\pi rx}{L}\right) \cos\left(\frac{2\pi px}{L}\right) dx = \begin{cases} L & \text{for } r = p = 0, \\ \frac{1}{2}L & \text{for } r = p > 0, \\ 0 & \text{for } r \neq p, \end{cases} \tag{12.2}$$

$$\int_{x_0}^{x_0+L} \sin\left(\frac{2\pi rx}{L}\right) \sin\left(\frac{2\pi px}{L}\right) dx = \begin{cases} 0 & \text{for } r = p = 0, \\ \frac{1}{2}L & \text{for } r = p > 0, \\ 0 & \text{for } r \neq p, \end{cases} \tag{12.3}$$

where r and p are integers greater than or equal to zero; these formulae are easily derived. A full discussion of why it is possible to expand a function as a sum of mutually orthogonal functions is given in chapter 17.

The Fourier series expansion of the function $f(x)$ is conventionally written

$$f(x) = \frac{a_0}{2} + \sum_{r=1}^{\infty} \left[a_r \cos\left(\frac{2\pi rx}{L}\right) + b_r \sin\left(\frac{2\pi rx}{L}\right) \right], \tag{12.4}$$

where a_0, a_r, b_r are constants called the *Fourier coefficients*. These coefficients are analogous to those in a power series expansion and the determination of their numerical values is the essential step in writing a function as a Fourier series.

This chapter continues with a discussion of how to find the Fourier coefficients for particular functions. We then discuss simplifications to the general Fourier series that may save considerable effort in calculations. This is followed by the alternative representation of a function as a complex Fourier series, and we conclude with a discussion of Parseval's theorem.

12.2 The Fourier coefficients

We have indicated that a series that satisfies the Dirichlet conditions may be written in the form (12.4). We now consider how to find the Fourier coefficients for any particular function. For a periodic function $f(x)$ of period L we will find that the Fourier coefficients are given by

$$a_r = \frac{2}{L} \int_{x_0}^{x_0+L} f(x) \cos\left(\frac{2\pi rx}{L}\right) dx, \tag{12.5}$$

$$b_r = \frac{2}{L} \int_{x_0}^{x_0+L} f(x) \sin\left(\frac{2\pi rx}{L}\right) dx, \tag{12.6}$$

where x_0 is arbitrary but is often taken as 0 or $-L/2$. The apparently arbitrary factor $\frac{1}{2}$ which appears in the a_0 term in (12.4) is included so that (12.5) may apply for $r = 0$ as well as $r > 0$. The relations (12.5) and (12.6) may be derived as follows.

Suppose the Fourier series expansion of $f(x)$ can be written as in (12.4),

$$f(x) = \frac{a_0}{2} + \sum_{r=1}^{\infty} \left[a_r \cos\left(\frac{2\pi rx}{L}\right) + b_r \sin\left(\frac{2\pi rx}{L}\right) \right].$$

Then, multiplying by $\cos(2\pi px/L)$, integrating over one full period in x and changing the order of the summation and integration, we get

$$\int_{x_0}^{x_0+L} f(x) \cos\left(\frac{2\pi px}{L}\right) dx = \frac{a_0}{2} \int_{x_0}^{x_0+L} \cos\left(\frac{2\pi px}{L}\right) dx$$

$$+ \sum_{r=1}^{\infty} a_r \int_{x_0}^{x_0+L} \cos\left(\frac{2\pi rx}{L}\right) \cos\left(\frac{2\pi px}{L}\right) dx$$

$$+ \sum_{r=1}^{\infty} b_r \int_{x_0}^{x_0+L} \sin\left(\frac{2\pi rx}{L}\right) \cos\left(\frac{2\pi px}{L}\right) dx.$$

$$(12.7)$$

We can now find the Fourier coefficients by considering (12.7) as p takes different values. Using the orthogonality conditions (12.1)–(12.3) of the previous section, we find that when $p = 0$ (12.7) becomes

$$\int_{x_0}^{x_0+L} f(x) dx = \frac{a_0}{2} L.$$

When $p \neq 0$ the only non-vanishing term on the RHS of (12.7) occurs when $r = p$, and so

$$\int_{x_0}^{x_0+L} f(x) \cos\left(\frac{2\pi rx}{L}\right) dx = \frac{a_r}{2} L.$$

The other Fourier coefficients b_r may be found by repeating the above process but multiplying by $\sin(2\pi px/L)$ instead of $\cos(2\pi px/L)$ (see exercise 12.2).

▶*Express the square-wave function illustrated in figure 12.2 as a Fourier series.*

Physically this might represent the input to an electrical circuit that switches between a high and a low state with time period T. The square wave may be represented by

$$f(t) = \begin{cases} -1 & \text{for } -\tfrac{1}{2}T \leq t < 0, \\ +1 & \text{for } 0 \leq t < \tfrac{1}{2}T. \end{cases}$$

In deriving the Fourier coefficients, we note firstly that the function is an odd function and so the series will contain only sine terms (this simplification is discussed further in the

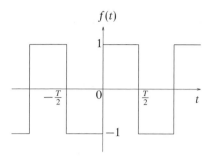

Figure 12.2   A square-wave function.

following section). To evaluate the coefficients in the sine series we use (12.6). Hence

$$b_r = \frac{2}{T} \int_{-T/2}^{T/2} f(t) \sin\left(\frac{2\pi rt}{T}\right) dt$$

$$= \frac{4}{T} \int_{0}^{T/2} \sin\left(\frac{2\pi rt}{T}\right) dt$$

$$= \frac{2}{\pi r} \left[1 - (-1)^r\right].$$

Thus the sine coefficients are zero if r is even and equal to $4/(\pi r)$ if r is odd. Hence the Fourier series for the square-wave function may be written as

$$f(t) = \frac{4}{\pi}\left(\sin \omega t + \frac{\sin 3\omega t}{3} + \frac{\sin 5\omega t}{5} + \cdots\right), \qquad (12.8)$$

where $\omega = 2\pi/T$ is called the *angular frequency*. ◄

12.3 Symmetry considerations

The example in the previous section employed the useful property that since the function to be represented was odd, all the cosine terms of the Fourier series were absent. It is often the case that the function we wish to express as a Fourier series has a particular symmetry, which we can exploit to reduce the calculational labour of evaluating Fourier coefficients. Functions that are symmetric or antisymmetric about the origin (i.e. even and odd functions respectively) admit particularly useful simplifications. Functions that are odd in x have no cosine terms (see section 12.1) and all the a-coefficients are equal to zero. Similarly, functions that are even in x have no sine terms and all the b-coefficients are zero. Since the Fourier series of odd or even functions contain only half the coefficients required for a general periodic function, there is a considerable reduction in the algebra needed to find a Fourier series.

The consequences of symmetry or antisymmetry of the function about the quarter period (i.e. about $L/4$) are a little less obvious. Furthermore, the results are not used as often as those above and the remainder of this section can be omitted on a first reading without loss of continuity. The following argument gives the required results.

Suppose that $f(x)$ has even or odd symmetry about $L/4$, i.e. $f(L/4 - x) = \pm f(x - L/4)$. For convenience, we make the substitution $s = x - L/4$ and hence $f(-s) = \pm f(s)$. We can now see that

$$b_r = \frac{2}{L} \int_{x_0}^{x_0+L} f(s) \sin\left(\frac{2\pi rs}{L} + \frac{\pi r}{2}\right) ds,$$

where the limits of integration have been left unaltered since f is, of course, periodic in s as well as in x. If we use the expansion

$$\sin\left(\frac{2\pi rs}{L} + \frac{\pi r}{2}\right) = \sin\left(\frac{2\pi rs}{L}\right) \cos\left(\frac{\pi r}{2}\right) + \cos\left(\frac{2\pi rs}{L}\right) \sin\left(\frac{\pi r}{2}\right),$$

we can immediately see that the trigonometric part of the integrand is an odd function of s if r is even and an even function of s if r is odd. Hence if $f(s)$ is even and r is even then the integral is zero, and if $f(s)$ is odd and r is odd then the integral is zero. Similar results can be derived for the Fourier a-coefficients and we conclude that

(i) if $f(x)$ is even about $L/4$ then $a_{2r+1} = 0$ and $b_{2r} = 0$,
(ii) if $f(x)$ is odd about $L/4$ then $a_{2r} = 0$ and $b_{2r+1} = 0$.

All the above results follow automatically when the Fourier coefficients are evaluated in any particular case, but prior knowledge of them will often enable some coefficients to be set equal to zero on inspection and so substantially reduce the computational labour. As an example, the square-wave function shown in figure 12.2 is (i) an odd function of t, so that all $a_r = 0$, and (ii) even about the point $t = T/4$, so that $b_{2r} = 0$. Thus we can say immediately that only sine terms of odd harmonics will be present and therefore will need to be calculated; this is confirmed in the expansion (12.8).

12.4 Discontinuous functions

The Fourier series expansion usually works well for functions that are discontinuous in the required range. However, the series itself does not produce a discontinuous function and we state without proof that the value of the expanded $f(x)$ at a discontinuity will be half-way between the upper and lower values. Expressing this more mathematically, at a point of finite discontinuity, x_d, the Fourier series converges to

$$\tfrac{1}{2} \lim_{\epsilon \to 0}[f(x_d + \epsilon) + f(x_d - \epsilon)].$$

At a discontinuity, the Fourier series representation of the function will overshoot its value. Although as more terms are included the overshoot moves in position

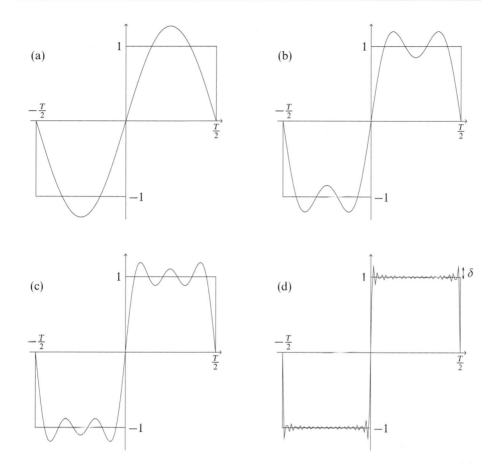

Figure 12.3 The convergence of a Fourier series expansion of a square-wave function, including (a) one term, (b) two terms, (c) three terms and (d) 20 terms. The overshoot δ is shown in (d).

arbitrarily close to the discontinuity, it never disappears even in the limit of an infinite number of terms. This behaviour is known as *Gibbs' phenomenon*. A full discussion is not pursued here but suffice it to say that the size of the overshoot is proportional to the magnitude of the discontinuity.

▶*Find the value to which the Fourier series of the square-wave function discussed in section 12.2 converges at $t = 0$.*

It can be seen that the function is discontinuous at $t = 0$ and, by the above rule, we expect the series to converge to a value half-way between the upper and lower values, in other words to converge to zero in this case. Considering the Fourier series of this function, (12.8), we see that all the terms are zero and hence the Fourier series converges to zero as expected. The Gibbs phenomenon for the square-wave function is shown in figure 12.3. ◀

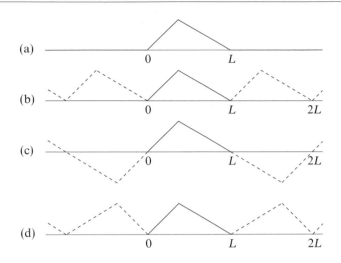

Figure 12.4 Possible periodic extensions of a function.

12.5 Non-periodic functions

We have already mentioned that a Fourier representation may sometimes be used for non-periodic functions. If we wish to find the Fourier series of a non-periodic function only within a fixed range then we may *continue* the function outside the range so as to make it periodic. The Fourier series of this periodic function would then correctly represent the non-periodic function in the desired range. Since we are often at liberty to extend the function in a number of ways, we can sometimes make it odd or even and so reduce the calculation required. Figure 12.4(b) shows the simplest extension to the function shown in figure 12.4(a). However, this extension has no particular symmetry. Figures 12.4(c), (d) show extensions as odd and even functions respectively with the benefit that only sine or cosine terms appear in the resulting Fourier series. We note that these last two extensions each give a function of period $2L$.

In view of the result of section 12.4, it must be added that the continuation must not be discontinuous at the end-points of the interval of interest; if it is the series will not converge to the required value there. This requirement that the series converges appropriately may reduce the choice of continuations. This is discussed further at the end of the following example.

▶ *Find the Fourier series of $f(x) = x^2$ for $0 < x \leq 2$.*

We must first make the function periodic. We do this by extending the range of interest to $-2 < x \leq 2$ in such a way that $f(x) = f(-x)$ and then letting $f(x + 4k) = f(x)$, where k is any integer. This is shown in figure 12.5. Now we have an even function of period 4. The Fourier series will faithfully represent $f(x)$ in the range, $-2 < x \leq 2$, although not outside it. Firstly we note that since we have made the specified function even in x by extending

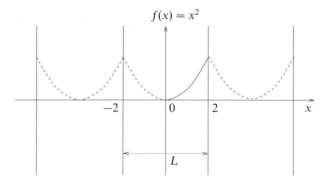

Figure 12.5 $f(x) = x^2$, $0 < x \le 2$, with the range extended to give periodicity.

the range, all the coefficients b_r will be zero. Now we apply (12.5) and (12.6) with $L = 4$ to determine the remaining coefficients:

$$a_r = \frac{2}{4} \int_{-2}^{2} x^2 \cos\left(\frac{2\pi r x}{4}\right) dx = \frac{4}{4} \int_{0}^{2} x^2 \cos\left(\frac{\pi r x}{2}\right) dx,$$

where the second equality holds because the function is even in x. Thus

$$a_r = \left[\frac{2}{\pi r} x^2 \sin\left(\frac{\pi r x}{2}\right)\right]_0^2 - \frac{4}{\pi r} \int_0^2 x \sin\left(\frac{\pi r x}{2}\right) dx$$

$$= \frac{8}{\pi^2 r^2} \left[x \cos\left(\frac{\pi r x}{2}\right)\right]_0^2 - \frac{8}{\pi^2 r^2} \int_0^2 \cos\left(\frac{\pi r x}{2}\right) dx$$

$$= \frac{16}{\pi^2 r^2} \cos \pi r$$

$$= \frac{16}{\pi^2 r^2} (-1)^r.$$

Since this expression for a_r has r^2 in its denominator, to evaluate a_0 we must return to the original definition,

$$a_r = \frac{2}{4} \int_{-2}^{2} f(x) \cos\left(\frac{\pi r x}{2}\right) dx.$$

From this we obtain

$$a_0 = \frac{2}{4} \int_{-2}^{2} x^2 \, dx = \frac{4}{4} \int_0^2 x^2 \, dx = \frac{8}{3}.$$

The final expression for $f(x)$ is then

$$x^2 = \frac{4}{3} + 16 \sum_{r=1}^{\infty} \frac{(-1)^r}{\pi^2 r^2} \cos\left(\frac{\pi r x}{2}\right) \qquad \text{for } 0 < x \le 2. \blacktriangleleft$$

We note that in the above example we could have extended the range so as to make the function odd. In other words we could have set $f(x) = -f(-x)$ and then made $f(x)$ periodic in such a way that $f(x + 4) = f(x)$. In this case the resulting Fourier series would be a series of just sine terms. However, although this will faithfully represent the function inside the required range, it does not converge to the correct values of $f(x) = \pm 4$ at $x = \pm 2$; it converges, instead, to zero, the average of the values at the two ends of the range.

12.6 Integration and differentiation

It is sometimes possible to find the Fourier series of a function by integration or differentiation of another Fourier series. If the Fourier series of $f(x)$ is integrated term by term then the resulting Fourier series converges to the integral of $f(x)$. Clearly, when integrating in such a way there is a constant of integration that must be found. If $f(x)$ is a continuous function of x for all x and $f(x)$ is also periodic then the Fourier series that results from differentiating term by term converges to $f'(x)$, provided that $f'(x)$ itself satisfies the Dirichlet conditions. These properties of Fourier series may be useful in calculating complicated Fourier series, since simple Fourier series may easily be evaluated (or found from standard tables) and often the more complicated series can then be built up by integration and/or differentiation.

▶ *Find the Fourier series of $f(x) = x^3$ for $0 < x \leq 2$.*

In the example discussed in the previous section we found the Fourier series for $f(x) = x^2$ in the required range. So, if we *integrate* this term by term, we obtain

$$\frac{x^3}{3} = \frac{4}{3}x + 32 \sum_{r=1}^{\infty} \frac{(-1)^r}{\pi^3 r^3} \sin\left(\frac{\pi r x}{2}\right) + c,$$

where c is, so far, an arbitrary constant. We have not yet found the Fourier series for x^3 because the term $\frac{4}{3}x$ appears in the expansion. However, by now *differentiating* the same initial expression for x^2 we obtain

$$2x = -8 \sum_{r=1}^{\infty} \frac{(-1)^r}{\pi r} \sin\left(\frac{\pi r x}{2}\right).$$

We can now write the full Fourier expansion of x^3 as

$$x^3 = -16 \sum_{r=1}^{\infty} \frac{(-1)^r}{\pi r} \sin\left(\frac{\pi r x}{2}\right) + 96 \sum_{r=1}^{\infty} \frac{(-1)^r}{\pi^3 r^3} \sin\left(\frac{\pi r x}{2}\right) + c.$$

Finally, we can find the constant, c, by considering $f(0)$. At $x = 0$, our Fourier expansion gives $x^3 = c$ since all the sine terms are zero, and hence $c = 0$. ◀

12.7 Complex Fourier series

As a Fourier series expansion in general contains both sine and cosine parts, it may be written more compactly using a complex exponential expansion. This simplification makes use of the property that $\exp(irx) = \cos rx + i \sin rx$. The complex Fourier series expansion is written

$$f(x) = \sum_{r=-\infty}^{\infty} c_r \exp\left(\frac{2\pi i r x}{L}\right), \tag{12.9}$$

where the Fourier coefficients are given by

$$c_r = \frac{1}{L} \int_{x_0}^{x_0+L} f(x) \exp\left(-\frac{2\pi i r x}{L}\right) dx. \tag{12.10}$$

This relation can be derived, in a similar manner to that of section 12.2, by multiplying (12.9) by $\exp(-2\pi i p x/L)$ before integrating and using the orthogonality relation

$$\int_{x_0}^{x_0+L} \exp\left(-\frac{2\pi i p x}{L}\right) \exp\left(\frac{2\pi i r x}{L}\right) dx = \begin{cases} L & \text{for } r = p, \\ 0 & \text{for } r \neq p. \end{cases}$$

The complex Fourier coefficients in (12.9) have the following relations to the real Fourier coefficients:

$$c_r = \tfrac{1}{2}(a_r - i b_r),$$
$$c_{-r} = \tfrac{1}{2}(a_r + i b_r). \tag{12.11}$$

Note that if $f(x)$ is real then $c_{-r} = c_r^*$, where the asterisk represents complex conjugation.

▶Find a complex Fourier series for $f(x) = x$ in the range $-2 < x < 2$.

Using (12.10), for $r \neq 0$,

$$c_r = \frac{1}{4} \int_{-2}^{2} x \exp\left(-\frac{\pi i r x}{2}\right) dx$$

$$= \left[-\frac{x}{2\pi i r} \exp\left(-\frac{\pi i r x}{2}\right)\right]_{-2}^{2} + \int_{-2}^{2} \frac{1}{2\pi i r} \exp\left(-\frac{\pi i r x}{2}\right) dx$$

$$= -\frac{1}{\pi i r} \left[\exp(-\pi i r) + \exp(\pi i r)\right] + \left[\frac{1}{r^2\pi^2} \exp\left(-\frac{\pi i r x}{2}\right)\right]_{-2}^{2}$$

$$= \frac{2i}{\pi r} \cos \pi r - \frac{2i}{r^2\pi^2} \sin \pi r = \frac{2i}{\pi r}(-1)^r. \tag{12.12}$$

For $r = 0$, we find $c_0 = 0$ and hence

$$x = \sum_{\substack{r=-\infty \\ r\neq 0}}^{\infty} \frac{2i(-1)^r}{r\pi} \exp\left(\frac{\pi i r x}{2}\right).$$

We note that the Fourier series derived for x in section 12.6 gives $a_r = 0$ for all r and

$$b_r = -\frac{4(-1)^r}{\pi r},$$

and so, using (12.11), we confirm that c_r and c_{-r} have the forms derived above. It is also apparent that the relationship $c_r^* = c_{-r}$ holds, as we expect since $f(x)$ is real. ◀

425

12.8 Parseval's theorem

Parseval's theorem gives a useful way of relating the Fourier coefficients to the function that they describe. Essentially a conservation law, it states that

$$\frac{1}{L}\int_{x_0}^{x_0+L} |f(x)|^2 dx = \sum_{r=-\infty}^{\infty} |c_r|^2$$

$$= \left(\tfrac{1}{2}a_0\right)^2 + \tfrac{1}{2}\sum_{r=1}^{\infty}(a_r^2 + b_r^2). \tag{12.13}$$

In a more memorable form, this says that the sum of the moduli squared of the complex Fourier coefficients is equal to the average value of $|f(x)|^2$ over one period. Parseval's theorem can be proved straightforwardly by writing $f(x)$ as a Fourier series and evaluating the required integral, but the algebra is messy. Therefore, we shall use an alternative method, for which the algebra is simple and which in fact leads to a more general form of the theorem.

Let us consider two functions $f(x)$ and $g(x)$, which are (or can be made) periodic with period L and which have Fourier series (expressed in complex form)

$$f(x) = \sum_{r=-\infty}^{\infty} c_r \exp\left(\frac{2\pi irx}{L}\right),$$

$$g(x) = \sum_{r=-\infty}^{\infty} \gamma_r \exp\left(\frac{2\pi irx}{L}\right),$$

where c_r and γ_r are the complex Fourier coefficients of $f(x)$ and $g(x)$ respectively. Thus

$$f(x)g^*(x) = \sum_{r=-\infty}^{\infty} c_r g^*(x) \exp\left(\frac{2\pi irx}{L}\right).$$

Integrating this equation with respect to x over the interval $(x_0, x_0 + L)$ and dividing by L, we find

$$\frac{1}{L}\int_{x_0}^{x_0+L} f(x)g^*(x)\,dx = \sum_{r=-\infty}^{\infty} c_r \frac{1}{L}\int_{x_0}^{x_0+L} g^*(x)\exp\left(\frac{2\pi irx}{L}\right) dx$$

$$= \sum_{r=-\infty}^{\infty} c_r \left[\frac{1}{L}\int_{x_0}^{x_0+L} g(x)\exp\left(\frac{-2\pi irx}{L}\right) dx\right]^*$$

$$= \sum_{r=-\infty}^{\infty} c_r \gamma_r^*,$$

where the last equality uses (12.10). Finally, if we let $g(x) = f(x)$ then we obtain Parseval's theorem (12.13). This result can be proved in a similar manner using the sine and cosine form of the Fourier series, but the algebra is slightly more complicated.

Parseval's theorem is sometimes used to sum series. However, if one is presented with a series to sum, it is not usually possible to decide which Fourier series should be used to evaluate it. Rather, useful summations are nearly always found serendipitously. The following example shows the evaluation of a sum by a Fourier series method.

▶ *Using Parseval's theorem and the Fourier series for $f(x) = x^2$ found in section 12.5, calculate the sum $\sum_{r=1}^{\infty} r^{-4}$.*

Firstly we find the average value of $[f(x)]^2$ over the interval $-2 < x \leq 2$:

$$\frac{1}{4} \int_{-2}^{2} x^4 \, dx = \frac{16}{5}.$$

Now we evaluate the right-hand side of (12.13):

$$\left(\tfrac{1}{2}a_0\right)^2 + \tfrac{1}{2} \sum_{1}^{\infty} a_r^2 + \tfrac{1}{2} \sum_{1}^{\infty} b_r^2 = \left(\tfrac{4}{3}\right)^2 + \tfrac{1}{2} \sum_{r=1}^{\infty} \frac{16^2}{\pi^4 r^4}.$$

Equating the two expression we find

$$\sum_{r=1}^{\infty} \frac{1}{r^4} = \frac{\pi^4}{90}. \quad ◀$$

12.9 Exercises

12.1 Prove the orthogonality relations stated in section 12.1.

12.2 Derive the Fourier coefficients b_r in a similar manner to the derivation of the a_r in section 12.2.

12.3 Which of the following functions of x could be represented by a Fourier series over the range indicated?

(a) $\tanh^{-1}(x)$, $-\infty < x < \infty$;
(b) $\tan x$, $-\infty < x < \infty$;
(c) $|\sin x|^{-1/2}$, $-\infty < x < \infty$;
(d) $\cos^{-1}(\sin 2x)$, $-\infty < x < \infty$;
(e) $x \sin(1/x)$, $-\pi^{-1} < x \leq \pi^{-1}$, cyclically repeated.

12.4 By moving the origin of t to the centre of an interval in which $f(t) = +1$, i.e. by changing to a new independent variable $t' = t - \tfrac{1}{4}T$, express the square-wave function in the example in section 12.2 as a cosine series. Calculate the Fourier coefficients involved (a) directly and (b) by changing the variable in result (12.8).

12.5 Find the Fourier series of the function $f(x) = x$ in the range $-\pi < x \leq \pi$. Hence show that

$$1 - \frac{1}{3} + \frac{1}{5} - \frac{1}{7} + \cdots = \frac{\pi}{4}.$$

12.6 For the function

$$f(x) = 1 - x, \qquad 0 \leq x \leq 1,$$

find (a) the Fourier sine series and (b) the Fourier cosine series. Which would be better for numerical evaluation? Relate your answer to the relevant periodic continuations.

12.7   For the continued functions used in exercise 12.6 and the derived corresponding series, consider (i) their derivatives and (ii) their integrals. Do they give meaningful equations? You will probably find it helpful to sketch all the functions involved.

12.8   The function $y(x) = x \sin x$ for $0 \le x \le \pi$ is to be represented by a Fourier series of period 2π that is either even or odd. By sketching the function and considering its derivative, determine which series will have the more rapid convergence. Find the full expression for the better of these two series, showing that the convergence $\sim n^{-3}$ and that alternate terms are missing.

12.9   Find the Fourier coefficients in the expansion of $f(x) = \exp x$ over the range $-1 < x < 1$. What value will the expansion have when $x = 2$?

12.10   By integrating term by term the Fourier series found in the previous question and using the Fourier series for $f(x) = x$ found in section 12.6, show that $\int \exp x\, dx = \exp x + c$. Why is it not possible to show that $d(\exp x)/dx = \exp x$ by differentiating the Fourier series of $f(x) = \exp x$ in a similar manner?

12.11   Consider the function $f(x) = \exp(-x^2)$ in the range $0 \le x \le 1$. Show how it should be continued to give as its Fourier series a series (the actual form is not wanted) (a) with only cosine terms, (b) with only sine terms, (c) with period 1 and (d) with period 2.

        Would there be any difference between the values of the last two series at (i) $x = 0$, (ii) $x = 1$?

12.12   Find, without calculation, which terms will be present in the Fourier series for the periodic functions $f(t)$, of period T, that are given in the range $-T/2$ to $T/2$ by:

        (a) $f(t) = 2$ for $0 \le |t| < T/4$, $f = 1$ for $T/4 \le |t| < T/2$;
        (b) $f(t) = \exp[-(t - T/4)^2]$;
        (c) $f(t) = -1$ for $-T/2 \le t < -3T/8$ and $3T/8 \le t < T/2, f(t) = 1$ for $-T/8 \le t < T/8$; the graph of f is completed by two straight lines in the remaining ranges so as to form a continuous function.

12.13   Consider the representation as a Fourier series of the displacement of a string lying in the interval $0 \le x \le L$ and fixed at its ends, when it is pulled aside by y_0 at the point $x = L/4$. Sketch the continuations for the region outside the interval that will

        (a) produce a series of period L,
        (b) produce a series that is antisymmetric about $x = 0$, and
        (c) produce a series that will contain only cosine terms.
        (d) What are (i) the periods of the series in (b) and (c) and (ii) the value of the 'a_0-term' in (c)?
        (e) Show that a typical term of the series obtained in (b) is

$$\frac{32y_0}{3n^2\pi^2} \sin\frac{n\pi}{4} \sin\frac{n\pi x}{L}.$$

12.14   Show that the Fourier series for the function $y(x) = |x|$ in the range $-\pi \le x < \pi$ is

$$y(x) = \frac{\pi}{2} - \frac{4}{\pi}\sum_{m=0}^{\infty}\frac{\cos(2m + 1)x}{(2m + 1)^2}.$$

        By integrating this equation term by term from 0 to x, find the function $g(x)$ whose Fourier series is

$$\frac{4}{\pi}\sum_{m=0}^{\infty}\frac{\sin(2m + 1)x}{(2m + 1)^3}.$$

428

Deduce the value of the sum S of the series

$$1 - \frac{1}{3^3} + \frac{1}{5^3} - \frac{1}{7^3} + \cdots .$$

12.15 Using the result of exercise 12.14, determine, as far as possible by inspection, the forms of the functions of which the following are the Fourier series:

(a)

$$\cos\theta + \frac{1}{9}\cos 3\theta + \frac{1}{25}\cos 5\theta + \cdots ;$$

(b)

$$\sin\theta + \frac{1}{27}\sin 3\theta + \frac{1}{125}\sin 5\theta + \cdots ;$$

(c)

$$\frac{L^2}{3} - \frac{4L^2}{\pi^2}\left[\cos\frac{\pi x}{L} - \frac{1}{4}\cos\frac{2\pi x}{L} + \frac{1}{9}\cos\frac{3\pi x}{L} - \cdots\right].$$

(You may find it helpful to first set $x = 0$ in the quoted result and so obtain values for $S_o = \sum(2m+1)^{-2}$ and other sums derivable from it.)

12.16 By finding a cosine Fourier series of period 2 for the function $f(t)$ that takes the form $f(t) = \cosh(t-1)$ in the range $0 \le t \le 1$, prove that

$$\sum_{n=1}^{\infty} \frac{1}{n^2\pi^2 + 1} = \frac{1}{e^2 - 1}.$$

Deduce values for the sums $\sum(n^2\pi^2 + 1)^{-1}$ over odd n and even n separately.

12.17 Find the (real) Fourier series of period 2 for $f(x) = \cosh x$ and $g(x) = x^2$ in the range $-1 \le x \le 1$. By integrating the series for $f(x)$ twice, prove that

$$\sum_{n=1}^{\infty} \frac{(-1)^{n+1}}{n^2\pi^2(n^2\pi^2 + 1)} = \frac{1}{2}\left(\frac{1}{\sinh 1} - \frac{5}{6}\right).$$

12.18 Express the function $f(x) = x^2$ as a Fourier sine series in the range $0 < x \le 2$ and show that it converges to zero at $x = \pm 2$.

12.19 Demonstrate explicitly for the square-wave function discussed in section 12.2 that Parseval's theorem (12.13) is valid. You will need to use the relationship

$$\sum_{m=0}^{\infty} \frac{1}{(2m+1)^2} = \frac{\pi^2}{8}.$$

Show that a filter that transmits frequencies only up to $8\pi/T$ will still transmit more than 90% of the power in such a square-wave voltage signal.

12.20 Show that the Fourier series for $|\sin\theta|$ in the range $-\pi \le \theta \le \pi$ is given by

$$|\sin\theta| = \frac{2}{\pi} - \frac{4}{\pi}\sum_{m=1}^{\infty} \frac{\cos 2m\theta}{4m^2 - 1}.$$

By setting $\theta = 0$ and $\theta = \pi/2$, deduce values for

$$\sum_{m=1}^{\infty} \frac{1}{4m^2 - 1} \quad \text{and} \quad \sum_{m=1}^{\infty} \frac{1}{16m^2 - 1}.$$

12.21   Find the complex Fourier series for the periodic function of period 2π defined in the range $-\pi \leq x \leq \pi$ by $y(x) = \cosh x$. By setting $x = 0$ prove that

$$\sum_{n=1}^{\infty} \frac{(-1)^n}{n^2 + 1} = \frac{1}{2} \left(\frac{\pi}{\sinh \pi} - 1 \right).$$

12.22   The repeating output from an electronic oscillator takes the form of a sine wave $f(t) = \sin t$ for $0 \leq t \leq \pi/2$; it then drops instantaneously to zero and starts again. The output is to be represented by a complex Fourier series of the form

$$\sum_{n=-\infty}^{\infty} c_n e^{4nti}.$$

Sketch the function and find an expression for c_n. Verify that $c_{-n} = c_n^*$. Demonstrate that setting $t = 0$ and $t = \pi/2$ produces differing values for the sum

$$\sum_{n=1}^{\infty} \frac{1}{16n^2 - 1}.$$

Determine the correct value and check it using the result of exercise 12.20.

12.23   Apply Parseval's theorem to the series found in the previous exercise and so derive a value for the sum of the series

$$\frac{17}{(15)^2} + \frac{65}{(63)^2} + \frac{145}{(143)^2} + \cdots + \frac{16n^2 + 1}{(16n^2 - 1)^2} + \cdots .$$

12.24   A string, anchored at $x = \pm L/2$, has a fundamental vibration frequency of $2L/c$, where c is the speed of transverse waves on the string. It is pulled aside at its centre point by a distance y_0 and released at time $t = 0$. Its subsequent motion can be described by the series

$$y(x, t) = \sum_{n=1}^{\infty} a_n \cos \frac{n\pi x}{L} \cos \frac{n\pi ct}{L}.$$

Find a general expression for a_n and show that only the odd harmonics of the fundamental frequency are present in the sound generated by the released string. By applying Parseval's theorem, find the sum S of the series $\sum_0^{\infty} (2m + 1)^{-4}$.

12.25   Show that Parseval's theorem for two real functions whose Fourier expansions have cosine and sine coefficients a_n, b_n and α_n, β_n takes the form

$$\frac{1}{L} \int_0^L f(x) g^*(x) \, dx = \frac{1}{4} a_0 \alpha_0 + \frac{1}{2} \sum_{n=1}^{\infty} (a_n \alpha_n + b_n \beta_n).$$

(a) Demonstrate that for $g(x) = \sin mx$ or $\cos mx$ this reduces to the definition of the Fourier coefficients.

(b) Explicitly verify the above result for the case in which $f(x) = x$ and $g(x)$ is the square-wave function, both in the interval $-1 \leq x \leq 1$.

12.26   An odd function $f(x)$ of period 2π is to be approximated by a Fourier sine series having only m terms. The error in this approximation is measured by the square deviation

$$E_m = \int_{-\pi}^{\pi} \left[f(x) - \sum_{n=1}^{m} b_n \sin nx \right]^2 dx.$$

By differentiating E_m with respect to the coefficients b_n, find the values of b_n that minimise E_m.

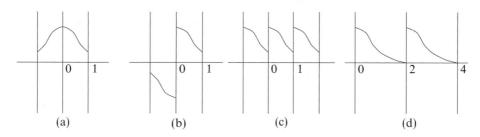

Figure 12.6   Continuations of $\exp(-x^2)$ in $0 \leq x \leq 1$ to give: (a) cosine terms only; (b) sine terms only; (c) period 1; (d) period 2.

Sketch the graph of the function $f(x)$, where

$$f(x) = \begin{cases} -x(\pi + x) & \text{for } -\pi \leq x < 0, \\ x(x - \pi) & \text{for } 0 \leq x < \pi. \end{cases}$$

If $f(x)$ is to be approximated by the first three terms of a Fourier sine series, what values should the coefficients have so as to minimise E_3? What is the resulting value of E_3?

12.10 Hints and answers

12.1   Note that the only integral of a sinusoid around a complete cycle of length L that is not zero is the integral of $\cos(2\pi nx/L)$ when $n = 0$.

12.3   Only (c). In terms of the Dirichlet conditions (section 12.1), the others fail as follows: (a) (i); (b) (ii); (d) (ii); (e) (iii).

12.5 $f(x) = 2\sum_1^\infty (-1)^{n+1} n^{-1} \sin nx$; set $x - \pi/2$.

12.7   (i) Series (a) from exercise 12.6 does not converge and cannot represent the function $y(x) = -1$. Series (b) reproduces the square-wave function of equation (12.8).

(ii) Series (a) gives the series for $y(x) = -x - \frac{1}{2}x^2 - \frac{1}{2}$ in the range $-1 \leq x \leq 0$ and for $y(x) = x - \frac{1}{2}x^2 - \frac{1}{2}$ in the range $0 \leq x \leq 1$. Series (b) gives the series for $y(x) = x + \frac{1}{2}x^2 + \frac{1}{2}$ in the range $-1 \leq x \leq 0$ and for $y(x) = x - \frac{1}{2}x^2 + \frac{1}{2}$ in the range $0 \leq x \leq 1$.

12.9 $f(x) = (\sinh 1)\left\{1 + 2\sum_1^\infty (-1)^n (1 + n^2\pi^2)^{-1}[\cos(n\pi x) - n\pi \sin(n\pi x)]\right\}$.
The series will converge to the same value as it does at $x = 0$, i.e. $f(0) = 1$.

12.11   See figure 12.6. (c) (i) $(1 + e^{-1})/2$, (ii) $(1 + e^{-1})/2$; (d) (i) $(1 + e^{-4})/2$, (ii) e^{-1}.

12.13   (d) (i) The periods are both $2L$; (ii) $y_0/2$.

12.15 $S_o = \pi^2/8$. If $S_e = \sum (2m)^{-2}$ then $S_e = \frac{1}{4}(S_e + S_o)$, yielding $S_o - S_e = \pi^2/12$ and $S_e + S_o = \pi^2/6$.
(a) $(\pi/4)(\pi/2 - |\theta|)$; (b) $(\pi\theta/4)(\pi/2 - |\theta|/2)$ from integrating (a). (c) Even function; average value $L^2/3$; $y(0) = 0$; $y(L) = L^2$; probably $y(x) = x^2$. Compare with the worked example in section 12.5.

12.17 $\cosh x = (\sinh 1)[1 + 2\sum_{n=1}^\infty (-1)^n (\cos n\pi x)/(n^2\pi^2 + 1)]$ and after integrating twice this form must be recovered. Use $x^2 = \frac{1}{3} + 4\sum (-1)^n (\cos n\pi x)/(n^2\pi^2)]$ to eliminate the quadratic term arising from the constants of integration; there is no linear term.

12.19 $C_{\pm(2m+1)} = \mp 2i/[(2m + 1)\pi]$; $\sum |C_n|^2 = (4/\pi^2) \times 2 \times (\pi^2/8)$; the values $n = \pm 1$, ± 3 contribute $> 90\%$ of the total.

12.21 $c_n = [(-1)^n \sinh \pi]/[\pi(1 + n^2)]$. Having set $x = 0$, separate out the $n = 0$ term and note that $(-1)^n = (-1)^{-n}$.

12.23 $(\pi^2 - 8)/16$.

12.25   (b) All a_n and α_n are zero; $b_n = 2(-1)^{n+1}/(n\pi)$ and $\beta_n = 4/(n\pi)$. You will need the result quoted in exercise 12.19.

13

Integral transforms

In the previous chapter we encountered the Fourier series representation of a periodic function in a fixed interval as a superposition of sinusoidal functions. It is often desirable, however, to obtain such a representation even for functions defined over an infinite interval and with no particular periodicity. Such a representation is called a *Fourier transform* and is one of a class of representations called *integral transforms*.

We begin by considering Fourier transforms as a generalisation of Fourier series. We then go on to discuss the properties of the Fourier transform and its applications. In the second part of the chapter we present an analogous discussion of the closely related *Laplace transform*.

13.1 Fourier transforms

The Fourier transform provides a representation of functions defined over an infinite interval and having no particular periodicity, in terms of a superposition of sinusoidal functions. It may thus be considered as a generalisation of the Fourier series representation of periodic functions. Since Fourier transforms are often used to represent time-varying functions, we shall present much of our discussion in terms of $f(t)$, rather than $f(x)$, although in some spatial examples $f(x)$ will be the more natural notation and we shall use it as appropriate. Our only requirement on $f(t)$ will be that $\int_{-\infty}^{\infty} |f(t)| \, dt$ is finite.

In order to develop the transition from Fourier series to Fourier transforms, we first recall that a function of period T may be represented as a complex Fourier series, cf. (12.9),

$$f(t) = \sum_{r=-\infty}^{\infty} c_r \, e^{2\pi i r t/T} = \sum_{r=-\infty}^{\infty} c_r \, e^{i\omega_r t}, \tag{13.1}$$

where $\omega_r = 2\pi r/T$. As the period T tends to infinity, the 'frequency quantum'

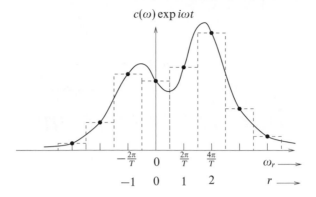

Figure 13.1 The relationship between the Fourier terms for a function of period T and the Fourier integral (the area below the solid line) of the function.

$\Delta\omega = 2\pi/T$ becomes vanishingly small and the spectrum of allowed frequencies ω_r becomes a continuum. Thus, the infinite sum of terms in the Fourier series becomes an integral, and the coefficients c_r become functions of the *continuous* variable ω, as follows.

We recall, cf. (12.10), that the coefficients c_r in (13.1) are given by

$$c_r = \frac{1}{T} \int_{-T/2}^{T/2} f(t)\, e^{-2\pi i r t/T}\, dt = \frac{\Delta\omega}{2\pi} \int_{-T/2}^{T/2} f(t)\, e^{-i\omega_r t}\, dt, \tag{13.2}$$

where we have written the integral in two alternative forms and, for convenience, made one period run from $-T/2$ to $+T/2$ rather than from 0 to T. Substituting from (13.2) into (13.1) gives

$$f(t) = \sum_{r=-\infty}^{\infty} \frac{\Delta\omega}{2\pi} \int_{-T/2}^{T/2} f(u)\, e^{-i\omega_r u}\, du\; e^{i\omega_r t}. \tag{13.3}$$

At this stage ω_r is still a discrete function of r equal to $2\pi r/T$.

The solid points in figure 13.1 are a plot of (say, the real part of) $c_r\, e^{i\omega_r t}$ as a function of r (or equivalently of ω_r) and it is clear that $(2\pi/T)c_r\, e^{i\omega_r t}$ gives the area of the rth broken-line rectangle. If T tends to ∞ then $\Delta\omega\ (= 2\pi/T)$ becomes infinitesimal, the width of the rectangles tends to zero and, from the mathematical definition of an integral,

$$\sum_{r=-\infty}^{\infty} \frac{\Delta\omega}{2\pi} g(\omega_r)\, e^{i\omega_r t} \;\rightarrow\; \frac{1}{2\pi} \int_{-\infty}^{\infty} g(\omega)\, e^{i\omega t}\, d\omega.$$

In this particular case

$$g(\omega_r) = \int_{-T/2}^{T/2} f(u)\, e^{-i\omega_r u}\, du,$$

and (13.3) becomes

$$f(t) = \frac{1}{2\pi} \int_{-\infty}^{\infty} d\omega \, e^{i\omega t} \int_{-\infty}^{\infty} du \, f(u) \, e^{-i\omega u}. \tag{13.4}$$

This result is known as *Fourier's inversion theorem*.

From it we may define the *Fourier transform* of $f(t)$ by

$$\widetilde{f}(\omega) = \frac{1}{\sqrt{2\pi}} \int_{-\infty}^{\infty} f(t) \, e^{-i\omega t} \, dt, \tag{13.5}$$

and its inverse by

$$f(t) = \frac{1}{\sqrt{2\pi}} \int_{-\infty}^{\infty} \widetilde{f}(\omega) \, e^{i\omega t} \, d\omega. \tag{13.6}$$

Including the constant $1/\sqrt{2\pi}$ in the definition of $\widetilde{f}(\omega)$ (whose mathematical existence as $T \to \infty$ is assumed here without proof) is clearly arbitrary, the only requirement being that the product of the constants in (13.5) and (13.6) should equal $1/(2\pi)$. Our definition is chosen to be as symmetric as possible.

▶ *Find the Fourier transform of the exponential decay function $f(t) = 0$ for $t < 0$ and $f(t) = A e^{-\lambda t}$ for $t \geq 0$ ($\lambda > 0$).*

Using the definition (13.5) and separating the integral into two parts,

$$\widetilde{f}(\omega) = \frac{1}{\sqrt{2\pi}} \int_{-\infty}^{0} (0) \, e^{-i\omega t} \, dt + \frac{A}{\sqrt{2\pi}} \int_{0}^{\infty} e^{-\lambda t} \, e^{-i\omega t} \, dt$$

$$= 0 + \frac{A}{\sqrt{2\pi}} \left[-\frac{e^{-(\lambda + i\omega)t}}{\lambda + i\omega} \right]_{0}^{\infty}$$

$$= \frac{A}{\sqrt{2\pi}(\lambda + i\omega)},$$

which is the required transform. It is clear that the multiplicative constant A does not affect the form of the transform, merely its amplitude. This transform may be verified by resubstitution of the above result into (13.6) to recover $f(t)$, but evaluation of the integral requires the use of complex-variable contour integration (chapter 24). ◀

13.1.1 The uncertainty principle

An important function that appears in many areas of physical science, either precisely or as an approximation to a physical situation, is the *Gaussian* or *normal* distribution. Its Fourier transform is of importance both in itself and also because, when interpreted statistically, it readily illustrates a form of *uncertainty principle*.

> ►*Find the Fourier transform of the normalised Gaussian distribution*
> $$f(t) = \frac{1}{\tau\sqrt{2\pi}} \exp\left(-\frac{t^2}{2\tau^2}\right), \qquad -\infty < t < \infty.$$

This Gaussian distribution is centred on $t = 0$ and has a root mean square deviation $\Delta t = \tau$. (Any reader who is unfamiliar with this interpretation of the distribution should refer to chapter 30.)

Using the definition (13.5), the Fourier transform of $f(t)$ is given by

$$\tilde{f}(\omega) = \frac{1}{\sqrt{2\pi}} \int_{-\infty}^{\infty} \frac{1}{\tau\sqrt{2\pi}} \exp\left(-\frac{t^2}{2\tau^2}\right) \exp(-i\omega t)\, dt$$

$$= \frac{1}{\sqrt{2\pi}} \int_{-\infty}^{\infty} \frac{1}{\tau\sqrt{2\pi}} \exp\left\{-\frac{1}{2\tau^2}\left[t^2 + 2\tau^2 i\omega t + (\tau^2 i\omega)^2 - (\tau^2 i\omega)^2\right]\right\} dt,$$

where the quantity $-(\tau^2 i\omega)^2/(2\tau^2)$ has been both added and subtracted in the exponent in order to allow the factors involving the variable of integration t to be expressed as a complete square. Hence the expression can be written

$$\tilde{f}(\omega) = \frac{\exp(-\frac{1}{2}\tau^2\omega^2)}{\sqrt{2\pi}} \left\{\frac{1}{\tau\sqrt{2\pi}} \int_{-\infty}^{\infty} \exp\left[-\frac{(t + i\tau^2\omega)^2}{2\tau^2}\right] dt\right\}.$$

The quantity inside the braces is the normalisation integral for the Gaussian and equals unity, although to show this strictly needs results from complex variable theory (chapter 24). That it is equal to unity can be made plausible by changing the variable to $s = t + i\tau^2\omega$ and assuming that the imaginary parts introduced into the integration path and limits (where the integrand goes rapidly to zero anyway) make no difference.

We are left with the result that

$$\tilde{f}(\omega) = \frac{1}{\sqrt{2\pi}} \exp\left(\frac{-\tau^2\omega^2}{2}\right), \tag{13.7}$$

which is another Gaussian distribution, centred on zero and with a root mean square deviation $\Delta\omega = 1/\tau$. It is interesting to note, and an important property, that the Fourier transform of a Gaussian is another Gaussian. ◄

In the above example the root mean square deviation in t was τ, and so it is seen that the deviations or 'spreads' in t and in ω are inversely related:

$$\Delta\omega\,\Delta t = 1,$$

independently of the value of τ. In physical terms, the narrower in time is, say, an electrical impulse the greater the spread of frequency components it must contain. Similar physical statements are valid for other pairs of Fourier-related variables, such as spatial position and wave number. In an obvious notation, $\Delta k\,\Delta x = 1$ for a Gaussian wave packet.

The uncertainty relations as usually expressed in quantum mechanics can be related to this if the de Broglie and Einstein relationships for momentum and energy are introduced; they are

$$p = \hbar k \qquad \text{and} \qquad E = \hbar\omega.$$

Here $\hbar$ is Planck's constant h divided by 2π. In a quantum mechanics setting $f(t)$

is a wavefunction and the distribution of the wave intensity in time is given by $|f|^2$ (also a Gaussian). Similarly, the intensity distribution in frequency is given by $|\tilde{f}|^2$. These two distributions have respective root mean square deviations of $\tau/\sqrt{2}$ and $1/(\sqrt{2}\tau)$, giving, after incorporation of the above relations,

$$\Delta E \,\Delta t = \hbar/2 \qquad \text{and} \qquad \Delta p \,\Delta x = \hbar/2.$$

The factors of $1/2$ that appear are specific to the Gaussian form, but any distribution $f(t)$ produces for the product $\Delta E \Delta t$ a quantity $\lambda \hbar$ in which λ is strictly positive (in fact, the Gaussian value of $1/2$ is the minimum possible).

13.1.2 Fraunhofer diffraction

We take our final example of the Fourier transform from the field of optics. The pattern of transmitted light produced by a partially opaque (or phase-changing) object upon which a coherent beam of radiation falls is called a *diffraction pattern* and, in particular, when the cross-section of the object is small compared with the distance at which the light is observed the pattern is known as a *Fraunhofer* diffraction pattern.

We will consider only the case in which the light is monochromatic with wavelength λ. The direction of the incident beam of light can then be described by the *wave vector* $\mathbf{k}$; the magnitude of this vector is given by the *wave number* $k = 2\pi/\lambda$ of the light. The essential quantity in a Fraunhofer diffraction pattern is the dependence of the observed amplitude (and hence intensity) on the angle θ between the viewing direction $\mathbf{k}'$ and the direction $\mathbf{k}$ of the incident beam. This is entirely determined by the spatial distribution of the amplitude and phase of the light at the object, the transmitted intensity in a particular direction $\mathbf{k}'$ being determined by the corresponding Fourier component of this spatial distribution.

As an example, we take as an object a simple two-dimensional screen of width $2Y$ on which light of wave number k is incident normally; see figure 13.2. We suppose that at the position $(0, y)$ the amplitude of the transmitted light is $f(y)$ per unit length in the y-direction ($f(y)$ may be complex). The function $f(y)$ is called an *aperture function*. Both the screen and beam are assumed infinite in the z-direction.

Denoting the unit vectors in the x- and y- directions by $\mathbf{i}$ and $\mathbf{j}$ respectively, the total light amplitude at a position $\mathbf{r}_0 = x_0\mathbf{i} + y_0\mathbf{j}$, with $x_0 > 0$, will be the superposition of all the (Huyghens') wavelets originating from the various parts of the screen. For large $r_0 \,(= |\mathbf{r}_0|)$, these can be treated as plane waves to give[§]

$$A(\mathbf{r}_0) = \int_{-Y}^{Y} \frac{f(y) \exp[i\mathbf{k}' \cdot (\mathbf{r}_0 - y\mathbf{j})]}{|\mathbf{r}_0 - y\mathbf{j}|} \, dy. \tag{13.8}$$

[§] This is the approach first used by Fresnel. For simplicity we have omitted from the integral a multiplicative inclination factor that depends on angle θ and decreases as θ increases.

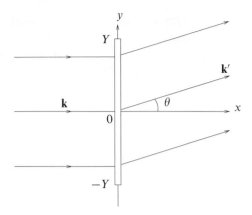

Figure 13.2   Diffraction grating of width $2Y$ with light of wavelength $2\pi/k$ being diffracted through an angle θ.

The factor $\exp[i\mathbf{k}' \cdot (\mathbf{r}_0 - y\mathbf{j})]$ represents the phase change undergone by the light in travelling from the point $y\mathbf{j}$ on the screen to the point $\mathbf{r}_0$, and the denominator represents the reduction in amplitude with distance. (Recall that the system is infinite in the z-direction and so the 'spreading' is effectively in two dimensions only.)

If the medium is the same on both sides of the screen then $\mathbf{k}' = k\cos\theta\,\mathbf{i} + k\sin\theta\,\mathbf{j}$, and if $r_0 \gg Y$ then expression (13.8) can be approximated by

$$A(\mathbf{r}_0) = \frac{\exp(i\mathbf{k}' \cdot \mathbf{r}_0)}{r_0} \int_{-\infty}^{\infty} f(y)\exp(-iky\sin\theta)\,dy. \tag{13.9}$$

We have used that $f(y) = 0$ for $|y| > Y$ to extend the integral to infinite limits. The intensity in the direction θ is then given by

$$I(\theta) = |A|^2 = \frac{2\pi}{r_0^2}|\tilde{f}(q)|^2, \tag{13.10}$$

where $q = k\sin\theta$.

▶*Evaluate $I(\theta)$ for an aperture consisting of two long slits each of width $2b$ whose centres are separated by a distance $2a$, $a > b$; the slits are illuminated by light of wavelength λ.*

The aperture function is plotted in figure 13.3. We first need to find $\tilde{f}(q)$:

$$\tilde{f}(q) = \frac{1}{\sqrt{2\pi}} \int_{-a-b}^{-a+b} e^{-iqx}\,dx + \frac{1}{\sqrt{2\pi}} \int_{a-b}^{a+b} e^{-iqx}\,dx$$

$$= \frac{1}{\sqrt{2\pi}}\left[-\frac{e^{-iqx}}{iq}\right]_{-a-b}^{-a+b} + \frac{1}{\sqrt{2\pi}}\left[-\frac{e^{-iqx}}{iq}\right]_{a-b}^{a+b}$$

$$= \frac{-1}{iq\sqrt{2\pi}}\left[e^{-iq(-a+b)} - e^{-iq(-a-b)} + e^{-iq(a+b)} - e^{-iq(a-b)}\right].$$

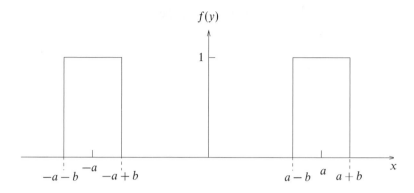

Figure 13.3   The aperture function $f(y)$ for two wide slits.

After some manipulation we obtain

$$\tilde{f}(q) = \frac{4\cos qa \sin qb}{q\sqrt{2\pi}}.$$

Now applying (13.10), and remembering that $q = (2\pi \sin \theta)/\lambda$, we find

$$I(\theta) = \frac{16\cos^2 qa \sin^2 qb}{q^2 r_0^2},$$

where r_0 is the distance from the centre of the aperture. ◀

13.1.3 The Dirac δ-function

Before going on to consider further properties of Fourier transforms we make a digression to discuss the Dirac δ-function and its relation to Fourier transforms. The δ-function is different from most functions encountered in the physical sciences but we will see that a rigorous mathematical definition exists; the utility of the δ-function will be demonstrated throughout the remainder of this chapter. It can be visualised as a very sharp narrow pulse (in space, time, density, etc.) which produces an integrated effect having a definite magnitude. The formal properties of the δ-function may be summarised as follows.

The Dirac δ-function has the property that

$$\delta(t) = 0 \quad \text{for } t \neq 0, \tag{13.11}$$

but its fundamental defining property is

$$\int f(t)\delta(t-a)\,dt = f(a), \tag{13.12}$$

provided the range of integration includes the point $t = a$; otherwise the integral equals zero. This leads immediately to two further useful results:

$$\int_{-a}^{b} \delta(t)\,dt = 1 \quad \text{for all } a, b > 0 \tag{13.13}$$

and

$$\int \delta(t - a)\,dt = 1, \tag{13.14}$$

provided the range of integration includes $t = a$.

Equation (13.12) can be used to derive further useful properties of the Dirac δ-function:

$$\delta(t) = \delta(-t), \tag{13.15}$$

$$\delta(at) = \frac{1}{|a|}\delta(t), \tag{13.16}$$

$$t\delta(t) = 0. \tag{13.17}$$

▶ *Prove that $\delta(bt) = \delta(t)/|b|$.*

Let us first consider the case where $b > 0$. It follows that

$$\int_{-\infty}^{\infty} f(t)\delta(bt)\,dt = \int_{-\infty}^{\infty} f\left(\frac{t'}{b}\right) \delta(t')\frac{dt'}{b} = \frac{1}{b}f(0) = \frac{1}{b}\int_{-\infty}^{\infty} f(t)\delta(t)\,dt,$$

where we have made the substitution $t' = bt$. But $f(t)$ is arbitrary and so we immediately see that $\delta(bt) = \delta(t)/b = \delta(t)/|b|$ for $b > 0$.

Now consider the case where $b = -c < 0$. It follows that

$$\int_{-\infty}^{\infty} f(t)\delta(bt)\,dt = \int_{\infty}^{-\infty} f\left(\frac{t'}{-c}\right) \delta(t') \left(\frac{dt'}{-c}\right) = \int_{-\infty}^{\infty} \frac{1}{c}f\left(\frac{t'}{-c}\right) \delta(t')\,dt'$$

$$= \frac{1}{c}f(0) = \frac{1}{|b|}f(0) = \frac{1}{|b|}\int_{-\infty}^{\infty} f(t)\delta(t)\,dt,$$

where we have made the substitution $t' = bt = -ct$. But $f(t)$ is arbitrary and so

$$\delta(bt) = \frac{1}{|b|}\delta(t),$$

for all b, which establishes the result. ◀

Furthermore, by considering an integral of the form

$$\int f(t)\delta(h(t))\,dt,$$

and making a change of variables to $z = h(t)$, we may show that

$$\delta(h(t)) = \sum_{i} \frac{\delta(t - t_i)}{|h'(t_i)|}, \tag{13.18}$$

where the t_i are those values of t for which $h(t) = 0$ and $h'(t)$ stands for dh/dt.

The derivative of the delta function, $\delta'(t)$, is defined by

$$\int_{-\infty}^{\infty} f(t)\delta'(t)\,dt = \left[f(t)\delta(t)\right]_{-\infty}^{\infty} - \int_{-\infty}^{\infty} f'(t)\delta(t)\,dt$$
$$= -f'(0), \tag{13.19}$$

and similarly for higher derivatives.

For many practical purposes, effects that are not strictly described by a δ-function may be analysed as such, if they take place in an interval much shorter than the response interval of the system on which they act. For example, the idealised notion of an impulse of magnitude J applied at time t_0 can be represented by

$$j(t) = J\delta(t - t_0). \tag{13.20}$$

Many physical situations are described by a δ-function in space rather than in time. Moreover, we often require the δ-function to be defined in more than one dimension. For example, the charge density of a point charge q at a point $\mathbf{r_0}$ may be expressed as a three-dimensional δ-function

$$\rho(\mathbf{r}) = q\delta(\mathbf{r} - \mathbf{r_0}) = q\delta(x - x_0)\delta(y - y_0)\delta(z - z_0), \tag{13.21}$$

so that a discrete 'quantum' is expressed as if it were a continuous distribution. From (13.21) we see that (as expected) the total charge enclosed in a volume V is given by

$$\int_V \rho(\mathbf{r})\,dV = \int_V q\delta(\mathbf{r} - \mathbf{r_0})\,dV = \begin{cases} q & \text{if } \mathbf{r_0} \text{ lies in } V, \\ 0 & \text{otherwise.} \end{cases}$$

Closely related to the Dirac δ-function is the *Heaviside* or *unit step function* $H(t)$, for which

$$H(t) = \begin{cases} 1 & \text{for } t > 0, \\ 0 & \text{for } t < 0. \end{cases} \tag{13.22}$$

This function is clearly discontinuous at $t = 0$ and it is usual to take $H(0) = 1/2$. The Heaviside function is related to the delta function by

$$H'(t) = \delta(t). \tag{13.23}$$

▶Prove relation (13.23).

Considering the integral

$$\int_{-\infty}^{\infty} f(t)H'(t)\, dt = \left[f(t)H(t) \right]_{-\infty}^{\infty} - \int_{-\infty}^{\infty} f'(t)H(t)\, dt$$

$$= f(\infty) - \int_{0}^{\infty} f'(t)\, dt$$

$$= f(\infty) - \left[f(t) \right]_{0}^{\infty} = f(0),$$

and comparing it with (13.12) when $a = 0$ immediately shows that $H'(t) = \delta(t)$. ◀

13.1.4 Relation of the δ-function to Fourier transforms

In the previous section we introduced the Dirac δ-function as a way of representing very sharp narrow pulses, but in no way related it to Fourier transforms. We now show that the δ-function can equally well be defined in a way that more naturally relates it to the Fourier transform.

Referring back to the Fourier inversion theorem (13.4), we have

$$f(t) = \frac{1}{2\pi} \int_{-\infty}^{\infty} d\omega\, e^{i\omega t} \int_{-\infty}^{\infty} du\, f(u)\, e^{-i\omega u}$$

$$= \int_{-\infty}^{\infty} du\, f(u) \left\{ \frac{1}{2\pi} \int_{-\infty}^{\infty} e^{i\omega(t-u)}\, d\omega \right\}.$$

Comparison of this with (13.12) shows that we may write the δ-function as

$$\delta(t - u) = \frac{1}{2\pi} \int_{-\infty}^{\infty} e^{i\omega(t-u)}\, d\omega. \tag{13.24}$$

Considered as a Fourier transform, this representation shows that a very narrow time peak at $t = u$ results from the superposition of a complete spectrum of harmonic waves, all frequencies having the same amplitude and all waves being in phase at $t = u$. This suggests that the δ-function may also be represented as the limit of the transform of a uniform distribution of unit height as the width of this distribution becomes infinite.

Consider the rectangular distribution of frequencies shown in figure 13.4(a). From (13.6), taking the inverse Fourier transform,

$$f_\Omega(t) = \frac{1}{\sqrt{2\pi}} \int_{-\Omega}^{\Omega} 1 \times e^{i\omega t}\, d\omega$$

$$= \frac{2\Omega}{\sqrt{2\pi}} \frac{\sin \Omega t}{\Omega t}. \tag{13.25}$$

This function is illustrated in figure 13.4(b) and it is apparent that, for large Ω, it becomes very large at $t = 0$ and also very narrow about $t = 0$, as we qualitatively

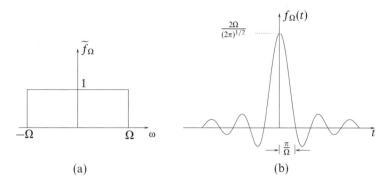

Figure 13.4   (a) A Fourier transform showing a rectangular distribution of frequencies between $\pm\Omega$; (b) the function of which it is the transform, which is proportional to $t^{-1}\sin\Omega t$.

expect and require. We also note that, in the limit $\Omega \to \infty$, $f_\Omega(t)$, as defined by the inverse Fourier transform, tends to $(2\pi)^{1/2}\delta(t)$ by virtue of (13.24). Hence we may conclude that the δ-function can also be represented by

$$\delta(t) = \lim_{\Omega \to \infty} \left(\frac{\sin \Omega t}{\pi t} \right). \tag{13.26}$$

Several other function representations are equally valid, e.g. the limiting cases of rectangular, triangular or Gaussian distributions; the only essential requirements are a knowledge of the area under such a curve and that undefined operations such as dividing by zero are not inadvertently carried out on the δ-function whilst some non-explicit representation is being employed.

We also note that the Fourier transform definition of the delta function, (13.24), shows that the latter is real since

$$\delta^*(t) = \frac{1}{2\pi} \int_{-\infty}^{\infty} e^{-i\omega t}\, d\omega = \delta(-t) = \delta(t).$$

Finally, the Fourier transform of a δ-function is simply

$$\widetilde{\delta}(\omega) = \frac{1}{\sqrt{2\pi}} \int_{-\infty}^{\infty} \delta(t)\, e^{-i\omega t}\, dt = \frac{1}{\sqrt{2\pi}}. \tag{13.27}$$

13.1.5 Properties of Fourier transforms

Having considered the Dirac δ-function, we now return to our discussion of the properties of Fourier transforms. As we would expect, Fourier transforms have many properties analogous to those of Fourier series in respect of the connection between the transforms of related functions. Here we list these properties without proof; they can be verified by working from the definition of the transform. As previously, we denote the Fourier transform of $f(t)$ by $\widetilde{f}(\omega)$ or $\mathscr{F}[f(t)]$.

(i) Differentiation:

$$\mathscr{F}[f'(t)] = i\omega\tilde{f}(\omega). \tag{13.28}$$

This may be extended to higher derivatives, so that

$$\mathscr{F}[f''(t)] = i\omega\mathscr{F}[f'(t)] = -\omega^2\tilde{f}(\omega),$$

and so on.

(ii) Integration:

$$\mathscr{F}\left[\int^t f(s)\,ds\right] = \frac{1}{i\omega}\tilde{f}(\omega) + 2\pi c\delta(\omega), \tag{13.29}$$

where the term $2\pi c\delta(\omega)$ represents the Fourier transform of the constant of integration associated with the indefinite integral.

(iii) Scaling:

$$\mathscr{F}[f(at)] = \frac{1}{a}\tilde{f}\left(\frac{\omega}{a}\right). \tag{13.30}$$

(iv) Translation:

$$\mathscr{F}[f(t+a)] = e^{ia\omega}\tilde{f}(\omega). \tag{13.31}$$

(v) Exponential multiplication:

$$\mathscr{F}\left[e^{\alpha t}f(t)\right] = \tilde{f}(\omega + i\alpha), \tag{13.32}$$

where α may be real, imaginary or complex.

▶*Prove relation (13.28).*

Calculating the Fourier transform of $f'(t)$ directly, we obtain

$$\mathscr{F}[f'(t)] = \frac{1}{\sqrt{2\pi}}\int_{-\infty}^{\infty} f'(t)\,e^{-i\omega t}\,dt$$

$$= \frac{1}{\sqrt{2\pi}}\left[e^{-i\omega t}f(t)\right]_{-\infty}^{\infty} + \frac{1}{\sqrt{2\pi}}\int_{-\infty}^{\infty} i\omega\,e^{-i\omega t}f(t)\,dt$$

$$= i\omega\tilde{f}(\omega),$$

if $f(t) \to 0$ at $t = \pm\infty$, as it must since $\int_{-\infty}^{\infty}|f(t)|\,dt$ is finite. ◀

To illustrate a use and also a proof of (13.32), let us consider an amplitude-modulated radio wave. Suppose a message to be broadcast is represented by $f(t)$. The message can be added electronically to a constant signal a of magnitude such that $a + f(t)$ is never negative, and then the sum can be used to modulate the amplitude of a carrier signal of frequency ω_c. Using a complex exponential notation, the transmitted amplitude is now

$$g(t) = A\,[a + f(t)]\,e^{i\omega_c t}. \tag{13.33}$$

Ignoring in the present context the effect of the term $Aa\exp(i\omega_c t)$, which gives a contribution to the transmitted spectrum only at $\omega = \omega_c$, we obtain for the new spectrum

$$\tilde{g}(\omega) = \frac{1}{\sqrt{2\pi}} A \int_{-\infty}^{\infty} f(t) e^{i\omega_c t} e^{-i\omega t} dt$$

$$= \frac{1}{\sqrt{2\pi}} A \int_{-\infty}^{\infty} f(t) e^{-i(\omega - \omega_c)t} dt$$

$$= A\tilde{f}(\omega - \omega_c), \tag{13.34}$$

which is simply a shift of the whole spectrum by the carrier frequency. The use of different carrier frequencies enables signals to be separated.

13.1.6 Odd and even functions

If $f(t)$ is odd or even then we may derive alternative forms of Fourier's inversion theorem, which lead to the definition of different transform pairs. Let us first consider an odd function $f(t) = -f(-t)$, whose Fourier transform is given by

$$\tilde{f}(\omega) = \frac{1}{\sqrt{2\pi}} \int_{-\infty}^{\infty} f(t) e^{-i\omega t} dt$$

$$= \frac{1}{\sqrt{2\pi}} \int_{-\infty}^{\infty} f(t)(\cos \omega t - i \sin \omega t) dt$$

$$= \frac{-2i}{\sqrt{2\pi}} \int_{0}^{\infty} f(t) \sin \omega t \, dt,$$

where in the last line we use the fact that $f(t)$ and $\sin \omega t$ are odd, whereas $\cos \omega t$ is even.

We note that $\tilde{f}(-\omega) = -\tilde{f}(\omega)$, i.e. $\tilde{f}(\omega)$ is an odd function of ω. Hence

$$f(t) = \frac{1}{\sqrt{2\pi}} \int_{-\infty}^{\infty} \tilde{f}(\omega) e^{i\omega t} d\omega = \frac{2i}{\sqrt{2\pi}} \int_{0}^{\infty} \tilde{f}(\omega) \sin \omega t \, d\omega$$

$$= \frac{2}{\pi} \int_{0}^{\infty} d\omega \, \sin \omega t \left\{ \int_{0}^{\infty} f(u) \sin \omega u \, du \right\}.$$

Thus we may define the *Fourier sine transform pair* for odd functions:

$$\tilde{f}_s(\omega) = \sqrt{\frac{2}{\pi}} \int_{0}^{\infty} f(t) \, \sin \omega t \, dt, \tag{13.35}$$

$$f(t) = \sqrt{\frac{2}{\pi}} \int_{0}^{\infty} \tilde{f}_s(\omega) \, \sin \omega t \, d\omega. \tag{13.36}$$

Note that although the Fourier sine transform pair was derived by considering an odd function $f(t)$ defined over all t, the definitions (13.35) and (13.36) only require $f(t)$ and $\tilde{f}_s(\omega)$ to be defined for positive t and ω respectively. For an

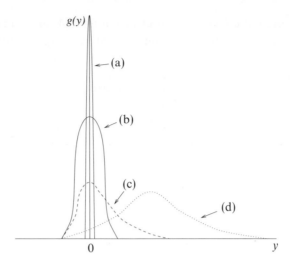

Figure 13.5   Resolution functions: (a) ideal δ-function; (b) typical unbiased resolution; (c) and (d) biases tending to shift observations to higher values than the true one.

even function, i.e. one for which $f(t) = f(-t)$, we can define the *Fourier cosine transform pair* in a similar way, but with $\sin \omega t$ replaced by $\cos \omega t$.

13.1.7 *Convolution and deconvolution*

It is apparent that any attempt to measure the value of a physical quantity is limited, to some extent, by the finite resolution of the measuring apparatus used. On the one hand, the physical quantity we wish to measure will be in general a function of an independent variable, x say, i.e. the true function to be measured takes the form $f(x)$. On the other hand, the apparatus we are using does not give the true output value of the function; a resolution function $g(y)$ is involved. By this we mean that the probability that an output value $y = 0$ will be recorded instead as being between y and $y+dy$ is given by $g(y)\, dy$. Some possible resolution functions of this sort are shown in figure 13.5. To obtain good results we wish the resolution function to be as close to a δ-function as possible (case (*a*)). A typical piece of apparatus has a resolution function of finite width, although if it is accurate the mean is centred on the true value (case (*b*)). However, some apparatus may show a bias that tends to shift observations to higher or lower values than the true ones (cases (*c*) and (*d*)), thereby exhibiting systematic error.

Given that the true distribution is $f(x)$ and the resolution function of our measuring apparatus is $g(y)$, we wish to calculate what the observed distribution $h(z)$ will be. The symbols x, y and z all refer to the same physical variable (e.g.

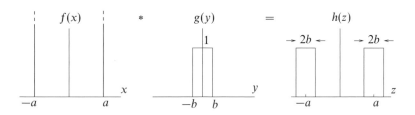

Figure 13.6   The convolution of two functions $f(x)$ and $g(y)$.

length or angle), but are denoted differently because the variable appears in the analysis in three different roles.

The probability that a true reading lying between x and $x + dx$, and so having probability $f(x) dx$ of being selected by the experiment, will be moved by the instrumental resolution by an amount $z - x$ into a small interval of width dz is $g(z - x) dz$. Hence the combined probability that the interval dx will give rise to an observation appearing in the interval dz is $f(x) dx g(z - x) dz$. Adding together the contributions from all values of x that can lead to an observation in the range z to $z + dz$, we find that the observed distribution is given by

$$h(z) = \int_{-\infty}^{\infty} f(x)g(z - x)\, dx. \tag{13.37}$$

The integral in (13.37) is called the *convolution* of the functions f and g and is often written $f * g$. The convolution defined above is commutative ($f * g = g * f$), associative and distributive. The observed distribution is thus the convolution of the true distribution and the experimental resolution function. The result will be that the observed distribution is broader and smoother than the true one and, if $g(y)$ has a bias, the maxima will normally be displaced from their true positions. It is also obvious from (13.37) that if the resolution is the ideal δ-function, $g(y) = \delta(y)$ then $h(z) = f(z)$ and the observed distribution is the true one.

It is interesting to note, and a very important property, that the convolution of any function $g(y)$ with a number of delta functions leaves a copy of $g(y)$ at the position of each of the delta functions.

▶*Find the convolution of the function $f(x) = \delta(x + a) + \delta(x - a)$ with the function $g(y)$ plotted in figure 13.6.*

Using the convolution integral (13.37)

$$h(z) = \int_{-\infty}^{\infty} f(x)g(z - x)\, dx = \int_{-\infty}^{\infty} [\delta(x + a) + \delta(x - a)]g(z - x)\, dx$$
$$= g(z + a) + g(z - a).$$

This convolution $h(z)$ is plotted in figure 13.6. ◀

Let us now consider the Fourier transform of the convolution (13.37); this is given by

$$\tilde{h}(k) = \frac{1}{\sqrt{2\pi}} \int_{-\infty}^{\infty} dz\, e^{-ikz} \left\{ \int_{-\infty}^{\infty} f(x)g(z-x)\, dx \right\}$$

$$= \frac{1}{\sqrt{2\pi}} \int_{-\infty}^{\infty} dx\, f(x) \left\{ \int_{-\infty}^{\infty} g(z-x)\, e^{-ikz}\, dz \right\}.$$

If we let $u = z - x$ in the second integral we have

$$\tilde{h}(k) = \frac{1}{\sqrt{2\pi}} \int_{-\infty}^{\infty} dx\, f(x) \left\{ \int_{-\infty}^{\infty} g(u)\, e^{-ik(u+x)}\, du \right\}$$

$$= \frac{1}{\sqrt{2\pi}} \int_{-\infty}^{\infty} f(x)\, e^{-ikx}\, dx \int_{-\infty}^{\infty} g(u)\, e^{-iku}\, du$$

$$= \frac{1}{\sqrt{2\pi}} \times \sqrt{2\pi}\, \tilde{f}(k) \times \sqrt{2\pi}\tilde{g}(k) = \sqrt{2\pi}\, \tilde{f}(k)\tilde{g}(k). \tag{13.38}$$

Hence the Fourier transform of a convolution $f * g$ is equal to the product of the separate Fourier transforms multiplied by $\sqrt{2\pi}$; this result is called the *convolution theorem*.

It may be proved similarly that the converse is also true, namely that the Fourier transform of the product $f(x)g(x)$ is given by

$$\mathcal{F}[f(x)g(x)] = \frac{1}{\sqrt{2\pi}} \tilde{f}(k) * \tilde{g}(k). \tag{13.39}$$

►*Find the Fourier transform of the function in figure 13.3 representing two wide slits by considering the Fourier transforms of* (i) *two δ-functions, at* $x = \pm a$, (ii) *a rectangular function of height 1 and width 2b centred on* $x = 0$.

(i) The Fourier transform of the two δ-functions is given by

$$\tilde{f}(q) = \frac{1}{\sqrt{2\pi}} \int_{-\infty}^{\infty} \delta(x-a)\, e^{-iqx}\, dx + \frac{1}{\sqrt{2\pi}} \int_{-\infty}^{\infty} \delta(x+a)\, e^{-iqx}\, dx$$

$$= \frac{1}{\sqrt{2\pi}} \left(e^{-iqa} + e^{iqa} \right) = \frac{2\cos qa}{\sqrt{2\pi}}.$$

(ii) The Fourier transform of the broad slit is

$$\tilde{g}(q) = \frac{1}{\sqrt{2\pi}} \int_{-b}^{b} e^{-iqx}\, dx = \frac{1}{\sqrt{2\pi}} \left[\frac{e^{-iqx}}{-iq} \right]_{-b}^{b}$$

$$= \frac{-1}{iq\sqrt{2\pi}} (e^{-iqb} - e^{iqb}) = \frac{2\sin qb}{q\sqrt{2\pi}}.$$

We have already seen that the convolution of these functions is the required function representing two wide slits (see figure 13.6). So, using the convolution theorem, the Fourier transform of the convolution is $\sqrt{2\pi}$ times the product of the individual transforms, i.e. $4\cos qa \sin qb/(q\sqrt{2\pi})$. This is, of course, the same result as that obtained in the example in subsection 13.1.2. ◄

The inverse of convolution, called *deconvolution*, allows us to find a true distribution $f(x)$ given an observed distribution $h(z)$ and a resolution function $g(y)$.

> ►*An experimental quantity $f(x)$ is measured using apparatus with a known resolution function $g(y)$ to give an observed distribution $h(z)$. How may $f(x)$ be extracted from the measured distribution?*

From the convolution theorem (13.38), the Fourier transform of the measured distribution is

$$\widetilde{h}(k) = \sqrt{2\pi}\,\widetilde{f}(k)\widetilde{g}(k),$$

from which we obtain

$$\widetilde{f}(k) = \frac{1}{\sqrt{2\pi}}\frac{\widetilde{h}(k)}{\widetilde{g}(k)}.$$

Then on inverse Fourier transforming we find

$$f(x) = \frac{1}{\sqrt{2\pi}}\mathscr{F}^{-1}\left[\frac{\widetilde{h}(k)}{\widetilde{g}(k)}\right].$$

In words, to extract the true distribution, we divide the Fourier transform of the observed distribution by that of the resolution function for each value of k and then take the inverse Fourier transform of the function so generated. ◄

This explicit method of extracting true distributions is straightforward for exact functions but, in practice, because of experimental and statistical uncertainties in the experimental data or because data over only a limited range are available, it is often not very precise, involving as it does three (numerical) transforms each requiring in principle an integral over an infinite range.

13.1.8 Correlation functions and energy spectra

The *cross-correlation* of two functions f and g is defined by

$$C(z) = \int_{-\infty}^{\infty} f^*(x)g(x+z)\,dx. \tag{13.40}$$

Despite the formal similarity between (13.40) and the definition of the convolution in (13.37), the use and interpretation of the cross-correlation and of the convolution are very different; the cross-correlation provides a quantitative measure of the similarity of two functions f and g as one is displaced through a distance z relative to the other. The cross-correlation is often notated as $C = f \otimes g$, and, like convolution, it is both associative and distributive. Unlike convolution, however, it is *not* commutative, in fact

$$[f \otimes g](z) = [g \otimes f]^*(-z). \tag{13.41}$$

> ▶ *Prove the Wiener–Kinchin theorem,*
> $$\widetilde{C}(k) = \sqrt{2\pi}\,[\widetilde{f}(k)]^*\widetilde{g}(k).\qquad(13.42)$$

Following a method similar to that for the convolution of f and g, let us consider the Fourier transform of (13.40):

$$
\begin{aligned}
\widetilde{C}(k) &= \frac{1}{\sqrt{2\pi}}\int_{-\infty}^{\infty} dz\,e^{-ikz}\left\{\int_{-\infty}^{\infty} f^*(x)g(x+z)\,dx\right\}\\
&= \frac{1}{\sqrt{2\pi}}\int_{-\infty}^{\infty} dx\,f^*(x)\left\{\int_{-\infty}^{\infty} g(x+z)\,e^{-ikz}\,dz\right\}.
\end{aligned}
$$

Making the substitution $u = x + z$ in the second integral we obtain

$$
\begin{aligned}
\widetilde{C}(k) &= \frac{1}{\sqrt{2\pi}}\int_{-\infty}^{\infty} dx\,f^*(x)\left\{\int_{-\infty}^{\infty} g(u)\,e^{-ik(u-x)}\,du\right\}\\
&= \frac{1}{\sqrt{2\pi}}\int_{-\infty}^{\infty} f^*(x)\,e^{ikx}\,dx\int_{-\infty}^{\infty} g(u)\,e^{-iku}\,du\\
&= \frac{1}{\sqrt{2\pi}}\times\sqrt{2\pi}\,[\widetilde{f}(k)]^*\times\sqrt{2\pi}\,\widetilde{g}(k) = \sqrt{2\pi}\,[\widetilde{f}(k)]^*\widetilde{g}(k).\ \blacktriangleleft
\end{aligned}
$$

Thus the Fourier transform of the cross-correlation of f and g is equal to the product of $[\widetilde{f}(k)]^*$ and $\widetilde{g}(k)$ multiplied by $\sqrt{2\pi}$. This a statement of the *Wiener–Kinchin theorem*. Similarly we can derive the converse theorem

$$\mathscr{F}\big[f^*(x)g(x)\big] = \frac{1}{\sqrt{2\pi}}\widetilde{f}\otimes\widetilde{g}.$$

If we now consider the special case where g is taken to be equal to f in (13.40) then, writing the LHS as $a(z)$, we have

$$a(z) = \int_{-\infty}^{\infty} f^*(x)f(x+z)\,dx;\qquad(13.43)$$

this is called the *auto-correlation function* of $f(x)$. Using the Wiener–Kinchin theorem (13.42) we see that

$$
\begin{aligned}
a(z) &= \frac{1}{\sqrt{2\pi}}\int_{-\infty}^{\infty}\widetilde{a}(k)\,e^{ikz}\,dk\\
&= \frac{1}{\sqrt{2\pi}}\int_{-\infty}^{\infty}\sqrt{2\pi}\,[\widetilde{f}(k)]^*\widetilde{f}(k)\,e^{ikz}\,dk,
\end{aligned}
$$

so that $a(z)$ is the inverse Fourier transform of $\sqrt{2\pi}\,|\widetilde{f}(k)|^2$, which is in turn called the *energy spectrum* of f.

13.1.9 Parseval's theorem

Using the results of the previous section we can immediately obtain *Parseval's theorem*. The most general form of this (also called the *multiplication theorem*) is obtained simply by noting from (13.42) that the cross-correlation (13.40) of two functions f and g can be written as

$$C(z) = \int_{-\infty}^{\infty} f^*(x)g(x+z)\,dx = \int_{-\infty}^{\infty} [\tilde{f}(k)]^* \tilde{g}(k)\, e^{ikz}\, dk. \qquad (13.44)$$

Then, setting $z = 0$ gives the multiplication theorem

$$\int_{-\infty}^{\infty} f^*(x)g(x)\,dx = \int [\tilde{f}(k)]^* \tilde{g}(k)\, dk. \qquad (13.45)$$

Specialising further, by letting $g = f$, we derive the most common form of Parseval's theorem,

$$\int_{-\infty}^{\infty} |f(x)|^2\, dx = \int_{-\infty}^{\infty} |\tilde{f}(k)|^2\, dk. \qquad (13.46)$$

When f is a physical amplitude these integrals relate to the total intensity involved in some physical process. We have already met a form of Parseval's theorem for Fourier series in chapter 12; it is in fact a special case of (13.46).

▶*The displacement of a damped harmonic oscillator as a function of time is given by*

$$f(t) = \begin{cases} 0 & \text{for } t < 0, \\ e^{-t/\tau} \sin \omega_0 t & \text{for } t \geq 0. \end{cases}$$

Find the Fourier transform of this function and so give a physical interpretation of Parseval's theorem.

Using the usual definition for the Fourier transform we find

$$\tilde{f}(\omega) = \int_{-\infty}^{0} 0 \times e^{-i\omega t}\, dt + \int_{0}^{\infty} e^{-t/\tau} \sin \omega_0 t\, e^{-i\omega t}\, dt.$$

Writing $\sin \omega_0 t$ as $(e^{i\omega_0 t} - e^{-i\omega_0 t})/2i$ we obtain

$$\tilde{f}(\omega) = 0 + \frac{1}{2i} \int_{0}^{\infty} \left[e^{-it(\omega-\omega_0-i/\tau)} - e^{-it(\omega+\omega_0-i/\tau)} \right] dt$$

$$= \frac{1}{2} \left[\frac{1}{\omega + \omega_0 - i/\tau} - \frac{1}{\omega - \omega_0 - i/\tau} \right],$$

which is the required Fourier transform. The physical interpretation of $|\tilde{f}(\omega)|^2$ is the energy content per unit frequency interval (i.e. the *energy spectrum*) whilst $|f(t)|^2$ is proportional to the sum of the kinetic and potential energies of the oscillator. Hence (to within a constant) Parseval's theorem shows the equivalence of these two alternative specifications for the total energy. ◀

13.1.10 Fourier transforms in higher dimensions

The concept of the Fourier transform can be extended naturally to more than one dimension. For instance we may wish to find the spatial Fourier transform of two- or three-dimensional functions of position. For example, in three dimensions we can define the Fourier transform of $f(x, y, z)$ as

$$\widetilde{f}(k_x, k_y, k_z) = \frac{1}{(2\pi)^{3/2}} \iiint f(x, y, z) \, e^{-ik_x x} e^{-ik_y y} e^{-ik_z z} \, dx \, dy \, dz, \tag{13.47}$$

and its inverse as

$$f(x, y, z) = \frac{1}{(2\pi)^{3/2}} \iiint \widetilde{f}(k_x, k_y, k_z) \, e^{ik_x x} e^{ik_y y} e^{ik_z z} \, dk_x \, dk_y \, dk_z. \tag{13.48}$$

Denoting the vector with components k_x, k_y, k_z by $\mathbf{k}$ and that with components x, y, z by $\mathbf{r}$, we can write the Fourier transform pair (13.47), (13.48) as

$$\widetilde{f}(\mathbf{k}) = \frac{1}{(2\pi)^{3/2}} \int f(\mathbf{r}) \, e^{-i\mathbf{k} \cdot \mathbf{r}} \, d^3\mathbf{r}, \tag{13.49}$$

$$f(\mathbf{r}) = \frac{1}{(2\pi)^{3/2}} \int \widetilde{f}(\mathbf{k}) \, e^{i\mathbf{k} \cdot \mathbf{r}} \, d^3\mathbf{k}. \tag{13.50}$$

From these relations we may deduce that the three-dimensional Dirac δ-function can be written as

$$\delta(\mathbf{r}) = \frac{1}{(2\pi)^3} \int e^{i\mathbf{k} \cdot \mathbf{r}} \, d^3\mathbf{k}. \tag{13.51}$$

Similar relations to (13.49), (13.50) and (13.51) exist for spaces of other dimensionalities.

> ►*In three-dimensional space a function $f(\mathbf{r})$ possesses spherical symmetry, so that $f(\mathbf{r}) = f(r)$. Find the Fourier transform of $f(\mathbf{r})$ as a one-dimensional integral.*

Let us choose spherical polar coordinates in which the vector $\mathbf{k}$ of the Fourier transform lies along the polar axis ($\theta = 0$). This we can do since $f(\mathbf{r})$ is spherically symmetric. We then have

$$d^3\mathbf{r} = r^2 \sin\theta \, dr \, d\theta \, d\phi \quad \text{and} \quad \mathbf{k} \cdot \mathbf{r} = kr \cos\theta,$$

where $k = |\mathbf{k}|$. The Fourier transform is then given by

$$\begin{aligned}
\widetilde{f}(\mathbf{k}) &= \frac{1}{(2\pi)^{3/2}} \int f(\mathbf{r}) \, e^{-i\mathbf{k} \cdot \mathbf{r}} \, d^3\mathbf{r} \\
&= \frac{1}{(2\pi)^{3/2}} \int_0^\infty dr \int_0^\pi d\theta \int_0^{2\pi} d\phi \, f(r) r^2 \sin\theta \, e^{-ikr\cos\theta} \\
&= \frac{1}{(2\pi)^{3/2}} \int_0^\infty dr \, 2\pi f(r) r^2 \int_0^\pi d\theta \, \sin\theta \, e^{-ikr\cos\theta}.
\end{aligned}$$

The integral over θ may be straightforwardly evaluated by noting that

$$\frac{d}{d\theta}(e^{-ikr\cos\theta}) = ikr \sin\theta \, e^{-ikr\cos\theta}.$$

Therefore

$$\begin{aligned}
\widetilde{f}(\mathbf{k}) &= \frac{1}{(2\pi)^{3/2}} \int_0^\infty dr \, 2\pi f(r) r^2 \left[\frac{e^{-ikr\cos\theta}}{ikr} \right]_{\theta=0}^{\theta=\pi} \\
&= \frac{1}{(2\pi)^{3/2}} \int_0^\infty 4\pi r^2 f(r) \left(\frac{\sin kr}{kr} \right) dr. \ ◄
\end{aligned}$$

452

A similar result may be obtained for two-dimensional Fourier transforms in which $f(\mathbf{r}) = f(\rho)$, i.e. $f(\mathbf{r})$ is independent of azimuthal angle ϕ. In this case, using the integral representation of the Bessel function $J_0(x)$ given at the very end of subsection 18.5.3, we find

$$\tilde{f}(\mathbf{k}) = \frac{1}{2\pi} \int_0^\infty 2\pi\rho f(\rho)J_0(k\rho)\, d\rho. \tag{13.52}$$

13.2 Laplace transforms

Often we are interested in functions $f(t)$ for which the Fourier transform does not exist because $f \nrightarrow 0$ as $t \to \infty$, and so the integral defining $\tilde{f}$ does not converge. This would be the case for the function $f(t) = t$, which does not possess a Fourier transform. Furthermore, we might be interested in a given function only for $t > 0$, for example when we are given its value at $t = 0$ in an initial-value problem. This leads us to consider the Laplace transform, $\bar{f}(s)$ or $\mathscr{L}[f(t)]$, of $f(t)$, which is defined by

$$\bar{f}(s) \equiv \int_0^\infty f(t)e^{-st}\, dt, \tag{13.53}$$

provided that the integral exists. We assume here that s is real, but complex values would have to be considered in a more detailed study. In practice, for a given function $f(t)$ there will be some real number s_0 such that the integral in (13.53) exists for $s > s_0$ but diverges for $s \leq s_0$.

Through (13.53) we define a *linear* transformation $\mathscr{L}$ that converts functions of the variable t to functions of a new variable s:

$$\mathscr{L}[af_1(t) + bf_2(t)] = a\mathscr{L}[f_1(t)] + b\mathscr{L}[f_2(t)] = a\bar{f}_1(s) + b\bar{f}_2(s). \tag{13.54}$$

> ►*Find the Laplace transforms of the functions* (i) $f(t) = 1$, (ii) $f(t) = e^{at}$, (iii) $f(t) = t^n$, *for* $n = 0, 1, 2, \ldots$.

(i) By direct application of the definition of a Laplace transform (13.53), we find

$$\mathscr{L}[1] = \int_0^\infty e^{-st}\, dt = \left[\frac{-1}{s}e^{-st}\right]_0^\infty = \frac{1}{s}, \quad \text{if } s > 0,$$

where the restriction $s > 0$ is required for the integral to exist.

(ii) Again using (13.53) directly, we find

$$\bar{f}(s) = \int_0^\infty e^{at}e^{-st}\, dt = \int_0^\infty e^{(a-s)t}\, dt$$

$$= \left[\frac{e^{(a-s)t}}{a-s}\right]_0^\infty = \frac{1}{s-a} \quad \text{if } s > a.$$

453

(iii) Once again using the definition (13.53) we have

$$\bar{f}_n(s) = \int_0^\infty t^n e^{-st} \, dt.$$

Integrating by parts we find

$$\bar{f}_n(s) = \left[\frac{-t^n e^{-st}}{s} \right]_0^\infty + \frac{n}{s} \int_0^\infty t^{n-1} e^{-st} \, dt$$

$$= 0 + \frac{n}{s} \bar{f}_{n-1}(s), \qquad \text{if } s > 0.$$

We now have a recursion relation between successive transforms and by calculating $\bar{f}_0$ we can infer $\bar{f}_1$, $\bar{f}_2$, etc. Since $t^0 = 1$, (i) above gives

$$\bar{f}_0 = \frac{1}{s}, \qquad \text{if } s > 0, \tag{13.55}$$

and

$$\bar{f}_1(s) = \frac{1}{s^2}, \qquad \bar{f}_2(s) = \frac{2!}{s^3}, \qquad \ldots, \qquad \bar{f}_n(s) = \frac{n!}{s^{n+1}} \qquad \text{if } s > 0.$$

Thus, in each case (i)–(iii), direct application of the definition of the Laplace transform (13.53) yields the required result. ◄

Unlike that for the Fourier transform, the inversion of the Laplace transform is not an easy operation to perform, since an explicit formula for $f(t)$, given $\bar{f}(s)$, is not straightforwardly obtained from (13.53). The general method for obtaining an inverse Laplace transform makes use of complex variable theory and is not discussed until chapter 25. However, progress can be made without having to find an *explicit* inverse, since we can prepare from (13.53) a 'dictionary' of the Laplace transforms of common functions and, when faced with an inversion to carry out, hope to find the given transform (together with its parent function) in the listing. Such a list is given in table 13.1.

When finding inverse Laplace transforms using table 13.1, it is useful to note that for all practical purposes the inverse Laplace transform is unique[§] and linear so that

$$\mathcal{L}^{-1}\left[a\bar{f}_1(s) + b\bar{f}_2(s) \right] = af_1(t) + bf_2(t). \tag{13.56}$$

In many practical problems the method of partial fractions can be useful in producing an expression from which the inverse Laplace transform can be found.

▶*Using table 13.1 find $f(t)$ if*

$$\bar{f}(s) = \frac{s+3}{s(s+1)}.$$

Using partial fractions $\bar{f}(s)$ may be written

$$\bar{f}(s) = \frac{3}{s} - \frac{2}{s+1}.$$

[§] This is not strictly true, since two functions can differ from one another at a finite number of isolated points but have the *same* Laplace transform.

$f(t)$	$\bar{f}(s)$	s_0
c	c/s	0
ct^n	$cn!/s^{n+1}$	0
$\sin bt$	$b/(s^2 + b^2)$	0
$\cos bt$	$s/(s^2 + b^2)$	0
e^{at}	$1/(s - a)$	a
$t^n e^{at}$	$n!/(s - a)^{n+1}$	a
$\sinh at$	$a/(s^2 - a^2)$	$\lvert a \rvert$
$\cosh at$	$s/(s^2 - a^2)$	$\lvert a \rvert$
$e^{at} \sin bt$	$b/[(s - a)^2 + b^2]$	a
$e^{at} \cos bt$	$(s - a)/[(s - a)^2 + b^2]$	a
$t^{1/2}$	$\frac{1}{2}(\pi/s^3)^{1/2}$	0
$t^{-1/2}$	$(\pi/s)^{1/2}$	0
$\delta(t - t_0)$	e^{-st_0}	0
$H(t - t_0) = \begin{cases} 1 & \text{for } t \geq t_0 \\ 0 & \text{for } t < t_0 \end{cases}$	e^{-st_0}/s	0

Table 13.1   Standard Laplace transforms. The transforms are valid for $s > s_0$.

Comparing this with the standard Laplace transforms in table 13.1, we find that the inverse transform of $3/s$ is 3 for $s > 0$ and the inverse transform of $2/(s + 1)$ is $2e^{-t}$ for $s > -1$, and so

$$f(t) = 3 - 2e^{-t}, \quad \text{if } s > 0. \blacktriangleleft$$

13.2.1 Laplace transforms of derivatives and integrals

One of the main uses of Laplace transforms is in solving differential equations. Differential equations are the subject of the next six chapters and we will return to the application of Laplace transforms to their solution in chapter 15. In the meantime we will derive the required results, i.e. the Laplace transforms of derivatives.

The Laplace transform of the first derivative of $f(t)$ is given by

$$\mathcal{L}\left[\frac{df}{dt}\right] = \int_0^\infty \frac{df}{dt} e^{-st}\, dt$$

$$= \left[f(t)e^{-st}\right]_0^\infty + s\int_0^\infty f(t)e^{-st}\, dt$$

$$= -f(0) + s\bar{f}(s), \quad \text{for } s > 0. \tag{13.57}$$

The evaluation relies on integration by parts and higher-order derivatives may be found in a similar manner.

▶ *Find the Laplace transform of d^2f/dt^2.*

Using the definition of the Laplace transform and integrating by parts we obtain

$$\mathcal{L}\left[\frac{d^2f}{dt^2}\right] = \int_0^\infty \frac{d^2f}{dt^2} e^{-st}\, dt$$

$$= \left[\frac{df}{dt}e^{-st}\right]_0^\infty + s\int_0^\infty \frac{df}{dt}e^{-st}\, dt$$

$$= -\frac{df}{dt}(0) + s[s\bar{f}(s) - f(0)], \qquad \text{for } s > 0,$$

where (13.57) has been substituted for the integral. This can be written more neatly as

$$\mathcal{L}\left[\frac{d^2f}{dt^2}\right] = s^2\bar{f}(s) - sf(0) - \frac{df}{dt}(0), \qquad \text{for } s > 0. \blacktriangleleft$$

In general the Laplace transform of the nth derivative is given by

$$\mathcal{L}\left[\frac{d^nf}{dt^n}\right] = s^n\bar{f} - s^{n-1}f(0) - s^{n-2}\frac{df}{dt}(0) - \cdots - \frac{d^{n-1}f}{dt^{n-1}}(0), \qquad \text{for } s > 0. \tag{13.58}$$

We now turn to integration, which is much more straightforward. From the definition (13.53),

$$\mathcal{L}\left[\int_0^t f(u)\, du\right] = \int_0^\infty dt\, e^{-st}\int_0^t f(u)\, du$$

$$= \left[-\frac{1}{s}e^{-st}\int_0^t f(u)\, du\right]_0^\infty + \int_0^\infty \frac{1}{s}e^{-st}f(t)\, dt.$$

The first term on the RHS vanishes at both limits, and so

$$\mathcal{L}\left[\int_0^t f(u)\, du\right] = \frac{1}{s}\mathcal{L}[f]. \tag{13.59}$$

13.2.2 Other properties of Laplace transforms

From table 13.1 it will be apparent that multiplying a function $f(t)$ by e^{at} has the effect on its transform that s is replaced by $s - a$. This is easily proved generally:

$$\mathcal{L}\left[e^{at}f(t)\right] = \int_0^\infty f(t)e^{at}e^{-st}\, dt$$

$$= \int_0^\infty f(t)e^{-(s-a)t}\, dt$$

$$= \bar{f}(s - a). \tag{13.60}$$

As it were, multiplying $f(t)$ by e^{at} moves the origin of s by an amount a.

We may now consider the effect of multiplying the Laplace transform $\bar{f}(s)$ by e^{-bs} ($b > 0$). From the definition (13.53),

$$e^{-bs}\bar{f}(s) = \int_0^\infty e^{-s(t+b)} f(t)\, dt$$

$$= \int_b^\infty e^{-sz} f(z-b)\, dz,$$

on putting $t + b = z$. Thus $e^{-bs}\bar{f}(s)$ is the Laplace transform of a function $g(t)$ defined by

$$g(t) = \begin{cases} 0 & \text{for } 0 < t \le b, \\ f(t-b) & \text{for } t > b. \end{cases}$$

In other words, the function f has been translated to 'later' t (larger values of t) by an amount b.

Further properties of Laplace transforms can be proved in similar ways and are listed below.

(i)
$$\mathcal{L}[f(at)] = \frac{1}{a}\bar{f}\left(\frac{s}{a}\right), \tag{13.61}$$

(ii)
$$\mathcal{L}[t^n f(t)] = (-1)^n \frac{d^n \bar{f}(s)}{ds^n}, \quad \text{for } n = 1, 2, 3, \ldots, \tag{13.62}$$

(iii)
$$\mathcal{L}\left[\frac{f(t)}{t}\right] = \int_s^\infty \bar{f}(u)\, du, \tag{13.63}$$

provided $\lim_{t\to 0}[f(t)/t]$ exists.

Related results may be easily proved.

▶ *Find an expression for the Laplace transform of $t\, d^2 f/dt^2$.*

From the definition of the Laplace transform we have

$$\mathcal{L}\left[t\frac{d^2 f}{dt^2}\right] = \int_0^\infty e^{-st} t \frac{d^2 f}{dt^2}\, dt$$

$$= -\frac{d}{ds}\int_0^\infty e^{-st}\frac{d^2 f}{dt^2}\, dt$$

$$= -\frac{d}{ds}[s^2\bar{f}(s) - sf(0) - f'(0)]$$

$$= -s^2\frac{d\bar{f}}{ds} - 2s\bar{f} + f(0). \blacktriangleleft$$

Finally we mention the convolution theorem for Laplace transforms (which is analogous to that for Fourier transforms discussed in subsection 13.1.7). If the functions f and g have Laplace transforms $\bar{f}(s)$ and $\bar{g}(s)$ then

$$\mathcal{L}\left[\int_0^t f(u)g(t-u)\, du\right] = \bar{f}(s)\bar{g}(s), \tag{13.64}$$

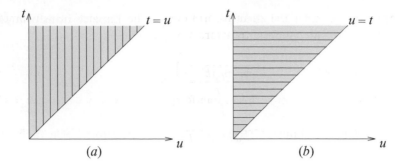

Figure 13.7 Two representations of the Laplace transform convolution (see text).

where the integral in the brackets on the LHS is the *convolution* of f and g, denoted by $f * g$. As in the case of Fourier transforms, the convolution defined above is commutative, i.e. $f * g = g * f$, and is associative and distributive. From (13.64) we also see that

$$\mathcal{L}^{-1}\left[\bar{f}(s)\bar{g}(s)\right] = \int_0^t f(u)g(t - u)\,du = f * g.$$

▶*Prove the convolution theorem (13.64) for Laplace transforms.*

From the definition (13.64),

$$\bar{f}(s)\bar{g}(s) = \int_0^\infty e^{-su}f(u)\,du \int_0^\infty e^{-sv}g(v)\,dv$$

$$= \int_0^\infty du \int_0^\infty dv\, e^{-s(u+v)}f(u)g(v).$$

Now letting $u + v = t$ changes the limits on the integrals, with the result that

$$\bar{f}(s)\bar{g}(s) = \int_0^\infty du\, f(u) \int_u^\infty dt\, g(t - u)\, e^{-st}.$$

As shown in figure 13.7(a) the shaded area of integration may be considered as the sum of vertical strips. However, we may instead integrate over this area by summing over horizontal strips as shown in figure 13.7(b). Then the integral can be written as

$$\bar{f}(s)\bar{g}(s) = \int_0^t du\, f(u) \int_0^\infty dt\, g(t - u)\, e^{-st}$$

$$= \int_0^\infty dt\, e^{-st} \left\{\int_0^t f(u)g(t - u)\,du\right\}$$

$$= \mathcal{L}\left[\int_0^t f(u)g(t - u)\,du\right]. \blacktriangleleft$$

458

The properties of the Laplace transform derived in this section can sometimes be useful in finding the Laplace transforms of particular functions.

▶ *Find the Laplace transform of $f(t) = t \sin bt$.*

Although we could calculate the Laplace transform directly, we can use (13.62) to give

$$\bar{f}(s) = (-1)\frac{d}{ds}\mathscr{L}\,[\sin bt] = -\frac{d}{ds}\left(\frac{b}{s^2 + b^2}\right) = \frac{2bs}{(s^2 + b^2)^2}, \qquad \text{for } s > 0. \blacktriangleleft$$

13.3 Concluding remarks

In this chapter we have discussed Fourier and Laplace transforms in some detail. Both are examples of *integral transforms*, which can be considered in a more general context.

A general integral transform of a function $f(t)$ takes the form

$$F(\alpha) = \int_a^b K(\alpha, t)f(t)\,dt, \tag{13.65}$$

where $F(\alpha)$ is the transform of $f(t)$ with respect to the *kernel* $K(\alpha, t)$, and α is the transform variable. For example, in the Laplace transform case $K(s, t) = e^{-st}$, $a = 0$, $b = \infty$.

Very often the inverse transform can also be written straightforwardly and we obtain a transform pair similar to that encountered in Fourier transforms. Examples of such pairs are

(i) the Hankel transform

$$F(k) = \int_0^\infty f(x)J_n(kx)x\,dx,$$

$$f(x) = \int_0^\infty F(k)J_n(kx)k\,dk,$$

where the J_n are Bessel functions of order n, and

(ii) the Mellin transform

$$F(z) = \int_0^\infty t^{z-1}f(t)\,dt,$$

$$f(t) = \frac{1}{2\pi i}\int_{-i\infty}^{i\infty} t^{-z}F(z)\,dz.$$

Although we do not have the space to discuss their general properties, the reader should at least be aware of this wider class of integral transforms.

13.4 Exercises

13.1   Find the Fourier transform of the function $f(t) = \exp(-|t|)$.

(a) By applying Fourier's inversion theorem prove that

$$\frac{\pi}{2}\exp(-|t|) = \int_0^\infty \frac{\cos \omega t}{1 + \omega^2}\, d\omega.$$

(b) By making the substitution $\omega = \tan\theta$, demonstrate the validity of Parseval's theorem for this function.

13.2   Use the general definition and properties of Fourier transforms to show the following.

(a) If $f(x)$ is periodic with period a then $\tilde{f}(k) = 0$, unless $ka = 2\pi n$ for integer n.
(b) The Fourier transform of $tf(t)$ is $id\tilde{f}(\omega)/d\omega$.
(c) The Fourier transform of $f(mt + c)$ is

$$\frac{e^{i\omega c/m}}{m}\tilde{f}\left(\frac{\omega}{m}\right).$$

13.3   Find the Fourier transform of $H(x-a)e^{-bx}$, where $H(x)$ is the Heaviside function.
13.4   Prove that the Fourier transform of the function $f(t)$ defined in the tf-plane by straight-line segments joining $(-T,0)$ to $(0,1)$ to $(T,0)$, with $f(t) = 0$ outside $|t| < T$, is

$$\tilde{f}(\omega) = \frac{T}{\sqrt{2\pi}}\mathrm{sinc}^2\left(\frac{\omega T}{2}\right),$$

where sinc x is defined as $(\sin x)/x$.

Use the general properties of Fourier transforms to determine the transforms of the following functions, graphically defined by straight-line segments and equal to zero outside the ranges specified:

(a) $(0,0)$ to $(0.5,1)$ to $(1,0)$ to $(2,2)$ to $(3,0)$ to $(4.5,3)$ to $(6,0)$;
(b) $(-2,0)$ to $(-1,2)$ to $(1,2)$ to $(2,0)$;
(c) $(0,0)$ to $(0,1)$ to $(1,2)$ to $(1,0)$ to $(2,-1)$ to $(2,0)$.

13.5   By taking the Fourier transform of the equation

$$\frac{d^2\phi}{dx^2} - K^2\phi = f(x),$$

show that its solution, $\phi(x)$, can be written as

$$\phi(x) = \frac{-1}{\sqrt{2\pi}}\int_{-\infty}^\infty \frac{e^{ikx}\tilde{f}(k)}{k^2 + K^2}\, dk,$$

where $\tilde{f}(k)$ is the Fourier transform of $f(x)$.

13.6   By differentiating the definition of the Fourier sine transform $\tilde{f}_s(\omega)$ of the function $f(t) = t^{-1/2}$ with respect to ω, and then integrating the resulting expression by parts, find an elementary differential equation satisfied by $\tilde{f}_s(\omega)$. Hence show that this function is its own Fourier sine transform, i.e. $\tilde{f}_s(\omega) = Af(\omega)$, where A is a constant. Show that it is also its own Fourier cosine transform. Assume that the limit as $x \to \infty$ of $x^{1/2}\sin\alpha x$ can be taken as zero.

13.7   Find the Fourier transform of the unit rectangular distribution

$$f(t) = \begin{cases} 1 & |t| < 1, \\ 0 & \text{otherwise.} \end{cases}$$

Determine the convolution of f with itself and, without further integration, deduce its transform. Deduce that

$$\int_{-\infty}^{\infty} \frac{\sin^2 \omega}{\omega^2} \, d\omega = \pi,$$

$$\int_{-\infty}^{\infty} \frac{\sin^4 \omega}{\omega^4} \, d\omega = \frac{2\pi}{3}.$$

13.8   Calculate the Fraunhofer spectrum produced by a diffraction grating, uniformly illuminated by light of wavelength $2\pi/k$, as follows. Consider a grating with $4N$ equal strips each of width a and alternately opaque and transparent. The aperture function is then

$$f(y) = \begin{cases} A & \text{for } (2n+1)a \le y < (2n+2)a, \quad -N \le n < N, \\ 0 & \text{otherwise.} \end{cases}$$

(a) Show, for diffraction at angle θ to the normal to the grating, that the required Fourier transform can be written

$$\tilde{f}(q) = (2\pi)^{-1/2} \sum_{r=-N}^{N-1} \exp(-2iarq) \int_{a}^{2a} A \exp(-iqu) \, du,$$

where $q = k \sin \theta$.

(b) Evaluate the integral and sum to show that

$$\tilde{f}(q) = (2\pi)^{-1/2} \exp(-iqa/2) \frac{A \sin(2qaN)}{q \cos(qa/2)},$$

and hence that the intensity distribution $I(\theta)$ in the spectrum is proportional to

$$\frac{\sin^2(2qaN)}{q^2 \cos^2(qa/2)}.$$

(c) For large values of N, the numerator in the above expression has very closely spaced maxima and minima as a function of θ and effectively takes its mean value, $1/2$, giving a low-intensity background. Much more significant peaks in $I(\theta)$ occur when $\theta = 0$ or the cosine term in the denominator vanishes. Show that the corresponding values of $|\tilde{f}(q)|$ are

$$\frac{2aNA}{(2\pi)^{1/2}} \quad \text{and} \quad \frac{4aNA}{(2\pi)^{1/2}(2m+1)\pi}, \quad \text{with } m \text{ integral.}$$

Note that the constructive interference makes the maxima in $I(\theta) \propto N^2$, not N. Of course, observable maxima only occur for $0 \le \theta \le \pi/2$.

13.9   By finding the complex Fourier *series* for its LHS show that either side of the equation

$$\sum_{n=-\infty}^{\infty} \delta(t + nT) = \frac{1}{T} \sum_{n=-\infty}^{\infty} e^{-2\pi nit/T}$$

can represent a periodic train of impulses. By expressing the function $f(t + nX)$, in which X is a constant, in terms of the Fourier *transform* $\tilde{f}(\omega)$ of $f(t)$, show that

$$\sum_{n=-\infty}^{\infty} f(t + nX) = \frac{\sqrt{2\pi}}{X} \sum_{n=-\infty}^{\infty} \tilde{f}\left(\frac{2n\pi}{X}\right) e^{2\pi nit/X}.$$

This result is known as the *Poisson summation formula*.

13.10 In many applications in which the frequency spectrum of an analogue signal is required, the best that can be done is to sample the signal $f(t)$ a finite number of times at fixed intervals, and then use a *discrete Fourier transform* F_k to estimate discrete points on the (true) frequency spectrum $\tilde{f}(\omega)$.

(a) By an argument that is essentially the converse of that given in section 13.1, show that, if N samples f_n, beginning at $t = 0$ and spaced τ apart, are taken, then $\tilde{f}(2\pi k/(N\tau)) \approx F_k\tau$ where

$$F_k = \frac{1}{\sqrt{2\pi}} \sum_{n=0}^{N-1} f_n e^{-2\pi nki/N}.$$

(b) For the function $f(t)$ defined by

$$f(t) = \begin{cases} 1 & \text{for } 0 \leq t < 1, \\ 0 & \text{otherwise,} \end{cases}$$

from which eight samples are drawn at intervals of $\tau = 0.25$, find a formula for $|F_k|$ and evaluate it for $k = 0, 1, \ldots, 7$.

(c) Find the exact frequency spectrum of $f(t)$ and compare the actual and estimated values of $\sqrt{2\pi}|\tilde{f}(\omega)|$ at $\omega = k\pi$ for $k = 0, 1, \ldots, 7$. Note the relatively good agreement for $k < 4$ and the lack of agreement for larger values of k.

13.11 For a function $f(t)$ that is non-zero only in the range $|t| < T/2$, the full frequency spectrum $\tilde{f}(\omega)$ can be constructed, in principle exactly, from values at discrete sample points $\omega = n(2\pi/T)$. Prove this as follows.

(a) Show that the coefficients of a complex Fourier *series* representation of $f(t)$ with period T can be written as

$$c_n = \frac{\sqrt{2\pi}}{T} \tilde{f}\left(\frac{2\pi n}{T}\right).$$

(b) Use this result to represent $f(t)$ as an infinite sum in the defining integral for $\tilde{f}(\omega)$, and hence show that

$$\tilde{f}(\omega) = \sum_{n=-\infty}^{\infty} \tilde{f}\left(\frac{2\pi n}{T}\right) \operatorname{sinc}\left(n\pi - \frac{\omega T}{2}\right),$$

where sinc x is defined as $(\sin x)/x$.

13.12 A signal obtained by sampling a function $x(t)$ at regular intervals T is passed through an electronic filter, whose response $g(t)$ to a unit δ-function input is represented in a tg-plot by straight lines joining $(0,0)$ to $(T, 1/T)$ to $(2T, 0)$ and is zero for all other values of t. The output of the filter is the convolution of the input, $\sum_{-\infty}^{\infty} x(t)\delta(t - nT)$, with $g(t)$.
Using the convolution theorem, and the result given in exercise 13.4, show that the output of the filter can be written

$$y(t) = \frac{1}{2\pi} \sum_{n=-\infty}^{\infty} x(nT) \int_{-\infty}^{\infty} \operatorname{sinc}^2\left(\frac{\omega T}{2}\right) e^{-i\omega[(n+1)T-t]} d\omega.$$

13.13 Find the Fourier transform specified in part (a) and then use it to answer part (b).

(a) Find the Fourier transform of

$$f(\gamma, p, t) = \begin{cases} e^{-\gamma t} \sin pt & t > 0, \\ 0 & t < 0, \end{cases}$$

where γ (> 0) and p are constant parameters.

(b) The current $I(t)$ flowing through a certain system is related to the applied voltage $V(t)$ by the equation

$$I(t) = \int_{-\infty}^{\infty} K(t - u)V(u)\,du,$$

where

$$K(\tau) = a_1 f(\gamma_1, p_1, \tau) + a_2 f(\gamma_2, p_2, \tau).$$

The function $f(\gamma, p, t)$ is as given in (a) and all the a_i, γ_i (> 0) and p_i are fixed parameters. By considering the Fourier transform of $I(t)$, find the relationship that must hold between a_1 and a_2 if the total net charge Q passed through the system (over a very long time) is to be zero for an arbitrary applied voltage.

13.14　Prove the equality

$$\int_0^\infty e^{-2at} \sin^2 at\,dt = \frac{1}{\pi} \int_0^\infty \frac{a^2}{4a^4 + \omega^4}\,d\omega.$$

13.15　A linear amplifier produces an output that is the convolution of its input and its response function. The Fourier transform of the response function for a particular amplifier is

$$\tilde{K}(\omega) = \frac{i\omega}{\sqrt{2\pi}(\alpha + i\omega)^2}.$$

Determine the time variation of its output $g(t)$ when its input is the Heaviside step function. (Consider the Fourier transform of a decaying exponential function and the result of exercise 13.2(b).)

13.16　In quantum mechanics, two equal-mass particles having momenta $\mathbf{p}_j = \hbar\mathbf{k}_j$ and energies $E_j = \hbar\omega_j$ and represented by plane wavefunctions $\phi_j = \exp[i(\mathbf{k}_j\cdot\mathbf{r}_j - \omega_j t)]$, $j = 1, 2$, interact through a potential $V = V(|\mathbf{r}_1 - \mathbf{r}_2|)$. In first-order perturbation theory the probability of scattering to a state with momenta and energies $\mathbf{p}_j', E_j'$ is determined by the modulus squared of the quantity

$$M = \iiint \psi_f^* V \psi_i\,d\mathbf{r}_1\,d\mathbf{r}_2\,dt.$$

The initial state, ψ_i, is $\phi_1\phi_2$ and the final state, ψ_f, is $\phi_1'\phi_2'$.

(a) By writing $\mathbf{r}_1 + \mathbf{r}_2 = 2\mathbf{R}$ and $\mathbf{r}_1 - \mathbf{r}_2 = \mathbf{r}$ and assuming that $d\mathbf{r}_1\,d\mathbf{r}_2 = d\mathbf{R}\,d\mathbf{r}$, show that M can be written as the product of three one-dimensional integrals.

(b) From two of the integrals deduce energy and momentum conservation in the form of δ-functions.

(c) Show that M is proportional to the Fourier transform of V, i.e. to $\tilde{V}(\mathbf{k})$ where $2\hbar\mathbf{k} = (\mathbf{p}_2 - \mathbf{p}_1) - (\mathbf{p}_2' - \mathbf{p}_1')$ or, alternatively, $\hbar\mathbf{k} = \mathbf{p}_1' - \mathbf{p}_1$.

13.17　For some ion–atom scattering processes, the potential V of the previous exercise may be approximated by $V = |\mathbf{r}_1 - \mathbf{r}_2|^{-1} \exp(-\mu|\mathbf{r}_1 - \mathbf{r}_2|)$. Show, using the result of the worked example in subsection 13.1.10, that the probability that the ion will scatter from, say, $\mathbf{p}_1$ to $\mathbf{p}_1'$ is proportional to $(\mu^2 + k^2)^{-2}$, where $k = |\mathbf{k}|$ and $\mathbf{k}$ is as given in part (c) of that exercise.

13.18 The equivalent duration and bandwidth, T_e and B_e, of a signal $x(t)$ are defined in terms of the latter and its Fourier transform $\tilde{x}(\omega)$ by

$$T_e = \frac{1}{x(0)} \int_{-\infty}^{\infty} x(t)\,dt,$$

$$B_e = \frac{1}{\tilde{x}(0)} \int_{-\infty}^{\infty} \tilde{x}(\omega)\,d\omega,$$

where neither $x(0)$ nor $\tilde{x}(0)$ is zero. Show that the product $T_e B_e = 2\pi$ (this is a form of uncertainty principle), and find the equivalent bandwidth of the signal

$$x(t) = \exp(-|t|/T).$$

For this signal, determine the fraction of the total energy that lies in the frequency range $|\omega| < B_e/4$. You will need the indefinite integral with respect to x of $(a^2 + x^2)^{-2}$, which is

$$\frac{x}{2a^2(a^2 + x^2)} + \frac{1}{2a^3}\tan^{-1}\frac{x}{a}.$$

13.19 Calculate directly the auto-correlation function $a(z)$ for the product $f(t)$ of the exponential decay distribution and the Heaviside step function,

$$f(t) = \frac{1}{\lambda}e^{-\lambda t}H(t).$$

Use the Fourier transform and energy spectrum of $f(t)$ to deduce that

$$\int_{-\infty}^{\infty} \frac{e^{i\omega z}}{\lambda^2 + \omega^2}\,d\omega = \frac{\pi}{\lambda}e^{-\lambda|z|}.$$

13.20 Prove that the cross-correlation $C(z)$ of the Gaussian and Lorentzian distributions

$$f(t) = \frac{1}{\tau\sqrt{2\pi}}\exp\left(-\frac{t^2}{2\tau^2}\right), \qquad g(t) = \left(\frac{a}{\pi}\right)\frac{1}{t^2 + a^2},$$

has as its Fourier transform the function

$$\frac{1}{\sqrt{2\pi}}\exp\left(-\frac{\tau^2\omega^2}{2}\right)\exp(-a|\omega|).$$

Hence show that

$$C(z) = \frac{1}{\tau\sqrt{2\pi}}\exp\left(\frac{a^2 - z^2}{2\tau^2}\right)\cos\left(\frac{az}{\tau^2}\right).$$

13.21 Prove the expressions given in table 13.1 for the Laplace transforms of $t^{-1/2}$ and $t^{1/2}$, by setting $x^2 = ts$ in the result

$$\int_0^{\infty} \exp(-x^2)\,dx = \tfrac{1}{2}\sqrt{\pi}.$$

13.22 Find the functions $y(t)$ whose Laplace transforms are the following:

(a) $1/(s^2 - s - 2)$;
(b) $2s/[(s + 1)(s^2 + 4)]$;
(c) $e^{-(\gamma+s)t_0}/[(s + \gamma)^2 + b^2]$.

13.23 Use the properties of Laplace transforms to prove the following without evaluating any Laplace integrals explicitly:

(a) $\mathscr{L}\left[t^{5/2}\right] = \frac{15}{8}\sqrt{\pi}s^{-7/2}$;
(b) $\mathscr{L}\left[(\sinh at)/t\right] = \tfrac{1}{2}\ln\left[(s + a)/(s - a)\right], \qquad s > |a|$;

(c) $\mathcal{L}[\sinh at \cos bt] = a(s^2 - a^2 + b^2)[(s-a)^2 + b^2]^{-1}[(s+a)^2 + b^2]^{-1}$.

13.24   Find the solution (the so-called *impulse response* or *Green's function*) of the equation

$$T\frac{dx}{dt} + x = \delta(t)$$

by proceeding as follows.

(a) Show by substitution that

$$x(t) = A(1 - e^{-t/T})H(t)$$

is a solution, for which $x(0) = 0$, of

$$T\frac{dx}{dt} + x = AH(t), \qquad (*)$$

where $H(t)$ is the Heaviside step function.

(b) Construct the solution when the RHS of $(*)$ is replaced by $AH(t - \tau)$, with $dx/dt = x = 0$ for $t < \tau$, and hence find the solution when the RHS is a rectangular pulse of duration τ.

(c) By setting $A = 1/\tau$ and taking the limit as $\tau \to 0$, show that the impulse response is $x(t) = T^{-1}e^{-t/T}$.

(d) Obtain the same result much more directly by taking the Laplace transform of each term in the original equation, solving the resulting algebraic equation and then using the entries in table 13.1.

13.25   This exercise is concerned with the limiting behaviour of Laplace transforms.

(a) If $f(t) = A + g(t)$, where A is a constant and the indefinite integral of $g(t)$ is bounded as its upper limit tends to ∞, show that

$$\lim_{s \to 0} s\bar{f}(s) = A.$$

(b) For $t > 0$, the function $y(t)$ obeys the differential equation

$$\frac{d^2y}{dt^2} + a\frac{dy}{dt} + by = c\cos^2 \omega t,$$

where a, b and c are positive constants. Find $\bar{y}(s)$ and show that $s\bar{y}(s) \to c/2b$ as $s \to 0$. Interpret the result in the t-domain.

13.26   By writing $f(x)$ as an integral involving the δ-function $\delta(\xi - x)$ and taking the Laplace transforms of both sides, show that the transform of the solution of the equation

$$\frac{d^4y}{dx^4} - y = f(x)$$

for which y and its first three derivatives vanish at $x = 0$ can be written as

$$\bar{y}(s) = \int_0^\infty f(\xi)\frac{e^{-s\xi}}{s^4 - 1}\, d\xi.$$

Use the properties of Laplace transforms and the entries in table 13.1 to show that

$$y(x) = \frac{1}{2}\int_0^x f(\xi)\,[\sinh(x - \xi) - \sin(x - \xi)]\, d\xi.$$

13.27 The function $f_a(x)$ is defined as unity for $0 < x < a$ and zero otherwise. Find its Laplace transform $\bar{f}_a(s)$ and deduce that the transform of $xf_a(x)$ is

$$\frac{1}{s^2}\left[1 - (1 + as)e^{-sa}\right].$$

Write $f_a(x)$ in terms of Heaviside functions and hence obtain an explicit expression for

$$g_a(x) = \int_0^x f_a(y)f_a(x - y)\, dy.$$

Use the expression to write $\bar{g}_a(s)$ in terms of the functions $\bar{f}_a(s)$ and $\bar{f}_{2a}(s)$, and their derivatives, and hence show that $\bar{g}_a(s)$ is equal to the square of $\bar{f}_a(s)$, in accordance with the convolution theorem.

13.28 Show that the Laplace transform of $f(t - a)H(t - a)$, where $a \geq 0$, is $e^{-as}\bar{f}(s)$ and that, if $g(t)$ is a periodic function of period T, $\bar{g}(s)$ can be written as

$$\frac{1}{1 - e^{-sT}}\int_0^T e^{-st}g(t)\, dt.$$

(a) Sketch the periodic function defined in $0 \leq t \leq T$ by

$$g(t) = \begin{cases} 2t/T & 0 \leq t < T/2, \\ 2(1 - t/T) & T/2 \leq t \leq T, \end{cases}$$

and, using the previous result, find its Laplace transform.

(b) Show, by sketching it, that

$$\frac{2}{T}[tH(t) + 2\sum_{n=1}^{\infty}(-1)^n(t - \tfrac{1}{2}nT)H(t - \tfrac{1}{2}nT)]$$

is another representation of $g(t)$ and hence derive the relationship

$$\tanh x = 1 + 2\sum_{n=1}^{\infty}(-1)^n e^{-2nx}.$$

13.5 Hints and answers

13.1 Note that the integrand has different analytic forms for $t < 0$ and $t \geq 0$. $(2/\pi)^{1/2}(1 + \omega^2)^{-1}$.

13.3 $(1/\sqrt{2\pi})[(b - ik)/(b^2 + k^2)]e^{-a(b+ik)}$.

13.5 Use or derive $\widetilde{\phi''}(k) = -k^2\tilde{\phi}(k)$ to obtain an algebraic equation for $\tilde{\phi}(k)$ and then use the Fourier inversion formula.

13.7 $(2/\sqrt{2\pi})(\sin\omega/\omega)$.
The convolution is $2 - |t|$ for $|t| < 2$, zero otherwise. Use the convolution theorem. $(4/\sqrt{2\pi})(\sin^2\omega/\omega^2)$.
Apply Parseval's theorem to f and to $f * f$.

13.9 The Fourier coefficient is T^{-1}, independent of n. Make the changes of variables $t \to \omega$, $n \to -n$ and $T \to 2\pi/X$ and apply the translation theorem.

13.11 (b) Recall that the infinite integral involved in defining $\tilde{f}(\omega)$ has a non-zero integrand only in $|t| < T/2$.

13.13 (a) $(1/\sqrt{2\pi})\{p/[(\gamma + i\omega)^2 + p^2]\}$.
(b) Show that $Q = \sqrt{2\pi}\tilde{I}(0)$ and use the convolution theorem. The required relationship is $a_1p_1/(\gamma_1^2 + p_1^2) + a_2p_2/(\gamma_2^2 + p_2^2) = 0$.

13.15 $\tilde{g}(\omega) = 1/[\sqrt{2\pi}(\alpha + i\omega)^2]$, leading to $g(t) = te^{-\alpha t}$.

466

13.17 $\tilde{V}(\mathbf{k}) \propto [-2\pi/(ik)] \int \{\exp[-(\mu - ik)r] - \exp[-(\mu + ik)r]\}\, dr.$

13.19   Note that the lower limit in the calculation of $a(z)$ is 0, for $z > 0$, and $|z|$, for $z < 0$. Auto-correlation $a(z) = [(1/(2\lambda^3)]\exp(-\lambda|z|).$

13.21   Prove the result for $t^{1/2}$ by integrating that for $t^{-1/2}$ by parts.

13.23   (a) Use (13.62) with $n = 2$ on $\mathscr{L}\left[\sqrt{t}\right]$; (b) use (13.63);
       (c) consider $\mathscr{L}\left[\exp(\pm at)\cos bt\right]$ and use the translation property, subsection 13.2.2.

13.25   (a) Note that $|\lim \int g(t)e^{-st}\, dt| \leq |\lim \int g(t)\, dt|.$
       (b) $(s^2 + as + b)\bar{y}(s) = \{c(s^2 + 2\omega^2)/[s(s^2 + 4\omega^2)]\} + (a + s)y(0) + y'(0).$
       For this damped system, at large t (corresponding to $s \to 0$) rates of change are negligible and the equation reduces to $by = c\cos^2 \omega t$. The average value of $\cos^2 \omega t$ is $\frac{1}{2}$.

13.27 $s^{-1}[1 - \exp(-sa)]$; $g_a(x) = x$ for $0 < x < a$, $g_a(x) = 2a - x$ for $a \leq x \leq 2a$, $g_a(x) = 0$ otherwise.

14

First-order ordinary differential equations

Differential equations are the group of equations that contain derivatives. Chapters 14–21 discuss a variety of differential equations, starting in this chapter and the next with those ordinary differential equations (ODEs) that have closed-form solutions. As its name suggests, an ODE contains only ordinary derivatives (no partial derivatives) and describes the relationship between these derivatives of the *dependent variable*, usually called y, with respect to the *independent variable*, usually called x. The solution to such an ODE is therefore a function of x and is written $y(x)$. For an ODE to have a closed-form solution, it must be possible to express $y(x)$ in terms of the standard elementary functions such as $\exp x$, $\ln x$, $\sin x$ etc. The solutions of some differential equations cannot, however, be written in closed form, but only as an infinite series; these are discussed in chapter 16.

Ordinary differential equations may be separated conveniently into different categories according to their general characteristics. The primary grouping adopted here is by the *order* of the equation. The order of an ODE is simply the order of the highest derivative it contains. Thus equations containing dy/dx, but no higher derivatives, are called first order, those containing d^2y/dx^2 are called second order and so on. In this chapter we consider first-order equations, and in the next, second- and higher-order equations.

Ordinary differential equations may be classified further according to *degree*. The degree of an ODE is the power to which the highest-order derivative is raised, after the equation has been rationalised to contain only integer powers of derivatives. Hence the ODE

$$\frac{d^3y}{dx^3} + x\left(\frac{dy}{dx}\right)^{3/2} + x^2y = 0,$$

is of third order and second degree, since after rationalisation it contains the term $(d^3y/dx^3)^2$.

The *general solution* to an ODE is the most general function $y(x)$ that satisfies the equation; it will contain *constants of integration* which may be determined by the application of some suitable *boundary conditions*. For example, we may be told that for a certain first-order differential equation, the solution $y(x)$ is equal to zero when the parameter x is equal to unity; this allows us to determine the value of the constant of integration. The *general solutions* to nth-order ODEs, which are considered in detail in the next chapter, will contain n (essential) arbitrary constants of integration and therefore we will need n boundary conditions if these constants are to be determined (see section 14.1). When the boundary conditions have been applied, and the constants found, we are left with a *particular solution* to the ODE, which obeys the given boundary conditions. Some ODEs of degree greater than unity also possess *singular solutions*, which are solutions that contain no arbitrary constants and cannot be found from the general solution; singular solutions are discussed in more detail in section 14.3. When any solution to an ODE has been found, it is always possible to check its validity by substitution into the original equation and verification that any given boundary conditions are met.

In this chapter, firstly we discuss various types of first-degree ODE and then go on to examine those higher-degree equations that can be solved in closed form. At the outset, however, we discuss the general form of the solutions of ODEs; this discussion is relevant to both first- and higher-order ODEs.

14.1 General form of solution

It is helpful when considering the general form of the solution of an ODE to consider the inverse process, namely that of obtaining an ODE from a given group of functions, each one of which is a solution of the ODE. Suppose the members of the group can be written as

$$y = f(x, a_1, a_2, \ldots, a_n), \tag{14.1}$$

each member being specified by a different set of values of the parameters a_i. For example, consider the group of functions

$$y = a_1 \sin x + a_2 \cos x; \tag{14.2}$$

here $n = 2$.

Since an ODE is required for which *any* of the group is a solution, it clearly must not contain any of the a_i. As there are n of the a_i in expression (14.1), we must obtain $n + 1$ equations involving them in order that, by elimination, we can obtain one final equation without them.

Initially we have only (14.1), but if this is differentiated n times, a total of $n + 1$ equations is obtained from which (in principle) all the a_i can be eliminated, to give one ODE satisfied by all the group. As a result of the n differentiations, $d^n y/dx^n$ will be present in one of the $n + 1$ equations and hence in the final equation, which will therefore be of nth order.

In the case of (14.2), we have

$$\frac{dy}{dx} = a_1 \cos x - a_2 \sin x,$$

$$\frac{d^2 y}{dx^2} = -a_1 \sin x - a_2 \cos x.$$

Here the elimination of a_1 and a_2 is trivial (because of the similarity of the forms of y and $d^2 y/dx^2$), resulting in

$$\frac{d^2 y}{dx^2} + y = 0,$$

a second-order equation.

Thus, to summarise, a group of functions (14.1) with n parameters satisfies an nth-order ODE in general (although in some degenerate cases an ODE of less than nth order is obtained). The intuitive converse of this is that the general solution of an nth-order ODE contains n arbitrary parameters (constants); for our purposes, this will be assumed to be valid although a totally general proof is difficult.

As mentioned earlier, external factors affect a system described by an ODE, by fixing the values of the dependent variables for particular values of the independent ones. These externally imposed (or *boundary*) conditions on the solution are thus the means of determining the parameters and so of specifying precisely which function is the required solution. It is apparent that the number of boundary conditions should match the number of parameters and hence the order of the equation, if a unique solution is to be obtained. Fewer independent boundary conditions than this will lead to a number of undetermined parameters in the solution, whilst an excess will usually mean that no acceptable solution is possible.

For an nth-order equation the required n boundary conditions can take many forms, for example the value of y at n different values of x, or the value of any $n-1$ of the n derivatives dy/dx, $d^2 y/dx^2$, ..., $d^n y/dx^n$ together with that of y, all for the same value of x, or many intermediate combinations.

14.2 First-degree first-order equations

First-degree first-order ODEs contain only dy/dx equated to some function of x and y, and can be written in either of two equivalent standard forms,

$$\frac{dy}{dx} = F(x, y), \qquad A(x, y)\, dx + B(x, y)\, dy = 0,$$

where $F(x, y) = -A(x, y)/B(x, y)$, and $F(x, y)$, $A(x, y)$ and $B(x, y)$ are in general functions of both x and y. Which of the two above forms is the more useful for finding a solution depends on the type of equation being considered. There

470

are several different types of first-degree first-order ODEs that are of interest in the physical sciences. These equations and their respective solutions are discussed below.

14.2.1 Separable-variable equations

A separable-variable equation is one which may be written in the conventional form

$$\frac{dy}{dx} = f(x)g(y), \tag{14.3}$$

where $f(x)$ and $g(y)$ are functions of x and y respectively, including cases in which $f(x)$ or $g(y)$ is simply a constant. Rearranging this equation so that the terms depending on x and on y appear on opposite sides (i.e. are separated), and integrating, we obtain

$$\int \frac{dy}{g(y)} = \int f(x)\,dx.$$

Finding the solution $y(x)$ that satisfies (14.3) then depends only on the ease with which the integrals in the above equation can be evaluated. It is also worth noting that ODEs that at first sight do not appear to be of the form (14.3) can sometimes be made separable by an appropriate factorisation.

▶Solve

$$\frac{dy}{dx} = x + xy.$$

Since the RHS of this equation can be factorised to give $x(1 + y)$, the equation becomes separable and we obtain

$$\int \frac{dy}{1+y} = \int x\,dx.$$

Now integrating both sides separately, we find

$$\ln(1 + y) = \frac{x^2}{2} + c,$$

and so

$$1 + y = \exp\left(\frac{x^2}{2} + c\right) = A\exp\left(\frac{x^2}{2}\right),$$

where c and hence A is an arbitrary constant. ◀

Solution method. *Factorise the equation so that it becomes separable. Rearrange it so that the terms depending on x and those depending on y appear on opposite sides and then integrate directly. Remember the constant of integration, which can be evaluated if further information is given.*

471

14.2.2 Exact equations

An *exact* first-degree first-order ODE is one of the form

$$A(x, y)\, dx + B(x, y)\, dy = 0 \quad \text{and for which} \quad \frac{\partial A}{\partial y} = \frac{\partial B}{\partial x}. \tag{14.4}$$

In this case $A(x, y)\, dx + B(x, y)\, dy$ is an exact differential, $dU(x, y)$ say (see section 5.3). In other words

$$A\, dx + B\, dy = dU = \frac{\partial U}{\partial x}\, dx + \frac{\partial U}{\partial y}\, dy,$$

from which we obtain

$$A(x, y) = \frac{\partial U}{\partial x}, \tag{14.5}$$

$$B(x, y) = \frac{\partial U}{\partial y}. \tag{14.6}$$

Since $\partial^2 U / \partial x \partial y = \partial^2 U / \partial y \partial x$ we therefore require

$$\frac{\partial A}{\partial y} = \frac{\partial B}{\partial x}. \tag{14.7}$$

If (14.7) holds then (14.4) can be written $dU(x, y) = 0$, which has the solution $U(x, y) = c$, where c is a constant and from (14.5) $U(x, y)$ is given by

$$U(x, y) = \int A(x, y)\, dx + F(y). \tag{14.8}$$

The function $F(y)$ can be found from (14.6) by differentiating (14.8) with respect to y and equating to $B(x, y)$.

►*Solve*

$$x\frac{dy}{dx} + 3x + y = 0.$$

Rearranging into the form (14.4) we have

$$(3x + y)\, dx + x\, dy = 0,$$

i.e. $A(x, y) = 3x + y$ and $B(x, y) = x$. Since $\partial A / \partial y = 1 = \partial B / \partial x$, the equation is exact, and by (14.8) the solution is given by

$$U(x, y) = \int (3x + y)\, dx + F(y) = c_1 \quad \Rightarrow \quad \frac{3x^2}{2} + yx + F(y) = c_1.$$

Differentiating $U(x, y)$ with respect to y and equating it to $B(x, y) = x$ we obtain $dF/dy = 0$, which integrates immediately to give $F(y) = c_2$. Therefore, letting $c = c_1 - c_2$, the solution to the original ODE is

$$\frac{3x^2}{2} + xy = c. \ \blacktriangleleft$$

Solution method. *Check that the equation is an exact differential using (14.7) then solve using (14.8). Find the function $F(y)$ by differentiating (14.8) with respect to y and using (14.6).*

14.2.3 Inexact equations: integrating factors

Equations that may be written in the form

$$A(x, y)\, dx + B(x, y)\, dy = 0 \quad \text{but for which} \quad \frac{\partial A}{\partial y} \neq \frac{\partial B}{\partial x} \qquad (14.9)$$

are known as inexact equations. However, the differential $A\, dx + B\, dy$ can always be made exact by multiplying by an *integrating factor* $\mu(x, y)$, which obeys

$$\frac{\partial(\mu A)}{\partial y} = \frac{\partial(\mu B)}{\partial x}. \qquad (14.10)$$

For an integrating factor that is a function of both x and y, i.e. $\mu = \mu(x, y)$, there exists no general method for finding it; in such cases it may sometimes be found by inspection. If, however, an integrating factor exists that is a function of either x or y alone then (14.10) can be solved to find it. For example, if we assume that the integrating factor is a function of x alone, i.e. $\mu = \mu(x)$, then (14.10) reads

$$\mu \frac{\partial A}{\partial y} = \mu \frac{\partial B}{\partial x} + B \frac{d\mu}{dx}.$$

Rearranging this expression we find

$$\frac{d\mu}{\mu} = \frac{1}{B} \left(\frac{\partial A}{\partial y} - \frac{\partial B}{\partial x} \right) dx = f(x)\, dx,$$

where we require $f(x)$ also to be a function of x only; indeed this provides a general method of determining whether the integrating factor μ is a function of x alone. This integrating factor is then given by

$$\mu(x) = \exp \left\{ \int f(x)\, dx \right\} \quad \text{where} \quad f(x) = \frac{1}{B} \left(\frac{\partial A}{\partial y} - \frac{\partial B}{\partial x} \right). \qquad (14.11)$$

Similarly, if $\mu = \mu(y)$ then

$$\mu(y) = \exp \left\{ \int g(y)\, dy \right\} \quad \text{where} \quad g(y) = \frac{1}{A} \left(\frac{\partial B}{\partial x} - \frac{\partial A}{\partial y} \right). \qquad (14.12)$$

▶*Solve*

$$\frac{dy}{dx} = -\frac{2}{y} - \frac{3y}{2x}.$$

Rearranging into the form (14.9), we have

$$(4x + 3y^2)\,dx + 2xy\,dy = 0, \tag{14.13}$$

i.e. $A(x, y) = 4x + 3y^2$ and $B(x, y) = 2xy$. Now

$$\frac{\partial A}{\partial y} = 6y, \qquad \frac{\partial B}{\partial x} = 2y,$$

so the ODE is not exact in its present form. However, we see that

$$\frac{1}{B}\left(\frac{\partial A}{\partial y} - \frac{\partial B}{\partial x}\right) = \frac{2}{x},$$

a function of x alone. Therefore an integrating factor exists that is also a function of x alone and, ignoring the arbitrary constant of integration, is given by

$$\mu(x) = \exp\left\{2\int\frac{dx}{x}\right\} = \exp(2\ln x) = x^2.$$

Multiplying (14.13) through by $\mu(x) = x^2$ we obtain

$$(4x^3 + 3x^2y^2)\,dx + 2x^3y\,dy = 4x^3\,dx + (3x^2y^2\,dx + 2x^3y\,dy) = 0.$$

By inspection this integrates immediately to give the solution $x^4 + y^2x^3 = c$, where c is a constant. ◀

Solution method. *Examine whether $f(x)$ and $g(y)$ are functions of only x or y respectively. If so, then the required integrating factor is a function of either x or y only, and is given by (14.11) or (14.12) respectively. If the integrating factor is a function of both x and y, then sometimes it may be found by inspection or by trial and error. In any case, the integrating factor μ must satisfy (14.10). Once the equation has been made exact, solve by the method of subsection 14.2.2.*

14.2.4 Linear equations

Linear first-order ODEs are a special case of inexact ODEs (discussed in the previous subsection) and can be written in the conventional form

$$\frac{dy}{dx} + P(x)y = Q(x). \tag{14.14}$$

Such equations can be made exact by multiplying through by an appropriate integrating factor in a similar manner to that discussed above. In this case, however, the integrating factor is always a function of x alone and may be expressed in a particularly simple form. An integrating factor $\mu(x)$ must be such that

$$\mu(x)\frac{dy}{dx} + \mu(x)P(x)y = \frac{d}{dx}\,[\mu(x)y] = \mu(x)Q(x), \tag{14.15}$$

474

which may then be integrated directly to give

$$\mu(x)y = \int \mu(x)Q(x)\,dx. \tag{14.16}$$

The required integrating factor $\mu(x)$ is determined by the first equality in (14.15), i.e.

$$\frac{d}{dx}(\mu y) = \mu\frac{dy}{dx} + \frac{d\mu}{dx}y = \mu\frac{dy}{dx} + \mu P y,$$

which immediately gives the simple relation

$$\frac{d\mu}{dx} = \mu(x)P(x) \quad \Rightarrow \quad \mu(x) = \exp\left\{\int P(x)\,dx\right\}. \tag{14.17}$$

▶Solve

$$\frac{dy}{dx} + 2xy = 4x.$$

The integrating factor is given immediately by

$$\mu(x) = \exp\left\{\int 2x\,dx\right\} = \exp x^2.$$

Multiplying through the ODE by $\mu(x) = \exp x^2$ and integrating, we have

$$y\exp x^2 = 4\int x\exp x^2\,dx = 2\exp x^2 + c.$$

The solution to the ODE is therefore given by $y = 2 + c\exp(-x^2)$. ◀

Solution method. *Rearrange the equation into the form (14.14) and multiply by the integrating factor $\mu(x)$ given by (14.17). The left- and right-hand sides can then be integrated directly, giving y from (14.16).*

14.2.5 Homogeneous equations

Homogeneous equation are ODEs that may be written in the form

$$\frac{dy}{dx} = \frac{A(x,y)}{B(x,y)} = F\left(\frac{y}{x}\right), \tag{14.18}$$

where $A(x,y)$ and $B(x,y)$ are homogeneous functions of the same degree. A function $f(x,y)$ is homogeneous of degree n if, for any λ, it obeys

$$f(\lambda x, \lambda y) = \lambda^n f(x,y).$$

For example, if $A = x^2 y - xy^2$ and $B = x^3 + y^3$ then we see that A and B are both homogeneous functions of degree 3. In general, for functions of the form of A and B, we see that for both to be homogeneous, and of the same degree, we require the sum of the powers in x and y in each term of A and B to be the same (in this example equal to 3). The RHS of a homogeneous ODE can be written as a function of y/x. The equation may then be solved by making the substitution $y = vx$, so that

$$\frac{dy}{dx} = v + x\frac{dv}{dx} = F(v).$$

This is now a separable equation and can be integrated directly to give

$$\int \frac{dv}{F(v) - v} = \int \frac{dx}{x}. \tag{14.19}$$

▶Solve

$$\frac{dy}{dx} = \frac{y}{x} + \tan\left(\frac{y}{x}\right).$$

Substituting $y = vx$ we obtain

$$v + x\frac{dv}{dx} = v + \tan v.$$

Cancelling v on both sides, rearranging and integrating gives

$$\int \cot v \, dv = \int \frac{dx}{x} = \ln x + c_1.$$

But

$$\int \cot v \, dv = \int \frac{\cos v}{\sin v} \, dv = \ln(\sin v) + c_2,$$

so the solution to the ODE is $y = x \sin^{-1} Ax$, where A is a constant. ◀

Solution method. *Check to see whether the equation is homogeneous. If so, make the substitution $y = vx$, separate variables as in (14.19) and then integrate directly. Finally replace v by y/x to obtain the solution.*

14.2.6 Isobaric equations

An isobaric ODE is a generalisation of the homogeneous ODE discussed in the previous section, and is of the form

$$\frac{dy}{dx} = \frac{A(x, y)}{B(x, y)}, \tag{14.20}$$

where the equation is dimensionally consistent if y and dy are each given a weight m relative to x and dx, i.e. if the substitution $y = vx^m$ makes it separable.

476

▶Solve
$$\frac{dy}{dx} = \frac{-1}{2yx}\left(y^2 + \frac{2}{x}\right).$$

Rearranging we have

$$\left(y^2 + \frac{2}{x}\right) dx + 2yx\, dy = 0.$$

Giving y and dy the weight m and x and dx the weight 1, the sums of the powers in each term on the LHS are $2m+1$, 0 and $2m+1$ respectively. These are equal if $2m+1 = 0$, i.e. if $m = -\frac{1}{2}$. Substituting $y = vx^m = vx^{-1/2}$, with the result that $dy = x^{-1/2}\,dv - \frac{1}{2}vx^{-3/2}\,dx$, we obtain

$$v\,dv + \frac{dx}{x} = 0,$$

which is separable and may be integrated directly to give $\frac{1}{2}v^2 + \ln x = c$. Replacing v by $y\sqrt{x}$ we obtain the solution $\frac{1}{2}y^2 x + \ln x = c$. ◀

Solution method. *Write the equation in the form $A\,dx + B\,dy = 0$. Giving y and dy each a weight m and x and dx each a weight 1, write down the sum of powers in each term. Then, if a value of m that makes all these sums equal can be found, substitute $y = vx^m$ into the original equation to make it separable. Integrate the separated equation directly, and then replace v by yx^{-m} to obtain the solution.*

14.2.7 Bernoulli's equation

Bernoulli's equation has the form

$$\frac{dy}{dx} + P(x)y = Q(x)y^n \qquad \text{where } n \neq 0 \text{ or } 1. \tag{14.21}$$

This equation is very similar in form to the linear equation (14.14), but is in fact non-linear due to the extra y^n factor on the RHS. However, the equation can be made linear by substituting $v = y^{1-n}$ and correspondingly

$$\frac{dy}{dx} = \left(\frac{y^n}{1-n}\right)\frac{dv}{dx}.$$

Substituting this into (14.21) and dividing through by y^n, we find

$$\frac{dv}{dx} + (1-n)P(x)v = (1-n)Q(x),$$

which is a linear equation and may be solved by the method described in subsection 14.2.4.

▶*Solve*

$$\frac{dy}{dx} + \frac{y}{x} = 2x^3 y^4.$$

If we let $v = y^{1-4} = y^{-3}$ then

$$\frac{dy}{dx} = -\frac{y^4}{3}\frac{dv}{dx}.$$

Substituting this into the ODE and rearranging, we obtain

$$\frac{dv}{dx} - \frac{3v}{x} = -6x^3,$$

which is linear and may be solved by multiplying through by the integrating factor (see subsection 14.2.4)

$$\exp\left\{-3\int\frac{dx}{x}\right\} = \exp(-3\ln x) = \frac{1}{x^3}.$$

This yields the solution

$$\frac{v}{x^3} = -6x + c.$$

Remembering that $v = y^{-3}$, we obtain $y^{-3} = -6x^4 + cx^3$. ◀

Solution method. *Rearrange the equation into the form (14.21) and make the substitution $v = y^{1-n}$. This leads to a linear equation in v, which can be solved by the method of subsection 14.2.4. Then replace v by y^{1-n} to obtain the solution.*

14.2.8 Miscellaneous equations

There are two further types of first-degree first-order equation that occur fairly regularly but do not fall into any of the above categories. They may be reduced to one of the above equations, however, by a suitable change of variable.

Firstly, we consider

$$\frac{dy}{dx} = F(ax + by + c), \tag{14.22}$$

where a, b and c are constants, i.e. x and y *only* appear on the RHS in the particular combination $ax + by + c$ and not in any other combination or by themselves. This equation can be solved by making the substitution $v = ax + by + c$, in which case

$$\frac{dv}{dx} = a + b\frac{dy}{dx} = a + bF(v), \tag{14.23}$$

which is separable and may be integrated directly.

▶Solve

$$\frac{dy}{dx} = (x+y+1)^2.$$

Making the substitution $v = x+y+1$, we obtain, as in (14.23),

$$\frac{dv}{dx} = v^2 + 1,$$

which is separable and integrates directly to give

$$\int \frac{dv}{1+v^2} = \int dx \quad \Rightarrow \quad \tan^{-1} v = x + c_1.$$

So the solution to the original ODE is $\tan^{-1}(x+y+1) = x+c_1$, where c_1 is a constant of integration. ◀

Solution method. *In an equation such as (14.22), substitute $v = ax+by+c$ to obtain a separable equation that can be integrated directly. Then replace v by $ax+by+c$ to obtain the solution.*

Secondly, we discuss

$$\frac{dy}{dx} = \frac{ax+by+c}{ex+fy+g}, \tag{14.24}$$

where a, b, c, e, f and g are all constants. This equation may be solved by letting $x = X + \alpha$ and $y = Y + \beta$, where α and β are constants found from

$$a\alpha + b\beta + c = 0 \tag{14.25}$$
$$e\alpha + f\beta + g = 0. \tag{14.26}$$

Then (14.24) can be written as

$$\frac{dY}{dX} = \frac{aX+bY}{eX+fY},$$

which is homogeneous and can be solved by the method of subsection 14.2.5. Note, however, that if $a/e = b/f$ then (14.25) and (14.26) are not independent and so cannot be solved uniquely for α and β. However, in this case, (14.24) reduces to an equation of the form (14.22), which was discussed above.

▶Solve

$$\frac{dy}{dx} = \frac{2x-5y+3}{2x+4y-6}.$$

Let $x = X + \alpha$ and $y = Y + \beta$, where α and β obey the relations

$$2\alpha - 5\beta + 3 = 0$$
$$2\alpha + 4\beta - 6 = 0,$$

which solve to give $\alpha = \beta = 1$. Making these substitutions we find

$$\frac{dY}{dX} = \frac{2X-5Y}{2X+4Y},$$

which is a homogeneous ODE and can be solved by substituting $Y = vX$ (see subsection 14.2.5) to obtain

$$\frac{dv}{dX} = \frac{2 - 7v - 4v^2}{X(2 + 4v)}.$$

This equation is separable, and using partial fractions we find

$$\int \frac{2 + 4v}{2 - 7v - 4v^2}\, dv = -\frac{4}{3} \int \frac{dv}{4v - 1} - \frac{2}{3} \int \frac{dv}{v + 2} = \int \frac{dX}{X},$$

which integrates to give

$$\ln X + \tfrac{1}{3} \ln(4v - 1) + \tfrac{2}{3} \ln(v + 2) = c_1,$$

or

$$X^3(4v - 1)(v + 2)^2 = \exp 3c_1.$$

Remembering that $Y = vX$, $x = X + 1$ and $y = Y + 1$, the solution to the original ODE is given by $(4y - x - 3)(y + 2x - 3)^2 = c_2$, where $c_2 = \exp 3c_1$. ◀

Solution method. *If in (14.24) $a/e \neq b/f$ then make the substitution $x = X + \alpha$, $y = Y + \beta$, where α and β are given by (14.25) and (14.26); the resulting equation is homogeneous and can be solved as in subsection 14.2.5. Substitute $v = Y/X$, $X = x - \alpha$ and $Y = y - \beta$ to obtain the solution. If $a/e = b/f$ then (14.24) is of the same form as (14.22) and may be solved accordingly.*

14.3 Higher-degree first-order equations

First-order equations of degree higher than the first do not occur often in the description of physical systems, since squared and higher powers of first-order derivatives usually arise from resistive or driving mechanisms, when an acceleration or other higher-order derivative is also present. They do sometimes appear in connection with geometrical problems, however.

Higher-degree first-order equations can be written as $F(x, y, dy/dx) = 0$. The most general standard form is

$$p^n + a_{n-1}(x, y)p^{n-1} + \cdots + a_1(x, y)p + a_0(x, y) = 0, \tag{14.27}$$

where for ease of notation we write $p = dy/dx$. If the equation can be solved for one of x, y or p then either an explicit or a parametric solution can sometimes be obtained. We discuss the main types of such equations below, including Clairaut's equation, which is a special case of an equation explicitly soluble for y.

14.3.1 *Equations soluble for p*

Sometimes the LHS of (14.27) can be factorised into the form

$$(p - F_1)(p - F_2) \cdots (p - F_n) = 0, \tag{14.28}$$

where $F_i = F_i(x, y)$. We are then left with solving the n first-degree equations $p = F_i(x, y)$. Writing the solutions to these first-degree equations as $G_i(x, y) = 0$, the general solution to (14.28) is given by the product

$$G_1(x, y)G_2(x, y) \cdots G_n(x, y) = 0. \tag{14.29}$$

▶Solve

$$(x^3 + x^2 + x + 1)p^2 - (3x^2 + 2x + 1)yp + 2xy^2 = 0. \tag{14.30}$$

This equation may be factorised to give

$$[(x + 1)p - y][(x^2 + 1)p - 2xy] = 0.$$

Taking each bracket in turn we have

$$(x + 1)\frac{dy}{dx} - y = 0,$$

$$(x^2 + 1)\frac{dy}{dx} - 2xy = 0,$$

which have the solutions $y - c(x + 1) = 0$ and $y - c(x^2 + 1) = 0$ respectively (see section 14.2 on first-degree first-order equations). Note that the arbitrary constants in these two solutions can be taken to be the same, since only one is required for a first-order equation. The general solution to (14.30) is then given by

$$[y - c(x + 1)] [y - c(x^2 + 1)] = 0. \blacktriangleleft$$

Solution method. *If the equation can be factorised into the form (14.28) then solve the first-order ODE $p - F_i = 0$ for each factor and write the solution in the form $G_i(x, y) = 0$. The solution to the original equation is then given by the product (14.29).*

14.3.2 Equations soluble for x

Equations that can be solved for x, i.e. such that they may be written in the form

$$x = F(y, p), \tag{14.31}$$

can be reduced to first-degree first-order equations in p by differentiating both sides with respect to y, so that

$$\frac{dx}{dy} = \frac{1}{p} = \frac{\partial F}{\partial y} + \frac{\partial F}{\partial p}\frac{dp}{dy}.$$

This results in an equation of the form $G(y, p) = 0$, which can be used together with (14.31) to eliminate p and give the general solution. Note that often a singular solution to the equation will be found at the same time (see the introduction to this chapter).

> ►*Solve*
>
> $$6y^2p^2 + 3xp - y = 0. \tag{14.32}$$

This equation can be solved for x explicitly to give $3x = (y/p) - 6y^2p$. Differentiating both sides with respect to y, we find

$$3\frac{dx}{dy} = \frac{3}{p} = \frac{1}{p} - \frac{y}{p^2}\frac{dp}{dy} - 6y^2\frac{dp}{dy} - 12yp,$$

which factorises to give

$$(1 + 6yp^2)\left(2p + y\frac{dp}{dy}\right) = 0. \tag{14.33}$$

Setting the factor containing dp/dy equal to zero gives a first-degree first-order equation in p, which may be solved to give $py^2 = c$. Substituting for p in (14.32) then yields the general solution of (14.32):

$$y^3 = 3cx + 6c^2. \tag{14.34}$$

If we now consider the first factor in (14.33), we find $6p^2y = -1$ as a possible solution. Substituting for p in (14.32) we find the singular solution

$$8y^3 + 3x^2 = 0.$$

Note that the singular solution contains no arbitrary constants and cannot be found from the general solution (14.34) by any choice of the constant c. ◄

Solution method. *Write the equation in the form (14.31) and differentiate both sides with respect to y. Rearrange the resulting equation into the form $G(y, p) = 0$, which can be used together with the original ODE to eliminate p and so give the general solution. If $G(y, p)$ can be factorised then the factor containing dp/dy should be used to eliminate p and give the general solution. Using the other factors in this fashion will instead lead to singular solutions.*

14.3.3 Equations soluble for y

Equations that can be solved for y, i.e. are such that they may be written in the form

$$y = F(x, p), \tag{14.35}$$

can be reduced to first-degree first-order equations in p by differentiating both sides with respect to x, so that

$$\frac{dy}{dx} = p = \frac{\partial F}{\partial x} + \frac{\partial F}{\partial p}\frac{dp}{dx}.$$

This results in an equation of the form $G(x, p) = 0$, which can be used together with (14.35) to eliminate p and give the general solution. An additional (singular) solution to the equation is also often found.

▶*Solve*

$$xp^2 + 2xp - y = 0. \tag{14.36}$$

This equation can be solved for y explicitly to give $y = xp^2 + 2xp$. Differentiating both sides with respect to x, we find

$$\frac{dy}{dx} = p = 2xp\frac{dp}{dx} + p^2 + 2x\frac{dp}{dx} + 2p,$$

which after factorising gives

$$(p+1)\left(p + 2x\frac{dp}{dx}\right) = 0. \tag{14.37}$$

To obtain the general solution of (14.36), we consider the factor containing dp/dx. This first-degree first-order equation in p has the solution $xp^2 = c$ (see subsection 14.3.1), which we then use to eliminate p from (14.36). Thus we find that the general solution to (14.36) is

$$(y \quad c)^2 - 4cx. \tag{14.38}$$

If instead, we set the other factor in (14.37) equal to zero, we obtain the very simple solution $p = -1$. Substituting this into (14.36) then gives

$$x + y = 0,$$

which is a singular solution to (14.36). ◀

Solution method. *Write the equation in the form (14.35) and differentiate both sides with respect to x. Rearrange the resulting equation into the form $G(x, p) = 0$, which can be used together with the original ODE to eliminate p and so give the general solution. If $G(x, p)$ can be factorised then the factor containing dp/dx should be used to eliminate p and give the general solution. Using the other factors in this fashion will instead lead to singular solutions.*

14.3.4 Clairaut's equation

Finally, we consider Clairaut's equation, which has the form

$$y = px + F(p) \tag{14.39}$$

and is therefore a special case of equations soluble for y, as in (14.35). It may be solved by a similar method to that given in subsection 14.3.3, but for Clairaut's equation the form of the general solution is particularly simple. Differentiating (14.39) with respect to x, we find

$$\frac{dy}{dx} = p = p + x\frac{dp}{dx} + \frac{dF}{dp}\frac{dp}{dx} \quad \Rightarrow \quad \frac{dp}{dx}\left(\frac{dF}{dp} + x\right) = 0. \tag{14.40}$$

Considering first the factor containing dp/dx, we find

$$\frac{dp}{dx} = \frac{d^2y}{dx^2} = 0 \quad \Rightarrow \quad y = c_1x + c_2. \tag{14.41}$$

Since $p = dy/dx = c_1$, if we substitute (14.41) into (14.39) we find $c_1 x + c_2 = c_1 x + F(c_1)$. Therefore the constant c_2 is given by $F(c_1)$, and the general solution to (14.39) is

$$y = c_1 x + F(c_1), \tag{14.42}$$

i.e. the general solution to Clairaut's equation can be obtained by replacing p in the ODE by the arbitrary constant c_1. Now, considering the second factor in (14.40), we also have

$$\frac{dF}{dp} + x = 0, \tag{14.43}$$

which has the form $G(x, p) = 0$. This relation may be used to eliminate p from (14.39) to give a singular solution.

▶ *Solve*

$$y = px + p^2. \tag{14.44}$$

From (14.42) the general solution is $y = cx + c^2$. But from (14.43) we also have $2p + x = 0 \Rightarrow p = -x/2$. Substituting this into (14.44) we find the singular solution $x^2 + 4y = 0$. ◀

Solution method. *Write the equation in the form (14.39), then the general solution is given by replacing p by some constant c, as shown in (14.42). Using the relation $dF/dp + x = 0$ to eliminate p from the original equation yields the singular solution.*

14.4 Exercises

14.1   A radioactive isotope decays in such a way that the number of atoms present at a given time, $N(t)$, obeys the equation

$$\frac{dN}{dt} = -\lambda N.$$

If there are initially N_0 atoms present, find $N(t)$ at later times.

14.2   Solve the following equations by separation of the variables:

(a) $y' - xy^3 = 0$;
(b) $y' \tan^{-1} x - y(1 + x^2)^{-1} = 0$;
(c) $x^2 y' + xy^2 = 4y^2$.

14.3   Show that the following equations either are exact or can be made exact, and solve them:

(a) $y(2x^2 y^2 + 1)y' + x(y^4 + 1) = 0$;
(b) $2xy' + 3x + y = 0$;
(c) $(\cos^2 x + y \sin 2x)y' + y^2 = 0$.

14.4   Find the values of α and β that make

$$dF(x, y) = \left(\frac{1}{x^2 + 2} + \frac{\alpha}{y} \right) dx + (xy^\beta + 1) \, dy$$

an exact differential. For these values solve $dF(x, y) = 0$.

484

14.5   By finding suitable integrating factors, solve the following equations:

(a) $(1 - x^2)y' + 2xy = (1 - x^2)^{3/2}$;
(b) $y' - y \cot x + \operatorname{cosec} x = 0$;
(c) $(x + y^3)y' = y$ (treat y as the independent variable).

14.6   By finding an appropriate integrating factor, solve

$$\frac{dy}{dx} = -\frac{2x^2 + y^2 + x}{xy}.$$

14.7   Find, in the form of an integral, the solution of the equation

$$\alpha \frac{dy}{dt} + y = f(t)$$

for a general function $f(t)$. Find the specific solutions for

(a) $f(t) = H(t)$,
(b) $f(t) = \delta(t)$,
(c) $f(t) = \beta^{-1} e^{-t/\beta} H(t)$ with $\beta < \alpha$.

For case (c), what happens if $\beta \to 0$?

14.8   A series electric circuit contains a resistance R, a capacitance C and a battery supplying a time-varying electromotive force $V(t)$. The charge q on the capacitor therefore obeys the equation

$$R \frac{dq}{dt} + \frac{q}{C} = V(t).$$

Assuming that initially there is no charge on the capacitor, and given that $V(t) = V_0 \sin \omega t$, find the charge on the capacitor as a function of time.

14.9   Using tangential–polar coordinates (see exercise 2.20), consider a particle of mass m moving under the influence of a force f directed towards the origin O. By resolving forces along the instantaneous tangent and normal and making use of the result of exercise 2.20 for the instantaneous radius of curvature, prove that

$$f = -mv \frac{dv}{dr} \qquad \text{and} \qquad mv^2 = fp \frac{dr}{dp}.$$

Show further that $h = mpv$ is a constant of the motion and that the law of force can be deduced from

$$f = \frac{h^2}{mp^3} \frac{dp}{dr}.$$

14.10   Use the result of exercise 14.9 to find the law of force, acting towards the origin, under which a particle must move so as to describe the following trajectories:

(a) A circle of radius a that passes through the origin;
(b) An equiangular spiral, which is defined by the property that the angle α between the tangent and the radius vector is constant along the curve.

14.11   Solve

$$(y - x)\frac{dy}{dx} + 2x + 3y = 0.$$

14.12   A mass m is accelerated by a time-varying force $\alpha \exp(-\beta t)v^3$, where v is its velocity. It also experiences a resistive force ηv, where η is a constant, owing to its motion through the air. The equation of motion of the mass is therefore

$$m \frac{dv}{dt} = \alpha \exp(-\beta t)v^3 - \eta v.$$

Find an expression for the velocity v of the mass as a function of time, given that it has an initial velocity v_0.

14.13 Using the results about Laplace transforms given in chapter 13 for df/dt and $tf(t)$, show, for a function $y(t)$ that satisfies

$$t\frac{dy}{dt} + (t - 1)y = 0 \qquad (*)$$

with $y(0)$ finite, that $\bar{y}(s) = C(1 + s)^{-2}$ for some constant C.

Given that

$$y(t) = t + \sum_{n=2}^{\infty} a_n t^n,$$

determine C and show that $a_n = (-1)^{n-1}/(n - 1)!$. Compare this result with that obtained by integrating $(*)$ directly.

14.14 Solve

$$\frac{dy}{dx} = \frac{1}{x + 2y + 1}.$$

14.15 Solve

$$\frac{dy}{dx} = -\frac{x + y}{3x + 3y - 4}.$$

14.16 If $u = 1 + \tan y$, calculate $d(\ln u)/dy$; hence find the general solution of

$$\frac{dy}{dx} = \tan x \cos y \,(\cos y + \sin y).$$

14.17 Solve

$$x(1 - 2x^2 y)\frac{dy}{dx} + y = 3x^2 y^2,$$

given that $y(1) = 1/2$.

14.18 A reflecting mirror is made in the shape of the surface of revolution generated by revolving the curve $y(x)$ about the x-axis. In order that light rays emitted from a point source at the origin are reflected back parallel to the x-axis, the curve $y(x)$ must obey

$$\frac{y}{x} = \frac{2p}{1 - p^2},$$

where $p = dy/dx$. By solving this equation for x, find the curve $y(x)$.

14.19 Find the curve with the property that at each point on it the sum of the intercepts on the x- and y-axes of the tangent to the curve (taking account of sign) is equal to 1.

14.20 Find a parametric solution of

$$x\left(\frac{dy}{dx}\right)^2 + \frac{dy}{dx} - y = 0$$

as follows.

(a) Write an equation for y in terms of $p = dy/dx$ and show that

$$p = p^2 + (2px + 1)\frac{dp}{dx}.$$

(b) Using p as the independent variable, arrange this as a linear first-order equation for x.

(c) Find an appropriate integrating factor to obtain

$$x = \frac{\ln p - p + c}{(1 - p)^2},$$

which, together with the expression for y obtained in (a), gives a parameterisation of the solution.

(d) Reverse the roles of x and y in steps (a) to (c), putting $dx/dy = p^{-1}$, and show that essentially the same parameterisation is obtained.

14.21 Using the substitutions $u = x^2$ and $v = y^2$, reduce the equation

$$xy \left(\frac{dy}{dx} \right)^2 - (x^2 + y^2 - 1)\frac{dy}{dx} + xy = 0$$

to Clairaut's form. Hence show that the equation represents a family of conics and the four sides of a square.

14.22 The action of the control mechanism on a particular system for an input $f(t)$ is described, for $t \geq 0$, by the coupled first-order equations:

$$\dot{y} + 4z = f(t),$$
$$\dot{z} - 2z = \dot{y} + \tfrac{1}{2}y.$$

Use Laplace transforms to find the response $y(t)$ of the system to a unit step input, $f(t) = H(t)$, given that $y(0) = 1$ and $z(0) = 0$.

Questions 23 to 31 are intended to give the reader practice in choosing an appropriate method. The level of difficulty varies within the set; if necessary, the hints may be consulted for an indication of the most appropriate approach.

14.23 Find the general solutions of the following:

(a) $\dfrac{dy}{dx} + \dfrac{xy}{a^2 + x^2} = x$; (b) $\dfrac{dy}{dx} = \dfrac{4y^2}{x^2} - y^2$.

14.24 Solve the following first-order equations for the boundary conditions given:

(a) $y' - (y/x) = 1,$ $y(1) = -1$;
(b) $y' - y \tan x = 1,$ $y(\pi/4) = 3$;
(c) $y' - y^2/x^2 = 1/4,$ $y(1) = 1$;
(d) $y' - y^2/x^2 = 1/4,$ $y(1) = 1/2$.

14.25 An electronic system has two inputs, to each of which a constant unit signal is applied, but starting at different times. The equations governing the system thus take the form

$$\dot{x} + 2y = H(t),$$
$$\dot{y} - 2x = H(t - 3).$$

Initially (at $t = 0$), $x = 1$ and $y = 0$; find $x(t)$ at later times.

14.26 Solve the differential equation

$$\sin x \frac{dy}{dx} + 2y \cos x = 1,$$

subject to the boundary condition $y(\pi/2) = 1$.

14.27 Find the complete solution of

$$\left(\frac{dy}{dx} \right)^2 - \frac{y}{x}\frac{dy}{dx} + \frac{A}{x} = 0,$$

where A is a positive constant.

14.28 Find the solution of

$$(5x + y - 7)\frac{dy}{dx} = 3(x + y + 1).$$

14.29 Find the solution $y = y(x)$ of

$$x\frac{dy}{dx} + y - \frac{y^2}{x^{3/2}} = 0,$$

subject to $y(1) = 1$.

14.30 Find the solution of

$$(2\sin y - x)\frac{dy}{dx} = \tan y,$$

if (a) $y(0) = 0$, and (b) $y(0) = \pi/2$.

14.31 Find the family of solutions of

$$\frac{d^2y}{dx^2} + \left(\frac{dy}{dx}\right)^2 + \frac{dy}{dx} = 0$$

that satisfy $y(0) = 0$.

14.5 Hints and answers

14.1 $N(t) = N_0 \exp(-\lambda t)$.

14.3 (a) exact, $x^2 y^4 + x^2 + y^2 = c$; (b) IF $= x^{-1/2}$, $x^{1/2}(x + y) = c$; (c) IF $= \sec^2 x$, $y^2 \tan x + y = c$.

14.5 (a) IF $= (1 - x^2)^{-2}$, $y = (1 - x^2)(k + \sin^{-1} x)$; (b) IF $= \operatorname{cosec} x$, leading to $y = k \sin x + \cos x$; (c) exact equation is $y^{-1}(dx/dy) - xy^{-2} = y$, leading to $x = y(k + y^2/2)$.

14.7 $y(t) = e^{-t/\alpha} \int^t \alpha^{-1} e^{t'/\alpha} f(t') dt'$; (a) $y(t) = 1 - e^{-t/\alpha}$; (b) $y(t) = \alpha^{-1} e^{-t/\alpha}$; (c) $y(t) = (e^{-t/\alpha} - e^{-t/\beta})/(\alpha - \beta)$. It becomes case (b).

14.9 Note that, if the angle between the tangent and the radius vector is α, then $\cos\alpha = dr/ds$ and $\sin\alpha = p/r$.

14.11 Homogeneous equation, put $y = vx$ to obtain $(1 - v)(v^2 + 2v + 2)^{-1} dv = x^{-1} dx$; write $1 - v$ as $2 - (1 + v)$, and $v^2 + 2v + 2$ as $1 + (1 + v)^2$; $A[x^2 + (x + y)^2] = \exp\left\{4\tan^{-1}[(x + y)/x]\right\}$.

14.13 $(1 + s)(d\bar{y}/ds) + 2\bar{y} = 0$. $C = 1$; use separation of variables to show directly that $y(t) = te^{-t}$.

14.15 The equation is of the form of (14.22), set $v = x + y$; $x + 3y + 2\ln(x + y - 2) = A$.

14.17 The equation is isobaric with weight $y = -2$; setting $y = vx^{-2}$ gives $v^{-1}(1 - v)^{-1}(1 - 2v) dv = x^{-1} dx$; $4xy(1 - x^2 y) = 1$.

14.19 The curve must satisfy $y = (1 - p^{-1})^{-1}(1 - x + px)$, which has solution $x = (p - 1)^{-2}$, leading to $y = (1 \pm \sqrt{x})^2$ or $x = (1 \pm \sqrt{y})^2$; the singular solution $p' = 0$ gives straight lines joining $(\theta, 0)$ and $(0, 1 - \theta)$ for any θ.

14.21 $v = qu + q/(q - 1)$, where $q = dv/du$. General solution $y^2 = cx^2 + c/(c - 1)$, hyperbolae for $c > 0$ and ellipses for $c < 0$. Singular solution $y = \pm(x \pm 1)$.

14.23 (a) Integrating factor is $(a^2 + x^2)^{1/2}$, $y = (a^2 + x^2)/3 + A(a^2 + x^2)^{-1/2}$; (b) separable, $y = x(x^2 + Ax + 4)^{-1}$.

14.25 Use Laplace transforms; $\bar{x}s(s^2 + 4) = s + s^2 - 2e^{-3s}$; $x(t) = \frac{1}{2}\sin 2t + \cos 2t - \frac{1}{2}H(t - 3) + \frac{1}{2}\cos(2t - 6)H(t - 3)$.

14.27 This is Clairaut's equation with $F(p) = A/p$. General solution $y = cx + A/c$; singular solution, $y = 2\sqrt{Ax}$.

14.29 Either Bernoulli's equation with $n = 2$ or an isobaric equation with $m = 3/2$; $y(x) = 5x^{3/2}/(2 + 3x^{5/2})$.

14.31   Show that $p = (Ce^x - 1)^{-1}$, where $p = dy/dx$; $y = \ln[C - e^{-x})/(C - 1)]$ or $\ln[D - (D - 1)e^{-x}]$ or $\ln(e^{-K} + 1 - e^{-x}) + K$.

15

Higher-order ordinary differential equations

Following on from the discussion of first-order ordinary differential equations (ODEs) given in the previous chapter, we now examine equations of second and higher order. Since a brief outline of the general properties of ODEs and their solutions was given at the beginning of the previous chapter, we will not repeat it here. Instead, we will begin with a discussion of various types of higher-order equation. This chapter is divided into three main parts. We first discuss linear equations with constant coefficients and then investigate linear equations with variable coefficients. Finally, we discuss a few methods that may be of use in solving general linear or non-linear ODEs. Let us start by considering some general points relating to *all* linear ODEs.

Linear equations are of paramount importance in the description of physical processes. Moreover, it is an empirical fact that, when put into mathematical form, many natural processes appear as higher-order linear ODEs, most often as second-order equations. Although we could restrict our attention to these second-order equations, the generalisation to nth-order equations requires little extra work, and so we will consider this more general case.

A linear ODE of general order n has the form

$$a_n(x)\frac{d^n y}{dx^n} + a_{n-1}(x)\frac{d^{n-1}y}{dx^{n-1}} + \cdots + a_1(x)\frac{dy}{dx} + a_0(x)y = f(x). \tag{15.1}$$

If $f(x) = 0$ then the equation is called *homogeneous*; otherwise it is *inhomogeneous*. The first-order linear equation studied in subsection 14.2.4 is a special case of (15.1). As discussed at the beginning of the previous chapter, the general solution to (15.1) will contain n arbitrary constants, which may be determined if n boundary conditions are also provided.

In order to solve any equation of the form (15.1), we must first find the general solution of the *complementary equation*, i.e. the equation formed by setting

$f(x) = 0$:

$$a_n(x)\frac{d^n y}{dx^n} + a_{n-1}(x)\frac{d^{n-1} y}{dx^{n-1}} + \cdots + a_1(x)\frac{dy}{dx} + a_0(x)y = 0. \tag{15.2}$$

To determine the general solution of (15.2), we must find n linearly independent functions that satisfy it. Once we have found these solutions, the general solution is given by a linear superposition of these n functions. In other words, if the n solutions of (15.2) are $y_1(x), y_2(x), \ldots, y_n(x)$, then the general solution is given by the linear superposition

$$y_c(x) = c_1 y_1(x) + c_2 y_2(x) + \cdots + c_n y_n(x), \tag{15.3}$$

where the c_m are arbitrary constants that may be determined if n boundary conditions are provided. The linear combination $y_c(x)$ is called the *complementary function* of (15.1).

The question naturally arises how we establish that any n individual solutions to (15.2) are indeed linearly independent. For n functions to be linearly independent over an interval, there must not exist *any* set of constants $c_1, c_2, \ldots, c_n$ such that

$$c_1 y_1(x) + c_2 y_2(x) + \cdots + c_n y_n(x) = 0 \tag{15.4}$$

over the interval in question, except for the trivial case $c_1 = c_2 = \cdots = c_n = 0$.

A statement equivalent to (15.4), which is perhaps more useful for the practical determination of linear independence, can be found by repeatedly differentiating (15.4), $n - 1$ times in all, to obtain n simultaneous equations for $c_1, c_2, \ldots, c_n$:

$$\begin{aligned} c_1 y_1(x) + c_2 y_2(x) + \cdots + c_n y_n(x) &= 0 \\ c_1 y_1'(x) + c_2 y_2'(x) + \cdots + c_n y_n'(x) &= 0 \\ &\vdots \\ c_1 y_1^{(n-1)}(x) + c_2 y_2^{(n-1)}(x) + \cdots + c_n y_n^{(n-1)}(x) &= 0, \end{aligned} \tag{15.5}$$

where the primes denote differentiation with respect to x. Referring to the discussion of simultaneous linear equations given in chapter 8, if the determinant of the coefficients of $c_1, c_2, \ldots, c_n$ is non-zero then the only solution to equations (15.5) is the trivial solution $c_1 = c_2 = \cdots = c_n = 0$. In other words, the n functions $y_1(x), y_2(x), \ldots, y_n(x)$ are linearly independent over an interval if

$$W(y_1, y_2, \ldots, y_n) = \begin{vmatrix} y_1 & y_2 & \cdots & y_n \\ y_1' & y_2' & & \vdots \\ \vdots & & \ddots & \vdots \\ y_1^{(n-1)} & \cdots & \cdots & y_n^{(n-1)} \end{vmatrix} \neq 0 \tag{15.6}$$

over that interval; $W(y_1, y_2, \ldots, y_n)$ is called the *Wronskian* of the set of functions. It should be noted, however, that the vanishing of the Wronskian does not guarantee that the functions are linearly dependent.

If the original equation (15.1) has $f(x) = 0$ (i.e. it is homogeneous) then of course the complementary function $y_c(x)$ in (15.3) is already the general solution. If, however, the equation has $f(x) \neq 0$ (i.e. it is inhomogeneous) then $y_c(x)$ is only one part of the solution. The general solution of (15.1) is then given by

$$y(x) = y_c(x) + y_p(x), \tag{15.7}$$

where $y_p(x)$ is the *particular integral*, which can be *any* function that satisfies (15.1) directly, provided it is linearly independent of $y_c(x)$. It should be emphasised for practical purposes that *any* such function, no matter how simple (or complicated), is equally valid in forming the general solution (15.7).

It is important to realise that the above method for finding the general solution to an ODE by superposing particular solutions assumes crucially that the ODE is linear. For non-linear equations, discussed in section 15.3, this method cannot be used, and indeed it is often impossible to find closed-form solutions to such equations.

15.1 Linear equations with constant coefficients

If the a_m in (15.1) are constants rather than functions of x then we have

$$a_n \frac{d^n y}{dx^n} + a_{n-1} \frac{d^{n-1} y}{dx^{n-1}} + \cdots + a_1 \frac{dy}{dx} + a_0 y = f(x). \tag{15.8}$$

Equations of this sort are very common throughout the physical sciences and engineering, and the method for their solution falls into two parts as discussed in the previous section, i.e. finding the complementary function $y_c(x)$ and finding the particular integral $y_p(x)$. If $f(x) = 0$ in (15.8) then we do not have to find a particular integral, and the complementary function is by itself the general solution.

15.1.1 Finding the complementary function $y_c(x)$

The complementary function must satisfy

$$a_n \frac{d^n y}{dx^n} + a_{n-1} \frac{d^{n-1} y}{dx^{n-1}} + \cdots + a_1 \frac{dy}{dx} + a_0 y = 0 \tag{15.9}$$

and contain n arbitrary constants (see equation (15.3)). The standard method for finding $y_c(x)$ is to try a solution of the form $y = Ae^{\lambda x}$, substituting this into (15.9). After dividing the resulting equation through by $Ae^{\lambda x}$, we are left with a polynomial equation in λ of order n; this is the *auxiliary equation* and reads

$$a_n \lambda^n + a_{n-1} \lambda^{n-1} + \cdots + a_1 \lambda + a_0 = 0. \tag{15.10}$$

In general the auxiliary equation has n roots, say $\lambda_1, \lambda_2, \ldots, \lambda_n$. In certain cases, some of these roots may be repeated and some may be complex. The three main cases are as follows.

(i) *All roots real and distinct.* In this case the n solutions to (15.9) are $\exp \lambda_m x$ for $m = 1$ to n. It is easily shown by calculating the Wronskian (15.6) of these functions that if all the λ_m are distinct then these solutions are linearly independent. We can therefore linearly superpose them, as in (15.3), to form the complementary function

$$y_c(x) = c_1 e^{\lambda_1 x} + c_2 e^{\lambda_2 x} + \cdots + c_n e^{\lambda_n x}. \tag{15.11}$$

(ii) *Some roots complex.* For the special (but usual) case that all the coefficients a_m in (15.9) are real, if one of the roots of the auxiliary equation (15.10) is complex, say $\alpha + i\beta$, then its complex conjugate $\alpha - i\beta$ is also a root. In this case we can write

$$c_1 e^{(\alpha + i\beta)x} + c_2 e^{(\alpha - i\beta)x} = e^{\alpha x}(d_1 \cos \beta x + d_2 \sin \beta x)$$

$$= A e^{\alpha x} \left\{ \begin{matrix} \sin \\ \cos \end{matrix} \right\} (\beta x + \phi), \tag{15.12}$$

where A and ϕ are arbitrary constants.

(iii) *Some roots repeated.* If, for example, λ_1 occurs k times ($k > 1$) as a root of the auxiliary equation, then we have not found n linearly independent solutions of (15.9); formally the Wronskian (15.6) of these solutions, having two or more identical columns, is equal to zero. We must therefore find $k - 1$ further solutions that are linearly independent of those already found and also of each other. By direct substitution into (15.9) we find that

$$x e^{\lambda_1 x}, \quad x^2 e^{\lambda_1 x}, \quad \ldots, \quad x^{k-1} e^{\lambda_1 x}$$

are also solutions, and by calculating the Wronskian it is easily shown that they, together with the solutions already found, form a linearly independent set of n functions. Therefore the complementary function is given by

$$y_c(x) = (c_1 + c_2 x + \cdots + c_k x^{k-1}) e^{\lambda_1 x} + c_{k+1} e^{\lambda_{k+1} x} + c_{k+2} e^{\lambda_{k+2} x} + \cdots + c_n e^{\lambda_n x}. \tag{15.13}$$

If more than one root is repeated the above argument is easily extended. For example, suppose as before that λ_1 is a k-fold root of the auxiliary equation and, further, that λ_2 is an l-fold root (of course, $k > 1$ and $l > 1$). Then, from the above argument, the complementary function reads

$$y_c(x) = (c_1 + c_2 x + \cdots + c_k x^{k-1}) e^{\lambda_1 x}$$

$$+ (c_{k+1} + c_{k+2} x + \cdots + c_{k+l} x^{l-1}) e^{\lambda_2 x}$$

$$+ c_{k+l+1} e^{\lambda_{k+l+1} x} + c_{k+l+2} e^{\lambda_{k+l+2} x} + \cdots + c_n e^{\lambda_n x}. \tag{15.14}$$

493

> ►*Find the complementary function of the equation*
>
> $$\frac{d^2y}{dx^2} - 2\frac{dy}{dx} + y = e^x. \tag{15.15}$$

Setting the RHS to zero, substituting $y = Ae^{\lambda x}$ and dividing through by $Ae^{\lambda x}$ we obtain the auxiliary equation

$$\lambda^2 - 2\lambda + 1 = 0.$$

The root $\lambda = 1$ occurs twice and so, although e^x is a solution to (15.15), we must find a further solution to the equation that is linearly independent of e^x. From the above discussion, we deduce that xe^x is such a solution, so that the full complementary function is given by the linear superposition

$$y_c(x) = (c_1 + c_2 x)e^x. \ \blacktriangleleft$$

Solution method. *Set the RHS of the ODE to zero (if it is not already so), and substitute $y = Ae^{\lambda x}$. After dividing through the resulting equation by $Ae^{\lambda x}$, obtain an nth-order polynomial equation in λ (the auxiliary equation, see (15.10)). Solve the auxiliary equation to find the n roots, $\lambda_1, \lambda_2, \ldots, \lambda_n$, say. If all these roots are real and distinct then $y_c(x)$ is given by (15.11). If, however, some of the roots are complex or repeated then $y_c(x)$ is given by (15.12) or (15.13), or the extension (15.14) of the latter, respectively.*

15.1.2 Finding the particular integral $y_p(x)$

There is no generally applicable method for finding the particular integral $y_p(x)$ but, for linear ODEs with constant coefficients and a simple RHS, $y_p(x)$ can often be found by inspection or by assuming a parameterised form similar to $f(x)$. The latter method is sometimes called the *method of undetermined coefficients*. If $f(x)$ contains only polynomial, exponential, or sine and cosine terms then, by assuming a trial function for $y_p(x)$ of similar form but one which contains a number of undetermined parameters and substituting this trial function into (15.9), the parameters can be found and $y_p(x)$ deduced. Standard trial functions are as follows.

(i) If $f(x) = ae^{rx}$ then try

$$y_p(x) = be^{rx}.$$

(ii) If $f(x) = a_1 \sin rx + a_2 \cos rx$ (a_1 or a_2 may be zero) then try

$$y_p(x) = b_1 \sin rx + b_2 \cos rx.$$

(iii) If $f(x) = a_0 + a_1 x + \cdots + a_N x^N$ (some a_m may be zero) then try

$$y_p(x) = b_0 + b_1 x + \cdots + b_N x^N.$$

(iv) If $f(x)$ is the sum or product of any of the above then try $y_p(x)$ as the sum or product of the corresponding individual trial functions.

It should be noted that this method fails if any term in the assumed trial function is also contained within the complementary function $y_c(x)$. In such a case the trial function should be multiplied by the smallest integer power of x such that it will then contain no term that already appears in the complementary function. The undetermined coefficients in the trial function can now be found by substitution into (15.8).

Three further methods that are useful in finding the particular integral $y_p(x)$ are those based on Green's functions, the variation of parameters, and a change in the dependent variable using knowledge of the complementary function. However, since these methods are also applicable to equations with variable coefficients, a discussion of them is postponed until section 15.2.

▶ *Find a particular integral of the equation*

$$\frac{d^2y}{dx^2} - 2\frac{dy}{dx} + y = e^x.$$

From the above discussion our first guess at a trial particular integral would be $y_p(x) = be^x$. However, since the complementary function of this equation is $y_c(x) = (c_1 + c_2x)e^x$ (as in the previous subsection), we see that e^x is already contained in it, as indeed is xe^x. Multiplying our first guess by the lowest integer power of x such that the result does not appear in $y_c(x)$, we therefore try $y_p(x) = bx^2e^x$. Substituting this into the ODE, we find that $b - 1/2$, so the particular integral is given by $y_p(x) = x^2e^x/2$. ◀

Solution method. *If the RHS of an ODE contains only functions mentioned at the start of this subsection then the appropriate trial function should be substituted into it, thereby fixing the undetermined parameters. If, however, the RHS of the equation is not of this form then one of the more general methods outlined in sub sections 15.2.3–15.2.5 should be used; perhaps the most straightforward of these is the variation-of-parameters method.*

15.1.3 Constructing the general solution $y_c(x) + y_p(x)$

As stated earlier, the full solution to the ODE (15.8) is found by adding together the complementary function and any particular integral. In order to illustrate further the material discussed in the last two subsections, let us find the general solution to a new example, starting from the beginning.

▶*Solve*

$$\frac{d^2 y}{dx^2} + 4y = x^2 \sin 2x. \tag{15.16}$$

First we set the RHS to zero and assume the trial solution $y = Ae^{\lambda x}$. Substituting this into (15.16) leads to the auxiliary equation

$$\lambda^2 + 4 = 0 \quad \Rightarrow \quad \lambda = \pm 2i. \tag{15.17}$$

Therefore the complementary function is given by

$$y_c(x) = c_1 e^{2ix} + c_2 e^{-2ix} = d_1 \cos 2x + d_2 \sin 2x. \tag{15.18}$$

We must now turn our attention to the particular integral $y_p(x)$. Consulting the list of standard trial functions in the previous subsection, we find that a first guess at a suitable trial function for this case should be

$$(ax^2 + bx + c)\sin 2x + (dx^2 + ex + f)\cos 2x. \tag{15.19}$$

However, we see that this trial function contains terms in $\sin 2x$ and $\cos 2x$, both of which already appear in the complementary function (15.18). We must therefore multiply (15.19) by the smallest integer power of x which ensures that none of the resulting terms appears in $y_c(x)$. Since multiplying by x will suffice, we finally assume the trial function

$$(ax^3 + bx^2 + cx)\sin 2x + (dx^3 + ex^2 + fx)\cos 2x. \tag{15.20}$$

Substituting this into (15.16) to fix the constants appearing in (15.20), we find the particular integral to be

$$y_p(x) = -\frac{x^3}{12}\cos 2x + \frac{x^2}{16}\sin 2x + \frac{x}{32}\cos 2x. \tag{15.21}$$

The general solution to (15.16) then reads

$$y(x) = y_c(x) + y_p(x)$$

$$= d_1 \cos 2x + d_2 \sin 2x - \frac{x^3}{12}\cos 2x + \frac{x^2}{16}\sin 2x + \frac{x}{32}\cos 2x. \blacktriangleleft$$

15.1.4 Linear recurrence relations

Before continuing our discussion of higher-order ODEs, we take this opportunity to introduce the discrete analogues of differential equations, which are called *recurrence relations* (or sometimes *difference equations*). Whereas a differential equation gives a prescription, in terms of current values, for the new value of a dependent variable at a point only infinitesimally far away, a recurrence relation describes how the next in a sequence of values u_n, defined only at (non-negative) integer values of the 'independent variable' n, is to be calculated.

In its most general form a recurrence relation expresses the way in which u_{n+1} is to be calculated from all the preceding values $u_0, u_1, \ldots, u_n$. Just as the most general differential equations are intractable, so are the most general recurrence relations, and we will limit ourselves to analogues of the types of differential equations studied earlier in this chapter, namely those that are linear, have constant coefficients and possess simple functions on the RHS. Such equations occur over a broad range of engineering and statistical physics as well as in the realms of finance, business planning and gambling! They form the basis of many numerical methods, particularly those concerned with the numerical solution of ordinary and partial differential equations.

A general recurrence relation is exemplified by the formula

$$u_{n+1} = \sum_{r=0}^{N-1} a_r u_{n-r} + k, \tag{15.22}$$

where N and the a_r are fixed and k is a constant or a simple function of n. Such an equation, involving terms of the series whose indices differ by up to N (ranging from $n-N+1$ to n), is called an Nth-order recurrence relation. It is clear that, given values for $u_0, u_1, \ldots, u_{N-1}$, this is a definitive scheme for generating the series and therefore has a unique solution.

Parallelling the nomenclature of differential equations, if the term not involving any u_n is absent, i.e. $k = 0$, then the recurrence relation is called *homogeneous*. The parallel continues with the form of the general solution of (15.22). If v_n is the general solution of the homogeneous relation, and w_n is *any* solution of the full relation, then

$$u_n = v_n + w_n$$

is the most general solution of the complete recurrence relation. This is straight-forwardly verified as follows:

$$u_{n+1} = v_{n+1} + w_{n+1}$$
$$= \sum_{r=0}^{N-1} a_r v_{n-r} + \sum_{r=0}^{N-1} a_r w_{n-r} + k$$
$$= \sum_{r=0}^{N-1} a_r (v_{n-r} + w_{n-r}) + k$$
$$= \sum_{r=0}^{N-1} a_r u_{n-r} + k.$$

Of course, if $k = 0$ then $w_n = 0$ for all n is a trivial particular solution and the complementary solution, v_n, is itself the most general solution.

First-order recurrence relations

First-order relations, for which $N = 1$, are exemplified by

$$u_{n+1} = au_n + k, \tag{15.23}$$

with u_0 specified. The solution to the homogeneous relation is immediate,

$$u_n = Ca^n,$$

and, if k is a constant, the particular solution is equally straightforward: $w_n = K$ for all n, provided K is chosen to satisfy

$$K = aK + k,$$

i.e. $K = k(1-a)^{-1}$. This will be sufficient unless $a = 1$, in which case $u_n = u_0 + nk$ is obvious by inspection.

Thus the general solution of (15.23) is

$$u_n = \begin{cases} Ca^n + k/(1-a) & a \neq 1, \\ u_0 + nk & a = 1. \end{cases} \tag{15.24}$$

If u_0 is specified for the case of $a \neq 1$ then C must be chosen as $C = u_0 - k/(1-a)$, resulting in the equivalent form

$$u_n = u_0 a^n + k \frac{1-a^n}{1-a}. \tag{15.25}$$

We now illustrate this method with a worked example.

▶ *A house-buyer borrows capital B from a bank that charges a fixed annual rate of interest $R\%$. If the loan is to be repaid over Y years, at what value should the fixed annual payments P, made at the end of each year, be set? For a loan over 25 years at 6%, what percentage of the first year's payment goes towards paying off the capital?*

Let u_n denote the outstanding debt at the end of year n, and write $R/100 = r$. Then the relevant recurrence relation is

$$u_{n+1} = u_n(1+r) - P$$

with $u_0 = B$. From (15.25) we have

$$u_n = B(1+r)^n - P \frac{1-(1+r)^n}{1-(1+r)}.$$

As the loan is to be repaid over Y years, $u_Y = 0$ and thus

$$P = \frac{Br(1+r)^Y}{(1+r)^Y - 1}.$$

The first year's interest is rB and so the fraction of the first year's payment going towards capital repayment is $(P - rB)/P$, which, using the above expression for P, is equal to $(1+r)^{-Y}$. With the given figures, this is (only) 23%. ◀

With only small modifications, the method just described can be adapted to handle recurrence relations in which the constant k in (15.23) is replaced by $k\alpha^n$, i.e. the relation is

$$u_{n+1} = au_n + k\alpha^n. \tag{15.26}$$

As for an inhomogeneous linear differential equation (see subsection 15.1.2), we may try as a potential particular solution a form which resembles the term that makes the equation inhomogeneous. Here, the presence of the term $k\alpha^n$ indicates that a particular solution of the form $u_n = A\alpha^n$ should be tried. Substituting this into (15.26) gives

$$A\alpha^{n+1} = aA\alpha^n + k\alpha^n,$$

from which it follows that $A = k/(\alpha - a)$ and that there is a particular solution having the form $u_n = k\alpha^n/(\alpha - a)$, provided $\alpha \neq a$. For the special case $\alpha = a$, the reader can readily verify that a particular solution of the form $u_n = An\alpha^n$ is appropriate. This mirrors the corresponding situation for linear differential equations when the RHS of the differential equation is contained in the complementary function of its LHS.

In summary, the general solution to (15.26) is

$$u_n = \begin{cases} C_1 a^n + k\alpha^n/(\alpha - a) & \alpha \neq a, \\ C_2 a^n + kn\alpha^{n-1} & \alpha = a, \end{cases} \tag{15.27}$$

with $C_1 = u_0 - k/(\alpha - a)$ and $C_2 = u_0$.

Second-order recurrence relations

We consider next recurrence relations that involve u_{n-1} in the prescription for u_{n+1} and treat the general case in which the intervening term, u_n, is also present. A typical equation is thus

$$u_{n+1} = au_n + bu_{n-1} + k. \tag{15.28}$$

As previously, the general solution of this is $u_n = v_n + w_n$, where v_n satisfies

$$v_{n+1} = av_n + bv_{n-1} \tag{15.29}$$

and w_n is *any* particular solution of (15.28); the proof follows the same lines as that given earlier.

We have already seen for a first-order recurrence relation that the solution to the homogeneous equation is given by terms forming a geometric series, and we consider a corresponding series of powers in the present case. Setting $v_n = A\lambda^n$ in (15.29) for some λ, as yet undetermined, gives the requirement that λ should satisfy

$$A\lambda^{n+1} = aA\lambda^n + bA\lambda^{n-1}.$$

Dividing through by $A\lambda^{n-1}$ (assumed non-zero) shows that λ could be either of the roots, λ_1 and λ_2, of

$$\lambda^2 - a\lambda - b = 0, \tag{15.30}$$

which is known as the *characteristic equation* of the recurrence relation.

That there are two possible series of terms of the form $A\lambda^n$ is consistent with the fact that two initial values (boundary conditions) have to be provided before the series can be calculated by repeated use of (15.28). These two values are sufficient to determine the appropriate coefficient A for each of the series. Since (15.29) is both linear and homogeneous, and is satisfied by both $v_n = A\lambda_1^n$ and $v_n = B\lambda_2^n$, its general solution is

$$v_n = A\lambda_1^n + B\lambda_2^n.$$

If the coefficients a and b are such that (15.30) has two equal roots, i.e. $a^2 = -4b$, then, as in the analogous case of repeated roots for differential equations (see subsection 15.1.1(iii)), the second term of the general solution is replaced by $Bn\lambda_1^n$ to give

$$v_n = (A + Bn)\lambda_1^n.$$

Finding a particular solution is straightforward if k is a constant: a trivial but adequate solution is $w_n = k(1 - a - b)^{-1}$ for all n. As with first-order equations, particular solutions can be found for other simple forms of k by trying functions similar to k itself. Thus particular solutions for the cases $k = Cn$ and $k = D\alpha^n$ can be found by trying $w_n = E + Fn$ and $w_n = G\alpha^n$ respectively.

▶Find the value of u_{16} if the series u_n satisfies

$$u_{n+1} + 4u_n + 3u_{n-1} = n$$

for $n \geq 1$, with $u_0 = 1$ and $u_1 = -1$.

We first solve the characteristic equation,

$$\lambda^2 + 4\lambda + 3 = 0,$$

to obtain the roots $\lambda = -1$ and $\lambda = -3$. Thus the complementary function is

$$v_n = A(-1)^n + B(-3)^n.$$

In view of the form of the RHS of the original relation, we try

$$w_n = E + Fn$$

as a particular solution and obtain

$$E + F(n + 1) + 4(E + Fn) + 3[E + F(n - 1)] = n,$$

yielding $F = 1/8$ and $E = 1/32$.
Thus the complete general solution is

$$u_n = A(-1)^n + B(-3)^n + \frac{n}{8} + \frac{1}{32},$$

and now using the given values for u_0 and u_1 determines A as $7/8$ and B as $3/32$. Thus

$$u_n = \frac{1}{32}\left[28(-1)^n + 3(-3)^n + 4n + 1\right].$$

Finally, substituting $n = 16$ gives $u_{16} = 4\,035\,633$, a value the reader may (or may not) wish to verify by repeated application of the initial recurrence relation. ◀

Higher-order recurrence relations

It will be apparent that linear recurrence relations of order $N > 2$ do not present any additional difficulty in principle, though two obvious practical difficulties are (i) that the characteristic equation is of order N and in general will not have roots that can be written in closed form and (ii) that a correspondingly large number of given values is required to determine the N otherwise arbitrary constants in the solution. The algebraic labour needed to solve the set of simultaneous linear equations that determines them increases rapidly with N. We do not give specific examples here, but some are included in the exercises at the end of the chapter.

15.1.5 Laplace transform method

Having briefly discussed recurrence relations, we now return to the main topic of this chapter, i.e. methods for obtaining solutions to higher-order ODEs. One such method is that of Laplace transforms, which is very useful for solving linear ODEs with constant coefficients. Taking the Laplace transform of such an equation transforms it into a purely *algebraic* equation in terms of the Laplace transform of the required solution. Once the algebraic equation has been solved for this Laplace transform, the general solution to the original ODE can be obtained by performing an inverse Laplace transform. One advantage of this method is that, for given boundary conditions, it provides the solution in just one step, instead of having to find the complementary function and particular integral separately.

In order to apply the method we need only two results from Laplace transform theory (see section 13.2). First, the Laplace transform of a function $f(x)$ is defined by

$$\bar{f}(s) \equiv \int_0^\infty e^{-sx} f(x)\, dx, \qquad (15.31)$$

from which we can derive the second useful relation. This concerns the Laplace transform of the nth derivative of $f(x)$:

$$\overline{f^{(n)}}(s) = s^n \bar{f}(s) - s^{n-1} f(0) - s^{n-2} f'(0) - \cdots - s f^{(n-2)}(0) - f^{(n-1)}(0), \qquad (15.32)$$

where the primes and superscripts in parentheses denote differentiation with respect to x. Using these relations, along with table 13.1, on p. 455, which gives Laplace transforms of standard functions, we are in a position to solve a linear ODE with constant coefficients by this method.

> ►*Solve*
> $$\frac{d^2y}{dx^2} - 3\frac{dy}{dx} + 2y = 2e^{-x}, \qquad (15.33)$$
> *subject to the boundary conditions* $y(0) = 2$, $y'(0) = 1$.

Taking the Laplace transform of (15.33) and using the table of standard results we obtain

$$s^2\bar{y}(s) - sy(0) - y'(0) - 3[s\bar{y}(s) - y(0)] + 2\bar{y}(s) = \frac{2}{s+1},$$

which reduces to

$$(s^2 - 3s + 2)\bar{y}(s) - 2s + 5 = \frac{2}{s+1}. \qquad (15.34)$$

Solving this algebraic equation for $\bar{y}(s)$, the Laplace transform of the required solution to (15.33), we obtain

$$\bar{y}(s) = \frac{2s^2 - 3s - 3}{(s+1)(s-1)(s-2)} = \frac{1}{3(s+1)} + \frac{2}{s-1} - \frac{1}{3(s-2)}, \qquad (15.35)$$

where in the final step we have used partial fractions. Taking the inverse Laplace transform of (15.35), again using table 13.1, we find the specific solution to (15.33) to be

$$y(x) = \tfrac{1}{3}e^{-x} + 2e^x - \tfrac{1}{3}e^{2x}. \ \blacktriangleleft$$

Note that if the boundary conditions in a problem are given as symbols, rather than just numbers, then the step involving partial fractions can often involve a considerable amount of algebra. The Laplace transform method is also very convenient for solving sets of *simultaneous* linear ODEs with constant coefficients.

> ►*Two electrical circuits, both of negligible resistance, each consist of a coil having self-inductance L and a capacitor having capacitance C. The mutual inductance of the two circuits is M. There is no source of e.m.f. in either circuit. Initially the second capacitor is given a charge CV_0, the first capacitor being uncharged, and at time $t = 0$ a switch in the second circuit is closed to complete the circuit. Find the subsequent current in the first circuit.*

Subject to the initial conditions $q_1(0) = \dot{q}_1(0) = \dot{q}_2(0) = 0$ and $q_2(0) = CV_0 = V_0/G$, say, we have to solve

$$L\ddot{q}_1 + M\ddot{q}_2 + Gq_1 = 0,$$
$$M\ddot{q}_1 + L\ddot{q}_2 + Gq_2 = 0.$$

On taking the Laplace transform of the above equations, we obtain

$$(Ls^2 + G)\bar{q}_1 + Ms^2\bar{q}_2 = sMV_0C,$$
$$Ms^2\bar{q}_1 + (Ls^2 + G)\bar{q}_2 = sLV_0C.$$

Eliminating $\bar{q}_2$ and rewriting as an equation for $\bar{q}_1$, we find

$$\bar{q}_1(s) = \frac{MV_0s}{[(L+M)s^2 + G][(L-M)s^2 + G]}$$

$$= \frac{V_0}{2G}\left[\frac{(L+M)s}{(L+M)s^2 + G} - \frac{(L-M)s}{(L-M)s^2 + G}\right].$$

Using table 13.1,

$$q_1(t) = \tfrac{1}{2} V_0 C (\cos \omega_1 t - \cos \omega_2 t),$$

where $\omega_1^2(L + M) = G$ and $\omega_2^2(L - M) = G$. Thus the current is given by

$$i_1(t) = \tfrac{1}{2} V_0 C (\omega_2 \sin \omega_2 t - \omega_1 \sin \omega_1 t). \blacktriangleleft$$

Solution method. *Perform a Laplace transform, as defined in (15.31), on the entire equation, using (15.32) to calculate the transform of the derivatives. Then solve the resulting algebraic equation for $\bar{y}(s)$, the Laplace transform of the required solution to the ODE. By using the method of partial fractions and consulting a table of Laplace transforms of standard functions, calculate the inverse Laplace transform. The resulting function $y(x)$ is the solution of the ODE that obeys the given boundary conditions.*

15.2 Linear equations with variable coefficients

There is no generally applicable method of solving equations with coefficients that are functions of x. Nevertheless, there are certain cases in which a solution is possible. Some of the methods discussed in this section are also useful in finding the general solution or particular integral for equations with constant coefficients that have proved impenetrable by the techniques discussed above.

15.2.1 The Legendre and Euler linear equations

Legendre's linear equation has the form

$$a_n(\alpha x + \beta)^n \frac{d^n y}{dx^n} + \cdots + a_1(\alpha x + \beta) \frac{dy}{dx} + a_0 y = f(x), \qquad (15.36)$$

where α, β and the a_n are constants and may be solved by making the substitution $\alpha x + \beta = e^t$. We then have

$$\frac{dy}{dx} = \frac{dt}{dx} \frac{dy}{dt} = \frac{\alpha}{\alpha x + \beta} \frac{dy}{dt}$$

$$\frac{d^2 y}{dx^2} = \frac{d}{dx} \frac{dy}{dx} = \frac{\alpha^2}{(\alpha x + \beta)^2} \left(\frac{d^2 y}{dt^2} - \frac{dy}{dt} \right)$$

and so on for higher derivatives. Therefore we can write the terms of (15.36) as

$$(\alpha x + \beta) \frac{dy}{dx} = \alpha \frac{dy}{dt},$$

$$(\alpha x + \beta)^2 \frac{d^2 y}{dx^2} = \alpha^2 \frac{d}{dt} \left(\frac{d}{dt} - 1 \right) y,$$

$$\vdots$$

$$(\alpha x + \beta)^n \frac{d^n y}{dx^n} = \alpha^n \frac{d}{dt} \left(\frac{d}{dt} - 1 \right) \cdots \left(\frac{d}{dt} - n + 1 \right) y.$$

(15.37)

Substituting equations (15.37) into the original equation (15.36), the latter becomes a linear ODE with constant coefficients, i.e.

$$a_n \alpha^n \frac{d}{dt}\left(\frac{d}{dt}-1\right)\cdots\left(\frac{d}{dt}-n+1\right)y + \cdots + a_1\alpha\frac{dy}{dt} + a_0 y = f\left(\frac{e^t - \beta}{\alpha}\right),$$

which can be solved by the methods of section 15.1.

A special case of Legendre's linear equation, for which $\alpha = 1$ and $\beta = 0$, is *Euler's equation*,

$$a_n x^n \frac{d^n y}{dx^n} + \cdots + a_1 x\frac{dy}{dx} + a_0 y = f(x); \tag{15.38}$$

it may be solved in a similar manner to the above by substituting $x = e^t$. If $f(x) = 0$ in (15.38) then substituting $y = x^\lambda$ leads to a simple algebraic equation in λ, which can be solved to yield the solution to (15.38). In the event that the algebraic equation for λ has repeated roots, extra care is needed. If λ_1 is a k-fold root $(k > 1)$ then the k linearly independent solutions corresponding to this root are $x^{\lambda_1}, x^{\lambda_1}\ln x, \ldots, x^{\lambda_1}(\ln x)^{k-1}$.

►*Solve*

$$x^2\frac{d^2 y}{dx^2} + x\frac{dy}{dx} - 4y = 0 \tag{15.39}$$

by both of the methods discussed above.

First we make the substitution $x = e^t$, which, after cancelling e^t, gives an equation with constant coefficients, i.e.

$$\frac{d}{dt}\left(\frac{d}{dt}-1\right)y + \frac{dy}{dt} - 4y = 0 \quad \Rightarrow \quad \frac{d^2 y}{dt^2} - 4y = 0. \tag{15.40}$$

Using the methods of section 15.1, the general solution of (15.40), and therefore of (15.39), is given by

$$y = c_1 e^{2t} + c_2 e^{-2t} = c_1 x^2 + c_2 x^{-2}.$$

Since the RHS of (15.39) is zero, we can reach the same solution by substituting $y = x^\lambda$ into (15.39). This gives

$$\lambda(\lambda - 1)x^\lambda + \lambda x^\lambda - 4x^\lambda = 0,$$

which reduces to

$$(\lambda^2 - 4)x^\lambda = 0.$$

This has the solutions $\lambda = \pm 2$, so we obtain again the general solution

$$y = c_1 x^2 + c_2 x^{-2}. \blacktriangleleft$$

Solution method. *If the ODE is of the Legendre form (15.36) then substitute $\alpha x + \beta = e^t$. This results in an equation of the same order but with constant coefficients, which can be solved by the methods of section 15.1. If the ODE is of the Euler form (15.38) with a non-zero RHS then substitute $x = e^t$; this again leads to an equation of the same order but with constant coefficients. If, however, $f(x) = 0$ in the Euler equation (15.38) then the equation may also be solved by substituting*

$y = x^\lambda$. *This leads to an algebraic equation whose solution gives the allowed values of λ; the general solution is then the linear superposition of these functions.*

15.2.2 Exact equations

Sometimes an ODE may be merely the derivative of another ODE of one order lower. If this is the case then the ODE is called exact. The nth-order linear ODE

$$a_n(x)\frac{d^n y}{dx^n} + \cdots + a_1(x)\frac{dy}{dx} + a_0(x)y = f(x), \tag{15.41}$$

is exact if the LHS can be written as a simple derivative, i.e. if

$$a_n(x)\frac{d^n y}{dx^n} + \cdots + a_0(x)y = \frac{d}{dx}\left[b_{n-1}(x)\frac{d^{n-1} y}{dx^{n-1}} + \cdots + b_0(x)y\right]. \tag{15.42}$$

It may be shown that, for (15.42) to hold, we require

$$a_0(x) - a_1'(x) + a_2''(x) - \cdots + (-1)^n a_n^{(n)}(x) = 0, \tag{15.43}$$

where the prime again denotes differentiation with respect to x. If (15.43) is satisfied then straightforward integration leads to a new equation of one order lower. If this simpler equation can be solved then a solution to the original equation is obtained. Of course, if the above process leads to an equation that is itself exact then the analysis can be repeated to reduce the order still further.

▶Solve

$$(1 - x^2)\frac{d^2 y}{dx^2} - 3x\frac{dy}{dx} - y = 1. \tag{15.44}$$

Comparing with (15.41), we have $a_2 = 1 - x^2$, $a_1 = -3x$ and $a_0 = -1$. It is easily shown that $a_0 - a_1' + a_2'' = 0$, so (15.44) is exact and can therefore be written in the form

$$\frac{d}{dx}\left[b_1(x)\frac{dy}{dx} + b_0(x)y\right] = 1. \tag{15.45}$$

Expanding the LHS of (15.45) we find

$$\frac{d}{dx}\left(b_1\frac{dy}{dx} + b_0 y\right) = b_1\frac{d^2 y}{dx^2} + (b_1' + b_0)\frac{dy}{dx} + b_0' y. \tag{15.46}$$

Comparing (15.44) and (15.46) we find

$$b_1 = 1 - x^2, \qquad b_1' + b_0 = -3x, \qquad b_0' = -1.$$

These relations integrate consistently to give $b_1 = 1 - x^2$ and $b_0 = -x$, so (15.44) can be written as

$$\frac{d}{dx}\left[(1 - x^2)\frac{dy}{dx} - xy\right] = 1. \tag{15.47}$$

Integrating (15.47) gives us directly the first-order linear ODE

$$\frac{dy}{dx} - \left(\frac{x}{1 - x^2}\right)y = \frac{x + c_1}{1 - x^2},$$

which can be solved by the method of subsection 14.2.4 and has the solution

$$y = \frac{c_1 \sin^{-1} x + c_2}{\sqrt{1 - x^2}} - 1. \blacktriangleleft$$

505

It is worth noting that, even if a higher-order ODE is not exact in its given form, it may sometimes be made exact by multiplying through by some suitable function, an *integrating factor*, cf. subsection 14.2.3. Unfortunately, no straightforward method for finding an integrating factor exists and one often has to rely on inspection or experience.

▶*Solve*

$$x(1-x^2)\frac{d^2y}{dx^2} - 3x^2\frac{dy}{dx} - xy = x. \tag{15.48}$$

It is easily shown that (15.48) is not exact, but we also see immediately that by multiplying it through by $1/x$ we recover (15.44), which is exact and is solved above. ◀

Another important point is that an ODE need not be linear to be exact, although no simple rule such as (15.43) exists if it is not linear. Nevertheless, it is often worth exploring the possibility that a non-linear equation is exact, since it could then be reduced in order by one and may lead to a soluble equation. This is discussed further in subsection 15.3.3.

Solution method. *For a linear ODE of the form (15.41) check whether it is exact using equation (15.43). If it is not then attempt to find an integrating factor which when multiplying the equation makes it exact. Once the equation is exact write the LHS as a derivative as in (15.42) and, by expanding this derivative and comparing with the LHS of the ODE, determine the functions $b_m(x)$ in (15.42). Integrate the resulting equation to yield another ODE, of one order lower. This may be solved or simplified further if the new ODE is itself exact or can be made so.*

15.2.3 Partially known complementary function

Suppose we wish to solve the nth-order linear ODE

$$a_n(x)\frac{d^ny}{dx^n} + \cdots + a_1(x)\frac{dy}{dx} + a_0(x)y = f(x), \tag{15.49}$$

and we happen to know that $u(x)$ is a solution of (15.49) when the RHS is set to zero, i.e. $u(x)$ is one part of the complementary function. By making the substitution $y(x) = u(x)v(x)$, we can transform (15.49) into an equation of order $n-1$ in dv/dx. This simpler equation may prove soluble.

In particular, if the original equation is of second order then we obtain a first-order equation in dv/dx, which may be soluble using the methods of section 14.2. In this way both the remaining term in the complementary function and the particular integral are found. This method therefore provides a useful way of calculating particular integrals for second-order equations with variable (or constant) coefficients.

> ►*Solve*
> $$\frac{d^2y}{dx^2} + y = \operatorname{cosec} x. \tag{15.50}$$

We see that the RHS does not fall into any of the categories listed in subsection 15.1.2, and so we are at an initial loss as to how to find the particular integral. However, the complementary function of (15.50) is

$$y_c(x) = c_1 \sin x + c_2 \cos x,$$

and so let us choose the solution $u(x) = \cos x$ (we could equally well choose $\sin x$) and make the substitution $y(x) = v(x)u(x) = v(x)\cos x$ into (15.50). This gives

$$\cos x \frac{d^2v}{dx^2} - 2\sin x \frac{dv}{dx} = \operatorname{cosec} x, \tag{15.51}$$

which is a first-order linear ODE in dv/dx and may be solved by multiplying through by a suitable integrating factor, as discussed in subsection 14.2.4. Writing (15.51) as

$$\frac{d^2v}{dx^2} - 2\tan x \frac{dv}{dx} = \frac{\operatorname{cosec} x}{\cos x}, \tag{15.52}$$

we see that the required integrating factor is given by

$$\exp\left\{-2\int \tan x \, dx\right\} = \exp\left[2\ln(\cos x)\right] = \cos^2 x.$$

Multiplying both sides of (15.52) by the integrating factor $\cos^2 x$ we obtain

$$\frac{d}{dx}\left(\cos^2 x \frac{dv}{dx}\right) - \cot x,$$

which integrates to give

$$\cos^2 x \frac{dv}{dx} = \ln(\sin x) + c_1.$$

After rearranging and integrating again, this becomes

$$v = \int \sec^2 x \ln(\sin x) \, dx + c_1 \int \sec^2 x \, dx$$
$$= \tan x \ln(\sin x) - x + c_1 \tan x + c_2.$$

Therefore the general solution to (15.50) is given by $y = uv = v\cos x$, i.e.

$$y = c_1 \sin x + c_2 \cos x + \sin x \ln(\sin x) - x\cos x,$$

which contains the full complementary function and the particular integral. ◄

Solution method. *If $u(x)$ is a known solution of the nth-order equation (15.49) with $f(x) = 0$, then make the substitution $y(x) = u(x)v(x)$ in (15.49). This leads to an equation of order $n - 1$ in dv/dx, which might be soluble.*

15.2.4 Variation of parameters

The method of variation of parameters proves useful in finding particular integrals for linear ODEs with variable (and constant) coefficients. However, it requires knowledge of the entire complementary function, not just of one part of it as in the previous subsection.

Suppose we wish to find a particular integral of the equation

$$a_n(x)\frac{d^n y}{dx^n} + \cdots + a_1(x)\frac{dy}{dx} + a_0(x)y = f(x), \tag{15.53}$$

and the complementary function $y_c(x)$ (the general solution of (15.53) with $f(x) = 0$) is

$$y_c(x) = c_1 y_1(x) + c_2 y_2(x) + \cdots + c_n y_n(x),$$

where the functions $y_m(x)$ are known. We now assume that a particular integral of (15.53) can be expressed in a form similar to that of the complementary function, but with the constants c_m replaced by functions of x, i.e. we assume a particular integral of the form

$$y_p(x) = k_1(x)y_1(x) + k_2(x)y_2(x) + \cdots + k_n(x)y_n(x). \tag{15.54}$$

This will no longer satisfy the complementary equation (i.e. (15.53) with the RHS set to zero) but might, with suitable choices of the functions $k_i(x)$, be made equal to $f(x)$, thus producing not a complementary function but a particular integral.

Since we have n arbitrary functions $k_1(x), k_2(x), \ldots, k_n(x)$, but only one restriction on them (namely the ODE), we may impose a further $n - 1$ constraints. We can choose these constraints to be as convenient as possible, and the simplest choice is given by

$$k_1'(x)y_1(x) + k_2'(x)y_2(x) + \cdots + k_n'(x)y_n(x) = 0$$
$$k_1'(x)y_1'(x) + k_2'(x)y_2'(x) + \cdots + k_n'(x)y_n'(x) = 0$$
$$\vdots \tag{15.55}$$
$$k_1'(x)y_1^{(n-2)}(x) + k_2'(x)y_2^{(n-2)}(x) + \cdots + k_n'(x)y_n^{(n-2)}(x) = 0$$
$$k_1'(x)y_1^{(n-1)}(x) + k_2'(x)y_2^{(n-1)}(x) + \cdots + k_n'(x)y_n^{(n-1)}(x) = \frac{f(x)}{a_n(x)},$$

where the primes denote differentiation with respect to x. The last of these equations is not a freely chosen constraint; given the previous $n - 1$ constraints and the original ODE, it must be satisfied.

This choice of constraints is easily justified (although the algebra is quite messy). Differentiating (15.54) with respect to x, we obtain

$$y_p' = k_1 y_1' + k_2 y_2' + \cdots + k_n y_n' + [k_1' y_1 + k_2' y_2 + \cdots + k_n' y_n],$$

where, for the moment, we drop the explicit x-dependence of these functions. Since we are free to choose our constraints as we wish, let us define the expression in parentheses to be zero, giving the first equation in (15.55). Differentiating again we find

$$y_p'' = k_1 y_1'' + k_2 y_2'' + \cdots + k_n y_n'' + [k_1' y_1' + k_2' y_2' + \cdots + k_n' y_n'].$$

Once more we can choose the expression in brackets to be zero, giving the second equation in (15.55). We can repeat this procedure, choosing the corresponding expression in each case to be zero. This yields the first $n-1$ equations in (15.55). The mth derivative of y_p for $m < n$ is then given by

$$y_p^{(m)} = k_1 y_1^{(m)} + k_2 y_2^{(m)} + \cdots + k_n y_n^{(m)}.$$

Differentiating y_p once more we find that its nth derivative is given by

$$y_p^{(n)} = k_1 y_1^{(n)} + k_2 y_2^{(n)} + \cdots + k_n y_n^{(n)} + [k_1' y_1^{(n-1)} + k_2' y_2^{(n-1)} + \cdots + k_n' y_n^{(n-1)}].$$

Substituting the expressions for $y_p^{(m)}$, $m = 0$ to n, into the original ODE (15.53), we obtain

$$\sum_{m=0}^{n} a_m [k_1 y_1^{(m)} + k_2 y_2^{(m)} + \cdots + k_n y_n^{(m)}] + a_n [k_1' y_1^{(n-1)} + k_2' y_2^{(n-1)} + \cdots + k_n' y_n^{(n-1)}] = f(x),$$

i.e.

$$\sum_{m=0}^{n} a_m \sum_{j=1}^{n} k_j y_j^{(m)} + a_n [k_1' y_1^{(n-1)} + k_2' y_2^{(n-1)} + \cdots + k_n' y_n^{(n-1)}] = f(x).$$

Rearranging the order of summation on the LHS, we find

$$\sum_{j=1}^{n} k_j [a_n y_j^{(n)} + \cdots + a_1 y_j' + a_0 y_j] + a_n [k_1' y_1^{(n-1)} + k_2' y_2^{(n-1)} + \cdots + k_n' y_n^{(n-1)}] = f(x). \tag{15.56}$$

But since the functions y_j are solutions of the complementary equation of (15.53) we have (for all j)

$$a_n y_j^{(n)} + \cdots + a_1 y_j' + a_0 y_j = 0.$$

Therefore (15.56) becomes

$$a_n [k_1' y_1^{(n-1)} + k_2' y_2^{(n-1)} + \cdots + k_n' y_n^{(n-1)}] = f(x),$$

which is the final equation given in (15.55).

Considering (15.55) to be a set of simultaneous equations in the set of unknowns $k_1'(x), k_2', \ldots, k_n'(x)$, we see that the determinant of the coefficients of these functions is equal to the Wronskian $W(y_1, y_2, \ldots, y_n)$, which is non-zero since the solutions $y_m(x)$ are linearly independent; see equation (15.6). Therefore (15.55) can be solved for the functions $k_m'(x)$, which in turn can be integrated, setting all constants of integration equal to zero, to give $k_m(x)$. The general solution to (15.53) is then given by

$$y(x) = y_c(x) + y_p(x) = \sum_{m=1}^{n} [c_m + k_m(x)] y_m(x).$$

Note that if the constants of integration are included in the $k_m(x)$ then, as well as finding the particular integral, we redefine the arbitrary constants c_m in the complementary function.

> ▶*Use the variation-of-parameters method to solve*
>
> $$\frac{d^2 y}{dx^2} + y = \operatorname{cosec} x, \tag{15.57}$$
>
> *subject to the boundary conditions $y(0) = y(\pi/2) = 0$.*

The complementary function of (15.57) is again

$$y_c(x) = c_1 \sin x + c_2 \cos x.$$

We therefore assume a particular integral of the form

$$y_p(x) = k_1(x) \sin x + k_2(x) \cos x,$$

and impose the additional constraints of (15.55), i.e.

$$k_1'(x) \sin x + k_2'(x) \cos x = 0,$$
$$k_1'(x) \cos x - k_2'(x) \sin x = \operatorname{cosec} x.$$

Solving these equations for $k_1'(x)$ and $k_2'(x)$ gives

$$k_1'(x) = \cos x \operatorname{cosec} x = \cot x,$$
$$k_2'(x) = -\sin x \operatorname{cosec} x = -1.$$

Hence, ignoring the constants of integration, $k_1(x)$ and $k_2(x)$ are given by

$$k_1(x) = \ln(\sin x),$$
$$k_2(x) = -x.$$

The general solution to the ODE (15.57) is therefore

$$y(x) = [c_1 + \ln(\sin x)] \sin x + (c_2 - x) \cos x,$$

which is identical to the solution found in subsection 15.2.3. Applying the boundary conditions $y(0) = y(\pi/2) = 0$ we find $c_1 = c_2 = 0$ and so

$$y(x) = \ln(\sin x) \sin x - x \cos x. \quad ◀$$

Solution method. *If the complementary function of (15.53) is known then assume a particular integral of the same form but with the constants replaced by functions of x. Impose the constraints in (15.55) and solve the resulting system of equations for the unknowns $k_1'(x), k_2', \ldots, k_n'(x)$. Integrate these functions, setting constants of integration equal to zero, to obtain $k_1(x), k_2(x), \ldots, k_n(x)$ and hence the particular integral.*

15.2.5 Green's functions

The Green's function method of solving linear ODEs bears a striking resemblance to the method of variation of parameters discussed in the previous subsection; it too requires knowledge of the entire complementary function in order to find the particular integral and therefore the general solution. The Green's function approach differs, however, since once the Green's function for a particular LHS of (15.1) and particular boundary conditions has been found, then the solution for *any* RHS (i.e. any $f(x)$) can be written down immediately, albeit in the form of an integral.

Although the Green's function method can be approached by considering the superposition of eigenfunctions of the equation (see chapter 17) and is also applicable to the solution of partial differential equations (see chapter 21), this section adopts a more utilitarian approach based on the properties of the Dirac delta function (see subsection 13.1.3) and deals only with the use of Green's functions in solving ODEs.

Let us again consider the equation

$$a_n(x)\frac{d^n y}{dx^n} + \cdots + a_1(x)\frac{dy}{dx} + a_0(x)y = f(x), \tag{15.58}$$

but for the sake of brevity we now denote the LHS by $\mathcal{L}y(x)$, i.e. as a linear differential operator acting on $y(x)$. Thus (15.58) now reads

$$\mathcal{L}y(x) = f(x). \tag{15.59}$$

Let us suppose that a function $G(x, z)$ (the *Green's function*) exists such that the general solution to (15.59), which obeys some set of imposed boundary conditions in the range $a \leq x \leq b$, is given by

$$y(x) = \int_a^b G(x, z)f(z)\,dz, \tag{15.60}$$

where z is an integration variable. If we apply the linear differential operator $\mathcal{L}$ to both sides of (15.60) and use (15.59) then we obtain

$$\mathcal{L}y(x) = \int_a^b [\mathcal{L}G(x, z)]\,f(z)\,dz = f(x). \tag{15.61}$$

Comparison of (15.61) with a standard property of the Dirac delta function (see subsection 13.1.3), namely

$$f(x) = \int_a^b \delta(x - z)f(z)\,dz,$$

for $a \leq x \leq b$, shows that for (15.61) to hold for any arbitrary function $f(x)$, we require (for $a \leq x \leq b$) that

$$\mathcal{L}G(x, z) = \delta(x - z), \tag{15.62}$$

i.e. the Green's function $G(x, z)$ must satisfy the original ODE with the RHS set equal to a delta function. $G(x, z)$ may be thought of physically as the response of a system to a unit impulse at $x = z$.

In addition to (15.62), we must impose two further sets of restrictions on $G(x, z)$. The first is the requirement that the general solution $y(x)$ in (15.60) obeys the boundary conditions. For *homogeneous* boundary conditions, in which $y(x)$ and/or its derivatives are required to be *zero* at specified points, this is most simply arranged by demanding that $G(x, z)$ itself obeys the boundary conditions when it is considered as a function of x alone; if, for example, we require $y(a) = y(b) = 0$ then we should also demand $G(a, z) = G(b, z) = 0$. Problems having inhomogeneous boundary conditions are discussed at the end of this subsection.

The second set of restrictions concerns the continuity or discontinuity of $G(x, z)$ and its derivatives at $x = z$ and can be found by integrating (15.62) with respect to x over the small interval $[z - \epsilon, \ z + \epsilon]$ and taking the limit as $\epsilon \to 0$. We then obtain

$$\lim_{\epsilon \to 0} \sum_{m=0}^{n} \int_{z-\epsilon}^{z+\epsilon} a_m(x) \frac{d^m G(x, z)}{dx^m} \, dx = \lim_{\epsilon \to 0} \int_{z-\epsilon}^{z+\epsilon} \delta(x - z) \, dx = 1. \qquad (15.63)$$

Since $d^n G/dx^n$ exists at $x = z$ but with value infinity, the $(n-1)$th-order derivative must have a finite discontinuity there, whereas all the lower-order derivatives, $d^m G/dx^m$ for $m < n - 1$, must be continuous at this point. Therefore the terms containing these derivatives cannot contribute to the value of the integral on the LHS of (15.63). Noting that, apart from an arbitrary additive constant, $\int (d^m G/dx^m) \, dx = d^{m-1} G/dx^{m-1}$, and integrating the terms on the LHS of (15.63) by parts we find

$$\lim_{\epsilon \to 0} \int_{z-\epsilon}^{z+\epsilon} a_m(x) \frac{d^m G(x, z)}{dx^m} \, dx = 0 \qquad (15.64)$$

for $m = 0$ to $n - 1$. Thus, since only the term containing $d^n G/dx^n$ contributes to the integral in (15.63), we conclude, after performing an integration by parts, that

$$\lim_{\epsilon \to 0} \left[a_n(x) \frac{d^{n-1} G(x, z)}{dx^{n-1}} \right]_{z-\epsilon}^{z+\epsilon} = 1. \qquad (15.65)$$

Thus we have the further n constraints that $G(x, z)$ and its derivatives up to order $n - 2$ are continuous at $x = z$ but that $d^{n-1} G/dx^{n-1}$ has a discontinuity of $1/a_n(z)$ at $x = z$.

Thus the properties of the Green's function $G(x, z)$ for an nth-order linear ODE may be summarised by the following.

(i) $G(x, z)$ obeys the original ODE but with $f(x)$ on the RHS set equal to a delta function $\delta(x - z)$.

(ii) When considered as a function of x alone $G(x, z)$ obeys the specified (homogeneous) boundary conditions on $y(x)$.

(iii) The derivatives of $G(x, z)$ with respect to x up to order $n-2$ are continuous at $x = z$, but the $(n-1)$th-order derivative has a discontinuity of $1/a_n(z)$ at this point.

▶ *Use Green's functions to solve*

$$\frac{d^2 y}{dx^2} + y = \operatorname{cosec} x, \tag{15.66}$$

subject to the boundary conditions $y(0) = y(\pi/2) = 0$.

From (15.62) we see that the Green's function $G(x, z)$ must satisfy

$$\frac{d^2 G(x, z)}{dx^2} + G(x, z) = \delta(x - z). \tag{15.67}$$

Now it is clear that for $x \neq z$ the RHS of (15.67) is zero, and we are left with the task of finding the general solution to the homogeneous equation, i.e. the complementary function. The complementary function of (15.67) consists of a linear superposition of $\sin x$ and $\cos x$ and *must* consist of different superpositions on either side of $x = z$, since its $(n-1)$th derivative (i.e. the first derivative in this case) is required to have a discontinuity there. Therefore we assume the form of the Green's function to be

$$G(x, z) = \begin{cases} A(z) \sin x + B(z) \cos x & \text{for } x < z, \\ C(z) \sin x + D(z) \cos x & \text{for } x > z. \end{cases}$$

Note that we have performed a similar (but not identical) operation to that used in the variation-of-parameters method, i.e. we have replaced the constants in the complementary function with functions (this time of z).

We must now impose the relevant restrictions on $G(x, z)$ in order to determine the functions $A(z), \ldots, D(z)$. The first of these is that $G(x, z)$ should itself obey the homogeneous boundary conditions $G(0, z) = G(\pi/2, z) = 0$. This leads to the conclusion that $B(z) = C(z) = 0$, so we now have

$$G(x, z) = \begin{cases} A(z) \sin x & \text{for } x < z, \\ D(z) \cos x & \text{for } x > z. \end{cases}$$

The second restriction is the continuity conditions given in equations (15.64), (15.65), namely that, for this second-order equation, $G(x, z)$ is continuous at $x = z$ and dG/dx has a discontinuity of $1/a_2(z) = 1$ at this point. Applying these two constraints we have

$$D(z) \cos z - A(z) \sin z = 0$$
$$-D(z) \sin z - A(z) \cos z = 1.$$

Solving these equations for $A(z)$ and $D(z)$, we find

$$A(z) = -\cos z, \qquad D(z) = -\sin z.$$

Thus we have

$$G(x, z) = \begin{cases} -\cos z \sin x & \text{for } x < z, \\ -\sin z \cos x & \text{for } x > z. \end{cases}$$

Therefore, from (15.60), the general solution to (15.66) that obeys the boundary conditions

$y(0) = y(\pi/2) = 0$ is given by

$$y(x) = \int_0^{\pi/2} G(x,z)\operatorname{cosec} z \, dz$$

$$= -\cos x \int_0^x \sin z \operatorname{cosec} z \, dz - \sin x \int_x^{\pi/2} \cos z \operatorname{cosec} z \, dz$$

$$= -x \cos x + \sin x \ln(\sin x),$$

which agrees with the result obtained in the previous subsections. ◄

As mentioned earlier, once a Green's function has been obtained for a given LHS and boundary conditions, it can be used to find a general solution for any RHS; thus, the solution of $d^2y/dx^2 + y = f(x)$, with $y(0) = y(\pi/2) = 0$, is given immediately by

$$y(x) = \int_0^{\pi/2} G(x,z)f(z) \, dz$$

$$= -\cos x \int_0^x \sin z \, f(z) \, dz - \sin x \int_x^{\pi/2} \cos z \, f(z) \, dz. \qquad (15.68)$$

As an example, the reader may wish to verify that if $f(x) = \sin 2x$ then (15.68) gives $y(x) = (-\sin 2x)/3$, a solution easily verified by direct substitution. In general, analytic integration of (15.68) for arbitrary $f(x)$ will prove intractable; then the integrals must be evaluated numerically.

Another important point is that although the Green's function method above has provided a general solution, it is also useful for finding a particular integral if the complementary function is known. This is easily seen since in (15.68) the constant integration limits 0 and $\pi/2$ lead merely to constant values by which the factors $\sin x$ and $\cos x$ are multiplied; thus the complementary function is reconstructed. The rest of the general solution, i.e. the particular integral, comes from the variable integration limit x. Therefore by changing $\int_x^{\pi/2}$ to $-\int^x$, and so dropping the constant integration limits, we can find just the particular integral. For example, a particular integral of $d^2y/dx^2 + y = f(x)$ that satisfies the above boundary conditions is given by

$$y_p(x) = -\cos x \int^x \sin z \, f(z) \, dz + \sin x \int^x \cos z \, f(z) \, dz.$$

A very important point to realise about the Green's function method is that a particular $G(x,z)$ applies to a given LHS of an ODE *and* the imposed boundary conditions, i.e. *the same equation with different boundary conditions will have a different Green's function*. To illustrate this point, let us consider again the ODE solved in (15.68), but with different boundary conditions.

> ►*Use Green's functions to solve*
> $$\frac{d^2 y}{dx^2} + y = f(x), \tag{15.69}$$
> *subject to the one-point boundary conditions* $y(0) = y'(0) = 0$.

We again require (15.67) to hold and so again we assume a Green's function of the form

$$G(x, z) = \begin{cases} A(z)\sin x + B(z)\cos x & \text{for } x < z, \\ C(z)\sin x + D(z)\cos x & \text{for } x > z. \end{cases}$$

However, we now require $G(x, z)$ to obey the boundary conditions $G(0, z) = G'(0, z) = 0$, which imply $A(z) = B(z) = 0$. Therefore we have

$$G(x, z) = \begin{cases} 0 & \text{for } x < z, \\ C(z)\sin x + D(z)\cos x & \text{for } x > z. \end{cases}$$

Applying the continuity conditions on $G(x, z)$ as before now gives

$$C(z)\sin z + D(z)\cos z = 0,$$
$$C(z)\cos z - D(z)\sin z = 1,$$

which are solved to give

$$C(z) = \cos z, \qquad D(z) = -\sin z.$$

So finally the Green's function is given by

$$G(x, z) = \begin{cases} 0 & \text{for } x < z, \\ \sin(x - z) & \text{for } x > z, \end{cases}$$

and the general solution to (15.69) that obeys the boundary conditions $y(0) = y'(0) = 0$ is

$$y(x) = \int_0^x G(x, z)f(z)\, dz$$
$$= \int_0^x \sin(x - z)f(z)\, dz. \blacktriangleleft$$

Finally, we consider how to deal with inhomogeneous boundary conditions such as $y(a) = \alpha$, $y(b) = \beta$ or $y(0) = y'(0) = \gamma$, where α, β, γ are non-zero. The simplest method of solution in this case is to make a change of variable such that the boundary conditions in the new variable, u say, are homogeneous, i.e. $u(a) = u(b) = 0$ or $u(0) = u'(0) = 0$ etc. For nth-order equations we generally require n boundary conditions to fix the solution, but these n boundary conditions can be of various types: we could have the n-point boundary conditions $y(x_m) = y_m$ for $m = 1$ to n, or the one-point boundary conditions $y(x_0) = y'(x_0) = \cdots = y^{(n-1)}(x_0) = y_0$, or something in between. In all cases a suitable change of variable is

$$u = y - h(x),$$

where $h(x)$ is an $(n-1)$th-order polynomial that obeys the boundary conditions.

For example, if we consider the second-order case with boundary conditions $y(a) = \alpha$, $y(b) = \beta$ then a suitable change of variable is

$$u = y - (mx + c),$$

where $y = mx + c$ is the straight line through the points (a, α) and (b, β), for which $m = (\alpha - \beta)/(a - b)$ and $c = (\beta a - \alpha b)/(a - b)$. Alternatively, if the boundary conditions for our second-order equation are $y(0) = y'(0) = \gamma$ then we would make the same change of variable, but this time $y = mx + c$ would be the straight line through $(0, \gamma)$ with slope γ, i.e. $m = c = \gamma$.

Solution method. *Require that the Green's function $G(x, z)$ obeys the original ODE, but with the RHS set to a delta function $\delta(x - z)$. This is equivalent to assuming that $G(x, z)$ is given by the complementary function of the original ODE, with the constants replaced by functions of z; these functions are different for $x < z$ and $x > z$. Now require also that $G(x, z)$ obeys the given homogeneous boundary conditions and impose the continuity conditions given in (15.64) and (15.65). The general solution to the original ODE is then given by (15.60). For inhomogeneous boundary conditions, make the change of dependent variable $u = y - h(x)$, where $h(x)$ is a polynomial obeying the given boundary conditions.*

15.2.6 Canonical form for second-order equations

In this section we specialise from nth-order linear ODEs with variable coefficients to those of order 2. In particular we consider the equation

$$\frac{d^2y}{dx^2} + a_1(x)\frac{dy}{dx} + a_0(x)y = f(x), \tag{15.70}$$

which has been rearranged so that the coefficient of d^2y/dx^2 is unity. By making the substitution $y(x) = u(x)v(x)$ we obtain

$$v'' + \left(\frac{2u'}{u} + a_1\right)v' + \left(\frac{u'' + a_1u' + a_0u}{u}\right)v = \frac{f}{u}, \tag{15.71}$$

where the prime denotes differentiation with respect to x. Since (15.71) would be much simplified if there were no term in v', let us choose $u(x)$ such that the first factor in parentheses on the LHS of (15.71) is zero, i.e.

$$\frac{2u'}{u} + a_1 = 0 \quad \Rightarrow \quad u(x) = \exp\left\{-\tfrac{1}{2}\int a_1(z)\,dz\right\}. \tag{15.72}$$

We then obtain an equation of the form

$$\frac{d^2v}{dx^2} + g(x)v = h(x), \tag{15.73}$$

where

$$g(x) = a_0(x) - \tfrac{1}{4}[a_1(x)]^2 - \tfrac{1}{2}a_1'(x)$$

$$h(x) = f(x) \exp \left\{ \tfrac{1}{2} \int a_1(z)\, dz \right\}.$$

Since (15.73) is of a simpler form than the original equation, (15.70), it may prove easier to solve.

▶*Solve*

$$4x^2 \frac{d^2 y}{dx^2} + 4x \frac{dy}{dx} + (x^2 - 1)y = 0. \tag{15.74}$$

Dividing (15.74) through by $4x^2$, we see that it is of the form (15.70) with $a_1(x) = 1/x$, $a_0(x) = (x^2 - 1)/4x^2$ and $f(x) = 0$. Therefore, making the substitution

$$y = vu = v \exp \left(-\int \frac{1}{2x}\, dx \right) = \frac{Av}{\sqrt{x}},$$

we obtain

$$\frac{d^2 v}{dx^2} + \frac{v}{4} = 0. \tag{15.75}$$

Equation (15.75) is easily solved to give

$$v = c_1 \sin \tfrac{1}{2}x + c_2 \cos \tfrac{1}{2}x,$$

so the solution of (15.74) is

$$y = \frac{v}{\sqrt{x}} = \frac{c_1 \sin \tfrac{1}{2}x + c_2 \cos \tfrac{1}{2}x}{\sqrt{x}}. \quad ◀$$

As an alternative to choosing $u(x)$ such that the coefficient of v' in (15.71) is zero, we could choose a different $u(x)$ such that the coefficient of v vanishes. For this to be the case, we see from (15.71) that we would require

$$u'' + a_1 u' + a_0 u = 0,$$

so $u(x)$ would have to be a solution of the original ODE with the RHS set to zero, i.e. part of the complementary function. If such a solution were known then the substitution $y = uv$ would yield an equation with no term in v, which could be solved by two straightforward integrations. This is a special (second-order) case of the method discussed in subsection 15.2.3.

Solution method. *Write the equation in the form (15.70), then substitute $y = uv$, where $u(x)$ is given by (15.72). This leads to an equation of the form (15.73), in which there is no term in dv/dx and which may be easier to solve. Alternatively, if part of the complementary function is known then follow the method of subsection 15.2.3.*

15.3 General ordinary differential equations

In this section, we discuss miscellaneous methods for simplifying general ODEs. These methods are applicable to both linear and non-linear equations and in some cases may lead to a solution. More often than not, however, finding a closed-form solution to a general non-linear ODE proves impossible.

15.3.1 Dependent variable absent

If an ODE does not contain the dependent variable y explicitly, but only its derivatives, then the change of variable $p = dy/dx$ leads to an equation of one order lower.

> ►*Solve*
> $$\frac{d^2y}{dx^2} + 2\frac{dy}{dx} = 4x \tag{15.76}$$

This is transformed by the substitution $p = dy/dx$ to the first-order equation

$$\frac{dp}{dx} + 2p = 4x. \tag{15.77}$$

The solution to (15.77) is then found by the method of subsection 14.2.4 and reads

$$p = \frac{dy}{dx} = ae^{-2x} + 2x - 1,$$

where a is a constant. Thus by direct integration the solution to the original equation, (15.76), is

$$y(x) = c_1 e^{-2x} + x^2 - x + c_2. \blacktriangleleft$$

An extension to the above method is appropriate if an ODE contains only derivatives of y that are of order m and greater. Then the substitution $p = d^m y/dx^m$ reduces the order of the ODE by m.

Solution method. *If the ODE contains only derivatives of y that are of order m and greater then the substitution $p = d^m y/dx^m$ reduces the order of the equation by m.*

15.3.2 Independent variable absent

If an ODE does not contain the independent variable x explicitly, except in d/dx, d^2/dx^2 etc., then as in the previous subsection we make the substitution $p = dy/dx$

but also write

$$\frac{d^2y}{dx^2} = \frac{dp}{dx} = \frac{dy}{dx}\frac{dp}{dy} = p\frac{dp}{dy}$$

$$\frac{d^3y}{dx^3} = \frac{d}{dx}\left(p\frac{dp}{dy}\right) = \frac{dy}{dx}\frac{d}{dy}\left(p\frac{dp}{dy}\right) = p^2\frac{d^2p}{dy^2} + p\left(\frac{dp}{dy}\right)^2, \qquad (15.78)$$

and so on for higher-order derivatives. This leads to an equation of one order lower.

▶Solve

$$1 + y\frac{d^2y}{dx^2} + \left(\frac{dy}{dx}\right)^2 = 0. \qquad (15.79)$$

Making the substitutions $dy/dx = p$ and $d^2y/dx^2 = p(dp/dy)$ we obtain the first-order ODE

$$1 + yp\frac{dp}{dy} + p^2 = 0,$$

which is separable and may be solved as in subsection 14.2.1 to obtain

$$(1 + p^2)y^2 = c_1.$$

Using $p = dy/dx$ we therefore have

$$p = \frac{dy}{dx} = \pm\sqrt{\frac{c_1^2 - y^2}{y^2}},$$

which may be integrated to give the general solution of (15.79); after squaring this reads

$$(x + c_2)^2 + y^2 = c_1^2. \quad◀$$

Solution method. *If the ODE does not contain x explicitly then substitute $p = dy/dx$, along with the relations for higher derivatives given in (15.78), to obtain an equation of one order lower, which may prove easier to solve.*

15.3.3 Non-linear exact equations

As discussed in subsection 15.2.2, an exact ODE is one that can be obtained by straightforward differentiation of an equation of one order lower. Moreover, the notion of exact equations is useful for both linear and non-linear equations, since an exact equation can be immediately integrated. It is possible, of course, that the resulting equation may itself be exact, so that the process can be repeated. In the non-linear case, however, there is no simple relation (such as (15.43) for the linear case) by which an equation can be shown to be exact. Nevertheless, a general procedure does exist and is illustrated in the following example.

> ►*Solve*
>
> $$2y\frac{d^3y}{dx^3} + 6\frac{dy}{dx}\frac{d^2y}{dx^2} = x. \qquad (15.80)$$

Directing our attention to the term on the LHS of (15.80) that contains the highest-order derivative, i.e. $2y\,d^3y/dx^3$, we see that it can be obtained by differentiating $2y\,d^2y/dx^2$ since

$$\frac{d}{dx}\left(2y\frac{d^2y}{dx^2}\right) = 2y\frac{d^3y}{dx^3} + 2\frac{dy}{dx}\frac{d^2y}{dx^2}. \qquad (15.81)$$

Rewriting the LHS of (15.80) using (15.81), we are left with $4(dy/dx)(d^2y/dy^2)$, which may itself be written as a derivative, i.e.

$$4\frac{dy}{dx}\frac{d^2y}{dx^2} = \frac{d}{dx}\left[2\left(\frac{dy}{dx}\right)^2\right]. \qquad (15.82)$$

Since, therefore, we can write the LHS of (15.80) as a sum of simple derivatives of other functions, (15.80) is exact. Integrating (15.80) with respect to x, and using (15.81) and (15.82), now gives

$$2y\frac{d^2y}{dx^2} + 2\left(\frac{dy}{dx}\right)^2 = \int x\,dx = \frac{x^2}{2} + c_1. \qquad (15.83)$$

Now we can repeat the process to find whether (15.83) is itself exact. Considering the term on the LHS of (15.83) that contains the highest-order derivative, i.e. $2y\,d^2y/dx^2$, we note that we obtain this by differentiating $2y\,dy/dx$, as follows:

$$\frac{d}{dx}\left(2y\frac{dy}{dx}\right) = 2y\frac{d^2y}{dx^2} + 2\left(\frac{dy}{dx}\right)^2.$$

The above expression already contains all the terms on the LHS of (15.83), so we can integrate (15.83) to give

$$2y\frac{dy}{dx} = \frac{x^3}{6} + c_1x + c_2.$$

Integrating once more we obtain the solution

$$y^2 = \frac{x^4}{24} + \frac{c_1x^2}{2} + c_2x + c_3. \quad ◄$$

It is worth noting that both linear equations (as discussed in subsection 15.2.2) and non-linear equations may sometimes be made exact by multiplying through by an appropriate integrating factor. Although no general method exists for finding such a factor, one may sometimes be found by inspection or inspired guesswork.

Solution method. *Rearrange the equation so that all the terms containing y or its derivatives are on the LHS, then check to see whether the equation is exact by attempting to write the LHS as a simple derivative. If this is possible then the equation is exact and may be integrated directly to give an equation of one order lower. If the new equation is itself exact the process can be repeated.*

15.3.4 *Isobaric or homogeneous equations*

It is straightforward to generalise the discussion of first-order isobaric equations given in subsection 14.2.6 to equations of general order n. An nth-order isobaric equation is one in which every term can be made dimensionally consistent upon giving y and dy each a weight m, and x and dx each a weight 1. Then the nth derivative of y with respect to x, for example, would have dimensions m in y and $-n$ in x. In the special case $m = 1$, for which the equation is dimensionally consistent, the equation is called homogeneous (not to be confused with linear equations with a zero RHS). If an equation is isobaric or homogeneous then the change in dependent variable $y = vx^m$ ($y = vx$ in the homogeneous case) followed by the change in independent variable $x = e^t$ leads to an equation in which the new independent variable t is absent except in the form d/dt.

▶*Solve*

$$x^3 \frac{d^2y}{dx^2} - (x^2 + xy)\frac{dy}{dx} + (y^2 + xy) = 0. \tag{15.84}$$

Assigning y and dy the weight m, and x and dx the weight 1, the weights of the five terms on the LHS of (15.84) are, from left to right: $m + 1$, $m + 1$, $2m$, $2m$, $m + 1$. For these weights all to be equal we require $m = 1$; thus (15.84) is a homogeneous equation. Since it is homogeneous we now make the substitution $y = vx$, which, after dividing the resulting equation through by x^3, gives

$$x\frac{d^2v}{dx^2} + (1 - v)\frac{dv}{dx} = 0. \tag{15.85}$$

Now substituting $x = e^t$ into (15.85) we obtain (after some working)

$$\frac{d^2v}{dt^2} - v\frac{dv}{dt} = 0, \tag{15.86}$$

which can be integrated directly to give

$$\frac{dv}{dt} = \tfrac{1}{2}v^2 + c_1. \tag{15.87}$$

Equation (15.87) is separable, and integrates to give

$$\tfrac{1}{2}t + d_2 = \int \frac{dv}{v^2 + d_1^2}$$

$$= \frac{1}{d_1}\tan^{-1}\left(\frac{v}{d_1}\right).$$

Rearranging and using $x = e^t$ and $y = vx$ we finally obtain the solution to (15.84) as

$$y = d_1 x \tan\left(\tfrac{1}{2}d_1 \ln x + d_1 d_2\right). \blacktriangleleft$$

Solution method. *Assume that y and dy have weight m, and x and dx weight 1, and write down the combined weights of each term in the ODE. If these weights can be made equal by assuming a particular value for m then the equation is isobaric (or homogeneous if $m = 1$). Making the substitution $y = vx^m$ followed by $x = e^t$ leads to an equation in which the new independent variable t is absent except in the form d/dt.*

15.3.5 *Equations homogeneous in* x *or* y *alone*

It will be seen that the intermediate equation (15.85) in the example of the previous subsection was simplified by the substitution $x = e^t$, in that this led to an equation in which the new independent variable t occurred only in the form d/dt; see (15.86). A closer examination of (15.85) reveals that it is dimensionally consistent in the independent variable x *taken alone*; this is equivalent to giving the dependent variable and its differential a weight $m = 0$. For any equation that is homogeneous in x alone, the substitution $x = e^t$ will lead to an equation that does not contain the new independent variable t except as d/dt. Note that the Euler equation of subsection 15.2.1 is a special, linear example of an equation homogeneous in x alone. Similarly, if an equation is homogeneous in y alone, then substituting $y = e^v$ leads to an equation in which the new dependent variable, v, occurs only in the form d/dv.

▶*Solve*

$$x^2 \frac{d^2 y}{dx^2} + x \frac{dy}{dx} + \frac{2}{y^3} = 0.$$

This equation is homogeneous in x alone, and on substituting $x = e^t$ we obtain

$$\frac{d^2 y}{dt^2} + \frac{2}{y^3} = 0,$$

which does not contain the new independent variable t except as d/dt. Such equations may often be solved by the method of subsection 15.3.2, but in this case we can integrate directly to obtain

$$\frac{dy}{dt} = \sqrt{2(c_1 + 1/y^2)}.$$

This equation is separable, and we find

$$\int \frac{dy}{\sqrt{2(c_1 + 1/y^2)}} = t + c_2.$$

By multiplying the numerator and denominator of the integrand on the LHS by y, we find the solution

$$\frac{\sqrt{c_1 y^2 + 1}}{\sqrt{2} c_1} = t + c_2.$$

Remembering that $t = \ln x$, we finally obtain

$$\frac{\sqrt{c_1 y^2 + 1}}{\sqrt{2} c_1} = \ln x + c_2. \ ◀$$

Solution method. *If the weight of x taken alone is the same in every term in the ODE then the substitution $x = e^t$ leads to an equation in which the new independent variable t is absent except in the form d/dt. If the weight of y taken alone is the same in every term then the substitution $y = e^v$ leads to an equation in which the new dependent variable v is absent except in the form d/dv.*

15.3.6 Equations having $y = Ae^x$ as a solution

Finally, we note that if any general (linear or non-linear) nth-order ODE is satisfied identically by assuming that

$$y = \frac{dy}{dx} = \cdots = \frac{d^n y}{dx^n} \tag{15.88}$$

then $y = Ae^x$ is a solution of that equation. This must be so because $y = Ae^x$ is a non-zero function that satisfies (15.88).

▶Find a solution of

$$(x^2 + x)\frac{dy}{dx}\frac{d^2 y}{dx^2} - x^2 y\frac{dy}{dx} - x\left(\frac{dy}{dx}\right)^2 = 0. \tag{15.89}$$

Setting $y = dy/dx = d^2 y/dx^2$ in (15.89), we obtain

$$(x^2 + x)y^2 - x^2 y^2 - xy^2 = 0,$$

which is satisfied identically. Therefore $y = Ae^x$ is a solution of (15.89); this is easily verified by directly substituting $y = Ae^x$ into (15.89). ◀

Solution method. *If the equation is satisfied identically by making the substitutions* $y = dy/dx = \cdots = d^n y/dx^n$ *then* $y = Ae^x$ *is a solution.*

15.4 Exercises

15.1 A simple harmonic oscillator, of mass m and natural frequency ω_0, experiences an oscillating driving force $f(t) = ma\cos\omega t$. Therefore, its equation of motion is

$$\frac{d^2 x}{dt^2} + \omega_0^2 x = a\cos\omega t,$$

where x is its position. Given that at $t = 0$ we have $x = dx/dt = 0$, find the function $x(t)$. Describe the solution if ω is approximately, but not exactly, equal to ω_0.

15.2 Find the roots of the auxiliary equation for the following. Hence solve them for the boundary conditions stated.

(a) $\dfrac{d^2 f}{dt^2} + 2\dfrac{df}{dt} + 5f = 0,$ with $f(0) = 1, f'(0) = 0.$

(b) $\dfrac{d^2 f}{dt^2} + 2\dfrac{df}{dt} + 5f = e^{-t}\cos 3t,$ with $f(0) = 0, f'(0) = 0.$

15.3 The theory of bent beams shows that at any point in the beam the 'bending moment' is given by K/ρ, where K is a constant (that depends upon the beam material and cross-sectional shape) and ρ is the radius of curvature at that point. Consider a light beam of length L whose ends, $x = 0$ and $x = L$, are supported at the same vertical height and which has a weight W suspended from its centre. Verify that at any point x ($0 \le x \le L/2$ for definiteness) the net magnitude of the bending moment (bending moment = force × perpendicular distance) due to the weight and support reactions, evaluated on either side of x, is $Wx/2$.

523

If the beam is only slightly bent, so that $(dy/dx)^2 \ll 1$, where $y = y(x)$ is the downward displacement of the beam at x, show that the beam profile satisfies the approximate equation

$$\frac{d^2y}{dx^2} = -\frac{Wx}{2K}.$$

By integrating this equation twice and using physically imposed conditions on your solution at $x = 0$ and $x = L/2$, show that the downward displacement at the centre of the beam is $WL^3/(48K)$.

15.4   Solve the differential equation

$$\frac{d^2f}{dt^2} + 6\frac{df}{dt} + 9f = e^{-t},$$

subject to the conditions $f = 0$ and $df/dt = \lambda$ at $t = 0$.

Find the equation satisfied by the positions of the turning points of $f(t)$ and hence, by drawing suitable sketch graphs, determine the number of turning points the solution has in the range $t > 0$ if (a) $\lambda = 1/4$, and (b) $\lambda = -1/4$.

15.5   The function $f(t)$ satisfies the differential equation

$$\frac{d^2f}{dt^2} + 8\frac{df}{dt} + 12f = 12e^{-4t}.$$

For the following sets of boundary conditions determine whether it has solutions, and, if so, find them:

(a) $f(0) = 0$, $f'(0) = 0$, $f(\ln\sqrt{2}) = 0$;
(b) $f(0) = 0$, $f'(0) = -2$, $f(\ln\sqrt{2}) = 0$.

15.6   Determine the values of α and β for which the following four functions are linearly dependent:

$$y_1(x) = x\cosh x + \sinh x,$$
$$y_2(x) = x\sinh x + \cosh x,$$
$$y_3(x) = (x + \alpha)e^x,$$
$$y_4(x) = (x + \beta)e^{-x}.$$

You will find it convenient to work with those linear combinations of the $y_i(x)$ that can be written the most compactly.

15.7   A solution of the differential equation

$$\frac{d^2y}{dx^2} + 2\frac{dy}{dx} + y = 4e^{-x}$$

takes the value 1 when $x = 0$ and the value e^{-1} when $x = 1$. What is its value when $x = 2$?

15.8   The two functions $x(t)$ and $y(t)$ satisfy the simultaneous equations

$$\frac{dx}{dt} - 2y = -\sin t,$$
$$\frac{dy}{dt} + 2x = 5\cos t.$$

Find explicit expressions for $x(t)$ and $y(t)$, given that $x(0) = 3$ and $y(0) = 2$. Sketch the solution trajectory in the xy-plane for $0 \le t < 2\pi$, showing that the trajectory crosses itself at $(0, 1/2)$ and passes through the points $(0, -3)$ and $(0, -1)$ in the negative x-direction.

15.9   Find the general solutions of

(a) $\dfrac{d^3 y}{dx^3} - 12\dfrac{dy}{dx} + 16y = 32x - 8,$

(b) $\dfrac{d}{dx}\left(\dfrac{1}{y}\dfrac{dy}{dx}\right) + (2a\coth 2ax)\left(\dfrac{1}{y}\dfrac{dy}{dx}\right) = 2a^2,$

where a is a constant.

15.10   Use the method of Laplace transforms to solve

(a) $\dfrac{d^2 f}{dt^2} + 5\dfrac{df}{dt} + 6f = 0,$ $f(0) = 1,\ f'(0) = -4,$

(b) $\dfrac{d^2 f}{dt^2} + 2\dfrac{df}{dt} + 5f = 0,$ $f(0) = 1,\ f'(0) = 0.$

15.11   The quantities $x(t),\ y(t)$ satisfy the simultaneous equations

$$\ddot{x} + 2n\dot{x} + n^2 x = 0,$$
$$\ddot{y} + 2n\dot{y} + n^2 y = \mu\dot{x},$$

where $x(0) = y(0) = \dot{y}(0) = 0$ and $\dot{x}(0) = \lambda$. Show that

$$y(t) = \tfrac{1}{2}\mu\lambda t^2 \left(1 - \tfrac{1}{3}nt\right)\exp(-nt).$$

15.12   Use Laplace transforms to solve, for $t \geq 0$, the differential equations

$$\ddot{x} + 2x + y = \cos t,$$
$$\ddot{y} + 2x + 3y = 2\cos t,$$

which describe a coupled system that starts from rest at the equilibrium position. Show that the subsequent motion takes place along a straight line in the xy-plane. Verify that the frequency at which the system is driven is equal to one of the resonance frequencies of the system; explain why there is *no* resonant behaviour in the solution you have obtained.

15.13   Two unstable isotopes A and B and a stable isotope C have the following decay rates per atom present: $A \to B$, $3\,\mathrm{s}^{-1}$; $A \to C$, $1\,\mathrm{s}^{-1}$; $B \to C$, $2\,\mathrm{s}^{-1}$. Initially a quantity v_0 of A is present, but there are no atoms of the other two types. Using Laplace transforms, find the amount of C present at a later time t.

15.14   For a lightly damped ($\gamma < \omega_0$) harmonic oscillator driven at its undamped resonance frequency ω_0, the displacement $x(t)$ at time t satisfies the equation

$$\frac{d^2 x}{dt^2} + 2\gamma\frac{dx}{dt} + \omega_0^2 x = F\sin\omega_0 t.$$

Use Laplace transforms to find the displacement at a general time if the oscillator starts from rest at its equilibrium position.

(a) Show that ultimately the oscillation has amplitude $F/(2\omega_0\gamma)$, with a phase lag of $\pi/2$ relative to the driving force per unit mass F.

(b) By differentiating the original equation, conclude that if $x(t)$ is expanded as a power series in t for small t, then the first non-vanishing term is $F\omega_0 t^3/6$. Confirm this conclusion by expanding your explicit solution.

15.15   The 'golden mean', which is said to describe the most aesthetically pleasing proportions for the sides of a rectangle (e.g. the ideal picture frame), is given by the limiting value of the ratio of successive terms of the Fibonacci series u_n, which is generated by

$$u_{n+2} = u_{n+1} + u_n,$$

with $u_0 = 0$ and $u_1 = 1$. Find an expression for the general term of the series and verify that the golden mean is equal to the larger root of the recurrence relation's characteristic equation.

15.16 In a particular scheme for numerically modelling one-dimensional fluid flow, the successive values, u_n, of the solution are connected for $n \geq 1$ by the difference equation

$$c(u_{n+1} - u_{n-1}) = d(u_{n+1} - 2u_n + u_{n-1}),$$

where c and d are positive constants. The boundary conditions are $u_0 = 0$ and $u_M = 1$. Find the solution to the equation, and show that successive values of u_n will have alternating signs if $c > d$.

15.17 The first few terms of a series u_n, starting with u_0, are $1, 2, 2, 1, 6, -3$. The series is generated by a recurrence relation of the form

$$u_n = Pu_{n-2} + Qu_{n-4},$$

where P and Q are constants. Find an expression for the general term of the series and show that, in fact, the series consists of two interleaved series given by

$$u_{2m} = \tfrac{2}{3} + \tfrac{1}{3}4^m,$$

$$u_{2m+1} = \tfrac{7}{3} - \tfrac{1}{3}4^m,$$

for $m = 0, 1, 2, \ldots$.

15.18 Find an explicit expression for the u_n satisfying

$$u_{n+1} + 5u_n + 6u_{n-1} = 2^n,$$

given that $u_0 = u_1 = 1$. Deduce that $2^n - 26(-3)^n$ is divisible by 5 for all non-negative integers n.

15.19 Find the general expression for the u_n satisfying

$$u_{n+1} = 2u_{n-2} - u_n$$

with $u_0 = u_1 = 0$ and $u_2 = 1$, and show that they can be written in the form

$$u_n = \frac{1}{5} - \frac{2^{n/2}}{\sqrt{5}} \cos\left(\frac{3\pi n}{4} - \phi\right),$$

where $\tan \phi = 2$.

15.20 Consider the seventh-order recurrence relation

$$u_{n+7} - u_{n+6} - u_{n+5} + u_{n+4} - u_{n+3} + u_{n+2} + u_{n+1} - u_n = 0.$$

Find the most general form of its solution, and show that:

(a) if only the four initial values $u_0 = 0$, $u_1 = 2$, $u_2 = 6$ and $u_3 = 12$, are specified, then the relation has one solution that cycles repeatedly through this set of four numbers;

(b) but if, in addition, it is required that $u_4 = 20$, $u_5 = 30$ and $u_6 = 42$ then the solution is unique, with $u_n = n(n+1)$.

15.21 Find the general solution of

$$x^2 \frac{d^2 y}{dx^2} - x \frac{dy}{dx} + y = x,$$

given that $y(1) = 1$ and $y(e) = 2e$.

15.22 Find the general solution of

$$(x+1)^2 \frac{d^2 y}{dx^2} + 3(x+1) \frac{dy}{dx} + y = x^2.$$

526

15.23 Prove that the general solution of

$$(x-2)\frac{d^2y}{dx^2} + 3\frac{dy}{dx} + \frac{4y}{x^2} = 0$$

is given by

$$y(x) = \frac{1}{(x-2)^2}\left[k\left(\frac{2}{3x} - \frac{1}{2} \right) + cx^2 \right].$$

15.24 Use the method of variation of parameters to find the general solutions of

(a) $\dfrac{d^2y}{dx^2} - y = x^n$, (b) $\dfrac{d^2y}{dx^2} - 2\dfrac{dy}{dx} + y = 2xe^x$.

15.25 Use the intermediate result of exercise 15.24(a) to find the Green's function that satisfies

$$\frac{d^2 G(x,\xi)}{dx^2} - G(x,\xi) = \delta(x-\xi) \qquad \text{with} \qquad G(0,\xi) = G(1,\xi) = 0.$$

15.26 Consider the equation

$$F(x,y) = x(x+1)\frac{d^2y}{dx^2} + (2-x^2)\frac{dy}{dx} - (2+x)y = 0.$$

(a) Given that $y_1(x) = 1/x$ is one of its solutions, find a second linearly independent one,
      (i) by setting $y_2(x) = y_1(x)u(x)$, and
      (ii) by noting the sum of the coefficients in the equation.
(b) Hence, using the variation of parameters method, find the general solution of

$$F(x,y) = (x+1)^2.$$

15.27 Show generally that if $y_1(x)$ and $y_2(x)$ are linearly independent solutions of

$$\frac{d^2y}{dx^2} + p(x)\frac{dy}{dx} + q(x)y = 0,$$

with $y_1(0) = 0$ and $y_2(1) = 0$, then the Green's function $G(x,\xi)$ for the interval $0 \le x, \xi \le 1$ and with $G(0,\xi) = G(1,\xi) = 0$ can be written in the form

$$G(x,\xi) = \begin{cases} y_1(x)y_2(\xi)/W(\xi) & 0 < x < \xi, \\ y_2(x)y_1(\xi)/W(\xi) & \xi < x < 1, \end{cases}$$

where $W(x) = W[y_1(x), y_2(x)]$ is the Wronskian of $y_1(x)$ and $y_2(x)$.

15.28 Use the result of the previous exercise to find the Green's function $G(x,\xi)$ that satisfies

$$\frac{d^2 G}{dx^2} + 3\frac{dG}{dx} + 2G = \delta(x-x),$$

in the interval $0 \le x, \xi \le 1$, with $G(0,\xi) = G(1,\xi) = 0$. Hence obtain integral expressions for the solution of

$$\frac{d^2y}{dx^2} + 3\frac{dy}{dx} + 2y = \begin{cases} 0 & 0 < x < x_0, \\ 1 & x_0 < x < 1, \end{cases}$$

distinguishing between the cases (a) $x < x_0$, and (b) $x > x_0$.

15.29   The equation of motion for a driven damped harmonic oscillator can be written

$$\ddot{x} + 2\dot{x} + (1 + \kappa^2)x = f(t),$$

with $\kappa \neq 0$. If it starts from rest with $x(0) = 0$ and $\dot{x}(0) = 0$, find the corresponding Green's function $G(t, \tau)$ and verify that it can be written as a function of $t - \tau$ only. Find the explicit solution when the driving force is the unit step function, i.e. $f(t) = H(t)$. Confirm your solution by taking the Laplace transforms of both it and the original equation.

15.30   Show that the Green's function for the equation

$$\frac{d^2y}{dx^2} + \frac{y}{4} = f(x),$$

subject to the boundary conditions $y(0) = y(\pi) = 0$, is given by

$$G(x, z) = \begin{cases} -2\cos\frac{1}{2}x \sin\frac{1}{2}z & 0 \le z \le x, \\ -2\sin\frac{1}{2}x \cos\frac{1}{2}z & x \le z \le \pi. \end{cases}$$

15.31   Find the Green's function $x = G(t, t_0)$ that solves

$$\frac{d^2x}{dt^2} + \alpha\frac{dx}{dt} = \delta(t - t_0)$$

under the initial conditions $x = dx/dt = 0$ at $t = 0$. Hence solve

$$\frac{d^2x}{dt^2} + \alpha\frac{dx}{dt} = f(t),$$

where $f(t) = 0$ for $t < 0$.
Evaluate your answer explicitly for $f(t) = Ae^{-at}$ $(t > 0)$.

15.32   Consider the equation

$$\frac{d^2y}{dx^2} + f(y) = 0,$$

where $f(y)$ can be any function.

(a) By multiplying through by dy/dx, obtain the general solution relating x and y.

(b) A mass m, initially at rest at the point $x = 0$, is accelerated by a force

$$f(x) = A(x_0 - x)\left[1 + 2\ln\left(1 - \frac{x}{x_0}\right)\right].$$

Its equation of motion is $m\,d^2x/dt^2 = f(x)$. Find x as a function of time, and show that ultimately the particle has travelled a distance x_0.

15.33   Solve

$$2y\frac{d^3y}{dx^3} + 2\left(y + 3\frac{dy}{dx}\right)\frac{d^2y}{dx^2} + 2\left(\frac{dy}{dx}\right)^2 = \sin x.$$

15.34   Find the general solution of the equation

$$x\frac{d^3y}{dx^3} + 2\frac{d^2y}{dx^2} = Ax.$$

15.35   Express the equation

$$\frac{d^2y}{dx^2} + 4x\frac{dy}{dx} + (4x^2 + 6)y = e^{-x^2}\sin 2x$$

in canonical form and hence find its general solution.

15.36   Find the form of the solutions of the equation

$$\frac{dy}{dx}\frac{d^3y}{dx^3} - 2\left(\frac{d^2y}{dx^2}\right)^2 + \left(\frac{dy}{dx}\right)^2 = 0$$

that have $y(0) = \infty$.
[You will need the result $\int^z \operatorname{cosech} u\, du = -\ln(\operatorname{cosech} z + \coth z)$.]

15.37   Consider the equation

$$x^p y'' + \frac{n+3-2p}{n-1} x^{p-1} y' + \left(\frac{p-2}{n-1}\right)^2 x^{p-2} y = y^n,$$

in which $p \neq 2$ and $n > -1$ but $n \neq 1$. For the boundary conditions $y(1) = 0$ and $y'(1) = \lambda$, show that the solution is $y(x) = v(x)x^{(p-2)/(n-1)}$, where $v(x)$ is given by

$$\int_0^{v(x)} \frac{dz}{\left[\lambda^2 + 2z^{n+1}/(n+1)\right]^{1/2}} = \ln x.$$

15.5 Hints and answers

15.1   The function is $a(\omega_0^2 - \omega^2)^{-1}(\cos\omega t - \cos\omega_0 t)$; for moderate t, $x(t)$ is a sine wave of linearly increasing amplitude $(t\sin\omega_0 t)/(2\omega_0)$; for large t it shows beats of maximum amplitude $2(\omega_0^2 - \omega^2)^{-1}$.

15.3   Ignore the term y'^2, compared with 1, in the expression for ρ. $y = 0$ at $x = 0$. From symmetry, $dy/dx = 0$ at $x - L/2$.

15.5   General solution $f(t) = Ae^{-6t} + Be^{-2t} - 3e^{-4t}$. (a) No solution, inconsistent boundary conditions; (b) $f(t) = 2e^{-6t} + e^{-2t} - 3e^{-4t}$.

15.7   The auxiliary equation has repeated roots and the RHS is contained in the complementary function. The solution is $y(x) = (A+Bx)e^{-x} + 2x^2 e^{-x}$. $y(2) = 5e^{-2}$.

15.9   (a) The auxiliary equation has roots 2, 2, -4; $(A+Bx)\exp 2x + C\exp(-4x) + 2x + 1$; (b) multiply through by $\sinh 2ax$ and note that $\int \operatorname{cosech} 2ax\, dx = (2a)^{-1}\ln(|\tanh ax|)$; $y = B(\sinh 2ax)^{1/2}(|\tanh ax|)^4$.

15.11   Use Laplace transforms; write $s(s+n)^{-4}$ as $(s+n)^{-3} - n(s+n)^{-4}$.

15.13 $\mathcal{L}|C(t)| = x_0(s+8)/[s(s+2)(s+4)]$, yielding $C(t) = x_0[1 + \frac{1}{2}\exp(-4t) - \frac{3}{2}\exp(-2t)]$.

15.15   The characteristic equation is $\lambda^2 - \lambda - 1 = 0$.
$u_n = [(1+\sqrt{5})^n - (1-\sqrt{5})^n]/(2^n\sqrt{5})$.

15.17   From u_4 and u_5, $P = 5, Q = -4$. $u_n = 3/2 - 5(-1)^n/6 + (-2)^n/4 + 2^n/12$.

15.19   The general solution is $A + B2^{n/2}\exp(i3\pi n/4) + C2^{n/2}\exp(i5\pi n/4)$. The initial values imply that $A = 1/5, B = (\sqrt{5}/10)\exp[i(n-\phi)]$ and $C = (\sqrt{5}/10)\exp[i(\pi+\phi)]$.

15.21   This is Euler's equation; setting $x = \exp t$ produces $d^2z/dt^2 - 2\,dz/dt + z = \exp t$, with complementary function $(A+Bt)\exp t$ and particular integral $t^2(\exp t)/2$; $y(x) = x + [x\ln x(1 + \ln x)]/2$.

15.23   After multiplication through by x^2 the coefficients are such that this is an exact equation. The resulting first-order equation, in standard form, needs an integrating factor $(x-2)^2/x^2$.

15.25   Given the boundary conditions, it is better to work with $\sinh x$ and $\sinh(1-x)$ than with $e^{\pm x}$; $G(x,\xi) = -[\sinh(1-\xi)\sinh x]/\sinh 1$ for $x < \xi$ and $-[\sinh(1-x)\sinh\xi]/\sinh 1$ for $x > \xi$.

15.27   Follow the method of subsection 15.2.5, but using general rather than specific functions.

15.29 $G(t,\tau) = 0$ for $t < \tau$ and $\kappa^{-1}e^{-(t-\tau)}\sin[\kappa(t-\tau)]$ for $t > \tau$. For a unit step input, $x(t) = (1+\kappa^2)^{-1}(1 - e^{-t}\cos\kappa t - \kappa^{-1}e^{-t}\sin\kappa t)$. Both transforms are equivalent to $s[(s+1)^2 + \kappa^2]\bar{x} = 1$.

15.31   Use continuity and the step condition on $\partial G/\partial t$ at $t = t_0$ to show that
$G(t, t_0) = \alpha^{-1}\{1 - \exp[\alpha(t_0 - t)]\}$ for $0 \le t_0 \le t$;
$x(t) = A(\alpha - a)^{-1}\{a^{-1}[1 - \exp(-at)] - \alpha^{-1}[1 - \exp(-\alpha t)]\}$.

15.33   The LHS of the equation is exact for two stages of integration and then needs
an integrating factor $\exp x$; $2y\, d^2y/dx^2 + 2y\, dy/dx + 2(dy/dx)^2$; $2y\, dy/dx + y^2 = d(y^2)/dx + y^2$; $y^2 = A\exp(-x) + Bx + C - (\sin x - \cos x)/2$.

15.35   Follow the method of subsection 15.2.6; $u(x) = e^{-x^2}$ and $v(x)$ satisfies $v'' + 4v = \sin 2x$, for which a particular integral is $(-x\cos 2x)/4$. The general solution is
$y(x) = [A\sin 2x + (B - \frac{1}{4}x)\cos 2x]e^{-x^2}$.

15.37   The equation is isobaric, with y of weight m, where $m + p - 2 = mn$; $v(x)$
satisfies $x^2v'' + xv' = v^n$. Set $x = e^t$ and $v(x) = u(t)$, leading to $u'' = u^n$ with
$u(0) = 0, u'(0) = \lambda$. Multiply both sides by u' to make the equation exact.

16

Series solutions of ordinary differential equations

In the previous chapter the solution of both homogeneous and non-homogeneous linear ordinary differential equations (ODEs) of order ≥ 2 was discussed. In particular we developed methods for solving some equations in which the coefficients were not constant but functions of the independent variable x. In each case we were able to write the solutions to such equations in terms of elementary functions, or as integrals. In general, however, the solutions of equations with variable coefficients cannot be written in this way, and we must consider alternative approaches.

In this chapter we discuss a method for obtaining solutions to linear ODEs in the form of convergent series. Such series can be evaluated numerically, and those occurring most commonly are named and tabulated. There is in fact no distinct borderline between this and the previous chapter, since solutions in terms of elementary functions may equally well be written as convergent series (i.e. the relevant Taylor series). Indeed, it is partly because some series occur so frequently that they are given special names such as $\sin x$, $\cos x$ or $\exp x$.

Since we shall be concerned principally with second-order linear ODEs in this chapter, we begin with a discussion of these equations, and obtain some general results that will prove useful when we come to discuss series solutions.

16.1 Second-order linear ordinary differential equations

Any homogeneous second-order linear ODE can be written in the form

$$y'' + p(x)y' + q(x)y = 0, \tag{16.1}$$

where $y' = dy/dx$ and $p(x)$ and $q(x)$ are given functions of x. From the previous chapter, we recall that the most general form of the solution to (16.1) is

$$y(x) = c_1 y_1(x) + c_2 y_2(x), \tag{16.2}$$

where $y_1(x)$ and $y_2(x)$ are *linearly independent* solutions of (16.1), and c_1 and c_2 are constants that are fixed by the boundary conditions (if supplied).

A full discussion of the linear independence of sets of functions was given at the beginning of the previous chapter, but for just two functions y_1 and y_2 to be linearly independent we simply require that y_2 is not a multiple of y_1. Equivalently, y_1 and y_2 must be such that the equation

$$c_1 y_1(x) + c_2 y_2(x) = 0$$

is *only* satisfied for $c_1 = c_2 = 0$. Therefore the linear independence of $y_1(x)$ and $y_2(x)$ can usually be deduced by inspection but in any case can always be verified by the evaluation of the Wronskian of the two solutions,

$$W(x) = \begin{vmatrix} y_1 & y_2 \\ y_1' & y_2' \end{vmatrix} = y_1 y_2' - y_2 y_1'. \tag{16.3}$$

If $W(x) \neq 0$ anywhere in a given interval then y_1 and y_2 are linearly independent in that interval.

An alternative expression for $W(x)$, of which we will make use later, may be derived by differentiating (16.3) with respect to x to give

$$W' = y_1 y_2'' + y_1' y_2' - y_2 y_1'' - y_2' y_1' = y_1 y_2'' - y_1'' y_2.$$

Since both y_1 and y_2 satisfy (16.1), we may substitute for y_1'' and y_2'' to obtain

$$W' = -y_1(py_2' + qy_2) + (py_1' + qy_1)y_2 = -p(y_1 y_2' - y_1' y_2) = -pW.$$

Integrating, we find

$$W(x) = C \exp\left\{ -\int^x p(u)\, du \right\}, \tag{16.4}$$

where C is a constant. We note further that in the special case $p(x) \equiv 0$ we obtain $W = \text{constant}$.

> ► *The functions $y_1 = \sin x$ and $y_2 = \cos x$ are both solutions of the equation $y'' + y = 0$. Evaluate the Wronskian of these two solutions, and hence show that they are linearly independent.*

The Wronskian of y_1 and y_2 is given by

$$W = y_1 y_2' - y_2 y_1' = -\sin^2 x - \cos^2 x = -1.$$

Since $W \neq 0$ the two solutions are linearly independent. We also note that $y'' + y = 0$ is a special case of (16.1) with $p(x) = 0$. We therefore expect, from (16.4), that W will be a constant, as is indeed the case. ◄

From the previous chapter we recall that, once we have obtained the general solution to the homogeneous second-order ODE (16.1) in the form (16.2), the general solution to the *inhomogeneous* equation

$$y'' + p(x)y' + q(x)y = f(x) \tag{16.5}$$

can be written as the sum of the solution to the homogeneous equation $y_c(x)$ (the complementary function) and *any* function $y_p(x)$ (the particular integral) that satisfies (16.5) and is linearly independent of $y_c(x)$. We have therefore

$$y(x) = c_1 y_1(x) + c_2 y_2(x) + y_p(x). \tag{16.6}$$

General methods for obtaining y_p, that are applicable to equations with variable coefficients, such as the variation of parameters or Green's functions, were discussed in the previous chapter. An alternative description of the Green's function method for solving inhomogeneous equations is given in the next chapter. For the present, however, we will restrict our attention to the solutions of homogeneous ODEs in the form of convergent series.

16.1.1 Ordinary and singular points of an ODE

So far we have implicitly assumed that $y(x)$ is a *real* function of a *real* variable x. However, this is not always the case, and in the remainder of this chapter we broaden our discussion by generalising to a *complex* function $y(z)$ of a *complex* variable z.

Let us therefore consider the second-order linear homogeneous ODE

$$y'' + p(z)y' + q(z) = 0, \tag{16.7}$$

where now $y' = dy/dz$; this is a straightforward generalisation of (16.1). A full discussion of complex functions and differentiation with respect to a complex variable z is given in chapter 24, but for the purposes of the present chapter we need not concern ourselves with many of the subtleties that exist. In particular, we may treat differentiation with respect to z in a way analogous to ordinary differentiation with respect to a real variable x.

In (16.7), if, at some point $z = z_0$, the functions $p(z)$ and $q(z)$ are finite and can be expressed as complex power series (see section 4.5), i.e.

$$p(z) = \sum_{n=0}^{\infty} p_n(z - z_0)^n, \qquad q(z) = \sum_{n=0}^{\infty} q_n(z - z_0)^n,$$

then $p(z)$ and $q(z)$ are said to be *analytic* at $z = z_0$, and this point is called an *ordinary point* of the ODE. If, however, $p(z)$ or $q(z)$, or both, diverge at $z = z_0$ then it is called a *singular point* of the ODE.

Even if an ODE is singular at a given point $z = z_0$, it may still possess a non-singular (finite) solution at that point. In fact the necessary and sufficient condition[§] for such a solution to exist is that $(z - z_0)p(z)$ and $(z - z_0)^2 q(z)$ are both analytic at $z = z_0$. Singular points that have this property are called *regular*

[§] See, for example, H. Jeffreys and B. S. Jeffreys, *Methods of Mathematical Physics*, 3rd edn (Cambridge: Cambridge University Press, 1966), p. 479.

singular points, whereas any singular point not satisfying both these criteria is termed an *irregular* or *essential* singularity.

▶*Legendre's equation has the form*

$$(1 - z^2)y'' - 2zy' + \ell(\ell + 1)y = 0, \tag{16.8}$$

where ℓ is a constant. Show that $z = 0$ is an ordinary point and $z = \pm 1$ are regular singular points of this equation.

Firstly, divide through by $1 - z^2$ to put the equation into our standard form (16.7):

$$y'' - \frac{2z}{1 - z^2}y' + \frac{\ell(\ell + 1)}{1 - z^2}y = 0.$$

Comparing this with (16.7), we identify $p(z)$ and $q(z)$ as

$$p(z) = \frac{-2z}{1 - z^2} = \frac{-2z}{(1 + z)(1 - z)}, \qquad q(z) = \frac{\ell(\ell + 1)}{1 - z^2} = \frac{\ell(\ell + 1)}{(1 + z)(1 - z)}.$$

By inspection, $p(z)$ and $q(z)$ are analytic at $z = 0$, which is therefore an ordinary point, but both diverge for $z = \pm 1$, which are thus singular points. However, at $z = 1$ we see that both $(z - 1)p(z)$ and $(z - 1)^2 q(z)$ are analytic and hence $z = 1$ is a regular singular point. Similarly, at $z = -1$ both $(z + 1)p(z)$ and $(z + 1)^2 q(z)$ are analytic, and it too is a regular singular point. ◀

So far we have assumed that z_0 is finite. However, we may sometimes wish to determine the nature of the point $|z| \to \infty$. This may be achieved straightforwardly by substituting $w = 1/z$ into the equation and investigating the behaviour at $w = 0$.

▶*Show that Legendre's equation has a regular singularity at $|z| \to \infty$.*

Letting $w = 1/z$, the derivatives with respect to z become

$$\frac{dy}{dz} = \frac{dy}{dw}\frac{dw}{dz} = -\frac{1}{z^2}\frac{dy}{dw} = -w^2\frac{dy}{dw},$$

$$\frac{d^2y}{dz^2} = \frac{dw}{dz}\frac{d}{dw}\left(\frac{dy}{dz}\right) = -w^2\left(-2w\frac{dy}{dw} - w^2\frac{d^2y}{dw^2}\right) = w^3\left(2\frac{dy}{dw} + w\frac{d^2y}{dw^2}\right).$$

If we substitute these derivatives into Legendre's equation (16.8) we obtain

$$\left(1 - \frac{1}{w^2}\right)w^3\left(2\frac{dy}{dw} + w\frac{d^2y}{dw^2}\right) + 2\frac{1}{w}w^2\frac{dy}{dw} + \ell(\ell + 1)y = 0,$$

which simplifies to give

$$w^2(w^2 - 1)\frac{d^2y}{dw^2} + 2w^3\frac{dy}{dw} + \ell(\ell + 1)y = 0.$$

Dividing through by $w^2(w^2 - 1)$ to put the equation into standard form, and comparing with (16.7), we identify $p(w)$ and $q(w)$ as

$$p(w) = \frac{2w}{w^2 - 1}, \qquad q(w) = \frac{\ell(\ell + 1)}{w^2(w^2 - 1)}.$$

At $w = 0$, $p(w)$ is analytic but $q(w)$ diverges, and so the point $|z| \to \infty$ is a singular point of Legendre's equation. However, since wp and w^2q are both analytic at $w = 0$, $|z| \to \infty$ is a regular singular point. ◀

Equation	Regular singularities	Essential singularities
Hypergeometric $z(1-z)y'' + [c - (a+b+1)z]y' - aby = 0$	$0, 1, \infty$	—
Legendre $(1-z^2)y'' - 2zy' + \ell(\ell+1)y = 0$	$-1, 1, \infty$	—
Associated Legendre $(1-z^2)y'' - 2zy' + \left[\ell(\ell+1) - \dfrac{m^2}{1-z^2}\right]y = 0$	$-1, 1, \infty$	—
Chebyshev $(1-z^2)y'' - zy' + v^2 y = 0$	$-1, 1, \infty$	—
Confluent hypergeometric $zy'' + (c-z)y' - ay = 0$	0	∞
Bessel $z^2 y'' + zy' + (z^2 - v^2)y = 0$	0	∞
Laguerre $zy'' + (1-z)y' + vy = 0$	0	∞
Associated Laguerre $zy'' + (m+1-z)y' + (v-m)y = 0$	0	∞
Hermite $y'' - 2zy' + 2vy = 0$	—	∞
Simple harmonic oscillator $y'' + \omega^2 y = 0$	—	∞

Table 16.1   Important second-order linear ODEs in the physical sciences and engineering.

Table 16.1 lists the singular points of several second-order linear ODEs that play important roles in the analysis of many problems in physics and engineering. A full discussion of the solutions to each of the equations in table 16.1 and their properties is left until chapter 18. We now discuss the general methods by which series solutions may be obtained.

16.2 Series solutions about an ordinary point

If $z = z_0$ is an ordinary point of (16.7) then it may be shown that *every* solution $y(z)$ of the equation is also analytic at $z = z_0$. From now on we will take z_0 as the origin, i.e. $z_0 = 0$. If this is not already the case, then a substitution $Z = z - z_0$ will make it so. Since every solution is analytic, $y(z)$ can be represented by a

535

power series of the form (see section 24.11)

$$y(z) = \sum_{n=0}^{\infty} a_n z^n. \tag{16.9}$$

Moreover, it may be shown that such a power series converges for $|z| < R$, where R is the radius of convergence and is equal to the distance from $z = 0$ to the nearest singular point of the ODE (see chapter 24). At the radius of convergence, however, the series may or may not converge (as shown in section 4.5).

Since every solution of (16.7) is analytic at an ordinary point, it is always possible to obtain two *independent* solutions (from which the general solution (16.2) can be constructed) of the form (16.9). The derivatives of y with respect to z are given by

$$y' = \sum_{n=0}^{\infty} n a_n z^{n-1} = \sum_{n=0}^{\infty} (n+1) a_{n+1} z^n, \tag{16.10}$$

$$y'' = \sum_{n=0}^{\infty} n(n-1) a_n z^{n-2} = \sum_{n=0}^{\infty} (n+2)(n+1) a_{n+2} z^n. \tag{16.11}$$

Note that, in each case, in the first equality the sum can still start at $n = 0$ since the first term in (16.10) and the first two terms in (16.11) are automatically zero. The second equality in each case is obtained by shifting the summation index so that the sum can be written in terms of coefficients of z^n. By substituting (16.9)–(16.11) into the ODE (16.7), and requiring that the coefficients of each power of z sum to zero, we obtain a *recurrence relation* expressing each a_n in terms of the previous a_r $(0 \leq r \leq n-1)$.

▶Find the series solutions, about $z = 0$, of
$$y'' + y = 0.$$

By inspection, $z = 0$ is an ordinary point of the equation, and so we may obtain two independent solutions by making the substitution $y = \sum_{n=0}^{\infty} a_n z^n$. Using (16.9) and (16.11) we find

$$\sum_{n=0}^{\infty} (n+2)(n+1) a_{n+2} z^n + \sum_{n=0}^{\infty} a_n z^n = 0,$$

which may be written as

$$\sum_{n=0}^{\infty} [(n+2)(n+1) a_{n+2} + a_n] z^n = 0.$$

For this equation to be satisfied we require that the coefficient of each power of z vanishes *separately*, and so we obtain the two-term recurrence relation

$$a_{n+2} = -\frac{a_n}{(n+2)(n+1)} \qquad \text{for } n \geq 0.$$

Using this relation, we can calculate, say, the even coefficients a_2, a_4, a_6 and so on, for a given a_0. Alternatively, starting with a_1, we obtain the odd coefficients a_3, a_5, etc. Two independent solutions of the ODE can be obtained by setting either $a_0 = 0$ or $a_1 = 0$. Firstly, if we set $a_1 = 0$ and choose $a_0 = 1$ then we obtain the solution

$$y_1(z) = 1 - \frac{z^2}{2!} + \frac{z^4}{4!} - \cdots = \sum_{n=0}^{\infty} \frac{(-1)^n}{(2n)!} z^{2n}.$$

Secondly, if we set $a_0 = 0$ and choose $a_1 = 1$ then we obtain a second, *independent*, solution

$$y_2(z) = z - \frac{z^3}{3!} + \frac{z^5}{5!} - \cdots = \sum_{n=0}^{\infty} \frac{(-1)^n}{(2n+1)!} z^{2n+1}.$$

Recognising these two series as $\cos z$ and $\sin z$, we can write the general solution as

$$y(z) = c_1 \cos z + c_2 \sin z,$$

where c_1 and c_2 are arbitrary constants that are fixed by boundary conditions (if supplied). We note that both solutions converge for all z, as might be expected since the ODE possesses no singular points (except $|z| \to \infty$). ◀

Solving the above example was quite straightforward and the resulting series were easily recognised and written in *closed form* (i.e. in terms of elementary functions); *this is not usually the case*. Another simplifying feature of the previous example was that we obtained a two-term recurrence relation relating a_{n+2} and a_n, so that the odd- and even-numbered coefficients were independent of one another. In general, the recurrence relation expresses a_n in terms of any number of the previous a_r $(0 \leq r \leq n - 1)$.

▶*Find the series solutions, about $z = 0$, of*

$$y'' - \frac{2}{(1 - z)^2} y = 0.$$

By inspection, $z = 0$ is an ordinary point, and therefore we may find two independent solutions by substituting $y = \sum_{n=0}^{\infty} a_n z^n$. Using (16.10) and (16.11), and multiplying through by $(1 - z)^2$, we find

$$(1 - 2z + z^2) \sum_{n=0}^{\infty} n(n-1)a_n z^{n-2} - 2\sum_{n=0}^{\infty} a_n z^n = 0,$$

which leads to

$$\sum_{n=0}^{\infty} n(n-1)a_n z^{n-2} - 2\sum_{n=0}^{\infty} n(n-1)a_n z^{n-1} + \sum_{n=0}^{\infty} n(n-1)a_n z^n - 2\sum_{n=0}^{\infty} a_n z^n = 0.$$

In order to write all these series in terms of the coefficients of z^n, we must shift the summation index in the first two sums, obtaining

$$\sum_{n=0}^{\infty}(n+2)(n+1)a_{n+2} z^n - 2\sum_{n=0}^{\infty}(n+1)n a_{n+1} z^n + \sum_{n=0}^{\infty}(n^2 - n - 2)a_n z^n = 0,$$

which can be written as

$$\sum_{n=0}^{\infty}(n+1)[(n+2)a_{n+2} - 2n a_{n+1} + (n-2)a_n]z^n = 0.$$

By demanding that the coefficients of each power of z vanish separately, we obtain the three-term recurrence relation

$$(n+2)a_{n+2} - 2na_{n+1} + (n-2)a_n = 0 \qquad \text{for } n \geq 0,$$

which determines a_n for $n \geq 2$ in terms of a_0 and a_1. Three-term (or more) recurrence relations are a nuisance and, in general, can be difficult to solve. This particular recurrence relation, however, has two straightforward solutions. One solution is $a_n = a_0$ for all n, in which case (choosing $a_0 = 1$) we find

$$y_1(z) = 1 + z + z^2 + z^3 + \cdots = \frac{1}{1-z}.$$

The other solution to the recurrence relation is $a_1 = -2a_0$, $a_2 = a_0$ and $a_n = 0$ for $n > 2$, so that (again choosing $a_0 = 1$) we obtain a *polynomial* solution to the ODE:

$$y_2(z) = 1 - 2z + z^2 = (1-z)^2.$$

The linear independence of y_1 and y_2 is obvious but can be checked by computing the Wronskian

$$W = y_1 y_2' - y_1' y_2 = \frac{1}{1-z}[-2(1-z)] - \frac{1}{(1-z)^2}(1-z)^2 = -3.$$

Since $W \neq 0$, the two solutions y_1 and y_2 are indeed linearly independent. The general solution of the ODE is therefore

$$y(z) = \frac{c_1}{1-z} + c_2(1-z)^2.$$

We observe that y_1 (and hence the general solution) is singular at $z = 1$, which is the singular point of the ODE nearest to $z = 0$, but the polynomial solution, y_2, is valid for all finite z. ◄

The above example illustrates the possibility that, in some cases, we may find that the recurrence relation leads to $a_n = 0$ for $n > N$, for one or both of the two solutions; we then obtain a *polynomial* solution to the equation. Polynomial solutions are discussed more fully in section 16.5, but one obvious property of such solutions is that they converge for all finite z. By contrast, as mentioned above, for solutions in the form of an infinite series the circle of convergence extends only as far as the singular point nearest to that about which the solution is being obtained.

16.3 Series solutions about a regular singular point

From table 16.1 we see that several of the most important second-order linear ODEs in physics and engineering have regular singular points in the finite complex plane. We must extend our discussion, therefore, to obtaining series solutions to ODEs about such points. In what follows we assume that the regular singular point about which the solution is required is at $z = 0$, since, as we have seen, if this is not already the case then a substitution of the form $Z = z - z_0$ will make it so.

If $z = 0$ is a regular singular point of the equation

$$y'' + p(z)y' + q(z)y = 0$$

then at least one of $p(z)$ and $q(z)$ is not analytic at $z = 0$, and in general we should not expect to find a power series solution of the form (16.9). We must therefore extend the method to include a more general form for the solution. In fact, it may be shown (Fuch's theorem) that there exists *at least one* solution to the above equation, of the form

$$y = z^\sigma \sum_{n=0}^\infty a_n z^n, \tag{16.12}$$

where the exponent σ is a number that may be real or complex and where $a_0 \neq 0$ (since, if it were otherwise, σ could be redefined as $\sigma + 1$ or $\sigma + 2$ or $\cdots$ so as to make $a_0 \neq 0$). Such a series is called a generalised power series or *Frobenius series*. As in the case of a simple power series solution, the radius of convergence of the Frobenius series is, in general, equal to the distance to the nearest singularity of the ODE.

Since $z = 0$ is a regular singularity of the ODE, it follows that $zp(z)$ and $z^2 q(z)$ are analytic at $z = 0$, so that we may write

$$zp(z) \equiv s(z) = \sum_{n=0}^\infty s_n z^n,$$

$$z^2 q(z) \equiv t(z) = \sum_{n=0}^\infty t_n z^n,$$

where we have defined the analytic functions $s(z)$ and $t(z)$ for later convenience. The original ODE therefore becomes

$$y'' + \frac{s(z)}{z} y' + \frac{t(z)}{z^2} y = 0.$$

Let us substitute the Frobenius series (16.12) into this equation. The derivatives of (16.12) with respect to z are given by

$$y' = \sum_{n=0}^\infty (n + \sigma) a_n z^{n+\sigma-1}, \tag{16.13}$$

$$y'' = \sum_{n=0}^\infty (n + \sigma)(n + \sigma - 1) a_n z^{n+\sigma-2}, \tag{16.14}$$

and we obtain

$$\sum_{n=0}^\infty (n + \sigma)(n + \sigma - 1) a_n z^{n+\sigma-2} + s(z) \sum_{n=0}^\infty (n + \sigma) a_n z^{n+\sigma-2} + t(z) \sum_{n=0}^\infty a_n z^{n+\sigma-2} = 0.$$

Dividing this equation through by $z^{\sigma-2}$, we find

$$\sum_{n=0}^\infty \left[(n + \sigma)(n + \sigma - 1) + s(z)(n + \sigma) + t(z) \right] a_n z^n = 0. \tag{16.15}$$

Setting $z = 0$, all terms in the sum with $n > 0$ vanish, implying that

$$[\sigma(\sigma - 1) + s(0)\sigma + t(0)]a_0 = 0,$$

which, since we require $a_0 \neq 0$, yields the *indicial equation*

$$\sigma(\sigma - 1) + s(0)\sigma + t(0) = 0. \tag{16.16}$$

This equation is a quadratic in σ and in general has two roots, the nature of which determines the forms of possible series solutions.

The two roots of the indicial equation, σ_1 and σ_2, are called the *indices* of the regular singular point. By substituting each of these roots into (16.15) in turn and requiring that the coefficients of each power of z vanish separately, we obtain a recurrence relation (for each root) expressing each a_n as a function of the previous a_r $(0 \leq r \leq n - 1)$. We will see that the larger root of the indicial equation always yields a solution to the ODE in the form of a Frobenius series (16.12). The form of the second solution depends, however, on the relationship between the two indices σ_1 and σ_2. There are three possible general cases: (i) distinct roots not differing by an integer; (ii) repeated roots; (iii) distinct roots differing by an integer (not equal to zero). Below, we discuss each of these in turn.

Before continuing, however, we note that, as was the case for solutions in the form of a simple power series, it is always worth investigating whether a Frobenius series found as a solution to a problem is summable in closed form or expressible in terms of known functions. We illustrate this point below, but the reader should avoid gaining the impression that this is always so or that, if one worked hard enough, a closed-form solution could always be found without using the series method. As mentioned earlier, this is *not* the case, and very often an infinite series solution is the best one can do.

16.3.1 Distinct roots not differing by an integer

If the roots of the indicial equation, σ_1 and σ_2, differ by an amount that is not an integer then the recurrence relations corresponding to each root lead to two linearly independent solutions of the ODE:

$$y_1(z) = z^{\sigma_1} \sum_{n=0}^{\infty} a_n z^n, \qquad y_2(z) = z^{\sigma_2} \sum_{n=0}^{\infty} b_n z^n,$$

with both solutions taking the form of a Frobenius series. The linear independence of these two solutions follows from the fact that y_2/y_1 is not a constant since $\sigma_1 - \sigma_2$ is not an integer. Because y_1 and y_2 are linearly independent, we may use them to construct the general solution $y = c_1 y_1 + c_2 y_2$.

We also note that this case includes complex conjugate roots where $\sigma_2 = \sigma_1^*$, since $\sigma_1 - \sigma_2 = \sigma_1 - \sigma_1^* = 2i \operatorname{Im} \sigma_1$ cannot be equal to a real integer.

▶ *Find the power series solutions about $z = 0$ of*
$$4zy'' + 2y' + y = 0.$$

Dividing through by $4z$ to put the equation into standard form, we obtain

$$y'' + \frac{1}{2z}y' + \frac{1}{4z}y = 0, \tag{16.17}$$

and on comparing with (16.7) we identify $p(z) = 1/(2z)$ and $q(z) = 1/(4z)$. Clearly $z = 0$ is a singular point of (16.17), but since $zp(z) = 1/2$ and $z^2q(z) = z/4$ are finite there, it is a regular singular point. We therefore substitute the Frobenius series $y = z^\sigma \sum_{n=0}^\infty a_n z^n$ into (16.17). Using (16.13) and (16.14), we obtain

$$\sum_{n=0}^\infty (n+\sigma)(n+\sigma-1)a_n z^{n+\sigma-2} + \frac{1}{2z}\sum_{n=0}^\infty (n+\sigma)a_n z^{n+\sigma-1} + \frac{1}{4z}\sum_{n=0}^\infty a_n z^{n+\sigma} = 0,$$

which, on dividing through by $z^{\sigma-2}$, gives

$$\sum_{n=0}^\infty \left[(n+\sigma)(n+\sigma-1) + \tfrac{1}{2}(n+\sigma) + \tfrac{1}{4}z \right] a_n z^n = 0. \tag{16.18}$$

If we set $z = 0$ then all terms in the sum with $n > 0$ vanish, and we obtain the indicial equation

$$\sigma(\sigma - 1) + \tfrac{1}{2}\sigma = 0,$$

which has roots $\sigma = 1/2$ and $\sigma = 0$. Since these roots do not differ by an integer, we expect to find two independent solutions to (16.17), in the form of Frobenius series.

Demanding that the coefficients of z^n vanish separately in (16.18), we obtain the recurrence relation

$$(n+\sigma)(n+\sigma-1)a_n + \tfrac{1}{2}(n+\sigma)a_n + \tfrac{1}{4}a_{n-1} = 0. \tag{16.19}$$

If we choose the larger root, $\sigma = 1/2$, of the indicial equation then (16.19) becomes

$$(4n^2 + 2n)a_n + a_{n-1} = 0 \quad \Rightarrow \quad a_n = \frac{-a_{n-1}}{2n(2n+1)}.$$

Setting $a_0 = 1$, we find $a_n = (-1)^n/(2n+1)!$, and so the solution to (16.17) is given by

$$y_1(z) = \sqrt{z}\sum_{n=0}^\infty \frac{(-1)^n}{(2n+1)!}z^n$$

$$= \sqrt{z} - \frac{(\sqrt{z})^3}{3!} + \frac{(\sqrt{z})^5}{5!} - \cdots = \sin\sqrt{z}.$$

To obtain the second solution we set $\sigma = 0$ (the smaller root of the indicial equation) in (16.19), which gives

$$(4n^2 - 2n)a_n + a_{n-1} = 0 \quad \Rightarrow \quad a_n = -\frac{a_{n-1}}{2n(2n-1)}.$$

Setting $a_0 = 1$ now gives $a_n = (-1)^n/(2n)!$, and so the second (independent) solution to (16.17) is

$$y_2(z) = \sum_{n=0}^\infty \frac{(-1)^n}{(2n)!}z^n = 1 - \frac{(\sqrt{z})^2}{2!} + \frac{(\sqrt{4})^4}{4!} - \cdots = \cos\sqrt{z}.$$

We may check that $y_1(z)$ and $y_2(z)$ are indeed linearly independent by computing the Wronskian as follows:

$$
\begin{aligned}
W &= y_1 y_2' - y_2 y_1' \\
&= \sin\sqrt{z}\left(-\frac{1}{2\sqrt{z}}\sin\sqrt{z}\right) - \cos\sqrt{z}\left(\frac{1}{2\sqrt{z}}\cos\sqrt{z}\right) \\
&= -\frac{1}{2\sqrt{z}}\left(\sin^2\sqrt{z} + \cos^2\sqrt{z}\right) = -\frac{1}{2\sqrt{z}} \neq 0.
\end{aligned}
$$

Since $W \neq 0$, the solutions $y_1(z)$ and $y_2(z)$ are linearly independent. Hence, the general solution to (16.17) is given by

$$
y(z) = c_1 \sin\sqrt{z} + c_2 \cos\sqrt{z}. \; \blacktriangleleft
$$

16.3.2 Repeated root of the indicial equation

If the indicial equation has a repeated root, so that $\sigma_1 = \sigma_2 = \sigma$, then obviously only one solution in the form of a Frobenius series (16.12) may be found as described above, i.e.

$$
y_1(z) = z^\sigma \sum_{n=0}^{\infty} a_n z^n.
$$

Methods for obtaining a second, linearly independent, solution are discussed in section 16.4.

16.3.3 Distinct roots differing by an integer

Whatever the roots of the indicial equation, the recurrence relation corresponding to the larger of the two always leads to a solution of the ODE. However, if the roots of the indicial equation differ by an integer then the recurrence relation corresponding to the smaller root may or may not lead to a second linearly independent solution, depending on the ODE under consideration. Note that for complex roots of the indicial equation, the 'larger' root is taken to be the one with the larger real part.

▶*Find the power series solutions about $z = 0$ of*

$$
z(z-1)y'' + 3zy' + y = 0. \tag{16.20}
$$

Dividing through by $z(z-1)$ to put the equation into standard form, we obtain

$$
y'' + \frac{3}{(z-1)}y' + \frac{1}{z(z-1)}y = 0, \tag{16.21}
$$

and on comparing with (16.7) we identify $p(z) = 3/(z-1)$ and $q(z) = 1/[z(z-1)]$. We immediately see that $z = 0$ is a singular point of (16.21), but since $zp(z) = 3z/(z-1)$ and $z^2 q(z) = z/(z-1)$ are finite there, it is a regular singular point and we expect to find at least one solution in the form of a Frobenius series. We therefore substitute $y = z^\sigma \sum_{n=0}^{\infty} a_n z^n$ into (16.21) and, using (16.13) and (16.14), we obtain

$$\sum_{n=0}^{\infty} (n+\sigma)(n+\sigma-1)a_n z^{n+\sigma-2} + \frac{3}{z-1} \sum_{n=0}^{\infty} (n+\sigma)a_n z^{n+\sigma-1}$$

$$+ \frac{1}{z(z-1)} \sum_{n=0}^{\infty} a_n z^{n+\sigma} = 0,$$

which, on dividing through by $z^{\sigma-2}$, gives

$$\sum_{n=0}^{\infty} \left[(n+\sigma)(n+\sigma-1) + \frac{3z}{z-1}(n+\sigma) + \frac{z}{z-1} \right] a_n z^n = 0.$$

Although we could use this expression to find the indicial equation and recurrence relations, the working is simpler if we now multiply through by $z-1$ to give

$$\sum_{n=0}^{\infty} [(z-1)(n+\sigma)(n+\sigma-1) + 3z(n+\sigma) + z] a_n z^n = 0. \tag{16.22}$$

If we set $z = 0$ then all terms in the sum with the exponent of z greater than zero vanish, and we obtain the indicial equation

$$\sigma(\sigma-1) = 0,$$

which has the roots $\sigma = 1$ and $\sigma = 0$. Since the roots differ by an integer (unity), it may not be possible to find two linearly independent solutions of (16.21) in the form of Frobenius series. We are guaranteed, however, to find one such solution corresponding to the larger root, $\sigma = 1$.

Demanding that the coefficients of z^n vanish separately in (16.22), we obtain the recurrence relation

$$(n-1+\sigma)(n-2+\sigma)a_{n-1} - (n+\sigma)(n+\sigma-1)a_n + 3(n-1+\sigma)a_{n-1} + a_{n-1} = 0,$$

which can be simplified to give

$$(n+\sigma-1)a_n = (n+\sigma)a_{n-1}. \tag{16.23}$$

On substituting $\sigma = 1$ into this expression, we obtain

$$a_n = \left(\frac{n+1}{n} \right) a_{n-1},$$

and on setting $a_0 = 1$ we find $a_n = n+1$; so one solution to (16.21) is given by

$$y_1(z) = z \sum_{n=0}^{\infty} (n+1)z^n = z(1 + 2z + 3z^2 + \cdots)$$

$$= \frac{z}{(1-z)^2}. \tag{16.24}$$

If we attempt to find a second solution (corresponding to the smaller root of the indicial equation) by setting $\sigma = 0$ in (16.23), we find

$$a_n = \left(\frac{n}{n-1} \right) a_{n-1}.$$

But we require $a_0 \neq 0$, so a_1 is formally infinite and the method fails. We discuss how to find a second linearly independent solution in the next section. ◀

One particular case is worth mentioning. If the point about which the solution is required, i.e. $z = 0$, is in fact an ordinary point of the ODE rather than a regular singular point, then substitution of the Frobenius series (16.12) leads to an indicial equation with roots $\sigma = 0$ and $\sigma = 1$. Although these roots differ by an integer (unity), the recurrence relations corresponding to the two roots yield two linearly independent power series solutions (one for each root), as expected from section 16.2.

16.4 Obtaining a second solution

Whilst attempting to construct solutions to an ODE in the form of Frobenius series about a regular singular point, we found in the previous section that when the indicial equation has a repeated root, or roots differing by an integer, we can (in general) find only one solution of this form. In order to construct the general solution to the ODE, however, we require two linearly independent solutions y_1 and y_2. We now consider several methods for obtaining a second solution in this case.

16.4.1 The Wronskian method

If y_1 and y_2 are two linearly independent solutions of the standard equation

$$y'' + p(z)y' + q(z)y = 0$$

then the Wronskian of these two solutions is given by $W(z) = y_1 y_2' - y_2 y_1'$. Dividing the Wronskian by y_1^2 we obtain

$$\frac{W}{y_1^2} = \frac{y_2'}{y_1} - \frac{y_1'}{y_1^2} y_2 = \frac{y_2'}{y_1} + \left[\frac{d}{dz} \left(\frac{1}{y_1} \right) \right] y_2 = \frac{d}{dz} \left(\frac{y_2}{y_1} \right),$$

which integrates to give

$$y_2(z) = y_1(z) \int^z \frac{W(u)}{y_1^2(u)}\, du.$$

Now using the alternative expression for $W(z)$ given in (16.4) with $C = 1$ (since we are not concerned with this normalising factor), we find

$$y_2(z) = y_1(z) \int^z \frac{1}{y_1^2(u)} \exp\left\{ -\int^u p(v)\, dv \right\} du. \tag{16.25}$$

Hence, given y_1, we can in principle compute y_2. Note that the lower limits of integration have been omitted. If constant lower limits are included then they merely lead to a constant times the first solution.

▶*Find a second solution to (16.21) using the Wronskian method.*

For the ODE (16.21) we have $p(z) = 3/(z-1)$, and from (16.24) we see that one solution to (16.21) is $y_1 = z/(1-z)^2$. Substituting for p and y_1 in (16.25) we have

$$
\begin{aligned}
y_2(z) &= \frac{z}{(1-z)^2} \int^z \frac{(1-u)^4}{u^2} \exp\left(-\int^u \frac{3}{v-1}\, dv\right) du \\
&= \frac{z}{(1-z)^2} \int^z \frac{(1-u)^4}{u^2} \exp\left[-3\ln(u-1)\right] du \\
&= \frac{z}{(1-z)^2} \int^z \frac{u-1}{u^2}\, du \\
&= \frac{z}{(1-z)^2} \left(\ln z + \frac{1}{z}\right).
\end{aligned}
$$

By calculating the Wronskian of y_1 and y_2 it is easily shown that, as expected, the two solutions are linearly independent. In fact, as the Wronskian has already been evaluated as $W(u) = \exp[-3\ln(u-1)]$, i.e. $W(z) = (z-1)^{-3}$, no calculation is needed. ◄

An alternative (but equivalent) method of finding a second solution is simply to assume that the second solution has the form $y_2(z) = u(z)y_1(z)$ for some function $u(z)$ to be determined (this method was discussed more fully in subsection 15.2.3). From (16.25), we see that the second solution derived from the Wronskian is indeed of this form. Substituting $y_2(z) = u(z)y_1(z)$ into the ODE leads to a first-order ODE in which u' is the dependent variable; this may then be solved.

16.4.2 The derivative method

The derivative method of finding a second solution begins with the derivation of a recurrence relation for the coefficients a_n in a Frobenius series solution, as in the previous section. However, rather than putting $\sigma = \sigma_1$ in this recurrence relation to evaluate the first series solution, we now keep σ as a variable parameter. This means that the computed a_n are functions of σ and the computed solution is now a function of z and σ:

$$
y(z, \sigma) = z^\sigma \sum_{n=0}^\infty a_n(\sigma) z^n. \tag{16.26}
$$

Of course, if we put $\sigma = \sigma_1$ in this, we obtain immediately the first series solution, but for the moment we leave σ as a parameter.

For brevity let us denote the differential operator on the LHS of our standard ODE (16.7) by $\mathcal{L}$, so that

$$
\mathcal{L} = \frac{d^2}{dz^2} + p(z)\frac{d}{dz} + q(z),
$$

and examine the effect of $\mathcal{L}$ on the series $y(z, \sigma)$ in (16.26). It is clear that the series $\mathcal{L}y(z, \sigma)$ will contain only a term in z^σ, since the recurrence relation defining the $a_n(\sigma)$ is such that these coefficients vanish for higher powers of z. But the coefficient of z^σ is simply the LHS of the indicial equation. Therefore, if the roots of the indicial equation are $\sigma = \sigma_1$ and $\sigma = \sigma_2$ then it follows that

$$\mathcal{L}y(z,\sigma) = a_0(\sigma - \sigma_1)(\sigma - \sigma_2)z^{\sigma}. \tag{16.27}$$

Therefore, as in the previous section, we see that for $y(z,\sigma)$ to be a solution of the ODE $\mathcal{L}y = 0$, σ must equal σ_1 or σ_2. For simplicity we shall set $a_0 = 1$ in the following discussion.

Let us first consider the case in which the two roots of the indicial equation are equal, i.e. $\sigma_2 = \sigma_1$. From (16.27) we then have

$$\mathcal{L}y(z,\sigma) = (\sigma - \sigma_1)^2 z^{\sigma}.$$

Differentiating this equation with respect to σ we obtain

$$\frac{\partial}{\partial\sigma}[\mathcal{L}y(z,\sigma)] = (\sigma - \sigma_1)^2 z^{\sigma}\ln z + 2(\sigma - \sigma_1)z^{\sigma},$$

which equals zero if $\sigma = \sigma_1$. But since $\partial/\partial\sigma$ and $\mathcal{L}$ are operators that differentiate with respect to different variables, we can reverse their order, implying that

$$\mathcal{L}\left[\frac{\partial}{\partial\sigma}y(z,\sigma)\right] = 0 \qquad \text{at } \sigma = \sigma_1.$$

Hence, the function in square brackets, evaluated at $\sigma = \sigma_1$ and denoted by

$$\left[\frac{\partial}{\partial\sigma}y(z,\sigma)\right]_{\sigma=\sigma_1}, \tag{16.28}$$

is also a solution of the original ODE $\mathcal{L}y = 0$, and is in fact the second linearly independent solution that we were looking for.

The case in which the roots of the indicial equation differ by an integer is slightly more complicated but can be treated in a similar way. In (16.27), since $\mathcal{L}$ differentiates with respect to z we may multiply (16.27) by any function of σ, say $\sigma - \sigma_2$, and take this function inside the operator $\mathcal{L}$ on the LHS to obtain

$$\mathcal{L}\left[(\sigma - \sigma_2)y(z,\sigma)\right] = (\sigma - \sigma_1)(\sigma - \sigma_2)^2 z^{\sigma}. \tag{16.29}$$

Therefore the function

$$[(\sigma - \sigma_2)y(z,\sigma)]_{\sigma=\sigma_2}$$

is also a solution of the ODE $\mathcal{L}y = 0$. However, it can be proved[§] that this function is a simple multiple of the first solution $y(z,\sigma_1)$, showing that it is not linearly independent and that we must find another solution. To do this we differentiate (16.29) with respect to σ and find

$$\frac{\partial}{\partial\sigma}\{\mathcal{L}[(\sigma - \sigma_2)y(z,\sigma)]\} = (\sigma - \sigma_2)^2 z^{\sigma} + 2(\sigma - \sigma_1)(\sigma - \sigma_2)z^{\sigma}$$
$$+ (\sigma - \sigma_1)(\sigma - \sigma_2)^2 z^{\sigma}\ln z,$$

[§] For a fuller discussion see, for example, K. F. Riley, *Mathematical Methods for the Physical Sciences* (Cambridge: Cambridge University Press, 1974), pp. 158–9.

which is equal to zero if $\sigma = \sigma_2$. As previously, since $\partial/\partial\sigma$ and $\mathcal{L}$ are operators that differentiate with respect to different variables, we can reverse their order to obtain

$$\mathcal{L}\left\{\frac{\partial}{\partial\sigma}[(\sigma - \sigma_2)y(z,\sigma)]\right\} = 0 \qquad \text{at } \sigma = \sigma_2,$$

and so the function

$$\left\{\frac{\partial}{\partial\sigma}[(\sigma - \sigma_2)y(z,\sigma)]\right\}_{\sigma=\sigma_2} \tag{16.30}$$

is also a solution of the original ODE $\mathcal{L}y = 0$, and is in fact the second linearly independent solution.

> ►*Find a second solution to (16.21) using the derivative method.*

From (16.23) the recurrence relation (with σ as a parameter) is given by

$$(n + \sigma - 1)a_n = (n + \sigma)a_{n-1}.$$

Setting $a_0 = 1$ we find that the coefficients have the particularly simple form $a_n(\sigma) = (\sigma + n)/\sigma$. We therefore consider the function

$$y(z, \sigma) = z^\sigma \sum_{n=0}^{\infty} a_n(\sigma)z^n = z^\sigma \sum_{n=0}^{\infty} \frac{\sigma + n}{\sigma}z^n.$$

The smaller root of the indicial equation for (16.21) is $\sigma_2 = 0$, and so from (16.30) a second, linearly independent, solution to the ODE is given by

$$\left\{\frac{\partial}{\partial\sigma}[\sigma y(z,\sigma)]\right\}_{\sigma=0} = \left\{\frac{\partial}{\partial\sigma}\left[z^\sigma\sum_{n=0}^{\infty}(\sigma + n)z^n\right]\right\}_{\sigma=0}.$$

The derivative with respect to σ is given by

$$\frac{\partial}{\partial\sigma}\left[z^\sigma\sum_{n=0}^{\infty}(\sigma + n)z^n\right] = z^\sigma \ln z \sum_{n=0}^{\infty}(\sigma + n)z^n + z^\sigma\sum_{n=0}^{\infty}z^n,$$

which on setting $\sigma = 0$ gives the second solution

$$y_2(z) = \ln z \sum_{n=0}^{\infty} nz^n + \sum_{n=0}^{\infty} z^n$$

$$= \frac{z}{(1-z)^2}\ln z + \frac{1}{1-z}$$

$$= \frac{z}{(1-z)^2}\left(\ln z + \frac{1}{z} - 1\right).$$

This second solution is the same as that obtained by the Wronskian method in the previous subsection except for the addition of some of the first solution. ◄

16.4.3 Series form of the second solution

Using any of the methods discussed above, we can find the general form of the second solution to the ODE. This form is most easily found, however, using the derivative method. Let us first consider the case where the two solutions of the indicial equation are equal. In this case a second solution is given by (16.28), which may be written as

$$y_2(z) = \left[\frac{\partial y(z,\sigma)}{\partial \sigma} \right]_{\sigma=\sigma_1}$$

$$= (\ln z) z^{\sigma_1} \sum_{n=0}^{\infty} a_n(\sigma_1) z^n + z^{\sigma_1} \sum_{n=1}^{\infty} \left[\frac{da_n(\sigma)}{d\sigma} \right]_{\sigma=\sigma_1} z^n$$

$$= y_1(z) \ln z + z^{\sigma_1} \sum_{n=1}^{\infty} b_n z^n, \tag{16.31}$$

where $b_n = [da_n(\sigma)/d\sigma]_{\sigma=\sigma_1}$. One could equally obtain the coefficients b_n by direct substitution of the form (16.31) into the original ODE.

In the case where the roots of the indicial equation differ by an integer (not equal to zero), then from (16.30) a second solution is given by

$$y_2(z) = \left\{ \frac{\partial}{\partial \sigma} [(\sigma - \sigma_2) y(z,\sigma)] \right\}_{\sigma=\sigma_2}$$

$$= \ln z \left[(\sigma - \sigma_2) z^{\sigma} \sum_{n=0}^{\infty} a_n(\sigma) z^n \right]_{\sigma=\sigma_2} + z^{\sigma_2} \sum_{n=0}^{\infty} \left[\frac{d}{d\sigma} (\sigma - \sigma_2) a_n(\sigma) \right]_{\sigma=\sigma_2} z^n.$$

But, as we mentioned in the previous section, $[(\sigma - \sigma_2) y(z,\sigma)]$ at $\sigma = \sigma_2$ is just a multiple of the first solution $y(z,\sigma_1)$. Therefore the second solution is of the form

$$y_2(z) = c y_1(z) \ln z + z^{\sigma_2} \sum_{n=0}^{\infty} b_n z^n, \tag{16.32}$$

where c is a constant. In some cases, however, c might be zero, and so the second solution would not contain the term in $\ln z$ and could be written simply as a Frobenius series. Clearly this corresponds to the case in which the substitution of a Frobenius series into the original ODE yields two solutions automatically. In either case, the coefficients b_n may also be found by direct substitution of the form (16.32) into the original ODE.

16.5 Polynomial solutions

We have seen that the evaluation of successive terms of a series solution to a differential equation is carried out by means of a recurrence relation. The form of the relation for a_n depends upon n, the previous values of a_r $(r < n)$ and the parameters of the equation. It may happen, as a result of this, that for some value of $n = N + 1$ the computed value a_{N+1} is zero and that all higher a_r also vanish. If this is so, and the corresponding solution of the indicial equation σ

is a positive integer or zero, then we are left with a finite polynomial of degree $N' = N + \sigma$ as a solution of the ODE:

$$y(z) = \sum_{n=0}^{N} a_n z^{n+\sigma}. \tag{16.33}$$

In many applications in theoretical physics (particularly in quantum mechanics) the termination of a potentially infinite series after a finite number of terms is of crucial importance in establishing physically acceptable descriptions and properties of systems. The condition under which such a termination occurs is therefore of considerable importance.

> ►*Find power series solutions about $z = 0$ of*
>
> $$y'' - 2zy' + \lambda y = 0. \tag{16.34}$$
>
> *For what values of λ does the equation possess a polynomial solution? Find such a solution for $\lambda = 4$.*

Clearly $z = 0$ is an ordinary point of (16.34) and so we look for solutions of the form $y = \sum_{n=0}^{\infty} a_n z^n$. Substituting this into the ODE and multiplying through by z^2 we find

$$\sum_{n=0}^{\infty} [n(n - 1) - 2z^2 n + \lambda z^2] a_n z^n = 0.$$

By demanding that the coefficients of each power of z vanish separately we derive the recurrence relation

$$n(n - 1)a_n - 2(n - 2)a_{n-2} + \lambda a_{n-2} = 0,$$

which may be rearranged to give

$$a_n = \frac{2(n - 2) - \lambda}{n(n - 1)} a_{n-2} \quad \text{for } n \geq 2. \tag{16.35}$$

The odd and even coefficients are therefore independent of one another, and two solutions to (16.34) may be derived. We either set $a_1 = 0$ and $a_0 = 1$ to obtain

$$y_1(z) = 1 - \lambda \frac{z^2}{2!} - \lambda(4 - \lambda)\frac{z^4}{4!} - \lambda(4 - \lambda)(8 - \lambda)\frac{z^6}{6!} - \cdots \tag{16.36}$$

or set $a_0 = 0$ and $a_1 = 1$ to obtain

$$y_2(z) = z + (2 - \lambda)\frac{z^3}{3!} + (2 - \lambda)(6 - \lambda)\frac{z^5}{5!} + (2 - \lambda)(6 - \lambda)(10 - \lambda)\frac{z^7}{7!} + \cdots .$$

Now, from the recurrence relation (16.35) (or in this case from the expressions for y_1 and y_2 themselves) we see that for the ODE to possess a polynomial solution we require $\lambda = 2(n - 2)$ for $n \geq 2$ or, more simply, $\lambda = 2n$ for $n \geq 0$, i.e. λ must be an even positive integer. If $\lambda = 4$ then from (16.36) the ODE has the polynomial solution

$$y_1(z) = 1 - \frac{4z^2}{2!} = 1 - 2z^2. \blacktriangleleft$$

A simpler method of obtaining finite polynomial solutions is to *assume* a solution of the form (16.33), where $a_N \neq 0$. Instead of starting with the lowest power of z, as we have done up to now, this time we start by considering the coefficient of the highest power z^N; such a power now exists because of our assumed form of solution.

> ►By assuming a polynomial solution find the values of λ in (16.34) for which such a solution exists.

We assume a polynomial solution to (16.34) of the form $y = \sum_{n=0}^{N} a_n z^n$. Substituting this form into (16.34) we find

$$\sum_{n=0}^{N} \left[n(n-1)a_n z^{n-2} - 2zn a_n z^{n-1} + \lambda a_n z^n \right] = 0.$$

Now, instead of starting with the lowest power of z, we start with the highest. Thus, demanding that the coefficient of z^N vanishes, we require $-2N + \lambda = 0$, i.e. $\lambda = 2N$, as we found in the previous example. By demanding that the coefficient of a general power of z is zero, the same recurrence relation as above may be derived and the solutions found. ◄

16.6 Exercises

16.1   Find two power series solutions about $z = 0$ of the differential equation

$$(1 - z^2)y'' - 3zy' + \lambda y = 0.$$

Deduce that the value of λ for which the corresponding power series becomes an Nth-degree polynomial $U_N(z)$ is $N(N+2)$. Construct $U_2(z)$ and $U_3(z)$.

16.2   Find solutions, as power series in z, of the equation

$$4zy'' + 2(1-z)y' - y = 0.$$

Identify one of the solutions and verify it by direct substitution.

16.3   Find power series solutions in z of the differential equation

$$zy'' - 2y' + 9z^5 y = 0.$$

Identify closed forms for the two series, calculate their Wronskian, and verify that they are linearly independent. Compare the Wronskian with that calculated from the differential equation.

16.4   Change the independent variable in the equation

$$\frac{d^2 f}{dz^2} + 2(z-a)\frac{df}{dz} + 4f = 0 \qquad (*)$$

from z to $x = z - \alpha$, and find two independent series solutions, expanded about $x = 0$, of the resulting equation. Deduce that the general solution of (*) is

$$f(z, \alpha) = A(z-\alpha)e^{-(z-\alpha)^2} + B \sum_{m=0}^{\infty} \frac{(-4)^m m!}{(2m)!} (z-\alpha)^{2m},$$

with A and B arbitrary constants.

16.5   Investigate solutions of Legendre's equation at one of its singular points as follows.

(a)   Verify that $z = 1$ is a regular singular point of Legendre's equation and that the indicial equation for a series solution in powers of $(z-1)$ has a double root at $\sigma = 0$.

(b)   Obtain the corresponding recurrence relation and show that a polynomial solution is obtained if ℓ is a positive integer.

550

(c) Determine the radius of convergence R of the $\sigma = 0$ series and relate it to the positions of the singularities of Legendre's equation.

16.6   Verify that $z = 0$ is a regular singular point of the equation

$$z^2 y'' - \tfrac{3}{2} z y' + (1 + z)y = 0,$$

and that the indicial equation has roots 2 and $1/2$. Show that the general solution is given by

$$y(z) = 6a_0 z^2 \sum_{n=0}^{\infty} \frac{(-1)^n (n+1) 2^{2n} z^n}{(2n+3)!}$$

$$+ b_0 \left(z^{1/2} + 2z^{3/2} - \frac{z^{1/2}}{4} \sum_{n=2}^{\infty} \frac{(-1)^n 2^{2n} z^n}{n(n-1)(2n-3)!} \right).$$

16.7   Use the derivative method to obtain, as a second solution of Bessel's equation for the case when $v = 0$, the following expression:

$$J_0(z) \ln z - \sum_{n=1}^{\infty} \frac{(-1)^n}{(n!)^2} \left(\sum_{r=1}^{n} \frac{1}{r} \right) \left(\frac{z}{2} \right)^{2n},$$

given that the first solution is $J_0(z)$, as specified by (18.79).

16.8   Consider a series solution of the equation

$$z y'' - 2y' + yz = 0 \qquad (*)$$

about its regular singular point.

(a) Show that its indicial equation has roots that differ by an integer but that the two roots nevertheless generate linearly independent solutions

$$y_1(z) = 3a_0 \sum_{n=1}^{\infty} \frac{(-1)^{n+1} \, 2nz^{2n+1}}{(2n+1)!},$$

$$y_2(z) = a_0 \sum_{n=0}^{\infty} \frac{(-1)^{n+1}(2n-1)z^{2n}}{(2n)!}.$$

(b) Show that $y_1(z)$ is equal to $3a_0(\sin z - z \cos z)$ by expanding the sinusoidal functions. Then, using the Wronskian method, find an expression for $y_2(z)$ in terms of sinusoids. You will need to write z^2 as $(z/\sin z)(z \sin z)$ and integrate by parts to evaluate the integral involved.

(c) Confirm that the two solutions are linearly independent by showing that their Wronskian is equal to $-z^2$, as would be expected from the form of $(*)$.

16.9   Find series solutions of the equation $y'' - 2zy' - 2y = 0$. Identify one of the series as $y_1(z) = \exp z^2$ and verify this by direct substitution. By setting $y_2(z) = u(z)y_1(z)$ and solving the resulting equation for $u(z)$, find an explicit form for $y_2(z)$ and deduce that

$$\int_0^x e^{-v^2} \, dv = e^{-x^2} \sum_{n=0}^{\infty} \frac{n!}{2(2n+1)!} (2x)^{2n+1}.$$

16.10   Solve the equation

$$z(1-z)\frac{d^2 y}{dz^2} + (1-z)\frac{dy}{dz} + \lambda y = 0$$

as follows.

(a) Identify and classify its singular points and determine their indices.

(b) Find one series solution in powers of z. Give a formal expression for a second linearly independent solution.

(c) Deduce the values of λ for which there is a polynomial solution $P_N(z)$ of degree N. Evaluate the first four polynomials, normalised in such a way that $P_N(0) = 1$.

16.11   Find the general power series solution about $z = 0$ of the equation

$$z\frac{d^2 y}{dz^2} + (2z - 3)\frac{dy}{dz} + \frac{4}{z}y = 0.$$

16.12   Find the radius of convergence of a series solution about the origin for the equation $(z^2 + az + b)y'' + 2y = 0$ in the following cases:

$$\text{(a) } a = 5, \ b = 6; \qquad \text{(b) } a = 5, \ b = 7.$$

Show that if a and b are real and $4b > a^2$, then the radius of convergence is always given by $b^{1/2}$.

16.13   For the equation $y'' + z^{-3}y = 0$, show that the origin becomes a regular singular point if the independent variable is changed from z to $x = 1/z$. Hence find a series solution of the form $y_1(z) = \sum_0^\infty a_n z^{-n}$. By setting $y_2(z) = u(z)y_1(z)$ and expanding the resulting expression for du/dz in powers of z^{-1}, show that $y_2(z)$ has the asymptotic form

$$y_2(z) = c\left[z + \ln z - \tfrac{1}{2} + O\left(\frac{\ln z}{z}\right)\right],$$

where c is an arbitrary constant.

16.14   Prove that the Laguerre equation,

$$z\frac{d^2 y}{dz^2} + (1 - z)\frac{dy}{dz} + \lambda y = 0,$$

has polynomial solutions $L_N(z)$ if λ is a non-negative integer N, and determine the recurrence relationship for the polynomial coefficients. Hence show that an expression for $L_N(z)$, normalised in such a way that $L_N(0) = N!$, is

$$L_N(z) = \sum_{n=0}^{N} \frac{(-1)^n (N!)^2}{(N-n)!(n!)^2} z^n.$$

Evaluate $L_3(z)$ explicitly.

16.15   The origin is an ordinary point of the Chebyshev equation,

$$(1 - z^2)y'' - zy' + m^2 y = 0,$$

which therefore has series solutions of the form $z^\sigma \sum_0^\infty a_n z^n$ for $\sigma = 0$ and $\sigma = 1$.

(a) Find the recurrence relationships for the a_n in the two cases and show that there exist polynomial solutions $T_m(z)$:

(i)   for $\sigma = 0$, when m is an even integer, the polynomial having $\tfrac{1}{2}(m+2)$ terms;

(ii) for $\sigma = 1$, when m is an odd integer, the polynomial having $\tfrac{1}{2}(m+1)$ terms.

(b) $T_m(z)$ is normalised so as to have $T_m(1) = 1$. Find explicit forms for $T_m(z)$ for $m = 0, 1, 2, 3$.

(c) Show that the corresponding non-terminating series solutions $S_m(z)$ have as their first few terms

$$S_0(z) = a_0 \left(z + \frac{1}{3!}z^3 + \frac{9}{5!}z^5 + \cdots \right),$$

$$S_1(z) = a_0 \left(1 - \frac{1}{2!}z^2 - \frac{3}{4!}z^4 - \cdots \right),$$

$$S_2(z) = a_0 \left(z - \frac{3}{3!}z^3 - \frac{15}{5!}z^5 - \cdots \right),$$

$$S_3(z) = a_0 \left(1 - \frac{9}{2!}z^2 + \frac{45}{4!}z^4 + \cdots \right).$$

16.16   Obtain the recurrence relations for the solution of Legendre's equation (18.1) in *inverse* powers of z, i.e. set $y(z) = \sum a_n z^{\sigma-n}$, with $a_0 \neq 0$. Deduce that, if ℓ is an integer, then the series with $\sigma = \ell$ will terminate and hence converge for all z, whilst the series with $\sigma = -(\ell + 1)$ does not terminate and hence converges only for $|z| > 1$.

16.7 Hints and answers

16.1   Note that $z = 0$ is an ordinary point of the equation.
For $\sigma = 0$, $a_{n+2}/a_n = [n(n+2) - \lambda]/[(n+1)(n+2)]$ and, correspondingly, for $\sigma = 1$, $U_2(z) = a_0(1 - 4z^2)$ and $U_3(z) = a_0(z - 2z^3)$.

16.3 $\sigma = 0$ and 3; $a_{6m}/a_0 = (-1)^m/(2m)!$ and $a_{6m}/a_0 = (-1)^m/(2m+1)!$, respectively. $y_1(z) = a_0 \cos z^3$ and $y_2(z) = a_0 \sin z^3$. The Wronskian is $+3a_0^2 z^2 \neq 0$.

16.5   (b) $a_{n+1}/a_n = [\ell(\ell+1) - n(n+1)]/[2(n+1)^2]$.
(c) $R = 2$, equal to the distance between $z = 1$ and the closest singularity at $z = -1$.

16.7   A typical term in the series for $y(\sigma, z)$ is $\dfrac{(-1)^n z^{2n}}{[(\sigma+2)(\sigma+4)\cdots(\sigma+2n)]^2}$.

16.9   The origin is an ordinary point. Determine the constant of integration by examining the behaviour of the related functions for small x.
$y_2(z) = (\exp z^2) \int_0^z \exp(-x^2)\, dx$.

16.11   Repeated roots $\sigma = 2$.

$$y(z) = az^2 - 4az^3 + 6bz^3 + \sum_{n=2}^{\infty} \frac{(n+1)(-2z)^{n+2}}{n!} \left\{ \frac{a}{4} + b\left[\ln z + g(n)\right] \right\},$$

where

$$g(n) = \frac{1}{n+1} - \frac{1}{n} - \frac{1}{n-1} - \cdots - \frac{1}{2} - 2.$$

16.13   The transformed equation is $xy'' + 2y' + y = 0$; $a_n = (-1)^n(n+1)^{-1}(n!)^{-2}a_0$; $du/dz = A[y_1(z)]^{-2}$.

16.15   (a) (i) $a_{n+2} = [a_n(n^2 - m^2)]/[(n+2)(n+1)]$,
(ii) $a_{n+2} = \{a_n[(n+1)^2 - m^2]\}/[(n+3)(n+2)]$; (b) 1, z, $2z^2 - 1$, $4z^3 - 3z$.

17

Eigenfunction methods for differential equations

In the previous three chapters we dealt with the solution of differential equations of order n by two methods. In one method, we found n independent solutions of the equation and then combined them, weighted with coefficients determined by the boundary conditions; in the other we found solutions in terms of series whose coefficients were related by (in general) an n-term recurrence relation and thence fixed by the boundary conditions. For both approaches the linearity of the equation was an important or essential factor in the utility of the method, and in this chapter our aim will be to exploit the superposition properties of linear differential equations even further.

We will be concerned with the solution of equations of the inhomogeneous form

$$\mathcal{L}y(x) = f(x), \tag{17.1}$$

where $f(x)$ is a prescribed or general function *and* the boundary conditions to be satisfied by the solution $y = y(x)$, for example at the limits $x = a$ and $x = b$, are given. The expression $\mathcal{L}y(x)$ stands for a linear differential operator $\mathcal{L}$ acting upon the function $y(x)$.

In general, unless $f(x)$ is both known and simple, it will not be possible to find particular integrals of (17.1), even if complementary functions can be found that satisfy $\mathcal{L}y = 0$. The idea is therefore to exploit the linearity of $\mathcal{L}$ by building up the required solution $y(x)$ as a *superposition*, generally containing an infinite number of terms, of some set of functions $\{y_i(x)\}$ that each individually satisfy the boundary conditions. Clearly this brings in a quite considerable complication but since, within reason, we may select the set of functions to suit ourselves, we can obtain sizeable compensation for this complication. Indeed, if the set chosen is one containing functions that, when acted upon by $\mathcal{L}$, produce particularly simple results then we can 'show a profit' on the operation. In particular, if the set consists of those functions y_i for which

$$\mathcal{L}y_i(x) = \lambda_i y_i(x), \tag{17.2}$$

where λ_i is a constant (and which satisfy the boundary conditions), then a distinct advantage may be obtained from the manoeuvre because all the differentiation will have disappeared from (17.1).

Equation (17.2) is clearly reminiscent of the equation satisfied by the *eigenvectors* $\mathbf{x}^i$ of a linear operator $\mathcal{A}$, namely

$$\mathcal{A}\mathbf{x}^i = \lambda_i \mathbf{x}^i, \tag{17.3}$$

where λ_i is a constant and is called the *eigenvalue* associated with $\mathbf{x}^i$. By analogy, in the context of differential equations a function $y_i(x)$ satisfying (17.2) is called an *eigenfunction* of the operator $\mathcal{L}$ (under the imposed boundary conditions) and λ_i is then called the eigenvalue associated with the eigenfunction $y_i(x)$. Clearly, the eigenfunctions $y_i(x)$ of $\mathcal{L}$ are only determined up to an arbitrary scale factor by (17.2).

Probably the most familiar equation of the form (17.2) is that which describes a simple harmonic oscillator, i.e.

$$\mathcal{L}y \equiv -\frac{d^2 y}{dt^2} = \omega^2 y, \quad \text{where } \mathcal{L} \equiv -d^2/dt^2. \tag{17.4}$$

Imposing the boundary condition that the solution is periodic with period T, the eigenfunctions in this case are given by $y_n(t) = A_n e^{i\omega_n t}$, where $\omega_n = 2\pi n/T$, $n = 0, \pm 1, \pm 2, \ldots$ and the A_n are constants. The eigenvalues are $\omega_n^2 = n^2 \omega_1^2 = n^2(2\pi/T)^2$. (Sometimes ω_n is referred to as the eigenvalue of this equation, but we will avoid such confusing terminology here.)

We may discuss a somewhat wider class of differential equations by considering a slightly more general form of (17.2), namely

$$\mathcal{L}y_i(x) = \lambda_i \rho(x) y_i(x), \tag{17.5}$$

where $\rho(x)$ is a *weight function*. In many applications $\rho(x)$ is unity for all x, in which case (17.2) is recovered; in general, though, it is a function determined by the choice of coordinate system used in describing a particular physical situation. The only requirement on $\rho(x)$ is that it is real and does not change sign in the range $a \leq x \leq b$, so that it can, without loss of generality, be taken to be non-negative throughout; of course, $\rho(x)$ must be the same function for all values of λ_i. A function $y_i(x)$ that satisfies (17.5) is called an eigenfunction of the operator $\mathcal{L}$ with respect to the weight function $\rho(x)$.

This chapter will not cover methods used to determine the eigenfunctions of (17.2) or (17.5), since we have discussed those in previous chapters, but, rather, will use the properties of the eigenfunctions to solve inhomogeneous equations of the form (17.1). We shall see later that the sets of eigenfunctions $y_i(x)$ of a particular class of operators called *Hermitian operators* (the operator in the simple harmonic oscillator equation is an example) have particularly useful properties and these will be studied in detail. It turns out that many of the interesting differential operators met within the physical sciences are Hermitian. Before continuing our discussion of the eigenfunctions of Hermitian operators, however, we will consider some properties of general sets of functions.

17.1 Sets of functions

In chapter 8 we discussed the definition of a vector space but concentrated on spaces of finite dimensionality. We consider now the *infinite*-dimensional space of all reasonably well-behaved functions $f(x)$, $g(x)$, $h(x)$, ... on the interval $a \leq x \leq b$. That these functions form a linear vector space is shown by noting the following properties. The set is closed under

(i) addition, which is commutative and associative, i.e.

$$f(x) + g(x) = g(x) + f(x),$$
$$[f(x) + g(x)] + h(x) = f(x) + [g(x) + h(x)],$$

(ii) multiplication by a scalar, which is distributive and associative, i.e.

$$\lambda [f(x) + g(x)] = \lambda f(x) + \lambda g(x),$$
$$\lambda [\mu f(x)] = (\lambda \mu)f(x),$$
$$(\lambda + \mu)f(x) = \lambda f(x) + \mu f(x).$$

Furthermore, in such a space

(iii) there exists a 'null vector' 0 such that $f(x) + 0 = f(x)$,
(iv) multiplication by unity leaves any function unchanged, i.e. $1 \times f(x) = f(x)$,
(v) each function has an associated negative function $-f(x)$ that is such that $f(x) + [-f(x)] = 0$.

By analogy with finite-dimensional vector spaces we now introduce a set of linearly independent basis functions $y_n(x)$, $n = 0, 1, \ldots, \infty$, such that *any* 'reasonable' function in the interval $a \leq x \leq b$ (i.e. it obeys the Dirichlet conditions discussed in chapter 12) can be expressed as the linear sum of these functions:

$$f(x) = \sum_{n=0}^{\infty} c_n y_n(x).$$

Clearly if a different set of linearly independent basis functions $u_n(x)$ is chosen then the function can be expressed in terms of the new basis,

$$f(x) = \sum_{n=0}^{\infty} d_n u_n(x),$$

where the d_n are a different set of coefficients. In each case, provided the basis functions are linearly independent, the coefficients are unique.

We may also define an *inner product* on our function space by

$$\langle f|g\rangle = \int_a^b f^*(x)g(x)\rho(x)\,dx, \tag{17.6}$$

where $\rho(x)$ is the weight function, which we require to be real and non-negative in the interval $a \leq x \leq b$. As mentioned above, $\rho(x)$ is often unity for all x. Two functions are said to be *orthogonal* (with respect to the weight function $\rho(x)$) on the interval $[a, b]$ if

$$\langle f|g\rangle = \int_a^b f^*(x)g(x)\rho(x)\,dx = 0, \tag{17.7}$$

and the *norm* of a function is defined as

$$\|f\| = \langle f|f\rangle^{1/2} = \left[\int_a^b f^*(x)f(x)\rho(x)\,dx\right]^{1/2} = \left[\int_a^b |f(x)|^2\rho(x)\,dx\right]^{1/2}. \tag{17.8}$$

It is also common practice to define a *normalised* function by $\hat{f} = f/\|f\|$, which has unit norm.

An infinite-dimensional vector space of functions, for which an inner product is defined, is called a *Hilbert space*. Using the concept of the inner product, we can choose a basis of linearly independent functions $\hat{\phi}_n(x)$, $n = 0, 1, 2, \ldots$ that are orthonormal, i.e. such that

$$\langle \hat{\phi}_i|\hat{\phi}_j\rangle = \int_a^b \hat{\phi}_i^*(x)\hat{\phi}_j(x)\rho(x)\,dx = \delta_{ij}. \tag{17.9}$$

If $y_n(x)$, $n = 0, 1, 2, \ldots$, are a linearly independent, but not orthonormal, basis for the Hilbert space then an orthonormal set of basis functions $\hat{\phi}_n$ may be produced (in a similar manner to that used in the construction of a set of orthogonal eigenvectors of an Hermitian matrix; see chapter 8) by the following procedure:

$$\phi_0 = y_0,$$
$$\phi_1 = y_1 - \hat{\phi}_0\langle\hat{\phi}_0|y_1\rangle,$$
$$\phi_2 = y_2 - \hat{\phi}_1\langle\hat{\phi}_1|y_2\rangle - \hat{\phi}_0\langle\hat{\phi}_0|y_2\rangle,$$
$$\vdots$$
$$\phi_n = y_n - \hat{\phi}_{n-1}\langle\hat{\phi}_{n-1}|y_n\rangle - \cdots - \hat{\phi}_0\langle\hat{\phi}_0|y_n\rangle,$$
$$\vdots$$

It is straightforward to check that each ϕ_n is orthogonal to all its predecessors ϕ_i, $i = 0, 1, 2, \ldots, n - 1$. This method is called *Gram–Schmidt orthogonalisation*. Clearly the functions ϕ_n form an orthogonal set, but in general they do not have unit norms.

> ▶*Starting from the linearly independent functions $y_n(x) = x^n$, $n = 0, 1, \ldots$, construct three orthonormal functions over the range $-1 < x < 1$, assuming a weight function of unity.*

The first unnormalised function ϕ_0 is simply equal to the first of the original functions, i.e.

$$\phi_0 = 1.$$

The normalisation is carried out by dividing by

$$\langle \phi_0 | \phi_0 \rangle^{1/2} = \left(\int_{-1}^{1} 1 \times 1 \, du \right)^{1/2} = \sqrt{2},$$

with the result that the first normalised function $\hat{\phi}_0$ is given by

$$\hat{\phi}_0 = \frac{\phi_0}{\sqrt{2}} = \sqrt{\tfrac{1}{2}}.$$

The second unnormalised function is found by applying the above Gram–Schmidt orthogonalisation procedure, i.e.

$$\phi_1 = y_1 - \hat{\phi}_0 \langle \hat{\phi}_0 | y_1 \rangle.$$

It can easily be shown that $\langle \hat{\phi}_0 | y_1 \rangle = 0$, and so $\phi_1 = x$. Normalising then gives

$$\hat{\phi}_1 = \phi_1 \left(\int_{-1}^{1} u \times u \, du \right)^{-1/2} = \sqrt{\tfrac{3}{2}} x.$$

The third unnormalised function is similarly given by

$$\begin{aligned} \phi_2 &= y_2 - \hat{\phi}_1 \langle \hat{\phi}_1 | y_2 \rangle - \hat{\phi}_0 \langle \hat{\phi}_0 | y_2 \rangle \\ &= x^2 - 0 - \tfrac{1}{3}, \end{aligned}$$

which, on normalising, gives

$$\hat{\phi}_2 = \phi_2 \left(\int_{-1}^{1} \left(u^2 - \tfrac{1}{3} \right)^2 du \right)^{-1/2} = \tfrac{1}{2} \sqrt{\tfrac{5}{2}} (3x^2 - 1).$$

By comparing the functions $\hat{\phi}_0$, $\hat{\phi}_1$ and $\hat{\phi}_2$ with the list in subsection 18.1.1, we see that this procedure has generated (multiples of) the first three Legendre polynomials. ◀

If a function is expressed in terms of an *orthonormal* basis $\hat{\phi}_n(x)$ as

$$f(x) = \sum_{n=0}^{\infty} c_n \hat{\phi}_n(x) \tag{17.10}$$

then the coefficients c_n are given by

$$c_n = \langle \hat{\phi}_n | f \rangle = \int_a^b \hat{\phi}_n^*(x) f(x) \rho(x) \, dx. \tag{17.11}$$

Note that this is true only if the basis is orthonormal.

17.1.1 *Some useful inequalities*

Since for a Hilbert space $\langle f|f \rangle \geq 0$, the inequalities discussed in subsection 8.1.3 hold. The proofs are not repeated here, but the relationships are listed for completeness.

(i) The Schwarz inequality states that

$$|\langle f|g \rangle| \leq \langle f|f \rangle^{1/2} \langle g|g \rangle^{1/2}, \qquad (17.12)$$

where the equality holds when $f(x)$ is a scalar multiple of $g(x)$, i.e. when they are linearly dependent.

(ii) The triangle inequality states that

$$\|f + g\| \leq \|f\| + \|g\|, \qquad (17.13)$$

where again equality holds when $f(x)$ is a scalar multiple of $g(x)$.

(iii) Bessel's inequality requires the introduction of an *orthonormal* basis $\hat{\phi}_n(x)$ so that any function $f(x)$ can be written as

$$f(x) = \sum_{n=0}^{\infty} c_n \hat{\phi}_n(x),$$

where $c_n = \langle \hat{\phi}_n|f \rangle$. Bessel's inequality then states that

$$\langle f|f \rangle \geq \sum_n |c_n|^2. \qquad (17.14)$$

The equality holds if the summation is over all the basis functions. If some values of n are omitted from the sum then the inequality results (unless, of course, the c_n happen to be zero for all values of n omitted, in which case the equality remains).

17.2 Adjoint, self-adjoint and Hermitian operators

Having discussed general sets of functions, we now return to the discussion of eigenfunctions of linear operators. We begin by introducing the *adjoint* of an operator $\mathcal{L}$, denoted by $\mathcal{L}^\dagger$, which is defined by

$$\int_a^b f^*(x) \left[\mathcal{L}g(x) \right] dx = \int_a^b [\mathcal{L}^\dagger f(x)]^* g(x) \, dx + \text{boundary terms}, \qquad (17.15)$$

where the boundary terms are evaluated at the end-points of the interval $[a, b]$. Thus, for any given linear differential operator $\mathcal{L}$, the adjoint operator $\mathcal{L}^\dagger$ can be found by repeated integration by parts.

An operator is said to be *self-adjoint* if $\mathcal{L}^\dagger = \mathcal{L}$. If, in addition, certain boundary conditions are met by the functions f and g on which a self-adjoint operator acts, or by the operator itself, such that the boundary terms in (17.15) vanish, then the operator is said to be *Hermitian* over the interval $a \leq x \leq b$. Thus, in this case,

$$\int_a^b f^*(x)\,[\mathcal{L}g(x)]\,dx = \int_a^b [\mathcal{L}f(x)]^*g(x)\,dx. \tag{17.16}$$

A little careful study will reveal the similarity between the definition of an Hermitian operator and the definition of an Hermitian matrix given in chapter 8.

> ►*Show that the linear operator $\mathcal{L} = d^2/dt^2$ is self-adjoint, and determine the required boundary conditions for the operator to be Hermitian over the interval t_0 to $t_0 + T$.*

Substituting into the LHS of the definition of the adjoint operator (17.15) and integrating by parts gives

$$\int_{t_0}^{t_0+T} f^* \frac{d^2g}{dt^2}\,dt = \left[f^* \frac{dg}{dt} \right]_{t_0}^{t_0+T} - \int_{t_0}^{t_0+T} \frac{df^*}{dt}\frac{dg}{dt}\,dt.$$

Integrating the second term on the RHS by parts once more yields

$$\int_{t_0}^{t_0+T} f^* \frac{d^2g}{dt^2}\,dt = \left[f^* \frac{dg}{dt} \right]_{t_0}^{t_0+T} + \left[-\frac{df^*}{dt} g \right]_{t_0}^{t_0+T} + \int_{t_0}^{t_0+T} g\,\frac{d^2f^*}{dt^2}\,dt,$$

which, by comparison with (17.15), proves that $\mathcal{L}$ is a self-adjoint operator. Moreover, from (17.16), we see that $\mathcal{L}$ is an Hermitian operator over the required interval provided

$$\left[f^* \frac{dg}{dt} \right]_{t_0}^{t_0+T} = \left[\frac{df^*}{dt} g \right]_{t_0}^{t_0+T}. \quad ◄$$

We showed in chapter 8 that the eigenvalues of Hermitian matrices are real and that their eigenvectors can be chosen to be orthogonal. Similarly, the eigenvalues of Hermitian operators are real and their eigenfunctions can be chosen to be orthogonal (we will prove these properties in the following section). Hermitian operators (or matrices) are often used in the formulation of quantum mechanics. The eigenvalues then give the possible measured values of an observable quantity such as energy or angular momentum, and the physical requirement that such quantities must be real is ensured by the reality of these eigenvalues. Furthermore, the infinite set of eigenfunctions of an Hermitian operator form a complete basis set over the relevant interval, so that it is possible to expand any function $y(x)$ obeying the appropriate conditions in an eigenfunction series over this interval:

$$y(x) = \sum_{n=0}^{\infty} c_n y_n(x), \tag{17.17}$$

where the choice of suitable values for the c_n will make the sum arbitrarily close to $y(x)$.[§] These useful properties provide the motivation for a detailed study of Hermitian operators.

[§] The proof of the completeness of the eigenfunctions of an Hermitian operator is beyond the scope of this book. The reader should refer, for example, to R. Courant and D. Hilbert, *Methods of Mathematical Physics* (New York: Interscience, 1953).

17.3 Properties of Hermitian operators

We now provide proofs of some of the useful properties of Hermitian operators. Again much of the analysis is similar to that for Hermitian matrices in chapter 8, although the present section stands alone. (Here, and throughout the remainder of this chapter, we will write out inner products in full. We note, however, that the inner product notation often provides a neat form in which to express results.)

17.3.1 Reality of the eigenvalues

Consider an Hermitian operator for which (17.5) is satisfied by at least two eigenfunctions $y_i(x)$ and $y_j(x)$, which have corresponding eigenvalues λ_i and λ_j, so that

$$\mathcal{L}y_i = \lambda_i \rho(x) y_i, \tag{17.18}$$

$$\mathcal{L}y_j = \lambda_j \rho(x) y_j, \tag{17.19}$$

where we have allowed for the presence of a weight function $\rho(x)$. Multiplying (17.18) by y_j^* and (17.19) by y_i^* and then integrating gives

$$\int_a^b y_j^* \mathcal{L}y_i \, dx = \lambda_i \int_a^b y_j^* y_i \rho \, dx, \tag{17.20}$$

$$\int_a^b y_i^* \mathcal{L}y_j \, dx = \lambda_j \int_a^b y_i^* y_j \rho \, dx. \tag{17.21}$$

Remembering that we have required $\rho(x)$ to be real, the complex conjugate of (17.20) becomes

$$\int_a^b y_j(\mathcal{L}y_i)^* \, dx = \lambda_i^* \int_a^b y_i^* y_j \rho \, dx, \tag{17.22}$$

and using the definition of an Hermitian operator (17.16) it follows that the LHS of (17.22) is equal to the LHS of (17.21). Thus

$$(\lambda_i^* - \lambda_j) \int_a^b y_i^* y_j \rho \, dx = 0. \tag{17.23}$$

If $i = j$ then $\lambda_i = \lambda_i^*$ (since $\int_a^b y_i^* y_i \rho \, dx \neq 0$), which is a statement that the eigenvalue λ_i is real.

17.3.2 Orthogonality and normalisation of the eigenfunctions

From (17.23), it is immediately apparent that two eigenfunctions y_i and y_j that correspond to different eigenvalues, i.e. such that $\lambda_i \neq \lambda_j$, satisfy

$$\int_a^b y_i^* y_j \rho \, dx = 0, \tag{17.24}$$

561

which is a statement of the orthogonality of y_i and y_j.

If one (or more) of the eigenvalues is degenerate, however, we have different eigenfunctions corresponding to the same eigenvalue, and the proof of orthogonality is not so straightforward. Nevertheless, an orthogonal set of eigenfunctions may be constructed using the *Gram–Schmidt orthogonalisation* method mentioned earlier in this chapter and used in chapter 8 to construct a set of orthogonal eigenvectors of an Hermitian matrix. We repeat the analysis here for completeness.

Suppose, for the sake of our proof, that λ_0 is k-fold degenerate, i.e.

$$\mathcal{L}y_i = \lambda_0 \rho y_i \quad \text{for } i = 0, 1, \ldots, k-1, \tag{17.25}$$

but that λ_0 is different from any of λ_k, λ_{k+1}, etc. Then any linear combination of these y_i is also an eigenfunction with eigenvalue λ_0 since

$$\mathcal{L}z \equiv \mathcal{L} \sum_{i=0}^{k-1} c_i y_i = \sum_{i=0}^{k-1} c_i \mathcal{L}y_i = \sum_{i=0}^{k-1} c_i \lambda_0 \rho y_i = \lambda_0 \rho z. \tag{17.26}$$

If the y_i defined in (17.25) are not already mutually orthogonal then consider the new eigenfunctions z_i constructed by the following procedure, in which each of the new functions z_i is to be normalised, to give $\hat{z}_i$, before proceeding to the construction of the next one (the normalisation can be carried out by dividing the eigenfunction z_i by $(\int_a^b z_i^* z_i \rho \, dx)^{1/2}$):

$$z_0 = y_0,$$

$$z_1 = y_1 - \left(\hat{z}_0 \int_a^b \hat{z}_0^* y_1 \rho \, dx \right),$$

$$z_2 = y_2 - \left(\hat{z}_1 \int_a^b \hat{z}_1^* y_2 \rho \, dx \right) - \left(\hat{z}_0 \int_a^b \hat{z}_0^* y_2 \rho \, dx \right),$$

$$\vdots$$

$$z_{k-1} = y_{k-1} - \left(\hat{z}_{k-2} \int_a^b \hat{z}_{k-2}^* y_{k-1} \rho \, dx \right) - \cdots - \left(\hat{z}_0 \int_a^b \hat{z}_0^* y_{k-1} \rho \, dx \right).$$

Each of the integrals is just a number and thus each new function z_i is, as can be shown from (17.26), an eigenvector of $\mathcal{L}$ with eigenvalue λ_0. It is straightforward to check that each z_i is orthogonal to all its predecessors. Thus, by this explicit construction we have shown that an orthogonal set of eigenfunctions of an Hermitian operator $\mathcal{L}$ can be obtained. Clearly the orthogonal set obtained, z_i, is not unique.

In general, since $\mathcal{L}$ is linear, the normalisation of its eigenfunctions $y_i(x)$ is arbitrary. It is often convenient, however, to work in terms of the normalised eigenfunctions $\hat{y}_i(x)$, so that $\int_a^b \hat{y}_i^* \hat{y}_i \rho \, dx = 1$. These therefore form an orthonormal set and we can write

$$\int_a^b \hat{y}_i^* \hat{y}_j \rho \, dx = \delta_{ij},$$ (17.27)

which is valid for all pairs of values i, j.

17.3.3 Completeness of the eigenfunctions

As noted earlier, the eigenfunctions of an Hermitian operator may be shown to form a complete basis set over the relevant interval. One may thus expand any (reasonable) function $y(x)$ obeying appropriate boundary conditions in an eigenfunction series over the interval, as in (17.17). Working in terms of the normalised eigenfunctions $\hat{y}_n(x)$, we may thus write

$$f(x) = \sum_n \hat{y}_n(x) \int_a^b \hat{y}_n^*(z) f(z) \rho(z) \, dz$$

$$= \int_a^b f(z) \rho(z) \sum_n \hat{y}_n(x) \hat{y}_n^*(z) \, dz.$$

Since this is true for any $f(x)$, we must have that

$$\rho(z) \sum_n \hat{y}_n(x) \hat{y}_n^*(z) = \delta(x - z).$$ (17.28)

This is called the *completeness* or *closure* property of the eigenfunctions. It defines a complete set. If the spectrum of eigenvalues of $\mathcal{L}$ is anywhere continuous then the eigenfunction $y_n(x)$ must be treated as $y(n, x)$ and an integration carried out over n.

We also note that the RHS of (17.28) is a δ-function and so is only non-zero when $z = x$; thus $\rho(z)$ on the LHS can be replaced by $\rho(x)$ if required, i.e.

$$\rho(z) \sum_n \hat{y}_n(x) \hat{y}_n^*(z) = \rho(x) \sum_n \hat{y}_n(x) \hat{y}_n^*(z).$$ (17.29)

17.3.4 Construction of real eigenfunctions

Recall that the eigenfunction y_i satisfies

$$\mathcal{L} y_i = \lambda_i \rho y_i$$ (17.30)

and that, for a real Hermitian operator, the complex conjugate of this gives

$$\mathcal{L} y_i^* = \lambda_i^* \rho y_i^* = \lambda_i \rho y_i^*,$$ (17.31)

where the last equality follows because the eigenvalues are real, i.e. $\lambda_i = \lambda_i^*$. Thus, y_i and y_i^* are eigenfunctions corresponding to the same eigenvalue and hence, because of the linearity of $\mathcal{L}$, at least one of $y_i^* + y_i$ and $i(y_i^* - y_i)$, which are

563

both real, is a non-zero eigenfunction corresponding to that eigenvalue. It follows that the eigenfunctions of a real Hermitian differential operator can always be made real by taking suitable linear combinations, though taking such linear combinations will only be necessary in cases where a particular λ is degenerate, i.e. corresponds to more than one linearly independent eigenfunction.

17.4 Sturm–Liouville equations

One of the most important applications of our discussion of Hermitian operators is to the study of *Sturm–Liouville equations*, which take the general form

$$p(x)\frac{d^2y}{dx^2} + r(x)\frac{dy}{dx} + q(x)y + \lambda\rho(x)y = 0, \quad \text{where } r(x) = \frac{dp(x)}{dx} \tag{17.32}$$

and p, q and r are real functions of x.[§] A variational approach to the Sturm–Liouville equation, which is useful in estimating the eigenvalues λ for a given set of boundary conditions on y, is discussed in chapter 22. For now, however, we concentrate on demonstrating that solutions of the Sturm–Liouville equation that satisfy appropriate boundary conditions are the eigenfunctions of an Hermitian operator.

It is clear that (17.32) can be written

$$\mathcal{L}y = \lambda\rho(x)y, \quad \text{where } \mathcal{L} \equiv -\left[p(x)\frac{d^2}{dx^2} + r(x)\frac{d}{dx} + q(x)\right]. \tag{17.33}$$

Using the condition that $r(x) = p'(x)$, it will be seen that the general Sturm–Liouville equation (17.32) can also be rewritten as

$$(py')' + qy + \lambda\rho y = 0, \tag{17.34}$$

where primes denote differentiation with respect to x. Using (17.33) this may also be written $\mathcal{L}y \equiv -(py')' - qy = \lambda\rho y$, which defines a more useful form for the Sturm–Liouville linear operator, namely

$$\mathcal{L} \equiv -\left[\frac{d}{dx}\left(p(x)\frac{d}{dx}\right) + q(x)\right]. \tag{17.35}$$

17.4.1 Hermitian nature of the Sturm–Liouville operator

As we now show, the linear operator of the Sturm–Liouville equation (17.35) is self-adjoint. Moreover, the operator is Hermitian over the range $[a, b]$ provided

---

[§] We note that sign conventions vary in this expression for the general Sturm–Liouville equation; some authors use $-\lambda\rho(x)y$ on the LHS of (17.32).

certain boundary conditions are met, namely that any two eigenfunctions y_i and y_j of (17.33) must satisfy

$$\left[y_i^* p y_j'\right]_{x=a} = \left[y_i^* p y_j'\right]_{x=b} \qquad \text{for all } i, j. \tag{17.36}$$

Rearranging (17.36), we can write

$$\left[y_i^* p y_j'\right]_{x=a}^{x=b} = 0 \tag{17.37}$$

as an equivalent statement of the required boundary conditions. These boundary conditions are in fact not too restrictive and are met, for instance, by the sets $y(a) = y(b) = 0$; $y(a) = y'(b) = 0$; $p(a) = p(b) = 0$ and by many other sets. It is important to note that in order to satisfy (17.36) and (17.37) one boundary condition must be specified at each end of the range.

> ▶*Prove that the Sturm–Liouville operator is Hermitian over the range $[a, b]$ and under the boundary conditions (17.37).*

Putting the Sturm–Liouville form $\mathcal{L}y = -(py')' - qy$ into the definition (17.16) of an Hermitian operator, the LHS may be written as a sum of two terms, i.e.

$$-\int_a^b \left[y_i^*(py_j')' + y_i^* q y_j\right] dx = -\int_a^b y_i^*(py_j')' \, dx - \int_a^b y_i^* q y_j \, dx.$$

The first term may be integrated by parts to give

$$-\left[y_i^* p y_j'\right]_a^b + \int_a^b (y_i^*)' p y_j' \, dx.$$

The boundary-value term in this is zero because of the boundary conditions, and so integrating by parts again yields

$$\left[(y_i^*)' p y_j\right]_a^b - \int_a^b ((y_i^*)' p)' y_j \, dx.$$

Again, the boundary-value term is zero, leaving us with

$$-\int_a^b \left[y_i^*(py_j')' + y_i^* q y_j\right] dx = -\int_a^b \left[y_j(p(y_i^*)')' + y_j q y_i^*\right] dx,$$

which proves that the Sturm–Liouville operator is Hermitian over the prescribed interval. ◀

It is also worth noting that, since $p(a) = p(b) = 0$ is a valid set of boundary conditions, many Sturm–Liouville equations possess a 'natural' interval $[a, b]$ over which the corresponding differential operator $\mathcal{L}$ is Hermitian *irrespective* of the boundary conditions satisfied by its eigenfunctions at $x = a$ and $x = b$ (the only requirement being that they are regular at these end-points).

17.4.2 Transforming an equation into Sturm–Liouville form

Many of the second-order differential equations encountered in physical problems are examples of the Sturm–Liouville equation (17.34). Moreover, *any* second-order

Equation	$p(x)$	$q(x)$	λ	$\rho(x)$
Hypergeometric	$x^c(1-x)^{a+b-c+1}$	0	$-ab$	$x^{c-1}(1-x)^{a+b-c}$
Legendre	$1-x^2$	0	$\ell(\ell+1)$	1
Associated Legendre	$1-x^2$	$-m^2/(1-x^2)$	$\ell(\ell+1)$	1
Chebyshev	$(1-x^2)^{1/2}$	0	ν^2	$(1-x^2)^{-1/2}$
Confluent hypergeometric	$x^c e^{-x}$	0	$-a$	$x^{c-1}e^{-x}$
Bessel*	x	$-\nu^2/x$	α^2	x
Laguerre	xe^{-x}	0	ν	e^{-x}
Associated Laguerre	$x^{m+1}e^{-x}$	0	ν	$x^m e^{-x}$
Hermite	e^{-x^2}	0	2ν	e^{-x^2}
Simple harmonic	1	0	ω^2	1

Table 17.1 The Sturm–Liouville form (17.34) for important ODEs in the physical sciences and engineering. The asterisk denotes that, for Bessel's equation, a change of variable $x \to x/\alpha$ is required to give the conventional normalisation used here, but is not needed for the transformation into Sturm–Liouville form.

differential equation of the form

$$p(x)y'' + r(x)y' + q(x)y + \lambda\rho(x)y = 0 \tag{17.38}$$

can be converted into Sturm–Liouville form by multiplying through by a suitable integrating factor, which is given by

$$F(x) = \exp\left\{\int^x \frac{r(u)-p'(u)}{p(u)}\,du\right\}. \tag{17.39}$$

It is easily verified that (17.38) then takes the Sturm–Liouville form,

$$[F(x)p(x)y']' + F(x)q(x)y + \lambda F(x)\rho(x)y = 0, \tag{17.40}$$

with a different, but still non-negative, weight function $F(x)\rho(x)$. Table 17.1 summarises the Sturm–Liouville form (17.34) for several of the equations listed in table 16.1. These forms can be determined using (17.39), as illustrated in the following example.

▶Put the following equations into Sturm–Liouville (SL) form:
  (i) $(1-x^2)y'' - xy' + \nu^2 y = 0$ (Chebyshev equation);
  (ii) $xy'' + (1-x)y' + \nu y = 0$ (Laguerre equation);
  (iii) $y'' - 2xy' + 2\nu y = 0$ (Hermite equation).

(i) From (17.39), the required integrating factor is

$$F(x) = \exp\left(\int^x \frac{u}{1-u^2}\,du\right) = \exp\left[-\tfrac{1}{2}\ln(1-x^2)\right] = (1-x^2)^{-1/2}.$$

Thus, the Chebyshev equation becomes

$$(1-x^2)^{1/2}y'' - x(1-x^2)^{-1/2}y' + \nu^2(1-x^2)^{-1/2}y = \left[(1-x^2)^{1/2}y'\right]' + \nu^2(1-x^2)^{-1/2}y = 0,$$

which is in SL form with $p(x) = (1-x^2)^{1/2}$, $q(x) = 0$, $\rho(x) = (1-x^2)^{-1/2}$ and $\lambda = \nu^2$.

(ii) From (17.39), the required integrating factor is

$$F(x) = \exp \left(\int^x -1 \, du \right) = \exp(-x).$$

Thus, the Laguerre equation becomes

$$xe^{-x}y'' + (1-x)e^{-x}y' + ve^{-x}y = (xe^{-x}y')' + ve^{-x}y = 0,$$

which is in SL form with $p(x) = xe^{-x}$, $q(x) = 0$, $\rho(x) = e^{-x}$ and $\lambda = v$.

(iii) From (17.39), the required integrating factor is

$$F(x) = \exp \left(\int^x -2u \, du \right) = \exp(-x^2).$$

Thus, the Hermite equation becomes

$$e^{-x^2}y'' - 2xe^{-x^2}y' + 2ve^{-x^2}y = (e^{-x^2}y')' + 2ve^{-x^2}y = 0,$$

which is in SL form with $p(x) = e^{-x^2}$, $q(x) = 0$, $\rho(x) = e^{-x^2}$ and $\lambda = 2v$. ◀

From the $p(x)$ entries in table 17.1, we may read off the natural interval over which the corresponding Sturm–Liouville operator (17.35) is Hermitian; in each case this is given by $[a, b]$, where $p(a) = p(b) = 0$. Thus, the natural interval for the Legendre equation, the associated Legendre equation and the Chebyshev equation is $[-1, 1]$; for the Laguerre and associated Laguerre equations the interval is $[0, \infty]$; and for the Hermite equation it is $[-\infty, \infty]$. In addition, from (17.37), one sees that for the simple harmonic equation one requires only that $[a, b] = [x_0, x_0 + 2\pi]$. We also note that, as required, the weight function in each case is finite and non-negative over the natural interval. Occasionally, a little more care is required when determining the conditions for a Sturm–Liouville operator of the form (17.35) to be Hermitian over some natural interval, as is illustrated in the following example.

▶*Express the hypergeometric equation,*

$$x(1-x)y'' + [c - (a+b+1)x]y' - aby = 0,$$

in Sturm–Liouville form. Hence determine the natural interval over which the resulting Sturm–Liouville operator is Hermitian and the corresponding conditions that one must impose on the parameters a, b and c.

As usual for an equation not already in SL form, we first determine the appropriate integrating factor. This is given, as in equation (17.39), by

$$F(x) = \exp\left[\int^x \frac{c - (a+b+1)u - 1 + 2u}{u(1-u)} \, du\right]$$

$$= \exp\left[\int^x \frac{c - 1 - (a+b-1)u}{u(1-u)} \, du\right]$$

$$= \exp\left[\int^x \left(\frac{c-1}{1-u} + \frac{c-1}{u} - \frac{a+b-1}{1-u}\right) du\right]$$

$$= \exp\left[(a+b-c)\ln(1-x) + (c-1)\ln x\right]$$

$$= x^{c-1}(1-x)^{a+b-c}.$$

When the equation is multiplied through by $F(x)$ it takes the form

$$\left[x^c(1-x)^{a+b-c+1}y'\right]' - abx^{c-1}(1-x)^{a+b-c}y = 0.$$

Now, for the corresponding Sturm–Liouville operator to be Hermitian, the conditions to be imposed are as follows.

(i) The boundary condition (17.37); if $c > 0$ and $a + b - c + 1 > 0$, this is satisfied automatically for $0 \le x \le 1$, which is thus the natural interval in this case.
(ii) The weight function $x^{c-1}(1-x)^{a+b-c}$ must be finite and not change sign in the interval $0 \le x \le 1$. This means that both exponents in it must be positive, i.e. $c - 1 > 0$ and $a + b - c > 0$.

Putting together the conditions on the parameters gives the double inequality $a + b > c > 1$. ◀

Finally, we consider Bessel's equation,

$$x^2 y'' + xy' + (x^2 - v^2)y = 0,$$

which may be converted into Sturm–Liouville form, but only in a somewhat unorthodox fashion. It is conventional first to divide the Bessel equation by x and then to change variables to $\bar{x} = x/\alpha$. In this case, it becomes

$$\bar{x}y''(\alpha\bar{x}) + y'(\alpha\bar{x}) - \frac{v^2}{\bar{x}}y(\alpha\bar{x}) + \alpha^2\bar{x}y(\alpha\bar{x}) = 0, \tag{17.41}$$

where a prime now indicates differentiation with respect to $\bar{x}$. Dropping the bars on the independent variable, we thus have

$$[xy'(\alpha x)]' - \frac{v^2}{x}y(\alpha x) + \alpha^2 xy(\alpha x) = 0, \tag{17.42}$$

which is in SL form with $p(x) = x$, $q(x) = -v^2/x$, $\rho(x) = x$ and $\lambda = \alpha^2$. It should be noted, however, that in this case the eigenvalue (actually its square root) appears in the argument of the dependent variable.

17.5 Superposition of eigenfunctions: Green's functions

We have already seen that if

$$\mathcal{L}y_n(x) = \lambda_n \rho(x) y_n(x), \tag{17.43}$$

where $\mathcal{L}$ is an Hermitian operator, then the eigenvalues λ_n are real and the eigenfunctions $y_n(x)$ are orthogonal (or can be made so). Let us assume that we know the eigenfunctions $y_n(x)$ of $\mathcal{L}$ that individually satisfy (17.43) and some imposed boundary conditions (for which $\mathcal{L}$ is Hermitian).

Now let us suppose we wish to solve the inhomogeneous differential equation

$$\mathcal{L}y(x) = f(x), \tag{17.44}$$

subject to the same boundary conditions. Since the eigenfunctions of $\mathcal{L}$ form a complete set, the full solution, $y(x)$, to (17.44) may be written as a superposition of eigenfunctions, i.e.

$$y(x) = \sum_{n=0}^{\infty} c_n y_n(x), \tag{17.45}$$

for some choice of the constants c_n. Making full use of the linearity of $\mathcal{L}$, we have

$$f(x) = \mathcal{L}y(x) = \mathcal{L}\left(\sum_{n=0}^{\infty} c_n y_n(x)\right) = \sum_{n=0}^{\infty} c_n \mathcal{L}y_n(x) = \sum_{n=0}^{\infty} c_n \lambda_n \rho(x) y_n(x). \tag{17.46}$$

Multiplying the first and last terms of (17.46) by y_j^* and integrating, we obtain

$$\int_a^b y_j^*(z)f(z)\,dz = \sum_{n=0}^{\infty} \int_a^b c_n \lambda_n y_j^*(z) y_n(z)\rho(z)\,dz, \tag{17.47}$$

where we have used z as the integration variable for later convenience. Finally, using the orthogonality condition (17.27), we see that the integrals on the RHS are zero unless $n = j$, and so obtain

$$c_n = \frac{1}{\lambda_n}\frac{\int_a^b y_n^*(z)f(z)\,dz}{\int_a^b y_n^*(z) y_n(z)\rho(z)\,dz}. \tag{17.48}$$

Thus, if we can find all the eigenfunctions of a differential operator then (17.48) can be used to find the weighting coefficients for the superposition, to give as the full solution

$$y(x) = \sum_{n=0}^{\infty} \frac{1}{\lambda_n}\frac{\int_a^b y_n^*(z)f(z)\,dz}{\int_a^b y_n^*(z) y_n(z)\rho(z)\,dz}\, y_n(x). \tag{17.49}$$

If we work with normalised eigenfunctions $\hat{y}_n(x)$, so that

$$\int_a^b \hat{y}_n^*(z)\hat{y}_n(z)\rho(z)\,dz = 1 \qquad \text{for all } n,$$

and we assume that we may interchange the order of summation and integration, then (17.49) can be written as

$$y(x) = \int_a^b \left\{ \sum_{n=0}^{\infty} \left[\frac{1}{\lambda_n} \hat{y}_n(x) \hat{y}_n^*(z) \right] \right\} f(z) \, dz.$$

The quantity in braces, which is a function of x and z only, is usually written $G(x, z)$, and is the *Green's function* for the problem. With this notation,

$$y(x) = \int_a^b G(x, z) f(z) \, dz, \tag{17.50}$$

where

$$G(x, z) = \sum_{n=0}^{\infty} \frac{1}{\lambda_n} \hat{y}_n(x) \hat{y}_n^*(z). \tag{17.51}$$

We note that $G(x, z)$ is determined entirely by the boundary conditions and the eigenfunctions $\hat{y}_n$, and hence by $\mathcal{L}$ itself, and that $f(z)$ depends purely on the RHS of the inhomogeneous equation (17.44). Thus, for a given $\mathcal{L}$ and boundary conditions we can establish, once and for all, a function $G(x, z)$ that will enable us to solve the inhomogeneous equation for *any* RHS. From (17.51) we also note that

$$G(x, z) = G^*(z, x). \tag{17.52}$$

We have already met the Green's function in the solution of second-order differential equations in chapter 15, as the function that satisfies the equation $\mathcal{L}[G(x, z)] = \delta(x - z)$ (and the boundary conditions). The formulation given above is an alternative, though equivalent, one.

▶ *Find an appropriate Green's function for the equation*
$$y'' + \tfrac{1}{4} y = f(x),$$
with boundary conditions $y(0) = y(\pi) = 0$. *Hence, solve for* (i) $f(x) = \sin 2x$ *and* (ii) $f(x) = x/2$.

One approach to solving this problem is to use the methods of chapter 15 and find a complementary function and particular integral. However, in order to illustrate the techniques developed in the present chapter we will use the superposition of eigenfunctions, which, as may easily be checked, produces the same solution.

The operator on the LHS of this equation is already Hermitian under the given boundary conditions, and so we seek its eigenfunctions. These satisfy the equation

$$y'' + \tfrac{1}{4} y = \lambda y.$$

This equation has the familiar solution

$$y(x) = A \sin \left(\sqrt{\tfrac{1}{4} - \lambda} \right) x + B \cos \left(\sqrt{\tfrac{1}{4} - \lambda} \right) x.$$

Now, the boundary conditions require that $B = 0$ and $\sin\left(\sqrt{\frac{1}{4} - \lambda}\right)\pi = 0$, and so

$$\sqrt{\tfrac{1}{4} - \lambda} = n, \qquad \text{where } n = 0, \pm 1, \pm 2, \ldots.$$

Therefore, the independent eigenfunctions that satisfy the boundary conditions are

$$y_n(x) = A_n \sin nx,$$

where n is any non-negative integer, and the corresponding eigenvalues are $\lambda_n = \frac{1}{4} - n^2$. The normalisation condition further requires

$$\int_0^\pi A_n^2 \sin^2 nx \, dx = 1 \quad \Rightarrow \quad A_n = \left(\frac{2}{\pi}\right)^{1/2}.$$

Comparison with (17.51) shows that the appropriate Green's function is therefore given by

$$G(x, z) = \frac{2}{\pi} \sum_{n=0}^\infty \frac{\sin nx \sin nz}{\frac{1}{4} - n^2}.$$

Case (i). Using (17.50), the solution with $f(x) = \sin 2x$ is given by

$$y(x) = \frac{2}{\pi} \int_0^\pi \left(\sum_{n=0}^\infty \frac{\sin nx \sin nz}{\frac{1}{4} - n^2}\right) \sin 2z \, dz = \frac{2}{\pi} \sum_{n=0}^\infty \frac{\sin nx}{\frac{1}{4} - n^2} \int_0^\pi \sin nz \sin 2z \, dz.$$

Now the integral is zero unless $n = 2$, in which case it is

$$\int_0^\pi \sin^2 2z \, dz = \frac{\pi}{2}.$$

Thus

$$y(x) = -\frac{2}{\pi} \frac{\sin 2x}{15/4} \frac{\pi}{2} = -\frac{4}{15} \sin 2x$$

is the full solution for $f(x) = \sin 2x$. This is, of course, exactly the solution found by using the methods of chapter 15.

Case (ii). The solution with $f(x) = x/2$ is given by

$$y(x) = \int_0^\pi \left(\frac{2}{\pi} \sum_{n=0}^\infty \frac{\sin nx \sin nz}{\frac{1}{4} - n^2}\right) \frac{z}{2} \, dz = \frac{1}{\pi} \sum_{n=0}^\infty \frac{\sin nx}{\frac{1}{4} - n^2} \int_0^\pi z \sin nz \, dz.$$

The integral may be evaluated by integrating by parts. For $n \neq 0$,

$$\int_0^\pi z \sin nz \, dz = \left[-\frac{z \cos nz}{n}\right]_0^\pi + \int_0^\pi \frac{\cos nz}{n} \, dz$$

$$= \frac{-\pi \cos n\pi}{n} + \left[\frac{\sin nz}{n^2}\right]_0^\pi$$

$$= -\frac{\pi(-1)^n}{n}.$$

For $n = 0$ the integral is zero, and thus

$$y(x) = \sum_{n=1}^\infty (-1)^{n+1} \frac{\sin nx}{n\left(\frac{1}{4} - n^2\right)},$$

is the full solution for $f(x) = x/2$. Using the methods of subsection 15.1.2, the solution is found to be $y(x) = 2x - 2\pi \sin(x/2)$, which may be shown to be equal to the above solution by expanding $2x - 2\pi \sin(x/2)$ as a Fourier sine series. ◀

17.6 A useful generalisation

Sometimes we encounter inhomogeneous equations of a form slightly more general than (17.1), given by

$$\mathcal{L}y(x) - \mu\rho(x)y(x) = f(x) \tag{17.53}$$

for some Hermitian operator $\mathcal{L}$, with y subject to the appropriate boundary conditions and μ a given (i.e. *fixed*) constant. To solve this equation we expand $y(x)$ and $f(x)$ in terms of the eigenfunctions $y_n(x)$ of the operator $\mathcal{L}$, which satisfy

$$\mathcal{L}y_n(x) = \lambda_n\rho(x)y_n(x).$$

Working in terms of the normalised eigenfunctions $\hat{y}_n(x)$, we first expand $f(x)$ as follows:

$$
\begin{aligned}
f(x) &= \sum_{n=0}^{\infty} \hat{y}_n(x) \int_a^b \hat{y}_n^*(z)f(z)\rho(z)\,dz \\
&= \int_a^b \rho(z) \sum_{n=0}^{\infty} \hat{y}_n(x)\hat{y}_n^*(z)f(z)\,dz.
\end{aligned}
\tag{17.54}
$$

Using (17.29) this becomes

$$
\begin{aligned}
f(x) &= \int_a^b \rho(x) \sum_{n=0}^{\infty} \hat{y}_n(x)\hat{y}_n^*(z)f(z)\,dz \\
&= \rho(x) \sum_{n=0}^{\infty} \hat{y}_n(x) \int_a^b \hat{y}_n^*(z)f(z)\,dz.
\end{aligned}
\tag{17.55}
$$

Next, we expand $y(x)$ as $y = \sum_{n=0}^{\infty} c_n\hat{y}_n(x)$ and seek the coefficients c_n. Substituting this and (17.55) into (17.53) we have

$$\rho(x)\sum_{n=0}^{\infty}(\lambda_n - \mu)c_n\hat{y}_n(x) = \rho(x)\sum_{n=0}^{\infty}\hat{y}_n(x)\int_a^b \hat{y}_n^*(z)f(z)\,dz,$$

from which we find that

$$c_n = \sum_{n=0}^{\infty} \frac{\int_a^b \hat{y}_n^*(z)f(z)\,dz}{\lambda_n - \mu}.$$

Hence the solution of (17.53) is given by

$$y = \sum_{n=0}^{\infty} c_n\hat{y}_n(x) = \sum_{n=0}^{\infty} \frac{\hat{y}_n(x)}{\lambda_n - \mu} \int_a^b \hat{y}_n^*(z)f(z)\,dz = \int_a^b \sum_{n=0}^{\infty} \frac{\hat{y}_n(x)\hat{y}_n^*(z)}{\lambda_n - \mu} f(z)\,dz.$$

From this we may identify the Green's function

$$G(x,z) = \sum_{n=0}^{\infty} \frac{\hat{y}_n(x)\hat{y}_n^*(z)}{\lambda_n - \mu}.$$

We note that if $\mu = \lambda_n$, i.e. if μ equals one of the eigenvalues of $\mathcal{L}$, then $G(x, z)$ becomes infinite and this method runs into difficulty. No solution then exists unless the RHS of (17.53) satisfies the relation

$$\int_a^b \hat{y}_n^*(x) f(x)\, dx = 0.$$

If the spectrum of eigenvalues of the operator $\mathcal{L}$ is anywhere continuous, the orthonormality and closure relationships of the normalised eigenfunctions become

$$\int_a^b \hat{y}_n^*(x) \hat{y}_m(x) \rho(x)\, dx = \delta(n - m),$$

$$\int_0^\infty \hat{y}_n^*(z) \hat{y}_n(x) \rho(x)\, dn = \delta(x - z).$$

Repeating the above analysis we then find that the Green's function is given by

$$G(x, z) = \int_0^\infty \frac{\hat{y}_n(x) \hat{y}_n^*(z)}{\lambda_n - \mu}\, dn.$$

17.7 Exercises

17.1 By considering $\langle h|h \rangle$, where $h = f + \lambda g$ with λ real, prove that, for two functions f and g,

$$\langle f|f \rangle \langle g|g \rangle \geq \tfrac{1}{4} [\langle f|g \rangle + \langle g|f \rangle]^2.$$

The function $y(x)$ is real and positive for all x. Its Fourier cosine transform $\tilde{y}_c(k)$ is defined by

$$\tilde{y}_c(k) = \int_{-\infty}^{\infty} y(x) \cos(kx)\, dx,$$

and it is given that $\tilde{y}_c(0) = 1$. Prove that

$$\tilde{y}_c(2k) \geq 2[\tilde{y}_c(k)]^2 - 1.$$

17.2 Write the homogeneous Sturm–Liouville eigenvalue equation for which $y(a) = y(b) = 0$ as

$$\mathcal{L}(y; \lambda) \equiv (py')' + qy + \lambda \rho y = 0,$$

where $p(x), q(x)$ and $\rho(x)$ are continuously differentiable functions. Show that if $z(x)$ and $F(x)$ satisfy $\mathcal{L}(z; \lambda) = F(x)$, with $z(a) = z(b) = 0$, then

$$\int_a^b y(x) F(x)\, dx = 0.$$

Demonstrate the validity of this general result by direct calculation for the specific case in which $p(x) = \rho(x) = 1$, $q(x) = 0$, $a = -1$, $b = 1$ and $z(x) = 1 - x^2$.

17.3 Consider the real eigenfunctions $y_n(x)$ of a Sturm–Liouville equation,

$$(py')' + qy + \lambda \rho y = 0, \qquad a \leq x \leq b,$$

in which $p(x), q(x)$ and $\rho(x)$ are continuously differentiable real functions and $p(x)$ does not change sign in $a \leq x \leq b$. Take $p(x)$ as positive throughout the interval, if necessary by changing the signs of all eigenvalues. For $a \leq x_1 \leq x_2 \leq b$, establish the identity

$$(\lambda_n - \lambda_m) \int_{x_1}^{x_2} \rho y_n y_m \, dx = \left[y_n \, p \, y_m' - y_m \, p \, y_n' \right]_{x_1}^{x_2}.$$

Deduce that if $\lambda_n > \lambda_m$ then $y_n(x)$ must change sign between two successive zeros of $y_m(x)$.

[The reader may find it helpful to illustrate this result by sketching the first few eigenfunctions of the system $y'' + \lambda y = 0$, with $y(0) = y(\pi) = 0$, and the Legendre polynomials $P_n(z)$ for $n = 2, 3, 4, 5$.]

17.4   Show that the equation

$$y'' + a\delta(x)y + \lambda y = 0,$$

with $y(\pm\pi) = 0$ and a real, has a set of eigenvalues λ satisfying

$$\tan(\pi\sqrt{\lambda}) = \frac{2\sqrt{\lambda}}{a}.$$

Investigate the conditions under which negative eigenvalues, $\lambda = -\mu^2$, with μ real, are possible.

17.5   Use the properties of Legendre polynomials to carry out the following exercises.

(a) Find the solution of $(1 - x^2)y'' - 2xy' + by = f(x)$, valid in the range $-1 \leq x \leq 1$ and finite at $x = 0$, in terms of Legendre polynomials.

(b) If $b = 14$ and $f(x) = 5x^3$, find the explicit solution and verify it by direct substitution.

[The first six Legendre polynomials are listed in Subsection 18.1.1.]

17.6   Starting from the linearly independent functions 1, x, x^2, x^3, ... , in the range $0 \leq x < \infty$, find the first three orthogonal functions ϕ_0, ϕ_1 and ϕ_2, with respect to the weight function $\rho(x) = e^{-x}$. By comparing your answers with the Laguerre polynomials generated by the recurrence relation (18.115), deduce the form of $\phi_3(x)$.

17.7   Consider the set of functions, $\{f(x)\}$, of the real variable x, defined in the interval $-\infty < x < \infty$, that $\to 0$ at least as quickly as x^{-1} as $x \to \pm\infty$. For unit weight function, determine whether each of the following linear operators is Hermitian when acting upon $\{f(x)\}$:

(a) $\dfrac{d}{dx} + x$;   (b) $-i\dfrac{d}{dx} + x^2$;   (c) $ix\dfrac{d}{dx}$;   (d) $i\dfrac{d^3}{dx^3}$.

17.8   A particle moves in a parabolic potential in which its natural angular frequency of oscillation is $\frac{1}{2}$. At time $t = 0$ it passes through the origin with velocity v. It is then suddenly subjected to an additional acceleration, of $+1$ for $0 \leq t \leq \pi/2$, followed by -1 for $\pi/2 < t \leq \pi$. At the end of this period it is again at the origin. Apply the results of the worked example in section 17.5 to show that

$$v = -\frac{8}{\pi} \sum_{m=0}^{\infty} \frac{1}{(4m+2)^2 - \frac{1}{4}} \approx -0.81.$$

17.9   Find an eigenfunction expansion for the solution, with boundary conditions $y(0) = y(\pi) = 0$, of the inhomogeneous equation

$$\frac{d^2 y}{dx^2} + \kappa y = f(x),$$

where κ is a constant and

$$f(x) = \begin{cases} x & 0 \le x \le \pi/2, \\ \pi - x & \pi/2 < x \le \pi. \end{cases}$$

17.10   Consider the following two approaches to constructing a Green's function.

(a) Find those eigenfunctions $y_n(x)$ of the self-adjoint linear differential operator d^2/dx^2 that satisfy the boundary conditions $y_n(0) = y_n(\pi) = 0$, and hence construct its Green's function $G(x,z)$.

(b) Construct the same Green's function using a method based on the complementary function of the appropriate differential equation and the boundary conditions to be satisfied at the position of the δ-function, showing that it is

$$G(x,z) = \begin{cases} x(z - \pi)/\pi & 0 \le x \le z, \\ z(x - \pi)/\pi & z \le x \le \pi. \end{cases}$$

(c) By expanding the function given in (b) in terms of the eigenfunctions $y_n(x)$, verify that it is the same function as that derived in (a).

17.11   The differential operator $\mathcal{L}$ is defined by

$$\mathcal{L}y = -\frac{d}{dx}\left(e^x \frac{dy}{dx}\right) - \tfrac{1}{4}e^x y.$$

Determine the eigenvalues λ_n of the problem

$$\mathcal{L}y_n = \lambda_n e^x y_n \qquad 0 < x < 1,$$

with boundary conditions

$$y(0) = 0, \qquad \frac{dy}{dx} + \tfrac{1}{2}y = 0 \quad \text{at} \quad x = 1.$$

(a) Find the corresponding unnormalised y_n, and also a weight function $\rho(x)$ with respect to which the y_n are orthogonal. Hence, select a suitable normalisation for the y_n.

(b) By making an eigenfunction expansion, solve the equation

$$\mathcal{L}y = -e^{x/2}, \qquad 0 < x < 1,$$

subject to the same boundary conditions as previously.

17.12   Show that the linear operator

$$\mathcal{L} \equiv \tfrac{1}{4}(1 + x^2)^2 \frac{d^2}{dx^2} + \tfrac{1}{2}x(1 + x^2)\frac{d}{dx} + a,$$

acting upon functions defined in $-1 \le x \le 1$ and vanishing at the end-points of the interval, is Hermitian with respect to the weight function $(1 + x^2)^{-1}$.

By making the change of variable $x = \tan(\theta/2)$, find two even eigenfunctions, $f_1(x)$ and $f_2(x)$, of the differential equation

$$\mathcal{L}u = \lambda u.$$

17.13   By substituting $x = \exp t$, find the normalised eigenfunctions $y_n(x)$ and the eigenvalues λ_n of the operator $\mathcal{L}$ defined by

$$\mathcal{L}y = x^2 y'' + 2xy' + \tfrac{1}{4}y, \qquad 1 \le x \le e,$$

with $y(1) = y(e) = 0$. Find, as a series $\sum a_n y_n(x)$, the solution of $\mathcal{L}y = x^{-1/2}$.

17.14   Express the solution of Poisson's equation in electrostatics,

$$\nabla^2 \phi(\mathbf{r}) = -\rho(\mathbf{r})/\epsilon_0,$$

where ρ is the non-zero charge density over a finite part of space, in the form of an integral and hence identify the Green's function for the ∇^2 operator.

17.15   In the quantum-mechanical study of the scattering of a particle by a potential, a Born-approximation solution can be obtained in terms of a function $y(\mathbf{r})$ that satisfies an equation of the form

$$(-\nabla^2 - K^2)y(\mathbf{r}) = F(\mathbf{r}).$$

Assuming that $y_k(\mathbf{r}) = (2\pi)^{-3/2} \exp(i\mathbf{k} \cdot \mathbf{r})$ is a suitably normalised eigenfunction of $-\nabla^2$ corresponding to eigenvalue k^2, find a suitable Green's function $G_K(\mathbf{r}, \mathbf{r}')$. By taking the direction of the vector $\mathbf{r} - \mathbf{r}'$ as the polar axis for a $\mathbf{k}$-space integration, show that $G_K(\mathbf{r}, \mathbf{r}')$ can be reduced to

$$\frac{1}{4\pi^2 |\mathbf{r} - \mathbf{r}'|} \int_{-\infty}^{\infty} \frac{w \sin w}{w^2 - w_0^2} \, dw,$$

where $w_0 = K|\mathbf{r} - \mathbf{r}'|$.
[This integral can be evaluated using a contour integration (chapter 24) to give $(4\pi |\mathbf{r} - \mathbf{r}'|)^{-1} \exp(iK|\mathbf{r} - \mathbf{r}'|)$.]

17.8 Hints and answers

17.1   Express the condition $\langle h|h \rangle \geq 0$ as a quadratic equation in λ and then apply the condition for no real roots, noting that $\langle f|g \rangle + \langle g|f \rangle$ is real. To put a limit on $\int y \cos^2 kx \, dx$, set $f = y^{1/2} \cos kx$ and $g = y^{1/2}$ in the inequality.

17.3   Follow an argument similar to that used for proving the reality of the eigenvalues, but integrate from x_1 to x_2, rather than from a to b. Take x_1 and x_2 as two successive zeros of $y_m(x)$ and note that, if the sign of y_m is α then the sign of $y'_m(x_1)$ is α whilst that of $y'_m(x_2)$ is $-\alpha$. Now assume that $y_n(x)$ does not change sign in the interval and has a constant sign β; show that this leads to a contradiction between the signs of the two sides of the identity.

17.5   (a) $y = \sum a_n P_n(x)$ with

$$a_n = \frac{n + 1/2}{b - n(n+1)} \int_{-1}^{1} f(z) P_n(z) \, dz;$$

(b) $5x^3 = 2P_3(x) + 3P_1(x)$, giving $a_1 = 1/4$ and $a_3 = 1$, leading to $y = 5(2x^3 - x)/4$.

17.7   (a) No, $\int gf^{*\prime} dx \neq 0$; (b) yes; (c) no, $i \int f^* g \, dx \neq 0$; (d) yes.

17.9   The normalised eigenfunctions are $(2/\pi)^{1/2} \sin nx$, with n an integer.
$y(x) = (4/\pi) \sum_{n \text{ odd}} [(-1)^{(n-1)/2} \sin nx]/[n^2(\kappa - n^2)]$.

17.11 $\lambda_n = (n + 1/2)^2 \pi^2$, $n = 0, 1, 2, \ldots$.
(a) Since $y_n(1) y'_m(1) \neq 0$, the Sturm–Liouville boundary conditions are not satisfied and the appropriate weight function has to be justified by inspection. The normalised eigenfunctions are $\sqrt{2} e^{-x/2} \sin[(n + 1/2)\pi x]$, with $\rho(x) = e^x$.
(b) $y(x) = (-2/\pi^3) \sum_{n=0}^{\infty} e^{-x/2} \sin[(n + 1/2)\pi x]/(n + 1/2)^3$.

17.13 $y_n(x) = \sqrt{2} x^{-1/2} \sin(n\pi \ln x)$ with $\lambda_n = -n^2 \pi^2$;

$$a_n = \begin{cases} -(n\pi)^{-2} \int_1^e \sqrt{2} x^{-1} \sin(n\pi \ln x) \, dx = -\sqrt{8}(n\pi)^{-3} & \text{for } n \text{ odd,} \\ 0 & \text{for } n \text{ even.} \end{cases}$$

17.15 Use the form of Green's function that is the integral over all eigenvalues of the 'outer product' of two eigenfunctions corresponding to the same eigenvalue, but with arguments $\mathbf{r}$ and $\mathbf{r}'$.

18

Special functions

In the previous two chapters, we introduced the most important second-order linear ODEs in physics and engineering, listing their regular and irregular singular points in table 16.1 and their Sturm–Liouville forms in table 17.1. These equations occur with such frequency that solutions to them, which obey particular commonly occurring boundary conditions, have been extensively studied and given special names. In this chapter, we discuss these so-called 'special functions' and their properties. In addition, we also discuss some special functions that are not derived from solutions of important second-order ODEs, namely the gamma function and related functions. These convenient functions appear in a number of contexts, and so in section 18.12 we gather together some of their properties, with a minimum of formal proofs.

18.1 Legendre functions

Legendre's differential equation has the form

$$(1 - x^2)y'' - 2xy' + \ell(\ell + 1)y = 0, \tag{18.1}$$

and has three regular singular points, at $x = -1, 1, \infty$. It occurs in numerous physical applications and particularly in problems with axial symmetry that involve the ∇^2 operator, when they are expressed in spherical polar coordinates. In normal usage the variable x in Legendre's equation is the cosine of the polar angle in spherical polars, and thus $-1 \le x \le 1$. The parameter ℓ is a given real number, and any solution of (18.1) is called a *Legendre function*.

In subsection 16.1.1, we showed that $x = 0$ is an ordinary point of (18.1), and so we expect to find two linearly independent solutions of the form $y = \sum_{n=0}^{\infty} a_n x^n$. Substituting, we find

$$\sum_{n=0}^{\infty} \left[n(n-1)a_n x^{n-2} - n(n-1)a_n x^n - 2na_n x^n + \ell(\ell+1)a_n x^n \right] = 0,$$

which on collecting terms gives

$$\sum_{n=0}^{\infty} \{(n+2)(n+1)a_{n+2} - [n(n+1) - \ell(\ell+1)]a_n\} x^n = 0.$$

The recurrence relation is therefore

$$a_{n+2} = \frac{[n(n+1) - \ell(\ell+1)]}{(n+1)(n+2)} a_n, \tag{18.2}$$

for $n = 0, 1, 2, \ldots$. If we choose $a_0 = 1$ and $a_1 = 0$ then we obtain the solution

$$y_1(x) = 1 - \ell(\ell+1)\frac{x^2}{2!} + (\ell-2)\ell(\ell+1)(\ell+3)\frac{x^4}{4!} - \cdots, \tag{18.3}$$

whereas on choosing $a_0 = 0$ and $a_1 = 1$ we find a second solution

$$y_2(x) = x - (\ell-1)(\ell+2)\frac{x^3}{3!} + (\ell-3)(\ell-1)(\ell+2)(\ell+4)\frac{x^5}{5!} - \cdots. \tag{18.4}$$

By applying the ratio test to these series (see subsection 4.3.2), we find that both series converge for $|x| < 1$, and so their radius of convergence is unity, which (as expected) is the distance to the nearest singular point of the equation. Since (18.3) contains only even powers of x and (18.4) contains only odd powers, these two solutions cannot be proportional to one another, and are therefore linearly independent. Hence, the general solution to (18.1) for $|x| < 1$ is

$$y(x) = c_1 y_1(x) + c_2 y_2(x).$$

18.1.1 Legendre functions for integer ℓ

In many physical applications the parameter ℓ in Legendre's equation (18.1) is an integer, i.e. $\ell = 0, 1, 2, \ldots$. In this case, the recurrence relation (18.2) gives

$$a_{\ell+2} = \frac{[\ell(\ell+1) - \ell(\ell+1)]}{(\ell+1)(\ell+2)} a_\ell = 0,$$

i.e. the series terminates and we obtain a polynomial solution of order ℓ. In particular, if ℓ is even, then $y_1(x)$ in (18.3) reduces to a polynomial, whereas if ℓ is odd the same is true of $y_2(x)$ in (18.4). These solutions (suitably normalised) are called the *Legendre polynomials* of order ℓ; they are written $P_\ell(x)$ and are valid for all finite x. It is conventional to normalise $P_\ell(x)$ in such a way that $P_\ell(1) = 1$, and as a consequence $P_\ell(-1) = (-1)^\ell$. The first few Legendre polynomials are easily constructed and are given by

$$P_0(x) = 1, \qquad\qquad P_1(x) = x,$$

$$P_2(x) = \tfrac{1}{2}(3x^2 - 1), \qquad P_3(x) = \tfrac{1}{2}(5x^3 - 3x),$$

$$P_4(x) = \tfrac{1}{8}(35x^4 - 30x^2 + 3), \quad P_5(x) = \tfrac{1}{8}(63x^5 - 70x^3 + 15x).$$

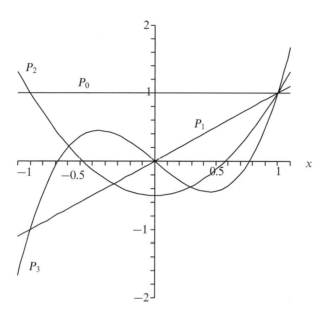

Figure 18.1 The first four Legendre polynomials.

The first four Legendre polynomials are plotted in figure 18.1.

Although, according to whether ℓ is an even or odd integer, respectively, either $y_1(x)$ in (18.3) or $y_2(x)$ in (18.4) terminates to give a multiple of the corresponding Legendre polynomial $P_\ell(x)$, the other series in each case does not terminate and therefore converges only for $|x| < 1$. According to whether ℓ is even or odd, we define *Legendre functions of the second kind* as $Q_\ell(x) = \alpha_\ell y_2(x)$ or $Q_\ell(x) = \beta_\ell y_1(x)$, respectively, where the constants α_ℓ and β_ℓ are conventionally taken to have the values

$$\alpha_\ell = \frac{(-1)^{\ell/2} 2^\ell [(\ell/2)!]^2}{\ell!} \qquad \text{for } \ell \text{ even,} \qquad (18.5)$$

$$\beta_\ell = \frac{(-1)^{(\ell+1)/2} 2^{\ell-1} \{[(\ell-1)/2]!\}^2}{\ell!} \qquad \text{for } \ell \text{ odd.} \qquad (18.6)$$

These normalisation factors are chosen so that the $Q_\ell(x)$ obey the same recurrence relations as the $P_\ell(x)$ (see subsection 18.1.2).

The general solution of Legendre's equation for *integer* ℓ is therefore

$$y(x) = c_1 P_\ell(x) + c_2 Q_\ell(x), \qquad (18.7)$$

where $P_\ell(x)$ is a polynomial of order ℓ, and so converges for all x, and $Q_\ell(x)$ is an infinite series that converges only for $|x| < 1.^§$

By using the Wronskian method, section 16.4, we may obtain closed forms for the $Q_\ell(x)$.

▶ *Use the Wronskian method to find a closed-form expression for $Q_0(x)$.*

From (16.25) a second solution to Legendre's equation (18.1), with $\ell = 0$, is

$$y_2(x) = P_0(x) \int^x \frac{1}{[P_0(u)]^2} \exp\left(\int^u \frac{2v}{1-v^2}\,dv\right)\,du$$

$$= \int^x \exp\left[-\ln(1-u^2)\right]\,du$$

$$= \int^x \frac{du}{(1-u^2)} = \tfrac{1}{2}\ln\left(\frac{1+x}{1-x}\right), \qquad (18.8)$$

where in the second line we have used the fact that $P_0(x) = 1$.

All that remains is to adjust the normalisation of this solution so that it agrees with (18.5). Expanding the logarithm in (18.8) as a Maclaurin series we obtain

$$y_2(x) = x + \frac{x^3}{3} + \frac{x^5}{5} + \cdots.$$

Comparing this with the expression for $Q_0(x)$, using (18.4) with $\ell = 0$ and the normalisation (18.5), we find that $y_2(x)$ is already correctly normalised, and so

$$Q_0(x) = \tfrac{1}{2}\ln\left(\frac{1+x}{1-x}\right).$$

Of course, we might have recognised the series (18.4) for $\ell = 0$, but to do so for larger ℓ would prove progressively more difficult. ◀

Using the above method for $\ell = 1$, we find

$$Q_1(x) = \tfrac{1}{2}x\ln\left(\frac{1+x}{1-x}\right) - 1.$$

Closed forms for higher-order $Q_\ell(x)$ may now be found using the recurrence relation (18.27) derived in the next subsection. The first few Legendre functions of the second kind are plotted in figure 18.2.

18.1.2 Properties of Legendre polynomials

As stated earlier, when encountered in physical problems the variable x in Legendre's equation is usually the cosine of the polar angle θ in spherical polar coordinates, and we then require the solution $y(x)$ to be regular at $x = \pm 1$, which corresponds to $\theta = 0$ or $\theta = \pi$. For this to occur we require the equation to have a polynomial solution, and so ℓ must be an integer. Furthermore, we also require

§ It is possible, in fact, to find a second solution in terms of an infinite series of *negative* powers of x that is finite for $|x| > 1$ (see exercise 16.16).

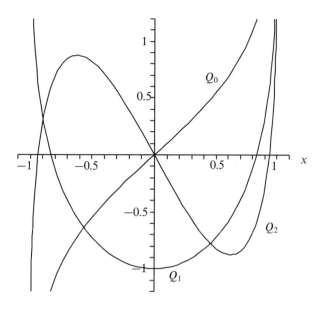

Figure 18.2   The first three Legendre functions of the second kind.

the coefficient c_2 of the function $Q_\ell(x)$ in (18.7) to be zero, since $Q_\ell(x)$ is singular at $x = \pm1$, with the result that the general solution is simply some multiple of the relevant Legendre polynomial $P_\ell(x)$. In this section we will study the properties of the Legendre polynomials $P_\ell(x)$ in some detail.

Rodrigues' formula

As an aid to establishing further properties of the Legendre polynomials we now develop Rodrigues' representation of these functions. Rodrigues' formula for the $P_\ell(x)$ is

$$P_\ell(x) = \frac{1}{2^\ell \ell!} \frac{d^\ell}{dx^\ell}(x^2 - 1)^\ell. \tag{18.9}$$

To prove that this is a representation we let $u = (x^2-1)^\ell$, so that $u' = 2\ell x(x^2-1)^{\ell-1}$ and

$$(x^2 - 1)u' - 2\ell xu = 0.$$

If we differentiate this expression $\ell + 1$ times using Leibnitz' theorem, we obtain

$$\left[(x^2 - 1)u^{(\ell+2)} + 2x(\ell + 1)u^{(\ell+1)} + \ell(\ell + 1)u^{(\ell)}\right] - 2\ell\left[xu^{(\ell+1)} + (\ell + 1)u^{(\ell)}\right] = 0,$$

which reduces to

$$(x^2 - 1)u^{(\ell+2)} + 2xu^{(\ell+1)} - \ell(\ell + 1)u^{(\ell)} = 0.$$

Changing the sign all through, we recover Legendre's equation (18.1) with $u^{(\ell)}$ as the dependent variable. Since, from (18.9), ℓ is an integer and $u^{(\ell)}$ is regular at $x = \pm 1$, we may make the identification

$$u^{(\ell)}(x) = c_\ell P_\ell(x), \tag{18.10}$$

for some constant c_ℓ that depends on ℓ. To establish the value of c_ℓ we note that the only term in the expression for the ℓth derivative of $(x^2 - 1)^\ell$ that does not contain a factor $x^2 - 1$, and therefore does not vanish at $x = 1$, is $(2x)^\ell \ell!(x^2 - 1)^0$. Putting $x = 1$ in (18.10) and recalling that $P_\ell(1) = 1$, therefore shows that $c_\ell = 2^\ell \ell!$, thus completing the proof of Rodrigues' formula (18.9).

▶ *Use Rodrigues' formula to show that*

$$I_\ell = \int_{-1}^{1} P_\ell(x)P_\ell(x)\,dx = \frac{2}{2\ell + 1}. \tag{18.11}$$

The result is trivially obvious for $\ell = 0$ and so we assume $\ell \geq 1$. Then, by Rodrigues' formula,

$$I_\ell = \frac{1}{2^{2\ell}(\ell!)^2} \int_{-1}^{1} \left[\frac{d^\ell(x^2 - 1)^\ell}{dx^\ell}\right] \left[\frac{d^\ell(x^2 - 1)^\ell}{dx^\ell}\right] dx.$$

Repeated integration by parts, with all boundary terms vanishing, reduces this to

$$I_\ell = \frac{(-1)^\ell}{2^{2\ell}(\ell!)^2} \int_{-1}^{1} (x^2 - 1)^\ell \frac{d^{2\ell}}{dx^{2\ell}}(x^2 - 1)^\ell \, dx$$

$$= \frac{(2\ell)!}{2^{2\ell}(\ell!)^2} \int_{-1}^{1} (1 - x^2)^\ell \, dx.$$

If we write

$$K_\ell = \int_{-1}^{1} (1 - x^2)^\ell \, dx,$$

then integration by parts (taking a factor 1 as the second part) gives

$$K_\ell = \int_{-1}^{1} 2\ell x^2(1 - x^2)^{\ell-1} \, dx.$$

Writing $2\ell x^2$ as $2\ell - 2\ell(1 - x^2)$ we obtain

$$K_\ell = 2\ell \int_{-1}^{1} (1 - x^2)^{\ell-1} \, dx - 2\ell \int_{-1}^{1} (1 - x^2)^\ell \, dx$$

$$= 2\ell K_{\ell-1} - 2\ell K_\ell$$

and hence the recurrence relation $(2\ell + 1)K_\ell = 2\ell K_{\ell-1}$. We therefore find

$$K_\ell = \frac{2\ell}{2\ell + 1} \frac{2\ell - 2}{2\ell - 1} \cdots \frac{2}{3} K_0 = 2^\ell \ell! \frac{2^\ell \ell!}{(2\ell + 1)!} 2 = \frac{2^{2\ell+1}(\ell!)^2}{(2\ell + 1)!},$$

which, when substituted into the expression for I_ℓ, establishes the required result. ◀

Mutual orthogonality

In section 17.4, we noted that Legendre's equation was of Sturm–Liouville form with $p = 1 - x^2$, $q = 0$, $\lambda = \ell(\ell + 1)$ and $\rho = 1$, and that its natural interval was $[-1, 1]$. Since the Legendre polynomials $P_\ell(x)$ are regular at the end-points $x = \pm 1$, they must be mutually orthogonal over this interval, i.e.

$$\int_{-1}^{1} P_\ell(x) P_k(x)\, dx = 0 \qquad \text{if } \ell \neq k. \tag{18.12}$$

Although this result follows from the general considerations of the previous chapter, it may also be proved directly, as shown in the following example.

▶*Prove directly that the Legendre polynomials $P_\ell(x)$ are mutually orthogonal over the interval $-1 < x < 1$.*

Since the $P_\ell(x)$ satisfy Legendre's equation we may write

$$\left[(1 - x^2) P_\ell'\right]' + \ell(\ell + 1) P_\ell = 0,$$

where $P_\ell' = dP_\ell/dx$. Multiplying through by P_k and integrating from $x = -1$ to $x = 1$, we obtain

$$\int_{-1}^{1} P_k \left[(1 - x^2) P_\ell'\right]' dx + \int_{-1}^{1} P_k \ell(\ell + 1) P_\ell\, dx = 0.$$

Integrating the first term by parts and noting that the boundary contribution vanishes at both limits because of the factor $1 - x^2$, we find

$$-\int_{-1}^{1} P_k'(1 - x^2) P_\ell'\, dx + \int_{-1}^{1} P_k \ell(\ell + 1) P_\ell\, dx = 0.$$

Now, if we reverse the roles of ℓ and k and subtract one expression from the other, we conclude that

$$[k(k + 1) - \ell(\ell + 1)] \int_{-1}^{1} P_k P_\ell\, dx = 0,$$

and therefore, since $k \neq \ell$, we must have the result (18.12). As a particular case, we note that if we put $k = 0$ we obtain

$$\int_{-1}^{1} P_\ell(x)\, dx = 0 \qquad \text{for } \ell \neq 0. \blacktriangleleft$$

As we discussed in the previous chapter, the mutual orthogonality (and completeness) of the $P_\ell(x)$ means that any reasonable function $f(x)$ (i.e. one obeying the Dirichlet conditions discussed at the start of chapter 12) can be expressed in the interval $|x| < 1$ as an infinite sum of Legendre polynomials,

$$f(x) = \sum_{\ell=0}^{\infty} a_\ell P_\ell(x), \tag{18.13}$$

where the coefficients a_ℓ are given by

$$a_\ell = \frac{2\ell + 1}{2} \int_{-1}^{1} f(x) P_\ell(x)\, dx. \tag{18.14}$$

▶*Prove the expression (18.14) for the coefficients in the Legendre polynomial expansion of a function $f(x)$.*

If we multiply (18.13) by $P_k(x)$ and integrate from $x = -1$ to $x = 1$ then we obtain

$$\int_{-1}^{1} P_k(x) f(x)\, dx = \sum_{\ell=0}^{\infty} a_\ell \int_{-1}^{1} P_k(x) P_\ell(x)\, dx$$

$$= a_k \int_{-1}^{1} P_k(x) P_k(x)\, dx = \frac{2a_k}{2k+1},$$

where we have used the orthogonality property (18.12) and the normalisation property (18.11). ◀

Generating function

A useful device for manipulating and studying sequences of functions or quantities labelled by an integer variable (here, the Legendre polynomials $P_\ell(x)$ labelled by ℓ) is a *generating function*. The generating function has perhaps its greatest utility in the area of probability theory (see chapter 30). However, it is also a great convenience in our present study.

The generating function for, say, a series of functions $f_n(x)$ for $n = 0, 1, 2, \dots$ is a function $G(x, h)$ containing, as well as x, a dummy variable h such that

$$G(x, h) = \sum_{n=0}^{\infty} f_n(x) h^n,$$

i.e. $f_n(x)$ is the coefficient of h^n in the expansion of G in powers of h. The utility of the device lies in the fact that sometimes it is possible to find a closed form for $G(x, h)$.

For our study of Legendre polynomials let us consider the functions $P_n(x)$ defined by the equation

$$G(x, h) = (1 - 2xh + h^2)^{-1/2} = \sum_{n=0}^{\infty} P_n(x) h^n. \tag{18.15}$$

As we show below, the functions so defined are identical to the Legendre polynomials and the function $(1 - 2xh + h^2)^{-1/2}$ is in fact the generating function for them. In the process we will also deduce several useful relationships between the various polynomials and their derivatives.

▶*Show that the functions $P_n(x)$ defined by (18.15) satisfy Legendre's equation*

In the following $dP_n(x)/dx$ will be denoted by P_n'. Firstly, we differentiate the defining equation (18.15) with respect to x and get

$$h(1 - 2xh + h^2)^{-3/2} = \sum P_n' h^n. \tag{18.16}$$

Also, we differentiate (18.15) with respect to h to yield

$$(x - h)(1 - 2xh + h^2)^{-3/2} = \sum n P_n h^{n-1}. \tag{18.17}$$

Equation (18.16) can then be written, using (18.15), as

$$h \sum P_n h^n = (1 - 2xh + h^2) \sum P'_n h^n,$$

and equating the coefficients of h^{n+1} we obtain the recurrence relation

$$P_n = P'_{n+1} - 2xP'_n + P'_{n-1}. \tag{18.18}$$

Equations (18.16) and (18.17) can be combined as

$$(x - h) \sum P'_n h^n = h \sum n P_n h^{n-1},$$

from which the coefficient of h^n yields a second recurrence relation,

$$xP'_n - P'_{n-1} = nP_n; \tag{18.19}$$

eliminating P'_{n-1} between (18.18) and (18.19) then gives the further result

$$(n + 1)P_n = P'_{n+1} - xP'_n. \tag{18.20}$$

If we now take the result (18.20) with n replaced by $n - 1$ and add x times (18.19) to it we obtain

$$(1 - x^2)P'_n = n(P_{n-1} - xP_n). \tag{18.21}$$

Finally, differentiating both sides with respect to x and using (18.19) again, we find

$$(1 - x^2)P''_n - 2xP'_n = n[(P'_{n-1} - xP'_n) - P_n]$$
$$= n(-nP_n - P_n) = -n(n + 1)P_n,$$

and so the P_n defined by (18.15) do indeed satisfy Legendre's equation. ◀

The above example shows that the functions $P_n(x)$ defined by (18.15) satisfy Legendre's equation with $\ell = n$ (an integer) and, also from (18.15), these functions are regular at $x = \pm 1$. Thus P_n must be some multiple of the nth Legendre polynomial. It therefore remains only to verify the normalisation. This is easily done at $x = 1$, when G becomes

$$G(1, h) = [(1 - h)^2]^{-1/2} = 1 + h + h^2 + \cdots,$$

and we can see that all the P_n so defined have $P_n(1) = 1$ as required, and are thus identical to the Legendre polynomials.

A particular use of the generating function (18.15) is in representing the inverse distance between two points in three-dimensional space in terms of Legendre polynomials. If two points $\mathbf{r}$ and $\mathbf{r}'$ are at distances r and r', respectively, from the origin, with $r' < r$, then

$$\frac{1}{|\mathbf{r} - \mathbf{r}'|} = \frac{1}{(r^2 + r'^2 - 2rr' \cos \theta)^{1/2}}$$
$$= \frac{1}{r[1 - 2(r'/r) \cos \theta + (r'/r)^2]^{1/2}}$$
$$= \frac{1}{r} \sum_{\ell=0}^{\infty} \left(\frac{r'}{r}\right)^\ell P_\ell(\cos \theta), \tag{18.22}$$

where θ is the angle between the two position vectors $\mathbf{r}$ and $\mathbf{r}'$. If $r' > r$, however,

r and r' must be exchanged in (18.22) or the series would not converge. This result may be used, for example, to write down the electrostatic potential at a point $\mathbf{r}$ due to a charge q at the point $\mathbf{r}'$. Thus, in the case $r' < r$, this is given by

$$V(\mathbf{r}) = \frac{q}{4\pi\epsilon_0 r} \sum_{\ell=0}^{\infty} \left(\frac{r'}{r}\right)^{\ell} P_{\ell}(\cos\theta).$$

We note that in the special case where the charge is at the origin, and $r' = 0$, only the $\ell = 0$ term in the series is non-zero and the expression reduces correctly to the familiar form $V(\mathbf{r}) = q/(4\pi\epsilon_0 r)$.

Recurrence relations

In our discussion of the generating function above, we derived several useful recurrence relations satisfied by the Legendre polynomials $P_n(x)$. In particular, from (18.18), we have the four-term recurrence relation

$$P'_{n+1} + P'_{n-1} = P_n + 2xP'_n.$$

Also, from (18.19)–(18.21), we have the three-term recurrence relations

$$P'_{n+1} = (n+1)P_n + xP'_n, \tag{18.23}$$
$$P'_{n-1} = -nP_n + xP'_n, \tag{18.24}$$
$$(1-x^2)P'_n = n(P_{n-1} - xP_n), \tag{18.25}$$
$$(2n+1)P_n = P'_{n+1} - P'_{n-1}, \tag{18.26}$$

where the final relation is obtained immediately by subtracting the second from the first. Many other useful recurrence relations can be derived from those given above and from the generating function.

▶*Prove the recurrence relation*
$$(n+1)P_{n+1} = (2n+1)xP_n - nP_{n-1}. \tag{18.27}$$

Substituting from (18.15) into (18.17), we find

$$(x-h)\sum P_n h^n = (1 - 2xh + h^2)\sum nP_n h^{n-1}.$$

Equating coefficients of h^n we obtain

$$xP_n - P_{n-1} = (n+1)P_{n+1} - 2xnP_n + (n-1)P_{n-1},$$

which on rearrangement gives the stated result. ◀

The recurrence relation derived in the above example is particularly useful in evaluating $P_n(x)$ for a given value of x. One starts with $P_0(x) = 1$ and $P_1(x) = x$ and iterates the recurrence relation until $P_n(x)$ is obtained.

18.2 Associated Legendre functions

The associated Legendre equation has the form

$$(1 - x^2)y'' - 2xy' + \left[\ell(\ell + 1) - \frac{m^2}{1 - x^2}\right] y = 0, \qquad (18.28)$$

which has three regular singular points at $x = -1, 1, \infty$ and reduces to Legendre's equation (18.1) when $m = 0$. It occurs in physical applications involving the operator ∇^2, when expressed in spherical polars. In such cases, $-\ell \leq m \leq \ell$ and m is restricted to integer values, which we will assume from here on. As was the case for Legendre's equation, in normal usage the variable x is the cosine of the polar angle in spherical polars, and thus $-1 \leq x \leq 1$. Any solution of (18.28) is called an *associated Legendre function*.

The point $x = 0$ is an ordinary point of (18.28), and one could obtain series solutions of the form $y = \sum_{n=0} a_n x^n$ in the same manner as that used for Legendre's equation. In this case, however, it is more instructive to note that if $u(x)$ is a solution of Legendre's equation (18.1), then

$$y(x) = (1 - x^2)^{|m|/2} \frac{d^{|m|}u}{dx^{|m|}} \qquad (18.29)$$

is a solution of the associated equation (18.28).

▶ *Prove that if $u(x)$ is a solution of Legendre's equation, then $y(x)$ given in (18.29) is a solution of the associated equation.*

For simplicity, let us begin by assuming that m is non-negative. Legendre's equation for u reads

$$(1 - x^2)u'' - 2xu' + \ell(\ell + 1)u = 0,$$

and, on differentiating this equation m times using Leibnitz' theorem, we obtain

$$(1 - x^2)v'' - 2x(m + 1)v' + (\ell - m)(\ell + m + 1)v = 0, \qquad (18.30)$$

where $v(x) = d^m u/dx^m$. On setting

$$y(x) = (1 - x^2)^{m/2}v(x),$$

the derivatives v' and v'' may be written as

$$v' = (1 - x^2)^{-m/2} \left(y' + \frac{mx}{1 - x^2} y\right),$$

$$v'' = (1 - x^2)^{-m/2} \left[y'' + \frac{2mx}{1 - x^2} y' + \frac{m}{1 - x^2} y + \frac{m(m + 2)x^2}{(1 - x^2)^2} y\right].$$

Substituting these expressions into (18.30) and simplifying, we obtain

$$(1 - x^2)y'' - 2xy' + \left[\ell(\ell + 1) - \frac{m^2}{1 - x^2}\right] y = 0,$$

which shows that y is a solution of the associated Legendre equation (18.28). Finally, we note that if m is negative, the value of m^2 is unchanged, and so a solution for positive m is also a solution for the corresponding negative value of m. ◀

From the two linearly independent series solutions to Legendre's equation given in (18.3) and (18.4), which we now denote by $u_1(x)$ and $u_2(x)$, we may obtain two linearly-independent series solutions, $y_1(x)$ and $y_2(x)$, to the associated equation by using (18.29). From the general discussion of the convergence of power series given in section 4.5.1, we see that both $y_1(x)$ and $y_2(x)$ will also converge for $|x| < 1$. Hence the general solution to (18.28) in this range is given by

$$y(x) = c_1 y_1(x) + c_2 y_2(x).$$

18.2.1 Associated Legendre functions for integer ℓ

If ℓ and m are both integers, as is the case in many physical applications, then the general solution to (18.28) is denoted by

$$y(x) = c_1 P_\ell^m(x) + c_2 Q_\ell^m(x), \tag{18.31}$$

where $P_\ell^m(x)$ and $Q_\ell^m(x)$ are associated Legendre functions of the first and second kind, respectively. For non-negative values of m, these functions are related to the ordinary Legendre functions for integer ℓ by

$$P_\ell^m(x) = (1 - x^2)^{m/2} \frac{d^m P_\ell}{dx^m}, \qquad Q_\ell^m(x) = (1 - x^2)^{m/2} \frac{d^m Q_\ell}{dx^m}. \tag{18.32}$$

We see immediately that, as required, the associated Legendre functions reduce to the ordinary Legendre functions when $m = 0$. Since it is m^2 that appears in the associated Legendre equation (18.28), the associated Legendre functions for negative m values must be proportional to the corresponding function for non-negative m. The constant of proportionality is a matter of convention. For the $P_\ell^m(x)$ it is usual to regard the definition (18.32) as being valid also for negative m values. Although differentiating a negative number of times is not defined, when $P_\ell(x)$ is expressed in terms of the Rodrigues' formula (18.9), this problem does not occur for $-\ell \le m \le \ell$.[§] In this case,

$$P_\ell^{-m}(x) = (-1)^m \frac{(\ell - m)!}{(\ell + m)!} P_\ell^m(x). \tag{18.33}$$

▶ *Prove the result (18.33).*

From (18.32) and the Rodrigues' formula (18.9) for the Legendre polynomials, we have

$$P_\ell^m(x) = \frac{1}{2^\ell \ell!} (1 - x^2)^{m/2} \frac{d^{\ell+m}}{dx^{\ell+m}} (x^2 - 1)^\ell,$$

and, without loss of generality, we may assume that m is non-negative. It is convenient to

---

[§] Some authors define $P_\ell^{-m}(x) = P_\ell^m(x)$, and similarly for the $Q_\ell^m(x)$, in which case m is replaced by $|m|$ in the definitions (18.32). It should be noted that, in this case, many of the results presented in this section also require m to be replaced by $|m|$.

write $(x^2 - 1) = (x+1)(x-1)$ and use Leibnitz' theorem to evaluate the derivative, which yields

$$P_\ell^m(x) = \frac{1}{2^\ell \ell!}(1-x^2)^{m/2} \sum_{r=0}^{\ell+m} \frac{(\ell+m)!}{r!(\ell+m-r)!} \frac{d^r(x+1)^\ell}{dx^r} \frac{d^{\ell+m-r}(x-1)^\ell}{dx^{\ell+m-r}}.$$

Considering the two derivative factors in a term in the summation, we note that the first is non-zero only for $r \leq \ell$ and the second is non-zero for $\ell + m - r \leq \ell$. Combining these conditions yields $m \leq r \leq \ell$. Performing the derivatives, we thus obtain

$$P_\ell^m(x) = \frac{1}{2^\ell \ell!}(1-x^2)^{m/2} \sum_{r=m}^{\ell} \frac{(\ell+m)!}{r!(\ell+m-r)!} \frac{\ell!(x+1)^{\ell-r}}{(\ell-r)!} \frac{\ell!(x-1)^{r-m}}{(r-m)!}$$

$$= (-1)^{m/2} \frac{\ell!(\ell+m)!}{2^\ell} \sum_{r=m}^{\ell} \frac{(x+1)^{\ell-r+\frac{m}{2}}(x-1)^{r-\frac{m}{2}}}{r!(\ell+m-r)!(\ell-r)!(r-m)!}. \qquad (18.34)$$

Repeating the above calculation for $P_\ell^{-m}(x)$ and identifying once more those terms in the sum that are non-zero, we find

$$P_\ell^{-m}(x) = (-1)^{-m/2} \frac{\ell!(\ell-m)!}{2^\ell} \sum_{r=0}^{\ell-m} \frac{(x+1)^{\ell-r-\frac{m}{2}}(x-1)^{r+\frac{m}{2}}}{r!(\ell-m-r)!(\ell-r)!(r+m)!}$$

$$= (-1)^{-m/2} \frac{\ell!(\ell-m)!}{2^\ell} \sum_{\bar{r}=m}^{\ell} \frac{(x+1)^{\ell-\bar{r}+\frac{m}{2}}(x-1)^{\bar{r}-\frac{m}{2}}}{(\bar{r}-m)!(\ell-\bar{r})!(\ell+m-\bar{r})!\bar{r}!}, \qquad (18.35)$$

where, in the second equality, we have rewritten the summation in terms of the new index $\bar{r} = r + m$. Comparing (18.34) and (18.35), we immediately arrive at the required result (18.33). ◄

Since $P_\ell(x)$ is a polynomial of order ℓ, we have $P_\ell^m(x) = 0$ for $|m| > \ell$. From its definition, it is clear that $P_\ell^m(x)$ is also a polynomial of order ℓ if m is even, but contains the factor $(1 - x^2)$ to a fractional power if m is odd. In either case, $P_\ell^m(x)$ is regular at $x = \pm 1$. The first few associated Legendre functions of the first kind are easily constructed and are given by (omitting the $m = 0$ cases)

$$P_1^1(x) = (1-x^2)^{1/2}, \qquad P_2^1(x) = 3x(1-x^2)^{1/2},$$

$$P_2^2(x) = 3(1-x^2), \qquad P_3^1(x) = \tfrac{3}{2}(5x^2 - 1)(1-x^2)^{1/2},$$

$$P_3^2(x) = 15x(1-x^2), \qquad P_3^3(x) = 15(1-x^2)^{3/2}.$$

Finally, we note that the associated Legendre functions of the second kind $Q_\ell^m(x)$, like $Q_\ell(x)$, are singular at $x = \pm 1$.

18.2.2 Properties of associated Legendre functions $P_\ell^m(x)$

When encountered in physical problems, the variable x in the associated Legendre equation (as in the ordinary Legendre equation) is usually the cosine of the polar angle θ in spherical polar coordinates, and we then require the solution $y(x)$ to be regular at $x = \pm 1$ (corresponding to $\theta = 0$ or $\theta = \pi$). For this to occur, we require ℓ to be an integer and the coefficient c_2 of the function $Q_\ell^m(x)$ in (18.31) to be zero, since $Q_\ell^m(x)$ is singular at $x = \pm 1$, with the result that the general solution is simply some multiple of one of the associated Legendre functions of the first kind, $P_\ell^m(x)$. We will study the further properties of these functions in the remainder of this subsection.

Mutual orthogonality

As noted in section 17.4, the associated Legendre equation is of Sturm–Liouville form $(py')' + qy + \lambda \rho y = 0$, with $p = 1 - x^2$, $q = -m^2/(1 - x^2)$, $\lambda = \ell(\ell + 1)$ and $\rho = 1$, and its natural interval is thus $[-1, 1]$. Since the associated Legendre functions $P_\ell^m(x)$ are regular at the end-points $x = \pm 1$, they must be mutually orthogonal over this interval for a fixed value of m, i.e.

$$\int_{-1}^{1} P_\ell^m(x) P_k^m(x) \, dx = 0 \qquad \text{if } \ell \neq k. \tag{18.36}$$

This result may also be proved directly in a manner similar to that used for demonstrating the orthogonality of the Legendre polynomials $P_\ell(x)$ in section 18.1.2. Note that the value of m must be the same for the two associated Legendre functions for (18.36) to hold. The normalisation condition when $\ell = k$ may be obtained using the Rodrigues' formula, as shown in the following example.

▶ Show that

$$I_{\ell m} \equiv \int_{-1}^{1} P_\ell^m(x) P_\ell^m(x) \, dx = \frac{2}{2\ell + 1} \frac{(\ell + m)!}{(\ell - m)!}. \tag{18.37}$$

From the definition (18.32) and the Rodrigues' formula (18.9) for $P_\ell(x)$, we may write

$$I_{\ell m} = \frac{1}{2^{2\ell}(\ell!)^2} \int_{-1}^{1} \left[(1 - x^2)^m \frac{d^{\ell+m}(x^2 - 1)^\ell}{dx^{\ell+m}} \right] \left[\frac{d^{\ell+m}(x^2 - 1)^\ell}{dx^{\ell+m}} \right] dx,$$

where the square brackets identify the factors to be used when integrating by parts. Performing the integration by parts $\ell + m$ times, and noting that all boundary terms vanish, we obtain

$$I_{\ell m} = \frac{(-1)^{\ell+m}}{2^{2\ell}(\ell!)^2} \int_{-1}^{1} (x^2 - 1)^\ell \frac{d^{\ell+m}}{dx^{\ell+m}} \left[(1 - x^2)^m \frac{d^{\ell+m}(x^2 - 1)^\ell}{dx^{\ell+m}} \right] dx.$$

Using Leibnitz' theorem, the second factor in the integrand may be written as

$$\frac{d^{\ell+m}}{dx^{\ell+m}} \left[(1 - x^2)^m \frac{d^{\ell+m}(x^2 - 1)^\ell}{dx^{\ell+m}} \right] = \sum_{r=0}^{\ell+m} \frac{(\ell + m)!}{r!(\ell + m - r)!} \frac{d^r(1 - x^2)^m}{dx^r} \frac{d^{2\ell+2m-r}(x^2 - 1)^\ell}{dx^{2\ell+2m-r}}.$$

Considering the two derivative factors in a term in the summation on the RHS, we see that the first is non-zero only for $r \leq 2m$, whereas the second is non-zero only for $2\ell + 2m - r \leq 2\ell$. Combining these conditions, we find that the only non-zero term in the sum is that for which $r = 2m$. Thus, we may write

$$I_{\ell m} = \frac{(-1)^{\ell+m}}{2^{2\ell}(\ell!)^2} \frac{(\ell + m)!}{(2m)!(\ell - m)!} \int_{-1}^{1} (1 - x^2)^\ell \frac{d^{2m}(1 - x^2)^m}{dx^{2m}} \frac{d^{2\ell}(1 - x^2)^\ell}{dx^{2\ell}} dx.$$

Since $d^{2\ell}(1-x^2)^\ell/dx^{2\ell} = (-1)^\ell(2\ell)!$, and noting that $(-1)^{2\ell+2m} = 1$, we have

$$I_{\ell m} = \frac{1}{2^{2\ell}(\ell!)^2}\frac{(2\ell)!(\ell+m)!}{(\ell-m)!}\int_{-1}^{1}(1-x^2)^\ell\,dx.$$

We have already shown in section 18.1.2 that

$$K_\ell \equiv \int_{-1}^{1}(1-x^2)^\ell\,dx = \frac{2^{2\ell+1}(\ell!)^2}{(2\ell+1)!},$$

and so we obtain the final result

$$I_{\ell m} = \frac{2}{2\ell+1}\frac{(\ell+m)!}{(\ell-m)!}. \quad \blacktriangleleft$$

The orthogonality and normalisation conditions, (18.36) and (18.37) respectively, mean that the associated Legendre functions $P_\ell^m(x)$, with m fixed, may be used in a similar way to the Legendre polynomials to expand any reasonable function $f(x)$ on the interval $|x| < 1$ in a series of the form

$$f(x) = \sum_{k=0}^{\infty} a_{m+k}P_{m+k}^m(x), \tag{18.38}$$

where, in this case, the coefficients are given by

$$a_\ell = \frac{2\ell+1}{2}\frac{(\ell-m)!}{(\ell+m)!}\int_{-1}^{1} f(x)P_\ell^m(x)\,dx.$$

We note that the series takes the form (18.38) because $P_\ell^m(x) = 0$ for $m > \ell$.

Finally, it is worth noting that the associated Legendre functions $P_\ell^m(x)$ must also obey a second orthogonality relationship. This has to be so because one may equally well write the associated Legendre equation (18.28) in Sturm–Liouville form $(py')'+qy+\lambda\rho y = 0$, with $p = 1-x^2, q = \ell(\ell+1), \lambda = -m^2$ and $\rho = (1-x^2)^{-1}$; once again the natural interval is $[-1, 1]$. Since the associated Legendre functions $P_\ell^m(x)$ are regular at the end-points $x = \pm 1$, they must therefore be mutually orthogonal with respect to the weight function $(1 - x^2)^{-1}$ over this interval for a fixed value of ℓ, i.e.

$$\int_{-1}^{1} P_\ell^m(x)P_\ell^k(x)(1-x^2)^{-1}\,dx = 0 \qquad \text{if } |m| \neq |k|. \tag{18.39}$$

One may also show straightforwardly that the corresponding normalisation condition when $m = k$ is given by

$$\int_{-1}^{1} P_\ell^m(x)P_\ell^m(x)(1-x^2)^{-1}\,dx = \frac{(\ell+m)!}{m(\ell-m)!}.$$

In solving physical problems, however, the orthogonality condition (18.39) is not of any practical use.

Generating function

The generating function for associated Legendre functions can be easily derived by combining their definition (18.32) with the generating function for the Legendre polynomials given in (18.15). We find that

$$G(x, h) = \frac{(2m)!(1 - x^2)^{m/2}}{2^m m!(1 - 2hx + h^2)^{m+1/2}} = \sum_{n=0}^{\infty} P_{n+m}^m(x)h^n. \tag{18.40}$$

> ▶Derive the expression (18.40) for the associated Legendre generating function.

The generating function (18.15) for the Legendre polynomials reads

$$\sum_{n=0}^{\infty} P_n h^n = (1 - 2xh + h^2)^{-1/2}.$$

Differentiating both sides of this result m times (assuming m to be non-negative), multiplying through by $(1 - x^2)^{m/2}$ and using the definition (18.32) of the associated Legendre functions, we obtain

$$\sum_{n=0}^{\infty} P_n^m h^n = (1 - x^2)^{m/2} \frac{d^m}{dx^m} (1 - 2xh + h^2)^{-1/2}.$$

Performing the derivatives on the RHS gives

$$\sum_{n=0}^{\infty} P_n^m h^n = \frac{1 \cdot 3 \cdot 5 \cdots (2m - 1)(1 - x^2)^{m/2} h^m}{(1 - 2xh + h^2)^{m+1/2}}.$$

Dividing through by h^m, re-indexing the summation on the LHS and noting that, quite generally,

$$1 \cdot 3 \cdot 5 \cdots (2r - 1) = \frac{1 \cdot 2 \cdot 3 \cdots 2r}{2 \cdot 4 \cdot 6 \cdots 2r} = \frac{(2r)!}{2^r r!},$$

we obtain the final result (18.40). ◀

Recurrence relations

As one might expect, the associated Legendre functions satisfy certain recurrence relations. Indeed, the presence of the two indices n and m means that a much wider range of recurrence relations may be derived. Here we shall content ourselves with quoting just four of the most useful relations:

$$P_n^{m+1} = \frac{2mx}{(1 - x^2)^{1/2}} P_n^m + [m(m - 1) - n(n + 1)]P_n^{m-1}, \tag{18.41}$$

$$(2n + 1)xP_n^m = (n + m)P_{n-1}^m + (n - m + 1)P_{n+1}^m, \tag{18.42}$$

$$(2n + 1)(1 - x^2)^{1/2}P_n^m = P_{n+1}^{m+1} - P_{n-1}^{m+1}, \tag{18.43}$$

$$2(1 - x^2)^{1/2}(P_n^m)' = P_n^{m+1} - (n + m)(n - m + 1)P_n^{m-1}. \tag{18.44}$$

We note that, by virtue of our adopted definition (18.32), these recurrence relations are equally valid for negative and non-negative values of m. These relations may be derived in a number of ways, such as using the generating function (18.40) or by differentiation of the recurrence relations for the Legendre polynomials $P_\ell(x)$.

▶ *Use the recurrence relation $(2n + 1)P_n = P'_{n+1} - P'_{n-1}$ for Legendre polynomials to derive the result (18.43).*

Differentiating the recurrence relation for the Legendre polynomials m times, we have

$$(2n + 1)\frac{d^m P_n}{dx^m} = \frac{d^{m+1}P_{n+1}}{dx^{m+1}} - \frac{d^{m+1}P_{n-1}}{dx^{m+1}}.$$

Multiplying through by $(1 - x^2)^{(m+1)/2}$ and using the definition (18.32) immediately gives the result (18.43). ◀

18.3 Spherical harmonics

The associated Legendre functions discussed in the previous section occur most commonly when obtaining solutions in spherical polar coordinates of Laplace's equation $\nabla^2 u = 0$ (see section 21.3.1). In particular, one finds that, for solutions that are finite on the polar axis, the angular part of the solution is given by

$$\Theta(\theta)\Phi(\phi) = P_\ell^m(\cos\theta)(C\cos m\phi + D\sin m\phi),$$

where ℓ and m are integers with $-\ell \leq m \leq \ell$. This general form is sufficiently common that particular functions of θ and ϕ called *spherical harmonics* are defined and tabulated. The spherical harmonics $Y_\ell^m(\theta, \phi)$ are defined by

$$Y_\ell^m(\theta, \phi) = (-1)^m \left[\frac{2\ell + 1}{4\pi}\frac{(\ell - m)!}{(\ell + m)!}\right]^{1/2} P_\ell^m(\cos\theta)\exp(im\phi). \qquad (18.45)$$

Using (18.33), we note that

$$Y_\ell^{-m}(\theta, \phi) = (-1)^m \left[Y_\ell^m(\theta, \phi)\right]^*,$$

where the asterisk denotes complex conjugation. The first few spherical harmonics $Y_\ell^m(\theta, \phi) \equiv Y_\ell^m$ are as follows:

$$Y_0^0 = \sqrt{\tfrac{1}{4\pi}}, \qquad\qquad Y_1^0 = \sqrt{\tfrac{3}{4\pi}}\cos\theta,$$

$$Y_1^{\pm 1} = \mp\sqrt{\tfrac{3}{8\pi}}\sin\theta\exp(\pm i\phi), \qquad Y_2^0 = \sqrt{\tfrac{5}{16\pi}}(3\cos^2\theta - 1),$$

$$Y_2^{\pm 1} = \mp\sqrt{\tfrac{15}{8\pi}}\sin\theta\cos\theta\exp(\pm i\phi), \qquad Y_2^{\pm 2} = \sqrt{\tfrac{15}{32\pi}}\sin^2\theta\exp(\pm 2i\phi).$$

Since they contain as their θ-dependent part the solution P_ℓ^m to the associated Legendre equation, the Y_ℓ^m are mutually orthogonal when integrated from -1 to $+1$ over $d(\cos\theta)$. Their mutual orthogonality with respect to ϕ $(0 \leq \phi \leq 2\pi)$ is even more obvious. The numerical factor in (18.45) is chosen to make the Y_ℓ^m an orthonormal set, i.e.

$$\int_{-1}^{1} \int_{0}^{2\pi} \left[Y_\ell^m(\theta, \phi) \right]^* Y_{\ell'}^{m'}(\theta, \phi) \, d\phi \, d(\cos\theta) = \delta_{\ell\ell'} \delta_{mm'}. \tag{18.46}$$

In addition, the spherical harmonics form a complete set in that any reasonable function (i.e. one that is likely to be met in a physical situation) of θ and ϕ can be expanded as a sum of such functions,

$$f(\theta, \phi) = \sum_{\ell=0}^{\infty} \sum_{m=-\ell}^{\ell} a_{\ell m} Y_\ell^m(\theta, \phi), \tag{18.47}$$

the constants $a_{\ell m}$ being given by

$$a_{\ell m} = \int_{-1}^{1} \int_{0}^{2\pi} \left[Y_\ell^m(\theta, \phi) \right]^* f(\theta, \phi) \, d\phi \, d(\cos\theta). \tag{18.48}$$

This is in exact analogy with a Fourier series and is a particular example of the general property of Sturm–Liouville solutions.

Aside from the orthonormality condition (18.46), the most important relationship obeyed by the Y_ℓ^m is the *spherical harmonic addition theorem*. This reads

$$P_\ell(\cos\gamma) = \frac{4\pi}{2\ell + 1} \sum_{m=-\ell}^{\ell} Y_\ell^m(\theta, \phi) [Y_\ell^m(\theta', \phi')]^*, \tag{18.49}$$

where (θ, ϕ) and (θ', ϕ') denote two different directions in our spherical polar coordinate system that are separated by an angle γ. In general, spherical trigonometry (or vector methods) shows that these angles obey the identity

$$\cos\gamma = \cos\theta \, \cos\theta' + \sin\theta \, \sin\theta' \, \cos(\phi - \phi'). \tag{18.50}$$

▶*Prove the spherical harmonic addition theorem (18.49).*

For the sake of brevity, it will be useful to denote the directions (θ, ϕ) and (θ', ϕ') by Ω and Ω', respectively. We will also denote the element of solid angle on the sphere by $d\Omega = d\phi \, d(\cos\theta)$. We begin by deriving the form of the closure relationship obeyed by the spherical harmonics. Using (18.47) and (18.48), and reversing the order of the summation and integration, we may write

$$f(\Omega) = \int_{4\pi} d\Omega' \, f(\Omega') \sum_{\ell m} Y_\ell^{m*}(\Omega') Y_\ell^m(\Omega),$$

where $\sum_{\ell m}$ is a convenient shorthand for the double summation in (18.47). Thus we may write the closure relationship for the spherical harmonics as

$$\sum_{\ell m} Y_\ell^m(\Omega) Y_\ell^{m*}(\Omega') = \delta(\Omega - \Omega'), \tag{18.51}$$

where $\delta(\Omega - \Omega')$ is a Dirac delta function with the properties that $\delta(\Omega - \Omega') = 0$ if $\Omega \neq \Omega'$ and $\int_{4\pi} \delta(\Omega) \, d\Omega = 1$.

Since $\delta(\Omega - \Omega')$ can depend only on the angle γ between the two directions Ω and Ω', we may also expand it in terms of a series of Legendre polynomials of the form

$$\delta(\Omega - \Omega') = \sum_\ell b_\ell P_\ell(\cos\gamma). \tag{18.52}$$

From (18.14), the coefficients in this expansion are given by

$$\begin{aligned} b_\ell &= \frac{2\ell + 1}{2} \int_{-1}^{1} \delta(\Omega - \Omega') P_\ell(\cos\gamma)\, d(\cos\gamma) \\ &= \frac{2\ell + 1}{4\pi} \int_{0}^{2\pi} \int_{-1}^{1} \delta(\Omega - \Omega') P_\ell(\cos\gamma)\, d(\cos\gamma)\, d\psi, \end{aligned}$$

where, in the second equality, we have introduced an additional integration over an azimuthal angle ψ about the direction Ω' (and γ is now the polar angle measured from Ω' to Ω). Since the rest of the integrand does not depend upon ψ, this is equivalent to multiplying it by $2\pi/2\pi$. However, the resulting double integral now has the form of a solid-angle integration over the whole sphere. Moreover, when $\Omega = \Omega'$, the angle γ separating the two directions is zero, and so $\cos\gamma = 1$. Thus, we find

$$b_\ell = \frac{2\ell + 1}{4\pi} P_\ell(1) = \frac{2\ell + 1}{4\pi},$$

and combining this expression with (18.51) and (18.52) gives

$$\sum_{\ell m} Y_\ell^m(\Omega) Y_\ell^{m*}(\Omega') = \sum_\ell \frac{2\ell + 1}{4\pi} P_\ell(\cos\gamma). \tag{18.53}$$

Comparing this result with (18.49), we see that, to complete the proof of the addition theorem, we now only need to show that the summations in ℓ on either side of (18.53) can be equated *term by term*.

That such a procedure is valid may be shown by considering an arbitrary rigid rotation of the coordinate axes, thereby defining new spherical polar coordinates $\bar\Omega$ on the sphere. Any given spherical harmonic $Y_\ell^m(\bar\Omega)$ in the new coordinates can be written as a linear combination of the spherical harmonics $Y_\ell^m(\Omega)$ of the old coordinates, *all* having the *same* value of ℓ. Thus,

$$Y_\ell^m(\bar\Omega) = \sum_{m'=-\ell}^{\ell} D_\ell^{mm'} Y_\ell^{m'}(\Omega),$$

where the coefficients $D_\ell^{mm'}$ depend on the rotation; note that in this expression Ω and $\bar\Omega$ refer to the same direction, but expressed in the two different coordinate systems. If we choose the polar axis of the new coordinate system to lie along the Ω' direction, then from (18.45), with m in that equation set equal to zero, we may write

$$P_\ell(\cos\gamma) = \sqrt{\frac{4\pi}{2\ell + 1}} Y_\ell^0(\bar\Omega) = \sum_{m'=-\ell}^{\ell} C_\ell^{0m'} Y_\ell^{m'}(\Omega)$$

for some set of coefficients C_ℓ^{0m} that depend on Ω'. Thus, we see that the equality (18.53) does indeed hold term by term in ℓ, thus proving the addition theorem (18.49). ◄

18.4 Chebyshev functions

Chebyshev's equation has the form

$$(1 - x^2)y'' - xy' + v^2 y = 0, \tag{18.54}$$

and has three regular singular points, at $x = -1, 1, \infty$. By comparing it with (18.1), we see that the Chebyshev equation is very similar in form to Legendre's equation. Despite this similarity, equation (18.54) does not occur very often in physical problems, though its solutions are of considerable importance in numerical analysis. The parameter v is a given real number, but in nearly all practical applications it takes an integer value. From here on we thus assume that $v = n$, where n is a non-negative integer. As was the case for Legendre's equation, in normal usage the variable x is the cosine of an angle, and so $-1 \le x \le 1$. Any solution of (18.54) is called a *Chebyshev function*.

The point $x = 0$ is an ordinary point of (18.54), and so we expect to find two linearly independent solutions of the form $y = \sum_{m=0}^{\infty} a_m x^m$. One could find the recurrence relations for the coefficients a_m in a similar manner to that used for Legendre's equation in section 18.1 (see exercise 16.15). For Chebyshev's equation, however, it is easier and more illuminating to take a different approach. In particular, we note that, on making the substitution $x = \cos\theta$, and consequently $d/dx = (-1/\sin\theta)\, d/d\theta$, Chebyshev's equation becomes (with $v = n$)

$$\frac{d^2y}{d\theta^2} + n^2 y = 0,$$

which is the simple harmonic equation with solutions $\cos n\theta$ and $\sin n\theta$. The corresponding linearly independent solutions of Chebyshev's equation are thus given by

$$T_n(x) = \cos(n\cos^{-1} x) \quad \text{and} \quad V_n(x) = \sin(n\cos^{-1} x). \tag{18.55}$$

It is straightforward to show that the $T_n(x)$ are *polynomials* of order n, whereas the $V_n(x)$ are *not* polynomials

▶ *Find explicit forms for the series expansions of $T_n(x)$ and $V_n(x)$.*

Writing $x = \cos\theta$, it is convenient first to form the complex superposition

$$\begin{aligned} T_n(x) + iV_n(x) &= \cos n\theta + i\sin n\theta \\ &= (\cos\theta + i\sin\theta)^n \\ &= \left(x + i\sqrt{1-x^2}\right)^n \qquad \text{for } |x| \le 1. \end{aligned}$$

Then, on expanding out the last expression using the binomial theorem, we obtain

$$T_n(x) = x^n - {}^nC_2 x^{n-2}(1-x^2) + {}^nC_4 x^{n-4}(1-x^2)^2 - \cdots, \tag{18.56}$$

$$V_n(x) = \sqrt{1-x^2}\left[{}^nC_1 x^{n-1} - {}^nC_3 x^{n-3}(1-x^2) + {}^nC_5 x^{n-5}(1-x^2)^2 - \cdots\right], \tag{18.57}$$

where ${}^nC_r = n!/[r!(n-r)!]$ is a binomial coefficient. We thus see that $T_n(x)$ is a polynomial of order n, but $V_n(x)$ is not a polynomial. ◀

It is conventional to define the additional functions

$$W_n(x) = (1-x^2)^{-1/2} T_{n+1}(x) \quad \text{and} \quad U_n(x) = (1-x^2)^{-1/2} V_{n+1}(x). \tag{18.58}$$

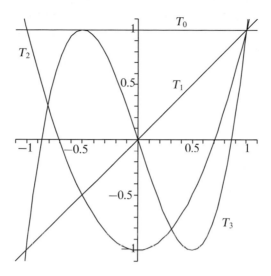

Figure 18.3   The first four Chebyshev polynomials of the first kind.

From (18.56) and (18.57), we see immediately that $U_n(x)$ is a *polynomial* of order n, but that $W_n(x)$ is *not* a polynomial. In practice, it is usual to work entirely in terms of $T_n(x)$ and $U_n(x)$, which are known, respectively, as *Chebyshev polynomials of the first and second kind*. In particular, we note that the general solution to Chebyshev's equation can be written in terms of these polynomials as

$$
y(x) = \begin{cases} c_1 T_n(x) + c_2 \sqrt{1 - x^2}\, U_{n-1}(x) & \text{for } n = 1, 2, 3, \ldots, \\ c_1 + c_2 \sin^{-1} x & \text{for } n = 0. \end{cases}
$$

The $n = 0$ solution could also be written as $d_1 + c_2 \cos^{-1} x$ with $d_1 = c_1 + \frac{1}{2}\pi c_2$.

The first few Chebyshev polynomials of the first kind are easily constructed and are given by

$$
\begin{aligned}
&T_0(x) = 1, &&T_1(x) = x, \\
&T_2(x) = 2x^2 - 1, &&T_3(x) = 4x^3 - 3x, \\
&T_4(x) = 8x^4 - 8x^2 + 1, &&T_5(x) = 16x^5 - 20x^3 + 5x.
\end{aligned}
$$

The functions $T_0(x)$, $T_1(x)$, $T_2(x)$ and $T_3(x)$ are plotted in figure 18.3. In general, the Chebyshev polynomials $T_n(x)$ satisfy $T_n(-x) = (-1)^n T_n(x)$, which is easily deduced from (18.56). Similarly, it is straightforward to deduce the following

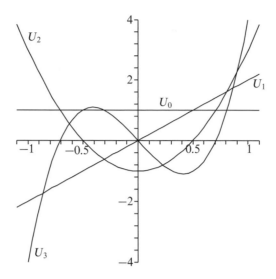

Figure 18.4   The first four Chebyshev polynomials of the second kind.

special values:

$$T_n(1) = 1, \qquad T_n(-1) = (-1)^n, \qquad T_{2n}(0) = (-1)^n, \qquad T_{2n+1}(0) = 0.$$

The first few Chebyshev polynomials of the second kind are also easily found and read

$$U_0(x) = 1, \qquad\qquad\qquad U_1(x) = 2x,$$

$$U_2(x) = 4x^2 - 1, \qquad\qquad U_3(x) = 8x^3 - 4x,$$

$$U_4(x) = 16x^4 - 12x^2 + 1, \qquad U_5(x) = 32x^5 - 32x^3 + 6x.$$

The functions $U_0(x)$, $U_1(x)$, $U_2(x)$ and $U_3(x)$ are plotted in figure 18.4. The Chebyshev polynomials $U_n(x)$ also satisfy $U_n(-x) = (-1)^n U_n(x)$, which may be deduced from (18.57) and (18.58), and have the special values:

$$U_n(1) = n + 1, \qquad U_n(-1) = (-1)^n(n+1), \qquad U_{2n}(0) = (-1)^n, \qquad U_{2n+1}(0) = 0.$$

▶Show that the Chebyshev polynomials $U_n(x)$ satisfy the differential equation

$$(1 - x^2)U_n''(x) - 3xU_n'(x) + n(n+2)U_n(x) = 0. \tag{18.59}$$

From (18.58), we have $V_{n+1} = (1 - x^2)^{1/2}U_n$ and these functions satisfy the Chebyshev equation (18.54) with $v = n + 1$, namely

$$(1 - x^2)V_{n+1}'' - xV_{n+1}' + (n+1)^2 V_{n+1} = 0. \tag{18.60}$$

598

Evaluating the first and second derivatives of V_{n+1}, we obtain

$$V'_{n+1} = (1 - x^2)^{1/2} U'_n - x(1 - x^2)^{-1/2} U_n$$
$$V''_{n+1} = (1 - x^2)^{1/2} U''_n - 2x(1 - x^2)^{-1/2} U'_n - (1 - x^2)^{-1/2} U_n - x^2(1 - x^2)^{-3/2} U_n.$$

Substituting these expressions into (18.60) and dividing through by $(1 - x^2)^{1/2}$, we find

$$(1 - x^2) U''_n - 3x U'_n - U_n + (n + 1)^2 U_n = 0,$$

which immediately simplifies to give the required result (18.59). ◀

18.4.1 Properties of Chebyshev polynomials

The Chebyshev polynomials $T_n(x)$ and $U_n(x)$ have their principal applications in numerical analysis. Their use in representing other functions over the range $|x| < 1$ plays an important role in numerical integration; Gauss–Chebyshev integration is of particular value for the accurate evaluation of integrals whose integrands contain factors $(1 - x^2)^{\pm 1/2}$. It is therefore worthwhile outlining some of their main properties.

Rodrigues' formula

The Chebyshev polynomials $T_n(x)$ and $U_n(x)$ may be expressed in terms of a Rodrigues' formula, in a similar way to that used for the Legendre polynomials discussed in section 18.1.2. For the Chebyshev polynomials, we have

$$T_n(x) = \frac{(-1)^n \sqrt{\pi} (1 - x^2)^{1/2}}{2^n (n - \frac{1}{2})!} \frac{d^n}{dx^n} (1 - x^2)^{n - \frac{1}{2}},$$

$$U_n(x) = \frac{(-1)^n \sqrt{\pi} (n + 1)}{2^{n+1} (n + \frac{1}{2})! (1 - x^2)^{1/2}} \frac{d^n}{dx^n} (1 - x^2)^{n + \frac{1}{2}}.$$

These Rodrigues' formulae may be proved in an analogous manner to that used in section 18.1.2 when establishing the corresponding expression for the Legendre polynomials.

Mutual orthogonality

In section 17.4, we noted that Chebyshev's equation could be put into Sturm–Liouville form with $p = (1 - x^2)^{1/2}$, $q = 0$, $\lambda = n^2$ and $\rho = (1 - x^2)^{-1/2}$, and its natural interval is thus $[-1, 1]$. Since the Chebyshev polynomials of the first kind, $T_n(x)$, are solutions of the Chebyshev equation and are regular at the end-points $x = \pm 1$, they must be mutually orthogonal over this interval with respect to the weight function $\rho = (1 - x^2)^{-1/2}$, i.e.

$$\int_{-1}^{1} T_n(x) T_m(x) (1 - x^2)^{-1/2} \, dx = 0 \qquad \text{if } n \neq m. \tag{18.61}$$

599

The normalisation, when $m = n$, is easily found by making the substitution $x = \cos\theta$ and using (18.55). We immediately obtain

$$\int_{-1}^{1} T_n(x)T_n(x)(1 - x^2)^{-1/2}\, dx = \begin{cases} \pi & \text{for } n = 0, \\ \pi/2 & \text{for } n = 1, 2, 3, \ldots. \end{cases} \tag{18.62}$$

The orthogonality and normalisation conditions mean that any (reasonable) function $f(x)$ can be expanded over the interval $|x| < 1$ in a series of the form

$$f(x) = \tfrac{1}{2}a_0 + \sum_{n=1}^{\infty} a_n T_n(x),$$

where the coefficients in the expansion are given by

$$a_n = \frac{2}{\pi}\int_{-1}^{1} f(x)T_n(x)(1 - x^2)^{-1/2}\, dx.$$

For the Chebyshev polynomials of the second kind, $U_n(x)$, we see from (18.58) that $(1 - x^2)^{1/2}U_n(x) = V_{n+1}(x)$ satisfies Chebyshev's equation (18.54) with $v = n + 1$. Thus, the orthogonality relation for the $U_n(x)$, obtained by replacing $T_i(x)$ by $V_{i+1}(x)$ in equation (18.61), reads

$$\int_{-1}^{1} U_n(x)U_m(x)(1 - x^2)^{1/2}\, dx = 0 \qquad \text{if } n \neq m.$$

The corresponding normalisation condition, when $n = m$, can again be found by making the substitution $x = \cos\theta$, as illustrated in the following example.

►*Show that*

$$I \equiv \int_{-1}^{1} U_n(x)U_n(x)(1 - x^2)^{1/2}\, dx = \frac{\pi}{2}.$$

From (18.58), we see that

$$I = \int_{-1}^{1} V_{n+1}(x)V_{n+1}(x)(1 - x^2)^{-1/2}\, dx,$$

which, on substituting $x = \cos\theta$, gives

$$I = \int_{\pi}^{0} \sin(n+1)\theta\, \sin(n+1)\theta\, \frac{1}{\sin\theta}\,(-\sin\theta)\, d\theta = \frac{\pi}{2}. \ ◄$$

The above orthogonality and normalisation conditions allow one to expand any (reasonable) function in the interval $|x| < 1$ in a series of the form

$$f(x) = \sum_{n=0}^{\infty} a_n U_n(x),$$

600

in which the coefficients a_n are given by

$$a_n = \frac{2}{\pi} \int_{-1}^{1} f(x)U_n(x)(1-x^2)^{1/2}\, dx.$$

Generating functions

The generating functions for the Chebyshev polynomials of the first and second kinds are given, respectively, by

$$G_{\mathrm{I}}(x,h) = \frac{1-xh}{1-2xh+h^2} = \sum_{n=0}^{\infty} T_n(x)h^n, \tag{18.63}$$

$$G_{\mathrm{II}}(x,h) = \frac{1}{1-2xh+h^2} = \sum_{n=0}^{\infty} U_n(x)h^n. \tag{18.64}$$

These prescriptions may be proved in a manner similar to that used in section 18.1.2 for the generating function of the Legendre polynomials. For the Chebyshev polynomials, however, the generating functions are of less practical use, since most of the useful results can be obtained more easily by taking advantage of the trigonometric forms (18.55), as illustrated below.

Recurrence relations

There exist many useful recurrence relationships for the Chebyshev polynomials $T_n(x)$ and $U_n(x)$. They are most easily derived by setting $x = \cos\theta$ and using (18.55) and (18.58) to write

$$T_n(x) = T_n(\cos\theta) = \cos n\theta, \tag{18.65}$$

$$U_n(x) = U_n(\cos\theta) = \frac{\sin(n+1)\theta}{\sin\theta}. \tag{18.66}$$

One may then use standard formulae for the trigonometric functions to derive a wide variety of recurrence relations. Of particular use are the trigonometric identities

$$\cos(n\pm1)\theta = \cos n\theta \cos\theta \mp \sin n\theta \sin\theta, \tag{18.67}$$

$$\sin(n\pm1)\theta = \sin n\theta \cos\theta \pm \cos n\theta \sin\theta. \tag{18.68}$$

▶*Show that the Chebyshev polynomials satisfy the recurrence relations*

$$T_{n+1}(x) - 2xT_n(x) + T_{n-1}(x) = 0, \tag{18.69}$$

$$U_{n+1}(x) - 2xU_n(x) + U_{n-1}(x) = 0. \tag{18.70}$$

Adding the result (18.67) with the plus sign to the corresponding result with a minus sign gives

$$\cos(n+1)\theta + \cos(n-1)\theta = 2\cos n\theta \cos\theta.$$

Using (18.65) and setting $x = \cos\theta$ immediately gives a rearrangement of the required result (18.69). Similarly, adding the plus and minus cases of result (18.68) gives

$$\sin(n+1)\theta + \sin(n-1)\theta = 2\sin n\theta \cos\theta.$$

Dividing through on both sides by $\sin\theta$ and using (18.66) yields (18.70). ◄

The recurrence relations (18.69) and (18.70) are extremely useful in the practical computation of Chebyshev polynomials. For example, given the values of $T_0(x)$ and $T_1(x)$ at some point x, the result (18.69) may be used iteratively to obtain the value of any $T_n(x)$ at that point; similarly, (18.70) may be used to calculate the value of any $U_n(x)$ at some point x, given the values of $U_0(x)$ and $U_1(x)$ at that point.

Further recurrence relations satisfied by the Chebyshev polynomials are

$$T_n(x) = U_n(x) - xU_{n-1}(x), \tag{18.71}$$

$$(1-x^2)U_n(x) = xT_{n+1}(x) - T_{n+2}(x), \tag{18.72}$$

which establish useful relationships between the two sets of polynomials $T_n(x)$ and $U_n(x)$. The relation (18.71) follows immediately from (18.68), whereas (18.72) follows from (18.67), with n replaced by $n+1$, on noting that $\sin^2\theta = 1 - x^2$. Additional useful results concerning the derivatives of Chebyshev polynomials may be obtained from (18.65) and (18.66), as illustrated in the following example.

►*Show that*

$$T_n'(x) = nU_{n-1}(x),$$

$$(1-x^2)U_n'(x) = xU_n(x) - (n+1)T_{n+1}(x).$$

These results are most easily derived from the expressions (18.65) and (18.66) by noting that $d/dx = (-1/\sin\theta)\,d/d\theta$. Thus,

$$T_n'(x) = -\frac{1}{\sin\theta}\frac{d(\cos n\theta)}{d\theta} = \frac{n\sin n\theta}{\sin\theta} = nU_{n-1}(x).$$

Similarly, we find

$$U_n'(x) = -\frac{1}{\sin\theta}\frac{d}{d\theta}\left[\frac{\sin(n+1)\theta}{\sin\theta}\right] = \frac{\sin(n+1)\theta\cos\theta}{\sin^3\theta} - \frac{(n+1)\cos(n+1)\theta}{\sin^2\theta}$$

$$= \frac{x\,U_n(x)}{1-x^2} - \frac{(n+1)T_{n+1}(x)}{1-x^2},$$

which rearranges immediately to yield the stated result. ◄

18.5 Bessel functions

Bessel's equation has the form

$$x^2y'' + xy' + (x^2 - v^2)y = 0, \tag{18.73}$$

which has a regular singular point at $x = 0$ and an essential singularity at $x = \infty$. The parameter v is a given number, which we may take as ≥ 0 with no loss of generality. The equation arises from physical situations similar to those involving Legendre's equation but when cylindrical, rather than spherical, polar coordinates are employed. The variable x in Bessel's equation is usually a multiple of a radial distance and therefore ranges from 0 to ∞.

We shall seek solutions to Bessel's equation in the form of infinite series. Writing (18.73) in the standard form used in chapter 16, we have

$$y'' + \frac{1}{x}y' + \left(1 - \frac{v^2}{x^2}\right)y = 0. \tag{18.74}$$

By inspection, $x = 0$ is a regular singular point; hence we try a solution of the form $y = x^\sigma \sum_{n=0}^{\infty} a_n x^n$. Substituting this into (18.74) and multiplying the resulting equation by $x^{2-\sigma}$, we obtain

$$\sum_{n=0}^{\infty} \left[(\sigma+n)(\sigma+n-1)+(\sigma+n)-v^2\right] a_n x^n + \sum_{n=0}^{\infty} a_n x^{n+2} = 0,$$

which simplifies to

$$\sum_{n=0}^{\infty} \left[(\sigma+n)^2 - v^2\right] a_n x^n + \sum_{n=0}^{\infty} a_n x^{n+2} = 0.$$

Considering the coefficient of x^0, we obtain the indicial equation

$$\sigma^2 - v^2 = 0,$$

and so $\sigma = \pm v$. For coefficients of higher powers of x we find

$$\left[(\sigma+1)^2 - v^2\right] a_1 = 0, \tag{18.75}$$

$$\left[(\sigma+n)^2 - v^2\right] a_n + a_{n-2} = 0 \quad \text{for } n \geq 2. \tag{18.76}$$

Substituting $\sigma = \pm v$ into (18.75) and (18.76), we obtain the recurrence relations

$$(1 \pm 2v)a_1 = 0, \tag{18.77}$$

$$n(n+2v)a_n + a_{n-2} = 0 \quad \text{for } n \geq 2. \tag{18.78}$$

We consider now the form of the general solution to Bessel's equation (18.73) for two cases: the case for which v is not an integer and that for which it is (including zero).

18.5.1 Bessel functions for non-integer v

If v is a non-integer then, in general, the two roots of the indicial equation, $\sigma_1 = v$ and $\sigma_2 = -v$, will not differ by an integer, and we may obtain two linearly independent solutions in the form of Frobenius series. Special considerations do arise, however, when $v = m/2$ for $m = 1, 3, 5, \ldots$, and $\sigma_1 - \sigma_2 = 2v = m$ is an (odd positive) integer. When this happens, we may always obtain a solution in the form of a Frobenius series corresponding to the larger root, $\sigma_1 = v = m/2$, as described above. However, for the smaller root, $\sigma_2 = -v = -m/2$, we must determine whether a second Frobenius series solution is possible by examining the recurrence relation (18.78), which reads

$$n(n - m)a_n + a_{n-2} = 0 \quad \text{for } n \geq 2.$$

Since m is an *odd* positive integer in this case, we can use this recurrence relation (starting with $a_0 \neq 0$) to calculate $a_2, a_4, a_6, \ldots$ in the knowledge that all these terms will remain finite. It is possible in this case, therefore, to find a second solution in the form of a Frobenius series, one that corresponds to the smaller root σ_2.

Thus, in general, for non-integer v we have from (18.77) and (18.78)

$$\begin{aligned}
a_n &= -\frac{1}{n(n \pm 2v)} a_{n-2} \quad \text{for } n = 2, 4, 6, \ldots, \\
&= 0 \quad \text{for } n = 1, 3, 5, \ldots.
\end{aligned}$$

Setting $a_0 = 1$ in each case, we obtain the two solutions

$$y_{\pm v}(x) = x^{\pm v} \left[1 - \frac{x^2}{2(2 \pm 2v)} + \frac{x^4}{2 \times 4(2 \pm 2v)(4 \pm 2v)} - \cdots \right].$$

It is customary, however, to set

$$a_0 = \frac{1}{2^{\pm v} \Gamma(1 \pm v)},$$

where $\Gamma(x)$ is the *gamma function*, described in subsection 18.12.1; it may be regarded as the generalisation of the factorial function to non-integer and/or negative arguments.[§] The two solutions of (18.73) are then written as $J_v(x)$ and $J_{-v}(x)$, where

$$\begin{aligned}
J_v(x) &= \frac{1}{\Gamma(v + 1)} \left(\frac{x}{2}\right)^v \left[1 - \frac{1}{v + 1} \left(\frac{x}{2}\right)^2 + \frac{1}{(v + 1)(v + 2)} \frac{1}{2!} \left(\frac{x}{2}\right)^4 - \cdots \right] \\
&= \sum_{n=0}^{\infty} \frac{(-1)^n}{n! \Gamma(v + n + 1)} \left(\frac{x}{2}\right)^{v+2n};
\end{aligned}$$

(18.79)

replacing v by $-v$ gives $J_{-v}(x)$. The functions $J_v(x)$ and $J_{-v}(x)$ are called *Bessel functions of the first kind, of order v*. Since the first term of each series is a finite non-zero multiple of x^v and x^{-v}, respectively, if v is not an integer then $J_v(x)$ and $J_{-v}(x)$ are linearly independent. This may be confirmed by calculating the Wronskian of these two functions. Therefore, for non-integer v the general solution of Bessel's equation (18.73) is given by

$$y(x) = c_1 J_v(x) + c_2 J_{-v}(x).$$

(18.80)

[§] In particular, $\Gamma(n + 1) = n!$ for $n = 0, 1, 2, \ldots$, and $\Gamma(n)$ is infinite if n is any integer ≤ 0.

We note that Bessel functions of half-integer order are expressible in closed form in terms of trigonometric functions, as illustrated in the following example.

> ► *Find the general solution of*
> $$x^2 y'' + xy' + (x^2 - \tfrac{1}{4})y = 0.$$

This is Bessel's equation with $v = 1/2$, so from (18.80) the general solution is simply

$$y(x) = c_1 J_{1/2}(x) + c_2 J_{-1/2}(x).$$

However, Bessel functions of half-integral order can be expressed in terms of trigonometric functions. To show this, we note from (18.79) that

$$J_{\pm 1/2}(x) = x^{\pm 1/2} \sum_{n=0}^{\infty} \frac{(-1)^n x^{2n}}{2^{2n \pm 1/2} n! \Gamma(1 + n \pm \tfrac{1}{2})}.$$

Using the fact that $\Gamma(x+1) = x\Gamma(x)$ and $\Gamma(\tfrac{1}{2}) = \sqrt{\pi}$, we find that, for $v = 1/2$,

$$
\begin{aligned}
J_{1/2}(x) &= \frac{(\tfrac{1}{2}x)^{1/2}}{\Gamma(\tfrac{3}{2})} - \frac{(\tfrac{1}{2}x)^{5/2}}{1!\,\Gamma(\tfrac{5}{2})} + \frac{(\tfrac{1}{2}x)^{9/2}}{2!\,\Gamma(\tfrac{7}{2})} - \cdots \\
&= \frac{(\tfrac{1}{2}x)^{1/2}}{(\tfrac{1}{2})\sqrt{\pi}} - \frac{(\tfrac{1}{2}x)^{5/2}}{1!(\tfrac{3}{2})(\tfrac{1}{2})\sqrt{\pi}} + \frac{(\tfrac{1}{2}x)^{9/2}}{2!(\tfrac{5}{2})(\tfrac{3}{2})(\tfrac{1}{2})\sqrt{\pi}} - \cdots \\
&= \frac{(\tfrac{1}{2}x)^{1/2}}{(\tfrac{1}{2})\sqrt{\pi}} \left(1 - \frac{x^2}{3!} + \frac{x^4}{5!} - \cdots \right) = \frac{(\tfrac{1}{2}x)^{1/2}}{(\tfrac{1}{2})\sqrt{\pi}} \frac{\sin x}{x} = \sqrt{\frac{2}{\pi x}} \sin x,
\end{aligned}
$$

whereas for $v = -1/2$ we obtain

$$
\begin{aligned}
J_{-1/2}(x) &= \frac{(\tfrac{1}{2}x)^{-1/2}}{\Gamma(\tfrac{1}{2})} - \frac{(\tfrac{1}{2}x)^{3/2}}{1!\,\Gamma(\tfrac{3}{2})} + \frac{(\tfrac{1}{2}x)^{7/2}}{2!\,\Gamma(\tfrac{5}{2})} - \cdots \\
&= \frac{(\tfrac{1}{2}x)^{-1/2}}{\sqrt{\pi}} \left(1 - \frac{x^2}{2!} + \frac{x^4}{4!} - \cdots \right) = \sqrt{\frac{2}{\pi x}} \cos x.
\end{aligned}
$$

Therefore the general solution we require is

$$y(x) = c_1 J_{1/2}(x) + c_2 J_{-1/2}(x) = c_1 \sqrt{\frac{2}{\pi x}} \sin x + c_2 \sqrt{\frac{2}{\pi x}} \cos x. \ ◄$$

18.5.2 Bessel functions for integer v

The definition of the Bessel function $J_v(x)$ given in (18.79) is, of course, valid for all values of v, but, as we shall see, in the case of integer v the general solution of Bessel's equation cannot be written in the form (18.80). Firstly, let us consider the case $v = 0$, so that the two solutions to the indicial equation are equal, and we clearly obtain only one solution in the form of a Frobenius series. From (18.79),

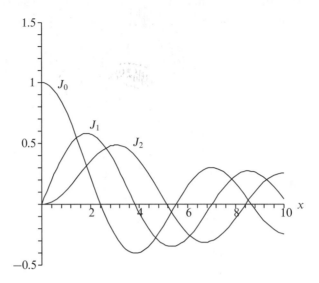

Figure 18.5 The first three integer-order Bessel functions of the first kind.

this is given by

$$J_0(x) = \sum_{n=0}^{\infty} \frac{(-1)^n x^{2n}}{2^{2n} n! \Gamma(1+n)}$$

$$= 1 - \frac{x^2}{2^2} + \frac{x^4}{2^2 4^2} - \frac{x^6}{2^2 4^2 6^2} + \cdots .$$

In general, however, if v is a positive integer then the solutions of the indicial equation differ by an integer. For the larger root, $\sigma_1 = v$, we may find a solution $J_v(x)$, for $v = 1, 2, 3, \ldots$, in the form of the Frobenius series given by (18.79). Graphs of $J_0(x)$, $J_1(x)$ and $J_2(x)$ are plotted in figure 18.5 for real x. For the smaller root, $\sigma_2 = -v$, however, the recurrence relation (18.78) becomes

$$n(n - m)a_n + a_{n-2} = 0 \quad \text{for } n \geq 2,$$

where $m = 2v$ is now an *even* positive integer, i.e. $m = 2, 4, 6, \ldots$. Starting with $a_0 \neq 0$ we may then calculate $a_2, a_4, a_6, \ldots$, but we see that when $n = m$ the coefficient a_n is formally infinite, and the method fails to produce a second solution in the form of a Frobenius series.

In fact, by replacing v by $-v$ in the definition of $J_v(x)$ given in (18.79), it can be shown that, for integer v,

$$J_{-v}(x) = (-1)^v J_v(x),$$

and hence that $J_v(x)$ and $J_{-v}(x)$ are linearly dependent. So, in this case, we cannot write the general solution to Bessel's equation in the form (18.80). One therefore defines the function

$$Y_v(x) = \frac{J_v(x)\cos v\pi - J_{-v}(x)}{\sin v\pi},$$
(18.81)

which is called a Bessel function of the *second kind* of order v (or, occasionally, a *Weber* or *Neumann* function). As Bessel's equation is linear, $Y_v(x)$ is clearly a solution, since it is just the weighted sum of Bessel functions of the first kind. Furthermore, for non-integer v it is clear that $Y_v(x)$ is linearly independent of $J_v(x)$. It may also be shown that the Wronskian of $J_v(x)$ and $Y_v(x)$ is non-zero for *all* values of v. Hence $J_v(x)$ and $Y_v(x)$ always constitute a pair of independent solutions.

▶ *If n is an integer, show that $Y_{n+1/2}(x) = (-1)^{n+1} J_{-n-1/2}(x)$.*

From (18.81), we have

$$Y_{n+1/2}(x) = \frac{J_{n+1/2}(x)\cos(n + \frac{1}{2})\pi - J_{-n-1/2}(x)}{\sin(n + \frac{1}{2})\pi}.$$

If n is an integer, $\cos(n + \frac{1}{2})\pi = 0$ and $\sin(n + \frac{1}{2})\pi = (-1)^n$, and so we immediately obtain $Y_{n+1/2}(x) = (-1)^{n+1} J_{-n-1/2}(x)$, as required. ◀

The expression (18.81) becomes an indeterminate form $0/0$ when v is an integer, however. This is so because for integer v we have $\cos v\pi = (-1)^v$ and $J_{-v}(x) = (-1)^v J_v(x)$. Nevertheless, this indeterminate form can be evaluated using l'Hôpital's rule (see chapter 4). Therefore, for integer v, we set

$$Y_v(x) = \lim_{\mu \to v} \left[\frac{J_\mu(x)\cos \mu\pi - J_{-\mu}(x)}{\sin \mu\pi} \right],$$
(18.82)

which gives a linearly independent second solution for this case. Thus, we may write the general solution of Bessel's equation, valid for *all* v, as

$$y(x) = c_1 J_v(x) + c_2 Y_v(x).$$
(18.83)

The functions $Y_0(x)$, $Y_1(x)$ and $Y_2(x)$ are plotted in figure 18.6

Finally, we note that, in some applications, it is convenient to work with complex linear combinations of Bessel functions of the first and second kinds given by

$$H_v^{(1)}(x) = J_v(x) + iY_v(x), \qquad H_v^{(2)}(x) = J_v(x) - iY_v(x);$$

these are called, respectively, *Hankel functions* of the first and second kind of order v.

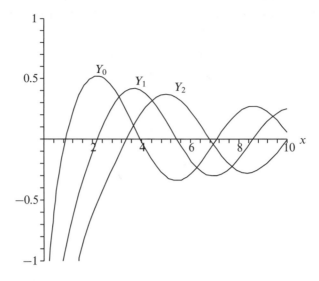

Figure 18.6 The first three integer-order Bessel functions of the second kind.

18.5.3 Properties of Bessel functions $J_v(x)$

In physical applications, we often require that the solution is regular at $x = 0$, but, from its definition (18.81) or (18.82), it is clear that $Y_v(x)$ is singular at the origin, and so in such physical situations the coefficient c_2 in (18.83) must be set to zero; the solution is then simply some multiple of $J_v(x)$. These Bessel functions of the first kind have various useful properties that are worthy of further discussion. Unless otherwise stated, the results presented in this section apply to Bessel functions $J_v(x)$ of integer and non-integer order.

Mutual orthogonality

In section 17.4, we noted that Bessel's equation (18.73) could be put into conventional Sturm–Liouville form with $p = x$, $q = -v^2/x$, $\lambda = \alpha^2$ and $\rho = x$, provided αx is the argument of y. From the form of p, we see that there is no natural interval over which one would expect the solutions of Bessel's equation corresponding to different eigenvalues λ (but fixed v) to be automatically orthogonal. Nevertheless, provided the Bessel functions satisfied appropriate boundary conditions, we would expect them to obey an orthogonality relationship over some interval $[a, b]$ of the form

$$\int_a^b x J_v(\alpha x) J_v(\beta x)\, dx = 0 \qquad \text{for } \alpha \neq \beta. \tag{18.84}$$

To determine the required boundary conditions for this result to hold, let us consider the functions $f(x) = J_v(\alpha x)$ and $g(x) = J_v(\beta x)$, which, as will be proved below, respectively satisfy the equations

$$x^2 f'' + x f' + (\alpha^2 x^2 - v^2)f = 0, \qquad (18.85)$$

$$x^2 g'' + x g' + (\beta^2 x^2 - v^2)g = 0. \qquad (18.86)$$

▶Show that $f(x) = J_v(\alpha x)$ satisfies (18.85).

If $f(x) = J_v(\alpha x)$ and we write $w = \alpha x$, then

$$\frac{df}{dx} = \alpha \frac{dJ_v(w)}{dw} \qquad \text{and} \qquad \frac{d^2 f}{dx^2} = \alpha^2 \frac{d^2 J_v(w)}{dw^2}.$$

When these expressions are substituted into (18.85), its LHS becomes

$$x^2 \alpha^2 \frac{d^2 J_v(w)}{dw^2} + x\alpha \frac{dJ_v(w)}{dw} + (\alpha^2 x^2 - v^2)J_v(w)$$

$$= w^2 \frac{d^2 J_v(w)}{dw^2} + w \frac{dJ_v(w)}{dw} + (w^2 - v^2)J_v(w).$$

But, from Bessel's equation itself, this final expression is equal to zero, thus verifying that $f(x)$ does satisfy (18.85). ◀

Now multiplying (18.86) by $f(x)$ and (18.85) by $g(x)$ and subtracting them gives

$$\frac{d}{dx}[x(fg' - gf')] = (\alpha^2 - \beta^2)xfg, \qquad (18.87)$$

where we have used the fact that

$$\frac{d}{dx}[x(fg' - gf')] = x(fg'' - gf'') + (fg' - gf').$$

By integrating (18.87) over any given range $x = a$ to $x = b$, we obtain

$$\int_a^b xf(x)g(x)\,dx = \frac{1}{\alpha^2 - \beta^2}\left[xf(x)g'(x) - xg(x)f'(x)\right]_a^b,$$

which, on setting $f(x) = J_v(\alpha x)$ and $g(x) = J_v(\beta x)$, becomes

$$\int_a^b xJ_v(\alpha x)J_v(\beta x)\,dx = \frac{1}{\alpha^2 - \beta^2}\left[\beta xJ_v(\alpha x)J_v'(\beta x) - \alpha xJ_v(\beta x)J_v'(\alpha x)\right]_a^b. \qquad (18.88)$$

If $\alpha \neq \beta$, and the interval $[a, b]$ is such that the expression on the RHS of (18.88) equals zero, then we obtain the orthogonality condition (18.84). This happens, for example, if $J_v(\alpha x)$ and $J_v(\beta x)$ vanish at $x = a$ and $x = b$, or if $J_v'(\alpha x)$ and $J_v'(\beta x)$ vanish at $x = a$ and $x = b$, or for many more general conditions. It should be noted that the boundary term is automatically zero at the point $x = 0$, as one might expect from the fact that the Sturm–Liouville form of Bessel's equation has $p(x) = x$.

If $\alpha = \beta$, the RHS of (18.88) takes the indeterminate form $0/0$. This may be evaluated using l'Hôpital's rule, or alternatively we may calculate the relevant integral directly.

▶ *Evaluate the integral*

$$\int_a^b J_\nu^2(\alpha x) x \, dx.$$

Ignoring the integration limits for the moment,

$$\int J_\nu^2(\alpha x) x \, dx = \frac{1}{\alpha^2} \int J_\nu^2(u) u \, du,$$

where $u = \alpha x$. Integrating by parts yields

$$I = \int J_\nu^2(u) u \, du = \tfrac{1}{2} u^2 J_\nu^2(u) - \int J_\nu(u) J_\nu'(u) u^2 \, du.$$

Now Bessel's equation (18.73) can be rearranged as

$$u^2 J_\nu(u) = \nu^2 J_\nu(u) - u J_\nu'(u) - u^2 J_\nu''(u),$$

which, on substitution into the expression for I, gives

$$I = \tfrac{1}{2} u^2 J_\nu^2(u) - \int J_\nu'(u) [\nu^2 J_\nu(u) - u J_\nu'(u) - u^2 J_\nu''(u)] \, du$$

$$= \tfrac{1}{2} u^2 J_\nu^2(u) - \tfrac{1}{2} \nu^2 J_\nu^2(u) + \tfrac{1}{2} u^2 [J_\nu'(u)]^2 + c.$$

Since $u = \alpha x$, the required integral is given by

$$\int_a^b J_\nu^2(\alpha x) x \, dx = \frac{1}{2} \left[\left(x^2 - \frac{\nu^2}{\alpha^2} \right) J_\nu^2(\alpha x) + x^2 [J_\nu'(\alpha x)]^2 \right]_a^b, \tag{18.89}$$

which gives the normalisation condition for Bessel functions of the first kind. ◀

Since the Bessel functions $J_\nu(x)$ possess the orthogonality property (18.88), we may expand any reasonable function $f(x)$, i.e. one obeying the Dirichlet conditions discussed in chapter 12, in the interval $0 \le x \le b$ as a sum of Bessel functions of a given (non-negative) order ν,

$$f(x) = \sum_{n=0}^\infty c_n J_\nu(\alpha_n x), \tag{18.90}$$

provided that the α_n are chosen such that $J_\nu(\alpha_n b) = 0$. The coefficients c_n are then given by

$$c_n = \frac{2}{b^2 J_{\nu+1}^2(\alpha_n b)} \int_0^b f(x) J_\nu(\alpha_n x) x \, dx. \tag{18.91}$$

The interval is taken to be $0 \le x \le b$, as then one need only ensure that the appropriate boundary condition is satisfied at $x = b$, since the boundary condition at $x = 0$ is met automatically.

▶*Prove the expression (18.91).*

If we multiply (18.90) by $xJ_\nu(\alpha_m x)$ and integrate from $x = 0$ to $x = b$ then we obtain

$$\int_0^b xJ_\nu(\alpha_m x)f(x)\,dx = \sum_{n=0}^\infty c_n \int_0^b xJ_\nu(\alpha_m x)J_\nu(\alpha_n x)\,dx$$

$$= c_m \int_0^b J_\nu^2(\alpha_m x)x\,dx$$

$$= \tfrac{1}{2}c_m b^2 J_\nu'^2(\alpha_m b) = \tfrac{1}{2}c_m b^2 J_{\nu+1}^2(\alpha_m b),$$

where in the last two lines we have used (18.88) with $\alpha_m = \alpha \neq \beta = \alpha_n$, (18.89), the fact that $J_\nu(\alpha_m b) = 0$ and (18.95), which is proved below. ◀

Recurrence relations

The recurrence relations enjoyed by Bessel functions of the first kind, $J_\nu(x)$, can be derived directly from the power series definition (18.79).

▶*Prove the recurrence relation*

$$\frac{d}{dx}[x^\nu J_\nu(x)] = x^\nu J_{\nu-1}(x). \tag{18.92}$$

From the power series definition (18.79) of $J_\nu(x)$ we obtain

$$\frac{d}{dx}[x^\nu J_\nu(x)] = \frac{d}{dx}\sum_{n=0}^\infty \frac{(-1)^n x^{2\nu+2n}}{2^{\nu+2n}n!\Gamma(\nu+n+1)}$$

$$= \sum_{n=0}^\infty \frac{(-1)^n x^{2\nu+2n-1}}{2^{\nu+2n-1}n!\Gamma(\nu+n)}$$

$$= x^\nu \sum_{n=0}^\infty \frac{(-1)^n x^{(\nu-1)+2n}}{2^{(\nu-1)+2n}n!\Gamma((\nu-1)+n+1)} = x^\nu J_{\nu-1}(x). ◀$$

It may similarly be shown that

$$\frac{d}{dx}[x^{-\nu}J_\nu(x)] = -x^{-\nu}J_{\nu+1}(x). \tag{18.93}$$

From (18.92) and (18.93) the remaining recurrence relations may be derived. Expanding out the derivative on the LHS of (18.92) and dividing through by $x^{\nu-1}$, we obtain the relation

$$xJ_\nu'(x) + \nu J_\nu(x) = xJ_{\nu-1}(x). \tag{18.94}$$

Similarly, by expanding out the derivative on the LHS of (18.93), and multiplying through by $x^{\nu+1}$, we find

$$xJ_\nu'(x) - \nu J_\nu(x) = -xJ_{\nu+1}(x). \tag{18.95}$$

Adding (18.94) and (18.95) and dividing through by x gives

$$J_{\nu-1}(x) - J_{\nu+1}(x) = 2J_\nu'(x). \tag{18.96}$$

Finally, subtracting (18.95) from (18.94) and dividing by x gives

$$J_{v-1}(x) + J_{v+1}(x) = \frac{2v}{x}J_v(x). \qquad (18.97)$$

▶Given that $J_{1/2}(x) = (2/\pi x)^{1/2}\sin x$ and that $J_{-1/2}(x) = (2/\pi x)^{1/2}\cos x$, express $J_{3/2}(x)$ and $J_{-3/2}(x)$ in terms of trigonometric functions.

From (18.95) we have

$$J_{3/2}(x) = \frac{1}{2x}J_{1/2}(x) - J'_{1/2}(x)$$

$$= \frac{1}{2x}\left(\frac{2}{\pi x}\right)^{1/2}\sin x - \left(\frac{2}{\pi x}\right)^{1/2}\cos x + \frac{1}{2x}\left(\frac{2}{\pi x}\right)^{1/2}\sin x$$

$$= \left(\frac{2}{\pi x}\right)^{1/2}\left(\frac{1}{x}\sin x - \cos x\right).$$

Similarly, from (18.94) we have

$$J_{-3/2}(x) = -\frac{1}{2x}J_{-1/2}(x) + J'_{-1/2}(x)$$

$$= -\frac{1}{2x}\left(\frac{2}{\pi x}\right)^{1/2}\cos x - \left(\frac{2}{\pi x}\right)^{1/2}\sin x - \frac{1}{2x}\left(\frac{2}{\pi x}\right)^{1/2}\cos x$$

$$= \left(\frac{2}{\pi x}\right)^{1/2}\left(-\frac{1}{x}\cos x - \sin x\right).$$

We see that, by repeated use of these recurrence relations, all Bessel functions $J_v(x)$ of half-integer order may be expressed in terms of trigonometric functions. From their definition (18.81), Bessel functions of the second kind, $Y_v(x)$, of half-integer order can be similarly expressed. ◀

Finally, we note that the relations (18.92) and (18.93) may be rewritten in integral form as

$$\int x^v J_{v-1}(x)\,dx = x^v J_v(x),$$

$$\int x^{-v} J_{v+1}(x)\,dx = -x^{-v} J_v(x).$$

If v is an integer, the recurrence relations of this section may be proved using the generating function for Bessel functions discussed below. It may be shown that Bessel functions of the second kind, $Y_v(x)$, also satisfy the recurrence relations derived above.

Generating function

The Bessel functions $J_v(x)$, where $v = n$ is an integer, can be described by a generating function in a way similar to that discussed for Legendre polynomials in subsection 18.1.2. The generating function for Bessel functions of integer order is given by

$$G(x, h) = \exp\left[\frac{x}{2}\left(h - \frac{1}{h}\right)\right] = \sum_{n=-\infty}^{\infty} J_n(x)h^n. \tag{18.98}$$

By expanding the exponential as a power series, it is straightforward to verify that the functions $J_n(x)$ defined by (18.98) are indeed Bessel functions of the first kind, as given by (18.79).

The generating function (18.98) is useful for finding, for Bessel functions of integer order, properties that can often be extended to the non-integer case. In particular, the Bessel function recurrence relations may be derived.

> ▶*Use the generating function to prove, for integer v, the recurrence relation (18.97), i.e.*
>
> $$J_{v-1}(x) + J_{v+1}(x) = \frac{2v}{x}J_v(x).$$

Differentiating $G(x, h)$ with respect to h we obtain

$$\frac{\partial G(x, h)}{\partial h} = \frac{x}{2}\left(1 + \frac{1}{h^2}\right)G(x, h) = \sum_{n=-\infty}^{\infty} nJ_n(x)h^{n-1},$$

which can be written using (18.98) again as

$$\frac{x}{2}\left(1 + \frac{1}{h^2}\right)\sum_{n=-\infty}^{\infty} J_n(x)h^n = \sum_{n=-\infty}^{\infty} nJ_n(x)h^{n-1}.$$

Equating coefficients of h^n we obtain

$$\frac{x}{2}[J_n(x) + J_{n+2}(x)] = (n + 1)J_{n+1}(x),$$

which, on replacing n by $v - 1$, gives the required recurrence relation. ◀

Integral representations

The generating function (18.98) is also useful for deriving *integral representations* of Bessel functions of integer order.

> ▶*Show that for integer n the Bessel function $J_n(x)$ is given by*
>
> $$J_n(x) = \frac{1}{\pi}\int_0^{\pi} \cos(n\theta - x\sin\theta)\, d\theta. \tag{18.99}$$

By expanding out the cosine term in the integrand in (18.99) we obtain the integral

$$I = \frac{1}{\pi}\int_0^{\pi} [\cos(x\sin\theta)\cos n\theta + \sin(x\sin\theta)\sin n\theta]\, d\theta. \tag{18.100}$$

Now, we may express $\cos(x\sin\theta)$ and $\sin(x\sin\theta)$ in terms of Bessel functions by setting $h = \exp i\theta$ in (18.98) to give

$$\exp\left[\frac{x}{2}(\exp i\theta - \exp(-i\theta))\right] = \exp(ix\sin\theta) = \sum_{m=-\infty}^{\infty} J_m(x)\exp im\theta.$$

Using de Moivre's theorem, $\exp i\theta = \cos\theta + i\sin\theta$, we then obtain

$$\exp(ix\sin\theta) = \cos(x\sin\theta) + i\sin(x\sin\theta) = \sum_{m=-\infty}^{\infty} J_m(x)(\cos m\theta + i\sin m\theta).$$

Equating the real and imaginary parts of this expression gives

$$\cos(x\sin\theta) = \sum_{m=-\infty}^{\infty} J_m(x)\cos m\theta,$$

$$\sin(x\sin\theta) = \sum_{m=-\infty}^{\infty} J_m(x)\sin m\theta.$$

Substituting these expressions into (18.100) then yields

$$I = \frac{1}{\pi} \sum_{m=-\infty}^{\infty} \int_0^\pi [J_m(x)\cos m\theta \cos n\theta + J_m(x)\sin m\theta \sin n\theta] \, d\theta.$$

However, using the orthogonality of the trigonometric functions [see equations (12.1)–(12.3)], we obtain

$$I = \frac{1}{\pi}\frac{\pi}{2}[J_n(x) + J_n(x)] = J_n(x),$$

which proves the integral representation (18.99). ◄

Finally, we mention the special case of the integral representation (18.99) for $n = 0$:

$$J_0(x) = \frac{1}{\pi} \int_0^\pi \cos(x\sin\theta)\,d\theta = \frac{1}{2\pi} \int_0^{2\pi} \cos(x\sin\theta)\,d\theta,$$

since $\cos(x\sin\theta)$ repeats itself in the range $\theta = \pi$ to $\theta = 2\pi$. However, $\sin(x\sin\theta)$ changes sign in this range and so

$$\frac{1}{2\pi} \int_0^{2\pi} \sin(x\sin\theta)\,d\theta = 0.$$

Using de Moivre's theorem, we can therefore write

$$J_0(x) = \frac{1}{2\pi} \int_0^{2\pi} \exp(ix\sin\theta)\,d\theta = \frac{1}{2\pi} \int_0^{2\pi} \exp(ix\cos\theta)\,d\theta.$$

There are in fact many other integral representations of Bessel functions; they can be derived from those given.

18.6 Spherical Bessel functions

When obtaining solutions of Helmholtz' equation $(\nabla^2 + k^2)u = 0$ in spherical polar coordinates (see section 21.3.2), one finds that, for solutions that are finite on the polar axis, the radial part $R(r)$ of the solution must satisfy the equation

$$r^2 R'' + 2rR' + [k^2 r^2 - \ell(\ell+1)]R = 0, \tag{18.101}$$

where ℓ is an integer. This equation looks very much like Bessel's equation and can in fact be reduced to it by writing $R(r) = r^{-1/2}S(r)$, in which case $S(r)$ then satisfies

$$r^2 S'' + rS' + \left[k^2 r^2 - (\ell + \tfrac{1}{2})^2 \right] S = 0.$$

On making the change of variable $x = kr$ and letting $y(x) = S(kr)$, we obtain

$$x^2 y'' + xy' + [x^2 - (\ell + \tfrac{1}{2})^2]y = 0,$$

where the primes now denote d/dx. This is Bessel's equation of order $\ell + \tfrac{1}{2}$ and has as its solutions $y(x) = J_{\ell+1/2}(x)$ and $Y_{\ell+1/2}(x)$. The general solution of (18.101) can therefore be written

$$R(r) = r^{-1/2}[c_1 J_{\ell+1/2}(kr) + c_2 Y_{\ell+1/2}(kr)],$$

where c_1 and c_2 are constants that may be determined from the boundary conditions on the solution. In particular, for solutions that are finite at the origin we require $c_2 = 0$.

The functions $x^{-1/2}J_{\ell+1/2}(x)$ and $x^{-1/2}Y_{\ell+1/2}(x)$, when suitably normalised, are called *spherical Bessel functions* of the first and second kind, respectively, and are denoted as follows:

$$j_\ell(x) = \sqrt{\frac{\pi}{2x}} J_{\ell+1/2}(x), \tag{18.102}$$

$$n_\ell(x) = \sqrt{\frac{\pi}{2x}} Y_{\ell+1/2}(x). \tag{18.103}$$

For integer ℓ, we also note that $Y_{\ell+1/2}(x) = (-1)^{\ell+1}J_{-\ell-1/2}(x)$, as discussed in section 18.5.2. Moreover, in section 18.5.1, we noted that Bessel functions of the first kind, $J_\nu(x)$, of half-integer order are expressible in closed form in terms of trigonometric functions. Thus, all spherical Bessel functions of both the first and second kinds may be expressed in such a form. In particular, using the results of the worked example in section 18.5.1, we find that

$$j_0(x) = \frac{\sin x}{x}, \tag{18.104}$$

$$n_0(x) = -\frac{\cos x}{x}. \tag{18.105}$$

Expressions for higher-order spherical Bessel functions are most easily obtained by using the recurrence relations for Bessel functions.

> ►*Show that the ℓth spherical Bessel function is given by*
>
> $$f_\ell(x) = (-1)^\ell x^\ell \left(\frac{1}{x}\frac{d}{dx}\right)^\ell f_0(x), \qquad (18.106)$$
>
> *where $f_\ell(x)$ denotes either $j_\ell(x)$ or $n_\ell(x)$.*

The recurrence relation (18.93) for Bessel functions of the first kind reads

$$J_{\nu+1}(x) = -x^\nu \frac{d}{dx}\left[x^{-\nu}J_\nu(x)\right].$$

Thus, on setting $\nu = \ell + \frac{1}{2}$ and rearranging, we find

$$x^{-1/2}J_{\ell+3/2}(x) = -x^\ell \frac{d}{dx}\left[\frac{x^{-1/2}J_{\ell+1/2}}{x^\ell}\right],$$

which on using (18.102) yields the recurrence relation

$$j_{\ell+1}(x) = -x^\ell \frac{d}{dx}[x^{-\ell}j_\ell(x)].$$

We now change $\ell + 1 \to \ell$ and iterate this result:

$$\begin{aligned}
j_\ell(x) &= -x^{\ell-1}\frac{d}{dx}[x^{-\ell+1}j_{\ell-1}(x)]\\
&= -x^{\ell-1}\frac{d}{dx}\left\{x^{-\ell+1}(-1)x^{\ell-2}\frac{d}{dx}\left[x^{-\ell+2}j_{\ell-2}(x)\right]\right\}\\
&= (-1)^2 \frac{x^\ell}{x}\frac{d}{dx}\left\{\frac{1}{x}\frac{d}{dx}\left[x^{-\ell+2}j_{\ell-2}(x)\right]\right\}\\
&= \cdots\\
&= (-1)^\ell x^\ell \left(\frac{1}{x}\frac{d}{dx}\right)^\ell j_0(x).
\end{aligned}$$

This is the expression for $j_\ell(x)$ as given in (18.106). One may prove the result (18.106) for $n_\ell(x)$ in an analogous manner by setting $\nu = \ell - \frac{1}{2}$ in the recurrence relation (18.92) for Bessel functions of the first kind and using the relationship $Y_{\ell+1/2}(x) = (-1)^{\ell+1}J_{-\ell-1/2}(x)$. ◄

Using result (18.106) and the expressions (18.104) and (18.105), one quickly finds, for example,

$$j_1(x) = \frac{\sin x}{x^2} - \frac{\cos x}{x}, \qquad j_2(x) = \left(\frac{3}{x^3} - \frac{1}{x}\right)\sin x - \frac{3\cos x}{x^2},$$

$$n_1(x) = -\frac{\cos x}{x^2} - \frac{\sin x}{x}, \qquad n_2(x) = -\left(\frac{3}{x^3} - \frac{1}{x}\right)\cos x - \frac{3\sin x}{x^2}.$$

Finally, we note that the orthogonality properties of the spherical Bessel functions follow directly from the orthogonality condition (18.88) for Bessel functions of the first kind.

18.7 Laguerre functions

Laguerre's equation has the form

$$xy'' + (1 - x)y' + \nu y = 0; \qquad (18.107)$$

it has a regular singularity at $x = 0$ and an essential singularity at $x = \infty$. The parameter v is a given real number, although it nearly always takes an integer value in physical applications. The Laguerre equation appears in the description of the wavefunction of the hydrogen atom. Any solution of (18.107) is called a *Laguerre function.*

Since the point $x = 0$ is a regular singularity, we may find at least one solution in the form of a Frobenius series (see section 16.3):

$$y(x) = \sum_{m=0}^{\infty} a_m x^{m+\sigma}. \tag{18.108}$$

Substituting this series into (18.107) and dividing through by $x^{\sigma-1}$, we obtain

$$\sum_{m=0}^{\infty} [(m + \sigma)(m + \sigma - 1) + (1 - x)(m + \sigma) + vx] a_m x^m = 0. \tag{18.109}$$

Setting $x = 0$, so that only the $m = 0$ term remains, we obtain the indicial equation $\sigma^2 = 0$, which trivially has $\sigma = 0$ as its repeated root. Thus, Laguerre's equation has only one solution of the form (18.108), and it, in fact, reduces to a simple power series. Substituting $\sigma = 0$ into (18.109) and demanding that the coefficient of x^{m+1} vanishes, we obtain the recurrence relation

$$a_{m+1} = \frac{m - v}{(m + 1)^2} a_m.$$

As mentioned above, in nearly all physical applications, the parameter v takes integer values. Therefore, if $v = n$, where n is a non-negative integer, we see that $a_{n+1} = a_{n+2} = \cdots = 0$, and so our solution to Laguerre's equation is a polynomial of order n. It is conventional to choose $a_0 = 1$, so that the solution is given by

$$L_n(x) = \frac{(-1)^n}{n!} \left[x^n - \frac{n^2}{1!} x^{n-1} + \frac{n^2(n-1)^2}{2!} x^{n-2} - \cdots + (-1)^n n! \right] \tag{18.110}$$

$$= \sum_{m=0}^{n} (-1)^m \frac{n!}{(m!)^2(n-m)!} x^m, \tag{18.111}$$

where $L_n(x)$ is called the nth *Laguerre polynomial.* We note in particular that $L_n(0) = 1$. The first few Laguerre polynomials are given by

$L_0(x) = 1,$ $\qquad\qquad 3!L_3(x) = -x^3 + 9x^2 - 18x + 6,$

$L_1(x) = -x + 1,$ $\qquad\quad 4!L_4(x) = x^4 - 16x^3 + 72x^2 - 96x + 24,$

$2!L_2(x) = x^2 - 4x + 2,$ $\quad 5!L_5(x) = -x^5 + 25x^4 - 200x^3 + 600x^2 - 600x + 120.$

The functions $L_0(x)$, $L_1(x)$, $L_2(x)$ and $L_3(x)$ are plotted in figure 18.7.

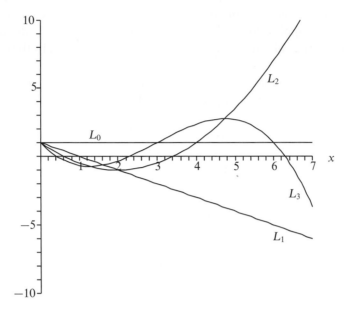

Figure 18.7   The first four Laguerre polynomials.

18.7.1 Properties of Laguerre polynomials

The Laguerre polynomials and functions derived from them are important in the analysis of the quantum mechanical behaviour of some physical systems. We therefore briefly outline their useful properties in this section.

Rodrigues' formula

The Laguerre polynomials can be expressed in terms of a Rodrigues' formula given by

$$L_n(x) = \frac{e^x}{n!} \frac{d^n}{dx^n} \left(x^n e^{-x} \right), \qquad (18.112)$$

which may be proved straightforwardly by calculating the nth derivative explicitly using Leibnitz' theorem and comparing the result with (18.111). This is illustrated in the following example.

▶*Prove that the expression (18.112) yields the nth Laguerre polynomial.*

Evaluating the nth derivative in (18.112) using Leibnitz' theorem, we find

$$L_n(x) = \frac{e^x}{n!} \sum_{r=0}^{n} {}^nC_r \frac{d^r x^n}{dx^r} \frac{d^{n-r} e^{-x}}{dx^{n-r}}$$

$$= \frac{e^x}{n!} \sum_{r=0}^{n} \frac{n!}{r!(n-r)!} \frac{n!}{(n-r)!} x^{n-r} (-1)^{n-r} e^{-x}$$

$$= \sum_{r=0}^{n} (-1)^{n-r} \frac{n!}{r!(n-r)!(n-r)!} x^{n-r}.$$

Relabelling the summation using the index $m = n - r$, we obtain

$$L_n(x) = \sum_{m=0}^{n} (-1)^m \frac{n!}{(m!)^2(n-m)!} x^m,$$

which is precisely the expression (18.111) for the nth Laguerre polynomial. ◀

Mutual orthogonality

In section 17.4, we noted that Laguerre's equation could be put into Sturm–Liouville form with $p = xe^{-x}$, $q = 0$, $\lambda = \nu$ and $\rho = e^{-x}$, and its natural interval is thus $[0, \infty]$. Since the Laguerre polynomials $L_n(x)$ are solutions of the equation and are regular at the end-points, they must be mutually orthogonal over this interval with respect to the weight function $\rho = e^{-x}$, i.e.

$$\int_0^\infty L_n(x)L_k(x)e^{-x}\,dx = 0 \qquad \text{if } n \neq k.$$

This result may also be proved directly using the Rodrigues' formula (18.112). Indeed, the normalisation, when $k = n$, is most easily found using this method.

▶*Show that*

$$I \equiv \int_0^\infty L_n(x)L_n(x)e^{-x}\,dx = 1. \tag{18.113}$$

Using the Rodrigues' formula (18.112), we may write

$$I = \frac{1}{n!} \int_0^\infty L_n(x) \frac{d^n}{dx^n}(x^n e^{-x})\,dx = \frac{(-1)^n}{n!} \int_0^\infty \frac{d^n L_n}{dx^n} x^n e^{-x}\,dx,$$

where, in the second equality, we have integrated by parts n times and used the fact that the boundary terms all vanish. When $d^n L_n/dx^n$ is evaluated using (18.111), only the derivative of the $m = n$ term survives and that has the value $[(-1)^n n! \, n!]/[(n!)^2 \, 0!] = (-1)^n$. Thus we have

$$I = \frac{1}{n!} \int_0^\infty x^n e^{-x}\,dx = 1,$$

where, in the second equality, we use the expression (18.153) defining the gamma function (see section 18.12). ◀

The above orthogonality and normalisation conditions allow us to expand any (reasonable) function in the interval $0 \leq x < \infty$ in a series of the form

$$f(x) = \sum_{n=0}^{\infty} a_n L_n(x),$$

in which the coefficients a_n are given by

$$a_n = \int_0^{\infty} f(x)L_n(x)e^{-x}\, dx.$$

We note that it is sometimes convenient to define the *orthonormal Laguerre functions* $\phi_n(x) = e^{-x/2}L_n(x)$, which may also be used to produce a series expansion of a function in the interval $0 \leq x < \infty$.

Generating function

The generating function for the Laguerre polynomials is given by

$$G(x, h) = \frac{e^{-xh/(1-h)}}{1-h} = \sum_{n=0}^{\infty} L_n(x)h^n. \tag{18.114}$$

We may prove this result by differentiating the generating function with respect to x and h, respectively, to obtain recurrence relations for the Laguerre polynomials, which may then be combined to show that the functions $L_n(x)$ in (18.114) do indeed satisfy Laguerre's equation (as discussed in the next subsection).

Recurrence relations

The Laguerre polynomials obey a number of useful recurrence relations. The three most important relations are as follows:

$$(n+1)L_{n+1}(x) = (2n+1-x)L_n(x) - nL_{n-1}(x), \tag{18.115}$$

$$L_{n-1}(x) = L'_{n-1}(x) - L'_n(x), \tag{18.116}$$

$$xL'_n(x) = nL_n(x) - nL_{n-1}(x). \tag{18.117}$$

The first two relations are easily derived from the generating function (18.114), and may be combined straightforwardly to yield the third result.

▶*Derive the recurrence relations (18.115) and (18.116).*

Differentiating the generating function (18.114) with respect to h, we find

$$\frac{\partial G}{\partial h} = \frac{(1-x-h)e^{-xh/(1-h)}}{(1-h)^3} = \sum nL_n h^{n-1}.$$

Thus, we may write

$$(1-x-h)\sum L_n h^n = (1-h)^2 \sum nL_n h^{n-1},$$

and, on equating coefficients of h^n on each side, we obtain

$$(1-x)L_n - L_{n-1} = (n+1)L_{n+1} - 2nL_n + (n-1)L_{n-1},$$

which trivially rearranges to give the recurrence relation (18.115).

To obtain the recurrence relation (18.116), we begin by differentiating the generating function (18.114) with respect to x, which yields

$$\frac{\partial G}{\partial x} = -\frac{he^{-xh/(1-h)}}{(1-h)^2} = \sum L_n' h^n,$$

and thus we have

$$-h \sum L_n h^n = (1-h) \sum L_n' h^n.$$

Equating coefficients of h^n on each side then gives

$$-L_{n-1} = L_n' - L_{n-1}',$$

which immediately simplifies to give (18.116). ◄

18.8 Associated Laguerre functions

The associated Laguerre equation has the form

$$xy'' + (m+1-x)y' + ny = 0; \tag{18.118}$$

it has a regular singularity at $x = 0$ and an essential singularity at $x = \infty$. We restrict our attention to the situation in which the parameters n and m are both non-negative integers, as is the case in nearly all physical problems. The associated Laguerre equation occurs most frequently in quantum-mechanical applications. Any solution of (18.118) is called an *associated Laguerre function*.

Solutions of (18.118) for non-negative integers n and m are given by the *associated Laguerre polynomials*

$$L_n^m(x) = (-1)^m \frac{d^m}{dx^m} L_{n+m}(x), \tag{18.119}$$

where $L_n(x)$ are the ordinary Laguerre polynomials.[§]

►*Show that the functions $L_n^m(x)$ defined in (18.119) are solutions of (18.118).*

Since the Laguerre polynomials $L_n(x)$ are solutions of Laguerre's equation (18.107), we have

$$xL_{n+m}'' + (1-x)L_{n+m}' + (n+m)L_{n+m} = 0.$$

Differentiating this equation m times using Leibnitz' theorem and rearranging, we find

$$xL_{n+m}^{(m+2)} + (m+1-x)L_{n+m}^{(m+1)} + nL_{n+m}^{(m)} = 0.$$

On multiplying through by $(-1)^m$ and setting $L_n^m = (-1)^m L_{n+m}^{(m)}$, in accord with (18.119), we obtain

$$x(L_n^m)'' + (m+1-x)(L_n^m)' + nL_n^m = 0,$$

which shows that the functions L_n^m are indeed solutions of (18.118). ◄

[§] Note that some authors define the associated Laguerre polynomials as $\mathcal{L}_n^m(x) = (d^m/dx^m)L_n(x)$, which is thus related to our expression (18.119) by $L_n^m(x) = (-1)^m \mathcal{L}_{n+m}^m(x)$.

In particular, we note that $L_n^0(x) = L_n(x)$. As discussed in the previous section, $L_n(x)$ is a polynomial of order n and so it follows that $L_n^m(x)$ is also. The first few associated Laguerre polynomials are easily found using (18.119):

$$L_0^m(x) = 1,$$
$$L_1^m(x) = -x + m + 1,$$
$$2!L_2^m(x) = x^2 - 2(m+2)x + (m+1)(m+2),$$
$$3!L_3^m(x) = -x^3 + 3(m+3)x^2 - 3(m+2)(m+3)x + (m+1)(m+2)(m+3).$$

Indeed, in the general case, one may show straightforwardly, from the definition (18.119) and the expression (18.111) for the ordinary Laguerre polynomials, that

$$L_n^m(x) = \sum_{k=0}^{n} (-1)^k \frac{(n+m)!}{k!(n-k)!(k+m)!} x^k. \tag{18.120}$$

18.8.1 Properties of associated Laguerre polynomials

The properties of the associated Laguerre polynomials follow directly from those of the ordinary Laguerre polynomials through the definition (18.119). We shall therefore only briefly outline the most useful results here.

Rodrigues' formula

A Rodrigues' formula for the associated Laguerre polynomials is given by

$$L_n^m(x) = \frac{e^x x^{-m}}{n!} \frac{d^n}{dx^n} (x^{n+m} e^{-x}). \tag{18.121}$$

It can be proved by evaluating the nth derivative using Leibnitz' theorem (see exercise 18.7).

Mutual orthogonality

In section 17.4, we noted that the associated Laguerre equation could be transformed into a Sturm–Liouville one with $p = x^{m+1} e^{-x}$, $q = 0$, $\lambda = n$ and $\rho = x^m e^{-x}$, and its natural interval is thus $[0, \infty]$. Since the associated Laguerre polynomials $L_n^m(x)$ are solutions of the equation and are regular at the end-points, those with the same m but differing values of the eigenvalue $\lambda = n$ must be mutually orthogonal over this interval with respect to the weight function $\rho = x^m e^{-x}$, i.e.

$$\int_0^\infty L_n^m(x) L_k^m(x) x^m e^{-x} \, dx = 0 \qquad \text{if } n \neq k.$$

This result may also be proved directly using the Rodrigues' formula (18.121), as may the normalisation condition when $k = n$.

> ▶*Show that*
>
> $$I \equiv \int_0^\infty L_n^m(x) L_n^m(x) x^m e^{-x}\, dx = \frac{(n+m)!}{n!}.$$ (18.122)

Using the Rodrigues' formula (18.121), we may write

$$I = \frac{1}{n!} \int_0^\infty L_n^m(x) \frac{d^n}{dx^n}(x^{n+m} e^{-x})\, dx = \frac{(-1)^n}{n!} \int_0^\infty \frac{d^n L_n^m}{dx^n} x^{n+m} e^{-x}\, dx,$$

where, in the second equality, we have integrated by parts n times and used the fact that the boundary terms all vanish. From (18.120) we see that $d^n L_n^m/dx^n = (-1)^n$. Thus we have

$$I = \frac{1}{n!} \int_0^\infty x^{n+m} e^{-x}\, dx = \frac{(n+m)!}{n!},$$

where, in the second equality, we use the expression (18.153) defining the gamma function (see section 18.12). ◀

The above orthogonality and normalisation conditions allow us to expand any (reasonable) function in the interval $0 \leq x < \infty$ in a series of the form

$$f(x) = \sum_{n=0}^\infty a_n L_n^m(x),$$

in which the coefficients a_n are given by

$$a_n = \frac{n!}{(n+m)!} \int_0^\infty f(x) L_n^m(x) x^m e^{-x}\, dx.$$

We note that it is sometimes convenient to define the *orthogonal associated Laguerre functions* $\phi_n^m(x) = x^{m/2} e^{-x/2} L_n^m(x)$, which may also be used to produce a series expansion of a function in the interval $0 \leq x < \infty$.

Generating function

The generating function for the associated Laguerre polynomials is given by

$$G(x,h) = \frac{e^{-xh/(1-h)}}{(1-h)^{m+1}} = \sum_{n=0}^\infty L_n^m(x) h^n.$$ (18.123)

This can be obtained by differentiating the generating function (18.114) for the ordinary Laguerre polynomials m times with respect to x, and using (18.119).

> ▶*Use the generating function (18.123) to obtain an expression for $L_n^m(0)$.*

From (18.123), we have

$$\sum_{n=0}^\infty L_n^m(0) h^n = \frac{1}{(1-h)^{m+1}}$$

$$= 1 + (m+1)h + \frac{(m+1)(m+2)}{2!} h^2 + \cdots + \frac{(m+1)(m+2)\cdots(m+n)}{n!} h^n + \cdots,$$

where, in the second equality, we have expanded the RHS using the binomial theorem. On equating coefficients of h^n, we immediately obtain

$$L_n^m(0) = \frac{(n+m)!}{n!m!}. \blacktriangleleft$$

Recurrence relations

The various recurrence relations satisfied by the associated Laguerre polynomials may be derived by differentiating the generating function (18.123) with respect to either or both of x and h, or by differentiating with respect to x the recurrence relations obeyed by the ordinary Laguerre polynomials, discussed in section 18.7.1. Of the many recurrence relations satisfied by the associated Laguerre polynomials, two of the most useful are as follows:

$$(n + 1)L_{n+1}^m(x) = (2n + m + 1 - x)L_n^m(x) - (n + m)L_{n-1}^m(x), \qquad (18.124)$$

$$x(L_n^m)'(x) = nL_n^m(x) - (n + m)L_{n-1}^m(x). \qquad (18.125)$$

For proofs of these relations the reader is referred to exercise 18.7.

18.9 Hermite functions

Hermite's equation has the form

$$y'' - 2xy' + 2vy = 0, \qquad (18.126)$$

and has an essential singularity at $x = \infty$. The parameter v is a given real number, although it nearly always takes an integer value in physical applications. The Hermite equation appears in the description of the wavefunction of the harmonic oscillator. Any solution of (18.126) is called a *Hermite function*.

Since $x = 0$ is an ordinary point of the equation, we may find two linearly independent solutions in the form of a power series (see section 16.2):

$$y(x) = \sum_{m=0}^{\infty} a_m x^m. \qquad (18.127)$$

Substituting this series into (18.107) yields

$$\sum_{m=0}^{\infty} [(m + 1)(m + 2)a_{m+2} + 2(v - m)a_m] x^m = 0.$$

Demanding that the coefficient of each power of x vanishes, we obtain the recurrence relation

$$a_{m+2} = -\frac{2(v - m)}{(m + 1)(m + 2)}a_m.$$

As mentioned above, in nearly all physical applications, the parameter v takes integer values. Therefore, if $v = n$, where n is a non-negative integer, we see that

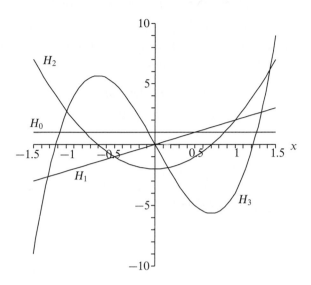

Figure 18.8   The first four Hermite polynomials.

$a_{n+2} = a_{n+4} = \cdots = 0$, and so one solution of Hermite's equation is a polynomial of order n. For even n, it is conventional to choose $a_0 = (-1)^{n/2} n! / (n/2)!$, whereas for odd n one takes $a_1 = (-1)^{(n-1)/2} 2n! / [\frac{1}{2}(n-1)]!$. These choices allow a general solution to be written as

$$H_n(x) = (2x)^n - n(n-1)(2x)^{n-1} + \frac{n(n-1)(n-2)(n-3)}{2!}(2x)^{n-4} - \cdots \quad (18.128)$$

$$= \sum_{m=0}^{[n/2]} (-1)^m \frac{n!}{m!(n-2m)!}(2x)^{n-2m}, \quad (18.129)$$

where $H_n(x)$ is called the nth *Hermite polynomial* and the notation $[n/2]$ denotes the integer part of $n/2$. We note in particular that $H_n(-x) = (-1)^n H_n(x)$. The first few Hermite polynomials are given by

$$H_0(x) = 1, \qquad\qquad H_3(x) = 8x^3 - 12x,$$

$$H_1(x) = 2x, \qquad\qquad H_4(x) = 16x^4 - 48x^2 + 12,$$

$$H_2(x) = 4x^2 - 2, \qquad H_5(x) = 32x^5 - 160x^3 + 120x.$$

The functions $H_0(x)$, $H_1(x)$, $H_2(x)$ and $H_3(x)$ are plotted in figure 18.8.

18.9.1 Properties of Hermite polynomials

The Hermite polynomials and functions derived from them are important in the analysis of the quantum mechanical behaviour of some physical systems. We therefore briefly outline their useful properties in this section.

Rodrigues' formula

The Rodrigues' formula for the Hermite polynomials is given by

$$H_n(x) = (-1)^n e^{x^2} \frac{d^n}{dx^n}(e^{-x^2}). \qquad (18.130)$$

This can be proved using Leibnitz' theorem.

▶Prove the Rodrigues' formula (18.130) for the Hermite polynomials.

Letting $u = e^{-x^2}$ and differentiating with respect to x, we quickly find that

$$u' + 2xu = 0.$$

Differentiating this equation $n + 1$ times using Leibnitz' theorem then gives

$$u^{(n+2)} + 2xu^{(n+1)} + 2(n+1)u^{(n)} = 0,$$

which, on introducing the new variable $v = (-1)^n u^{(n)}$, reduces to

$$v'' + 2xv' + 2(n+1)v = 0. \qquad (18.131)$$

Now letting $y = e^{x^2}v$, we may write the derivatives of v as

$$v' = e^{-x^2}(y' - 2xy),$$
$$v'' = e^{-x^2}(y'' - 4xy' + 4x^2y - 2y).$$

Substituting these expressions into (18.131), and dividing through by e^{-x^2}, finally yields Hermite's equation,

$$y'' - 2xy + 2ny = 0,$$

thus demonstrating that $y = (-1)^n e^{x^2} d^n(e^{-x^2})/dx^n$ is indeed a solution. Moreover, since this solution is clearly a polynomial of order n, it must be some multiple of $H_n(x)$. The normalisation is easily checked by noting that, from (18.130), the highest-order term is $(2x)^n$, which agrees with the expression (18.128). ◀

Mutual orthogonality

We saw in section 17.4 that Hermite's equation could be cast in Sturm–Liouville form with $p = e^{-x^2}$, $q = 0$, $\lambda = 2n$ and $\rho = e^{-x^2}$, and its natural interval is thus $[-\infty, \infty]$. Since the Hermite polynomials $H_n(x)$ are solutions of the equation and are regular at the end-points, they must be mutually orthogonal over this interval with respect to the weight function $\rho = e^{-x^2}$, i.e.

$$\int_{-\infty}^{\infty} H_n(x)H_k(x)e^{-x^2}\,dx = 0 \qquad \text{if } n \neq k.$$

This result may also be proved directly using the Rodrigues' formula (18.130). Indeed, the normalisation, when $k = n$, is most easily found in this way.

►*Show that*

$$I \equiv \int_{-\infty}^{\infty} H_n(x)H_n(x)e^{-x^2}\,dx = 2^n n!\sqrt{\pi}. \tag{18.132}$$

Using the Rodrigues' formula (18.130), we may write

$$I = (-1)^n \int_{0}^{\infty} H_n(x)\frac{d^n}{dx^n}(e^{-x^2})\,dx = \int_{-\infty}^{\infty} \frac{d^n H_n}{dx^n}e^{-x^2}\,dx,$$

where, in the second equality, we have integrated by parts n times and used the fact that the boundary terms all vanish. From (18.128) we see that $d^n H_n/dx^n = 2^n n!$. Thus we have

$$I = 2^n n! \int_{-\infty}^{\infty} e^{-x^2}\,dx = 2^n n!\sqrt{\pi},$$

where, in the second equality, we use the standard result for the area under a Gaussian (see section 6.4.2). ◄

The above orthogonality and normalisation conditions allow any (reasonable) function in the interval $-\infty \le x < \infty$ to be expanded in a series of the form

$$f(x) = \sum_{n=0}^{\infty} a_n H_n(x),$$

in which the coefficients a_n are given by

$$a_n = \frac{1}{2^n n!\sqrt{\pi}} \int_{-\infty}^{\infty} f(x)H_n(x)e^{-x^2}\,dx.$$

We note that it is sometimes convenient to define the *orthogonal Hermite functions* $\phi_n(x) = e^{-x^2/2}H_n(x)$; they also may be used to produce a series expansion of a function in the interval $-\infty \le x < \infty$. Indeed, $\phi_n(x)$ is proportional to the wavefunction of a particle in the nth energy level of a quantum harmonic oscillator.

Generating function

The generating function equation for the Hermite polynomials reads

$$G(x,h) = e^{2hx-h^2} = \sum_{n=0}^{\infty} \frac{H_n(x)}{n!}h^n, \tag{18.133}$$

a result that may be proved using the Rodrigues' formula (18.130).

►*Show that the functions $H_n(x)$ in (18.133) are the Hermite polynomials.*

It is often more convenient to write the generating function (18.133) as

$$G(x,h) = e^{x^2}e^{-(x-h)^2} = \sum_{n=0}^{\infty} \frac{H_n(x)}{n!}h^n.$$

Differentiating this form k times with respect to h gives

$$\sum_{n=k}^{\infty} \frac{H_n}{(n-k)!} h^{n-k} = \frac{\partial^k G}{\partial h^k} = e^{x^2} \frac{\partial^k}{\partial h^k} e^{-(x-h)^2} = (-1)^k e^{x^2} \frac{\partial^k}{\partial x^k} e^{-(x-h)^2}.$$

Relabelling the summation on the LHS using the new index $m = n - k$, we obtain

$$\sum_{m=0}^{\infty} \frac{H_{m+k}}{m!} h^m = (-1)^k e^{x^2} \frac{\partial^k}{\partial x^k} e^{-(x-h)^2}.$$

Setting $h = 0$ in this equation, we find

$$H_k(x) = (-1)^k e^{x^2} \frac{d^k}{dx^k} (e^{-x^2}),$$

which is the Rodrigues' formula (18.130) for the Hermite polynomials. ◄

The generating function (18.133) is also useful for determining special values of the Hermite polynomials. In particular, it is straightforward to show that $H_{2n}(0) = (-1)^n (2n)!/n!$ and $H_{2n+1}(0) = 0$.

Recurrence relations

The two most useful recurrence relations satisfied by the Hermite polynomials are given by

$$H_{n+1}(x) = 2x H_n(x) - 2n H_{n-1}(x), \tag{18.134}$$

$$H_n'(x) = 2n H_{n-1}(x). \tag{18.135}$$

The first relation provides a simple iterative way of evaluating the nth Hermite polynomials at some point $x = x_0$, given the values of $H_0(x)$ and $H_1(x)$ at that point. For proofs of these recurrence relations, see exercise 18.5.

18.10 Hypergeometric functions

The hypergeometric equation has the form

$$x(1-x)y'' + [c - (a+b+1)x]y' - aby = 0, \tag{18.136}$$

and has three regular singular points, at $x = 0, 1, \infty$, but no essential singularities. The parameters a, b and c are given real numbers.

In our discussions of Legendre functions, associated Legendre functions and Chebyshev functions in sections 18.1, 18.2 and 18.4, respectively, it was noted that in each case the corresponding second-order differential equation had three regular singular points, at $x = -1, 1, \infty$, and no essential singularities. The hypergeometric equation can, in fact, be considered as the 'canonical form' for second-order differential equations with this number of singularities. It may be shown[§] that,

---

[§] See, for example, J. Mathews and R. L. Walker, *Mathematical Methods of Physics*, 2nd edn (Reading MA: Addision–Wesley, 1971).

by making appropriate changes of the independent and dependent variables, any second-order differential equation with three regular singularities and an ordinary point at infinity can be transformed into the hypergeometric equation (18.136) with the singularities at $= -1$, 1 and ∞. As we discuss below, this allows Legendre functions, associated Legendre functions and Chebyshev functions, for example, to be written as particular cases of *hypergeometric functions*, which are the solutions to (18.136).

Since the point $x = 0$ is a regular singularity of (18.136), we may find at least one solution in the form of a Frobenius series (see section 16.3):

$$y(x) = \sum_{n=0}^{\infty} a_n x^{n+\sigma}. \tag{18.137}$$

Substituting this series into (18.136) and dividing through by $x^{\sigma-1}$, we obtain

$$\sum_{n=0}^{\infty} \{(1-x)(n+\sigma)(n+\sigma-1) + [c - (a+b+1)x](n+\sigma) - abx\} a_n x^n = 0. \tag{18.138}$$

Setting $x = 0$, so that only the $n = 0$ term remains, we obtain the indicial equation $\sigma(\sigma - 1) + c\sigma = 0$, which has the roots $\sigma = 0$ and $\sigma = 1 - c$. Thus, provided c is not an integer, one can obtain two linearly independent solutions of the hypergeometric equation in the form (18.137).

For $\sigma = 0$ the corresponding solution is a simple power series. Substituting $\sigma = 0$ into (18.138) and demanding that the coefficient of x^n vanishes, we find the recurrence relation

$$n[(n-1) + c]a_n - [(n-1)(a+b+n-1) + ab]a_{n-1} = 0, \tag{18.139}$$

which, on simplifying and replacing n by $n + 1$, yields the recurrence relation

$$a_{n+1} = \frac{(a+n)(b+n)}{(n+1)(c+n)}a_n. \tag{18.140}$$

It is conventional to make the simple choice $a_0 = 1$. Thus, provided c is not a negative integer or zero, we may write the solution as follows:

$$F(a, b, c; x) = 1 + \frac{ab}{c}\frac{x}{1!} + \frac{a(a+1)b(b+1)}{c(c+1)}\frac{x^2}{2!} + \cdots \tag{18.141}$$

$$= \frac{\Gamma(c)}{\Gamma(a)\Gamma(b)} \sum_{n=0}^{\infty} \frac{\Gamma(a+n)\Gamma(b+n)}{\Gamma(c+n)} \frac{x^n}{n!}, \tag{18.142}$$

where $F(a, b, c; x)$ is known as the *hypergeometric function* or *hypergeometric series*, and in the second equality we have used the property (18.154) of the gamma function.[§] It is straightforward to show that the hypergeometric series converges in the range $|x| < 1$. It also converges at $x = 1$ if $c > a + b$ and at $x = -1$ if $c > a + b - 1$. We also note that $F(a, b, c; x)$ is symmetric in the parameters a and b, i.e. $F(a, b, c; x) = F(b, a, c; x)$.

The hypergeometric function $y(x) = F(a, b, c; x)$ is clearly not the general solution to the hypergeometric equation (18.136), since we must also consider the second root of the indicial equation. Substituting $\sigma = 1 - c$ into (18.138) and demanding that the coefficient of x^n vanishes, we find that we must have

$$n(n + 1 - c)a_n - [(n - c)(a + b + n - c) + ab]a_{n-1} = 0,$$

which, on comparing with (18.139) and replacing n by $n + 1$, yields the recurrence relation

$$a_{n+1} = \frac{(a - c + 1 + n)(b - c + 1 + n)}{(n + 1)(2 - c + n)}a_n.$$

We see that this recurrence relation has the same form as (18.140) if one makes the replacements $a \to a - c + 1$, $b \to b - c + 1$ and $c \to 2 - c$. Thus, provided c, $a - b$ and $c - a - b$ are all non-integers, the general solution to the hypergeometric equation, valid for $|x| < 1$, may be written as

$$y(x) = AF(a, b, c; x) + Bx^{1-c}F(a - c + 1, b - c + 1, 2 - c; x),$$

$$(18.143)$$

where A and B are arbitrary constants to be fixed by the boundary conditions on the solution. If the solution is to be regular at $x = 0$, one requires $B = 0$.

18.10.1 Properties of hypergeometric functions

Since the hypergeometric equation is so general in nature, it is not feasible to present a comprehensive account of the hypergeometric functions. Nevertheless, we outline here some of their most important properties.

Special cases

As mentioned above, the general nature of the hypergeometric equation allows us to write a large number of elementary functions in terms of the hypergeometric functions $F(a, b, c; x)$. Such identifications can be made from the series expansion (18.142) directly, or by transformation of the hypergeometric equation into a more familiar equation, the solutions to which are already known. Some particular examples of well known special cases of the hypergeometric function are as follows:

[§] We note that it is also common to denote the hypergeometric function by $_2F_1(a, b, c; x)$. This slightly odd-looking notation is meant to signify that, in the coefficient of each power of x, there are two parameters (a and b) in the numerator and one parameter (c) in the denominator.

$$F(a, b, b; x) = (1 - x)^{-a},$$

$$F(\tfrac{1}{2}, \tfrac{1}{2}, \tfrac{3}{2}; x^2) = x^{-1} \sin^{-1} x,$$

$$F(1, 1, 2; -x) = x^{-1} \ln(1 + x),$$

$$F(\tfrac{1}{2}, 1, \tfrac{3}{2}; -x^2) = x^{-1} \tan^{-1} x,$$

$$\lim_{m \to \infty} F(1, m, 1; x/m) = e^x,$$

$$F(\tfrac{1}{2}, 1, \tfrac{3}{2}; x^2) = \tfrac{1}{2} x^{-1} \ln[(1 + x)/(1 - x)],$$

$$F(\tfrac{1}{2}, -\tfrac{1}{2}, \tfrac{1}{2}; \sin^2 x) = \cos x,$$

$$F(m + 1, -m, 1; (1 - x)/2) = P_m(x),$$

$$F(\tfrac{1}{2}, p, p; \sin^2 x) = \sec x,$$

$$F(m, -m, \tfrac{1}{2}; (1 - x)/2) = T_m(x),$$

where m is an integer, $P_m(x)$ is the mth Legendre polynomial and $T_m(x)$ is the mth Chebyshev polynomial of the first kind. Some of these results are proved in exercise 18.11.

▶ *Show that $F(m, -m, \tfrac{1}{2}; (1 - x)/2) = T_m(x)$.*

Let us prove this result by transforming the hypergeometric equation. The form of the result suggests that we should make the substitution $x = (1 - z)/2$ into (18.136), in which case $d/dx = -2d/dz$. Thus, letting $u(z) = y(x)$ and setting $a = m$, $b = -m$ and $c = 1/2$, (18.136) becomes

$$\frac{(1 - z)}{2} \frac{(1 + z)}{2} (-2)^2 \frac{d^2 u}{dz^2} + \left[\tfrac{1}{2} - (m - m + 1) \frac{1 - z}{2} \right] (-2) \frac{du}{dz} - (m)(-m)u = 0.$$

On simplifying, we obtain

$$(1 - z^2) \frac{d^2 u}{dz^2} - z \frac{du}{dz} + m^2 u = 0,$$

which has the form of Chebyshev's equation, (18.54). This equation has $u(z) = T_m(z)$ as its power series solution, and so $F(m, -m, \tfrac{1}{2}; (1 - z)/2)$ and $T_m(z)$ are equal to within a normalisation factor. On comparing the expressions (18.141) and (18.56) at $x = 0$, i.e. at $z = 1$, we see that they both have value 1. Hence, the normalisations already agree and we obtain the required result. ◀

Integral representation

One of the most useful representations for the hypergeometric functions is in terms of an integral, which may be derived using the properties of the gamma and beta functions discussed in section 18.12. The integral representation reads

$$F(a, b, c; x) = \frac{\Gamma(c)}{\Gamma(b)\Gamma(c - b)} \int_0^1 t^{b-1} (1 - t)^{c-b-1} (1 - tx)^{-a} \, dt, \tag{18.144}$$

and requires $c > b > 0$ for the integral to converge.

▶ *Prove the result (18.144).*

From the series expansion (18.142), we have

$$F(a, b, c; x) = \frac{\Gamma(c)}{\Gamma(a)\Gamma(b)} \sum_{n=0}^{\infty} \frac{\Gamma(a + n)\Gamma(b + n)}{\Gamma(c + n)} \frac{x^n}{n!}$$

$$= \frac{\Gamma(c)}{\Gamma(a)\Gamma(b)\Gamma(c - b)} \sum_{n=0}^{\infty} \Gamma(a + n) B(b + n, c - b) \frac{x^n}{n!},$$

where in the second equality we have used the expression (18.165) relating the gamma and beta functions. Using the definition (18.162) of the beta function, we then find

$$F(a,b,c;x) = \frac{\Gamma(c)}{\Gamma(a)\Gamma(b)\Gamma(c-b)} \sum_{n=0}^{\infty} \Gamma(a+n)\frac{x^n}{n!} \int_0^1 t^{b+n-1}(1-t)^{c-b-1}\,dt$$

$$= \frac{\Gamma(c)}{\Gamma(b)\Gamma(c-b)} \int_0^1 dt\, t^{b-1}(1-t)^{c-b-1} \sum_{n=0}^{\infty} \frac{\Gamma(a+n)}{\Gamma(a)}\frac{(tx)^n}{n!},$$

where in the second equality we have rearranged the expression and reversed the order of integration and summation. Finally, one recognises the sum over n as being equal to $(1-tx)^{-a}$, and so we obtain the final result (18.144). ◀

The integral representation may be used to prove a wide variety of properties of the hypergeometric functions. As a simple example, on setting $x = 1$ in (18.144), and using properties of the beta function discussed in section 18.12.2, one quickly finds that, provided c is not a negative integer or zero and $c > a + b$,

$$F(a,b,c;1) = \frac{\Gamma(c)\Gamma(c-a-b)}{\Gamma(c-a)\Gamma(c-b)}.$$

Relationships between hypergeometric functions

There exist a great many relationships between hypergeometric functions with different arguments. These are most easily derived by making use of the integral representation (18.144) or the series form (18.141). It is not feasible to list all the relationships here, so we simply note two useful examples, which read

$$F(a,b,c;x) = (1-x)^{c-a-b} F(c-a, c-b, c; x), \tag{18.145}$$

$$F'(a,b,c;x) = \frac{ab}{c} F(a+1, b+1, c+1; x), \tag{18.146}$$

where the prime in the second relation denotes d/dx. The first result follows straightforwardly from the integral representation using the substitution $t = (1-u)/(1-ux)$, whereas the second result may be proved more easily from the series expansion.

In addition to the above results, one may also derive relationships between $F(a,b,c;x)$ and any two of the six 'contiguous functions' $F(a \pm 1, b, c; x)$, $F(a, b \pm 1, c; x)$ and $F(a, b, c \pm 1; x)$. These 'contiguous relations' serve as the recurrence relations for the hypergeometric functions. An example of such a relationship is

$$(c-a)F(a-1,b,c;x)+(2a-c-ax+bx)F(a,b,c;x)+a(x-1)F(a+1,b,c;x) = 0.$$

Repeated application of such relationships allows one to express $F(a+l, b+m, c+n; x)$, where l, m, n are integers (with $c+n$ not equalling a negative integer or zero), as a linear combination of $F(a,b,c;x)$ and one of its contiguous functions.

18.11 Confluent hypergeometric functions

The confluent hypergeometric equation has the form

$$xy'' + (c - x)y' - ay = 0; \tag{18.147}$$

it has a regular singularity at $x = 0$ and an essential singularity at $x = \infty$. This equation can be obtained by merging two of the singularities of the ordinary hypergeometric equation (18.136). The parameters a and c are given real numbers.

▶*Show that setting $x = z/b$ in the hypergeometric equation, and letting $b \to \infty$, yields the confluent hypergeometric equation.*

Substituting $x = z/b$ into (18.136), with $d/dx = b\,d/dz$, and letting $u(z) = y(x)$, we obtain

$$bz \left(1 - \frac{z}{b}\right) \frac{d^2u}{dz^2} + [bc - (a + b + 1)z]\frac{du}{dz} - abu = 0,$$

which clearly has regular singular points at $z = 0$, b and ∞. If we now merge the last two singularities by letting $b \to \infty$, we obtain

$$zu'' + (c - z)u' - au = 0,$$

where the primes denote d/dz. Hence $u(z)$ must satisfy the confluent hypergeometric equation. ◀

In our discussion of Bessel, Laguerre and associated Laguerre functions, it was noted that the corresponding second-order differential equation in each case had a single regular singular point at $x = 0$ and an essential singularity at $x = \infty$. From table 16.1, we see that this is also true for the confluent hypergeometric equation. Indeed, this equation can be considered as the 'canonical form' for second-order differential equations with this pattern of singularities. Consequently, as we mention below, the Bessel, Laguerre and associated Laguerre functions can all be written in terms of the *confluent hypergeometric functions*, which are the solutions of (18.147).

The solutions of the confluent hypergeometric equation are obtained from those of the ordinary hypergeometric equation by again letting $x \to x/b$ and carrying out the limiting process $b \to \infty$. Thus, from (18.141) and (18.143), two linearly independent solutions of (18.147) are (when c is not an integer)

$$y_1(x) = 1 + \frac{a}{c}\frac{x}{1!} + \frac{a(a+1)}{c(c+1)}\frac{z^2}{2!} + \cdots \equiv M(a, c; x), \tag{18.148}$$

$$y_2(x) = x^{1-c} M(a - c + 1, 2 - c; x), \tag{18.149}$$

where $M(a, c; x)$ is called the *confluent hypergeometric function* (or *Kummer function*).[§] It is worth noting, however, that $y_1(x)$ is singular when $c = 0, -1, -2, \ldots$ and $y_2(x)$ is singular when $c = 2, 3, 4, \ldots$. Thus, it is conventional to take the

---

[§] We note that an alternative notation for the confluent hypergeometric function is $_1F_1(a, c; x)$.

second solution to (18.147) as a linear combination of (18.148) and (18.149) given by

$$U(a,c;x) \equiv \frac{\pi}{\sin \pi c} \left[\frac{M(a,c;x)}{\Gamma(a-c+1)\Gamma(c)} - x^{1-c} \frac{M(a-c+1,2-c;x)}{\Gamma(a)\Gamma(2-c)} \right].$$

This has a well behaved limit as c approaches an integer.

18.11.1 Properties of confluent hypergeometric functions

The properties of confluent hypergeometric functions can be derived from those of ordinary hypergeometric functions by letting $x \to x/b$ and taking the limit $b \to \infty$, in the same way as both the equation and its solution were derived. A general procedure of this sort is called a *confluence* process.

Special cases

The general nature of the confluent hypergeometric equation allows one to write a large number of elementary functions in terms of the confluent hypergeometric functions $M(a,c;x)$. Once again, such identifications can be made from the series expansion (18.148) directly, or by transformation of the confluent hypergeometric equation into a more familiar equation for which the solutions are already known. Some particular examples of well known special cases of the confluent hypergeometric function are as follows:

$$M(a,a;x) = e^x,$$

$$M(1,2;2x) = \frac{e^x \sinh x}{x},$$

$$M(-n,1;x) = L_n(x),$$

$$M(-n,m+1;x) = \frac{n!m!}{(n+m)!} L_n^m(x),$$

$$M(-n,\tfrac{1}{2};x^2) = \frac{(-1)^n n!}{(2n)!} H_{2n}(x),$$

$$M(-n,\tfrac{3}{2};x^2), = \frac{(-1)^n n!}{2(2n+1)!} \frac{H_{2n+1}(x)}{x},$$

$$M(v+\tfrac{1}{2},2v+1;2ix) = v!e^{ix}(\tfrac{x}{2})^{-v} J_v(x), \quad M(\tfrac{1}{2},\tfrac{3}{2};-x^2) = \frac{\sqrt{\pi}}{2x}\,\mathrm{erf}(x),$$

where n and m are integers, $L_n^m(x)$ is an associated Legendre polynomial, $H_n(x)$ is a Hermite polynomial, $J_v(x)$ is a Bessel function and $\mathrm{erf}(x)$ is the error function discussed in section 18.12.4.

Integral representation

Using the integral representation (18.144) of the ordinary hypergeometric function, exchanging a and b and carrying out the process of confluence gives

$$M(a,c,x) = \frac{\Gamma(c)}{\Gamma(a)\Gamma(c-a)} \int_0^1 e^{tx} t^{a-1}(1-t)^{c-a-1}\, dt,$$

$$(18.150)$$

which converges provided $c > a > 0$.

> ►*Prove the result (18.150).*

Since $F(a, b, c; x)$ is unchanged by swapping a and b, we may write its integral representation (18.144) as

$$F(a, b, c; x) = \frac{\Gamma(c)}{\Gamma(a)\Gamma(c - a)} \int_0^1 t^{a-1}(1 - t)^{c-a-1}(1 - tx)^{-b} \, dt.$$

Setting $x = z/b$ and taking the limit $b \to \infty$, we obtain

$$M(a, c; z) = \frac{\Gamma(c)}{\Gamma(a)\Gamma(c - a)} \int_0^1 t^{a-1}(1 - t)^{c-a-1} \lim_{b \to \infty} \left(1 - \frac{tz}{b}\right)^{-b} dt.$$

Since the limit is equal to e^{tz}, we obtain result (18.150). ◄

Relationships between confluent hypergeometric functions

A large number of relationships exist between confluent hypergeometric functions with different arguments. These are straightforwardly derived using the integral representation (18.150) or the series form (18.148). Here, we simply note two useful examples, which read

$$M(a, c; x) = e^x M(c - a, c; -x), \tag{18.151}$$

$$M'(a, c; x) = \frac{a}{c} M(a + 1, c + 1; x), \tag{18.152}$$

where the prime in the second relation denotes d/dx. The first result follows straightforwardly from the integral representation, and the second result may be proved from the series expansion (see exercise 18.19).

In an analogous manner to that used for the ordinary hypergeometric functions, one may also derive relationships between $M(a, c; x)$ and any two of the four 'contiguous functions' $M(a \pm 1, c; x)$ and $M(a, c \pm 1; x)$. These serve as the recurrence relations for the confluent hypergeometric functions. An example of such a relationship is

$$(c - a)M(a - 1, c; x) + (2a - c + x)M(a, c; x) - aM(a + 1, c; x) = 0.$$

18.12 The gamma function and related functions

Many times in this chapter, and often throughout the rest of the book, we have made mention of the gamma function and related functions such as the beta and error functions. Although not derived as the solutions of important second-order ODEs, these convenient functions appear in a number of contexts, and so here we gather together some of their properties. This final section should be regarded merely as a reference containing some useful relations obeyed by these functions; a minimum of formal proofs is given.

18.12.1 The gamma function

The *gamma function* $\Gamma(n)$ is defined by

$$\Gamma(n) = \int_0^\infty x^{n-1} e^{-x} \, dx, \tag{18.153}$$

which converges for $n > 0$, where in general n is a real number. Replacing n by $n+1$ in (18.153) and integrating the RHS by parts, we find

$$\Gamma(n+1) = \int_0^\infty x^n e^{-x} \, dx$$

$$= \left[-x^n e^{-x} \right]_0^\infty + \int_0^\infty n x^{n-1} e^{-x} \, dx$$

$$= n \int_0^\infty x^{n-1} e^{-x} \, dx,$$

from which we obtain the important result

$$\Gamma(n+1) = n\Gamma(n). \tag{18.154}$$

From (18.153), we see that $\Gamma(1) = 1$, and so, if n is a positive integer,

$$\Gamma(n+1) = n!. \tag{18.155}$$

In fact, equation (18.155) serves as a definition of the factorial function even for non-integer n. For negative n the factorial function is defined by

$$n! = \frac{(n+m)!}{(n+m)(n+m-1)\cdots(n+1)}, \tag{18.156}$$

where m is any positive integer that makes $n + m > 0$. Different choices of m $(> -n)$ do not lead to different values for $n!$. A plot of the gamma function is given in figure 18.9, where it can be seen that the function is infinite for negative integer values of n, in accordance with (18.156). For an extension of the factorial function to complex arguments, see exercise 18.15.

By letting $x = y^2$ in (18.153), we immediately obtain another useful representation of the gamma function given by

$$\Gamma(n) = 2 \int_0^\infty y^{2n-1} e^{-y^2} \, dy. \tag{18.157}$$

Setting $n = \frac{1}{2}$ we find the result

$$\Gamma\left(\tfrac{1}{2}\right) = 2 \int_0^\infty e^{-y^2} \, dy = \int_{-\infty}^\infty e^{-y^2} \, dy = \sqrt{\pi},$$

where we have used the standard integral discussed in section 6.4.2. From this result, $\Gamma(n)$ for half-integral n can be found using (18.154). Some immediately derivable factorial values of half integers are

$$\left(-\tfrac{3}{2}\right)! = -2\sqrt{\pi}, \quad \left(-\tfrac{1}{2}\right)! = \sqrt{\pi}, \quad \left(\tfrac{1}{2}\right)! = \tfrac{1}{2}\sqrt{\pi}, \quad \left(\tfrac{3}{2}\right)! = \tfrac{3}{4}\sqrt{\pi}.$$

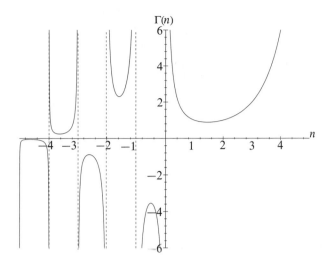

Figure 18.9   The gamma function $\Gamma(n)$.

Moreover, it may be shown for non-integral n that the gamma function satisfies the important identity

$$\Gamma(n)\Gamma(1-n) = \frac{\pi}{\sin n\pi}. \tag{18.158}$$

This is proved for a restricted range of n in the next section, once the beta function has been introduced.

It can also be shown that the gamma function is given by

$$\Gamma(n+1) = \sqrt{2\pi n}\, n^n e^{-n}\left(1 + \frac{1}{12n} + \frac{1}{288n^2} - \frac{139}{51\,840n^3} + \cdots\right) = n!, \tag{18.159}$$

which is known as *Stirling's asymptotic series*. For large n the first term dominates, and so

$$n! \approx \sqrt{2\pi n}\, n^n e^{-n}; \tag{18.160}$$

this is known as *Stirling's approximation*. This approximation is particularly useful in statistical thermodynamics, when arrangements of a large number of particles are to be considered.

▶*Prove Stirling's approximation $n! \approx \sqrt{2\pi n}\, n^n e^{-n}$ for large n.*

From (18.153), the extended definition of the factorial function (which is valid for $n > -1$) is given by

$$n! = \int_0^\infty x^n e^{-x}\, dx = \int_0^\infty e^{n\ln x - x}\, dx. \tag{18.161}$$

637

If we let $x = n + y$, then

$$\ln x = \ln n + \ln \left(1 + \frac{y}{n}\right)$$

$$= \ln n + \frac{y}{n} - \frac{y^2}{2n^2} + \frac{y^3}{3n^3} - \cdots.$$

Substituting this result into (18.161), we obtain

$$n! = \int_{-n}^{\infty} \exp \left[n \left(\ln n + \frac{y}{n} - \frac{y^2}{2n^2} + \cdots\right) - n - y\right] dy.$$

Thus, when n is sufficiently large, we may approximate $n!$ by

$$n! \approx e^{n \ln n - n} \int_{-\infty}^{\infty} e^{-y^2/(2n)} \, dy = e^{n \ln n - n} \sqrt{2\pi n} = \sqrt{2\pi n} \; n^n e^{-n},$$

which is Stirling's approximation (18.160). ◀

18.12.2 The beta function

The *beta function* is defined by

$$B(m, n) = \int_{0}^{1} x^{m-1} (1 - x)^{n-1} \, dx, \tag{18.162}$$

which converges for $m > 0$, $n > 0$, where m and n are, in general, real numbers. By letting $x = 1 - y$ in (18.162) it is easy to show that $B(m, n) = B(n, m)$. Other useful representations of the beta function may be obtained by suitable changes of variable. For example, putting $x = (1 + y)^{-1}$ in (18.162), we find that

$$B(m, n) = \int_{0}^{\infty} \frac{y^{n-1} \, dy}{(1 + y)^{m+n}}. \tag{18.163}$$

Alternatively, if we let $x = \sin^2 \theta$ in (18.162), we obtain immediately

$$B(m, n) = 2 \int_{0}^{\pi/2} \sin^{2m-1} \theta \, \cos^{2n-1} \theta \, d\theta. \tag{18.164}$$

The beta function may also be written in terms of the gamma function as

$$B(m, n) = \frac{\Gamma(m)\Gamma(n)}{\Gamma(m + n)}. \tag{18.165}$$

▶ *Prove the result (18.165).*

Using (18.157), we have

$$\Gamma(n)\Gamma(m) = 4 \int_{0}^{\infty} x^{2n-1} e^{-x^2} \, dx \int_{0}^{\infty} y^{2m-1} e^{-y^2} \, dy$$

$$= 4 \int_{0}^{\infty} \int_{0}^{\infty} x^{2n-1} y^{2m-1} e^{-(x^2+y^2)} \, dx \, dy.$$

Changing variables to plane polar coordinates (ρ, ϕ) given by $x = \rho \cos \phi$, $y = \rho \sin \phi$, we obtain

$$\Gamma(n)\Gamma(m) = 4 \int_0^{\pi/2} \int_0^\infty \rho^{2(m+n-1)} e^{-\rho^2} \sin^{2m-1} \phi \cos^{2n-1} \phi \, \rho \, d\rho \, d\phi$$

$$= 4 \int_0^{\pi/2} \sin^{2m-1} \phi \cos^{2n-1} \phi \, d\phi \int_0^\infty \rho^{2(m+n)-1} e^{-\rho^2} \, d\rho$$

$$= B(m, n)\Gamma(m + n),$$

where in the last line we have used the results (18.157) and (18.164). ◄

The result (18.165) is useful in proving the identity (18.158) satisfied by the gamma function, since

$$\Gamma(n)\Gamma(1 - n) = B(1 - n, n) = \int_0^\infty \frac{y^{n-1} \, dy}{1 + y},$$

where, in the second equality, we have used the integral representation (18.163). For $0 < n < 1$ this integral can be evaluated using contour integration and has the value $\pi/(\sin n\pi)$ (see exercise 24.19), thereby proving result (18.158) for this range of n. Extensions to other ranges require more sophisticated methods.

18.12.3 The incomplete gamma function

In the definition (18.153) of the gamma function, we may divide the range of integration into two parts and write

$$\Gamma(n) = \int_0^x u^{n-1} e^{-u} \, du + \int_x^\infty u^{n-1} e^{-u} \, du \equiv \gamma(n, x) + \Gamma(n, x),$$

$$(18.166)$$

whereby we have defined the *incomplete gamma functions* $\gamma(n, x)$ and $\Gamma(n, x)$, respectively. The choice of which of these two functions to use is merely a matter of convenience.

► *Show that if n is a positive integer*

$$\Gamma(n, x) = (n - 1)! e^{-x} \sum_{k=0}^{n-1} \frac{x^k}{k!}.$$

From (18.166), on integrating by parts we find

$$\Gamma(n, x) = \int_x^\infty u^{n-1} e^{-u} du = x^{n-1} e^{-x} + (n - 1) \int_x^\infty u^{n-2} e^{-u} du$$

$$= x^{n-1} e^{-x} + (n - 1)\Gamma(n - 1, x),$$

which is valid for arbitrary n. If n is an integer, however, we obtain

$$\Gamma(n, x) = e^{-x} [x^{n-1} + (n - 1)x^{n-2} + (n - 1)(n - 2)x^{n-3} + \cdots + (n - 1)!]$$

$$= (n - 1)! e^{-x} \sum_{k=0}^{n-1} \frac{x^k}{k!},$$

which is the required result. ◀

We note that is it conventional to define, in addition, the functions

$$P(a, x) \equiv \frac{\gamma(a, x)}{\Gamma(a)}, \qquad Q(a, x) \equiv \frac{\Gamma(a, x)}{\Gamma(a)},$$

which are also often called incomplete gamma functions; it is clear that $Q(a, x) = 1 - P(a, x)$.

18.12.4 The error function

Finally, we mention the *error function*, which is encountered in probability theory and in the solutions of some partial differential equations. The error function is related to the incomplete gamma function by $\mathrm{erf}(x) = \gamma(\frac{1}{2}, x^2)/\sqrt{\pi}$ and is thus given by

$$\mathrm{erf}(x) = \frac{2}{\sqrt{\pi}} \int_0^x e^{-u^2}\, du = 1 - \frac{2}{\sqrt{\pi}} \int_x^\infty e^{-u^2}\, du. \tag{18.167}$$

From this definition we can easily see that

$$\mathrm{erf}(0) = 0, \qquad \mathrm{erf}(\infty) = 1, \qquad \mathrm{erf}(-x) = -\mathrm{erf}(x).$$

By making the substitution $y = \sqrt{2}u$ in (18.167), we find

$$\mathrm{erf}(x) = \sqrt{\frac{2}{\pi}} \int_0^{\sqrt{2}x} e^{-y^2/2}\, dy.$$

The cumulative probability function $\Phi(x)$ for the standard Gaussian distribution (discussed in section 30.9.1) may be written in terms of the error function as follows:

$$\begin{aligned}
\Phi(x) &= \frac{1}{\sqrt{2\pi}} \int_{-\infty}^x e^{-y^2/2}\, dy \\
&= \frac{1}{2} + \frac{1}{\sqrt{2\pi}} \int_0^x e^{-y^2/2}\, dy \\
&= \frac{1}{2} + \frac{1}{2}\mathrm{erf}\left(\frac{x}{\sqrt{2}}\right).
\end{aligned}$$

It is also sometimes useful to define the *complementary error function*

$$\mathrm{erfc}(x) = 1 - \mathrm{erf}(x) = \frac{2}{\sqrt{\pi}} \int_x^\infty e^{-u^2}\, du = \frac{\Gamma(\frac{1}{2}, x^2)}{\sqrt{\pi}}. \tag{18.168}$$

18.13 Exercises

18.1   Use the explicit expressions

$$Y_0^0 = \sqrt{\tfrac{1}{4\pi}}, \qquad\qquad Y_1^0 = \sqrt{\tfrac{3}{4\pi}}\cos\theta,$$

$$Y_1^{\pm 1} = \mp\sqrt{\tfrac{3}{8\pi}}\sin\theta\exp(\pm i\phi), \qquad Y_2^0 = \sqrt{\tfrac{5}{16\pi}}(3\cos^2\theta - 1),$$

$$Y_2^{\pm 1} = \mp\sqrt{\tfrac{15}{8\pi}}\sin\theta\cos\theta\exp(\pm i\phi), \qquad Y_2^{\pm 2} = \sqrt{\tfrac{15}{32\pi}}\sin^2\theta\exp(\pm 2i\phi),$$

to verify for $\ell = 0, 1, 2$ that

$$\sum_{m=-\ell}^{\ell} |Y_\ell^m(\theta,\phi)|^2 = \frac{2\ell+1}{4\pi},$$

and so is independent of the values of θ and ϕ. This is true for any ℓ, but a general proof is more involved. This result helps to reconcile intuition with the apparently arbitrary choice of polar axis in a general quantum mechanical system.

18.2   Express the function

$$f(\theta,\phi) = \sin\theta[\sin^2(\theta/2)\cos\phi + i\cos^2(\theta/2)\sin\phi] + \sin^2(\theta/2)$$

as a sum of spherical harmonics.

18.3   Use the generating function for the Legendre polynomials $P_n(x)$ to show that

$$\int_0^1 P_{2n+1}(x)\,dx = (-1)^n \frac{(2n)!}{2^{2n+1}n!(n+1)!}$$

and that, except for the case $n = 0$,

$$\int_0^1 P_{2n}(x)\,dx = 0.$$

18.4   Carry through the following procedure as a proof of the result

$$I_n = \int_{-1}^1 P_n(z)P_n(z)\,dz = \frac{2}{2n+1}.$$

(a) Square both sides of the generating-function definition of the Legendre polynomials,

$$(1 - 2zh + h^2)^{-1/2} = \sum_{n=0}^{\infty} P_n(z)h^n.$$

(b) Express the RHS as a sum of powers of h, obtaining expressions for the coefficients.

(c) Integrate the RHS from -1 to 1 and use the orthogonality property of the Legendre polynomials.

(d) Similarly integrate the LHS and expand the result in powers of h.

(e) Compare coefficients.

18.5   The Hermite polynomials $H_n(x)$ may be defined by

$$\Phi(x,h) = \exp(2xh - h^2) = \sum_{n=0}^{\infty} \frac{1}{n!}H_n(x)h^n.$$

Show that

$$\frac{\partial^2 \Phi}{\partial x^2} - 2x\frac{\partial\Phi}{\partial x} + 2h\frac{\partial\Phi}{\partial h} = 0,$$

and hence that the $H_n(x)$ satisfy the Hermite equation

$$y'' - 2xy' + 2ny = 0,$$

where n is an integer ≥ 0.

Use Φ to prove that

(a) $H_n'(x) = 2nH_{n-1}(x)$,
(b) $H_{n+1}(x) - 2xH_n(x) + 2nH_{n-1}(x) = 0$.

18.6   A charge $+2q$ is situated at the origin and charges of $-q$ are situated at distances $\pm a$ from it along the polar axis. By relating it to the generating function for the Legendre polynomials, show that the electrostatic potential Φ at a point (r, θ, ϕ) with $r > a$ is given by

$$\Phi(r, \theta, \phi) = \frac{2q}{4\pi\epsilon_0 r} \sum_{s=1}^{\infty} \left(\frac{a}{r}\right)^{2s} P_{2s}(\cos\theta).$$

18.7   For the associated Laguerre polynomials, carry through the following exercises.

(a) Prove the Rodrigues' formula

$$L_n^m(x) = \frac{e^x x^{-m}}{n!} \frac{d^n}{dx^n}(x^{n+m}e^{-x}),$$

taking the polynomials to be defined by

$$L_n^m(x) = \sum_{k=0}^{n}(-1)^k \frac{(n+m)!}{k!(n-k)!(k+m)!}x^k.$$

(b) Prove the recurrence relations

$$(n+1)L_{n+1}^m(x) = (2n+m+1-x)L_n^m(x) - (n+m)L_{n-1}^m(x),$$

$$x(L_n^m)'(x) = nL_n^m(x) - (n+m)L_{n-1}^m(x),$$

but this time taking the polynomial as defined by

$$L_n^m(x) = (-1)^m \frac{d^m}{dx^m}L_{n+m}(x)$$

or the generating function.

18.8   The quantum mechanical wavefunction for a one-dimensional simple harmonic oscillator in its nth energy level is of the form

$$\psi(x) = \exp(-x^2/2)H_n(x),$$

where $H_n(x)$ is the nth Hermite polynomial. The generating function for the polynomials is

$$G(x, h) = e^{2hx - h^2} = \sum_{n=0}^{\infty} \frac{H_n(x)}{n!}h^n.$$

(a) Find $H_i(x)$ for $i = 1, 2, 3, 4$.
(b) Evaluate by direct calculation

$$\int_{-\infty}^{\infty} e^{-x^2} H_p(x)H_q(x)\,dx,$$

(i) for $p = 2$, $q = 3$; (ii) for $p = 2$, $q = 4$; (iii) for $p = q = 3$. Check your answers against the expected values $2^p p!\sqrt{\pi}\,\delta_{pq}$.

[You will find it convenient to use

$$\int_{-\infty}^{\infty} x^{2n} e^{-x^2} \, dx = \frac{(2n)! \sqrt{\pi}}{2^{2n} n!}$$

for integer $n \geq 0$.]

18.9 By initially writing $y(x)$ as $x^{1/2} f(x)$ and then making subsequent changes of variable, reduce Stokes' equation,

$$\frac{d^2 y}{dx^2} + \lambda x y = 0,$$

to Bessel's equation. Hence show that a solution that is finite at $x = 0$ is a multiple of $x^{1/2} J_{1/3}(\frac{2}{3} \sqrt{\lambda x^3})$.

18.10 By choosing a suitable form for h in their generating function,

$$G(z, h) = \exp \left[\frac{z}{2} \left(h - \frac{1}{h} \right) \right] = \sum_{n=-\infty}^{\infty} J_n(z) h^n,$$

show that integral representations of the Bessel functions of the first kind are given, for integral m, by

$$J_{2m}(z) = \frac{(-1)^m}{2\pi} \int_0^{2\pi} \cos(z \cos \theta) \cos 2m\theta \, d\theta \qquad m \geq 1,$$

$$J_{2m+1}(z) = \frac{(-1)^m}{2\pi} \int_0^{2\pi} \sin(z \cos \theta) \cos(2m+1)\theta \, d\theta \qquad m \geq 0.$$

18.11 Identify the series for the following hypergeometric functions, writing them in terms of better known functions:

(a) $F(a, b, b; z)$,
(b) $F(1, 1, 2; -x)$,
(c) $F(\frac{1}{2}, 1, \frac{3}{2}; -x^2)$,
(d) $F(\frac{1}{2}, \frac{1}{2}, \frac{3}{2}; x^2)$,
(e) $F(-a, a, \frac{1}{2}; \sin^2 x)$; this is a much more difficult exercise.

18.12 By making the substitution $z = (1 - x)/2$ and suitable choices for a, b and c, convert the hypergeometric equation,

$$z(1 - z) \frac{d^2 u}{dz^2} + [c - (a + b + 1)z] \frac{du}{dz} - abu = 0,$$

into the Legendre equation,

$$(1 - x^2) \frac{d^2 y}{dx^2} - 2x \frac{dy}{dx} + \ell(\ell + 1)y = 0.$$

Hence, using the hypergeometric series, generate the Legendre polynomials $P_\ell(x)$ for the integer values $\ell = 0, 1, 2, 3$. Comment on their normalisations.

18.13 Find a change of variable that will allow the integral

$$I = \int_1^{\infty} \frac{\sqrt{u - 1}}{(u + 1)^2} \, du$$

to be expressed in terms of the beta function, and so evaluate it.

18.14 Prove that, if m and n are both greater than -1, then

$$I = \int_0^{\infty} \frac{u^m}{(au^2 + b)^{(m+n+2)/2}} \, du = \frac{\Gamma[\frac{1}{2}(m+1)] \Gamma[\frac{1}{2}(n+1)]}{2a^{(m+1)/2} b^{(n+1)/2} \Gamma[\frac{1}{2}(m+n+2)]}.$$

Deduce the value of

$$J = \int_0^\infty \frac{(u+2)^2}{(u^2+4)^{5/2}} \, du.$$

18.15 The complex function $z!$ is defined by

$$z! = \int_0^\infty u^z e^{-u} \, du \qquad \text{for Re } z > -1.$$

For Re $z \leq -1$ it is defined by

$$z! = \frac{(z+n)!}{(z+n)(z+n-1)\cdots(z+1)},$$

where n is any (positive) integer $> -\text{Re } z$. Being the ratio of two polynomials, $z!$ is analytic everywhere in the finite complex plane except at the poles that occur when z is a negative integer.

(a) Show that the definition of $z!$ for Re $z \leq -1$ is independent of the value of n chosen.
(b) Prove that the residue of $z!$ at the pole $z = -m$, where m is an integer > 0, is $(-1)^{m-1}/(m-1)!$.

18.16 For $-1 < \text{Re } z < 1$, use the definition and value of the beta function to show that

$$z!(-z)! = \int_0^\infty \frac{u^z}{(1+u)^2} \, du.$$

Contour integration gives the value of the integral on the RHS of the above equation as $\pi z \operatorname{cosec} \pi z$. Use this to deduce the value of $(-\frac{1}{2})!$.

18.17 The integral

$$I = \int_{-\infty}^\infty \frac{e^{-k^2}}{k^2+a^2} \, dk, \qquad (*)$$

in which $a > 0$, occurs in some statistical mechanics problems. By first considering the integral

$$J = \int_0^\infty e^{iu(k+ia)} \, du,$$

and a suitable variation of it, show that $I = (\pi/a) \exp(a^2) \operatorname{erfc}(a)$, where $\operatorname{erfc}(x)$ is the complementary error function.

18.18 Consider two series expansions of the error function as follows.

(a) Obtain a series expansion of the error function $\operatorname{erf}(x)$ in ascending powers of x. How many terms are needed to give a value correct to four significant figures for $\operatorname{erf}(1)$?
(b) Obtain an asymptotic expansion that can be used to estimate $\operatorname{erfc}(x)$ for large x (> 0) in the form of a series

$$\operatorname{erfc}(x) = R(x) = e^{-x^2} \sum_{n=0}^\infty \frac{a_n}{x^n}.$$

Consider what bounds can be put on the estimate and at what point the infinite series should be terminated in a practical estimate. In particular, estimate $\operatorname{erfc}(1)$ and test the answer for compatibility with that in part (a).

18.19 For the functions $M(a,c;z)$ that are the solutions of the confluent hypergeometric equation,

644

(a) use their series representation to prove that

$$b \frac{d}{dz} M(a,c;z) = a M(a+1,c+1;z);$$

(b) use an integral representation to prove that

$$M(a,c;z) = e^z M(c-a,c;-z).$$

18.20 The Bessel function $J_\nu(z)$ can be considered as a special case of the solution $M(a,c;z)$ of the confluent hypergeometric equation, the connection being

$$\lim_{a \to \infty} \frac{M(a,\nu+1;-z/a)}{\Gamma(\nu+1)} = z^{-\nu/2} J_\nu(2\sqrt{z}).$$

Prove this equality by writing each side in terms of an infinite series and showing that the series are the same.

18.21 Find the differential equation satisfied by the function $y(x)$ defined by

$$y(x) = Ax^{-n} \int_0^x e^{-t} t^{n-1} \, dt \equiv Ax^{-n} \gamma(n,x),$$

and, by comparing it with the confluent hypergeometric function, express y as a multiple of the solution $M(a,c;z)$ of that equation. Determine the value of A that makes y equal to M.

18.22 Show, from its definition, that the Bessel function of the second kind, and of integral order ν, can be written as

$$Y_\nu(z) = \frac{1}{\pi} \left[\frac{\partial J_\mu(z)}{\partial \mu} - (-1)^\nu \frac{\partial J_{-\mu}(z)}{\partial \mu} \right]_{\mu=\nu}.$$

Using the explicit series expression for $J_\mu(z)$, show that $\partial J_\mu(z)/\partial \mu$ can be written as

$$J_\nu(z) \ln \left(\frac{z}{2} \right) + g(\nu,z),$$

and deduce that $Y_\nu(z)$ can be expressed as

$$Y_\nu(z) = \frac{2}{\pi} J_\nu(z) \ln \left(\frac{z}{2} \right) + h(\nu,z),$$

where $h(\nu,z)$, like $g(\nu,z)$, is a power series in z.

18.23 Prove two of the properties of the incomplete gamma function $P(a,x^2)$ as follows.

(a) By considering its form for a suitable value of a, show that the error function can be expressed as a particular case of the incomplete gamma function.

(b) The Fresnel integrals, of importance in the study of the diffraction of light, are given by

$$C(x) = \int_0^x \cos \left(\frac{\pi}{2} t^2 \right) dt, \qquad S(x) = \int_0^x \sin \left(\frac{\pi}{2} t^2 \right) dt.$$

Show that they can be expressed in terms of the error function by

$$C(x) + iS(x) = A \operatorname{erf} \left[\frac{\sqrt{\pi}}{2} (1-i)x \right],$$

where A is a (complex) constant, which you should determine. Hence express $C(x) + iS(x)$ in terms of the incomplete gamma function.

18.24   The solutions $y(x, a)$ of the equation

$$\frac{d^2 y}{dx^2} - (\tfrac{1}{4}x^2 + a)y = 0 \qquad (*)$$

are known as parabolic cylinder functions.

(a) If $y(x, a)$ is a solution of $(*)$, determine which of the following are also solutions: (i) $y(a, -x)$, (ii) $y(-a, x)$, (iii) $y(a, ix)$ and (iv) $y(-a, ix)$.

(b) Show that one solution of $(*)$, even in x, is

$$y_1(x, a) = e^{-x^2/4} M(\tfrac{1}{2}a + \tfrac{1}{4}, \tfrac{1}{2}, \tfrac{1}{2}x^2),$$

where $M(\alpha, c, z)$ is the confluent hypergeometric function satisfying

$$z\frac{d^2 M}{dz^2} + (c - z)\frac{dM}{dz} - \alpha M = 0.$$

You may assume (or prove) that a second solution, odd in x, is given by $y_2(x, a) = xe^{-x^2/4} M(\tfrac{1}{2}a + \tfrac{3}{4}, \tfrac{3}{2}, \tfrac{1}{2}x^2)$.

(c) Find, as an infinite series, an explicit expression for $e^{x^2/4} y_1(x, a)$.

(d) Using the results from part (a), show that $y_1(x, a)$ can also be written as

$$y_1(x, a) = e^{x^2/4} M(-\tfrac{1}{2}a + \tfrac{1}{4}, \tfrac{1}{2}, -\tfrac{1}{2}x^2).$$

(e) By making a suitable choice for a deduce that

$$1 + \sum_{n=1}^{\infty} \frac{b_n \, x^{2n}}{(2n)!} = e^{x^2/2}\left(1 + \sum_{n=1}^{\infty} \frac{(-1)^n \, b_n \, x^{2n}}{(2n)!}\right),$$

where $b_n = \prod_{r=1}^{n}(2r - \tfrac{3}{2})$.

18.14 Hints and answers

18.1   Note that taking the square of the modulus eliminates all mention of ϕ.

18.3   Integrate both sides of the generating function definition from $x = 0$ to $x = 1$, and then expand the resulting term, $(1 + h^2)^{1/2}$, using a binomial expansion. Show that $^{1/2}C_m$ can be written as $[(-1)^{m-1}(2m - 2)!]/[2^{2m-1}m!(m - 1)!]$.

18.5   Prove the stated equation using the explicit closed form of the generating function. Then substitute the series and require the coefficient of each power of h to vanish.
(b) Differentiate result (a) and then use (a) again to replace the derivatives.

18.7   (a) Write the result of using Leibnitz' theorem on the product of x^{n+m} and e^{-x} as a finite sum, evaluate the separated derivatives, and then re-index the summation.
(b) For the first recurrence relation, differentiate the generating function with respect to h and then use the generating function again to replace the exponential. Equating coefficients of h^n then yields the result. For the second, differentiate the corresponding relationship for the ordinary Laguerre polynomials m times.

18.9 $x^2 f'' + xf' + (\lambda x^3 - \tfrac{1}{4})f = 0$. Then, in turn, set $x^{3/2} = u$, and $\tfrac{2}{3}\lambda^{1/2}u = v$; then v satisfies Bessel's equation with $\nu = \tfrac{1}{3}$.

18.11 (a) $(1 - z)^{-a}$. (b) $x^{-1}\ln(1 + x)$. (c) Compare the calculated coefficients with those of $\tan^{-1} x$. $F(\tfrac{1}{2}, 1, \tfrac{3}{2}; -x^2) = x^{-1}\tan^{-1} x$. (d) $x^{-1}\sin^{-1} x$. (e) Note that a term containing x^{2n} can only arise from the first $n + 1$ terms of an expansion in powers of $\sin^2 x$; make a few trials. $F(-a, a, \tfrac{1}{2}; \sin^2 x) = \cos 2ax$.

18.13 Looking for $f(x) = u$ such that $u + 1$ is an inverse power of x with $f(0) = \infty$ and $f(1) = 1$ leads to $f(x) = 2x^{-1} - 1$. $I = B(\tfrac{1}{2}, \tfrac{3}{2})/\sqrt{2} = \pi/(2\sqrt{2})$.

18.15 (a) Show that the ratio of two definitions based on m and n, with $m > n > -\text{Re } z$, is unity, independent of the actual values of m and n.
(b) Consider the limit as $z \to -m$ of $(z + m)z!$, with the definition of $z!$ based on n where $n > m$.

18.17 Express the integrand in partial fractions and use J, as given, and $J' = \int_0^\infty \exp[-iu(k - ia)] \, du$ to express I as the sum of two double integral expressions. Reduce them using the standard Gaussian integral, and then make a change of variable $2v = u + 2a$.

18.19 (b) Using the representation

$$M(a, b; z) = \frac{\Gamma(b)}{\Gamma(b - a)\Gamma(a)} \int_0^1 e^{zt} t^{a-1} (1 - t)^{b-a-1} \, dt$$

allows the equality to be established, without actual integration, by changing the integration variable to $s = 1 - t$.

18.21 Calculate $y'(x)$ and $y''(x)$ and then eliminate $x^{-1}e^{-x}$ to obtain $xy'' + (n + 1 + x)y' + ny = 0$; $M(n, n + 1; -x)$. Comparing the expansion of the hypergeometric series with the result of term by term integration of the expansion of the integrand shows that $A = n$.

18.23 (a) If the dummy variable in the incomplete gamma function is t, make the change of variable $y = +\sqrt{t}$. Now choose a so that $2(a - 1) + 1 = 0$; $\text{erf}(x) = P(\frac{1}{2}, x^2)$.
(b) Change the integration variable u in the standard representation of the RHS to s, given by $u = \frac{1}{2}\sqrt{\pi}(1 - i)s$, and note that $(1 - i)^2 = -2i$. $A = (1 + i)/2$. From part (a), $C(x) + iS(x) = \frac{1}{2}(1 + i)P(\frac{1}{2}, -\frac{1}{2}\pi i x^2)$.

19

Quantum operators

Although the previous chapter was principally concerned with the use of linear operators and their eigenfunctions in connection with the solution of given differential equations, it is of interest to study the properties of the operators themselves and determine which of them follow purely from the nature of the operators, without reference to specific forms of eigenfunctions.

19.1 Operator formalism

The results we will obtain in this chapter have most of their applications in the field of quantum mechanics and our descriptions of the methods will reflect this. In particular, when we discuss a function ψ that depends upon variables such as space coordinates and time, and possibly also on some non-classical variables, ψ will usually be a quantum-mechanical wavefunction that is being used to describe the state of a physical system. For example, the value of $|\psi|^2$ for a particular set of values of the variables is interpreted in quantum mechanics as being the probability that the system's variables have that set of values.

To this end, we will be no more specific about the functions involved than attaching just enough labels to them that a particular function, or a particular set of functions, is identified. A convenient notation for this kind of approach is that already hinted at, but not specifically stated, in subsection 17.1, where the definition of an inner product is given. This notation, often called the Dirac notation, denotes a state whose wavefunction is ψ by $|\psi\rangle$; since ψ belongs to a vector space of functions, $|\psi\rangle$ is known as a *ket vector*. Ket vectors, or simply kets, must not be thought of as completely analogous to physical vectors. Quantum mechanics associates the same physical state with $ke^{i\theta}|\psi\rangle$ as it does with $|\psi\rangle$ for all real k and θ and so there is no loss of generality in taking k as 1 and θ as 0. On the other hand, the combination $c_1|\psi_1\rangle + c_2|\psi_2\rangle$, where $|\psi_1\rangle$ and $|\psi_2\rangle$

represent different states, is a ket that represents a continuum of different states as the complex numbers c_1 and c_2 are varied.

If we need to specify a state more closely – say we know that it corresponds to a plane wave with a wave number whose magnitude is k – then we indicate this with a label; the corresponding ket vector would be written as $|k\rangle$. If we also knew the direction of the wave then $|\mathbf{k}\rangle$ would be the appropriate form. Clearly, in general, the more labels we include, the more precisely the corresponding state is specified.

The Dirac notation for the Hermitian conjugate (dual vector) of the ket vector $|\psi\rangle$ is written as $\langle\psi|$ and is known as a *bra vector*; the wavefunction describing this state is ψ^*, the complex conjugate of ψ. The inner product of two wavefunctions $\int \psi^*\phi\,dv$ is then denoted by $\langle\psi|\phi\rangle$ or, more generally if a non-unit weight function ρ is involved, by

$$\langle\psi|\rho|\phi\rangle, \quad \text{evaluated as} \quad \int \psi^*(\mathbf{r})\phi(\mathbf{r})\rho(\mathbf{r})\,d\mathbf{r}. \tag{19.1}$$

Given the (contrived) names for the two sorts of vectors, an inner product like $\langle\psi|\phi\rangle$ becomes a particular type of 'bra(c)ket'. Despite its somewhat whimsical construction, this type of quantity has a fundamental role to play in the interpretation of quantum theory, because expectation values, probabilities and transition rates are all expressed in terms of them. For physical states the inner product of the corresponding ket with itself, with or without an explicit weight function, is non-zero, and it is usual to take

$$\langle\psi|\psi\rangle = 1.$$

Although multiplying a ket vector by a constant does not change the state described by the vector, acting upon it with a more general linear operator A results (in general) in a ket describing a different state. For example, if ψ is a state that is described in one-dimensional x-space by the wavefunction $\psi(x) = \exp(-x^2)$ and A is the differential operator $\partial/\partial x$, then

$$|\psi_1\rangle = A|\psi\rangle \equiv |A\psi\rangle$$

is the ket associated with the state whose wavefunction is $\psi_1(x) = -2x\exp(-x^2)$, clearly a different state. This allows us to attach a meaning to an expression such as $\langle\phi|A|\psi\rangle$ through the equation

$$\langle\phi|A|\psi\rangle = \langle\phi|\psi_1\rangle, \tag{19.2}$$

i.e. it is the inner product of $|\psi_1\rangle$ and $|\phi\rangle$. We have already used this notation in equation (19.1), but there the effect of the operator A was merely multiplication by a weight function.

If it should happen that the effect of an operator acting upon a particular ket is to produce a scalar multiple of that ket, i.e.

$$A|\psi\rangle = \lambda|\psi\rangle, \qquad (19.3)$$

then, just as for matrices and differential equations, $|\psi\rangle$ is called an *eigenket* or, more usually, an *eigenstate* of A, with corresponding eigenvalue λ; to mark this special property the state will normally be denoted by $|\lambda\rangle$, rather than by the more general $|\psi\rangle$. Taking the Hermitian conjugate of this ket vector eigenequation gives a bra vector equation,

$$\langle\psi|A^\dagger = \lambda^*\langle\psi|. \qquad (19.4)$$

It should be noted that the complex conjugate of the eigenvalue appears in this equation. Should the action of A on $|\psi\rangle$ produce an unphysical state (usually one whose wavefunction is identically zero, and is therefore unacceptable as a quantum-mechanical wavefunction because of the required probability interpretation) we denote the result either by 0 or by the ket vector $|\emptyset\rangle$ according to context. Formally, $|\emptyset\rangle$ can be considered as an eigenket of any operator, but one for which the eigenvalue is always zero.

If an operator A is Hermitian ($A^\dagger = A$) then its eigenvalues are real and the eigenstates can be chosen to be orthogonal; this can be shown in the same way as in chapter 17 (but using a different notation). As indicated there, the reality of their eigenvalues is one reason why Hermitian operators form the basis of measurement in quantum mechanics; in that formulation of physics, the eigenvalues of an operator are the *only* possible values that can be obtained when a measurement of the physical quantity corresponding to the operator is made. Actual individual measurements must always result in real values, even if they are combined in a complex form ($x + iy$ or $re^{i\theta}$) for final presentation or analysis, and using only Hermitian operators ensures this. The proof of the reality of the eigenvalues using the Dirac notation is given below in a worked example.

In the same notation the Hermitian property of an operator A is represented by the double equality

$$\langle A\,\phi|\psi\rangle = \langle\phi|A|\psi\rangle = \langle\phi|A\,\psi\rangle.$$

It should be remembered that the definition of an Hermitian operator involves specifying boundary conditions that the wavefunctions considered must satisfy. Typically, they are that the wavefunctions vanish for large values of the spatial variables upon which they depend; this deals with most physical systems since they are nearly all formally infinite in extent. Some model systems require the wavefunction to be periodic or to vanish at finite values of a spatial variable.

Depending on the nature of the physical system, the eigenvalues of a particular linear operator may be discrete, part of a continuum, or a mixture of both. For example, the energy levels of the bound proton–electron system (the hydrogen atom) are discrete, but if the atom is ionised and the electron is free, the energy spectrum of the system is continuous. This system has discrete negative and continuous positive eigenvalues for the operator corresponding to the total energy (the Hamiltonian).

▶ *Using the Dirac notation, show that the eigenvalues of an Hermitian operator are real.*

Let $|a\rangle$ be an eigenstate of Hermitian operator A corresponding to eigenvalue a, then

$$A|a\rangle = a|a\rangle,$$
$$\Rightarrow \quad \langle a|A|a\rangle = \langle a|a|a\rangle = a\langle a|a\rangle,$$

and

$$\langle a|A^\dagger = a^*\langle a|,$$
$$\Rightarrow \quad \langle a|A^\dagger|a\rangle = a^*\langle a|a\rangle,$$
$$\langle a|A|a\rangle = a^*\langle a|a\rangle, \quad \text{since } A \text{ is Hermitian.}$$

Hence,

$$(a - a^*)\langle a|a\rangle = 0,$$
$$\Rightarrow \quad a = a^*, \text{ since } \langle a|a\rangle \neq 0.$$

Thus a is real. ◀

It is not our intention to describe the complete axiomatic basis of quantum mechanics, but rather to show what can be learned about linear operators, and in particular about their eigenvalues, without recourse to explicit wavefunctions on which the operators act.

Before we proceed to do that, we close this subsection with a number of results, expressed in Dirac notation, that the reader should verify by inspection or by following the lines of argument sketched in the statements. Where a sum over a complete set of eigenvalues is shown, it should be replaced by an integral for those parts of the eigenvalue spectrum that are continuous. With the notation that $|a_n\rangle$ is an eigenstate of Hermitian operator A with non-degenerate eigenvalue a_n (or, if a_n is k-fold degenerate, then a set of k mutually orthogonal eigenstates has been constructed and the states relabelled), we have the following results.

$$A|a_n\rangle = a_n|a_n\rangle,$$
$$\langle a_m|a_n\rangle = \delta_{mn} \quad \text{(orthonormality of eigenstates)}, \tag{19.5}$$
$$A(c_n|a_n\rangle + c_m|a_m\rangle) = c_n a_n|a_n\rangle + c_m a_m|a_m\rangle \quad \text{(linearity).} \tag{19.6}$$

The definitions of the sum and product of two operators are

$$(A + B)|\psi\rangle \equiv A|\psi\rangle + B|\psi\rangle, \tag{19.7}$$
$$AB|\psi\rangle \equiv A(B|\psi\rangle) \quad (\neq BA|\psi\rangle \text{ in general}), \tag{19.8}$$
$$\Rightarrow \quad A^p|a_n\rangle = a_n^p|a_n\rangle. \tag{19.9}$$

651

If $A|a_n\rangle = a|a_n\rangle$ for all $N_1 \leq n \leq N_2$, then

$$|\psi\rangle = \sum_{n=N_1}^{N_2} d_n|a_n\rangle \text{ satisfies } A|\psi\rangle = a|\psi\rangle \text{ for any set of } d_i.$$

For a general state $|\psi\rangle$,

$$|\psi\rangle = \sum_{n=0}^{\infty} c_n|a_n\rangle, \text{ where } c_n = \langle a_n|\psi\rangle. \tag{19.10}$$

This can also be expressed as the operator identity,

$$1 = \sum_{n=0}^{\infty} |a_n\rangle\langle a_n|, \tag{19.11}$$

in the sense that

$$|\psi\rangle = 1\,|\psi\rangle = \sum_{n=0}^{\infty} |a_n\rangle\langle a_n|\psi\rangle = \sum_{n=0}^{\infty} c_n|a_n\rangle.$$

It also follows that

$$1 = \langle\psi|\psi\rangle = \left(\sum_{m=0}^{\infty} c_m^*\langle a_m|\right)\left(\sum_{n=0}^{\infty} c_n|a_n\rangle\right) = \sum_{m,n}^{\infty} c_m^* c_n \delta_{mn} = \sum_{n=0}^{\infty} |c_n|^2. \tag{19.12}$$

Similarly, the expectation value of the physical variable corresponding to A is

$$\langle\psi|A|\psi\rangle = \sum_{m,n}^{\infty} c_m^*\langle a_m|A|a_n\rangle c_n = \sum_{m,n}^{\infty} c_m^*\langle a_m|a_n|a_n\rangle c_n$$

$$= \sum_{m,n}^{\infty} c_m^* c_n a_n \delta_{mn} = \sum_{n=0}^{\infty} |c_n|^2 a_n. \tag{19.13}$$

19.1.1 Commutation and commutators

As has been noted above, the product AB of two linear operators may or may not be equal to the product BA. That is

$$AB|\psi\rangle \text{ is not necessarily equal to } BA|\psi\rangle.$$

If A and B are both purely multiplicative operators, multiplication by $f(\mathbf{r})$ and $g(\mathbf{r})$ say, then clearly the order of the operations is immaterial, the result $|f(\mathbf{r})g(\mathbf{r})\psi\rangle$ being obtained in both cases. However, consider a case in which A is the differential operator $\partial/\partial x$ and B is the operator 'multiply by x'. Then the wavefunction describing $AB|\psi\rangle$ is

$$\frac{\partial}{\partial x}(x\psi(x)) = \psi(x) + x\frac{\partial\psi}{\partial x},$$

whilst that for $BA|\psi\rangle$ is simply

$$x\frac{\partial\psi}{\partial x},$$

which is not the same.

If the result

$$AB|\psi\rangle = BA|\psi\rangle$$

is true for *all* ket vectors $|\psi\rangle$, then A and B are said to *commute*; otherwise they are non-commuting operators.

A convenient way to express the commutation properties of two linear operators is to define their *commutator*, $[A, B]$, by

$$[A, B]|\psi\rangle \equiv AB|\psi\rangle - BA|\psi\rangle. \tag{19.14}$$

Clearly two operators that commute have a zero commutator. But, for the example given above we have that

$$\left[\frac{\partial}{\partial x}, x\right]\psi(x) = \left(\psi(x) + x\frac{\partial\psi}{\partial x}\right) - \left(x\frac{\partial\psi}{\partial x}\right) = \psi(x) = 1 \times \psi$$

or, more simply, that

$$\left[\frac{\partial}{\partial x}, x\right] = 1; \tag{19.15}$$

in words, the commutator of the differential operator $\partial/\partial x$ and the multiplicative operator x is the multiplicative operator 1. It should be noted that the order of the linear operators is important and that

$$[A, B] = -[B, A]. \tag{19.16}$$

Clearly any linear operator commutes with itself and some other obvious zero commutators (when operating on wavefunctions with 'reasonable' properties) are:

$[A, I]$, where I is the identity operator;

$[A^n, A^m]$, for any positive integers n and m;

$[A, p(A)]$, where $p(x)$ is any polynomial in x;

$[A, c]$, where A is any linear operator and c is any constant;

$[f(x), g(x)]$, where the functions are mutliplicative;

$[A(x), B(y)]$, where the operators act on different variables, with

$\left[\dfrac{\partial}{\partial x}, \dfrac{\partial}{\partial y}\right]$ as a specific example.

Simple identities amongst commutators include the following:

$$[A, B + C] = [A, B] + [A, C],\qquad(19.17)$$

$$[A + B, C] = [A, C] + [B, C],\qquad(19.18)$$

$$[A, BC] = ABC - BCA + BAC - BAC$$

$$= (AB - BA)C + B(AC - CA)$$

$$= [A, B]C + B[A, C],\qquad(19.19)$$

$$[AB, C] = A[B, C] + [A, C]B.\qquad(19.20)$$

▶*If A and B are two linear operators that both commute with their commutator, prove that $[A, B^n] = nB^{n-1}[A, B]$ and that $[A^n, B] = nA^{n-1}[A, B]$.*

Define C_n by $C_n = [A, B^n]$. We aim to find a reduction formula for C_n:

$$C_n = \left[A, B\, B^{n-1}\right]$$

$$= [A, B]\, B^{n-1} + B\left[A, B^{n-1}\right], \text{ using } (19.19),$$

$$= B^{n-1}[A, B] + B\left[A, B^{n-1}\right], \text{ since } [[A, B], B] = 0,$$

$$= B^{n-1}[A, B] + BC_{n-1}, \text{ the required reduction formula,}$$

$$= B^{n-1}[A, B] + B\{B^{n-2}[A, B] + BC_{n-2}\}, \text{ applying the formula,}$$

$$= 2B^{n-1}[A, B] + B^2 C_{n-2}$$

$$= \cdots$$

$$= nB^{n-1}[A, B] + B^n C_0.$$

However, $C_0 = [A, I] = 0$ and so $C_n = nB^{n-1}[A, B]$.

Using equation (19.16) and interchanging A and B in the result just obtained, we find

$$[A^n, B] = -[B, A^n] = -nA^{n-1}[B, A] = nA^{n-1}[A, B],$$

as stated in the question. ◀

As the power of a linear operator can be defined, so can its exponential; this situation parallels that for matrices, which are of course a particular set of operators that act upon state functions represented by vectors. The definition follows that for the exponential of a scalar or matrix, namely

$$\exp A = \sum_{n=0}^{\infty} \frac{A^n}{n!}.\qquad(19.21)$$

Related functions of A, such as $\sin A$ and $\cos A$, can be defined in a similar way.

Since any linear operator commutes with itself, when two functions of it are combined in some way, the result takes a form similar to that for the corresponding functions of scalar quantities. Consider, for example, the function $f(A)$ defined by $f(A) = 2 \sin A \cos A$. Expressing $\sin A$ and $\cos A$ in terms of their defining series, we have

$$f(A) = 2 \sum_{m=0}^{\infty} \frac{(-1)^m A^{2m+1}}{(2m+1)!} \sum_{n=0}^{\infty} \frac{(-1)^n A^{2n}}{(2n)!}.$$

Writing $m + n$ as r and replacing n by s, we have

$$f(A) = 2 \sum_{r=0}^{\infty} A^{2r+1} \left(\sum_{s=0}^{r} \frac{(-1)^{r-s}}{(2r-2s+1)!} \frac{(-1)^s}{(2s)!} \right)$$

$$= 2 \sum_{r=0}^{\infty} (-1)^r c_r A^{2r+1},$$

where

$$c_r = \sum_{s=0}^{r} \frac{1}{(2r-2s+1)!\,(2s)!} = \frac{1}{(2r+1)!} \sum_{s=0}^{r} {}^{2r+1}C_{2s}.$$

By adding the binomial expansions of $2^{2r+1} = (1+1)^{2r+1}$ and $0 = (1-1)^{2r+1}$, it can easily be shown that

$$2^{2r+1} = 2 \sum_{s=0}^{r} {}^{2r+1}C_{2s} \quad \Rightarrow \quad c_r = \frac{2^{2r}}{(2r+1)!}.$$

It then follows that

$$2 \sin A \cos A = 2 \sum_{r=0}^{\infty} \frac{(-1)^r A^{2r+1} 2^{2r}}{(2r+1)!} = \sum_{r=0}^{\infty} \frac{(-1)^r (2A)^{2r+1}}{(2r+1)!} = \sin 2A,$$

a not unexpected result.

However, if two (or more) linear operators that do not commute are involved, combining functions of them is more complicated and the results less intuitively obvious. We take as a particular case the product of two exponential functions and, even then, take the simplified case in which each linear operator commutes with their commutator (so that we may use the results from the previous worked example).

▶*If A and B are two linear operators that both commute with their commutator, show that*

$$\exp(A) \exp(B) = \exp(A + B + \tfrac{1}{2}[A, B]).$$

We first find the commutator of A and $\exp \lambda B$, where λ is a scalar quantity introduced for later algebraic convenience:

$$[A, e^{\lambda B}] = \left[A, \sum_{n=0}^{\infty} \frac{(\lambda B)^n}{n!}\right] = \sum_{n=0}^{\infty} \frac{\lambda^n}{n!} [A, B^n]$$

$$= \sum_{n=0}^{\infty} \frac{\lambda^n}{n!} nB^{n-1} [A, B], \text{ using the earlier result,}$$

$$= \sum_{n=1}^{\infty} \frac{\lambda^n}{n!} nB^{n-1} [A, B]$$

$$= \lambda \sum_{m=0}^{\infty} \frac{\lambda^m B^m}{m!} [A, B], \text{ writing } m = n - 1,$$

$$= \lambda e^{\lambda B} [A, B].$$

Now consider the derivative with respect to λ of the function

$$f(\lambda) = e^{\lambda A} e^{\lambda B} e^{-\lambda(A+B)}.$$

In the following calculation we use the fact that the derivative of $e^{\lambda C}$ is $Ce^{\lambda C}$; this is the same as $e^{\lambda C} C$, since any two functions of the same operator commute. Differentiating the three-factor product gives

$$\frac{df}{d\lambda} = e^{\lambda A} A e^{\lambda B} e^{-\lambda(A+B)} + e^{\lambda A} e^{\lambda B} B e^{-\lambda(A+B)} + e^{\lambda A} e^{\lambda B} (-A - B) e^{-\lambda(A+B)}$$

$$= e^{\lambda A} (e^{\lambda B} A + \lambda e^{\lambda B} [A, B]) e^{-\lambda(A+B)} + e^{\lambda A} e^{\lambda B} B e^{-\lambda(A+B)}$$
$$\quad - e^{\lambda A} e^{\lambda B} A e^{-\lambda(A+B)} - e^{\lambda A} e^{\lambda B} B e^{-\lambda(A+B)}$$

$$= e^{\lambda A} \lambda e^{\lambda B} [A, B] e^{-\lambda(A+B)}$$

$$= \lambda [A, B] f(\lambda).$$

In the second line we have used the result obtained above to replace $Ae^{\lambda B}$, and in the last line have used the fact that $[A, B]$ commutes with each of A and B, and hence with any function of them.

Integrating this scalar differential equation with respect to λ and noting that $f(0) = 1$, we obtain

$$\ln f = \tfrac{1}{2} \lambda^2 [A, B] \quad \Rightarrow \quad e^{\lambda A} e^{\lambda B} e^{-\lambda(A+B)} = f(\lambda) = e^{\frac{1}{2}\lambda^2 [A, B]}.$$

Finally, post-multiplying both sides of the equation by $e^{\lambda(A+B)}$ and setting $\lambda = 1$ yields

$$e^A e^B = e^{\frac{1}{2}[A, B] + A + B}. \quad \blacktriangleleft$$

19.2 Physical examples of operators

We now turn to considering some of the specific linear operators that play a part in the description of physical systems. In particular, we will examine the properties of some of those that appear in the quantum-mechanical description of the physical world.

As stated earlier, the operators corresponding to physical observables are restricted to Hermitian operators (which have real eigenvalues) as this ensures the reality of predicted values for experimentally measured quantities. The two basic quantum-mechanical operators are those corresponding to position $\mathbf{r}$ and momentum $\mathbf{p}$. One prescription for making the transition from classical to quantum mechanics is to express classical quantities in terms of these two variables in Cartesian coordinates and then make the component by component substitutions

$$\mathbf{r} \rightarrow \text{multiplicative operator } \mathbf{r} \quad \text{and} \quad \mathbf{p} \rightarrow \text{differential operator } -i\hbar\nabla.$$
$$(19.22)$$

This generates the quantum operators corresponding to the classical quantities. For the sake of completeness, we should add that if the classical quantity contains a product of factors whose corresponding operators A and B do not commute, then the operator $\frac{1}{2}(AB + BA)$ is to be substituted for the product.

The substitutions (19.22) invoke operators that are closely connected with the two that we considered at the start of the previous subsection, namely x and $\partial/\partial x$. One, x, corresponds exactly to the x-component of the prescribed quantum position operator; the other, however, has been multiplied by the imaginary constant $-i\hbar$, where $\hbar$ is the Planck constant divided by 2π. This has the (subtle) effect of converting the differential operator into the x-component of an *Hermitian* operator; this is easily verified using integration by parts to show that it satisfies equation (17.16). Without the extra imaginary factor (which changes sign under complex conjugation) the two sides of the equation differ by a minus sign.

Making the differential operator Hermitian does not change in any essential way its commutation properties, and the commutation relation of the two basic quantum operators reads

$$[p_x, x] = \left[-i\hbar\frac{\partial}{\partial x}, x\right] = -i\hbar.$$
$$(19.23)$$

Corresponding results hold when x is replaced, in both operators, by y or z. However, it should be noted that if different Cartesian coordinates appear in the two operators then the operators commute, i.e.

$$[p_x, y] = [p_x, z] = [p_y, x] = [p_y, z] = [p_z, x] = [p_z, y] = 0.$$
$$(19.24)$$

As an illustration of the substitution rules, we now construct the Hamiltonian (the quantum-mechanical energy operator) H for a particle of mass m moving in a potential $V(x, y, z)$ when it has one of its allowed energy values, i.e its energy is E_n, where $H|\psi_n\rangle = E_n|\psi_n\rangle$. This latter equation when expressed in a particular coordinate system is the Schrödinger equation for the particle. In terms of position and momentum, the total classical energy of the particle is given by

$$E = \frac{p^2}{2m} + V(x, y, z) = \frac{p_x^2 + p_y^2 + p_z^2}{2m} + V(x, y, z).$$

Substituting $-i\hbar\partial/\partial x$ for p_x (and similarly for p_y and p_z) in the first term on the

RHS gives

$$\frac{(-i\hbar)^2}{2m}\frac{\partial}{\partial x}\frac{\partial}{\partial x} + \frac{(-i\hbar)^2}{2m}\frac{\partial}{\partial y}\frac{\partial}{\partial y} + \frac{(-i\hbar)^2}{2m}\frac{\partial}{\partial z}\frac{\partial}{\partial z}.$$

The potential energy V, being a function of position only, becomes a purely multiplicative operator, thus creating the full expression for the Hamiltonian,

$$H = -\frac{\hbar^2}{2m}\left(\frac{\partial^2}{\partial x^2} + \frac{\partial^2}{\partial y^2} + \frac{\partial^2}{\partial z^2}\right) + V(x, y, z),$$

and giving the corresponding Schrödinger equation as

$$H\psi_n = -\frac{\hbar^2}{2m}\left(\frac{\partial^2\psi_n}{\partial x^2} + \frac{\partial^2\psi_n}{\partial y^2} + \frac{\partial^2\psi_n}{\partial z^2}\right) + V(x, y, z)\psi_n = E_n\psi_n.$$

We are not so much concerned in this section with solving such differential equations, but with the commutation properties of the operators from which they are constructed. To this end, we now turn our attention to the topic of angular momentum, the operators for which can be constructed in a straightforward manner from the two basic sets.

19.2.1 Angular momentum operators

As required by the substitution rules, we start by expressing angular momentum in terms of the classical quantities $\mathbf{r}$ and $\mathbf{p}$, namely $\mathbf{L} = \mathbf{r} \times \mathbf{p}$ with Cartesian components

$$L_z = xp_y - yp_x, \quad L_x = yp_z - zp_y, \quad L_y = zp_x - xp_z.$$

Making the substitutions (19.22) yields as the corresponding quantum-mechanical operators

$$L_z = -i\hbar\left(x\frac{\partial}{\partial y} - y\frac{\partial}{\partial x}\right),$$

$$L_x = -i\hbar\left(y\frac{\partial}{\partial z} - z\frac{\partial}{\partial y}\right), \tag{19.25}$$

$$L_y = -i\hbar\left(z\frac{\partial}{\partial x} - x\frac{\partial}{\partial z}\right).$$

It should be noted that for xp_y, say, x and $\partial/\partial y$ commute, and there is no ambiguity about the way it is to be carried into its quantum form. Further, since the operators corresponding to each of its factors commute and are Hermitian, the operator corresponding to the product is Hermitian. This was shown directly for matrices in exercise 8.7, and can be verified using equation (17.16).

The first question that arises is whether or not these three operators commute.

Consider first

$$L_xL_y = -\hbar^2 \left(y\frac{\partial}{\partial z} - z\frac{\partial}{\partial y} \right) \left(z\frac{\partial}{\partial x} - x\frac{\partial}{\partial z} \right)$$

$$= -\hbar^2 \left(y\frac{\partial}{\partial x} + yz\frac{\partial^2}{\partial z\partial x} - yx\frac{\partial^2}{\partial z^2} - z^2\frac{\partial^2}{\partial y\partial x} + zx\frac{\partial^2}{\partial y\partial z} \right).$$

Now consider

$$L_yL_x = -\hbar^2 \left(z\frac{\partial}{\partial x} - x\frac{\partial}{\partial z} \right) \left(y\frac{\partial}{\partial z} - z\frac{\partial}{\partial y} \right)$$

$$= -\hbar^2 \left(zy\frac{\partial^2}{\partial x\partial z} - z^2\frac{\partial^2}{\partial x\partial y} - xy\frac{\partial^2}{\partial z^2} + x\frac{\partial}{\partial y} + xz\frac{\partial^2}{\partial z\partial y} \right).$$

These two expressions are *not* the same. The difference between them, i.e. the commutator of L_x and L_y, is given by

$$\left[L_x, L_y \right] = L_xL_y - L_yL_x = \hbar^2 \left(x\frac{\partial}{\partial y} - y\frac{\partial}{\partial x} \right) = i\hbar L_z. \tag{19.26}$$

This, and two similar results obtained by permuting x, y and z cyclically, summarise the commutation relationships between the quantum operators corresponding to the three Cartesian components of angular momentum:

$$\begin{aligned}
\left[L_x, L_y \right] &= i\hbar L_z, \\
\left[L_y, L_z \right] &= i\hbar L_x, \\
\left[L_z, L_x \right] &= i\hbar L_y.
\end{aligned} \tag{19.27}$$

As well as its separate components of angular momentum, the total angular momentum associated with a particular state $|\psi\rangle$ is a physical quantity of interest. This is measured by the operator corresponding to the sum of squares of its components,

$$L^2 = L_x^2 + L_y^2 + L_z^2. \tag{19.28}$$

This is an Hermitian operator, as each term in it is the product of two Hermitian operators that (trivially) commute. It might seem natural to want to 'take the square root' of this operator, but such a process is undefined and we will not pursue the matter.

We next show that, although no two of its components commute, the total angular momentum operator does commute with each of its components. In the proof we use some of the properties (19.17) to (19.20) and result (19.27). We begin with

$$[L^2, L_z] = [L_x^2 + L_y^2 + L_z^2, L_z]$$
$$= L_x[L_x, L_z] + [L_x, L_z]L_x$$
$$+ L_y[L_y, L_z] + [L_y, L_z]L_y + [L_z^2, L_z]$$
$$= L_x(-i\hbar)L_y + (-i\hbar)L_yL_x + L_y(i\hbar)L_x + (i\hbar)L_xL_y + 0$$
$$= 0.$$

Thus operators L^2 and L_z commute and, continuing in the same way, it can be shown that

$$[L^2, L_x] = [L^2, L_y] = [L^2, L_z] = 0. \tag{19.29}$$

Eigenvalues of the angular momentum operators

We will now use the commutation relations for L^2 and its components to find the eigenvalues of L^2 and L_z, without reference to any specific wavefunction. In other words, the eigenvalues of the operators follow from the structure of their commutators. There is nothing particular about L_z, and L_x or L_y could equally well have been chosen, though, in general, it is not possible to find states that are simultaneously eigenstates of two or more of L_x, L_y and L_z.

To help with the calculation, it is convenient to define the two operators

$$U \equiv L_x + iL_y \quad \text{and} \quad D \equiv L_x - iL_y.$$

These operators are not Hermitian; they are in fact Hermitian conjugates, in that $U^\dagger = D$ and $D^\dagger = U$, but they do not represent measurable physical quantities. We first note their multiplication and commutation properties:

$$UD = (L_x + iL_y)(L_x - iL_y) = L_x^2 + L_y^2 + i[L_y, L_x]$$
$$= L^2 - L_z^2 + \hbar L_z, \tag{19.30}$$
$$DU = (L_x - iL_y)(L_x + iL_y) = L_x^2 + L_y^2 - i[L_y, L_x]$$
$$= L^2 - L_z^2 - \hbar L_z, \tag{19.31}$$
$$[L_z, U] = [L_z, L_x] + i[L_z, L_y] = i\hbar L_y + \hbar L_x = \hbar U, \tag{19.32}$$
$$[L_z, D] = [L_z, L_x] - i[L_z, L_y] = i\hbar L_y - \hbar L_x = -\hbar D. \tag{19.33}$$

In the same way as was shown for matrices, it can be demonstrated that if two operators commute they have a common set of eigenstates. Since L^2 and L_z commute they possess such a set; let one of the set be $|\psi\rangle$ with

$$L^2|\psi\rangle = a|\psi\rangle \quad \text{and} \quad L_z|\psi\rangle = b|\psi\rangle.$$

Now consider the state $|\psi'\rangle = U|\psi\rangle$ and the actions of L^2 and L_z upon it.

Consider first $L^2|\psi'\rangle$, recalling that L^2 commutes with both L_x and L_y and hence with U:

$$L^2|\psi'\rangle = L^2 U|\psi\rangle = UL^2|\psi\rangle = Ua|\psi\rangle = aU|\psi\rangle = a|\psi'\rangle.$$

Thus, $|\psi'\rangle$ is also an eigenstate of L^2, corresponding to the same eigenvalue as $|\psi\rangle$. Now consider the action of L_z:

$$\begin{aligned}
L_z|\psi'\rangle &= L_z U|\psi\rangle \\
&= (UL_z + \hbar U)|\psi\rangle, \text{ using } [L_z, U] = \hbar U, \\
&= Ub|\psi\rangle + \hbar U|\psi\rangle \\
&= (b + \hbar)U|\psi\rangle \\
&= (b + \hbar)|\psi'\rangle.
\end{aligned}$$

Thus, $|\psi'\rangle$ is also an eigenstate of L_z, but with eigenvalue $b + \hbar$.

In summary, the effect of U acting upon $|\psi\rangle$ is to produce a new state that has the same eigenvalue for L^2 and is still an eigenstate of L_z, though with that eigenvalue increased by $\hbar$. An exactly analogous calculation shows that the effect of D acting upon $|\psi\rangle$ is to produce another new state, one that also has the same eigenvalue for L^2 and is also still an eigenstate of L_z, though with the eigenvalue decreased by $\hbar$ in this case. For these reasons, U and D are usually known as *ladder* operators.

It is clear that, by starting from any arbitrary eigenstate and repeatedly applying either U or D, we could generate a series of eigenstates, all of which have the eigenvalue a for L^2, but increment in their L_z eigenvalues by $\pm\hbar$. However, we also have the physical requirement that, for real values of the z-component, its square cannot exceed the square of the total angular momentum, i.e. $b^2 \leq a$. Thus b has a maximum value c that satisfies

$$c^2 \leq a \quad \text{but} \quad (c + \hbar)^2 > a;$$

let the corresponding eigenstate be $|\psi_u\rangle$ with $L_z|\psi_u\rangle = c|\psi_u\rangle$. Now it is still true that

$$L_z U|\psi_u\rangle = (c + \hbar)U|\psi_u\rangle,$$

and, to make this compatible with the physical constraint, we must have that $U|\psi_u\rangle$ is the zero ket vector $|\emptyset\rangle$. Now, using result (19.31), we have

$$\begin{aligned}
DU|\psi_u\rangle &= (L^2 - L_z^2 - \hbar L_z)|\psi_u\rangle, \\
\Rightarrow \quad 0|\emptyset\rangle = D|\emptyset\rangle &= (a^2 - c^2 - \hbar c)|\psi_u\rangle, \\
\Rightarrow \quad a &= c(c + \hbar).
\end{aligned}$$

This gives the relationship between a and c. We now establish the possible forms for c.

If we start with eigenstate $|\psi_u\rangle$, which has the highest eigenvalue c for L_z, and operate repeatedly on it with the (down) ladder operator D, we will generate a state $|\psi_d\rangle$ which, whilst still an eigenstate of L^2 with eigenvalue a, has the lowest physically possible value, d say, for the eigenvalue of L_z. If this happens after n operations we will have that $d = c - n\hbar$ and

$$L_z|\psi_d\rangle = (c - n\hbar)|\psi_d\rangle.$$

Arguing in the same way as previously that $D|\psi_d\rangle$ must be an unphysical ket vector, we conclude that

$$
\begin{aligned}
0|\emptyset\rangle = U|\emptyset\rangle &= UD|\psi_d\rangle \\
&= (L^2 - L_z^2 + \hbar L_z)|\psi_d\rangle, \text{ using (19.30),} \\
&= [a - (c - n\hbar)^2 + \hbar(c - n\hbar)]|\psi_d\rangle \\
\Rightarrow \quad a &= (c - n\hbar)^2 - \hbar(c - n\hbar).
\end{aligned}
$$

Equating the two results for a gives

$$c^2 + c\hbar = c^2 - 2cn\hbar + n^2\hbar^2 - c\hbar + n\hbar^2,$$
$$2c(n+1) = n(n+1)\hbar,$$
$$c = \tfrac{1}{2}n\hbar.$$

Since n is necessarily integral, c is an integer multiple of $\tfrac{1}{2}\hbar$. This result is valid irrespective of which eigenstate $|\psi\rangle$ we started with, though the actual value of the integer n depends on $|\psi_u\rangle$ and hence upon $|\psi\rangle$.

Denoting $\tfrac{1}{2}n$ by ℓ we can say that the possible eigenvalues of the operator L_z, and hence the possible results of a measurement of the z-component of the angular momentum of a system, are given by

$$\ell\hbar, \quad (\ell - 1)\hbar, \quad (\ell - 2)\hbar, \quad \ldots, \quad -\ell\hbar.$$

The value of a for all $2\ell + 1$ of the corresponding states,

$$|\psi_u\rangle, \quad D|\psi_u\rangle, \quad D^2|\psi_u\rangle, \quad \ldots, \quad D^{2\ell}|\psi_u\rangle,$$

is $\ell(\ell + 1)\hbar^2$.

The similarity of form between this eigenvalue and that appearing in Legendre's equation is not an accident. It is intimately connected with the facts (i) that L^2 is a measure of the rotational kinetic energy of a particle in a system centred on the origin, and (ii) that in spherical polar coordinates L^2 has the same form as the angle-dependent part of ∇^2, which, as we have seen, is itself proportional to the quantum-mechanical kinetic energy operator. Legendre's equation and the associated Legendre equation arise naturally when $\nabla^2\psi = f(r)$ is solved in spherical polar coordinates using the method of separation of variables discussed in chapter 21.

The derivation of the eigenvalues $\ell(\ell + 1)\hbar^2$ and $m\hbar$, with $-\ell \leq m \leq \ell$, depends only on the commutation relationships between the corresponding operators. Any other set of four operators with the same commutation structure would result in the same eigenvalue spectrum. In fact, quantum mechanically, orbital angular momentum is restricted to cases in which n is even and so ℓ is an integer; this is in accord with the requirement placed on ℓ if solutions to $\nabla^2\psi = f(r)$ that are finite on the polar axis are to be obtained. The non-classical notion of internal angular momentum (spin) for a particle provides a set of operators that are able to take both integral and half-integral multiples of $\hbar$ as their eigenvalues.

We have already seen that, for a state $|\ell,m\rangle$ that has a z-component of angular momentum $m\hbar$, the state $U|\ell,m\rangle$ is one with its z-component of angular momentum equal to $(m+1)\hbar$. But the new state ket vector so produced is not necessarily normalised so as to make $\langle\ell,m+1|\ell,m+1\rangle = 1$. We will conclude this discussion of angular momentum by calculating the coefficients μ_m and ν_m in the equations

$$U|\ell,m\rangle = \mu_m|\ell,m+1\rangle \quad \text{and} \quad D|\ell,m\rangle = \nu_m|\ell,m-1\rangle$$

on the basis that $\langle\ell,r|\ell,r\rangle = 1$ for all ℓ and r.

To do so, we consider the inner product $I = \langle\ell,m|DU|\ell,m\rangle$, evaluated in two different ways. We have already noted that U and D are Hermitian conjugates and so I can be written as

$$I = \langle\ell,m|U^\dagger U|\ell,m\rangle = \mu_m^*\langle\ell,m|\ell,m\rangle\mu_m = |\mu_m|^2.$$

But, using equation (19.31), it can also be expressed as

$$\begin{aligned}
I &= \langle\ell,m|L^2 - L_z^2 - \hbar L_z|\ell,m\rangle \\
&= \langle\ell,m|\ell(\ell+1)\hbar^2 - m^2\hbar^2 - m\hbar^2|\ell,m\rangle \\
&= [\ell(\ell+1)\hbar^2 - m^2\hbar^2 - m\hbar^2]\langle\ell,m|\ell,m\rangle \\
&= [\ell(\ell+1) - m(m+1)]\hbar^2.
\end{aligned}$$

Thus we are required to have

$$|\mu_m|^2 = [\ell(\ell+1) - m(m+1)]\hbar^2,$$

but can choose that all μ_m are real and non-negative (recall that $|m| \leq \ell$). A similar calculation can be used to calculate ν_m. The results are summarised in the equations

$$U|\ell,m\rangle = \sqrt{\ell(\ell+1) - m(m+1)}\,\hbar|\ell,m+1\rangle, \tag{19.34}$$

$$D|\ell,m\rangle = \sqrt{\ell(\ell+1) - m(m-1)}\,\hbar|\ell,m-1\rangle. \tag{19.35}$$

It can easily be checked that $U|\ell,\ell\rangle = |\emptyset\rangle = D|\ell,-\ell\rangle$.

19.2.2 Uncertainty principles

The next topic we explore is the quantitative consequences of a non-zero commutator for two quantum (Hermitian) operators that correspond to physical variables.

As previously noted, the expectation value in a state $|\psi\rangle$ of the physical quantity A corresponding to the operator A is $E[A] = \langle\psi|A|\psi\rangle$. Any one measurement of A can only yield one of the eigenvalues of A. But if repeated measurements could be made on a large number of identical systems, a discrete or continuous range of values would be obtained. It is a natural extension of normal data analysis to measure the uncertainty in the value of A by the observed variance in the measured values of A, denoted by $(\Delta A)^2$ and calculated as the average value of $(A - E[A])^2$. The expected value of this variance for the state $|\psi\rangle$ is given by $\langle\psi|(A - E[A])^2|\psi\rangle$.

We now give a mathematical proof that there is a theoretical lower limit for the product of the uncertainties in any two physical quantities, and we start by proving a result similar to the Schwarz inequality. Let $|u\rangle$ and $|v\rangle$ be any two state vectors and let λ be any *real* scalar. Then consider the vector $|w\rangle = |u\rangle + \lambda|v\rangle$ and, in particular, note that

$$0 \leq \langle w\,|\,w\rangle = \langle u\,|\,u\rangle + \lambda(\langle u\,|\,v\rangle + \langle v\,|\,u\rangle) + \lambda^2\langle v\,|\,v\rangle.$$

This is a quadratic inequality in λ and therefore the quadratic equation formed by equating the RHS to zero must have no real roots. The coefficient of λ is $(\langle u\,|\,v\rangle + \langle v\,|\,u\rangle) = 2\,\mathrm{Re}\,\langle u\,|\,v\rangle$ and its square is thus ≥ 0. The condition for no real roots of the quadratic is therefore

$$0 \leq (\langle u\,|\,v\rangle + \langle v\,|\,u\rangle)^2 \leq 4\langle u\,|\,u\rangle\,\langle v\,|\,v\rangle. \tag{19.36}$$

This result will now be applied to state vectors constructed from $|\psi\rangle$, the state vector of the particular system for which we wish to establish a relationship between the uncertainties in the two physical variables corresponding to (Hermitian) operators A and B. We take

$$|u\rangle = (A - E[A])\,|\psi\rangle \quad \text{and} \quad |v\rangle = i(B - E[B])\,|\psi\rangle. \tag{19.37}$$

Then

$$\langle u\,|\,u\rangle = \langle\psi\,|(A - E[A])^2|\psi\rangle = (\Delta A)^2,$$
$$\langle v\,|\,v\rangle = \langle\psi\,|(B - E[B])^2|\psi\rangle = (\Delta B)^2.$$

Further,

$$\begin{aligned}
\langle u\,|\,v\rangle &= \langle\psi\,|\,(A - E[A])i(B - E[B])\,|\psi\rangle \\
&= i\langle\psi\,|\,AB\,|\psi\rangle - iE[A]\langle\psi\,|\,B\,|\psi\rangle - iE[B]\langle\psi\,|\,A\,|\psi\rangle + iE[A]E[B]\langle\psi\,|\,\psi\rangle \\
&= i\langle\psi\,|\,AB\,|\psi\rangle - iE[A]E[B].
\end{aligned}$$

In the second line, we have moved expectation values, which are purely numbers, out of the inner products and used the normalisation condition $\langle \psi | \psi \rangle = 1$. Similarly

$$\langle v \,|\, u \rangle = -i\langle \psi \,|\, BA \,|\, \psi \rangle + iE[A]E[B].$$

Adding these two results gives

$$\langle u \,|\, v \rangle + \langle v \,|\, u \rangle = i\langle \psi \,|\, AB - BA \,|\, \psi \rangle,$$

and substitution into (19.36) yields

$$0 \le (i\langle \psi \,|\, AB - BA \,|\, \psi \rangle)^2 \le 4(\Delta A)^2(\Delta B)^2$$

At first sight, the middle term of this inequality might appear to be negative, but this is not so. Since A and B are Hermitian, $AB - BA$ is anti-Hermitian, as is easily demonstrated. Since i is also anti-Hermitian, the quantity in the parentheses in the middle term is real and its square non-negative. Rearranging the equation and expressing it in terms of the commutator of A and B gives the generalised form of the *Uncertainty Principle*. For any particular state $|\psi\rangle$ of a system, this provides the quantitative relationship between the minimum value that the product of the uncertainties in A and B can have and the expectation value, in that state, of their commutator,

$$(\Delta A)^2(\Delta B)^2 \ge \tfrac{1}{4}|\langle \psi \,|\, [A, B] \,|\, \psi \rangle|^2. \tag{19.38}$$

Immediate observations include the following:

  (i) If A and B commute there is no absolute restriction on the accuracy with which the corresponding physical quantities may be known. That is not to say that ΔA and ΔB will always be zero, only that they may be.
 (ii) If the commutator of A and B is a constant, $k \ne 0$, then the RHS of equation (19.38) is necessarily equal to $\tfrac{1}{4}|k|^2$, whatever the form of $|\psi\rangle$, and it is not possible to have $\Delta A = \Delta B = 0$.
(iii) Since the RHS depends upon $|\psi\rangle$, it is possible, even for two operators that do not commute, for the lower limit of $(\Delta A)^2(\Delta B)^2$ to be zero. This will occur if the commutator $[A, B]$ is itself an operator whose expectation value in the particular state $|\psi\rangle$ happens to be zero.

To illustrate the third of these, we might consider the components of angular momentum discussed in the previous subsection. There, in equation (19.27), we found that the commutator of the operators corresponding to the x- and y-components of angular momentum is non-zero; in fact, it has the value $i\hbar L_z$. This means that if the state $|\psi\rangle$ of a system happened to be such that $\langle \psi | L_z | \psi \rangle = 0$, as it would if, for example, it were the eigenstate of L_z, $|\psi\rangle = |\ell, 0\rangle$, then there would be no fundamental reason why the physical values of both L_x and L_y should not be known exactly. Indeed, if the state were spherically symmetric, and hence formally an eigenstate of L^2 with $\ell = 0$, all three components of angular momentum could be (and are) known to be zero.

> ▶*Working in one dimension, show that the minimum value of the product $\Delta p_x \times \Delta x$ for a particle is $\frac{1}{2}\hbar$. Find the form of the wavefunction that attains this minimum value for a particle whose expectation values for position and momentum are $\bar{x}$ and $\bar{p}$, respectively.*

We have already seen, in (19.23) that the commutator of p_x and x is $-i\hbar$, a constant. Therefore, irrespective of the actual form of $|\psi\rangle$, the RHS of (19.38) is $\frac{1}{4}\hbar^2$ (see observation (ii) above). Thus, since all quantities are positive, taking the square roots of both sides of the equation shows directly that

$$\Delta p_x \times \Delta x \geq \tfrac{1}{2}\hbar.$$

Returning to the derivation of the Uncertainty Principle, we see that the inequality becomes an equality only when

$$(\langle u \,|\, v \rangle + \langle v \,|\, u \rangle)^2 = 4\langle u \,|\, u \rangle \langle v \,|\, v \rangle.$$

The RHS of this equality has the value $4\|u\|^2\|v\|^2$ and so, by virtue of Schwarz's inequality, we have

$$
\begin{aligned}
4\|u\|^2\|v\|^2 &= (\langle u \,|\, v \rangle + \langle v \,|\, u \rangle)^2 \\
&\leq (|\langle u \,|\, v \rangle| + |\langle v \,|\, u \rangle|)^2 \\
&\leq (\|u\| \, \|v\| + \|v\| \, \|u\|)^2 \\
&= 4\|u\|^2\|v\|^2.
\end{aligned}
$$

Since the LHS is less than or equal to something that has the same value as itself, all of the inequalities are, in fact, equalities. Thus $\langle u|v \rangle = \|u\| \, \|v\|$, showing that $|u\rangle$ and $|v\rangle$ are parallel vectors, i.e. $|u\rangle = \mu|v\rangle$ for some scalar μ.

We now transform this condition into a constraint that the wavefunction $\psi = \psi(x)$ must satisfy. Recalling the definitions (19.37) of $|u\rangle$ and $|v\rangle$ in terms of $|\psi\rangle$, we have

$$\left(-i\hbar\frac{d}{dx} - \bar{p} \right)\psi = \mu i(x - \bar{x})\psi,$$

$$\frac{d\psi}{dx} + \frac{1}{\hbar}[\mu(x - \bar{x}) - i\bar{p}]\psi = 0.$$

The IF for this equation is $\exp\left[\dfrac{\mu(x - \bar{x})^2}{2\hbar} - \dfrac{i\bar{p}x}{\hbar} \right]$, giving

$$\frac{d}{dx}\left\{ \psi \exp\left[\frac{\mu(x - \bar{x})^2}{2\hbar} - \frac{i\bar{p}x}{\hbar} \right] \right\} = 0,$$

which, in turn, leads to

$$\psi(x) = A\exp\left[-\frac{\mu(x - \bar{x})^2}{2\hbar} \right]\exp\left(\frac{i\bar{p}x}{\hbar} \right).$$

From this it is apparent that the minimum uncertainty product $\Delta p_x \times \Delta x$ is obtained when the probability density $|\psi(x)|^2$ has the form of a Gaussian distribution centred on $\bar{x}$. The value of μ is not fixed by this consideration and it could be anything (positive); a large value for μ would yield a small value for Δx but a correspondingly large one for Δp_x. ◀

19.2.3 Annihilation and creation operators

As a final illustration of the use of operator methods in physics we consider their application to the quantum mechanics of a simple harmonic oscillator (s.h.o.). Although we will start with the conventional description of a one-dimensional oscillator, using its position and momentum, we will recast the description in terms of two operators and their commutator and show that many important conclusions can be reached from studying these alone.

The Hamiltonian for a particle of mass m with momentum p moving in a one-dimensional parabolic potential $V(x) = \frac{1}{2}kx^2$ is

$$H = \frac{p^2}{2m} + \frac{1}{2}kx^2 = \frac{p^2}{2m} + \frac{1}{2}m\omega^2 x^2,$$

where its classical frequency of oscillation ω is given by $\omega^2 = k/m$. We recall that the corresponding operators, p and x, do not commute and that $[p, x] = -i\hbar$.

In analogy with the ladder operators used when discussing angular momentum, we define two new operators:

$$A \equiv \sqrt{\frac{m\omega}{2}}\, x + \frac{ip}{\sqrt{2m\omega}} \quad \text{and} \quad A^\dagger \equiv \sqrt{\frac{m\omega}{2}}\, x - \frac{ip}{\sqrt{2m\omega}}. \tag{19.39}$$

Since both x and p are Hermitian, A and $A^\dagger$ are Hermitian conjugates, though neither is Hermitian and they do not represent physical quantities that can be measured.

Now consider the two products $A^\dagger A$ and $AA^\dagger$:

$$A^\dagger A = \frac{m\omega}{2}\, x^2 - \frac{ipx}{2} + \frac{ixp}{2} + \frac{p^2}{2m\omega} = \frac{H}{\omega} - \frac{i}{2}[p, x] = \frac{H}{\omega} - \frac{\hbar}{2},$$

$$AA^\dagger = \frac{m\omega}{2}\, x^2 + \frac{ipx}{2} - \frac{ixp}{2} + \frac{p^2}{2m\omega} = \frac{H}{\omega} + \frac{i}{2}[p, x] = \frac{H}{\omega} + \frac{\hbar}{2}.$$

From these it follows that

$$H = \tfrac{1}{2}\omega(A^\dagger A + AA^\dagger) \tag{19.40}$$

and that

$$\left[A, A^\dagger\right] = \hbar. \tag{19.41}$$

Further,

$$[H, A] = \left[\tfrac{1}{2}\omega(A^\dagger A + AA^\dagger), A\right]$$
$$= \tfrac{1}{2}\omega\left(A^\dagger 0 + \left[A^\dagger, A\right]A + A\left[A^\dagger, A\right] + 0\,A^\dagger\right)$$
$$= \tfrac{1}{2}\omega(-\hbar A - A\hbar) = -\hbar\omega A. \tag{19.42}$$

Similarly,

$$\left[H, A^\dagger\right] = \hbar\omega A^\dagger \tag{19.43}$$

Before we apply these relationships to the question of the energy spectrum of the s.h.o., we need to prove one further result. This is that if B is an Hermitian operator then $\langle \psi \,|\, B^2 \,|\, \psi \rangle \geq 0$ for *any* $|\psi\rangle$. The proof, which involves introducing an arbitrary complete set of orthonormal base states $|\phi_i\rangle$ and using equation (19.11), is as follows:

$$\langle \psi \mid B^2 \mid \psi \rangle = \langle \psi \mid B \times 1 \times B \mid \psi \rangle$$

$$= \sum_i \langle \psi \mid B \mid \phi_i \rangle \langle \phi_i \mid B \mid \psi \rangle$$

$$= \sum_i \langle \psi \mid B \mid \phi_i \rangle \left(\langle \phi_i \mid B \mid \psi \rangle^* \right)^*$$

$$= \sum_i \langle \psi \mid B \mid \phi_i \rangle \left(\langle \psi \mid B^\dagger \mid \phi_i \rangle \right)^*$$

$$= \sum_i \langle \psi \mid B \mid \phi_i \rangle \langle \psi \mid B \mid \phi_i \rangle^*, \qquad \text{since } B \text{ is Hermitian,}$$

$$= \sum_i | \langle \psi \mid B \mid \phi_i \rangle |^2 \geq 0.$$

We note, for future reference, that the Hamiltonian H for the s.h.o. is the sum of two terms each of this form and therefore conclude that $\langle \psi | H | \psi \rangle \geq 0$ for all $|\psi\rangle$.

The energy spectrum of the simple harmonic oscillator

Let the normalised ket vector $|n\rangle$ (or $|E_n\rangle$) denote the nth energy state of the s.h.o. with energy E_n. Then it must be an eigenstate of the (Hermitian) Hamiltonian H and satisfy

$$H|n\rangle = E_n|n\rangle \text{ with } \langle m|n\rangle = \delta_{mn}.$$

Now consider the state $A|n\rangle$ and the effect of H upon it:

$$HA|n\rangle = AH|n\rangle - \hbar\omega A|n\rangle, \qquad \text{using (19.42),}$$

$$= AE_n|n\rangle - \hbar\omega A|n\rangle$$

$$= (E_n - \hbar\omega)A|n\rangle.$$

Thus $A|n\rangle$ is an eigenstate of H corresponding to energy $E_n - \hbar\omega$ and must be some multiple of the normalised ket vector $|E_n - \hbar\omega\rangle$, i.e.

$$A| E_n \rangle \equiv A|n\rangle = c_n|E_n - \hbar\omega\rangle,$$

where c_n is not necessarily of unit modulus. Clearly, A is an operator that generates a new state that is lower in energy by $\hbar\omega$; it can thus be compared to the operator D, which has a similar effect in the context of the z-component of angular momentum. Because it possesses the property of reducing the energy of the state by $\hbar\omega$, which, as we will see, is one quantum of excitation energy for the oscillator, the operator A is called an *annihilation operator*. Repeated application of A, m times say, will produce a state whose energy is $m\hbar\omega$ lower than that of the original:

$$A^m|E_n\rangle = c_n c_{n-1} \cdots c_{n-m+1}|E_n - m\hbar\omega\rangle. \tag{19.44}$$

In a similar way it can be shown that $A^\dagger$ parallels the operator U of our angular momentum discussion and creates an additional quantum of energy each time it is applied:

$$(A^\dagger)^m|E_n\rangle = d_n d_{n+1} \cdots d_{n+m-1}|E_n + m\hbar\omega\rangle. \tag{19.45}$$

It is therefore known as a *creation operator*.

As noted earlier, the expectation value of the oscillator's energy operator $\langle\psi|H|\psi\rangle$ must be non-negative, and therefore it must have a lowest value. Let this be E_0, with corresponding eigenstate $|0\rangle$. Since the energy-lowering property of A applies to any eigenstate of H, in order to avoid a contradiction we must have that $A|0\rangle = |\emptyset\rangle$. It then follows from (19.40) that

$$\begin{aligned}
H|0\rangle &= \tfrac{1}{2}\omega(A^\dagger A + AA^\dagger)|0\rangle \\
&= \tfrac{1}{2}\omega A^\dagger A|0\rangle + \tfrac{1}{2}\omega(A^\dagger A + \hbar)|0\rangle, \qquad \text{using (19.41),} \\
&= 0 + 0 + \tfrac{1}{2}\hbar\omega|0\rangle.
\end{aligned} \tag{19.46}$$

This shows that the commutator structure of the operators and the form of the Hamiltonian imply that the lowest energy (its ground-state energy) is $\tfrac{1}{2}\hbar\omega$; this is a result that has been derived without explicit reference to the corresponding wavefunction. This non-zero lowest value for the energy, known as the zero-point energy of the oscillator, and the discrete values for the allowed energy states are quantum-mechanical in origin; classically such an oscillator could have any non-negative energy, including zero.

Working back from this result, we see that the energy levels of the s.h.o. are $\tfrac{1}{2}\hbar\omega$, $\tfrac{3}{2}\hbar\omega$, $\tfrac{5}{2}\hbar\omega$,..., $(m + \tfrac{1}{2})\hbar\omega$,..., and that the corresponding (unnormalised) ket vectors can be written as

$$|0\rangle, \qquad A^\dagger|0\rangle, \qquad (A^\dagger)^2|0\rangle, \qquad \ldots, \qquad (A^\dagger)^m|0\rangle, \qquad \ldots.$$

This notation, and elaborations of it, are often used in the quantum treatment of classical fields such as the electromagnetic field. Thus, as the reader should verify, $A(A^\dagger)^3 A^2 A^\dagger A(A^\dagger)^4|0\rangle$ is a state with energy $\tfrac{9}{2}\hbar\omega$, whilst $A(A^\dagger)^3 A^5 A^\dagger A(A^\dagger)^4|0\rangle$ is not a physical state at all.

The normalisation of the eigenstates

In order to make quantitative calculations using the previous results we need to establish the values of the c_n and d_n that appear in equations (19.44) and (19.45). To do this, we first establish the operator recurrence relation

$$A^m(A^\dagger)^m = A^{m-1}(A^\dagger)^m A + m\hbar A^{m-1}(A^\dagger)^{m-1}. \tag{19.47}$$

The proof, which makes repeated use of $\left[A, A^{\dagger} \right] = \hbar$, is as follows:

$$
\begin{aligned}
A^{m}(A^{\dagger})^{m} &= A^{m-1}AA^{\dagger}(A^{\dagger})^{m-1} \\
&= A^{m-1}(A^{\dagger}A + \hbar)(A^{\dagger})^{m-1} \\
&= A^{m-1}A^{\dagger}A(A^{\dagger})^{m-1} + \hbar A^{m-1}(A^{\dagger})^{m-1} \\
&= A^{m-1}A^{\dagger}(A^{\dagger}A + \hbar)(A^{\dagger})^{m-2} + \hbar A^{m-1}(A^{\dagger})^{m-1} \\
&= A^{m-1}(A^{\dagger})^{2}A(A^{\dagger})^{m-2} + A^{m-1}A^{\dagger}\hbar(A^{\dagger})^{m-2} + \hbar A^{m-1}(A^{\dagger})^{m-1} \\
&= A^{m-1}(A^{\dagger})^{2}(A^{\dagger}A + \hbar)(A^{\dagger})^{m-3} + 2\hbar A^{m-1}(A^{\dagger})^{m-1} \\
&\;\;\vdots \\
&= A^{m-1}(A^{\dagger})^{m}A + m\hbar A^{m-1}(A^{\dagger})^{m-1}.
\end{aligned}
$$

Now we take the expectation values in the ground state $|0\rangle$ of both sides of this operator equation and note that the first term on the RHS is zero since it contains the term $A|0\rangle$. The non-vanishing terms are

$$
\langle 0 | A^{m}(A^{\dagger})^{m} | 0 \rangle = m\hbar \langle 0 | A^{m-1}(A^{\dagger})^{m-1} | 0 \rangle.
$$

The LHS is the square of the norm of $(A^{\dagger})^{m}|0\rangle$, and, from equation (19.45), it is equal to

$$
|d_{0}|^{2}|d_{1}|^{2} \cdots |d_{m-1}|^{2} \langle 0 | 0 \rangle.
$$

Similarly, the RHS is equal to

$$
m\hbar \, |d_{0}|^{2}|d_{1}|^{2} \cdots |d_{m-2}|^{2} \langle 0 | 0 \rangle.
$$

It follows that $|d_{m-1}|^{2} = m\hbar$ and, taking all coefficients as real, $d_{m} = \sqrt{(m+1)\hbar}$. Thus the correctly normalised state of energy $(n + \frac{1}{2})\hbar$, obtained by repeated application of $A^{\dagger}$ to the ground state, is given by

$$
|n\rangle = \frac{(A^{\dagger})^{n}}{(n! \, \hbar^{n})^{1/2}} \, |0\rangle. \tag{19.48}
$$

To evaluate the c_{n}, we note that, from the commutator of A and $A^{\dagger}$,

$$
\begin{aligned}
\left[A, A^{\dagger} \right] |n\rangle &= AA^{\dagger}|n\rangle - A^{\dagger}A|n\rangle \\
\hbar |n\rangle &= \sqrt{(n+1)\hbar}\, A \, |n+1\rangle - c_{n} A^{\dagger} \, |n-1\rangle \\
&= \sqrt{(n+1)\hbar}\, c_{n+1} \, |n\rangle - c_{n}\sqrt{n\hbar} \, |n\rangle, \\
\hbar &= \sqrt{(n+1)\hbar}\, c_{n+1} - c_{n}\sqrt{n\hbar},
\end{aligned}
$$

which has the obvious solution $c_{n} = \sqrt{n\hbar}$. To summarise:

$$
c_{n} = \sqrt{n\hbar} \quad \text{and} \quad d_{n} = \sqrt{(n+1)\hbar}. \tag{19.49}
$$

We end this chapter with another worked example. This one illustrates how the operator formalism that we have developed can be used to obtain results that would involve a number of non-trivial integrals if tackled using explicit wavefunctions.

▶Given that the first-order change in the ground-state energy of a quantum system when it is perturbed by a small additional term H' in the Hamiltonian is $\langle 0|H'|0\rangle$, find the first-order change in the energy of a simple harmonic oscillator in the presence of an additional potential $V'(x) = \lambda x^3 + \mu x^4$.

From the definitions of A and $A^\dagger$, equation (19.39), we can write

$$x = \frac{1}{\sqrt{2m\omega}}(A + A^\dagger) \quad \Rightarrow \quad H' = \frac{\lambda}{(2m\omega)^{3/2}}(A + A^\dagger)^3 + \frac{\mu}{(2m\omega)^2}(A + A^\dagger)^4.$$

We now compute successive values of $(A + A^\dagger)^n |0\rangle$ for $n = 1, 2, 3, 4$, remembering that

$$A|n\rangle = \sqrt{n\hbar}\,|n-1\rangle \quad \text{and} \quad A^\dagger|n\rangle = \sqrt{(n+1)\hbar}\,|n+1\rangle :$$

$$(A + A^\dagger)|0\rangle = 0 + \hbar^{1/2}|1\rangle,$$
$$(A + A^\dagger)^2|0\rangle = \hbar|0\rangle + \sqrt{2}\,\hbar|2\rangle,$$
$$(A + A^\dagger)^3|0\rangle = 0 + \hbar^{3/2}|1\rangle + 2\hbar^{3/2}|1\rangle + \sqrt{6}\,\hbar^{3/2}|3\rangle$$
$$= 3\hbar^{3/2}|1\rangle + \sqrt{6}\,\hbar^{3/2}|3\rangle,$$
$$(A + A^\dagger)^4|0\rangle = 3\hbar^2|0\rangle + \sqrt{18}\,\hbar^2|2\rangle + \sqrt{18}\,\hbar^2|2\rangle + \sqrt{24}\,\hbar^2|4\rangle.$$

To find the energy shift we need to form the inner product of each of these state vectors with $|0\rangle$. But $|0\rangle$ is orthogonal to all $|n\rangle$ if $n \neq 0$. Consequently, the term $\langle 0|(A + A^\dagger)^3|0\rangle$ in the expectation value is zero, and in the expression for $\langle 0|(A + A^\dagger)^4|0\rangle$ only the first term is non-zero; its value is $3\hbar^2$. The perturbation energy is thus given by

$$\langle 0|H'|0\rangle = \frac{3\mu\hbar^2}{(2m\omega)^2}.$$

It could have been anticipated on symmetry grounds that the expectation of λx^3, an odd function of x, would be zero, but the calculation gives this result automatically. The contribution of the quadratic term in the perturbation would have been much harder to anticipate! ◀

19.3 Exercises

19.1 Show that the commutator of two operators that correspond to two physical observables cannot itself correspond to another physical observable.

19.2 By expressing the operator L_z, corresponding to the z-component of angular momentum, in spherical polar coordinates (r, θ, ϕ), show that the angular momentum of a particle about the polar axis cannot be known at the same time as its azimuthal position around that axis.

19.3 In quantum mechanics, the time dependence of the state function $|\psi\rangle$ of a system is given, as a further postulate, by the equation

$$i\hbar \frac{\partial}{\partial t}|\psi\rangle = H|\psi\rangle,$$

where H is the Hamiltonian of the system. Use this to find the time dependence of the expectation value $\langle A \rangle$ of an operator A that itself has no explicit time dependence. Hence show that operators that commute with the Hamiltonian correspond to the classical 'constants of the motion'.

For a particle of mass m moving in a one-dimensional potential $V(x)$, prove Ehrenfest's theorem:

$$\frac{d\langle p_x \rangle}{dt} = -\left\langle \frac{dV}{dx} \right\rangle \quad \text{and} \quad \frac{d\langle x \rangle}{dt} = \frac{\langle p_x \rangle}{m}.$$

19.4   Show that the Pauli matrices

$$S_x = \tfrac{1}{2}\hbar \begin{pmatrix} 0 & 1 \\ 1 & 0 \end{pmatrix}, \quad S_y = \tfrac{1}{2}\hbar \begin{pmatrix} 0 & -i \\ i & 0 \end{pmatrix}, \quad S_z = \tfrac{1}{2}\hbar \begin{pmatrix} 1 & 0 \\ 0 & -1 \end{pmatrix},$$

which are used as the operators corresponding to intrinsic spin of $\tfrac{1}{2}\hbar$ in non-relativistic quantum mechanics, satisfy $S_x^2 = S_y^2 = S_z^2 = \tfrac{1}{4}\hbar^2 I$, and have the same commutation properties as the components of orbital angular momentum. Deduce that any state $|\psi\rangle$ represented by the column vector $(a, \; b)^{\mathrm{T}}$ is an eigenstate of S^2 with eigenvalue $3\hbar^2/4$.

19.5   Find closed-form expressions for $\cos C$ and $\sin C$, where C is the matrix

$$C = \begin{pmatrix} 1 & 1 \\ 1 & -1 \end{pmatrix}.$$

Demonstrate that the 'expected' relationships

$$\cos^2 C + \sin^2 C = I \quad \text{and} \quad \sin 2C = 2 \sin C \cos C$$

are valid.

19.6   Operators A and B anticommute. Evaluate $(A + B)^{2n}$ for a few values of n and hence propose an expression for c_{nr} in the expansion

$$(A + B)^{2n} = \sum_{r=0}^{n} c_{nr} A^{2n-2r} B^{2r}.$$

Prove your proposed formula for general values of n, using the method of induction.

Show that

$$\cos(A + B) = \sum_{n=0}^{\infty} \sum_{r=0}^{n} d_{nr} A^{2n-2r} B^{2r},$$

where the d_{nr} are constants whose values you should determine.

By taking as A the matrix $A = \begin{pmatrix} 0 & 1 \\ 1 & 0 \end{pmatrix}$, confirm that your answer is consistent with that obtained in exercise 19.5.

19.7   Expressed in terms of the annihilation and creation operators A and $A^\dagger$ discussed in the text, a system has an unperturbed Hamiltonian $H_0 = \hbar \omega A^\dagger A$. The system is disturbed by the addition of a perturbing Hamiltonian $H_1 = g\hbar\omega(A + A^\dagger)$, where g is real. Show that the effect of the perturbation is to move the whole energy spectrum of the system down by $g^2 \hbar \omega$.

19.8   For a system of N electrons in their ground state $|0\rangle$, the Hamiltonian is

$$H = \sum_{n=1}^{N} \frac{p_{x_n}^2 + p_{y_n}^2 + p_{z_n}^2}{2m} + \sum_{n=1}^{N} V(x_n, y_n, z_n).$$

Show that $\left[p_{x_n}^2, x_n \right] = -2i\hbar p_{x_n}$, and hence that the expectation value of the double commutator $[[x, H], x]$, where $x = \sum_{n=1}^{N} x_n$, is given by

$$\langle 0 | \, [[x, H], x] \, | 0 \rangle = \frac{N\hbar^2}{m}.$$

Now evaluate the expectation value using the eigenvalue properties of H, namely $H|r\rangle = E_r|r\rangle$, and deduce the *sum rule for oscillation strengths*,

$$\sum_{r=0}^{\infty}(E_r - E_0)|\langle r\,|\,x\,|\,0\rangle\,|^2 = \frac{N\hbar^2}{2m}.$$

19.9   By considering the function

$$F(\lambda) = \exp(\lambda A)B\exp(-\lambda A),$$

where A and B are linear operators and λ is a parameter, and finding its derivatives with respect to λ, prove that

$$e^A B e^{-A} = B + [A, B] + \frac{1}{2!}[A, [A, B]] + \frac{1}{3!}[A, [A, [A, B]]] + \cdots.$$

Use this result to express

$$\exp\left(\frac{iL_x\theta}{\hbar}\right) L_y \exp\left(\frac{-iL_x\theta}{\hbar}\right)$$

as a linear combination of the angular momentum operators L_x, L_y and L_z.

19.10   For a system containing more than one particle, the total angular momentum J and its components are represented by operators that have completely analogous commutation relations to those for the operators for a single particle, i.e. J^2 has eigenvalue $j(j + 1)\hbar^2$ and J_z has eigenvalue $m_j\hbar$ for the state $|j, m_j\rangle$. The usual orthonormality relationship $\langle j', m'_j\,|\,j, m_j\rangle = \delta_{j'j}\,\delta_{m'_j m_j}$ is also valid.

A system consists of two (distinguishable) particles A and B. Particle A is in an $\ell = 3$ state and can have state functions of the form $|A, 3, m_A\rangle$, whilst B is in an $\ell = 2$ state with possible state functions $|B, 2, m_B\rangle$. The range of possible values for j is $|3 - 2| \le j \le |3 + 2|$, i.e. $1 \le j \le 5$, and the overall state function can be written as

$$|j, m_j\rangle = \sum_{m_A + m_B = m_j} C_{m_A\,m_B}^{j\,m_j}|A, 3, m_A\rangle\,|B, 2, m_B\rangle.$$

The numerical coefficients $C_{m_A\,m_B}^{J\,m_j}$ are known as *Clebsch–Gordon* coefficients.

Assume (as can be shown) that the ladder operators $U(AB)$ and $D(AB)$ for the system can be written as $U(A) + U(B)$ and $D(A) + D(B)$, respectively, and that they lead to relationships equivalent to (19.34) and (19.35) with ℓ replaced by j and m by m_j.

(a) Apply the operators to the (obvious) relationship

$$|AB, 5, 5\rangle = |A, 3, 3\rangle\,|B, 2, 2\rangle$$

to show that

$$|AB, 5, 4\rangle = \sqrt{\tfrac{6}{10}}\,|A, 3, 2\rangle\,|B, 2, 2\rangle + \sqrt{\tfrac{4}{10}}\,|A, 3, 3\rangle\,|B, 2, 1\rangle.$$

(b) Find, to within an overall sign, the real coefficients c and d in the expansion

$$|AB, 4, 4\rangle = c|A, 3, 2\rangle\,|B, 2, 2\rangle + d|A, 3, 3\rangle\,|B, 2, 1\rangle$$

by requiring it to be orthogonal to $|AB, 5, 4\rangle$. Check your answer by considering $U(AB)|AB, 4, 4\rangle$.

(c) Find, to within an overall sign, and as efficiently as possible, an expression for $|AB, 4, -3\rangle$ as a sum of products of the form $|A, 3, m_A\rangle\,|B, 2, m_B\rangle$.

19.4 Hints and answers

19.1   Show that the commutator is anti-Hermitian.

19.3   Use the Hermitian conjugate of the given equation to obtain the time dependence of $\langle\psi|$. The rate of change of $\langle\psi|A|\psi\rangle$ is $i\langle\psi|[H,A]|\psi\rangle$. Note that $[H,p_x] = [V,p_x]$ and $[H,x] = [p_x^2,x]/2m$.

19.5   Show that $C^2 = 2I$.

$$\cos C = \cos\sqrt{2}\begin{pmatrix} 1 & 0 \\ 0 & 1 \end{pmatrix}, \quad \sin C = \frac{\sin\sqrt{2}}{\sqrt{2}}\begin{pmatrix} 1 & 1 \\ 1 & -1 \end{pmatrix}.$$

19.7   Express the total Hamiltonian in terms of $B = A + gI$ and determine the value of $[B, B^{\dagger}]$.

19.9   Show that, if $F^{(n)}$ is the nth derivative of $F(\lambda)$, then $F^{(n+1)} = [A, F^{(n)}]$. Use a Taylor series in λ to evaluate $F(1)$, using derivatives evaluated at $\lambda = 0$. Successively reduce the level of nesting of each multiple commutator by using the result of evaluating the previous term. The given expression reduces to $\cos\theta\, L_y - \sin\theta\, L_z$.

Partial differential equations: general and particular solutions

In this chapter and the next the solution of differential equations of types typically encountered in the physical sciences and engineering is extended to situations involving more than one independent variable. A partial differential equation (PDE) is an equation relating an unknown function (the dependent variable) of two or more variables to its partial derivatives with respect to those variables. The most commonly occurring independent variables are those describing position and time, and so we will couch our discussion and examples in notation appropriate to them.

As in other chapters we will focus our attention on the equations that arise most often in physical situations. We will restrict our discussion, therefore, to linear PDEs, i.e. those of first degree in the dependent variable. Furthermore, we will discuss primarily second-order equations. The solution of first-order PDEs will necessarily be involved in treating these, and some of the methods discussed can be extended without difficulty to third- and higher-order equations. We shall also see that many ideas developed for ordinary differential equations (ODEs) can be carried over directly into the study of PDEs.

In this chapter we will concentrate on general solutions of PDEs in terms of arbitrary functions and the particular solutions that may be derived from them in the presence of boundary conditions. We also discuss the existence and uniqueness of the solutions to PDEs under given boundary conditions.

In the next chapter the methods most commonly used in practice for obtaining solutions to PDEs subject to given boundary conditions will be considered. These methods include the separation of variables, integral transforms and Green's functions. This division of material is rather arbitrary and has been made only to emphasise the general usefulness of the latter methods. In particular, it will be readily apparent that some of the results of the present chapter are in fact solutions in the form of separated variables, but arrived at by a different approach.

20.1 Important partial differential equations

Most of the important PDEs of physics are second-order and linear. In order to gain familiarity with their general form, some of the more important ones will now be briefly discussed. These equations apply to a wide variety of different physical systems.

Since, in general, the PDEs listed below describe three-dimensional situations, the independent variables are $\mathbf{r}$ and t, where $\mathbf{r}$ is the position vector and t is time. The actual variables used to specify the position vector $\mathbf{r}$ are dictated by the coordinate system in use. For example, in Cartesian coordinates the independent variables of position are x, y and z, whereas in spherical polar coordinates they are r, θ and ϕ. The equations may be written in a coordinate-independent manner, however, by the use of the Laplacian operator ∇^2.

20.1.1 The wave equation

The wave equation

$$\nabla^2 u = \frac{1}{c^2}\frac{\partial^2 u}{\partial t^2} \tag{20.1}$$

describes as a function of position and time the displacement from equilibrium, $u(\mathbf{r}, t)$, of a vibrating string or membrane or a vibrating solid, gas or liquid. The equation also occurs in electromagnetism, where u may be a component of the electric or magnetic field in an electromagnetic wave or the current or voltage along a transmission line. The quantity c is the speed of propagation of the waves.

> ▶ *Find the equation satisfied by small transverse displacements $u(x, t)$ of a uniform string of mass per unit length ρ held under a uniform tension T, assuming that the string is initially located along the x-axis in a Cartesian coordinate system.*

Figure 20.1 shows the forces acting on an elemental length Δs of the string. If the tension T in the string is uniform along its length then the net upward vertical force on the element is

$$\Delta F = T \sin\theta_2 - T \sin\theta_1.$$

Assuming that the angles θ_1 and θ_2 are both small, we may make the approximation $\sin\theta \approx \tan\theta$. Since at any point on the string the slope $\tan\theta = \partial u/\partial x$, the force can be written

$$\Delta F = T\left[\frac{\partial u(x + \Delta x, t)}{\partial x} - \frac{\partial u(x, t)}{\partial x}\right] \approx T\frac{\partial^2 u(x, t)}{\partial x^2}\,\Delta x,$$

where we have used the definition of the partial derivative to simplify the RHS.

This upward force may be equated, by Newton's second law, to the product of the mass of the element and its upward acceleration. The element has a mass $\rho\,\Delta s$, which is approximately equal to $\rho\,\Delta x$ if the vibrations of the string are small, and so we have

$$\rho\,\Delta x\,\frac{\partial^2 u(x, t)}{\partial t^2} = T\frac{\partial^2 u(x, t)}{\partial x^2}\,\Delta x.$$

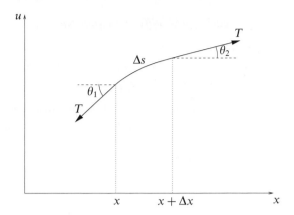

Figure 20.1   The forces acting on an element of a string under uniform tension T.

Dividing both sides by Δx we obtain, for the vibrations of the string, the one-dimensional wave equation

$$\frac{\partial^2 u}{\partial x^2} = \frac{1}{c^2}\frac{\partial^2 u}{\partial t^2},$$

where $c^2 = T/\rho$. ◄

The longitudinal vibrations of an elastic rod obey a very similar equation to that derived in the above example, namely

$$\frac{\partial^2 u}{\partial x^2} = \frac{\rho}{E}\frac{\partial^2 u}{\partial t^2};$$

here ρ is the mass per unit volume and E is Young's modulus.

The wave equation can be generalised slightly. For example, in the case of the vibrating string, there could also be an external upward vertical force $f(x,t)$ per unit length acting on the string at time t. The transverse vibrations would then satisfy the equation

$$T\frac{\partial^2 u}{\partial x^2} + f(x,t) = \rho\frac{\partial^2 u}{\partial t^2},$$

which is clearly of the form 'upward force per unit length = mass per unit length × upward acceleration'.

Similar examples, but involving two or three spatial dimensions rather than one, are provided by the equation governing the transverse vibrations of a stretched membrane subject to an external vertical force density $f(x,y,t)$,

$$T\left(\frac{\partial^2 u}{\partial x^2} + \frac{\partial^2 u}{\partial y^2}\right) + f(x,y,t) = \rho(x,y)\frac{\partial^2 u}{\partial t^2},$$

where ρ is the mass per unit area of the membrane and T is the tension.

20.1.2 The diffusion equation

The diffusion equation

$$\kappa \nabla^2 u = \frac{\partial u}{\partial t} \tag{20.2}$$

describes the temperature u in a region containing no heat sources or sinks; it also applies to the diffusion of a chemical that has a concentration $u(\mathbf{r}, t)$. The constant κ is called the diffusivity. The equation is clearly second order in the three spatial variables, but first order in time.

▶*Derive the equation satisfied by the temperature $u(\mathbf{r}, t)$ at time t for a material of uniform thermal conductivity k, specific heat capacity s and density ρ. Express the equation in Cartesian coordinates.*

Let us consider an arbitrary volume V lying within the solid and bounded by a surface S (this may coincide with the surface of the solid if so desired). At any point in the solid the rate of heat flow per unit area in any given direction $\hat{\mathbf{r}}$ is proportional to minus the component of the temperature gradient in that direction and so is given by $(-k\nabla u) \cdot \hat{\mathbf{r}}$. The total flux of heat *out* of the volume V per unit time is given by

$$-\frac{dQ}{dt} = \int_S (-k\nabla u) \cdot \hat{\mathbf{n}} \, dS$$
$$= \int_V \nabla \cdot (-k\nabla u) \, dV, \tag{20.3}$$

where Q is the total heat energy in V at time t and $\hat{\mathbf{n}}$ is the outward-pointing unit normal to S; note that we have used the divergence theorem to convert the surface integral into a volume integral.

We can also express Q as a volume integral over V,

$$Q = \int_V s\rho u \, dV,$$

and its rate of change is then given by

$$\frac{dQ}{dt} = \int_V s\rho \frac{\partial u}{\partial t} \, dV, \tag{20.4}$$

where we have taken the derivative with respect to time inside the integral (see section 5.12).

Comparing (20.3) and (20.4), and remembering that the volume V is arbitrary, we obtain the three-dimensional diffusion equation

$$\kappa \nabla^2 u = \frac{\partial u}{\partial t},$$

where the diffusion coefficient $\kappa = k/(s\rho)$. To express this equation in Cartesian coordinates, we simply write ∇^2 in terms of x, y and z to obtain

$$\kappa \left(\frac{\partial^2 u}{\partial x^2} + \frac{\partial^2 u}{\partial y^2} + \frac{\partial^2 u}{\partial z^2} \right) = \frac{\partial u}{\partial t}. \quad ◀$$

The diffusion equation just derived can be generalised to

$$k\nabla^2 u + f(\mathbf{r}, t) = s\rho \frac{\partial u}{\partial t}.$$

The second term, $f(\mathbf{r}, t)$, represents a varying density of heat sources throughout the material but is often not required in physical applications. In the most general case, k, s and ρ may depend on position $\mathbf{r}$, in which case the first term becomes $\nabla \cdot (k\nabla u)$. However, in the simplest application the heat flow is one-dimensional with no heat sources, and the equation becomes (in Cartesian coordinates)

$$\frac{\partial^2 u}{\partial x^2} = \frac{s\rho}{k}\frac{\partial u}{\partial t}.$$

20.1.3 Laplace's equation

Laplace's equation,

$$\nabla^2 u = 0, \tag{20.5}$$

may be obtained by setting $\partial u/\partial t = 0$ in the diffusion equation (20.2), and describes (for example) the *steady-state* temperature distribution in a solid in which there are no heat sources – i.e. the temperature distribution after a long time has elapsed.

Laplace's equation also describes the gravitational potential in a region containing no matter or the electrostatic potential in a charge-free region. Further, it applies to the flow of an incompressible fluid with no sources, sinks or vortices; in this case u is the velocity potential, from which the velocity is given by $v = \nabla u$.

20.1.4 Poisson's equation

Poisson's equation,

$$\nabla^2 u = \rho(\mathbf{r}), \tag{20.6}$$

describes the same physical situations as Laplace's equation, but in regions containing matter, charges or sources of heat or fluid. The function $\rho(\mathbf{r})$ is called the source density and in physical applications usually contains some multiplicative physical constants. For example, if u is the electrostatic potential in some region of space, in which case ρ is the density of electric charge, then $\nabla^2 u = -\rho(\mathbf{r})/\epsilon_0$, where ϵ_0 is the permittivity of free space. Alternatively, u might represent the gravitational potential in some region where the matter density is given by ρ; then $\nabla^2 u = 4\pi G\rho(\mathbf{r})$, where G is the gravitational constant.

20.1.5 Schrödinger's equation

The Schrödinger equation

$$-\frac{\hbar^2}{2m}\nabla^2 u + V(\mathbf{r})u = i\hbar\frac{\partial u}{\partial t}, \tag{20.7}$$

describes the quantum mechanical wavefunction $u(\mathbf{r}, t)$ of a non-relativistic particle of mass m; $\hbar$ is Planck's constant divided by 2π. Like the diffusion equation it is second order in the three spatial variables and first order in time.

20.2 General form of solution

Before turning to the methods by which we may hope to solve PDEs such as those listed in the previous section, it is instructive, as for ODEs in chapter 14, to study how PDEs may be formed from a set of possible solutions. Such a study can provide an indication of how equations obtained not from possible solutions but from physical arguments might be solved.

For definiteness let us suppose we have a set of functions involving two independent variables x and y. Without further specification this is of course a very wide set of functions, and we could not expect to find a useful equation that they all satisfy. However, let us consider a type of function $u_i(x, y)$ in which x and y appear in a particular way, such that u_i can be written as a function (however complicated) *of a single variable p*, itself a simple function of x and y.

Let us illustrate this by considering the three functions

$$u_1(x, y) = x^4 + 4(x^2y + y^2 + 1),$$
$$u_2(x, y) = \sin x^2 \cos 2y + \cos x^2 \sin 2y,$$
$$u_3(x, y) = \frac{x^2 + 2y + 2}{3x^2 + 6y + 5}.$$

These are all fairly complicated functions of x and y and a single differential equation of which each one is a solution is not obvious. However, if we observe that in fact each can be expressed as a function of the variable $p = x^2 + 2y$ alone (with no other x or y involved) then a great simplification takes place. Written in terms of p the above equations become

$$u_1(x, y) = (x^2 + 2y)^2 + 4 = p^2 + 4 = f_1(p),$$
$$u_2(x, y) = \sin(x^2 + 2y) = \sin p = f_2(p),$$
$$u_3(x, y) = \frac{(x^2 + 2y) + 2}{3(x^2 + 2y) + 5} = \frac{p + 2}{3p + 5} = f_3(p).$$

Let us now form, for each u_i, the partial derivatives $\partial u_i/\partial x$ and $\partial u_i/\partial y$. In each case these are (writing both the form for general p and the one appropriate to our particular case, $p = x^2 + 2y$)

$$\frac{\partial u_i}{\partial x} = \frac{df_i(p)}{dp}\frac{\partial p}{\partial x} = 2xf_i',$$
$$\frac{\partial u_i}{\partial y} = \frac{df_i(p)}{dp}\frac{\partial p}{\partial y} = 2f_i',$$

for $i = 1, 2, 3$. All reference to the form of f_i can be eliminated from these

680

equations by cross-multiplication, obtaining

$$\frac{\partial p}{\partial y}\frac{\partial u_i}{\partial x} = \frac{\partial p}{\partial x}\frac{\partial u_i}{\partial y},$$

or, for our specific form, $p = x^2 + 2y$,

$$\frac{\partial u_i}{\partial x} = x\frac{\partial u_i}{\partial y}. \tag{20.8}$$

It is thus apparent that not only are the three functions u_1, u_2 u_3 solutions of the PDE (20.8) but so also is *any arbitrary function $f(p)$* of which the argument p has the form $x^2 + 2y$.

20.3 General and particular solutions

In the last section we found that the first-order PDE (20.8) has as a solution *any* function of the variable $x^2 + 2y$. This points the way for the solution of PDEs of other orders, as follows. It is *not* generally true that an nth-order PDE can always be considered as resulting from the elimination of n arbitrary *functions* from its solution (as opposed to the elimination of n arbitrary *constants* for an nth-order ODE, see section 14.1). However, given specific PDEs we can try to solve them by seeking combinations of variables in terms of which the solutions may be expressed as arbitrary functions. Where this is possible we may expect n combinations to be involved in the solution.

Naturally, the exact functional form of the solution for any particular situation must be determined by some set of boundary conditions. For instance, if the PDE contains two independent variables x and y then for complete determination of its solution the boundary conditions will take a form equivalent to specifying $u(x, y)$ along a suitable continuum of points in the xy-plane (usually along a line).

We now discuss the general and particular solutions of first- and second-order PDEs. In order to simplify the algebra, we will restrict our discussion to equations containing just two independent variables x and y. Nevertheless, the method presented below may be extended to equations containing several independent variables.

20.3.1 First-order equations

Although most of the PDEs encountered in physical contexts are second order (i.e. they contain $\partial^2 u/\partial x^2$ or $\partial^2 u/\partial x\partial y$, etc.), we now discuss first-order equations to illustrate the general considerations involved in the form of the solution and in satisfying any boundary conditions on the solution.

The most general first-order linear PDE (containing two independent variables) is of the form

$$A(x, y)\frac{\partial u}{\partial x} + B(x, y)\frac{\partial u}{\partial y} + C(x, y)u = R(x, y), \tag{20.9}$$

where $A(x, y)$, $B(x, y)$, $C(x, y)$ and $R(x, y)$ are given functions. Clearly, if either $A(x, y)$ or $B(x, y)$ is zero then the PDE may be solved straightforwardly as a first-order linear ODE (as discussed in chapter 14), the only modification being that the arbitrary constant of integration becomes an *arbitrary function* of x or y respectively.

> ▶ *Find the general solution $u(x, y)$ of*
> $$x\frac{\partial u}{\partial x} + 3u = x^2.$$

Dividing through by x we obtain

$$\frac{\partial u}{\partial x} + \frac{3u}{x} = x,$$

which is a linear equation with integrating factor (see subsection 14.2.4)

$$\exp\left(\int \frac{3}{x}\, dx\right) = \exp(3\ln x) = x^3.$$

Multiplying through by this factor we find

$$\frac{\partial}{\partial x}(x^3 u) = x^4,$$

which, on integrating with respect to x, gives

$$x^3 u = \frac{x^5}{5} + f(y),$$

where $f(y)$ is an *arbitrary function* of y. Finally, dividing through by x^3, we obtain the solution

$$u(x, y) = \frac{x^2}{5} + \frac{f(y)}{x^3}. \quad ◀$$

When the PDE contains partial derivatives with respect to both independent variables then, of course, we cannot employ the above procedure but must seek an alternative method. Let us for the moment restrict our attention to the special case in which $C(x, y) = R(x, y) = 0$ and, following the discussion of the previous section, look for solutions of the form $u(x, y) = f(p)$ where p is some, at present unknown, combination of x and y. We then have

$$\frac{\partial u}{\partial x} = \frac{df(p)}{dp}\frac{\partial p}{\partial x},$$
$$\frac{\partial u}{\partial y} = \frac{df(p)}{dp}\frac{\partial p}{\partial y},$$

which, when substituted into the PDE (20.9), give

$$\left[A(x,y)\frac{\partial p}{\partial x} + B(x,y)\frac{\partial p}{\partial y} \right] \frac{df(p)}{dp} = 0.$$

This removes all reference to the actual form of the function $f(p)$ if, for non-trivial p, we have

$$A(x,y)\frac{\partial p}{\partial x} + B(x,y)\frac{\partial p}{\partial y} = 0. \tag{20.10}$$

Let us now consider the necessary condition for $f(p)$ to remain constant as x and y vary; this is that p itself remains constant. Thus for f to remain constant implies that x and y must vary in such a way that

$$dp = \frac{\partial p}{\partial x}\,dx + \frac{\partial p}{\partial y}\,dy = 0. \tag{20.11}$$

The forms of (20.10) and (20.11) are very alike and become the same if we require that

$$\frac{dx}{A(x,y)} = \frac{dy}{B(x,y)}. \tag{20.12}$$

By integrating this expression the form of p can be found.

▶*For*

$$x\frac{\partial u}{\partial x} - 2y\frac{\partial u}{\partial y} = 0, \tag{20.13}$$

find (i) *the solution that takes the value* $2y+1$ *on the line* $x = 1$, *and* (ii) *a solution that has the value 4 at the point* $(1,1)$.

If we seek a solution of the form $u(x,y) = f(p)$, we deduce from (20.12) that $u(x,y)$ will be constant along lines of (x,y) that satisfy

$$\frac{dx}{x} = \frac{dy}{-2y},$$

which on integrating gives $x = cy^{-1/2}$. Identifying the constant of integration c with $p^{1/2}$ (to avoid fractional powers), we conclude that $p = x^2y$. Thus the general solution of the PDE (20.13) is

$$u(x,y) = f(x^2y),$$

where f is an arbitrary function.

We must now find the particular solutions that obey each of the imposed boundary conditions. For boundary condition (i) a little thought shows that the particular solution required is

$$u(x,y) = 2(x^2y) + 1 = 2x^2y + 1. \tag{20.14}$$

For boundary condition (ii) some obviously acceptable solutions are

$$u(x,y) = x^2y + 3,$$
$$u(x,y) = 4x^2y,$$
$$u(x,y) = 4.$$

Each is a valid solution (the freedom of choice of form arises from the fact that u is specified at only one point $(1, 1)$, and not along a continuum (say), as in boundary condition (i)). All three are particular examples of the general solution, which may be written, for example, as

$$u(x, y) = x^2 y + 3 + g(x^2 y),$$

where $g = g(x^2 y) = g(p)$ is an arbitrary function subject only to $g(1) = 0$. For this example, the forms of g corresponding to the particular solutions listed above are $g(p) = 0$, $g(p) = 3p - 3$, $g(p) = 1 - p$. ◄

As mentioned above, in order to find a solution of the form $u(x, y) = f(p)$ we require that the original PDE contains no term in u, but only terms containing its partial derivatives. If a term in u is present, so that $C(x, y) \neq 0$ in (20.9), then the procedure needs some modification, since we cannot simply divide out the dependence on $f(p)$ to obtain (20.10). In such cases we look instead for a solution of the form $u(x, y) = h(x, y)f(p)$. We illustrate this method in the following example.

►*Find the general solution of*

$$x \frac{\partial u}{\partial x} + 2 \frac{\partial u}{\partial y} - 2u = 0. \tag{20.15}$$

We seek a solution of the form $u(x, y) = h(x, y)f(p)$, with the consequence that

$$\frac{\partial u}{\partial x} = \frac{\partial h}{\partial x} f(p) + h \frac{df(p)}{dp} \frac{\partial p}{\partial x},$$

$$\frac{\partial u}{\partial y} = \frac{\partial h}{\partial y} f(p) + h \frac{df(p)}{dp} \frac{\partial p}{\partial y}.$$

Substituting these expressions into the PDE (20.15) and rearranging, we obtain

$$\left(x \frac{\partial h}{\partial x} + 2 \frac{\partial h}{\partial y} - 2h \right) f(p) + \left(x \frac{\partial p}{\partial x} + 2 \frac{\partial p}{\partial y} \right) h \frac{df(p)}{dp} = 0.$$

The first factor in parentheses is just the original PDE with u replaced by h. Therefore, if h is *any* solution of the PDE, *however simple*, this term will vanish, to leave

$$\left(x \frac{\partial p}{\partial x} + 2 \frac{\partial p}{\partial y} \right) h \frac{df(p)}{dp} = 0,$$

from which, as in the previous case, we obtain

$$x \frac{\partial p}{\partial x} + 2 \frac{\partial p}{\partial y} = 0.$$

From (20.11) and (20.12) we see that $u(x, y)$ will be constant along lines of (x, y) that satisfy

$$\frac{dx}{x} = \frac{dy}{2},$$

which integrates to give $x = c \exp(y/2)$. Identifying the constant of integration c with p we find $p = x \exp(-y/2)$. Thus the general solution of (20.15) is

$$u(x, y) = h(x, y)f(x \exp(-\tfrac{1}{2}y)),$$

where $f(p)$ is any arbitrary function of p and $h(x, y)$ is any solution of (20.15).

If we take, for example, $h(x, y) = \exp y$, which clearly satisfies (20.15), then the general solution is

$$u(x, y) = (\exp y)f(x \exp(-\tfrac{1}{2}y)).$$

Alternatively, $h(x, y) = x^2$ also satisfies (20.15) and so the general solution to the equation can also be written

$$u(x, y) = x^2 g(x \exp(-\tfrac{1}{2}y)),$$

where g is an arbitrary function of p; clearly $g(p) = f(p)/p^2$. ◀

20.3.2 Inhomogeneous equations and problems

Let us discuss in a more general form the particular solutions of (20.13) found in the second example of the previous subsection. It is clear that, so far as this equation is concerned, if $u(x, y)$ is a solution then so is any multiple of $u(x, y)$ or any linear sum of separate solutions $u_1(x, y) + u_2(x, y)$. However, when it comes to fitting the boundary conditions this is not so.

For example, although $u(x, y)$ in (20.14) satisfies the PDE and the boundary condition $u(1, y) = 2y + 1$, the function $u_1(x, y) = 4u(x, y) = 8xy + 4$, whilst satisfying the PDE, takes the value $8y + 4$ on the line $x = 1$ and so does not satisfy the required boundary condition. Likewise the function $u_2(x, y) = u(x, y) + f_1(x^2 y)$, for arbitrary f_1, satisfies (20.13) but takes the value $u_2(1, y) = 2y + 1 + f_1(y)$ on the line $x = 1$, and so is not of the required form unless f_1 is identically zero.

Thus we see that when treating the superposition of solutions of PDEs two considerations arise, one concerning the equation itself and the other connected to the boundary conditions. The *equation* is said to be homogeneous if the fact that $u(x, y)$ is a solution implies that $\lambda u(x, y)$, for any constant λ, is also a solution. However, the *problem* is said to be homogeneous if, in addition, the boundary conditions are such that if they are satisfied by $u(x, y)$ then they are also satisfied by $\lambda u(x, y)$. The last requirement itself is referred to as that of *homogeneous boundary conditions*.

For example, the PDE (20.13) is homogeneous but the general first-order equation (20.9) would not be homogeneous unless $R(x, y) = 0$. Furthermore, the boundary condition (i) imposed on the solution of (20.13) in the previous subsection is not homogeneous though, in this case, the boundary condition

$$u(x, y) = 0 \quad \text{on the line } y = 4x^{-2}$$

would be, since $u(x, y) = \lambda(x^2 y - 4)$ satisfies this condition for any λ and, being a function of $x^2 y$, satisfies (20.13).

The reason for discussing the homogeneity of PDEs and their boundary conditions is that in linear PDEs there is a close parallel to the complementary-function and particular-integral property of ODEs. The general solution of an inhomogeneous problem can be written as the sum of *any* particular solution of the problem and the general solution of the corresponding homogeneous problem (as for ODEs, we require that the particular solution is not already contained in the general solution of the homogeneous problem). Thus, for example, the general solution of

$$\frac{\partial u}{\partial x} - x\frac{\partial u}{\partial y} + au = f(x, y),$$ (20.16)

subject to, say, the boundary condition $u(0, y) = g(y)$, is given by

$$u(x, y) = v(x, y) + w(x, y),$$

where $v(x, y)$ is any solution (however simple) of (20.16) such that $v(0, y) = g(y)$ and $w(x, y)$ is the general solution of

$$\frac{\partial w}{\partial x} - x\frac{\partial w}{\partial y} + aw = 0,$$ (20.17)

with $w(0, y) = 0$. If the boundary conditions are sufficiently specified then the only possible solution of (20.17) will be $w(x, y) \equiv 0$ and $v(x, y)$ will be the complete solution by itself.

Alternatively, we may begin by finding the general solution of the inhomogeneous equation (20.16) *without* regard for any boundary conditions; it is just the sum of the general solution to the homogeneous equation and a particular integral of (20.16), both without reference to the boundary conditions. The boundary conditions can then be used to find the appropriate particular solution from the general solution.

We will not discuss at length general methods of obtaining particular integrals of PDEs but merely note that some of those methods available for ordinary differential equations can be suitably extended.[§]

▶*Find the general solution of*

$$y\frac{\partial u}{\partial x} - x\frac{\partial u}{\partial y} = 3x.$$ (20.18)

Hence find the most general particular solution (i) which satisfies $u(x, 0) = x^2$ and (ii) which has the value $u(x, y) = 2$ at the point $(1, 0)$.

This equation is inhomogeneous, and so let us first find the general solution of (20.18) without regard for any boundary conditions. We begin by looking for the solution of the corresponding homogeneous equation ((20.18) but with the RHS equal to zero) of the form $u(x, y) = f(p)$. Following the same procedure as that used in the solution of (20.13) we find that $u(x, y)$ will be constant along lines of (x, y) that satisfy

$$\frac{dx}{y} = \frac{dy}{-x} \quad \Rightarrow \quad \frac{x^2}{2} + \frac{y^2}{2} = c.$$

Identifying the constant of integration c with $p/2$, we find that the general solution of the

---

[§] See for example H. T. H. Piaggio, *An Elementary Treatise on Differential Equations and their Applications* (London: G. Bell and Sons, Ltd, 1954), pp. 175 ff.

homogeneous equation is $u(x,y) = f(x^2 + y^2)$ for arbitrary function f. Now by inspection a particular integral of (20.18) is $u(x,y) = -3y$, and so the general solution to (20.18) is

$$u(x,y) = f(x^2 + y^2) - 3y.$$

Boundary condition (i) requires $u(x,0) = f(x^2) = x^2$, i.e. $f(z) = z$, and so the particular solution in this case is

$$u(x,y) = x^2 + y^2 - 3y.$$

Similarly, boundary condition (ii) requires $u(1,0) = f(1) = 2$. One possibility is $f(z) = 2z$, and if we make this choice, then one way of writing the most general particular solution is

$$u(x,y) = 2x^2 + 2y^2 - 3y + g(x^2 + y^2),$$

where g is any arbitrary function for which $g(1) = 0$. Alternatively, a simpler choice would be $f(z) = 2$, leading to

$$u(x,y) = 2 - 3y + g(x^2 + y^2). \blacktriangleleft$$

Although we have discussed the solution of inhomogeneous problems only for first-order equations, the general considerations hold true for linear PDEs of higher order.

20.3.3 Second-order equations

As noted in section 20.1, second-order linear PDEs are of great importance in describing the behaviour of many physical systems. As in our discussion of first-order equations, for the moment we shall restrict our discussion to equations with just two independent variables; extensions to a greater number of independent variables are straightforward.

The most general second-order linear PDE (containing two independent variables) has the form

$$A\frac{\partial^2 u}{\partial x^2} + B\frac{\partial^2 u}{\partial x \partial y} + C\frac{\partial^2 u}{\partial y^2} + D\frac{\partial u}{\partial x} + E\frac{\partial u}{\partial y} + Fu = R(x,y), \qquad (20.19)$$

where $A, B, \ldots, F$ and $R(x,y)$ are given functions of x and y. Because of the nature of the solutions to such equations, they are usually divided into three classes, a division of which we will make further use in subsection 20.6.2. The equation (20.19) is called *hyperbolic* if $B^2 > 4AC$, *parabolic* if $B^2 = 4AC$ and *elliptic* if $B^2 < 4AC$. Clearly, if A, B and C are functions of x and y (rather than just constants) then the equation might be of different types in different parts of the xy-plane.

Equation (20.19) obviously represents a very large class of PDEs, and it is usually impossible to find closed-form solutions to most of these equations. Therefore, for the moment we shall consider only homogeneous equations, with $R(x,y) = 0$, and make the further (greatly simplifying) restriction that, throughout the remainder of this section, $A, B, \ldots, F$ are not functions of x and y but merely constants.

We now tackle the problem of solving some types of second-order PDE with constant coefficients by seeking solutions that are arbitrary functions of particular combinations of independent variables, just as we did for first-order equations.

Following the discussion of the previous section, we can hope to find such solutions only if all the terms of the equation involve the same total number of differentiations, i.e. all terms are of the same order, although the number of differentiations with respect to the individual independent variables may be different. This means that in (20.19) we require the constants D, E and F to be identically zero (we have, of course, already assumed that $R(x, y)$ is zero), so that we are now considering only equations of the form

$$A\frac{\partial^2 u}{\partial x^2} + B\frac{\partial^2 u}{\partial x \partial y} + C\frac{\partial^2 u}{\partial y^2} = 0, \tag{20.20}$$

where A, B and C are constants. We note that both the one-dimensional wave equation,

$$\frac{\partial^2 u}{\partial x^2} - \frac{1}{c^2}\frac{\partial^2 u}{\partial t^2} = 0,$$

and the two-dimensional Laplace equation,

$$\frac{\partial^2 u}{\partial x^2} + \frac{\partial^2 u}{\partial y^2} = 0,$$

are of this form, but that the diffusion equation,

$$\kappa\frac{\partial^2 u}{\partial x^2} - \frac{\partial u}{\partial t} = 0,$$

is not, since it contains a first-order derivative.

Since all the terms in (20.20) involve two differentiations, by assuming a solution of the form $u(x, y) = f(p)$, where p is some unknown function of x and y (or t), we may be able to obtain a common factor $d^2f(p)/dp^2$ as the only appearance of f on the LHS. Then, because of the zero RHS, all reference to the form of f can be cancelled out.

We can gain some guidance on suitable forms for the combination $p = p(x, y)$ by considering $\partial u/\partial x$ when u is given by $u(x, y) = f(p)$, for then

$$\frac{\partial u}{\partial x} = \frac{df(p)}{dp}\frac{\partial p}{\partial x}.$$

Clearly differentiation of this equation with respect to x (or y) will not lead to a single term on the RHS, containing f only as $d^2f(p)/dp^2$, unless the factor $\partial p/\partial x$ is a constant so that $\partial^2 p/\partial x^2$ and $\partial^2 p/\partial x \partial y$ are necessarily zero. This shows that p must be a linear function of x. In an exactly similar way p must also be a linear function of y, i.e. $p = ax + by$.

If we assume a solution of (20.20) of the form $u(x, y) = f(ax+by)$, and evaluate the terms ready for substitution into (20.20), we obtain

$$\frac{\partial u}{\partial x} = a\frac{df(p)}{dp}, \qquad \frac{\partial u}{\partial y} = b\frac{df(p)}{dp},$$

$$\frac{\partial^2 u}{\partial x^2} = a^2\frac{d^2 f(p)}{dp^2}, \qquad \frac{\partial^2 u}{\partial x \partial y} = ab\frac{d^2 f(p)}{dp^2}, \qquad \frac{\partial^2 u}{\partial y^2} = b^2\frac{d^2 f(p)}{dp^2},$$

which on substitution give

$$\left(Aa^2 + Bab + Cb^2\right)\frac{d^2 f(p)}{dp^2} = 0. \tag{20.21}$$

This is the form we have been seeking, since now a solution independent of the form of f can be obtained if we require that a and b satisfy

$$Aa^2 + Bab + Cb^2 = 0.$$

From this quadratic, two values for the ratio of the two constants a and b are obtained,

$$b/a = [-B \pm (B^2 - 4AC)^{1/2}]/2C.$$

If we denote these two ratios by λ_1 and λ_2 then *any* functions of the two variables

$$p_1 = x + \lambda_1 y, \qquad p_2 = x + \lambda_2 y$$

will be solutions of the original equation (20.20). The omission of the constant factor a from p_1 and p_2 is of no consequence since this can always be absorbed into the particular form of any chosen function; only the *relative* weighting of x and y in p is important.

Since p_1 and p_2 are in general different, we can thus write the general solution of (20.20) as

$$u(x, y) = f(x + \lambda_1 y) + g(x + \lambda_2 y), \tag{20.22}$$

where f and g are arbitrary functions.

Finally, we note that the alternative solution $d^2 f(p)/dp^2 = 0$ to (20.21) leads only to the trivial solution $u(x, y) = kx + ly + m$, for which all second derivatives are individually zero.

▶ *Find the general solution of the one-dimensional wave equation*

$$\frac{\partial^2 u}{\partial x^2} - \frac{1}{c^2}\frac{\partial^2 u}{\partial t^2} = 0.$$

This equation is (20.20) with $A = 1$, $B = 0$ and $C = -1/c^2$, and so the values of λ_1 and λ_2 are the solutions of

$$1 - \frac{\lambda^2}{c^2} = 0,$$

namely $\lambda_1 = -c$ and $\lambda_2 = c$. This means that arbitrary functions of the quantities

$$p_1 = x - ct, \qquad p_2 = x + ct$$

will be satisfactory solutions of the equation and that the general solution will be

$$u(x,t) = f(x - ct) + g(x + ct), \tag{20.23}$$

where f and g are arbitrary functions. This solution is discussed further in section 20.4. ◀

The method used to obtain the general solution of the wave equation may also be applied straightforwardly to Laplace's equation.

▶ *Find the general solution of the two-dimensional Laplace equation*

$$\frac{\partial^2 u}{\partial x^2} + \frac{\partial^2 u}{\partial y^2} = 0. \tag{20.24}$$

Following the established procedure, we look for a solution that is a function $f(p)$ of $p = x + \lambda y$, where from (20.24) λ satisfies

$$1 + \lambda^2 = 0.$$

This requires that $\lambda = \pm i$, and satisfactory variables p are $p = x \pm iy$. The general solution required is therefore, in terms of arbitrary functions f and g,

$$u(x, y) = f(x + iy) + g(x - iy). \blacktriangleleft$$

It will be apparent from the last two examples that the nature of the appropriate linear combination of x and y depends upon whether $B^2 > 4AC$ or $B^2 < 4AC$. This is exactly the same criterion as determines whether the PDE is hyperbolic or elliptic. Hence as a general result, hyperbolic and elliptic equations of the form (20.20), given the restriction that the constants A, B and C are real, have as solutions functions whose arguments have the form $x + \alpha y$ and $x + i\beta y$ respectively, where α and β themselves are real.

The one case not covered by this result is that in which $B^2 = 4AC$, i.e. a parabolic equation. In this case λ_1 and λ_2 are not different and only one suitable combination of x and y results, namely

$$u(x, y) = f(x - (B/2C)y).$$

To find the second part of the general solution we try, in analogy with the corresponding situation for ordinary differential equations, a solution of the form

$$u(x, y) = h(x, y)g(x - (B/2C)y).$$

Substituting this into (20.20) and using $A = B^2/4C$ results in

$$\left(A \frac{\partial^2 h}{\partial x^2} + B \frac{\partial^2 h}{\partial x \partial y} + C \frac{\partial^2 h}{\partial y^2} \right) g = 0.$$

Therefore we require $h(x, y)$ to be any solution of the original PDE. There are several simple solutions of this equation, but as only one is required we take the simplest non-trivial one, $h(x, y) = x$, to give the general solution of the parabolic equation

$$u(x, y) = f(x - (B/2C)y) + xg(x - (B/2C)y). \tag{20.25}$$

We could, of course, have taken $h(x, y) = y$, but this only leads to a solution that is already represented by (20.25).

> ▶ *Solve*
>
> $$\frac{\partial^2 u}{\partial x^2} + 2\frac{\partial^2 u}{\partial x \partial y} + \frac{\partial^2 u}{\partial y^2} = 0,$$
>
> *subject to the boundary conditions $u(0, y) = 0$ and $u(x, 1) = x^2$.*

From our general result, functions of $p = x + \lambda y$ will be solutions provided

$$1 + 2\lambda + \lambda^2 = 0,$$

i.e. $\lambda = -1$ and the equation is parabolic. The general solution is therefore

$$u(x, y) = f(x - y) + xg(x - y).$$

The boundary condition $u(0, y) = 0$ implies $f(p) \equiv 0$, whilst $u(x, 1) = x^2$ yields

$$xg(x - 1) = x^2,$$

which gives $g(p) = p + 1$, Therefore the particular solution required is

$$u(x, y) = x(p + 1) = x(x - y + 1). \blacktriangleleft$$

To reinforce the material discussed above we will now give alternative derivations of the general solutions (20.22) and (20.25) by expressing the original PDE in terms of new variables before solving it. The actual solution will then become almost trivial; but, of course, it will be recognised that suitable new variables could hardly have been guessed if it were not for the work already done. This does not detract from the validity of the derivation to be described, only from the likelihood that it would be discovered by inspection.

We start again with (20.20) and change to new variables

$$\zeta = x + \lambda_1 y, \qquad \eta = x + \lambda_2 y.$$

With this change of variables, we have from the chain rule that

$$\frac{\partial}{\partial x} = \frac{\partial}{\partial \zeta} + \frac{\partial}{\partial \eta},$$

$$\frac{\partial}{\partial y} = \lambda_1 \frac{\partial}{\partial \zeta} + \lambda_2 \frac{\partial}{\partial \eta}.$$

Using these and the fact that

$$A + B\lambda_i + C\lambda_i^2 = 0 \qquad \text{for } i = 1, 2,$$

equation (20.20) becomes

$$[2A + B(\lambda_1 + \lambda_2) + 2C\lambda_1\lambda_2]\frac{\partial^2 u}{\partial \zeta \partial \eta} = 0.$$

Then, providing the factor in brackets does not vanish, for which the required condition is easily shown to be $B^2 \neq 4AC$, we obtain

$$\frac{\partial^2 u}{\partial \zeta \partial \eta} = 0,$$

which has the successive integrals

$$\frac{\partial u}{\partial \eta} = F(\eta), \quad u(\zeta, \eta) = f(\eta) + g(\zeta).$$

This solution is just the same as (20.22),

$$u(x, y) = f(x + \lambda_2 y) + g(x + \lambda_1 y).$$

If the equation is parabolic (i.e. $B^2 = 4AC$), we instead use the new variables

$$\zeta = x + \lambda y, \qquad \eta = x,$$

and recalling that $\lambda = -(B/2C)$ we can reduce (20.20) to

$$A \frac{\partial^2 u}{\partial \eta^2} = 0.$$

Two straightforward integrations give as the general solution

$$u(\zeta, \eta) = \eta g(\zeta) + f(\zeta),$$

which in terms of x and y has exactly the form of (20.25),

$$u(x, y) = xg(x + \lambda y) + f(x + \lambda y).$$

Finally, as hinted at in subsection 20.3.2 with reference to first-order linear PDEs, some of the methods used to find particular integrals of linear ODEs can be suitably modified to find particular integrals of PDEs of higher order. In simple cases, however, an appropriate solution may often be found by inspection.

> ►*Find the general solution of*
>
> $$\frac{\partial^2 u}{\partial x^2} + \frac{\partial^2 u}{\partial y^2} = 6(x + y).$$

Following our previous methods and results, the complementary function is

$$u(x, y) = f(x + iy) + g(x - iy),$$

and only a particular integral remains to be found. By inspection a particular integral of the equation is $u(x, y) = x^3 + y^3$, and so the general solution can be written

$$u(x, y) = f(x + iy) + g(x - iy) + x^3 + y^3. \quad ◄$$

20.4 The wave equation

We have already found that the general solution of the one-dimensional wave equation is

$$u(x,t) = f(x - ct) + g(x + ct), \tag{20.26}$$

where f and g are arbitrary functions. However, the equation is of such general importance that further discussion will not be out of place.

Let us imagine that $u(x,t) = f(x-ct)$ represents the displacement of a string at time t and position x. It is clear that all positions x and times t for which $x - ct =$ constant will have the same instantaneous displacement. But $x - ct =$ constant is exactly the relation between the time and position of an observer travelling with speed c along the positive x-direction. Consequently this moving observer sees a constant displacement of the string, whereas to a stationary observer, the initial profile $u(x,0)$ moves with speed c along the x-axis as if it were a rigid system. Thus $f(x - ct)$ represents a wave form of constant shape travelling along the positive x-axis with speed c, the actual form of the wave depending upon the function f. Similarly, the term $g(x + ct)$ is a constant wave form travelling with speed c in the negative x-direction. The general solution (20.23) represents a superposition of these.

If the functions f and g are the same then the complete solution (20.23) represents identical progressive waves going in opposite directions. This may result in a wave pattern whose profile does not progress, described as a *standing wave*. As a simple example, suppose both $f(p)$ and $g(p)$ have the form[§]

$$f(p) = g(p) = A\cos(kp + \epsilon).$$

Then (20.23) can be written as

$$\begin{aligned}
u(x,t) &= A[\cos(kx - kct + \epsilon) + \cos(kx + kct + \epsilon)] \\
&= 2A\cos(kct)\cos(kx + \epsilon).
\end{aligned}$$

The important thing to notice is that the shape of the wave pattern, given by the factor in x, is the same at all times but that its amplitude $2A\cos(kct)$ depends upon time. At some points x that satisfy

$$\cos(kx + \epsilon) = 0$$

there is no displacement at any time; such points are called *nodes*.

So far we have not imposed any boundary conditions on the solution (20.26). The problem of finding a solution to the wave equation that satisfies given boundary conditions is normally treated using the method of separation of variables

---

[§] In the usual notation, k is the wave number ($= 2\pi/$wavelength) and $kc = \omega$, the angular frequency of the wave.

discussed in the next chapter. Nevertheless, we now consider *D'Alembert's solution* $u(x,t)$ of the wave equation subject to initial conditions (boundary conditions) in the following general form:

initial displacement, $u(x,0) = \phi(x);$ initial velocity, $\dfrac{\partial u(x,0)}{\partial t} = \psi(x).$

The functions $\phi(x)$ and $\psi(x)$ are given and describe the displacement and velocity of each part of the string at the (arbitrary) time $t = 0$.

It is clear that what we need are the particular forms of the functions f and g in (20.26) that lead to the required values at $t = 0$. This means that

$$\phi(x) = u(x,0) = f(x - 0) + g(x + 0), \tag{20.27}$$

$$\psi(x) = \frac{\partial u(x,0)}{\partial t} = -cf'(x - 0) + cg'(x + 0), \tag{20.28}$$

where it should be noted that $f'(x - 0)$ stands for $df(p)/dp$ evaluated, after the differentiation, at $p = x - c \times 0$; likewise for $g'(x + 0)$.

Looking on the above two left-hand sides as functions of $p = x \pm ct$, but everywhere evaluated at $t = 0$, we may integrate (20.28) between an arbitrary (and irrelevant) lower limit p_0 and an indefinite upper limit p to obtain

$$\frac{1}{c} \int_{p_0}^{p} \psi(q)\, dq + K = -f(p) + g(p),$$

the constant of integration K depending on p_0. Comparing this equation with (20.27), with x replaced by p, we can establish the forms of the functions f and g as

$$f(p) = \frac{\phi(p)}{2} - \frac{1}{2c} \int_{p_0}^{p} \psi(q)\, dq - \frac{K}{2}, \tag{20.29}$$

$$g(p) = \frac{\phi(p)}{2} + \frac{1}{2c} \int_{p_0}^{p} \psi(q)\, dq + \frac{K}{2}. \tag{20.30}$$

Adding (20.29) with $p = x - ct$ to (20.30) with $p = x + ct$ gives as the solution to the original problem

$$u(x,t) = \frac{1}{2} [\phi(x - ct) + \phi(x + ct)] + \frac{1}{2c} \int_{x-ct}^{x+ct} \psi(q)\, dq, \tag{20.31}$$

in which we notice that all dependence on p_0 has disappeared.

Each of the terms in (20.31) has a fairly straightforward physical interpretation. In each case the factor $1/2$ represents the fact that only half a displacement profile that starts at any particular point on the string travels towards any other position x, the other half travelling away from it. The first term $\frac{1}{2}\phi(x - ct)$ arises from the initial displacement at a distance ct to the left of x; this travels forward arriving at x at time t. Similarly, the second contribution is due to the initial displacement at a distance ct to the right of x. The interpretation of the final term is a little less obvious. It can be viewed as representing the accumulated transverse displacement at position x due to the passage past x of all parts of the initial motion whose effects can reach x within a time t, both backward and forward travelling.

The extension to the three-dimensional wave equation of solutions of the type we have so far encountered presents no serious difficulty. In Cartesian coordinates the three-dimensional wave equation is

$$\frac{\partial^2 u}{\partial x^2} + \frac{\partial^2 u}{\partial y^2} + \frac{\partial^2 u}{\partial z^2} - \frac{1}{c^2}\frac{\partial^2 u}{\partial t^2} = 0. \tag{20.32}$$

In close analogy with the one-dimensional case we try solutions that are functions of linear combinations of all four variables,

$$p = lx + my + nz + \mu t.$$

It is clear that a solution $u(x, y, z, t) = f(p)$ will be acceptable provided that

$$\left(l^2 + m^2 + n^2 - \frac{\mu^2}{c^2}\right)\frac{d^2 f(p)}{dp^2} - 0.$$

Thus, as in the one-dimensional case, f can be arbitrary provided that

$$l^2 + m^2 + n^2 = \mu^2/c^2.$$

Using an obvious normalisation, we take $\mu = \pm c$ and l, m, n as three numbers such that

$$l^2 + m^2 + n^2 = 1.$$

In other words (l, m, n) are the Cartesian components of a unit vector $\hat{\mathbf{n}}$ that points along the direction of propagation of the wave. The quantity p can be written in terms of vectors as the scalar expression $p = \hat{\mathbf{n}} \cdot \mathbf{r} \pm ct$, and the general solution of (20.32) is then

$$u(x, y, z, t) = u(\mathbf{r}, t) = f(\hat{\mathbf{n}} \cdot \mathbf{r} - ct) + g(\hat{\mathbf{n}} \cdot \mathbf{r} + ct), \tag{20.33}$$

where $\hat{\mathbf{n}}$ is *any* unit vector. It would perhaps be more transparent to write $\hat{\mathbf{n}}$ explicitly as one of the arguments of u.

20.5 The diffusion equation

One important class of second-order PDEs, which we have not yet considered in detail, is that in which the second derivative with respect to one variable appears, but only the first derivative with respect to another (usually time). This is exemplified by the one-dimensional diffusion equation

$$\kappa\frac{\partial^2 u(x, t)}{\partial x^2} = \frac{\partial u}{\partial t}, \tag{20.34}$$

695

in which κ is a constant with the dimensions length$^2 \times$ time^{-1}. The physical constants that go to make up κ in a particular case depend upon the nature of the process (e.g. solute diffusion, heat flow, etc.) and the material being described.

With (20.34) we cannot hope to repeat successfully the method of subsection 20.3.3, since now $u(x, t)$ is differentiated a different number of times on the two sides of the equation; any attempted solution in the form $u(x, t) = f(p)$ with $p = ax + bt$ will lead only to an equation in which the form of f cannot be cancelled out. Clearly we must try other methods.

Solutions may be obtained by using the standard method of separation of variables discussed in the next chapter. Alternatively, a simple solution is also given if both sides of (20.34), as it stands, are separately set equal to a constant α (say), so that

$$\frac{\partial^2 u}{\partial x^2} = \frac{\alpha}{\kappa}, \qquad \frac{\partial u}{\partial t} = \alpha.$$

These equations have the general solutions

$$u(x, t) = \frac{\alpha}{2\kappa}x^2 + xg(t) + h(t) \quad \text{and} \quad u(x, t) = \alpha t + m(x)$$

respectively and may be made compatible with each other if $g(t)$ is taken as constant, $g(t) = g$ (where g could be zero), $h(t) = \alpha t$ and $m(x) = (\alpha/2\kappa)x^2 + gx$. An acceptable solution is thus

$$u(x, t) = \frac{\alpha}{2\kappa}x^2 + gx + \alpha t + \text{constant}. \tag{20.35}$$

Let us now return to seeking solutions of equations by combining the independent variables in particular ways. Having seen that a linear combination of x and t will be of no value, we must search for other possible combinations. It has been noted already that κ has the dimensions length$^2 \times$ time^{-1} and so the combination of variables

$$\eta = \frac{x^2}{\kappa t}$$

will be dimensionless. Let us see if we can satisfy (20.34) with a solution of the form $u(x, t) = f(\eta)$. Evaluating the necessary derivatives we have

$$\frac{\partial u}{\partial x} = \frac{df(\eta)}{d\eta}\frac{\partial \eta}{\partial x} = \frac{2x}{\kappa t}\frac{df(\eta)}{d\eta},$$

$$\frac{\partial^2 u}{\partial x^2} = \frac{2}{\kappa t}\frac{df(\eta)}{d\eta} + \left(\frac{2x}{\kappa t}\right)^2\frac{d^2 f(\eta)}{d\eta^2},$$

$$\frac{\partial u}{\partial t} = -\frac{x^2}{\kappa t^2}\frac{df(\eta)}{d\eta}.$$

Substituting these expressions into (20.34) we find that the new equation can be written entirely in terms of η,

$$4\eta \frac{d^2 f(\eta)}{d\eta^2} + (2+\eta) \frac{df(\eta)}{d\eta} = 0.$$

This is a straightforward ODE, which can be solved as follows. Writing $f'(\eta) = df(\eta)/d\eta$, etc., we have

$$\frac{f''(\eta)}{f'(\eta)} = -\frac{1}{2\eta} - \frac{1}{4}$$

$$\Rightarrow \quad \ln[\eta^{1/2} f'(\eta)] = -\frac{\eta}{4} + c$$

$$\Rightarrow \quad f'(\eta) = \frac{A}{\eta^{1/2}} \exp\left(\frac{-\eta}{4}\right)$$

$$\Rightarrow \quad f(\eta) = A \int_{\eta_0}^{\eta} \mu^{-1/2} \exp\left(\frac{-\mu}{4}\right) d\mu.$$

If we now write this in terms of a slightly different variable

$$\zeta = \frac{\eta^{1/2}}{2} = \frac{x}{2(\kappa t)^{1/2}},$$

then $d\zeta = \frac{1}{4}\eta^{-1/2} d\eta$, and the solution to (20.34) is given by

$$u(x,t) = f(\eta) = g(\zeta) = B \int_{\zeta_0}^{\zeta} \exp(-v^2) \, dv. \tag{20.36}$$

Here B is a constant and it should be noticed that x and t appear on the RHS only in the indefinite upper limit ζ, and then only in the combination $xt^{-1/2}$. If ζ_0 is chosen as zero then $u(x,t)$ is, to within a constant factor,[§] the error function $\mathrm{erf}[x/2(\kappa t)^{1/2}]$, which is tabulated in many reference books. Only non-negative values of x and t are to be considered here, so that $\zeta \geq \zeta_0$.

Let us try to determine what kind of (say) temperature distribution and flow this represents. For definiteness we take $\zeta_0 = 0$. Firstly, since $u(x,t)$ in (20.36) depends only upon the product $xt^{-1/2}$, it is clear that all points x at times t such that $xt^{-1/2}$ has the same value have the same temperature. Put another way, at any specific time t the region having a particular temperature has moved along the positive x-axis a distance proportional to the square root of t. This is a typical *diffusion* process.

Notice that, on the one hand, at $t = 0$ the variable $\zeta \to \infty$ and u becomes quite independent of x (except perhaps at $x = 0$); the solution then represents a uniform spatial temperature distribution. On the other hand, at $x = 0$ we have that $u(x,t)$ is identically zero for all t.

[§] Take $B = 2\pi^{-1/2}$ to give the usual error function normalised in such a way that $\mathrm{erf}(\infty) = 1$. See section 18.12.4.

▶An infrared laser delivers a pulse of (heat) energy E to a point P on a large insulated sheet of thickness b, thermal conductivity k, specific heat s and density ρ. The sheet is initially at a uniform temperature. If $u(r, t)$ is the excess temperature a time t later, at a point that is a distance r ($\gg b$) from P, then show that a suitable expression for u is

$$u(r, t) = \frac{\alpha}{t} \exp\left(-\frac{r^2}{2\beta t}\right),$$ (20.37)

where α and β are constants. (Note that we use r instead of ρ to denote the radial coordinate in plane polars so as to avoid confusion with the density.)

Further, (i) show that $\beta = 2k/(s\rho)$; (ii) demonstrate that the excess heat energy in the sheet is independent of t, and hence evaluate α; and (iii) prove that the total heat flow past any circle of radius r is E.

The equation to be solved is the heat diffusion equation

$$k\nabla^2 u(\mathbf{r}, t) = s\rho \frac{\partial u(\mathbf{r}, t)}{\partial t}.$$

Since we only require the solution for $r \gg b$ we can treat the problem as two-dimensional with obvious circular symmetry. Thus only the r-derivative term in the expression for $\nabla^2 u$ is non-zero, giving

$$\frac{k}{r} \frac{\partial}{\partial r}\left(r \frac{\partial u}{\partial r}\right) = s\rho \frac{\partial u}{\partial t},$$ (20.38)

where now $u(\mathbf{r}, t) = u(r, t)$.

(i) Substituting the given expression (20.37) into (20.38) we obtain

$$\frac{2k\alpha}{\beta t^2}\left(\frac{r^2}{2\beta t} - 1\right)\exp\left(-\frac{r^2}{2\beta t}\right) = \frac{s\rho\alpha}{t^2}\left(\frac{r^2}{2\beta t} - 1\right)\exp\left(-\frac{r^2}{2\beta t}\right),$$

from which we find that (20.37) is a solution, provided $\beta = 2k/(s\rho)$.
(ii) The excess heat in the system at any time t is

$$b\rho s \int_0^\infty u(r, t) 2\pi r\, dr = 2\pi b\rho s\alpha \int_0^\infty \frac{r}{t}\exp\left(-\frac{r^2}{2\beta t}\right) dr$$
$$= 2\pi b\rho s\alpha\beta.$$

The excess heat is therefore independent of t and so must be equal to the total heat input E, implying that

$$\alpha = \frac{E}{2\pi b\rho s\beta} = \frac{E}{4\pi bk}.$$

(iii) The total heat flow past a circle of radius r is

$$-2\pi rbk \int_0^\infty \frac{\partial u(r, t)}{\partial r} dt = -2\pi rbk \int_0^\infty \frac{E}{4\pi bkt}\left(\frac{-r}{\beta t}\right)\exp\left(-\frac{r^2}{2\beta t}\right) dt$$
$$= E\left[\exp\left(-\frac{r^2}{2\beta t}\right)\right]_0^\infty = E \quad \text{for all } r.$$

As we would expect, all the heat energy E deposited by the laser will eventually flow past a circle of any given radius r. ◀

20.6 Characteristics and the existence of solutions

So far in this chapter we have discussed how to find general solutions to various types of first- and second-order linear PDE. Moreover, given a set of boundary conditions we have shown how to find the particular solution (or class of solutions) that satisfies them. For first-order equations, for example, we found that if the value of $u(x, y)$ is specified along some curve in the xy-plane then the solution to the PDE is in general unique, but that if $u(x, y)$ is specified at only a single point then the solution is not unique: there exists a class of particular solutions all of which satisfy the boundary condition. In this section and the next we make more rigorous the notion of the respective types of boundary condition that cause a PDE to have a unique solution, a class of solutions, or no solution at all.

20.6.1 First-order equations

Let us consider the general first-order PDE (20.9) but now write it as

$$A(x, y)\frac{\partial u}{\partial x} + B(x, y)\frac{\partial u}{\partial y} = F(x, y, u). \tag{20.39}$$

Suppose we wish to solve this PDE subject to the boundary condition that $u(x, y) = \phi(s)$ is specified along some curve C in the xy-plane that is described parametrically by the equations $x = x(s)$ and $y = y(s)$, where s is the arc length along C. The variation of u along C is therefore given by

$$\frac{du}{ds} = \frac{\partial u}{\partial x}\frac{dx}{ds} + \frac{\partial u}{\partial y}\frac{dy}{ds} = \frac{d\phi}{ds}. \tag{20.40}$$

We may then solve the two (inhomogeneous) simultaneous linear equations (20.39) and (20.40) for $\partial u/\partial x$ and $\partial u/\partial y$, *unless* the determinant of the coefficients vanishes (see section 8.18), i.e. unless

$$\begin{vmatrix} dx/ds & dy/ds \\ A & B \end{vmatrix} = 0.$$

At each point in the xy-plane this equation determines a set of curves called *characteristic curves* (or just *characteristics*), which thus satisfy

$$B\frac{dx}{ds} - A\frac{dy}{ds} = 0,$$

or, multiplying through by ds/dx and dividing through by A,

$$\frac{dy}{dx} = \frac{B(x, y)}{A(x, y)}. \tag{20.41}$$

However, we have already met (20.41) in subsection 20.3.1 on first-order PDEs, where solutions of the form $u(x, y) = f(p)$, where p is some combination of x and y, were discussed. Comparing (20.41) with (20.12) we see that the characteristics are merely those curves along which p is constant.

Since the partial derivatives $\partial u/\partial x$ and $\partial u/\partial y$ may be evaluated provided the boundary curve C does *not* lie along a characteristic, defining $u(x, y) = \phi(s)$ along C is sufficient to specify the solution to the original problem (equation plus boundary conditions) near the curve C, in terms of a Taylor expansion about C. Therefore the characteristics can be considered as the curves along which information about the solution $u(x, y)$ 'propagates'. This is best understood by using an example.

> ▶*Find the general solution of*
>
> $$x\frac{\partial u}{\partial x} - 2y\frac{\partial u}{\partial y} = 0 \qquad (20.42)$$
>
> *that takes the value $2y + 1$ on the line $x = 1$ between $y = 0$ and $y = 1$.*

We solved this problem in subsection 20.3.1 for the case where $u(x, y)$ takes the value $2y + 1$ along the *entire* line $x = 1$. We found then that the general solution to the equation (ignoring boundary conditions) is of the form

$$u(x, y) = f(p) = f(x^2 y),$$

for some arbitrary function f. Hence the characteristics of (20.42) are given by $x^2 y = c$ where c is a constant; some of these curves are plotted in figure 20.2 for various values of c. Furthermore, we found that the particular solution for which $u(1, y) = 2y + 1$ *for all y* was given by

$$u(x, y) = 2x^2 y + 1.$$

In the present case the value of $x^2 y$ is fixed by the boundary conditions only between $y = 0$ and $y = 1$. However, since the characteristics are curves along which $x^2 y$, and hence $f(x^2 y)$, remains constant, the solution is determined everywhere along any characteristic that intersects the line segment denoting the boundary conditions. Thus $u(x, y) = 2x^2 y + 1$ is the particular solution that holds in the shaded region in figure 20.2 (corresponding to $0 \leq c \leq 1$).

Outside this region, however, the solution is not precisely specified, and any function of the form

$$u(x, y) = 2x^2 y + 1 + g(x^2 y)$$

will satisfy both the equation and the boundary condition, provided $g(p) = 0$ for $0 \leq p \leq 1$. ◀

In the above example the boundary curve was not itself a characteristic and furthermore it crossed each characteristic *once only*. For a general boundary curve C this may not be the case. Firstly, if C is itself a characteristic (or is just a single point) then information about the solution cannot 'propagate' away from C, and so the solution remains unspecified everywhere except on C.

The second possibility is that C (although not a characteristic itself) crosses some characteristics more than once, as in figure 20.3. In this case specifying the value of $u(x, y)$ along the curve PQ determines the solution along all the characteristics that intersect it. Therefore, also specifying $u(x, y)$ along QR can *overdetermine* the problem solution and generally results in there being no solution.

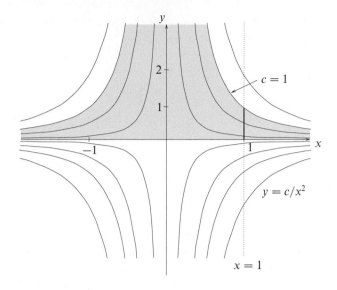

Figure 20.2   The characteristics of equation (20.42). The shaded region shows where the solution to the equation is defined, given the imposed boundary condition at $x = 1$ between $y = 0$ and $y = 1$, shown as a bold vertical line.

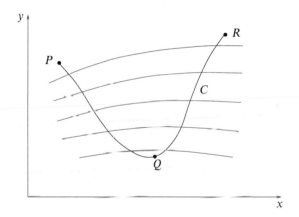

Figure 20.3   A boundary curve C that crosses characteristics more than once.

20.6.2 Second-order equations

The concept of characteristics can be extended naturally to second- (and higher-) order equations. In this case let us write the general second-order linear PDE (20.19) as

$$A(x, y)\frac{\partial^2 u}{\partial x^2} + B(x, y)\frac{\partial^2 u}{\partial x \partial y} + C(x, y)\frac{\partial^2 u}{\partial y^2} = F\left(x, y, u, \frac{\partial u}{\partial x}, \frac{\partial u}{\partial y}\right). \qquad (20.43)$$

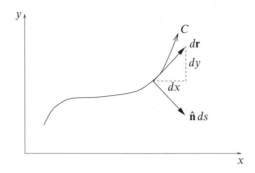

Figure 20.4 A boundary curve C and its tangent and unit normal at a given point.

For second-order equations we might expect that relevant boundary conditions would involve specifying u, or some of its first derivatives, or both, along a suitable set of boundaries bordering or enclosing the region over which a solution is sought. Three common types of boundary condition occur and are associated with the names of Dirichlet, Neumann and Cauchy. They are as follows.

(i) *Dirichlet*: The value of u is specified at each point of the boundary.
(ii) *Neumann*: The value of $\partial u/\partial n$, the *normal derivative* of u, is specified at each point of the boundary. Note that $\partial u/\partial n = \nabla u \cdot \hat{\mathbf{n}}$, where $\hat{\mathbf{n}}$ is the normal to the boundary at each point.
(iii) *Cauchy*: Both u and $\partial u/\partial n$ are specified at each point of the boundary.

Let us consider for the moment the solution of (20.43) subject to the Cauchy boundary conditions, i.e. u and $\partial u/\partial n$ are specified along some boundary curve C in the xy-plane defined by the parametric equations $x = x(s)$, $y = y(s)$, s being the arc length along C (see figure 20.4). Let us suppose that along C we have $u(x, y) = \phi(s)$ and $\partial u/\partial n = \psi(s)$. At any point on C the vector $d\mathbf{r} = dx\,\mathbf{i} + dy\,\mathbf{j}$ is a tangent to the curve and $\hat{\mathbf{n}}\,ds = dy\,\mathbf{i} - dx\,\mathbf{j}$ is a vector normal to the curve. Thus on C we have

$$\frac{\partial u}{\partial s} \equiv \nabla u \cdot \frac{d\mathbf{r}}{ds} = \frac{\partial u}{\partial x}\frac{dx}{ds} + \frac{\partial u}{\partial y}\frac{dy}{ds} = \frac{d\phi(s)}{ds},$$
$$\frac{\partial u}{\partial n} \equiv \nabla u \cdot \hat{\mathbf{n}} = \frac{\partial u}{\partial x}\frac{dy}{ds} - \frac{\partial u}{\partial y}\frac{dx}{ds} = \psi(s).$$

These two equations may then be solved straightforwardly for the first partial derivatives $\partial u/\partial x$ and $\partial u/\partial y$ along C. Using the chain rule to write

$$\frac{d}{ds} = \frac{dx}{ds}\frac{\partial}{\partial x} + \frac{dy}{ds}\frac{\partial}{\partial y},$$

702

we may differentiate the two first derivatives $\partial u/\partial x$ and $\partial u/\partial y$ along the boundary to obtain the pair of equations

$$\frac{d}{ds}\left(\frac{\partial u}{\partial x}\right) = \frac{dx}{ds}\frac{\partial^2 u}{\partial x^2} + \frac{dy}{ds}\frac{\partial^2 u}{\partial x \partial y},$$

$$\frac{d}{ds}\left(\frac{\partial u}{\partial y}\right) = \frac{dx}{ds}\frac{\partial^2 u}{\partial x \partial y} + \frac{dy}{ds}\frac{\partial^2 u}{\partial y^2}.$$

We may now solve these two equations, together with the original PDE (20.43), for the second partial derivatives of u, *except* where the determinant of their coefficients equals zero,

$$\begin{vmatrix} A & B & C \\ \dfrac{dx}{ds} & \dfrac{dy}{ds} & 0 \\ 0 & \dfrac{dx}{ds} & \dfrac{dy}{ds} \end{vmatrix} = 0.$$

Expanding out the determinant,

$$A\left(\frac{dy}{ds}\right)^2 - B\left(\frac{dx}{ds}\right)\left(\frac{dy}{ds}\right) + C\left(\frac{dx}{ds}\right)^2 = 0.$$

Multiplying through by $(ds/dx)^2$ we obtain

$$A\left(\frac{dy}{dx}\right)^2 - B\frac{dy}{dx} + C = 0, \tag{20.44}$$

which is the ODE for the curves in the xy-plane along which the second partial derivatives of u *cannot* be found.

As for the first-order case, the curves satisfying (20.44) are called characteristics of the original PDE. These characteristics have tangents at each point given by (when $A \neq 0$)

$$\frac{dy}{dx} = \frac{B \pm \sqrt{B^2 - 4AC}}{2A}. \tag{20.45}$$

Clearly, when the original PDE is hyperbolic ($B^2 > 4AC$), equation (20.45) defines two families of real curves in the xy-plane; when the equation is parabolic ($B^2 = 4AC$) it defines one family of real curves; and when the equation is elliptic ($B^2 < 4AC$) it defines two families of complex curves. Furthermore, when A, B and C are constants, rather than functions of x and y, the equations of the characteristics will be of the form $x + \lambda y = \text{constant}$, which is reminiscent of the form of solution discussed in subsection 20.3.3.

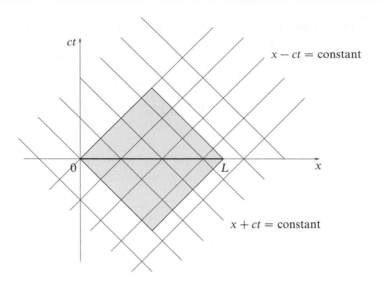

Figure 20.5   The characteristics for the one-dimensional wave equation. The shaded region indicates the region over which the solution is determined by specifying Cauchy boundary conditions at $t = 0$ on the line segment $x = 0$ to $x = L$.

▶Find the characteristics of the one-dimensional wave equation

$$\frac{\partial^2 u}{\partial x^2} - \frac{1}{c^2}\frac{\partial^2 u}{\partial t^2} = 0.$$

This is a hyperbolic equation with $A = 1$, $B = 0$ and $C = -1/c^2$. Therefore from (20.44) the characteristics are given by

$$\left(\frac{dx}{dt}\right)^2 = c^2,$$

and so the characteristics are the straight lines $x - ct = $ constant and $x + ct = $ constant. ◀

The characteristics of second-order PDEs can be considered as the curves along which *partial* information about the solution $u(x, y)$ 'propagates'. Consider a point in the space that has the independent variables as its coordinates; unless both of the two characteristics that pass through the point intersect the curve along which the boundary conditions are specified, the solution will not be determined at that point. In particular, if the equation is hyperbolic, so that we obtain two families of real characteristics in the xy-plane, then Cauchy boundary conditions propagate partial information concerning the solution along the characteristics, belonging to each family, that intersect the boundary curve C. The solution u is then specified in the region common to these two families of characteristics. For instance, the characteristics of the hyperbolic one-dimensional wave equation in the last example are shown in figure 20.5. By specifying Cauchy boundary

Equation type	Boundary	Conditions
hyperbolic	open	Cauchy
parabolic	open	Dirichlet or Neumann
elliptic	closed	Dirichlet or Neumann

Table 20.1   The appropriate boundary conditions for different types of partial differential equation.

conditions u and $\partial u/\partial t$ on the line segment $t = 0$, $x = 0$ to L, the solution is specified in the shaded region.

As in the case of first-order PDEs, however, problems can arise. For example, if for a hyperbolic equation the boundary curve intersects any characteristic more than once then Cauchy conditions along C can overdetermine the problem, resulting in there being no solution. In this case either the boundary curve C must be altered, or the boundary conditions on the offending parts of C must be relaxed to Dirichlet or Neumann conditions.

The general considerations involved in deciding which boundary conditions are appropriate for a particular problem are complex, and we do not discuss them any further here.[§] We merely note that whether the various types of boundary condition are appropriate (in that they give a solution that is unique, sometimes to within a constant, and is well defined) depends upon the type of second-order equation under consideration and on whether the region of solution is bounded by a closed or an open curve (or a surface if there are more than two independent variables). Note that part of a closed boundary may be at infinity if conditions are imposed on u or $\partial u/\partial n$ there.

It may be shown that the appropriate boundary-condition and equation-type pairings are as given in table 20.1.

For example, Laplace's equation $\nabla^2 u = 0$ is elliptic and thus requires either Dirichlet or Neumann boundary conditions on a closed boundary which, as we have already noted, may be at infinity if the behaviour of u is specified there (most often u or $\partial u/\partial n \rightarrow 0$ at infinity).

20.7 Uniqueness of solutions

Although we have merely stated the appropriate boundary types and conditions for which, in the general case, a PDE has a unique, well-defined solution, sometimes to within an additive constant, it is often important to be able to prove that a unique solution is obtained.

[§] For a discussion the reader is referred, for example, to P. M. Morse and H. Feshbach, *Methods of Theoretical Physics, Part I* (New York: McGraw-Hill, 1953), chap. 6.

As an important example let us consider Poisson's equation in three dimensions,

$$\nabla^2 u(\mathbf{r}) = \rho(\mathbf{r}), \tag{20.46}$$

with either Dirichlet or Neumann conditions on a closed boundary appropriate to such an elliptic equation; for brevity, in (20.46), we have absorbed any physical constants into ρ. We aim to show that, to within an unimportant constant, the solution of (20.46) is *unique* if either the potential u or its normal derivative $\partial u/\partial n$ is specified on all surfaces bounding a given region of space (including, if necessary, a hypothetical spherical surface of indefinitely large radius on which u or $\partial u/\partial n$ is prescribed to have an arbitrarily small value). Stated more formally this is as follows.

Uniqueness theorem. *If u is real and its first and second partial derivatives are continuous in a region V and on its boundary S, and $\nabla^2 u = \rho$ in V and either $u = f$ or $\partial u/\partial n = g$ on S, where ρ, f and g are prescribed functions, then u is unique (at least to within an additive constant).*

▶*Prove the uniqueness theorem for Poisson's equation.*

Let us suppose on the contrary that two solutions $u_1(\mathbf{r})$ and $u_2(\mathbf{r})$ both satisfy the conditions given above, and denote their difference by the function $w = u_1 - u_2$. We then have

$$\nabla^2 w = \nabla^2 u_1 - \nabla^2 u_2 = \rho - \rho = 0,$$

so that w satisfies Laplace's equation in V. Furthermore, since either $u_1 = f = u_2$ or $\partial u_1/\partial n = g = \partial u_2/\partial n$ on S, we must have either $w = 0$ or $\partial w/\partial n = 0$ on S.

If we now use Green's first theorem, (11.19), for the case where both scalar functions are taken as w we have

$$\int_V \left[w\nabla^2 w + (\nabla w) \cdot (\nabla w) \right] dV = \int_S w \frac{\partial w}{\partial n} \, dS.$$

However, either condition, $w = 0$ or $\partial w/\partial n = 0$, makes the RHS vanish whilst the first term on the LHS vanishes since $\nabla^2 w = 0$ in V. Thus we are left with

$$\int_V |\nabla w|^2 \, dV = 0.$$

Since $|\nabla w|^2$ can never be negative, this can only be satisfied if

$$\nabla w = \mathbf{0},$$

i.e. if w, and hence $u_1 - u_2$, is a constant in V.

If Dirichlet conditions are given then $u_1 \equiv u_2$ on (some part of) S and hence $u_1 = u_2$ everywhere in V. For Neumann conditions, however, u_1 and u_2 can differ throughout V by an arbitrary (but unimportant) constant. ◀

The importance of this uniqueness theorem lies in the fact that if a solution to Poisson's (or Laplace's) equation that fits the given set of Dirichlet or Neumann conditions can be found by any means whatever, then that solution is the correct one, since only one exists. This result is the mathematical justification for the *method of images*, which is discussed more fully in the next chapter.

706

We also note that often the same general method, used in the above example for proving the uniqueness theorem for Poisson's equation, can be employed to prove the uniqueness (or otherwise) of solutions to other equations and boundary conditions.

20.8 Exercises

20.1   Determine whether the following can be written as functions of $p = x^2 + 2y$ only, and hence whether they are solutions of (20.8):

(a) $x^2(x^2 - 4) + 4y(x^2 - 2) + 4(y^2 - 1)$;
(b) $x^4 + 2x^2y + y^2$;
(c) $[x^4 + 4x^2y + 4y^2 + 4]/[2x^4 + x^2(8y + 1) + 8y^2 + 2y]$.

20.2   Find partial differential equations satisfied by the following functions $u(x, y)$ for all arbitrary functions f and all arbitrary constants a and b:

(a) $u(x, y) = f(x^2 - y^2)$;
(b) $u(x, y) = (x - a)^2 + (y - b)^2$;
(c) $u(x, v) = y^n f(y/x)$;
(d) $u(x, y) = f(x + ay)$.

20.3   Solve the following partial differential equations for $u(x, y)$ with the boundary conditions given:

(a) $x\dfrac{\partial u}{\partial x} + xy = u$, $u = 2y$ on the line $x = 1$;

(b) $1 + x\dfrac{\partial u}{\partial y} = xu$, $u(x, 0) = x$.

20.4   Find the most general solutions $u(x, y)$ of the following equations, consistent with the boundary conditions stated:

(a) $y\dfrac{\partial u}{\partial x} - x\dfrac{\partial u}{\partial y} = 0$, $u(x, 0) = 1 + \sin x$;

(b) $i\dfrac{\partial u}{\partial x} = 3\dfrac{\partial u}{\partial y}$, $u = (4 + 3i)x^2$ on the line $x = y$;

(c) $\sin x \sin y\dfrac{\partial u}{\partial x} + \cos x \cos y\dfrac{\partial u}{\partial y} = 0$, $u = \cos 2y$ on $x + y = \pi/2$;

(d) $\dfrac{\partial u}{\partial x} + 2x\dfrac{\partial u}{\partial y} = 0$, $u = 2$ on the parabola $y = x^2$.

20.5   Find solutions of

$$\frac{1}{x}\frac{\partial u}{\partial x} + \frac{1}{y}\frac{\partial u}{\partial y} = 0$$

for which (a) $u(0, y) = y$ and (b) $u(1, 1) = 1$.

20.6   Find the most general solutions $u(x, y)$ of the following equations consistent with the boundary conditions stated:

(a) $y\dfrac{\partial u}{\partial x} - x\dfrac{\partial u}{\partial y} = 3x$, $u = x^2$ on the line $y = 0$;

707

(b) $y\dfrac{\partial u}{\partial x} - x\dfrac{\partial u}{\partial y} = 3x, \quad u(1,0) = 2;$

(c) $y^2\dfrac{\partial u}{\partial x} + x^2\dfrac{\partial u}{\partial y} = x^2 y^2(x^3 + y^3),$ no boundary conditions.

20.7   Solve

$$\sin x\dfrac{\partial u}{\partial x} + \cos x\dfrac{\partial u}{\partial y} = \cos x$$

subject to (a) $u(\pi/2, y) = 0$ and (b) $u(\pi/2, y) = y(y+1).$

20.8   A function $u(x, y)$ satisfies

$$2\dfrac{\partial u}{\partial x} + 3\dfrac{\partial u}{\partial y} = 10,$$

and takes the value 3 on the line $y = 4x$. Evaluate $u(2, 4).$

20.9   If $u(x, y)$ satisfies

$$\dfrac{\partial^2 u}{\partial x^2} - 3\dfrac{\partial^2 u}{\partial x \partial y} + 2\dfrac{\partial^2 u}{\partial y^2} = 0$$

and $u = -x^2$ and $\partial u/\partial y = 0$ for $y = 0$ and all x, find the value of $u(0, 1).$

20.10   Consider the partial differential equation

$$\dfrac{\partial^2 u}{\partial x^2} - 3\dfrac{\partial^2 u}{\partial x \partial y} + 2\dfrac{\partial^2 u}{\partial y^2} = 0. \qquad (*)$$

(a) Find the function $u(x, y)$ that satisfies $(*)$ and the boundary condition $u = \partial u/\partial y = 1$ when $y = 0$ for all x. Evaluate $u(0, 1).$

(b) In which region of the xy-plane would u be determined if the boundary condition were $u = \partial u/\partial y = 1$ when $y = 0$ for all $x > 0$?

20.11   In those cases in which it is possible to do so, evaluate $u(2, 2)$, where $u(x, y)$ is the solution of

$$2y\dfrac{\partial u}{\partial x} - x\dfrac{\partial u}{\partial y} = xy(2y^2 - x^2)$$

that satisfies the (separate) boundary conditions given below.

(a) $u(x, 1) = x^2$ for all $x.$
(b) $u(x, 1) = x^2$ for $x \geq 0.$
(c) $u(x, 1) = x^2$ for $0 \leq x \leq 3.$
(d) $u(x, 0) = x$ for $x \geq 0.$
(e) $u(x, 0) = x$ for all $x.$
(f) $u(1, \sqrt{10}) = 5.$
(g) $u(\sqrt{10}, 1) = 5.$

20.12   Solve

$$6\dfrac{\partial^2 u}{\partial x^2} - 5\dfrac{\partial^2 u}{\partial x \partial y} + \dfrac{\partial^2 u}{\partial y^2} = 14,$$

subject to $u = 2x + 1$ and $\partial u/\partial y = 4 - 6x$, both on the line $y = 0.$

20.13   By changing the independent variables in the previous exercise to

$$\xi = x + 2y \quad \text{and} \quad \eta = x + 3y,$$

show that it must be possible to write $14(x^2 + 5xy + 6y^2)$ in the form

$$f_1(x + 2y) + f_2(x + 3y) - (x^2 + y^2),$$

and determine the forms of $f_1(z)$ and $f_2(z).$

20.14   Solve

$$\frac{\partial^2 u}{\partial x \partial y} + 3\frac{\partial^2 u}{\partial y^2} = x(2y + 3x).$$

20.15   Find the most general solution of $\partial^2 u/\partial x^2 + \partial^2 u/\partial y^2 = x^2 y^2$.

20.16   An infinitely long string on which waves travel at speed c has an initial displacement

$$y(x) = \begin{cases} \sin(\pi x/a), & -a \le x \le a, \\ 0, & |x| > a. \end{cases}$$

It is released from rest at time $t = 0$, and its subsequent displacement is described by $y(x, t)$.

By expressing the initial displacement as one explicit function incorporating Heaviside step functions, find an expression for $y(x, t)$ at a general time $t > 0$. In particular, determine the displacement as a function of time (a) at $x = 0$, (b) at $x = a$, and (c) at $x = a/2$.

20.17   The non-relativistic Schrödinger equation (20.7) is similar to the diffusion equation in having different orders of derivatives in its various terms; this precludes solutions that are arbitrary functions of particular linear combinations of variables. However, since exponential functions do not change their forms under differentiation, solutions in the form of exponential functions of combinations of the variables may still be possible.

Consider the Schrödinger equation for the case of a constant potential, i.e. for a free particle, and show that it has solutions of the form $A \exp(lx + my + nz + \lambda t)$, where the only requirement is that

$$-\frac{\hbar^2}{2m}\left(l^2 + m^2 + n^2\right) = i\hbar\lambda.$$

In particular, identify the equation and wavefunction obtained by taking λ as $-iE/\hbar$, and l, m and n as $ip_x/\hbar, ip_y/\hbar$ and $ip_z/\hbar$, respectively, where E is the energy and $\mathbf{p}$ the momentum of the particle; these identifications are essentially the content of the de Broglie and Einstein relationships.

20.18   Like the Schrödinger equation of the previous exercise, the equation describing the transverse vibrations of a rod,

$$a^4 \frac{\partial^4 u}{\partial x^4} + \frac{\partial^2 u}{\partial t^2} = 0,$$

has different orders of derivatives in its various terms. Show, however, that it has solutions of exponential form, $u(x, t) = A \exp(\lambda x + i\omega t)$, provided that the relation $a^4 \lambda^4 = \omega^2$ is satisfied.

Use a linear combination of such allowed solutions, expressed as the sum of sinusoids and hyperbolic sinusoids of λx, to describe the transverse vibrations of a rod of length L clamped at both ends. At a clamped point both u and $\partial u/\partial x$ must vanish; show that this implies that $\cos(\lambda L) \cosh(\lambda L) = 1$, thus determining the frequencies ω at which the rod can vibrate.

20.19   An incompressible fluid of density ρ and negligible viscosity flows with velocity v along a thin, straight, perfectly light and flexible tube, of cross-section A which is held under tension T. Assume that small transverse displacements u of the tube are governed by

$$\frac{\partial^2 u}{\partial t^2} + 2v\frac{\partial^2 u}{\partial x \partial t} + \left(v^2 - \frac{T}{\rho A}\right)\frac{\partial^2 u}{\partial x^2} = 0.$$

(a) Show that the general solution consists of a superposition of two waveforms travelling with different speeds.

(b) The tube initially has a small transverse displacement $u = a \cos kx$ and is suddenly released from rest. Find its subsequent motion.

20.20 A sheet of material of thickness w, specific heat capacity c and thermal conductivity k is isolated in a vacuum, but its two sides are exposed to fluxes of radiant heat of strengths J_1 and J_2. Ignoring short-term transients, show that the temperature difference between its two surfaces is steady at $(J_2 - J_1)w/2k$, whilst their average temperature increases at a rate $(J_2 + J_1)/cw$.

20.21 In an electrical cable of resistance R and capacitance C, each per unit length, voltage signals obey the equation $\partial^2 V / \partial x^2 = RC \partial V / \partial t$. This has solutions of the form given in (20.36) and also of the form $V = Ax + D$.

(a) Find a combination of these that represents the situation after a steady voltage V_0 is applied at $x = 0$ at time $t = 0$.
(b) Obtain a solution describing the propagation of the voltage signal resulting from the application of the signal $V = V_0$ for $0 < t < T$, $V = 0$ otherwise, to the end $x = 0$ of an infinite cable.
(c) Show that for $t \gg T$ the maximum signal occurs at a value of x proportional to $t^{1/2}$ and has a magnitude proportional to t^{-1}.

20.22 The daily and annual variations of temperature at the surface of the earth may be represented by sine-wave oscillations, with equal amplitudes and periods of 1 day and 365 days respectively. Assume that for (angular) frequency ω the temperature at depth x in the earth is given by $u(x, t) = A \sin(\omega t + \mu x) \exp(-\lambda x)$, where λ and μ are constants.

(a) Use the diffusion equation to find the values of λ and μ.
(b) Find the ratio of the depths below the surface at which the two amplitudes have dropped to $1/20$ of their surface values.
(c) At what time of year is the soil coldest at the greater of these depths, assuming that the smoothed annual variation in temperature at the surface has a minimum on February 1st?

20.23 Consider each of the following situations in a qualitative way and determine the equation type, the nature of the boundary curve and the type of boundary conditions involved:

(a) a conducting bar given an initial temperature distribution and then thermally isolated;
(b) two long conducting concentric cylinders, on each of which the voltage distribution is specified;
(c) two long conducting concentric cylinders, on each of which the charge distribution is specified;
(d) a semi-infinite string, the end of which is made to move in a prescribed way.

20.24 *This example gives a formal demonstration that the type of a second-order PDE (elliptic, parabolic or hyperbolic) cannot be changed by a new choice of independent variable. The algebra is somewhat lengthy, but straightforward.*
If a change of variable $\xi = \xi(x, y)$, $\eta = \eta(x, y)$ is made in (20.19), so that it reads

$$A' \frac{\partial^2 u}{\partial \xi^2} + B' \frac{\partial^2 u}{\partial \xi \partial \eta} + C' \frac{\partial^2 u}{\partial \eta^2} + D' \frac{\partial u}{\partial \xi} + E' \frac{\partial u}{\partial \eta} + F'u = R'(\xi, \eta),$$

show that

$$B'^2 - 4A'C' = (B^2 - 4AC) \left[\frac{\partial(\xi, \eta)}{\partial(x, y)} \right]^2.$$

Hence deduce the conclusion stated above.

20.25   The Klein–Gordon equation (which is satisfied by the quantum-mechanical wave-function $\Phi(\mathbf{r})$ of a relativistic spinless particle of non-zero mass m) is

$$\nabla^2 \Phi - m^2 \Phi = 0.$$

Show that the solution for the scalar field $\Phi(\mathbf{r})$ in any volume V bounded by a surface S is unique if either Dirichlet or Neumann boundary conditions are specified on S.

20.9 Hints and answers

20.1   (a) Yes, $p^2 - 4p - 4$; (b) no, $(p - y)^2$; (c) yes, $(p^2 + 4)/(2p^2 + p)$.

20.3   Each equation is effectively an ordinary differential equation, but with a function of the non-integrated variable as the constant of integration;
(a) $u = xy(2 - \ln x)$; (b) $u = x^{-1}(1 - e^y) + xe^y$.

20.5   (a) $(y^2 - x^2)^{1/2}$; (b) $1 + f(y^2 - x^2)$, where $f(0) = 0$.

20.7 $u = y + f(y - \ln(\sin x))$; (a) $u = \ln(\sin x)$; (b) $u = y + [y - \ln(\sin x)]^2$.

20.9   General solution is $u(x, y) = f(x + y) + g(x + y/2)$. Show that $2p = -g'(p)/2$, and hence $g(p) = k - 2p^2$, whilst $f(p) = p^2 - k$, leading to $u(x, y) = -x^2 + y^2/2$; $u(0, 1) = 1/2$.

20.11 $p = x^2 + 2y^2$; $u(x, y) = f(p) + x^2y^2/2$.

    (a) $u(x, y) = (x^2 + 2y^2 + x^2y^2 - 2)/2$; $u(2, 2) = 13$. The line $y = 1$ cuts each characteristic in zero or two distinct points, but this causes no difficulty with the given boundary conditions.

    (b) As in (a).

    (c) The solution is defined over the space between the ellipses $p = 2$ and $p = 11$; $(2, 2)$ lies on $p = 12$, and so $u(2, 2)$ is undetermined.

    (d) $u(x, y) = (x^2 + 2y^2)^{1/2} + x^2y^2/2$; $u(2, 2) = 8 + \sqrt{12}$.

    (e) The line $y = 0$ cuts each characteristic in two distinct points. No differentiable form of $f(p)$ gives $f(\pm a) = \pm a$ respectively, and so there is no solution.

    (f) The solution is only specified on $p = 21$, and so $u(2, 2)$ is undetermined.

    (g) The solution is specified on $p = 12$, and so $u(2, 2) = 5 + \frac{1}{2}(4)(4) = 13$.

20.13   The equation becomes $\partial^2 f/\partial \xi \partial \eta = -14$, with solution $f(\xi, \eta) = f(\xi) + g(\eta) - 14\xi\eta$, which can be compared with the answer from the previous question; $f_1(z) = 10z^2$ and $f_2(z) = 5z^2$.

20.15 $u(x, y) - f(x + iy) + g(x - iy) + (1/12)x^4(y^2 - (1/15)x^2)$. In the last term, x and y may be interchanged. There are (infinitely) many other possibilities for the specific PI, e.g. $[15x^2y^2(x^2 + y^2) - (x^6 + y^6)]/360$.

20.17 $E = p^2/(2m)$, the relationship between energy and momentum for a non-relativistic particle; $u(\mathbf{r}, t) = A\exp[i(\mathbf{p} \cdot \mathbf{r} - Et)/\hbar]$, a plane wave of wave number $\mathbf{k} = \mathbf{p}/\hbar$ and angular frequency $\omega = E/\hbar$ travelling in the direction $\mathbf{p}/p$.

20.19   (a) $c = v \pm \alpha$ where $\alpha^2 = T/\rho A$;
(b) $u(x, t) = a\cos[k(x - vt)]\cos(k\alpha t) - (v a/\alpha)\sin[k(x - vt)]\sin(k\alpha t)$.

20.21   (a) $V_0\left[1 - (2/\sqrt{\pi}) \int^{\frac{1}{2}x(CR/t)^{1/2}} \exp(-v^2)\, dv\right]$;

(b) consider the input as equivalent to V_0 applied at $t = 0$ and continued and $-V_0$ applied at $t = T$ and continued;

$$V(x, t) = \frac{2V_0}{\sqrt{\pi}} \int_{\frac{1}{2}x(CR/t)^{1/2}}^{\frac{1}{2}x[CR/(t-T)]^{1/2}} \exp\left(-v^2\right)\, dv;$$

(c) For $t \gg T$, maximum at $x = [2t/(CR)]^{1/2}$ with value $\dfrac{V_0 T \exp(-\frac{1}{2})}{(2\pi)^{1/2}t}$.

711

20.23   (a) Parabolic, open, Dirichlet $u(x,0)$ given, Neumann $\partial u/\partial x = 0$ at $x = \pm L/2$
             for all t;
         (b) elliptic, closed, Dirichlet;
         (c) elliptic, closed, Neumann $\partial u/\partial n = \sigma/\epsilon_0$;
         (d) hyperbolic, open, Cauchy.

20.25   Follow an argument similar to that in section 20.7 and argue that the additional
         term $\int m^2|w|^2\,dV$ must be zero, and hence that $w = 0$ everywhere.

Partial differential equations: separation of variables and other methods

In the previous chapter we demonstrated the methods by which general solutions of some partial differential equations (PDEs) may be obtained in terms of arbitrary functions. In particular, solutions containing the independent variables in definite combinations were sought, thus reducing the effective number of them.

In the present chapter we begin by taking the opposite approach, namely that of trying to keep the independent variables as separate as possible, using the method of separation of variables. We then consider integral transform methods by which one of the independent variables may be eliminated, at least from differential coefficients. Finally, we discuss the use of Green's functions in solving inhomogeneous problems.

21.1 Separation of variables: the general method

Suppose we seek a solution $u(x, y, z, t)$ to some PDE (expressed in Cartesian coordinates). Let us attempt to obtain one that has the product form[§]

$$u(x, y, z, t) = X(x)Y(y)Z(z)T(t). \tag{21.1}$$

A solution that has this form is said to be *separable* in x, y, z and t, and seeking solutions of this form is called the method of *separation of variables*.

As simple examples we may observe that, of the functions

(i) $xyz^2 \sin bt$,   (ii) $xy + zt$,   (iii) $(x^2 + y^2)z \cos \omega t$,

(i) is completely separable, (ii) is inseparable in that no single variable can be separated out from it and written as a multiplicative factor, whilst (iii) is separable in z and t but not in x and y.

[§] It should be noted that the conventional use here of upper-case (capital) letters to denote the functions of the corresponding lower-case variable is intended to enable an easy correspondence between a function and its argument to be made.

When seeking PDE solutions of the form (21.1), we are requiring not that there is no connection at all between the functions X, Y, Z and T (for example, certain parameters may appear in two or more of them), but only that X does not depend upon y, z, t, that Y does not depend on x, z, t, and so on.

For a general PDE it is likely that a separable solution is impossible, but certainly some common and important equations do have useful solutions of this form, and we will illustrate the method of solution by studying the three-dimensional wave equation

$$\nabla^2 u(\mathbf{r}) = \frac{1}{c^2} \frac{\partial^2 u(\mathbf{r})}{\partial t^2}. \tag{21.2}$$

We will work in Cartesian coordinates for the present and assume a solution of the form (21.1); the solutions in alternative coordinate systems, e.g. spherical or cylindrical polars, are considered in section 21.3. Expressed in Cartesian coordinates (21.2) takes the form

$$\frac{\partial^2 u}{\partial x^2} + \frac{\partial^2 u}{\partial y^2} + \frac{\partial^2 u}{\partial z^2} = \frac{1}{c^2} \frac{\partial^2 u}{\partial t^2}; \tag{21.3}$$

substituting (21.1) gives

$$\frac{d^2 X}{dx^2} YZT + X \frac{d^2 Y}{dy^2} ZT + XY \frac{d^2 Z}{dz^2} T = \frac{1}{c^2} XYZ \frac{d^2 T}{dt^2},$$

which can also be written as

$$X'' YZT + XY'' ZT + XYZ'' T = \frac{1}{c^2} XYZT'', \tag{21.4}$$

where in each case the primes refer to the *ordinary* derivative with respect to the independent variable upon which the function depends. This emphasises the fact that each of the functions X, Y, Z and T has only one independent variable and thus its only derivative is its total derivative. For the same reason, in each term in (21.4) three of the four functions are unaltered by the partial differentiation and behave exactly as constant multipliers.

If we now divide (21.4) throughout by $u = XYZT$ we obtain

$$\frac{X''}{X} + \frac{Y''}{Y} + \frac{Z''}{Z} = \frac{1}{c^2} \frac{T''}{T}. \tag{21.5}$$

This form shows the particular characteristic that is the basis of the method of separation of variables, namely that of the four terms the first is a function of x only, the second of y only, the third of z only and the RHS a function of t only and yet there is an equation connecting them. This can only be so for all x, y, z and t if *each* of the terms does not in fact, despite appearances, depend upon the corresponding independent variable but *is equal to a constant*, the four constants being such that (21.5) is satisfied.

Since there is only one equation to be satisfied and four constants involved, there is considerable freedom in the values they may take. For the purposes of our illustrative example let us make the choice of $-l^2$, $-m^2$, $-n^2$, for the first three constants. The constant associated with $c^{-2}T''/T$ must then have the value $-\mu^2 = -(l^2 + m^2 + n^2)$.

Having recognised that each term of (21.5) is individually equal to a constant (or parameter), we can now replace (21.5) by four separate ordinary differential equations (ODEs):

$$\frac{X''}{X} = -l^2, \quad \frac{Y''}{Y} = -m^2, \quad \frac{Z''}{Z} = -n^2, \quad \frac{1}{c^2}\frac{T''}{T} = -\mu^2. \qquad (21.6)$$

The important point to notice is not the simplicity of the equations (21.6) (the corresponding ones for a general PDE are usually far from simple) but that, by the device of assuming a separable solution, a *partial* differential equation (21.3), containing derivatives with respect to the four independent variables all in one equation, has been reduced to four *separate ordinary* differential equations (21.6). The ordinary equations are connected through four constant parameters that satisfy an algebraic relation. These constants are called *separation constants*.

The general solutions of the equations (21.6) can be deduced straightforwardly and are

$$X(x) = A\exp(ilx) + B\exp(-ilx),$$
$$Y(y) = C\exp(imy) + D\exp(-imy),$$
$$Z(z) = E\exp(inz) + F\exp(-inz),$$
$$T(t) = G\exp(ic\mu t) + H\exp(-ic\mu t),$$
$$\qquad (21.7)$$

where $A, B, \ldots, H$ are constants, which may be determined if boundary conditions are imposed on the solution. Depending on the geometry of the problem and any boundary conditions, it is sometimes more appropriate to write the solutions (21.7) in the alternative form

$$X(x) = A'\cos lx + B'\sin lx,$$
$$Y(y) = C'\cos my + D'\sin my,$$
$$Z(z) = E'\cos nz + F'\sin nz,$$
$$T(t) = G'\cos(c\mu t) + H'\sin(c\mu t),$$
$$\qquad (21.8)$$

for some different set of constants $A', B', \ldots, H'$. Clearly the choice of how best to represent the solution depends on the problem being considered.

As an example, suppose that we take as particular solutions the four functions

$$X(x) = \exp(ilx), \qquad Y(y) = \exp(imy),$$
$$Z(z) = \exp(inz), \qquad T(t) = \exp(-ic\mu t).$$

This gives a particular solution of the original PDE (21.3)

$$u(x, y, z, t) = \exp(ilx)\exp(imy)\exp(inz)\exp(-ic\mu t)$$
$$= \exp[i(lx + my + nz - c\mu t)],$$

which is a special case of the solution (20.33) obtained in the previous chapter and represents a plane wave of unit amplitude propagating in a direction given by the vector with components l, m, n in a Cartesian coordinate system. In the conventional notation of wave theory, l, m and n are the components of the wave-number vector $\mathbf{k}$, whose magnitude is given by $k = 2\pi/\lambda$, where λ is the wavelength of the wave; $c\mu$ is the angular frequency ω of the wave. This gives the equation in the form

$$u(x, y, z, t) = \exp[i(k_x x + k_y y + k_z z - \omega t)]$$
$$= \exp[i(\mathbf{k} \cdot \mathbf{r} - \omega t)],$$

and makes the exponent dimensionless.

The method of separation of variables can be applied to many commonly occurring PDEs encountered in physical applications.

> ▶*Use the method of separation of variables to obtain for the one-dimensional diffusion equation*
>
> $$\kappa \frac{\partial^2 u}{\partial x^2} = \frac{\partial u}{\partial t}, \qquad (21.9)$$
>
> *a solution that tends to zero as $t \to \infty$ for all x.*

Here we have only two independent variables x and t and we therefore assume a solution of the form

$$u(x, t) = X(x)T(t).$$

Substituting this expression into (21.9) and dividing through by $u = XT$ (and also by κ) we obtain

$$\frac{X''}{X} = \frac{T'}{\kappa T}.$$

Now, arguing exactly as above that the LHS is a function of x only and the RHS is a function of t only, we conclude that each side must equal a constant, which, anticipating the result and noting the imposed boundary condition, we will take as $-\lambda^2$. This gives us two ordinary equations,

$$X'' + \lambda^2 X = 0, \qquad (21.10)$$
$$T' + \lambda^2 \kappa T = 0, \qquad (21.11)$$

which have the solutions

$$X(x) = A\cos\lambda x + B\sin\lambda x,$$
$$T(t) = C\exp(-\lambda^2\kappa t).$$

Combining these to give the assumed solution $u = XT$ yields (absorbing the constant C into A and B)

$$u(x, t) = (A\cos\lambda x + B\sin\lambda x)\exp(-\lambda^2\kappa t). \qquad (21.12)$$

In order to satisfy the boundary condition $u \to 0$ as $t \to \infty$, $\lambda^2 \kappa$ must be > 0. Since κ is real and > 0, this implies that λ is a real non-zero number and that the solution is sinusoidal in x and is not a disguised hyperbolic function; this was our reason for choosing the separation constant as $-\lambda^2$. ◄

As a final example we consider Laplace's equation in Cartesian coordinates; this may be treated in a similar manner.

▶Use the method of separation of variables to obtain a solution for the two-dimensional Laplace equation,

$$\frac{\partial^2 u}{\partial x^2} + \frac{\partial^2 u}{\partial y^2} = 0. \tag{21.13}$$

If we assume a solution of the form $u(x,y) = X(x)Y(y)$ then, following the above method, and taking the separation constant as λ^2, we find

$$X'' = \lambda^2 X, \qquad Y'' = -\lambda^2 Y.$$

Taking λ^2 as > 0, the general solution becomes

$$u(x,y) = (A \cosh \lambda x + B \sinh \lambda x)(C \cos \lambda y + D \sin \lambda y). \tag{21.14}$$

An alternative form, in which the exponentials are written explicitly, may be useful for other geometries or boundary conditions:

$$u(x,y) = [A \exp \lambda x + B \exp(-\lambda x)](C \cos \lambda y + D \sin \lambda y), \tag{21.15}$$

with different constants A and B.

If $\lambda^2 < 0$ then the roles of x and y interchange. The particular combination of sinusoidal and hyperbolic functions and the values of λ allowed will be determined by the geometrical properties of any specific problem, together with any prescribed or necessary boundary conditions. ◄

We note here that a particular case of the solution (21.14) links up with the 'combination' result $u(x,y) = f(x + iy)$ of the previous chapter (equations (20.24) and following), namely that if $A = B$ and $D = iC$ then the solution is the same as $f(p) = AC \exp \lambda p$ with $p = x + iy$.

21.2 Superposition of separated solutions

It will be noticed in the previous two examples that there is considerable freedom in the values of the separation constant λ, the only essential requirement being that λ has the *same* value in both parts of the solution, i.e. the part depending on x and the part depending on y (or t). This is a general feature for solutions in separated form, which, if the original PDE has n independent variables, will contain $n - 1$ separation constants. All that is required in general is that we associate the correct function of one independent variable with the appropriate functions of the others, the correct function being the one with the same values of the separation constants.

If the original PDE is linear (as are the Laplace, Schrödinger, diffusion and wave equations) then mathematically acceptable solutions can be formed by

717

superposing solutions corresponding to different allowed values of the separation constants. To take a two-variable example: if

$$u_{\lambda_1}(x, y) = X_{\lambda_1}(x)Y_{\lambda_1}(y)$$

is a solution of a linear PDE obtained by giving the separation constant the value λ_1, then the superposition

$$u(x, y) = a_1 X_{\lambda_1}(x)Y_{\lambda_1}(y) + a_2 X_{\lambda_2}(x)Y_{\lambda_2}(y) + \cdots = \sum_i a_i X_{\lambda_i}(x)Y_{\lambda_i}(y) \tag{21.16}$$

is also a solution for any constants a_i, provided that the λ_i are the allowed values of the separation constant λ given the imposed boundary conditions. Note that if the boundary conditions allow any of the separation constants to be zero then the form of the general solution is normally different and must be deduced by returning to the separated ordinary differential equations. We will encounter this behaviour in section 21.3.

The value of the superposition approach is that a boundary condition, say that $u(x, y)$ takes a particular form $f(x)$ when $y = 0$, might be met by choosing the constants a_i such that

$$f(x) = \sum_i a_i X_{\lambda_i}(x)Y_{\lambda_i}(0).$$

In general, this will be possible provided that the functions $X_{\lambda_i}(x)$ form a complete set – as do the sinusoidal functions of Fourier series or the spherical harmonics discussed in subsection 18.3.

> ▶ *A semi-infinite rectangular metal plate occupies the region $0 \leq x \leq \infty$ and $0 \leq y \leq b$ in the xy-plane. The temperature at the far end of the plate and along its two long sides is fixed at $0\,°C$. If the temperature of the plate at $x = 0$ is also fixed and is given by $f(y)$, find the steady-state temperature distribution $u(x,y)$ of the plate. Hence find the temperature distribution if $f(y) = u_0$, where u_0 is a constant.*

The physical situation is illustrated in figure 21.1. With the notation we have used several times before, the two-dimensional heat diffusion equation satisfied by the temperature $u(x, y, t)$ is

$$\kappa \left(\frac{\partial^2 u}{\partial x^2} + \frac{\partial^2 u}{\partial y^2} \right) = \frac{\partial u}{\partial t},$$

with $\kappa = k/(s\rho)$. In this case, however, we are asked to find the steady-state temperature, which corresponds to $\partial u/\partial t = 0$, and so we are led to consider the (two-dimensional) Laplace equation

$$\frac{\partial^2 u}{\partial x^2} + \frac{\partial^2 u}{\partial y^2} = 0.$$

We saw that assuming a separable solution of the form $u(x, y) = X(x)Y(y)$ led to solutions such as (21.14) or (21.15), or equivalent forms with x and y interchanged. In the current problem we have to satisfy the boundary conditions $u(x, 0) = 0 = u(x, b)$ and so a solution that is sinusoidal in y seems appropriate. Furthermore, since we require $u(\infty, y) = 0$ it is best to write the x-dependence of the solution explicitly in terms of

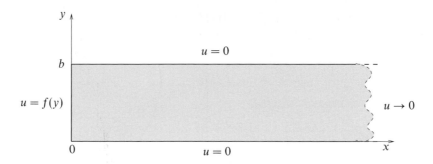

Figure 21.1   A semi-infinite metal plate whose edges are kept at fixed temperatures.

exponentials rather than of hyperbolic functions. We therefore write the separable solution in the form (21.15) as

$$u(x, y) = [A \exp \lambda x + B \exp(-\lambda x)](C \cos \lambda y + D \sin \lambda y).$$

Applying the boundary conditions, we see firstly that $u(\infty, y) = 0$ implies $A = 0$ if we take $\lambda > 0$. Secondly, since $u(x, 0) = 0$ we may set $C = 0$, which, if we absorb the constant D into B, leaves us with

$$u(x, y) = B \exp(-\lambda x) \sin \lambda y.$$

But, using the condition $u(x, b) = 0$, we require $\sin \lambda b = 0$ and so λ must be equal to $n\pi/b$, where n is any positive integer.

Using the principle of superposition (21.16), the general solution satisfying the given boundary conditions can therefore be written

$$u(x, y) = \sum_{n=1}^{\infty} B_n \exp(-n\pi x/b) \sin(n\pi y/b), \tag{21.17}$$

for some constants B_n. Notice that in the sum in (21.17) we have omitted negative values of n since they would lead to exponential terms that diverge as $x \to \infty$. The $n = 0$ term is also omitted since it is identically zero. Using the remaining boundary condition $u(0, y) = f(y)$ we see that the constants B_n must satisfy

$$f(y) = \sum_{n=1}^{\infty} B_n \sin(n\pi y/b). \tag{21.18}$$

This is clearly a Fourier sine series expansion of $f(y)$ (see chapter 12). For (21.18) to hold, however, the continuation of $f(y)$ outside the region $0 \le y \le b$ must be an odd periodic function with period $2b$ (see figure 21.2). We also see from figure 21.2 that if the original function $f(y)$ does not equal zero at either of $y = 0$ and $y = b$ then its continuation has a discontinuity at the corresponding point(s); nevertheless, as discussed in chapter 12, the Fourier series will converge to the mid-points of these jumps and hence tend to zero in this case. If, however, the top and bottom edges of the plate were held not at $0\,°C$ but at some other non-zero temperature, then, in general, the final solution would possess discontinuities at the corners $x = 0$, $y = 0$ and $x = 0$, $y = b$.

Bearing in mind these technicalities, the coefficients B_n in (21.18) are given by

$$B_n = \frac{2}{b} \int_0^b f(y) \sin\left(\frac{n\pi y}{b}\right) dy. \tag{21.19}$$

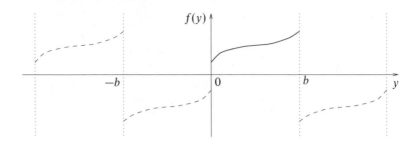

Figure 21.2 The continuation of $f(y)$ for a Fourier sine series.

Therefore, if $f(y) = u_0$ (i.e. the temperature of the side at $x = 0$ is constant along its length), (21.19) becomes

$$B_n = \frac{2}{b} \int_0^b u_0 \sin\left(\frac{n\pi y}{b}\right) dy$$

$$= \left[-\frac{2u_0}{b} \frac{b}{n\pi} \cos\left(\frac{n\pi y}{b}\right)\right]_0^b$$

$$= -\frac{2u_0}{n\pi}[(-1)^n - 1] = \begin{cases} 4u_0/n\pi & \text{for } n \text{ odd,} \\ 0 & \text{for } n \text{ even.} \end{cases}$$

Therefore the required solution is

$$u(x,y) = \sum_{n \text{ odd}} \frac{4u_0}{n\pi} \exp\left(-\frac{n\pi x}{b}\right) \sin\left(\frac{n\pi y}{b}\right). \blacktriangleleft$$

In the above example the boundary conditions meant that one term in each part of the separable solution could be immediately discarded, making the problem much easier to solve. Sometimes, however, a little ingenuity is required in writing the separable solution in such a way that certain parts can be neglected immediately.

> ►*Suppose that the semi-infinite rectangular metal plate in the previous example is replaced by one that in the x-direction has finite length a. The temperature of the right-hand edge is fixed at $0\,°C$ and all other boundary conditions remain as before. Find the steady-state temperature in the plate.*

As in the previous example, the boundary conditions $u(x,0) = 0 = u(x,b)$ suggest a solution that is sinusoidal in y. In this case, however, we require $u = 0$ on $x = a$ (rather than at infinity) and so a solution in which the x-dependence is written in terms of hyperbolic functions, such as (21.14), rather than exponentials is more appropriate. Moreover, since the constants in front of the hyperbolic functions are, at this stage, arbitrary, we may write the separable solution in the most convenient way that ensures that the condition $u(a, y) = 0$ is straightforwardly satisfied. We therefore write

$$u(x,y) = [A \cosh \lambda(a - x) + B \sinh \lambda(a - x)](C \cos \lambda y + D \sin \lambda y).$$

Now the condition $u(a, y) = 0$ is easily satisfied by setting $A = 0$. As before the conditions $u(x,0) = 0 = u(x,b)$ imply $C = 0$ and $\lambda = n\pi/b$ for integer n. Superposing the solutions for different n we then obtain

$$u(x, y) = \sum_{n=1}^{\infty} B_n \sinh[n\pi(a - x)/b] \sin(n\pi y/b), \qquad (21.20)$$

for some constants B_n. We have omitted negative values of n in the sum (21.20) since the relevant terms are already included in those obtained for positive n. Again the $n = 0$ term is identically zero. Using the final boundary condition $u(0, y) = f(y)$ as above we find that the constants B_n must satisfy

$$f(y) = \sum_{n=1}^{\infty} B_n \sinh(n\pi a/b) \sin(n\pi y/b),$$

and, remembering the caveats discussed in the previous example, the B_n are therefore given by

$$B_n = \frac{2}{b \sinh(n\pi a/b)} \int_0^b f(y) \sin(n\pi y/b) \, dy. \qquad (21.21)$$

For the case where $f(y) = u_0$, following the working of the previous example gives (21.21) as

$$B_n = \frac{4u_0}{n\pi \sinh(n\pi a/b)} \quad \text{for } n \text{ odd}, \qquad B_n = 0 \quad \text{for } n \text{ even}. \qquad (21.22)$$

The required solution is thus

$$u(x, y) = \sum_{n \text{ odd}} \frac{4u_0}{n\pi \sinh(n\pi a/b)} \sinh[n\pi(a - x)/b] \sin\left(n\pi y/b\right).$$

We note that, as required, in the limit $a \to \infty$ this solution tends to the solution of the previous example. ◀

Often the principle of superposition can be used to write the solution to problems with more complicated boundary conditions as the sum of solutions to problems that each satisfy only some part of the boundary condition but when added together satisfy all the conditions.

▶ Find the steady-state temperature in the (finite) rectangular plate of the previous example, subject to the boundary conditions $u(x, b) = 0$, $u(a, y) = 0$ and $u(0, y) = f(y)$ as before, but now, in addition, $u(x, 0) = g(x)$.

Figure 21.3(c) shows the imposed boundary conditions for the metal plate. Although we could find a solution to this problem using the methods presented above, we can arrive at the answer almost immediately by using the principle of superposition and the result of the previous example.

Let us suppose the required solution $u(x, y)$ is made up of two parts:

$$u(x, y) = v(x, y) + w(x, y),$$

where $v(x, y)$ is the solution satisfying the boundary conditions shown in figure 21.3(a),

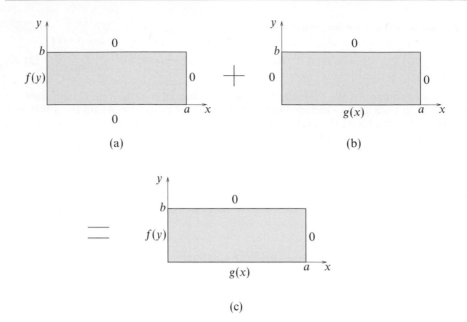

Figure 21.3   Superposition of boundary conditions for a metal plate.

whilst $w(x, y)$ is the solution satisfying the boundary conditions in figure 21.3(b). It is clear that $v(x, y)$ is simply given by the solution to the previous example,

$$v(x, y) = \sum_{n \text{ odd}} B_n \sinh \left[\frac{n\pi(a - x)}{b} \right] \sin \left(\frac{n\pi y}{b} \right),$$

where B_n is given by (21.21). Moreover, by symmetry, $w(x, y)$ must be of the same form as $v(x, y)$ but with x and a interchanged with y and b, respectively, and with $f(y)$ in (21.21) replaced by $g(x)$. Therefore the required solution can be written down immediately without further calculation as

$$u(x, y) = \sum_{n \text{ odd}} B_n \sinh \left[\frac{n\pi(a - x)}{b} \right] \sin \left(\frac{n\pi y}{b} \right) + \sum_{n \text{ odd}} C_n \sinh \left[\frac{n\pi(b - y)}{a} \right] \sin \left(\frac{n\pi x}{a} \right),$$

the B_n being given by (21.21) and the C_n by

$$C_n = \frac{2}{a \sinh(n\pi b/a)} \int_0^a g(x) \sin(n\pi x/a) \, dx.$$

Clearly, this method may be extended to cases in which three or four sides of the plate have non-zero boundary conditions. ◀

As a final example of the usefulness of the principle of superposition we now consider a problem that illustrates how to deal with inhomogeneous boundary conditions by a suitable change of variables.

722

▶*A bar of length L is initially at a temperature of $0\,^\circ$C. One end of the bar ($x = 0$) is held at $0\,^\circ$C and the other is supplied with heat at a constant rate per unit area of H. Find the temperature distribution within the bar after a time t.*

With our usual notation, the heat diffusion equation satisfied by the temperature $u(x,t)$ is

$$\kappa \frac{\partial^2 u}{\partial x^2} = \frac{\partial u}{\partial t},$$

with $\kappa = k/(s\rho)$, where k is the thermal conductivity of the bar, s is its specific heat capacity and ρ is its density.

The boundary conditions can be written as

$$u(x,0) = 0, \qquad u(0,t) = 0, \qquad \frac{\partial u(L,t)}{\partial x} = \frac{H}{k},$$

the last of which is inhomogeneous. In general, inhomogeneous boundary conditions can cause difficulties and it is usual to attempt a transformation of the problem into an equivalent homogeneous one. To this end, let us assume that the solution to our problem takes the form

$$u(x,t) = v(x,t) + w(x),$$

where the function $w(x)$ is to be suitably determined. In terms of v and w the problem becomes

$$\kappa\left(\frac{\partial^2 v}{\partial x^2} + \frac{d^2 w}{dx^2}\right) = \frac{\partial v}{\partial t},$$
$$v(x,0) + w(x) = 0,$$
$$v(0,t) + w(0) = 0,$$
$$\frac{\partial v(L,t)}{\partial x} + \frac{dw(L)}{dx} = \frac{H}{k}.$$

There are several ways of choosing $w(x)$ so as to make the new problem straightforward. Using some physical insight, however, it is clear that ultimately (at $t = \infty$), when all transients have died away, the end $x = L$ will attain a temperature u_0 such that $ku_0/L = H$ and there will be a constant temperature gradient $u(x,\infty) = u_0 x/L$. We therefore choose

$$w(x) = \frac{Hx}{k}.$$

Since the second derivative of $w(x)$ is zero, v satisfies the diffusion equation and the boundary conditions on v are now given by

$$v(x,0) = -\frac{Hx}{k}, \qquad v(0,t) = 0, \qquad \frac{\partial v(L,t)}{\partial x} = 0,$$

which are homogeneous in x.

From (21.12) a separated solution for the one-dimensional diffusion equation is

$$v(x,t) = (A\cos\lambda x + B\sin\lambda x)\exp(-\lambda^2 \kappa t),$$

corresponding to a separation constant $-\lambda^2$. If we restrict λ to be real then all these solutions are transient ones decaying to zero as $t \to \infty$. These are just what is required to add to $w(x)$ to give the correct solution as $t \to \infty$. In order to satisfy $v(0,t) = 0$, however, we require $A = 0$. Furthermore, since

$$\frac{\partial v}{\partial x} = B\exp(-\lambda^2 \kappa t)\lambda\cos\lambda x,$$

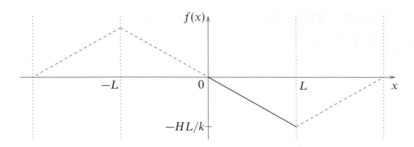

Figure 21.4 The appropriate continuation for a Fourier series containing only sine terms.

in order to satisfy $\partial v(L, t)/\partial x = 0$ we require $\cos \lambda L = 0$, and so λ is restricted to the values

$$\lambda = \frac{n\pi}{2L},$$

where n is an odd non-negative integer, i.e. $n = 1, 3, 5, \ldots$.

Thus, to satisfy the boundary condition $v(x, 0) = -Hx/k$, we must have

$$\sum_{n \text{ odd}} B_n \sin\left(\frac{n\pi x}{2L}\right) = -\frac{Hx}{k},$$

in the range $x = 0$ to $x = L$. In this case we must be more careful about the continuation of the function $-Hx/k$, for which the Fourier sine series is required. We want a series that is odd in x (sine terms only) and continuous as $x = 0$ and $x = L$ (no discontinuities, since the series must converge at the end-points). This leads to a continuation of the function as shown in figure 21.4, with a period of $L' = 4L$. Following the discussion of section 12.3, since this continuation is odd about $x = 0$ and even about $x = L'/4 = L$ it can indeed be expressed as a Fourier sine series containing only odd-numbered terms.

The corresponding Fourier series coefficients are found to be

$$B_n = \frac{-8HL}{k\pi^2} \frac{(-1)^{(n-1)/2}}{n^2} \qquad \text{for } n \text{ odd},$$

and thus the final formula for $u(x, t)$ is

$$u(x, t) = \frac{Hx}{k} - \frac{8HL}{k\pi^2} \sum_{n \text{ odd}} \frac{(-1)^{(n-1)/2}}{n^2} \sin\left(\frac{n\pi x}{2L}\right) \exp\left(-\frac{kn^2\pi^2 t}{4L^2 s\rho}\right),$$

giving the temperature for all positions $0 \le x \le L$ and for all times $t \ge 0$. ◀

We note that in all the above examples the boundary conditions restricted the separation constant(s) to an infinite number of *discrete* values, usually integers. If, however, the boundary conditions allow the separation constant(s) λ to take a *continuum* of values then the summation in (21.16) is replaced by an integral over λ. This is discussed further in connection with integral transform methods in section 21.4.

21.3 Separation of variables in polar coordinates

So far we have considered the solution of PDEs only in Cartesian coordinates, but many systems in two and three dimensions are more naturally expressed in some form of polar coordinates, in which full advantage can be taken of any inherent symmetries. For example, the potential associated with an isolated point charge has a very simple expression, $q/(4\pi\epsilon_0 r)$, when polar coordinates are used, but involves all three coordinates and square roots when Cartesians are employed. For these reasons we now turn to the separation of variables in plane polar, cylindrical polar and spherical polar coordinates.

Most of the PDEs we have considered so far have involved the operator ∇^2, e.g. the wave equation, the diffusion equation, Schrödinger's equation and Poisson's equation (and of course Laplace's equation). It is therefore appropriate that we recall the expressions for ∇^2 when expressed in polar coordinate systems. From chapter 10, in plane polars, cylindrical polars and spherical polars, respectively, we have

$$\nabla^2 = \frac{1}{\rho}\frac{\partial}{\partial\rho}\left(\rho\frac{\partial}{\partial\rho}\right) + \frac{1}{\rho^2}\frac{\partial^2}{\partial\phi^2}, \tag{21.23}$$

$$\nabla^2 = \frac{1}{\rho}\frac{\partial}{\partial\rho}\left(\rho\frac{\partial}{\partial\rho}\right) + \frac{1}{\rho^2}\frac{\partial^2}{\partial\phi^2} + \frac{\partial^2}{\partial z^2}, \tag{21.24}$$

$$\nabla^2 = \frac{1}{r^2}\frac{\partial}{\partial r}\left(r^2\frac{\partial}{\partial r}\right) + \frac{1}{r^2\sin\theta}\frac{\partial}{\partial\theta}\left(\sin\theta\frac{\partial}{\partial\theta}\right) + \frac{1}{r^2\sin^2\theta}\frac{\partial^2}{\partial\phi^2}. \tag{21.25}$$

Of course the first of these may be obtained from the second by taking z to be identically zero.

21.3.1 Laplace's equation in polar coordinates

The simplest of the equations containing ∇^2 is Laplace's equation,

$$\nabla^2 u(\mathbf{r}) = 0. \tag{21.26}$$

Since it contains most of the essential features of the other more complicated equations, we will consider its solution first.

Laplace's equation in plane polars

Suppose that we need to find a solution of (21.26) that has a prescribed behaviour on the circle $\rho = a$ (e.g. if we are finding the shape taken up by a circular drumskin when its rim is slightly deformed from being planar). Then we may seek solutions of (21.26) that are separable in ρ and ϕ (measured from some arbitrary radius as $\phi = 0$) and hope to accommodate the boundary condition by examining the solution for $\rho = a$.

Thus, writing $u(\rho, \phi) = P(\rho)\Phi(\phi)$ and using the expression (21.23), Laplace's equation (21.26) becomes

$$\frac{\Phi}{\rho} \frac{\partial}{\partial \rho} \left(\rho \frac{\partial P}{\partial \rho} \right) + \frac{P}{\rho^2} \frac{\partial^2 \Phi}{\partial \phi^2} = 0.$$

Now, employing the same device as previously, that of dividing through by $u = P\Phi$ and multiplying through by ρ^2, results in the separated equation

$$\frac{\rho}{P} \frac{\partial}{\partial \rho} \left(\rho \frac{\partial P}{\partial \rho} \right) + \frac{1}{\Phi} \frac{\partial^2 \Phi}{\partial \phi^2} = 0.$$

Following our earlier argument, since the first term on the RHS is a function of ρ only, whilst the second term depends only on ϕ, we obtain the two *ordinary* equations

$$\frac{\rho}{P} \frac{d}{d\rho} \left(\rho \frac{dP}{d\rho} \right) = n^2, \tag{21.27}$$

$$\frac{1}{\Phi} \frac{d^2 \Phi}{d\phi^2} = -n^2, \tag{21.28}$$

where we have taken the separation constant to have the form n^2 for later convenience; for the present, n is a general (complex) number.

Let us first consider the case in which $n \neq 0$. The second equation, (21.28), then has the general solution

$$\Phi(\phi) = A \exp(in\phi) + B \exp(-in\phi). \tag{21.29}$$

Equation (21.27), on the other hand, is the homogeneous equation

$$\rho^2 P'' + \rho P' - n^2 P = 0,$$

which must be solved either by trying a power solution in ρ or by making the substitution $\rho = \exp t$ as described in subsection 15.2.1 and so reducing it to an equation with constant coefficients. Carrying out this procedure we find

$$P(\rho) = C\rho^n + D\rho^{-n}. \tag{21.30}$$

Returning to the solution (21.29) of the azimuthal equation (21.28), we can see that if Φ, and hence u, is to be single-valued and so not change when ϕ increases by 2π then n must be an integer. Mathematically, other values of n are permissible, but for the description of real physical situations it is clear that this limitation must be imposed. Having thus restricted the possible values of n in one part of the solution, the same limitations must be carried over into the radial part, (21.30). Thus we may write a particular solution of the two-dimensional Laplace equation as

$$u(\rho, \phi) = (A \cos n\phi + B \sin n\phi)(C\rho^n + D\rho^{-n}),$$

where A, B, C, D are arbitrary constants and n is any integer.

We have not yet, however, considered the solution when $n = 0$. In this case, the solutions of the separated ordinary equations (21.28) and (21.27), respectively, are easily shown to be

$$\Phi(\phi) = A\phi + B,$$
$$P(\rho) = C \ln \rho + D.$$

But, in order that $u = P\Phi$ is single-valued, we require $A = 0$, and so the solution for $n = 0$ is simply (absorbing B into C and D)

$$u(\rho, \phi) = C \ln \rho + D.$$

Superposing the solutions for the different allowed values of n, we can write the general solution to Laplace's equation in plane polars as

$$u(\rho, \phi) = (C_0 \ln \rho + D_0) + \sum_{n=1}^{\infty} (A_n \cos n\phi + B_n \sin n\phi)(C_n \rho^n + D_n \rho^{-n}), \tag{21.31}$$

where n can take only integer values. Negative values of n have been omitted from the sum since they are already included in the terms obtained for positive n. We note that, since $\ln \rho$ is singular at $\rho = 0$, whenever we solve Laplace's equation in a region containing the origin, C_0 must be identically zero.

▶*A circular drumskin has a supporting rim at $\rho = a$. If the rim is twisted so that it is displaced vertically by a small amount $\epsilon(\sin \phi + 2 \sin 2\phi)$, where ϕ is the azimuthal angle with respect to a given radius, find the resulting displacement $u(\rho, \phi)$ over the entire drumskin.*

The transverse displacement of a circular drumskin is usually described by the two-dimensional wave equation. In this case, however, there is no time dependence and so $u(\rho, \phi)$ solves the two-dimensional Laplace equation, subject to the imposed boundary condition.

Referring to (21.31), since we wish to find a solution that is finite everywhere inside $\rho = a$, we require $C_0 = 0$ and $D_n = 0$ for all $n > 0$. Now the boundary condition at the rim requires

$$u(a, \phi) = D_0 + \sum_{n=1}^{\infty} C_n a^n (A_n \cos n\phi + B_n \sin n\phi) = \epsilon(\sin \phi + 2 \sin 2\phi).$$

Firstly we see that we require $D_0 = 0$ and $A_n = 0$ for all n. Furthermore, we must have $C_1 B_1 a = \epsilon$, $C_2 B_2 a^2 = 2\epsilon$ and $B_n = 0$ for $n > 2$. Hence the appropriate shape for the drumskin (valid over the whole skin, not just the rim) is

$$u(\rho, \phi) = \frac{\epsilon \rho}{a} \sin \phi + \frac{2\epsilon \rho^2}{a^2} \sin 2\phi = \frac{\epsilon \rho}{a} \left(\sin \phi + \frac{2\rho}{a} \sin 2\phi \right). \blacktriangleleft$$

Laplace's equation in cylindrical polars

Passing to three dimensions, we now consider the solution of Laplace's equation in cylindrical polar coordinates,

$$\frac{1}{\rho}\frac{\partial}{\partial \rho}\left(\rho\frac{\partial u}{\partial \rho}\right) + \frac{1}{\rho^2}\frac{\partial^2 u}{\partial \phi^2} + \frac{\partial^2 u}{\partial z^2} = 0. \tag{21.32}$$

We note here that, even when considering a cylindrical physical system, if there is no dependence of the physical variables on z (i.e. along the length of the cylinder) then the problem may be treated using two-dimensional plane polars, as discussed above.

For the more general case, however, we proceed as previously by trying a solution of the form

$$u(\rho, \phi, z) = P(\rho)\Phi(\phi)Z(z),$$

which, on substitution into (21.32) and division through by $u = P\Phi Z$, gives

$$\frac{1}{P\rho}\frac{d}{d\rho}\left(\rho\frac{dP}{d\rho}\right) + \frac{1}{\Phi\rho^2}\frac{d^2\Phi}{d\phi^2} + \frac{1}{Z}\frac{d^2Z}{dz^2} = 0.$$

The last term depends only on z, and the first and second (taken together) depend only on ρ and ϕ. Taking the separation constant to be k^2, we find

$$\frac{1}{Z}\frac{d^2Z}{dz^2} = k^2,$$

$$\frac{1}{P\rho}\frac{d}{d\rho}\left(\rho\frac{dP}{d\rho}\right) + \frac{1}{\Phi\rho^2}\frac{d^2\Phi}{d\phi^2} + k^2 = 0.$$

The first of these equations has the straightforward solution

$$Z(z) = E\exp(-kz) + F\exp kz.$$

Multiplying the second equation through by ρ^2, we obtain

$$\frac{\rho}{P}\frac{d}{d\rho}\left(\rho\frac{dP}{d\rho}\right) + \frac{1}{\Phi}\frac{d^2\Phi}{d\phi^2} + k^2\rho^2 = 0,$$

in which the second term depends only on Φ and the other terms depend only on ρ. Taking the second separation constant to be m^2, we find

$$\frac{1}{\Phi}\frac{d^2\Phi}{d\phi^2} = -m^2, \tag{21.33}$$

$$\rho\frac{d}{d\rho}\left(\rho\frac{dP}{d\rho}\right) + (k^2\rho^2 - m^2)P = 0. \tag{21.34}$$

The equation in the azimuthal angle ϕ has the very familiar solution

$$\Phi(\phi) = C\cos m\phi + D\sin m\phi.$$

728

As in the two-dimensional case, single-valuedness of u requires that m is an integer. However, in the particular case $m = 0$ the solution is

$$\Phi(\phi) = C\phi + D.$$

This form is appropriate to a solution with axial symmetry ($C = 0$) or one that is multivalued, but manageably so, such as the magnetic scalar potential associated with a current I (in which case $C = I/(2\pi)$ and D is arbitrary).

Finally, the ρ-equation (21.34) may be transformed into Bessel's equation of order m by writing $\mu = k\rho$. This has the solution

$$P(\rho) = AJ_m(k\rho) + BY_m(k\rho).$$

The properties of these functions were investigated in chapter 16 and will not be pursued here. We merely note that $Y_m(k\rho)$ is singular at $\rho = 0$, and so, when seeking solutions to Laplace's equation in cylindrical coordinates within some region containing the $\rho = 0$ axis, we require $B = 0$.

The complete separated-variable solution in cylindrical polars of Laplace's equation $\nabla^2 u = 0$ is thus given by

$$u(\rho, \phi, z) = [AJ_m(k\rho) + BY_m(k\rho)][C\cos m\phi + D\sin m\phi][E\exp(-kz) + F\exp kz]. \tag{21.35}$$

Of course we may use the principle of superposition to build up more general solutions by adding together solutions of the form (21.35) for all allowed values of the separation constants k and m.

▶ *A semi-infinite solid cylinder of radius a has its curved surface held at $0\,°C$ and its base held at a temperature T_0. Find the steady-state temperature distribution in the cylinder.*

The physical situation is shown in figure 21.5. The steady-state temperature distribution $u(\rho, \phi, z)$ must satisfy Laplace's equation subject to the imposed boundary conditions. Let us take the cylinder to have its base in the $z = 0$ plane and to extend along the positive z-axis. From (21.35), in order that u is finite everywhere in the cylinder we immediately require $B = 0$ and $F = 0$. Furthermore, since the boundary conditions, and hence the temperature distribution, are axially symmetric, we require $m = 0$, and so the general solution must be a superposition of solutions of the form $J_0(k\rho)\exp(-kz)$ for all allowed values of the separation constant k.

The boundary condition $u(a, \phi, z) = 0$ restricts the allowed values of k, since we must have $J_0(ka) = 0$. The zeros of Bessel functions are given in most books of mathematical tables, and we find that, to two decimal places,

$$J_0(x) = 0 \quad \text{for } x = 2.40, \ 5.52, \ 8.65, \ \ldots.$$

Writing the allowed values of k as k_n for $n = 1, 2, 3, \ldots$ (so, for example, $k_1 = 2.40/a$), the required solution takes the form

$$u(\rho, \phi, z) = \sum_{n=1}^{\infty} A_n J_0(k_n\rho)\exp(-k_n z).$$

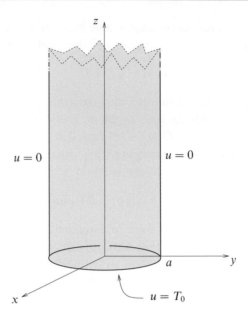

Figure 21.5 A uniform metal cylinder whose curved surface is kept at $0\,^\circ$C and whose base is held at a temperature T_0.

By imposing the remaining boundary condition $u(\rho, \phi, 0) = T_0$, the coefficients A_n can be found in a similar way to Fourier coefficients but this time by exploiting the orthogonality of the Bessel functions, as discussed in chapter 16. From this boundary condition we require

$$u(\rho, \phi, 0) = \sum_{n=1}^{\infty} A_n J_0(k_n \rho) = T_0.$$

If we multiply this expression by $\rho J_0(k_r \rho)$ and integrate from $\rho = 0$ to $\rho = a$, and use the orthogonality of the Bessel functions $J_0(k_n \rho)$, then the coefficients are given by (18.91) as

$$A_n = \frac{2T_0}{a^2 J_1^2(k_n a)} \int_0^a J_0(k_n \rho) \rho \, d\rho. \tag{21.36}$$

The integral on the RHS can be evaluated using the recurrence relation (18.92) of chapter 16,

$$\frac{d}{dz}[z J_1(z)] = z J_0(z),$$

which on setting $z = k_n \rho$ yields

$$\frac{1}{k_n} \frac{d}{d\rho}[k_n \rho J_1(k_n \rho)] = k_n \rho J_0(k_n \rho).$$

Therefore the integral in (21.36) is given by

$$\int_0^a J_0(k_n \rho) \rho \, d\rho = \left[\frac{1}{k_n} \rho J_1(k_n \rho) \right]_0^a = \frac{1}{k_n} a J_1(k_n a),$$

and the coefficients A_n may be expressed as

$$A_n = \frac{2T_0}{a^2 J_1^2(k_n a)} \left[\frac{a J_1(k_n a)}{k_n} \right] = \frac{2T_0}{k_n a J_1(k_n a)}.$$

The steady-state temperature in the cylinder is then given by

$$u(\rho, \phi, z) = \sum_{n=1}^{\infty} \frac{2T_0}{k_n a J_1(k_n a)} J_0(k_n \rho) \exp(-k_n z). \blacktriangleleft$$

We note that if, in the above example, the base of the cylinder were not kept at a uniform temperature T_0, but instead had some fixed temperature distribution $T(\rho, \phi)$, then the solution of the problem would become more complicated. In such a case, the required temperature distribution $u(\rho, \phi, z)$ is in general *not* axially symmetric, and so the separation constant m is not restricted to be zero but may take any integer value. The solution will then take the form

$$u(\rho, \phi, z) = \sum_{m=0}^{\infty} \sum_{n=1}^{\infty} J_m(k_{nm}\rho)(C_{nm} \cos m\phi + D_{nm} \sin m\phi) \exp(-k_{nm} z),$$

where the separation constants k_{nm} are such that $J_m(k_{nm}a) = 0$, i.e. $k_{nm}a$ is the nth zero of the mth-order Bessel function At the base of the cylinder we would then require

$$u(\rho, \phi, 0) = \sum_{m=0}^{\infty} \sum_{n=1}^{\infty} J_m(k_{nm}\rho)(C_{nm} \cos m\phi + D_{nm} \sin m\phi) = T(\rho, \phi).$$

$$(21.37)$$

The coefficients C_{nm} could be found by multiplying (21.37) by $J_q(k_{rq}\rho) \cos q\phi$, integrating with respect to ρ and ϕ over the base of the cylinder and exploiting the orthogonality of the Bessel functions and of the trigonometric functions. The D_{nm} could be found in a similar way by multiplying (21.37) by $J_q(k_{rq}\rho) \sin q\phi$.

Laplace's equation in spherical polars

We now come to an equation that is very widely applicable in physical science, namely $\nabla^2 u = 0$ in spherical polar coordinates:

$$\frac{1}{r^2} \frac{\partial}{\partial r} \left(r^2 \frac{\partial u}{\partial r} \right) + \frac{1}{r^2 \sin\theta} \frac{\partial}{\partial \theta} \left(\sin\theta \frac{\partial u}{\partial \theta} \right) + \frac{1}{r^2 \sin^2\theta} \frac{\partial^2 u}{\partial \phi^2} = 0. \qquad (21.38)$$

Our method of procedure will be as before; we try a solution of the form

$$u(r, \theta, \phi) = R(r)\Theta(\theta)\Phi(\phi).$$

Substituting this in (21.38), dividing through by $u = R\Theta\Phi$ and multiplying by r^2, we obtain

$$\frac{1}{R} \frac{d}{dr} \left(r^2 \frac{dR}{dr} \right) + \frac{1}{\Theta \sin\theta} \frac{d}{d\theta} \left(\sin\theta \frac{d\Theta}{d\theta} \right) + \frac{1}{\Phi \sin^2\theta} \frac{d^2\Phi}{d\phi^2} = 0. \qquad (21.39)$$

The first term depends only on r and the second and third terms (taken together) depend only on θ and ϕ. Thus (21.39) is equivalent to the two equations

$$\frac{1}{R}\frac{d}{dr}\left(r^2\frac{dR}{dr}\right) = \lambda, \tag{21.40}$$

$$\frac{1}{\Theta\sin\theta}\frac{d}{d\theta}\left(\sin\theta\frac{d\Theta}{d\theta}\right) + \frac{1}{\Phi\sin^2\theta}\frac{d^2\Phi}{d\phi^2} = -\lambda. \tag{21.41}$$

Equation (21.40) is a homogeneous equation,

$$r^2\frac{d^2R}{dr^2} + 2r\frac{dR}{dr} - \lambda R = 0,$$

which can be reduced, by the substitution $r = \exp t$ (and writing $R(r) = S(t)$), to

$$\frac{d^2S}{dt^2} + \frac{dS}{dt} - \lambda S = 0.$$

This has the straightforward solution

$$S(t) = A\exp\lambda_1 t + B\exp\lambda_2 t,$$

and so the solution to the radial equation is

$$R(r) = Ar^{\lambda_1} + Br^{\lambda_2},$$

where $\lambda_1 + \lambda_2 = -1$ and $\lambda_1\lambda_2 = -\lambda$. We can thus take λ_1 and λ_2 as given by ℓ and $-(\ell+1)$; λ then has the form $\ell(\ell+1)$. (It should be noted that at this stage nothing has been either assumed or proved about whether ℓ is an integer.)

Hence we have obtained some information about the first factor in the separated-variable solution, which will now have the form

$$u(r,\theta,\phi) = \left[Ar^\ell + Br^{-(\ell+1)}\right]\Theta(\theta)\Phi(\phi), \tag{21.42}$$

where Θ and Φ must satisfy (21.41) with $\lambda = \ell(\ell+1)$.

The next step is to take (21.41) further. Multiplying through by $\sin^2\theta$ and substituting for λ, it too takes a separated form:

$$\left[\frac{\sin\theta}{\Theta}\frac{d}{d\theta}\left(\sin\theta\frac{d\Theta}{d\theta}\right) + \ell(\ell+1)\sin^2\theta\right] + \frac{1}{\Phi}\frac{d^2\Phi}{d\phi^2} = 0. \tag{21.43}$$

Taking the separation constant as m^2, the equation in the azimuthal angle ϕ has the same solution as in cylindrical polars, namely

$$\Phi(\phi) = C\cos m\phi + D\sin m\phi.$$

As before, single-valuedness of u requires that m is an integer; for $m = 0$ we again have $\Phi(\phi) = C\phi + D$.

Having settled the form of $\Phi(\phi)$, we are left only with the equation satisfied by $\Theta(\theta)$, which is

$$\frac{\sin\theta}{\Theta}\frac{d}{d\theta}\left(\sin\theta\frac{d\Theta}{d\theta}\right) + \ell(\ell+1)\sin^2\theta = m^2. \tag{21.44}$$

A change of independent variable from θ to $\mu = \cos\theta$ will reduce this to a form for which solutions are known, and of which some study has been made in chapter 16. Putting

$$\mu = \cos\theta, \qquad \frac{d\mu}{d\theta} = -\sin\theta, \qquad \frac{d}{d\theta} = -(1-\mu^2)^{1/2}\frac{d}{d\mu},$$

the equation for $M(\mu) \equiv \Theta(\theta)$ reads

$$\frac{d}{d\mu}\left[(1-\mu^2)\frac{dM}{d\mu}\right] + \left[\ell(\ell+1) - \frac{m^2}{1-\mu^2}\right]M = 0. \tag{21.45}$$

This equation is the *associated Legendre equation*, which was mentioned in sub-section 18.2 in the context of Sturm–Liouville equations.

We recall that for the case $m = 0$, (21.45) reduces to Legendre's equation, which was studied at length in chapter 16, and has the solution

$$M(\mu) = EP_\ell(\mu) + FQ_\ell(\mu). \tag{21.46}$$

We have not solved (21.45) explicitly for general m, but the solutions were given in subsection 18.2 and are the associated Legendre functions $P_\ell^m(\mu)$ and $Q_\ell^m(\mu)$, where

$$P_\ell^m(\mu) = (1-\mu^2)^{|m|/2}\frac{d^{|m|}}{d\mu^{|m|}}P_\ell(\mu), \tag{21.47}$$

and similarly for $Q_\ell^m(\mu)$. We then have

$$M(\mu) = EP_\ell^m(\mu) + FQ_\ell^m(\mu); \tag{21.48}$$

here m must be an integer, $0 \le |m| \le \ell$. We note that if we require solutions to Laplace's equation that are finite when $\mu = \cos\theta = \pm 1$ (i.e. on the polar axis where $\theta = 0, \pi$), then we must have $F = 0$ in (21.46) and (21.48) since $Q_\ell^m(\mu)$ diverges at $\mu = \pm 1$.

It will be remembered that one of the important conditions for obtaining finite polynomial solutions of Legendre's equation is that ℓ is an integer ≥ 0. This condition therefore applies also to the solutions (21.46) and (21.48) and is reflected back into the radial part of the general solution given in (21.42).

Now that the solutions of each of the three ordinary differential equations governing R, Θ and Φ have been obtained, we may assemble a complete separated-variable solution of Laplace's equation in spherical polars. It is

$$u(r, \theta, \phi) = (Ar^\ell + Br^{-(\ell+1)})(C \cos m\phi + D \sin m\phi)[EP_\ell^m(\cos \theta) + FQ_\ell^m(\cos \theta)],$$

$$(21.49)$$

where the three bracketted factors are connected only through the *integer* parameters ℓ and m, $0 \leq |m| \leq \ell$. As before, a general solution may be obtained by superposing solutions of this form for the allowed values of the separation constants ℓ and m. As mentioned above, if the solution is required to be finite on the polar axis then $F = 0$ for all ℓ and m.

▶*An uncharged conducting sphere of radius a is placed at the origin in an initially uniform electrostatic field E. Show that it behaves as an electric dipole.*

The uniform field, taken in the direction of the polar axis, has an electrostatic potential

$$u = -Ez = -Er \cos \theta,$$

where u is arbitrarily taken as zero at $z = 0$. This satisfies Laplace's equation $\nabla^2 u = 0$, as must the potential v when the sphere is present; for large r the asymptotic form of v must still be $-Er \cos \theta$.

Since the problem is clearly axially symmetric, we have immediately that $m = 0$, and since we require v to be finite on the polar axis we must have $F = 0$ in (21.49). Therefore the solution must be of the form

$$v(r, \theta, \phi) = \sum_{\ell=0}^{\infty} (A_\ell r^\ell + B_\ell r^{-(\ell+1)}) P_\ell(\cos \theta).$$

Now the $\cos \theta$-dependence of v for large r indicates that the (θ, ϕ)-dependence of $v(r, \theta, \phi)$ is given by $P_1^0(\cos \theta) = \cos \theta$. Thus the r-dependence of v must also correspond to an $\ell = 1$ solution, and the most general such solution (outside the sphere, i.e. for $r \geq a$) is

$$v(r, \theta, \phi) = (A_1 r + B_1 r^{-2}) P_1(\cos \theta).$$

The asymptotic form of v for large r immediately gives $A_1 = -E$ and so yields the solution

$$v(r, \theta, \phi) = \left(-Er + \frac{B_1}{r^2} \right) \cos \theta.$$

Since the sphere is conducting, it is an equipotential region and so v must not depend on θ for $r = a$. This can only be the case if $B_1/a^2 = Ea$, thus fixing B_1. The final solution is therefore

$$v(r, \theta, \phi) = -Er \left(1 - \frac{a^3}{r^3} \right) \cos \theta.$$

Since a dipole of moment p gives rise to a potential $p/(4\pi\epsilon_0 r^2)$, this result shows that the sphere behaves as a dipole of moment $4\pi\epsilon_0 a^3 E$, because of the charge distribution induced on its surface; see figure 21.6. ◀

Often the boundary conditions are not so easily met, and it is necessary to use the mutual orthogonality of the associated Legendre functions (and the trigonometric functions) to obtain the coefficients in the general solution.

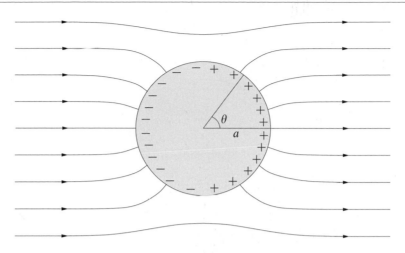

Figure 21.6   Induced charge and field lines associated with a conducting sphere placed in an initially uniform electrostatic field.

▶*A hollow split conducting sphere of radius a is placed at the origin. If one half of its surface is charged to a potential v_0 and the other half is kept at zero potential, find the potential v inside and outside the sphere.*

Let us choose the top hemisphere to be charged to v_0 and the bottom hemisphere to be at zero potential, with the plane in which the two hemispheres meet perpendicular to the polar axis; this is shown in figure 21.7. The boundary condition then becomes

$$v(a, \theta, \phi) = \begin{cases} v_0 & \text{for } 0 < \theta < \pi/2 \quad (0 < \cos\theta < 1), \\ 0 & \text{for } \pi/2 < \theta < \pi \quad (-1 < \cos\theta < 0). \end{cases} \tag{21.50}$$

The problem is clearly axially symmetric and so we may set $m = 0$. Also, we require the solution to be finite on the polar axis and so it cannot contain $Q_\ell(\cos\theta)$. Therefore the general form of the solution to (21.38) is

$$v(r, \theta, \phi) = \sum_{\ell=0}^{\infty}(A_\ell r^\ell + B_\ell r^{-(\ell+1)})P_\ell(\cos\theta). \tag{21.51}$$

Inside the sphere (for $r < a$) we require the solution to be finite at the origin and so $B_\ell = 0$ for all ℓ in (21.51). Imposing the boundary condition at $r = a$ we must then have

$$v(a, \theta, \phi) = \sum_{\ell=0}^{\infty} A_\ell a^\ell P_\ell(\cos\theta),$$

where $v(a, \theta, \phi)$ is also given by (21.50). Exploiting the mutual orthogonality of the Legendre polynomials, the coefficients in the Legendre polynomial expansion are given by (18.14) as (writing $\mu = \cos\theta$)

$$A_\ell a^\ell = \frac{2\ell + 1}{2}\int_{-1}^{1} v(a, \theta, \phi)P_\ell(\mu)d\mu$$

$$= \frac{2\ell + 1}{2}v_0\int_{0}^{1} P_\ell(\mu)d\mu,$$

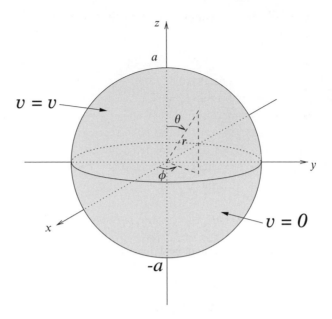

Figure 21.7 A hollow split conducting sphere with its top half charged to a potential v_0 and its bottom half at zero potential.

where in the last line we have used (21.50). The integrals of the Legendre polynomials are easily evaluated (see exercise 17.3) and we find

$$A_0 = \frac{v_0}{2}, \qquad A_1 = \frac{3v_0}{4a}, \qquad A_2 = 0, \qquad A_3 = -\frac{7v_0}{16a^3}, \qquad \cdots,$$

so that the required solution inside the sphere is

$$v(r, \theta, \phi) = \frac{v_0}{2}\left[1 + \frac{3r}{2a}P_1(\cos\theta) - \frac{7r^3}{8a^3}P_3(\cos\theta) + \cdots\right].$$

Outside the sphere (for $r > a$) we require the solution to be bounded as r tends to infinity and so in (21.51) we must have $A_\ell = 0$ for all ℓ. In this case, by imposing the boundary condition at $r = a$ we require

$$v(a, \theta, \phi) = \sum_{\ell=0}^{\infty} B_\ell a^{-(\ell+1)} P_\ell(\cos\theta),$$

where $v(a, \theta, \phi)$ is given by (21.50). Following the above argument the coefficients in the expansion are given by

$$B_\ell a^{-(\ell+1)} = \frac{2\ell+1}{2} v_0 \int_0^1 P_\ell(\mu)d\mu,$$

so that the required solution outside the sphere is

$$v(r, \theta, \phi) = \frac{v_0 a}{2r}\left[1 + \frac{3a}{2r}P_1(\cos\theta) - \frac{7a^3}{8r^3}P_3(\cos\theta) + \cdots\right]. \blacktriangleleft$$

In the above example, on the equator of the sphere (i.e. at $r = a$ and $\theta = \pi/2$) the potential is given by

$$v(a, \pi/2, \phi) = v_0/2,$$

i.e. mid-way between the potentials of the top and bottom hemispheres. This is so because a Legendre polynomial expansion of a function behaves in the same way as a Fourier series expansion, in that it converges to the average of the two values at any discontinuities present in the original function.

If the potential on the surface of the sphere had been given as a function of θ and ϕ, then we would have had to consider a double series summed over ℓ and m (for $-\ell \leq m \leq \ell$), since, in general, the solution would not have been axially symmetric.

Finally, we note in general that, when obtaining solutions of Laplace's equation in spherical polar coordinates, one finds that, for solutions that are finite on the polar axis, the angular part of the solution is given by

$$\Theta(\theta)\Phi(\phi) = P_\ell^m(\cos\theta)(C\cos m\phi + D\sin m\phi),$$

where ℓ and m are integers with $-\ell \leq m \leq \ell$. This general form is sufficiently common that particular functions of θ and ϕ called *spherical harmonics* are defined and tabulated (see section 18.3).

21.3.2 Other equations in polar coordinates

The development of the solutions of $\nabla^2 u = 0$ carried out in the previous subsection can be employed to solve other equations in which the ∇^2 operator appears. Since we have discussed the general method in some depth already, only an outline of the solutions will be given here.

Let us first consider the wave equation

$$\nabla^2 u = \frac{1}{c^2}\frac{\partial^2 u}{\partial t^2}, \tag{21.52}$$

and look for a separated solution of the form $u = F(\mathbf{r})T(t)$, so that initially we are separating only the spatial and time dependences. Substituting this form into (21.52) and taking the separation constant as k^2 we obtain

$$\nabla^2 F + k^2 F = 0, \qquad \frac{d^2 T}{dt^2} + k^2 c^2 T = 0. \tag{21.53}$$

The second equation has the simple solution

$$T(t) = A\exp(i\omega t) + B\exp(-i\omega t), \tag{21.54}$$

where $\omega = kc$; this may also be expressed in terms of sines and cosines, of course. The first equation in (21.53) is referred to as *Helmholtz's equation*; we discuss it below.

We may treat the diffusion equation

$$\kappa \nabla^2 u = \frac{\partial u}{\partial t}$$

in a similar way. Separating the spatial and time dependences by assuming a solution of the form $u = F(\mathbf{r})T(t)$, and taking the separation constant as k^2, we find

$$\nabla^2 F + k^2 F = 0, \qquad \frac{dT}{dt} + k^2 \kappa T = 0.$$

Just as in the case of the wave equation, the spatial part of the solution satisfies Helmholtz's equation. It only remains to consider the time dependence, which has the simple solution

$$T(t) = A \, \exp(-k^2 \kappa t).$$

Helmholtz's equation is clearly of central importance in the solutions of the wave and diffusion equations. It can be solved in polar coordinates in much the same way as Laplace's equation, and indeed reduces to Laplace's equation when $k = 0$. Therefore, we will merely sketch the method of its solution in each of the three polar coordinate systems.

Helmholtz's equation in plane polars

In two-dimensional plane polar coordinates, Helmholtz's equation takes the form

$$\frac{1}{\rho} \frac{\partial}{\partial \rho} \left(\rho \frac{\partial F}{\partial \rho} \right) + \frac{1}{\rho^2} \frac{\partial^2 F}{\partial \phi^2} + k^2 F = 0.$$

If we try a separated solution of the form $F(\mathbf{r}) = P(\rho)\Phi(\phi)$, and take the separation constant as m^2, we find

$$\frac{d^2 \Phi}{d\phi^2} + m^2 \phi = 0,$$

$$\frac{d^2 P}{d\rho^2} + \frac{1}{\rho} \frac{dP}{d\rho} + \left(k^2 - \frac{m^2}{\rho^2} \right) P = 0.$$

As for Laplace's equation, the angular part has the familiar solution (if $m \neq 0$)

$$\Phi(\phi) = A \cos m\phi + B \sin m\phi,$$

or an equivalent form in terms of complex exponentials. The radial equation differs from that found in the solution of Laplace's equation, but by making the substitution $\mu = k\rho$ it is easily transformed into Bessel's equation of order m (discussed in chapter 16), and has the solution

$$P(\rho) = C J_m(k\rho) + D Y_m(k\rho),$$

where Y_m is a Bessel function of the second kind, which is infinite at the origin and is not to be confused with a spherical harmonic (these are written with a superscript as well as a subscript).

Putting the two parts of the solution together we have

$$F(\rho, \phi) = [A \cos m\phi + B \sin m\phi][C J_m(k\rho) + D Y_m(k\rho)]. \qquad (21.55)$$

Clearly, for solutions of Helmholtz's equation that are required to be finite at the origin, we must set $D = 0$.

▶Find the four lowest frequency modes of oscillation of a circular drumskin of radius a whose circumference is held fixed in a plane.

The transverse displacement $u(\mathbf{r}, t)$ of the drumskin satisfies the two-dimensional wave equation

$$\nabla^2 u = \frac{1}{c^2} \frac{\partial^2 u}{\partial t^2},$$

with $c^2 = T/\sigma$, where T is the tension of the drumskin and σ is its mass per unit area. From (21.54) and (21.55) a separated solution of this equation, in plane polar coordinates, that is finite at the origin is

$$u(\rho, \phi, t) = J_m(k\rho)(A \cos m\phi + B \sin m\phi) \exp(\pm i\omega t),$$

where $\omega = kc$. Since we require the solution to be single-valued we must have m as an integer. Furthermore, if the drumskin is clamped at its outer edge $\rho = a$ then we also require $u(a, \phi, t) = 0$. Thus we need

$$J_m(ka) = 0,$$

which in turn restricts the allowed values of k. The zeros of Bessel functions can be obtained from most books of tables; the first few are

$$J_0(x) = 0 \quad \text{for } x \approx 2.40, \ 5.52, \ 8.65, \ldots,$$
$$J_1(x) = 0 \quad \text{for } x \approx 3.83, \ 7.02, \ 10.17, \ldots,$$
$$J_2(x) = 0 \quad \text{for } x \approx 5.14, \ 8.42, \ 11.62 \ldots.$$

The smallest value of x for which any of the Bessel functions is zero is $x \approx 2.40$, which occurs for $J_0(x)$. Thus the lowest-frequency mode has $k = 2.40/a$ and angular frequency $\omega = 2.40c/a$. Since $m = 0$ for this mode, the shape of the drumskin is

$$u \propto J_0\left(2.40\frac{\rho}{a}\right);$$

this is illustrated in figure 21.8.

Continuing in the same way, the next three modes are given by

$$\omega = 3.83\frac{c}{a}, \quad u \propto J_1\left(3.83\frac{\rho}{a}\right)\cos\phi, \quad J_1\left(3.83\frac{\rho}{a}\right)\sin\phi;$$
$$\omega = 5.14\frac{c}{a}, \quad u \propto J_2\left(5.14\frac{\rho}{a}\right)\cos 2\phi, \quad J_2\left(5.14\frac{\rho}{a}\right)\sin 2\phi;$$
$$\omega = 5.52\frac{c}{a}, \quad u \propto J_0\left(5.52\frac{\rho}{a}\right).$$

These modes are also shown in figure 21.8. We note that the second and third frequencies have *two* corresponding modes of oscillation; these frequencies are therefore two-fold degenerate. ◀

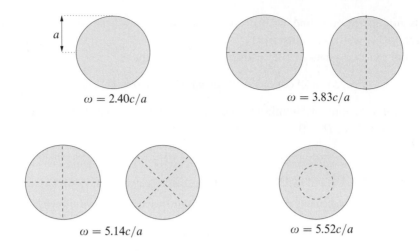

$\omega = 2.40c/a$ $\omega = 3.83c/a$

$\omega = 5.14c/a$ $\omega = 5.52c/a$

Figure 21.8   The modes of oscillation with the four lowest frequencies for a circular drumskin of radius a. The dashed lines indicate the nodes, where the displacement of the drumskin is always zero.

Helmholtz's equation in cylindrical polars

Generalising the above method to three-dimensional cylindrical polars is straightforward, and following a similar procedure to that used for Laplace's equation we find the separated solution of Helmholtz's equation takes the form

$$F(\rho, \phi, z) = \left[A J_m \left(\sqrt{k^2 - \alpha^2} \, \rho \right) + B Y_m \left(\sqrt{k^2 - \alpha^2} \, \rho \right) \right]$$
$$\times (C \cos m\phi + D \sin m\phi)[E \exp(i\alpha z) + F \exp(-i\alpha z)],$$

where α and m are separation constants. We note that the angular part of the solution is the same as for Laplace's equation in cylindrical polars.

Helmholtz's equation in spherical polars

In spherical polars, we find again that the angular parts of the solution $\Theta(\theta)\Phi(\phi)$ are identical to those of Laplace's equation in this coordinate system, i.e. they are the spherical harmonics $Y_\ell^m(\theta, \phi)$, and so we shall not discuss them further.

The radial equation in this case is given by

$$r^2 R'' + 2r R' + [k^2 r^2 - \ell(\ell + 1)] R = 0, \tag{21.56}$$

which has an additional term $k^2 r^2 R$ compared with the radial equation for the Laplace solution. The equation (21.56) looks very much like Bessel's equation. In fact, by writing $R(r) = r^{-1/2} S(r)$ and making the change of variable $\mu = kr$, it can be reduced to Bessel's equation of order $\ell + \frac{1}{2}$, which has as its solutions $S(\mu) = J_{\ell+1/2}(\mu)$ and $Y_{\ell+1/2}(\mu)$ (see section 18.6). The separated solution to

Helmholtz's equation in spherical polars is thus

$$F(r,\theta,\phi) = r^{-1/2}[AJ_{\ell+1/2}(kr) + BY_{\ell+1/2}(kr)](C\cos m\phi + D\sin m\phi)$$
$$\times[EP_\ell^m(\cos\theta) + FQ_\ell^m(\cos\theta)]. \tag{21.57}$$

For solutions that are finite at the origin we require $B = 0$, and for solutions that are finite on the polar axis we require $F = 0$. It is worth mentioning that the solutions proportional to $r^{-1/2}J_{\ell+1/2}(kr)$ and $r^{-1/2}Y_{\ell+1/2}(kr)$, when suitably normalised, are called *spherical Bessel functions* of the first and second kind, respectively, and are denoted by $j_\ell(kr)$ and $n_\ell(\mu)$ (see section 18.6).

As mentioned at the beginning of this subsection, the separated solution of the wave equation in spherical polars is the product of a time-dependent part (21.54) and a spatial part (21.57). It will be noticed that, although this solution corresponds to a solution of definite frequency $\omega = kc$, the zeros of the radial function $j_\ell(kr)$ are not equally spaced in r, except for the case $\ell = 0$ involving $j_0(kr)$, and so there is no precise wavelength associated with the solution.

To conclude this subsection, let us mention briefly the Schrödinger equation for the electron in a hydrogen atom, the nucleus of which is taken at the origin and is assumed massive compared with the electron. Under these circumstances the Schrödinger equation is

$$-\frac{\hbar^2}{2m}\nabla^2 u - \frac{e^2}{4\pi\epsilon_0}\frac{u}{r} = i\hbar\frac{\partial u}{\partial t}.$$

For a 'stationary-state' solution, for which the energy is a constant E and the time-dependent factor T in u is given by $T(t) = A\exp(-iEt/\hbar)$, the above equation is similar to, but not quite the same as, the Helmholtz equation.[§] However, as with the wave equation, the angular parts of the solution are identical to those for Laplace's equation and are expressed in terms of spherical harmonics.

The important point to note is that for *any* equation involving ∇^2, provided θ and ϕ do not appear in the equation other than as part of ∇^2, a separated-variable solution in spherical polars will always lead to spherical harmonic solutions. This is the case for the Schrödinger equation describing an atomic electron in a central potential $V(r)$.

21.3.3 Solution by expansion

It is sometimes possible to use the uniqueness theorem discussed in the previous chapter, together with the results of the last few subsections, in which Laplace's equation (and other equations) were considered in polar coordinates, to obtain solutions of such equations appropriate to particular physical situations.

[§] For the solution by series of the r-equation in this case the reader may consult, for example, L. Schiff, *Quantum Mechanics* (New York: McGraw-Hill, 1955), p. 82.

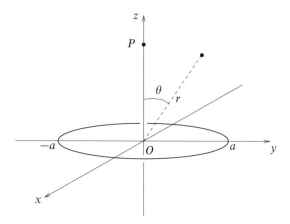

Figure 21.9   The polar axis Oz is taken as normal to the plane of the ring of matter and passing through its centre.

We will illustrate the method for Laplace's equation in spherical polars and first assume that the required solution of $\nabla^2 u = 0$ can be written as a superposition in the normal way:

$$u(r, \theta, \phi) = \sum_{\ell=0}^{\infty} \sum_{m=-\ell}^{\ell} (Ar^{\ell} + Br^{-(\ell+1)})P_{\ell}^{m}(\cos\theta)(C\cos m\phi + D\sin m\phi).$$
(21.58)

Here, all the constants A, B, C, D may depend upon ℓ and m, and we have assumed that the required solution is finite on the polar axis. As usual, boundary conditions of a physical nature will then fix or eliminate some of the constants; for example, u finite at the origin implies all $B = 0$, or axial symmetry implies that only $m = 0$ terms are present.

The essence of the method is then to find the remaining constants by determining u at values of r, θ, ϕ for which it can be evaluated *by other means*, e.g. by direct calculation on an axis of symmetry. Once the remaining constants have been fixed by these special considerations to have particular values, the uniqueness theorem can be invoked to establish that they must have these values in general.

▶*Calculate the gravitational potential at a general point in space due to a uniform ring of matter of radius a and total mass M.*

Everywhere except on the ring the potential $u(\mathbf{r})$ satisfies the Laplace equation, and so if we use polar coordinates with the normal to the ring as polar axis, as in figure 21.9, a solution of the form (21.58) can be assumed.

We expect the potential $u(r, \theta, \phi)$ to tend to zero as $r \to \infty$, and also to be finite at $r = 0$. At first sight this might seem to imply that all A and B, and hence u, must be identically zero, an unacceptable result. In fact, what it means is that different expressions must apply to different regions of space. On the ring itself we no longer have $\nabla^2 u = 0$ and so it is not surprising that the form of the expression for u changes there. Let us therefore take two separate regions.

In the region $r > a$

(i) we must have $u \to 0$ as $r \to \infty$, implying that all $A = 0$, *and*
(ii) the system is axially symmetric and so only $m = 0$ terms appear.

With these restrictions we can write as a trial form

$$u(r, \theta, \phi) = \sum_{\ell=0}^{\infty} B_\ell r^{-(\ell+1)} P_\ell^0(\cos \theta). \tag{21.59}$$

The constants B_ℓ are still to be determined; this we do by calculating *directly* the potential where this can be done simply – in this case, on the polar axis.

Considering a point P on the polar axis at a distance $z \ (> a)$ from the plane of the ring (taken as $\theta = \pi/2$), all parts of the ring are at a distance $(z^2 + a^2)^{1/2}$ from it. The potential at P is thus straightforwardly

$$u(z, 0, \phi) - -\frac{GM}{(z^2 + a^2)^{1/2}}, \tag{21.60}$$

where G is the gravitational constant. This must be the same as (21.59) for the particular values $r = z$, $\theta - 0$, and ϕ undefined. Since $P_\ell^0(\cos \theta) = P_\ell(\cos \theta)$ with $P_\ell(1) = 1$, putting $r = z$ in (21.59) gives

$$u(z, 0, \phi) = \sum_{\ell=0}^{\infty} \frac{B_\ell}{z^{\ell+1}}. \tag{21.61}$$

However, expanding (21.60) for $z > a$ (as it applies to this region of space) we obtain

$$u(z, 0, \phi) = -\frac{GM}{z} \left[1 - \frac{1}{2} \left(\frac{a}{z}\right)^2 + \frac{3}{8} \left(\frac{a}{z}\right)^4 - \cdots \right],$$

which on comparison with (21.61) gives[§]

$$B_0 - -GM,$$

$$B_{2\ell} = -\frac{GMa^{2\ell}(-1)^\ell(2\ell - 1)!!}{2^\ell \ell!} \qquad \text{for } \ell \geq 1, \tag{21.62}$$

$$B_{2\ell+1} = 0.$$

We now conclude the argument by saying that if a solution for a general point (r, θ, ϕ) exists at all, which of course we very much expect on physical grounds, then it must be (21.59) with the B_ℓ given by (21.62). This is so because thus defined it is a function with no arbitrary constants and which satisfies all the boundary conditions, and the uniqueness theorem states that there is only one such function. The expression for the potential in the region $r > a$ is therefore

$$u(r, \theta, \phi) = -\frac{GM}{r} \left[1 + \sum_{\ell=1}^{\infty} \frac{(-1)^\ell(2\ell - 1)!!}{2^\ell \ell!} \left(\frac{a}{r}\right)^{2\ell} P_{2\ell}(\cos \theta) \right].$$

The expression for $r < a$ can be found in a similar way. The finiteness of u at $r = 0$ and the axial symmetry give

$$u(r, \theta, \phi) = \sum_{\ell=0}^{\infty} A_\ell r^\ell P_\ell^0(\cos \theta).$$

[§] $(2\ell - 1)!! = 1 \times 3 \times \cdots \times (2\ell - 1)$.

Comparing this expression for $r = z$, $\theta = 0$ with the $z < a$ expansion of (21.60), which is valid for any z, establishes $A_{2\ell+1} = 0$, $A_0 = -GM/a$ and

$$A_{2\ell} = -\frac{GM}{a^{2\ell+1}} \frac{(-1)^\ell (2\ell - 1)!!}{2^\ell \ell!},$$

so that the final expression valid, and convergent, for $r < a$ is thus

$$u(r, \theta, \phi) = -\frac{GM}{a} \left[1 + \sum_{\ell=1}^{\infty} \frac{(-1)^\ell (2\ell - 1)!!}{2^\ell \ell!} \left(\frac{r}{a} \right)^{2\ell} P_{2\ell}(\cos \theta) \right].$$

It is easy to check that the solution obtained has the expected physical value for large r and for $r = 0$ and is continuous at $r = a$. ◄

21.3.4 Separation of variables for inhomogeneous equations

So far our discussion of the method of separation of variables has been limited to the solution of homogeneous equations such as the Laplace equation and the wave equation. The solutions of inhomogeneous PDEs are usually obtained using the Green's function methods to be discussed below in section 21.5. However, as a final illustration of the usefulness of the separation of variables, we now consider its application to the solution of inhomogeneous equations.

Because of the added complexity in dealing with inhomogeneous equations, we shall restrict our discussion to the solution of Poisson's equation,

$$\nabla^2 u = \rho(\mathbf{r}), \tag{21.63}$$

in spherical polar coordinates, although the general method can accommodate other coordinate systems and equations. In physical problems the RHS of (21.63) usually contains some multiplicative constant(s). If u is the electrostatic potential in some region of space in which ρ is the density of electric charge then $\nabla^2 u = -\rho(\mathbf{r})/\epsilon_0$. Alternatively, u might represent the gravitational potential in some region where the matter density is given by ρ, so that $\nabla^2 u = 4\pi G\rho(\mathbf{r})$.

We will simplify our discussion by assuming that the required solution u is finite on the polar axis and also that the system possesses axial symmetry about that axis – in which case ρ does not depend on the azimuthal angle ϕ. The key to the method is then to assume a separated form for both the solution u and the density term ρ.

From the discussion of Laplace's equation, for systems with axial symmetry only $m = 0$ terms appear, and so the angular part of the solution can be expressed in terms of Legendre polynomials $P_\ell(\cos \theta)$. Since these functions form an orthogonal set let us expand both u and ρ in terms of them:

$$u = \sum_{\ell=0}^{\infty} R_\ell(r) P_\ell(\cos \theta), \tag{21.64}$$

$$\rho = \sum_{\ell=0}^{\infty} F_\ell(r) P_\ell(\cos \theta), \tag{21.65}$$

where the coefficients $R_\ell(r)$ and $F_\ell(r)$ in the Legendre polynomial expansions are functions of r. Since in any particular problem ρ is given, we can find the coefficients $F_\ell(r)$ in the expansion in the usual way (see subsection 18.1.2). It then only remains to find the coefficients $R_\ell(r)$ in the expansion of the solution u.

Writing ∇^2 in spherical polars and substituting (21.64) and (21.65) into (21.63) we obtain

$$\sum_{\ell=0}^{\infty} \left[\frac{P_\ell(\cos\theta)}{r^2} \frac{d}{dr}\left(r^2 \frac{dR_\ell}{dr}\right) + \frac{R_\ell}{r^2 \sin\theta} \frac{d}{d\theta}\left(\sin\theta \frac{dP_\ell(\cos\theta)}{d\theta}\right) \right] = \sum_{\ell=0}^{\infty} F_\ell(r) P_\ell(\cos\theta).$$

(21.66)

However, if, in equation (21.44) of our discussion of the angular part of the solution to Laplace's equation, we set $m = 0$ we conclude that

$$\frac{1}{\sin\theta} \frac{d}{d\theta}\left(\sin\theta \frac{dP_\ell(\cos\theta)}{d\theta}\right) = -\ell(\ell+1)P_\ell(\cos\theta).$$

Substituting this into (21.66), we find that the LHS is greatly simplified and we obtain

$$\sum_{\ell=0}^{\infty} \left[\frac{1}{r^2} \frac{d}{dr}\left(r^2 \frac{dR_\ell}{dr}\right) - \frac{\ell(\ell+1)R_\ell}{r^2} \right] P_\ell(\cos\theta) = \sum_{\ell=0}^{\infty} F_\ell(r) P_\ell(\cos\theta).$$

This relation is most easily satisfied by equating terms on both sides for each value of ℓ separately, so that for $\ell = 0, 1, 2, \ldots$ we have

$$\frac{1}{r^2} \frac{d}{dr}\left(r^2 \frac{dR_\ell}{dr}\right) - \frac{\ell(\ell+1)R_\ell}{r^2} = F_\ell(r). \qquad (21.67)$$

This is an ODE in which $F_\ell(r)$ is given, and it can therefore be solved for $R_\ell(r)$. The solution to Poisson's equation, u, is then obtained by making the superposition (21.64).

▶In a certain system, the electric charge density ρ is distributed as follows:

$$\rho = \begin{cases} Ar\cos\theta & \text{for } 0 \leq r < a, \\ 0 & \text{for } r \geq a. \end{cases}$$

Find the electrostatic potential inside and outside the charge distribution, given that both the potential and its radial derivative are continuous everywhere.

The electrostatic potential u satisfies

$$\nabla^2 u = \begin{cases} -(A/\epsilon_0)r\cos\theta & \text{for } 0 \leq r < a, \\ 0 & \text{for } r \geq a. \end{cases}$$

For $r < a$ the RHS can be written $-(A/\epsilon_0)rP_1(\cos\theta)$, and the coefficients in (21.65) are simply $F_1(r) = -(Ar/\epsilon_0)$ and $F_\ell(r) = 0$ for $\ell \neq 1$. Therefore we need only calculate $R_1(r)$, which satisfies (21.67) for $\ell = 1$:

$$\frac{1}{r^2} \frac{d}{dr}\left(r^2 \frac{dR_1}{dr}\right) - \frac{2R_1}{r^2} = -\frac{Ar}{\epsilon_0}.$$

This can be rearranged to give

$$r^2 R_1'' + 2r R_1' - 2R_1 = -\frac{Ar^3}{\epsilon_0},$$

where the prime denotes differentiation with respect to r. The LHS is homogeneous and the equation can be reduced by the substitution $r = \exp t$, and writing $R_1(r) = S(t)$, to

$$\ddot{S} + \dot{S} - 2S = -\frac{A}{\epsilon_0} \exp 3t, \tag{21.68}$$

where the dots indicate differentiation with respect to t.

This is an inhomogeneous second-order ODE with constant coefficients and can be straightforwardly solved by the methods of subsection 15.2.1 to give

$$S(t) = c_1 \exp t + c_2 \exp(-2t) - \frac{A}{10\epsilon_0} \exp 3t.$$

Recalling that $r = \exp t$ we find

$$R_1(r) = c_1 r + c_2 r^{-2} - \frac{A}{10\epsilon_0} r^3.$$

Since we are interested in the region $r < a$ we must have $c_2 = 0$ for the solution to remain finite. Thus inside the charge distribution the electrostatic potential has the form

$$u_1(r, \theta, \phi) = \left(c_1 r - \frac{A}{10\epsilon_0} r^3 \right) P_1(\cos\theta). \tag{21.69}$$

Outside the charge distribution (for $r \geq a$), however, the electrostatic potential obeys Laplace's equation, $\nabla^2 u = 0$, and so given the symmetry of the problem and the requirement that $u \to \infty$ as $r \to \infty$ the solution must take the form

$$u_2(r, \theta, \phi) = \sum_{\ell=0}^{\infty} \frac{B_\ell}{r^{\ell+1}} P_\ell(\cos\theta). \tag{21.70}$$

We can now use the boundary conditions at $r = a$ to fix the constants in (21.69) and (21.70). The requirement of continuity of the potential and its radial derivative at $r = a$ imply that

$$u_1(a, \theta, \phi) = u_2(a, \theta, \phi),$$
$$\frac{\partial u_1}{\partial r}(a, \theta, \phi) = \frac{\partial u_2}{\partial r}(a, \theta, \phi).$$

Clearly $B_\ell = 0$ for $\ell \neq 1$; carrying out the necessary differentiations and setting $r = a$ in (21.69) and (21.70) we obtain the simultaneous equations

$$c_1 a - \frac{A}{10\epsilon_0} a^3 = \frac{B_1}{a^2},$$
$$c_1 - \frac{3A}{10\epsilon_0} a^2 = -\frac{2B_1}{a^3},$$

which may be solved to give $c_1 = Aa^2/(6\epsilon_0)$ and $B_1 = Aa^5/(15\epsilon_0)$. Since $P_1(\cos\theta) = \cos\theta$, the electrostatic potentials inside and outside the charge distribution are given, respectively, by

$$u_1(r, \theta, \phi) = \frac{A}{\epsilon_0} \left(\frac{a^2 r}{6} - \frac{r^3}{10} \right) \cos\theta, \qquad u_2(r, \theta, \phi) = \frac{Aa^5}{15\epsilon_0} \frac{\cos\theta}{r^2}. \blacktriangleleft$$

21.4 Integral transform methods

In the method of separation of variables our aim was to keep the independent variables in a PDE as separate as possible. We now discuss the use of integral transforms in solving PDEs, a method by which one of the independent variables can be eliminated from the differential coefficients. It will be assumed that the reader is familiar with Laplace and Fourier transforms and their properties, as discussed in chapter 13.

The method consists simply of transforming the PDE into one containing derivatives with respect to a smaller number of variables. Thus, if the original equation has just two independent variables, it may be possible to reduce the PDE into a soluble ODE. The solution obtained can then (where possible) be transformed back to give the solution of the original PDE. As we shall see, boundary conditions can usually be incorporated in a natural way.

Which sort of transform to use, and the choice of the variable(s) with respect to which the transform is to be taken, is a matter of experience; we illustrate this in the example below. In practice, transforms can be taken with respect to each variable in turn, and the transformation that affords the greatest simplification can be pursued further.

> ▶ *A semi-infinite tube of constant cross-section contains initially pure water. At time $t = 0$, one end of the tube is put into contact with a salt solution and maintained at a concentration u_0. Find the total amount of salt that has diffused into the tube after time t, if the diffusion constant is κ.*

The concentration $u(x, t)$ at time t and distance x from the end of the tube satisfies the diffusion equation

$$\kappa \frac{\partial^2 u}{\partial x^2} = \frac{\partial u}{\partial t}, \tag{21.71}$$

which has to be solved subject to the boundary conditions $u(0, t) = u_0$ for all t and $u(x, 0) = 0$ for all $x > 0$.

Since we are interested only in $t > 0$, the use of the Laplace transform is suggested. Furthermore, it will be recalled from chapter 13 that one of the major virtues of Laplace transformations is the possibility they afford of replacing derivatives of functions by simple multiplication by a scalar. If the derivative with respect to time were so removed, equation (21.71) would contain only differentiation with respect to a single variable. Let us therefore take the Laplace transform of (21.71) with respect to t:

$$\int_0^\infty \kappa \frac{\partial^2 u}{\partial x^2} \exp(-st) \, dt = \int_0^\infty \frac{\partial u}{\partial t} \exp(-st) \, dt.$$

On the LHS the (double) differentiation is with respect to x, whereas the integration is with respect to the independent variable t. Therefore the derivative can be taken outside the integral. Denoting the Laplace transform of $u(x, t)$ by $\bar{u}(x, s)$ and using result (13.57) to rewrite the transform of the derivative on the RHS (or by integrating directly by parts), we obtain

$$\kappa \frac{\partial^2 \bar{u}}{\partial x^2} = s\bar{u}(x, s) - u(x, 0).$$

But from the boundary condition $u(x, 0) = 0$ the last term on the RHS vanishes, and the solution is immediate:

$$\bar{u}(x,s) = A \exp\left(\sqrt{\frac{s}{\kappa}}\, x\right) + B \exp\left(-\sqrt{\frac{s}{\kappa}}\, x\right),$$

where the constants A and B may depend on s.

We require $u(x,t) \to 0$ as $x \to \infty$ and so we must also have $\bar{u}(\infty, s) = 0$; consequently we require that $A = 0$. The value of B is determined by the need for $u(0, t) = u_0$ and hence that

$$\bar{u}(0,s) = \int_0^\infty u_0 \exp(-st)\, dt = \frac{u_0}{s}.$$

We thus conclude that the appropriate expression for the Laplace transform of $u(x,t)$ is

$$\bar{u}(x,s) = \frac{u_0}{s} \exp\left(-\sqrt{\frac{s}{\kappa}}\, x\right). \tag{21.72}$$

To obtain $u(x,t)$ from this result requires the inversion of this transform – a task that is generally difficult and requires a contour integration. This is discussed in chapter 24, but for completeness we note that the solution is

$$u(x,t) = u_0 \left[1 - \mathrm{erf}\left(\frac{x}{\sqrt{4\kappa t}}\right)\right],$$

where $\mathrm{erf}(x)$ is the error function discussed in section 18.12.4. (The more complete sets of mathematical tables list this inverse Laplace transform.)

In the present problem, however, an alternative method is available. Let $w(t)$ be the amount of salt that has diffused into the tube in time t; then

$$w(t) = \int_0^\infty u(x,t)\, dx,$$

and its transform is given by

$$
\begin{aligned}
\bar{w}(s) &= \int_0^\infty dt \, \exp(-st) \int_0^\infty u(x,t)\, dx \\
&= \int_0^\infty dx \int_0^\infty u(x,t) \exp(-st)\, dt \\
&= \int_0^\infty \bar{u}(x,s)\, dx.
\end{aligned}
$$

Substituting for $\bar{u}(x,s)$ from (21.72) into the last integral and integrating, we obtain

$$\bar{w}(s) = u_0 \kappa^{1/2} s^{-3/2}.$$

This expression is much simpler to invert, and referring to the table of standard Laplace transforms (table 13.1) we find

$$w(t) = 2(\kappa/\pi)^{1/2} u_0 t^{1/2},$$

which is thus the required expression for the amount of diffused salt at time t. ◄

The above example shows that in some circumstances the use of a Laplace transformation can greatly simplify the solution of a PDE. However, it will have been observed that (as with ODEs) the easy elimination of some derivatives is usually paid for by the introduction of a difficult inverse transformation. This problem, although still present, is less severe for Fourier transformations.

> ►*An infinite metal bar has an initial temperature distribution $f(x)$ along its length. Find the temperature distribution at a later time t.*

We are interested in values of x from $-\infty$ to ∞, which suggests Fourier transformation with respect to x. Assuming that the solution obeys the boundary conditions $u(x, t) \to 0$ and $\partial u/\partial x \to 0$ as $|x| \to \infty$, we may Fourier-transform the one-dimensional diffusion equation (21.71) to obtain

$$\frac{\kappa}{\sqrt{2\pi}} \int_{-\infty}^{\infty} \frac{\partial^2 u(x, t)}{\partial x^2} \exp(-ikx)\, dx = \frac{1}{\sqrt{2\pi}} \frac{\partial}{\partial t} \int_{-\infty}^{\infty} u(x, t) \exp(-ikx)\, dx,$$

where on the RHS we have taken the partial derivative with respect to t outside the integral. Denoting the Fourier transform of $u(x, t)$ by $\tilde{u}(k, t)$, and using equation (13.28) to rewrite the Fourier transform of the second derivative on the LHS, we then have

$$-\kappa k^2 \tilde{u}(k, t) = \frac{\partial \tilde{u}(k, t)}{\partial t}.$$

This first-order equation has the simple solution

$$\tilde{u}(k, t) = \tilde{u}(k, 0) \exp(-\kappa k^2 t),$$

where the initial conditions give

$$\tilde{u}(k, 0) = \frac{1}{\sqrt{2\pi}} \int_{-\infty}^{\infty} u(x, 0) \exp(-ikx)\, dx$$

$$= \frac{1}{\sqrt{2\pi}} \int_{-\infty}^{\infty} f(x) \exp(-ikx)\, dx = \tilde{f}(k).$$

Thus we may write the Fourier transform of the solution as

$$\tilde{u}(k, t) = \tilde{f}(k) \exp(-\kappa k^2 t) = \sqrt{2\pi}\, \tilde{f}(k) \tilde{G}(k, t), \tag{21.73}$$

where we have defined the function $\tilde{G}(k, t) = (\sqrt{2\pi})^{-1} \exp(-\kappa k^2 t)$. Since $\tilde{u}(k, t)$ can be written as the product of two Fourier transforms, we can use the convolution theorem, subsection 13.1.7, to write the solution as

$$u(x, t) = \int_{-\infty}^{\infty} G(x - x', t) f(x')\, dx',$$

where $G(x, t)$ is the Green's function for this problem (see subsection 15.2.5). This function is the inverse Fourier transform of $\tilde{G}(k, t)$ and is thus given by

$$G(x, t) = \frac{1}{2\pi} \int_{-\infty}^{\infty} \exp(-\kappa k^2 t) \exp(ikx)\, dk$$

$$= \frac{1}{2\pi} \int_{-\infty}^{\infty} \exp\left[-\kappa t\left(k^2 - \frac{ix}{\kappa t}k\right)\right] dk.$$

Completing the square in the integrand we find

$$G(x, t) = \frac{1}{2\pi} \exp\left(-\frac{x^2}{4\kappa t}\right) \int_{-\infty}^{\infty} \exp\left[-\kappa t\left(k - \frac{ix}{2\kappa t}\right)^2\right] dk$$

$$= \frac{1}{2\pi} \exp\left(-\frac{x^2}{4\kappa t}\right) \int_{-\infty}^{\infty} \exp\left(-\kappa t k'^2\right) dk'$$

$$= \frac{1}{\sqrt{4\pi\kappa t}} \exp\left(-\frac{x^2}{4\kappa t}\right),$$

where in the second line we have made the substitution $k' = k - ix/(2\kappa t)$, and in the last

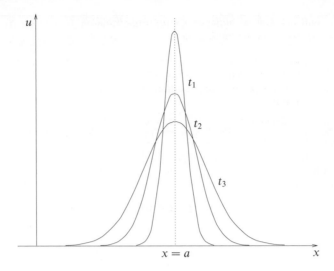

Figure 21.10   Diffusion of heat from a point source in a metal bar: the curves show the temperature u at position x for various times $t_1 < t_2 < t_3$. The area under the curves remains constant, since the total heat energy is conserved.

line we have used the standard result for the integral of a Gaussian, given in subsection 6.4.2. (Strictly speaking the change of variable from k to k' shifts the path of integration off the real axis, since k' is complex for real k, and so results in a complex integral, as will be discussed in chapter 24. Nevertheless, in this case the path of integration can be shifted back to the real axis without affecting the value of the integral.)

Thus the temperature in the bar at a later time t is given by

$$u(x, t) = \frac{1}{\sqrt{4\pi\kappa t}} \int_{-\infty}^{\infty} \exp\left[-\frac{(x - x')^2}{4\kappa t}\right] f(x')\, dx', \qquad (21.74)$$

which may be evaluated (numerically if necessary) when the form of $f(x)$ is given. ◀

As we might expect from our discussion of Green's functions in chapter 15, we see from (21.74) that, if the initial temperature distribution is $f(x) = \delta(x - a)$, i.e. a 'point' source at $x = a$, then the temperature distribution at later times is simply given by

$$u(x, t) = G(x - a, t) = \frac{1}{\sqrt{4\pi\kappa t}} \exp\left[-\frac{(x - a)^2}{4\kappa t}\right].$$

The temperature at several later times is illustrated in figure 21.10, which shows that the heat diffuses out from its initial position; the width of the Gaussian increases as $\sqrt{t}$, a dependence on time which is characteristic of diffusion processes.

The reader may have noticed that in both examples using integral transforms the solutions have been obtained in closed form – albeit in one case in the form of an integral. This differs from the infinite series solutions usually obtained via the separation of variables. It should be noted that this behaviour is a result of the infinite range in x, rather than of the transform method itself. In fact the method of separation of variables would yield the same solutions, since in the infinite-range case the separation constant is not restricted to take on an infinite set of discrete values but may have any real value, with the result that the sum over λ becomes an integral, as mentioned at the end of section 21.2.

►*An infinite metal bar has an initial temperature distribution $f(x)$ along its length. Find the temperature distribution at a later time t using the method of separation of variables.*

This is the same problem as in the previous example, but we now seek a solution by separating variables. From (21.12) a separated solution for the one-dimensional diffusion equation is given by

$$u(x,t) = [A \exp(i\lambda x) + B \exp(-i\lambda x)] \exp(-\kappa\lambda^2 t),$$

where $-\lambda^2$ is the separation constant. Since the bar is infinite we do not require the solution to take a given form at any finite value of x (for instance at $x = 0$) and so there is no restriction on λ other than its being real. Therefore instead of the superposition of such solutions in the form of a sum over allowed values of λ we have an integral over all λ,

$$u(x,t) = \frac{1}{\sqrt{2\pi}} \int_{-\infty}^{\infty} A(\lambda) \exp(-\kappa\lambda^2 t) \exp(i\lambda x)\, d\lambda, \tag{21.75}$$

where in taking λ from $-\infty$ to ∞ we need include only one of the complex exponentials, we have taken a factor $1/\sqrt{2\pi}$ out of $A(\lambda)$ for convenience. We can see from (21.75) that the expression for $u(x,t)$ has the form of an inverse Fourier transform (where λ is the transform variable). Therefore, Fourier-transforming both sides and using the Fourier inversion theorem, we find

$$\widetilde{u}(\lambda, t) = A(\lambda) \exp(-\kappa\lambda^2 t).$$

Now, the initial boundary condition requires

$$u(x,0) = \frac{1}{\sqrt{2\pi}} \int_{-\infty}^{\infty} A(\lambda) \exp(i\lambda x)\, d\lambda = f(x),$$

from which, using the Fourier inversion theorem once more, we see that $A(\lambda) = \widetilde{f}(\lambda)$. Therefore we have

$$\widetilde{u}(\lambda, t) = \widetilde{f}(\lambda) \exp(-\kappa\lambda^2 t),$$

which is identical to (21.73) in the previous example (but with k replaced by λ), and hence leads to the same result. ◄

21.5 Inhomogeneous problems – Green's functions

In chapters 15 and 17 we encountered Green's functions and found them a useful tool for solving inhomogeneous linear ODEs. We now discuss their usefulness in solving inhomogeneous linear PDEs.

For the sake of brevity we shall again denote a linear PDE by

$$\mathcal{L}u(\mathbf{r}) = \rho(\mathbf{r}), \tag{21.76}$$

where $\mathcal{L}$ is a linear partial differential operator. For example, in Laplace's equation we have $\mathcal{L} = \nabla^2$, whereas for Helmholtz's equation $\mathcal{L} = \nabla^2 + k^2$. Note that we have not specified the dimensionality of the problem, and (21.76) may, for example, represent Poisson's equation in two or three (or more) dimensions. The reader will also notice that for the sake of simplicity we have not included any time dependence in (21.76). Nevertheless, the following discussion can be generalised to include it.

As we discussed in subsection 20.3.2, a problem is inhomogeneous if the fact that $u(\mathbf{r})$ is a solution does *not* imply that any constant multiple $\lambda u(\mathbf{r})$ is also a solution. This inhomogeneity may derive from either the PDE itself or from the boundary conditions imposed on the solution.

In our discussion of Green's function solutions of inhomogeneous ODEs (see subsection 15.2.5) we dealt with inhomogeneous boundary conditions by making a suitable change of variable such that in the new variable the boundary conditions were homogeneous. In an analogous way, as illustrated in the final example of section 21.2, it is usually possible to make a change of variables in PDEs to transform between inhomogeneity of the boundary conditions and inhomogeneity of the equation. Therefore let us assume for the moment that the boundary conditions imposed on the solution $u(\mathbf{r})$ of (21.76) are homogeneous. This most commonly means that if we seek a solution to (21.76) in some region V then on the surface S that bounds V the solution obeys the conditions $u(\mathbf{r}) = 0$ or $\partial u / \partial n = 0$, where $\partial u / \partial n$ is the normal derivative of u at the surface S.

We shall discuss the extension of the Green's function method to the direct solution of problems with inhomogeneous boundary conditions in subsection 21.5.2, but we first highlight how the Green's function approach to solving ODEs can be simply extended to PDEs for homogeneous boundary conditions.

21.5.1 Similarities to Green's functions for ODEs

As in the discussion of ODEs in chapter 15, we may consider the Green's function for a system described by a PDE as the response of the system to a 'unit impulse' or 'point source'. Thus if we seek a solution to (21.76) that satisfies some homogeneous boundary conditions on $u(\mathbf{r})$ then the Green's function $G(\mathbf{r}, \mathbf{r}_0)$ for the problem is a solution of

$$\mathcal{L}G(\mathbf{r}, \mathbf{r}_0) = \delta(\mathbf{r} - \mathbf{r}_0), \tag{21.77}$$

where $\mathbf{r}_0$ lies in V. The Green's function $G(\mathbf{r}, \mathbf{r}_0)$ must also satisfy the imposed (homogeneous) boundary conditions.

It is understood that in (21.77) the $\mathcal{L}$ operator expresses differentiation with respect to $\mathbf{r}$ as opposed to $\mathbf{r}_0$. Also, $\delta(\mathbf{r} - \mathbf{r}_0)$ is the Dirac delta function (see chapter 13) of dimension appropriate to the problem; it may be thought of as representing a unit-strength point source at $\mathbf{r} = \mathbf{r}_0$.

Following an analogous argument to that given in subsection 15.2.5 for ODEs, if the boundary conditions on $u(\mathbf{r})$ are homogeneous then a solution to (21.76) that satisfies the imposed boundary conditions is given by

$$u(\mathbf{r}) = \int G(\mathbf{r}, \mathbf{r}_0) \rho(\mathbf{r}_0) \, dV(\mathbf{r}_0), \tag{21.78}$$

where the integral on $\mathbf{r}_0$ is over some appropriate 'volume'. In two or more dimensions, however, the task of finding directly a solution to (21.77) that satisfies the imposed boundary conditions on S can be a difficult one, and we return to this in the next subsection.

An alternative approach is to follow a similar argument to that presented in chapter 17 for ODEs and so to construct the Green's function for (21.76) as a superposition of eigenfunctions of the operator $\mathcal{L}$, provided $\mathcal{L}$ is Hermitian. By analogy with an ordinary differential operator, a partial differential operator is Hermitian if it satisfies

$$\int_V v^*(\mathbf{r}) \mathcal{L} w(\mathbf{r}) \, dV = \left[\int_V w^*(\mathbf{r}) \mathcal{L} v(\mathbf{r}) \, dV \right]^*,$$

where the asterisk denotes complex conjugation and v and w are arbitrary functions obeying the imposed (homogeneous) boundary condition on the solution of $\mathcal{L} u(\mathbf{r}) = 0$.

The eigenfunctions $u_n(\mathbf{r})$, $n = 0, 1, 2, \ldots$, of $\mathcal{L}$ satisfy

$$\mathcal{L} u_n(\mathbf{r}) = \lambda_n u_n(\mathbf{r}),$$

where λ_n are the corresponding eigenvalues, which are all real for an Hermitian operator $\mathcal{L}$. Furthermore, each eigenfunction must obey any imposed (homogeneous) boundary conditions. Using an argument analogous to that given in chapter 17, the Green's function for the problem is given by

$$G(\mathbf{r}, \mathbf{r}_0) = \sum_{n=0}^{\infty} \frac{u_n(\mathbf{r}) u_n^*(\mathbf{r}_0)}{\lambda_n}. \tag{21.79}$$

From (21.79) we see immediately that the Green's function (irrespective of how it is found) enjoys the property

$$G(\mathbf{r}, \mathbf{r}_0) = G^*(\mathbf{r}_0, \mathbf{r}).$$

Thus, if the Green's function is real then it is symmetric in its two arguments.

Once the Green's function has been obtained, the solution to (21.76) is again given by (21.78). For PDEs this approach can become very cumbersome, however, and so we shall not pursue it further here.

21.5.2 General boundary-value problems

As mentioned above, often inhomogeneous boundary conditions can be dealt with by making an appropriate change of variables, such that the boundary conditions in the new variables are homogeneous although the equation itself is generally inhomogeneous. In this section, however, we extend the use of Green's functions to problems with inhomogeneous boundary conditions (and equations). This provides a more consistent and intuitive approach to the solution of such *boundary-value problems*.

For definiteness we shall consider Poisson's equation

$$\nabla^2 u(\mathbf{r}) = \rho(\mathbf{r}), \tag{21.80}$$

but the material of this section may be extended to other linear PDEs of the form (21.76). Clearly, Poisson's equation reduces to Laplace's equation for $\rho(\mathbf{r}) = 0$ and so our discussion is equally applicable to this case.

We wish to solve (21.80) in some region V bounded by a surface S, which may consist of several disconnected parts. As stated above, we shall allow the possibility that the boundary conditions on the solution $u(\mathbf{r})$ may be inhomogeneous on S, although as we shall see this method reduces to those discussed above in the special case that the boundary conditions are in fact homogeneous.

The two common types of inhomogeneous boundary condition for Poisson's equation are (as discussed in subsection 20.6.2):

  (i) Dirichlet conditions, in which $u(\mathbf{r})$ is specified on S, and
  (ii) Neumann conditions, in which $\partial u/\partial n$ is specified on S.

In general, specifying *both* Dirichlet *and* Neumann conditions on S overdetermines the problem and leads to there being no solution.

The specification of the surface S requires some further comment, since S may have several disconnected parts. If we wish to solve Poisson's equation inside some closed surface S then the situation is straightforward and is shown in figure 21.11(a). If, however, we wish to solve Poisson's equation in the gap between two closed surfaces (for example in the gap between two concentric conducting cylinders) then the volume V is bounded by a surface S that has two disconnected parts S_1 and S_2, as shown in figure 21.11(b); the direction of the normal to the surface is always taken as pointing *out* of the volume V. A similar situation arises when we wish to solve Poisson's equation *outside* some closed surface S_1. In this case the volume V is infinite but is treated formally by taking the surface S_2 as a large sphere of radius R and letting R tend to infinity.

In order to solve (21.80) subject to either Dirichlet or Neumann boundary conditions on S, we will remind ourselves of Green's second theorem, equation (11.20), which states that, for two scalar functions $\phi(\mathbf{r})$ and $\psi(\mathbf{r})$ defined in some volume V bounded by a surface S,

$$\int_V (\phi\nabla^2\psi - \psi\nabla^2\phi)\, dV = \int_S (\phi\nabla\psi - \psi\nabla\phi)\cdot\hat{\mathbf{n}}\, dS, \tag{21.81}$$

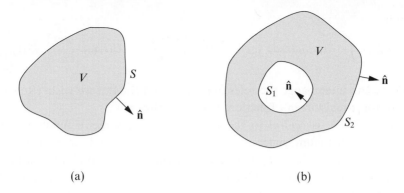

Figure 21.11 Surfaces used for solving Poisson's equation in different regions V.

where on the RHS it is common to write, for example, $\nabla\psi \cdot \hat{\mathbf{n}}\,dS$ as $(\partial\psi/\partial n)\,dS$. The expression $\partial\psi/\partial n$ stands for $\nabla\psi \cdot \hat{\mathbf{n}}$, the rate of change of ψ in the direction of the unit outward normal $\hat{\mathbf{n}}$ to the surface S.

The Green's function for Poisson's equation (21.80) must satisfy

$$\nabla^2 G(\mathbf{r}, \mathbf{r}_0) = \delta(\mathbf{r} - \mathbf{r}_0), \tag{21.82}$$

where $\mathbf{r}_0$ lies in V. (As mentioned above, we may think of $G(\mathbf{r}, \mathbf{r}_0)$ as the solution to Poisson's equation for a unit-strength point source located at $\mathbf{r} = \mathbf{r}_0$.) Let us for the moment impose no boundary conditions on $G(\mathbf{r}, \mathbf{r}_0)$.

If we now let $\phi = u(\mathbf{r})$ and $\psi = G(\mathbf{r}, \mathbf{r}_0)$ in Green's theorem (21.81) then we obtain

$$\int_V \left[u(\mathbf{r})\nabla^2 G(\mathbf{r}, \mathbf{r}_0) - G(\mathbf{r}, \mathbf{r}_0)\,\nabla^2 u(\mathbf{r}) \right]\,dV(\mathbf{r})$$
$$= \int_S \left[u(\mathbf{r})\frac{\partial G(\mathbf{r}, \mathbf{r}_0)}{\partial n} - G(\mathbf{r}, \mathbf{r}_0)\frac{\partial u(\mathbf{r})}{\partial n} \right]\,dS(\mathbf{r}),$$

where we have made explicit that the volume and surface integrals are with respect to $\mathbf{r}$. Using (21.80) and (21.82) the LHS can be simplified to give

$$\int_V \left[u(\mathbf{r})\delta(\mathbf{r} - \mathbf{r}_0) - G(\mathbf{r}, \mathbf{r}_0)\rho(\mathbf{r}) \right]\,dV(\mathbf{r})$$
$$= \int_S \left[u(\mathbf{r})\frac{\partial G(\mathbf{r}, \mathbf{r}_0)}{\partial n} - G(\mathbf{r}, \mathbf{r}_0)\frac{\partial u(\mathbf{r})}{\partial n} \right]\,dS(\mathbf{r}). \tag{21.83}$$

Since $\mathbf{r}_0$ lies within the volume V,

$$\int_V u(\mathbf{r})\delta(\mathbf{r} - \mathbf{r}_0)\,dV(\mathbf{r}) = u(\mathbf{r}_0),$$

and thus on rearranging (21.83) the solution to Poisson's equation (21.80) can be written as

$$u(\mathbf{r}_0) = \int_V G(\mathbf{r}, \mathbf{r}_0) \rho(\mathbf{r}) \, dV(\mathbf{r}) + \int_S \left[u(\mathbf{r}) \frac{\partial G(\mathbf{r}, \mathbf{r}_0)}{\partial n} - G(\mathbf{r}, \mathbf{r}_0) \frac{\partial u(\mathbf{r})}{\partial n} \right] dS(\mathbf{r}).$$
(21.84)

Clearly, we can interchange the roles of $\mathbf{r}$ and $\mathbf{r}_0$ in (21.84) if we wish. (Remember also that, for a real Green's function, $G(\mathbf{r}, \mathbf{r}_0) = G(\mathbf{r}_0, \mathbf{r})$.)

Equation (21.84) is *central* to the extension of the Green's function method to problems with inhomogeneous boundary conditions, and we next discuss its application to both Dirichlet and Neumann boundary-value problems. But, before doing so, we also note that if the boundary condition on S is in fact homogeneous, so that $u(\mathbf{r}) = 0$ or $\partial u(\mathbf{r})/\partial n = 0$ on S, then demanding that the Green's function $G(\mathbf{r}, \mathbf{r}_0)$ also obeys the same boundary condition causes the surface integral in (21.84) to vanish, and we are left with the familiar form of solution given in (21.78). The extension of (21.84) to a PDE other than Poisson's equation is discussed in exercise 21.28.

21.5.3 Dirichlet problems

In a Dirichlet problem we require the solution $u(\mathbf{r})$ of Poisson's equation (21.80) to take specific values on some surface S that bounds V, i.e. we require that $u(\mathbf{r}) = f(\mathbf{r})$ on S where f is a given function.

If we seek a Green's function $G(\mathbf{r}, \mathbf{r}_0)$ for this problem it must clearly satisfy (21.82), but we are free to choose the boundary conditions satisfied by $G(\mathbf{r}, \mathbf{r}_0)$ in such a way as to make the solution (21.84) as simple as possible. From (21.84), we see that by choosing

$$G(\mathbf{r}, \mathbf{r}_0) = 0 \quad \text{for } \mathbf{r} \text{ on } S \tag{21.85}$$

the second term in the surface integral vanishes. Since $u(\mathbf{r}) = f(\mathbf{r})$ on S, (21.84) then becomes

$$u(\mathbf{r}_0) = \int_V G(\mathbf{r}, \mathbf{r}_0) \rho(\mathbf{r}) \, dV(\mathbf{r}) + \int_S f(\mathbf{r}) \frac{\partial G(\mathbf{r}, \mathbf{r}_0)}{\partial n} \, dS(\mathbf{r}). \tag{21.86}$$

Thus we wish to find the *Dirichlet Green's function* that

  (i) satisfies (21.82) and hence is singular at $\mathbf{r} = \mathbf{r}_0$, and
  (ii) obeys the boundary condition $G(\mathbf{r}, \mathbf{r}_0) = 0$ for $\mathbf{r}$ on S.

In general, it is difficult to obtain this function directly, and so it is useful to separate these two requirements. We therefore look for a solution of the form

$$G(\mathbf{r}, \mathbf{r}_0) = F(\mathbf{r}, \mathbf{r}_0) + H(\mathbf{r}, \mathbf{r}_0),$$

where $F(\mathbf{r}, \mathbf{r}_0)$ satisfies (21.82) and has the required singular character at $\mathbf{r} = \mathbf{r}_0$ but does not necessarily obey the boundary condition on S, whilst $H(\mathbf{r}, \mathbf{r}_0)$ satisfies the corresponding homogeneous equation (i.e. Laplace's equation) inside V but is adjusted in such a way that the sum $G(\mathbf{r}, \mathbf{r}_0)$ equals zero on S. The Green's function $G(\mathbf{r}, \mathbf{r}_0)$ is still a solution of (21.82) since

$$\nabla^2 G(\mathbf{r}, \mathbf{r}_0) = \nabla^2 F(\mathbf{r}, \mathbf{r}_0) + \nabla^2 H(\mathbf{r}, \mathbf{r}_0) = \nabla^2 F(\mathbf{r}, \mathbf{r}_0) + 0 = \delta(\mathbf{r} - \mathbf{r}_0).$$

The function $F(\mathbf{r}, \mathbf{r}_0)$ is called the *fundamental solution* and will clearly take different forms depending on the dimensionality of the problem. Let us first consider the fundamental solution to (21.82) in three dimensions.

> ► *Find the fundamental solution to Poisson's equation in three dimensions that tends to zero as $|\mathbf{r}| \to \infty$.*

We wish to solve

$$\nabla^2 F(\mathbf{r}, \mathbf{r}_0) = \delta(\mathbf{r} - \mathbf{r}_0) \tag{21.87}$$

in three dimensions, subject to the boundary condition $F(\mathbf{r}, \mathbf{r}_0) \to 0$ as $|\mathbf{r}| \to \infty$. Since the problem is spherically symmetric about $\mathbf{r}_0$, let us consider a large sphere S of radius R centred on $\mathbf{r}_0$, and integrate (21.87) over the enclosed volume V. We then obtain

$$\int_V \nabla^2 F(\mathbf{r}, \mathbf{r}_0) \, dV = \int_V \delta(\mathbf{r} - \mathbf{r}_0) \, dV = 1, \tag{21.88}$$

since V encloses the point $\mathbf{r}_0$. However, using the divergence theorem,

$$\int_V \nabla^2 F(\mathbf{r}, \mathbf{r}_0) \, dV = \int_S \nabla F(\mathbf{r}, \mathbf{r}_0) \cdot \hat{\mathbf{n}} \, dS, \tag{21.89}$$

where $\hat{\mathbf{n}}$ is the unit normal to the large sphere S at any point.

Since the problem is spherically symmetric about $\mathbf{r}_0$, we expect that

$$F(\mathbf{r}, \mathbf{r}_0) = F(|\mathbf{r} - \mathbf{r}_0|) = F(r),$$

i.e. that F has the same value everywhere on S. Thus, evaluating the surface integral in (21.89) and equating it to unity from (21.88), we have[§]

$$4\pi r^2 \frac{dF}{dr}\bigg|_{r=R} = 1.$$

Integrating this expression we obtain

$$F(r) = -\frac{1}{4\pi r} + \text{constant},$$

but, since we require $F(\mathbf{r}, \mathbf{r}_0) \to 0$ as $|\mathbf{r}| \to \infty$, the constant must be zero. The fundamental solution in three dimensions is consequently given by

$$F(\mathbf{r}, \mathbf{r}_0) = -\frac{1}{4\pi|\mathbf{r} - \mathbf{r}_0|}. \tag{21.90}$$

This is clearly also the full Green's function for Poisson's equation subject to the boundary condition $u(\mathbf{r}) \to 0$ as $|\mathbf{r}| \to \infty$. ◄

Using (21.90) we can write down the solution of Poisson's equation to find,

---

[§] A vertical bar to the right of an expression is a common alternative to enclosing the expression in square brackets; as usual, the subscript shows the value of the variable at which the expression is to be evaluated.

for example, the electrostatic potential $u(\mathbf{r})$ due to some distribution of electric charge $\rho(\mathbf{r})$. The electrostatic potential satisfies

$$\nabla^2 u(\mathbf{r}) = -\frac{\rho}{\epsilon_0},$$

where $u(\mathbf{r}) \to 0$ as $|\mathbf{r}| \to \infty$. Since the boundary condition on the surface at infinity is homogeneous the surface integral in (21.86) vanishes, and using (21.90) we recover the familiar solution

$$u(\mathbf{r}_0) = \int \frac{\rho(\mathbf{r})}{4\pi\epsilon_0 |\mathbf{r} - \mathbf{r}_0|} \, dV(\mathbf{r}), \tag{21.91}$$

where the volume integral is over all space.

We can develop an analogous theory in two dimensions. As before the fundamental solution satisfies

$$\nabla^2 F(\mathbf{r}, \mathbf{r}_0) = \delta(\mathbf{r} - \mathbf{r}_0), \tag{21.92}$$

where $\delta(\mathbf{r} - \mathbf{r}_0)$ is now the two-dimensional delta function. Following an analogous method to that used in the previous example, we find the fundamental solution in two dimensions to be given by

$$F(\mathbf{r}, \mathbf{r}_0) = \frac{1}{2\pi} \ln |\mathbf{r} - \mathbf{r}_0| + \text{constant}. \tag{21.93}$$

From the form of the solution we see that in two dimensions we cannot apply the condition $F(\mathbf{r}, \mathbf{r}_0) \to 0$ as $|\mathbf{r}| \to \infty$, and in this case the constant does not necessarily vanish.

We now return to the task of constructing the full Dirichlet Green's function. To do so we wish to add to the fundamental solution a solution of the homogeneous equation (in this case Laplace's equation) such that $G(\mathbf{r}, \mathbf{r}_0) = 0$ on S, as required by (21.86) and its attendant conditions. The appropriate Green's function is constructed by adding to the fundamental solution 'copies' of itself that represent 'image' sources at different locations *outside* V. Hence this approach is called the *method of images*.

In summary, if we wish to solve Poisson's equation in some region V subject to Dirichlet boundary conditions on its surface S then the procedure and argument are as follows.

(i) To the single source $\delta(\mathbf{r} - \mathbf{r}_0)$ inside V add image sources *outside* V

$$\sum_{n=1}^{N} q_n \delta(\mathbf{r} - \mathbf{r}_n) \quad \text{with } \mathbf{r}_n \text{ outside } V,$$

where the positions $\mathbf{r}_n$ and the strengths q_n of the image sources are to be determined as described in step (iii) below.

(ii) Since all the image sources lie outside V, the fundamental solution corresponding to each source satisfies Laplace's equation *inside* V. Thus we may add the fundamental solutions $F(\mathbf{r}, \mathbf{r}_n)$ corresponding to each image source to that corresponding to the single source inside V, obtaining the Green's function

$$G(\mathbf{r}, \mathbf{r}_0) = F(\mathbf{r}, \mathbf{r}_0) + \sum_{n=1}^{N} q_n F(\mathbf{r}, \mathbf{r}_n).$$

(iii) Now adjust the positions $\mathbf{r}_n$ and strengths q_n of the image sources so that the required boundary conditions are satisfied on S. For a Dirichlet Green's function we require $G(\mathbf{r}, \mathbf{r}_0) = 0$ for $\mathbf{r}$ on S.

(iv) The solution to Poisson's equation subject to the Dirichlet boundary condition $u(\mathbf{r}) = f(\mathbf{r})$ on S is then given by (21.86).

In general it is very difficult to find the correct positions and strengths for the images, i.e. to make them such that the boundary conditions on S are satisfied. Nevertheless, it is possible to do so for certain problems that have simple geometry. In particular, for problems in which the boundary S consists of straight lines (in two dimensions) or planes (in three dimensions), positions of the image points can be deduced simply by imagining the boundary lines or planes to be mirrors in which the single source in V (at $\mathbf{r}_0$) is reflected.

> ▶ *Solve Laplace's equation $\nabla^2 u = 0$ in three dimensions in the half-space $z > 0$, given that $u(\mathbf{r}) = f(\mathbf{r})$ on the plane $z = 0$.*

The surface S bounding V consists of the xy-plane and the surface at infinity. Therefore, the Dirichlet Green's function for this problem must satisfy $G(\mathbf{r}, \mathbf{r}_0) = 0$ on $z = 0$ and $G(\mathbf{r}, \mathbf{r}_0) \to 0$ as $|\mathbf{r}| \to \infty$. Thus it is clear in this case that we require one image source at a position $\mathbf{r}_1$ that is the reflection of $\mathbf{r}_0$ in the plane $z = 0$, as shown in figure 21.12 (so that $\mathbf{r}_1$ lies in $z < 0$, outside the region in which we wish to obtain a solution). It is also clear that the strength of this image should be -1.

Therefore by adding the fundamental solutions corresponding to the original source and its image we obtain the Green's function

$$G(\mathbf{r}, \mathbf{r}_0) = -\frac{1}{4\pi|\mathbf{r} - \mathbf{r}_0|} + \frac{1}{4\pi|\mathbf{r} - \mathbf{r}_1|}, \tag{21.94}$$

where $\mathbf{r}_1$ is the reflection of $\mathbf{r}_0$ in the plane $z = 0$, i.e. if $\mathbf{r}_0 = (x_0, y_0, z_0)$ then $\mathbf{r}_1 = (x_0, y_0, -z_0)$. Clearly $G(\mathbf{r}, \mathbf{r}_0) \to 0$ as $|\mathbf{r}| \to \infty$ as required. Also $G(\mathbf{r}, \mathbf{r}_0) = 0$ on $z = 0$, and so (21.94) is the desired Dirichlet Green's function.

The solution to Laplace's equation is then given by (21.86) with $\rho(\mathbf{r}) = 0$,

$$u(\mathbf{r}_0) = \int_S f(\mathbf{r}) \frac{\partial G(\mathbf{r}, \mathbf{r}_0)}{\partial n} \, dS(\mathbf{r}). \tag{21.95}$$

Clearly the surface at infinity makes no contribution to this integral. The outward-pointing unit vector normal to the xy-plane is simply $\hat{\mathbf{n}} = -\mathbf{k}$ (where $\mathbf{k}$ is the unit vector in the z-direction), and so

$$\frac{\partial G(\mathbf{r}, \mathbf{r}_0)}{\partial n} = -\frac{\partial G(\mathbf{r}, \mathbf{r}_0)}{\partial z} = -\mathbf{k} \cdot \nabla G(\mathbf{r}, \mathbf{r}_0).$$

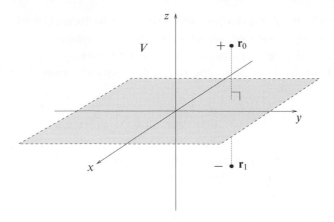

Figure 21.12 The arrangement of images for solving Laplace's equation in the half-space $z > 0$.

We may evaluate this normal derivative by writing the Green's function (21.94) explicitly in terms of x, y and z (and x_0, y_0 and z_0) and calculating the partial derivative with respect to z directly. It is usually quicker, however, to use the fact that[§]

$$\nabla |\mathbf{r} - \mathbf{r}_0| = \frac{\mathbf{r} - \mathbf{r}_0}{|\mathbf{r} - \mathbf{r}_0|}; \tag{21.96}$$

thus

$$\nabla G(\mathbf{r}, \mathbf{r}_0) = \frac{\mathbf{r} - \mathbf{r}_0}{4\pi |\mathbf{r} - \mathbf{r}_0|^3} - \frac{\mathbf{r} - \mathbf{r}_1}{4\pi |\mathbf{r} - \mathbf{r}_1|^3}.$$

Since $\mathbf{r}_0 = (x_0, y_0, z_0)$ and $\mathbf{r}_1 = (x_0, y_0, -z_0)$ the normal derivative is given by

$$-\frac{\partial G(\mathbf{r}, \mathbf{r}_0)}{\partial z} = -\mathbf{k} \cdot \nabla G(\mathbf{r}, \mathbf{r}_0)$$

$$= -\frac{z - z_0}{4\pi |\mathbf{r} - \mathbf{r}_0|^3} + \frac{z + z_0}{4\pi |\mathbf{r} - \mathbf{r}_1|^3}.$$

Therefore on the surface $z = 0$, and writing out the dependence on x, y and z explicitly, we have

$$-\frac{\partial G(\mathbf{r}, \mathbf{r}_0)}{\partial z}\bigg|_{z=0} = \frac{2z_0}{4\pi [(x - x_0)^2 + (y - y_0)^2 + z_0^2]^{3/2}}.$$

Inserting this expression into (21.95) we obtain the solution

$$u(x_0, y_0, z_0) = \frac{z_0}{2\pi} \int_{-\infty}^{\infty} \int_{-\infty}^{\infty} \frac{f(x, y)}{[(x - x_0)^2 + (y - y_0)^2 + z_0^2]^{3/2}} \, dx \, dy. \blacktriangleleft$$

An analogous procedure may be applied in two-dimensional problems. For

[§] Since $|\mathbf{r} - \mathbf{r}_0|^2 = (\mathbf{r} - \mathbf{r}_0) \cdot (\mathbf{r} - \mathbf{r}_0)$ we have $\nabla |\mathbf{r} - \mathbf{r}_0|^2 = 2(\mathbf{r} - \mathbf{r}_0)$, from which we obtain

$$\nabla (|\mathbf{r} - \mathbf{r}_0|^2)^{1/2} = \frac{1}{2} \frac{2(\mathbf{r} - \mathbf{r}_0)}{(|\mathbf{r} - \mathbf{r}_0|^2)^{1/2}} = \frac{\mathbf{r} - \mathbf{r}_0}{|\mathbf{r} - \mathbf{r}_0|}.$$

Note that this result holds in two *and* three dimensions.

example, in solving Poisson's equation in two dimensions in the half-space $x > 0$ we again require just one image charge, of strength $q_1 = -1$, at a position $\mathbf{r}_1$ that is the reflection of $\mathbf{r}_0$ in the line $x = 0$. Since we require $G(\mathbf{r}, \mathbf{r}_0) = 0$ when $\mathbf{r}$ lies on $x = 0$, the constant in (21.93) must equal zero, and so the Dirichlet Green's function is

$$G(\mathbf{r}, \mathbf{r}_0) = \frac{1}{2\pi} \left(\ln |\mathbf{r} - \mathbf{r}_0| - \ln |\mathbf{r} - \mathbf{r}_1| \right).$$

Clearly $G(\mathbf{r}, \mathbf{r}_0)$ tends to zero as $|\mathbf{r}| \to \infty$. If, however, we wish to solve the two-dimensional Poisson equation in the quarter space $x > 0$, $y > 0$, then more image points are required.

> ▶ A line charge in the z-direction of charge density λ is placed at some position $\mathbf{r}_0$ in the quarter-space $x > 0$, $y > 0$. Calculate the force per unit length on the line charge due to the presence of thin earthed plates along $x = 0$ and $y = 0$.

Here we wish to solve Poisson's equation,

$$\nabla^2 u = -\frac{\lambda}{\epsilon_0} \delta(\mathbf{r} - \mathbf{r}_0),$$

in the quarter space $x > 0$, $y > 0$. It is clear that we require three image line charges with positions and strengths as shown in figure 21.13 (all of which lie outside the region in which we seek a solution). The boundary condition that the electrostatic potential u is zero on $x = 0$ and $y = 0$ (shown as the 'curve' C in figure 21.13) is then automatically satisfied, and so this system of image charges is directly equivalent to the original situation of a single line charge in the presence of the earthed plates along $x = 0$ and $y = 0$. Thus the electrostatic potential is simply equal to the Dirichlet Green's function

$$u(\mathbf{r}) = G(\mathbf{r}, \mathbf{r}_0) = -\frac{\lambda}{2\pi\epsilon_0} \left(\ln |\mathbf{r} - \mathbf{r}_0| - \ln |\mathbf{r} - \mathbf{r}_1| + \ln |\mathbf{r} - \mathbf{r}_2| - \ln |\mathbf{r} - \mathbf{r}_3| \right),$$

which equals zero on C and on the 'surface' at infinity.

The force on the line charge at $\mathbf{r}_0$, therefore, is simply that due to the three line charges at $\mathbf{r}_1$, $\mathbf{r}_2$ and $\mathbf{r}_3$. The elecrostatic potential due to a line charge at $\mathbf{r}_i$, $i = 1$, 2 or 3, is given by the fundamental solution

$$u_i(\mathbf{r}) = \mp\frac{\lambda}{2\pi\epsilon_0} \ln |\mathbf{r} - \mathbf{r}_i| + c,$$

the upper or lower sign being taken according to whether the line charge is positive or negative, respectively. Therefore the force per unit length on the line charge at $\mathbf{r}_0$, due to the one at $\mathbf{r}_i$, is given by

$$-\lambda \nabla u_i(\mathbf{r}) \Big|_{\mathbf{r}=\mathbf{r}_0} = \pm\frac{\lambda^2}{2\pi\epsilon_0} \frac{\mathbf{r}_0 - \mathbf{r}_i}{|\mathbf{r}_0 - \mathbf{r}_i|^2}.$$

Adding the contributions from the three image charges shown in figure 21.13, the total force experienced by the line charge at $\mathbf{r}_0$ is given by

$$\mathbf{F} = \frac{\lambda^2}{2\pi\epsilon_0} \left(-\frac{\mathbf{r}_0 - \mathbf{r}_1}{|\mathbf{r}_0 - \mathbf{r}_1|^2} + \frac{\mathbf{r}_0 - \mathbf{r}_2}{|\mathbf{r}_0 - \mathbf{r}_2|^2} - \frac{\mathbf{r}_0 - \mathbf{r}_3}{|\mathbf{r}_0 - \mathbf{r}_3|^2} \right),$$

where, from the figure, $\mathbf{r}_0 - \mathbf{r}_1 = 2y_0\mathbf{j}$, $\mathbf{r}_0 - \mathbf{r}_2 = 2x_0\mathbf{i} + 2y_0\mathbf{j}$ and $\mathbf{r}_0 - \mathbf{r}_3 = 2x_0\mathbf{i}$. Thus, in

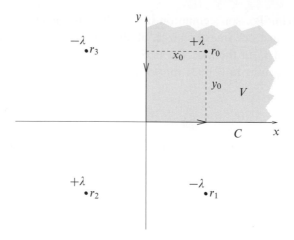

Figure 21.13 The arrangement of images for finding the force on a line charge situated in the (two-dimensional) quarter-space $x > 0$, $y > 0$, when the planes $x = 0$ and $y = 0$ are earthed.

terms of x_0 and y_0, the total force on the line charge due to the charge induced on the plates is given by

$$\mathbf{F} = \frac{\lambda^2}{2\pi\epsilon_0}\left(-\frac{1}{2y_0}\mathbf{j} + \frac{2x_0\,\mathbf{i} + 2y_0\,\mathbf{j}}{4x_0^2 + 4y_0^2} - \frac{1}{2x_0}\mathbf{i}\right)$$

$$= -\frac{\lambda^2}{4\pi\epsilon_0(x_0^2 + y_0^2)}\left(\frac{y_0^2}{x_0}\mathbf{i} + \frac{x_0^2}{y_0}\mathbf{j}\right). \ \blacktriangleleft$$

Further generalisations are possible. For instance, solving Poisson's equation in the two-dimensional strip $-\infty < x < \infty$, $0 < y < b$ requires an infinite series of image points.

So far we have considered problems in which the boundary S consists of straight lines (in two dimensions) or planes (in three dimensions), in which simple reflections of the source at $\mathbf{r}_0$ in these boundaries fix the positions of the image points. For more complicated (curved) boundaries this is no longer possible, and finding the appropriate position(s) and strength(s) of the image source(s) requires further work.

▶*Use the method of images to find the Dirichlet Green's function for solving Poisson's equation outside a sphere of radius a centred at the origin.*

We need to find a solution of Poisson's equation valid outside the sphere of radius a. Since an image point $\mathbf{r}_1$ cannot lie in this region, it must be located within the sphere. The Green's function for this problem is therefore

$$G(\mathbf{r}, \mathbf{r}_0) = -\frac{1}{4\pi|\mathbf{r} - \mathbf{r}_0|} - \frac{q}{4\pi|\mathbf{r} - \mathbf{r}_1|},$$

where $|\mathbf{r}_0| > a$, $|\mathbf{r}_1| < a$ and q is the strength of the image which we have yet to determine. Clearly, $G(\mathbf{r}, \mathbf{r}_0) \to 0$ on the surface at infinity.

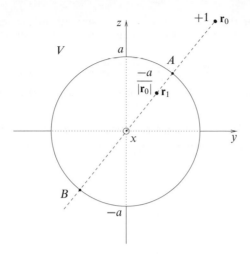

Figure 21.14 The arrangement of images for solving Poisson's equation outside a sphere of radius a centred at the origin. For a charge $+1$ at $\mathbf{r}_0$, the image point $\mathbf{r}_1$ is given by $(a/|\mathbf{r}_0|)^2\mathbf{r}_0$ and the strength of the image charge is $-a/|\mathbf{r}_0|$.

By symmetry we expect the image point $\mathbf{r}_1$ to lie on the same radial line as the original source, $\mathbf{r}_0$, as shown in figure 21.14, and so $\mathbf{r}_1 = k\mathbf{r}_0$ where $k < 1$. However, for a Dirichlet Green's function we require $G(\mathbf{r} - \mathbf{r}_0) = 0$ on $|\mathbf{r}| = a$, and the form of the Green's function suggests that we need

$$|\mathbf{r} - \mathbf{r}_0| \propto |\mathbf{r} - \mathbf{r}_1| \quad \text{for all } |\mathbf{r}| = a. \tag{21.97}$$

Referring to figure 21.14, if this relationship is to hold over the whole surface of the sphere, then it must certainly hold for the points A and B. We thus require

$$\frac{|\mathbf{r}_0| - a}{a - |\mathbf{r}_1|} = \frac{|\mathbf{r}_0| + a}{a + |\mathbf{r}_1|},$$

which reduces to $|\mathbf{r}_1| = a^2/|\mathbf{r}_0|$. Therefore the image point must be located at the position

$$\mathbf{r}_1 = \frac{a^2}{|\mathbf{r}_0|^2}\mathbf{r}_0.$$

It may now be checked that, for this location of the image point, (21.97) is satisfied over the whole sphere. Using the geometrical result

$$|\mathbf{r} - \mathbf{r}_1|^2 = |\mathbf{r}|^2 - \frac{2a^2}{|\mathbf{r}_0|^2}\mathbf{r} \cdot \mathbf{r}_0 + \frac{a^4}{|\mathbf{r}_0|^2}$$

$$= \frac{a^2}{|\mathbf{r}_0|^2}\left(|\mathbf{r}_0|^2 - 2\mathbf{r} \cdot \mathbf{r}_0 + a^2\right) \quad \text{for } |\mathbf{r}| = a, \tag{21.98}$$

we see that, on the surface of the sphere,

$$|\mathbf{r} - \mathbf{r}_1| = \frac{a}{|\mathbf{r}_0|}|\mathbf{r} - \mathbf{r}_0| \quad \text{for } |\mathbf{r}| = a. \tag{21.99}$$

Therefore, in order that $G = 0$ at $|\mathbf{r}| = a$, the strength of the image charge must be $-a/|\mathbf{r}_0|$. Consequently, the Dirichlet Green's function for the exterior of the sphere is

$$G(\mathbf{r}, \mathbf{r}_0) = -\frac{1}{4\pi|\mathbf{r} - \mathbf{r}_0|} + \frac{a/|\mathbf{r}_0|}{4\pi\,|\mathbf{r} - (a^2/|\mathbf{r}_0|^2)\mathbf{r}_0|}.$$

For a less formal treatment of the same problem see exercise 21.22. ◄

If we seek solutions to Poisson's equation in the *interior* of a sphere then the above analysis still holds, but $\mathbf{r}$ and $\mathbf{r}_0$ are now inside the sphere and the image $\mathbf{r}_1$ lies outside it.

For two-dimensional Dirichlet problems outside the circle $|\mathbf{r}| = a$, we are led by arguments similar to those employed previously to use the same image point as in the three-dimensional case, namely

$$\mathbf{r}_1 = \frac{a^2}{|\mathbf{r}_0|^2}\mathbf{r}_0. \tag{21.100}$$

As illustrated below, however, it is usually necessary to take the image strength as -1 in two-dimensional problems.

> ►*Solve Laplace's equation in the two-dimensional region $|\mathbf{r}| \leq a$, subject to the boundary condition $u = f(\phi)$ on $|\mathbf{r}| = a$.*

In this case we wish to find the Dirichlet Green's function in the interior of a disc of radius a, so the image charge must lie outside the disc. Taking the strength of the image to be -1, we have

$$G(\mathbf{r}, \mathbf{r}_0) = \frac{1}{2\pi}\ln|\mathbf{r} - \mathbf{r}_0| - \frac{1}{2\pi}\ln|\mathbf{r} - \mathbf{r}_1| + c,$$

where $\mathbf{r}_1 = (a^2/|\mathbf{r}_0|^2)\mathbf{r}_0$ lies outside the disc, and c is a constant that includes the strength of the image charge and does not necessarily equal zero.

Since we require $G(\mathbf{r}, \mathbf{r}_0) = 0$ when $|\mathbf{r}| = a$, the value of the constant c is determined, and the Dirichlet Green's function for this problem is given by

$$G(\mathbf{r}, \mathbf{r}_0) = \frac{1}{2\pi}\left(\ln|\mathbf{r} - \mathbf{r}_0| - \ln\left|\mathbf{r} - \frac{a^2}{|\mathbf{r}_0|^2}\mathbf{r}_0\right| - \ln\frac{|\mathbf{r}_0|}{a}\right). \tag{21.101}$$

Using plane polar coordinates, the solution to the boundary-value problem can be written as a line integral around the circle $\rho = a$:

$$u(\mathbf{r}_0) = \int_C f(\mathbf{r})\frac{\partial G(\mathbf{r}, \mathbf{r}_0)}{\partial n}\,dl$$

$$= \int_0^{2\pi} f(\mathbf{r})\left.\frac{\partial G(\mathbf{r}, \mathbf{r}_0)}{\partial \rho}\right|_{\rho=a} a\,d\phi. \tag{21.102}$$

The normal derivative of the Green's function (21.101) is given by

$$\frac{\partial G(\mathbf{r}, \mathbf{r}_0)}{\partial \rho} = \frac{\mathbf{r}}{|\mathbf{r}|}\cdot\nabla G(\mathbf{r}, \mathbf{r}_0)$$

$$= \frac{\mathbf{r}}{2\pi|\mathbf{r}|}\cdot\left(\frac{\mathbf{r} - \mathbf{r}_0}{|\mathbf{r} - \mathbf{r}_0|^2} - \frac{\mathbf{r} - \mathbf{r}_1}{|\mathbf{r} - \mathbf{r}_1|^2}\right). \tag{21.103}$$

Using the fact that $\mathbf{r}_1 = (a^2/|\mathbf{r}_0|^2)\mathbf{r}_0$ and the geometrical result (21.99), we find that

$$\left.\frac{\partial G(\mathbf{r}, \mathbf{r}_0)}{\partial \rho}\right|_{\rho=a} = \frac{a^2 - |\mathbf{r}_0|^2}{2\pi a|\mathbf{r} - \mathbf{r}_0|^2}.$$

In plane polar coordinates, $\mathbf{r} = \rho\cos\phi\,\mathbf{i} + \rho\sin\phi\,\mathbf{j}$ and $\mathbf{r}_0 = \rho_0\cos\phi_0\,\mathbf{i} + \rho_0\sin\phi_0\,\mathbf{j}$, and so

$$\left.\frac{\partial G(\mathbf{r}, \mathbf{r}_0)}{\partial \rho}\right|_{\rho=a} = \left(\frac{1}{2\pi a}\right)\frac{a^2 - \rho_0^2}{a^2 + \rho_0^2 - 2a\rho_0\cos(\phi - \phi_0)}.$$

On substituting into (21.102), we obtain

$$u(\rho_0, \phi_0) = \frac{1}{2\pi}\int_0^{2\pi} \frac{(a^2 - \rho_0^2)f(\phi)\,d\phi}{a^2 + \rho_0^2 - 2a\rho_0\cos(\phi - \phi_0)}, \tag{21.104}$$

which is the solution to the problem. ◀

21.5.4 Neumann problems

In a Neumann problem we require the normal derivative of the solution of Poisson's equation to take on specific values on some surface S that bounds V, i.e. we require $\partial u(\mathbf{r})/\partial n = f(\mathbf{r})$ on S, where f is a given function. As we shall see, much of our discussion of Dirichlet problems can be immediately taken over into the solution of Neumann problems.

As we proved in section 20.7 of the previous chapter, specifying Neumann boundary conditions determines the relevant solution of Poisson's equation to within an (unimportant) additive constant. Unlike Dirichlet conditions, Neumann conditions impose a self-consistency requirement. In order for a solution u to exist, it is necessary that the following consistency condition holds:

$$\int_S f\,dS = \int_S \nabla u \cdot \hat{\mathbf{n}}\,dS = \int_V \nabla^2 u\,dV = \int_V \rho\,dV, \tag{21.105}$$

where we have used the divergence theorem to convert the surface integral into a volume integral. As a physical example, the integral of the normal component of an electric field over a surface bounding a given volume cannot be chosen arbitrarily when the charge inside the volume has already been specified (Gauss's theorem).

Let us again consider (21.84), which is central to our discussion of Green's functions in inhomogeneous problems. It reads

$$u(\mathbf{r}_0) = \int_V G(\mathbf{r}, \mathbf{r}_0)\rho(\mathbf{r})\,dV(\mathbf{r}) + \int_S \left[u(\mathbf{r})\frac{\partial G(\mathbf{r}, \mathbf{r}_0)}{\partial n} - G(\mathbf{r}, \mathbf{r}_0)\frac{\partial u(\mathbf{r})}{\partial n}\right]dS(\mathbf{r}).$$

As always, the Green's function must obey

$$\nabla^2 G(\mathbf{r}, \mathbf{r}_0) = \delta(\mathbf{r} - \mathbf{r}_0),$$

where $\mathbf{r}_0$ lies in V. In the solution of Dirichlet problems in the previous subsection, we chose the Green's function to obey the boundary condition $G(\mathbf{r}, \mathbf{r}_0) = 0$ on S

and, in a similar way, we might wish to choose $\partial G(\mathbf{r}, \mathbf{r}_0)/\partial n = 0$ in the solution of Neumann problems. However, in general this is *not* permitted since the Green's function must obey the consistency condition

$$\int_S \frac{\partial G(\mathbf{r}, \mathbf{r}_0)}{\partial n}\, dS = \int_S \nabla G(\mathbf{r}, \mathbf{r}_0) \cdot \hat{\mathbf{n}}\, dS = \int_V \nabla^2 G(\mathbf{r}, \mathbf{r}_0)\, dV = 1.$$

The simplest permitted boundary condition is therefore

$$\frac{\partial G(\mathbf{r}, \mathbf{r}_0)}{\partial n} = \frac{1}{A} \qquad \text{for } \mathbf{r} \text{ on } S,$$

where A is the area of the surface S; this defines a *Neumann Green's function*.

If we require $\partial u(\mathbf{r})/\partial n = f(\mathbf{r})$ on S, the solution to Poisson's equation is given by

$$u(\mathbf{r}_0) = \int_V G(\mathbf{r}, \mathbf{r}_0)\rho(\mathbf{r})\, dV(\mathbf{r}) + \frac{1}{A} \int_S u(\mathbf{r})\, dS(\mathbf{r}) - \int_S G(\mathbf{r}, \mathbf{r}_0)f(\mathbf{r})\, dS(\mathbf{r})$$

$$= \int_V G(\mathbf{r}, \mathbf{r}_0)\rho(\mathbf{r})\, dV(\mathbf{r}) + \langle u(\mathbf{r})\rangle_S - \int_S G(\mathbf{r}, \mathbf{r}_0)f(\mathbf{r})\, dS(\mathbf{r}), \qquad (21.106)$$

where $\langle u(\mathbf{r})\rangle_S$ is the average of u over the surface S and is a freely specifiable constant. For Neumann problems in which the volume V is bounded by a surface S at infinity, we do not need the $\langle u(\mathbf{r})\rangle_S$ term. For example, if we wish to solve a Neumann problem outside the unit sphere centred at the origin then $r > a$ is the region V throughout which we require the solution; this region may be considered as being bounded by two disconnected surfaces, the surface of the sphere and a surface at infinity. By requiring that $u(\mathbf{r}) \to 0$ as $|\mathbf{r}| \to \infty$, the term $\langle u(\mathbf{r})\rangle_S$ becomes zero.

As mentioned above, much of our discussion of Dirichlet problems can be taken over into the solution of Neumann problems. In particular, we may use the method of images to find the appropriate Neumann Green's function.

> ▶*Solve Laplace's equation in the two-dimensional region $|\mathbf{r}| \le a$ subject to the boundary condition $\partial u/\partial n = f(\phi)$ on $|\mathbf{r}| = a$, with $\int_0^{2\pi} f(\phi)\, d\phi = 0$ as required by the consistency condition (21.105).*

Let us assume, as in Dirichlet problems with this geometry, that a single image charge is placed outside the circle at

$$\mathbf{r}_1 = \frac{a^2}{|\mathbf{r}_0|^2}\mathbf{r}_0,$$

where $\mathbf{r}_0$ is the position of the source inside the circle (see equation (21.100)). Then, from (21.99), we have the useful geometrical result

$$|\mathbf{r} - \mathbf{r}_1| = \frac{a}{|\mathbf{r}_0|}|\mathbf{r} - \mathbf{r}_0| \qquad \text{for } |\mathbf{r}| = a. \qquad (21.107)$$

Leaving the strength q of the image as a parameter, the Green's function has the form

$$G(\mathbf{r}, \mathbf{r}_0) = \frac{1}{2\pi} \left(\ln|\mathbf{r} - \mathbf{r}_0| + q \ln|\mathbf{r} - \mathbf{r}_1| + c \right). \qquad (21.108)$$

Using plane polar coordinates, the radial (i.e. normal) derivative of this function is given by

$$\frac{\partial G(\mathbf{r}, \mathbf{r}_0)}{\partial \rho} = \frac{\mathbf{r}}{|\mathbf{r}|} \cdot \nabla G(\mathbf{r}, \mathbf{r}_0)$$

$$= \frac{\mathbf{r}}{2\pi |\mathbf{r}|} \cdot \left[\frac{\mathbf{r} - \mathbf{r}_0}{|\mathbf{r} - \mathbf{r}_0|^2} + \frac{q(\mathbf{r} - \mathbf{r}_1)}{|\mathbf{r} - \mathbf{r}_1|^2} \right].$$

Using (21.107), on the perimeter of the circle $\rho = a$ the radial derivative takes the form

$$\frac{\partial G(\mathbf{r}, \mathbf{r}_0)}{\partial \rho}\bigg|_{\rho=a} = \frac{1}{2\pi |\mathbf{r}|} \left[\frac{|\mathbf{r}|^2 - \mathbf{r} \cdot \mathbf{r}_0}{|\mathbf{r} - \mathbf{r}_0|^2} + \frac{q|\mathbf{r}|^2 - q(a^2/|\mathbf{r}_0|^2)\mathbf{r} \cdot \mathbf{r}_0}{(a^2/|\mathbf{r}_0|^2)|\mathbf{r} - \mathbf{r}_0|^2} \right]$$

$$= \frac{1}{2\pi a} \frac{1}{|\mathbf{r} - \mathbf{r}_0|^2} \left[|\mathbf{r}|^2 + q|\mathbf{r}_0|^2 - (1 + q)\mathbf{r} \cdot \mathbf{r}_0 \right],$$

where we have set $|\mathbf{r}|^2 = a^2$ in the second term on the RHS, but not in the first. If we take $q = 1$, the radial derivative simplifies to

$$\frac{\partial G(\mathbf{r}, \mathbf{r}_0)}{\partial \rho}\bigg|_{\rho=a} = \frac{1}{2\pi a},$$

or $1/L$, where L is the circumference, and so (21.108) with $q = 1$ is the required Neumann Green's function.

Since $\rho(\mathbf{r}) = 0$, the solution to our boundary-value problem is now given by (21.106) as

$$u(\mathbf{r}_0) = \langle u(\mathbf{r}) \rangle_C - \int_C G(\mathbf{r}, \mathbf{r}_0) f(\mathbf{r}) \, dl(\mathbf{r}),$$

where the integral is around the circumference of the circle C. In plane polar coordinates $\mathbf{r} = \rho \cos \phi \, \mathbf{i} + \rho \sin \phi \, \mathbf{j}$ and $\mathbf{r}_0 = \rho_0 \cos \phi_0 \, \mathbf{i} + \rho_0 \sin \phi_0 \, \mathbf{j}$, and again using (21.107) we find that on C the Green's function is given by

$$G(\mathbf{r}, \mathbf{r}_0)|_{\rho=a} = \frac{1}{2\pi} \left[\ln |\mathbf{r} - \mathbf{r}_0| + \ln \left(\frac{a}{|\mathbf{r}_0|} |\mathbf{r} - \mathbf{r}_0| \right) + c \right]$$

$$= \frac{1}{2\pi} \left(\ln |\mathbf{r} - \mathbf{r}_0|^2 + \ln \frac{a}{|\mathbf{r}_0|} + c \right)$$

$$= \frac{1}{2\pi} \left\{ \ln \left[a^2 + \rho_0^2 - 2a\rho_0 \cos(\phi - \phi_0) \right] + \ln \frac{a}{\rho_0} + c \right\}. \quad (21.109)$$

Since $dl = a \, d\phi$ on C, the solution to the problem is given by

$$u(\rho_0, \phi_0) = \langle u \rangle_C - \frac{a}{2\pi} \int_0^{2\pi} f(\phi) \ln[a^2 + \rho_0^2 - 2a\rho_0 \cos(\phi - \phi_0)] \, d\phi.$$

The contributions of the final two terms terms in the Green's function (21.109) vanish because $\int_0^{2\pi} f(\phi) \, d\phi = 0$. The average value of u around the circumference, $\langle u \rangle_C$, is a freely specifiable constant as we would expect for a Neumann problem. This result should be compared with the result (21.104) for the corresponding Dirichlet problem, but it should be remembered that in the one case $f(\phi)$ is a potential, and in the other the gradient of a potential. ◄

21.6 Exercises

21.1   Solve the following first-order partial differential equations by separating the variables:

$$\text{(a)} \quad \frac{\partial u}{\partial x} - x \frac{\partial u}{\partial y} = 0; \qquad \text{(b)} \quad x \frac{\partial u}{\partial x} - 2y \frac{\partial u}{\partial y} = 0.$$

21.2 A cube, made of material whose conductivity is k, has as its six faces the planes $x = \pm a$, $y = \pm a$ and $z = \pm a$, and contains no internal heat sources. Verify that the temperature distribution

$$u(x, y, z, t) = A \cos \frac{\pi x}{a} \sin \frac{\pi z}{a} \exp \left(-\frac{2\kappa \pi^2 t}{a^2} \right)$$

obeys the appropriate diffusion equation. Across which faces is there heat flow? What is the direction and rate of heat flow at the point $(3a/4, a/4, a)$ at time $t = a^2/(\kappa \pi^2)$?

21.3 The wave equation describing the transverse vibrations of a stretched membrane under tension T and having a uniform surface density ρ is

$$T \left(\frac{\partial^2 u}{\partial x^2} + \frac{\partial^2 u}{\partial y^2} \right) = \rho \frac{\partial^2 u}{\partial t^2}.$$

Find a separable solution appropriate to a membrane stretched on a frame of length a and width b, showing that the natural angular frequencies of such a membrane are given by

$$\omega^2 = \frac{\pi^2 T}{\rho} \left(\frac{n^2}{a^2} + \frac{m^2}{b^2} \right),$$

where n and m are any positive integers.

21.4 Schrödinger's equation for a non-relativistic particle in a constant potential region can be taken as

$$-\frac{\hbar^2}{2m} \left(\frac{\partial^2 u}{\partial x^2} + \frac{\partial^2 u}{\partial y^2} + \frac{\partial^2 u}{\partial z^2} \right) = i\hbar \frac{\partial u}{\partial t}.$$

(a) Find a solution, separable in the four independent variables, that can be written in the form of a plane wave,

$$\psi(x, y, z, t) = A \exp[i(\mathbf{k} \cdot \mathbf{r} - \omega t)].$$

Using the relationships associated with de Broglie ($\mathbf{p} = \hbar \mathbf{k}$) and Einstein ($E = \hbar \omega$), show that the separation constants must be such that

$$p_x^2 + p_y^2 + p_z^2 = 2mE.$$

(b) Obtain a different separable solution describing a particle confined to a box of side a (ψ must vanish at the walls of the box). Show that the energy of the particle can only take the quantised values

$$E = \frac{\hbar^2 \pi^2}{2ma^2} (n_x^2 + n_y^2 + n_z^2),$$

where n_x, n_y and n_z are integers.

21.5 Denoting the three terms of ∇^2 in spherical polars by ∇_r^2, ∇_θ^2, ∇_ϕ^2 in an obvious way, evaluate $\nabla_r^2 u$, etc. for the two functions given below and verify that, in each case, although the individual terms are not necessarily zero their sum $\nabla^2 u$ is zero. Identify the corresponding values of ℓ and m.

(a) $u(r, \theta, \phi) = \left(Ar^2 + \dfrac{B}{r^3} \right) \dfrac{3 \cos^2 \theta - 1}{2}$.

(b) $u(r, \theta, \phi) = \left(Ar + \dfrac{B}{r^2} \right) \sin \theta \exp i\phi$.

21.6 Prove that the expression given in equation (21.47) for the associated Legendre function $P_\ell^m(\mu)$ satisfies the appropriate equation, (21.45), as follows.

(a) Evaluate $dP_\ell^m(\mu)/d\mu$ and $d^2P_\ell^m(\mu)/d\mu^2$, using the forms given in (21.47), and substitute them into (21.45).
(b) Differentiate Legendre's equation m times using Leibnitz' theorem.
(c) Show that the equations obtained in (a) and (b) are multiples of each other, and hence that the validity of (b) implies that of (a).

21.7 Continue the analysis of exercise 10.20, concerned with the flow of a very viscous fluid past a sphere, to find the full expression for the stream function $\psi(r, \theta)$. At the surface of the sphere $r = a$, the velocity field $\mathbf{u} = \mathbf{0}$, whilst far from the sphere $\psi \simeq (Ur^2 \sin^2 \theta)/2$.

Show that $f(r)$ can be expressed as a superposition of powers of r, and determine which powers give acceptable solutions. Hence show that

$$\psi(r, \theta) = \frac{U}{4}\left(2r^2 - 3ar + \frac{a^3}{r}\right)\sin^2 \theta.$$

21.8 The motion of a very viscous fluid in the two-dimensional (wedge) region $-\alpha < \phi < \alpha$ can be described, in (ρ, ϕ) coordinates, by the (biharmonic) equation

$$\nabla^2\nabla^2\psi \equiv \nabla^4\psi = 0,$$

together with the boundary conditions $\partial\psi/\partial\phi = 0$ at $\phi = \pm\alpha$, which represent the fact that there is no radial fluid velocity close to either of the bounding walls because of the viscosity, and $\partial\psi/\partial\rho = \pm\rho$ at $\phi = \pm\alpha$, which impose the condition that azimuthal flow increases linearly with r along any radial line. Assuming a solution in separated-variable form, show that the full expression for ψ is

$$\psi(\rho, \phi) = \frac{\rho^2}{2}\frac{\sin 2\phi - 2\phi \cos 2\alpha}{\sin 2\alpha - 2\alpha \cos 2\alpha}.$$

21.9 A circular disc of radius a is heated in such a way that its perimeter $\rho = a$ has a steady temperature distribution $A + B\cos^2 \phi$, where ρ and ϕ are plane polar coordinates and A and B are constants. Find the temperature $T(\rho, \phi)$ everywhere in the region $\rho < a$.

21.10 Consider possible solutions of Laplace's equation inside a circular domain as follows.

(a) Find the solution in plane polar coordinates ρ, ϕ, that takes the value $+1$ for $0 < \phi < \pi$ and the value -1 for $-\pi < \phi < 0$, when $\rho = a$.
(b) For a point (x, y) on or inside the circle $x^2 + y^2 = a^2$, identify the angles α and β defined by

$$\alpha = \tan^{-1}\frac{y}{a+x} \quad \text{and} \quad \beta = \tan^{-1}\frac{y}{a-x}.$$

Show that $u(x, y) = (2/\pi)(\alpha + \beta)$ is a solution of Laplace's equation that satisfies the boundary conditions given in (a).
(c) Deduce a Fourier series expansion for the function

$$\tan^{-1}\frac{\sin\phi}{1+\cos\phi} + \tan^{-1}\frac{\sin\phi}{1-\cos\phi}.$$

21.11 The free transverse vibrations of a thick rod satisfy the equation

$$a^4\frac{\partial^4 u}{\partial x^4} + \frac{\partial^2 u}{\partial t^2} = 0.$$

Obtain a solution in separated-variable form and, for a rod clamped at one end,

$x = 0$, and free at the other, $x = L$, show that the angular frequency of vibration ω satisfies

$$\cosh\left(\frac{\omega^{1/2}L}{a}\right) = -\sec\left(\frac{\omega^{1/2}L}{a}\right).$$

[At a clamped end both u and $\partial u/\partial x$ vanish, whilst at a free end, where there is no bending moment, $\partial^2 u/\partial x^2$ and $\partial^3 u/\partial x^3$ are both zero.]

21.12 A membrane is stretched between two concentric rings of radii a and b ($b > a$). If the smaller ring is transversely distorted from the planar configuration by an amount $c|\phi|$, $-\pi \le \phi \le \pi$, show that the membrane then has a shape given by

$$u(\rho, \phi) = \frac{c\pi}{2}\frac{\ln(b/\rho)}{\ln(b/a)} - \frac{4c}{\pi}\sum_{m \text{ odd}}\frac{a^m}{m^2(b^{2m} - a^{2m})}\left(\frac{b^{2m}}{\rho^m} - \rho^m\right)\cos m\phi.$$

21.13 A string of length L, fixed at its two ends, is plucked at its mid-point by an amount A and then released. Prove that the subsequent displacement is given by

$$u(x, t) = \sum_{n=0}^{\infty}\frac{8A(-1)^n}{\pi^2(2n+1)^2}\sin\left[\frac{(2n+1)\pi x}{L}\right]\cos\left[\frac{(2n+1)\pi ct}{L}\right],$$

where, in the usual notation, $c^2 = T/\rho$.

Find the total kinetic energy of the string when it passes through its unplucked position, by calculating it in each mode (each n) and summing, using the result

$$\sum_0^{\infty}\frac{1}{(2n+1)^2} = \frac{\pi^2}{8}.$$

Confirm that the total energy is equal to the work done in plucking the string initially.

21.14 Prove that the potential for $\rho < a$ associated with a vertical split cylinder of radius a, the two halves of which ($\cos\phi > 0$ and $\cos\phi < 0$) are maintained at equal and opposite potentials $\pm V$, is given by

$$u(\rho, \phi) = \frac{4V}{\pi}\sum_{n=0}^{\infty}\frac{(-1)^n}{2n+1}\left(\frac{\rho}{a}\right)^{2n+1}\cos(2n+1)\phi.$$

21.15 A conducting spherical shell of radius a is cut round its equator and the two halves connected to voltages of $+V$ and $-V$. Show that an expression for the potential at the point (r, θ, ϕ) anywhere inside the two hemispheres is

$$u(r, \theta, \phi) = V\sum_{n=0}^{\infty}\frac{(-1)^n(2n)!(4n+3)}{2^{2n+1}n!(n+1)!}\left(\frac{r}{a}\right)^{2n+1}P_{2n+1}(\cos\theta).$$

[This is the spherical polar analogue of the previous question.]

21.16 A slice of biological material of thickness L is placed into a solution of a radioactive isotope of constant concentration C_0 at time $t = 0$. For a later time t find the concentration of radioactive ions at a depth x inside one of its surfaces if the diffusion constant is κ.

21.17 Two identical copper bars are each of length a. Initially, one is at $0\,°C$ and the other is at $100\,°C$; they are then joined together end to end and thermally isolated. Obtain in the form of a Fourier series an expression $u(x, t)$ for the temperature at any point a distance x from the join at a later time t. Bear in mind the heat flow conditions at the free ends of the bars.

Taking $a = 0.5\,\text{m}$ estimate the time it takes for one of the free ends to attain a temperature of $55\,°C$. The thermal conductivity of copper is $3.8 \times 10^2\,\text{J}\,\text{m}^{-1}\,\text{K}^{-1}\,\text{s}^{-1}$, and its specific heat capacity is $3.4 \times 10^6\,\text{J}\,\text{m}^{-3}\,\text{K}^{-1}$.

21.18   A sphere of radius a and thermal conductivity k_1 is surrounded by an infinite medium of conductivity k_2 in which far away the temperature tends to T_∞. A distribution of heat sources $q(\theta)$ embedded in the sphere's surface establish steady temperature fields $T_1(r,\theta)$ inside the sphere and $T_2(r,\theta)$ outside it. It can be shown, by considering the heat flow through a small volume that includes part of the sphere's surface, that

$$k_1 \frac{\partial T_1}{\partial r} - k_2 \frac{\partial T_2}{\partial r} = q(\theta) \quad \text{on} \quad r = a.$$

Given that

$$q(\theta) = \frac{1}{a} \sum_{n=0}^{\infty} q_n P_n(\cos \theta),$$

find complete expressions for $T_1(r,\theta)$ and $T_2(r,\theta)$. What is the temperature at the centre of the sphere?

21.19   Using result (21.74) from the worked example in the text, find the general expression for the temperature $u(x,t)$ in the bar, given that the temperature distribution at time $t = 0$ is $u(x,0) = \exp(-x^2/a^2)$.

21.20   Working in *spherical* polar coordinates $\mathbf{r} = (r,\theta,\phi)$, but for a system that has azimuthal symmetry around the polar axis, consider the following gravitational problem.

(a)   Show that the gravitational potential due to a uniform disc of radius a and mass M, centred at the origin, is given for $r < a$ by

$$-\frac{2GM}{a} \left[1 - \frac{r}{a} P_1(\cos \theta) + \frac{1}{2} \left(\frac{r}{a}\right)^2 P_2(\cos \theta) - \frac{1}{8} \left(\frac{r}{a}\right)^4 P_4(\cos \theta) + \cdots \right],$$

and for $r > a$ by

$$-\frac{GM}{r} \left[1 - \frac{1}{4} \left(\frac{a}{r}\right)^2 P_2(\cos \theta) + \frac{1}{8} \left(\frac{a}{r}\right)^4 P_4(\cos \theta) - \cdots \right],$$

where the polar axis is normal to the plane of the disc.

(b)   Reconcile the presence of a term $P_1(\cos \theta)$, which is odd under $\theta \to \pi - \theta$, with the symmetry with respect to the plane of the disc of the physical system.

(c)   Deduce that the gravitational field near an infinite sheet of matter of constant density ρ per unit area is $2\pi G\rho$.

21.21   In the region $-\infty < x,\, y < \infty$ and $-t \leq z \leq t$, a charge-density wave $\rho(\mathbf{r}) = A \cos qx$, in the x-direction, is represented by

$$\rho(\mathbf{r}) = \frac{e^{iqx}}{\sqrt{2\pi}} \int_{-\infty}^{\infty} \tilde{\rho}(\alpha) e^{i\alpha z} \, d\alpha.$$

The resulting potential is represented by

$$V(\mathbf{r}) = \frac{e^{iqx}}{\sqrt{2\pi}} \int_{-\infty}^{\infty} \tilde{V}(\alpha) e^{i\alpha z} \, d\alpha.$$

Determine the relationship between $\tilde{V}(\alpha)$ and $\tilde{\rho}(\alpha)$, and hence show that the potential at the point $(0,0,0)$ is

$$\frac{A}{\pi\epsilon_0} \int_{-\infty}^{\infty} \frac{\sin kt}{k(k^2 + q^2)} \, dk.$$

21.22   Point charges q and $-qa/b$ (with $a < b$) are placed, respectively, at a point P, a distance b from the origin O, and a point Q between O and P, a distance a^2/b from O. Show, by considering similar triangles QOS and SOP, where S is any point on the surface of the sphere centred at O and of radius a, that the net potential anywhere on the sphere due to the two charges is zero.

Use this result (backed up by the uniqueness theorem) to find the force with which a point charge q placed a distance b from the centre of a spherical conductor of radius a ($< b$) is attracted to the sphere (i) if the sphere is earthed, and (ii) if the sphere is uncharged and insulated.

21.23   Find the Green's function $G(\mathbf{r}, \mathbf{r}_0)$ in the half-space $z > 0$ for the solution of $\nabla^2 \Phi = 0$ with Φ specified in cylindrical polar coordinates (ρ, ϕ, z) on the plane $z = 0$ by

$$\Phi(\rho, \phi, z) = \begin{cases} 1 & \text{for } \rho \leq 1, \\ 1/\rho & \text{for } \rho > 1. \end{cases}$$

Determine the variation of $\Phi(0, 0, z)$ along the z-axis.

21.24   Electrostatic charge is distributed in a sphere of radius R centred on the origin. Determine the form of the resultant potential $\phi(\mathbf{r})$ at distances much greater than R, as follows.

(a)   Express in the form of an integral over all space the solution of

$$\nabla^2 \phi = -\frac{\rho(\mathbf{r})}{\epsilon_0}.$$

(b)   Show that, for $r \gg r'$,

$$|\mathbf{r} - \mathbf{r}'| = r - \frac{\mathbf{r} \cdot \mathbf{r}'}{r} + O\left(\frac{1}{r}\right).$$

(c)   Use results (a) and (b) to show that $\phi(\mathbf{r})$ has the form

$$\phi(\mathbf{r}) = \frac{M}{r} + \frac{\mathbf{d} \cdot \mathbf{r}}{r^3} + O\left(\frac{1}{r^3}\right).$$

Find expressions for M and $\mathbf{d}$, and identify them physically.

21.25   Find, in the form of an infinite series, the Green's function of the ∇^2 operator for the Dirichlet problem in the region $-\infty < x < \infty$, $-\infty < y < \infty$, $-c \leq z \leq c$.

21.26   Find the Green's function for the three-dimensional Neumann problem

$$\nabla^2 \phi = 0 \quad \text{for } z > 0 \qquad \text{and} \qquad \frac{\partial \phi}{\partial z} = f(x, y) \quad \text{on } z = 0.$$

Determine $\phi(x, y, z)$ if

$$f(x, y) = \begin{cases} \delta(y) & \text{for } |x| < a, \\ 0 & \text{for } |x| \geq a. \end{cases}$$

21.27   Determine the Green's function for the Klein–Gordon equation in a half-space as follows.

(a)   By applying the divergence theorem to the volume integral

$$\int_V \left[\phi(\nabla^2 - m^2)\psi - \psi(\nabla^2 - m^2)\phi\right] dV,$$

obtain a Green's function expression, as the sum of a volume integral and a surface integral, for the function $\phi(\mathbf{r}')$ that satisfies

$$\nabla^2 \phi - m^2 \phi = \rho$$

in V and takes the specified form $\phi = f$ on S, the boundary of V. The Green's function, $G(\mathbf{r}, \mathbf{r}')$, to be used satisfies

$$\nabla^2 G - m^2 G = \delta(\mathbf{r} - \mathbf{r}')$$

and vanishes when $\mathbf{r}$ is on S.

(b) When V is all space, $G(\mathbf{r}, \mathbf{r}')$ can be written as $G(t) = g(t)/t$, where $t = |\mathbf{r} - \mathbf{r}'|$ and $g(t)$ is bounded as $t \to \infty$. Find the form of $G(t)$.

(c) Find $\phi(\mathbf{r})$ in the half-space $x > 0$ if $\rho(\mathbf{r}) = \delta(\mathbf{r} - \mathbf{r}_1)$ and $\phi = 0$ both on $x = 0$ and as $r \to \infty$.

21.28 Consider the PDE $\mathcal{L}u(\mathbf{r}) = \rho(\mathbf{r})$, for which the differential operator $\mathcal{L}$ is given by

$$\mathcal{L} = \nabla \cdot [p(\mathbf{r})\nabla] + q(\mathbf{r}),$$

where $p(\mathbf{r})$ and $q(\mathbf{r})$ are functions of position. By proving the generalised form of Green's theorem,

$$\int_V (\phi \mathcal{L}\psi - \psi \mathcal{L}\phi)\, dV = \oint_S p(\phi \nabla \psi - \psi \nabla \phi) \cdot \hat{\mathbf{n}}\, dS,$$

show that the solution of the PDE is given by

$$u(\mathbf{r}_0) = \int_V G(\mathbf{r}, \mathbf{r}_0)\rho(\mathbf{r})\, dV(\mathbf{r}) + \oint_S p(\mathbf{r}) \left[u(\mathbf{r}) \frac{\partial G(\mathbf{r}, \mathbf{r}_0)}{\partial n} - G(\mathbf{r}, \mathbf{r}_0) \frac{\partial u(\mathbf{r})}{\partial n} \right] dS(\mathbf{r}),$$

where $G(\mathbf{r}, \mathbf{r}_0)$ is the Green's function satisfying $\mathcal{L}G(\mathbf{r}, \mathbf{r}_0) = \delta(\mathbf{r} - \mathbf{r}_0)$.

21.7 Hints and answers

21.1 (a) $C \exp[\lambda(x^2 + 2y)]$; (b) $C(x^2 y)^\lambda$.

21.3 $u(x, y, t) = \sin(n\pi x/a)\sin(m\pi y/b)(A \sin \omega t + B \cos \omega t)$.

21.5 (a) $6u/r^2$, $-6u/r^2$, 0, $\ell = 2$ (or -3), $m = 0$;
(b) $2u/r^2$, $(\cot^2 \theta - 1)u/r^2$; $-u/(r^2 \sin^2 \theta)$, $\ell = 1$ (or -2), $m = \pm 1$.

21.7 Solutions of the form r^ℓ give ℓ as $-1, 1, 2, 4$. Because of the asymptotic form of ψ, an r^4 term cannot be present. The coefficients of the three remaining terms are determined by the two boundary conditions $\mathbf{u} = \mathbf{0}$ on the sphere and the form of ψ for large r.

21.9 Express $\cos^2 \phi$ in terms of $\cos 2\phi$; $T(\rho, \phi) = A + B/2 + (B\rho^2/2a^2)\cos 2\phi$.

21.11 $(A \cos mx + B \sin mx + C \cosh mx + D \sinh mx)\cos(\omega t + \epsilon)$, with $m^4 a^4 = \omega^2$.

21.13 $E_n = 16\rho A^2 c^2 / [(2n + 1)^2 \pi^2 L]$; $E = 2\rho c^2 A^2 / L = \int_0^A [2Tv/(\frac{1}{2}L)]\, dv$.

21.15 Note that the boundary value function is a square wave that is *symmetric* in ϕ.

21.17 Since there is no heat flow at $x = \pm a$, use a series of period $4a$, $u(x, 0) = 100$ for $0 < x \leq 2a$, $u(x, 0) = 0$ for $-2a \leq x < 0$.

$$u(x, t) = 50 + \frac{200}{\pi} \sum_{n=0}^{\infty} \frac{1}{2n+1} \sin\left[\frac{(2n+1)\pi x}{2a}\right] \exp\left[-\frac{k(2n+1)^2 \pi^2 t}{4a^2 s}\right].$$

Taking only the $n = 0$ term gives $t \approx 2300$ s.

21.19 $u(x, t) = [a/(a^2 + 4\kappa t)^{1/2}] \exp[-x^2/(a^2 + 4\kappa t)]$.

21.21 Fourier-transform Poisson's equation to show that $\tilde{\rho}(\alpha) = \epsilon_0(\alpha^2 + q^2)\tilde{V}(\alpha)$.

21.23 Follow the worked example that includes result (21.95). For part of the explicit integration, substitute $\rho = z \tan \alpha$.

$$\Phi(0, 0, z) = \frac{z(1 + z^2)^{1/2} - z^2 + (1 + z^2)^{1/2} - 1}{z(1 + z^2)^{1/2}}.$$

21.25   The terms in $G(\mathbf{r}, \mathbf{r}_0)$ that are additional to the fundamental solution are

$$\frac{1}{4\pi} \sum_{n=2}^{\infty} (-1)^n \left\{ \left[(x - x_0)^2 + (y - y_0)^2 + (z + (-1)^n z_0 - nc)^2 \right]^{-1/2} \right.$$

$$\left. + \left[(x - x_0)^2 + (y - y_0)^2 + (z + (-1)^n z_0 + nc)^2 \right]^{-1/2} \right\}.$$

21.27   (a) As given in equation (21.86), but with $\mathbf{r}_0$ replaced by $\mathbf{r}'$.
         (b) Move the origin to $\mathbf{r}'$ and integrate the defining Green's equation to obtain

$$4\pi t^2 \frac{dG}{dt} - m^2 \int_0^t G(t') 4\pi t'^2 \, dt' = 1,$$

leading to $G(t) = [-1/(4\pi t)]e^{-mt}$.
         (c) $\phi(\mathbf{r}) = [-1/(4\pi)](p^{-1}e^{-mp} - q^{-1}e^{-mq})$, where $p = |\mathbf{r} - \mathbf{r}_1|$ and $q = |\mathbf{r} - \mathbf{r}_2|$ with $\mathbf{r}_1 = (x_1, y_1, z_1)$ and $\mathbf{r}_2 = (-x_1, y_1, z_1)$.

22

Calculus of variations

In chapters 2 and 5 we discussed how to find stationary values of functions of a single variable $f(x)$, of several variables $f(x, y, \dots)$ and of constrained variables, where $x, y, \dots$ are subject to the n constraints $g_i(x, y, \dots) = 0$, $i = 1, 2, \dots, n$. In all these cases the forms of the functions f and g_i were known, and the problem was one of finding the appropriate values of the variables x, y etc.

We now turn to a different kind of problem in which we are interested in bringing about a particular condition for a given expression (usually maximising or minimising it) by varying the *functions* on which the expression depends. For instance, we might want to know in what shape a fixed length of rope should be arranged so as to enclose the largest possible area, or in what shape it will hang when suspended under gravity from two fixed points. In each case we are concerned with a general maximisation or minimisation criterion by which the function $y(x)$ that satisfies the given problem may be found.

The calculus of variations provides a method for finding the function $y(x)$. The problem must first be expressed in a mathematical form, and the form most commonly applicable to such problems is an *integral*. In each of the above questions, the quantity that has to be maximised or minimised by an appropriate choice of the function $y(x)$ may be expressed as an integral involving $y(x)$ and the variables describing the geometry of the situation.

In our example of the rope hanging from two fixed points, we need to find the shape function $y(x)$ that minimises the gravitational potential energy of the rope. Each elementary piece of the rope has a gravitational potential energy proportional both to its vertical height above an arbitrary zero level and to the length of the piece. Therefore the total potential energy is given by an integral for the whole rope of such elementary contributions. The particular function $y(x)$ for which the value of this integral is a minimum will give the shape assumed by the hanging rope.

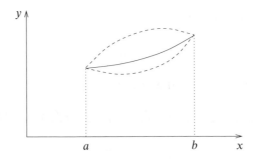

Figure 22.1 Possible paths for the integral (22.1). The solid line is the curve along which the integral is assumed stationary. The broken curves represent small variations from this path.

So in general we are led by this type of question to study the value of an integral whose integrand has a specified form in terms of a certain function and its derivatives, and to study how that value changes when the form of the function is varied. Specifically, we aim to find the function that makes the integral *stationary*, i.e. the function that makes the value of the integral a local maximum or minimum. Note that, unless stated otherwise, y' is used to denote dy/dx throughout this chapter. We also assume that all the functions we need to deal with are sufficiently smooth and differentiable.

22.1 The Euler–Lagrange equation

Let us consider the integral

$$I = \int_a^b F(y, y', x)\, dx, \tag{22.1}$$

where a, b and the form of the function F are fixed by given considerations, e.g. the physics of the problem, but the curve $y(x)$ is to be chosen so as to make stationary the value of I, which is clearly a function, or more accurately a *functional*, of this curve, i.e. $I = I[y(x)]$. Referring to figure 22.1, we wish to find the function $y(x)$ (given, say, by the solid line) such that first-order small changes in it (for example the two broken lines) will make only second-order changes in the value of I.

Writing this in a more mathematical form, let us suppose that $y(x)$ is the function required to make I stationary and consider making the replacement

$$y(x) \rightarrow y(x) + \alpha\eta(x), \tag{22.2}$$

where the parameter α is small and $\eta(x)$ is an arbitrary function with sufficiently amenable mathematical properties. For the value of I to be stationary with respect to these variations, we require

$$\left.\frac{dI}{d\alpha}\right|_{\alpha=0} = 0 \quad \text{for all } \eta(x). \tag{22.3}$$

Substituting (22.2) into (22.1) and expanding as a Taylor series in α we obtain

$$I(y, \alpha) = \int_a^b F(y + \alpha\eta, y' + \alpha\eta', x)\, dx$$
$$= \int_a^b F(y, y', x)\, dx + \int_a^b \left(\frac{\partial F}{\partial y}\alpha\eta + \frac{\partial F}{\partial y'}\alpha\eta'\right) dx + O(\alpha^2).$$

With this form for $I(y, \alpha)$ the condition (22.3) implies that for all $\eta(x)$ we require

$$\delta I = \int_a^b \left(\frac{\partial F}{\partial y}\eta + \frac{\partial F}{\partial y'}\eta'\right) dx = 0,$$

where δI denotes the first-order variation in the value of I due to the variation (22.2) in the function $y(x)$. Integrating the second term by parts this becomes

$$\left[\eta \frac{\partial F}{\partial y'}\right]_a^b + \int_a^b \left[\frac{\partial F}{\partial y} - \frac{d}{dx}\left(\frac{\partial F}{\partial y'}\right)\right] \eta(x)\, dx = 0. \tag{22.4}$$

In order to simplify the result we will assume, for the moment, that the end-points are fixed, i.e. not only a and b are given but also $y(a)$ and $y(b)$. This restriction means that we require $\eta(a) = \eta(b) = 0$, in which case the first term on the LHS of (22.4) equals zero at both end-points. Since (22.4) must be satisfied for arbitrary $\eta(x)$, it is easy to see that we require

$$\frac{\partial F}{\partial y} = \frac{d}{dx}\left(\frac{\partial F}{\partial y'}\right). \tag{22.5}$$

This is known as the *Euler–Lagrange* (EL) equation, and is a differential equation for $y(x)$, since the function F is known.

22.2 Special cases

In certain special cases a first integral of the EL equation can be obtained for a general form of F.

22.2.1 *F does not contain y explicitly*

In this case $\partial F/\partial y = 0$, and (22.5) can be integrated immediately giving

$$\frac{\partial F}{\partial y'} = \text{constant}. \tag{22.6}$$

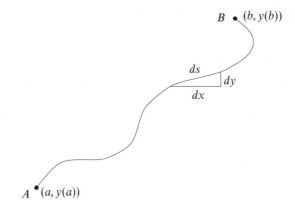

Figure 22.2   An arbitrary path between two fixed points.

▶*Show that the shortest curve joining two points is a straight line.*

Let the two points be labelled A and B and have coordinates $(a, y(a))$ and $(b, y(b))$ respectively (see figure 22.2). Whatever the shape of the curve joining A to B, the length of an element of path ds is given by

$$ds = \left[(dx)^2 + (dy)^2\right]^{1/2} = (1 + y'^2)^{1/2}dx,$$

and hence the total path length along the curve is given by

$$L = \int_a^b (1 + y'^2)^{1/2} \, dx. \tag{22.7}$$

We must now apply the results of the previous section to determine that path which makes L stationary (clearly a minimum in this case). Since the integral does not contain y (or indeed x) explicitly, we may use (22.6) to obtain

$$k = \frac{\partial F}{\partial y'} = \frac{y'}{(1 + y'^2)^{1/2}}.$$

where k is a constant. This is easily rearranged and integrated to give

$$y = \frac{k}{(1 - k^2)^{1/2}} x + c,$$

which, as expected, is the equation of a straight line in the form $y = mx + c$, with $m = k/(1 - k^2)^{1/2}$. The value of m (or k) can be found by demanding that the straight line passes through the points A and B and is given by $m = [y(b) - y(a)]/(b - a)$. Substituting the equation of the straight line into (22.7) we find that, again as expected, the total path length is given by

$$L^2 = [y(b) - y(a)]^2 + (b - a)^2. \blacktriangleleft$$

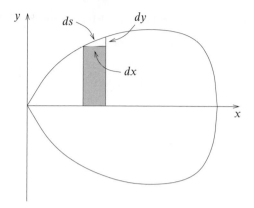

Figure 22.3   A convex closed curve that is symmetrical about the x-axis.

22.2.2 F does not contain x explicitly

In this case, multiplying the E.L equation (22.5) by y' and using

$$\frac{d}{dx}\left(y'\frac{\partial F}{\partial y'}\right) = y'\frac{d}{dx}\left(\frac{\partial F}{\partial y'}\right) + y''\frac{\partial F}{\partial y'}$$

we obtain

$$y'\frac{\partial F}{\partial y} + y''\frac{\partial F}{\partial y'} = \frac{d}{dx}\left(y'\frac{\partial F}{\partial y'}\right).$$

But since F is a function of y and y' only, and not explicitly of x, the LHS of this equation is just the total derivative of F, namely dF/dx. Hence, integrating we obtain

$$F - y'\frac{\partial F}{\partial y'} = \text{constant.} \qquad (22.8)$$

> ►*Find the closed convex curve of length l that encloses the greatest possible area.*

Without any loss of generality we can assume that the curve passes through the origin and can further suppose that it is symmetric with respect to the x-axis; this assumption is not essential. Using the distance s along the curve, measured from the origin, as the independent variable and y as the dependent one, we have the boundary conditions $y(0) = y(l/2) = 0$. The element of area shown in figure 22.3 is then given by

$$dA = y\,dx = y\left[(ds)^2 - (dy)^2\right]^{1/2},$$

and the total area by

$$A = 2\int_0^{l/2} y(1 - y'^2)^{1/2}\,ds; \qquad (22.9)$$

here y' stands for dy/ds rather than dy/dx. Since the integrand does not contain s explicitly,

779

we can use (22.8) to obtain a first integral of the EL equation for y, namely

$$y(1 - y'^2)^{1/2} + yy'^2(1 - y'^2)^{-1/2} = k,$$

where k is a constant. On rearranging this gives

$$ky' = \pm(k^2 - y^2)^{1/2},$$

which, using $y(0) = 0$, integrates to

$$y/k = \sin(s/k). \tag{22.10}$$

The other end-point, $y(l/2) = 0$, fixes the value of k as $l/(2\pi)$ to yield

$$y = \frac{l}{2\pi} \sin \frac{2\pi s}{l}.$$

From this we obtain $dy = \cos(2\pi s/l)\, ds$ and since $(ds)^2 = (dx)^2 + (dy)^2$ we find also that $dx = \pm \sin(2\pi s/l)\, ds$. This in turn can be integrated and, using $x(0) = 0$, gives x in terms of s as

$$x - \frac{l}{2\pi} = -\frac{l}{2\pi} \cos \frac{2\pi s}{l}.$$

We thus obtain the expected result that x and y lie on the circle of radius $l/(2\pi)$ given by

$$\left(x - \frac{l}{2\pi}\right)^2 + y^2 = \frac{l^2}{4\pi^2}.$$

Substituting the solution (22.10) into the expression for the total area (22.9), it is easily verified that $A = l^2/(4\pi)$. A much quicker derivation of this result is possible using plane polar coordinates. ◀

The previous two examples have been carried out in some detail, even though the answers are more easily obtained in other ways, expressly so that the method is transparent and the way in which it works can be filled in mentally at almost every step. The next example, however, does not have such an intuitively obvious solution.

▶ *Two rings, each of radius a, are placed parallel with their centres $2b$ apart and on a common normal. An open-ended axially symmetric soap film is formed between them (see figure 22.4). Find the shape assumed by the film.*

Creating the soap film requires an energy γ per unit area (numerically equal to the surface tension of the soap solution). So the stable shape of the soap film, i.e. the one that minimises the energy, will also be the one that minimises the surface area (neglecting gravitational effects).

It is obvious that any convex surface, shaped such as that shown as the broken line in figure 22.4(a), cannot be a minimum but it is not clear whether some shape intermediate between the cylinder shown by solid lines in (a), with area $4\pi ab$ (or twice this for the double surface of the film), and the form shown in (b), with area approximately $2\pi a^2$, will produce a lower total area than both of these extremes. If there is such a shape (e.g. that in figure 22.4(c)), then it will be that which is the best compromise between two requirements, the need to minimise the ring-to-ring distance measured on the film surface (a) and the need to minimise the average waist measurement of the surface (b).

We take cylindrical polar coordinates as in figure 22.4(c) and let the radius of the soap film at height z be $\rho(z)$ with $\rho(\pm b) = a$. Counting only one side of the film, the element of

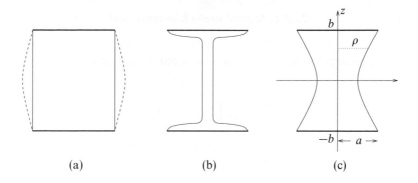

Figure 22.4   Possible soap films between two parallel circular rings.

surface area between z and $z + dz$ is

$$dS = 2\pi\rho \left[(dz)^2 + (d\rho)^2\right]^{1/2},$$

so the total surface area is given by

$$S = 2\pi \int_{-b}^{b} \rho(1 + \rho'^2)^{1/2} \, dz. \qquad (22.11)$$

Since the integrand does not contain z explicitly, we can use (22.8) to obtain an equation for ρ that minimises S, i.e.

$$\rho(1 + \rho'^2)^{1/2} - \rho\rho'^2(1 + \rho'^2)^{-1/2} = k,$$

where k is a constant. Multiplying through by $(1 + \rho'^2)^{1/2}$, rearranging to find an explicit expression for ρ' and integrating we find

$$\cosh^{-1} \frac{\rho}{k} = \frac{z}{k} + c.$$

where c is the constant of integration. Using the boundary conditions $\rho(\pm b) = a$, we require $c = 0$ and k such that $a/k = \cosh b/k$ (if b/a is too large, no such k can be found). Thus the curve that minimises the surface area is

$$\rho/k = \cosh(z/k),$$

and in profile the soap film is a catenary (see section 22.4) with the minimum distance from the axis equal to k. ◀

22.3 Some extensions

It is quite possible to relax many of the restrictions we have imposed so far. For example, we can allow end-points that are constrained to lie on given curves rather than being fixed, or we can consider problems with several dependent and/or independent variables or higher-order derivatives of the dependent variable. Each of these extensions is now discussed.

22.3.1 Several dependent variables

Here we have $F = F(y_1, y_1', y_2, y_2', \ldots, y_n, y_n', x)$ where each $y_i = y_i(x)$. The analysis in this case proceeds as before, leading to n separate but simultaneous equations for the $y_i(x)$,

$$\frac{\partial F}{\partial y_i} = \frac{d}{dx}\left(\frac{\partial F}{\partial y_i'}\right), \qquad i = 1, 2, \ldots, n. \tag{22.12}$$

22.3.2 Several independent variables

With n independent variables, we need to extremise multiple integrals of the form

$$I = \int \int \cdots \int F\left(y, \frac{\partial y}{\partial x_1}, \frac{\partial y}{\partial x_2}, \ldots, \frac{\partial y}{\partial x_n}, x_1, x_2, \ldots, x_n\right) dx_1 \, dx_2 \cdots dx_n.$$

Using the same kind of analysis as before, we find that the extremising function $y = y(x_1, x_2, \ldots, x_n)$ must satisfy

$$\frac{\partial F}{\partial y} = \sum_{i=1}^{n} \frac{\partial}{\partial x_i}\left(\frac{\partial F}{\partial y_{x_i}}\right), \tag{22.13}$$

where y_{x_i} stands for $\partial y / \partial x_i$.

22.3.3 Higher-order derivatives

If in (22.1) $F = F(y, y', y'', \ldots, y^{(n)}, x)$ then using the same method as before and performing repeated integration by parts, it can be shown that the required extremising function $y(x)$ satisfies

$$\frac{\partial F}{\partial y} - \frac{d}{dx}\left(\frac{\partial F}{\partial y'}\right) + \frac{d^2}{dx^2}\left(\frac{\partial F}{\partial y''}\right) - \cdots + (-1)^n \frac{d^n}{dx^n}\left(\frac{\partial F}{\partial y^{(n)}}\right) = 0, \tag{22.14}$$

provided that $y = y' = \cdots = y^{(n-1)} = 0$ at both end-points. If y, or any of its derivatives, is not zero at the end-points then a corresponding contribution or contributions will appear on the RHS of (22.14).

22.3.4 Variable end-points

We now discuss the very important generalisation to variable end-points. Suppose, as before, we wish to find the function $y(x)$ that extremises the integral

$$I = \int_a^b F(y, y', x) \, dx,$$

but this time we demand only that the lower end-point is fixed, while we allow $y(b)$ to be arbitrary. Repeating the analysis of section 22.1, we find from (22.4)

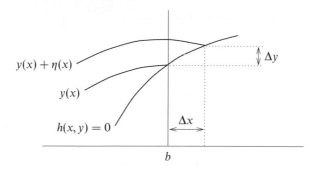

Figure 22.5 Variation of the end-point b along the curve $h(x, y) = 0$.

that we require

$$\left[\eta \frac{\partial F}{\partial y'} \right]_a^b + \int_a^b \left[\frac{\partial F}{\partial y} - \frac{d}{dx} \left(\frac{\partial F}{\partial y'} \right) \right] \eta(x)\, dx = 0. \tag{22.15}$$

Obviously the EL equation (22.5) must still hold for the second term on the LHS to vanish. Also, since the lower end-point is fixed, i.e. $\eta(a) = 0$, the first term on the LHS automatically vanishes at the lower limit. However, in order that it also vanishes at the upper limit, we require in addition that

$$\left. \frac{\partial F}{\partial y'} \right|_{x=b} = 0. \tag{22.16}$$

Clearly if both end-points may vary then $\partial F/\partial y'$ must vanish at both ends.

An interesting and more general case is where the lower end-point is again fixed at $x = a$, but the upper end-point is free to lie anywhere on the curve $h(x, y) = 0$. Now in this case, the variation in the value of I due to the arbitrary variation (22.2) is given to first order by

$$\delta I = \left[\frac{\partial F}{\partial y'} \eta \right]_a^b + \int_a^b \left(\frac{\partial F}{\partial y} - \frac{d}{dx} \frac{\partial F}{\partial y'} \right) \eta\, dx + F(b)\Delta x, \tag{22.17}$$

where Δx is the displacement in the x-direction of the upper end-point, as indicated in figure 22.5, and $F(b)$ is the value of F at $x = b$. In order for (22.17) to be valid, we of course require the displacement Δx to be small.

From the figure we see that $\Delta y = \eta(b) + y'(b)\Delta x$. Since the upper end-point must lie on $h(x, y) = 0$ we also require that, at $x = b$,

$$\frac{\partial h}{\partial x}\Delta x + \frac{\partial h}{\partial y}\Delta y = 0,$$

which on substituting our expression for Δy and rearranging becomes

$$\left(\frac{\partial h}{\partial x} + y' \frac{\partial h}{\partial y} \right) \Delta x + \frac{\partial h}{\partial y} \eta = 0. \tag{22.18}$$

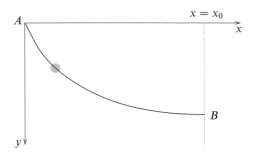

Figure 22.6   A frictionless wire along which a small bead slides. We seek the shape of the wire that allows the bead to travel from the origin O to the line $x = x_0$ in the least possible time.

Now, from (22.17) the condition $\delta I = 0$ requires, besides the EL equation, that at $x = b$, the other two contributions cancel, i.e.

$$F\Delta x + \frac{\partial F}{\partial y'}\eta = 0. \tag{22.19}$$

Eliminating Δx and η between (22.18) and (22.19) leads to the condition that at the end-point

$$\left(F - y'\frac{\partial F}{\partial y'}\right)\frac{\partial h}{\partial y} - \frac{\partial F}{\partial y'}\frac{\partial h}{\partial x} = 0. \tag{22.20}$$

In the special case where the end-point is free to lie anywhere on the vertical line $x = b$, we have $\partial h/\partial x = 1$ and $\partial h/\partial y = 0$. Substituting these values into (22.20), we recover the end-point condition given in (22.16).

▶ *A frictionless wire in a vertical plane connects two points A and B, A being higher than B. Let the position of A be fixed at the origin of an xy-coordinate system, but allow B to lie anywhere on the vertical line $x = x_0$ (see figure 22.6). Find the shape of the wire such that a bead placed on it at A will slide under gravity to B in the shortest possible time.*

This is a variant of the famous brachistochrone (shortest time) problem, which is often used to illustrate the calculus of variations. Conservation of energy tells us that the particle speed is given by

$$v = \frac{ds}{dt} = \sqrt{2gy},$$

where s is the path length along the wire and g is the acceleration due to gravity. Since the element of path length is $ds = (1 + y'^2)^{1/2}dx$, the total time taken to travel to the line $x = x_0$ is given by

$$t = \int_{x=0}^{x=x_0} \frac{ds}{v} = \frac{1}{\sqrt{2g}}\int_0^{x_0}\sqrt{\frac{1 + y'^2}{y}}\,dx.$$

Because the integrand does not contain x explicitly, we can use (22.8) with the specific form $F = \sqrt{1 + y'^2}/\sqrt{y}$ to find a first integral; on simplification this yields

$$\left[y(1 + y'^2)\right]^{1/2} = k,$$

where k is a constant. Letting $a = k^2$ and solving for y' we find

$$y' = \frac{dy}{dx} = \sqrt{\frac{a-y}{y}},$$

which on substituting $y = a \sin^2 \theta$ integrates to give

$$x = \frac{a}{2}(2\theta - \sin 2\theta) + c.$$

Thus the parametric equations of the curve are given by

$$x = b(\phi - \sin \phi) + c, \qquad y = b(1 - \cos \phi),$$

where $b = a/2$ and $\phi = 2\theta$; they define a cycloid, the curve traced out by a point on the rim of a wheel of radius b rolling along the x-axis. We must now use the end-point conditions to determine the constants b and c. Since the curve passes through the origin, we see immediately that $c = 0$. Now since $y(x_0)$ is arbitrary, i.e. the upper end-point can lie anywhere on the curve $x = x_0$, the condition (22.20) reduces to (22.16), so that we also require

$$\left. \frac{\partial F}{\partial y'} \right|_{x=x_0} = \left. \frac{y'}{\sqrt{y(1+y'^2)}} \right|_{x=x_0} = 0,$$

which implies that $y' = 0$ at $x = x_0$. In words, the tangent to the cycloid at B must be parallel to the x-axis; this requires $\pi b = x_0$. ◀

22.4 Constrained variation

Just as the problem of finding the stationary values of a function $f(x, y)$ subject to the constraint $g(x, y) = $ constant is solved by means of Lagrange's undetermined multipliers (see chapter 5), so the corresponding problem in the calculus of variations is solved by an analogous method.

Suppose that we wish to find the stationary values of

$$I = \int_a^b F(y, y', x) \, dx,$$

subject to the constraint that the value of

$$J = \int_a^b G(y, y', x) \, dx$$

is held constant. Following the method of Lagrange undetermined multipliers let us define a new functional

$$K = I + \lambda J = \int_a^b (F + \lambda G) \, dx,$$

and find its *unconstrained* stationary values. Repeating the analysis of section 22.1 we find that we require

$$\frac{\partial F}{\partial y} - \frac{d}{dx}\left(\frac{\partial F}{\partial y'}\right) + \lambda \left[\frac{\partial G}{\partial y} - \frac{d}{dx}\left(\frac{\partial G}{\partial y'}\right) \right] = 0,$$

785

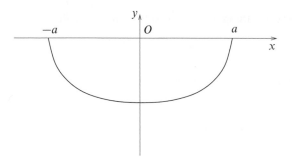

Figure 22.7   A uniform rope with fixed end-points suspended under gravity.

which, together with the original constraint $J = $ constant, will yield the required solution $y(x)$.

This method is easily generalised to cases with more than one constraint by the introduction of more Lagrange multipliers. If we wish to find the stationary values of an integral I subject to the multiple constraints that the values of the integrals J_i be held constant for $i = 1, 2, \ldots, n$, then we simply find the unconstrained stationary values of the new integral

$$K = I + \sum_{1}^{n} \lambda_i J_i.$$

▶*Find the shape assumed by a uniform rope when suspended by its ends from two points at equal heights.*

We will solve this problem using x (see figure 22.7) as the independent variable. Let the rope of length $2L$ be suspended between the points $x = \pm a$, $y = 0$ $(L > a)$ and have uniform linear density ρ. We then need to find the stationary value of the rope's gravitational potential energy,

$$I = -\rho g \int y \, ds = -\rho g \int_{-a}^{a} y(1 + y'^2)^{1/2} \, dx,$$

with respect to small changes in the form of the rope but subject to the constraint that the total length of the rope remains constant, i.e.

$$J = \int ds = \int_{-a}^{a} (1 + y'^2)^{1/2} \, dx = 2L.$$

We thus define a new integral (omitting the factor -1 from I for brevity)

$$K = I + \lambda J = \int_{-a}^{a} (\rho g y + \lambda)(1 + y'^2)^{1/2} \, dx$$

and find its stationary values. Since the integrand does not contain the independent variable x explicitly, we can use (22.8) to find the first integral:

$$(\rho g y + \lambda)\left(1 + y'^2\right)^{1/2} - (\rho g y + \lambda)\left(1 + y'^2\right)^{-1/2} y'^2 = k,$$

786

where k is a constant; this reduces to

$$y'^2 = \left(\frac{\rho g y + \lambda}{k}\right)^2 - 1.$$

Making the substitution $\rho g y + \lambda = k \cosh z$, this can be integrated easily to give

$$\frac{k}{\rho g} \cosh^{-1}\left(\frac{\rho g y + \lambda}{k}\right) = x + c,$$

where c is the constant of integration.

We now have three unknowns, λ, k and c, that must be evaluated using the two end conditions $y(\pm a) = 0$ and the constraint $J = 2L$. The end conditions give

$$\cosh \frac{\rho g (a + c)}{k} = \frac{\lambda}{k} = \cosh \frac{\rho g (-a + c)}{k},$$

and since $a \neq 0$, these imply $c = 0$ and $\lambda/k = \cosh(\rho g a/k)$. Putting $c = 0$ into the constraint, in which $y' = \sinh(\rho g x/k)$, we obtain

$$2L = \int_{-a}^{a} \left[1 + \sinh^2\left(\frac{\rho g x}{k}\right)\right]^{1/2} dx$$

$$= \frac{2k}{\rho g} \sinh\left(\frac{\rho g a}{k}\right).$$

Collecting together the values for the constants, the form adopted by the rope is therefore

$$y(x) = \frac{k}{\rho g}\left[\cosh\left(\frac{\rho g x}{k}\right) - \cosh\left(\frac{\rho g a}{k}\right)\right],$$

where k is the solution of $\sinh(\rho g a/k) = \rho g L/k$. This curve is known as a catenary. ◀

22.5 Physical variational principles

Many results in both classical and quantum physics can be expressed as variational principles, and it is often when expressed in this form that their physical meaning is most clearly understood. Moreover, once a physical phenomenon has been written as a variational principle, we can use all the results derived in this chapter to investigate its behaviour. It is usually possible to identify conserved quantities, or symmetries of the system of interest, that otherwise might be found only with considerable effort. From the wide range of physical variational principles we will select two examples from familiar areas of classical physics, namely geometric optics and mechanics.

22.5.1 Fermat's principle in optics

Fermat's principle in geometrical optics states that a ray of light travelling in a region of variable refractive index follows a path such that the total optical path length (physical length × refractive index) is stationary.

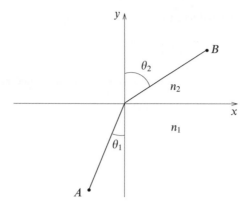

Figure 22.8   Path of a light ray at the plane interface between media with refractive indices n_1 and n_2, where $n_2 < n_1$.

▶*From Fermat's principle deduce Snell's law of refraction at an interface.*

Let the interface be at $y = $ constant (see figure 22.8) and let it separate two regions with refractive indices n_1 and n_2 respectively. On a ray the element of physical path length is $ds = (1 + y'^2)^{1/2}dx$, and so for a ray that passes through the points A and B, the total optical path length is

$$P = \int_A^B n(y)(1 + y'^2)^{1/2} \, dx.$$

Since the integrand does not contain the independent variable x explicitly, we use (22.8) to obtain a first integral, which, after some rearrangement, reads

$$n(y)\left(1 + y'^2\right)^{-1/2} = k,$$

where k is a constant. Recalling that y' is the tangent of the angle ϕ between the instantaneous direction of the ray and the x-axis, this *general* result, which is not dependent on the configuration presently under consideration, can be put in the form

$$n \cos \phi = \text{constant}$$

along a ray, even though n and ϕ vary individually.

   For our particular configuration n is constant in each medium and therefore so is y'. Thus the rays travel in straight lines in each medium (as anticipated in figure 22.8, but not assumed in our analysis), and since k is constant along the *whole* path we have $n_1 \cos \phi_1 = n_2 \cos \phi_2$, or in terms of the conventional angles in the figure

$$n_1 \sin \theta_1 = n_2 \sin \theta_2. \; ◀$$

22.5.2   *Hamilton's principle in mechanics*

Consider a mechanical system whose configuration can be uniquely defined by a number of coordinates q_i (usually distances and angles) together with time t and which experiences only forces derivable from a potential. Hamilton's principle

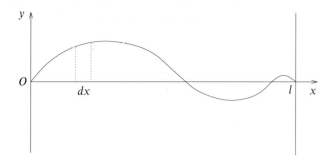

Figure 22.9 Transverse displacement on a taut string that is fixed at two points a distance l apart.

states that in moving from one configuration at time t_0 to another at time t_1 the motion of such a system is such as to make

$$\mathcal{L} = \int_{t_0}^{t_1} L(q_1, q_2 \ldots, q_n, \dot{q}_1, \dot{q}_2, \ldots, \dot{q}_n, t) \, dt \qquad (22.21)$$

stationary. The *Lagrangian* L is defined, in terms of the kinetic energy T and the potential energy V (with respect to some reference situation), by $L = T - V$. Here V is a function of the q_i only, not of the $\dot{q}_i$. Applying the EL equation to $\mathcal{L}$ we obtain *Lagrange's equations*,

$$\frac{\partial L}{\partial q_i} = \frac{d}{dt}\left(\frac{\partial L}{\partial \dot{q}_i}\right), \qquad i = 1, 2, \ldots, n.$$

▶*Using Hamilton's principle derive the wave equation for small transverse oscillations of a taut string.*

In this example we are in fact considering a generalisation of (22.21) to a case involving one isolated independent coordinate t, together with a *continuum* in which the q_i become the continuous variable x. The expressions for T and V therefore become integrals over x rather than sums over the label i.

If ρ and τ are the local density and tension of the string, both of which may depend on x, then, referring to figure 22.9, the kinetic and potential energies of the string are given by

$$T = \int_0^l \frac{\rho}{2}\left(\frac{\partial y}{\partial t}\right)^2 dx, \qquad V = \int_0^l \frac{\tau}{2}\left(\frac{\partial y}{\partial x}\right)^2 dx$$

and (22.21) becomes

$$\mathcal{L} = \frac{1}{2}\int_{t_0}^{t_1} dt \int_0^l \left[\rho\left(\frac{\partial y}{\partial t}\right)^2 - \tau\left(\frac{\partial y}{\partial x}\right)^2\right] dx.$$

Using (22.13) and the fact that y does not appear explicitly, we obtain

$$\frac{\partial}{\partial t}\left(\rho\frac{\partial y}{\partial t}\right) - \frac{\partial}{\partial x}\left(\tau\frac{\partial y}{\partial x}\right) = 0.$$

If, in addition, ρ and τ do not depend on x or t then

$$\frac{\partial^2 y}{\partial x^2} = \frac{1}{c^2}\frac{\partial^2 y}{\partial t^2},$$

where $c^2 = \tau/\rho$. This is the wave equation for small transverse oscillations of a taut uniform string. ◄

22.6 General eigenvalue problems

We have seen in this chapter that the problem of finding a curve that makes the value of a given integral stationary when the integral is taken along the curve results, in each case, in a differential equation for the curve. It is not a great extension to ask whether this may be used to solve differential equations, by setting up a suitable variational problem and then seeking ways other than the Euler equation of finding or estimating stationary solutions.

We shall be concerned with differential equations of the form $\mathcal{L}y = \lambda\rho(x)y$, where the differential operator $\mathcal{L}$ is self-adjoint, so that $\mathcal{L} = \mathcal{L}^\dagger$ (with appropriate boundary conditions on the solution y) and $\rho(x)$ is some weight function, as discussed in chapter 17. In particular, we will concentrate on the Sturm–Liouville equation as an explicit example, but much of what follows can be applied to other equations of this type.

We have already discussed the solution of equations of the Sturm–Liouville type in chapter 17 and the same notation will be used here. In this section, however, we will adopt a variational approach to estimating the eigenvalues of such equations.

Suppose we search for stationary values of the integral

$$I = \int_a^b \left[p(x)y'^2(x) - q(x)y^2(x)\right] dx, \tag{22.22}$$

with $y(a) = y(b) = 0$ and p and q any sufficiently smooth and differentiable functions of x. However, in addition we impose a normalisation condition

$$J = \int_a^b \rho(x)y^2(x)\,dx = \text{constant}. \tag{22.23}$$

Here $\rho(x)$ is a positive weight function defined in the interval $a \le x \le b$, but which may in particular cases be a constant.

Then, as in section 22.4, we use undetermined Lagrange multipliers,[§] and

---

[§] We use $-\lambda$, rather than λ, so that the final equation (22.24) appears in the conventional Sturm–Liouville form.

consider $K = I - \lambda J$ given by

$$K = \int_a^b \left[py'^2 - (q + \lambda \rho)y^2 \right] dx.$$

On application of the EL equation (22.5) this yields

$$\frac{d}{dx}\left(p\frac{dy}{dx} \right) + qy + \lambda \rho y = 0, \tag{22.24}$$

which is exactly the Sturm–Liouville equation (17.34), with eigenvalue λ. Now, since both I and J are quadratic in y and its derivative, finding stationary values of K is equivalent to finding stationary values of I/J. This may also be shown by considering the functional $\Lambda = I/J$, for which

$$\begin{aligned}
\delta\Lambda &= (\delta I/J) - (I/J^2)\delta J \\
&= (\delta I - \Lambda\delta J)/J \\
&= \delta K/J.
\end{aligned}$$

Hence, extremising Λ is equivalent to extremising K. Thus we have the important result that *finding functions y that make I/J stationary is equivalent to finding functions y that are solutions of the Sturm–Liouville equation; the resulting value of I/J equals the corresponding eigenvalue of the equation.*

Of course this does not tell us how to find such a function y and, naturally, to have to do this by solving (22.24) directly defeats the purpose of the exercise. We will see in the next section how some progress can be made. It is worth recalling that the functions $p(x)$, $q(x)$ and $\rho(x)$ can have many different forms, and so (22.24) represents quite a wide variety of equations.

We now recall some properties of the solutions of the Sturm–Liouville equation. The eigenvalues λ_i of (22.24) are real and will be assumed non-degenerate (for simplicity). We also assume that the corresponding eigenfunctions have been made real, so that normalised eigenfunctions $y_i(x)$ satisfy the orthogonality relation (as in (17.24))

$$\int_a^b y_i y_j \rho \, dx = \delta_{ij}. \tag{22.25}$$

Further, we take the boundary condition in the form

$$\left[y_i p y_j' \right]_{x=a}^{x=b} = 0; \tag{22.26}$$

this can be satisfied by $y(a) = y(b) = 0$, but also by many other sets of boundary conditions.

791

▶*Show that*

$$\int_a^b \left(y'_j p y'_i - y_j q y_i \right) dx = \lambda_i \delta_{ij}.$$ (22.27)

Let y_i be an eigenfunction of (22.24), corresponding to a particular eigenvalue λ_i, so that

$$\left(p y'_i \right)' + (q + \lambda_i \rho) y_i = 0.$$

Multiplying this through by y_j and integrating from a to b (the first term by parts) we obtain

$$\left[y_j \left(p y'_i \right) \right]_a^b - \int_a^b y'_j (p y'_i) \, dx + \int_a^b y_j (q + \lambda_i \rho) y_i \, dx = 0.$$ (22.28)

The first term vanishes by virtue of (22.26), and on rearranging the other terms and using (22.25), we find the result (22.27). ◀

We see at once that, if the function $y(x)$ minimises I/J, i.e. satisfies the Sturm–Liouville equation, then putting $y_i = y_j = y$ in (22.25) and (22.27) yields J and I respectively on the left-hand sides; thus, as mentioned above, the minimised value of I/J is just the eigenvalue λ, introduced originally as the undetermined multiplier.

▶*For a function y satisfying the Sturm–Liouville equation verify that, provided (22.26) is satisfied, $\lambda = I/J$.*

Firstly, we multiply (22.24) through by y to give

$$y(py')' + qy^2 + \lambda \rho y^2 = 0.$$

Now integrating this expression by parts we have

$$\left[ypy' \right]_a^b - \int_a^b \left(py'^2 - qy^2 \right) dx + \lambda \int_a^b \rho y^2 \, dx = 0.$$

The first term on the LHS is zero, the second is simply $-I$ and the third is λJ. Thus $\lambda = I/J$. ◀

22.7 Estimation of eigenvalues and eigenfunctions

Since the eigenvalues λ_i of the Sturm–Liouville equation are the stationary values of I/J (see above), it follows that any evaluation of I/J must yield a value that lies between the lowest and highest eigenvalues of the corresponding Sturm–Liouville equation, i.e.

$$\lambda_{\min} \le \frac{I}{J} \le \lambda_{\max},$$

where, depending on the equation under consideration, either $\lambda_{\min} = -\infty$ and

λ_{max} is finite, or $\lambda_{\text{max}} = \infty$ and λ_{min} is finite. Notice that here we have departed from direct consideration of the minimising problem and made a statement about a calculation in which no actual minimisation is necessary.

Thus, as an example, for an equation with a finite lowest eigenvalue λ_0 any evaluation of I/J provides an upper bound on λ_0. Further, we will now show that the estimate λ obtained is a better estimate of λ_0 than the estimated (guessed) function y is of y_0, the true eigenfunction corresponding to λ_0. The sense in which 'better' is used here will be clear from the final result.

Firstly, we expand the estimated or *trial function* y in terms of the complete set y_i:

$$y = y_0 + c_1 y_1 + c_2 y_2 + \cdots,$$

where, if a good trial function has been guessed, the c_i will be small. Using (22.25) we have immediately that $J = 1 + \sum_i |c_i|^2$. The other required integral is

$$I = \int_a^b \left[p \left(y_0' + \sum_i c_i y_i' \right)^2 - q \left(y_0 + \sum_i c_i y_i \right)^2 \right] dx.$$

On multiplying out the squared terms, all the cross terms vanish because of (22.27) to leave

$$\lambda = \frac{I}{J}$$

$$= \frac{\lambda_0 + \sum_i |c_i|^2 \lambda_i}{1 + \sum_j |c_j|^2}$$

$$= \lambda_0 + \sum_i |c_i|^2 (\lambda_i - \lambda_0) + \mathrm{O}(c^4).$$

Hence λ differs from λ_0 by a term second order in the c_i, even though y differed from y_0 by a term first order in the c_i; this is what we aimed to show. We notice incidentally that, since $\lambda_0 < \lambda_i$ for all i, λ is shown to be necessarily $\geq \lambda_0$, with equality only if all $c_i = 0$, i.e. if $y \equiv y_0$.

The method can be extended to the second and higher eigenvalues by imposing, in addition to the original constraints and boundary conditions, a restriction of the trial functions to only those that are orthogonal to the eigenfunctions corresponding to lower eigenvalues. (Of course, this requires complete or nearly complete knowledge of these latter eigenfunctions.) An example is given at the end of the chapter (exercise 22.25).

We now illustrate the method we have discussed by considering a simple example, one for which, as on previous occasions, the answer is obvious.

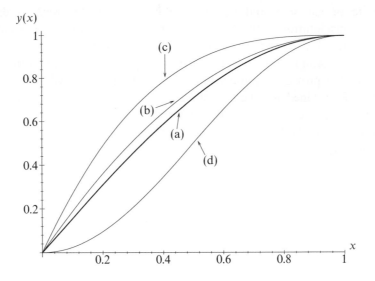

Figure 22.10 Trial solutions used to estimate the lowest eigenvalue λ of $-y'' = \lambda y$ with $y(0) = y'(1) = 0$. They are: (a) $y = \sin(\pi x/2)$, the exact result; (b) $y = 2x - x^2$; (c) $y = x^3 - 3x^2 + 3x$; (d) $y = \sin^2(\pi x/2)$.

▶ *Estimate the lowest eigenvalue of the equation*

$$-\frac{d^2y}{dx^2} = \lambda y, \qquad 0 \le x \le 1, \tag{22.29}$$

with boundary conditions

$$y(0) = 0, \qquad y'(1) = 0. \tag{22.30}$$

We need to find the lowest value λ_0 of λ for which (22.29) has a solution $y(x)$ that satisfies (22.30). The exact answer is of course $y = A\sin(x\pi/2)$ and $\lambda_0 = \pi^2/4 \approx 2.47$.

Firstly we note that the Sturm–Liouville equation reduces to (22.29) if we take $p(x) = 1$, $q(x) = 0$ and $\rho(x) = 1$ and that the boundary conditions satisfy (22.26). Thus we are able to apply the previous theory.

We will use three trial functions so that the effect on the estimate of λ_0 of making better or worse 'guesses' can be seen. One further preliminary remark is relevant, namely that the estimate is independent of any constant multiplicative factor in the function used. This is easily verified by looking at the form of I/J. We normalise each trial function so that $y(1) = 1$, purely in order to facilitate comparison of the various function shapes.

Figure 22.10 illustrates the trial functions used, curve (a) being the exact solution $y = \sin(\pi x/2)$. The other curves are (b) $y(x) = 2x - x^2$, (c) $y(x) = x^3 - 3x^2 + 3x$, and (d) $y(x) = \sin^2(\pi x/2)$. The choice of trial function is governed by the following considerations:

  (i) the boundary conditions (22.30) *must* be satisfied.
 (ii) a 'good' trial function ought to mimic the correct solution as far as possible, but it may not be easy to guess even the general shape of the correct solution in some cases.
(iii) the evaluation of I/J should be as simple as possible.

It is easily verified that functions (b), (c) and (d) all satisfy (22.30) but, so far as mimicking the correct solution is concerned, we would expect from the figure that (b) would be superior to the other two. The three evaluations are straightforward, using (22.22) and (22.23):

$$\lambda_b = \frac{\int_0^1 (2 - 2x)^2\, dx}{\int_0^1 (2x - x^2)^2\, dx} = \frac{4/3}{8/15} = 2.50$$

$$\lambda_c = \frac{\int_0^1 (3x^2 - 6x + 3)^2\, dx}{\int_0^1 (x^3 - 3x^2 + 3x)^2\, dx} = \frac{9/5}{9/14} = 2.80$$

$$\lambda_d = \frac{\int_0^1 (\pi^2/4)\sin^2(\pi x)\, dx}{\int_0^1 \sin^4(\pi x/2)\, dx} = \frac{\pi^2/8}{3/8} = 3.29.$$

We expected all evaluations to yield estimates greater than the lowest eigenvalue, 2.47, and this is indeed so. From these trials alone we are able to say (only) that $\lambda_0 \leq 2.50$. As expected, the best approximation (b) to the true eigenfunction yields the lowest, and therefore the best, upper bound on λ_0. ◄

We may generalise the work of this section to other differential equations of the form $\mathcal{L}y = \lambda \rho y$, where $\mathcal{L} = \mathcal{L}^\dagger$. In particular, one finds

$$\lambda_{\min} < \frac{I}{J} \leq \lambda_{\max},$$

where I and J are now given by

$$I = \int_a^b y^*(\mathcal{L}y)\, dx \qquad \text{and} \qquad J = \int_a^b \rho y^* y\, dx. \tag{22.31}$$

It is straightforward to show that, for the special case of the Sturm–Liouville equation, for which

$$\mathcal{L}y = -(py')' - qy,$$

the expression for I in (22.31) leads to (22.22).

22.8 Adjustment of parameters

Instead of trying to estimate λ_0 by selecting a large number of different trial functions, we may also use trial functions that include one or more parameters which themselves may be adjusted to give the lowest value to $\lambda = I/J$ and hence the best estimate of λ_0. The justification for this method comes from the knowledge that no matter what form of function is chosen, nor what values are assigned to the parameters, provided the boundary conditions are satisfied λ can never be less than the required λ_0.

To illustrate this method an example from quantum mechanics will be used. The time-independent Schrödinger equation is formally written as the eigenvalue equation $H\psi = E\psi$, where H is a linear operator, ψ the wavefunction describing a quantum mechanical system and E the energy of the system. The energy operator H is called the Hamiltonian and for a particle of mass m moving in a one-dimensional harmonic oscillator potential is given by

$$H = -\frac{\hbar^2}{2m}\frac{d^2}{dx^2} + \frac{kx^2}{2},$$ (22.32)

where $\hbar$ is Planck's constant divided by 2π.

> ▶*Estimate the ground-state energy of a quantum harmonic oscillator.*

Using (22.32) in $H\psi = E\psi$, the Schrödinger equation is

$$-\frac{\hbar^2}{2m}\frac{d^2\psi}{dx^2} + \frac{kx^2}{2}\psi = E\psi, \qquad -\infty < x < \infty.$$ (22.33)

The boundary conditions are that ψ should vanish as $x \to \pm\infty$. Equation (22.33) is a form of the Sturm–Liouville equation in which $p = \hbar^2/(2m)$, $q = -kx^2/2$, $\rho = 1$ and $\lambda = E$; it can be solved by the methods developed previously, e.g. by writing the eigenfunction ψ as a power series in x.

However, our purpose here is to illustrate variational methods and so we take as a trial wavefunction $\psi = \exp(-\alpha x^2)$, where α is a positive parameter whose value we will choose later. This function certainly $\to 0$ as $x \to \pm\infty$ and is convenient for calculations. Whether it approximates the true wave function is unknown, but if it does not our estimate will still be valid, although the upper bound will be a poor one.

With $y = \exp(-\alpha x^2)$ and therefore $y' = -2\alpha x \exp(-\alpha x^2)$, the required estimate is

$$E = \lambda = \frac{\int_{-\infty}^{\infty}[(\hbar^2/2m)4\alpha^2 x^2 + (k/2)x^2]e^{-2\alpha x^2}\,dx}{\int_{-\infty}^{\infty}e^{-2\alpha x^2}\,dx} = \frac{\hbar^2\alpha}{2m} + \frac{k}{8\alpha}.$$ (22.34)

This evaluation is easily carried out using the reduction formula

$$I_n = \frac{n-1}{4\alpha}I_{n-2}, \quad \text{for integrals of the form} \quad I_n = \int_{-\infty}^{\infty} x^n e^{-2\alpha x^2}\,dx.$$ (22.35)

So, we have obtained the estimate (22.34), involving the parameter α, for the oscillator's ground-state energy, i.e. the lowest eigenvalue of H. In line with our previous discussion we now minimise λ with respect to α. Putting $d\lambda/d\alpha = 0$ (clearly a minimum), yields $\alpha = (km)^{1/2}/(2\hbar)$, which in turn gives as the minimum value for λ

$$E = \frac{\hbar}{2}\left(\frac{k}{m}\right)^{1/2} = \frac{\hbar\omega}{2},$$ (22.36)

where we have put $(k/m)^{1/2}$ equal to the classical angular frequency ω.

The method thus leads to the conclusion that the ground-state energy E_0 is $\leq \frac{1}{2}\hbar\omega$. In fact, as is well known, the equality sign holds, $\frac{1}{2}\hbar\omega$ being just the zero-point energy of a quantum mechanical oscillator. Our estimate gives the exact value because $\psi(x) = \exp(-\alpha x^2)$ is the correct functional form for the ground state wavefunction and the particular value of α that we have found is that needed to make ψ an eigenfunction of H with eigenvalue $\frac{1}{2}\hbar\omega$. ◀

An alternative but equivalent approach to this is developed in the exercises that follow, as is an extension of this particular problem to estimating the second-lowest eigenvalue (see exercise 22.25).

22.9 Exercises

22.1 A surface of revolution, whose equation in cylindrical polar coordinates is $\rho = \rho(z)$, is bounded by the circles $\rho = a$, $z = \pm c$ $(a > c)$. Show that the function that makes the surface integral $I = \int \rho^{-1/2} \, dS$ stationary with respect to small variations is given by $\rho(z) = k + z^2/(4k)$, where $k = [a \pm (a^2 - c^2)^{1/2}]/2$.

22.2 Show that the lowest value of the integral

$$\int_A^B \frac{(1 + y'^2)^{1/2}}{y} \, dx,$$

where A is $(-1, 1)$ and B is $(1, 1)$, is $2 \ln(1 + \sqrt{2})$. Assume that the Euler–Lagrange equation gives a minimising curve.

22.3 The refractive index n of a medium is a function only of the distance r from a fixed point O. Prove that the equation of a light ray, assumed to lie in a plane through O, travelling in the medium satisfies (in plane polar coordinates)

$$\frac{1}{r^2} \left(\frac{dr}{d\phi} \right)^2 = \frac{r^2 \, n^2(r)}{a^2 \, n^2(a)} - 1,$$

where a is the distance of the ray from O at the point at which $dr/d\phi = 0$.

If $n = [1 + (\alpha^2/r^2)]^{1/2}$ and the ray starts and ends far from O, find its deviation (the angle through which the ray is turned), if its minimum distance from O is a.

22.4 The Lagrangian for a π-meson is given by

$$L(\mathbf{x}, t) = \tfrac{1}{2}(\dot{\phi}^2 - |\nabla\phi|^2 - \mu^2\phi^2),$$

where μ is the meson mass and $\phi(\mathbf{x}, t)$ is its wavefunction. Assuming Hamilton's principle, find the wave equation satisfied by ϕ.

22.5 Prove the following results about general systems.

(a) For a system described in terms of coordinates q_i and t, show that if t does not appear explicitly in the expressions for x, y and z $(x = x(q_i, t)$, etc.) then the kinetic energy T is a homogeneous quadratic function of the $\dot{q}_i$ (it may also involve the q_i). Deduce that $\sum_i \dot{q}_i(\partial T/\partial \dot{q}_i) = 2T$.

(b) Assuming that the forces acting on the system are derivable from a potential V, show, by expressing dT/dt in terms of q_i and $\dot{q}_i$, that $d(T + V)/dt = 0$.

22.6 For a system specified by the coordinates q and t, show that the equation of motion is unchanged if the Lagrangian $L(q, \dot{q}, t)$ is replaced by

$$L_1 = L + \frac{d\phi(q, t)}{dt},$$

where ϕ is an arbitrary function. Deduce that the equation of motion of a particle that moves in one dimension subject to a force $-dV(x)/dx$ (x being measured from a point O) is unchanged if O is forced to move with a constant velocity v (x still being measured from O).

22.7 In cylindrical polar coordinates, the curve $(\rho(\theta), \theta, \alpha\rho(\theta))$ lies on the surface of the cone $z = \alpha\rho$. Show that geodesics (curves of minimum length joining two points) on the cone satisfy

$$\rho^4 = c^2[\beta^2\rho'^2 + \rho^2],$$

where c is an arbitrary constant, but β has to have a particular value. Determine the form of $\rho(\theta)$ and hence find the equation of the shortest path on the cone between the points $(R, -\theta_0, \alpha R)$ and $(R, \theta_0, \alpha R)$.

[You will find it useful to determine the form of the derivative of $\cos^{-1}(u^{-1})$.]

22.8 Derive the differential equations for the plane-polar coordinates, r and ϕ, of a particle of unit mass moving in a field of potential $V(r)$. Find the form of V if the path of the particle is given by $r = a \sin \phi$.

22.9 You are provided with a line of length $\pi a/2$ and negligible mass and some lead shot of total mass M. Use a variational method to determine how the lead shot must be distributed along the line if the loaded line is to hang in a circular arc of radius a when its ends are attached to two points at the same height. Measure the distance s along the line from its centre.

22.10 Extend the result of subsection 22.2.2 to the case of several dependent variables $y_i(x)$, showing that, if x does not appear explicitly in the integrand, then a first integral of the Euler–Lagrange equations is

$$F - \sum_{i=1}^{n} y_i' \frac{\partial F}{\partial y_i'} = \text{constant.}$$

22.11 A general result is that light travels through a variable medium by a path which minimises the travel time (this is an alternative formulation of Fermat's principle). With respect to a particular cylindrical polar coordinate system (ρ, ϕ, z), the speed of light $v(\rho, \phi)$ is independent of z. If the path of the light is parameterised as $\rho = \rho(z)$, $\phi = \phi(z)$, use the result of the previous exercise to show that

$$v^2(\rho'^2 + \rho^2\phi'^2 + 1)$$

is constant along the path.

For the particular case when $v = v(\rho) = b(a^2 + \rho^2)^{1/2}$, show that the two Euler–Lagrange equations have a common solution in which the light travels along a helical path given by $\phi = Az + B$, $\rho = C$, provided that A has a particular value.

22.12 Light travels in the vertical xz-plane through a slab of material which lies between the planes $z = z_0$ and $z = 2z_0$, and in which the speed of light $v(z) = c_0z/z_0$. Using the alternative formulation of Fermat's principle, given in the previous question, show that the ray paths are arcs of circles.

Deduce that, if a ray enters the material at $(0, z_0)$ at an angle to the vertical, $\pi/2 - \theta$, of more than $30°$, then it does not reach the far side of the slab.

22.13 A dam of capacity V (less than $\pi b^2 h/2$) is to be constructed on level ground next to a long straight wall which runs from $(-b, 0)$ to $(b, 0)$. This is to be achieved by joining the ends of a new wall, of height h, to those of the existing wall. Show that, in order to minimise the length L of new wall to be built, it should form part of a circle, and that L is then given by

$$\int_{-b}^{b} \frac{dx}{(1 - \lambda^2 x^2)^{1/2}},$$

where λ is found from

$$\frac{V}{hb^2} = \frac{\sin^{-1} \mu}{\mu^2} - \frac{(1 - \mu^2)^{1/2}}{\mu}$$

and $\mu = \lambda b$.

22.14 In the brachistochrone problem of subsection 22.3.4 show that if the upper endpoint can lie anywhere on the curve $h(x, y) = 0$, then the curve of quickest descent $y(x)$ meets $h(x, y) = 0$ at right angles.

22.15 The Schwarzchild metric for the static field of a non-rotating spherically symmetric black hole of mass M is given by

$$(ds)^2 = c^2 \left(1 - \frac{2GM}{c^2 r}\right) (dt)^2 - \frac{(dr)^2}{1 - 2GM/(c^2 r)} - r^2 (d\theta)^2 - r^2 \sin^2 \theta \, (d\phi)^2.$$

Considering only motion confined to the plane $\theta = \pi/2$, and assuming that the path of a small test particle is such as to make $\int ds$ stationary, find two first integrals of the equations of motion. From their Newtonian limits, in which GM/r, $\dot{r}^2$ and $r^2\dot{\phi}^2$ are all $\ll c^2$, identify the constants of integration.

22.16   Use result (22.27) to evaluate

$$ J = \int_{-1}^{1} (1 - x^2)P'_m(x)P'_n(x)\,dx, $$

where $P_m(x)$ is a Legendre polynomial of order m.

22.17   Determine the minimum value that the integral

$$ J = \int_{0}^{1} [x^4(y'')^2 + 4x^2(y')^2]\,dx $$

can have, given that y is not singular at $x = 0$ and that $y(1) = y'(1) = 1$. Assume that the Euler–Lagrange equation gives the lower limit, and verify retrospectively that your solution makes the first term on the LHS of equation (22.15) vanish.

22.18   Show that $y'' - xy + \lambda x^2 y = 0$ has a solution for which $y(0) = y(1) = 0$ and $\lambda \le 147/4$.

22.19   Find an appropriate, but simple, trial function and use it to estimate the lowest eigenvalue λ_0 of Stokes' equation,

$$ \frac{d^2y}{dx^2} + \lambda xy = 0, \qquad \text{with } y(0) = y(\pi) = 0. $$

Explain why your estimate must be strictly greater than λ_0.

22.20   Estimate the lowest eigenvalue, λ_0, of the equation

$$ \frac{d^2y}{dx^2} - x^2y + \lambda y = 0, \qquad y(-1) = y(1) = 0, $$

using a quadratic trial function.

22.21   A drumskin is stretched across a fixed circular rim of radius a. Small transverse vibrations of the skin have an amplitude $z(\rho, \phi, t)$ that satisfies

$$ \nabla^2 z = \frac{1}{c^2}\frac{\partial^2 z}{\partial t^2} $$

in plane polar coordinates. For a normal mode independent of azimuth, $z = Z(\rho)\cos\omega t$, find the differential equation satisfied by $Z(\rho)$. By using a trial function of the form $a^\nu - \rho^\nu$, with adjustable parameter ν, obtain an estimate for the lowest normal mode frequency.
  [The exact answer is $(5.78)^{1/2}c/a$.]

22.22   Consider the problem of finding the lowest eigenvalue, λ_0, of the equation

$$ (1 + x^2)\frac{d^2y}{dx^2} + 2x\frac{dy}{dx} + \lambda y = 0, \qquad y(\pm 1) = 0. $$

(a) Recast the problem in variational form, and derive an approximation λ_1 to λ_0 by using the trial function $y_1(x) = 1 - x^2$.
(b) Show that an improved estimate λ_2 is obtained by using $y_2(x) = \cos(\pi x/2)$.
(c) Prove that the estimate $\lambda(\gamma)$ obtained by taking $y_1(x) + \gamma y_2(x)$ as the trial function is

$$ \lambda(\gamma) = \frac{64/15 + 64\gamma/\pi - 384\gamma/\pi^3 + (\pi^2/3 + 1/2)\gamma^2}{16/15 + 64\gamma/\pi^3 + \gamma^2}. $$

Investigate $\lambda(\gamma)$ numerically as γ is varied, or, more simply, show that $\lambda(-1.80) = 3.668$, an improvement on both λ_1 and λ_2.

22.23 For the boundary conditions given below, obtain a functional $\Lambda(y)$ whose stationary values give the eigenvalues of the equation

$$(1+x)\frac{d^2y}{dx^2} + (2+x)\frac{dy}{dx} + \lambda y = 0, \qquad y(0) = 0, \; y'(2) = 0.$$

Derive an approximation to the lowest eigenvalue λ_0 using the trial function $y(x) = xe^{-x/2}$. For what value(s) of γ would

$$y(x) = xe^{-x/2} + \beta \sin \gamma x$$

be a suitable trial function for attempting to obtain an improved estimate of λ_0?

22.24 This is an alternative approach to the example in section 22.8. Using the notation of that section, the expectation value of the energy of the state ψ is given by $\int \psi^* H \psi \, dv$. Denote the eigenfunctions of H by ψ_i, so that $H\psi_i = E_i \psi_i$, and, since H is self-adjoint (Hermitian), $\int \psi_j^* \psi_i \, dv = \delta_{ij}$.

(a) By writing any function ψ as $\sum c_j \psi_j$ and following an argument similar to that in section 22.7, show that

$$E = \frac{\int \psi^* H \psi \, dv}{\int \psi^* \psi \, dv} \geq E_0,$$

the energy of the lowest state. This is the Rayleigh–Ritz principle.

(b) Using the same trial function as in section 22.8, $\psi = \exp(-\alpha x^2)$, show that the same result is obtained.

22.25 This is an extension to section 22.8 and the previous question. With the ground-state (i.e. the lowest-energy) wavefunction as $\exp(-\alpha x^2)$, take as a trial function the orthogonal wave function $x^{2n+1} \exp(-\alpha x^2)$, using the integer n as a variable parameter. Use either Sturm–Liouville theory or the Rayleigh–Ritz principle to show that the energy of the second lowest state of a quantum harmonic oscillator is $\leq 3\hbar\omega/2$.

22.26 The Hamiltonian H for the hydrogen atom is

$$-\frac{\hbar^2}{2m}\nabla^2 - \frac{q^2}{4\pi\epsilon_0 r}.$$

For a spherically symmetric state, as may be assumed for the ground state, the only relevant part of ∇^2 is that involving differentiation with respect to r.

(a) Define the integrals J_n by

$$J_n = \int_0^\infty r^n e^{-2\beta r} \, dr$$

and show that, for a trial wavefunction of the form $\exp(-\beta r)$ with $\beta > 0$, $\int \psi^* H \psi \, dv$ and $\int \psi^* \psi \, dv$ (see exercise 22.24(a)) can be expressed as $aJ_1 - bJ_2$ and cJ_2 respectively, where a, b and c are factors which you should determine.

(b) Show that the estimate of E is minimised when $\beta = mq^2/(4\pi\epsilon_0\hbar^2)$.

(c) Hence find an upper limit for the ground-state energy of the hydrogen atom. In fact, $\exp(-\beta r)$ is the correct form for the wavefunction and the limit gives the actual value.

22.27 The upper and lower surfaces of a film of liquid, which has surface energy per unit area (surface tension) γ and density ρ, have equations $z = p(x)$ and $z = q(x)$, respectively. The film has a given volume V (per unit depth in the y-direction) and lies in the region $-L < x < L$, with $p(0) = q(0) = p(L) = q(L) = 0$. The total energy (per unit depth) of the film consists of its surface energy and its gravitational energy, and is expressed by

$$ E = \frac{\rho g}{2} \int_{-L}^{L} (p^2 - q^2)\, dx + \gamma \int_{-L}^{L} \left[(1 + p'^2)^{1/2} + (1 + q'^2)^{1/2} \right] dx. $$

(a) Express V in terms of p and q.

(b) Show that, if the total energy is minimised, p and q must satisfy

$$ \frac{p'^2}{(1 + p'^2)^{1/2}} - \frac{q'^2}{(1 + q'^2)^{1/2}} = \text{constant}. $$

(c) As an approximate solution, consider the equations

$$ p = a(L - |x|), \qquad q = b(L - |x|), $$

where a and b are sufficiently small that a^3 and b^3 can be neglected compared with unity. Find the values of a and b that minimise E.

22.28 A particle of mass m moves in a one-dimensional potential well of the form

$$ V(x) = -\mu \frac{\hbar^2 \alpha^2}{m} \operatorname{sech}^2 \alpha x, $$

where μ and α are positive constants. As in exercise 22.26, the expectation value $\langle E \rangle$ of the energy of the system is $\int \psi^* H \psi \, dx$, where the self-adjoint operator H is given by $-(\hbar^2/2m)d^2/dx^2 + V(x)$. Using trial wavefunctions of the form $y = A \operatorname{sech} \beta x$, show the following:

(a) for $\mu = 1$, there is an exact eigenfunction of H, with a corresponding $\langle E \rangle$ of half of the maximum depth of the well;

(b) for $\mu = 6$, the 'binding energy' of the ground state is at least $10\hbar^2\alpha^2/(3m)$.

[You will find it useful to note that for $u, v \geq 0$, $\operatorname{sech} u \operatorname{sech} v \geq \operatorname{sech}(u + v)$.]

22.29 The Sturm–Liouville equation can be extended to two independent variables, x and z, with little modification. In equation (22.22), y'^2 is replaced by $(\nabla y)^2$ and the integrals of the various functions of $y(x, z)$ become two-dimensional, i.e. the infinitesimal is $dx \, dz$.

The vibrations of a trampoline 4 units long and 1 unit wide satisfy the equation

$$ \nabla^2 y + k^2 y = 0. $$

By taking the simplest possible permissible polynomial as a trial function, show that the lowest mode of vibration has $k^2 \leq 10.63$ and, by direct solution, that the actual value is 10.49.

22.10 Hints and answers

22.1 Note that the integrand, $2\pi \rho^{1/2}(1 + \rho'^2)^{1/2}$, does not contain z explicitly.

22.3 $I = \int n(r)[r^2 + (dr/d\phi)^2]^{1/2} \, d\phi$. Take axes such that $\phi = 0$ when $r = \infty$. If $\beta = (\pi - \text{deviation angle})/2$ then $\beta = \phi$ at $r = a$, and the equation reduces to

$$ \frac{\beta}{(a^2 + \alpha^2)^{1/2}} = \int_{-\infty}^{\infty} \frac{dr}{r(r^2 - a^2)^{1/2}}, $$

which can be evaluated by putting $r = a(y + y^{-1})/2$, or successively $r = a \cosh \psi$, $y = \exp \psi$ to yield a deviation of $\pi[(a^2 + \alpha^2)^{1/2} - a]/a$.

22.5 (a) $\partial x/\partial t = 0$ and so $\dot{x} = \sum_i \dot{q}_i \partial x/\partial q_i$; (b) use

$$\sum_i \dot{q}_i \frac{d}{dt}\left(\frac{\partial T}{\partial \dot{q}_i}\right) = \frac{d}{dt}(2T) - \sum_i \ddot{q}_i \frac{\partial T}{\partial \dot{q}_i}.$$

22.7 Use result (22.8); $\beta^2 = 1 + \alpha^2$.
Put $\rho = uc$ to obtain $d\theta/du = \beta/[u(u^2 - 1)^{1/2}]$. Remember that $\cos^{-1}$ is a multivalued function; $\rho(\theta) = [R\cos(\theta_0/\beta)]/[\cos(\theta/\beta)]$.

22.9 $-\lambda y'(1 - y'^2)^{-1/2} = 2gP(s)$, $y = y(s)$, $P(s) = \int_0^s \rho(s')\, ds'$. The solution, $y = -a\cos(s/a)$, and $2P(\pi a/4) = M$ together give $\lambda = -gM$. The required $\rho(s)$ is given by $[M/(2a)]\sec^2(s/a)$.

22.11 Note that the ϕ E–L equation is automatically satisfied if $v \neq v(\phi)$. $A = 1/a$.

22.13 Circle is $\lambda^2 x^2 + [\lambda y + (1 - \lambda^2 b^2)^{1/2}]^2 = 1$. Use the fact that $\int y\, dx = V/h$ to determine the condition on λ.

22.15 Denoting $(ds)^2/(dt)^2$ by f^2, the Euler–Lagrange equation for ϕ gives $r^2\dot{\phi} = Af$, where A corresponds to the angular momentum of the particle. Use the result of exercise 22.10 to obtain $c^2 - (2GM/r) = Bf$, where, to first order in small quantities,

$$cB = c^2 - \frac{GM}{r} + \frac{1}{2}(\dot{r}^2 + r^2\dot{\phi}^2),$$

which reads 'total energy = rest mass + gravitational energy + radial and azimuthal kinetic energy'.

22.17 Convert the equation to the usual form, by writing $y'(x) = u(x)$, and obtain $x^2 u'' + 4xu' - 4u = 0$ with general solution $Ax^{-4} + Bx$. Integrating a second time and using the boundary conditions gives $y(x) = (1 + x^2)/2$ and $J = 1$; $\eta(1) = 0$, since $y'(1)$ is fixed, and $\partial F/\partial u' = 2x^4 u' = 0$ at $x = 0$.

22.19 Using $y = \sin x$ as a trial function shows that $\lambda_0 \leq 2/\pi$. The estimate must be $> \lambda_0$ since the trial function does not satisfy the original equation.

22.21 $Z'' + \rho^{-1}Z' + (\omega/c)^2 Z = 0$, with $Z(a) = 0$ and $Z'(0) = 0$; this is an SL equation with $p = \rho$, $q = 0$ and weight function ρ/c^2. Estimate of $\omega^2 = [c^2 v/(2a^2)][0.5 - 2(v+2)^{-1} + (2v+2)^{-1}]^{-1}$, which minimises to $c^2(2+\sqrt{2})^2/(2a^2) = 5.83c^2/a^2$ when $v = \sqrt{2}$.

22.23 Note that the original equation is not self-adjoint; it needs an integrating factor of e^x. $\Lambda(y) = [\int_0^2 (1 + x)e^x y'^2\, dx]/[\int_0^2 e^x y^2\, dx]$; $\lambda_0 \leq 3/8$. Since $y'(2)$ must equal 0, $\gamma = (\pi/2)(n + \frac{1}{2})$ for some integer n.

22.25 $E_1 \leq (\hbar\omega/2)(8n^2 + 12n + 3)/(4n + 1)$, which has a minimum value $3\hbar\omega/2$ when integer $n = 0$.

22.27 (a) $V = \int_{-L}^{L} (p - q)\, dx$. (c) Use $V = (a - b)L^2$ to eliminate b from the expression for E; now the minimisation is with respect to a alone. The values for a and b are $\pm V/(2L^2) - V\rho g/(6\gamma)$.

22.29 The SL equation has $p = 1$, $q = 0$, and $\rho = 1$.
Use $u(x, z) = x(4 - x)z(1 - z)$ as a trial function; numerator = 1088/90, denominator = 512/450. Direct solution $k^2 = 17\pi^2/16$.

23

Integral equations

It is not unusual in the analysis of a physical system to encounter an equation in which an unknown but required function $y(x)$, say, appears under an integral sign. Such an equation is called an *integral equation*, and in this chapter we discuss several methods for solving the more straightforward examples of such equations.

Before embarking on our discussion of methods for solving various integral equations, we begin with a warning that many of the integral equations met in practice cannot be solved by the elementary methods presented here but must instead be solved numerically, usually on a computer. Nevertheless, the regular occurrence of several simple types of integral equation that may be solved analytically is sufficient reason to explore these equations more fully.

We shall begin this chapter by discussing how a differential equation can be transformed into an integral equation and by considering the most common types of linear integral equation. After introducing the operator notation and considering the existence of solutions for various types of equation, we go on to discuss elementary methods of obtaining closed-form solutions of simple integral equations. We then consider the solution of integral equations in terms of infinite series and conclude by discussing the properties of integral equations with Hermitian kernels, i.e. those in which the integrands have particular symmetry properties.

23.1 Obtaining an integral equation from a differential equation

Integral equations occur in many situations, partly because we may always rewrite a differential equation as an integral equation. It is sometimes advantageous to make this transformation, since questions concerning the existence of a solution are more easily answered for integral equations (see section 23.3), and, furthermore, an integral equation can incorporate automatically any boundary conditions on the solution.

We shall illustrate the principles involved by considering the differential equation

$$y''(x) = f(x, y), \tag{23.1}$$

where $f(x, y)$ can be any function of x and y but not of $y'(x)$. Equation (23.1) thus represents a large class of linear and non-linear second-order differential equations.

We can convert (23.1) into the corresponding integral equation by first integrating with respect to x to obtain

$$y'(x) = \int_0^x f(z, y(z)) \, dz + c_1.$$

Integrating once more, we find

$$y(x) = \int_0^x du \int_0^u f(z, y(z)) \, dz + c_1 x + c_2.$$

Provided we do not change the region in the uz-plane over which the double integral is taken, we can reverse the order of the two integrations. Changing the integration limits appropriately, we find

$$y(x) = \int_0^x f(z, y(z)) \, dz \int_z^x du + c_1 x + c_2 \tag{23.2}$$

$$= \int_0^x (x - z) f(z, y(z)) \, dz + c_1 x + c_2; \tag{23.3}$$

this is a non-linear (for general $f(x, y)$) *Volterra* integral equation.

It is straightforward to incorporate any boundary conditions on the solution $y(x)$ by fixing the constants c_1 and c_2 in (23.3). For example, we might have the one-point boundary condition $y(0) = a$ and $y'(0) = b$, for which it is clear that we must set $c_1 = b$ and $c_2 = a$.

23.2 Types of integral equation

From (23.3), we can see that even a relatively simple differential equation such as (23.1) can lead to a corresponding integral equation that is non-linear. In this chapter, however, we will restrict our attention to *linear* integral equations, which have the general form

$$g(x)y(x) = f(x) + \lambda \int_a^b K(x, z) y(z) \, dz. \tag{23.4}$$

In (23.4), $y(x)$ is the unknown function, while the functions $f(x)$, $g(x)$ and $K(x, z)$ are assumed known. $K(x, z)$ is called the *kernel* of the integral equation. The integration limits a and b are also assumed known, and may be constants or functions of x, and λ is a known constant or parameter.

In fact, we shall be concerned with various special cases of (23.4), which are known by particular names. Firstly, if $g(x) = 0$ then the unknown function $y(x)$ appears only under the integral sign, and (23.4) is called a linear integral equation *of the first kind*. Alternatively, if $g(x) = 1$, so that $y(x)$ appears twice, once inside the integral and once outside, then (23.4) is called a linear integral equation *of the second kind*. In either case, if $f(x) = 0$ the equation is called *homogeneous*, otherwise *inhomogeneous*.

We can distinguish further between different types of integral equation by the form of the integration limits a and b. If these limits are fixed constants then the equation is called a *Fredholm* equation. If, however, the upper limit $b = x$ (i.e. it is variable) then the equation is called a *Volterra* equation; such an equation is analogous to one with fixed limits but for which the kernel $K(x, z) = 0$ for $z > x$. Finally, we note that any equation for which either (or both) of the integration limits is infinite, or for which $K(x, z)$ becomes infinite in the range of integration, is called a *singular* integral equation.

23.3 Operator notation and the existence of solutions

There is a close correspondence between linear integral equations and the matrix equations discussed in chapter 8. However, the former involve linear, integral relations between functions in an infinite-dimensional function space (see chapter 17), whereas the latter specify linear relations among vectors in a finite-dimensional vector space.

Since we are restricting our attention to linear integral equations, it will be convenient to introduce the linear integral operator $\mathcal{K}$, whose action on an arbitrary function y is given by

$$\mathcal{K}y = \int_a^b K(x, z)y(z) \, dz. \tag{23.5}$$

This is analogous to the introduction in chapters 16 and 17 of the notation $\mathcal{L}$ to describe a linear differential operator. Furthermore, we may define the Hermitian conjugate $\mathcal{K}^\dagger$ by

$$\mathcal{K}^\dagger y = \int_a^b K^*(z, x)y(z) \, dz,$$

where the asterisk denotes complex conjugation and we have reversed the order of the arguments in the kernel.

It is clear from (23.5) that $\mathcal{K}$ is indeed linear. Moreover, since $\mathcal{K}$ operates on the infinite-dimensional space of (reasonable) functions, we may make an obvious analogy with matrix equations and consider the action of $\mathcal{K}$ on a function f as that of a matrix on a column vector (both of infinite dimension).

When written in operator form, the integral equations discussed in the previous section resemble equations familiar from linear algebra. For example, the inhomogeneous Fredholm equation of the first kind may be written as

$$0 = f + \lambda \mathcal{K} y,$$

which has the unique solution $y = -\mathcal{K}^{-1} f / \lambda$, provided that $f \neq 0$ and the inverse operator $\mathcal{K}^{-1}$ exists.

Similarly, we may write the corresponding Fredholm equation of the second kind as

$$y = f + \lambda \mathcal{K} y. \tag{23.6}$$

In the homogeneous case, where $f = 0$, this reduces to $y = \lambda \mathcal{K} y$, which is reminiscent of an eigenvalue problem in linear algebra (except that λ appears on the other side of the equation) and, similarly, only has solutions for at most a countably infinite set of *eigenvalues* λ_i. The corresponding solutions y_i are called the eigenfunctions.

In the inhomogeneous case ($f \neq 0$), the solution to (23.6) can be written symbolically as

$$y = (1 - \lambda \mathcal{K})^{-1} f,$$

again provided that the inverse operator exists. It may be shown that, in general, (23.6) does possess a unique solution if $\lambda \neq \lambda_i$, i.e. when λ does not equal one of the eigenvalues of the corresponding homogeneous equation.

When λ does equal one of these eigenvalues, (23.6) may have either many solutions or no solution at all, depending on the form of f. If the function f is orthogonal to *every* eigenfunction of the equation

$$g = \lambda^* \mathcal{K}^\dagger g \tag{23.7}$$

that belongs to the eigenvalue λ^*, i.e.

$$\langle g | f \rangle = \int_a^b g^*(x) f(x) \, dx = 0$$

for every function g obeying (23.7), then it can be shown that (23.6) has many solutions. Otherwise the equation has no solution. These statements are discussed further in section 23.7, for the special case of integral equations with Hermitian kernels, i.e. those for which $\mathcal{K} = \mathcal{K}^\dagger$.

23.4 Closed-form solutions

In certain very special cases, it may be possible to obtain a closed-form solution of an integral equation. The reader should realise, however, when faced with an integral equation, that in general it will not be soluble by the simple methods presented in this section but must instead be solved using (numerical) iterative methods, such as those outlined in section 23.5.

23.4.1 Separable kernels

The most straightforward integral equations to solve are Fredholm equations with *separable* (or *degenerate*) kernels. A kernel is separable if it has the form

$$K(x,z) = \sum_{i=1}^{n} \phi_i(x)\psi_i(z), \qquad (23.8)$$

where $\phi_i(x)$ are $\psi_i(z)$ are respectively functions of x only and of z only and the number of terms in the sum, n, is finite.

Let us consider the solution of the (inhomogeneous) Fredholm equation of the second kind,

$$y(x) = f(x) + \lambda \int_a^b K(x,z)y(z)\,dz, \qquad (23.9)$$

which has a separable kernel of the form (23.8). Writing the kernel in its separated form, the functions $\phi_i(x)$ may be taken outside the integral over z to obtain

$$y(x) = f(x) + \lambda \sum_{i=1}^{n} \phi_i(x) \int_a^b \psi_i(z)y(z)\,dz.$$

Since the integration limits a and b are constant for a Fredholm equation, the integral over z in each term of the sum is just a constant. Denoting these constants by

$$c_i = \int_a^b \psi_i(z)y(z)\,dz, \qquad (23.10)$$

the solution to (23.9) is found to be

$$y(x) = f(x) + \lambda \sum_{i=1}^{n} c_i \phi_i(x), \qquad (23.11)$$

where the constants c_i can be evaluated by substituting (23.11) into (23.10).

> ►*Solve the integral equation*
>
> $$y(x) = x + \lambda \int_0^1 (xz + z^2)y(z)\,dz. \qquad (23.12)$$

The kernel for this equation is $K(x,z) = xz + z^2$, which is clearly separable, and using the notation in (23.8) we have $\phi_1(x) = x$, $\phi_2(x) = 1$, $\psi_1(z) = z$ and $\psi_2(z) = z^2$. From (23.11) the solution to (23.12) has the form

$$y(x) = x + \lambda(c_1 x + c_2),$$

where the constants c_1 and c_2 are given by (23.10) as

$$c_1 = \int_0^1 z[z + \lambda(c_1 z + c_2)]\,dz = \tfrac{1}{3} + \tfrac{1}{3}\lambda c_1 + \tfrac{1}{2}\lambda c_2,$$

$$c_2 = \int_0^1 z^2[z + \lambda(c_1 z + c_2)]\,dz = \tfrac{1}{4} + \tfrac{1}{4}\lambda c_1 + \tfrac{1}{3}\lambda c_2.$$

These two simultaneous linear equations may be straightforwardly solved for c_1 and c_2 to give

$$c_1 = \frac{24 + \lambda}{72 - 48\lambda - \lambda^2} \quad \text{and} \quad c_2 = \frac{18}{72 - 48\lambda - \lambda^2},$$

so that the solution to (23.12) is

$$y(x) = \frac{(72 - 24\lambda)x + 18\lambda}{72 - 48\lambda - \lambda^2}. \blacktriangleleft$$

In the above example, we see that (23.12) has a (finite) unique solution provided that λ is not equal to either root of the quadratic in the denominator of $y(x)$. The roots of this quadratic are in fact the *eigenvalues* of the corresponding homogeneous equation, as mentioned in the previous section. In general, if the separable kernel contains n terms, as in (23.8), there will be n such eigenvalues, although they may not all be different.

Kernels consisting of trigonometric (or hyperbolic) functions of sums or differences of x and z are also often separable.

▶ *Find the eigenvalues and corresponding eigenfunctions of the homogeneous Fredholm equation*

$$y(x) = \lambda \int_0^\pi \sin(x + z)\, y(z)\, dz. \tag{23.13}$$

The kernel of this integral equation can be written in separated form as

$$K(x, z) = \sin(x + z) = \sin x \cos z + \cos x \sin z,$$

so, comparing with (23.8), we have $\phi_1(x) = \sin x$, $\phi_2(x) = \cos x$, $\psi_1(z) = \cos z$ and $\psi_2(z) = \sin z$.

Thus, from (23.11), the solution to (23.13) has the form

$$y(x) = \lambda(c_1 \sin x + c_2 \cos x),$$

where the constants c_1 and c_2 are given by

$$c_1 = \lambda \int_0^\pi \cos z\, (c_1 \sin z + c_2 \cos z)\, dz = \frac{\lambda \pi}{2} c_2, \tag{23.14}$$

$$c_2 = \lambda \int_0^\pi \sin z\, (c_1 \sin z + c_2 \cos z)\, dz = \frac{\lambda \pi}{2} c_1. \tag{23.15}$$

Combining these two equations we find $c_1 = (\lambda \pi / 2)^2 c_1$, and, assuming that $c_1 \neq 0$, this gives $\lambda = \pm 2/\pi$, the two eigenvalues of the integral equation (23.13).

By substituting each of the eigenvalues back into (23.14) and (23.15), we find that the eigenfunctions corresponding to the eigenvalues $\lambda_1 = 2/\pi$ and $\lambda_2 = -2/\pi$ are given respectively by

$$y_1(x) = A(\sin x + \cos x) \quad \text{and} \quad y_2(x) = B(\sin x - \cos x), \tag{23.16}$$

where A and B are arbitrary constants. ◀

23.4.2 Integral transform methods

If the kernel of an integral equation can be written as a function of the difference $x - z$ of its two arguments, then it is called a *displacement* kernel. An integral equation having such a kernel, and which also has the integration limits $-\infty$ to ∞, may be solved by the use of Fourier transforms (chapter 13).

If we consider the following integral equation with a displacement kernel,

$$y(x) = f(x) + \lambda \int_{-\infty}^{\infty} K(x - z)y(z)\, dz, \qquad (23.17)$$

the integral over z clearly takes the form of a convolution (see chapter 13). Therefore, Fourier-transforming (23.17) and using the convolution theorem, we obtain

$$\tilde{y}(k) = \tilde{f}(k) + \sqrt{2\pi}\lambda\tilde{K}(k)\tilde{y}(k),$$

which may be rearranged to give

$$\tilde{y}(k) = \frac{\tilde{f}(k)}{1 - \sqrt{2\pi}\lambda\tilde{K}(k)}. \qquad (23.18)$$

Taking the inverse Fourier transform, the solution to (23.17) is given by

$$y(x) = \frac{1}{\sqrt{2\pi}} \int_{-\infty}^{\infty} \frac{\tilde{f}(k)\exp(ikx)}{1 - \sqrt{2\pi}\lambda\tilde{K}(k)}\, dk.$$

If we can perform this inverse Fourier transformation then the solution can be found explicitly; otherwise it must be left in the form of an integral.

▶ *Find the Fourier transform of the function*

$$g(x) = \begin{cases} 1 & \textit{if } |x| \le a, \\ 0 & \textit{if } |x| > a. \end{cases}$$

Hence find an explicit expression for the solution of the integral equation

$$y(x) = f(x) + \lambda \int_{-\infty}^{\infty} \frac{\sin(x - z)}{x - z}y(z)\, dz. \qquad (23.19)$$

Find the solution for the special case $f(x) = (\sin x)/x$.

The Fourier transform of $g(x)$ is given directly by

$$\tilde{g}(k) = \frac{1}{\sqrt{2\pi}} \int_{-a}^{a} \exp(-ikx)\, dx = \left[\frac{1}{\sqrt{2\pi}} \frac{\exp(-ikx)}{(-ik)} \right]_{-a}^{a} = \sqrt{\frac{2}{\pi}} \frac{\sin ka}{k}. \qquad (23.20)$$

The kernel of the integral equation (23.19) is $K(x - z) = [\sin(x - z)]/(x - z)$. Using (23.20), it is straightforward to show that the Fourier transform of the kernel is

$$\tilde{K}(k) = \begin{cases} \sqrt{\pi/2} & \text{if } |k| \le 1, \\ 0 & \text{if } |k| > 1. \end{cases} \qquad (23.21)$$

Thus, using (23.18), we find the Fourier transform of the solution to be

$$\tilde{y}(k) = \begin{cases} \tilde{f}(k)/(1 - \pi\lambda) & \text{if } |k| \leq 1, \\ \tilde{f}(k) & \text{if } |k| > 1. \end{cases} \tag{23.22}$$

Inverse Fourier-transforming, and writing the result in a slightly more convenient form, the solution to (23.19) is given by

$$y(x) = f(x) + \left(\frac{1}{1 - \pi\lambda} - 1\right) \frac{1}{\sqrt{2\pi}} \int_{-1}^{1} \tilde{f}(k) \exp(ikx) \, dk$$

$$= f(x) + \frac{\pi\lambda}{1 - \pi\lambda} \frac{1}{\sqrt{2\pi}} \int_{-1}^{1} \tilde{f}(k) \exp(ikx) \, dk. \tag{23.23}$$

It is clear from (23.22) that when $\lambda = 1/\pi$, which is the only eigenvalue of the corresponding homogeneous equation to (23.19), the solution becomes infinite, as we would expect.

For the special case $f(x) = (\sin x)/x$, the Fourier transform $\tilde{f}(k)$ is identical to that in (23.21), and the solution (23.23) becomes

$$y(x) = \frac{\sin x}{x} + \left(\frac{\pi\lambda}{1 - \pi\lambda}\right) \frac{1}{\sqrt{2\pi}} \int_{-1}^{1} \sqrt{\frac{\pi}{2}} \exp(ikx) \, dk$$

$$= \frac{\sin x}{x} + \left(\frac{\pi\lambda}{1 - \pi\lambda}\right) \frac{1}{2} \left[\frac{\exp(ikx)}{ix}\right]_{k=-1}^{k=1}$$

$$= \frac{\sin x}{x} + \left(\frac{\pi\lambda}{1 - \pi\lambda}\right) \frac{\sin x}{x} = \left(\frac{1}{1 - \pi\lambda}\right) \frac{\sin x}{x}. \blacktriangleleft$$

If, instead, the integral equation (23.17) had integration limits 0 and x (so making it a Volterra equation) then its solution could be found, in a similar way, by using the convolution theorem for Laplace transforms (see chapter 13). We would find

$$\bar{y}(s) = \frac{\bar{f}(s)}{1 - \lambda\bar{K}(s)},$$

where s is the Laplace transform variable. Often one may use the dictionary of Laplace transforms given in table 13.1 to invert this equation and find the solution $y(x)$. In general, however, the evaluation of inverse Laplace transform integrals is difficult, since (in principle) it requires a contour integration; see chapter 24.

As a final example of the use of Fourier transforms in solving integral equations, we mention equations that have integration limits $-\infty$ and ∞ and a kernel of the form

$$K(x, z) = \exp(-ixz).$$

Consider, for example, the inhomogeneous Fredholm equation

$$y(x) = f(x) + \lambda \int_{-\infty}^{\infty} \exp(-ixz) \, y(z) \, dz. \tag{23.24}$$

The integral over z is clearly just (a multiple of) the Fourier transform of $y(z)$, so we can write

$$y(x) = f(x) + \sqrt{2\pi}\lambda\tilde{y}(x). \tag{23.25}$$

If we now take the Fourier transform of (23.25) but continue to denote the independent variable by x (i.e. rather than k, for example), we obtain

$$\tilde{y}(x) = \tilde{f}(x) + \sqrt{2\pi}\lambda y(-x). \tag{23.26}$$

Substituting (23.26) into (23.25) we find

$$y(x) = f(x) + \sqrt{2\pi}\lambda\left[\tilde{f}(x) + \sqrt{2\pi}\lambda y(-x)\right],$$

but on making the change $x \to -x$ and substituting back in for $y(-x)$, this gives

$$y(x) = f(x) + \sqrt{2\pi}\lambda\tilde{f}(x) + 2\pi\lambda^2\left[f(-x) + \sqrt{2\pi}\lambda\tilde{f}(-x) + 2\pi\lambda^2 y(x)\right].$$

Thus the solution to (23.24) is given by

$$y(x) = \frac{1}{1 - (2\pi)^2\lambda^4}\left[f(x) + (2\pi)^{1/2}\lambda\tilde{f}(x) + 2\pi\lambda^2 f(-x) + (2\pi)^{3/2}\lambda^3\tilde{f}(-x)\right]. \tag{23.27}$$

Clearly, (23.24) possesses a unique solution provided $\lambda \neq \pm 1/\sqrt{2\pi}$ or $\pm i/\sqrt{2\pi}$; these are easily shown to be the eigenvalues of the corresponding homogeneous equation (for which $f(x) \equiv 0$).

▶Solve the integral equation

$$y(x) = \exp\left(-\frac{x^2}{2}\right) + \lambda\int_{-\infty}^{\infty}\exp(-ixz)\,y(z)\,dz, \tag{23.28}$$

where λ is a real constant. Show that the solution is unique unless λ has one of two particular values. Does a solution exist for either of these two values of λ?

Following the argument given above, the solution to (23.28) is given by (23.27) with $f(x) = \exp(-x^2/2)$. In order to write the solution explicitly, however, we must calculate the Fourier transform of $f(x)$. Using equation (13.7), we find $\tilde{f}(k) = \exp(-k^2/2)$, from which we note that $f(x)$ has the special property that its functional form is identical to that of its Fourier transform. Thus, the solution to (23.28) is given by

$$y(x) = \frac{1}{1 - (2\pi)^2\lambda^4}\left[1 + (2\pi)^{1/2}\lambda + 2\pi\lambda^2 + (2\pi)^{3/2}\lambda^3\right]\exp\left(-\frac{x^2}{2}\right). \tag{23.29}$$

Since λ is restricted to be real, the solution to (23.28) will be unique unless $\lambda = \pm 1/\sqrt{2\pi}$, at which points (23.29) becomes infinite. In order to find whether solutions exist for either of these values of λ we must return to equations (23.25) and (23.26).

Let us first consider the case $\lambda = +1/\sqrt{2\pi}$. Putting this value into (23.25) and (23.26), we obtain

$$y(x) = f(x) + \tilde{y}(x), \tag{23.30}$$
$$\tilde{y}(x) = \tilde{f}(x) + y(-x). \tag{23.31}$$

Substituting (23.31) into (23.30) we find

$$y(x) = f(x) + \tilde{f}(x) + y(-x),$$

but on changing x to $-x$ and substituting back in for $y(-x)$, this gives

$$y(x) = f(x) + \tilde{f}(x) + f(-x) + \tilde{f}(-x) + y(x).$$

Thus, in order for a solution to exist, we require that the function $f(x)$ obeys

$$f(x) + \tilde{f}(x) + f(-x) + \tilde{f}(-x) = 0.$$

This is satisfied if $f(x) = -\tilde{f}(x)$, i.e. if the functional form of $f(x)$ is minus the form of its Fourier transform. We may repeat this analysis for the case $\lambda = -1/\sqrt{2\pi}$, and, in a similar way, we find that this time we require $f(x) = \tilde{f}(x)$.

In our case $f(x) = \exp(-x^2/2)$, for which, as we mentioned above, $f(x) = \tilde{f}(x)$. Therefore, (23.28) possesses no solution when $\lambda = +1/\sqrt{2\pi}$ but has many solutions when $\lambda = -1/\sqrt{2\pi}$. ◄

A similar approach to the above may be taken to solve equations with kernels of the form $K(x, y) = \cos xy$ or $\sin xy$, either by considering the integral over y in each case as the real or imaginary part of the corresponding Fourier transform or by using Fourier cosine or sine transforms directly.

23.4.3 Differentiation

A closed-form solution to a Volterra equation may sometimes be obtained by differentiating the equation to obtain the corresponding differential equation, which may be easier to solve.

▶Solve the integral equation

$$y(x) = x - \int_0^x xz^2 y(z)\,dz. \tag{23.32}$$

Dividing through by x, we obtain

$$\frac{y(x)}{x} = 1 - \int_0^x z^2 y(z)\,dz,$$

which may be differentiated with respect to x to give

$$\frac{d}{dx}\left[\frac{y(x)}{x}\right] = -x^2 y(x) = -x^3 \left[\frac{y(x)}{x}\right].$$

This equation may be integrated straightforwardly, and we find

$$\ln\left[\frac{y(x)}{x}\right] = -\frac{x^4}{4} + c,$$

where c is a constant of integration. Thus the solution to (23.32) has the form

$$y(x) = Ax \exp\left(-\frac{x^4}{4}\right), \tag{23.33}$$

where A is an arbitrary constant.

Since the original integral equation (23.32) contains no arbitrary constants, neither should its solution. We may calculate the value of the constant, A, by substituting the solution (23.33) back into (23.32), from which we find $A = 1$. ◄

23.5 Neumann series

As mentioned above, most integral equations met in practice will not be of the simple forms discussed in the last section and so, in general, it is not possible to find closed-form solutions. In such cases, we might try to obtain a solution in the form of an infinite series, as we did for differential equations (see chapter 16).

Let us consider the equation

$$y(x) = f(x) + \lambda \int_a^b K(x,z)y(z)\,dz, \qquad (23.34)$$

where either both integration limits are constants (for a Fredholm equation) or the upper limit is variable (for a Volterra equation). Clearly, if λ were small then a crude (but reasonable) approximation to the solution would be

$$y(x) \approx y_0(x) = f(x),$$

where $y_0(x)$ stands for our 'zeroth-order' approximation to the solution (and is not to be confused with an eigenfunction).

Substituting this crude guess under the integral sign in the original equation, we obtain what should be a better approximation:

$$y_1(x) = f(x) + \lambda \int_a^b K(x,z)y_0(z)\,dz = f(x) + \lambda \int_a^b K(x,z)f(z)\,dz,$$

which is first order in λ. Repeating the procedure once more results in the second-order approximation

$$y_2(x) = f(x) + \lambda \int_a^b K(x,z)y_1(z)\,dz$$

$$= f(x) + \lambda \int_a^b K(x,z_1)f(z_1)\,dz_1 + \lambda^2 \int_a^b dz_1 \int_a^b K(x,z_1)K(z_1,z_2)f(z_2)\,dz_2.$$

It is clear that we may continue this process to obtain progressively higher-order approximations to the solution. Introducing the functions

$$K_1(x,z) = K(x,z),$$

$$K_2(x,z) = \int_a^b K(x,z_1)K(z_1,z)\,dz_1,$$

$$K_3(x,z) = \int_a^b dz_1 \int_a^b K(x,z_1)K(z_1,z_2)K(z_2,z)\,dz_2,$$

and so on, which obey the recurrence relation

$$K_n(x,z) = \int_a^b K(x,z_1)K_{n-1}(z_1,z)\,dz_1,$$

813

we may write the nth-order approximation as

$$y_n(x) = f(x) + \sum_{m=1}^{n} \lambda^m \int_a^b K_m(x, z) f(z) \, dz. \tag{23.35}$$

The solution to the original integral equation is then given by $y(x) = \lim_{n \to \infty} y_n(x)$, *provided the infinite series converges.* Using (23.35), this solution may be written as

$$y(x) = f(x) + \lambda \int_a^b R(x, z; \lambda) f(z) \, dz, \tag{23.36}$$

where the *resolvent kernel* $R(x, z; \lambda)$ is given by

$$R(x, z; \lambda) = \sum_{m=0}^{\infty} \lambda^m K_{m+1}(x, z). \tag{23.37}$$

Clearly, the resolvent kernel, and hence the series solution, will converge provided λ is sufficiently small. In fact, it may be shown that the series converges in some domain of $|\lambda|$ provided the original kernel $K(x, z)$ is bounded in such a way that

$$|\lambda|^2 \int_a^b dx \int_a^b |K(x, z)|^2 \, dz < 1. \tag{23.38}$$

▶ *Use the Neumann series method to solve the integral equation*

$$y(x) = x + \lambda \int_0^1 xz y(z) \, dz. \tag{23.39}$$

Following the method outlined above, we begin with the crude approximation $y(x) \approx y_0(x) = x$. Substituting this under the integral sign in (23.39), we obtain the next approximation

$$y_1(x) = x + \lambda \int_0^1 xz y_0(z) \, dz = x + \lambda \int_0^1 xz^2 dz = x + \frac{\lambda x}{3},$$

Repeating the procedure once more, we obtain

$$y_2(x) = x + \lambda \int_0^1 xz y_1(z) \, dz$$

$$= x + \lambda \int_0^1 xz \left(z + \frac{\lambda z}{3} \right) dz = x + \left(\frac{\lambda}{3} + \frac{\lambda^2}{9} \right) x.$$

For this simple example, it is easy to see that by continuing this process the solution to (23.39) is obtained as

$$y(x) = x + \left[\frac{\lambda}{3} + \left(\frac{\lambda}{3} \right)^2 + \left(\frac{\lambda}{3} \right)^3 + \cdots \right] x.$$

Clearly the expression in brackets is an infinite geometric series with first term $\lambda/3$ and common ratio $\lambda/3$. Thus, *provided* $|\lambda| < 3$, this infinite series converges to the value $\lambda/(3 - \lambda)$, and the solution to (23.39) is

$$y(x) = x + \frac{\lambda x}{3 - \lambda} = \frac{3x}{3 - \lambda}. \tag{23.40}$$

Finally, we note that the requirement that $|\lambda| < 3$ may also be derived very easily from the condition (23.38). ◄

23.6 Fredholm theory

In the previous section, we found that a solution to the integral equation (23.34) can be obtained as a Neumann series of the form (23.36), where the resolvent kernel $R(x, z; \lambda)$ is written as an infinite power series in λ. This solution is valid provided the infinite series converges.

A related, but more elegant, approach to the solution of integral equations using infinite series was found by Fredholm. We will not reproduce Fredholm's analysis here, but merely state the results we need. Essentially, *Fredholm theory* provides a formula for the resolvent kernel $R(x, z; \lambda)$ in (23.36) in terms of the ratio of two infinite series:

$$R(x, z; \lambda) = \frac{D(x, z; \lambda)}{d(\lambda)}. \tag{23.41}$$

The numerator and denominator in (23.41) are given by

$$D(x, z; \lambda) = \sum_{n=0}^{\infty} \frac{(-1)^n}{n!} D_n(x, z) \lambda^n, \tag{23.42}$$

$$d(\lambda) = \sum_{n=0}^{\infty} \frac{(-1)^n}{n!} d_n \lambda^n, \tag{23.43}$$

where the functions $D_n(x, z)$ and the constants d_n are found from recurrence relations as follows. We start with

$$D_0(x, z) = K(x, z) \qquad \text{and} \qquad d_0 = 1, \tag{23.44}$$

where $K(x, z)$ is the kernel of the original integral equation (23.34). The higher-order coefficients of λ in (23.43) and (23.42) are then obtained from the two recurrence relations

$$d_n = \int_a^b D_{n-1}(x, x)\, dx, \tag{23.45}$$

$$D_n(x, z) = K(x, z)d_n - n \int_a^b K(x, z_1)D_{n-1}(z_1, z)\, dz_1. \tag{23.46}$$

Although the formulae for the resolvent kernel appear complicated, they are often simple to apply. Moreover, for the Fredholm solution the power series (23.42) and (23.43) are both guaranteed to converge for all values of λ, unlike

Neumann series, which converge only if the condition (23.38) is satisfied. Thus the Fredholm method leads to a unique, non-singular solution, provided that $d(\lambda) \neq 0$. In fact, as we might suspect, the solutions of $d(\lambda) = 0$ give the eigenvalues of the homogeneous equation corresponding to (23.34), i.e. with $f(x) \equiv 0$.

▶ *Use Fredholm theory to solve the integral equation (23.39).*

Using (23.36) and (23.41), the solution to (23.39) can be written in the form

$$y(x) = x + \lambda \int_0^1 R(x, z; \lambda) z \, dz = x + \lambda \int_0^1 \frac{D(x, z; \lambda)}{d(\lambda)} z \, dz. \tag{23.47}$$

In order to find the form of the resolvent kernel $R(x, z; \lambda)$, we begin by setting

$$D_0(x, z) = K(x, z) = xz \qquad \text{and} \qquad d_0 = 1$$

and use the recurrence relations (23.45) and (23.46) to obtain

$$d_1 = \int_0^1 D_0(x, x) \, dx = \int_0^1 x^2 \, dx = \frac{1}{3},$$

$$D_1(x, z) = \frac{xz}{3} - \int_0^1 xz_1^2 z \, dz_1 = \frac{xz}{3} - xz \left[\frac{z_1^3}{3} \right]_0^1 = 0.$$

Applying the recurrence relations again we find that $d_n = 0$ and $D_n(x, z) = 0$ for $n > 1$. Thus, from (23.42) and (23.43), the numerator and denominator of the resolvent respectively are given by

$$D(x, z; \lambda) = xz \qquad \text{and} \qquad d(\lambda) = 1 - \frac{\lambda}{3}.$$

Substituting these expressions into (23.47), we find that the solution to (23.39) is given by

$$y(x) = x + \lambda \int_0^1 \frac{xz^2}{1 - \lambda/3} \, dz$$

$$= x + \lambda \left[\frac{x}{1 - \lambda/3} \frac{z^3}{3} \right]_0^1 = x + \frac{\lambda x}{3 - \lambda} = \frac{3x}{3 - \lambda},$$

which, as expected, is the same as the solution (23.40) found by constructing a Neumann series. ◀

23.7 Schmidt–Hilbert theory

The Schmidt–Hilbert (SH) theory of integral equations may be considered as analogous to the Sturm–Liouville (SL) theory of differential equations, discussed in chapter 17, and is concerned with the properties of integral equations with *Hermitian* kernels. An Hermitian kernel enjoys the property

$$K(x, z) = K^*(z, x), \tag{23.48}$$

and it is clear that a special case of (23.48) occurs for a real kernel that is also symmetric with respect to its two arguments.

Let us begin by considering the homogeneous integral equation

$$y = \lambda \mathcal{K} y,$$

where the integral operator $\mathcal{K}$ has an Hermitian kernel. As discussed in section 23.3, in general this equation will have solutions only for $\lambda = \lambda_i$, where the λ_i are the eigenvalues of the integral equation, the corresponding solutions y_i being the eigenfunctions of the equation.

By following similar arguments to those presented in chapter 17 for SL theory, it may be shown that the eigenvalues λ_i of an Hermitian kernel are real and that the corresponding eigenfunctions y_i belonging to different eigenvalues are orthogonal and form a complete set. If the eigenfunctions are suitably normalised, we have

$$\langle y_i | y_j \rangle = \int_a^b y_i^*(x) y_j(x)\, dx = \delta_{ij}. \tag{23.49}$$

If an eigenvalue is degenerate then the eigenfunctions corresponding to that eigenvalue can be made orthogonal by the Gram–Schmidt procedure, in a similar way to that discussed in chapter 17 in the context of SL theory.

Like SL theory, SH theory does not provide a method of obtaining the eigenvalues and eigenfunctions of any particular homogeneous integral equation with an Hermitian kernel; for this we have to turn to the methods discussed in the previous sections of this chapter. Rather, SH theory is concerned with the general properties of the solutions to such equations. Where SH theory becomes applicable, however, is in the solution of inhomogeneous integral equations with Hermitian kernels for which the eigenvalues and eigenfunctions of the corresponding homogeneous equation are already known.

Let us consider the inhomogeneous equation

$$y = f + \lambda \mathcal{K} y, \tag{23.50}$$

where $\mathcal{K} = \mathcal{K}^\dagger$ and for which we know the eigenvalues λ_i and normalised eigenfunctions y_i of the corresponding homogeneous problem. The function f may or may not be expressible solely in terms of the eigenfunctions y_i, and to accommodate this situation we write the unknown solution y as $y = f + \sum_i a_i y_i$, where the a_i are expansion coefficients to be determined.

Substituting this into (23.50), we obtain

$$f + \sum_i a_i y_i = f + \lambda \sum_i \frac{a_i y_i}{\lambda_i} + \lambda \mathcal{K} f, \tag{23.51}$$

where we have used the fact that $y_i = \lambda_i \mathcal{K} y_i$. Forming the inner product of both sides of (23.51) with y_j, we find

$$\sum_i a_i \langle y_j | y_i \rangle = \lambda \sum_i \frac{a_i}{\lambda_i} \langle y_j | y_i \rangle + \lambda \langle y_j | \mathcal{K} f \rangle. \tag{23.52}$$

Since the eigenfunctions are orthonormal and $\mathcal{K}$ is an Hermitian operator, we have that both $\langle y_j | y_i \rangle = \delta_{ij}$ and $\langle y_j | \mathcal{K} f \rangle = \langle \mathcal{K} y_j | f \rangle = \lambda_j^{-1} \langle y_j | f \rangle$. Thus the coefficients a_j are given by

$$a_j = \frac{\lambda \lambda_j^{-1} \langle y_j | f \rangle}{1 - \lambda \lambda_j^{-1}} = \frac{\lambda \langle y_j | f \rangle}{\lambda_j - \lambda}, \tag{23.53}$$

and the solution is

$$y = f + \sum_i a_i y_i = f + \lambda \sum_i \frac{\langle y_i | f \rangle}{\lambda_i - \lambda} y_i. \tag{23.54}$$

This also shows, incidentally, that a formal representation for the resolvent kernel is

$$R(x, z; \lambda) = \sum_i \frac{y_i(x) y_i^*(z)}{\lambda_i - \lambda}. \tag{23.55}$$

If f *can* be expressed as a linear superposition of the y_i, i.e. $f = \sum_i b_i y_i$, then $b_i = \langle y_i | f \rangle$ and the solution can be written more briefly as

$$y = \sum_i \frac{b_i}{1 - \lambda \lambda_i^{-1}} y_i. \tag{23.56}$$

We see from (23.54) that the inhomogeneous equation (23.50) has a unique solution provided $\lambda \neq \lambda_i$, i.e. when λ is not equal to one of the eigenvalues of the corresponding homogeneous equation. However, if λ does equal one of the eigenvalues λ_j then, in general, the coefficients a_j become singular and no (finite) solution exists.

Returning to (23.53), we notice that even if $\lambda = \lambda_j$ a non-singular solution to the integral equation is still possible provided that the function f is orthogonal to every eigenfunction corresponding to the eigenvalue λ_j, i.e.

$$\langle y_j | f \rangle = \int_a^b y_j^*(x) f(x) \, dx = 0.$$

The following worked example illustrates the case in which f can be expressed in terms of the y_i. One in which it cannot is considered in exercise 23.14.

> ▶ *Use Schmidt–Hilbert theory to solve the integral equation*
>
> $$y(x) = \sin(x + \alpha) + \lambda \int_0^\pi \sin(x + z) y(z) \, dz. \tag{23.57}$$

It is clear that the kernel $K(x, z) = \sin(x + z)$ is real and symmetric in x and z and is thus Hermitian. In order to solve this inhomogeneous equation using SH theory, however, we must first find the eigenvalues and eigenfunctions of the corresponding homogeneous equation.

In fact, we have considered the solution of the corresponding homogeneous equation (23.13) already, in subsection 23.4.1, where we found that it has two eigenvalues $\lambda_1 = 2/\pi$ and $\lambda_2 = -2/\pi$, with eigenfunctions given by (23.16). The normalised eigenfunctions are

$$y_1(x) = \frac{1}{\sqrt{\pi}}(\sin x + \cos x) \quad \text{and} \quad y_2(x) = \frac{1}{\sqrt{\pi}}(\sin x - \cos x) \tag{23.58}$$

and are easily shown to obey the orthonormality condition (23.49).

Using (23.54), the solution to the inhomogeneous equation (23.57) has the form

$$y(x) = a_1 y_1(x) + a_2 y_2(x), \tag{23.59}$$

where the coefficients a_1 and a_2 are given by (23.53) with $f(x) = \sin(x + \alpha)$. Therefore, using (23.58),

$$a_1 = \frac{1}{1 - \pi\lambda/2} \int_0^\pi \frac{1}{\sqrt{\pi}}(\sin z + \cos z)\sin(z + \alpha)\,dz = \frac{\sqrt{\pi}}{2 - \pi\lambda}(\cos\alpha + \sin\alpha),$$

$$a_2 = \frac{1}{1 + \pi\lambda/2} \int_0^\pi \frac{1}{\sqrt{\pi}}(\sin z - \cos z)\sin(z + \alpha)\,dz = \frac{\sqrt{\pi}}{2 + \pi\lambda}(\cos\alpha - \sin\alpha).$$

Substituting these expressions for a_1 and a_2 into (23.59) and simplifying, we find that the solution to (23.57) is given by

$$y(x) = \frac{1}{1 - (\pi\lambda/2)^2}\left[\sin(x + \alpha) + (\pi\lambda/2)\cos(x - \alpha)\right]. \blacktriangleleft$$

23.8 Exercises

23.1 Solve the integral equation

$$\int_0^\infty \cos(xv)y(v)\,dv = \exp(-x^2/2)$$

for the function $y = y(x)$ for $x > 0$. Note that for $x < 0$, $y(x)$ can be chosen as is most convenient.

23.2 Solve

$$\int_0^\infty f(t)\exp(-st)\,dt = \frac{a}{a^2 + s^2}.$$

23.3 Convert

$$f(x) = \exp x + \int_0^x (x - y)f(y)\,dy$$

into a differential equation, and hence show that its solution is

$$(\alpha + \beta x)\exp x + \gamma \exp(-x),$$

where α, β and γ are constants that should be determined.

23.4 Use the fact that its kernel is separable, to solve for $y(x)$ the integral equation

$$y(x) = A\cos(x + a) + \lambda \int_0^\pi \sin(x + z)y(z)\,dz.$$

[This equation is an inhomogeneous extension of the homogeneous Fredholm equation (23.13), and is similar to equation (23.57).]

23.5 Solve for $\phi(x)$ the integral equation

$$\phi(x) = f(x) + \lambda \int_0^1 \left[\left(\frac{x}{y} \right)^n + \left(\frac{y}{x} \right)^n \right] \phi(y)\, dy,$$

where $f(x)$ is bounded for $0 < x < 1$ and $-\frac{1}{2} < n < \frac{1}{2}$, expressing your answer in terms of the quantities $F_m = \int_0^1 f(y) y^m\, dy$.

(a) Give the explicit solution when $\lambda = 1$.
(b) For what values of λ are there no solutions unless $F_{\pm n}$ are in a particular ratio? What is this ratio?

23.6 Consider the inhomogeneous integral equation

$$f(x) = g(x) + \lambda \int_a^b K(x, y) f(y)\, dy,$$

for which the kernel $K(x, y)$ is real, symmetric and continuous in $a \le x \le b$, $a \le y \le b$.

(a) If λ is one of the eigenvalues λ_i of the homogeneous equation

$$f_i(x) = \lambda_i \int_a^b K(x, y) f_i(y)\, dy,$$

prove that the inhomogeneous equation can only a have non-trivial solution if $g(x)$ is orthogonal to the corresponding eigenfunction $f_i(x)$.

(b) Show that the only values of λ for which

$$f(x) = \lambda \int_0^1 xy(x + y) f(y)\, dy$$

has a non-trivial solution are the roots of the equation

$$\lambda^2 + 120\lambda - 240 = 0.$$

(c) Solve

$$f(x) = \mu x^2 + \int_0^1 2xy(x + y) f(y)\, dy.$$

23.7 The kernel of the integral equation

$$\psi(x) = \lambda \int_a^b K(x, y) \psi(y)\, dy$$

has the form

$$K(x, y) = \sum_{n=0}^{\infty} h_n(x) g_n(y),$$

where the $h_n(x)$ form a complete orthonormal set of functions over the interval $[a, b]$.

(a) Show that the eigenvalues λ_i are given by

$$|\mathsf{M} - \lambda^{-1}\mathsf{I}| = 0,$$

where M is the matrix with elements

$$M_{kj} = \int_a^b g_k(u) h_j(u)\, du.$$

If the corresponding solutions are $\psi^{(i)}(x) = \sum_{n=0}^{\infty} a_n^{(i)} h_n(x)$, find an expression for $a_n^{(i)}$.

(b) Obtain the eigenvalues and eigenfunctions over the interval $[0, 2\pi]$ if

$$K(x, y) = \sum_{n=1}^{\infty} \frac{1}{n} \cos nx \cos ny.$$

23.8   By taking its Laplace transform, and that of $x^n e^{-ax}$, obtain the explicit solution of

$$f(x) = e^{-x} \left[x + \int_0^x (x - u)e^u f(u)\, du \right].$$

Verify your answer by substitution.

23.9   For $f(t) = \exp(-t^2/2)$, use the relationships of the Fourier transforms of $f'(t)$ and $tf(t)$ to that of $f(t)$ itself to find a simple differential equation satisfied by $\tilde{f}(\omega)$, the Fourier transform of $f(t)$, and hence determine $\tilde{f}(\omega)$ to within a constant. Use this result to solve the integral equation

$$\int_{-\infty}^{\infty} e^{-t(t-2x)/2} h(t)\, dt = e^{3x^2/8}$$

for $h(t)$.

23.10   Show that the equation

$$f(x) = x^{-1/3} + \lambda \int_0^{\infty} f(y) \exp(-xy)\, dy$$

has a solution of the form $Ax^\alpha + Bx^\beta$. Determine the values of α and β, and show that those of A and B are

$$\frac{1}{1 - \lambda^2 \Gamma(\tfrac{1}{3}) \Gamma(\tfrac{2}{3})} \quad \text{and} \quad \frac{\lambda \Gamma(\tfrac{2}{3})}{1 - \lambda^2 \Gamma(\tfrac{1}{3}) \Gamma(\tfrac{2}{3})},$$

where $\Gamma(z)$ is the gamma function.

23.11   At an international 'peace' conference a large number of delegates are seated around a circular table with each delegation sitting near its allies and diametrically opposite the delegation most bitterly opposed to it. The position of a delegate is denoted by θ, with $0 \leq \theta \leq 2\pi$. The fury $f(\theta)$ felt by the delegate at θ is the sum of his own natural hostility $h(\theta)$ and the influences on him of each of the other delegates; a delegate at position ϕ contributes an amount $K(\theta - \phi)f(\phi)$. Thus

$$f(\theta) = h(\theta) + \int_0^{2\pi} K(\theta - \phi)f(\phi)\, d\phi.$$

Show that if $K(\psi)$ takes the form $K(\psi) = k_0 + k_1 \cos \psi$ then

$$f(\theta) = h(\theta) + p + q \cos \theta + r \sin \theta$$

and evaluate p, q and r. A positive value for k_1 implies that delegates tend to placate their opponents but upset their allies, whilst negative values imply that they calm their allies but infuriate their opponents. A walkout will occur if $f(\theta)$ exceeds a certain threshold value for some θ. Is this more likely to happen for positive or for negative values of k_1?

23.12   By considering functions of the form $h(x) = \int_0^x (x - y)f(y)\, dy$, show that the solution $f(x)$ of the integral equation

$$f(x) = x + \tfrac{1}{2} \int_0^1 |x - y| f(y)\, dy$$

satisfies the equation $f''(x) = f(x)$.

By examining the special cases $x = 0$ and $x = 1$, show that

$$f(x) = \frac{2}{(e+3)(e+1)}[(e+2)e^x - ee^{-x}].$$

23.13   The operator $\mathcal{M}$ is defined by

$$\mathcal{M}f(x) \equiv \int_{-\infty}^{\infty} K(x,y)f(y)\,dy,$$

where $K(x,y) = 1$ inside the square $|x| < a, |y| < a$ and $K(x,y) = 0$ elsewhere. Consider the possible eigenvalues of $\mathcal{M}$ and the eigenfunctions that correspond to them; show that the only possible eigenvalues are 0 and $2a$ and determine the corresponding eigenfunctions. Hence find the general solution of

$$f(x) = g(x) + \lambda \int_{-\infty}^{\infty} K(x,y)f(y)\,dy.$$

23.14   For the integral equation

$$y(x) = x^{-3} + \lambda \int_{a}^{b} x^2 z^2 y(z)\,dz,$$

show that the resolvent kernel is $5x^2z^2/[5 - \lambda(b^5 - a^5)]$ and hence solve the equation. For what range of λ is the solution valid?

23.15   Use Fredholm theory to show that, for the kernel

$$K(x,z) = (x+z)\exp(x-z)$$

over the interval $[0,1]$, the resolvent kernel is

$$R(x,z;\lambda) = \frac{\exp(x-z)[(x+z) - \lambda(\frac{1}{2}x + \frac{1}{2}z - xz - \frac{1}{3})]}{1 - \lambda - \frac{1}{12}\lambda^2},$$

and hence solve

$$y(x) = x^2 + 2\int_{0}^{1}(x+z)\exp(x-z)\,y(z)\,dz,$$

expressing your answer in terms of I_n, where $I_n = \int_{0}^{1} u^n \exp(-u)\,du$.

23.16   This exercise shows that following formal theory is not necessarily the best way to get practical results!

(a) Determine the eigenvalues $\lambda_{\pm}$ of the kernel $K(x,z) = (xz)^{1/2}(x^{1/2} + z^{1/2})$ and show that the corresponding eigenfunctions have the forms

$$y_{\pm}(x) = A_{\pm}(\sqrt{2}x^{1/2} \pm \sqrt{3}x),$$

where $A_{\pm}^2 = 5/(10 \pm 4\sqrt{6})$.

(b) Use Schmidt–Hilbert theory to solve

$$y(x) = 1 + \frac{5}{2}\int_{0}^{1} K(x,z)y(z)\,dz.$$

(c) As will have been apparent, the algebra involved in the formal method used in (b) is long and error-prone, and it is in fact much more straightforward to use a trial function $1 + \alpha x^{1/2} + \beta x$. Check your answer by doing so.

23.9 Hints and answers

23.1   Define $y(-x) = y(x)$ and use the cosine Fourier transform inversion theorem; $y(x) = (2/\pi)^{1/2} \exp(-x^2/2)$.

23.3 $f''(x) - f(x) = \exp x$; $\alpha = 3/4$, $\beta = 1/2$, $\gamma = 1/4$.

23.5   (a) $\phi(x) = f(x) - (1 + 2n)F_n x^n - (1 - 2n)F_{-n}x^{-n}$. (b) There are no solutions for $\lambda = [1 \pm (1 - 4n^2)^{-1/2}]^{-1}$ unless $F_{\pm n} = 0$ or $F_n/F_{-n} = \mp[(1 - 2n)/(1 + 2n)]^{1/2}$.

23.7   (a) $a_n^{(i)} = \int_a^b h_n(x)\psi^{(i)}(x)\,dx$; (b) use $(1/\sqrt{\pi})\cos nx$ and $(1/\sqrt{\pi})\sin nx$; M is diagonal; eigenvalues $\lambda_k = k/\pi$ with eigenfunctions $\psi^{(k)}(x) = (1/\sqrt{\pi})\cos kx$.

23.9 $d\tilde{f}/d\omega = -\omega\tilde{f}$, leading to $\tilde{f}(\omega) = Ae^{-\omega^2/2}$. Rearrange the integral as a convolution and deduce that $\tilde{h}(\omega) = Be^{-3\omega^2/2}$; $h(t) = Ce^{-t^2/6}$, where resubstitution and Gaussian normalisation show that $C = \sqrt{2/(3\pi)}$.

23.11 $p = k_0 H/(1 - 2\pi k_0)$, $q = k_1 H_c/(1 - \pi k_1)$ and $r = k_1 H_s/(1 - \pi k_1)$, where $H = \int_0^{2\pi} h(z)\,dz$, $H_c = \int_0^{2\pi} h(z)\cos z\,dz$, and $H_s = \int_0^{2\pi} h(z)\sin z\,dz$. Positive values of $k_1(\approx \pi^{-1})$ are most likely to cause a conference breakdown.

23.13   For eigenvalue 0 : $f(x) = 0$ for $|x| < a$ or $f(x)$ is such that $\int_{-a}^a f(y)\,dy = 0$. For eigenvalue $2a$: $f(x) = \mu S(x, a)$ with μ a constant and $S(x, a) \equiv [H(a + x) - H(x - a)]$, where $H(z)$ is the Heaviside step function. Take $f(x) = g(x) + cGS(x, a)$, where $G = \int_{-a}^a g(z)\,dz$. Show that $c = \lambda/(1 - 2a\lambda)$.

23.15 $y(x) = x^2 - (3I_3 x + I_2)\exp x$.

Complex variables

Throughout this book references have been made to results derived from the theory of complex variables. This theory thus becomes an integral part of the mathematics appropriate to physical applications. Indeed, so numerous and widespread are these applications that the whole of the next chapter is devoted to a systematic presentation of some of the more important ones. This current chapter develops the general theory on which these applications are based. The difficulty with it, from the point of view of a book such as the present one, is that the underlying basis has a distinctly pure mathematics flavour.

Thus, to adopt a comprehensive rigorous approach would involve a large amount of groundwork in analysis, for example formulating precise definitions of continuity and differentiability, developing the theory of sets and making a detailed study of boundedness. Instead, we will be selective and pursue only those parts of the formal theory that are needed to establish the results used in the next chapter and elsewhere in this book.

In this spirit, the proofs that have been adopted for some of the standard results of complex variable theory have been chosen with an eye to simplicity rather than sophistication. This means that in some cases the imposed conditions are more stringent than would be strictly necessary if more sophisticated proofs were used; where this happens the less restrictive results are usually stated as well. The reader who is interested in a fuller treatment should consult one of the many excellent textbooks on this fascinating subject.[§]

One further concession to 'hand-waving' has been made in the interests of keeping the treatment to a moderate length. In several places phrases such as 'can be made as small as we like' are used, rather than a careful treatment in terms of 'given $\epsilon > 0$, there exists a $\delta > 0$ such that'. In the authors' experience, some

---

[§] For example, K. Knopp, *Theory of Functions, Part I* (New York: Dover, 1945); E. G. Phillips, *Functions of a Complex Variable with Applications* 7th edn (Edinburgh: Oliver and Boyd, 1951); E. C. Titchmarsh, *The Theory of Functions* (Oxford: Oxford University Press, 1952).

students are more at ease with the former type of statement, despite its lack of precision, whilst others, those who would contemplate only the latter, are usually well able to supply it for themselves.

24.1 Functions of a complex variable

The quantity $f(z)$ is said to be a function of the complex variable z if to every value of z in a certain domain R (a region of the Argand diagram) there corresponds one or more values of $f(z)$. Stated like this $f(z)$ could be any function consisting of a real and an imaginary part, each of which is, in general, itself a function of x and y. If we denote the real and imaginary parts of $f(z)$ by u and v, respectively, then

$$f(z) = u(x, y) + iv(x, y).$$

In this chapter, however, we will be primarily concerned with functions that are single-valued, so that to each value of z there corresponds just one value of $f(z)$, and are differentiable in a particular sense, which we now discuss.

A function $f(z)$ that is single-valued in some domain R is *differentiable* at the point z in R if the *derivative*

$$f'(z) = \lim_{\Delta z \to 0} \left[\frac{f(z + \Delta z) - f(z)}{\Delta z} \right] \tag{24.1}$$

exists and is unique, in that its value does not depend upon the direction in the Argand diagram from which Δz tends to zero.

▶ *Show that the function $f(z) = x^2 - y^2 + i2xy$ is differentiable for all values of z.*

Considering the definition (24.1), and taking $\Delta z = \Delta x + i\Delta y$, we have

$$\frac{f(z + \Delta z) - f(z)}{\Delta z}$$

$$= \frac{(x + \Delta x)^2 - (y + \Delta y)^2 + 2i(x + \Delta x)(y + \Delta y) - x^2 + y^2 - 2ixy}{\Delta x + i\Delta y}$$

$$= \frac{2x\Delta x + (\Delta x)^2 - 2y\Delta y - (\Delta y)^2 + 2i(x\Delta y + y\Delta x + \Delta x\Delta y)}{\Delta x + i\Delta y}$$

$$= 2x + i2y + \frac{(\Delta x)^2 - (\Delta y)^2 + 2i\Delta x\Delta y}{\Delta x + i\Delta y}.$$

Now, in whatever way Δx and Δy are allowed to tend to zero (e.g. taking $\Delta y = 0$ and letting $\Delta x \to 0$ or vice versa), the last term on the RHS will tend to zero and the unique limit $2x + i2y$ will be obtained. Since z was arbitrary, $f(z)$ with $u = x^2 - y^2$ and $v = 2xy$ is differentiable at all points in the (finite) complex plane. ◀

We note that the above working can be considerably reduced by recognising that, since $z = x + iy$, we can write $f(z)$ as

$$f(z) = x^2 - y^2 + 2ixy = (x + iy)^2 = z^2.$$

We then find that

$$f'(z) = \lim_{\Delta z \to 0} \left[\frac{(z + \Delta z)^2 - z^2}{\Delta z} \right] = \lim_{\Delta z \to 0} \left[\frac{(\Delta z)^2 + 2z\Delta z}{\Delta z} \right]$$

$$= \left(\lim_{\Delta z \to 0} \Delta z \right) + 2z = 2z,$$

from which we see immediately that the limit both exists and is independent of the way in which $\Delta z \to 0$. Thus we have verified that $f(z) = z^2$ is differentiable for all (finite) z. We also note that the derivative is analogous to that found for real variables.

Although the definition of a differentiable function clearly includes a wide class of functions, the concept of differentiability is restrictive and, indeed, some functions are not differentiable at any point in the complex plane.

▶ *Show that the function $f(z) = 2y + ix$ is not differentiable anywhere in the complex plane.*

In this case $f(z)$ cannot be written simply in terms of z, and so we must consider the limit (24.1) in terms of x and y explicitly. Following the same procedure as in the previous example we find

$$\frac{f(z + \Delta z) - f(z)}{\Delta z} = \frac{2y + 2\Delta y + ix + i\Delta x - 2y - ix}{\Delta x + i\Delta y}$$

$$= \frac{2\Delta y + i\Delta x}{\Delta x + i\Delta y}.$$

In this case the limit will clearly depend on the direction from which $\Delta z \to 0$. Suppose $\Delta z \to 0$ along a line through z of slope m, so that $\Delta y = m\Delta x$, then

$$\lim_{\Delta z \to 0} \left[\frac{f(z + \Delta z) - f(z)}{\Delta z} \right] = \lim_{\Delta x, \Delta y \to 0} \left[\frac{2\Delta y + i\Delta x}{\Delta x + i\Delta y} \right] = \frac{2m + i}{1 + im}.$$

This limit is dependent on m and hence on the direction from which $\Delta z \to 0$. Since this conclusion is independent of the value of z, and hence true for all z, $f(z) = 2y + ix$ is nowhere differentiable. ◄

A function that is single-valued and differentiable at all points of a domain R is said to be *analytic* (or *regular*) in R. A function may be analytic in a domain except at a finite number of points (or an infinite number if the domain is infinite); in this case it is said to be analytic except at these points, which are called the *singularities* of $f(z)$. In our treatment we will not consider cases in which an infinite number of singularities occur in a finite domain.

▶*Show that the function $f(z) = 1/(1-z)$ is analytic everywhere except at $z = 1$.*

Since $f(z)$ is given explicitly as a function of z, evaluation of the limit (24.1) is somewhat easier. We find

$$f'(z) = \lim_{\Delta z \to 0} \left[\frac{f(z + \Delta z) - f(z)}{\Delta z} \right]$$

$$= \lim_{\Delta z \to 0} \left[\frac{1}{\Delta z} \left(\frac{1}{1 - z - \Delta z} - \frac{1}{1 - z} \right) \right]$$

$$= \lim_{\Delta z \to 0} \left[\frac{1}{(1 - z - \Delta z)(1 - z)} \right] = \frac{1}{(1 - z)^2},$$

independently of the way in which $\Delta z \to 0$, provided $z \neq 1$. Hence $f(z)$ is analytic everywhere except at the singularity $z = 1$. ◀

24.2 The Cauchy–Riemann relations

From examining the previous examples, it is apparent that for a function $f(z)$ to be differentiable and hence analytic there must be some particular connection between its real and imaginary parts u and v.

By considering a general function we next establish what this connection must be. If the limit

$$L = \lim_{\Delta z \to 0} \left[\frac{f(z + \Delta z) - f(z)}{\Delta z} \right] \tag{24.2}$$

is to exist and be unique, in the way required for differentiability, then any two specific ways of letting $\Delta z \to 0$ must produce the same limit. In particular, moving parallel to the real axis and moving parallel to the imaginary axis must do so. This is certainly a necessary condition, although it may not be sufficient.

If we let $f(z) = u(x, y) + iv(x, y)$ and $\Delta z = \Delta x + i\Delta y$ then we have

$$f(z + \Delta z) = u(x + \Delta x, \ y + \Delta y) + iv(x + \Delta x, \ y + \Delta y),$$

and the limit (24.2) is given by

$$L = \lim_{\Delta x, \Delta y \to 0} \left[\frac{u(x + \Delta x, y + \Delta y) + iv(x + \Delta x, y + \Delta y) - u(x, y) - iv(x, y)}{\Delta x + i\Delta y} \right].$$

If we first suppose that Δz is purely real, so that $\Delta y = 0$, we obtain

$$L = \lim_{\Delta x \to 0} \left[\frac{u(x + \Delta x, y) - u(x, y)}{\Delta x} + i \frac{v(x + \Delta x, y) - v(x, y)}{\Delta x} \right] = \frac{\partial u}{\partial x} + i \frac{\partial v}{\partial x}, \tag{24.3}$$

provided each limit exists at the point z. Similarly, if Δz is taken as purely imaginary, so that $\Delta x = 0$, we find

$$L = \lim_{\Delta y \to 0} \left[\frac{u(x, y + \Delta y) - u(x, y)}{i\Delta y} + i \frac{v(x, y + \Delta y) - v(x, y)}{i\Delta y} \right] = \frac{1}{i} \frac{\partial u}{\partial y} + \frac{\partial v}{\partial y}. \tag{24.4}$$

For f to be differentiable at the point z, expressions (24.3) and (24.4) must be identical. It follows from equating real and imaginary parts that *necessary* conditions for this are

$$\frac{\partial u}{\partial x} = \frac{\partial v}{\partial y} \qquad \text{and} \qquad \frac{\partial v}{\partial x} = -\frac{\partial u}{\partial y}. \tag{24.5}$$

These two equations are known as the *Cauchy–Riemann relations*.

We can now see why for the earlier examples (i) $f(z) = x^2 - y^2 + i2xy$ might be differentiable and (ii) $f(z) = 2y + ix$ could not be.

(i) $u = x^2 - y^2$, $v = 2xy$:

$$\frac{\partial u}{\partial x} = 2x = \frac{\partial v}{\partial y} \qquad \text{and} \qquad \frac{\partial v}{\partial x} = 2y = -\frac{\partial u}{\partial y},$$

(ii) $u = 2y$, $v = x$:

$$\frac{\partial u}{\partial x} = 0 = \frac{\partial v}{\partial y} \qquad \text{but} \qquad \frac{\partial v}{\partial x} = 1 \neq -2 = -\frac{\partial u}{\partial y}.$$

It is apparent that for $f(z)$ to be analytic something more than the existence of the partial derivatives of u and v with respect to x and y is required; this something is that they satisfy the Cauchy–Riemann relations.

We may enquire also as to the *sufficient* conditions for $f(z)$ to be analytic in R. It can be shown[§] that a sufficient condition is that the four partial derivatives exist, *are continuous* and satisfy the Cauchy–Riemann relations. It is the additional requirement of continuity that makes the difference between the necessary conditions and the sufficient conditions.

►*In which domain(s) of the complex plane is $f(z) = |x| - i|y|$ an analytic function?*

Writing $f = u + iv$ it is clear that both $\partial u/\partial y$ and $\partial v/\partial x$ are zero in all four quadrants and hence that the second Cauchy–Riemann relation in (24.5) is satisfied everywhere.

Turning to the first Cauchy–Riemann relation, in the first quadrant $(x > 0, y > 0)$ we have $f(z) = x - iy$ so that

$$\frac{\partial u}{\partial x} = 1, \qquad \frac{\partial v}{\partial y} = -1,$$

which clearly violates the first relation in (24.5). Thus $f(z)$ is not analytic in the first quadrant.

Following a similar argument for the other quadrants, we find

$$\frac{\partial u}{\partial x} = -1 \quad \text{or} \quad +1 \quad \text{for } x < 0 \text{ and } x > 0, \text{ respectively,}$$

$$\frac{\partial v}{\partial y} = -1 \quad \text{or} \quad +1 \quad \text{for } y > 0 \text{ and } y < 0, \text{ respectively.}$$

Therefore $\partial u/\partial x$ and $\partial v/\partial y$ are equal, and hence $f(z)$ is analytic only in the second and fourth quadrants. ◄

[§] See, for example, any of the references given on page 824.

Since x and y are related to z and its complex conjugate z^* by

$$x = \frac{1}{2}(z + z^*) \quad \text{and} \quad y = \frac{1}{2i}(z - z^*), \tag{24.6}$$

we may formally regard any function $f = u + iv$ as a function of z and z^*, rather than x and y. If we do this and examine $\partial f / \partial z^*$ we obtain

$$\begin{aligned}
\frac{\partial f}{\partial z^*} &= \frac{\partial f}{\partial x}\frac{\partial x}{\partial z^*} + \frac{\partial f}{\partial y}\frac{\partial y}{\partial z^*} \\
&= \left(\frac{\partial u}{\partial x} + i\frac{\partial v}{\partial x}\right)\left(\frac{1}{2}\right) + \left(\frac{\partial u}{\partial y} + i\frac{\partial v}{\partial y}\right)\left(-\frac{1}{2i}\right) \\
&= \frac{1}{2}\left(\frac{\partial u}{\partial x} - \frac{\partial v}{\partial y}\right) + \frac{i}{2}\left(\frac{\partial v}{\partial x} + \frac{\partial u}{\partial y}\right).
\end{aligned} \tag{24.7}$$

Now, if f is analytic then the Cauchy–Riemann relations (24.5) must be satisfied, and these immediately give that $\partial f / \partial z^*$ is identically zero. Thus we conclude that if f is analytic then f cannot be a function of z^* and any expression representing an analytic function of z can contain x and y only in the combination $x + iy$, *not* in the combination $x - iy$.

We conclude this section by discussing some properties of analytic functions that are of great practical importance in theoretical physics. These can be obtained simply from the requirement that the Cauchy–Riemann relations must be satisfied by the real and imaginary parts of an analytic function.

The most important of these results can be obtained by differentiating the first Cauchy–Riemann relation with respect to one independent variable, and the second with respect to the other independent variable, to obtain the two chains of equalities

$$\frac{\partial}{\partial x}\left(\frac{\partial u}{\partial x}\right) = \frac{\partial}{\partial x}\left(\frac{\partial v}{\partial y}\right) = \frac{\partial}{\partial y}\left(\frac{\partial v}{\partial x}\right) = -\frac{\partial}{\partial y}\left(\frac{\partial u}{\partial y}\right),$$

$$\frac{\partial}{\partial x}\left(\frac{\partial v}{\partial x}\right) = -\frac{\partial}{\partial x}\left(\frac{\partial u}{\partial y}\right) = -\frac{\partial}{\partial y}\left(\frac{\partial u}{\partial x}\right) = -\frac{\partial}{\partial y}\left(\frac{\partial v}{\partial y}\right).$$

Thus both u and v are *separately* solutions of Laplace's equation in two dimensions, i.e.

$$\frac{\partial^2 u}{\partial x^2} + \frac{\partial^2 u}{\partial y^2} = 0 \quad \text{and} \quad \frac{\partial^2 v}{\partial x^2} + \frac{\partial^2 v}{\partial y^2} = 0. \tag{24.8}$$

We will make significant use of this result in the next chapter.

A further useful result concerns the two families of curves $u(x, y) = $ constant and $v(x, y) = $ constant, where u and v are the real and imaginary parts of any analytic function $f = u + iv$. As discussed in chapter 10, the vector normal to the curve $u(x, y) = $ constant is given by

$$\nabla u = \frac{\partial u}{\partial x}\mathbf{i} + \frac{\partial u}{\partial y}\mathbf{j}, \tag{24.9}$$

where $\mathbf{i}$ and $\mathbf{j}$ are the unit vectors along the x- and y-axes, respectively. A similar expression exists for ∇v, the normal to the curve $v(x, y) = \text{constant}$. Taking the scalar product of these two normal vectors, we obtain

$$\nabla u \cdot \nabla v = \frac{\partial u}{\partial x}\frac{\partial v}{\partial x} + \frac{\partial u}{\partial y}\frac{\partial v}{\partial y}$$

$$= -\frac{\partial u}{\partial x}\frac{\partial u}{\partial y} + \frac{\partial u}{\partial y}\frac{\partial u}{\partial x} = 0,$$

where in the last line we have used the Cauchy–Riemann relations to rewrite the partial derivatives of v as partial derivatives of u. Since the scalar product of the normal vectors is zero, they must be orthogonal, and the curves $u(x, y) = \text{constant}$ and $v(x, y) = \text{constant}$ must therefore intersect at *right angles*.

▶*Use the Cauchy–Riemann relations to show that, for any analytic function $f = u + iv$, the relation $|\nabla u| = |\nabla v|$ must hold.*

From (24.9) we have

$$|\nabla u|^2 = \nabla u \cdot \nabla u = \left(\frac{\partial u}{\partial x}\right)^2 + \left(\frac{\partial u}{\partial y}\right)^2.$$

Using the Cauchy–Riemann relations to write the partial derivatives of u in terms of those of v, we obtain

$$|\nabla u|^2 = \left(\frac{\partial v}{\partial y}\right)^2 + \left(\frac{\partial v}{\partial x}\right)^2 = |\nabla v|^2,$$

from which the result $|\nabla u| = |\nabla v|$ follows immediately. ◀

24.3 Power series in a complex variable

The theory of power series in a real variable was considered in chapter 4, which also contained a brief discussion of the natural extension of this theory to a series such as

$$f(z) = \sum_{n=0}^{\infty} a_n z^n, \tag{24.10}$$

where z is a complex variable and the a_n are, in general, complex. We now consider complex power series in more detail.

Expression (24.10) is a power series about the origin and may be used for general discussion, since a power series about any other point z_0 can be obtained by a change of variable from z to $z - z_0$. If z were written in its modulus and argument form, $z = r \exp i\theta$, expression (24.10) would become

$$f(z) = \sum_{n=0}^{\infty} a_n r^n \exp(in\theta). \tag{24.11}$$

This series is absolutely convergent if

$$\sum_{n=0}^{\infty} |a_n| r^n, \tag{24.12}$$

which is a series of positive real terms, is convergent. Thus tests for the absolute convergence of real series can be used in the present context, and of these the most appropriate form is based on the Cauchy root test. With the *radius of convergence R* defined by

$$\frac{1}{R} = \lim_{n \to \infty} |a_n|^{1/n}, \tag{24.13}$$

the series (24.10) is absolutely convergent if $|z| < R$ and divergent if $|z| > R$. If $|z| = R$ then no particular conclusion may be drawn, and this case must be considered separately, as discussed in subsection 4.5.1.

A circle of radius R centred on the origin is called the *circle of convergence* of the series $\sum a_n z^n$. The cases $R = 0$ and $R = \infty$ correspond, respectively, to convergence at the origin only and convergence everywhere. For R finite the convergence occurs in a restricted part of the z-plane (the Argand diagram). For a power series about a general point z_0, the circle of convergence is, of course, centred on that point.

> ▶*Find the parts of the z-plane for which the following series are convergent:*
>
> (i) $\displaystyle\sum_{n=0}^{\infty} \frac{z^n}{n!},$ (ii) $\displaystyle\sum_{n=0}^{\infty} n! z^n,$ (iii) $\displaystyle\sum_{n=1}^{\infty} \frac{z^n}{n}.$

(i) Since $(n!)^{1/n}$ behaves like n as $n \to \infty$ we find $\lim(1/n!)^{1/n} = 0$. Hence $R = \infty$ and the series is convergent for all z. (ii) Correspondingly, $\lim(n!)^{1/n} = \infty$. Thus $R = 0$ and the series converges only at $z = 0$. (iii) As $n \to \infty$, $(n)^{1/n}$ has a lower limit of 1 and hence $\lim(1/n)^{1/n} = 1/1 = 1$. Thus the series is absolutely convergent if the condition $|z| < 1$ is satisfied. ◀

Case (iii) in the above example provides a good illustration of the fact that on its circle of convergence a power series may or may not converge. For this particular series, the circle of convergence is $|z| = 1$, so let us consider the convergence of the series at two different points on this circle. Taking $z = 1$, the series becomes

$$\sum_{n=1}^{\infty} \frac{1}{n} = 1 + \frac{1}{2} + \frac{1}{3} + \frac{1}{4} + \cdots,$$

which is easily shown to diverge (by, for example, grouping terms, as discussed in subsection 4.3.2). Taking $z = -1$, however, the series is given by

$$\sum_{n=1}^{\infty} \frac{(-1)^n}{n} = -1 + \frac{1}{2} - \frac{1}{3} + \frac{1}{4} - \cdots,$$

which is an alternating series whose terms decrease in magnitude and which therefore converges.

The ratio test discussed in subsection 4.3.2 may also be employed to investigate the absolute convergence of a complex power series. A series is absolutely convergent if

$$\lim_{n \to \infty} \frac{|a_{n+1}||z|^{n+1}}{|a_n||z|^n} = \lim_{n \to \infty} \frac{|a_{n+1}||z|}{|a_n|} < 1 \tag{24.14}$$

and hence the radius of convergence R of the series is given by

$$\frac{1}{R} = \lim_{n \to \infty} \frac{|a_{n+1}|}{|a_n|}.$$

For instance, in case (i) of the previous example, we have

$$\frac{1}{R} = \lim_{n \to \infty} \frac{n!}{(n+1)!} = \lim_{n \to \infty} \frac{1}{n+1} = 0.$$

Thus the series is absolutely convergent for all (finite) z, confirming the previous result.

Before turning to particular power series, we conclude this section by stating the important result[§] that *the power series $\sum_0^\infty a_n z^n$ has a sum that is an analytic function of z inside its circle of convergence.*

As a corollary to the above theorem, it may further be shown that if $f(z) = \sum a_n z^n$ then, inside the circle of convergence of the series,

$$f'(z) = \sum_{n=0}^{\infty} n a_n z^{n-1}.$$

Repeated application of this result demonstrates that any power series can be differentiated any number of times inside its circle of convergence.

24.4 Some elementary functions

In the example at the end of the previous section it was shown that the function $\exp z$ *defined* by

$$\exp z = \sum_{n=0}^{\infty} \frac{z^n}{n!} \tag{24.15}$$

is convergent for all z of finite modulus and is thus, by the discussion of the previous section, an analytic function over the whole z-plane.[¶] Like its

---

[§] For a proof see, for example, K. F. Riley, *Mathematical Methods for the Physical Sciences* (Cambridge: Cambridge University Press, 1974), p. 446.

[¶] Functions that are analytic in the *whole* z-plane are usually called *integral* or *entire* functions.

real-variable counterpart it is called the *exponential function*; also like its real counterpart it is equal to its own derivative.

The multiplication of two exponential functions results in a further exponential function, in accordance with the corresponding result for real variables.

▶*Show that* $\exp z_1 \exp z_2 = \exp(z_1 + z_2)$.

From the series expansion (24.15) of $\exp z_1$ and a similar expansion for $\exp z_2$, it is clear that the coefficient of $z_1^r z_2^s$ in the corresponding series expansion of $\exp z_1 \exp z_2$ is simply $1/(r!s!)$.

But, from (24.15) we also have

$$\exp(z_1 + z_2) = \sum_{n=0}^{\infty} \frac{(z_1 + z_2)^n}{n!}.$$

In order to find the coefficient of $z_1^r z_2^s$ in this expansion, we clearly have to consider the term in which $n = r + s$, namely

$$\frac{(z_1 + z_2)^{r+s}}{(r+s)!} = \frac{1}{(r+s)!} \left({}^{r+s}C_0 z_1^{r+s} + \cdots + {}^{r+s}C_s z_1^r z_2^s + \cdots + {}^{r+s}C_{r+s} z_2^{r+s} \right).$$

The coefficient of $z_1^r z_2^s$ in this is given by

$${}^{r+s}C_s \frac{1}{(r+s)!} = \frac{(r+s)!}{s!r!} \frac{1}{(r+s)!} = \frac{1}{r!s!}.$$

Thus, since the corresponding coefficients on the two sides are equal, and all the series involved are absolutely convergent for all z, we can conclude that $\exp z_1 \exp z_2 = \exp(z_1 + z_2)$. ◀

As an extension of (24.15) we may also define the complex exponent of a real number $a > 0$ by the equation

$$a^z = \exp(z \ln a), \tag{24.16}$$

where $\ln a$ is the natural logarithm of a. The particular case $a = e$ and the fact that $\ln e = 1$ enable us to write $\exp z$ interchangeably with e^z. If z is real then the definition agrees with the familiar one.

The result for $z = iy$,

$$\exp iy = \cos y + i \sin y, \tag{24.17}$$

has been met already in equation (3.23). Its immediate extension is

$$\exp z = (\exp x)(\cos y + i \sin y). \tag{24.18}$$

As z varies over the complex plane, the modulus of $\exp z$ takes all real positive values, except that of 0. However, two values of z that differ by $2\pi ki$, for any integer k, produce the same value of $\exp z$, as given by (24.18), and so $\exp z$ is periodic with period $2\pi i$. If we denote $\exp z$ by t, then the strip $-\pi < y \le \pi$ in the z-plane corresponds to the whole of the t-plane, except for the point $t = 0$.

The sine, cosine, sinh and cosh functions of a complex variable are defined from the exponential function exactly as are those for real variables. The functions derived from them (e.g. tan and tanh), the identities they satisfy and their derivative properties are also just as for real variables. In view of this we will not give them further attention here.

The inverse function of $\exp z$ is given by w, the solution of

$$\exp w = z. \tag{24.19}$$

This inverse function was discussed in chapter 3, but we mention it again here for completeness. By virtue of the discussion following (24.18), w is not uniquely defined and is indeterminate to the extent of any integer multiple of $2\pi i$. If we express z as

$$z = r \exp i\theta,$$

where r is the (real) modulus of z and θ is its argument $(-\pi < \theta \leq \pi)$, then multiplying z by $\exp(2ik\pi)$, where k is an integer, will result in the same complex number z. Thus we may write

$$z = r \exp[i(\theta + 2k\pi)],$$

where k is an integer. If we denote w in (24.19) by

$$w = \operatorname{Ln} z = \ln r + i(\theta + 2k\pi), \tag{24.20}$$

where $\ln r$ is the natural logarithm (to base e) of the real positive quantity r, then $\operatorname{Ln} z$ is an infinitely multivalued function of z. Its *principal value*, denoted by $\ln z$, is obtained by taking $k = 0$ so that its argument lies in the range $-\pi$ to π. Thus

$$\ln z = \ln r + i\theta, \quad \text{with } -\pi < \theta \leq \pi. \tag{24.21}$$

Now that the logarithm of a complex variable has been defined, definition (24.16) of a general power can be extended to cases other than those in which a is real and positive. If t ($\neq 0$) and z are both complex, then the zth power of t is defined by

$$t^z = \exp(z \operatorname{Ln} t). \tag{24.22}$$

Since $\operatorname{Ln} t$ is multivalued, so is this definition. Its principal value is obtained by giving $\operatorname{Ln} t$ its principal value, $\ln t$.

If t ($\neq 0$) is complex but z is real and equal to $1/n$, then (24.22) provides a definition of the nth root of t. Because of the multivaluedness of $\operatorname{Ln} t$, there will be more than one nth root of any given t.

▶*Show that there are exactly n distinct nth roots of t.*

From (24.22) the nth roots of t are given by

$$t^{1/n} = \exp\left(\frac{1}{n}\operatorname{Ln} t\right).$$

On the RHS let us write t as follows:

$$t = r \exp[i(\theta + 2k\pi)],$$

where k is an integer. We then obtain

$$t^{1/n} = \exp\left[\frac{1}{n}\ln r + i\frac{(\theta + 2k\pi)}{n}\right]$$

$$= r^{1/n}\exp\left[i\frac{(\theta + 2k\pi)}{n}\right],$$

where $k = 0, 1, \ldots, n-1$; for other values of k we simply recover the roots already found. Thus t has n distinct nth roots. ◀

24.5 Multivalued functions and branch cuts

In the definition of an analytic function, one of the conditions imposed was that the function is single-valued. However, as shown in the previous section, the logarithmic function, a complex power and a complex root are all multivalued. Nevertheless, it happens that the properties of analytic functions can still be applied to these and other multivalued functions of a complex variable provided that suitable care is taken. This care amounts to identifying the *branch points* of the multivalued function $f(z)$ in question. If z is varied in such a way that its path in the Argand diagram forms a closed curve that encloses a branch point, then, in general, $f(z)$ will not return to its original value.

For definiteness let us consider the multivalued function $f(z) = z^{1/2}$ and express z as $z = r \exp i\theta$. From figure 24.1(a), it is clear that, as the point z traverses any closed contour C that does not enclose the origin, θ will return to its original value after one complete circuit. However, for any closed contour C' that does enclose the origin, after one circuit $\theta \to \theta + 2\pi$ (see figure 24.1(b)). Thus, for the function $f(z) = z^{1/2}$, after one circuit

$$r^{1/2}\exp(i\theta/2) \;\to\; r^{1/2}\exp[i(\theta + 2\pi)/2] = -r^{1/2}\exp(i\theta/2).$$

In other words, the value of $f(z)$ changes around any closed loop enclosing the origin; in this case $f(z) \to -f(z)$. Thus $z = 0$ is a branch point of the function $f(z) = z^{1/2}$.

We note in this case that if any closed contour enclosing the origin is traversed *twice* then $f(z) = z^{1/2}$ returns to its original value. The number of loops around a branch point required for any given function $f(z)$ to return to its original value depends on the function in question, and for some functions (e.g. $\operatorname{Ln} z$, which also has a branch point at the origin) the original value is never recovered.

In order that $f(z)$ may be treated as single-valued, we may define a *branch cut* in the Argand diagram. A branch cut is a line (or curve) in the complex plane and may be regarded as an artificial barrier that we must not cross. Branch cuts are positioned in such a way that we are prevented from making a complete

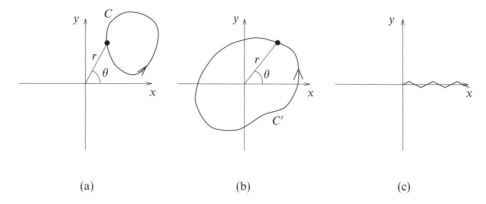

Figure 24.1 (a) A closed contour not enclosing the origin; (b) a closed contour enclosing the origin; (c) a possible branch cut for $f(z) = z^{1/2}$.

circuit around any one branch point, and so the function in question remains single-valued.

For the function $f(z) = z^{1/2}$, we may take as a branch cut any curve starting at the origin $z = 0$ and extending out to $|z| = \infty$ in any direction, since all such curves would equally well prevent us from making a closed loop around the branch point at the origin. It is usual, however, to take the cut along either the real or the imaginary axis. For example, in figure 24.1(c), we take the cut as the positive real axis. By agreeing not to cross this cut, we restrict θ to lie in the range $0 \leq \theta < 2\pi$, and so keep $f(z)$ single-valued.

These ideas are easily extended to functions with more than one branch point.

▶*Find the branch points of $f(z) = \sqrt{z^2 + 1}$, and hence sketch suitable arrangements of branch cuts.*

We begin by writing $f(z)$ as

$$f(z) = \sqrt{z^2 + 1} = \sqrt{(z - i)(z + i)}.$$

As shown above, the function $g(z) = z^{1/2}$ has a branch point at $z = 0$. Thus we might expect $f(z)$ to have branch points at values of z that make the expression under the square root equal to zero, i.e. at $z = i$ and $z = -i$.

As shown in figure 24.2(a), we use the notation

$$z - i = r_1 \exp i\theta_1 \qquad \text{and} \qquad z + i = r_2 \exp i\theta_2.$$

We can therefore write $f(z)$ as

$$f(z) = \sqrt{r_1 r_2} \exp(i\theta_1/2) \exp(i\theta_2/2) = \sqrt{r_1 r_2} \exp\left[i(\theta_1 + \theta_2)/2\right].$$

Let us now consider how $f(z)$ changes as we make one complete circuit around various closed loops C in the Argand diagram. If C encloses

(i) neither branch point, then $\theta_1 \to \theta_1$, $\theta_2 \to \theta_2$ and so $f(z) \to f(z)$;

(ii) $z = i$ but not $z = -i$, then $\theta_1 \to \theta_1 + 2\pi$, $\theta_2 \to \theta_2$ and so $f(z) \to -f(z)$;

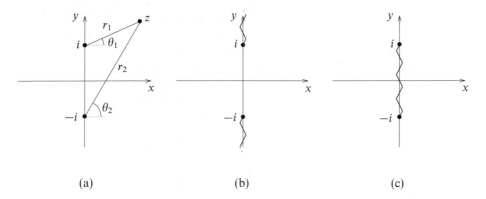

Figure 24.2 (a) Coordinates used in the analysis of the branch points of $f(z) = (z^2 + 1)^{1/2}$; (b) one possible arrangement of branch cuts; (c) another possible branch cut, which is finite.

(iii) $z = -i$ but not $z = i$, then $\theta_1 \to \theta_1$, $\theta_2 \to \theta_2 + 2\pi$ and so $f(z) \to -f(z)$;

(iv) both branch points, then $\theta_1 \to \theta_1 + 2\pi$, $\theta_2 \to \theta_2 + 2\pi$ and so $f(z) \to f(z)$.

Thus, as expected, $f(z)$ changes value around loops containing either $z = i$ or $z = -i$ (but not both). We must therefore choose branch cuts that prevent us from making a complete loop around either branch point; one suitable choice is shown in figure 24.2(b).

For this $f(z)$, however, we have noted that after traversing a loop containing *both* branch points the function returns to its original value. Thus we may choose an alternative, *finite*, branch cut that allows this possibility but still prevents us from making a complete loop around just one of the points. A suitable cut is shown in figure 24.2(c). ◀

24.6 Singularities and zeros of complex functions

A singular point of a complex function $f(z)$ is any point in the Argand diagram at which $f(z)$ fails to be analytic. We have already met one sort of singularity, the branch point, and in this section we will consider other types of singularity as well as discuss the zeros of complex functions.

If $f(z)$ has a singular point at $z = z_0$ but is analytic at all points in some neighbourhood containing z_0 but no other singularities, then $z = z_0$ is called an *isolated singularity*. (Clearly, branch points are not isolated singularities.)

The most important type of isolated singularity is the *pole*. If $f(z)$ has the form

$$f(z) = \frac{g(z)}{(z - z_0)^n},\tag{24.23}$$

where n is a positive integer, $g(z)$ is analytic at all points in some neighbourhood containing $z = z_0$ and $g(z_0) \neq 0$, then $f(z)$ has a *pole of order n* at $z = z_0$. An alternative (though equivalent) definition is that

$$\lim_{z \to z_0} [(z - z_0)^n f(z)] = a,\tag{24.24}$$

where a is a finite, non-zero complex number. We note that if the above limit is equal to zero, then $z = z_0$ is a pole of order less than n, or $f(z)$ is analytic there; if the limit is infinite then the pole is of an order greater than n. It may also be shown that if $f(z)$ has a pole at $z = z_0$, then $|f(z)| \to \infty$ as $z \to z_0$ from any direction in the Argand diagram.[§] If no finite value of n can be found such that (24.24) is satisfied, then $z = z_0$ is called an *essential singularity*.

▶*Find the singularities of the functions*

$$\text{(i)} \ f(z) = \frac{1}{1-z} - \frac{1}{1+z}, \qquad \text{(ii)} \ f(z) = \tanh z.$$

(i) If we write $f(z)$ as

$$f(z) = \frac{1}{1-z} - \frac{1}{1+z} = \frac{2z}{(1-z)(1+z)},$$

we see immediately from either (24.23) or (24.24) that $f(z)$ has poles of order 1 (or *simple poles*) at $z = 1$ and $z = -1$.

(ii) In this case we write

$$f(z) = \tanh z = \frac{\sinh z}{\cosh z} = \frac{\exp z - \exp(-z)}{\exp z + \exp(-z)}.$$

Thus $f(z)$ has a singularity when $\exp z = -\exp(-z)$ or, equivalently, when

$$\exp z = \exp[i(2n + 1)\pi] \exp(-z),$$

where n is any integer. Equating the arguments of the exponentials we find $z = (n + \frac{1}{2})\pi i$, for integer n.

Furthermore, using l'Hôpital's rule (see chapter 4) we have

$$\lim_{z \to (n+\frac{1}{2})\pi i} \left\{ \frac{[z - (n + \frac{1}{2})\pi i] \sinh z}{\cosh z} \right\}$$

$$= \lim_{z \to (n+\frac{1}{2})\pi i} \left\{ \frac{[z - (n + \frac{1}{2})\pi i] \cosh z + \sinh z}{\sinh z} \right\} = 1.$$

Therefore, from (24.24), each singularity is a simple pole. ◀

Another type of singularity exists at points for which the value of $f(z)$ takes an indeterminate form such as $0/0$ but $\lim_{z \to z_0} f(z)$ exists and is independent of the direction from which z_0 is approached. Such points are called *removable singularities*.

▶*Show that $f(z) = (\sin z)/z$ has a removable singularity at $z = 0$.*

It is clear that $f(z)$ takes the indeterminate form $0/0$ at $z = 0$. However, by expanding $\sin z$ as a power series in z, we find

$$f(z) = \frac{1}{z} \left(z - \frac{z^3}{3!} + \frac{z^5}{5!} - \cdots \right) = 1 - \frac{z^2}{3!} + \frac{z^4}{5!} - \cdots .$$

[§] Although perhaps intuitively obvious, this result really requires formal demonstration by analysis.

Thus $\lim_{z \to 0} f(z) = 1$ independently of the way in which $z \to 0$, and so $f(z)$ has a removable singularity at $z = 0$. ◀

An expression common in mathematics, but which we have so far avoided using explicitly in this chapter, is 'z tends to infinity'. For a real variable such as $|z|$ or R, 'tending to infinity' has a reasonably well defined meaning. For a complex variable needing a two-dimensional plane to represent it, the meaning is not intrinsically well defined. However, it is convenient to have a unique meaning and this is provided by the following *definition*: the behaviour of $f(z)$ *at infinity* is given by that of $f(1/\xi)$ at $\xi = 0$, where $\xi = 1/z$.

> ▶*Find the behaviour at infinity of* (i) $f(z) = a + bz^{-2}$, (ii) $f(z) = z(1 + z^2)$ *and* (iii) $f(z) = \exp z$.

(i) $f(z) = a + bz^{-2}$: on putting $z = 1/\xi$, $f(1/\xi) = a + b\xi^2$, which is analytic at $\xi = 0$; thus f is analytic at $z = \infty$.
(ii) $f(z) = z(1 + z^2)$: $f(1/\xi) = 1/\xi + 1/\xi^3$; thus f has a pole of order 3 at $z = \infty$.
(iii) $f(z) = \exp z$: $f(1/\zeta) = \sum_0^\infty (n!)^{-1}\xi^{-n}$; thus f has an essential singularity at $z = \infty$. ◀

We conclude this section by briefly mentioning the *zeros* of a complex function. As the name suggests, if $f(z_0) = 0$ then $z = z_0$ is called a zero of the function $f(z)$. Zeros are classified in a similar way to poles, in that if

$$f(z) = (z - z_0)^n g(z),$$

where n is a positive integer and $g(z_0) \neq 0$, then $z = z_0$ is called a *zero of order* n of $f(z)$. If $n = 1$ then $z = z_0$ is called a *simple zero*. It may further be shown that if $z = z_0$ is a zero of order n of $f(z)$ then it is also a pole of order n of the function $1/f(z)$.

We will return in section 24.11 to the classification of zeros and poles in terms of their series expansions.

24.7 Conformal transformations

We now turn our attention to the subject of transformations, by which we mean a change of coordinates from the complex variable $z = x + iy$ to another, say $w = r + is$, by means of a prescribed formula:

$$w = g(z) = r(x, y) + is(x, y).$$

Under such a transformation, or *mapping*, the Argand diagram for the z-variable is transformed into one for the w-variable, although the complete z-plane might be mapped onto only a part of the w-plane, or onto the whole of the w-plane, or onto some or all of the w-plane covered more than once.

We shall consider only those mappings for which w and z are related by a function $w = g(z)$ and its inverse $z = h(w)$ with both functions analytic, except possibly at a few isolated points; such mappings are called *conformal*. Their

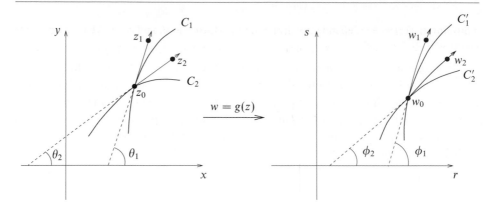

Figure 24.3 Two curves C_1 and C_2 in the z-plane, which are mapped onto C_1' and C_2' in the w-plane.

important properties are that, except at points at which $g'(z)$, and hence $h'(z)$, is zero or infinite:

(i) continuous lines in the z-plane transform into continuous lines in the w-plane;

(ii) the angle between two intersecting curves in the z-plane equals the angle between the corresponding curves in the w-plane;

(iii) the magnification, as between the z-plane and the w-plane, of a small line element in the neighbourhood of any particular point is independent of the direction of the element;

(iv) any analytic function of z transforms to an analytic function of w and vice versa.

Result (i) is immediate, and results (ii) and (iii) can be justified by the following argument. Let two curves C_1 and C_2 pass through the point z_0 in the z-plane and let z_1 and z_2 be two points on their respective tangents at z_0, each a distance ρ from z_0. The same prescription with w replacing z describes the transformed situation; however, the transformed tangents may not be straight lines and the distances of w_1 and w_2 from w_0 have not yet been shown to be equal. This situation is illustrated in figure 24.3.

In the z-plane z_1 and z_2 are given by

$$z_1 - z_0 = \rho \exp i\theta_1 \qquad \text{and} \qquad z_2 - z_0 = \rho \exp i\theta_2.$$

The corresponding descriptions in the w-plane are

$$w_1 - w_0 = \rho_1 \exp i\phi_1 \qquad \text{and} \qquad w_2 - w_0 = \rho_2 \exp i\phi_2.$$

The angles θ_i and ϕ_i are clear from figure 24.3. The transformed angles ϕ_i are those made with the r-axis by the tangents to the transformed curves at their point of intersection. Since any finite-length tangent may be curved, w_i is more strictly given by $w_i - w_0 = \rho_i \exp i(\phi_i + \delta\phi_i)$, where $\delta\phi_i \to 0$ as $\rho_i \to 0$, i.e. as $\rho \to 0$.

Now since $w = g(z)$, where g is analytic, we have

$$\lim_{z_1 \to z_0} \left(\frac{w_1 - w_0}{z_1 - z_0} \right) = \lim_{z_2 \to z_0} \left(\frac{w_2 - w_0}{z_2 - z_0} \right) = \frac{dg}{dz}\Big|_{z=z_0},$$

which may be written as

$$\lim_{\rho \to 0} \left\{ \frac{\rho_1}{\rho} \exp[i(\phi_1 + \delta\phi_1 - \theta_1)] \right\} = \lim_{\rho \to 0} \left\{ \frac{\rho_2}{\rho} \exp[i(\phi_2 + \delta\phi_2 - \theta_2)] \right\} = g'(z_0). \tag{24.25}$$

Comparing magnitudes and phases (i.e. arguments) in the equalities (24.25) gives the stated results (ii) and (iii) and adds quantitative information to them, namely that for *small* line elements

$$\frac{\rho_1}{\rho} \approx \frac{\rho_2}{\rho} \approx |g'(z_0)|, \tag{24.26}$$

$$\phi_1 - \theta_1 \approx \phi_2 - \theta_2 \approx \arg g'(z_0). \tag{24.27}$$

For strict comparison with result (ii), (24.27) must be written as $\theta_1 - \theta_2 = \phi_1 - \phi_2$, with an ordinary equality sign, since the angles are only defined in the limit $\rho \to 0$ when (24.27) becomes a true identity. We also see from (24.26) that the linear magnification factor is $|g'(z_0)|$; similarly, small areas are magnified by $|g'(z_0)|^2$.

Since in the neighbourhoods of corresponding points in a transformation angles are preserved and magnifications are independent of direction, it follows that small plane figures are transformed into figures of the same shape, but, in general, ones that are magnified and rotated (though not distorted). However, we also note that at points where $g'(z) = 0$, the angle $\arg g'(z)$ through which line elements are rotated is undefined; these are called *critical points* of the transformation.

The final result (iv) is perhaps the most important property of conformal transformations. If $f(z)$ is an analytic function of z and $z = h(w)$ is also analytic, then $F(w) = f(h(w))$ is analytic in w. Its importance lies in the further conclusions it allows us to draw from the fact that, since f is analytic, the real and imaginary parts of $f = \phi + i\psi$ are necessarily solutions of

$$\frac{\partial^2 \phi}{\partial x^2} + \frac{\partial^2 \phi}{\partial y^2} = 0 \quad \text{and} \quad \frac{\partial^2 \psi}{\partial x^2} + \frac{\partial^2 \psi}{\partial y^2} = 0. \tag{24.28}$$

Since the transformation property ensures that $F = \Phi + i\Psi$ is also analytic, we can conclude that its real and imaginary parts must themselves satisfy Laplace's equation in the w-plane:

$$\frac{\partial^2 \Phi}{\partial r^2} + \frac{\partial^2 \Phi}{\partial s^2} = 0 \quad \text{and} \quad \frac{\partial^2 \Psi}{\partial r^2} + \frac{\partial^2 \Psi}{\partial s^2} = 0. \tag{24.29}$$

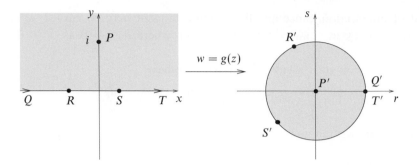

Figure 24.4   Transforming the upper half of the z-plane into the interior of the unit circle in the w-plane, in such a way that $z = i$ is mapped onto $w = 0$ and the points $x = \pm\infty$ are mapped onto $w = 1$.

Further, suppose that (say) Re $f(z) = \phi$ is constant over a boundary C in the z-plane; then Re $F(w) = \Phi$ is constant over C in the z-plane. But this is the same as saying that Re $F(w)$ is constant over the boundary C' in the w-plane, C' being the curve into which C is transformed by the conformal transformation $w = g(z)$. This result is exploited extensively in the next chapter to solve Laplace's equation for a variety of two-dimensional geometries.

Examples of useful conformal transformations are numerous. For instance, $w = z + b$, $w = (\exp i\phi)z$ and $w = az$ correspond, respectively, to a translation by b, a rotation through an angle ϕ and a stretching (or contraction) in the radial direction (for a real). These three examples can be combined into the general linear transformation $w = az + b$, where, in general, a and b are complex. Another example is the inversion mapping $w = 1/z$, which maps the interior of the unit circle to the exterior and vice versa. Other, more complicated, examples also exist.

▶*Show that if the point z_0 lies in the upper half of the z-plane then the transformation*

$$w = (\exp i\phi)\frac{z - z_0}{z - z_0^*}$$

maps the upper half of the z-plane into the interior of the unit circle in the w-plane. Hence find a similar transformation that maps the point $z = i$ onto $w = 0$ and the points $x = \pm\infty$ onto $w = 1$.

Taking the modulus of w, we have

$$|w| = \left|(\exp i\phi)\frac{z - z_0}{z - z_0^*}\right| = \left|\frac{z - z_0}{z - z_0^*}\right|.$$

However, since the complex conjugate z_0^* is the reflection of z_0 in the real axis, if z and z_0 both lie in the upper half of the z-plane then $|z - z_0| \leq |z - z_0^*|$; thus $|w| \leq 1$, as required. We also note that (i) the equality holds only when z lies on the real axis, and so this axis is mapped onto the boundary of the unit circle in the w-plane; (ii) the point z_0 is mapped onto $w = 0$, the origin of the w-plane.

By fixing the images of two points in the z-plane, the constants z_0 and ϕ can also be fixed. Since we require the point $z = i$ to be mapped onto $w = 0$, we have immediately

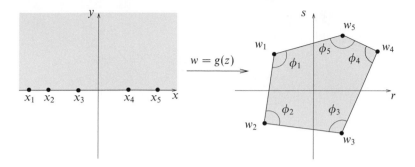

Figure 24.5 Transforming the upper half of the z-plane into the interior of a polygon in the w-plane, in such a way that the points $x_1, x_2, \ldots, x_n$ are mapped onto the vertices $w_1, w_2, \ldots, w_n$ of the polygon with interior angles $\phi_1, \phi_2, \ldots, \phi_n$.

$z_0 = i$. By further requiring $z = \pm\infty$ to be mapped onto $w = 1$, we find $1 = w = \exp i\phi$ and so $\phi = 0$. The required transformation is therefore

$$w = \frac{z - i}{z + i},$$

and is illustrated in figure 24.4. ◄

We conclude this section by mentioning the rather curious *Schwarz–Christoffel* transformation.[§] Suppose, as shown in figure 24.5, that we are interested in a (finite) number of points $x_1, x_2, \ldots, x_n$ on the real axis in the z-plane. Then by means of the transformation

$$w = \left\{ A \int_0^z (\xi - x_1)^{(\phi_1/\pi) - 1} (\xi - x_2)^{(\phi_2/\pi) - 1} \cdots (\xi - x_n)^{(\phi_n/\pi) - 1} \, d\xi \right\} + B, \quad (24.30)$$

we may map the upper half of the z-plane onto the interior of a closed polygon in the w-plane having n vertices $w_1, w_2, \ldots, w_n$ (which are the images of $x_1, x_2, \ldots, x_n$) with corresponding interior angles $\phi_1, \phi_2, \ldots, \phi_n$, as shown in figure 24.5. The real axis in the z-plane is transformed into the boundary of the polygon itself. The constants A and B are complex in general and determine the position, size and orientation of the polygon. It is clear from (24.30) that $dw/dz = 0$ at $x = x_1, x_2, \ldots, x_n$, and so the transformation is not conformal at these points.

There are various subtleties associated with the use of the Schwarz–Christoffel transformation. For example, if one of the points on the real axis in the z-plane (usually x_n) is taken at infinity, then the corresponding factor in (24.30) (i.e. the one involving x_n) is not present. In this case, the point(s) $x = \pm\infty$ are considered as one point, since they transform to a single vertex of the polygon in the w-plane.

[§] Strictly speaking, the use of this transformation requires an understanding of complex integrals, which are discussed in section 24.8.

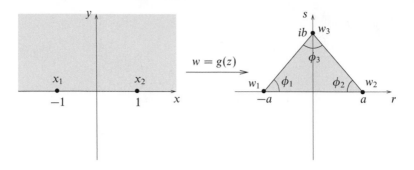

Figure 24.6 Transforming the upper half of the z-plane into the interior of a triangle in the w-plane.

We can also map the upper half of the z-plane into an infinite *open* polygon by considering it as the limiting case of some closed polygon.

▶*Find a transformation that maps the upper half of the z-plane into the triangular region shown in figure 24.6 in such a way that the points $x_1 = -1$ and $x_2 = 1$ are mapped into the points $w = -a$ and $w = a$, respectively, and the point $x_3 = \pm\infty$ is mapped into $w = ib$. Hence find a transformation that maps the upper half of the z-plane into the region $-a < r < a$, $s > 0$ of the w-plane, as shown in figure 24.7.*

Let us denote the angles at w_1 and w_2 in the w-plane by $\phi_1 = \phi_2 = \phi$, where $\phi = \tan^{-1}(b/a)$. Since x_3 is taken at infinity, we may omit the corresponding factor in (24.30) to obtain

$$w = \left\{ A \int_0^z (\xi + 1)^{(\phi/\pi)-1}(\xi - 1)^{(\phi/\pi)-1} \, d\xi \right\} + B$$

$$= \left\{ A \int_0^z (\xi^2 - 1)^{(\phi/\pi)-1} \, d\xi \right\} + B. \tag{24.31}$$

The required transformation may then be found by fixing the constants A and B as follows. Since the point $z = 0$ lies on the line segment $x_1 x_2$, it will be mapped onto the line segment $w_1 w_2$ in the w-plane, and by symmetry must be mapped onto the point $w = 0$. Thus setting $z = 0$ and $w = 0$ in (24.31) we obtain $B = 0$. An expression for A can be found in the form of an integral by setting (for example) $z = 1$ and $w = a$ in (24.31).

We may consider the region in the w-plane in figure 24.7 to be the limiting case of the triangular region in figure 24.6 with the vertex w_3 at infinity. Thus we may use the above, but with the angles at w_1 and w_2 set to $\phi = \pi/2$. From (24.31), we obtain

$$w = A \int_0^z \frac{d\xi}{\sqrt{\xi^2 - 1}} = iA \sin^{-1} z.$$

By setting $z = 1$ and $w = a$, we find $iA = 2a/\pi$, so the required transformation is

$$w = \frac{2a}{\pi} \sin^{-1} z. \ \blacktriangleleft$$

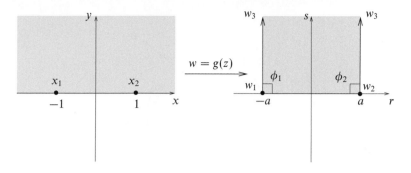

Figure 24.7   Transforming the upper half of the z-plane into the interior of the region $-a < r < a$, $s > 0$ in the w-plane.

24.8 Complex integrals

Corresponding to integration with respect to a real variable, it is possible to define integration with respect to a complex variable between two complex limits. Since the z-plane is two-dimensional there is clearly greater freedom and hence ambiguity in what is meant by a complex integral. If a complex function $f(z)$ is single-valued and continuous in some region R in the complex plane, then we can define the complex integral of $f(z)$ between two points A and B along some curve in R; its value will depend, in general, upon the path taken between A and B (see figure 24.8). However, we will find that for some paths that are different but bear a particular relationship to each other the value of the integral does *not* depend upon which of the paths is adopted.

Let a particular path C be described by a continuous (real) parameter t ($\alpha \leq t \leq \beta$) that gives successive positions on C by means of the equations

$$x = x(t), \qquad y = y(t), \tag{24.32}$$

with $t = \alpha$ and $t = \beta$ corresponding to the points A and B, respectively. Then the integral along path C of a continuous function $f(z)$ is written

$$\int_C f(z)\,dz \tag{24.33}$$

and can be given explicitly as a sum of real integrals as follows:

$$\begin{aligned}
\int_C f(z)\,dz &= \int_C (u + iv)(dx + i\,dy) \\
&= \int_C u\,dx - \int_C v\,dy + i\int_C u\,dy + i\int_C v\,dx \\
&= \int_\alpha^\beta u\frac{dx}{dt}\,dt - \int_\alpha^\beta v\frac{dy}{dt}\,dt + i\int_\alpha^\beta u\frac{dy}{dt}\,dt + i\int_\alpha^\beta v\frac{dx}{dt}\,dt.
\end{aligned} \tag{24.34}$$

845

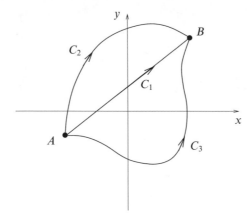

Figure 24.8   Some alternative paths for the integral of a function $f(z)$ between A and B.

The question of when such an integral exists will not be pursued, except to state that a sufficient condition is that dx/dt and dy/dt are continuous.

▶ *Evaluate the complex integral of $f(z) = z^{-1}$ along the circle $|z| = R$, starting and finishing at $z = R$.*

The path C_1 is parameterised as follows (figure 24.9(a)):

$$z(t) = R\cos t + iR\sin t, \qquad 0 \le t \le 2\pi,$$

whilst $f(z)$ is given by

$$f(z) = \frac{1}{x+iy} = \frac{x-iy}{x^2+y^2}.$$

Thus the real and imaginary parts of $f(z)$ are

$$u = \frac{x}{x^2+y^2} = \frac{R\cos t}{R^2} \qquad \text{and} \qquad v = \frac{-y}{x^2+y^2} = -\frac{R\sin t}{R^2}.$$

Hence, using expression (24.34),

$$\int_{C_1} \frac{1}{z}\, dz = \int_0^{2\pi} \frac{\cos t}{R}(-R\sin t)\, dt - \int_0^{2\pi} \left(\frac{-\sin t}{R}\right) R\cos t\, dt$$

$$+ i\int_0^{2\pi} \frac{\cos t}{R} R\cos t\, dt + i\int_0^{2\pi} \left(\frac{-\sin t}{R}\right)(-R\sin t)\, dt \qquad (24.35)$$

$$= 0 + 0 + i\pi + i\pi = 2\pi i. \ \blacktriangleleft$$

With a bit of experience, the reader may be able to evaluate integrals like the LHS of (24.35) directly without having to write them as four separate real integrals. In the present case,

$$\int_{C_1} \frac{dz}{z} = \int_0^{2\pi} \frac{-R\sin t + iR\cos t}{R\cos t + iR\sin t}\, dt = \int_0^{2\pi} i\, dt = 2\pi i. \qquad (24.36)$$

846

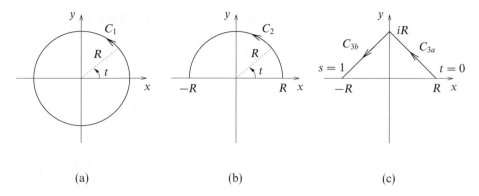

Figure 24.9 Different paths for an integral of $f(z) = z^{-1}$. See the text for details.

This very important result will be used many times later, and the following should be carefully noted: (i) its value, (ii) that this value is independent of R.

In the above example the contour was closed, and so it began and ended at the same point in the Argand diagram. We can evaluate complex integrals along open paths in a similar way.

▶ *Evaluate the complex integral of* $f(z) = z^{-1}$ *along the following paths (see figure 24.9):*

  (i) *the contour* C_2 *consisting of the semicircle* $|z| = R$ *in the half-plane* $y \geq 0$,

  (ii) *the contour* C_3 *made up of the two straight lines* C_{3a} *and* C_{3b}.

(i) This is just as in the previous example, except that now $0 \leq t \leq \pi$. With this change, we have from (24.35) or (24.36) that

$$\int_{C_2} \frac{dz}{z} = \pi i. \tag{24.37}$$

(ii) The straight lines that make up the contour C_3 may be parameterised as follows:

$$C_{3a}, \qquad z = (1-t)R + itR \qquad \text{for } 0 \leq t \leq 1;$$
$$C_{3b}, \qquad z = -sR + i(1-s)R \qquad \text{for } 0 \leq s \leq 1.$$

With these parameterisations the required integrals may be written

$$\int_{C_3} \frac{dz}{z} = \int_0^1 \frac{-R + iR}{R + t(-R + iR)} \, dt + \int_0^1 \frac{-R - iR}{iR + s(-R - iR)} \, ds. \tag{24.38}$$

If we could take over from real-variable theory that, for real t, $\int (a+bt)^{-1} \, dt = b^{-1} \ln(a+bt)$ even if a and b are complex, then these integrals could be evaluated immediately. However, to do this would be presuming to some extent what we wish to show, and so the evaluation

847

must be made in terms of entirely real integrals. For example, the first is given by

$$\int_0^1 \frac{-R+iR}{R(1-t)+itR}\, dt = \int_0^1 \frac{(-1+i)(1-t-it)}{(1-t)^2+t^2}\, dt$$

$$= \int_0^1 \frac{2t-1}{1-2t+2t^2}\, dt + i\int_0^1 \frac{1}{1-2t+2t^2}\, dt$$

$$= \frac{1}{2}\left[\ln(1-2t+2t^2)\right]_0^1 + \frac{i}{2}\left[2\tan^{-1}\left(\frac{t-\frac{1}{2}}{\frac{1}{2}}\right)\right]_0^1$$

$$= 0 + \frac{i}{2}\left[\frac{\pi}{2}-\left(-\frac{\pi}{2}\right)\right] = \frac{\pi i}{2}.$$

The second integral on the RHS of (24.38) can also be shown to have the value $\pi i/2$. Thus

$$\int_{C_3} \frac{dz}{z} = \pi i. \blacktriangleleft$$

Considering the results of the preceding two examples, which have common integrands and limits, some interesting observations are possible. Firstly, the two integrals from $z = R$ to $z = -R$, along C_2 and C_3, respectively, have the same value, even though the paths taken are different. It also follows that if we took a closed path C_4, given by C_2 from R to $-R$ and C_3 traversed backwards from $-R$ to R, then the integral round C_4 of z^{-1} would be zero (both parts contributing equal and opposite amounts). This is to be compared with result (24.36), in which closed path C_1, beginning and ending at the same place as C_4, yields a value $2\pi i$.

It is not true, however, that the integrals along the paths C_2 and C_3 are equal for any function $f(z)$, or, indeed, that their values are independent of R in general.

▶Evaluate the complex integral of $f(z) = \mathrm{Re}\, z$ along the paths C_1, C_2 and C_3 shown in figure 24.9.

(i) If we take $f(z) = \mathrm{Re}\, z$ and the contour C_1 then

$$\int_{C_1} \mathrm{Re}\, z\, dz = \int_0^{2\pi} R\cos t(-R\sin t + iR\cos t)\, dt = i\pi R^2.$$

(ii) Using C_2 as the contour,

$$\int_{C_2} \mathrm{Re}\, z\, dz = \int_0^{\pi} R\cos t(-R\sin t + iR\cos t)\, dt = \tfrac{1}{2}i\pi R^2.$$

(iii) Finally the integral along $C_3 = C_{3a} + C_{3b}$ is given by

$$\int_{C_3} \mathrm{Re}\, z\, dz = \int_0^1 (1-t)R(-R+iR)\, dt + \int_0^1 (-sR)(-R-iR)\, ds$$

$$= \tfrac{1}{2}R^2(-1+i) + \tfrac{1}{2}R^2(1+i) = iR^2. \blacktriangleleft$$

The results of this section demonstrate that the value of an integral between the same two points may depend upon the path that is taken between them but, at the same time, suggest that, under some circumstances, the value is independent of the path. The general situation is summarised in the result of the next section, namely Cauchy's theorem, which is the cornerstone of the integral calculus of complex variables.

Before discussing Cauchy's theorem, however, we note an important result concerning complex integrals that will be of some use later. Let us consider the integral of a function $f(z)$ along some path C. If M is an upper bound on the value of $|f(z)|$ on the path, i.e. $|f(z)| \leq M$ on C, and L is the length of the path C, then

$$\left| \int_C f(z)\, dz \right| \leq \int_c |f(z)||dz| \leq M \int_C dl = ML. \tag{24.39}$$

It is straightforward to verify that this result does indeed hold for the complex integrals considered earlier in this section.

24.9 Cauchy's theorem

Cauchy's theorem states that if $f(z)$ is an analytic function, and $f'(z)$ is continuous at each point within and on a closed contour C, then

$$\oint_C f(z)\, dz = 0. \tag{24.40}$$

In this statement and from now on we denote an integral around a closed contour by $\oint_C$.

To prove this theorem we will need the two-dimensional form of the divergence theorem, known as Green's theorem in a plane (see section 11.3). This says that if p and q are two functions with continuous first derivatives within and on a closed contour C (bounding a domain R) in the xy-plane, then

$$\iint_R \left(\frac{\partial p}{\partial x} + \frac{\partial q}{\partial y} \right) dxdy = \oint_C (p\, dy - q\, dx). \tag{24.41}$$

With $f(z) = u + iv$ and $dz = dx + i\, dy$, this can be applied to

$$I = \oint_C f(z)\, dz = \oint_C (u\, dx - v\, dy) + i \oint_C (v\, dx + u\, dy)$$

to give

$$I = \iint_R \left[\frac{\partial(-u)}{\partial y} + \frac{\partial(-v)}{\partial x} \right] dx\, dy + i \iint_R \left[\frac{\partial(-v)}{\partial y} + \frac{\partial u}{\partial x} \right] dx\, dy. \tag{24.42}$$

Now, recalling that $f(z)$ is analytic and therefore that the Cauchy–Riemann relations (24.5) apply, we see that each integrand is identically zero and thus I is also zero; this proves Cauchy's theorem.

In fact, the conditions of the above proof are more stringent than they need be. The continuity of $f'(z)$ is not necessary for the proof of Cauchy's theorem,

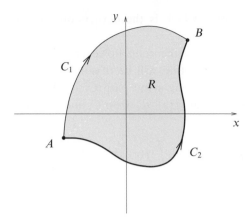

Figure 24.10 Two paths C_1 and C_2 enclosing a region R.

analyticity of $f(z)$ within and on C being sufficient. However, the proof then becomes more complicated and is too long to be given here.[§]

The connection between Cauchy's theorem and the zero value of the integral of z^{-1} around the composite path C_4 discussed towards the end of the previous section is apparent: the function z^{-1} is analytic in the two regions of the z-plane enclosed by contours (C_2 and C_{3a}) and (C_2 and C_{3b}).

▶Suppose two points A and B in the complex plane are joined by two different paths C_1 and C_2. Show that if $f(z)$ is an analytic function on each path and in the region enclosed by the two paths, then the integral of $f(z)$ is the same along C_1 and C_2.

The situation is shown in figure 24.10. Since $f(z)$ is analytic in R, it follows from Cauchy's theorem that we have

$$\int_{C_1} f(z)\,dz - \int_{C_2} f(z)\,dz = \oint_{C_1 - C_2} f(z)\,dz = 0,$$

since $C_1 - C_2$ forms a closed contour enclosing R. Thus we immediately obtain

$$\int_{C_1} f(z)\,dz = \int_{C_2} f(z)\,dz,$$

and so the values of the integrals along C_1 and C_2 are equal. ◀

An important application of Cauchy's theorem is in proving that, in some cases, it is possible to deform a closed contour C into another contour γ in such a way that the integrals of a function $f(z)$ around each of the contours have the same value.

[§] The reader may refer to almost any book that is devoted to complex variables and the theory of functions.

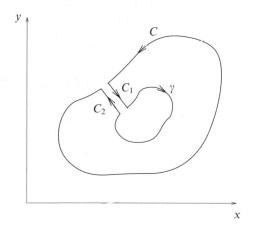

Figure 24.11   The contour used to prove the result (24.43).

▶Consider two closed contours C and γ in the Argand diagram, γ being sufficiently small that it lies completely within C. Show that if the function $f(z)$ is analytic in the region between the two contours then

$$\oint_C f(z)\, dz = \oint_\gamma f(z)\, dz. \tag{24.43}$$

To prove this result we consider a contour as shown in figure 24.11. The two close parallel lines C_1 and C_2 join γ and C, which are 'cut' to accommodate them. The new contour Γ so formed consists of C, C_1, γ and C_2.

Within the area bounded by Γ, the function $f(z)$ is analytic, and therefore, by Cauchy's theorem (24.40),

$$\oint_\Gamma f(z)\, dz = 0. \tag{24.44}$$

Now the parts C_1 and C_2 of Γ are traversed in opposite directions, and in the limit lie on top of each other, and so their contributions to (24.44) cancel. Thus

$$\oint_C f(z)\, dz + \oint_\gamma f(z)\, dz = 0. \tag{24.45}$$

The sense of the integral round γ is opposite to the conventional (anticlockwise) one, and so by traversing γ in the usual sense, we establish the result (24.43). ◀

A sort of converse of Cauchy's theorem is known as *Morera's theorem*, which states that if $f(z)$ is a continuous function of z in a closed domain R bounded by a curve C and, further, $\oint_C f(z)\, dz = 0$, then $f(z)$ is analytic in R.

24.10 Cauchy's integral formula

Another very important theorem in the theory of complex variables is *Cauchy's integral formula*, which states that if $f(z)$ is analytic within and on a closed contour C and z_0 is a point within C then

$$f(z_0) = \frac{1}{2\pi i} \oint_C \frac{f(z)}{z - z_0} \, dz. \tag{24.46}$$

This formula is saying that the value of an analytic function anywhere inside a closed contour is uniquely determined by its values on the contour[§] and that the specific expression (24.46) can be given for the value at the interior point.

We may prove Cauchy's integral formula by using (24.43) and taking γ to be a circle centred on the point $z = z_0$, of small enough radius ρ that it all lies inside C. Then, since $f(z)$ is analytic inside C, the integrand $f(z)/(z - z_0)$ is analytic in the space between C and γ. Thus, from (24.43), the integral around γ has the same value as that around C.

We then use the fact that any point z on γ is given by $z = z_0 + \rho \exp i\theta$ (and so $dz = i\rho \exp i\theta \, d\theta$). Thus the value of the integral around γ is given by

$$I = \oint_\gamma \frac{f(z)}{z - z_0} \, dz = \int_0^{2\pi} \frac{f(z_0 + \rho \exp i\theta)}{\rho \exp i\theta} i\rho \exp i\theta \, d\theta$$

$$= i \int_0^{2\pi} f(z_0 + \rho \exp i\theta) \, d\theta.$$

If the radius of the circle γ is now shrunk to zero, i.e. $\rho \to 0$, then $I \to 2\pi i f(z_0)$, thus establishing the result (24.46).

An extension to Cauchy's integral formula can be made, yielding an integral expression for $f'(z_0)$:

$$f'(z_0) = \frac{1}{2\pi i} \int_C \frac{f(z)}{(z - z_0)^2} \, dz, \tag{24.47}$$

under the same conditions as previously stated.

▶*Prove Cauchy's integral formula for $f'(z_0)$ given in (24.47).*

To show this, we use the definition of a derivative and (24.46) itself to evaluate

$$f'(z_0) = \lim_{h \to 0} \frac{f(z_0 + h) - f(z_0)}{h}$$

$$= \lim_{h \to 0} \left[\frac{1}{2\pi i} \oint_C \frac{f(z)}{h} \left(\frac{1}{z - z_0 - h} - \frac{1}{z - z_0} \right) dz \right]$$

$$= \lim_{h \to 0} \left[\frac{1}{2\pi i} \oint_C \frac{f(z)}{(z - z_0 - h)(z - z_0)} \, dz \right]$$

$$= \frac{1}{2\pi i} \oint_C \frac{f(z)}{(z - z_0)^2} \, dz,$$

which establishes result (24.47). ◀

[§] The similarity between this and the uniqueness theorem for the Laplace equation with Dirichlet boundary conditions (see chapter 20) is apparent.

Further, it may be proved by induction that the nth derivative of $f(z)$ is also given by a Cauchy integral,

$$f^{(n)}(z_0) = \frac{n!}{2\pi i} \oint_C \frac{f(z)\, dz}{(z-z_0)^{n+1}}. \tag{24.48}$$

Thus, if the value of the analytic function is known on C then not only may the value of the function at any interior point be calculated, but also the values of *all* its derivatives.

The observant reader will notice that (24.48) may also be obtained by the formal device of differentiating under the integral sign with respect to z_0 in Cauchy's integral formula (24.46):

$$
\begin{aligned}
f^{(n)}(z_0) &= \frac{1}{2\pi i} \oint_C \frac{\partial^n}{\partial z_0^n} \left[\frac{f(z)}{(z-z_0)} \right]\, dz \\
&= \frac{n!}{2\pi i} \oint_C \frac{f(z)\, dz}{(z-z_0)^{n+1}}.
\end{aligned}
$$

▶*Suppose that $f(z)$ is analytic inside and on a circle C of radius R centred on the point $z = z_0$. If $|f(z)| \le M$ on the circle, where M is some constant, show that*

$$|f^{(n)}(z_0)| \le \frac{Mn!}{R^n}. \tag{24.49}$$

From (24.48) we have

$$|f^{(n)}(z_0)| = \frac{n!}{2\pi} \left| \oint_C \frac{f(z)\, dz}{(z-z_0)^{n+1}} \right|,$$

and on using (24.39) this becomes

$$|f^{(n)}(z_0)| \le \frac{n!}{2\pi} \frac{M}{R^{n+1}} 2\pi R = \frac{Mn!}{R^n}.$$

This result is known as *Cauchy's inequality*. ◀

We may use Cauchy's inequality to prove *Liouville's theorem*, which states that if $f(z)$ is analytic and bounded for all z then f is a constant. Setting $n = 1$ in (24.49) and letting $R \to \infty$, we find $|f'(z_0)| = 0$ and hence $f'(z_0) = 0$. Since $f(z)$ is analytic for all z, we may take z_0 as any point in the z-plane and thus $f'(z) = 0$ for all z; this implies $f(z) = $ constant. Liouville's theorem may be used in turn to prove the *fundamental theorem of algebra* (see exercise 24.9).

24.11 Taylor and Laurent series

Following on from (24.48), we may establish *Taylor's theorem* for functions of a complex variable. If $f(z)$ is analytic inside and on a circle C of radius R centred on the point $z = z_0$, and z is a point inside C, then

$$f(z) = \sum_{n=0}^{\infty} a_n (z-z_0)^n, \tag{24.50}$$

where a_n is given by $f^{(n)}(z_0)/n!$. The Taylor expansion is valid inside the region of analyticity and, for any particular z_0, can be shown to be unique.

To prove Taylor's theorem (24.50), we note that, since $f(z)$ is analytic inside and on C, we may use Cauchy's formula to write $f(z)$ as

$$f(z) = \frac{1}{2\pi i} \oint_C \frac{f(\xi)}{\xi - z} \, d\xi, \tag{24.51}$$

where ξ lies on C. Now we may expand the factor $(\xi - z)^{-1}$ as a geometric series in $(z - z_0)/(\xi - z_0)$,

$$\frac{1}{\xi - z} = \frac{1}{\xi - z_0} \sum_{n=0}^{\infty} \left(\frac{z - z_0}{\xi - z_0} \right)^n,$$

so (24.51) becomes

$$\begin{aligned}
f(z) &= \frac{1}{2\pi i} \oint_C \frac{f(\xi)}{\xi - z_0} \sum_{n=0}^{\infty} \left(\frac{z - z_0}{\xi - z_0} \right)^n d\xi \\
&= \frac{1}{2\pi i} \sum_{n=0}^{\infty} (z - z_0)^n \oint_C \frac{f(\xi)}{(\xi - z_0)^{n+1}} \, d\xi \\
&= \frac{1}{2\pi i} \sum_{n=0}^{\infty} (z - z_0)^n \frac{2\pi i f^{(n)}(z_0)}{n!},
\end{aligned} \tag{24.52}$$

where we have used Cauchy's integral formula (24.48) for the derivatives of $f(z)$. Cancelling the factors of $2\pi i$, we thus establish the result (24.50) with $a_n = f^{(n)}(z_0)/n!$.

▶*Show that if $f(z)$ and $g(z)$ are analytic in some region R, and $f(z) = g(z)$ within some subregion S of R, then $f(z) = g(z)$ throughout R.*

It is simpler to consider the (analytic) function $h(z) = f(z) - g(z)$, and to show that because $h(z) = 0$ in S it follows that $h(z) = 0$ throughout R.

If we choose a point $z = z_0$ in S, then we can expand $h(z)$ in a Taylor series about z_0,

$$h(z) = h(z_0) + h'(z_0)(z - z_0) + \tfrac{1}{2} h''(z_0)(z - z_0)^2 + \cdots,$$

which will converge inside some circle C that extends at least as far as the nearest part of the boundary of R, since $h(z)$ is analytic in R. But since z_0 lies in S, we have

$$h(z_0) = h'(z_0) = h''(z_0) = \cdots = 0,$$

and so $h(z) = 0$ inside C. We may now expand about a new point, which can lie anywhere within C, and repeat the process. By continuing this procedure we may show that $h(z) = 0$ throughout R.

This result is called the *identity theorem* and, in fact, the equality of $f(z)$ and $g(z)$ throughout R follows from their equality along any curve of non-zero length in R, or even at a countably infinite number of points in R. ◀

So far we have assumed that $f(z)$ is analytic inside and on the (circular) contour C. If, however, $f(z)$ has a singularity inside C at the point $z = z_0$, then it cannot be expanded in a Taylor series. Nevertheless, suppose that $f(z)$ has a pole of order p at $z = z_0$ but is analytic at every other point inside and on C. Then the function $g(z) = (z - z_0)^p f(z)$ is analytic at $z = z_0$, and so may be expanded as a Taylor series about $z = z_0$:

$$g(z) = \sum_{n=0}^{\infty} b_n (z - z_0)^n. \tag{24.53}$$

Thus, for all z inside C, $f(z)$ will have a power series representation of the form

$$f(z) = \frac{a_{-p}}{(z - z_0)^p} + \cdots + \frac{a_{-1}}{z - z_0} + a_0 + a_1(z - z_0) + a_2(z - z_0)^2 + \cdots, \tag{24.54}$$

with $a_{-p} \neq 0$. Such a series, which is an extension of the Taylor expansion, is called a *Laurent series*. By comparing the coefficients in (24.53) and (24.54), we see that $a_n = b_{n+p}$. Now, the coefficients b_n in the Taylor expansion of $g(z)$ are seen from (24.52) to be given by

$$b_n = \frac{g^{(n)}(z_0)}{n!} = \frac{1}{2\pi i} \oint \frac{g(z)}{(z - z_0)^{n+1}} \, dz,$$

and so for the coefficients a_n in (24.54) we have

$$a_n = \frac{1}{2\pi i} \oint \frac{g(z)}{(z - z_0)^{n+1+p}} \, dz = \frac{1}{2\pi i} \oint \frac{f(z)}{(z - z_0)^{n+1}} \, dz,$$

an expression that is valid for both positive and negative n.

The terms in the Laurent series with $n \geq 0$ are collectively called the *analytic part*, whilst the remainder of the series, consisting of terms in inverse powers of $z - z_0$, is called the *principal part*. Depending on the nature of the point $z = z_0$, the principal part may contain an infinite number of terms, so that

$$f(z) = \sum_{n=-\infty}^{+\infty} a_n (z - z_0)^n. \tag{24.55}$$

In this case we would expect the principal part to converge only for $|(z - z_0)^{-1}|$ less than some constant, i.e. *outside* some circle centred on z_0. However, the analytic part will converge *inside* some (different) circle also centred on z_0. If the latter circle has the greater radius then the Laurent series will converge in the region R *between* the two circles (see figure 24.12); otherwise it does not converge at all.

In fact, it may be shown that any function $f(z)$ that is analytic in a region R between two such circles C_1 and C_2 centred on $z = z_0$ can be expressed as a Laurent series about z_0 that converges in R. We note that, depending on the nature of the point $z = z_0$, the inner circle may be a point (when the principal part contains only a finite number of terms) and the outer circle may have an infinite radius.

We may use the Laurent series of a function $f(z)$ about any point $z = z_0$ to

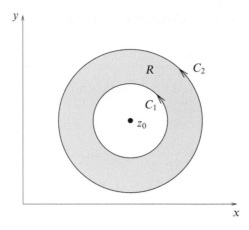

Figure 24.12   The region of convergence R for a Laurent series of $f(z)$ about a point $z = z_0$ where $f(z)$ has a singularity.

classify the nature of that point. If $f(z)$ is actually analytic at $z = z_0$, then in (24.55) all a_n for $n < 0$ must be zero. It may happen that not only are all a_n zero for $n < 0$ but a_0, a_1, ..., a_{m-1} are all zero as well. In this case, the first non-vanishing term in (24.55) is $a_m(z - z_0)^m$, with $m > 0$, and $f(z)$ is then said to have a *zero of order m* at $z = z_0$.

If $f(z)$ is not analytic at $z = z_0$, then two cases arise, as discussed above (p is here taken as positive):

    (i) it is possible to find an integer p such that $a_{-p} \neq 0$ but $a_{-p-k} = 0$ for all integers $k > 0$;

    (ii) it is not possible to find such a lowest value of $-p$.

In case (i), $f(z)$ is of the form (24.54) and is described as having a *pole of order p* at $z = z_0$; the value of a_{-1} (not a_{-p}) is called the *residue* of $f(z)$ at the pole $z = z_0$, and will play an important part in later applications.

For case (ii), in which the negatively decreasing powers of $z - z_0$ do not terminate, $f(z)$ is said to have an *essential singularity*. These definitions should be compared with those given in section 24.6.

▶*Find the Laurent series of*

$$f(z) = \frac{1}{z(z - 2)^3}$$

about the singularities $z = 0$ and $z = 2$ (separately). Hence verify that $z = 0$ is a pole of order 1 and $z = 2$ is a pole of order 3, and find the residue of $f(z)$ at each pole.

To obtain the Laurent series about $z = 0$, we make the factor in parentheses in the denominator take the form $(1 - \alpha z)$, where α is some constant, and thus obtain

$$f(z) = -\frac{1}{8z(1 - z/2)^3}$$

$$= -\frac{1}{8z}\left[1 + (-3)\left(-\frac{z}{2}\right) + \frac{(-3)(-4)}{2!}\left(-\frac{z}{2}\right)^2 + \frac{(-3)(-4)(-5)}{3!}\left(-\frac{z}{2}\right)^3 + \cdots\right]$$

$$= -\frac{1}{8z} - \frac{3}{16} - \frac{3z}{16} - \frac{5z^2}{32} - \cdots.$$

Since the lowest power of z is -1, the point $z = 0$ is a pole of order 1. The residue of $f(z)$ at $z = 0$ is simply the coefficient of z^{-1} in the Laurent expansion about that point and is equal to $-1/8$.

The Laurent series about $z = 2$ is most easily found by letting $z = 2 + \xi$ (or $z - 2 = \xi$) and substituting into the expression for $f(z)$ to obtain

$$f(z) = \frac{1}{(2 + \xi)\xi^3} = \frac{1}{2\xi^3(1 + \xi/2)}$$

$$= \frac{1}{2\xi^3}\left[1 - \left(\frac{\xi}{2}\right) + \left(\frac{\xi}{2}\right)^2 - \left(\frac{\xi}{2}\right)^3 + \left(\frac{\xi}{2}\right)^4 - \cdots\right]$$

$$= \frac{1}{2\xi^3} - \frac{1}{4\xi^2} + \frac{1}{8\xi} - \frac{1}{16} + \frac{\xi}{32} - \cdots$$

$$= \frac{1}{2(z - 2)^3} - \frac{1}{4(z - 2)^2} + \frac{1}{8(z - 2)} - \frac{1}{16} + \frac{z - 2}{32} - \cdots.$$

From this series we see that $z = 2$ is a pole of order 3 and that the residue of $f(z)$ at $z = 2$ is $1/8$. ◄

As we shall see in the next few sections, finding the residue of a function at a singularity is of crucial importance in the evaluation of complex integrals. Specifically, formulae exist for calculating the residue of a function at a particular (singular) point $z = z_0$ without having to expand the function explicitly as a Laurent series about z_0 and identify the coefficient of $(z - z_0)^{-1}$. The type of formula generally depends on the nature of the singularity at which the residue is required.

►*Suppose that $f(z)$ has a pole of order m at the point $z = z_0$. By considering the Laurent series of $f(z)$ about z_0, derive a general expression for the residue $R(z_0)$ of $f(z)$ at $z = z_0$. Hence evaluate the residue of the function*

$$f(z) = \frac{\exp iz}{(z^2 + 1)^2}$$

at the point $z = i$.

If $f(z)$ has a pole of order m at $z = z_0$, then its Laurent series about this point has the form

$$f(z) = \frac{a_{-m}}{(z - z_0)^m} + \cdots + \frac{a_{-1}}{(z - z_0)} + a_0 + a_1(z - z_0) + a_2(z - z_0)^2 + \cdots,$$

which, on multiplying both sides of the equation by $(z - z_0)^m$, gives

$$(z - z_0)^m f(z) = a_{-m} + a_{-m+1}(z - z_0) + \cdots + a_{-1}(z - z_0)^{m-1} + \cdots.$$

Differentiating both sides $m - 1$ times, we obtain

$$\frac{d^{m-1}}{dz^{m-1}}[(z - z_0)^m f(z)] = (m - 1)! \, a_{-1} + \sum_{n=1}^{\infty} b_n (z - z_0)^n,$$

for some coefficients b_n. In the limit $z \to z_0$, however, the terms in the sum disappear, and after rearranging we obtain the formula

$$R(z_0) = a_{-1} = \lim_{z \to z_0} \left\{ \frac{1}{(m - 1)!} \frac{d^{m-1}}{dz^{m-1}}[(z - z_0)^m f(z)] \right\}, \qquad (24.56)$$

which gives the value of the residue of $f(z)$ at the point $z = z_0$.

If we now consider the function

$$f(z) = \frac{\exp iz}{(z^2 + 1)^2} = \frac{\exp iz}{(z + i)^2 (z - i)^2},$$

we see immediately that it has poles of order 2 (*double* poles) at $z = i$ and $z = -i$. To calculate the residue at (for example) $z = i$, we may apply the formula (24.56) with $m = 2$. Performing the required differentiation, we obtain

$$\frac{d}{dz}[(z - i)^2 f(z)] = \frac{d}{dz} \left[\frac{\exp iz}{(z + i)^2} \right]$$

$$= \frac{1}{(z + i)^4}[(z + i)^2 i \exp iz - 2(\exp iz)(z + i)].$$

Setting $z = i$, we find the residue is given by

$$R(i) = \frac{1}{1!} \frac{1}{16} \left(-4ie^{-1} - 4ie^{-1} \right) = -\frac{i}{2e}. \blacktriangleleft$$

An important special case of (24.56) occurs when $f(z)$ has a *simple pole* (a pole of order 1) at $z = z_0$. Then the residue at z_0 is given by

$$R(z_0) = \lim_{z \to z_0} [(z - z_0)f(z)]. \qquad (24.57)$$

If $f(z)$ has a simple pole at $z = z_0$ and, as is often the case, has the form $g(z)/h(z)$, where $g(z)$ is analytic and non-zero at z_0 and $h(z_0) = 0$, then (24.57) becomes

$$R(z_0) = \lim_{z \to z_0} \frac{(z - z_0)g(z)}{h(z)} = g(z_0) \lim_{z \to z_0} \frac{(z - z_0)}{h(z)}$$

$$= g(z_0) \lim_{z \to z_0} \frac{1}{h'(z)} = \frac{g(z_0)}{h'(z_0)}, \qquad (24.58)$$

where we have used l'Hôpital's rule. This result often provides the simplest way of determining the residue at a simple pole.

24.12 Residue theorem

Having seen from Cauchy's theorem that the value of an integral round a closed contour C is zero if the integrand is analytic inside the contour, it is natural to ask what value it takes when the integrand is not analytic inside C. The answer to this is contained in the residue theorem, which we now discuss.

Suppose the function $f(z)$ has a pole of order m at the point $z = z_0$, and so can be written as a Laurent series about z_0 of the form

$$f(z) = \sum_{n=-m}^{\infty} a_n(z - z_0)^n. \qquad (24.59)$$

Now consider the integral I of $f(z)$ around a closed contour C that encloses $z = z_0$, but no other singular points. Using Cauchy's theorem, this integral has the same value as the integral around a circle γ of radius ρ centred on $z = z_0$, since $f(z)$ is analytic in the region between C and γ. On the circle we have $z = z_0 + \rho \exp i\theta$ (and $dz = i\rho \exp i\theta \, d\theta$), and so

$$I = \oint_{\gamma} f(z) \, dz$$

$$= \sum_{n=-m}^{\infty} a_n \oint (z - z_0)^n \, dz$$

$$= \sum_{n=-m}^{\infty} a_n \int_0^{2\pi} i\rho^{n+1} \exp[i(n + 1)\theta] \, d\theta.$$

For every term in the series with $n \neq -1$, we have

$$\int_0^{2\pi} i\rho^{n+1} \exp[i(n + 1)\theta] \, d\theta = \left[\frac{i\rho^{n+1} \exp[i(n + 1)\theta]}{i(n + 1)} \right]_0^{2\pi} = 0,$$

but for the $n = -1$ term we obtain

$$\int_0^{2\pi} i \, d\theta = 2\pi i.$$

Therefore only the term in $(z - z_0)^{-1}$ contributes to the value of the integral around γ (and therefore C), and I takes the value

$$I = \oint_C f(z) \, dz = 2\pi i a_{-1}. \qquad (24.60)$$

Thus the integral around any closed contour containing a single pole of general order m (or, by extension, an essential singularity) is equal to $2\pi i$ times the residue of $f(z)$ at $z = z_0$.

If we extend the above argument to the case where $f(z)$ is continuous within and on a closed contour C and analytic, except for a finite number of poles, within C, then we arrive at the *residue theorem*

$$\oint_C f(z) \, dz = 2\pi i \sum_j R_j, \qquad (24.61)$$

where $\sum_j R_j$ is the sum of the residues of $f(z)$ at its poles within C.

The method of proof is indicated by figure 24.13, in which (a) shows the original contour C referred to in (24.61) and (b) shows a contour C' giving the same value

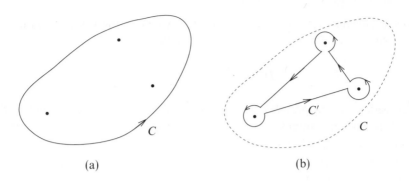

(a)   (b)

Figure 24.13   The contours used to prove the residue theorem: (a) the original contour; (b) the contracted contour encircling each of the poles.

to the integral, because f is analytic between C and C'. Now the contribution to the C' integral from the polygon (a triangle for the case illustrated) joining the small circles is zero, since f is also analytic inside C'. Hence the whole value of the integral comes from the circles and, by result (24.60), each of these contributes $2\pi i$ times the residue at the pole it encloses. All the circles are traversed in their positive sense if C is thus traversed and so the residue theorem follows. Formally, Cauchy's theorem (24.40) is a particular case of (24.61) in which C encloses no poles.

Finally we prove another important result, for later use. Suppose that $f(z)$ has a simple pole at $z = z_0$ and so may be expanded as the Laurent series

$$f(z) = \phi(z) + a_{-1}(z - z_0)^{-1},$$

where $\phi(z)$ is analytic within some neighbourhood surrounding z_0. We wish to find an expression for the integral I of $f(z)$ along an *open* contour C, which is the arc of a circle of radius ρ centred on $z = z_0$ given by

$$|z - z_0| = \rho, \qquad \theta_1 \le \arg(z - z_0) \le \theta_2, \tag{24.62}$$

where ρ is chosen small enough that no singularity of f, other than $z = z_0$, lies within the circle. Then I is given by

$$I = \int_C f(z)\, dz = \int_C \phi(z)\, dz + a_{-1} \int_C (z - z_0)^{-1}\, dz.$$

If the radius of the arc C is now allowed to tend to zero, then the first integral tends to zero, since the path becomes of zero length and ϕ is analytic and therefore continuous along it. On C, $z = \rho e^{i\theta}$ and hence the required expression for I is

$$I = \lim_{\rho \to 0} \int_C f(z)\, dz = \lim_{\rho \to 0} \left(a_{-1} \int_{\theta_1}^{\theta_2} \frac{1}{\rho e^{i\theta}} i\rho e^{i\theta}\, d\theta \right) = ia_{-1}(\theta_2 - \theta_1). \tag{24.63}$$

We note that result (24.60) is a special case of (24.63) in which θ_2 is equal to $\theta_1 + 2\pi$.

24.13 Definite integrals using contour integration

The remainder of this chapter is devoted to methods of applying contour integration and the residue theorem to various types of definite integral. However, three applications of contour integration, in which obtaining a value for the integral is not the prime purpose of the exercise, have been postponed until chapter 25. They are the location of the zeros of a complex polynomial, the evaluation of the sums of certain infinite series and the determination of inverse Laplace transforms.

For the integral evaluations considered here, not much preamble is given since, for this material, the simplest explanation is felt to be via a series of worked examples that can be used as models.

24.13.1 Integrals of sinusoidal functions

Suppose that an integral of the form

$$\int_0^{2\pi} F(\cos\theta, \sin\theta)\, d\theta \tag{24.64}$$

is to be evaluated. It can be made into a contour integral around the unit circle C by writing $z = \exp i\theta$, and hence

$$\cos\theta = \tfrac{1}{2}(z + z^{-1}), \qquad \sin\theta = -\tfrac{1}{2}i(z - z^{-1}), \qquad d\theta = -iz^{-1}\, dz. \tag{24.65}$$

This contour integral can then be evaluated using the residue theorem, provided the transformed integrand has only a finite number of poles inside the unit circle and none on it.

> ▶ *Evaluate*
>
> $$I = \int_0^{2\pi} \frac{\cos 2\theta}{a^2 + b^2 - 2ab\cos\theta}\, d\theta, \qquad b > a > 0. \tag{24.66}$$

By de Moivre's theorem (section 3.4),

$$\cos n\theta = \tfrac{1}{2}(z^n + z^{-n}). \tag{24.67}$$

Using $n = 2$ in (24.67) and straightforward substitution for the other functions of θ in (24.66) gives

$$I = \frac{i}{2ab} \oint_C \frac{z^4 + 1}{z^2(z - a/b)(z - b/a)}\, dz.$$

Thus there are two poles inside C, a double pole at $z = 0$ and a simple pole at $z = a/b$ (recall that $b > a$).

We could find the residue of the integrand at $z = 0$ by expanding the integrand as a Laurent series in z and identifying the coefficient of z^{-1}. Alternatively, we may use the formula (24.56) with $m = 2$. Choosing the latter method and denoting the integrand by $f(z)$, we have

$$\frac{d}{dz}[z^2 f(z)] = \frac{d}{dz}\left[\frac{z^4 + 1}{(z - a/b)(z - b/a)}\right]$$

$$= \frac{(z - a/b)(z - b/a)4z^3 - (z^4 + 1)[(z - a/b) + (z - b/a)]}{(z - a/b)^2(z - b/a)^2}.$$

Now setting $z = 0$ and applying (24.56), we find

$$R(0) = \frac{a}{b} + \frac{b}{a}.$$

For the simple pole at $z = a/b$, equation (24.57) gives the residue as

$$R(a/b) = \lim_{z \to (a/b)} \left[(z - a/b)f(z)\right] = \frac{(a/b)^4 + 1}{(a/b)^2(a/b - b/a)}$$

$$= -\frac{a^4 + b^4}{ab(b^2 - a^2)}.$$

Therefore by the residue theorem

$$I = 2\pi i \times \frac{i}{2ab}\left[\frac{a^2 + b^2}{ab} - \frac{a^4 + b^4}{ab(b^2 - a^2)}\right] = \frac{2\pi a^2}{b^2(b^2 - a^2)}. \blacktriangleleft$$

24.13.2 Some infinite integrals

We next consider the evaluation of an integral of the form

$$\int_{-\infty}^{\infty} f(x)\,dx,$$

where $f(z)$ has the following properties:

(i) $f(z)$ is analytic in the upper half-plane, $\text{Im}\,z \geq 0$, except for a finite number of poles, none of which is on the real axis;

(ii) on a semicircle Γ of radius R (figure 24.14), R times the maximum of $|f|$ on Γ tends to zero as $R \to \infty$ (a sufficient condition is that $zf(z) \to 0$ as $|z| \to \infty$);

(iii) $\int_{-\infty}^{0} f(x)\,dx$ and $\int_{0}^{\infty} f(x)\,dx$ both exist.

Since

$$\left|\int_{\Gamma} f(z)\,dz\right| \leq 2\pi R \times (\text{maximum of } |f| \text{ on } \Gamma),$$

condition (ii) ensures that the integral along Γ tends to zero as $R \to \infty$, after which it is obvious from the residue theorem that the required integral is given by

$$\int_{-\infty}^{\infty} f(x)\,dx = 2\pi i \times (\text{sum of the residues at poles with } \text{Im}\,z > 0).$$

$$(24.68)$$

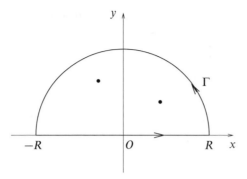

Figure 24.14   A semicircular contour in the upper half-plane.

▶ *Evaluate*

$$I = \int_0^\infty \frac{dx}{(x^2 + a^2)^4}, \qquad \text{where } a \text{ is real.}$$

The complex function $(z^2 + a^2)^{-4}$ has poles of order 4 at $z = \pm ai$, of which only $z = ai$ is in the upper half-plane. Conditions (ii) and (iii) are clearly satisfied. For higher-order poles, formula (24.56) for evaluating residues can be tiresome to apply. So, instead, we put $z = ai + \xi$ and expand for small ξ to obtain[§]

$$\frac{1}{(z^2 + a^2)^4} = \frac{1}{(2ai\xi + \xi^2)^4} = \frac{1}{(2ai\xi)^4}\left(1 - \frac{i\xi}{2a}\right)^{-4}.$$

The coefficient of ξ^{-1} is given by

$$\frac{1}{(2a)^4}\frac{(-4)(-5)(-6)}{3!}\left(\frac{-i}{2a}\right)^3 = \frac{-5i}{32a^7},$$

and hence by the residue theorem

$$\int_{-\infty}^\infty \frac{dx}{(x^2 + a^2)^4} = \frac{10\pi}{32a^7},$$

and so $I = 5\pi/(32a^7)$. ◀

Condition (i) of the previous method required there to be no poles of the integrand on the real axis, but in fact simple poles on the real axis can be accommodated by indenting the contour as shown in figure 24.15. The indentation at the pole $z = z_0$ is in the form of a semicircle γ of radius ρ in the upper half-plane, thus excluding the pole from the interior of the contour.

[§] This illustrates another useful technique for determining residues.

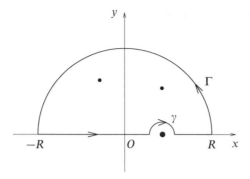

Figure 24.15   An indented contour used when the integrand has a simple pole on the real axis.

What is then obtained from a contour integration, apart from the contributions for Γ and γ, is called the *principal value of the integral*, defined as $\rho \to 0$ by

$$P \int_{-R}^{R} f(x)\, dx \equiv \int_{-R}^{z_0-\rho} f(x)\, dx + \int_{z_0+\rho}^{R} f(x)\, dx.$$

The remainder of the calculation goes through as before, but the contribution from the semicircle, γ, must be included. Result (24.63) of section 24.12 shows that since only a simple pole is involved its contribution is

$$-ia_{-1}\pi, \tag{24.69}$$

where a_{-1} is the residue at the pole and the minus sign arises because γ is traversed in the clockwise (negative) sense.

We defer giving an example of an indented contour until we have established *Jordan's lemma*; we will then work through an example illustrating both. Jordan's lemma enables infinite integrals involving sinusoidal functions to be evaluated.

For a function $f(z)$ of a complex variable z, if

 (i) *$f(z)$ is analytic in the upper half-plane except for a finite number of poles in $\operatorname{Im} z > 0$,*
 (ii) *the maximum of $|f(z)| \to 0$ as $|z| \to \infty$ in the upper half-plane,*
 (iii) *$m > 0$,*

then

$$I_\Gamma = \int_\Gamma e^{imz} f(z)\, dz \to 0 \qquad as\ R \to \infty, \tag{24.70}$$

where Γ is the same semicircular contour as in figure 24.14.

Note that this condition (ii) is less stringent than the earlier condition (ii) (see the start of this section), since we now only require $M(R) \to 0$ and not $RM(R) \to 0$, where M is the maximum[§] of $|f(z)|$ on $|z| = R$.

[§] More strictly, the least upper bound.

The proof of the lemma is straightforward once it has been observed that, for $0 \leq \theta \leq \pi/2$,

$$1 \geq \frac{\sin \theta}{\theta} \geq \frac{2}{\pi}. \tag{24.71}$$

Then, since on Γ we have $|\exp(imz)| = |\exp(-mR \sin \theta)|$,

$$I_\Gamma \leq \int_\Gamma |e^{imz} f(z)| \, |dz| \leq MR \int_0^\pi e^{-mR \sin \theta} \, d\theta = 2MR \int_0^{\pi/2} e^{-mR \sin \theta} \, d\theta.$$

Thus, using (24.71),

$$I_\Gamma \leq 2MR \int_0^{\pi/2} e^{-mR(2\theta/\pi)} \, d\theta = \frac{\pi M}{m} \left(1 - e^{-mR} \right) < \frac{\pi M}{m};$$

hence, as $R \to \infty$, I_Γ tends to zero since M tends to zero.

> ▶ Find the principal value of
> $$\int_{-\infty}^{\infty} \frac{\cos mx}{x - a} \, dx, \qquad \text{for a real, } m > 0.$$

Consider the function $(z - a)^{-1} \exp(imz)$; although it has no poles in the upper half-plane it does have a simple pole at $z = a$, and further $|(z - a)^{-1}| \to 0$ as $|z| \to \infty$. We will use a contour like that shown in figure 24.15 and apply the residue theorem. Symbolically,

$$\int_{-R}^{a-\rho} + \int_\gamma + \int_{a+\rho}^R + \int_\Gamma = 0. \tag{24.72}$$

Now as $R \to \infty$ and $\rho \to 0$ we have $\int_\Gamma \to 0$, by Jordan's lemma, and from (24.68) and (24.69) we obtain

$$P \int_{-\infty}^{\infty} \frac{e^{imx}}{x - a} \, dx - i\pi a_{-1} = 0, \tag{24.73}$$

where a_{-1} is the residue of $(z - a)^{-1} \exp(imz)$ at $z = a$, which is $\exp(ima)$. Then taking the real and imaginary parts of (24.73) gives

$$P \int_{-\infty}^{\infty} \frac{\cos mx}{x - a} \, dx = -\pi \sin ma, \qquad \text{as required,}$$

$$P \int_{-\infty}^{\infty} \frac{\sin mx}{x - a} \, dx = \pi \cos ma, \qquad \text{as a bonus.} \ ◀$$

24.13.3 Integrals of multivalued functions

We have discussed briefly some of the properties and difficulties associated with certain multivalued functions such as $z^{1/2}$ or $\text{Ln} \, z$. It was mentioned that one method of managing such functions is by means of a 'cut plane'. A similar technique can be used with advantage to evaluate some kinds of infinite integral involving real functions for which the corresponding complex functions are multivalued. A typical contour employed for functions with a single branch point

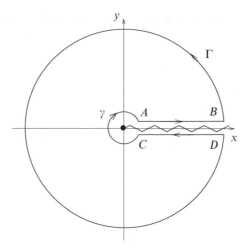

Figure 24.16   A typical cut-plane contour for use with multivalued functions that have a single branch point located at the origin.

located at the origin is shown in figure 24.16. Here Γ is a large circle of radius R and γ is a small one of radius ρ, both centred on the origin. Eventually we will let $R \to \infty$ and $\rho \to 0$.

The success of the method is due to the fact that because the integrand is multivalued, its values along the two lines AB and CD joining $z = \rho$ to $z = R$ are *not* equal and opposite although both are related to the corresponding real integral. Again an example provides the best explanation.

▶ *Evaluate*

$$I = \int_0^\infty \frac{dx}{(x + a)^3 x^{1/2}}, \qquad a > 0.$$

We consider the integrand $f(z) = (z + a)^{-3} z^{-1/2}$ and note that $|zf(z)| \to 0$ on the two circles as $\rho \to 0$ and $R \to \infty$. Thus the two circles make no contribution to the contour integral.

The only pole of the integrand inside the contour is at $z = -a$ (and is of order 3). To determine its residue we put $z = -a + \xi$ and expand (noting that $(-a)^{1/2}$ equals $a^{1/2} \exp(i\pi/2) = ia^{1/2}$):

$$\frac{1}{(z + a)^3 z^{1/2}} = \frac{1}{\xi^3 ia^{1/2}(1 - \xi/a)^{1/2}}$$

$$= \frac{1}{i\xi^3 a^{1/2}} \left(1 + \frac{1}{2}\frac{\xi}{a} + \frac{3}{8}\frac{\xi^2}{a^2} + \cdots \right).$$

The residue is thus $-3i/(8a^{5/2})$.

The residue theorem (24.61) now gives

$$\int_{AB} + \int_\Gamma + \int_{DC} + \int_\gamma = 2\pi i \left(\frac{-3i}{8a^{5/2}} \right).$$

We have seen that $\int_\Gamma$ and $\int_\gamma$ vanish, and if we denote z by x along the line AB then it has the value $z = x \exp 2\pi i$ along the line DC (note that $\exp 2\pi i$ must not be set equal to 1 until after the substitution for z has been made in $\int_{DC}$). Substituting these expressions,

$$\int_0^\infty \frac{dx}{(x+a)^3 x^{1/2}} + \int_\infty^0 \frac{dx}{[x \exp 2\pi i + a]^3 x^{1/2} \exp(\frac{1}{2} 2\pi i)} = \frac{3\pi}{4a^{5/2}}.$$

Thus

$$\left(1 - \frac{1}{\exp \pi i}\right) \int_0^\infty \frac{dx}{(x+a)^3 x^{1/2}} = \frac{3\pi}{4a^{5/2}}$$

and

$$I = \frac{1}{2} \times \frac{3\pi}{4a^{5/2}}. \blacktriangleleft$$

Several other examples of integrals of multivalued functions around a variety of contours are included in the exercises that follow.

24.14 Exercises

24.1 Find an analytic function of $z = x + iy$ whose imaginary part is

$$(y \cos y + x \sin y) \exp x.$$

24.2 Find a function $f(z)$, analytic in a suitable part of the Argand diagram, for which

$$\text{Re } f = \frac{\sin 2x}{\cosh 2y - \cos 2x}.$$

Where are the singularities of $f(z)$?

24.3 Find the radii of convergence of the following Taylor series:

$$(a) \sum_{n=2}^\infty \frac{z^n}{\ln n}, \quad (b) \sum_{n=1}^\infty \frac{n! z^n}{n^n},$$

$$(c) \sum_{n=1}^\infty z^n n^{\ln n}, \quad (d) \sum_{n=1}^\infty \left(\frac{n+p}{n}\right)^{n^2} z^n, \text{ with } p \text{ real.}$$

24.4 Find the Taylor series expansion about the origin of the function $f(z)$ defined by

$$f(z) = \sum_{r=1}^\infty (-1)^{r+1} \sin\left(\frac{pz}{r}\right),$$

where p is a constant. Hence verify that $f(z)$ is a convergent series for all z.

24.5 Determine the types of singularities (if any) possessed by the following functions at $z = 0$ and $z = \infty$:

(a) $(z-2)^{-1}$, (b) $(1+z^3)/z^2$, (c) $\sinh(1/z)$,
(d) e^z/z^3, (e) $z^{1/2}/(1+z^2)^{1/2}$.

24.6 Identify the zeros, poles and essential singularities of the following functions:

(a) $\tan z$, (b) $[(z-2)/z^2] \sin[1/(1-z)]$, (c) $\exp(1/z)$,
(d) $\tan(1/z)$, (e) $z^{2/3}$.

24.7   Find the real and imaginary parts of the functions (i) z^2, (ii) e^z, and (iii) $\cosh \pi z$. By considering the values taken by these parts on the boundaries of the region $0 \le x, y \le 1$, determine the solution of Laplace's equation in that region that satisfies the boundary conditions

$$\phi(x, 0) = 0, \qquad\qquad \phi(0, y) = 0,$$
$$\phi(x, 1) = x, \qquad\qquad \phi(1, y) = y + \sin \pi y.$$

24.8   Show that the transformation

$$w = \int_0^z \frac{1}{(\zeta^3 - \zeta)^{1/2}} \, d\zeta$$

transforms the upper half-plane into the interior of a square that has one corner at the origin of the w-plane and sides of length L, where

$$L = \int_0^{\pi/2} \operatorname{cosec}^{1/2} \theta \, d\theta.$$

24.9   The *fundamental theorem of algebra* states that, for a complex polynomial $p_n(z)$ of degree n, the equation $p_n(z) = 0$ has precisely n complex roots. By applying Liouville's theorem (see the end of section 24.10) to $f(z) = 1/p_n(z)$, prove that $p_n(z) = 0$ has at least one complex root. Factor out that root to obtain $p_{n-1}(z)$ and, by repeating the process, prove the above theorem.

24.10   Show that, if a is a positive real constant, the function $\exp(iaz^2)$ is analytic and $\to 0$ as $|z| \to \infty$ for $0 < \arg z \le \pi/4$. By applying Cauchy's theorem to a suitable contour prove that

$$\int_0^\infty \cos(ax^2) \, dx = \sqrt{\frac{\pi}{8a}}.$$

24.11   The function

$$f(z) = (1 - z^2)^{1/2}$$

of the complex variable z is defined to be real and positive on the real axis in the range $-1 < x < 1$. Using cuts running along the real axis for $1 < x < +\infty$ and $-\infty < x < -1$, show how $f(z)$ is made single-valued and evaluate it on the upper and lower sides of both cuts.

Use these results and a suitable contour in the complex z-plane to evaluate the integral

$$I = \int_1^\infty \frac{dx}{x(x^2 - 1)^{1/2}}.$$

Confirm your answer by making the substitution $x = \sec \theta$.

24.12   By considering the real part of

$$\int \frac{-iz^{n-1} \, dz}{1 - a(z + z^{-1}) + a^2},$$

where $z = \exp i\theta$ and n is a non-negative integer, evaluate

$$\int_0^\pi \frac{\cos n\theta}{1 - 2a\cos\theta + a^2} \, d\theta$$

for a real and > 1.

24.13   Prove that if $f(z)$ has a simple zero at z_0, then $1/f(z)$ has residue $1/f'(z_0)$ there. Hence evaluate

$$\int_{-\pi}^\pi \frac{\sin\theta}{a - \sin\theta} \, d\theta,$$

where a is real and > 1.

24.14   Prove that, for $\alpha > 0$, the integral

$$\int_0^\infty \frac{t \sin \alpha t}{1 + t^2} \, dt$$

has the value $(\pi/2) \exp(-\alpha)$.

24.15   Prove that

$$\int_0^\infty \frac{\cos mx}{4x^4 + 5x^2 + 1} \, dx = \frac{\pi}{6} \left(4e^{-m/2} - e^{-m} \right) \qquad \text{for } m > 0.$$

24.16   Show that the principal value of the integral

$$\int_{-\infty}^\infty \frac{\cos(x/a)}{x^2 - a^2} \, dx$$

is $-(\pi/a) \sin 1$.

24.17   The following is an alternative (and roundabout!) way of evaluating the Gaussian integral.

     (a)   Prove that the integral of $[\exp(i\pi z^2)]\mathrm{cosec}\,\pi z$ around the parallelogram with corners $\pm 1/2 \pm R \exp(i\pi/4)$ has the value $2i$.

     (b)   Show that the parts of the contour parallel to the real axis do not contribute when $R \to \infty$.

     (c)   Evaluate the integrals along the other two sides by putting $z' = r \exp(i\pi/4)$ and working in terms of $z' + \frac{1}{2}$ and $z' - \frac{1}{2}$. Hence, by letting $R \to \infty$ show that

$$\int_{-\infty}^\infty e^{-\pi r^2} \, dr = 1.$$

24.18   By applying the residue theorem around a wedge-shaped contour of angle $2\pi/n$, with one side along the real axis, prove that the integral

$$\int_0^\infty \frac{dx}{1 + x^n},$$

where n is real and ≥ 2, has the value $(\pi/n)\mathrm{cosec}\,(\pi/n)$.

24.19   Using a suitable cut plane, prove that if α is real and $0 < \alpha < 1$ then

$$\int_0^\infty \frac{x^{-\alpha}}{1 + x} \, dx$$

has the value $\pi \, \mathrm{cosec} \, \pi\alpha$.

24.20   Show that

$$\int_0^\infty \frac{\ln x}{x^{3/4}(1 + x)} \, dx = -\sqrt{2}\pi^2.$$

24.21   By integrating a suitable function around a large semicircle in the upper half-plane and a small semicircle centred on the origin, determine the value of

$$I = \int_0^\infty \frac{(\ln x)^2}{1 + x^2} \, dx$$

and deduce, as a by-product of your calculation, that

$$\int_0^\infty \frac{\ln x}{1 + x^2} \, dx = 0.$$

24.22   The equation of an ellipse in plane polar coordinates r, θ, with one of its foci at the origin, is

$$\frac{l}{r} = 1 - \epsilon \cos \theta,$$

where l is a length (that of the latus rectum) and ϵ $(0 < \epsilon < 1)$ is the eccentricity of the ellipse. Express the area of the ellipse as an integral around the unit circle in the complex plane, and show that the only singularity of the integrand inside the circle is a double pole at $z_0 = \epsilon^{-1} - (\epsilon^{-2} - 1)^{1/2}$.

By setting $z = z_0 + \xi$ and expanding the integrand in powers of ξ, find the residue at z_0 and hence show that the area is equal to $\pi l^2 (1 - \epsilon^2)^{-3/2}$.

[In terms of the semi-axes a and b of the ellipse, $l = b^2/a$ and $\epsilon^2 = (a^2 - b^2)/a^2$.]

24.15 Hints and answers

24.1 $\partial u / \partial y = -(\exp x)(y \cos y + x \sin y + \sin y);\ z \exp z$.

24.3   (a) 1; (b) e; (c) 1; (d) e^{-p}.

24.5   (a) Analytic, analytic; (b) double pole, single pole; (c) essential singularity, analytic; (d) triple pole, essential singularity; (e) branch point, branch point.

24.7   (i) $x^2 - y^2, 2xy$; (ii) $e^x \cos y, e^x \sin y$; (iii) $\cosh \pi x \cos \pi y, \sinh \pi x \sin \pi y$;
$\phi(x, y) = xy + (\sinh \pi x \sin \pi y)/\sinh \pi$.

24.9   Assume that $p_r(x)$ $(r = n, n-1, \ldots, 1)$ has no roots and then argue by the method of contradiction.

24.11   With $0 \leq \theta_1 < 2\pi$ and $-\pi < \theta_2 \leq \pi$, $f(z) = (r_1 r_2)^{1/2} \exp[i(\theta_1 + \theta_2 - \pi)]$. The four values are $\pm i(x^2 - 1)^{1/2}$, with the plus sign corresponding to points near the cut that lie in the second and fourth quadrants. $I = \pi/2$.

24.13   The only pole inside the unit circle is at $z = ia - i(a^2 - 1)^{1/2}$; the residue is given by $-(i/2)(a^2 - 1)^{-1/2}$; the integral has value $2\pi[a(a^2 - 1)^{-1/2} - 1]$.

24.15   Factorise the denominator, showing that the relevant simple poles are at $i/2$ and i.

24.17   (a) The only pole is at the origin with residue π^{-1};
(b) each is $O[\exp(-\pi R^2 \mp \sqrt{2\pi}R)]$;
(c) the sum of the integrals is $2i \int_{-R}^{R} \exp(-\pi r^2)\, dr$.

24.19   Use a contour like that shown in figure 24.16.

24.21   Note that $\rho \ln^n \rho \to 0$ as $\rho \to 0$ for all n. When z is on the negative real axis, $(\ln z)^2$ contains three terms; one of the corresponding integrals is a standard form. The residue at $z = i$ is $i\pi^2/8$; $I = \pi^3/8$.

25

Applications of complex variables

In chapter 24, we developed the basic theory of the functions of a complex variable, $z = x + iy$, studied their analyticity (differentiability) properties and derived a number of results concerned with values of contour integrals in the complex plane. In this current chapter we will show how some of those results and properties can be exploited to tackle problems arising directly from physical situations or from apparently unrelated parts of mathematics.

In the former category will be the use of the differential properties of the real and imaginary parts of a function of a complex variable to solve problems involving Laplace's equation in two dimensions, whilst an example of the latter might be the summation of certain types of infinite series. Other applications, such as the Bromwich inversion formula for Laplace transforms, appear as mathematical problems that have their origins in physical applications; the Bromwich inversion enables us to extract the spatial or temporal response of a system to an initial input from the representation of that response in 'frequency space' – or, more correctly, imaginary frequency space.

Other topics that will be considered are the location of the (complex) zeros of a polynomial, the approximate evaluation of certain types of contour integrals using the methods of steepest descent and stationary phase, and the so-called 'phase-integral' solutions to some differential equations. For each of these a brief introduction is given at the start of the relevant section and to repeat them here would be pointless. We will therefore move on to our first topic of complex potentials.

25.1 Complex potentials

Towards the end of section 24.2 of the previous chapter it was shown that the real and the imaginary parts of an analytic function of z are separately solutions of Laplace's equation in two dimensions. Analytic functions thus offer a possible way

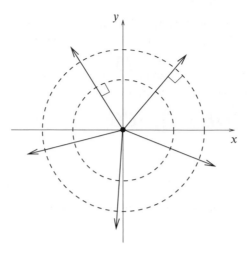

Figure 25.1 The equipotentials (dashed circles) and field lines (solid lines) for a line charge perpendicular to the z-plane.

of solving some two-dimensional physical problems describable by a potential satisfying $\nabla^2 \phi = 0$. The general method is known as that of *complex potentials*.

We also found that if $f = u + iv$ is an analytic function of z then any curve $u = $ constant intersects any curve $v = $ constant at right angles. In the context of solutions of Laplace's equation, this result implies that the real and imaginary parts of $f(z)$ have an additional connection between them, for if the set of contours on which one of them is a constant represents the equipotentials of a system then the contours on which the other is constant, being orthogonal to each of the first set, must represent the *corresponding* field lines or stream lines, depending on the context. The analytic function f is the complex potential. It is conventional to use ϕ and ψ (rather than u and v) to denote the real and imaginary parts of a complex potential, so that $f = \phi + i\psi$.

As an example, consider the function

$$f(z) = \frac{-q}{2\pi\epsilon_0} \ln z \qquad (25.1)$$

in connection with the physical situation of a line charge of strength q per unit length passing through the origin, perpendicular to the z-plane (figure 25.1). Its real and imaginary parts are

$$\phi = \frac{-q}{2\pi\epsilon_0} \ln |z|, \qquad \psi = \frac{-q}{2\pi\epsilon_0} \arg z. \qquad (25.2)$$

The contours in the z-plane of $\phi = $ constant are concentric circles and those of $\psi = $ constant are radial lines. As expected, these are orthogonal sets, but in addition they are, respectively, the equipotentials and electric field lines appropriate to the field produced by the line charge. The minus sign is needed in (25.1) because the value of ϕ must decrease with increasing distance from the origin.

Suppose we make the choice that the real part ϕ of the analytic function f gives the conventional potential function; ψ could equally well be selected. Then we may consider how the direction and magnitude of the field are related to f.

▶*Show that for any complex (electrostatic) potential $f(z)$ the strength of the electric field is given by $E = |f'(z)|$ and that its direction makes an angle of $\pi - \arg[f'(z)]$ with the x-axis.*

Because $\phi = $ constant is an equipotential, the field has components

$$E_x = -\frac{\partial \phi}{\partial x} \quad \text{and} \quad E_y = -\frac{\partial \phi}{\partial y}. \tag{25.3}$$

Since f is analytic, (i) we may use the Cauchy–Riemann relations (24.5) to change the second of these, obtaining

$$E_x = -\frac{\partial \phi}{\partial x} \quad \text{and} \quad E_y = \frac{\partial \psi}{\partial x}; \tag{25.4}$$

(ii) the direction of differentiation at a point is immaterial and so

$$\frac{df}{dz} = \frac{\partial f}{\partial x} = \frac{\partial \phi}{\partial x} + i\frac{\partial \psi}{\partial x} = -E_x + iE_y. \tag{25.5}$$

From these it can be seen that the field at a point is given in magnitude by $E = |f'(z)|$ and that it makes an angle with the x-axis given by $\pi - \arg[f'(z)]$. ◀

It will be apparent from the above that much of physical interest can be calculated by working directly in terms of f and z. In particular, the electric field vector $\mathbf{E}$ may be represented, using (25.5) above, by the quantity

$$\mathcal{E} = E_x + iE_y = -[f'(z)]^*.$$

Complex potentials can be used in two-dimensional fluid mechanics problems in a similar way. If the flow is stationary (i.e. the velocity of the fluid does not depend on time) and irrotational, and the fluid is both incompressible and non-viscous, then the velocity of the fluid can be described by $\mathbf{V} = \nabla\phi$, where ϕ is the velocity potential and satisfies $\nabla^2\phi = 0$. If, for a complex potential $f = \phi + i\psi$, the real part ϕ is taken to represent the velocity potential then the curves $\psi = $ constant will be the streamlines of the flow. In a direct parallel with the electric field, the velocity may be represented in terms of the complex potential by

$$\mathcal{V} = V_x + iV_y = [f'(z)]^*,$$

the difference of a minus sign reflecting the same difference between the definitions of $\mathbf{E}$ and $\mathbf{V}$. The speed of the flow is equal to $|f'(z)|$. Points where $f'(z) = 0$, and thus the velocity is zero, are called *stagnation points* of the flow.

Analogously to the electrostatic case, a line *source* of fluid at $z = z_0$, perpendicular to the z-plane (i.e. a point from which fluid is emerging at a constant rate), is described by the complex potential

$$f(z) = k \ln(z - z_0),$$

where k is the strength of the source. A sink is similarly represented, but with k replaced by $-k$. Other simple examples are as follows.

(i) The flow of a fluid at a constant speed V_0 and at an angle α to the x-axis is described by $f(z) = V_0(\exp i\alpha)z$.

(ii) Vortex flow, in which fluid flows azimuthally in an anticlockwise direction around some point z_0, the speed of the flow being inversely proportional to the distance from z_0, is described by $f(z) = -ik \ln(z - z_0)$, where k is the strength of the vortex. For a clockwise vortex k is replaced by $-k$.

► *Verify that the complex potential*

$$f(z) = V_0 \left(z + \frac{a^2}{z} \right)$$

is appropriate to a circular cylinder of radius a placed so that it is perpendicular to a uniform fluid flow of speed V_0 parallel to the x-axis.

Firstly, since $f(z)$ is analytic except at $z = 0$, both its real and imaginary parts satisfy Laplace's equation in the region exterior to the cylinder. Also $f(z) \to V_0 z$ as $z \to \infty$, so that Re $f(z) \to V_0 x$, which is appropriate to a uniform flow of speed V_0 in the x-direction far from the cylinder.

Writing $z = r \exp i\theta$ and using de Moivre's theorem we have

$$f(z) = V_0 \left[r \exp i\theta + \frac{a^2}{r} \exp(-i\theta) \right]$$

$$= V_0 \left(r + \frac{a^2}{r} \right) \cos\theta + iV_0 \left(r - \frac{a^2}{r} \right) \sin\theta.$$

Thus we see that the streamlines of the flow described by $f(z)$ are given by

$$\psi = V_0 \left(r - \frac{a^2}{r} \right) \sin\theta = \text{constant}.$$

In particular, $\psi = 0$ on $r = a$, independently of the value of θ, and so $r = a$ must be a streamline. Since there can be no flow of fluid across streamlines, $r = a$ must correspond to a boundary along which the fluid flows tangentially. Thus $f(z)$ is a solution of Laplace's equation that satisfies all the physical boundary conditions of the problem, and so, by the uniqueness theorem, it is the appropriate complex potential. ◄

By a similar argument, the complex potential $f(z) = -E(z - a^2/z)$ (note the minus signs) is appropriate to a conducting circular cylinder of radius a placed perpendicular to a uniform electric field $\mathbf{E}$ in the x-direction.

The real and imaginary parts of a complex potential $f = \phi + i\psi$ have another interesting relationship in the context of Laplace's equation in electrostatics or fluid mechanics. Let us choose ϕ as the conventional potential, so that ψ represents the stream function (or electric field, depending on the application), and consider

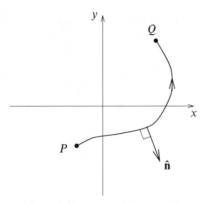

Figure 25.2   A curve joining the points P and Q. Also shown is $\hat{\mathbf{n}}$, the unit vector normal to the curve.

the difference in the values of ψ at any two points P and Q connected by some path C, as shown in figure 25.2. This difference is given by

$$\psi(Q) - \psi(P) = \int_P^Q d\psi = \int_P^Q \left(\frac{\partial \psi}{\partial x} \, dx + \frac{\partial \psi}{\partial y} \, dy \right),$$

which, on using the Cauchy–Riemann relations, becomes

$$\psi(Q) - \psi(P) = \int_P^Q \left(-\frac{\partial \phi}{\partial y} \, dx + \frac{\partial \phi}{\partial x} \, dy \right)$$

$$= \int_P^Q \nabla \phi \cdot \hat{\mathbf{n}} \, ds \; = \; \int_P^Q \frac{\partial \phi}{\partial n} \, ds,$$

where $\hat{\mathbf{n}}$ is the vector unit normal to the path C and s is the arc length along the path; the last equality is written in terms of the normal derivative $\partial \phi / \partial n \equiv \nabla \phi \cdot \hat{\mathbf{n}}$.

Now suppose that in an electrostatics application, the path C is the surface of a conductor; then

$$\frac{\partial \phi}{\partial n} = -\frac{\sigma}{\epsilon_0},$$

where σ is the surface charge density per unit length normal to the xy-plane. Therefore $-\epsilon_0[\psi(Q) - \psi(P)]$ is equal to the charge per unit length normal to the xy-plane on the surface of the conductor between the points P and Q. Similarly, in fluid mechanics applications, if the density of the fluid is ρ and its velocity is $\mathbf{V}$ then

$$\rho[\psi(Q) - \psi(P)] = \rho \int_P^Q \nabla \phi \cdot \hat{\mathbf{n}} \, ds = \rho \int_P^Q \mathbf{V} \cdot \hat{\mathbf{n}} \, ds$$

is equal to the mass flux between P and Q per unit length perpendicular to the xy-plane.

> ▶ A conducting circular cylinder of radius a is placed with its centre line passing through the origin and perpendicular to a uniform electric field **E** in the x-direction. Find the charge per unit length induced on the half of the cylinder that lies in the region x < 0.

As mentioned immediately following the previous example, the appropriate complex potential for this problem is $f(z) = -E(z - a^2/z)$. Writing $z = r \exp i\theta$ this becomes

$$f(z) = -E \left[r \exp i\theta - \frac{a^2}{r} \exp(-i\theta) \right]$$

$$= -E \left(r - \frac{a^2}{r} \right) \cos\theta - iE \left(r + \frac{a^2}{r} \right) \sin\theta,$$

so that on $r = a$ the imaginary part of f is given by

$$\psi = -2Ea \sin\theta.$$

Therefore the induced charge q per unit length on the left half of the cylinder, between $\theta = \pi/2$ and $\theta = 3\pi/2$, is given by

$$q = 2\epsilon_0 Ea[\sin(3\pi/2) - \sin(\pi/2)] = -4\epsilon_0 Ea. \blacktriangleleft$$

25.2 Applications of conformal transformations

In section 24.7 of the previous chapter it was shown that, under a conformal transformation $w = g(z)$ from $z = x + iy$ to a new variable $w = r + is$, if a solution of Laplace's equation in some region R of the xy-plane can be found as the real or imaginary part of an analytic function[§] of z, then the same expression put in terms of r and s will be a solution of Laplace's equation in the corresponding region R' of the w-plane, and vice versa. In addition, if the solution is constant over the boundary C of the region R in the xy-plane, then the solution in the w-plane will take the same constant value over the corresponding curve C' that bounds R'.

Thus, from any two-dimensional solution of Laplace's equation for a particular geometry, typified by those discussed in the previous section, further solutions for other geometries can be obtained by making conformal transformations. From the physical point of view the given geometry is usually complicated, and so the solution is sought by transforming to a simpler one. However, working from simpler to more complicated situations can provide useful experience and make it more likely that the reverse procedure can be tackled successfully.

[§] In fact, the original solution in the xy-plane need not be given explicitly as the real or imaginary part of an analytic function. Any solution of $\nabla^2 \phi = 0$ in the xy-plane is carried over into another solution of $\nabla^2 \phi = 0$ in the new variables by a conformal transformation, and vice versa.

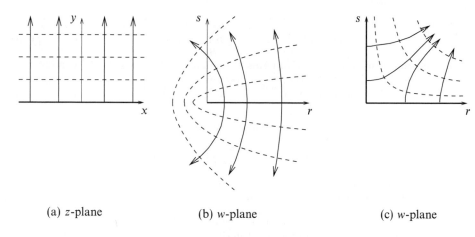

(a) z-plane   (b) w-plane   (c) w-plane

Figure 25.3   The equipotential lines (broken) and field lines (solid) (a) for an infinite charged conducting plane at $y = 0$, where $z = x + iy$, and after the transformations (b) $w = z^2$ and (c) $w = z^{1/2}$ of the situation shown in (a).

> ▶ *Find the complex electrostatic potential associated with an infinite charged conducting plate $y = 0$, and thus obtain those associated with*
>   (i) *a semi-infinite charged conducting plate ($r > 0$, $s = 0$);*
>   (ii) *the inside of a right-angled charged conducting wedge ($r > 0$, $s = 0$ and $r = 0$, $s > 0$).*

Figure 25.3(a) shows the equipotentials (broken lines) and field lines (solid lines) for the infinite charged conducting plane $y = 0$. Suppose that we elect to make the real part of the complex potential coincide with the conventional electrostatic potential. If the plate is charged to a potential V then clearly

$$\phi(x, y) = V - ky, \tag{25.6}$$

where k is related to the charge density σ by $k = \sigma/\epsilon_0$, since physically the electric field $\mathbf{E}$ has components $(0, \sigma/\epsilon_0)$ and $\mathbf{E} = -\nabla\phi$.

Thus what is needed is an analytic function of z, of which the real part is $V - ky$. This can be obtained by inspection, but we may proceed formally and use the Cauchy–Riemann relations to obtain the imaginary part $\psi(x, y)$ as follows:

$$\frac{\partial\psi}{\partial y} = \frac{\partial\phi}{\partial x} = 0 \quad \text{and} \quad \frac{\partial\psi}{\partial x} = -\frac{\partial\phi}{\partial y} = k.$$

Hence $\psi = kx + c$ and, absorbing c into V, the required complex potential is

$$f(z) = V - ky + ikx = V + ikz. \tag{25.7}$$

(i) Now consider the transformation

$$w = g(z) = z^2. \tag{25.8}$$

This satisfies the criteria for a conformal mapping (except at $z = 0$) and carries the upper half of the z-plane into the entire w-plane; the equipotential plane $y = 0$ goes into the half-plane $r > 0$, $s = 0$.

By the general results proved, $f(z)$, when expressed in terms of r and s, will give a complex potential whose real part will be constant on the half-plane in question; we deduce that

$$F(w) = f(z) = V + ikz = V + ikw^{1/2} \qquad (25.9)$$

is the required potential. Expressed in terms of r, s and $\rho = (r^2 + s^2)^{1/2}$, $w^{1/2}$ is given by

$$w^{1/2} = \rho^{1/2} \left[\left(\frac{\rho + r}{2\rho} \right)^{1/2} + i \left(\frac{\rho - r}{2\rho} \right)^{1/2} \right], \qquad (25.10)$$

and, in particular, the electrostatic potential is given by

$$\Phi(r, s) = \operatorname{Re} F(w) = V - \frac{k}{\sqrt{2}} \left[(r^2 + s^2)^{1/2} - r \right]^{1/2}. \qquad (25.11)$$

The corresponding equipotentials and field lines are shown in figure 25.3(b). Using results (25.3)–(25.5), the magnitude of the electric field is

$$|\mathbf{E}| = |F'(w)| = |\tfrac{1}{2} ikw^{-1/2}| = \tfrac{1}{2} k(r^2 + s^2)^{-1/4}.$$

(ii) A transformation 'converse' to that used in (i),

$$w = g(z) = z^{1/2},$$

has the effect of mapping the upper half of the z-plane into the first quadrant of the w-plane and the conducting plane $y = 0$ into the wedge $r > 0$, $s = 0$ and $r = 0$, $s > 0$.

The complex potential now becomes

$$\begin{aligned} F(w) &= V + ikw^2 \\ &= V + ik[(r^2 - s^2) + 2irs], \qquad (25.12) \end{aligned}$$

showing that the electrostatic potential is $V - 2krs$ and that the electric field has components

$$\mathbf{E} = (2ks, 2kr). \qquad (25.13)$$

Figure 25.3(c) indicates the approximate equipotentials and field lines. (Note that, in both transformations, $g'(z)$ is either 0 or ∞ at the origin, and so neither transformation is conformal there. Consequently there is no violation of result (ii), given at the start of section 24.7, concerning the angles between intersecting lines.) ◄

The *method of images*, discussed in section 21.5, can be used in conjunction with conformal transformations to solve some problems involving Laplace's equation in two dimensions.

►*A wedge of angle π/α with its vertex at $z = 0$ is formed by two semi-infinite conducting plates, as shown in figure 25.4(a). A line charge of strength q per unit length is positioned at $z = z_0$, perpendicular to the z-plane. By considering the transformation $w = z^\alpha$, find the complex electrostatic potential for this situation.*

Let us consider the action of the transformation $w = z^\alpha$ on the lines defining the positions of the conducting plates. The plate that lies along the positive x-axis is mapped onto the positive r-axis in the w-plane, whereas the plate that lies along the direction $\exp(i\pi/\alpha)$ is mapped into the negative r-axis, as shown in figure 25.4(b). Similarly the line charge at z_0 is mapped onto the point $w_0 = z_0^\alpha$.

From figure 25.4(b), we see that in the w-plane the problem can be solved by introducing a second line charge of opposite sign at the point w_0^*, so that the potential $\Phi = 0$ along the r-axis. The complex potential for such an arrangement is simply

$$F(w) = -\frac{q}{2\pi\epsilon_0} \ln(w - w_0) + \frac{q}{2\pi\epsilon_0} \ln(w - w_0^*).$$

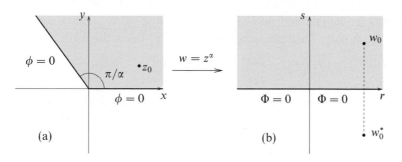

Figure 25.4 (a) An infinite conducting wedge with interior angle π/α and a line charge at $z = z_0$; (b) after the transformation $w = z^\alpha$, with an additional image charge placed at $w = w_0^*$.

Substituting $w = z^\alpha$ into the above shows that the required complex potential in the original z-plane is

$$f(z) = \frac{q}{2\pi\epsilon_0} \ln\left(\frac{z^\alpha - z_0^{*\alpha}}{z^\alpha - z_0^\alpha}\right). \blacktriangleleft$$

It should be noted that the appearance of a complex conjugate in the final expression is not in conflict with the general requirement that the complex potential be analytic. It is z^* that must not appear; here, $z_0^{*\alpha}$ is no more than a parameter of the problem.

25.3 Location of zeros

The residue theorem, relating the value of a closed contour integral to the sum of the residues at the poles enclosed by the contour, was discussed in the previous chapter. One important practical use of an extension to the theorem is that of locating the zeros of functions of a complex variable. The location of such zeros has a particular application in electrical network and general oscillation theory, since the complex zeros of certain functions (usually polynomials) give the system parameters (usually frequencies) at which system instabilities occur. As the basis of a method for locating these zeros we next prove three important theorems.

(i) If $f(z)$ has poles as its only singularities inside a closed contour C and is not zero at any point on C then

$$\oint_C \frac{f'(z)}{f(z)}\, dz = 2\pi i \sum_j (N_j - P_j). \tag{25.14}$$

Here N_j is the order of the jth zero of $f(z)$ enclosed by C. Similarly P_j is the order of the jth pole of $f(z)$ inside C.

To prove this we note that, at each position z_j, $f(z)$ can be written as

$$f(z) = (z - z_j)^{m_j}\phi(z), \tag{25.15}$$

879

where $\phi(z)$ is analytic and non-zero at $z = z_j$ and m_j is positive for a zero and negative for a pole. Then the integrand $f'(z)/f(z)$ takes the form

$$\frac{f'(z)}{f(z)} = \frac{m_j}{z - z_j} + \frac{\phi'(z)}{\phi(z)}. \tag{25.16}$$

Since $\phi(z_j) \neq 0$, the second term on the RHS is analytic; thus the integrand has a simple pole at $z = z_j$, with residue m_j. For zeros $m_j = N_j$ and for poles $m_j = -P_j$, and thus (25.14) follows from the residue theorem.

(ii) If $f(z)$ is analytic inside C and not zero at any point on it then

$$2\pi \sum_j N_j = \Delta_C[\arg f(z)], \tag{25.17}$$

where $\Delta_C[x]$ denotes the variation in x around the contour C.

Since f is analytic, there are no P_j; further, since

$$\frac{f'(z)}{f(z)} = \frac{d}{dz}[\operatorname{Ln} f(z)], \tag{25.18}$$

equation (25.14) can be written

$$2\pi i \sum N_j = \oint_C \frac{f'(z)}{f(z)}\, dz = \Delta_C[\operatorname{Ln} f(z)]. \tag{25.19}$$

However,

$$\Delta_C[\operatorname{Ln} f(z)] = \Delta_C[\ln |f(z)|] + i\Delta_C[\arg f(z)], \tag{25.20}$$

and, since C is a closed contour, $\ln |f(z)|$ must return to its original value; so the real term on the RHS is zero. Comparison of (25.19) and (25.20) then establishes (25.17), which is known as the *principle of the argument*.

(iii) If $f(z)$ and $g(z)$ are analytic within and on a closed contour C and $|g(z)| < |f(z)|$ on C then $f(z)$ and $f(z) + g(z)$ have the same number of zeros inside C; this is *Rouché's theorem*.

With the conditions given, neither $f(z)$ nor $f(z) + g(z)$ can have a zero on C. So, applying theorem (ii) with an obvious notation,

$$\begin{aligned} 2\pi \sum_j N_j(f+g) &= \Delta_C[\arg(f+g)] \\ &= \Delta_C[\arg f] + \Delta_C[\arg(1 + g/f)] \\ &= 2\pi \sum_k N_k(f) + \Delta_C[\arg(1 + g/f)]. \end{aligned} \tag{25.21}$$

Further, since $|g| < |f|$ on C, $1 + g/f$ always lies *within* a unit circle centred on $z = 1$; thus its argument *always* lies in the range $-\pi/2 < \arg(1 + g/f) < \pi/2$ and cannot change by any multiple of 2π. It must therefore return to its original value when z returns to its starting point having traversed C. Hence the second term on the RHS of (25.21) is zero and the theorem is established.

The importance of Rouché's theorem is that for some functions, in particular polynomials, only the behaviour of a single term in the function need be considered if the contour is chosen appropriately. For example, for a polynomial, $f(z) + g(z) = \sum_0^N b_i z^i$, only the properties of its largest power, taken as $f(z)$, need be investigated if a circular contour is chosen with radius R sufficiently large that, on the contour, the magnitude of the largest power term, $|b_N R^N|$, is greater than the sum of the magnitudes of all other terms. It is obvious that $f(z) = b_N z^N$ has N zeros inside $|z| = R$ (all at the origin); consequently, $f + g$ also has N zeros inside the same circle.

The corresponding situation, in which only the properties of the polynomial's smallest power, again taken as $f(z)$, need be investigated is a circular contour with a radius R chosen sufficiently *small* that, on the contour, the magnitude of the smallest power term (usually the constant term in a polynomial) is greater than the sum of the magnitudes of all other terms. Then, a similar argument to that given above shows that, since $f(z) = b_0$ has no zeros inside $|z| = R$, neither does $f + g$.

A weak form of the *maximum-modulus theorem* may also be deduced. This states that if $f(z)$ is analytic within and on a simple closed contour C then $|f(z)|$ attains its maximum value on the boundary of C. The proof is as follows.

Let $|f(z)| \leq M$ on C with equality at at least one point of C. Now suppose that there is a point $z = a$ inside C such that $|f(a)| > M$. Then the function $h(z) \equiv f(a)$ is such that $|h(z)| > |-f(z)|$ on C, and thus, by Rouché's theorem, $h(z)$ and $h(z) - f(z)$ have the same number of zeros inside C. But $h(z) (\equiv f(a))$ has no zeros inside C, and, again by Rouché's theorem, this would imply that $f(a) - f(z)$ has no zeros in C. However, $f(a) - f(z)$ clearly has a zero at $z = a$, and so we have a contradiction; the assumption of the existence of a point $z = a$ inside C such that $|f(a)| > M$ must be invalid. This establishes the theorem.

The stronger form of the maximum-modulus theorem, which we do not prove, states, in addition, that the maximum value of $f(z)$ is not attained at any interior point except for the case where $f(z)$ is a constant.

▶ *Show that the four zeros of* $h(z) = z^4 + z + 1$ *occur one in each quadrant of the Argand diagram and that all four lie between the circles* $|z| = 2/3$ *and* $|z| = 3/2$.

Putting $z = x$ and $z = iy$ shows that no zeros occur on the real or imaginary axes. They must therefore occur in conjugate pairs, as can be shown by taking the complex conjugate of $h(z) = 0$.

Now take C as the contour $OXYO$ shown in figure 25.5 and consider the changes $\Delta[\arg h]$ in the argument of $h(z)$ as z traverses C.

(i) OX: $\arg h$ is everywhere zero, since h is real, and thus $\Delta_{OX}[\arg h] = 0$.
(ii) XY: $z = R \exp i\theta$ and so $\arg h$ changes by an amount

$$\Delta_{XY}[\arg h] = \Delta_{XY}[\arg z^4] + \Delta_{XY}[\arg(1 + z^{-3} + z^{-4})]$$
$$= \Delta_{XY}[\arg R^4 e^{4i\theta}] + \Delta_{XY}\{\arg[1 + O(R^{-3})]\}$$
$$= 2\pi + O(R^{-3}). \tag{25.22}$$

881

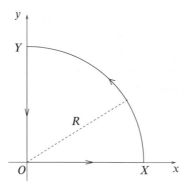

Figure 25.5   A contour for locating the zeros of a polynomial that occur in the first quadrant of the Argand diagram.

(iii) YO: $z = iy$ and so $\arg h = \tan^{-1} y/(y^4 + 1)$, which starts at $O(R^{-3})$ and finishes at 0 as y goes from large R to 0. It never reaches $\pi/2$ because $y^4 + 1 = 0$ has no real positive root. Thus $\Delta_{YO}[\arg h] = 0$.

Hence for the complete contour $\Delta_C[\arg h] = 0 + 2\pi + 0 + O(R^{-3})$, and, if R is allowed to tend to infinity, we deduce from (25.17) that $h(z)$ has one zero in the first quadrant. Furthermore, since the roots occur in conjugate pairs, a second root must lie in the fourth quadrant, and the other pair must lie in the second and third quadrants.

To show that the zeros lie within the given annulus in the z-plane we apply Rouché's theorem, as follows.

(i) With C as $|z| = 3/2$, $f = z^4$, $g = z + 1$. Now $|f| = 81/16$ on C and $|g| \leq 1 + |z| < 5/2 < 81/16$. Thus, since $z^4 = 0$ has four roots inside $|z| = 3/2$, so also does $z^4 + z + 1 = 0$.

(ii) With C as $|z| = 2/3$, $f = 1$, $g = z^4 + z$. Now $f = 1$ on C and $|g| \leq |z^4| + |z| = 16/81 + 2/3 = 70/81 < 1$. Thus, since $f = 0$ has no roots inside $|z| = 2/3$, neither does $1 + z + z^4 = 0$.

Hence the four zeros of $h(z) = z^4 + z + 1$ occur one in each quadrant and all lie between the circles $|z| = 2/3$ and $|z| = 3/2$. ◀

A further technique useful for locating the zeros of functions is explained in exercise 25.8.

25.4 Summation of series

We now turn to an application of contour integration which at first sight might seem to lie in an unrelated area of mathematics, namely the summation of infinite series. Sometimes a real infinite series with index n, say, can be summed with the help of a suitable complex function that has poles on the real axis at the various positions $z = n$ with the corresponding residues at those poles equal to the values of the terms of the series. A worked example provides the best explanation of how the technique is applied; other examples will be found in the exercises.

▶*By considering*

$$\oint_C \frac{\pi \cot \pi z}{(a+z)^2} \, dz,$$

where a is not an integer and C is a circle of large radius, evaluate

$$\sum_{n=-\infty}^{\infty} \frac{1}{(a+n)^2}.$$

The integrand has (i) simple poles at $z = $ integer n, for $-\infty < n < \infty$, due to the factor $\cot \pi z$ and (ii) a double pole at $z = -a$.

(i) To find the residue of $\cot \pi z$, put $z = n + \xi$ for small ξ:

$$\cot \pi z = \frac{\cos(n\pi + \xi\pi)}{\sin(n\pi + \xi\pi)} \approx \frac{\cos n\pi}{(\cos n\pi)\xi\pi} = \frac{1}{\xi\pi}.$$

The residue of the integrand at $z = n$ is thus $\pi(a+n)^{-2}\pi^{-1}$.

(ii) Putting $z = -a + \xi$ for small ξ and determining the coefficient of ξ^{-1} gives[§]

$$\frac{\pi \cot \pi z}{(a+z)^2} = \frac{\pi}{\xi^2} \cot(-a\pi + \xi\pi)$$

$$= \frac{\pi}{\xi^2} \left\{ \cot(-a\pi) + \xi \left[\frac{d}{dz}(\cot \pi z) \right]_{z=-a} + \cdots \right\},$$

so that the residue at the double pole $z = -a$ is given by

$$\pi [-\pi \csc^2 \pi z]_{z=-a} = -\pi^2 \csc^2 \pi a.$$

Collecting together these results to express the residue theorem gives

$$I = \oint_C \frac{\pi \cot \pi z}{(a+z)^2} \, dz = 2\pi i \left[\sum_{n=-N}^{N} \frac{1}{(a+n)^2} - \pi^2 \csc^2 \pi a \right], \tag{25.23}$$

where N equals the integer part of R. But as the radius R of C tends to ∞, $\cot \pi z \to \mp i$ (depending on whether Im z is greater or less than zero, respectively). Thus

$$I < k \int \frac{dz}{(a+z)^2},$$

which tends to 0 as $R \to \infty$. Thus $I \to 0$ as R (and hence N) $\to \infty$, and (25.23) establishes the result

$$\sum_{n=-\infty}^{\infty} \frac{1}{(a+n)^2} = \frac{\pi^2}{\sin^2 \pi a}. \quad ◀$$

Series with alternating signs in the terms, i.e. $(-1)^n$, can also be attempted in this way but using $\csc \pi z$ rather than $\cot \pi z$, since the former has residue $(-1)^n \pi^{-1}$ at $z = n$ (see exercise 25.11).

[§] This again illustrates one of the techniques for determining residues.

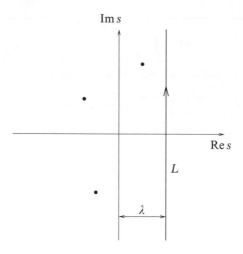

Figure 25.6   The integration path of the inverse Laplace transform is along the infinite line L. The quantity λ must be positive and large enough for all poles of the integrand to lie to the left of L.

25.5 Inverse Laplace transform

As a further example of the use of contour integration we now discuss a method whereby the process of Laplace transformation, discussed in chapter 13, can be inverted.

It will be recalled that the Laplace transform $\bar{f}(s)$ of a function $f(x)$, $x \geq 0$, is given by

$$\bar{f}(s) = \int_0^\infty e^{-sx} f(x)\, dx, \qquad \text{Re } s > s_0. \tag{25.24}$$

In chapter 13, functions $f(x)$ were deduced from the transforms by means of a prepared dictionary. However, an explicit formula for an unknown inverse may be written in the form of an integral. It is known as the *Bromwich integral* and is given by

$$f(x) = \frac{1}{2\pi i} \int_{\lambda - i\infty}^{\lambda + i\infty} e^{sx} \bar{f}(s)\, ds, \qquad \lambda > 0, \tag{25.25}$$

where s is treated as a complex variable and the integration is along the line L indicated in figure 25.6. The position of the line is dictated by the requirements that λ is positive and that all singularities of $\bar{f}(s)$ lie to the left of the line.

That (25.25) really is the unique inverse of (25.24) is difficult to show for general functions and transforms, but the following verification should at least make it

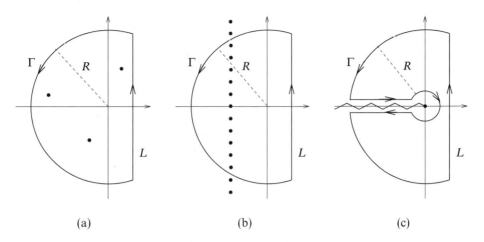

(a)   (b)   (c)

Figure 25.7 Some contour completions for the integration path L of the inverse Laplace transform. For details of when each is appropriate see the main text.

plausible:

$$f(x) = \frac{1}{2\pi i} \int_{\lambda-i\infty}^{\lambda+i\infty} ds\, e^{sx} \int_0^\infty e^{-su} f(u)\, du, \qquad \text{Re}(s) > 0, \text{ i.e. } \lambda > 0,$$

$$= \frac{1}{2\pi i} \int_0^\infty du\, f(u) \int_{\lambda-i\infty}^{\lambda+i\infty} e^{s(x-u)}\, ds$$

$$= \frac{1}{2\pi i} \int_0^\infty du\, f(u) \int_{-\infty}^\infty e^{\lambda(x-u)} e^{ip(x-u)} i\, dp, \qquad \text{putting } s = \lambda + ip,$$

$$= \frac{1}{2\pi} \int_0^\infty f(u) e^{\lambda(x-u)} 2\pi \delta(x-u)\, du$$

$$= \begin{cases} f(x) & x \geq 0, \\ 0 & x < 0. \end{cases} \tag{25.26}$$

Our main purpose here is to demonstrate the use of contour integration. To employ it in the evaluation of the line integral (25.25), the path L must be made part of a closed contour in such a way that the contribution from the completion either vanishes or is simply calculable.

A typical completion is shown in figure 25.7(a) and would be appropriate if $\bar{f}(s)$ had a finite number of poles. For more complicated cases, in which $\bar{f}(s)$ has an infinite sequence of poles but all to the left of L as in figure 25.7(b), a sequence of circular-arc completions that pass between the poles must be used and $f(x)$ is obtained as a series. If $\bar{f}(s)$ is a multivalued function then a cut plane is needed and a contour such as that shown in figure 25.7(c) might be appropriate.

We consider here only the simple case in which the contour in figure 25.7(a) is used; we refer the reader to the exercises at the end of the chapter for others.

Ideally, we would like the contribution to the integral from the circular arc Γ to tend to zero as its radius $R \to \infty$. Using a modified version of Jordan's lemma, it may be shown that this is indeed the case if there exist constants $M > 0$ and $\alpha > 0$ such that on Γ

$$|\bar{f}(s)| \le \frac{M}{R^\alpha}.$$

Moreover, this condition always holds when $\bar{f}(s)$ has the form

$$\bar{f}(s) = \frac{P(s)}{Q(s)},$$

where $P(s)$ and $Q(s)$ are polynomials and the degree of $Q(s)$ is greater than that of $P(s)$.

When the contribution from the part-circle Γ tends to zero as $R \to \infty$, we have from the residue theorem that the inverse Laplace transform (25.25) is given simply by

$$f(t) = \sum \left(\text{residues of } \bar{f}(s)e^{sx} \text{ at all poles} \right). \tag{25.27}$$

> ► *Find the function $f(x)$ whose Laplace transform is*
>
> $$\bar{f}(s) = \frac{s}{s^2 - k^2},$$
>
> *where k is a constant.*

It is clear that $\bar{f}(s)$ is of the form required for the integral over the circular arc Γ to tend to zero as $R \to \infty$, and so we may use the result (25.27) directly. Now

$$\bar{f}(s)e^{sx} = \frac{se^{sx}}{(s-k)(s+k)},$$

and thus has simple poles at $s = k$ and $s = -k$. Using (24.57) the residues at each pole can be easily calculated as

$$R(k) = \frac{ke^{kx}}{2k} \quad \text{and} \quad R(-k) = \frac{ke^{-kx}}{2k}.$$

Thus the inverse Laplace transform is given by

$$f(x) = \tfrac{1}{2}\left(e^{kx} + e^{-kx} \right) = \cosh kx.$$

This result may be checked by computing the forward transform of $\cosh kx$. ◄

Sometimes a little more care is required when deciding in which half-plane to close the contour C.

> ► *Find the function $f(x)$ whose Laplace transform is*
>
> $$\bar{f}(s) = \frac{1}{s}(e^{-as} - e^{-bs}),$$
>
> *where a and b are fixed and positive, with $b > a$.*

From (25.25) we have the integral

$$f(x) = \frac{1}{2\pi i} \int_{\lambda - i\infty}^{\lambda + i\infty} \frac{e^{(x-a)s} - e^{(x-b)s}}{s} \, ds. \tag{25.28}$$

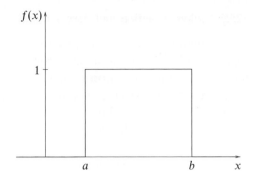

Figure 25.8   The result of the Laplace inversion of $\bar{f}(s) = s^{-1}(e^{-as} - e^{-bs})$ with $b > a$.

Now, despite appearances to the contrary, the integrand has no poles, as may be confirmed by expanding the exponentials as Taylor series about $s = 0$. Depending on the value of x, several cases arise.

(i) For $x < a$ both exponentials in the integrand will tend to zero as Re $s \to \infty$. Thus we may close L with a circular arc Γ in the *right* half-plane (λ can be as small as desired), and we observe that $s \times$ integrand tends to zero everywhere on Γ as $R \to \infty$. With no poles enclosed and no contribution from Γ, the integral along L must also be zero. Thus

$$f(x) = 0 \qquad \text{for } x < a. \tag{25.29}$$

(ii) For $x > b$ the exponentials in the integrand will tend to zero as Re $s \to -\infty$, and so we may close L in the left half-plane, as in figure 25.7(a). Again the integral around Γ vanishes for infinite R, and so, by the residue theorem,

$$f(x) = 0 \qquad \text{for } x > b. \tag{25.30}$$

(iii) For $a < x < b$ the two parts of the integrand behave in different ways and have to be treated separately:

$$I_1 - I_2 \equiv \frac{1}{2\pi i} \int_L \frac{e^{(x-a)s}}{s} \, ds - \frac{1}{2\pi i} \int_L \frac{e^{(x-b)s}}{s} \, ds.$$

The integrand of I_1 then vanishes in the far left-hand half-plane, but does now have a (simple) pole at $s = 0$. Closing L in the left half-plane, and using the residue theorem, we obtain

$$I_1 = \text{residue at } s = 0 \text{ of } s^{-1}e^{(x-a)s} = 1. \tag{25.31}$$

The integrand of I_2, however, vanishes in the far right-hand half-plane (and also has a simple pole at $s = 0$) and is evaluated by a circular-arc completion in that half-plane. Such a contour encloses no poles and leads to $I_2 = 0$.

Thus, collecting together results (25.29)–(25.31) we obtain

$$f(x) = \begin{cases} 0 & \text{for } x < a, \\ 1 & \text{for } a < x < b, \\ 0 & \text{for } x > b, \end{cases}$$

as shown in figure 25.8. ◀

25.6 Stokes' equation and Airy integrals

Much of the analysis of situations occurring in physics and engineering is concerned with what happens at a boundary within or surrounding a physical system. Sometimes the existence of a boundary imposes conditions on the behaviour of variables describing the state of the system; obvious examples include the zero displacement at its end-points of an anchored vibrating string and the zero potential contour that must coincide with a grounded electrical conductor.

More subtle are the effects at internal boundaries, where the same non-vanishing variable has to describe the situation on either side of the boundary but its behaviour is quantitatively, or even *qualitatively*, different in the two regions. In this section we will study an equation, Stokes' equation, whose solutions have this latter property; as well as solutions written as series in the usual way, we will find others expressed as complex integrals.

The Stokes' equation can be written in several forms, e.g.

$$\frac{d^2y}{dx^2} + \lambda xy = 0; \quad \frac{d^2y}{dx^2} + xy = 0; \quad \frac{d^2y}{dx^2} = xy.$$

We will adopt the last of these, but write it as

$$\frac{d^2y}{dz^2} = zy \tag{25.32}$$

to emphasis that its complex solutions are valid for a complex independent variable z, though this also means that particular care has to be exercised when examining their behaviour in different parts of the complex z-plane. The other forms of Stokes' equation can all be reduced to that of (25.32) by suitable (complex) scaling of the independent variable.

25.6.1 The solutions of Stokes' equation

It will be immediately apparent that, even for z restricted to be real and denoted by x, the behaviour of the solutions to (25.32) will change markedly as x passes through $x = 0$. For positive x they will have similar characteristics to the solutions of $y'' = k^2 y$, where k is real; these have monotonic exponential forms, either increasing or decreasing. On the other hand, when x is negative the solutions will be similar to those of $y'' + k^2 y = 0$, i.e. oscillatory functions of x. This is just the sort of behaviour shown by the wavefunction describing light diffracted by a sharp edge or by the quantum wavefunction describing a particle near to the boundary of a region which it is classically forbidden to enter on energy grounds. Other examples could be taken from the propagation of electromagnetic radiation in an ion plasma or wave-guide.

Let us examine in a bit more detail the behaviour of plots of possible solutions $y(z)$ of Stokes' equation in the region near $z = 0$ and, in particular, what may

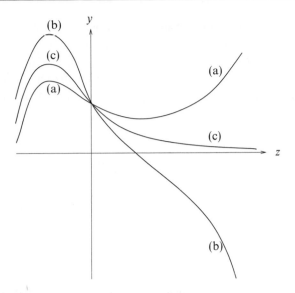

Figure 25.9 Behaviour of the solutions $y(z)$ of Stokes' equation near $z = 0$ for various values of $\lambda = -y'(0)$. (a) with λ small, (b) with λ large and (c) with λ appropriate to the Airy function Ai(z).

happen in the region $z > 0$. For definiteness and ease of illustration (see figure 25.9), let us suppose that both y and z, and hence the derivatives of y, are real and that $y(0)$ is positive; if it were negative, our conclusions would not be changed since equation (25.32) is invariant under $y(z) \to -y(z)$. The only difference would be that all plots of $y(z)$ would be reflected in the z-axis.

We first note that d^2y/dx^2, and hence also the curvature of the plot, has the same sign as z, i.e. it has positive curvature when $z > 0$, for so long as $y(z)$ remains positive there. What will happen to the plot for $z > 0$ therefore depends crucially on the value of $y'(0)$. If this slope is positive or only slightly negative the positive curvature will carry the plot, either immediately or ultimately, further away from the z-axis. On the other hand, if $y'(0)$ is negative but sufficiently large in magnitude, the plot will cross the $y = 0$ line; if this happens the sign of the curvature reverses and again the plot will be carried ever further from the z-axis, only this time towards large negative values.

Between these two extremes it seems at least plausible that there is a particular negative value of $y'(0)$ that leads to a plot that approaches the z-axis asymptotically, never crosses it (and so always has positive curvature), and has a slope that, whilst always negative, tends to zero in magnitude. There is such a solution, known as Ai(z), whose properties we will examine further in the following subsections. The three cases are illustrated in figure 25.9.

The behaviour of the solutions of (25.32) in the region $z < 0$ is more straightforward, in that, whatever the sign of y at any particular point z, the curvature always has the opposite sign. Consequently the curve always bends towards the z-axis, crosses it, and then bends towards the axis again. Thus the curve exhibits oscillatory behaviour. Furthermore, as $-z$ increases, the curvature for any given $|y|$ gets larger; as a consequence, the oscillations become increasingly more rapid and their amplitude decreases.

25.6.2 Series solution of Stokes' equation

Obtaining a series solution of Stokes' equation presents no particular difficulty when the methods of chapter 16 are used. The equation, written in the form

$$\frac{d^2 y}{dz^2} - zy = 0,$$

has no singular points except at $z = \infty$. Every other point in the z-plane is an ordinary point and so two linearly independent series expansions about it (formally with indicial values $\sigma = 0$ and $\sigma = 1$) can be found. Those about $z = 0$ take the forms $\sum_0^\infty a_n z^n$ and $\sum_0^\infty b_n z^{n+1}$. The corresponding recurrence relations are

$$(n+3)(n+2)a_{n+3} = a_n \quad \text{and} \quad (n+4)(n+3)b_{n+3} = b_n,$$

and the two series (with $a_0 = b_0 = 1$) take the forms

$$y_1(z) = 1 + \frac{z^3}{(3)(2)} + \frac{z^6}{(6)(5)(3)(2)} + \cdots,$$

$$y_2(z) = z + \frac{z^4}{(4)(3)} + \frac{z^7}{(7)(6)(4)(3)} + \cdots.$$

The ratios of successive terms for the two series are thus

$$\frac{a_{n+3}z^{n+3}}{a_n z^n} = \frac{z^3}{(n+3)(n+2)} \quad \text{and} \quad \frac{b_{n+3}z^{n+4}}{b_n z^{n+1}} = \frac{z^3}{(n+4)(n+3)}.$$

It follows from the ratio test that both series are absolutely convergent for all z. A similar argument shows that the series for their derivatives are also absolutely convergent for all z. Any solution of the Stokes' equation is representable as a superposition of the two series and so is analytic for all finite z; it is therefore an integral function with its only singularity at infinity.

25.6.3 Contour integral solutions

We now move on to another form of solution of the Stokes' equation (25.32), one that takes the form of a contour integral in which z appears as a parameter in the integrand. Consider the contour integral

$$y(z) = \int_a^b f(t) \exp(zt) \, dt, \tag{25.33}$$

in which a, b and $f(t)$ are all yet to be chosen. Note that the contour is in the complex t-plane and that the path from a to b can be distorted as required so long as no poles of the integrand are trapped between an original path and its distortion.

Substitution of (25.33) into (25.32) yields

$$\int_a^b t^2 f(t) \exp(zt) \, dt = \int_a^b z \, f(t) \exp(zt) \, dt$$

$$= [f(t) \exp(zt)]_a^b - \int_a^b \frac{df(t)}{dt} \exp(zt) \, dt.$$

If we could choose the limits a and b so that the end-point contributions vanish then Stokes' equation would be satisfied by (25.33), provided $f(t)$ satisfies

$$\frac{df(t)}{dt} + t^2 f(t) = 0 \quad \Rightarrow \quad f(t) = A \exp(-\tfrac{1}{3} t^3), \tag{25.34}$$

where A is any constant.

To make the end-point contributions vanish we must choose a and b such that $\exp(-\tfrac{1}{3} t^3 + zt) = 0$ for both values of t. This can only happen if $|a| \to \infty$ and $|b| \to \infty$ and, even then, only if $\mathrm{Re}\,(t^3)$ is positive. This condition is satisfied if

$$2n\pi - \tfrac{1}{2}\pi < 3 \arg(t) < 2n\pi + \tfrac{1}{2}\pi \text{ for some integer } n.$$

Thus a and b must each be at infinity in one of the three shaded areas shown in figure 25.10, but clearly not in the same area as this would lead to a zero value for the contour integral. This leaves three contours (marked C_1, C_2 and C_3 in the figure) that start and end in different sectors. However, only two of them give rise to independent integrals since the path $C_2 + C_3$ is equivalent to (can be distorted into) the path C_1.

The two integral functions given particular names are

$$\mathrm{Ai}(z) = \frac{1}{2\pi i} \int_{C_1} \exp(-\tfrac{1}{3} t^3 + zt) \, dt \tag{25.35}$$

and

$$\mathrm{Bi}(z) = \frac{1}{2\pi} \int_{C_2} \exp(-\tfrac{1}{3} t^3 + zt) \, dt - \frac{1}{2\pi} \int_{C_3} \exp(-\tfrac{1}{3} t^3 + zt) \, dt. \tag{25.36}$$

Stokes' equation is unchanged if the independent variable is changed from z to ζ, where $\zeta = \exp(2\pi i/3)z \equiv \Omega z$. This is also true for the repeated change $z \to \Omega\zeta = \Omega^2 z$. The same changes of variable, rotations of the complex plane through $2\pi/3$ or $4\pi/3$, carry the three contours C_1, C_2 and C_3 into each other,

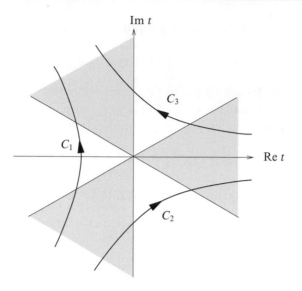

Figure 25.10   The contours used in the complex t-plane to define the functions Ai(z) and Bi(z).

though sometimes the sense of traversal is reversed. Consequently there are relationships connecting Ai and Bi when the rotated variables are used as their arguments. As two examples,

$$\text{Ai}(z) + \Omega\text{Ai}(\Omega z) + \Omega^2\text{Ai}(\Omega^2 z) = 0, \tag{25.37}$$

$$\text{Bi}(z) = i[\Omega^2\text{Ai}(\Omega^2 z) - \Omega\text{Ai}(\Omega z)] = e^{-\pi i/6}\text{Ai}(ze^{-2\pi i/3}) + e^{\pi i/6}\text{Ai}(ze^{2\pi i/3}). \tag{25.38}$$

Since the only requirements for the integral paths is that they start and end in the correct sectors, we can distort path C_1 so that it lies on the imaginary axis for virtually its whole length and just to the left of the axis at its two ends. This enables us to obtain an alternative expression for Ai(z), as follows.

Setting $t = is$, where s is real and $-\infty < s < \infty$, converts the integral representation of Ai(z) to

$$\text{Ai}(z) = \frac{1}{2\pi}\int_{-\infty}^{\infty}\exp[i(\tfrac{1}{3}s^3 + zs)]\,ds.$$

Now, the exponent in this integral is an odd function of s and so the imaginary part of the integrand contributes nothing to the integral. What is left is therefore

$$\text{Ai}(z) = \frac{1}{\pi}\int_0^{\infty}\cos(\tfrac{1}{3}s^3 + zs)\,ds. \tag{25.39}$$

This form shows explicitly that when z is real, so is Ai(z).

This same representation can also be used to justify the association of the contour integral (25.35) with the particular solution of Stokes' equation that decays monotonically to zero for real $z > 0$ as $|z| \to \infty$. As discussed in subsection 25.6.1, all solutions except the one called Ai(z) tend to $\pm\infty$ as z (real) takes on increasingly large positive values and so their asymptotic forms reflect this. In a worked example in subsection 25.8.2 we use the method of steepest descents (a saddle-point method) to show that the function defined by (25.39) has exactly the characteristic asymptotic property expected of Ai(z) (see page 911). It follows that it is the same function as Ai(z), up to a real multiplicative constant.

The choice of definition (25.36) as the other named solution Bi(z) of Stokes' equation is a less obvious one. However, it is made on the basis of its behaviour for negative real values of z. As discussed earlier, Ai(z) oscillates almost sinusoidally in this region, except for a relatively slow increase in frequency and an even slower decrease in amplitude as $-z$ increases. The solution Bi(z) is chosen to be the particular function that exhibits the same behaviour as Ai(z) except that it is in quadrature with Ai, i.e. it is $\pi/2$ out of phase with it. Specifically, as $x \to -\infty$,

$$\mathrm{Ai}(x) \sim \frac{1}{\sqrt{2\pi}x^{1/4}} \sin\left(\frac{2|x|^{3/2}}{3} + \frac{\pi}{4}\right), \tag{25.40}$$

$$\mathrm{Bi}(x) \sim \frac{1}{\sqrt{2\pi}x^{1/4}} \cos\left(\frac{2|x|^{3/2}}{3} + \frac{\pi}{4}\right). \tag{25.41}$$

There is a close parallel between this choice and that of taking sine and cosine functions as the basic independent solutions of the simple harmonic oscillator equation. Plots of Ai(z) and Bi(z) for real z are shown in figure 25.11.

> ►By choosing a suitable contour for C_1 in (25.35), express Ai(0) in terms of the gamma function.

With z set equal to zero, (25.35) takes the form

$$\mathrm{Ai}(0) = \frac{1}{2\pi i} \int_{C_1} \exp(-\tfrac{1}{3}t^3)\, dt.$$

We again use the freedom to choose the specific line of the contour so as to make the actual integration as simple as possible.

Here we consider C_1 as made up of two straight-line segments: one along the line $\arg t = 4\pi/3$, starting at infinity in the correct sector and ending at the origin; the other starting at the origin and going to infinity along the line $\arg t = 2\pi/3$, thus ending in the correct final sector. On each, we set $\tfrac{1}{3}t^3 = s$, where s is real and positive on both lines. Then $dt = e^{4\pi i/3}(3s)^{-2/3}\, ds$ on the first segment and $dt = e^{2\pi i/3}(3s)^{-2/3}\, ds$ on the second.

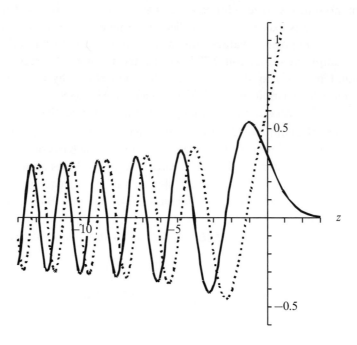

Figure 25.11   The functions Ai(z) (full line) and Bi(z) (broken line) for real z.

Then we have

$$
\begin{aligned}
\mathrm{Ai}(0) &= \frac{1}{2\pi i}\int_{\infty}^{0} e^{-s}e^{4\pi i/3}(3s)^{-2/3}\,ds + \frac{1}{2\pi i}\int_{0}^{\infty} e^{-s}e^{2\pi i/3}(3s)^{-2/3}\,ds \\
&= \frac{3^{-2/3}}{2\pi i}\int_{0}^{\infty} e^{-s}(-e^{4\pi i/3} + e^{2\pi i/3})s^{-2/3}\,ds \\
&= \frac{\sqrt{3}i\,3^{-2/3}}{2\pi i}\int_{0}^{\infty} e^{-s}s^{-2/3}\,ds \\
&= \frac{3^{-1/6}}{2\pi}\Gamma(\tfrac{1}{3}),
\end{aligned}
$$

where we have used the standard integral defining the gamma function in the last line. ◄

Finally in this subsection we should mention that the Airy functions and their derivatives are closely related to Bessel functions of orders $\pm\tfrac{1}{3}$ and $\pm\tfrac{2}{3}$ and that there exist many representations, both as linear combinations and as indefinite integrals, of one in terms of the other.[§]

25.7 WKB methods

Throughout this book we have had many occasions on which it has been necessary to solve the equation

$$\frac{d^2 y}{dx^2} + k_0^2 f(x) y = 0 \qquad (25.42)$$

when the notionally general function $f(x)$ has been, in fact, a constant, usually the unit function $f(x) = 1$. Then the solutions have been elementary and of the form $A \sin k_0 x$ or $A \cos k_0 x$ with arbitrary but constant amplitude A.

Explicit solutions of (25.42) for a non-constant $f(x)$ are only possible in a limited number of cases, but, as we will show, some progress can be made if $f(x)$ is a slowly varying function of x, in the sense that it does not change much in a range of x of the order of k_0^{-1}.

We will also see that it is possible to handle situations in which $f(x)$ is complex; this enables us to deal with, for example, the passage of waves through an absorbing medium. Developing such solutions will involve us in finding the integrals of some complex quantities, integrals that will behave differently in the various parts of the complex plane – hence their inclusion in this chapter.

25.7.1 Phase memory

Before moving on to the formal development of WKB methods[¶] we discuss the concept of *phase memory* which is the underlying idea behind them.

Let us first suppose that $f(x)$ is real, positive and essentially constant over a range of x and define $n(x)$ as the positive square root of $f(x)$; $n(x)$ is then also real, positive and essentially constant over the same range of x. We adopt this notation so that the connection can be made with the description of an electromagnetic wave travelling through a medium of dielectric constant $f(x)$ and, consequently, refractive index $n(x)$. The quantity $y(x)$ would be the electric or magnetic field of the wave. For this simplified case, in which we can omit the

---

[§] These relationships and many other properties of the Airy functions can be found in, for example, M. Abramowitz and I. A. Stegun (eds), *Handbook of Mathematical Functions* (New York: Dover, 1965) pp. 446–50.

[¶] So called because they were used, independently, by Wentzel, Kramers and Brillouin to tackle certain wave-mechanical problems in 1926, though they had earlier been studied in some depth by Jeffreys and used as far back as the first half of the nineteenth century by Green.

x-dependence of $n(x)$, the solution would be (as usual)

$$y(x) = A \exp(-ik_0 n x), \tag{25.43}$$

with both A and n constant. The quantity $k_0 n x$ would be real and would be called the 'phase' of the wave; it increases linearly with x.

As a first variation on this simple picture, we may allow $f(x)$ to be complex, though, for the moment, still constant. Then $n(x)$ is still a constant, albeit a complex one:

$$n = \mu + iv.$$

The solution is formally the same as before; however, whilst it still exhibits oscillatory behaviour, the amplitude of the oscillations either grows or declines, depending upon the sign of v:

$$y(x) = A \exp(-ik_0 n x) = A \exp[-ik_0(\mu + iv)x] = A \exp(k_0 v x) \exp(-ik_0 \mu x).$$

This solution with v negative is the appropriate description for a wave travelling in a uniform absorbing medium. The quantity $k_0(\mu + iv)x$ is usually called the *complex phase* of the wave.

We now allow $f(x)$, and hence $n(x)$, to be both complex and varying with position, though, as we have noted earlier, there will be restrictions on how rapidly $f(x)$ may vary if valid solutions are to be obtained. The obvious extension of solution (25.43) to the present case would be

$$y(x) = A \exp[-ik_0 n(x)x], \tag{25.44}$$

but direct substitution of this into equation (25.42) gives

$$y'' + k_0^2 n^2 y = -k_0^2 (n'^2 x^2 + 2nn'x)y - ik_0(n''x + 2n')y.$$

Clearly the RHS can only be zero, as is required by the equation, if n' and n'' are both very small, or if some unlikely relationship exists between them.

To try to improve on this situation, we consider how the phase ϕ of the solution changes as the wave passes through an infinitesimal thickness dx of the medium. The infinitesimal (complex) phase change $d\phi$ for this is clearly $k_0 n(x)\,dx$, and therefore will be

$$\Delta\phi = k_0 \int_0^x n(u)\,du$$

for a finite thickness x of the medium. This suggests that an improvement on (25.44) might be

$$y(x) = A \exp\left(-ik_0 \int_0^x n(u)\,du\right). \tag{25.45}$$

This is still not an exact solution as now

$$y''(x) + k_0^2 n^2(x)y(x) = -ik_0 n'(x)y(x).$$

This still requires $k_0 n'(x)$ to be small (compared with, say, $k_0^2 n^2(x)$), but is some improvement (not least in complexity!) on (25.44) and gives some measure of the conditions under which the solution might be a suitable approximation.

The integral in equation (25.45) embodies what is sometimes referred to as the *phase memory* approach; it expresses the notion that the phase of the wave-like solution is the cumulative effect of changes it undergoes as it passes through the medium. If the medium were uniform the overall change would be proportional to nx, as in (25.43); the extent to which it is not uniform is reflected in the amount by which the integral differs from nx.

The condition for solution (25.45) to be a reasonable approximation can be written as $n'k_0^{-1} \ll n^2$ or, in words, the change in n over an x-range of k_0^{-1} should be small compared with n^2. For light in an optical medium, this means that the refractive index n, which is of the order of unity, must change very little over a distance of a few wavelengths.

For some purposes the above approximation is adequate, but for others further refinement is needed. This comes from considering solutions that are still wave-like but have amplitudes, as well as phases, that vary with position. These are the WKB solutions developed and studied in the next three subsections.

25.7.2 *Constructing the WKB solutions*

Having formulated the notion of phase memory, we now construct the WKB solutions of the general equation (25.42), in which $f(x)$ can now be both position-dependent and complex. As we have already seen, it is the possibility of a complex phase that permits the existence of wave-like solutions with varying amplitudes. Since $n(x)$ is calculated as the square root of $f(x)$, there is an ambiguity in its overall sign. In physical applications this is normally resolved unambiguously by considerations such as the inevitable increase in entropy of the system, but, so far as dealing with purely mathematical questions is concerned, the ambiguity must be borne in mind.

The process we adopt is an iterative one based on the assumption that the second derivative of the complex phase with respect to x is very small and can be approximated at each stage of the iteration. So we start with equation (25.42) and look for a solution of the form

$$y(x) = A \exp[i\phi(x)], \tag{25.46}$$

where A is a constant. When this is substituted into (25.42) the equation becomes

$$\left[-\left(\frac{d\phi}{dx}\right)^2 + i\frac{d^2\phi}{dx^2} + k_0^2 n^2(x) \right] y(x) = 0. \tag{25.47}$$

Setting the quantity in square brackets to zero produces a non-linear equation for which there is no obvious solution for a general $n(x)$. However, on the assumption that $d^2\phi/dx^2$ is small, an iterative solution can be found.

As a first approximation ϕ'' is ignored, and the solution

$$\frac{d\phi}{dx} \approx \pm k_0 n(x)$$

is obtained. From this, differentiation gives an approximate value for

$$\frac{d^2\phi}{dx^2} \approx \pm k_0 \frac{dn}{dx},$$

which can be substituted into equation (25.47) to give, as a second approximation for $d\phi/dx$, the expression

$$\frac{d\phi}{dx} \approx \pm \left[k_0^2 n^2(x) \pm ik_0 \frac{dn}{dx} \right]^{1/2}$$

$$= \pm k_0 n \left(1 \pm \frac{i}{2k_0 n^2} \frac{dn}{dx} + \cdots \right)$$

$$\approx \pm k_0 n + \frac{i}{2n} \frac{dn}{dx}.$$

This can now be integrated to give an approximate expression for $\phi(x)$ as follows:

$$\phi(x) = \pm k_0 \int_{x_0}^{x} n(u)\,du + \frac{i}{2} \ln[n(x)], \tag{25.48}$$

where the constant of integration has been formally incorporated into the lower limit x_0 of the integral. Now, noting that $\exp(i\frac{1}{2}i\ln n) = n^{-1/2}$, substitution of (25.48) into equation (25.46) gives

$$y_\pm(x) = \frac{A}{n^{1/2}} \exp\left[\pm ik_0 \int_{x_0}^{x} n(u)\,du \right] \tag{25.49}$$

as two independent WKB solutions of the original equation (25.42). This result is essentially the same as that in (25.45) except that the amplitude has been divided by $\sqrt{n(x)}$, i.e. by $[f(x)]^{1/4}$. Since $f(x)$ may be complex, this may introduce an additional x-dependent phase into the solution as well as the more obvious change in amplitude.

▶*Find two independent WKB solutions of Stokes' equation in the form*

$$\frac{d^2 y}{dx^2} + \lambda xy = 0, \text{ with } \lambda \text{ real and } > 0.$$

The form of the equation is the same as that in (25.42) with $f(x) = x$, and therefore $n(x) = x^{1/2}$. The WKB solutions can be read off immediately using (25.49), so long as we remember that although $f(x)$ is real, it has four fourth roots and that therefore the constant appearing in a solution can be complex. Two independent WKB solutions are

$$y_\pm(x) = \frac{A_\pm}{|x|^{1/4}} \exp\left[\pm i\sqrt{\lambda} \int^{x} \sqrt{u}\,du \right] = \frac{A_\pm}{|x|^{1/4}} \exp\left[\pm i\frac{2\sqrt{\lambda}}{3} x^{3/2} \right].$$

$$\tag{25.50}$$

The precise combination of these two solutions that is required for any particular problem has to be determined from the problem. ◀

When Stokes' equation is applied more generally to functions of a complex variable, i.e. the real variable x is replaced by the complex variable z, it has solutions whose type of behaviour depends upon where z lies in the complex plane. For the particular case $\lambda = -1$, when Stokes' equation takes the form

$$\frac{d^2 y}{dz^2} = zy$$

and the two WKB solutions (with the inverse fourth root written explicitly) are

$$y_{1,2}(z) = \frac{A_{1,2}}{z^{1/4}} \exp\left[\mp \frac{2}{3} z^{3/2}\right], \tag{25.51}$$

one of the solutions, $\text{Ai}(z)$ (see section 25.6), has the property that it is real whenever z is real, whether positive or negative. For negative real z it has sinusoidal behaviour, but it becomes an evanescent wave for real positive z.

Since the function $z^{3/2}$ has a branch point at $z = 0$ and therefore has an abrupt (complex) change in its argument there, it is clear that neither of the two functions in (25.51), nor any fixed combination of them, can be equal to $\text{Ai}(z)$ for all values of z. More explicitly, for z real and positive, $\text{Ai}(z)$ is proportional to $y_1(z)$, which is real and has the form of a decaying exponential function, whilst for z real and negative, when $z^{3/2}$ is purely imaginary and $y_1(z)$ and $y_2(z)$ are both oscillatory, it is clear that $\text{Ai}(z)$ must contain both y_1 and y_2 with equal amplitudes.

The actual combinations of $y_1(z)$ and $y_2(z)$ needed to coincide with these two asymptotic forms of $\text{Ai}(z)$ are as follows.

For z real and > 0, $\qquad c_1 y_1(z) = \frac{1}{2\sqrt{\pi} z^{1/4}} \exp\left[-\frac{2}{3} z^{3/2}\right].$ $\qquad$ (25.52)

For z real and < 0, $\qquad c_2 [y_1(z) e^{i\pi/4} - y_2(z) e^{-i\pi/4}]$

$$= \frac{1}{\sqrt{\pi} (-z)^{1/4}} \sin\left[\frac{2}{3}(-z)^{3/2} + \frac{\pi}{4}\right]. \tag{25.53}$$

Therefore it must be the case that the constants used to form $\text{Ai}(z)$ from the solutions (25.51) change as z moves from one part of the complex plane to another. In fact, the changes occur for particular values of the argument of z; these boundaries are therefore radial lines in the complex plane and are known as *Stokes lines*. For Stokes' equation they occur when $\arg z$ is equal to 0, $2\pi/3$ or $4\pi/3$.

The general occurrence of a change in the arbitrary constants used to make up a solution, as its argument crosses certain boundaries in the complex plane, is known as the Stokes phenomenon and is discussed further in subsection 25.7.4.

> ►*Apply the WKB method to the problem of finding the quantum energy levels E of a particle of mass m bound in a symmetrical one-dimensional potential well V(x) that has only a single minimum. The relevant Schrödinger equation is*
>
> $$-\frac{\hbar^2}{2m}\frac{d^2\psi}{dx^2} + V(x)\psi = E\psi.$$
>
> *Relate the problem close to each of the classical 'turning points', $x = \pm a$ at which $E - V(x) = 0$, to Stokes' equation and assume that it is appropriate to use the solution Ai(x) given in equations (25.52) and (25.53) at $x = a$. Show that if the general WKB solution in the 'classically allowed' region $-a < x < a$ is to match such Airy solutions at both turning points, then*
>
> $$\int_{-a}^{a} k(x)\,dx = (n + \tfrac{1}{2})\pi,$$
>
> *where $k^2(x) = 2m[E - V(x)]/\hbar^2$ and $n = 0, 1, 2, \ldots$.*
>
> *For a symmetric potential $V(x) = V_0 x^{2s}$, where s is a positive integer, show that in this approximation the energy of the nth level is given by $E_n = c_s(n + \tfrac{1}{2})^{2s/(s+1)}$, where c_s is a constant depending on s but not upon n.*

We start by multiplying the equation through by $2m/\hbar^2$, writing $2m[E - V(x)]/\hbar^2$ as $k^2(x)$, and rearranging the equation to read

$$\frac{d^2\psi}{dx^2} + k^2(x)\psi = 0, \tag{25.54}$$

noting that, with E and $V(x)$ given, the equation $E = V(a)$ determines the value of a and that $k(a) = 0$.

For $-a < x < a$, where $k^2(x)$ is positive, the form of the WKB solutions are given directly by (25.49) as

$$\psi_\pm = \frac{C}{\sqrt{k(x)}}\exp\left[\pm i\int^x k(u)\,du\right].$$

Just beyond the turning point $x = a$, where

$$E - V(x) = 0 - V'(a)(x - a) + O[(x - a)^2],$$

equation (25.54) can be approximated by

$$\frac{d^2\psi}{dx^2} - \frac{2mV'(a)}{\hbar^2}(x - a)\psi = 0. \tag{25.55}$$

This, in turn, can be reduced to Stokes' equation by first setting $x - a = \mu z$ and $\psi(x) \equiv y(z)$, so converting it into

$$\frac{1}{\mu^2}\frac{d^2y}{dz^2} - \frac{2\mu m V'(a)}{\hbar^2}zy = 0,$$

and then choosing $\mu = [\hbar^2/2mV'(a)]^{1/3}$. The equation then reads

$$\frac{d^2y}{dz^2} = zy.$$

Since the solution must be evanescent for $x > a$, i.e. for $z > 0$, we assume that the appropriate solution there is Ai(z); this implies that, for z small and negative (just inside the classically allowed region), the solution has the form given by (25.53), namely

$$\frac{A}{(-z)^{1/4}}\sin\left[\frac{2}{3}(-z)^{3/2} + \frac{\pi}{4}\right],$$

for some constant A. This form is only valid for negative z close to $z = 0$ and is not appropriate within the well as a whole, where the approximation (25.55) leading to Stokes' equation is not valid. However, it does allow us to determine the correct combination of the WKB solutions found earlier for the proper continuation inside the well of the solution found for $z > 0$. This is

$$\psi_1(x) = \frac{A}{\sqrt{k(x)}} \sin \left(\int_x^a k(u)\, du + \frac{\pi}{4} \right).$$

A similar argument gives the continuation inside the well of the evanescent solution required in the region $x < -a$ as

$$\psi_2(x) = \frac{B}{\sqrt{k(x)}} \sin \left(\int_{-a}^x k(u)\, du + \frac{\pi}{4} \right).$$

However, for a consistent solution to the problem, these two functions must match, both in magnitude and slope, at any arbitrary point x inside the well. We therefore require both of the equalities

$$\frac{A}{\sqrt{k(x)}} \sin \left(\int_x^a k(u)\, du + \frac{\pi}{4} \right) = \frac{B}{\sqrt{k(x)}} \sin \left(\int_{-a}^x k(u)\, du + \frac{\pi}{4} \right) \quad \text{(i)}$$

and

$$-\frac{1}{2} \frac{Ak'}{\sqrt{k^3(x)}} \sin \left(\int_x^a k(u)\, du + \frac{\pi}{4} \right) + \frac{A}{\sqrt{k(x)}} [-k(x)] \cos \left(\int_x^a k(u)\, du + \frac{\pi}{4} \right)$$

$$= -\frac{1}{2} \frac{Bk'}{\sqrt{k^3(x)}} \sin \left(\int_{-a}^x k(u)\, du + \frac{\pi}{4} \right) + \frac{B}{\sqrt{k(x)}} [k(x)] \cos \left(\int_{-a}^x k(u)\, du + \frac{\pi}{4} \right). \quad \text{(ii)}$$

The general condition for the validity of the WKB solutions is that the derivatives of the function appearing in the phase integral are small in some sense (see subsection 25.7.3 for a more general discussion); here, if $k'/\sqrt{k^3} \ll k/\sqrt{k}$, i.e. $k' \ll k^2$, then we can ignore the k' terms in equation (ii) above. In fact, for this particular situation, this approximation is not needed since the first of the equalities, equation (i), ensures that the k'-dependent terms in the second equality (ii) cancel. Either way, we are left with a pair of homogeneous equations for A and B. For them to give consistent values for the ratio A/B, it must be that

$$\frac{A}{\sqrt{k(x)}} \sin \left(\int_x^a k(u)\, du + \frac{\pi}{4} \right) \times \frac{B}{\sqrt{k(x)}} [k(x)] \cos \left(\int_{-a}^x k(u)\, du + \frac{\pi}{4} \right)$$

$$= \frac{A}{\sqrt{k(x)}} [-k(x)] \cos \left(\int_x^a k(u)\, du + \frac{\pi}{4} \right) \times \frac{B}{\sqrt{k(x)}} \sin \left(\int_{-a}^x k(u)\, du + \frac{\pi}{4} \right).$$

This condition reduces to

$$\sin \left[\left(\int_x^a k(u)\, du + \frac{\pi}{4} \right) + \left(\int_{-a}^x k(u)\, du + \frac{\pi}{4} \right) \right] = 0,$$

$$\sin \left[\left(\int_{-a}^a k(u)\, du + \frac{\pi}{2} \right) \right] = 0,$$

$$\Rightarrow \int_{-a}^a k(u)\, du = (n + \tfrac{1}{2})\pi.$$

Since $k(x) > 0$ in the range $-a < x < a$, n may take the values $0, 1, 2, \ldots$.

If $V(x)$ has the form $V(x) = V_0 x^{2s}$ then, for the nth allowed energy level, $E_n = V_0 a_n^{2s}$ and

$$k^2(x) = \frac{2m}{\hbar^2} (E_n - V_0 x^{2s}).$$

901

The result just proved gives

$$\int_{-a_n}^{a_n} \frac{\sqrt{2mV_0}}{\hbar} (a_n^{2s} - x^{2s})^{1/2} \, dx = (n + \tfrac{1}{2})\pi.$$

Writing $x = va_n$ shows that the integral is proportional to $a_n^{s+1} I_s$, where I_s is the integral between -1 and $+1$ of $(1 - v^{2s})^{1/2}$ and does not depend upon n. Thus $E_n \propto a_n^{2s}$ and $a_n^{s+1} \propto (n + \tfrac{1}{2})$, implying that $E_n \propto (n + \tfrac{1}{2})^{2s/s+1}$.

Although not asked for, we note that the above result indicates that, for a simple harmonic oscillator, for which $s = 1$, the energy levels $[E_n \sim (n + \tfrac{1}{2})]$ are equally spaced, whilst for very large s, corresponding to a square well, the energy levels vary as n^2. Both of these results agree with what is found from detailed analyses of the individual cases. ◀

25.7.3 Accuracy of the WKB solutions

We may also ask when we can expect the WKB solutions to the Stokes' equation to be reasonable approximations. Although our final form for the WKB solutions is not exactly that used when the condition $|n'k_0^{-1}| \ll |n^2|$ was derived, it should give the same order of magnitude restriction as a more careful analysis. For the derivation of (25.51), $k_0^2 = -1$, $n(z) = [f(z)]^{1/2} = z^{1/2}$, and the criterion becomes $\tfrac{1}{2}|z^{-1/2}| \ll |z|$, or, in round terms, $|z|^3 \gg 1$.

For the more general equation, typified by (25.42), the condition for the validity of the WKB solutions can usually be satisfied by making some quantity, often $|z|$, sufficiently large. Alternatively, a parameter such as k_0 can be made large enough that the validity criterion is satisfied to any pre-specified level. However, from a practical point of view, natural physical parameters cannot be varied at will, and requiring z to be large may well reduce the value of the method to virtually zero. It is normally more useful to try to obtain an improvement on a WKB solution by multiplying it by a series whose terms contain increasing inverse powers of the variable, so that the result can be applied successfully for moderate, and not just excessively large, values of the variable.

We do not have the space to discuss the properties and pitfalls of such asymptotic expansions in any detail, but exercise 25.18 will provide the reader with a model of the general procedure. A few particular points that should be noted are given as follows.

(i) If the multiplier is analytic as $z \to \infty$, then it will be represented by a series that is convergent for $|z|$ greater than some radius of convergence R.

(ii) If the multiplier is not analytic as $z \to \infty$, as is usually the case, then the multiplier series eventually diverges and there is a z-dependent optimal number of terms that the series should contain in order to give the best accuracy.

(iii) For a fixed value of $\arg z$, the asymptotic expansion of the multiplier is unique. However, the same asymptotic expansion can represent more than

902

one function and the same function may need different expansions for different values of $\arg z$.

Finally in this subsection we note that, although the form of equation (25.42) may appear rather restrictive, in that it contains no term in y', the results obtained so far can be applied to an equation such as

$$\frac{d^2y}{dz^2} + P(z)\frac{dy}{dz} + Q(z)y = 0. \qquad (25.56)$$

To make this possible, a change of either the dependent or the independent variable is made. For the former we write

$$Y(z) = y(z)\exp\left(\frac{1}{2}\int^z P(u)\,du\right) \quad \Rightarrow \quad \frac{d^2Y}{dz^2} + \left(Q - \frac{1}{4}P^2 - \frac{1}{2}\frac{dP}{dz}\right)Y = 0,$$

whilst for the latter we introduce a new independent variable ζ defined by

$$\frac{d\zeta}{dz} = \exp\left(-\int^z P(u)\,du\right) \quad \Rightarrow \quad \frac{d^2y}{d\zeta^2} + Q\left(\frac{dz}{d\zeta}\right)^2 y = 0.$$

In either case, equation (25.56) is reduced to the form of (25.42), though it will be clear that the two sets of WKB solutions (which are, of course, only approximations) will not be the same.

25.7.4 The Stokes phenomenon

As we saw in subsection 25.7.2, the combination of WKB solutions of a differential equation required to reproduce the asymptotic form of the accurate solution $y(z)$ of the same equation, varies according to the region of the z-plane in which z lies. We now consider this behaviour, known as the Stokes phenomenon, in a little more detail.

Let $y_1(z)$ and $y_2(z)$ be the two WKB solutions of a second-order differential equation. Then any solution $Y(z)$ of the same equation can be written asymptotically as

$$Y(z) \sim A_1 y_1(z) + A_2 y_2(z), \qquad (25.57)$$

where, although we will be considering (abrupt) changes in them, we will continue to refer to A_1 and A_2 as constants, as they are within any one region. In order to produce the required change in the linear combination, as we pass over a Stokes line from one region of the z-plane to another, one of the constants must change (relative to the other) as the border between the regions is crossed.

At first sight, this may seem impossible without causing a discernible discontinuity in the representation of $Y(z)$. However, we must recall that the WKB solutions are approximations, and that, as they contain a phase integral, for certain values of $\arg z$ the phase $\phi(z)$ will be purely imaginary and the factors

$\exp[\pm i\phi(z)]$ will be purely real. What is more, one such factor, known as the *dominant* term, will be exponentially large, whilst the other (the *subdominant* term) will be exponentially small. A *Stokes line* is precisely where this happens.

We can now see how the change takes place without an observable discontinuity occurring. Suppose that $y_1(z)$ is very large and $y_2(z)$ is very small on a Stokes line. Then a finite change in A_2 will have a negligible effect on $Y(z)$; in fact, Stokes showed, for some particular cases, that the change is less than the uncertainty in $y_1(z)$ arising from the approximations made in deriving it. Since the solution with any particular asymptotic form is determined in a region bounded by two Stokes lines to within an overall multiplicative constant and the original equation is linear, the change in A_2 when one of the Stokes lines is crossed must be proportional to A_1, i.e. A_2 changes to $A_2 + SA_1$, where S is a constant (the *Stokes constant*) characteristic of the particular line but independent of A_1 and A_2. It should be emphasised that, at a Stokes line, if the dominant term is not present in a solution, then the multiplicative constant in the subdominant term *cannot* change as the line is crossed.

As an example, consider the Bessel function $J_0(z)$ of zero order. It is single-valued, differentiable everywhere, and can be written as a series in powers of z^2. It is therefore an integral even function of z. However, its asymptotic approximations for two regions of the z-plane, $\text{Re } z > 0$ and z real and negative, are given by

$$J_0(z) \sim \frac{1}{\sqrt{2\pi}} \frac{1}{\sqrt{z}} \left(e^{iz} e^{-i\pi/4} + e^{-iz} e^{i\pi/4} \right), \quad |\arg(z)| < \tfrac{1}{2}\pi, \ |\arg(z^{-1/2})| < \tfrac{1}{4}\pi,$$

$$J_0(z) \sim \frac{1}{\sqrt{2\pi}} \frac{1}{\sqrt{z}} \left(e^{iz} e^{3i\pi/4} + e^{-iz} e^{i\pi/4} \right), \quad \arg(z) = \pi, \ \arg(z^{-1/2}) = -\tfrac{1}{2}\pi.$$

We note in passing that neither of these expressions is naturally single-valued, and a prescription for taking the square root has to be given. Equally, neither is an even function of z. For our present purpose the important point to note is that, for both expressions, on the line $\arg z = \pi/2$ both z-dependent exponents become real. For large $|z|$ the second term in each expression is large; this is the dominant term, and its multiplying constant $e^{i\pi/4}$ is the same in both expressions. Contrarywise, the first term in each expression is small, and its multiplying constant does change, from $e^{-i\pi/4}$ to $e^{3i\pi/4}$, as $\arg z$ passes through $\pi/2$ whilst increasing from 0 to π. It is straightforward to calculate the Stokes constant for this Stokes line as follows:

$$S = \frac{A_2(\text{new}) - A_2(\text{old})}{A_1} = \frac{e^{3i\pi/4} - e^{-i\pi/4}}{e^{i\pi/4}} = e^{i\pi/2} - e^{-i\pi/2} = 2i.$$

If we had moved (in the negative sense) from $\arg z = 0$ to $\arg z = -\pi$, the relevant Stokes line would have been $\arg z = -\pi/2$. There the first term in each expression is dominant, and it would have been the constant $e^{i\pi/4}$ in the second term that would have changed. The final argument of $z^{-1/2}$ would have been $+\pi/2$.

Finally, we should mention that the lines in the z-plane on which the exponents in the WKB solutions are purely imaginary, and the two solutions have equal amplitudes, are usually called the *anti-Stokes lines*. For the general Bessel's equation they are the real positive and real negative axes.

25.8 Approximations to integrals

In this section we will investigate a method of finding approximations to the values or forms of certain types of infinite integrals. The class of integrals to be considered is that containing integrands that are, or can be, represented by exponential functions of the general form $g(z)\exp[f(z)]$. The exponents $f(z)$ may be complex, and so integrals of sinusoids can be handled as well as those with more obvious exponential properties. We will be using the analyticity properties of the functions of a complex variable to move the integration path to a part of the complex plane where a general integrand can be approximated well by a standard form; the standard form is then integrated explicitly.

The particular standard form to be employed is that of a Gaussian function of a real variable, for which the integral between infinite limits is well known. This form will be generated by expressing $f(z)$ as a Taylor series expansion about a point z_0, at which the linear term in the expansion vanishes, i.e. where $f'(z) = 0$. Then, apart from a constant multiplier, the exponential function will behave like $\exp[\frac{1}{2}f''(z_0)(z - z_0)^2]$ and, by choosing an appropriate direction for the contour to take as it passes through the point, this can be made into a normal Gaussian function of a real variable and its integral may then be found.

25.8.1 Level lines and saddle points

Before we can discuss the method outlined above in more detail, a number of observations about functions of a complex variable and, in particular, about the properties of the exponential function need to be made. For a general analytic function,

$$f(z) = \phi(x, y) + i\psi(x, y), \tag{25.58}$$

of the complex variable $z = x + iy$, we recall that, not only do both ϕ and ψ satisfy Laplace's equation, but $\nabla\phi$ and $\nabla\psi$ are orthogonal. This means that the lines on which one of ϕ and ψ is constant are exactly the lines on which the other is changing most rapidly.

Let us apply these observations to the function

$$h(z) \equiv \exp[f(z)] = \exp(\phi)\exp(i\psi), \tag{25.59}$$

recalling that the functions ϕ and ψ are themselves real. The magnitude of $h(z)$, given by $\exp(\phi)$, is constant on the lines of constant ϕ, which are known as the

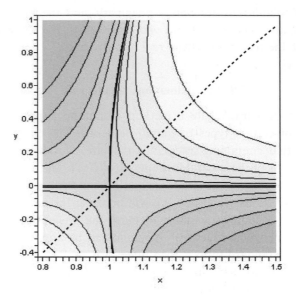

Figure 25.12   A greyscale plot with associated contours of the value of $|h(z)|$, where $h(z) = \exp[i(z^3 + 6z^2 - 15z + 8)]$, in the neighbourhood of one of its saddle points; darker shading corresponds to larger magnitudes. The plot also shows the two level lines (thick solid lines) through the saddle and part of the line of steepest descents (dashed line) passing over it. At the saddle point, the angle between the line of steepest descents and a level line is $\pi/4$.

level lines of the function. It follows that the direction in which the magnitude of $h(z)$ changes most rapidly at any point z is in a direction perpendicular to the level line passing through that point. This is therefore the line through z on which the phase of $h(z)$, namely $\psi(z)$, is constant. Lines of constant phase are therefore sometimes referred to as *lines of steepest descent* (or steepest ascent).

We further note that $|h(z)|$ can never be negative and that neither ϕ nor ψ can have a finite maximum at any point at which $f(z)$ is analytic. This latter observation follows from the fact that at a maximum of, say, $\phi(x, y)$, both $\partial^2\phi/\partial x^2$ and $\partial^2\phi/\partial y^2$ would have to be negative; if this were so, Laplace's equation could not be satisfied, leading to a contradiction. A similar argument shows that a minimum of either ϕ or ψ is not possible wherever $f(z)$ is analytic. A more positive conclusion is that, since the two unmixed second partial derivatives $\partial^2\phi/\partial x^2$ and $\partial^2\phi/\partial y^2$ must have opposite signs, the only possible conclusion about a point at which $\nabla\phi$ is defined and equal to zero is that the point is a saddle point of $h(z)$. An example of a saddle point is shown as a greyscale plot in figure 25.12 and, more pictorially, in figure 5.2.

From the observations contained in the two previous paragraphs, we deduce that a path that follows the lines of steepest descent (or ascent) can never form a closed loop. On such a path, ϕ, and hence $|h(z)|$, must continue to decrease (increase) until the path meets a singularity of $f(z)$. It also follows that if a level line of $h(z)$ forms a closed loop in the complex plane, then the loop must enclose a singularity of $f(z)$. This may (if $\phi \to \infty$) or may not (if $\phi \to -\infty$) produce a singularity in $h(z)$.

We now turn to the study of the behaviour of $h(z)$ at a saddle point and how this enables us to find an approximation to the integral of $h(z)$ along a contour that can be deformed to pass through the saddle point. At a saddle point z_0, at which $f'(z_0) = 0$, both $\nabla\phi$ and $\nabla\psi$ are zero, and consequently the magnitude and phase of $h(z)$ are both stationary. The Taylor expansion of $f(z)$ at such a point takes the form

$$f(z) = f(z_0) + 0 + \frac{1}{2!}f''(z_0)(z - z_0)^2 + O(z - z_0)^3. \tag{25.60}$$

We assume that $f''(z_0) \neq 0$ and write it explicitly as $f''(z_0) \equiv Ae^{i\alpha}$, thus defining the real quantities A and α. If it happens that $f''(z_0) = 0$, then two or more saddle points coalesce and the Taylor expansion must be continued until the first non-vanishing term is reached; we will not consider this case further, though the general method of proceeding will be apparent from what follows. If we also abbreviate the (in general) complex quantity $f(z_0)$ to f_0, then (25.60) takes the form

$$f(z) = f_0 + \tfrac{1}{2}Ae^{i\alpha}(z - z_0)^2 + O(z - z_0)^3. \tag{25.61}$$

To study the implications of this approximation for $h(z)$, we write $z - z_0$ as $\rho\,e^{i\theta}$ with ρ and θ both real. Then

$$|h(z)| = |\exp(f_0)|\,\exp[\tfrac{1}{2}A\rho^2\cos(2\theta + \alpha) + O(\rho^3)]. \tag{25.62}$$

This shows that there are four values of θ for which $|h(z)|$ is independent of ρ (to second order). These therefore correspond to two crossing level lines given by

$$\theta = \tfrac{1}{2}\left(\pm\tfrac{1}{2}\pi - \alpha\right) \text{ and } \theta = \tfrac{1}{2}\left(\pm\tfrac{3}{2}\pi - \alpha\right). \tag{25.63}$$

The two level lines cross at right angles to each other. It should be noted that the continuations of the two level lines away from the saddle are not straight in general. At the saddle they have to satisfy (25.63), but away from it the lines must take whatever directions are needed to make $\nabla\phi = 0$. In figure 25.12 one of the level lines ($|h| = 1$) has a continuation ($y = 0$) that is straight; the other does not and bends away from its initial direction $x = 1$.

So far as the phase of $h(z)$ is concerned, we have

$$\arg[h(z)] = \arg(f_0) + \tfrac{1}{2}A\rho^2\sin(2\theta + \alpha) + O(\rho^3),$$

which shows that there are four other directions (two lines crossing at right angles) in which the phase of $h(z)$ is independent of ρ. They make angles of $\pi/4$ with the level lines through z_0 and are given by

$$\theta = -\tfrac{1}{2}\alpha, \quad \theta = \tfrac{1}{2}(\pm\pi - \alpha), \quad \theta = \pi - \tfrac{1}{2}\alpha.$$

From our previous discussion it follows that these four directions will be the lines of steepest descent (or ascent) on moving away from the saddle point. In particular, the two directions for which the term $\cos(2\theta + \alpha)$ in (25.62) is *negative* will be the directions in which $|h(z)|$ *decreases* most rapidly from its value at the saddle point. These two directions are antiparallel, and a steepest descents path following them is a smooth locally straight line passing the saddle point. It is known as the *line of steepest descents* (l.s.d.) through the saddle point. Note that 'descents' is plural as on this line the value of $|h(z)|$ decreases on *both* sides of the saddle. This is the line which we will make the path of the contour integral of $h(z)$ follow. Part of a typical l.s.d. is indicated by the dashed line in figure 25.12.

25.8.2 Steepest descents method

To help understand how an integral along the line of steepest descents can be handled in a mechanical way, it is instructive to consider the case where the function $f(z) = -\beta z^2$ and $h(z) = \exp(-\beta z^2)$. The saddle point is situated at $z = z_0 = 0$, with $f_0 = f(z_0) = 1$ and $f''(z_0) = -2\beta$, implying that $A = 2|\beta|$ and $\alpha = \pm\pi + \arg\beta$, with the $\pm$ sign chosen to put α in the range $0 \le \alpha < 2\pi$. Then the l.s.d. is determined by the requirement that $\sin(2\theta+\alpha) = 0$ whilst $\cos(2\theta+\alpha)$ is negative; together these imply that, for the l.s.d., $\theta = -\tfrac{1}{2}\arg\beta$ or $\theta = \pi - \tfrac{1}{2}\arg\beta$.

Since the Taylor series for $f(z) = -\beta z^2$ terminates after three terms, expansion (25.61) for this particular function is not an approximation to $h(z)$, but is exact. Consequently, a contour integral starting and ending in regions of the complex plane where the function tends to zero and following the l.s.d. through the saddle point at $z = 0$ will not only have a straight-line path, but will yield an exact result. Setting $z = te^{-\frac{1}{2}\arg\beta}$ will reduce the integral to that of a Gaussian function:

$$e^{-\frac{1}{2}\arg\beta} \int_{-\infty}^{\infty} e^{-|\beta|t^2}\, dt = e^{-\frac{1}{2}\arg\beta} \sqrt{\frac{\pi}{|\beta|}}.$$

The saddle-point method for a more general function aims to simulate this approach by deforming the integration contour C and forcing it to pass through a saddle point $z = z_0$, where, whatever the function, the leading z-dependent term in the exponent will be a quadratic function of $z - z_0$, thus turning the integrand into one that can be approximated by a Gaussian.

The path well away from the saddle point may be changed in any convenient way so long as it remains within the relevant sectors, as determined by the end-points of C. By a 'sector' we mean a region of the complex plane, any part of which can be reached from any other part of the same region without crossing any of the continuations to infinity of the level lines that pass through the saddle. In practical applications the start- and end-points of the path are nearly always at singularities of $f(z)$ with Re $f(z) \to -\infty$ and $|h(z)| \to 0$.

We now set out the complete procedure for the simplest form of integral evaluation that uses a method of steepest descents. Extensions, such as including higher terms in the Taylor expansion or having to pass through more than one saddle point in order to have appropriate termination points for the contour, can be incorporated, but the resulting calculations tend to be long and complicated, and we do not have space to pursue them in a book such as this one.

As our general integrand we take a function of the form $g(z)h(z)$, where, as before, $h(z) = \exp[f(z)]$. The function $g(z)$ should neither vary rapidly nor have zeros or singularities close to any saddle point used to evaluate the integral. Rapidly varying factors should be incorporated in the exponent, usually in the form of a logarithm. Provided $g(z)$ satisfies these criteria, it is sufficient to treat it as a constant multiplier when integrating, assigning to it its value at the saddle point, $g(z_0)$.

Incorporating this and retaining only the first two non-vanishing terms in equation (25.61) gives the integrand as

$$g(z_0) \exp(f_0) \exp[\tfrac{1}{2} A e^{i\alpha} (z - z_0)^2]. \tag{25.64}$$

From the way in which it was defined, it follows that on the l.s.d. the imaginary part of $f(z)$ is constant ($= \text{Im } f_0$) and that the final exponent in (25.64) is either zero (at z_0) or negative. We can therefore write it as $-s^2$, where s is real. Further, since $\exp[f(z)] \to 0$ at the start- and end-points of the contour, we must have that s runs from $-\infty$ to $+\infty$, the sense of s being chosen so that it is negative approaching the saddle and positive when leaving it.

Making this change of variable,

$$\tfrac{1}{2} A e^{i\alpha}(z - z_0)^2 = -s^2, \text{ with } dz = \pm\sqrt{\frac{2}{A}} \exp[\tfrac{1}{2} i(\pi - \alpha)] \, ds, \tag{25.65}$$

allows us to express the contribution to the integral from the neighbourhood of the saddle point as

$$\pm g(z_0) \exp(f_0) \sqrt{\frac{2}{A}} \exp[\tfrac{1}{2} i(\pi - \alpha)] \int_{-\infty}^{\infty} \exp(-s^2) \, ds.$$

The simple saddle-point approximation assumes that this is the only contribution, and gives as the value of the contour integral

$$\int_C g(z) \exp[f(z)] \, dz = \pm\sqrt{\frac{2\pi}{A}} \, g(z_0) \exp(f_0) \exp[\tfrac{1}{2} i(\pi - \alpha)], \tag{25.66}$$

where we have used the standard result that $\int_{-\infty}^{\infty} \exp(-s^2) \, ds = \sqrt{\pi}$. The overall $\pm$

sign is determined by the direction θ in the complex plane in which the distorted contour passes through the saddle point. If $-\frac{1}{2}\pi < \theta \leq \frac{1}{2}\pi$, then the positive sign is taken; if not, then the negative sign is appropriate. In broad terms, if the integration path through the saddle is in the direction of an increasing real part for z, then the overall sign is positive.

Formula (25.66) is the main result from a steepest descents approach to evaluating a contour integral of the type considered, in the sense that it is the leading term in any more refined calculation of the same integral. As can be seen, it is as an 'omnibus' formula, the various components of which can be found by considering a number of separate, less-complicated, calculations.

Before presenting a worked example that generates a substantial result, useful in another connection, it is instructive to consider an integral that can be simply and exactly evaluated by other means and then apply the saddle-point result to it. Of course, the steepest descents method will appear heavy-handed, but our purpose is to show it in action and to try to see why it works.

Consider the real integral

$$I = \int_{-\infty}^{\infty} \exp(10t - t^2) \, dt.$$

This can be evaluated directly by making the substitution $s = t - 5$ as follows:

$$I = \int_{-\infty}^{\infty} \exp(10t - t^2) \, dt = \int_{-\infty}^{\infty} \exp(25 - s^2) \, ds = e^{25} \int_{-\infty}^{\infty} \exp(-s^2) \, ds = \sqrt{\pi} e^{25}.$$

The saddle-point approach to the same problem is to consider the integral as a contour integral in the complex plane, but one that lies along the real axis. The saddle points of the integrand occur where $f'(t) = 10 - 2t = 0$; there is thus a single saddle point at $t = t_0 = 5$. This is on the real axis, and no distortion of the contour is necessary. The value f_0 of the exponent is $f(5) = 50 - 25 = 25$, whilst its second derivative at the saddle point is $f''(5) = -2$. Thus, $A = 2$ and $\alpha = \pi$. The contour clearly passes through the saddle point in the direction $\theta = 0$, i.e. in the positive sense on the real axis, and so the overall sign must be $+$. Since $g(t_0)$ is formally unity, we have all the ingredients needed for substitution in formula (25.66), which reads

$$I = +\sqrt{\frac{2\pi}{2}} \, 1 \, \exp(25) \, \exp[\tfrac{1}{2} i(\pi - \pi)] = \sqrt{\pi} e^{25}.$$

As it happens, this is exactly the same result as that obtained by accurate calculation. This would not normally be the case, but here it is, because of the quadratic nature of $10t - t^2$; all of its derivatives beyond the second are identically zero and no approximation of the exponent is involved.

Given the very large value of the integrand at the saddle point itself, the reader may wonder whether there really is a saddle there. However, evaluating the integrand at points lying on a line through the saddle point perpendicular to the l.s.d., i.e. on the imaginary t-axis, provides some reassurance. Whether μ is positive or negative,

$$h(5 + i\mu) = \exp(50 + 10i\mu - 25 - 10i\mu + \mu^2) = \exp(25 + \mu^2).$$

This is greater than $h(5)$ for all μ and increases as $|\mu|$ increases, showing that the integration path really does lie at a minimum of $h(t)$ for a traversal in this direction.

We now give a fully worked solution to a problem that could not be easily tackled by elementary means.

▶ *Apply the saddle-point method to the function defined by*

$$F(x) = \frac{1}{\pi} \int_0^\infty \cos(\tfrac{1}{3}s^3 + xs)\,ds$$

to show that its form for large positive real x is one that tends asymptotically to zero, hence enabling $F(x)$ to be identified with the Airy function, Ai(x).

We first express the integral as an exponential function and then make the change of variable $s = x^{1/2}t$ to bring it into the canonical form $\int g(t)\exp[f(t)]\,dt$ as follows:

$$F(x) = \frac{1}{\pi} \int_0^\infty \cos(\tfrac{1}{3}s^3 + xs)\,ds$$

$$= \frac{1}{2\pi} \int_{-\infty}^\infty \exp[i(\tfrac{1}{3}s^3 + xs)]\,ds$$

$$= \frac{1}{2\pi} \int_{-\infty}^\infty x^{1/2} \exp[ix^{3/2}(\tfrac{1}{3}t^3 + t)]\,dt.$$

We now seek to find an approximate expression for this contour integral by deforming its path along the real t-axis into one passing over a saddle point of the integrand. Considered as a function of t, the multiplying factor $x^{1/2}/2\pi$ is a constant, and any effects due to the proximity of its zeros and singularities to any saddle point do not arise.

The saddle points are situated where

$$0 = f'(t) = ix^{3/2}(t^2 + 1) \quad \Rightarrow \quad t = \pm i.$$

For reasons discussed later, we choose to use the saddle point at $t = t_0 = i$. At this point,

$$f(i) = ix^{3/2}(-\tfrac{1}{3}i + i) = -\tfrac{2}{3}x^{3/2} \text{ and } Ae^{i\alpha} \equiv f''(i) = ix^{3/2}(2i) = -2x^{3/2},$$

and so $A = 2x^{3/2}$ and $\alpha = \pi$.

Now, expanding $f(t)$ around $t = i$ by setting $t = i + \rho e^{i\theta}$, we have

$$f(t) = f(i) + 0 + \frac{1}{2!}f''(i)(t - i)^2 + O[(t - i)^3]$$

$$= -\frac{2}{3}x^{3/2} + \frac{1}{2}2x^{3/2}e^{i\pi}\rho^2 e^{2i\theta} + O(\rho^3).$$

For the l.s.d. contour that crosses the saddle point we need the second term in this last line to decrease as ρ increases. This happens if $\pi + 2\theta = \pm\pi$, i.e. if $\theta = 0$ or $\theta = -\pi$ (or $+\pi$); thus, the l.s.d. through the saddle is oriented parallel to the real t-axis. Given the initial contour direction, the deformed contour should approach the saddle point from the direction $\theta = -\pi$ and leave it along the line $\theta = 0$. Since $-\pi/2 < 0 \leq \pi/2$, the overall sign of the 'omnibus' approximation formula is determined as positive.

Finally, putting the various values into the formula yields

$$F(x) \sim + \left(\frac{2\pi}{A}\right)^{1/2} g(i) \exp[f(i)] \exp[\tfrac{1}{2}i(\pi - \alpha)]$$

$$= + \left(\frac{2\pi}{2x^{3/2}}\right)^{1/2} \frac{x^{1/2}}{2\pi} \exp\left(-\frac{2}{3}x^{3/2}\right) \exp[\tfrac{1}{2}i(\pi - \pi)]$$

$$= \frac{1}{2\sqrt{\pi}x^{1/4}} \exp\left(-\frac{2}{3}x^{3/2}\right).$$

This is the leading term in the asymptotic expansion of $F(x)$, which, as shown in equation (25.39), is a particular contour integral solution of Stokes' equation. The fact that it tends to zero in a monotonic way as $x \to +\infty$ allows it to be identified with the Airy function, Ai(x).

We may ask why the saddle point at $t = -i$ was not used. The answer to this is as follows. Of course, any path that starts and ends in the right sectors will suffice, but if another saddle point exists close to the one used, then the Taylor expansion actually employed is likely to be less effective than if there were no other saddle points or if there were only distant ones.

An investigation of the same form as that used at $t = +i$ shows that the saddle at $t = -i$ is higher by a factor of $\exp(\tfrac{4}{3}x^{3/2})$ and that its l.s.d. is orientated parallel to the imaginary t-axis. Thus a path that went through it would need to go via a region of largish negative imaginary t, over the saddle at $t = -i$, and then, when it reached the col at $t = +i$, bend sharply and follow part of the same l.s.d. as considered earlier. Thus the contribution from the $t = -i$ saddle would be incomplete and roughly half of that from the $t = +i$ saddle would still have to be included. The more serious error would come from the first of these, as, clearly, the part of the path that lies in the plane Re $t = 0$ is not symmetric and is far from Guassian-like on the side nearer the origin. The Gaussian-path approximation used will therefore not be a good one, and, what is more, the resulting error will be magnified by a factor $\exp(\tfrac{4}{3}x^{3/2})$ compared with the best estimate. So, both on the grounds of simplicity and because the effect of the other (neglected) saddle point is likely to be less severe, we choose to use the one at $t = +i$. ◄

25.8.3 Stationary phase method

In the previous subsection we showed how to use the saddle points of an exponential function of a complex variable to evaluate approximately a contour integral of that function. This was done by following the lines of steepest descent that passed through the saddle point; these are lines on which the phase of the exponential is constant but its amplitude is varying at the maximum possible rate for that function. We now introduce an alternative method, one that entirely reverses the roles of amplitude and phase. To see how such an alternative approach might work, it is useful to study how the integral of an exponential function of a complex variablecan be represented as the sum of infinitesimal vectors in the complex plane.

We start by studying the familiar integral

$$I_0 = \int_{-\infty}^{\infty} \exp(-z^2)\,dz, \tag{25.67}$$

which we already know has the value $\sqrt{\pi}$ when z is real. This choice of demonstration model is not accidental, but is motivated by the fact that, as we have already shown, in the neighbourhood of a saddle point all exponential integrands can be approximated by a Gaussian function of this form.

The same integral can also be thought of as an integral in the complex plane, in which the integration contour happens to be along the real axis. Since the integrand is analytic, the contour could be distorted into any other that had the same end-points, $z = -\infty$ and $z = +\infty$, both on the real axis.

As a particular possibility, we consider an arc of a circle of radius R centred on $z = 0$. It is easily shown that $\cos 2\theta \geq 1 + 4\theta/\pi$ for $-\pi/4 < \theta \leq 0$, where θ is measured from the positive real z-axis and $-\pi < \theta \leq \pi$. It follows from writing $z = R e^{i\theta}$ on the arc that, if the arc is confined to the region $-\pi/4 < \theta \leq 0$ (actually, $|\theta| < \pi/4$ is sufficient), then the integral of $\exp(-z^2)$ tends to zero as $R \to \infty$ anywhere on the arc. A similar result holds for an arc confined to the region $||\theta| - \pi| < \pi/4$. We also note for future use that, for $\pi/4 < \theta < 3\pi/4$ or $-\pi/4 > \theta > -3\pi/4$, the integrand $\exp(-z^2)$ grows without limit as $R \to \infty$, and that the larger R is, the more precipitous is the 'drop or rise' in its value on crossing the four radial lines $\theta = \pm\pi/4$ and $\theta = \pm 3\pi/4$.

Now consider a contour that consists of an arc at infinity running from $\theta = \pi$ to $\theta = \pi - \alpha$ joined to a straight line, $\theta = -\alpha$, which passes through $z = 0$ and continues to infinity, where it in turn joins an arc at infinity running from $\theta = -\alpha$ to $\theta = 0$. This contour has the same start- and end-points as that used in I_0, and so the integral of $\exp(-z^2)$ along it must also have the value $\sqrt{\pi}$. As the contributions to the integral from the arcs vanish, provided $\alpha < \pi/4$, it follows that the integral of $\exp(-z^2)$ along the infinite line $\theta = -\alpha$ is $\sqrt{\pi}$. If we now take α arbitrarily close to $\pi/4$, we may substitute $z = s \exp(-i\pi/4)$ into (25.67) and obtain

$$\sqrt{\pi} = \int_{-\infty}^{\infty} \exp(-z^2)\, dz$$

$$= \exp(-i\pi/4) \int_{-\infty}^{\infty} \exp(is^2)\, ds \tag{25.68}$$

$$= \sqrt{2\pi} \exp(-i\pi/4) \left[\int_0^{\infty} \cos(\tfrac{1}{2}\pi u^2)\, du + i \int_0^{\infty} \sin(\tfrac{1}{2}\pi u^2)\, du \right]. \tag{25.69}$$

The final line was obtained by making a scale change $s = \sqrt{\pi/2}\, u$. This enables the two integrals to be identified with the Fresnel integrals $C(x)$ and $S(x)$,

$$C(x) = \int_0^x \cos(\tfrac{1}{2}\pi u^2)\, du \text{ and } S(x) = \int_0^x \sin(\tfrac{1}{2}\pi u^2)\, du,$$

mentioned on page 645. Equation (25.69) can be rewritten as

$$\frac{(1+i)\sqrt{\pi}}{\sqrt{2}} = \sqrt{2\pi}\, [\, C(\infty) + iS(\infty)\,],$$

from which it follows that $C(\infty) = S(\infty) = \frac{1}{2}$. Clearly, $C(-\infty) = S(-\infty) = -\frac{1}{2}$.

We are now in a position to examine these two equivalent ways of evaluating I_0 in terms of sums of infinitesimal vectors in the complex plane. When the integral $\int_{-\infty}^{\infty} \exp(-z^2) \, dz$ is evaluated as a real integral, or a complex one along the real z-axis, each element dz generates a vector of length $\exp(-z^2) \, dz$ in an Argand diagram, usually called the *amplitude–phase diagram* for the integral. For this integration, whilst all vector contributions lie along the real axis, they do differ in magnitude, starting vanishingly small, growing to a maximum length of $1 \times dz$, and then reducing until they are again vanishingly small. At any stage, their vector sum (in this case, the same as their algebraic sum) is a measure of the indefinite integral

$$I(x) = \int_{-\infty}^{x} \exp(-z^2) \, dz. \tag{25.70}$$

The total length of the vector sum when $x \to \infty$ is, of course, $\sqrt{\pi}$, and it should not be overlooked that the sum is a vector parallel to (actually coinciding with) the real axis in the amplitude–phase diagram. Formally this indicates that the integral is real. This 'ordinary' view of evaluating the integral generates the same amplitude–phase diagram as does the method of steepest descents. This is because for this particular integrand the l.s.d. never leaves the real axis.

Now consider the same integral evaluated using the form of equation (25.69). Here, each contribution, as the integration variable goes from u to $u + du$, is of the form

$$g(u) \, du = \cos(\tfrac{1}{2}\pi u^2) \, du + i \sin(\tfrac{1}{2}\pi u^2) \, du.$$

As infinitesimal vectors in the amplitude–phase diagram, *all* $g(u) \, du$ have the *same* magnitude du, but their directions change continuously. Near $u = 0$, where u^2 is small, the change is slow and each vector element is approximately equal to $\sqrt{2\pi} \exp(-i\pi/4) \, du$; these contributions are all in phase and add up to a significant vector contribution in the direction $\theta = -\pi/4$. This is illustrated by the central part of the curve in part (b) of figure 25.13, in which the amplitude–phase diagram for the 'ordinary' integration, discussed above, is drawn as part (a).

Part (b) of the figure also shows that the vector representing the indefinite integral (25.70) initially (s large and negative) spirals out, in a clockwise sense, from around the point $0 + i0$ in the amplitude–phase diagram and ultimately (s large and positive) spirals in, in an anticlockwise direction, to the point $\sqrt{\pi} + i0$. The total curve is called a Cornu spiral. In physical applications, such as the diffraction of light at a straight edge, the relevant limits of integration are typically $-\infty$ and some finite value x. Then, as can be seen, the resulting vector sum is complex in general, with its magnitude (the distance from $0 + i0$ to the point on the spiral corresponding to $z = x$) growing steadily for $x < 0$ but showing oscillations when $x > 0$.

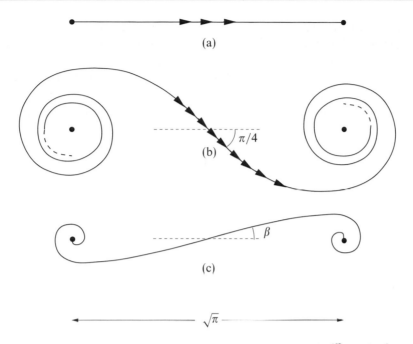

Figure 25.13   Amplitude–phase diagrams for the integral $\int_{-\infty}^{\infty}\exp(-z^2)\,dz$ using different contours in the complex z-plane. (a) Using the real axis, as in the steepest descents method. (b) Using the level line $z = u\exp(-\tfrac{1}{4}i\pi)$ that passes through the saddle point, as in the stationary phase method. (c) Using a path that makes a positive angle β ($< \pi/4$) with the z-axis.

The final curve, 25.13(c), shows the amplitude-phase diagram corresponding to an integration path that is along a line making a positive angle β ($0 < \beta < \pi/4$) with the real z-axis. In this case, the constituent infinitesimal vectors vary in both length and direction. Note that the curve passes through its centre point with the positive gradient $\tan\beta$ and that the directions of the spirals around the winding points are reversed as compared with case (b).

It is important to recognise that, although the three paths illustrated (and the infinity of other similar paths not illustrated) each produce a different phase–amplitude diagram, the vectors joining the initial and final points in the diagrams are *all the same*. For this particular integrand they are all (i) parallel to the positive real axis, showing that the integral is real and giving its sign, and (ii) of length $\sqrt{\pi}$, giving its magnitude.

What is apparent from figure 25.13(b), is that, because of the rapidly varying phase at either end of the spiral, the contributions from the infinitesimal vectors in those regions largely cancel each other. It is only in the central part of the spiral where the individual contributions are all nearly in phase that a substantial net contribution arises. If, on this part of the contour, where the phase is virtually

915

stationary, the magnitude of any factor, $g(z)$, multiplying the exponential function, $\exp[f(z)] \sim \exp[Ae^{i\alpha}(z-z_0)^2]$, is at least comparable to its magnitude elsewhere, then this result can be used to obtain an approximation to the value of the integral of $h(z) = g(z)\exp[f(z)]$. This is the basis of the method of stationary phase.

Returning to the behaviour of a function $\exp[f(z)]$ at one of its saddle points, we can now see how the considerations of the previous paragraphs can be applied there. We already know, from equation (25.62) and the discussion immediately following it, that in the equation

$$h(z) \approx g(z_0)\exp(f_0)\,\exp\{\tfrac{1}{2}A\rho^2[\cos(2\theta+\alpha)+i\sin(2\theta+\alpha)]\}$$

$$(25.71)$$

the second exponent is purely imaginary on a level line, and equal to zero at the saddle point itself. What is more, since $\nabla\psi = 0$ at the saddle, the phase is stationary there; on one level line it is a maximum and on the other it is a minimum. As there are two level lines through a saddle point, a path on which the amplitude of the integrand is constant could go straight on at the saddle point or it could turn through a right angle. For the moment we assume that it runs continuously through the saddle.

On the level line for which the phase at the saddle point is a minimum, we can write the phase of $h(z)$ as approximately

$$\arg g(z_0) + \operatorname{Im} f_0 + v^2,$$

where v is real, $iv^2 = \tfrac{1}{2}Ae^{i\alpha}(z-z_0)^2$ and, as previously, $Ae^{i\alpha} = f''(z_0)$. Then

$$e^{i\pi/4}\,dv = \pm\sqrt{\frac{A}{2}}\,e^{i\alpha/2}\,dz, \qquad (25.72)$$

leading to an approximation to the integral of

$$\int h(z)\,dz \approx \pm g(z_0)\exp(f_0)\int_{-\infty}^{\infty}\exp(iv^2)\sqrt{\frac{A}{2}}\,\exp[i(\tfrac{1}{4}\pi - \tfrac{1}{2}\alpha)]\,dv$$

$$= \pm g(z_0)\exp(f_0)\sqrt{\pi}\,\exp(i\pi/4)\sqrt{\frac{A}{2}}\,\exp[i(\tfrac{1}{4}\pi - \tfrac{1}{2}\alpha)]$$

$$= \pm\sqrt{\frac{2\pi}{A}}\,g(z_0)\exp(f_0)\exp[\tfrac{1}{2}i(\pi - \alpha)]. \qquad (25.73)$$

Result (25.68) was used to obtain the second line above. The $\pm$ ambiguity is again resolved by the direction θ of the contour; it is positive if $-3\pi/4 < \theta \le \pi/4$; otherwise, it is negative.

What we have ignored in obtaining result (25.73) is that we have integrated along a level line and that therefore the integrand has the same magnitude far from the saddle as it has at the saddle itself. This could be dismissed by referring to the fact that contributions to the integral from the ends of the Cornu spiral are self-cancelling, as discussed previously. However, the ends of the contour *must* be in regions where the integrand is vanishingly small, and so at each end of the level line we need to add a further section of path that makes the contour terminate correctly.

Fortunately, this can be done without adding to the value of the integral. This is because, as noted in the second paragraph following equation (25.67), far from the saddle the level line will be at a finite height up a 'precipitous cliff' that separates the region where the integrand grows without limit from the one where it tends to zero. To move down the cliff-face into the zero-level valley requires an ever smaller step the further we move away from the saddle; as the integrand is finite, the contribution to the integral is vanishingly small. In figure 25.12, this additional piece of path length might, for example, correspond to the infinitesimal move from a point on the large positive x-axis (where $h(z)$ has value 1) to a point just above it (where $h(z) \approx 0$).

Now that formula (25.73) has been justified, we may note that it is exactly the same as that for the method of steepest descents, equation (25.66). A similar calculation using the level line on which the phase is a maximum also reproduces the steepest-descents formula. It would appear that 'all roads lead to Rome'. However, as we explain later, some roads are more difficult than others. Where a problem involves using more than one saddle point, if the steepest-descents approach is tractable, it will usually be the more straight forward to apply.

Typical amplitude-phase diagrams for an integration along a level line that goes straight through the saddle are shown in parts (a) and (b) of figure 25.14. The value of the integral is given, in both magnitude and phase, by the vector **v** joining the initial to the final winding points and, of course, is the same in both cases. Part (a) corresponds to the case of the phase being a minimum at the saddle; the vector path crosses **v** at an angle of $-\pi/4$. When a path on which the phase at the saddle is a maximum is used, the Cornu spiral is as in part (b) of the figure; then the vector path crosses **v** at an angle of $+\pi/4$. As can be seen, the two spirals are mirror images of each other.

Clearly a straight-through level line path will start and end in different zero-level valleys. For one that turns through a right angle at the saddle point, the end-point could be in a different valley (for a function such as $\exp(-z^2)$, there is only one other) or in the same one. In the latter case the integral will give a zero value, unless a singularity of $h(z)$ happens to have been enclosed by the contour. Parts (c) and (d) of figure 25.14 illustrate the phase–amplitude diagrams for these two cases. In (c) the path turns through a right angle ($+\pi/2$, as it happens) at the saddle point, but finishes up in a different valley from that in which it started. In (d) it also turns through a right angle but returns to the same valley, albeit close to the other precipice from that near its starting point. This makes no difference and the result is zero, the two half spirals in the diagram producing resultants that cancel.

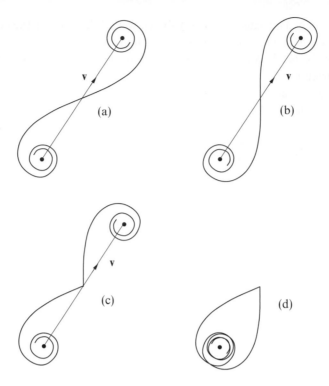

Figure 25.14   Amplitude–phase diagrams for stationary phase integration. (a) Using a straight-through path on which the phase is a minimum. (b) Using a straight-through path on which the phase is a maximum. (c) Using a level line that turns through $+\pi/2$ at the saddle point but starts and finishes in different valleys. (d) Using a level line that turns through a right angle but finishes in the same valley as it started. In cases (a), (b) and (c) the integral value is represented by **v** (see text). In case (d) the integral has value zero.

We do not have the space to consider cases with two or more saddle points, but even more care is needed with the stationary phase approach than when using the steepest-descents method. At a saddle point there is only one l.s.d. but there are two level lines. If more than one saddle point is required to reach the appropriate end-point of an integration, or an intermediate zero-level valley has to be used, then care is needed in linking the corresponding level lines in such a way that the links do not make a significant, but unknown, contribution to the integral. Yet more complications can arise if a level line through one saddle point crosses a line of steepest ascent through a second saddle.

We conclude this section with a worked example that has direct links to the two preceding sections of this chapter.

▶*In the worked example in subsection 25.8.2 the function*

$$F(x) = \frac{1}{\pi} \int_0^\infty \cos(\tfrac{1}{3}s^3 + xs)\, ds \qquad (*)$$

was shown to have the properties associated with the Airy function, Ai(x), when $x > 0$. Use the stationary phase method to show that, for $x < 0$ and $-x$ sufficiently large,

$$F(x) \sim \frac{1}{\sqrt{\pi}(-x)^{1/4}} \sin\left[\frac{2}{3}(-x)^{3/2} + \frac{\pi}{4}\right],$$

in accordance with equation (25.53) for Ai(z).

Since the cosine function is an even function and its argument in $(*)$ is purely real, we may consider $F(x)$ as the real part of

$$G(x) = \frac{1}{2\pi} \int_{-\infty}^\infty \exp[\, i(\tfrac{1}{3}s^3 + xs)\,].$$

This is of the standard form for a saddle-point approach with $g(s) = 1/2\pi$ and $f(s) = i(\tfrac{1}{3}s^3 + xs)$. The latter has $f'(s) = 0$ when $s^2 = -x$. Since $x < 0$ there are two saddle points at $s = +\sqrt{-x}$ and $s = -\sqrt{-x}$. These are both on the real axis separated by a distance $2\sqrt{-x}$.

If $-x$ is sufficiently large, the Gaussian-like stationary phase integrals can be treated separately and their contributions simply added. In terms of a phase–amplitude diagram, the Cornu spiral from the first saddle will have effectively reached its final winding point before the spiral from the second saddle begins. The second spiral therefore takes the final point of the first as its starting point; the vector representing its net contribution need not be in the same direction as that arising from the first spiral, and in general it will not be.

Near the saddle at $s = +\sqrt{-x}$ the form of $f(s)$ is, in the usual notation,

$$f(s) = f_0 + \tfrac{1}{2}Ae^{i\alpha}(s - s_0)^2$$
$$= -\frac{2i}{3}(-x)^{3/2} + \frac{1}{2}\,2\sqrt{-x}\,e^{i\pi/2}\,(\rho\,e^{i\theta})^2$$
$$= -\frac{2i}{3}(-x)^{3/2} + \sqrt{-x}\,e^{i\pi/2}\,\rho^2(\cos 2\theta + i\sin 2\theta).$$

For the exponent to be purely imaginary requires $\sin 2\theta = 0$, implying that the level lines are given by $\theta = 0, \pi/2, \pi$ or $3\pi/2$. The same conclusions hold at the saddle at $s = -\sqrt{-x}$, which differs only in that the sign of f_0 is reversed and $\alpha = 3\pi/2$ rather than $\pi/2$; $\exp(i\alpha)$ is imaginary in both cases. Thus the obvious path is one that approaches both saddles from the direction $\theta = \pi$ and leaves them in the direction $\theta = 0$. As $-3\pi/4 < 0 < \pi/4$, the $\pm$ choice is resolved as positive at both saddles.

Next we calculate the approximate values of the integrals from equation (25.73). At $s = +\sqrt{-x}$ it is

$$+\sqrt{\frac{2\pi}{2\sqrt{-x}}}\,\frac{1}{2\pi}\,\exp\left[-\frac{2i}{3}(-x)^{3/2}\right]\exp\left[\frac{i}{2}\left(\pi - \frac{\pi}{2}\right)\right]$$

$$= +\frac{1}{2\sqrt{\pi}(-x)^{1/4}}\,\exp\left[-i\left(\frac{2}{3}(-x)^{3/2} - \frac{\pi}{4}\right)\right].$$

The corresponding contribution from the saddle at $s = -\sqrt{-x}$ is

$$+\sqrt{\frac{2\pi}{2\sqrt{-x}}}\,\frac{1}{2\pi}\,\exp\left[\frac{2i}{3}(-x)^{3/2}\right]\exp\left[\frac{i}{2}\left(\pi - \frac{3\pi}{2}\right)\right],$$

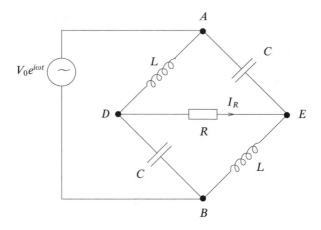

Figure 25.15   The inductor–capacitor–resistor network for exercise 25.1.

which can also be simplified, and gives

$$+\frac{1}{2\sqrt{\pi}(-x)^{1/4}} \exp\left[i\left(\frac{2}{3}(-x)^{3/2} - \frac{\pi}{4}\right)\right].$$

Adding the two contributions and taking the real part of the sum, though this is not necessary here because the sum is real anyway, we obtain

$$F(x) = \frac{2}{2\sqrt{\pi}(-x)^{1/4}} \cos\left(\frac{2}{3}(-x)^{3/2} - \frac{\pi}{4}\right)$$

$$= \frac{1}{\sqrt{\pi}(-x)^{1/4}} \sin\left(\frac{2}{3}(-x)^{3/2} + \frac{\pi}{4}\right),$$

in agreement with the asymptotic form given in (25.53). ◀

25.9 Exercises

25.1   In the method of complex impedances for a.c. circuits, an inductance L is represented by a complex impedance $Z_L = i\omega L$ and a capacitance C by $Z_C = 1/(i\omega C)$. Kirchhoff's circuit laws,

$$\sum_i I_i = 0 \text{ at a node and } \sum_i Z_i I_i = \sum_j V_j \text{ around any closed loop,}$$

are then applied as if the circuit were a d.c. one.

   Apply this method to the a.c. bridge connected as in figure 25.15 to show that if the resistance R is chosen as $R = (L/C)^{1/2}$ then the amplitude of the current, I_R, through it is independent of the angular frequency ω of the applied a.c. voltage $V_0 e^{i\omega t}$.

   Determine how the phase of I_R, relative to that of the voltage source, varies with the angular frequency ω.

25.2   A long straight fence made of conducting wire mesh separates two fields and stands one metre high. Sometimes, on fine days, there is a vertical electric field over flat open countryside. Well away from the fence the strength of the field is E_0. By considering the effect of the transformation $w = (1-z^2)^{1/2}$ on the real and imaginary z-axes, find the strengths of the field (a) at a point one metre directly above the fence, (b) at ground level one metre to the side of the fence, and (c) at a point that is level with the top of the fence but one metre to the side of it. What is the direction of the field in case (c)?

25.3 For the function

$$f(z) = \ln\left(\frac{z+c}{z-c}\right),$$

where c is real, show that the real part u of f is constant on a circle of radius $c\,\mathrm{cosech}\,u$ centred on the point $z = c\coth u$. Use this result to show that the electrical capacitance per unit length of two parallel cylinders of radii a, placed with their axes $2d$ apart, is proportional to $[\cosh^{-1}(d/a)]^{-1}$.

25.4 Find a complex potential in the z-plane appropriate to a physical situation in which the half-plane $x > 0$, $y = 0$ has zero potential and the half-plane $x < 0$, $y = 0$ has potential V.

By making the transformation $w = a(z + z^{-1})/2$, with a real and positive, find the electrostatic potential associated with the half-plane $r > a$, $s = 0$ and the half-plane $r < -a$, $s = 0$ at potentials 0 and V, respectively.

25.5 By considering in turn the transformations

$$z = \tfrac{1}{2}c(w + w^{-1}) \quad \text{and} \quad w = \exp\zeta,$$

where $z = x + iy$, $w = r\exp i\theta$, $\zeta = \xi + i\eta$ and c is a real positive constant, show that $z = c\cosh\zeta$ maps the strip $\xi \ge 0$, $0 \le \eta \le 2\pi$, onto the whole z-plane. Which curves in the z-plane correspond to the lines $\xi = $ constant and $\eta = $ constant? Identify those corresponding to $\xi = 0$, $\eta = 0$ and $\eta = 2\pi$.

The electric potential ϕ of a charged conducting strip $-c \le x \le c$, $y = 0$, satisfies

$$\phi \sim -k\ln(x^2 + y^2)^{1/2} \text{ for large values of } (x^2 + y^2)^{1/2},$$

with ϕ constant on the strip. Show that $\phi = \mathrm{Re}[-k\cosh^{-1}(z/c)]$ and that the magnitude of the electric field near the strip is $k(c^2 - x^2)^{-1/2}$.

25.6 For the equation $8z^3 + z + 1 = 0$:

(a) show that all three roots lie between the circles $|z| = 3/8$ and $|z| = 5/8$;
(b) find the approximate location of the real root, and hence deduce that the complex ones lie in the first and fourth quadrants and have moduli greater than 0.5.

25.7 Use contour integration to answer the following questions about the complex zeros of a polynomial equation.

(a) Prove that $z^8 + 3z^3 + 7z + 5$ has two zeros in the first quadrant.
(b) Find in which quadrants the zeros of $2z^3 + 7z^2 + 10z + 6$ lie. Try to locate them.

25.8 The following is a method of determining the number of zeros of an nth-degree polynomial $f(z)$ inside the contour C given by $|z| = R$:

(a) put $z = R(1 + it)/(1 - it)$, with $t = \tan(\theta/2)$, in the range $-\infty \le t \le \infty$;
(b) obtain $f(z)$ as

$$\frac{A(t) + iB(t)}{(1 - it)^n}\frac{(1 + it)^n}{(1 + it)^n};$$

(c) it follows that $\arg f(z) = \tan^{-1}(B/A) + n\tan^{-1}t$;
(d) and that $\Delta_C[\arg f(z)] = \Delta_C[\tan^{-1}(B/A)] + n\pi$;
(e) determine $\Delta_C[\tan^{-1}(B/A)]$ by evaluating $\tan^{-1}(B/A)$ at $t = \pm\infty$ and finding the discontinuities in B/A by inspection or using a sketch graph.

Then, by the principle of the argument, the number of zeros inside C is given by the integer $(2\pi)^{-1}\Delta_C[\arg f(z)]$.

It can be shown that the zeros of $z^4 + z + 1$ lie one in each quadrant. Use the above method to show that the zeros in the second and third quadrants have $|z| < 1$.

25.9   Prove that

$$\sum_{-\infty}^{\infty} \frac{1}{n^2 + \frac{3}{4}n + \frac{1}{8}} = 4\pi.$$

Carry out the summation numerically, say between -4 and 4, and note how much of the sum comes from values near the poles of the contour integration.

25.10   This exercise illustrates a method of summing some infinite series.

(a) Determine the residues at all the poles of the function

$$f(z) = \frac{\pi \cot \pi z}{a^2 + z^2},$$

where a is a positive real constant.

(b) By evaluating, in two different ways, the integral I of $f(z)$ along the straight line joining $-\infty - ia/2$ and $+\infty - ia/2$, show that

$$\sum_{n=1}^{\infty} \frac{1}{a^2 + n^2} = \frac{\pi \coth \pi a}{2a} - \frac{1}{2a^2}.$$

(c) Deduce the value of $\sum_1^{\infty} n^{-2}$.

25.11   By considering the integral of

$$\left(\frac{\sin \alpha z}{\alpha z}\right)^2 \frac{\pi}{\sin \pi z}, \qquad \alpha < \frac{\pi}{2},$$

around a circle of large radius, prove that

$$\sum_{m=1}^{\infty} (-1)^{m-1} \frac{\sin^2 m\alpha}{(m\alpha)^2} = \frac{1}{2}.$$

25.12   Use the Bromwich inversion, and contours similar to that shown in figure 25.7(a), to find the functions of which the following are the Laplace transforms:

(a) $s(s^2 + b^2)^{-1}$;
(b) $n!(s - a)^{-(n+1)}$, with n a positive integer and $s > a$;
(c) $a(s^2 - a^2)^{-1}$, with $s > |a|$.

Compare your answers with those given in a table of standard Laplace transforms.

25.13   Find the function $f(t)$ whose Laplace transform is

$$\bar{f}(s) = \frac{e^{-s} - 1 + s}{s^2}.$$

25.14   A function $f(t)$ has the Laplace transform

$$F(s) = \frac{1}{2i} \ln \left(\frac{s + i}{s - i}\right),$$

the complex logarithm being defined by a finite branch cut running along the imaginary axis from $-i$ to i.

(a) Convince yourself that, for $t > 0$, $f(t)$ can be expressed as a closed contour integral that encloses only the branch cut.

(b) Calculate $F(s)$ on either side of the branch cut, evaluate the integral and hence determine $f(t)$.

(c) Confirm that the derivative with respect to s of the Laplace transform integral of your answer is the same as that given by dF/ds.

25.15 Use the contour in figure 25.7(c) to show that the function with Laplace transform $s^{-1/2}$ is $(\pi x)^{-1/2}$.

[For an integrand of the form $r^{-1/2}\exp(-rx)$ change variable to $t = r^{1/2}$.]

25.16 Transverse vibrations of angular frequency ω on a string stretched with constant tension T are described by $u(x,t) = y(x)e^{-i\omega t}$, where

$$\frac{d^2y}{dx^2} + \frac{\omega^2 m(x)}{T}\,y(x) = 0.$$

Here, $m(x) = m_0 f(x)$ is the mass per unit length of the string and, in the general case, is a function of x. Find the first-order W.K.B. solution for $y(x)$.

Due to imperfections in its manufacturing process, a particular string has a small periodic variation in its linear density of the form $m(x) = m_0[1 + \epsilon\sin(2\pi x/L)]$, where $\epsilon \ll 1$. A progressive wave (i.e. one in which no energy is lost) travels in the positive x-direction along the string. Show that its amplitude fluctuates by $\pm\frac{1}{4}\epsilon$ of its value A_0 at $x = 0$ and that, to first order in ϵ, the phase of the wave is

$$\frac{\epsilon\,\omega\,L}{2\pi}\sqrt{\frac{m_0}{T}}\,\sin^2\frac{\pi x}{L}$$

ahead of what it would be if the string were uniform, with $m(x) = m_0$.

25.17 The equation

$$\frac{d^2y}{dz^2} + \left(v + \frac{1}{2} - \frac{1}{4}z^2\right)y = 0,$$

sometimes called the Weber–Hermite equation, has solutions known as parabolic cylinder functions. Find, to within (possibly complex) multiplicative constants, the two W.K.B. solutions of this equation that are valid for large $|z|$. In each case, determine the leading term and show that the multiplicative correction factor is of the form $1 + O(v^2/z^2)$.

Identify the Stokes and anti-Stokes lines for the equation. On which of the Stokes lines is the W.K.B. solution that tends to zero for z large, real and negative, the dominant solution?

25.18 A W.K.B. solution of Bessel's equation of order zero,

$$\frac{d^2y}{dz^2} + \frac{1}{z}\frac{dy}{dz} + y = 0, \qquad (*)$$

valid for large $|z|$ and $-\pi/2 < \arg z < 3\pi/2$, is $y(z) = Az^{-1/2}e^{iz}$. Obtain an improvement on this by finding a multiplier of $y(z)$ in the form of an asymptotic expansion in inverse powers of z as follows.

(a) Substitute for $y(z)$ in $(*)$ and show that the equation is satisfied to $O(z^{-5/2})$.

(b) Now replace the constant A by $A(z)$ and find the equation that must be satisfied by $A(z)$. Look for a solution of the form $A(z) = z^\sigma \sum_{n=0}^\infty a_n z^{-n}$, where $a_0 = 1$. Show that $\sigma = 0$ is the only acceptable solution to the indicial equation and obtain a recurrence relation for the a_n.

(c) To within a (complex) constant, the expression $y(z) = A(z)z^{-1/2}e^{iz}$ is the asymptotic expansion of the Hankel function $H_0^{(1)}(z)$. Show that it is a divergent expansion for all values of z and estimate, in terms of z, the value of N such that $\sum_{n=0}^{N} a_n z^{-n-1/2}e^{iz}$ gives the best estimate of $H_0^{(1)}(z)$.

25.19 The function $h(z)$ of the complex variable z is defined by the integral

$$h(z) = \int_{-i\infty}^{i\infty} \exp(t^2 - 2zt)\,dt.$$

(a) Make a change of integration variable, $t = iu$, and evaluate $h(z)$ using a standard integral. Is your answer valid for all finite z?

(b) Evaluate the integral using the method of steepest descents, considering in particular the cases (i) z is real and positive, (ii) z is real and negative and (iii) z is purely imaginary and equal to $i\beta$, where β is real. In each case sketch the corresponding contour in the complex t-plane.

(c) Evaluate the integral for the same three cases as specified in part (b) using the method of stationary phases. To determine an appropriate contour that passes through a saddle point $t = t_0$, write $t = t_0 + (u + iv)$ and apply the criterion for determining a level line. Sketch the relevant contour in each case, indicating what freedom there is to distort it.

Comment on the accuracy of the results obtained using the approximate methods adopted in (b) and (c).

25.20 Use the method of steepest descents to show that an approximate value for the integral

$$F(z) = \int_{-\infty}^{\infty} \exp[iz(\tfrac{1}{5}t^5 + t)]\,dt,$$

where z is real and positive, is

$$\left(\frac{2\pi}{z}\right)^{1/2} \exp(-\beta z)\cos(\beta z - \tfrac{1}{8}\pi),$$

where $\beta = 4/(5\sqrt{2})$.

25.21 The stationary phase approximation to an integral of the form

$$F(v) = \int_a^b g(t)e^{ivf(t)}\,dt, \qquad |v| \gg 1,$$

where $f(t)$ is a real function of t and $g(t)$ is a slowly varying function (when compared with the argument of the exponential), can be written as

$$F(v) \sim \left(\frac{2\pi}{|v|}\right)^{1/2} \sum_{n=1}^{N} \frac{g(t_n)}{\sqrt{A_n}} \exp\left\{i\left[vf(t_n) + \frac{\pi}{4}\,\mathrm{sgn}\left(vf''(t_n)\right)\right]\right\},$$

where the t_n are the N stationary points of $f(t)$ that lie in $a < t_1 < t_2 < \cdots < t_N < b$, $A_n = |f''(t_n)|$, and $\mathrm{sgn}(x)$ is the sign of x.

Use this result to find an approximation, valid for large positive values of v, to the integral

$$F(v, z) = \int_{-\infty}^{\infty} \frac{1}{1+t^2} \cos[(2t^3 - 3zt^2 - 12z^2t)v]\,dt,$$

where z is a real positive parameter.

25.22 The Bessel function $J_v(z)$ is given for $|\arg z| < \tfrac{1}{2}\pi$ by the integral around a contour C of the function

$$g(z) = \frac{1}{2\pi i}\,t^{-(v+1)}\exp\left[\frac{z}{2}\left(t - \frac{1}{t}\right)\right].$$

The contour starts and ends along the negative real t-axis and encircles the origin in the positive sense. It can be considered to be made up of two contours. One of them, C_2, starts at $t = -\infty$, runs through the third quadrant to the point

$t = -i$ and then approaches the origin in the fourth quadrant in a curve that is ultimately antiparallel to the positive real axis. The other contour, C_1, is the mirror image of this in the real axis; it is confined to the upper half-plane, passes through $t = i$ and is antiparallel to the real t-axis at both of its extremities. The contribution to $J_\nu(z)$ from the curve C_k is $\frac{1}{2}H_\nu^{(k)}$, the function $H_\nu^{(k)}$ being known as a Hankel function.

Using the method of steepest descents, establish the leading term in an asymptotic expansion for $H_\nu^{(1)}$ for z real, large and positive. Deduce, without detailed calculation, the corresponding result for $H_\nu^{(2)}$. Hence establish the asymptotic form of $J_\nu(z)$ for the same range of z.

25.23   Use the method of steepest descents to find an asymptotic approximation, valid for z large, real and positive, to the function defined by

$$F_\nu(z) = \int_C \exp(-iz\sin t + i\nu t)\,dt,$$

where ν is real and non-negative and C is a contour that starts at $t = -\pi + i\infty$ and ends at $t = -i\infty$.

25.10 Hints and answers

25.1   Apply Kirchhoff's laws to three independent loops, say $ADBA$, $ADEA$ and $DBED$. Eliminate other currents from the equations to obtain $I_R = \omega_0 C V_0[(\omega_0^2 - \omega^2 - 2i\omega\omega_0)/(\omega_0^2 + \omega^2)]$, where $\omega_0^2 = (LC)^{-1}$; $|I_R| = \omega_0 C V_0$; the phase of I_R is $\tan^{-1}[(-2\omega\omega_0)/(\omega_0^2 - \omega^2)]$.

25.3   Set $c\coth u_1 = -d$, $c\coth u_2 = +d$, $|c\operatorname{cosech} u| = a$ and note that the capacitance is proportional to $(u_2 - u_1)^{-1}$.

25.5 $\xi = $ constant, ellipses $x^2(a+1)^{-2} + y^2(a-1)^{-2} = c^2/(4a^2)$; $\eta = $ constant, hyperbolae $x^2(\cos\alpha)^{-2} - y^2(\sin\alpha)^{-2} = c^2$. The curves are the cuts $-c \le x \le c$, $y = 0$ and $|x| \ge c$, $y = 0$. The curves for $\eta = 2\pi$ are the same as those for $\eta = 0$.

25.7   (a) For a quarter-circular contour enclosing the first quadrant, the change in the argument of the function is $0 + 8(\pi/2) + 0$ (since $y^8 + 5 = 0$ has no real roots); (b) one negative real zero; a conjugate pair in the second and third quadrants, $-\frac{3}{2}, -1 \pm i$.

25.9   Evaluate

$$\int \frac{\pi\cot\pi z}{\left(\frac{1}{2} + z\right)\left(\frac{1}{4} + z\right)}\,dz$$

around a large circle centred on the origin; residue at $z = -1/2$ is 0; residue at $z = -1/4$ is $4\pi\cot(-\pi/4)$.

25.11   The behaviour of the integrand for large $|z|$ is $|z|^{-2}\exp[(2\alpha - \pi)|z|]$. The residue at $z = \pm m$, for each integer m, is $\sin^2(m\alpha)(-1)^m/(m\alpha)^2$. The contour contributes nothing.
Required summation $= [\text{total sum} - (m = 0 \text{ term})]/2$.

25.13   Note that $\bar{f}(s)$ has no pole at $s = 0$. For $t < 0$ close the Bromwich contour in the right half-plane, and for $t > 1$ in the left half-plane. For $0 < t < 1$ the integrand has to be split into separate terms containing e^{-s} and $s - 1$ and the completions made in the right and left half-planes, respectively. The last of these completed contours now contains a second-order pole at $s = 0$. $f(t) = 1 - t$ for $0 < t < 1$, but is 0 otherwise.

25.15 $\int_\Gamma$ and $\int_\gamma$ tend to 0 as $R \to \infty$ and $\rho \to 0$. Put $s = r\exp i\pi$ and $s = r\exp(-i\pi)$ on the two sides of the cut and use $\int_0^\infty \exp(-t^2 x)\,dt = \frac{1}{2}(\pi/x)^{1/2}$. There are no poles inside the contour.

925

25.17 Use the binomial theorem to expand, in inverse powers of z, both the square root in the exponent and the fourth root in the multiplier, working to $O(z^{-2})$. The leading terms are $y_1(z) = Ce^{-z^2/4}z^\nu$ and $y_2(z) = De^{z^2/4}z^{-(\nu+1)}$. Stokes lines: $\arg z = 0, \pi/2, \pi, 3\pi/2$; anti-Stokes lines: $\arg z = (2n+1)\pi/4$ for $n = 0, 1, 2, 3$. y_1 is dominant on $\arg z = \pi/2$ or $3\pi/2$.

25.19 (a) $i\sqrt{\pi}e^{-z^2}$, valid for all z, including $i\sqrt{\pi}\exp(\beta^2)$ in case (iii).
(b) The same values as in (a). The (only) saddle point, at $t_0 = z$, is traversed in the direction $\theta = +\frac{1}{2}\pi$ in all cases, though the path in the complex t-plane varies with each case.
(c) The same values as in (a). The level lines are $v = \pm u$. In cases (i) and (ii) the contour turns through a right angle at the saddle point.
All three methods give *exact* answers in this case of a quadratic exponent.

25.21 Saddle points at $t_1 = -z$ and $t_2 = 2z$ with $f_1'' = -18z$ and $f_2'' = 18z$. Approximation is

$$\left(\frac{\pi}{9zv}\right)^{1/2}\left[\frac{\cos(7vz^3 - \frac{1}{4}\pi)}{1+z^2} + \frac{\cos(20vz^3 - \frac{1}{4}\pi)}{1+4z^2}\right].$$

25.23 Saddle point at $t_0 = \cos^{-1}(v/z)$ is traversed in the direction $\theta = -\frac{1}{4}\pi$. $F_v(z) \approx (2\pi/z)^{1/2}\exp[i(z - \frac{1}{2}v\pi - \frac{1}{4}\pi)]$.

26

Tensors

It may seem obvious that the quantitative description of physical processes cannot depend on the coordinate system in which they are represented. However, we may turn this argument around: since physical results must indeed be independent of the choice of coordinate system, what does this imply about the nature of the quantities involved in the description of physical processes? The study of these implications and of the classification of physical quantities by means of them forms the content of the present chapter.

Although the concepts presented here may be applied, with little modification, to more abstract spaces (most notably the four-dimensional space–time of special or general relativity), we shall restrict our attention to our familiar three-dimensional Euclidean space. This removes the need to discuss the properties of differentiable manifolds and their tangent and dual spaces. The reader who is interested in these more technical aspects of tensor calculus in general spaces, and in particular their application to general relativity, should consult one of the many excellent textbooks on the subject.[§]

Before the presentation of the main development of the subject, we begin by introducing the summation convention, which will prove very useful in writing tensor equations in a more compact form. We then review the effects of a change of basis in a vector space; such spaces were discussed in chapter 8. This is followed by an investigation of the rotation of Cartesian coordinate systems, and finally we broaden our discussion to include more general coordinate systems and transformations.

[§] For example, R. D'Inverno, *Introducing Einstein's Relativity* (Oxford: Oxford University Press, 1992); J. Foster and J. D. Nightingale, *A Short Course in General Relativity* (New York: Springer, 2006); B. F. Schutz, *A First Course in General Relativity* (Cambridge; Cambridge University Press 1985).

26.1 Some notation

Before proceeding further, we introduce the *summation convention* for subscripts, since its use looms large in the work of this chapter. The convention is that any *lower-case* alphabetic subscript that appears *exactly* twice in any term of an expression is understood to be summed over all the values that a subscript in that position can take (unless the contrary is specifically stated). The subscripted quantities may appear in the numerator and/or the denominator of a term in an expression. This naturally implies that any such pair of repeated subscripts must occur only in subscript positions that have the same range of values. Sometimes the ranges of values have to be specified but usually they are apparent from the context.

The following simple examples illustrate what is meant (in the three-dimensional case):

(i) $a_i x_i$ stands for $a_1 x_1 + a_2 x_2 + a_3 x_3$;

(ii) $a_{ij} b_{jk}$ stands for $a_{i1} b_{1k} + a_{i2} b_{2k} + a_{i3} b_{3k}$;

(iii) $a_{ij} b_{jk} c_k$ stands for $\sum_{j=1}^{3} \sum_{k=1}^{3} a_{ij} b_{jk} c_k$;

(iv) $\dfrac{\partial v_i}{\partial x_i}$ stands for $\dfrac{\partial v_1}{\partial x_1} + \dfrac{\partial v_2}{\partial x_2} + \dfrac{\partial v_3}{\partial x_3}$;

(v) $\dfrac{\partial^2 \phi}{\partial x_i \partial x_i}$ stands for $\dfrac{\partial^2 \phi}{\partial x_1^2} + \dfrac{\partial^2 \phi}{\partial x_2^2} + \dfrac{\partial^2 \phi}{\partial x_3^2}$.

Subscripts that are summed over are called *dummy subscripts* and the others *free subscripts*. It is worth remarking that when introducing a dummy subscript into an expression, care should be taken not to use one that is already present, either as a free or as a dummy subscript. For example, $a_{ij} b_{jk} c_{kl}$ cannot, and must not, be replaced by $a_{ij} b_{jj} c_{jl}$ or by $a_{il} b_{lk} c_{kl}$, but could be replaced by $a_{im} b_{mk} c_{kl}$ or by $a_{im} b_{mn} c_{nl}$. Naturally, free subscripts must not be changed at all unless the working calls for it.

Furthermore, as we have done throughout this book, we will make frequent use of the Kronecker delta δ_{ij}, which is defined by

$$\delta_{ij} = \begin{cases} 1 & \text{if } i = j, \\ 0 & \text{otherwise.} \end{cases}$$

When the summation convention has been adopted, the main use of δ_{ij} is to replace one subscript by another in certain expressions. Examples might include

$$b_j \delta_{ij} = b_i,$$

and

$$a_{ij} \delta_{jk} = a_{ij} \delta_{kj} = a_{ik}. \tag{26.1}$$

In the second of these the dummy index shared by both terms on the left-hand side (namely j) has been replaced by the free index carried by the Kronecker delta (namely k), and the delta symbol has disappeared. In matrix language, (26.1) can be written as $\mathsf{AI} = \mathsf{A}$, where A is the matrix with elements a_{ij} and I is the unit matrix having the same dimensions as A.

In some expressions we may use the Kronecker delta to replace indices in a number of different ways, e.g.

$$a_{ij}b_{jk}\delta_{ki} = a_{ij}b_{ji} \quad \text{or} \quad a_{kj}b_{jk},$$

where the two expressions on the RHS are totally equivalent to one another.

26.2 Change of basis

In chapter 8 some attention was given to the subject of changing the basis set (or coordinate system) in a vector space and it was shown that, under such a change, different types of quantity behave in different ways. These results are given in section 8.15, but are summarised below for convenience, using the summation convention. Although throughout this section we will remind the reader that we are using this convention, it will simply be assumed in the remainder of the chapter.

If we introduce a set of basis vectors $\mathbf{e}_1, \mathbf{e}_2, \mathbf{e}_3$ into our familiar three-dimensional (vector) space, then we can describe any vector $\mathbf{x}$ in terms of its components x_1, x_2, x_3 with respect to this basis:

$$\mathbf{x} = x_1\mathbf{e}_1 + x_2\mathbf{e}_2 + x_3\mathbf{e}_3 = x_i\mathbf{e}_i,$$

where we have used the summation convention to write the sum in a more compact form. If we now introduce a new basis $\mathbf{e}'_1, \mathbf{e}'_2, \mathbf{e}'_3$ related to the old one by

$$\mathbf{e}'_j = S_{ij}\mathbf{e}_i \quad \text{(sum over } i\text{)}, \tag{26.2}$$

where the coefficient S_{ij} is the ith component of the vector $\mathbf{e}'_j$ with respect to the unprimed basis, then we may write $\mathbf{x}$ with respect to the new basis as

$$\mathbf{x} = x'_1\mathbf{e}'_1 + x'_2\mathbf{e}'_2 + x'_3\mathbf{e}'_3 = x'_i\mathbf{e}'_i \quad \text{(sum over } i\text{)}.$$

If we denote the matrix with elements S_{ij} by S, then the components x'_i and x_i in the two bases are related by

$$x'_i = (\mathsf{S}^{-1})_{ij}x_j \quad \text{(sum over } j\text{)},$$

where, using the summation convention, there is an implicit sum over j from $j = 1$ to $j = 3$. In the special case where the transformation is a rotation of the coordinate axes, the transformation matrix S is orthogonal and we have

$$x'_i = (\mathsf{S}^{\mathsf{T}})_{ij}x_j = S_{ji}x_j \quad \text{(sum over } j\text{)}. \tag{26.3}$$

Scalars behave differently under transformations, however, since they remain unchanged. For example, the value of the scalar product of two vectors $\mathbf{x} \cdot \mathbf{y}$ (which is just a number) is unaffected by the transformation from the unprimed to the primed basis. Different again is the behaviour of linear operators. If a linear operator $\mathcal{A}$ is represented by some matrix A in a given coordinate system then in the new (primed) coordinate system it is represented by a new matrix, $A' = S^{-1}AS$.

In this chapter we develop a general formulation to describe and classify these different types of behaviour under a change of basis (or coordinate transformation). In the development, the generic name *tensor* is introduced, and certain scalars, vectors and linear operators are described respectively as tensors of zeroth, first and second order (the *order* – or *rank* – corresponds to the number of subscripts needed to specify a particular element of the tensor). Tensors of third and fourth order will also occupy some of our attention.

26.3 Cartesian tensors

We begin our discussion of tensors by considering a particular class of coordinate transformation – namely rotations – and we shall confine our attention strictly to the rotation of Cartesian coordinate systems. Our object is to study the properties of various types of mathematical quantities, and their associated physical interpretations, when they are described in terms of Cartesian coordinates and the axes of the coordinate system are rigidly rotated from a basis $\mathbf{e}_1, \mathbf{e}_2, \mathbf{e}_3$ (lying along the Ox_1, Ox_2 and Ox_3 axes) to a new one $\mathbf{e}'_1, \mathbf{e}'_2, \mathbf{e}'_3$ (lying along the Ox'_1, Ox'_2 and Ox'_3 axes).

Since we shall be more interested in how the components of a vector or linear operator are changed by a rotation of the axes than in the relationship between the two sets of basis vectors $\mathbf{e}_i$ and $\mathbf{e}'_i$, let us define the transformation matrix L as the inverse of the matrix S in (26.2). Thus, from (26.2), the components of a position vector $\mathbf{x}$, in the old and new bases respectively, are related by

$$x'_i = L_{ij}x_j. \tag{26.4}$$

Because we are considering only rigid rotations of the coordinate axes, the transformation matrix L will be orthogonal, i.e. such that $L^{-1} = L^T$. Therefore the inverse transformation is given by

$$x_i = L_{ji}x'_j. \tag{26.5}$$

The orthogonality of L also implies relations among the elements of L that express the fact that $LL^T = L^TL = I$. In subscript notation they are given by

$$L_{ik}L_{jk} = \delta_{ij} \quad \text{and} \quad L_{ki}L_{kj} = \delta_{ij}. \tag{26.6}$$

Furthermore, in terms of the basis vectors of the primed and unprimed Cartesian

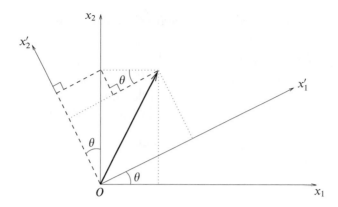

Figure 26.1   Rotation of Cartesian axes by an angle θ about the x_3-axis. The three angles marked θ and the parallels (broken lines) to the primed axes show how the first two equations of (26.7) are constructed.

coordinate systems, the transformation matrix is given by

$$L_{ij} = \mathbf{e}'_i \cdot \mathbf{e}_j.$$

We note that the product of two rotations is also a rotation. For example, suppose that $x'_i = L_{ij}x_j$ and $x''_i = M_{ij}x'_j$; then the composite rotation is described by

$$x''_i = M_{ij}x'_j = M_{ij}L_{jk}x_k = (\mathsf{ML})_{ik}x_k,$$

corresponding to the matrix ML.

▶*Find the transformation matrix* L *corresponding to a rotation of the coordinate axes through an angle θ about the $\mathbf{e}_3$-axis (or x_3-axis), as shown in figure 26.1.*

Taking $\mathbf{x}$ as a position vector — the most obvious choice – we see from the figure that the components of $\mathbf{x}$ with respect to the new (primed) basis are given in terms of the components in the old (unprimed) basis by

$$\begin{aligned}
x'_1 &= x_1 \cos\theta + x_2 \sin\theta, \\
x'_2 &= -x_1 \sin\theta + x_2 \cos\theta, \\
x'_3 &= x_3.
\end{aligned}$$

(26.7)

The (orthogonal) transformation matrix is thus

$$\mathsf{L} = \begin{pmatrix} \cos\theta & \sin\theta & 0 \\ -\sin\theta & \cos\theta & 0 \\ 0 & 0 & 1 \end{pmatrix}.$$

The inverse equations are

$$\begin{aligned}
x_1 &= x'_1 \cos\theta - x'_2 \sin\theta, \\
x_2 &= x'_1 \sin\theta + x'_2 \cos\theta, \\
x_3 &= x'_3,
\end{aligned}$$

(26.8)

in line with (26.5). ◀

931

26.4 First- and zero-order Cartesian tensors

Using the above example as a guide, we may consider any set of three quantities v_i, which are directly or indirectly functions of the coordinates x_i and possibly involve some constants, and ask how their values are changed by any rotation of the Cartesian axes. The specific question to be answered is whether the specific forms v_i' in the new variables can be obtained from the old ones v_i using (26.4),

$$v_i' = L_{ij}v_j. \tag{26.9}$$

If so, the v_i are said to form the components of a *vector* or *first-order Cartesian tensor* **v**. By definition, the position coordinates are themselves the components of such a tensor. The first-order tensor **v** does not change under rotation of the coordinate axes; nevertheless, since the basis set does change, from $\mathbf{e}_1, \mathbf{e}_2, \mathbf{e}_3$ to $\mathbf{e}_1', \mathbf{e}_2', \mathbf{e}_3'$, the components of **v** must also change. The changes must be such that

$$\mathbf{v} = v_i\mathbf{e}_i = v_i'\mathbf{e}_i' \tag{26.10}$$

is unchanged.

Since the transformation (26.9) is orthogonal, the components of any such first-order Cartesian tensor also obey a relation that is the inverse of (26.9),

$$v_i = L_{ji}v_j'. \tag{26.11}$$

We now consider explicit examples. In order to keep the equations to reasonable proportions, the examples will be restricted to the x_1x_2-plane, i.e. there are no components in the x_3-direction. Three-dimensional cases are no different in principle – but much longer to write out.

▶ *Which of the following pairs (v_1, v_2) form the components of a first-order Cartesian tensor in two dimensions?*:

$$\text{(i) } (x_2, -x_1), \qquad \text{(ii) } (x_2, x_1), \qquad \text{(iii) } (x_1^2, x_2^2).$$

We shall consider the rotation discussed in the previous example, and to save space we denote $\cos\theta$ by c and $\sin\theta$ by s.

(i) Here $v_1 = x_2$ and $v_2 = -x_1$, referred to the old axes. In terms of the new coordinates they will be $v_1' = x_2'$ and $v_2' = -x_1'$, i.e.

$$\begin{aligned} v_1' &= x_2' &= -sx_1 + cx_2 \\ v_2' &= -x_1' &= -cx_1 - sx_2. \end{aligned} \tag{26.12}$$

Now if we start again and evaluate v_1' and v_2' as given by (26.9) we find that

$$\begin{aligned} v_1' &= L_{11}v_1 + L_{12}v_2 = cx_2 + s(-x_1) \\ v_2' &= L_{21}v_1 + L_{22}v_2 = -s(x_2) + c(-x_1). \end{aligned} \tag{26.13}$$

The expressions for v_1' and v_2' in (26.12) and (26.13) are the same whatever the values of θ (i.e. for *all* rotations) and thus by definition (26.9) the pair $(x_2, -x_1)$ *is* a first-order Cartesian tensor.

(ii) Here $v_1 = x_2$ and $v_2 = x_1$. Following the same procedure,

$$v_1' = x_2' = -sx_1 + cx_2$$
$$v_2' = x_1' = cx_1 + sx_2.$$

But, by (26.9), for a Cartesian tensor we must have

$$v_1' = cv_1 + sv_2 = cx_2 + sx_1$$

$$v_2' = (-s)v_1 + cv_2 = -sx_2 + cx_1.$$

These two sets of expressions do not agree and thus the pair (x_2, x_1) is not a first-order Cartesian tensor.

(iii) $v_1 = x_1^2$ and $v_2 = x_2^2$. As in (ii) above, considering the first component alone is sufficient to show that this pair is also not a first-order tensor. Evaluating v_1' directly gives

$$v_1' = x_1'^2 = c^2 x_1^2 + 2csx_1x_2 + s^2 x_2^2,$$

whilst (26.9) requires that

$$v_1' = cv_1 + sv_2 = cx_1^2 + sx_2^2,$$

which is quite different. ◀

There are many physical examples of first-order tensors (i.e. vectors) that will be familiar to the reader. As a straightforward one, we may take the set of Cartesian components of the momentum of a particle of mass m, $(m\dot{x}_1, m\dot{x}_2, m\dot{x}_3)$. This set transforms in all essentials as (x_1, x_2, x_3), since the other operations involved, multiplication by a number and differentiation with respect to time, are quite unaffected by any orthogonal transformation of the axes. Similarly, acceleration and force are represented by the components of first-order tensors.

Other more complicated vectors involving the position coordinates more than once, such as the angular momentum of a particle of mass m, namely $\mathbf{J} = \mathbf{x} \times \mathbf{p} = m(\mathbf{x} \times \dot{\mathbf{x}})$, are also first-order tensors. That this is so is less obvious in component form than for the earlier examples, but may be verified by writing out the components of $\mathbf{J}$ explicitly or by appealing to the quotient law to be discussed in section 26.7 and using the Cartesian tensor ϵ_{ijk} from section 26.8.

Having considered the effects of rotations on vector-like sets of quantities we may consider quantities that are unchanged by a rotation of axes. In our previous nomenclature these have been called *scalars* but we may also describe them as *tensors of zero order*. They contain only one element (formally, the number of subscripts needed to identify a particular element is zero); the most obvious non-trivial example associated with a rotation of axes is the square of the distance of a point from the origin, $r^2 = x_1^2 + x_2^2 + x_3^2$. In the new coordinate system it will have the form $r'^2 = x_1'^2 + x_2'^2 + x_3'^2$, which for any rotation has the same value as $x_1^2 + x_2^2 + x_3^2$.

In fact any scalar product of two first-order tensors (vectors) is a zero-order tensor (scalar), as might be expected since it can be written in a coordinate-free way as $\mathbf{u} \cdot \mathbf{v}$.

▶*By considering the components of the vectors $\mathbf{u}$ and $\mathbf{v}$ with respect to two Cartesian coordinate systems (related by a rotation), show that the scalar product $\mathbf{u} \cdot \mathbf{v}$ is invariant under rotation.*

In the original (unprimed) system the scalar product is given in terms of components by $u_i v_i$ (summed over i), and in the rotated (primed) system by

$$u_i' v_i' = L_{ij} u_j L_{ik} v_k = L_{ij} L_{ik} u_j v_k = \delta_{jk} u_j v_k = u_j v_j,$$

where we have used the orthogonality relation (26.6). Since the resulting expression in the rotated system is the same as that in the original system, the scalar product is indeed invariant under rotations. ◀

The above result leads directly to the identification of many physically important quantities as zero-order tensors. Perhaps the most immediate of these is energy, either as potential energy or as an energy density (e.g. $\mathbf{F} \cdot d\mathbf{r}$, $e\mathbf{E} \cdot d\mathbf{r}$, $\mathbf{D} \cdot \mathbf{E}$, $\mathbf{B} \cdot \mathbf{H}$, $\boldsymbol{\mu} \cdot \mathbf{B}$), but others, such as the angle between two directed quantities, are important. In fact, in most analyses of physical situations it is a scalar quantity (such as energy) that is to be determined. Such quantities are *invariant* under a rotation of axes and so it is possible to work with the most convenient set of axes and still have confidence in the results.

Complementing the way in which a zero-order tensor was obtained from two first-order tensors, so a first-order tensor can be obtained from a zero-order tensor (i.e. a scalar). We show this by taking a specific example, that of the electric field $\mathbf{E} = -\nabla\phi$; this is derived from a scalar, the electrostatic potential ϕ, and has components

$$E_i = -\frac{\partial \phi}{\partial x_i}. \tag{26.14}$$

Clearly, $\mathbf{E}$ *is* a first-order tensor, but we may prove this more formally by considering the behaviour of its components (26.14) under a rotation of the coordinate axes, since the components of the electric field E_i' are then given by

$$E_i' = \left(-\frac{\partial \phi}{\partial x_i}\right)' = -\frac{\partial \phi'}{\partial x_i'} = -\frac{\partial x_j}{\partial x_i'}\frac{\partial \phi}{\partial x_j} = L_{ij} E_j, \tag{26.15}$$

where (26.5) has been used to evaluate $\partial x_j / \partial x_i'$. Now (26.15) is in the form (26.9), thus confirming that the components of the electric field do behave as the components of a first-order tensor.

▶ *If v_i are the components of a first-order tensor, show that $\nabla \cdot \mathbf{v} = \partial v_i/\partial x_i$ is a zero-order tensor.*

In the rotated coordinate system $\nabla \cdot \mathbf{v}$ is given by

$$\left(\frac{\partial v_i}{\partial x_i}\right)' = \frac{\partial v_i'}{\partial x_i'} = \frac{\partial x_j}{\partial x_i'}\frac{\partial}{\partial x_j}(L_{ik}v_k) = L_{ij}L_{ik}\frac{\partial v_k}{\partial x_j},$$

since the elements L_{ij} are not functions of position. Using the orthogonality relation (26.6) we then find

$$\frac{\partial v_i'}{\partial x_i'} = L_{ij}L_{ik}\frac{\partial v_k}{\partial x_j} = \delta_{jk}\frac{\partial v_k}{\partial x_j} = \frac{\partial v_j}{\partial x_j}.$$

Hence $\partial v_i/\partial x_i$ is invariant under rotation of the axes and is thus a zero-order tensor; this was to be expected since it can be written in a coordinate-free way as $\nabla \cdot \mathbf{v}$. ◀

26.5 Second- and higher-order Cartesian tensors

Following on from scalars with no subscripts and vectors with one subscript, we turn to sets of quantities that require two subscripts to identify a particular element of the set. Let these quantities by denoted by T_{ij}.

Taking (26.9) as a guide we define a *second-order Cartesian tensor* as follows: the T_{ij} form the components of such a tensor if, under the same conditions as for (26.9),

$$T_{ij}' = L_{ik}L_{jl}T_{kl} \tag{26.16}$$

and

$$T_{ij} = L_{ki}L_{lj}T_{kl}'. \tag{26.17}$$

At the same time we may define a Cartesian tensor of general order as follows. The set of expressions $T_{ij\cdots k}$ form the components of a Cartesian tensor if, for all rotations of the axes of coordinates given by (26.4) and (26.5), subject to (26.6), the expressions using the new coordinates, $T_{ij\cdots k}'$ are given by

$$T_{ij\cdots k}' = L_{ip}L_{jq}\cdots L_{kr}T_{pq\cdots r} \tag{26.18}$$

and

$$T_{ij\cdots k} = L_{pi}L_{qj}\cdots L_{rk}T_{pq\cdots r}'. \tag{26.19}$$

It is apparent that in three dimensions, an Nth-order Cartesian tensor has 3^N components.

Since a second-order tensor has two subscripts, it is natural to display its components in matrix form. The notation $[T_{ij}]$ is used, as well as T, to denote the matrix having T_{ij} as the element in the ith row and jth column.[§]

We may think of a second-order tensor **T** as a geometrical entity in a similar way to that in which we viewed linear operators (which transform one vector into

---

[§] We can also denote the column matrix containing the elements v_i of a vector by $[v_i]$.

another, without reference to any coordinate system) and consider the matrix containing its components as a representation of the tensor with respect to a particular coordinate system. Moreover, the matrix $T = [T_{ij}]$, containing the components of a second-order tensor, behaves in the same way under orthogonal transformations $T' = LTL^T$ as a linear operator.

However, not all linear operators are second-order tensors. More specifically, the two subscripts in a second-order tensor must refer to the same coordinate system. In particular, this means that any linear operator that transforms a vector into a vector in a different vector space cannot be a second-order tensor. Thus, although the elements L_{ij} of the transformation matrix are written with two subscripts, they cannot be the components of a tensor since the two subscripts each refer to a different coordinate system.

As examples of sets of quantities that are readily shown to be second-order tensors we consider the following.

(i) *The outer product of two vectors.* Let u_i and v_i, $i = 1, 2, 3$, be the components of two vectors $\mathbf{u}$ and $\mathbf{v}$, and consider the set of quantities T_{ij} defined by

$$T_{ij} = u_i v_j. \tag{26.20}$$

The set T_{ij} are called the components of the the *outer product* of $\mathbf{u}$ and $\mathbf{v}$. Under rotations the components T_{ij} become

$$T'_{ij} = u'_i v'_j = L_{ik} u_k L_{jl} v_l = L_{ik} L_{jl} u_k v_l = L_{ik} L_{jl} T_{kl}, \tag{26.21}$$

which shows that they do transform as the components of a second-order tensor. Use has been made in (26.21) of the fact that u_i and v_i are the components of first-order tensors.

The outer product of two vectors is often denoted, without reference to any coordinate system, as

$$\mathbf{T} = \mathbf{u} \otimes \mathbf{v}. \tag{26.22}$$

(This is not to be confused with the vector product of two vectors, which is itself a vector and is discussed in chapter 7.) The expression (26.22) gives the basis to which the components T_{ij} of the second-order tensor refer: since $\mathbf{u} = u_i \mathbf{e}_i$ and $\mathbf{v} = v_i \mathbf{e}_i$, we may write the tensor $\mathbf{T}$ as

$$\mathbf{T} = u_i \mathbf{e}_i \otimes v_j \mathbf{e}_j = u_i v_j \mathbf{e}_i \otimes \mathbf{e}_j = T_{ij} \mathbf{e}_i \otimes \mathbf{e}_j. \tag{26.23}$$

Moreover, as for the case of first-order tensors (see equation (26.10)) we note that the quantities T'_{ij} are the components of the *same* tensor $\mathbf{T}$, but referred to a different coordinate system, i.e.

$$\mathbf{T} = T_{ij} \mathbf{e}_i \otimes \mathbf{e}_j = T'_{ij} \mathbf{e}'_i \otimes \mathbf{e}'_j.$$

These concepts can be extended to higher-order tensors.

(ii) *The gradient of a vector.* Suppose v_i represents the components of a vector; let us consider the quantities generated by forming the derivatives of each v_i, $i = 1, 2, 3$, with respect to each x_j, $j = 1, 2, 3$, i.e.

$$T_{ij} = \frac{\partial v_i}{\partial x_j}.$$

These nine quantities form the components of a second-order tensor, as can be seen from the fact that

$$T'_{ij} = \frac{\partial v'_i}{\partial x'_j} = \frac{\partial (L_{ik} v_k)}{\partial x_l} \frac{\partial x_l}{\partial x'_j} = L_{ik} \frac{\partial v_k}{\partial x_l} L_{jl} = L_{ik} L_{jl} T_{kl}.$$

In coordinate-free language the tensor $\mathbf{T}$ may be written as $\mathbf{T} = \nabla \mathbf{v}$ and hence gives meaning to the concept of the gradient of a vector, a quantity that was not discussed in the chapter on vector calculus (chapter 10).

A test of whether any given set of quantities forms the components of a second-order tensor can always be made by direct substitution of the x'_i in terms of the x_i, followed by comparison with the right-hand side of (26.16). This procedure is extremely laborious, however, and it is almost always better to try to recognise the set as being expressible in one of the forms just considered, or to make alternative tests based on the quotient law of section 26.7 below.

▶*Show that the T_{ij} given by*

$$\mathbf{T} = [T_{ij}] = \begin{pmatrix} x_2^2 & -x_1 x_2 \\ -x_1 x_2 & x_1^2 \end{pmatrix} \qquad (26.24)$$

are the components of a second-order tensor.

Again we consider a rotation θ about the $\mathbf{e}_3$-axis. Carrying out the direct evaluation first we obtain, using (26.7),

$$T'_{11} = x'_2{}^2 - s^2 x_1^2 - 2sc x_1 x_2 + c^2 x_2^2,$$
$$T'_{12} = -x'_1 x'_2 = sc x_1^2 + (s^2 - c^2) x_1 x_2 - sc x_2^2,$$
$$T'_{21} = -x'_1 x'_2 = sc x_1^2 + (s^2 - c^2) x_1 x_2 - sc x_2^2,$$
$$T'_{22} = x'_1{}^2 = c^2 x_1^2 + 2sc x_1 x_2 + s^2 x_2^2.$$

Now, evaluating the right-hand side of (26.16),

$$T'_{11} = cc x_2^2 + cs(-x_1 x_2) + sc(-x_1 x_2) + ss x_1^2,$$
$$T'_{12} = c(-s) x_2^2 + cc(-x_1 x_2) + s(-s)(-x_1 x_2) + sc x_1^2,$$
$$T'_{21} = (-s) c x_2^2 + (-s) s(-x_1 x_2) + cc(-x_1 x_2) + cs x_1^2,$$
$$T'_{22} = (-s)(-s) x_2^2 + (-s) c(-x_1 x_2) + c(-s)(-x_1 x_2) + cc x_1^2.$$

After reorganisation, the corresponding expressions are seen to be the same, showing, as required, that the T_{ij} are the components of a second-order tensor.

The same result could be inferred much more easily, however, by noting that the T_{ij} are in fact the components of the outer product of the vector $(x_2, -x_1)$ with itself. That $(x_2, -x_1)$ is indeed a vector was established by (26.12) and (26.13). ◀

Physical examples involving second-order tensors will be discussed in the later sections of this chapter, but we might note here that, for example, magnetic susceptibility and electrical conductivity are described by second-order tensors.

26.6 The algebra of tensors

Because of the similarity of first- and second-order tensors to column vectors and matrices, it would be expected that similar types of algebraic operation can be carried out with them and so provide ways of constructing new tensors from old ones. In the remainder of this chapter, instead of referring to the T_{ij} (say) as the *components* of a second-order tensor $\mathbf{T}$, we may sometimes simply refer to T_{ij} as the tensor. It should always be remembered, however, that the T_{ij} are in fact just the components of $\mathbf{T}$ in a given coordinate system and that T'_{ij} refers to the components of the *same* tensor $\mathbf{T}$ in a different coordinate system.

The addition and subtraction of tensors follows an obvious definition; namely that if $V_{ij\cdots k}$ and $W_{ij\cdots k}$ are (the components of) tensors of the same order, then their sum and difference, $S_{ij\cdots k}$ and $D_{ij\cdots k}$ respectively, are given by

$$S_{ij\cdots k} = V_{ij\cdots k} + W_{ij\cdots k},$$
$$D_{ij\cdots k} = V_{ij\cdots k} - W_{ij\cdots k},$$

for each set of values $i, j, \ldots, k$. That $S_{ij\cdots k}$ and $D_{ij\cdots k}$ are the components of tensors follows immediately from the linearity of a rotation of coordinates.

It is equally straightforward to show that if the $T_{ij\cdots k}$ are the components of a tensor, then so is the set of quantities formed by interchanging the order of (a pair of) indices, e.g. $T_{ji\cdots k}$.

If $T_{ji\cdots k}$ is found to be identical with $T_{ij\cdots k}$ then $T_{ij\cdots k}$ is said to be *symmetric* with respect to its first two subscripts (or simply 'symmetric', for second-order tensors). If, however, $T_{ji\cdots k} = -T_{ij\cdots k}$ for every element then it is an *antisymmetric* tensor. An arbitrary tensor is neither symmetric nor antisymmetric but can always be written as the sum of a symmetric tensor $S_{ij\cdots k}$ and an antisymmetric tensor $A_{ij\cdots k}$:

$$T_{ij\cdots k} = \tfrac{1}{2}(T_{ij\cdots k} + T_{ji\cdots k}) + \tfrac{1}{2}(T_{ij\cdots k} - T_{ji\cdots k})$$
$$= S_{ij\cdots k} + A_{ij\cdots k}.$$

Of course these properties are valid for any pair of subscripts.

In (26.20) in the previous section we had an example of a kind of 'multiplication' of two tensors, thereby producing a tensor of higher order – in that case two first-order tensors were multiplied to give a second-order tensor. Inspection of (26.21) shows that there is nothing particular about the orders of the tensors involved and it follows as a general result that the outer product of an Nth-order tensor with an Mth-order tensor will produce an $(M + N)$th-order tensor.

An operation that produces the opposite effect – namely, generates a tensor of smaller rather than larger order – is known as *contraction* and consists of making two of the subscripts equal and summing over all values of the equalised subscripts.

▶*Show that the process of contraction of an Nth-order tensor produces another tensor, of order N − 2.*

Let $T_{ij\ldots l\ldots m\ldots k}$ be the components of an Nth-order tensor, then

$$T'_{ij\ldots l\ldots m\ldots k} = \underbrace{L_{ip}L_{jq}\cdots L_{lr}\cdots L_{ms}\cdots L_{kn}}_{N \text{ factors}}\, T_{pq\ldots r\ldots s\ldots n}.$$

Thus if, for example, we make the two subscripts l and m equal and sum over all values of these subscripts, we obtain

$$\begin{aligned}
T'_{ij\ldots l\ldots l\ldots k} &= L_{ip}L_{jq}\cdots L_{lr}\cdots L_{ls}\cdots L_{kn}T_{pq\ldots r\ldots s\ldots n}\\
&= L_{ip}L_{jq}\cdots \delta_{rs}\cdots L_{kn}T_{pq\ldots r\ldots s\ldots n}\\
&= \underbrace{L_{ip}L_{jq}\cdots L_{kn}}_{(N-2) \text{ factors}}\, T_{pq\ldots r\ldots r\ldots n},
\end{aligned}$$

showing that $T_{ij\ldots l\ldots l\ldots k}$ are the components of a (different) Cartesian tensor of order $N - 2$. ◀

For a second-rank tensor, the process of contraction is the same as taking the trace of the corresponding matrix. The trace T_{ii} itself is thus a zero-order tensor (or scalar) and hence invariant under rotations, as was noted in chapter 8.

The process of taking the scalar product of two vectors can be recast into tensor language as forming the outer product $T_{ij} = u_i v_j$ of two first-order tensors **u** and **v** and then contracting the second-order tensor **T** so formed, to give $T_{ii} = u_i v_i$, a scalar (invariant under a rotation of axes).

As yet another example of a familiar operation that is a particular case of a contraction, we may note that the multiplication of a column vector $[u_i]$ by a matrix $[B_{ij}]$ to produce another column vector $[v_i]$,

$$B_{ij}u_j = v_i,$$

can be looked upon as the contraction T_{ijj} of the third-order tensor T_{ijk} formed from the outer product of B_{ij} and u_k.

26.7 The quotient law

The previous paragraph appears to give a heavy-handed way of describing a familiar operation, but it leads us to ask whether it has a converse. To put the question in more general terms: if we know that **B** and **C** are tensors and also that

$$A_{pq\ldots k\ldots m}B_{ij\ldots k\ldots n} = C_{pq\ldots mij\ldots n}, \tag{26.25}$$

does this imply that the $A_{pq\ldots k\ldots m}$ also form the components of a tensor $\mathbf{A}$? Here $\mathbf{A}$, $\mathbf{B}$ and $\mathbf{C}$ are respectively of Mth, Nth and $(M+N-2)$th order and it should be noted that the subscript k that has been contracted may be any of the subscripts in $\mathbf{A}$ and $\mathbf{B}$ independently.

The *quotient law* for tensors states that if (26.25) holds in all rotated coordinate frames then the $A_{pq\ldots k\ldots m}$ do indeed form the components of a tensor $\mathbf{A}$. To prove it for general M and N is no more difficult regarding the ideas involved than to show it for specific M and N, but this does involve the introduction of a large number of subscript symbols. We will therefore take the case $M = N = 2$, but it will be readily apparent that the principle of the proof holds for general M and N.

We thus start with (say)

$$A_{pk}B_{ik} = C_{pi}, \tag{26.26}$$

where B_{ik} and C_{pi} are arbitrary second-order tensors. Under a rotation of coordinates the set A_{pk} (tensor or not) transforms into a new set of quantities that we will denote by A'_{pk}. We thus obtain in succession the following steps, using (26.16), (26.17) and (26.6):

$$
\begin{aligned}
A'_{pk}B'_{ik} &= C'_{pi} && \text{(transforming (26.26))},\\
&= L_{pq}L_{ij}C_{qj} && \text{(since } \mathbf{C} \text{ is a tensor)},\\
&= L_{pq}L_{ij}A_{ql}B_{jl} && \text{(from (26.26))},\\
&= L_{pq}L_{ij}A_{ql}L_{mj}L_{nl}B'_{mn} && \text{(since } \mathbf{B} \text{ is a tensor)},\\
&= L_{pq}L_{nl}A_{ql}B'_{in} && \text{(since } L_{ij}L_{mj} = \delta_{im}).
\end{aligned}
$$

Now k on the left and n on the right are dummy subscripts and thus we may write

$$(A'_{pk} - L_{pq}L_{kl}A_{ql})B'_{ik} = 0. \tag{26.27}$$

Since B_{ik}, and hence B'_{ik}, is an arbitrary tensor, we must have

$$A'_{pk} = L_{pq}L_{kl}A_{ql},$$

showing that the A'_{pk} are given by the general formula (26.18) and hence that the A_{pk} are the components of a second-order tensor. By following an analogous argument, the same result (26.27) and deduction could be obtained if (26.26) were replaced by

$$A_{pk}B_{ki} = C_{pi},$$

i.e. the contraction being now with respect to a different pair of indices.

Use of the quotient law to test whether a given set of quantities is a tensor is generally much more convenient than making a direct substitution. A particular way in which it is applied is by contracting the given set of quantities, having

N subscripts, with an arbitrary Nth-order tensor (i.e. one having independently variable components) and determining whether the result is a scalar.

> ► *Use the quotient law to show that the elements of* T, *equation (26.24), are the components of a second-order tensor.*

The outer product $x_i x_j$ is a second-order tensor. Contracting this with the T_{ij} given in (26.24) we obtain

$$T_{ij} x_i x_j = x_2^2 x_1^2 - x_1 x_2 x_1 x_2 - x_1 x_2 x_2 x_1 + x_1^2 x_2^2 = 0,$$

which is clearly invariant (a zeroth-order tensor). Hence by the quotient theorem T_{ij} must also be a tensor. ◄

26.8 The tensors δ_{ij} and ϵ_{ijk}

In many places throughout this book we have encountered and used the two-subscript quantity δ_{ij} defined by

$$\delta_{ij} = \begin{cases} 1 & \text{if } i = j, \\ 0 & \text{otherwise.} \end{cases}$$

Let us now also introduce the three-subscript *Levi–Civita symbol* ϵ_{ijk}, the value of which is given by

$$\epsilon_{ijk} = \begin{cases} +1 & \text{if } i, j, k \text{ is an even permutation of } 1, 2, 3, \\ -1 & \text{if } i, j, k \text{ is an odd permutation of } 1, 2, 3, \\ 0 & \text{otherwise.} \end{cases}$$

We will now show that δ_{ij} and ϵ_{ijk} are respectively the components of a second- and a third-order Cartesian tensor. Notice that the coordinates x_i do not appear explicitly in the components of these tensors, their components consisting entirely of 0 and ± 1.

In passing, we also note that ϵ_{ijk} is totally antisymmetric, i.e. it changes sign under the interchange of any pair of subscripts. In fact ϵ_{ijk}, or any scalar multiple of it, is the *only* three-subscript quantity with this property.

Treating δ_{ij} first, the proof that it is a second-order tensor is straightforward since if, from (26.16), we consider the equation

$$\delta'_{kl} = L_{ki} L_{lj} \delta_{ij} = L_{ki} L_{li} = \delta_{kl},$$

we see that the transformation of δ_{ij} generates the same expression (a pattern of 0's and 1's) as does the definition of δ'_{ij} in the transformed coordinates. Thus δ_{ij} transforms according to the appropriate tensor transformation law and is therefore a second-order tensor.

Turning now to ϵ_{ijk}, we have to consider the quantity

$$\epsilon'_{lmn} = L_{li} L_{mj} L_{nk} \epsilon_{ijk}. \tag{26.28}$$

Let us begin, however, by noting that we may use the Levi–Civita symbol to write an expression for the determinant of a 3×3 matrix A,

$$|A|\epsilon_{lmn} = A_{li}A_{mj}A_{nk}\epsilon_{ijk}, \qquad (26.29)$$

which may be shown to be equivalent to the Laplace expansion (see chapter 8).[§] Indeed many of the properties of determinants discussed in chapter 8 can be proved very efficiently using this expression (see exercise 26.9).

▶*Evaluate the determinant of the matrix*

$$A = \begin{pmatrix} 2 & 1 & -3 \\ 3 & 4 & 0 \\ 1 & -2 & 1 \end{pmatrix}.$$

Setting $l = 1$, $m = 2$ and $n = 3$ in (26.29) we find

$$\begin{aligned} |A| &= \epsilon_{ijk}A_{1i}A_{2j}A_{3k} \\ &= (2)(4)(1) - (2)(0)(-2) - (1)(3)(1) + (-3)(3)(-2) \\ &\quad + (1)(0)(1) - (-3)(4)(1) = 35, \end{aligned}$$

which may be verified using the Laplace expansion method. ◀

We can now show that the ϵ_{ijk} are in fact the components of a third-order tensor. Using (26.29) with the general matrix A replaced by the specific transformation matrix L, we can rewrite the RHS of (26.28) in terms of $|L|$

$$\epsilon'_{lmn} = L_{li}L_{mj}L_{nk}\epsilon_{ijk} = |L|\epsilon_{lmn}.$$

Since L is orthogonal its determinant has the value unity, and so $\epsilon'_{lmn} = \epsilon_{lmn}$. Thus we see that ϵ'_{lmn} has exactly the properties of ϵ_{ijk} but with i, j, k replaced by l, m, n, i.e. it is the same as the expression ϵ_{ijk} written using the new coordinates. This shows that ϵ_{ijk} is a third-order Cartesian tensor.

In addition to providing a convenient notation for the determinant of a matrix, δ_{ij} and ϵ_{ijk} can be used to write many of the familiar expressions of vector algebra and calculus as contracted tensors. For example, provided we are using right-handed Cartesian coordinates, the vector product $\mathbf{a} = \mathbf{b} \times \mathbf{c}$ has as its ith component $a_i = \epsilon_{ijk}b_jc_k$; this should be contrasted with the outer product $\mathbf{T} = \mathbf{b} \otimes \mathbf{c}$, which is a second-order tensor having the components $T_{ij} = b_ic_j$.

[§] This may be readily extended to an $N \times N$ matrix A, i.e.

$$|A|\epsilon_{i_1 i_2 \cdots i_N} = A_{i_1 j_1}A_{i_2 j_2} \cdots A_{i_N j_N}\epsilon_{j_1 j_2 \cdots j_N},$$

where $\epsilon_{i_1 i_2 \cdots i_N}$ equals 1 if $i_1 i_2 \cdots i_N$ is an even permutation of $1, 2, \ldots, N$ and equals -1 if it is an odd permutation; otherwise it equals zero.

> ►*Write the following as contracted Cartesian tensors:* $\mathbf{a} \cdot \mathbf{b}$, $\nabla^2 \phi$, $\nabla \times \mathbf{v}$, $\nabla(\nabla \cdot \mathbf{v})$, $\nabla \times (\nabla \times \mathbf{v})$, $(\mathbf{a} \times \mathbf{b}) \cdot \mathbf{c}$.

The corresponding (contracted) tensor expressions are readily seen to be as follows:

$$\mathbf{a} \cdot \mathbf{b} = a_i b_i = \delta_{ij} a_i b_j,$$

$$\nabla^2 \phi = \frac{\partial^2 \phi}{\partial x_i \partial x_i} = \delta_{ij} \frac{\partial^2 \phi}{\partial x_i \partial x_j},$$

$$(\nabla \times \mathbf{v})_i = \epsilon_{ijk} \frac{\partial v_k}{\partial x_j},$$

$$[\nabla(\nabla \cdot \mathbf{v})]_i = \frac{\partial}{\partial x_i}\left(\frac{\partial v_j}{\partial x_j}\right) = \delta_{jk} \frac{\partial^2 v_j}{\partial x_i \partial x_k},$$

$$[\nabla \times (\nabla \times \mathbf{v})]_i = \epsilon_{ijk} \frac{\partial}{\partial x_j}\left(\epsilon_{klm} \frac{\partial v_m}{\partial x_l}\right) = \epsilon_{ijk}\epsilon_{klm} \frac{\partial^2 v_m}{\partial x_j \partial x_l},$$

$$(\mathbf{a} \times \mathbf{b}) \cdot \mathbf{c} = \delta_{ij} c_i \epsilon_{jkl} a_k b_l = \epsilon_{ikl} c_i a_k b_l. ◄$$

An important relationship between the ϵ- and δ- tensors is expressed by the identity

$$\epsilon_{ijk}\epsilon_{klm} = \delta_{il}\delta_{jm} - \delta_{im}\delta_{jl}. \tag{26.30}$$

To establish the validity of this identity between two fourth-order tensors (the LHS is a once-contracted sixth-order tensor) we consider the various possible cases.

The RHS of (26.30) has the values

$$+1 \text{ if } i = l \text{ and } j = m \neq i, \tag{26.31}$$

$$-1 \text{ if } i = m \text{ and } j = l \neq i, \tag{26.32}$$

$$0 \text{ for any other set of subscript values } i, j, l, m. \tag{26.33}$$

In each product on the LHS k has the same value in both factors and for a non-zero contribution none of i, l, j, m can have the same value as k. Since there are only three values, 1, 2 and 3, that any of the subscripts may take, the only non-zero possibilities are $i = l$ and $j = m$ or vice versa but not all four subscripts equal (since then each ϵ factor is zero, as it would be if $i = j$ or $l = m$). This reproduces (26.33) for the LHS of (26.30) and also the conditions (26.31) and (26.32). The values in (26.31) and (26.32) are also reproduced in the LHS of (26.30) since

(i) if $i = l$ and $j = m$, $\epsilon_{ijk} = \epsilon_{lmk} = \epsilon_{klm}$ and, whether ϵ_{ijk} is $+1$ or -1, the product of the two factors is $+1$; and

(ii) if $i = m$ and $j = l$, $\epsilon_{ijk} = \epsilon_{mlk} = -\epsilon_{klm}$ and thus the product $\epsilon_{ijk}\epsilon_{klm}$ (no summation) has the value -1.

This concludes the establishment of identity (26.30).

A useful application of (26.30) is in obtaining alternative expressions for vector quantities that arise from the vector product of a vector product.

▶*Obtain an alternative expression for* $\nabla \times (\nabla \times \mathbf{v})$.

As shown in the previous example, $\nabla \times (\nabla \times \mathbf{v})$ can be expressed in tensor form as

$$[\nabla \times (\nabla \times \mathbf{v})]_i = \epsilon_{ijk}\epsilon_{klm}\frac{\partial^2 v_m}{\partial x_j \partial x_l}$$

$$= (\delta_{il}\delta_{jm} - \delta_{im}\delta_{jl})\frac{\partial^2 v_m}{\partial x_j \partial x_l}$$

$$= \frac{\partial}{\partial x_i}\left(\frac{\partial v_j}{\partial x_j}\right) - \frac{\partial^2 v_i}{\partial x_j \partial x_j}$$

$$= [\nabla(\nabla \cdot \mathbf{v})]_i - \nabla^2 v_i,$$

where in the second line we have used the identity (26.30). This result has already been mentioned in chapter 10 and the reader is referred there for a discussion of its applicability. ◀

By examining the various possibilities, it is straightforward to verify that, more generally,

$$\epsilon_{ijk}\epsilon_{pqr} = \begin{vmatrix} \delta_{ip} & \delta_{iq} & \delta_{ir} \\ \delta_{jp} & \delta_{jq} & \delta_{jr} \\ \delta_{kp} & \delta_{kq} & \delta_{kr} \end{vmatrix} \tag{26.34}$$

and it is easily seen that (26.30) is a special case of this result. From (26.34) we can derive alternative forms of (26.30), for example,

$$\epsilon_{ijk}\epsilon_{ilm} = \delta_{jl}\delta_{km} - \delta_{jm}\delta_{kl}. \tag{26.35}$$

The pattern of subscripts in these identities is most easily remembered by noting that the subscripts on the first δ on the RHS are those that immediately follow (cyclically, if necessary) the common subscript, here i, in each ϵ-term on the LHS; the remaining combinations of j, k, l, m as subscripts in the other δ-terms on the RHS can then be filled in automatically.

Contracting (26.35) by setting $j = l$ (say) we obtain, since $\delta_{kk} = 3$ when using the summation convention,

$$\epsilon_{ijk}\epsilon_{ijm} = 3\delta_{km} - \delta_{km} = 2\delta_{km},$$

and by contracting once more, setting $k = m$, we further find that

$$\epsilon_{ijk}\epsilon_{ijk} = 6. \tag{26.36}$$

26.9 Isotropic tensors

It will have been noticed that, unlike most of the tensors discussed (except for scalars), δ_{ij} and ϵ_{ijk} have the property that all their components have values that are the same whatever rotation of axes is made, i.e. the component values are independent of the transformation L_{ij}. Specifically, δ_{11} has the value 1 in all coordinate frames, whereas for a general second-order tensor $\mathbf{T}$ all we know is that if $T_{11} = f_{11}(x_1, x_2, x_3)$ then $T'_{11} = f_{11}(x'_1, x'_2, x'_3)$. Tensors with the former property are called *isotropic* (or *invariant*) tensors.

It is important to know the most general form that an isotropic tensor can take, since the description of the physical properties, e.g. the conductivity, magnetic susceptibility or tensile strength, of an isotropic medium (i.e. a medium having the same properties whichever way it is orientated) involves an isotropic tensor. In the previous section it was shown that δ_{ij} and ϵ_{ijk} are second- and third-order isotropic tensors; we will now show that, to within a scalar multiple, they are the only such isotropic tensors.

Let us begin with isotropic second-order tensors. Suppose T_{ij} is an isotropic tensor; then, by definition, for *any* rotation of the axes we must have that

$$T_{ij} = T'_{ij} = L_{ik}L_{jl}T_{kl} \tag{26.37}$$

for each of the nine components.

First consider a rotation of the axes by $2\pi/3$ about the $(1, 1, 1)$ direction; this takes Ox_1, Ox_2, Ox_3 into Ox'_2, Ox'_3, Ox'_1 respectively. For this rotation $L_{13} = 1$, $L_{21} = 1$, $L_{32} = 1$ and all other $L_{ij} = 0$. This requires that $T_{11} = T'_{11} = T_{33}$. Similarly $T_{12} = T'_{12} = T_{31}$. Continuing in this way, we find:

(a) $T_{11} = T_{22} = T_{33}$;

(b) $T_{12} = T_{23} = T_{31}$;

(c) $T_{21} = T_{32} = T_{13}$.

Next, consider a rotation of the axes (from their original position) by $\pi/2$ about the Ox_3-axis. In this case $L_{12} = -1$, $L_{21} = 1$, $L_{33} = 1$ and all other $L_{ij} = 0$. Amongst other relationships, we must have from (26.37) that:

$$T_{13} = (-1) \times 1 \times T_{23};$$
$$T_{23} = 1 \times 1 \times T_{13}.$$

Hence $T_{13} = T_{23} = 0$ and therefore, by parts (b) and (c) above, each element $T_{ij} = 0$ except for T_{11}, T_{22} and T_{33}, which are all the same. This shows that $T_{ij} = \lambda\delta_{ij}$.

▶*Show that $\lambda\epsilon_{ijk}$ is the only isotropic third-order Cartesian tensor.*

The general line of attack is as above and so only a minimum of explanation will be given.

$$T_{ijk} = T'_{ijk} = L_{il}L_{jm}L_{kn}T_{lmn} \quad \text{(in all, there are 27 elements).}$$

Rotate about the $(1, 1, 1)$ direction: this is equivalent to making subscript permutations $1 \to 2 \to 3 \to 1$. We find

(a) $T_{111} = T_{222} = T_{333}$,

(b) $T_{112} = T_{223} = T_{331}$ (and two similar sets),

(c) $T_{123} = T_{231} = T_{312}$ (and a set involving odd permutations of 1, 2, 3).

Rotate by $\pi/2$ about the Ox_3-axis: $L_{12} = -1$, $L_{21} = 1$, $L_{33} = 1$, the other $L_{ij} = 0$.

(d) $T_{111} = (-1) \times (-1) \times (-1) \times T_{222} = -T_{222}$,
(e) $T_{112} = (-1) \times (-1) \times 1 \times T_{221}$,
(f) $T_{221} = 1 \times 1 \times (-1) \times T_{112}$,
(g) $T_{123} = (-1) \times 1 \times 1 \times T_{213}$.

Relations (a) and (d) show that elements with all subscripts the same are zero. Relations (e), (f) and (b) show that all elements with repeated subscripts are zero. Relations (g) and (c) show that $T_{123} = T_{231} = T_{312} = -T_{213} = -T_{321} = -T_{132}$.

In total, T_{ijk} differs from ϵ_{ijk} by at most a scalar factor, but since ϵ_{ijk} (and hence $\lambda\epsilon_{ijk}$) has already been shown to be an isotropic tensor, T_{ijk} must be the most general third-order isotropic Cartesian tensor. ◄

Using exactly the same procedures as those employed for δ_{ij} and ϵ_{ijk}, it may be shown that the only isotropic first-order tensor is the trivial one with all elements zero.

26.10 Improper rotations and pseudotensors

So far we have considered rigid rotations of the coordinate axes described by an orthogonal matrix L with $|L| = +1$, (26.4). Strictly speaking such transformations are called *proper rotations*. We now broaden our discussion to include transformations that are still described by an orthogonal matrix L but for which $|L| = -1$; these are called *improper rotations*.

This kind of transformation can always be considered as an *inversion* of the coordinate axes through the origin represented by the equation

$$x_i' = -x_i, \qquad (26.38)$$

combined with a proper rotation. The transformation may be looked upon alternatively as one that changes an initially right-handed coordinate system into a left-handed one; any prior or subsequent proper rotation will not change this state of affairs. The most obvious example of a transformation with $|L| = -1$ is the matrix corresponding to (26.38) itself; in this case $L_{ij} = -\delta_{ij}$.

As we have emphasised in earlier chapters, any real physical vector **v** may be considered as a geometrical object (i.e. an arrow in space), which can be referred to independently of any coordinate system and whose direction and magnitude cannot be altered merely by describing it in terms of a different coordinate system. Thus the components of **v** transform as $v_i' = L_{ij}v_j$ under *all* rotations (proper and improper).

We can define another type of object, however, whose components may also be labelled by a single subscript but which transforms as $v_i' = L_{ij}v_j$ under proper rotations and as $v_i' = -L_{ij}v_j$ (note the minus sign) under improper rotations. In this case, the v_i are not strictly the components of a true first-order Cartesian tensor but instead are said to form the components of a first-order Cartesian *pseudotensor* or *pseudovector*.

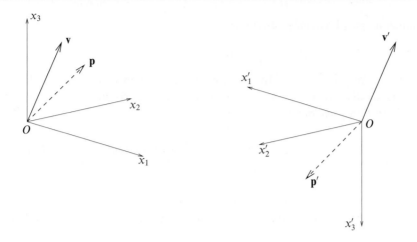

Figure 26.2 The behaviour of a vector **v** and a pseudovector **p** under a reflection through the origin of the coordinate system x_1, x_2, x_3 giving the new system x'_1, x'_2, x'_3.

It is important to realise that a pseudovector (as its name suggests) is not a geometrical object in the usual sense. In particular, it should *not* be considered as a real physical arrow in space, since its direction is reversed by an *improper* transformation of the coordinate axes (such as an inversion through the origin). This is illustrated in figure 26.2, in which the pseudovector **p** is shown as a broken line to indicate that it is not a real physical vector.

Corresponding to vectors and pseudovectors, zeroth-order objects may be divided into scalars and pseudoscalars — the latter being invariant under rotation but changing sign on reflection.

We may also extend the notion of scalars and pseudoscalars, vectors and pseudovectors, to objects with two or more subscripts. For two subscripts, as defined previously, any quantity with components that transform as $T'_{ij} = L_{ik}L_{jl}T_{kl}$ under *all* rotations (proper and improper) is called a second-order Cartesian tensor. If, however, $T'_{ij} = L_{ik}L_{jl}T_{kl}$ under proper rotations but $T'_{ij} = -L_{ik}L_{jl}T_{kl}$ under improper ones (which include reflections), then the T_{ij} are the components of a second-order Cartesian pseudotensor. In general the components of Cartesian pseudotensors of arbitrary order transform as

$$T'_{ij\cdots k} = |\mathsf{L}|L_{il}L_{jm}\cdots L_{kn}T_{lm\cdots n}, \tag{26.39}$$

where $|\mathsf{L}|$ is the determinant of the transformation matrix.

For example, from (26.29) we have that

$$|\mathsf{L}|\epsilon_{ijk} = L_{il}L_{jm}L_{kn}\epsilon_{lmn},$$

but since $|\mathsf{L}| = \pm 1$ we may rewrite this as

$$\epsilon_{ijk} = |\mathsf{L}|L_{il}L_{jm}L_{kn}\epsilon_{lmn}.$$

From this expression, we see that although ϵ_{ijk} behaves as a tensor under proper rotations, as discussed in section 26.8, it should properly be regarded as a third-order Cartesian *pseudo*tensor.

▶*If b_j and c_k are the components of vectors, show that the quantities $a_i = \epsilon_{ijk}b_jc_k$ form the components of a pseudovector.*

In a new coordinate system we have

$$
\begin{aligned}
a'_i &= \epsilon'_{ijk}b'_jc'_k \\
&= |\mathsf{L}|L_{il}L_{jm}L_{kn}\epsilon_{lmn}L_{jp}b_pL_{kq}c_q \\
&= |\mathsf{L}|L_{il}\epsilon_{lmn}\delta_{mp}\delta_{nq}b_pc_q \\
&= |\mathsf{L}|L_{il}\epsilon_{lmn}b_mc_n \\
&= |\mathsf{L}|L_{il}a_l,
\end{aligned}
$$

from which we see immediately that the quantities a_i form the components of a pseudovector. ◀

The above example is worth some further comment. If we denote the vectors with components b_j and c_k by $\mathbf{b}$ and $\mathbf{c}$ respectively then, as mentioned in section 26.8, the quantities $a_i = \epsilon_{ijk}b_jc_k$ are the components of the real vector $\mathbf{a} = \mathbf{b} \times \mathbf{c}$, *provided that we are using a right-handed Cartesian coordinate system.* However, in a coordinate system that is left-handed the quantities $a'_i = \epsilon'_{ijk}b'_jc'_k$ are *not* the components of the physical vector $\mathbf{a} = \mathbf{b} \times \mathbf{c}$, which has, instead, the components $-a'_i$. It is therefore important to note the handedness of a coordinate system before attempting to write in component form the vector relation $\mathbf{a} = \mathbf{b} \times \mathbf{c}$ (which is true without reference to any coordinate system).

It is worth noting that, although pseudotensors can be useful mathematical objects, the description of the real physical world must usually be in terms of tensors (i.e. scalars, vectors, etc.).[§] For example, the temperature or density of a gas must be a scalar quantity (rather than a pseudoscalar), since its value does not change when the coordinate system used to describe it is inverted through the origin. Similarly, velocity, magnetic field strength or angular momentum can only be described by a vector, and not by a pseudovector.

At this point, it may be useful to make a brief comment on the distinction between *active* and *passive* transformations of a physical system, as this difference often causes confusion. In this chapter, we are concerned solely with passive trans-

---

[§] In fact the quantum-mechanical description of elementary particles, such as electrons, protons and neutrons, requires the introduction of a new kind of mathematical object called a *spinor*, which is not a scalar, vector, or more general tensor. The study of spinors, however, falls beyond the scope of this book.

formations, for which the physical system of interest is left unaltered, and only the coordinate system used to describe it is changed. In an active transformation, however, the system itself is altered.

As an example, let us consider a particle of mass m that is located at a position $\mathbf{x}$ relative to the origin O and hence has velocity $\dot{\mathbf{x}}$. The angular momentum of the particle about O is thus $\mathbf{J} = m(\mathbf{x} \times \dot{\mathbf{x}})$. If we merely invert the Cartesian coordinates used to describe this system through O, neither the magnitude nor direction of any of these vectors will be changed, since they may be considered simply as arrows in space that are independent of the coordinates used to describe them. If, however, we perform the analogous active transformation on the system, by inverting the position vector of the particle through O, then it is clear that the direction of particle's velocity will also be reversed, since it is simply the time derivative of the position vector, but that the direction of its angular momentum vector remains unaltered. This suggests that vectors can be divided into two categories, as follows: *polar* vectors (such as position and velocity), which reverse direction under an active inversion of the physical system through the origin, and *axial* vectors (such as angular momentum), which remain unchanged. It should be emphasised that at no point in this discussion have we used the concept of a pseudovector to describe a real physical quantity.§

26.11 Dual tensors

Although pseudotensors are not themselves appropriate for the description of physical phenomena, they are sometimes needed; for example, we may use the pseudotensor ϵ_{ijk} to associate with every *antisymmetric* second-order tensor A_{ij} (in three dimensions) a pseudovector p_i given by

$$p_i = \tfrac{1}{2}\epsilon_{ijk} A_{jk}; \tag{26.40}$$

p_i is called the *dual* of A_{ij}. Thus if we denote the antisymmetric tensor $\mathbf{A}$ by the matrix

$$\mathbf{A} = [A_{ij}] = \begin{pmatrix} 0 & A_{12} & -A_{31} \\ -A_{12} & 0 & A_{23} \\ A_{31} & -A_{23} & 0 \end{pmatrix}$$

then the components of its dual pseudovector are $(p_1, p_2, p_3) = (A_{23}, A_{31}, A_{12})$.

---

§ The scalar product of a polar vector and an axial vector is a pseudoscalar. It was the experimental detection of the dependence of the angular distribution of electrons of (polar vector) momentum $\mathbf{p}_e$ emitted by polarised nuclei of (axial vector) spin $\mathbf{J}_N$ upon the pseudoscalar quantity $\mathbf{J}_N \cdot \mathbf{p}_e$ that established the existence of the non-conservation of parity in β-decay.

▶ *Using (26.40), show that $A_{ij} = \epsilon_{ijk}p_k$.*

By contracting both sides of (26.40) with ϵ_{ijk}, we find

$$\epsilon_{ijk}p_k = \tfrac{1}{2}\epsilon_{ijk}\epsilon_{klm}A_{lm}.$$

Using the identity (26.30) then gives

$$\epsilon_{ijk}p_k = \tfrac{1}{2}(\delta_{il}\delta_{jm} - \delta_{im}\delta_{jl})A_{lm}$$
$$= \tfrac{1}{2}(A_{ij} - A_{ji}) = \tfrac{1}{2}(A_{ij} + A_{ij}) = A_{ij},$$

where in the last line we use the fact that $A_{ij} = -A_{ji}$. ◀

By a simple extension, we may associate a dual pseudoscalar s with every totally antisymmetric third-rank tensor A_{ijk}, i.e. one that is antisymmetric with respect to the interchange of every possible pair of subscripts; s is given by

$$s = \frac{1}{3!}\epsilon_{ijk}A_{ijk}. \tag{26.41}$$

Since A_{ijk} is a totally antisymmetric three-subscript quantity, we expect it to equal some multiple of ϵ_{ijk} (since this is the only such quantity). In fact $A_{ijk} = s\epsilon_{ijk}$, as can be proved by substituting this expression into (26.41) and using (26.36).

26.12 Physical applications of tensors

In this section some physical applications of tensors will be given. First-order tensors are familiar as vectors and so we will concentrate on second-order tensors, starting with an example taken from mechanics.

Consider a collection of rigidly connected point particles of which the αth, which has mass $m^{(\alpha)}$ and is positioned at $\mathbf{r}^{(\alpha)}$ with respect to an origin O, is typical. Suppose that the rigid assembly is rotating about an axis through O with angular velocity $\boldsymbol{\omega}$.

The angular momentum $\mathbf{J}$ about O of the assembly is given by

$$\mathbf{J} = \sum_\alpha \left(\mathbf{r}^{(\alpha)} \times \mathbf{p}^{(\alpha)}\right).$$

But $\mathbf{p}^{(\alpha)} = m^{(\alpha)}\dot{\mathbf{r}}^{(\alpha)}$ and $\dot{\mathbf{r}}^{(\alpha)} = \boldsymbol{\omega} \times \mathbf{r}^{(\alpha)}$, for any α, and so in subscript form the components of $\mathbf{J}$ are given by

$$J_i = \sum_\alpha m^{(\alpha)}\epsilon_{ijk}x_j^{(\alpha)}\dot{x}_k^{(\alpha)}$$
$$= \sum_\alpha m^{(\alpha)}\epsilon_{ijk}x_j^{(\alpha)}\epsilon_{klm}\omega_l x_m^{(\alpha)}$$
$$= \sum_\alpha m^{(\alpha)}(\delta_{il}\delta_{jm} - \delta_{im}\delta_{jl})x_j^{(\alpha)}x_m^{(\alpha)}\omega_l$$
$$= \sum_\alpha m^{(\alpha)}\left[\left(r^{(\alpha)}\right)^2\delta_{il} - x_i^{(\alpha)}x_l^{(\alpha)}\right]\omega_l \equiv I_{il}\omega_l, \tag{26.42}$$

where I_{il} is a symmetric second-order Cartesian tensor (by the quotient rule, see section 26.7, since **J** and ω are vectors). The tensor is called the *inertia tensor* at O of the assembly and depends only on the distribution of masses in the assembly and not upon the direction or magnitude of ω.

A more realistic situation obtains if a continuous rigid body is considered. In this case, $m^{(\alpha)}$ must be replaced everywhere by $\rho(\mathbf{r})\,dx\,dy\,dz$ and all summations by integrations over the volume of the body. Written out in full in Cartesians, the inertia tensor for a continuous body would have the form

$$\mathsf{I} = [I_{ij}] = \begin{pmatrix} \int(y^2+z^2)\rho\,dV & -\int xy\rho\,dV & -\int xz\rho\,dV \\ -\int xy\rho\,dV & \int(z^2+x^2)\rho\,dV & -\int yz\rho\,dV \\ -\int xz\rho\,dV & -\int yz\rho\,dV & \int(x^2+y^2)\rho\,dV \end{pmatrix},$$

where $\rho = \rho(x,y,z)$ is the mass distribution and dV stands for $dx\,dy\,dz$; the integrals are to be taken over the whole body. The diagonal elements of this tensor are called the *moments of inertia* and the off-diagonal elements without the minus signs are known as the *products of inertia*.

▶*Show that the kinetic energy of the rotating system is given by $T = \frac{1}{2}I_{jl}\omega_j\omega_l$.*

By an argument parallel to that already made for **J**, the kinetic energy is given by

$$T = \frac{1}{2}\sum_\alpha m^{(\alpha)}\left(\dot{\mathbf{r}}^{(\alpha)}\cdot\dot{\mathbf{r}}^{(\alpha)}\right)$$

$$= \frac{1}{2}\sum_\alpha m^{(\alpha)}\epsilon_{ijk}\omega_j x_k^{(\alpha)}\epsilon_{ilm}\omega_l x_m^{(\alpha)}$$

$$= \frac{1}{2}\sum_\alpha m^{(\alpha)}(\delta_{jl}\delta_{km} - \delta_{jm}\delta_{kl})x_k^{(\alpha)}x_m^{(\alpha)}\omega_j\omega_l$$

$$= \frac{1}{2}\sum_\alpha m^{(\alpha)}\left[\delta_{jl}\left(r^{(\alpha)}\right)^2 - x_j^{(\alpha)}x_l^{(\alpha)}\right]\omega_j\omega_l$$

$$= \frac{1}{2}I_{jl}\omega_j\omega_l.$$

Alternatively, since $J_j = I_{jl}\omega_l$ we may write the kinetic energy of the rotating system as $T = \frac{1}{2}J_j\omega_j$. ◀

The above example shows that the kinetic energy of the rotating body can be expressed as a scalar obtained by twice contracting ω with the inertia tensor. It also shows that the moment of inertia of the body about a line given by the unit vector $\hat{\mathbf{n}}$ is $I_{jl}\hat{n}_j\hat{n}_l$ (or $\hat{\mathbf{n}}^{\mathsf{T}}\mathsf{I}\hat{\mathbf{n}}$ in matrix form).

Since **I** ($\equiv I_{jl}$) is a real symmetric second-order tensor, it has associated with it three mutually perpendicular directions that are its *principal axes* and have the following properties (proved in chapter 8):

(i) with each axis is associated a principal moment of inertia λ_μ, $\mu = 1, 2, 3$;

(ii) when the rotation of the body is about one of these axes, the angular velocity and the angular momentum are parallel and given by

$$\mathbf{J} = \mathsf{I}\omega = \lambda_\mu\omega,$$

i.e. ω is an eigenvector of **I** with eigenvalue λ_μ;

(iii) referred to these axes as coordinate axes, the inertia tensor is diagonal with diagonal entries $\lambda_1, \lambda_2, \lambda_3$.

Two further examples of physical quantities represented by second-order tensors are magnetic susceptibility and electrical conductivity. In the first case we have (in standard notation)

$$M_i = \chi_{ij} H_j, \tag{26.43}$$

and in the second case

$$j_i = \sigma_{ij} E_j. \tag{26.44}$$

Here $\mathbf{M}$ is the magnetic moment per unit volume and $\mathbf{j}$ the current density (current per unit perpendicular area). In both cases we have on the left-hand side a vector and on the right-hand side the contraction of a set of quantities with another vector. Each set of quantities must therefore form the components of a second-order tensor.

For isotropic media $\mathbf{M} \propto \mathbf{H}$ and $\mathbf{j} \propto \mathbf{E}$, but for anisotropic materials such as crystals the susceptibility and conductivity may be different along different crystal axes, making χ_{ij} and σ_{ij} general second-order tensors, although they are usually symmetric.

▶The electrical conductivity $\boldsymbol{\sigma}$ in a crystal is measured by an observer to have components as shown:

$$[\sigma_{ij}] = \begin{pmatrix} 1 & \sqrt{2} & 0 \\ \sqrt{2} & 3 & 1 \\ 0 & 1 & 1 \end{pmatrix}. \tag{26.45}$$

Show that there is one direction in the crystal along which no current can flow. Does the current flow equally easily in the two perpendicular directions?

The current density in the crystal is given by $j_i = \sigma_{ij} E_j$, where σ_{ij}, relative to the observer's coordinate system, is given by (26.45). Since $[\sigma_{ij}]$ is a symmetric matrix, it possesses three mutually perpendicular eigenvectors (or principal axes) with respect to which the conductivity tensor is diagonal, with diagonal entries $\lambda_1, \lambda_2, \lambda_3$, the eigenvalues of $[\sigma_{ij}]$.

As discussed in chapter 8, the eigenvalues of $[\sigma_{ij}]$ are given by $|\sigma - \lambda \mathbf{I}| = 0$. Thus we require

$$\begin{vmatrix} 1 - \lambda & \sqrt{2} & 0 \\ \sqrt{2} & 3 - \lambda & 1 \\ 0 & 1 & 1 - \lambda \end{vmatrix} = 0,$$

from which we find

$$(1 - \lambda)[(3 - \lambda)(1 - \lambda) - 1] - 2(1 - \lambda) = 0.$$

This simplifies to give $\lambda = 0, 1, 4$ so that, with respect to its principal axes, the conductivity tensor has components σ'_{ij} given by

$$[\sigma'_{ij}] = \begin{pmatrix} 4 & 0 & 0 \\ 0 & 1 & 0 \\ 0 & 0 & 0 \end{pmatrix}.$$

Since $j'_i = \sigma'_{ij} E'_j$, we see immediately that along one of the principal axes there is no current flow and along the two perpendicular directions the current flows are not equal. ◀

We can extend the idea of a second-order tensor that relates two vectors to a situation where two physical second-order tensors are related by a fourth-order tensor. The most common occurrence of such relationships is in the theory of elasticity. This is not the place to give a detailed account of elasticity theory, but suffice it to say that the local deformation of an elastic body at any interior point P can be described by a second-order symmetric tensor e_{ij} called the *strain tensor*. It is given by

$$e_{ij} = \frac{1}{2}\left(\frac{\partial u_i}{\partial x_j} + \frac{\partial u_j}{\partial x_i}\right),$$

where $\mathbf{u}$ is the displacement vector describing the strain of a small volume element whose unstrained position relative to the origin is $\mathbf{x}$. Similarly we can describe the stress in the body at P by the second-order symmetric *stress tensor* p_{ij}; the quantity p_{ij} is the x_j-component of the stress vector acting across a plane through P whose normal lies in the x_i-direction. A generalisation of Hooke's law then relates the stress and strain tensors by

$$p_{ij} = c_{ijkl}e_{kl} \tag{26.46}$$

where c_{ijkl} is a fourth-order Cartesian tensor.

▶*Assuming that the most general fourth-order isotropic tensor is*

$$c_{ijkl} = \lambda\delta_{ij}\delta_{kl} + \eta\delta_{ik}\delta_{jl} + \nu\delta_{il}\delta_{jk}, \tag{26.47}$$

find the form of (26.46) for an isotropic medium having Young's modulus E and Poisson's ratio σ.

For an isotropic medium we must have an isotropic tensor for c_{ijkl}, and so we assume the form (26.47). Substituting this into (26.46) yields

$$p_{ij} = \lambda\delta_{ij}e_{kk} + \eta e_{ij} + \nu e_{ji}.$$

But e_{ij} is symmetric, and if we write $\eta + \nu = 2\mu$, then this takes the form

$$p_{ij} = \lambda e_{kk}\delta_{ij} + 2\mu e_{ij},$$

in which λ and μ are known as *Lamé constants*. It will be noted that if $e_{ij} = 0$ for $i \neq j$ then the same is true of p_{ij}, i.e. the principal axes of the stress and strain tensors coincide.

Now consider a simple tension in the x_1-direction, i.e. $p_{11} = S$ but all other $p_{ij} = 0$. Then denoting e_{kk} (summed over k) by θ we have, in addition to $e_{ij} = 0$ for $i \neq j$, the three equations

$$S = \lambda\theta + 2\mu e_{11},$$
$$0 = \lambda\theta + 2\mu e_{22},$$
$$0 = \lambda\theta + 2\mu e_{33}.$$

Adding them gives

$$S = \theta(3\lambda + 2\mu).$$

Substituting for θ from this into the first of the three, and recalling that Young's modulus is defined by $S = E e_{11}$, gives E as

$$E = \frac{\mu(3\lambda + 2\mu)}{\lambda + \mu}. \tag{26.48}$$

Further, Poisson's ratio is defined as $\sigma = -e_{22}/e_{11}$ (or $-e_{33}/e_{11}$) and is thus

$$\sigma = \left(\frac{1}{e_{11}}\right)\frac{\lambda\theta}{2\mu} = \left(\frac{1}{e_{11}}\right)\left(\frac{\lambda}{2\mu}\right)\frac{Ee_{11}}{3\lambda+2\mu} = \frac{\lambda}{2(\lambda+\mu)}. \tag{26.49}$$

Solving (26.48) and (26.49) for λ and μ gives finally

$$p_{ij} = \frac{\sigma E}{(1+\sigma)(1-2\sigma)}e_{kk}\delta_{ij} + \frac{E}{(1+\sigma)}e_{ij}. \blacktriangleleft$$

26.13 Integral theorems for tensors

In chapter 11, we discussed various integral theorems involving vector and scalar fields. Most notably, we considered the divergence theorem, which states that, for any vector field $\mathbf{a}$,

$$\int_V \nabla \cdot \mathbf{a}\, dV = \oint_S \mathbf{a}\cdot\hat{\mathbf{n}}\, dS, \tag{26.50}$$

where S is the surface enclosing the volume V and $\hat{\mathbf{n}}$ is the outward-pointing unit normal to S at each point.

Writing (26.50) in subscript notation, we have

$$\int_V \frac{\partial a_k}{\partial x_k}\, dV = \oint_S a_k\hat{n}_k\, dS. \tag{26.51}$$

Although we shall not prove it rigorously, (26.51) can be extended in an obvious manner to relate integrals of *tensor fields*, rather than just vector fields, over volumes and surfaces, with the result

$$\int_V \frac{\partial T_{ij\cdots k\cdots m}}{\partial x_k}\, dV = \oint_S T_{ij\cdots k\cdots m}\hat{n}_k\, dS.$$

This form of the divergence theorem for general tensors can be very useful in vector calculus manipulations.

> ▶ *A vector field $\mathbf{a}$ satisfies $\nabla\cdot\mathbf{a}=0$ inside some volume V and $\mathbf{a}\cdot\hat{\mathbf{n}}=0$ on the boundary surface S. By considering the divergence theorem applied to $T_{ij}=x_i a_j$, show that $\int_V \mathbf{a}\, dV = \mathbf{0}$.*

Applying the divergence theorem to $T_{ij}=x_i a_j$ we find

$$\int_V \frac{\partial T_{ij}}{\partial x_j}\, dV = \int_V \frac{\partial(x_i a_j)}{\partial x_j}\, dV = \oint_S x_i a_j\hat{n}_j\, dS = 0,$$

since $a_j\hat{n}_j = 0$. By expanding the volume integral we obtain

$$\int_V \frac{\partial(x_i a_j)}{\partial x_j}\, dV = \int_V \frac{\partial x_i}{\partial x_j}a_j\, dV + \int_V x_i\frac{\partial a_j}{\partial x_j}\, dV$$

$$= \int_V \delta_{ij}a_j\, dV$$

$$= \int_V a_i\, dV = 0,$$

where in going from the first to the second line we used $\partial x_i/\partial x_j = \delta_{ij}$ and $\partial a_j/\partial x_j = 0$. ◀

The other integral theorems discussed in chapter 11 can be extended in a similar way. For example, written in tensor notation Stokes' theorem states that, for a vector field a_i,

$$\int_S \epsilon_{ijk} \frac{\partial a_k}{\partial x_j} \hat{n}_i \, dS = \oint_C a_k \, dx_k.$$

For a general tensor field this has the straightforward extension

$$\int_S \epsilon_{ijk} \frac{\partial T_{lm\cdots k\cdots n}}{\partial x_j} \hat{n}_i \, dS = \oint_C T_{lm\cdots k\cdots n} \, dx_k.$$

26.14 Non-Cartesian coordinates

So far we have restricted our attention to the study of tensors when they are described in terms of Cartesian coordinates and the axes of coordinates are rigidly rotated, sometimes together with an inversion of axes through the origin. In the remainder of this chapter we shall extend the concepts discussed in the previous sections by considering arbitrary coordinate transformations from one general coordinate system to another. Although this generalisation brings with it several complications, we shall find that many of the properties of Cartesian tensors are still valid for more general tensors. Before considering general coordinate transformations, however, we begin by reminding ourselves of some properties of general curvilinear coordinates, as discussed in chapter 10.

The position of an arbitrary point P in space may be expressed in terms of the three curvilinear coordinates u_1, u_2, u_3. We saw in chapter 10 that if $\mathbf{r}(u_1, u_2, u_3)$ is the position vector of the point P then at P there exist two sets of basis vectors

$$\mathbf{e}_i = \frac{\partial \mathbf{r}}{\partial u_i} \qquad \text{and} \qquad \boldsymbol{\epsilon}_i = \nabla u_i, \tag{26.52}$$

where $i = 1, 2, 3$. In general, the vectors in each set neither are of unit length nor form an orthogonal basis. However, the sets $\mathbf{e}_i$ and $\boldsymbol{\epsilon}_i$ are reciprocal systems of vectors and so

$$\mathbf{e}_i \cdot \boldsymbol{\epsilon}_j = \delta_{ij}. \tag{26.53}$$

In the context of general tensor analysis, it is more usual to denote the second set of vectors $\boldsymbol{\epsilon}_i$ in (26.52) by $\mathbf{e}^i$, the index being placed as a superscript to distinguish it from the (different) vector $\mathbf{e}_i$, which is a member of the first set in (26.52). Although this positioning of the index may seem odd (not least because of the possibility of confusion with powers) it forms part of a slight modification to the summation convention that we will adopt for the remainder of this chapter. This is as follows: any lower-case alphabetic index that appears exactly twice in any term of an expression, *once as a subscript and once as a superscript*, is to be summed over all the values that an index in that position can take (unless the contrary is specifically stated). All other aspects of the summation convention remain unchanged.

With the introduction of superscripts, the reciprocity relation (26.53) should be rewritten so that both sides of (26.54) have one subscript and one superscript, i.e. as

$$\mathbf{e}_i \cdot \mathbf{e}^j = \delta_i^j. \tag{26.54}$$

The alternative form of the Kronecker delta is defined in a similar way to previously, i.e. it equals unity if $i = j$ and is zero otherwise.

For similar reasons it is usual to denote the curvilinear coordinates themselves by u^1, u^2, u^3, with the index raised, so that

$$\mathbf{e}_i = \frac{\partial \mathbf{r}}{\partial u^i} \qquad \text{and} \qquad \mathbf{e}^i = \nabla u^i. \tag{26.55}$$

From the first equality we see that we may consider a superscript that appears in the denominator of a partial derivative as a subscript.

Given the two bases $\mathbf{e}_i$ and $\mathbf{e}^i$, we may write a general vector $\mathbf{a}$ equally well in terms of either basis as follows:

$$\mathbf{a} = a^1 \mathbf{e}_1 + a^2 \mathbf{e}_2 + a^3 \mathbf{e}_3 = a^i \mathbf{e}_i;$$
$$\mathbf{a} = a_1 \mathbf{e}^1 + a_2 \mathbf{e}^2 + a_3 \mathbf{e}^3 = a_i \mathbf{e}^i.$$

The a^i are called the *contravariant* components of the vector $\mathbf{a}$ and the a_i the *covariant* components, the position of the index (either as a subscript or superscript) serving to distinguish between them. Similarly, we may call the $\mathbf{e}_i$ the covariant basis vectors and the $\mathbf{e}^i$ the contravariant ones.

> ▶*Show that the contravariant and covariant components of a vector* $\mathbf{a}$ *are given by* $a^i = \mathbf{a} \cdot \mathbf{e}^i$ *and* $a_i = \mathbf{a} \cdot \mathbf{e}_i$ *respectively.*

For the contravariant components, we find

$$\mathbf{a} \cdot \mathbf{e}^i = a^j \mathbf{e}_j \cdot \mathbf{e}^i = a^j \delta_j^i = a^i,$$

where we have used the reciprocity relation (26.54). Similarly, for the covariant components,

$$\mathbf{a} \cdot \mathbf{e}_i = a_j \mathbf{e}^j \cdot \mathbf{e}_i = a_j \delta_i^j = a_i. \; ◀$$

The reason that the notion of contravariant and covariant components of a vector (and the resulting superscript notation) was not introduced earlier is that for Cartesian coordinate systems the two sets of basis vectors $\mathbf{e}_i$ and $\mathbf{e}^i$ are identical and, hence, so are the components of a vector with respect to either basis. Thus, for Cartesian coordinates, we may speak simply of the components of the vector and there is no need to differentiate between contravariance and covariance, or to introduce superscripts to make a distinction between them.

If we consider the components of higher-order tensors in non-Cartesian co-ordinates, there are even more possibilities. As an example, let us consider a second-order tensor **T**. Using the outer product notation in (26.23), we may write **T** in three different ways:

$$\mathbf{T} = T^{ij}\mathbf{e}_i \otimes \mathbf{e}_j = T^i{}_j\mathbf{e}_i \otimes \mathbf{e}^j = T_{ij}\mathbf{e}^i \otimes \mathbf{e}^j,$$

where T^{ij}, $T^i{}_j$ and T_{ij} are called the *contravariant, mixed* and *covariant* components of **T** respectively. It is important to remember that these three sets of quantities form the components of the *same* tensor **T** but refer to different (tensor) bases made up from the basis vectors of the coordinate system. Again, if we are using Cartesian coordinates then all three sets of components are identical.

We may generalise the above equation to higher-order tensors. Components carrying only superscripts or only subscripts are referred to as the contravariant and covariant components respectively; all others are called mixed components.

26.15 The metric tensor

Any particular curvilinear coordinate system is completely characterised at each point in space by the nine quantities

$$g_{ij} = \mathbf{e}_i \cdot \mathbf{e}_j, \tag{26.56}$$

which, as we will show, are the covariant components of a symmetric second-order tensor **g** called the *metric tensor*.

Since an infinitesimal vector displacement can be written as $d\mathbf{r} = du^i\mathbf{e}_i$, we find that the square of the infinitesimal arc length $(ds)^2$ can be written in terms of the metric tensor as

$$(ds)^2 = d\mathbf{r} \cdot d\mathbf{r} = du^i\mathbf{e}_i \cdot du^j\mathbf{e}_j = g_{ij}\,du^i du^j. \tag{26.57}$$

It may further be shown that the volume element dV is given by

$$dV = \sqrt{g}\,du^1\,du^2\,du^3, \tag{26.58}$$

where g is the determinant of the matrix $[g_{ij}]$, which has the covariant components of the metric tensor as its elements.

If we compare equations (26.57) and (26.58) with the analogous ones in section 10.10 then we see that in the special case where the coordinate system is orthogonal (so that $\mathbf{e}_i \cdot \mathbf{e}_j = 0$ for $i \neq j$) the metric tensor can be written in terms of the coordinate-system scale factors h_i, $i = 1, 2, 3$ as

$$g_{ij} = \begin{cases} h_i^2 & i = j, \\ 0 & i \neq j. \end{cases}$$

Its determinant is then given by $g = h_1^2 h_2^2 h_3^2$.

> ►*Calculate the elements g_{ij} of the metric tensor for cylindrical polar coordinates. Hence find the square of the infinitesimal arc length $(ds)^2$ and the volume dV for this coordinate system.*

As discussed in section 10.9, in cylindrical polar coordinates $(u^1, u^2, u^3) = (\rho, \phi, z)$ and so the position vector $\mathbf{r}$ of any point P may be written

$$\mathbf{r} = \rho \cos \phi \, \mathbf{i} + \rho \sin \phi \, \mathbf{j} + z \, \mathbf{k}.$$

From this we obtain the (covariant) basis vectors:

$$\mathbf{e}_1 = \frac{\partial \mathbf{r}}{\partial \rho} = \cos \phi \, \mathbf{i} + \sin \phi \, \mathbf{j};$$

$$\mathbf{e}_2 = \frac{\partial \mathbf{r}}{\partial \phi} = -\rho \sin \phi \, \mathbf{i} + \rho \cos \phi \, \mathbf{j};$$

$$\mathbf{e}_3 = \frac{\partial \mathbf{r}}{\partial z} = \mathbf{k}. \tag{26.59}$$

Thus the components of the metric tensor $[g_{ij}] = [\mathbf{e}_i \cdot \mathbf{e}_j]$ are found to be

$$\mathsf{G} = [g_{ij}] = \begin{pmatrix} 1 & 0 & 0 \\ 0 & \rho^2 & 0 \\ 0 & 0 & 1 \end{pmatrix}, \tag{26.60}$$

from which we see that, as expected for an orthogonal coordinate system, the metric tensor is diagonal, the diagonal elements being equal to the squares of the scale factors of the coordinate system.

From (26.57), the square of the infinitesimal arc length in this coordinate system is given by

$$(ds)^2 = g_{ij} \, du^i \, du^j = (d\rho)^2 + \rho^2 (d\phi)^2 + (dz)^2,$$

and, using (26.58), the volume element is found to be

$$dV = \sqrt{g} \, du^1 \, du^2 \, du^3 = \rho \, d\rho \, d\phi \, dz.$$

These expressions are identical to those derived in section 10.9. ◄

We may also express the scalar product of two vectors in terms of the metric tensor:

$$\mathbf{a} \cdot \mathbf{b} = a^i \mathbf{e}_i \cdot b^j \mathbf{e}_j = g_{ij} a^i b^j, \tag{26.61}$$

where we have used the contravariant components of the two vectors. Similarly, using the covariant components, we can write the same scalar product as

$$\mathbf{a} \cdot \mathbf{b} = a_i \mathbf{e}^i \cdot b_j \mathbf{e}^j = g^{ij} a_i b_j, \tag{26.62}$$

where we have defined the nine quantities $g^{ij} = \mathbf{e}^i \cdot \mathbf{e}^j$. As we shall show, they form the contravariant components of the metric tensor $\mathbf{g}$ and are, in general, different from the quantities g_{ij}. Finally, we could express the scalar product in terms of the contravariant components of one vector and the covariant components of the other,

$$\mathbf{a} \cdot \mathbf{b} = a_i \mathbf{e}^i \cdot b^j \mathbf{e}_j = a_i b^j \delta^i_j = a_i b^i, \tag{26.63}$$

where we have used the reciprocity relation (26.54). Similarly, we could write

$$\mathbf{a} \cdot \mathbf{b} = a^i \mathbf{e}_i \cdot b_j \mathbf{e}^j = a^i b_j \delta_i^j = a^i b_i. \tag{26.64}$$

By comparing the four alternative expressions (26.61)–(26.64) for the scalar product of two vectors we can deduce one of the most useful properties of the quantities g_{ij} and g^{ij}. Since $g_{ij} a^i b^j = a^i b_i$ holds for any arbitrary vector components a^i, it follows that

$$g_{ij} b^j = b_i,$$

which illustrates the fact that the covariant components g_{ij} of the metric tensor can be used to *lower an index*. In other words, it provides a means of obtaining the covariant components of a vector from its contravariant components. By a similar argument, we have

$$g^{ij} b_j = b^i,$$

so that the contravariant components g^{ij} can be used to perform the reverse operation of *raising an index*.

It is straightforward to show that the contravariant and covariant basis vectors, $\mathbf{e}^i$ and $\mathbf{e}_i$ respectively, are related in the same way as other vectors, i.e. by

$$\mathbf{e}^i = g^{ij} \mathbf{e}_j \quad \text{and} \quad \mathbf{e}_i = g_{ij} \mathbf{e}^j.$$

We also note that, since $\mathbf{e}_i$ and $\mathbf{e}^i$ are reciprocal systems of vectors in three-dimensional space (see chapter 7), we may write

$$\mathbf{e}^i = \frac{\mathbf{e}_j \times \mathbf{e}_k}{\mathbf{e}_i \cdot (\mathbf{e}_j \times \mathbf{e}_k)},$$

for the combination of subscripts $i, j, k = 1, 2, 3$ and its cyclic permutations. A similar expression holds for $\mathbf{e}_i$ in terms of the $\mathbf{e}^i$-basis. Moreover, it may be shown that $|\mathbf{e}_1 \cdot (\mathbf{e}_2 \times \mathbf{e}_3)| = \sqrt{g}$.

> ▶*Show that the matrix $[g^{ij}]$ is the inverse of the matrix $[g_{ij}]$. Hence calculate the contravariant components g^{ij} of the metric tensor in cylindrical polar coordinates.*

Using the index-lowering and index-raising properties of g_{ij} and g^{ij} on an arbitrary vector $\mathbf{a}$, we find

$$\delta_k^i a^k = a^i = g^{ij} a_j = g^{ij} g_{jk} a^k.$$

But, since $\mathbf{a}$ is arbitrary, we must have

$$g^{ij} g_{jk} = \delta_k^i. \tag{26.65}$$

Denoting the matrix $[g_{ij}]$ by G and $[g^{ij}]$ by $\hat{\mathsf{G}}$, equation (26.65) can be written in matrix form as $\hat{\mathsf{G}}\mathsf{G} = \mathsf{I}$, where I is the unit matrix. Hence G and $\hat{\mathsf{G}}$ are inverse matrices of each other.

Thus, by inverting the matrix G in (26.60), we find that the elements g^{ij} are given in cylindrical polar coordinates by

$$\hat{G} = [g^{ij}] = \begin{pmatrix} 1 & 0 & 0 \\ 0 & 1/\rho^2 & 0 \\ 0 & 0 & 1 \end{pmatrix}. \blacktriangleleft$$

So far we have not considered the components of the metric tensor g^i_j with one subscript and one superscript. By analogy with (26.56), these mixed components are given by

$$g^i_j = \mathbf{e}^i \cdot \mathbf{e}_j = \delta^i_j,$$

and so the components of g^i_j are identical to those of δ^i_j. We may therefore consider the δ^i_j to be the mixed components of the metric tensor $\mathbf{g}$.

26.16 General coordinate transformations and tensors

We now discuss the concept of general transformations from one coordinate system, u^1, u^2, u^3, to another, u'^1, u'^2, u'^3. We can describe the coordinate transform using the three equations

$$u'^i = u'^i(u^1, u^2, u^3),$$

for $i = 1, 2, 3$, in which the new coordinates u'^i can be arbitrary functions of the old ones u^i rather than just represent linear orthogonal transformations (rotations) of the coordinate axes. We shall assume also that the transformation can be inverted, so that we can write the old coordinates in terms of the new ones as

$$u^i = u^i(u'^1, u'^2, u'^3),$$

As an example, we may consider the transformation from spherical polar to Cartesian coordinates, given by

$$x = r \sin\theta \cos\phi,$$
$$y = r \sin\theta \sin\phi,$$
$$z = r \cos\theta,$$

which is clearly not a linear transformation.

The two sets of basis vectors in the new coordinate system, u'^1, u'^2, u'^3, are given as in (26.55) by

$$\mathbf{e}'_i = \frac{\partial \mathbf{r}}{\partial u'^i} \qquad \text{and} \qquad \mathbf{e}'^i = \nabla u'^i. \tag{26.66}$$

Considering the first set, we have from the chain rule that

$$\frac{\partial \mathbf{r}}{\partial u^j} = \frac{\partial u'^i}{\partial u^j} \frac{\partial \mathbf{r}}{\partial u'^i},$$

so that the basis vectors in the old and new coordinate systems are related by

$$\mathbf{e}_j = \frac{\partial u'^i}{\partial u^j}\mathbf{e}'_i. \tag{26.67}$$

Now, since we can write any arbitrary vector $\mathbf{a}$ in terms of either basis as

$$\mathbf{a} = a'^i\mathbf{e}'_i = a^j\mathbf{e}_j = a^j\frac{\partial u'^i}{\partial u^j}\mathbf{e}'_i,$$

it follows that the contravariant components of a vector must transform as

$$a'^i = \frac{\partial u'^i}{\partial u^j}a^j. \tag{26.68}$$

In fact, we use this relation as the defining property for a set of quantities a^i to form the contravariant components of a vector.

▶*Find an expression analogous to (26.67) relating the basis vectors $\mathbf{e}^i$ and $\mathbf{e}'^i$ in the two coordinate systems. Hence deduce the way in which the covariant components of a vector change under a coordinate transformation.*

If we consider the second set of basis vectors in (26.66), $\mathbf{e}'^i = \nabla u'^i$, we have from the chain rule that

$$\frac{\partial u^j}{\partial x} = \frac{\partial u^j}{\partial u'^i}\frac{\partial u'^i}{\partial x}$$

and similarly for $\partial u^j/\partial y$ and $\partial u^j/\partial z$. So the basis vectors in the old and new coordinate systems are related by

$$\mathbf{e}^j = \frac{\partial u^j}{\partial u'^i}\mathbf{e}'^i. \tag{26.69}$$

For any arbitrary vector $\mathbf{a}$,

$$\mathbf{a} = a'_i\mathbf{e}'^i = a_j\mathbf{e}^j = a_j\frac{\partial u^j}{\partial u'^i}\mathbf{e}'^i$$

and so the covariant components of a vector must transform as

$$a'_i = \frac{\partial u^j}{\partial u'^i}a_j. \tag{26.70}$$

Analogously to the contravariant case (26.68), we take this result as the defining property of the covariant components of a vector. ◀

We may compare the transformation laws (26.68) and (26.70) with those for a first-order Cartesian tensor under a rigid rotation of axes. Let us consider a rotation of Cartesian axes x^i through an angle θ about the 3-axis to a new set x'^i, $i = 1, 2, 3$, as given by (26.7) and the inverse transformation (26.8). It is straightforward to show that

$$\frac{\partial x^j}{\partial x'^i} = \frac{\partial x'^i}{\partial x^j} = L_{ij},$$

where the elements L_{ij} are given by

$$\mathsf{L} = \begin{pmatrix} \cos\theta & \sin\theta & 0 \\ -\sin\theta & \cos\theta & 0 \\ 0 & 0 & 1 \end{pmatrix}.$$

Thus (26.68) and (26.70) agree with our earlier definition in the special case of a rigid rotation of Cartesian axes.

Following on from (26.68) and (26.70), we proceed in a similar way to define general tensors of higher rank. For example, the contravariant, mixed and covariant components, respectively, of a second-order tensor must transform as follows:

contravariant components, $\quad T'^{ij} = \dfrac{\partial u'^i}{\partial u^k}\dfrac{\partial u'^j}{\partial u^l}T^{kl}$;

mixed components, $\quad T'^i{}_j = \dfrac{\partial u'^i}{\partial u^k}\dfrac{\partial u^l}{\partial u'^j}T^k{}_l$;

covariant components, $\quad T'_{ij} = \dfrac{\partial u^k}{\partial u'^i}\dfrac{\partial u^l}{\partial u'^j}T_{kl}.$

It is important to remember that these quantities form the components of the *same* tensor $\mathbf{T}$ but refer to different tensor bases made up from the basis vectors of the different coordinate systems. For example, in terms of the contravariant components we may write

$$\mathbf{T} = T^{ij}\mathbf{e}_i \otimes \mathbf{e}_j = T'^{ij}\mathbf{e}'_i \otimes \mathbf{e}'_j.$$

We can clearly go on to define tensors of higher order, with arbitrary numbers of covariant (subscript) and contravariant (superscript) indices, by demanding that their components transform as follows:

$$T'^{ij\cdots k}{}_{lm\cdots n} = \frac{\partial u'^i}{\partial u^a}\frac{\partial u'^j}{\partial u^b}\cdots\frac{\partial u'^k}{\partial u^c}\frac{\partial u^d}{\partial u'^l}\frac{\partial u^e}{\partial u'^m}\cdots\frac{\partial u^f}{\partial u'^n}T^{ab\cdots c}{}_{de\cdots f}. \qquad (26.71)$$

Using the revised summation convention described in section 26.14, the algebra of general tensors is completely analogous to that of the Cartesian tensors discussed earlier. For example, as with Cartesian coordinates, the Kronecker delta is a tensor provided it is written as the mixed tensor δ^i_j since

$$\delta'^i_j = \frac{\partial u'^i}{\partial u^k}\frac{\partial u^l}{\partial u'^j}\delta^k_l = \frac{\partial u'^i}{\partial u^k}\frac{\partial u^k}{\partial u'^j} = \frac{\partial u'^i}{\partial u'^j} = \delta^i_j,$$

where we have used the chain rule to justify the third equality. This also shows that δ^i_j is isotropic. As discussed at the end of section 26.15, the δ^i_j can be considered as the mixed components of the metric tensor $\mathbf{g}$.

> ▶*Show that the quantities* $g_{ij} = \mathbf{e}_i \cdot \mathbf{e}_j$ *form the covariant components of a second-order tensor.*

In the new (primed) coordinate system we have

$$g'_{ij} = \mathbf{e}'_i \cdot \mathbf{e}'_j,$$

but using (26.67) for the inverse transformation, we have

$$\mathbf{e}'_i = \frac{\partial u^k}{\partial u'^i} \mathbf{e}_k,$$

and similarly for $\mathbf{e}'_j$. Thus we may write

$$g'_{ij} = \frac{\partial u^k}{\partial u'^i} \frac{\partial u^l}{\partial u'^j} \mathbf{e}_k \cdot \mathbf{e}_l = \frac{\partial u^k}{\partial u'^i} \frac{\partial u^l}{\partial u'^j} g_{kl},$$

which shows that the g_{ij} are indeed the covariant components of a second-order tensor (the metric tensor $\mathbf{g}$). ◀

A similar argument to that used in the above example shows that the quantities g^{ij} form the contravariant components of a second-order tensor which transforms according to

$$g'^{ij} = \frac{\partial u'^i}{\partial u^k} \frac{\partial u'^j}{\partial u^l} g^{kl}.$$

In the previous section we discussed the use of the components g_{ij} and g^{ij} in the raising and lowering of indices in contravariant and covariant vectors. This can be extended to tensors of arbitrary rank. In general, contraction of a tensor with g_{ij} will convert the contracted index from being contravariant (superscript) to covariant (subscript), i.e. it is lowered. This can be repeated for as many indices are required. For example,

$$T_{ij} = g_{ik} T^k{}_j = g_{ik} g_{jl} T^{kl}. \tag{26.72}$$

Similarly contraction with g^{ij} raises an index, i.e.

$$T^{ij} = g^{ik} T_k{}^j = g^{ik} g^{jl} T_{kl}. \tag{26.73}$$

That (26.72) and (26.73) are mutually consistent may be shown by using the fact that $g^{ik} g_{kj} = \delta^i_j$.

26.17 Relative tensors

In section 26.10 we introduced the concept of pseudotensors in the context of the rotation (proper or improper) of a set of Cartesian axes. Generalising to arbitrary coordinate transformations leads to the notion of a *relative tensor*.

For an arbitrary coordinate transformation from one general coordinate system

u^i to another u'^i, we may define the Jacobian of the transformation (see chapter 6) as the determinant of the transformation matrix $[\partial u'^i/\partial u^j]$: this is usually denoted by

$$J = \left| \frac{\partial u'}{\partial u} \right|.$$

Alternatively, we may interchange the primed and unprimed coordinates to obtain $|\partial u/\partial u'| = 1/J$: unfortunately this also is often called the Jacobian of the transformation.

Using the Jacobian J, we define a relative tensor of weight w as one whose components transform as follows:

$$T'^{ij\cdots k}_{\quad lm\cdots n} = \frac{\partial u'^i}{\partial u^a} \frac{\partial u'^j}{\partial u^b} \cdots \frac{\partial u'^k}{\partial u^c} \frac{\partial u^d}{\partial u'^l} \frac{\partial u^e}{\partial u'^m} \cdots \frac{\partial u^f}{\partial u'^n} T^{ab\cdots c}_{\quad de\cdots f} \left| \frac{\partial u}{\partial u'} \right|^w. \tag{26.74}$$

Comparing this expression with (26.71), we see that a true (or *absolute*) general tensor may be considered as a relative tensor of weight $w = 0$. If $w = -1$, on the other hand, the relative tensor is known as a general *pseudotensor* and if $w = 1$ as a *tensor density*.

It is worth comparing (26.74) with the definition (26.39) of a Cartesian pseudotensor. For the latter, we are concerned only with its behaviour under a rotation (proper or improper) of Cartesian axes, for which the Jacobian $J = \pm 1$. Thus, general relative tensors of weight $w = -1$ and $w = 1$ would both satisfy the definition (26.39) of a Cartesian pseudotensor.

> ▶ *If the g_{ij} are the covariant components of the metric tensor, show that the determinant g of the matrix $[g_{ij}]$ is a relative scalar of weight $w = 2$.*

The components g_{ij} transform as

$$g'_{ij} = \frac{\partial u^k}{\partial u'^i} \frac{\partial u^l}{\partial u'^j} g_{kl}.$$

Defining the matrices $\mathsf{U} = [\partial u^i/\partial u'^j]$, $\mathsf{G} = [g_{ij}]$ and $\mathsf{G}' = [g'_{ij}]$, we may write this expression as

$$\mathsf{G}' = \mathsf{U}^\mathsf{T} \mathsf{G} \mathsf{U}.$$

Taking the determinant of both sides, we obtain

$$g' = |\mathsf{U}|^2 g = \left| \frac{\partial u}{\partial u'} \right|^2 g,$$

which shows that g is a relative scalar of weight $w = 2$. ◀

From the discussion in section 26.8, it can be seen that ϵ_{ijk} is a covariant relative tensor of weight -1. We may also define the contravariant tensor ϵ^{ijk}, which is numerically equal to ϵ_{ijk} but is a relative tensor of weight $+1$.

If two relative tensors have weights w_1 and w_2 respectively then, from (26.74), the outer product of the two tensors, or any contraction of them, is a relative tensor of weight $w_1 + w_2$. As a special case, we may use ϵ_{ijk} and ϵ^{ijk} to construct pseudovectors from antisymmetric tensors and vice versa, in an analogous way to that discussed in section 26.11.

For example, if the A^{ij} are the contravariant components of an antisymmetric tensor ($w = 0$) then

$$p_i = \tfrac{1}{2}\epsilon_{ijk}A^{jk}$$

are the covariant components of a pseudovector ($w = -1$), since ϵ_{ijk} has weight $w = -1$. Similarly, we may show that

$$A^{ij} = \epsilon^{ijk}p_k.$$

26.18 Derivatives of basis vectors and Christoffel symbols

In Cartesian coordinates, the basis vectors $\mathbf{e}_i$ are constant and so their derivatives with respect to the coordinates vanish. In a general coordinate system, however, the basis vectors $\mathbf{e}_i$ and $\mathbf{e}^i$ are functions of the coordinates. Therefore, in order that we may differentiate general tensors we must consider the derivatives of the basis vectors.

First consider the derivative $\partial \mathbf{e}_i/\partial u^j$. Since this is itself a vector, it can be written as a linear combination of the basis vectors $\mathbf{e}_k$, $k = 1, 2, 3$. If we introduce the symbol $\Gamma^k{}_{ij}$ to denote the coefficients in this combination, we have

$$\frac{\partial \mathbf{e}_i}{\partial u^j} = \Gamma^k{}_{ij}\mathbf{e}_k. \tag{26.75}$$

The coefficient $\Gamma^k{}_{ij}$ is the kth component of the vector $\partial \mathbf{e}_i/\partial u^j$. Using the reciprocity relation $\mathbf{e}^i \cdot \mathbf{e}_j = \delta^i_j$, these 27 numbers are given (at each point in space) by

$$\Gamma^k{}_{ij} = \mathbf{e}^k \cdot \frac{\partial \mathbf{e}_i}{\partial u^j}. \tag{26.76}$$

Furthermore, by differentiating the reciprocity relation $\mathbf{e}^i \cdot \mathbf{e}_j = \delta^i_j$ with respect to the coordinates, and using (26.76), it is straightforward to show that the derivatives of the contravariant basis vectors are given by

$$\frac{\partial \mathbf{e}^i}{\partial u^j} = -\Gamma^i{}_{kj}\mathbf{e}^k. \tag{26.77}$$

The symbol $\Gamma^k{}_{ij}$ is called a *Christoffel symbol* (of the second kind), but, despite appearances to the contrary, these quantities do *not* form the components of a third-order tensor. It is clear from (26.76) that in Cartesian coordinates $\Gamma^k{}_{ij} = 0$ for all values of the indices i, j and k.

> ►*Using (26.76), deduce the way in which the quantities $\Gamma^k{}_{ij}$ transform under a general coordinate transformation, and hence show that they do not form the components of a third-order tensor.*

In a new coordinate system

$$\Gamma'^k{}_{ij} = \mathbf{e}'^k \cdot \frac{\partial \mathbf{e}'_i}{\partial u'^j},$$

but from (26.69) and (26.67) respectively we have, on reversing primed and unprimed variables,

$$\mathbf{e}'^k = \frac{\partial u'^k}{\partial u^n}\mathbf{e}^n \quad \text{and} \quad \mathbf{e}'_i = \frac{\partial u^l}{\partial u'^i}\mathbf{e}_l.$$

Therefore in the new coordinate system the quantities $\Gamma'^k{}_{ij}$ are given by

$$
\begin{aligned}
\Gamma'^k{}_{ij} &= \frac{\partial u'^k}{\partial u^n}\mathbf{e}^n \cdot \frac{\partial}{\partial u'^j}\left(\frac{\partial u^l}{\partial u'^i}\mathbf{e}_l\right) \\
&= \frac{\partial u'^k}{\partial u^n}\mathbf{e}^n \cdot \left(\frac{\partial^2 u^l}{\partial u'^j \partial u'^i}\mathbf{e}_l + \frac{\partial u^l}{\partial u'^i}\frac{\partial \mathbf{e}_l}{\partial u'^j}\right) \\
&= \frac{\partial u'^k}{\partial u^n}\frac{\partial^2 u^l}{\partial u'^j \partial u'^i}\mathbf{e}^n \cdot \mathbf{e}_l + \frac{\partial u'^k}{\partial u^n}\frac{\partial u^l}{\partial u'^i}\frac{\partial u^m}{\partial u'^j}\mathbf{e}^n \cdot \frac{\partial \mathbf{e}_l}{\partial u^m} \\
&= \frac{\partial u'^k}{\partial u^l}\frac{\partial^2 u^l}{\partial u'^j \partial u'^i} + \frac{\partial u'^k}{\partial u^n}\frac{\partial u^l}{\partial u'^i}\frac{\partial u^m}{\partial u'^j}\Gamma^n{}_{lm},
\end{aligned}
\tag{26.78}
$$

where in the last line we have used (26.76) and the reciprocity relation $\mathbf{e}^n \cdot \mathbf{e}_l = \delta^n_l$. From (26.78), because of the presence of the first term on the right-hand side, we conclude immediately that the $\Gamma^k{}_{ij}$ do not form the components of a third-order tensor. ◄

In a given coordinate system, in principle we may calculate the $\Gamma^k{}_{ij}$ using (26.76). In practice, however, it is often quicker to use an alternative expression, which we now derive, for the Christoffel symbol in terms of the metric tensor g_{ij} and its derivatives with respect to the coordinates.

Firstly we note that the Christoffel symbol $\Gamma^k{}_{ij}$ is symmetric with respect to the interchange of its two subscripts i and j. This is easily shown: since

$$\frac{\partial \mathbf{e}_i}{\partial u^j} = \frac{\partial^2 \mathbf{r}}{\partial u^j \partial u^i} = \frac{\partial^2 \mathbf{r}}{\partial u^i \partial u^j} = \frac{\partial \mathbf{e}_j}{\partial u^i},$$

it follows from (26.75) that $\Gamma^k{}_{ij}\mathbf{e}_k = \Gamma^k{}_{ji}\mathbf{e}_k$. Taking the scalar product with $\mathbf{e}^l$ and using the reciprocity relation $\mathbf{e}_k \cdot \mathbf{e}^l = \delta^l_k$ gives immediately that

$$\Gamma^l{}_{ij} = \Gamma^l{}_{ji}.$$

To obtain an expression for $\Gamma^k{}_{ij}$ we then use $g_{ij} = \mathbf{e}_i \cdot \mathbf{e}_j$ and consider the derivative

$$
\begin{aligned}
\frac{\partial g_{ij}}{\partial u^k} &= \frac{\partial \mathbf{e}_i}{\partial u^k} \cdot \mathbf{e}_j + \mathbf{e}_i \cdot \frac{\partial \mathbf{e}_j}{\partial u^k} \\
&= \Gamma^l{}_{ik}\mathbf{e}_l \cdot \mathbf{e}_j + \mathbf{e}_i \cdot \Gamma^l{}_{jk}\mathbf{e}_l \\
&= \Gamma^l{}_{ik}g_{lj} + \Gamma^l{}_{jk}g_{il},
\end{aligned}
\tag{26.79}
$$

where we have used the definition (26.75). By cyclically permuting the free indices i, j, k in (26.79), we obtain two further equivalent relations,

$$\frac{\partial g_{jk}}{\partial u^i} = \Gamma^l_{ji} g_{lk} + \Gamma^l_{ki} g_{jl} \qquad (26.80)$$

and

$$\frac{\partial g_{ki}}{\partial u^j} = \Gamma^l_{kj} g_{li} + \Gamma^l_{ij} g_{kl}. \qquad (26.81)$$

If we now add (26.80) and (26.81) together and subtract (26.79) from the result, we find

$$\frac{\partial g_{jk}}{\partial u^i} + \frac{\partial g_{ki}}{\partial u^j} - \frac{\partial g_{ij}}{\partial u^k} = \Gamma^l_{ji} g_{lk} + \Gamma^l_{ki} g_{jl} + \Gamma^l_{kj} g_{li} + \Gamma^l_{ij} g_{kl} - \Gamma^l_{ik} g_{lj} - \Gamma^l_{jk} g_{il}$$
$$= 2\Gamma^l_{ij} g_{kl},$$

where we have used the symmetry properties of both Γ^l_{ij} and g_{ij}. Contracting both sides with g^{mk} leads to the required expression for the Christoffel symbol in terms of the metric tensor and its derivatives, namely

$$\Gamma^m_{ij} = \tfrac{1}{2} g^{mk} \left(\frac{\partial g_{jk}}{\partial u^i} + \frac{\partial g_{ki}}{\partial u^j} - \frac{\partial g_{ij}}{\partial u^k} \right). \qquad (26.82)$$

▶*Calculate the Christoffel symbols Γ^m_{ij} for cylindrical polar coordinates.*

We may use either (26.75) or (26.82) to calculate the Γ^m_{ij} for this simple coordinate system. In cylindrical polar coordinates $(u^1, u^2, u^3) = (\rho, \phi, z)$, the basis vectors $\mathbf{e}_i$ are given by (26.59). It is straightforward to show that the only derivatives of those vectors with respect to the coordinates that are non-zero are

$$\frac{\partial \mathbf{e}_\rho}{\partial \phi} = \frac{1}{\rho} \mathbf{e}_\phi, \qquad \frac{\partial \mathbf{e}_\phi}{\partial \rho} = \frac{1}{\rho} \mathbf{e}_\phi, \qquad \frac{\partial \mathbf{e}_\phi}{\partial \phi} = -\rho \mathbf{e}_\rho.$$

Thus, from (26.75), we have immediately that

$$\Gamma^2_{12} = \Gamma^2_{21} = \frac{1}{\rho} \qquad \text{and} \qquad \Gamma^1_{22} = -\rho. \qquad (26.83)$$

Alternatively, using (26.82) and the fact that $g_{11} = 1$, $g_{22} = \rho^2$, $g_{33} = 1$ and the other components are zero, we see that the only three non-zero Christoffel symbols are indeed $\Gamma^2_{12} = \Gamma^2_{21}$ and Γ^1_{22}. These are given by

$$\Gamma^2_{12} = \Gamma^2_{21} = \frac{1}{2g_{22}} \frac{\partial g_{22}}{\partial u^1} = \frac{1}{2\rho^2} \frac{\partial}{\partial \rho} (\rho^2) = \frac{1}{\rho},$$

$$\Gamma^1_{22} = -\frac{1}{2g_{11}} \frac{\partial g_{22}}{\partial u^1} = -\frac{1}{2} \frac{\partial}{\partial \rho} (\rho^2) = -\rho,$$

which agree with the expressions found directly from (26.75) and given in (26.83). ◀

26.19 Covariant differentiation

For Cartesian tensors we noted that the derivative of a scalar is a (covariant) vector. This is also true for *general* tensors, as may be shown by considering the differential of a scalar

$$d\phi = \frac{\partial \phi}{\partial u^i} \, du^i.$$

Since the du^i are the components of a contravariant vector and $d\phi$ is a scalar, we have by the quotient law, discussed in section 26.7, that the quantities $\partial \phi / \partial u^i$ must form the components of a covariant vector. As a second example, if the contravariant components in Cartesian coordinates of a vector $\mathbf{v}$ are v^i, then the quantities $\partial v^i / \partial x^j$ form the components of a second-order tensor.

However, it is straightforward to show that in non-Cartesian coordinates differentiation of the components of a general tensor, other than a scalar, with respect to the coordinates does *not* in general result in the components of another tensor.

▶*Show that, in general coordinates, the quantities $\partial v^i / \partial u^j$ do not form the components of a tensor.*

We may show this directly by considering

$$\left(\frac{\partial v^i}{\partial u^j} \right)' = \frac{\partial v'^i}{\partial u'^j} = \frac{\partial u^k}{\partial u'^j} \frac{\partial v'^i}{\partial u^k}$$

$$= \frac{\partial u^k}{\partial u'^j} \frac{\partial}{\partial u^k} \left(\frac{\partial u'^i}{\partial u^l} v^l \right)$$

$$= \frac{\partial u^k}{\partial u'^j} \frac{\partial u'^i}{\partial u^l} \frac{\partial v^l}{\partial u^k} + \frac{\partial u^k}{\partial u'^j} \frac{\partial^2 u'^i}{\partial u^k \partial u^l} v^l. \tag{26.84}$$

The presence of the second term on the right-hand side of (26.84) shows that the $\partial v^i / \partial x^j$ do not form the components of a second-order tensor. This term arises because the 'transformation matrix' $[\partial u'^i / \partial u^j]$ changes as the position in space at which it is evaluated is changed. This is not true in Cartesian coordinates, for which the second term vanishes and $\partial v^i / \partial x^j$ is a second-order tensor. ◀

We may, however, use the Christoffel symbols discussed in the previous section to define a new *covariant* derivative of the components of a tensor that does result in the components of another tensor.

Let us first consider the derivative of a vector $\mathbf{v}$ with respect to the coordinates. Writing the vector in terms of its contravariant components $\mathbf{v} = v^i \mathbf{e}_i$, we find

$$\frac{\partial \mathbf{v}}{\partial u^j} = \frac{\partial v^i}{\partial u^j} \mathbf{e}_i + v^i \frac{\partial \mathbf{e}_i}{\partial u^j}, \tag{26.85}$$

where the second term arises because, in general, the basis vectors $\mathbf{e}_i$ are not constant (this term vanishes in Cartesian coordinates). Using (26.75) we write

$$\frac{\partial \mathbf{v}}{\partial u^j} = \frac{\partial v^i}{\partial u^j}\mathbf{e}_i + v^i \Gamma^k_{\;ij}\mathbf{e}_k.$$

Since i and k are dummy indices in the last term on the right-hand side, we may interchange them to obtain

$$\frac{\partial \mathbf{v}}{\partial u^j} = \frac{\partial v^i}{\partial u^j}\mathbf{e}_i + v^k \Gamma^i_{\;kj}\mathbf{e}_i = \left(\frac{\partial v^i}{\partial u^j} + v^k \Gamma^i_{\;kj}\right)\mathbf{e}_i. \qquad (26.86)$$

The reason for the interchanging the dummy indices, as shown in (26.86), is that we may now factor out $\mathbf{e}_i$. The quantity in parentheses is called the *covariant derivative*, for which the standard notation is

$$v^i_{\;;j} \equiv \frac{\partial v^i}{\partial u^j} + \Gamma^i_{\;kj}v^k, \qquad (26.87)$$

the semicolon subscript denoting covariant differentiation. A similar short-hand notation also exists for the partial derivatives, a comma being used for these instead of a semicolon; for example, $\partial v^i/\partial u^j$ is denoted by $v^i_{\;,j}$. In Cartesian coordinates all the $\Gamma^i_{\;kj}$ are zero, and so the covariant derivative reduces to the simple partial derivative $\partial v^i/\partial u^j$.

Using the short-hand semicolon notation, the derivative of a vector may be written in the very compact form

$$\frac{\partial \mathbf{v}}{\partial u^j} = v^i_{\;,j}\mathbf{e}_i$$

and, by the quotient rule (section 26.7), it is clear that the $v^i_{\;,j}$ are the (mixed) components of a second-order tensor. This may also be verified directly, using the transformation properties of $\partial v^i/\partial u^j$ and $\Gamma^i_{\;kj}$ given in (26.84) and (26.78) respectively.

In general, we may regard the $v^i_{\;,j}$ as the mixed components of a second-order tensor called the covariant derivative of $\mathbf{v}$ and denoted by $\nabla \mathbf{v}$. In Cartesian coordinates, the components of this tensor are just $\partial v^i/\partial x^j$.

▶*Calculate $v^i_{\;;i}$ in cylindrical polar coordinates.*

Contracting (26.87) we obtain

$$v^i_{\;;i} = \frac{\partial v^i}{\partial u^i} + \Gamma^i_{\;ki}v^k.$$

Now from (26.83) we have

$$\Gamma^i_{\;1i} = \Gamma^1_{\;11} + \Gamma^2_{\;12} + \Gamma^3_{\;13} = 1/\rho,$$
$$\Gamma^i_{\;2i} = \Gamma^1_{\;21} + \Gamma^2_{\;22} + \Gamma^3_{\;23} = 0,$$
$$\Gamma^i_{\;3i} = \Gamma^1_{\;31} + \Gamma^2_{\;32} + \Gamma^3_{\;33} = 0,$$

and so

$$
\begin{aligned}
v^i{}_{;i} &= \frac{\partial v^\rho}{\partial \rho} + \frac{\partial v^\phi}{\partial \phi} + \frac{\partial v^z}{\partial z} + \frac{1}{\rho} v^\rho \\
&= \frac{1}{\rho} \frac{\partial}{\partial \rho} (\rho v^\rho) + \frac{\partial v^\phi}{\partial \phi} + \frac{\partial v^z}{\partial z}.
\end{aligned}
$$

This result is identical to the expression for the divergence of a vector field in cylindrical polar coordinates given in section 10.9. This is discussed further in section 26.20. ◀

So far we have considered only the covariant derivative of the contravariant components v^i of a vector. The corresponding result for the covariant components v_i may be found in a similar way, by considering the derivative of $\mathbf{v} = v_i \mathbf{e}^i$ and using (26.77) to obtain

$$
v_{i;j} = \frac{\partial v_i}{\partial u^j} - \Gamma^k{}_{ij} v_k. \tag{26.88}
$$

Comparing the expressions (26.87) and (26.88) for the covariant derivative of the contravariant and covariant components of a vector respectively, we see that there are some similarities and some differences. It may help to remember that the index with respect to which the covariant derivative is taken (j in this case), is also the last subscript on the Christoffel symbol; the remaining indices can then be arranged in only one way without raising or lowering them. It only remains to note that for a covariant index (subscript) the Christoffel symbol carries a minus sign, whereas for a contravariant index (superscript) the sign is positive.

Following a similar procedure to that which led to equation (26.87), we may obtain expressions for the covariant derivatives of higher-order tensors.

▶*By considering the derivative of the second-order tensor* $\mathbf{T}$ *with respect to the coordinate* u^k, *find an expression for the covariant derivative* $T^{ij}{}_{;k}$ *of its contravariant components.*

Expressing $\mathbf{T}$ in terms of its contravariant components, we have

$$
\begin{aligned}
\frac{\partial \mathbf{T}}{\partial u^k} &= \frac{\partial}{\partial u^k} (T^{ij} \mathbf{e}_i \otimes \mathbf{e}_j) \\
&= \frac{\partial T^{ij}}{\partial u^k} \mathbf{e}_i \otimes \mathbf{e}_j + T^{ij} \frac{\partial \mathbf{e}_i}{\partial u^k} \otimes \mathbf{e}_j + T^{ij} \mathbf{e}_i \otimes \frac{\partial \mathbf{e}_j}{\partial u^k}.
\end{aligned}
$$

Using (26.75), we can rewrite the derivatives of the basis vectors in terms of Christoffel symbols to obtain

$$
\frac{\partial \mathbf{T}}{\partial u^k} = \frac{\partial T^{ij}}{\partial u^k} \mathbf{e}_i \otimes \mathbf{e}_j + T^{ij} \Gamma^l{}_{ik} \mathbf{e}_l \otimes \mathbf{e}_j + T^{ij} \mathbf{e}_i \otimes \Gamma^l{}_{jk} \mathbf{e}_l.
$$

Interchanging the dummy indices i and l in the second term and j and l in the third term on the right-hand side, this becomes

$$
\frac{\partial \mathbf{T}}{\partial u^k} = \left(\frac{\partial T^{ij}}{\partial u^k} + \Gamma^i{}_{lk} T^{lj} + \Gamma^j{}_{lk} T^{il} \right) \mathbf{e}_i \otimes \mathbf{e}_j,
$$

where the expression in parentheses is the required covariant derivative

$$T^{ij}{}_{;k} = \frac{\partial T^{ij}}{\partial u^k} + \Gamma^i{}_{lk} T^{lj} + \Gamma^j{}_{lk} T^{il}. \tag{26.89}$$

Using (26.89), the derivative of the tensor $\mathbf{T}$ with respect to u^k can now be written in terms of its contravariant components as

$$\frac{\partial \mathbf{T}}{\partial u^k} = T^{ij}{}_{;k} \mathbf{e}_i \otimes \mathbf{e}_j. \blacktriangleleft$$

Results similar to (26.89) may be obtained for the the covariant derivatives of the mixed and covariant components of a second-order tensor. Collecting these results together, we have

$$T^{ij}{}_{;k} = T^{ij}{}_{,k} + \Gamma^i{}_{lk} T^{lj} + \Gamma^j{}_{lk} T^{il},$$
$$T^i{}_{j;k} = T^i{}_{j,k} + \Gamma^i{}_{lk} T^l{}_j - \Gamma^l{}_{jk} T^i{}_l,$$
$$T_{ij;k} = T_{ij,k} - \Gamma^l{}_{ik} T_{lj} - \Gamma^l{}_{jk} T_{il},$$

where we have used the comma notation for partial derivatives. The position of the indices in these expressions is very systematic: for each contravariant index (superscript) on the LHS we add a term on the RHS containing a Christoffel symbol with a plus sign, and for every covariant index (subscript) we add a corresponding term with a minus sign. This is extended straightforwardly to tensors with an arbitrary number of contravariant and covariant indices.

We note that the quantities $T^{ij}{}_{;k}$, $T^i{}_{j;k}$ and $T_{ij;k}$ are the components of the *same* third-order tensor $\nabla \mathbf{T}$ with respect to different tensor bases, i.e.

$$\nabla \mathbf{T} = T^{ij}{}_{;k} \mathbf{e}_i \otimes \mathbf{e}_j \otimes \mathbf{e}^k = T^i{}_{j;k} \mathbf{e}_i \otimes \mathbf{e}^j \otimes \mathbf{e}^k = T_{ij;k} \mathbf{e}^i \otimes \mathbf{e}^j \otimes \mathbf{e}^k.$$

We conclude this section by considering briefly the covariant derivative of a scalar. The covariant derivative differs from the simple partial derivative with respect to the coordinates only because the basis vectors of the coordinate system change with position in space (hence for Cartesian coordinates there is no difference). However, a scalar ϕ does not depend on the basis vectors at all and so its covariant derivative must be the same as its partial derivative, i.e.

$$\phi_{;j} = \frac{\partial \phi}{\partial u^j} = \phi_{,j}. \tag{26.90}$$

26.20 Vector operators in tensor form

In section 10.10 we used vector calculus methods to find expressions for vector differential operators, such as grad, div, curl and the Laplacian, in general *orthogonal* curvilinear coordinates, taking cylindrical and spherical polars as particular examples. In this section we use the framework of general tensors that we have developed to obtain, in tensor form, expressions for these operators that are valid in *all* coordinate systems, whether orthogonal or not.

971

In order to compare the results obtained here with those given in section 10.10 for orthogonal coordinates, it is necessary to remember that here we are working with the (in general) non-unit basis vectors $\mathbf{e}_i = \partial \mathbf{r}/\partial u^i$ or $\mathbf{e}^i = \nabla u^i$. Thus the components of a vector $\mathbf{v} = v^i \mathbf{e}_i$ are not the same as the components $\hat{v}^i$ appropriate to the corresponding unit basis $\hat{\mathbf{e}}_i$. In fact, if the scale factors of the coordinate system are h_i, $i = 1, 2, 3$, then $v^i = \hat{v}^i/h_i$ (no summation over i).

As mentioned in section 26.15, for an orthogonal coordinate system with scale factors h_i we have

$$
g_{ij} = \begin{cases} h_i^2 & \text{if } i = j, \\ 0 & \text{otherwise} \end{cases} \qquad \text{and} \qquad g^{ij} = \begin{cases} 1/h_i^2 & \text{if } i = j, \\ 0 & \text{otherwise,} \end{cases}
$$

and so the determinant g of the matrix $[g_{ij}]$ is given by $g = h_1^2 h_2^2 h_3^2$.

Gradient

The gradient of a scalar ϕ is given by

$$
\nabla \phi = \phi_{;i} \mathbf{e}^i = \frac{\partial \phi}{\partial u^i} \mathbf{e}^i, \tag{26.91}
$$

since the covariant derivative of a scalar is the same as its partial derivative.

Divergence

Replacing the partial derivatives that occur in Cartesian coordinates with covariant derivatives, the divergence of a vector field $\mathbf{v}$ in a general coordinate system is given by

$$
\nabla \cdot \mathbf{v} = v^i_{;i} = \frac{\partial v^i}{\partial u^i} + \Gamma^i_{ki} v^k.
$$

Using the expression (26.82) for the Christoffel symbol in terms of the metric tensor, we find

$$
\Gamma^i_{ki} = \tfrac{1}{2} g^{il} \left(\frac{\partial g_{il}}{\partial u^k} + \frac{\partial g_{kl}}{\partial u^i} - \frac{\partial g_{ki}}{\partial u^l} \right) = \tfrac{1}{2} g^{il} \frac{\partial g_{il}}{\partial u^k}. \tag{26.92}
$$

The last two terms have cancelled because

$$
g^{il} \frac{\partial g_{kl}}{\partial u^i} = g^{li} \frac{\partial g_{ki}}{\partial u^l} = g^{il} \frac{\partial g_{ki}}{\partial u^l},
$$

where in the first equality we have interchanged the dummy indices i and l, and in the second equality have used the symmetry of the metric tensor.

We may simplify (26.92) still further by using a result concerning the derivative of the determinant of a matrix whose elements are functions of the coordinates.

▶*Suppose* $\mathsf{A} = [a_{ij}]$, $\mathsf{B} = [b^{ij}]$ *and that* $\mathsf{B} = \mathsf{A}^{-1}$. *By considering the determinant* $a = |\mathsf{A}|$, *show that*

$$\frac{\partial a}{\partial u^k} = ab^{ji}\frac{\partial a_{ij}}{\partial u^k}.$$

If we denote the cofactor of the element a_{ij} by Δ^{ij} then the elements of the inverse matrix are given by (see chapter 8)

$$b^{ij} = \frac{1}{a}\Delta^{ji}. \tag{26.93}$$

However, the determinant of A is given by

$$a = \sum_j a_{ij}\Delta^{ij},$$

in which we have *fixed* i and written the sum over j explicitly, for clarity. Partially differentiating both sides with respect to a_{ij}, we then obtain

$$\frac{\partial a}{\partial a_{ij}} = \Delta^{ij}, \tag{26.94}$$

since a_{ij} does not occur in any of the cofactors Δ^{ij}.

Now, if the a_{ij} depend on the coordinates then so will the determinant a and, by the chain rule, we have

$$\frac{\partial a}{\partial u^k} = \frac{\partial a}{\partial a_{ij}}\frac{\partial a_{ij}}{\partial u^k} = \Delta^{ij}\frac{\partial a_{ij}}{\partial u^k} = ab^{ji}\frac{\partial a_{ij}}{\partial u^k}, \tag{26.95}$$

in which we have used (26.93) and (26.94). ◀

Applying the result (26.95) to the determinant g of the metric tensor, and remembering both that $g^{ik}g_{kj} = \delta^i_j$ and that g^{ij} is symmetric, we obtain

$$\frac{\partial g}{\partial u^k} = gg^{ij}\frac{\partial g_{ij}}{\partial u^k}. \tag{26.96}$$

Substituting (26.96) into (26.92) we find that the expression for the Christoffel symbol can be much simplified to give

$$\Gamma^i_{ki} = \frac{1}{2g}\frac{\partial g}{\partial u^k} = \frac{1}{\sqrt{g}}\frac{\partial\sqrt{g}}{\partial u^k}.$$

Thus finally we obtain the expression for the divergence of a vector field in a general coordinate system as

$$\nabla \cdot \mathbf{v} = v^i{}_{;i} = \frac{1}{\sqrt{g}}\frac{\partial}{\partial u^j}(\sqrt{g}v^j). \tag{26.97}$$

Laplacian

If we replace $\mathbf{v}$ by $\nabla\phi$ in $\nabla \cdot \mathbf{v}$ then we obtain the Laplacian $\nabla^2\phi$. From (26.91), we have

$$v_i\mathbf{e}^i = \mathbf{v} = \nabla\phi = \frac{\partial\phi}{\partial u^i}\mathbf{e}^i,$$

and so the covariant components of $\mathbf{v}$ are given by $v_i = \partial\phi/\partial u^i$. In (26.97), however, we require the contravariant components v^i. These may be obtained by raising the index using the metric tensor, to give

$$v^j = g^{jk}v_k = g^{jk}\frac{\partial\phi}{\partial u^k}.$$

Substituting this into (26.97) we obtain

$$\nabla^2\phi = \frac{1}{\sqrt{g}}\frac{\partial}{\partial u^j}\left(\sqrt{g}g^{jk}\frac{\partial\phi}{\partial u^k}\right). \tag{26.98}$$

▶ *Use (26.98) to find the expression for $\nabla^2\phi$ in an orthogonal coordinate system with scale factors h_i, $i = 1, 2, 3$.*

For an orthogonal coordinate system $\sqrt{g} = h_1 h_2 h_3$; further, $g^{ij} = 1/h_i^2$ if $i = j$ and $g^{ij} = 0$ otherwise. Therefore, from (26.98) we have

$$\nabla^2\phi = \frac{1}{h_1 h_2 h_3}\frac{\partial}{\partial u^j}\left(\frac{h_1 h_2 h_3}{h_j^2}\frac{\partial\phi}{\partial u^j}\right),$$

which agrees with the results of section 10.10. ◀

Curl

The special vector form of the curl of a vector field exists only in three dimensions. We therefore consider a more general form valid in higher-dimensional spaces as well. In a general space the operation curl $\mathbf{v}$ is defined by

$$(\text{curl } \mathbf{v})_{ij} = v_{i;j} - v_{j;i},$$

which is an antisymmetric covariant tensor.

In fact the difference of derivatives can be simplified, since

$$v_{i;j} - v_{j;i} = \frac{\partial v_i}{\partial u^j} - \Gamma^l{}_{ij}v_l - \frac{\partial v_j}{\partial u^i} + \Gamma^l{}_{ji}v_l$$

$$= \frac{\partial v_i}{\partial u^j} - \frac{\partial v_j}{\partial u^i},$$

where the Christoffel symbols have cancelled because of their symmetry properties. Thus curl $\mathbf{v}$ can be written in terms of partial derivatives as

$$(\text{curl } \mathbf{v})_{ij} = \frac{\partial v_i}{\partial u^j} - \frac{\partial v_j}{\partial u^i}.$$

Generalising slightly the discussion of section 26.17, in three dimensions we may associate with this antisymmetric second-order tensor a vector with contravariant components,

$$(\nabla \times \mathbf{v})^i = -\frac{1}{2\sqrt{g}}\epsilon^{ijk}(\text{curl } \mathbf{v})_{jk}$$

$$= -\frac{1}{2\sqrt{g}}\epsilon^{ijk}\left(\frac{\partial v_j}{\partial u^k} - \frac{\partial v_k}{\partial u^j}\right) = \frac{1}{\sqrt{g}}\epsilon^{ijk}\frac{\partial v_k}{\partial u^j};$$

this is the analogue of the expression in Cartesian coordinates discussed in section 26.8.

26.21 Absolute derivatives along curves

In section 26.19 we discussed how to differentiate a general tensor with respect to the coordinates and introduced the covariant derivative. In this section we consider the slightly different problem of calculating the derivative of a tensor along a curve $\mathbf{r}(t)$ that is parameterised by some variable t.

Let us begin by considering the derivative of a vector $\mathbf{v}$ along the curve. If we introduce an arbitrary coordinate system u^i with basis vectors $\mathbf{e}_i$, $i = 1, 2, 3$, then we may write $\mathbf{v} = v^i \mathbf{e}_i$ and so obtain

$$\frac{d\mathbf{v}}{dt} = \frac{dv^i}{dt} \mathbf{e}_i + v^i \frac{d\mathbf{e}_i}{dt}$$
$$= \frac{dv^i}{dt} \mathbf{e}_i + v^i \frac{\partial \mathbf{e}_i}{\partial u^k} \frac{du^k}{dt} ;$$

here the chain rule has been used to rewrite the last term on the right-hand side. Using (26.75) to write the derivatives of the basis vectors in terms of Christoffel symbols, we obtain

$$\frac{d\mathbf{v}}{dt} = \frac{dv^i}{dt} \mathbf{e}_i + \Gamma^j{}_{ik} v^i \frac{du^k}{dt} \mathbf{e}_j.$$

Interchanging the dummy indices i and j in the last term, we may factor out the basis vector and find

$$\frac{d\mathbf{v}}{dt} = \left(\frac{dv^i}{dt} + \Gamma^i{}_{jk} v^j \frac{du^k}{dt} \right) \mathbf{e}_i.$$

The term in parentheses is called the *absolute* (or *intrinsic*) derivative of the components v^i along the curve $\mathbf{r}(t)$ and is usually denoted by

$$\frac{\delta v^i}{\delta t} \equiv \frac{dv^i}{dt} + \Gamma^i{}_{jk} v^j \frac{du^k}{dt} = v^i{}_{;k} \frac{du^k}{dt}.$$

With this notation, we may write

$$\frac{d\mathbf{v}}{dt} = \frac{\delta v^i}{\delta t} \mathbf{e}_i = v^i{}_{;k} \frac{du^k}{dt} \mathbf{e}_i. \tag{26.99}$$

Using the same method, the absolute derivative of the covariant components v_i of a vector is given by

$$\frac{\delta v_i}{\delta t} \equiv v_{i;k} \frac{du^k}{dt}.$$

Similarly, the absolute derivatives of the contravariant, mixed and covariant

975

components of a second-order tensor $\mathbf{T}$ are

$$\frac{\delta T^{ij}}{\delta t} \equiv T^{ij}_{\ ;k}\frac{du^k}{dt},$$

$$\frac{\delta T^i_{\ j}}{\delta t} \equiv T^i_{\ j;k}\frac{du^k}{dt},$$

$$\frac{\delta T_{ij}}{\delta t} \equiv T_{ij;k}\frac{du^k}{dt}.$$

The derivative of $\mathbf{T}$ along the curve $\mathbf{r}(t)$ may then be written in terms of, for example, its contravariant components as

$$\frac{d\mathbf{T}}{dt} = \frac{\delta T^{ij}}{\delta t}\mathbf{e}_i \otimes \mathbf{e}_j = T^{ij}_{\ ;k}\frac{du^k}{dt}\mathbf{e}_i \otimes \mathbf{e}_j.$$

26.22 Geodesics

As an example of the use of the absolute derivative, we conclude this chapter with a brief discussion of geodesics. A geodesic in real three-dimensional space is a straight line, which has two equivalent defining properties. Firstly, it is the curve of shortest length between two points and, secondly, it is the curve whose tangent vector always points in the same direction (along the line). Although in this chapter we have considered explicitly only our familiar three-dimensional space, much of the mathematical formalism developed can be generalised to more abstract spaces of higher dimensionality in which the familiar ideas of Euclidean geometry are no longer valid. It is often of interest to find geodesic curves in such spaces by using the defining properties of straight lines in Euclidean space.

We shall not consider these more complicated spaces explicitly but will determine the equation that a geodesic in Euclidean three-dimensional space (i.e. a straight line) must satisfy, deriving it in a sufficiently general way that our method may be applied with little modification to finding the equations satisfied by geodesics in more abstract spaces.

Let us consider a curve $\mathbf{r}(s)$, parameterised by the arc length s from some point on the curve, and choose as our defining property for a geodesic that its tangent vector $\mathbf{t} = d\mathbf{r}/ds$ always points in the same direction everywhere on the curve, i.e.

$$\frac{d\mathbf{t}}{ds} = \mathbf{0}. \tag{26.100}$$

Alternatively, we could exploit the property that the distance between two points is a minimum along a geodesic and use the calculus of variations (see chapter 22); this would lead to the same final result (26.101).

If we now introduce an arbitrary coordinate system u^i with basis vectors $\mathbf{e}_i$, $i = 1, 2, 3$, then we may write $\mathbf{t} = t^i\mathbf{e}_i$, and from (26.99) we find

$$\frac{d\mathbf{t}}{ds} = t^i_{\ ;k}\frac{du^k}{ds}\mathbf{e}_i = \mathbf{0}.$$

Writing out the covariant derivative, we obtain

$$\left(\frac{dt^i}{ds} + \Gamma^i{}_{jk} t^j \frac{du^k}{ds} \right) \mathbf{e}_i = \mathbf{0}.$$

But, since $t^j = du^j/ds$, it follows that the equation satisfied by a geodesic is

$$\frac{d^2 u^i}{ds^2} + \Gamma^i{}_{jk} \frac{du^j}{ds} \frac{du^k}{ds} = 0. \tag{26.101}$$

▶ *Find the equations satisfied by a geodesic (straight line) in cylindrical polar coordinates.*

From (26.83), the only non-zero Christoffel symbols are $\Gamma^1{}_{22} = -\rho$ and $\Gamma^2{}_{12} = \Gamma^2{}_{21} = 1/\rho$. Thus the required geodesic equations are

$$\frac{d^2 u^1}{ds^2} + \Gamma^1{}_{22} \frac{du^2}{ds} \frac{du^2}{ds} = 0 \qquad \Rightarrow \qquad \frac{d^2 \rho}{ds^2} - \rho \left(\frac{d\phi}{ds} \right)^2 = 0,$$

$$\frac{d^2 u^2}{ds^2} + 2\Gamma^2{}_{12} \frac{du^1}{ds} \frac{du^2}{ds} = 0 \qquad \Rightarrow \qquad \frac{d^2 \phi}{ds^2} + \frac{2}{\rho} \frac{d\rho}{ds} \frac{d\phi}{ds} = 0,$$

$$\frac{d^2 u^3}{ds^2} = 0 \qquad \Rightarrow \qquad \frac{d^2 z}{ds^2} = 0. \ ◀$$

26.23 Exercises

26.1   Use the basic definition of a Cartesian tensor to show the following.

(a) That for any general, but fixed, ϕ,

$$(u_1, u_2) = (x_1 \cos \phi - x_2 \sin \phi, \ x_1 \sin \phi + x_2 \cos \phi)$$

are the components of a first-order tensor in two dimensions.

(b) That

$$\begin{pmatrix} x_2^2 & x_1 x_2 \\ x_1 x_2 & x_1^2 \end{pmatrix}$$

is not a tensor of order 2. To establish that a single element does not transform correctly is sufficient.

26.2   The components of two vectors, $\mathbf{A}$ and $\mathbf{B}$, and a second-order tensor, $\mathbf{T}$, are given in one coordinate system by

$$A = \begin{pmatrix} 1 \\ 0 \\ 0 \end{pmatrix}, \qquad B = \begin{pmatrix} 0 \\ 1 \\ 0 \end{pmatrix}, \qquad T = \begin{pmatrix} 2 & \sqrt{3} & 0 \\ \sqrt{3} & 4 & 0 \\ 0 & 0 & 2 \end{pmatrix}.$$

In a second coordinate system, obtained from the first by rotation, the components of $\mathbf{A}$ and $\mathbf{B}$ are

$$A' = \frac{1}{2} \begin{pmatrix} \sqrt{3} \\ 0 \\ 1 \end{pmatrix}, \qquad B' = \frac{1}{2} \begin{pmatrix} -1 \\ 0 \\ \sqrt{3} \end{pmatrix}.$$

Find the components of $\mathbf{T}$ in this new coordinate system and hence evaluate, with a minimum of calculation,

$$T_{ij} T_{ji}, \qquad T_{ki} T_{jk} T_{ij}, \qquad T_{ik} T_{mn} T_{ni} T_{km}.$$

26.3 In section 26.3 the transformation matrix for a rotation of the coordinate axes was derived, and this approach is used in the rest of the chapter. An alternative view is that of taking the coordinate axes as fixed and rotating the components of the system; this is equivalent to reversing the signs of all rotation angles.

Using this alternative view, determine the matrices representing (a) a positive rotation of $\pi/4$ about the x-axis and (b) a rotation of $-\pi/4$ about the y-axis. Determine the initial vector $\mathbf{r}$ which, when subjected to (a) followed by (b), finishes at $(3, 2, 1)$.

26.4 Show how to decompose the Cartesian tensor T_{ij} into three tensors,

$$T_{ij} = U_{ij} + V_{ij} + S_{ij},$$

where U_{ij} is symmetric and has zero trace, V_{ij} is isotropic and S_{ij} has only three independent components.

26.5 Use the quotient law discussed in section 26.7 to show that the array

$$\begin{pmatrix} y^2 + z^2 - x^2 & -2xy & -2xz \\ -2yx & x^2 + z^2 - y^2 & -2yz \\ -2zx & -2zy & x^2 + y^2 - z^2 \end{pmatrix}$$

forms a second-order tensor.

26.6 Use tensor methods to establish the following vector identities:

(a) $(\mathbf{u} \times \mathbf{v}) \times \mathbf{w} = (\mathbf{u} \cdot \mathbf{w})\mathbf{v} - (\mathbf{v} \cdot \mathbf{w})\mathbf{u}$;
(b) $\operatorname{curl}(\phi\mathbf{u}) = \phi \operatorname{curl}\mathbf{u} + (\operatorname{grad}\phi) \times \mathbf{u}$;
(c) $\operatorname{div}(\mathbf{u} \times \mathbf{v}) = \mathbf{v} \cdot \operatorname{curl}\mathbf{u} - \mathbf{u} \cdot \operatorname{curl}\mathbf{v}$;
(d) $\operatorname{curl}(\mathbf{u} \times \mathbf{v}) = (\mathbf{v} \cdot \operatorname{grad})\mathbf{u} - (\mathbf{u} \cdot \operatorname{grad})\mathbf{v} + \mathbf{u}\operatorname{div}\mathbf{v} - \mathbf{v}\operatorname{div}\mathbf{u}$;
(e) $\operatorname{grad}\frac{1}{2}(\mathbf{u} \cdot \mathbf{u}) = \mathbf{u} \times \operatorname{curl}\mathbf{u} + (\mathbf{u} \cdot \operatorname{grad})\mathbf{u}$.

26.7 Use result (e) of the previous question and the general divergence theorem for tensors to show that, for a vector field $\mathbf{A}$,

$$\int_S \left[\mathbf{A}(\mathbf{A} \cdot d\mathbf{S}) - \tfrac{1}{2}A^2 d\mathbf{S}\right] = \int_V [\mathbf{A}\operatorname{div}\mathbf{A} - \mathbf{A} \times \operatorname{curl}\mathbf{A}]\, dV,$$

where S is the surface enclosing volume V.

26.8 A column matrix a has components a_x, a_y, a_z and A is the matrix with elements $A_{ij} = -\epsilon_{ijk}a_k$.

(a) What is the relationship between column matrices b and c if Ab = c?
(b) Find the eigenvalues of A and show that a is one of its eigenvectors. Explain why this must be so.

26.9 Equation (26.29),

$$|A|\epsilon_{lmn} = A_{li}A_{mj}A_{nk}\epsilon_{ijk},$$

is a more general form of the expression (8.47) for the determinant of a 3×3 matrix A. The latter could have been written as

$$|A| = \epsilon_{ijk}A_{i1}A_{j2}A_{k3},$$

whilst the former removes the explicit mention of $1, 2, 3$ at the expense of an additional Levi–Civita symbol. As stated in the footnote on p. 942, (26.29) can be readily extended to cover a general $N \times N$ matrix.

Use the form given in (26.29) to prove properties (i), (iii), (v), (vi) and (vii) of determinants stated in subsection 8.9.1. Property (iv) is obvious by inspection. For definiteness take $N = 3$, but convince yourself that your methods of proof would be valid for any positive integer N.

26.10   A symmetric second-order Cartesian tensor is defined by

$$T_{ij} = \delta_{ij} - 3x_i x_j.$$

Evaluate the following surface integrals, each taken over the surface of the unit sphere:

$$(a) \int T_{ij}\, dS; \quad (b) \int T_{ik} T_{kj}\, dS; \quad (c) \int x_i T_{jk}\, dS.$$

26.11   Given a non-zero vector $\mathbf{v}$, find the value that should be assigned to α to make

$$P_{ij} = \alpha v_i v_j \quad \text{and} \quad Q_{ij} = \delta_{ij} - \alpha v_i v_j$$

into parallel and orthogonal projection tensors, respectively, i.e. tensors that satisfy, respectively, $P_{ij}v_j = v_i$, $P_{ij}u_j = 0$ and $Q_{ij}v_j = 0$, $Q_{ij}u_j = u_i$, for any vector $\mathbf{u}$ that is orthogonal to $\mathbf{v}$.

Show, in particular, that Q_{ij} is unique, i.e. that if another tensor T_{ij} has the same properties as Q_{ij} then $(Q_{ij} - T_{ij})w_j = 0$ for *any* vector $\mathbf{w}$.

26.12   In four dimensions, define second-order antisymmetric tensors, F_{ij} and Q_{ij}, and a first-order tensor, S_i, as follows:

(a) $F_{23} = H_1$, $Q_{23} = B_1$ and their cyclic permutations;
(b) $F_{i4} = -D_i$, $Q_{i4} = E_i$ for $i = 1, 2, 3$;
(c) $S_4 = \rho$, $S_i = J_i$ for $i = 1, 2, 3$.

Then, taking x_4 as t and the other symbols to have their usual meanings in electromagnetic theory, show that the equations $\sum_j \partial F_{ij}/\partial x_j = S_i$ and $\partial Q_{jk}/\partial x_i + \partial Q_{ki}/\partial x_j + \partial Q_{ij}/\partial x_k = 0$ reproduce Maxwell's equations. In the latter i, j, k is any set of three subscripts selected from 1, 2, 3, 4, but chosen in such a way that they are all different.

26.13   In a certain crystal the unit cell can be taken as six identical atoms lying at the corners of a regular octahedron. Convince yourself that these atoms can also be considered as lying at the centres of the faces of a cube and hence that the crystal has cubic symmetry. Use this result to prove that the conductivity tensor for the crystal, σ_{ij}, must be isotropic.

26.14   Assuming that the current density $\mathbf{j}$ and the electric field $\mathbf{E}$ appearing in equation (26.44) are first-order Cartesian tensors, show explicitly that the electrical conductivity tensor σ_{ij} transforms according to the law appropriate to a second-order tensor.

The rate W at which energy is dissipated per unit volume, as a result of the current flow, is given by $\mathbf{E} \cdot \mathbf{j}$. Determine the limits between which W must lie for a given value of $|\mathbf{E}|$ as the direction of $\mathbf{E}$ is varied.

26.15   In a certain system of units, the electromagnetic stress tensor M_{ij} is given by

$$M_{ij} = E_i E_j + B_i B_j - \tfrac{1}{2}\delta_{ij}(E_k E_k + B_k B_k),$$

where the electric and magnetic fields, $\mathbf{E}$ and $\mathbf{B}$, are first-order tensors. Show that M_{ij} is a second-order tensor.

Consider a situation in which $|\mathbf{E}| = |\mathbf{B}|$, but the directions of $\mathbf{E}$ and $\mathbf{B}$ are not parallel. Show that $\mathbf{E} \pm \mathbf{B}$ are principal axes of the stress tensor and find the corresponding principal values. Determine the third principal axis and its corresponding principal value.

26.16   A rigid body consists of four particles of masses $m, 2m, 3m, 4m$, respectively situated at the points (a, a, a), $(a, -a, -a)$, $(-a, a, -a)$, $(-a, -a, a)$ and connected together by a light framework.

(a)   Find the inertia tensor at the origin and show that the principal moments of inertia are $20ma^2$ and $(20 \pm 2\sqrt{5})ma^2$.

979

(b) Find the principal axes and verify that they are orthogonal.

26.17   A rigid body consists of eight particles, each of mass m, held together by light rods. In a certain coordinate frame the particles are at positions

$$\pm a(3, 1, -1), \quad \pm a(1, -1, 3), \quad \pm a(1, 3, -1), \quad \pm a(-1, 1, 3).$$

Show that, when the body rotates about an axis through the origin, if the angular velocity and angular momentum vectors are parallel then their ratio must be $40ma^2$, $64ma^2$ or $72ma^2$.

26.18   The paramagnetic tensor χ_{ij} of a body placed in a magnetic field, in which its energy density is $-\frac{1}{2}\mu_0 \mathbf{M} \cdot \mathbf{H}$ with $M_i = \sum_j \chi_{ij} H_j$, is

$$\begin{pmatrix} 2k & 0 & 0 \\ 0 & 3k & k \\ 0 & k & 3k \end{pmatrix}.$$

Assuming depolarizing effects are negligible, find how the body will orientate itself if the field is horizontal, in the following circumstances:

(a) the body can rotate freely;

(b) the body is suspended with the $(1, 0, 0)$ axis vertical;

(c) the body is suspended with the $(0, 1, 0)$ axis vertical.

26.19   A block of wood contains a number of thin soft-iron nails (of constant permeability). A unit magnetic field directed eastwards induces a magnetic moment in the block having components $(3, 1, -2)$, and similar fields directed northwards and vertically upwards induce moments $(1, 3, -2)$ and $(-2, -2, 2)$ respectively. Show that all the nails lie in parallel planes.

26.20   For tin, the conductivity tensor is diagonal, with entries a, a, and b when referred to its crystal axes. A single crystal is grown in the shape of a long wire of length L and radius r, the axis of the wire making polar angle θ with respect to the crystal's 3-axis. Show that the resistance of the wire is $L(\pi r^2 ab)^{-1} \left(a\cos^2\theta + b\sin^2\theta\right)$.

26.21   By considering an isotropic body subjected to a uniform hydrostatic pressure (no shearing stress), show that the bulk modulus k, defined by the ratio of the pressure to the fractional decrease in volume, is given by $k = E/[3(1 - 2\sigma)]$ where E is Young's modulus and σ is Poisson's ratio.

26.22   For an isotropic elastic medium under dynamic stress, at time t the displacement u_i and the stress tensor p_{ij} satisfy

$$p_{ij} = c_{ijkl} \left(\frac{\partial u_k}{\partial x_l} + \frac{\partial u_l}{\partial x_k} \right) \quad \text{and} \quad \frac{\partial p_{ij}}{\partial x_j} = \rho \frac{\partial^2 u_i}{\partial t^2},$$

where c_{ijkl} is the isotropic tensor given in equation (26.47) and ρ is a constant. Show that both $\nabla \cdot \mathbf{u}$ and $\nabla \times \mathbf{u}$ satisfy wave equations and find the corresponding wave speeds.

26.23   A fourth-order tensor T_{ijkl} has the properties

$$T_{jikl} = -T_{ijkl}, \qquad T_{ijlk} = -T_{ijkl}.$$

Prove that for any such tensor there exists a second-order tensor K_{mn} such that

$$T_{ijkl} = \epsilon_{ijm}\epsilon_{kln}K_{mn}$$

and give an explicit expression for K_{mn}. Consider two (separate) special cases, as follows.

(a) Given that T_{ijkl} is isotropic and $T_{ijji} = 1$, show that T_{ijkl} is uniquely determined and express it in terms of Kronecker deltas.

(b) If now T_{ijkl} has the additional property

$$T_{klij} = -T_{ijkl},$$

show that T_{ijkl} has only three linearly independent components and find an expression for T_{ijkl} in terms of the vector

$$V_i = -\tfrac{1}{4}\epsilon_{jkl}T_{ijkl}.$$

26.24   Working in cylindrical polar coordinates ρ, ϕ, z, parameterise the straight line (geodesic) joining $(1, 0, 0)$ to $(1, \pi/2, 1)$ in terms of s, the distance along the line. Show by substitution that the geodesic equations, derived at the end of section 26.22, are satisfied.

26.25   In a general coordinate system u^i, $i = 1, 2, 3$, in three-dimensional Euclidean space, a volume element is given by

$$dV = |\mathbf{e}_1 \, du^1 \cdot (\mathbf{e}_2 \, du^2 \times \mathbf{e}_3 \, du^3)|.$$

Show that an alternative form for this expression, written in terms of the determinant g of the metric tensor, is given by

$$dV = \sqrt{g} \, du^1 \, du^2 \, du^3.$$

Show that, under a general coordinate transformation to a new coordinate system u'^i, the volume element dV remains unchanged, i.e. show that it is a scalar quantity.

26.26   By writing down the expression for the square of the infinitesimal arc length $(ds)^2$ in spherical polar coordinates, find the components g_{ij} of the metric tensor in this coordinate system. Hence, using (26.97), find the expression for the divergence of a vector field $\mathbf{v}$ in spherical polars. Calculate the Christoffel symbols (of the second kind) $\Gamma^i{}_{jk}$ in this coordinate system.

26.27   Find an expression for the second covariant derivative $v_{i;jk} \equiv (v_{i;j})_{;k}$ of a vector v_i (see (26.88)). By interchanging the order of differentiation and then subtracting the two expressions, we define the components $R^l{}_{ijk}$ of the *Riemann tensor* as

$$v_{i;jk} - v_{i;kj} \equiv R^l{}_{ijk}v_l.$$

Show that in a general coordinate system u^i these components are given by

$$R^l{}_{ijk} = \frac{\partial \Gamma^l{}_{ik}}{\partial u^j} - \frac{\partial \Gamma^l{}_{ij}}{\partial u^k} + \Gamma^m{}_{ik}\Gamma^l{}_{mj} - \Gamma^m{}_{ij}\Gamma^l{}_{mk}.$$

By first considering Cartesian coordinates, show that all the components $R^l{}_{ijk} \equiv 0$ for *any* coordinate system in three-dimensional Euclidean space. In such a space, therefore, we may change the order of the covariant derivatives without changing the resulting expression.

26.28   A curve $\mathbf{r}(t)$ is parameterised by a scalar variable t. Show that the length of the curve between two points, A and B, is given by

$$L = \int_A^B \sqrt{g_{ij} \frac{du^i}{dt} \frac{du^j}{dt}}\, dt.$$

Using the calculus of variations (see chapter 22), show that the curve $\mathbf{r}(t)$ that minimises L satisfies the equation

$$\frac{d^2 u^i}{dt^2} + \Gamma^i_{\ jk} \frac{du^j}{dt} \frac{du^k}{dt} = \frac{\ddot{s}}{\dot{s}} \frac{du^i}{dt},$$

where s is the arc length along the curve, $\dot{s} = ds/dt$ and $\ddot{s} = d^2 s/dt^2$. Hence, show that if the parameter t is of the form $t = as + b$, where a and b are constants, then we recover the equation for a geodesic (26.101).
         [A parameter which, like t, is the sum of a linear transformation of s and a translation is called an *affine* parameter.]

26.29   We may define Christoffel symbols of the first kind by

$$\Gamma_{ijk} = g_{il} \Gamma^l_{\ jk}.$$

Show that these are given by

$$\Gamma_{kij} = \frac{1}{2} \left(\frac{\partial g_{ik}}{\partial u^j} + \frac{\partial g_{jk}}{\partial u^i} - \frac{\partial g_{ij}}{\partial u^k} \right).$$

By permuting indices, verify that

$$\frac{\partial g_{ij}}{\partial u^k} = \Gamma_{ijk} + \Gamma_{jik}.$$

Using the fact that $\Gamma^l_{\ jk} = \Gamma^l_{\ kj}$, show that

$$g_{ij;k} \equiv 0,$$

i.e. that the covariant derivative of the metric tensor is identically zero in all coordinate systems.

26.24 Hints and answers

26.1   (a) $u_1' = x_1 \cos(\phi - \theta) - x_2 \sin(\phi - \theta)$, etc.;
        (b) $u_{11}' = s^2 x_1^2 - 2scx_1 x_2 + c^2 x_2^2 \neq c^2 x_2^2 + csx_1 x_2 + scx_1 x_2 + s^2 x_1^2$.
26.3   (a) $(1/\sqrt{2})(\sqrt{2}, 0, 0; 0, 1, -1; 0, 1, 1)$. (b) $(1/\sqrt{2})(1, 0, -1; 0, \sqrt{2}, 0; 1, 0, 1)$.
        $\mathbf{r} = (2\sqrt{2},\ -1 + \sqrt{2},\ -1 - \sqrt{2})^\mathrm{T}$.
26.5   Twice contract the array with the outer product of (x, y, z) with itself, thus obtaining the expression $-(x^2 + y^2 + z^2)^2$, which is an invariant and therefore a scalar.
26.7   Write $A_j(\partial A_i / \partial x_j)$ as $\partial(A_i A_j)/\partial x_j - A_i(\partial A_j / \partial x_j)$.
26.9   (i) Write out the expression for $|A^\mathrm{T}|$, contract both sides of the equation with ϵ_{lmn} and pick out the expression for $|A|$ on the RHS. Note that $\epsilon_{lmn}\epsilon_{lmn}$ is a numerical scalar.
        (iii) Each non-zero term on the RHS contains any particular row index once and only once. The same can be said for the Levi–Civita symbol on the LHS. Thus interchanging two rows is equivalent to interchanging two of the subscripts of ϵ_{lmn}, and thereby reversing its sign. Consequently, the magnitude of $|A|$ remains the same but its sign is changed.
        (v) If, say, $A_{pi} = \lambda A_{pj}$, for some particular pair of values i and j and all p then,

in the (multiple) summation on the RHS, each A_{nk} appears multiplied by (with no summation over i and j)

$$\epsilon_{ijk}A_{li}A_{mj} + \epsilon_{jik}A_{lj}A_{mi} = \epsilon_{ijk}\lambda A_{lj}A_{mj} + \epsilon_{jik}A_{lj}\lambda A_{mj} = 0,$$

since $\epsilon_{ijk} = -\epsilon_{jik}$. Consequently, grouped in this way all terms are zero and $|\mathsf{A}| = 0$.
(vi) Replace A_{mj} by $A_{mj} + \lambda A_{lj}$ and note that $\lambda A_{li}A_{lj}A_{nk}\epsilon_{ijk} = 0$ by virtue of result (v).
(vii) If $\mathsf{C} = \mathsf{AB}$,

$$|\mathsf{C}|\epsilon_{lmn} = A_{lx}B_{xi}A_{my}B_{yj}A_{nz}B_{zk}\epsilon_{ijk}.$$

Contract this with ϵ_{lmn} and show that the RHS is equal to $\epsilon_{xyz}|\mathsf{A}^{\mathrm{T}}|\epsilon_{xyz}|\mathsf{B}|$. It then follows from result (i) that $|\mathsf{C}| = |\mathsf{A}||\mathsf{B}|$.

26.11 $\alpha = |\mathbf{v}|^{-2}$. Note that the most general vector has components $w_i = \lambda v_i + \mu u_i^{(1)} + v u_i^{(2)}$, where both $\mathbf{u}^{(1)}$ and $\mathbf{u}^{(2)}$ are orthogonal to $\mathbf{v}$.

26.13   Construct the orthogonal transformation matrix S for the symmetry operation of (say) a rotation of $2\pi/3$ about a body diagonal and, setting $\mathsf{L} = \mathsf{S}^{-1} = \mathsf{S}^{\mathrm{T}}$, construct $\sigma' = \mathsf{L}\sigma\mathsf{L}^{\mathrm{T}}$ and require $\sigma' = \sigma$. Repeat the procedure for (say) a rotation of $\pi/2$ about the x_3-axis. These together show that $\sigma_{11} = \sigma_{22} = \sigma_{33}$ and that all other $\sigma_{ij} = 0$. Further symmetry requirements do not provide any additional constraints.

26.15   The transformation of δ_{ij} has to be included; the principal values are $\pm\mathbf{E}\cdot\mathbf{B}$. The third axis is in the direction $\pm\mathbf{B}\times\mathbf{E}$ with principal value $-|\mathbf{E}|^2$.

26.17   The principal moments give the required ratios.

26.19   The principal permeability, in direction $(1,1,2)$, has value 0. Thus all the nails lie in planes to which this is the normal.

26.21   Take $p_{11} = p_{22} = p_{33} = -p$, and $p_{ij} = e_{ij} = 0$ for $i \ne j$, leading to $-p = (\lambda + 2\mu/3)e_{ii}$. The fractional volume change is e_{ii}; λ and μ are as defined in (26.46) and the worked example that follows it.

26.23   Consider $Q_{pq} = \epsilon_{pij}\epsilon_{qkl}T_{ijkl}$ and show that $K_{mn} = Q_{mn}/4$ has the required property. (a) Argue from the isotropy of T_{ijkl} and ϵ_{ijk} for that of K_{mn} and hence that it must be a multiple of δ_{mn}. Show that the multiplier is uniquely determined and that $T_{ijkl} = (\delta_{il}\delta_{jk} - \delta_{ik}\delta_{jl})/6$.
(b) By relabelling dummy subscripts and using the stated antisymmetry property, show that $K_{nm} = -K_{mn}$. Show that $-2V_i = \epsilon_{mn}K_{mn}$ and hence that $K_{mn} = \epsilon_{imn}V_i$. $T_{ijkl} = \epsilon_{kli}V_j - \epsilon_{klj}V_i$.

26.25   Use $|\mathbf{e}_1\cdot(\mathbf{e}_2\times\mathbf{e}_3)| = \sqrt{g}$.
Recall that $\sqrt{g'} = |\partial u/\partial u'|\sqrt{g}$ and $du'^1\,du'^2\,du'^3 = |\partial u'/\partial u|\,du^1\,du^2\,du^3$.

26.27 $(v_{i;j})_{;k} = (v_{i;j})_{,k} - \Gamma^l_{ik}v_{l;j} - \Gamma^l_{jk}v_{i;l}$ and $v_{i;j} = v_{i,j} - \Gamma^m_{ij}v_m$. If all components of a tensor equal zero in one coordinate system, then they are zero in all coordinate systems.

26.29   Use $g_{il}g^{ln} = \delta^n_i$ and $g_{ij} = g_{ji}$. Show that

$$g_{ij;k} = \left(\frac{\partial g_{ij}}{\partial u^k} - \Gamma_{jik} - \Gamma_{ijk}\right)\mathbf{e}^i\otimes\mathbf{e}^j$$

and then use the earlier result.

27

Numerical methods

It happens frequently that the end product of a calculation or piece of analysis is one or more algebraic or differential equations, or an integral that cannot be evaluated in closed form or in terms of tabulated or pre-programmed functions. From the point of view of the physical scientist or engineer, who needs numerical values for prediction or comparison with experiment, the calculation or analysis is thus incomplete.

With the ready availability of standard packages on powerful computers for the numerical solution of equations, both algebraic and differential, and for the evaluation of integrals, in principle there is no need for the investigator to do anything other than turn to them. However, it should be a part of every engineer's or scientist's repertoire to have some understanding of the kinds of procedure that are being put into practice within those packages. The present chapter indicates (at a simple level) some of the ways in which analytically intractable problems can be tackled using numerical methods.

In the restricted space available in a book of this nature, it is clearly not possible to give anything like a full discussion, even of the elementary points that will be made in this chapter. The limited objective adopted is that of explaining and illustrating by simple examples some of the basic principles involved. In many cases, the examples used can be solved in closed form anyway, but this 'obviousness' of the answers should not detract from their illustrative usefulness, and it is hoped that their transparency will help the reader to appreciate some of the inner workings of the methods described.

The student who proposes to study complicated sets of equations or make repeated use of the same procedures by, for example, writing computer programs to carry out the computations, will find it essential to acquire a good under-standing of topics hardly mentioned here. Amongst these are the sensitivity of the adopted procedures to errors introduced by the limited accuracy with which a numerical value can be stored in a computer (rounding errors) and to the errors introduced as a result of approximations made in setting up the numerical procedures (truncation errors). For this scale of application, books specifically devoted to numerical analysis, data analysis and computer programming should be consulted.

So far as is possible, the method of presentation here is that of indicating and discussing in a qualitative way the main steps in the procedure, and then of following this with an elementary worked example. The examples have been restricted in complexity to a level at which they can be carried out with a pocket calculator. Naturally it will not be possible for the student to check all the numerical values presented, unless he or she has a programmable calculator or computer readily available, and even then it might be tedious to do so. However, it is advisable to check the initial step and at least one step in the middle of each repetitive calculation given in the text, so that how the symbolic equations are used with actual numbers is understood. Clearly the intermediate step should be chosen to be at a point in the calculation at which the changes are still sufficiently large that they can be detected by whatever calculating device is used.

Where alternative methods for solving the same type of problem are discussed, for example in finding the roots of a polynomial equation, we have usually taken the same example to illustrate each method. This could give the mistaken impression that the methods are very restricted in applicability, but it is felt by the authors that using the same examples repeatedly has sufficient advantages, in terms of illustrating the *relative* characteristics of competing methods, to justify doing so. Once the principles are clear, little is to be gained by using new examples each time, and, in fact, having some prior knowledge of the 'correct answer' should allow the reader to judge the efficiency and dangers of particular methods as the successive steps are followed through.

One other point remains to be mentioned. Here, in contrast with every other chapter of this book, the value of a large selection of exercises is not clear cut. The reader with sufficient computing resources to tackle them can easily devise algebraic or differential equations to be solved, or functions to be integrated (which perhaps have arisen in other contexts). Further, the solutions of these problems will be self-checking, for the most part. Consequently, although a number of exercises are included, no attempt has been made to test the full range of ideas treated in this chapter.

27.1 Algebraic and transcendental equations

The problem of finding the real roots of an equation of the form $f(x) = 0$, where $f(x)$ is an algebraic or transcendental function of x, is one that can sometimes be treated numerically, even if explicit solutions in closed form are not feasible.

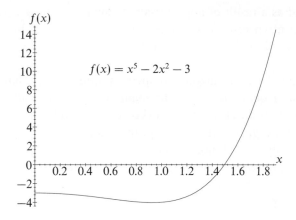

Figure 27.1 A graph of the function $f(x) = x^5 - 2x^2 - 3$ for x in the range $0 \le x \le 1.9$.

Examples of the types of equation mentioned are the quartic equation,

$$ax^4 + bx + c = 0,$$

and the transcendental equation,

$$x - 3 \tanh x = 0.$$

The latter type is characterised by the fact that it contains, in effect, a polynomial of infinite order on the LHS.

We will discuss four methods that, in various circumstances, can be used to obtain the real roots of equations of the above types. In all cases we will take as the specific equation to be solved the fifth-order polynomial equation

$$f(x) \equiv x^5 - 2x^2 - 3 = 0. \tag{27.1}$$

The reasons for using the same equation each time were discussed in the introduction to this chapter.

For future reference, and so that the reader may follow some of the calculations leading to the evaluation of the real root of (27.1), a graph of $f(x)$ in the range $0 \le x \le 1.9$ is shown in figure 27.1.

Equation (27.1) is one for which no solution can be found in closed form, that is in the form $x = a$, where a does not explicitly contain x. The general scheme to be employed will be an iterative one in which successive approximations to a real root of (27.1) will be obtained, each approximation, it is to be hoped, being better than the preceding one; certainly, we require that the approximations converge and that they have as their limit the sought-for root. Let us denote the required root by ξ and the values of successive approximations by $x_1, x_2, \ldots, x_n, \ldots$. Then, for any particular method to be successful,

$$\lim_{n \to \infty} x_n = \xi, \quad \text{where } f(\xi) = 0. \tag{27.2}$$

However, success as defined here is not the only criterion. Since, in practice, only a finite number of iterations will be possible, it is important that the values of x_n be close to that of ξ for all $n > N$, where N is a relatively low number; exactly how low it is naturally depends on the computing resources available and the accuracy required in the final answer.

So that the reader may assess the progress of the calculations that follow, we record that to nine significant figures the real root of equation (27.1) has the value

$$\xi = 1.495\,106\,40. \tag{27.3}$$

We now consider in turn four methods for determining the value of this root.

27.1.1 Rearrangement of the equation

If equation (27.1), $f(x) = 0$, can be recast into the form

$$x = \phi(x), \tag{27.4}$$

where $\phi(x)$ is a *slowly* varying function of x, then an iteration scheme

$$x_{n+1} = \phi(x_n) \tag{27.5}$$

will often produce a fair approximation to the root ξ after a few iterations, as follows. Clearly, $\xi = \phi(\xi)$, since $f(\xi) = 0$; thus, when x_n is close to ξ, the next approximation, x_{n+1}, will differ little from x_n, the actual size of the difference giving an order-of-magnitude indication of the inaccuracy in x_{n+1} (when compared with ξ).

In the present case, the equation can be written

$$x = (2x^2 + 3)^{1/5}. \tag{27.6}$$

Because of the presence of the one-fifth power, the RHS is rather insensitive to the value of x used to compute it, and so the form (27.6) fits the general requirements for the method to work satisfactorily. It remains only to choose a starting approximation. It is easy to see from figure 27.1 that the value $x = 1.5$ would be a good starting point, but, so that the behaviour of the procedure at values some way from the actual root can be studied, we will make a poorer choice, $x_1 = 1.7$.

With this starting value and the general recurrence relationship

$$x_{n+1} = (2x_n^2 + 3)^{1/5}, \tag{27.7}$$

n	x_n	$f(x_n)$
1	1.7	5.42
2	1.544 18	1.01
3	1.506 86	2.28×10^{-1}
4	1.497 92	5.37×10^{-2}
5	1.495 78	1.28×10^{-2}
6	1.495 27	3.11×10^{-3}
7	1.495 14	7.34×10^{-4}
8	1.495 12	1.76×10^{-4}

Table 27.1 Successive approximations to the root of (27.1) using the method of rearrangement.

n	A_n	$f(A_n)$	B_n	$f(B_n)$	x_n	$f(x_n)$
1	1.0	−4.0000	1.7	5.4186	1.2973	−2.6916
2	1.2973	−2.6916	1.7	5.4186	1.4310	−1.0957
3	1.4310	−1.0957	1.7	5.4186	1.4762	−0.3482
4	1.4762	−0.3482	1.7	5.4186	1.4897	−0.1016
5	1.4897	−0.1016	1.7	5.4186	1.4936	−0.0289
6	1.4936	−0.0289	1.7	5.4186	1.4947	−0.0082

Table 27.2 Successive approximations to the root of (27.1) using linear interpolation.

successive values can be found. These are recorded in table 27.1. Although not strictly necessary, the value of $f(x_n) \equiv x_n^5 - 2x_n^2 - 3$ is also shown at each stage.

It will be seen that x_7 and all later x_n agree with the precise answer (27.3) to within one part in 10^4. However, $f(x_n)$ and $x_n - \xi$ are both reduced by a factor of only about 4 for each iteration; thus a large number of iterations would be needed to produce a very accurate answer. The factor 4 is, of course, specific to this particular problem and would be different for a different equation. The successive values of x_n are shown in graph (a) of figure 27.2.

27.1.2 Linear interpolation

In this approach two values, A_1 and B_1, of x are chosen with $A_1 < B_1$ and such that $f(A_1)$ and $f(B_1)$ have opposite signs. The chord joining the two points $(A_1, f(A_1))$ and $(B_1, f(B_1))$ is then notionally constructed, as illustrated in graph (b) of figure 27.2, and the value x_1 at which the chord cuts the x-axis is determined by the *interpolation formula*

$$x_n = \frac{A_n f(B_n) - B_n f(A_n)}{f(B_n) - f(A_n)}, \qquad (27.8)$$

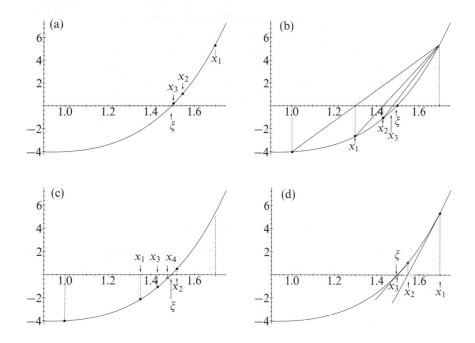

Figure 27.2 Graphical illustrations of the iteration methods discussed in the text: (a) rearrangement; (b) linear interpolation; (c) binary chopping; (d) Newton–Raphson.

with $n = 1$. Next, $f(x_1)$ is evaluated and the process repeated after replacing either A_1 or B_1 by x_1, according to whether $f(x_1)$ has the same sign as $f(A_1)$ or $f(B_1)$, respectively. In figure 27.2(b), A_1 is the one replaced.

As can be seen in the particular example that we are considering, with this method there is a tendency, if the curvature of $f(x)$ is of constant sign near the root, for one of the two ends of the successive chords to remain unchanged.

Starting with the initial values $A_1 = 1$ and $B_1 = 1.7$, the results of the first five iterations using (27.8) are given in table 27.2 and indicated in graph (b) of figure 27.2. As with the rearrangement method, the improvement in accuracy, as measured by $f(x_n)$ and $x_n - \xi$, is a fairly constant factor at each iteration (approximately 3 in this case), and for our particular example there is little to choose between the two. Both tend to their limiting value of ξ monotonically, from either higher or lower values, and this makes it difficult to estimate limits within which ξ can safely be presumed to lie. The next method to be described gives at any stage a range of values within which ξ is *known* to lie.

n	A_n	$f(A_n)$	B_n	$f(B_n)$	x_n	$f(x_n)$
1	1.0000	−4.0000	1.7000	5.4186	1.3500	−2.1610
2	1.3500	−2.1610	1.7000	5.4186	1.5250	0.5968
3	1.3500	−2.1610	1.5250	0.5968	1.4375	−0.9946
4	1.4375	−0.9946	1.5250	0.5968	1.4813	−0.2573
5	1.4813	−0.2573	1.5250	0.5968	1.5031	0.1544
6	1.4813	−0.2573	1.5031	0.1544	1.4922	−0.0552
7	1.4922	−0.0552	1.5031	0.1544	1.4977	0.0487
8	1.4922	−0.0552	1.4977	0.0487	1.4949	−0.0085

Table 27.3   Successive approximations to the root of (27.1) using binary chopping.

27.1.3 Binary chopping

Again two values of x, A_1 and B_1, that straddle the root are chosen, such that $A_1 < B_1$ and $f(A_1)$ and $f(B_1)$ have opposite signs. The interval between them is then halved by forming

$$x_n = \tfrac{1}{2}(A_n + B_n), \tag{27.9}$$

with $n = 1$, and $f(x_1)$ is evaluated. It should be noted that x_1 is determined solely by A_1 and B_1, and not by the values of $f(A_1)$ and $f(B_1)$ as in the linear interpolation method. Now x_1 is used to replace either A_1 or B_1, depending on which of $f(A_1)$ or $f(B_1)$ has the same sign as $f(x_1)$, i.e. if $f(A_1)$ and $f(x_1)$ have the same sign then x_1 replaces A_1. The process is then repeated to obtain x_2, x_3, etc.

This has been carried through in table 27.3 for our standard equation (27.1) and is illustrated in figure 27.2(c). The entries have been rounded to four places of decimals. It is suggested that the reader follows through the sequential replacements of the A_n and B_n in the table and correlates the first few of these with graph (c) of figure 27.2.

Clearly, the accuracy with which ξ is known in this approach increases by only a factor of 2 at each step, but this accuracy is predictable at the outset of the calculation and (unless $f(x)$ has very violent behaviour near $x = \xi$) a range of x in which ξ lies can be safely stated at any stage. At the stage reached in the last row of table 27.3 it may be stated that $1.4949 < \xi < 1.4977$. Thus binary chopping gives a simple approximation method (it involves less multiplication than linear interpolation, for example) that is predictable and relatively safe, although its convergence is slow.

27.1.4 Newton–Raphson method

The Newton–Raphson (NR) procedure is somewhat similar to the interpolation method, but, as will be seen, has one distinct advantage over the latter. Instead

n	x_n	$f(x_n)$
1	1.7	5.42
2	1.545 01	1.03
3	1.498 87	7.20×10^{-2}
4	1.495 13	4.49×10^{-4}
5	1.495 106 40	2.6×10^{-8}
6	1.495 106 40	—

Table 27.4   Successive approximations to the root of (27.1) using the Newton–Raphson method.

of (notionally) constructing the chord between two points on the curve of $f(x)$ against x, the tangent to the curve is notionally constructed at each successive value of x_n, and the next value, x_{n+1}, is taken as the point at which the tangent cuts the axis $f(x) = 0$. This is illustrated in graph (d) of figure 27.2.

If the nth value is x_n, the tangent to the curve of $f(x)$ at that point has slope $f'(x_n)$ and passes through the point $x = x_n$, $y = f(x_n)$. Its equation is thus

$$y(x) = (x - x_n)f'(x_n) + f(x_n). \tag{27.10}$$

The value of x at which $y = 0$ is then taken as x_{n+1}; thus the condition $y(x_{n+1}) = 0$ yields, from (27.10), the iteration scheme

$$x_{n+1} = x_n - \frac{f(x_n)}{f'(x_n)}. \tag{27.11}$$

This is the *Newton–Raphson iteration formula*. Clearly, if x_n is close to ζ then x_{n+1} is close to x_n, as it should be. It is also apparent that if any of the x_n comes close to a stationary point of f, so that $f'(x_n)$ is close to zero, the scheme is not going to work well.

For our standard example, (27.11) becomes

$$x_{n+1} = x_n - \frac{x_n^5 - 2x_n^2 - 3}{5x_n^4 - 4x_n} = \frac{4x_n^5 - 2x_n^2 + 3}{5x_n^4 - 4x_n}. \tag{27.12}$$

Again taking a starting value of $x_1 = 1.7$, we obtain in succession the entries in table 27.4. The different values are given to an increasing number of decimal places as the calculation proceeds; $f(x_n)$ is also recorded.

It is apparent that this method is unlike the previous ones in that the increase in accuracy of the answer is not constant throughout the iterations but improves dramatically as the required root is approached. Away from the root the behaviour of the series is less satisfactory, and from its geometrical interpretation it can be seen that if, for example, there were a maximum or minimum near the root then the series could oscillate between values on either side of it (instead of 'homing in' on the root). The reason for the good convergence near the root is discussed in the next section.

Of the four methods mentioned, no single one is ideal, and, in practice, some mixture of them is usually to be preferred. The particular combination of methods selected will depend a great deal on how easily the progress of the calculation may be monitored, but some combination of the first three methods mentioned, followed by the NR scheme if great accuracy were required, would be suitable for most situations.

27.2 Convergence of iteration schemes

For iteration schemes in which x_{n+1} can be expressed as a differentiable function of x_n, for example the rearrangement or NR methods of the previous section, a partial analysis of the conditions necessary for a successful scheme can be made as follows.

Suppose the general iteration formula is expressed as

$$x_{n+1} = F(x_n) \tag{27.13}$$

((27.7) and (27.12) are examples). Then the sequence of values $x_1, x_2, \ldots, x_n, \ldots$ is required to converge to the value ξ that satisfies both

$$f(\xi) = 0 \quad \text{and} \quad \xi = F(\xi). \tag{27.14}$$

If the error in the solution at the nth stage is ϵ_n, i.e. $x_n = \xi + \epsilon_n$, then

$$\xi + \epsilon_{n+1} = x_{n+1} = F(x_n) = F(\xi + \epsilon_n). \tag{27.15}$$

For the iteration process to converge, a decreasing error is required, i.e. $|\epsilon_{n+1}| < |\epsilon_n|$. To see what this implies about F, we expand the right-hand term of (27.15) by means of a Taylor series and use (27.14) to replace (27.15) by

$$\xi + \epsilon_{n+1} = \xi + \epsilon_n F'(\xi) + \tfrac{1}{2}\epsilon_n^2 F''(\xi) + \cdots. \tag{27.16}$$

This shows that, for small ϵ_n,

$$\epsilon_{n+1} \approx F'(\xi)\epsilon_n$$

and that a necessary (but not sufficient) condition for convergence is that

$$|F'(\xi)| < 1. \tag{27.17}$$

It should be noted that this is a condition on $F'(\xi)$ and not on $f'(\xi)$, which may have any finite value. Figure 27.3 illustrates in a graphical way how the convergence proceeds for the case $0 < F'(\xi) < 1$.

Equation (27.16) suggests that if $F(x)$ can be chosen so that $F'(\xi) = 0$ then the ratio $|\epsilon_{n+1}/\epsilon_n|$ could be made very small, of order ϵ_n in fact. To go even further, if it can be arranged that the first few derivatives of F vanish at $x = \xi$ then the

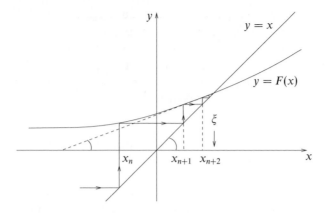

Figure 27.3 Illustration of the convergence of the iteration scheme $x_{n+1} = F(x_n)$ when $0 < F'(\xi) < 1$, where $\xi = F(\xi)$. The line $y = x$ makes an angle $\pi/4$ with the axes. The broken line makes an angle $\tan^{-1} F'(\xi)$ with the x-axis.

convergence, once x_n has become close to ξ, could be very rapid indeed. If the first $N - 1$ derivatives of F vanish at $x = \xi$, i.e.

$$F'(\xi) = F''(\xi) = \cdots = F^{(N-1)}(\xi) = 0 \qquad (27.18)$$

and consequently

$$\epsilon_{n+1} = O(\epsilon_n^N), \qquad (27.19)$$

then the scheme is said to have *Nth-order convergence*.

This is the explanation of the significant difference in convergence between the NR scheme and the others discussed (judged by reference to (27.19), so that the differentiability of the function F is not a prerequisite). The NR procedure has second-order convergence, as is shown by the following analysis. Since

$$F(x) = x - \frac{f(x)}{f'(x)},$$

$$F'(x) = 1 - \frac{f'(x)}{f'(x)} + \frac{f(x)f''(x)}{[f'(x)]^2}$$

$$= \frac{f(x)f''(x)}{[f'(x)]^2}.$$

Now, provided $f'(\xi) \neq 0$, it follows that $F'(\xi) = 0$ because $f(x) = 0$ at $x = \xi$.

n	x_{n+1}	ϵ_n
1	8.5	4.5
2	5.191	1.19
3	4.137	1.4×10^{-1}
4	4.002\,257	2.3×10^{-3}
5	4.000\,000\,637	6.4×10^{-7}
6	4	—

Table 27.5 Successive approximations to $\sqrt{16}$ using the iteration scheme (27.20).

▶*The following is an iteration scheme for finding the square root of X:*

$$x_{n+1} = \frac{1}{2}\left(x_n + \frac{X}{x_n}\right). \tag{27.20}$$

Show that it has second-order convergence and illustrate its efficiency by finding, say, $\sqrt{16}$ starting with a very poor guess, $\sqrt{16} = 1$.

If this scheme does converge to ξ then ξ will satisfy

$$\xi = \frac{1}{2}\left(\xi + \frac{X}{\xi}\right) \quad \Rightarrow \quad \xi^2 = X,$$

as required. The iteration function F is given by

$$F(x) = \frac{1}{2}\left(x + \frac{X}{x}\right),$$

and so, since $\xi^2 = X$,

$$F'(\xi) = \frac{1}{2}\left(1 - \frac{X}{x^2}\right)_{x=\xi} = 0,$$

whilst

$$F''(\xi) = \left(\frac{X}{x^3}\right)_{x=\xi} = \frac{1}{\xi} \neq 0.$$

Thus the procedure has second-order, but not third-order, convergence.

We now show the procedure in action. Table 27.5 gives successive values of x_n and of ϵ_n, the difference between x_n and the true value, 4. As we can see, the scheme is crude initially, but once x_n gets close to ξ, it homes in on the true value extremely rapidly. ◀

27.3 Simultaneous linear equations

As we saw in chapter 8, many situations in physical science can be described approximately or exactly by a set of N simultaneous linear equations in N

variables (unknowns), x_i, $i = 1, 2, \ldots, N$. The equations take the general form

$$A_{11}x_1 + A_{12}x_2 + \cdots + A_{1N}x_N = b_1,$$
$$A_{21}x_1 + A_{22}x_2 + \cdots + A_{2N}x_N = b_2, \qquad (27.21)$$
$$\vdots$$
$$A_{N1}x_1 + A_{N2}x_2 + \cdots + A_{NN}x_N = b_N,$$

where the A_{ij} are constants and form the elements of a square matrix A. The b_i are given and form a column matrix b. If A is non-singular then (27.21) can be solved for the x_i using the inverse of A, according to the formula

$$x = A^{-1}b.$$

This approach was discussed at length in chapter 8 and will not be considered further here.

27.3.1 Gaussian elimination

We follow instead a continuation of one of the earliest techniques acquired by a student of algebra, namely the solving of simultaneous equations (initially only two in number) by the successive elimination of all the variables but one. This (known as *Gaussian elimination*) is achieved by using, at each stage, one of the equations to obtain an explicit expression for one of the remaining x_i in terms of the others and then substituting for that x_i in all other remaining equations. Eventually a single linear equation in just one of the unknowns is obtained. This is then solved and the result is resubstituted in previously derived equations (in reverse order) to establish values for all the x_i.

This method is probably very familiar to the reader, and so a specific example to illustrate this alone seems unnecessary. Instead, we will show how a calculation along such lines might be arranged so that the errors due to the inherent lack of precision in any calculating equipment do not become excessive. This can happen if the value of N is large and particularly (and we will merely state this) if the elements $A_{11}, A_{22}, \ldots, A_{NN}$ on the leading diagonal of the matrix in (27.21) are small compared with the off-diagonal elements.

The process to be described is known as *Gaussian elimination with interchange*. The only, but essential, difference from straightforward elimination is that before each variable x_i is eliminated, the equations are reordered to put the largest (in modulus) remaining coefficient of x_i on the leading diagonal.

We will take as an illustration a straightforward three-variable example, which can in fact be solved perfectly well without any interchange since, with simple numbers and only two eliminations to perform, rounding errors do not have a chance to build up. However, the important thing is that the reader should appreciate how this would apply in (say) a computer program for a 1000-variable case, perhaps with unforeseeable zeros or very small numbers appearing on the leading diagonal.

▶*Solve the simultaneous equations*

$$
\begin{array}{llrll}
\text{(a)} & x_1 & +6x_2 & -4x_3 & = 8, \\
\text{(b)} & 3x_1 & -20x_2 & +x_3 & = 12, \\
\text{(c)} & -x_1 & +3x_2 & +5x_3 & = 3.
\end{array}
\tag{27.22}
$$

Firstly, we interchange rows (a) and (b) to bring the term $3x_1$ onto the leading diagonal. In the following, we label the important equations (I), (II), (III), and the others alphabetically. A general (i.e. variable) label will be denoted by j.

$$
\begin{array}{llrll}
\text{(I)} & 3x_1 & -20x_2 & +x_3 & = 12, \\
\text{(d)} & x_1 & +6x_2 & -4x_3 & = 8, \\
\text{(e)} & -x_1 & +3x_2 & +5x_3 & = 3.
\end{array}
$$

For $(j) = $ (d) and (e), replace row (j) by

$$\text{row } (j) - \frac{a_{j1}}{3} \times \text{row (I)},$$

where a_{j1} is the coefficient of x_1 in row (j), to give the two equations

$$\text{(II)} \quad \left(6 + \tfrac{20}{3}\right)x_2 \ + \left(-4 - \tfrac{1}{3}\right)x_3 \ = 8 - \tfrac{12}{3},$$

$$\text{(f)} \quad \left(3 - \tfrac{20}{3}\right)x_2 \ \ + \left(5 + \tfrac{1}{3}\right)x_3 \ = 3 + \tfrac{12}{3}.$$

Now $|6 + \tfrac{20}{3}| > |3 - \tfrac{20}{3}|$ and so no interchange is required before the next elimination. To eliminate x_2, replace row (f) by

$$\text{row (f)} - \frac{\left(-\tfrac{11}{3}\right)}{\tfrac{38}{3}} \times \text{row (II)}.$$

This gives

$$\text{(III)} \qquad \left[\tfrac{16}{3} + \tfrac{11}{38} \times \tfrac{(-13)}{3}\right] x_3 = 7 + \tfrac{11}{38} \times 4.$$

Collecting together and tidying up the final equations, we have

$$
\begin{array}{llrll}
\text{(I)} & 3x_1 & -20x_2 & +x_3 & = 12, \\
\text{(II)} & & 38x_2 & -13x_3 & = 12, \\
\text{(III)} & & & x_3 & = 2.
\end{array}
$$

Starting with (III) and working backwards, it is now a simple matter to obtain

$$x_1 = 10, \qquad x_2 = 1, \qquad x_3 = 2. \ \blacktriangleleft$$

27.3.2 Gauss–Seidel iteration

In the example considered in the previous subsection an explicit way of solving a set of simultaneous equations was given, the accuracy obtainable being limited only by the rounding errors in the calculating facilities available, and the calculation was planned to minimise these. However, in some situations it may be that only an approximate solution is needed. If, for a large number of variables, this is the case then an iterative method may produce a satisfactory degree of precision with less calculation. Such a method, known as *Gauss–Seidel iteration*, is based upon the following analysis.

The problem is again that of finding the components of the column matrix x that satisfies

$$Ax = b \tag{27.23}$$

when A and b are a given matrix and column matrix, respectively.

The steps of the Gauss–Seidel scheme are as follows.

(i) Rearrange the equations (usually by simple division on both sides of each equation) so that all diagonal elements of the new matrix C are unity, i.e. (27.23) becomes

$$Cx = d, \tag{27.24}$$

where $C = I - F$, and F has zeros as its diagonal elements.

(ii) Step (i) produces

$$Fx + d = Ix = x, \tag{27.25}$$

and this forms the basis of an iteration scheme,

$$x_{n+1} = Fx_n + d, \tag{27.26}$$

where x_n is the nth approximation to the required solution vector ξ.

(iii) To improve the convergence, the matrix F, which has zeros on its leading diagonal, can be written as the sum of two matrices L and U that have non-zero elements only below and above the leading diagonal, respectively:

$$L_{ij} = \begin{cases} F_{ij} & \text{if } i > j, \\ 0 & \text{otherwise,} \end{cases}$$
$$ \tag{27.27}$$
$$U_{ij} = \begin{cases} F_{ij} & \text{if } i < j, \\ 0 & \text{otherwise.} \end{cases}$$

This allows the latest values of the components of x to be used at each stage and an improved form of (27.26) to be obtained:

$$x_{n+1} = Lx_{n+1} + Ux_n + d. \tag{27.28}$$

To see why this is possible, we note, for example, that when calculating, say, the fourth component of x_{n+1}, its first three components are already known, and, because of the structure of L, these are the only ones needed to evaluate the fourth component of Lx_{n+1}.

n	x_1	x_2	x_3
1	2	2	2
2	4	0.1	1.34
3	12.76	1.381	2.323
4	9.008	0.867	1.881
5	10.321	1.042	2.039
6	9.902	0.987	1.988
7	10.029	1.004	2.004

Table 27.6 Successive approximations to the solution of simultaneous equations (27.29) using the Gauss–Seidel iteration method.

▶*Obtain an approximate solution to the simultaneous equations*

$$\begin{array}{rrrr} x_1 & +6x_2 & -4x_3 & = 8, \\ 3x_1 & -20x_2 & +x_3 & = 12, \\ -x_1 & +3x_2 & +5x_3 & = 3. \end{array} \qquad (27.29)$$

These are the same equations as were solved in subsection 27.3.1.

Divide the equations by 1, −20 and 5, respectively, to give

$$x_1 + 6x_2 - 4x_3 = 8,$$
$$-0.15x_1 + x_2 - 0.05x_3 = -0.6,$$
$$-0.2x_1 + 0.6x_2 + x_3 = 0.6.$$

Thus, set out in matrix form, (27.28) is, in this case, given by

$$\begin{pmatrix} x_1 \\ x_2 \\ x_3 \end{pmatrix}_{n+1} = \begin{pmatrix} 0 & 0 & 0 \\ 0.15 & 0 & 0 \\ 0.2 & -0.6 & 0 \end{pmatrix} \begin{pmatrix} x_1 \\ x_2 \\ x_3 \end{pmatrix}_{n+1}$$
$$+ \begin{pmatrix} 0 & -6 & 4 \\ 0 & 0 & 0.05 \\ 0 & 0 & 0 \end{pmatrix} \begin{pmatrix} x_1 \\ x_2 \\ x_3 \end{pmatrix}_{n} + \begin{pmatrix} 8 \\ -0.6 \\ 0.6 \end{pmatrix}.$$

Suppose initially ($n = 1$) we guess each component to have the value 2. Then the successive sets of values of the three quantities generated by this scheme are as shown in table 27.6. Even with the rather poor initial guess, a close approximation to the exact result, $x_1 = 10$, $x_2 = 1$, $x_3 = 2$, is obtained in only a few iterations. ◀

27.3.3 Tridiagonal matrices

Although for the solution of most matrix equations $\mathsf{A}\mathbf{x} = \mathbf{b}$ the number of operations required increases rapidly with the size $N \times N$ of the matrix (roughly as N^3), for one particularly simple kind of matrix the computing required increases only linearly with N. This type often occurs in physical situations in which objects in an ordered set interact only with their nearest neighbours and is one in which only the leading diagonal and the diagonals immediately above and below it contain non-zero entries. Such matrices are known as tridiagonal matrices. They may also be used in numerical approximations to the solutions of certain types of differential equation.

A typical matrix equation involving a tridiagonal matrix is as follows:

$$
\begin{pmatrix}
b_1 & c_1 & & & & \\
a_2 & b_2 & c_2 & & & \\
& a_3 & b_3 & c_3 & & \\
& & \ddots & \ddots & \ddots & \\
& & & a_{N-1} & b_{N-1} & c_{N-1} \\
& & & & a_N & b_N
\end{pmatrix}
\begin{pmatrix}
x_1 \\ x_2 \\ x_3 \\ \vdots \\ x_{N-1} \\ x_N
\end{pmatrix}
=
\begin{pmatrix}
y_1 \\ y_2 \\ y_3 \\ \vdots \\ y_{N-1} \\ y_N
\end{pmatrix}
\tag{27.30}
$$

So as to keep the entries in the matrix as free from subscripts as possible, we have used a, b and c to indicate subdiagonal, leading diagonal and superdiagonal elements, respectively. As a consequence, we have had to change the notation for the column matrix on the RHS from b to (say) y.

In such an equation the first and last rows involve x_1 and x_N, respectively, and so the solution could be found by letting x_1 be unknown and then solving in turn each row of the equation in terms of x_1, and finally determining x_1 by requiring the next-to-last line to generate for x_N an equation compatible with that given by the last line. However, if the matrix is large this becomes a very cumbersome operation, and a simpler method is to assume a form of solution

$$x_{i-1} = \theta_{i-1} x_i + \phi_{i-1}. \tag{27.31}$$

Since the ith line of the matrix equation is

$$a_i x_{i-1} + b_i x_i + c_i x_{i+1} = y_i,$$

we must have, by substituting for x_{i-1}, that

$$(a_i \theta_{i-1} + b_i)x_i + c_i x_{i+1} = y_i - a_i \phi_{i-1}.$$

This is also in the form of (27.31), but with i replaced by $i+1$. Thus the recurrence formulae for θ_i and ϕ_i are

$$\theta_i = \frac{-c_i}{a_i \theta_{i-1} + b_i}, \qquad \phi_i = \frac{y_i - a_i \phi_{i-1}}{a_i \theta_{i-1} + b_i}, \tag{27.32}$$

provided the denominator does not vanish for any i. From the first of the matrix equations it follows that $\theta_1 = -c_1/b_1$ and $\phi_1 = y_1/b_1$. The equations may now be solved for the x_i in two stages without carrying through an unknown quantity. First, all the θ_i and ϕ_i are generated using (27.32) and the values of θ_1 and ϕ_1, then, as a second stage, (27.31) is used to evaluate the x_i, starting with x_N ($= \phi_N$) and working backwards.

> ►Solve the following tridiagonal matrix equation, in which only non-zero elements are shown:
>
> $$\begin{pmatrix} 1 & 2 & & & & \\ -1 & 2 & 1 & & & \\ & 2 & -1 & 2 & & \\ & & 3 & 1 & 1 & \\ & & & 3 & 4 & 2 \\ & & & & -2 & 2 \end{pmatrix} \begin{pmatrix} x_1 \\ x_2 \\ x_3 \\ x_4 \\ x_5 \\ x_6 \end{pmatrix} = \begin{pmatrix} 4 \\ 3 \\ -3 \\ 10 \\ 7 \\ -2 \end{pmatrix}. \qquad (27.33)$$

The solution is set out in table 27.7, in which the arrows indicate the general flow of the calculation. First, the columns of a_i, b_i, c_i and y_i are filled in from the original equation (27.33) and then the recurrence relations (27.32) are used to fill in the successive rows starting at the top; on each row we work from left to right as far as and including the ϕ_i column. Finally, the bottom entry in the the x_i column is set equal to the bottom entry in the completed ϕ_i column and the rest of the x_i column is completed by using (27.31) and working up from the bottom. Thus the solution is $x_1 = 2$; $x_2 = 1$; $x_3 = 3$; $x_4 = -1$; $x_5 =$

	a_i	b_i	c_i		$a_i\theta_{i-1} + b_i$	θ_i	y_i	$a_i\phi_{i-1}$	ϕ_i	x_i	
↓	0	1	2	→	1	-2	4	0	4	2	↑
↓	-1	2	1	→	4	$-1/4$	3	-4	$7/4$	1	↑
↓	2	-1	2	→	$-3/2$	$4/3$	-3	$7/2$	$13/3$	3	↑
↓	3	1	1	→	5	$-1/5$	10	13	$-3/5$	-1	↑
↓	3	4	2	→	$17/5$	$-10/17$	7	$-9/5$	$44/17$	2	↑
↓	-2	2	0	→	$54/17$	0	-2	$-88/17$	1 →	1	↑

Table 27.7   The solution of tridiagonal matrix equation (27.33). The arrows indicate the general flow of the calculation, as described in the text.

2; $x_6 = 1$. ◄

27.4 Numerical integration

As noted at the start of this chapter, with modern computers and computer packages – some of which will present solutions in algebraic form, where that is possible – the inability to find a closed-form expression for an integral no longer presents a problem. But, just as for the solution of algebraic equations, it is extremely important that scientists and engineers should have some idea of the procedures on which such packages are based. In this section we discuss some of the more elementary methods used to evaluate integrals numerically and at the same time indicate the basis of more sophisticated procedures.

The standard integral evaluation has the form

$$I = \int_a^b f(x)\,dx, \qquad (27.34)$$

where the integrand $f(x)$ may be given in analytic or tabulated form, but for the cases under consideration no closed-form expression for I can be obtained. All

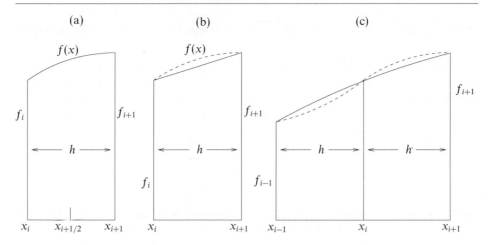

Figure 27.4 (a) Definition of nomenclature. (b) The approximation in using the trapezium rule; $f(x)$ is indicated by the broken curve. (c) Simpson's rule approximation; $f(x)$ is indicated by the broken curve. The solid curve is part of the approximating parabola.

numerical evaluations of I are based on regarding I as the area under the curve of $f(x)$ between the limits $x = a$ and $x = b$ and attempting to estimate that area.

The simplest methods of doing this involve dividing up the interval $a \leq x \leq b$ into N equal sections, each of length $h = (b - a)/N$. The dividing points are labelled x_i, with $x_0 = a$, $x_N = b$, i running from 0 to N. The point x_i is a distance ih from a. The central value of x in a strip $(x = x_i + h/2)$ is denoted for brevity by $x_{i+1/2}$, and for the same reason $f(x_i)$ is written as f_i. This nomenclature is indicated graphically in figure 27.4(a).

So that we may compare later estimates of the area under the curve with the true value, we next obtain an exact expression for I, even though we cannot evaluate it. To do this we need to consider only one strip, say that between x_i and x_{i+1}. For this strip the area is, using Taylor's expansion,

$$
\int_{-h/2}^{h/2} f(x_{i+1/2} + y)\, dy = \int_{-h/2}^{h/2} \sum_{n=0}^{\infty} f^{(n)}(x_{i+1/2}) \frac{y^n}{n!}\, dy
$$

$$
= \sum_{n=0}^{\infty} f_{i+1/2}^{(n)} \int_{-h/2}^{h/2} \frac{y^n}{n!}\, dy
$$

$$
= \sum_{n \text{ even}}^{\infty} f_{i+1/2}^{(n)} \frac{2}{(n+1)!} \left(\frac{h}{2}\right)^{n+1}. \tag{27.35}
$$

It should be noted that, in this exact expression, only the even derivatives of f survive the integration and all derivatives are evaluated at $x_{i+1/2}$. Clearly other exact expressions are possible, e.g. the integral of $f(x_i + y)$ over the range $0 \leq y \leq h$, but we will find (27.35) the most useful for our purposes.

Although the preceding discussion has implicitly assumed that both of the limits a and b are finite, with the consequence that N is finite, the general method can be adapted to treat some cases in which one of the limits is infinite. It is sufficient to consider one infinite limit, as an integral with limits $-\infty$ and ∞ can be considered as the sum of two integrals, each with one infinite limit.

Consider the integral

$$I = \int_a^\infty f(x)\,dx,$$

where a is chosen large enough that the integrand is monotonically decreasing for $x > a$ and falls off more quickly than x^{-2}. The change of variable $t = 1/x$ converts this integral into

$$I = \int_0^{1/a} \frac{1}{t^2} f\left(\frac{1}{t}\right)\,dt.$$

It is now an integral over a finite range and the methods indicated earlier can be applied to it. The value of the integrand at the lower end of the t-range is zero.

In a similar vein, integrals with an upper limit of ∞ and an integrand that is known to behave asymptotically as $g(x)e^{-\alpha x}$, where $g(x)$ is a smooth function, can be converted into an integral over a finite range by setting $x = -\alpha^{-1} \ln \alpha t$. Again, the lower limit, a, for this part of the integral should be positive and chosen beyond the last turning point of $g(x)$. The part of the integral for $x < a$ is treated in the normal way. However, it should be added that if the asymptotic form of the integrand is known to be a linear or quadratic (decreasing) exponential then there are better ways of estimating it numerically; these are discussed in subsection 27.4.3 on Gaussian integration.

We now turn to practical ways of approximating I, given the values of f_i, or a means to calculate them, for $i = 0, 1, \ldots, N$.

27.4.1 Trapezium rule

In this simple case the area shown in figure 27.4(a) is approximated as shown in figure 27.4(b), i.e. by a trapezium. The area A_i of the trapezium is

$$A_i = \tfrac{1}{2}(f_i + f_{i+1})h, \tag{27.36}$$

and if such contributions from all strips are added together then the estimate of the total, and hence of I, is

$$I(\text{estim.}) = \sum_{i=0}^{N-1} A_i = \frac{h}{2}(f_0 + 2f_1 + 2f_2 + \cdots + 2f_{N-1} + f_N). \tag{27.37}$$

This provides a very simple expression for estimating integral (27.34); its accuracy is limited only by the extent to which h can be made very small (and hence N very large) without making the calculation excessively long. Clearly the estimate provided is exact only if $f(x)$ is a linear function of x.

The error made in calculating the area of the strip when the trapezium rule is used may be estimated as follows. The values used are f_i and f_{i+1}, as in (27.36). These can be expressed accurately in terms of $f_{i+1/2}$ and its derivatives by the Taylor series

$$f_{i+1/2\pm1/2} = f_{i+1/2} \pm \frac{h}{2}f'_{i+1/2} + \frac{1}{2!}\left(\frac{h}{2}\right)^2 f''_{i+1/2} \pm \frac{1}{3!}\left(\frac{h}{2}\right)^3 f^{(3)}_{i+1/2} + \cdots .$$

Thus

$$A_i(\text{estim.}) = \tfrac{1}{2}h(f_i + f_{i+1})$$

$$= h\left[f_{i+1/2} + \frac{1}{2!}\left(\frac{h}{2}\right)^2 f''_{i+1/2} + O(h^4) \right],$$

whilst, from the first few terms of the exact result (27.35),

$$A_i(\text{exact}) = hf_{i+1/2} + \frac{2}{3!}\left(\frac{h}{2}\right)^3 f''_{i+1/2} + O(h^5).$$

Thus the error $\Delta A_i = A_i(\text{estim.}) - A_i(\text{exact})$ is given by

$$\Delta A_i = \left(\tfrac{1}{8} - \tfrac{1}{24}\right) h^3 f''_{i+1/2} + O(h^5)$$

$$\approx \tfrac{1}{12} h^3 f''_{i+1/2}.$$

The total error in $I(\text{estim.})$ is thus given approximately by

$$\Delta I(\text{estim.}) \approx \tfrac{1}{12} n h^3 \langle f'' \rangle = \tfrac{1}{12}(b-a)h^2\langle f'' \rangle, \qquad (27.38)$$

where $\langle f'' \rangle$ represents an average value for the second derivative of f over the interval a to b.

▶ *Use the trapezium rule with $h = 0.5$ to evaluate*

$$I = \int_0^2 (x^2 - 3x + 4)\,dx,$$

and, by evaluating the integral exactly, examine how well (27.38) estimates the error.

With $h = 0.5$, we will need five values of $f(x) = x^2 - 3x + 4$ for use in formula (27.37). They are $f(0) = 4$, $f(0.5) = 2.75$, $f(1) = 2$, $f(1.5) = 1.75$ and $f(2) = 2$. Putting these into (27.37) gives

$$I(\text{estim.}) = \frac{0.5}{2}(4 + 2 \times 2.75 + 2 \times 2 + 2 \times 1.75 + 2) = 4.75.$$

The exact value is

$$I(\text{exact}) = \left[\frac{x^3}{3} - \frac{3x^2}{2} + 4x \right]_0^2 = 4\tfrac{2}{3}.$$

The difference between the estimate of the integral and the exact answer is $1/12$. Equation (27.38) estimates this error as $2 \times 0.25 \times \langle f'' \rangle / 12$. Our (deliberately chosen!) integrand is one for which $\langle f'' \rangle$ can be evaluated trivially. Because $f(x)$ is a quadratic function of x, its second derivative is constant, and equal to 2 in this case. Thus $\langle f'' \rangle$ has value 2 and (27.38) estimates the error as $1/12$; that the estimate is exactly right should be no surprise since the Taylor expansion for a quadratic polynomial about any point always terminates after three terms and so no higher-order terms in h have been ignored in (27.38). ◄

27.4.2 Simpson's rule

Whereas the trapezium rule makes a linear interpolation of f, Simpson's rule effectively mimics the local variation of $f(x)$ using parabolas. The strips are treated two at a time (figure 27.4(c)) and therefore their number, N, should be made even.

In the neighbourhood of x_i, for i odd, it is supposed that $f(x)$ can be adequately represented by a quadratic form,

$$f(x_i + y) = f_i + ay + by^2. \tag{27.39}$$

In particular, applying this to $y = \pm h$ yields two expressions involving b

$$f_{i+1} = f(x_i + h) = f_i + ah + bh^2,$$
$$f_{i-1} = f(x_i - h) = f_i - ah + bh^2;$$

thus

$$bh^2 = \tfrac{1}{2}(f_{i+1} + f_{i-1} - 2f_i).$$

Now, in the representation (27.39), the area of the double strip from x_{i-1} to x_{i+1} is given by

$$A_i(\text{estim.}) = \int_{-h}^{h} (f_i + ay + by^2)\, dy = 2hf_i + \tfrac{2}{3}bh^3.$$

Substituting for bh^2 then yields, for the estimated area,

$$A_i(\text{estim.}) = 2hf_i + \tfrac{2}{3}h \times \tfrac{1}{2}(f_{i+1} + f_{i-1} - 2f_i)$$
$$= \tfrac{1}{3}h(4f_i + f_{i+1} + f_{i-1}),$$

an expression involving only given quantities. It should be noted that the values of neither b nor a need be calculated.

For the full integral,

$$I(\text{estim.}) = \tfrac{1}{3}h \left(f_0 + f_N + 4 \sum_{m \text{ odd}} f_m + 2 \sum_{m \text{ even}} f_m \right). \tag{27.40}$$

It can be shown, by following the same procedure as in the trapezium rule case, that the error in the estimated area is approximately

$$\Delta I(\text{estim.}) \approx \frac{(b-a)}{180} h^4 \langle f^{(4)} \rangle.$$

27.4.3 Gaussian integration

In the cases considered in the previous two subsections, the function f was mimicked by linear and quadratic functions. These yield exact answers if f itself is a linear or quadratic function (respectively) of x. This process could be continued by increasing the order of the polynomial mimicking-function so as to increase the accuracy with which more complicated functions f could be numerically integrated. However, the same effect can be achieved with less effort by not insisting upon equally spaced points x_i.

The detailed analysis of such methods of numerical integration, in which the integration points are not equally spaced and the weightings given to the values at each point do not fall into a few simple groups, is too long to be given in full here. Suffice it to say that the methods are based upon mimicking the given function with a weighted sum of mutually orthogonal polynomials. The polynomials, $F_n(x)$, are chosen to be orthogonal with respect to a particular weight function $w(x)$, i.e.

$$\int_a^b F_n(x)F_m(x)w(x)\,dx = k_n\delta_{nm},$$

where k_n is some constant that may depend upon n. Often the weight function is unity and the polynomials are mutually orthogonal in the most straightforward sense; this is the case for Gauss–Legendre integration for which the appropriate polynomials are the Legendre polynomials, $P_n(x)$. This particular scheme is discussed in more detail below.

Other schemes cover cases in which one or both of the integral limits a and b are not finite. For example, if the limits are 0 and ∞ and the integrand contains a negative exponential function $e^{-\alpha x}$, a simple change of variable can cast it into a form for which Gauss–Laguerre integration would be particularly well suited. This form of quadrature is based upon the Laguerre polynomials, for which the appropriate weight function is $w(x) = e^{-x}$. Advantage is taken of this, and the handling of the exponential factor in the integrand is effectively carried out analytically. If the other factors in the integrand can be well mimicked by low-order polynomials, then a Gauss–Laguerre integration using only a modest number of points gives accurate results.

If we also add that the integral over the range $-\infty$ to ∞ of an integrand containing an explicit factor $\exp(-\beta x^2)$ may be conveniently calculated using a scheme based on the Hermite polynomials, the reader will appreciate the close connection between the various Gaussian quadrature schemes and the sets of eigenfunctions discussed in chapter 18. As noted above, the Gauss–Legendre scheme, which we discuss next, is just such a scheme, though its weight function, being unity throughout the range, is not explicitly displayed in the integrand.

Gauss–Legendre quadrature can be applied to integrals over any finite range though the Legendre polynomials $P_\ell(x)$ on which it is based are only defined

1005

and orthogonal over the interval $-1 \leq x \leq 1$, as discussed in subsection 18.1.2. Therefore, in order to use their properties, the integral between limits a and b in (27.34) has to be changed to one between the limits -1 and $+1$. This is easily done with a change of variable from x to z given by

$$z = \frac{2x - b - a}{b - a},$$

so that I becomes

$$I = \frac{b - a}{2} \int_{-1}^{1} g(z) \, dz, \tag{27.41}$$

in which $g(z) \equiv f(x)$.

The n integration points x_i for an n-point Gauss–Legendre integration are given by the zeros of $P_n(x)$, i.e. the x_i are such that $P_n(x_i) = 0$. The integrand $g(x)$ is mimicked by the $(n-1)$th-degree polynomial

$$G(x) = \sum_{i=1}^{n} \frac{P_n(x)}{(x - x_i)P_n'(x_i)} g(x_i),$$

which coincides with $g(x)$ at each of the points x_i, $i = 1, 2, \ldots, n$. To see this it should be noted that

$$\lim_{x \to x_k} \frac{P_n(x)}{(x - x_i)P_n'(x_i)} = \delta_{ik}.$$

It then follows, to the extent that $g(x)$ is well reproduced by $G(x)$, that

$$\int_{-1}^{1} g(x) \, dx \approx \sum_{i=1}^{n} \frac{g(x_i)}{P_n'(x_i)} \int_{-1}^{1} \frac{P_n(x)}{x - x_i} \, dx. \tag{27.42}$$

The expression

$$w(x_i) \equiv \frac{1}{P_n'(x_i)} \int_{-1}^{1} \frac{P_n(x)}{x - x_i} \, dx$$

can be shown, using the properties of Legendre polynomials, to be equal to

$$w_i = \frac{2}{(1 - x_i^2)|P_n'(x_i)|^2},$$

which is thus the weighting to be attached to the factor $g(x_i)$ in the sum (27.42). The latter then becomes

$$\int_{-1}^{1} g(x) \, dx \approx \sum_{i=1}^{n} w_i g(x_i). \tag{27.43}$$

In fact, because of the particular properties of Legendre polynomials, it can be shown that (27.43) integrates exactly any polynomial of degree up to $2n - 1$. The error in the approximate equality is of the order of the $2n$th derivative of g, and so, provided $g(x)$ is a reasonably smooth function, the approximation is a good one.

Taking 3-point integration as an example, the three x_i are the zeros of $P_3(x) = \frac{1}{2}(5x^3 - 3x)$, namely 0 and $\pm 0.774\,60$, and the corresponding weights are

$$\frac{2}{1 \times \left(-\frac{3}{2}\right)^2} = \frac{8}{9} \quad \text{and} \quad \frac{2}{(1 - 0.6) \times \left(\frac{6}{2}\right)^2} = \frac{5}{9}.$$

Table 27.8 gives the integration points (in the range $-1 \le x_i \le 1$) and the corresponding weights w_i for a selection of n-point Gauss–Legendre schemes.

> ▶ *Using a 3-point formula in each case, evaluate the integral*
>
> $$I = \int_0^1 \frac{1}{1 + x^2}\, dx,$$
>
> (i) *using the trapezium rule*, (ii) *using Simpson's rule*, (iii) *using Gaussian integration. Also evaluate the integral analytically and compare the results.*

(i) Using the trapezium rule, we obtain

$$I = \frac{1}{2} \times \frac{1}{2} \left[f(0) + 2f\left(\tfrac{1}{2}\right) + f(1) \right]$$
$$= \frac{1}{4} \left[1 + \tfrac{8}{5} + \tfrac{1}{2} \right] = 0.7750.$$

(ii) Using Simpson's rule, we obtain

$$I = \frac{1}{3} \times \frac{1}{2} \left[f(0) + 4f\left(\tfrac{1}{2}\right) + f(1) \right]$$
$$= \frac{1}{6} \left[1 + \tfrac{16}{5} + \tfrac{1}{2} \right] = 0.7833.$$

(iii) Using Gaussian integration, we obtain

$$I = \frac{1 - 0}{2} \int_{-1}^1 \frac{dz}{1 + \frac{1}{4}(z + 1)^2}$$
$$= \frac{1}{2} \left\{ 0.555\,56 \left[f(-0.774\,60) + f(0.774\,60) \right] + 0.888\,89 f(0) \right\}$$
$$= \frac{1}{2} \left\{ 0.555\,56 \left[0.987\,458 + 0.559\,503 \right] + 0.888\,89 \times 0.8 \right\}$$
$$= 0.785\,27.$$

(iv) Exact evaluation gives

$$I = \int_0^1 \frac{dx}{1 + x^2} = \left[\tan^{-1} x \right]_0^1 = \frac{\pi}{4} = 0.785\,40.$$

In practice, a compromise has to be struck between the accuracy of the result achieved and the calculational labour that goes into obtaining it. ◀

Further Gaussian quadrature procedures, ones that utilise the properties of the Chebyshev polynomials, are available for integrals over finite ranges when the integrands involve factors of the form $(1 - x^2)^{\pm 1/2}$. In the same way as decreasing linear and quadratic exponentials are handled through the weight functions in Gauss–Laguerre and Gauss–Hermite quadrature, respectively, the square root

Gauss–Legendre integration

$$\int_{-1}^{1} f(x)\,dx = \sum_{i=1}^{n} w_i f(x_i)$$

$\pm x_i$	w_i	$\pm x_i$	w_i
$n = 2$		$n = 9$	
0.57735 02692	1.00000 00000	0.00000 00000	0.33023 93550
		0.32425 34234	0.31234 70770
$n = 3$		0.61337 14327	0.26061 06964
0.00000 00000	0.88888 88889	0.83603 11073	0.18064 81607
0.77459 66692	0.55555 55556	0.96816 02395	0.08127 43884
$n = 4$		$n = 10$	
0.33998 10436	0.65214 51549	0.14887 43390	0.29552 42247
0.86113 63116	0.34785 48451	0.43339 53941	0.26926 67193
		0.67940 95683	0.21908 63625
$n = 5$		0.86506 33667	0.14945 13492
0.00000 00000	0.56888 88889	0.97390 65285	0.06667 13443
0.53846 93101	0.47862 86705		
0.90617 98459	0.23692 68851	$n = 12$	
		0.12523 34085	0.24914 70458
$n = 6$		0.36783 14990	0.23349 25365
0.23861 91861	0.46791 39346	0.58731 79543	0.20316 74267
0.66120 93865	0.36076 15730	0.76990 26742	0.16007 83285
0.93246 95142	0.17132 44924	0.90411 72564	0.10693 93260
		0.98156 06342	0.04717 53364
$n = 7$			
0.00000 00000	0.41795 91837	$n = 20$	
0.40584 51514	0.38183 00505	0.07652 65211	0.15275 33871
0.74153 11856	0.27970 53915	0.22778 58511	0.14917 29865
0.94910 79123	0.12948 49662	0.37370 60887	0.14209 61093
		0.51086 70020	0.13168 86384
$n = 8$		0.63605 36807	0.11819 45320
0.18343 46425	0.36268 37834	0.74633 19065	0.10193 01198
0.52553 24099	0.31370 66459	0.83911 69718	0.08327 67416
0.79666 64774	0.22238 10345	0.91223 44283	0.06267 20483
0.96028 98565	0.10122 85363	0.96397 19272	0.04060 14298
		0.99312 85992	0.01761 40071

Table 27.8 The integration points and weights for a number of n-point Gauss–Legendre integration formulae. The points are given as $\pm x_i$ and the contributions from both $+x_i$ and $-x_i$ must be included. However, the contribution from any point $x_i = 0$ must be counted only once.

factor is treated accurately in Gauss–Chebyshev integration. Thus

$$\int_{-1}^{1} \frac{f(x)}{\sqrt{1-x^2}}\, dx \approx \sum_{i=1}^{n} w_i f(x_i), \tag{27.44}$$

where the integration points x_i are the zeros of the Chebyshev polynomials of the first kind $T_n(x)$ and w_i are the corresponding weights. Fortunately, both sets are analytic and can be written compactly for all n as

$$x_i = \cos \frac{(i-\frac{1}{2})\pi}{n}, \qquad w_i = \frac{\pi}{n} \qquad \text{for } i = 1, \dots, n. \tag{27.45}$$

Note that, for any given n, all points are weighted equally and that no special action is required to deal with the integrable singularities at $x = \pm 1$; they are dealt with automatically through the weight function.

For integrals involving factors of the form $(1-x^2)^{1/2}$, the corresponding formula, based on Chebyshev polynomials of the second kind $U_n(x)$, is

$$\int_{-1}^{1} f(x)\sqrt{1-x^2}\, dx \approx \sum_{i=1}^{n} w_i f(x_i), \tag{27.46}$$

with integration points and weights given, for $i = 1, \dots, n$, by

$$x_i = \cos \frac{i\pi}{n+1}, \qquad w_i = \frac{\pi}{n+1} \sin^2 \frac{i\pi}{n+1}. \tag{27.47}$$

For discussions of the many other schemes available, as well as their relative merits, the reader is referred to books devoted specifically to the theory of numerical analysis. There, details of integration points and weights, as well as quantitative estimates of the error involved in replacing an integral by a finite sum, will be found. Table 27.9 gives the points and weights for a selection of Gauss–Laguerre and Gauss–Hermite schemes.[§]

27.4.4 Monte Carlo methods

Surprising as it may at first seem, random numbers may be used to carry out numerical integration. The random element comes in principally when selecting the points at which the integrand is evaluated, and naturally does not extend to the actual values of the integrand!

For the most part we will continue to use as our model one-dimensional integrals between finite limits, as typified by equation (27.34). Extensions to cover infinite or multidimensional integrals will be indicated briefly at the end of the section. It should be noted here, however, that Monte Carlo methods – the name

---

[§] They, and those presented in table 27.8 for Gauss–Legendre integration, are taken from the much more comprehensive sets to be found in M. Abramowitz and I. A. Stegun (eds), *Handbook of Mathematical Functions* (New York: Dover, 1965).

Gauss–Laguerre and Gauss–Hermite integration

$$\int_0^\infty e^{-x} f(x)\, dx = \sum_{i=1}^n w_i f(x_i) \qquad \int_{-\infty}^\infty e^{-x^2} f(x)\, dx = \sum_{i=1}^n w_i f(x_i)$$

x_i	w_i	$\pm x_i$	w_i
$n = 2$		**$n = 2$**	
0.58578 64376	0.85355 33906	0.70710 67812	0.88622 69255
3.41421 35624	0.14644 66094		
		$n = 3$	
$n = 3$		0.00000 00000	1.18163 59006
0.41577 45568	0.71109 30099	1.22474 48714	0.29540 89752
2.29428 03603	0.27851 77336		
6.28994 50829	0.01038 92565	**$n = 4$**	
		0.52464 76233	0.80491 40900
$n = 4$		1.65068 01239	0.08131 28354
0.32254 76896	0.60315 41043		
1.74576 11012	0.35741 86924	**$n = 5$**	
4.53662 02969	0.03888 79085	0.00000 00000	0.94530 87205
9.39507 09123	0.00053 92947	0.95857 24646	0.39361 93232
		2.02018 28705	0.01995 32421
$n = 5$			
0.26356 03197	0.52175 56106	**$n = 6$**	
1.41340 30591	0.39866 68111	0.43607 74119	0.72462 95952
3.59642 57710	0.07594 24497	1.33584 90740	0.15706 73203
7.08581 00059	0.00361 17587	2.35060 49737	0.00453 00099
12.6408 00844	0.00002 33700		
		$n = 7$	
$n = 6$		0.00000 00000	0.81026 46176
0.22284 66042	0.45896 46740	0.81628 78829	0.42560 72526
1.18893 21017	0.41700 08308	1.67355 16288	0.05451 55828
2.99273 63261	0.11337 33821	2.65196 13568	0.00097 17812
5.77514 35691	0.01039 91975		
9.83746 74184	0.00026 10172	**$n = 8$**	
15.9828 73981	0.00000 08985	0.38118 69902	0.66114 70126
		1.15719 37124	0.20780 23258
$n = 7$		1.98165 67567	0.01707 79830
0.19304 36766	0.40931 89517	2.93063 74203	0.00019 96041
1.02666 48953	0.42183 12779		
2.56787 67450	0.14712 63487	**$n = 9$**	
4.90035 30845	0.02063 35145	0.00000 00000	0.72023 52156
8.18215 34446	0.00107 40101	0.72355 10188	0.43265 15590
12.7341 80292	0.00001 58655	1.46855 32892	0.08847 45274
19.3957 27862	0.00000 00317	2.26658 05845	0.00494 36243
		3.19099 32018	0.00003 96070

Table 27.9 The integration points and weights for a number of n-point Gauss–Laguerre and Gauss–Hermite integration formulae. Where the points are given as $\pm x_i$, the contributions from both $+x_i$ and $-x_i$ must be included. However, the contribution from any point $x_i = 0$ must be counted only once.

has become attached to methods based on randomly generated numbers – in many ways come into their own when used on multidimensional integrals over regions with complicated boundaries.

It goes without saying that in order to use random numbers for calculational purposes a supply of them must be available. There was a time when they were provided in book form as a two-dimensional array of random digits in the range 0 to 9. The user could generate the successive digits of a random number of any desired length by selecting their positions in the table in any predetermined and systematic way. Nowadays all computers and nearly all pocket calculators offer a function which supplies a sequence of decimal numbers, ξ, that, for all practical purposes, are randomly and uniformly chosen in the range $0 \le \xi < 1$. The maximum number of significant figures available in each random number depends on the precision of the generating device. We will defer the details of how these numbers are produced until later in this subsection, where it will also be shown how random numbers distributed in a prescribed way can be generated.

All integrals of the general form shown in equation (27.34) can, by a suitable change of variable, be brought to the form

$$\theta = \int_0^1 f(x)\,dx, \tag{27.48}$$

and we will use this as our standard model.

All approaches to integral evaluation based on random numbers proceed by estimating a quantity whose expectation value is equal to the sought-for value θ. The estimator t must be unbiased, i.e. we must have $E[t] = \theta$, and the method must provide some measure of the likely error in the result. The latter will appear generally as the variance of the estimate, with its usual statistical interpretation, and not as a band in which the true answer is known to lie with certainty.

The various approaches really differ from each other only in the degree of sophistication employed to keep the variance of the estimate of θ small. The overall efficiency of any particular method has to take into account not only the variance of the estimate but also the computing and book-keeping effort required to achieve it.

We do not have the space to describe even the most elementary methods in full detail, but the main thrust of each approach should be apparent to the reader from the brief descriptions that follow.

Crude Monte Carlo

The most straightforward application is one in which the random numbers are used to pick sample points at which $f(x)$ is evaluated. These values are then averaged:

$$t = \frac{1}{n} \sum_{i=1}^{n} f(\xi_i).$$ (27.49)

Stratified sampling

Here the range of x is broken up into k subranges,

$$0 = \alpha_0 < \alpha_1 < \cdots < \alpha_k = 1,$$

and crude Monte Carlo evaluation is carried out in each subrange. The estimate $E[t]$ is then calculated as

$$E[t] = \sum_{j=1}^{k} \sum_{i=1}^{n_j} \frac{\alpha_j - \alpha_{j-1}}{n_j} f\left(\alpha_{j-1} + \xi_{ij}(\alpha_j - \alpha_{j-1})\right).$$ (27.50)

This is an unbiased estimator of θ with variance

$$\sigma_t^2 = \sum_{j=1}^{k} \frac{\alpha_j - \alpha_{j-1}}{n_j} \int_{\alpha_{j-1}}^{\alpha_j} [f(x)]^2 \, dx - \sum_{j=1}^{k} \frac{1}{n_j} \left[\int_{\alpha_{j-1}}^{\alpha_j} f(x) \, dx\right]^2.$$

This variance can be made less than that for crude Monte Carlo, whilst using the same total number of random numbers, $n = \sum n_j$, if the differences between the average values of $f(x)$ in the various subranges are significantly greater than the variations in f within each subrange. It is easier administratively to make all subranges equal in length, but better, if it can be managed, to make them such that the variations in f are approximately equal in all the individual subranges.

Importance sampling

Although we cannot integrate $f(x)$ analytically – we would not be using Monte Carlo methods if we could – if we can find another function $g(x)$ that *can* be integrated analytically and mimics the shape of f then the variance in the estimate of θ can be reduced significantly compared with that resulting from the use of crude Monte Carlo evaluation.

Firstly, if necessary the function g must be renormalised, so that $G(x) = \int_0^x g(y)dy$ has the property $G(1) = 1$. Clearly, it also has the property $G(0) = 0$. Then, since

$$\theta = \int_0^1 \frac{f(x)}{g(x)} \, dG(x),$$

it follows that finding the expectation value of $f(\eta)/g(\eta)$ using a random number η, distributed in such a way that $\xi = G(\eta)$ is uniformly distributed on $(0, 1)$, is equivalent to estimating θ. This involves being able to find the inverse function of G; a discussion of how to do this is given towards the end of this subsection. If $g(\eta)$ mimics $f(\eta)$ well, $f(\eta)/g(\eta)$ will be nearly constant and the estimation will have a very small variance. Further, any error in inverting the relationship between η and ξ will not be important since $f(\eta)/g(\eta)$ will be largely independent of the value of η.

As an example, consider the function $f(x) = [\tan^{-1}(x)]^{1/2}$, which is not analytically integrable over the range $(0, 1)$ but is well mimicked by the easily integrated function $g(x) = x^{1/2}(1 - x^2/6)$. The ratio of the two varies from 1.00 to 1.06 as x varies from 0 to 1. The integral of g over this range is $0.619\,048$, and so it has to be renormalised by the factor $1.615\,38$. The value of the integral of $f(x)$ from 0 to 1 can then be estimated by averaging the value of

$$\frac{[\tan^{-1}(\eta)]^{1/2}}{(1.615\,38)\,\eta^{1/2}(1 - \tfrac{1}{6}\eta^2)}$$

for random variables η which are such that $G(\eta)$ is uniformly distributed on $(0, 1)$. Using batches of as few as ten random numbers gave a value 0.630 for θ, with standard deviation 0.003. The corresponding result for crude Monte Carlo, using the same random numbers, was 0.634 ± 0.065. The increase in precision is obvious, though the additional labour involved would not be justified for a single application.

Control variates

The control-variate method is similar to, but not the same as, importance sampling. Again, an analytically integrable function that mimics $f(x)$ in shape has to be found. The function, known as the control variate, is first scaled so as to match f as closely as possible in magnitude and then its integral is found in closed form. If we denote the scaled control variate by $h(x)$, then the estimate of θ is computed as

$$t = \int_0^1 [f(x) - h(x)]\,dx + \int_0^1 h(x)\,dx. \tag{27.51}$$

The first integral in (27.51) is evaluated using (crude) Monte Carlo, whilst the second is known analytically. Although the first integral should have been rendered small by the choice of $h(x)$, it is its variance that matters. The method relies on the following result (see equation (30.136)):

$$V[t - t'] = V[t] + V[t'] - 2\,\mathrm{Cov}[t, t'],$$

and on the fact that if t estimates θ whilst t' estimates θ' using the same random numbers, then the covariance of t and t' can be larger than the variance of t', and indeed will be so if the integrands producing θ and θ' are highly correlated.

To evaluate the same integral as was estimated previously using importance sampling, we take as $h(x)$ the function $g(x)$ used there, before it was renormalised. Again using batches of ten random numbers, the estimated value for θ was found to be 0.629 ± 0.004, a result almost identical to that obtained using importance sampling, in both value and precision. Since we knew already that $f(x)$ and $g(x)$ diverge monotonically by about 6% as x varies over the range $(0, 1)$, we could have made a small improvement to our control variate by scaling it by 1.03 before using it in equation (27.51).

Antithetic variates

As a final example of a method that improves on crude Monte Carlo, and one that is particularly useful when monotonic functions are to be integrated, we mention the use of antithetic variates. This method relies on finding two estimates t and t' of θ that are strongly anticorrelated (i.e. $\text{Cov}[t, t']$ is large and negative) and using the result

$$V[\tfrac{1}{2}(t + t')] = \tfrac{1}{4} V[t] + \tfrac{1}{4} V[t'] + \tfrac{1}{2} \text{Cov}[t, t'].$$

For example, the use of $\tfrac{1}{2}[f(\xi) + f(1 - \xi)]$ instead of $f(\xi)$ involves only twice as many evaluations of f, and no more random variables, but generally gives an improvement in precision significantly greater than this. For the integral of $f(x) = [\tan^{-1}(x)]^{1/2}$, using as previously a batch of ten random variables, an estimate of 0.623 ± 0.018 was found. This is to be compared with the crude Monte Carlo result, 0.634 ± 0.065, obtained using the same number of random variables.

For a fuller discussion of these methods, and of theoretical estimates of their efficiencies, the reader is referred to more specialist treatments. For practical implementation schemes, a book dedicated to scientific computing should be consulted.[§]

Hit or miss method

We now come to the approach that, in spirit, is closest to the activities that gave Monte Carlo methods their name. In this approach, one or more straightforward yes/no decisions are made on the basis of numbers drawn at random – the end result of each trial is either a hit or a miss! In this section we are concerned with numerical integration, but the general Monte Carlo approach, in which one estimates a physical quantity that is hard or impossible to calculate directly by simulating the physical processes that determine it, is widespread in modern science. For example, the calculation of the efficiencies of detector arrays in experiments to study elementary particle interactions are nearly always carried out in this way. Indeed, in a normal experiment, far more simulated interactions are generated in computers than ever actually occur when the experiment is taking real data.

As was noted in chapter 2, the process of evaluating a one-dimensional integral $\int_a^b f(x)dx$ can be regarded as that of finding the area between the curve $y = f(x)$

[§] e.g. W. H. Press, S. A. Teukolsky, W. T. Vetterling and B. P. Flannery, *Numerical Recipes in C: The Art of Scientific Computing*, 2nd edn (Cambridge: Cambridge University Press, 1992).

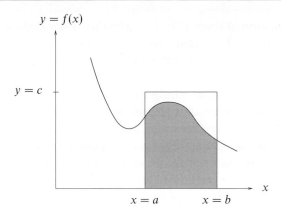

Figure 27.5 A simple rectangular figure enclosing the area (shown shaded) which is equal to $\int_a^b f(x)\,dx$.

and the x-axis in the range $a \le x \le b$. It may not be possible to do this analytically, but if, as shown in figure 27.5, we can enclose the curve in a simple figure whose area can be found trivially then the ratio of the required (shaded) area to that of the bounding figure, $c(b-a)$, is the same as the probability that a randomly selected point inside the boundary will lie below the line.

In order to accommodate cases in which $f(x)$ can be negative in part of the x-range, we treat a slightly more general case. Suppose that, for $a \le x \le b$, $f(x)$ is bounded and known to lie in the range $A \le f(x) \le B$; then the transformation

$$z = \frac{x-a}{b-a}$$

will reduce the integral $\int_a^b f(x)\,dx$ to the form

$$A(b-a) + (B-A)(b-a) \int_0^1 h(z)\,dz, \qquad (27.52)$$

where

$$h(z) = \frac{1}{B-A} \left[f\left((b-a)z + a\right) - A \right].$$

In this form z lies in the range $0 \le z \le 1$ and $h(z)$ lies in the range $0 \le h(z) \le 1$, i.e. both are suitable for simulation using the standard random-number generator. It should be noted that, for an efficient estimation, the bounds A and B should be drawn as tightly as possible – preferably, but not necessarily, they should be equal to the minimum and maximum values of f in the range. The reason for this is that random numbers corresponding to values which $f(x)$ cannot reach add nothing to the estimation but do increase its variance.

It only remains to estimate the final integral on the RHS of equation (27.52). This we do by selecting pairs of random numbers, ξ_1 and ξ_2, and testing whether

$h(\xi_1) > \xi_2$. The fraction of times that this inequality is satisfied estimates the value of the integral (without the scaling factors $(B - A)(b - a)$) since the expectation value of this fraction is the ratio of the area below the curve $y = h(z)$ to the area of a unit square.

To illustrate the evaluation of multiple integrals using Monte Carlo techniques, consider the relatively elementary problem of finding the volume of an irregular solid bounded by planes, say an octahedron. In order to keep the description brief, but at the same time illustrate the general principles involved, let us suppose that the octahedron has two vertices on each of the three Cartesian axes, one on either side of the origin for each axis. Denote those on the x-axis by $x_1 (< 0)$ and $x_2 (> 0)$, and similarly for the y- and z-axes. Then the whole of the octahedron can be enclosed by the rectangular parallelepiped

$$x_1 \le x \le x_2, \quad y_1 \le y \le y_2, \quad z_1 \le z \le z_2.$$

Any point in the octahedron lies inside or on the parallelepiped, but any point in the parallelepiped may or may not lie inside the octahedron.

The equation of the plane containing the three vertex points $(x_i, 0, 0), (0, y_j, 0)$ and $(0, 0, z_k)$ is

$$\frac{x}{x_i} + \frac{y}{y_j} + \frac{z}{z_k} = 1 \qquad \text{for } i, j, k = 1, 2, \tag{27.53}$$

and the condition that any general point (x, y, z) lies on the same side of the plane as the origin is that

$$\frac{x}{x_i} + \frac{y}{y_j} + \frac{z}{z_k} - 1 \le 0. \tag{27.54}$$

For the point to be inside or on the octahedron, equation (27.54) must therefore be satisfied for *all eight* of the sets of i, j and k given in (27.53).

Thus an estimate of the volume of the octahedron can be made by generating random numbers ξ from the usual uniform distribution and then using them in sets of three, according to the following scheme.

With integer m labelling the mth set of three random numbers, calculate

$$x = x_1 + \xi_{3m-2}(x_2 - x_1),$$
$$y = y_1 + \xi_{3m-1}(y_2 - y_1),$$
$$z = z_1 + \xi_{3m}(z_2 - z_1).$$

Define a variable n_m as 1 if (27.54) is satisfied for all eight combinations of i, j, k values and as 0 otherwise. The volume V can then be estimated using $3M$ random numbers from the formula

$$\frac{V}{(x_2 - x_1)(y_2 - y_1)(z_2 - z_1)} = \frac{1}{M} \sum_{m=1}^{M} n_m.$$

It will be seen that, by replacing each n_m in the summation by $f(x, y, z)n_m$, this procedure could be extended to estimate the integral of the function f over the volume of the solid. The method has special value if f is too complicated to have analytic integrals with respect to x, y and z or if the limits of any of these integrals are determined by anything other than the simplest combinations of the other variables. If large values of f are known to be concentrated in particular regions of the integration volume, then some form of stratified sampling should be used.

It will be apparent that this general method can be extended to integrals of general functions, bounded but not necessarily continuous, over volumes with complicated bounding surfaces and, if appropriate, in more than three dimensions.

Random number generation

Earlier in this subsection we showed how to evaluate integrals using sequences of numbers that we took to be distributed uniformly on the interval $0 \leq \xi < 1$. In reality the sequence of numbers is not truly random, since each is generated in a mechanistic way from its predecessor and eventually the sequence will repeat itself. However, the cycle is so long that in practice this is unlikely to be a problem, and the reproducibility of the sequence can even be turned to advantage when checking the accuracy of the rest of a calculational program. Much research has gone into the best ways to produce such 'pseudo-random' sequences of numbers. We do not have space to pursue them here and will limit ourselves to one recipe that works well in practice.

Given any particular starting (integer) value x_0, the following algorithm will generate a full cycle of m values for ξ_i, uniformly distributed on $0 \leq \xi_i < 1$, before repeats appear:

$$x_i = ax_{i-1} + c \quad (\text{mod } m); \qquad \xi_i = \frac{x_i}{m}.$$

Here c is an odd integer and a has the form $a = 4k + 1$, with k an integer. For practical reasons, in computers and calculators m is taken as a (fairly high) power of 2, typically the 32nd power.

The uniform distribution can be used to generate random numbers y distributed according to a more general probability distribution $f(y)$ on the range $a \leq y \leq b$ if the inverse of the indefinite integral of f can be found, either analytically or by means of a look-up table. In other words, if

$$F(y) = \int_a^y f(t)\, dt,$$

for which $F(a) = 0$ and $F(b) = 1$, then $F(y)$ is uniformly distributed on $(0, 1)$. This approach is not limited to finite a and b; a could be $-\infty$ and b could be ∞.

The procedure is thus to select a random number ξ from a uniform distribution on $(0, 1)$ and then take as the random number y the value of $F^{-1}(\xi)$. We now illustrate this with a worked example.

> ▶ *Find an explicit formula that will generate a random number y distributed on $(-\infty, \infty)$ according to the Cauchy distribution*
>
> $$f(y)\, dy = \left(\frac{a}{\pi}\right) \frac{dy}{a^2 + y^2},$$
>
> *given a random number ξ uniformly distributed on $(0, 1)$.*

The first task is to determine the indefinite integral:

$$F(y) = \int_{-\infty}^{y} \left(\frac{a}{\pi}\right) \frac{dt}{a^2 + t^2} = \frac{1}{\pi} \tan^{-1} \frac{y}{a} + \frac{1}{2}.$$

Now, if y is distributed as we wish then $F(y)$ is uniformly distributed on $(0, 1)$. This follows from the fact that the derivative of $F(y)$ is $f(y)$. We therefore set $F(y)$ equal to ξ and obtain

$$\xi = \frac{1}{\pi} \tan^{-1} \frac{y}{a} + \frac{1}{2},$$

yielding

$$y = a \tan[\pi(\xi - \tfrac{1}{2})].$$

This explicit formula shows how to change a random number ξ drawn from a population uniformly distributed on $(0, 1)$ into a random number y distributed according to the Cauchy distribution. ◀

Look-up tables operate as described below for cumulative distributions $F(y)$ that are non-invertible, i.e. $F^{-1}(y)$ cannot be expressed in closed form. They are especially useful if many random numbers are needed but great sampling accuracy is not essential. The method for an N-entry table can be summarised as follows. Define w_m by $F(w_m) = m/N$ for $m = 1, 2, \ldots, N$, and store a table of

$$y(m) = \tfrac{1}{2}(w_m + w_{m-1}).$$

As each random number y is needed, calculate k as the integral part of $N\xi$ and take y as given by $y(k)$.

Normally, such a look-up table would have to be used for generating random numbers with a Gaussian distribution, as the cumulative integral of a Gaussian is non-invertible. It would be, in essence, table 30.3, with the roles of argument and value interchanged. In this particular case, an alternative, based on the central limit theorem, can be considered.

With ξ_i generated in the usual way, i.e. uniformly distributed on the interval $0 \le \xi < 1$, the random variable

$$y = \sum_{i=1}^{n} \xi_i - \tfrac{1}{2} n \tag{27.55}$$

is normally distributed with mean 0 and variance $n/12$ when n is large. This approach does produce a continuous spectrum of possible values for y, but needs many values of ξ_i for each value of y and is a very poor approximation if the wings of the Gaussian distribution have to be sampled accurately. For nearly all practical purposes a Gaussian look-up table is to be preferred.

27.5 Finite differences

It will have been noticed that earlier sections included several equations linking sequential values of f_i and the derivatives of f evaluated at one of the x_i. In this section, by way of preparation for the numerical treatment of differential equations, we establish these relationships in a more systematic way.

Again we consider a set of values f_i of a function $f(x)$ evaluated at equally spaced points x_i, their separation being h. As before, the basis for our discussion will be a Taylor series expansion, but on this occasion about the point x_i:

$$f_{i\pm1} = f_i \pm hf_i' + \frac{h^2}{2!}f_i'' \pm \frac{h^3}{3!}f_i^{(3)} + \cdots . \tag{27.56}$$

In this section, and subsequently, we denote the nth derivative evaluated at x_i by $f_i^{(n)}$.

From (27.56), three different expressions that approximate $f_i^{(1)}$ can be derived. The first of these, obtained by subtracting the $\pm$ equations, is

$$f_i^{(1)} \equiv \left(\frac{df}{dx}\right)_{x_i} = \frac{f_{i+1} - f_{i-1}}{2h} - \frac{h^2}{3!}f_i^{(3)} - \cdots . \tag{27.57}$$

The quantity $(f_{i+1} - f_{i-1})/(2h)$ is known as the central difference approximation to $f_i^{(1)}$ and can be seen from (27.57) to be in error by approximately $(h^2/6)f_i^{(3)}$.

An alternative approximation, obtained from (27.56+) alone, is given by

$$f_i^{(1)} \equiv \left(\frac{df}{dx}\right)_{x_i} = \frac{f_{i+1} - f_i}{h} - \frac{h}{2!}f_i^{(2)} - \cdots . \tag{27.58}$$

The *forward difference* approximation, $(f_{i+1} - f_i)/h$, is clearly a poorer approximation, since it is in error by approximately $(h/2)f_i^{(2)}$ as compared with $(h^2/6)f_i^{(3)}$. Similarly, the backward difference $(f_i - f_{i-1})/h$ obtained from (27.56−) is not as good as the central difference; the sign of the error is reversed in this case.

This type of differencing approximation can be continued to the higher derivatives of f in an obvious manner. By adding the two equations (27.56$\pm$), a central difference approximation to $f_i^{(2)}$ can be obtained:

$$f_i^{(2)} \equiv \left(\frac{d^2f}{dx^2}\right) \approx \frac{f_{i+1} - 2f_i + f_{i-1}}{h^2} . \tag{27.59}$$

The error in this approximation (also known as the second difference of f) is easily shown to be about $(h^2/12)f_i^{(4)}$.

Of course, if the function $f(x)$ is a sufficiently simple polynomial in x, all derivatives beyond a particular one will vanish and there is no error in taking the differences to obtain the derivatives.

> ▶The following is copied from the tabulation of a second-degree polynomial $f(x)$ at values of x from 1 to 12 inclusive:
>
> $$2, \ 2, \ ?, \ 8, \ 14, \ 22, \ 32, \ 46, \ ?, \ 74, \ 92, \ 112.$$
>
> The entries marked ? were illegible and in addition one error was made in transcription. Complete and correct the table. Would your procedure have worked if the copying error had been in $f(6)$?

Write out the entries again in row (a) below, and where possible calculate first differences in row (b) and second differences in row (c). Denote the jth entry in row (n) by $(n)_j$.

(a) 2	2	?	8	14	22	32	46	?	74	92	112
(b)	0	?	?	6	8	10	14	?	?	18	20
(c)		?	?	?	2	2	4	?	?	?	2

Because the polynomial is second-degree, the second differences $(c)_j$, which are proportional to $d^2 f/dx^2$, should be constant, and clearly the constant should be 2. That is, $(c)_6$ should equal 2 and $(b)_7$ should equal 12 (not 14). Since all the $(c)_j = 2$, we can conclude that $(b)_2 = 2$, $(b)_3 = 4$, $(b)_8 = 14$, and $(b)_9 = 16$. Working these changes back to row (a) shows that $(a)_3 = 4$, $(a)_8 = 44$ (not 46), and $(a)_9 = 58$.

The entries therefore should read

$$\text{(a) } 2, \ 2, \ \mathbf{4}, \ 8, \ 14, \ 22, \ 32, \ \mathbf{44}, \ \mathbf{58}, \ 74, \ 92, \ 112,$$

where the amended entries are shown in bold type.

It is easily verified that if the error were in $f(6)$ no two computable entries in row (c) would be equal, and it would not be clear what the correct common entry should be. Nevertheless, trial and error might arrive at a self-consistent scheme. ◀

27.6 Differential equations

For the remaining sections of this chapter our attention will be on the solution of differential equations by numerical methods. Some of the general difficulties of applying numerical methods to differential equations will be all too apparent. Initially we consider only the simplest kind of equation – one of first order, typically represented by

$$\frac{dy}{dx} = f(x, y), \tag{27.60}$$

where y is taken as the dependent variable and x the independent one. If this equation can be solved analytically then that is the best course to adopt. But sometimes it is not possible to do so and a numerical approach becomes the only one available. In fact, most of the examples that we will use can be solved easily by an explicit integration, but, for the purposes of illustration, this is an advantage rather than the reverse since useful comparisons can then be made between the numerically derived solution and the exact one.

x	h							y(exact)
	0.01	0.1	0.5	1.0	1.5	2	3	
0	(1)	(1)	(1)	(1)	(1)	(1)	(1)	(1)
0.5	0.605	0.590	0.500	0	−0.500	−1	−2	0.607
1.0	0.366	0.349	0.250	0	0.250	1	4	0.368
1.5	0.221	0.206	0.125	0	−0.125	−1	−8	0.223
2.0	0.134	0.122	0.063	0	0.063	1	16	0.135
2.5	0.081	0.072	0.032	0	−0.032	−1	−32	0.082
3.0	0.049	0.042	0.016	0	0.016	1	64	0.050

Table 27.10 The solution y of differential equation (27.61) using the Euler forward difference method for various values of h. The exact solution is also shown.

27.6.1 Difference equations

Consider the differential equation

$$\frac{dy}{dx} = -y, \qquad y(0) = 1, \tag{27.61}$$

and the possibility of solving it numerically by approximating dy/dx by a finite difference along the lines indicated in section 27.5. We start with the forward difference

$$\left(\frac{dy}{dx}\right)_{x_i} \simeq \frac{y_{i+1} - y_i}{h}, \tag{27.62}$$

where we use the notation of section 27.5 but with f replaced by y. In this particular case, it leads to the recurrence relation

$$y_{i+1} = y_i + h\left(\frac{dy}{dx}\right)_i = y_i - hy_i = (1 - h)y_i. \tag{27.63}$$

Thus, since $y_0 = y(0) = 1$ is given, $y_1 = y(0 + h) = y(h)$ can be calculated, and so on (this is the *Euler* method). Table 27.10 shows the values of $y(x)$ obtained if this is done using various values of h and for selected values of x. The exact solution, $y(x) = \exp(-x)$, is also shown.

It is clear that to maintain anything like a reasonable accuracy only very small steps, h, can be used. Indeed, if h is taken to be too large, not only is the accuracy bad but, as can be seen, for $h > 1$ the calculated solution oscillates (when it should be monotonic), and for $h > 2$ it diverges. Equation (27.63) is of the form $y_{i+1} = \lambda y_i$, and a necessary condition for non-divergence is $|\lambda| < 1$, i.e. $0 < h < 2$, though in no way does this ensure accuracy.

Part of this difficulty arises from the poor approximation (27.62); its right-hand side is a closer approximation to dy/dx evaluated at $x = x_i + h/2$ than to dy/dx at $x = x_i$. This is the result of using a forward difference rather than the

1021

x	y(estim.)	y(exact)
−0.5	(1.648)	—
0	(1.000)	(1.000)
0.5	0.648	0.607
1.0	0.352	0.368
1.5	0.296	0.223
2.0	0.056	0.135
2.5	0.240	0.082
3.0	−0.184	0.050

Table 27.11   The solution of differential equation (27.61) using the Milne central difference method with $h = 0.5$ and accurate starting values.

more accurate, but of course still approximate, central difference. A more accurate method based on central differences (*Milne's method*) gives the recurrence relation

$$y_{i+1} = y_{i-1} + 2h \left(\frac{dy}{dx} \right)_i \qquad (27.64)$$

in general and, in this particular case,

$$y_{i+1} = y_{i-1} - 2hy_i. \qquad (27.65)$$

An additional difficulty now arises, since two initial values of y are needed. The second must be estimated by other means (e.g. by using a Taylor series, as discussed later), but for illustration purposes we will take the accurate value, $y(-h) = \exp h$, as the value of y_{-1}. If h is taken as, say, 0.5 and (27.65) is applied repeatedly, then the results shown in table 27.11 are obtained.

Although some improvement in the early values of the calculated $y(x)$ is noticeable, as compared with the corresponding ($h = 0.5$) column of table 27.10, this scheme soon runs into difficulties, as is obvious from the last two rows of the table.

Some part of this poor performance is not really attributable to the approximations made in estimating dy/dx but to the form of the equation itself and hence of its solution. *Any* rounding error occurring in the evaluation effectively introduces into y some contamination by the solution of

$$\frac{dy}{dx} = +y.$$

This equation has the solution $y(x) = \exp x$ and so grows without limit; ultimately it will dominate the sought-for solution and thus render the calculations totally inaccurate.

We have only illustrated, rather than analysed, some of the difficulties associated with simple finite-difference iteration schemes for first-order differential equations, but they may be summarised as (i) insufficiently precise approximations to the derivatives and (ii) inherent instability due to rounding errors.

27.6.2 Taylor series solutions

Since a Taylor series expansion is exact if all its terms are included, and the limits of convergence are not exceeded, we may seek to use one to evaluate y_1, y_2, etc. for an equation

$$\frac{dy}{dx} = f(x, y), \tag{27.66}$$

when the initial value $y(x_0) = y_0$ is given.

The Taylor series is

$$y(x + h) = y(x) + hy'(x) + \frac{h^2}{2!}y''(x) + \frac{h^3}{3!}y^{(3)}(x) + \cdots . \tag{27.67}$$

In the present notation, at the point $x = x_i$ this is written

$$y_{i+1} = y_i + hy_i^{(1)} + \frac{h^2}{2!}y_i^{(2)} + \frac{h^3}{3!}y_i^{(3)} + \cdots . \tag{27.68}$$

But, for the required solution $y(x)$, we know that

$$y_i^{(1)} = \left(\frac{dy}{dx}\right)_{x_i} = f(x_i, y_i), \tag{27.69}$$

and the value of the second derivative at $x = x_i$, $y = y_i$ can be obtained from it:

$$y_i^{(2)} = \frac{\partial f}{\partial x} + \frac{\partial f}{\partial y}\frac{dy}{dx} = \frac{\partial f}{\partial x} + f\frac{\partial f}{\partial y}. \tag{27.70}$$

This process can be continued for the third and higher derivatives, all of which are to be evaluated at (x_i, y_i).

Having obtained expressions for the derivatives $y_i^{(n)}$ in (27.67), two alternative ways of proceeding are open to us:

(i) equation (27.68) is used to evaluate y_{i+1}, the whole process is repeated to obtain y_{i+2}, and so on;
(ii) equation (27.68) is applied several times but using a different value of h each time, and so the corresponding values of $y(x + h)$ are obtained.

It is clear that, on the one hand, approach (i) does not require so many terms of (27.67) to be kept, but, on the other hand, the $y_i(n)$ have to be recalculated at each step. With approach (ii), fairly accurate results for y may be obtained for values of x close to the given starting value, but for large values of h a large number of terms of (27.67) must be kept. As an example of approach (ii) we solve the following problem.

x	y(estim.)	y(exact)
0	1.0000	1.0000
0.1	1.2346	1.2346
0.2	1.5619	1.5625
0.3	2.0331	2.0408
0.4	2.7254	2.7778
0.5	3.7500	4.0000

Table 27.12   The solution of differential equation (27.71) using a Taylor series.

▶*Find the numerical solution of the equation*

$$\frac{dy}{dx} = 2y^{3/2}, \qquad y(0) = 1, \tag{27.71}$$

for $x = 0.1$ *to* 0.5 *in steps of* 0.1. *Compare it with the exact solution obtained analytically.*

Since the right-hand side of the equation does not contain x explicitly, (27.70) is greatly simplified and the calculation becomes a repeated application of

$$y_i^{(n+1)} = \frac{\partial y^{(n)}}{\partial y}\frac{dy}{dx} = f\frac{\partial y^{(n)}}{\partial y}.$$

The necessary derivatives and their values at $x = 0$, where $y = 1$, are given below:

$$y(0) = 1 \qquad\qquad\qquad\qquad 1$$
$$y' = 2y^{3/2} \qquad\qquad\qquad\qquad 2$$
$$y'' = (3/2)(2y^{1/2})(2y^{3/2}) = 6y^2 \qquad 6$$
$$y^{(3)} = (12y)2y^{3/2} = 24y^{5/2} \qquad 24$$
$$y^{(4)} = (60y^{3/2})2y^{3/2} = 120y^3 \qquad 120$$
$$y^{(5)} = (360y^2)2y^{3/2} = 720y^{7/2} \qquad 720$$

Thus the Taylor expansion of the solution about the origin (in fact a Maclaurin series) is

$$y(x) = 1 + 2x + \frac{6}{2!}x^2 + \frac{24}{3!}x^3 + \frac{120}{4!}x^4 + \frac{720}{5!}x^5 + \cdots .$$

Hence, y(estim.) $= 1 + 2x + 3x^2 + 4x^3 + 5x^4 + 6x^5$. Values calculated from this are given in table 27.12. Comparison with the exact values shows that using the first six terms gives a value that is correct to one part in 100, up to $x = 0.3$. ◀

27.6.3 Prediction and correction

An improvement in the accuracy obtainable using difference methods is possible if steps are taken, sometimes retrospectively, to allow for inaccuracies in approximating derivatives by differences. We will describe only the simplest schemes of this kind and begin with a *prediction* method, usually called the *Adams method*.

1024

The forward difference estimate of y_{i+1}, namely

$$y_{i+1} = y_i + h \left(\frac{dy}{dx} \right)_i = y_i + hf(x_i, y_i), \tag{27.72}$$

would give exact results if y were a linear function of x in the range $x_i \leq x \leq x_i + h$. The idea behind the Adams method is to allow some relaxation of this and suppose that y can be adequately approximated by a parabola over the interval $x_{i-1} \leq x \leq x_{i+1}$. In the same interval, dy/dx can then be approximated by a linear function:

$$f(x, y) = \frac{dy}{dx} \approx a + b(x - x_i) \quad \text{for } x_i - h \leq x \leq x_i + h.$$

The values of a and b are fixed by the calculated values of f at x_{i-1} and x_i, which we may denote by f_{i-1} and f_i:

$$a = f_i, \qquad b = \frac{f_i - f_{i-1}}{h}.$$

Thus

$$y_{i+1} - y_i \approx \int_{x_i}^{x_i+h} \left[f_i + \frac{(f_i - f_{i-1})}{h}(x - x_i) \right] dx,$$

which yields

$$y_{i+1} = y_i + hf_i + \tfrac{1}{2}h(f_i - f_{i-1}). \tag{27.73}$$

The last term of this expression is seen to be a correction to result (27.72). That it is, in some sense, the second-order correction,

$$\tfrac{1}{2}h^2 y^{(2)}_{i-1/2},$$

to a first-order formula is apparent.

Such a procedure requires, in addition to a value for y_0, a value for either y_1 or y_{-1}, so that f_1 or f_{-1} can be used to initiate the iteration. This has to be obtained by other methods, e.g. a Taylor series expansion.

Improvements to simple difference formulae can also be obtained by using *correction* methods. In these, a rough prediction of the value y_{i+1} is made first, and then this is used in a better formula, not originally usable since it, in turn, requires a value of y_{i+1} for its evaluation. The value of y_{i+1} is then recalculated, using this better formula.

Such a scheme based on the forward difference formula might be as follows:

(i) predict y_{i+1} using $y_{i+1} = y_i + hf_i$;

(ii) calculate f_{i+1} using this value;

(iii) recalculate y_{i+1} using $y_{i+1} = y_i + h(f_i + f_{i+1})/2$. Here $(f_i + f_{i+1})/2$ has replaced the f_i used in (i), since it better represents the average value of dy/dx in the interval $x_i \leq x \leq x_i + h$.

Steps (ii) and (iii) can be iterated to improve further the approximation to the average value of dy/dx, but this will not compensate for the omission of higher-order derivatives in the forward difference formula.

Many more complex schemes of prediction and correction, in most cases combining the two in the same process, have been devised, but the reader is referred to more specialist texts for discussions of them. However, because it offers some clear advantages, one group of methods will be set out explicitly in the next subsection. This is the general class of schemes known as Runge–Kutta methods.

27.6.4 Runge–Kutta methods

The Runge–Kutta method of integrating

$$\frac{dy}{dx} = f(x, y) \tag{27.74}$$

is a step-by-step process of obtaining an approximation for y_{i+1} by starting from the value of y_i. Among its advantages are that no functions other than f are used, no subsidiary differentiation is needed and no additional starting values need be calculated.

To be set against these advantages is the fact that f is evaluated using somewhat complicated arguments and that this has to be done several times for each increase in the value of i. However, once a procedure has been established, for example on a computer, the method usually gives good results.

The basis of the method is to simulate the (accurate) Taylor series for $y(x_i + h)$, not by calculating all the higher derivatives of y at the point x_i but by taking a particular combination of the values of the first derivative of y evaluated at a number of carefully chosen points. Equation (27.74) is used to evaluate these derivatives. The accuracy can be made to be up to whatever power of h is desired, but, naturally, the greater the accuracy, the more complex the calculation, and, in any case, rounding errors cannot ultimately be avoided.

The setting up of the calculational scheme may be illustrated by considering the particular case in which second-order accuracy in h is required. To second order, the Taylor expansion is

$$y_{i+1} = y_i + hf_i + \frac{h^2}{2}\left(\frac{df}{dx}\right)_{x_i}, \tag{27.75}$$

where

$$\left(\frac{df}{dx}\right)_{x_i} = \left(\frac{\partial f}{\partial x} + f\frac{\partial f}{\partial y}\right)_{x_i} \equiv \frac{\partial f_i}{\partial x} + f_i\frac{\partial f_i}{\partial y},$$

the last step being merely the definition of an abbreviated notation.

We assume that this can be simulated by a form

$$y_{i+1} = y_i + \alpha_1 h f_i + \alpha_2 h f(x_i + \beta_1 h, \ y_i + \beta_2 h f_i), \tag{27.76}$$

which in effect uses a weighted mean of the value of dy/dx at x_i and its value at some point yet to be determined. The object is to choose values of α_1, α_2, β_1 and β_2 such that (27.76) coincides with (27.75) up to the coefficient of h^2.

Expanding the function f in the last term of (27.76) in a Taylor series of its own, we obtain

$$f(x_i + \beta_1 h, \ y_i + \beta_2 h f_i) = f(x_i, y_i) + \beta_1 h \frac{\partial f_i}{\partial x} + \beta_2 h f_i \frac{\partial f_i}{\partial y} + O(h^2).$$

Putting this result into (27.76) and rearranging in powers of h, we obtain

$$y_{i+1} = y_i + (\alpha_1 + \alpha_2) h f_i + \alpha_2 h^2 \left(\beta_1 \frac{\partial f_i}{\partial x} + \beta_2 f_i \frac{\partial f_i}{\partial y} \right). \tag{27.77}$$

Comparing this with (27.75) shows that there is, in fact, some freedom remaining in the choice of the α's and β's. In terms of an arbitrary α_1 ($\neq 1$),

$$\alpha_2 = 1 - \alpha_1, \qquad \beta_1 = \beta_2 = \frac{1}{2(1 - \alpha_1)}.$$

One possible choice is $\alpha_1 = 0.5$, giving $\alpha_2 = 0.5$, $\beta_1 = \beta_2 = 1$. In this case the procedure (equation (27.76)) can be summarised by

$$y_{i+1} = y_i + \tfrac{1}{2}(a_1 + a_2), \tag{27.78}$$

where

$$a_1 = h f(x_i, y_i),$$
$$a_2 = h f(x_i + h, \ y_i + a_1).$$

Similar schemes giving higher-order accuracy in h can be devised. Two such schemes, given without derivation, are as follows.

(i) To order h^3,

$$y_{i+1} = y_i + \tfrac{1}{6}(b_1 + 4b_2 + b_3), \tag{27.79}$$

where

$$b_1 = h f(x_i, y_i),$$
$$b_2 = h f(x_i + \tfrac{1}{2}h, \ y_i + \tfrac{1}{2}b_1),$$
$$b_3 = h f(x_i + h, \ y_i + 2b_2 - b_1).$$

(ii) To order h^4,

$$y_{i+1} = y_i + \tfrac{1}{6}(c_1 + 2c_2 + 2c_3 + c_4), \qquad (27.80)$$

where

$$c_1 = hf(x_i, y_i),$$
$$c_2 = hf(x_i + \tfrac{1}{2}h, \; y_i + \tfrac{1}{2}c_1),$$
$$c_3 = hf(x_i + \tfrac{1}{2}h, \; y_i + \tfrac{1}{2}c_2),$$
$$c_4 = hf(x_i + h, \; y_i + c_3).$$

27.6.5 Isoclines

The final method to be described for first-order differential equations is not so much numerical as graphical, but since it is sometimes useful it is included here. The method, known as that of *isoclines*, involves sketching for a number of values of a parameter c those curves (the isoclines) in the xy-plane along which $f(x, y) = c$, i.e. those curves along which dy/dx is a constant of known value. It should be noted that isoclines are not generally straight lines. Since a straight line of slope dy/dx at and through any particular point is a tangent to the curve $y = y(x)$ at that point, small elements of straight lines, with slopes appropriate to the isoclines they cut, effectively form the curve $y = y(x)$.

Figure 27.6 illustrates in outline the method as applied to the solution of

$$\frac{dy}{dx} = -2xy. \qquad (27.81)$$

The thinner curves (rectangular hyperbolae) are a selection of the isoclines along which $-2xy$ is constant and equal to the corresponding value of c. The small cross lines on each curve show the slopes $(= c)$ that solutions of (27.81) must have if they cross the curve. The thick line is the solution for which $y = 1$ at $x = 0$; it takes the slope dictated by the value of c on each isocline it crosses. The analytic solution with these properties is $y(x) = \exp(-x^2)$.

27.7 Higher-order equations

So far the discussion of numerical solutions of differential equations has been in terms of one dependent and one independent variable related by a first-order equation. It is straightforward to carry out an extension to the case of several dependent variables $y_{[r]}$ governed by R first-order equations:

$$\frac{dy_{[r]}}{dx} = f_{[r]}(x, y_{[1]}, y_{[2]}, \ldots, y_{[R]}), \qquad r = 1, 2, \ldots, R.$$

We have enclosed the label r in brackets so that there is no confusion between, say, the second dependent variable $y_{[2]}$ and the value y_2 of a variable y at the

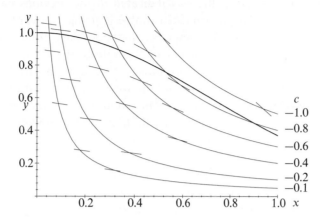

Figure 27.6 The isocline method. The cross lines on each isocline show the slopes that solutions of $dy/dx = -2xy$ must have at the points where they cross the isoclines. The heavy line is the solution with $y(0) = 1$, namely $\exp(-x^2)$.

second calculational point x_2. The integration of these equations by the methods discussed in the previous section presents no particular difficulty, provided that all the equations are advanced through each particular step before any of them is taken through the following step.

Higher-order equations in one dependent and one independent variable can be reduced to a set of simultaneous equations, provided that they can be written in the form

$$\frac{d^R y}{dx^R} = f(x, y, y', \ldots, y^{(R-1)}), \tag{27.82}$$

where R is the order of the equation. To do this, a new set of variables $p_{[r]}$ is defined by

$$p_{[r]} = \frac{d^r y}{dx^r}, \qquad r = 1, 2, \ldots, R - 1. \tag{27.83}$$

Equation (27.82) is then equivalent to the following set of simultaneous first-order equations:

$$\frac{dy}{dx} = p_{[1]},$$

$$\frac{dp_{[r]}}{dx} = p_{[r+1]}, \qquad r = 1, 2, \ldots, R - 2, \tag{27.84}$$

$$\frac{dp_{[R-1]}}{dx} = f(x, y, p_{[1]}, \ldots, p_{[R-1]}).$$

1029

These can then be treated in the way indicated in the previous paragraph. The extension to more than one dependent variable is straightforward.

In practical problems it often happens that boundary conditions applicable to a higher-order equation consist not of the values of the function and all its derivatives at one particular point but of, say, the values of the function at two separate end-points. In these cases a solution cannot be found using an explicit step-by-step 'marching' scheme, in which the solutions at successive values of the independent variable are calculated using solution values previously found. Other methods have to be tried.

One obvious method is to treat the problem as a 'marching one', but to use a number of (intelligently guessed) initial values for the derivatives at the starting point. The aim is then to find, by interpolation or some other form of iteration, those starting values for the derivatives that will produce the given value of the function at the finishing point.

In some cases the problem can be reduced by a differencing scheme to a matrix equation. Such a case is that of a second-order equation for $y(x)$ with constant coefficients and given values of y at the two end-points. Consider the second-order equation

$$y'' + 2ky' + \mu y = f(x), \tag{27.85}$$

with the boundary conditions

$$y(0) = A, \qquad y(1) = B.$$

If (27.85) is replaced by a central difference equation,

$$\frac{y_{i+1} - 2y_i + y_{i-1}}{h^2} + 2k\frac{y_{i+1} - y_{i-1}}{2h} + \mu y_i = f(x_i),$$

we obtain from it the recurrence relation

$$(1 + kh)y_{i+1} + (\mu h^2 - 2)y_i + (1 - kh)y_{i-1} = h^2 f(x_i).$$

For $h = 1/(N - 1)$ this is in exactly the form of the $N \times N$ tridiagonal matrix equation (27.30), with

$$b_1 = b_N = 1, \qquad c_1 = a_N = 0,$$

$$a_i = 1 - kh, \qquad b_i = \mu h^2 - 2, \qquad c_i = 1 + kh, \qquad i = 2, 3, \ldots, N - 1,$$

and y_1 replaced by A, y_N by B and y_i by $h^2 f(x_i)$ for $i = 2, 3, \ldots, N - 1$. The solutions can be obtained as in (27.31) and (27.32).

27.8 Partial differential equations

The extension of previous methods to partial differential equations, thus involving two or more independent variables, proceeds in a more or less obvious way. Rather than an interval divided into equal steps by the points at which solutions to the equations are to be found, a mesh of points in two or more dimensions has to be set up and all the variables given an increased number of subscripts.

Considerations of the stability, accuracy and feasibility of particular calculational schemes are the same as for the one-dimensional case in principle, but in practice are too complicated to be discussed here.

Rather than note generalities that we are unable to pursue in any quantitative way, we will conclude this chapter by indicating in outline how two familiar partial differential equations of physical science can be set up for numerical solution. The first of these is Laplace's equation in two dimensions,

$$\frac{\partial^2 \phi}{\partial x^2} + \frac{\partial^2 \phi}{\partial y^2} = 0, \tag{27.86}$$

the value of ϕ being given on the perimeter of a closed domain.

A grid with spacings Δx and Δy in the two directions is first chosen, so that, for example, x_i stands for the point $x_0 + i\Delta x$ and $\phi_{i,j}$ for the value $\phi(x_i, y_j)$. Next, using a second central difference formula, (27.86) is turned into

$$\frac{\phi_{i+1,j} - 2\phi_{i,j} + \phi_{i-1,j}}{(\Delta x)^2} + \frac{\phi_{i,j+1} - 2\phi_{i,j} + \phi_{i,j-1}}{(\Delta y)^2} = 0, \tag{27.87}$$

for $i = 0, 1, \ldots, N$ and $j = 0, 1, \ldots, M$. If $(\Delta x)^2 = \lambda(\Delta y)^2$ then this becomes the recurrence relationship

$$\phi_{i+1,j} + \phi_{i-1,j} + \lambda(\phi_{i,j+1} + \phi_{i,j-1}) = 2(1 + \lambda)\phi_{i,j}. \tag{27.88}$$

The boundary conditions in their simplest form (i.e. for a rectangular domain) mean that

$$\phi_{0,j}, \quad \phi_{N,j}, \quad \phi_{i,0}, \quad \phi_{i,M} \tag{27.89}$$

have predetermined values. Non-rectangular boundaries can be accommodated, either by more complex boundary-value prescriptions or by using non-Cartesian coordinates.

To find a set of values satisfying (27.88), an initial guess of a complete set of values for the $\phi_{i,j}$ is made, subject to the requirement that the quantities listed in (27.89) have the given fixed values; those values that are not on the boundary are then adjusted iteratively in order to try to bring about condition (27.88) everywhere. Clearly one scheme is to set $\lambda = 1$ and recalculate each $\phi_{i,j}$ as the mean of the four current values at neighbouring grid-points, using (27.88) directly, and then to iterate this recalculation until no value of ϕ changes significantly after a complete cycle through all values of i and j. This procedure is the simplest of such 'relaxation' methods; for a slightly more sophisticated scheme see exercise 27.26 at the end of this chapter. The reader is referred to specialist books for fuller accounts of how this approach can be made faster and more accurate.

Our final example is based upon the one-dimensional diffusion equation for the temperature ϕ of a system:

$$\frac{\partial \phi}{\partial t} = \kappa \frac{\partial^2 \phi}{\partial x^2}. \tag{27.90}$$

If $\phi_{i,j}$ stands for $\phi(x_0 + i\Delta x, \ t_0 + j\Delta t)$ then a forward difference representation of the time derivative and a central difference representation for the spatial derivative lead to the following relationship:

$$\frac{\phi_{i,j+1} - \phi_{i,j}}{\Delta t} = \kappa \frac{\phi_{i+1,j} - 2\phi_{i,j} + \phi_{i-1,j}}{(\Delta x)^2}. \tag{27.91}$$

This allows the construction of an explicit scheme for generating the temperature distribution at later times, given that it is known at some earlier time:

$$\phi_{i,j+1} = \alpha(\phi_{i+1,j} + \phi_{i-1,j}) + (1 - 2\alpha)\phi_{i,j}, \tag{27.92}$$

where $\alpha = \kappa \Delta t / (\Delta x)^2$.

Although this scheme is explicit, it is not a good one because of the asymmetric way in which the differences are formed. However, the effect of this can be minimised if we study and correct for the errors introduced in the following way. Taylor's series for the time variable gives

$$\phi_{i,j+1} = \phi_{i,j} + \Delta t \frac{\partial \phi_{i,j}}{\partial t} + \frac{(\Delta t)^2}{2!} \frac{\partial^2 \phi_{i,j}}{\partial t^2} + \cdots, \tag{27.93}$$

using the same notation as previously. Thus the first correction term to the LHS of (27.91) is

$$-\frac{\Delta t}{2} \frac{\partial^2 \phi_{i,j}}{\partial t^2}. \tag{27.94}$$

The first term omitted on the RHS of the same equation is, by a similar argument,

$$-\kappa \frac{2(\Delta x)^2}{4!} \frac{\partial^4 \phi_{i,j}}{\partial x^4}. \tag{27.95}$$

But, using the fact that ϕ satisfies (27.90), we obtain

$$\frac{\partial^2 \phi}{\partial t^2} = \frac{\partial}{\partial t}\left(\kappa \frac{\partial^2 \phi}{\partial x^2}\right) = \kappa \frac{\partial^2}{\partial x^2}\left(\frac{\partial \phi}{\partial t}\right) = \kappa^2 \frac{\partial^4 \phi}{\partial x^4}, \tag{27.96}$$

and so, to this accuracy, the two errors (27.94) and (27.95) can be made to cancel if α is chosen such that

$$-\frac{\kappa^2 \Delta t}{2} = -\frac{2\kappa(\Delta x)^2}{4!}, \qquad \text{i.e. } \alpha = \frac{1}{6}.$$

27.9 Exercises

27.1 Use an iteration procedure to find the root of the equation $40x = \exp x$ to four significant figures.

27.2 Using the Newton–Raphson procedure find, correct to three decimal places, the root nearest to 7 of the equation $4x^3 + 2x^2 - 200x - 50 = 0$.

27.3 Show the following results about rearrangement schemes for polynomial equations.

(a) That if a polynomial equation $g(x) \equiv x^m - f(x) = 0$, where $f(x)$ is a polynomial of degree less than m and for which $f(0) \neq 0$, is solved using a rearrangement iteration scheme $x_{n+1} = [f(x_n)]^{1/m}$, then, in general, the scheme will have only first-order convergence.

(b) By considering the cubic equation

$$x^3 - ax^2 + 2abx - (b^3 + ab^2) = 0$$

for arbitrary non-zero values of a and b, demonstrate that, in special cases, the same rearrangement scheme can give second- (or higher-) order convergence.

27.4 The square root of a number N is to be determined by means of the iteration scheme

$$x_{n+1} = x_n \left[1 - \left(N - x_n^2\right) f(N)\right].$$

Determine how to choose $f(N)$ so that the process has second-order convergence. Given that $\sqrt{7} \approx 2.65$, calculate $\sqrt{7}$ as accurately as a single application of the formula will allow.

27.5 Solve the following set of simultaneous equations using Gaussian elimination (including interchange where it is formally desirable):

$$x_1 + 3x_2 + 4x_3 + 2x_4 = 0,$$
$$2x_1 + 10x_2 - 5x_3 + x_4 = 6,$$
$$4x_2 + 3x_3 + 3x_4 = 20,$$
$$-3x_1 + 6x_2 + 12x_3 - 4x_4 = 16.$$

27.6 The following table of values of a polynomial $p(x)$ of low degree contains an error. Identify and correct the erroneous value and extend the table up to $x = 1.2$.

x	$p(x)$	x	$p(x)$
0.0	0.000	0.5	0.165
0.1	0.011	0.6	0.216
0.2	0.040	0.7	0.245
0.3	0.081	0.8	0.256
0.4	0.128	0.9	0.243

27.7 Simultaneous linear equations that result in tridiagonal matrices can sometimes be treated as three-term recurrence relations, and their solution may be found in a similar manner to that described in chapter 15. Consider the tridiagonal simultaneous equations

$$x_{i-1} + 4x_i + x_{i+1} = 3(\delta_{i+1,0} - \delta_{i-1,0}), \quad i = 0, \pm 1, \pm 2, \ldots .$$

Prove that, for $i > 0$, the equations have a general solution of the form $x_i = \alpha p^i + \beta q^i$, where p and q are the roots of a certain quadratic equation. Show that a similar result holds for $i < 0$. In each case express x_0 in terms of the arbitrary constants $\alpha, \beta, \ldots$.

Now impose the condition that x_i is bounded as $i \to \pm\infty$ and obtain a unique solution.

1033

27.8 A possible rule for obtaining an approximation to an integral is the *mid-point rule*, given by

$$\int_{x_0}^{x_0+\Delta x} f(x)\,dx = \Delta x\, f(x_0 + \tfrac{1}{2}\Delta x) + O(\Delta x^3).$$

Writing h for Δx, and evaluating all derivatives at the mid-point of the interval $(x, x+\Delta x)$, use a Taylor series expansion to find, up to $O(h^5)$, the coefficients of the higher-order errors in both the trapezium and mid-point rules. Hence find a linear combination of these two rules that gives $O(h^5)$ accuracy for each step Δx.

27.9 Although it can easily be shown, by direct calculation, that

$$\int_0^\infty e^{-x} \cos(kx)\,dx = \frac{1}{1+k^2},$$

the form of the integrand is appropriate for Gauss–Laguerre numerical integration. Using a 5-point formula, investigate the range of values of k for which the formula gives accurate results. At about what value of k do the results become inaccurate at the 1% level?

27.10 Using the points and weights given in table 27.9, answer the following questions.

(a) A table of unnormalised Hermite polynomials $H_n(x)$ has been spattered with ink blots and gives $H_5(x)$ as $32x^5 - ?x^3 + 120x$ and $H_4(x)$ as $?x^4 - ?x^2 + 12$, where the coefficients marked ? cannot be read. What should they read?

(b) What is the value of the integral

$$I = \int_{-\infty}^{\infty} \frac{e^{-2x^2}}{4x^2 + 3x + 1}\,dx,$$

as given by a 7-point integration routine?

27.11 Consider the integrals I_p defined by

$$I_p = \int_{-1}^{1} \frac{x^{2p}}{\sqrt{1-x^2}}\,dx.$$

(a) By setting $x = \sin\theta$ and using the results given in exercise 2.42, show that I_p has the value

$$I_p = 2\,\frac{2p-1}{2p}\,\frac{2p-3}{2p-2}\,\cdots\,\frac{1}{2}\,\frac{\pi}{2}.$$

(b) Evaluate I_p for $p = 1, 2, \ldots, 6$ using 5- and 6-point Gauss–Chebyshev integration (conveniently run on a spreadsheet such as *Excel*) and compare the results with those in (a). In particular, show that, as expected, the 5-point scheme first fails to be accurate when the order of the polynomial numerator $(2p)$ exceeds $(2 \times 5) - 1 = 9$. Likewise, verify that the 6-point scheme evaluates I_5 accurately but is in error for I_6.

27.12 In normal use only a single application of n-point Gaussian quadrature is made, using a value of n that is estimated from experience to be 'safe'. However, it is instructive to examine what happens when n is changed in a controlled way.

(a) Evaluate the integral

$$I_n = \int_2^5 \sqrt{7x - x^2 - 10}\,dx$$

using n-point Gauss–Legendre formulae for $n = 2, 3, \ldots, 6$. Estimate (to 4 s.f.) the value I_∞ you would obtain for very large n and compare it with the result I obtained by exact integration. Explain why the variation of I_n with n is monotonically decreasing.

(b) Try to repeat the processes described in (a) for the integrals

$$J_n = \int_2^5 \frac{1}{\sqrt{7x - x^2 - 10}} \, dx.$$

Why is it very difficult to estimate J_∞?

27.13 Given a random number η uniformly distributed on $(0, 1)$, determine the function $\xi = \xi(\eta)$ that would generate a random number ξ distributed as

(a) 2ξ on $0 \le \xi < 1$,

(b) $\frac{3}{2}\sqrt{\xi}$ on $0 \le \xi < 1$,

(c) $\frac{\pi}{4a} \cos \frac{\pi\xi}{2a}$ on $-a \le \xi < a$,

(d) $\frac{1}{2} \exp(-|\xi|)$ on $-\infty < \xi < \infty$.

27.14 A, B and C are three circles of unit radius with centres in the xy-plane at $(1, 2), (2.5, 1.5)$ and $(2, 3)$, respectively. Devise a hit or miss Monte Carlo calculation to determine the size of the area that lies outside C but inside A and B, as well as inside the square centred on $(2, 2.5)$, that has sides of length 2 parallel to the coordinate axes. You should choose your sampling region so as to make the estimation as efficient as possible. Take the random number distribution to be uniform on $(0, 1)$ and determine the inequalities that have to be tested using the random numbers chosen.

27.15 Use a Taylor series to solve the equation

$$\frac{dy}{dx} + xy = 0, \qquad y(0) = 1,$$

evaluating $y(x)$ for $x = 0.0$ to 0.5 in steps of 0.1.

27.16 Consider the application of the predictor–corrector method described near the end of subsection 27.6.3 to the equation

$$\frac{dy}{dx} = x + y, \qquad y(0) = 0.$$

Show, by comparison with a Taylor series expansion, that the expression obtained for y_{i+1} in terms of x_i and y_i by applying the three steps indicated (without any repeat of the last two) is correct to $O(h^2)$. Using steps of $h = 0.1$ compute the value of $y(0.3)$ and compare it with the value obtained by solving the equation analytically.

27.17 A more refined form of the Adams predictor–corrector method for solving the first-order differential equation

$$\frac{dy}{dx} = f(x, y)$$

is known as the Adams–Moulton–Bashforth scheme. At any stage (say the nth) in an Nth-order scheme, the values of x and y at the previous N solution points are first used to *predict* the value of y_{n+1}. This approximate value of y at the next solution point, x_{n+1}, denoted by $\bar{y}_{n+1}$, is then used together with those at the previous $N - 1$ solution points to make a more refined (*corrected*) estimation of $y(x_{n+1})$. The calculational procedure for a third-order scheme is summarised by the two following two equations:

$$\bar{y}_{n+1} = y_n + h(a_1 f_n + a_2 f_{n-1} + a_3 f_{n-2}) \qquad \text{(predictor)},$$
$$y_{n+1} = y_n + h(b_1 f(x_{n+1}, \bar{y}_{n+1}) + b_2 f_n + b_3 f_{n-1}) \qquad \text{(corrector)}.$$

(a) Find Taylor series expansions for f_{n-1} and f_{n-2} in terms of the function $f_n = f(x_n, y_n)$ and its derivatives at x_n.

1035

(b) Substitute them into the predictor equation and, by making that expression for $\bar{y}_{n+1}$ coincide with the true Taylor series for y_{n+1} up to order h^3, establish simultaneous equations that determine the values of a_1, a_2 and a_3.

(c) Find the Taylor series for f_{n+1} and substitute it and that for f_{n-1} into the corrector equation. Make the corrected prediction for y_{n+1} coincide with the true Taylor series by choosing the weights b_1, b_2 and b_3 appropriately.

(d) The values of the numerical solution of the differential equation

$$\frac{dy}{dx} = \frac{2(1+x)y + x^{3/2}}{2x(1+x)}$$

at three values of x are given in the following table:

x	0.1	0.2	0.3
$y(x)$	0.030 628	0.084 107	0.150 328

Use the above predictor–corrector scheme to find the value of $y(0.4)$ and compare your answer with the accurate value, 0.225 577.

27.18   If $dy/dx = f(x, y)$ then show that

$$\frac{d^2 f}{dx^2} = \frac{\partial^2 f}{\partial x^2} + 2f\frac{\partial^2 f}{\partial x \partial y} + f^2\frac{\partial^2 f}{\partial y^2} + \frac{\partial f}{\partial x}\frac{\partial f}{\partial y} + f\left(\frac{\partial f}{\partial y}\right)^2.$$

Hence verify, by substitution and the subsequent expansion of arguments in Taylor series of their own, that the scheme given in (27.79) coincides with the Taylor expansion (27.68), i.e.

$$y_{i+1} = y_i + hy_i^{(1)} + \frac{h^2}{2!}y_i^{(2)} + \frac{h^3}{3!}y_i^{(3)} + \cdots,$$

up to terms in h^3.

27.19   To solve the ordinary differential equation

$$\frac{du}{dt} = f(u, t)$$

for $f = f(t)$, the explicit two-step finite difference scheme

$$u_{n+1} = \alpha u_n + \beta u_{n-1} + h(\mu f_n + v f_{n-1})$$

may be used. Here, in the usual notation, h is the time step, $t_n = nh$, $u_n = u(t_n)$ and $f_n = f(u_n, t_n)$; $\alpha, \beta, \mu,$ and v are constants.

(a) A particular scheme has $\alpha = 1, \beta = 0, \mu = 3/2$ and $v = -1/2$. By considering Taylor expansions about $t = t_n$ for both u_{n+j} and f_{n+j}, show that this scheme gives errors of order h^3.

(b) Find the values of α, β, μ and v that will give the greatest accuracy.

27.20   Set up a finite difference scheme to solve the ordinary differential equation

$$x\frac{d^2\phi}{dx^2} + \frac{d\phi}{dx} = 0$$

in the range $1 \leq x \leq 4$, subject to the boundary conditions $\phi(1) = 2$ and $d\phi/dx = 2$ at $x = 4$. Using N equal increments, Δx, in x, obtain the general difference equation and state how the boundary conditions are incorporated into the scheme. Setting Δx equal to the (crude) value 1, obtain the relevant simultaneous equations and so obtain rough estimates for $\phi(2), \phi(3)$ and $\phi(4)$.

Finally, solve the original equation analytically and compare your numerical estimates with the accurate values.

27.21 Write a computer program that would solve, for a range of values of λ, the differential equation

$$\frac{dy}{dx} = \frac{1}{\sqrt{x^2 + \lambda y^2}}, \qquad y(0) = 1,$$

using a third-order Runge–Kutta scheme. Consider the difficulties that might arise when $\lambda < 0$.

27.22 Use the isocline approach to sketch the family of curves that satisfies the non-linear first-order differential equation

$$\frac{dy}{dx} = \frac{a}{\sqrt{x^2 + y^2}}.$$

27.23 For some problems, numerical or algebraic experimentation may suggest the form of the complete solution. Consider the problem of numerically integrating the first-order wave equation

$$\frac{\partial u}{\partial t} + A \frac{\partial u}{\partial x} = 0,$$

in which A is a positive constant. A finite difference scheme for this partial differential equation is

$$\frac{u(p, n+1) - u(p, n)}{\Delta t} + A \frac{u(p, n) - u(p-1, n)}{\Delta x} = 0,$$

where $x = p\Delta x$ and $t = n\Delta t$, with p any integer and n a non-negative integer. The initial values are $u(0, 0) = 1$ and $u(p, 0) = 0$ for $p \neq 0$.

(a) Carry the difference equation forward in time for two or three steps and attempt to identify the pattern of solution. Establish the criterion for the method to be numerically stable.
(b) Suggest a general form for $u(p, n)$, expressing it in generator function form, i.e. as '$u(p, n)$ is the coefficient of s^p in the expansion of $G(n, s)$'.
(c) Using your form of solution (or that given in the answers!), obtain an explicit general expression for $u(p, n)$ and verify it by direct substitution into the difference equation.
(d) An analytic solution of the original PDE indicates that an initial disturbance propagates undistorted. Under what circumstances would the difference scheme reproduce that behaviour?

27.24 In exercise 27.23 the difference scheme for solving

$$\frac{\partial u}{\partial t} + \frac{\partial u}{\partial x} = 0,$$

in which A has been set equal to unity, was one-sided in both space (x) and time (t). A more accurate procedure (known as the Lax–Wendroff scheme) is

$$\frac{u(p, n+1) - u(p, n)}{\Delta t} + \frac{u(p+1, n) - u(p-1, n)}{2\Delta x}$$
$$= \frac{\Delta t}{2} \left[\frac{u(p+1, n) - 2u(p, n) + u(p-1, n)}{(\Delta x)^2} \right].$$

(a) Establish the orders of accuracy of the two finite difference approximations on the LHS of the equation.
(b) Establish the accuracy with which the expression in the brackets approximates $\partial^2 u / \partial x^2$.
(c) Show that the RHS of the equation is such as to make the whole difference scheme accurate to second order in both space and time.

27.25   Laplace's equation,

$$\frac{\partial^2 V}{\partial x^2} + \frac{\partial^2 V}{\partial y^2} = 0,$$

is to be solved for the region and boundary conditions shown in figure 27.7.

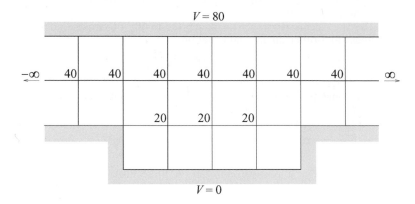

V = 80

$-\infty$ 40   40   40   40   40   40   40 ∞

20   20   20

V = 0

Figure 27.7   Region, boundary values and initial guessed solution values.

Starting from the given initial guess for the potential values V, and using the simplest possible form of relaxation, obtain a better approximation to the actual solution. Do not aim to be more accurate than ± 0.5 units, and so terminate the process when subsequent changes would be no greater than this.

27.26   Consider the solution, $\phi(x, y)$, of Laplace's equation in two dimensions using a relaxation method on a square grid with common spacing h. As in the main text, denote $\phi(x_0 + ih, \ y_0 + jh)$ by $\phi_{i,j}$. Further, define $\phi_{i,j}^{m,n}$ by

$$\phi_{i,j}^{m,n} \equiv \frac{\partial^{m+n}\phi}{\partial x^m \, \partial y^n}$$

evaluated at $(x_0 + ih, \ y_0 + jh)$.

(a) Show that

$$\phi_{i,j}^{4,0} + 2\phi_{i,j}^{2,2} + \phi_{i,j}^{0,4} = 0.$$

(b) Working up to terms of order h^5, find Taylor series expansions, expressed in terms of the $\phi_{i,j}^{m,n}$, for

$$S_{\pm,0} = \phi_{i+1,j} + \phi_{i-1,j},$$
$$S_{0,\pm} = \phi_{i,j+1} + \phi_{i,j-1}.$$

(c) Find a corresponding expansion, to the same order of accuracy, for $\phi_{i\pm1,j+1} + \phi_{i\pm1,j-1}$ and hence show that

$$S_{\pm,\pm} = \phi_{i+1,j+1} + \phi_{i+1,j-1} + \phi_{i-1,j+1} + \phi_{i-1,j-1}$$

has the form

$$4\phi_{i,j}^{0,0} + 2h^2(\phi_{i,j}^{2,0} + \phi_{i,j}^{0,2}) + \frac{h^4}{6}(\phi_{i,j}^{4,0} + 6\phi_{i,j}^{2,2} + \phi_{i,j}^{0,4}).$$

1038

(d) Evaluate the expression $4(S_{\pm,0}+S_{0,\pm})+S_{\pm,\pm}$ and hence deduce that a possible relaxation scheme, good to the fifth order in h, is to recalculate each $\phi_{i,j}$ as the weighted mean of the current values of its four nearest neighbours (each with weight $\frac{1}{5}$) and its four next-nearest neighbours (each with weight $\frac{1}{20}$).

27.27   The Schrödinger equation for a quantum mechanical particle of mass m moving in a one-dimensional harmonic oscillator potential $V(x) = kx^2/2$ is

$$-\frac{\hbar^2}{2m}\frac{d^2\psi}{dx^2} + \frac{kx^2\psi}{2} = E\psi.$$

For physically acceptable solutions, the wavefunction $\psi(x)$ must be finite at $x = 0$, tend to zero as $x \to \pm\infty$ and be normalised, so that $\int |\psi|^2\,dx = 1$. In practice, these constraints mean that only certain (quantised) values of E, the energy of the particle, are allowed. The allowed values fall into two groups: those for which $\psi(0) = 0$ and those for which $\psi(0) \neq 0$.

Show that if the unit of length is taken as $[\hbar^2/(mk)]^{1/4}$ and the unit of energy is taken as $\hbar(k/m)^{1/2}$, then the Schrödinger equation takes the form

$$\frac{d^2\psi}{dy^2} + (2E' - y^2)\psi = 0.$$

Devise an outline computerised scheme, using Runge–Kutta integration, that will enable you to:

(a) determine the three lowest allowed values of E;
(b) tabulate the normalised wavefunction corresponding to the lowest allowed energy.

You should consider explicitly:

(i)   the variables to use in the numerical integration;
(ii)   how starting values near $y = 0$ are to be chosen;
(iii) how the condition on ψ as $y \to \pm\infty$ is to be implemented;
(iv) how the required values of E are to be extracted from the results of the integration;
(v)   how the normalisation is to be carried out.

27.10 Hints and answers

27.1   5.370.
27.3   (a) $\xi \neq 0$ and $f'(\xi) \neq 0$ in general; (b) $\xi = b$, but $f'(b) = 0$ whilst $f(b) \neq 0$.
27.5   Interchange is formally needed for the first two steps, though in this case no error will result if it is not carried out; $x_1 = -12$, $x_2 = 2$, $x_3 = -1$, $x_4 = 5$.
27.7   The quadratic equation is $z^2 + 4z + 1 = 0$; $\alpha + \beta - 3 = x_0 = \alpha' + \beta' + 3$.
        With $p = -2 + \sqrt{3}$ and $q = -2 - \sqrt{3}$, β must be zero for $i > 0$ and α' must be zero for $i < 0$; $x_i = 3(-2 + \sqrt{3})^i$ for $i > 0$, $x_i = 0$ for $i = 0$, $x_i = -3(-2 - \sqrt{3})^i$ for $i < 0$.
27.9   The error is 1% or less for $|k|$ less than about 1.1.
27.11   Exact values (6 s.f.) for $p = 1, 2, \dots, 6$ are $1.570\,796$, $1.178\,097$, $0.981\,748$, $0.859\,029$, $0.773\,126$, $0.708\,699$. The Gauss–Chebyshev integration is in error by about 1% when $n = p$.
27.13   Listed below are the relevant indefinite integrals $F(y)$ of the distributions together with the functions $\xi = \xi(\eta)$:

(a) y^2, $\xi = \sqrt{\eta}$;
(b) $y^{3/2}$, $\xi = \eta^{2/3}$;

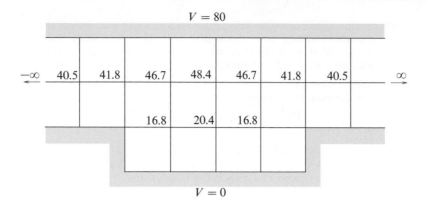

$$V = 80$$

$-\infty$ 40.5 41.8 46.7 48.4 46.7 41.8 40.5 ∞

16.8 20.4 16.8

$$V = 0$$

Figure 27.8   The solution to exercise 27.25.

(c) $\frac{1}{2}\{\sin[\pi y/(2a)] + 1\}$, $\xi = (2a/\pi)\sin^{-1}(2\eta - 1)$;

(d) $\frac{1}{2}\exp y$ for $y \le 0$, $\frac{1}{2}[2 - \exp(-y)]$ for $y > 0$; $\xi = \ln 2\eta$ for $0 < \eta \le \frac{1}{2}$, $\xi = -\ln[2(1 - \eta)]$ for $\frac{1}{2} < \eta < 1$.

27.15 $1 - x^2/2 + x^4/8 - x^6/48$; 1.0000, 0.9950, 0.9802, 0.9560, 0.9231, 0.8825; exact solution $y = \exp(-x^2/2)$.

27.17   (b) $a_1 = 23/12$, $a_2 = -4/3$, $a_3 = 5/12$.
(c) $b_1 = 5/12$, $b_2 = 2/3$, $b_3 = -1/12$.
(d) $\bar{y}(0.4) = 0.224\,582$, $y(0.4) = 0.225\,527$ after correction.

27.19   (a) The error is $5h^3 u_n^{(3)}/12 + O(h^4)$.
(b) $\alpha = -4$, $\beta = 5$, $\mu = 4$ and $v = 2$

27.21   For λ positive the solutions are (boringly) monotonic functions of x. With $y(0)$ given, there are no real solutions at all for *any* negative λ!

27.23   (a) Setting $A\Delta t = c\Delta x$ gives, for example, $u(0,2) = (1 - c)^2$, $u(1,2) = 2c(1 - c)$, $u(2,2) = c^2$. For stability, $0 < c < 1$.
(b) $G(n,s) = [(1 - c) + cs]^n$ for $0 \le p \le n$.
(c) $[n!(1 - c)^{n-p}c^p]/[p!(n - p)!]$.
(d) When $c = 1$ and the difference equation becomes $u(p, n + 1) = u(p - 1, n)$.

27.25   See figure 27.8.

27.27   If $x = \alpha y$ then

$$\frac{d^2\psi}{dy^2} - \alpha^4 \frac{mk}{\hbar^2} y^2\psi + \alpha^2 \frac{2mE}{\hbar^2}\psi = 0.$$

Solutions will be either symmetric or antisymmetric with $\psi(0) \ne 0$ but $\psi'(0) = 0$ for the former and vice versa for the latter. Integration to a largish but finite value of y followed by an interpolation procedure to estimate the values of E that lead to $\psi(\infty) = 0$ needs to be incorporated. Simple numerical integration such as Simpson's rule will suffice for the normalisation integral. The solutions should be $\lambda = 1, 3, 5, \ldots$.

28

Group theory

For systems that have some degree of symmetry, full exploitation of that symmetry is desirable. Significant physical results can sometimes be deduced simply by a study of the symmetry properties of the system under investigation. Consequently it becomes important, for such a system, to identify all those operations (rotations, reflections, inversions) that carry the system into a physically indistinguishable copy of itself.

The study of the properties of the complete set of such operations forms one application of *group theory*. Though this is the aspect of most interest to the physical scientist, group theory itself is a much larger subject and of great importance in its own right. Consequently we leave until the next chapter any direct applications of group theoretical results and concentrate on building up the general mathematical properties of groups.

28.1 Groups

As an example of symmetry properties, let us consider the sets of operations, such as rotations, reflections, and inversions, that transform physical objects, for example molecules, into physically indistinguishable copies of themselves, so that only the labelling of identical components of the system (the atoms) changes in the process. For differently shaped molecules there are different sets of operations, but in each case it is a well-defined set, and with a little practice all members of each set can be identified.

As simple examples, consider (*a*) the hydrogen molecule, and (*b*) the ammonia molecule illustrated in figure 28.1. The hydrogen molecule consists of two atoms H of hydrogen and is carried into itself by any of the following operations:

   (i) any rotation about its long axis;
   (ii) rotation through π about an axis perpendicular to the long axis and passing through the point M that lies midway between the atoms;

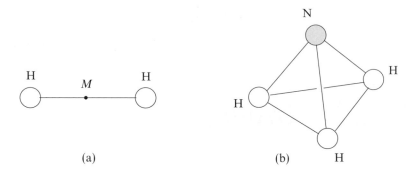

Figure 28.1   (a) The hydrogen molecule, and (b) the ammonia molecule.

(iii) inversion through the point M;

(iv) reflection in the plane that passes through M and has its normal parallel to the long axis.

These operations collectively form the set of symmetry operations for the hydrogen molecule.

The somewhat more complex ammonia molecule consists of a tetrahedron with an equilateral triangular base at the three corners of which lie hydrogen atoms H, whilst a nitrogen atom N is sited at the fourth vertex of the tetrahedron. The set of symmetry operations on this molecule is limited to rotations of $\pi/3$ and $2\pi/3$ about the axis joining the centroid of the equilateral triangle to the nitrogen atom, and reflections in the three planes containing that axis and each of the hydrogen atoms in turn. However, if the nitrogen atom could be replaced by a fourth hydrogen atom, and all interatomic distances equalised in the process, the number of symmetry operations would be greatly increased.

Once *all* the possible operations in any particular set have been identified, it must follow that the result of applying two such operations in succession will be identical to that obtained by the sole application of some third (usually different) operation in the set – for if it were not, a new member of the set would have been found, contradicting the assumption that all members have been identified.

Such observations introduce two of the main considerations relevant to deciding whether a set of objects, here the rotation, reflection and inversion operations, qualifies as a *group* in the mathematically tightly defined sense. These two considerations are (i) whether there is some law for combining two members of the set, and (ii) whether the result of the combination is also a member of the set. The obvious rule of combination has to be that the second operation is carried out on the system that results from application of the first operation, and we have already seen that the second requirement is satisfied by the inclusion of all such operations in the set. However, for a set to qualify as a group, more than these two conditions have to be satisfied, as will now be made clear.

28.1.1 Definition of a group

A group G is a set of elements $\{X, Y, \ldots\}$, together with a rule for combining them that associates with each ordered pair X, Y a 'product' or combination law $X \bullet Y$ for which the following conditions must be satisfied.

   (i) For *every* pair of elements X, Y that belongs to G, the product $X \bullet Y$ also belongs to G. (This is known as the *closure property* of the group.)

  (ii) For all triples X, Y, Z the *associative law* holds; in symbols,

$$X \bullet (Y \bullet Z) = (X \bullet Y) \bullet Z. \tag{28.1}$$

 (iii) There exists a unique element I, belonging to G, with the property that

$$I \bullet X = X = X \bullet I \tag{28.2}$$

for *all* X belonging to G. This element I is known as the *identity element* of the group.

 (iv) For every element X of G, there exists an element X^{-1}, also belonging to G, such that

$$X^{-1} \bullet X = I = X \bullet X^{-1}. \tag{28.3}$$

X^{-1} is called the *inverse* of X.

An alternative notation in common use is to write the elements of a group G as the set $\{G_1, G_2, \ldots\}$ or, more briefly, as $\{G_i\}$, a typical element being denoted by G_i.

It should be noticed that, as given, the nature of the operation $\bullet$ is not stated. It should also be noticed that the more general term *element*, rather than *operation*, has been used in this definition. We will see that the general definition of a group allows as elements not only sets of operations on an object but also sets of numbers, of functions and of other objects, provided that the interpretation of $\bullet$ is appropriately defined.

In one of the simplest examples of a group, namely the group of all integers under addition, the operation $\bullet$ is taken to be ordinary addition. In this group the role of the identity I is played by the integer 0, and the inverse of an integer X is $-X$. That requirements (i) and (ii) are satisfied by the integers under addition is trivially obvious. A second simple group, under ordinary multiplication, is formed by the two numbers 1 and -1; in this group, closure is obvious, 1 is the identity element, and each element is its own inverse.

It will be apparent from these two examples that the number of elements in a group can be either finite or infinite. In the former case the group is called a *finite group* and the number of elements it contains is called the *order* of the group, which we will denote by g; an alternative notation is $|G|$ but has obvious dangers if matrices are involved. In the notation in which $\mathcal{G} = \{G_1, G_2, \ldots, G_n\}$ the order of the group is clearly n.

As we have noted, for the integers under addition zero is the identity. For the group of rotations and reflections, the operation of doing nothing, i.e. the null operation, plays this role. This latter identification may seem artificial, but it is an operation, albeit trivial, which does leave the system in a physically indistinguishable state, and needs to be included. One might add that without it the set of operations would not form a group and none of the powerful results we will derive later in this and the next chapter could be justifiably applied to give deductions of physical significance.

In the examples of rotations and reflections mentioned earlier, $\bullet$ has been taken to mean that the left-hand operation is carried out on the system that results from application of the right-hand operation. Thus

$$Z = X \bullet Y \tag{28.4}$$

means that the effect on the system of carrying out Z is the same as would be obtained by first carrying out Y and then carrying out X. The order of the operations should be noted; it is arbitrary in the first instance but, once chosen, must be adhered to. The choice we have made is dictated by the fact that most of our applications involve the effect of rotations and reflections on functions of space coordinates, and it is usual, and our practice in the rest of this book, to write operators acting on functions to the left of the functions.

It will be apparent that for the above-mentioned group, integers under ordinary addition, it is true that

$$Y \bullet X = X \bullet Y \tag{28.5}$$

for all pairs of integers X, Y. If any two particular elements of a group satisfy (28.5), they are said to *commute* under the operation $\bullet$; if all pairs of elements in a group satisfy (28.5), then the group is said to be *Abelian*. The set of all integers forms an infinite Abelian group under (ordinary) addition.

As we show below, requirements (iii) and (iv) of the definition of a group are over-demanding (but self-consistent), since in each of equations (28.2) and (28.3) the second equality can be deduced from the first by using the associativity required by (28.1). The mathematical steps in the following arguments are all very simple, but care has to be taken to make sure that nothing that has not yet been proved is used to justify a step. For this reason, and to act as a model in logical deduction, a reference in Roman numerals to the previous result, or to the group definition used, is given over each equality sign. Such explicit detailed referencing soon becomes tiresome, but it should always be available if needed.

▶ *Using only the first equalities in (28.2) and (28.3), deduce the second ones.*

Consider the expression $X^{-1} \bullet (X \bullet X^{-1})$;

$$X^{-1} \bullet (X \bullet X^{-1}) \overset{\text{(ii)}}{=} (X^{-1} \bullet X) \bullet X^{-1} \overset{\text{(iv)}}{=} I \bullet X^{-1}$$
$$\overset{\text{(iii)}}{=} X^{-1}. \tag{28.6}$$

But X^{-1} belongs to $\mathcal{G}$, and so from (iv) there is an element U in $\mathcal{G}$ such that

$$U \bullet X^{-1} = I. \tag{v}$$

Form the product of U with the first and last expressions in (28.6) to give

$$U \bullet (X^{-1} \bullet (X \bullet X^{-1})) = U \bullet X^{-1} \overset{\text{(v)}}{=} I. \tag{28.7}$$

Transforming the left-hand side of this equation gives

$$U \bullet (X^{-1} \bullet (X \bullet X^{-1})) \overset{\text{(ii)}}{=} (U \bullet X^{-1}) \bullet (X \bullet X^{-1})$$
$$\overset{\text{(v)}}{=} I \bullet (X \bullet X^{-1})$$
$$\overset{\text{(iii)}}{=} X \bullet X^{-1}. \tag{28.8}$$

Comparing (28.7), (28.8) shows that

$$X \bullet X^{-1} = I, \tag{iv'}$$

i.e. the second equality in group definition (iv). Similarly

$$X \bullet I \overset{\text{(iv)}}{=} X \bullet (X^{-1} \bullet X) \overset{\text{(ii)}}{=} (X \bullet X^{-1}) \bullet X$$
$$\overset{\text{(iv)'}}{=} I \bullet X$$
$$\overset{\text{(iii)}}{=} X. \tag{iii'}$$

i.e. the second equality in group definition (iii). ◀

The uniqueness of the identity element I can also be demonstrated rather than assumed. Suppose that I', belonging to $\mathcal{G}$, also has the property

$$I' \bullet X = X = X \bullet I' \qquad \text{for all } X \text{ belonging to } \mathcal{G}.$$

Take X as I, then

$$I' \bullet I = I. \tag{28.9}$$

Further, from (iii'),

$$X = X \bullet I \qquad \text{for all } X \text{ belonging to } \mathcal{G},$$

and setting $X = I'$ gives

$$I' = I' \bullet I. \tag{28.10}$$

It then follows from (28.9), (28.10) that $I = I'$, showing that in any particular group the identity element is unique.

In a similar way it can be shown that the inverse of any particular element is unique. If U and V are two postulated inverses of an element X of $\mathcal{G}$, by considering the product

$$U \bullet (X \bullet V) = (U \bullet X) \bullet V,$$

it can be shown that $U = V$. The proof is left to the reader.

Given the uniqueness of the inverse of any particular group element, it follows that

$$(U \bullet V \bullet \cdots \bullet Y \bullet Z) \bullet (Z^{-1} \bullet Y^{-1} \bullet \cdots \bullet V^{-1} \bullet U^{-1})$$

$$= (U \bullet V \bullet \cdots \bullet Y) \bullet (Z \bullet Z^{-1}) \bullet (Y^{-1} \bullet \cdots \bullet V^{-1} \bullet U^{-1})$$

$$= (U \bullet V \bullet \cdots \bullet Y) \bullet (Y^{-1} \bullet \cdots \bullet V^{-1} \bullet U^{-1})$$

$$\vdots$$

$$= I,$$

where use has been made of the associativity and of the two equations $Z \bullet Z^{-1} = I$ and $I \bullet X = X$. Thus the inverse of a product is the product of the inverses in reverse order, i.e.

$$(U \bullet V \bullet \cdots \bullet Y \bullet Z)^{-1} = (Z^{-1} \bullet Y^{-1} \bullet \cdots \bullet V^{-1} \bullet U^{-1}). \tag{28.11}$$

Further elementary results that can be obtained by arguments similar to those above are as follows.

(i) Given any pair of elements X, Y belonging to $\mathcal{G}$, there exist unique elements U, V, also belonging to $\mathcal{G}$, such that

$$X \bullet U = Y \qquad \text{and} \qquad V \bullet X = Y.$$

Clearly $U = X^{-1} \bullet Y$, and $V = Y \bullet X^{-1}$, and they can be shown to be unique. This result is sometimes called the *division axiom*.

(ii) The *cancellation law* can be stated as follows. If

$$X \bullet Y = X \bullet Z$$

for some X belonging to $\mathcal{G}$, then $Y = Z$. Similarly,

$$Y \bullet X = Z \bullet X$$

implies the same conclusion.

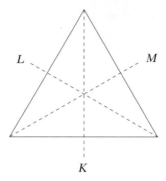

Figure 28.2 Reflections in the three perpendicular bisectors of the sides of an equilateral triangle take the triangle into itself.

(iii) Forming the product of each element of G with a fixed element X of G simply permutes the elements of G; this is often written symbolically as $G \bullet X = G$. If this were not so, and $X \bullet Y$ and $X \bullet Z$ were not different even though Y and Z were, application of the cancellation law would lead to a contradiction. This result is called the *permutation law*.

In any finite group of order g, any element X when combined with itself to form successively $X^2 = X \bullet X$, $X^3 = X \bullet X^2$, ... will, after at most $g - 1$ such combinations, produce the group identity I. Of course X^2, X^3, ... are some of the original elements of the group, and not new ones. If the actual number of combinations needed is $m - 1$, i.e. $X^m = I$, then m is called the *order of the element* X in G. The order of the identity of a group is always 1, and that of any other element of a group that is its own inverse is always 2.

> ► *Determine the order of the group of (two-dimensional) rotations and reflections that take a plane equilateral triangle into itself and the order of each of the elements. The group is usually known as 3m (to physicists and crystallographers) or C_{3v} (to chemists).*

There are two (clockwise) rotations, by $2\pi/3$ and $4\pi/3$, about an axis perpendicular to the plane of the triangle. In addition, reflections in the perpendicular bisectors of the three sides (see figure 28.2) have the defining property. To these must be added the identity operation. Thus in total there are six distinct operations and so $g = 6$ for this group. To reproduce the identity operation either of the rotations has to be applied three times, whilst any of the reflections has to be applied just twice in order to recover the original situation. Thus each rotation element of the group has order 3, and each reflection element has order 2. ◄

A so-called *cyclic group* is one for which all members of the group can be generated from just one element X (say). Thus a cyclic group of order g can be written as

$$G = \{I, X, X^2, X^3, \ldots, X^{g-1}\}.$$

It is clear that cyclic groups are always Abelian and that each element, apart from the identity, has order g, the order of the group itself.

28.1.2 Further examples of groups

In this section we consider some sets of objects, each set together with a law of combination, and investigate whether they qualify as groups and, if not, why not.

We have already seen that the integers form a group under ordinary addition, but it is immediately apparent that (even if zero is excluded) they do *not* do so under ordinary multiplication. Unity must be the identity of the set, but the requisite inverse of any integer n, namely $1/n$, does not belong to the set of integers for any n other than unity.

Other infinite sets of quantities that do form groups are the sets of all real numbers, or of all complex numbers, under addition, and of the same two sets excluding 0 under multiplication. All these groups are Abelian.

Although subtraction and division are normally considered the obvious counterparts of the operations of (ordinary) addition and multiplication, they are not acceptable operations for use within groups since the associative law, (28.1), does not hold. Explicitly,

$$X - (Y - Z) \neq (X - Y) - Z,$$
$$X \div (Y \div Z) \neq (X \div Y) \div Z.$$

From within the field of all non-zero complex numbers we can select just those that have unit modulus, i.e. are of the form $e^{i\theta}$ where $0 \leq \theta < 2\pi$, to form a group under multiplication, as can easily be verified:

$$
\begin{aligned}
e^{i\theta_1} \times e^{i\theta_2} &= e^{i(\theta_1 + \theta_2)} && \text{(closure),} \\
e^{i0} &= 1 && \text{(identity),} \\
e^{i(2\pi - \theta)} \times e^{i\theta} &= e^{i2\pi} \equiv e^{i0} = 1 && \text{(inverse).}
\end{aligned}
$$

Closely related to the above group is the set of 2×2 rotation matrices that take the form

$$M(\theta) = \begin{pmatrix} \cos\theta & -\sin\theta \\ \sin\theta & \cos\theta \end{pmatrix}$$

where, as before, $0 \leq \theta < 2\pi$. These form a group when the law of combination is that of matrix multiplication. The reader can easily verify that

$$
\begin{aligned}
M(\theta)M(\phi) &= M(\theta + \phi) && \text{(closure),} \\
M(0) &= I_2 && \text{(identity),} \\
M(2\pi - \theta) &= M^{-1}(\theta) && \text{(inverse).}
\end{aligned}
$$

Here I_2 is the unit 2×2 matrix.

28.2 Finite groups

Whilst many properties of physical systems (e.g. angular momentum) are related to the properties of infinite, and, in particular, continuous groups, the symmetry properties of crystals and molecules are more intimately connected with those of finite groups. We therefore concentrate in this section on finite sets of objects that can be combined in a way satisfying the group postulates.

Although it is clear that the set of all integers does not form a group under ordinary multiplication, restricted sets can do so if the operation involved is multiplication (mod N) for suitable values of N; this operation will be explained below.

As a simple example of a group with only four members, consider the set S defined as follows:

$$S = \{1, 3, 5, 7\} \quad \text{under multiplication (mod 8)}.$$

To find the product (mod 8) of any two elements, we multiply them together in the ordinary way, and then divide the answer by 8, treating the remainder after doing so as the product of the two elements. For example, $5 \times 7 = 35$, which on dividing by 8 gives a remainder of 3. Clearly, since $Y \times Z = Z \times Y$, the full set of different products is

$$
\begin{aligned}
&1 \times 1 = 1, \quad 1 \times 3 = 3, \quad 1 \times 5 = 5, \quad 1 \times 7 = 7, \\
&3 \times 3 = 1, \quad 3 \times 5 = 7, \quad 3 \times 7 = 5, \\
&5 \times 5 = 1, \quad 5 \times 7 = 3, \\
&7 \times 7 = 1.
\end{aligned}
$$

The first thing to notice is that each multiplication produces a member of the original set, i.e. the set is closed. Obviously the element 1 takes the role of the identity, i.e. $1 \times Y = Y$ for all members Y of the set. Further, for each element Y of the set there is an element Z (equal to Y, as it happens, in this case) such that $Y \times Z = 1$, i.e. each element has an inverse. These observations, together with the associativity of multiplication (mod 8), show that the set S is an Abelian group of order 4.

It is convenient to present the results of combining any two elements of a group in the form of multiplication tables – akin to those which used to appear in elementary arithmetic books before electronic calculators were invented! Written in this much more compact form the above example is expressed by table 28.1. Although the order of the two elements being combined does not matter here because the group is Abelian, we adopt the convention that if the product in a general multiplication table is written $X \bullet Y$ then X is taken from the left-hand column and Y is taken from the top row. Thus the bold '**7**' in the table is the result of 3×5, rather than of 5×3.

Whilst it would make no difference to the basic information content in a table to present the rows and columns with their headings in random orders, it is

	1	3	5	7
1	1	3	5	7
3	3	1	7	5
5	5	7	1	3
7	7	5	3	1

Table 28.1 The table of products for the elements of the group $S = \{1, 3, 5, 7\}$ under multiplication (mod 8).

usual to list the elements in the same order in both the vertical and horizontal headings in any one table. The actual order of the elements in the common list, whilst arbitrary, is normally chosen to make the table have as much symmetry as possible. This is initially a matter of convenience, but, as we shall see later, some of the more subtle properties of groups are revealed by putting next to each other elements of the group that are alike in certain ways.

Some simple general properties of group multiplication tables can be deduced immediately from the fact that each row or column constitutes the elements of the group.

  (i) Each element appears once and only once in each row or column of the table; this must be so since $\mathcal{G} \bullet X = \mathcal{G}$ (the permutation law) holds.
 (ii) The inverse of any element Y can be found by looking along the row in which Y appears in the left-hand column (the Yth row), and noting the element Z at the head of the column (the Zth column) in which the identity appears as the table entry. An immediate corollary is that whenever the identity appears on the leading diagonal, it indicates that the corresponding header element is of order 2 (unless it happens to be the identity itself).
(iii) For any Abelian group the multiplication table is symmetric about the leading diagonal.

To get used to the ideas involved in using group multiplication tables, we now consider two more sets of integers under multiplication (mod N):

$$S' = \{1, 5, 7, 11\} \quad \text{under multiplication (mod 24), and}$$
$$S'' = \{1, 2, 3, 4\} \quad \text{under multiplication (mod 5).}$$

These have group multiplication tables 28.2(a) and (b) respectively, as the reader should verify.

If tables 28.1 and 28.2(a) for the groups S and S' are compared, it will be seen that they have essentially the same structure, i.e if the elements are written as $\{I, A, B, C\}$ in both cases, then the two tables are each equivalent to table 28.3.

For S, $I = 1$, $A = 3$, $B = 5$, $C = 7$ and the law of combination is multiplication (mod 8), whilst for S', $I = 1$, $A = 5$, $B = 7$, $C = 11$ and the law of combination

(a)

	1	5	7	11
1	1	5	7	11
5	5	1	11	7
7	7	11	1	5
11	11	7	5	1

(b)

	1	2	3	4
1	1	2	3	4
2	2	4	1	3
3	3	1	4	2
4	4	3	2	1

Table 28.2 (a) The multiplication table for the group $S' = \{1,5,7,11\}$ under multiplication (mod 24). (b) The multiplication table for the group $S'' = \{1,2,3,4\}$ under multiplication (mod 5).

	I	A	B	C
I	I	A	B	C
A	A	I	C	B
B	B	C	I	A
C	C	B	A	I

Table 28.3 The common structure exemplified by tables 28.1 and 28.2(a).

	1	i	-1	$-i$
1	1	i	-1	$-i$
i	i	-1	$-i$	1
-1	1	$-i$	1	i
$-i$	$-i$	1	i	-1

Table 28.4 The group table for the set $\{1, i, -1, -i\}$ under ordinary multiplication of complex numbers.

is multiplication (mod 24). However, the really important point is that the two groups S and S' have equivalent group multiplication tables – they are said to be *isomorphic*, a matter to which we will return more formally in section 28.5.

> ►*Determine the behaviour of the set of four elements*
>
> $$\{1, i, -1, -i\}$$
>
> *under the ordinary multiplication of complex numbers. Show that they form a group and determine whether the group is isomorphic to either of the groups S (itself isomorphic to S') and S'' defined above.*

That the elements form a group under the associative operation of complex multiplication is immediate; there is an identity (1), each possible product generates a member of the set and each element has an inverse (1, $-i$, -1, i, respectively). The group table has the form shown in table 28.4.

We now ask whether this table can be made to look like table 28.3, which is the standardised form of the tables for S and S'. Since the identity element of the group (1) will have to be represented by I, and '1' only appears on the leading diagonal twice whereas I appears on the leading diagonal four times in table 28.3, it is clear that no

1051

	1	i	-1	$-i$
1	1	i	-1	$-i$
i	i	-1	$-i$	1
-1	-1	$-i$	1	i
$-i$	$-i$	1	i	-1

	1	2	4	3
1	1	2	4	3
2	2	4	3	1
4	4	3	1	2
3	3	1	2	4

Table 28.5   A comparison between tables 28.4 and 28.2(b), the latter with its columns reordered.

	I	A	B	C
I	I	A	B	C
A	A	B	C	I
B	B	C	I	A
C	C	I	A	B

Table 28.6   The common structure exemplified by tables 28.4 and 28.2(b), the latter with its columns reordered.

amount of relabelling (or, equivalently, no allocation of the symbols A, B, C, amongst i, -1, $-i$) can bring table 28.4 into the form of table 28.3. We conclude that the group $\{1, i, -1, -i\}$ is not isomorphic to $\mathcal{S}$ or $\mathcal{S}'$. An alternative way of stating the observation is to say that the group contains only one element of order 2 whilst a group corresponding to table 28.3 contains three such elements.

However, if the rows and columns of table 28.2(b) – in which the identity does appear twice on the diagonal and which therefore has the potential to be equivalent to table 28.4 – are rearranged by making the heading order 1, 2, 4, 3 then the two tables can be compared in the forms shown in table 28.5. They can thus be seen to have the same structure, namely that shown in table 28.6.

We therefore conclude that the group of four elements $\{1, i, -1, -i\}$ under ordinary multiplication of complex numbers is isomorphic to the group $\{1, 2, 3, 4\}$ under multiplication (mod 5). ◄

What we have done does not prove it, but the two tables 28.3 and 28.6 are in fact the only possible tables for a group of order 4, i.e. a group containing exactly four elements.

28.3 Non-Abelian groups

So far, all the groups for which we have constructed multiplication tables have been based on some form of arithmetic multiplication, a commutative operation, with the result that the groups have been Abelian and the tables symmetric about the leading diagonal. We now turn to examples of groups in which some non-commutation occurs. It should be noted, in passing, that non-commutation *cannot* occur *throughout* a group, as the identity always commutes with any element in its group.

As a first example we consider again as elements of a group the two-dimensional operations which transform an equilateral triangle into itself (see the end of subsection 28.1.1). It has already been shown that there are six such operations: the null operation, two rotations (by $2\pi/3$ and $4\pi/3$ about an axis perpendicular to the plane of the triangle) and three reflections in the perpendicular bisectors of the three sides. To abbreviate we will denote these operations by symbols as follows.

(i) I is the null operation.

(ii) R and R' are (clockwise) rotations by $2\pi/3$ and $4\pi/3$ respectively.

(iii) K, L, M are reflections in the three lines indicated in figure 28.2.

Some products of the operations of the form $X \bullet Y$ (where it will be recalled that the symbol $\bullet$ means that the second operation X is carried out on the system resulting from the application of the first operation Y) are easily calculated:

$$R \bullet R = R', \qquad R' \bullet R' = R, \qquad R \bullet R' = I = R' \bullet R$$
$$K \bullet K = L \bullet L = M \bullet M = I. \tag{28.12}$$

Others, such as $K \bullet M$, are more difficult, but can be found by a little thought, or by making a model triangle or drawing a sequence of diagrams such as those following.

showing that $K \bullet M = R'$. In the same way,

shows that $M \bullet K = R$, and

shows that $R \bullet L = K$.

Proceeding in this way we can build up the complete multiplication table (table 28.7). In fact, it is not necessary to draw any more diagrams, as all remaining products can be deduced algebraically from the three found above and

	I	R	R'	K	L	M
I	I	R	R'	K	L	M
R	R	R'	I	M	K	L
R'	R'	I	R	L	M	K
K	K	L	M	I	R	R'
L	L	M	K	R'	I	R
M	M	K	L	R	R'	I

Table 28.7   The group table for the two-dimensional symmetry operations on an equilateral triangle.

the more self-evident results given in (28.12). A number of things may be noticed about this table.

(i) It is *not* symmetric about the leading diagonal, indicating that some pairs of elements in the group do not commute.

(ii) There is some symmetry within the 3×3 blocks that form the four quarters of the table. This occurs because we have elected to put similar operations close to each other when choosing the order of table headings – the two rotations (or three if I is viewed as a rotation by $0\pi/3$) are next to each other, and the three reflections also occupy adjacent columns and rows. We will return to this later.

That two groups of the same order may be isomorphic carries over to non-Abelian groups. The next two examples are each concerned with sets of six objects; they will be shown to form groups that, although very different in nature from the rotation–reflection group just considered, are isomorphic to it.

We consider first the set $\mathcal{M}$ of six orthogonal 2×2 matrices given by

$$I = \begin{pmatrix} 1 & 0 \\ 0 & 1 \end{pmatrix} \qquad A = \begin{pmatrix} -\frac{1}{2} & \frac{\sqrt{3}}{2} \\ -\frac{\sqrt{3}}{2} & -\frac{1}{2} \end{pmatrix} \qquad B = \begin{pmatrix} -\frac{1}{2} & \frac{-\sqrt{3}}{2} \\ \frac{\sqrt{3}}{2} & -\frac{1}{2} \end{pmatrix}$$

$$C = \begin{pmatrix} -1 & 0 \\ 0 & 1 \end{pmatrix} \qquad D = \begin{pmatrix} \frac{1}{2} & -\frac{\sqrt{3}}{2} \\ -\frac{\sqrt{3}}{2} & -\frac{1}{2} \end{pmatrix} \qquad E = \begin{pmatrix} \frac{1}{2} & \frac{\sqrt{3}}{2} \\ \frac{\sqrt{3}}{2} & -\frac{1}{2} \end{pmatrix}$$

$$(28.13)$$

the combination law being that of ordinary matrix multiplication. Here we use italic, rather than the sans serif used for matrices elsewhere, to emphasise that the matrices are group elements.

Although it is tedious to do so, it can be checked that the product of any two of these matrices, in either order, is also in the set. However, the result is generally different in the two cases, as matrix multiplication is non-commutative. The matrix I clearly acts as the identity element of the set, and during the checking for closure it is found that the inverse of each matrix is contained in the set, I, C, D and E being their own inverses. The group table is shown in table 28.8.

	I	A	B	C	D	E
I	I	A	B	C	D	E
A	A	B	I	E	C	D
B	B	I	A	D	E	C
C	C	D	E	I	A	B
D	D	E	C	B	I	A
E	E	C	D	A	B	I

Table 28.8   The group table, under matrix multiplication, for the set $\mathcal{M}$ of six orthogonal 2×2 matrices given by (28.13).

The similarity to table 28.7 is striking. If $\{R, R', K, L, M\}$ of that table are replaced by $\{A, B, C, D, E\}$ respectively, the two tables are identical, without even the need to reshuffle the rows and columns. The two groups, one of reflections and rotations of an equilateral triangle, the other of matrices, are isomorphic.

Our second example of a group isomorphic to the same rotation–reflection group is provided by a set of functions of an undetermined variable x. The functions are as follows:

$$f_1(x) = x, \qquad f_2(x) = 1/(1-x), \qquad f_3(x) = (x-1)/x,$$

$$f_4(x) = 1/x, \qquad f_5(x) = 1-x, \qquad f_6(x) = x/(x-1),$$

and the law of combination is

$$f_i(x) \bullet f_j(x) = f_i(f_j(x)),$$

i.e. the function on the right acts as the argument of the function on the left to produce a new function of x. It should be emphasised that it is the functions that are the elements of the group. The variable x is the 'system' on which they act, and plays much the same role as the triangle does in our first example of a non-Abelian group.

To show an explicit example, we calculate the product $f_6 \bullet f_3$. The product will be the function of x obtained by evaluating $y/(y-1)$, when y is set equal to $(x-1)/x$. Explicitly

$$f_6(f_3) = \frac{(x-1)/x}{(x-1)/x \; - \; 1} = 1 - x = f_5(x).$$

Thus $f_6 \bullet f_3 = f_5$. Further examples are

$$f_2 \bullet f_2 = \frac{1}{1 - 1/(1-x)} = \frac{x-1}{x} = f_3,$$

and

$$f_6 \bullet f_6 = \frac{x/(x-1)}{x/(x-1) \; - \; 1} = x = f_1. \tag{28.14}$$

The multiplication table for this set of six functions has all the necessary properties to show that they form a group. Further, if the symbols $f_1, f_2, f_3, f_4, f_5, f_6$ are replaced by I, A, B, C, D, E respectively the table becomes identical to table 28.8. This justifies our earlier claim that this group of functions, with argument substitution as the law of combination, is isomorphic to the group of reflections and rotations of an equilateral triangle.

28.4 Permutation groups

The operation of rearranging n distinct objects amongst themselves is called a *permutation* of degree n, and since many symmetry operations on physical systems can be viewed in that light, the properties of permutations are of interest. For example, the symmetry operations on an equilateral triangle, to which we have already given much attention, can be considered as the six possible rearrangements of the marked corners of the triangle amongst three fixed points in space, much as in the diagrams used to compute table 28.7. In the same way, the symmetry operations on a cube can be viewed as a rearrangement of its corners amongst eight points in space, albeit with many constraints, or, with fewer complications, as a rearrangement of its body diagonals in space. The details will be left until we review the possible finite groups more systematically.

The notations and conventions used in the literature to describe permutations are very varied and can easily lead to confusion. We will try to avoid this by using letters $a, b, c, \ldots$ (rather than numbers) for the objects that are rearranged by a permutation and by adopting, before long, a 'cycle notation' for the permutations themselves. It is worth emphasising that it is the *permutations*, i.e. the acts of rearranging, and not the objects themselves (represented by letters) that form the elements of permutation groups. The complete group of all permutations of degree n is usually denoted by S_n or Σ_n. The number of possible permutations of degree n is $n!$, and so this is the order of S_n.

Suppose the ordered set of six distinct objects $\{a\ b\ c\ d\ e\ f\}$ is rearranged by some process into $\{b\ e\ f\ a\ d\ c\}$; then we can represent this mathematically as

$$\theta\{a\ b\ c\ d\ e\ f\} = \{b\ e\ f\ a\ d\ c\},$$

where θ is a permutation of degree 6. The permutation θ can be denoted by [2 5 6 1 4 3], since the first object, a, is replaced by the second, b, the second object, b, is replaced by the fifth, e, the third by the sixth, f, etc. The equation can then be written more explicitly as

$$\theta\{a\ b\ c\ d\ e\ f\} = [2\ 5\ 6\ 1\ 4\ 3]\{a\ b\ c\ d\ e\ f\} = \{b\ e\ f\ a\ d\ c\}.$$

If ϕ is a second permutation, also of degree 6, then the obvious interpretation of the product $\phi \bullet \theta$ of the two permutations is

$$\phi \bullet \theta\{a\ b\ c\ d\ e\ f\} = \phi(\theta\{a\ b\ c\ d\ e\ f\}).$$

1056

Suppose that ϕ is the permutation [4 5 3 6 2 1]; then

$$\phi \bullet \theta\{a\ b\ c\ d\ e\ f\} = [4\ 5\ 3\ 6\ 2\ 1][2\ 5\ 6\ 1\ 4\ 3]\{a\ b\ c\ d\ e\ f\}$$
$$= [4\ 5\ 3\ 6\ 2\ 1]\{b\ e\ f\ a\ d\ c\}$$
$$= \{a\ d\ f\ c\ e\ b\}$$
$$= [1\ 4\ 6\ 3\ 5\ 2]\{a\ b\ c\ d\ e\ f\}.$$

Written in terms of the permutation notation this result is

$$[4\ 5\ 3\ 6\ 2\ 1][2\ 5\ 6\ 1\ 4\ 3] = [1\ 4\ 6\ 3\ 5\ 2].$$

A concept that is very useful for working with permutations is that of decomposition into cycles. The cycle notation is most easily explained by example. For the permutation θ given above:

> the 1st object, a, has been replaced by the 2nd, b;
> the 2nd object, b, has been replaced by the 5th, e;
> the 5th object, e, has been replaced by the 4th, d;
> the 4th object, d, has been replaced by the 1st, a.

This brings us back to the beginning of a closed cycle, which is conveniently represented by the notation (1 2 5 4), in which the successive replacement positions are enclosed, in sequence, in parentheses. Thus (1 2 5 4) means 2nd → 1st, 5th → 2nd, 4th → 5th, 1st → 4th. It should be noted that the object initially in the first listed position replaces that in the final position indicated in the bracket – here 'a' is put into the fourth position by the permutation. Clearly the cycle (5 4 1 2), or any other that involved the same numbers in the same relative order, would have exactly the same meaning and effect. The remaining two objects, c and f, are interchanged by θ or, more formally, are rearranged according to a cycle of length 2, a *transposition*, represented by (3 6). Thus the complete representation (specification) of θ is

$$\theta = (1\ 2\ 5\ 4)(3\ 6).$$

The positions of objects that are unaltered by a permutation are either placed by themselves in a pair of parentheses or omitted altogether. The former is recommended as it helps to indicate how many objects are involved – important when the object in the last position is unchanged, or the permutation is the identity, which leaves all objects unaltered in position! Thus the identity permutation of degree 6 is

$$I = (1)(2)(3)(4)(5)(6),$$

though in practice it is often shortened to (1).

It will be clear that the cycle representation is unique, to within the internal absolute ordering of the numbers in each bracket as already noted, and that each number appears once and only once in the representation of any particular permutation.

The *order of any permutation* of degree n within the group S_n can be read off from the cyclic representation and is given by the lowest common multiple (LCM) of the lengths of the cycles. Thus I has order 1, as it must, and the permutation θ discussed above has order 4 (the LCM of 4 and 2).

Expressed in cycle notation our second permutation ϕ is (3)(1 4 6)(2 5), and the product $\phi \bullet \theta$ is calculated as

$$(3)(1\ 4\ 6)(2\ 5) \bullet (1\ 2\ 5\ 4)(3\ 6)\{a\ b\ c\ d\ e\ f\} = (3)(1\ 4\ 6)(2\ 5)\{b\ e\ f\ a\ d\ c\}$$
$$= \{a\ d\ f\ c\ e\ b\}$$
$$= (1)(5)(2\ 4\ 3\ 6)\{a\ b\ c\ d\ e\ f\}.$$

i.e. expressed as a relationship amongst the elements of the group of permutations of degree 6 (not yet proved as a group, but reasonably anticipated), this result reads

$$(3)(1\ 4\ 6)(2\ 5) \bullet (1\ 2\ 5\ 4)(3\ 6) = (1)(5)(2\ 4\ 3\ 6).$$

We note, for practice, that ϕ has order 6 (the LCM of 1, 3, and 2) and that the product $\phi \bullet \theta$ has order 4.

The number of elements in the group S_n of all permutations of degree n is $n!$ and clearly increases very rapidly as n increases. Fortunately, to illustrate the essential features of permutation groups it is sufficient to consider the case $n = 3$, which involves only six elements. They are as follows (with labelling which the reader will by now recognise as anticipatory):

$$I = (1)(2)(3) \quad A = (1\ 2\ 3) \quad B = (1\ 3\ 2)$$
$$C = (1)(2\ 3) \quad D = (3)(1\ 2) \quad E = (2)(1\ 3)$$

It will be noted that A and B have order 3, whilst C, D and E have order 2. As perhaps anticipated, their combination products are exactly those corresponding to table 28.8, I, C, D and E being their own inverses. For example, putting in all steps explicitly,

$$D \bullet C\{a\ b\ c\} = (3)(1\ 2) \bullet (1)(2\ 3)\{a\ b\ c\}$$
$$= (3)(12)\{a\ c\ b\}$$
$$= \{c\ a\ b\}$$
$$= (3\ 2\ 1)\{a\ b\ c\}$$
$$= (1\ 3\ 2)\{a\ b\ c\}$$
$$= B\{a\ b\ c\}.$$

In brief, the six permutations belonging to S_3 form yet another non-Abelian group isomorphic to the rotation–reflection symmetry group of an equilateral triangle.

28.5 Mappings between groups

Now that we have available a range of groups that can be used as examples, we return to the study of more general group properties. From here on, when there is no ambiguity we will write the product of two elements, $X \bullet Y$, simply as XY, omitting the explicit combination symbol. We will also continue to use 'multiplication' as a loose generic name for the combination process between elements of a group.

If $\mathcal{G}$ and $\mathcal{G}'$ are two groups, we can study the effect of a *mapping*

$$\Phi : \mathcal{G} \to \mathcal{G}'$$

of $\mathcal{G}$ onto $\mathcal{G}'$. If X is an element of $\mathcal{G}$ we denote its *image* in $\mathcal{G}'$ under the mapping Φ by $X' = \Phi(X)$.

A technical term that we have already used is *isomorphic*. We will now define it formally. Two groups $\mathcal{G} = \{X, Y, \ldots\}$ and $\mathcal{G}' = \{X', Y', \ldots\}$ are said to be *isomorphic* if there is a one-to-one correspondence

$$X \leftrightarrow X', \ Y \leftrightarrow Y', \ \ldots$$

between their elements such that

$$XY = Z \qquad \text{implies} \qquad X'Y' = Z'$$

and vice versa.

In other words, isomorphic groups have the same (multiplication) structure, although they may differ in the nature of their elements, combination law and notation. Clearly if groups $\mathcal{G}$ and $\mathcal{G}'$ are isomorphic, and $\mathcal{G}$ and $\mathcal{G}''$ are isomorphic, then it follows that $\mathcal{G}'$ and $\mathcal{G}''$ are isomorphic. We have already seen an example of four groups (of functions of x, of orthogonal matrices, of permutations and of the symmetries of an equilateral triangle) that are isomorphic, all having table 28.8 as their multiplication table.

Although our main interest is in isomorphic relationships between groups, the wider question of mappings of one set of elements onto another is of some importance, and we start with the more general notion of a homomorphism.

Let $\mathcal{G}$ and $\mathcal{G}'$ be two groups and Φ a mapping of $\mathcal{G} \to \mathcal{G}'$. If for every pair of elements X and Y in $\mathcal{G}$

$$(XY)' = X'Y'$$

then Φ is called a homomorphism, and $\mathcal{G}'$ is said to be a homomorphic image of $\mathcal{G}$.

The essential defining relationship, expressed by $(XY)' = X'Y'$, is that the same result is obtained whether the product of two elements is formed first and the image then taken or the images are taken first and the product then formed.

Three immediate consequences of the above definition are proved as follows.

(i) If I is the identity of $\mathcal{G}$ then $IX = X$ for all X in $\mathcal{G}$. Consequently

$$X' = (IX)' = I'X',$$

for all X' in $\mathcal{G}'$. Thus I' is the identity in $\mathcal{G}'$. In words, the identity element of $\mathcal{G}$ maps into the identity element of $\mathcal{G}'$.

(ii) Further,

$$I' = (XX^{-1})' = X'(X^{-1})'.$$

That is, $(X^{-1})' = (X')^{-1}$. In words, the image of an inverse is the same element in $\mathcal{G}'$ as the inverse of the image.

(iii) If element X in $\mathcal{G}$ is of order m, i.e. $I = X^m$, then

$$I' = (X^m)' = (XX^{m-1})' = X'(X^{m-1})' = \cdots = \underbrace{X'X' \cdots X'}_{m \text{ factors}}.$$

In words, the image of an element has the same order as the element.

What distinguishes an isomorphism from the more general homomorphism are the requirements that in an isomorphism:

(I) different elements in $\mathcal{G}$ must map into different elements in $\mathcal{G}'$ (whereas in a homomorphism several elements in $\mathcal{G}$ may have the same image in $\mathcal{G}'$), that is, $x' = y'$ must imply $x = y$;

(II) any element in $\mathcal{G}'$ must be the image of some element in $\mathcal{G}$.

An immediate consequence of (I) and result (iii) for homomorphisms is that isomorphic groups each have the same number of elements of any given order.

For a general homomorphism, the set of elements of $\mathcal{G}$ whose image in $\mathcal{G}'$ is I' is called the *kernel* of the homomorphism; this is discussed further in the next section. In an isomorphism the kernel consists of the identity I alone. To illustrate both this point and the general notion of a homomorphism, consider a mapping between the additive group of real numbers $\mathfrak{R}$ and the multiplicative group of complex numbers with unit modulus, $U(1)$. Suppose that the mapping $\mathfrak{R} \to U(1)$ is

$$\Phi : x \to e^{ix};$$

then this is a homomorphism since

$$(x + y)' \to e^{i(x+y)} = e^{ix}e^{iy} = x'y'.$$

However, it is not an isomorphism because many (an infinite number) of the elements of $\mathfrak{R}$ have the same image in $U(1)$. For example, $\pi, 3\pi, 5\pi, \ldots$ in $\mathfrak{R}$ all have the image -1 in $U(1)$ and, furthermore, all elements of $\mathfrak{R}$ of the form $2\pi n$, where n is an integer, map onto the identity element in $U(1)$. The latter set forms the kernel of the homomorphism.

	I	A	B	C	D	E
I	**I**	**A**	**B**	C	D	E
A	**A**	**B**	**I**	E	C	D
B	**B**	**I**	**A**	D	E	C
C	C	D	E	I	A	B
D	D	E	C	B	I	A
E	E	C	D	A	B	I

(a)

	I	A	B	C
I	**I**	**A**	B	C
A	**A**	**I**	C	B
B	B	C	I	A
C	C	B	A	I

(b)

Table 28.9   Reproduction of (a) table 28.8 and (b) table 28.3 with the relevant subgroups shown in bold.

For the sake of completeness, we add that a homomorphism for which (I) above holds is said to be a *monomorphism* (or an isomorphism *into*), whilst a homomorphism for which (II) holds is called an *epimorphism* (or an isomorphism *onto*). If, in either case, the other requirement is met as well then the monomorphism or epimorphism is also an isomorphism.

Finally, if the initial and final groups are the same, $G = G'$, then the isomorphism $G \to G'$ is termed an *automorphism*.

28.6 Subgroups

More detailed inspection of tables 28.8 and 28.3 shows that not only do the complete tables have the properties associated with a group multiplication table (see section 28.2) but so do the upper left corners of each table taken on their own. The relevant parts are shown in bold in the tables 28.9(a) and (b).

This observation immediately prompts the notion of a *subgroup*. A subgroup of a group G can be formally defined as any non-empty subset $\mathcal{H} = \{H_i\}$ of G, the elements of which themselves behave as a group under the same rule of combination as applies in G itself. As for all groups, the order of the subgroup is equal to the number of elements it contains; we will denote it by h or $|\mathcal{H}|$.

Any group G contains two trivial subgroups:

(i) G itself;
(ii) the set $\mathcal{I}$ consisting of the identity element alone.

All other subgroups of G are termed *proper subgroups*. In a group with multiplication table 28.8 the elements $\{I, A, B\}$ form a proper subgroup, as do $\{I, A\}$ in a group with table 28.3 as its group table.

Some groups have no proper subgroups. For example, the so-called *cyclic groups*, mentioned at the end of subsection 28.1.1, have no subgroups other than the whole group or the identity alone. Tables 28.10(a) and (b) show the multiplication tables for two of these groups. Table 28.6 is also the group table for a cyclic group, that of order 4.

(a)

	I	A	B
I	I	A	B
A	A	B	I
B	B	I	A

(b)

	I	A	B	C	D
I	I	A	B	C	D
A	A	B	C	D	I
B	B	C	D	I	A
C	C	D	I	A	B
D	D	I	A	B	C

Table 28.10   The group tables of two cyclic groups, of orders 3 and 5. They have no proper subgroups.

It will be clear that for a cyclic group G repeated combination of any element with itself generates all other elements of G, before finally reproducing itself. So, for example, in table 28.10(b), starting with (say) D, repeated combination with itself produces, in turn, C, B, A, I and finally D again. As noted earlier, in any cyclic group G every element, apart from the identity, is of order g, the order of the group itself.

The two tables shown are for groups of orders 3 and 5. It will be proved in subsection 28.7.2 that the order of any group is a multiple of the order of any of its subgroups (Lagrange's theorem), i.e. in our general notation, g is a multiple of h. It thus follows that a group of order p, where p is any prime, must be cyclic and cannot have any proper subgroups. The groups for which tables 28.10(a) and (b) are the group tables are two such examples. Groups of non-prime order may (table 28.3) or may not (table 28.6) have proper subgroups.

As we have seen, repeated multiplication of an element X (not the identity) by itself will generate a subgroup $\{X, X^2, X^3, \ldots\}$. The subgroup will clearly be Abelian, and if X is of order m, i.e. $X^m = I$, the subgroup will have m distinct members. If m is less than g – though, in view of Lagrange's theorem, m must be a factor of g – the subgroup will be a proper subgroup. We can deduce, in passing, that the order of any element of a group is an exact divisor of the order of the group.

Some obvious properties of the subgroups of a group G, which can be listed without formal proof, are as follows.

(i) The identity element of G belongs to every subgroup $\mathcal{H}$.
(ii) If element X belongs to a subgroup $\mathcal{H}$, so does X^{-1}.
(iii) The set of elements in G that belong to every subgroup of G themselves form a subgroup, though this may consist of the identity alone.

Properties of subgroups that need more explicit proof are given in the following sections, though some need the development of new concepts before they can be established. However, we can begin with a theorem, applicable to all homomorphisms, not just isomorphisms, that requires no new concepts.

Let $\Phi : G \to G'$ be a homomorphism of G into G'; then

    (i) the set of elements $\mathcal{H}'$ in $\mathcal{G}'$ that are images of the elements of $\mathcal{G}$ forms a subgroup of $\mathcal{G}'$;

    (ii) the set of elements $\mathcal{K}$ in $\mathcal{G}$ that are mapped onto the identity I' in $\mathcal{G}'$ forms a subgroup of $\mathcal{G}$.

As indicated in the previous section, the subgroup $\mathcal{K}$ is called the *kernel* of the homomorphism.

To prove (i), suppose Z and W belong to $\mathcal{H}'$, with $Z = X'$ and $W = Y'$, where X and Y belong to $\mathcal{G}$. Then

$$ZW = X'Y' = (XY)'$$

and therefore belongs to $\mathcal{H}'$, and

$$Z^{-1} = (X')^{-1} = (X^{-1})'$$

and therefore belongs to $\mathcal{H}'$. These two results, together with the fact that I' belongs to $\mathcal{H}'$, are enough to establish result (i).

To prove (ii), suppose X and Y belong to $\mathcal{K}$; then

$$(XY)' = X'Y' = I'I' = I' \qquad \text{(closure)},$$

$$I' - (XX^{-1})' = X'(X^{-1})' = I'(X^{-1})' = (X^{-1})'$$

and therefore X^{-1} belongs to $\mathcal{K}$. These two results, together with the fact that I belongs to $\mathcal{K}$, are enough to establish (ii). An illustration of this result is provided by the mapping Φ of $\mathfrak{R} \to U(1)$ considered in the previous section. Its kernel consists of the set of real numbers of the form $2\pi n$, where n is an integer; it forms a subgroup of $\mathcal{R}$, the additive group of real numbers.

In fact the kernel $\mathcal{K}$ of a homomorphism is a *normal* subgroup of $\mathcal{G}$. The defining property of such a subgroup is that for every element X in $\mathcal{G}$ and every element Y in the subgroup, XYX^{-1} belongs to the subgroup. This property is easily verified for the kernel $\mathcal{K}$, since

$$(XYX^{-1})' = X'Y'(X^{-1})' = X'I'(X^{-1})' = X'(X^{-1})' = I'.$$

Anticipating the discussion of subsection 28.7.2, the cosets of a normal subgroup themselves form a group (see exercise 28.16).

28.7 Subdividing a group

We have already noted, when looking at the (arbitrary) order of headings in a group table, that some choices appear to make the table more orderly than do others. In the following subsections we will identify ways in which the elements of a group can be divided up into sets with the property that the members of any one set are more like the other members of the set, in some particular regard, than they are like any element that does not belong to the set. We will find that these divisions will be such that the group is *partitioned*, i.e. the elements will be divided into sets in such a way that each element of the group belongs to one, and only one, such set.

We note in passing that the subgroups of a group do *not* form such a partition, not least because the identity element is in every subgroup, rather than being in precisely one. In other words, despite the nomenclature, a group is not simply the aggregate of its proper subgroups.

28.7.1 Equivalence relations and classes

We now specify in a more mathematical manner what it means for two elements of a group to be 'more like' one another than like a third element, as mentioned in section 28.2. Our introduction will apply to any set, whether a group or not, but our main interest will ultimately be in two particular applications to groups. We start with the formal definition of an equivalence relation.

An *equivalence relation* on a set S is a relationship $X \sim Y$, between two elements X and Y belonging to S, in which the definition of the symbol $\sim$ must satisfy the requirements of

    (i) reflexivity, $X \sim X$;
    (ii) symmetry, $X \sim Y$ implies $Y \sim X$;
    (iii) transitivity, $X \sim Y$ and $Y \sim Z$ imply $X \sim Z$.

Any particular two elements either satisfy or do not satisfy the relationship.

The general notion of an equivalence relation is very straightforward, and the requirements on the symbol $\sim$ seem undemanding; but not all relationships qualify. As an example within the topic of groups, if it meant 'has the same order as' then clearly all the requirements would be satisfied. However, if it meant 'commutes with' then it would not be an equivalence relation, since although A commutes with I, and I commutes with C, this does not necessarily imply that A commutes with C, as is obvious from table 28.8.

It may be shown that an equivalence relation on S divides up S into *classes* C_i such that:

    (i) X and Y belong to the same class if, and only if, $X \sim Y$;
    (ii) every element W of S belongs to exactly one class.

This may be shown as follows. Let X belong to S, and define the subset S_X of S to be the set of all elements U of S such that $X \sim U$. Clearly by reflexivity X belongs to S_X. Suppose first that $X \sim Y$, and let Z be any element of S_Y. Then $Y \sim Z$, and hence by transitivity $X \sim Z$, which means that Z belongs to S_X. Conversely, since the symmetry law gives $Y \sim X$, if Z belongs to S_X then this implies that Z belongs to $\mathcal{S}_Y$. These two results together mean that the two subsets $\mathcal{S}_X$ and $\mathcal{S}_Y$ have the same members and hence are equal.

Now suppose that $\mathcal{S}_X$ equals $\mathcal{S}_Y$. Since Y belongs to $\mathcal{S}_Y$ it also belongs to $\mathcal{S}_X$ and hence $X \sim Y$. This completes the proof of (i), once the distinct subsets of type $\mathcal{S}_X$ are identified as the classes $\mathcal{C}_i$. Statement (ii) is an immediate corollary, the class in question being identified as $\mathcal{S}_W$.

The most important property of an equivalence relation is as follows.

Two different subsets $\mathcal{S}_X$ and $\mathcal{S}_Y$ can have no element in common, and the collection of all the classes $\mathcal{C}_i$ is a 'partition' of S, i.e. every element in S belongs to one, and only one, of the classes.

To prove this, suppose $\mathcal{S}_X$ and $\mathcal{S}_Y$ have an element Z in common; then $X \sim Z$ and $Y \sim Z$ and so by the symmetry and transitivity laws $X \sim Y$. By the above theorem this implies $\mathcal{S}_X$ equals $\mathcal{S}_Y$. But this contradicts the fact that $\mathcal{S}_X$ and $\mathcal{S}_Y$ are different subsets. Hence $\mathcal{S}_X$ and $\mathcal{S}_Y$ can have no element in common.

Finally, if the elements of S are used in turn to define subsets and hence classes in S, every element U is in the subset $\mathcal{S}_U$ that is either a class already found or constitutes a new one. It follows that the classes exhaust S, i.e. every element is in some class.

Having established the general properties of equivalence relations, we now turn to two specific examples of such relationships, in which the general set S has the more specialised properties of a group $\mathcal{G}$ and the equivalence relation $\sim$ is chosen in such a way that the relatively transparent general results for equivalence relations can be used to derive powerful, but less obvious, results about the properties of groups.

28.7.2 Congruence and cosets

As the first application of equivalence relations we now prove Lagrange's theorem which is stated as follows.

Lagrange's theorem. If $\mathcal{G}$ is a finite group of order g and $\mathcal{H}$ is a subgroup of $\mathcal{G}$ of order h then g is a multiple of h.

We take as the definition of $\sim$ that, given X and Y belonging to $\mathcal{G}$, $X \sim Y$ if $X^{-1}Y$ belongs to $\mathcal{H}$. This is the same as saying that $Y = XH_i$ for some element H_i belonging to $\mathcal{H}$; technically X and Y are said to be left-congruent with respect to $\mathcal{H}$.

This defines an equivalence relation, since it has the following properties.

(i) Reflexivity: $X \sim X$, since $X^{-1}X = I$ and I belongs to any subgroup.
(ii) Symmetry: $X \sim Y$ implies that $X^{-1}Y$ belongs to $\mathcal{H}$ and so, therefore, does its inverse, since $\mathcal{H}$ is a group. But $(X^{-1}Y)^{-1} = Y^{-1}X$ and, as this belongs to $\mathcal{H}$, it follows that $Y \sim X$.

(iii) Transitivity: $X \sim Y$ and $Y \sim Z$ imply that $X^{-1}Y$ and $Y^{-1}Z$ belong to $\mathcal{H}$ and so, therefore, does their product $(X^{-1}Y)(Y^{-1}Z) = X^{-1}Z$, from which it follows that $X \sim Z$.

With $\sim$ proved as an equivalence relation, we can immediately deduce that it divides $\mathcal{G}$ into disjoint (non-overlapping) classes. For this particular equivalence relation the classes are called the *left cosets* of $\mathcal{H}$. Thus each element of $\mathcal{G}$ is in one and only one left coset of $\mathcal{H}$. The left coset containing any particular X is usually written $X\mathcal{H}$, and denotes the set of elements of the form XH_i (one of which is X itself since $\mathcal{H}$ contains the identity element); it must contain h different elements, since if it did not, and two elements were equal,

$$XH_i = XH_j,$$

we could deduce that $H_i = H_j$ and that $\mathcal{H}$ contained fewer than h elements.

From our general results about equivalence relations it now follows that the left cosets of $\mathcal{H}$ are a 'partition' of $\mathcal{G}$ into a number of sets each containing h members. Since there are g members of $\mathcal{G}$ and each must be in just one of the sets, it follows that g is a multiple of h. This concludes the proof of Lagrange's theorem.

The number of left cosets of $\mathcal{H}$ in $\mathcal{G}$ is known as the *index* of $\mathcal{H}$ in $\mathcal{G}$ and is written $[\mathcal{G} : \mathcal{H}]$; numerically the index $= g/h$. For the record we note that, for the trivial subgroup $\mathcal{I}$, which contains only the identity element, $[\mathcal{G} : \mathcal{I}] = g$ and that, for a subgroup $\mathcal{J}$ of subgroup $\mathcal{H}$, $[\mathcal{G} : \mathcal{H}][\mathcal{H} : \mathcal{J}] = [\mathcal{G} : \mathcal{J}]$.

The validity of *Lagrange's theorem* was established above using the far-reaching properties of equivalence relations. However, for this specific purpose there is a more direct and self-contained proof, which we now give.

Let X be some particular element of a finite group $\mathcal{G}$ of order g, and $\mathcal{H}$ be a subgroup of $\mathcal{G}$ of order h, with typical element Y_i. Consider the set of elements

$$X\mathcal{H} \equiv \{XY_1, XY_2, \ldots, XY_h\}.$$

This set contains h distinct elements, since if any two were equal, i.e. $XY_i = XY_j$ with $i \neq j$, this would contradict the cancellation law. As we have already seen, the set is called a left coset of $\mathcal{H}$.

We now prove three simple results.

- *Two cosets are either disjoint or identical.* Suppose cosets $X_1\mathcal{H}$ and $X_2\mathcal{H}$ have an element in common, i.e. $X_1Y_1 = X_2Y_2$ for some Y_1, Y_2 in $\mathcal{H}$. Then $X_1 = X_2Y_2Y_1^{-1}$, and since Y_1 and Y_2 both belong to $\mathcal{H}$ so does $Y_2Y_1^{-1}$; thus X_1 belongs to the left coset $X_2\mathcal{H}$. Similarly X_2 belongs to the left coset $X_1\mathcal{H}$. Consequently, either the two cosets are identical or it was wrong to assume that they have an element in common.

- *Two cosets $X_1\mathcal{H}$ and $X_2\mathcal{H}$ are identical if, and only if, $X_2^{-1}X_1$ belongs to $\mathcal{H}$.* If $X_2^{-1}X_1$ belongs to $\mathcal{H}$ then $X_1 = X_2Y_i$ for some i, and

$$X_1\mathcal{H} = X_2Y_i\mathcal{H} = X_2\mathcal{H},$$

since by the permutation law $Y_i\mathcal{H} = \mathcal{H}$. Thus the two cosets are identical.

Conversely, suppose $X_1\mathcal{H} = X_2\mathcal{H}$. Then $X_2^{-1}X_1\mathcal{H} = \mathcal{H}$. But one element of $\mathcal{H}$ (on the left of the equation) is I; thus $X_2^{-1}X_1$ must also be an element of $\mathcal{H}$ (on the right). This proves the stated result.

- *Every element of $\mathcal{G}$ is in some left coset $X\mathcal{H}$.* This follows trivially since $\mathcal{H}$ contains I, and so the element X_i is in the coset $X_i\mathcal{H}$.

The final step in establishing Lagrange's theorem is, as previously, to note that each coset contains h elements, that the cosets are disjoint and that every one of the g elements in $\mathcal{G}$ appears in one and only one distinct coset. It follows that $g = kh$ for some integer k.

As noted earlier, Lagrange's theorem justifies our statement that any group of order p, where p is prime, must be cyclic and cannot have any proper subgroups: since any subgroup must have an order that divides p, this can only be 1 or p, corresponding to the two trivial subgroups $\mathcal{I}$ and the whole group.

It may be helpful to see an example worked through explicitly, and we again use the same six-element group.

▶*Find the left cosets of the proper subgroup $\mathcal{H}$ of the group $\mathcal{G}$ that has table 28.8 as its multiplication table.*

The subgroup consists of the set of elements $\mathcal{H} = \{I, A, B\}$. We note in passing that it has order 3, which, as required by Lagrange's theorem, is a divisor of 6, the order of $\mathcal{G}$. As in all cases, $\mathcal{H}$ itself provides the first (left) coset, formally the coset

$$I\mathcal{H} = \{II, IA, IB\} = \{I, A, B\}.$$

We continue by choosing an element not already selected, C say, and form

$$C\mathcal{H} = \{CI, CA, CB\} = \{C, D, E\}.$$

These two cosets of $\mathcal{H}$ exhaust $\mathcal{G}$, and are therefore the only cosets, the index of $\mathcal{H}$ in $\mathcal{G}$ being equal to 2.

This completes the example, but it is useful to demonstrate that it would not have mattered if we had taken D, say, instead of I to form a first coset

$$D\mathcal{H} = \{DI, DA, DB\} = \{D, E, C\},$$

and then, from previously unselected elements, picked B, say:

$$B\mathcal{H} = \{BI, BA, BB\} = \{B, I, A\}.$$

The same two cosets would have resulted. ◀

It will be noticed that the cosets are the same groupings of the elements of $\mathcal{G}$ which we earlier noted as being the choice of adjacent column and row headings that give the multiplication table its 'neatest' appearance. Furthermore, if $\mathcal{H}$ is a *normal* subgroup of $\mathcal{G}$ then its (left) cosets themselves form a group (see exercise 28.16).

28.7.3 Conjugates and classes

Our second example of an equivalence relation is concerned with those elements X and Y of a group $\mathcal{G}$ that can be connected by a transformation of the form $Y = G_i^{-1} X G_i$, where G_i is an (appropriate) element of $\mathcal{G}$. Thus $X \sim Y$ if there exists an element G_i of $\mathcal{G}$ such that $Y = G_i^{-1} X G_i$. Different pairs of elements X and Y will, in general, require different group elements G_i. Elements connected in this way are said to be *conjugates*.

We first need to establish that this does indeed define an equivalence relation, as follows.

(i) Reflexivity: $X \sim X$, since $X = I^{-1} X I$ and I belongs to the group.

(ii) Symmetry: $X \sim Y$ implies $Y = G_i^{-1} X G_i$ and therefore $X = (G_i^{-1})^{-1} Y G_i^{-1}$. Since G_i belongs to $\mathcal{G}$ so does G_i^{-1}, and it follows that $Y \sim X$.

(iii) Transitivity: $X \sim Y$ and $Y \sim Z$ imply $Y = G_i^{-1} X G_i$ and $Z = G_j^{-1} Y G_j$ and therefore $Z = G_j^{-1} G_i^{-1} X G_i G_j = (G_i G_j)^{-1} X (G_i G_j)$. Since G_i and G_j belong to $\mathcal{G}$ so does $G_i G_j$, from which it follows that $X \sim Z$.

These results establish conjugacy as an equivalence relation and hence show that it divides $\mathcal{G}$ into classes, two elements being in the same class if, and only if, they are conjugate.

Immediate corollaries are:

(i) If Z is in the class containing I then

$$Z = G_i^{-1} I G_i = G_i^{-1} G_i = I.$$

Thus, since any conjugate of I can be shown to be I, the identity must be in a class by itself.

(ii) If X is in a class by itself then

$$Y = G_i^{-1} X G_i$$

must imply that $Y = X$. But

$$X = G_i G_i^{-1} X G_i G_i^{-1}$$

for any G_i, and so

$$X = G_i (G_i^{-1} X G_i) G_i^{-1} = G_i Y G_i^{-1} = G_i X G_i^{-1},$$

i.e. $X G_i = G_i X$ for all G_i.

Thus commutation with all elements of the group is a necessary (and sufficient) condition for any particular group element to be in a class by itself. In an Abelian group each element is in a class by itself.

(iii) In any group G the set S of elements in classes by themselves is an Abelian subgroup (known as the *centre* of G). We have shown that I belongs to S, and so if, further, $X G_i = G_i X$ and $Y G_i = G_i Y$ for all G_i belonging to G then:

(a) $(XY)G_i = X G_i Y = G_i(XY)$, i.e. the closure of S, and

(b) $X G_i = G_i X$ implies $X^{-1} G_i = G_i X^{-1}$, i.e. the inverse of X belongs to S.

Hence S is a group, and clearly Abelian.

Yet again for illustration purposes, we use the six-element group that has table 28.8 as its group table.

▶*Find the conjugacy classes of the group G having table 28.8 as its multiplication table.*

As always, I is in a class by itself, and we need consider it no further.

Consider next the results of forming $X^{-1}AX$, as X runs through the elements of G.

$$
\begin{array}{llllll}
I^{-1}AI & A^{-1}AA & B^{-1}AB & C^{-1}AC & D^{-1}AD & E^{-1}AE \\
= IA & = IA & = AI & = CE & = DC & = ED \\
= A & = A & = A & = B & = B & = B
\end{array}
$$

Only A and B are generated. It is clear that $\{A, B\}$ is one of the conjugacy classes of G. This can be verified by forming all elements $X^{-1}BX$; again only A and B appear.

We now need to pick an element not in the two classes already found. Suppose we pick C. Just as for A, we compute $X^{-1}CX$, as X runs through the elements of G. The calculations can be done directly using the table and give the following:

$$
\begin{array}{lllllll}
X & : I & A & B & C & D & E \\
X^{-1}CX & : C & E & D & C & E & D
\end{array}
$$

Thus C, D and E belong to the same class. The group is now exhausted, and so the three conjugacy classes are

$$\{I\}, \qquad \{A, B\}, \qquad \{C, D, E\}. \blacktriangleleft$$

In the case of this small and simple, but non-Abelian, group, only the identity is in a class by itself (i.e. only I commutes with all other elements). It is also the only member of the centre of the group.

Other areas from which examples of conjugacy classes can be taken include permutations and rotations. Two permutations can only be (but are not necessarily) in the same class if their cycle specifications have the same structure. For example, in S_5 the permutations (1 3 5)(2)(4) and (2 5 3)(1)(4) could be in the same class as each other but not in the class that contains (1 5)(2 4)(3). An example of permutations with the same cycle structure yet in different conjugacy classes is given in exercise 29. 10.

In the case of the continuous rotation group, rotations by the same angle θ about any two axes labelled i and j are in the same class, because the group contains a rotation that takes the first axis into the second. Without going into mathematical details, a rotation about axis i can be represented by the operator $R_i(\theta)$, and the two rotations are connected by a relationship of the form

$$R_j(\theta) = \phi_{ij}^{-1} R_i(\theta)\phi_{ij},$$

in which ϕ_{ij} is the member of the full continuous rotation group that takes axis i into axis j.

28.8 Exercises

28.1   For each of the following sets, determine whether they form a group under the operation indicated (where it is relevant you may assume that matrix multiplication is associative):

(a) the integers (mod 10) under addition;
(b) the integers (mod 10) under multiplication;
(c) the integers $1, 2, 3, 4, 5, 6$ under multiplication (mod 7);
(d) the integers $1, 2, 3, 4, 5$ under multiplication (mod 6);
(e) all matrices of the form

$$\begin{pmatrix} a & a-b \\ 0 & b \end{pmatrix},$$

where a and b are integers (mod 5) and $a \neq 0 \neq b$, under matrix multiplication;
(f) those elements of the set in (e) that are of order 1 or 2 (taken together);
(g) all matrices of the form

$$\begin{pmatrix} 1 & 0 & 0 \\ a & 1 & 0 \\ b & c & 1 \end{pmatrix},$$

where a, b, c are integers, under matrix multiplication.

28.2   Which of the following relationships between X and Y are equivalence relations? Give a proof of your conclusions in each case:

(a) X and Y are integers and $X - Y$ is odd;
(b) X and Y are integers and $X - Y$ is even;
(c) X and Y are people and have the same postcode;
(d) X and Y are people and have a parent in common;
(e) X and Y are people and have the same mother;
(f) X and Y are $n \times n$ matrices satisfying $Y = PXQ$, where P and Q are elements of a group $\mathcal{G}$ of $n \times n$ matrices.

28.3   Define a binary operation $\bullet$ on the set of real numbers by

$$x \bullet y = x + y + rxy,$$

where r is a non-zero real number. Show that the operation $\bullet$ is associative.
        Prove that $x \bullet y = -r^{-1}$ if, and only if, $x = -r^{-1}$ or $y = -r^{-1}$. Hence prove that the set of all real numbers excluding $-r^{-1}$ forms a group under the operation $\bullet$.

28.4   Prove that the relationship $X \sim Y$, defined by $X \sim Y$ if Y can be expressed in the form

$$Y = \frac{aX + b}{cX + d},$$

with a, b, c and d as integers, is an equivalence relation on the set of real numbers $\Re$. Identify the class that contains the real number 1.

28.5   The following is a 'proof' that reflexivity is an unnecessary axiom for an equivalence relation.

Because of symmetry $X \sim Y$ implies $Y \sim X$. Then by transitivity $X \sim Y$ and $Y \sim X$ imply $X \sim X$. Thus symmetry and transitivity imply reflexivity, which therefore need not be separately required.

Demonstrate the flaw in this proof using the set consisting of all real numbers plus the number i. Show by investigating the following specific cases that, whether or not reflexivity actually holds, it cannot be deduced from symmetry and transitivity alone.

(a) $X \sim Y$ if $X + Y$ is real.
(b) $X \sim Y$ if XY is real.

28.6   Prove that the set $\mathcal{M}$ of matrices

$$A = \begin{pmatrix} a & b \\ 0 & c \end{pmatrix},$$

where a, b, c are integers (mod 5) and $a \neq 0 \neq c$, form a non-Abelian group under matrix multiplication.

Show that the subset containing elements of $\mathcal{M}$ that are of order 1 or 2 do not form a proper subgroup of $\mathcal{M}$,

(a) using Lagrange's theorem,
(b) by direct demonstration that the set is not closed.

28.7 S is the set of all 2×2 matrices of the form

$$A = \begin{pmatrix} w & x \\ y & z \end{pmatrix}, \qquad \text{where } wz - xy = 1.$$

Show that S is a group under matrix multiplication. Which element(s) have order 2? Prove that an element A has order 3 if $w + z + 1 = 0$.

28.8   Show that, under matrix multiplication, matrices of the form

$$M(a_0, \mathbf{a}) = \begin{pmatrix} a_0 + a_1 i & -a_2 + a_3 i \\ a_2 + a_3 i & a_0 - a_1 i \end{pmatrix},$$

where a_0 and the components of column matrix $\mathbf{a} = (a_1 \; a_2 \; a_3)^{\mathrm{T}}$ are real numbers satisfying $a_0^2 + |\mathbf{a}|^2 = 1$, constitute a group. Deduce that, under the transformation $\mathbf{z} \to M\mathbf{z}$, where $\mathbf{z}$ is any column matrix, $|\mathbf{z}|^2$ is invariant.

28.9   If $\mathcal{A}$ is a group in which every element other than the identity, I, has order 2, prove that $\mathcal{A}$ is Abelian. Hence show that if X and Y are distinct elements of $\mathcal{A}$, neither being equal to the identity, then the set $\{I, X, Y, XY\}$ forms a subgroup of $\mathcal{A}$.

Deduce that if $\mathcal{B}$ is a group of order $2p$, with p a prime greater than 2, then $\mathcal{B}$ must contain an element of order p.

28.10   The group of rotations (excluding reflections and inversions) in three dimensions that take a cube into itself is known as the group 432 (or O in the usual chemical notation). Show by each of the following methods that this group has 24 elements.

(a) Identify the distinct relevant axes and count the number of qualifying rotations about each.

(b) The orientation of the cube is determined if the directions of two of its body diagonals are given. Consider the number of distinct ways in which one body diagonal can be chosen to be 'vertical', say, and a second diagonal made to lie along a particular direction.

28.11 Identify the eight symmetry operations on a square. Show that they form a group $\mathcal{D}_4$ (known to crystallographers as $4mm$ and to chemists as C_{4v}) having one element of order 1, five of order 2 and two of order 4. Find its proper subgroups and the corresponding cosets.

28.12 If $\mathcal{A}$ and $\mathcal{B}$ are two groups, then their direct product, $\mathcal{A} \times \mathcal{B}$, is defined to be the set of ordered pairs (X, Y), with X an element of $\mathcal{A}$, Y an element of $\mathcal{B}$ and multiplication given by $(X, Y)(X', Y') = (XX', YY')$. Prove that $\mathcal{A} \times \mathcal{B}$ is a group.

Denote the cyclic group of order n by C_n and the symmetry group of a regular n-sided figure (an n-gon) by $\mathcal{D}_n$ – thus $\mathcal{D}_3$ is the symmetry group of an equilateral triangle, as discussed in the text.

(a) By considering the orders of each of their elements, show (i) that $C_2 \times C_3$ is isomorphic to C_6, and (ii) that $C_2 \times \mathcal{D}_3$ is isomorphic to $\mathcal{D}_6$.

(b) Are any of $\mathcal{D}_4$, C_8, $C_2 \times C_4$, $C_2 \times C_2 \times C_2$ isomorphic?

28.13 Find the group $\mathcal{G}$ generated under matrix multiplication by the matrices

$$ A = \begin{pmatrix} 0 & 1 \\ 1 & 0 \end{pmatrix}, \qquad B = \begin{pmatrix} 0 & i \\ i & 0 \end{pmatrix}. $$

Determine its proper subgroups, and verify for each of them that its cosets exhaust $\mathcal{G}$.

28.14 Show that if p is prime then the set of rational number pairs (a, b), excluding $(0, 0)$, with multiplication defined by

$$ (a, b) \bullet (c, d) = (e, f), \quad \text{where} \quad (a + b\sqrt{p})(c + d\sqrt{p}) = e + f\sqrt{p}, $$

forms an Abelian group. Show further that the mapping $(a, b) \to (a, -b)$ is an automorphism.

28.15 Consider the following mappings between a permutation group and a cyclic group.

(a) Denote by A_n the subset of the permutation group S_n that contains all the even permutations. Show that A_n is a subgroup of S_n.

(b) List the elements of S_3 in cycle notation and identify the subgroup A_3.

(c) For each element X of S_3, let $p(X) = 1$ if X belongs to A_3 and $p(X) = -1$ if it does not. Denote by C_2 the multiplicative cyclic group of order 2. Determine the images of each of the elements of S_3 for the following four mappings:

$$ \begin{aligned} \Phi_1 &: S_3 \to C_2 & X &\to p(X), \\ \Phi_2 &: S_3 \to C_2 & X &\to -p(X), \\ \Phi_3 &: S_3 \to A_3 & X &\to X^2, \\ \Phi_4 &: S_3 \to S_3 & X &\to X^3. \end{aligned} $$

(d) For each mapping, determine whether the kernel $\mathcal{K}$ is a subgroup of S_3 and, if so, whether the mapping is a homomorphism.

28.16 For the group $\mathcal{G}$ with multiplication table 28.8 and proper subgroup $\mathcal{H} = \{I, A, B\}$, denote the coset $\{I, A, B\}$ by $\mathcal{C}_1$ and the coset $\{C, D, E\}$ by $\mathcal{C}_2$. Form the set of all possible products of a member of $\mathcal{C}_1$ with itself, and denote this by $\mathcal{C}_1\mathcal{C}_1$.

Similarly compute C_2C_2, C_1C_2 and C_2C_1. Show that each product coset is equal to C_1 or to C_2, and that a 2×2 multiplication table can be formed, demonstrating that C_1 and C_2 are themselves the elements of a group of order 2. A subgroup like $\mathcal{H}$ whose cosets themselves form a group is a *normal subgroup*.

28.17 The group of all non-singular $n \times n$ matrices is known as the general linear group $GL(n)$ and that with only real elements as $GL(n, \mathbf{R})$. If $\mathbf{R}^*$ denotes the multiplicative group of non-zero real numbers, prove that the mapping $\Phi : GL(n, \mathbf{R}) \to \mathbf{R}^*$, defined by $\Phi(\mathsf{M}) = \det \mathsf{M}$, is a homomorphism.

Show that the kernel $\mathcal{K}$ of Φ is a subgroup of $GL(n, \mathbf{R})$. Determine its cosets and show that they themselves form a group.

28.18 The group of reflection–rotation symmetries of a square is known as $\mathcal{D}_4$; let X be one of its elements. Consider a mapping $\Phi : \mathcal{D}_4 \to S_4$, the permutation group on four objects, defined by $\Phi(X) =$ the permutation induced by X on the set $\{x, y, d, d'\}$, where x and y are the two principal axes, and d and d' the two principal diagonals, of the square. For example, if R is a rotation by $\pi/2$, $\Phi(R) = (12)(34)$. Show that $\mathcal{D}_4$ is mapped onto a subgroup of S_4 and, by constructing the multiplication tables for $\mathcal{D}_4$ and the subgroup, prove that the mapping is a homomorphism.

28.19 Given that matrix M is a member of the multiplicative group $GL(3, \mathbf{R})$, determine, for each of the following additional constraints on M (applied separately), whether the subset satisfying the constraint is a subgroup of $GL(3, \mathbf{R})$:

(a) $\mathsf{M}^T = \mathsf{M}$;

(b) $\mathsf{M}^T \mathsf{M} = \mathsf{I}$;

(c) $|\mathsf{M}| = 1$;

(d) $M_{ij} = 0$ for $j > i$ and $M_{ii} \neq 0$.

28.20 The elements of the quaternion group, $\mathcal{Q}$, are the set

$$\{1, -1, i, -i, j, -j, k, -k\},$$

with $i^2 = j^2 = k^2 = -1$, $ij = k$ and its cyclic permutations, and $ji = -k$ and its cyclic permutations. Find the proper subgroups of $\mathcal{Q}$ and the corresponding cosets. Show that all of the subgroups are normal. Show that $\mathcal{Q}$ cannot be isomorphic to the group $4mm$ (C_{4v}) considered in exercise 28.11.

28.21 Show that $\mathcal{D}_4$, the group of symmetries of a square, has two isomorphic subgroups of order 4. Show further that there exists a two-to-one homomorphism from the quaternion group $\mathcal{Q}$, of exercise 28.20, onto one (and hence either) of these two subgroups, and determine its kernel.

28.22 Show that the matrices

$$M(\theta, x, y) = \begin{pmatrix} \cos\theta & -\sin\theta & x \\ \sin\theta & \cos\theta & y \\ 0 & 0 & 1 \end{pmatrix},$$

where $0 \leq \theta < 2\pi$, $-\infty < x < \infty$, $-\infty < y < \infty$, form a group under matrix multiplication.

Show that those $M(\theta, x, y)$ for which $\theta = 0$ form a subgroup and identify its cosets. Show that the cosets themselves form a group.

28.23 Find (a) all the proper subgroups and (b) all the conjugacy classes of the symmetry group of a regular pentagon.

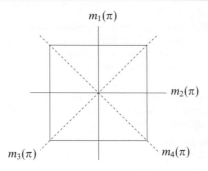

Figure 28.3 The notation for exercise 28.11.

28.9 Hints and answers

28.1 § (a) Yes, (b) no, there is no inverse for 2, (c) yes, (d) no, 2×3 is not in the set, (e) yes, (f) yes, they form a subgroup of order 4, $[1,0;0,1]$ $[4,0;0,4]$ $[1,2;0,4]$ $[4,3;0,1]$, (g) yes.

28.3 $x \bullet (y \bullet z) = x + y + z + r(xy + xz + yz) + r^2 xyz = (x \bullet y) \bullet z$. Show that assuming $x \bullet y = -r^{-1}$ leads to $(rx + 1)(ry + 1) = 0$. The inverse of x is $x^{-1} = -x/(1 + rx)$; show that this is not equal to $-r^{-1}$.

28.5 (a) Consider both $X = i$ and $X \neq i$. Here, $i \nsim i$. (b) In this case $i \sim i$, but the conclusion cannot be deduced from the other axioms. In both cases i is in a class by itself and no Y, as used in the false proof, can be found.

28.7 † Use $|AB| = |A||B| = 1 \times 1 = 1$ to prove closure. The inverse has $w \leftrightarrow z$, $x \leftrightarrow -x$, $y \leftrightarrow -y$, giving $|A^{-1}| = 1$, i.e. it is in the set. The only element of order 2 is $-I$; A^2 can be simplified to $[-(w + 1), -x; -y, -(z + 1)]$.

28.9 If $XY = Z$, show that $Y = XZ$ and $X = ZY$, then form YX. Note that the elements of $\mathcal{B}$ can only have orders 1, 2 or p. Suppose they all have order 1 or 2; then using the earlier result, whilst noting that 4 does not divide $2p$, leads to a contradiction.

28.11 Using the notation indicated in figure 28.3, R being a rotation of $\pi/2$ about an axis perpendicular to the square, we have: I has order 1; R^2, m_1, m_2, m_3, m_4 have order 2; R, R^3 have order 4.
subgroup $\{I, R, R^2, R^3\}$ has cosets $\{I, R, R^2, R^3\}$, $\{m_1, m_2, m_3, m_4\}$;
subgroup $\{I, R^2, m_1, m_2\}$ has cosets $\{I, R^2, m_1, m_2\}$, $\{R, R^3, m_3, m_4\}$;
subgroup $\{I, R^2, m_3, m_4\}$ has cosets $\{I, R^2, m_3, m_4\}$, $\{R, R^3, m_1, m_2\}$;
subgroup $\{I, R^2\}$ has cosets $\{I, R^2\}$, $\{R, R^3\}$, $\{m_1, m_2\}$, $\{m_3, m_4\}$;
subgroup $\{I, m_1\}$ has cosets $\{I, m_1\}$, $\{R, m_3\}$, $\{R^2, m_2\}$, $\{R^3, m_4\}$;
subgroup $\{I, m_2\}$ has cosets $\{I, m_2\}$, $\{R, m_4\}$, $\{R^2, m_1\}$, $\{R^3, m_3\}$;
subgroup $\{I, m_3\}$ has cosets $\{I, m_3\}$, $\{R, m_2\}$, $\{R^2, m_4\}$, $\{R^3, m_1\}$;
subgroup $\{I, m_4\}$ has cosets $\{I, m_4\}$, $\{R, m_1\}$, $\{R^2, m_3\}$, $\{R^3, m_2\}$.

28.13 $\mathcal{G} = \{I, A, B, B^2, B^3, AB, AB^2, AB^3\}$. The proper subgroups are as follows: $\{I, A\}$, $\{I, B^2\}$, $\{I, AB^2\}$, $\{I, B, B^2, B^3\}$, $\{I, B^2, AB, AB^3\}$.

28.15 (b) $A_3 = \{(1), (123), (132)\}$.
(d) For Φ_1, $\mathcal{K} = \{(1), (123), (132)\}$ is a subgroup.
For Φ_2, $\mathcal{K} = \{(23), (13), (12)\}$ is not a subgroup because it has no identity element.
For Φ_3, $\mathcal{K} = \{(1), (23), (13), (12)\}$ is not a subgroup because it is not closed.

§ Where matrix elements are given as a list, the convention used is [row 1; row 2;...], individual entries in each row being separated by commas.

For Φ_4, $\mathcal{K} = \{(1), (123), (132)\}$ is a subgroup.
Only Φ_1 is a homomorphism; Φ_4 fails because, for example, $[(23)(13)]' \neq (23)'(13)'$.

28.17 Recall that, for any pair of matrices P and Q, $|PQ| = |P||Q|$. $\mathcal{K}$ is the set of all matrices with unit determinant. The cosets of $\mathcal{K}$ are the sets of matrices whose determinants are equal; $\mathcal{K}$ itself is the identity in the group of cosets.

28.19 (a) No, because the set is not closed, (b) yes, (c) yes, (d) yes.

28.21 Each subgroup contains the identity, a rotation by π, and two reflections. The homomorphism is $\pm 1 \to I$, $\pm i \to R^2$, $\pm j \to m_x$, $\pm k \to m_y$ with kernel $\{1, -1\}$.

28.23 There are 10 elements in all: I, rotations R^i ($i = 1, 4$) and reflections m_j ($j = 1, 5$).
(a) There are five proper subgroups of order 2, $\{I, m_j\}$ and one proper subgroup of order 5, $\{I, R, R^2, R^3, R^4\}$.
(b) Four conjugacy classes, $\{I\}, \{R, R^4\}, \{R^2, R^3\}, \{m_1, m_2, m_3, m_4, m_5\}$.

29

Representation theory

As indicated at the start of the previous chapter, significant conclusions can often be drawn about a physical system simply from the study of its symmetry properties. That chapter was devoted to setting up a formal mathematical basis, group theory, with which to describe and classify such properties; the current chapter shows how to implement the consequences of the resulting classifications and obtain concrete physical conclusions about the system under study. The connection between the two chapters is akin to that between working with coordinate-free vectors, each denoted by a single symbol, and working with a coordinate system in which the same vectors are expressed in terms of components.

The 'coordinate systems' that we will choose will be ones that are expressed in terms of matrices; it will be clear that ordinary numbers would not be sufficient, as they make no provision for any non-commutation amongst the elements of a group. Thus, in this chapter the group elements will be *represented* by matrices that have the same commutation relations as the members of the group, whatever the group's original nature (symmetry operations, functional forms, matrices, permutations, etc.). For some abstract groups it is difficult to give a written description of the elements and their properties without recourse to such representations. Most of our applications will be concerned with representations of the groups that consist of the symmetry operations on molecules containing two or more identical atoms.

Firstly, in section 29.1, we use an elementary example to demonstrate the kind of conclusions that can be reached by arguing purely on symmetry grounds. Then in sections 29.2–29.10 we develop the formal side of representation theory and establish general procedures and results. Finally, these are used in section 29.11 to tackle a variety of problems drawn from across the physical sciences.

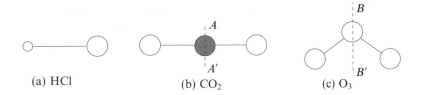

Figure 29.1 Three molecules, (a) hydrogen chloride, (b) carbon dioxide and (c) ozone, for which symmetry considerations impose varying degrees of constraint on their possible electric dipole moments.

29.1 Dipole moments of molecules

Some simple consequences of *symmetry* can be demonstrated by considering whether a permanent electric dipole moment can exist in any particular molecule; three simple molecules, hydrogen chloride, carbon dioxide and ozone, are illustrated in figure 29.1. Even if a molecule is electrically neutral, an electric dipole moment will exist in it if the centres of gravity of the positive charges (due to protons in the atomic nuclei) and of the negative charges (due to the electrons) do not coincide.

For hydrogen chloride there is no reason why they should coincide; indeed, the normal picture of the binding mechanism in this molecule is that the electron from the hydrogen atom moves its average position from that of its proton nucleus to somewhere between the hydrogen and chlorine nuclei. There is no compensating movement of positive charge, and a net dipole moment is to be expected – and is found experimentally.

For the linear molecule carbon dioxide it seems obvious that it cannot have a dipole moment, because of its symmetry. Putting this rather more rigorously, we note that any rotation about the long axis of the molecule leaves it totally unchanged; consequently, any component of a permanent electric dipole perpendicular to that axis must be zero (a non-zero component would rotate although no physical change had taken place in the molecule). That only leaves the possibility of a component parallel to the axis. However, a rotation of π radians about the axis AA' shown in figure 29.1(b) carries the molecule into itself, as does a reflection in a plane through the carbon atom and perpendicular to the molecular axis (i.e. one with its normal parallel to the axis). In both cases the two oxygen atoms change places but, as they are identical, the molecule is indistinguishable from the original. Either 'symmetry operation' would reverse the sign of any dipole component directed parallel to the molecular axis; this can only be compatible with the indistinguishability of the original and final systems if the parallel component is zero. Thus on symmetry grounds carbon dioxide cannot have a permanent electric dipole moment.

Finally, for ozone, which is angular rather than linear, symmetry does not place such tight constraints. A dipole-moment component parallel to the axis BB' (figure 29.1(c)) is possible, since there is no symmetry operation that reverses the component in that direction and at the same time carries the molecule into an indistinguishable copy of itself. However, a dipole moment perpendicular to BB' is not possible, since a rotation of π about BB' would both reverse any such component and carry the ozone molecule into itself – two contradictory conclusions unless the component is zero.

In summary, symmetry requirements appear in the form that some or all components of permanent electric dipoles in molecules are forbidden; they do not show that the other components do exist, only that they may. The greater the symmetry of the molecule, the tighter the restrictions on potentially non-zero components of its dipole moment.

In section 23.11 other, more complicated, physical situations will be analysed using results derived from representation theory. In anticipation of these results, and since it may help the reader to understand where the developments in the next nine sections are leading, we make here a broad, powerful, but rather formal, statement as follows.

If a physical system is such that after the application of particular rotations or reflections (or a combination of the two) the final system is indistinguishable from the original system then its behaviour, and hence the functions that describe its behaviour, must have the corresponding property of invariance when subjected to the same rotations and reflections.

29.2 Choosing an appropriate formalism

As mentioned in the introduction to this chapter, the elements of a finite group $\mathcal{G}$ can be *represented* by matrices; this is done in the following way. A suitable column matrix u, known as a *basis vector*,[§] is chosen and is written in terms of its components u_i, the *basis functions*, as $u = (u_1 \ u_2 \ \cdots \ u_n)^T$. The u_i may be of a variety of natures, e.g. numbers, coordinates, functions or even a set of labels, though for any one basis vector they will all be of the same kind.

Once chosen, the basis vector can be used to generate an n-dimensional *representation* of the group as follows. An element X of the group is selected and its effect on each basis function u_i is determined. If the action of X on u_1 is to produce u_1', etc. then the set of equations

$$u_i' = X u_i \tag{29.1}$$

[§] This usage of the term *basis vector* is not exactly the same as that introduced in subsection 8.1.1.

generates a new column matrix $u' = (u'_1 \; u'_2 \; \cdots \; u'_n)^T$. Having established u and u' we can determine the $n \times n$ matrix, $M(X)$ say, that connects them by

$$u' = M(X)u. \tag{29.2}$$

It may seem natural to use the matrix $M(X)$ so generated as the representative matrix of the element X; in fact, because we have already chosen the convention whereby $Z = XY$ implies that the effect of applying element Z is the same as that of first applying Y and then applying X to the result, one further step has to be taken. So that the representative matrices $D(X)$ may follow the same convention, i.e.

$$D(Z) = D(X)D(Y),$$

and at the same time respect the normal rules of matrix multiplication, it is necessary to take the *transpose* of $M(X)$ as the representative matrix $D(X)$. Explicitly,

$$D(X) = M^T(X) \tag{29.3}$$

and (29.2) becomes

$$u' = D^T(X)u. \tag{29.4}$$

Thus the procedure for determining the matrix $D(X)$ that represents the group element X in a representation based on basis vector u is summarised by equations (29.1)–(29.4).[§]

This procedure is then repeated for each element X of the group, and the resulting set of $n \times n$ matrices $D = \{D(X)\}$ is said to be the n-dimensional representation of G having u as its basis. The need to take the transpose of each matrix $M(X)$ is not of any fundamental significance, since the only thing that really matters is whether the matrices $D(X)$ have the appropriate multiplication properties – and, as defined, they do.

In cases in which the basis functions are labels, the actions of the group elements are such as to cause rearrangements of the labels. Correspondingly the matrices $D(X)$ contain only '1's and '0's as entries; each row and each column contains a single '1'.

[§] An alternative procedure in which a row vector is used as the basis vector is possible. Defining equations of the form $u^T X = u^T D(X)$ are used, and no additional transpositions are needed to define the representative matrices. However, row-matrix equations are cumbersome to write out and in all other parts of this book we have adopted the convention of writing operators (here the group element) to the left of the object on which they operate (here the basis vector).

> ▶ *For the group S_3 of permutations on three objects, which has group multiplication table 28.8 on p. 1055, with (in cycle notation)*
>
> $$I = (1)(2)(3), \quad A = (1\,2\,3), \quad B = (1\,3\,2$$
> $$C = (1)(2\,3), \quad D = (3)(1\,2), \quad E = (2)(1\,3),$$
>
> *use as the components of a basis vector the ordered letter triplets*
>
> $$u_1 = \{P\,Q\,R\}, \quad u_2 = \{Q\,R\,P\}, \quad u_3 = \{R\,P\,Q\},$$
> $$u_4 = \{P\,R\,Q\}, \quad u_5 = \{Q\,P\,R\}, \quad u_6 = \{R\,Q\,P\}.$$
>
> *Generate a six-dimensional representation $D = \{D(X)\}$ of the group and confirm that the representative matrices multiply according to table 28.8, e.g.*
>
> $$D(C)D(B) = D(E).$$

It is immediate that the identity permutation $I = (1)(2)(3)$ leaves all u_i unchanged, i.e. $u_i' = u_i$ for all i. The representative matrix $D(I)$ is thus I_6, the 6×6 unit matrix.

We next take X as the permutation $A = (1\,2\,3)$ and, using (29.1), let it act on each of the components of the basis vector:

$$u_1' = Au_1 = (1\,2\,3)\{P\,Q\,R\} = \{Q\,R\,P\} = u_2$$
$$u_2' = Au_2 = (1\,2\,3)\{Q\,R\,P\} = \{R\,P\,Q\} = u_3$$
$$\vdots \qquad\qquad\qquad\qquad \vdots$$
$$u_6' = Au_6 = (1\,2\,3)\{R\,Q\,P\} = \{Q\,P\,R\} = u_5.$$

The matrix $M(A)$ has to be such that $\mathbf{u}' = M(A)\mathbf{u}$ (here dots replace zeros to aid readability):

$$\mathbf{u}' = \begin{pmatrix} u_2 \\ u_3 \\ u_1 \\ u_6 \\ u_4 \\ u_5 \end{pmatrix} = \begin{pmatrix} \cdot & 1 & \cdot & \cdot & \cdot & \cdot \\ \cdot & \cdot & 1 & \cdot & \cdot & \cdot \\ 1 & \cdot & \cdot & \cdot & \cdot & \cdot \\ \cdot & \cdot & \cdot & \cdot & \cdot & 1 \\ \cdot & \cdot & \cdot & 1 & \cdot & \cdot \\ \cdot & \cdot & \cdot & \cdot & 1 & \cdot \end{pmatrix} \begin{pmatrix} u_1 \\ u_2 \\ u_3 \\ u_4 \\ u_5 \\ u_6 \end{pmatrix} \equiv M(A)\mathbf{u}.$$

$D(A)$ is then equal to $M^T(A)$.

The other $D(X)$ are determined in a similar way. In general, if

$$Xu_i = u_j,$$

then $[M(X)]_{ij} = 1$, leading to $[D(X)]_{ji} = 1$ and $[D(X)]_{jk} = 0$ for $k \neq i$. For example,

$$Cu_3 = (1)(23)\{R\,P\,Q\} = \{R\,Q\,P\} = u_6$$

implies that $[D(C)]_{63} = 1$ and $[D(C)]_{6k} = 0$ for $k = 1, 2, 4, 5, 6$. When calculated in full

$$D(C) = \begin{pmatrix} \cdot & \cdot & \cdot & 1 & \cdot & \cdot \\ \cdot & \cdot & \cdot & \cdot & 1 & \cdot \\ \cdot & \cdot & \cdot & \cdot & \cdot & 1 \\ 1 & \cdot & \cdot & \cdot & \cdot & \cdot \\ \cdot & 1 & \cdot & \cdot & \cdot & \cdot \\ \cdot & \cdot & 1 & \cdot & \cdot & \cdot \end{pmatrix}, \quad D(B) = \begin{pmatrix} \cdot & 1 & \cdot & \cdot & \cdot & \cdot \\ \cdot & \cdot & 1 & \cdot & \cdot & \cdot \\ 1 & \cdot & \cdot & \cdot & \cdot & \cdot \\ \cdot & \cdot & \cdot & \cdot & \cdot & 1 \\ \cdot & \cdot & \cdot & 1 & \cdot & \cdot \\ \cdot & \cdot & \cdot & \cdot & 1 & \cdot \end{pmatrix},$$

$$D(E) = \begin{pmatrix} \cdot & \cdot & \cdot & \cdot & \cdot & 1 \\ \cdot & \cdot & \cdot & 1 & \cdot & \cdot \\ \cdot & \cdot & \cdot & \cdot & 1 & \cdot \\ \cdot & 1 & \cdot & \cdot & \cdot & \cdot \\ \cdot & \cdot & 1 & \cdot & \cdot & \cdot \\ 1 & \cdot & \cdot & \cdot & \cdot & \cdot \end{pmatrix},$$

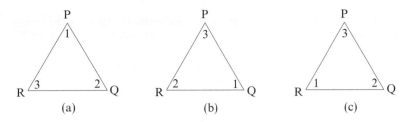

Figure 29.2   Diagram (a) shows the definition of the basis vector, (b) shows the effect of applying a clockwise rotation of $2\pi/3$ and (c) shows the effect of applying a reflection in the mirror axis through Q.

from which it can be verified that $D(C)D(B) = D(E)$. ◀

Whilst a representation obtained in this way necessarily has the same dimension as the order of the group it represents, there are, in general, square matrices of both smaller and larger dimensions that can be used to represent the group, though their existence may be less obvious.

One possibility that arises when the group elements are symmetry operations on an object whose position and orientation can be referred to a space coordinate system is called the *natural representation*. In it the representative matrices $D(X)$ describe, in terms of a fixed coordinate system, what happens to a coordinate system that moves with the object when X is applied. There is usually some redundancy of the coordinates used in this type of representation, since interparticle distances are fixed and fewer than $3N$ coordinates, where N is the number of identical particles, are needed to specify uniquely the object's position and orientation. Subsection 29.11.1 gives an example that illustrates both the advantages and disadvantages of the natural representation. We continue here with an example of a natural representation that has no such redundancy.

▶ *Use the fact that the group considered in the previous worked example is isomorphic to the group of two-dimensional symmetry operations on an equilateral triangle to generate a three-dimensional representation of the group.*

Label the triangle's corners as 1, 2, 3 and three fixed points in space as P, Q, R, so that initially corner 1 lies at point P, 2 lies at point Q, and 3 at point R. We take P, Q, R as the components of the basis vector.

In figure 29.2, (a) shows the initial configuration and also, formally, the result of applying the identity I to the triangle; it is therefore described by the basis vector, $(P \quad Q \quad R)^{T}$.

Diagram (b) shows the the effect of a clockwise rotation by $2\pi/3$, corresponding to element A in the previous example; the new column matrix is $(Q \quad R \quad P)^{T}$.

Diagram (c) shows the effect of a typical mirror reflection – the one that leaves the corner at point Q unchanged (element D in table 28.8 and the previous example); the new column matrix is now $(R \quad Q \quad P)^{T}$.

In similar fashion it can be concluded that the column matrix corresponding to element B, rotation by $4\pi/3$, is $(R \quad P \quad Q)^{T}$, and that the other two reflections C and E result in column matrices $(P \quad R \quad Q)^T$ and $(Q \quad P \quad R)^T$ respectively. The forms of the representative matrices $M^{nat}(X)$, (29.2), are now determined by equations such as, for element E,

$$\begin{pmatrix} Q \\ P \\ R \end{pmatrix} = \begin{pmatrix} 0 & 1 & 0 \\ 1 & 0 & 0 \\ 0 & 0 & 1 \end{pmatrix} \begin{pmatrix} P \\ Q \\ R \end{pmatrix}$$

implying that

$$D^{nat}(E) = \begin{pmatrix} 0 & 1 & 0 \\ 1 & 0 & 0 \\ 0 & 0 & 1 \end{pmatrix}^T = \begin{pmatrix} 0 & 1 & 0 \\ 1 & 0 & 0 \\ 0 & 0 & 1 \end{pmatrix}.$$

In this way the complete representation is obtained as

$$D^{nat}(I) = \begin{pmatrix} 1 & 0 & 0 \\ 0 & 1 & 0 \\ 0 & 0 & 1 \end{pmatrix}, \quad D^{nat}(A) = \begin{pmatrix} 0 & 0 & 1 \\ 1 & 0 & 0 \\ 0 & 1 & 0 \end{pmatrix}, \quad D^{nat}(B) = \begin{pmatrix} 0 & 1 & 0 \\ 0 & 0 & 1 \\ 1 & 0 & 0 \end{pmatrix},$$

$$D^{nat}(C) = \begin{pmatrix} 1 & 0 & 0 \\ 0 & 0 & 1 \\ 0 & 1 & 0 \end{pmatrix}, \quad D^{nat}(D) = \begin{pmatrix} 0 & 0 & 1 \\ 0 & 1 & 0 \\ 1 & 0 & 0 \end{pmatrix}, \quad D^{nat}(E) = \begin{pmatrix} 0 & 1 & 0 \\ 1 & 0 & 0 \\ 0 & 0 & 1 \end{pmatrix}.$$

It should be emphasised that although the group contains six elements this representation is three-dimensional. ◀

We will concentrate on matrix representations of *finite* groups, particularly rotation and reflection groups (the so-called crystal point groups). The general ideas carry over to infinite groups, such as the continuous rotation groups, but in a book such as this, which aims to cover many areas of applicable mathematics, some topics can only be mentioned and not explored. We now give the formal definition of a representation.

Definition. *A representation* $D = \{D(X)\}$ *of a group* $\mathcal{G}$ *is an assignment of a non-singular square* $n \times n$ *matrix* $D(X)$ *to each element* X *belonging to* $\mathcal{G}$, *such that*

(i) $D(I) = I_n$, *the unit* $n \times n$ *matrix*,

(ii) $D(X)D(Y) = D(XY)$ *for any two elements* X *and* Y *belonging to* $\mathcal{G}$, *i.e. the matrices multiply in the same way as the group elements they represent*.

As mentioned previously, a representation by $n \times n$ matrices is said to be an *n-dimensional representation* of $\mathcal{G}$. The dimension n is not to be confused with g, the order of the group, which gives the number of matrices needed in the representation, though they might not all be different.

A consequence of the two defining conditions for a representation is that the matrix associated with the inverse of X is the inverse of the matrix associated with X. This follows immediately from setting $Y = X^{-1}$ in (ii):

$$D(X)D(X^{-1}) = D(XX^{-1}) = D(I) = I_n;$$

hence

$$D(X^{-1}) = [D(X)]^{-1}.$$

As an example, the four-element Abelian group that consists of the set $\{1, i, -1, -i\}$ under ordinary multiplication has a two-dimensional representation based on the column matrix $(1 \quad i)^T$:

$$D(1) = \begin{pmatrix} 1 & 0 \\ 0 & 1 \end{pmatrix}, \qquad D(i) = \begin{pmatrix} 0 & -1 \\ 1 & 0 \end{pmatrix},$$

$$D(-1) = \begin{pmatrix} -1 & 0 \\ 0 & -1 \end{pmatrix}, \qquad D(-i) = \begin{pmatrix} 0 & 1 \\ -1 & 0 \end{pmatrix}.$$

The reader should check that $D(i)D(-i) = D(1)$, $D(i)D(i) = D(-1)$ etc., i.e. that the matrices do have exactly the same multiplication properties as the elements of the group. Having done so, the reader may also wonder why anybody would bother with the representative matrices, when the original elements are so much simpler to handle! As we will see later, once some general properties of matrix representations have been established, the analysis of large groups, both Abelian and non-Abelian, can be reduced to routine, almost cookbook, procedures.

An n-dimensional representation of $\mathcal{G}$ is a homomorphism of $\mathcal{G}$ into the set of invertible $n \times n$ matrices (i.e. $n \times n$ matrices that have inverses or, equivalently, have non-zero determinants); this set is usually known as the general linear group and denoted by $GL(n)$. In general the same matrix may represent more than one element of $\mathcal{G}$; if, however, all the matrices representing the elements of $\mathcal{G}$ are *different* then the representation is said to be *faithful*, and the homomorphism becomes an isomorphism onto a subgroup of $GL(n)$.

A trivial but important representation is $D(X) = I_n$ for all elements X of $\mathcal{G}$. Clearly both of the defining relationships are satisfied, and there is no restriction on the value of n. However, such a representation is not a faithful one.

To sum up, in the context of a rotation–reflection group, the transposes of the set of $n \times n$ matrices $D(X)$ that make up a representation D may be thought of as describing what happens to an n-component basis vector of coordinates, $(x \quad y \quad \cdots)^T$, or of functions, $(\Psi_1 \quad \Psi_2 \quad \cdots)^T$, the Ψ_i themselves being functions of coordinates, when the group operation X is carried out on each of the coordinates or functions. For example, to return to the symmetry operations on an equilateral triangle, the clockwise rotation by $2\pi/3$, R, carries the three-dimensional basis vector $(x \quad y \quad z)^T$ into the column matrix

$$\begin{pmatrix} -\frac{1}{2}x + \frac{\sqrt{3}}{2}y \\ -\frac{\sqrt{3}}{2}x - \frac{1}{2}y \\ z \end{pmatrix}$$

whilst the two-dimensional basis vector of functions $(r^2 \quad 3z^2 - r^2)^T$ is unaltered, as neither r nor z is changed by the rotation. The fact that z is unchanged by any of the operations of the group shows that the components x, y, z actually divide (i.e. are 'reducible', to anticipate a more formal description) into two sets:

one comprises z, which is unchanged by any of the operations, and the other comprises x, y, which change as a pair into linear combinations of themselves. This is an important observation to which we return in section 29.4.

29.3 Equivalent representations

If D is an n-dimensional representation of a group G, and Q is any fixed invertible $n \times n$ matrix ($|Q| \neq 0$), then the set of matrices defined by the similarity transformation

$$D_Q(X) = Q^{-1}D(X)Q \tag{29.5}$$

also forms a representation D_Q of G, said to be *equivalent* to D. We can see from a comparison with the definition in section 29.2 that they do form a representation:

(i) $D_Q(I) = Q^{-1}D(I)Q = Q^{-1}I_n Q = I_n$,
(ii) $D_Q(X)D_Q(Y) = Q^{-1}D(X)QQ^{-1}D(Y)Q = Q^{-1}D(X)D(Y)Q$
$ = Q^{-1}D(XY)Q = D_Q(XY).$

Since we can always transform between equivalent representations using a non-singular matrix Q, we will consider such representations to be one and the same.

Despite the similarity of words and manipulations to those of subsection 28.7.1, that two representations are equivalent does not constitute an 'equivalence relation' – for example, the reflexive property does not hold for a general fixed matrix Q. However, if Q were not fixed, but simply restricted to belonging to a set of matrices that themselves form a group, then (29.5) would constitute an equivalence relation.

The general invertible matrix Q that appears in the definition (29.5) of equivalent matrices describes changes arising from a change in the coordinate system (i.e. in the set of basis functions). As before, suppose that the effect of an operation X on the basis functions is expressed by the action of M(X) (which is equal to $D^T(X)$) on the corresponding basis vector:

$$u' = M(X)u = D^T(X)u. \tag{29.6}$$

A change of basis would be given by $u_Q = Qu$ and $u'_Q = Qu'$, and we may write

$$u'_Q = Qu' = QM(X)u = QD^T(X)Q^{-1}u_Q. \tag{29.7}$$

This is of the same form as (29.6), i.e.

$$u'_Q = D^T_{Q^T}(X)u_Q, \tag{29.8}$$

where $D_{Q^T}(X) = (Q^T)^{-1}D(X)Q^T$ is related to $D(X)$ by a similarity transformation. Thus $D_{Q^T}(X)$ represents the same linear transformation as $D(X)$, but with respect to a new basis vector u_Q; this supports our contention that representations connected by similarity transformations should be considered as the *same* representation.

▶ *For the four-element Abelian group consisting of the set* $\{1, i, -1, -i\}$ *under ordinary multiplication, discussed near the end of section 29.2, change the basis vector from* $u = (1 \quad i)^T$ *to* $u_Q = (3 - i \quad 2i - 5)^T$. *Find the real transformation matrix* Q. *Show that the transformed representative matrix for element* i, $D_{Q^T}(i)$, *is given by*

$$D_{Q^T}(i) = \begin{pmatrix} 17 & -29 \\ 10 & -17 \end{pmatrix}$$

and verify that $D^T_{Q^T}(i)u_Q = iu_Q$.

Firstly, we solve the matrix equation

$$\begin{pmatrix} 3 - i \\ 2i - 5 \end{pmatrix} = \begin{pmatrix} a & b \\ c & d \end{pmatrix} \begin{pmatrix} 1 \\ i \end{pmatrix},$$

with a, b, c, d real. This gives Q and hence Q^{-1} as

$$Q = \begin{pmatrix} 3 & -1 \\ -5 & 2 \end{pmatrix}, \qquad Q^{-1} = \begin{pmatrix} 2 & 1 \\ 5 & 3 \end{pmatrix}.$$

Following (29.7) we now find the transpose of $D_{Q^T}(i)$ as

$$QD^T(i)Q^{-1} = \begin{pmatrix} 3 & -1 \\ -5 & 2 \end{pmatrix} \begin{pmatrix} 0 & 1 \\ -1 & 0 \end{pmatrix} \begin{pmatrix} 2 & 1 \\ 5 & 3 \end{pmatrix} = \begin{pmatrix} 17 & 10 \\ -29 & -17 \end{pmatrix}$$

and hence $D_{Q^T}(i)$ is as stated. Finally,

$$D^T_{Q^T}(i)u_Q = \begin{pmatrix} 17 & 10 \\ -29 & -17 \end{pmatrix} \begin{pmatrix} 3 - i \\ 2i - 5 \end{pmatrix} = \begin{pmatrix} 1 + 3i \\ -2 - 5i \end{pmatrix}$$

$$= i \begin{pmatrix} 3 - i \\ 2i - 5 \end{pmatrix} = iu_Q,$$

as required. ◀

Although we will not prove it, it can be shown that any finite representation of a finite group of linear transformations that preserve spatial length (or, in quantum mechanics, preserve the magnitude of a wavefunction) is equivalent to a representation in which all the matrices are unitary (see chapter 8) and so from now on we will consider only *unitary representations*.

29.4 Reducibility of a representation

We have seen already that it is possible to have more than one representation of any particular group. For example, the group $\{1, i, -1, -i\}$ under ordinary multiplication has been shown to have a set of 2×2 matrices, and a set of four unit $n \times n$ matrices I_n, as two of its possible representations.

Consider two or more representations, $D^{(1)}$, $D^{(2)}$, $\ldots$, $D^{(N)}$, which may be of different dimensions, of a group $\mathcal{G}$. Now combine the matrices $D^{(1)}(X)$, $D^{(2)}(X)$, $\ldots$, $D^{(N)}(X)$ that correspond to element X of $\mathcal{G}$ into a larger *block-diagonal* matrix:

$$D(X) = \begin{pmatrix} D^{(1)}(X) & & & \\ & D^{(2)}(X) & & \\ & & \ddots & \\ & & & D^{(N)}(X) \end{pmatrix} \qquad (29.9)$$

Then $D = \{D(X)\}$ is the matrix representation of the group obtained by combining the basis vectors of $D^{(1)}$, $D^{(2)}$, $\ldots$, $D^{(N)}$ into one larger basis vector. If, knowingly or unknowingly, we had started with this larger basis vector and found the matrices of the representation D to have the form shown in (29.9), or to have a form that can be transformed into this by a similarity transformation (29.5) (using, of course, the *same* matrix Q for each of the matrices $D(X)$) then we would say that D is *reducible* and that each matrix $D(X)$ can be written as the *direct sum* of smaller representations:

$$D(X) = D^{(1)}(X) \oplus D^{(2)}(X) \oplus \cdots \oplus D^{(N)}(X).$$

It may be that some or all of the matrices $D^{(1)}(X)$, $D^{(2)}(X)$, $\ldots$, $D^{(N)}$ themselves can be further reduced – i.e. written in block diagonal form. For example, suppose that the representation $D^{(1)}$, say, has a basis vector $(x \quad y \quad z)^{\mathrm{T}}$; then, for the symmetry group of an equilateral triangle, whilst x and y are mixed together for at least one of the operations X, z is never changed. In this case the 3×3 representative matrix $D^{(1)}(X)$ can itself be written in block diagonal form as a

2×2 matrix and a 1×1 matrix. The direct-sum matrix $D(X)$ can now be written

$$
D(X) = \begin{pmatrix} \begin{matrix} a & b \\ c & d \end{matrix} & & & & \mathbf{0} \\ & 1 & & & \\ & & D^{(2)}(X) & & \\ & & & \ddots & \\ \mathbf{0} & & & & D^{(N)}(X) \end{pmatrix} \tag{29.10}
$$

but the first two blocks can be reduced no further.

When all the other representations $D^{(2)}(X)$, ... have been similarly treated, what remains is said to be *irreducible* and has the characteristic of being block diagonal, with blocks that individually cannot be reduced further. The blocks are known as the *irreducible representations of* $\mathcal{G}$, often abbreviated to the *irreps of* $\mathcal{G}$, and we denote them by $\hat{D}^{(i)}$. They form the building blocks of representation theory, and it is their properties that are used to analyse any given physical situation which is invariant under the operations that form the elements of $\mathcal{G}$. Any representation can be written as a linear combination of irreps.

If, however, the initial choice u of basis vector for the representation D is arbitrary, as it is in general, then it is unlikely that the matrices $D(X)$ will assume obviously block diagonal forms (it should be noted, though, that since the matrices are square, even a matrix with non-zero entries only in the extreme top right and bottom left positions is technically block diagonal). In general, it will be possible to reduce them to block diagonal matrices with more than one block; this reduction corresponds to a transformation Q to a new basis vector u_Q, as described in section 29.3.

In any particular representation D, each constituent irrep $\hat{D}^{(i)}$ may appear any number of times, or not at all, subject to the obvious restriction that the sum of all the irrep dimensions must add up to the dimension of D itself. Let us say that $\hat{D}^{(i)}$ appears m_i times. The general expansion of D is then written

$$
D = m_1\hat{D}^{(1)} \oplus m_2\hat{D}^{(2)} \oplus \cdots \oplus m_N\hat{D}^{(N)}, \tag{29.11}
$$

where if $\mathcal{G}$ is finite so is N.

This is such an important result that we shall now restate the situation in somewhat different language. When the set of matrices that forms a representation of a particular group of symmetry operations has been brought to irreducible form, the implications are as follows.

(i) Those components of the basis vector that correspond to rows in the representation matrices with a single-entry block, i.e. a 1×1 block, are unchanged by the operations of the group. Such a coordinate or function is said to transform according to a one-dimensional irrep of $\mathcal{G}$. In the example given in (29.10), that the entry on the third row forms a 1×1 block implies that the third entry in the basis vector $(x \quad y \quad z \quad \cdots)^{\mathrm{T}}$, namely z, is invariant under the two-dimensional symmetry operations on an equilateral triangle in the xy-plane.

(ii) If, in any of the g matrices of the representation, the largest-sized block located on the row or column corresponding to a particular coordinate (or function) in the basis vector is $n \times n$, then that coordinate (or function) is mixed by the symmetry operations with $n - 1$ others and is said to transform according to an n-dimensional irrep of $\mathcal{G}$. Thus in the matrix (29.10), x is the first entry in the complete basis vector; the first row of the matrix contains two non-zero entries, as does the first column, and so x is part of a two-component basis vector whose components are mixed by the symmetry operations of $\mathcal{G}$. The other component is y.

The result (29.11) may also be formulated in terms of the more abstract notion of vector spaces (chapter 8). The set of g matrices that forms an n-dimensional representation D of the group $\mathcal{G}$ can be thought of as acting on column matrices corresponding to vectors in an n-dimensional vector space V spanned by the basis functions of the representation. If there exists a *proper subspace* W of V, such that if a vector whose column matrix is w belongs to W then the vector whose column matrix is D(X)w also belongs to W, for all X belonging to $\mathcal{G}$, then it follows that D is reducible. We say that the subspace W is invariant under the actions of the elements of $\mathcal{G}$. With D unitary, the orthogonal complement $W_\perp$ of W, i.e. the vector space V remaining when the subspace W has been removed, is also invariant, and all the matrices D(X) split into two blocks acting separately on W and $W_\perp$. Both W and $W_\perp$ may contain further invariant subspaces, in which case the matrices will be split still further.

As a concrete example of this approach, consider in plane polar coordinates ρ, ϕ the effect of rotations about the polar axis on the infinite-dimensional vector space V of all functions of ϕ that satisfy the Dirichlet conditions for expansion as a Fourier series (see section 12.1). We take as our basis functions the set $\{\sin m\phi, \cos m\phi\}$ for integer values $m = 0, 1, 2, \ldots$; this is an infinite-dimensional representation ($n = \infty$) and, since a rotation about the polar axis can be through any angle α ($0 \leq \alpha < 2\pi$), the group $\mathcal{G}$ is a subgroup of the continuous rotation group and has its order g formally equal to infinity.

Now, for some k, consider a vector w in the space W_k spanned by $\{\sin k\phi, \cos k\phi\}$, say $w = a\sin k\phi + b\cos k\phi$. Under a rotation by α about the polar axis, $a\sin k\phi$ becomes $a\sin k(\phi + \alpha)$, which can be written as $a\cos k\alpha \sin k\phi + a\sin k\alpha \cos k\phi$, i.e as a linear combination of $\sin k\phi$ and $\cos k\phi$; similarly $\cos k\phi$ becomes another linear combination of the same two functions. The newly generated vector w', whose column matrix w' is given by $\mathsf{w}' = \mathsf{D}(\alpha)\mathsf{w}$, therefore belongs to W_k for any α and we can conclude that W_k is an invariant irreducible two-dimensional subspace of V. It follows that $\mathsf{D}(\alpha)$ is reducible and that, since the result holds for every k, in its reduced form $\mathsf{D}(\alpha)$ has an infinite series of identical 2×2 blocks on its leading diagonal; each block will have the form

$$\begin{pmatrix} \cos\alpha & -\sin\alpha \\ \sin\alpha & \cos\alpha \end{pmatrix}.$$

We note that the particular case $k = 0$ is special, in that then $\sin k\phi = 0$ and $\cos k\phi = 1$, for all ϕ; consequently the first 2×2 block in $\mathsf{D}(\alpha)$ is reducible further and becomes two single-entry blocks.

A second illustration of the connection between the behaviour of vector spaces under the actions of the elements of a group and the form of the matrix representation of the group is provided by the vector space spanned by the spherical harmonics $Y_{\ell m}(\theta, \phi)$. This contains subspaces, corresponding to the different values of ℓ, that are invariant under the actions of the elements of the full three-dimensional rotation group; the corresponding matrices are block-diagonal, and those entries that correspond to the part of the basis containing $Y_{\ell m}(\theta, \phi)$ form a $(2\ell + 1) \times (2\ell + 1)$ block.

To illustrate further the irreps of a group, we return again to the group $\mathcal{G}$ of two-dimensional rotation and reflection symmetries of an equilateral triangle, or equivalently the permutation group S_3; this may be shown, using the methods of section 29.7 below, to have three irreps. Firstly, we have already seen that the set $\mathcal{M}$ of six orthogonal 2×2 matrices given in section (28.3), equation (28.13), is isomorphic to $\mathcal{G}$. These matrices therefore form not only a representation of $\mathcal{G}$, but a faithful one. It should be noticed that, although $\mathcal{G}$ contains six elements, the matrices are only 2×2. However, they contain no invariant 1×1 sub-block (which for 2×2 matrices would require them all to be diagonal) and neither can all the matrices be made block-diagonal by the *same* similarity transformation; they therefore form a two-dimensional irrep of $\mathcal{G}$.

Secondly, as previously noted, every group has one (unfaithful) irrep in which every element is represented by the 1×1 matrix I_1, or, more simply, 1.

Thirdly an (unfaithful) irrep of $\mathcal{G}$ is given by assignment of the one-dimensional set of six 'matrices' $\{1, 1, 1, -1, -1, -1\}$ to the symmetry operations $\{I, R, R', K, L, M\}$ respectively, or to the group elements $\{I, A, B, C, D, E\}$ respectively; see section 28.3. In terms of the permutation group S_3, 1 corresponds to even permutations and -1 to odd permutations, 'odd' or 'even' referring to the number of simple pair interchanges to which a permutation is equivalent. That these assignments are in accord with the group multiplication table 28.8 should be checked.

Thus the three irreps of the group $\mathcal{G}$ (i.e. the group $3m$ or C_{3v} or S_3), are, using the conventional notation A_1, A_2, E (see section 29.8), as follows:

$$
\begin{array}{r|cccccc}
 & \multicolumn{6}{c}{\text{Element}} \\
 & I & A & B & C & D & E \\
\hline
A_1 & 1 & 1 & 1 & 1 & 1 & 1 \\
\text{Irrep} \quad A_2 & 1 & 1 & 1 & -1 & -1 & -1 \\
E & M_I & M_A & M_B & M_C & M_D & M_E
\end{array}
\tag{29.12}
$$

where

$$
M_I = \begin{pmatrix} 1 & 0 \\ 0 & 1 \end{pmatrix}, \qquad
M_A = \begin{pmatrix} -\frac{1}{2} & \frac{\sqrt{3}}{2} \\ -\frac{\sqrt{3}}{2} & -\frac{1}{2} \end{pmatrix}, \qquad
M_B = \begin{pmatrix} -\frac{1}{2} & -\frac{\sqrt{3}}{2} \\ \frac{\sqrt{3}}{2} & -\frac{1}{2} \end{pmatrix},
$$

$$
M_C = \begin{pmatrix} -1 & 0 \\ 0 & 1 \end{pmatrix}, \qquad
M_D = \begin{pmatrix} \frac{1}{2} & -\frac{\sqrt{3}}{2} \\ -\frac{\sqrt{3}}{2} & -\frac{1}{2} \end{pmatrix}, \qquad
M_E = \begin{pmatrix} \frac{1}{2} & \frac{\sqrt{3}}{2} \\ \frac{\sqrt{3}}{2} & -\frac{1}{2} \end{pmatrix}.
$$

29.5 The orthogonality theorem for irreducible representations

We come now to the central theorem of representation theory, a theorem that justifies the relatively routine application of certain procedures to determine the restrictions that are inherent in physical systems that have some degree of rotational or reflection symmetry. The development of the theorem is long and quite complex when presented in its entirety, and the reader will have to refer elsewhere for the proof.[§]

The theorem states that, in a certain sense, the irreps of a group $\mathcal{G}$ are as orthogonal as possible, as follows. If, for each irrep, the elements in any one position in each of the g matrices are used to make up g-component column matrices then

(i) any two such column matrices coming from different irreps are orthogonal;

(ii) any two such column matrices coming from different positions in the matrices of the same irrep are orthogonal.

This orthogonality is in addition to the irreps' being in the form of orthogonal (unitary) matrices and thus each comprising mutually orthogonal rows and columns.

[§] See, e.g., H. F. Jones, *Groups, Representations and Physics* (Bristol: Institute of Physics, 1998); J. F. Cornwell, *Group Theory in Physics*, vol 2 (London: Academic Press, 1984); J-P. Serre, *Linear Representations of Finite Groups* (New York: Springer, 1977).

More mathematically, if we denote the entry in the ith row and jth column of a matrix $D(X)$ by $[D(X)]_{ij}$, and $\hat{D}^{(\lambda)}$ and $\hat{D}^{(\mu)}$ are two irreps of $\mathcal{G}$ having dimensions n_λ and n_μ respectively, then

$$\sum_X \left[\hat{D}^{(\lambda)}(X)\right]_{ij}^* \left[\hat{D}^{(\mu)}(X)\right]_{kl} = \frac{g}{n_\lambda}\delta_{ik}\delta_{jl}\delta_{\lambda\mu}. \tag{29.13}$$

This rather forbidding-looking equation needs some further explanation.

Firstly, the asterisk indicates that the complex conjugate should be taken if necessary, though all our representations so far have involved only real matrix elements. Each Kronecker delta function on the right-hand side has the value 1 if its two subscripts are equal and has the value 0 otherwise. Thus the right-hand side is only non-zero if $i = k$, $j = l$ and $\lambda = \mu$, all at the same time.

Secondly, the summation over the group elements X means that g contributions have to be added together, each contribution being a product of entries drawn from the representative matrices in the two irreps $\hat{D}^{(\lambda)} = \{\hat{D}^{(\lambda)}(X)\}$ and $\hat{D}^{(\mu)} = \{\hat{D}^{(\mu)}(X)\}$. The g contributions arise as X runs over the g elements of $\mathcal{G}$.

Thus, putting these remarks together, the summation will produce zero if either

(i) the matrix elements are not taken from exactly the same position in every matrix, including cases in which it is not possible to do so because the irreps $\hat{D}^{(\lambda)}$ and $\hat{D}^{(\mu)}$ have different dimensions, or

(ii) even if $\hat{D}^{(\lambda)}$ and $\hat{D}^{(\mu)}$ do have the same dimensions and the matrix elements are from the same positions in every matrix, they are different irreps, i.e. $\lambda \neq \mu$.

Some numerical illustrations based on the irreps A_1, A_2 and E of the group $3m$ (or C_{3v} or S_3) will probably provide the clearest explanation (see (29.12)).

(a) Take $i = j = k = l = 1$, with $\hat{D}^{(\lambda)} = A_1$ and $\hat{D}^{(\mu)} = A_2$. Equation (29.13) then reads

$$1(1) + 1(1) + 1(1) + 1(-1) + 1(-1) + 1(-1) = 0,$$

as expected, since $\lambda \neq \mu$.

(b) Take (i, j) as $(1, 2)$ and (k, l) as $(2, 2)$, corresponding to different matrix positions within the same irrep $\hat{D}^{(\lambda)} = \hat{D}^{(\mu)} = E$. Substituting in (29.13) gives

$$0(1) + \left(-\tfrac{\sqrt{3}}{2}\right)\left(-\tfrac{1}{2}\right) + \left(\tfrac{\sqrt{3}}{2}\right)\left(-\tfrac{1}{2}\right) + 0(1) + \left(-\tfrac{\sqrt{3}}{2}\right)\left(-\tfrac{1}{2}\right) + \left(\tfrac{\sqrt{3}}{2}\right)\left(-\tfrac{1}{2}\right) = 0.$$

(c) Take (i, j) as $(1, 2)$, and (k, l) as $(1, 2)$, corresponding to the same matrix positions within the same irrep $\hat{D}^{(\lambda)} = \hat{D}^{(\mu)} = E$. Substituting in (29.13) gives

$$0(0) + \left(-\tfrac{\sqrt{3}}{2}\right)\left(-\tfrac{\sqrt{3}}{2}\right) + \left(\tfrac{\sqrt{3}}{2}\right)\left(\tfrac{\sqrt{3}}{2}\right) + 0(0) + \left(-\tfrac{\sqrt{3}}{2}\right)\left(-\tfrac{\sqrt{3}}{2}\right) + \left(\tfrac{\sqrt{3}}{2}\right)\left(\tfrac{\sqrt{3}}{2}\right) = \tfrac{6}{2}.$$

(d) No explicit calculation is needed to see that if $i = j = k = l = 1$, with $\hat{D}^{(\lambda)} = \hat{D}^{(\mu)} = A_1$ (or A_2), then each term in the sum is either 1^2 or $(-1)^2$ and the total is 6, as predicted by the right-hand side of (29.13) since $g = 6$ and $n_\lambda = 1$.

29.6 Characters

The actual matrices of general representations and irreps are cumbersome to work with, and they are not unique since there is always the freedom to change the coordinate system, i.e. the components of the basis vector (see section 29.3), and hence the entries in the matrices. However, one thing that does not change for a matrix under such an equivalence (similarity) transformation – i.e. under a change of basis – is the trace of the matrix. This was shown in chapter 8, but is repeated here. The trace of a matrix A is the sum of its diagonal elements,

$$\mathrm{Tr}\, A = \sum_{i=1}^{n} A_{ii}$$

or, using the summation convention (section 26.1), simply A_{ii}. Under a similarity transformation, again using the summation convention,

$$
\begin{aligned}
[D_Q(X)]_{ii} &= [Q^{-1}]_{ij}[D(X)]_{jk}[Q]_{ki} \\
&= [D(X)]_{jk}[Q]_{ki}[Q^{-1}]_{ij} \\
&= [D(X)]_{jk}[I]_{kj} \\
&= [D(X)]_{jj},
\end{aligned}
$$

showing that the traces of equivalent matrices are equal.

This fact can be used to greatly simplify work with representations, though with some partial loss of the information content of the full matrices. For example, using trace values alone it is not possible to distinguish between the two groups known as $4mm$ and $\bar{4}2m$, or as C_{4v} and D_{2d} respectively, even though the two groups are not isomorphic. To make use of these simplifications we now define the characters of a representation.

Definition. *The characters $\chi(D)$ of a representation D of a group $\mathcal{G}$ are defined as the traces of the matrices $D(X)$, one for each element X of $\mathcal{G}$.*

At this stage there will be g characters, but, as we noted in subsection 28.7.3, elements A, B of $\mathcal{G}$ in the same conjugacy class are connected by equations of the form $B = X^{-1}AX$. It follows that their matrix representations are connected by corresponding equations of the form $D(B) = D(X^{-1})D(A)D(X)$, and so by the argument just given their representations will have equal traces and hence equal characters. Thus *elements in the same conjugacy class have the same characters,*

$3m$	I	A, B	C, D, E	
A_1	1	1	1	z; z^2; $x^2 + y^2$
A_2	1	1	-1	R_z
E	2	-1	0	(x, y); (xz, yz); (R_x, R_y); $(x^2 - y^2, 2xy)$

Table 29.1   The character table for the irreps of group $3m$ (C_{3v} or S_3). The right-hand column lists some common functions that transform according to the irrep against which each is shown (see text).

though, in general, these will vary from one representation to another. However, it might also happen that two or more conjugacy classes have the same characters in a representation – indeed, in the trivial irrep A_1, see (29.12), every element inevitably has the character 1.

For the irrep A_2 of the group $3m$, the classes $\{I\}$, $\{A, B\}$ and $\{C, D, E\}$ have characters 1, 1 and -1, respectively, whilst they have characters 2, -1 and 0 respectively in irrep E.

We are thus able to draw up a *character table* for the group $3m$ as shown in table 29.1. This table holds in compact form most of the important information on the behaviour of functions under the two-dimensional rotational and reflection symmetries of an equilateral triangle, i.e. under the elements of group $3m$. The entry under I for any irrep gives the dimension of the irrep, since it is equal to the trace of the unit matrix whose dimension is equal to that of the irrep. In other words, for the λth irrep $\chi^{(\lambda)}(I) = n_\lambda$, where n_λ is its dimension.

In the extreme right-hand column we list some common functions of Cartesian coordinates that transform, under the group $3m$, according to the irrep on whose line they are listed. Thus, as we have seen, z, z^2, and $x^2 + y^2$ are all unchanged by the group operations (though x and y individually are affected) and so are listed against the one-dimensional irrep A_1. Each of the pairs (x, y), (xz, yz), and $(x^2 - y^2, 2xy)$, however, is mixed as a pair by some of the operations, and so these pairs are listed against the two-dimensional irrep E: each pair forms a basis set for this irrep.

The quantities R_x, R_y and R_z refer to rotations about the indicated axes; they transform in the same way as the corresponding components of angular momentum $\mathbf{J}$, and their behaviour can be established by examining how the components of $\mathbf{J} = \mathbf{r} \times \mathbf{p}$ transform under the operations of the group. To do this explicitly is beyond the scope of this book. However, it can be noted that R_z, being listed opposite the one-dimensional A_2, is unchanged by I and by the rotations A and B but changes sign under the mirror reflections C, D, and E, as would be expected.

29.6.1 Orthogonality property of characters

Some of the most important properties of characters can be deduced from the orthogonality theorem (29.13),

$$\sum_X \left[\hat{\mathsf{D}}^{(\lambda)}(X)\right]_{ij}^* \left[\hat{\mathsf{D}}^{(\mu)}(X)\right]_{kl} = \frac{g}{n_\lambda} \delta_{ik} \delta_{jl} \delta_{\lambda\mu}.$$

If we set $j = i$ and $l = k$, so that both factors in any particular term in the summation refer to diagonal elements of the representative matrices, and then sum both sides over i and k, we obtain

$$\sum_X \sum_{i=1}^{n_\lambda} \sum_{k=1}^{n_\mu} \left[\hat{\mathsf{D}}^{(\lambda)}(X)\right]_{ii}^* \left[\hat{\mathsf{D}}^{(\mu)}(X)\right]_{kk} = \frac{g}{n_\lambda} \sum_{i=1}^{n_\lambda} \sum_{k=1}^{n_\mu} \delta_{ik} \delta_{ik} \delta_{\lambda\mu}.$$

Expressed in term of characters, this reads

$$\sum_X \left[\chi^{(\lambda)}(X)\right]^* \chi^{(\mu)}(X) = \frac{g}{n_\lambda} \sum_{i=1}^{n_\lambda} \delta_{ii}^2 \delta_{\lambda\mu} = \frac{g}{n_\lambda} \sum_{i=1}^{n_\lambda} 1 \times \delta_{\lambda\mu} = g\delta_{\lambda\mu}. \tag{29.14}$$

In words, the (g-component) 'vectors' formed from the characters of the various irreps of a group are mutually orthogonal, but each one has a squared magnitude (the sum of the squares of its components) equal to the order of the group.

Since, as noted in the previous subsection, group elements in the same class have the same characters, (29.14) can be written as a sum over classes rather than elements. If c_i denotes the number of elements in class C_i and X_i any element of C_i, then

$$\sum_i c_i \left[\chi^{(\lambda)}(X_i)\right]^* \chi^{(\mu)}(X_i) = g\delta_{\lambda\mu}. \tag{29.15}$$

Although we do not prove it here, there also exists a 'completeness' relation for characters. It makes a statement about the products of characters for a fixed pair of group elements, X_1 and X_2, when the products are summed over all possible irreps of the group. This is the converse of the summation process defined by (29.14). The completeness relation states that

$$\sum_\lambda \left[\chi^{(\lambda)}(X_1)\right]^* \chi^{(\lambda)}(X_2) = \frac{g}{c_1} \delta_{C_1 C_2}, \tag{29.16}$$

where element X_1 belongs to conjugacy class C_1 and X_2 belongs to C_2. Thus the sum is zero unless X_1 and X_2 belong to the same class. For table 29.1 we can verify that these results are valid.

(i) For $\hat{\mathsf{D}}^{(\lambda)} = \hat{\mathsf{D}}^{(\mu)} = \mathrm{A}_1$ or A_2, (29.15) reads

$$1(1) + 2(1) + 3(1) = 6,$$

whilst for $\hat{\mathsf{D}}^{(\lambda)} = \hat{\mathsf{D}}^{(\mu)} = E$, it gives

$$1(2^2) + 2(1) + 3(0) = 6.$$

(ii) For $\hat{\mathsf{D}}^{(\lambda)} = A_2$ and $\hat{\mathsf{D}}^{(\mu)} = E$, say, (29.15) reads

$$1(1)(2) + 2(1)(-1) + 3(-1)(0) = 0.$$

(iii) For $X_1 = A$ and $X_2 = D$, say, (29.16) reads

$$1(1) + 1(-1) + (-1)(0) = 0,$$

whilst for $X_1 = C$ and $X_2 = E$, both of which belong to class C_3 for which $c_3 = 3$,

$$1(1) + (-1)(-1) + (0)(0) = 2 = \frac{6}{3}.$$

29.7 Counting irreps using characters

The expression of a general representation $\mathsf{D} = \{D(X)\}$ in terms of irreps, as given in (29.11), can be simplified by going from the full matrix form to that of characters. Thus

$$\mathsf{D}(X) = m_1 \hat{\mathsf{D}}^{(1)}(X) \oplus m_2 \hat{\mathsf{D}}^{(2)}(X) \oplus \cdots \oplus m_N \hat{\mathsf{D}}^{(N)}(X)$$

becomes, on taking the trace of both sides,

$$\chi(X) = \sum_{\lambda=1}^{N} m_\lambda \chi^{(\lambda)}(X). \tag{29.17}$$

Given the characters of the irreps of the group G to which the elements X belong, and the characters of the representation $\mathsf{D} = \{\mathsf{D}(X)\}$, the g equations (29.17) can be solved as simultaneous equations in the m_λ, either by inspection or by multiplying both sides by $[\chi^{(\mu)}(X)]^*$ and summing over X, making use of (29.14) and (29.15), to obtain

$$m_\mu = \frac{1}{g} \sum_X [\chi^{(\mu)}(X)]^* \chi(X) = \frac{1}{g} \sum_i c_i [\chi^{(\mu)}(X_i)]^* \chi(X_i). \tag{29.18}$$

That an unambiguous formula can be given for each m_λ, once the *character set* (the set of characters of each of the group elements or, equivalently, of each of the conjugacy classes) of D is known, shows that, for any particular group, two representations with the same characters are equivalent. This strongly suggests something that can be shown, namely, *the number of irreps = the number of conjugacy classes*. The argument is as follows. Equation (29.17) is a set of simultaneous equations for N unknowns, the m_λ, some of which may be zero. The value of N is equal to the number of irreps of G. There are g different values of X, but the number of *different* equations is only equal to the number of distinct conjugacy classes, since any two elements of $\mathcal{G}$ in the same class have the same character set and therefore generate the same equation. For a unique solution to simultaneous equations in N unknowns, exactly N independent equations are needed. Thus N is also the number of classes, establishing the stated result.

> ►*Determine the irreps contained in the representation of the group 3m in the vector space spanned by the functions* x^2, y^2, xy.

We first note that although these functions are not orthogonal they form a basis set for a representation, since they are linearly independent quadratic forms in x and y and any other quadratic form can be written (uniquely) in terms of them. We must establish how they transform under the symmetry operations of group $3m$. We need to do so only for a representative element of each conjugacy class, and naturally we take the simplest in each case.

The first class contains only I (as always) and clearly $D(I)$ is the 3×3 unit matrix.

The second class contains the rotations, A and B, and we choose to find $D(A)$. Since, under A,

$$x \rightarrow -\frac{1}{2}x + \frac{\sqrt{3}}{2}y \quad \text{and} \quad y \rightarrow -\frac{\sqrt{3}}{2}x - \frac{1}{2}y,$$

it follows that

$$x^2 \rightarrow \tfrac{1}{4}x^2 - \tfrac{\sqrt{3}}{2}xy + \tfrac{3}{4}y^2, \qquad y^2 \rightarrow \tfrac{3}{4}x^2 + \tfrac{\sqrt{3}}{2}xy + \tfrac{1}{4}y^2 \tag{29.19}$$

and

$$xy \rightarrow \tfrac{\sqrt{3}}{4}x^2 - \tfrac{1}{2}xy - \tfrac{\sqrt{3}}{4}y^2. \tag{29.20}$$

Hence $D(A)$ can be deduced and is given below.

The third and final class contains the reflections, C, D and E; of these C is much the easiest to deal with. Under C, $x \rightarrow -x$ and $y \rightarrow y$, causing xy to change sign but leaving x^2 and y^2 unaltered. The three matrices needed are thus

$$D(I) = I_3, \quad D(C) = \begin{pmatrix} 1 & 0 & 0 \\ 0 & 1 & 0 \\ 0 & 0 & -1 \end{pmatrix}, \quad D(A) = \begin{pmatrix} \frac{1}{4} & \frac{3}{4} & -\frac{\sqrt{3}}{2} \\ \frac{3}{4} & \frac{1}{4} & \frac{\sqrt{3}}{2} \\ \frac{\sqrt{3}}{4} & -\frac{\sqrt{3}}{4} & -\frac{1}{2} \end{pmatrix};$$

their traces are respectively 3, 1 and 0.

It should be noticed that much more work has been done here than is necessary, since the traces can be computed immediately from the effects of the symmetry operations on the basis functions. All that is needed is the weight of each basis function in the transformed expression for that function; these are clearly 1, 1, 1 for I, and $\frac{1}{4}, \frac{1}{4}, -\frac{1}{2}$ for A, from (29.19) and (29.20), and 1, 1, -1 for C, from the observations made just above the displayed matrices. The traces are then the sums of these weights. The off-diagonal elements of the matrices need not be found, nor need the matrices be written out.

From (29.17) we now need to find a superposition of the characters of the irreps that gives representation D in the bottom line of table 29.2.

By inspection it is obvious that $D = A_1 \oplus E$, but we can use (29.18) formally:

$$m_{A_1} = \tfrac{1}{6}[1(1)(3) + 2(1)(0) + 3(1)(1)] = 1,$$
$$m_{A_2} = \tfrac{1}{6}[1(1)(3) + 2(1)(0) + 3(-1)(1)] = 0,$$
$$m_E = \tfrac{1}{6}[1(2)(3) + 2(-1)(0) + 3(0)(1)] = 1.$$

Thus A_1 and E appear once each in the reduction of D, and A_2 not at all. Table 29.1 gives the further information, not needed here, that it is the combination $x^2 + y^2$ that transforms as a one-dimensional irrep and the pair $(x^2 - y^2, 2xy)$ that forms a basis of the two-dimensional irrep, E. ◄

Irrep	I	Classes AB	CDE
A_1	1	1	1
A_2	1	1	-1
E	2	-1	0
D	3	0	1

Table 29.2   The characters of the irreps of the group $3m$ and of the representation D, which must be a superposition of some of them.

29.7.1 Summation rules for irreps

The first summation rule for irreps is a simple restatement of (29.14), with μ set equal to λ; it then reads

$$\sum_X \left[\chi^{(\lambda)}(X)\right]^* \chi^{(\lambda)}(X) = g.$$

In words, the sum of the squares (modulus squared if necessary) of the characters of an irrep taken over all elements of the group adds up to the order of the group. For group $3m$ (table 29.1), this takes the following explicit forms:

$$\text{for } A_1, \qquad 1(1^2) + 2(1^2) + 3(1^2) = 6;$$
$$\text{for } A_2, \qquad 1(1^2) + 2(1^2) + 3(-1)^2 = 6;$$
$$\text{for } E, \qquad 1(2^2) + 2(-1)^2 + 3(0^2) = 6.$$

We next prove a theorem that is concerned not with a summation within an irrep but with a summation over irreps.

Theorem. *If n_μ is the dimension of the μth irrep of a group $\mathcal{G}$ then*

$$\sum_\mu n_\mu^2 = g,$$

where g is the order of the group.

   Proof. Define a representation of the group in the following way. Rearrange the rows of the multiplication table of the group so that whilst the elements in a particular order head the columns, their inverses in the same order head the rows. In this arrangement of the $g \times g$ table, the leading diagonal is entirely occupied by the identity element. Then, for each element X of the group, take as representative matrix the multiplication-table array obtained by replacing X by 1 and all other element symbols by 0. The matrices $D^{\text{reg}}(X)$ so obtained form the *regular representation* of $\mathcal{G}$; they are each $g \times g$, have a single non-zero entry '1' in each row and column and (as will be verified by a little experimentation) have

	I	A	B
I	I	A	B
A	A	B	I
B	B	I	A

(a)

	I	A	B
I	I	A	B
B	B	I	A
A	A	B	I

(b)

Table 29.3 (a) The multiplication table of the cyclic group of order 3, and (b) its reordering used to generate the regular representation of the group.

the same multiplication structure as the group $\mathcal{G}$ itself, i.e. they form a faithful representation of $\mathcal{G}$.

Although not part of the proof, a simple example may help to make these ideas more transparent. Consider the cyclic group of order 3. Its multiplication table is shown in table 29.3(a) (a repeat of table 28.10(a) of the previous chapter), whilst table 29.3(b) shows the same table reordered so that the columns are still labelled in the order I, A, B but the rows are now labelled in the order $I^{-1} = I$, $A^{-1} = B$, $B^{-1} = A$. The three matrices of the regular representation are then

$$\mathsf{D}^{\mathrm{reg}}(I) = \begin{pmatrix} 1 & 0 & 0 \\ 0 & 1 & 0 \\ 0 & 0 & 1 \end{pmatrix}, \quad \mathsf{D}^{\mathrm{reg}}(A) = \begin{pmatrix} 0 & 1 & 0 \\ 0 & 0 & 1 \\ 1 & 0 & 0 \end{pmatrix}, \quad \mathsf{D}^{\mathrm{reg}}(B) = \begin{pmatrix} 0 & 0 & 1 \\ 1 & 0 & 0 \\ 0 & 1 & 0 \end{pmatrix}.$$

An alternative, more mathematical, definition of the regular representation of a group is

$$\left[\mathsf{D}^{\mathrm{reg}}(G_k)\right]_{ij} = \begin{cases} 1 & \text{if } G_k G_j = G_i, \\ 0 & \text{otherwise.} \end{cases}$$

We now return to the proof. With the construction given, the regular representation has characters as follows:

$$\chi^{\mathrm{reg}}(I) = g, \qquad \chi^{\mathrm{reg}}(X) = 0 \quad \text{if } X \neq I.$$

We now apply (29.18) to $\mathsf{D}^{\mathrm{reg}}$ to obtain for the number m_μ of times that the irrep $\hat{\mathsf{D}}^{(\mu)}$ appears in $\mathsf{D}^{\mathrm{reg}}$ (see 29.11))

$$m_\mu = \frac{1}{g} \sum_X \left[\chi^{(\mu)}(X)\right]^* \chi^{\mathrm{reg}}(X) = \frac{1}{g} \left[\chi^{(\mu)}(I)\right]^* \chi^{\mathrm{reg}}(I) = \frac{1}{g} n_\mu g = n_\mu.$$

Thus an irrep $\hat{\mathsf{D}}^{(\mu)}$ of dimension n_μ appears n_μ times in $\mathsf{D}^{\mathrm{reg}}$, and so by counting the total number of basis functions, or by considering $\chi^{\mathrm{reg}}(I)$, we can conclude that

$$\sum_{\mu} n_{\mu}^2 = g. \tag{29.21}$$

This completes the proof.

As before, our standard demonstration group $3m$ provides an illustration. In this case we have seen already that there are two one-dimensional irreps and one two-dimensional irrep. This is in accord with (29.21) since

$$1^2 + 1^2 + 2^2 = 6, \quad \text{which is the order } g \text{ of the group.}$$

Another straightforward application of the relation (29.21), to the group with multiplication table 29.3(a), yields immediate results. Since $g = 3$, none of its irreps can have dimension 2 or more, as $2^2 = 4$ is too large for (29.21) to be satisfied. Thus all irreps must be one-dimensional and there must be three of them (consistent with the fact that each element is in a class of its own, and that there are therefore three classes). The three irreps are the sets of 1×1 matrices (numbers)

$$A_1 = \{1, 1, 1\} \qquad A_2 = \{1, \omega, \omega^2\} \qquad A_2^* = \{1, \omega^2, \omega\},$$

where $\omega = \exp(2\pi i/3)$; since the matrices are 1×1, the same set of nine numbers would be, of course, the entries in the character table for the irreps of the group. The fact that the numbers in each irrep are all cube roots of unity is discussed below. As will be noticed, two of these irreps are complex – an unusual occurrence in most applications – and form a complex conjugate pair of one-dimensional irreps. In practice, they function much as a two-dimensional irrep, but this is to be ignored for formal purposes such as theorems.

A further property of characters can be derived from the fact that all elements in a conjugacy class have the same order. Suppose that the element X has order m, i.e. $X^m = I$. This implies for a representation D of dimension n that

$$[\mathsf{D}(X)]^m = I_n. \tag{29.22}$$

Representations equivalent to D are generated as before by using similarity transformations of the form

$$\mathsf{D}_\mathsf{Q}(X) = \mathsf{Q}^{-1}\mathsf{D}(X)\mathsf{Q}.$$

In particular, if we choose the columns of Q to be the eigenvectors of D(X) then, as discussed in chapter 8,

$$\mathsf{D}_\mathsf{Q}(X) = \begin{pmatrix} \lambda_1 & 0 & \cdots & 0 \\ 0 & \lambda_2 & & \vdots \\ \vdots & & \ddots & 0 \\ 0 & \cdots & 0 & \lambda_n \end{pmatrix}$$

where the λ_i are the eigenvalues of $D(X)$. Therefore, from (29.22), we have that

$$
\begin{pmatrix}
\lambda_1^m & 0 & \cdots & 0 \\
0 & \lambda_2^m & & \vdots \\
\vdots & & \ddots & 0 \\
0 & \cdots & 0 & \lambda_n^m
\end{pmatrix}
=
\begin{pmatrix}
1 & 0 & \cdots & 0 \\
0 & 1 & & \vdots \\
\vdots & & \ddots & 0 \\
0 & \cdots & 0 & 1
\end{pmatrix}.
$$

Hence all the eigenvalues λ_i are mth roots of unity, and so $\chi(X)$, the trace of $D(X)$, is the sum of n of these. In view of the implications of Lagrange's theorem (section 28.6 and subsection 28.7.2), the only values of m allowed are the divisors of the order g of the group.

29.8 Construction of a character table

In order to decompose representations into irreps on a routine basis using characters, it is necessary to have available a character table for the group in question. Such a table gives, for each irrep μ of the group, the character $\chi^{(\mu)}(X)$ of the class to which group element X belongs. To construct such a table the following properties of a group, established earlier in this chapter, may be used:

(i) the number of classes equals the number of irreps;
(ii) the 'vector' formed by the characters from a given irrep is orthogonal to the 'vector' formed by the characters from a different irrep;
(iii) $\sum_{\mu} n_{\mu}^2 = g$, where n_{μ} is the dimension of the μth irrep and g is the order of the group;
(iv) the identity irrep (one-dimensional with all characters equal to 1) is present for every group;
(v) $\sum_X \left| \chi^{(\mu)}(X) \right|^2 = g$.
(vi) $\chi^{(\mu)}(X)$ is the sum of n_{μ} mth roots of unity, where m is the order of X.

▶*Construct the character table for the group 4mm (or C_{4v}) using the properties of classes, irreps and characters so far established.*

The group 4mm is the group of two-dimensional symmetries of a square, namely rotations of 0, $\pi/2$, π and $3\pi/2$ and reflections in the mirror planes parallel to the coordinate axes and along the main diagonals. These are illustrated in figure 29.3. For this group there are eight elements:

• the identity, I;
• rotations by $\pi/2$ and $3\pi/2$, R and R';
• a rotation by π, Q ;
• four mirror reflections m_x, m_y, m_d and $m_{d'}$.

Requirements (i) to (iv) at the start of this section put tight constraints on the possible character sets, as the following argument shows.

The group is non-Abelian (clearly $Rm_x \neq m_x R$), and so there are fewer than eight classes, and hence fewer than eight irreps. But requirement (iii), with $g = 8$, then implies

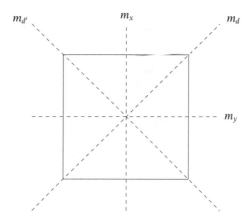

Figure 29.3   The mirror planes associated with 4*mm*, the group of two-dimensional symmetries of a square.

that at least one irrep has dimension 2 or greater. However, there can be no irrep with dimension 3 or greater, since $3^2 > 8$, nor can there be more than one two-dimensional irrep, since $2^2 + 2^2 = 8$ would rule out a contribution to the sum in (iii) of 1^2 from the identity irrep, and this must be present. Thus the only possibility is one two-dimensional irrep and, to make the sum in (iii) correct, four one-dimensional irreps.

Therefore using (i) we can now deduce that there are five classes. This same conclusion can be reached by evaluating $X^{-1}YX$ for every pair of elements in $\mathcal{G}$, as in the description of conjugacy classes given in the previous chapter. However, it is tedious to do so and certainly much longer than the above. The five classes are I, Q, $\{R, R'\}$, $\{m_x, m_y\}$, $\{m_d, m_{d'}\}$.

It is straightforward to show that only I and Q commute with every element of the group, so they are the only elements in classes of their own. Each other class must have at least 2 members, but, as there are three classes to accommodate $8 - 2 = 6$ elements, there must be exactly 2 in each class. This does not pair up the remaining 6 elements, but does say that the five classes have 1, 1, 2, 2, and 2 elements. Of course, if we had started by dividing the group into classes, we would know the number of elements in each class directly.

We cannot entirely ignore the group structure (though it sometimes happens that the results are independent of the group structure – for example, all non-Abelian groups of order 8 have the same character table!); thus we need to note in the present case that $m_i^2 = I$ for $i = x$, y, d or d' and, as can be proved directly, $Rm_i = m_i R'$ for the same four values of label i. We also recall that for any pair of elements X and Y, $D(XY) = D(X)D(Y)$. We may conclude the following for the one-dimensional irreps.

(a) In view of result (vi), $\chi(m_i) = D(m_i) = \pm 1$.
(b) Since $R^4 = I$, result (vi) requires that $\chi(R)$ is one of 1, i, -1, $-i$. But, since $D(R)D(m_i) = D(m_i)D(R')$, and the $D(m_i)$ are just numbers, $D(R) = D(R')$. Further

$$D(R)D(R) = D(R)D(R') = D(RR') = D(I) = 1,$$

and so $D(R) = \pm 1 = D(R')$.
(c) $D(Q) = D(RR) = D(R)D(R) = 1$.

If we add this to the fact that the characters of the identity irrep A_1 are all unity then we can fill in those entries in character table 29.4 shown in bold.

Suppose now that the three missing entries in a one-dimensional irrep are p, q and r, where each can only be ± 1. Then, allowing for the numbers in each class, orthogonality

$4mm$	I	Q	R, R'	m_x, m_y	$m_d, m_{d'}$
A_1	**1**	**1**	**1**	**1**	**1**
A_2	**1**	**1**	1	-1	-1
B_1	**1**	**1**	-1	1	-1
B_2	**1**	**1**	-1	-1	1
E	**2**	-2	0	0	0

Table 29.4   The character table deduced for the group $4mm$. For an explanation of the entries in bold see the text.

with the characters of A_1 requires that

$$1(1)(1) + 1(1)(1) + 2(1)(p) + 2(1)(q) + 2(1)(r) = 0.$$

The only possibility is that two of p, q, and r equal -1 and the other equals $+1$. This can be achieved in three different ways, corresponding to the need to find three further different one-dimensional irreps. Thus the first four lines of entries in character table 29.4 can be completed. The final line can be completed by requiring it to be orthogonal to the other four. Property (v) has not been used here though it could have replaced part of the argument given. ◀

29.9 Group nomenclature

The nomenclature of published character tables, as we have said before, is erratic and sometimes unfortunate; for example, often E is used to represent, not only a two-dimensional irrep, but also the identity operation, where we have used I. Thus the symbol E might appear in both the column and row headings of a table, though with quite different meanings in the two cases. In this book we use roman capitals to denote irreps.

One-dimensional irreps are regularly denoted by A and B, B being used if a rotation about the principal axis of $2\pi/n$ has character -1. Here n is the highest integer such that a rotation of $2\pi/n$ is a symmetry operation of the system, and the principal axis is the one about which this occurs. For the group of operations on a square, $n = 4$, the axis is the perpendicular to the square and the rotation in question is R. The names for the group, $4mm$ and C_{4v}, derive from the fact that here n is equal to 4. Similarly, for the operations on an equilateral triangle, $n = 3$ and the group names are $3m$ and C_{3v}, but because the rotation by $2\pi/3$ has character $+1$ in all its one-dimensional irreps (see table 29.1), only A appears in the irrep list.

Two-dimensional irreps are denoted by E, as we have already noted, and three-dimensional irreps by T, although in many cases the symbols are modified by primes and other alphabetic labels to denote variations in behaviour from one irrep to another in respect of mirror reflections and parity inversions. In the study of molecules, alternative names based on molecular angular momentum properties are common. It is beyond the scope of this book to list all these variations, or to give a large selection of character tables; our aim is to demonstrate and justify the use of those found in the literature specifically dedicated to crystal physics or molecular chemistry.

Variations in notation are not restricted to the naming of groups and their irreps, but extend to the symbols used to identify a typical element, and hence all members, of a conjugacy class in a group. In physics these are usually of the types n_z, $\bar{n}_z$ or m_x. The first of these denotes a rotation of $2\pi/n$ about the z-axis, and the second the same thing followed by parity inversion (all vectors $\mathbf{r}$ go to $-\mathbf{r}$), whilst the third indicates a mirror reflection in a plane, in this case the plane $x = 0$.

Typical chemistry symbols for classes are NC_n, NC_n^2, NC_n^x, NS_n, σ_v, σ^{xy}. Here the first symbol N, where it appears, shows that there are N elements in the class (a useful feature). The subscript n has the same meaning as in the physics notation, but σ rather than m is used for a mirror reflection, subscripts v, d or h or superscripts xy, xz or yz denoting the various orientations of the relevant mirror planes. Symmetries involving parity inversions are denoted by S; thus S_n is the chemistry analogue of $\bar{n}$. None of what is said in this and the previous paragraph should be taken as definitive, but merely as a warning of common variations in nomenclature and as an initial guide to corresponding entities. Before using any set of group character tables, the reader should ensure that he or she understands the precise notation being employed.

29.10 Product representations

In quantum mechanical investigations we are often faced with the calculation of what are called matrix elements. These normally take the form of integrals over all space of the product of two or more functions whose analytic forms depend on the microscopic properties (usually angular momentum and its components) of the electrons or nuclei involved. For 'bonding' calculations involving 'overlap integrals' there are usually two functions involved, whilst for transition probabilities a third function, giving the spatial variation of the interaction Hamiltonian, also appears under the integral sign.

If the environment of the microscopic system under investigation has some symmetry properties, then sometimes these can be used to establish, without detailed evaluation, that the multiple integral must have zero value. We now express the essential content of these ideas in group theoretical language.

Suppose we are given an integral of the form

$$J = \int \Psi \phi \, d\tau \quad \text{or} \quad J = \int \Psi \xi \phi \, d\tau$$

to be evaluated over all space in a situation in which the physical system is invariant under a particular group $\mathcal{G}$ of symmetry operations. For the integral to be non-zero the integrand must be invariant under each of these operations. In group theoretical language, *the integrand must transform as the identity, the one-dimensional representation* A_1 *of* G; more accurately, some non-vanishing part of the integrand must do so.

An alternative way of saying this is that if under the symmetry operations of G the integrand transforms according to a representation D and D does not contain A_1 amongst its irreps then the integral J is necessarily zero. It should be noted that the converse is not true; J may be zero even if A_1 is present, since the integral, whilst showing the required invariance, may still have the value zero.

It is evident that we need to establish how to find the irreps that go to make up a representation of a double or triple product when we already know the irreps according to which the factors in the product transform. The method is established by the following theorem.

Theorem. *For each element of a group the character in a product representation is the product of the corresponding characters in the separate representations.*

Proof. Suppose that $\{u_i\}$ and $\{v_j\}$ are two sets of basis functions, that transform under the operations of a group G according to representations $D^{(\lambda)}$ and $D^{(\mu)}$ respectively. Denote by u and v the corresponding basis vectors and let X be an element of the group. Then the functions generated from u_i and v_j by the action of X are calculated as follows, using (29.1) and (29.4):

$$u'_i = Xu_i = \left[\left(D^{(\lambda)}(X)\right)^T u\right]_i = \left[D^{(\lambda)}(X)\right]_{ii} u_i + \sum_{l \neq i} \left[\left(D^{(\lambda)}(X)\right)^T\right]_{il} u_l,$$

$$v'_j = Xv_j = \left[\left(D^{(\mu)}(X)\right)^T v\right]_j = \left[D^{(\mu)}(X)\right]_{jj} v_j + \sum_{m \neq j} \left[\left(D^{(\mu)}(X)\right)^T\right]_{jm} v_m.$$

Here $[D(X)]_{ij}$ is just a single element of the matrix $D(X)$ and $[D(X)]_{kk} = [D^T(X)]_{kk}$ is simply a diagonal element from the matrix – the repeated subscript does not indicate summation. Now, if we take as basis functions for a product representation $D^{prod}(X)$ the products $w_k = u_i v_j$ (where the $n_\lambda n_\mu$ various possible pairs of values i, j are labelled by k), we have also that

$$w'_k = Xw_k = Xu_i v_j = (Xu_i)(Xv_j)$$
$$= \left[D^{(\lambda)}(X)\right]_{ii} \left[D^{(\mu)}(X)\right]_{jj} u_i v_j + \text{terms not involving the product } u_i v_j.$$

This is to be compared with

$$w'_k = Xw_k = \left[\left(D^{prod}(X)\right)^T w\right]_k = \left[D^{prod}(X)\right]_{kk} w_k + \sum_{n \neq k} \left[\left(D^{prod}(X)\right)^T\right]_{kn} w_n,$$

where $D^{prod}(X)$ is the product representation matrix for element X of the group. The comparison shows that

$$\left[D^{prod}(X)\right]_{kk} = \left[D^{(\lambda)}(X)\right]_{ii} \left[D^{(\mu)}(X)\right]_{jj}.$$

1104

It follows that

$$\chi^{\text{prod}}(X) = \sum_{k=1}^{n_\lambda n_\mu} \left[\mathbf{D}^{\text{prod}}(X) \right]_{kk}$$

$$= \sum_{i=1}^{n_\lambda} \sum_{j=1}^{n_\mu} \left[\mathbf{D}^{(\lambda)}(X) \right]_{ii} \left[\mathbf{D}^{(\mu)}(X) \right]_{jj}$$

$$= \left\{ \sum_{i=1}^{n_\lambda} \left[\mathbf{D}^{(\lambda)}(X) \right]_{ii} \right\} \left\{ \sum_{j=1}^{n_\mu} \left[\mathbf{D}^{(\mu)}(X) \right]_{jj} \right\}$$

$$= \chi^{(\lambda)}(X)\, \chi^{(\mu)}(X). \tag{29.23}$$

This proves the theorem, and a similar argument leads to the corresponding result for integrands in the form of a product of three or more factors.

An immediate corollary is that *an integral whose integrand is the product of two functions transforming according to two different irreps is necessarily zero*. To see this, we use (29.18) to determine whether irrep A_1 appears in the product character set $\chi^{\text{prod}}(X)$:

$$m_{A_1} = \frac{1}{g} \sum_X \left[\chi^{(A_1)}(X) \right]^* \chi^{\text{prod}}(X) = \frac{1}{g} \sum_X \chi^{\text{prod}}(X) = \frac{1}{g} \sum_X \chi^{(\lambda)}(X) \chi^{(\mu)}(X).$$

We have used the fact that $\chi^{(A_1)}(X) = 1$ for all X but now note that, by virtue of (29.14), the expression on the right of this equation is equal to zero unless $\lambda = \mu$.

Any complications due to non-real characters have been ignored – in practice, they are handled automatically as it is usually $\Psi^* \phi$, rather than $\Psi \phi$, that appears in integrands, though many functions are real in any case, and nearly all characters are.

Equation (29.23) is a general result for integrands but, specifically in the context of chemical bonding, it implies that for the possibility of bonding to exist, the two quantum wavefunctions must transform according to the same irrep. This is discussed further in the next section.

29.11 Physical applications of group theory

As we indicated at the start of chapter 28 and discussed in a little more detail at the beginning of the present chapter, some physical systems possess symmetries that allow the results of the present chapter to be used in their analysis. We consider now some of the more common sorts of problem in which these results find ready application.

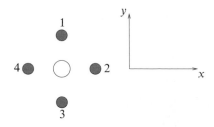

Figure 29.4   A molecule consisting of four atoms of iodine and one of manganese.

29.11.1 Bonding in molecules

We have just seen that whether chemical bonding can take place in a molecule is strongly dependent upon whether the wavefunctions of the two atoms forming a bond transform according to the same irrep. Thus it is sometimes useful to be able to find a wavefunction that does transform according to a particular irrep of a group of transformations. This can be done if the characters of the irrep are known and a sensible starting point can be guessed. We state without proof that starting from any n-dimensional basis vector $\Psi \equiv (\Psi_1 \ \Psi_2 \ \cdots \ \Psi_n)^{\mathrm{T}}$, where $\{\Psi_i\}$ is a set of wavefunctions, the new vector $\Psi^{(\lambda)} \equiv (\Psi_1^{(\lambda)} \ \Psi_2^{(\lambda)} \ \cdots \ \Psi_n^{(\lambda)})^{\mathrm{T}}$ generated by

$$\Psi_i^{(\lambda)} = \sum_X \chi^{(\lambda)^*}(X) X \Psi_i \tag{29.24}$$

will transform according to the λth irrep. If the randomly chosen Ψ happens not to contain any component that transforms in the desired way then the $\Psi^{(\lambda)}$ so generated is found to be a zero vector and it is necessary to select a new starting vector. An illustration of the use of this 'projection operator' is given in the next example.

> ►*Consider a molecule made up of four iodine atoms lying at the corners of a square in the xy-plane, with a manganese atom at its centre, as shown in figure 29.4. Investigate whether the molecular orbital given by the superposition of p-state (angular momentum l = 1) atomic orbitals*
>
> $$\Psi_1 = \Psi_y(\mathbf{r} - \mathbf{R}_1) + \Psi_x(\mathbf{r} - \mathbf{R}_2) - \Psi_y(\mathbf{r} - \mathbf{R}_3) - \Psi_x(\mathbf{r} - \mathbf{R}_4)$$
>
> *can bond to the d-state atomic orbitals of the manganese atom described by either (i) $\phi_1 = (3z^2 - r^2)f(r)$ or (ii) $\phi_2 = (x^2 - y^2)f(r)$, where f(r) is a function of r and so is unchanged by any of the symmetry operations of the molecule. Such linear combinations of atomic orbitals are known as ring orbitals.*

We have eight basis functions, the atomic orbitals $\Psi_x(N)$ and $\Psi_y(N)$, where $N = 1, 2, 3, 4$ and indicates the position of an iodine atom. Since the wavefunctions are those of p-states they have the forms $xf(r)$ or $yf(r)$ and lie in the directions of the x- and y-axes shown in the figure. Since r is not changed by any of the symmetry operations, $f(r)$ can be treated as a constant. The symmetry group of the system is $4mm$, whose character table is table 29.4.

Case (i). The manganese atomic orbital $\phi_1 = (3z^2 - r^2)f(r)$, lying at the centre of the molecule, is not affected by any of the symmetry operations since z and r are unchanged by them. It clearly transforms according to the identity irrep A_1. We therefore need to know which combination of the iodine orbitals $\Psi_x(N)$ and $\Psi_y(N)$, if any, also transforms according to A_1.

We use the projection operator (29.24). If we choose $\Psi_x(1)$ as the arbitrary one-dimensional starting vector, we unfortunately obtain zero (as the reader may wish to verify), but $\Psi_y(1)$ is found to generate a new non-zero one-dimensional vector transforming according to A_1. The results of acting on $\Psi_y(1)$ with the various symmetry elements X can be written down by inspection (see the discussion in section 29.2). So, for example, the $\Psi_y(1)$ orbital centred on iodine atom 1 and aligned along the positive y-axis is changed by the anticlockwise rotation of $\pi/2$ produced by R' into an orbital centred on atom 4 and aligned along the negative x-axis; thus $R'\Psi_y(1) = -\Psi_x(4)$. The complete set of group actions on $\Psi_y(1)$ is:

$$I, \ \Psi_y(1); \qquad Q, \ -\Psi_y(3); \qquad R, \ \Psi_x(2); \qquad R', \ -\Psi_x(4);$$

$$m_x, \ \Psi_y(1); \qquad m_y, \ -\Psi_y(3); \qquad m_d, \ \Psi_x(2); \qquad m_{d'}, \ -\Psi_x(4).$$

Now $\chi^{(A_1)}(X) = 1$ for all X, so (29.24) states that the sum of the above results for $X\Psi_y(1)$, all with weight 1, gives a vector (here, since the irrep is one-dimensional, just a wavefunction) that transforms according to A_1 and is therefore capable of forming a chemical bond with the manganese wavefunction ϕ_1. It is

$$\Psi^{(A_1)} = 2[\Psi_y(1) - \Psi_y(3) + \Psi_x(2) - \Psi_x(4)],$$

though, of course, the factor 2 is irrelevant. This is precisely the ring orbital Ψ_1 given in the problem, but here it is generated rather than guessed beforehand.

Case (ii). The atomic orbital $\phi_2 = (x^2 - y^2)f(r)$ behaves as follows under the action of typical conjugacy class members:

$$I, \ \phi_2; \qquad Q, \ \phi_2; \qquad R, \ (y^2 - x^2)f(r) = -\phi_2; \qquad m_x, \ \phi_2; \qquad m_d, \ -\phi_2.$$

From this we see that ϕ_2 transforms as a one-dimensional irrep, but, from table 29.4, that irrep is B_1 not A_1 (the irrep according to which Ψ_1 transforms, as already shown). Thus ϕ_2 and Ψ_1 cannot form a bond. ◄

The original question did not ask for the the ring orbital to which ϕ_2 may bond, but it can be generated easily by using the values of $X\Psi_y(1)$ calculated in case (i) and now weighting them according to the characters of B_1:

$$\Psi^{(B_1)} = \Psi_y(1) - \Psi_y(3) + (-1)\Psi_x(2) - (-1)\Psi_x(4)$$
$$+ \Psi_y(1) - \Psi_y(3) + (-1)\Psi_x(2) - (-1)\Psi_x(4)$$
$$= 2[\Psi_y(1) - \Psi_x(2) - \Psi_y(3) + \Psi_x(4)].$$

Now we will find the other irreps of $4mm$ present in the space spanned by the basis functions $\Psi_x(N)$ and $\Psi_y(N)$; at the same time this will illustrate the important point that since we are working with characters we are only interested in the diagonal elements of the representative matrices. This means (section 29.2) that if we work in the natural representation D^{nat} we need consider only those functions that transform, wholly or partially, into themselves. Since we have no need to write out the matrices explicitly, their size (8×8) is no drawback. All the irreps spanned by the basis functions $\Psi_x(N)$ and $\Psi_y(N)$ can be determined by considering the actions of the group elements upon them, as follows.

(i) Under I all eight basis functions are unchanged, and $\chi(I) = 8$.

(ii) The rotations R, R' and Q change the value of N in every case and so all diagonal elements of the natural representation are zero and $\chi(R) = \chi(Q) = 0$.

(iii) m_x takes x into $-x$ and y into y and, for $N = 1$ and 3, leaves N unchanged, with the consequences (remember the forms of $\Psi_x(N)$ and $\Psi_y(N)$) that

$$\Psi_x(1) \to -\Psi_x(1), \quad \Psi_x(3) \to -\Psi_x(3),$$
$$\Psi_y(1) \to \Psi_y(1), \quad \Psi_y(3) \to \Psi_y(3).$$

Thus $\chi(m_x)$ has four non-zero contributions, -1, -1, 1 and 1, together with four zero contributions. The total is thus zero.

(iv) m_d and $m_{d'}$ leave no atom unchanged and so $\chi(m_d) = 0$.

The character set of the natural representation is thus 8, 0, 0, 0, 0, which, either by inspection or by applying formula (29.18), shows that

$$\mathsf{D}^{\mathrm{nat}} = A_1 \oplus A_2 \oplus B_1 \oplus B_2 \oplus 2E,$$

i.e. that all possible irreps are present. We have constructed previously the combinations of $\Psi_x(N)$ and $\Psi_y(N)$ that transform according to A_1 and B_1. The others can be found in the same way.

29.11.2 Matrix elements in quantum mechanics

In section 29.10 we outlined the procedure for determining whether a matrix element that involves the product of three factors as an integrand is necessarily zero. We now illustrate this with a specific worked example.

▶*Determine whether a 'dipole' matrix element of the form*

$$J = \int \Psi_{d_1} x \Psi_{d_2} \, d\tau,$$

where Ψ_{d_1} and Ψ_{d_2} are d-state wavefunctions of the forms $xyf(r)$ and $(x^2 - y^2)g(r)$ respectively, can be non-zero (i) in a molecule with symmetry C_{3v} (or 3m), such as ammonia, and (ii) in a molecule with symmetry C_{4v} (or 4mm), such as the MnI_4 molecule considered in the previous example.

We will need to make reference to the character tables of the two groups. The table for C_{3v} is table 29.1 (section 29.6); that for C_{4v} is reproduced as table 29.5 from table 29.4 but with the addition of another column showing how some common functions transform.

We make use of (29.23), extended to the product of three functions. No attention need be paid to $f(r)$ and $g(r)$ as they are unaffected by the group operations.

Case (i). From the character table 29.1 for C_{3v}, we see that each of xy, x and $x^2 - y^2$ forms part of a basis set transforming according to the two-dimensional irrep E. Thus we may fill in the array of characters (using chemical notation for the classes, except that we continue to use I rather than E) as shown in table 29.6. The last line is obtained by

4mm	I	Q	R, R'	m_x, m_y	$m_d, m_{d'}$	
A_1	1	1	1	1	1	$z; z^2; x^2 + y^2$
A_2	1	1	1	-1	-1	R_z
B_1	1	1	-1	1	-1	$x^2 - y^2$
B_2	1	1	-1	-1	1	xy
E	2	-2	0	0	0	$(x, y); (xz, yz); (R_x, R_y)$

Table 29.5   The character table for the irreps of group $4mm$ (or C_{4v}). The right-hand column lists some common functions, or, for the two-dimensional irrep E, pairs of functions, that transform according to the irrep against which they are shown.

Function	Irrep	Classes		
		I	$2C_3$	$3\sigma_v$
xy	E	2	-1	0
x	E	2	-1	0
$x^2 - y^2$	E	2	-1	0
product		8	-1	0

Table 29.6   The character sets, for the group C_{3v} (or $3mm$), of three functions and of their product $x^2 y(x^2 - y^2)$.

Function	Irrep	Classes				
		I	C_2	$2C_6$	$2\sigma_v$	$2\sigma_d$
xy	B_2	1	1	-1	-1	1
x	E	2	-2	0	0	0
$x^2 \quad y^2$	B_1	1	1	-1	1	-1
product		2	2	0	0	0

Table 29.7   The character sets, for the group C_{4v} (or $4mm$), of three functions, and of their product $x^2 y(x^2 - y^2)$.

multiplying together the corresponding characters for each of the three elements. Now, by inspection, or by applying (29.18), i.e.

$$m_{A_1} = \tfrac{1}{6}[1(1)(8) + 2(1)(-1) + 3(1)(0)] = 1,$$

we see that irrep A_1 does appear in the reduced representation of the product, and so J is not necessarily zero.

   Case (ii). From table 29.5 we find that, under the group C_{4v}, xy and $x^2 - y^2$ transform as irreps B_2 and B_1 respectively and that x is part of a basis set transforming as E. Thus the calculation table takes the form of table 29.7 (again, chemical notation for the classes has been used).

   Here inspection is sufficient, as the product is exactly that of irrep E and irrep A_1 is certainly not present. Thus J is necessarily zero and the dipole matrix element vanishes. ◀

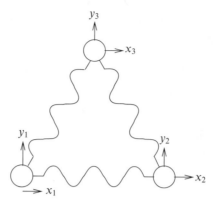

Figure 29.5   An equilateral array of masses and springs.

29.11.3 Degeneracy of normal modes

As our final area for illustrating the usefulness of group theoretical results we consider the normal modes of a vibrating system (see chapter 9). This analysis has far-reaching applications in physics, chemistry and engineering. For a given system, normal modes that are related by some symmetry operation have the same frequency of vibration; the modes are said to be *degenerate*. It can be shown that such modes span a vector space that transforms according to some irrep of the group $\mathcal{G}$ of symmetry operations of the system. Moreover, the degeneracy of the modes equals the dimension of the irrep. As an illustration, we consider the following example.

▶*Investigate the possible vibrational modes of the equilateral triangular arrangement of equal masses and springs shown in figure 29.5. Demonstrate that two are degenerate.*

Clearly the symmetry group is that of the symmetry operations on an equilateral triangle, namely $3m$ (or C_{3v}), whose character table is table 29.1. As on a previous occasion, it is most convenient to use the natural representation D^{nat} of this group (it almost always saves having to write out matrices explicitly) acting on the six-dimensional vector space $(x_1, y_1, x_2, y_2, x_3, y_3)$. In this example the natural and regular representations coincide, but this is not usually the case.

We note that in table 29.1 the second class contains the rotations A (by $\pi/3$) and B (by $2\pi/3$), also known as R and R'. This class is known as 3_z in crystallographic notation, or C_3 in chemical notation, as explained in section 29.9. The third class contains C, D, E, the three mirror reflections.

Clearly $\chi(I) = 6$. Since all position labels are changed by a rotation, $\chi(3_z) = 0$. For the mirror reflections the simplest representative class member to choose is the reflection m_y in the plane containing the y_3-axis, since then only label 3 is unchanged; under m_y, $x_3 \to -x_3$ and $y_3 \to y_3$, leading to the conclusion that $\chi(m_y) = 0$. Thus the character set is 6, 0, 0.

Using (29.18) and the character table 29.1 shows that

$$D^{nat} = A_1 \oplus A_2 \oplus 2E.$$

However, we have so far allowed x_i, y_i to be completely general, and we must now identify and remove those irreps that do not correspond to vibrations. These will be the irreps corresponding to bodily translations of the triangle and to its rotation without relative motion of the three masses.

Bodily translations are linear motions of the centre of mass, which has coordinates

$$x = (x_1 + x_2 + x_3)/3 \quad \text{and} \quad y = (y_1 + y_2 + y_3)/3.$$

Table 29.1 shows that such a coordinate pair (x, y) transforms according to the two-dimensional irrep E; this accounts for one of the two such irreps found in the natural representation.

It can be shown that, as stated in table 29.1, planar bodily rotations of the triangle – rotations about the z-axis, denoted by R_z – transform as irrep A_2. Thus, when the linear motions of the centre of mass, and pure rotation about it, are removed from our reduced representation, we are left with $E \oplus A_1$. So, E and A_1 must be the irreps corresponding to the internal vibrations of the triangle – one doubly degenerate mode and one non-degenerate mode.

The physical interpretation of this is that two of the normal modes of the system have the same frequency and one normal mode has a different frequency (barring accidental coincidences for other reasons). It may be noted that in quantum mechanics the energy quantum of a normal mode is proportional to its frequency. ◄

In general, group theory does not tell us what the frequencies are, since it is entirely concerned with the symmetry of the system and not with the values of masses and spring constants. However, using this type of reasoning, the results from representation theory can be used to predict the degeneracies of atomic energy levels and, given a perturbation whose Hamiltonian (energy operator) has some degree of symmetry, the extent to which the perturbation will resolve the degeneracy. Some of these ideas are explored a little further in the next section and in the exercises.

29.11.4 Breaking of degeneracies

If a physical system has a high degree of symmetry, invariant under a group G of reflections and rotations, say, then, as implied above, it will normally be the case that some of its eigenvalues (of energy, frequency, angular momentum etc.) are degenerate. However, if a perturbation that is invariant only under the operations of the elements of a smaller symmetry group (a subgroup of G) is added, some of the original degeneracies may be broken. The results derived from representation theory can be used to decide the extent of the degeneracy-breaking.

The normal procedure is to use an N-dimensional basis vector, consisting of the N degenerate eigenfunctions, to generate an N-dimensional representation of the symmetry group of the perturbation. This representation is then decomposed into irreps. In general, eigenfunctions that transform according to different irreps no longer share the same frequency of vibration. We illustrate this with the following example.

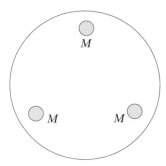

Figure 29.6   A circular drumskin loaded with three symmetrically placed masses.

►*A circular drumskin has three equal masses placed on it at the vertices of an equilateral triangle, as shown in figure 29.6. Determine which degenerate normal modes of the drumskin can be split in frequency by this perturbation.*

When no masses are present the normal modes of the drum-skin are either non-degenerate or two-fold degenerate (see chapter 21). The degenerate eigenfunctions Ψ of the nth normal mode have the forms

$$J_n(kr)(\cos n\theta)e^{\pm i\omega t} \quad \text{or} \quad J_n(kr)(\sin n\theta)e^{\pm i\omega t}.$$

Therefore, as explained above, we need to consider the two-dimensional vector space spanned by $\Psi_1 = \sin n\theta$ and $\Psi_2 = \cos n\theta$. This will generate a two-dimensional representation of the group $3m$ (or C_{3v}), the symmetry group of the perturbation. Taking the easiest element from each of the three classes (identity, rotations, and reflections) of group $3m$, we have

$$I\Psi_1 = \Psi_1, \quad I\Psi_2 = \Psi_2,$$
$$A\Psi_1 = \sin\left[n\left(\theta - \tfrac{2}{3}\pi\right)\right] = \left(\cos\tfrac{2}{3}n\pi\right)\Psi_1 - \left(\sin\tfrac{2}{3}n\pi\right)\Psi_2,$$
$$A\Psi_2 = \cos\left[n\left(\theta - \tfrac{2}{3}\pi\right)\right] = \left(\cos\tfrac{2}{3}n\pi\right)\Psi_2 + \left(\sin\tfrac{2}{3}n\pi\right)\Psi_1,$$
$$C\Psi_1 = \sin[n(\pi - \theta)] = -(\cos n\pi)\Psi_1,$$
$$C\Psi_2 = \cos[n(\pi - \theta)] = (\cos n\pi)\Psi_2.$$

The three representative matrices are therefore

$$\mathsf{D}(I) = \mathsf{I}_2, \quad \mathsf{D}(A) = \begin{pmatrix} \cos\tfrac{2}{3}n\pi & -\sin\tfrac{2}{3}n\pi \\ \sin\tfrac{2}{3}n\pi & \cos\tfrac{2}{3}n\pi \end{pmatrix}, \quad \mathsf{D}(C) = \begin{pmatrix} -\cos n\pi & 0 \\ 0 & \cos n\pi \end{pmatrix}.$$

The characters of this representation are $\chi(I) = 2$, $\chi(A) = 2\cos(2n\pi/3)$ and $\chi(C) = 0$. Using (29.18) and table 29.1, we find that

$$m_{A_1} = \tfrac{1}{6}\left(2 + 4\cos\tfrac{2}{3}n\pi\right) = m_{A_2}$$
$$m_E = \tfrac{1}{6}\left(4 - 4\cos\tfrac{2}{3}n\pi\right).$$

Thus

$$\mathsf{D} = \begin{cases} A_1 \oplus A_2 & \text{if } n = 3,\ 6,\ 9,\ \ldots, \\ E & \text{otherwise.} \end{cases}$$

Hence the normal modes $n = 3,\ 6,\ 9,\ \ldots$ each transform under the operations of $3m$

as the sum of two one-dimensional irreps and, using the reasoning given in the previous example, are therefore split in frequency by the perturbation. For other values of n the representation is irreducible and so the degeneracy cannot be split. ◄

29.12 Exercises

29.1 A group $\mathcal{G}$ has four elements I, X, Y and Z, which satisfy $X^2 = Y^2 = Z^2 = XYZ = I$. Show that $\mathcal{G}$ is Abelian and hence deduce the form of its character table.

Show that the matrices

$$D(I) = \begin{pmatrix} 1 & 0 \\ 0 & 1 \end{pmatrix}, \qquad D(X) = \begin{pmatrix} -1 & 0 \\ 0 & -1 \end{pmatrix},$$

$$D(Y) = \begin{pmatrix} -1 & -p \\ 0 & 1 \end{pmatrix}, \qquad D(Z) = \begin{pmatrix} 1 & p \\ 0 & -1 \end{pmatrix},$$

where p is a real number, form a representation D of $\mathcal{G}$. Find its characters and decompose it into irreps.

29.2 Using a square whose corners lie at coordinates $(\pm 1, \pm 1)$, form a natural representation of the dihedral group $\mathcal{D}_4$. Find the characters of the representation, and, using the information (and class order) in table 29.4 (p. 110?), express the representation in terms of irreps.

Now form a representation in terms of eight 2×2 orthogonal matrices, by considering the effect of each of the elements of $\mathcal{D}_4$ on a general vector (x, y). Confirm that this representation is one of the irreps found using the natural representation.

29.3 The quaternion group $\mathcal{Q}$ (see exercise 28.20) has eight elements $\{\pm 1, \pm i, \pm j, \pm k\}$ obeying the relations

$$i^2 = j^2 = k^2 = -1, \quad ij = k = -ji.$$

Determine the conjugacy classes of $\mathcal{Q}$ and deduce the dimensions of its irreps. Show that $\mathcal{Q}$ is homomorphic to the four-element group $\mathcal{V}$, which is generated by two distinct elements a and b with $a^2 = b^2 = (ab)^2 = I$. Find the one-dimensional irreps of $\mathcal{V}$ and use these to help determine the full character table for $\mathcal{Q}$.

29.4 Construct the character table for the irreps of the permutation group S_4 as follows.

(a) By considering the possible forms of its cycle notation, determine the number of elements in each conjugacy class of the permutation group S_4, and show that S_4 has five irreps. Give the logical reasoning that shows they must consist of two three-dimensional, one two-dimensional, and two one-dimensional irreps.

(b) By considering the odd and even permutations in the group S_4, establish the characters for one of the one-dimensional irreps.

(c) Form a natural matrix representation of 4×4 matrices based on a set of objects $\{a, b, c, d\}$, which may or may not be equal to each other, and, by selecting one example from each conjugacy class, show that this natural representation has characters 4, 2, 1, 0, 0. In the four-dimensional vector space in which each of the four coordinates takes on one of the four values a, b, c or d, the one-dimensional subspace consisting of the four points with coordinates of the form $\{a, a, a, a\}$ is invariant under the permutation group and hence transforms according to the invariant irrep A_1. The remaining three-dimensional subspace is irreducible; use this and the characters deduced above to establish the characters for one of the three-dimensional irreps, T_1.

1113

(d) Complete the character table using orthogonality properties, and check the summation rule for each irrep. You should obtain table 29.8.

Irrep	Typical element and class size				
	(1)	(12)	(123)	(1234)	(12)(34)
	1	6	8	6	3
A_1	1	1	1	1	1
A_2	1	−1	1	−1	1
E	2	0	−1	0	2
T_1	3	1	0	−1	−1
T_2	3	−1	0	1	−1

Table 29.8   The character table for the permutation group S_4.

29.5   In exercise 28.10, the group of pure rotations taking a cube into itself was found to have 24 elements. The group is isomorphic to the permutation group S_4, considered in the previous question, and hence has the same character table, once corresponding classes have been established. By counting the number of elements in each class, make the correspondences below (the final two cannot be decided purely by counting, and should be taken as given).

Permutation class type	Symbol (physics)	Action
(1)	I	none
(123)	3	rotations about a body diagonal
(12)(34)	2_z	rotation of π about the normal to a face
(1234)	4_z	rotations of $\pm\pi/2$ about the normal to a face
(12)	2_d	rotation of π about an axis through the centres of opposite edges

Reformulate the character table 29.8 in terms of the elements of the rotation symmetry group (432 or O) of a cube and use it when answering exercises 29.7 and 29.8.

29.6   Consider a regular hexagon orientated so that two of its vertices lie on the x-axis. Find matrix representations of a rotation R through $2\pi/6$ and a reflection m_y in the y-axis by determining their effects on vectors lying in the xy-plane . Show that a reflection m_x in the x-axis can be written as $m_x = m_y R^3$, and that the 12 elements of the symmetry group of the hexagon are given by R^n or $R^n m_y$.

Using the representations of R and m_y as generators, find a two-dimensional representation of the symmetry group, C_6, of the regular hexagon. Is it a faithful representation?

29.7   In a certain crystalline compound, a thorium atom lies at the centre of a regular octahedron of six sulphur atoms at positions $(\pm a, 0, 0)$, $(0, \pm a, 0)$, $(0, 0, \pm a)$. These can be considered as being positioned at the centres of the faces of a cube of side $2a$. The sulphur atoms produce at the site of the thorium atom an electric field that has the same symmetry group as a cube (432 or O).

The five degenerate d-electron orbitals of the thorium atom can be expressed, relative to any arbitrary polar axis, as

$$(3\cos^2\theta - 1)f(r), \qquad e^{\pm i\phi}\sin\theta\cos\theta f(r), \qquad e^{\pm 2i\phi}\sin^2\theta f(r).$$

A rotation about that polar axis by an angle ϕ' effectively changes ϕ to $\phi - \phi'$.

Use this to show that the character of the rotation in a representation based on the orbital wavefunctions is given by

$$1 + 2\cos\phi' + 2\cos 2\phi'$$

and hence that the characters of the representation, in the order of the symbols given in exercise 29.5, is 5, -1, 1, -1, 1. Deduce that the five-fold degenerate level is split into two levels, a doublet and a triplet.

29.8 Sulphur hexafluoride is a molecule with the same structure as the crystalline compound in exercise 29.7, except that a sulphur atom is now the central atom. The following are the forms of some of the electronic orbitals of the sulphur atom, together with the irreps according to which they transform under the symmetry group 432 (or O).

$$
\begin{array}{ll}
\Psi_s = f(r) & A_1 \\
\Psi_{p_1} = z f(r) & T_1 \\
\Psi_{d_1} = (3z^2 - r^2)f(r) & E \\
\Psi_{d_2} = (x^2 - y^2)f(r) & E \\
\Psi_{d_3} = xyf(r) & T_2
\end{array}
$$

The function x transforms according to the irrep T_1. Use the above data to determine whether dipole matrix elements of the form $J = \int \phi_1 x \phi_2 \, d\tau$ can be non-zero for the following pairs of orbitals ϕ_1, ϕ_2 in a sulphur hexafluoride molecule: (a) Ψ_{d1}, Ψ_s; (b) Ψ_{d1}, Ψ_{p1}; (c) Ψ_{d2}, Ψ_{d1}; (d) Ψ_s, Ψ_{d3}; (e) Ψ_{p1}, Ψ_s.

29.9 The hydrogen atoms in a methane molecule CH_4 form a perfect tetrahedron with the carbon atom at its centre. The molecule is most conveniently described mathematically by placing the hydrogen atoms at the points $(1, 1, 1)$, $(1, -1, -1)$, $(-1, 1, -1)$ and $(-1, -1, 1)$. The symmetry group to which it belongs, the tetrahedral group ($\bar{4}3m$ or T_d), has classes typified by I, 3, 2_z, m_d and $\bar{4}_z$, where the first three are as in exercise 29.5, m_d is a reflection in the mirror plane $x - y = 0$ and $\bar{4}_z$ is a rotation of $\pi/2$ about the z-axis followed by an inversion in the origin. A reflection in a mirror plane can be considered as a rotation of π about an axis perpendicular to the plane, followed by an inversion in the origin.

The character table for the group $\bar{4}3m$ is very similar to that for the group 432, and has the form shown in table 29.9.

Irreps	Typical element and class size					Functions transforming according to irrep
	I	3	2_z	$\bar{4}_z$	m_d	
	1	8	3	6	6	
A_1	1	1	1	1	1	$x^2 + y^2 + z^2$
A_2	1	1	1	-1	-1	
E	2	-1	2	0	0	$(x^2 - y^2, 3z^2 - r^2)$
T_1	3	0	-1	1	-1	(R_x, R_y, R_z)
T_2	3	0	-1	-1	1	(x, y, z); (xy, yz, zx)

Table 29.9 The character table for group $\bar{4}3m$.

By following the steps given below, determine how many different internal vibration frequencies the CH_4 molecule has.

(a) Consider a representation based on the twelve coordinates x_i, y_i, z_i for $i = 1, 2, 3, 4$. For those hydrogen atoms that transform into themselves, a rotation through an angle θ about an axis parallel to one of the coordinate axes gives rise in the natural representation to the diagonal elements 1 for the corresponding coordinate and $2\cos\theta$ for the two orthogonal coordinates. If the rotation is followed by an inversion then these entries are multiplied by -1. Atoms not transforming into themselves give a zero diagonal contribution. Show that the characters of the natural representation are 12, 0, 0, 0, 2 and hence that its expression in terms of irreps is

$$A_1 \oplus E \oplus T_1 \oplus 2T_2.$$

(b) The irreps of the bodily translational and rotational motions are included in this expression and need to be identified and removed. Show that when this is done it can be concluded that there are three different internal vibration frequencies in the CH_4 molecule. State their degeneracies and check that they are consistent with the expected number of normal coordinates needed to describe the internal motions of the molecule.

29.10 Investigate the properties of an alternating group and construct its character table as follows.

(a) The set of even permutations of four objects (a proper subgroup of S_4) is known as the *alternating group* A_4. List its twelve members using cycle notation.

(b) Assume that all permutations with the same cycle structure belong to the same conjugacy class. Show that this leads to a contradiction, and hence demonstrates that, even if two permutations have the same cycle structure, they do not necessarily belong to the same class.

(c) By evaluating the products

$$p_1 = (123)(4) \bullet (12)(34) \bullet (132)(4) \quad \text{and} \quad p_2 = (132)(4) \bullet (12)(34) \bullet (123)(4)$$

deduce that the three elements of A_4 with structure of the form $(12)(34)$ belong to the same class.

(d) By evaluating products of the form $(1\alpha)(\beta\gamma) \bullet (123)(4) \bullet (1\alpha)(\beta\gamma)$, where α, β, γ are various combinations of 2, 3, 4, show that the class to which $(123)(4)$ belongs contains at least four members. Show the same for $(124)(3)$.

(e) By combining results (b), (c) and (d) deduce that A_4 has exactly four classes, and determine the dimensions of its irreps.

(f) Using the orthogonality properties of characters and noting that elements of the form $(124)(3)$ have order 3, find the character table for A_4.

29.11 Use the results of exercise 28.23 to find the character table for the dihedral group $\mathcal{D}_5$, the symmetry group of a regular pentagon.

29.12 Demonstrate that equation (29.24) does, indeed, generate a set of vectors transforming according to an irrep λ, by sketching and superposing drawings of an equilateral triangle of springs and masses, based on that shown in figure 29.5.

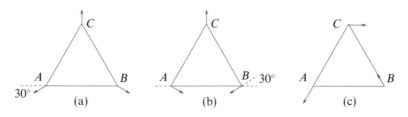

Figure 29.7 The three normal vibration modes of the equilateral array. Mode (a) is known as the 'breathing mode'. Modes (b) and (c) transform according to irrep E and have equal vibrational frequencies.

(a) Make an initial sketch showing an arbitrary small mass displacement from, say, vertex C. Draw the results of operating on this initial sketch with each of the symmetry elements of the group $3m$ (C_{3v}).

(b) Superimpose the results, weighting them according to the characters of irrep A_1 (table 29.1 in section 29.6) and verify that the resultant is a symmetrical arrangement in which all three masses move symmetrically towards (or away from) the centroid of the triangle. The mode is illustrated in figure 29.7(a).

(c) Start again, this time considering a displacement δ of C parallel to the x-axis. Form a similar superposition of sketches weighted according to the characters of irrep E (note that the reflections are not needed). The resultant contains some bodily displacement of the triangle, since this also transforms according to E. Show that the displacement of the centre of mass is $\bar{x} = \delta$, $\bar{y} = 0$. Subtract this out, and verify that the remainder is of the form shown in figure 29.7(c).

(d) Using an initial displacement parallel to the y-axis, and an analogous procedure, generate the remaining normal mode, degenerate with that in (c) and shown in figure 29.7(b).

29.13   Further investigation of the crystalline compound considered in exercise 29.7 shows that the octahedron is not quite perfect but is elongated along the $(1, 1, 1)$ direction with the sulphur atoms at positions $\pm(a+\delta, \delta, \delta)$, $\pm(\delta, a+\delta, \delta)$, $\pm(\delta, \delta, a+\delta)$, where $\delta \ll a$. This structure is invariant under the (crystallographic) symmetry group 32 with three two-fold axes along directions typified by $(1, -1, 0)$. The latter axes, which are perpendicular to the $(1, 1, 1)$ direction, are axes of two-fold symmetry for the perfect octahedron. The group 32 is really the three-dimensional version of the group $3m$ and has the same character table as table 29.1 (section 29.6). Use this to show that, when the distortion of the octahedron is included, the doublet found in exercise 29.7 is unsplit but the triplet breaks up into a singlet and a doublet.

29.13   Hints and answers

29.1   There are four classes and hence four one-dimensional irreps, which must have entries as follows: 1, 1, 1, 1,   1, 1, -1, -1;   1, -1, 1, -1;   1, -1, -1, 1. The characters of D are 2, -2, 0, 0 and so the irreps present are the last two of these.

29.3   There are five classes $\{1\}, \{-1\}, \{\pm i\}, \{\pm j\}, \{\pm k\}$; there are four one-dimensional irreps and one two-dimensional irrep. Show that $ab = ba$. The homomorphism is $\pm 1 \rightarrow I$, $\pm i \rightarrow a$, $\pm j \rightarrow b$, $\pm k \rightarrow ab$. V is Abelian and hence has four one-dimensional irreps.

In the class order given above, the characters for Q are as follows:

$\hat{D}^{(1)}$, 1, 1, 1, 1, 1; $\hat{D}^{(2)}$, 1, 1, 1, -1, -1; $\hat{D}^{(3)}$, 1, 1, -1, 1, -1;

$\hat{D}^{(4)}$, 1, 1, -1, -1, 1; $\hat{D}^{(5)}$, 2, -2, 0, 0, 0.

29.5   Note that the fourth and fifth classes each have 6 members.

29.7   The five basis functions of the representation are multiplied by 1, $e^{-i\phi'}$, $e^{+i\phi'}$, $e^{-2i\phi'}$, $e^{+2i\phi'}$ as a result of the rotation. The character is the sum of these for rotations of 0, $2\pi/3$, π, $\pi/2$, π; $D^{\text{rep}} = E + T_2$.

29.9   (b) The bodily translation has irrep T_2 and the rotation has irrep T_1. The irreps of the internal vibrations are A_1, E, T_2, with respective degeneracies 1, 2, 3, making six internal coordinates (12 in total, minus three translational, minus three rotational).

29.11   There are four classes and hence four irreps, which can only be the identity irrep, one other one-dimensional irrep, and two two-dimensional irreps. In the class order $\{I\}, \{R, R^4\}, \{R^2, R^3\}, \{m_i\}$ the second one-dimensional irrep must (because of orthogonality) have characters 1, 1, 1, -1. The summation rules and orthogonality require the other two character sets to be 2, $(-1 + \sqrt{5})/2$, $(-1 - \sqrt{5})/2$, 0 and 2, $(-1 - \sqrt{5})/2$, $(-1 + \sqrt{5})/2$, 0. Note that R has order 5 and that, e.g., $(-1 + \sqrt{5})/2 = \exp(2\pi i/5) + \exp(8\pi i/5)$.

29.13   The doublet irrep E (characters 2, -1, 0) appears in both 432 and 32 and so is unsplit. The triplet T_1 (characters 3, 0, 1) splits under 32 into doublet E (characters 2, -1, 0) and singlet A_1 (characters 1, 1, 1).

Probability

All scientists will know the importance of experiment and observation and, equally, be aware that the results of some experiments depend to a degree on chance. For example, in an experiment to measure the heights of a random sample of people, we would not be in the least surprised if all the heights were found to be different; but, if the experiment were repeated often enough, we would expect to find some sort of regularity in the results. Statistics, which is the subject of the next chapter, is concerned with the analysis of real experimental data of this sort. First, however, we discuss probability. To a pure mathematician, probability is an entirely theoretical subject based on axioms. Although this axiomatic approach is important, and we discuss it briefly, an approach to probability more in keeping with its eventual applications in statistics is adopted here.

We first discuss the terminology required, with particular reference to the convenient graphical representation of experimental results as Venn diagrams. The concepts of random variables and distributions of random variables are then introduced. It is here that the connection with statistics is made; we assert that the results of many experiments are random variables and that those results have some sort of regularity, which is represented by a distribution. Precise definitions of a random variable and a distribution are then given, as are the defining equations for some important distributions. We also derive some useful quantities associated with these distributions.

30.1 Venn diagrams

We call a single performance of an experiment a *trial* and each possible result an *outcome*. The *sample space* S of the experiment is then the set of all possible outcomes of an individual trial. For example, if we throw a six-sided die then there are six possible outcomes that together form the sample space of the experiment. At this stage we are not concerned with how likely a particular outcome might

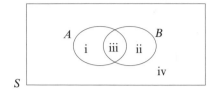

Figure 30.1 A Venn diagram.

be (we will return to the probability of an outcome in due course) but rather will concentrate on the classification of possible outcomes. It is clear that some sample spaces are finite (e.g. the outcomes of throwing a die) whilst others are infinite (e.g. the outcomes of measuring people's heights). Most often, one is not interested in individual outcomes but in whether an outcome belongs to a given subset A (say) of the sample space S; these subsets are called *events*. For example, we might be interested in whether a person is taller or shorter than 180 cm, in which case we divide the sample space into just two events: namely, that the outcome (height measured) is (i) greater than 180 cm or (ii) less than 180 cm.

A common graphical representation of the outcomes of an experiment is the *Venn diagram*. A Venn diagram usually consists of a rectangle, the interior of which represents the sample space, together with one or more closed curves inside it. The interior of each closed curve then represents an event. Figure 30.1 shows a typical Venn diagram representing a sample space S and two events A and B. Every possible outcome is assigned to an appropriate region; in this example there are four regions to consider (marked i to iv in figure 30.1):

   (i) outcomes that belong to event A but not to event B;
  (ii) outcomes that belong to event B but not to event A;
 (iii) outcomes that belong to both event A and event B;
 (iv) outcomes that belong to neither event A nor event B.

▶*A six-sided die is thrown. Let event A be 'the number obtained is divisible by 2' and event B be 'the number obtained is divisible by 3'. Draw a Venn diagram to represent these events.*

It is clear that the outcomes 2, 4, 6 belong to event A and that the outcomes 3, 6 belong to event B. Of these, 6 belongs to both A and B. The remaining outcomes, 1, 5, belong to neither A nor B. The appropriate Venn diagram is shown in figure 30.2. ◀

In the above example, one outcome, 6, is divisible by both 2 and 3 and so belongs to both A and B. This outcome is placed in region iii of figure 30.1, which is called the *intersection* of A and B and is denoted by $A \cap B$ (see figure 30.3(a)). If no events lie in the region of intersection then A and B are said to be *mutually exclusive* or *disjoint*. In this case, often the Venn diagram is drawn so that the closed curves representing the events A and B do not overlap, so as to make

1120

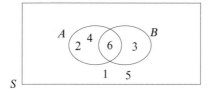

Figure 30.2 The Venn diagram for the outcomes of the die-throwing trials described in the worked example.

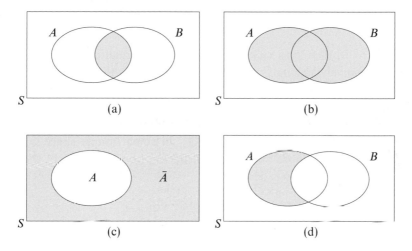

Figure 30.3 Venn diagrams: the shaded regions show (a) $A \cap B$, the intersection of two events A and B, (b) $A \cup B$, the union of events A and B, (c) the complement $\bar{A}$ of an event A, (d) $A - B$, those outcomes in A that do not belong to B.

graphically explicit the fact that A and B are disjoint. It is not necessary, however, to draw the diagram in this way, since we may simply assign zero outcomes to the shaded region in figure 30.3(a). An event that contains no outcomes is called the *empty event* and denoted by $\emptyset$. The event comprising all the elements that belong to either A or B, or to both, is called the *union* of A and B and is denoted by $A \cup B$ (see figure 30.3(b)). In the previous example, $A \cup B = \{2, 3, 4, 6\}$. It is sometimes convenient to talk about those outcomes that do *not* belong to a particular event. The set of outcomes that do not belong to A is called the *complement* of A and is denoted by $\bar{A}$ (see figure 30.3(c)); this can also be written as $\bar{A} = S - A$. It is clear that $A \cup \bar{A} = S$ and $A \cap \bar{A} = \emptyset$.

The above notation can be extended in an obvious way, so that $A - B$ denotes the outcomes in A that do not belong to B. It is clear from figure 30.3(d) that $A - B$ can also be written as $A \cap \bar{B}$. Finally, when *all* the outcomes in event B (say) also belong to event A, but A may contain, in addition, outcomes that do

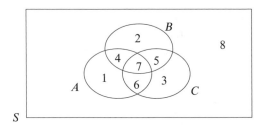

Figure 30.4   The general Venn diagram for three events is divided into eight regions.

not belong to B, then B is called a *subset* of A, a situation that is denoted by $B \subset A$; alternatively, one may write $A \supset B$, which states that A *contains* B. In this case, the closed curve representing the event B is often drawn lying completely within the closed curve representing the event A.

The operations $\cup$ and $\cap$ are extended straightforwardly to more than two events. If there exist n events $A_1, A_2, \ldots, A_n$, in some sample space S, then the event consisting of all those outcomes that belong to *one or more* of the A_i is the *union* of $A_1, A_2, \ldots, A_n$ and is denoted by

$$A_1 \cup A_2 \cup \cdots \cup A_n. \tag{30.1}$$

Similarly, the event consisting of all the outcomes that belong to *every one* of the A_i is called the *intersection* of $A_1, A_2, \ldots, A_n$ and is denoted by

$$A_1 \cap A_2 \cap \cdots \cap A_n. \tag{30.2}$$

If, for *any* pair of values i, j with $i \neq j$,

$$A_i \cap A_j = \emptyset \tag{30.3}$$

then the events A_i and A_j are said to be *mutually exclusive* or *disjoint*.

Consider three events A, B and C with a Venn diagram such as is shown in figure 30.4. It will be clear that, in general, the diagram will be divided into eight regions and they will be of four different types. Three regions correspond to a single event; three regions are each the intersection of exactly two events; one region is the three-fold intersection of all three events; and finally one region corresponds to none of the events. Let us now consider the numbers of different regions in a general n-event Venn diagram.

For one-event Venn diagrams there are two regions, for the two-event case there are four regions and, as we have just seen, for the three-event case there are eight. In the general n-event case there are 2^n regions, as is clear from the fact that any particular region R lies either inside or outside the closed curve of any particular event. With two choices (inside or outside) for each of n closed curves, there are 2^n different possible combinations with which to characterise R. Once n

gets beyond three it becomes impossible to draw a simple two-dimensional Venn diagram, but this does not change the results.

The 2^n regions will break down into $n+1$ types, with the numbers of each type as follows[§]

$$\text{no events,} \quad {}^nC_0 = 1;$$
$$\text{one event but no intersections,} \quad {}^nC_1 = n;$$
$$\text{two-fold intersections,} \quad {}^nC_2 = \tfrac{1}{2}n(n-1);$$
$$\text{three-fold intersections,} \quad {}^nC_3 = \tfrac{1}{3!}n(n-1)(n-2);$$
$$\vdots$$
$$\text{an } n\text{-fold intersection,} \quad {}^nC_n = 1.$$

That this makes a total of 2^n can be checked by considering the binomial expansion

$$2^n = (1+1)^n = 1 + n + \tfrac{1}{2}n(n-1) + \cdots + 1.$$

Using Venn diagrams, it is straightforward to show that the operations $\cap$ and $\cup$ obey the following algebraic laws:

commutativity, $\quad A \cap B = B \cap A, \quad A \cup B = B \cup A;$

associativity, $\quad (A \cap B) \cap C = A \cap (B \cap C), \quad (A \cup B) \cup C = A \cup (B \cup C);$

distributivity, $\quad A \cap (B \cup C) = (A \cap B) \cup (A \cap C),$
$\quad\quad\quad\quad\quad A \cup (B \cap C) = (A \cup B) \cap (A \cup C);$

idempotency, $\quad A \cap A = A, \quad A \cup A = A.$

▶*Show that* (i) $A \cup (A \cap B) = A \cap (A \cup B) = A$, (ii) $(A - D) \cup (A \cap B) = A$.

(i) Using the distributivity and idempotency laws above, we see that

$$A \cup (A \cap B) = (A \cup A) \cap (A \cup B) = A \cap (A \cup B).$$

By sketching a Venn diagram it is immediately clear that both expressions are equal to A. Nevertheless, we here proceed in a more formal manner in order to deduce this result algebraically. Let us begin by writing

$$X = A \cup (A \cap B) = A \cap (A \cup B), \quad\quad\quad (30.4)$$

from which we want to deduce a simpler expression for the event X. Using the first equality in (30.4) and the algebraic laws for $\cap$ and $\cup$, we may write

$$A \cap X = A \cap [A \cup (A \cap B)]$$
$$= (A \cap A) \cup [A \cap (A \cap B)]$$
$$= A \cup (A \cap B) = X.$$

[§] The symbols ${}^nC_i$, for $i = 0, 1, 2, \ldots, n$, are a convenient notation for combinations; they and their properties are discussed in chapter 1.

Since $A \cap X = X$ we must have $X \subset A$. Now, using the second equality in (30.4) in a similar way, we find

$$A \cup X = A \cup [A \cap (A \cup B)]$$
$$= (A \cup A) \cap [A \cup (A \cup B)]$$
$$= A \cap (A \cup B) = X,$$

from which we deduce that $A \subset X$. Thus, since $X \subset A$ and $A \subset X$, we must conclude that $X = A$.

(ii) Since we do not know how to deal with compound expressions containing a minus sign, we begin by writing $A - B = A \cap \bar{B}$ as mentioned above. Then, using the distributivity law, we obtain

$$(A - B) \cup (A \cap B) = (A \cap \bar{B}) \cup (A \cap B)$$
$$= A \cap (\bar{B} \cup B)$$
$$= A \cap S = A.$$

In fact, this result, like the first one, can be proved trivially by drawing a Venn diagram. ◄

Further useful results may be derived from Venn diagrams. In particular, it is simple to show that the following rules hold:

(i)   if $A \subset B$ then $\bar{A} \supset \bar{B}$;
(ii) $\overline{A \cup B} = \bar{A} \cap \bar{B}$;
(iii) $\overline{A \cap B} = \bar{A} \cup \bar{B}$.

Statements (ii) and (iii) are known jointly as *de Morgan's laws* and are sometimes useful in simplifying logical expressions.

► *There exist two events A and B such that*
$$\overline{(X \cup A)} \cup \overline{(X \cup \bar{A})} = B.$$
Find an expression for the event X in terms of A and B.

We begin by taking the complement of both sides of the above expression: applying de Morgan's laws we obtain

$$\bar{B} = (X \cup A) \cap (X \cup \bar{A}).$$

We may then use the algebraic laws obeyed by $\cap$ and $\cup$ to yield

$$\bar{B} = X \cup (A \cap \bar{A}) = X \cup \emptyset = X.$$

Thus, we find that $X = \bar{B}$. ◄

30.2 Probability

In the previous section we discussed Venn diagrams, which are graphical representations of the possible outcomes of experiments. We did not, however, give any indication of how likely each outcome or event might be when any particular experiment is performed. Most experiments show some regularity. By this we mean that the relative frequency of an event is approximately the same on each occasion that a set of trials is performed. For example, if we throw a die N

times then we expect that a six will occur approximately $N/6$ times (assuming, of course, that the die is not biased). The regularity of outcomes allows us to define the *probability*, $\Pr(A)$, as the expected relative frequency of event A in a large number of trials. More quantitatively, if an experiment has a total of n_S outcomes in the sample space S, and n_A of these outcomes correspond to the event A, then the probability that event A will occur is

$$\Pr(A) = \frac{n_A}{n_S}. \tag{30.5}$$

30.2.1 Axioms and theorems

From (30.5) we may deduce the following properties of the probability $\Pr(A)$.

(i) For any event A in a sample space S,

$$0 \le \Pr(A) \le 1. \tag{30.6}$$

If $\Pr(A) = 1$ then A is a certainty; if $\Pr(A) = 0$ then A is an impossibility.

(ii) For the entire sample space S we have

$$\Pr(S) = \frac{n_S}{n_S} = 1, \tag{30.7}$$

which simply states that we are certain to obtain one of the possible outcomes.

(iii) If A and B are two events in S then, from the Venn diagrams in figure 30.3, we see that

$$n_{A\cup B} = n_A + n_B - n_{A\cap B}, \tag{30.8}$$

the final subtraction arising because the outcomes in the intersection of A and B are counted twice when the outcomes of A are added to those of B. Dividing both sides of (30.8) by n_S, we obtain the *addition rule* for probabilities

$$\Pr(A \cup B) = \Pr(A) + \Pr(B) - \Pr(A \cap B). \tag{30.9}$$

However, if A and B are *mutually exclusive* events ($A \cap B = \emptyset$) then $\Pr(A \cap B) = 0$ and we obtain the special case

$$\Pr(A \cup B) = \Pr(A) + \Pr(B). \tag{30.10}$$

(iv) If $\bar{A}$ is the complement of A then $\bar{A}$ and A are mutually exclusive events. Thus, from (30.7) and (30.10) we have

$$1 = \Pr(S) = \Pr(A \cup \bar{A}) = \Pr(A) + \Pr(\bar{A}),$$

from which we obtain the *complement law*

$$\Pr(\bar{A}) = 1 - \Pr(A). \tag{30.11}$$

This is particularly useful for problems in which evaluating the probability of the complement is easier than evaluating the probability of the event itself.

▶*Calculate the probability of drawing an ace or a spade from a pack of cards.*

Let A be the event that an ace is drawn and B the event that a spade is drawn. It immediately follows that $\Pr(A) = \frac{4}{52} = \frac{1}{13}$ and $\Pr(B) = \frac{13}{52} = \frac{1}{4}$. The intersection of A and B consists of only the ace of spades and so $\Pr(A \cap B) = \frac{1}{52}$. Thus, from (30.9)

$$\Pr(A \cup B) = \tfrac{1}{13} + \tfrac{1}{4} - \tfrac{1}{52} = \tfrac{4}{13}.$$

In this case it is just as simple to recognise that there are 16 cards in the pack that satisfy the required condition (13 spades plus three other aces) and so the probability is $\frac{16}{52}$. ◀

The above theorems can easily be extended to a greater number of events. For example, if $A_1, A_2, \dots, A_n$ are mutually exclusive events then (30.10) becomes

$$\Pr(A_1 \cup A_2 \cup \cdots \cup A_n) = \Pr(A_1) + \Pr(A_2) + \cdots + \Pr(A_n). \qquad (30.12)$$

Furthermore, if $A_1, A_2, \dots, A_n$ (whether mutually exclusive or not) *exhaust* S, i.e. are such that $A_1 \cup A_2 \cup \cdots \cup A_n = S$, then

$$\Pr(A_1 \cup A_2 \cup \cdots \cup A_n) = \Pr(S) = 1. \qquad (30.13)$$

▶*A biased six-sided die has probabilities $\frac{1}{2}p,\ p,\ p,\ p,\ p,\ 2p$ of showing 1, 2, 3, 4, 5, 6 respectively. Calculate p.*

Given that the individual events are mutually exclusive, (30.12) can be applied to give

$$\Pr(1 \cup 2 \cup 3 \cup 4 \cup 5 \cup 6) = \tfrac{1}{2}p + p + p + p + p + 2p = \tfrac{13}{2}p.$$

The union of all possible outcomes on the LHS of this equation is clearly the sample space, S, and so

$$\Pr(S) = \tfrac{13}{2}p.$$

Now using (30.7),

$$\tfrac{13}{2}p = \Pr(S) = 1 \quad \Rightarrow \quad p = \tfrac{2}{13}. \ ◀$$

When the possible outcomes of a trial correspond to more than two events, and those events are *not* mutually exclusive, the calculation of the probability of the union of a number of events is more complicated, and the generalisation of the addition law (30.9) requires further work. Let us begin by considering the union of three events A_1, A_2 and A_3, which need not be mutually exclusive. We first define the event $B = A_2 \cup A_3$ and, using the addition law (30.9), we obtain

$$\Pr(A_1 \cup A_2 \cup A_3) = \Pr(A_1 \cup B) = \Pr(A_1) + \Pr(B) - \Pr(A_1 \cap B). \qquad (30.14)$$

However, we may write $\Pr(A_1 \cap B)$ as

$$\begin{aligned}
\Pr(A_1 \cap B) &= \Pr[A_1 \cap (A_2 \cup A_3)] \\
&= \Pr[(A_1 \cap A_2) \cup (A_1 \cap A_3)] \\
&= \Pr(A_1 \cap A_2) + \Pr(A_1 \cap A_3) - \Pr(A_1 \cap A_2 \cap A_3).
\end{aligned}$$

Substituting this expression, and that for $\Pr(B)$ obtained from (30.9), into (30.14) we obtain the probability addition law for three general events,

$$\begin{aligned}
\Pr(A_1 \cup A_2 \cup A_3) &= \Pr(A_1) + \Pr(A_2) + \Pr(A_3) - \Pr(A_2 \cap A_3) - \Pr(A_1 \cap A_3) \\
&\quad - \Pr(A_1 \cap A_2) + \Pr(A_1 \cap A_2 \cap A_3). \tag{30.15}
\end{aligned}$$

▶*Calculate the probability of drawing from a pack of cards one that is an ace or is a spade or shows an even number (2, 4, 6, 8, 10).*

If, as previously, A is the event that an ace is drawn, $\Pr(A) = \frac{4}{52}$. Similarly the event B, that a spade is drawn, has $\Pr(B) = \frac{13}{52}$. The further possibility C, that the card is even (but not a picture card) has $\Pr(C) = \frac{20}{52}$. The two-fold intersections have probabilities

$$\Pr(A \cap B) = \frac{1}{52}, \quad \Pr(A \cap C) = 0, \quad \Pr(B \cap C) = \frac{5}{52}.$$

There is no three-fold intersection as events A and C are mutually exclusive. Hence

$$\Pr(A \cup B \cup C) = \frac{1}{52} [(4 + 13 + 20) - (1 + 0 + 5) + (0)] = \frac{31}{52}.$$

The reader should identify the 31 cards involved. ◀

When the probabilities are combined to calculate the probability for the union of the n general events, the result, which may be proved by induction upon n (see the answer to exercise 30.4), is

$$\begin{aligned}
\Pr(A_1 \cup A_2 \cup \cdots \cup A_n) &= \sum_i \Pr(A_i) - \sum_{i,j} \Pr(A_i \cap A_j) + \sum_{i,j,k} \Pr(A_i \cap A_j \cap A_k) \\
&\quad - \cdots + (-1)^{n+1} \Pr(A_1 \cap A_2 \cap \cdots \cap A_n). \tag{30.16}
\end{aligned}$$

Each summation runs over all possible sets of subscripts, except those in which any two subscripts in a set are the same. The number of terms in the summation of probabilities of m-fold intersections of the n events is given by ^nC_m (as discussed in section 30.1). Equation (30.9) is a special case of (30.16) in which $n = 2$ and only the first two terms on the RHS survive. We now illustrate this result with a worked example that has $n = 4$ and includes a four-fold intersection.

1127

> ► Find the probability of drawing from a pack a card that has at least one of the following properties:
>
> A, it is an ace;
> B, it is a spade;
> C, it is a black honour card (ace, king, queen, jack or 10);
> D, it is a black ace.

Measuring all probabilities in units of $\frac{1}{52}$, the single-event probabilities are

$$\Pr(A) = 4, \qquad \Pr(B) = 13, \qquad \Pr(C) = 10, \qquad \Pr(D) = 2.$$

The two-fold intersection probabilities, measured in the same units, are

$$\Pr(A \cap B) = 1, \qquad \Pr(A \cap C) = 2, \qquad \Pr(A \cap D) = 2,$$
$$\Pr(B \cap C) = 5, \qquad \Pr(B \cap D) = 1, \qquad \Pr(C \cap D) = 2.$$

The three-fold intersections have probabilities

$$\Pr(A \cap B \cap C) = 1, \qquad \Pr(A \cap B \cap D) = 1, \qquad \Pr(A \cap C \cap D) = 2, \qquad \Pr(B \cap C \cap D) = 1.$$

Finally, the four-fold intersection, requiring all four conditions to hold, is satisfied only by the ace of spades, and hence (again in units of $\frac{1}{52}$)

$$\Pr(A \cap B \cap C \cap D) = 1.$$

Substituting in (30.16) gives

$$P = \frac{1}{52} [(4 + 13 + 10 + 2) - (1 + 2 + 2 + 5 + 1 + 2) + (1 + 1 + 2 + 1) - (1)] = \frac{20}{52}. \blacktriangleleft$$

We conclude this section on basic theorems by deriving a useful general expression for the probability $\Pr(A \cap B)$ that two events A and B both occur in the case where A (say) is the union of a set of n *mutually exclusive* events A_i. In this case

$$A \cap B = (A_1 \cap B) \cup \cdots \cup (A_n \cap B),$$

where the events $A_i \cap B$ are also mutually exclusive. Thus, from the addition law (30.12) for mutually exclusive events, we find

$$\Pr(A \cap B) = \sum_i \Pr(A_i \cap B). \tag{30.17}$$

Moreover, in the special case where the events A_i *exhaust* the sample space S, we have $A \cap B = S \cap B = B$, and we obtain the *total probability law*

$$\Pr(B) = \sum_i \Pr(A_i \cap B). \tag{30.18}$$

30.2.2 Conditional probability

So far we have defined only probabilities of the form 'what is the probability that event A happens?'. In this section we turn to *conditional probability*, the probability that a particular event occurs *given* the occurrence of another, possibly related, event. For example, we may wish to know the probability of event B, drawing an ace from a pack of cards from which one has already been removed, given that event A, the card already removed was itself an ace, has occurred.

We denote this probability by $\Pr(B|A)$ and may obtain a formula for it by considering the total probability $\Pr(A \cap B) = \Pr(B \cap A)$ that both A and B will occur. This may be written in two ways, i.e.

$$\Pr(A \cap B) = \Pr(A)\Pr(B|A)$$
$$= \Pr(B)\Pr(A|B).$$

From this we obtain

$$\Pr(A|B) = \frac{\Pr(A \cap B)}{\Pr(B)} \tag{30.19}$$

and

$$\Pr(B|A) = \frac{\Pr(B \cap A)}{\Pr(A)}. \tag{30.20}$$

In terms of Venn diagrams, we may think of $\Pr(B|A)$ as the probability of B in the reduced sample space defined by A. Thus, if two events A and B are mutually exclusive then

$$\Pr(A|B) = 0 = \Pr(B|A). \tag{30.21}$$

When an experiment consists of drawing objects at random from a given set of objects, it is termed *sampling a population*. We need to distinguish between two different ways in which such a *sampling experiment* may be performed. After an object has been drawn at random from the set it may either be put aside or returned to the set before the next object is randomly drawn. The former is termed 'sampling without replacement', the latter 'sampling with replacement'.

▶*Find the probability of drawing two aces at random from a pack of cards (i) when the first card drawn is replaced at random into the pack before the second card is drawn, and (ii) when the first card is put aside after being drawn.*

Let A be the event that the first card is an ace, and B the event that the second card is an ace. Now

$$\Pr(A \cap B) = \Pr(A)\Pr(B|A),$$

and for both (i) and (ii) we know that $\Pr(A) = \frac{4}{52} = \frac{1}{13}$.

(i) If the first card is replaced in the pack before the next is drawn then $\Pr(B|A) = \Pr(B) = \frac{4}{52} = \frac{1}{13}$, since A and B are independent events. We then have

$$\Pr(A \cap B) = \Pr(A)\Pr(B) = \frac{1}{13} \times \frac{1}{13} = \frac{1}{169}.$$

(ii) If the first card is put aside and the second then drawn, A and B are not independent and $\Pr(B|A) = \frac{3}{51}$, with the result that

$$\Pr(A \cap B) = \Pr(A)\Pr(B|A) = \frac{1}{13} \times \frac{3}{51} = \frac{1}{221}. \ ◀$$

Two events A and B are *statistically independent* if $\Pr(A|B) = \Pr(A)$ (or equivalently if $\Pr(B|A) = \Pr(B)$). In words, the probability of A given B is then the same as the probability of A regardless of whether B occurs. For example, if we throw a coin and a die at the same time, we would normally expect that the probability of throwing a six was independent of whether a head was thrown. If A and B are statistically independent then it follows that

$$\Pr(A \cap B) = \Pr(A)\Pr(B). \tag{30.22}$$

In fact, on the basis of intuition and experience, (30.22) may be regarded as the *definition* of the statistical independence of two events.

The idea of statistical independence is easily extended to an arbitrary number of events $A_1, A_2, \ldots, A_n$. The events are said to be (mutually) independent if

$$\Pr(A_i \cap A_j) = \Pr(A_i)\Pr(A_j),$$
$$\Pr(A_i \cap A_j \cap A_k) = \Pr(A_i)\Pr(A_j)\Pr(A_k),$$
$$\vdots$$
$$\Pr(A_1 \cap A_2 \cap \cdots \cap A_n) = \Pr(A_1)\Pr(A_2)\cdots\Pr(A_n),$$

for all combinations of indices i, j and k for which no two indices are the same. Even if all n events are not mutually independent, any two events for which $\Pr(A_i \cap A_j) = \Pr(A_i)\Pr(A_j)$ are said to be *pairwise independent*.

We now derive two results that often prove useful when working with conditional probabilities. Let us suppose that an event A is the union of n *mutually exclusive* events A_i. If B is some other event then from (30.17) we have

$$\Pr(A \cap B) = \sum_i \Pr(A_i \cap B).$$

Dividing both sides of this equation by $\Pr(B)$, and using (30.19), we obtain

$$\Pr(A|B) = \sum_i \Pr(A_i|B), \tag{30.23}$$

which is the *addition law for conditional probabilities*.

Furthermore, if the set of mutually exclusive events A_i exhausts the sample space S then, from the *total probability law* (30.18), the probability $\Pr(B)$ of some event B in S can be written as

$$\Pr(B) = \sum_i \Pr(A_i)\Pr(B|A_i). \tag{30.24}$$

▶*A collection of traffic islands connected by a system of one-way roads is shown in figure 30.5. At any given island a car driver chooses a direction at random from those available. What is the probability that a driver starting at O will arrive at B?*

In order to leave O the driver must pass through one of A_1, A_2, A_3 or A_4, which thus form a complete set of mutually exclusive events. Since at each island (including O) the driver chooses a direction at random from those available, we have that $\Pr(A_i) = \frac{1}{4}$ for

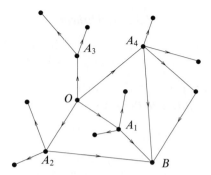

Figure 30.5   A collection of traffic islands connected by one-way roads.

$i = 1, 2, 3, 4$. From figure 30.5, we see also that

$$\Pr(B|A_1) = \tfrac{1}{3}, \quad \Pr(B|A_2) = \tfrac{1}{3}, \quad \Pr(B|A_3) = 0, \quad \Pr(B|A_4) = \tfrac{2}{4} = \tfrac{1}{2}.$$

Thus, using the total probability law (30.24), we find that the probability of arriving at B is given by

$$\Pr(B) = \sum_i \Pr(A_i)\Pr(B|A_i) = \tfrac{1}{4}\left(\tfrac{1}{3} + \tfrac{1}{3} + 0 + \tfrac{1}{2}\right) = \tfrac{7}{24}. \quad \blacktriangleleft$$

Finally, we note that the concept of conditional probability may be straightforwardly extended to several compound events. For example, in the case of three events A, B, C, we may write $\Pr(A \cap B \cap C)$ in several ways, e.g.

$$\Pr(A \cap B \cap C) = \Pr(C)\Pr(A \cap B|C)$$
$$= \Pr(B \cap C)\Pr(A|B \cap C)$$
$$= \Pr(C)\Pr(B|C)\Pr(A|B \cap C).$$

> ▶ *Suppose $\{A_i\}$ is a set of mutually exclusive events that exhausts the sample space S. If B and C are two other events in S, show that*
>
> $$\Pr(B|C) = \sum_i \Pr(A_i|C)\Pr(B|A_i \cap C).$$

Using (30.19) and (30.17), we may write

$$\Pr(C)\Pr(B|C) = \Pr(B \cap C) = \sum_i \Pr(A_i \cap B \cap C). \tag{30.25}$$

Each term in the sum on the RHS can be expanded as an appropriate product of conditional probabilities,

$$\Pr(A_i \cap B \cap C) = \Pr(C)\Pr(A_i|C)\Pr(B|A_i \cap C).$$

Substituting this form into (30.25) and dividing through by $\Pr(C)$ gives the required result. ◀

30.2.3 Bayes' theorem

In the previous section we saw that the probability that both an event A and a related event B will occur can be written either as $\Pr(A)\Pr(B|A)$ or $\Pr(B)\Pr(A|B)$. Hence

$$\Pr(A)\Pr(B|A) = \Pr(B)\Pr(A|B),$$

from which we obtain *Bayes' theorem,*

$$\Pr(A|B) = \frac{\Pr(A)}{\Pr(B)}\Pr(B|A). \qquad (30.26)$$

This theorem clearly shows that $\Pr(B|A) \neq \Pr(A|B)$, unless $\Pr(A) = \Pr(B)$. It is sometimes useful to rewrite $\Pr(B)$, if it is not known directly, as

$$\Pr(B) = \Pr(A)\Pr(B|A) + \Pr(\bar{A})\Pr(B|\bar{A})$$

so that Bayes' theorem becomes

$$\Pr(A|B) = \frac{\Pr(A)\Pr(B|A)}{\Pr(A)\Pr(B|A) + \Pr(\bar{A})\Pr(B|\bar{A})}. \qquad (30.27)$$

▶*Suppose that the blood test for some disease is reliable in the following sense: for people who are infected with the disease the test produces a positive result in 99.99% of cases; for people not infected a positive test result is obtained in only 0.02% of cases. Furthermore, assume that in the general population one person in 10 000 people is infected. A person is selected at random and found to test positive for the disease. What is the probability that the individual is actually infected?*

Let A be the event that the individual is infected and B be the event that the individual tests positive for the disease. Using Bayes' theorem the probability that a person who tests positive is actually infected is

$$\Pr(A|B) = \frac{\Pr(A)\Pr(B|A)}{\Pr(A)\Pr(B|A) + \Pr(\bar{A})\Pr(B|\bar{A})}.$$

Now $\Pr(A) = 1/10000 = 1 - \Pr(\bar{A})$, and we are told that $\Pr(B|A) = 9999/10000$ and $\Pr(B|\bar{A}) = 2/10000$. Thus we obtain

$$\Pr(A|B) = \frac{1/10000 \times 9999/10000}{(1/10000 \times 9999/10000) + (9999/10000 \times 2/10000)} = \frac{1}{3}.$$

Thus, there is only a one in three chance that a person chosen at random, who tests positive for the disease, is actually infected.

At a first glance, this answer may seem a little surprising, but the reason for the counter-intuitive result is that the probability that a randomly selected person is not infected is 9999/10000, which is very high. Thus, the 0.02% chance of a positive test for an uninfected person becomes significant. ◀

We note that (30.27) may be written in a more general form if S is not simply divided into A and $\bar{A}$ but, rather, into *any* set of mutually exclusive events A_i that exhaust S. Using the total probability law (30.24), we may then write

$$\Pr(B) = \sum_i \Pr(A_i)\Pr(B|A_i),$$

so that Bayes' theorem takes the form

$$\Pr(A|B) = \frac{\Pr(A)\Pr(B|A)}{\sum_i \Pr(A_i)\Pr(B|A_i)}, \tag{30.28}$$

where the event A need not coincide with any of the A_i.

As a final point, we comment that sometimes we are concerned only with the *relative* probabilities of two events A and C (say), given the occurrence of some other event B. From (30.26) we then obtain a different form of Bayes' theorem,

$$\frac{\Pr(A|B)}{\Pr(C|B)} = \frac{\Pr(A)\Pr(B|A)}{\Pr(C)\Pr(B|C)}, \tag{30.29}$$

which does not contain $\Pr(B)$ at all.

30.3 Permutations and combinations

In equation (30.5) we defined the probability of an event A in a sample space S as

$$\Pr(A) = \frac{n_A}{n_S},$$

where n_A is the number of outcomes belonging to event A and n_S is the total number of possible outcomes. It is therefore necessary to be able to count the number of possible outcomes in various common situations.

30.3.1 Permutations

Let us first consider a set of n objects that are all different. We may ask in how many ways these n objects may be arranged, i.e. how many *permutations* of these objects exist. This is straightforward to deduce, as follows: the object in the first position may be chosen in n different ways, that in the second position in $n-1$ ways, and so on until the final object is positioned. The number of possible arrangements is therefore

$$n(n-1)(n-2)\cdots(1) = n! \tag{30.30}$$

Generalising (30.30) slightly, let us suppose we choose only k ($< n$) objects from n. The number of possible permutations of these k objects selected from n is given by

$$\underbrace{n(n-1)(n-2)\cdots(n-k+1)}_{k \text{ factors}} = \frac{n!}{(n-k)!} \equiv {}^nP_k. \tag{30.31}$$

In calculating the number of permutations of the various objects we have so far assumed that the objects are sampled *without replacement* – i.e. once an object has been drawn from the set it is put aside. As mentioned previously, however, we may instead replace each object before the next is chosen. The number of permutations of k objects from n *with replacement* may be calculated very easily since the first object can be chosen in n different ways, as can the second, the third, etc. Therefore the number of permutations is simply n^k. This may also be viewed as the number of permutations of k objects from n where repetitions are allowed, i.e. each object may be used as often as one likes.

> ►*Find the probability that in a group of k people at least two have the same birthday (ignoring 29 February).*

It is simplest to begin by calculating the probability that no two people share a birthday, as follows. Firstly, we imagine each of the k people in turn pointing to their birthday on a year planner. Thus, we are sampling the 365 days of the year 'with replacement' and so the total number of possible outcomes is $(365)^k$. Now (for the moment) we assume that no two people share a birthday and imagine the process being repeated, except that as each person points out their birthday it is crossed off the planner. In this case, we are sampling the days of the year 'without replacement', and so the possible number of outcomes for which all the birthdays are different is

$$^{365}P_k = \frac{365!}{(365-k)!}.$$

Hence the probability that all the birthdays are different is

$$p = \frac{365!}{(365-k)! \, 365^k}.$$

Now using the complement rule (30.11), the probability q that two or more people have the same birthday is simply

$$q = 1 - p = 1 - \frac{365!}{(365-k)! \, 365^k}.$$

This expression may be conveniently evaluated using Stirling's approximation for $n!$ when n is large, namely

$$n! \sim \sqrt{2\pi n} \left(\frac{n}{e}\right)^n,$$

to give

$$q \approx 1 - e^{-k} \left(\frac{365}{365-k}\right)^{365-k+0.5}.$$

It is interesting to note that if $k = 23$ the probability is a little greater than a half that at least two people have the same birthday, and if $k = 50$ the probability rises to 0.970. This can prove a good bet at a party of non-mathematicians! ◄

So far we have assumed that all n objects are different (or *distinguishable*). Let us now consider n objects of which n_1 are identical and of type 1, n_2 are identical and of type 2, $\ldots, n_m$ are identical and of type m (clearly $n = n_1 + n_2 + \cdots + n_m$). From (30.30) the number of permutations of these n objects is again $n!$. However, the number of *distinguishable* permutations is only

$$\frac{n!}{n_1!n_2!\cdots n_m!},\tag{30.32}$$

since the ith group of identical objects can be rearranged in $n_i!$ ways without changing the distinguishable permutation.

►*A set of snooker balls consists of a white, a yellow, a green, a brown, a blue, a pink, a black and 15 reds. How many distinguishable permutations of the balls are there?*

In total there are 22 balls, the 15 reds being indistinguishable. Thus from (30.32) the number of distinguishable permutations is

$$\frac{22!}{(1!)(1!)(1!)(1!)(1!)(1!)(1!)(15!)} = \frac{22!}{15!} = 859\,541\,760. \blacktriangleleft$$

30.3.2 Combinations

We now consider the number of *combinations* of various objects when their order is immaterial. Assuming all the objects to be distinguishable, from (30.31) we see that the number of permutations of k objects chosen from n is $^nP_k = n!/(n-k)!$. Now, since we are no longer concerned with the order of the chosen objects, which can be internally arranged in $k!$ different ways, the number of combinations of k objects from n is

$$\frac{n!}{(n-k)!k!} \equiv {}^nC_k \equiv \begin{pmatrix} n \\ k \end{pmatrix} \quad \text{for } 0 \le k \le n,\tag{30.33}$$

where, as noted in chapter 1, ^nC_k is called the *binomial coefficient* since it also appears in the binomial expansion for positive integer n, namely

$$(a+b)^n = \sum_{k=0}^{n} {}^nC_k a^k b^{n-k}.\tag{30.34}$$

►*A hand of 13 playing cards is dealt from a well-shuffled pack of 52. What is the probability that the hand contains two aces?*

Since the order of the cards in the hand is immaterial, the total number of distinct hands is simply equal to the number of combinations of 13 objects drawn from 52, i.e. $^{52}C_{13}$. However, the number of hands containing two aces is equal to the number of ways, $^4C_2$, in which the two aces can be drawn from the four available, multiplied by the number of ways, $^{48}C_{11}$, in which the remaining 11 cards in the hand can be drawn from the 48 cards that are not aces. Thus the required probability is given by

$$\frac{^4C_2 \,^{48}C_{11}}{^{52}C_{13}} = \frac{4!}{2!2!} \frac{48!}{11!37!} \frac{13!39!}{52!}$$

$$= \frac{(3)(4)}{2} \frac{(12)(13)(38)(39)}{(49)(50)(51)(52)} = 0.213 \blacktriangleleft$$

Another useful result that may be derived using the binomial coefficients is the number of ways in which n distinguishable objects can be divided into m piles, with n_i objects in the ith pile, $i = 1, 2, \ldots, m$ (the ordering of objects within each pile being unimportant). This may be straightforwardly calculated as follows. We may choose the n_1 objects in the first pile from the original n objects in $^nC_{n_1}$ ways. The n_2 objects in the second pile can then be chosen from the $n - n_1$ remaining objects in $^{n-n_1}C_{n_2}$ ways, etc. We may continue in this fashion until we reach the $(m - 1)$th pile, which may be formed in $^{n-n_1-\cdots-n_{m-2}}C_{n_{m-1}}$ ways. The remaining objects then form the mth pile and so can only be 'chosen' in one way. Thus the total number of ways of dividing the original n objects into m piles is given by the product

$$
\begin{aligned}
N &= {}^nC_{n_1} \, {}^{n-n_1}C_{n_2} \cdots {}^{n-n_1-\cdots-n_{m-2}}C_{n_{m-1}} \\
&= \frac{n!}{n_1!(n-n_1)!} \frac{(n-n_1)!}{n_2!(n-n_1-n_2)!} \cdots \frac{(n-n_1-n_2-\cdots-n_{m-2})!}{n_{m-1}!(n-n_1-n_2-\cdots-n_{m-2}-n_{m-1})!} \\
&= \frac{n!}{n_1!(n-n_1)!} \frac{(n-n_1)!}{n_2!(n-n_1-n_2)!} \cdots \frac{(n-n_1-n_2-\cdots-n_{m-2})!}{n_{m-1}!n_m!} \\
&= \frac{n!}{n_1!n_2!\cdots n_m!}.
\end{aligned}
\tag{30.35}
$$

These numbers are called *multinomial coefficients* since (30.35) is the coefficient of $x_1^{n_1} x_2^{n_2} \cdots x_m^{n_m}$ in the multinomial expansion of $(x_1 + x_2 + \cdots + x_m)^n$, i.e. for positive integer n

$$
(x_1 + x_2 + \cdots + x_m)^n = \sum_{\substack{n_1,n_2,\ldots,n_m \\ n_1+n_2+\cdots+n_m=n}} \frac{n!}{n_1!n_2!\cdots n_m!} x_1^{n_1} x_2^{n_2} \cdots x_m^{n_m}.
$$

For the case $m = 2$, $n_1 = k$, $n_2 = n - k$, (30.35) reduces to the binomial coefficient ^nC_k. Furthermore, we note that the multinomial coefficient (30.35) is identical to the expression (30.32) for the number of distinguishable permutations of n objects, n_i of which are identical and of type i (for $i = 1, 2, \ldots, m$ and $n_1 + n_2 + \cdots + n_m = n$). A few moments' thought should convince the reader that the two expressions (30.35) and (30.32) must be identical.

> ►*In the card game of bridge, each of four players is dealt 13 cards from a full pack of 52. What is the probability that each player is dealt an ace?*

From (30.35), the total number of distinct bridge dealings is $52!/(13!13!13!13!)$. However, the number of ways in which the four aces can be distributed with one in each hand is $4!/(1!1!1!1!) = 4!$; the remaining 48 cards can then be dealt out in $48!/(12!12!12!12!)$ ways. Thus the probability that each player receives an ace is

$$
4! \frac{48!}{(12!)^4} \frac{(13!)^4}{52!} = \frac{24(13)^4}{(49)(50)(51)(52)} = 0.105. \; ◄
$$

As in the case of permutations we might ask how many combinations of k objects can be chosen from n *with replacement* (repetition). To calculate this, we may imagine the n (distinguishable) objects set out on a table. Each combination of k objects can then be made by pointing to k of the n objects in turn (with repetitions allowed). These k equivalent selections distributed amongst n different but re-choosable objects are strictly analogous to the placing of k indistinguishable 'balls' in n different boxes with no restriction on the number of balls in each box. A particular selection in the case $k = 7$, $n = 5$ may be symbolised as

$$\text{xxx}|\ |\text{x}|\text{xx}|\text{x}.$$

This denotes three balls in the first box, none in the second, one in the third, two in the fourth and one in the fifth. We therefore need only consider the number of (distinguishable) ways in which k crosses and $n-1$ vertical lines can be arranged, i.e. the number of permutations of $k + n - 1$ objects of which k are identical crosses and $n - 1$ are identical lines. This is given by (30.33) as

$$\frac{(k + n - 1)!}{k!(n - 1)!} = {}^{n+k-1}C_k. \tag{30.36}$$

We note that this expression also occurs in the binomial expansion for negative integer powers. If n is a positive integer, it is straightforward to show that (see chapter 1)

$$(a + b)^{-n} = \sum_{k=0}^{\infty} (-1)^k \, {}^{n+k-1}C_k a^{-n-k} b^k,$$

where a is taken to be larger than b in magnitude.

▶ *A system contains a number N of (non-interacting) particles, each of which can be in any of the quantum states of the system. The structure of the set of quantum states is such that there exist R energy levels with corresponding energies E_i and degeneracies g_i (i.e. the ith energy level contains g_i quantum states). Find the numbers of distinct ways in which the particles can be distributed among the quantum states of the system such that the ith energy level contains n_i particles, for $i = 1, 2, \ldots, R$, in the cases where the particles are*

    (i) *distinguishable with no restriction on the number in each state;*
    (ii) *indistinguishable with no restriction on the number in each state;*
    (iii) *indistinguishable with a maximum of one particle in each state;*
    (iv) *distinguishable with a maximum of one particle in each state.*

It is easiest to solve this problem in two stages. Let us first consider distributing the N particles among the R energy levels, *without* regard for the individual degenerate quantum states that comprise each level. If the particles are *distinguishable* then the number of distinct arrangements with n_i particles in the ith level, $i = 1, 2, \ldots, R$, is given by (30.35) as

$$\frac{N!}{n_1! n_2! \cdots n_R!}.$$

If, however, the particles are *indistinguishable* then clearly there exists only one distinct arrangement having n_i particles in the ith level, $i = 1, 2, \ldots, R$. If we suppose that there exist w_i ways in which the n_i particles in the ith energy level can be distributed among the g_i degenerate states, then it follows that the number of distinct ways in which the N

particles can be distributed among all R quantum states of the system, with n_i particles in the ith level, is given by

$$
W\{n_i\} = \begin{cases} \dfrac{N!}{n_1!n_2!\cdots n_R!} \displaystyle\prod_{i=1}^{R} w_i & \text{for distinguishable particles,} \\[4mm] \displaystyle\prod_{i=1}^{R} w_i & \text{for indistinguishable particles.} \end{cases} \tag{30.37}
$$

It therefore remains only for us to find the appropriate expression for w_i in each of the cases (i)–(iv) above.

Case (i). If there is no restriction on the number of particles in each quantum state, then in the ith energy level each particle can reside in any of the g_i degenerate quantum states. Thus, if the particles are distinguishable then the number of distinct arrangements is simply $w_i = g_i^{n_i}$. Thus, from (30.37),

$$
W\{n_i\} = \frac{N!}{n_1!n_2!\cdots n_R!} \prod_{i=1}^{R} g_i^{n_i} = N! \prod_{i=1}^{R} \frac{g_i^{n_i}}{n_i!}.
$$

Such a system of particles (for example atoms or molecules in a classical gas) is said to obey Maxwell–Boltzmann statistics.

Case (ii). If the particles are indistinguishable and there is no restriction on the number in each state then, from (30.36), the number of distinct arrangements of the n_i particles among the g_i states in the ith energy level is

$$
w_i = \frac{(n_i + g_i - 1)!}{n_i!(g_i - 1)!}.
$$

Substituting this expression in (30.37), we obtain

$$
W\{n_i\} = \prod_{i=1}^{R} \frac{(n_i + g_i - 1)!}{n_i!(g_i - 1)!}.
$$

Such a system of particles (for example a gas of photons) is said to obey Bose–Einstein statistics.

Case (iii). If a maximum of one particle can reside in each of the g_i degenerate quantum states in the ith energy level then the number of particles in each state is either 0 or 1. Since the particles are indistinguishable, w_i is equal to the number of distinct arrangements in which n_i states are occupied and $g_i - n_i$ states are unoccupied; this is given by

$$
w_i = {}^{g_i}C_{n_i} = \frac{g_i!}{n_i!(g_i - n_i)!}.
$$

Thus, from (30.37), we have

$$
W\{n_i\} = \prod_{i=1}^{R} \frac{g_i!}{n_i!(g_i - n_i)!}.
$$

Such a system is said to obey Fermi–Dirac statistics, and an example is provided by an electron gas.

Case (iv). Again, the number of particles in each state is either 0 or 1. If the particles are distinguishable, however, each arrangement identified in case (iii) can be reordered in $n_i!$ different ways, so that

$$
w_i = {}^{g_i}P_{n_i} = \frac{g_i!}{(g_i - n_i)!}.
$$

Substituting this expression into (30.37) gives

$$W\{n_i\} = N! \prod_{i=1}^{R} \frac{g_i!}{n_i!(g_i - n_i)!}.$$

Such a system of particles has the names of no famous scientists attached to it, since it appears that it never occurs in nature. ◄

30.4 Random variables and distributions

Suppose an experiment has an outcome sample space S. A real variable X that is defined for all possible outcomes in S (so that a real number – not necessarily unique – is assigned to each possible outcome) is called a *random variable* (RV). The outcome of the experiment may already be a real number and hence a random variable, e.g. the number of heads obtained in 10 throws of a coin, or the sum of the values if two dice are thrown. However, more arbitrary assignments are possible, e.g. the assignment of a 'quality' rating to each successive item produced by a manufacturing process. Furthermore, assuming that a probability can be assigned to all possible outcomes in a sample space S, it is possible to assign a *probability distribution* to any random variable. Random variables may be divided into two classes, discrete and continuous, and we now examine each of these in turn.

30.4.1 Discrete random variables

A random variable X that takes only discrete values $x_1, x_2, \ldots, x_n$, with probabilities $p_1, p_2, \ldots, p_n$, is called a discrete random variable. The number of values n for which X has a non-zero probability is finite or at most countably infinite. As mentioned above, an example of a discrete random variable is the number of heads obtained in 10 throws of a coin. If X is a discrete random variable, we can define a *probability function* (PF) $f(x)$ that assigns probabilities to all the distinct values that X can take, such that

$$f(x) = \Pr(X = x) = \begin{cases} p_i & \text{if } x = x_i, \\ 0 & \text{otherwise.} \end{cases} \tag{30.38}$$

A typical PF (see figure 30.6) thus consists of spikes, at *valid values* of X, whose height at x corresponds to the probability that $X = x$. Since the probabilities must sum to unity, we require

$$\sum_{i=1}^{n} f(x_i) = 1. \tag{30.39}$$

We may also define the *cumulative probability function* (CPF) of X, $F(x)$, whose value gives the probability that $X \leq x$, so that

$$F(x) = \Pr(X \leq x) = \sum_{x_i \leq x} f(x_i). \tag{30.40}$$

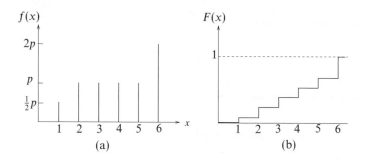

Figure 30.6 (a) A typical probability function for a discrete distribution, that for the biased die discussed earlier. Since the probabilities must sum to unity we require $p = 2/13$. (b) The cumulative probability function for the same discrete distribution. (Note that a different scale has been used for (b).)

Hence $F(x)$ is a step function that has upward jumps of p_i at $x = x_i$, $i = 1, 2, \ldots, n$, and is constant between possible values of X. We may also calculate the probability that X lies between two limits, l_1 and l_2 $(l_1 < l_2)$; this is given by

$$\Pr(l_1 < X \le l_2) = \sum_{l_1 < x_i \le l_2} f(x_i) = F(l_2) - F(l_1), \tag{30.41}$$

i.e. it is the sum of all the probabilities for which x_i lies within the relevant interval.

▶*A bag contains seven red balls and three white balls. Three balls are drawn at random and not replaced. Find the probability function for the number of red balls drawn.*

Let X be the number of red balls drawn. Then

$$\Pr(X = 0) = f(0) = \frac{3}{10} \times \frac{2}{9} \times \frac{1}{8} = \frac{1}{120},$$
$$\Pr(X = 1) = f(1) = \frac{3}{10} \times \frac{2}{9} \times \frac{7}{8} \times 3 = \frac{7}{40},$$
$$\Pr(X = 2) = f(2) = \frac{3}{10} \times \frac{7}{9} \times \frac{6}{8} \times 3 = \frac{21}{40},$$
$$\Pr(X = 3) = f(3) = \frac{7}{10} \times \frac{6}{9} \times \frac{5}{8} = \frac{7}{24}.$$

It should be noted that $\sum_{i=0}^{3} f(i) = 1$, as expected. ◀

30.4.2 Continuous random variables

A random variable X is said to have a *continuous* distribution if X is defined for a continuous range of values between given limits (often $-\infty$ to ∞). An example of a continuous random variable is the height of a person drawn from a population, which can take *any* value (within limits!). We can define the *probability density function* (PDF) $f(x)$ of a continuous random variable X such that

$$\Pr(x < X \le x + dx) = f(x)\,dx,$$

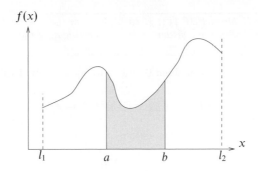

Figure 30.7   The probability density function for a continuous random variable X that can take values only between the limits l_1 and l_2. The shaded area under the curve gives $\Pr(a < X \leq b)$, whereas the total area under the curve, between the limits l_1 and l_2, is equal to unity.

i.e. $f(x)\,dx$ is the probability that X lies in the interval $x < X \leq x + dx$. Clearly $f(x)$ must be a real function that is everywhere ≥ 0. If X can take only values between the limits l_1 and l_2 then, in order for the sum of the probabilities of all possible outcomes to be equal to unity, we require

$$\int_{l_1}^{l_2} f(x)\,dx = 1.$$

Often X can take any value between $-\infty$ and ∞ and so

$$\int_{-\infty}^{\infty} f(x)\,dx = 1.$$

The probability that X lies in the interval $a < X \leq b$ is then given by

$$\Pr(a < X \leq b) = \int_{a}^{b} f(x)\,dx, \tag{30.42}$$

i.e. $\Pr(a < X \leq b)$ is equal to the area under the curve of $f(x)$ between these limits (see figure 30.7).

We may also define the cumulative probability function $F(x)$ for a continuous random variable by

$$F(x) = \Pr(X \leq x) = \int_{l_1}^{x} f(u)\,du, \tag{30.43}$$

where u is a (dummy) integration variable. We can then write

$$\Pr(a < X \leq b) = F(b) - F(a).$$

From (30.43) it is clear that $f(x) = dF(x)/dx$.

> ►*A random variable X has a PDF $f(x)$ given by Ae^{-x} in the interval $0 < x < \infty$ and zero elsewhere. Find the value of the constant A and hence calculate the probability that X lies in the interval $1 < X \leq 2$.*

We require the integral of $f(x)$ between 0 and ∞ to equal unity. Evaluating this integral, we find

$$\int_0^\infty Ae^{-x}\,dx = \left[-Ae^{-x}\right]_0^\infty = A,$$

and hence $A = 1$. From (30.42), we then obtain

$$\Pr(1 < X \leq 2) = \int_1^2 f(x)\,dx = \int_1^2 e^{-x}\,dx = -e^{-2} - (-e^{-1}) = 0.23. \quad ◄$$

It is worth mentioning here that a *discrete* RV can in fact be treated as continuous and assigned a corresponding probability density function. If X is a discrete RV that takes only the values $x_1, x_2, \ldots, x_n$ with probabilities $p_1, p_2, \ldots, p_n$ then we may describe X as a continuous RV with PDF

$$f(x) = \sum_{i=1}^n p_i \delta(x - x_i), \tag{30.44}$$

where $\delta(x)$ is the Dirac delta function discussed in subsection 13.1.3. From (30.42) and the fundamental property of the delta function (13.12), we see that

$$\Pr(a < X \leq b) = \int_a^b f(x)\,dx,$$

$$= \sum_{i=1}^n p_i \int_a^b \delta(x - x_i)\,dx = \sum_i p_i,$$

where the final sum extends over those values of i for which $a < x_i \leq b$.

30.4.3 Sets of random variables

It is common in practice to consider two or more random variables simultaneously. For example, one might be interested in both the height and weight of a person drawn at random from a population. In the general case, these variables may depend on one another and are described by *joint probability density functions*; these are discussed fully in section 30.11. We simply note here that if we have (say) two random variables X and Y then by analogy with the single-variable case we define their joint probability density function $f(x, y)$ in such a way that, if X and Y are discrete RVs,

$$\Pr(X = x_i, \ Y = y_j) = f(x_i, y_j),$$

or, if X and Y are continuous RVs,

$$\Pr(x < X \leq x + dx, \ y < Y \leq y + dy) = f(x, y)\,dx\,dy.$$

In many circumstances, however, random variables do not depend on one another, i.e. they are *independent*. As an example, for a person drawn at random from a population, we might expect height and IQ to be independent random variables. Let us suppose that X and Y are two random variables with probability density functions $g(x)$ and $h(y)$ respectively. In mathematical terms, X and Y are independent RVs if their joint probability density function is given by $f(x, y) = g(x)h(y)$. Thus, for independent RVs, if X and Y are both discrete then

$$\Pr(X = x_i, \ Y = y_j) = g(x_i)h(y_j)$$

or, if X and Y are both continuous, then

$$\Pr(x < X \le x + dx, \ y < Y \le y + dy) = g(x)h(y)\,dx\,dy.$$

The important point in each case is that the RHS is simply the product of the individual probability density functions (compare with the expression for $\Pr(A \cap B)$ in (30.22) for statistically independent events A and B). By a simple extension, one may also consider the case where one of the random variables is discrete and the other continuous. The above discussion may also be trivially extended to any number of independent RVs X_i, $i = 1, 2, \ldots, N$.

▶ *The independent random variables X and Y have the PDFs $g(x) = e^{-x}$ and $h(y) = 2e^{-2y}$ respectively. Calculate the probability that X lies in the interval $1 < X \le 2$ and Y lies in the interval $0 < Y \le 1$.*

Since X and Y are independent RVs, the required probability is given by

$$\Pr(1 < X \le 2, \ 0 < Y \le 1) = \int_1^2 g(x)\,dx \ \int_0^1 h(y)\,dy$$

$$= \int_1^2 e^{-x}\,dx \ \int_0^1 2e^{-2y}\,dy$$

$$= \left[-e^{-x}\right]_1^2 \times \left[-e^{-2y}\right]_0^1 - 0.23 \times 0.86 = 0.20. \ ◀$$

30.5 Properties of distributions

For a single random variable X, the probability density function $f(x)$ contains all possible information about how the variable is distributed. However, for the purposes of comparison, it is conventional and useful to characterise $f(x)$ by certain of its properties. Most of these standard properties are defined in terms of *averages* or *expectation values*. In the most general case, the expectation value $E[g(X)]$ of any function $g(X)$ of the random variable X is defined as

$$E[g(X)] = \begin{cases} \sum_i g(x_i)f(x_i) & \text{for a discrete distribution,} \\ \int g(x)f(x)\,dx & \text{for a continuous distribution,} \end{cases} \quad (30.45)$$

where the sum or integral is over all allowed values of X. It is assumed that the series is absolutely convergent or that the integral exists, as the case may be. From its definition it is straightforward to show that the expectation value has the following properties:

(i) if a is a constant then $E[a] = a$;
(ii) if a is a constant then $E[ag(X)] = aE[g(X)]$;
(iii) if $g(X) = s(X) + t(X)$ then $E[g(X)] = E[s(X)] + E[t(X)]$.

It should be noted that the expectation value is not a function of X but is instead a number that depends on the form of the probability density function $f(x)$ and the function $g(x)$. Most of the standard quantities used to characterise $f(x)$ are simply the expectation values of various functions of the random variable X. We now consider these standard quantities.

30.5.1 Mean

The property most commonly used to characterise a probability distribution is its *mean*, which is defined simply as the expectation value $E[X]$ of the variable X itself. Thus, the mean is given by

$$E[X] = \begin{cases} \sum_i x_i f(x_i) & \text{for a discrete distribution,} \\ \int xf(x)\,dx & \text{for a continuous distribution.} \end{cases} \tag{30.46}$$

The alternative notations μ and $\langle x \rangle$ are also commonly used to denote the mean. If in (30.46) the series is not absolutely convergent, or the integral does not exist, we say that the distribution does not have a mean, but this is very rare in physical applications.

> ▶ *The probability of finding a* 1s *electron in a hydrogen atom in a given infinitesimal volume* dV *is* $\psi^* \psi \, dV$, *where the quantum mechanical wavefunction* ψ *is given by*
>
> $$\psi = Ae^{-r/a_0}.$$
>
> *Find the value of the real constant* A *and thereby deduce the mean distance of the electron from the origin.*

Let us consider the random variable $R = $ 'distance of the electron from the origin'. Since the 1s orbital has no θ- or ϕ-dependence (it is spherically symmetric), we may consider the infinitesimal volume element dV as the spherical shell with inner radius r and outer radius $r + dr$. Thus, $dV = 4\pi r^2 \, dr$ and the PDF of R is simply

$$\Pr(r < R \leq r + dr) \equiv f(r)\,dr = 4\pi r^2 A^2 e^{-2r/a_0} \, dr.$$

The value of A is found by requiring the total probability (i.e. the probability that the electron is *somewhere*) to be unity. Since R must lie between zero and infinity, we require that

$$A^2 \int_0^\infty e^{-2r/a_0} 4\pi r^2 \, dr = 1.$$

Integrating by parts we find $A = 1/(\pi a_0^3)^{1/2}$. Now, using the definition of the mean (30.46), we find

$$E[R] = \int_0^\infty r f(r)\, dr = \frac{4}{a_0^3} \int_0^\infty r^3 e^{-2r/a_0}\, dr.$$

The integral on the RHS may be integrated by parts and takes the value $3a_0^4/8$; consequently we find that $E[R] = 3a_0/2$. ◀

30.5.2 Mode and median

Although the mean discussed in the last section is the most common measure of the 'average' of a distribution, two other measures, which do not rely on the concept of expectation values, are frequently encountered.

The *mode* of a distribution is the value of the random variable X at which the probability (density) function $f(x)$ has its greatest value. If there is more than one value of X for which this is true then each value may equally be called the mode of the distribution.

The *median* M of a distribution is the value of the random variable X at which the cumulative probability function $F(x)$ takes the value $\frac{1}{2}$, i.e. $F(M) = \frac{1}{2}$. Related to the median are the lower and upper quartiles Q_l and Q_u of the PDF, which are defined such that

$$F(Q_l) = \tfrac{1}{4}, \qquad F(Q_u) = \tfrac{3}{4}.$$

Thus the median and lower and upper quartiles divide the PDF into four regions each containing one quarter of the probability. Smaller subdivisions are also possible, e.g. the nth percentile, P_n, of a PDF is defined by $F(P_n) = n/100$.

▶ *Find the mode of the PDF for the distance from the origin of the electron whose wavefunction was given in the previous example.*

We found in the previous example that the PDF for the electron's distance from the origin was given by

$$f(r) = \frac{4r^2}{a_0^3} e^{-2r/a_0}. \tag{30.47}$$

Differentiating $f(r)$ with respect to r, we obtain

$$\frac{df}{dr} = \frac{8r}{a_0^3} \left(1 - \frac{r}{a_0}\right) e^{-2r/a_0}.$$

Thus $f(r)$ has turning points at $r = 0$ and $r = a_0$, where $df/dr = 0$. It is straightforward to show that $r = 0$ is a minimum and $r = a_0$ is a maximum. Moreover, it is also clear that $r = a_0$ is a global maximum (as opposed to just a local one). Thus the mode of $f(r)$ occurs at $r = a_0$. ◀

30.5.3 Variance and standard deviation

The *variance* of a distribution, $V[X]$, also written σ^2, is defined by

$$V[X] = E\left[(X - \mu)^2\right] = \begin{cases} \sum_j (x_j - \mu)^2 f(x_j) & \text{for a discrete distribution,} \\ \int (x - \mu)^2 f(x)\,dx & \text{for a continuous distribution.} \end{cases}$$

(30.48)

Here μ has been written for the expectation value $E[X]$ of X. As in the case of the mean, unless the series and the integral in (30.48) converge the distribution does not have a variance. From the definition (30.48) we may easily derive the following useful properties of $V[X]$. If a and b are constants then

(i) $V[a] = 0$,
(ii) $V[aX + b] = a^2 V[X]$.

The variance of a distribution is always positive; its positive square root is known as the *standard deviation* of the distribution and is often denoted by σ. Roughly speaking, σ measures the spread (about $x = \mu$) of the values that X can assume.

▶*Find the standard deviation of the PDF for the distance from the origin of the electron whose wavefunction was discussed in the previous two examples.*

Inserting the expression (30.47) for the PDF $f(r)$ into (30.48), the variance of the random variable R is given by

$$V[R] = \int_0^\infty (r - \mu)^2 \frac{4r^2}{a_0^3} e^{-2r/a_0}\,dr = \frac{4}{a_0^3} \int_0^\infty (r^4 - 2r^3\mu + r^2\mu^2) e^{-2r/a_0}\,dr,$$

where the mean $\mu = E[R] = 3a_0/2$. Integrating each term in the integrand by parts we obtain

$$V[R] = 3a_0^2 - 3\mu a_0 + \mu^2 = \frac{3a_0^2}{4}.$$

Thus the standard deviation of the distribution is $\sigma = \sqrt{3}a_0/2$. ◀

We may also use the definition (30.48) to derive the *Bienaymé–Chebyshev inequality*, which provides a useful upper limit on the probability that random variable X takes values outside a given range centred on the mean. Let us consider the case of a continuous random variable, for which

$$\Pr(|X - \mu| \geq c) = \int_{|x-\mu|\geq c} f(x)\,dx,$$

where the integral on the RHS extends over all values of x satisfying the inequality

$|x - \mu| \geq c$. From (30.48), we find that

$$\sigma^2 \geq \int_{|x-\mu| \geq c} (x - \mu)^2 f(x) \, dx \geq c^2 \int_{|x-\mu| \geq c} f(x) \, dx. \qquad (30.49)$$

The first inequality holds because both $(x - \mu)^2$ and $f(x)$ are non-negative for all x, and the second inequality holds because $(x - \mu)^2 \geq c^2$ over the range of integration. However, the RHS of (30.49) is simply equal to $c^2 \Pr(|X - \mu| \geq c)$, and thus we obtain the required inequality

$$\Pr(|X - \mu| \geq c) \leq \frac{\sigma^2}{c^2}.$$

A similar derivation may be carried through for the case of a discrete random variable. Thus, for *any* distribution $f(x)$ that possesses a variance we have, for example,

$$\Pr(|X - \mu| \geq 2\sigma) \leq \frac{1}{4} \quad \text{and} \quad \Pr(|X - \mu| \geq 3\sigma) \leq \frac{1}{9}.$$

30.5.4 Moments

The mean (or expectation) of X is sometimes called the *first moment* of X, since it is defined as the sum or integral of the probability density function multiplied by the first power of x. By a simple extension the kth moment of a distribution is defined by

$$\mu_k \equiv E[X^k] = \begin{cases} \sum_j x_j^k f(x_j) & \text{for a discrete distribution,} \\ \int x^k f(x) \, dx & \text{for a continuous distribution.} \end{cases} \qquad (30.50)$$

For notational convenience, we have introduced the symbol μ_k to denote $E[X^k]$, the kth moment of the distribution. Clearly, the mean of the distribution is then denoted by μ_1, often abbreviated simply to μ, as in the previous subsection, as this rarely causes confusion.

A useful result that relates the second moment, the mean and the variance of a distribution is proved using the properties of the expectation operator:

$$\begin{aligned} V[X] &= E\left[(X - \mu)^2\right] \\ &= E\left[X^2 - 2\mu X + \mu^2\right] \\ &= E\left[X^2\right] - 2\mu E[X] + \mu^2 \\ &= E\left[X^2\right] - 2\mu^2 + \mu^2 \\ &= E\left[X^2\right] - \mu^2. \end{aligned} \qquad (30.51)$$

In alternative notations, this result can be written

$$\langle (x - \mu)^2 \rangle = \langle x^2 \rangle - \langle x \rangle^2 \quad \text{or} \quad \sigma^2 = \mu_2 - \mu_1^2.$$

▶*A biased die has probabilities $p/2, p, p, p, p, 2p$ of showing 1, 2, 3, 4, 5, 6 respectively. Find (i) the mean, (ii) the second moment and (iii) the variance of this probability distribution.*

By demanding that the sum of the probabilities equals unity we require $p = 2/13$. Now, using the definition of the mean (30.46) for a discrete distribution,

$$E[X] = \sum_j x_j f(x_j) = 1 \times \tfrac{1}{2}p + 2 \times p + 3 \times p + 4 \times p + 5 \times p + 6 \times 2p$$

$$= \frac{53}{2}p = \frac{53}{2} \times \frac{2}{13} = \frac{53}{13}.$$

Similarly, using the definition of the second moment (30.50),

$$E[X^2] = \sum_j x_j^2 f(x_j) = 1^2 \times \tfrac{1}{2}p + 2^2 p + 3^2 p + 4^2 p + 5^2 p + 6^2 \times 2p$$

$$= \frac{253}{2}p = \frac{253}{13}.$$

Finally, using the definition of the variance (30.48), with $\mu = 53/13$, we obtain

$$V[X] = \sum_j (x_j - \mu)^2 f(x_j)$$

$$= (1 - \mu)^2 \tfrac{1}{2}p + (2 - \mu)^2 p + (3 - \mu)^2 p + (4 - \mu)^2 p + (5 - \mu)^2 p + (6 - \mu)^2 2p$$

$$= \left(\frac{3120}{169}\right) p = \frac{480}{169}.$$

It is easy to verify that $V[X] = E[X^2] - (E[X])^2$. ◀

In practice, to calculate the moments of a distribution it is often simpler to use the moment generating function discussed in subsection 30.7.2. This is particularly true for higher-order moments, where direct evaluation of the sum or integral in (30.50) can be somewhat laborious.

30.5.5 Central moments

The variance $V[X]$ is sometimes called the *second central moment* of the distribution, since it is defined as the sum or integral of the probability density function multiplied by the *second* power of $x - \mu$. The origin of the term 'central' is that by subtracting μ from x before squaring we are considering the moment about the mean of the distribution, rather than about $x = 0$. Thus the kth *central* moment of a distribution is defined as

$$\nu_k \equiv E\left[(X - \mu)^k\right] = \begin{cases} \sum_j (x_j - \mu)^k f(x_j) & \text{for a discrete distribution,} \\ \int (x - \mu)^k f(x)\, dx & \text{for a continuous distribution.} \end{cases} \tag{30.52}$$

It is convenient to introduce the notation ν_k for the kth central moment. Thus $V[X] \equiv \nu_2$ and we may write (30.51) as $\nu_2 = \mu_2 - \mu_1^2$. Clearly, the first central moment of a distribution is always zero since, for example in the continuous case,

$$\nu_1 = \int (x - \mu) f(x)\, dx = \int x f(x)\, dx - \mu \int f(x)\, dx = \mu - (\mu \times 1) = 0.$$

We note that the notation μ_k and ν_k for the moments and central moments respectively is not universal. Indeed, in some books their meanings are reversed.

We can write the kth central moment of a distribution in terms of its kth and lower-order moments by expanding $(X - \mu)^k$ in powers of X. We have already noted that $\nu_2 = \mu_2 - \mu_1^2$, and similar expressions may be obtained for higher-order central moments. For example,

$$
\begin{aligned}
\nu_3 &= E\left[(X - \mu_1)^3\right] \\
&= E\left[X^3 - 3\mu_1 X^2 + 3\mu_1^2 X - \mu_1^3\right] \\
&= \mu_3 - 3\mu_1\mu_2 + 3\mu_1^2\mu_1 - \mu_1^3 \\
&= \mu_3 - 3\mu_1\mu_2 + 2\mu_1^3.
\end{aligned}
\tag{30.53}
$$

In general, it is straightforward to show that

$$
\nu_k = \mu_k - {}^kC_1\mu_{k-1}\mu_1 + \cdots + (-1)^r\,{}^kC_r\mu_{k-r}\mu_1^r + \cdots + (-1)^{k-1}({}^kC_{k-1} - 1)\mu_1^k.
\tag{30.54}
$$

Once again, direct evaluation of the sum or integral in (30.52) can be rather tedious for higher moments, and it is usually quicker to use the moment generating function (see subsection 30.7.2), from which the central moments can be easily evaluated as well.

> ▶ *The PDF for a Gaussian distribution (see subsection 30.9.1) with mean μ and variance σ^2 is given by*
> $$ f(x) = \frac{1}{\sigma\sqrt{2\pi}} \exp\left[-\frac{(x - \mu)^2}{2\sigma^2}\right]. $$
> *Obtain an expression for the kth central moment of this distribution.*

As an illustration, we will perform this calculation by evaluating the integral in (30.52) directly. Thus, the kth central moment of $f(x)$ is given by

$$
\begin{aligned}
\nu_k &= \int_{-\infty}^{\infty} (x - \mu)^k f(x)\,dx \\
&= \frac{1}{\sigma\sqrt{2\pi}} \int_{-\infty}^{\infty} (x - \mu)^k \exp\left[-\frac{(x - \mu)^2}{2\sigma^2}\right] dx \\
&= \frac{1}{\sigma\sqrt{2\pi}} \int_{-\infty}^{\infty} y^k \exp\left(-\frac{y^2}{2\sigma^2}\right) dy,
\end{aligned}
\tag{30.55}
$$

where in the last line we have made the substitution $y = x - \mu$. It is clear that if k is odd then the integrand is an odd function of y and hence the integral equals zero. Thus, $\nu_k = 0$ if k is odd. When k is even, we could calculate ν_k by integrating by parts to obtain a reduction formula, but it is more elegant to consider instead the standard integral (see subsection 6.4.2)

$$
I = \int_{-\infty}^{\infty} \exp(-\alpha y^2)\,dy = \pi^{1/2}\alpha^{-1/2},
$$

and differentiate it repeatedly with respect to α (see section 5.12). Thus, we obtain

$$\frac{dI}{d\alpha} = -\int_{-\infty}^{\infty} y^2 \exp(-\alpha y^2)\, dy = -\tfrac{1}{2}\pi^{1/2}\alpha^{-3/2}$$

$$\frac{d^2 I}{d\alpha^2} = \int_{-\infty}^{\infty} y^4 \exp(-\alpha y^2)\, dy = (\tfrac{1}{2})(\tfrac{3}{2})\pi^{1/2}\alpha^{-5/2}$$

$$\vdots$$

$$\frac{d^n I}{d\alpha^n} = (-1)^n \int_{-\infty}^{\infty} y^{2n} \exp(-\alpha y^2)\, dy = (-1)^n (\tfrac{1}{2})(\tfrac{3}{2}) \cdots (\tfrac{1}{2}(2n-1))\pi^{1/2}\alpha^{-(2n+1)/2}.$$

Setting $\alpha = 1/(2\sigma^2)$ and substituting the above result into (30.55), we find (for k even)

$$\nu_k = (\tfrac{1}{2})(\tfrac{3}{2}) \cdots (\tfrac{1}{2}(k-1))(2\sigma^2)^{k/2} = (1)(3) \cdots (k-1)\sigma^k. \;\blacktriangleleft$$

One may also characterise a probability distribution $f(x)$ using the closely related *normalised* and dimensionless central moments

$$\gamma_k \equiv \frac{\nu_k}{\nu_2^{k/2}} = \frac{\nu_k}{\sigma^k}.$$

From this set, γ_3 and γ_4 are more commonly called, respectively, the *skewness* and *kurtosis* of the distribution. The skewness γ_3 of a distribution is zero if it is symmetrical about its mean. If the distribution is skewed to values of x smaller than the mean then $\gamma_3 < 0$. Similarly $\gamma_3 > 0$ if the distribution is skewed to higher values of x.

From the above example, we see that the kurtosis of the Gaussian distribution (subsection 30.9.1) is given by

$$\gamma_4 = \frac{\nu_4}{\nu_2^2} = \frac{3\sigma^4}{\sigma^4} = 3.$$

It is therefore common practice to define the *excess kurtosis* of a distribution as $\gamma_4 - 3$. A positive value of the excess kurtosis implies a relatively narrower peak and wider wings than the Gaussian distribution with the same mean and variance. A negative excess kurtosis implies a wider peak and shorter wings.

Finally, we note here that one can also describe a probability density function $f(x)$ in terms of its *cumulants*, which are again related to the central moments. However, we defer the discussion of cumulants until subsection 30.7.4, since their definition is most easily understood in terms of generating functions.

30.6 Functions of random variables

Suppose X is some random variable for which the probability density function $f(x)$ is known. In many cases, we are more interested in a related random variable $Y = Y(X)$, where $Y(X)$ is some function of X. What is the probability density function $g(y)$ for the new random variable Y? We now discuss how to obtain this function.

30.6.1 Discrete random variables

If X is a discrete RV that takes only the values x_i, $i = 1, 2, \ldots, n$, then Y must also be discrete and takes the values $y_i = Y(x_i)$, although some of these values may be identical. The probability function for Y is given by

$$g(y) = \begin{cases} \sum_j f(x_j) & \text{if } y = y_i, \\ 0 & \text{otherwise,} \end{cases} \tag{30.56}$$

where the sum extends over those values of j for which $y_i = Y(x_j)$. The simplest case arises when the function $Y(X)$ possesses a single-valued inverse $X(Y)$. In this case, only one x-value corresponds to each y-value, and we obtain a closed-form expression for $g(y)$ given by

$$g(y) = \begin{cases} f(x(y_i)) & \text{if } y = y_i, \\ 0 & \text{otherwise.} \end{cases}$$

If $Y(X)$ does not possess a single-valued inverse then the situation is more complicated and it may not be possible to obtain a closed-form expression for $g(y)$. Nevertheless, whatever the form of $Y(X)$, one can always use (30.56) to obtain the numerical values of the probability function $g(y)$ at $y = y_i$.

30.6.2 Continuous random variables

If X is a continuous RV, then so too is the new random variable $Y = Y(X)$. The probability that Y lies in the range y to $y + dy$ is given by

$$g(y)\,dy = \int_{dS} f(x)\,dx, \tag{30.57}$$

where dS corresponds to all values of x for which Y lies in the range y to $y + dy$. Once again the simplest case occurs when $Y(X)$ possesses a single-valued inverse $X(Y)$. In this case, we may write

$$g(y)\,dy = \left| \int_{x(y)}^{x(y+dy)} f(x')\,dx' \right| = \int_{x(y)}^{x(y) + \left| \frac{dx}{dy} \right| dy} f(x')\,dx',$$

from which we obtain

$$g(y) = f(x(y)) \left| \frac{dx}{dy} \right|. \tag{30.58}$$

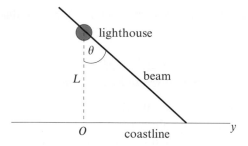

Figure 30.8   The illumination of a coastline by the beam from a lighthouse.

> ►*A lighthouse is situated at a distance L from a straight coastline, opposite a point O, and sends out a narrow continuous beam of light simultaneously in opposite directions. The beam rotates with constant angular velocity. If the random variable Y is the distance along the coastline, measured from O, of the spot that the light beam illuminates, find its probability density function.*

The situation is illustrated in figure 30.8. Since the light beam rotates at a constant angular velocity, θ is distributed uniformly between $-\pi/2$ and $\pi/2$, and so $f(\theta) = 1/\pi$. Now $y = L\tan\theta$, which possesses the single-valued inverse $\theta = \tan^{-1}(y/L)$, provided that θ lies between $-\pi/2$ and $\pi/2$. Since $dy/d\theta = L\sec^2\theta = L(1 + \tan^2\theta) = L[1 + (y/L)^2]$, from (30.58) we find

$$g(y) = \frac{1}{\pi}\left|\frac{d\theta}{dy}\right| = \frac{1}{\pi L[1 + (y/L)^2]} \qquad \text{for } -\infty < y < \infty.$$

A distribution of this form is called a *Cauchy distribution* and is discussed in subsection 30.9.5. ◄

If $Y(X)$ does not possess a single-valued inverse then we encounter complications, since there exist several intervals in the X-domain for which Y lies between y and $y + dy$. This is illustrated in figure 30.9, which shows a function $Y(X)$ such that $X(Y)$ is a double-valued function of Y. Thus the range y to $y + dy$ corresponds to X's being either in the range x_1 to $x_1 + dx_1$ or in the range x_2 to $x_2 + dx_2$. In general, it may not be possible to obtain an expression for $g(y)$ in closed form, although the distribution may always be obtained numerically using (30.57). However, a closed-form expression may be obtained in the case where there exist single-valued functions $x_1(y)$ and $x_2(y)$ giving the two values of x that correspond to any given value of y. In this case,

$$g(y)\,dy = \left|\int_{x_1(y)}^{x_1(y+dy)} f(x)\,dx\right| + \left|\int_{x_2(y)}^{x_2(y+dy)} f(x)\,dx\right|,$$

from which we obtain

$$g(y) = f(x_1(y))\left|\frac{dx_1}{dy}\right| + f(x_2(y))\left|\frac{dx_2}{dy}\right|. \tag{30.59}$$

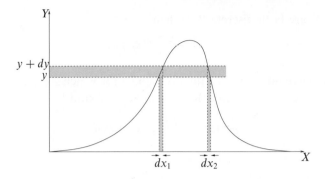

Figure 30.9   Illustration of a function $Y(X)$ whose inverse $X(Y)$ is a double-valued function of Y. The range y to $y + dy$ corresponds to X being either in the range x_1 to $x_1 + dx_1$ or in the range x_2 to $x_2 + dx_2$.

This result may be generalised straightforwardly to the case where the range y to $y + dy$ corresponds to more than two x-intervals.

▶The random variable X is Gaussian distributed (see subsection 30.9.1) with mean μ and variance σ^2. Find the PDF of the new variable $Y = (X - \mu)^2/\sigma^2$.

It is clear that $X(Y)$ is a double-valued function of Y. However, in this case, it is straightforward to obtain single-valued functions giving the two values of x that correspond to a given value of y; these are $x_1 = \mu - \sigma\sqrt{y}$ and $x_2 = \mu + \sigma\sqrt{y}$, where $\sqrt{y}$ is taken to mean the positive square root. The PDF of X is given by

$$f(x) = \frac{1}{\sigma\sqrt{2\pi}} \exp\left[-\frac{(x-\mu)^2}{2\sigma^2}\right].$$

Since $dx_1/dy = -\sigma/(2\sqrt{y})$ and $dx_2/dy = \sigma/(2\sqrt{y})$, from (30.59) we obtain

$$g(y) = \frac{1}{\sigma\sqrt{2\pi}} \exp(-\tfrac{1}{2}y)\left|\frac{\sigma}{2\sqrt{y}}\right| + \frac{1}{\sigma\sqrt{2\pi}} \exp(-\tfrac{1}{2}y)\left|\frac{\sigma}{2\sqrt{y}}\right|$$

$$= \frac{1}{2\sqrt{\pi}}(\tfrac{1}{2}y)^{-1/2} \exp(-\tfrac{1}{2}y).$$

As we shall see in subsection 30.9.3, this is the gamma distribution $\gamma(\tfrac{1}{2}, \tfrac{1}{2})$. ◀

30.6.3 Functions of several random variables

We may extend our discussion further, to the case in which the new random variable is a function of *several* other random variables. For definiteness, let us consider the random variable $Z = Z(X, Y)$, which is a function of two other RVs X and Y. Given that these variables are described by the joint probability density function $f(x, y)$, we wish to find the probability density function $p(z)$ of the variable Z.

If X and Y are both discrete RVs then

$$p(z) = \sum_{i,j} f(x_i, y_j), \tag{30.60}$$

where the sum extends over all values of i and j for which $Z(x_i, y_j) = z$. Similarly, if X and Y are both continuous RVs then $p(z)$ is found by requiring that

$$p(z)\,dz = \iint_{dS} f(x, y)\,dx\,dy, \tag{30.61}$$

where dS is the infinitesimal area in the xy-plane lying between the curves $Z(x, y) = z$ and $Z(x, y) = z + dz$.

▶*Suppose X and Y are independent continuous random variables in the range $-\infty$ to ∞, with PDFs $g(x)$ and $h(y)$ respectively. Obtain expressions for the PDFs of $Z = X + Y$ and $W = XY$.*

Since X and Y are independent RVs, their joint PDF is simply $f(x, y) = g(x)h(y)$. Thus, from (30.61), the PDF of the sum $Z = X + Y$ is given by

$$p(z)\,dz = \int_{-\infty}^{\infty} dx\, g(x) \int_{z-x}^{z+dz-x} dy\, h(y)$$

$$= \left(\int_{-\infty}^{\infty} g(x)h(z-x)\,dx \right) dz.$$

Thus $p(z)$ is the *convolution* of the PDFs of g and h (i.e. $p = g * h$, see subsection 13.1.7). In a similar way, the PDF of the product $W = XY$ is given by

$$q(w)\,dw = \int_{-\infty}^{\infty} dx\, g(x) \int_{w/|x|}^{(w+dw)/|x|} dy\, h(y)$$

$$= \left(\int_{-\infty}^{\infty} g(x)h(w/x)\,\frac{dx}{|x|} \right) dw \blacktriangleleft$$

The prescription (30.61) is readily generalised to functions of n random variables $Z = Z(X_1, X_2, \ldots, X_n)$, in which case the infinitesimal 'volume' element dS is the region in $x_1 x_2 \cdots x_n$-space between the (hyper)surfaces $Z(x_1, x_2, \ldots, x_n) = z$ and $Z(x_1, x_2, \ldots, x_n) = z + dz$. In practice, however, the integral is difficult to evaluate, since one is faced with the complicated geometrical problem of determining the limits of integration. Fortunately, an alternative (and powerful) technique exists for evaluating integrals of this kind. One eliminates the geometrical problem by integrating over *all* values of the variables x_i *without* restriction, while shifting the constraint on the variables to the integrand. This is readily achieved by multiplying the integrand by a function that equals unity in the infinitesimal region dS and zero elsewhere. From the discussion of the Dirac delta function in subsection 13.1.3, we see that $\delta(Z(x_1, x_2, \ldots, x_n) - z)\,dz$ satisfies these requirements, and so in the most general case we have

$$p(z) = \iint \cdots \int f(x_1, x_2, \ldots, x_n)\delta(Z(x_1, x_2, \ldots, x_n) - z)\,dx_1 dx_2 \ldots dx_n, \tag{30.62}$$

where the range of integration is over all possible values of the variables x_i. This integral is most readily evaluated by substituting in (30.62) the Fourier integral representation of the Dirac delta function discussed in subsection 13.1.4, namely

$$\delta(Z(x_1, x_2, \ldots, x_n) - z) = \frac{1}{2\pi} \int_{-\infty}^{\infty} e^{ik(Z(x_1, x_2, \ldots, x_n) - z)} \, dk. \qquad (30.63)$$

This is best illustrated by considering a specific example.

> ▶A general one-dimensional random walk consists of n independent steps, each of which can be of a different length and in either direction along the x-axis. If $g(x)$ is the PDF for the (positive or negative) displacement X along the x-axis achieved in a single step, obtain an expression for the PDF of the total displacement S after n steps.

The total displacement S is simply the algebraic sum of the displacements X_i achieved in each of the n steps, so that

$$S - X_1 + X_2 + \cdots + X_n.$$

Since the random variables X_i are independent and have the same PDF $g(x)$, their joint PDF is simply $g(x_1)g(x_2)\cdots g(x_n)$. Substituting this into (30.62), together with (30.63), we obtain

$$p(s) = \int_{-\infty}^{\infty} \int_{-\infty}^{\infty} \cdots \int_{-\infty}^{\infty} g(x_1)g(x_2)\cdots g(x_n) \frac{1}{2\pi} \int_{-\infty}^{\infty} e^{ik[(x_1 + x_2 + \cdots + x_n) - s]} \, dk \, dx_1 dx_2 \cdots dx_n$$

$$= \frac{1}{2\pi} \int_{-\infty}^{\infty} dk \, e^{-iks} \left(\int_{-\infty}^{\infty} g(x)e^{ikx} \, dx \right)^n. \qquad (30.64)$$

It is convenient to define the *characteristic function* $C(k)$ of the variable X as

$$C(k) = \int_{\infty}^{\infty} g(x)e^{ikx} \, dx,$$

which is simply related to the Fourier transform of $g(x)$. Then (30.64) may be written as

$$p(s) = \frac{1}{2\pi} \int_{-\infty}^{\infty} e^{-iks} [C(k)]^n \, dk.$$

Thus $p(s)$ can be found by evaluating two Fourier integrals. Characteristic functions will be discussed in more detail in subsection 30.7.3. ◀

30.6.4 Expectation values and variances

In some cases, one is interested only in the expectation value or the variance of the new variable Z rather than in its full probability density function. For definiteness, let us consider the random variable $Z = Z(X, Y)$, which is a function of two RVs X and Y with a known joint distribution $f(x, y)$; the results we will obtain are readily generalised to more (or fewer) variables.

It is clear that $E[Z]$ and $V[Z]$ can be obtained, in principle, by first using the methods discussed above to obtain $p(z)$ and then evaluating the appropriate sums or integrals. The intermediate step of calculating $p(z)$ is not necessary, however, since it is straightforward to obtain expressions for $E[Z]$ and $V[Z]$ in terms of the variables X and Y. For example, if X and Y are continuous RVs then the expectation value of Z is given by

$$E[Z] = \int zp(z)\,dz = \int\int Z(x,y)f(x,y)\,dx\,dy. \tag{30.65}$$

An analogous result exists for discrete random variables.

Integrals of the form (30.65) are often difficult to evaluate. Nevertheless, we may use (30.65) to derive an important general result concerning expectation values. If X and Y are *any* two random variables and a and b are arbitrary constants then by letting $Z = aX + bY$ we find

$$E[aX + bY] = aE[X] + bE[Y].$$

Furthermore, we may use this result to obtain an *approximate* expression for the expectation value $E[Z(X,Y)]$ of any arbitrary function of X and Y. Letting $\mu_X = E[X]$ and $\mu_Y = E[Y]$, and provided $Z(X,Y)$ can be reasonably approximated by the linear terms of its Taylor expansion about the point (μ_X, μ_Y), we have

$$Z(X,Y) \approx Z(\mu_X,\mu_Y) + \left(\frac{\partial Z}{\partial X}\right)(X-\mu_X) + \left(\frac{\partial Z}{\partial Y}\right)(Y-\mu_Y), \tag{30.66}$$

where the partial derivatives are evaluated at $X = \mu_X$ and $Y = \mu_Y$. Taking the expectation values of both sides, we find

$$E[Z(X,Y)] \approx Z(\mu_X,\mu_Y) + \left(\frac{\partial Z}{\partial X}\right)(E[X]-\mu_X) + \left(\frac{\partial Z}{\partial Y}\right)(E[Y]-\mu_Y) = Z(\mu_X,\mu_Y),$$

which gives the approximate result $E[Z(X,Y)] \approx Z(\mu_X,\mu_Y)$.

By analogy with (30.65), the variance of $Z = Z(X,Y)$ is given by

$$V[Z] = \int (z-\mu_Z)^2 p(z)\,dz = \int\int [Z(x,y)-\mu_Z]^2 f(x,y)\,dx\,dy, \tag{30.67}$$

where $\mu_Z = E[Z]$. We may use this expression to derive a second useful result. If X and Y are two *independent* random variables, so that $f(x,y) = g(x)h(y)$, and a, b and c are constants then by setting $Z = aX + bY + c$ in (30.67) we obtain

$$V[aX + bY + c] = a^2 V[X] + b^2 V[Y]. \tag{30.68}$$

From (30.68) we also obtain the important special case

$$V[X + Y] = V[X - Y] = V[X] + V[Y].$$

Provided X and Y are indeed independent random variables, we may obtain an approximate expression for $V[Z(X,Y)]$, for any arbitrary function $Z(X,Y)$, in a similar manner to that used in approximating $E[Z(X,Y)]$ above. Taking the variance of both sides of (30.66), and using (30.68), we find

$$V[Z(X, Y)] \approx \left(\frac{\partial Z}{\partial X}\right)^2 V[X] + \left(\frac{\partial Z}{\partial Y}\right)^2 V[Y], \qquad (30.69)$$

the partial derivatives being evaluated at $X = \mu_X$ and $Y = \mu_Y$.

30.7 Generating functions

As we saw in chapter 16, when dealing with particular sets of functions f_n, each member of the set being characterised by a different non-negative integer n, it is sometimes possible to summarise the whole set by a single function of a dummy variable (say t), called a generating function. The relationship between the generating function and the nth member f_n of the set is that if the generating function is expanded as a power series in t then f_n is the coefficient of t^n. For example, in the expansion of the generating function $G(z, t) = (1 - 2zt + t^2)^{-1/2}$, the coefficient of t^n is the nth Legendre polynomial $P_n(z)$, i.e.

$$G(z, t) = (1 - 2zt + t^2)^{-1/2} = \sum_{n=0}^{\infty} P_n(z) t^n.$$

We found that many useful properties of, and relationships between, the members of a set of functions could be established using the generating function and other functions obtained from it, e.g. its derivatives.

Similar ideas can be used in the area of probability theory, and two types of generating function can be usefully defined, one more generally applicable than the other. The more restricted of the two, applicable only to discrete integral distributions, is called a probability generating function; this is discussed in the next section. The second type, a moment generating function, can be used with both discrete and continuous distributions and is considered in subsection 30.7.2. From the moment generating function, we may also construct the closely re-lated characteristic and cumulant generating functions; these are discussed in subsections 30.7.3 and 30.7.4 respectively.

30.7.1 Probability generating functions

As already indicated, probability generating functions are restricted in applicability to integer distributions, of which the most common (the binomial, the Poisson and the geometric) are considered in this and later subsections. In such distributions a random variable may take only non-negative integer values. The actual possible values may be finite or infinite in number, but, for formal purposes, all integers, $0, 1, 2, \ldots$ are considered possible. If only a finite number of integer values can occur in any particular case then those that cannot occur are included but are assigned zero probability.

If, as previously, the probability that the random variable X takes the value x_n is $f(x_n)$, then

$$\sum_n f(x_n) = 1.$$

In the present case, however, only non-negative integer values of x_n are possible, and we can, without ambiguity, write the probability that X takes the value n as f_n, with

$$\sum_{n=0}^{\infty} f_n = 1. \tag{30.70}$$

We may now define the *probability generating function* $\Phi_X(t)$ by

$$\Phi_X(t) \equiv \sum_{n=0}^{\infty} f_n t^n. \tag{30.71}$$

It is immediately apparent that $\Phi_X(t) = E[t^X]$ and that, by virtue of (30.70), $\Phi_X(1) = 1$.

Probably the simplest example of a probability generating function (PGF) is provided by the random variable X defined by

$$X = \begin{cases} 1 & \text{if the outcome of a single trial is a 'success',} \\ 0 & \text{if the trial ends in 'failure'.} \end{cases}$$

If the probability of success is p and that of failure $q\ (= 1 - p)$ then

$$\Phi_X(t) = qt^0 + pt^1 + 0 + 0 + \cdots = q + pt. \tag{30.72}$$

This type of random variable is discussed much more fully in subsection 30.8.1. In a similar but slightly more complicated way, a Poisson-distributed integer variable with mean λ (see subsection 30.8.4) has a PGF

$$\Phi_X(t) = \sum_{n=0}^{\infty} \frac{e^{-\lambda}\lambda^n}{n!} t^n = e^{-\lambda} e^{\lambda t}. \tag{30.73}$$

We note that, as required, $\Phi_X(1) = 1$ in both cases.

Useful results will be obtained from this kind of approach only if the summation (30.71) can be carried out explicitly in particular cases and the functions derived from $\Phi_X(t)$ can be shown to be related to meaningful parameters. Two such relationships can be obtained by differentiating (30.71) with respect to t. Taking the first derivative we find

$$\frac{d\Phi_X(t)}{dt} = \sum_{n=0}^{\infty} n f_n t^{n-1} \quad \Rightarrow \quad \Phi'_X(1) = \sum_{n=0}^{\infty} n f_n = E[X], \tag{30.74}$$

and differentiating once more we obtain

$$\frac{d^2\Phi_X(t)}{dt^2} = \sum_{n=0}^{\infty} n(n-1)f_n t^{n-2} \quad \Rightarrow \quad \Phi_X''(1) = \sum_{n=0}^{\infty} n(n-1)f_n = E[X(X-1)].$$
(30.75)

Equation (30.74) shows that $\Phi_X'(1)$ gives the mean of X. Using both (30.75) and (30.51) allows us to write

$$\begin{aligned}
\Phi_X''(1) + \Phi_X'(1) - \left[\Phi_X'(1)\right]^2 &= E[X(X-1)] + E[X] - (E[X])^2 \\
&= E\left[X^2\right] - E[X] + E[X] - (E[X])^2 \\
&= E\left[X^2\right] - (E[X])^2 \\
&= V[X],
\end{aligned}$$
(30.76)

and so express the variance of X in terms of the derivatives of its probability generating function.

> ►*A random variable X is given by the number of trials needed to obtain a first success when the chance of success at each trial is constant and equal to p. Find the probability generating function for X and use it to determine the mean and variance of X.*

Clearly, at least one trial is needed, and so $f_0 = 0$. If n (≥ 1) trials are needed for the first success, the first $n-1$ trials must have resulted in failure. Thus

$$\Pr(X = n) = q^{n-1}p, \qquad n \geq 1,$$
(30.77)

where $q = 1 - p$ is the probability of failure in each individual trial.

The corresponding probability generating function is thus

$$\begin{aligned}
\Phi_X(t) &= \sum_{n=0}^{\infty} f_n t^n = \sum_{n=1}^{\infty} (q^{n-1}p)t^n \\
&= \frac{p}{q} \sum_{n=1}^{\infty} (qt)^n = \frac{p}{q} \times \frac{qt}{1-qt} = \frac{pt}{1-qt},
\end{aligned}$$
(30.78)

where we have used the result for the sum of a geometric series, given in chapter 4, to obtain a closed-form expression for $\Phi_X(t)$. Again, as must be the case, $\Phi_X(1) = 1$.

To find the mean and variance of X we need to evaluate $\Phi_X'(1)$ and $\Phi_X''(1)$. Differentiating (30.78) gives

$$\Phi_X'(t) = \frac{p}{(1-qt)^2} \quad \Rightarrow \quad \Phi_X'(1) = \frac{p}{p^2} = \frac{1}{p},$$

$$\Phi_X''(t) = \frac{2pq}{(1-qt)^3} \quad \Rightarrow \quad \Phi_X''(1) = \frac{2pq}{p^3} = \frac{2q}{p^2}.$$

Thus, using (30.74) and (30.76),

$$E[X] = \Phi_X'(1) = \frac{1}{p},$$

$$\begin{aligned}
V[X] &= \Phi_X''(1) + \Phi_X'(1) - [\Phi_X'(1)]^2 \\
&= \frac{2q}{p^2} + \frac{1}{p} - \frac{1}{p^2} = \frac{q}{p^2}.
\end{aligned}$$

A distribution with probabilities of the general form (30.77) is known as a *geometric distribution* and is discussed in subsection 30.8.2. This form of distribution is common in 'waiting time' problems (subsection 30.9.3). ◄

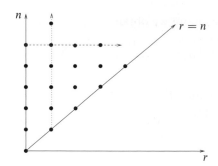

Figure 30.10   The pairs of values of n and r used in the evaluation of $\Phi_{X+Y}(t)$.

Sums of random variables

We now turn to considering the sum of two or more independent random variables, say X and Y, and denote by S_2 the random variable

$$S_2 = X + Y.$$

If $\Phi_{S_2}(t)$ is the PGF for S_2, the coefficient of t^n in its expansion is given by the probability that $X + Y = n$ and is thus equal to the sum of the probabilities that $X = r$ and $Y = n - r$ for all values of r in $0 \leq r \leq n$. Since such outcomes for different values of r are mutually exclusive, we have

$$\Pr(X + Y = n) = \sum_{r=0}^{n} \Pr(X = r)\Pr(Y = n - r). \tag{30.79}$$

Multiplying both sides of (30.79) by t^n and summing over all values of n enables us to express this relationship in terms of probability generating functions as follows:

$$\Phi_{X+Y}(t) = \sum_{n=0}^{\infty} \Pr(X + Y = n)t^n = \sum_{n=0}^{\infty}\sum_{r=0}^{n} \Pr(X = r)t^r \Pr(Y = n - r)t^{n-r}$$

$$= \sum_{r=0}^{\infty}\sum_{n=r}^{\infty} \Pr(X = r)t^r \Pr(Y = n - r)t^{n-r}.$$

The change in summation order is justified by reference to figure 30.10, which illustrates that the summations are over exactly the same pairs of values of n and r, but with the first (inner) summation over the points in a column rather than over the points in a row. Now, setting $n = r + s$ gives the final result,

$$\Phi_{X+Y}(t) = \sum_{r=0}^{\infty} \Pr(X = r)t^r \sum_{s=0}^{\infty} \Pr(Y = s)t^s$$

$$= \Phi_X(t)\Phi_Y(t), \tag{30.80}$$

i.e. the PGF of the sum of two independent random variables is equal to the product of their individual PGFs. The same result can be deduced in a less formal way by noting that if X and Y are independent then

$$E\left[t^{X+Y}\right] = E\left[t^X\right] E\left[t^Y\right].$$

Clearly result (30.80) can be extended to more than two random variables by writing $S_3 = S_2 + Z$ etc., to give

$$\Phi_{\left(\sum_{i=1}^n X_i\right)}(t) = \prod_{i=1}^n \Phi_{X_i}(t), \tag{30.81}$$

and, further, if all the X_i have the same probability distribution,

$$\Phi_{\left(\sum_{i=1}^n X_i\right)}(t) = [\Phi_X(t)]^n. \tag{30.82}$$

This latter result has immediate application in the deduction of the PGF for the binomial distribution from that for a single trial, equation (30.72).

Variable-length sums of random variables

As a final result in the theory of probability generating functions we show how to calculate the PGF for a sum of N random variables, all with the same probability distribution, when the value of N is itself a random variable but one with a known probability distribution. In symbols, we wish to find the distribution of

$$S_N = X_1 + X_2 + \cdots + X_N, \tag{30.83}$$

where N is a random variable with $\Pr(N = n) = h_n$ and PGF $\chi_N(t) = \sum h_n t^n$.

The probability ξ_k that $S_N = k$ is given by a sum of conditional probabilities, namely[§]

$$\xi_k = \sum_{n=0}^\infty \Pr(N = n) \Pr(X_0 + X_1 + X_2 + \cdots + X_n = k)$$

$$= \sum_{n=0}^\infty h_n \times \text{coefficient of } t^k \text{ in } [\Phi_X(t)]^n.$$

Multiplying both sides of this equation by t^k and summing over all k, we obtain

---

[§] Formally $X_0 = 0$ has to be included, since $\Pr(N = 0)$ may be non-zero.

an expression for the PGF $\Xi_S(t)$ of S_N:

$$\Xi_S(t) = \sum_{k=0}^{\infty} \xi_k t^k = \sum_{k=0}^{\infty} t^k \sum_{n=0}^{\infty} h_n \times \text{coefficient of } t^k \text{ in } [\Phi_X(t)]^n$$

$$= \sum_{n=0}^{\infty} h_n \sum_{k=0}^{\infty} t^k \times \text{coefficient of } t^k \text{ in } [\Phi_X(t)]^n$$

$$= \sum_{n=0}^{\infty} h_n [\Phi_X(t)]^n$$

$$= \chi_N(\Phi_X(t)). \qquad (30.84)$$

In words, the PGF of the sum S_N is given by the compound function $\chi_N(\Phi_X(t))$ obtained by substituting $\Phi_X(t)$ for t in the PGF for the number of terms N in the sum. We illustrate this with the following example.

> ▶ *The probability distribution for the number of eggs in a clutch is Poisson distributed with mean λ, and the probability that each egg will hatch is p (and is independent of the size of the clutch). Use the results stated in (30.72) and (30.73) to show that the PGF (and hence the probability distribution) for the number of chicks that hatch corresponds to a Poisson distribution having mean λp.*

The number of chicks that hatch is given by a sum of the form (30.83) in which $X_i = 1$ if the ith chick hatches and $X_i = 0$ if it does not. As given by (30.72), $\Phi_X(t)$ is thus $(1-p)+pt$. The value of N is given by a Poisson distribution with mean λ; thus, from (30.73), in the terminology of our previous discussion,

$$\chi_N(t) = e^{-\lambda} e^{\lambda t}.$$

We now substitute these forms into (30.84) to obtain

$$\Xi_S(t) = \exp(-\lambda) \exp[\lambda \Phi_X(t)]$$
$$= \exp(-\lambda) \exp\{\lambda[(1-p)+pt]\}$$
$$= \exp(-\lambda p) \exp(\lambda p t).$$

But this is exactly the PGF of a Poisson distribution with mean λp.

That this implies that the probability is Poisson distributed is intuitively obvious since, in the expansion of the PGF as a power series in t, every coefficient will be precisely that implied by such a distribution. A solution of the same problem by direct calculation appears in the answer to exercise 30.29. ◄

30.7.2 Moment generating functions

As we saw in section 30.5 a probability function is often expressed in terms of its moments. This leads naturally to the second type of generating function, a *moment generating function*. For a random variable X, and a real number t, the moment generating function (MGF) is defined by

$$M_X(t) = E\left[e^{tX}\right] = \begin{cases} \sum_i e^{tx_i} f(x_i) & \text{for a discrete distribution,} \\ \int e^{tx} f(x)\, dx & \text{for a continuous distribution.} \end{cases} \qquad (30.85)$$

The MGF will exist for all values of t provided that X is bounded and always exists at the point $t = 0$ where $M(0) = E(1) = 1$.

It will be apparent that the PGF and the MGF for a random variable X are closely related. The former is the expectation of t^X whilst the latter is the expectation of e^{tX}:

$$\Phi_X(t) = E\left[t^X\right], \qquad M_X(t) = E\left[e^{tX}\right].$$

The MGF can thus be obtained from the PGF by replacing t by e^t, and vice versa. The MGF has more general applicability, however, since it can be used with both continuous and discrete distributions whilst the PGF is restricted to non-negative integer distributions.

As its name suggests, the MGF is particularly useful for obtaining the moments of a distribution, as is easily seen by noting that

$$E\left[e^{tX}\right] = E\left[1 + tX + \frac{t^2 X^2}{2!} + \cdots\right]$$

$$= 1 + E[X]t + E\left[X^2\right]\frac{t^2}{2!} + \cdots.$$

Assuming that the MGF exists for all t around the point $t = 0$, we can deduce that the moments of a distribution are given in terms of its MGF by

$$E[X^n] = \left.\frac{d^n M_X(t)}{dt^n}\right|_{t=0}. \tag{30.86}$$

Similarly, by substitution in (30.51), the variance of the distribution is given by

$$V[X] = M_X''(0) - \left[M_X'(0)\right]^2, \tag{30.87}$$

where the prime denotes differentiation with respect to t.

▶ *The MGF for the Gaussian distribution (see the end of subsection 30.9.1) is given by*
$$M_X(t) = \exp\left(\mu t + \tfrac{1}{2}\sigma^2 t^2\right).$$
Find the expectation and variance of this distribution.

Using (30.86),

$$M_X'(t) = \left(\mu + \sigma^2 t\right)\exp\left(\mu t + \tfrac{1}{2}\sigma^2 t^2\right) \qquad \Rightarrow \qquad E[X] = M_X'(0) = \mu,$$

$$M_X''(t) = \left[\sigma^2 + (\mu + \sigma^2 t)^2\right]\exp\left(\mu t + \tfrac{1}{2}\sigma^2 t^2\right) \qquad \Rightarrow \qquad M_X''(0) = \sigma^2 + \mu^2.$$

Thus, using (30.87),

$$V[X] = \sigma^2 + \mu^2 - \mu^2 = \sigma^2.$$

That the mean is found to be μ and the variance σ^2 justifies the use of these symbols in the Gaussian distribution. ◀

The moment generating function has several useful properties that follow from its definition and can be employed in simplifying calculations.

Scaling and shifting

If $Y = aX + b$, where a and b are arbitrary constants, then

$$M_Y(t) = E\left[e^{tY}\right] = E\left[e^{t(aX+b)}\right] = e^{bt}E\left[e^{atX}\right] = e^{bt}M_X(at). \tag{30.88}$$

This result is often useful for obtaining the *central* moments of a distribution. If the MFG of X is $M_X(t)$ then the variable $Y = X - \mu$ has the MGF $M_Y(t) = e^{-\mu t}M_X(t)$, which clearly generates the central moments of X, i.e.

$$E[(X - \mu)^n] = E[Y^n] = M_Y^{(n)}(0) = \left(\frac{d^n}{dt^n}[e^{-\mu t}M_X(t)]\right)_{t=0}.$$

Sums of random variables

If $X_1, X_2, \ldots, X_N$ are independent random variables and $S_N = X_1 + X_2 + \cdots + X_N$ then

$$M_{S_N}(t) = E\left[e^{tS_N}\right] = E\left[e^{t(X_1+X_2+\cdots+X_N)}\right] = E\left[\prod_{i=1}^{N} e^{tX_i}\right].$$

Since the X_i are *independent*,

$$M_{S_N}(t) = \prod_{i=1}^{N} E\left[e^{tX_i}\right] = \prod_{i=1}^{N} M_{X_i}(t). \tag{30.89}$$

In words, the MGF of the sum of N independent random variables is the product of their individual MGFs. By combining (30.89) with (30.88), we obtain the more general result that the MGF of $S_N = c_1X_1 + c_2X_2 + \cdots + c_NX_N$ (where the c_i are constants) is given by

$$M_{S_N}(t) = \prod_{i=1}^{N} M_{X_i}(c_i t). \tag{30.90}$$

Variable-length sums of random variables

Let us consider the sum of N independent random variables X_i ($i = 1, 2, \ldots, N$), all with the same probability distribution, and let us suppose that N is itself a random variable with a known distribution. Following the notation of section 30.7.1,

$$S_N = X_1 + X_2 + \cdots + X_N,$$

where N is a random variable with $\Pr(N = n) = h_n$ and probability generating function $\chi_N(t) = \sum h_n t^n$. For definiteness, let us assume that the X_i are continuous RVs (an analogous discussion can be given in the discrete case). Thus, the probability that value of S_N lies in the interval s to $s + ds$ is given by[§]

$$\Pr(s < S_N \leq s + ds) = \sum_{n=0}^{\infty} \Pr(N = n) \Pr(s < X_0 + X_1 + X_2 \cdots + X_n \leq s + ds).$$

Write $\Pr(s < S_N \leq s + ds)$ as $f_N(s)\,ds$ and $\Pr(s < X_0 + X_1 + X_2 \cdots + X_n \leq s + ds)$ as $f_n(s)\,ds$. The kth moment of the PDF $f_N(s)$ is given by

$$\mu^k = \int s^k f_N(s)\,ds = \int s^k \sum_{n=0}^{\infty} \Pr(N = n) f_n(s)\,ds$$

$$= \sum_{n=0}^{\infty} \Pr(N = n) \int s^k f_n(s)\,ds$$

$$= \sum_{n=0}^{\infty} h_n \times (k! \times \text{coefficient of } t^k \text{ in } [M_X(t)]^n)$$

Thus the MGF of S_N is given by

$$M_{S_N}(t) = \sum_{k=0}^{\infty} \frac{\mu^k}{k!} t^k = \sum_{n=0}^{\infty} h_n \sum_{k=0}^{\infty} t^k \times \text{coefficient of } t^k \text{ in } [M_X(t)]^n$$

$$= \sum_{n=0}^{\infty} h_n [M_X(t)]^n$$

$$= \chi_N(M_X(t)).$$

In words, the MGF of the sum S_N is given by the compound function $\chi_N(M_X(t))$ obtained by substituting $M_X(t)$ for t in the PGF for the number of terms N in the sum.

Uniqueness

If the MGF of the random variable X_1 is identical to that for X_2 then the probability distributions of X_1 and X_2 are identical. This is intuitively reasonable although a rigorous proof is complicated,[¶] and beyond the scope of this book.

30.7.3 Characteristic function

The *characteristic function* (CF) of a random variable X is defined as

$$C_X(t) = E\left[e^{itX}\right] = \begin{cases} \sum_j e^{itx_j} f(x_j) & \text{for a discrete distribution,} \\ \int e^{itx} f(x)\,dx & \text{for a continuous distribution} \end{cases} \tag{30.91}$$

[§] As in the previous section, X_0 has to be formally included, since $\Pr(N = 0)$ may be non-zero.

[¶] See, for example, P. A. Moran, *An Introduction to Probability Theory* (New York: Oxford Science Publications, 1984).

so that $C_X(t) = M_X(it)$, where $M_X(t)$ is the MGF of X. Clearly, the characteristic function and the MGF are very closely related and can be used interchangeably. Because of the formal similarity between the definitions of $C_X(t)$ and $M_X(t)$, the characteristic function possesses analogous properties to those listed in the previous section for the MGF, with only minor modifications. Indeed, by substituting *it* for t in any of the relations obeyed by the MGF and noting that $C_X(t) = M_X(it)$, we obtain the corresponding relationship for the characteristic function. Thus, for example, the moments of X are given in terms of the derivatives of $C_X(t)$ by

$$E[X^n] = (-i)^n C_X^{(n)}(0).$$

Similarly, if $Y = aX + b$ then $C_Y(t) = e^{ibt} C_X(at)$.

Whether to describe a random variable by its characteristic function or by its MGF is partly a matter of personal preference. However, the use of the CF does have some advantages. Most importantly, the replacement of the exponential e^{tX} in the definition of the MGF by the complex oscillatory function e^{itX} in the CF means that in the latter we avoid any difficulties associated with convergence of the relevant sum or integral. Furthermore, when X is a continous RV, we see from (30.91) that $C_X(t)$ is related to the Fourier transform of the PDF $f(x)$. As a consequence of Fourier's inversion theorem, we may obtain $f(x)$ from $C_X(t)$ by performing the inverse transform

$$f(x) = \frac{1}{2\pi} \int_{-\infty}^{\infty} C_X(t) e^{-itx}\, dt.$$

30.7.4 Cumulant generating function

As mentioned at the end of subsection 30.5.5, we may also describe a probability density function $f(x)$ in terms of its *cumulants*. These quantities may be expressed in terms of the moments of the distribution and are important in sampling theory, which we discuss in the next chapter. The cumulants of a distribution are best defined in terms of its cumulant generating function (CGF), given by $K_X(t) = \ln M_X(t)$ where $M_X(t)$ is the MGF of the distribution. If $K_X(t)$ is expanded as a power series in t then the kth cumulant κ_k of $f(x)$ is the coefficient of $t^k/k!$:

$$K_X(t) = \ln M_X(t) \equiv \kappa_1 t + \kappa_2 \frac{t^2}{2!} + \kappa_3 \frac{t^3}{3!} + \cdots . \tag{30.92}$$

Since $M_X(0) = 1$, $K_X(t)$ contains no constant term.

▶*Find all the cumulants of the Gaussian distribution discussed in the previous example.*

The moment generating function for the Gaussian distribution is $M_X(t) = \exp\left(\mu t + \frac{1}{2}\sigma^2 t^2\right)$. Thus, the cumulant generating function has the simple form

$$K_X(t) = \ln M_X(t) = \mu t + \frac{1}{2}\sigma^2 t^2.$$

Comparing this expression with (30.92), we find that $\kappa_1 = \mu$, $\kappa_2 = \sigma^2$ and all other cumulants are equal to zero. ◄

We may obtain expressions for the cumulants of a distribution in terms of its moments by differentiating (30.92) with respect to t to give

$$\frac{dK_X}{dt} = \frac{1}{M_X}\frac{dM_X}{dt}.$$

Expanding each term as power series in t and cross-multiplying, we obtain

$$\left(\kappa_1 + \kappa_2 t + \kappa_3\frac{t^2}{2!} + \cdots\right)\left(1 + \mu_1 t + \mu_2\frac{t^2}{2!} + \cdots\right) = \left(\mu_1 + \mu_2 t + \mu_3\frac{t^2}{2!} + \cdots\right),$$

and, on equating coefficients of like powers of t on each side, we find

$$\mu_1 = \kappa_1,$$
$$\mu_2 = \kappa_2 + \kappa_1\mu_1,$$
$$\mu_3 = \kappa_3 + 2\kappa_2\mu_1 + \kappa_1\mu_2,$$
$$\mu_4 = \kappa_4 + 3\kappa_3\mu_1 + 3\kappa_2\mu_2 + \kappa_1\mu_3,$$
$$\vdots$$
$$\mu_k = \kappa_k + {}^{k-1}C_1\kappa_{k-1}\mu_1 + \cdots + {}^{k-1}C_r\kappa_{k-r}\mu_r + \cdots + \kappa_1\mu_{k-1}.$$

Solving these equations for the κ_k, we obtain (for the first four cumulants)

$$\kappa_1 = \mu_1,$$
$$\kappa_2 = \mu_2 - \mu_1^2 = v_2,$$
$$\kappa_3 = \mu_3 - 3\mu_2\mu_1 + 2\mu_1^3 = v_3,$$
$$\kappa_4 = \mu_4 - 4\mu_3\mu_1 + 12\mu_2\mu_1^2 - 3\mu_2^2 - 6\mu_1^4 = v_4 - 3v_2^2. \tag{30.93}$$

Higher-order cumulants may be calculated in the same way but become increasingly lengthy to write out in full.

The principal property of cumulants is their additivity, which may be proved by combining (30.92) with (30.90). If $X_1, X_2, \ldots, X_N$ are independent random variables and $K_{X_i}(t)$ for $i = 1, 2, \ldots, N$ is the CGF for X_i then the CGF of $S_N = c_1 X_1 + c_2 X_2 + \cdots + c_N X_N$ (where the c_i are constants) is given by

$$K_{S_N}(t) = \sum_{i=1}^{N} K_{X_i}(c_i t).$$

Cumulants also have the useful property that, under a change of origin $X \to X + a$ the first cumulant undergoes the change $\kappa_1 \to \kappa_1 + a$ but all higher-order cumulants remain unchanged. Under a change of scale $X \to bX$, cumulant κ_r undergoes the change $\kappa_r \to b^r\kappa_r$.

Distribution	Probability law $f(x)$	MGF	$E[X]$	$V[X]$
binomial	$^nC_x p^x q^{n-x}$	$(pe^t + q)^n$	np	npq
negative binomial	$^{r+x-1}C_x p^r q^x$	$\left(\dfrac{p}{1-qe^t}\right)^r$	$\dfrac{rq}{p}$	$\dfrac{rq}{p^2}$
geometric	$q^{x-1}p$	$\dfrac{pe^t}{1-qe^t}$	$\dfrac{1}{p}$	$\dfrac{q}{p^2}$
hypergeometric	$\dfrac{(Np)!(Nq)!n!(N-n)!}{x!(Np-x)!(n-x)!(Nq-n+x)!N!}$		np	$\dfrac{N-n}{N-1}npq$
Poisson	$\dfrac{\lambda^x}{x!}e^{-\lambda}$	$e^{\lambda(e^t-1)}$	λ	λ

Table 30.1 Some important discrete probability distributions.

30.8 Important discrete distributions

Having discussed some general properties of distributions, we now consider the more important discrete distributions encountered in physical applications. These are discussed in detail below, and summarised for convenience in table 30.1; we refer the reader to the relevant section below for an explanation of the symbols used.

30.8.1 The binomial distribution

Perhaps the most important discrete probability distribution is the *binomial distribution*. This distribution describes processes that consist of a number of independent identical *trials* with two possible outcomes, A and $B = \bar{A}$. We may call these outcomes 'success' and 'failure' respectively. If the probability of a success is $\Pr(A) = p$ then the probability of a failure is $\Pr(B) = q = 1 - p$. If we perform n trials then the discrete random variable

$$X = \text{number of times } A \text{ occurs}$$

can take the values $0, 1, 2, \ldots, n$; its distribution amongst these values is described by the *binomial distribution*.

We now calculate the probability that in n trials we obtain x successes (and so $n-x$ failures). One way of obtaining such a result is to have x successes followed by $n-x$ failures. Since the trials are assumed independent, the probability of this is

$$\underbrace{pp\cdots p}_{x \text{ times}} \times \underbrace{qq\cdots q}_{n-x \text{ times}} = p^x q^{n-x}.$$

This is, however, just one permutation of x successes and $n-x$ failures. The total

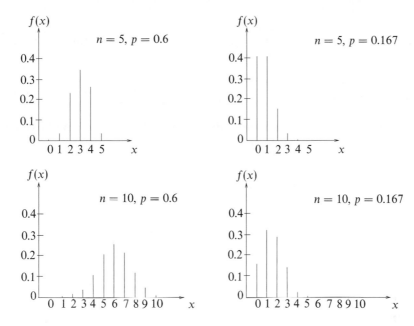

Figure 30.11   Some typical binomial distributions with various combinations of parameters n and p.

number of permutations of n objects, of which x are identical and of type 1 and $n - x$ are identical and of type 2, is given by (30.33) as

$$\frac{n!}{x!(n-x)!} \equiv {}^nC_x.$$

Therefore, the total probability of obtaining x successes from n trials is

$$f(x) = \Pr(X = x) = {}^nC_x\, p^x q^{n-x} = {}^nC_x\, p^x (1-p)^{n-x}, \tag{30.94}$$

which is the *binomial probability distribution formula*. When a random variable X follows the binomial distribution for n trials, with a probability of success p, we write $X \sim \mathrm{Bin}(n, p)$. Then the random variable X is often referred to as a binomial *variate*. Some typical binomial distributions are shown in figure 30.11.

> ▶*If a single six-sided die is rolled five times, what is the probability that a six is thrown exactly three times?*

Here the number of 'trials' $n = 5$, and we are interested in the random variable

$$X = \text{number of sixes thrown.}$$

Since the probability of a 'success' is $p = \frac{1}{6}$, the probability of obtaining exactly three sixes in five throws is given by (30.94) as

$$\Pr(X = 3) = \frac{5!}{3!(5-3)!} \left(\frac{1}{6}\right)^3 \left(\frac{5}{6}\right)^{(5-3)} = 0.032. \ \blacktriangleleft$$

For evaluating binomial probabilities a useful result is the binomial recurrence formula

$$\Pr(X = x + 1) = \frac{p}{q}\left(\frac{n - x}{x + 1}\right)\Pr(X = x),\tag{30.95}$$

which enables successive probabilities $\Pr(X = x + k)$, $k = 1, 2, \ldots$, to be calculated once $\Pr(X = x)$ is known; it is often quicker to use than (30.94).

▶ *The random variable X is distributed as $X \sim \text{Bin}(3, \frac{1}{2})$. Evaluate the probability function $f(x)$ using the binomial recurrence formula.*

The probability $\Pr(X = 0)$ may be calculated using (30.94) and is

$$\Pr(X = 0) = {}^{3}C_0 \left(\tfrac{1}{2}\right)^{0} \left(\tfrac{1}{2}\right)^{3} = \tfrac{1}{8}.$$

The ratio $p/q = \tfrac{1}{2}/\tfrac{1}{2} = 1$ in this case and so, using the binomial recurrence formula (30.95), we find

$$\Pr(X = 1) = 1 \times \frac{3 - 0}{0 + 1} \times \frac{1}{8} = \frac{3}{8},$$

$$\Pr(X = 2) = 1 \times \frac{3 - 1}{1 + 1} \times \frac{3}{8} = \frac{3}{8},$$

$$\Pr(X = 3) = 1 \times \frac{3 - 2}{2 + 1} \times \frac{3}{8} = \frac{1}{8},$$

results which may be verified by direct application of (30.94). ◀

We note that, as required, the binomial distribution satifies

$$\sum_{x=0}^{n} f(x) = \sum_{x=0}^{n} {}^{n}C_x\, p^x q^{n-x} = (p + q)^n = 1.$$

Furthermore, from the definitions of $E[X]$ and $V[X]$ for a discrete distribution, we may show that for the binomial distribution $E[X] = np$ and $V[X] = npq$. The direct summations involved are, however, rather cumbersome and these results are obtained much more simply using the moment generating function.

The moment generating function for the binomial distribution

To find the MGF for the binomial distribution we consider the binomial random variable X to be the sum of the random variables X_i, $i = 1, 2, \ldots, n$, which are defined by

$$X_i = \begin{cases} 1 & \text{if a 'success' occurs on the ith trial,} \\ 0 & \text{if a 'failure' occurs on the ith trial.} \end{cases}$$

Thus

$$M_i(t) = E\left[e^{tX_i}\right] = e^{0t} \times \Pr(X_i = 0) + e^{1t} \times \Pr(X_i = 1)$$
$$= 1 \times q + e^t \times p$$
$$= pe^t + q.$$

From (30.89), it follows that the MGF for the binomial distribution is given by

$$M(t) = \prod_{i=1}^{n} M_i(t) = (pe^t + q)^n. \tag{30.96}$$

We can now use the moment generating function to derive the mean and variance of the binomial distribution. From (30.96)

$$M'(t) = npe^t(pe^t + q)^{n-1},$$

and from (30.86)

$$E[X] = M'(0) = np(p + q)^{n-1} = np,$$

where the last equality follows from $p + q = 1$.

Differentiating with respect to t once more gives

$$M''(t) = e^t(n-1)np^2(pe^t + q)^{n-2} + e^t np(pe^t + q)^{n-1},$$

and from (30.86)

$$E[X^2] = M''(0) = n^2 p^2 - np^2 + np.$$

Thus, using (30.87)

$$V[X] = M''(0) - \left[M'(0)\right]^2 = n^2 p^2 - np^2 + np - n^2 p^2 = np(1-p) = npq.$$

Multiple binomial distributions

Suppose X and Y are two *independent* random variables, both of which are described by binomial distributions with a common probability of success p, but with (in general) different numbers of trials n_1 and n_2, so that $X \sim \text{Bin}(n_1, p)$ and $Y \sim \text{Bin}(n_2, p)$. Now consider the random variable $Z = X + Y$. We could calculate the probability distribution of Z directly using (30.60), but it is much easier to use the MGF (30.96).

Since X and Y are independent random variables, the MGF $M_Z(t)$ of the new variable $Z = X + Y$ is given simply by the product of the individual MGFs $M_X(t)$ and $M_Y(t)$. Thus, we obtain

$$M_Z(t) = M_X(t)M_Y(t) = (pe^t + q)^{n_1}(pe^t + q)^{n_2} = (pe^t + q)^{n_1+n_2},$$

which we recognise as the MGF of $Z \sim \text{Bin}(n_1 + n_2, p)$. Hence Z is also described by a binomial distribution.

This result may be extended to any number of binomial distributions. If X_i,

$i = 1, 2, \ldots, N$, is distributed as $X_i \sim \text{Bin}(n_i, p)$ then $Z = X_1 + X_2 + \cdots + X_N$ is distributed as $Z \sim \text{Bin}(n_1 + n_2 + \cdots + n_N, p)$, as would be expected since the result of $\sum_i n_i$ trials cannot depend on how they are split up. A similar proof is also possible using either the probability or cumulant generating functions.

Unfortunately, no equivalent simple result exists for the probability distribution of the *difference* $Z = X - Y$ of two binomially distributed variables.

30.8.2 *The geometric and negative binomial distributions*

A special case of the binomial distribution occurs when instead of the number of successes we consider the discrete random variable

$$X = \text{number of trials required to obtain the first success.}$$

The probability that x trials are required in order to obtain the first success, is simply the probability of obtaining $x - 1$ failures followed by one success. If the probability of a success on each trial is p, then for $x > 0$

$$f(x) = \Pr(X = x) = (1 - p)^{x-1} p = q^{x-1} p,$$

where $q = 1 - p$. This distribution is sometimes called the *geometric distribution*. The probability generating function for this distribution is given in (30.78). By replacing t by e^t in (30.78) we immediately obtain the MGF of the geometric distribution

$$M(t) = \frac{pe^t}{1 - qe^t},$$

from which its mean and variance are found to be

$$E[X] = \frac{1}{p}, \qquad V[X] = \frac{q}{p^2}.$$

Another distribution closely related to the binomial is the negative binomial distribution. This describes the probability distribution of the random variable

$$X = \text{number of failures before the } r\text{th success.}$$

One way of obtaining x failures before the rth success is to have $r - 1$ successes followed by x failures followed by the rth success, for which the probability is

$$\underbrace{pp \cdots p}_{r-1 \text{ times}} \times \underbrace{qq \cdots q}_{x \text{ times}} \times p = p^r q^x.$$

However, the first $r + x - 1$ factors constitute just one permutation of $r - 1$ successes and x failures. The total number of permutations of these $r + x - 1$ objects, of which $r - 1$ are identical and of type 1 and x are identical and of type

2, is $^{r+x-1}C_x$. Therefore, the total probability of obtaining x failures before the rth success is

$$f(x) = \Pr(X = x) = {}^{r+x-1}C_x p^r q^x,$$

which is called the *negative binomial distribution* (see the related discussion on p. 1137). It is straightforward to show that the MGF of this distribution is

$$M(t) = \left(\frac{p}{1 - qe^t}\right)^r,$$

and that its mean and variance are given by

$$E[X] = \frac{rq}{p} \quad \text{and} \quad V[X] = \frac{rq}{p^2}.$$

30.8.3 The hypergeometric distribution

In subsection 30.8.1 we saw that the probability of obtaining x successes in n *independent* trials was given by the binomial distribution. Suppose that these n 'trials' actually consist of drawing at random n balls, from a set of N such balls of which M are red and the rest white. Let us consider the random variable $X = $ number of red balls drawn.

On the one hand, if the balls are drawn *with replacement* then the trials are independent and the probability of drawing a red ball is $p = M/N$ each time. Therefore, the probability of drawing x red balls in n trials is given by the binomial distribution as

$$\Pr(X = x) = \frac{n!}{x!(n-x)!}p^x(1-p)^{n-x}.$$

On the other hand, if the balls are drawn *without replacement* the trials are not independent and the probability of drawing a red ball depends on how many red balls have already been drawn. We can, however, still derive a general formula for the probability of drawing x red balls in n trials, as follows.

The number of ways of drawing x red balls from M is ^MC_x, and the number of ways of drawing $n - x$ white balls from $N - M$ is $^{N-M}C_{n-x}$. Therefore, the total number of ways to obtain x red balls in n trials is $^MC_x \, ^{N-M}C_{n-x}$. However, the total number of ways of drawing n objects from N is simply ^NC_n. Hence the probability of obtaining x red balls in n trials is

$$\Pr(X = x) = \frac{^MC_x \, ^{N-M}C_{n-x}}{^NC_n}$$

$$= \frac{M!}{x!(M-x)!}\frac{(N-M)!}{(n-x)!(N-M-n+x)!}\frac{n!(N-n)!}{N!}, \quad (30.97)$$

$$= \frac{(Np)!(Nq)!\,n!(N-n)!}{x!(Np-x)!(n-x)!(Nq-n+x)!\,N!}, \quad (30.98)$$

where in the last line $p = M/N$ and $q = 1 - p$. This is called the *hypergeometric distribution*.

By performing the relevant summations directly, it may be shown that the hypergeometric distribution has mean

$$E[X] = n\frac{M}{N} = np$$

and variance

$$V[X] = \frac{nM(N-M)(N-n)}{N^2(N-1)} = \frac{N-n}{N-1}npq.$$

> ▶*In the UK National Lottery each participant chooses six different numbers between 1 and 49. In each weekly draw six numbered winning balls are subsequently drawn. Find the probabilities that a participant chooses $0, 1, 2, 3, 4, 5, 6$ winning numbers correctly.*

The probabilities are given by a hypergeometric distribution with N (the total number of balls) $= 49$, M (the number of winning balls drawn) $= 6$, and n (the number of numbers chosen by each participant) $= 6$. Thus, substituting in (30.97), we find

$$\Pr(0) = \frac{^{6}C_0 \,^{43}C_6}{^{49}C_6} = \frac{1}{2.29}, \quad \Pr(1) = \frac{^{6}C_1 \,^{43}C_5}{^{49}C_6} = \frac{1}{2.42},$$

$$\Pr(2) = \frac{^{6}C_2 \,^{43}C_4}{^{49}C_6} = \frac{1}{7.55}, \quad \Pr(3) = \frac{^{6}C_3 \,^{43}C_3}{^{49}C_6} = \frac{1}{56.6},$$

$$\Pr(4) = \frac{^{6}C_4 \,^{43}C_2}{^{49}C_6} = \frac{1}{1032}, \quad \Pr(5) = \frac{^{6}C_5 \,^{43}C_1}{^{49}C_6} = \frac{1}{54\,200},$$

$$\Pr(6) = \frac{^{6}C_6 \,^{43}C_0}{^{49}C_6} = \frac{1}{13.98 \times 10^6}.$$

It can easily be seen that

$$\sum_{i=0}^{6} \Pr(i) = 0.44 + 0.41 + 0.13 + 0.02 + \mathrm{O}(10^{-3}) = 1,$$

as expected. ◀

Note that if the number of trials (balls drawn) is small compared with N, M and $N - M$ then not replacing the balls is of little consequence, and we may approximate the hypergeometric distribution by the binomial distribution (with $p = M/N$); this is much easier to evaluate.

30.8.4 The Poisson distribution

We have seen that the binomial distribution describes the number of successful outcomes in a certain number of trials n. The Poisson distribution also describes the probability of obtaining a given number of successes but for situations in which the number of 'trials' cannot be enumerated; rather it describes the situation in which discrete events occur in a continuum. Typical examples of discrete random variables X described by a Poisson distribution are the number of telephone calls received by a switchboard in a given interval, or the number of stars above a certain brightness in a particular area of the sky. Given a mean rate of occurrence λ of these events in the relevant interval or area, the Poisson distribution gives the probability $\Pr(X = x)$ that exactly x events will occur.

We may derive the form of the Poisson distribution as the limit of the binomial distribution when the number of trials $n \to \infty$ and the probability of 'success' $p \to 0$, in such a way that $np = \lambda$ remains finite. Thus, in our example of a telephone switchboard, suppose we wish to find the probability that exactly x calls are received during some time interval, given that the mean number of calls in such an interval is λ. Let us begin by dividing the time interval into a large number, n, of equal shorter intervals, in each of which the probability of receiving a call is p. As we let $n \to \infty$ then $p \to 0$, but since we require the mean number of calls in the interval to equal λ, we must have $np = \lambda$. The probability of x successes in n trials is given by the binomial formula as

$$\Pr(X = x) = \frac{n!}{x!(n - x)!} p^x (1 - p)^{n-x}. \tag{30.99}$$

Now as $n \to \infty$, with x finite, the ratio of the n-dependent factorials in (30.99) behaves asymptotically as a power of n, i.e.

$$\lim_{n \to \infty} \frac{n!}{(n - x)!} = \lim_{n \to \infty} n(n - 1)(n - 2) \cdots (n - x + 1) \sim n^x.$$

Also

$$\lim_{n \to \infty} \lim_{p \to 0} (1 - p)^{n-x} = \lim_{p \to 0} \frac{(1 - p)^{\lambda/p}}{(1 - p)^x} = \frac{e^{-\lambda}}{1}.$$

Thus, using $\lambda = np$, (30.99) tends to the *Poisson distribution*

$$f(x) = \Pr(X = x) = \frac{e^{-\lambda} \lambda^x}{x!}, \tag{30.100}$$

which gives the probability of obtaining exactly x calls in the given time interval. As we shall show below, λ is the mean of the distribution. Events following a Poisson distribution are usually said to occur randomly in time.

Alternatively we may derive the Poisson distribution directly, without considering a limit of the binomial distribution. Let us again consider our example of a telephone switchboard. Suppose that the probability that x calls have been received in a time interval t is $P_x(t)$. If the average number of calls received in a unit time is λ then in a further small time interval Δt the probability of receiving a call is $\lambda \Delta t$, provided Δt is short enough that the probability of receiving two or more calls in this small interval is negligible. Similarly the probability of receiving no call during the same small interval is simply $1 - \lambda \Delta t$.

Thus, for $x > 0$, the probability of receiving exactly x calls in the total interval

$t + \Delta t$ is given by

$$P_x(t + \Delta t) = P_x(t)(1 - \lambda \Delta t) + P_{x-1}(t)\lambda \Delta t.$$

Rearranging the equation, dividing through by Δt and letting $\Delta t \to 0$, we obtain the differential recurrence equation

$$\frac{dP_x(t)}{dt} = \lambda P_{x-1}(t) - \lambda P_x(t). \tag{30.101}$$

For $x = 0$ (i.e. no calls received), however, (30.101) simplifies to

$$\frac{dP_0(t)}{dt} = -\lambda P_0(t),$$

which may be integrated to give $P_0(t) = P_0(0)e^{-\lambda t}$. But since the probability $P_0(0)$ of receiving no calls in a zero time interval must equal unity, we have $P_0(t) = e^{-\lambda t}$. This expression for $P_0(t)$ may then be substituted back into (30.101) with $x = 1$ to obtain a differential equation for $P_1(t)$ that has the solution $P_1(t) = \lambda t e^{-\lambda t}$. We may repeat this process to obtain expressions for $P_2(t), P_3(t), \ldots, P_x(t)$, and we find

$$P_x(t) = \frac{(\lambda t)^x}{x!} e^{-\lambda t}. \tag{30.102}$$

By setting $t = 1$ in (30.102), we again obtain the Poisson distribution (30.100) for obtaining exactly x calls in a unit time interval.

If a discrete random variable is described by a Poisson distribution of mean λ then we write $X \sim \text{Po}(\lambda)$. As it must be, the sum of the probabilities is unity:

$$\sum_{x=0}^{\infty} \Pr(X = x) = e^{-\lambda} \sum_{x=0}^{\infty} \frac{\lambda^x}{x!} = e^{-\lambda} e^{\lambda} = 1.$$

From (30.100) we may also derive the *Poisson recurrence formula*,

$$\Pr(X = x + 1) = \frac{\lambda}{x + 1} \Pr(X = x) \quad \text{for } x = 0, 1, 2, \ldots, \tag{30.103}$$

which enables successive probabilities to be calculated easily once one is known.

▶*A person receives on average one e-mail message per half-hour interval. Assuming that the e-mails are received randomly in time, find the probabilities that in any particular hour 0, 1, 2, 3, 4, 5 messages are received.*

Let X = number of e-mails received per hour. Clearly the mean number of e-mails per hour is two, and so X follows a Poisson distribution with $\lambda = 2$, i.e.

$$\Pr(X = x) = \frac{2^x}{x!} e^{-2}.$$

Thus $\Pr(X = 0) = e^{-2} = 0.135$, $\Pr(X = 1) = 2e^{-2} = 0.271$, $\Pr(X = 2) = 2^2 e^{-2}/2! = 0.271$, $\Pr(X = 3) = 2^3 e^{-2}/3! = 0.180$, $\Pr(X = 4) = 2^4 e^{-2}/4! = 0.090$, $\Pr(X = 5) = 2^5 e^{-2}/5! = 0.036$. These results may also be calculated using the recurrence formula (30.103). ◀

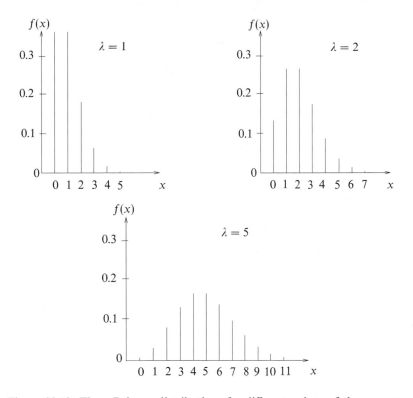

Figure 30.12 Three Poisson distributions for different values of the parameter λ.

The above example illustrates the point that a Poisson distribution typically rises and then falls. It either has a maximum when x is equal to the integer part of λ or, if λ happens to be an integer, has equal maximal values at $x = \lambda - 1$ and $x = \lambda$. The Poisson distribution always has a long 'tail' towards higher values of X but the higher the value of the mean the more symmetric the distribution becomes. Typical Poisson distributions are shown in figure 30.12. Using the definitions of mean and variance, we may show that, for the Poisson distribution, $E[X] = \lambda$ and $V[X] = \lambda$. Nevertheless, as in the case of the binomial distribution, performing the relevant summations directly is rather tiresome, and these results are much more easily proved using the MGF.

The moment generating function for the Poisson distribution

The MGF of the Poisson distribution is given by

$$M_X(t) = E\left[e^{tX}\right] = \sum_{x=0}^{\infty} \frac{e^{tx}e^{-\lambda}\lambda^x}{x!} = e^{-\lambda}\sum_{x=0}^{\infty}\frac{(\lambda e^t)^x}{x!} = e^{-\lambda}e^{\lambda e^t} = e^{\lambda(e^t-1)}$$

$$(30.104)$$

from which we obtain

$$M_X'(t) = \lambda e^t e^{\lambda(e^t-1)},$$
$$M_X''(t) = (\lambda^2 e^{2t} + \lambda e^t)e^{\lambda(e^t-1)}.$$

Thus, the mean and variance of the Poisson distribution are given by

$$E[X] = M_X'(0) = \lambda \quad \text{and} \quad V[X] = M_X''(0) - [M_X'(0)]^2 = \lambda.$$

The Poisson approximation to the binomial distribution

Earlier we derived the Poisson distribution as the limit of the binomial distribution when $n \to \infty$ and $p \to 0$ in such a way that $np = \lambda$ remains finite, where λ is the mean of the Poisson distribution. It is not surprising, therefore, that the Poisson distribution is a very good approximation to the binomial distribution for large n (≥ 50, say) and small p (≤ 0.1, say). Moreover, it is easier to calculate as it involves fewer factorials.

> ▶ *In a large batch of light bulbs, the probability that a bulb is defective is 0.5%. For a sample of 200 bulbs taken at random, find the approximate probabilities that 0, 1 and 2 of the bulbs respectively are defective.*

Let the random variable X = number of defective bulbs in a sample. This is distributed as $X \sim \text{Bin}(200, 0.005)$, implying that $\lambda = np = 1.0$. Since n is large and p small, we may approximate the distribution as $X \sim \text{Po}(1)$, giving

$$\Pr(X = x) \approx e^{-1}\frac{1^x}{x!},$$

from which we find $\Pr(X = 0) \approx 0.37$, $\Pr(X = 1) \approx 0.37$, $\Pr(X = 2) \approx 0.18$. For comparison, it may be noted that the exact values calculated from the binomial distribution are identical to those found here to two decimal places. ◀

Multiple Poisson distributions

Mirroring our discussion of multiple binomial distributions in subsection 30.8.1, let us suppose X and Y are two *independent* random variables, both of which are described by Poisson distributions with (in general) different means, so that $X \sim \text{Po}(\lambda_1)$ and $Y \sim \text{Po}(\lambda_2)$. Now consider the random variable $Z = X + Y$. We may calculate the probability distribution of Z directly using (30.60), but we may derive the result much more easily by using the moment generating function (or indeed the probability or cumulant generating functions).

Since X and Y are independent RVs, the MGF for Z is simply the product of the individual MGFs for X and Y. Thus, from (30.104),

$$M_Z(t) = M_X(t)M_Y(t) = e^{\lambda_1(e^t-1)}e^{\lambda_2(e^t-1)} = e^{(\lambda_1+\lambda_2)(e^t-1)},$$

which we recognise as the MGF of $Z \sim \text{Po}(\lambda_1 + \lambda_2)$. Hence Z is also Poisson distributed and has mean $\lambda_1 + \lambda_2$. Unfortunately, no such simple result holds for the *difference* $Z = X - Y$ of two independent Poisson variates. A closed-form expression for the PDF of this Z does exist, but it is a rather complicated combination of exponentials and a modified Bessel function.[§]

> ► *Two types of e-mail arrive independently and at random: external e-mails at a mean rate of one every five minutes and internal e-mails at a rate of two every five minutes. Calculate the probability of receiving two or more e-mails in any two-minute interval.*

Let

$$X = \text{number of external e-mails per two-minute interval,}$$
$$Y = \text{number of internal e-mails per two-minute interval.}$$

Since we expect on average one external e-mail and two internal e-mails every five minutes we have $X \sim \text{Po}(0.4)$ and $Y \sim \text{Po}(0.8)$. Letting $Z = X + Y$ we have $Z \sim \text{Po}(0.4 + 0.8) = \text{Po}(1.2)$. Now

$$\text{Pr}(Z \geq 2) = 1 - \text{Pr}(Z < 2) = 1 - \text{Pr}(Z = 0) - \text{Pr}(Z = 1)$$

and

$$\text{Pr}(Z = 0) = e^{-1.2} = 0.301,$$
$$\text{Pr}(Z = 1) = e^{-1.2}\frac{1.2}{1} = 0.361.$$

Hence $\text{Pr}(Z \geq 2) = 1 - 0.301 - 0.361 = 0.338.$ ◄

The above result can be extended, of course, to any number of Poisson processes, so that if $X_i = \text{Po}(\lambda_i)$, $i = 1, 2, \ldots, n$ then the random variable $Z = X_1 + X_2 + \cdots + X_n$ is distributed as $Z \sim \text{Po}(\lambda_1 + \lambda_2 + \cdots + \lambda_n)$.

30.9 Important continuous distributions

Having discussed the most commonly encountered discrete probability distributions, we now consider some of the more important continuous probability distributions. These are summarised for convenience in table 30.2; we refer the reader to the relevant subsection below for an explanation of the symbols used.

30.9.1 The Gaussian distribution

By far the most important continuous probability distribution is the *Gaussian* or *normal* distribution. The reason for its importance is that a great many random variables of interest, in all areas of the physical sciences and beyond, are described either exactly or approximately by a Gaussian distribution. Moreover, the Gaussian distribution can be used to approximate other, more complicated, probability distributions.

[§] For a derivation see, for example, M. P. Hobson and A. N. Lasenby, *Monthly Notices of the Royal Astronomical Society*, **298**, 905 (1998).

Distribution	Probability law $f(x)$	MGF	$E[X]$	$V[X]$
Gaussian	$\dfrac{1}{\sigma\sqrt{2\pi}}\exp\left[-\dfrac{(x-\mu)^2}{2\sigma^2}\right]$	$\exp(\mu t + \tfrac{1}{2}\sigma^2 t^2)$	μ	σ^2
exponential	$\lambda e^{-\lambda x}$	$\left(\dfrac{\lambda}{\lambda-t}\right)$	$\dfrac{1}{\lambda}$	$\dfrac{1}{\lambda^2}$
gamma	$\dfrac{\lambda}{\Gamma(r)}(\lambda x)^{r-1}e^{-\lambda x}$	$\left(\dfrac{\lambda}{\lambda-t}\right)^{r}$	$\dfrac{r}{\lambda}$	$\dfrac{r}{\lambda^2}$
chi-squared	$\dfrac{1}{2^{n/2}\Gamma(n/2)}x^{(n/2)-1}e^{-x/2}$	$\left(\dfrac{1}{1-2t}\right)^{n/2}$	n	$2n$
uniform	$\dfrac{1}{b-a}$	$\dfrac{e^{bt}-e^{at}}{(b-a)t}$	$\dfrac{a+b}{2}$	$\dfrac{(b-a)^2}{12}$

Table 30.2   Some important continuous probability distributions.

The probability density function for a Gaussian distribution of a random variable X, with mean $E[X] = \mu$ and variance $V[X] = \sigma^2$, takes the form

$$f(x) = \frac{1}{\sigma\sqrt{2\pi}}\exp\left[-\frac{1}{2}\left(\frac{x-\mu}{\sigma}\right)^2\right]. \tag{30.105}$$

The factor $1/\sqrt{2\pi}$ arises from the normalisation of the distribution,

$$\int_{-\infty}^{\infty} f(x)\,dx = 1;$$

the evaluation of this integral is discussed in subsection 6.4.2. The Gaussian distribution is symmetric about the point $x = \mu$ and has the characteristic 'bell' shape shown in figure 30.13. The width of the curve is described by the standard deviation σ: if σ is large then the curve is broad, and if σ is small then the curve is narrow (see the figure). At $x = \mu \pm \sigma$, $f(x)$ falls to $e^{-1/2} \approx 0.61$ of its peak value; these points are points of inflection, where $d^2 f/dx^2 = 0$. When a random variable X follows a Gaussian distribution with mean μ and variance σ^2, we write $X \sim N(\mu, \sigma^2)$.

The effects of changing μ and σ are only to shift the curve along the x-axis or to broaden or narrow it, respectively. Thus all Gaussians are equivalent in that a change of origin and scale can reduce them to a standard form. We therefore consider the random variable $Z = (X - \mu)/\sigma$, for which the PDF takes the form

$$\phi(z) = \frac{1}{\sqrt{2\pi}}\exp\left(-\frac{z^2}{2}\right), \tag{30.106}$$

which is called the *standard Gaussian distribution* and has mean $\mu = 0$ and variance $\sigma^2 = 1$. The random variable Z is called the *standard variable*.

From (30.105) we can define the cumulative probability function for a Gaussian

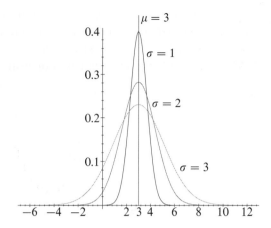

Figure 30.13 The Gaussian or normal distribution for mean $\mu = 3$ and various values of the standard deviation σ.

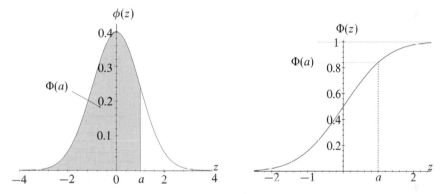

Figure 30.14 On the left, the standard Gaussian distribution $\phi(z)$; the shaded area gives $\Pr(Z < a) = \Phi(a)$. On the right, the cumulative probability function $\Phi(z)$ for a standard Gaussian distribution $\phi(z)$.

distribution as

$$F(x) = \Pr(X < x) = \frac{1}{\sigma\sqrt{2\pi}} \int_{-\infty}^{x} \exp\left[-\frac{1}{2}\left(\frac{u-\mu}{\sigma}\right)^2\right] du,$$

(30.107)

where u is a (dummy) integration variable. Unfortunately, this (indefinite) integral cannot be evaluated analytically. It is therefore standard practice to tabulate values of the cumulative probability function for the standard Gaussian distribution (see figure 30.14), i.e.

$$\Phi(z) = \Pr(Z < z) = \frac{1}{\sqrt{2\pi}} \int_{-\infty}^{z} \exp\left(-\frac{u^2}{2}\right) du.$$

(30.108)

It is usual only to tabulate $\Phi(z)$ for $z > 0$, since it can be seen easily, from figure 30.14 and the symmetry of the Gaussian distribution, that $\Phi(-z) = 1 - \Phi(z)$; see table 30.3. Using such a table it is then straightforward to evaluate the probability that Z lies in a given range of z-values. For example, for a and b constant,

$$\Pr(Z < a) = \Phi(a),$$
$$\Pr(Z > a) = 1 - \Phi(a),$$
$$\Pr(a < Z \leq b) = \Phi(b) - \Phi(a).$$

Remembering that $Z = (X - \mu)/\sigma$ and comparing (30.107) and (30.108), we see that

$$F(x) = \Phi\left(\frac{x - \mu}{\sigma}\right),$$

and so we may also calculate the probability that the original random variable X lies in a given x-range. For example,

$$\Pr(a < X \leq b) = \frac{1}{\sigma\sqrt{2\pi}} \int_a^b \exp\left[-\frac{1}{2}\left(\frac{u - \mu}{\sigma}\right)^2\right] du \qquad (30.109)$$
$$= F(b) - F(a) \qquad (30.110)$$
$$= \Phi\left(\frac{b - \mu}{\sigma}\right) - \Phi\left(\frac{a - \mu}{\sigma}\right). \qquad (30.111)$$

▶*If X is described by a Gaussian distribution of mean μ and variance σ^2, calculate the probabilities that X lies within 1σ, 2σ and 3σ of the mean.*

From (30.111)

$$\Pr(\mu - n\sigma < X \leq \mu + n\sigma) = \Phi(n) - \Phi(-n) = \Phi(n) - [1 - \Phi(n)],$$

and so from table 30.3

$$\Pr(\mu - \sigma < X \leq \mu + \sigma) = 2\Phi(1) - 1 = 0.6826 \approx 68.3\%,$$
$$\Pr(\mu - 2\sigma < X \leq \mu + 2\sigma) = 2\Phi(2) - 1 = 0.9544 \approx 95.4\%,$$
$$\Pr(\mu - 3\sigma < X \leq \mu + 3\sigma) = 2\Phi(3) - 1 = 0.9974 \approx 99.7\%.$$

Thus we expect X to be distributed in such a way that about two thirds of the values will lie between $\mu - \sigma$ and $\mu + \sigma$, 95% will lie within 2σ of the mean and 99.7% will lie within 3σ of the mean. These limits are called the one-, two- and three-sigma limits respectively; it is particularly important to note that they are independent of the actual values of the mean and variance. ◀

There are many other ways in which the Gaussian distribution may be used. We now illustrate some of the uses in more complicated examples.

$\Phi(z)$	.00	.01	.02	.03	.04	.05	.06	.07	.08	.09
0.0	.5000	.5040	.5080	.5120	.5160	.5199	.5239	.5279	.5319	.5359
0.1	.5398	.5438	.5478	.5517	.5557	.5596	.5636	.5675	.5714	.5753
0.2	.5793	.5832	.5871	.5910	.5948	.5987	.6026	.6064	.6103	.6141
0.3	.6179	.6217	.6255	.6293	.6331	.6368	.6406	.6443	.6480	.6517
0.4	.6554	.6591	.6628	.6664	.6700	.6736	.6772	.6808	.6844	.6879
0.5	.6915	.6950	.6985	.7019	.7054	.7088	.7123	.7157	.7190	.7224
0.6	.7257	.7291	.7324	.7357	.7389	.7422	.7454	.7486	.7517	.7549
0.7	.7580	.7611	.7642	.7673	.7704	.7734	.7764	.7794	.7823	.7852
0.8	.7881	.7910	.7939	.7967	.7995	.8023	.8051	.8078	.8106	.8133
0.9	.8159	.8186	.8212	.8238	.8264	.8289	.8315	.8340	.8365	.8389
1.0	.8413	.8438	.8461	.8485	.8508	.8531	.8554	.8577	.8599	.8621
1.1	.8643	.8665	.8686	.8708	.8729	.8749	.8770	.8790	.8810	.8830
1.2	.8849	.8869	.8888	.8907	.8925	.8944	.8962	.8980	.8997	.9015
1.3	.9032	.9049	.9066	.9082	.9099	.9115	.9131	.9147	.9162	.9177
1.4	.9192	.9207	.9222	.9236	.9251	.9265	.9279	.9292	.9306	.9319
1.5	.9332	.9345	.9357	.9370	.9382	.9394	.9406	.9418	.9429	.9441
1.6	.9452	.9463	.9474	.9484	.9495	.9505	.9515	.9525	.9535	.9545
1.7	.9554	.9564	.9573	.9582	.9591	.9599	.9608	.9616	.9625	.9633
1.8	.9641	.9649	.9656	.9664	.9671	.9678	.9686	.9693	.9699	.9706
1.9	.9713	.9719	.9726	.9732	.9738	.9744	.9750	.9756	.9761	.9767
2.0	.9772	.9778	.9783	.9788	.9793	.9798	.9803	.9808	.9812	.9817
2.1	.9821	.9826	.9830	.9834	.9838	.9842	.9846	.9850	.9854	.9857
2.2	.9861	.9864	.9868	.9871	.9875	.9878	.9881	.9884	.9887	.9890
2.3	.9893	.9896	.9898	.9901	.9904	.9906	.9909	.9911	.9913	.9916
2.4	.9918	.9920	.9922	.9925	.9927	.9929	.9931	.9932	.9934	.9936
2.5	.9938	.9940	.9941	.9943	.9945	.9946	.9948	.9949	.9951	.9952
2.6	.9953	.9955	.9956	.9957	.9959	.9960	.9961	.9962	.9963	.9964
2.7	.9965	.9966	.9967	.9968	.9969	.9970	.9971	.9972	.9973	.9974
2.8	.9974	.9975	.9976	.9977	.9977	.9978	.9979	.9979	.9980	.9981
2.9	.9981	.9982	.9982	.9983	.9984	.9984	.9985	.9985	.9986	.9986
3.0	.9987	.9987	.9987	.9988	.9988	.9989	.9989	.9989	.9990	.9990
3.1	.9990	.9991	.9991	.9991	.9992	.9992	.9992	.9992	.9993	.9993
3.2	.9993	.9993	.9994	.9994	.9994	.9994	.9994	.9995	.9995	.9995
3.3	.9995	.9995	.9995	.9996	.9996	.9996	.9996	.9996	.9996	.9997
3.4	.9997	.9997	.9997	.9997	.9997	.9997	.9997	.9997	.9997	.9998

Table 30.3 The cumulative probability function $\Phi(z)$ for the standard Gaussian distribution, as given by (30.108). The units and the first decimal place of z are specified in the column under $\Phi(z)$ and the second decimal place is specified by the column headings. Thus, for example, $\Phi(1.23) = 0.8907$.

> ►*Sawmill A produces boards whose distribution of lengths is well approximated by a Gaussian with mean 209.4 cm and standard deviation 5.0 cm. A board is accepted if it is longer than 200 cm but is rejected otherwise. Show that 3% of boards are rejected.*
> *Sawmill B produces boards of the same standard deviation but of mean length 210.1 cm. Find the proportion of boards rejected if they are drawn at random from the outputs of A and B in the ratio 3 : 1.*

Let X = length of boards from A, so that $X \sim N(209.4, \, (5.0)^2)$ and

$$\Pr(X < 200) = \Phi\left(\frac{200 - \mu}{\sigma}\right) = \Phi\left(\frac{200 - 209.4}{5.0}\right) = \Phi(-1.88).$$

But, since $\Phi(-z) = 1 - \Phi(z)$ we have, using table 30.3,

$$\Pr(X < 200) = 1 - \Phi(1.88) = 1 - 0.9699 = 0.0301,$$

i.e. 3.0% of boards are rejected.

Now let Y = length of boards from B, so that $Y \sim N(210.1, \, (5.0)^2)$ and

$$\Pr(Y < 200) = \Phi\left(\frac{200 - 210.1}{5.0}\right) = \Phi(-2.02)$$
$$= 1 - \Phi(2.02)$$
$$= 1 - 0.9783 = 0.0217.$$

Therefore, when taken alone, only 2.2% of boards from B are rejected. If, however, boards are drawn at random from A and B in the ratio 3 : 1 then the proportion rejected is

$$\tfrac{1}{4}(3 \times 0.030 + 1 \times 0.022) = 0.028 = 2.8\%. \; \blacktriangleleft$$

We may sometimes work backwards to derive the mean and standard deviation of a population that is known to be Gaussian distributed.

> ►*The time taken for a computer 'packet' to travel from Cambridge UK to Cambridge MA is Gaussian distributed. 6.8% of the packets take over 200 ms to make the journey, and 3.0% take under 140 ms. Find the mean and standard deviation of the distribution.*

Let X = journey time in ms; we are told that $X \sim N(\mu, \sigma^2)$ where μ and σ are unknown. Since 6.8% of journey times are longer than 200 ms,

$$\Pr(X > 200) = 1 - \Phi\left(\frac{200 - \mu}{\sigma}\right) = 0.068,$$

from which we find

$$\Phi\left(\frac{200 - \mu}{\sigma}\right) = 1 - 0.068 = 0.932.$$

Using table 30.3, we have therefore

$$\frac{200 - \mu}{\sigma} = 1.49. \tag{30.112}$$

Also, 3.0% of journey times are under 140 ms, so

$$\Pr(X < 140) = \Phi\left(\frac{140 - \mu}{\sigma}\right) = 0.030.$$

Now using $\Phi(-z) = 1 - \Phi(z)$ gives

$$\Phi\left(\frac{\mu - 140}{\sigma}\right) = 1 - 0.030 = 0.970.$$

Using table 30.3 again, we find

$$\frac{\mu - 140}{\sigma} = 1.88. \tag{30.113}$$

Solving the simultaneous equations (30.112) and (30.113) gives $\mu = 173.5$, $\sigma = 17.8$. ◀

The moment generating function for the Gaussian distribution

Using the definition of the MGF (30.85),

$$M_X(t) = E\left[e^{tX}\right] = \int_{-\infty}^{\infty} \frac{1}{\sigma\sqrt{2\pi}} \exp\left[tx - \frac{(x-\mu)^2}{2\sigma^2}\right] dx$$
$$= c\exp\left(\mu t + \tfrac{1}{2}\sigma^2 t^2\right),$$

where the final equality is established by completing the square in the argument of the exponential and writing

$$c = \int_{-\infty}^{\infty} \frac{1}{\sigma\sqrt{2\pi}} \exp\left\{-\frac{[x - (\mu + \sigma^2 t)]^2}{2\sigma^2}\right\} dx.$$

However, the final integral is simply the normalisation integral for the Gaussian distribution, and so $c = 1$ and the MGF is given by

$$M_X(t) = \exp\left(\mu t + \tfrac{1}{2}\sigma^2 t^2\right). \tag{30.114}$$

We showed in subsection 30.7.2 that this MGF leads to $E[X] = \mu$ and $V[X] = \sigma^2$, as required.

Gaussian approximation to the binomial distribution

We may consider the Gaussian distribution as the limit of the binomial distribution when the number of trials $n \to \infty$ but the probability of a success p remains finite, so that $np \to \infty$ also. (This contrasts with the Poisson distribution, which corresponds to the limit $n \to \infty$ and $p \to 0$ with $np = \lambda$ remaining finite.) In other words, a Gaussian distribution results when an experiment with a finite probability of success is repeated a large number of times. We now show how this Gaussian limit arises.

The binomial probability function gives the probability of x successes in n trials as

$$f(x) = \frac{n!}{x!(n-x)!} p^x (1-p)^{n-x}.$$

Taking the limit as $n \to \infty$ (and $x \to \infty$) we may approximate the factorials by Stirling's approximation

$$n! \sim \sqrt{2\pi n} \left(\frac{n}{e}\right)^n$$

x	$f(x)$ (binomial)	$f(x)$ (Gaussian)
0	0.0001	0.0001
1	0.0016	0.0014
2	0.0106	0.0092
3	0.0425	0.0395
4	0.1115	0.1119
5	0.2007	0.2091
6	0.2508	0.2575
7	0.2150	0.2091
8	0.1209	0.1119
9	0.0403	0.0395
10	0.0060	0.0092

Table 30.4   Comparison of the binomial distribution for $n = 10$ and $p = 0.6$ with its Gaussian approximation.

to obtain

$$f(x) \approx \frac{1}{\sqrt{2\pi n}} \left(\frac{x}{n}\right)^{-x-1/2} \left(\frac{n-x}{n}\right)^{-n+x-1/2} p^x (1-p)^{n-x}$$

$$= \frac{1}{\sqrt{2\pi n}} \exp\left[-\left(x + \tfrac{1}{2}\right) \ln \frac{x}{n} - \left(n - x + \tfrac{1}{2}\right) \ln \frac{n-x}{n} \right.$$

$$\left. + x \ln p + (n - x)\ln(1 - p) \right].$$

By expanding the argument of the exponential in terms of $y = x - np$, where $1 \ll y \ll np$ and keeping only the dominant terms, it can be shown that

$$f(x) \approx \frac{1}{\sqrt{2\pi n}} \frac{1}{\sqrt{p(1-p)}} \exp\left[-\frac{1}{2} \frac{(x - np)^2}{np(1-p)} \right],$$

which is of Gaussian form with $\mu = np$ and $\sigma = \sqrt{np(1-p)}$.

Thus we see that the *value* of the Gaussian *probability density function* $f(x)$ is a good approximation to the *probability* of obtaining x successes in n trials. This approximation is actually very good even for relatively small n. For example, if $n = 10$ and $p = 0.6$ then the Gaussian approximation to the binomial distribution is (30.105) with $\mu = 10 \times 0.6 = 6$ and $\sigma = \sqrt{10 \times 0.6(1 - 0.6)} = 1.549$. The probability functions $f(x)$ for the binomial and associated Gaussian distributions for these parameters are given in table 30.4, and it can be seen that the Gaussian approximation is a good one.

Strictly speaking, however, since the Gaussian distribution is continuous and the binomial distribution is discrete, we should use the integral of $f(x)$ for the Gaussian distribution in the calculation of approximate binomial probabilities. More specifically, we should apply a *continuity correction* so that the discrete integer x in the binomial distribution becomes the interval $[x - 0.5, x + 0.5]$ in the Gaussian distribution. Explicitly,

$$\Pr(X = x) \approx \frac{1}{\sigma\sqrt{2\pi}} \int_{x-0.5}^{x+0.5} \exp\left[-\frac{1}{2}\left(\frac{u-\mu}{\sigma}\right)^2\right] du.$$

The Gaussian approximation is particularly useful for estimating the binomial probability that X lies between the (integer) values x_1 and x_2,

$$\Pr(x_1 < X \leq x_2) \approx \frac{1}{\sigma\sqrt{2\pi}} \int_{x_1-0.5}^{x_2+0.5} \exp\left[-\frac{1}{2}\left(\frac{u-\mu}{\sigma}\right)^2\right] du.$$

▶ *A manufacturer makes computer chips of which 10% are defective. For a random sample of 200 chips, find the approximate probability that more than 15 are defective.*

We first define the random variable

$$X = \text{number of defective chips in the sample,}$$

which has a binomial distribution $X \sim \text{Bin}(200, 0.1)$. Therefore, the mean and variance of this distribution are

$$E[X] = 200 \times 0.1 = 20 \quad \text{and} \quad V[X] = 200 \times 0.1 \times (1 - 0.1) = 18,$$

and we may approximate the binomial distribution with a Gaussian distribution such that $X \sim N(20, 18)$. The standard variable is

$$Z = \frac{X - 20}{\sqrt{18}},$$

and so, using $X = 15.5$ to allow for the continuity correction,

$$\Pr(X > 15.5) = \Pr\left(Z > \frac{15.5 - 20}{\sqrt{18}}\right) = \Pr(Z > -1.06)$$
$$= \Pr(Z < 1.06) = 0.86. \quad \blacktriangleleft$$

Gaussian approximation to the Poisson distribution

We first met the Poisson distribution as the limit of the binomial distribution for $n \to \infty$ and $p \to 0$, taken in such a way that $np = \lambda$ remains finite. Further, in the previous subsection, we considered the Gaussian distribution as the limit of the binomial distribution when $n \to \infty$ but p remains finite, so that $np \to \infty$ also. It should come as no surprise, therefore, that the Gaussian distribution can also be used to approximate the Poisson distribution when the mean λ becomes large. The probability function for the Poisson distribution is

$$f(x) = e^{-\lambda}\frac{\lambda^x}{x!},$$

which, on taking the logarithm of both sides, gives

$$\ln f(x) = -\lambda + x \ln \lambda - \ln x!. \tag{30.115}$$

Stirling's approximation for large x gives

$$x! \approx \sqrt{2\pi x} \left(\frac{x}{e}\right)^x$$

implying that

$$\ln x! \approx \ln \sqrt{2\pi x} + x \ln x - x,$$

which, on substituting into (30.115), yields

$$\ln f(x) \approx -\lambda + x \ln \lambda - (x \ln x - x) - \ln \sqrt{2\pi x}.$$

Since we expect the Poisson distribution to peak around $x = \lambda$, we substitute $\epsilon = x - \lambda$ to obtain

$$\ln f(x) \approx -\lambda + (\lambda + \epsilon) \left\{\ln \lambda - \ln \left[\lambda \left(1 + \frac{\epsilon}{\lambda}\right)\right]\right\} + (\lambda + \epsilon) - \ln \sqrt{2\pi(\lambda + \epsilon)}.$$

Using the expansion $\ln(1 + z) = z - z^2/2 + \cdots$, we find

$$\ln f(x) \approx \epsilon - (\lambda + \epsilon) \left(\frac{\epsilon}{\lambda} - \frac{\epsilon^2}{2\lambda^2}\right) - \ln \sqrt{2\pi\lambda} - \frac{1}{2} \left(\frac{\epsilon}{\lambda} - \frac{\epsilon^2}{2\lambda^2}\right)$$

$$\approx -\frac{\epsilon^2}{2\lambda} - \ln \sqrt{2\pi\lambda},$$

when only the dominant terms are retained, after using the fact that ϵ is of the order of the standard deviation of x, i.e. of order $\lambda^{1/2}$. On exponentiating this result we obtain

$$f(x) \approx \frac{1}{\sqrt{2\pi\lambda}} \exp \left[-\frac{(x - \lambda)^2}{2\lambda}\right],$$

which is the Gaussian distribution with $\mu = \lambda$ and $\sigma^2 = \lambda$.

The larger the value of λ, the better is the Gaussian approximation to the Poisson distribution; the approximation is reasonable even for $\lambda = 5$, but $\lambda \geq 10$ is safer. As in the case of the Gaussian approximation to the binomial distribution, a continuity correction is necessary since the Poisson distribution is discrete.

> ►*E-mail messages are received by an author at an average rate of one per hour. Find the probability that in a day the author receives 24 messages or more.*

We first define the random variable

$$X = \text{number of messages received in a day}.$$

Thus $E[X] = 1 \times 24 = 24$, and so $X \sim \text{Po}(24)$. Since $\lambda > 10$ we may approximate the Poisson distribution by $X \sim \text{N}(24, 24)$. Now the standard variable is

$$Z = \frac{X - 24}{\sqrt{24}},$$

and, using the continuity correction, we find

$$\Pr(X > 23.5) = \Pr \left(Z > \frac{23.5 - 24}{\sqrt{24}}\right)$$

$$= \Pr(Z > -0.102) = \Pr(Z < 0.102) = 0.54. ◄$$

In fact, almost all probability distributions tend towards a Gaussian when the numbers involved become large – that this should happen is required by the central limit theorem, which we discuss in section 30.10.

Multiple Gaussian distributions

Suppose X and Y are *independent* Gaussian-distributed random variables, so that $X \sim N(\mu_1, \sigma_1^2)$ and $Y \sim N(\mu_2, \sigma_2^2)$. Let us now consider the random variable $Z = X + Y$. The PDF for this random variable may be found directly using (30.61), but it is easier to use the MGF. From (30.114), the MGFs of X and Y are

$$M_X(t) = \exp\left(\mu_1 t + \tfrac{1}{2}\sigma_1^2 t^2\right), \qquad M_Y(t) = \exp\left(\mu_2 t + \tfrac{1}{2}\sigma_2^2 t^2\right).$$

Using (30.89), since X and Y are independent RVs, the MGF of $Z = X + Y$ is simply the product of $M_X(t)$ and $M_Y(t)$. Thus, we have

$$M_Z(t) = M_X(t)M_Y(t) = \exp\left(\mu_1 t + \tfrac{1}{2}\sigma_1^2 t^2\right)\exp\left(\mu_2 t + \tfrac{1}{2}\sigma_2^2 t^2\right)$$
$$= \exp\left[(\mu_1 + \mu_2)t + \tfrac{1}{2}(\sigma_1^2 + \sigma_2^2)t^2\right],$$

which we recognise as the MGF for a Gaussian with mean $\mu_1 + \mu_2$ and variance $\sigma_1^2 + \sigma_2^2$. Thus, Z is also Gaussian distributed: $Z \sim N(\mu_1 + \mu_2,\ \sigma_1^2 + \sigma_2^2)$.

A similar calculation may be performed to calculate the PDF of the random variable $W = X - Y$. If we introduce the variable $\tilde{Y} = -Y$ then $W = X + \tilde{Y}$, where $\tilde{Y} \sim N(-\mu_1,\ \sigma_1^2)$. Thus, using the result above, we find $W \sim N(\mu_1 - \mu_2,\ \sigma_1^2 + \sigma_2^2)$.

> ▶*An executive travels home from her office every evening. Her journey consists of a train ride, followed by a bicycle ride. The time spent on the train is Gaussian distributed with mean 52 minutes and standard deviation 1.8 minutes, while the time for the bicycle journey is Gaussian distributed with mean 8 minutes and standard deviation 2.6 minutes. Assuming these two factors are independent, estimate the percentage of occasions on which the whole journey takes more than 65 minutes.*

We first define the random variables

$$X = \text{time spent on train}, \qquad Y = \text{time spent on bicycle},$$

so that $X \sim N(52, (1.8)^2)$ and $Y \sim N(8, (2.6)^2)$. Since X and Y are independent, the total journey time $T = X + Y$ is distributed as

$$T \sim N(52 + 8,\ (1.8)^2 + (2.6)^2) = N(60, (3.16)^2).$$

The standard variable is thus

$$Z = \frac{T - 60}{3.16},$$

and the required probability is given by

$$\Pr(T > 65) = \Pr\left(Z > \frac{65 - 60}{3.16}\right) = \Pr(Z > 1.58) = 1 - 0.943 = 0.057.$$

Thus the total journey time exceeds 65 minutes on 5.7% of occasions. ◀

The above results may be extended. For example, if the random variables X_i, $i = 1, 2, \ldots, n$, are distributed as $X_i \sim N(\mu_i, \sigma_i^2)$ then the random variable $Z = \sum_i c_i X_i$ (where the c_i are constants) is distributed as $Z \sim N(\sum_i c_i \mu_i, \sum_i c_i^2 \sigma_i^2)$.

30.9.2 The log-normal distribution

If the random variable X follows a Gaussian distribution then the variable $Y = e^X$ is described by a *log-normal* distribution. Clearly, if X can take values in the range $-\infty$ to ∞, then Y will lie between 0 and ∞. The probability density function for Y is found using the result (30.58). It is

$$g(y) = f(x(y)) \left| \frac{dx}{dy} \right| = \frac{1}{\sigma \sqrt{2\pi}} \frac{1}{y} \exp\left[-\frac{(\ln y - \mu)^2}{2\sigma^2} \right].$$

We note that μ and σ^2 are not the mean and variance of the log-normal distribution, but rather the parameters of the corresponding Gaussian distribution for X. The mean and variance of Y, however, can be found straightforwardly using the MGF of X, which reads $M_X(t) = E[e^{tX}] = \exp(\mu t + \frac{1}{2}\sigma^2 t^2)$. Thus, the mean of Y is given by

$$E[Y] = E[e^X] = M_X(1) = \exp(\mu + \tfrac{1}{2}\sigma^2),$$

and the variance of Y reads

$$V[Y] = E[Y^2] - (E[Y])^2 = E[e^{2X}] - (E[e^X])^2$$
$$= M_X(2) - [M_X(1)]^2 = \exp(2\mu + \sigma^2)[\exp(\sigma^2) - 1].$$

In figure 30.15, we plot some examples of the log-normal distribution for various values of the parameters μ and σ^2.

30.9.3 The exponential and gamma distributions

The exponential distribution with positive parameter λ is given by

$$f(x) = \begin{cases} \lambda e^{-\lambda x} & \text{for } x > 0, \\ 0 & \text{for } x \le 0 \end{cases} \tag{30.116}$$

and satisfies $\int_{-\infty}^{\infty} f(x)\,dx = 1$ as required. The exponential distribution occurs naturally if we consider the distribution of the length of intervals between successive events in a Poisson process or, equivalently, the distribution of the interval (i.e. the waiting time) before the first event. If the average number of events per unit interval is λ then on average there are λx events in interval x, so that from the Poisson distribution the probability that there will be no events in this interval is given by

$$\Pr(\text{no events in interval } x) = e^{-\lambda x}.$$

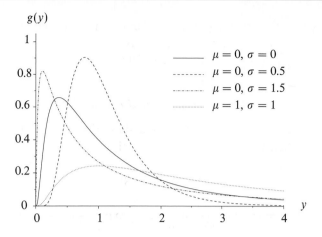

Figure 30.15 The PDF $g(y)$ for the log-normal distribution for various values of the parameters μ and σ.

The probability that an event occurs in the next infinitesimal interval $[x, x + dx]$ is given by $\lambda\, dx$, so that

$$\text{Pr(the first event occurs in interval } [x, x + dx]) = e^{-\lambda x}\lambda\, dx.$$

Hence the required probability density function is given by

$$f(x) = \lambda e^{-\lambda x}.$$

The expectation and variance of the exponential distribution can be evaluated as $1/\lambda$ and $(1/\lambda)^2$ respectively. The MGF is given by

$$M(t) = \frac{\lambda}{\lambda\ t}. \tag{30.117}$$

We may generalise the above discussion to obtain the PDF for the interval between every rth event in a Poisson process or, equivalently, the interval (waiting time) before the rth event. We begin by using the Poisson distribution to give

$$\text{Pr}(r - 1 \text{ events occur in interval } x) = e^{-\lambda x}\frac{(\lambda x)^{r-1}}{(r - 1)!},$$

from which we obtain

$$\text{Pr}(r\text{th event occurs in the interval } [x, x + dx]) = e^{-\lambda x}\frac{(\lambda x)^{r-1}}{(r - 1)!}\lambda\, dx.$$

Thus the required PDF is

$$f(x) = \frac{\lambda}{(r - 1)!}(\lambda x)^{r-1}e^{-\lambda x}, \tag{30.118}$$

which is known as the *gamma distribution* of order r with parameter λ. Although our derivation applies only when r is a positive integer, the gamma distribution is

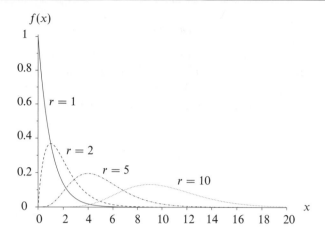

Figure 30.16   The PDF $f(x)$ for the gamma distributions $\gamma(\lambda, r)$ with $\lambda = 1$ and $r = 1, 2, 5, 10$.

defined for all positive r by replacing $(r-1)!$ by $\Gamma(r)$ in (30.118); see section 18.12 for a discussion of the gamma function $\Gamma(x)$. If a random variable X is described by a gamma distribution of order r with parameter λ, we write $X \sim \gamma(\lambda, r)$; we note that the exponential distribution is the special case $\gamma(\lambda, 1)$. The gamma distribution $\gamma(\lambda, r)$ is plotted in figure 30.16 for $\lambda = 1$ and $r = 1, 2, 5, 10$. For large r, the gamma distribution tends to the Gaussian distribution whose mean and variance are specified by (30.120) below.

The MGF for the gamma distribution is obtained from that for the exponential distribution, by noting that we may consider the interval between every rth event in a Poisson process as the sum of r intervals between successive events. Thus the rth-order gamma variate is the sum of r independent exponentially distributed random variables. From (30.117) and (30.90), the MGF of the gamma distribution is therefore given by

$$M(t) = \left(\frac{\lambda}{\lambda - t} \right)^r, \tag{30.119}$$

from which the mean and variance are found to be

$$E[X] = \frac{r}{\lambda}, \qquad V[X] = \frac{r}{\lambda^2}. \tag{30.120}$$

We may also use the above MGF to prove another useful theorem regarding multiple gamma distributions. If $X_i \sim \gamma(\lambda, r_i)$, $i = 1, 2, \ldots, n$, are independent gamma variates then the random variable $Y = X_1 + X_2 + \cdots + X_n$ has MGF

$$M(t) = \prod_{i=1}^{n} \left(\frac{\lambda}{\lambda - t} \right)^{r_i} = \left(\frac{\lambda}{\lambda - t} \right)^{r_1 + r_2 + \cdots + r_n}. \tag{30.121}$$

Thus Y is also a gamma variate, distributed as $Y \sim \gamma(\lambda, r_1 + r_2 + \cdots + r_n)$.

30.9.4 The chi-squared distribution

In subsection 30.6.2, we showed that if X is Gaussian distributed with mean μ and variance σ^2, such that $X \sim N(\mu, \sigma^2)$, then the random variable $Y = (x - \mu)^2/\sigma^2$ is distributed as the gamma distribution $Y \sim \gamma(\frac{1}{2}, \frac{1}{2})$. Let us now consider n independent Gaussian random variables $X_i \sim N(\mu_i, \sigma_i^2)$, $i = 1, 2, \ldots, n$, and define the new variable

$$\chi_n^2 = \sum_{i=1}^{n} \frac{(X_i - \mu_i)^2}{\sigma_i^2}. \tag{30.122}$$

Using the result (30.121) for multiple gamma distributions, χ_n^2 must be distributed as the gamma variate $\chi_n^2 \sim \gamma(\frac{1}{2}, \frac{1}{2}n)$, which from (30.118) has the PDF

$$f(\chi_n^2) = \frac{\frac{1}{2}}{\Gamma(\frac{1}{2}n)} (\tfrac{1}{2}\chi_n^2)^{(n/2)-1} \exp(-\tfrac{1}{2}\chi_n^2)$$

$$= \frac{1}{2^{n/2}\Gamma(\frac{1}{2}n)} (\chi_n^2)^{(n/2)-1} \exp(-\tfrac{1}{2}\chi_n^2). \tag{30.123}$$

This is known as the *chi-squared distribution* of order n and has numerous applications in statistics (see chapter 31). Setting $\lambda = \frac{1}{2}$ and $r = \frac{1}{2}n$ in (30.120), we find that

$$E[\chi_n^2] = n, \qquad V[\chi_n^2] = 2n.$$

An important generalisation occurs when the n Gaussian variables X_i are *not* linearly independent but are instead required to satisfy a linear constraint of the form

$$c_1 X_1 + c_2 X_2 + \cdots + c_n X_n = 0, \tag{30.124}$$

in which the constants c_i are not all zero. In this case, it may be shown (see exercise 30.40) that the variable χ_n^2 defined in (30.122) is still described by a chi-squared distribution, but one of order $n - 1$. Indeed, this result may be trivially extended to show that if the n Gaussian variables X_i satisfy m linear constraints of the form (30.124) then the variable χ_n^2 defined in (30.122) is described by a chi-squared distribution of order $n - m$.

30.9.5 The Cauchy and Breit–Wigner distributions

A random variable X (in the range $-\infty$ to ∞) that obeys the *Cauchy distribution* is described by the PDF

$$f(x) = \frac{1}{\pi} \frac{1}{1 + x^2}.$$

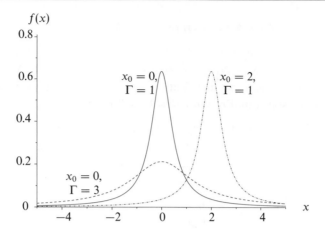

Figure 30.17 The PDF $f(x)$ for the Breit–Wigner distribution for different values of the parameters x_0 and Γ.

This is a special case of the *Breit–Wigner distribution*

$$f(x) = \frac{1}{\pi} \frac{\frac{1}{2}\Gamma}{\frac{1}{4}\Gamma^2 + (x - x_0)^2},$$

which is encountered in the study of nuclear and particle physics. In figure 30.17, we plot some examples of the Breit–Wigner distribution for several values of the parameters x_0 and Γ.

We see from the figure that the peak (or mode) of the distribution occurs at $x = x_0$. It is also straightforward to show that the parameter Γ is equal to the width of the peak at half the maximum height. Although the Breit–Wigner distribution is symmetric about its peak, it does not formally possess a mean since the integrals $\int_{-\infty}^{0} xf(x)\,dx$ and $\int_{0}^{\infty} xf(x)\,dx$ both diverge. Similar divergences occur for all higher moments of the distribution.

30.9.6 The uniform distribution

Finally we mention the very simple, but common, *uniform distribution*, which describes a continuous random variable that has a constant PDF over its allowed range of values. If the limits on X are a and b then

$$f(x) = \begin{cases} 1/(b - a) & \text{for } a \leq x \leq b, \\ 0 & \text{otherwise.} \end{cases}$$

The MGF of the uniform distribution is found to be

$$M(t) = \frac{e^{bt} - e^{at}}{(b - a)t},$$

and its mean and variance are given by

$$E[X] = \frac{a+b}{2}, \qquad V[X] = \frac{(b-a)^2}{12}.$$

30.10 The central limit theorem

In subsection 30.9.1 we discussed approximating the binomial and Poisson distributions by the Gaussian distribution when the number of trials is large. We now discuss why the Gaussian distribution is so common and therefore so important. The *central limit theorem* may be stated as follows.

Central limit theorem. *Suppose that X_i, $i = 1, 2, \ldots, n$, are independent random variables, each of which is described by a probability density function $f_i(x)$ (these may all be different) with a mean μ_i and a variance σ_i^2. The random variable $Z = \left(\sum_i X_i \right) / n$, i.e. the 'mean' of the X_i, has the following properties:*

   (i) *its expectation value is given by $E[Z] = \left(\sum_i \mu_i \right) / n$;*
   (ii) *its variance is given by $V[Z] = \left(\sum_i \sigma_i^2 \right) / n^2$;*
   (iii) *as $n \to \infty$ the probability function of Z tends to a Gaussian with corresponding mean and variance.*

We note that for the theorem to hold, the probability density functions $f_i(x)$ must possess formal means and variances. Thus, for example, if any of the X_i were described by a Cauchy distribution then the theorem would not apply.

Properties (i) and (ii) of the theorem are easily proved, as follows. Firstly

$$E[Z] = \frac{1}{n}(E[X_1] + E[X_2] + \cdots + E[X_n]) = \frac{1}{n}(\mu_1 + \mu_2 + \cdots + \mu_n) = \frac{\sum_i \mu_i}{n},$$

a result which does *not* require that the X_i are *independent* random variables. If $\mu_i = \mu$ for all i then this becomes

$$E[Z] = \frac{n\mu}{n} = \mu.$$

Secondly, if the X_i *are* independent, it follows from an obvious extension of (30.68) that

$$V[Z] = V\left[\frac{1}{n}(X_1 + X_2 + \cdots + X_n) \right]$$

$$= \frac{1}{n^2}(V[X_1] + V[X_2] + \cdots + V[X_n]) = \frac{\sum_i \sigma_i^2}{n^2}.$$

Let us now consider property (iii), which is the reason for the ubiquity of the Gaussian distribution and is most easily proved by considering the moment generating function $M_Z(t)$ of Z. From (30.90), this MGF is given by

$$M_Z(t) = \prod_{i=1}^{n} M_{X_i}\left(\frac{t}{n} \right),$$

where $M_{X_i}(t)$ is the MGF of $f_i(x)$. Now

$$M_{X_i}\left(\frac{t}{n}\right) = 1 + \frac{t}{n}E[X_i] + \frac{1}{2}\frac{t^2}{n^2}E[X_i^2] + \cdots$$

$$= 1 + \mu_i\frac{t}{n} + \frac{1}{2}(\sigma_i^2 + \mu_i^2)\frac{t^2}{n^2} + \cdots,$$

and as n becomes large

$$M_{X_i}\left(\frac{t}{n}\right) \approx \exp\left(\frac{\mu_i t}{n} + \frac{1}{2}\sigma_i^2\frac{t^2}{n^2}\right),$$

as may be verified by expanding the exponential up to terms including $(t/n)^2$. Therefore

$$M_Z(t) \approx \prod_{i=1}^{n}\exp\left(\frac{\mu_i t}{n} + \frac{1}{2}\sigma_i^2\frac{t^2}{n^2}\right) = \exp\left(\frac{\sum_i \mu_i}{n}t + \frac{1}{2}\frac{\sum_i \sigma_i^2}{n^2}t^2\right).$$

Comparing this with the form of the MGF for a Gaussian distribution, (30.114), we can see that the probability density function $g(z)$ of Z tends to a Gaussian distribution with mean $\sum_i \mu_i/n$ and variance $\sum_i \sigma_i^2/n^2$. In particular, if we consider Z to be the mean of n *independent* measurements of the *same* random variable X (so that $X_i = X$ for $i = 1, 2, \ldots, n$) then, as $n \to \infty$, Z has a Gaussian distribution with mean μ and variance σ^2/n.

We may use the central limit theorem to derive an analogous result to (iii) above for the *product* $W = X_1 X_2 \cdots X_n$ of the n independent random variables X_i. Provided the X_i only take values between zero and infinity, we may write

$$\ln W = \ln X_1 + \ln X_2 + \cdots + \ln X_n,$$

which is simply the sum of n new random variables $\ln X_i$. Thus, provided these new variables each possess a formal mean and variance, the PDF of $\ln W$ will tend to a Gaussian in the limit $n \to \infty$, and so the product W will be described by a log-normal distribution (see subsection 30.9.2).

30.11 Joint distributions

As mentioned briefly in subsection 30.4.3, it is common in the physical sciences to consider simultaneously two or more random variables that are not independent, in general, and are thus described by *joint probability density functions*. We will return to the subject of the interdependence of random variables after first presenting some of the general ways of characterising joint distributions. We will concentrate mainly on *bivariate* distributions, i.e. distributions of only two random variables, though the results may be extended readily to multivariate distributions. The subject of multivariate distributions is large and a detailed study is beyond the scope of this book; the interested reader should therefore consult one of the many specialised texts. However, we do discuss the multinomial and multivariate Gaussian distributions, in section 30.15.

The first thing to note when dealing with bivariate distributions is that the distinction between discrete and continuous distributions may not be as clear as for the single variable case; the random variables can both be discrete, or both continuous, or one discrete and the other continuous. In general, for the random variables X and Y, the joint distribution will take an infinite number of values unless both X and Y have only a finite number of values. In this chapter we will consider only the cases where X and Y are either both discrete or both continuous random variables.

30.11.1 Discrete bivariate distributions

In direct analogy with the one-variable (univariate) case, if X is a discrete random variable that takes the values $\{x_i\}$ and Y one that takes the values $\{y_j\}$ then the probability function of the joint distribution is defined as

$$f(x, y) = \begin{cases} \Pr(X = x_i,\ Y = y_j) & \text{for } x = x_i,\ y = y_j, \\ 0 & \text{otherwise.} \end{cases}$$

We may therefore think of $f(x, y)$ as a set of spikes at valid points in the xy-plane, whose height at (x_i, y_i) represents the probability of obtaining $X = x_i$ and $Y = y_j$. The normalisation of $f(x, y)$ implies

$$\sum_i \sum_j f(x_i, y_j) = 1, \tag{30.125}$$

where the sums over i and j take all valid pairs of values. We can also define the cumulative probability function

$$F(x, y) = \sum_{x_i \leq x} \sum_{y_j \leq y} f(x_i, y_j), \tag{30.126}$$

from which it follows that the probability that X lies in the range $[a_1, a_2]$ and Y lies in the range $[b_1, b_2]$ is given by

$$\Pr(a_1 < X \leq a_2,\ b_1 < Y \leq b_2) = F(a_2, b_2) - F(a_1, b_2) - F(a_2, b_1) + F(a_1, b_1).$$

Finally, we define X and Y to be *independent* if we can write their joint distribution in the form

$$f(x, y) = f_X(x) f_Y(y), \tag{30.127}$$

i.e. as the product of two univariate distributions.

30.11.2 Continuous bivariate distributions

In the case where both X and Y are continuous random variables, the PDF of the joint distribution is defined by

$$f(x, y)\, dx\, dy = \Pr(x < X \leq x + dx, \ y < Y \leq y + dy),$$

(30.128)

so $f(x, y)\, dx\, dy$ is the probability that x lies in the range $[x, x + dx]$ and y lies in the range $[y, y + dy]$. It is clear that the two-dimensional function $f(x, y)$ must be everywhere non-negative and that normalisation requires

$$\int_{-\infty}^{\infty} \int_{-\infty}^{\infty} f(x, y)\, dx\, dy = 1.$$

It follows further that

$$\Pr(a_1 < X \leq a_2, \ b_1 < Y \leq b_2) = \int_{b_1}^{b_2} \int_{a_1}^{a_2} f(x, y)\, dx\, dy.$$

(30.129)

We can also define the cumulative probability function by

$$F(x, y) = \Pr(X \leq x, \ Y \leq y) = \int_{-\infty}^{x} \int_{-\infty}^{y} f(u, v)\, du\, dv,$$

from which we see that (as for the discrete case),

$$\Pr(a_1 < X \leq a_2, \ b_1 < Y \leq b_2) = F(a_2, b_2) - F(a_1, b_2) - F(a_2, b_1) + F(a_1, b_1).$$

Finally we note that the definition of independence (30.127) for discrete bivariate distributions also applies to continuous bivariate distributions.

> ▶ *A flat table is ruled with parallel straight lines a distance D apart, and a thin needle of length $l < D$ is tossed onto the table at random. What is the probability that the needle will cross a line?*

Let θ be the angle that the needle makes with the lines, and let x be the distance from the centre of the needle to the nearest line. Since the needle is tossed 'at random' onto the table, the angle θ is uniformly distributed in the interval $[0, \pi]$, and the distance x is uniformly distributed in the interval $[0, D/2]$. Assuming that θ and x are independent, their joint distribution is just the product of their individual distributions, and is given by

$$f(\theta, x) = \frac{1}{\pi} \frac{1}{D/2} = \frac{2}{\pi D}.$$

The needle will cross a line if the distance x of its centre from that line is less than $\frac{1}{2} l \sin \theta$. Thus the required probability is

$$\frac{2}{\pi D} \int_0^{\pi} \int_0^{\frac{1}{2} l \sin \theta} dx\, d\theta = \frac{2}{\pi D} \frac{l}{2} \int_0^{\pi} \sin \theta\, d\theta = \frac{2l}{\pi D}.$$

This gives an experimental (but cumbersome) method of determining π. ◀

30.11.3 Marginal and conditional distributions

Given a bivariate distribution $f(x, y)$, we may be interested only in the probability function for X *irrespective of the value of Y* (or vice versa). This *marginal* distribution of X is obtained by summing or integrating, as appropriate, the joint probability distribution over all allowed values of Y. Thus, the marginal distribution of X (for example) is given by

$$f_X(x) = \begin{cases} \sum_j f(x, y_j) & \text{for a discrete distribution,} \\ \int f(x, y)\, dy & \text{for a continuous distribution.} \end{cases} \tag{30.130}$$

It is clear that an analogous definition exists for the marginal distribution of Y.

Alternatively, one might be interested in the probability function of X *given that Y takes some specific value of $Y = y_0$*, i.e. $\Pr(X = x | Y = y_0)$. This *conditional* distribution of X is given by

$$g(x) = \frac{f(x, y_0)}{f_Y(y_0)},$$

where $f_Y(y)$ is the marginal distribution of Y. The division by $f_Y(y_0)$ is necessary in order that $g(x)$ is properly normalised.

30.12 Properties of joint distributions

The probability density function $f(x, y)$ contains all the information on the joint probability distribution of two random variables X and Y. In a similar manner to that presented for univariate distributions, however, it is conventional to characterise $f(x, y)$ by certain of its properties, which we now discuss. Once again, most of these properties are based on the concept of expectation values, which are defined for joint distributions in an analogous way to those for single-variable distributions (30.46). Thus, the expectation value of any function $g(X, Y)$ of the random variables X and Y is given by

$$E[g(X, Y)] = \begin{cases} \sum_i \sum_j g(x_i, y_j) f(x_i, y_j) & \text{for the discrete case,} \\ \int_{-\infty}^{\infty} \int_{-\infty}^{\infty} g(x, y) f(x, y)\, dx\, dy & \text{for the continuous case.} \end{cases}$$

30.12.1 Means

The means of X and Y are defined respectively as the expectation values of the variables X and Y. Thus, the mean of X is given by

$$E[X] = \mu_X = \begin{cases} \sum_i \sum_j x_i f(x_i, y_j) & \text{for the discrete case,} \\ \int_{-\infty}^{\infty} \int_{-\infty}^{\infty} x f(x, y)\, dx\, dy & \text{for the continuous case.} \end{cases} \tag{30.131}$$

$E[Y]$ is obtained in a similar manner.

▶*Show that if X and Y are independent random variables then $E[XY] = E[X]E[Y]$.*

Let us consider the case where X and Y are continuous random variables. Since X and Y are independent $f(x, y) = f_X(x)f_Y(y)$, so that

$$E[XY] = \int_{-\infty}^{\infty} \int_{-\infty}^{\infty} xy f_X(x) f_Y(y)\, dx\, dy = \int_{-\infty}^{\infty} x f_X(x)\, dx \int_{-\infty}^{\infty} y f_Y(y)\, dy = E[X]E[Y].$$

An analogous proof exists for the discrete case. ◀

30.12.2 Variances

The definitions of the variances of X and Y are analogous to those for the single-variable case (30.48), i.e. the variance of X is given by

$$V[X] = \sigma_X^2 = \begin{cases} \sum_i \sum_j (x_i - \mu_X)^2 f(x_i, y_j) & \text{for the discrete case,} \\ \int_{-\infty}^{\infty} \int_{-\infty}^{\infty} (x - \mu_X)^2 f(x, y)\, dx\, dy & \text{for the continuous case.} \end{cases} \quad (30.132)$$

Equivalent definitions exist for the variance of Y.

30.12.3 Covariance and correlation

Means and variances of joint distributions provide useful information about their marginal distributions, but we have not yet given any indication of how to measure the relationship between the two random variables. Of course, it may be that the two random variables are independent, but often this is not so. For example, if we measure the heights and weights of a sample of people we would not be surprised to find a tendency for tall people to be heavier than short people and vice versa. We will show in this section that two functions, the *covariance* and the *correlation*, can be defined for a bivariate distribution and that these are useful in characterising the relationship between the two random variables.

The *covariance* of two random variables X and Y is defined by

$$\text{Cov}[X, Y] = E[(X - \mu_X)(Y - \mu_Y)], \quad (30.133)$$

where μ_X and μ_Y are the expectation values of X and Y respectively. Clearly related to the covariance is the *correlation* of the two random variables, defined by

$$\text{Corr}[X, Y] = \frac{\text{Cov}[X, Y]}{\sigma_X \sigma_Y}, \quad (30.134)$$

where σ_X and σ_Y are the standard deviations of X and Y respectively. It can be shown that the correlation function lies between -1 and $+1$. If the value assumed is negative, X and Y are said to be *negatively correlated*, if it is positive they are said to be *positively correlated* and if it is zero they are said to be *uncorrelated*. We will now justify the use of these terms.

One particularly useful consequence of its definition is that the covariance of two *independent* variables, X and Y, is zero. It immediately follows from (30.134) that their correlation is also zero, and this justifies the use of the term 'uncorrelated' for two such variables. To show this extremely important property we first note that

$$
\begin{aligned}
\text{Cov}[X, Y] &= E[(X - \mu_X)(Y - \mu_Y)] \\
&= E[XY - \mu_X Y - \mu_Y X + \mu_X \mu_Y] \\
&= E[XY] - \mu_X E[Y] - \mu_Y E[X] + \mu_X \mu_Y \\
&= E[XY] - \mu_X \mu_Y.
\end{aligned}
\tag{30.135}
$$

Now, if X and Y are independent then $E[XY] = E[X]E[Y] = \mu_X \mu_Y$ and so $\text{Cov}[X, Y] = 0$. It is important to note that the converse of this result is not necessarily true; two variables dependent on each other can still be uncorrelated. In other words, it is possible (and not uncommon) for two variables X and Y to be described by a joint distribution $f(x, y)$ that *cannot* be factorised into a product of the form $g(x)h(y)$, but for which $\text{Corr}[X, Y] = 0$. Indeed, from the definition (30.133), we see that for any joint distribution $f(x, y)$ that is symmetric in x about μ_X (or similarly in y) we have $\text{Corr}[X, Y] = 0$.

We have already asserted that if the correlation of two random variables is positive (negative) they are said to be positively (negatively) correlated. We have also stated that the correlation lies between -1 and $+1$. The terminology suggests that if the two RVs are identical (i.e. $X = Y$) then they are completely correlated and that their correlation should be $+1$. Likewise, if $X = -Y$ then the functions are completely anticorrelated and their correlation should be -1. Values of the correlation function between these extremes show the existence of some degree of correlation. In fact it is not necessary that $X = Y$ for $\text{Corr}[X, Y] = 1$; it is sufficient that Y is a linear function of X, i.e. $Y = aX + b$ (with a positive). If a is negative then $\text{Corr}[X, Y] = -1$. To show this we first note that $\mu_Y = a\mu_X + b$. Now

$$
Y = aX + b = aX + \mu_Y - a\mu_X \quad \Rightarrow \quad Y - \mu_Y = a(X - \mu_X),
$$

and so using the definition of the covariance (30.133)

$$
\text{Cov}[X, Y] = aE[(X - \mu_X)^2] = a\sigma_X^2.
$$

It follows from the properties of the variance (subsection 30.5.3) that $\sigma_Y = |a|\sigma_X$ and so, using the definition (30.134) of the correlation,

$$
\text{Corr}[X, Y] = \frac{a\sigma_X^2}{|a|\sigma_X^2} = \frac{a}{|a|},
$$

which is the stated result.

It should be noted that, even if the possibilities of X and Y being non-zero are mutually exclusive, $\text{Corr}[X, Y]$ need not have value ± 1.

> ►*A biased die gives probabilities* $\frac{1}{2}p$, p, p, p, p, $2p$ *of throwing* 1, 2, 3, 4, 5, 6 *respectively.*
> *If the random variable* X *is the number shown on the die and the random variable* Y *is*
> *defined as* X^2, *calculate the covariance and correlation of* X *and* Y.

We have already calculated in subsections 30.2.1 and 30.5.4 that

$$p = \frac{2}{13}, \quad E[X] = \frac{53}{13}, \quad E[X^2] = \frac{253}{13}, \quad V[X] = \frac{480}{169}.$$

Using (30.135), we obtain

$$\text{Cov}[X, Y] = \text{Cov}[X, X^2] = E[X^3] - E[X]E[X^2].$$

Now $E[X^3]$ is given by

$$E[X^3] = 1^3 \times \tfrac{1}{2}p + (2^3 + 3^3 + 4^3 + 5^3)p + 6^3 \times 2p$$
$$= \frac{1313}{2}p = 101,$$

and the covariance of X and Y is given by

$$\text{Cov}[X, Y] = 101 - \frac{53}{13} \times \frac{253}{13} = \frac{3660}{169}.$$

The correlation is defined by $\text{Corr}[X, Y] = \text{Cov}[X, Y]/\sigma_X \sigma_Y$. The standard deviation of Y may be calculated from the definition of the variance. Letting $\mu_Y = E[X^2] = \frac{253}{13}$ gives

$$\sigma_Y^2 = \frac{p}{2}\left(1^2 - \mu_Y\right)^2 + p\left(2^2 - \mu_Y\right)^2 + p\left(3^2 - \mu_Y\right)^2 + p\left(4^2 - \mu_Y\right)^2$$
$$+ p\left(5^2 - \mu_Y\right)^2 + 2p\left(6^2 - \mu_Y\right)^2$$
$$= \frac{187\,356}{169}p = \frac{28\,824}{169}.$$

We deduce that

$$\text{Corr}[X, Y] = \frac{3660}{169}\sqrt{\frac{169}{28\,824}}\sqrt{\frac{169}{480}} \approx 0.984.$$

Thus the random variables X and Y display a strong degree of positive correlation, as we would expect. ◄

We note that the covariance of X and Y occurs in various expressions. For example, if X and Y are *not* independent then

$$V[X + Y] = E\left[(X + Y)^2\right] - (E[X + Y])^2$$
$$= E\left[X^2\right] + 2E[XY] + E\left[Y^2\right] - \{(E[X])^2 + 2E[X]E[Y] + (E[Y])^2\}$$
$$= V[X] + V[Y] + 2(E[XY] - E[X]E[Y])$$
$$= V[X] + V[Y] + 2\,\text{Cov}[X, Y].$$

More generally, we find (for a, b and c constant)

$$V[aX + bY + c] = a^2 V[X] + b^2 V[Y] + 2ab\ \text{Cov}[X, Y].$$

$$(30.136)$$

Note that if X and Y are in fact independent then $\text{Cov}[X, Y] = 0$ and we recover the expression (30.68) in subsection 30.6.4.

We may use (30.136) to obtain an approximate expression for $V[f(X, Y)]$ for any arbitrary function f, even when the random variables X and Y are correlated. Approximating $f(X, Y)$ by the linear terms of its Taylor expansion about the point (μ_X, μ_Y), we have

$$f(X, Y) \approx f(\mu_X, \mu_Y) + \left(\frac{\partial f}{\partial X} \right) (X - \mu_X) + \left(\frac{\partial f}{\partial Y} \right) (Y - \mu_Y),$$

$$(30.137)$$

where the partial derivatives are evaluated at $X = \mu_X$ and $Y = \mu_Y$. Taking the variance of both sides, and using (30.136), we find

$$V[f(X, Y)] \approx \left(\frac{\partial f}{\partial X} \right)^2 V[X] + \left(\frac{\partial f}{\partial Y} \right)^2 V[Y] + 2 \left(\frac{\partial f}{\partial X} \right) \left(\frac{\partial f}{\partial Y} \right) \text{Cov}[X, Y].$$

$$(30.138)$$

Clearly, if $\text{Cov}[X, Y] = 0$, we recover the result (30.69) derived in subsection 30.6.4. We note that (30.138) is exact if $f(X, Y)$ is linear in X and Y.

For several variables X_i, $i = 1, 2, \ldots, n$, we can define the symmetric (positive definite) *covariance matrix* whose elements are

$$V_{ij} = \text{Cov}[X_i, X_j],$$

$$(30.139)$$

and the symmetric (positive definite) *correlation matrix*

$$\rho_{ij} = \text{Corr}[X_i, X_j].$$

The diagonal elements of the covariance matrix are the variances of the variables, whilst those of the correlation matrix are unity. For several variables, (30.138) generalises to

$$V[f(X_1, X_2, \ldots, X_n)] \approx \sum_i \left(\frac{\partial f}{\partial X_i} \right)^2 V[X_i] + \sum_i \sum_{j \neq i} \left(\frac{\partial f}{\partial X_i} \right) \left(\frac{\partial f}{\partial X_j} \right) \text{Cov}[X_i, X_j],$$

where the partial derivatives are evaluated at $X_i = \mu_{X_i}$.

▶A card is drawn at random from a normal 52-card pack and its identity noted. The card is replaced, the pack shuffled and the process repeated. Random variables W, X, Y, Z are defined as follows:

$W = 2$ if the drawn card is a heart; $W = 0$ otherwise.
$X = 4$ if the drawn card is an ace, king, or queen; $X = 2$ if the card is a jack or ten; $X = 0$ otherwise.
$Y = 1$ if the drawn card is red; $Y = 0$ otherwise.
$Z = 2$ if the drawn card is black and an ace, king or queen; $Z = 0$ otherwise.

Establish the correlation matrix for W, X, Y, Z.

The means of the variables are given by

$$\mu_W = 2 \times \tfrac{1}{4} = \tfrac{1}{2}, \quad \mu_X = \left(4 \times \tfrac{3}{13}\right) + \left(2 \times \tfrac{2}{13}\right) = \tfrac{16}{13},$$
$$\mu_Y = 1 \times \tfrac{1}{2} = \tfrac{1}{2}, \quad \mu_Z = 2 \times \tfrac{6}{52} = \tfrac{3}{13}.$$

The variances, calculated from $\sigma_U^2 = V[U] = E\left[U^2\right] - (E[U])^2$, where $U = W, X, Y$ or Z, are

$$\sigma_W^2 = \left(4 \times \tfrac{1}{4}\right) - \left(\tfrac{1}{2}\right)^2 = \tfrac{3}{4}, \quad \sigma_X^2 = \left(16 \times \tfrac{3}{13}\right) + \left(4 \times \tfrac{2}{13}\right) - \left(\tfrac{16}{13}\right)^2 = \tfrac{472}{169},$$
$$\sigma_Y^2 = \left(1 \times \tfrac{1}{2}\right) - \left(\tfrac{1}{2}\right)^2 = \tfrac{1}{4}, \quad \sigma_Z^2 = \left(4 \times \tfrac{6}{52}\right) - \left(\tfrac{3}{13}\right)^2 = \tfrac{69}{169}.$$

The covariances are found by first calculating $E[WX]$ etc. and then forming $E[WX] - \mu_W \mu_X$ etc.

$$E[WX] = 2(4)\left(\tfrac{3}{52}\right) + 2(2)\left(\tfrac{2}{52}\right) = \tfrac{8}{13}, \quad \text{Cov}[W, X] = \tfrac{8}{13} - \tfrac{1}{2}\left(\tfrac{16}{13}\right) = 0,$$

$$E[WY] = 2(1)\left(\tfrac{1}{4}\right) = \tfrac{1}{2}, \quad\quad\quad\quad \text{Cov}[W, Y] = \tfrac{1}{2} - \tfrac{1}{2}\left(\tfrac{1}{2}\right) = \tfrac{1}{4},$$

$$E[WZ] = 0, \quad\quad\quad\quad\quad\quad\quad\quad \text{Cov}[W, Z] = 0 - \tfrac{1}{2}\left(\tfrac{3}{13}\right) = -\tfrac{3}{26},$$

$$E[XY] = 4(1)\left(\tfrac{6}{52}\right) + 2(1)\left(\tfrac{4}{52}\right) = \tfrac{8}{13}, \quad \text{Cov}[X, Y] = \tfrac{8}{13} - \tfrac{16}{13}\left(\tfrac{1}{2}\right) = 0,$$

$$E[XZ] = 4(2)\left(\tfrac{6}{52}\right) = \tfrac{12}{13}, \quad\quad\quad \text{Cov}[X, Z] = \tfrac{12}{13} - \tfrac{16}{13}\left(\tfrac{3}{13}\right) = \tfrac{108}{169},$$

$$E[YZ] = 0, \quad\quad\quad\quad\quad\quad\quad\quad \text{Cov}[Y, Z] = 0 - \tfrac{1}{2}\left(\tfrac{3}{13}\right) = -\tfrac{3}{26}.$$

The correlations $\text{Corr}[W, X]$ and $\text{Corr}[X, Y]$ are clearly zero; the remainder are given by

$$\text{Corr}[W, Y] = \tfrac{1}{4}\left(\tfrac{3}{4} \times \tfrac{1}{4}\right)^{-1/2} = 0.577,$$

$$\text{Corr}[W, Z] = -\tfrac{3}{26}\left(\tfrac{3}{4} \times \tfrac{69}{169}\right)^{-1/2} = -0.209,$$

$$\text{Corr}[X, Z] = \tfrac{108}{169}\left(\tfrac{472}{169} \times \tfrac{69}{169}\right)^{-1/2} = 0.598,$$

$$\text{Corr}[Y, Z] = -\tfrac{3}{26}\left(\tfrac{1}{4} \times \tfrac{69}{169}\right)^{-1/2} = -0.361.$$

Finally, then, we can write down the correlation matrix:

$$\rho = \begin{pmatrix} 1 & 0 & 0.58 & -0.21 \\ 0 & 1 & 0 & 0.60 \\ 0.58 & 0 & 1 & -0.36 \\ -0.21 & 0.60 & -0.36 & 1 \end{pmatrix}.$$

As would be expected, X is uncorrelated with either W or Y, colour and face-value being two independent characteristics. Positive correlations are to be expected between W and Y and between X and Z; both correlations are fairly strong. Moderate anticorrelations exist between Z and both W and Y, reflecting the fact that it is impossible for W and Y to be positive if Z is positive. ◄

Finally, let us suppose that the random variables X_i, $i = 1, 2, \ldots, n$, are related to a second set of random variables $Y_k = Y_k(X_1, X_2, \ldots, X_n)$, $k = 1, 2, \ldots, m$. By expanding each Y_k as a Taylor series as in (30.137) and inserting the resulting expressions into the definition of the covariance (30.133), we find that the elements of the covariance matrix for the Y_k variables are given by

$$\text{Cov}[Y_k, Y_l] \approx \sum_i \sum_j \left(\frac{\partial Y_k}{\partial X_i} \right) \left(\frac{\partial Y_l}{\partial X_j} \right) \text{Cov}[X_i, X_j].$$
(30.140)

It is straightforward to show that this relation is exact if the Y_k are linear combinations of the X_i. Equation (30.140) can then be written in matrix form as

$$V_Y = SV_X S^T,$$
(30.141)

where V_Y and V_X are the covariance matrices of the Y_k and X_l variables respectively and S is the rectangular $m \times n$ matrix with elements $S_{ki} = \partial Y_k / \partial X_i$.

30.13 Generating functions for joint distributions

It is straightforward to generalise the discussion of generating function in section 30.7 to joint distributions. For a multivariate distribution $f(X_1, X_2, \ldots, X_n)$ of non-negative integer random variables X_i, $i = 1, 2, \ldots, n$, we define the probability generating function to be

$$\Phi(t_1, t_2, \ldots, t_n) = E[t_1^{X_1} t_2^{X_2} \cdots t_n^{X_n}].$$

As in the single-variable case, we may also define the closely related moment generating function, which has wider applicability since it is not restricted to non-negative integer random variables but can be used with any set of discrete or continuous random variables X_i ($i = 1, 2, \ldots, n$). The MGF of the multivariate distribution $f(X_1, X_2, \ldots, X_n)$ is defined as

$$M(t_1, t_2, \ldots, t_n) = E[e^{t_1 X_1} e^{t_2 X_2} \cdots e^{t_n X_n}] = E[e^{t_1 X_1 + t_2 X_2 + \cdots + t_n X_n}]$$
(30.142)

and may be used to evaluate (joint) moments of $f(X_1, X_2, \ldots, X_n)$. By performing a derivation analogous to that presented for the single-variable case in subsection 30.7.2, it can be shown that

$$E[X_1^{m_1} X_2^{m_2} \cdots X_n^{m_n}] = \frac{\partial^{m_1 + m_2 + \cdots + m_n} M(0, 0, \ldots, 0)}{\partial t_1^{m_1} \partial t_2^{m_2} \cdots \partial t_n^{m_n}}.$$
(30.143)

Finally we note that, by analogy with the single-variable case, the characteristic function and the cumulant generating function of a multivariate distribution are defined respectively as

$$C(t_1, t_2, \ldots, t_n) = M(it_1, it_2, \ldots, it_n) \quad \text{and} \quad K(t_1, t_2, \ldots, t_n) = \ln M(t_1, t_2, \ldots, t_n).$$

▶*Suppose that the random variables X_i, $i = 1, 2, \ldots, n$, are described by the PDF*

$$f(\mathbf{x}) = f(x_1, x_2, \ldots, x_n) = N \exp(-\tfrac{1}{2} \mathbf{x}^{\mathrm{T}} \mathsf{A} \mathbf{x}),$$

where the column vector $\mathbf{x} = (x_1 \quad x_2 \quad \cdots \quad x_n)^{\mathrm{T}}$, A is an $n \times n$ symmetric matrix and N is a normalisation constant such that

$$\int_\infty f(\mathbf{x}) \, d^n \mathbf{x} \equiv \int_{-\infty}^{\infty} \int_{-\infty}^{\infty} \cdots \int_{-\infty}^{\infty} f(x_1, x_2, \ldots, x_n) \, dx_1 \, dx_2 \cdots dx_n = 1.$$

Find the MGF of $f(\mathbf{x})$.

From (30.142), the MGF is given by

$$M(t_1, t_2, \ldots, t_n) = N \int_\infty \exp(-\tfrac{1}{2} \mathbf{x}^{\mathrm{T}} \mathsf{A} \mathbf{x} + \mathbf{t}^{\mathrm{T}} \mathbf{x}) \, d^n \mathbf{x}, \tag{30.144}$$

where the column vector $\mathbf{t} = (t_1 \quad t_2 \quad \cdots \quad t_n)^{\mathrm{T}}$. In order to evaluate this multiple integral, we begin by noting that

$$\mathbf{x}^{\mathrm{T}} \mathsf{A} \mathbf{x} - 2\mathbf{t}^{\mathrm{T}} \mathbf{x} = (\mathbf{x} - \mathsf{A}^{-1}\mathbf{t})^{\mathrm{T}} \mathsf{A} (\mathbf{x} - \mathsf{A}^{-1}\mathbf{t}) - \mathbf{t}^{\mathrm{T}} \mathsf{A}^{-1} \mathbf{t},$$

which is the matrix equivalent of 'completing the square'. Using this expression in (30.144) and making the substitution $\mathbf{y} = \mathbf{x} - \mathsf{A}^{-1}\mathbf{t}$, we obtain

$$M(t_1, t_2, \ldots, t_n) = c \exp(\tfrac{1}{2} \mathbf{t}^{\mathrm{T}} \mathsf{A}^{-1} \mathbf{t}), \tag{30.145}$$

where the constant c is given by

$$c = N \int_\infty \exp(-\tfrac{1}{2} \mathbf{y}^{\mathrm{T}} \mathsf{A} \mathbf{y}) \, d^n \mathbf{y}.$$

From the normalisation condition for N, we see that $c = 1$, as indeed it must be in order that $M(0, 0, \ldots, 0) = 1$. ◀

30.14 Transformation of variables in joint distributions

Suppose the random variables X_i, $i = 1, 2, \ldots, n$, are described by the multivariate PDF $f(x_1, x_2 \ldots, x_n)$. If we wish to consider random variables Y_j, $j = 1, 2, \ldots, m$, related to the X_i by $Y_j = Y_j(X_1, X_2, \ldots, X_m)$ then we may calculate $g(y_1, y_2, \ldots, y_m)$, the PDF for the Y_j, in a similar way to that in the univariate case by demanding that

$$|f(x_1, x_2 \ldots, x_n) \, dx_1 \, dx_2 \cdots dx_n| = |g(y_1, y_2, \ldots, y_m) \, dy_1 \, dy_2 \cdots dy_m|.$$

From the discussion of changing the variables in multiple integrals given in chapter 6 it follows that, in the special case where $n = m$,

$$g(y_1, y_2, \ldots, y_m) = f(x_1, x_2 \ldots, x_n)|J|,$$

1206

where

$$J \equiv \frac{\partial(x_1, x_2 \ldots, x_n)}{\partial(y_1, y_2, \ldots, y_n)} = \begin{vmatrix} \dfrac{\partial x_1}{\partial y_1} & \cdots & \dfrac{\partial x_n}{\partial y_1} \\ \vdots & \ddots & \vdots \\ \dfrac{\partial x_1}{\partial y_n} & \cdots & \dfrac{\partial x_n}{\partial y_n} \end{vmatrix},$$

is the Jacobian of the x_i with respect to the y_j.

> ▶ *Suppose that the random variables X_i, $i = 1, 2, \ldots, n$, are independent and Gaussian distributed with means μ_i and variances σ_i^2 respectively. Find the PDF for the new variables $Z_i = (X_i - \mu_i)/\sigma_i$, $i = 1, 2, \ldots, n$. By considering an elemental spherical shell in $\mathbf{Z}$-space, find the PDF of the chi-squared random variable $\chi_n^2 = \sum_{i=1}^n Z_i^2$.*

Since the X_i are independent random variables,

$$f(x_1, x_2, \ldots, x_n) = f(x_1) f(x_2) \cdots f(x_n) = \frac{1}{(2\pi)^{n/2} \sigma_1 \sigma_2 \cdots \sigma_n} \exp\left[-\sum_{i=1}^n \frac{(x_i - \mu_i)^2}{2\sigma_i^2} \right].$$

To derive the PDF for the variables Z_i, we require

$$|f(x_1, x_2, \ldots, x_n) \, dx_1 \, dx_2 \cdots dx_n| = |g(z_1, z_2, \ldots, z_n) \, dz_1 \, dz_2 \cdots dz_n|,$$

and, noting that $dz_i = dx_i/\sigma_i$, we obtain

$$g(z_1, z_2, \ldots, z_n) = \frac{1}{(2\pi)^{n/2}} \exp\left(-\frac{1}{2} \sum_{i=1}^n z_i^2 \right).$$

Let us now consider the random variable $\chi_n^2 = \sum_{i=1}^n Z_i^2$, which we may regard as the square of the distance from the origin in the n-dimensional $\mathbf{Z}$-space. We now require that

$$g(z_1, z_2, \ldots, z_n) \, dz_1 \, dz_2 \cdots dz_n = h(\chi_n^2) d\chi_n^2.$$

If we consider the infinitesimal volume $dV = dz_1 \, dz_2 \cdots dz_n$ to be that enclosed by the n-dimensional spherical shell of radius χ_n and thickness $d\chi_n$ then we may write $dV = A\chi_n^{n-1} d\chi_n$, for some constant A. We thus obtain

$$h(\chi_n^2) d\chi_n^2 \propto \exp(-\tfrac{1}{2}\chi_n^2)\chi_n^{n-1} d\chi_n \propto \exp(-\tfrac{1}{2}\chi_n^2)\chi_n^{n-2} d\chi_n^2,$$

where we have used the fact that $d\chi_n^2 = 2\chi_n \, d\chi_n$. Thus we see that the PDF for χ_n^2 is given by

$$h(\chi_n^2) = B \exp(-\tfrac{1}{2}\chi_n^2)\chi_n^{n-2},$$

for some constant B. This constant may be determined from the normalisation condition

$$\int_0^\infty h(\chi_n^2) \, d\chi_n^2 = 1$$

and is found to be $B = [2^{n/2}\Gamma(\tfrac{1}{2}n)]^{-1}$. This is the nth-order chi-squared distribution discussed in subsection 30.9.4. ◀

30.15 Important joint distributions

In this section we will examine two important multivariate distributions, the *multinomial distribution*, which is an extension of the binomial distribution, and the *multivariate Gaussian distribution*.

30.15.1 The multinomial distribution

The binomial distribution describes the probability of obtaining x 'successes' from n independent trials, where each trial has only two possible outcomes. This may be generalised to the case where each trial has k possible outcomes with respective probabilities p_1, p_2, ..., p_k. If we consider the random variables X_i, $i = 1, 2, \ldots, n$, to be the number of outcomes of type i in n trials then we may calculate their joint probability function

$$f(x_1, x_2, \ldots, x_k) = \Pr(X_1 = x_1, \ X_2 = x_2, \ \ldots, \ X_k = x_k),$$

where we must have $\sum_{i=1}^{k} x_i = n$. In n trials the probability of obtaining x_1 outcomes of type 1, followed by x_2 outcomes of type 2 etc. is given by

$$p_1^{x_1} p_2^{x_2} \cdots p_k^{x_k}.$$

However, the number of distinguishable permutations of this result is

$$\frac{n!}{x_1! x_2! \cdots x_k!},$$

and thus

$$f(x_1, x_2, \ldots, x_k) = \frac{n!}{x_1! x_2! \cdots x_k!} p_1^{x_1} p_2^{x_2} \cdots p_k^{x_k}. \tag{30.146}$$

This is the *multinomial probability distribution*.

If $k = 2$ then the multinomial distribution reduces to the familiar binomial distribution. Although in this form the binomial distribution appears to be a function of two random variables, it must be remembered that, in fact, since $p_2 = 1 - p_1$ and $x_2 = n - x_1$, the distribution of X_1 is entirely determined by the parameters p and n. That X_1 has a *binomial* distribution is shown by remembering that it represents the number of objects of a particular type obtained from sampling with replacement, which led to the original definition of the binomial distribution. In fact, any of the random variables X_i has a binomial distribution, i.e. the marginal distribution of each X_i is binomial with parameters n and p_i. It immediately follows that

$$E[X_i] = np_i \qquad \text{and} \qquad V[X_i]^2 = np_i(1 - p_i). \tag{30.147}$$

> ▶*At a village fête patrons were invited, for a 10 p entry fee, to pick without looking six tickets from a drum containing equal large numbers of red, blue and green tickets. If five or more of the tickets were of the same colour a prize of 100 p was awarded. A consolation award of 40 p was made if two tickets of each colour were picked. Was a good time had by all?*

In this case, all types of outcome (red, blue and green) have the same probabilities. The probability of obtaining any given combination of tickets is given by the multinomial distribution with $n = 6$, $k = 3$ and $p_i = \frac{1}{3}$, $i = 1, 2, 3$.

(i) The probability of picking six tickets of the same colour is given by

$$\text{Pr (six of the same colour)} = 3 \times \frac{6!}{6!0!0!} \left(\frac{1}{3}\right)^6 \left(\frac{1}{3}\right)^0 \left(\frac{1}{3}\right)^0 = \frac{1}{243}.$$

The factor of 3 is present because there are three different colours.
(ii) The probability of picking five tickets of one colour and one ticket of another colour is

$$\text{Pr(five of one colour; one of another)} = 3 \times 2 \times \frac{6!}{5!1!0!} \left(\frac{1}{3}\right)^5 \left(\frac{1}{3}\right)^1 \left(\frac{1}{3}\right)^0 = \frac{4}{81}.$$

The factors of 3 and 2 are included because there are three ways to choose the colour of the five matching tickets, and then two ways to choose the colour of the remaining ticket.
(iii) Finally, the probability of picking two tickets of each colour is

$$\text{Pr (two of each colour)} = \frac{6!}{2!2!2!} \left(\frac{1}{3}\right)^2 \left(\frac{1}{3}\right)^2 \left(\frac{1}{3}\right)^2 = \frac{10}{81}.$$

Thus the expected return to any patron was, in pence,

$$100 \left(\frac{1}{243} + \frac{4}{81}\right) + \left(40 \times \frac{10}{81}\right) = 10.29$$

A good time was had by all but the stallholder! ◀

30.15.2 The multivariate Gaussian distribution

A particularly interesting multivariate distribution is provided by the generalisation of the Gaussian distribution to multiple random variables X_i, $i = 1, 2, \ldots, n$. If the expectation value of X_i is $E(X_i) = \mu_i$ then the general form of the PDF is given by

$$f(x_1, x_2, \ldots, x_n) = N \exp\left[-\frac{1}{2} \sum_i \sum_j a_{ij}(x_i - \mu_i)(x_j - \mu_j)\right],$$

where $a_{ij} = a_{ji}$ and N is a normalisation constant that we give below. If we write the column vectors $\mathbf{x} = (x_1 \quad x_2 \quad \cdots \quad x_n)^{\mathrm{T}}$ and $\mu = (\mu_1 \quad \mu_2 \quad \cdots \quad \mu_n)^{\mathrm{T}}$, and denote the matrix with elements a_{ij} by A then

$$f(\mathbf{x}) = f(x_1, x_2, \ldots, x_n) = N \exp\left[-\frac{1}{2}(\mathbf{x} - \mu)^{\mathrm{T}} \mathsf{A}(\mathbf{x} - \mu)\right],$$

where A is symmetric. Using the same method as that used to derive (30.145) it is straightforward to show that the MGF of $f(\mathbf{x})$ is given by

$$M(t_1, t_2, \ldots, t_n) = \exp\left(\mu^{\mathrm{T}}\mathbf{t} + \frac{1}{2}\mathbf{t}^{\mathrm{T}}\mathsf{A}^{-1}\mathbf{t}\right),$$

where the column matrix $\mathbf{t} = (t_1 \quad t_2 \quad \cdots \quad t_n)^{\mathrm{T}}$. From the MGF, we find that

$$E[X_i X_j] = \frac{\partial^2 M(0, 0, \ldots, 0)}{\partial t_i \partial t_j} = \mu_i \mu_j + (\mathsf{A}^{-1})_{ij},$$

and thus, using (30.135), we obtain

$$\text{Cov}[X_i, X_j] = E[(X_i - \mu_i)(X_j - \mu_j)] = (A^{-1})_{ij}.$$

Hence A is equal to the inverse of the covariance matrix V of the X_i, see (30.139). Thus, with the correct normalisation, $f(x)$ is given by

$$f(x) = \frac{1}{(2\pi)^{n/2}(\det V)^{1/2}} \exp\left[-\tfrac{1}{2}(x-\mu)^T V^{-1}(x-\mu)\right].$$

$$(30.148)$$

►*Evaluate the integral*

$$I = \int_\infty \exp\left[-\tfrac{1}{2}(x-\mu)^T V^{-1}(x-\mu)\right] d^n x,$$

where V is a symmetric matrix, and hence verify the normalisation in (30.148).

We begin by making the substitution $y = x - \mu$ to obtain

$$I = \int_\infty \exp(-\tfrac{1}{2}y^T V^{-1}y) \, d^n y.$$

Since V is a symmetric matrix, it may be diagonalised by an orthogonal transformation to the new set of variables $y' = S^T y$, where S is the orthogonal matrix with the normalised eigenvectors of V as its columns (see section 8.16). In this new basis, the matrix V becomes

$$V' = S^T V S = \text{diag}(\lambda_1, \lambda_2, \ldots, \lambda_n),$$

where the λ_i are the eigenvalues of V. Also, since S is orthogonal, $\det S = \pm 1$, and so

$$d^n y = |\det S| \, d^n y' = d^n y'.$$

Thus we can write I as

$$I = \int_{-\infty}^{\infty} \int_{-\infty}^{\infty} \cdots \int_{-\infty}^{\infty} \exp\left(-\sum_{i=1}^{n} \frac{y_i'^2}{2\lambda_i}\right) dy_1' \, dy_2' \cdots dy_n'$$

$$= \prod_{i=1}^{n} \int_{-\infty}^{\infty} \exp\left(-\frac{y_i'^2}{2\lambda_i}\right) dy_i' = (2\pi)^{n/2}(\lambda_1 \lambda_2 \cdots \lambda_n)^{1/2}, \qquad (30.149)$$

where we have used the standard integral $\int_{-\infty}^{\infty} \exp(-\alpha y^2) \, dy = (\pi/\alpha)^{1/2}$ (see subsection 6.4.2). From section 8.16, however, we note that the product of eigenvalues in (30.149) is equal to det V. Thus we finally obtain

$$I = (2\pi)^{n/2}(\det V)^{1/2},$$

and hence the normalisation in (30.148) ensures that $f(x)$ integrates to unity. ◄

The above example illustrates some important points concerning the multivariate Gaussian distribution. In particular, we note that the Y_i' are *independent* Gaussian variables with mean zero and variance λ_i. Thus, given a general set of n Gaussian variables x with means μ and covariance matrix V, one can always perform the above transformation to obtain a new set of variables y', which are linear combinations of the old ones and are distributed as independent Gaussians with zero mean and variances λ_i.

This result is extremely useful in proving many of the properties of the multivariate Gaussian. For example, let us consider the quadratic form (multiplied by 2) appearing in the exponent of (30.148) and write it as χ_n^2, i.e.

$$\chi_n^2 = (\mathbf{x} - \boldsymbol{\mu})^{\mathsf{T}} \mathsf{V}^{-1} (\mathbf{x} - \boldsymbol{\mu}). \qquad (30.150)$$

From (30.149), we see that we may also write it as

$$\chi_n^2 = \sum_{i=1}^{n} \frac{y_i'^2}{\lambda_i},$$

which is the sum of n independent Gaussian variables with mean zero and unit variance. Thus, as our notation implies, the quantity χ_n^2 is distributed as a chi-squared variable of order n. As illustrated in exercise 30.40, if the variables X_i are required to satisfy m linear constraints of the form $\sum_{i=1}^{n} c_i X_i = 0$ then χ_n^2 defined in (30.150) is distributed as a chi-squared variable of order $n - m$.

30.16 Exercises

30.1   By shading or numbering Venn diagrams, determine which of the following are valid relationships between events. For those that are, prove the relationship using de Morgan's laws.

(a) $\overline{(\bar{X} \cup Y)} = X \cap \bar{Y}$.
(b) $\bar{X} \cup \bar{Y} = \overline{(X \cup Y)}$.
(c) $(X \cup Y) \cap Z = (X \cup Z) \cap Y$.
(d) $X \cup \overline{(Y \cap Z)} = (X \cup \bar{Y}) \cap \bar{Z}$.
(e) $X \cup \overline{(Y \cap Z)} = (X \cup \bar{Y}) \cup \bar{Z}$.

30.2   Given that events X, Y and Z satisfy

$$(X \cap Y) \cup (Z \cap X) \cup \overline{(\bar{X} \cup \bar{Y})} = \overline{(\bar{Z} \cup \bar{Y})} \cup \{[(\bar{Z} \cup \bar{X}) \cup (\bar{X} \cap Z)] \cap Y\},$$

prove that $X \supset Y$, and that either $X \cap Z = \emptyset$ or $Y \supset Z$.

30.3 A and B each have two unbiased four-faced dice, the four faces being numbered 1, 2, 3, 4. Without looking, B tries to guess the sum x of the numbers on the bottom faces of A's two dice after they have been thrown onto a table. If the guess is correct B receives x^2 euros, but if not he loses x euros.

Determine B's expected gain per throw of A's dice when he adopts each of the following strategies:

(a) he selects x at random in the range $2 \le x \le 8$;
(b) he throws his own two dice and guesses x to be whatever they indicate;
(c) he takes your advice and always chooses the same value for x. Which number would you advise?

30.4   Use the method of induction to prove equation (30.16), the probability addition law for the union of n general events.

30.5   Two duellists, A and B, take alternate shots at each other, and the duel is over when a shot (fatal or otherwise!) hits its target. Each shot fired by A has a probability α of hitting B, and each shot fired by B has a probability β of hitting A. Calculate the probabilities P_1 and P_2, defined as follows, that A will win such a duel: P_1, A fires the first shot; P_2, B fires the first shot.

If they agree to fire simultaneously, rather than alternately, what is the probability P_3 that A will win, i.e. hit B without being hit himself?

30.6 $X_1, X_2, \ldots, X_n$ are independent, identically distributed, random variables drawn from a uniform distribution on $[0, 1]$. The random variables A and B are defined by

$$A = \min(X_1, X_2, \ldots, X_n), \qquad B = \max(X_1, X_2, \ldots, X_n).$$

For any fixed k such that $0 \le k \le \frac{1}{2}$, find the probability, p_n, that both

$$A \le k \qquad \text{and} \qquad B \ge 1 - k.$$

Check your general formula by considering directly the cases (a) $k = 0$, (b) $k = \frac{1}{2}$, (c) $n = 1$ and (d) $n = 2$.

30.7 A tennis tournament is arranged on a straight knockout basis for 2^n players, and for each round, except the final, opponents for those still in the competition are drawn at random. The quality of the field is so even that in any match it is equally likely that either player will win. Two of the players have surnames that begin with 'Q'. Find the probabilities that they play each other

(a) in the final,
(b) at some stage in the tournament.

30.8 This exercise shows that the odds are hardly ever 'evens' when it comes to dice rolling.

(a) Gamblers A and B each roll a fair six-faced die, and B wins if his score is strictly greater than A's. Show that the odds are 7 to 5 in A's favour.
(b) Calculate the probabilities of scoring a total T from two rolls of a fair die for $T = 2, 3, \ldots, 12$. Gamblers C and D each roll a fair die twice and score respective totals T_C and T_D, D winning if $T_D > T_C$. Realising that the odds are not equal, D insists that C should increase her stake for each game. C agrees to stake £1.10 per game, as compared to D's £1.00 stake. Who will show a profit?

30.9 An electronics assembly firm buys its microchips from three different suppliers; half of them are bought from firm X, whilst firms Y and Z supply 30% and 20%, respectively. The suppliers use different quality-control procedures and the percentages of defective chips are 2%, 4% and 4% for X, Y and Z, respectively. The probabilities that a defective chip will fail two or more assembly-line tests are 40%, 60% and 80%, respectively, whilst all defective chips have a 10% chance of escaping detection. An assembler finds a chip that fails only one test. What is the probability that it came from supplier X?

30.10 As every student of probability theory will know, Bayesylvania is awash with natives, not all of whom can be trusted to tell the truth, and lost, and apparently somewhat deaf, travellers who ask the same question several times in an attempt to get directions to the nearest village.

One such traveller finds himself at a T-junction in an area populated by the Asciis and Bisciis in the ratio 11 to 5. As is well known, the Biscii always lie, but the Ascii tell the truth three quarters of the time, giving independent answers to all questions, even to immediately repeated ones.

(a) The traveller asks one particular native twice whether he should go to the left or to the right to reach the local village. Each time he is told 'left'. Should he take this advice, and, if he does, what are his chances of reaching the village?
(b) The traveller then asks the same native the same question a third time, and for a third time receives the answer 'left'. What should the traveller do now? Have his chances of finding the village been altered by asking the third question?

30.11   A boy is selected at random from amongst the children belonging to families with n children. It is known that he has at least two sisters. Show that the probability that he has $k - 1$ brothers is

$$\frac{(n-1)!}{(2^{n-1} - n)(k-1)!(n-k)!},$$

for $1 \leq k \leq n - 2$ and zero for other values of k. Assume that boys and girls are equally likely.

30.12   Villages A, B, C and D are connected by overhead telephone lines joining AB, AC, BC, BD and CD. As a result of severe gales, there is a probability p (the same for each link) that any particular link is broken.

(a) Show that the probability that a call can be made from A to B is

$$1 - p^2 - 2p^3 + 3p^4 - p^5.$$

(b) Show that the probability that a call can be made from D to A is

$$1 - 2p^2 - 2p^3 + 5p^4 - 2p^5.$$

30.13   A set of $2N + 1$ rods consists of one of each integer length $1, 2, \ldots, 2N, 2N + 1$. Three, of lengths a, b and c, are selected, of which a is the longest. By considering the possible values of b and c, determine the number of ways in which a non-degenerate triangle (i.e. one of non-zero area) can be formed (i) if a is even, and (ii) if a is odd. Combine these results appropriately to determine the total number of non-degenerate triangles that can be formed using three of the $2N + 1$ rods, and hence show that the probability that such a triangle can be formed from a random selection (without replacement) of three rods is

$$\frac{(N-1)(4N+1)}{2(4N^2 - 1)}.$$

30.14   A certain marksman never misses his target, which consists of a disc of unit radius with centre O. The probability that any given shot will hit the target within a distance t of O is t^2, for $0 \leq t \leq 1$. The marksman fires n independent shots at the target, and the random variable Y is the radius of the smallest circle with centre O that encloses all the shots. Determine the PDF for Y and hence find the expected area of the circle.

The shot that is furthest from O is now rejected and the corresponding circle determined for the remaining $n - 1$ shots. Show that its expected area is

$$\frac{n-1}{n+1}\pi.$$

30.15   The duration (in minutes) of a telephone call made from a public call-box is a random variable T. The probability density function of T is

$$f(t) = \begin{cases} 0 & t < 0, \\ \frac{1}{2} & 0 \leq t < 1, \\ ke^{-2t} & t \geq 1, \end{cases}$$

where k is a constant. To pay for the call, 20 pence has to be inserted at the beginning, and a further 20 pence after each subsequent half-minute. Determine by how much the average cost of a call exceeds the cost of a call of average length charged at 40 pence per minute.

30.16   Kittens from different litters do not get on with each other, and fighting breaks out whenever two kittens from different litters are present together. A cage initially contains x kittens from one litter and y from another. To quell the

1213

fighting, kittens are removed at random, one at a time, until peace is restored. Show, by induction, that the expected number of kittens finally remaining is

$$N(x, y) = \frac{x}{y+1} + \frac{y}{x+1}.$$

30.17 If the scores in a cup football match are equal at the end of the normal period of play, a 'penalty shoot-out' is held in which each side takes up to five shots (from the penalty spot) alternately, the shoot-out being stopped if one side acquires an unassailable lead (i.e. has a lead greater than its opponents have shots remaining). If the scores are still level after the shoot-out a 'sudden death' competition takes place.

In sudden death each side takes one shot and the competition is over if one side scores and the other does not; if both score, or both fail to score, a further shot is taken by each side, and so on. Team 1, which takes the first penalty, has a probability p_1, which is independent of the player involved, of scoring and a probability q_1 $(= 1 - p_1)$ of missing; p_2 and q_2 are defined likewise.

Define $\Pr(i : x, y)$ as the probability that team i has scored x goals after y attempts, and let $f(M)$ be the probability that the shoot-out terminates after a *total* of M shots.

(a) Prove that the probability that 'sudden death' will be needed is

$$f(11+) = \sum_{r=0}^{5} ({}^5C_r)^2 (p_1 p_2)^r (q_1 q_2)^{5-r}.$$

(b) Give reasoned arguments (preferably without first looking at the expressions involved) which show that

$$f(M = 2N) = \sum_{r=0}^{2N-6} \left\{ \begin{array}{l} p_2 \Pr(1 : r, N) \Pr(2 : 5 - N + r, N - 1) \\ + q_2 \Pr(1 : 6 - N + r, N) \Pr(2 : r, N - 1) \end{array} \right\}$$

for $N = 3, 4, 5$ and

$$f(M = 2N + 1) = \sum_{r=0}^{2N-5} \left\{ \begin{array}{l} p_1 \Pr(1 : 5 - N + r, N) \Pr(2 : r, N) \\ + q_1 \Pr(1 : r, N) \Pr(2 : 5 - N + r, N) \end{array} \right\}$$

for $N = 3, 4$.

(c) Give an explicit expression for $\Pr(i : x, y)$ and hence show that if the teams are so well matched that $p_1 = p_2 = 1/2$ then

$$f(2N) = \sum_{r=0}^{2N-6} \left(\frac{1}{2^{2N}} \right) \frac{N!(N-1)!6}{r!(N-r)!(6-N+r)!(2N-6-r)!},$$

$$f(2N+1) = \sum_{r=0}^{2N-5} \left(\frac{1}{2^{2N}} \right) \frac{(N!)^2}{r!(N-r)!(5-N+r)!(2N-5-r)!}.$$

(d) Evaluate these expressions to show that, expressing $f(M)$ in units of 2^{-8}, we have

M	6	7	8	9	10	11+
$f(M)$	8	24	42	56	63	63

Give a simple explanation of why $f(10) = f(11+)$.

30.18 A particle is confined to the one-dimensional space $0 \leq x \leq a$, and classically it can be in any small interval dx with equal probability. However, quantum mechanics gives the result that the probability distribution is proportional to $\sin^2(n\pi x/a)$, where n is an integer. Find the variance in the particle's position in both the classical and quantum-mechanical pictures, and show that, although they differ, the latter tends to the former in the limit of large n, in agreement with the correspondence principle of physics.

30.19 A continuous random variable X has a probability density function $f(x)$; the corresponding cumulative probability function is $F(x)$. Show that the random variable $Y = F(X)$ is uniformly distributed between 0 and 1.

30.20 For a non-negative integer random variable X, in addition to the probability generating function $\Phi_X(t)$ defined in equation (30.71), it is possible to define the probability generating function

$$\Psi_X(t) = \sum_{n=0}^{\infty} g_n t^n,$$

where g_n is the probability that $X > n$.

(a) Prove that Φ_X and Ψ_X are related by

$$\Psi_X(t) = \frac{1 - \Phi_X(t)}{1 - t}.$$

(b) Show that $E[X]$ is given by $\Psi_X(1)$ and that the variance of X can be expressed as $2\Psi'_X(1) + \Psi_X(1) - [\Psi_X(1)]^2$.

(c) For a particular random variable X, the probability that $X > n$ is equal to α^{n+1}, with $0 < \alpha < 1$. Use the results in (b) to show that $V[X] = \alpha(1 - \alpha)^{-2}$.

30.21 This exercise is about interrelated binomial trials.

(a) In two sets of binomial trials T and t, the probabilities that a trial has a successful outcome are P and p, respectively, with corresponding probabilities of failure of $Q = 1 - P$ and $q = 1 - p$. One 'game' consists of a trial T, followed, if T is successful, by a trial t and then a further trial T. The two trials continue to alternate until one of the T-trials fails, at which point the game ends. The score S for the game is the total number of successes in the t-trials. Find the PGF for S and use it to show that

$$E[S] = \frac{Pp}{Q}, \qquad V[S] = \frac{Pp(1 - Pq)}{Q^2}.$$

(b) Two normal unbiased six-faced dice A and B are rolled alternately starting with A; if A shows a 6 the experiment ends. If B shows an odd number no points are scored, if it shows a 2 or a 4 then one point is scored, whilst if it records a 6 then two points are awarded. Find the average and standard deviation of the score for the experiment and show that the latter is the greater.

30.22 Use the formula obtained in subsection 30.8.2 for the moment generating function of the geometric distribution to determine the CGF, $K_n(t)$, for the number of trials needed to record n successes. Evaluate the first four cumulants, and use them to confirm the stated results for the mean and variance, and to show that the distribution has skewness and kurtosis given, respectively, by

$$\frac{2 - p}{\sqrt{n(1 - p)}} \qquad \text{and} \qquad 3 + \frac{6 - 6p + p^2}{n(1 - p)}.$$

30.23 A point P is chosen at random on the circle $x^2 + y^2 = 1$. The random variable X denotes the distance of P from $(1, 0)$. Find the mean and variance of X and the probability that X is greater than its mean.

30.24 As assistant to a celebrated and imperious newspaper proprietor, you are given the job of running a lottery, in which each of his five million readers will have an equal independent chance, p, of winning a million pounds; you have the job of choosing p. However, if nobody wins it will be bad for publicity, whilst if more than two readers do so, the prize cost will more than offset the profit from extra circulation – in either case you will be sacked! Show that, however you choose p, there is more than a 40% chance you will soon be clearing your desk.

30.25 The number of errors needing correction on each page of a set of proofs follows a Poisson distribution of mean μ. The cost of the first correction on any page is α and that of each subsequent correction on the same page is β. Prove that the average cost of correcting a page is

$$\alpha + \beta(\mu - 1) - (\alpha - \beta)e^{-\mu}.$$

30.26 In the game of Blackball, at each turn Muggins draws a ball at random from a bag containing five white balls, three red balls and two black balls; after being recorded, the ball is replaced in the bag. A white ball earns him \$1, whilst a red ball gets him \$2; in either case, he also has the option of leaving with his current winnings or of taking a further turn on the same basis. If he draws a black ball the game ends and he loses all he may have gained previously. Find an expression for Muggins' expected return if he adopts the strategy of drawing up to n balls, provided he has not been eliminated by then.

Show that, as the entry fee to play is \$3, Muggins should be dissuaded from playing Blackball, but, if that cannot be done, what value of n would you advise him to adopt?

30.27 Show that, for large r, the value at the maximum of the PDF for the gamma distribution of order r with parameter λ is approximately $\lambda / \sqrt{2\pi(r-1)}$.

30.28 A husband and wife decide that their family will be complete when it includes two boys and two girls – but that this would then be enough! The probability that a new baby will be a girl is p. Ignoring the possibility of identical twins, show that the expected size of their family is

$$2 \left(\frac{1}{pq} - 1 - pq \right),$$

where $q = 1 - p$.

30.29 The probability distribution for the number of eggs in a clutch is $\text{Po}(\lambda)$, and the probability that each egg will hatch is p (independently of the size of the clutch). Show by direct calculation that the probability distribution for the number of chicks that hatch is $\text{Po}(\lambda p)$ and so justify the assumptions made in the worked example at the end of subsection 30.7.1.

30.30 A shopper buys 36 items at random in a supermarket, where, because of the sales tax imposed, the final digit (the number of pence) in the price is uniformly and randomly distributed from 0 to 9. Instead of adding up the bill exactly, she rounds each item to the nearest 10 pence, rounding up or down with equal probability if the price ends in a '5'. Should she suspect a mistake if the cashier asks her for 23 pence more than she estimated?

30.31 Under EU legislation on harmonisation, all kippers are to weigh 0.2000 kg, and vendors who sell underweight kippers must be fined by their government. The weight of a kipper is normally distributed, with a mean of 0.2000 kg and a standard deviation of 0.0100 kg. They are packed in cartons of 100 and large quantities of them are sold.

Every day, a carton is to be selected at random from each vendor and tested according to one of the following schemes, which have been approved for the purpose.

(a) The entire carton is weighed, and the vendor is fined 2500 euros if the average weight of a kipper is less than 0.1975 kg.
(b) Twenty five kippers are selected at random from the carton; the vendor is fined 100 euros if the average weight of a kipper is less than 0.1980 kg.
(c) Kippers are removed one at a time, at random, until one has been found that weighs *more* than 0.2000 kg; the vendor is fined $4n(n-1)$ euros, where n is the number of kippers removed.

Which scheme should the Chancellor of the Exchequer be urging his government to adopt?

30.32 In a certain parliament, the government consists of 75 New Socialites and the opposition consists of 25 Preservatives. Preservatives never change their mind, always voting against government policy without a second thought; New Socialites vote randomly, but with probability p that they will vote for their party leader's policies.

Following a decision by the New Socialites' leader to drop certain manifesto commitments, N of his party decide to vote consistently with the opposition. The leader's advisors reluctantly admit that an election must be called if N is such that, at any vote on government policy, the chance of a simple majority in favour would be less than 80%. Given that $p = 0.8$, estimate the lowest value of N that would precipitate an election.

30.33 A practical-class demonstrator sends his twelve students to the storeroom to collect apparatus for an experiment, but forgets to tell each which type of component to bring. There are three types, A, B and C, held in the stores (in large numbers) in the proportions 20%, 30% and 50%, respectively, and each student picks a component at random. In order to set up one experiment, one unit each of A and B and two units of C are needed. Let $\Pr(N)$ be the probability that at least N experiments can be set up.

(a) Evaluate $\Pr(3)$.
(b) Find an expression for $\Pr(N)$ in terms of k_1 and k_2, the numbers of components of types A and B respectively selected by the students. Show that $\Pr(2)$ can be written in the form

$$\Pr(2) = (0.5)^{12} \sum_{i=2}^{6} {}^{12}C_i \, (0.4)^i \sum_{j=2}^{8-i} {}^{12-i}C_j \, (0.6)^j.$$

(c) By considering the conditions under which no experiments can be set up, show that $\Pr(1) = 0.9145$.

30.34 The random variables X and Y take integer values, x and y, both ≥ 1, and such that $2x + y \leq 2a$, where a is an integer greater than 1. The joint probability within this region is given by

$$\Pr(X = x, Y = y) = c(2x + y),$$

where c is a constant, and it is zero elsewhere.

Show that the marginal probability $\Pr(X = x)$ is

$$\Pr(X = x) = \frac{6(a - x)(2x + 2a + 1)}{a(a - 1)(8a + 5)},$$

and obtain expressions for $\Pr(Y = y)$, (a) when y is even and (b) when y is odd. Show further that

$$E[Y] = \frac{6a^2 + 4a + 1}{8a + 5}.$$

[You will need the results about series involving the natural numbers given in subsection 4.2.5.]

30.35　The continuous random variables X and Y have a joint PDF proportional to $xy(x - y)^2$ with $0 \leq x \leq 1$ and $0 \leq y \leq 1$. Find the marginal distributions for X and Y and show that they are negatively correlated with correlation coefficient $-\frac{2}{3}$.

30.36　A discrete random variable X takes integer values $n = 0, 1, \ldots, N$ with probabilities p_n. A second random variable Y is defined as $Y = (X - \mu)^2$, where μ is the expectation value of X. Prove that the covariance of X and Y is given by

$$\text{Cov}[X, Y] = \sum_{n=0}^{N} n^3 p_n - 3\mu \sum_{n=0}^{N} n^2 p_n + 2\mu^3.$$

Now suppose that X takes all of its possible values with equal probability, and hence demonstrate that two random variables can be uncorrelated, even though one is defined in terms of the other.

30.37　Two continuous random variables X and Y have a joint probability distribution

$$f(x, y) = A(x^2 + y^2),$$

where A is a constant and $0 \leq x \leq a, 0 \leq y \leq a$. Show that X and Y are negatively correlated with correlation coefficient $-15/73$. By sketching a rough contour map of $f(x, y)$ and marking off the regions of positive and negative correlation, convince yourself that this (perhaps counter-intuitive) result is plausible.

30.38　A continuous random variable X is uniformly distributed over the interval $[-c, c]$. A sample of $2n + 1$ values of X is selected at random and the random variable Z is defined as the *median* of that sample. Show that Z is distributed over $[-c, c]$ with probability density function

$$f_n(z) = \frac{(2n + 1)!}{(n!)^2 (2c)^{2n+1}} (c^2 - z^2)^n.$$

Find the variance of Z.

30.39　Show that, as the number of trials n becomes large but $np_i = \lambda_i$, $i = 1, 2, \ldots, k - 1$, remains finite, the multinomial probability distribution (30.146),

$$M_n(x_1, x_2, \ldots, x_k) = \frac{n!}{x_1! x_2! \cdots x_k!} p_1^{x_1} p_2^{x_2} \cdots p_k^{x_k},$$

can be approximated by a multiple Poisson distribution with $k - 1$ factors:

$$M_n'(x_1, x_2, \ldots, x_{k-1}) = \prod_{i=1}^{k-1} \frac{e^{-\lambda_i} \lambda_i^{x_i}}{x_i!}.$$

(Write $\sum_i^{k-1} p_i = \delta$ and express all terms involving subscript k in terms of n and δ, either exactly or approximately. You will need to use $n! \approx n^\epsilon [(n - \epsilon)!]$ and $(1 - a/n)^n \approx e^{-a}$ for large n.)

(a)　Verify that the terms of M_n' when summed over all values of $x_1, x_2, \ldots, x_{k-1}$ add up to unity.

(b)　If $k = 7$ and $\lambda_i = 9$ for all $i = 1, 2, \ldots, 6$, estimate, using the appropriate Gaussian approximation, the chance that at least three of $x_1, x_2, \ldots, x_6$ will be 15 or greater.

30.40　The variables X_i, $i = 1, 2, \ldots, n$, are distributed as a multivariate Gaussian, with means μ_i and a covariance matrix V. If the X_i are required to satisfy the linear constraint $\sum_{i=1}^{n} c_i X_i = 0$, where the c_i are constants (and not all equal to zero), show that the variable

$$\chi_n^2 = (\mathbf{x} - \mu)^T V^{-1}(\mathbf{x} - \mu)$$

follows a chi-squared distribution of order $n - 1$.

30.17 Hints and answers

30.1   (a) Yes, (b) no, (c) no, (d) no, (e) yes.

30.3   Show that, if $p_x/16$ is the probability that the total will be x, then the corresponding gain is $[p_x(x^2 + x) - 16x]/16$. (a) A loss of 0.36 euros; (b) a gain of $27/64$ euros; (c) a gain of $46/16$ euros, provided he takes your advice and guesses '6' each time.

30.5 $P_1 = \alpha(\alpha + \beta - \alpha\beta)^{-1}$; $P_2 = \alpha(1 - \beta)(\alpha + \beta - \alpha\beta)^{-1}$; $P_3 = P_2$.

30.7   If p_r is the probability that before the rth round both players are still in the tournament (and therefore have not met each other), show that

$$p_{r+1} = \frac{1}{4} \frac{2^{n+1-r} - 2}{2^{n+1-r} - 1} p_r \quad \text{and hence that} \quad p_r = \left(\frac{1}{2}\right)^{r-1} \frac{2^{n+1-r} - 1}{2^n - 1}.$$

(a)   The probability that they meet in the final is $p_n = 2^{-(n-1)}(2^n - 1)^{-1}$.

(b)   The probability that they meet at some stage in the tournament is given by the sum $\sum_{r=1}^{n} p_r(2^{n+1-r} - 1)^{-1} = 2^{-(n-1)}$.

30.9   The relative probabilities are $X : Y : Z = 50 : 36 : 8$ (in units of 10^{-4}); $25/47$.

30.11   Take A_j as the event that a family consists of j boys and $n - j$ girls, and B as the event that the boy has at least two sisters. Apply Bayes' theorem.

30.13   (i) For a even, the number of ways is $1 + 3 + 5 + \cdots + (a - 3)$, and (ii) for a odd it is $2 + 4 + 6 + \cdots + (a - 3)$. Combine the results for $a = 2m$ and $a = 2m + 1$, with m running from 2 to N, to show that the total number of non-degenerate triangles is given by $N(4N + 1)(N - 1)/6$. The number of possible selections of a set of three rods is $(2N + 1)(2N)(2N - 1)/6$.

30.15   Show that $k = e^2$ and that the average duration of a call is 1 minute. Let p_n be the probability that the call ends during the interval $0.5(n - 1) \le t < 0.5n$ and $c_n = 20n$ be the corresponding cost. Prove that $p_1 = p_2 = \frac{1}{4}$ and that $p_n = \frac{1}{2}e^2(e - 1)e^{-n}$, for $n \ge 3$. It follows that the average cost is

$$E[C] = \frac{30}{2} + 20\frac{e^2(e - 1)}{2}\sum_{n=3}^{\infty} n e^{-n}.$$

The arithmetico-geometric series has sum $(3e^{-1} - 2e^{-2})/(e - 1)^2$ and the total charge is $5(e + 1)/(e - 1) = 10.82$ pence more than the 40 pence a uniform rate would cost.

30.17   (a) The scores must be equal, at r each, after five attempts each.

(b) M can only be even if team 2 gets too far ahead (or drops too far behind) to be caught (or catch up), with conditional probability p_2 (or q_2). Conversely, M can only be odd as a result of a final action by team 1.

(c) $\Pr(i : x, y) = {}^y C_x p_i^x q_i^{y-x}$.

(d) If the match is still alive at the tenth kick, team 2 is just as likely to lose it as to take it into sudden death.

30.19   Show that $dY/dX = f$ and use $g(y) = f(x)|dx/dy|$.

30.21   (a)   Use result (30.84) to show that the PGF for S is $Q/(1 - Pq - Ppt)$. Then use equations (30.74) and (30.76).

(b) The PGF for the score is $6/(21 - 10t - 5t^2)$ and the average score is $10/3$. The variance is $145/9$ and the standard deviation is 4.01.

30.23   Mean $= 4/\pi$. Variance $= 2 - (16/\pi^2)$. Probability that X exceeds its mean $= 1 - (2/\pi)\sin^{-1}(2/\pi) = 0.561$.

30.25   Consider, separately, 0, 1 and ≥ 2 errors on a page.

30.27   Show that the maximum occurs at $x = (r - 1)/\lambda$, and then use Stirling's approximation to find the maximum value.

30.29 $\Pr(k \text{ chicks hatching}) = \sum_{n=k}^{\infty} \text{Po}(n, \lambda)\,\text{Bin}(n, p)$.

30.31   There is not much to choose between the schemes. In (a) the critical value of the standard variable is -2.5 and the average fine would be 15.5 euros. For (b) the corresponding figures are -1.0 and 15.9 euros. Scheme (c) is governed by a geometric distribution with $p = q = \frac{1}{2}$, and leads to an expected fine of $\sum_{n=1}^{\infty} 4n(n - 1)(\frac{1}{2})^n$. The sum can be evaluated by differentiating the result $\sum_{n=1}^{\infty} p^n = p/(1 - p)$ with respect to p, and gives the expected fine as 16 euros.

30.33   (a) $[12!(0.5)^6(0.3)^3(0.2)^3]/(6!\,3!\,3!) = 0.0624$.

30.35   You will need to establish the normalisation constant for the distribution (36), the common mean value $(3/5)$ and the common standard deviation $(3/10)$. The marginal distributions are $f(x) = 3x(6x^2 - 8x + 3)$, and the same function of y. The covariance has the value $-3/50$, yielding a correlation of $-2/3$.

30.37 $A = 3/(24a^4)$; $\mu_X = \mu_Y = 5a/8$; $\sigma_X^2 = \sigma_Y^2 = 73a^2/960$; $E[XY] = 3a^2/8$; $\text{Cov}[X, Y] = -a^2/64$.

30.39   (b) With the continuity correction $\Pr(x_i \geq 15) = 0.0334$. The probability that at least three are 15 or greater is 7.5×10^{-4}.

Statistics

In this chapter, we turn to the study of statistics, which is concerned with the analysis of experimental data. In a book of this nature we cannot hope to do justice to such a large subject; indeed, many would argue that statistics belongs to the realm of experimental science rather than in a mathematics textbook. Nevertheless, physical scientists and engineers are regularly called upon to perform a statistical analysis of their data and to present their results in a statistical context. Therefore, we will concentrate on this aspect of a much more extensive subject.[§]

31.1 Experiments, samples and populations

We may regard the product of any experiment as a set of N measurements of some quantity x or set of quantities $x, y, \ldots, z$. This set of measurements constitutes the *data*. Each measurement (or *data item*) consists accordingly of a single number x_i or a set of numbers $(x_i, y_i, \ldots, , z_i)$, where $i = 1, \ldots, , N$. For the moment, we will assume that each data item is a single number, although our discussion can be extended to the more general case.

As a result of inaccuracies in the measurement process, or because of intrinsic variability in the quantity x being measured, one would expect the N measured values $x_1, x_2, \ldots, x_N$ to be different each time the experiment is performed. We may

[§] There are, in fact, two separate schools of thought concerning statistics: the frequentist approach and the Bayesian approach. Indeed, which of these approaches is the more fundamental is still a matter of heated debate. Here we shall concentrate primarily on the more traditional frequentist approach (despite the preference of some of the authors for the Bayesian viewpoint!). For a fuller discussion of the frequentist approach one could refer to, for example, A. Stuart and K. Ord, *Kendall's Advanced Theory of Statistics, vol. 1* (London: Edward Arnold, 1994) or J. F. Kenney and E. S. Keeping, *Mathematics of Statistics* (New York: Van Nostrand, 1954). For a discussion of the Bayesian approach one might consult, for example, D. S. Sivia, *Data Analysis: A Bayesian Tutorial* (Oxford: Oxford University Press, 1996).

therefore consider the x_i as a set of N random variables. In the most general case, these random variables will be described by some N-dimensional joint probability density function $P(x_1, x_2, \ldots, x_N)$.[§] In other words, an experiment consisting of N measurements is considered as a single random *sample* from the joint distribution (or *population*) $P(\mathbf{x})$, where $\mathbf{x}$ denotes a point in the N-dimensional data space having coordinates $(x_1, x_2, \ldots, x_N)$.

The situation is simplified considerably if the sample values x_i are *independent*. In this case, the N-dimensional joint distribution $P(\mathbf{x})$ factorises into the product of N one-dimensional distributions,

$$P(\mathbf{x}) = P(x_1)P(x_2) \cdots P(x_N). \tag{31.1}$$

In the general case, each of the one-dimensional distributions $P(x_i)$ may be different. A typical example of this occurs when N independent measurements are made of some quantity x but the accuracy of the measuring procedure varies between measurements.

It is often the case, however, that each sample value x_i is drawn independently from the *same* population. In this case, $P(\mathbf{x})$ is of the form (31.1), but, in addition, $P(x_i)$ has the same form for each value of i. The measurements $x_1, x_2, \ldots, x_N$ are then said to form a *random sample of size N* from the one-dimensional population $P(x)$. This is the most common situation met in practice and, unless stated otherwise, we will assume from now on that this is the case.

31.2 Sample statistics

Suppose we have a set of N measurements $x_1, x_2, \ldots, x_N$. Any function of these measurements (that contains no unknown parameters) is called a *sample statistic*, or often simply a *statistic*. Sample statistics provide a means of characterising the data. Although the resulting characterisation is inevitably incomplete, it is useful to be able to describe a set of data in terms of a few pertinent numbers. We now discuss the most commonly used sample statistics.

[§] In this chapter, we will adopt the common convention that $P(x)$ denotes the particular probability density function that applies to its argument, x. This obviates the need to use a different letter for the PDF of each new variable. For example, if X and Y are random variables with different PDFs, then properly one should denote these distributions by $f(x)$ and $g(y)$, say. In our shorthand notation, these PDFs are denoted by $P(x)$ and $P(y)$, where it is understood that the functional form of the PDF may be different in each case.

188.7	204.7	193.2	169.0
168.1	189.8	166.3	200.0

Table 31.1   Experimental data giving eight measurements of the round trip time in milliseconds for a computer 'packet' to travel from Cambridge UK to Cambridge MA.

31.2.1 Averages

The simplest number used to characterise a sample is the *mean*, which for N values x_i, $i = 1, 2, \ldots, N$, is defined by

$$\bar{x} = \frac{1}{N} \sum_{i=1}^{N} x_i. \tag{31.2}$$

In words, the *sample mean* is the sum of the sample values divided by the number of values in the sample.

▶ *Table 31.1 gives eight values for the round trip time in milliseconds for a computer 'packet' to travel from Cambridge UK to Cambridge MA. Find the sample mean.*

Using (31.2) the sample mean in milliseconds is given by

$$\bar{x} = \tfrac{1}{8}(188.7 + 204.7 + 193.2 + 169.0 + 168.1 + 189.8 + 166.3 + 200.0)$$
$$= \frac{1479.8}{8} = 184.975.$$

Since the sample values in table 31.1 are quoted to an accuracy of one decimal place, it is usual to quote the mean to the same accuracy, i.e. as $\bar{x} = 185.0$. ◀

Strictly speaking the mean given by (31.2) is the *arithmetic mean* and this is by far the most common definition used for a mean. Other definitions of the mean are possible, though less common, and include

(i) the *geometric mean*,

$$\bar{x}_g = \left(\prod_{i=1}^{N} x_i \right)^{1/N}, \tag{31.3}$$

(ii) the *harmonic mean*,

$$\bar{x}_h = \frac{N}{\sum_{i=1}^{N} 1/x_i}, \tag{31.4}$$

(iii) the *root mean square*,

$$\bar{x}_{rms} = \left(\frac{\sum_{i=1}^{N} x_i^2}{N} \right)^{1/2}. \tag{31.5}$$

It should be noted that, $\bar{x}$, $\bar{x}_h$ and $\bar{x}_{rms}$ would remain well defined even if some sample values were negative, but the value of $\bar{x}_g$ could then become complex. The geometric mean should not be used in such cases.

▶*Calculate $\bar{x}_g$, $\bar{x}_h$ and $\bar{x}_{rms}$ for the sample given in table 31.1.*

The geometric mean is given by (31.3) to be

$$\bar{x}_g = (188.7 \times 204.7 \times \cdots \times 200.0)^{1/8} = 184.4.$$

The harmonic mean is given by (31.4) to be

$$\bar{x}_h = \frac{8}{(1/188.7) + (1/204.7) + \cdots + (1/200.0)} = 183.9.$$

Finally, the root mean square is given by (31.5) to be

$$\bar{x}_{rms} = \left[\tfrac{1}{8}(188.7^2 + 204.7^2 + \cdots + 200.0^2)\right]^{1/2} = 185.5. \blacktriangleleft$$

Two other measures of the 'average' of a sample are its *mode* and *median*. The mode is simply the most commonly occurring value in the sample. A sample may possess several modes, however, and thus it can be misleading in such cases to use the mode as a measure of the average of the sample. The median of a sample is the halfway point when the sample values x_i ($i = 1, 2, \ldots, N$) are arranged in ascending (or descending) order. Clearly, this depends on whether the size of the sample, N, is odd or even. If N is odd then the median is simply equal to $x_{(N+1)/2}$, whereas if N is even the median of the sample is usually taken to be $\tfrac{1}{2}(x_{N/2} + x_{(N/2)+1})$.

▶*Find the mode and median of the sample given in table 31.1.*

From the table we see that each sample value occurs exactly once, and so any value may be called the mode of the sample.

To find the sample median, we first arrange the sample values in ascending order and obtain

166.3, 168.1, 169.0, 188.7, 189.8, 193.2, 200.0, 204.7.

Since the number of sample values $N = 8$, which is even, the median of the sample is

$$\tfrac{1}{2}(x_4 + x_5) = \tfrac{1}{2}(188.7 + 189.8) = 189.25. \blacktriangleleft$$

31.2.2 Variance and standard deviation

The variance and standard deviation both give a measure of the spread of values in a sample about the sample mean $\bar{x}$. The *sample variance* is defined by

$$s^2 = \frac{1}{N} \sum_{i=1}^{N} (x_i - \bar{x})^2, \tag{31.6}$$

and the *sample standard deviation* is the positive square root of the sample variance, i.e.

$$s = \sqrt{\frac{1}{N}\sum_{i=1}^{N}(x_i - \bar{x})^2}.$$ (31.7)

►*Find the sample variance and sample standard deviation of the data given in table 31.1.*

We have already found that the sample mean is 185.0 to one decimal place. However, when the mean is to be used in the subsequent calculation of the sample variance it is better to use the most accurate value available. In this case the exact value is 184.975, and so using (31.6),

$$s^2 = \frac{1}{8}\left[(188.7 - 184.975)^2 + \cdots + (200.0 - 184.975)^2\right]$$

$$= \frac{1608.36}{8} = 201.0,$$

where once again we have quoted the result to one decimal place. The sample standard deviation is then given by $s = \sqrt{201.0} = 14.2$. As it happens, in this case the difference between the true mean and the rounded value is very small compared with the variation of the individual readings about the mean and using the rounded value has a negligible effect; however, this would not be so if the difference were comparable to the sample standard deviation. ◄

Using the definition (31.7), it is clear that in order to calculate the standard deviation of a sample we must first calculate the sample mean. This requirement can be avoided, however, by using an alternative form for s^2. From (31.6), we see that

$$s^2 = \frac{1}{N}\sum_{i=1}^{N}(x_i - \bar{x})^2$$

$$= \frac{1}{N}\sum_{i=1}^{N}x_i^2 - \frac{1}{N}\sum_{i=1}^{N}2x_i\bar{x} + \frac{1}{N}\sum_{i=1}^{N}\bar{x}^2$$

$$= \overline{x^2} - 2\bar{x}^2 + \bar{x}^2 = \overline{x^2} - \bar{x}^2$$

We may therefore write the sample variance s^2 as

$$s^2 = \overline{x^2} - \bar{x}^2 = \frac{1}{N}\sum_{i=1}^{N}x_i^2 - \left(\frac{1}{N}\sum_{i=1}^{N}x_i\right)^2,$$ (31.8)

from which the sample standard deviation is found by taking the positive square root. Thus, by evaluating the quantities $\sum_{i=1}^{N}x_i$ and $\sum_{i=1}^{N}x_i^2$ for our sample, we can calculate the sample mean and sample standard deviation at the same time.

▶ *Calculate* $\sum_{i=1}^{N} x_i$ *and* $\sum_{i=1}^{N} x_i^2$ *for the data given in table 31.1 and hence find the mean and standard deviation of the sample.*

From table 31.1, we obtain

$$\sum_{i=1}^{N} x_i = 188.7 + 204.7 + \cdots + 200.0 = 1479.8,$$

$$\sum_{i=1}^{N} x_i^2 = (188.7)^2 + (204.7)^2 + \cdots + (200.0)^2 = 275\,334.36.$$

Since $N = 8$, we find as before (quoting the final results to one decimal place)

$$\bar{x} = \frac{1479.8}{8} = 185.0, \qquad s = \sqrt{\frac{275\,334.36}{8} - \left(\frac{1479.8}{8}\right)^2} = 14.2. \; \blacktriangleleft$$

31.2.3 Moments and central moments

By analogy with our discussion of probability distributions in section 30.5, the sample mean and variance may also be described respectively as the first moment and second central moment of the sample. In general, for a sample x_i, $i = 1, 2, \ldots, N$, we define the rth moment m_r and rth central moment n_r as

$$m_r = \frac{1}{N} \sum_{i=1}^{N} x_i^r, \tag{31.9}$$

$$n_r = \frac{1}{N} \sum_{i=1}^{N} (x_i - m_1)^r. \tag{31.10}$$

Thus the sample mean $\bar{x}$ and variance s^2 may also be written as m_1 and n_2 respectively. As is common practice, we have introduced a notation in which a sample statistic is denoted by the Roman letter corresponding to whichever Greek letter is used to describe the corresponding population statistic. Thus, we use m_r and n_r to denote the rth moment and central moment of a sample, since in section 30.5 we denoted the rth moment and central moment of a population by μ_r and ν_r respectively.

This notation is particularly useful, since the rth central moment of a sample, m_r, may be expressed in terms of the rth- and lower-order sample moments n_r in a way exactly analogous to that derived in subsection 30.5.5 for the corresponding population statistics. As discussed in the previous section, the sample variance is given by $s^2 = \overline{x^2} - \bar{x}^2$ but this may also be written as $n_2 = m_2 - m_1^2$, which is to be compared with the corresponding relation $\nu_2 = \mu_2 - \mu_1^2$ derived in subsection 30.5.3 for population statistics. This correspondence also holds for higher-order central moments of the sample. For example,

$$n_3 = \frac{1}{N} \sum_{i=1}^{N} (x_i - m_1)^3$$

$$= \frac{1}{N} \sum_{i=1}^{N} (x_i^3 - 3m_1 x_i^2 + 3m_1^2 x_i - m_1^3)$$

$$= m_3 - 3m_1 m_2 + 3m_1^2 m_1 - m_1^3$$

$$= m_3 - 3m_1 m_2 + 2m_1^3, \tag{31.11}$$

which may be compared with equation (30.53) in the previous chapter.

Mirroring our discussion of the normalised central moments γ_r of a population in subsection 30.5.5, we can also describe a sample in terms of the dimensionless quantities

$$g_k = \frac{n_k}{n_2^{k/2}} = \frac{n_k}{s^k};$$

g_3 and g_4 are called the sample skewness and kurtosis. Likewise, it is common to define the *excess* kurtosis of a sample by $g_4 - 3$.

31.2.4 Covariance and correlation

So far we have assumed that each data item of the sample consists of a single number. Now let us suppose that each item of data consists of a pair of numbers, so that the sample is given by (x_i, y_i), $i = 1, 2, \ldots, N$.

We may calculate the sample means, $\bar{x}$ and $\bar{y}$, and sample variances, s_x^2 and s_y^2, of the x_i and y_i values individually but these statistics do not provide any measure of the relationship between the x_i and y_i. By analogy with our discussion in subsection 30.12.3 we measure any interdependence between the x_i and y_i in terms of the *sample covariance*, which is given by

$$V_{xy} = \frac{1}{N} \sum_{i=1}^{N} (x_i - \bar{x})(y_i - \bar{y})$$

$$= \overline{(x - \bar{x})(y - \bar{y})}$$

$$= \overline{xy} - \bar{x}\bar{y}. \tag{31.12}$$

Writing out the last expression in full, we obtain the form most useful for calculations, which reads

$$V_{xy} = \frac{1}{N} \left(\sum_{i=1}^{N} x_i y_i \right) - \frac{1}{N^2} \left(\sum_{i=1}^{N} x_i \right) \left(\sum_{i=1}^{N} y_i \right).$$

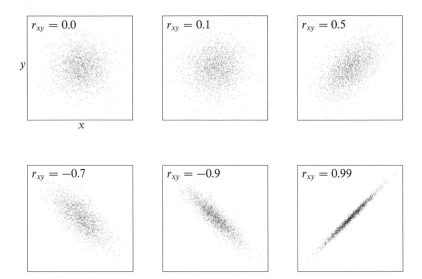

Figure 31.1   Scatter plots for two-dimensional data samples of size $N = 1000$, with various values of the correlation r. No scales are plotted, since the value of r is unaffected by shifts of origin or changes of scale in x and y.

We may also define the closely related *sample correlation* by

$$r_{xy} = \frac{V_{xy}}{s_x s_y},$$

which can take values between -1 and $+1$. If the x_i and y_i are independent then $V_{xy} = 0 = r_{xy}$, and from (31.12) we see that $\overline{xy} = \bar{x}\bar{y}$. It should also be noted that the value of r_{xy} is not altered by shifts in the origin or by changes in the scale of the x_i or y_i. In other words, if $x' = ax + b$ and $y' = cy + d$, where a, b, c, d are constants, then $r_{x'y'} = r_{xy}$. Figure 31.1 shows scatter plots for several two-dimensional random samples x_i, y_i of size $N = 1000$, each with a different value of r_{xy}.

▶*Ten UK citizens are selected at random and their heights and weights are found to be as follows (to the nearest cm or kg respectively):*

Person	A	B	C	D	E	F	G	H	I	J
Height (cm)	194	168	177	180	171	190	151	169	175	182
Weight (kg)	75	53	72	80	75	75	57	67	46	68

Calculate the sample correlation between the heights and weights.

In order to find the sample correlation, we begin by calculating the following sums (where x_i are the heights and y_i are the weights)

$$\sum_i x_i = 1757, \qquad \sum_i y_i = 668,$$

$$\sum_i x_i^2 = 310\,041, \qquad \sum_i y_i^2 = 45\,746, \qquad \sum_i x_i y_i = 118\,029.$$

The sample consists of $N = 10$ pairs of numbers, so the means of the x_i and of the y_i are given by $\bar{x} = 175.7$ and $\bar{y} = 66.8$. Also, $\overline{xy} = 11\,802.9$. Similarly, the standard deviations of the x_i and y_i are calculated, using (31.8), as

$$s_x = \sqrt{\frac{310\,041}{10} - \left(\frac{1757}{10}\right)^2} = 11.6,$$

$$s_y = \sqrt{\frac{45\,746}{10} - \left(\frac{668}{10}\right)^2} = 10.6.$$

Thus the sample correlation is given by

$$r_{xy} = \frac{\overline{xy} - \bar{x}\bar{y}}{s_x s_y} = \frac{11\,802.9 - (175.7)(66.8)}{(11.6)(10.6)} = 0.54.$$

Thus there is a moderate positive correlation between the heights and weights of the people measured. ◄

It is straightforward to generalise the above discussion to data samples of arbitrary dimension, the only complication being one of notation. We choose to denote the ith data item from an n-dimensional sample as $(x_i^{(1)}, x_i^{(2)}, \ldots, x_i^{(n)})$, where the bracketed superscript runs from 1 to n and labels the elements within a given data item whereas the subscript i runs from 1 to N and labels the data items within the sample. In this n-dimensional case, we can define the *sample covariance matrix* whose elements are

$$V_{kl} = \overline{x^{(k)} x^{(l)}} - \overline{x^{(k)}}\; \overline{x^{(l)}}$$

and the *sample correlation matrix* with elements

$$r_{kl} = \frac{V_{kl}}{s_k s_l}.$$

Both these matrices are clearly symmetric but are *not* necessarily positive definite.

31.3 Estimators and sampling distributions

In general, the population $P(x)$ from which a sample $x_1, x_2, \ldots, x_N$ is drawn is *unknown*. The *central aim* of statistics is to use the sample values x_i to infer certain properties of the unknown population $P(x)$, such as its mean, variance and higher moments. To keep our discussion in general terms, let us denote the various parameters of the population by $a_1, a_2, \ldots$, or collectively by $\mathbf{a}$. Moreover, we make the dependence of the population on the values of these quantities explicit by writing the population as $P(x|\mathbf{a})$. For the moment, we are assuming that the sample values x_i are independent and drawn from the same (one-dimensional) population $P(x|\mathbf{a})$, in which case

$$P(\mathbf{x}|\mathbf{a}) = P(x_1|\mathbf{a}) P(x_2|\mathbf{a}) \cdots P(x_N|\mathbf{a}).$$

Suppose, we wish to *estimate* the value of one of the quantities $a_1, a_2, \ldots$, which we will denote simply by a. Since the sample values x_i provide our only source of information, any estimate of a must be some function of the x_i, i.e. some sample statistic. Such a statistic is called an *estimator* of a and is usually denoted by $\hat{a}(\mathbf{x})$, where $\mathbf{x}$ denotes the sample elements $x_1, x_2, \ldots, x_N$.

Since an estimator $\hat{a}$ is a function of the sample values of the random variables $x_1, x_2, \ldots, x_N$, it too must be a random variable. In other words, if a number of random samples, each of the same size N, are taken from the (one-dimensional) population $P(x|a)$ then the value of the estimator $\hat{a}$ will vary from one sample to the next and in general will not be equal to the true value a. This variation in the estimator is described by its *sampling distribution* $P(\hat{a}|a)$. From section 30.14, this is given by

$$P(\hat{a}|\mathbf{a}) \, d\hat{a} = P(\mathbf{x}|\mathbf{a}) \, d^N\mathbf{x},$$

where $d^N\mathbf{x}$ is the infinitesimal 'volume' in $\mathbf{x}$-space lying between the 'surfaces' $\hat{a}(\mathbf{x}) = \hat{a}$ and $\hat{a}(\mathbf{x}) = \hat{a} + d\hat{a}$. The form of the sampling distribution generally depends upon the estimator under consideration and upon the form of the population from which the sample was drawn, including, as indicated, the true values of the quantities $\mathbf{a}$. It is also usually dependent on the sample size N.

▶*The sample values $x_1, x_2, \ldots, x_N$ are drawn independently from a Gaussian distribution with mean μ and variance σ. Suppose that we choose the sample mean $\bar{x}$ as our estimator $\hat{\mu}$ of the population mean. Find the sampling distributions of this estimator.*

The sample mean $\bar{x}$ is given by

$$\bar{x} = \frac{1}{N}(x_1 + x_2 + \cdots + x_N),$$

where the x_i are independent random variables distributed as $x_i \sim N(\mu, \sigma^2)$. From our discussion of multiple Gaussian distributions on page 1189, we see immediately that $\bar{x}$ will also be Gaussian distributed as $N(\mu, \sigma^2/N)$. In other words, the sampling distribution of $\bar{x}$ is given by

$$P(\bar{x}|\mu, \sigma) = \frac{1}{\sqrt{2\pi\sigma^2/N}} \exp\left[-\frac{(\bar{x}-\mu)^2}{2\sigma^2/N}\right]. \tag{31.13}$$

Note that the variance of this distribution is σ^2/N. ◀

31.3.1 Consistency, bias and efficiency of estimators

For any particular quantity a, we may in fact define any number of different estimators, each of which will have its own sampling distribution. The quality of a given estimator $\hat{a}$ may be assessed by investigating certain properties of its sampling distribution $P(\hat{a}|\mathbf{a})$. In particular, an estimator $\hat{a}$ is usually judged on the three criteria of *consistency*, *bias* and *efficiency*, each of which we now discuss.

Consistency

An estimator $\hat{a}$ is *consistent* if its value tends to the true value a in the large-sample limit, i.e.

$$\lim_{N \to \infty} \hat{a} = a.$$

Consistency is usually a minimum requirement for a useful estimator. An equivalent statement of consistency is that in the limit of large N the sampling distribution $P(\hat{a}|\mathbf{a})$ of the estimator must satisfy

$$\lim_{N \to \infty} P(\hat{a}|\mathbf{a}) \to \delta(\hat{a} - a).$$

Bias

The expectation value of an estimator $\hat{a}$ is given by

$$E[\hat{a}] = \int \hat{a}P(\hat{a}|\mathbf{a})\,d\hat{a} = \int \hat{a}(\mathbf{x})P(\mathbf{x}|\mathbf{a})\,d^N\mathbf{x}, \tag{31.14}$$

where the second integral extends over all possible values that can be taken by the sample elements $x_1, x_2, \ldots, x_N$. This expression gives the expected mean value of $\hat{a}$ from an infinite number of samples, each of size N. The *bias* of an estimator $\hat{a}$ is then defined as

$$b(\mathbf{a}) = E[\hat{a}] - a. \tag{31.15}$$

We note that the bias b does not depend on the measured sample values $x_1, x_2, \ldots, x_N$. In general, though, it will depend on the sample size N, the functional form of the estimator $\hat{a}$ and, as indicated, on the true properties $\mathbf{a}$ of the population, including the true value of a itself. If $b = 0$ then $\hat{a}$ is called an *unbiased* estimator of a.

> ▶An estimator $\hat{a}$ is biased in such a way that $E[\hat{a}] = a + b(a)$, where the bias $b(a)$ is given by $(b_1 - 1)a + b_2$ and b_1 and b_2 are known constants. Construct an unbiased estimator of a.

Let us first write $E[\hat{a}]$ in the clearer form

$$E[\hat{a}] = a + (b_1 - 1)a + b_2 = b_1 a + b_2.$$

The task of constructing an unbiased estimator is now trivial, and an appropriate choice is $\hat{a}' = (\hat{a} - b_2)/b_1$, which (as required) has the expectation value

$$E[\hat{a}'] = \frac{E[\hat{a}] - b_2}{b_1} = a. \blacktriangleleft$$

Efficiency

The variance of an estimator is given by

$$V[\hat{a}] = \int (\hat{a} - E[\hat{a}])^2 P(\hat{a}|\mathbf{a})\,d\hat{a} = \int (\hat{a}(\mathbf{x}) - E[\hat{a}])^2 P(\mathbf{x}|\mathbf{a})\,d^N\mathbf{x} \tag{31.16}$$

and describes the spread of values $\hat{a}$ about $E[\hat{a}]$ that would result from a large number of samples, each of size N. An estimator with a smaller variance is said to be more *efficient* than one with a larger variance. As we show in the next section, for any given quantity a of the population there exists a theoretical *lower limit* on the variance of *any* estimator $\hat{a}$. This result is known as *Fisher's inequality* (or the *Cramér–Rao inequality*) and reads

$$V[\hat{a}] \geq \left(1 + \frac{\partial b}{\partial a}\right)^2 \bigg/ E\left[-\frac{\partial^2 \ln P}{\partial a^2}\right], \qquad (31.17)$$

where P stands for the population $P(\mathbf{x}|\mathbf{a})$ and b is the bias of the estimator. Denoting the quantity on the RHS of (31.17) by $V_{\min}$, the *efficiency* e of an estimator is defined as

$$e = V_{\min}/V[\hat{a}].$$

An estimator for which $e = 1$ is called a *minimum-variance* or *efficient* estimator. Otherwise, if $e < 1$, $\hat{a}$ is called an *inefficient* estimator.

It should be noted that, in general, there is no unique 'optimal' estimator $\hat{a}$ for a particular property a. To some extent, there is always a trade-off between bias and efficiency. One must often weigh the relative merits of an unbiased, inefficient estimator against another that is more efficient but slightly biased. Nevertheless, a common choice is the *best unbiased estimator* (BUE), which is simply the unbiased estimator $\hat{a}$ having the smallest variance $V[\hat{a}]$.

Finally, we note that some qualities of estimators are related. For example, suppose that $\hat{a}$ is an unbiased estimator, so that $E[\hat{a}] = a$ and $V[\hat{a}] \to 0$ as $N \to \infty$. Using the Bienaymé–Chebyshev inequality discussed in subsection 30.5.3, it follows immediately that $\hat{a}$ is also a consistent estimator. Nevertheless, it does *not* follow that a consistent estimator is unbiased.

> ▶ *The sample values $x_1, x_2, \ldots, x_N$ are drawn independently from a Gaussian distribution with mean μ and variance σ. Show that the sample mean $\bar{x}$ is a consistent, unbiased, minimum-variance estimator of μ.*

We found earlier that the sampling distribution of $\bar{x}$ is given by

$$P(\bar{x}|\mu, \sigma) = \frac{1}{\sqrt{2\pi\sigma^2/N}} \exp\left[-\frac{(\bar{x} - \mu)^2}{2\sigma^2/N}\right],$$

from which we see immediately that $E[\bar{x}] = \mu$ and $V[\bar{x}] = \sigma^2/N$. Thus $\bar{x}$ is an unbiased estimator of μ. Moreover, since it is also true that $V[\bar{x}] \to 0$ as $N \to \infty$, $\bar{x}$ is a consistent estimator of μ.

In order to determine whether $\bar{x}$ is a minimum-variance estimator of μ, we must use Fisher's inequality (31.17). Since the sample values x_i are independent and drawn from a Gaussian of mean μ and standard deviation σ, we have

$$\ln P(\mathbf{x}|\mu, \sigma) = -\frac{1}{2} \sum_{i=1}^{N} \left[\ln(2\pi\sigma^2) + \frac{(x_i - \mu)^2}{\sigma^2}\right],$$

and, on differentiating twice with respect to μ, we find

$$\frac{\partial^2 \ln P}{\partial \mu^2} = -\frac{N}{\sigma^2}.$$

This is independent of the x_i and so its expectation value is also equal to $-N/\sigma^2$. With b set equal to zero in (31.17), Fisher's inequality thus states that, for *any* unbiased estimator $\hat{\mu}$ of the population mean,

$$V[\hat{\mu}] \geq \frac{\sigma^2}{N}.$$

Since $V[\bar{x}] = \sigma^2/N$, the sample mean $\bar{x}$ is a minimum-variance estimator of μ. ◄

31.3.2 Fisher's inequality

As mentioned above, Fisher's inequality provides a lower limit on the variance of *any* estimator $\hat{a}$ of the quantity a; it reads

$$V[\hat{a}] \geq \left(1 + \frac{\partial b}{\partial a}\right)^2 \bigg/ E\left[-\frac{\partial^2 \ln P}{\partial a^2}\right], \tag{31.18}$$

where P stands for the population $P(\mathbf{x}|a)$ and b is the bias of the estimator. We now present a proof of this inequality. Since the derivation is somewhat complicated, and many of the details are unimportant, this section can be omitted on a first reading. Nevertheless, some aspects of the proof will be useful when the efficiency of maximum-likelihood estimators is discussed in section 31.5.

►*Prove Fisher's inequality (31.18).*

The normalisation of $P(\mathbf{x}|a)$ is given by

$$\int P(\mathbf{x}|a)\, d^N\mathbf{x} = 1, \tag{31.19}$$

where $d^N\mathbf{x} = dx_1 dx_2 \cdots dx_N$ and the integral extends over all the allowed values of the sample items x_i. Differentiating (31.19) with respect to the parameter a, we obtain

$$\int \frac{\partial P}{\partial a}\, d^N\mathbf{x} = \int \frac{\partial \ln P}{\partial a} P\, d^N\mathbf{x} = 0. \tag{31.20}$$

We note that the second integral is simply the expectation value of $\partial \ln P/\partial a$, where the average is taken over all possible samples x_i, $i = 1, 2, \ldots, N$. Further, by equating the two expressions for $\partial E[\hat{a}]/\partial a$ obtained by differentiating (31.15) and (31.14) with respect to a we obtain, dropping the functional dependencies, a second relationship,

$$1 + \frac{\partial b}{\partial a} = \int \hat{a}\frac{\partial P}{\partial a}\, d^N\mathbf{x} = \int \hat{a}\frac{\partial \ln P}{\partial a} P\, d^N\mathbf{x}. \tag{31.21}$$

Now, multiplying (31.20) by $\alpha(a)$, where $\alpha(a)$ is *any* function of a, and subtracting the result from (31.21), we obtain

$$\int [\hat{a} - \alpha(a)]\frac{\partial \ln P}{\partial a} P\, d^N\mathbf{x} = 1 + \frac{\partial b}{\partial a}.$$

At this point we must invoke the Schwarz inequality proved in subsection 8.1.3. The proof is trivially extended to multiple integrals and shows that for two real functions, $g(\mathbf{x})$ and $h(\mathbf{x})$,

$$\left(\int g^2(\mathbf{x}) \, d^N\mathbf{x} \right) \left(\int h^2(\mathbf{x}) \, d^N\mathbf{x} \right) \geq \left(\int g(\mathbf{x}) h(\mathbf{x}) \, d^N\mathbf{x} \right)^2. \tag{31.22}$$

If we now let $g = [\hat{a} - \alpha(a)]\sqrt{P}$ and $h = (\partial \ln P / \partial a)\sqrt{P}$, we find

$$\left\{ \int [\hat{a} - \alpha(a)]^2 P \, d^N\mathbf{x} \right\} \left[\int \left(\frac{\partial \ln P}{\partial a} \right)^2 P \, d^N\mathbf{x} \right] \geq \left(1 + \frac{\partial b}{\partial a} \right)^2.$$

On the LHS, the factor in braces represents the expected spread of $\hat{a}$-values around the point $\alpha(a)$. The minimum value that this integral may take occurs when $\alpha(a) = E[\hat{a}]$. Making this substitution, we recognise the integral as the variance $V[\hat{a}]$, and so obtain the result

$$V[\hat{a}] \geq \left(1 + \frac{\partial b}{\partial a} \right)^2 \left[\int \left(\frac{\partial \ln P}{\partial a} \right)^2 P \, d^N\mathbf{x} \right]^{-1}. \tag{31.23}$$

We note that the factor in brackets is the expectation value of $(\partial \ln P / \partial a)^2$.

Fisher's inequality is, in fact, often quoted in the form (31.23). We may recover the form (31.18) by noting that on differentiating (31.20) with respect to a we obtain

$$\int \left(\frac{\partial^2 \ln P}{\partial a^2} P + \frac{\partial \ln P}{\partial a} \frac{\partial P}{\partial a} \right) d^N\mathbf{x} = 0.$$

Writing $\partial P / \partial a$ as $(\partial \ln P / \partial a)P$ and rearranging we find that

$$\int \left(\frac{\partial \ln P}{\partial a} \right)^2 P \, d^N\mathbf{x} = -\int \frac{\partial^2 \ln P}{\partial a^2} P \, d^N\mathbf{x}.$$

Substituting this result in (31.23) gives

$$V[\hat{a}] \geq -\left(1 + \frac{\partial b}{\partial a} \right)^2 \left[\int \frac{\partial^2 \ln P}{\partial a^2} P \, d^N\mathbf{x} \right]^{-1}.$$

Since the factor in brackets is the expectation value of $\partial^2 \ln P / \partial a^2$, we have recovered result (31.18). ◄

31.3.3 Standard errors on estimators

For a given sample $x_1, x_2, \ldots, x_N$, we may calculate the value of an estimator $\hat{a}(\mathbf{x})$ for the quantity a. It is also necessary, however, to give some measure of the statistical uncertainty in this estimate. One way of characterising this uncertainty is with the standard deviation of the sampling distribution $P(\hat{a}|a)$, which is given simply by

$$\sigma_{\hat{a}} = (V[\hat{a}])^{1/2}. \tag{31.24}$$

If the estimator $\hat{a}(\mathbf{x})$ were calculated for a large number of samples, each of size N, then the standard deviation of the resulting $\hat{a}$ values would be given by (31.24). Consequently, $\sigma_{\hat{a}}$ is called the *standard error* on our estimate.

In general, however, the standard error $\sigma_{\hat{a}}$ depends on the true values of some or all of the quantities **a** and they may be unknown. When this occurs, one must substitute estimated values of any unknown quantities into the expression for $\sigma_{\hat{a}}$ in order to obtain an estimated standard error $\hat{\sigma}_{\hat{a}}$. One then quotes the result as

$$a = \hat{a} \pm \hat{\sigma}_{\hat{a}}.$$

> ▶ *Ten independent sample values x_i, $i = 1, 2, \ldots, 10$, are drawn at random from a Gaussian distribution with standard deviation $\sigma = 1$. The sample values are as follows (to two decimal places):*
>
> | 2.22 | 2.56 | 1.07 | 0.24 | 0.18 | 0.95 | 0.73 | −0.79 | 2.09 | 1.81 |
>
> *Estimate the population mean μ, quoting the standard error on your result.*

We have shown in the final worked example of subsection 31.3.1 that, in this case, $\bar{x}$ is a consistent, unbiased, minimum-variance estimator of μ and has variance $V[\bar{x}] = \sigma^2/N$. Thus, our estimate of the population mean with its associated standard error is

$$\hat{\mu} = \bar{x} \pm \frac{\sigma}{\sqrt{N}} = 1.11 \pm 0.32.$$

If the true value of σ had not been known, we would have needed to use an estimated value $\hat{\sigma}$ in the expression for the standard error. Useful basic estimators of σ are discussed in subsection 31.4.2. ◀

It should be noted that the above approach is most meaningful for unbiased estimators. In this case, $E[\hat{a}] = a$ and so $\sigma_{\hat{a}}$ describes the spread of $\hat{a}$-values about the true value a. For a biased estimator, however, the spread about the true value a is given by the *root mean square error* $\epsilon_{\hat{a}}$, which is defined by

$$\begin{aligned}
\epsilon_{\hat{a}}^2 &= E[(\hat{a} - a)^2] \\
&= E[(\hat{a} - E[\hat{a}])^2] + (E[\hat{a}] - a)^2 \\
&= V[\hat{a}] + b(\mathbf{a})^2.
\end{aligned}$$

We see that $\epsilon_{\hat{a}}^2$ is the sum of the variance of $\hat{a}$ and the square of the bias and so can be interpreted as the sum of squares of statistical and systematic errors. For a biased estimator, it is often more appropriate to quote the result as

$$a = \hat{a} \pm \epsilon_{\hat{a}}.$$

As above, it may be necessary to use estimated values $\hat{\mathbf{a}}$ in the expression for the root mean square error and thus to quote only an estimate $\hat{\epsilon}_{\hat{a}}$ of the error.

31.3.4 Confidence limits on estimators

An alternative (and often equivalent) way of quoting a statistical error is with a *confidence interval*. Let us assume that, *other* than the quantity of interest a, the quantities **a** have known fixed values. Thus we denote the sampling distribution

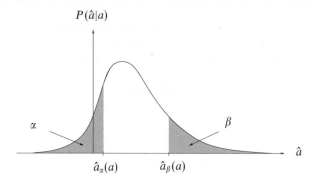

Figure 31.2   The sampling distribution $P(\hat{a}|a)$ of some estimator $\hat{a}$ for a given value of a. The shaded regions indicate the two probabilities $\Pr(\hat{a} < \hat{a}_\alpha(a)) = \alpha$ and $\Pr(\hat{a} > \hat{a}_\beta(a)) = \beta$.

of $\hat{a}$ by $P(\hat{a}|a)$. For any particular value of a, one can determine the two values $\hat{a}_\alpha(a)$ and $\hat{a}_\beta(a)$ such that

$$\Pr(\hat{a} < \hat{a}_\alpha(a)) = \int_{-\infty}^{\hat{a}_\alpha(a)} P(\hat{a}|a)\,d\hat{a} = \alpha, \tag{31.25}$$

$$\Pr(\hat{a} > \hat{a}_\beta(a)) = \int_{\hat{a}_\beta(a)}^{\infty} P(\hat{a}|a)\,d\hat{a} = \beta. \tag{31.26}$$

This is illustrated in figure 31.2. Thus, for any particular value of a, the probability that the estimator $\hat{a}$ lies within the limits $\hat{a}_\alpha(a)$ and $\hat{a}_\beta(a)$ is given by

$$\Pr(\hat{a}_\alpha(a) < \hat{a} < \hat{a}_\beta(a)) = \int_{\hat{a}_\alpha(a)}^{\hat{a}_\beta(a)} P(\hat{a}|a)\,d\hat{a} = 1 - \alpha - \beta.$$

Now, let us suppose that from our sample $x_1, x_2, \ldots, x_N$, we actually obtain the value $\hat{a}_{\text{obs}}$ for our estimator. If $\hat{a}$ is a good estimator of a then we would expect $\hat{a}_\alpha(a)$ and $\hat{a}_\beta(a)$ to be monotonically increasing functions of a (i.e. $\hat{a}_\alpha$ and $\hat{a}_\beta$ *both* change in the *same* sense as a when the latter is varied). Assuming this to be the case, we can uniquely define the two numbers a_- and a_+ by the relationships

$$\hat{a}_\alpha(a_+) = \hat{a}_{\text{obs}} \quad \text{and} \quad \hat{a}_\beta(a_-) = \hat{a}_{\text{obs}}.$$

From (31.25) and (31.26) it follows that

$$\Pr(a_+ < a) = \alpha \quad \text{and} \quad \Pr(a_- > a) = \beta,$$

which when taken together imply

$$\Pr(a_- < a < a_+) = 1 - \alpha - \beta. \tag{31.27}$$

Thus, from our estimate $\hat{a}_{\text{obs}}$, we have determined two values a_- and a_+ such that this interval contains the true value of a with probability $1 - \alpha - \beta$. It should be emphasised that a_- and a_+ are random variables. If a large number of samples,

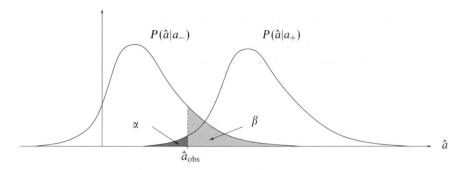

Figure 31.3   An illustration of how the observed value of the estimator, $\hat{a}_{\text{obs}}$, and the given values α and β determine the two confidence limits a_- and a_+, which are such that $\hat{a}_\alpha(a_+) = \hat{a}_{\text{obs}} = \hat{a}_\beta(a_-)$.

each of size N, were analysed then the interval $[a_-, a_+]$ would contain the true value a on a fraction $1 - \alpha - \beta$ of the occasions.

The interval $[a_-, a_+]$ is called a *confidence interval* on a at the *confidence level* $1 - \alpha - \beta$. The values a_- and a_+ themselves are called respectively the *lower confidence limit* and the *upper confidence limit* at this confidence level. In practice, the confidence level is often quoted as a percentage. A convenient way of presenting our results is

$$\int_{-\infty}^{\hat{a}_{\text{obs}}} P(\hat{a}|a_+)\, d\hat{a} = \alpha, \tag{31.28}$$

$$\int_{\hat{a}_{\text{obs}}}^{\infty} P(\hat{a}|a_-)\, d\hat{a} = \beta. \tag{31.29}$$

The confidence limits may then be found by solving these equations for a_- and a_+ either analytically or numerically. The situation is illustrated graphically in figure 31.3.

Occasionally one might not combine the results (31.28) and (31.29) but use either one or the other to provide a *one-sided* confidence interval on a. Whenever the results are combined to provide a *two-sided* confidence interval, however, the interval is *not* specified uniquely by the confidence level $1 - \alpha - \beta$. In other words, there are generally an infinite number of intervals $[a_-, a_+]$ for which (31.27) holds. To specify a unique interval, one often chooses $\alpha = \beta$, resulting in the *central confidence interval* on a. All cases can be covered by calculating the quantities $c = \hat{a} - a_-$ and $d = a_+ - \hat{a}$ and quoting the result of an estimate as

$$a = \hat{a}_{-c}^{+d}.$$

So far we have assumed that the quantities **a** other than the quantity of interest a are known in advance. If this is not the case then the construction of confidence limits is considerably more complicated. This is discussed in subsection 31.3.6.

31.3.5 Confidence limits for a Gaussian sampling distribution

An important special case occurs when the sampling distribution is Gaussian; if the mean is a and the standard deviation is $\sigma_{\hat{a}}$ then

$$P(\hat{a}|a, \sigma_{\hat{a}}) = \frac{1}{\sqrt{2\pi\sigma_{\hat{a}}^2}} \exp\left[-\frac{(\hat{a}-a)^2}{2\sigma_{\hat{a}}^2}\right]. \tag{31.30}$$

For almost any (consistent) estimator $\hat{a}$, the sampling distribution will tend to this form in the large-sample limit $N \to \infty$, as a consequence of the central limit theorem. For a sampling distribution of the form (31.30), the above procedure for determining confidence intervals becomes straightforward. Suppose, from our sample, we obtain the value $\hat{a}_{\text{obs}}$ for our estimator. In this case, equations (31.28) and (31.29) become

$$\Phi\left(\frac{\hat{a}_{\text{obs}} - a_+}{\sigma_{\hat{a}}}\right) = \alpha,$$

$$1 - \Phi\left(\frac{\hat{a}_{\text{obs}} - a_-}{\sigma_{\hat{a}}}\right) = \beta,$$

where $\Phi(z)$ is the cumulative probability function for the standard Gaussian distribution, discussed in subsection 30.9.1. Solving these equations for a_- and a_+ gives

$$a_- = \hat{a}_{\text{obs}} - \sigma_{\hat{a}}\Phi^{-1}(1-\beta), \tag{31.31}$$

$$a_+ = \hat{a}_{\text{obs}} + \sigma_{\hat{a}}\Phi^{-1}(1-\alpha); \tag{31.32}$$

we have used the fact that $\Phi^{-1}(\alpha) = -\Phi^{-1}(1-\alpha)$ to make the equations symmetric. The value of the inverse function $\Phi^{-1}(z)$ can be read off directly from table 30.3, given in subsection 30.9.1. For the normally used central confidence interval one has $\alpha = \beta$. In this case, we see that quoting a result using the standard error, as

$$a = \hat{a} \pm \sigma_{\hat{a}}, \tag{31.33}$$

is equivalent to taking $\Phi^{-1}(1-\alpha) = 1$. From table 30.3, we find $\alpha = 1 - 0.8413 = 0.1587$, and so this corresponds to a confidence level of $1 - 2(0.1587) \approx 0.683$. Thus, the standard error limits give the 68.3% central confidence interval.

▶ *Ten independent sample values x_i, $i = 1, 2, \ldots, 10$, are drawn at random from a Gaussian distribution with standard deviation $\sigma = 1$. The sample values are as follows (to two decimal places):*

| 2.22 | 2.56 | 1.07 | 0.24 | 0.18 | 0.95 | 0.73 | −0.79 | 2.09 | 1.81 |

Find the 90% central confidence interval on the population mean μ.

Our estimator $\hat{\mu}$ is the sample mean $\bar{x}$. As shown towards the end of section 31.3, the sampling distribution of $\bar{x}$ is Gaussian with mean $E[\bar{x}]$ and variance $V[\bar{x}] = \sigma^2/N$. Since $\sigma = 1$ in this case, the standard error is given by $\sigma_{\hat{x}} = \sigma/\sqrt{N} = 0.32$. Moreover, in subsection 31.3.3, we found the mean of the above sample to be $\bar{x} = 1.11$.

For the 90% central confidence interval, we require $\alpha = \beta = 0.05$. From table 30.3, we find

$$\Phi^{-1}(1 - \alpha) = \Phi^{-1}(0.95) = 1.65,$$

and using (31.31) and (31.32) we obtain

$$a_- = \bar{x} - 1.65\sigma_{\bar{x}} = 1.11 - (1.65)(0.32) = 0.58,$$
$$a_+ = \bar{x} + 1.65\sigma_{\bar{x}} = 1.11 + (1.65)(0.32) = 1.64.$$

Thus, the 90% central confidence interval on μ is $[0.58, 1.64]$. For comparison, the true value used to create the sample was $\mu = 1$. ◄

In the case where the standard error $\sigma_{\hat{a}}$ in (31.33) is not known in advance, one must use a value $\hat{\sigma}_{\hat{a}}$ estimated from the sample. In principle, this complicates somewhat the construction of confidence intervals, since properly one should consider the two-dimensional joint sampling distribution $P(\hat{a}, \hat{\sigma}_{\hat{a}}|a)$. Nevertheless, in practice, provided $\hat{\sigma}_{\hat{a}}$ is a fairly good estimate of $\sigma_{\hat{a}}$ the above procedure may be applied with reasonable accuracy. In the special case where the sample values x_i are drawn from a Gaussian distribution with unknown μ and σ, it is in fact possible to obtain *exact* confidence intervals on the mean μ, for a sample of any size N, using Student's t-distribution. This is discussed in subsection 31.7.5.

31.3.6 Estimation of several quantities simultaneously

Suppose one uses a sample $x_1, x_2, \ldots, x_N$ to calculate the values of several estimators $\hat{a}_1, \hat{a}_2, \ldots, \hat{a}_M$ (collectively denoted by $\hat{\mathbf{a}}$) of the quantities $a_1, a_2, \ldots, a_M$ (collectively denoted by $\mathbf{a}$) that describe the population from which the sample was drawn. The joint sampling distribution of these estimators is an M-dimensional PDF $P(\hat{\mathbf{a}}|\mathbf{a})$ given by

$$P(\hat{\mathbf{a}}|\mathbf{a})\, d^M\hat{\mathbf{a}} = P(\mathbf{x}|\mathbf{a})\, d^N\mathbf{x}.$$

►*Sample values $x_1, x_2, \ldots, x_N$ are drawn independently from a Gaussian distribution with mean μ and standard deviation σ. Suppose we choose the sample mean $\bar{x}$ and sample standard deviation s respectively as estimators $\hat{\mu}$ and $\hat{\sigma}$. Find the joint sampling distribution of these estimators.*

Since each data value x_i in the sample is assumed to be independent of the others, the joint probability distribution of sample values is given by

$$P(\mathbf{x}|\mu, \sigma) = (2\pi\sigma^2)^{-N/2} \exp\left[-\frac{\sum_i (x_i - \mu)^2}{2\sigma^2}\right].$$

We may rewrite the sum in the exponent as follows:

$$\sum_i (x_i - \mu)^2 = \sum_i (x_i - \bar{x} + \bar{x} - \mu)^2$$
$$= \sum_i (x_i - \bar{x})^2 + 2(\bar{x} - \mu) \sum_i (x_i - \bar{x}) + \sum_i (\bar{x} - \mu)^2$$
$$= Ns^2 + N(\bar{x} - \mu)^2,$$

where in the last line we have used the fact that $\sum_i(x_i - \bar{x}) = 0$. Hence, for given values of μ and σ, the sampling distribution is in fact a function only of the sample mean $\bar{x}$ and the standard deviation s. Thus the sampling distribution of $\bar{x}$ and s must satisfy

$$P(\bar{x}, s|\mu, \sigma)\, d\bar{x}\, ds = (2\pi\sigma^2)^{-N/2} \exp\left\{-\frac{N[(\bar{x} - \mu)^2 + s^2]}{2\sigma^2}\right\} dV, \qquad (31.34)$$

where $dV = dx_1\, dx_2 \cdots dx_N$ is an element of volume in the sample space which yields simultaneously values of $\bar{x}$ and s that lie within the region bounded by $[\bar{x}, \bar{x} + d\bar{x}]$ and $[s, s + ds]$. Thus our only remaining task is to express dV in terms of $\bar{x}$ and s and their differentials.

Let S be the point in sample space representing the sample $(x_1, x_2, \ldots, x_N)$. For given values of $\bar{x}$ and s, we require the sample values to satisfy both the condition

$$\sum_i x_i = N\bar{x},$$

which defines an $(N - 1)$-dimensional hyperplane in the sample space, and the condition

$$\sum_i (x_i - \bar{x})^2 = Ns^2,$$

which defines an $(N - 1)$-dimensional hypersphere. Thus S is constrained to lie in the intersection of these two hypersurfaces, which is itself an $(N - 2)$-dimensional hypersphere. Now, the volume of an $(N - 2)$-dimensional hypersphere is proportional to s^{N-1}. It follows that the volume dV between two concentric $(N - 2)$-dimensional hyperspheres of radius $\sqrt{N}s$ and $\sqrt{N}(s + ds)$ and two $(N - 1)$-dimensional hyperplanes corresponding to $\bar{x}$ and $\bar{x} + d\bar{x}$ is

$$dV = As^{N-2}\, ds\, d\bar{x},$$

where A is some constant. Thus, substituting this expression for dV into (31.34), we find

$$P(\bar{x}, s|\mu, \sigma) = C_1 \exp\left[-\frac{N(\bar{x} - \mu)^2}{2\sigma^2}\right] C_2 s^{N-2} \exp\left(-\frac{Ns^2}{2\sigma^2}\right) = P(\bar{x}|\mu, \sigma)P(s|\sigma), \qquad (31.35)$$

where C_1 and C_2 are constants. We have written $P(\bar{x}, s|\mu, \sigma)$ in this form to show that it separates naturally into two parts, one depending only on $\bar{x}$ and the other only on s. Thus, $\bar{x}$ and s are *independent* variables. Separate normalisations of the two factors in (31.35) require

$$C_1 = \left(\frac{N}{2\pi\sigma^2}\right)^{1/2} \qquad \text{and} \qquad C_2 = 2\left(\frac{N}{2\sigma^2}\right)^{(N-1)/2} \frac{1}{\Gamma\left(\frac{1}{2}(N-1)\right)},$$

where the calculation of C_2 requires the use of the gamma function, discussed in section 18.12. ◀

The *marginal* sampling distribution of any one of the estimators $\hat{a}_i$ is given simply by

$$P(\hat{a}_i|\mathbf{a}) = \int \cdots \int P(\hat{\mathbf{a}}|\mathbf{a})\, d\hat{a}_1 \cdots d\hat{a}_{i-1} d\hat{a}_{i+1} \cdots d\hat{a}_M,$$

and the expectation value $E[\hat{a}_i]$ and variance $V[\hat{a}_i]$ of $\hat{a}_i$ are again given by (31.14) and (31.16) respectively. By analogy with the one-dimensional case, the standard error $\sigma_{\hat{a}_i}$ on the estimator $\hat{a}_i$ is given by the positive square root of $V[\hat{a}_i]$. With several estimators, however, it is usual to quote their full covariance matrix. This $M \times M$ matrix has elements

$$V_{ij} = \text{Cov}[\hat{a}_i, \hat{a}_j] = \int (\hat{a}_i - E[\hat{a}_i])(\hat{a}_j - E[\hat{a}_j])P(\hat{\mathbf{a}}|\mathbf{a}) \, d^M \hat{\mathbf{a}}$$

$$= \int (\hat{a}_i - E[\hat{a}_i])(\hat{a}_j - E[\hat{a}_j])P(\mathbf{x}|\mathbf{a}) \, d^N \mathbf{x}.$$

Fisher's inequality can be generalised to the multi-dimensional case. Adapting the proof given in subsection 31.3.2, one may show that, in the case where the estimators are efficient and have zero bias, the elements of the *inverse* of the covariance matrix are given by

$$(V^{-1})_{ij} = E\left[-\frac{\partial^2 \ln P}{\partial a_i \partial a_j}\right], \tag{31.36}$$

where P denotes the population $P(\mathbf{x}|\mathbf{a})$ from which the sample is drawn. The quantity on the RHS of (31.36) is the element F_{ij} of the so-called *Fisher matrix* F of the estimators.

> ►*Calculate the covariance matrix of the estimators $\bar{x}$ and s in the previous example.*

As shown in (31.35), the joint sampling distribution $P(\bar{x}, s|\mu, \sigma)$ factorises, and so the estimators $\bar{x}$ and s are independent. Thus, we conclude immediately that

$$\text{Cov}[\bar{x}, s] = 0.$$

Since we have already shown in the worked example at the end of subsection 31.3.1 that $V[\bar{x}] = \sigma^2/N$, it only remains to calculate $V[s]$. From (31.35), we find

$$E[s^r] = C_2 \int_0^\infty s^{N-2+r} \exp\left(-\frac{Ns^2}{2\sigma^2}\right) ds = \left(\frac{2}{N}\right)^{r/2} \frac{\Gamma\left(\frac{1}{2}(N-1+r)\right)}{\Gamma\left(\frac{1}{2}(N-1)\right)} \sigma^r,$$

where we have evaluated the integral using the definition of the gamma function given in the Appendix. Thus, the expectation value of the sample standard deviation is

$$E[s] = \left(\frac{2}{N}\right)^{1/2} \frac{\Gamma\left(\frac{1}{2}N\right)}{\Gamma\left(\frac{1}{2}(N-1)\right)} \sigma, \tag{31.37}$$

and its variance is given by

$$V[s] = E[s^2] - (E[s])^2 = \frac{\sigma^2}{N}\left\{N - 1 - 2\left[\frac{\Gamma\left(\frac{1}{2}N\right)}{\Gamma\left(\frac{1}{2}(N-1)\right)}\right]^2\right\}$$

We note, in passing, that (31.37) shows that s is a *biased* estimator of σ. ◄

The idea of a confidence interval can also be extended to the case where several quantities are estimated simultaneously but then the practical construction of an interval is considerably more complicated. The general approach is to construct an M-dimensional *confidence region R* in **a**-space. By analogy with the one-dimensional case, for a given confidence level of (say) $1 - \alpha$, one first constructs a region $\hat{R}$ in $\hat{\mathbf{a}}$-space, such that

$$\iint_{\hat{R}} P(\hat{\mathbf{a}}|\mathbf{a})\, d^M\hat{\mathbf{a}} = 1 - \alpha.$$

A common choice for such a region is that bounded by the 'surface' $P(\hat{\mathbf{a}}|\mathbf{a}) = $ constant. By considering all possible values $\mathbf{a}$ and the values of $\hat{\mathbf{a}}$ lying within the region $\hat{R}$, one can construct a $2M$-dimensional region in the combined space $(\hat{\mathbf{a}}, \mathbf{a})$. Suppose now that, from our sample $\mathbf{x}$, the values of the estimators are $\hat{a}_{i,\text{obs}}$, $i = 1, 2, \ldots, M$. The intersection of the M 'hyperplanes' $\hat{a}_i = \hat{a}_{i,\text{obs}}$ with the $2M$-dimensional region will determine an M-dimensional region which, when projected onto $\mathbf{a}$-space, will determine a confidence limit R at the confidence level $1 - \alpha$. It is usually the case that this confidence region has to be evaluated numerically.

The above procedure is clearly rather complicated in general and a simpler approximate method that uses the likelihood function is discussed in subsection 31.5.5. As a consequence of the central limit theorem, however, in the large-sample limit, $N \to \infty$, the joint sampling distribution $P(\hat{\mathbf{a}}|\mathbf{a})$ will tend, in general, towards the multivariate Gaussian

$$P(\hat{\mathbf{a}}|\mathbf{a}) = \frac{1}{(2\pi)^{M/2}|\mathsf{V}|^{1/2}} \exp\left[-\tfrac{1}{2}Q(\hat{\mathbf{a}}, \mathbf{a})\right], \tag{31.38}$$

where V is the covariance matrix of the estimators and the quadratic form Q is given by

$$Q(\hat{\mathbf{a}}, \mathbf{a}) = (\hat{\mathbf{a}} - \mathbf{a})^{\mathrm{T}} \mathsf{V}^{-1} (\hat{\mathbf{a}} - \mathbf{a}).$$

Moreover, in the limit of large N, the inverse covariance matrix tends to the Fisher matrix F given in (31.36), i.e. $\mathsf{V}^{-1} \to \mathsf{F}$.

For the Gaussian sampling distribution (31.38), the process of obtaining confidence intervals is greatly simplified. The surfaces of constant $P(\hat{\mathbf{a}}|\mathbf{a})$ correspond to surfaces of constant $Q(\hat{\mathbf{a}}, \mathbf{a})$, which have the shape of M-dimensional ellipsoids in $\hat{\mathbf{a}}$-space, centred on the true values $\mathbf{a}$. In particular, let us suppose that the ellipsoid $Q(\hat{\mathbf{a}}, \mathbf{a}) = c$ (where c is some constant) contains a fraction $1 - \alpha$ of the total probability. Now suppose that, from our sample $\mathbf{x}$, we obtain the values $\hat{\mathbf{a}}_{\text{obs}}$ for our estimators. Because of the obvious symmetry of the quadratic form Q with respect to $\mathbf{a}$ and $\hat{\mathbf{a}}$, it is clear that the ellipsoid $Q(\mathbf{a}, \hat{\mathbf{a}}_{\text{obs}}) = c$ in $\mathbf{a}$-space that is centred on $\hat{\mathbf{a}}_{\text{obs}}$ should contain the true values $\mathbf{a}$ with probability $1 - \alpha$. Thus $Q(\mathbf{a}, \hat{\mathbf{a}}_{\text{obs}}) = c$ defines our required confidence region R at this confidence level. This is illustrated in figure 31.4 for the two-dimensional case.

It remains only to determine the constant c corresponding to the confidence level $1 - \alpha$. As discussed in subsection 30.15.2, the quantity $Q(\hat{\mathbf{a}}, \mathbf{a})$ is distributed as a χ^2 variable of order M. Thus, the confidence region corresponding to the

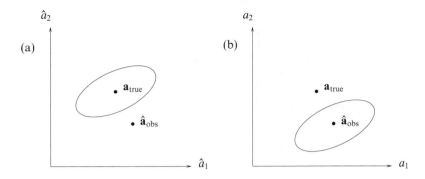

Figure 31.4 (a) The ellipse $Q(\hat{\mathbf{a}}, \mathbf{a}) = c$ in $\hat{\mathbf{a}}$-space. (b) The ellipse $Q(\mathbf{a}, \hat{\mathbf{a}}_{\text{obs}}) = c$ in $\mathbf{a}$-space that corresponds to a confidence region R at the level $1 - \alpha$, when c satisfies (31.39).

confidence level $1 - \alpha$ is given by $Q(\mathbf{a}, \hat{\mathbf{a}}_{\text{obs}}) = c$, where the constant c satisfies

$$\int_0^c P(\chi_M^2)\, d(\chi_M^2) = 1 - \alpha, \tag{31.39}$$

and $P(\chi_M^2)$ is the chi-squared PDF of order M, discussed in subsection 30.9.4. This integral may be evaluated numerically to determine the constant c. Alternatively, some reference books tabulate the values of c corresponding to given confidence levels and various values of M. A representative selection of values of c is given in table 31.2; there the number of degrees of freedom is denoted by the more usual n, rather than M.

31.4 Some basic estimators

In many cases, one does not know the functional form of the population from which a sample is drawn. Nevertheless, in a case where the sample values $x_1, x_2, \ldots, x_N$ are each drawn *independently* from a one-dimensional population $P(x)$, it is possible to construct some basic estimators for the moments and central moments of $P(x)$. In this section, we investigate the estimating properties of the common sample statistics presented in section 31.2. In fact, expectation values and variances of these sample statistics can be calculated *without* prior knowledge of the functional form of the population; they depend only on the sample size N and certain moments and central moments of $P(x)$.

31.4.1 Population mean μ

Let us suppose that the parent population $P(x)$ has mean μ and variance σ^2. An obvious estimator $\hat{\mu}$ of the population mean is the sample mean $\bar{x}$. Provided μ and σ^2 are both finite, we may apply the central limit theorem directly to obtain

%	99	95	10	5	2.5	1	0.5	0.1
$n = 1$	$1.57\,10^{-4}$	$3.93\,10^{-3}$	2.71	3.84	5.02	6.63	7.88	10.83
2	$2.01\,10^{-2}$	0.103	4.61	5.99	7.38	9.21	10.60	13.81
3	0.115	0.352	6.25	7.81	9.35	11.34	12.84	16.27
4	0.297	0.711	7.78	9.49	11.14	13.28	14.86	18.47
5	0.554	1.15	9.24	11.07	12.83	15.09	16.75	20.52
6	0.872	1.64	10.64	12.59	14.45	16.81	18.55	22.46
7	1.24	2.17	12.02	14.07	16.01	18.48	20.28	24.32
8	1.65	2.73	13.36	15.51	17.53	20.09	21.95	26.12
9	2.09	3.33	14.68	16.92	19.02	21.67	23.59	27.88
10	2.56	3.94	15.99	18.31	20.48	23.21	25.19	29.59
11	3.05	4.57	17.28	19.68	21.92	24.73	26.76	31.26
12	3.57	5.23	18.55	21.03	23.34	26.22	28.30	32.91
13	4.11	5.89	19.81	22.36	24.74	27.69	29.82	34.53
14	4.66	6.57	21.06	23.68	26.12	29.14	31.32	36.12
15	5.23	7.26	22.31	25.00	27.49	30.58	32.80	37.70
16	5.81	7.96	23.54	26.30	28.85	32.00	34.27	39.25
17	6.41	8.67	24.77	27.59	30.19	33.41	35.72	40.79
18	7.01	9.39	25.99	28.87	31.53	34.81	37.16	42.31
19	7.63	10.12	27.20	30.14	32.85	36.19	38.58	43.82
20	8.26	10.85	28.41	31.41	34.17	37.57	40.00	45.31
21	8.90	11.59	29.62	32.67	35.48	38.93	41.40	46.80
22	9.54	12.34	30.81	33.92	36.78	40.29	42.80	48.27
23	10.20	13.09	32.01	35.17	38.08	41.64	44.18	49.73
24	10.86	13.85	33.20	36.42	39.36	42.98	45.56	51.18
25	11.52	14.61	34.38	37.65	40.65	44.31	46.93	52.62
30	14.95	18.49	40.26	43.77	46.98	50.89	53.67	59.70
40	22.16	26.51	51.81	55.76	59.34	63.69	66.77	73.40
50	29.71	34.76	63.17	67.50	71.42	76.15	79.49	86.66
60	37.48	43.19	74.40	79.08	83.30	88.38	91.95	99.61
70	45.44	51.74	85.53	90.53	95.02	100.4	104.2	112.3
80	53.54	60.39	96.58	101.9	106.6	112.3	116.3	124.8
90	61.75	69.13	107.6	113.1	118.1	124.1	128.3	137.2
100	70.06	77.93	118.5	124.3	129.6	135.8	140.2	149.4

Table 31.2 The tabulated values are those which a variable distributed as χ^2 with n degrees of freedom exceeds with the given percentage probability. For example, a variable having a χ^2 distribution with 14 degrees of freedom takes values in excess of 21.06 on 10% of occasions.

exact expressions, valid for samples of any size N, for the expectation value and variance of $\bar{x}$. From parts (i) and (ii) of the central limit theorem, discussed in section 30.10, we immediately obtain

$$E[\bar{x}] = \mu, \qquad V[\bar{x}] = \frac{\sigma^2}{N}. \qquad (31.40)$$

Thus we see that $\bar{x}$ is an unbiased estimator of μ. Moreover, we note that the standard error in $\bar{x}$ is $\sigma/\sqrt{N}$, and so the sampling distribution of $\bar{x}$ becomes more tightly centred around μ as the sample size N increases. Indeed, since $V[\bar{x}] \to 0$ as $N \to \infty$, $\bar{x}$ is also a consistent estimator of μ.

In the limit of large N, we may in fact obtain an *approximate* form for the full sampling distribution of $\bar{x}$. Part (iii) of the central limit theorem (see section 30.10) tells us immediately that, for large N, the sampling distribution of $\bar{x}$ is given approximately by the Gaussian form

$$P(\bar{x}|\mu, \sigma) \approx \frac{1}{\sqrt{2\pi\sigma^2/N}} \exp\left[-\frac{(\bar{x} - \mu)^2}{2\sigma^2/N}\right].$$

Note that this does *not* depend on the form of the original parent population. If, however, the parent population is in fact Gaussian then this result is *exact* for samples of *any* size N (as is immediately apparent from our discussion of multiple Gaussian distributions in subsection 30.9.1).

31.4.2 Population variance σ^2

An estimator for the population variance σ^2 is not so straightforward to define as one for the mean. Complications arise because, in many cases, the true mean of the population μ is not known. Nevertheless, let us begin by considering the case where in fact μ is known. In this event, a useful estimator is

$$\widehat{\sigma^2} = \frac{1}{N}\sum_{i=1}^{N}(x_i - \mu)^2 = \left(\frac{1}{N}\sum_{i=1}^{N}x_i^2\right) - \mu^2. \qquad (31.41)$$

▶*Show that $\widehat{\sigma^2}$ is an unbiased and consistent estimator of the population variance σ^2.*

The expectation value of $\widehat{\sigma^2}$ is given by

$$E[\widehat{\sigma^2}] = \frac{1}{N}E\left[\sum_{i=1}^{N}x_i^2\right] - \mu^2 = E[x_i^2] - \mu^2 = \mu_2 - \mu^2 = \sigma^2,$$

from which we see that the estimator is unbiased. The variance of the estimator is

$$V[\widehat{\sigma^2}] = \frac{1}{N^2}V\left[\sum_{i=1}^{N}x_i^2\right] + V[\mu^2] = \frac{1}{N}V[x_i^2] = \frac{1}{N}(\mu_4 - \mu_2^2),$$

in which we have used that fact that $V[\mu^2] = 0$ and $V[x_i^2] = E[x_i^4] - (E[x_i^2])^2 = \mu_4 - \mu_2^2$, where μ_r is the rth population moment. Since $\widehat{\sigma^2}$ is unbiased and $V[\widehat{\sigma^2}] \to 0$ as $N \to \infty$, showing that it is also a consistent estimator of σ^2, the result is established. ◄

If the true mean of the population is unknown, however, a natural alternative is to replace μ by $\bar{x}$ in (31.41), so that our estimator is simply the sample variance s^2 given by

$$s^2 = \frac{1}{N} \sum_{i=1}^{N} x_i^2 - \left(\frac{1}{N} \sum_{i=1}^{N} x_i \right)^2.$$

In order to determine the properties of this estimator, we must calculate $E[s^2]$ and $V[s^2]$. This task is straightforward but lengthy. However, for the investigation of the properties of a *central* moment of the sample, there exists a useful trick that simplifies the calculation. We can assume, with no loss of generality, that the mean μ_1 of the population from which the sample is drawn is equal to zero. With this assumption, the population central moments, v_r, are identical to the corresponding moments μ_r, and we may perform our calculation in terms of the latter. At the end, however, we replace μ_r by v_r in the final result and so obtain a general expression that is valid even in cases where $\mu_1 \neq 0$.

►*Calculate $E[s^2]$ and $V[s^2]$ for a sample of size N.*

The expectation value of the sample variance s^2 for a sample of size N is given by

$$E[s^2] = \frac{1}{N} E \left[\sum_i x_i^2 \right] - \frac{1}{N^2} E \left[\left(\sum_i x_i \right)^2 \right]$$

$$= \frac{1}{N} N E[x_i^2] - \frac{1}{N^2} E \left[\sum_i x_i^2 + \sum_{\substack{i,j \\ j \neq i}} x_i x_j \right]. \tag{31.42}$$

The number of terms in the double summation in (31.42) is $N(N-1)$, so we find

$$E[s^2] = E[x_i^2] - \frac{1}{N^2} (N E[x_i^2] + N(N-1) E[x_i x_j]).$$

Now, since the sample elements x_i and x_j are independent, $E[x_i x_j] = E[x_i] E[x_j] = 0$, assuming the mean μ_1 of the parent population to be zero. Denoting the rth moment of the population by μ_r, we thus obtain

$$E[s^2] = \mu_2 - \frac{\mu_2}{N} = \frac{N-1}{N} \mu_2 = \frac{N-1}{N} \sigma^2, \tag{31.43}$$

where in the last line we have used the fact that the population mean is zero, and so $\mu_2 = v_2 = \sigma^2$. However, the final result is also valid in the case where $\mu_1 \neq 0$.

Using the above method, we can also find the variance of s^2, although the algebra is rather heavy going. The variance of s^2 is given by

$$V[s^2] = E[s^4] - (E[s^2])^2, \tag{31.44}$$

where $E[s^2]$ is given by (31.43). We therefore need only consider how to calculate $E[s^4]$, where s^4 is given by

$$s^4 = \left[\frac{\sum_i x_i^2}{N} - \left(\frac{\sum_i x_i}{N} \right)^2 \right]^2$$

$$= \frac{(\sum_i x_i^2)^2}{N^2} - 2 \frac{(\sum_i x_i^2)(\sum_i x_i)^2}{N^3} + \frac{(\sum_i x_i)^4}{N^4}. \tag{31.45}$$

We will consider in turn each of the three terms on the RHS. In the first term, the sum $(\sum_i x_i^2)^2$ can be written as

$$\left(\sum_i x_i^2 \right)^2 = \sum_i x_i^4 + \sum_{\substack{i,j \\ j \neq i}} x_i^2 x_j^2,$$

where the first sum contains N terms and the second contains $N(N-1)$ terms. Since the sample elements x_i and x_j are assumed independent, we have $E[x_i^2 x_j^2] = E[x_i^2]E[x_j^2] = \mu_2^2$, and so

$$E\left[\left(\sum_i x_i^2 \right)^2 \right] = N\mu_4 + N(N-1)\mu_2^2.$$

Turning to the second term on the RHS of (31.45),

$$\left(\sum_i x_i^2 \right) \left(\sum_i x_i \right)^2 = \sum_i x_i^4 + \sum_{\substack{i,j \\ j \neq i}} x_i^3 x_j + \sum_{\substack{i,j \\ j \neq i}} x_i^2 x_j^2 + \sum_{\substack{i,j,k \\ k \neq j \neq i}} x_i^2 x_j x_k.$$

Since the mean of the population has been assumed to equal zero, the expectation values of the second and fourth sums on the RHS vanish. The first and third sums contain N and $N(N-1)$ terms respectively, and so

$$E\left[\left(\sum_i x_i^2 \right) \left(\sum_i x_i \right)^2 \right] = N\mu_4 + N(N-1)\mu_2^2.$$

Finally, we consider the third term on the RHS of (31.45), and write

$$\left(\sum_i x_i \right)^4 = \sum_i x_i^4 + \sum_{\substack{i,j \\ j \neq i}} x_i^3 x_j + \sum_{\substack{i,j \\ j \neq i}} x_i^2 x_j^2 + \sum_{\substack{i,j,k \\ k \neq j \neq i}} x_i^2 x_j x_k + \sum_{\substack{i,j,k,l \\ l \neq k \neq j \neq i}} x_i x_j x_k x_l.$$

The expectation values of the second, fourth and fifth sums are zero, and the first and third sums contain N and $3N(N-1)$ terms respectively (for the third sum, there are $N(N-1)/2$ ways of choosing i and j, and the multinomial coefficient of $x_i^2 x_j^2$ is $4!/(2!2!) = 6$). Thus

$$E\left[\left(\sum_i x_i \right)^4 \right] = N\mu_4 + 3N(N-1)\mu_2^2.$$

Collecting together terms, we therefore obtain

$$E[s^4] = \frac{(N-1)^2}{N^3} \mu_4 + \frac{(N-1)(N^2-2N+3)}{N^3} \mu_2^2, \tag{31.46}$$

which, together with the result (31.43), may be substituted into (31.44) to obtain finally

$$V[s^2] = \frac{(N-1)^2}{N^3} \mu_4 - \frac{(N-1)(N-3)}{N^3} \mu_2^2$$

$$= \frac{N-1}{N^3} [(N-1)v_4 - (N-3)v_2^2], \tag{31.47}$$

where in the last line we have used again the fact that, since the population mean is zero, $\mu_r = v_r$. However, result (31.47) holds even when the population mean is not zero. ◄

From (31.43), we see that s^2 is a *biased* estimator of σ^2, although the bias becomes negligible for large N. However, it immediately follows that an unbiased estimator of σ^2 is given simply by

$$\widehat{\sigma^2} = \frac{N}{N-1}s^2, \tag{31.48}$$

where the multiplicative factor $N/(N-1)$ is often called *Bessel's correction*. Thus in terms of the sample values x_i, $i = 1, 2, \ldots, N$, an unbiased estimator of the population variance σ^2 is given by

$$\widehat{\sigma^2} = \frac{1}{N-1}\sum_{i=1}^{N}(x_i - \bar{x})^2. \tag{31.49}$$

Using (31.47), we find that the variance of the estimator $\widehat{\sigma^2}$ is

$$V[\widehat{\sigma^2}] = \left(\frac{N}{N-1}\right)^2 V[s^2] = \frac{1}{N}\left(v_4 - \frac{N-3}{N-1}v_2^2\right),$$

where v_r is the rth central moment of the parent population. We note that, since $E[\widehat{\sigma^2}] = \sigma^2$ and $V[\widehat{\sigma^2}] \to 0$ as $N \to \infty$, the statistic $\widehat{\sigma^2}$ is also a consistent estimator of the population variance.

31.4.3 Population standard deviation σ

The standard deviation σ of a population is defined as the positive square root of the population variance σ^2 (as, indeed, our notation suggests). Thus, it is common practice to take the positive square root of the variance estimator as our estimator for σ. Thus, we take

$$\hat{\sigma} = \left(\widehat{\sigma^2}\right)^{1/2}, \tag{31.50}$$

where $\widehat{\sigma^2}$ is given by either (31.41) or (31.48), depending on whether the population mean μ is known or unknown. Because of the square root in the definition of $\hat{\sigma}$, it is not possible in either case to obtain an exact expression for $E[\hat{\sigma}]$ and $V[\hat{\sigma}]$. Indeed, although in each case the estimator is the positive square root of an unbiased estimator of σ^2, it is *not* itself an unbiased estimator of σ. However, the bias does becomes negligible for large N.

►*Obtain approximate expressions for $E[\hat{\sigma}]$ and $V[\hat{\sigma}]$ for a sample of size N in the case where the population mean μ is unknown.*

As the population mean is unknown, we use (31.50) and (31.48) to write our estimator in the form

$$\hat{\sigma} = \left(\frac{N}{N-1}\right)^{1/2} s,$$

where s is the sample standard deviation. The expectation value of this estimator is given by

$$E[\hat{\sigma}] = \left(\frac{N}{N-1}\right)^{1/2} E[(s^2)^{1/2}] \approx \left(\frac{N}{N-1}\right)^{1/2} (E[s^2])^{1/2} = \sigma.$$

An approximate expression for the variance of $\hat{\sigma}$ may be found using (31.47) and is given by

$$V[\hat{\sigma}] = \frac{N}{N-1} V[(s^2)^{1/2}] \approx \frac{N}{N-1} \left[\frac{d}{d(s^2)}(s^2)^{1/2}\right]^2_{s^2=E[s^2]} V[s^2]$$

$$\approx \frac{N}{N-1} \left[\frac{1}{4s^2}\right]_{s^2=E[s^2]} V[s^2].$$

Using the expressions (31.43) and (31.47) for $E[s^2]$ and $V[s^2]$ respectively, we obtain

$$V[\hat{\sigma}] \approx \frac{1}{4Nv_2}\left(v_4 - \frac{N-3}{N-1}v_2^2\right). \quad \blacktriangleleft$$

31.4.4 Population moments μ_r

We may straightforwardly generalise our discussion of estimation of the population mean $\mu \, (= \mu_1)$ in subsection 31.4.1 to the estimation of the rth population moment μ_r. An obvious choice of estimator is the rth sample moment m_r. The expectation value of m_r is given by

$$E[m_r] = \frac{1}{N}\sum_{i=1}^{N} E[x_i^r] = \frac{N\mu_r}{N} = \mu_r,$$

and so it is an unbiased estimator of μ_r.

The variance of m_r may be found in a similar manner, although the calculation is a little more complicated. We find that

$$V[m_r] = E[(m_r - \mu_r)^2]$$

$$= \frac{1}{N^2}E\left[\left(\sum_i x_i^r - N\mu_r\right)^2\right]$$

$$= \frac{1}{N^2}E\left[\sum_i x_i^{2r} + \sum_i\sum_{j\neq i}x_i^r x_j^r - 2N\mu_r\sum_i x_i^r + N^2\mu_r^2\right]$$

$$= \frac{1}{N}\mu_{2r} - \mu_r^2 + \frac{1}{N^2}\sum_i\sum_{j\neq i}E[x_i^r x_j^r]. \qquad (31.51)$$

However, since the sample values x_i are assumed to be independent, we have

$$E[x_i^r x_j^r] = E[x_i^r]E[x_j^r] = \mu_r^2. \tag{31.52}$$

The number of terms in the sum on the RHS of (31.51) is $N(N-1)$, and so we find

$$V[m_r] = \frac{1}{N}\mu_{2r} - \mu_r^2 + \frac{N-1}{N}\mu_r^2 = \frac{\mu_{2r} - \mu_r^2}{N}. \tag{31.53}$$

Since $E[m_r] = \mu_r$ and $V[m_r] \to 0$ as $N \to \infty$, the rth sample moment m_r is also a consistent estimator of μ_r.

▶ *Find the covariance of the sample moments m_r and m_s for a sample of size N.*

We obtain the covariance of the sample moments m_r and m_s in a similar manner to that used above to obtain the variance of m_r. From the definition of covariance, we have

$$\text{Cov}[m_r, m_s] = E[(m_r - \mu_r)(m_s - \mu_s)]$$

$$= \frac{1}{N^2}E\left[\left(\sum_i x_i^r - N\mu_r\right)\left(\sum_j x_j^s - N\mu_s\right)\right]$$

$$= \frac{1}{N^2}E\left[\sum_i x_i^{r+s} + \sum_i\sum_{j\neq i} x_i^r x_j^s - N\mu_r\sum_j x_j^s - N\mu_s\sum_i x_i^r + N^2\mu_r\mu_s\right]$$

Assuming the x_i to be independent, we may again use result (31.52) to obtain

$$\text{Cov}[m_r, m_s] = \frac{1}{N^2}[N\mu_{r+s} + N(N-1)\mu_r\mu_s - N^2\mu_r\mu_s - N^2\mu_s\mu_r + N^2\mu_r\mu_s]$$

$$= \frac{1}{N}\mu_{r+s} + \frac{N-1}{N}\mu_r\mu_s - \mu_r\mu_s$$

$$= \frac{\mu_{r+s} - \mu_r\mu_s}{N}.$$

We note that by setting $r = s$, we recover the expression (31.53) for $V[m_r]$. ◀

31.4.5 Population central moments v_r

We may generalise the discussion of estimators for the second central moment v_2 (or equivalently σ^2) given in subsection 31.4.2 to the estimation of the rth central moment v_r. In particular, we saw in that subsection that our choice of estimator for v_2 depended on whether the population mean μ_1 is known; the same is true for the estimation of v_r.

Let us first consider the case in which μ_1 is known. From (30.54), we may write v_r as

$$v_r = \mu_r - {}^rC_1\mu_{r-1}\mu_1 + \cdots + (-1)^k\,{}^rC_k\mu_{r-k}\mu_1^k + \cdots + (-1)^{r-1}({}^rC_{r-1} - 1)\mu_1^r.$$

If μ_1 is known, a suitable estimator is obviously

$$\hat{v}_r = m_r - {}^rC_1 m_{r-1}\mu_1 + \cdots + (-1)^k\,{}^rC_k m_{r-k}\mu_1^k + \cdots + (-1)^{r-1}({}^rC_{r-1} - 1)\mu_1^r,$$

where m_r is the rth sample moment. Since μ_1 and the binomial coefficients are (known) constants, it is immediately clear that $E[\hat{v}_r] = v_r$, and so $\hat{v}_r$ is an unbiased estimator of v_r. It is also possible to obtain an expression for $V[\hat{v}_r]$, though the calculation is somewhat lengthy.

In the case where the population mean μ_1 is *not* known, the situation is more complicated. We saw in subsection 31.4.2 that the second sample moment n_2 (or s^2) is *not* an unbiased estimator of v_2 (or σ^2). Similarly, the rth *central* moment of a sample, n_r, is not an unbiased estimator of the rth population central moment v_r. However, in all cases the bias becomes negligible in the limit of large N.

As we also found in the same subsection, there are complications in calculating the expectation and variance of n_2; these complications increase considerably for general r. Nevertheless, we have derived already in this chapter *exact* expressions for the expectation value of the first few sample central moments, which are valid for samples of any size N. From (31.40), (31.43) and (31.46), we find

$$E[n_1] = 0,$$

$$E[n_2] = \frac{N-1}{N} v_2,$$ (31.54)

$$E[n_2^2] = \frac{N-1}{N^3} [(N-1)v_4 + (N^2 - 2N + 3)v_2^2].$$

By similar arguments it can be shown that

$$E[n_3] = \frac{(N-1)(N-2)}{N^2} v_3,$$ (31.55)

$$E[n_4] = \frac{N-1}{N^3} [(N^2 - 3N + 3)v_4 + 3(2N - 3)v_2^2].$$ (31.56)

From (31.54) and (31.55), we see that unbiased estimators of v_2 and v_3 are

$$\hat{v}_2 = \frac{N}{N-1} n_2,$$ (31.57)

$$\hat{v}_3 = \frac{N^2}{(N-1)(N-2)} n_3,$$ (31.58)

where (31.57) simply re-establishes our earlier result that $\widehat{\sigma^2} = Ns^2/(N-1)$ is an unbiased estimator of σ^2.

Unfortunately, the pattern that appears to be emerging in (31.57) and (31.58) is *not* continued for higher r, as is seen immediately from (31.56). Nevertheless, in the limit of large N, the bias becomes negligible, and often one simply takes $\hat{v}_r = n_r$. For large N, it may be shown that

$$E[n_r] \approx v_r$$

$$V[n_r] \approx \frac{1}{N}(v_{2r} - v_r^2 + r^2 v_2 v_{r-1}^2 - 2r v_{r-1} v_{r+1})$$

$$\text{Cov}[n_r, n_s] \approx \frac{1}{N}(v_{r+s} - v_r v_s + rs v_2 v_{r-1} v_{s-1} - r v_{r-1} v_{s+1} - s v_{s-1} v_{r+1})$$

31.4.6 *Population covariance* Cov[x, y] *and correlation* Corr[x, y]

So far we have assumed that each of our N independent samples consists of a single number x_i. Let us now extend our discussion to a situation in which each sample consists of two numbers x_i, y_i, which we may consider as being drawn randomly from a two-dimensional population $P(x, y)$. In particular, we now consider estimators for the population covariance $\text{Cov}[x, y]$ and for the correlation $\text{Corr}[x, y]$.

When μ_x and μ_y are *known*, an appropriate estimator of the population covariance is

$$\widehat{\text{Cov}}[x, y] = \overline{xy} - \mu_x\mu_y = \left(\frac{1}{N} \sum_{i=1}^{N} x_i y_i \right) - \mu_x\mu_y. \tag{31.59}$$

This estimator is unbiased since

$$E\left[\widehat{\text{Cov}}[x, y] \right] = \frac{1}{N} E\left[\sum_{i=1}^{N} x_i y_i \right] - \mu_x\mu_y = E[x_i y_i] - \mu_x\mu_y = \text{Cov}[x, y].$$

Alternatively, if μ_x and μ_y are *unknown*, it is natural to replace μ_x and μ_y in (31.59) by the sample means $\bar{x}$ and $\bar{y}$ respectively, in which case we recover the sample covariance $V_{xy} = \overline{xy} - \bar{x}\bar{y}$ discussed in subsection 31.2.4. This estimator is biased but an unbiased estimator of the population covariance is obtained by forming

$$\widehat{\text{Cov}}[x, y] = \frac{N}{N-1} V_{xy}. \tag{31.60}$$

► *Calculate the expectation value of the sample covariance* V_{xy} *for a sample of size N.*

The sample covariance is given by

$$V_{xy} = \left(\frac{1}{N} \sum_i x_i y_i \right) - \left(\frac{1}{N} \sum_i x_i \right) \left(\frac{1}{N} \sum_j y_j \right).$$

Thus its expectation value is given by

$$E[V_{xy}] = \frac{1}{N} E\left[\sum_i x_i y_i \right] - \frac{1}{N^2} E\left[\left(\sum_i x_i \right) \left(\sum_j x_j \right) \right]$$

$$= E[x_i y_i] - \frac{1}{N^2} E\left[\sum_i x_i y_i + \sum_{\substack{i,j \\ j \neq i}} x_i y_j \right]$$

Since the number of terms in the double sum on the RHS is $N(N-1)$, we have

$$E[V_{xy}] = E[x_i y_i] - \frac{1}{N^2}(NE[x_i y_i] + N(N-1)E[x_i y_j])$$

$$= E[x_i y_i] - \frac{1}{N^2}(NE[x_i y_i] + N(N-1)E[x_i]E[y_j])$$

$$= E[x_i y_i] - \frac{1}{N}\left(E[x_i y_i] + (N-1)\mu_x \mu_y\right) = \frac{N-1}{N}\text{Cov}[x, y],$$

where we have used the fact that, since the samples are independent, $E[x_i y_j] = E[x_i]E[y_j]$. ◀

It is possible to obtain expressions for the variances of the estimators (31.59) and (31.60) but these quantities depend upon higher moments of the population $P(x, y)$ and are extremely lengthy to calculate.

Whether the means μ_x and μ_y are known or unknown, an estimator of the population correlation $\text{Corr}[x, y]$ is given by

$$\widehat{\text{Corr}}[x, y] = \frac{\widehat{\text{Cov}}[x, y]}{\hat{\sigma}_x \hat{\sigma}_y}, \tag{31.61}$$

where $\widehat{\text{Cov}}[x, y]$, $\hat{\sigma}_x$ and $\hat{\sigma}_y$ are the appropriate estimators of the population co-variance and standard deviations. Although this estimator is only asymptotically unbiased, i.e. for large N, it is widely used because of its simplicity. Once again the variance of the estimator depends on the higher moments of $P(x, y)$ and is difficult to calculate.

In the case in which the means μ_x and μ_y are unknown, a suitable (but biased) estimator is

$$\widehat{\text{Corr}}[x, y] = \frac{N}{N-1}\frac{V_{xy}}{s_x s_y} = \frac{N}{N-1}r_{xy}, \tag{31.62}$$

where s_x and s_y are the sample standard deviations of the x_i and y_i respectively and r_{xy} is the sample correlation. In the special case when the parent population $P(x, y)$ is Gaussian, it may be shown that, if $\rho = \text{Corr}[x, y]$,

$$E[r_{xy}] = \rho - \frac{\rho(1-\rho^2)}{2N} + O(N^{-2}), \tag{31.63}$$

$$V[r_{xy}] = \frac{1}{N}(1-\rho^2)^2 + O(N^{-2}), \tag{31.64}$$

from which the expectation value and variance of the estimator $\widehat{\text{Corr}}[x, y]$ may be found immediately.

We note finally that our discussion may be extended, without significant alteration, to the general case in which each data item consists of n numbers $x_i, y_i, \ldots, z_i$.

31.4.7 A worked example

To conclude our discussion of basic estimators, we reconsider the set of experimental data given in subsection 31.2.4. We carry the analysis as far as calculating the standard errors in the estimated population parameters, including the population correlation.

▶ *Ten UK citizens are selected at random and their heights and weights are found to be as follows (to the nearest cm or kg respectively):*

Person	A	B	C	D	E	F	G	H	I	J
Height (cm)	194	168	177	180	171	190	151	169	175	182
Weight (kg)	75	53	72	80	75	75	57	67	46	68

Estimate the means, μ_x and μ_y, and standard deviations, σ_x and σ_y, of the two-dimensional joint population from which the sample was drawn, quoting the standard error on the estimate in each case. Estimate also the correlation $\mathrm{Corr}[x, y]$ of the population, and quote the standard error on the estimate under the assumption that the population is a multivariate Gaussian.

In subsection 31.2.4, we calculated various sample statistics for these data. In particular, we found that for our sample of size $N = 10$,

$$\bar{x} = 175.7, \qquad \bar{y} = 66.8,$$

$$s_x = 11.6, \qquad s_y = 10.6, \qquad r_{xy} = 0.54.$$

Let us begin by estimating the means μ_x and μ_y. As discussed in subsection 31.4.1, the sample mean is an unbiased, consistent estimator of the population mean. Moreover, the standard error on $\bar{x}$ (say) is $\sigma_x/\sqrt{N}$. In this case, however, we do not know the true value of σ_x and we must estimate it using $\hat{\sigma}_x = \sqrt{N/(N-1)}s_x$. Thus, our estimates of μ_x and μ_y, with associated standard errors, are

$$\hat{\mu}_x = \bar{x} \pm \frac{s_x}{\sqrt{N-1}} = 175.7 \pm 3.9,$$

$$\hat{\mu}_y = \bar{y} \pm \frac{s_y}{\sqrt{N-1}} = 66.8 \pm 3.5.$$

We now turn to estimating σ_x and σ_y. As just mentioned, our estimate of σ_x (say) is $\hat{\sigma}_x = \sqrt{N/(N-1)}s_x$. Its variance (see the final line of subsection 31.4.3) is given approximately by

$$V[\hat{\sigma}] \approx \frac{1}{4Nv_2}\left(v_4 - \frac{N-3}{N-1}v_2^2\right).$$

Since we do not know the true values of the population central moments v_2 and v_4, we must use their estimated values in this expression. We may take $\hat{v}_2 = \hat{\sigma}_x^2 = (\hat{\sigma})^2$, which we have already calculated. It still remains, however, to estimate v_4. As implied near the end of subsection 31.4.5, it is acceptable to take $\hat{v}_4 = n_4$. Thus for the x_i and y_i values, we have

$$(\hat{v}_4)_x = \frac{1}{N}\sum_{i=1}^{N}(x_i - \bar{x})^4 = 53\,411.6$$

$$(\hat{v}_4)_y = \frac{1}{N}\sum_{i=1}^{N}(y_i - \bar{y})^4 = 27\,732.5$$

Substituting these values into (31.50), we obtain

$$\hat{\sigma}_x = \left(\frac{N}{N-1}\right)^{1/2} s_x \pm (\hat{V}[\hat{\sigma}_x])^{1/2} = 12.2 \pm 6.7, \tag{31.65}$$

$$\hat{\sigma}_y = \left(\frac{N}{N-1}\right)^{1/2} s_y \pm (\hat{V}[\hat{\sigma}_y])^{1/2} = 11.2 \pm 3.6. \tag{31.66}$$

Finally, we estimate the population correlation $\text{Corr}[x, y]$, which we shall denote by ρ. From (31.62), we have

$$\hat{\rho} = \frac{N}{N-1} r_{xy} = 0.60.$$

Under the *assumption* that the sample was drawn from a two-dimensional Gaussian population $P(x, y)$, the variance of our estimator is given by (31.64). Since we do not know the true value of ρ, we must use our estimate $\hat{\rho}$. Thus, we find that the standard error $\Delta\rho$ in our estimate is given approximately by

$$\Delta\rho \approx \frac{10}{9}\left(\frac{1}{10}\right)[1 - (0.60)^2]^2 = 0.05. \blacktriangleleft$$

31.5 Maximum-likelihood method

The population from which the sample $x_1, x_2, \ldots, x_N$ is drawn is, in general, *unknown*. In the previous section, we assumed that the sample values were independent and drawn from a one-dimensional population $P(x)$, and we considered basic estimators of the moments and central moments of $P(x)$. We did *not*, however, assume a particular functional form for $P(x)$. We now discuss the process of *data modelling*, in which a specific form is assumed for the population.

In the most general case, it will not be known whether the sample values are independent, and so let us consider the full joint population $P(\mathbf{x})$, where $\mathbf{x}$ is the point in the N-dimensional data space with coordinates $x_1, x_2, \ldots, x_N$. We then adopt the *hypothesis H* that the probability distribution of the sample values has some particular functional form $L(\mathbf{x}; \mathbf{a})$, dependent on the values of some set of *parameters* a_i, $i = 1, 2, \ldots, m$. Thus, we have

$$P(\mathbf{x}|\mathbf{a}, H) = L(\mathbf{x}; \mathbf{a}),$$

where we make explicit the conditioning on both the assumed functional form and on the parameter values. $L(\mathbf{x}; \mathbf{a})$ is called the *likelihood function*. Hypotheses of this type form the basis of *data modelling* and *parameter estimation*. One proposes a particular model for the underlying population and then attempts to estimate from the sample values $x_1, x_2, \ldots, x_N$ the values of the parameters $\mathbf{a}$ defining this model.

▶*A company measures the duration (in minutes) of the N intervals x_i, $i = 1, 2, \ldots, N$ between successive telephone calls received by its switchboard. Suppose that the sample values x_i are drawn independently from the distribution $P(x|\tau) = (1/\tau)\exp(-x/\tau)$, where τ is the mean interval between calls. Calculate the likelihood function $L(\mathbf{x}; \tau)$.*

Since the sample values are independent and drawn from the stated distribution, the

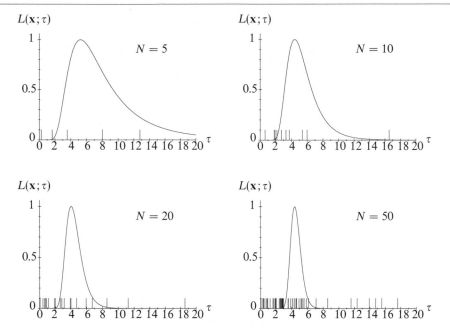

Figure 31.5 Examples of the likelihood function (31.67) for samples of different size N. In each case, the true value of the parameter is $\tau = 4$ and the sample values x_i are indicated by the short vertical lines. For the purposes of illustration, in each case the likelihood function is normalised so that its maximum value is unity.

likelihood is given by

$$
\begin{aligned}
L(\mathbf{x};\tau) &= P(x_i|\tau)P(x_2|\tau)\cdots P(x_N|\tau) \\
&= \frac{1}{\tau}\exp\left(-\frac{x_1}{\tau}\right)\frac{1}{\tau}\exp\left(-\frac{x_2}{\tau}\right)\cdots\frac{1}{\tau}\exp\left(-\frac{x_N}{\tau}\right) \\
&= \frac{1}{\tau^N}\exp\left[-\frac{1}{\tau}(x_1 + x_2 + \cdots + x_N)\right].
\end{aligned}
\tag{31.67}
$$

which is to be considered as a function of τ, given that the sample values x_i are fixed. ◄

The likelihood function (31.67) depends on just a single parameter τ. Plots of the likelihood function, considered as a function of τ, are shown in figure 31.5 for samples of different size N. The true value of the parameter τ used to generate the sample values was 4. In each case, the sample values x_i are indicated by the short vertical lines. For the purposes of illustration, the likelihood function in each case has been scaled so that its maximum value is unity (this is, in fact, common practice). We see that when the sample size is small, the likelihood function is very broad. As N increases, however, the function becomes narrower (its width is inversely proportional to $\sqrt{N}$) and tends to a Gaussian-like shape, with its peak centred on 4, the true value of τ. We discuss these properties of the likelihood function in more detail in subsection 31.5.6.

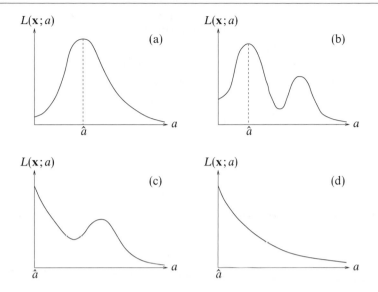

Figure 31.6   Typical shapes of one-dimensional likelihood functions $L(\mathbf{x};a)$ encountered in practice, when, for illustration purposes, it is assumed that the parameter a is restricted to the range zero to infinity. The ML estimator in the various cases occurs at: (a) the only stationary point; (b) one of several stationary points; (c) an end-point of the allowed parameter range that is not a stationary point (although stationary points do exist); (d) an end-point of the allowed parameter range in which no stationary point exists.

31.5.1 The maximum-likelihood estimator

Since the likelihood function $L(\mathbf{x};a)$ gives the probability density associated with any particular set of values of the parameters $\mathbf{a}$, our best estimate $\hat{\mathbf{a}}$ of these parameters is given by the values of $\mathbf{a}$ for which $L(\mathbf{x};a)$ is a maximum. This is called the *maximum-likelihood estimator* (or ML estimator).

In general, the likelihood function can have a complicated shape when considered as a function of $\mathbf{a}$, particularly when the dimensionality of the space of parameters $a_1, a_2, \ldots, a_M$ is large. It may be that the values of some parameters are either known or assumed in advance, in which case the effective dimensionality of the likelihood function is reduced accordingly. However, even when the likelihood depends on just a single parameter a (either intrinsically or as the result of assuming particular values for the remaining parameters), its form may be complicated when the sample size N is small. Frequently occurring shapes of one-dimensional likelihood functions are illustrated in figure 31.6, where we have assumed, for definiteness, that the allowed range of the parameter a is zero to infinity. In each case, the ML estimate $\hat{a}$ is also indicated. Of course, the 'shape' of higher-dimensional likelihood functions may be considerably more complicated.

In many simple cases, however, the likelihood function $L(\mathbf{x};\mathbf{a})$ has a single maximum that occurs at a stationary point (the likelihood function is then termed *unimodal*). In this case, the ML estimators of the parameters a_i, $i = 1, 2, \ldots, M$, may be found *without* evaluating the full likelihood function $L(\mathbf{x}; \mathbf{a})$. Instead, one simply solves the M simultaneous equations

$$\left. \frac{\partial L}{\partial a_i} \right|_{\mathbf{a}=\hat{\mathbf{a}}} = 0 \qquad \text{for } i = 1, 2, \ldots, M. \tag{31.68}$$

Since $\ln z$ is a monotonically increasing function of z (and therefore has the same stationary points), it is often more convenient, in fact, to maximise the *log-likelihood function*, $\ln L(\mathbf{x}; \mathbf{a})$, with respect to the a_i. Thus, one may, as an alternative, solve the equations

$$\left. \frac{\partial \ln L}{\partial a_i} \right|_{\mathbf{a}=\hat{\mathbf{a}}} = 0 \qquad \text{for } i = 1, 2, \ldots, M. \tag{31.69}$$

Clearly, (31.68) and (31.69) will lead to the same ML estimates $\hat{\mathbf{a}}$ of the parameters. In either case, it is, of course, prudent to check that the point $\mathbf{a} = \hat{\mathbf{a}}$ is a local maximum.

▶ *Find the ML estimate of the parameter τ in the previous example, in terms of the measured values x_i, $i = 1, 2, \ldots, N$.*

From (31.67), the log-likelihood function in this case is given by

$$\ln L(\mathbf{x}; \tau) = \sum_{i=1}^{N} \ln \left(\frac{1}{\tau} e^{-x_i/\tau} \right) = - \sum_{i=1}^{N} \left(\ln \tau + \frac{x_i}{\tau} \right). \tag{31.70}$$

Differentiating with respect to the parameter τ and setting the result equal to zero, we find

$$\frac{\partial \ln L}{\partial \tau} = - \sum_{i=1}^{N} \left(\frac{1}{\tau} - \frac{x_i}{\tau^2} \right) = 0.$$

Thus the ML estimate of the parameter τ is given by

$$\hat{\tau} = \frac{1}{N} \sum_{i=1}^{N} x_i, \tag{31.71}$$

which is simply the sample mean of the N measured intervals. ◀

In the previous example we assumed that the sample values x_i were drawn independently from the *same* parent distribution. The ML method is more flexible than this restriction might seem to imply, and it can equally well be applied to the common case in which the samples x_i are independent but each is drawn from a *different* distribution.

> ►*In an experiment, N independent measurements x_i of some quantity are made. Suppose that the random measurement error on the ith sample value is Gaussian distributed with mean zero and known standard deviation σ_i. Calculate the ML estimate of the true value μ of the quantity being measured.*

As the measurements are independent, the likelihood factorises:

$$L(\mathbf{x};\mu,\{\sigma_k\}) = \prod_{i=1}^{N} P(x_i|\mu,\sigma_i),$$

where $\{\sigma_k\}$ denotes collectively the set of known standard deviations $\sigma_1,\sigma_2,\ldots,\sigma_N$. The individual distributions are given by

$$P(x_i|\mu,\sigma_i) = \frac{1}{\sqrt{2\pi\sigma_i^2}} \exp\left[-\frac{(x_i-\mu)^2}{2\sigma_i^2}\right].$$

and so the full log-likelihood function is given by

$$\ln L(\mathbf{x};\mu,\{\sigma_k\}) = -\frac{1}{2}\sum_{i=1}^{N}\left[\ln(2\pi\sigma_i^2) + \frac{(x_i-\mu)^2}{\sigma_i^2}\right].$$

Differentiating this expression with respect to μ and setting the result equal to zero, we find

$$\frac{\partial \ln L}{\partial \mu} = \sum_{i=1}^{N} \frac{x_i-\mu}{\sigma_i^2} = 0,$$

from which we obtain the ML estimator

$$\hat{\mu} = \frac{\sum_{i=1}^{N}(x_i/\sigma_i^2)}{\sum_{i=1}^{N}(1/\sigma_i^2)}. \tag{31.72}$$

This estimator is commonly used when averaging data with different *statistical weights* $w_i = 1/\sigma_i^2$. We note that when all the variances σ_i^2 have the same value the estimator reduces to the sample mean of the data x_i. ◄

There is, in fact, no requirement in the ML method that the sample values be independent. As an illustration, we shall generalise the above example to a case in which the measurements x_i are not all independent. This would occur, for example, if these measurements were based at least in part on the same data.

> ►*In an experiment N measurements x_i of some quantity are made. Suppose that the random measurement errors on the samples are drawn from a joint Gaussian distribution with mean zero and known covariance matrix V. Calculate the ML estimate of the true value μ of the quantity being measured.*

From (30.148), the likelihood in this case is given by

$$L(\mathbf{x};\mu,\mathsf{V}) = \frac{1}{(2\pi)^{N/2}|\mathsf{V}|^{1/2}} \exp\left[-\tfrac{1}{2}(\mathbf{x}-\mu\mathbf{1})^{\mathrm{T}}\mathsf{V}^{-1}(\mathbf{x}-\mu\mathbf{1})\right],$$

where $\mathbf{x}$ is the column matrix with components $x_1,x_2,\ldots,x_N$ and $\mathbf{1}$ is the column matrix with all components equal to unity. Thus, the log-likelihood function is given by

$$\ln L(\mathbf{x};\mu,\mathsf{V}) = -\tfrac{1}{2}\left[N\ln(2\pi) + \ln|\mathsf{V}| + (\mathbf{x}-\mu\mathbf{1})^{\mathrm{T}}\mathsf{V}^{-1}(\mathbf{x}-\mu\mathbf{1})\right].$$

Differentiating with respect to μ and setting the result equal to zero gives

$$\frac{\partial \ln L}{\partial \mu} = \mathbf{1}^{\mathrm{T}} \mathsf{V}^{-1}(\mathbf{x} - \mu \mathbf{1}) = 0.$$

Thus, the ML estimator is given by

$$\hat{\mu} = \frac{\mathbf{1}^{\mathrm{T}} \mathsf{V}^{-1} \mathbf{x}}{\mathbf{1}^{\mathrm{T}} \mathsf{V}^{-1} \mathbf{1}} = \frac{\sum_{i,j} (V^{-1})_{ij} x_j}{\sum_{i,j} (V^{-1})_{ij}}.$$

In the case of uncorrelated errors in measurement, $(V^{-1})_{ij} = \delta_{ij}/\sigma_i^2$ and our estimator reduces to that given in (31.72). ◀

In all the examples considered so far, the likelihood function has been effectively one-dimensional, either instrinsically or under the assumption that the values of all but one of the parameters are known in advance. As the following example involving two parameters shows, the application of the ML method to the estimation of several parameters simultaneously is straightforward.

> ▶*In an experiment N measurements x_i of some quantity are made. Suppose the random error on each sample value is drawn independently from a Gaussian distribution of mean zero but unknown standard deviation σ (which is the same for each measurement). Calculate the ML estimates of the true value μ of the quantity being measured and the standard deviation σ of the random errors.*

In this case the log-likelihood function is given by

$$\ln L(\mathbf{x}; \mu, \sigma) = -\frac{1}{2} \sum_{i=1}^{N} \left[\ln(2\pi\sigma^2) + \frac{(x_i - \mu)^2}{\sigma^2} \right].$$

Taking partial derivatives of $\ln L$ with respect to μ and σ and setting the results equal to zero at the joint estimate $\hat{\mu}, \hat{\sigma}$, we obtain

$$\sum_{i=1}^{N} \frac{x_i - \hat{\mu}}{\hat{\sigma}^2} = 0, \tag{31.73}$$

$$\sum_{i=1}^{N} \frac{(x_i - \hat{\mu})^2}{\hat{\sigma}^3} - \sum_{i=1}^{N} \frac{1}{\hat{\sigma}} = 0. \tag{31.74}$$

In principle, one should solve these two equations simultaneously for $\hat{\mu}$ and $\hat{\sigma}$, but in this case we notice that the first is solved immediately by

$$\hat{\mu} = \frac{1}{N} \sum_{i=1}^{N} x_i = \bar{x},$$

where $\bar{x}$ is the sample mean. Substituting this result into the second equation, we find

$$\hat{\sigma} = \sqrt{\frac{1}{N} \sum_{i=1}^{N} (x_i - \bar{x})^2} = s,$$

where s is the sample standard deviation. As shown in subsection 31.4.3, s is a biased estimator of σ. The reason why the ML method may produce a biased estimator is discussed in the next subsection. ◀

31.5.2 Transformation invariance and bias of ML estimators

An extremely useful property of ML estimators is that they are *invariant* to parameter transformations. Suppose that, instead of estimating some parameter a of the assumed population, we wish to estimate some function $\alpha(a)$ of the parameter. The ML estimator $\hat{\alpha}(a)$ is given by the value assumed by the function $\alpha(a)$ at the maximum point of the likelihood, which is simply equal to $\alpha(\hat{a})$. Thus, we have the very convenient property

$$\hat{\alpha}(a) = \alpha(\hat{a}).$$

We do not have to worry about the distinction between estimating a and estimating a function of a. This is *not* true, in general, for other estimation procedures.

▶ *A company measures the duration (in minutes) of the N intervals x_i, $i = 1, 2, \dots, N$, between successive telephone calls received by its switchboard. Suppose that the sample values x_i are drawn independently from the distribution $P(x|\tau) = (1/\tau)\exp(-x/\tau)$. Find the ML estimate of the parameter $\lambda = 1/\tau$.*

This is the same problem as the first one considered in subsection 31.5.1. In terms of the new parameter λ, the log-likelihood function is given by

$$\ln L(\mathbf{x}; \lambda) = \sum_{i=1}^{N} \ln(\lambda e^{-\lambda x_i}) = \sum_{i=1}^{N}(\ln \lambda - \lambda x_i).$$

Differentiating with respect to λ and setting the result equal to zero, we have

$$\frac{\partial \ln L}{\partial \lambda} = \sum_{i=1}^{N} \left(\frac{1}{\lambda} - x_i \right) = 0.$$

Thus, the ML estimator of the parameter λ is given by

$$\hat{\lambda} = \left(\frac{1}{N} \sum_{i=1}^{N} x_i \right)^{-1} = \bar{x}^{-1}. \tag{31.75}$$

Referring back to (31.71), we see that, as expected, the ML estimators of λ and τ are related by $\hat{\lambda} = 1/\hat{\tau}$. ◀

Although this invariance property is useful it also means that, in general, ML estimators may be *biased*. In particular, one must be aware of the fact that even if $\hat{a}$ is an unbiased ML estimator of a it does *not* follow that the estimator $\hat{\alpha}(a)$ is also unbiased. In the limit of large N, however, the bias of ML estimators always tends to zero. As an illustration, it is straightforward to show (see exercise 31.8) that the ML estimators $\hat{\tau}$ and $\hat{\lambda}$ in the above example have expectation values

$$E[\hat{\tau}] = \tau \quad \text{and} \quad E[\hat{\lambda}] = \frac{N}{N-1}\lambda. \tag{31.76}$$

In fact, since $\hat{\tau} = \bar{x}$ and the sample values are independent, the first result follows immediately from (31.40). Thus, $\hat{\tau}$ is unbiased, but $\hat{\lambda} = 1/\hat{\tau}$ is biased, albeit that the bias tends to zero for large N.

31.5.3 Efficiency of ML estimators

We showed in subsection 31.3.2 that Fisher's inequality puts a lower limit on the variance $V[\hat{a}]$ of any estimator of the parameter a. Under our hypothesis H on p. 1255, the functional form of the population is given by the likelihood function, i.e. $P(\mathbf{x}|\mathbf{a}, H) = L(\mathbf{x}; \mathbf{a})$. Thus, if this hypothesis is correct, we may replace P by L in Fisher's inequality (31.18), which then reads

$$V[\hat{a}] \geq \left(1 + \frac{\partial b}{\partial a}\right)^2 \bigg/ E\left[-\frac{\partial^2 \ln L}{\partial a^2}\right],$$

where b is the bias in the estimator $\hat{a}$. We usually denote the RHS by V_{min}.

An important property of ML estimators is that *if* there exists an efficient estimator $\hat{a}_{\text{eff}}$, i.e. one for which $V[\hat{a}_{\text{eff}}] = V_{\text{min}}$, then it *must* be the ML estimator or some function thereof. This is easily shown by replacing P by L in the proof of Fisher's inequality given in subsection 31.3.2. In particular, we note that the equality in (31.22) holds only if $h(\mathbf{x}) = cg(\mathbf{x})$, where c is a constant. Thus, if an efficient estimator $\hat{a}_{\text{eff}}$ exists, this is equivalent to demanding that

$$\frac{\partial \ln L}{\partial a} = c[\hat{a}_{\text{eff}} - \alpha(a)].$$

Now, the ML estimator $\hat{a}_{\text{ML}}$ is given by

$$\left.\frac{\partial \ln L}{\partial a}\right|_{a=\hat{a}_{\text{ML}}} = 0 \quad \Rightarrow \quad c[\hat{a}_{\text{eff}} - \alpha(\hat{a}_{\text{ML}})] = 0,$$

which, in turn, implies that $\hat{a}_{\text{eff}}$ must be some function of $\hat{a}_{\text{ML}}$.

▶*Show that the ML estimator $\hat{\tau}$ given in (31.71) is an efficient estimator of the parameter τ.*

As shown in (31.70), the log-likelihood function in this case is

$$\ln L(\mathbf{x}; \tau) = -\sum_{i=1}^{N} \left(\ln \tau + \frac{x_i}{\tau}\right).$$

Differentiating twice with respect to τ, we find

$$\frac{\partial^2 \ln L}{\partial \tau^2} = \sum_{i=1}^{N} \left(\frac{1}{\tau^2} - \frac{2x_i}{\tau^3}\right) = \frac{N}{\tau^2}\left(1 - \frac{2}{\tau N}\sum_{i=1}^{N} x_i\right), \tag{31.77}$$

and so the expectation value of this expression is

$$E\left[\frac{\partial^2 \ln L}{\partial \tau^2}\right] = \frac{N}{\tau^2}\left(1 - \frac{2}{\tau}E[x_i]\right) = -\frac{N}{\tau^2},$$

where we have used the fact that $E[x] = \tau$. Setting $b = 0$ in (31.18), we thus find that for *any* unbiased estimator of τ,

$$V[\hat{\tau}] \geq \frac{\tau^2}{N}.$$

From (31.76), we see that the ML estimator $\hat{\tau} = \sum_i x_i/N$ is unbiased. Moreover, using the fact that $V[x] = \tau^2$, it follows immediately from (31.40) that $V[\hat{\tau}] = \tau^2/N$. Thus $\hat{\tau}$ is a minimum-variance estimator of τ. ◀

31.5.4 Standard errors and confidence limits on ML estimators

The ML method provides a procedure for obtaining a particular set of estimators $\hat{\mathbf{a}}_{\mathrm{ML}}$ for the parameters $\mathbf{a}$ of the assumed population $P(\mathbf{x}|\mathbf{a})$. As for any other set of estimators, the associated standard errors, covariances and confidence intervals can be found as described in subsections 31.3.3 and 31.3.4.

> ▶*A company measures the duration (in minutes) of the 10 intervals x_i, $i = 1, 2, \ldots, 10$, between successive telephone calls made to its switchboard to be as follows:*
>
> $\qquad$ 0.43 $\quad$ 0.24 $\quad$ 3.03 $\quad$ 1.93 $\quad$ 1.16 $\quad$ 8.65 $\quad$ 5.33 $\quad$ 6.06 $\quad$ 5.62 $\quad$ 5.22.
>
> *Supposing that the sample values are drawn independently from the probability distribution $P(x|\tau) = (1/\tau)\exp(-x/\tau)$, find the ML estimate of the mean τ and quote an estimate of the standard error on your result.*

As shown in (31.71) the (unbiased) ML estimator $\hat{\tau}$ in this case is simply the sample mean $\bar{x} = 3.77$. Also, as shown in subsection 31.5.3, $\hat{\tau}$ is a minimum-variance estimator with $V[\hat{\tau}] = \tau^2/N$. Thus, the standard error in $\hat{\tau}$ is simply

$$\sigma_{\hat{\tau}} = \frac{\tau}{\sqrt{N}}. \tag{31.78}$$

Since we do not know the true value of τ, however, we must instead quote an estimate $\hat{\sigma}_{\hat{\tau}}$ of the standard error, obtained by substituting our estimate $\hat{\tau}$ for τ in (31.78). Thus, we quote our final result as

$$\tau = \hat{\tau} \pm \frac{\hat{\tau}}{\sqrt{N}} = 3.77 \pm 1.19. \tag{31.79}$$

For comparison, the true value used to create the sample was $\tau = 4$. ◀

For the particular problem considered in the above example, it is in fact possible to derive the full sampling distribution of the ML estimator $\hat{\tau}$ using characteristic functions, and it is given by

$$P(\hat{\tau}|\tau) = \frac{N^N}{(N-1)!} \frac{\hat{\tau}^{N-1}}{\tau^N} \exp\left(-\frac{N\hat{\tau}}{\tau}\right), \tag{31.80}$$

where N is the size of the sample. This function is plotted in figure 31.7 for the case $\tau = 4$ and $N = 10$, which pertains to the above example. Knowledge of the analytic form of the sampling distribution allows one to place *confidence limits* on the estimate $\hat{\tau}$ obtained, as discussed in subsection 31.3.4.

> ▶*Using the sample values in the above example, obtain the 68% central confidence interval on the value of τ.*

For the sample values given, our observed value of the ML estimator is $\hat{\tau}_{\mathrm{obs}} = 3.77$. Thus, from (31.28) and (31.29), the 68% central confidence interval $[\tau_-, \tau_+]$ on the value of τ is found by solving the equations

$$\int_{-\infty}^{\hat{\tau}_{\mathrm{obs}}} P(\hat{\tau}|\tau_+)\, d\hat{\tau} = 0.16,$$

$$\int_{\hat{\tau}_{\mathrm{obs}}}^{\infty} P(\hat{\tau}|\tau_-)\, d\hat{\tau} = 0.16,$$

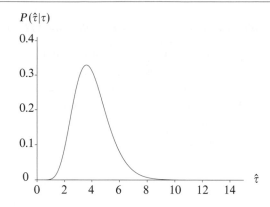

Figure 31.7   The sampling distribution $P(\hat{\tau}|\tau)$ for the estimator $\hat{\tau}$ for the case $\tau = 4$ and $N = 10$.

where $P(\hat{\tau}|\tau)$ is given by (31.80) with $N = 10$. The above integrals can be evaluated analytically but the calculations are rather cumbersome. It is much simpler to evaluate them by numerical integration, from which we find $[\tau_-, \tau_+] = [2.86, 5.46]$. Alternatively, we could quote the estimate and its 68% confidence interval as

$$\tau = 3.77\,^{+1.69}_{-0.91}.$$

Thus we see that the 68% central confidence interval is not symmetric about the estimated value, and differs from the standard error calculated above. This is a result of the (non-Gaussian) shape of the sampling distribution $P(\hat{\tau}|\tau)$, apparent in figure 31.7. ◄

In many problems, however, it is not possible to derive the full sampling distribution of an ML estimator $\hat{a}$ in order to obtain its confidence intervals. Indeed, one may not even be able to obtain an analytic formula for its standard error $\sigma_{\hat{a}}$. This is particularly true when one is estimating several parameter $\hat{\mathbf{a}}$ simultaneously, since the joint sampling distribution will be, in general, very complicated. Nevertheless, as we discuss below, the likelihood function $L(\mathbf{x}; \mathbf{a})$ *itself* can be used very simply to obtain standard errors and confidence intervals. The justification for this has its roots in the *Bayesian* approach to statistics, as opposed to the more traditional *frequentist* approach we have adopted here. We now give a brief discussion of the Bayesian viewpoint on parameter estimation.

31.5.5 The Bayesian interpretation of the likelihood function

As stated at the beginning of section 31.5, the likelihood function $L(\mathbf{x}; \mathbf{a})$ is defined by

$$P(\mathbf{x}|\mathbf{a}, H) = L(\mathbf{x}; \mathbf{a}),$$

where H denotes our hypothesis of an assumed functional form. Now, using *Bayes' theorem* (see subsection 30.2.3), we may write

$$P(\mathbf{a}|\mathbf{x}, H) = \frac{P(\mathbf{x}|\mathbf{a}, H)P(\mathbf{a}|H)}{P(\mathbf{x}|H)}, \tag{31.81}$$

which provides us with an expression for the probability distribution $P(\mathbf{a}|\mathbf{x}, H)$ of the parameters $\mathbf{a}$, given the (fixed) data $\mathbf{x}$ and our hypothesis H, in terms of other quantities that we may assign. The various terms in (31.81) have special formal names, as follows.

- The quantity $P(\mathbf{a}|H)$ on the RHS is the *prior* probability, which represents our state of knowledge of the parameter values (given the hypothesis H) *before* we have analysed the data.
- This probability is modified by the experimental data $\mathbf{x}$ through the *likelihood* $P(\mathbf{x}|\mathbf{a}, H)$.
- When appropriately normalised by the *evidence* $P(\mathbf{x}|H)$, this yields the *posterior* probability $P(\mathbf{a}|\mathbf{x}, H)$, which is the quantity of interest.
- The posterior encodes *all* our inferences about the values of the parameters $\mathbf{a}$. Strictly speaking, from a Bayesian viewpoint, this *entire function*, $P(\mathbf{a}|\mathbf{x}, H)$, is the 'answer' to a parameter estimation problem.

Given a particular hypothesis, the (normalising) evidence factor $P(\mathbf{x}|H)$ is unimportant, since it does not depend explicitly upon the parameter values $\mathbf{a}$. Thus, it is often omitted and one considers only the proportionality relation

$$P(\mathbf{a}|\mathbf{x}, H) \propto P(\mathbf{x}|\mathbf{a}, H)P(\mathbf{a}|H). \tag{31.82}$$

If necessary, the posterior distribution can be normalised empirically, by requiring that it integrates to unity, i.e. $\int P(\mathbf{a}|\mathbf{x}, H) \, d^m\mathbf{a} = 1$, where the integral extends over all values of the parameters $a_1, a_2, \ldots, a_m$.

The prior $P(\mathbf{a}|H)$ in (31.82) should reflect our entire knowledge concerning the values of the parameters $\mathbf{a}$, *before* the analysis of the current data $\mathbf{x}$. For example, there may be some physical reason to require some or all of the parameters to lie in a given range. If we are largely ignorant of the values of the parameters, we often indicate this by choosing a *uniform* (or very broad) prior,

$$P(\mathbf{a}|H) = \text{constant},$$

in which case the posterior distribution is simply proportional to the likelihood. In this case, we thus have

$$P(\mathbf{a}|\mathbf{x}, H) \propto L(\mathbf{x}; \mathbf{a}). \tag{31.83}$$

In other words, if we assume a uniform prior then we can identify the posterior distribution (up to a normalising factor) with $L(\mathbf{x}; \mathbf{a})$, considered as a function of the parameters $\mathbf{a}$.

Thus, a Bayesian statistician considers the ML estimates $\hat{a}_{\mathrm{ML}}$ of the parameters to be the values that maximise the posterior $P(\mathbf{a}|\mathbf{x}, H)$ under the assumption of a uniform prior. More importantly, however, a Bayesian would *not* calculate the standard error or confidence interval on this estimate using the (classical) method employed in subsection 31.3.4. Instead, a far more straightforward approach is adopted. Let us assume, for the moment, that one is estimating just a single parameter a. Using (31.83), we may determine the values a_- and a_+ such that

$$\Pr(a < a_-|\mathbf{x}, H) = \int_{-\infty}^{a_-} L(\mathbf{x}; a)\, da = \alpha,$$

$$\Pr(a > a_+|\mathbf{x}, H) = \int_{a_+}^{\infty} L(\mathbf{x}; a)\, da = \beta.$$

where it is assumed that the likelihood has been normalised in such a way that $\int L(\mathbf{x}; a)\, da = 1$. Combining these equations gives

$$\Pr(a_- \leq a < a_+|\mathbf{x}, H) = \int_{a_-}^{a_+} L(\mathbf{x}; a)\, da = 1 - \alpha - \beta, \tag{31.84}$$

and $[a_-, a_+]$ is the *Bayesian confidence interval* on the value of a at the confidence level $1 - \alpha - \beta$. As in the case of classical confidence intervals, one often quotes the central confidence interval, for which $\alpha = \beta$. Another common choice (where possible) is to use the two values a_- and a_+ satisfying (31.84), for which $L(\mathbf{x}; a_-) = L(\mathbf{x}; a_+)$.

It should be understood that a frequentist would consider the Bayesian confidence interval as an *approximation* to the (classical) confidence interval discussed in subsection 31.3.4. Conversely, a Bayesian would consider the confidence interval defined in (31.84) to be the more meaningful. In fact, the difference between the Bayesian and classical confidence intervals is rather subtle. The classical confidence interval is defined in such a way that if one took a large number of samples each of size N and constructed the confidence interval in each case then the proportion of cases in which the true value of a would be contained within the interval is $1 - \alpha - \beta$. For the Bayesian confidence interval, one does not rely on the frequentist concept of a large number of repeated samples. Instead, its meaning is that, given the single sample $\mathbf{x}$ (and our hypothesis H for the functional form of the population), the probability that a lies within the interval $[a_-, a_+]$ is $1 - \alpha - \beta$.

By adopting the Bayesian viewpoint, the likelihood function $L(\mathbf{x}; a)$ may also be used to obtain an approximation $\hat{\sigma}_{\hat{a}}$ to the standard error in the ML estimator; the approximation is given by

$$\hat{\sigma}_{\hat{a}} = \left(-\frac{\partial^2 \ln L}{\partial a^2}\bigg|_{a=\hat{a}} \right)^{-1/2}. \tag{31.85}$$

Clearly, if $L(\mathbf{x}; a)$ were a Gaussian centred on $a = \hat{a}$ then $\hat{\sigma}_{\hat{a}}$ would be its standard deviation. Indeed, in this case, the resulting 'one-sigma' limits would constitute a

$L(\mathbf{x}; \tau)$

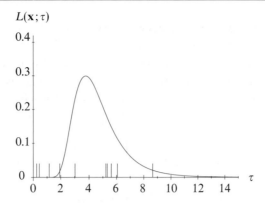

Figure 31.8   The likelihood function $L(\mathbf{x}; \tau)$ (normalised to unit area) for the sample values given in the worked example in subsection 31.5.4 and indicated here by short vertical lines.

68.3% Bayesian central confidence interval. Even when $L(\mathbf{x}; a)$ is not Gaussian, however, (31.85) is often used as a measure of the standard error.

▶ *For the sample data given in subsection 31.5.4, use the likelihood function to estimate the standard error $\hat{\sigma}_{\hat{\tau}}$ in the ML estimator $\hat{\tau}$ and obtain the Bayesian 68% central confidence interval on τ.*

We showed in (31.67) that the likelihood function in this case is given by

$$L(\mathbf{x}; \tau) = \frac{1}{\tau^N} \exp\left[-\frac{1}{\tau}(x_1 + x_2 + \cdots + x_N) \right].$$

where x_i, $i = 1, 2, \ldots, N$, denotes the sample value and $N = 10$. This likelihood function is plotted in figure 31.8, after normalising (numerically) to unit area. The short vertical lines in the figure indicate the sample values. We see that the likelihood function peaks at the ML estimate $\hat{\tau} = 3.77$ that we found in subsection 31.5.4. Also, from (31.77), we have

$$\frac{\partial^2 \ln L}{\partial \tau^2} = \frac{N^2}{\tau}\left(1 - \frac{2}{\tau N}\sum_{i=1}^{N} x_i \right).$$

Remembering that $\hat{\tau} = \sum_i x_i/N$, our estimate of the standard error in $\hat{\tau}$ is

$$\hat{\sigma}_{\hat{\tau}} = \left(-\frac{\partial^2 \ln L}{\partial \tau^2}\bigg|_{\tau=\hat{\tau}} \right)^{-1/2} = \frac{\hat{\tau}}{\sqrt{N}} = 1.19,$$

which is precisely the estimate of the standard error we obtained in subsection 31.5.4. It should be noted, however, that in general we would not expect the two estimates of standard error made by the different methods to be identical.

In order to calculate the Bayesian 68% central confidence interval, we must determine the values a_- and a_+ that satisfy (31.84) with $\alpha = \beta = 0.16$. In this case, the calculation can be performed analytically but is somewhat tedious. It is trivial, however, to determine a_- and a_+ numerically and we find the confidence interval to be $[3.16, 6.20]$. Thus we can quote our result with 68% central confidence limits as

$$\tau = 3.77^{+2.43}_{-0.61}.$$

By comparing this result with that given towards the end of subsection 31.5.4, we see that, as we might expect, the Bayesian and classical confidence intervals differ somewhat. ◀

The above discussion is generalised straightforwardly to the estimation of several parameters $a_1, a_2, \ldots, a_M$ simultaneously. The elements of the inverse of the covariance matrix of the ML estimators can be approximated by

$$(V^{-1})_{ij} = -\left.\frac{\partial^2 \ln L}{\partial a_i \partial a_j}\right|_{\mathbf{a}=\hat{\mathbf{a}}}. \tag{31.86}$$

From (31.36), we see that (at least for unbiased estimators) the expectation value of (31.86) is equal to the element F_{ij} of the Fisher matrix.

The construction of a multi-dimensional *Bayesian confidence region* is also straightforward. For a given confidence level $1 - \alpha$ (say), it is most common to construct the confidence region as the M-dimensional region R in $\mathbf{a}$-space, bounded by the 'surface' $L(\mathbf{x}; \mathbf{a}) = \text{constant}$, for which

$$\int_R L(\mathbf{x}; \mathbf{a}) \, d^M\mathbf{a} = 1 - \alpha,$$

where it is assumed that $L(\mathbf{x}; \mathbf{a})$ is normalised to unit volume. Moreover, we see from (31.83) that (assuming a uniform prior probability) we may obtain the *marginal* posterior distribution for any parameter a_i simply by integrating the likelihood function $L(\mathbf{x}; \mathbf{a})$ over the other parameters:

$$P(a_i | \mathbf{x}, H) = \int \cdots \int L(\mathbf{x}; \mathbf{a}) \, da_1 \cdots da_{i-1} da_{i+1} \cdots da_M.$$

Here the integral extends over all possible values of the parameters, and again is it assumed that the likelihood function is normalised in such a way that $\int L(\mathbf{x}; \mathbf{a}) \, d^M\mathbf{a} = 1$. This marginal distribution can then be used as above to determine Bayesian confidence intervals on each a_i separately.

▶ *Ten independent sample values x_i, $i = 1, 2, \ldots, 10$, are drawn at random from a Gaussian distribution with unknown mean μ and standard deviation σ. The sample values are as follows (to two decimal places):*

| 2.22 | 2.56 | 1.07 | 0.24 | 0.18 | 0.95 | 0.73 | −0.79 | 2.09 | 1.81 |

Find the Bayesian 95% central confidence intervals on μ and σ separately.

The likelihood function in this case is

$$L(\mathbf{x}; \mu, \sigma) = (2\pi\sigma^2)^{-N/2} \exp\left[-\frac{1}{2\sigma^2} \sum_{i=1}^{N} (x_i - \mu)^2\right]. \tag{31.87}$$

Assuming uniform priors on μ and σ (over their natural ranges of $-\infty \to \infty$ and $0 \to \infty$ respectively), we may identify this likelihood function with the posterior probability, as in (31.83). Thus, the marginal posterior distribution on μ is given by

$$P(\mu | \mathbf{x}, H) \propto \int_0^\infty \frac{1}{\sigma^N} \exp\left[-\frac{1}{2\sigma^2} \sum_{i=1}^{N} (x_i - \mu)^2\right] d\sigma.$$

By substituting $\sigma = 1/u$ (so that $d\sigma = -du/u^2$) and integrating by parts either $(N-2)/2$ or $(N-3)/2$ times, we find

$$P(\mu|\mathbf{x}, H) \propto \left[N(\bar{x} - \mu)^2 + Ns^2\right]^{-(N-1)/2},$$

where we have used the fact that $\sum_i (x_i - \mu)^2 = N(\bar{x} - \mu)^2 + Ns^2$, $\bar{x}$ being the sample mean and s^2 the sample variance. We may now obtain the 95% central confidence interval by finding the values μ_- and μ_+ for which

$$\int_{-\infty}^{\mu_-} P(\mu|\mathbf{x}, H)\, d\mu = 0.025 \qquad \text{and} \qquad \int_{\mu_+}^{\infty} P(\mu|\mathbf{x}, H)\, d\mu = 0.025.$$

The normalisation of the posterior distribution and the values μ_- and μ_+ are easily obtained by numerical integration. Substituting in the appropriate values $N = 10$, $\bar{x} = 1.11$ and $s = 1.01$, we find the required confidence interval to be $[0.29, 1.97]$.

To obtain a confidence interval on σ, we must first obtain the corresponding marginal posterior distribution. From (31.87), again using the fact that $\sum_i (x_i - \mu)^2 = N(\bar{x} - \mu)^2 + Ns^2$, this is given by

$$P(\sigma|\mathbf{x}, H) \propto \frac{1}{\sigma^N} \exp\left(-\frac{Ns^2}{2\sigma^2}\right) \int_{-\infty}^{\infty} \exp\left[-\frac{N(\bar{x} - \mu)^2}{2\sigma^2}\right] d\mu.$$

Noting that the integral of a one-dimensional Gaussian is proportional to σ, we conclude that

$$P(\sigma|\mathbf{x}, H) \propto \frac{1}{\sigma^{N-1}} \exp\left(-\frac{Ns^2}{2\sigma^2}\right).$$

The 95% central confidence interval on σ can then be found in an analogous manner to that on μ, by solving numerically the equations

$$\int_0^{\sigma_-} P(\sigma|\mathbf{x}, H)\, d\sigma = 0.025 \qquad \text{and} \qquad \int_{\sigma_+}^{\infty} P(\sigma|\mathbf{x}, H)\, d\sigma = 0.025.$$

We find the required interval to be $[0.76, 2.16]$. ◀

31.5.6 Behaviour of ML estimators for large N

As mentioned in subsection 31.3.6, in the large-sample limit $N \to \infty$, the sampling distribution of a set of (consistent) estimators $\hat{\mathbf{a}}$, whether ML or not, will tend, in general, to a multivariate Gaussian centred on the true values $\mathbf{a}$. This is a direct consequence of the central limit theorem. Similarly, in the limit $N \to \infty$ the likelihood function $L(\mathbf{x}; \mathbf{a})$ *also* tends towards a multivariate Gaussian but one centred on the ML estimate(s) $\hat{\mathbf{a}}$. Thus ML estimators are always *asymptotically consistent*. This limiting process was illustrated for the one-dimensional case by figure 31.5.

Thus, as N becomes large, the likelihood function tends to the form

$$L(\mathbf{x}; \mathbf{a}) = L_{\max} \exp\left[-\tfrac{1}{2} Q(\mathbf{a}, \hat{\mathbf{a}})\right],$$

where Q denotes the quadratic form

$$Q(\mathbf{a}, \hat{\mathbf{a}}) = (\mathbf{a} - \hat{\mathbf{a}})^{\mathrm{T}} \mathsf{V}^{-1} (\mathbf{a} - \hat{\mathbf{a}})$$

and the matrix V^{-1} is given by

$$\left(V^{-1}\right)_{ij} = -\left.\frac{\partial^2 \ln L}{\partial a_i \partial a_j}\right|_{\mathbf{a}=\hat{\mathbf{a}}}.$$

Moreover, in the limit of large N, this matrix tends to the Fisher matrix given in (31.36), i.e. $V^{-1} \to F$. Hence ML estimators are *asymptotically minimum-variance*.

Comparison of the above results with those in subsection 31.3.6 shows that the large-sample limit of the likelihood function $L(\mathbf{x}; \mathbf{a})$ has the same form as the large-sample limit of the joint estimator sampling distribution $P(\hat{\mathbf{a}}|\mathbf{a})$. The only difference is that $P(\hat{\mathbf{a}}|\mathbf{a})$ is centred in $\hat{\mathbf{a}}$-space on the true values $\hat{\mathbf{a}} = \mathbf{a}$ whereas $L(\mathbf{x}; \mathbf{a})$ is centred in $\mathbf{a}$-space on the ML estimates $\mathbf{a} = \hat{\mathbf{a}}$. From figure 31.4 and its accompanying discussion, we therefore conclude that, in the large-sample limit, the Bayesian and classical confidence limits on the parameters *coincide*.

31.5.7 *Extended maximum-likelihood method*

It is sometimes the case that the number of data items N in our sample is itself a random variable. Such experiments are typically those in which data are collected for a certain period of time during which events occur at random in some way, as opposed to those in which a prearranged number of data items are collected. In particular, let us consider the case where the sample values $x_1, x_2, \ldots, x_N$ are drawn independently from some distribution $P(x|\mathbf{a})$ and the sample size N is a random variable described by a Poisson distribution with mean λ, i.e. $N \sim \text{Po}(\lambda)$. The likelihood function in this case is given by

$$L(\mathbf{x}, N; \lambda, \mathbf{a}) = \frac{\lambda^N}{N!} e^{-\lambda} \prod_{i=1}^{N} P(x_i|\mathbf{a}), \tag{31.88}$$

and is often called the *extended likelihood function*. The function $L(\mathbf{x}; \lambda, \mathbf{a})$ can be used as before to estimate parameter values or obtain confidence intervals. Two distinct cases arise in the use of the extended likelihood function, depending on whether the Poisson parameter λ is a function of the parameters $\mathbf{a}$ or is an independent parameter.

Let us first consider the case in which λ is a function of the parameters $\mathbf{a}$. From (31.88), we can write the extended log-likelihood function as

$$\ln L = N \ln \lambda(\mathbf{a}) - \lambda(\mathbf{a}) + \sum_{i=1}^{N} \ln P(x_i|\mathbf{a}) = -\lambda(\mathbf{a}) + \sum_{i=1}^{N} \ln[\lambda(\mathbf{a})P(x_i|\mathbf{a})].$$

where we have ignored terms not depending on $\mathbf{a}$. The ML estimates $\hat{\mathbf{a}}$ of the parameters can then be found in the usual way, and the ML estimate of the Poisson parameter is simply $\hat{\lambda} = \lambda(\hat{\mathbf{a}})$. The errors on our estimators $\hat{\mathbf{a}}$ will be, in general, smaller than those obtained in the usual likelihood approach, since our estimate includes information from the value of N as well as the sample values x_i.

The other possibility is that λ is an independent parameter and not a function of the parameters $\mathbf{a}$. In this case, the extended log-likelihood function is

$$\ln L = N \ln \lambda - \lambda + \sum_{i=1}^{N} \ln P(x_i|\mathbf{a}), \qquad (31.89)$$

where we have omitted terms not depending on λ or $\mathbf{a}$. Differentiating with respect to λ and setting the result equal to zero, we find that the ML estimate of λ is simply

$$\hat{\lambda} = N.$$

By differentiating (31.89) with respect to the parameters a_i and setting the results equal to zero, we obtain the usual ML estimates $\hat{a}_i$ of their values. In this case, however, the errors in our estimates will be larger, in general, than those in the standard likelihood approach, since they must include the effect of statistical uncertainty in the parameter λ.

31.6 The method of least squares

The method of least squares is, in fact, just a special case of the method of maximum likelihood. Nevertheless, it is so widely used as a method of parameter estimation that it has acquired a special name of its own. At the outset, let us suppose that a data sample consists of a set of pairs (x_i, y_i), $i = 1, 2, \ldots, N$. For example, these data might correspond to the temperature y_i measured at various points x_i along some metal rod.

For the moment, we will suppose that the x_i are known exactly, whereas there exists a measurement error (or *noise*) n_i on each of the values y_i. Moreover, let us assume that the true value of y at any position x is given by some function $y = f(x; \mathbf{a})$ that depends on the M unknown parameters $\mathbf{a}$. Then

$$y_i = f(x_i; \mathbf{a}) + n_i.$$

Our aim is to estimate the values of the parameters $\mathbf{a}$ from the data sample.

Bearing in mind the central limit theorem, let us suppose that the n_i are drawn from a *Gaussian* distribution with no systematic bias and hence zero mean. In the most general case the measurement errors n_i might *not* be independent but be described by an N-dimensional multivariate Gaussian with non-trivial covariance matrix N, whose elements $N_{ij} = \text{Cov}[n_i, n_j]$ we assume to be known. Under these assumptions it follows from (30.148), that the likelihood function is

$$L(\mathbf{x}, \mathbf{y}; \mathbf{a}) = \frac{1}{(2\pi)^{N/2}|\mathsf{N}|^{1/2}} \exp\left[-\tfrac{1}{2}\chi^2(\mathbf{a})\right],$$

where the quantity denoted by χ^2 is given by the quadratic form

$$\chi^2(\mathbf{a}) = \sum_{i,j=1}^{N} [y_i - f(x_i;\mathbf{a})](\mathbf{N}^{-1})_{ij}[y_j - f(x_j;\mathbf{a})] = (\mathbf{y} - \mathbf{f})^{\mathrm{T}}\mathbf{N}^{-1}(\mathbf{y} - \mathbf{f}).$$
(31.90)

In the last equality, we have rewritten the expression in matrix notation by defining the column vector $\mathbf{f}$ with elements $f_i = f(x_i;\mathbf{a})$. We note that in the (common) special case in which the measurement errors n_i are *independent*, their covariance matrix takes the diagonal form $\mathbf{N} = \mathrm{diag}(\sigma_1^2, \sigma_2^2, \ldots, \sigma_N^2)$, where σ_i is the standard deviation of the measurement error n_i. In this case, the expression (31.90) for χ^2 reduces to

$$\chi^2(\mathbf{a}) = \sum_{i=1}^{N} \left[\frac{y_i - f(x_i;\mathbf{a})}{\sigma_i} \right]^2 .$$

The least-squares (LS) estimators $\hat{\mathbf{a}}_{\mathrm{LS}}$ of the parameter values are defined as those that minimise the value of $\chi^2(\mathbf{a})$; they are usually determined by solving the M equations

$$\left. \frac{\partial \chi^2}{\partial a_i} \right|_{\mathbf{a}=\hat{\mathbf{a}}_{\mathrm{LS}}} = 0 \qquad \text{for } i = 1, 2, \ldots, M.$$
(31.91)

Clearly, if the measurement errors n_i are indeed Gaussian distributed, as assumed above, then the LS and ML estimators of the parameters $\mathbf{a}$ coincide. Because of its relative simplicity, the method of least squares is often applied to cases in which the n_i are not Gaussian distributed. The resulting estimators $\hat{\mathbf{a}}_{\mathrm{LS}}$ are *not* the ML estimators, and the best that can be said in justification is that the method is an obviously sensible procedure for parameter estimation that has stood the test of time.

Finally, we note that the method of least squares is easily extended to the case in which each measurement y_i depends on several variables, which we denote by $\mathbf{x}_i$. For example, y_i might represent the temperature measured at the (three-dimensional) position $\mathbf{x}_i$ in a room. In this case, the data is modelled by a function $y = f(\mathbf{x}_i;\mathbf{a})$, and the remainder of the above discussion carries through unchanged.

31.6.1 Linear least squares

We have so far made no restriction on the form of the function $f(x;\mathbf{a})$. It so happens, however, that, for a model in which $f(x;\mathbf{a})$ is a *linear* function of the parameters $a_1, a_2, \ldots, a_M$, one can always obtain analytic expressions for the LS estimators $\hat{\mathbf{a}}_{\mathrm{LS}}$ and their variances. The general form of this kind of model is

$$f(x;\mathbf{a}) = \sum_{i=1}^{M} a_i h_i(x),$$
(31.92)

where $\{h_1(x), h_2(x), \ldots, h_M(x)\}$ is some set of linearly independent fixed functions of x, often called the *basis functions*. Note that the functions $h_i(x)$ themselves may be highly non-linear functions of x. The 'linear' nature of the model (31.92) refers only to its dependence on the *parameters* a_i. Furthermore, in this case, it may be shown that the LS estimators $\hat{a}_i$ have zero bias and are minimum-variance, irrespective of the probability density function from which the measurement errors n_i are drawn.

In order to obtain analytic expressions for the LS estimators $\hat{\mathbf{a}}_{LS}$, it is convenient to write (31.92) in the form

$$f(x_i; \mathbf{a}) = \sum_{j=1}^{M} R_{ij} a_j, \tag{31.93}$$

where $R_{ij} = h_j(x_i)$ is an element of the *response matrix* R of the experiment. The expression for χ^2 given in (31.90) can then be written, in matrix notation, as

$$\chi^2(\mathbf{a}) = (\mathbf{y} - \mathsf{R}\mathbf{a})^\mathsf{T} \mathsf{N}^{-1} (\mathbf{y} - \mathsf{R}\mathbf{a}). \tag{31.94}$$

The LS estimates of the parameters $\mathbf{a}$ are now found, as shown in (31.91), by differentiating (31.94) with respect to the a_i and setting the resulting expressions equal to zero. Denoting by $\nabla \chi^2$ the vector with elements $\partial \chi^2 / \partial a_i$, we find

$$\nabla \chi^2 = -2 \mathsf{R}^\mathsf{T} \mathsf{N}^{-1} (\mathbf{y} - \mathsf{R}\mathbf{a}). \tag{31.95}$$

This can be verified by writing out the expression (31.94) in component form and differentiating directly.

▶*Verify result (31.95) by formulating the calculation in component form.*

To make the derivation less cumbersome, let us adopt the summation convention discussed in section 26.1, in which it is understood that any subscript that appears *exactly* twice in any term of an expression is to be summed over all the values that a subscript in that position can take. Thus, writing (31.94) in component form, we have

$$\chi^2(\mathbf{a}) = (y_i - R_{ik}a_k)(N^{-1})_{ij}(y_j - R_{jl}a_l).$$

Differentiating with respect to a_p gives

$$\frac{\partial \chi^2}{\partial a_p} = -R_{ik}\delta_{kp}(N^{-1})_{ij}(y_j - R_{jl}a_l) + (y_i - R_{ik}a_k)(N^{-1})_{ij}(-R_{jl}\delta_{lp})$$

$$= -R_{ip}(N^{-1})_{ij}(y_j - R_{jl}a_l) - (y_i - R_{ik}a_k)(N^{-1})_{ij}R_{jp}, \tag{31.96}$$

where δ_{ij} is the Kronecker delta symbol discussed in section 26.1. By swapping the indices i and j in the second term on the RHS of (31.96) and using the fact that the matrix N^{-1} is symmetric, we obtain

$$\frac{\partial \chi^2}{\partial a_p} = -2R_{ip}(N^{-1})_{ij}(y_j - R_{jk}a_k)$$

$$= -2(R^\mathsf{T})_{pi}(N^{-1})_{ij}(y_j - R_{jk}a_k). \tag{31.97}$$

If we denote the vector with components $\partial \chi^2 / \partial a_p$, $p = 1, 2, \ldots, M$, by $\nabla \chi^2$ and write the RHS of (31.97) in matrix notation, we recover the result (31.95). ◀

Setting the expression (31.95) equal to zero at $\mathbf{a} = \hat{\mathbf{a}}$, we find

$$-2R^T N^{-1} y + 2R^T N^{-1} R\hat{\mathbf{a}} = 0.$$

Provided the matrix $R^T N^{-1} R$ is not singular, we may solve this equation for $\hat{\mathbf{a}}$ to obtain

$$\hat{\mathbf{a}} = (R^T N^{-1} R)^{-1} R^T N^{-1} y \equiv Sy, \tag{31.98}$$

thus defining the $M \times N$ matrix S. It follows that the LS estimates $\hat{a}_i$, $i = 1, 2, \ldots, M$, are linear functions of the original measurements y_j, $j = 1, 2, \ldots, N$. Moreover, using the error propagation formula (30.141) derived in subsection 30.12.3, we find that the covariance matrix of the estimators $\hat{a}_i$ is given by

$$V \equiv \text{Cov}[\hat{a}_i, \hat{a}_j] = SNS^T = (R^T N^{-1} R)^{-1}. \tag{31.99}$$

The two equations (31.98) and (31.99) contain the complete method of least squares. In particular, we note that, if one calculates the LS estimates using (31.98) then one has already obtained their covariance matrix (31.99).

▶ *Prove result (31.99).*

Using the definition of S given in (31.98), the covariance matrix (31.99) becomes

$$V = SNS^T$$
$$= [(R^T N^{-1} R)^{-1} R^T N^{-1}] N [(R^T N^{-1} R)^{-1} R^T N^{-1}]^T.$$

Using the result $(AB \cdots C)^T = C^T \cdots B^T A^T$ for the transpose of a product of matrices and noting that, for any non-singular matrix, $(A^{-1})^T = (A^T)^{-1}$ we find

$$V = (R^T N^{-1} R)^{-1} R^T N^{-1} N (N^T)^{-1} R[(R^T N^{-1} R)^T]^{-1}$$
$$= (R^T N^{-1} R)^{-1} R^T N^{-1} R (R^T N^{-1} R)^{-1}$$
$$= (R^T N^{-1} R)^{-1},$$

where we have also used the fact that N is symmetric and so $N^T = N$. ◀

It is worth noting that one may also write the elements of the (inverse) covariance matrix as

$$(V^{-1})_{ij} = \frac{1}{2} \left(\frac{\partial^2 \chi^2}{\partial a_i \partial a_j} \right)_{\mathbf{a} = \hat{\mathbf{a}}},$$

which is the same as the Fisher matrix (31.36) in cases where the measurement errors are Gaussian distributed (and so the log-likelihood is $\ln L = -\chi^2/2$). This proves, at least for this case, our earlier statement that the LS estimators are minimum-variance. In fact, since $f(x; \mathbf{a})$ is linear in the parameters $\mathbf{a}$, one can write χ^2 *exactly* as

$$\chi^2(\mathbf{a}) = \chi^2(\hat{\mathbf{a}}) + \frac{1}{2} \sum_{i,j=1}^{M} \left(\frac{\partial^2 \chi^2}{\partial a_i \partial a_j} \right)_{\mathbf{a} = \hat{\mathbf{a}}} (a_i - \hat{a}_i)(a_j - \hat{a}_j),$$

which is quadratic in the parameters a_i. Hence the form of the likelihood function

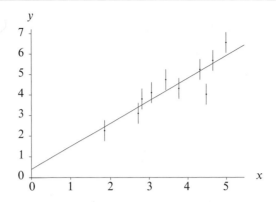

Figure 31.9   A set of data points with error bars indicating the uncertainty $\sigma = 0.5$ on the y-values. The straight line is $y = \hat{m}x + \hat{c}$, where $\hat{m}$ and $\hat{c}$ are the least-squares estimates of the slope and intercept.

$L \propto \exp(-\chi^2/2)$ is Gaussian. From the discussions of subsections 31.3.6 and 31.5.6, it follows that the 'surfaces' $\chi^2(\mathbf{a}) = c$, where c is a constant, bound ellipsoidal *confidence regions* for the parameters a_i. The relationship between the value of the constant c and the confidence level is given by (31.39).

▶*An experiment produces the following data sample pairs* (x_i, y_i):

x_i:	1.85	2.72	2.81	3.06	3.42	3.76	4.31	4.47	4.64	4.99
y_i:	2.26	3.10	3.80	4.11	4.74	4.31	5.24	4.03	5.69	6.57

where the x_i-values are known exactly but each y_i-value is measured only to an accuracy of $\sigma = 0.5$. Assuming the underlying model for the data to be a straight line $y = mx + c$, find the LS estimates of the slope m and intercept c and quote the standard error on each estimate.

The data are plotted in figure 31.9, together with error bars indicating the uncertainty in the y_i-values. Our model of the data is a straight line, and so we have

$$f(x; c, m) = c + mx.$$

In the language of (31.92), our basis functions are $h_1(x) = 1$ and $h_2(x) = x$ and our model parameters are $a_1 = c$ and $a_2 = m$. From (31.93) the elements of the response matrix are $R_{ij} = h_j(x_i)$, so that

$$\mathsf{R} = \begin{pmatrix} 1 & x_1 \\ 1 & x_2 \\ \vdots & \vdots \\ 1 & x_N \end{pmatrix}, \tag{31.100}$$

where x_i are the data values and $N = 10$ in our case. Further, since the standard deviation on each measurement error is σ, we have $\mathsf{N} = \sigma^2\mathsf{I}$, where I is the $N \times N$ identity matrix. Because of this simple form for N, the expression (31.98) for the LS estimates reduces to

$$\hat{\mathbf{a}} = \sigma^2(\mathsf{R}^\mathsf{T}\mathsf{R})^{-1}\frac{1}{\sigma^2}\mathsf{R}^\mathsf{T}\mathbf{y} = (\mathsf{R}^\mathsf{T}\mathsf{R})^{-1}\mathsf{R}^\mathsf{T}\mathbf{y}. \tag{31.101}$$

Note that we cannot expand the inverse in the last line, since R itself is not square and hence does not possess an inverse. Inserting the form for R in (31.100) into the expression (31.101), we find

$$\begin{pmatrix} \hat{c} \\ \hat{m} \end{pmatrix} = \begin{pmatrix} \sum_i 1 & \sum_i x_i \\ \sum_i x_i & \sum_i x_i^2 \end{pmatrix}^{-1} \begin{pmatrix} \sum_i y_i \\ \sum_i x_i y_i \end{pmatrix}$$

$$= \frac{1}{N(\overline{x^2} - \bar{x}^2)} \begin{pmatrix} \overline{x^2} & -\bar{x} \\ -\bar{x} & 1 \end{pmatrix} \begin{pmatrix} N\bar{y} \\ N\overline{xy} \end{pmatrix}.$$

We thus obtain the LS estimates

$$\hat{m} = \frac{\overline{xy} - \bar{x}\,\bar{y}}{\overline{x^2} - \bar{x}^2} \qquad \text{and} \qquad \hat{c} = \frac{\overline{x^2}\bar{y} - \bar{x}\,\overline{xy}}{\overline{x^2} - \bar{x}^2} = \bar{y} - \hat{m}\bar{x}, \tag{31.102}$$

where the last expression for $\hat{c}$ shows that the best-fit line passes through the 'centre of mass' $(\bar{x}, \bar{y})$ of the data sample. To find the standard errors on our results, we must calculate the covariance matrix of the estimators. This is given by (31.99), which in our case reduces to

$$V = \sigma^2 (R^T R)^{-1} = \frac{\sigma^2}{N(\overline{x^2} - \bar{x}^2)} \begin{pmatrix} \overline{x^2} & -\bar{x} \\ -\bar{x} & 1 \end{pmatrix}. \tag{31.103}$$

The standard error on each estimator is simply the positive square root of the corresponding diagonal element, i.e. $\sigma_{\hat{c}} = \sqrt{V_{11}}$ and $\sigma_{\hat{m}} = \sqrt{V_{22}}$, and the covariance of the estimators $\hat{m}$ and $\hat{c}$ is given by $\text{Cov}[\hat{c}, \hat{m}] = V_{12} = V_{21}$. Inserting the data sample averages and moments into (31.102) and (31.103), we find

$$c = \hat{c} \pm \sigma_{\hat{c}} = 0.40 \pm 0.62 \qquad \text{and} \qquad m = \hat{m} \pm \sigma_{\hat{m}} = 1.11 \pm 0.17.$$

The 'best-fit' straight line $y = \hat{m}x + \hat{c}$ is plotted in figure 31.9. For comparison, the true values used to create the data were $m = 1$ and $c = 1$. ◄

The extension of the method to fitting data to a higher-order polynomial, such as $f(x; \mathbf{a}) = a_1 + a_2 x + a_3 x^2$, is obvious. However, as the order of the polynomial increases the matrix inversions become rather complicated. Indeed, even when the matrices are inverted numerically, the inversion is prone to numerical instabilities. A better approach is to replace the basis functions $h_m(x) = x^m$, $m = 1, 2, \ldots, M$, with a set of polynomials that are 'orthogonal over the data', i.e. such that

$$\sum_{i=1}^N h_l(x_i) h_m(x_i) = 0 \qquad \text{for } l \neq m.$$

Such a set of polynomial basis functions can always be found by using the Gram–Schmidt orthogonalisation procedure presented in section 17.1. The details of this approach are beyond the scope of our discussion but we note that, in this case, the matrix $R^T R$ is diagonal and may be inverted easily.

31.6.2 Non-linear least squares

If the function $f(x; \mathbf{a})$ is *not* linear in the parameters $\mathbf{a}$ then, in general, it is not possible to obtain an explicit expression for the LS estimates $\hat{\mathbf{a}}$. Instead, one must use an iterative (numerical) procedure, which we now outline. In practice, however, such problems are best solved using one of the many commercially available software packages.

One begins by making a first guess $\mathbf{a}^0$ for the values of the parameters. At this point in parameter space, the components of the gradient $\nabla \chi^2$ will not be equal to zero, in general (unless one makes a very lucky guess!). Thus, for at least some values of i, we have

$$\left. \frac{\partial \chi^2}{\partial a_i} \right|_{\mathbf{a}=\mathbf{a}^0} \neq 0.$$

Our aim is to find a small increment $\delta \mathbf{a}$ in the values of the parameters, such that

$$\left. \frac{\partial \chi^2}{\partial a_i} \right|_{\mathbf{a}=\mathbf{a}^0+\delta \mathbf{a}} = 0 \qquad \text{for all } i. \tag{31.104}$$

If our first guess $\mathbf{a}^0$ were sufficiently close to the true (local) minimum of χ^2, we could find the required increment $\delta \mathbf{a}$ by expanding the LHS of (31.104) as a Taylor series about $\mathbf{a} = \mathbf{a}^0$, keeping only the zeroth-order and first-order terms:

$$\left. \frac{\partial \chi^2}{\partial a_i} \right|_{\mathbf{a}=\mathbf{a}^0+\delta \mathbf{a}} \approx \left. \frac{\partial \chi^2}{\partial a_i} \right|_{\mathbf{a}=\mathbf{a}^0} + \sum_{j=1}^{M} \left. \frac{\partial^2 \chi^2}{\partial a_i \partial a_j} \right|_{\mathbf{a}=\mathbf{a}^0} \delta a_j. \tag{31.105}$$

Setting this expression to zero, we find that the increments δa_j may be found by solving the set of M linear equations

$$\sum_{j=1}^{M} \left. \frac{\partial^2 \chi^2}{\partial a_i \partial a_j} \right|_{\mathbf{a}=\mathbf{a}^0} \delta a_j = - \left. \frac{\partial \chi^2}{\partial a_i} \right|_{\mathbf{a}=\mathbf{a}^0}.$$

It most cases, however, our first guess $\mathbf{a}^0$ will not be sufficiently close to the true minimum for (31.105) to be an accurate approximation, and consequently (31.104) will not be satisfied. In this case, $\mathbf{a}^1 = \mathbf{a}^0 + \delta \mathbf{a}$ is (hopefully) an improved guess at the parameter values; the whole process is then repeated until convergence is achieved.

It is worth noting that, when one is estimating several parameters $\mathbf{a}$, the function $\chi^2(\mathbf{a})$ may be *very* complicated. In particular, it may possess numerous local extrema. The procedure outlined above will converge to the local extremum 'nearest' to the first guess $\mathbf{a}^0$. Since, in fact, we are interested only in the local minimum that has the absolute lowest value of $\chi^2(\mathbf{a})$, it is clear that a large part of solving the problem is to make a 'good' first guess.

31.7 Hypothesis testing

So far we have concentrated on using a data sample to obtain a number or a set of numbers. These numbers may be estimated values for the moments or central moments of the population from which the sample was drawn or, more generally, the values of some parameters $\mathbf{a}$ in an assumed model for the data. Sometimes, however, one wishes to use the data to give a 'yes' or 'no' answer to a particular question. For example, one might wish to know whether some assumed model does, in fact, provide a good fit to the data, or whether two parameters have the same value.

31.7.1 Simple and composite hypotheses

In order to use data to answer questions of this sort, the question must be posed precisely. This is done by first asserting that some *hypothesis* is true. The hypothesis under consideration is traditionally called the *null hypothesis* and is denoted by H_0. In particular, this usually specifies some form $P(\mathbf{x}|H_0)$ for the probability density function from which the data $\mathbf{x}$ are drawn. If the hypothesis determines the PDF uniquely, then it is said to be a *simple hypothesis*. If, however, the hypothesis determines the functional form of the PDF but not the values of certain parameters $\mathbf{a}$ on which it depends then it is called a *composite hypothesis*.

One decides whether to *accept* or *reject* the null hypothesis H_0 by performing some *statistical test*, as described below in subsection 31.7.2. In fact, formally one uses a statistical test to decide between the null hypothesis H_0 and the *alternative hypothesis* H_1. We define the latter to be the complement $\overline{H}_0$ of the null hypothesis *within some restricted hypothesis space known (or assumed) in advance*. Hence, rejection of H_0 implies acceptance of H_1, and vice versa.

As an example, let us consider the case in which a sample $\mathbf{x}$ is drawn from a Gaussian distribution with a known variance σ^2 but with an unknown mean μ. If one adopts the null hypothesis H_0 that $\mu = 0$, which we write as $H_0 : \mu = 0$, then the corresponding alternative hypothesis must be $H_1 : \mu \neq 0$. Note that, in this case, H_0 is a simple hypothesis whereas H_1 is a composite hypothesis. If, however, one adopted the null hypothesis $H_0 : \mu < 0$ then the alternative hypothesis would be $H_1 : \mu \geq 0$, so that both H_0 and H_1 would be composite hypotheses. Very occasionally both H_0 and H_1 will be simple hypotheses. In our illustration, this would occur, for example, if one knew in advance that the mean μ of the Gaussian distribution were equal to either zero or unity. In this case, if one adopted the null hypothesis $H_0 : \mu = 0$ then the alternative hypothesis would be $H_1 : \mu = 1$.

31.7.2 Statistical tests

In our discussion of hypothesis testing we will restrict our attention to cases in which the null hypothesis H_0 is *simple* (see above). We begin by constructing a *test statistic* $t(\mathbf{x})$ from the data sample. Although, in general, the test statistic need not be just a (scalar) number, and could be a multi-dimensional (vector) quantity, we will restrict our attention to the former case. Like any statistic, $t(\mathbf{x})$ will be a

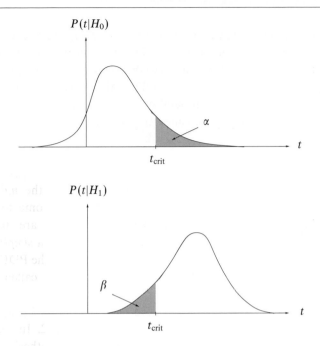

Figure 31.10   The sampling distributions $P(t|H_0)$ and $P(t|H_1)$ of a test statistic t. The shaded areas indicate the (one-tailed) regions for which $\Pr(t > t_{\text{crit}}|H_0) = \alpha$ and $\Pr(t < t_{\text{crit}}|H_1) = \beta$ respectively.

random variable. Moreover, given the simple null hypothesis H_0 concerning the PDF from which the sample was drawn, we may determine (in principle) the sampling distribution $P(t|H_0)$ of the test statistic. A typical example of such a sampling distribution is shown in figure 31.10. One defines for t a *rejection region* containing some fraction α of the total probability. For example, the (one-tailed) rejection region could consist of values of t greater than some value t_{crit}, for which

$$\Pr(t > t_{\text{crit}}|H_0) = \int_{t_{\text{crit}}}^{\infty} P(t|H_0)\, dt = \alpha; \qquad (31.106)$$

this is indicated by the shaded region in the upper half of figure 31.10. Equally, a (one-tailed) rejection region could consist of values of t less than some value t_{crit}. Alternatively, one could define a (two-tailed) rejection region by two values t_1 and t_2 such that $\Pr(t_1 < t < t_2|H_0) = \alpha$. In all cases, if the observed value of t lies in the rejection region then H_0 is *rejected* at *significance level* α; otherwise H_0 is *accepted* at this same level.

It is clear that there is a probability α of rejecting the null hypothesis H_0 even if it is true. This is called an *error of the first kind*. Conversely, an *error of the second kind* occurs when the hypothesis H_0 is accepted even though it is false (in which case H_1 is true). The probability β (say) that such an error will occur is, in general, difficult to calculate, since the alternative hypothesis H_1 is often composite. Nevertheless, in the case where H_1 is a simple hypothesis, it is straightforward (in principle) to calculate β. Denoting the corresponding sampling distribution of t by $P(t|H_1)$, the probability β is the integral of $P(t|H_1)$ over the *complement* of the rejection region, called the *acceptance region*. For example, in the case corresponding to (31.106) this probability is given by

$$\beta = \Pr(t < t_{\text{crit}}|H_1) = \int_{-\infty}^{t_{\text{crit}}} P(t|H_1)\,dt.$$

This is illustrated in figure 31.10. The quantity $1 - \beta$ is called the *power* of the statistical test to reject the wrong hypothesis.

31.7.3 The Neyman–Pearson test

In the case where H_0 and H_1 are both simple hypotheses, the *Neyman–Pearson lemma* (which we shall not prove) allows one to determine the 'best' rejection region and test statistic to use.

We consider first the choice of rejection region. Even in the general case, in which the test statistic **t** is a multi-dimensional (vector) quantity, the Neyman–Pearson lemma states that, for a given significance level α, the rejection region for H_0 giving the highest power for the test is the region of t-space for which

$$\frac{P(\mathbf{t}|H_0)}{P(\mathbf{t}|H_1)} > c, \tag{31.107}$$

where c is some constant determined by the required significance level.

In the case where the test statistic t is a simple scalar quantity, the Neyman–Pearson lemma is also useful in deciding which such statistic is the 'best' in the sense of having the maximum power for a given significance level α. From (31.107), we can see that the best statistic is given by the *likelihood ratio*

$$t(\mathbf{x}) = \frac{P(\mathbf{x}|H_0)}{P(\mathbf{x}|H_1)}. \tag{31.108}$$

and that the corresponding rejection region for H_0 is given by $t < t_{\text{crit}}$. In fact, it is clear that any statistic $u = f(t)$ will be equally good, provided that $f(t)$ is a monotonically increasing function of t. The rejection region is then $u < f(t_{\text{crit}})$. Alternatively, one may use any test statistic $v = g(t)$ where $g(t)$ is a monotonically decreasing function of t; in this case the rejection region becomes $v > g(t_{\text{crit}})$. To construct such statistics, however, one must know $P(\mathbf{x}|H_0)$ and $P(\mathbf{x}|H_1)$ explicitly, and such cases are rare.

> ▶*Ten independent sample values x_i, $i = 1, 2, \ldots, 10$, are drawn at random from a Gaussian distribution with standard deviation $\sigma = 1$. The mean μ of the distribution is known to equal either zero or unity. The sample values are as follows:*
>
> | 2.22 | 2.56 | 1.07 | 0.24 | 0.18 | 0.95 | 0.73 | −0.79 | 2.09 | 1.81 |
>
> *Test the null hypothesis $H_0 : \mu = 0$ at the 10% significance level.*

The restricted nature of the hypothesis space means that our null and alternative hypotheses are $H_0 : \mu = 0$ and $H_1 : \mu = 1$ respectively. Since H_0 and H_1 are both simple hypotheses, the best test statistic is given by the likelihood ratio (31.108). Thus, denoting the means by μ_0 and μ_1, we have

$$
t(\mathbf{x}) = \frac{\exp\left[-\frac{1}{2}\sum_i (x_i - \mu_0)^2\right]}{\exp\left[-\frac{1}{2}\sum_i (x_i - \mu_1)^2\right]} = \frac{\exp\left[-\frac{1}{2}\sum_i (x_i^2 - 2\mu_0 x_i + \mu_0^2)\right]}{\exp\left[-\frac{1}{2}\sum_i (x_i^2 - 2\mu_1 x_i + \mu_1^2)\right]}
$$
$$
= \exp\left[(\mu_0 - \mu_1)\sum_i x_i - \tfrac{1}{2}N(\mu_0^2 - \mu_1^2)\right].
$$

Inserting the values $\mu_0 = 0$ and $\mu_1 = 1$, yields $t = \exp(-N\bar{x} + \frac{1}{2}N)$, where $\bar{x}$ is the sample mean. Since $-\ln t$ is a monotonically decreasing function of t, however, we may equivalently use as our test statistic

$$
v = -\frac{1}{N}\ln t + \tfrac{1}{2} = \bar{x},
$$

where we have divided by the sample size N and added $\frac{1}{2}$ for convenience. Thus we may take the sample mean as our test statistic. From (31.13), we know that the sampling distribution of the sample mean under our null hypothesis H_0 is the Gaussian distribution $N(\mu_0, \sigma^2/N)$, where $\mu_0 = 0$, $\sigma^2 = 1$ and $N = 10$. Thus $\bar{x} \sim N(0, 0.1)$.

Since $\bar{x}$ is a monotonically decreasing function of t, our best rejection region for a given significance α is $\bar{x} > \bar{x}_{\text{crit}}$, where $\bar{x}_{\text{crit}}$ depends on α. Thus, in our case, $\bar{x}_{\text{crit}}$ is given by

$$
\alpha = 1 - \Phi\left(\frac{\bar{x}_{\text{crit}} - \mu_0}{\sigma}\right) = 1 - \Phi(\sqrt{10}\,\bar{x}_{\text{crit}}),
$$

where $\Phi(z)$ is the cumulative distribution function for the standard Gaussian. For a 10% significance level we have $\alpha = 0.1$ and, from table 30.3 in subsection 30.9.1, we find $\bar{x}_{\text{crit}} = 0.405$. Thus the rejection region on $\bar{x}$ is

$$
\bar{x} > 0.405.
$$

From the sample, we deduce that $\bar{x} = 1.11$, and so we can clearly reject the null hypothesis $H_0 : \mu = 0$ at the 10% significance level. It can, in fact, be rejected at a higher significance level. As revealed on p. 1239, the data was generated using $\mu = 1$. ◀

31.7.4 The generalised likelihood-ratio test

If the null hypothesis H_0 or the alternative hypothesis H_1 is composite (or both are composite) then the corresponding distributions $P(\mathbf{x}|H_0)$ and $P(\mathbf{x}|H_1)$ are not uniquely determined, in general, and so we cannot use the Neyman–Pearson lemma to obtain the 'best' test statistic t. Nevertheless, in many cases, there still exists a general procedure for constructing a test statistic t which has useful properties and which reduces to the Neyman–Pearson statistic (31.108) in the special case where H_0 and H_1 are both simple hypotheses.

Consider the quite general, and commonly occurring, case in which the data sample $\mathbf{x}$ is drawn from a population $P(\mathbf{x}|\mathbf{a})$ with a known (or assumed) functional form but depends on the unknown values of some parameters $a_1, a_2, \ldots, a_M$. Moreover, suppose we wish to test the null hypothesis H_0 that the parameter values $\mathbf{a}$ lie in some subspace S of the full parameter space $\mathcal{A}$. In other words, on the basis of the sample $\mathbf{x}$ it is desired to test the null hypothesis $H_0 : (a_1, a_2, \ldots, a_M$ lies in $S)$ against the alternative hypothesis $H_1 : (a_1, a_2, \ldots, a_M$ lies in $\overline{S})$, where $\overline{S}$ is $\mathcal{A} - S$.

Since the functional form of the population is known, we may write down the likelihood function $L(\mathbf{x}; \mathbf{a})$ for the sample. Ordinarily, the likelihood will have a maximum as the parameters $\mathbf{a}$ are varied over the entire parameter space $\mathcal{A}$. This is the usual maximum-likelihood estimate of the parameter values, which we denote by $\hat{\mathbf{a}}$. If, however, the parameter values are allowed to vary only over the subspace S then the likelihood function will be maximised at the point $\hat{\mathbf{a}}_S$, which may or may not coincide with the global maximum $\hat{\mathbf{a}}$. Now, let us take as our test statistic the *generalised likelihood ratio*

$$t(\mathbf{x}) = \frac{L(\mathbf{x}; \hat{\mathbf{a}}_S)}{L(\mathbf{x}; \hat{\mathbf{a}})}, \qquad (31.109)$$

where $L(\mathbf{x}; \hat{\mathbf{a}}_S)$ is the maximum value of the likelihood function in the subspace S and $L(\mathbf{x}; \hat{\mathbf{a}})$ is its maximum value in the entire parameter space $\mathcal{A}$. It is clear that t is a function of the sample values only and must lie between 0 and 1.

We will concentrate on the special case where H_0 is the simple hypothesis $H_0 : \mathbf{a} = \mathbf{a}_0$. The subspace S then consists of only the single point $\mathbf{a}_0$. Thus (31.109) becomes

$$t(\mathbf{x}) = \frac{L(\mathbf{x}; \mathbf{a}_0)}{L(\mathbf{x}; \hat{\mathbf{a}})}, \qquad (31.110)$$

and the sampling distribution $P(t|H_0)$ can be determined (in principle). As in the previous subsection, the best rejection region for a given significance α is simply $t < t_{\text{crit}}$, where the value t_{crit} depends on α. Moreover, as before, an equivalent procedure is to use as a test statistic $u = f(t)$, where $f(t)$ is any monotonically increasing function of t; the corresponding rejection region is then $u < f(t_{\text{crit}})$. Similarly, one may use a test statistic $v = g(t)$, where $g(t)$ is any monotonically decreasing function of t; the rejection region then becomes $v > g(t_{\text{crit}})$. Finally, we note that if H_1 is also a simple hypothesis $H_1 : \mathbf{a} = \mathbf{a}_1$, then (31.110) reduces to the Neyman–Pearson test statistic (31.108).

> ▶ *Ten independent sample values x_i, $i = 1, 2, \ldots, 10$, are drawn at random from a Gaussian distribution with standard deviation $\sigma = 1$. The sample values are as follows:*
>
> 2.22   2.56   1.07   0.24   0.18   0.95   0.73   −0.79   2.09   1.81
>
> *Test the null hypothesis $H_0 : \mu = 0$ at the 10% significance level.*

We must test the (simple) null hypothesis $H_0 : \mu = 0$ against the (composite) alternative hypothesis $H_1 : \mu \neq 0$. Thus, the subspace S is the single point $\mu = 0$, whereas A is the entire μ-axis. The likelihood function is

$$L(\mathbf{x}; \mu) = \frac{1}{(2\pi)^{N/2}} \exp\left[-\tfrac{1}{2}\sum_i (x_i - \mu)^2\right],$$

which has its global maximum at $\mu = \bar{x}$. The test statistic t is then given by

$$t(\mathbf{x}) = \frac{L(\mathbf{x}; 0)}{L(\mathbf{x}; \bar{x})} = \frac{\exp\left[-\tfrac{1}{2}\sum_i x_i^2\right]}{\exp\left[-\tfrac{1}{2}\sum_i (x_i - \bar{x})^2\right]} = \exp\left(-\tfrac{1}{2}N\bar{x}^2\right).$$

It is in fact more convenient to consider the test statistic

$$v = -2\ln t = N\bar{x}^2.$$

Since $-2\ln t$ is a monotonically decreasing function of t, the rejection region now becomes $v > v_{\mathrm{crit}}$, where

$$\int_{v_{\mathrm{crit}}}^{\infty} P(v|H_0)\, dv = \alpha, \tag{31.111}$$

α being the significance level of the test. Thus it only remains to determine the sampling distribution $P(v|H_0)$. Under the null hypothesis H_0, we expect $\bar{x}$ to be Gaussian distributed, with mean zero and variance $1/N$. Thus, from subsection 30.9.4, v will follow a *chi-squared* distribution of order 1. Substituting the appropriate form for $P(v|H_0)$ in (31.111) and setting $\alpha = 0.1$, we find by numerical integration (or from table 31.2) that $v_{\mathrm{crit}} = N\bar{x}^2_{\mathrm{crit}} = 2.71$. Since $N = 10$, the rejection region on $\bar{x}$ at the 10% significance level is thus

$$\bar{x} < -0.52 \qquad \text{and} \qquad \bar{x} > 0.52.$$

As noted before, for this sample $\bar{x} = 1.11$, and so we may reject the null hypothesis $H_0 : \mu = 0$ at the 10% significance level. ◀

The above example illustrates the general situation that if the maximum-likelihood estimates $\hat{\mathbf{a}}$ of the parameters fall in or near the subspace S then the sample will be considered consistent with H_0 and the value of t will be near unity. If $\hat{\mathbf{a}}$ is distant from S then the sample will not be in accord with H_0 and ordinarily t will have a small (positive) value.

It is clear that in order to prescribe the rejection region for t, or for a related statistic u or v, it is necessary to know the sampling distribution $P(t|H_0)$. If H_0 is simple then one can in principle determine $P(t|H_0)$, although this may prove difficult in practice. Moreover, if H_0 is composite, then it may not be possible to obtain $P(t|H_0)$, even in principle. Nevertheless, a useful approximate form for $P(t|H_0)$ exists in the large-sample limit. Consider the null hypothesis

$$H_0 : (a_1 = a_1^0, a_2 = a_2^0, \ldots, a_R = a_R^0), \quad \text{where } R \leq M$$

and the a_i^0 are fixed numbers. (In fact, we may fix the values of any subset containing R of the M parameters.) If H_0 is true then it follows from our discussion in subsection 31.5.6 (although we shall not prove it) that, when the sample size N is large, the quantity $-2\ln t$ follows approximately a *chi-squared* distribution of order R.

31.7.5 Student's t-test

Student's t-test is just a special case of the generalised likelihood ratio test applied to a sample $x_1, x_2, \ldots, x_N$ drawn independently from a Gaussian distribution for which *both* the mean μ and variance σ^2 are unknown, and for which one wishes to distinguish between the hypotheses

$$H_0 : \mu = \mu_0, \quad 0 < \sigma^2 < \infty, \qquad \text{and} \qquad H_1 : \mu \neq \mu_0, \quad 0 < \sigma^2 < \infty,$$

where μ_0 is a given number. Here, the parameter space $\mathcal{A}$ is the half-plane $-\infty < \mu < \infty$, $0 < \sigma^2 < \infty$, whereas the subspace $\mathcal{S}$ characterised by the null hypothesis H_0 is the line $\mu = \mu_0$, $0 < \sigma^2 < \infty$.

The likelihood function for this situation is given by

$$L(\mathbf{x}; \mu, \sigma^2) = \frac{1}{(2\pi\sigma^2)^{N/2}} \exp\left[-\frac{\sum_i (x_i - \mu)^2}{2\sigma^2}\right].$$

On the one hand, as shown in subsection 31.5.1, the values of μ and σ^2 that maximise L in $\mathcal{A}$ are $\mu = \bar{x}$ and $\sigma^2 = s^2$, where $\bar{x}$ is the sample mean and s^2 is the sample variance. On the other hand, to maximise L in the subspace $\mathcal{S}$ we set $\mu = \mu_0$, and the only remaining parameter is σ^2; the value of σ^2 that maximises L is then easily found to be

$$\widehat{\sigma^2} = \frac{1}{N} \sum_{i=1}^{N} (x_i - \mu_0)^2.$$

To retain, in due course, the standard notation for Student's t-test, in this section we will denote the generalised likelihood ratio by λ (rather than t); it is thus given by

$$\begin{aligned}
\lambda(\mathbf{x}) &= \frac{L(\mathbf{x}; \mu_0, \widehat{\sigma^2})}{L(\mathbf{x}; \bar{x}, s^2)} \\
&= \frac{[(2\pi/N)\sum_i (x_i - \mu_0)^2]^{-N/2} \exp(-N/2)}{[(2\pi/N)\sum_i (x_i - \bar{x})^2]^{-N/2} \exp(-N/2)} = \left[\frac{\sum_i (x_i - \bar{x})^2}{\sum_i (x_i - \mu_0)^2}\right]^{N/2}.
\end{aligned} \tag{31.112}$$

Normally, our next step would be to find the sampling distribution of λ under the assumption that H_0 were true. It is more conventional, however, to work in terms of a related test statistic t, which was first devised by William Gossett, who wrote under the pen name of 'Student'.

1284

The sum of squares in the denominator of (31.112) may be put into the form

$$\sum_i (x_i - \mu_0)^2 = N(\bar{x} - \mu_0)^2 + \sum_i (x_i - \bar{x})^2.$$

Thus, on dividing the numerator and denominator in (31.112) by $\sum_i (x_i - \bar{x})^2$ and rearranging, the generalised likelihood ratio λ can be written

$$\lambda = \left(1 + \frac{t^2}{N-1}\right)^{-N/2},$$

where we have defined the new variable

$$t = \frac{\bar{x} - \mu_0}{s/\sqrt{N-1}}. \tag{31.113}$$

Since t^2 is a monotonically decreasing function of λ, the corresponding rejection region is $t^2 > c$, where c is a positive constant depending on the required significance level α. It is conventional, however, to use t itself as our test statistic, in which case our rejection region becomes two-tailed and is given by

$$t < -t_{\text{crit}} \qquad \text{and} \qquad t > t_{\text{crit}}, \tag{31.114}$$

where t_{crit} is the positive square root of the constant c.

The definition (31.113) and the rejection region (31.114) form the basis of Student's t-test. It only remains to determine the sampling distribution $P(t|H_0)$. At the outset, it is worth noting that if we write the expression (31.113) for t in terms of the standard estimator $\hat{\sigma} = \sqrt{Ns^2/(N-1)}$ of the standard deviation then we obtain

$$t = \frac{\bar{x} - \mu_0}{\hat{\sigma}/\sqrt{N}}. \tag{31.115}$$

If, in fact, we knew the true value of σ and used it in this expression for t then it is clear from our discussion in section 31.3 that t would follow a Gaussian distribution with mean 0 and variance 1, i.e. $t \sim N(0,1)$. When σ is not known, however, we have to use our estimate $\hat{\sigma}$ in (31.115), with the result that t is no longer distributed as the standard Gaussian. As one might expect from the central limit theorem, however, the distribution of t does tend towards the standard Gaussian for large values of N.

As noted earlier, the exact distribution of t, valid for any value of N, was first discovered by William Gossett. From (31.35), if the hypothesis H_0 is true then the joint sampling distribution of $\bar{x}$ and s is given by

$$P(\bar{x}, s|H_0) = Cs^{N-2} \exp\left(-\frac{Ns^2}{2\sigma^2}\right) \exp\left[-\frac{N(\bar{x} - \mu_0)^2}{2\sigma^2}\right], \tag{31.116}$$

where C is a normalisation constant. We can use this result to obtain the joint sampling distribution of s and t by demanding that

$$P(\bar{x}, s|H_0)\, d\bar{x}\, ds = P(t, s|H_0)\, dt\, ds.$$

Using (31.113) to substitute for $\bar{x} - \mu_0$ in (31.116), and noting that $d\bar{x} = (s/\sqrt{N-1})\,dt$, we find

$$P(\bar{x}, s|H_0)\,d\bar{x}\,ds = As^{N-1} \exp\left[-\frac{Ns^2}{2\sigma^2}\left(1 + \frac{t^2}{N-1}\right)\right]\,dt\,ds,$$

where A is another normalisation constant. In order to obtain the sampling distribution of t alone, we must integrate $P(t, s|H_0)$ with respect to s over its allowed range, from 0 to ∞. Thus, the required distribution of t alone is given by

$$P(t|H_0) = \int_0^\infty P(t, s|H_0)\,ds = A\int_0^\infty s^{N-1} \exp\left[-\frac{Ns^2}{2\sigma^2}\left(1 + \frac{t^2}{N-1}\right)\right]\,ds.$$
(31.117)

To carry out this integration, we set $y = s\{1 + [t^2/(N-1)]\}^{1/2}$, which on substitution into (31.117) yields

$$P(t|H_0) = A\left(1 + \frac{t^2}{N-1}\right)^{-N/2}\int_0^\infty y^{N-1}\exp\left(-\frac{Ny^2}{2\sigma^2}\right)\,dy.$$

Since the integral over y does not depend on t, it is simply a constant. We thus find that that the sampling distribution of the variable t is

$$P(t|H_0) = \frac{1}{\sqrt{(N-1)\pi}}\frac{\Gamma\left(\frac{1}{2}N\right)}{\Gamma\left(\frac{1}{2}(N-1)\right)}\left(1 + \frac{t^2}{N-1}\right)^{-N/2},$$
(31.118)

where we have used the condition $\int_{-\infty}^\infty P(t|H_0)\,dt = 1$ to determine the normalisation constant (see exercise 31.18).

The distribution (31.118) is called *Student's t-distribution with $N-1$ degrees of freedom*. A plot of Student's t-distribution is shown in figure 31.11 for various values of N. For comparison, we also plot the standard Gaussian distribution, to which the t-distribution tends for large N. As is clear from the figure, the t-distribution is symmetric about $t = 0$. In table 31.3 we list some critical points of the cumulative probability function $C_n(t)$ of the t-distribution, which is defined by

$$C_n(t) = \int_{-\infty}^t P(t'|H_0)\,dt',$$

where $n = N - 1$ is the number of degrees of freedom. Clearly, $C_n(t)$ is analogous to the cumulative probability function $\Phi(z)$ of the Gaussian distribution, discussed in subsection 30.9.1. For comparison purposes, we also list the critical points of $\Phi(z)$, which corresponds to the t-distribution for $N = \infty$.

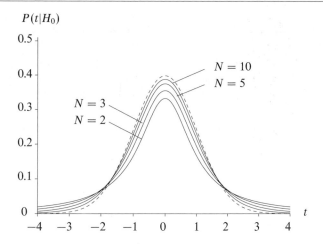

Figure 31.11   Student's t-distribution for various values of N. The broken curve shows the standard Gaussian distribution for comparison.

> ► *Ten independent sample values x_i, $i = 1, 2, \ldots, 10$, are drawn at random from a Gaussian distribution with unknown mean μ and unknown standard deviation σ. The sample values are as follows:*
>
> | 2.22 | 2.56 | 1.07 | 0.24 | 0.18 | 0.95 | 0.73 | −0.79 | 2.09 | 1.81 |
>
> *Test the null hypothesis $H_0 : \mu = 0$ at the 10% significance level.*

For our null hypothesis, $\mu_0 = 0$. Since for this sample $\bar{x} = 1.11$, $s = 1.01$ and $N = 10$, it follows from (31.113) that

$$t = \frac{\bar{x}}{s/\sqrt{N-1}} = 3.33.$$

The rejection region for t is given by (31.114) where t_{crit} is such that

$$C_{N-1}(t_{\text{crit}}) = 1 - \alpha/2,$$

and α is the required significance of the test. In our case $\alpha = 0.1$ and $N = 10$, and from table 31.3 we find $t_{\text{crit}} = 1.83$. Thus our rejection region for H_0 at the 10% significance level is

$$t < -1.83 \qquad \text{and} \qquad t > 1.83.$$

For our sample $t = 3.30$ and so we can clearly reject the null hypothesis $H_0 : \mu = 0$ at this level. ◄

It is worth noting the connection between the t-test and the classical confidence interval on the mean μ. The central confidence interval on μ at the confidence level $1 - \alpha$ is the set of values for which

$$-t_{\text{crit}} < \frac{\bar{x} - \mu}{s/\sqrt{N-1}} < t_{\text{crit}},$$

$C_n(t)$	0.5	0.6	0.7	0.8	0.9	0.950	0.975	0.990	0.995	0.999
$n = 1$	0.00	0.33	0.73	1.38	3.08	6.31	12.7	31.8	63.7	318.3
2	0.00	0.29	0.62	1.06	1.89	2.92	4.30	6.97	9.93	22.3
3	0.00	0.28	0.58	0.98	1.64	2.35	3.18	4.54	5.84	10.2
4	0.00	0.27	0.57	0.94	1.53	2.13	2.78	3.75	4.60	7.17
5	0.00	0.27	0.56	0.92	1.48	2.02	2.57	3.37	4.03	5.89
6	0.00	0.27	0.55	0.91	1.44	1.94	2.45	3.14	3.71	5.21
7	0.00	0.26	0.55	0.90	1.42	1.90	2.37	3.00	3.50	4.79
8	0.00	0.26	0.55	0.89	1.40	1.86	2.31	2.90	3.36	4.50
9	0.00	0.26	0.54	0.88	1.38	1.83	2.26	2.82	3.25	4.30
10	0.00	0.26	0.54	0.88	1.37	1.81	2.23	2.76	3.17	4.14
11	0.00	0.26	0.54	0.88	1.36	1.80	2.20	2.72	3.11	4.03
12	0.00	0.26	0.54	0.87	1.36	1.78	2.18	2.68	3.06	3.93
13	0.00	0.26	0.54	0.87	1.35	1.77	2.16	2.65	3.01	3.85
14	0.00	0.26	0.54	0.87	1.35	1.76	2.15	2.62	2.98	3.79
15	0.00	0.26	0.54	0.87	1.34	1.75	2.13	2.60	2.95	3.73
16	0.00	0.26	0.54	0.87	1.34	1.75	2.12	2.58	2.92	3.69
17	0.00	0.26	0.53	0.86	1.33	1.74	2.11	2.57	2.90	3.65
18	0.00	0.26	0.53	0.86	1.33	1.73	2.10	2.55	2.88	3.61
19	0.00	0.26	0.53	0.86	1.33	1.73	2.09	2.54	2.86	3.58
20	0.00	0.26	0.53	0.86	1.33	1.73	2.09	2.53	2.85	3.55
25	0.00	0.26	0.53	0.86	1.32	1.71	2.06	2.49	2.79	3.46
30	0.00	0.26	0.53	0.85	1.31	1.70	2.04	2.46	2.75	3.39
40	0.00	0.26	0.53	0.85	1.30	1.68	2.02	2.42	2.70	3.31
50	0.00	0.26	0.53	0.85	1.30	1.68	2.01	2.40	2.68	3.26
100	0.00	0.25	0.53	0.85	1.29	1.66	1.98	2.37	2.63	3.17
200	0.00	0.25	0.53	0.84	1.29	1.65	1.97	2.35	2.60	3.13
∞	0.00	0.25	0.52	0.84	1.28	1.65	1.96	2.33	2.58	3.09

Table 31.3   The confidence limits t of the cumulative probability function $C_n(t)$ for Student's t-distribution with n degrees of freedom. For example, $C_5(0.92) = 0.8$. The row $n = \infty$ is also the corresponding result for the standard Gaussian distribution.

where t_{crit} satisfies $C_{N-1}(t_{\mathrm{crit}}) = \alpha/2$. Thus the required confidence interval is

$$\bar{x} - \frac{t_{\mathrm{crit}}s}{\sqrt{N-1}} < \mu < \bar{x} + \frac{t_{\mathrm{crit}}s}{\sqrt{N-1}}.$$

Hence, in the above example, the 90% classical central confidence interval on μ is

$$0.49 < \mu < 1.73.$$

The t-distribution may also be used to compare different samples from Gaussian distributions. In particular, let us consider the case where we have two independent samples of sizes N_1 and N_2, drawn respectively from Gaussian distributions with a common variance σ^2 but with possibly different means μ_1 and μ_2. On the basis of the samples, one wishes to distinguish between the hypotheses

$$H_0 : \mu_1 = \mu_2, \quad 0 < \sigma^2 < \infty \qquad \text{and} \qquad H_1 : \mu_1 \neq \mu_2, \quad 0 < \sigma^2 < \infty.$$

In other words, we wish to test the null hypothesis that the samples are drawn from populations having the same mean. Suppose that the measured sample means and standard deviations are $\bar{x}_1$, $\bar{x}_2$ and s_1, s_2 respectively. In an analogous way to that presented above, one may show that the generalised likelihood ratio can be written as

$$\lambda = \left(1 + \frac{t^2}{N_1 + N_2 - 2}\right)^{-(N_1+N_2)/2}.$$

In this case, the variable t is given by

$$t = \frac{\bar{w} - \omega}{\hat{\sigma}} \left(\frac{N_1 N_2}{N_1 + N_2}\right)^{1/2}, \tag{31.119}$$

where $\bar{w} = \bar{x}_1 - \bar{x}_2$, $\omega = \mu_1 - \mu_2$ and

$$\hat{\sigma} = \left[\frac{N_1 s_1^2 + N_2 s_2^2}{N_1 + N_2 - 2}\right]^{1/2}.$$

It is straightforward (albeit with complicated algebra) to show that the variable t in (31.119) follows Student's t-distribution with $N_1 + N_2 - 2$ degrees of freedom, and so we may use an appropriate form of Student's t-test to investigate the null hypothesis $H_0 : \mu_1 = \mu_2$ (or equivalently $H_0 : \omega = 0$). As above, the t-test can be used to place a confidence interval on $\omega = \mu_1 - \mu_2$.

> ▶ Suppose that two classes of students take the same mathematics examination and the following percentage marks are obtained:
>
> | Class 1: | 66 | 62 | 34 | 55 | 77 | 80 | 55 | 60 | 69 | 47 | 50 |
> | Class 2: | 64 | 90 | 76 | 56 | 81 | 72 | 70 |
>
> Assuming that the two sets of examinations marks are drawn from Gaussian distributions with a common variance, test the hypothesis $H_0 : \mu_1 = \mu_2$ at the 5% significance level. Use your result to obtain the 95% classical central confidence interval on $\omega = \mu_1 - \mu_2$.

We begin by calculating the mean and standard deviation of each sample. The number of values in each sample is $N_1 = 11$ and $N_2 = 7$ respectively, and we find

$$\bar{x}_1 = 59.5, \quad s_1 = 12.8 \quad \text{and} \quad \bar{x}_2 = 72.7, \quad s_2 = 10.3,$$

leading to $\bar{w} = \bar{x}_1 - \bar{x}_2 = -13.2$ and $\hat{\sigma} = 12.6$. Setting $\omega = 0$ in (31.119), we thus find $t = -2.17$.

The rejection region for H_0 is given by (31.114), where t_{crit} satisfies

$$C_{N_1+N_2-2}(t_{\text{crit}}) = 1 - \alpha/2, \tag{31.120}$$

where α is the required significance level of the test. In our case we set $\alpha = 0.05$, and from table 31.3 with $n = 16$ we find that $t_{\text{crit}} = 2.12$. The rejection region is therefore

$$t < -2.12 \qquad \text{and} \qquad t > 2.12.$$

Since $t = -2.17$ for our samples, we can reject the null hypothesis $H_0 : \mu_1 = \mu_2$, although only by a small margin. (Indeed, it is easily shown that one cannot reject H_0 at the 2% significance level). The 95% central confidence interval on $\omega = \mu_1 - \mu_2$ is given by

$$\bar{w} - \hat{\sigma} t_{\text{crit}} \left(\frac{N_1 + N_2}{N_1 N_2} \right)^{1/2} < \quad \omega \quad < \bar{w} + \hat{\sigma} t_{\text{crit}} \left(\frac{N_1 + N_2}{N_1 N_2} \right)^{1/2},$$

where t_{crit} is given by (31.120). Thus, we find

$$-26.1 < \omega < -0.28,$$

which, as expected, does not (quite) contain $\omega = 0$. ◀

In order to apply Student's t-test in the above example, we had to make the assumption that the samples were drawn from Gaussian distributions possessing a common variance, which is clearly unjustified a $priori$. We can, however, perform another test on the data to investigate whether the additional hypothesis $\sigma_1^2 = \sigma_2^2$ is reasonable; this test is discussed in the next subsection. If this additional test shows that the hypothesis $\sigma_1^2 = \sigma_2^2$ may be accepted (at some suitable significance level), then we may indeed use the analysis in the above example to infer that the null hypothesis $H_0 : \mu_1 = \mu_2$ may be rejected at the 5% significance level. If, however, we find that the additional hypothesis $\sigma_1^2 = \sigma_2^2$ must be rejected, then we can only infer from the above example that the hypothesis that the two samples were drawn from the same Gaussian distribution may be rejected at the 5% significance level.

Throughout the above discussion, we have assumed that samples are drawn from a Gaussian distribution. Although this is true for many random variables, in practice it is usually impossible to know a $priori$ whether this is case. It can be shown, however, that Student's t-test remains reasonably accurate even if the sampled distribution(s) differ considerably from a Gaussian. Indeed, for sampled distributions that differ only slightly from a Gaussian form, the accuracy of the test is remarkably good. Nevertheless, when applying the t-test, it is always important to remember that the assumption of a Gaussian parent population is central to the method.

31.7.6 Fisher's F-test

Having concentrated on tests for the mean μ of a Gaussian distribution, we now consider tests for its standard deviation σ. Before discussing Fisher's F-test for comparing the standard deviations of two samples, we begin by considering the case when an independent sample $x_1, x_2, \ldots, x_N$ is drawn from a Gaussian

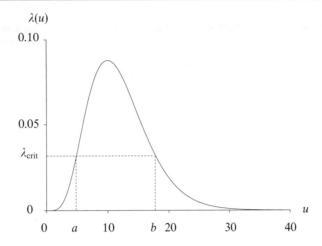

Figure 31.12 The sampling distribution $P(u|H_0)$ for $N = 10$; this is a chi-squared distribution for $N - 1$ degrees of freedom.

distribution with unknown μ and σ, and we wish to distinguish between the two hypotheses

$$H_0 : \sigma^2 = \sigma_0^2, \quad -\infty < \mu < \infty \quad \text{and} \quad H_1 : \sigma^2 \neq \sigma_0^2, \quad -\infty < \mu < \infty,$$

where σ_0^2 is a given number. Here, the parameter space $\mathcal{A}$ is the half-plane $-\infty < \mu < \infty$, $0 < \sigma^2 < \infty$, whereas the subspace S characterised by the null hypothesis H_0 is the line $\sigma^2 = \sigma_0^2$, $-\infty < \mu < \infty$.

The likelihood function for this situation is given by

$$L(\mathbf{x}; \mu, \sigma^2) = \frac{1}{(2\pi\sigma^2)^{N/2}} \exp\left[-\frac{\sum_i (x_i - \mu)^2}{2\sigma^2}\right].$$

The maximum of L in $\mathcal{A}$ occurs at $\mu = \bar{x}$ and $\sigma^2 = s^2$, whereas the maximum of L in S is at $\mu = \bar{x}$ and $\sigma^2 = \sigma_0^2$. Thus, the generalised likelihood ratio is given by

$$\lambda(\mathbf{x}) = \frac{L(\mathbf{x}; \bar{x}, \sigma_0^2)}{L(\mathbf{x}; \bar{x}, s^2)} = \left(\frac{u}{N}\right)^{N/2} \exp\left[-\tfrac{1}{2}(u - N)\right],$$

where we have introduced the variable

$$u = \frac{Ns^2}{\sigma_0^2} = \frac{\sum_i (x_i - \bar{x})^2}{\sigma_0^2}. \tag{31.121}$$

An example of this distribution is plotted in figure 31.12 for $N = 10$. From the figure, we see that the rejection region $\lambda < \lambda_{\text{crit}}$ corresponds to a two-tailed rejection region on u given by

$$0 < u < a \quad \text{and} \quad b < u < \infty,$$

where a and b are such that $\lambda_{\text{crit}}(a) = \lambda_{\text{crit}}(b)$, as shown in figure 31.12. In practice, however, it is difficult to determine a and b for a given significance level α, so a slightly different rejection region, which we now describe, is usually adopted.

The sampling distribution $P(u|H_0)$ may be found straightforwardly from the sampling distribution of s given in (31.35). Let us first determine $P(s^2|H_0)$ by demanding that

$$P(s|H_0)\,ds = P(s^2|H_0)\,d(s^2),$$

from which we find

$$P(s^2|H_0) = \frac{P(s|H_0)}{2s} = \left(\frac{N}{2\sigma_0^2}\right)^{(N-1)/2} \frac{(s^2)^{(N-3)/2}}{\Gamma\left(\frac{1}{2}(N-1)\right)} \exp\left(-\frac{Ns^2}{2\sigma_0^2}\right). \qquad (31.122)$$

Thus, the sampling distribution of $u = Ns^2/\sigma_0^2$ is given by

$$P(u|H_0) = \frac{1}{2^{(N-1)/2}\Gamma\left(\frac{1}{2}(N-1)\right)} u^{(N-3)/2} \exp\left(-\tfrac{1}{2}u\right).$$

We note, in passing, that the distribution of u is precisely that of an $(N-1)$th-order chi-squared variable (see subsection 30.9.4), i.e. $u \sim \chi_{N-1}^2$. Although it does not give quite the best test, one then takes the rejection region to be

$$0 < u < a \qquad \text{and} \qquad b < u < \infty,$$

with a and b chosen such that the two tails have *equal areas*; the advantage of this choice is that tabulations of the chi-squared distribution make the size of this region relatively easy to estimate. Thus, for a given significance level α, we have

$$\int_0^a P(u|H_0)\,du = \alpha/2 \qquad \text{and} \qquad \int_b^\infty P(u|H_0)\,du = \alpha/2.$$

▶*Ten independent sample values x_i, $i = 1, 2, \ldots, 10$, are drawn at random from a Gaussian distribution with unknown mean μ and standard deviation σ. The sample values are as follows:*

$$2.22 \quad 2.56 \quad 1.07 \quad 0.24 \quad 0.18 \quad 0.95 \quad 0.73 \quad -0.79 \quad 2.09 \quad 1.81$$

Test the null hypothesis $H_0 : \sigma^2 = 2$ at the 10% significance level.

For our null hypothesis $\sigma_0^2 = 2$. Since for this sample $s = 1.01$ and $N = 10$, from (31.121) we have $u = 5.10$. For $\alpha = 0.1$ we find, either numerically or using table 31.2, that $a = 3.33$ and $b = 16.92$. Thus, our rejection region is

$$0 < u < 3.33 \qquad \text{and} \qquad 16.92 < u < \infty.$$

The value $u = 5.10$ from our sample does not lie in the rejection region, and so we cannot reject the null hypothesis $H_0 : \sigma^2 = 2$. ◀

We now turn to Fisher's F-test. Let us suppose that two independent samples of sizes N_1 and N_2 are drawn from Gaussian distributions with means and variances μ_1, σ_1^2 and μ_2, σ_2^2 respectively, and we wish to distinguish between the two hypotheses

$$H_0 : \sigma_1^2 = \sigma_2^2 \quad \text{and} \quad H_1 : \sigma_1^2 \neq \sigma_2^2.$$

In this case, the generalised likelihood ratio is found to be

$$\lambda = \frac{(N_1 + N_2)^{(N_1+N_2)/2}}{N_1^{N_1/2} N_2^{N_2/2}} \frac{\left[F(N_1 - 1)/(N_2 - 1)\right]^{N_1/2}}{\left[1 + F(N_1 - 1)/(N_2 - 1)\right]^{(N_1+N_2)/2}},$$

where F is given by the variance ratio

$$F = \frac{N_1 s_1^2/(N_1 - 1)}{N_2 s_2^2/(N_2 - 1)} \equiv \frac{u^2}{v^2} \tag{31.123}$$

and s_1 and s_2 are the standard deviations of the two samples. On plotting λ as a function of F, it is apparent that the rejection region $\lambda < \lambda_{\text{crit}}$ corresponds to a two-tailed test on F. Nevertheless, as will shall see below, by defining the fraction (31.123) appropriately, it is customary to make a one-tailed test on F.

The distribution of F may be obtained in a reasonably straightforward manner by making use of the distribution of the sample variance s^2 given in (31.122). Under our null hypothesis H_0, the two Gaussian distributions share a common variance, which we denote by σ^2. Changing the variable in (31.122) from s^2 to u^2 we find that u^2 has the sampling distribution

$$P(u^2|H_0) = \left(\frac{N - 1}{2\sigma^2}\right)^{(N-1)/2} \frac{1}{\Gamma\left(\frac{1}{2}(N - 1)\right)} (u^2)^{(N-3)/2} \exp\left[-\frac{(N - 1)u^2}{2\sigma^2}\right].$$

Since u^2 and v^2 are independent, their joint distribution is simply the product of their individual distributions and is given by

$$P(u^2|H_0)P(v^2|H_0) = A(u^2)^{(N_1-3)/2}(v^2)^{(N_2-3)/2} \exp\left[-\frac{(N_1 - 1)u^2 + (N_2 - 1)v^2}{2\sigma^2}\right],$$

where the constant A is given by

$$A = \frac{(N_1 - 1)^{(N_1-1)/2}(N_2 - 1)^{(N_2-1)/2}}{2^{(N_1+N_2-2)/2}\sigma^{(N_1+N_2-2)}\Gamma\left(\frac{1}{2}(N_1 - 1)\right)\Gamma\left(\frac{1}{2}(N_2 - 1)\right)}. \tag{31.124}$$

Now, for fixed v we have $u^2 = Fv^2$ and $d(u^2) = v^2 dF$. Thus, the joint sampling distribution $P(v^2, F|H_0)$ is obtained by requiring that

$$P(v^2, F|H_0) \, d(v^2) \, dF = P(u^2|H_0) P(v^2|H_0) \, d(u^2) \, d(v^2).$$

(31.125)

In order to find the distribution of F alone, we now integrate $P(v^2, F|H_0)$ with respect to v^2 from 0 to ∞, from which we obtain

$P(F|H_0)$

$$= \left(\frac{N_1 - 1}{N_2 - 1} \right)^{(N_1-1)/2} \frac{F^{(N_1-3)/2}}{B \left(\frac{1}{2}(N_1 - 1), \frac{1}{2}(N_2 - 1) \right)} \left(1 + \frac{N_1 - 1}{N_2 - 1} F \right)^{-(N_1+N_2-2)/2},$$

(31.126)

where $B \left(\frac{1}{2}(N_1 - 1), \frac{1}{2}(N_2 - 1) \right)$ is the beta function defined in subsection 18.12.2. $P(F|H_0)$ is called the *F-distribution* (or occasionally the *Fisher distribution*) with $(N_1 - 1, N_2 - 1)$ degrees of freedom.

▶*Evaluate the integral $\int_0^\infty P(v^2, F|H_0) \, d(v^2)$ to obtain result (31.126).*

From (31.125), we have

$$P(F|H_0) = A F^{(N_1-3)/2} \int_0^\infty (v^2)^{(N_1+N_2-4)/2} \exp \left\{ -\frac{[(N_1 - 1)F + (N_2 - 1)]v^2}{2\sigma^2} \right\} d(v^2).$$

Making the substitution $x = [(N_1 - 1)F + (N_2 - 1)]v^2/(2\sigma^2)$, we obtain

$$P(F|H_0) = A \left[\frac{2\sigma^2}{(N_1 - 1)F + (N_2 - 1)} \right]^{(N_1+N_2-2)/2} F^{(N_1-3)/2} \int_0^\infty x^{(N_1+N_2-4)/2} e^{-x} \, dx$$

$$= A \left[\frac{2\sigma^2}{(N_1 - 1)F + (N_2 - 1)} \right]^{(N_1+N_2-2)/2} F^{(N_1-3)/2} \Gamma \left(\frac{1}{2}(N_1 + N_2 - 2) \right),$$

where in the last line we have used the definition of the gamma function given in section 18.12. Using the further result (18.165), which expresses the beta function in terms of the gamma function, and the expression for A given in (31.124), we see that $P(F|H_0)$ is indeed given by (31.126). ◀

As it does not matter whether the ratio F given in (31.123) is defined as u^2/v^2 or as v^2/u^2, it is conventional to put the larger sample variance on the top, so that F is always greater than or equal to unity. A large value of F indicates that the sample variances u^2 and v^2 are very different whereas a value of F close to unity means that they are very similar. Therefore, for a given significance α, it is

$C_{n_1,n_2}(F)$	$n_1 = 1$	2	3	4	5	6	7	8
$n_2 = 1$	161	200	216	225	230	234	237	239
2	18.5	19.0	19.2	19.2	19.3	19.3	19.4	19.4
3	10.1	9.55	9.28	9.12	9.01	8.94	8.89	8.85
4	7.71	6.94	6.59	6.39	6.26	6.16	6.09	6.04
5	6.61	5.79	5.41	5.19	5.05	4.95	4.88	4.82
6	5.99	5.14	4.76	4.53	4.39	4.28	4.21	4.15
7	5.59	4.74	4.35	4.12	3.97	3.87	3.79	3.73
8	5.32	4.46	4.07	3.84	3.69	3.58	3.50	3.44
9	5.12	4.26	3.86	3.63	3.48	3.37	3.29	3.23
10	4.96	4.10	3.71	3.48	3.33	3.22	3.14	3.07
20	4.35	3.49	3.10	2.87	2.71	2.60	2.51	2.45
30	4.17	3.32	2.92	2.69	2.53	2.42	2.33	2.27
40	4.08	3.23	2.84	2.61	2.45	2.34	2.25	2.18
50	4.03	3.18	2.79	2.56	2.40	2.29	2.20	2.13
100	3.94	3.09	2.70	2.46	2.31	2.19	2.10	2.03
∞	3.84	3.00	2.60	2.37	2.21	2.10	2.01	1.94

	$n_1 = 9$	10	20	30	40	50	100	∞
$n_2 = 1$	241	242	248	250	251	252	253	254
2	19.4	19.4	19.4	19.5	19.5	19.5	19.5	19.5
3	8.81	8.79	8.66	8.62	8.59	8.58	8.55	8.53
4	6.00	5.96	5.80	5.75	5.72	5.70	5.66	5.63
5	4.77	4.74	4.56	4.50	4.46	4.44	4.41	4.37
6	4.10	4.06	3.87	3.81	3.77	3.75	3.71	3.67
7	3.68	3.64	3.44	3.38	3.34	3.32	3.27	3.23
8	3.39	3.35	3.15	3.08	3.04	3.02	2.97	2.93
9	3.18	3.14	2.94	2.86	2.83	2.80	2.76	2.71
10	3.02	2.98	2.77	2.70	2.66	2.64	2.59	2.54
20	2.39	2.35	2.12	2.04	1.99	1.97	1.91	1.84
30	2.21	2.16	1.93	2.69	1.79	1.76	1.70	1.62
40	2.12	2.08	1.84	1.74	1.69	1.66	1.59	1.51
50	2.07	2.03	1.78	1.69	1.63	1.60	1.52	1.44
100	1.97	1.93	1.68	1.57	1.52	1.48	1.39	1.28
∞	1.88	1.83	1.57	1.46	1.39	1.35	1.24	1.00

Table 31.4   Values of F for which the cumulative probability function $C_{n_1,n_2}(F)$ of the F-distribution with (n_1, n_2) degrees of freedom has the value 0.95. For example, for $n_1 = 10$ and $n_2 = 6$, $C_{n_1,n_2}(4.06) = 0.95$.

customary to define the rejection region on F as $F > F_{\text{crit}}$, where

$$C_{n_1,n_2}(F_{\text{crit}}) = \int_1^{F_{\text{crit}}} P(F|H_0)\, dF = \alpha,$$

and $n_1 = N_1 - 1$ and $n_2 = N_2 - 1$ are the numbers of degrees of freedom. Table 31.4 lists values of F_{crit} corresponding to the 5% significance level (i.e. $\alpha = 0.05$) for various values of n_1 and n_2.

1295

▶Suppose that two classes of students take the same mathematics examination and the following percentage marks are obtained:

Class 1:	66	62	34	55	77	80	55	60	69	47	50
Class 2:	64	90	76	56	81	72	70				

Assuming that the two sets of examinations marks are drawn from Gaussian distributions, test the hypothesis $H_0 : \sigma_1^2 = \sigma_2^2$ at the 5% significance level.

The variances of the two samples are $s_1^2 = (12.8)^2$ and $s_2^2 = (10.3)^2$ and the sample sizes are $N_1 = 11$ and $N_2 = 7$. Thus, we have

$$u^2 = \frac{N_1 s_1^2}{N_1 - 1} = 180.2 \quad \text{and} \quad v^2 = \frac{N_2 s_2^2}{N_2 - 1} = 123.8,$$

where we have taken u^2 to be the larger value. Thus, $F = u^2/v^2 = 1.46$ to two decimal places. Since the first sample contains eleven values and the second contains seven values, we take $n_1 = 10$ and $n_2 = 6$. Consulting table 31.4, we see that, at the 5% significance level, $F_{\text{crit}} = 4.06$. Since our value lies comfortably below this, we conclude that there is no statistical evidence for rejecting the hypothesis that the two samples were drawn from Gaussian distributions with a common variance. ◀

It is also common to define the variable $z = \frac{1}{2} \ln F$, the distribution of which can be found straightforwardly from (31.126). This is a useful change of variable since it can be shown that, for large values of n_1 and n_2, the variable z is distributed approximately as a Gaussian with mean $\frac{1}{2}(n_2^{-1} - n_1^{-1})$ and variance $\frac{1}{2}(n_2^{-1} + n_1^{-1})$.

31.7.7 Goodness of fit in least-squares problems

We conclude our discussion of hypothesis testing with an example of a goodness-of-fit test. In section 31.6, we discussed the use of the method of least squares in estimating the best-fit values of a set of parameters $\mathbf{a}$ in a given model $y = f(x; \mathbf{a})$ for a data set (x_i, y_i), $i = 1, 2, \ldots, N$. We have not addressed, however, the question of whether the best-fit model $y = f(x; \hat{\mathbf{a}})$ does, in fact, provide a good fit to the data. In other words, we have not considered thus far how to verify that the functional form f of our assumed model is indeed correct. In the language of hypothesis testing, we wish to distinguish between the two hypotheses

$$H_0 : \text{model is correct} \quad \text{and} \quad H_1 : \text{model is incorrect}.$$

Given the vague nature of the alternative hypothesis H_1, we clearly cannot use the generalised likelihood-ratio test. Nevertheless, it is still possible to test the null hypothesis H_0 at a given significance level α.

The least-squares estimates of the parameters $\hat{a}_1, \hat{a}_2, \ldots, \hat{a}_M$, as discussed in section 31.6, are those values that minimise the quantity

$$\chi^2(\mathbf{a}) = \sum_{i,j=1}^{N} [y_i - f(x_i; \mathbf{a})](\mathsf{N}^{-1})_{ij}[y_j - f(x_j; \mathbf{a})] = (\mathbf{y} - \mathbf{f})^{\mathrm{T}} \mathsf{N}^{-1} (\mathbf{y} - \mathbf{f}).$$

In the last equality, we rewrote the expression in matrix notation by defining the column vector f with elements $f_i = f(x_i; \mathbf{a})$. The value $\chi^2(\hat{\mathbf{a}})$ at this minimum can be used as a statistic to test the null hypothesis H_0, as follows. The N quantities $y_i - f(x_i; \mathbf{a})$ are Gaussian distributed. However, provided the function $f(x_j; \mathbf{a})$ is linear in the parameters $\mathbf{a}$, the equations (31.98) that determine the least-squares estimate $\hat{\mathbf{a}}$ constitute a set of M linear constraints on these N quantities. Thus, as discussed in subsection 30.15.2, the sampling distribution of the quantity $\chi^2(\hat{\mathbf{a}})$ will be a *chi-squared distribution with $N - M$ degrees of freedom* (d.o.f), which has the expectation value and variance

$$E[\chi^2(\hat{\mathbf{a}})] = N - M \quad \text{and} \quad V[\chi^2(\hat{\mathbf{a}})] = 2(N - M).$$

Thus we would expect the value of $\chi^2(\hat{\mathbf{a}})$ to lie typically in the range $(N - M) \pm \sqrt{2(N - M)}$. A value lying outside this range may suggest that the assumed model for the data is incorrect. A very small value of $\chi^2(\hat{\mathbf{a}})$ is usually an indication that the model has too many free parameters and has 'over-fitted' the data. More commonly, the assumed model is simply incorrect, and this usually results in a value of $\chi^2(\hat{\mathbf{a}})$ that is larger than expected.

One can choose to perform either a one-tailed or a two-tailed test on the value of $\chi^2(\hat{\mathbf{a}})$. It is usual, for a given significance level α, to define the one-tailed rejection region to be $\chi^2(\hat{\mathbf{a}}) > k$, where the constant k satisfies

$$\int_k^\infty P(\chi_n^2) \, d\chi_n^2 = \alpha \tag{31.127}$$

and $P(\chi_n^2)$ is the PDF of the chi-squared distribution with $n = N - M$ degrees of freedom (see subsection 30.9.4).

▶An experiment produces the following data sample pairs (x_i, y_i):

| x_i: | 1.85 | 2.72 | 2.81 | 3.06 | 3.42 | 3.76 | 4.31 | 4.47 | 4.64 | 4.99 |
| y_i: | 2.26 | 3.10 | 3.80 | 4.11 | 4.74 | 4.31 | 5.24 | 4.03 | 5.69 | 6.57 |

where the x_i-values are known exactly but each y_i-value is measured only to an accuracy of $\sigma = 0.5$. At the one-tailed 5% significance level, test the null hypothesis H_0 that the underlying model for the data is a straight line $y = mx + c$.

These data are the same as those investigated in section 31.6 and plotted in figure 31.9. As shown previously, the least squares estimates of the slope m and intercept c are given by

$$\hat{m} = 1.11 \quad \text{and} \quad \hat{c} = 0.4. \tag{31.128}$$

Since the error on each y_i-value is drawn independently from a Gaussian distribution with standard deviation σ, we have

$$\chi^2(\mathbf{a}) = \sum_{i=1}^N \left[\frac{y_i - f(x_i; \mathbf{a})}{\sigma} \right]^2 = \sum_{i=1}^N \left[\frac{y_i - mx_i - c}{\sigma} \right]^2. \tag{31.129}$$

Inserting the values (31.128) into (31.129), we obtain $\chi^2(\hat{m}, \hat{c}) = 11.5$. In our case, the number of data points is $N = 10$ and the number of fitted parameters is $M = 2$. Thus, the number of degrees of freedom is $n = N - M = 8$. Setting $n = 8$ and $\alpha = 0.05$ in (31.127) we find from table 31.2 that $k = 15.51$. Hence our rejection region is

$$\chi^2(\hat{m}, \hat{c}) > 15.51.$$

Since above we found $\chi^2(\hat{m}, \hat{c}) = 11.5$, we cannot reject the null hypothesis that the underlying model for the data is a straight line $y = mx + c$. ◀

As mentioned above, our analysis is only valid if the function $f(x; \mathbf{a})$ is linear in the parameters $\mathbf{a}$. Nevertheless, it is so convenient that it is sometimes applied in non-linear cases, provided the non-linearity is not too severe.

31.8 Exercises

31.1   A group of students uses a pendulum experiment to measure g, the acceleration of free fall, and obtains the following values (in m s^{-2}): 9.80, 9.84, 9.72, 9.74, 9.87, 9.77, 9.28, 9.86, 9.81, 9.79, 9.82. What would you give as the best value and standard error for g as measured by the group?

31.2   Measurements of a certain quantity gave the following values: 296, 316, 307, 278, 312, 317, 314, 307, 313, 306, 320, 309. Within what limits would you say there is a 50% chance that the correct value lies?

31.3   The following are the values obtained by a class of 14 students when measuring a physical quantity x: 53.8, 53.1, 56.9, 54.7, 58.2, 54.1, 56.4, 54.8, 57.3, 51.0, 55.1, 55.0, 54.2, 56.6.

(a) Display these results as a histogram and state what you would give as the best value for x.

(b) Without calculation, estimate how much reliance could be placed upon your answer to (a).

(c) Databooks give the value of x as 53.6 with negligible error. Are the data obtained by the students in conflict with this?

31.4   Two physical quantities x and y are connected by the equation

$$y^{1/2} = \frac{x}{ax^{1/2} + b},$$

and measured pairs of values for x and y are as follows:

x:	10	12	16	20
y:	409	196	114	94

Determine the best values for a and b by graphical means, and (either by hand or by using a built-in calculator routine) by a least-squares fit to an appropriate straight line.

31.5   Measured quantities x and y are known to be connected by the formula

$$y = \frac{ax}{x^2 + b},$$

where a and b are constants. Pairs of values obtained experimentally are

x:	2.0	3.0	4.0	5.0	6.0
y:	0.32	0.29	0.25	0.21	0.18

Use these data to make best estimates of the values of y that would be obtained for (a) $x = 7.0$, and (b) $x = -3.5$. As measured by fractional error, which estimate is likely to be the more accurate?

31.6 Prove that the sample mean is the best *linear unbiased estimator* of the population mean μ as follows.

(a) If the real numbers $a_1, a_2, \ldots, a_n$ satisfy the constraint $\sum_{i=1}^{n} a_i = C$, where C is a given constant, show that $\sum_{i=1}^{n} a_i^2$ is minimised by $a_i = C/n$ for all i.

(b) Consider the linear estimator $\hat{\mu} = \sum_{i=1}^{n} a_i x_i$. Impose the conditions (i) that it is *unbiased* and (ii) that it is as *efficient* as possible.

31.7 A population contains individuals of k types in equal proportions. A quantity X has mean μ_i amongst individuals of type i and variance σ^2, which has the same value for all types. In order to estimate the mean of X over the whole population, two schemes are considered; each involves a total sample size of nk. In the first the sample is drawn randomly from the whole population, whilst in the second (*stratified sampling*) n individuals are randomly selected from each of the k types.

Show that in both cases the estimate has expectation

$$\mu = \frac{1}{k} \sum_{i=1}^{k} \mu_i,$$

but that the variance of the first scheme exceeds that of the second by an amount

$$\frac{1}{k^2 n} \sum_{i=1}^{k} (\mu_i - \mu)^2.$$

31.8 Carry through the following proofs of statements made in subsections 31.5.2 and 31.5.3 about the ML estimators $\hat{\tau}$ and $\hat{\lambda}$.

(a) Find the expectation values of the ML estimators $\hat{\tau}$ and $\hat{\lambda}$ given, respectively, in (31.71) and (31.75). Hence verify equations (31.76), which show that, even though an ML estimator is unbiased, it does not follow that functions of it are also unbiased.

(b) Show that $E[\hat{\tau}^2] = (N+1)\tau^2/N$ and hence prove that $\hat{\tau}$ is a minimum-variance estimator of τ.

31.9 Each of a series of experiments consists of a large, but unknown, number n ($\gg 1$) of trials in each of which the probability of success p is the same, but also unknown. In the ith experiment, $i = 1, 2, \ldots, N$, the total number of successes is x_i ($\gg 1$). Determine the log-likelihood function.

Using Stirling's approximation to $\ln(n - x)$, show that

$$\frac{d \ln(n - x)}{dn} \approx \frac{1}{2(n - x)} + \ln(n - x),$$

and hence evaluate $\partial(^n C_x)/\partial n$.

By finding the (coupled) equations determining the ML estimators $\hat{p}$ and $\hat{n}$, show that, to order n^{-1}, they must satisfy the simultaneous 'arithmetic' and 'geometric' mean constraints

$$\hat{n}\hat{p} = \frac{1}{N} \sum_{i=1}^{N} x_i \quad \text{and} \quad (1 - \hat{p})^N = \prod_{i=1}^{N} \left(1 - \frac{x_i}{\hat{n}}\right).$$

31.10   This exercise is intended to illustrate the dangers of applying formalised estimator techniques to distributions that are not well behaved in a statistical sense.

The following are five sets of 10 values, all drawn from the same Cauchy distribution with parameter a.

(i)	4.81	−1.24	1.30	−0.23	2.98
	−1.13	−8.32	2.62	−0.79	−2.85
(ii)	0.07	1.54	0.38	−2.76	−8.82
	1.86	−4.75	4.81	1.14	−0.66
(iii)	0.72	4.57	0.86	−3.86	0.30
	−2.00	2.65	−17.44	−2.26	−8.83
(iv)	−0.15	202.76	−0.21	−0.58	−0.14
	0.36	0.44	3.36	−2.96	5.51
(v)	0.24	−3.33	−1.30	3.05	3.99
	1.59	−7.76	0.91	2.80	−6.46

Ignoring the fact that the Cauchy distribution does not have a finite variance (or even a formal mean), show that $\hat{a}$, the ML estimator of a, has to satisfy

$$s(\hat{a}) = \sum_{i=1}^{10} \frac{1}{1 + x_i^2/\hat{a}^2} = 5. \quad (*)$$

Using a programmable calculator, spreadsheet or computer, find the value of $\hat{a}$ that satisfies $(*)$ for each of the data sets and compare it with the value $a = 1.6$ used to generate the data. Form an opinion regarding the variance of the estimator.

Show further that if it is assumed that $(E[\hat{a}])^2 = E[\hat{a}^2]$, then $E[\hat{a}] = v_2^{1/2}$, where v_2 is the second (central) moment of the distribution, which for the Cauchy distribution is infinite!

31.11   According to a particular theory, two dimensionless quantities X and Y have equal values. Nine measurements of X gave values of 22, 11, 19, 19, 14, 27, 8, 24 and 18, whilst seven measured values of Y were 11, 14, 17, 14, 19, 16 and 14. Assuming that the measurements of both quantities are Gaussian distributed with a common variance, are they consistent with the theory? An alternative theory predicts that $Y^2 = \pi^2 X$; are the data consistent with this proposal?

31.12   On a certain (testing) steeplechase course there are 12 fences to be jumped, and any horse that falls is not allowed to continue in the race. In a season of racing a total of 500 horses started the course and the following numbers fell at each fence:

Fence:	1	2	3	4	5	6	7	8	9	10	11	12
Falls:	62	75	49	29	33	25	30	17	19	11	15	12

Use this data to determine the overall probability of a horse's falling at a fence, and test the hypothesis that it is the same for all horses and fences as follows.

(a) Draw up a table of the expected number of falls at each fence on the basis of the hypothesis.

(b) Consider for each fence i the standardised variable

$$z_i = \frac{\text{estimated falls} - \text{actual falls}}{\text{standard deviation of estimated falls}},$$

and use it in an appropriate χ^2 test.

(c) Show that the data indicates that the odds against all fences being equally testing are about 40 to 1. Identify the fences that are significantly easier or harder than the average.

31.13   A similar technique to that employed in exercise 31.12 can be used to test correlations between characteristics of sampled data. To illustrate this consider the following problem.

During an investigation into possible links between mathematics and classical music, pupils at a school were asked whether they had preferences (a) between mathematics and english, and (b) between classical and pop music. The results are given below.

	Classical	None	Pop
Mathematics	23	13	14
None	17	17	36
English	30	10	40

By computing tables of expected numbers, based on the assumption that no correlations exist, and calculating the relevant values of χ^2, determine whether there is any evidence for

(a) a link between academic and musical tastes, and
(b) a claim that pupils either had preferences in both areas or had no preference.

You will need to consider the appropriate value for the number of degrees of freedom to use when applying the χ^2 test.

31.14   Three candidates X, Y and Z were standing for election to a vacant seat on their college's Student Committee. The members of the electorate (current first-year students, consisting of 150 men and 105 women) were each allowed to cross out the name of the candidate they least wished to be elected, the other two candidates then being credited with one vote each. The following data are known.

(a) X received 100 votes from men, whilst Y received 65 votes from women.
(b) Z received five more votes from men than X received from women.
(c) The total votes cast for X and Y were equal.

Analyse this data in such a way that a χ^2 test can be used to determine whether voting was other than random (i) amongst men and (ii) amongst women.

31.15   A particle detector consisting of a shielded scintillator is being tested by placing it near a particle source whose intensity can be controlled by the use of absorbers. It might register counts even in the absence of particles from the source because of the cosmic ray background.

The number of counts n registered in a fixed time interval as a function of the source strength s is given in as:

source strength s:	0	1	2	3	4	5	6
counts n:	6	11	20	42	44	62	61

At any given source strength, the number of counts is expected to be Poisson distributed with mean

$$n = a + bs,$$

where a and b are constants. Analyse the data for a fit to this relationship and obtain the best values for a and b together with their standard errors.

(a) How well is the cosmic ray background determined?
(b) What is the value of the correlation coefficient between a and b? Is this consistent with what would happen if the cosmic ray background were imagined to be negligible?
(c) Do the data fit the expected relationship well? Is there any evidence that the reported data 'are too good a fit'?

31.16 The function $y(x)$ is known to be a quadratic function of x. The following table gives the measured values and uncorrelated standard errors of y measured at various values of x (in which there is negligible error):

x	1	2	3	4	5
$y(x)$	3.5 ± 0.5	2.0 ± 0.5	3.0 ± 0.5	6.5 ± 1.0	10.5 ± 1.0

Construct the response matrix R using as basis functions 1, x, x^2. Calculate the matrix $R^T N^{-1} R$ and show that its inverse, the covariance matrix V, has the form

$$V = \frac{1}{9184} \begin{pmatrix} 12\,592 & -9708 & 1580 \\ -9708 & 8413 & -1461 \\ 1580 & -1461 & 269 \end{pmatrix}.$$

Use this matrix to find the best values, and their uncertainties, for the coefficients of the quadratic form for $y(x)$.

31.17 The following are the values and standard errors of a physical quantity $f(\theta)$ measured at various values of θ (in which there is negligible error):

θ	0	$\pi/6$	$\pi/4$	$\pi/3$
$f(\theta)$	3.72 ± 0.2	1.98 ± 0.1	-0.06 ± 0.1	-2.05 ± 0.1

θ	$\pi/2$	$2\pi/3$	$3\pi/4$	π
$f(\theta)$	-2.83 ± 0.2	1.15 ± 0.1	3.99 ± 0.2	9.71 ± 0.4

Theory suggests that f should be of the form $a_1 + a_2 \cos\theta + a_3 \cos 2\theta$. Show that the normal equations for the coefficients a_i are

$$481.3a_1 + 158.4a_2 - 43.8a_3 = 284.7,$$
$$158.4a_1 + 218.8a_2 + 62.1a_3 = -31.1,$$
$$-43.8a_1 + 62.1a_2 + 131.3a_3 = 368.4.$$

(a) If you have matrix inversion routines available on a computer, determine the best values and variances for the coefficients a_i and the correlation between the coefficients a_1 and a_2.

(b) If you have only a calculator available, solve for the values using a Gauss–Seidel iteration and start from the approximate solution $a_1 = 2$, $a_2 = -2$, $a_3 = 4$.

31.18 Prove that the expression given for the Student's t-distribution in equation (31.118) is correctly normalised.

31.19 Verify that the F-distribution $P(F)$ given explicitly in equation (31.126) is symmetric between the two data samples, i.e. that it retains the same form but with N_1 and N_2 interchanged, if F is replaced by $F' = F^{-1}$. Symbolically, if $P'(F')$ is the distribution of F' and $P(F) = \eta(F, N_1, N_2)$, then $P'(F') = \eta(F', N_2, N_1)$.

31.20 It is claimed that the two following sets of values were obtained (a) by randomly drawing from a normal distribution that is $N(0, 1)$ and then (b) randomly assigning each reading to one of two sets A and B:

Set A:	-0.314	0.603	-0.551	-0.537	-0.160	-1.635	0.719
	0.610	0.482	-1.757	0.058			
Set B:	-0.691	1.515	-1.642	-1.736	1.224	1.423	1.165

Make tests, including t- and F-tests, to establish whether there is any evidence that either claims is, or both claims are, false.

31.9 Hints and answers

31.1 Note that the reading of 9.28 m s^{-2} is clearly in error, and should not be used in the calculation; 9.80 ± 0.02 m s^{-2}.

31.3 (a) 55.1. (b) Note that two thirds of the readings lie within ± 2 of the mean and that 14 readings are being used. This gives a standard error in the mean ≈ 0.6. (c) Student's t has a value of about 2.5 for 13 d.o.f. (degrees of freedom), and therefore it is likely at the 3% significance level that the data are in conflict with the accepted value.

31.5 Plot or calculate a least-squares fit of either x^2 versus x/y or xy versus y/x to obtain $a \approx 1.19$ and $b \approx 3.4$. (a) 0.16; (b) -0.27. Estimate (b) is the more accurate because, using the fact that $y(-x) = -y(x)$, it is effectively obtained by *inter*polation rather than *extra*polation.

31.7 Recall that, because of the equal proportions of each type, the *expected* numbers of each type in the first scheme is n. Show that the variance of the estimator for the second scheme is $\sigma^2/(kn)$. When calculating that for the first scheme, recall that $\overline{x_i^2} = \mu_i^2 + \sigma^2$ and note that μ_i^2 can be written as $(\mu_i - \mu + \mu)^2$.

31.9 The log-likelihood function is

$$\ln L = \sum_{i=1}^{N} \ln {}^nC_{x_i} + \sum_{i=1}^{N} x_i \ln p + \left(Nn - \sum_{i=1}^{N} x_i \right) \ln(1-p);$$

$$\frac{\partial({}^nC_x)}{\partial n} \approx \ln\left(\frac{n}{n-x} \right) - \frac{x}{2n(n-x)}.$$

Ignore the second term on the RHS of the above to obtain

$$\sum_{i=1}^{N} \ln\left(\frac{n}{n-x_i} \right) + N\ln(1-p) = 0.$$

31.11 $\bar{X} = 18.0 \pm 2.2$, $\bar{Y} = 15.0 \pm 1.1$. $\hat{\sigma} - 4.92$ giving $t = 1.21$ for 14 d.o.f., and is significant only at the 75% level. Thus there is no significant disagreement between the data and the theory. For the second theory, only the mean values can be tested as Y^2 will not be Gaussian distributed. The difference in the means is $\bar{Y}^2 - \pi^2 \bar{X} = 47 \pm 36$ and is only significantly different from zero at the 82% level. Again the data is consistent with the proposed theory.

31.13 Consider how many entries may be chosen freely in the table if all row and column totals are to match the observed values. It should be clear that for an $m \times n$ table the number of degrees of freedom is $(m-1)(n-1)$.

(a) In order to make the fractions expressing each preference or lack of preference correct, the expected distribution, if there were no correlation, must be

	Classical	None	Pop
Mathematics	17.5	10	22.5
None	24.5	14	31.5
English	28	16	36

This gives a χ^2 of 12.3 for four d.o.f., making it less than 2% likely that no correlation exists.

(b) The expected distribution, if there were no correlation, is

	Music preference	No music preference
Academic preference	104	26
No academic preference	56	14

This gives a χ^2 of 1.2 for one d.o.f and no evidence for the claim.

31.15 As the distribution at each value of s is Poisson, the best estimate of the measurement error is the square root of the number of counts, i.e. $\sqrt{n(s)}$. Linear regression gives $a = 4.3 \pm 2.1$ and $b = 10.06 \pm 0.94$.

(a) The cosmic ray background must be present, since $n(0) \neq 0$ but its value of about 4 is uncertain to within a factor 2.

(b) The correlation coefficient between a and b is -0.63. Yes; if a were reduced towards zero then b would have to be increased to compensate.

(c) Yes, $\chi^2 = 4.9$ for five d.o.f., which is almost exactly the 'expected' value, neither too good nor too bad.

31.17 $a_1 = 2.02 \pm 0.06$, $a_2 = -2.99 \pm 0.09$, $a_3 = 4.90 \pm 0.10$; $r_{12} = -0.60$.

31.19 Note that $|dF| = |dF'/F'^2|$ and write

$$1 + \frac{N_1 - 1}{(N_2 - 1)F'} \quad \text{as} \quad \left[\frac{N_1 - 1}{(N_2 - 1)F'} \right] \left[1 + \frac{(N_2 - 1)F'}{N_1 - 1} \right].$$

Index

Where the discussion of a topic runs over two consecutive pages, reference is made only to the first of these. For discussions spread over three or more pages the first and last page numbers are given; these references are usually to the major treatment of the corresponding topic. Isolated refences to a topic, including those appearing on consecutive pages, are listed individually. Some long topics are split, e.g. 'Fourier transforms' and 'Fourier transforms, examples'. The letter 'n' after a page number indicates that the topic is discussed in a footnote on the relevant page.